Energy and the Environment in the 21st Century

edited by
Jefferson W. Tester
David O. Wood
Nancy A. Ferrari

with
János M. Beér
William J. Duggan
Leon R. Glicksman
Kent F. Hansen
Howard J. Herzog
John B. Heywood
Thomas H. Lee
Adel F. Sarofim
Malcolm A. Weiss
David C. White

The MIT Press
Cambridge, Massachusetts
London, England

This book was set in Linotron Baskerville by DEKR Corporation and was printed and bound in the United States of America.

Library of Congress Cataloging-in-Publication Data

Energy and the environment in the 21st century : proceedings of the conference held at the Massachusetts Institute of Technology, Cambridge, MA, March 26–28, 1990 / edited by Jefferson W. Tester, David O. Wood, Nancy A. Ferrari.
p. cm.
Includes bibliographical references and index.
ISBN 0-262-20078-3
1. Energy development—Environmental aspects—Congresses.
2. Energy consumption—Environmental aspects—Congresses.
I. Tester, Jefferson W. II. Wood, David O. III. Ferrari, Nancy A.
IV. Massachusetts Institute of Technology.
TD195.E49E515 1991
333.79'14—dc20 91-32
CIP

Energy and the Environment in the 21st Century

Proceedings of the Conference held at the Massachusetts Institute of Technology Cambridge, Massachusetts

March 26–28, 1990

Energy Laboratory
Massachusetts Institute of Technology
Cambridge, MA

Principal Editors

Jefferson W. Tester
Conference Co-chair
Professor of Chemical Engineering
Director, Energy Laboratory

David O. Wood
Conference Co-chair
Director, Center for Energy Policy Research

Nancy A. Ferrari
Energy Laboratory

Conference Organizers and Energy/Environmental Technology and Policy Study Group

János M. Beér
Professor of Chemical Engineering
Science Director, Combustion Research Facility

William J. Duggan
Administrative Officer, Energy Laboratory

Leon R. Glicksman
Professor of Building Technology

Kent F. Hansen
Professor of Nuclear Engineering

Howard J. Herzog
Research Engineer, Energy Laboratory

John B. Heywood
Professor of Mechanical Engineering
Director, Sloan Automotive Laboratory

Thomas H. Lee
Professor of Electrical Engineering and Computer Science, Emeritus

Adel F. Sarofim
Professor of Chemical Engineering

Malcolm A. Weiss
Research Staff, Energy Laboratory

David C. White
Professor of Electrical Engineering

The conference was sponsored in part by:

Sloan Foundation
American Academy of Arts and Sciences
Exxon Foundation

Session Chairs

Please note: All session chairs are from the Massachusetts Institute of Technology unless otherwise indicated.

Plenary Session I:

Paul E. Gray
President, Massachusetts Institute of Technology

Jefferson W. Tester
Professor of Chemical Engineering
Director, Energy Laboratory

David O. Wood
Director, Center for Energy Policy Research

Plenary Session II:

David O. Wood
Director, Center for Energy Policy Research

John M. Deutch
Provost, Massachusetts Institute of Technology

Session A—Transportation Systems:

Malcolm A. Weiss
Research Staff, Energy Laboratory

John B. Heywood
Professor of Mechanical Engineering
Director, Sloan Automotive Laboratory

Session B—Industrial Processes:

Jefferson W. Tester
Professor of Chemical Engineering
Director, Energy Laboratory

Howard J. Herzog
Research Engineer, Energy Laboratory

Session B-1—Industrial Process and Materials Manufacture:

Jefferson W. Tester
Professor of Chemical Engineering
Director, Energy Laboratory

James Wei
Professor of Chemical Engineering

Session B-2—Metals, Minerals, and Ceramics Manufacture:

Donald R. Sadoway
Professor of Materials Science and Engineering

Session B-3—Chemicals, Petroleum, and Pulp and Paper Manufacture

Kenneth A. Smith
Professor of Chemical Engineering
Associate Provost and Vice President for Research

Session C—Building Systems:

Leon R. Glicksman
Professor of Architecture

Session D—Electric Power Systems:

David C. White
Professor of Electrical Engineering

Session D-1—Electric Power for Developing Nations:

Dietmar Winje
Professor and Head of Division, Energy Engineering and Resource Economics
Technical University of Berlin

Session D-2—Fossil-Fuel Based Electric Power Technologies:

János M. Beér
Professor of Chemical Engineering
Science Director, Combustion Research Facility

Session D-3—Nuclear-Based Electric Power Technologies:

Kent F. Hansen
Professor of Nuclear Engineering

Session D-4—Alternative-Energy Based Electric Power Systems:

Jon G. McGowan
Professor of Mechanical Engineering
University of Massachusetts, Amherst

Session D-5—Electric Power End-Use and Network Efficiency:

David C. White
Professor of Electrical Engineering

Session E—Economics and Policy:

David O. Wood
Director, Center for Energy Policy Research

Session F—Advanced Energy Supply Technologies:

Robert Duffield
Energy Laboratory

Ronald Davidson
Professor of Physics
Associate Director, Plasma Fusion Laboratory

Table of Contents

Preface

As we enter the 21st century, policymakers, researchers, industrial leaders, and individuals the world over are becoming increasingly concerned about how to deal with the environmental consequences of energy use. Recognizing that concern, in late 1988 the Energy Laboratory's Energy/Environmental Technology and Policy Study Group began to plan a major international conference at MIT. One important motivation for holding such a conference was to help us better understand the roles of energy technology and policy in the context of our research and teaching responsibilities. At the same time, we wanted to provide an open forum for serious discussion of environmental problems and solutions, in the spirit of the Workshop on Alternative Energy Strategies (WAES), a major international meeting held at MIT 16 years earlier. We hoped to achieve a comprehensive treatment of energy services worldwide with quantitative discussions of how to minimize the environmental impacts of providing these services. And we wanted to adopt a conference structure where experts from industry, academia, and government would be invited to address the interrelated issues of energy technology and environmental policy.

After much discussion, we agreed that the specific objectives of the conference should be to:

- Evaluate short- and long-term projections of the environmental effects of energy supply, production, conversion, and utilization from national and international perspectives;
- Review the interaction of energy technologies and policies and their effects on local, regional, and global environmental systems;
- Examine the roles of technology and policy in providing the energy needed for the social and economic well-being of a growing world population while ensuring a safe and sustainable physical environment;
- Relate our understanding of energy technology and environmental system linkages to methods of internalizing the actual environmental costs of energy production and use at local, regional, national, and international levels; and
- Define a science, engineering, and policy research agenda that can address critical energy technology issues.

The result of this planning was the three-day conference "Energy and the Environment in the 21st Century," which was held at MIT, March 26–28, 1990, and hosted by the MIT Energy Laboratory and the MIT Center for Energy Policy Research. More than 600 registrants and hundreds of other interested observers attended.

The conference opened with a plenary session (I) at which governmental, industrial, and academic speakers considered environmental science, demographics, economics, policy, and energy technology issues. Concurrent sessions were then devoted to specific end-use areas, namely, transportation systems (Session A), industrial processes (Session B), building systems (Session C), and electric power systems (Session D). Each session analyzed in depth the environmental impacts of current technologies and opportunities for reducing those impacts, for example, through conservation, improved energy efficiency, re-engineering of processes and materials, fuel switching, and use of new energy supply technologies. Technology and economic concerns relevant to both developed and developing countries were considered, and critical policy options and appropriate research and development priorities were discussed.

Additional special concurrent sessions covered economics and policy (Session E) and advanced energy supply technologies (Session F). Session E dealt with economic and policy initiatives for reducing greenhouse gas emissions; Session F focused on nonconventional energy supply technologies with long-term potential, such as solar photovoltaic, hot dry rock geothermal, magnetohydrodynamic, and fusion energy.

The conference closed with a plenary session (II) at which forum members shared their perspectives on policy strategies for the efficient and equitable management of energy resources to create a sustainable global environment.

Speakers presenting invited papers at the conference were asked to submit manuscripts to be reviewed and edited for this proceedings volume. Almost all of the more than 80 papers presented appear here. They provide an up-to-date, in-depth analysis of a broad spectrum of current and new energy technologies and their environmental impacts. Every effort has been made to maintain a consistent format within each paper, though not necessarily from paper to paper. Major findings are outlined in the conference summary that follows this preface and the summaries that precede the papers in each session. We have also provided a detailed index to make the volume useful as a reference book on the general subject of energy and the environment. The complete conference program appears as Appendix A, and an energy unit conversion table appears as Appendix B.

The papers included in this volume reflect the views and opinions of the authors and do not necessarily represent the official positions of an author's sponsoring organization, or the Massachusetts Institute of Technology, the Energy Laboratory, or the Center for Energy Policy Research. Specific questions and requests for additional information should be directed to the authors.

A primary motivation for organizing and hosting the conference at MIT was to provide an accessible forum at which students, faculty, and the general public could participate in discussions and debates unbiased by underlying special interests. Therefore, partial funding for the conference was provided by the Energy Laboratory's endowed funds. We are grateful to the American Academy of Arts and Sciences, the Sloan Foundation, and the Exxon Foundation for providing additional support.

On behalf of the conference study group and organizing committee and the conference chairs, we would like to express special thanks to assistant editors Nancy Lattanzio, Laura Holgate, and Chris Herrick, associate editor Susan Innes, and production assistant Jay Corbett for their exceptional efforts toward producing these proceedings. In addition, we thank our colleagues in the Energy Laboratory at MIT, in particular, Nancy Stauffer for providing writing assistance; Joni Bubluski for preparing promotional materials; Betty Sheridan and Barbara Johnson for coordinating the complex arrangements; Anne Carbone, Sue Chan, Karen Luxton, and Kathleen Ross for their innumerable contributions toward making the conference and its proceedings a reality; and Gayle Fitzgerald and her associates at MIT's Conference Services for providing much-needed assistance in the planning and implementing of conference arrangements.

Jefferson W. Tester
David O. Wood
Nancy A. Ferrari

Summary and Perspectives on Energy and the Environment in the 21st Century

Nancy W. Stauffer
Nancy A. Ferrari
David O. Wood
Jefferson W. Tester
Energy Laboratory and
Center for Energy Policy Research
Massachusetts Institute of Technology
Cambridge, MA

In 1988, the Energy Laboratory convened the Energy/Environmental Technology and Policy Study Group, bringing together a dozen of the Laboratory's research faculty and staff people with expertise in a range of fields relating to energy. Our mandate was to design and develop an ongoing program of research that would examine interrelationships among energy activities and environmental effects and would explore new roles that technology and policy can play in achieving national and international goals. Discussions and research led to development of several themes basic to the design of such a research program. Those themes were likewise foremost in our minds as we planned one of our activities, the conference "Energy and the Environment in the 21st Century." As described below, presentations, discussions, and debate at that conference provided considerable insight and information that have helped us expand and clarify our initial ideas.

The availability of energy resources and technologies to convert them into usable forms has been critical to world economic development for many decades. The history of the industrial revolution begins with James Watt's invention of an energy-conversion technology, the steam engine, and is written largely in terms of the continued invention and application of increasingly sophisticated energy-conversion and end-use technologies that increase the economic productivity of human effort.

But economic growth fueled by the application of energy-conversion technologies has had a serious negative effect—the use of environmental resources (air, water, land) as depositories for waste products. The regenerative capacities of those environmental resources are now being stressed by the sheer volume of waste produced by an ever-expanding world population.

The close relationship between energy supply and use and environmental damage leads many to conclude that technology itself is the villain. In our opinion, that view is short-sighted and misses the essential role of technology in solving environmental problems. Indeed, we cannot change the environmental consequences of energy use without changing the mix and characteristics of the technologies that produce, convert, distribute, and use energy. The world needs another "technological revolution" if it is to ensure the efficient and sustainable use of the energy, environmental, and other resources needed for economic and social well-being.

One view suggests that the potential for technological response in the short run is limited by the characteristics of capital equipment now in place. The potential is greater in the longer run because that equipment can be replaced. Discussion at the conference supported the notion that, on the supply side, near-term gains through technology are unlikely. But it did point to conservation opportunities on the demand side—to areas where gains may be achievable through more efficient end-use technologies. In both the industrial and the buildings sector, for example, better technology is available but not yet implemented, for economic or institutional reasons. In other cases, significant efficiency and emissions improvements could result from additional fine-tuning of existing technologies, for example, the internal combustion engine.

The technological response for the long term could involve a number of new supply technologies that promise to be more environmentally benign and more efficient than those in use today. Examples discussed in Sessions D and F include advanced fossil-fuel systems using magnetohydrodynamic power generation or fuel cells; new generations of nuclear reactors; renewables such as wind, solar (thermal and photovoltaic), and geothermal (hydrothermal and hot dry rock) systems; and nuclear fusion. An objective assessment of these technologies will be needed to establish priorities for allocating funds for research, development, and demonstration projects.

The other critical factor in accomplishing the necessary technological revolution is, of course, policy. Technology and policy are intertwined in ways that change over time. As mentioned above, in the near term, our actions are largely limited to adjusting the characteristics and utilization of existing technologies. Thus, in the near term, our policy options are limited by technological possibilities. In the long term, our objective is to develop and deploy new and improved technologies.

Thus, in the long term, policy can determine the rate and direction of technology development. An effective energy/environmental plan will include both policies that create incentives and controls to increase energy efficiency and policies that encourage innovative thinkers to begin working on more environmentally benign technologies for future use.

One recurrent message for both policymakers and technology researchers is to adopt a broad view. On the policy side, we must be careful not to focus on a single environmental problem. Environmental problems are so interconnected that any policy that affects one tends to affect them all. For example, policies directed toward reducing acid rain may encourage the installation of scrubbers to permit the continued use of coal—an outcome inconsistent with the goal of reducing carbon dioxide emissions so as to control global climate change. We must also realize that technological advances alone cannot solve our energy and environmental problems. Patterns of energy demand and forms and amounts of effluents are directly influenced by interactions among population change, resource use and availability, and level of technological capabilities. To achieve energy and environmental sustainability, policymakers must also consider goals and outlooks for population growth and levels of economic activity.

We should also take a broad view as we look for promising approaches on the technology side. For example, we should not focus solely on making vehicles more fuel efficient but should also consider ways of making the entire transportation system more efficient or perhaps ways of restructuring demand. In the electric power area, we need to consider not single generation technologies but combinations of technologies plus regulatory constraints and load-management programs. And when reviewing which new technologies to pursue for the longer term, our analysis should include not only their projected availability and economic competitiveness but also the time and costs required for their development, their simplicity and operability, and the environmental impacts and costs associated with the entire fuel cycle.

Who are the key actors needed to achieve this second technological revolution? The three primary actors and their respective roles are not surprising. Government will set policy; universities and industry will refine existing technologies, invent new ones, and train future technologists and policymakers; and industry will further develop and implement the technology advances. The unusual factor here is the degree to which interaction among the actors must occur, largely because of the interconnectedness of energy/environmental technology and policy. Industry must make the adjustments in the short run and implement the new technologies for the longer term. Therefore, industry must provide information to both the policymakers and the university researchers as to what is possible. Indeed, industrial representatives should be intimately involved in the choice and design of new technology so that it is well suited to their needs.

Also critical to this undertaking are the developing nations. Models assessing various types of emissions from various sources suggest that emissions from now-developing nations can be expected to increase significantly by 2025. Indeed, the developing nations will become more important emitters than the developed nations. As world leaders consider international policies to reduce emissions, they need to give special attention to the developing nations for two reasons. First, those nations should not suffer because of the behavior of the developed nations in the past. And second, the economic penalty from emissions reduction policies will vary considerably from region to region and may be particularly high in developing nations. Cooperative international action is needed to ensure that developing nations achieve energy systems that are efficient and environmentally benign. Research in the developed nations should place a high priority on new technologies that are appropriate for use in developing nations. Because electricity consumption in the developing nations is likely to be high, attractive options include generators based on renewable, nonfossil-based resources and possibly small, easy-to-operate, ultrasafe nuclear reactors.

A critical question that has been hotly debated at the conference and elsewhere is, What level of action is appropriate now to deal with environmental damage that may occur in the future? Answering that question is difficult largely because we do not know what the future will bring. Mathematical models described at the conference can assess increased emissions—their generation, properties, lifetimes, interactions, and impacts. Other models attempt to analyze and predict changes in climate due to those emissions. Yet other studies consider how such climate change may alter various components of our world and how we may be able to adapt and cope. In all cases, the most striking theme is the difficulty and the enormous uncertainty involved in the modeling and analysis process and hence also in the results. (See Session E and Plenary Session II for further discussion.)

The greatest disagreement about the appropriate level of action arose at the conference in discussions about global warming—a problem that involves an unprecedented degree of uncertainty and scales of both time and space that are outside our experience with other environmental issues. At one extreme, conference participants recommended an immediate attack on sources of global warming, using the strongest measures that all nations would accept and focusing all our resources on developing and disseminating the needed technologies worldwide. At the other extreme, participants argued that we should take no action because the harm caused by the level of climate change predicted for the next 100 years would not be significant.

Not surprisingly, the majority of the participants advocated a moderate approach. They believed we should work to lessen uncertainty regarding climate change and that we should increase our ability to respond and adapt by developing new drought-resistant cultivars and low-pollution energy conversion devices. In addition, they warned against letting scientific uncertainty stymie the political process, suggesting instead the adoption of conservative regulations that can be adjusted as scientific understanding grows. And they recommended policies that make sense whether we are on the brink of a climate catastrophe or not—that is, policies that can provide multiple benefits, for instance, in terms of other environmental problems, long-term economic growth, and budget and trade concerns.

As we consider what steps to take on both policy and technology fronts, it is necessary to look in more detail at specific end-use sectors, notably, transportation systems, industrial processes, building systems, and electric power systems. The issues and opportunities vary considerably from sector to sector. For example, energy and environmental problems related to transportation systems (Session A) are difficult to alleviate, largely because of the increasing worldwide demand for transportation services and the technical merits of current petroleum-based technologies. Possible efficiency and emissions improvements were noted at the conference; but participants disagreed about the extent, cost, and performance debits of those improvements. In this sector, industry's role in advancing technology is vital. But federal support is required for the long-term basic research needed to accompany such advances.

Achieving across-the-board efficiency improvement in the area of industrial processes (Session B) is difficult because the potential for gain varies widely from industry to industry. In some industries, existing equipment could be used more effectively; or new, more efficient equipment or processes could be employed but are not for economic reasons. Other industries depend on processes that are inherently energy intensive and difficult to change. In those cases, recycling is critical; and research is needed to improve the quality of recycled products and to design products specifically for ease of recycling. As agreed at the conference, economics remains the most important factor motivating efficiency improvements. In many cases, the adoption of new technologies is impeded by the long life of capital equipment. Historically, growing industries such as petroleum refining and polyethylene production have invested substantial financial resources in implementing energy conservation measures. How quickly other industries will follow their lead will depend both on economic factors and on policy developments including regulations and standards.

Developing and implementing more energy-efficient practices in building systems (Session C) also present special challenges. Research budgets are low; and the industry is fragmented, risk averse, and subject to complex regulations. Nevertheless, conference participants cited some promising trends, including new legislative initiatives, expanded federal funding, and emergence of promising new technologies. Although buildings have become more energy efficient during the past decade, the gains resulted from application of existing technologies. Additional gains will require a concerted effort by industry, government, and researchers to develop innovative concepts and to encourage their widespread adoption. Participants agreed that it will be cheaper to develop and implement more efficient building technologies to save energy than to develop and use cleaner technologies to supply energy.

As a major consumer of the world's primary energy resources, electric power systems (Session D) are a growing source of environmental concern. All the options for generating power involve certain problems. Participants agreed that coal will probably continue as the dominant fuel for electricity generation, with natural gas coming into greater use in the next two decades. Therefore, it is imperative that we implement clean coal technologies, improve control systems, and develop better operating and maintenance programs. The benefits of retrofitting today's plants must be weighed against the benefits of accelerated deployment of advanced clean coal technology. Nuclear power generation offers many environmental advantages; and some new, safer designs look promising. But revitalization of the nuclear option will require reformed licensing procedures, technical and political solutions to the waste-disposal problem, improved nuclear plant performance, and renewed public confidence in nuclear power. The role of renewable energy technologies in generating electric power should be explored and expanded. Further development of renewables is encouraged by the success of currently operating geothermal plants and wind and photovoltaic systems. Finally, policies creating appropriate economic incentives are needed to ensure the efficiency of both end-use equipment and electric power networks.

Throughout discussions at the conference and within our study group, one underlying theme was evident: We need to adopt a new perspective as we tackle our energy/environmental future. We should view that future not as a series of isolated problems to be solved but as a set of interconnected trade-offs that must be made between having sufficient energy and bearing the related environmental and other costs. Those trade-offs must assign, at least implicitly, a relative importance to having diverse energy services, a clean environment, national and international security, acceptable economic costs, a reasonable "quality of life," and so on. After thus establishing our priorities, we can pursue the technological and policy measures needed to achieve the most important objectives. Finally, we must be aware that solving the world's energy and environmental problems depends critically on the talents and contributions of future generations of engineers and scientists. The education and training of those future generations are central to achieving environmentally and economically sustainable energy systems.

Introductory Address
Energy and the Environment in the 21st Century

Richard D. Morgenstern
US Environmental Protection Agency
Washington, DC

Thank you, President Gray. I am pleased to be here today representing Administrator Bill Reilly and the Environmental Protection Agency at this important meeting on "Energy and the Environment in the 21st Century."

No one will fault the organizers of this conference for poor timing, poised as we are for the final segment of the Clean Air Act debate. This legislation will affect the decisions of utility and industrial SO_2 emitters, as well as all of us driving cars over the course of the coming decade and beyond. It is also not long before negotiations are expected to begin on a framework convention for global climate change—which may ultimately affect the basic energy decisions made by individuals, corporations, and governments in every country on earth over the next century and beyond.

Many of the problems posed by the energy/environment conflict have classically fallen into the category of what economists call externalities—beyond the system of rational private sector decisionmaking. For those of us who worked on the so-called energy crisis of the 1970s, many of the issues déjà vu, as Yogi Berra would say, all over again. My own perspective is one of a senior civil servant and former economics professor now plying his trade in the meeting rooms and hallways of the Washington establishment.

Conservation was a good idea back in the 1970s and it's a good idea now. Nonfossil fuel energy technologies—and I mean a full range of such technologies—were good ideas then, often in need of field testing and commercial development, and they remain so today. Reducing tropospheric pollution made good sense for local environments, and now appears an especially wise policy. Similarly, preserving wetlands and saving beaches have had long-standing appeal. What distinguishes these issues today is a growing awareness of the fragility of our environment and the need for both local and global actions. We sometimes have a hard time selling the connection between the micro and macro in this arena: as one Washington wag recently noted, "It is kind of hard to imagine that some guy in Italy, using an underarm spray deodorant, can increase the risk of my getting a sunburn."

The crucial difference between the discussions of these issues 15 years ago and now stems from the enhanced appreciation of what that lonely *Voyager* spacecraft confirmed: namely that we have but one planet on which to stake human existence, and that our lot cannot be separated from those of other global inhabitants. As *Landsat* photos and models developed by many in this great university confirm, environmental pollution increasingly constitutes a global phenomenon, for which global solutions will be required.

Consistent with its mandate, the Environmental Protection Agency (EPA) looks broadly at the energy/environment connection. To set the context for the issues before this meeting, I want to take a moment and tell you about the experiences and frustrations of what I believe are two particularly relevant issues: acid rain and stratospheric ozone depletion. Each of them follows William James's observations about new ideas in science: first, they are treated as preposterous notions that are highly improbable; then they are embraced as perhaps true, but not that important; and finally, they become viewed as mainstream, and persons who first debunked the ideas, often scurry to appear to have invented or anticipated them in the first place. Whether or not this model fits the global climate issues, of course, remains to be seen.

In the case of acid rain, it has been more than a decade since the issue first started receiving regular attention in *Science* magazine. And it will be another decade before the full reductions are made. The debate has been very contentious, with strong vested—including major regional—interests lining up on all sides and the crafting of compromises that have struck many as odd and sometimes wasteful. In the end, a heavily market-oriented solution is being adopted. Even today, there is still a chorus of folks, including some serious scientists, who question the state of our knowledge of both the problem and the proposed solutions.

In the case of chlorofluorocarbons (CFCs), which have now been controlled for their adverse effects as ozone depletors and carcinogens rather than as greenhouse gases, we have an issue that was debated for a while then pushed to the back burner as new information indicated the problem was less immediate. It was the dogged, pioneering work of a handful of atmospheric scientists, mostly in the United States, who helped forge an international scientific consensus on the role of

CFCs in the depletion of the stratospheric ozone layer. Based on this scientific convergence, political leaders, activists and hard-working policy analysts pushed for controls. Ultimately, the issue was given a real boost by those abstract graphics that graced several national news magazines and showed the seasonal depletion of ozone in the Antarctic. Today even the skeptics generally applaud decisions to make major reductions of CFCs, and the major producers have taken the lead in calling for rapid phaseout. Critics—especially those who are uncomfortable with comparisons to global climate change—are quick to point out that we're dealing with a set of chemicals produced by a small number of companies, in a small number of countries, without which humankind survived quite well until 50 years ago. Further, they note, the controlling of CFCs may, despite the recently enacted taxes, turn out to be quite profitable to the producers.

With that background, my remarks today will focus on the issue of climate change and the connection to the energy system. The official US position is referred to as the "no regrets" strategy. Secretary of State Baker has defined this to mean ". . . that while we are pursuing the serious scientific research that is critical to any responsible approach we're also hedging our bets in an economically sound way." The "hedges" he refers to entail, for example, phasing out CFCs, encouraging reforestation, supporting the Clean Air Act, and promoting energy conservation. Some European nations are calling for quick action while others, including Japan, the USSR, China, and the US are reluctant to move ahead too quickly. We also have a major international study process under way—the Intergovernmental Panel on Climate Change (IPCC)—organized under the auspices of United Nations Environmental Program (UNEP) and World Meteorological Organization (WMO), which is slated to report at the end of the summer on the science, impacts, and policy options. President Bush has offered to host the initial negotiating session for a framework convention—which is likely to occur early next year.

With respect to a framework convention on global climate change, let me be clear that when we discuss emission reductions we're not excluding adaptation as an option. As John Firor's recent article reminded us, even if we froze emissions today, some warming would be inevitable. Some argue that adaptation should be the only policy response. I will not question that here today. I will confine my remarks to the topic of emission reduction, for which all agree a long lead time is involved.

I would define the issue as whether or not, and if so, how, the nations of the world should move to the point of making commitments—with targets and timetables—beyond the barebones framework convention now contemplated, and beyond the "no regrets" strategy, to attain significant emission reductions. The answer, I think, depends on three issues which I would now like to discuss—science, technology, and economics.

First, the science. World population has increased threefold since 1900. Industrial production has increased by a factor of 50, four fifths of it since 1950. Consumption of fossil fuel has increased by a factor of 30, and the atmospheric concentration of carbon dioxide has increased by about 25 percent since pre-industrial times. In the tropics, 10 trees are cut for each one planted. Human activity is clearly changing features of the earth, including the chemical composition of the earth's atmosphere—probably, as most scientists would agree, warming the earth's climate.

There is, of course, much uncertainty over the rate and magnitude of this warming. The rate and magnitude of temperature change depend largely on the reactions of the earth's geophysical and biological feedback mechanisms including changes in atmospheric levels of water vapor, snow and ice cover, and the effects of clouds—all of which are poorly understood and only crudely captured in our models.

According to the National Academy of Sciences, significant global climate change is at least as likely to occur in the next century as not. Based on a series of studies dating to the late 1970s, the Academy has estimated that an effective doubling of carbon dioxide levels (or its equivalent, taking into account other gases) could increase global average temperatures by 1.5 to 4.5°C. Moreover, doubling of CO_2 is only a point on a rising trend. With the exception of CFCs, greenhouse gas emissions will continue to rise at a rapid rate for the foreseeable future. In fact, an effective doubling of pre-industrial CO_2 concentrations is expected to occur before the year 2050—and based on some recent projections perhaps as early as 2020 or 2030.

Skeptics question the significance of the projected temperature changes. While the difference in mean annual temperature between Boston and Washington is only 3.3°C, the average global temperature could increase as much in the next century as it has in the 18,000 years since the last ice age.

Some have detected a type of Hegelian swing in the science community. If most of the science I've cited so far would be considered the "thesis," then we're clearly in the "antithesis" phase. Increasingly, the work of the National Academy has been called into question by those who see much less warming over the next century and some, a vocal minority, who see an actual cooling. It would be one thing if this scientific debate were carried out in the laboratories of the great universities, but it takes on quite a different meaning when the debate is played out on the front page of the *New York Times*.

What is needed is a grand synthesis. Unlike the stratospheric ozone debate that was ultimately subject to reasonable verification, the climate debate is unlikely to be similarly edified, at least in the near term. Clearly, more research is called for, but we should not expect real world verification of warming very soon. What we are left with are incremental improvements in our ability to understand and model climate and additional reviews of the existing evidence. The next significant review on the horizon is the science panel of the IPCC. Composed of eminent experts from around the world and headed by John Houghton, head of the British Meteorological Service, this panel is expected to report at the end of the summer. Indications are that this report might indeed serve as a synthesis, or at least a set of conclusions that are generally accepted by the international community.

Turning to the issue of technology development, let me share with you, if you can imagine, an outline of a 100-year plan for responding to climate change. Roughly speaking, the first 50 years would be years of transition, when environmentally friendly technologies are developed and introduced. The second 50 years would be a period of environmental recovery.

The first decade would be devoted to reducing the uncertainties surrounding climate change and improving energy efficiency through the increased use of available technology. The second decade would be characterized by a significant increase in the use of nonfossil fuels, including improved nuclear power plants. The third decade would see the spread of third generation CFCs, CO_2 fixation technologies, and revolutionary low-energy production processes. The fourth decade would witness advances in carbon absorption, the use of biotechnology to reverse desertification, significant net gains from reforestation, and enhanced oceanic sinks. Only in the fifth decade would we address the "more exotic technologies"—fusion, orbiting solar power plants, magma electricity generation, and other new forms of energy that, several hundred years after their introduction, might render fossil fuels unnecessary.

For those of you who are unnerved by this plan, let me assure you that it is not the product of some young EPA researcher gone mad. Rather, this 100-year plan was recently unveiled in Washington by the Japanese Ministry of Technology and Industry.

Recognizing that the United States does not function like Japan, the question arises as to the appropriate role of government in encouraging the development of long-term technological responses to climate change. I would suggest that the government should provide a stable economic environment in which incentives for research, development, and demonstration are not solely concentrated on the short term. And, as in the case of CFCs, government must stand ready to act even when all the technological solutions are not worked out completely. Such technology forcing, as it is called, is controversial and certainly not costless, but it is sometimes necessary in matters of this sort. And, because of the fundamentally international nature of the issue, governments should facilitate technology transfer through means that do not compromise legitimate commercial interests. Indeed, technology transfer will be key to the ultimate success of any global emission limiting strategy.

The third topic that I would like to touch upon is the need for careful analysis of the costs and economic impacts of possible climate change and alternative measures for responding to it. The most striking aspect here is the huge gap in the results between the so-called "top down" models, which typically use econometric techniques—and generally calculate relatively higher costs for reducing CO_2 emissions—and the so-called "bottom-up" approaches which typically focus on process engineering analyses for specific sectors or technologies and generally calculate lower costs.

Both of these approaches have merits, yet neither has sufficient believability that a policymaker would want to go too far out on a limb with the results. One has only to recall the rancorous cost debates on acid rain or what we now realize were exaggerated cost estimates developed by the CFC industry to realize that we're hitting a raw nerve. The search in Washington, and in capitals around the world, is to push the frontier on the cost issue—to integrate these modeling approaches and try to develop some consensus on the true costs of greenhouse gas reductions. Experts at this meeting can play a major role in this research and we welcome their input.

A related economic issue is the question of how to value the well-being of future generations. The practice of discounting is traditionally used in both the private and public sectors to evaluate the relative values of different cost or benefit streams over time. Clearly, this is a reasonable approach but the issue is more complex when decisions regarding the climate our grandchildren and great-grandchildren may inherit is compared with those decisions involving 10- or 20-year time horizons. That is not to say that a zero discount rate is the right answer by any means but, by the same token, it's hardly credible to use the 10 percent or the 7 percent real rates that many economists glibly endorse.

Recognizing that the environmental consequences of global warming is an area of significant uncertainty, work done at EPA and elsewhere suggests that the impacts could be dramatic. Millions of acres of coastal wetlands could be lost. Fertile agricultural states could become arid and unproductive. Essential climatic conditions for thousands of species of flora could move hundreds of miles north, with some species unable to follow quickly enough to avoid extinction. Eventually, new wetlands would be created, other states would become more fertile, and New Englanders might come to prefer flowering dogwoods in the spring to colorful maples in the fall. Such ecological renewal, however, may take considerable time to work itself out.

The question I raise is how do we value such large scale ecological changes? Do we simple tote up the measurable economic costs and benefits and essentially punt on the so-called "softer" effects? More research, including economic research, is needed to try to understand and, ultimately, to place values on environmental resources, particularly large systems, that may be threatened by global warming.

The final economic issue I will touch upon concerns the use of market mechanisms, including taxes, subsidies, and emissions trading schemes. While economists have long touted the virtues of these approaches, there is now a much greater openness among environmentalists and policymakers to using such mechanisms to achieve environmental objectives. Perhaps the clearest example of this is the proposed revision to the Clean Air Act (CAA) regarding trading of SO_2 emissions. Most observers believe that the use of economic incentives in the CAA has allowed a consensus to emerge around a more stringent environmental goal than would have been the case if economic incentives had not been used. Earlier this month, the Ways and Means Committee held hearings on the prospects for environmental fees and taxes—clearly a first. It is noteworthy that several environmental groups supported revenue neutral gasoline or carbon taxes, which, combined with the previously announced interest of at least one domestic automaker, has the makings of a curious coalition.

The idea of trading emissions reductions, among gases and/or nations, opens up two important possibilities. The first is the obvious opportunity to achieve any overall reduction target at lowest cost. The second, and perhaps more important internationally, is the possibility that trading would result in the movement of financial and technological resources from the developed to the developing world as countries such as the US purchase cheaper reductions in countries where the marginal cost of abatement is not as high. It is probable that any international agreement to limit emissions would include the possibility of trading, and may even establish mechanisms for facilitating it.

Conclusion

As we confront the challenge of anthropogenically induced global warming we see that the spilling forth of emission into the atmosphere may be altering life as we know it, and may be threatening the economic activity that has generated those same emissions. The pervasive nature of the emissions and the fact that they are controllable only through the concerted actions of dozens of sovereign governments and, ultimately, billions of individuals make this undertaking daunting at best. Yet humanity should not be suicidal as a species. Over the next few years we may well see concerted international action to restrain growth of greenhouse gas emissions. This would shape the environment for energy investment and could accelerate the commercialization of less polluting technologies. If we are to avoid a bumpy ride on this transition, it is essential that experts from various disciplines collaborate to design flexible approaches which will ignite rather than dampen the technological innovations and entrepreneurial initiatives essential to the resolution of these issues.

We should take from our experience with acid rain the realization that actions affecting the economic well-being of major groups require patience and ultimately a measure of good will to resolve. From the CFC experience we should take hope that the global community will be able to achieve a meaningful agreement for responding to common threats, and that the technical and business communities will be able to develop the necessary technologies and market them to global consumers at an acceptable price.

The program these next few days provides a much needed blend of interdisciplinary collaboration and augurs well for the long trip we are beginning. Thank you.

Keynote Address
Energy and the Environment in the 21st Century

United States Senator Albert Gore, Jr. (TN)
Washington, DC

Thank you very much. I'm very glad to be here. I congratulate the organizers of this conference for gathering a distinguished group of scientists who are really the leaders in these related fields. It's obvious from the roster that you've really got an outstanding lineup.

I just returned late last night from one of the most interesting experiences I have ever had, going under the north polar ice cap in a nuclear submarine with scientists who are using a new instrument for measuring ice thickness. As a member of the Armed Services Committee, I have been in negotiations with the US Navy for several months now to find ways to release the 30 years' worth of data on ice thickness. We're making real progress. It was interesting to me that trips under the ice cap were once primarily training missions designed to increase our proficiency in countering the threat from Russian submarine missile launchers. And while that is the continuing focus of these missions, the importance placed by the American people on understanding the polar regions and the climatic system of the entire earth is now overtaking the threat from the Soviet Union as a source of greatest concern. Of course, partly that is due not only to increased concern over climate change but also to a change in the political context within which this conference and all discussions of climate change now take place.

We have witnessed a rather extraordinary political earthquake in the last six months. That phrase "political earthquake" has a special meaning for me. My science subcommittee has, among other things, jurisdiction over real earthquake events and how we respond to them. Recently, I chaired a hearing where a scientist from the Soviet Union, their leading expert on the Armenian earthquake, joined a US earthquake expert from San Francisco and described exactly how tectonic plates create earthquakes. With two large plates of the earth pressing against one another, we see the pressure build up—actually we don't see it, it's obscured from view on the surface—but deep down the pressure builds up until one moves over the other, submerging it, sending out shock waves that knock down the buildings.

History is made not just of names, dates, and places, but also of large ideas that move through history the way those tectonic plates move through the crust of the earth. In Europe for the last 45 years, we have seen a large idea called democracy pressing against one called communism along the fault line that runs right through Berlin. And even though the changes have not been visible on the surface of the political landscape, beneath the surface in the hearts and minds of the men and women in the communist world, some very profound shifts have taken place, and just in the last few months. Suddenly, democracy moved over communism, submerging it, sending out shock waves that knocked down the Berlin Wall and all of the political structures in Eastern Europe.

I happened to be in San Francisco a few days after the World Series earthquake there. Of course if a novelist had written a plot that included a World Series between San Francisco and Oakland with an earthquake knocking out the bridge connecting the cities during the middle of the pre-game warm-up ceremonies, with families separated, the publisher would have rejected it out of hand as completely improbable. In any event, a few days after that earthquake, I saw in the Marina district of San Francisco some buildings that were still standing but had been knocked off their foundations. The police were deployed to prevent the occupants from going back inside the buildings because they weren't safe to use any more. The remaining political structures in the Soviet Union and throughout the communist world, I hope and pray in China as well, are like those buildings. They are still standing, but they're leaning a little bit. They've been knocked off their foundations and they're not safe to use any more. It really has been inspiring to hear young men and women, especially on university campuses throughout the world, quoting Thomas Jefferson, singing "We Shall Overcome," believing we shall overcome, and calling for a different way of governing their nations.

The environment has played a very large role in producing this political change in the communist world. I returned recently from eight days in the Soviet Union and found to my surprise that half of the newly elected members of the Supreme Soviet were elected on environmental platforms, and in Eastern Europe the most potent source of resistance and protest has been from environmentalists. Consider this description of an area around a Romanian town, provided by a reporter for the *New York Times*: "For about fifteen miles around, every growing thing in this once gentle valley looks as if it has been dipped in ink. Trees and bushes are black. The grass is stained. The houses and streets look like the inside of a chimney. Even the sheep on the hillsides are dingy grey. Two area factories there dump 30,000 pounds, tons of poison each year making this town with 6,000 residents one of the most polluted places in all of Europe." In the Ukraine, one of the 15 republics, or is it 14 this week, they dump more particulates into the air than eight times that emitted in the entire United States. At the Aral Sea, which Soviet environmentalists say is a worse catastrophe than Chernobyl, the fourth largest inland body of water in the world now has two thirds of its water gone because of a diversion scheme, and the fishing fleet is now 20 miles from the shoreline, sitting in the middle of what is now a desert, and the dead seabed that has been left now is caked with salt laced with pesticides. The regional climate has been changed and windstorms now lift 40 million tons of pesticide-laden salt into the air each year. It's been found in Iran and Turkey, as far north as the Arctic Circle.

These changes and the political changes which have just taken place in Eastern Europe illustrate something about change itself. We get used to slow, gradual change and we're often surprised when a sudden systemic change takes place. It's the same in science. When I was running for President, one interviewer asked, "What is your favorite color? What is your favorite movie? What is your favorite book?" I responded to the last question, "*The Structure of Scientific Revolution* by Thomas Kuhn." The reporter said, "No, I mean really, what is your favorite book?"

Twenty-three years ago, as a student at that other university down the street, one of my professors was Roger Revelle, teaching a course on population, but he had recently come from his work with the International Geophysical Year where he hired Dr. Keeling, the young researcher, and Roger Revelle was really the one person who first insisted upon measuring CO_2 in the atmosphere. And 23 years ago he presented to his undergraduates the first readouts from Mauna Loa, and you only have to see a few of those loops to start wondering how high up they're going to go. And I've watched those loops continue for the last 23 years and the pattern seems pretty clear to me. There is a sudden and systemic change that is taking place right before our eyes. It is not just the greenhouse effect or the hole in the ozone layer or the disappearance of the rain forest. It is a deeper change of which these other problems are manifestations.

What has changed fundamentally is the relationship between humankind and the ecological system of this planet. There are three causes of the change: the population explosion, the scientific and technological revolution, and a very old pattern of thinking that allows us to tolerate environmental vandalism on a global scale. But we are still in the stage right now where we're trying to decide among ourselves whether we really have a problem. I find that to be a source of puzzlement, really, because the pattern seems so obvious. I think one reason we have this debate over whether the problem is real is that we have difficulty telling the difference between sudden systemic change and slow gradual change, because our frame of reference is the experience of a single lifetime and what seems slow and gradual in the context of a single lifetime can be sudden and systemic and dramatic in the context of a geological age. People 500 years ago used to think with great conviction that the earth was flat, because their frame of reference was their own experience. You walk outside and look at the earth and it sure looks flat. One has to expand the context and enlarge the frame of reference in order to conceive of a new paradigm, a new model, a new shape for the earth. You can't see the roundness of the earth in your own experience just walking around. If you walk outside and it's a nice day you're tempted to conclude the environment's OK, because you have to have a very large frame of reference in order to envision the global environment and the threats that it faces from the present trend in human activities. I like to use an analogy to explain this—an old science experiment. If you place a frog into boiling water, it will jump out. But if the same frog is put in a pot of lukewarm water and it is slowly brought to a boil, the frog will just sit there until it is rescued. In the years I've told that story, it has become increasingly important to rescue the frog in the middle of the story. The point of the story of course is the frog's nervous system needs a sharp contrast in order to get the message. The slow gradual change in the temperature of the water obscures the danger that comes when a certain threshold is reached and there is a phase change in the pot of water and its consequences for the frog.

We're like that frog in the sense that the earth's environment is nearing a kind of boiling point and we're sitting dead in the water trying to figure out whether anything significant is happening. We faced the same thing with the emergence of nuclear weapons. Decades ago scientists and policymakers, mothers and fathers with children and grandchildren, people concerned about what then seemed like an imminent nuclear holocaust, together created a clock like a timer on a bomb ticking toward the unthinkable. Armageddon would come at 12 o'clock high. Over the years, the hands on that clock have been moved to within minutes of midnight. Now, as they seem to be moving backward, it's time to think about a new clock. My purpose here is to start that new clock and to amplify the sound of its ticking, to make us aware as a nation, and as a nation among nations, of the crisis and the opportunity before us, to make us aware of the urgent need for action, of the policies we must consider before this clock strikes, and of what it will take to move its hands backwards away from an environmental midnight. Still, however, the scientific community itself is engaged in this debate about whether the problem is real.

Economists have what they call "externalities"—things which don't fit neatly into the system of supply and demand and prices that one is dealing with at the time are shuffled into a big catch-all category called "externalities." I think scientists do also. As Thomas Kuhn illustrates, the shift from one side of the paradigm to another is marked by a phase in which the number of scientific externalities—observations that do not fit within the old paradigm—increases rapidly. For a while they can be dealt with by just labeling them "externalities," let's don't worry about them. Then the volume builds up to the point where the old paradigm is strained to the breaking point. Someone discovers a new paradigm that integrates and explains and contains and places in context all of the externalities from the old paradigm, and the new paradigm is adopted.

We are in the middle of a shift now in the paradigm we use to explain our relationship to the earth. One idea is pressing against another, one paradigm is exerting pressure on the other. And one reason why it is taking such a time to convince people is that those who are using the old paradigm to explain our relationship to the earth don't see the mosaic pattern contained in the reports coming in from around the world.

Let me try to put it in different terms. When I was a young reporter in Nashville, there was a weatherman just starting with one of the TV stations there named Pat Sajak. I watch his show now, "Wheel of Fortune," and I'm one of the few men in America who watches "Wheel of Fortune" mainly for Pat Sajak. His sidekick Vanna White turns up white cards that reveal letters underneath and the object for the contestants is to guess the well-known phrase with as few letters showing as possible. And I saw it recently when something happened that's not unusual on that show. Just enough letters were showing so that the audience had guessed the phrase and began to get frustrated that the contestants didn't see it yet. The contestants were locked in their competition and had tight minds and they just didn't see it. I think that's where we are on the crisis in the global environment. I think people at the grassroots level are beginning to see it pretty clearly and beginning to feel frustration that so many world political leaders just don't see it yet.

What are the letters that have been turned up that they have seen to put together this pattern? Well, the Exxon *Valdez* was one of them. The hole in the ozone layer above Antarctica was another and, as recent reports have shown, a smaller hole above the Arctic opened up last year. Hypodermic needles washing up on the beaches as children play in the surf, that's a card turned up. The garbage barge taking an 18-month tour of the Caribbean, then returning to New York no better for having made that long, hot journey. Hazardous waste being shipped in the same tanker trucks as orange juice on the return trip, that's a card turned up. Dead seals washing up on the beaches of the North Sea. Record temperatures in the last decade. The hottest January this year in the history of recorded January temperatures.

I don't know if you've ever tried to put together a jigsaw puzzle with your family around the dining room table, and at first the pieces are hard to place, and after a lot of them are in place the rest can be fitted in fairly easily and the process picks up speed. That's where we are. People are putting the pieces together themselves. Sometimes the scientific community can focus so much on the details, insisting quite properly on the degree of certainty that is customary in using the scientific method before reaching a "conclusion," that the overall pattern is obscured.

I remember and often cite an experience I had as a young student studying geography, when one of my classmates pointed to a map of the world in front of the room and pointed to the outline of South America, if you can see it in your mind's eye sticking out into the South Atlantic, and then pointed to the outline of Africa indented on the other side of the South Atlantic, and if you've ever seen those lines up close you've been struck by how closely they match, and my classmate asked the question to the teacher, "Did South America and Africa ever fit together like a jigsaw puzzle?" And the teacher confidently replied, "Of course not, that is ridiculous!" When the first geographers proposed that idea seriously, it didn't fit the old paradigm. And the scientific community said, "No, that's ridiculous!" There were debates up through the early 1960s when leading proponents of continental drift and tectonic plates were laughed off the stage of major scientific meetings. Scientists worked out the details of what would be necessary for continents to move and concluded it couldn't happen.

Yogi Berra once said, "What gets us into trouble is not what we don't know, it's what we know for sure that just ain't so."

There is an assumption today coming from the old paradigm used to describe our relationship to the earth that "just ain't so." The assumption is that the earth is so vast and nature so powerful, human civilization cannot possibly overwhelm any of its important interactions. That just ain't so. Continents can move, and do. We can profoundly alter the forces of nature and we are doing so right now. Flying over the tundra yesterday, I looked out at the edge of the horizon and saw a sight every single one of us has seen. The band of blue and the darkness above. From space the astronauts have taken pictures showing that same bright blue band and the stark contrast it poses to the vastness of the earth itself. The thickness of the atmosphere is one one-thousandth the diameter of the earth. Now there are some specialists here who will have a definitional problem on the atmosphere. But if you go up to half the atmosphere's weight, it's about three and half miles and then 30 miles to the top of the stratosphere. The total volume of the earth's atmosphere is small enough that we now have the capacity to profoundly alter its makeup. The air in this auditorium here, which we're breathing into our lungs, has six times as many chlorine atoms in each lung-full as it did 40 years ago, or three billion years ago in this space. It's remarkable to think that in just 40 years we could take one family of chemicals and alter the concentration of chlorine atoms in the atmosphere of the entire earth by 600 percent, evidence that we can indeed change the makeup of the atmosphere. Now chlorine atoms in that concentration don't hurt human health, at least directly, as far as we know, but they do burn a hole in the ozone layer and thin it out all over the earth. Similarly with carbon dioxide, we are changing the concentrations dramatically.

In Antarctica, where I visited a year and a half ago, the Soviets and the French have produced what to me is the single most compelling bit of evidence from the ice cores at Vastok. If you go back over 160,000 years and look at the concentrations of CO_2, they fluctuated between 200 and 300 parts per million—200, or roughly 180 to 200, parts per million during the last ice age and the ice age before that, and 300 parts per million during the period of great warming between the ice ages. We're now at 353 parts per million, and within 37 years, some say 75 years, we're going to 600 parts per million. The same record from Vastok shows that temperatures went up and down in lockstep with the CO_2 fluctuations.

Somebody will show that graph during your conferences here. I'd like you to perform a mental exercise when you see that graph. Take the CO_2 line and extend it up to 600 and see if it doesn't jar your thinking about what's appropriate in our interaction with the ecological system of the earth.

The one trend that to me shows this dramatic change more clearly than any other, the change in our relationship to the earth, is the population graph. If you draw a graph of population that defines human population growth from the beginning of the species until the present time, first you have to decide where to start. I don't want to start a religious argument, certainly not at MIT, because some people think it was only a few thousand years ago. Some say a million years ago. Again, you have a definitional problem. But if you start a million years ago at the edge of this stage with two people, and you know who they were, and you draw a line of human population to the other end of the stage, the first 500,000 years showed virtually no change. Then over the next 350,000 there's still virtually no change. In between the last two ice ages you had very small change. But then after the last ice age, about 45,000 years ago, you start getting some change. Fifteen thousand years ago, when agriculture appeared, you start getting some noticeable growth in the human population. By the time of Christopher Columbus's voyage 500 years ago, there were 500,000,000 people on earth. By the time Thomas Jefferson wrote the Declaration of Independence, there were 1 billion people on earth. By the end of World War II, there were 2 billion people on earth, a little more. Now we've come a million years, tens of thousands of human lifetimes, to 2 billion people. In the last 40 years we've gone from 2 billion to 5 billion. And in the next 40 years, we're going from 5 billion to [more than] 6 billion. Now that trend tells us something. Looking at it in its totality do you notice anything different about this part of the pattern?

If only population followed that pattern it wouldn't be a problem, but look at the graphs on methane and ask about oxidation of the atmosphere. You can go back 10 thousand years on methane and find exactly the same concentrations until our century, roughly, and then since World War II the figures on increases have been phenomenal. Look at the figures on loss of ozone in the stratosphere. The pattern is exactly the same. The loss of rain forest land and forest land generally, the pattern is exactly the same. Soil erosion—we've lost, well, 50 percent of the topsoil in Iowa has flowed past Memphis in the last hundred years. The pattern is exactly the same worldwide. The loss of living species, there the flat part of the line goes back not one million years but 65 million years all the way out of the auditorium, across the street and to the next building, and then suddenly in our century it jumps straight up at a rate 1,000 times greater than ever before. Depending on how you count the number of species in the rain forest, we could lose as many as half of all the living species and half of all the genetic information on earth within the lifetimes of people in this auditorium.

Now I said there were three causes of this problem. Let me take them one by one, because there are solutions in each of the three areas. First, let's take technology and especially energy. Technology shapes thinking. Marshall McLuhan said, "The medium is the message," but somebody else put it more pointedly in saying that if the only tool you have is a hammer, every problem begins to look like a nail. By choosing technologies we crystallize thought into substance and that substance contains an aura of potentials, of potential and possibilities, and our thinking begins to take the form of the technologies we use to modify the world around us. People fought wars for thousands of years, but when nuclear weapons were invented and the technology of war changed, we had to begin thinking about war differently. Similarly, we have gained sustenance by exploiting the earth and its resources for thousands of years. But in this century, the same scientific and technological revolution that produced nuclear weapons has now given us new technologies for exploiting the earth, which when taken together have consequences so different from previous patterns of exploitation that we have to begin thinking differently about our relationship to the earth and the means by which we exploit the earth to gain sustenance. We could have a much larger population, comfortably, on this earth, if we had different technologies and a different way of thinking about our relationship to the earth. We need new technologies and we can't wait 100 years to develop them.

I have proposed a Strategic Environment Initiative, consciously drawing upon the Strategic Defense Initiative (SDI) as a precedent, not because I like SDI, I don't, but I've seen what it has accomplished in focusing effort and resources, and we need that level of commitment, that level of resources or more, and that kind of focus to develop new technologies to substitute for the ones that have been associated with environmental destruction. The United States should take the lead in accelerating and disseminating these new technologies around the world. We need, secondly, centers of training throughout the Third World to develop a cadre of individuals and nations that can use these new technologies. Third, we need new market signals to accurately measure the consequences of the technologies and the fuels that we use. National income accounting, for example, should be changed in order to reflect the depreciation of natural resources. We need to accurately measure the cost of fuels that inject CO_2 and other pollutants into the atmosphere, and I've called for a CO_2 tax. We need to encourage conservation and efficiency in all of its forms. Next, I propose an Environmental Security Trust Fund to penalize technologies which help to destroy the environment and subsidize the purchase of alternatives and substitutes which allow us to achieve sustainable growth without destroying the environment. Next, we need a virgin materials fee to level the playing field between recyclable commodities and those based on this continuing pattern of unsustainable consumption. We need to ban certain chemicals: chlorofluorocarbons are first on the list but others will follow. Next, we need to begin planning now for systemic and structural changes to investigate substitutes for the internal combustion engine in uses which are important to our society. We all take 4,000 pounds of metal with us everywhere we go and it's hard to think about not doing that. But if we are going to continue that pattern we need new technologies to make it possible.

The second cause, after technology, is population. We need to have birth control technologies widely available throughout the world but we need to focus on politics and government as a technology itself and understand what the experts on population have been telling us about how to stabilize world population. The availability of birth control is a prerequisite, but much more is needed. We need a global effort to lift child survival rates above the threshold where decisions about family size are affected. The studies indicate that there are two prerequisites for the demographic transition. The first is a child survival rate that makes parents comfortable with the notion that two children will have a high likelihood of living to adulthood to take care of them in their old age. The second prerequisite is a level of education, particularly among women, to empower them to participate in the choice. It is not presently imaginable, but we need a global Marshall Plan, fueled with Japanese and European capital as well as our own, to focus on lifting child survival rates and education levels all over the earth, to the threshold necessary to accomplish the demographic transition. Thirty-seven thousand children under the age of five die every single day because of starvation and simple maladies like diarrhea and as they die, the population explodes in the countries where they live. We must also focus on the links between this issue and social justice. Debt for nature swaps are good ideas. Robert MacNamara has called the flow of money from the Third World to the industrial world, which now exceeds all of the aid in the opposite direction, a blood transfusion from the sick to the healthy.

After technology and population, comes the third and final cause of the problem and the area where most of the change is needed and that is in the area of thinking. We human beings are a kind of technology. In the debate years ago about appropriate technology, questions were asked like, "Is a nuclear power plant appropriate for a Third World country?" Now the question is, "When God created human beings did God choose an appropriate technology for dominion over the earth?" The jury is still out and we must provide the answer. I saw an episode of "Saturday Night Live" not long ago where they used a make-believe commercial for a product called "The Catapulto," a fairly large consumer device that you buy and put in your backyard. It looks like one of the old medieval catapults, but it's a miniature version, spring loaded, very powerful, and according to this commercial, when you take out your garbage instead of worrying about landfills or incineration, just put it in the Catapulto, and pull the lever and throw it over into your neighbor's yard a few houses away. It was a very funny skit. We laugh at that, but that's what our civilization is doing. That's why the hazardous waste is in the orange juice trucks. That's why the garbage barge is taking its journey. And if we're not putting it into the next city or the next state, we're putting it into the future. What has to change is our definition of the word "backyard." In political terms, if a landfill is proposed near somebody's backyard, every other issue becomes secondary, and that's the main focus of concern. When the air we breathe has 600 percent more chlorine atoms, it illustrates the fact that now the entire globe is our backyard. I believe we are very close to a change in political awareness that will give these global environmental issues the same kind of political significance that a landfill siting controversy does at the local level. We have lost our "eco-librium" if you will, and people are attempting to force changes which will allow us to regain it.

Psychologists have a clinical term they call denial.

When an addict lives his life around a pattern that is focused on consumption of a controlled substance or a substance that he abuses, the first reaction when confronted with the pattern is to deny that it exists. Part of our problem in changing the pattern of thinking about our relationship to the earth has to do with this phenomenon of denial. We are addicted to a pattern of consuming the earth's resources at an unsustainable rate and we refuse to recognize the destructive aspects of the pattern. We turn to science for help but science sometimes is part of the problem. The scientific method of observing the earth, experimenting with it and learning from the results of the experiment is part of a very old pattern of thinking that defines us as separate from the earth. We are witnessing a kind of Heisenberg Principle writ large. In developing new science and new technologies, we have changed our relationship to the earth upon which we are experimenting. A philosopher named Ivan Illich said, "We are searching for a language to describe the shadow our future throws." We are part of the web of life, not separate from it. The percentage of salt in our bloodstreams is precisely the same as the percentage of salt in the world's oceans.

What is needed are new lines on the graph I drew across the stage. One that will define a dramatic systemic change in public awareness and take it straight up, and then a change in leadership to help us confront the problem on a global scale. One scientist who has been active in this field for many years said to me recently after listening to this list of proposed solutions I've just offered to you—a global Marshall Plan, a Strategic Environment Initiative, an Environmental Security Trust Fund, and other proposals—and he said, "You know, Al, as I listen to that, I have to tell you, it's unlikely that we're going to be able to confront this problem." And I said, "Wait a minute. What if I had asked you six months ago, how likely is it that within the next few months every single nation in Eastern Europe will abandon communism, pull down the statues of Lenin, embrace democracy and capitalism and leave the Soviet orbit for the west. Would you have said that's unlikely?" He said, "Yes, I would have." What changed was their thinking in Eastern Europe. What was unlikely became likely, what was impossible became imperative, what was unimaginable was done. The Berlin Wall came down in their minds before the first chisel hit the first stone.

As we shift from one paradigm to another and change our way of thinking about the relationship between humankind and the ecological system of this planet, as we change in the political context the definition of the backyard we want to protect and preserve for future generations, then what is today unimaginable will become imaginable. What seems impossible will be done. The obstacles, I concede, appear immovable, but so did the Berlin Wall. As we begin to confront this task, let us take some comfort from the certain knowledge that "We Shall Overcome." Thank you very much.

Plenary Session I

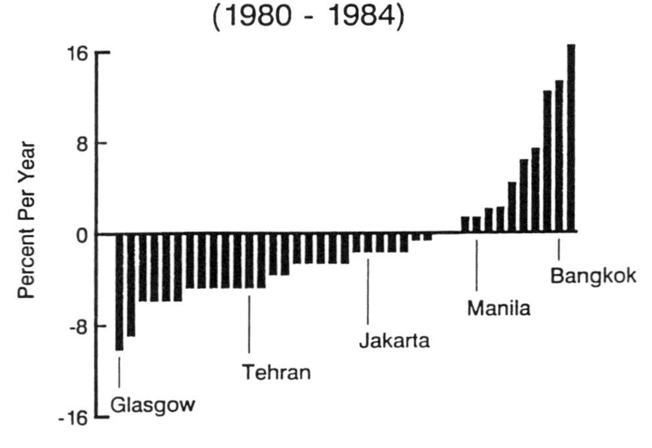

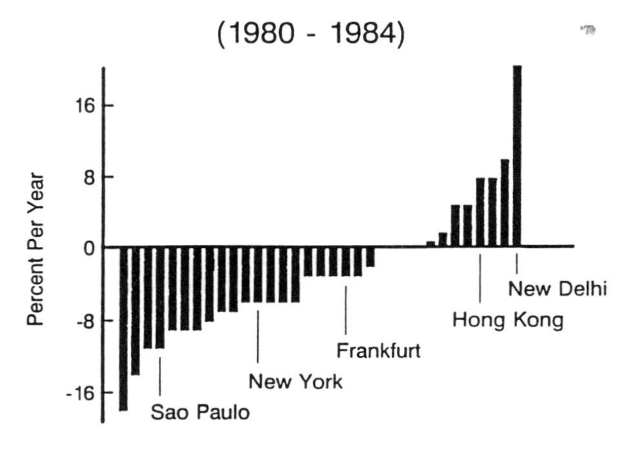

Figure 1. Trends in annual averages of air pollutant concentrations measured at multiple sites within cities throughout the world for the period 1980–84. Each bar represents a city, several of which are identified. Few cities in the developing world have been extensively monitored and many therefore do not appear in this figure; many would be expected to have high levels of particulate matter, especially if they make extensive use of high-sulfur coal for heating and energy generation.

developing countries are not yet monitoring concentrations of atmospheric trace species, but it seems likely that most of them presently have increasing levels of emissions and decreasing air quality.

The other pollutant for which substantial data are available is sulfur dioxide (SO_2). In more than a quarter of the cities, the annual average SO_2 concentrations exceed that thought to have a high potential for severe health and corrosion impacts, 16–24 parts per billion. The difference between the combined site averages in the 54 cities surveyed is nearly a factor of 100. The half-dozen cities with the highest annual averages include locations in Europe, Asia, and South America.

Trends in the ambient annual average levels of SO_2 are shown in Figure 1b. Of 33 cities, 27 have downward or stationary trends, reflecting strong efforts to reduce emissions of sulfur gases. However, 6 cities have upward trends, including locations in Asia, Australasia, and Europe.

A more detailed overview is possible for the United States, since the concentrations of a number of atmospheric trace species have been monitored regularly for more than a decade at many state or federally supervised locations. The information has been collected and summarized on a national basis since 1975 and is presented in a form in which typical concentrations, their ranges throughout the differing monitoring sites, and their long-term trends can be readily seen.

The US urban trends in ozone over a five-year period are shown in Figure 2a. (The data are for the annual second highest daily maximum one-hour concentration, since that is the way the air quality legislation was written.) The 1981–85 period is chosen because it was one in which major efforts were made to limit emissions of volatile organic compounds and of nitrogen oxides, two chemical precursors of atmospheric ozone. The chemical relationships are complex, and it is apparent that, over this period at least, reductions in precursor emissions had little significant effect on ozone peak values.

Perhaps the most dramatic evidence for improvement in an air quality component as a result of legislation is that for airborne lead in US urban areas. Figure 2b shows the ambient lead data for the past decade. It is easy to see that the substantial reduction in leaded gasoline use that has occurred during this period has resulted in sharply decreasing atmospheric lead concentrations. As of 1983, the mean lead level was about 3 $\mu g/m^3$.

The message from these examples of trends in local and regional air quality is clearly mixed. In many cities of the world, air quality is improving. In others, however, deterioration is taking place. Where vigorous action to reduce emissions has occurred, definite improvements are generally noted, as in the case of lead in US cities. Ozone remains the exception, its complex chemical generation mechanisms precluding improvements

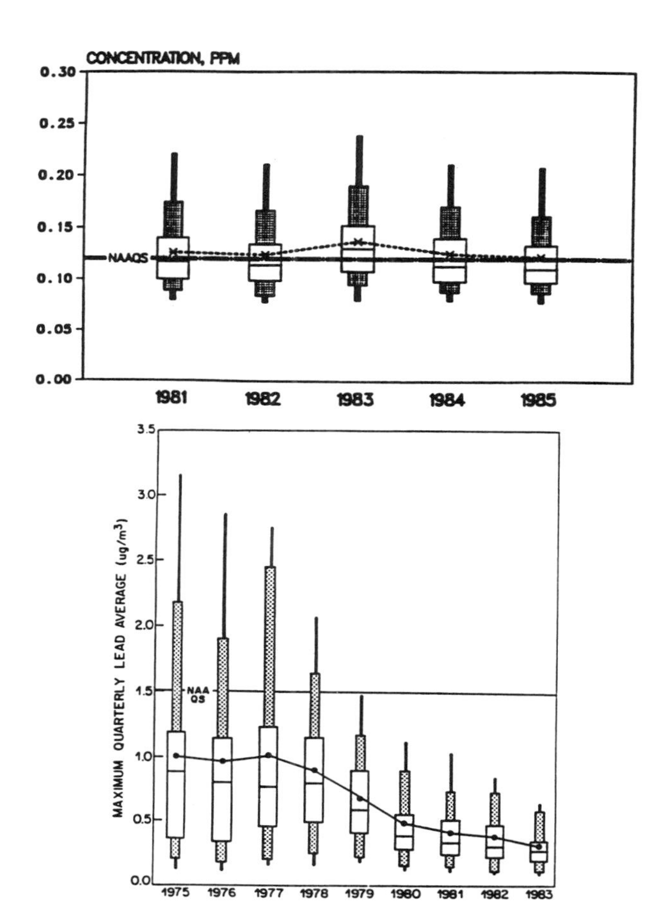

Figure 2. Boxplots of trends in US urban air constituents in annual mean concentrations. The mean (dot) and median (bar) values are indicated within the box, whose top and bottom boundaries are the upper and lower quartiles in the data. The shaded boxes extend to the 10th and 90th percentiles. The top and bottom bars extend to the 5th and 95th percentiles. (a) Ozone, annual second highest daily maximum one-hour levels, 1981–1985, 523 sites; (b) Lead, maximum quarterly average concentrations, 61 sites, 1975–1983. NAAQS indicates the national ambient air quality standard (EPA, 1985).

by relatively simple emissions reductions of some of its precursor chemicals.

Space and Time Scales

Air quality is often discussed in fragmentary fashion, as if the sources for one atmospheric species bore no relationship to the sources for another, and as if the air quality problems on a local or regional scale bore little relationship to problems on a global scale. These assumptions, generally implicit rather than explicit, are easily demonstrated to be false. Figure 3 illustrates the point. To construct the figure, we have taken mean atmospheric lifetimes for a few species of concern, permitted the species to be advected at a typical wind speed of 5 m/s, and computed the mean distance that each species travels before it is removed from the atmosphere by the predominant chemical or physical loss process. On the Figure we indicate as well a few distances of interest: Newark, New Jersey, to Boston, Massachusetts, probably the upper limit of what one might term the regional scale; San Diego, California, to Boston, probably the upper limit of continental scale distances; and the earth's circumference, the ultimate global scale indicator. Nitrogen oxides, which react rapidly, are clearly best characterized as locally acting trace constituents. Sulfur dioxide is regional in spatial scale. Lead and hydrogen chloride are subcontinental or continental in scale, while the scale for carbon monoxide (CO) is of the order of the hemispheric scale. Methane (CH_4) and the chlorofluorocarbon $CFCl_3$ (CFC-11) clearly exceed the global scale. The atmospheric concentration patterns of the species are consistent with their typical spatial scales, NO_x being the most variable of the species in Figure 3, and $CFCl_3$ the least.

To modify the concentration of any of these species, emissions must be reduced at the sources. At this conceptual stage it becomes evident that atmospheric impacts that sort themselves neatly on an intellectual level do not do so on a practical level. This is demonstrated by Table 1, which shows whether a particular atmospheric species is emitted by a specific source. As can be seen, most sources (electrical energy generating sources

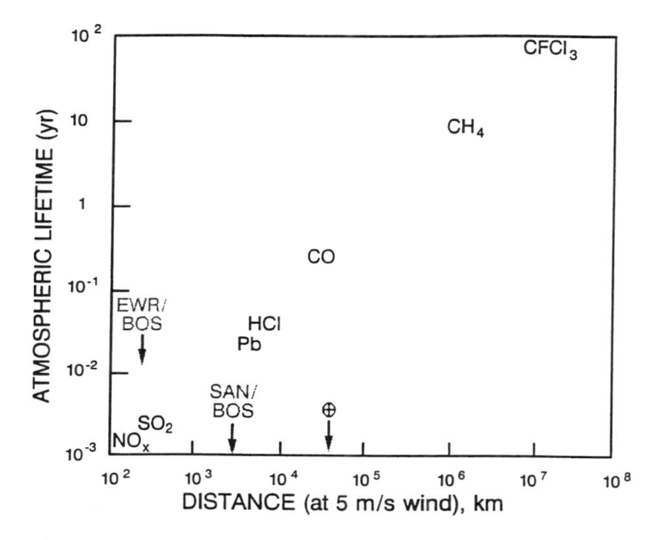

Figure 3. Time and space scales for selected species emitted from anthropogenic activities. The symbol ⊕ indicates the earth's circumference.

Table 1. Substances emitted from specific sources.

Emittant	Coal	Petroleum	Natural Gas
CO_2	x	x	x
CO	x	x	
NO_x	x	x	x
SO_2	x	x	
CH_4	x	x	x
HCl	x		
Pb			
$CFCl_3$			

are emphasized here) emit a spectrum of products, with impacts on all spatial and time scales.

A Synoptic Framework for Ensemble Assessment

A synoptic approach to the assessment of environmental impacts is one that views impacts from a common framework so that they may be compared with each other and so that a perspective may be reached concerning the multiplicity of impacts. The design for such a framework (Crutzen and Graedel, 1986; Graedel, 1989) seeks to establish the causal relationships between "critical environmental properties" and potential sources of environmental change. Critical environmental properties, in the sense used here, are those few attributes of the environment that parties involved in the assessment agree to be crucially important. Which environmental properties are valued in a particular case will depend upon specific social, political, and environmental circumstances, but in any case the list represents an explicit value judgment defining the terms in which users are willing to describe the environmental aspects of "what kind of planet they want." One of the clearest lessons from assessment experience of the last decade is that unless some definite—and preferably short—list of critical environmental properties is specified as a focus for assessment, the subsequent analysis is unlikely to be very useful for comprehensive policymaking.

For purposes of the present discussion, we have selected such a set of critical atmospheric properties (see Table 2). The goal is to understand the relationships between these critical properties of the atmosphere and the natural fluctuations and human activities that might be sources of significant change in them. Recent advances in understanding of atmospheric chemistry and its interactions with climate and biosphere now begin to allow specification of such relationships in terms of fundamental biological, chemical, and physical processes. Present knowledge regarding the critical atmospheric properties affected by changes in specific atmospheric chemicals is given quantitative expression in Figure 4, where both direct and indirect chemical effects are indicated. Thus, changes in ozone concentrations are shown to directly affect the critical atmospheric component "ultraviolet energy absorption" because the ozone molecules themselves have the ultimate impact. Halocarbons and nitrous oxide, though surely relevant to ultraviolet energy absorption, are shown to affect this critical atmospheric component indirectly, because their action occurs by changing the concentration of ozone. It is important to note from Figure 4 that a significant number of chemicals are involved in multiple impacts. Sources of disturbance or international policies that affect these chemicals must therefore be assessed in terms of multiple kinds of impacts on the atmosphere. Present knowledge regarding the sources of emission of specific atmospheric chemicals is given qualitative expression in Figure 5. The Figure shows that a number of atmospheric chemicals are affected by many different kinds of sources.

Figures 4 and 5 can be combined to provide a synoptic framework for atmospheric assessment. One begins with a critical atmospheric component like "precipitation acidification" and its immediate chemical causes (Figure 4), and identifies the sources responsible for initiating those interactions (Figure 5). The result is Figure 6, a matrix that shows the impact of each potential source of atmospheric change on each critical atmospheric component. The assessment is qualitative, reflecting to some extent the present state of knowledge. It also includes estimates of the reliability of that knowledge, an important component of such an assessment effort.

The simplest atmospheric impact assessments involve only a single cell of the matrix. A typical example is the study of the impacts of a single source, such as a new coal-fired power station, on a single critical atmospheric component, such as precipitation acidification (location *a* in Figure 6). More complex atmospheric assessments have addressed the question of aggregate impacts across different kinds of sources. A contemporary example is the study of the net impact on the earth's thermal radiation budget caused by chemical perturbations due to fossil fuel combustion, biomass burning, land-use changes, and industrialization (e.g., locations *b* in Figure 6). Even more useful for the purposes of policy and management are assessments of the impacts of a single source on several critical atmospheric properties. The simple study noted above would fall into this category if the impacts of coal combustion were assessed not only on acidification, but also on photochemical oxidant production, materials corrosion, visibility degradation, etc. (e.g., locations *c* in Figure 6). If desired, the columns could be summed to give the net impact of the ensemble of sources on each critical property. (Examples of column assessments include National Research Council [1983] and Bolin et al. [1986].) Similarly, the rows could

Table 2. Definitions of critical atmospheric properties.

Critical Properties	Definition
Ultraviolet energy absorption	This property reflects the ability, especially of stratospheric ozone, to absorb ultraviolet solar radiation, thus shielding the earth's surface from its deleterious effects. This property is commonly addressed in discussions of the "stratospheric ozone problem."
Radiation budget alteration	This property reflects the complicated processes through which the atmosphere transmits much of the energy arriving from the sun at visible wavelengths while absorbing much of the energy radiated from the earth and its atmosphere at infrared wavelengths. The balance of these energy flows, interacting with the hydrological cycle, exerts considerable influence on the earth's temperature. This property is commonly addressed in discussions of the "greenhouse problem."
Photochemical oxidant formation	This property reflects the oxidizing potential of the atmosphere, produced by a variety of highly reactive gases, especially tropospheric ozone. In this paper, the emphasis is on local-scale oxidants that are often implicated in smog-related problems, such as asthma, crop damage, and degradation of works of art.
Precipitation acidity	This property reflects the acid-base balance of the atmosphere, as represented in the chemical composition of rain, snow, and fog.
Visibility degradation	Visibility is reduced when light of visible wavelengths is absorbed or scattered by gases or particles in the atmosphere.
Materials corrosion	This property reflects the ability of the atmosphere to corrode materials exposed to it, often through the chloridation or sulfurization of marble, masonry, iron, aluminum, copper, and other materials.
Oxidation (cleansing) efficiency of the atmosphere	This property reflects the efficiency with which gases emitted into the atmosphere are broken down by reactions with the hydroxyl radical ($OH\cdot$) or other oxidizers. Increasing concentrations of CO, CH_4, and perhaps NO_x are thought likely to decrease oxidation efficiency by reacting with and removing $OH\cdot$.

be summed to give the net effect of each source on the ensemble of properties. (Examples of row assessments include National Research Council [1979] and National Research Council [1981].)

Figure 6 shows that the sources of most general concern, as indicated by their impact ratings, are almost wholly anthropogenic in nature: fossil fuel combustion, biomass combustion, and industrial processes. Emissions from crop production, especially CH_4 from rice paddies, and from estuaries near heavily populated areas may have future impacts. Some sources (the oceans, domestic animals, and vegetation) have sufficiently small effects on atmospheric processes that they need not cause much concern, even if their emissions should increase.

Among the most troublesome interactions between development and environment are those that involve cumulative impacts. In general, cumulative impacts become important when sources of perturbation to the environment are grouped sufficiently closely in space or time that they exceed the natural system's ability to remove or dissipate the resultant disturbance. The basic data required to structure such assessments are the characteristic time and space scales of the atmospheric constituents and development activities. For example, perturbations to gases with very long lifetimes accumulate over decades to centuries around the world as a whole. Today's perturbations to those gases will still be affecting the atmosphere decades hence, and perturbations occurring anywhere in the world will affect the atmosphere everywhere in the world. Long-lived gases tend to be radiatively active, thus giving the "greenhouse" syndrome its long-term, global-scale character. At the other extreme, heavy hydrocarbons and coarse particles, being short-lived, drop out of the atmosphere in a matter of hours, normally traveling a few hundred kilometers or less from their sources. The atmospheric properties of visibility reduction and photochemical oxidant formation associated with these chemicals thus take on their acute, relatively local character. Species with moderate atmospheric lifetimes include a group of chemicals associated with the acidification of precipitation, all with characteristic scales of a few days and a couple of thousand kilometers.

These concerns, together with the evidence that the concentrations of many chemical compounds in the biosphere are increasing, indicate that additional extensions to the conceptual framework are needed to provide it with spatial and temporal dimensions. The effort described earlier was focused on present-day impacts. One can envision, however, preparing separate versions of the framework shown in Figure 6 for global, regional, and local interactions, and for different epochs in time.

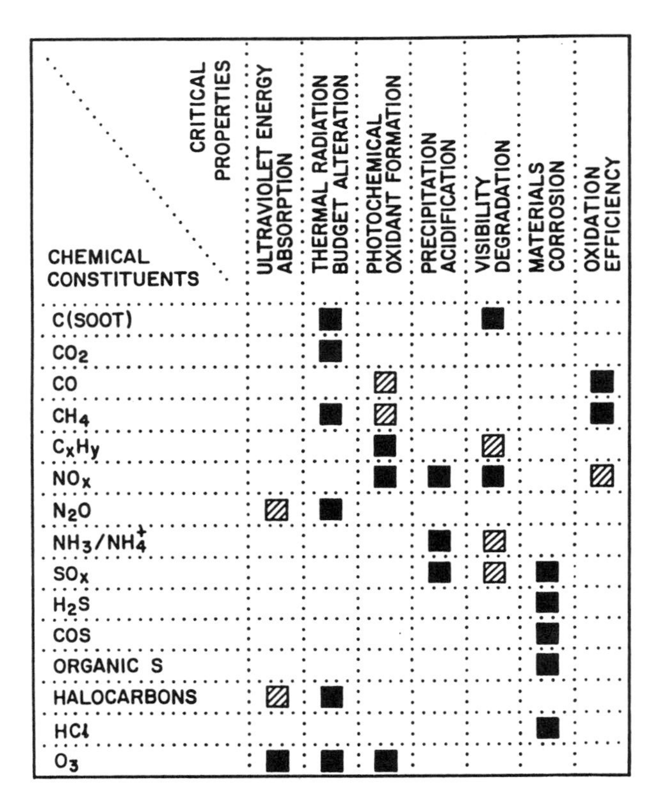

Figure 4. Major impacts of atmospheric chemistry on critical atmospheric properties. The squares indicate that the listed chemical is expected to have a significant impact on the listed property of the atmosphere. Solid squares indicate that the impact is direct, hatched squares that it is indirect. Definitions of the atmospheric properties are given in Table 2.

As suggested in Figure 7, this might consist of multiple versions of Figure 6 created to reflect "slices of time" through the evolving conditions at 25- or 50-year intervals. The result, it is hoped, will help to put the changing character of interactions between human activities and the environment into a truly synoptic historical and geographical perspective. This historical dimension may require assessment of what kind of human activity was being carried out at specific times in the past, and how much and what kind might be conducted in the future.

Extending the Assessment to Different Regimes

It is important to recall at this point that the previously presented assessments have been limited to impacts of development on the atmosphere. In fact, other regimes in the biospheric system are important as well. For ex-

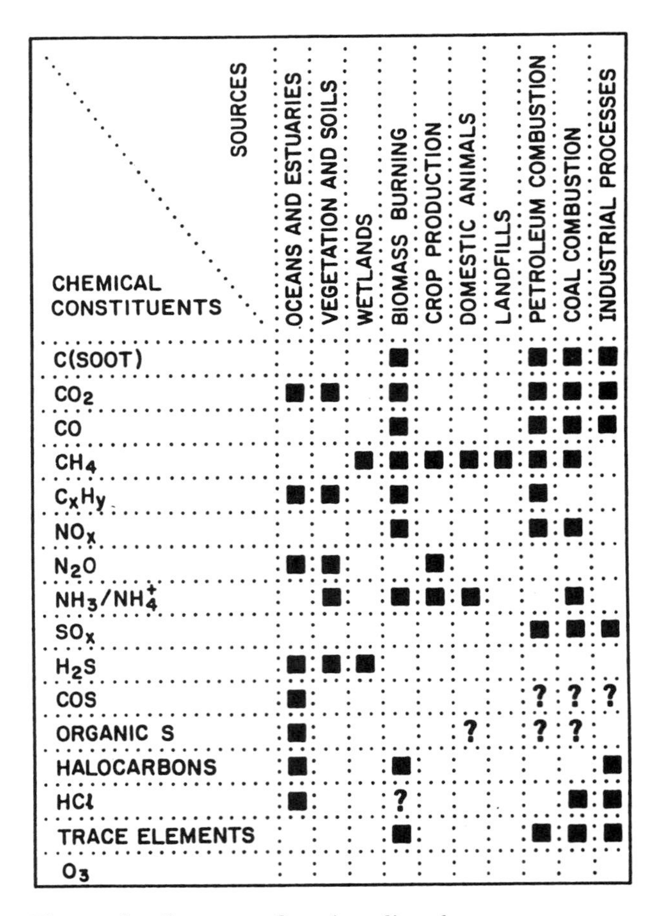

Figure 5. Sources of major disturbances to atmospheric chemistry. The squares indicate that the listed source is expected to exert a significant effect on the listed chemical. A question mark indicates that emission of a significant amount of a compound from a particular source is uncertain.

ample, concerns for acidification impacts through time are based not on the accumulation of relevant chemicals in the atmosphere, but rather on their accumulation in other media, such as soil and water. Expanding synoptic assessment frameworks to contend with additional environmental and developmental dimensions is a major part of sustainable development research. Beginnings have been made for water systems (Douglas, 1976) and soil systems (Harnoz, 1986), but much is yet to be done.

A relatively straightforward addition to the above framework would be one or more critical environmental properties that reflect the role of atmospheric chemicals as direct fertilizers or toxins for plants. Such a modification would allow the integrated treatment of such phenomena as the stimulation of plant growth by carbon dioxide and its inhibition by sulfur oxides—both products of fossil fuel combustion. Somewhat more ambi-

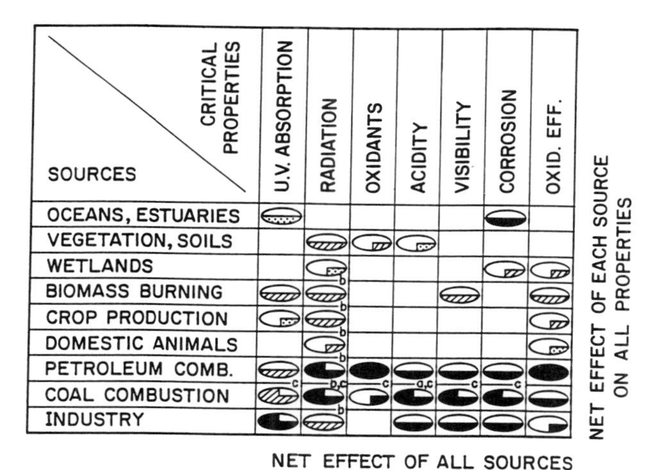

Figure 6. An initial ensemble assessment of impacts on the global atmosphere. The critical atmospheric properties defined in Table 2 are listed as the column headings of the matrix. The sources of disturbances to these properties are listed as row headings. Cell entries assess the relative impact of each source on each component and the relative scientific certainty of the assessment. Column totals would, in principle, represent the net effect of all sources on each critical atmospheric property. Row totals would indicate the net effect of each source on all critical atmospheric properties. These totals are envisioned as judgmental, qualitative assessments rather than as literal, quantitative summations. The significance of the letters in the cells is described in the text (Crutzen and Graedel, 1986).

tiously, the approach could be expanded beyond its present chemical focus to include the appropriate physical and biological processes, and the sources of disturbance to them. Ultimately, the need is for a qualitative framework that puts in perspective the impacts of human activities and natural fluctuations, not just on the atmospheric environment, but also on soils, water, and the biosphere as a whole.

Although a comprehensive treatment has yet to be made for other regimes, it is possible to envision one way in which such an assessment might be accomplished. The first step would be to select regimes of interest other than the atmosphere, such as oceans or soils. The second step would be to use the same analytical approach taken for the atmosphere to determine the matrix elements on a source-impact diagram similar to that of Figure 6. A possible example for soils is shown in Figure 8. The matrix elements are then summed in some appropriate

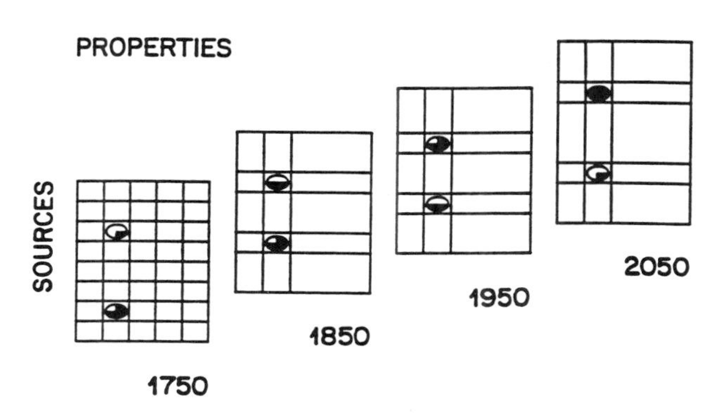

Figure 7. A history of disturbances to the atmosphere, as might be expressed through a time series of source-impact matrices such as that shown in Figure 6. This display suggests how two of the matrix elements would be evaluated at each of the "time slices," perhaps becoming more or less significant with time, as indicated here. The full assessment would include such an evaluation for each individual matrix element at each time slice.

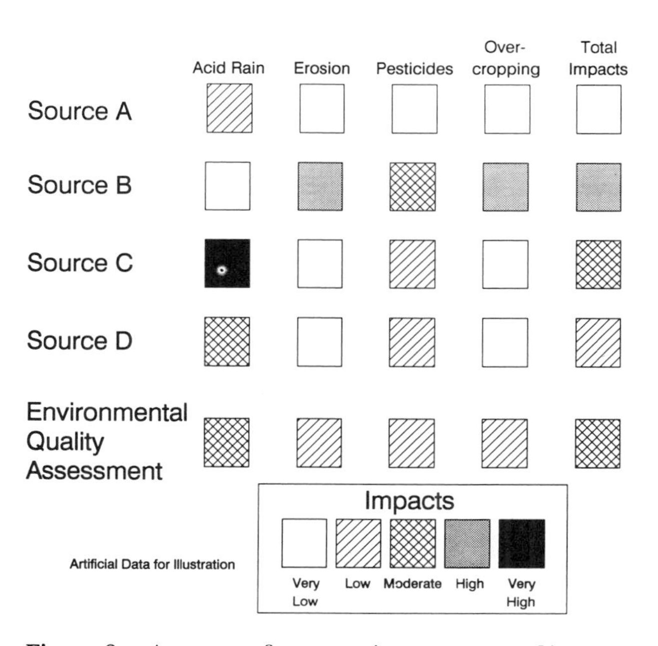

Figure 8. A concept for synoptic assessment of impacts on global soils. The data are artificial, and appear for didactic purposes only. This figure for soil impacts is analogous to Figure 6 for atmospheric impacts.

fashion to give a total regime impact for each source; these appear in the right column of Figure 8. The same sequence of operations is repeated for each regime of interest, not only for the present and perhaps for the past, but for future times as well.

In the final ensemble display, the total impact columns from each regime display are extracted and combined, producing a display of the type shown in Figure 9. A sequence of such diagrams would be used to illustrate progress or retrogression over time.

Emissions Projections

The comprehensive assessment techniques presented in Figures 4 through 9 have not been utilized in any consistent and rigorous way thus far, not even in the case of the atmosphere. To indicate the potential usefulness of some of these approaches, however, we have completed a simpler operation: that of assessing for the present epoch, and projecting, under a specified set of development assumptions, the emissions (rather than the impacts) of a number of trace gases as a consequence of electrical energy generation. We selected seven countries, spanning the range of highly to slightly developed, searched out their present populations and electrical generating capacities, and made the crucial but perhaps not unrealistic assumption that by the year 2025 every citizen in each country would be provided with electrical energy at half the rate consumed by today's average Italian. Some of the pertinent figures involved in making this assessment are given in Table 3. Two items are of special note. One is the great disparity in energy use per capita, a multiplier of about 100 separating the high and low extremes. The second is the enormous population increases anticipated in several of the countries, populations between two and four times those of today being anticipated for India, Kenya, and Nigeria only 35 years from the present.

The computation of emissions intensities as a consequence of electrical energy generation in each of the countries is performed by solving the following equation for each species i:

$$E_i = \left(\sum_j [M_j \cdot \epsilon_{i,j}] \right) \Big/ A$$

where M_j is the annual use of fuel j in the country; $\epsilon_{i,j}$

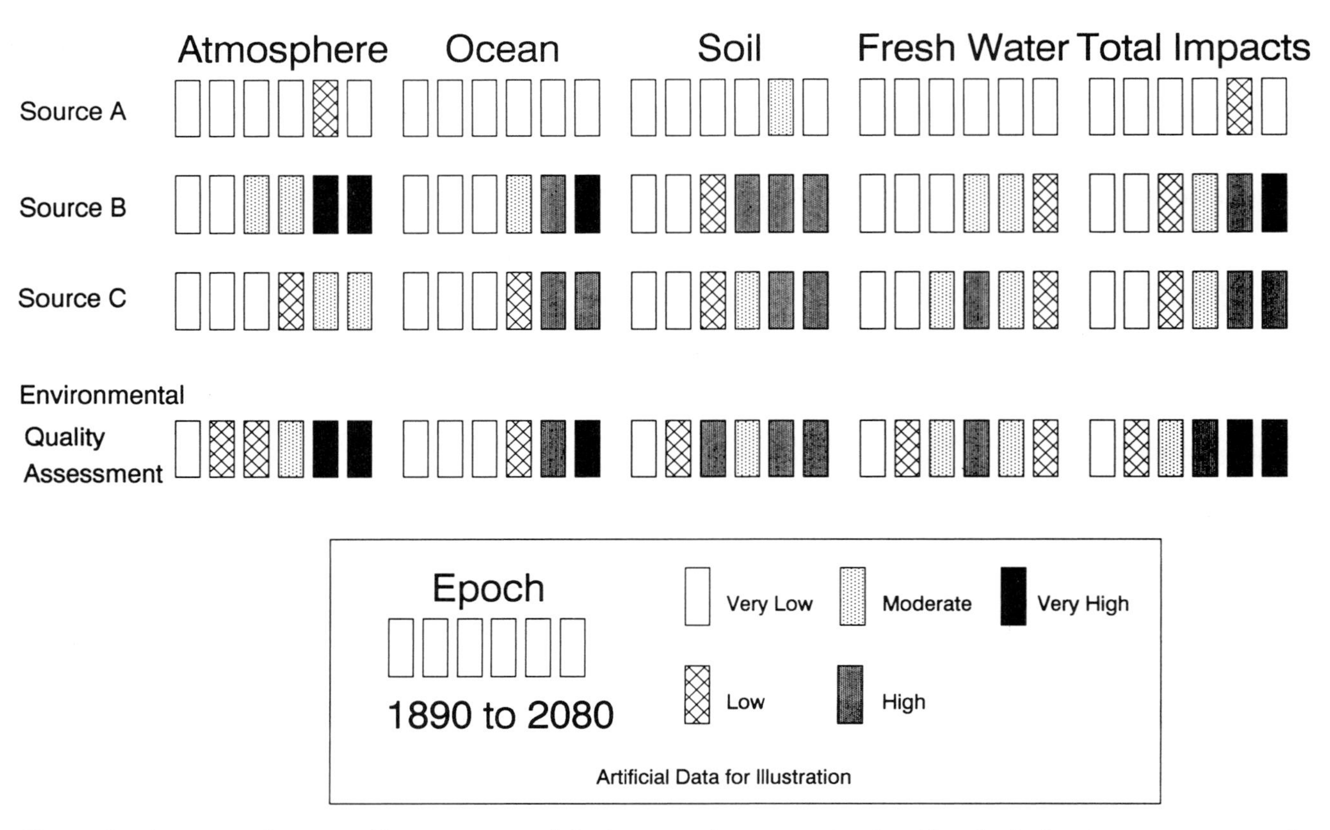

Figure 9. A concept for ensemble assessment of the impacts of sources related to biospheric development on several different environmental regimes. The data are artificial, and appear for didactic purposes only. This figure is an extension across regimes of Figure 8.

Table 3. Selected country statistics.*

Country	1984 Energy Use (Billion kWh)	1984 Population (Millions)	1984 Energy Use (10^3 kWh/person)	2025 Population (Millions)
USA	2413	237.2	10.17	296
Italy	161	57.0	2.82	69
India	138	746.4	0.18	1445
China	327	1034.9	0.32	1493
Nigeria	7.5	88.1	0.09	303
Brazil	152	134.4	1.13	235
Kenya	1.7	19.4	0.09	80

* Energy use data are from the *World Almanac* (1986); population data and projections are from Keyfitz (1989).

is the emission factor (grams of emittant i per gram of fuel j); and A is the country area. Electricity is generated almost entirely by a mix of five processes: coal combustion, petroleum combustion, natural gas combustion, nuclear power, and hydropower. The latter two processes emit no species to the atmosphere. The year 2025 assessment requires adopting projections of population growth, per capita energy consumption, and the fractions of the needed electrical energy supplied by the different energy sources. We assume that hydropower will provide as much electricity as it is now providing, but no more, that nuclear power will provide for each country 150 percent of its current capacity by 2025, and that supplies of petroleum and natural gas will begin to diminish and to increase in cost by 2025.

To display the results of this exercise, we reflect the emissions intensities for the countries on a world map with the intensities hatched according to a logarithmic scale. The display for CO_2, a product of all combustion of fossil fuel, is given in Figure 10. Carbon dioxide is of particular concern, of course, as a major component of the greenhouse effect. The 1980 diagram indicates that Italy has the highest CO_2 emissions intensity among the seven countries, even though the United States consumes significantly more energy per capita. India and China have intermediate emissions intensities (but note again that the scale is logarithmic, so the actual values are rather low), and the emissions intensities of the African and South American countries are low indeed.

The same analysis for the year 2025 gives quite different results. Italy retains its high ranking, but shares it with India and Nigeria. Kenya and China are projected to have emissions intensities nearly as high. The United States and Brazil are high as well, but the intensity is restrained in those countries by the relatively large land areas per capita. It is clear that if actual emissions intensities prove to be anything like those anticipated by the scenario used here, the populous countries of Asia and Africa will be producing a large portion of the planet's CO_2 emissions by the end of the first quarter of the 21st century.

The maps for NO_x, largely a product of high temperature petroleum combustion, are shown in Figure 11. NO_x, as we have seen, has impacts on a local scale, especially the generation of photochemical smog. The 1980 analysis shows Italy and the United States at the high end of the NO_x emissions intensities, with the other countries being moderately or very low. The 2025 picture is quite different, with increased energy demand pushing NO_x emissions in India, Nigeria, Kenya, and China to relatively high levels. In those countries, smog during sunlit summer months may become severe.

Finally, Figure 12 depicts emissions intensities in the seven countries for SO_2. Sulfur dioxide is implicated in acidic precipitation and reduced visibility, among other effects, and is generated by the combustion of sulfur-containing fossil fuel, both coal and petroleum. The 1980 assessment shows Italy with high emissions intensity, most of the other countries at modest levels, and the African countries at low levels. In 2025, with much energy being supplied by readily available coal, the prediction is for very high SO_2 emissions intensities in Italy, Nigeria, and India and for high intensities in China and Kenya. Brazil and the United States will be somewhat lower, aided by their large land areas. There is definite potential in much of Asia and Africa, particularly, for increases in corrosion and degradation of monuments, sculptures, and buildings, for acidification of lakes, and for decreases in visibility.

An alternative display for the information in Figures 10 through 12 is given in Figure 13, which is a variation of Figure 9 with impacts being replaced by emissions intensities and the seven countries serving as the sources. On this display, the transition of the developing countries from modest emissions sources to major emissions sources over the next 35 years is clearly evident for emittants having a variety of impacts and spatial scales. The principal message of this analysis is that increased energy use, especially if it involves the combustion of large amounts of coal, seems likely to produce global impacts from CO_2 emissions, regional impacts from SO_2 (and other) emissions, and local impacts from NO_x (and other) emissions. Although the impacts will be most severe in the countries with the highest emissions intensities, some level of impacts will be detectable all across the globe.

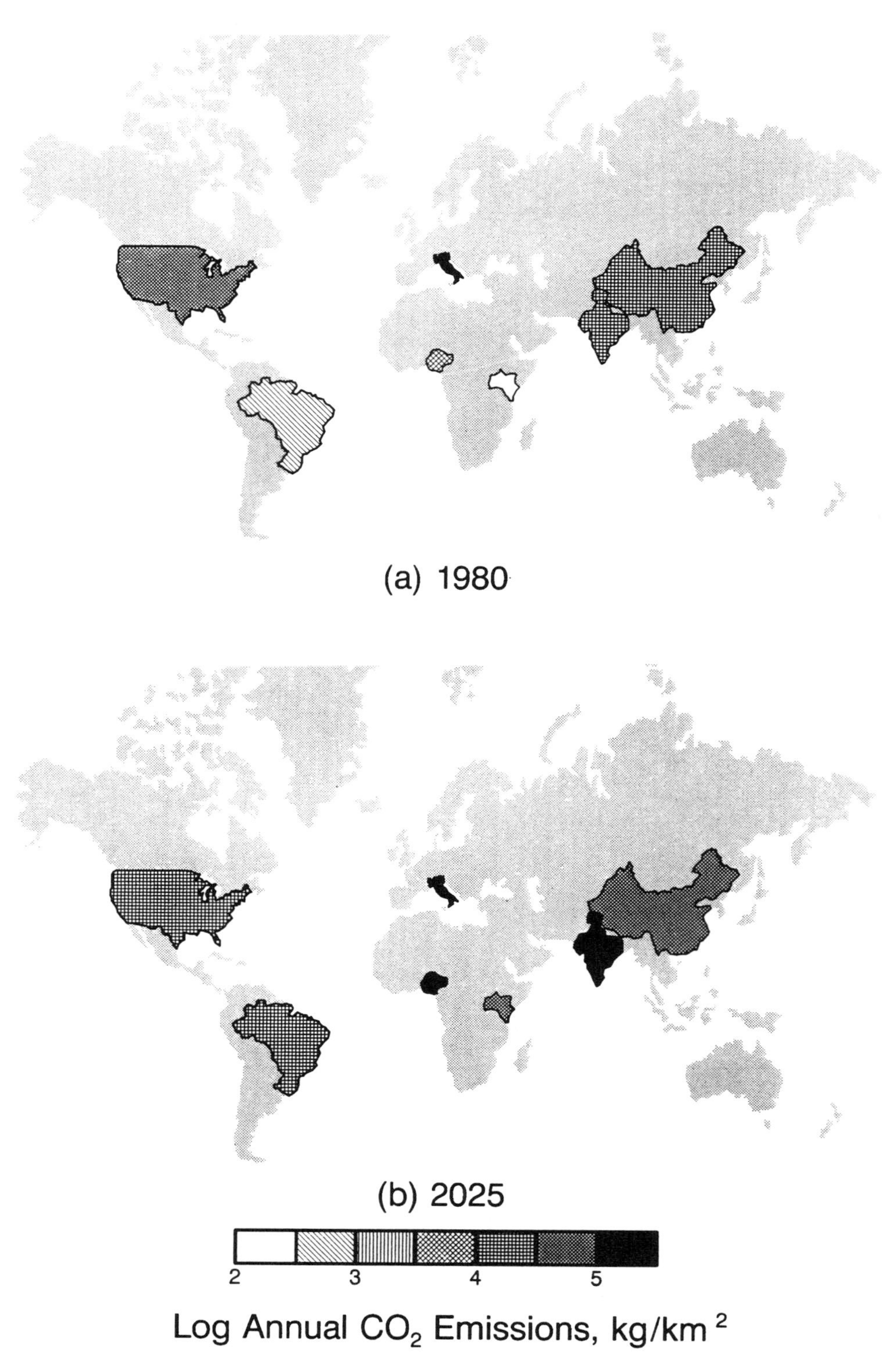

Figure 10. Carbon dioxide emissions intensities as a result of electrical energy generation for selected countries for 1984 and 2025, assuming that the supply of per capita energy in 2025 is equal to that of Italy in 1984.

Figure 11. Nitrogen oxide emissions intensities as a result of electrical energy generation for selected countries for 1984 and 2025, assuming that the supply of per capita energy in 2025 is equal to that of Italy in 1984.

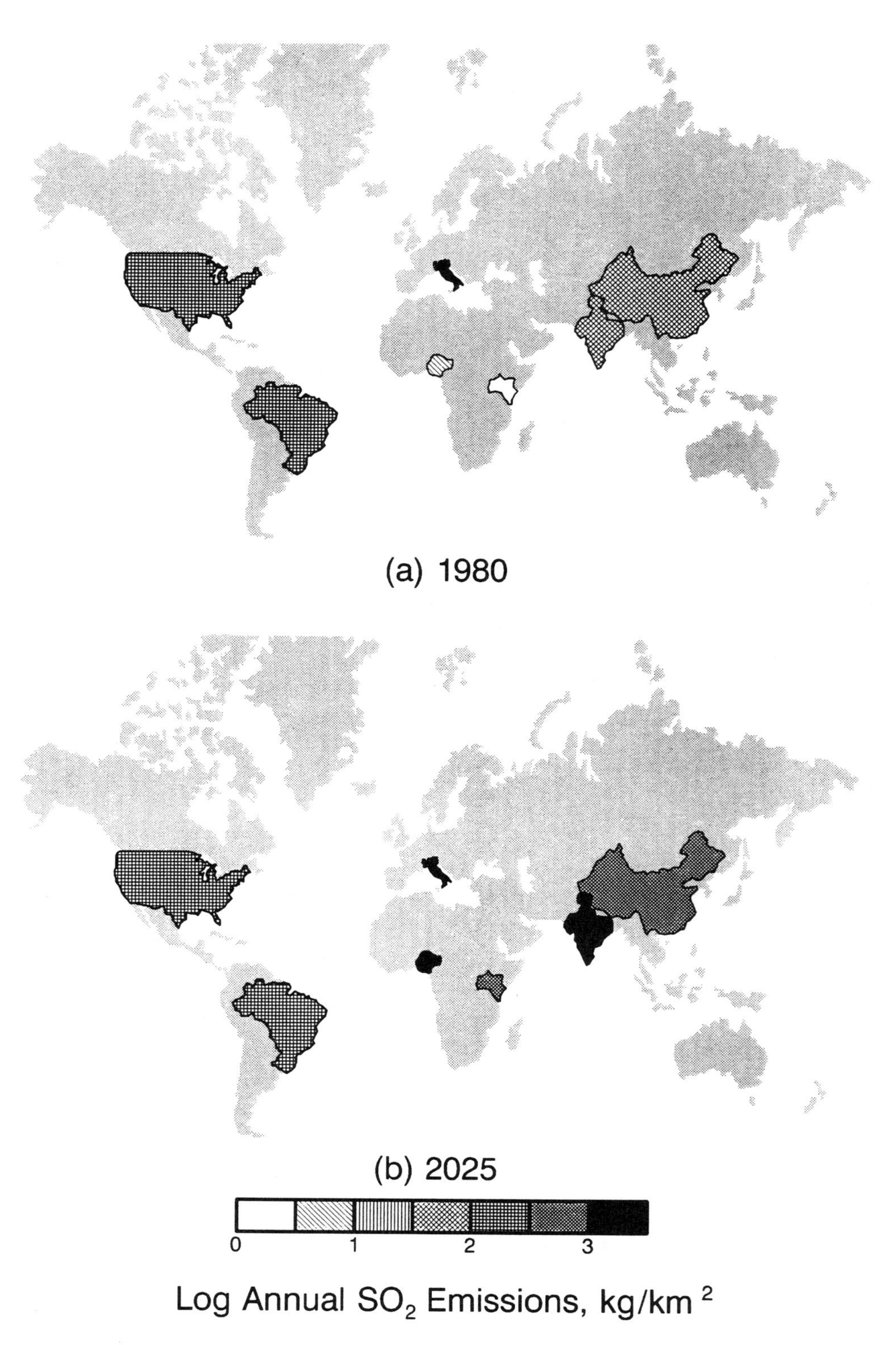

Figure 12. Sulfur dioxide emissions intensities as a result of electrical energy generation for selected countries for 1984 and 2025, assuming that the supply of per capita energy in 2025 is equal to that of Italy in 1984.

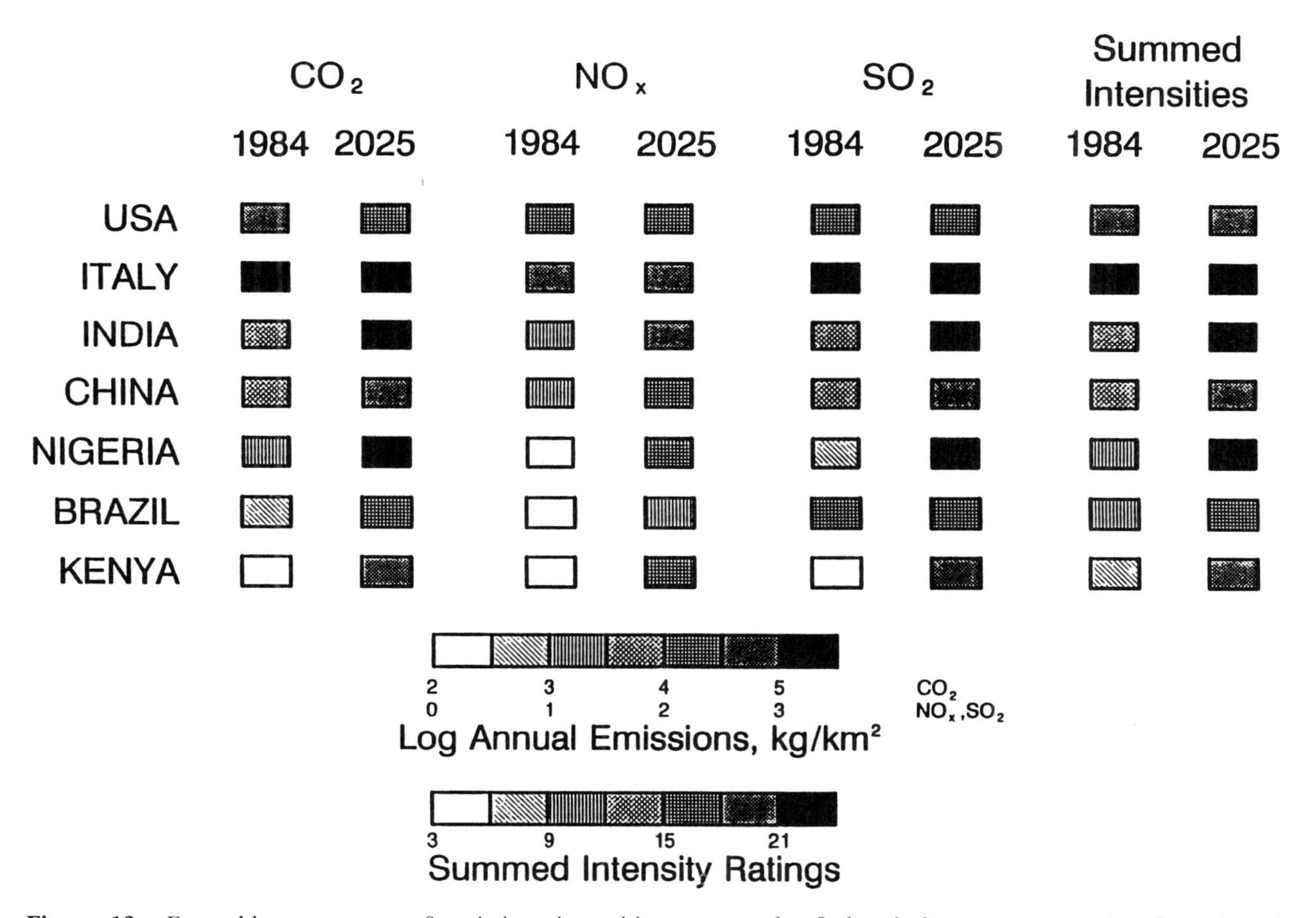

Figure 13. Ensemble assessment of emissions intensities as a result of electrical energy generation for selected countries in 1984 and 2025.

Conclusions

Energy generation involving the combustion of fossil fuels results in the generation and emission of a number of trace species to the atmosphere. These species are of various reactivities (hence lifetimes) and manifest their impacts on all spatial scales, from local to global. Such an assessment does not mean that development should not occur, but that its impacts on planetary processes should be studied and minimized.

The spectrum of impacts indicates that neither individual trace species nor individual impacts should be treated in isolation, but that the ensemble of emittants and impacts should be analyzed in a single concerted approach. This paper indicates one possible technique for such an analysis. Many variations of this approach are possible, but the recognition of the concept itself is paramount. Only through integrated approaches can world development be kept on a sustainable course.

References

Bolin, B., B. R. Döös, J. Jäger, and R. A. Warrick, *The Greenhouse Effect, Climate Change, and Ecosystems,* SCOPE Monograph 29, John Wiley, New York, NY (1986).

Crutzen, P. J., and T. E. Graedel, "The Role of Atmospheric Chemistry in Environment-Development Interactions," in *Sustainable Development of the Biosphere,* W. C. Clark and R. E. Munn, eds., Cambridge Univ. Press, pp. 213–250 (1986).

Douglas, I., "Urban Hydrology," *Geographical Journal,* 142, 65–72 (1976).

Environmental Protection Agency (EPA), *National Air Quality and Emission Trends Report,* EPA-450/4-84-029, Research Triangle Park, NC (1985).

Global Environment Monitoring System, *Assessment of Urban Air Quality,* United Nations Environment Program, Geneva, Switzerland (1988).

Graedel, T. E., "Regional Environmental Forces: A Methodology for Assessment and Prediction," in *Energy: Production, Consumption, and Consequences,* National Academy Press, Washington, DC, pp. 85–110 (1989).

Harnoz, Z., *Perspectives for Developing the Land Component of the Biosphere Program,* Collaborative Publication 86–22, International Institute for Applied Systems Analysis, Laxenburg, Austria (1986).

Keyfitz, N., "The Growing Human Population," *Scientific American,* 261 (3), 118–126 (1989).

National Research Council, *Nuclear and Alternative Energy Systems,* National Academy Press, Washington, DC (1979).

National Research Council, *Atmosphere-Biosphere Interactions: Towards a Better Understanding of Ecological Consequences of Fossil Fuel Combustion,* National Academy Press, Washington, DC (1981).

National Research Council, *Changing Climate,* National Academy Press, Washington, DC (1983).

World Almanac, Newspaper Enterprise Association, New York, NY (1986).

Global Atmospheric Chemistry and Global Pollution

Ronald G. Prinn
Center for Global Change Science
Massachusetts Institute of Technology
Cambridge, MA

Introduction: Global Atmospheric Chemistry

The global atmosphere is a chemically complex evolving system with fundamental chemical connections to the oceans, the solid earth, and, most importantly, the biota. Much of the key activity occurs in either the troposphere (between the surface and 8–12 km altitude) or the stratosphere (between the top of troposphere and 50 km altitude). In the troposphere the most important cycles in this system begin with emission from the surface of several key reactive species (Figure 1). Water vapor (H_2O) evaporated from the oceans and evapotranspired from the land is attacked in the atmosphere by electronically excited oxygen atoms $O(^1D)$ to produce two OH free radicals. These OH radicals are arguably the single most important oxidizing (cleansing) agent for the atmosphere. They are removed on a time scale of about one second by reaction with 1) a wide range of natural hydrocarbons emitted by soil, water, and enteric microorganisms or vegetation (e.g., methane (CH_4), isoprene, terpenes), 2) a wide range of anthropogenic hydrocarbons (e.g., CH_4, unburnt fuel hydrocarbons); and 3) carbon monoxide and sulfur dioxide of both industrial and natural origin. As a result of these reactions, either H atoms or organic free radicals (denoted R in Figure 1) are produced that combine with atmospheric oxygen to form hydroperoxy (HO_2) or organoperoxy (RO_2) free radicals. At this point nitrogen oxides emitted into the air as a result of combustion or produced from N_2 and O_2 in lightning play a key role (Crutzen, 1979). Nitric oxide (NO) can react with HO_2 or RO_2 to produce nitrogen dioxide (NO_2) and regenerate OH radicals from HO_2. The NO_2 can then either combine with OH to form nitric acid vapor (HNO_3) and potentially nitric acid rain, or it can be dissociated by near-ultraviolet solar radiation to regenerate NO and produce O atoms. These O atoms then combine with atmospheric O_2 to produce ozone (O_3). For example,

$$OH + CO \rightarrow H + CO_2$$
$$H + O_2 \rightarrow HO_2$$
$$HO_2 + NO \rightarrow OH + NO_2$$
$$\text{ultraviolet} + NO_2 \rightarrow NO + O$$
$$O_2 + O \rightarrow O_3$$

This ozone in turn can be dissociated by solar photons in a very narrow band of ultraviolet wavelengths (290–310 nanometers, which can penetrate through the ozone layer above) to produce O_2 and the excited oxygen atoms $O(^1D)$ with which we began our discussion of the cycle above.

The chemical cycles involving the stratosphere are also strongly influenced by surface emissions (Figure 1). Nitrous oxide (N_2O) from natural and industrial sources, methane, industrial chlorofluorocarbons like $CFCl_3$ and CF_2Cl_2, and water vapor are transported from the surface up into the stratosphere where photochemical reactions driven by solar ultraviolet radiation can decompose them to form a number of very reactive species (NO, NO_2, Cl, ClO, OH, HO_2) that can catalytically destroy ozone. For example, the Cl derived from the chlorofluorocarbons can destroy stratospheric ozone by the reactions

$$Cl + O_3 \rightarrow ClO + O_2$$
$$\text{ultraviolet} + O_3 \rightarrow O_2 + O$$
$$ClO + O \rightarrow Cl + O_2$$

without being itself consumed (Molina and Rowland, 1974; Rowland and Molina, 1975). Because stratospheric ozone is produced essentially exclusively by ultraviolet dissociation of O_2, its rate of global production is set externally by the supply of ultraviolet radiation from the sun. The global rate of destruction of ozone depends, however, on the supply of catalysts, which is rapidly rising due to continued chlorofluorocarbon emissions at the surface. Removal of the catalysts from the stratosphere occurs through their (partially reversible) conversion to harmless reservoir species (HNO_3, $ClONO_2$, HCl, etc., in Figure 1), many of which can be transported down again to the troposphere ultimately to rain out or deposit at the surface.

The Antarctic ozone hole phenomenon (Figure 2) provides a dramatic illustration of the fact that the global chemical system is changing. This phenomenon can occur because temperatures in the meteorologically isolated air mass lying between 10 and 30 km altitude

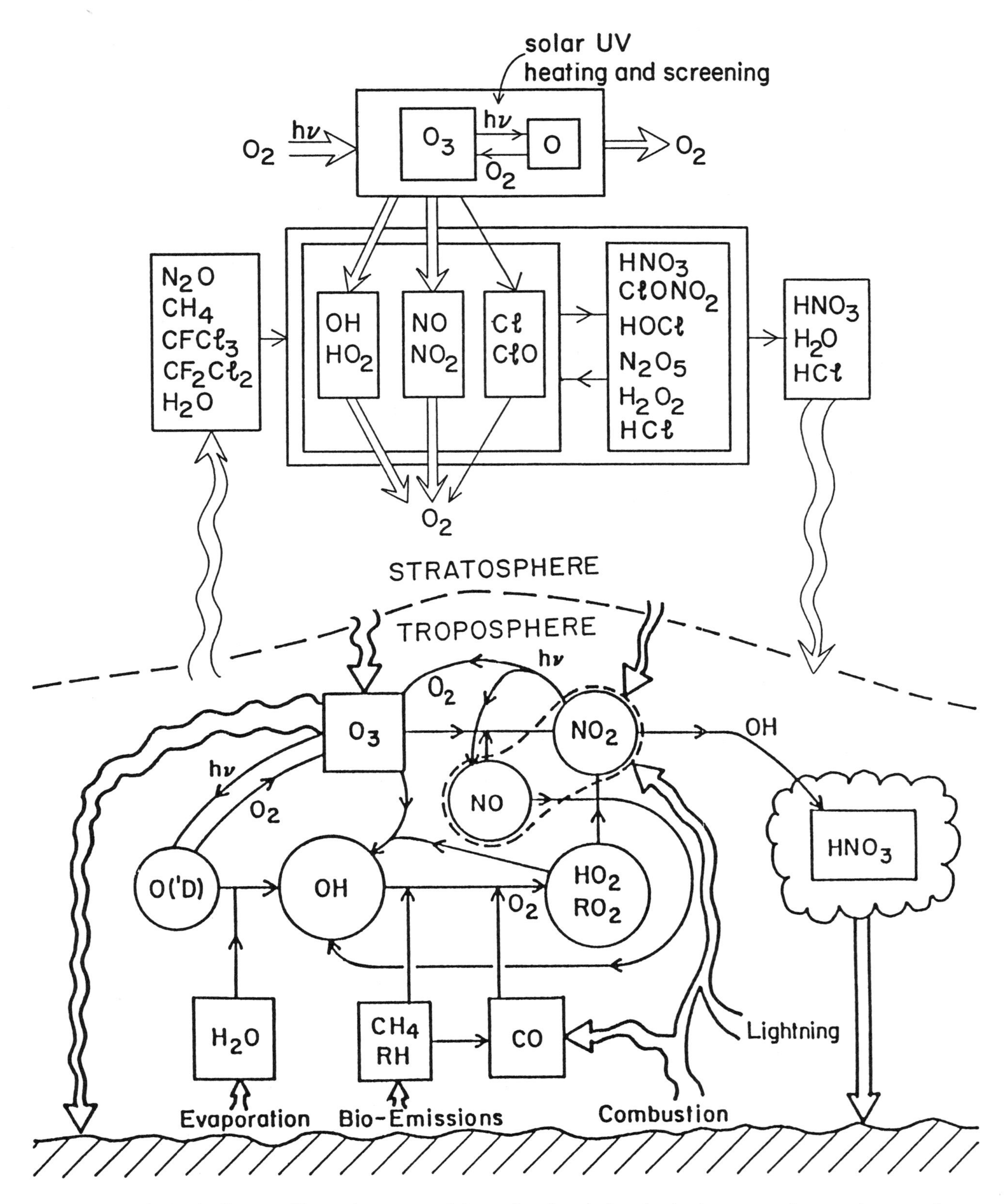

Figure 1. Schematic diagram illustrating some of the major chemical cycles in the troposphere and stratosphere.

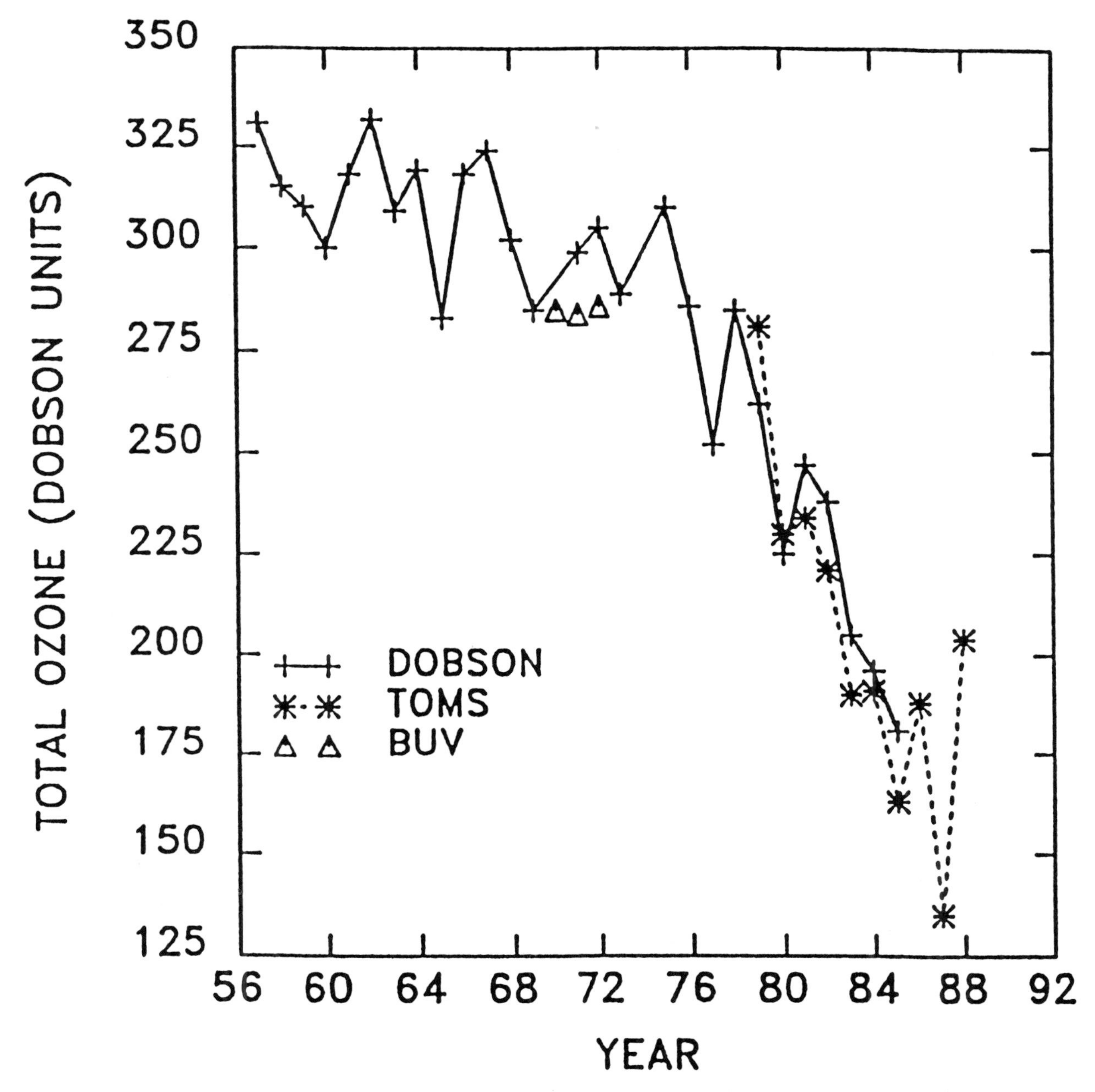

Figure 2. October monthly mean total ozone measurements over Halley Bay as measured by Halley Bay Dobson and BUV (Backscatter Ultraviolet) instrument on Nimbus 4. The polar minimum values from TOMS for October are also shown. Dobson = 2.7×10^{16} molecule cm^{-2}. From Schoeberl *et al.* (1989) and Stolarski (1988).

region in the Antarctic winter become so low (less than 190 K) that ice crystals condense in this region. Decomposition of some of the catalyst reservoir species occurs on and within these ice crystals, producing ClO in the early Antarctic spring. For example,

$$\text{sunlight} + ClONO_2(\text{ice}) + HCl(\text{ice}) \rightarrow Cl + Cl + HNO_3(\text{ice})$$
$$2Cl + 2O_3 \rightarrow 2ClO + 2O_2$$

The ClO levels become very high because HNO_3 remains in the ice and is therefore not available to contribute to the conversion of ClO back to its relatively harmless reservoir compound $ClONO_2$. The scarcity in the Antarctic spring of ultraviolet radiation capable of decomposing O_3 means that different catalytic cycles operate here than the one given earlier. For example, the reaction sequence

$$2ClO \rightarrow ClOOCl$$
$$\text{sunlight} + ClOOCl \rightarrow 2Cl + O_2$$
$$2(Cl + O_3) \rightarrow 2(ClO + O_2)$$

serves to efficiently destroy O_3 without requiring O_3 photodissociation and without destroying ClO (Molina *et al.*, 1987).

Several of the major players in the chemistry of the atmosphere are also major contributors to the greenhouse effect. In particular, CH_4, N_2O, $CFCl_3$, and CF_2Cl_2 all are strong absorbers and emitters of infrared radiation at wavelengths that the other two major greenhouse gases, H_2O and carbon dioxide (CO_2), do not absorb. As a result of this, their contribution per molecule to the greenhouse effect is very much greater than that of CO_2. Indeed, model calculations have shown that if current rates of increase of the greenhouse gases are extrapolated into the future the predicted greenhouse warming due to CH_4, N_2O, $CFCl_3$, and CF_2Cl_2 will equal that due to CO_2 by about the year 2030 (Ramanathan *et al.*, 1985; Dickinson and Cicerone, 1986).

Long-lived Trace Gases: Global Pollution

Trace gases whose lifetimes in the atmosphere exceed the one- to two-year time for mixing between Northern and Southern Hemispheric midlatitudes can influence the chemical and radiative properties of the atmosphere irrespective of the distribution and localization of their surface sources. Local emissions of these long-lived trace gases therefore cause global pollution. Several of the major trace gases discussed in the Introduction above are in this category, and, significantly, these long-lived trace gases are observed to be increasing today at rates which, projected into the future, are predicted to lead to significant enhancement of the earth's greenhouse effect and depletion of the global stratospheric ozone layer.

The long-lived trace gases can be conveniently divided into three categories based on their chemical reactivity. Some of these long-lived species are very inert (lifetimes exceeding 70 years) because they can be destroyed only by mixing up into the stratosphere above 25 kilometers altitude. There they can be dissociated by solar ultraviolet photons with wavelengths less than 250 nanometers. In this first category are the industrial chlorofluorocarbons $CFCl_3$, CF_2Cl_2, and $CF_2ClCFCl_2$ which are widely used as refrigerants, propellants, plastic foaming agents, and solvents and which have lifetimes of around 75, 180, and 140 years, respectively, at the present time (Golombek and Prinn, 1986, 1989). Measurements made in the ALE/GAGE (Atmospheric Lifetime Experiment/Global Atmospheric Gases Experiment) network of automated stations (located in Ireland, Oregon, Barbados, Samoa, Tasmania; Prinn, 1988, Prinn *et al.*, 1983) indicate that $CFCl_3$, CF_2Cl_2, and $CF_2ClCFCl_2$ are currently increasing at rates of about 5 percent, 5 percent, and 10 percent per year, respectively, over the globe (see Figures 3 through 5). Also in this class is nitrous oxide (N_2O), which has a present-day lifetime of about 170 years and is increasing at 0.2 to 0.3 percent per year (Figure 6) and has both natural sources (soils, ocean) and human-made sources (disturbed and fertilized soil, combustion). Due to the approximately three-year transport time between the surface and the middle stratosphere, the larger the rate of increase for a gas in this class, the smaller the ratio of its stratospheric and tropospheric contents, and thus the longer its present-day lifetime relative to its steady-state (zero increase) lifetime (Golombek and Prinn, 1986). As discussed earlier, all of these gases, when dissociated in the stratosphere, yield reactive species that catalytically destroy ozone and play a major role in the processes causing the Antarctic ozone hole. Destruction of stratospheric ozone in turn increases ultraviolet dosages at the surface, affecting living organisms and influencing the oxidation process in the lower atmosphere by accelerating photodissociation of O_3 to produce $O(^1D)$.

These very inert gases pose special environmental problems associated with their very long lifetimes. For example, according to ALE/GAGE network measurements, $CFCl_3$ and CF_2Cl_2 were building up in the atmosphere in 1988 at the respective rates of 206 and 307 kilotons per year. These values result from subtracting their small atmospheric loss rates of 82 and 55 kilotons per year from their large surface emission rates of 288 and 362 kilotons per year, respectively. Halving their emission rates, which should occur through implementation of the "Montreal Protocol on Substances that Deplete the Ozone Layer," will not decrease their rates of buildup to zero. This requires emissions to equal loss rates, which in turn requires decreases in $CFCl_3$ and CF_2Cl_2 emissions by factors of about 3 and 6, respectively, after taking into account their decreasing lifetimes as they approach a steady-state.

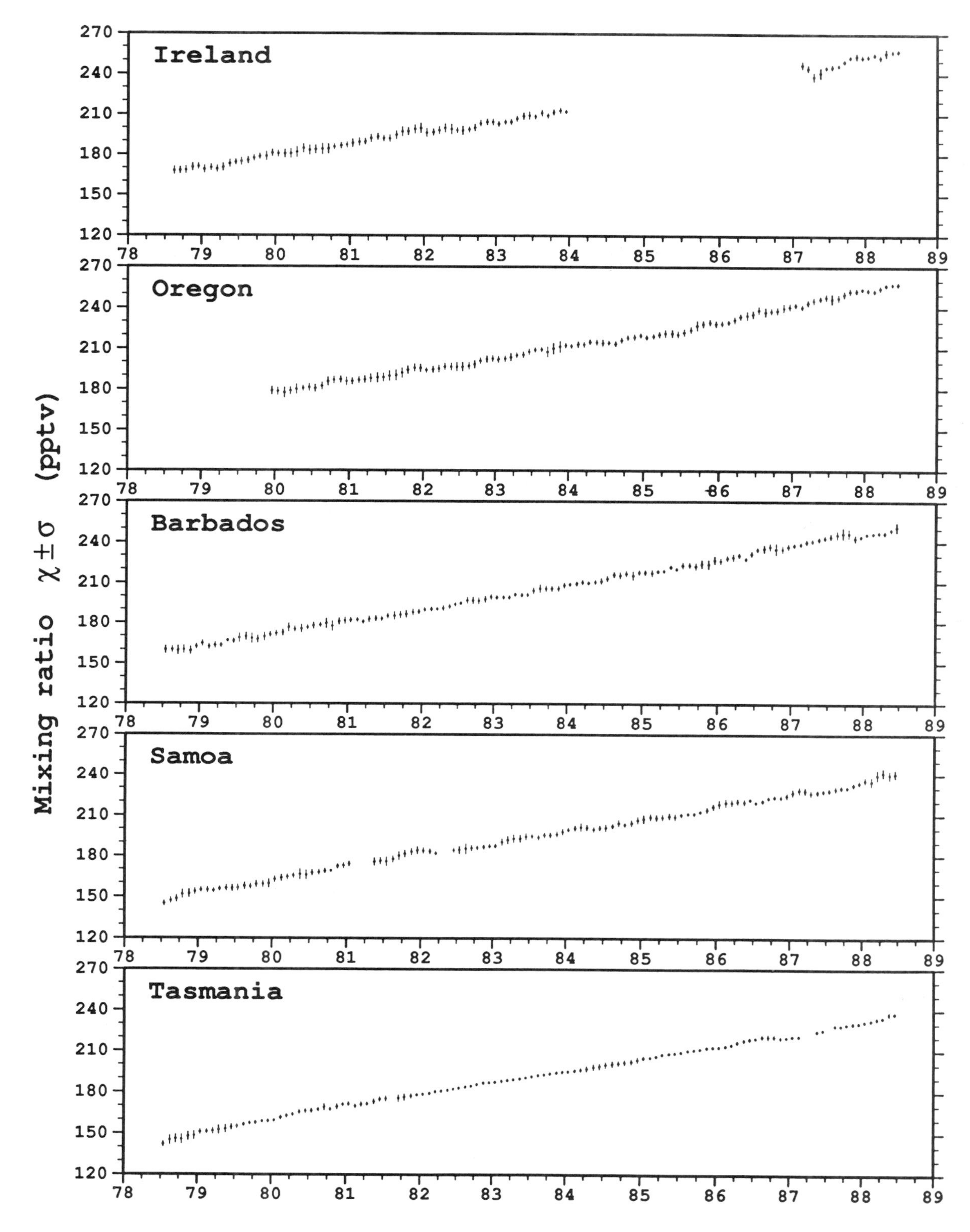

Figure 3. Monthly mean mixing ratios and standard deviations for $CFCl_3$ (CFC-11) determined from high frequency real-time measurements from 1978 to 1988 at the five stations of the ALE/GAGE network (pptv = parts in 10^{12} by volume). From Prinn *et al.* (1983) and Cunnold *et al.* (1986).

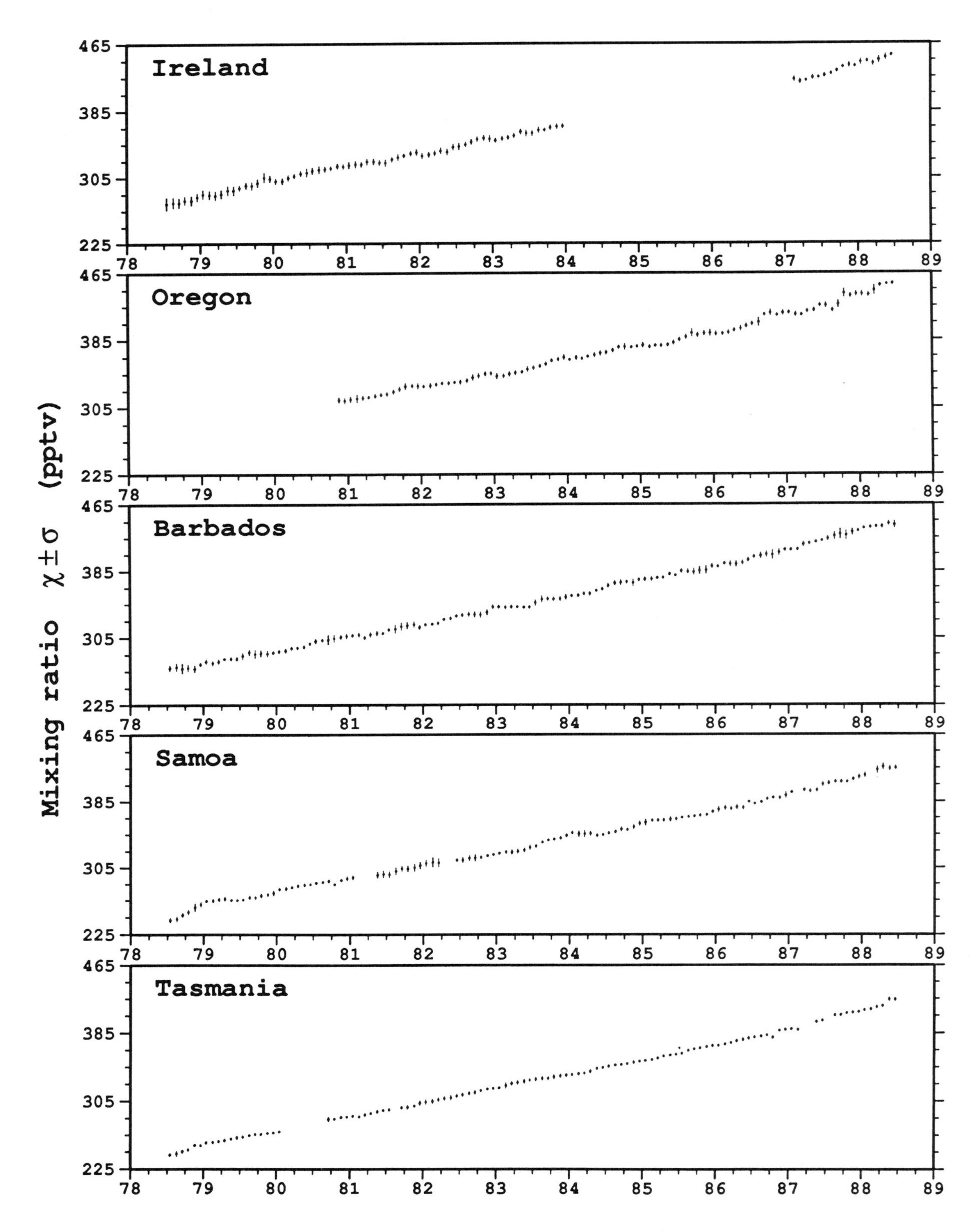

Figure 4. As in Figure 3, but for CF_2Cl_2 (CFC-12). From Cunnold *et al.* (1986).

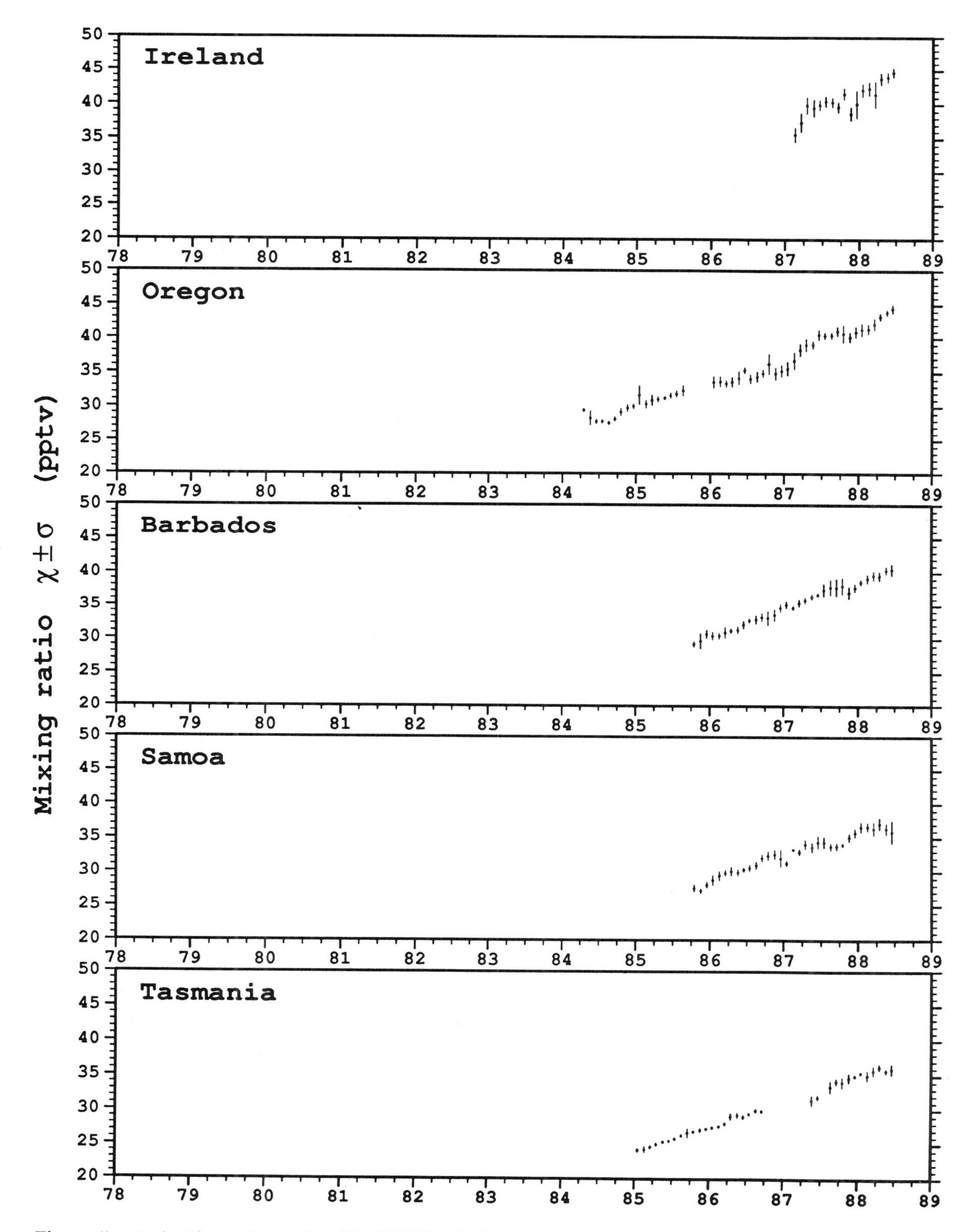

Figure 5. As in Figure 3, but for $CF_2ClCFCl_2$ (CFC-113). From Fraser *et al.* (1990).

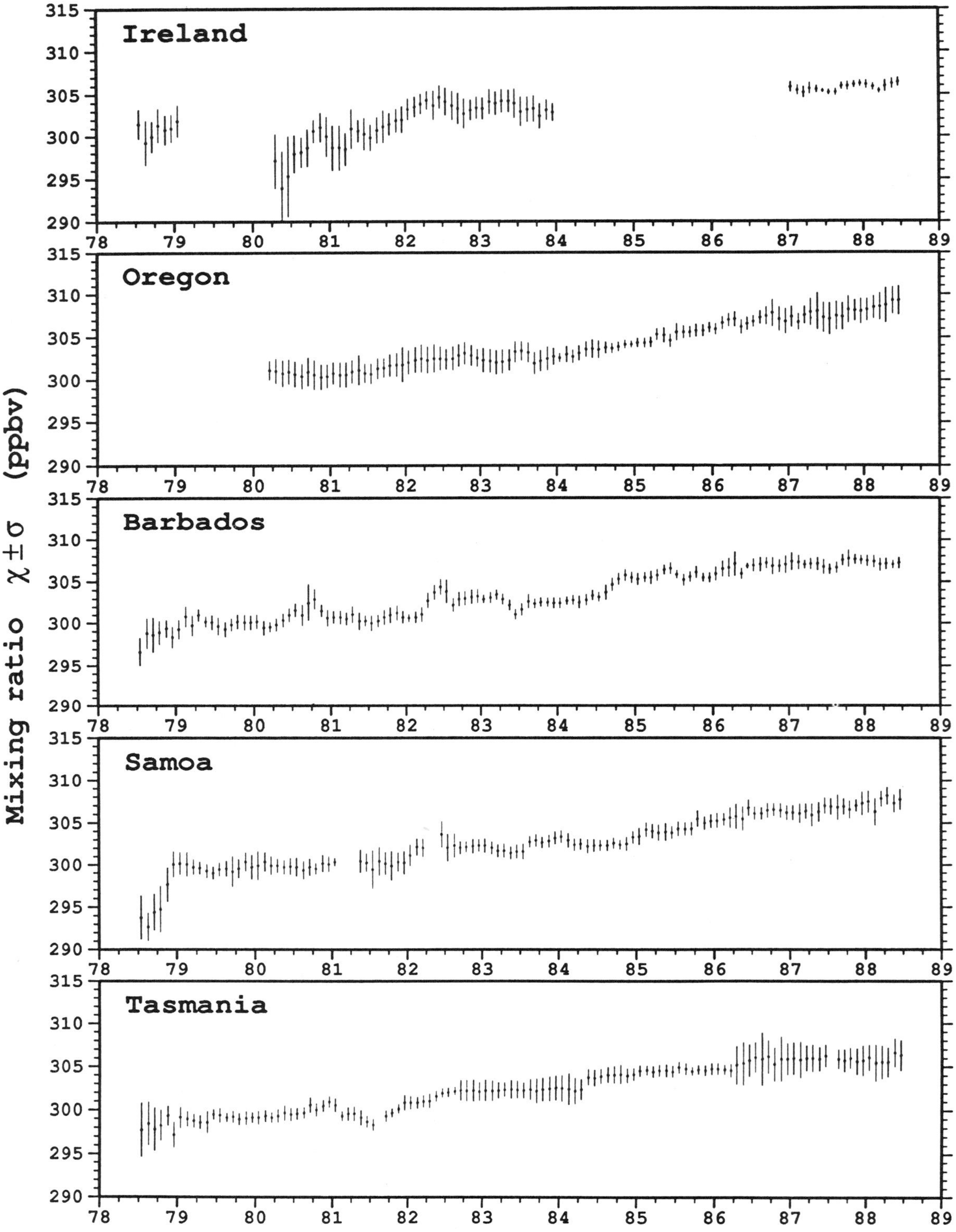

Figure 6. As in Figure 3, but for N_2O (ppbv = parts in 10^9 by volume). From Prinn *et al.* (1990).

A second category of long-lived species is less inert (lifetimes typically less than 15 years) because they can be destroyed by reaction with hydroxyl free radicals (OH) in the lower atmosphere. In this category is atmospheric methane (CH_4) with a lifetime of 9.6 years (Prinn *et al.*, 1987), which is increasing today at about 1 percent per year (Figure 7). Methane (like nitrous oxide) has both natural sources (wetlands, tundra) and human-controlled sources (cattle, combustion, rice agriculture) (Ehhalt, 1988). Also in this category is the widely used industrial solvent and cleaning agent methyl chloroform (CH_3CCl_3) with a lifetime of 6.3 years (Prinn *et al.*, 1987), which is increasing globally at about 4 percent per year (Figure 8). Finally in this category are the hydrochlorofluorocarbons such as $CHClF_2$ (16-year present-day lifetime), currently increasing at about 10 percent per year, and $CHCl_2CF_3$ (1.5-year steady-state lifetime) and CH_2FCF_3 (15-year steady-state lifetime), proposed as industrial replacements for the chlorofluorocarbons $CFCl_3$ and CF_2Cl_2.

The concentrations of OH radicals are determined to a significant extent by short-lived species whose life cycles begin and end on local and regional scales (e.g., NO, NO_2, O_3, CO, hydrocarbons, etc., in Figure 1). Thus the global atmospheric chemistry of the long-lived species in the second category above is linked in very important ways to the local and regional scale atmospheric chemistry of the short-lived species that determine hydroxyl radical levels. One important implication of this linkage is that the positive trends in species like CH_4 may in part be due to a decrease in OH radicals and thus a decrease in CH_4 loss rates rather than exclu-

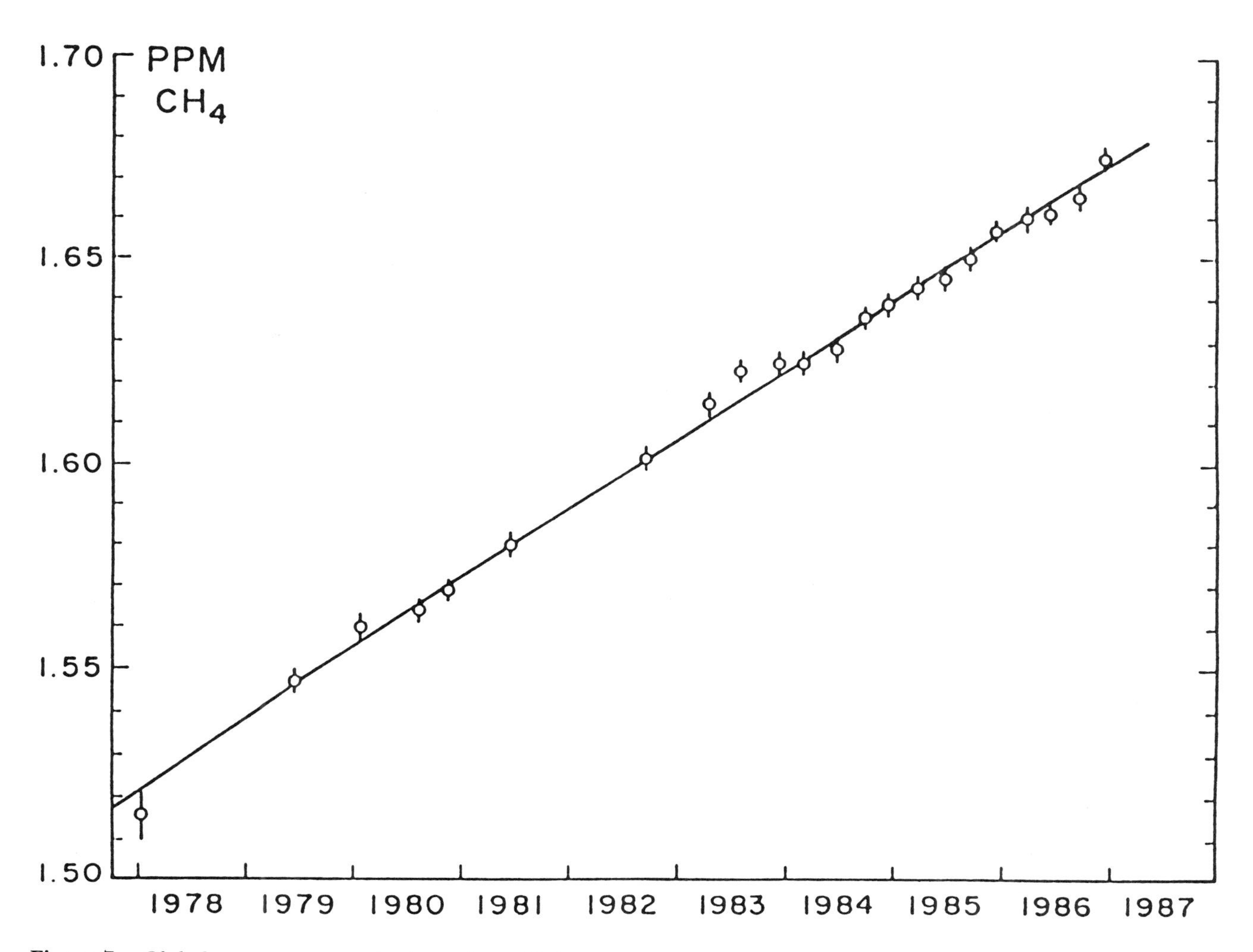

Figure 7. Global average tropospheric methane (CH_4) concentrations measured over the period 1978 to 1987 (ppm = parts in 10^6 by volume). From Blake and Rowland (1988).

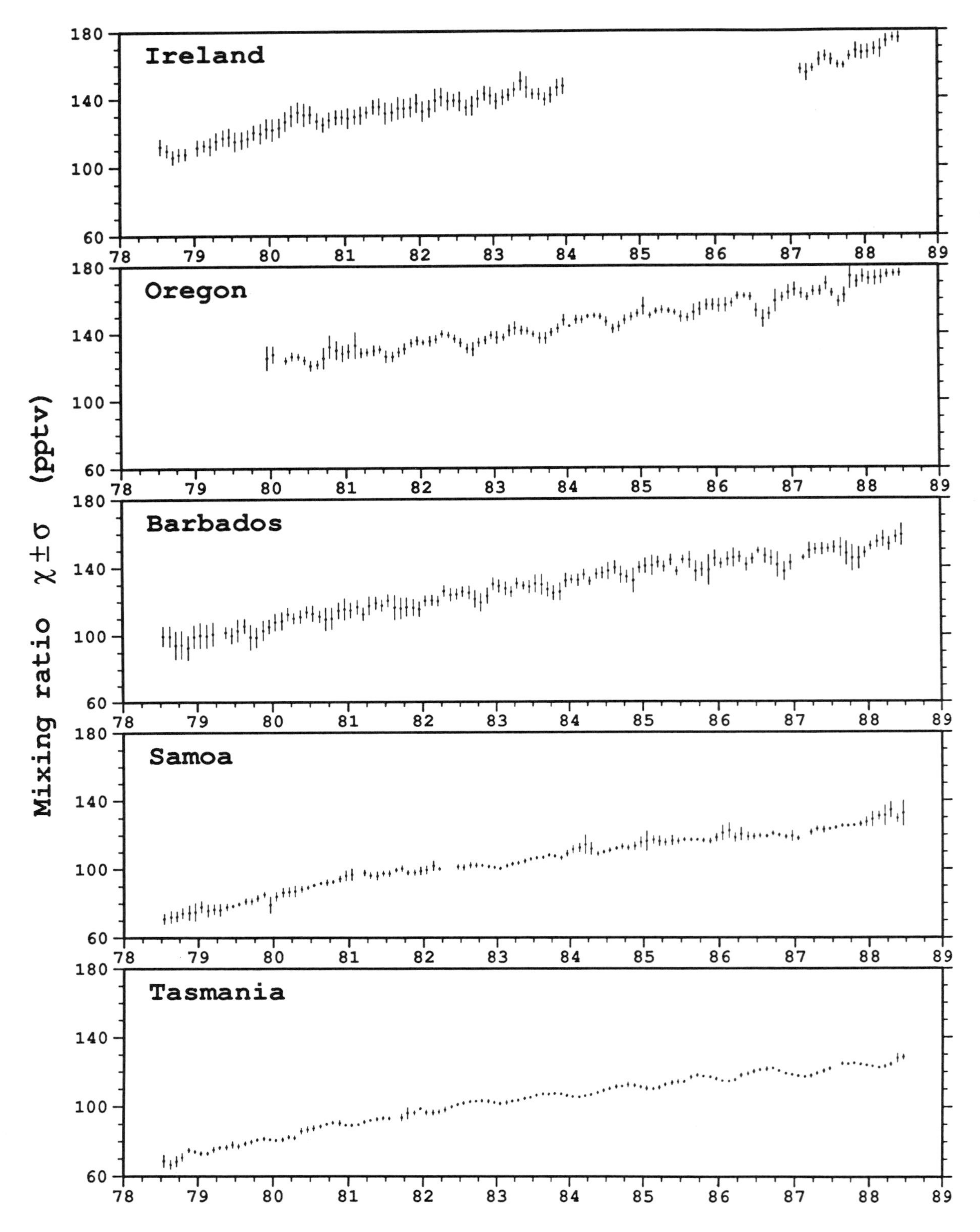

Figure 8. As in Figure 3, but for CH_3CCl_3. From Prinn *et al.* (1987).

sively to an increase in CH_4 emission rates. Another implication is that the destruction of long-lived gases in this second category is weighted heavily to the hot tropical lower atmosphere about whose local and regional chemistry we are only beginning to be informed (Donahue and Prinn, 1990). In this respect, it is important to note that while $CHCl_2CF_3$ (which is the proposed replacement for $CFCl_3$) is predicted to have a much smaller influence on the ozone layer and climate than $CFCl_3$ (for equal emissions of the two compounds), the removal of $CHCl_2CF_3$ occurs predominantly in the tropical lower troposphere and not in the midlatitude continental regions where most of it will be emitted (Prinn and Golombek, 1990). The potential effects of any harmful chemicals produced from $CHCl_2CF_3$ destruction in the tropics therefore need to be carefully assessed before large-scale production begins.

Carbon dioxide (CO_2) defines a third category of long-lived gases. It is essentially chemically inert in the atmosphere and is cycled between atmosphere, upper ocean, green vegetation, and soils on decadal time scales and between the upper ocean and deep ocean on a time scale of a few hundred years. The deep ocean acts as the major sink for CO_2 in this picture. Recent estimates of the approximate fluxes between the various CO_2 reservoirs and the sizes of these reservoirs are illustrated in Figure 9. Depending on the size of the unknown regrowth flux X in Figure 9, the time scale for loss of atmospheric CO_2 to the deep ocean is 160 to 240 years. Combustion of fossil fuel and deforestation are injecting sufficient amounts of additional carbon dioxide into these cycles to cause atmospheric CO_2 to have increased by about 26 percent between 1750 and the present and to be increasing today at about 0.34 percent per year (Figure 10). This increase in CO_2, along with the increases in CH_4, $CFCl_3$, CF_2Cl_2, and other long-lived gases, is the basis for current projections of a possible global warming over the next century.

By drilling deep into the Antarctic and Greenland ice caps and analyzing the ancient air trapped in the ice it has recently been possible to determine the variations in atmospheric composition that occurred during the glacial and interglacial periods over the last 150,000 years. Remarkably, the variations in concentrations of both methane and carbon dioxide are correlated with variations of temperature (Barnola *et al.*, 1987; Cicerone and Oremland, 1988). In particular, CH_4 and CO_2 are generally low during the ice ages and high during the warm interglacial epochs. This suggests the existence of a biogeochemical-climate feedback whereby increases in temperature or greenhouse gas concentrations lead to increases in greenhouse gas emissions or global warming, respectively.

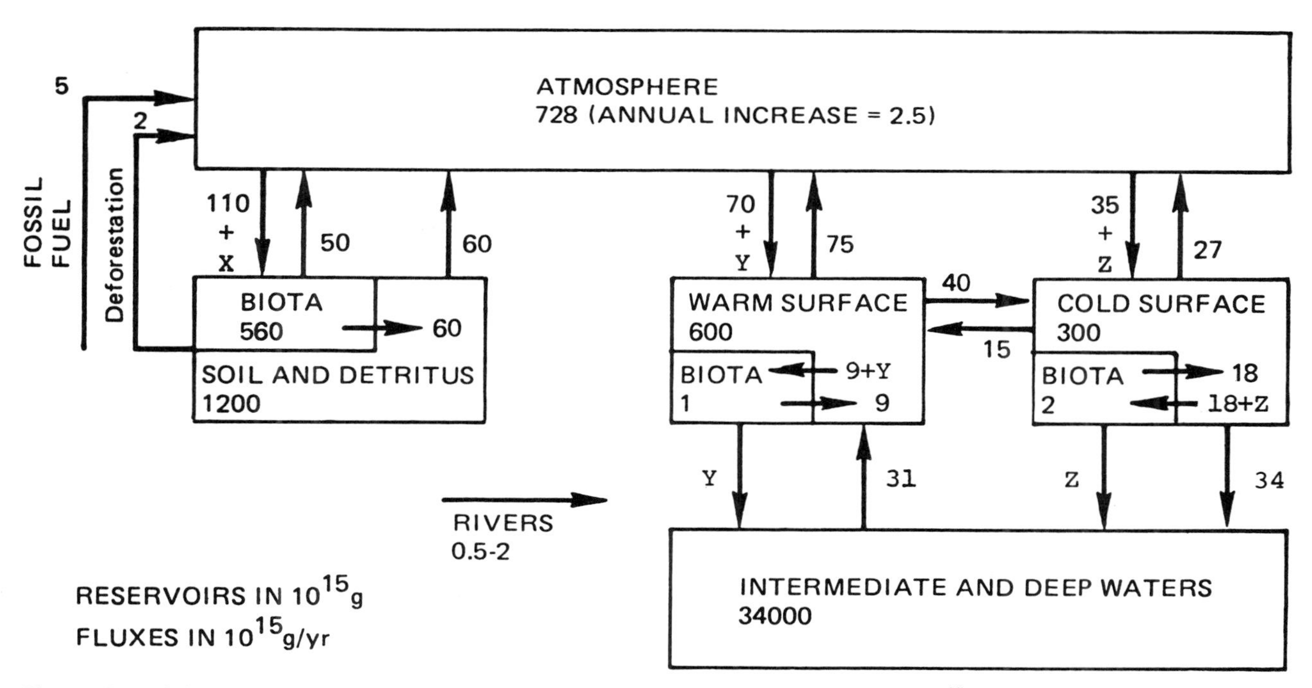

Figure 9. Global cycle of carbon dioxide. Content of each box given in units of 10^{15} grams. Fluxes between boxes given in units of 10^{15} grams per year. The sum of the biological sinks X, Y, and Z is constrained to be 1.5×10^{15} grams per year to balance the budget. From Committee on Earth Sciences (1985) with update of fluxes based on Oeschger and Seigenthaler (1988).

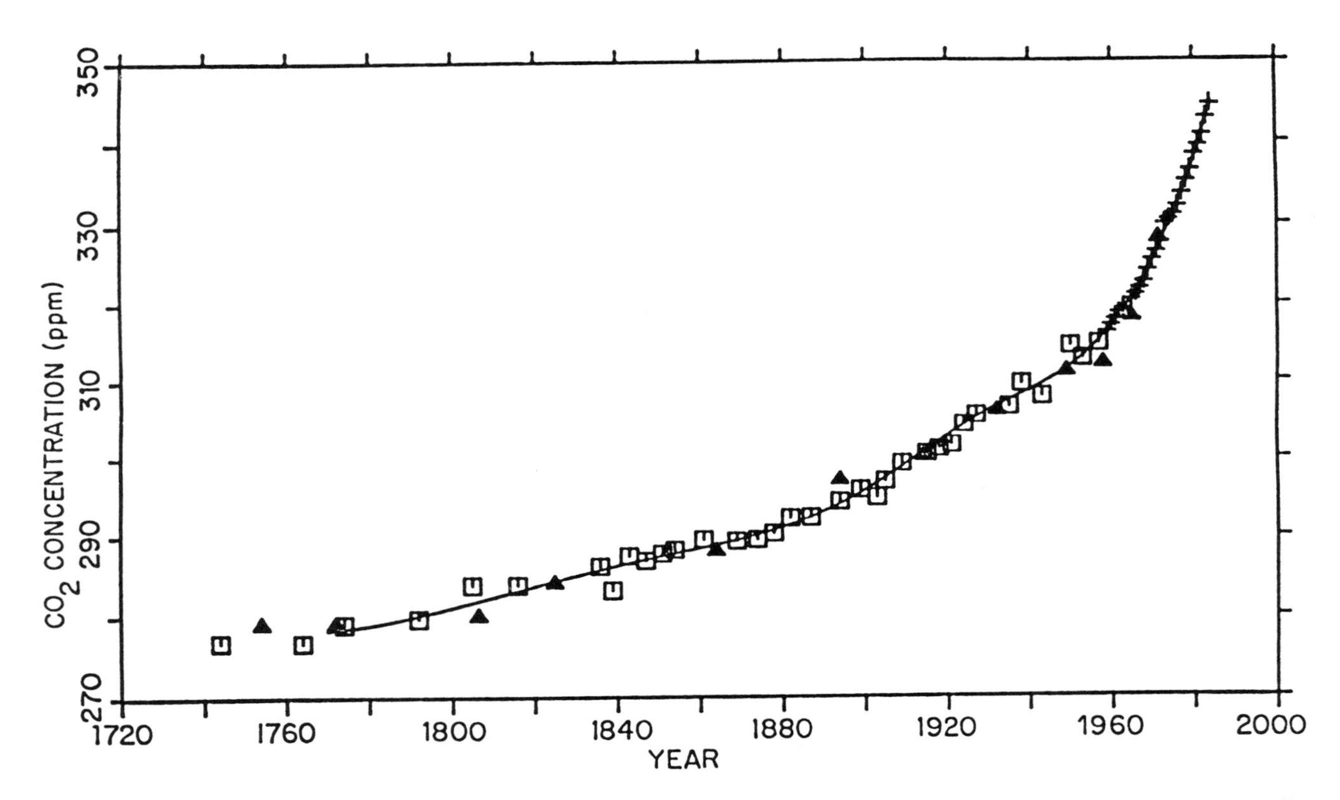

Figure 10. Carbon dioxide concentration in the atmosphere. Crosses are atmospheric measurements by C.D. Keeling in Hawaii. Squares and triangles are measurements in ancient air trapped in Antarctic ice cores (after Oeschger and Siegenthaler, 1988). Units are parts in 10^6 by volume.

Future Challenges: A Global Atmospheric Chemistry Program

One of the urgent needs in addressing global environmental change is to understand better the chemical, physical, and biological processes that determine atmospheric trace gas changes of the type discussed above and to use this understanding to explain the past and predict the future evolution of the atmosphere.

The International Global Atmospheric Chemistry (IGAC) Programme has been created recently in response to the growing international concern about these rapid atmospheric chemical changes and their potential impact on humankind. This program, while emphasizing atmospheric composition and chemistry, recognizes that the earth's atmosphere, oceans, land, and biota collectively form an interacting system that determines the global environment and its susceptibility to change. The International Geosphere-Biosphere Programme (IGBP) is a wide-ranging interdisciplinary international undertaking that addresses all major aspects of this interactive system. The IGAC program is intended to be a vital contributor to the broader interdisciplinary program of IGBP, providing the important atmospheric chemistry component and recognizing its linkages with the biosphere and human activities.

The overall goal of IGAC is to measure, understand, and thereby predict changes now and over the next century in the chemistry of the global atmosphere with particular emphasis on changes affecting the oxidizing power of the atmosphere, the impact of atmospheric composition on climate, and the interactions of atmospheric chemistry with the biota. This goal is broad and encompasses several urgent environmental issues, including the increasing acidity of rainfall, the depletion of stratospheric ozone, the greenhouse warming due to accumulation of trace gases, and the biological damage from increased oxidant levels. The IGAC program, which already involves over 100 scientists from 20 nations, should enable us to attain this goal within a time frame useful for meaningful predictions.

Acknowledgments

I thank Gail Rodriguez for TEXing. The author's research is supported by NASA Grant NAGW-474 and NAGW-732 and NSF Grant 87-10102 to MIT.

References

Barnola, J., D. Raynaud, Y. Korotkevich, and C. Lorius, 1987. "Vostok ice core: a 160,000-year record of atmospheric CO_2." *Nature* 329, pp. 408–414.

Blake, D., and S. Rowland, 1988. "Continuing world-wide increase in tropospheric methane, 1978 to 1987." *Science* 239, pp. 1129–1131.

Cicerone, R., and R. Oremland, 1988. "Biogeochemical aspects of atmospheric methane." *Biogeochemical Cycles* 2, pp. 299–327.

Committee on Earth Sciences, 1985. *A strategy for earth sciences from space in the 1980s and 1990s, part II: atmosphere and interactions with the solid earth, oceans, and biota.* National Academy Press, Washington, DC, 149 pps.

Crutzen, P., 1979. "The role of NO and NO_2 in the chemistry of the troposphere and stratosphere." *Ann. Rev. Earth Planet. Sci.* 7, pp. 443–472.

Cunnold, D., R. Prinn, R. Rasmussen, P. Simmonds, F. Alyea, C. Cardelino, A. Crawford, P. Fraser, and R. Rosen, 1986. "Atmospheric lifetime and annual release estimates for $CFCl_3$ and CF_2Cl_2 for 5 years of ALE data." *J. Geophys. Res.* 91, pp. 10797–10817.

Dickinson, R., and Cicerone, R., 1986. "Future global warming from atmospheric trace gases." *Nature* 319, pp. 109–115.

Donahue, N., and R. Prinn, 1990. "Non-methane hydrocarbon chemistry in the remote marine boundary layer." *J. Geophys. Res.*, in press.

Ehhalt, D.H., 1988. "How has the atmospheric concentration of CH_4 changed?" In *The Changing Atmosphere* (ed. F. Rowland and I. Isaksen, J. Wiley & Sons, Chichester) pp. 25–32.

Fraser, P., R. Prinn, R. Rasmussen, P. Simmonds, F. Alyea, A. Crawford, and D. Cunnold, 1990. "Global distributions and trends of $CFCl_2CF_2Cl$ (CFC-113) determined from GAGE data." Manuscript in preparation.

Golombek, A., and R. Prinn, 1986. "A global three-dimensional model of the circulation and chemistry of $CFCl_3$, CF_2Cl_2, CH_3CCl_3, CCl_4, and N_2O." *J. Geophys. Res.*, 91, pp. 3985–4001.

Golombek, A. and R. Prinn, 1989. "Global 3-dimensional model calculations of the budgets and present-day atmospheric lifetimes of $CF_2ClCFCl_2$ and $CHClF_2$." *Geophys. Res. Lett.* 16, pp. 1153–1156.

Molina, M., T. Tso, L. Molina, and F. Wang, 1987. "Antarctic stratospheric chemistry of chlorine nitrate, hydrogen chloride, and ice: release of active chlorine." *Science* 238, pp. 1253–1257.

Molina, M.J., and F.S. Rowland, 1974. "Stratospheric sink for chlorofluoromethanes: chlorine catalysed destruction of ozone." *Nature* 249, pp. 810–812.

Oeschger, H., and U. Seigenthaler, 1988. "How has the atmospheric concentration of CO_2 changed?" In *The Changing Atmosphere* (ed. F. Rowland and I. Isaksen, J. Wiley & Sons, Chichester) pp. 5–23.

Prinn, R., D. Cunnold, R. Rasmussen, P. Simmonds, F. Alyea, A. Crawford, P. Fraser, and R. Rosen, 1990. "Atmospheric trends and emissions of nitrous oxide deduced from ten years of ALE/GAGE data." *J. Geophys. Res.*, in press.

Prinn, R., 1988. "How have the atmospheric concentrations of the halocarbons changed?" In *The Changing Atmosphere* (ed. F. Rowland and I. Isaksen, J. Wiley & Sons, Chichester) pp. 33–48.

Prinn, R., D. Cunnold, R. Rasmussen, P. Simmonds, F. Alyea, A. Crawford, P. Fraser, and R. Rosen, 1987. "Atmospheric trends in methyl chloroform and the global average for the hydroxyl radical." *Science* 238, pp. 945–950.

Prinn, R., P. Simmonds, R. Rasmussen, R. Rosen, F. Alyea, C. Cardelino, A. Crawford, D. Cunnold, P. Fraser, and J. Lovelock, 1983. "The atmospheric lifetime experiment, I: introduction, instrumentation and overview." *J. Geophys. Res.* 88, pp. 8353–8368.

Prinn, R., and A. Golombek, 1990. "Global atmospheric chemistry of CFC-123." *Nature* 344, pp. 47–49.

Ramanathan, V., R. Cicerone, H. Singh, and J. Kiehl, 1985. "Trace gas trends and their potential role in climate change." *J. Geophys. Res.* 90, pp. 5547–5566.

Rowland, F., and Molina, M., 1975. "Chlorofluoromethanes in the environment." *Rev. Geophys. Space Phys.* 13, pp. 1–35.

Schoeberl, M., R. Stolarski, and A. Krueger, 1989. "The 1988 Antarctic ozone depletion: comparison with previous year depletions." *Geophys. Res. Lett.* 16, pp. 377–380.

Stolarski, R., 1988. "Changes in ozone over the Antarctic." In *The Changing Atmosphere* (ed. F. Rowland and I. Isaksen, J. Wiley & Sons, Chichester) pp. 105–119.

Prediction of Future Climate Change*

Stephen H. Schneider
National Center for Atmospheric Research**
Boulder, CO

Why Build a Model?

To predict the result of some event in nature it is common to build and perform an experiment. But what if the issues are very complex or the scale of the experiment unmanageably large? To forecast the effect of human pollution on climate poses just such a dilemma, for this uncontrolled experiment is now being performed on "laboratory earth." How then can we be anticipatory, if no meaningful physical experiment can be performed? While nothing can provide certain answers, we can turn to a surrogate lab, not a room with test tubes and Bunsen burners, but a small box with transistors and microchips. We can build mathematical models of the earth and perform our experiments in computers.

Mathematical models translate conceptual ideas into quantitative statements. There are a range of models, from those that simply treat one or two processes in detail to large-scale, multi-process simulation models. Such models usually are not faithful simulators of the full complexity of reality, of course, but they can tell us the logical consequences of explicit sets of plausible assumptions. To me, that certainly is a big step beyond pure conception—or to put it more crudely, modeling is a major advance over "hand-waving" forecasts of global changes.

The kinds of problems that can be studied by climate models include the downstream atmospheric effects of unusual ocean surface temperature patterns (e.g., so-called "El Niño" events), climatic effects of volcanic explosions, ice age-interglacial sequences, ancient climates, the climatic effects of human pollutants such as carbon dioxide (CO_2), and even the climatic aftereffects of nuclear war. We will consider a few examples of these later.

* Adapted in part from "Climate Modeling," *Scientific American* 256 No. 5, 1987.

** The National Center for Atmospheric Research is sponsored by the National Science Foundation.

Basic Elements of Models

To simulate the climate, a modeler needs to decide which components of the climatic system to include and which variables to involve. For example, if we choose to simulate the long-term sequence of glacials and interglacials, our model needs to include as explicitly as possible the effects of all the important interacting components of the climatic system operating over the past million years or so. Besides the atmosphere, these include the ice masses, upper and deep oceans, and the up and down motions of the earth's crust. Even life influences the climate and thus must be included too; plants, for example, can affect the chemical composition of the air and seas as well as the reflectivity (i.e., albedo) or water-cycling character of the land. These mutually interacting subsystems form part of the internal components of the model. On the other hand, if we are only interested in modeling very short-term weather events—say, over a single week—then our model can ignore any changes in the glaciers, deep oceans, land shapes, and forests, since these variables obviously change little over one week's time. For short-term weather, only the atmosphere itself needs to be part of the model's internal climatic system.

The slowly varying factors such as oceans or glaciers are said to be external to the internal part of the climatic system being modeled. Modelers also refer to external factors as boundary conditions, since they form boundaries for the internal model components. These boundaries are not always physical ones, such as the oceans, which are at the bottom of the atmosphere, but can also be mass or momentum or energy fluxes across physical boundaries. An example is the solar radiation impinging on the earth. Solar radiation is often referred to by climatic modelers as a *boundary forcing function* of the model for two reasons: the energy output from the sun is not an interactive, internal component of the climatic system of the model; and the energy from the sun forces the climate toward a certain temperature distribution.

We could restrict a model to predict only a globally averaged temperature that never changes its value over time (i.e., is in equilibrium). This very simple model would consist of an internal part that, when averaged over all of the atmosphere, oceans, biosphere, and glaciers, would describe two characteristics: the average reflectivity of the earth and its average greenhouse properties. The boundary condition for such a model would be merely the incoming solar energy. Such a model is called *zero dimensional*, since it collapses the east-west, north-south, and up-down space dimensions of the actual world into one point that represents some global average of all earth-atmosphere system temperatures in all places. It also collapses all three-dimensional processes into a global average—which may, in some cases such as heat transport by winds, completely neglect that process. If our zero-dimensional model were expanded to resolve temperature or winds at different latitudes and longitudes and heights, then it would be three-dimensional. The *resolution* of a model refers to the number of dimensions included and to the amount of spatial detail with which each dimension is explicitly treated.

Modelers speak of a hierarchy of models that ranges from simple earth-averaged, time-independent, temperature models to high-resolution, three-dimensional, time-dependent models known as general circulation models (GCMs). While three-dimensional, time-dependent models are usually more dynamically accurate, they are very computer intensive. Thus, it is sometimes necessary to run physically, chemically, and (somewhat) biologically comprehensive models at lower resolution. Choosing the optimum combination of factors is an intuitive art that trades off completeness and (the modelers hope) accuracy for tractability and economy. Moreover, the theoretical feasibility of long-range simulation must also be evaluated; in other words, some problems are inherently unpredictable. For those problems where predictability is not ruled out in principle, such a tradeoff between accuracy and economy is not "scientific" per se, but rather is a value judgment, based on the weighing of many factors. Making this judgment depends strongly on the problem the model is to address.

The Problems of Parameterizations and Climatic Feedback Mechanisms

Clouds, being very bright, reflect a large fraction of sunlight back to space, thereby helping control the earth's temperature. Thus, predicting the changing amount of cloudiness over time is essential to reliable climate simulation. But most individual clouds are smaller than even the smallest area represented by the smallest resolved element (i.e., "grid box") of a global climate or weather prediction model. A single thunderstorm is typically a few kilometers in size, not a few hundred—the size of many "high-resolution" global model grids. Therefore, no global climate model available now (or likely to be available in the next few decades) can explicitly resolve every individual cloud. These important climatic elements are therefore called *sub-grid-scale phenomena*. Yet, even though we cannot explicitly treat all individual clouds, we can deal with their collective effects on the grid-scale climate. The method for doing so is known as *parameterization*, a contraction for "parametric representation." Instead of solving for sub-grid-scale details, which is impractical, we search for a relationship between climatic variables we do resolve (for example, those whose variations occur over larger areas than the grid size) and those we do not resolve. For instance, climatic modelers have examined years of data on the humidity of the atmosphere averaged over large areas and have related these values to cloudiness averaged over that area. It is typical to choose an area the size of a numerical model's grid—a few hundred kilometers on a side. While it is not possible to find a perfect correspondence between these averaged variables, reasonable relationships have been found in a wide variety of circumstances. These relationships typically require a few factors, or *parameters*, some of which are derived empirically from observed data, not computed from first principles. The parameterization method applies to almost all simulation models, whether dealing with physical, biological, or even social systems. The most important parameterizations affect processes called *feedback mechanisms*. This concept is well known outside of computer-modeling circles. The word feedback is vernacular. As the term implies, information can be "fed back" to you that will possibly alter your behavior.

So it is in the climate system. Processes interact to modify the overall climatic state. Suppose, for example, a cold snap brings on a high albedo snow cover that tends to reduce the amount of solar heat absorbed, subsequently intensifying the cold. This interactive process is known to climatologists as the *snow-and-ice/albedo/temperature feedback mechanism*. Its destabilizing, *positive-feedback* effect is becoming well understood and has been incorporated into the parameterizations of most climatic models. Unfortunately, other potentially important feedback mechanisms are not usually as well understood. One of the most difficult ones is so-called *cloud feedback*, which could be either a positive or negative feedback process, depending on circumstances.

Model Verification

Given the uncertainties associated with model parameterizations and feedbacks, then, can we have confidence in model predictions? At least several methods can be used, and none by itself is sufficient. First, we must check overall model-simulation skill against the real climate for today's conditions to see if the control experiment is reliable. The seasonal cycle is one good test. Figure 1, from the work of Syukuro Manabe and R.J. Stouffer at the Geophysical Fluid Dynamics Laboratory at Princeton, NJ, shows how remarkably well a three-dimensional global circulation model can simulate the regional distribution of the seasonal cycle of surface air temperature—a well-understood climate change that is, when averaged over the earth, larger than ice age-interglacial changes! The seasonal-cycle simulation is a necessary test of what can be called "fast physics," like cloud formations, but it does not tell us how well the model simulates slow changes in ice cover or soil organic matter or deep ocean temperatures, since these variables do not change much over a seasonal cycle, though they do influence long-term trends.

A second method of verification is to test in isolation individual physical subcomponents of the model (such as its parameterizations) directly against real data and/or more highly resolved process models. This still is not a guarantee that the net effect of all interacting physical subcomponents has been properly treated, but it is an important test. For example, the upward infrared radiation emitted from the earth to space can be measured from satellites or calculated in a climate model. If this quantity is subtracted from the emitted upward infrared radiation at the earth's surface, then the difference between these quantities, G, can be identified as the "greenhouse effect," as in Figure 2 from Raval and Ramanathan (1989). Although radiative processes may not be the only ones operative in nature or in the general circulation model (GCM) results for G, the close agreement among the satellite results (line labeled ERBE on Figure 2), a GCM (labeled CCM), and line-by-line radiative transfer calculations on Figure 2 give strong evidence that this physical subcomponent—known as the "water-vapor/greenhouse feedback"—is well modeled at grid scale. Another just-emerging set of validation tests of internal model processes is to compare model-generated and observed statistics of grid-point daily variability (e.g., Mearns et al., in press; Rind et al., 1989). These have provided some examples of excellent agreements between model and observed variability (e.g., daily temperature variance) as well as examples of poorer agreement (e.g., daily relative humidity variances).

Third, some researchers express more confidence a priori in a model whose internal makeup includes more spatial resolution or physical detail, believing that "more is better." In some cases and for some problems this is true, but by no means for all. The "optimal" level of complexity depends upon the problem we are trying to solve and the resources available to the task.

All three methods must constantly be used and reused as models evolve if we are to improve the credibility of their predictions. And to these we can add a fourth method: the model's ability to simulate the very different climates of the ancient earth or even those of other planets. Perhaps the most perplexing question, though, is whether we should ever consider our confidence in models' forecasts sufficient reason to alter our present social policies—on CO_2-producing activities, for example. Viewed from this angle, the seemingly academic field of mathematical climate modeling could become a fundamental tool for assessing public policy.

Sensitivity and Scenario Analysis

Although the ultimate goal of any forecasting simulation may be to produce a single, accurate time-series projection of some evolving variable, a lesser goal may still be quite useful and certainly is more realizable: to specify plausible scenarios of various uncertain or unpredictable variables and then to evaluate the sensitivity of some predicted variable to either different scenarios or different model assumptions.

For example, in order to predict the societal impact of climatic changes from increasing concentrations of certain trace gases like CO_2, it is first necessary to invoke behavioral assumptions about future population, economic, and technological trends. (Such factors are external to the climatic forecast model, of course, but must be forecast nonetheless.) Although these may be impossible to forecast with confidence, a set of plausible scenarios can be derived. The differential consequences for the climatic forecast of each of these scenarios can then be evaluated.

Another sort of sensitivity analysis involves building a model that incorporates important, but highly uncertain, variables into its internal structure. Key internal factors, such as the cloud feedback or vertical oceanic mixing parameters, can be varied over a plausible range of values in order to help determine which internal processes have the most importance for the sensitivity of the climate to, say, CO_2 buildup. Even though one cannot be certain which of the simulations is most realistic, sensitivity analyses can 1) help set up a priority list for further work on uncertain internal model elements and 2) help to estimate the plausible range of climatic futures to which society may have to adapt over the next several decades. Given these plausible futures, some of us might choose to avoid a low-to-moderate probability, high-consequence outcome associated with some specific scenario. Indeed public policy often seeks to avoid plausible, high-cost scenarios. So too, do people purchase insurance. On the other hand, more risk-prone people might prefer extra scientific certainty before asking society to invest present resources to hedge against uncertain, even if plausible, climatic futures. At a minimum, cross-sensitivity analysis in which the response of some forecast variables to multiple variations in uncertain internal and/or external parameters allows us to examine

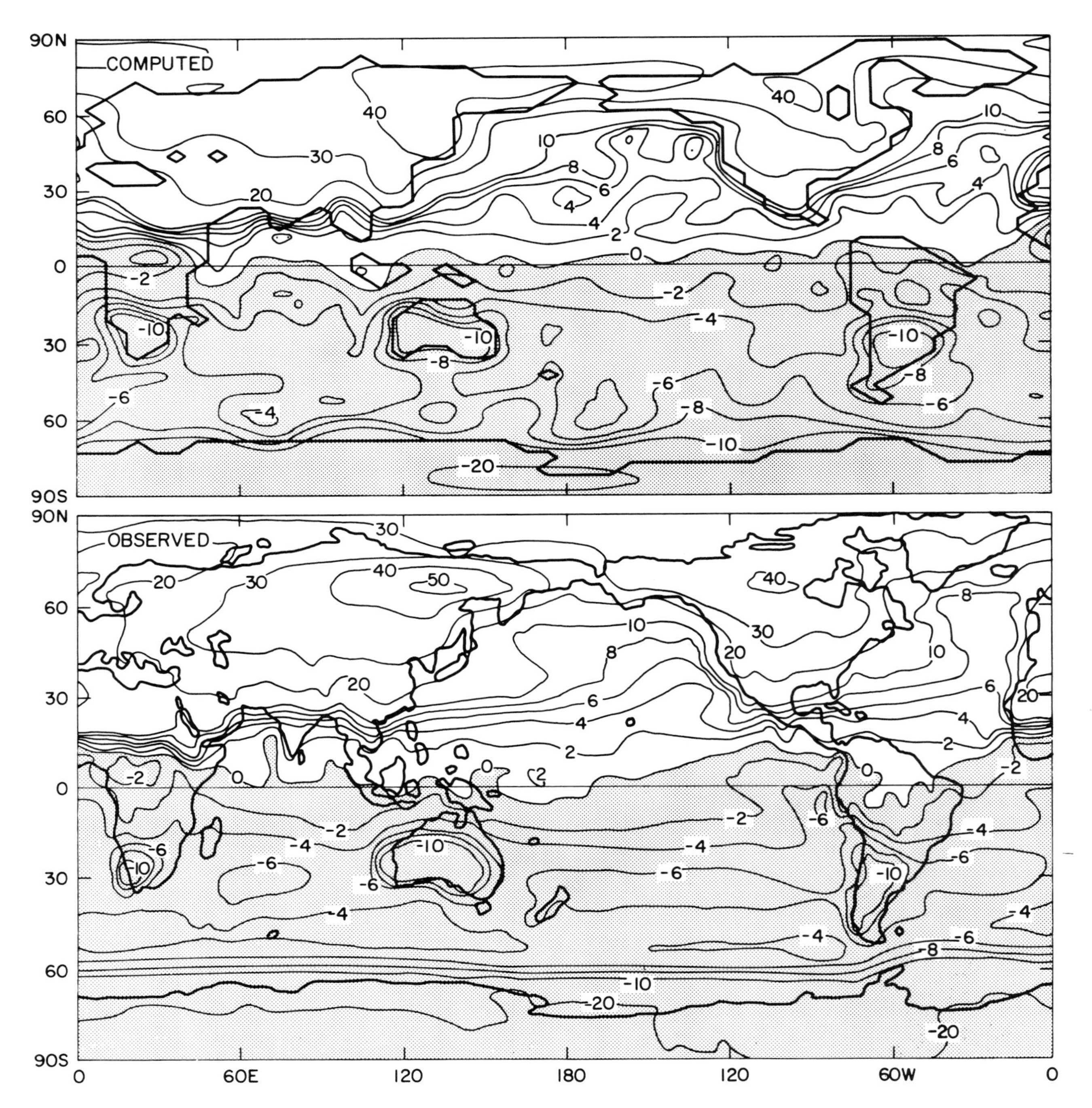

Figure 1. A three-dimensional climate model has been used to compute the winter to summer temperature differences all over the globe. The model's performance can be verified against the observed data shown below. This verification exercise shows that the model quite impressively reproduces many of the features of the seasonal cycle. These seasonal temperature extremes are mostly larger than those occurring between ice ages and interglacials or for any plausible future carbon dioxide change. (Source: S. Manabe and R. J. Stouffer, 1980, "Sensitivity of a global climate model to an increase of CO_2 concentration in the atmosphere," *Journal of Geophysical Research* 85:5529-5554.)

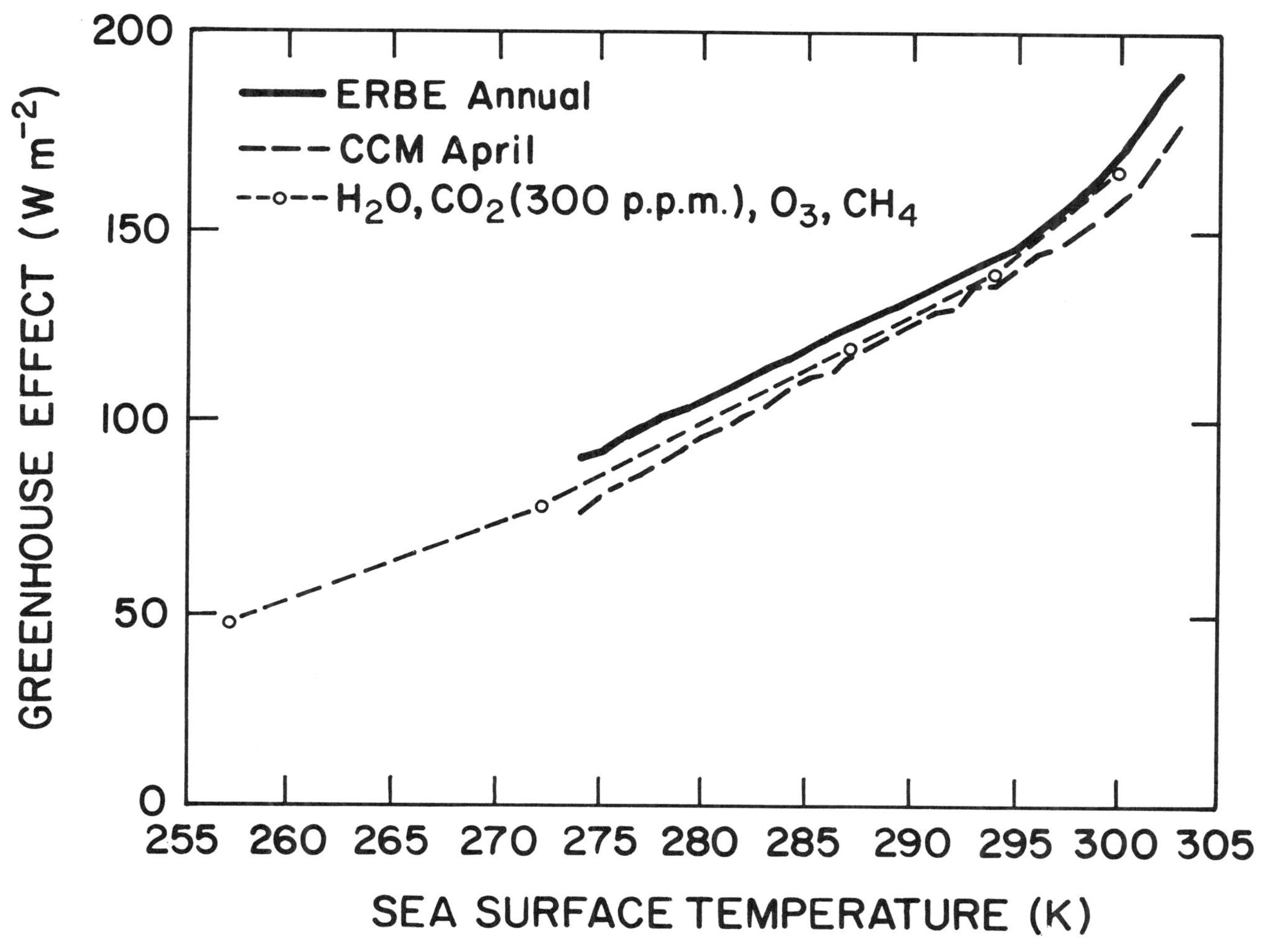

Figure 2. Comparison of greenhouse effect, heat trapping parameter, G, and surface temperature, obtained from three sources—bold line: ERBE annual values, obtained by averaging April, July, and October 1985 and January 1986 satellite measurements; thick dashed line, three-dimensional climate-model simulations for a perpetual April simulation (National Center for Atmospheric Research Community Climate Model); thin dashed line, line-by-line radiation-model calculations by Dr. A. Arking using CO_2, O_3 and CH_4. The line-by-line model results come close to the CCM and the ERBE values. (Source: Raval, A. and V. Ramanathan 1989, "Observational determination of the greenhouse effect," *Nature* 342, 758).

quantitatively the differential consequences of explicit sets of plausible assumptions.

In any case, even if we cannot produce a reliable single forecast of some future variables, we might be able to provide much more credible sensitivity analyses, which can have practical applications in helping us to investigate a range of probabilities and consequences of plausible scenarios. Such predictions may simply be the best "forecasts" that honest natural or social scientists can provide to inform society on a plausible range of alternative futures of complex systems. How to react to such information, of course, is in the realm of values and politics.

Let us proceed then to several examples of this process.

Climatic Model Results

Ancient Paleoclimates

To investigate future climatic changes, let us first turn not to the present, but to the ancient times. If the same basic models that we use to estimate effects of greenhouse gas increases into the next century can be applied to the ancient climatic changes and a reasonable simulation is obtained, then both scientific explanation of the ancient paleoclimate and some verification of the model's ability to reproduce radically different future climatic periods can be obtained.

Three-dimensional atmospheric circulation model studies which explicitly resolve land and sea and also explicitly calculate atmospheric motions have been applied to the sequence of climatic changes from the past ice age millennium, 18,000 years ago to the present. One of the most successful paleoclimatic simulations to date was performed by John Kutzbach and several colleagues at the University of Wisconsin in Madison (e.g., Kutzbach and Guetter, 1986). Kutzbach attempted to explain the warmest period in recent climatic history, the so-called "climatic optimum" that occurred between about 5,000 to 9,000 years ago. It was a time when summertime northern continental temperatures were probably several degrees warmer than at present and monsoon rainfall was more intense throughout Africa and Asia. Kutzbach found that the optimum could be explained simply by the fact that the tilt of the earth's axis (its obliquity) was slightly greater then than now. Also, the orbit was such that the earth was closer to the sun (i.e., perihelion) in June rather than in January as it now is. These variations, these slight perturbations in the earth's orbit, do not make substantial changes in the annual amount of solar radiation received on the earth, but do change substantially the difference between winter and summer heating periods. About 5 percent more solar heat over much of the Northern Hemisphere summer and a comparable amount less in winter occurred 9,000 years ago compared to the present. This change was sufficient in Kutzbach's simulations to substantially alter mid-continental warming in the summer months, which led to enhanced Asian and African monsoonal rainfall and river runoff in the models. His results matched quite well with considerable amounts of paleoclimatic evidence gathered by an international team of scientists (COHMAP, 1988), and are helping to explain one of the important mysteries in the paleoclimatic record—the climatic optimum. However, Kutzbach's simulations for 12,000 years ago were too warm, perhaps because he neglected the effects of blowing dust on cooling the climate. Clearly, more research is needed.

Forecasting Global Warming into the 21st Century

The most important question surrounding the greenhouse gas controversy is simply: What will be the regional distribution of climatic changes associated with significant increases in CO_2 and other trace greenhouse gases? [Other trace gases such as chlorofluorocarbons, methane, nitrogen oxides, or ozone can, all taken together, have a greenhouse effect comparable to CO_2 over the next century (e.g., Dickinson and Cicerone, 1986)]. To investigate such possibilities one needs a model with regional resolution. It needs to include processes such as the hydrologic cycle and the storage of moisture in the soils—since these factors are so critical to both climatic change and its impacts on agriculture and water supplies. To investigate plausible climatic scenarios, modelers have typically run "equilibrium simulations" in which instantaneously increased values of carbon dioxide are imposed at an initial time and held fixed while the model is allowed to approach a new equilibrium. Syukuro Manabe and R. J. Stouffer (1980) for example, in one of the most widely quoted results (shown in Figure 3), find a summer "dry zone" in the middle of North America, as well as increased moistness in some of the monsoon belts. All of this was from a doubling of CO_2 held fixed over time. The model was allowed to run several decades of simulated time in order to reach equilibrium. But Manabe's team used an artificially constructed ocean that consisted of a uniform mixed layer depth of 70 meters with no heat flow to or from the deep ocean below. While this shallow "sea" allows the seasonal cycle to be satisfactorily simulated (see Figure 1), a purely mixed layer ocean does not include the important processes whereby water is transported horizontally from the tropics toward the poles or vertically between the mixed layer and the abyssal depths. The latter processes slow the approach toward thermal equilibrium of the surface waters, and certainly would affect the transient evolution of the surface temperature changes in the ocean due to the actual time-evolving increase of trace greenhouse gases.

Thus, during the transient phase of warming the surface temperature increases from latitude to latitude and land to sea could have a different pattern than at equilibrium. This, in turn, could cause significantly different climatic anomalies during the transient phase than would be inferred from equilibrium sensitivity tests to fixed increases in trace gases.

To answer this transient response question more reliably, it has been proposed that three-dimensional atmospheric models be coupled to fairly realistic three-dimensional oceanic models. To date, only a handful of such model experiments have been run (e.g., Washington et al., 1989; Stouffer et al., 1989), but none over the century or two time scale needed to adequately address this important issue. One reason for detailing the CO_2 transient/regional climate anomaly example here is to exemplify the need for various sensitivity experiments across a hierarchy of models. Such methods are especially essential during the development phase of modeling. In 1980, Starley Thompson and I ran quasi-one-dimensional models to examine the potential importance of this transient issue (e.g., Schneider and Thompson, 1981). However, these models could not provide reliable simulations because of their physical simplicity. But the one-dimensional models were economically efficient enough to be able to be run over the century time span needed to explore the CO_2 transient issue. Our findings suggested that it is imperative to run high-resolution, coupled atmospheric, oceanic, cryospheric,

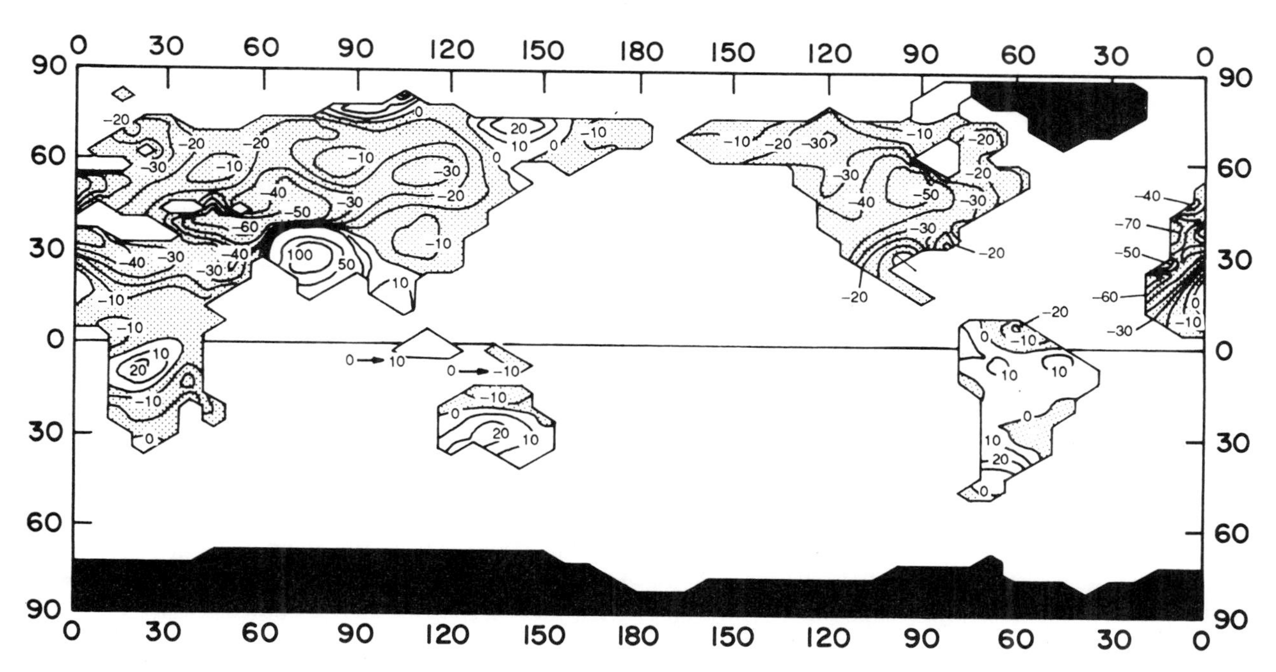

Figure 3. CO_2-induced change in soil moisture expressed as a percentage of soil moisture obtained from a computer model with doubled CO_2 compared to a control run with normal CO_2 amounts. Note the nonuniform response of this ecologically important variable to the uniform change in CO_2. (Source: S. Manabe and R. Wetherald, "Reduction in Summer Soil Wetness Induced by an Increase in Atmospheric Carbon Dioxide," *Science* 232, 626, 1986).

and land surface sub-model models if regional, time-evolving scenarios of climatic changes are to have any hope of credibility. On the other hand, the more complex three-dimensional general circulation models are not yet at an adequate phase of development to have trustworthy coupled atmosphere/ocean models that are both well verified and economical enough to be run over the hundred years needed for greenhouse gas-transient simulations. Thus, the simple model helps to identify and bound potentially important problems, provide some quantitative sensitivity studies, and help set priorities for three-dimensional coupled model research or needed observational programs over the next few decades. Since the agricultural and other environmental impacts of increasing greenhouse gases depend on the specific regional and seasonal distribution of climatic change, resolution of the transient climate response debate, among others, is critical for climatic impact assessment and ultimate policy responses to the advent or prospect of increasing greenhouse gases.

The Current Global Warming Debate: Science or Media Hype?

Before discussing any potential management responses to such projections, let me first point out that not all knowledgeable scientists are in agreement as to the probability that such changes will occur. In fact, if one has followed the very noisy, often polemical debate in the media recently, one might get the (I believe false) impression that there are but two radically opposed schools of thought about global warming: 1) that climatic changes will be so severe, so sudden, and so certain that major species extinction events will intensify, sea-level rise will create tens of millions of environmental refugees, millions to perhaps billions of people will starve, and devastated ecosystems are a virtual certainty or, alternatively, 2) there is nothing but uncertainty about global warming, no evidence that the 20th century has done what the modelers have predicted, and the people arguing for change are just "environmental extremists"—thus, there is no need for any management response to an event that is improbable and in no case should any such responses interfere with the "free market" and bankrupt the nation (for example, see the cover story of *Forbes Magazine*, Brookes, 1989). Unfortunately, while such a highly charged and polarized debate makes entertaining opinion page reading or viewing for the ratings dominated media, it provides a very poor description of the reality of the actual scientific debate or the broad consensus on basic issues within the scientific community. In my opinion, the "end of the world" or "nothing to worry about" are the two *least likely* cases, with almost any scenario in between having a higher probability of occurrence.

Figure 4 shows a projection of global warming possibilities into the 21st century drawn by an international group of scientists that was convened by the well-established International Council of Scientific Unions. It shows warming from a very moderate additional one-half degree C (0.9°F) up to a catastrophic 5°C (9°F) or greater warming before the end of the next century. I do not hesitate to call the latter extreme catastrophic because that is the magnitude of warming that occurred between about 15,000 years ago and 5,000 years ago—the end of the last ice age to our present interglacial epoch. It took nature some 5,000 to 10,000 years to accomplish that warming, and it was accompanied by a 100-meter (330-foot) or so rise in sea level, thousands-of-kilometers migration of forest species, radically altered habitats, species extinctions, species evolution, and other major environmental changes. Indeed, the ice age to interglacial transition revamped the ecological face of the planet. If the mid-to-upper part of the scenarios occurred as rapidly as they are projected on Figure 4, then indeed they would justify the substantial concern that many scientists have over the prospect of global warming. On the other hand, the many unknown factors which make most scientists, myself included, hesitant to make anything other than qualitative or intuitive probabilistic kinds of forecasts could well suggest a 21st century climate in which global temperature change would be only a degree or so. While even that seemingly small change might be serious for certain life forms (for example, some plant or animal species living near the tops of mountains that would be driven to extinction with even a small warming), by and large changes of less than a degree or so taking place over a century or more would clearly add less stress to natural (and certainly to human) systems than would changes of several degrees taking place in 50 years or less. Indeed, the rate of change may be a most critically important factor to the adaptive capacity of both humans and natural systems, particularly the latter, since ecosystems do not have the option of planting new seeds to match the new climate the way our farmers do.

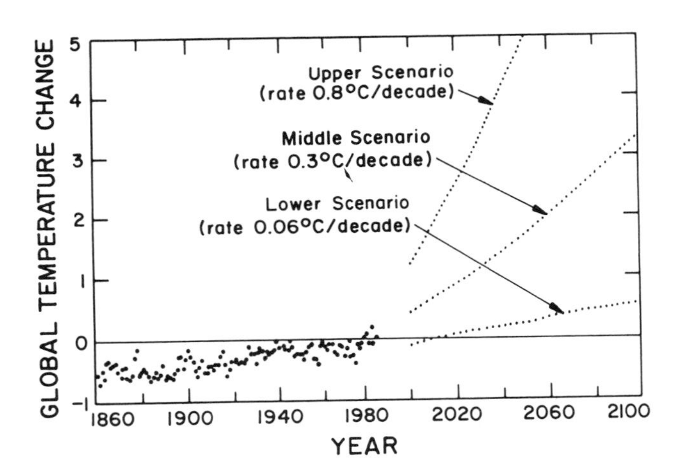

Figure 4. Three scenarios for global temperature change to the year 2100 derived from combining uncertainties in future trace greenhouse gas projections with uncertainties of modeling the climatic response to those projections. Sustained global temperature changes beyond 2°C (3.6°F) would be unprecedented during the era of human civilization. The middle to upper range represents climatic change at a pace 10 to 100 times faster than typical long-term natural average rates of change. [Source: Jaeger, J. April 1988, "Developing Policies for Responding to Climatic Change: A Summary of the Discussions and Recommendations of the Workshops Held in Villach 28 September to 2 October 1987" (WCIP-1, WMO/TD-No. 225)]

Critics of immediate policy responses to global warming are quick to point out the many uncertainties that could reduce the average projections made by climate models (such as the middle line on Figure 4). Indeed, most climate modelers include similar caveats in their papers, and many of us do resent somewhat the implications of some critics that they are the ones who are responsibly pointing out these uncertainties to the public, whereas the modelers are somehow deliberately suppressing uncertainties in order to overstate the issue (e.g., see *Detroit News*, 1989, and Schneider, 1989a). Many critiques (for example, see George C. Marshall Institute, 1989) somehow forget to stress that the sword of uncertainty has two blades: that is, uncertainties in physical or biological processes which make it possible for the present generation of models to have overestimated future warming effects are just as likely to have caused the models to have underestimated change.

The public policy dilemma is how to act even though we will not know in detail what will happen—it is my opinion that the scientific community will not be able to provide definitive information over the next decade, or perhaps two, about the precise timing and magnitude of century-long climate changes, especially if research efforts remain at current levels. Public policymakers will have to address how much information is "enough" to warrant action and what kinds of measures can be taken to deal with the plausible range of environmental changes. Unfortunately, the probability of such changes cannot be estimated by definitive analytical methods. Rather, we will have to rely on the intuition of experts, which is why a highly confusing and polarized media debate can be paralyzing to anticipatory management. Fortunately, making such scientific judgments is indeed the purpose of deliberative bodies such as the US National Academy of Sciences (NAS) or the International Council of Scientific Unions. The NAS regularly convenes a spectrum of experts to provide the best estimates of the probabilities of various scenarios of change. These people can deliberate to a considerable extent away from the confusion of noisy media debates in which extreme opposites are typically pitted. Half a dozen such assessments over the past 10 years have all reaffirmed the plausibility of unprecedented climate change building into the next 50 to 100 years.

Let me briefly summarize the arguments of critics, some of whom challenge these assessments. Many critics contend that the warming trend of the past century of about 0.5°C (e.g., Jones and Wigley, 1989) is suspect because the thermometer record is not very reliable (e.g., Ellsaesser, 1984). Of course, the scientists who produce such records say the same thing (e.g., Karl and Jones, 1988), but many are not certain whether the needed corrections will necessarily reduce the trend as opposed to the other direction. Moreover, some critics cite the neglect of ocean temperature data collected from buckets dropped over the sides of schooners in the Victorian or pre-Victorian times, since some of those records suggest that the 1850s may have seen ocean temperatures nearly as warm as the present (e.g., Bottomley et al., 1990; Lindzen, 1990). The reason those kinds of pre-1900 ocean temperature data are typically discounted in most assessments is that they were collected over only a few percent or so of the earth's surface (see Figure 5). In addition, the measurements simply are not reliable. The most reliable statement that seems reasonable to infer from the available temperature records is that a warming of some 0.5°C (0.9°F) has occurred globally over the past 100 years (e.g., IPCC, 1990). If greenhouse gas pollution were the *only* cause of that warming trend, then this is broadly consistent with the middle of the lower one half of the projected range of warming made by climate models as seen in Figure 4. Does this mean that nature has already told us that future global warming will be one half of what most models typically project? Unfortunately, we have only been accurately measuring the energy output of the sun from space over the past 10 years or so and have very little knowledge of the precise quantitative nature of this or other factors that could have influenced the temperature trends this century. Without accounting precisely for such factors, even accurate temperature data over the past century could not tell us very much about how the earth has responded to the pollution injected since pre-industrial times (e.g., Gilliland and Schneider, 1984; Wigley and Raper, 1990). Furthermore, although some critics have suggested that a century-long heating of the sun by a few tenths of a percent could account for the 20th century warming of the earth (e.g., George C. Marshall Institute, 1989), these critics often forget to mention that it is equally likely, since it was essentially unmeasured, that the sun could have cooled down by that amount, thereby having damped any greenhouse effect that otherwise would have been in the record, thus fooling us into thinking that the global warming from pollution to date is one half of what may have actually occurred. Quite simply, the critics cannot have it both ways—what we do not know could increase or decrease our estimates.

Finally, the principal reason that advocates of concern over the prospects of global warming—and I am

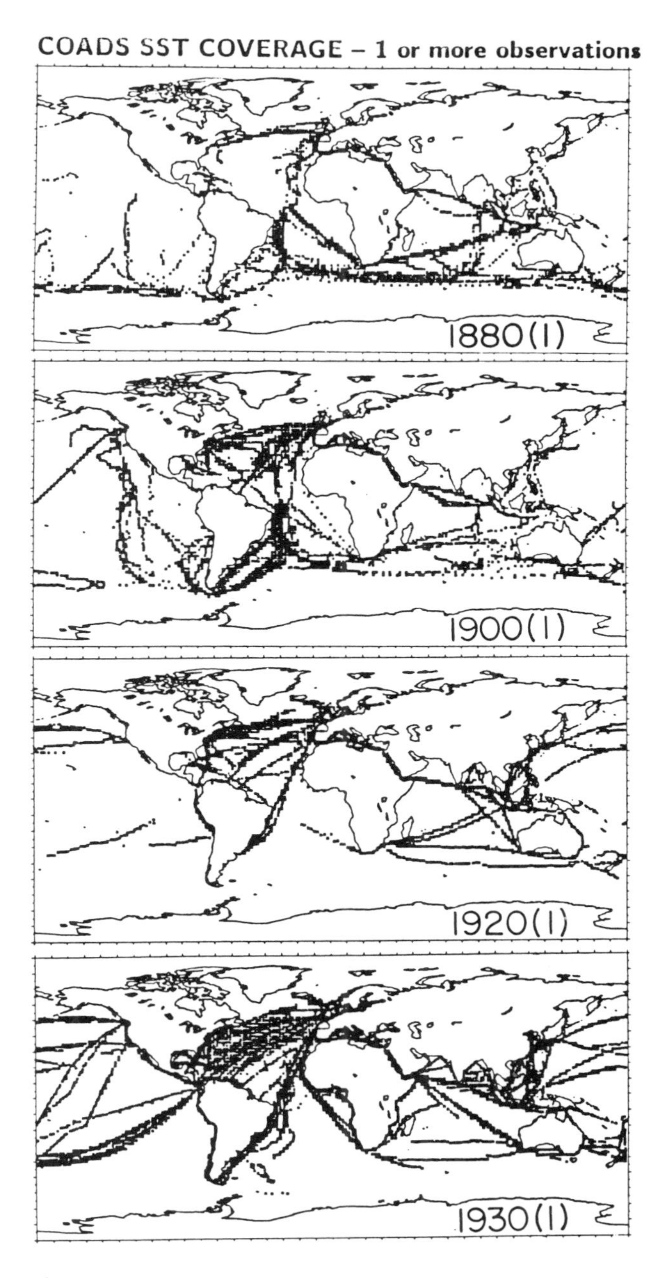

Figure 5. Ship tracks over which one or more sea surface temperature measurements were made during January of four different years. Coverage before 1900 is so sparse and measurement accuracy so questionable that very little reliability can be attached to temperature trend reconstructions before the 20th century. (Source: Dennis Shea and K. Trenberth, National Center for Atmospheric Research, private communication).

unabashedly one of them (for example, in my recent book, *Global Warming: Are We Entering the Greenhouse Century?* Schneider, 1989b)—stand before groups such as Congressional committees with our concerns is not based solely on speculative theory. Rather, our concern is because the models we use to foreshadow the future have already been validated to a considerable degree, although not to the full satisfaction of any responsible scientist. For example, we know from observations of nature, a point often neglected by the critics of global warming, that the last ice age, which was about 5°C (9°F) colder than the present era, also had carbon dioxide levels about 25 percent less than the pre-industrial values. Methane, another very potent greenhouse gas, also was reduced by nearly a factor of two relative to pre-industrial values. Ice cores in Antarctica have shown us, since these cores contain gas bubbles that are records of the atmospheric composition going back over 150,000 years, that the previous interglacial age some 125,000 to 130,000 years ago had CO_2 and methane levels comparable to that in the preindustrial times of the present interglacial. The nearly simultaneous change in these greenhouse gases and in planetary temperature over geological epochs (see Figure 6) is roughly what one would expect based on projections from today's generation of computer models. However, we still cannot assert that this greenhouse gas-temperature coincidence is proof that our models are quantitatively correct, since other factors were operating during the ice age-interglacial cycles. The best we can say is that the evidence is strong, but circumstantial.

Other forms of validation of the sensitivity of climate models to radiation changes involve the very large changes that occur between winter and summer or during ancient times, as was discussed earlier. If there were a gross error (e.g., a factor of 10) in estimating how cloudiness changes, for example, that could amplify or damp the models' projections to hypothetical scenarios of increased CO_2, the models would simply be incapable of capturing the seasonal cycle of surface temperature to anywhere near the degree of precision they appear to. Figure 1 showed that, indeed, the seasonal cycle is typically well simulated. Cloudiness changes would occur very rapidly relative to the time-frame of a season, but effects of the deep ocean would not. Therefore, the seasonal cycle test is not a direct validation of all potential global warming changes, but again, is strong circumstantial evidence. It is not a test of the reliability of a model forecast of the regional change in storm tracks, for example, which is so critical to land management. On the other hand, the seasonal test does give us, say, a factor of two to three confidence that the magnitude of global temperature change projected in Figure 4 has a good chance of being correct—indeed, that is why I often refer to the prospect for global warming in the 21st century as "coin-flipping odds of unprecedented change," since a sustained global change of more than

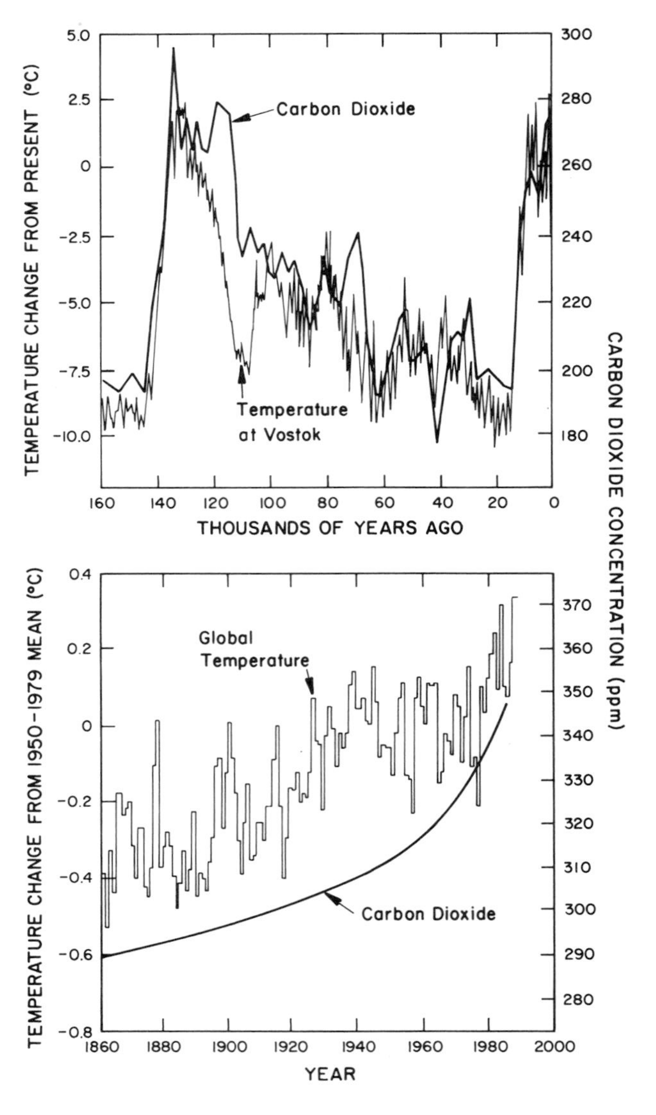

Figure 6. Carbon dioxide and temperature are very closely correlated over the past 160,000 years (top) and, to a lesser extent, over the past 100 years (bottom). The long-term record, based on evidence from Antarctica, shows how the local temperature and atmospheric carbon dioxide rose nearly in step as an ice age ended about 130,000 years ago, fell almost in synchrony at the onset of a new glacial period and rose again as the ice retreated about 10,000 years ago. The recent temperature record shows a slight global warming as traced by workers at the Climatic Research Unit of the University of East Anglia. Whether the accompanying buildup of carbon dioxide in the atmosphere caused the half-degree warming is hotly debated. (Source: Data for top: Laboratory de Glaciologie et de Geophysique de l'environment, France. Data for bottom: Climatic Research Unit, University of East Anglia, UK).

2°C (3.6°F) would be unprecedented during the past 10,000 year era of human civilization's development.

One final comment about the scientific discussions, often not well represented in the public debate, is that it is not true that the bulk of the scientific community is in constant controversy and disarray about the basic science of the greenhouse effect. The "greenhouse effect," the heat-trapping properties of the atmosphere and its gases and particles, is well understood and well validated. Indeed, it is as good a theory as there is in the atmospheric sciences. It explains, for example, the very hot conditions under the thick atmosphere of Venus and the very cold conditions under the thin, weak "greenhouse" of Mars. It explains the thousands of laboratory observations of the transfer of radiant energy through various gases, the millions of aircraft and balloon observations of the earth's temperature structure and its radiative fluxes, and the literally billions and billions of satellite observations of the same quantities. Moreover, very recently A. Raval and V. Ramanathan used satellite observations to study the very important water vapor–greenhouse feedback mechanism, a process that is central to most models' estimates of some 3° plus or minus 1.5°C equilibrium warming from doubling CO_2. They conclude that "The greenhouse effect is found to increase significantly with sea surface temperature. The rate of increase gives compelling evidence for the positive feedback between surface temperature, water vapour and the greenhouse effect; the magnitude of the feedback is consistent with that predicted by climate models." (Raval and Ramanathan, 1989). In other words, the heat-trapping capacity of the atmosphere is well understood and well measured on earth, and much of the sometimes polemical debate in the media over the greenhouse effect has little basis in reality. The empirical confirmation of the natural greenhouse effect, which is consistent with the greenhouse effect of climate models, stands in stark contrast to the theoretical arguments on physical principles of Ellsaesser (1984) or Lindzen (1990) that negative temperature–water vapor feedback processes in parts of the tropics will reduce present model estimates of global warming by a factor of four.

It is well known that the 25 percent increase in carbon dioxide documented since the industrial revolution, the 100 percent increase in methane since the industrial revolution, and the introduction of synthetic chemicals such as chlorofluorocarbons (also responsible for stratospheric ozone depletion) since the 1950s should have trapped roughly two extra watts of radiant energy over every square meter of earth. That part is well accepted by most climate specialists. However, what is less well accepted is how to translate that two watts of heating into "X" degrees of temperature change, since this involves assumptions about how that heating will be distributed among surface temperature rises, evaporation increases, cloudiness changes, ice changes and so forth. The factor of two to three uncertainty in global temperature rise projections as cited in typical National Academy of Sciences' reports reflects a legitimate estimate of uncertainty held by most in the scientific community. Indeed, recent modifications of the British Meteorological Office climate model to attempt to mimic the effects of cloud droplets halved its model's sensitivity to doubled CO_2—but it is still well within the often-cited 1.5–4.5°C range. Of course, the authors of the study, Mitchell, Senior, and Ingram (1989), wisely pointed out that "although the revised cloud scheme is more detailed, it is not necessarily more accurate than the less sophisticated scheme." I have never seen this forthright and important caveat quoted by any of the global warming critics who cite the British work as a reason to lower our concern by a factor of two or so. Finally, as stated earlier, prediction of the detailed regional distribution of climatic anomalies, that is, where and when it will be wetter and drier, how many floods might occur in the spring in California or forest fires in Wyoming in August, is simply highly speculative, although some plausible scenarios can be given. Some such scenarios are given in Table 1, from the National Academy of Sciences' 1987 assessment.

Although climatic models are far from fully verified for future simulations, the seasonal and paleoclimatic simulations are strong evidence that state-of-the-art climatic models already have considerable skills. An awareness of just what models are and what they can and cannot do is probably the best we can ask of the public and its representatives. Then, the tough policy problem is how to apply the society's values in facing the future given the possible outcomes that climatic models foretell. Modelers will continue to develop and refine new models by turning to larger computers to run them and more observations to improve and verify them. We must ask the indulgence of society to recognize that immediate, definitive answers are not likely, as coupling of higher-resolution atmosphere and ocean (i.e., land surface and chemistry) submodels will take a decade or more to develop. In essence, what climate models and their applications typify is a growing class of problems not unique to climate but also familiar in other disciplines: nuclear waste disposal, safety of food additives or drugs, efficacy of strategic defense and so forth. These are problems for which guaranteed "scientific" answers cannot be obtained—except by performing the experiment on ourselves. To deal with these complex socio-technical problems requires a new understanding of the central role of uncertainty and the willingness to deal with probabilistic estimation. It also calls for more modeling—provided the context of that modeling is understood and parallel observational programs for validation are concurrently pursued. Efficient functioning of society will depend upon our understanding of both the utility and limitations of models. However, in some senses more than economic efficiency is at stake, as the very survival of some societies—particularly in low lying

Table 1. Possible Climate Changes from Doubling of CO_2.

Large Stratospheric Cooling (virtually certain). Reduced ozone concentrations in the upper stratosphere will lead to reduced absorption of solar ultraviolet radiation and therefore less heating. Increases in tne stratospheric concentration of carbon dioxide and other radiatively active trace gases will increase the radiation of heat from the stratosphere. The combination of decreased heating and increased cooling will lead to a major lowering of temperatures in the upper stratosphere.
Global-Mean Surface Warming (very probable). For a doubling of atmospheric carbon dioxide (or its radiative equivalent from all the greenhouse gases), the *long-term* global-mean surface warming is expected to be in the range of 1.5-4.5°C. The most significant uncertainty arises from the effects of clouds. Of course, the *actual* rate of warming over the next century will be governed by the growth rate of greenhouse gases, natural fluctuations in the climate system, and the detailed response of the slowly responding parts of the climate system, i.e., oceans and glacial ice.
Global-Mean Precipitation Increase (very probable). Increased heating of the surface will lead to increased evaporation and, therefore, to greater global mean precipitation. Despite this increase in global average precipitation, some individual regions might well experience decreases in rainfall.
Reduction of Sea Ice (very probable). As the climate warms, total sea ice is expected to be reduced.
Polar Winter Surface Warming (very probable). As the sea ice boundary is shifted poleward, the models predict a dramatically enhanced surface warming in winter polar regions. The greater fraction of open water and thinner sea ice will probably lead to warming of the polar surface air by as much as three times the global mean warming.
Summer Continental Dryness/Warming (likely in the long term). Several studies have predicted a marked long-term drying of the soil moisture over some mid-latitude interior continental regions during summer. This dryness is mainly caused by an earlier termination of snowmelt and rainy periods, and an earlier onset of the spring-to-summer reduction of soil wetness. Of course, these simulations of long-term equilibrium conditions may not offer a reliable guide to trends over the next few decades of changing atmospheric composition and changing climate.
High-Latitude Precipitation Increase (probable). As the climate warms, the increased poleward penetration of warm, moist air should increase the average annual precipitation in high latitudes.
Rise in Global Mean Sea Level (probable). A rise in mean sea level is generally expected due to thermal expansion of sea water in the warmer future climate. Far less certain is the contribution due to melting or calving of land ice.

(Source: National Academy of Sciences, 1987)

coastal areas—will depend on present decisions to deal with plausible climatic futures.

If the public is totally ignorant of the nature, use, or validity of climatic (or many other kinds of) models, then public policymaking based on model results will be haphazard at best. In this case, the decisionmaking process tends to be dominated by special interests or a technically trained elite.

Adaptation or Prevention?

Public responses to the advent or prospect of global climate change typically come in two categories: adaptation and prevention. With regard to *adaptation*, flexibility of adaptive measures needs to be considered now. Indeed, if the entire global warming debate comes out closer to the views of the present critics and only small changes occur, what would be lost, for example, by improving the flexibility of water supply systems? After all, nature will continue to give us wet and dry years. The 100-year flood will happen sometime, as will the 100-year drought—with or without global warming, which of course, could change the odds of such extremes. Therefore, increasing management flexibility will pay dividends (e.g., see Waggoner, 1990) even in (what I believe to be) the unlikely event that global warming proves to be minimal. This situation provides a metaphor to buying insurance. Only a foolish or desperate person would over- or underinvest in insurance. Here, unlike the insurance metaphor, where a premium is "wasted" if one does not collect on any damages, we actually can get benefits for our investment in flexibility as we buy insurance against the prospect of rapid climate change at the same time. Of course, no one gets a return on investment without making an investment, and that is true not just in the private sector, but obviously in the public sector as well.

The other category of management response is *prevention*. That simply means slowing down the rate at which the gases injected into the atmosphere that can modify the climate are produced. The principal way to do that falls primarily in the jurisdiction of energy use and production managers. However, since deforestation is an important component of CO_2 production, halting deforestation can help to reduce the atmospheric buildup of CO_2. Thus, prevention is also in the purview of such land management agencies. For example, domestic animals are a major source of methane, and such livestock graze extensively on public lands. Therefore, solutions to methane emission controls will involve consideration of the use of public lands for this purpose. As another example, microbial communities in soils decompose dead organic matter into greenhouse gases, such as carbon dioxide, methane, or nitrous oxide. Since these microbes typically increase their metabolic activity when the soil warms, deforestation or grazing which removes vegetation cover and results in warmer soils could enhance the production of these greenhouse gases (e.g., Lashof, 1989). These issues are contentious, but nonetheless need to be examined for their potential effectiveness as emission control strategies.

The best strategies should have high leverage—that is, help to solve more than one problem with a single investment. It is my personal view that such high leverage or "tie-in" strategies are the best approach to dealing with prevention, or at least delaying the rate of the buildup of greenhouse gases, and thus the prospect of rapid global warming (e.g., Schneider, 1989b). It is obvious that the things to do first are the things that make the most sense regardless of whether global warming materializes. Energy efficiency, as is often mentioned, is the single most important "tie-in" strategy. Using energy efficiently will not only reduce the prospect of rapid climate change, but also reduce acid rain (itself a threat to many natural lands), reduce air pollution in cities, reduce balance of payment deficits in energy importing countries by reducing dependence on foreign supplies of energy, and in the long run improve product competitiveness by reducing the energy components of manufactured products (unfortunately about twice as high for the United States as for our more efficient competitors, Japan, Italy, or West Germany) (e.g., Chandler, 1988).

I believe that flexibility for adaptation and high-leverage strategies for prevention are not premature policy actions. If we are to manage a future with increasing uncertainty in environmental conditions effectively, it is imperative that management flexibility be increased. *Ideology is the enemy of flexibility.* It is incumbent on all of us—individuals, businesses, and governments—to rethink any of our ideologically rigid positions in order to fashion ways to enhance the flexibility of management. This can provide the opportunity to manage the potential consequences of serious climate change more effectively, as well as improve the capacity of present systems to deal with the natural variability of the environment, itself a well-demonstrated threat to many of our abilities or resources.

References

Bottomley, M., Folland, C.K., Hsiung, J., Newell, R.E., and Parker, D.E., 1990. "Global Ocean Surface Temperature Atlas," United Kingdom Meteorological Office, Bracknell.

Brookes, W.T., 1989. "The Global Warming Panic," *Forbes*, 25 December 1989, 96–102.

Chandler, W.U., Geller, H.S. and Ledbetter, N.R., 1988. *Energy Efficiency: A New Agenda*, GW Press, Springfield, VA.

COHMAP Members (P.M. Anderson et al.), 1988. "Climatic Changes of the Last 18,000 Years: Observations and Model Simulations," *Science* 241, 1043–52.

Detroit News, 1989. "Loads of Media Coverage," Editorial, 22 November 1989.

Dickinson, R.E. and Cicerone, R.J., 1986. "Future Global Warming from Atmospheric Trace Gases," *Nature* 319, 109–115.

Ellsaesser, H.W., 1984. "The Climatic Effect of CO_2: A Different View," *Atmos. Environ.* 18, 431–434.

Gilliland, R.L. and Schneider, S.H., 1984. "Volcanic, CO_2 and Solar Forcing of Northern and Southern Hemisphere Surface Air Temperatures," *Nature* 310, 38–41.

Intergovernmental Panel on Climate Change (IPCC), (2nd Draft) March 1990. *Scientific Assessment of Climate Change*, World Meteorological Organization, Geneva.

Jaeger, J., April 1988. "Developing Policies for Responding to Climatic Change: A Summary of the Discussions and Recommendations of the Workshops Held in Villach 28 September to 2 October 1987," (WCIP-1, WMO/TD-No. 225).

Jones, P.D. and Wigley, T.M.L., 1990. "Global Warming Trends," *Scientific American* 263, No. 2, August, 84–91.

Karl, T.R. and Jones, P.D. 1989. "Urban Bias in Area-averaged Surface Air Temperature Trends," *Bull. Amer. Meteor. Soc.* 70: 265–270.

Kutzbach, J.E. and Guetter, P.J., 1986. "The Influence of Changing Orbital Parameters and Surface Boundary Conditions on Climate Simulations for the Past 18,000 Years," *J. Atmos. Sci.* 43, 1726–1759.

Lashof, D.A., 1989. "The Dynamic Greenhouse: Feedback Processes That May Influence Future Concentrations of Atmospheric Trace Gases and Climatic Change," *Climatic Change* 14, 213–242.

Lindzen, R., 1990. "Some Coolness Concerning Global Warming," *Bull. Amer. Meteor. Soc.* 77, 288–299.

Manabe, S. and Stouffer, R.J., 1980. "Sensitivity of a Global Climate Model to an Increase of CO_2 Concentration in the Atmosphere," *Journal of Geophysical Research*, 85: 5529–5554.

Manabe, S. and Wetherald, R., 1986. "Reduction in Summer Soil Wetness Induced by an Increase in Atmospheric Carbon Dioxide," *Science*, 232, 626.

George C. Marshall Institute, 1989. *Scientific Perspectives on the Greenhouse Problem*, Washington, DC, 37 pp.

Mearns, L.O., S.H. Schneider, S.L. Thompson and L.R. McDaniel. "Analysis of Climate Variability in General Circulation Models: Comparison with Observations and Changes in Variability in $2xCO_2$ Experiments." *J. Geophys. Res.* (in press).

Mitchell, J.F.B., W.J. Ingram and C.A. Senior, 1989. "CO_2 and Climate: A Missing Feedback?" *Nature* 341, 132–134.

National Academy of Sciences, 1987. *Current Issues in Atmospheric Change*, National Academy Press, Washington, DC.

Raval, A. and Ramanathan, V., 1989. "Observational Determination of the Greenhouse Effect," *Nature* 342, 758.

Rind, D., R. Goldberg and R. Ruedy. 1989. "Change in Climate Variability in the 21st Century," *Climatic Change* 14: 5–37.

Schneider, S.H., 1989a. "News Plays Fast and Loose with the Facts," *Detroit News*, 5 December 1989.

Schneider, S.H., 1989b. *Global Warming: Are We Entering the Greenhouse Century*? Sierra Club Books, San Francisco.

Schneider, S.H., and Thompson, S.L., 1981. "Atmospheric CO_2 and Climate: Importance of the Transient Response," *J. Atmos. Sci.* 37, 895–900.

Shea, D. and Trenberth, K., National Center for Atmospheric Research, private communication.

Stouffer, R.J., Manabe, S. and K. Bryan. 1989. "Interhemispheric Asymmetry in Climate Response to a Gradual Increase of Atmospheric CO_2," *Nature* 342, 660–662.

Waggoner, P.E. (ed.), 1990. *Climate Change and US Water Resources*, John Wiley and Sons, NY.

Washington, W.M. and Meehl, G.A., 1989. "Climate Sensitivity Due to Increased CO_2: Experiments with a Coupled Atmosphere and Ocean General Circulation Model," *Climate Dynamics* 4, 1–38.

Wigley, T.M.L. and S.C.B. Raper, 1990. "Natural Variability of the Climate System and Detection of the Greenhouse Effect," *Nature* 344, 324-327.

The Interface of Environmental Science, Technology, and Policy

John H. Gibbons
Office of Technology Assessment
United States Congress
Washington, DC

The successes and excesses of the Industrial Revolution led to the earliest US energy policies—subsidies to encourage production, and anti-trust legislation to maintain some competitiveness in the field. Later, in the 1970s and early 1980s, security concerns engaged our interest in energy policy. When MIT established its Energy Laboratory, US consumers and politicians feared OPEC's domination of world energy markets—we sensed price gouging and foresaw no end to the escalations; we also knew they could interrupt oil supplies with very little effort. Emerging awareness of environmental externalities associated with energy use (particularly coal and nuclear power) captured some attention in those days, but national security was the prevalent concern.

Since then, new research, new policies, and the simple passage of time have led to some important changes. We learned about the large, long-run elasticity of demand to energy price; we successfully demonstrated that there is no fixed, long-term relationship between energy and economic growth; the basis of oil production broadened beyond the Middle East; we established the Strategic Petroleum Reserve (SPR); we relied on technology to increase supplies—nuclear power now delivers nearly 20 percent of our electricity—and decrease demand growth—still near 1973 levels despite a 40 percent increase in GNP.

That's a lot of progress, but there are signs of nonsustainability. Energy efficiency is slipping; oil imports have reached levels equivalent to those in existence just prior to the embargo and account for about one third of our negative balance of payments; hopes for discoveries of new giant domestic oil fields go unfilled and petroleum production is falling. We're increasingly troubled by the intimate links between energy use and environmental problems, from nuclear waste, to urban ozone, to acid rain, to global warming, and by the paucity of progress in dealing with such problems.

Though security concerns remain, environmental and economic concerns play an ever larger role in energy policy. In fact, many observers view the central, national issues of energy and environment as two sides of the same policy coin—the policy being to provide the goods and services of a strong, competitive, industrial economy with as little cost (including externalities) as possible.

Throughout the seventies and eighties, when national security was our chief energy "worry," policymakers acted on public concerns without scientific certainty to guide them. We spent billions to defend our energy interests around the globe, including the Persian Gulf, with no certainty that the expenditures would avoid war or that failure to spend would make interruptions more likely. We made the decision to fill the SPR without knowing where future oil prices would go, and daily, quietly, pay a huge price for this insurance against the havoc of oil supply disruptions. We spent millions to assure nuclear safety in worst case scenarios that likely will never occur. Our leaders took the best available information and combined it with experience to devise policy. And these policies are justified as hedges against particular risks, not on the basis of secondary gains.

Scientific uncertainty remains the rule even though emphasis has shifted from security to environment. We still have few firm measures of the health and environmental effects of air pollution, but we do know the effects that individual pollutants can have on the body. We cannot yet detect the signal of anthropogenic climate change in the midst of El Niño, volcanic eruptions, and other natural phenomena. But we know the greenhouse effect is real and insidious, in that by the time the effects are inarguably observed, it will be too late to ameliorate them. Perhaps instinct has not yet coupled with knowledge to persuade policymakers to insure against such contingencies. Indeed, energy and environmental policy seems to invoke a "show me" attitude in many people. Today's leaders have seen war, seen the effects of nuclear radiation, but health risks are many, and massive changes in climate occurred long before any of us were born. As René Dubos once observed, the tragedy of humankind is that we are both able and willing to adapt to almost any change that comes slowly, even if such change is for the worse.

Uncertainty and paradox drive the scientific process—force the next question, the next experiment. They need not and do not stymie the political process. Scientific understanding and consensus greatly aid policymaking, but the many structures, functions, and levels of government normally militate against "one right answer" regardless of scientific opinion. The very nature of the political process is that decisions must be made despite uncertainty and sharp differences of opinion;

that there are times when decisions must be made, with or without full analysis; and that the business of policymaking is a *process*, not a unique event in time. More careful systems studies, modeling, and scenario analysis can lead us toward greater coherence, consistency, and equity, but policy is made and tuned incrementally, based on the tools we have available at a particular time. A strategy that combines an initially conservative position (that reflects scientific uncertainty), followed by actions that have acceptable costs based on that position and continued pursuit of greater scientific understanding, and followed further by policy updates that reflect superior knowledge appears to have worked in several situations. For instance, we set a standard for worker radiation exposure because that permitted us to go forward with nuclear applications in industry, electric utilities, and medicine, but we continued research and eventually tightened the standard. On the other hand, initial fears and restrictions of environmental release of genetically engineered organisms have been weakened after continuing research and experience provided justification. Such a strategic approach to energy and environmental policy is sensible, for although it is a political tarbaby we cannot let go of, the rewards for attentiveness are many.

I have previously postulated a goal of "least cost provision of goods and services" for energy and environmental policy. A realistic approach to this goal requires us to lift our sights above the historical focus on expanding supplies, cleaning up associated messes, and, almost as an aside, making consumption more efficient. That focus ignores our planetary limits, and as Kenneth Boulding points out, "The only people who believe in infinite expansion in a finite system are economists and other madmen." Technology is not the Holy Grail; it cannot provide everlasting life to dwindling supplies and a rapidly degrading environment in the face of indefinitely expanding population and economic activity. To achieve energy and environmental sustainability, we will have to integrate historical goals of production and efficiency with the goals and outlooks for population growth and level of economic activity. We must think on a global rather than a regional or national scale, for international commerce and governance will greatly affect our own ability to deal with these problems.

Hopefully, we will have begun to take Boulding's admonition to heart long before our technological options of improved efficiency and more benign supplies run out. Those options, however, are not ultimate solutions, but mechanisms for realizing two goals: first, to mitigate damage while new, long-term solutions are devised and population growth is stabilized; and second, to create a situation where, under future conditions of a given amount of energy use, the corresponding amount of environmental impact will be minimized (see Figure 1).

What does state-of-the-art science and technology offer today's energy and environmental policymaker? On the supply side, nuclear fission power remains attractive to some, in large part because of the paucity of alternatives. Many energy planners favor increased reliance on nuclear power, since its use does not emit carbon dioxide or the pollutants that cause acid rain. Current generation light-water reactors, however, are increasingly expensive to build and require elaborately engineered safeguards. Accidents at Three Mile Island and Chernobyl have tarnished public perception of reactor safety and reliability. Much of the public doubts that adequate radioactive management and disposal techniques exist or can be developed or that a remedy for proliferation of weapons-grade nuclear materials has been found. Such doubts hinge not on lack of technology or scientific understanding, per se, but on lack of confidence in the individuals and institutions that are necessarily involved. These doubts are much tougher to resolve than technical problems. As a consequence of these handicaps, the growing attractiveness of other energy sources, and the slow-down of demand growth, US utilities have not ordered a new nuclear plant since 1978.

Advanced reactor designs could help restore the promise of nuclear energy. "Passively stable" reactors could be built that would prevent runaway chain reactions without relying on an external control system. Standardized designs could improve quality control, reduce construction costs, licensing complexities, and downtime. Perhaps most important, successful demonstration of radioactive waste-disposal capabilities will be critical to widespread public acceptance of nuclear power.

New electric-generation technologies promise greater efficiency around the globe. Fluidized-bed combustion, in which burning coal is suspended (fluidized) and mixed with crushed limestone in a stream of air, can reduce emissions of pollutants. Some analysts feel that a very promising future option for electric-power generation is the aeroderivative turbine, which is based on jet engine designs and burns gas. With additional refinement, this technology could raise electric conversion efficiency from today's 33 percent to more than 45 percent.

Solar energy, though often viewed as an expensive "soft path," is, in fact, a very hard "high-tech" path. It is one of a precious few potential successors to fossil fuels, which we now view as far more limited than we did a few years ago. Solar energy markets continue to expand, and unlike nuclear power, the price of solar energy is projected to continue to drop (see Figure 2). Electricity produced by photovoltaic cells, which directly convert sunlight to electricity, now costs 30 cents per kilowatt-hour and is already a common power source for remote power, calculators, watches, and satellites. These small-scale applications and niche markets help to sustain the industry as the technology develops, but photovoltaic cells remain more expensive than conven-

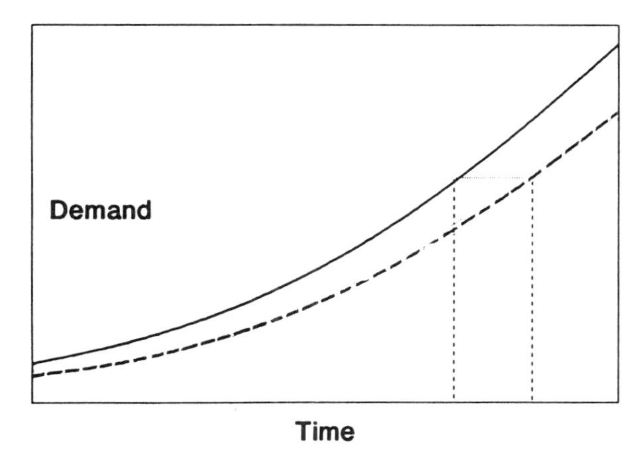

The change in efficiency of energy use translates to only delaying for a few years the inevitable growth of demand.

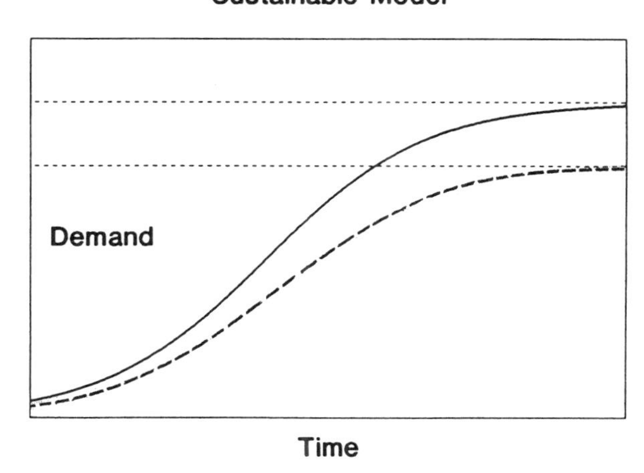

The change in efficiency of energy use translates to a permanent lowering of demand and, therefore, pollution for a given level of economic activity.

Figure 1. Alternative growth models.

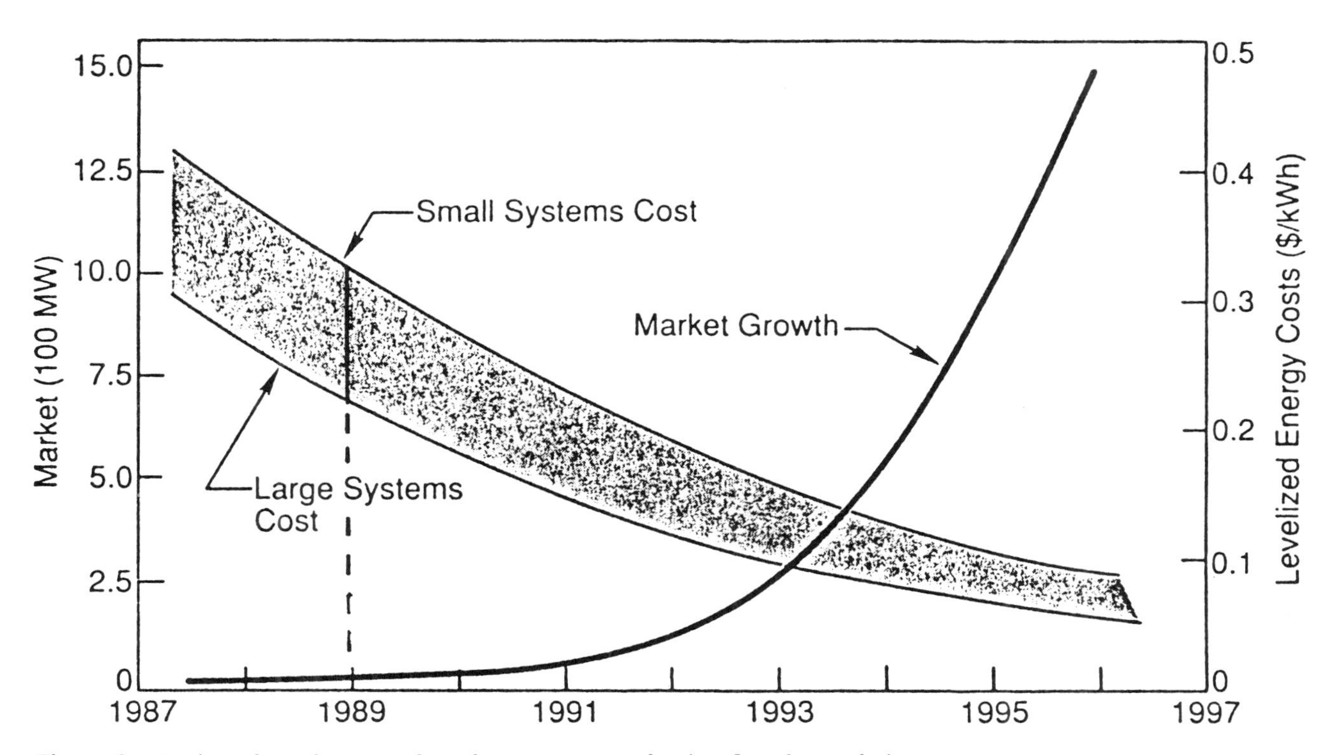

Figure 2. Projected market growth and energy cost reduction for photovoltaics.

tional electricity generation for most applications. Further advances in microelectronic and semiconductors promise to increase efficiency and further reduce costs.

Solar-power generation is also taking place on a larger scale at plants that convert solar energy to heat. In these solar-central thermal systems, mirrors or lenses focus sunlight onto a receiver containing a fluid that then conducts heat to a conventional electric generator. According to its owners, the Luz plant in California can produce electricity for eight cents per kilowatt-hour, compared to three cents per kilowatt-hour for electricity generated by conventional combustion turbines. Research on advanced lightweight mirrors and better heat-transfer fluids, such as molten salts, may improve these results. Hydroelectric power, biomass, wind power, and geothermal energy are all potential replacements for our current fuels (Gibbons, Blair, et al., 1989).

The successes of the past two decades have not exhausted the potential for contributions from energy efficiency to energy and environmental policy. The energy productivity of the US economy gained more than 24 percent between 1973 and 1986 (see Figure 3), partly through efficiency gains and partly because of changes in the mix of goods and services produced. Throughout the 1970s, actions to improve energy efficiency occurred primarily in industry, transportation, commercial buildings, and residences when fuel was a very significant operating cost and, therefore, when the payback for such an investment could be realized very quickly. Actions were often accelerated even further by public policy incentives. Some of the actions involved changes in patterns of energy use, such as lowering thermostats, but most involved investments in technology, either retrofits of existing technology (e.g., insulating existing homes) or new investments in technology (e.g., energy-efficient new construction or automobiles with improved mileage). More than a decade later, returns from these investments can often still be seen. Although there was usually sufficient incentive to replace capital with more energy efficient technology on a life-cycle cost basis, the initial capital cost was often enough of a disincentive to defer the investment until the existing capital approached the end of its useful life.

By contrast, energy efficiency gains in the 1980s, although often just as significant as those in the 1970s, have frequently been realized as incidental benefits to other investments aimed at improving the competitiveness of US products in world markets. Energy efficiency

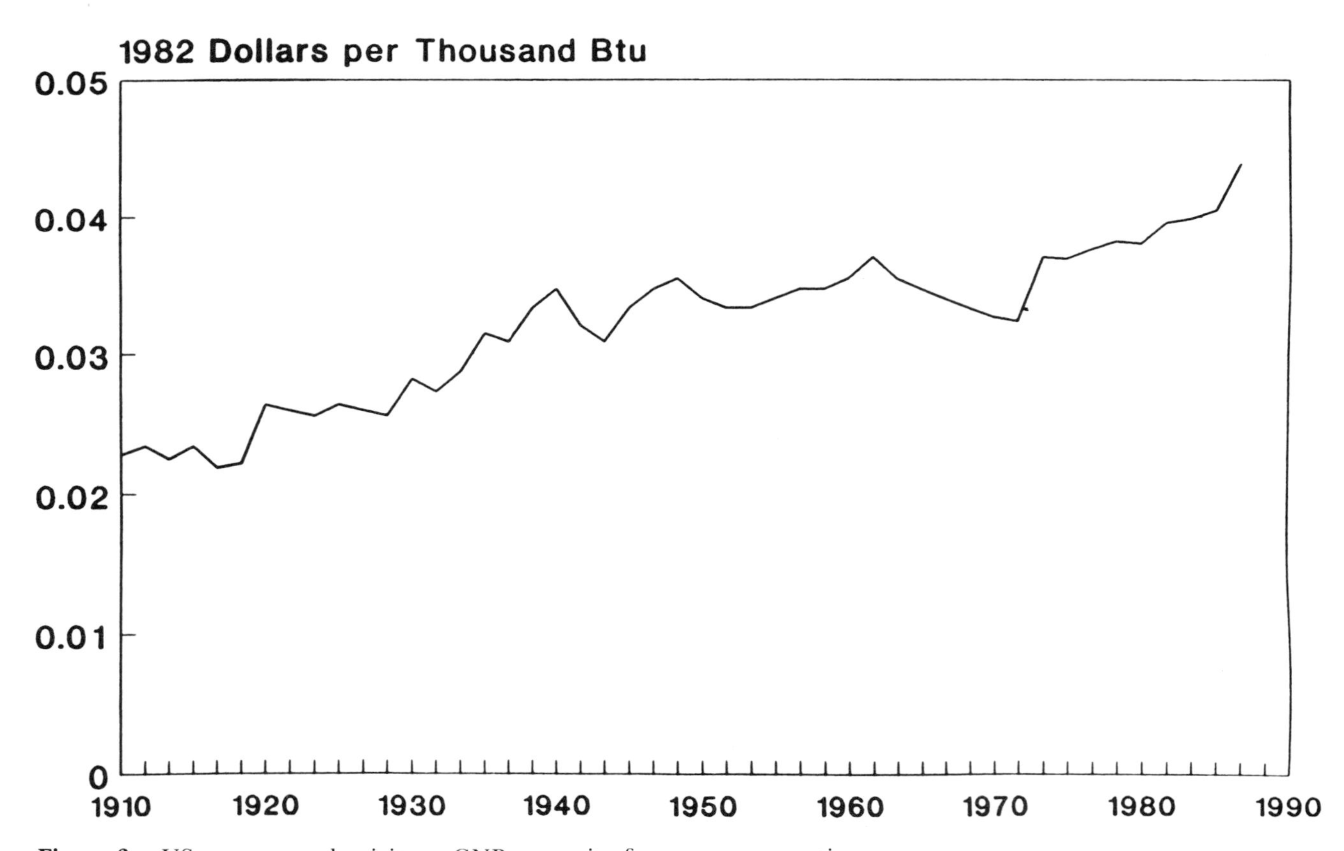

Figure 3. US energy productivity as GNP per unit of energy consumption.

investments in both the 1970s and the 1980s were, and generally continue to be, easier and more cost effective than funding new sources (Gibbons and Blair, 1990).

Many more opportunities for energy efficiency investment exist. The buildings sector of the global economy holds many opportunities for improved energy efficiency. In 1985 buildings in industrialized countries consumed 37 quads, almost equal to OPEC's production. New condensing furnaces could significantly reduce this demand. Because they reabsorb much heat from exhaust gases, condensing furnaces need 28 percent less fuel and emit fewer pollutants into the atmosphere than do conventional gas furnaces. Systems for controlling the indoor environment can monitor outdoor and indoor temperatures, sunlight, and the location of people, and then provide light and conditioned air where needed. These systems typically can provide energy savings of about 20 percent. A combination of improved lamps, reflectors, and daytime lighting can cut the consumption of energy for lighting by 75 percent or more.

Advanced building materials can sharply reduce loss of heat through windows, doors, and walls. In "superinsulated" homes, where normal insulation is doubled and forms an airtight seal in walls, heat radiating from people, light, stoves, and other appliances alone can warm the house. In comparison with the average home built in the US, some superinsulated homes in Minnesota require 68 percent less heat; for some residences in Sweden, the savings is 89 percent.

In industry, sensors and controls, advanced heat-recovery systems, and friction-reducing technologies can decrease energy consumption. A great opportunity for improving efficiency is cogeneration—the combined production of heat and electricity. Only a third of the energy from the steam produced by a boiler in a conventional electric-power plant is converted to electricity; in a cogeneration plant, much of the energy remaining in the used steam serves as a heat source for other industrial processes.

Other efficiency measures are specific to each industry. In the paper industry, automated process control, greater process speeds, and high-pressure rollers can boost efficiencies significantly. Advanced processes in the steel industry offer energy savings of at least 40 percent in US plants. In developing nations, efficiencies could be improved still further: China and India use four times as much energy to make a ton of steel as does Japan (Fulkerson, Auerbach, et al., 1988).

Transportation in industrialized and less developed countries constitutes the largest and most rapidly growing drain on the world's oil reserves and is a major threat to the environment. Cars and light trucks consume more than one out of every three barrels of oil and contribute 15 percent of carbon dioxide emissions in the US. During the past 15 years, new cars and trucks have become markedly more efficient through strategies such as increased use of light materials, improved engine design, installation of radial tires to reduce rolling resistance, and the redesigning of exteriors to decrease aerodynamic drag. Further gains in vehicle efficiency could come from a variety of technologies such as continuously variable transmissions and direct-injection diesel engines.

Capitalizing on many of these opportunities to the extent needed for environmental protection may require policy intervention. Many of the energy efficiency improvements in the 1970s and 1980s are attributable to market response, but the market isn't perfect. The driver, for instance, doesn't pay the national security costs of protecting oil supplies. The homeowner doesn't pay the costs of strip mining damage, acid rain, or global warming entailed in generating electricity. The US Congress recently passed legislation imposing minimum efficiency standards on all new appliances. This measure was necessary because builders, who seek to minimize initial costs, were eschewing cost-effective technologies to the detriment of building occupants, who are more interested in life-cycle costs. We may also need new tax policies to complement such regulatory policies.

Recent research (Chandler and Nicholls, 1990) compared the effectiveness of a $1 per gallon gasoline tax with a "carbon" tax applied to coal, oil, and natural gas ($1.73/GJ; $1.43/GJ; and $1.00/GJ, respectively) that would raise the same initial revenues as the gasoline tax. The effect of the two taxes on energy consumption following the adjustment to higher prices would be quite different. Under the carbon tax, US consumption would decline by 9 quads, about an 11 percent drop from current total consumption of 80 quads. By contrast, the gasoline tax would reduce US energy consumption about 4 quads. The carbon tax would reduce carbon emissions by approximately 190 million tons, the gasoline tax by 85 million tons. The carbon tax is therefore estimated to be two to three times more effective in reducing carbon emissions. The carbon emissions reduction under the carbon tax would exceed the total annual carbon emissions of present-day Poland, the world's sixth largest generator of carbon emissions. Adjusted revenues from the carbon and gasoline taxes would total $97 and $79 billion, respectively.

Neither a gasoline tax nor a carbon tax would be a simple step for policymakers. It would require attention to possible inflationary effects, varying regional impacts, and regressive taxation of the poor. But such problems should be soluble, and if instituted over a number of years, the impacts would be minimal because of substitution capabilities. If a gasoline tax or carbon tax were introduced as a tax shift, inflation and macroeconomic impacts should be zero. Either a gasoline tax or a carbon tax would effectively reduce US carbon emissions. Since a gasoline tax mainly affects oil use, while a carbon tax would affect total fossil fuel use, the taxes could be applied in separate or complementary fashion. The choice of "no new taxes" versus a carbon tax versus a gasoline tax depends on one's belief regarding the need to take action. If one believes that the evidence clearly calls for priority action to reduce the risks from carbon emissions, then the choice should be a carbon tax. However, if one believes that only actions that increase public welfare or help achieve complementary public goals are justified in reducing the risk of climatic change, then the gasoline tax would be the preferable choice since it would also help cut our trade deficit and reduce air pollution. If one believes that the importance of sticking to election promises outweighs the importance of adjusting policies to meet changing global challenges, then clearly "no new tax" is the right choice.

These technologies and policy options only deal with half the energy and environment equation, however, and with only a small fraction of the world. Energy-related environmental degradation can be symbolized as the product of a relatively straightforward identity:

$$\text{Pollution} \equiv \left[\frac{\text{Pollution}}{\text{Energy}}\right]\left[\frac{\text{Energy}}{\text{GNP}}\right] \times \left[\frac{\text{GNP}}{\text{Population}}\right]\left[\text{Population}\right]$$

To the extent we have made any effort to correct energy-related environmental degradation, we have concentrated our efforts on reducing the amount of pollution produced per unit of energy, with technologies like scrubbers, and on reducing the amount of energy required to produce a unit of GNP, with technologies that improve the efficiency of conversion and utilization. Even though substantial potential to improve the economy and environment remains in those first two terms of the identity, we would be fools not to pay attention to the last two terms as well. What is an adequate and sustainable standard of living? For how many people must we, can we, provide it?

The importance of integrating energy and environmental policy with global economic and security policy becomes especially evident here. For instance, our response to the changes in governance in China, the USSR, and Eastern Europe will be central to energy and environmental policy.

The path of industrial development in China could have a greater effect on the atmospheric accumulation of carbon dioxide than that of any other nation. China's critical role stems from its huge and growing population, its tendency toward energy-intensive processes, its poor energy efficiency, and its massive reliance on coal. Between 1980 and 1986, China's manufacturing sector grew by 12 percent a year, the fastest growth in any large nation in the world. The average energy intensity of China's industrial sector has dropped, but it remains higher than the intensity of any other developing nation. Indeed, the potential for improved efficiency is China's chief future energy resource (Meyers, 1988).

Achieving that potential will require large transfers of technology and capital from the industrialized world, transfers that are threatened by events like those at Tiananmen Square last June. It will also require reform of China's energy-pricing policy. Coal in China is priced at one quarter the international level. Folk wisdom has it that "one ton of coal could not even buy a ton of sand; one barrel of oil could not even buy a bottle of liquor."

Industrialized countries with aging infrastructure will also have a major impact on future energy consumption and carbon emissions. Energy intensity in the Soviet Union is twice the average of nations belonging to the OECD and shows no sign of improvement. The new policies of perestroika and glasnost, which encourage efficiency, market-oriented systems, and global cooperation hold great promise for the global economy and environment. As the USSR, China, and other centrally planned economies move toward a more rational price system, they will soon recognize, however, that "free world" market prices still do not reflect major external costs. Because we now realize that these costs may include extraordinary global environmental problems, it may be up to more industrialized countries to encourage—through technology transfer, subsidies, or loans—policies or technologies that take these countries beyond the levels of efficiency justifiable on the basis of free-market prices. Such policies would require unprecedented levels of international cooperation. However, helping the developing and restructuring nations build a high-level of energy efficiency and a low level of carbon intensity into their new industrial infrastructure may be far less expensive than taking equivalent corrective actions in our own economy.

Both domestically and internationally, we must devise policies that are equitable. The developing world's share of energy consumption is small but inefficient, and demand is rapidly growing. If demand growth continues at present rates in the developing world, their output of CO_2 will match that of the industrialized world within a few decades. With the help of the industrialized world, however, developing countries could apply technical solutions that would promote economic growth while keeping energy-demand growth relatively low. One important analysis (Goldemberg, Johansson, et al., 1987) shows that application of the best energy technology available today could provide a developing nation with a mid-1970s European level of energy services, while increasing energy consumption by only 20 percent over average consumption of a developing country in 1980. This model also confirms that industrialized countries could continue economic growth but consume less energy than they do today.

Why should less developed countries worry at all about saving energy when their prime concern is generating economic growth, which includes increasing the availability of energy services? The answer is that energy efficiency can reconcile the simultaneous goals of development and environmental protection. Thus technological progress can result in providing goods and services with less energy, capital, and environmental impacts.

But environmental problems result from resource flows, and the greater the economic activity and population, the greater the flow. Many developing nations are experiencing unsupportable population growth (see Figure 4). But UNICEF and others have shown (UNICEF, 1984; Chandler, 1985) that encouraging female literacy, breast feeding, and primary maternal and child health care, and coupling those techniques with family planning technologies, can make a significant difference. The developing countries have the fastest growing rates of resource consumption not only because their per capita consumption rates have been so low, but also because their populations are growing so fast. We cannot continue merely to ignore or wave our hands at population growth, especially since we know there are very effective, relatively inexpensive ways to control it. Attending to the unmet need and desire of people to have smaller families would be a humane, relatively inexpensive, and major step.

Policymakers face some challenging tasks in the decades ahead. Polls show that US citizens and people around the world are concerned about the environment and that they're willing to pay a substantial price to keep it clean. Science and technology will play a critical role in determining whether a "clean" environment is a realistic goal and what it will cost us to obtain. Policymakers need data, they need carefully drawn scenarios, they need attractive new technologies. They also need help interpreting available information and disseminating it to the public. Our government is founded on the principle that an informed people can make wise choices for their current and future welfare.

Good analysis is an essential element of good policy in today's world. But technical analysis is not the only component of policy, and policymakers need not—indeed cannot—wait for the final word to be uttered in a technical area before they act. Policy is an exquisite blend of many factors, including international tensions, economic factors, and cultural values as well as technical know-how. To ignore any factor, to weigh one or the other too heavily, to be paralyzed by the absence of absolute truth, is to abandon the responsibilities of leadership.

Analysis and experience have shown that investments in energy efficiency, new, nonfossil energy supplies, and population control can avert major, costly economic and environmental problems. It is now up to technologists and policymakers to develop an implementation plan for these measures that will contribute to sustainable development around the globe and mitigate future, potential damages such as climate change while we generate greater certainty about anthropogenic contributions to this phenomenon.

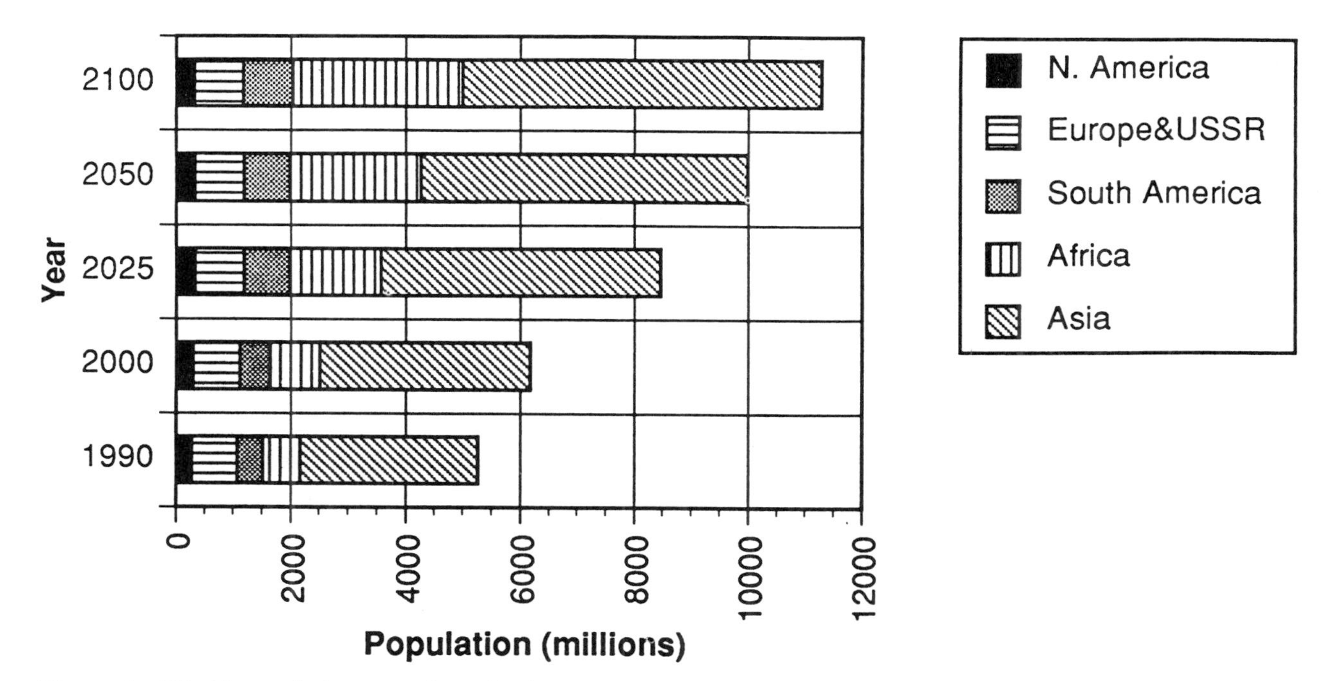

Figure 4. World population growth projections.
Source: Bulatao, Rudolfo A., Bos, Eduard, Stephens, Patience W., and Vu, My T., *Europe, Middle East, and Africa (EMN) Region Population Projections, 1989–90 Edition* (Washington, DC: World Bank, 1989).

Acknowledgments

The author is grateful to his associates, Holly Gwin, Peter Blair, and Gretchen Kolsrud, for their assistance in preparing this paper. The views expressed in this paper are those of the author and not necessarily those of the Office of Technology Assessment.

References

Chandler, W.U., *Investing in Children,* Worldwatch Institute, Washington, DC (1985).

Chandler, W.U., and Nicholls, A.K., *Assessing Carbon Emissions Control Strategies: A Carbon Tax or a Gasoline Tax?* American Council for An Energy-Efficient Economy, Washington, DC (1990).

Fulkerson, W., Auerbach, S.I., et al., *Energy Technology R&D: What Could Make a Difference?* Oak Ridge National Laboratory, Oak Ridge, TN (1988).

Gibbons, J.H., and Blair, P.D., "Energy Efficiency: Its Potential and Limits to the Year 2000," in Helm, J.L., ed., *Energy: Production, Consumption, and Consequences,* National Academy Press, Washington, DC (1990).

Gibbons, J.H., Blair, P.D., et al., "Strategies for Energy Use," *Scientific American* 260(9), pp. 136–143 (1989).

Goldemberg, J., Johansson, T.B., et al., *Energy for Development,* World Resources Institute, Washington, DC (1987).

Meyers, S., ed., *Energy Markets and the Future of Energy Demand: Proceedings of the Chinese-American Symposium,* Lawrence Berkeley Laboratory, Berkeley, CA (1988).

United Nations Children's Fund (UNICEF), *State of the World's Children 1985,* Oxford University Press, New York, NY (1984).

Energy and Environment: Strategic Perspectives on Policy Design*

William C. Clark
John F. Kennedy School of Government
Harvard University
Cambridge, MA

Introduction

Serious debate on the policy implications of interactions between energy use and environmental quality has now been under way for at least two decades. The focus of attention has shifted frequently, ranging on the energy side from security of supply to management of demand, and on the environmental side from oil pollution of local coastlines to thermal alteration of the global climate. Policy studies through the years have mirrored these shifts in substantive focus, with emphasis changing from the estimation of resource adequacy to the assessment of environmental limits (Baumgartner and Midttun, 1987). While each of these perspectives has produced insights, there has been a certain faddishness to the energy-environment debate, and a related failure to develop sustained management strategies.

As we approach the 21st century, the convergence of a number of trends, events, and findings provides a unique opportunity to shape a truly comprehensive and lasting strategy for managing energy-environment interactions. New energy and environment studies by a number of national and international organizations have recognized and explored the prospects for such strategies (OECD, 1990; EEC, 1989; US DOE, 1990). The challenge of this decade is to seize the full potential of the opportunity before us, rather than letting it once more slip away as a result of narrow preoccupation with a few currently popular environmental problems, energy technologies, or institutional reforms. In seeking to meet the challenge, it may help to address anew four sets of questions that have endured over the past decades of debate and that continue to provide a useful framework for thinking about the design of strategies for managing interactions between energy and environment:

- *What* relation has the energy-environment debate to larger questions of social, economic, and political development? What energy services and environmental properties merit priority consideration in the design of development policy?
- *How* can societies intervene to manage the interactions of energy and environment? How can a rich and balanced menu of intervention options be constructed that will help different societies, characterized by different values, problems, and opportunities, to shape management strategies that work for them?
- *Who* is responsible for which elements of the interactions between environment and development? Who endures which consequences of those interactions? Who must cooperate if management strategies are to be effective?
- *When* should particular management actions be undertaken, given the highly uncertain and changeable nature of interactions between energy and environment? When is waiting for new information or technology a rational policy?

A commitment to addressing such questions in national and international policy debates will not guarantee more effective strategies for the management of interactions between energy and environment. But it may help to avoid a preoccupation with the scene under this year's lamppost, and in so doing to broaden our perspective on the problems and prospects before us. This paper sketches a few of the many important issues that such a wide ranging reconnaissance would surely encounter.

What Are the Larger Contexts of the Energy-Environment Debate?

The energy-environment debate has for much of the last half century been posed as a conflict of economic growth versus environmental preservation. Fortunately, this unproductive and ultimately false dichotomy has begun to be superseded during the last decade by a growing consensus on the need for "sustainable development."

The United Nations' World Commission on Environment and Development (WCED), chaired by Norway's Gro Harlem Brundtland, characterized sustainable development as paths of social, economic, and political

* This paper was prepared while the author was Jean Monnet Visiting Professor at the European University Institute in Firenze, Italy. A longer version will be published by the Institute.

progress that meet "the needs of the present without compromising the ability of future generations to meet their own needs" (WCED, 1987). The need to shepherd the environmental resources on which all human development depends is implicit in the Brundtland Commission's statement. But so is a recognition that protecting people from poverty is a necessary component of protecting the planet from people.

An Earth Transformed by Human Action

The environmental dimensions of the sustainable development challenge are only beginning to be fully appreciated. This is especially true for the large spatial scales and long temporal spans that are important for the understanding of interactions between energy use and environment. A century ago, George Perkins Marsh titled one of the first efforts to assess environmental changes on a global scale *The Earth as Modified by Human Action* (March, 1864). Today, the explosive growth of the human population and its economic activities has left the planet not just modified, but fundamentally transformed (Kates, Turner and Clark, 1990). Since the beginning of the 18th century, people have cleared the earth of 6 million square kilometers of forest. Sediment loads have risen threefold in major river systems, and eightfold or more in smaller basins supporting the centers of human civilization. Irrigation schemes and impoundments have increased the withdrawals of water from its natural circulation by 30- to 40-fold, to a present total of perhaps 3600 cubic kilometers a year.

Substantial changes in the planet's chemistry have also taken place. The carbon dioxide concentration in the atmosphere has risen by 25 percent, while that of methane has doubled. The human-induced flows of major elements such as sulfur and nitrogen are now of comparable magnitude to natural flows at the global scale, and vastly exceed natural fluxes in heavily developed regions. Trace metals, many of which can be toxic to life, have also had their global flows dramatically increased by human activity: lead by a factor of 18, cadmium by a factor of five, and zinc by a factor of three. More than 70,000 chemicals previously unknown to nature have been synthesized. Several of these—for example DDT and the CFCs—have been shown to affect the environment significantly, even at very low concentrations. Although some of these changes are a result of long established trends, some of which appear to be decelerating, many are rapidly intensifying phenomena of the latter half of the 20th century (Figure 1).

The human dimensions of the sustainable development challenge are equally daunting, if more familiar (Clark, 1989). Over the past 300 years, the number of people on earth has increased by a factor of eight; average life expectancy has at least doubled. During the same period, human economic activity has become in-

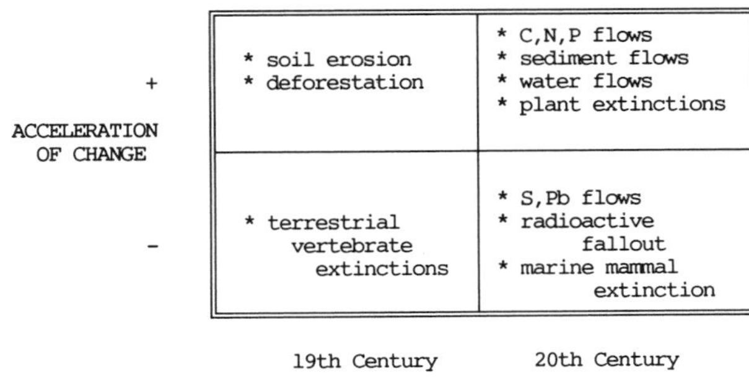

Figure 1. Timing of human transformation of the biosphere. The figure classifies global patterns of human induced change in the biosphere. "Recency" is determined by tallying the cumulative human induced change in each listed component from the times of earliest human civilization to the present. Each component is then classified according to whether it reached 50 percent of its total cumulative change to date in the 19th or 20th century. "Acceleration" is determined by comparing the average rate of human induced change for each component in the 1980s with the rate in the 1950s. Where the 1980s rate is higher than the 1950s rate, "acceleration" is scored "+", otherwise it is scored "−". Data are from Kates, Turner and Clark (1990).

creasingly global, with demands for goods and services in one part of the planet being met with supplies from half a world away. The volume of goods exchanged in international trade has increased by a factor of 800 or more and now represents more than a third of the world's total economic product. Since the middle of the last century, agricultural activities have converted nine million square kilometers of the earth's surface into permanent croplands; energy use has risen by a factor of 80.

As a result of this growth, the earth today is inhabited by more than five billion humans, who appropriate 40 percent of the organic material produced on its land surface. Each year, the human species produces double its weight in steel and, in the process, consumes the equivalent of two tons of coal for every man, woman, and child on the planet. The distribution of these people, their impacts on the environment, and the benefits they receive from development vary tremendously. At one extreme, the richest 15 percent of the world's human population consumes more than one third of the fertilizer and more than one half of the energy used by the species. At the other extreme, perhaps one quarter of the population goes hungry during at least some season of the year. More than a third live in countries where the mortality rate for young children is greater than one in ten. The vast majority exist on per capita incomes below the official poverty level in the United States.

As we look to the future, it is encouraging that the growth rate of human population is declining virtually everywhere. But even if trends responsible for this decline continue, the next century will in all likelihood see yet one more doubling of the number of people trying to extract a living from the earth. Nearly all of that increase will take place in today's poorer countries. According to the Brundtland Commission, a five- to tenfold increase in world economic activity will be necessary over the next 50 years to meet even the most basic needs and aspirations of this burgeoning population. The implications of this desperately needed economic growth for the already stressed planetary environment are at least problematic and may be overwhelming. The following simple numerical example, drawn from the unpublished work of Yoichi Kaya and a manuscript by Yoshiki Ogawa (1990), starkly illustrates the difficulties involved.

Tradeoffs in Sustainable Development

The emission of a pollutant like carbon dioxide (X) from a given area (A) can, as a matter of definition, be characterized as the product of the population (P) per unit area, times the income ($) per capita, times the pollution (X) emissions per unit economic value produced:

$$(X/A) \equiv (P/A) * (\$/P) * (X/\$). \qquad (1)$$

From equation (1), a comparable equation characterizing the rates of change in each of the terms can be derived. For purposes here, the result can be simplified as follows:

$$(X/A)' \approx (P/A)' + (\$/P)' + (X/\$)', \qquad (2)$$

where the ′ sign indicates rate of change (e.g., percent per year) in each respective term. Thus, the rate of change in emissions for an area equals the rate of change in population for the area, plus the rate of change in per capita income for that population, plus the rate of change in emissions intensity, all expressed as percent per year. Replacing these general terms with specific numbers raises the sobering prospect that sustainable development may often involve tradeoffs that leave both "people" advocates and environment advocates unhappy.

Consider as an example the planet's challenge in dealing with the emissions of carbon dioxide that are responsible in part for the risk of climate warming. Global carbon dioxide emissions have been growing at an average rate of a bit less than 2 percent per year for the last two decades. Some participants in the current debate have called for "freezing" or reducing emissions, (i.e., bringing their growth rate to zero or less). Any serious attempt to reduce the risks of carbon dioxide–induced climate change (whatever those risks may be) will at a minimum have to reduce the rate of emissions growth to significantly less than its historical value (i.e., make $(X/A)'$ less than 2 percent per year).

Turning to the right hand side of the equation, let us see how such reduced pollution rates might be achieved. The world population growth rate, the first term on the right side of equation (2), has only recently dropped below 2 percent. The most optimistic forecasts assume an average of no less than 1 percent per year increase in population over the next 30 to 40 years (World Bank, 1989). Looking to the second term in equation (2), the Brundtland Commission has argued that growth in per capita income of at least 3 percent per year over the same period will be necessary to relieve the worst ravages of world poverty (WCED, 1987). As for the third term, Kaya has shown that the maximum rates of improvement in emissions intensity that have been sustained over large areas and multiple decades are on the order of −1 percent per year. France achieved perhaps −2 percent per year during the peak decade of its conversion to nuclear power. Pulling these numbers together gives equation (3):

$$\begin{matrix} (X/A)' & \neq & (P/A)' & + & (\$/P)' & + & (X/\$)' \\ <2\% & & 1\% & & 3\% & & -1\%. \end{matrix} \qquad (3)$$

To make this account balance would require that within the next decade or so the entire world undertake draconian measures to bring population growth to zero, or engage successfully and continuously in the kind of technological revolution that the French used to convert most of their central generating capacity to nuclear power, or reduce by a full percentage point its objectives for growth of individual well-being. It should be noted that such a percentage point drop in rates of per capita income growth would condemn an entire additional generation of the world's poor to lives of poverty that could already have been significantly mitigated at the higher growth rate.

These are not attractive alternatives. Since equation (1) is merely an accounting identity, not a causal model, there may be ways around the choices implied—say by learning how to increase incomes in the developing countries through measures that simultaneously decrease the carbon intensity of their economies. But such tricks will not be easy, given the massive scale and momentum of the factors involved. At the very least, the long-term, large-scale relationships among population growth, economic well-being, technical change, and environmental degradation place some very tight constraints on the design of strategies for balancing concerns for energy and environment. And we must face the real prospect that some additional, though bounded, degradation of the global environment may be a necessary component of strategies for sustainable development.

Energy Use as One of Many Sources of Environmental Degradation

Energy-related pollution and resource use are among the most important human activities responsible for degradation of the environment. But the design of effective management strategies requires that energy's contribution to the overall problems of sustainable development be seen in the context of other major sources of environmental transformation.

A comprehensive review of the sources of environmental transformation, and energy's role as one of those sources, is beyond the scope of this paper. Interested readers will find partial treatments in Brooks and Hollander (1979), OECD (1983), Clark and Munn (1986), WCED (1987), Turner et al. (1990), and the new studies cited in the introduction to this paper. In general, it is clear that for most of the world throughout most of human history, agricultural alterations of landscapes have been the route through which societies have most significantly altered the environment. Manufacturing—the processing of materials—has played an increasing role as an agent of environmental change over the last hundred years. The impact of energy-related activities, though locally important for centuries, has achieved global importance only relatively recently.

In seeking to understand the complex relations of energy-related and other sources of environmental change, it is useful to focus first on the relatively simple case of the atmosphere. Figure 2, drawn from the work of Paul Crutzen and Thomas Graedel (1986), lists a number of components of the atmospheric environment that are of present concern to society. Most of these "valued environmental components"—stratospheric ozone, global climate, acid precipitation, materials corrosion, etc.—are indeed the subject of contemporary policy debate. The figure also gives the results of scientific research showing which components are influenced by which chemical emissions. Note that, in general, each component is affected by multiple chemical emissions. Figure 3 summarizes research tracing each of the chemical species listed in Figure 2 back to the natural and human processes responsible for its emission. It shows that most chemical species of importance for the atmospheric environment have multiple sources of emissions responsible for changing their concentrations. In particular, for none of those species listed are energy activities the only source of change in the contemporary environment. Finally, Figure 4 combines and simplifies the previous two figures to show how various human and natural processes affect valued components of the atmospheric environment, via the intermediary of specific chemical emissions.

Two aspects of Figure 4 are worth emphasizing. First, looking down the columns of the figure, it is clear that no single source—including energy activities such as fossil fuel combustion—is the sole determinant of change in any of the valued environmental components. Second, looking across the rows, it is clear that most sources affect more than one valued environmental component, while several—including energy activities—affect multiple components. In practical terms, this means that any management strategy targeted at energy activities is likely to have significant impacts across a wide range of valued environmental components. Useful comparisons of such strategies should logically evaluate costs and benefits across the relevant range of components. Unfortunately, most analysis is directed to either single cells in Figure 4, or to integrated "column" assessments of the differential contribution of various sources to single environmental components. The "row" assessments most needed for policy design are virtually nonexistent. Approaches that enable a more comprehensive and realistic accounting of the environmental benefits secured by alternative energy policies are needed.

The preceding atmospheric focus provides an important but hardly comprehensive perspective on the question of energy-environment relationships. To obtain a broader if more tentative indication of the contribution made by energy-related activities to contemporary environmental problems, I summarize below relevant aspects of a global analysis of environmental

VALUED ENVIRONMENTAL COMPONENT	TRACE CONSTITUENTS							
	CO	CO_2	CH_4	NO_x	N_2O	SO_x	CFC	Rad
Greenhouse effect		+	+		+	-	+	
Stratospheric ozone		+/-	+/-	+/-	+/-		-	
Acid deposition				+		+		
Photochemical smog				+				
Material corrosion						+		
Visibility				+		+		
Self-cleansing of atmosphere	-		+/-	+				

Figure 2. Atmospheric trace constituents and their impacts. The figure lists only principal effects of changes in trace constituent concentrations on valued environmental components. "+" indicates that an increase in the trace gas leads to an increase of the component (e.g., increased CFCs lead to increase in greenhouse effect); "−" indicates decrease of the component (e.g., increased CFCs lead to decrease in stratospheric ozone). "+/−" indicates that changes in the specified trace gas concentration can affect the component either positively or negatively depending on the absolute concentrations of the specified or other gases. No entry indicates no known major impact. Abbreviations in trace gases are "CFC" for chlorofluorocarbons in the stratosphere; "Rad" for radionuclides. Source is Graedel and Crutzen (1989).

SOURCE	TRACE CONSTITUENTS							
	CO	CO_2	CH_4	NO_x	N_2O	SO_x	CFC	Rad
Natural Processes								
oceans		+						
wetlands		+	+					
vegetation		+						
miscellaneous								+
Land Use								
deforestation		+			+			
biomass burning	+	+		+	+			
ruminant cult.			+					
rice cultivation			+					
N - fertilization					+			
landfills		+	+					
Manufacturing								
ore smelting		+				+		
pig iron prod.	+	+						
CFC use							+	
coke production		+	+					
Energy								
fossil fuel p & d			+					
coal combustion	+	+		+	+	+		+
oil combustion		+		+	+	+		
gas combustion		+		+	+	+		
nuclear energy								+

Figure 3. Principal sources of atmospheric trace constituents. The figure lists only principal contemporary sources of changes in trace constituent concentrations. Abbreviations in trace gases are "CFC" for chlorofluorocarbons in the stratosphere; "Rad" for radionuclides. Abbreviations in sources are "p & d" for production and distribution. Sources for data are Crutzen and Graedel (1986) and Darmstadter et al. (1987).

hazards that I am conducting with several colleagues (Norberg-Bohm et al., 1990).

Figure 5 lists 28 such hazards that cause societies serious concern in at least some parts of the world. For each hazard, we identified the human activities contributing to it. As shown in the figure, energy-related activities have a potentially significant direct impact on a bit less than half of the hazards. In addition, using a variety of measures, we attempted to characterize the relative seriousness of each hazard in terms of consequences for people, ecosystems, and material welfare. Figure 5 shows those hazards ranking in the top quartile of current and future consequences for India and the United States. Looking first at current consequences, it can be seen that virtually every top-ranked hazard in the United States has energy activities as one of its significant contributors, while very few of India's top ranked problems are directly related to energy use. Looking to the future, the contrast is less evident, with slightly more than half of the top ranked hazards in both countries showing energy activities as a potentially contributing factor.

It should be emphasized that the results reported in Figure 5 are tentative and likely to change as the research continues. Nonetheless, even these preliminary findings suggest some potentially useful perspectives on the broader context of the energy-environment debate:

- The relative importance of energy activities as a source of environmental degradation differs significantly among components of the environment. Although many of the world's environmental problems are directly related to energy activities, many are not. Quality of atmospheric and biotic components of the environment is perhaps most intimately tied to energy activities, while quality of other environmental components like the land are less directly affected.
- The relative importance of energy activities as a source of environmental degradation differs significantly among countries today. For many developed countries, energy-related activities may contribute to most of their serious environmental problems; for many developing countries, energy may rank far behind activities affecting land use and sanitation as a source of degradation.
- The relative importance of energy activities as a source of environmental degradation is likely to be different in the future than it is today. There is some suggestion that energy activities will take on a less dominant role as source of major problems in the developed countries, at the same time they gain prominence in developing regions.

How Can the Range of Management Options Be Expanded?

In this section I attempt to translate the broad perspective on energy-environment interactions developed above into an operational framework for the design of management strategies. The advantages of addressing environmental degradation through preventative source reduction rather than restorative impact mitigation are manifest. Physical, economic, and moral arguments have been articulated in support of preventative approaches, often resulting in their embodiment as official policy for environmental management.

As illustrated above, however, source reduction may often appear to entail unattractive tradeoffs with other highly desirable goals. It may be at least temporarily infeasible on economic or political grounds. Nations may reasonably differ in the priority they accord to reducing particular domestic sources of pollution, even when those pollutants affect other nations. Finally, even in the most optimistic of circumstances, total source reduction will rarely be feasible or desirable. Some environmental degradation is likely to occur for the foreseeable future as human societies continue their centuries-long transformation of the planet.

For all the reasons cited above, a dogmatic preoccupation with source reduction policies is unlikely to promote the most effective long run strategies for managing interactions between energy and environment. A premature focus on single pollutants or technologies will likewise benefit neither people nor the environment.

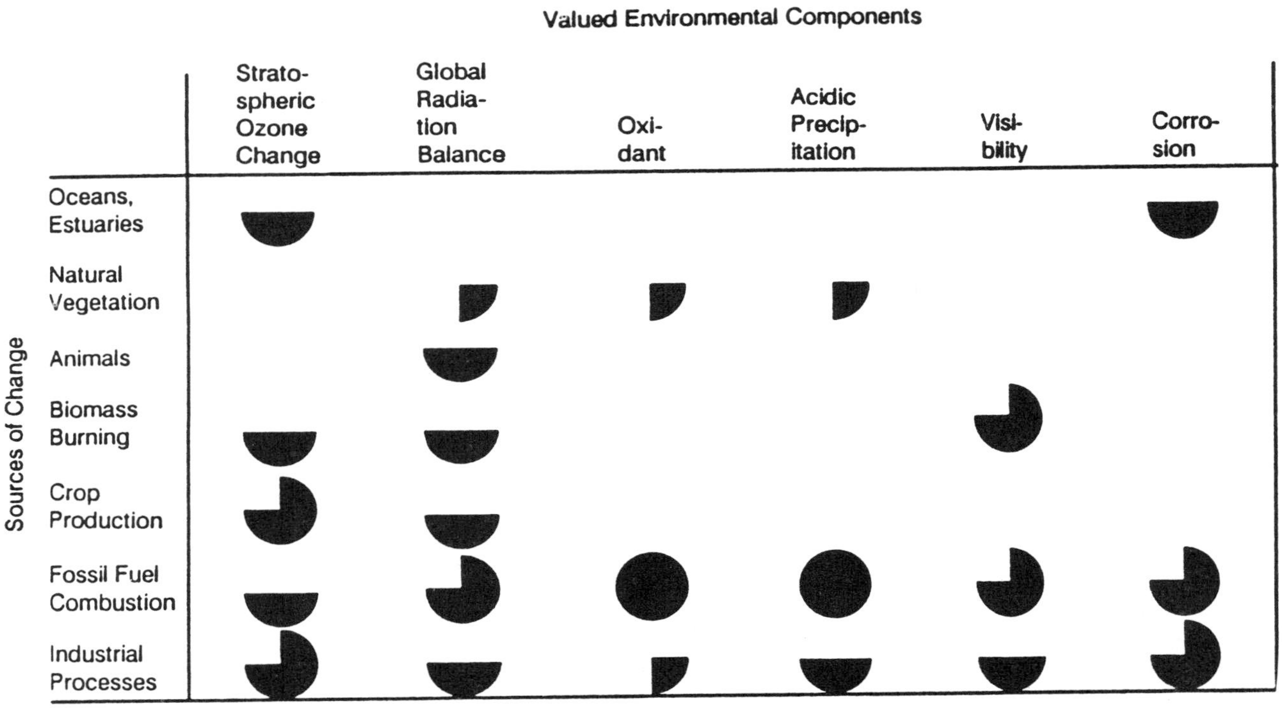

Valued Environmental Components

Strato-spheric Ozone Change | Global Radia-tion Balance | Oxi-dant | Acidic Precip-itation | Visi-bility | Corro-sion

Sources of Change

Oceans, Estuaries
Natural Vegetation
Animals
Biomass Burning
Crop Production
Fossil Fuel Combustion
Industrial Processes

Figure 4. Sources of changes in valued components of the atmosphere. Full "pie" indicates major influence, partial "pie" proportionally less influence, no entry relatively insignificant influence. Source: Crutzen and Graedel (1986).

Instead, a concerted effort is needed to expand the range of management options that can be called upon by different societies, characterized by different values, problems and opportunities, to shape energy strategies that work for them. Enriching the agenda of options for intervention has been one of the greatest contributions that policy analysts have made to other areas of management; there is no reason to expect that the potential contribution is any less in the field of energy-environment studies.

A useful first step in expanding the range of options for management in other technology-based sectors is the development of a systematic causal taxonomy of hazard. Panel A of Figure 6 presents one such taxonomy, based on work of the Center for Technology, Environment and Development at Clark University (Kates et al., 1985) but modified to incorporate the concept of "valued environmental components" (VECs) addressed above (Norberg-Bohm et al., 1990). As an example, the figure suggests how the hazard of change in one *valued environmental component* (e.g., climate) can be understood in terms of the *material releases* (e.g., CO_2) that affect it, the *human activities* (e.g., fossil fuel combustion) that cause those releases, the *human wants* (e.g., energy services) that create a demand for those activities, and the *human needs* (e.g., shelter) that ultimately underlie particular wants. Following the impacts of climate change "downstream" to the right, the figure also highlights *exposure* to the changed VEC (e.g., coastal dwellers encountering rising sea level), and *consequences* of that exposure (e.g., loss of housing). The full "taxonomy" of any environmental problem is, of course, more complex than this simple illustration. But that full complexity can readily be incorporated in the framework just presented, as suggested by the fuller treatment of the climate change problem given in Figure 7.

As noted earlier, most of the discussion of policies for managing energy-environment interactions has focused towards the left hand side of the taxonomy, particularly on the reduction of per capita *needs* for energy through conservation measures, or changing the *human activities* (i.e., the energy technologies) that are involved in meeting those needs. In principle, however, managerial interventions can be applied at every link in the causal chain (Schelling, 1983; Clark, 1985). Figure 6, panel B, illustrates an example in which we could manage the hazard of climate change through modifying *needs* (e.g., by disseminating knowledge and means of family planning), *wants* (e.g., by encouraging energy conserving consumer habits), *activities* (e.g., by shifting from fossil to nuclear fuels), *releases* (e.g., by capturing released CO_2 through reforestation projects), the *VEC* climate itself (e.g., by cooling our work environments with air conditioning), *exposure* (e.g., by building sea

Hazards	Energy Source	Current India	Current USA	Future India	Future USA
WATER QUALITY					
Biological contaminants		*			
Metals and toxics	*		*		*
Eutrophication					
Sedimentation					
Ocean pollution	*				
LAND (the Lithosphere)					
Soil - chemical damage		*			
Soil - physical damage		*		*	
Quantity of arable land					
BIOTA					
Animal habitat	*			*	*
Pure food supplies	*	*	*		*
Rate of gene mutation	*				
ATMOSPHERE					
Stratospheric ozone					*
Climate change	*			*	
Acid rain	*		*		
Photochemical oxidant	*		*		
Toxic air pollutants	*		*		*
THE HUMAN ENVIRONMENT					
Indoor air quality - radon			*		
Indoor air quality - other	*		*		
Chemicals in workplace					
Radiation (other than radon)	*			*	*
Accidental chemical releases					
RENEWABLE RESOURCES					
Stock of fisheries					
Stock of wildlife				*	*
Forestry reserves	*				
Groundwater resources					
NATURAL ENVIRONMENTAL HAZARDS					
Floods					
Droughts		*			
Pest epidemics		*			

Figure 5. Energy and the most serious environmental hazards. The "Hazard" column lists 28 potentially most serious environmental hazards that often appear in national and international assessments. The "Energy Source" column indicates which hazards can be significantly affected by energy related activities. The next four columns indicate the hazards that score in the top quartile of seriousness when evaluated for current and future impacts in India and the US (Norberg-Bohm et al., 1990).

walls), or *consequences* (e.g., by paying compensation to groups adversely affected).

As the illustrations suggest, it is not difficult—guided by the causal structure—to imagine examples of managerial options that could be applied at all points in the hazard chain. But further structure and stimulus for the exercise of imagination can be obtained by distinguishing among the kinds of management that might be employed. One approach that has proven useful in other fields focuses separate attention on three "decision orders"—the *technological, organizational,* and *behavioral*—as means of bringing about management goals (Montgomery, 1974). Figure 8 suggests how these distinctions might help to structure thinking about managing the hazards of climate change due to greenhouse gas emissions.

Technological options provide new ways of doing things. Examples relevant to the greenhouse effect cited in Figure 8 range from making available birth control techniques through developing safe nuclear power to designing new crops. *Organizational* options alter the relationships of authority and responsibility among human actors. In the energy-environment interplay, examples include returning US government support to family planning efforts, adoption of international protocols regulating carbon dioxide emissions, and the establishment of compensation principles and payments between those most responsible for, and those most impacted by, climate change. Finally, *behavioral* options change the things that people want to do. Encouraging energy conservation, improving the social acceptability of intentional climate modification, and abandonment of areas threatened by rising sea level are all behavioral options that might be invoked as part of a strategy for responding to the hazards of climate change.

Some of the options illustrated in Figure 8 are self-evident, others are highlighted by the two-way categorization of options provided by the causal structure of hazard on the one hand and the decision orders approach to managerial means on the other. The point of using the examples is not to argue that some should be adopted while others should be ignored in strategies for managing climate change. Rather, it is to show that systematic and creative thinking about the range of possible options that might be applied to the management of energy-environment systems is both possible and worthwhile. For just as the causal structure helps to avoid the common preoccupation with source reduction measures, so the decision order structure can help to compensate for prevalent biases in favor of or against management through technological fixes, regulatory fiat, or exhortation.

By encouraging systematic articulation of the full range of possible managerial means and points of intervention in the hazard chain, the menu of options available for consideration in the policy debate can be significantly enriched. The fact that some of these options may at first seem unwise, uneconomical, or impractical is largely irrelevant. It is a fundamental finding of policy studies that what is thought to be possible is strongly influenced by what is thought to be desirable, and vice versa (March and Olsen, 1976; Wildavsky, 1979). On "big" problems for which significant analysis is justified—precisely the sort of problems encountered in the energy-environment debate—it will therefore almost always be worthwhile to expand the range of options available for consideration before spending too much effort deciding which option is "best." Once a rich menu has been created and explored, then different actors in the energy policy debate—individuals, firms, states, and others—can begin assembling the strategies that make most sense for them. It is at this latter stage of the policy design process, when choice among means for accomplishing the same ends is the central issue, that the classic

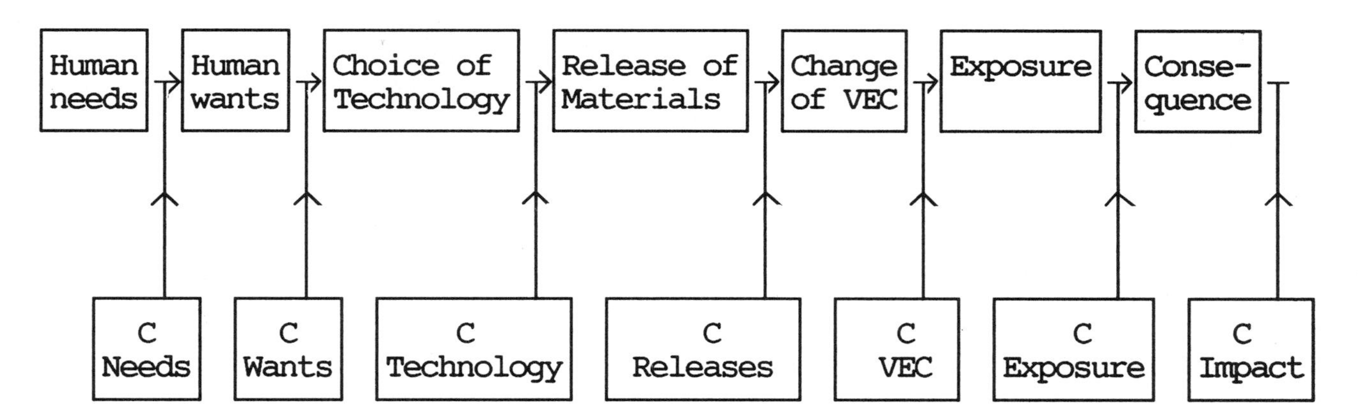

Figure 6. A taxonomy of environmental hazard and management.

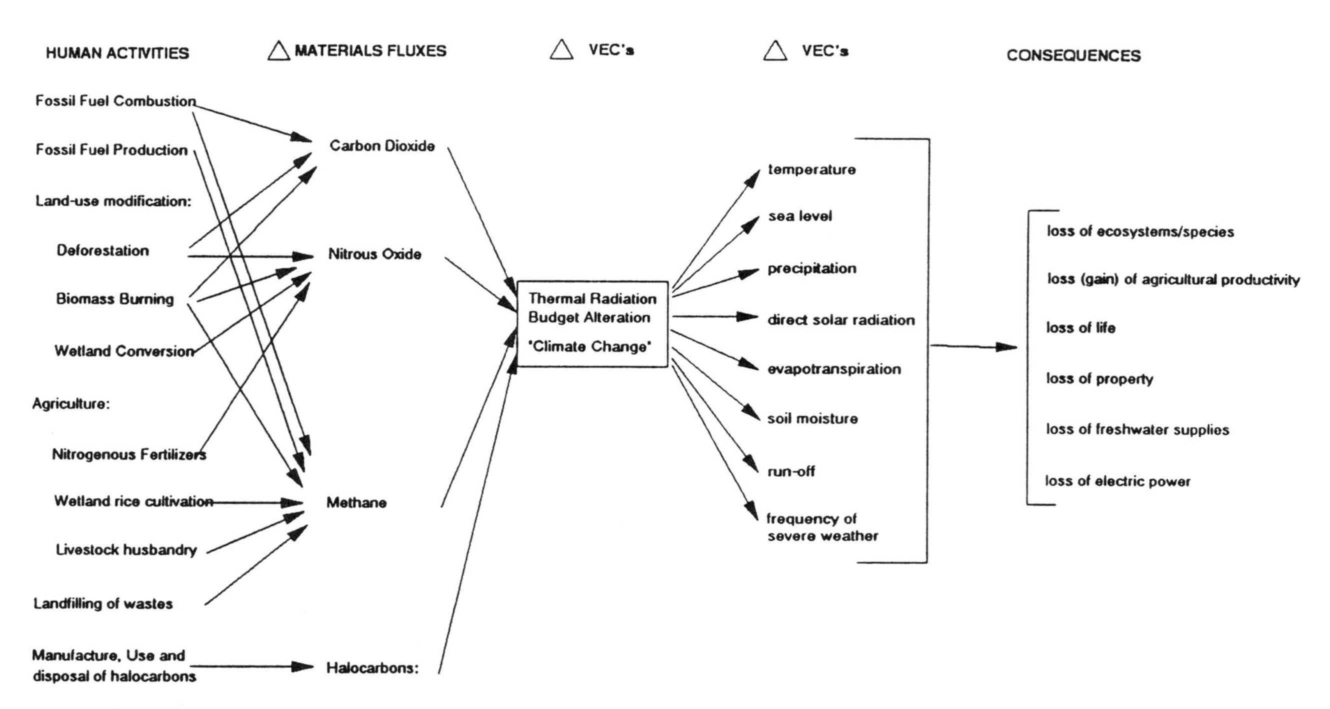

Figure 7. Hazard taxonomy for global climate change (Source: Norberg-Bohm et al., 1990).

Locus of Management	Decision Order: Technological	Organizational	Behavioral
Change Needs	*birth control	*return US gvt support to family planning	*birth control information
Change Wants	*improve energy efficiency	*eliminate subsidies forest clearing	*encourage energy conservation
Change Activity	*safe nuclear *cheap solar *convert to gas from coal	*ban CFCs *debt-nature *natl regulation of C emissions	*switch to low carbon fuels for heating
Change Releases	*prime biology pump in ocean *scrub stack gases *plant trees	*aerosols *orbital umbrellas *weather modification	*encourage social acceptance of intentional weather modification
Change Environment	*air condition *sea walls *water transfer	*coastal zoning	*migrate to better climate
Change Consequence	*improve crops *help trees migrate	*compensation payments *open trade	*abandon flooded cities *open new ports *change diet *encourage water conservation

Figure 8. A taxonomy of hazard management through decision orders. Row headings are from the taxonomy of management options introduced in Figure 6. Column headings are the decision orders discussed in the text. Cell entries from Schelling (1983), Clark (1985), and other sources.

tools of economic analysis become most useful and, indeed, indispensable. But to rely on such tools exclusively or prematurely poses a risk analogous to searching with a powerful magnifying glass under a convenient lamppost, when the real need is for broad reconnaissance of the surrounding shadows. The approaches that have been sketched in this section are neither more nor less than heuristic devices to assist in that initial exploratory reconnaissance.

Who Plays What Roles in the Management of Energy-Environment Interactions?

This section focuses on the actors involved in causing, experiencing, and managing interactions between energy use and environment. Its emphasis is on the special problems of negotiation, cooperation, and equity that arise in connection with the increasingly transboundary and multigenerational character of those interactions. Special attention is paid to asymmetries between those actors who reap the benefits of energy development and those who endure the costs of associated environmental degradation.

Scales of Interaction

The first critical issue in sorting out the relations among various actors in the energy-environment debate is one of scale. If the only parties affected by the environmental degradation that can accompany energy activities were the same parties directly benefiting from those activities, then the design of management strategies would be relatively straightforward. In reality, however, there can be significant separation in space and time of those who reap the benefits of energy use, and those who bear the environmental costs of that use. Nowhere is this situation more pronounced that in the atmospheric environment.

Figure 9 shows the characteristic time and space scales of various chemical species shown earlier to provide the linkages between energy use and changes in valued components of the atmospheric environment (compare Figures 2 and 3; the data for Figure 9 are also from Crutzen and Graedel, 1986). In the sense used here, the characteristic time scale is the time it takes for a substantial fraction (viz., two thirds) of a given quantity of the chemical emitted to the atmosphere to be lost through chemical reaction or physical removal. The characteristic spatial scale is simply the mean east-west distance that a given molecule can be expected to move in that characteristic time as a result of normal air mass circulation.

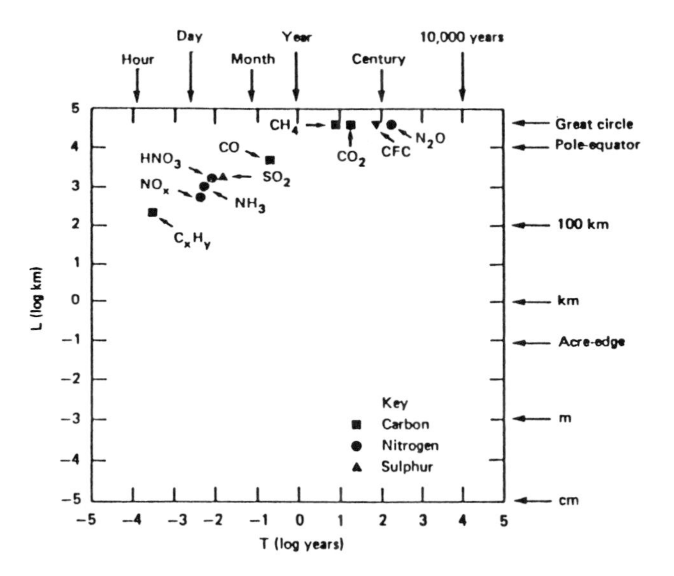

Figure 9. Space and time scales for trace gases in the atmosphere. The horizontal axis gives a typical residence time for each chemical species in the free troposphere (units are log years). Vertical axis gives characteristic spatial scale for distance (east-west) traveled in characteristic time (units are log kilometers). Data are from Crutzen and Graedel (1986).

Several major greenhouse gases—CO_2, N_2O, CFCs—have atmospheric lifetimes of 100 years or more, a sufficient time to assure that wherever they are emitted, they will spread around the entire globe. Our grandchildren will still be dealing with the consequences of today's emissions of each of these gases. Methane—also a greenhouse gas, but one that also plays a significant role in the atmosphere's self-cleansing capacity—has a lifetime of perhaps a decade. This is enough to assure thorough mixing within the northern or southern hemispheres, if somewhat incomplete mixing between them. Carbon monoxide—a combustion product that generally decreases the atmosphere's self-cleansing ability—has a lifetime of several months and thus a continental scale distribution. The oxides of nitrogen and sulfur involved in acid rain have significantly shorter lifetimes—on the order of days. Even this short interval is enough to assure that they are carried hundreds to thousands of kilometers from their source of emissions. Since a typical country in the modern world has dimensions on the order of a thousand kilometers, this gives NO_x and SO_x their well-known transboundary character. The heavier hydrocarbons, ozone, particulates, and certain aerosols have lifetimes sufficiently short to render them largely local phenomena, generally associated with air pollution in particular urban or industrial areas.

The policy implications of this short review of time and space scales are simple but of fundamental importance. Basic physical and chemical properties of the pollutants released through energy use dictate that managing energy-environment relationships has inherent multinational and multigenerational dimensions. As the total volume of pollutants increases, so will the already significant scales over which the users of energy and the users of environment find themselves linked to one another. It is to the problems of fairness associated with these transnational, transgenerational linkages that I turn next.

The Fair Allocation of Pollution Rights and Responsibilities

In view of the scale arguments advanced above, any energy-environment strategy must confront problems of fair allocation of rights to emissions and responsibilities for resulting environmental problems. Allocation issues arise, though often implicitly, in policies involving fixed percentage reductions in emissions, licensing of pollution rights, tradable pollution permits, polluter-pays compensation arrangements, or similar schemes. Questions of rights and responsibilities are clearly of importance for even relatively local pollution problems. But they become increasingly relevant and complex in the case of the long-lived, transboundary emissions that are associated with modern energy development.

Rights and responsibilities associated with long-lived, transboundary pollutants have historically been treated within accounting frameworks focused on total national emissions in a given base year. Such is the case, for example, in the European Community's Large Combustion Plant directive to limit acid emissions and the Montreal Protocol to limit CFC emissions. In principle, however, a large number of other accounting frameworks are possible, each with different implications for which actors appear to have the greatest responsibility. For example, possibilities such as emissions per capita, per unit land area, or per unit economic production have been proposed. Moreover, for those gases with especially long lifetimes it is plausible that historical and even future emissions could be considered in fair allocation schemes.

The choice of accounting frameworks for assessing emissions associated with energy activities therefore has strong strategic as well as technical dimensions. Who appears responsible for mitigating the risks of environmental degradation is strongly influenced by what we choose to count in assessing emissions. Who seems to have emitted "too much" and who still has emission "credit" is likewise affected by how our accounts are kept. Other things being equal, big countries may see advantages in accounts based on emissions per unit area, late developing countries may prefer an assessment that takes into account historical emissions, and so on. The design of accounting frameworks is thus an important aspect of the continuing international effort to balance energy and environmental issues in strategies of sustainable development.

In keeping with their strategic importance, accounting frameworks for pollutant emissions should not merely reflect technical preferences, convenience, or predilections of environmental accountants. Neither should choices be made merely in the service of special national or sectoral interests seeking to impose their own preferences on the international community. In the long run, no arbitrary or narrowly based choice of accounts is likely to receive the broad based support that will be needed to effectively manage the environmental risks of energy development. Needed is an approach to emissions accounting that is transparent in its implications, and that can be critically adopted, rejected, or bargained over by interested parties in the energy-environment debate. More ambitiously, it can be argued that all parties to those debates have a long run obligation to seek accounting frameworks that are as fair as possible in allocating rights for emissions and responsibility for the risks they entail among the peoples, nations, and generations sharing this planet.

The Case of Carbon Dioxide Emissions

An illustration of the issues that might be involved in the design and evaluation of fair accounting schemes is provided by the recent work of Susan Subak on carbon dioxide emissions (Subak and Clark, 1990). In essence, this work explores a range of accounts defined by various possible answers to questions of account numerator (i.e., what is to be counted as an emission?) and account denominator (i.e., emissions per what?).

In recognition of the role that nation states are likely to play in the negotiation of any international agreement on regulation of carbon dioxide emissions, all the accounts are computed so as to reflect a national frame of reference. Within this frame, three of the possible denominators were investigated: emissions per capita (national emissions/national population), emissions per unit area (national emissions/national land area), and emissions per nation (i.e., national emissions). The numerator chosen for evaluation was total carbon dioxide emissions from all human sources, in particular fossil fuel combustion and net deforestation. In recognition of the long lifetime of CO_2 in the atmosphere, accounts were computed both for current emissions (i.e., mean for 1980 to 1986) and for a century's cumulative emissions (i.e., total for 1860 to 1986).

The two numerators and three denominators described above produce a total of six accounts. Data were developed to allow evaluation of each account for 132 countries (see Figure 10). Figure 11 lists the highest ranked nations for each of the accounts. Clearly, the selection of accounting framework *does* make a difference in which countries appear most responsible for the risk of climate change. Using the traditional definition of current CO_2 emissions per nation produces the ranks shown in the first column. As expected, big countries and heavily industrialized countries—today's military and economic superpowers—dominate the top of the list. Other accounts, however, suggest other groupings of interests, and thus other possible coalitions in promoting or blocking action to restrict the emission of carbon dioxide to the atmosphere. As Subak has pointed out, for example, developing countries appear least accountable under cumulative accounts in general, and under cumulative per capita accounts in particular. As a group, they would appear most accountable under—and perhaps be expected to oppose—current per area accounts. Not all countries in these broad economic groupings fare alike, however. Big sparsely settled areas in the Middle East and North Africa receive very low rankings in the current per area accounts. And several Latin American countries appear high on the current release per capita list.

The immediate conclusion to be drawn from this preliminary analysis is that all energy pollution accounts have strategic implications, and no account is likely to serve the strategic interests of all countries equally. This may make it easier to promote an international discussion on the a priori fairness of alternative accounting systems. For example, the three denominators shown in Figure 11 place the rights and obligations of states, individuals, and "nature" respectively at the center of the debate. Similarly, a cumulative account numerator will tend to cast the debate on rights and obligations in a multiple generation context, while a current account confines notions of justice to the here and now. Which of these schemes "ought" to be adopted by the international community in support of its negotiations over the environmental risks of energy use is beyond the scope of this paper. But the scope for input from scholars concerned with environmental ethics (e.g., Sagoff, 1988) as well as scholars concerned with environmental realpolitik (e.g., Grubb, 1990) is as broad as the need for such input is great.

When Should What Sorts of Management Actions Be Undertaken?

This final section of the paper examines the all important question of timing in the management of interactions between energy and environment. In view of the highly uncertain and changeable nature of those interactions, an adaptive strategy is required that stresses planning for contingencies, effective learning from experience, and rapid incorporation of new understanding into management practice.

Uncertainty and (In)Action

Our scientific and technical understanding of the complex relationships between energy use and environmental change is and will continue to be uncertain and incomplete. The stakes involved in managing those relationships are enormous. As Funtowicz and Ravetz (1990) have pointed out, this combination of high uncertainty and decision stakes pushes efforts to formulate energy-environment management policies out of the realm of technical analysis and into that of total social assessment. Uncertainty becomes not just an expression of ignorance, but a strategic asset to be employed in advocating particular policies preferred for other, more fundamental reasons. We should thus not be surprised when, as has happened so frequently in debates bearing on energy-environment policy, uncertainty about the future is cited as justification both for immediate action, and for postponing action indefinitely.

A particularly striking example is provided by recent debates about the risks of climate change and the possible need for policies to restrict the emission of greenhouse gases from, inter alia, the use of fossil fuels. The range of uncertainty in assessments of future climate changes that might result from continued emissions of greenhouse gases is shown in Figure 12. The numbers reported in the figure reflect the subjective

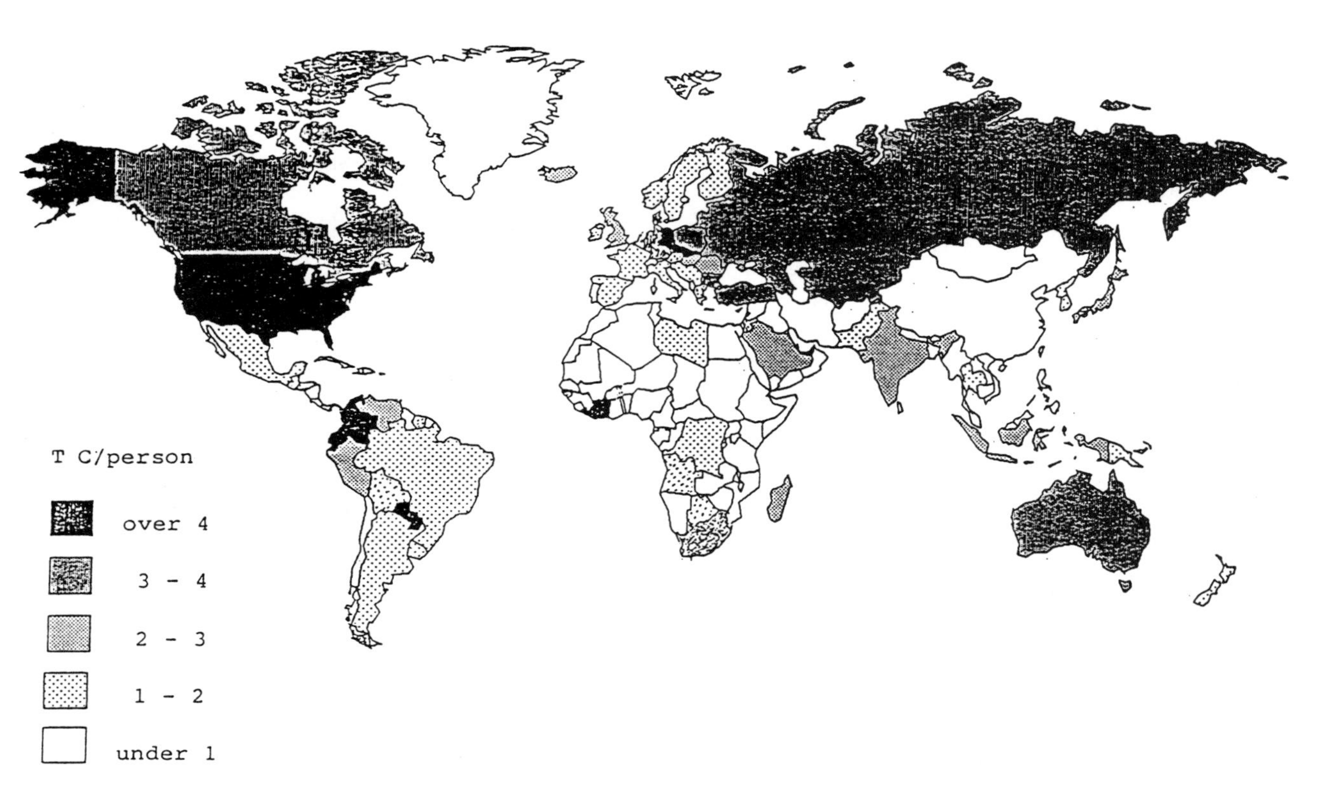

Figure 10. Distribution of current carbon dioxide emissions per capita. Data shown are averages for total current (1980–1986) human emissions resulting from fossil fuel use and forest clearing. Source: Subak and Clark (1990).

judgments of an international team of experts in climate and energy use assembled at Villach, Austria, in 1987 (Jäger, 1988). These experts concluded that based on present understanding, the actual future climate could do anything from warming at rates no faster than those experienced over the past century, to warming at rates so fast as to produce a global climate hotter than ever experienced in the course of human civilization within their own lifetimes.

How has the policy debate responded to this vast scientific uncertainty? Two instructive extremes are provided by the formal positions of the United States and the Federal Republic of Germany (FRG). To caricature complex reality only slightly, the United States' so-called "no regrets" strategy reflects a preoccupation with the lower bound of possibilities shown in Figure 12. Its objective is to adopt measures that will be beneficial even in the unlikely but not impossible event that minimal global warming occurs. Virtually no attention is paid to the regrets that might be in store under such a minimalist strategy should the climate, as is likely, actually follow a more vigorous warming path. The FRG's so-called "Vorsorge" or prevention strategy, in contrast, reflects a preoccupation with the upper bound of possibilities shown in Figure 12. Its objective is to adopt measures that will provide protection even in the unlikely but not impossible event that maximal global warming occurs. Virtually no attention is paid to the regrets that may be in store under such a maximalist strategy should the climate, as is likely, actually follow a less vigorous warming path. These contrasting national responses to the scientific uncertainties of the greenhouse effect clearly have less to do with different technical assessments than with different domestic politics. (The Americans have Mr. Sununu, while the Germans have die Grünen.) The question is not whether politics can or should be removed from the determination of strategies for energy-environment management. Rather, it is whether analysis can help the political process to strategize more effectively and efficiently.

COUNTRY	Current emissions per NATION	CAPITA	AREA	Cumulative emissions per NATION	CAPITA	AREA
United States	1	1	16	1	4	11
Soviet Union	2	9	28	2	10	18
China	3	33	26	5	25	20
Brazil	4	22	31	10	18	24
Japan	5	19	4	8	16	6
Germany, West	6	12	3	4	7	3
United Kingdom	7	13	5	3	5	2
Mexico	8	25	22	20	21	22
India	9	35	29	6	23	14
Columbia	10	5	17	99	99	99
Indonesia	11	31	24	16	22	19
Poland	12	11	8	13	99	8
France	13	21	13	9	9	9
Italy	14	24	9	15	17	10
Germany, East	15	3	2	11	3	4
Canada	16	10	34	7	1	23
South Africa	17	15	23	19	14	16
Czechoslovakia	18	6	7	18	8	7
Nigeria	19	32	25	99	99	99
Korea, South	20	26	6	99	99	99
Ivory Coast	21	2	14	99	99	99
Australia	22	7	35	12	2	25
Rumania	23	17	12	23	15	12
Spain	24	27	18	24	20	15
Thailand	25	30	20	21	19	13
Ecuador	26	4	15	99	12	99
Venezuela	27	14	27	99	99	99
Peru	28	18	30	99	99	99
Zaire	29	28	32	99	99	99
Korea, North	30	23	10	99	99	99
Netherlands	31	16	1	22	11	5
Argentina	32	29	33	14	13	21
Malaysia	33	20	19	99	99	99
Belgium	99	99	99	17	6	1
Pakistan	99	99	99	25	24	17
Bulgaria	99	8	11	99	99	99
Philippines	99	34	21	99	99	99

Figure 11. Rankings of carbon dioxide emissions by six accounts. Numbers indicate the rank order of the given country, out of 132 countries, in terms of its score in the given account. "1" signifies the highest score, "10" the tenth highest score. "99" indicates an insignificant score lower than for all countries given other numbers. All accounts are for total CO_2 emissions from all human sources. Current accounts are for average emissions over the period 1980 to 1986. Cumulative accounts are for total emissions over period 1860 to 1986. For further explanations of accounts, see text. Data are from calculations performed by Susan Subak (Subak and Clark, 1990).

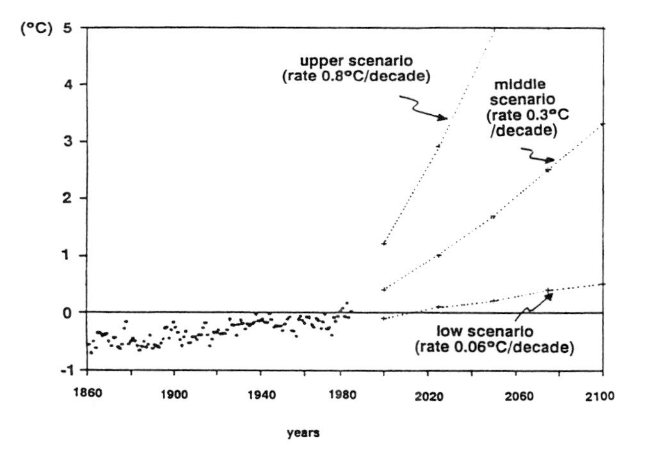

Figure 12. Scenarios of climate change. Shown are scenarios of changes in globally averaged surface temperatures that might result from continued emissions of greenhouse gases to the atmosphere. Dots to the left of 1985 are actual annual earth surface temperatures, scaled relative to 1985 temperature. Scenarios are constructed such that there is judged to be a chance of nine tenths that the actual climate will lie between the outer bounds, and a chance of one half that it will lie above or below the middle scenario. The range reflects uncertainties in both the functioning of climate, and the future course of human emissions. Source: Jäger (1988).

The Costs of Static Strategies

A first step in the direction of making analysis more useful is to insist upon the dynamic character of virtually all issues involving energy-environment tradeoffs. Both the "no regrets" and the "Vorsorge" responses caricatured above—as well as most strategies presently debated for addressing issues of acid rain, nuclear waste disposal and the like—reflect essentially static views of the decision process. Typically, a set of current forecasts, options, and uncertainties is taken as given, and a decision is made regarding the best course of action. Further research to reduce the uncertainties may be more or less centrally endorsed in such decisions. Significant attention is paid to just what sorts of new knowledge might realistically be expected to improve decision making capabilities, and to promote effective revision of initial policies. In contrast, the reality of energy-environment interactions is that problems, options, and knowledge are all changing on long and comparable time scales. There is rarely any realistic possibility of a static or even strictly sequential approach that first assembles new knowledge, then determines appropriate action, and thus changes the environmental impacts of energy choices to make them more to our liking.

The irrelevance of static or sequential decision making to most energy-environment affairs can be illustrated by examining some of the key time scales involved in management of the greenhouse effect. These are summarized in Figure 13. The fundamental time scales are set by the growth of energy use and the environment's response to the resulting greenhouse gas emissions (panel A of Figure 13). According to the scenarios presented earlier in Figure 12, it could take from 15 to 150 years for the climate to warm by 1° C; twice as long for it to warm by 2° C. The one degree level has been proposed as a threshold beyond which significant ecosystem effects might be expected (Bach, 1988). The two degree level would make the climate warmer than it has been throughout the history of human civilization, and is thus on a priori grounds a number to be viewed with respect.

Obviously, it would be desirable to have more certain information before initiating expensive steps to re-

Phenomenon	Years (best estimate and range)
A) The climate system:[a]	
Global warming of 1° C:	35 (15 - 150)
Global warming of 2° C:	70 (30 - 300)
Natural withdrawal of CO_2 from atmosphere	90 (80 - 100)
B) Knowledge: Scientific consensus on predictions of[b]	
Global temperature	3 (0 - 5)
Global sea level	10 (5 - 20)
Regional moisture	20 (10 - 50)
Regional socio-economic impacts	20 (10 - 50)
C) Response: Actions and adjustments[c]	
New industrial processes	15 (10 - 20)
Fuel substitution in energy systems	25 (10 - 40)
Shifts in location of major industries	50 (20 - 80)
Shifts in distribution of national influence	45 (30 - 60)

Figure 13. Timing in the management of global climate change.

Notes:[a]Time required for actual warming to occur, taken from Figure 12 (Jäger, 1988).

[b]Time required for scientific consensus to emerge (Schneider et al., 1990 for natural science; author's estimate for impact studies).

[c]Time required for, or relevant to, response actions and adjustments (Clark, 1987). For "industrial processes" this is the time it has historically taken to shift an entire industrial sector (e.g., steel) from one fundamental production process (e.g., basic oxygen) to another.

duce the risk of climate change. But as can be seen in panel B of Figure 13, it could easily take 5 to 20 years before a scientific consensus is reached on even the global dimensions of climate change. Agreement on the regional changes in moisture, runoff, etc., that are central to impact assessment could take 10 to 50 years—with luck (Schneider et al., 1990). Comparing the time scales of actual climate change and of improvements in our ability to predict such change, we find they are comparable. It is not at all inconceivable that significant climate changes could happen before we could predict them with any confidence, leaving much of our knowledge-gathering activities with roughly the utility of yesterday's weather forecast.

In addition, even if we did decide that there was adequate information to act, significant action itself takes time. Historical experience suggests that even under strong incentives, it takes on the order of 15 years to switch from one fundamental production process (e.g., basic oxygen process steel making) to newer processes throughout a national industrial sector. Major fuel switches in energy systems have taken even longer—on the order of 25 years (Clark, 1987). Add these response times to the times necessary to build scientific consensus, and we face a minimum of 50 years before actions driven by consensual scientific knowledge could be expected to significantly affect the course of climate change. But in 50 years, even the median climate scenario of Figure 12 would have long exceeded one degree of warming and be well on its way to an unprecedented two degrees warmer global climate. Moreover, as already noted, most greenhouse gases have very long atmospheric lifetimes: it would take almost a century for a significant amount of excess carbon dioxide to be removed from the atmosphere through natural processes *after* major reductions in human emissions were accomplished.

A strategy of first waiting for consensual scientific knowledge, then—if the knowledge so indicates—initiating major technological programs to reduce greenhouse gas emissions, and finally waiting for nature to remove excess greenhouse gases from the atmosphere could be expected to exert control on climate change by the year 2100—providing no surprises or irreversible changes occur. Conversely, a preoccupation with avoiding the risks associated with the "worst-case" scenario of climate change shown in Figure 12 could lock us into industrial processes, energy systems, and sea wall schemes that could severely constrain our technological options for dealing with other environment and development problems for half a century. Honorable people can differ over which choices associated with static or sequential strategies are most defensible. While they do so, however, it does seem wise to consider what dynamically conceived adaptive strategies might have to offer the policy debate.

The Dynamics of Usable Knowledge

The comparable time scales of problem development, knowledge development, and policy response described above strongly suggest that we will often have to experience the environmental problems of energy use, learn about those problems, and attempt to manage them all at the same time. To the extent that we wish to structure these interactions as a matter of formal policy, associated strategies will have to be dynamic and adaptive. Questions that immediately arise from such an adaptive perspective include the following: How should we apportion our attention between "best estimate" forecasts of environmental impacts and "extreme" estimates of not-impossible outcomes? What options for dealing with the environmental impacts of energy use need to be implemented immediately? Which should be held in reserve as events unfold? While waiting, what research and development activities will be most useful to pursue?

Even asking these questions in a disciplined and sustained manner has the potential for improving much contemporary policy debate. Serious efforts to provide answers can be extremely useful, as demonstrated by recent work of Gary Yohe on the value of information in dealing with global climate change (Yohe, 1990). Yohe adapts a number of classic approaches from economics and decision theory to perform the following analyses:

1. rank alternative response options given current scientific understanding (e.g., the scenarios of Figure 12);
2. determine the optimal timing of implementing those options given current scientific understanding;
3. show how the rankings and timing might change in the future as a result of additional information;
4. calculate the value of different kinds of new information from the perspective of the changes in assessments that they would cause;
5. recommend the sorts of research that might generate the most valuable new forms of information.

Based on a detailed application of these approaches to the problems of responding to sea level rise resulting from climate warming, Yohe advances a number of findings that are likely to have broad relevance to the formulation of policy for energy and environment.

First, he shows that dynamic policies need not be policies that retain total flexibility to respond to changing conditions. For example, the "perfectly contingent" option of adding sand as needed to keep a valuable recreational island above rising sea level is shown to be inferior to the "perfectly discrete" option of building a protective dike in anticipation of rising sea level. Yohe gives two reasons for this finding that suggest a more general relevance. One is a simple matter of costs. Where contingent policies must be initiated early and continued indefinitely, there is room for discrete policies involving a later, one-time expense to be cheaper under plausible discounting schemes. Another is that discrete options may often be treated as insurance policies: a little early investment in creating the technical and institutional capacity to respond (e.g., to build a dike) can then suffice until, if ever, changing information indicates that the time for exercising the option is at hand. Conversely, where the discrete option requires relatively early, large, and continuing investments, then the contingent response may be expected to dominate. The general analysis obviously does not answer the question of what to do in particular cases, but it does suggest the kinds of questions and estimates that would be required to increase the rationality of such particular decisions.

Yohe also points out that the value of future information about environmental change—whether provided by better predictions, or just by watching how the environment in fact changes—is diminished to the extent that society has access to cheap contingent responses or "insurance" options of the sort described above. The systematic expansion of the range of options available for consideration, as discussed earlier, is therefore especially important in cases where research and monitoring seems likely to be extremely expensive, too late, or inconclusive. Conversely, research and monitoring become increasingly valuable where contingency responses are scarce or discrete options require expensive preparation.

What kind of information is likely to be most valuable for structuring efficient responses to environmental change? This question is particularly relevant given the present and past policies of many governments to fund research rather than immediate action on environmental problems relating to energy use. Significantly, Yohe shows that under a wide range of conditions the crucial information is that which allows us to decide as early as possible whether we are in fact on the "extreme" trajectories of the sort shown in the upper bound of Figure 12. Research that refines the "best estimate" may be scientifically more satisfying, but is significantly less useful to society. The central relevance for policy of better information about the nature and likelihood of extreme trajectories and events has been stressed by many analysts (e.g., Clark, 1985; Waggoner, 1990). A strong implication is that the seriousness of any policy of "getting more information" before acting can be judged according to how resolutely the recommended information getting focuses on the assessment of extreme scenarios. Unfortunately, nearly all governments and most scientists are disinclined to explore such scenarios, preferring—for very different reasons—to focus exclusively on "best estimates" and modest spreads around them. The social consequences of such willful myopia may be substantial. The need for systematic thinking about the nature and probability of the "not impossible," as well as the "most likely," is correspondingly great (Clark, 1990). A commitment to such thinking as a matter of national and international policy would not be a bad place to begin the formulation of energy and environment strategies for the 21st century.

Bibliography

Bach, W. *Modelling the climatic effects of trace gases: reduction strategy and options for a low risk policy.* Paper presented at the World Congress on Climate and Development. (Hamburg). 1988.

Baumgartner, T. and A. Midttun. *The politics of energy forecasting.* (Oxford: Clarendon Press). 1987.

Brooks, H. "The typology of surprises in technology, institutions and development" in W. C. Clark and R. E. Munn, eds. *Sustainable development of the biosphere* (Cambridge: Cambridge University Press). pp. 455–473. 1986.

Brooks, H., and J. M. Hollander. "United States energy alternatives to 2010 and beyond: The CONAES study." *Annual Review of Energy* 4:1–70. 1979.

Clark, W.C. *On the practical implications of the carbon dioxide question.* WP-85-43. (Laxenburg, Austria: International Institute for Applied Systems Analysis). 1985.

Clark, W. C. "Scale relationships in the interactions of climate, ecosystems, and societies" in K. C. Land and S. H. Schneider, eds. *Forecasting in the Social and Natural Sciences.* (Dordrecht, Netherlands: Reidel). pp. 337–378. 1987.

Clark, W.C. "Managing planet earth." *Scientific American* 261 (3): 46–54. 1989.

Clark, W.C. *Visions of the 21st century: Conventional wisdom and other surprises in the global interactions of population, technology and environment.* Science, Technology and Public Policy Program Discussion Paper 89–90. (Cambridge, MA: Kennedy School of Government, Harvard University). 1990.

Clark, W. C. and R. E. Munn, eds. *Sustainable development of the biosphere.* (Cambridge: Cambridge University Press). 1986.

Crutzen, P. J., and T. E. Graedel. "The role of atmospheric chemistry in environment-development interactions" in Clark, W. C. and R. E. Munn, eds. *Sustainable development of the biosphere.* (Cambridge: Cambridge University Press). pp. 213–250. 1986.

EEC (European Economic Community). *Energy and environment.* [Communication from the European Commission to the Council, November] (Brussels: EEC). 1989.

Funtowicz, S.O. and J.R. Ravetz. *Uncertainty and quality in science for policy.* (Dordrecht: Kluwer). 1990.

Graedel, T. and P. Crutzen. "The changing atmosphere," *Scientific American* 261 (3): 58–68. 1989.

Grubb, M. *The greenhouse effect: negotiating targets.* (London: Royal Institute of International Affairs). 1990.

Jäger, J. *Developing policies for responding to climate change.* [Summary of the discussions and recommendations made at workshops held in Villach, Austria and Bellagio, Italy, World Meteorological Organization]. World Climate Program Publ. No. 1 WMO/TD-No. 225. (Geneva: WMO). 1988.

Kates, R. W., C. Hohenemser and J. X. Kasperson, eds. *Perilous progress: Managing the hazards of technology.* (Boulder, CO: Westview Press). 1985.

Kates, R.W., B.L. Turner II, and W.C. Clark. "The great transformation." Chapter 1 in B.L. Turner II et al. (eds). *The earth as transformed by human action.* (New York: Cambridge Univ. Press). 1990.

March, J. G., and J. P. Olsen. *Ambiguity and choice in organizations.* (Bergen, Norway: Universitetsforlaget). 1976.

Marsh, G. P. *The earth as modified by human action.* (New York: Scribner). [Reprinted 1965, Cambridge, MA: Harvard University Press]. 1864.

Montgomery, J.D. *Technology and civic life: making and implementing development decisions.* (Cambridge, MA: MIT Press). 1974.

Norberg-Bohm, V., W.C. Clark, M. Koehler, J. Marrs. *Comparing environmental hazards: the development and evaluation of a method based on a causal taxonomy of environmental hazards.* Progress Report, Ver. 1.1. Global environmental policy program Discussion Paper. (Cambridge, MA: Kennedy School of Government, Harvard University). 1990.

OECD (Organization for Economic Cooperation and Development). *Environment effects of energy systems: the OECD COMPASS Projects.* (Paris: OECD). 1983.

OECD (Organization for Economic Cooperation and Development). *Energy and environment: policy overview.* (Paris: OECD). 1990.

Ogawa, Yoshiki. *Energy consumption and global warming.* (Japan: Institute of Energy Economics). 1990.

Sagoff, M. *The economy of the earth: philosophy, law and the environment.* (Cambridge: Cambridge Univ. Press). 1988.

Schelling, T. "Climatic change: Implications for welfare and policy" in National Research Council. *Changing climate.* (Washington, DC: National Academy Press). pp. 449–482 1983.

Schneider, S.H., L.O. Mearns, and P.H. Gleick. "Prospects for climate change" in P. Waggoner, ed. *Climate change and US water resources.* (New York: John Wiley). 1990.

Subak, S. and W.C. Clark. *Accounts for the greenhouse gases: towards the design of fair assessments.* (Stockholm: Stockholm Environment Institute). 1990.

Turner, B.L. II et al. (eds). *The earth as transformed by human action.* (New York: Cambridge Univ. Press). 1990.

US DOE (United States Department of Energy). *A national energy plan for the United States of America.* (Washington, DC: DOE). 1990.

Waggoner, P. ed. *Climate change and US water resources.* (New York: John Wiley).

WCED (U.N. World Commission on Environment and Development, the Brundtland Commission). *Our common future.* (New York: Oxford University Press). 1987.

Wildavsky, A. *Speaking truth to power: The art and craft of policy analysis.* (Boston, MA: Little, Brown). 1979.

World Bank. *World Development Report.* (New York: Oxford University Press). 1989.

Yohe, G. *Uncertainty, climate change and the economic value of information: An economic methodology for evaluating the timing and relative efficacy of alternative responses to climate change.* (Stockholm: Stockholm Environment Institute). 1990.

Energy and Environmental Policies in Developed and Developing Countries

José Goldemberg
University of São Paulo
São Paulo, Brazil

Abstract

Energy is responsible for more than half of the environmental problems facing humankind. In addition to other problems caused by the need to ensure a steady supply of energy—essential for civilization as we know it today—pollutants associated with energy production and use constitute an increased threat to the environment. On a global scale, industrialized countries are no longer the dominant contributor to such problems, and pollution produced in centrally planned economies (CPEs) and the lesser developed countries (LDCs) is beginning to dominate the picture. Solutions will involve a degree of cooperation between the "rich" and the "poor" nations based upon the enlightened self-interest of both, not merely on good will and philanthropy. After analyzing the proposed solutions in the Organization for Economic Cooperation and Development countries (OECD) and other countries, the discussion turns to the actions needed to steer development of the LDCs (and to some extent CPEs) in a direction that will not hinder economic growth, but will be environmentally non-destructive.

Introduction

Energy is responsible for an important part of the environmental problems facing humankind today. Extraction of coal from under the ground used to be a risky and polluting business even in the 15th century, but the burning of such coal represented only a minor nuisance to the environment. Today, the burning of immense quantities of coal, oil, gas, and fuelwood represents a serious environmental threat; the sulfur and nitrogen oxides which result from the burning of fossil fuels are the main causes of smog, acid rain, and a variety of other disturbances in our environment.

Carbon dioxide is not a pollutant in the usual sense because it cannot be filtered or easily captured; the burning of carbonaceous materials is responsible for 57 percent of the global greenhouse effect (the warming of the earth as a whole) (see Figure 1) in contrast to problems caused by acid rain which are geographically localized although their effects may cross national boundaries.

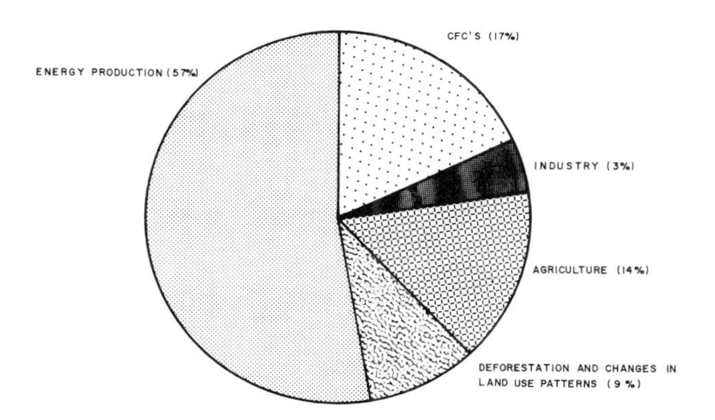

Figure 1. Contributions to the greenhouse effect.

On the other hand, energy is absolutely essential for development and huge amounts are needed to achieve the level of amenities available to people living in the industrialized nations. An average US citizen consumes 7.2 tons of petroleum equivalent (TEP) per year. In contrast, an average person living in the developing countries has to make due with 10 times less energy. Differences in climate among industrialized and developing countries cannot account for a difference of an order of magnitude in energy consumption. (See Table 1 for indicators of energy consumption per capita and per unit of production for various countries.)

From 1975 to 1985, Organization for Economic Cooperation and Development (OECD) countries made important progress in halting energy growth. The fact that energy consumption is growing again in some of these countries—mainly in the US—is, however, a cause of grave concern. This growth is due largely to personal transportation and the heating and cooling of residential and commercial buildings.

In the LDCs (lesser developed countries, including China), however, energy consumption has been growing linearly for a long time (and not exponentially!) and shows no indication of saturation (see Figure 2). This is also true for the Soviet Union and Eastern European

Table 1. Indicators of energy consumption per capita and per unit of product (1986).

	Total Population (millions)	Total Energy Consumption (MTEP)	TEP/Capita	TEP/US$1,000	GDP (US$)/capita
OECD	818	3,855	4.71	0.38	12,119
USA			7.2	0.41	17,480
Japan			3.2	0.20	12,840
FRG			4.5	0.30	12,080
France			3.6	0.28	10,720
England			3.8	0.46	8,870
Italy			2.5	0.24	8,550
LDCs	3,830	1,915	0.50	0.59	680
Brazil			0.8	0.55	1,810
India			0.2	0.80	290
Mexico			1.2	0.78	1,860
Korea			1.4	0.60	2,370
China			0.6	2.12	330
CPEs	475	2,040	4.30	0.61	6,749
USSR*			4.9	0.62	7,340
GDR*			5.9	0.56	10,235
Czechoslovakia*			4.8	0.55	8,500
Total	5,123	7,810			

* Toe/1000 US$ refers to 1985.

(Sources: British Petroleum Statistical Review of Energy, World Bank Report on World Development, 1988; Handbook of Economic Statistics (CIA), 1987.)

centrally planned economies (CPEs). The political leadership of LDCs, and their populations, in general accept the idea that it is necessary to grow at all costs repeating (if needed) the mistakes made in the past by industrialized countries. In addition, in a number of LDCs intense deforestation is taking place due to changing land use patterns. Some of the deforestation—mainly in Africa—is linked to energy uses such as cooking. Deforestation of any kind contributes significantly to CO_2 emissions, justified by some as an inevitable result of development. There are therefore different and even conflicting energy and environmental policies in the industrialized and developing countries.

Clearly, the problem faced here is global—although the northern hemisphere will probably be more affected by a warming of the globe than the southern hemisphere. No isolated region of the world or group of countries can reduce emissions so as to offset the emissions of the other groups, as can be seen in Figure 2.

The reductions of emissions needed to stabilize atmospheric concentrations at the present levels are the following:

1. 50 percent emission reductions in carbon dioxide (CO_2)
2. 10 percent of methane (CH_4)
3. 75 percent of chlorofluorocarbons (CFCs)
4. 80 percent of nitrous oxide (N_2O)

These are ambitious targets. The only proposal seriously considered so far is a 20 percent reduction in the greenhouse gas emissions of industrialized countries to be achieved by year 2005. This corresponds to the recommendations made by the Toronto Conference on Climate Change in 1988. [1] Even that target will not be easy to reach.

Energy and the Environment in Developed Countries

Between 1973 and 1986 there was no net primary-energy consumption growth in the US, while the gross domestic product (GDP) increased 36 percent. The ratio of primary energy use to GDP (i.e., the energy intensity of the US economy) declined, therefore, at an average rate of 2.4 percent per year in this period. In the same period, primary energy use in OECD countries increased 4 percent while GDP increased 38 percent, so that the energy intensity of these countries decreased at an average 2.2 percent per year. Figure 3 shows the data for the OECD countries indicating also a sharp drop in oil imports. These oil imports were replaced by other energy sources or made unnecessary due to an improved efficiency in energy use.

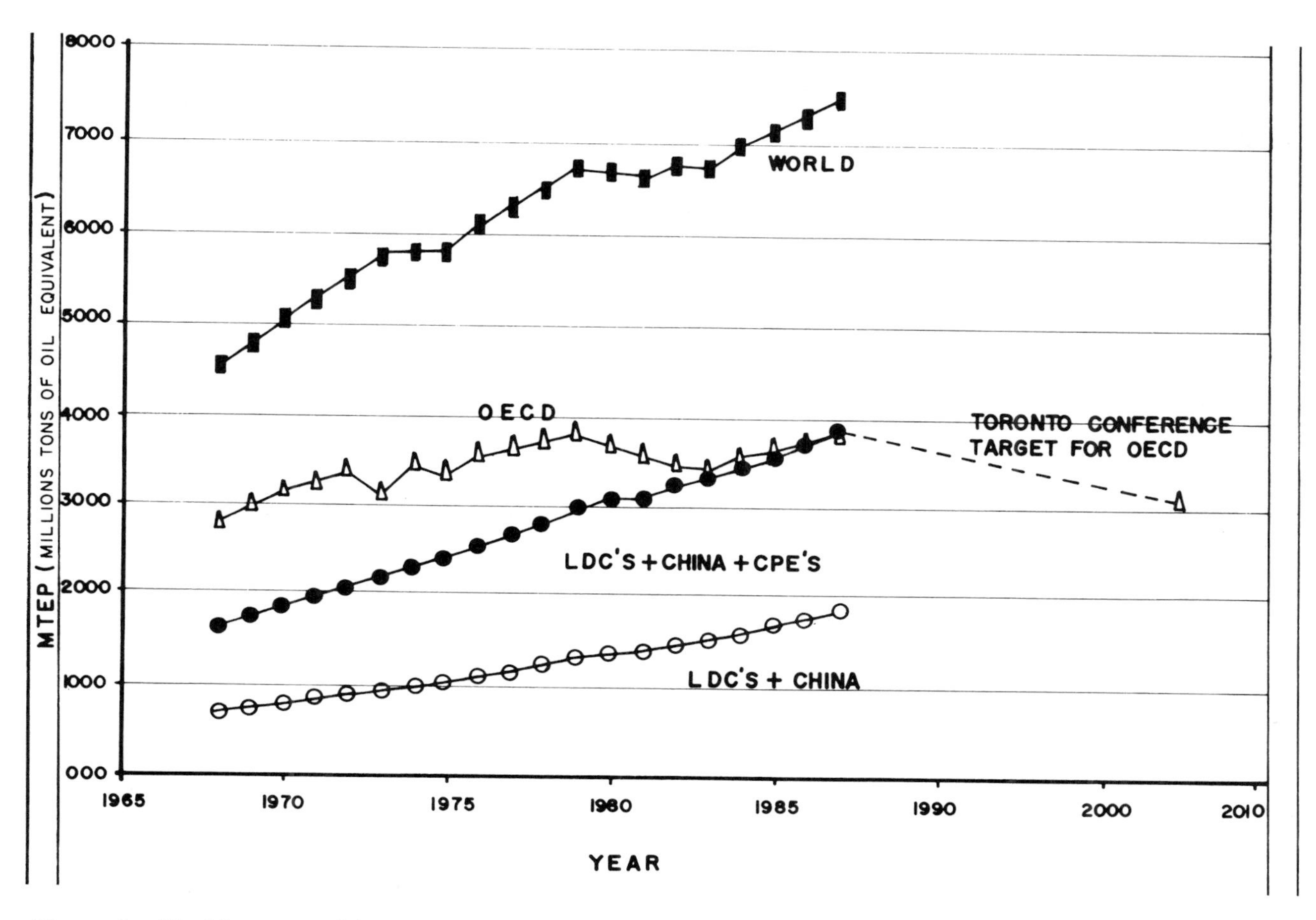

Figure 2. World commercial energy consumption, 1968–1987 (OECD—Organization for Economic Cooperation and Development countries; LDC—lesser developed countries; CPE—centrally planned economies).

The decoupling of GDP increase and energy use is considered a major new feature of the economies of industrialized countries that has forced a reanalysis of the role played by energy in modern societies. Such decoupling should not, in reality, be considered a novelty, because long-term studies of the evolution of the energy/GDP ratio (E/GDP) over time for a number of countries shows that this ratio has been decreasing steadily—except for a period when the heavy industrial infrastructure was put into place.

The data indicates [2] that latecomers in the development process also follow this pattern with less accentuated peaks (Figure 4); they do not have to reach high values of the E/GDP ratio even in the initial stages of industrialization because they benefit from modern methods of manufacturing and more efficient systems of transportation.

As pointed out by Drucker, Strout and Williams et al. [3–5], a saturation of consumer goods consumption has been achieved in industrialized societies, economic activity has been moving in the direction of services and not heavy industry, and there has been revolution in materials clearly indicating a shift toward the use of less energy-intensive materials. Given these trends, which started before the oil crisis of 1973, the increase in oil prices only accelerated structural changes in the industrialized countries.

Particularly Strout [4] shows that consumption of 10 energy intensive materials in 52 countries (wood pulp, paper and paperboard, chemical fertilizers, hydraulic cement, steel products, copper, lead, zinc, aluminum, and tin) are closely correlated with GDP up to incomes of $2,000 (in 1970 US dollars). Above that, the strong coupling of these indicators disappears. This means that in more affluent societies, where most of the infrastructure is already in place the structural shifts mentioned above strongly dominate the picture (see Figure 5).

If one concentrates on the industrial sector of the US economy, the energy intensity of this sector de-

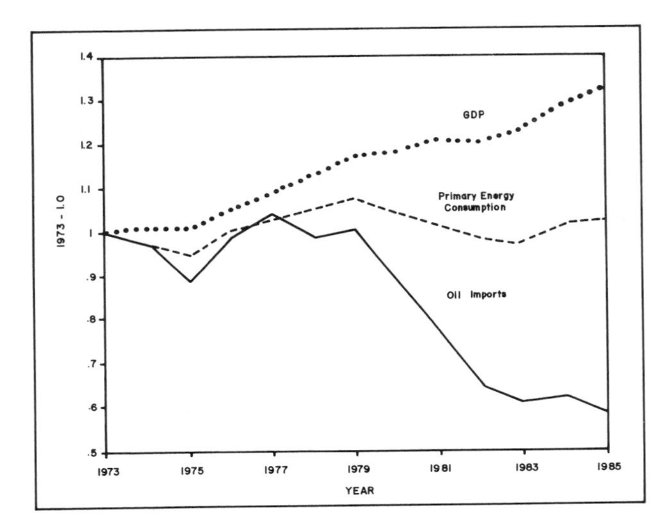

Figure 3. Primary energy consumption, net oil imports, and gross domestic product for the Organization for Economic Cooperation and Development countries, relative to the values of these quantities in 1973.

creased 3.5 percent per year in the period 1973–1986—faster than the overall energy intensity. This decrease was due to two equally important reasons [5]:

1. a shift from energy-intensive manufacture of materials to lighter materials, leading to a reduction of energy of 1.6 percent per year; and
2. a reduction in the energy intensity of many industrial sectors accounting for 1.9 percent per year.

A third reason is the import of energy intensive materials from developing countries to areas to which heavy industry is migrating. This seems to be of secondary importance today but will certainly become more important in the future. In 1986, the net imports of energy embodied in goods and services traded by the US reached about 3 percent of total energy use in the US [6]. In addition, it is estimated that 25 percent of the primary metals industry of the US will be shut down in 10 to 20 years.

The sum of all this is that the decline in energy intensity in the developed countries is taking place for a variety of reasons and the fast pace of technological advances in goods and materials (coupled with greater attention to efficient energy use) leads to an accentuated decline in energy intensity over the years indicated as in Figure 6.

Environmental considerations are bound to accelerate such trends and "energy conservation" (a synonym for efficient energy use) will play a very important role in the future. The political leadership in the OECD countries, pressured by public opinion, will probably adopt public policies designed to cut CO_2 emissions by imposing new taxes on emissions or introducing new mandatory regulations such as the efficiency standards imposed by the US Congress in 1975. These standards were remarkably successful in increasing the fuel performance of American cars by more than 50 percent. The energy efficieny of new cars has increased from 14 miles per gallon in 1973 to 28 miles per gallon in 1987.

The generally stated goal is to impose by international agreement (as done for the production of CFCs through the Montreal Protocol) a 2 percent per year reduction in energy consumption, leading to a 20 percent reduction by the year 2005. One half of that reduction could come from improved efficiency in energy generation and the other half by improved end-use efficiency.

The cost of such actions is strongly argued in some of the industrialized countries today and explains the resistance of the United States, Japan, and the United Kingdom in accepting the establishment of a worldwide carbon tax, the main burden of which would fall on their shoulders as the main CO_2 emitters. For example, it has been argued [7] that the moderate carbon taxes proposed by US Environmental Protection Agency (EPA) ($25/ton on coal, $6/bbl on oil and $0.50/mcf on gas) would:

- raise the average tariff of electricity by 20 percent (from present rate of 58 mills/kwh),
- raise the price of gasoline by 17 to 20 cents/gallon, and
- raise the average residential heating costs $85–$100. (Figures above are in US dollars)

Because unilateral action taken by any of the leading industrialized countries could indeed put them at an economic disadvantage relative to the others, multilateral action is needed. This is a typical example of an important action needed—and recognized by many—where market forces will probably not do it alone. The imposition of a carbon tax might increase costs and require government regulations, but will most likely be enacted once people feeling the effects of CO_2 emissions—such as those in the Los Angeles area—become willing to pay for it. In addition, other actions, such as an international agreement among leading car manufacturers to increase fuel performance, might have far-reaching effects. The developed countries can certainly lower energy consumption without affecting trends of development. It is possible for them to reach the Toronto Conference recommendations by year 2005.

Evidence to this effect can be found in a Canadian study [8] completed in the spring of 1989. The study concluded that it is feasible for Canada to meet the Toronto Conference target of a 20 percent reduction in CO_2 emissions by 2005 and that energy efficiency measures are the most cost-effective means of moving toward this target. The study also determined that energy efficiency measures could provide three fourths of the reductions, that fuel switching was needed for only one

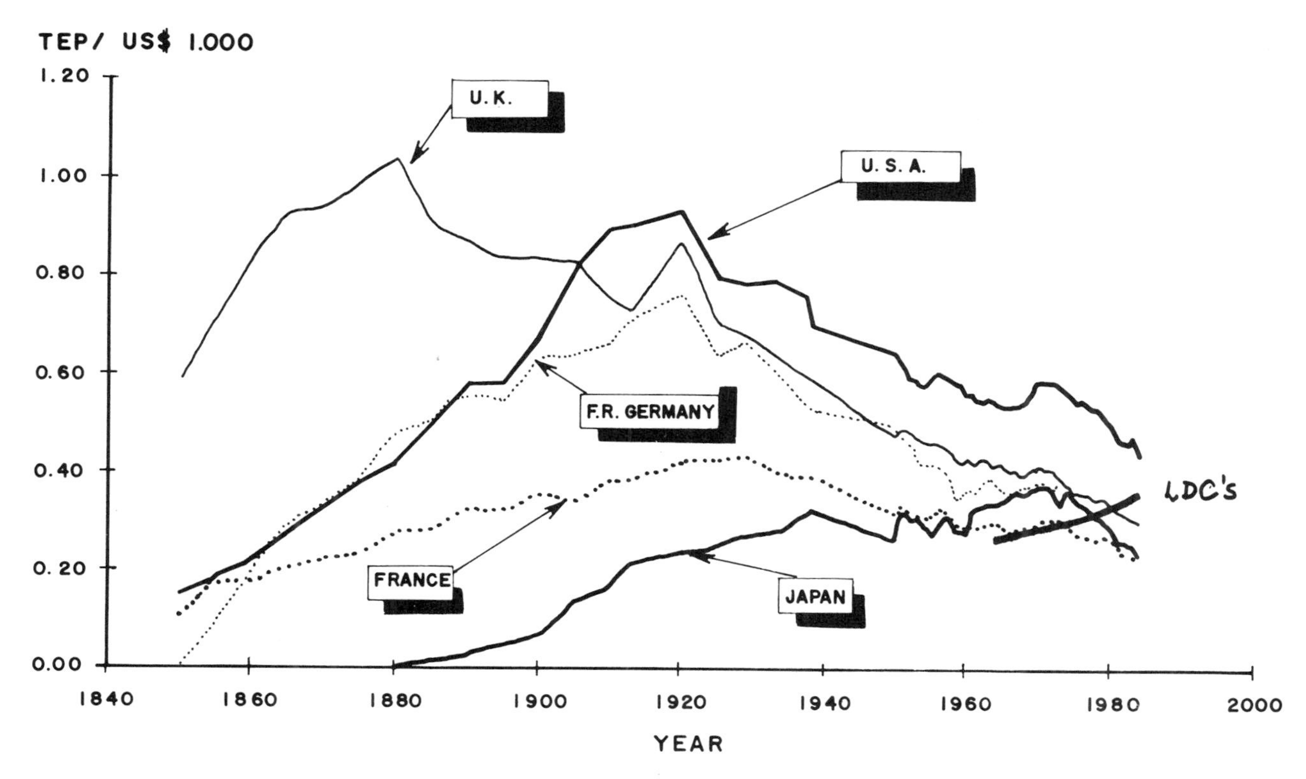

Figure 4. Evolution of the energy intensities in different countries.

fourth of total reductions, and that implementing the proposed strategy would save Canadians $100 billion through lower energy bills.

Energy and the Environment in Developing Countries

In developing countries, public sensitivity regarding environmental preservation is weak. Until 1972 the political leadership in many of these countries held a widespread perception that development should be pursued at all costs. They believed the eradication of misery and underdevelopment would bring with it a better environment. After the 1972 Stockholm World Conference on Environment some progress was made through efforts to introduce environmental concerns in many of the projects being implemented in LDCs. Particularly, the World Bank has become more sensitive to these actions and exhaustive studies of environmental impacts are required in many cases to avoid repeating horrors of the past.

It is worth mentioning that pollution can be categorized in the following ways:

1. *local pollution* produced by poor people due to the lack of sanitation and waste disposal. Although less damaging to the atmosphere, it seriously affects the living conditions of the poor.
2. *regional pollution* due to electricity generation from burning coal and oil, and automobile and freight transportation. It severely affects the atmosphere of large cities (such as Los Angeles and Mexico City) and leads to long-distance and transboundary pollution in the forms of smog and acid rain.
3. *global and atmospheric pollution*, a recent problem linked to large-scale energy production and consumption, deforestation, and CFC production. The "greenhouse effect" due to such emissions is the first global threat to the environment faced by humankind. The only practical way to cope with this problem is to reduce the amount of CO_2 released into the atmosphere—by reducing energy consumption—and to sequester CO_2 from the atmosphere, which can be done with reforestation.

By and large, it is recognized that GDP has to grow and that in the initial stages of development—and many of the LDCs are still in those early stages—energy has to grow as fast as GDP and even faster when establishing an infrastructure of roads, bridges, heavy industry, and houses.

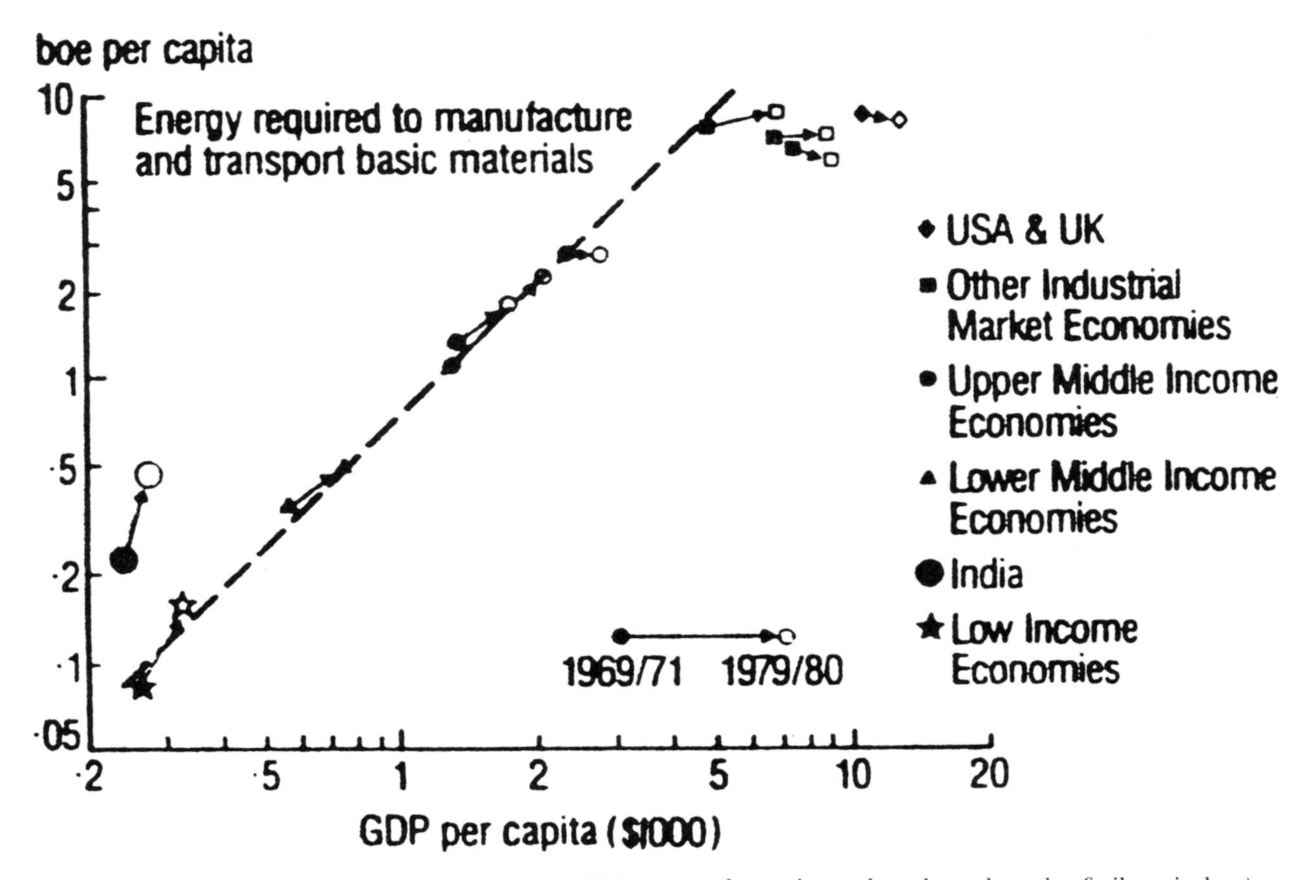

Figure 5. Energy use in materials consumption (GDP—gross domestic product; boe—barrels of oil equivalent).

Annual growth rates of 5 to 10 percent are always looked upon as the goal, although only a few LDCs have succeeded at that. Even more modest growth rates require large investments, but a 4 to 5 percent per year growth in energy consumption has been achieved in some of the "better off" LDCs. Coupled with a growing population, the evolution of energy consumption (as shown in Figure 2) indicates more or less *linear* growth. This is encouraging since an exponential growth could have been expected. If present trends continue, in about 20 years the total energy consumption in LDCs will reach the present level of consumption in industrialized countries.

Because the population of LDCs is presently three times that of industrialized countries and continues to grow, consumption per capita in LDCs would still be appreciably lower. Such prospects have led to fears that some industrial countries will try to impede the progress of developing countries in order to prevent additional threats to the environment. In Figure 2, only commercial energy consumption is considered; in many LDCs noncommercial consumption (wood fuel, charcoal, agricultural wastes) can represent a large portion of energy resources. Overall their contribution is on the order of 30 percent and declining as development takes place and fossil fuels replace noncommercial sources [9]. Development might also increase the growth rate of energy consumption in LDCs in the near future because the energy intensity of such countries is increasing as shown by Gibbons [10] (Figure 6).

The contribution of LDCs to environmental damage—particularly CO_2 and other greenhouse gas emissions—is increasing and might jeopardize all efforts made by industrialized countries to reduce these emissions. In addition to the trends pointed out above, China is planning to install some 100 gigawatts of coal-burning electricity generation within the next 10 to 15 years and 100 million refrigerators that use CFCs.

Urgent actions are needed to engage LDCs in the task of reducing global environmental degradation. This has become even more urgent because energy consumption has resumed growth in some OECD countries as pointed out above.

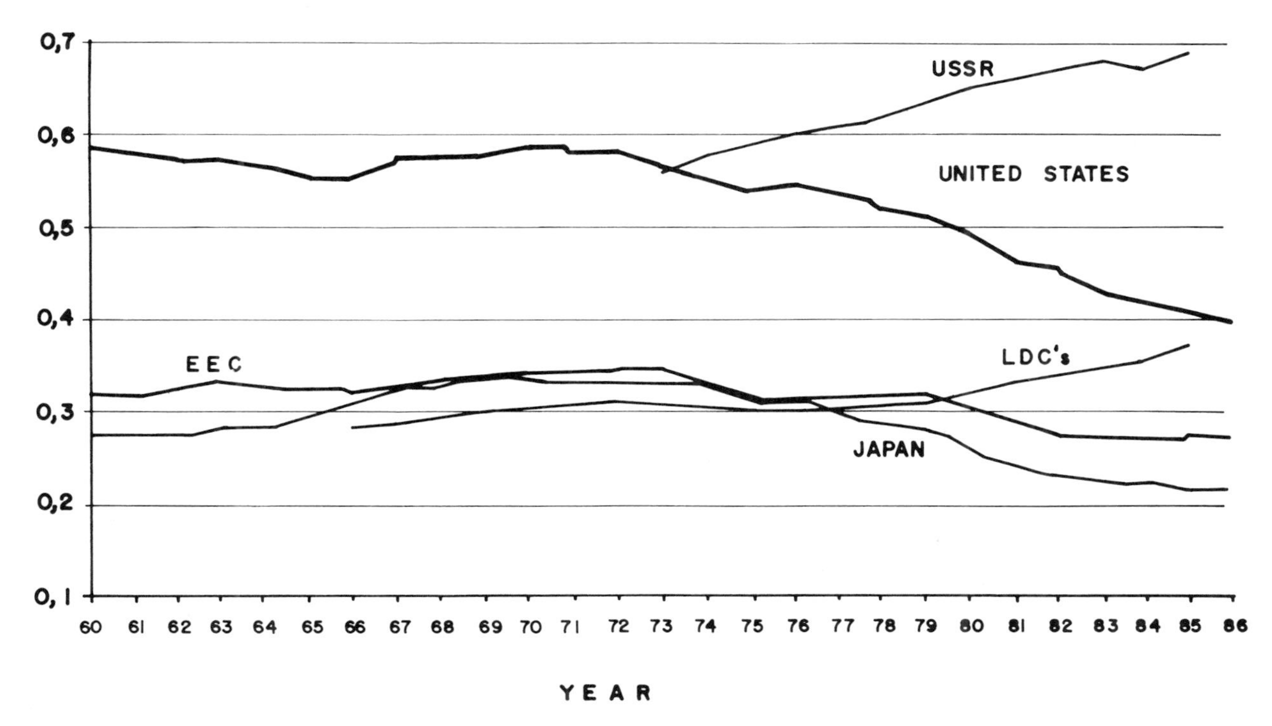

Figure 6. Energy intensity (EEC—European Economic Community).

The Deforestation Issue

Deforestation is taking place in many countries around the world—mainly in tropical forests—with a net contribution of some 30 percent of the total emitted by fossil fuel burning. The present knowledge of deforestation rates in LDCs is admittedly rather shaky, but Table 3 shows the countries that are the leading contributors.

Deforestation is taking place for a variety of reasons such as the production of hydroelectricity (which requires the formation of artificial lakes) or the production of fuelwood and charcoal for cooking or industrial purposes.

It's mainly in Africa that deforestation is intimately related to energy production. The total amount of fuelwood consumed in Africa (in the forms of charcoal and fuelwood) is approximately 300 million tons per year—used mostly for domestic purposes. Table 4 describes population and energy consumption in Africa. To obtain this fuelwood an estimated 2 million hectares of land are deforested each year with grim consequences that include soil erosion, loss of species, and local climate change. On a global scale, the burning of 300 million tons of fuelwood contributes some 4 percent to world CO_2 emissions. At least one half of the deforestation takes place to increase agricultural land and pastures, and for the extraction of valuable timber (mainly in the tropical forests of the Amazonia and Indonesia).

From the LDCs viewpoint such actions are necessary and legitimate in order to promote development. Yet, it has been pointed out repeatedly that deforestation of tropical forests, generally speaking, will only result in short-term gains because most of these forests are rather unique ecological systems unable to sustain agriculture and pastures for extended periods of time. Deforestation of the Amazonias forest, for example, has become predatory (except perhaps in the fringes of the forest) and the Brazilian government has been strongly pressured (both internally and externally) to stop further clearing of the forest before a full ecological and economic zoning of the area is conducted. [11]

Short-term concerns are frequently much stronger than long-term concerns and this is certainly the case for local and global environmental preservation as seen by less developed countries.

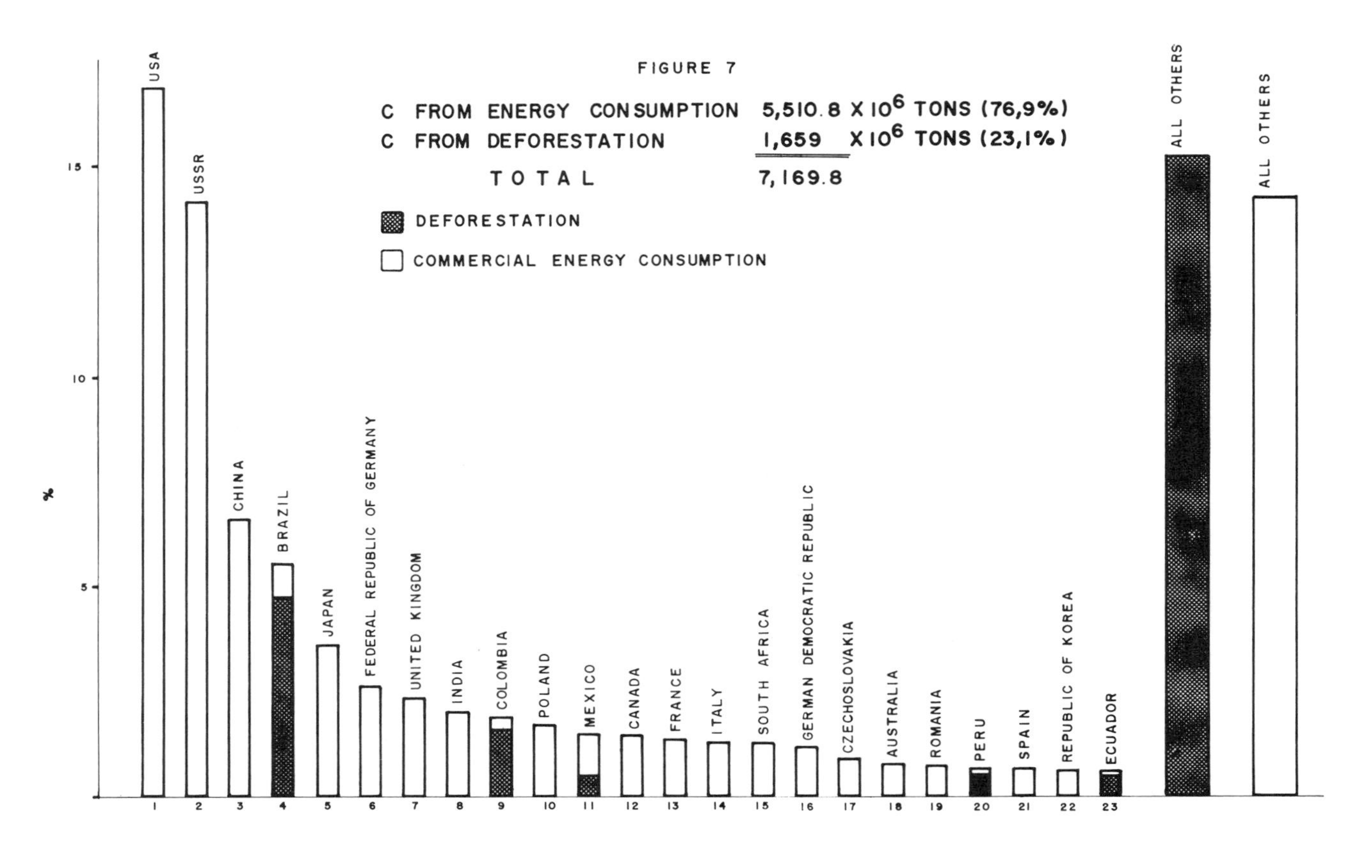

Figure 7. World deforestation and commercial energy consumption.

Proposed Solutions

The actions needed to reduce greenhouse gas emissions to levels that would stabilize the atmospheric composition are quite different for industrialized and nonindustrialized countries. A recent study prepared by McKinsey and Company, Inc. for the Ministerial Conference on Atmospheric Pollution and Climatic Change (November 7–8, 1989, Noordwijk, the Netherlands) considered a variety of actions to be taken in the OECD countries and other areas of the world. [12]

Contribution of the OECD Countries

In 1985, the OECD countries were responsible for the emission of 4.9 Gigatons of CO_2 equivalent per year (contributions of gases other than CO_2 converted in "equivalent" amounts with respect to global warming). Low-cost or profitable actions (in energy conservation) could reduce these emissions by 1.4 Gton per year through phasing out CFCs and by 0.4 Gton per year through efficiency improvements in fossil fuel use. The final result would be a reduction 3.9 Gtons/year greater than the EPA targets.

Contribution from Other Countries

The additional 3.9 Gton/year reduction would result from:

1. Eastern Europe mainly by switching from coal- to gas-fired plants for electricity and through energy efficient industrial processes, residential heating, and lighting (a 0.9 Gton/year reduction) and
2. actions in the rest of the world (mainly LDCs) that promote use of biomass, energy conservation, and most important *reforestation*—in addition to phasing out CFCs (a 3.0 Gton/year reduction).

McKinsey and Co. estimate the costs for such actions as the following:

Continued CFC funding. Will initially cost $150–200 million per year, and then $600–700 million per year should a 100 percent phase out be pursued.

Expanded forest management funding. Will cost up to a maximum of $10–15 billion per year. Reforestation has the capacity to yield a net carbon emission

Table 2. Total carbon emissions (× 10^6 tons).

			Percent	CO_2/capita
1	United States	1,201.6	16.9	5.00
2	USSR	1,010.8	14.1	3.59
3	China	554.3	7.7	0.53
4	Brazil*	391.5	5.5	2.83
5	Japan	256.1	3.6	2.11
6	Federal Republic of Germany	186.3	2.6	3.07
7	United Kingdom	166.2	2.3	2.94
8	India	144.3	2.0	0.19
9	Colombia*	135.8	1.9	4.68
10	Poland	124.5	1.7	3.32
11	Mexico*	106.6	1.5	1.33
12	Canada	105.2	1.5	4.09
13	France	98.4	1.4	1.79
14	Italy	94.9	1.3	1.65
15	South Africa	92.5	1.3	2.78
16	German Democratic Republic	92.3	1.3	5.50
17	Czechoslovakia	65.8	0.9	4.21
18	Australia	61.3	0.8	3.85
19	Romania	55.8	0.8	2.41
20	Peru*	50.8	0.7	2.56
21	Spain	49.8	0.7	1.28
22	Republic of Korea	45.4	0.6	1.08
23	Ecuador*	43.6	0.6	2.20
24	All Others	1,106	15.4	
25	Deforestation from all others	1,082	15.1	
World Total		**7,169.8**	**100.0**	

* Deforestation included

Table 3. Carbon emissions from deforestation (million tons).

Brazil	336
Indonesia	192
Colombia	123
Ivory Coast	101
Thailand	95
Laos	85
Nigeria	60
Philippine Islands	57
Burma	51
Peru	45
Ecuador	40
Vietnam	36
Zaire	35
Mexico	33
India	33
Total	1,659

reduction of 2.5 Gton/year (mostly in tropical countries) at an estimated cost of $10–15 billion per year or $1,000 per hectare.

Funding of fossil fuel energy conservation in LDCs. Concessional funds needed may be of the same order of magnitude as those for forest management ($10–15 billion per year). A similar amount might be needed to continue energy conservation efforts in OECD countries.

Actions Needed

Total international investments needed to stabilize the atmosphere are on the order of $30–45 billion per year. The challenge is to obtain such resources—who pays and who benefits?

One idea is to establish a carbon tax of $1.00 per barrel of oil equivalent (or $6.00 per ton of coal equivalent). Such a tax would collect $50 billion per year from the regions listed in Table 5. The purpose of the carbon tax should not be to discourage consumption by increasing the cost of energy. There is evidence that this method is hardly effective. The proposed tax should in reality be considered a way to raise the resources needed to overcome initial resistance to energy efficiency measures in industrialized countries and to help the LDCs overcome their problems.

One serious problem with this proposal is that most funds would be used on a concessional basis, that is, as grants and not loans. This is not the usual mechanism on which international public banks operate (World Bank, IDB, Asian Development Bank, etc.).

The resources would have to be applied essentially in the following areas:

1. Leapfrogging as a development policy;
2. Transfer to LDCs of the technologies needed for environmentally sound development;
3. Reforestation.

"Leapfrogging" as a Development Policy

Strout [4] has already stressed the fact that as developing countries reach a given level of development the consumption of basic materials levels off (at GDP per capita) on the order of $2,000 per year (1970 US dollars). This is a rather high income for many LDCs. However, this would occur as the natural result of market forces without undue interference of public policies. Now that a greater environmental concern exists the leveling off would be expected to take place at lower levels of income, if the adoption of the modern available technology is accelerated.

This outcome was expected by some even before

Table 4. Population and consumption of energy in Africa (1985).

	Population (millions)	Oil derivatives kep/capita	Electricity kWh/capita	Fuelwood kg/capita	Fuelwood (total) $\times 10^6$ tons
UMA*	56.4	274	498	206	11.6
Egypt	46.8	302	574	n.d.	—
Sahel-Sudan	97.0	25	34	682	66.2
Atlantic Coast	142.5	79	87	610	86.9
East Africa	103.4	42	99	810	83.7
SADCC**	70.6	45	252	928	65.6
Total	516.7				313.9

* Union of the Arabian Maghreb
** Southern African Development Coordination Conference
(Source: Jean Marie Martin, Énergie Internationale 1989–1990.)

Table 5. Commercial energy consumption and shares of carbon tax (1987).

Area	Energy consumed (MTEP)	Percent
OECD	3,855	49.4
US	1,849	24.0
Western Europe	1,296	17.0
LDC	1,915	24.5
China	700	9.0
CPE	2,040	26.1
USSR	1,444	18.0
Total	7,811	100.0

1 TEP ≅ 6 barrels oil equivalent

the emphasis on environmental concerns. As pointed out by Malenbaum [13]:

> . . . in contrast to earlier views the present prospect is that "intensity of use" for most materials will not continue to expand in poor nations until high levels of economic development are in fact achieved. Rather the effects of technological advance elsewhere are spread to them at a relatively early stage of economic growth.

It would be up to international banks (such as the World Bank and the Export-Import Bank of the United States), as well as regional banks, to preferentially finance industrial and related projects that are energy efficient according to the standards of industrialized nations. With this as basic policy, expensive retrofitting could be avoided and savings in the consumption of energy could be encouraged. For example, if a new factory to manufacture refrigerators is installed in Brazil, it should be based on models that consume at most 400 kwh/year instead of the less efficient 800 kwh/year models now common in that country. If 1,500,000 new models are sold each year, this will represent savings of 400 million kwh per year (representing investments in 35,000 kw of generating capacity and saving at least $70 million in additional investments). The cost of the more efficient refrigerators will probably not exceed $10–20 per refrigerator.

Similar plans could be applied to automobiles and other devices essential for the development in LDCs [14].

Transfer to LDCs of the Technologies Needed for Environmentally Sound Development

The importance of providing LDCs with environmentally sound technologies is shown by the refusal of some LDCs to join the Montreal Protocol and reduce CFCs use. It is argued by some LDCs—mainly India—that substitutes for CFCs are expensive and adoption of the costly replacements will "punish" India. The only solution is to transfer to LDCs (free of charge) the technology needed to phase out CFCs, since this is in the self-interest of industrialized countries. Internal mechanisms to compensate industries for the costs involved would have to be established in the industrialized countries.

Reforestation

Again, it is hard to see the economic benefits to LDCs of engaging in large reforestation programs, although they are justified by the need to recapture CO_2 from the atmosphere. The areas to be reforested are of the order of 10–15 million hectares as mentioned above and it would be difficult to obtain an economic return from such areas. Most of it will have to be done as an ecologically oriented reforestation and therefore financing will have to be done on a concessional basis.

In the particular case of the African countries where deforestation is directly linked to energy uses such as cooking, the adoption of modern cooking methods using kerosene and gas (superior fuels from all viewpoints) is the appropriate strategy. In the author's opinion, the effort to optimize fuelwood cooking stoves is an entirely mistaken strategy. The sooner rural populations are introduced to modern cooking devices the better in terms of both standard of living and environmental conservation.

Conclusion

Although substantial, the sums of money needed to stabilize the atmosphere are not larger than what is currently spent by the industrialized countries in official development aid (ODA) ($50 billion or 0.4 percent of their total gross domestic products). It seems most reasonable to increase such spending—not simply for the philanthropic reasons generally behind official development aid, but to stabilize the atmosphere for the good of *all* countries.

Appendix: Note on Energy Conservation in the Soviet Union

The potential for energy conservation in the Soviet Union is generally underestimated [15] (and other Eastern European countries). The industrial sector accounts for 70 percent of primary energy and 80 percent of electricity consumption in the Soviet Union. In this sector, the ratio of energy consumed per unit of goods produced (E/GNP in TEP/US$1,000) is 40 percent larger than in the United States.

In addition, the ratio E/GNP of OECD countries decreased by an average of 2.2 percent per year in the period 1973–1986 while it increased by 1.2 percent per year in the Soviet Union during this period (see Figure 6).

Because per capita consumption is low in the residential sector and there is a relatively small number of automobiles, energy consumption in these sectors is low and efficiency should increase by incorporating new standards and technologies which are available in the OECD countries.

The reduction of energy demand in the industrial sector could come from structural shifts (as in the US), combined with standard energy conservation measures that are well in line with the ongoing efforts to modernize the Soviet economy.

References

1. Conference on the Changing Atmosphere: Implications for Global Security. Toronto, June 1988.
2. Jean Marie Martin, "L'intensité energetique de l'activité economique: les evolutions de très long periodes livrentelles des enseignements utiles?" *Eco.Soc.* 49, 27(1988).
3. P. F. Drucker, "The Changed World Economy" *Foreign Affairs* 768–91 (1986).
4. A. M. Strout, "Energy-intensive materials and the developing countries" in *Materials and Society* 9(3), 281–330 (1985).
5. R. H. Williams, E. D. Larson and M. H. Ross, "Materials, Affluence and Industrial Energy Use" *Ann. Rev. Energy* 12, 99–144 (1987).
6. J. M. Roop, "The trade effects on energy use in the US economy: an input-output analysis" Presented at the 8th Ann. Int. Assoc. En. Econ. N.Am. Conf., Cambridge, Mass. (1986).
7. "Implications of global climate policies" Center for Strategic and International Studies, Washington, DC, June 27 (1989).
8. "Study on the reduction of energy-related greenhouse gas emissions" Prepared by DPA Group Inc. and CH_4 International Company for the Department of Energy, Mines and Resources of Canada, March 1989.
9. J. Goldemberg, T. B. Johansson, A. K. N. Reddy, and R. H. Williams, *Energy for a Sustainable World* Wiley Eastern Limited, 1988.
10. J. H. Gibbons, P. D. Blair, and H. L. Gwin, "Strategies for energy use" *Scientific American*, Vol. 261 (3) 86–93, September 1989.
11. "Amazonia: facts, problems and solutions" International Symposium on the Amazon. University of São Paulo, São Paulo, Brazil, July 1989 (Coordinator: J. Goldemberg).
12. Background paper on funding mechanisms prepared by McKinsey and Company Inc. for the Ministerial Conference on Atmospheric Pollution and Climate Change. Noordwijk, the Netherlands Nov. 6–7 (1989).
13. W. Malenbaum, "World Demand for Raw Materials in 1985 and 2000" E/MJ Mining Informational Services, McGraw Hill, Inc. New York (1987).
14. L. Schipper, "Efficient household electricity usage in Indonesia" Lawrence Berkeley Laboratory, Berkeley, CA, draft paper, January 1989.
15. R. H. Williams, "Decoupling energy and economic growth in the Soviet Union" Center for Energy and Environmental Studies, Princeton, NJ, draft paper, Oct. 12 (1989).

Population and the Global Environment

Nazli Choucri
Department of Political Science
Technology and Development Program
Massachusetts Institute of Technology
Cambridge, MA

Introduction

Anthropogenic sources of global environmental change can be traced to three interdependent factors: 1) trends in human activities and policies, such as population growth, economic growth, and the legitimization of activities that generate environmental degradation, 2) technological and industrial development, and 3) prevailing patterns of natural resource use, including energy consumption, deforestation, and land and water use. Technological and industrial contributions to global environmental change are obviously crucial, as are patterns of energy and resource use. The most critical influence on the environment, however, is worldwide demographic change, which generates environmental effects both directly and through its impact on resource use patterns and applications of technology.

The record of population growth is well documented, as are the differentials among states and the implications for per capita claims on the world's resources. The population characteristics of the globe—and of the constituent nations—are among the most robust features of the world we live in. This paper examines the demographic dimension of global change, reviewing the global record, future projections, and implications for modeling global population/environment relationships.

The environmental consequences of population growth are rooted in the fact that every human being requires some minimal amount of basic resources (food, water, air, living space) and that the total resources required by a society increase with population size. The "population nexus" refers to the interactions or dynamic convergence of population, resource needs, and levels of knowledge and technology, including organizational and mechanical skills. Technological change provides new resources (or new uses for existing resources) and may also increase demand for resource uses.

The developmental process is generally one of increased output, enhanced productivity, improved standards of living—and greater environmental degradation. The economic concept of demand is restricted to "willingness to purchase"; the political concept of demand means claims on the political system, claims on governance (including economic, political, social, and other claims or benefits), and claims on environments. The larger the population and the higher the rate of growth, the greater are the social, economic, political, and other demands engendered.

The environmental consequences of population are, therefore, intimately tied to levels of knowledge and skills (technology) and patterns of resource use (especially consumption of energy in all its forms). In the most basic sense numbers do matter, as do the demands generated by populations and their ability to meet those demands. For example, a global population at the technological level of Bangladesh—with per capita carbon emissions at less than 1 percent of the per capita level of the United States—would have a different impact on both the local and global environments than it would at a higher level of industrialization.

Trends in Global Population

Global Record

The global demographic record for the past 250 years is fairly well established, subject only to regional differences and narrow uncertainties in the aggregates. The size of the present population, though not precisely enumerated, is probably between 5.1 and 5.3 billion.

The doubling time for the first billion human beings on earth took 120 years, while the doubling time for the second billion took 47 years. Until recently the trend has been upward; by the late 1960s and early 1970s we observed the first discernible decline in the global rates (departures from the peak of just over 2 percent per year in the 1960s). Since then, the global aggregates have shown a small but downward adjustment. The declines of the 1970s—noteworthy in a historical context—were not sufficient to affect total aggregates since the population still grew at a rate of 220,000 persons per day.

Figure 1 shows the historical record and average growth, in terms of average annual rate as well as overall numbers.

The historical record shows that population growth is intimately related to patterns of energy use and advances in technology and, by extension, to patterns of environmental degradation. The intense interaction among population variables prevents simple assignment of responsibility for global environmental change. Nonetheless, the clear association between the growth of human population and economic activities, on the one hand, and the generation of greenhouse gases (carbon dioxide, methane, nitrous oxide, the chlorofluorocarbons, and others) on the other, illustrates the demographic sources of atmospheric alterations.

Demographic Transition

Populations are generally described in terms of their stage in the "demographic transition." Any given growth rate can be the result of different combinations of births and deaths. The transition refers to adjustments in the birth and death rates and to the theory that describes these adjustments. Specifically, demographic transition means the change from high and proximate levels of mortality and fertility to low levels and rates close to replacement. Demographic transition consists of four stages: 1) high mortality and fertility (life expectancy less than 45 years and total fertility rate (TFR) at six or more), 2) mortality declines earlier than fertility, which is also declining, 3) acceleration of both mortality and fertility declines, and 4) low mortality and low fertility (life expectancy greater than 65 and TFR less than three). Overall, the transition indicates declines in death rates preceding declines in birth rates, and a great variation in the time path of the transition across countries and regions.

While the dynamic processes of the transition are specified in terms of adjustments in fertility and mortality, there is a wide range of views and evidence concerning the determinants of births and deaths. Human beings can die at any age, but fertility, by contrast, is limited to the reproductive age groups. Aggregate changes in global fertility rates—from 4.99 to 3.64 between 1950–55 and 1980–85—mask great variation across regions and countries, as do the forces that shape births and deaths.

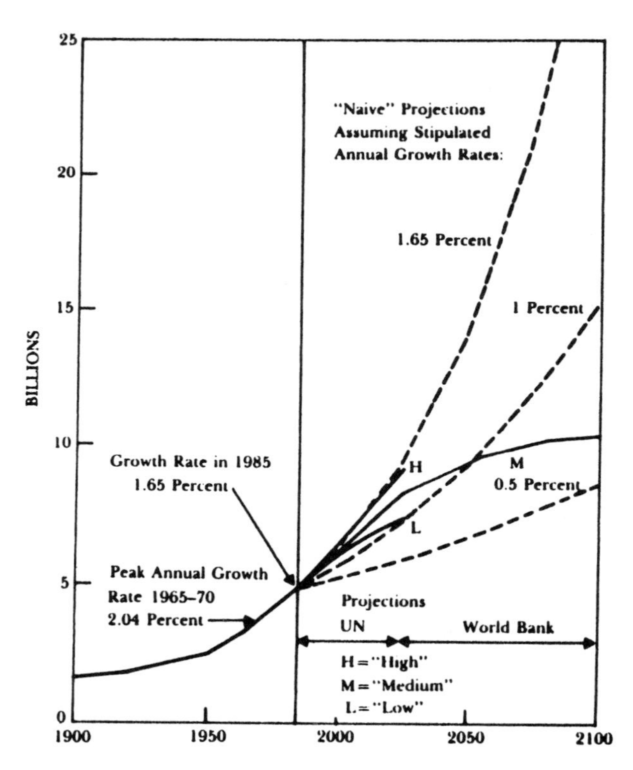

Figure 1. World population growth: 1900–2100 (1900–85: estimates; 1985–2100: projections and extrapolations). Source: Paul Demeny, "The World Demographic Situation," in *World Population and U.S. Policy*, Jane Menken, editor (New York, NY: W. W. Norton, 1986), p. 35.

Differences in Levels and Rates of Growth

Global aggregates obscure significant regional and national differences. The unequal distributions of human population across sovereign states are important in the formulation and design of population policies, since the locus of decisionmaking still resides with states (and state systems).

In 1985, 76 percent of the world's population was classified as "less developed," and its rate of population growth was highest. The less developed countries more than doubled their total population between 1950 and 1985. The importance of China, however, obscures the regional variation among the developing countries. China's 50 percent reduction in fertility rate during this period was due to the combination of strong population control policy (rise in marriage age and one child per family established in 1979) and notable famines. Although the present population level is not much higher than that required for replacement (2.1 births per woman), it takes several decades before fertility decline is reflected in the age structure and in attendant fertility patterns.

The effects of the demographic transition from 1950 to 1985 translate into the average annual population increases shown in Figure 2. Clearly, the *natural* increase in population—the difference between births and deaths—varies extensively across regions. While the global rate of natural increase in 1980–85 stood at 16.6/1000, the highest rate was in South Asia (21.7/1000),

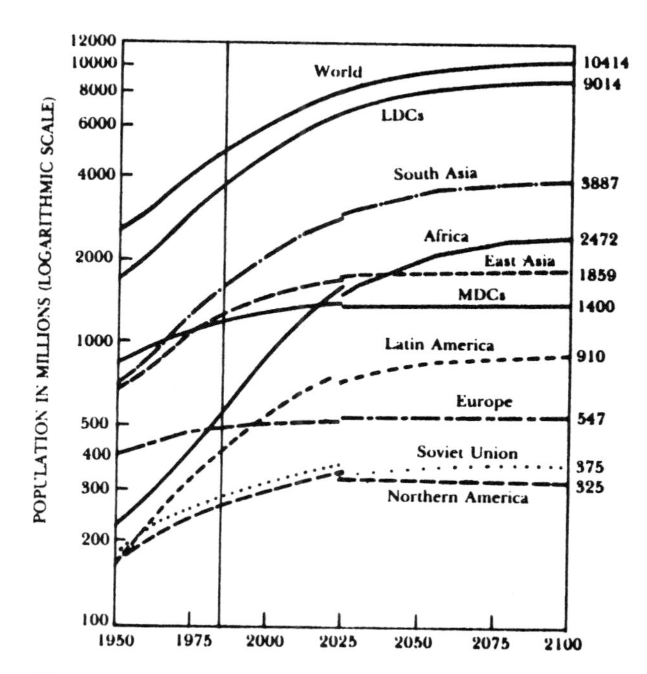

Figure 2. World population growth: 1950–2100 (1950–85: estimates; 1985–2100: projections). Source: Paul Demeny, "The World Demographic Situation," in *World Population and U.S. Policy,* Jane Menken, editor (New York, NY: W. W. Norton, 1986), p. 65.

and the lowest in Europe (3.0/1000). For the less developed countries as a whole, the average rate of natural increase was 20.2/1000.

Composition and Distribution

Demographic changes and variations among countries are due to a combination of socioeconomic conditions, development strategies, and population policy (i.e., toward fertility control, international migration, and provision of health and related services). Select socioeconomic indicators are *inversely* related to fertility (i.e., education, female employment). To the extent that they are the subject of government policy, these indicators can have a profound effect on fertility.

Urbanization

As a global trend, urbanization has been on the rise. It is a powerful determinant of energy use and directly reflects income changes and changes in the composition of the labor force.

Forty-one percent of the global population is urban; with an annual average increase of 2.84 percent, the lesser developed countries have expanded their urban populations fourfold over a 35-year period. In addition, the urban trend has implications for infrastructure demand, the need to expand built environments, and the demand for energy associated with increased conglomeration of populations. In 1950 the 15 largest urban concentrations were located in the developed countries; by 1980, 9 of the 15 most concentrated urban centers of population were in the developing states; 4 of these were in China and India.

International Migration

The movement of people across territorial boundaries is small relative to the total global population. Only 50 million people, or 1 percent of the world's population, are now living in countries in which they were not born. In some cases the number of migrants is large relative to the population in the country of destination (80 percent in the Gulf countries of the Middle East); in others the proportion is small, but the total numbers are high (the US has the largest number of persons born abroad—14 million—but this represents only 6.2 percent of the total population). Regardless of scale or scope, international migration is generally not neutral with respect to economic performance or patterns of resource use. In many regions of the world international migration facilitates or accelerates economic and industrial development and the provision of services. For example, the post–World War II reconstruction program in Europe was greatly aided by immigration from North Africa, Turkey, Yugoslavia, and other areas; the entire oil industry in the Middle East is built on the large-scale migration of foreign workers; agriculture in California draws heavily on Mexican workers; and the list goes on.

Population Policy

Narrowly construed, population policy refers to family planning programs designed to influence fertility rates. More broadly, however, population policy refers to any formal effort to influence the levels, rates, composition, and/or distribution of a society's demographic characteristics. Demographers have generally preferred to adopt the restrictive definition, while planners and designers of social policy have preferred the broader definition. Traditionally, the lines of policy debate have been sharply drawn between those who believe that family planning programs are essential in order to influence fertility rates and those who argue that economic development is both necessary and sufficient for this purpose (and that, at most, fertility control programs may facilitate the decline in the birth rate). The strong interdependence of fertility behavior and socioeconomic development, however, makes it difficult to untangle the effects of family planning programs and evaluate the contending views. Nonetheless, the historical record over the past 20 years provides sufficient evidence to suggest that population control programs do have a significant impact on fertility rates and that demographic and economic processes are so highly interconnected as to seriously question the utility of socioeconomic models that do not explicitly endogenize the demographic processes.

The policy/fertility record is as follows: by 1986, countries accounting for 78 percent of the population of developing nations had adopted policies designed to reduce their fertility rate. The 30 percent decline in fertility in these countries since 1950 was accompanied by a sharp increase in the use of fertility inhibiting technologies. Concurrently, strong modernization programs induced socioeconomic development. The significant intervening variables between socioeconomic development and fertility are not disputed (i.e., age at marriage, income, level of education of women, and employment of females). The availability of fertility control technology adds a powerful impact—a sort of multiplier effect. However, the relationship between fertility and socioeconomic variables, such as employment and education, varies significantly over time and across countries (see, for example, Cochrane, 1986). Furthermore, there are no good quantitative estimates of the specific responsiveness of fertility changes to specific socioeconomic variables (Mauldin, 1989:77). These problems complicate simple causal inferences.

Statistical uncertainties aside, however, the dominant role of the public sector—in setting policies, extending supporting services, and providing fertility control technologies—is crucial; so is the role of international institutions and external sources of both technology and finance. While the manipulability of fertility is clear, the causal sequence and the time frame over which adjustments take place is less clear.

Global Prospects

Projections

The relatively unambiguous demographic record of the past two centuries contrasts sharply with the remarkable diversity in prevailing projections of future trends. While the "dynamics" of aggregate demographic change are determined only by fertility, mortality, and their interactions, the demographic system is not "closed"; these individual elements are strongly influenced by technological, social, and other factors. Figure 1 above showed the record of population growth since 1900 and the future projections to 2100—the "naive" as well as the alternative projections. On balance, the most "reasonable" view is that by 2100 the global population may stabilize at 10 billion human beings. But it may also stabilize at 14 billion; and the differential is not insignificant.

Age Distribution

The age structure of the population projected by the UN for 2020 compared to data from 1985 shows that aging will be commensurate with the long-term stabilization of population—when the global population has effectively made its demographic transition (United Nations, 1989:54). An expansion of the age group 60 and over from 8.8 percent to 14.3 percent of the global total will involve important socioeconomic adjustments. The implications for labor, employment, productivity, and the provision of health and other social services, though of increasing concern, have yet to be fully recognized. While the overall dependency rate drops by only 3 percent, the shift away from the cohort 14 years of age or younger implies a gradually declining fertility rate characteristic of stabilization.

Again, global aggregates obscure regional and national differences. For example, while the developing states will retain a youthful distribution of population well into the next century, the industrial states are rapidly aging. Population growth in the United States and the Soviet Union is slowing, both countries are aging, and the effects on the labor force are already apparent (Torrey and Kingkade, 1990).

Doubling Time

Population doubling time given current rates of growth is a stark summary indicator of regional differences. The population of Africa could double in roughly 40 years, while that of Europe may take over 240 years to double (see Figure 3). The global shifts in the distribution of population are obviously not neutral with respect to energy use or other socioeconomic factors.

As indicated in Figure 1, World Bank "medium" projections anticipate a global population of 6.1 billion in the year 2000, rising to 8.2 billion by 2025, and to 10.4 billion by 2100. This growth involves major shifts in the regional distribution of population between now and then but no significant changes subsequently. By 2100 Europe will account for 5.2 percent of the global population (half the 1985 level), North America will account for 3.1 percent (compared to 5.5 percent in 1985), and the less developed regions will account for 87 percent of the world population. The implications for Europe are noteworthy, even graphic: "Europe is literally melting away like snow in the sun from 15 percent of the world population in 1950 to one-third that relative share in 2025" (United Nations, 1989:8).

This view of future population, based on projections such as those of the United Nations, the World

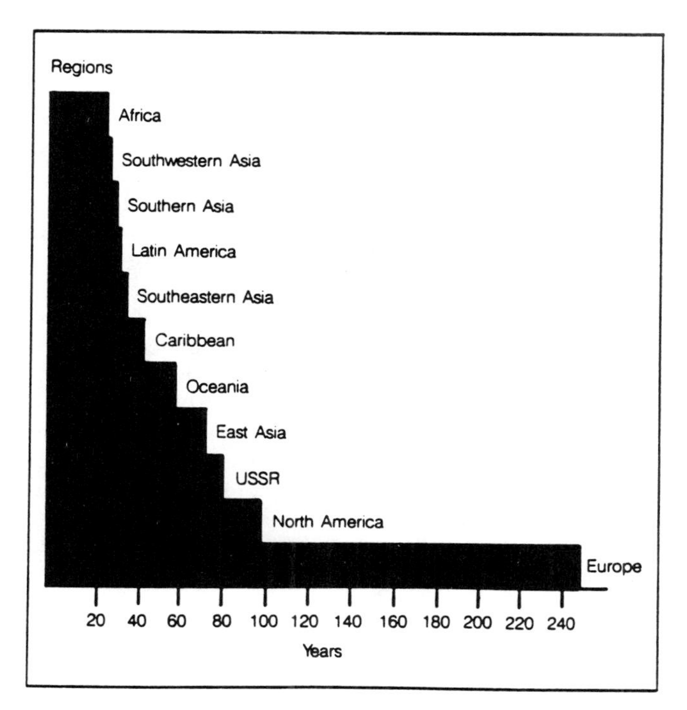

Figure 3. Population doubling times by region, at current rates of growth. Source: World Resources Institute, *World Resources 1987*. (Washington, DC: World Resources Institute, 1987), p. 12.

Bank, and the US Bureau of the Census, is generally used as exogenous input in models of energy and/or environment. This view is built exclusively on the fertility and mortality relationship and makes no provision for the influence of energy and environmental variables or interactions on future population patterns. As such, it is strictly a demographic projection. It does not incorporate population/environment relationships or population/energy use interactions, either directly or indirectly.

Methods: Approaches and Problems

Population projections are based on extrapolations from past trends or on identification of a point in the future at which replacement level fertility will be attained and from which intervening rates of fertility can then be derived by interpolation to the base period. The historical record has informed the theory of demographic transition as well as the determinants of fertility, mortality, and migration. Unfortunately, as Keyfitz notes, there are no clear theoretical explanations from which to derive a behavioral model for future population (Keyfitz, 1983).

Projection Method

The most common method for population projection is known as the "component projection method." It is based on 1) age/sex survival rates to determine survivors in each age group; 2) an age-specific fertility rate applied to the total number of females in each reproductive age group (births are distributed according to the assumed sex ratio at birth; based on survival probabilities, the number of survivors from calculated births is calculated for each time period); and 3) the number of migrants, which is added to or subtracted from the projected additions to population (i.e., survivors) from steps 1 and 2 (IIASA, 1989:16, 41, summarizing conventional procedures). In this sense the projection system draws only on births, deaths, and migration:

$$P_{t+1} = P_t (B_t - D_t + M_t)$$

where P_t = initial population size and structure
B_t = birth rate (fertility)
D_t = death rate (mortality)
M_t = net migration

Sources of error or bias are accordingly contained in estimates of initial population size, age/sex distribution, fertility rates, mortality rates, and migration rates. Among the most common sources of error (or variability) are the growth rate assumptions and the initial estimates of population parameters for developing countries (especially for those large countries—China, India, and others—that have major effects on global demographic trends).

The component method is basically computational rather than behavioral or causal, since it does not determine B_t, D_t, and M_t as functions of socioeconomic or policy variables. Similarity in logic, computation, and initialization is responsible for similarities in projections. Citing Keyfitz (1983), the International Institute for Applied Systems Analysis (IIASA) analysis pointed to the strong possibility that global population projections are not independent, and suggested that agreement therefore should not constitute validation. Differences among projections, determined exclusively by differences in initial condition parameters, cannot be construed as differences in theory or in explanatory power.

The importance of population projections lies not in their precision (or even rough accuracy) but in the sensitivity of socioeconomic, energy, or environmental variables to levels, rates, and distribution of population. Hence, improvements in projections need be delineated only to the extent that they generate discernible effects whose implications may be significant for these systems and/or demographic factors themselves. Among the possible strategies for endogenizing the demographic parameters in population projections are 1) making explicit the relationships of income/fertility and education/fertility, 2) making provisions for the population policy/fertility relationship, and 3) incorporating the migration/ fertility relationship.

Population as a Driving Factor

Population variables and projections are widely used as inputs in energy models and, more recently, in models of CO_2 emissions. Uniformly using population factors (usually size and rate of growth) as inputs "driving" the system assumes that system behavior has no effect on demographic patterns. Energy models and energy/environment models commonly treat population (i.e., size, rate of growth, labor force, labor productivity) as exogenous—influencing but not influenced by the relationships modeled. This practice in essence presumes the complete insensitivity of population to socioeconomic, energy, or other variables in the system modeled. In the short run this is probably a useful approach. Over the longer run, however, population becomes a significant modifier of the natural environment and thus is critical to models of long-term processes (more than 10 years). Endogenizing population is complicated by the lack of a comprehensive theory of population/energy/environment interactions. The processes of growth and development are powerful intervening dynamics shaping both sides of the relationship—demographic and energy/environment. Moreover, the level and type of technology (both organizational and mechanical knowledge and skills) significantly affects growth and development.

Recent reviews of major energy and energy/environment models undertaken by IIASA (1989) and by the Stanford University Energy Modeling Forum (EMF 6 and EMF 11) illustrate the continued practice of using population variables as exogenous factors. The impact of energy on the environment is most readily traced via the fuel mix used to generate energy-based services. Demographic factors enter directly in the demand side; over time shifts in economic and other activities affect fertility, mortality, and migration, which in turn influence the demand for energy almost immediately. The adjustment period may vary across regions and countries as well as over time, but the interactivity can be delineated based on pieces of theory, analysis, projection from demographic and energy/economy analysis, and the nascent energy/environment literature. Clearly, no model can or should do everything; nonetheless, the practice of shaping modeling decisions and defining system boundaries based on past practices may be distorting in a world where the signals are pointing to greater interactions rather than enhanced autonomy of social processes.

The draft report to Congress entitled *Policy Options for Stabilizing Global Climate Change* (Environmental Protection Agency, February 1989) has among its objectives improved definition of the potential effects of global climate change and identification of the options available for influencing the composition of the atmosphere and the rate of alterations. The scenarios examined express population as an exogenous factor. Drawing upon the US Bureau of the Census and World Bank projections, the scenarios differ on the rate at which the global population is projected to stabilize. The major feedback specified is from atmospheric gas concentration and temperature change (inputs) to select emissions forecasting modules (intervening processes). As a strategy shaped by expediency, this approach is certainly reasonable. However, it obscures the variety of demographic processes (and their interconnections) that affect patterns of emissions of the trace gases. It also obscures regional variations in the processes that influence the production of trace gases as the mix of human activities differs at different levels of development. Among the limitations of the study, as identified in the Report, is its failure to address the population and the demographic processes, reflecting the conventional exogenous treatment of population without exploring, or partially testing, the effects of this practice.

Illustrating Recursive Processes

One of the most useful frameworks for exploring CO_2 development options is the Edmonds and Reilly model, which provides five detailed commercial energy/demand/supply and CO_2 balances to 2100 (Edmonds and Reilly, 1983; 1984; 1985). Population and GNP projections drive the energy demand. Technological change affects energy demand, but no feedback effect is set from technological change to GNP, labor, or population. Over the time frame of the model (to 2100), it is reasonable to expect that such feedback effects might provide some useful insights beyond those based on exogenous population projections. This expectation has already been reinforced by sensitivity analysis of this model, showing the salience of labor productivity, change in energy efficiency, and income of less developed countries.

These concerns are relevant to other intertemporal energy/environment models. There is a case for endogenizing demographic factors whenever they are influenced *over time* by the processes modeled, such as energy use, environmental degradation, emissions of trace gases, and so forth. In the short run (five years or less) such interactions may not be significant. However, there are exceptions, most notably deforestation, soil erosion, land degradation, and expansion of noncommercial energy use. Without specifying the time frame, the United Nations Fund for Population Activities (UNFPA) starkly states that population growth alone may account for 80 percent of the loss of forest cover (Sadik, 1990:11). In the longer run (over decades, even a century) there is no historical or empirical basis upon which to presume the separation of demographic from energy or environmental processes. Climate affects water and soil conditions, and shifts in temperature would, in all likelihood, influence sea level, precipitation patterns, and the distribution of land and water resources. Even the brief sketch provided by the UNFPA illustrates the potentially alarming interactions of population and environment (Sadik, 1990: 10–12).

Population, Development, and the Global Environment

The current state of population projection and of demographic theory is not designed to address resource (energy) availability, differentials among populations in levels of economic performance, knowledge and skills, or overall technological capability—or their impacts on fertility and mortality. In large part this is due to the inherent complexity of these issues and the tradition of social inquiry that eschews integrative approaches in favor of detailed foci on particulars, in this case fertility and mortality. Such considerations notwithstanding, the fact remains that these variables are crucial intervening factors shaping the impacts of population on social and natural environments. Thus, demographic factors, however compelling, if considered alone, tell us only a part of the story; the *interactions* among population characteristics, technological change, and patterns of resource use (especially energy) define the nature of the population effects at any point in time. For the globe as a whole as well as for individual states, these interactions shape the effects of humans on both the natural and the social environments.

Differences in Development

Though states all over the world generate many of the same effluents—carbon dioxide (CO_2), carbon monoxide (CO), methane (CH_4), nitrous oxide (N_2O), chlorofluorocarbons (CFCs), and others—they tend to do so in different ways and in different amounts, according to size, population, geographical location (climate zone), and level of technology and industrialization. It remains a matter of conjecture whether forms of government and political regimes influence patterns of emissions. Table 1 shows the basic patterns of carbon emissions for the top contributing states.

The most advanced industrial countries, characterized by high levels and rates of technological development combined with access to resources, generate high levels of consumption, high levels of CO_2 emissions, extensive wastes, and high levels of other greenhouse gases, with attendant effects on the natural environment. There are differences among the industrial states, of course, but the basic trends of high per capita emissions rates are broadly similar. The dominant case is the United States, with a GNP per capita of $16,757 and roughly five metric tons of carbon emissions per capita (1986 figures). Despite differences in energy efficiency, the industrial states, by their very nature, rank high on emissions rates per capita compared to the global average.

The developing or industrializing countries are considerably more varied in size, level of economic growth, and level and type of environmental degradation—but they all produce remarkably lower carbon emissions than the global average. At one end of the spectrum are states with large or growing populations and relatively limited basic technology, contributing low carbon emissions per capita. As these countries industrialize, however, their large populations will make them significant contributors. For example, China's emissions per capita are 0.5 metric tons per year, but on the aggregate China is already the world's third largest contributor of global CO_2. Given China's population and its 20 percent share of the global use of coal, its industrialization process will expand its emissions.

At the other end of the spectrum are those states with low density, limited basic technology, and limited resource access. They are typically poor, with close to subsistence levels of development. Chad is a typical case, with a per capita income of $160 and carbon emissions less than 0.5 percent of the US level. These states will not affect global balances markedly in the foreseeable future.

In this context the oil-rich countries of the developing world, with sparse population and a rich resource base, are distinctive because of their rapid rates of environmental degradation and generation of a wide range of effluents. All these states have remarkably high levels of carbon emissions per capita, due in part to petroleum production and in part to rate of industrialization and infrastructure development. Saudi Arabia, for example, produces 2,584 kg of carbon per capita (1986).

Projecting Environmental Impacts

Globally, the industrial states obviously generate more effluents and affect the global balances more than do the developing countries. Over time, however, with greater industrialization worldwide the major sources of emissions and effluents will be significantly more widely distributed than they are at present.

In terms of the global carbon budget, the stabilization level of the future world population (estimated at 10 billion to 14 billion people) will have radically different impacts depending on the level of development worldwide. Depending on whether the global population converges at the level of development of Bangla-

Table 1. Major Contributors of Carbon Emissions to the Atmosphere, 1986.

Total CO_2 Emissions

	Carbon (1) (Thous. mt)	Population (Millions)	Carbon/Pop (Kg)	GNP/Pop ($/pers.)
1 United States	1,201,624	241.6	4,973.61	17,480
2 Soviet Union	1,010,804	281.1	3,595.89	8,384
3 China	554,349	1,054.0	525.95	300
4 Brazil	388,521	138.4	2,807.23	1,810
5 Japan	256,084	121.5	2,107.68	12,840
6 Indonesia	220,127	166.4	1,322.88	490
7 West Germany	186,269	60.9	3,058.60	12,080
8 India	177,326	781.4	226.93	290
9 United Kingdom	166,195	56.7	2,931.13	8,870
10 Colombia	135,831	29.0	4,683.83	1,230
Rest of World	2,918,870	1,986		
Top 10 Share	60%	60%		
World Total	7,216,000	4,917		

CO_2 Emissions from Deforestation

	Carbon (Mill. mt)	Pop (Millions)	Carbon/Pop (Kg)
1 Brazil	336	138.4	2,428
2 Indonesia	192	166.4	1,154
3 Colombia	123	29.0	4,241
4 Ivory Coast	101	10.7	9,439
5 Thailand	95	52.6	1,806
6 Laos	85	3.7	22,973
7 Nigeria	60	103.1	582
8 Philippines	57	57.3	995
9 Burma	51	38.0	1,342
10 Peru	45	19.8	2,273
11 Ecuador	40	9.6	4,167
12 Vietnam	36	63.3	569
13 Zaire	35	31.7	1,104
14 Mexico	33	80.2	411
15 India	33	781.4	42
Rest of World	337	3,729	
Top 15 Share	80%	32%	
Total	1659	4,917	

CO_2 Emissions from Energy Use

	Carbon (Thous. mt)	Pop (Millions)	Carbon/Pop (Kg)
1 United States	1,191,764	241.6	4,933
2 Soviet Union	992,421	281.1	3,530
3 China	532,388	1,054.0	505
4 Japan	246,394	121.5	2,028
5 West Germany	182,666	60.9	2,999
6 United Kingdom	164,373	56.7	2,899
7 India	139,971	781.4	179
8 Poland	122,329	37.5	3,262
9 Canada	103,834	25.6	4,056
10 France	95,162	55.4	1,718
11 South Africa	91,664	32.3	2,838
12 East Germany	90,731	16.6	5,466
13 Italy	90,103	57.2	1,575
14 Mexico	70,787	80.2	883
15 Czechoslovakia	64,430	15.5	4,157
Rest of World	1,194,984	2,000	
Top 15 Share	78%	59%	
Total	5,374,000	4,917	

Sources: Marland et al. (1989); World Bank (1989); Houghton et al. (1987); Central Intelligence Agency (various years).

desh (with near-trivial levels of carbon emissions), or at the level of Iran (the country that demarcates the global median in CO_2 emissions per capita), or at that of Italy (close to the global per capita average), there will be significant differences in the environment.

If we were to imagine a global population today at the level of development of the United States (roughly 5 metric tons of carbon per capita), global emissions would be three-and-one-half times current levels. Table 2 illustrates an alternative future by sketching two cases. Case 1 assumes a future world population of 10 billion persons and shows the projected carbon emissions for the world at the levels of development of various countries, in million tons and as a percent of the 1986 global totals. Case 2 shows what the total emissions would be in 2010 if the world had the population growth rates of these same countries, assuming constant emissions at the per capita world average in 1986.

Case 1 shows that with a future global population of 10 billion at the level of affluence of the United States today, and with present technology, global carbon emissions could be about seven times greater than 1986 levels. Case 2 shows that even if emissions were held constant at the 1986 average per capita level, the population

Table 2. Two Sketches.

	Total Global Carbon In Million Tons For the Year 2010	% of 1986 Global Total
Case 1. Assumes future world population of 10 billion		
Global per capita carbon emissions of:		
Bangladesh (30 kg.)	301	4
India (227 kg.)	2,269	31
China (526 kg.)	5,260	73
Iran (684 kg.)*	6,839	95
Italy (1,659 kg.)**	16,593	230
Japan (2,108 kg.)	21,077	292
Brazil (2,807 kg.)	28,072	389
US (4,974 kg.)	49,736	689
Case 2. Assumes world carbon per capita constant at global average (1,468 kg)		
Global population growth rates of:		
Bangladesh (2.6%)	13,361	185
India (2.2%)	12,165	169
China (1.2%)	9,608	133
Iran (2.8%)	14,000	194
Italy (0.3%)	7,754	107
Japan (0.7%)	8,531	118
Brazil (2.2%)	12,165	169
US (1.0%)	9,162	127

*Iran is at the median of carbon emissions per capita.
**Italy is close to the average global carbon emissions per capita.
Sources: Marland et al. (1989); World Bank (1989).

growth rate alone (under different scenarios) would have a substantial impact on the global level. This can be seen by noting the differential impacts of lower population growth rates (e.g., Italy and Japan) compared to those of higher growth rates (e.g., Bangladesh and Iran).

All of this presumes "no-surprise" futures and the prevalence of current patterns of industrialization and technological development. The cases are therefore only illustrative. They show that demographic factors are indeed compelling but that technological change and patterns of resource use have significant impacts as well. Together they highlight the dilemmas posed by the population nexus.

Challenges to Theory and Policy

Given the impacts of population on the environment and the interactions of population change with socioeconomic development, a central priority must be to improve both theory and policy on population.

The absence of a dynamic "causal" theory of population interactions with socioeconomic conditions may explain the relatively simple treatment of population in energy models, or models of aspects of global environmental change. The omission is serious—not only in terms of direct population effects and feedback but also with respect to the more complex population/resource/technology interactions which provide the context and significance of demographic factors in any particular state or social environment.

Contemporary demographic analysis—beyond the demographic transition—needs to address ways in which populations adapt or fail to adapt to their environments (Coleman, 1986:14). Also important is the need to relate more explicitly and precisely the interactions among population, energy, and environment. At least five tasks can be identified:

- Uncoupling processes of demographic responses (fertility, mortality, and migration trends) from those of demographic pressures (size and rates of growth);
- Analyzing the population feedback systems and how they vary in different demographic contexts and with different population nexuses;
- Accounting for the existence of complex demography/society linkages, such as warfare, infanticide, natural disasters, and disease, etc., that could substantially alter prevailing patterns of energy/environment interactions;
- Specifying the population/energy/environment interactions for different states at different levels of development;
- Identifying alternative forms of "sustainable" population/environment relationships in different demographic contexts.

Policy

Theory aside, there still remains the crucial issue of managing the global population, taking into account regional and national demographic characteristics, and framing a viable global policy. Population policy, in the broadest sense, has become a highly political issue, necessitating the articulation of underlying norms to help guide policy formulation. Since states at different levels of development generate different energy demand patterns and different forms of effluents—with different implications for the local and the global environment—the norms and principles for policy formation must respond to these differences.

A global approach to policy for managing population parameters can only be based on voluntary compliance internationally; coercion is simply not an option. Five principles, together, frame the basis for compliance.

These are 1) legitimacy: intervention strategies must be viewed as legitimate by all actors; 2) equity: interventions must be fair and appropriate; 3) consensus: policy must be adopted through procedures predicated on volition, not coercion; 4) universality: coverage must be global, encompassing all sovereign states; and 5) efficacy: implementation must be effective and not necessarily efficient.

The fact remains, however, that even if population policy worldwide were strengthened substantially, the most optimistic scenario still projects a future population of at least 10 billion people (rather than 14 billion or higher). In other words, strong population policy will only make the current projection more viable; it cannot shift the trend sufficiently to bring the number down below 10 billion by 2010.

Whatever the long-term management strategies for anthropogenic sources of global environmental change, the population factor is a crucial component. Population policy alone, however effective or comprehensive, is not sufficient to generate the adjustments, but it is necessary. The population nexus as a whole—the interaction of population, resources, and technological change—must become the focus of global policy.

Acknowledgments

I am grateful to Diane Beth Hyman, W. Parker Mauldin, and Jan Sundgren for comments on an earlier version of this paper.

References

Bilsborrow, R. E., "The Demographics of Macro-economic-demographic Models," *Population Bulletin of the United Nations,* United Nations, Department of International Economic and Social Affairs, no. 26, pp. 39–83 (1989).

Bongaarts, J., W. P. Mauldin, and J. Phillips, "The Demographic Impact of Family Planning Programs," paper prepared for the Meeting on Population and Development, Development Assistance Committee of OECD, April 1990, Paris.

Central Intelligence Agency, *The World Factbook,* US Government Printing Office, Washington, DC (various years).

Choucri, N., and R. C. North, "Lateral Pressure in International Relations: Concept and Theory," in M. Midlarsky, ed., *Handbook of War Studies,* Unwin Hyman, Inc., Winchester, MA, pp. 289–326 (1989).

Choucri, N. (ed.), *Multidisciplinary Perspectives on Population and Conflict,* Syracuse University Press, Syracuse, NY (1984).

Choucri, N., and C. Heye, "Simulation Models," in J. P. Weyant and T. A. Kuczmowski, eds., *Planning in a Risky Environment: A Handbook of Energy/Economy Modeling,* Pergamon Press, Oxford, England, in press.

Choucri, N., and R. C. North, *Nations in Conflict: National Growth and International Violence,* W. H. Freeman and Company, San Francisco, CA (1975).

Cochrane, S. H., *Fertility and Education: What Do We Really Know?* Johns Hopkins Press, Baltimore, MD (1979).

Coleman, D., "Population Regulation: A Long-Range View," in D. Coleman, and R. Schofield, eds., *The State of Population Theory: Forward from Malthus,* Basil Blackwell, Inc., New York, NY (1986).

Coleman, D., and R. Schofield, eds., *The State of Population Theory: Forward from Malthus,* Basil Blackwell, Inc., New York, NY (1986).

Demeny, P., "The World Demographic Situation," in J. Menken, ed., *World Population and U.S. Policy,* W. W. Norton, New York, NY (1986), pp. 27–66.

Edmonds, J., and J. Reilly, "Global Energy and CO_2 to the Year 2050," *Energy Journal* 4(3), p. 21 (1983).

Edmonds, J., and J. Reilly, "The IEA/ORAU Long-Term Global Energy-CO_2 Model," USDOE No. DE-AC05-84OR21400, Washington, DC (1984).

Edmonds, J., and J. Reilly, *Global Energy: Assessing the Future,* Oxford University Press, New York, NY (1985).

Energy Modeling Forum, "World Oil," EMF Report 6, Stanford University Modeling Forum, Stanford, CA (February 1982).

Environmental Protection Agency, *Policy Options for Stabilizing Global Climate Change,* draft report to Congress (February 1989).

Houghton, R.A., et al., "The Flux of Carbon from Terrestrial Ecosystems to the Atmosphere in 1980 Due to Changes in Land Use: Geographic Distribution of the Global Flux," *Tellus* 39B, 1987, pp. 122–139.

International Institute for Environment and Development and World Resources Institute, *World Resources 1987,* Basic Books, New York, NY (1987).

International Institute for Environment and Development and World Resources Institute, *World Resources 1990–91,* Oxford University Press, New York, NY (1990).

Keyfitz, N., *Can Knowledge Improve Forecasts?* Publication No. RR-83-05, International Institute for Applied Systems Analysis, Laxenburg, Austria (1983), cited in Toth et al. (1989).

Keyfitz, N., "The Growing Human Population," *Scientific American* 261(3), pp. 118–135 (September 1989).

Marland, G., et al., *Estimates of CO_2 Emissions from Fossil Fuel Burning and Cement Manufacturing, Based on the United Nations Energy Statistics and the U.S. Bureau of Mines Cement Manufacturing Data,* Oak Ridge National Laboratory, Oak Ridge, TN (1989).

Mauldin, W. P., "The Effectiveness of Family-planning Programmes," *Population Bulletin of the United Nations,* United Nations, Department of International Economic and Social Affairs, no. 27, pp. 69–94 (1989).

North, R. C., *War, Peace, Survival,* Westview Press, Boulder, CO (1990).

Sadik, N., *The State of World Population 1990,* United Nations Population Fund, New York, NY (1990).

Torrey, B. B., and W. W. Kingkade, "Population Dynamics of the United States and the Soviet Union," *Science* 247, pp. 1548–1552 (March 30, 1990).

Toth, F. L., E. Hizsnyik, and W. C. Clark, eds., *Scenarios of Socioeconomic Development for Studies of Global Environmental Change: A Critical Review,* International Institute for Applied Systems Analysis, Laxenburg, Austria (1989).

United Nations, *Population Bulletin of the United Nations No. 26,*

United Nations Publication E.89.XIII.6, New York, NY (1989).
United Nations, *Population Bulletin of the United Nations No. 27,* United Nations Publication E.89.XIII.7, New York, NY (1989).
United Nations, *World Population at the Turn of the Century,* United Nations Population Studies No. 111, Publication E.89.XIII.2, New York, NY (1989).
World Bank, *World Development Report 1989,* Oxford University Press, New York, NY (1989).

Economic Policy in the Face of Global Warming*

William D. Nordhaus
Department of Economics
Yale University
New Haven, CT

Introduction

Recent studies have identified four major global environmental risks. These are the increasing evidence of widespread damage from acid rain; the appearance of the Antarctic "ozone hole," interpreted by some as the harbinger of global ozone depletion that threatens to remove the shield from harmful ultraviolet radiation; deforestation, especially in the tropical rain forests, which may upset the global ecological balance and deplete genetic resources; and increases in certain atmospheric gases that may produce significant climatic changes over the next century through the "greenhouse effect."

These four global environmental issues have many common features, although the specific scientific and policy questions differ. Each is scientifically complex and controversial. All have costs and benefits that transcend national boundaries. This paper is confined entirely to the problem of global warming both because it is likely to have the most important economic impacts and because the science and economics are most mature here.

In what follows, I first review the theory and evidence on the greenhouse effect. I then present evidence on the impacts of greenhouse warming, the costs of stabilizing climate, and the kinds of adaptations that might be available. In the final section, I review current policy and sketch a set of policy recommendations for the near term.

Scientific Issues

The Greenhouse Effect and Greenhouse Gases

Scientists have suspected for almost two centuries that changing the chemical composition of the atmosphere would alter our planet's climate. In the first careful numerical calculations, S. A. Arrhenius estimated in 1896 that a doubling of the atmospheric concentrations of carbon dioxide (CO_2) would increase global mean temperature by 4°–6°C.[1]

What causes the greenhouse effect? The atmosphere consists of several "radiatively active" gases that absorb radiation at different points of the spectrum. The "greenhouse gases" are transparent to incoming solar radiation but absorb significant amounts of outgoing radiation. The net absorption of radiation produces a happy result, raising the earth's temperature about 33°C (59°F). The greenhouse effect helps explain the hot temperatures on Venus along with the frigid conditions of Mars.

The concern about the greenhouse effect occurs today because human activities are currently raising atmospheric concentrations of greenhouse gases and threatening a significant and undesirable climate change. The major greenhouse gases (GHGs) are carbon dioxide, methane, nitrous oxides, and chlorofluorocarbons (CFCs). Scientific monitoring has firmly established the buildup of the major GHGs. Table 1 shows the important GHGs, recent and projected concentrations, and the past and estimated future growth rates of major GHGs.

Not all greenhouse gases are created equal. GHGs have different radiative properties and different lifetimes. Table 2 shows the important greenhouse gases, their "instantaneous" and "total" contribution to global warming,[2] and the industries in which the emissions originate. Carbon dioxide is the major contributor to global warming, with most CO_2 emissions coming from the combustion of fossil fuels. Of these fossil fuels, natural gas has 58 percent as much CO_2 emissions per unit energy as coal, and petroleum has 81 percent as much CO_2 per unit energy as coal. The second most important source of GHG emissions is the CFCs, which are small in volume but have a warming potential almost 20,000 times as powerful as CO_2 per unit of volume.

One difficulty in analyzing the damages from climate change and the costs of slowing climate change is to find a common unit of measurement. In this paper, I will translate all impacts and costs into a common unit, the "CO_2 equivalent" of GHG emissions. By working in this metric, we can attempt to ensure cost effectiveness of policies in different sectors.

* Helpful comments were provided by Henry Aaron, Jesse Ausubel, Clark Bullard, Robert Chen, William Clark, John Perry, Thomas Schelling, Charles Schultze, and Aaron Wildavsky.

Table 1. Estimated concentrations of important greenhouse gases.

	Atmospheric Concentration of Greenhouse Gas				
	Level [parts per billion]			Growth [percent per year]	
Greenhouse Gas	1850	1986	2100[a]	1850–1986	1986–2100
Carbon Dioxide (000)	290	348	630	0.16	0.52
Methane	880	1675	3100	0.56	0.54
Nitrogen Oxides	285	340	380	0.15	0.10
Chlorofluorocarbons[b]	0	0.62	2.90	—	1.37

[a] Projected from EPA [1989].
[b] Includes only major sources, CFC-11 and CFC-12.
Sources: Wuebbels and Edmunds [1988] and EPA [1989].

Table 2. Estimated contribution of different greenhouse gases to global warming for concentration changes, 1985–2100.

A. Sources by Chemical Compound

Greenhouse Gas	Relative Contribution: Instantaneous	Relative Contribution: Total	Source of Emission
CO_2	76.1%	94.7%	Largely from combustion of fossil fuels.
Methane	9.6	0.8	Unknown. From a wide variety of biological and agricultural activities.
CFCs	11.6	3.3	Wholly industrial, from both aerosols and non-aerosols.
Nitrous Oxides	2.7	1.2	From fertilizers and energy use.

Source: Nordhaus [1990a]. Instantaneous contribution measures the impact of concentration change at the instant of release. "Total" impact estimates the relative contribution to global warming over indefinite future.

B. Sources by Economic Activity

Greenhouse Gas	Relative Contribution: Instantaneous	Relative Contribution: Total	Source of Emission
Energy	62.8%	76.2%	CO_2 Emissions, Nitrous Oxides, Methane
Agriculture	20.6	19.8	CO_2, Methane, Nitrous Oxides
Industry	0.7	0.1	Methane
Natural	4.3	0.7	Methane, Nitrous Oxides
Other	11.6	3.3	CFCs

Source: Nordhaus [1990]. Estimates of emission sources are from EPA [1989a], vol. I, Chapter II. Note that sources are highly uncertain for methane and nitrous oxides.

Climate Models and Forecasted Climate Change

No one disputes the buildup of greenhouse gases. To project climate changes many years into the future, however, requires use of climate models that trace out the effect of a changing radiative balance upon major climatic variables. Because we are heading into uncharted waters, the models cannot rely upon historical experience but must extrapolate beyond current observations.

Climate models are mathematical representations of important variables such as temperature, humidity, winds, soil moisture, and sea ice. Large "general-circulation models" or GCMs simulate changes in weather, in steps of a few minutes, over a century or more. The largest models use 500 kilometer square grids through several layers of the atmosphere. Such models are unfortunately extremely expensive to run, and a single CO_2 scenario might take the largest supercomputer up to a year to calculate.

Two important features of the results emerge from existing studies. First, the central estimates of the *equilibrium* impact of a CO_2-equivalent doubling have changed little since the earliest calculations. The last thorough review by the National Academy of Sciences concluded in 1983:[3]

> When it is assumed that the CO_2 content of the atmosphere is doubled and statistical thermal equilibrium is achieved, all models predict a global surface warming. [GCMs] indicate global warming [from CO_2 doubling] to be in the range between about 1.5 and 4.5° C.

Second, though short-run weather forecasting has improved dramatically in recent years, model estimates of the impact of CO_2 doubling are not converging. A few recent model runs are outside the 1.5°–4.5° C range cited above.

GCMs produce a trove of other interesting numerical results on predicted future climates. Table 3 reports an attempt to characterize the results and uncertainties about future climate change. Most experts believe that mean temperature will rise and that the warmer climate will increase precipitation and runoff. Some models foresee hotter and drier climates in mid-continental regions, such as the US Midwest. Forecasting climate changes at particular locations (such as in California or Texas or in my favorite ski area) has proven intractable, and many climate modelers do not expect to be able to forecast regional climates accurately in the foreseeable future.

Consistency with the Historical Temperature Record

A major challenge is to validate the model predictions. There are several possible ways to test the models, but the proof of the pudding will ultimately occur when and if global temperatures actually begin to rise.

Historical records indicate that global mean temperature has increased about 0.5° C since the 1880s. Whether the observed temperature record is consistent with the predictions of climate models is a hotly debated question. Some authors have used statistical techniques to test for the presence of a "greenhouse signal" in the upward trend of temperature over the last century. The hypothesis that the climate is a trendless, white-noise process can be rejected at a high level of confidence. Still, a great deal of evidence suggests that climatic variables fluctuate over periods of a century or more. Unfortunately, we do not know enough about the background trends and cycles to know whether the warming in recent years is normal climatic fluctuation or something new and different.[4] To date, statistical analysis of the historical record has lagged far behind construction of new and more refined GCMs. But the historical record is an important, independent source of evidence about the pace of global warming.

Uncertainties about Future Climate Change

Coping with greenhouse warming is particularly nettlesome because of the uncertainties about its magnitude and timing that cascade from stage to stage. The first uncertainty concerns the rate of economic growth over the next century. Other questions involve the rate of emission of GHGs per unit of economic activity; the rate of atmospheric retention of different GHGs; the equilibrium relationship between increased concentrations in GHGs and climate change; the speed with which actual climate will move to the new equilibrium; and the extent to which climate would change were humans not meddling in the climate.

Table 3. Range of estimates from climate models about equilibrium impact on major variables of doubling of CO_2.

Variable	Projection of probable global average change	Distribution of regional change	Confidence in prediction:[c] Global average	Confidence in prediction:[c] Regional average
Temperature	+2 to +5°C	−3 to +10°C	High	Medium
Sea Level[a]	+10 to +100 cm		High	
Precipitation	+7 to +15%	−20 to +20%	High	Low
Soil Moisture	???[b]	−50 to +50%	???[b]	Medium
Runoff	Increase	−50 to +50%	Medium	Low
Severe Storms	???[b]	???[b]	???[b]	???[b]

[a] Increases in sea level are the average of the global rate. Sea level rise in particular locations will be higher or lower than this figure depending upon local geological conditions.
[b] No basis for forecast of this variable.
[c] The "confidence in prediction" is a subjective estimate of experts of their confidence that the range of estimates provided is accurate. These estimates are based upon formal models, historical analogy, and other experience.

Source: Adopted from L. Mearns, P. H. Gleich, and S. H. Schneider, "Prospects for Climate Change," in Paul Waggoner, ed., *Climate Change and US Water Resources*, New York, Wiley, forthcoming, 1990.

Although uncertainty breeds frustration, a rational response to climate change must take it into account. To estimate the uncertainty about future climate change, I have combined estimates of uncertainty about future climate change from a number of different sources. Estimates of the distribution of emissions of different GHGs are derived from a prior study by Nordhaus and Yohe [1983]. For other GHGs, we use estimates of emissions and concentrations of CFCs, methane, and other GHGs from EPA [1989]. The assumed impact of rising GHGs follows the consensus of modelers that a CO_2 doubling would in equilibrium raise global temperatures by 3° C (with a standard deviation of 1° C). Finally, I derive a lag structure of actual temperature behind the equilibrium temperature using evidence from GCMs along with statistical evidence from the historical temperature record.

Figure 1 presents the estimated range of greenhouse warming over the period 1850 to 2000 and compares the estimated range with an index of mean surface temperature over the last century.[5] The calculated median estimate of realized warming in 1990 is around 0.7° C from 1800 and about 0.6° C from the beginning of the temperature record in 1880. This estimate approximates reasonably well the actual temperature increase of 0.5° C shown by the highly volatile series in Figure 1.

Figure 2 shows the estimated range of greenhouse warming over the next century. My estimate of the most likely global temperature rise is 1.8° C from 1800 to 2050 and 3.3° C from 1800 to 2100. This calculation indicates that the chances are one in four that temperature change from 1800 to 2050 will be either less than 1.5° C or greater than 2.2° C.

Another question is when the canonical 3° C warming is estimated to occur; this question is important because many economic studies investigate the impact of a 3° C warming climate scenario. My calculations indicate that the average temperature will rise 3° from 1800 to 2090; there is a one-in-four chance that this much warming could occur by 2075 but less than a one-in-twenty chance of a 3° warming before 2050. On the optimistic side, these calculations suggest that there is almost one chance in two that global mean temperature will rise less than 3° C between 1800 and 2100.

It should be emphasized that projecting climate change is hazardous at best. But these estimates suggest that a 3° C warming is most likely to occur around a century from now and is quite unlikely before the middle of the next century.

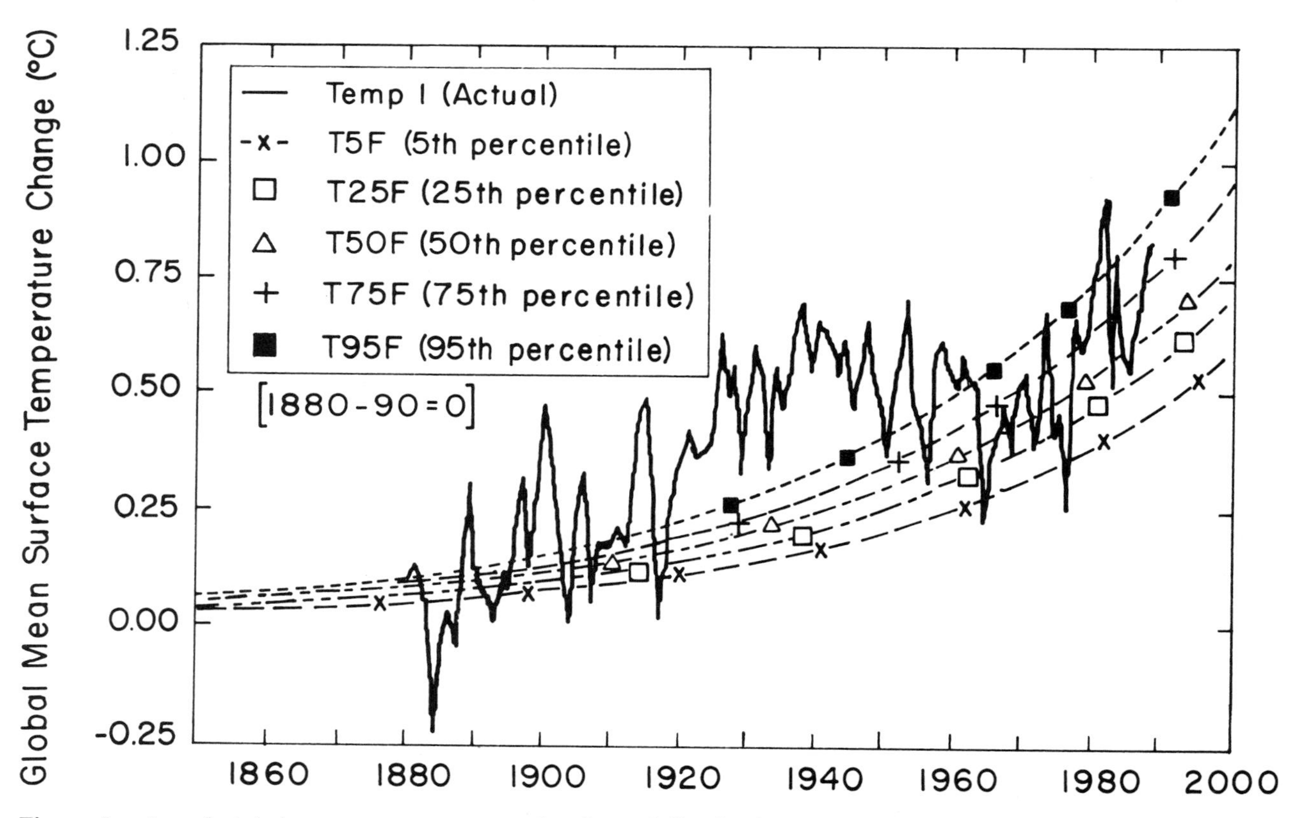

Figure 1. Actual global mean temperature and estimated distribution of greenhouse warming, 1850–2000.

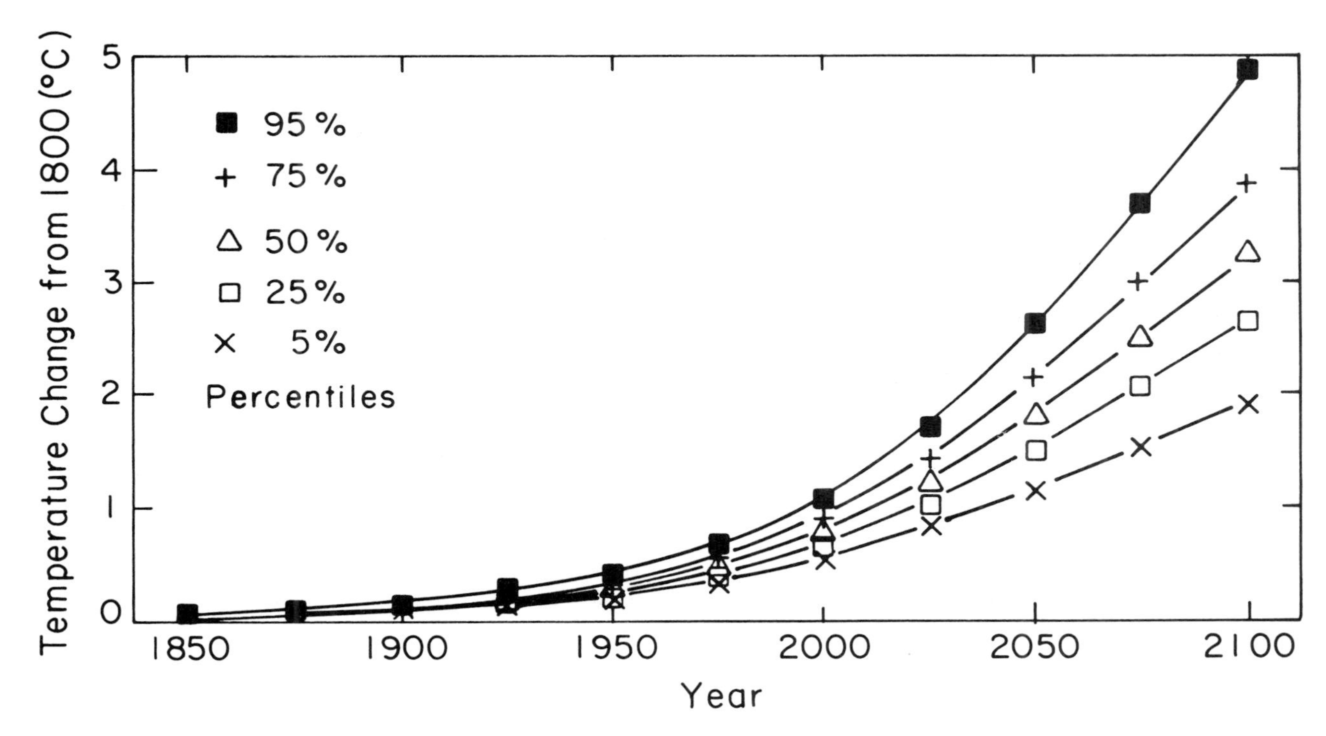

Figure 2. Change in temperature from 1800: distribution of possible outcomes.

Impacts of Climate Change

We now move from the *terra infirma* of climate change to the *terra incognita* of social and economic impacts of climate change. This section first describes the effects of greenhouse warming upon the economy, then presents some estimates of what measures to slow greenhouse warming would cost, and finally addresses the issue of potential adaptations to greenhouse warming.

Impacts of Greenhouse Warming

To appreciate the impacts of greenhouse warming, a number of general remarks are useful. To begin with, it must be recognized that human societies thrive in a wide variety of climatic zones. Climate variables like temperature or humidity have little effect upon the net value of economic activity in advanced countries. Indeed, in part due to technological changes like air conditioning, migration patterns in the United States have favored warmer regions.

At the same time, although most analyses focus primarily upon globally averaged surface temperature, this variable is not the most important for impacts. Variables like precipitation or water levels and extremes of droughts or freezes are likely to be more important. Mean temperature is chosen because it is a useful *index* of climate change that tends to be associated with most other important changes. A related point involves the size of projected climate changes in comparison to the day-to-day changes we normally experience. The variations in weather that we experience in our daily lives will swamp the likely changes over the next century. The change in temperature between 8:00 and 9:00 am on an April morning is normally greater than the expected change from 1990 to 2090. Few people are likely to notice the CO_2 signal amidst the noisy pandemonium of their daily lives.

Economic Effects of Climate Change: US

Climate change is likely to have different effects on different sectors.[6] In general, those sectors of the economy that have a significant interaction with unmanaged ecosystems—that is, are heavily dependent upon naturally occurring rainfall, runoff, or temperatures—may be significantly affected by climate change. Agriculture, forestry, and coastal activities fall in this category. Most of the US economy has little *direct* interaction with climate, and the impacts of climate change are likely to be very small in these sectors. For example, cardiovascular surgery and microprocessor fabrication are undertaken in carefully controlled environments and are unlikely to be directly affected by climate change.

Table 4 presents a breakdown of US national income, organized by the sensitivity of the sector to greenhouse warming.[7] Approximately 3 percent of US national output originates in climate-sensitive sectors and another 10 percent in sectors modestly sensitive to climatic change. About 87 percent of national output comes from sectors that are negligibly affected by climate change. These measures of output may understate the impact of climate change on well-being because they omit important nonmarket activities—especially leisure activities—that may be more sensitive to climatic change than measured output.

What are the likely effects of climate change on individual sectors? The following is a synopsis of recent studies.

Agriculture. Agriculture is the most climate-sensitive of the major sectors. Studies suggest that greenhouse warming will reduce yields in many crops. On the other hand, the associated fertilization effect of higher levels of CO_2 will tend to raise yields. After a careful review, a recent National Academy of Sciences report stated, "Thus, we do not regard the hypothesized CO_2-induced climate changes as a major direct threat to American agriculture over the next few decades."[8] The Environmental Protection Agency (EPA) found that the value of US agricultural output is likely to rise or fall by as much as $10 billion annually depending on the magnitude of the climate change.[9] It should be noted, however, that no studies of the impact of climate change on agriculture allow for a full adaptation of farming to the changes, including changed cropping patterns, land use changes, and migration of labor and capital. Current estimates of the costs of climate change therefore probably overestimate the costs (or underestimate the benefits) of future changes.

Sea-level rise. Most studies indicate a gradual rise in average sea level over the next century, with a consensus estimate until recently of a sea-level rise of 70 cm over the next century (recent evidence has cut this estimate sharply). For the consensus estimate, EPA projects the costs of sea-level rise for the US to be: land loss of around 6,000 square miles, protection costs (by levees and dikes) of high-value property, and miscellaneous protection of open coasts. The total capital outlay is in

Table 4. Breakdown of economic activity by vulnerability to climatic change, 1981.

	National Income		
Sector	Value (billions)	Percent of Total	
Total National Income	2414.1	100.0	100.0
Potentially Severely Impacted			3.1
Farms	67.1	2.8	
Forestry, fisheries, other	7.7	0.3	
Moderate Potential Impact			10.1
Construction	109.1	4.5	
Water transportation	6.3	0.3	
Energy and Utilities[1]			
Energy (electric, gas, oil)	45.9	1.9	
Water and sanitary	5.7	0.2	
Real Estate[2]			
Land-rent component	51.2	2.1	
Hotels, lodging, recreation	25.4	1.1	
Negligible Effect			86.8
Mining	45.1	1.9	
Manufacturing	581.3	24.1	
Other transportation and communication	132.6	5.5	
Finance, insurance, and balance real estate	274.8	11.4	
Trade	349.4	14.5	
Other services	325.2	13.5	
Government services	337.0	14.0	
Rest of World	50.3	2.1	

[1] National income in electric, gas, sanitary industry is subdivided on the basis of consumption of major components.
[2] Estimate of land-rent component is drawn from two sources: national balance sheet data on values of land and structures and from surveys of housing prices. Estimate assumes that 25 percent of non-labor income in real estate industry is from land rents.

Source: Data are based on the US National Accounts, Survey of Current Business, July 1984.

the order of $100 billion,[10] which is approximately 0.1 percent of cumulative gross private domestic investment over the period 1985 to 2050.

Energy. Greenhouse warming will increase the demand for space cooling and decrease the demand for space heating. The net impact of CO_2 doubling is estimated to be less than $1 billion at 1981 levels of national income.

Other marketed goods and services. Many other sectors are likely to be affected, although numerical estimates of the effects are not available. The forest products industry may benefit from CO_2 fertilization.[11] Water systems (such as runoff in rivers or the length of ice-free periods) may be significantly affected, but the costs are likely to be determined more by the rate of climate change than the new equilibrium climate. Construction in temperate climates will be favorably affected because of a longer period of warm weather. The impact upon recreation and water transportation is mixed depending upon the initial climate. Cold regions may gain while hot regions may lose; investments in water skiing will appreciate while those in snow skiing will depreciate. But for the bulk of the economy—manufacturing, mining, utilities, finance, trade, and most service industries—it is difficult to find major direct impacts of the projected climate changes over the next 50 to 75 years.

Nonmarketed goods and services. Many valuable goods and services escape the net of the national income accounts and might affect the calculations. Among the areas of importance are human health, biological diversity, amenity values of everyday life and leisure, and environmental quality. Some people will place a high moral, aesthetic, or environmental value on preventing climate change, but I know of no comprehensive estimates of what people are willing to pay to stop greenhouse warming. One study projects important gains for the US from modest increases in average temperature.[12] I am aware of no studies that point to major nonmarket costs, but further analysis will be required to decide whether these omitted sectors will affect the overall assessment of the cost of greenhouse warming.

In sum, the economic impact upon the US economy of the climatic changes induced by a doubling of CO_2 concentrations is likely to be small. The best guess today is that the impact, in terms of those variables that have been quantified, is likely to be around one fourth of 1 percent of national income by the middle or later part of the next century. However, current studies omit many potentially important effects so this estimate has a large margin of error.

Economic Effects of Climate Change: Outside the US[13]

To date, studies for other countries are fragmentary, and no general conclusions are possible at this time. Existing evidence suggests that other advanced industrial countries are likely to experience modest impacts similar to those of the United States. Detailed studies for the Netherlands, as well as a less comprehensive study for six large regions (the US, Europe, Brazil, China, Australia, and the USSR) found that the overall impact of a CO_2-equivalent doubling will be small and probably difficult to detect over a half-century or more.[14]

On the other hand, small countries that are heavily dependent on coastal activities or suffer major climate change may be severely affected. Studies suggest that Bangladesh may be inundated and that the Maldives will disappear over the long run, but the timing of these impacts is conjectural. Particular concerns arise where activities cannot easily migrate in response to climate change. Such situations include natural reserves (like Yosemite) or populations limited to small areas (like south-sea islanders).

Developing countries are probably more vulnerable to greenhouse warming than are advanced countries, particularly those poor countries living on the ragged edge of subsistence with few resources to divert to dealing with climate change. However, most poor countries are heavily dependent upon agriculture, so the benefits of CO_2 fertilization might offset the damages from climate change. Much more work needs to be done on the impacts of climate change on developing countries.

These reflections lead to a surprising conclusion: our best guess is that CO_2-induced climate change will produce a combination of gains and losses with no strong presumption of substantial net economic damages. This conclusion should not be interpreted as a brief *in favor of* climate change. Rather, it suggests that those who paint a bleak picture of desert earth devoid of fruitful economic activity may be exaggerating the injuries and neglecting the benefits of climate change.

Policies to Cope with the Threat of Global Warming

In response to the threat of global warming, a wide variety of responses is available (see Table 5). A first option, taking steps to slow or prevent greenhouse warming, has received the greatest public attention. Most policy discussion has focused on reducing energy consumption or switching to nonfossil fuels, while some have suggested reforestation to remove CO_2 from the atmosphere. One important goal of policy should be cost effectiveness—structuring policies to get the maximal reduction in harmful climatic change for a given level of expenditure.

A second option is to offset greenhouse warming through climatic engineering. Suggestions over the last

Table 5. Alternative responses to the threat of greenhouse warming.

1. Slow or prevent greenhouse warming: reduce emissions and concentrations of greenhouse gases.
   a. Reduce energy consumption
   b. Reduce GHG emissions per unit of energy consumption or GNP:
      - Shift to low-CO_2 fuels
      - Divert CO_2 from entering atmosphere
      - Shift to substitutes for CFCs
   c. Remove greenhouse gases from atmosphere
      - Grow and pickle trees

2. Offset climatic effects.
   a. Climatic engineering
      - Paint roads and roofs white, put particles into stratosphere

3. Adapt to warmer climate.
   a. Decentralized/market adaptations
      - Movement of population and capital to new temperate zones
      - Corn belt migrates toward Canada and Siberia
   b. Central/governmental policies
      - Build dikes to prevent ocean's invasion
      - Land-use regulations
      - Research on drought-tolerant crops

few years include painting roads and roofs white or shooting particulate matter into the stratosphere to cool the earth. Many climatologists fault these proposals, arguing in effect that "you shouldn't fool with Mother Nature," but the proposals have received insufficient analysis.

A final option is to adapt to the warmer climate. Adaptation could take place gradually on a decentralized basis through the automatic response of people and institutions or through markets as the climate warms and the oceans rise. In addition, governments could prevent harmful climatic impacts by land-use regulations or investments in research on living in a warmer climate.

Preventive Policies

The major policy question surrounding the greenhouse effect is whether steps should be taken in the near term to prevent global warming. Whether preventive action should be taken depends on the costs of preventing GHG emissions relative to the damages that the GHGs would cause if they continue unchecked.

Knowledge of the costs of slowing climate change is rudimentary. This section will review the costs of slowing greenhouse warming through reduction of emissions and atmospheric concentrations of greenhouse gases (strategy 1 in Table 5). These examples—reducing CFC emissions, reducing CO_2 emissions, and reforestation—are not the only options, but they have been studied most intensively.

In calculating the cost of preventive measures, we measure costs in terms of tons of CO_2 equivalent. Measures that cost up to $5 per ton of CO_2 equivalent are inexpensive; at this cost, global warming could be stopped dead in its tracks at a total cost of less than $40 billion per year (about 0.2 percent of global income). Costs near $10 to $50 per ton CO_2 equivalent are expensive but manageable (costing 0.5 to 2.5 percent of global income). Measures in excess of $100 per ton of CO_2 are extremely expensive.

Reducing CFC Emissions. A first strategy involves reducing emissions of chlorofluorocarbons (CFCs) into the atmosphere. This step is particularly important because CFCs are extremely powerful greenhouse gases. It is currently believed that new chemical substitutes for the two most important CFCs can be found that will significantly reduce greenhouse warming. A rough estimate is that these substitutes can reduce warming at a cost less than $5 per ton of CO_2 equivalent. This policy is extremely cost effective, bringing a significant reduction of warming at modest cost.

Reducing CO_2 Emissions. Any major reduction in GHGs will require a significant reduction in CO_2 emissions, more than 95 percent of which come from the energy sector and deforestation. Carbon dioxide emissions can be reduced through increases in energy efficiency, decreases in final energy services, substitution of less GHG-intensive fossil fuels for more GHG-intensive fossil fuels, substitution of nonfossil fuels for fossil fuels, and technological change that allows new techniques of production along with new products and services.

Because energy interacts with the economy in so many ways, good estimates of the costs of reducing CO_2 emissions require complex models of the energy system. Such models must incorporate the behavior of both producers and consumers along with consideration of each of the possible methods for reducing emissions. I have reviewed a number of studies,[15] and I will now discuss the current estimates of the long-run costs of reducing CO_2 emissions.[18]

The studies of the *long-run* costs of CO_2 reduction that I have reviewed lead to two major conclusions. First, the cost of reducing CO_2 emissions is low for small curbs. Reductions of up to 10 percent of CO_2 emissions from the energy sector can be attained at an average cost of around $10 per ton of CO_2 reduced. With current global annual emissions of around 6 billion tons of CO_2, a 10 percent reduction would cost around $6 billion annually.

The second conclusion is that the cost of reducing CO_2 emissions grows rapidly and becomes extreme for substantial reductions. I estimate that, in the long run with today's energy technologies, the marginal cost of a 50 percent reduction in CO_2 emissions is approximately $130 per ton CO_2. In other words, inducing producers and consumers to reduce their CO_2 emissions by one half would require a carbon tax (or the regulatory equivalent thereof) of \$130 per ton of CO_2, which would generate annual taxes of around \$400 billion. The *total* resource cost of a 50 percent reduction in CO_2 emissions is about \$180 billion annually, slightly less than 1 percent of world output at current price and output levels.

The incremental costs of reducing CO_2 emissions rise rapidly because no substitutes currently exist for many uses of fossil fuels. For example, a major reduction in CO_2 emissions from transportation would require either that people travel less or that fewer goods are transported, both of which would be quite costly.

Forests. Several studies have proposed using trees as a method of removing carbon from the atmosphere. Among the major proposals are slowing the deforestation of tropical forests; reforesting open land, thereby increasing the amount of carbon locked into the biosphere; a "tree bounty," which subsidizes the sequestration of wood in durable products; and a "tree pickling" program, in which trees are stored indefinitely.

No detailed study of the economics of tropical forests has been undertaken here. However, deforestation may be adding 0.5 billion to 3 billion tons of carbon per year to CO_2 emissions (this amounting to between 5 and 30 percent of total GHG emissions). Much deforestation is uneconomical in tropical regions even without invoking greenhouse considerations. If so, the cessation of uneconomical deforestation can significantly and inexpensively slow greenhouse warming.

I estimate that the three other reforestation options can remove carbon from the atmosphere at modest costs; however, they can contribute only marginally to reducing atmospheric concentrations.

Summary on Costs of Prevention. As shown in Figure 3, it appears that a significant fraction of GHG emissions—perhaps one sixth—can be eliminated at relatively low cost. The most cost effective policies to slow greenhouse warming include curbs on CFC production and preventing uneconomical deforestation. Putting all the low-cost options together, I estimate that around one sixth of CO_2-equivalent emissions can be reduced at an average cost of \$4 per ton of CO_2 equivalent, for a total cost of about \$6 billion annually.

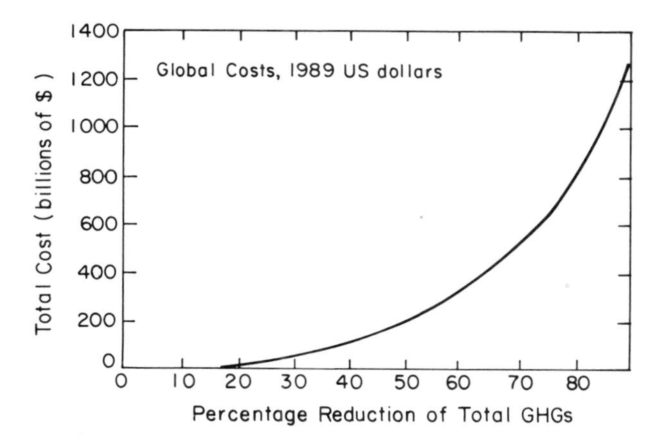

Figure 3. Costs of reducing GHG emissions.

After the low-cost options have been exhausted, further reductions in GHG emissions will require curbing CO_2 emissions, say through taxes or regulations on the carbon content of fuels. But curbing GHG emissions rapidly hits diminishing returns: a 50 percent reduction in GHG emissions in the long run will cost about \$200 billion annually, which is about 1 percent of global output. Attempts to restrict GHG emissions severely in a short period would be even more costly.

Adaptation Policies

Faced with the prospect of changing climate, societies may decide to adapt. The most important adaptations are those taken by *private* agents—consumers and businesses, for example. Decentralized adaptations—population migration, relocation of capital, land reclamation, and technological change—will occur more or less automatically in response to changing relative incomes, prices, and environmental conditions.

Governments also play an important role by ensuring that the legal and economic structure is conducive to adaptation, particularly by making sure that the environmental or climatic changes get reliably translated into the price and income signals that will induce private adaptation. Fulfilling this role may prove difficult because so many of the impacts of climate change are not properly priced. For example, greenhouse warming may alter runoff patterns of major rivers.[17] Because water is allocated in such an archaic way, there is no guarantee that it will be efficiently allocated when water availability changes. Governments can improve adaptation by introducing general allocational devices (such as water auctions) that will channel resources to their highest-value uses. Use of land near sea coasts and in flood plains poses similar issues.

Speeds of Adjustment in Prevention and Adaptation. Adaptation and prevention are often treated as symmetrical policies. They differ in one crucial respect, however. While preventive policies must generally precede global warming, adaptation policies can occur simultaneously. This distinction is crucial here, for cause precedes effect by a half century or more. To stabilize climate, immediate action is necessary; adaptations can wait for many years. This contrast underscores one of the major obstacles to responding intelligently to threatening climate change—the long time scale involved.[18]

A common mistake in thinking about this issue is to impose a slowly changing climate upon today's world and to ignore the inevitable evolution that will take place over the coming decades. If it takes 80 years or more for CO_2 doubling, as suggested earlier, adaptations will be spread over a similar period. Yet social and economic structures change enormously over such a time. Recall how much the world has changed since 1910. That was the age of empires, when the Ottoman, Austrian, and Czarist regimes ruled much of Eurasia. The map of Europe has been redrawn three times since 1910 and is again being restructured as this is written. The power density of the US was about 1.5 horsepower per capita as opposed to 130 horsepower per capita today; one sixth of horsepower was horses, and the 21 million horses were the major polluters. Unheard of were air conditioning, nuclear power, electronics, and computers.

This catalogue makes clear how foolish it would be to prescribe adaptive steps *now* to smooth the transition to climate changes over the next century. The time scale of most adaptations is much shorter than the time scale of climate change. Carbon dioxide doubling will take place over the next century. By contrast financial markets adjust in minutes, product prices in weeks, labor markets in a few years, and the economic "long run" is usually reckoned at no more than two decades. To adapt now would be akin to building a Maginot Line in 1935 to cope with military threats of the 1990s.

These considerations suggest that it would be unwise to undertake costly adaptive policies unless they satisfy one of three criteria: (1) they have such long leadtimes that they must be undertaken now to be effective; (2) they have a clear presumption of being economical even in the absence of climate change; or (3) the penalty for delay is extremely high. By this criterion, it is difficult to enumerate any adaptive measures other than the general maxim to promote a healthy economy, to strive to internalize most external effects to ensure an appropriate reponse to changing climatic signals, and to raise the national savings rate to provide the investments needed for changing infrastructure.

Policies to Combat Global Warming

The discussion up to now indicates that it is difficult to uncover any major economic costs of climate change. At the same time, the prospect that the climate may change in a catastrophic fashion might justify steps to slow climate change. What should the United States do now to respond to the threat of global warming over the next century?

A Cost-Benefit Approach

On economic grounds alone, any policy that promises incremental benefits worth more than its incremental costs is worth undertaking. As mentioned above, it is useful to define policies as carbon taxes that penalize emissions of greenhouse gases in proportion to their global warming potential. These "taxes" are in a sense a metaphor for explicit government steps to reduce GHG emissions through energy or gasoline taxes, CFC bans or regulatory limits, prohibitions on tree cutting, taxes on carbon emissions, or energy-efficiency standards.

Using the estimates of damages outlined above, and assuming a low discount rate on future damages from climate change, I calculate that an efficient policy would impose a penalty on GHG emissions of around $5 per ton CO_2 equivalent.[19] This level of penalty would produce a total reduction of about 12 percent of GHG emissions, including a large reduction in CFCs and a small reduction in CO_2 emissions. As Table 6 shows, such a tax amounts to $3.50 per ton tax on coal, 58 cents per barrel on oil, and 1.4 cents per gallon on gasoline. US revenues from a $5 per ton carbon tax would amount to about $10 billion annually. Table 6 also shows the impact of a severe restraint—$100 per ton CO_2—which would be close to the tax required to reduce CO_2 emissions by one half. The "high-tax" strategy would clearly have a significant impact upon the US economy.

It is useful to compare these costs with historical events or regulatory programs. A low-cost program for slowing global warming (say one associated with the low-tax proposal in Table 6) would impose a burden equivalent to a major US regulation, such as those on drinking water, noise, or surface mining.[20]

The more stringent program to cut GHG emissions by one half (associated with the high-tax scenario in Table 6) would impose annual costs of around 1 percent of world output. This can be compared to the costs of *all* environmental, health, and safety regulations in the United States, which were estimated to cost 1 to 3 percent of GNP.[21] Another parallel is with the impact of the energy price increases of the 1970s. Dale Jorgenson estimates that the energy-price increase lowered US output growth by 0.2 percent per year, or a total of about 3 percent, since 1973. Charles Schultze found similar estimates for the impact of the first oil shock.

Realistic Complications

A simple economic cost-benefit analysis is a useful starting point for considering appropriate actions, but it omits two practical issues that complicate policy enormously: discounting and uncertainty.

Should We Discount Future Climate Damages? The prospect of future climate change raises the question of how to discount the costs of future climate change in making current decisions. This issue is particularly thorny because the atmospheric residence time of CO_2 is hundreds of years.

In part, the issue of discounting is an ethical question, reflecting the relative valuation of well-being of

Table 6. Illustrative measures of impact of different carbon taxes.

Sector of Impact	Level of Stringency of GHG Reductions	
	Low Tax	High Tax
Tax Effect		
Tax on CO_2 equivalent (per ton carbon)	$5.00	$100
Impact on fossil fuels prices (1989 prices):		
Coal Price		
Per metric ton	$3.50	$70
Percentage increase	10%	205%
Oil Price		
Per barrel	$0.58	$11.65
Percentage increase	2.8%	55%
Gasoline Price		
Per gallon	1.4 cents	28 cents
Percentage increase	1.2%	23.3%
Overall Impacts		
Estimated reduction of GHG emissions (CO_2 equivalent)	13%	45%
Total tax revenues, US (billions)	$10	$196
Estimated global net economic benefits (+) or costs (−), billions of $ per year, 1989 global economy[a]	$12	−$96

Note: Figures do not take into account the reduction in GHG emissions in response to carbon tax; that is, they are "without feedback." These estimates are drawn from Nordhaus [1990].

[a] Assumes a discount rate equal to 1 percent in excess of the growth of output and damages from a CO_2 doubling equal to 1 percent of global output.

current and future generations. But a discount rate on climate change cannot be chosen arbitrarily and without regard to other decisions. A discount rate close to the return on capital in most countries—say 8 percent per year or more—would imply that we should forget about climate change for a few decades.

On the other hand, a low discount rate—say 4 percent per year or less—would give considerable weight today to climate changes in the late 21st century. But such a low discount rate implies that other investment opportunities have been largely exhausted—hardly an attractive assumption in a capital-starved world. A low discount rate on climate change along with a high return on capital is simply inconsistent. Faced with the dilemma of a low social discount rate and a high return on capital, the efficient policy would be to invest heavily in high-return capital now and then use the fruits of those investments to slow climate change in the future.

Uncertainty. Clearly, global warming is rife with uncertainty—about future emissions paths, about the GHG-climate linkage, about the timing of climate change, about the impacts of climate upon flora and fauna, about the costs of slowing climate change, and even about the speed with which we can reduce the uncertainties. How should we proceed in the face of uncertainty? Like generals or environmentalists who assume the worst case?[22] Or like cigarette manufacturers who assert that unproved equals untrue?

One approach would be to take a "certainty equivalent" or "best guess" analysis, ignore uncertainty and the costs of decisionmaking, and plunge ahead. The cost-benefit analysis performed above embodies this approach. It is appropriate as long as the risks are symmetrical and when the uncertainties are unlikely to be resolved in the foreseeable future. Unfortunately, neither of these conditions is likely to be satisfied for the greenhouse effect.

Risk asymmetry. Virtually all observers agree that the uncertainties of climate change are asymmetrical; we are likely to be increasingly averse to climate change the larger the change. To go from a 2° to a 4° C warming is much more alarming than to move from a 0° to 2° C warming. The greater the warming, the further we move from our current climate and the greater the potential for unforeseen events. Moreover, it is the extreme events—droughts and hurricanes, heat waves and freezes, river flooding and lake freezing—that produce major economic losses. As probability distributions shift, the frequency of extreme events increases (or decreases) proportionately more than the change in the mean. Whether the increases in unpleasant extremes (like droughts in the corn belt) will be greater or less than the increases in pleasant extremes (like frost-free winters in the citrus belt) is, like most questions about climate change, unanswered.

In addition, climatologists generally think that the chance of unpleasant surprises rises as the magnitude and pace of climate change increases. We must go back 5 million to 15 million years to find a climate equivalent to what we are likely to produce over the next 100 years;

the concentrations of GHGs in the next century will exceed levels previously observed. Climate systems are complex, and we do not know whether they may have multiple locally stable equilibria. It is sobering to remember that the Antarctic "ozone hole" was a complete surprise.

Among the kinds of responses that have been suggested and cannot be ruled out are: major shifts of glaciers, leading to a rise in sea level of 20 feet or more in a few centuries; dramatic changes in ocean currents, such as displacement of those in the North Atlantic, that would lead to a major shift in climates of Atlantic coastal communities; large-scale desertification of the grain belts of the world; and the possibility that climate changes will upset the delicate balance of bugs, viruses, and humans as the tropical climates that are so hospitable to spawning and spreading new diseases move poleward. No one has demonstrated that these impacts *will* occur. Rather, it seems likely that unexpected and unwelcome phenomena, like the Antarctic ozone hole, will occur more frequently under conditions of more rapid climatic change.

Learning. The threat of an unforeseen calamity argues for more aggressive action than a plain-vanilla cost-benefit analysis would suggest. However, the possibility of resolving uncertainties about climate change argues for postponing action until our knowledge is more secure. Most scientists believe that research can improve our understanding about the timing, extent, and impact of climate change. Improved understanding could sharpen our calculations about appropriate policies. The best investment today may be in *learning* about climate change rather than in *preventing* it.

Putting this proposition concretely, we could easily make serious mistakes in attempting to prevent climate change. Imagine that a massive nuclear-power program had been mandated 20 years ago only to find that the technology was expensive and unacceptable. Learning to cope with the threat of climate change includes not only improving our estimates of the consequences of climate change but performing research and development on inexpensive and reliable ways of slowing climate change.

Policies to Slow Greenhouse Warming

I conclude by suggesting the direction that policy should follow in slowing greenhouse warming and by comparing the idealized approach with current policy in the United States and other countries.

A Framework for Policy. In designing policies to slow global warming, we must first take into account that this is a *global* issue. Efficient policies will involve steps by all countries to restrict GHG emissions. In order to induce international cooperation, the United States and other rich nations may need to subsidize actions by poor nations (say to slow deforestation or to phase out CFC use). While unilateral action may be better than nothing, concerted action is better still.

Given the identified costs of global warming, the world would be well advised to take three modest steps to slow global warming while avoiding precipitous and ill-designed actions that it may later regret. A first set of measures should improve our understanding of greenhouse warming. Such steps would include augmented monitoring of the global environment; analyses of past climatic records, as well as intensive analysis of the environmental and economic impacts of climate change, past and future; and analyses of potential steps to slow climate change. Understanding of climate change has improved enormously over the last two decades, and further research will help to sharpen our pencils for the tough decisions to be made in the future.

Second, countries should support research and development on new technologies that will slow climate change—particularly on energy technologies that have low GHG emissions per unit of output. Too little is invested in these technologies because of a "double externality"; private returns are less than social returns both because the fruits of R&D are available to those who spent nothing on research and because the benefits of GHG reductions are currently worth nothing in the marketplace.

Energy technologies that replace fossil-fuel use require greater government support than they currently receive. Inherently safe nuclear power, solar energy, as well as energy conservation, are particularly promising targets for government R&D support.

A third and more involved policy is to identify and accelerate the myriad otherwise-sensible measures that would tend to slow global warming. Many steps could contribute to slowing global warming at little or no economic cost. These steps include efforts to strengthen international agreements that severely restrict CFCs, moves to slow or curb uneconomical deforestation, and steps to slow the growth of uneconomical use of fossil fuels, say through higher taxes on gasoline, on hydrocarbons, or on all fossil fuels.

Given the world's agenda of unsolved problems, these three steps should suffice for today. For those concerned with pressing further in reducing long-term risks—or should new evidence emerge to indicate that the risks are greater than is now believed—one more step might be worth taking.

Fourth, a set of global environmental taxes or their equivalent could be imposed on the CO_2-equivalent emissions of greenhouse gases, particularly on CO_2 emissions from the combustion of fossil fuels. The analysis in this study suggests that a GHG tax in the order of $5 per ton of CO_2 equivalent would be a reasonable response to the future costs of climate change. A carbon tax would be preferable to regulatory interventions because taxes provide incentives to minimize the costs of attaining a given level of GHG reduction while regulations often do not. To reap the maximum advantage from a carbon tax, an international agreement should be concluded so that the tax or restraint applies in all major countries.

Some would argue that carbon taxes fall in category three above as sensible economic policy. Consumption of fossil fuels has many negative spillovers besides the greenhouse effect, such as local pollution, traffic congestion, wear and tear on roads, accidents, and so forth. In addition to slowing global warming, carbon taxes would restrain consumption of fossil fuels, encourage R&D on nonfossil fuels, favor fuel switching to low-GHG fuels like methane, lower oil imports, reduce the trade and budget deficits, and raise the national saving rate. Indeed, in the tax kingdom, carbon taxes are the rara avis that increases rather than reduces economic efficiency.

While these arguments for a carbon tax are persuasive, I do not recommend it today. Negotiating a global carbon tax would be a daunting task even for a President who likes taxes and has no other problems to handle. Reducing the risks of climate change is a worthwhile objective, but humanity faces many other risks and worthy goals. This point can be seen in terms of the demand for investment. World saving today is distributed among a long list of uses—factories and equipment, training and education, health and hospitals, transportation and communications, research and development, housing and environmental protection, population control and curing drug dependency. Reducing GHG emissions is yet another investment, an investment in preventing the damages from climate change. Given the agenda of urgent needs, it is difficult to see how we can justify a much larger share of investment devoted to this area than the modest proposal laid out above.

Current Policy Initiatives. How does this framework for policy compare with the approach taken by the Bush administration? Political concerns about greenhouse warming burst on the scene in part because an unusual heat wave in the summer of 1988 coincided with a growing scientific consensus on the gravity of greenhouse warming. Candidate Bush promised to move ahead vigorously on environmental issues and endorsed action to slow global warming. Other heads of government joined with President Bush at the Paris Summit in July 1989 in recognizing the need for international cooperation to solve global environmental problems.

The administration's policy on global warming was finally unveiled in February 1990.[23] Current policies and proposals can be divided into our four categories as follows:

1. In the category of expanding knowledge, the administration proposes a major expansion in its funding for "global climate change." These programs are largely scientific and, at $1.03 billion for FY 1991, represent a 57 percent increase in outlays over the prior year.
2. Federal support for energy R&D is designated as an important part of the solution. Table 7 shows a breakdown of the administration's energy R&D budget for fiscal year 1991. Although the total budget of the Department of Energy is designated for a modest increase, civilian R&D programs will decline sharply, with a major cut in fossil energy programs and little change in low-GHG technologies after correcting for inflation. The reduction in conservation is puzzling given the importance of increased energy efficiency.
3. The US government has taken the lead in the most important single initiative on global warming—policies to phase out the most powerful CFCs. The relevant agreement is contained in the Montreal Protocol of 1987, which commits signatories to cut production in half by 1998. (Note that the phaseout of CFCs was undertaken in order to prevent ozone depletion; the impact upon global warming is a welcome but serendipitous side effect.) Even with the Montreal Protocol in place, however, the climatic impact of CFC emissions is projected to rise over the next century.
4. The Bush administration does not propose to go much beyond these three categories. It proposes a "plant-a-tree" program, with proposed outlays of $175 million annually, to plant a billion trees each year. I estimate this program will reduce global emissions of GHGs by about 0.01 percent.

On the whole, policy to date represents a reasonable if cautious response to the threat of global warming. The Bush administration should be commended for following a reasoned approach and avoiding measures that would lock the economy onto a path that could not respond flexibly to new information or emerging technologies—if only all policies were so well informed!

Final Thoughts

In conclusion, the United States and other major countries would be well served by continuing to take steps in the three areas outlined above—improving knowledge, investing in R&D in new technologies, and tilting away from greenhouse gases. Pursuing this approach, we will be prepared for whatever developments unfold in the future—for a tightening of the screws if the threat of global warming accelerates or for a relaxation of policy if science or technology were to alleviate our concerns.

However, behind all these suggestions lies one major piece of advice: The threat of climate change is uncertain. It may be large, and might conceivably be devastating. But we face many threats. And don't forget

Table 7. Energy research and development funds in FY 1991 budget, Budget Authority.

	In millions		
	FY 1990	FY 1991	Percent Change
Total DOE Budget	$16,423	$17,480	6.4
Total, Civilian Applied Research and Development	1,618	1,382	−14.6
R&D on Technologies with low GHG Emissions:	735	762	3.6
Conservation	194	183	−5.9
Solar, Renewables, Geothermal, Other	138	175	27.3
Nuclear (reactor programs only)	83	79	−5.5
Magnetic Fusion	320	325	1.6
R&D on Technologies with high GHG Emissions:	883	621	−29.7
Coal	829	566	−31.7
"Clean Coal"	554	456	−17.7
Other coal	275	110	−59.9
Oil & Gas	54	54	0.6
Basic Research in the DOE	2,079	2,355	13.3
Basic Physics	872	952	9.2
Superconducting Super Collider	219	318	45.2
Biological, Environmental, etc.	418	436	4.4
Basic Energy Science	570	649	13.8
National Science Foundation			
Total budget	1,651	1,853	12.2

Source: US Department of Energy Posture Statement and Fiscal Year 1991 Budget Overview, January 1990 and Budget of the United States Government, Fiscal Year 1991.

that humans have the capacity to do great harm through ill-designed schemes, as the socialist experiment of the last four decades clearly shows. Gather information, move cautiously, use efficient methods to cope with climate change, and fashion policies flexibly so that you can respond quickly to new information.

Notes

1. A short history of scientific concerns about the greenhouse effect is provided in Ausubel [1983].
2. The traditional estimate of the relative importance of different greenhouse gases uses the *instantaneous* contribution of a gas to global warming (in °C). The traditional estimate has the defect that GHGs differ in their lifetimes and chemical transformations. In order to calculate the *total* contribution of each GHG, I have estimated the sum of the instantaneous contributions over the indefinite future (in °C-years).
3. National Research Council [1983], p. 28.
4. One eminent climatologist stated that he had "99 percent" confidence that the warming of the 1980s was associated with the greenhouse effect (see Hansen [1988]). By contrast, four other respected scientists wrote that "no conclusion about the magnitude of the greenhouse effect in the next century can be drawn from the 0.5° C warming that has occurred in the last 100 years" (Marshall Institute [1989], p. 8).
5. A description of the procedure for calculating the uncertainties is outlined in Nordhaus [1990b].
6. The most careful studies of the impact of greenhouse warming have been conducted for the United States, and this review will therefore concentrate here. The most comprehensive is a recent study by the US EPA [1989a]. Although the studies reviewed here use different assumptions, we should envisage the estimated impacts as occurring late in the second half of the next century.
7. "National income" is total national output measured at factor costs. It equals GNP less indirect business taxes and depreciation.
8. NRC [1983], p. 45.
9. EPA [1989a].
10. EPA [1989a].
11. Binkley [1998].
12. See National Research Council [1978].
13. I reiterate that the studies reviewed here represent "best-guess" scenarios of climate change. They omit uncertainties and possible non-linearities, a topic that we examine in the last section.
14. See Coolfont Workshop [1989].
15. These studies include estimates from specific technologies (such as CO_2 scrubbing and substitution of methane for oil and coal); econometric or elasticity studies (often using highly aggregated models); and mathematical programming or optimization approaches (which often use activity-analysis specifications of energy technologies). For all three of these approaches, one can estimate the cost of reducing CO_2 emissions as a function of the *penalty* or *tax* imposed upon those who emit CO_2.

16. Note that these estimates are of the long-run cost—that is, the cost after the capital stock has fully adjusted. Attempts to reduce CO_2 emissions in the short run would be much more expensive. Also, these costs do not include any adjustments for unmeasured or external environmental, health, or economic effects.
17. See Ravelle and Waggoner [1983].
18. Many of the issues in this section are developed at length in a superb essay by Schelling [1983].
19. The analysis draws upon Nordhaus [1990]. More precisely, it assumes that the discount rate on goods and services exceeds the growth rate of the economy by 1 percent per year. If the damage from a doubling of CO_2 is 0.25 percent of total output, then the efficient CO_2 tax is $3.2 per ton CO_2 equivalent; if the damage is 1 percent of output, the efficient tax is $12.7 per ton. We choose $5 as an illustrative intermediate figure.
20. See Litan and Nordhaus [1983], Chapter 2.
21. *Id.*
22. For a nontechnical and very imaginative treatment of uncertainties, see Schneider [1989a].
23. The major political statement is found in President Bush's speech to the IPCC (the Intergovernmental Panel on Climate Change) at Georgetown University on February 5, 1990. The 1990 *Economic Report of the President* contains an extensive discussion of the scientific, economic, and policy issues. The details of the budget changes are contained in the US Budget for FY 1991.

Bibliography

Ausubel, Jesse H. [1983]. "Historical Note," in National Research Council [1983], pp. 488–491.

Ausubel, Jesse H. and Asit Biswas, editors [1980]. *Climatic Constraints and Human Activities*, Pergamon Press, Oxford, 1980.

Ausubel, Jesse H. and William D. Nordhaus [1983]. "A Review of Estimates of Future Carbon Dioxide Emissions," in National Research Council [1983], pp. 153–185.

Binkley, Clark S. [1988]. "A Case Study of the Effects of CO_2-Induced Climatic Warming on Forest Growth and the Forest Sector: B. Economic Effects on the World's Forest Sector," in M. L. Parry, T. R. Carter, and N. T. Konijn, eds., *The Impacts of Climatic Variations on Agriculture*, Dordrecht, Netherlands, Kluwer Academic Publishers, 1988, pp. 197–218.

Coolfont Workshop [1989]. *Climate Impact Response Functions: Report of a Workshop Held at Coolfont, West Virginia*, September 11–14, 1989, National Climate Program Office, Washington, DC.

Edmunds, J. A. and J. M. Reilly [1983]. "Global Energy and CO_2 to the Year 2050," *The Energy Journal*, vol. 4, pp. 21–47.

EPA [1989a]. US Environmental Protection Agency, *The Potential Effects of Global Climate Change on the United States: Report to Congress*, EPA-230-05-89-050, December 1989.

EPA [1989]. US Environmental Protection Agency, *Policy Options for Stabilizing Global Climate*, Draft Report to Congress, February 1989.

Hansen, James [1988]. Testimony before the Senate Energy Committee, June 23, 1988.

Hansen, James and Sergej Lebedeff [1987]. "Global Trends of Measured Surface Air Temperature," *Journal of Geophysical Research*, vol. 92, No. D11, November 20, 1987, pp. 13, 345-13, 372.

Kates, Robert W., Jesse H. Ausubel, and Mimi Berberian [1985]. *Climate Impact Assessment: Studies of the Interaction of Climate and Society (SCOPE 27)*, Wiley, New York, 1985.

Litan, Robert and William D. Nordhaus [1983]. *Reforming Federal Regulation*, Chapter 2, Yale University Press, New Haven, CT.

Marshall Institute [1989]. Frederick Seitz, Karl Bendelsen, Robert Jastrow, and William A. Nierenberg, *Scientific Perspectives on the Greenhouse Problem: Executive Summary*, George C. Marshall Institute, Washington, DC, 1989.

McElroy, Michael B. and Ross J. Salawitch [1989]. "Changing Global Composition of the Global Stratosphere," *Science*, vol. 243, February 10, 1989, pp. 763–71.

National Research Council (NRC) [1978]. *International Perspectives on the Study of Climate and Society*, National Academy Press, Washington, DC, 1978.

National Research Council (NRC) [1983]. *Changing Climate*, National Academy Press, Washington, DC, 1983.

National Research Council (NRC) [1987]. *Responding to Changes in Sea Level: Engineering Implications*, National Academy Press, Washington, DC, 1987.

Nordhaus, William D. [1979]. *The Efficient Use of Energy Resources*, Yale University Press, New Haven, CT, 1979.

——— [1990], "To Slow or Not to Slow: The Economics of the Greenhouse Effect," paper presented to the annual meetings of the American Association for the Advancement of Science, New Orleans, February 1990.

——— [1990a]. "Contribution of Different Greenhouse Gases to Global Warming: A New Technique for Measuring Impact," processed, February, 1990.

——— [1990b]. "Uncertainty about Future Climate Change: Estimates of Probably Likely Paths," processed, January, 1990.

Nordhaus, William D. and Jesse Ausubel [1983]. "A Review of Estimates of Future Carbon Dioxide Emissions," in National Research Council-National Academy of Sciences, *Changing Climate*, National Academy Press, 1983.

Nordhaus, William D. and Gary Yohe [1983a]. "Future Carbon Dioxide Emissions from Fossil Fuels," in National Research Council-National Academy of Sciences, *Changing Climate*, National Academy Press, 1983.

Raiffa, H. [1968]. *Decision Analysis: Introductory Lectures on Choices under Uncertainty*, Reading, MA, Addison-Wesley, 1968.

Ramanathan, V., R. J. Cicerone, H. B. Singh, and J. T. Kiehl [1985]. "Trace Gas Trends and Their Potential Role in Climate Change," *Journal of Geophysical Research*, vol. 90, pp. 5547–5566.

Ramanathan, V., et al. [1987]. "Climate-Chemical Interactions and Effects of Changing Atmospheric Trace Gases," *Review of Geophysics*, vol. 25, 1987, pp. 1441–1482.

Ravelle, Roger R. and Paul E. Waggoner [1983]. "Effects of a Carbon Dioxide-Induced Climatic Change on Water Supplied in the Western United States," in National Research Council [1983], pp. 419–32.

Schelling, Thomas C. [1983]. "Climatic Change: Implications for Welfare and Policy," in National Research Council [1983], pp. 449–82.

Schneider, Stephen H. [1989]. "The Greenhouse Effect: Science and Policy," *Science*, vol. 243, February 10, 1989, pp. 771–81.

——— [1989a]. *Global Warming: Are We Entering the Greenhouse Century?*, Sierra Club Books, San Francisco, CA.

Kerr, Richard A. [1989]. "Bringing Down the Sea Level Rise," *Science*, vol. 246, December 22, 1989, p. 1563.

Stouffer, R. J., S. Manabe, and K. Bryan, "Interhemispheric Asymmetry in Climate Response to a Gradual Increase of Atmospheric CO_2," *Nature*, vol. 342, December 7, 1989, pp. 660–662.

Wagonner, Paul E., ed. [1989]. *Climate Change and U.S. Water Resources*, Wiley, New York, 1989, in press.

Wuebbles, Donald J. and Jae Edmonds [1988]. *A Primer on Greenhouse Gases*, prepared for the Department of Energy, DOE/NBB-0083, March 1988.

Energy Technology: Problems and Solutions

Paul E. Gray
Jefferson W. Tester
David O. Wood
Massachusetts Institute of Technology
Cambridge, MA

Introduction

The course of the industrial revolution has been greatly influenced by the availability of energy to support the development and practical application of scientific advances and inventions. Increasingly flexible energy forms—most importantly, electricity—have both stimulated and harnessed the technological fruits of invention and have produced enormous increases in world economic productivity.

Over most of this period, however, the environmental costs of energy production and use have been of relatively little interest to the world's citizens; but that is changing. Since the 1950s, concerns about environmental costs, especially in developed countries, have been manifested in a broad range of laws and regulations designed to control and limit environmental pollution and damage. Developing countries have shared in the general concern, but have been unable to divert economic resources from increasing economic productivity and growth to the task of mitigating the environmental consequences of that development. Until recently, most environmental policy initiatives in developed countries have focused on intranational issues; that is also changing with increasing concern about the global consequences of energy production and use, most importantly ozone depletion and the potential of climate change.

While energy and environmental policy were closely linked in the 1950s and 1960s—e.g., in the development of US new source performance standards for stationary combustion—the OPEC oil embargo of the 1970s diverted policy attention to energy security. The falling energy prices of the 1980s, however, coupled with the dramatic increase in the flexibility of world energy supplies, the coincident emergence of the ozone depletion issue, and the potential for global climate change, have dramatically refocused public attention on the inextricable interaction between energy and environmental systems. Characterizing energy and environmental system linkages is now widely appreciated as a key ingredient in understanding and designing policies to internalize the environmental costs of energy production and use at the local, regional, national, international, and global levels.

Central to the design of effective energy/environmental policies is the recognition that these policies will shape, and be shaped by, the near- and long-term technological opportunities to influence energy production and use consistent with environmental goals. Almost without exception, every perspective on addressing the environmental consequences of energy production and use involves the re-engineering of existing technologies and/or the development of new technologies to accomplish environmental goals, consistent with world objectives of social, economic, and environmental well-being.

At the grass roots level, individuals are asking the question, "What can I do to relieve stress on the environment and reduce pollution and other forms of environmental degradation?" Some responses seem self-evident: turn the thermostat down (or up); recycle; change to more efficient light bulbs and appliances; buy smaller, more fuel-efficient cars; car-pool or use mass transportation. Other responses are more difficult to assess: lobby for increased use of nuclear power to reduce greenhouse gas emissions, or lobby against nuclear power to reduce risks of potential nuclear accidents and the need for radioactive waste disposal; promote legislation to increase efficiency standards for energy-using industrial and consumer equipment, or support fuel and/or emissions taxes to provide economic incentives to reduce energy use and the generation of environmentally damaging emissions; support international protocols setting targets for the reduction of environmentally damaging emissions, or promote a more integrated approach to global economic, social, and environmental development.

The development of responses to such questions involves the consideration of human and social values and attitudes, and ethical and moral issues, as well as technical and economic factors. Satisfactory answers—responsible energy futures for both the nation and the world—must account for and respect both technical and social considerations. We want to emphasize that constructive solutions to the world's energy and environmental problems depend critically on the talents and contributions of future generations of engineers and scientists. Therefore, the education and training of these future generations are central to achieving environmentally and economically sustainable energy systems.

Our main purpose in this paper is to reflect on technology that is likely to affect the future development of environmentally acceptable energy resources. While we make no claim to completeness in coverage, it is our intention to be reasonably broad in this overview.

Scope of the Problem

For this discussion, we have included two complementary approaches to reducing environmental stress: conservation and energy efficiency improvements to lower demand and new, more environmentally benign energy technologies to replace our existing capital stock of energy supply and service equipment.

We begin by reviewing the evolutional growth of energy use worldwide to establish a basis for change. Then we frame the major issues and constraints and establish criteria for evaluating conservation and new energy technologies. Finally, we suggest how energy technology could evolve to reduce or limit the negative environmental consequences of providing the energy services needed to sustain and improve the quality of human life on earth.

During the 20th century, worldwide annual consumption of energy increased 15-fold from about 20 quads in 1900 (1 quad = 10^{15} BTU ≃ 10^{18} joules = 36 × 10^6 tons of coal = 180 × 10^6 bbl of oil) to over 300 quads in 1990 (see Figure 1a). Part of this essentially exponential growth was driven by the increase in the world's population from 1.5 to 5.3 billion, or about 3.5-fold during this period. On average, the annual per capita energy consumption worldwide has increased more than 4-fold from about 13 to about 60 quads per billion people (or the equivalent from about 480 to about 2200 kg coal consumed per person per year).

By plotting the total and per capita consumption and population data on semi-logarithmic coordinates as shown in Figure 1b, similar, generally exponential trends appear with obvious perturbations due to world wars and other major political-economic events such as the Middle East oil embargo in 1973. The quasi-exponential growth of energy consumption can be attributed to man's ingenuity and the subsequent development of technology to implement new ideas and improve our standard of living. Energy is required in the manufacture of the resulting new products and to provide power for operating many of them.

The amount of energy consumed in the developed countries of the world underscores the effects of technology implementation. For example, the US is by far the highest per capita consumer of energy in the world today at an equivalent 320 quads per billion (80 quads for 250 million people), almost six times the current world's average. Part of this is attributable to the large geographic size of the US and to its climate, but much

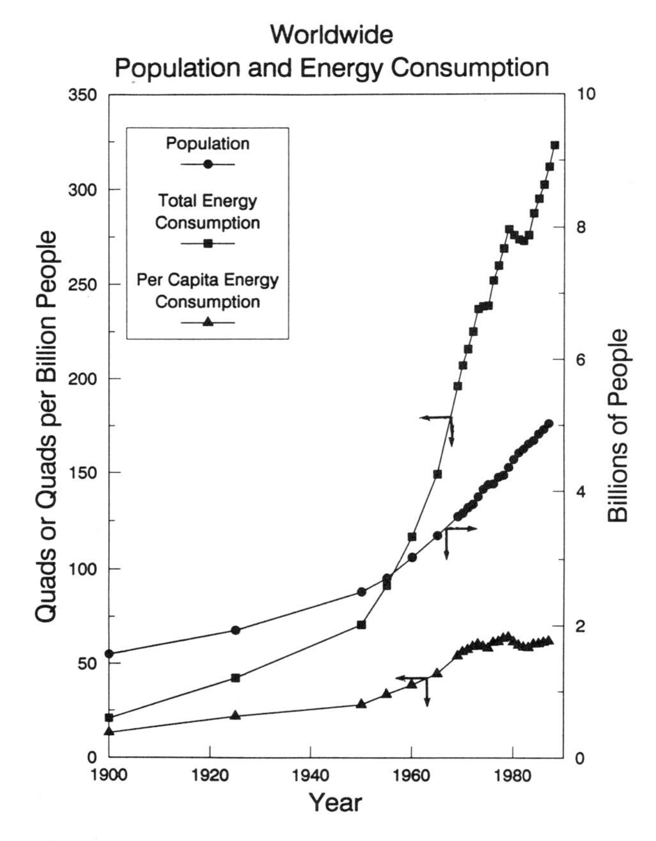

Figure 1a (linear coordinates). Historical data and projections of annual worldwide primary energy consumption (total and per capita) and world population. Sources: World Energy Conference (1974, 1987), *UN Yearbook of World Energy Statistics* (1980, 1982), *World Almanac* (1990), *BP Statistical Review of World Energy* (July 1989).

of it correlates directly with our standard of living and the development of technology, and the fact that energy as a commodity is relatively cheap. In Figures 2a and 2b, current annual per capita energy consumption levels are plotted for several countries. Most notable of course is the vast difference between the developed and developing countries. At first glance, these comparisons suggest that the US has a long way to go in the conservation area, but this is a complex matter to which we will return later.

Recognizing the quantitative dimensions of the energy consumption situation is key to understanding the problem. With changing world demographics and the less developed countries increasingly dominating our total population and struggling to improve their standards of living, total energy consumption may very well increase as fast or faster than the population, even in the face of massive conservation efforts.

For the first 73 years of this century, with few ex-

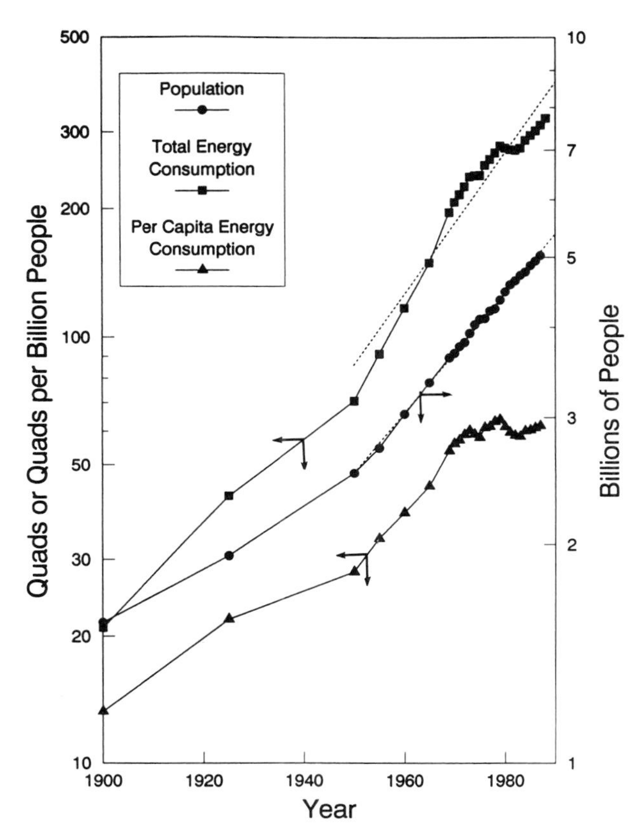

Figure 1b (semilogarithmic coordinates). Historical data and projections of annual worldwide primary energy consumption (total and per capita) and world population. Sources: World Energy Conference (1974, 1987), *UN Yearbook of World Energy Statistics* (1980, 1982), *World Almanac* (1990), *BP Statistical Review of World Energy* (July 1989).

ceptions, the energy intensity of manufacturing a new product was not a concern. Exceptions include petrochemicals and primary metals production. For other industries, fuel requirements or costs for supplying electricity or heat were of much less importance than the capital costs for plant construction or the raw material costs required to manufacture the product. Consequently, not much attention was given to making processes more energy efficient. With the OPEC oil embargo in 1973, world oil prices increased dramatically, forcing new interest in energy efficiency and conservation. Much of this enthusiasm dwindled in the mid to late 1980s as world oil prices fell, creating the impression among many people that energy efficiency was again of lesser importance. For example, US energy consumption in the last two years has shown a 4 percent increase and is now slightly over 80 quads per year.

The recent concerns over every aspect of our environment—locally, with tropospheric ozone, chemical

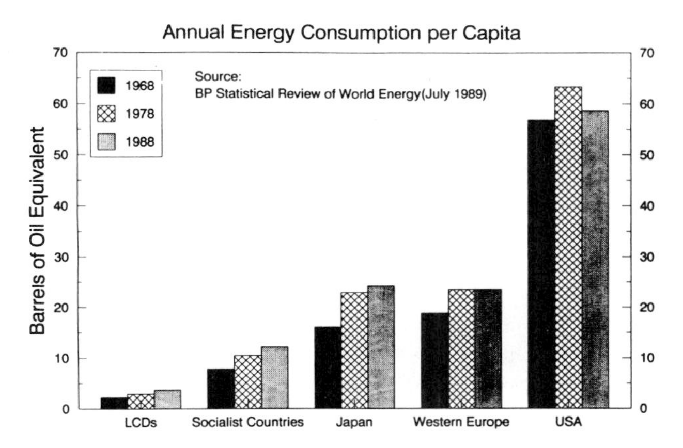

Figure 2a. Annual per capita energy consumption by country for 1968, 1978, and 1988. Source: *BP Statistical Review of World Energy* (July 1989).

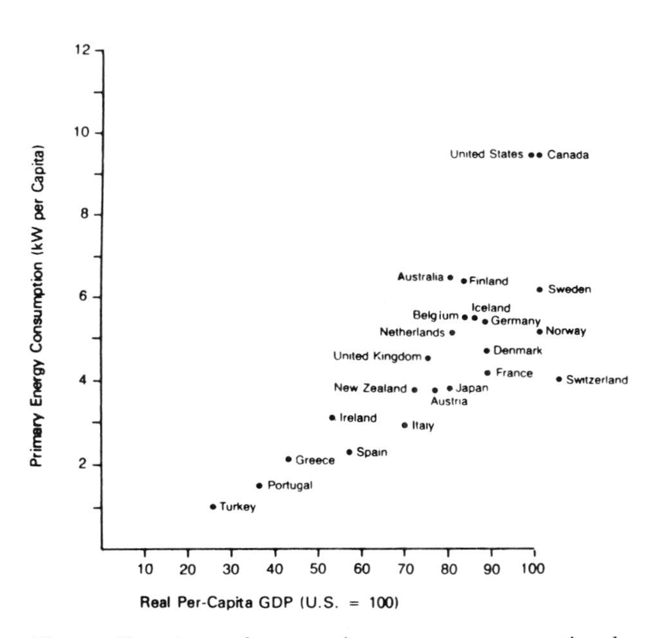

Figure 2b. Annual per capita energy consumption by country in 1982 versus per capita GDP (US=100). Source: Goldemberg et al. (1988).

releases, ground water pollutants, and particulates; regionally, with acid rain; and globally, with pollution of the oceans, buildup of greenhouse gases, and stratospheric ozone depletion by CFCs—have restimulated public interest in energy supply and end use.

Many have looked to the science and engineering community for technical solutions to these important local, regional, and global environmental problems—for "a technical fix," if you will. Others seek social solutions as a "moral fix" and insist on greater frugality with energy, increased conservation, and better stewardship for the earth. In this arena, as in so many areas of human endeavor, satisfactory solutions will require both sound technology and understanding of the social and human contexts.

Most of the environmental problems that we face today are fundamentally linked to worldwide energy supply and use, particularly our strong reliance on fossil fuels and combustion processes as the primary energy source. Figures 3a and 3b show the distribution of world energy demand in 1987 as reported by White and Golomb (1989) from data provided by the World Energy Conference and by a British Petroleum statistical survey. Note that out of a total of 333 quads consumed, 293 quads, or almost 90 percent, is provided by fossil fuel combustion.

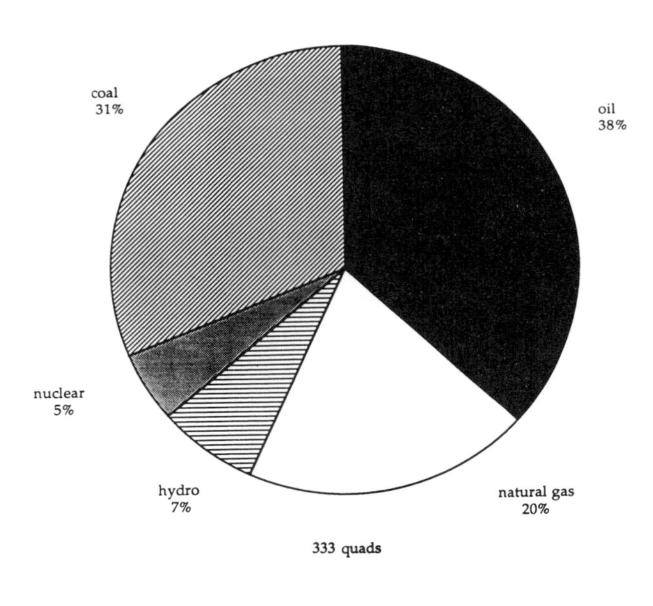

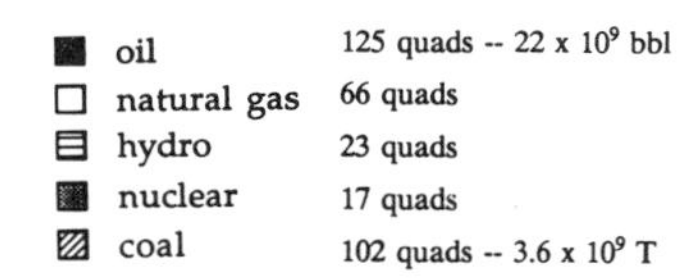

Figure 3a (overall). Distribution of world energy demand. Source: White and Golomb (1989) and *BP Statistical Review of World Energy* (July 1988).

The next critical factor that will affect what solutions are viable relates to how this energy is used. Figure 4 displays the US consumption of energy in four primary end use sectors:

- Transportation (air, land, water)
- Industrial Processes (primary materials and chemicals manufacture)
- Building Systems (residential and commercial space conditioning)
- Electric Power Systems (energy supplied by fossil, fission, and renewable resources)

Roughly speaking, each sector accounts for about one fourth of the total energy consumed. Each one has a specific set of constraints that limits the type and magnitude of technical changes or corrections that could reduce environmental effects, either by increased efficiency or by changing the mode of energy supply. Before we can evaluate new options and identify where to focus future energy technology research and development (R&D), it is necessary to cite a few of the likely major constraints and limitations that face us in making improvements in the next 50 to 100 years. These include the following:

- Liquid fuels (gasoline, kerosene, methanol, etc.) will be required for most transportation systems to ensure adequate range and performance. Consequently, fossil fuels will inevitably play a major role as a primary energy source.
- Current public resistance to nuclear fission in the US and several other countries, based primarily on concerns over safety risks and radioactive waste issues, has brought nuclear power developments to a standstill. Fission cannot play a major role in replacing fossil fuels in these countries unless these attitudes change.
- Thermal energy to electricity cycles limited by second-law efficiencies will be the dominant mode of producing electric power.
- Alternatives to fossil and fission energy systems (fusion, solar, geothermal, and other renewables) will require, first, technological breakthroughs of varying degrees, and, second, significant reductions in costs or an infusion of economic incentives to stimulate their substitution as major energy supply systems. In the near term, low market prices for conventional petroleum fuels will further compound these economic requirements. Investment in synthetic fuels and nonfossil alternative energy systems will be sluggish unless government incentives and/or environmental damage costs are factored in.
- Because of load demands and economic factors, new electric power generation plants added in developed and developing countries will have capacities in the 10 to 500 MWe range rather than in the 1000 to 1500

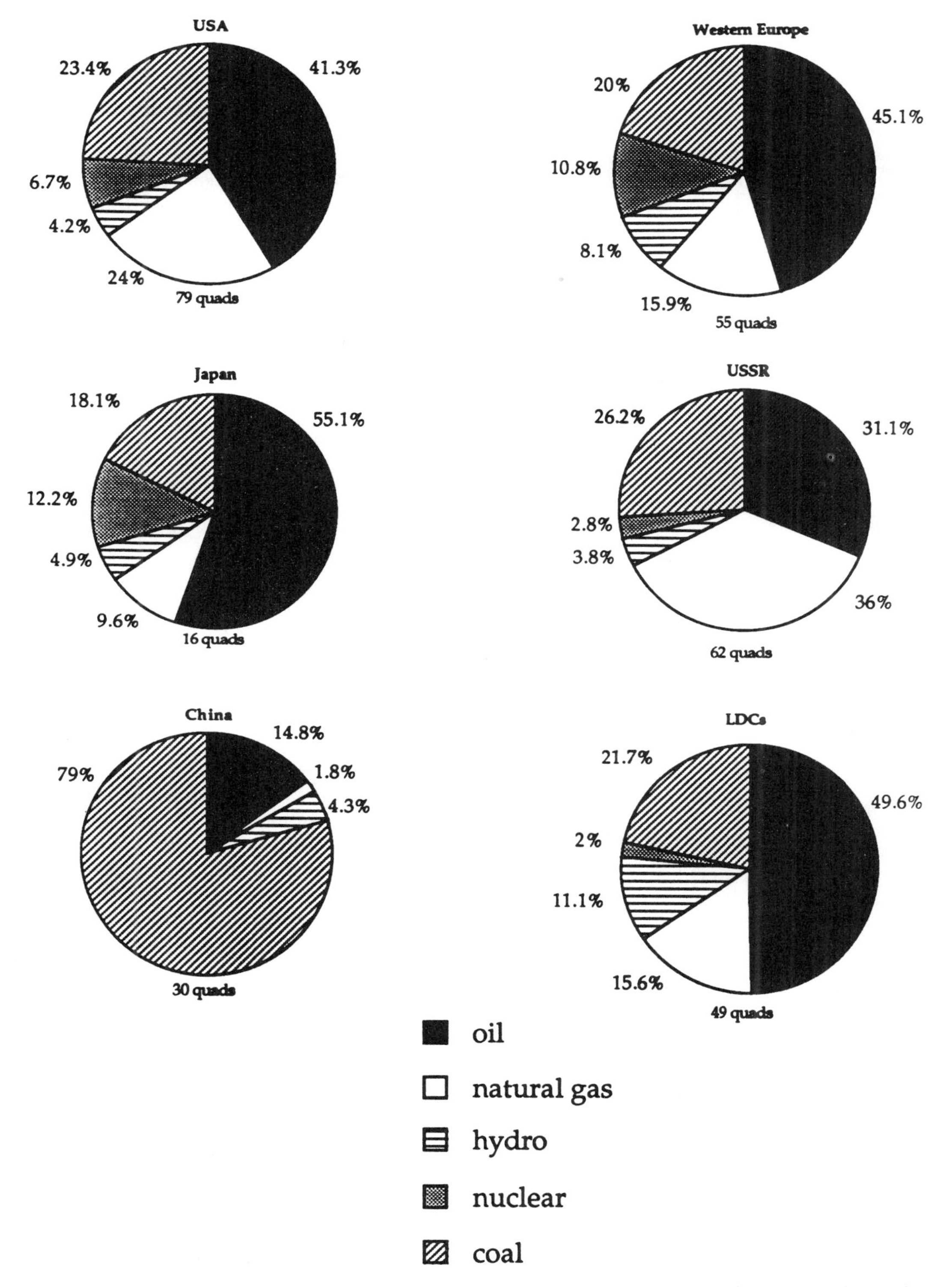

Figure 3b (by region). Distribution of world energy demand. Source: *Review of World Energy* (July 1988).

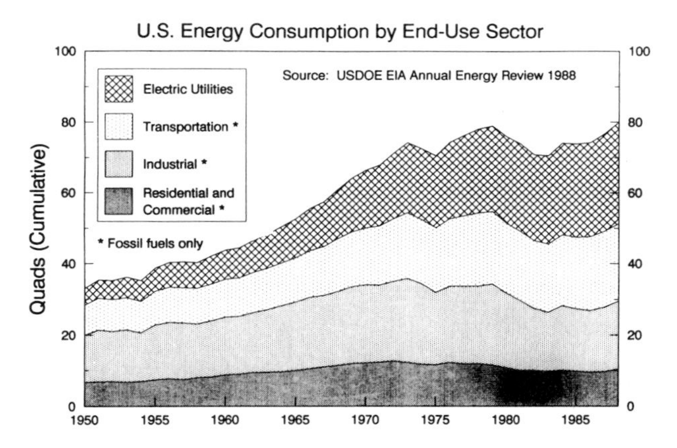

1988 U.S. Energy Consumption by End-Use Sector

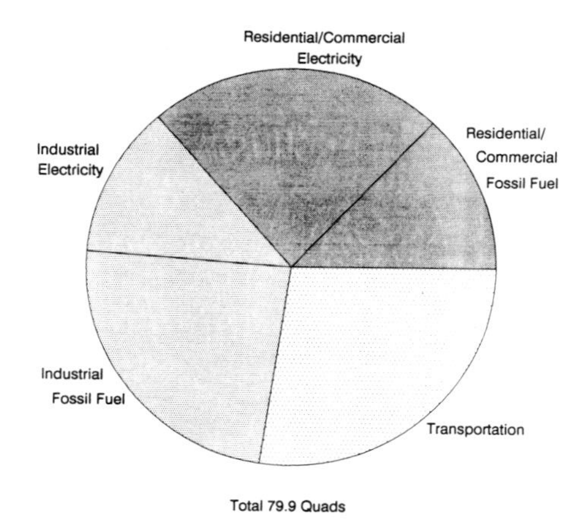

Figure 4. US consumption of energy in the residential/commercial, transportation, and industrial sectors. Source: USDOE (September 1989).

MWe capacity range typical of power plants built in the 1970s and 1980s.

- Materials lifecycle issues such as durability, biodegradability, recyclability, toxicity, and energy intensity in production will strongly influence the survival of new and existing materials in the marketplace.
- Public policy will increasingly attempt to internalize environmental costs of particular energy production and use components into overall capital and operating costs. Policy instruments to accomplish this objective will include both efficiency and emissions standards for energy producing and using technologies, and fiscal measures such as performance and emissions taxes.

Apart from these constraints, we can still pose the question as to what characteristics are desirable for a 21st century energy supply system. At the very least, this would help us construct criteria for comparing one alternative with another as a means of establishing national or international goals and priorities for developing new energy technologies. Desirable characteristics would certainly include:

- A large and globally well-distributed resource.
- Availability on an uninterruptible basis (not subject to changes in weather, etc.).
- Safe and simple designs to build and operate.
- Minimal environmental impact from "cradle to grave," including minimal effluents (CO_2, NO_x, SO_x, particulates, etc.) and minimal wastes during resource recovery, conversion, and production.
- Adaptable for electricity, heating, and cogeneration needs under baseload and peaking conditions.
- Technology easily transferable to other countries without sacrifice of security or economic balance.
- Producible at competitive costs.

In the context of this paper, we discuss a few examples of energy technologies to illustrate how closely they meet these criteria. Clearly, no single new energy source is likely to meet all of them but the degree that one does should influence how vigorously it should be pursued. Ultimately this translates into how much R&D funding it receives.

The world seeks a combination of energy supply and energy service technologies that simultaneously satisfy conditions of energy efficiency, economic acceptability, environmental quality, and social equity and well-being. Scientists and engineers have been enormously productive in inventing and developing technologies for increasing our standard of living and improving economic wealth. Unfortunately, much of our technological productivity has involved stressing or manipulating our environment without sufficient ecological accountability.

For a long time, increasing the height of a smokestack at a large, coal-fired power plant was thought to be sufficient for reducing local environmental problems. In the future, complete removal of SO_x, NO_x, and particulates may be required to meet environmental and health concerns. Increased capital costs for pollution abatement equipment and associated losses in energy efficiency provide incentives for introducing more environmentally efficient alternative technologies to replace our current stock of fossil and nuclear plants as they wear out. Even if demand is flat in a given region, a real opportunity exists to reduce emissions through replacement. In addition, these same new energy technologies could be implemented as demand grows in developing countries of the world.

Conservation and End Use Efficiency Improvement

It is widely held that improving end use efficiency is the best way to mitigate the environmental consequences of energy use. This view is based on analyses of the technical possibilities for improving current efficiency levels, and the economic costs of such improvements compared with the alternative of expanding energy supply. In this section, we will review some of the most promising possibilities, including improvements of current end use technologies as well as those that may be possible via fundamental redesign and reorganization of the technologies and systems providing energy related services. We will attempt to identify the important opportunities for improving efficiencies (see Table 2 below) but cannot provide any detailed or comparative evaluation.

Both the historical and prospective importance of efficiency relative to other factors influencing environmental quality may be illustrated using the following growth accounting relationship whose recent application for analysis of environmental effects has been popularized by Kaya (1990) and Kaya et al. (1989), among others. Suppose we assume that environmental degradation is correlated with some specific emission, say CO_2. These emissions depend upon the mix and characteristics of fossil fuel supplies and conversion technologies, the efficiency of energy use in producing output, the level of output and the size of the population. An identity relating these factors is

$$CO_2 \equiv \frac{CO_2}{\text{Energy}} * \frac{\text{Energy}}{\text{Output}} * \frac{\text{Output}}{\text{Population}} * \text{Population}$$

The first-order approximation to this identity relates the growth rates in these variables. Thus,

$$c = \frac{c}{e} + \frac{e}{o} + \frac{o}{p} + p$$

where all growth rates are in percent per year, and

c = growth rate in CO_2,
e = total fossil energy consumption per year,
o = annual output expressed as GDP,
c/e = growth rate in carbon intensity of energy supply,
e/o = growth rate in energy efficiency,
o/p = growth rate in economic well-being, and
p = growth rate in population.

In Table 1(a), we calibrate the growth rates given above using data for the United States (1950–85). The increasing importance of efficiency improvements in reducing CO_2 emissions growth in the US is very clear. Over the period 1950–85, efficiency improvements (−0.99 percent per year) were almost twice as important as CO_2 intensity of the energy used (−0.53 percent per year) in reducing the growth rate in CO_2 emissions. Further, a very substantial part of that contribution comes in the post-1970 period (−2.05 percent per year), due to the stimulus of higher energy prices on investments in efficiency improvements.

These historical data also illustrate the central importance of growth in economic well-being and population to environmental degradation. For example, note

Table 1. Historical and Prospective Annual Growth Rates for US CO_2 Emissions and Contributing Factors.

Period	c	c/e	e/o	o/p	p
		(a)			
1950–1985	1.61	−0.53	−0.99	1.84	1.29
1950–1970	2.64	−0.64	−0.20	1.99	1.49
1970–1985	0.23	−0.39	−2.05	1.64	1.03
		(b)			
1985–2000	−0.53	−0.28	−3.25	2.26	0.74
2000–2025	−2.16	−1.62	−3.03	2.18	0.32
2025–2050	−1.35	−1.59	−1.76	2.06	−0.06
2050–2075	−0.89	−0.85	−1.55	1.53	−0.03
2075–2100	−1.15	−0.72	−1.43	0.99	0.01

Sources: (a) Historical data based on Jorgenson and Wilcoxen (1990) for CO_2; Alterman (1985, 1988) for fossil fuel consumption; *Economic Report of the President* (1990) for Gross Domestic Product and Population; (b) based on EPA (1989) results for the scenario "rapidly changing world with policy."

that the annual growth rate in CO_2 emissions would have been some 94 percent higher (3.13 percent per year instead of 1.61 percent per year) over the 1950–85 period if there had been no improvements in carbon intensity or efficiency. Clearly, population growth and improved economic welfare have been major contributing factors to growth in CO_2 emissions for the US.

Next, let us consider the prospective contributions of efficiency improvements. We begin by extending the growth accounting analysis of Table 1(a) to interpret the results for one scenario from EPA (1989), the so-called "rapidly changing world with policy" scenario that has so influenced the international discussion of policies for addressing potential global climate change. This scenario includes a broad set of policies to stimulate both end use efficiency and the production and use of nonfossil fuel sources. Note that the EPA projections (Table 1(b)) generally continue the historical trends. For example, efficiency improvements continue to be the most important source of reductions in the growth of CO_2 emissions. The effect of policy is to accelerate both improvements in efficiency and carbon intensity over historical rates, resulting in negative CO_2 emissions growth rates. Finally, economic well-being and population growth continue to be important factors, but with steadily diminishing effects for the US.

We now turn to a more detailed analysis of the sources of efficiency improvements. International comparisons of energy consumption per unit of national output are frequently employed in assessing both historical trends and prospective improvements in energy service efficiency. These performance indices, although seemingly transparent, are inherently complex since they integrate demographic, social, economic, and technological effects.

Returning to our earlier discussion of Figure 2b, we note that while the magnitudes of aggregate primary energy consumption (E) per unit of national output, expressed as gross domestic product (GDP), differ markedly between Canada and the US versus Germany and Japan, all four countries experienced significant declines in their E/GDP ratios for the period 1972–86, as shown in Figure 5. While many factors account for the differences in the E/GDP levels (most importantly energy prices, but also geographic size, population levels and distribution, industrial product mix, living area per family, climate, and cultural factors), it is clear that the same forces operated in all four countries to cause energy service end users to increase their efficiency of consumption, at least through the early 1980s. Key questions for the future are (1) what are the technical possibilities for continuing energy efficiency improvements, and (2) how competitive are these new technologies likely to be?

As a case study, the historical record of efficiency improvements for the United States is worth reviewing. A recent US Department of Energy (DOE) report from the Offices of Policy, Planning, and Analysis and of Conservation and Renewable Energy (September 1989) documents the magnitude of efficiency improvement in the areas of transportation, residential and commercial buildings, and industrial processes. Dramatic reductions in energy use for the US from 1973 to 1986 have led to a nearly constant primary energy consumption level. For example, Figure 6 shows the overall effect of conservation from 1972 to 1986 while Figure 7 gives a breakdown for residential, commercial, industrial, and transportation sectors. The degree of conservation savings for transportation is smaller than the other sectors, reflecting a 45 percent increase in vehicle-miles traveled between 1972 and 1986. Savings categories in the other sectors are indicated in Figure 7 as well, where a number of factors had a significant impact.

One must keep in mind that the USDOE study only considers conservation in a limited time frame following the Middle East oil embargo in 1973. The magnitude of conservation gains are based on a projection of energy use assuming 1972 efficiencies and subsequent energy prices. This treatment overstates the level of conservation improvements since 1973. As can be seen from Figure 5, efficiency gains have been occurring for a long time due to technological improvements, even though

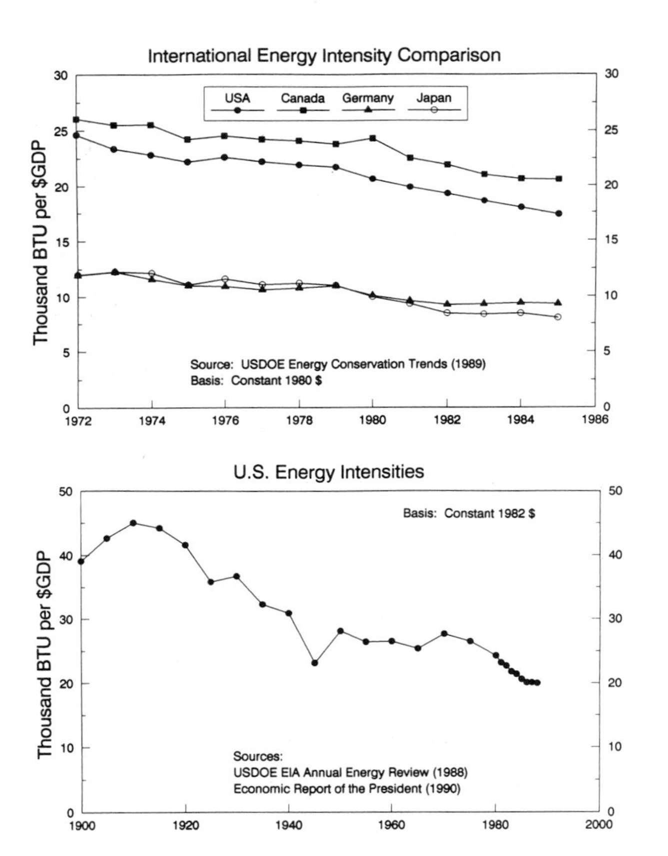

Figure 5. International energy intensity comparison for the US, Japan, Germany, and Canada. Also shown are US energy intensities for the period 1900–90. Sources: *US Energy Data Book* (1984), *Economic Report of the President* (January 1990), USDOE (September 1989).

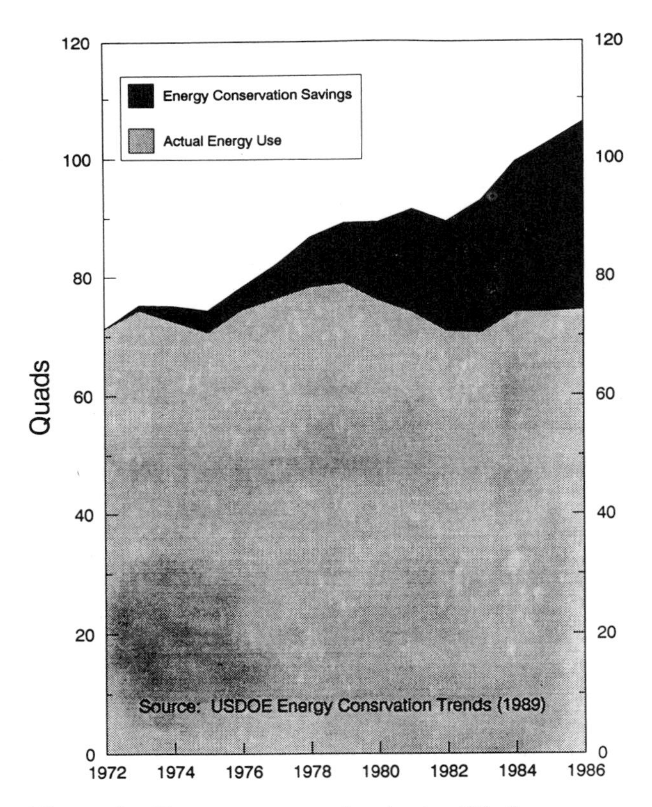

Figure 6. Energy conservation in the US. Source: USDOE (September 1989).

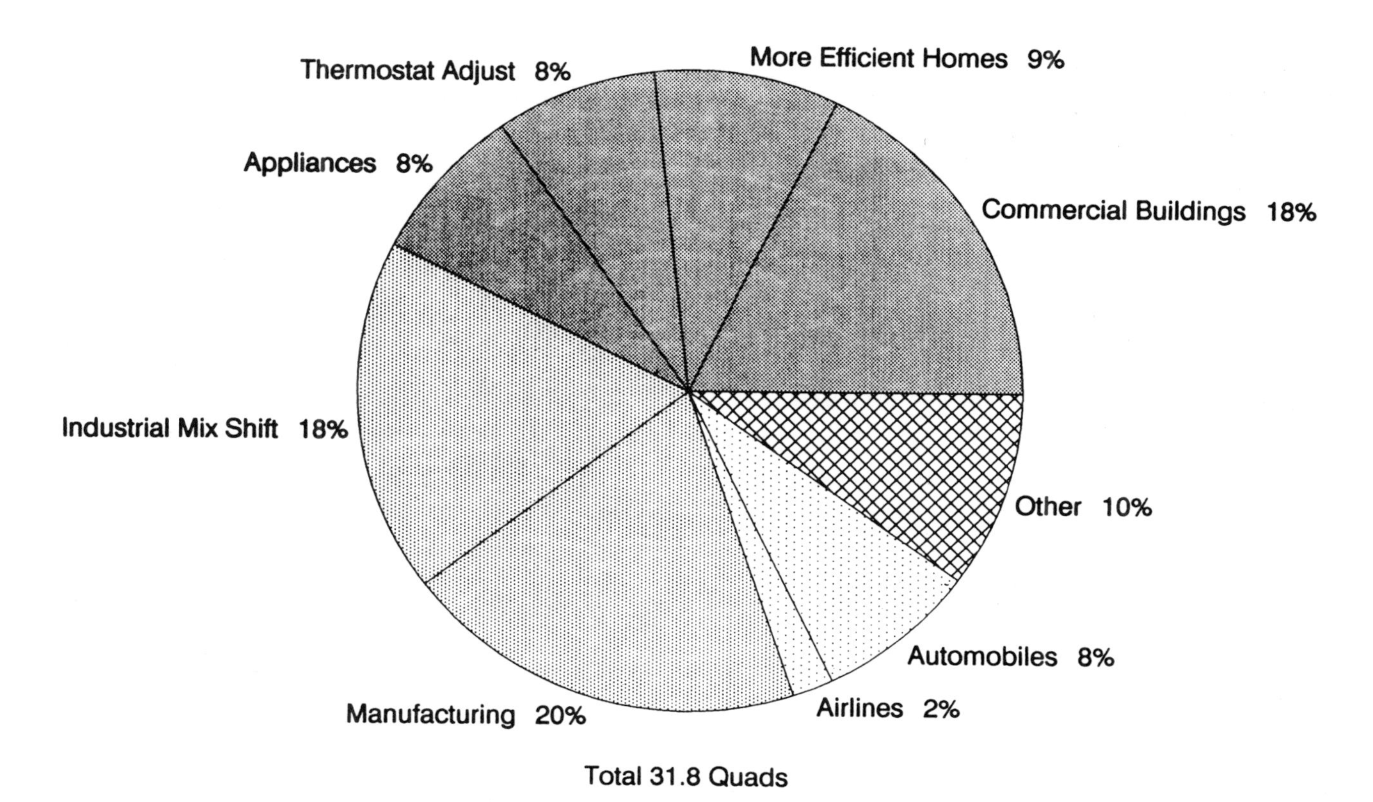

Figure 7. End-use contributions to US energy savings. Source: USDOE (September 1989).

real energy prices in the US have, on average, dropped at least since the 1920s.

While the historical record is impressive, it remains an open question as to the future potential for new conservation and energy efficiency improvements. Many studies find evidence that investments in end use energy efficiency are well below cost effective levels. The recent study by Carlsmith et al. (1990) surveys these studies and provides a new estimate based on the most recent data. Carlsmith and his colleagues employ a family of economic-engineering models to estimate detailed energy consumption for two scenarios for the period 1990–2010. The scenarios are 1) a "where we are headed" scenario reflecting continuance of current trends and policies, e.g., existing appliance and automobile efficiency performance standards, and 2) a "cost effective efficiency" scenario based on life cycle cost minimization. Both scenarios reflect Energy Information Administration (EIA) assumptions about economic growth, population, and energy prices and policies. Carlsmith estimates that the energy savings in 2010 from cost effective efficiency investments amount to about 13.8 quads, or about 15.7 percent of the "where we are headed" estimate of 101.7 quads. Using EIA energy price estimates, this implies a saving of $185 billion (1989 dollars) to energy consumers in 2010.

These results indicate that significant reductions in energy consumption are possible even with current policies. Other studies surveyed by Carlsmith indicate that substantially more savings are technically possible, and would occur with increased energy prices and/or more aggressive performance standards. Of particular interest are Fulkerson et al. (1989), Johansson et al. (1989), and Goldemberg et al. (1988). In the remainder of this section, we focus on several particularly attractive technical possibilities for improving energy efficiency. See Table 2 for a partial list of promising areas of conservation technology.

Two categories of technological developments must be considered in evaluating opportunities to improve end use energy efficiency. These include 1) technologies that actually provide energy services (e.g., automobiles, appliances, buildings, and industrial processes), and 2) technologies that may radically change our economic and social organization (e.g., microelectronics, telecommunications, and advanced transportation systems).

An example of end use service technology improvement would be energy use and conservation as they relate to the production of primary materials. The per capita consumption of primary materials (steel, concrete, paper, aluminum, etc.) and chemicals (petroleum, plastics, etc.) is very important in projecting worldwide energy demand, particularly for the developing countries of the world. Figure 8 shows the growth in demand for

Table 2. Some Promising Areas for Technology to Improve End Use Efficiency and Conservation

End Use Sector	Specific Opportunities
Transportation	advanced high efficiency automobile and aircraft engines continuously variable transmissions levitated high speed trains improved scheduling and control downsizing and weight reductions fuel cell/electric powered cars aerodynamic vehicle design
Building Systems	building envelope insulations window insulation heat pumps advanced space conditioning control systems high efficiency lighting improved refrigeration systems air to air heat exchangers innovative furnace and water heater design improved cooking appliances
Industrial Processes	
General	energy integration (pinch technology) cogeneration re-engineering processes for energy efficiency advanced separation methods variable speed drives/motors advanced/automated process control
Materials Manufacture	improved durability materials substitution • non-portland-based cements • ceramics • superconductors • composites advanced recycling technology for • paper • plastics • aluminum and steel • other primary metals

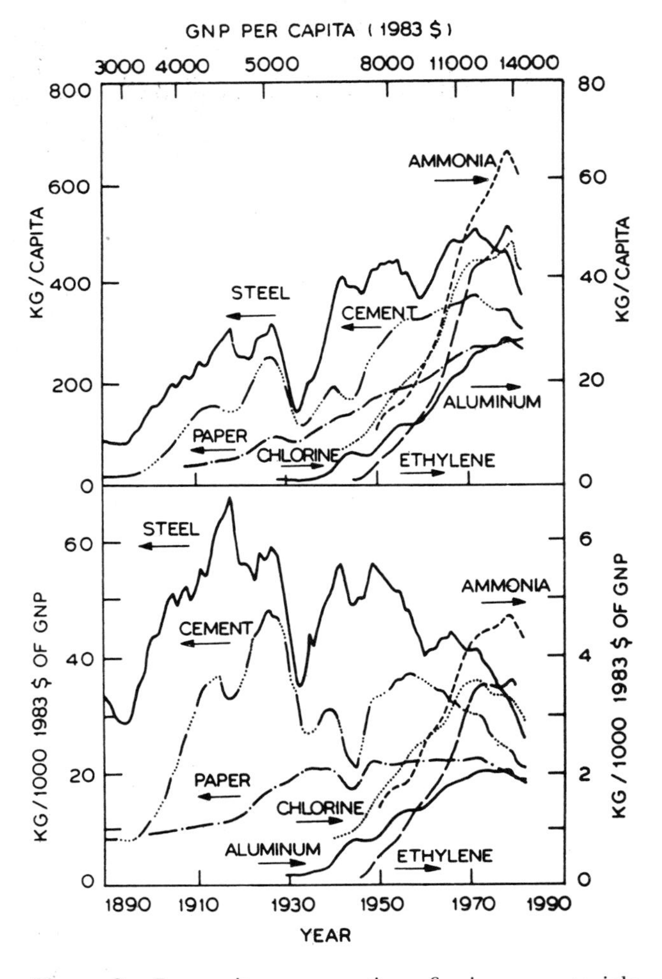

Figure 8. Per capita consumption of primary materials in the US. Source: Goldemberg et al. (1988).

the US. Additional gains in energy efficiency are envisioned by re-engineered processes for manufacturing primary materials and chemicals. Terms such as "heat integration," "waste minimization," and "recyclability" have been added to the vocabularies of our practicing engineers as viable methods for energy intensity reduction. Materials technology, in general, can contribute significantly to providing new products that have better life cycle costs. Improved durability reduces energy intensity and environmental impact related to the manufacture as well as the disposal of the product. Recycling of ferrous metal, lead, and aluminum products has been remarkably successful and there are signs that recycling of plastics such as polystyrene and polyethylene will have similar success.

The ultimate role of composites and ceramics is still unclear, particularly in regard to replacement of commodity materials such as wood, concrete, steel, and aluminum. Current production processes for many of these advanced materials are very energy intensive and the tonnage produced is relatively small. Likewise, a growth in the application of high-temperature superconducting materials may result in advanced high efficiency mass transportation systems such as transcontinental, high-speed, magnetically levitated and propelled trains. Nonetheless, the energy requirements to produce the materials and to construct these systems may significantly offset their efficiency gains. Similar concerns have been raised about the energy requirements for widespread development of silicon-based photovoltaic systems. Again, the whole energy cycle needs to be examined.

Perhaps even greater possibilities for increased energy efficiency exist with new technologies that fundamentally affect our economic and social organization. For example, major opportunities can result from better electronic communication and control systems. As the population grows, many of the world's cities have become "unlivable" or at least less livable. Proposals for helping to deurbanize and debottleneck our cities by improved transportation, communication, and control systems need to be carefully considered. Our productivity can only increase by reductions in commuting and travel time. To put this issue in perspective, just imagine how inefficient the offices of the world would be today without our telephones, electronic copiers, word processors/PCs and, of course, FAX machines. How these devices will evolve and multiply in the future is largely driven by society's needs. One must not underestimate the future role of advanced communication and control systems in making us more energy efficient. For example, substantial direct savings are possible in the transportation sector, both for better propulsion systems and for a reduction in vehicle miles traveled.

While important technology improvements in energy service efficiencies are possible, the rate at which these new technologies may penetrate into practice is more problematic. For example, the USDOE report on conservation points out that much of the efficiency improvement of the 1970s and early 1980s in the US is beginning to be "undone" by continuing low oil, gas, coal, and electricity prices. This suggests that end users may have a very low demand for efficiency increases and implies that new technologies and technical possibilities will penetrate only very slowly into the marketplace unless their costs are well below existing technologies or unless performance standards are imposed. We have seen above, for example, that significant technical opportunities for conservation exist in the areas of residential insulations, lighting, space conditioning, and appliances. Success in achieving the socially desirable rate of commercial development and adoption of these technical opportunities will depend on the economic efficiency of the relevant markets. While beyond our present scope, it is important to point out that studies such as Carlsmith et al. (1990) suggest that the relevant markets are not efficient due, for example, to asymmetries in information between producers and consumers of buildings and durable goods, market power of durable goods producers, and environmental externalities. To the extent that specific market failures can be identified and evaluated, they must be the subject of appropriate policy intervention. We will turn to this issue in the section on Regulation and Policy. For now, it is important to note that investments of national financial resources for research, development, and demonstration (RD&D) of new energy service technologies must be consistent with the expectation that end users will demand these new technologies.

Seeking Solutions through New Energy Technology

The technological options for reducing environmental impact in the face of growing worldwide hunger for energy can be roughly divided into two broad categories. First there are the "high tech" solutions such as second generation, passively safe, modular fission reactors, fusion reactors, photovoltaic systems, and the like. Then there are the "low tech" solutions such as better stoves for cooking with solid fuels, more efficient furnaces and refrigerators, laws limiting deforestation or increasing car pooling, reducing population growth, etc. None of these should be dismissed out of hand until a careful evaluation has been made of their potential impact for solving the problem. Discussing these options is, of course, the main agenda item of this conference.

In order to provide an overview of the topics covered in energy services sessions of the conference, we would like to discuss a few specific technology opportunities that could help mitigate the environmental problems we have. This discussion is by no means complete and is not intended to endorse our favorites. In practice, a sustainable global environment requires multiple solutions, each matched to achieve optimal performance between the source of supply and the end use. We must not only improve the efficiency of fossil systems and their associated combustion steps to reduce emissions to acceptable levels, but we must also seek out alternative, non-combustion-based technologies as replacements. New technologies certainly would include solar photovoltaics, biomass, geothermal, and other renewables as well as advanced fission and fusion energy supply systems. In any case, each technology will have its most appropriate application—no single energy supply option can meet all current or projected demands for electricity, transportation fuels, process and space heating and air conditioning.

Fossil Energy Systems. A key opportunity for reducing emissions is the redesign of combustion and exhaust gas cleanup processes for stationary systems like electric power plants (see Session D-2). For example, emphasis is currently being placed on developing clean coal technologies to mitigate the release of SO_x, NO_x, and particulates. Improved conversion efficiencies by using combined cycle, cogeneration systems are already being implemented by utilities. More efficient processing of synthetic fuels, improved utilization of remote natural gas, and the direct conversion of the chemical energy contained in hydrocarbons to electricity in fuel cells represent some of the other opportunities for improving fossil energy systems. Dealing with carbon dioxide release is more difficult because the capture and storage of carbon dioxide is very energy intensive and potentially environmentally damaging itself.

Fission Reactors. Greater use of nuclear energy as a replacement for fossil fuels can reduce atmospheric pollution on local, regional, and global scales since no CO_2, NO_x, SO_x, or particulates are emitted from a nuclear power plant. (The mining of uranium does, of course, result in some emissions on a small scale.)

Unfortunately, it has become a commonplace to assert that the nuclear industry in the United States is now dead, that its death was probably a suicide, and that the public is both passionate and unified in its determination to see that it stays buried. The present state of affairs needs no explication. Three Mile Island and Chernobyl cannot be expunged from our collective consciousness. Seabrook and Shoreham are real-time examples of the depth of the conviction held by our political system, perhaps even by a majority of the public, about the risks and benefits of nuclear power. In addition to concerns about safety and reliability, waste management, proliferation, and other risks associated with nuclear technology are important issues that increase public skepticism.

Certainly, we have made mistakes in the past, both in managing nuclear technology and in the ways we have addressed public concerns about nuclear energy. We must be willing to learn from these mistakes, to explore different approaches to the design of nuclear energy plants, and to improve public awareness and understanding of these issues if nuclear energy is to play a role in our future.

Today, nuclear power in the United States, while providing 20 percent of our total electricity, is neither acceptable politically nor attractive economically in the form in which it has been employed during the past 30 years. The American public is enormously skeptical of nuclear power, and there is reason for skepticism. Light-water reactors (LWRs), based on those used in nuclear-powered submarines, have been scaled up to a size that requires multiple, independent, redundant, active safety systems in order to prevent the release of radioactive materials in the event of an accident that causes loss of cooling. This "defense-in-depth" has been effective: it worked even at Three Mile Island, where there was only a very small release of radioactivity to the environment despite partial melting of the core. Still, public opinion has not altered.

Experience with nuclear power programs is very different in different nations. France and Japan have large, successful, and growing programs. Conversely, in Sweden and West Germany, as in the United States, there is a de facto freeze on further development. The reasons for the differences are complex and beyond the scope of this paper. However, there is no profound difference in the technology used in these nations. They all use the basic LWR designs. Thus, the benefits and the problems are essentially the same. It is unlikely that we will see cultural changes that would lead to worldwide acceptance of the existing technology. Rather we must require the technology to change so as to expand the level of public acceptance. New second-generation nuclear energy plants must meet firm requirements:

- Plants must be able to pass a tough realistic test to assure that even a total cooling system failure will not compromise safety and will not result in a melt-down of the fuel or a dangerous release of radioactivity.
- Plants must be designed so that their processes and materials make them passively safe without the active intervention of operators.
- The political, social, and technical problems of radioactive fuel waste disposal must be solved.
- New technical and institutional concepts will be needed to avoid misuse of nuclear fuel and technology for the production of weapons grade materials.

Technological advances in reactor design concepts and in materials in the past 20 years make all of these requirements possible today.

Advanced nuclear reactors will produce less power: 80 to 600 MWe compared with 1000 to 1500 MWe for reactors built in the 1970s. This smaller scale offers significant advantages. First, because higher standards of quality can be achieved, these plants will be modular and largely factory-built, rather than being custom-built to unique designs on the site. Second, their modular character and standard design makes possible licensing by type rather than individually. This approach is similar to that used for commercial aircraft, which are certified and licensed by class or type, not by individual aircraft. Third, new reactors could be added gradually at a power plant site as energy demand increases. Although contrary to conventional wisdom regarding economies of scale, these advantages should translate into cost savings for the customer. Since construction times will be shorter (possibly with even lower unit costs), incurred interest debt on invested capital will be substantially reduced.

Engineers now believe it is possible to design and build a nuclear reactor that will safely tolerate the simultaneous failure of its control system, its cooling system, and even loss of the coolant itself. That is not the only measure of safety, but it is clearly a hurdle which nuclear reactors must clear to overcome public skepticism. Different kinds of nuclear reactors have different levels of safety. For example, Lawrence M. Lidsky, a professor in MIT's nuclear engineering department, has identified four levels of safety classifications:

- Level zero: No hazardous materials or confined energy sources.
- Level one: No need for active systems in the event of subsystem failure. Immune to major structural failure and operator error.
- Level two: No need for active systems in the event of

subsystem failure. Reduced immunity to major structural failure or operator error.

- Level three: Positive response required in the event of subsystem malfunction or operator error. Defense-in-depth. No immunity to major structural failure.

The more than 100 nuclear power plants currently operating in the United States are all safety level three, light water reactors that were designed 20 to 40 years ago. All of the advanced second-generation nuclear power plants under consideration are safer; some are considered at level one and others at level two. Level zero is not practically attainable for a nuclear power plant, since it involves hazardous material and a confined energy source. Absolute safety without risk cannot be guaranteed, whether you are designing a bathtub or a nuclear reactor.

Six advanced reactor concepts are currently under active study for the next generation of nuclear power plants. None have been developed to the stage of a full-scale demonstration reactor, but all are based on experience with other similar reactors.

The advanced light water reactor concepts are the AP600 (Advanced Pressurized light water reactor, 600 MWe), conceived by Westinghouse; the SBWR (Simplified Boiling Water light water reactor, also 600 MWe), conceived by General Electric; and PIUS (Process Inherent Ultimate Safety reactor), a 500 to 640 MWe reactor conceived by ASEA Brown Boveri (ABB) of Sweden.

The helium gas-cooled reactors are the MHTGR (Modular High Temperature Gas-cooled Reactor), a 540 MWe plant using four 135 MWe reactors, conceived by GA Technologies and a consortium under USDOE leadership, and the MGR-GT (Modular Gas-Cooled Reactor Gas Turbine), a stand-alone 90 MWe reactor, conceived by MIT Professors Lidsky and David D. Lanning, whose work was supported by a consortium of American and Japanese firms.

The advanced concept for sodium liquid metal cooled reactors is the Integrated Fast Reactor being investigated by General Electric and the Argonne National Laboratory. The associated fuel cycle has the potential for separating the long-lived from the short-lived components of high-level nuclear waste, recycling the former back into the reactor as nuclear fuel and disposing of the latter in a geologic repository. Modifying the fuel cycle in this way reduces the period during which the hazard of the buried waste is large. This makes the stable containment of such waste in a repository more viable, and hence more acceptable to the public.

Although each concept has certain advantages and disadvantages, evidently several good ideas exist that deserve further consideration and research funding, particularly in view of the need for viable nonfossil primary energy supply technologies. The key issue will finally be public acceptance, but we must first move to the next level of demonstrating the technologies.

Solar, Wind, Hydro, Biomass, and Other Renewables. Although popular with many environmentalists because they have minimal emissions, solar, wind, hydro, and other renewables do have their limitations. For example, the desirable characteristic of "availability on an uninterruptible basis" introduced earlier in this paper cannot be met for these sources without a storage system of some type. In the case of wind or hydro, pumped storage can be used. Solar electric will also need backup systems on the grid to provide continuous power on demand. In some situations it may be very attractive to use a "cogeneration" concept with combined supply systems; for example, a solar, biomass, nuclear combination to provide both reliability and minimal environmental impact in a nonfossil electricity supply system of the future. Land and water use and the consumption of raw materials and energy needed in the siting and production of components and system hardware are among the major environmental impacts of renewable systems. For example, biomass farming requires land, fertilizer, and water and may displace food production needs. Siting hydroelectric dams, wind mills, and solar collectors presents impacts that must be dealt with on an increasing scale as these energy sources expand.

When substituted for fossil energy supply systems, renewables would, of course, have a very positive effect on reducing greenhouse gases such as CO_2. Unfortunately, the costs of using renewables for supplying primary baseload energy have only been competitive with fossil fuels in a few specific cases. Ultimately, costs need to be reduced and/or policy and regulatory incentives provided to increase the contribution of renewables to significant levels worldwide. There are, nonetheless, enormous opportunities for renewables that deserve careful consideration. (See in this volume specific alternative energy papers by Darkazalli, DiPippo, Otte, and McGowan; and for general treatment of solar and other renewables see also Hartley and Schueler.)

Geothermal Heat Mining Systems. In reviewing the desirable characteristics for energy supply systems cited earlier, some geothermal energy concepts being proposed meet these criteria as well as or better than nuclear (fission and fusion), solar photovoltaics, and other renewables like wind and wave power. Hot dry rock (HDR) and magma geothermal energy systems, which employ a closed loop heat mining concept to extract the thermal energy contained in the accessible portion of the earth's crust, may be particularly attractive as a fossil fuel replacement because of their simplicity and low environmental impact. The current concept being pursued for HDR involves drilling holes in impermeable, crystalline rock to depths of three to seven km (10,000–25,000 ft), where temperatures range from 200–300°C, and then fracturing the rock at depth using hydraulic techniques to create artificial permeability within the reservoir. Heat is extracted by pumping water through the fractured system in a closed loop connected to the surface. HDR technology has been under development by the Los Alamos National Laboratory under USDOE sponsorship and by the Camborne School of Mines in the United Kingdom under UKDOE sponsorship and on a smaller scale in Germany, Japan, France, and the USSR (see Armstead and Tester (1987) and the paper by Brown, Potter, and Myers herein for details). Magma systems are being evaluated by Sandia National Laboratories, also under USDOE sponsorship.

Included below is a summary of points made in a recent report on HDR technology by Tester, Brown, and Potter (1989):

- The HDR resource is large and well distributed globally with a technically accessible worldwide resource base greater than 10^7 quads. This magnitude is particularly significant in view of our current annual consumption of about 320 quads. Even without any recharge and only modest recovery levels, HDR could meet the world's primary energy needs for a long time into the future. The high grade HDR resource, defined in terms of high geothermal gradients above 60°C/km, represents about 10 percent of the total, while mid-grade with gradients of 30 to 60°C/km represents about 30 percent, and low grade with gradients below 30°C/km represents about 60 percent.
- HDR systems are inherently simple in design and safe to operate with no danger of explosion or catastrophic failure or leakage of harmful contaminants. They are adaptable for base load or peaking supply of electricity, space or process heat supply, and cogeneration applications.
- HDR can provide power on an uninterruptible basis not subject to changes in the weather. The optimal scale of individual HDR power plants is from 20 to 100 MWe—in line with current utility plans for added capacity and very suitable for remote applications. Furthermore, the technology for building these power plants already exists.
- Implementation technology for HDR is easily transferable to developing countries without compromise of strategic or economic balance.
- With essentially no effluents (no CO_2, SO_x, or NO_x) or waste, HDR systems are environmentally benign.
- Based on reasonable assumptions regarding reservoir formation and performance and current costs for well drilling and power plant construction, numerous economic projections suggest that electricity produced from high grade HDR resources would be competitive in today's marketplace at breakeven prices of five to six cents/kWh. Low- to mid-grade resources will require improved technology to reduce development costs before they would become commercially competitive at current electricity prices.

Given all these positive attributes, our description of HDR may seem overly optimistic. Certainly the field demonstration of a commercial-scale system has yet to be achieved in either a technical or economic sense. Two major technical barriers remain as constraints to commercial development: 1) establishing methods of creating reservoirs of sufficient size and lifetime to achieve economic fluid production rates for all HDR grades and 2) reducing the high cost of drilling wells deep into the earth to permit economic utilization of low and mid-grade HDR resources. As for any new energy source, the R&D effort (and budget) required to overcome these constraints needs to be balanced against the downstream benefits, be they purely economic or environmental. In the view of some experts, the number and magnitude of the technical barriers for HDR are relatively small and manageable in comparison to other long-term options.

Fusion Technology. If ever there was, in the media's eye, a silver bullet, "cold fusion" certainly fit the bill. According to the first news releases, it was "simple, safe, and easy to implement." Unfortunately, all the media attention surrounding the controversy over the veracity of the cold fusion experiments has overshadowed the quality work that has gone into "hot" plasma fusion research over the last 45 years. Here the potential energy payoff is so great and the scientific and political motivation so strong that a very large and productive research effort is already in place. Very few informed individuals question the logic of having a viable scientific program to examine fusion as an alternative energy source, although some members of Congress are impatient with the slow rate of progress toward significant energy production. However, the experimental systems employed today are large, complex, and costly, suggesting perhaps that a more technically and financially consolidated international research effort should be considered. Several fusion concepts probably should be pursued but to try to have parallel efforts in different countries may be more than we can afford.

This aspect of system complexity needs to be discussed further. As mentioned earlier, a desirable characteristic of a primary energy supply system is simplicity. This feature is particularly important if the technology is to be easily transferred to developing countries. For example, light water reactors (LWRs) of today's vintage are complex enough to tax the infrastructure of even the largest utilities in the developed world. To operate these systems reliably—at least subject to safety and safeguard regulations imposed by public agencies like the NRC—is at least an order of magnitude more intricate than comparably sized oil- or coal-fired plants. One can only imagine that fusion-fired electric plants based on a Tokamak (contained plasma) design, even in their second or third generation, would be more complicated, albeit possibly safer, than LWRs. These operational issues will undoubtedly be as important to the ultimate success of fusion power as meeting all the required technical and economic objectives.

Advanced fission, fusion, and solar photovoltaic

concepts have received substantial R&D support, not only because they are viable alternatives to fossil fuels, but also because they each have a relatively large group of advocates. Hot dry rock (HDR), magma, and some renewables, on the other hand, have much smaller constituencies and may not have received a level of R&D funding commensurate with their potential in comparison to the other advanced technologies mentioned. We recognize that this is a sensitive issue because with limited funding not all options can be pursued simultaneously. What is needed is a careful and objective evaluation of each new energy technology on a common basis so that national and international priorities can be established in the context of remaining technical and economic requirements for each alternative. In the end, a balanced program is needed to pursue several nonfossil energy alternatives weighing the probability that a specific technology will succeed against projections of its R&D costs and times required for commercialization.

Transportation Systems. The issue of energy supply for transportation systems is very much centered on the need for liquid petroleum fuels (for example, see Longwell et al., *Fuels to Drive Our Future,* National Academy Press, 1990, as well as papers in the Transportation Systems session of this conference). The need for liquid fuels raises the specter of worldwide oil distribution and supply among net consuming and producing nations along with its economic and political implications. Important topics range from identifying alternative liquid fuels such as synthetic fuels from oil shale and coal, methanol from natural gas or ethanol from biomass to advanced technology possibilities such as electric or hydrogen-powered vehicles for both individual and mass transportation systems to increase energy efficiency and reduce environmental stress. The interconnectedness of local, regional, and global environmental and health effects need to be carefully considered to avoid conflicts.

Regulation and Policy

We now turn to a brief consideration of policy issues posed by the need to better manage global energy and environmental resources. Perhaps the first point to make is that governments, industries, and citizens of developed countries have a broad and increasing experience with energy and environmental policies at local, regional, national, and international levels. International cooperation is evidenced by the cooperative agreements embodied in the programs of the International Energy Agency, and by the 1988 Montreal Protocol limiting the manufacture of CFCs. Further, all developed countries have national energy and environmental policies. For example, the US maintains a strategic petroleum reserve, supports a diversified program of energy technology research and development, and has a comprehensive program of automobile and appliance efficiency standards, all components of its energy policy. In addition, we have a fairly comprehensive set of environmental policies, mostly in the form of performance standards and emissions trading programs for air pollution, and performance standards for water pollution and for land use. Less developed countries are more varied in their energy/environmental policies, but environmental issues especially are of increasing concern.

While there is considerable experience to draw on in the design and evaluation of energy/environmental policy, at least two new elements must be addressed as we contemplate the 21st century and beyond. First, it is becoming increasingly important to recognize the policy implications of the interdependency of energy and environmental systems. In this regard, energy/environmental policymakers may be well served by studying the paradigms and concepts of ecologists who have long appreciated the importance of understanding the interdependencies of ecosystems. For example, current US air quality policies promote the installation of scrubbers on stationary combustion facilities, thereby creating an incentive to extend the life of these facilities. Life extension of coal-fired facilities, however, prolongs their contribution to greenhouse gases, most importantly CO_2, and so may be a problem in designing policies to address potential global climate change. Another example is the US synthetic fuels program, which in the 1970s had an ambitious objective of increasing our domestic production of petroleum-like products, thereby reducing national dependence on foreign sources. Synthetic fuels production of oil shale or coal, however, could have serious consequences for increased greenhouse gases, as well as other environmental impacts, and so major investments in these facilities would have presented a complication for policymakers concerned with global climate change.

Second, the potential for conflict between near-term and long-term energy/environmental objectives and policies must be recognized and addressed. Broadly speaking, in the near term our objectives are to adjust the characteristics and utilization of existing technologies to achieve socially efficient levels of investment in energy conservation consistent with near-term environmental goals. Policy instruments include an array of economic incentives and performance standards that induce changes in the characteristics and utilization of energy producing and using technologies by imposing economic costs, either explicitly or through lost opportunities. That such policies are required is supported by a number of comprehensive studies. For example, Carlsmith et al. (1990) have demonstrated that market failures are distorting the cost effective investments in energy conservation. As noted above, these distortions are estimated to cost US energy consumers about $185 billion (1989 dollars) in the year 2010. Clearly there are possibilities here for the design and implementation of policies involving economic incentives and performance standards to eliminate these distortions.

It is very important, however, to keep in mind that these policies, designed to address near-term objectives of improving energy efficiency, are not necessarily consistent with our longer term objective of developing and deploying new technologies that internalize the environmental consequences of the entire "fuel cycle" (including production, conversion, and end use). For example, setting standards for energy efficiency may provide only a modest incentive for entrepreneurs to strive to revolutionize the delivery of end use services in a way that internalizes, or dramatically reduces, environmental effects. In fact, performance standards may provide little of the economic incentive to compensate for the risks associated either with pioneering new technologies or with the reorganization of social and economic activity. The longer-term objective of socially efficient management and use of environmental resources may require policies that substantially increase the economic burden of not contributing to this objective, consistent with the time for "revolutions" in technology and in organization. Thus, the assertion that any policy that increases energy efficiency also contributes to our environmental management objectives may be very misleading, or even incorrect. Policymakers must carefully formulate near- and long-term objectives, and must design policy "constellations" that balance these potentially competing objectives.

It is beyond our present scope to pursue in any detail the design and analysis of policies to correct these *market failures*. We do, however, want to make a number of observations concerning issues that must be addressed in designing, analyzing, and implementing energy/environmental policies for today, the 21st century, and beyond. First, we need to quantify carefully the environmental costs associated with existing energy supply and end use systems, and the social benefits associated with investments to mitigate these costs. Here the environmental economists and engineers must provide information for policymakers to evaluate in choosing efficient and equitable targets for reducing the environmental costs of human activity. Having done this we need to estimate costs for the research, development, and demonstration needed to bring new and re-engineered energy supply and end use technologies to the marketplace. Important criteria for establishing R&D priorities for new technologies include estimates of the technological breakthroughs required and the time needed for them to happen. For example, what are the technical and economic requirements for fusion energy versus photovoltaics or hot dry rock geothermal? These are difficult and complex questions that will require objective leadership to establish energy policies and strategies.

Second, government and industry in this country have had a largely adversarial relationship when it comes to policies concerning environmental and economic consequences of technological development. Reliance has been on regulation rather than on cooperation. This stems largely from the lack of clearly articulated and agreed upon standards for performance, safety, cleanliness, or risk. Without such criteria, it is not surprising that continual conflict and misunderstanding persist between groups and individuals with differing concerns. That, taken with the lack of technical and scientific knowledge at high levels of decisionmaking in the legislative and executive branches of government, and in the public itself, has meant that we do not have consistent, well-thought-out and clearly articulated environmental and energy policies.

The creators of technological developments and the policymakers thus have a responsibility to thoroughly explore the multiple consequences of new developments. They need to develop guidelines and policies for sustainable development that reflect concern for the long-term, global implications of large-scale technologies in particular, and that support the innovation of less intrusive, more adaptable technologies at all levels.

Third, the greenhouse effect and the threat of global warming is the most dramatic and current example of the double-edged quality of technological development. Resolution of this global problem in the context of existing local and regional environmental problems will require a concerted technological, political, economic, and social collaboration on an international scale that has never been seen before.

Fourth, solutions to the dilemma of increasing energy use in the face of environmental degradation or collapse must occur on many fronts—there is no single answer. For example, conservation is an important component, but it can only go so far in curbing growth in energy consumption without eroding our standard of living and quality of life as world population rises and energy and resource needs increase. The limits of conservation are somewhat difficult to define since there is inherent synergism between improving efficiency by conservation and replacing or altering energy supplies to minimize environmental stress. For example, switching from a coal-fired steam boiler to a gas-fired combined cycle cogeneration system not only lowers CO_2, SO_x and NO_x emissions, but also increases the overall utilization energy efficiency in a thermodynamic sense. However, even with increased conservation and efficiency improvements, the average level of energy consumption per capita for the world as a whole will undoubtedly increase at some finite rate, though, as shown in Figure 1, the rate will probably be less than the rate of population increase.

Alternative nonfossil energy systems are needed, as are improved environmental control technologies to reduce SO_x and NO_x emissions in fossil fuel systems. Prioritizing our approaches is critical and will require national and international leadership. In part, the public blames science and technology for generating the problems associated with energy use, including smog, acid rain, and toxic wastes. Engineers and scientists will have to provide solutions that carefully knit together technical, economic, and moral perspectives. Those of us who develop, promote, and apply technological innovation have a moral responsibility to explore and consider, to the fullest extent possible, the environmental consequences of any innovation. New ideas are to be judged not on their technical merits alone—but also on their impact on the environment from "cradle to grave." We must foster the ethical responsibility of good stewardship for the earth.

To implement these technological innovations will require not only that new energy systems provide environmental improvements at costs comparable to, or below, existing technologies but that they also be acceptable to consumers. RD&D policymakers need to consider the ultimate marketability of a new technology as well as its technical attributes.

Finally, we feel strongly that an important function of policy must be to motivate young people, and especially engineers and scientists, to accept with enthusiasm the challenges of internalizing the environmental consequences of human activity. While policies can provide the incentives and standards to achieve environmental goals, more is needed. Leaders must be bold and persuasive in providing young people with both the ideals and the knowledge to create sustainable future economic/energy/environmental systems in the 21st century and beyond. We now discuss some of these educational needs and opportunities.

Educational Implications

Although our primary purpose in writing this paper has been to reflect on technological aspects of future environmentally conscious energy supply and service systems, we would like to make some observations about the implications for educational institutions engaged in the preparation of future generations of engineers.

Technological developments do not exist in a vacuum, as has been noted several times above, and not all of their consequences—either fortunate or unfortunate—can be anticipated. But the imperative to understand the implications of a development in its broadest and most encompassing terms is a professional responsibility of the engineer, one that must be incorporated into the task from the outset.

Now what does this mean for engineering and engineering education today? We need to stimulate interest in young people to help us develop technical solutions. We need to restructure our education programs. Of course engineering students should be expected to obtain a firm foundation in the sciences basic to their technical field, and to begin to develop a working knowledge of current technology in their field of interest. Beyond that, there are a number of things we could do:

- Instruction in the humanities, arts, and social sciences should be designed to give the engineering student some understanding of societies and cultures, of the complex relationships between society and technology, and of human values and relationships. Engineering is, obviously, a socially derived and culturally influenced activity, and engineers cannot function effectively without being steeped in those contexts.
- While all engineers should have an appreciation for and sensitivity to the social environment in which they operate, some engineers—who might be called "interface engineers"—will work directly on issues of application, impact, and implementation in a broader context. They need direct engagement in their education with these issues. These students should tackle subjects and engage in research on topics that directly address the political, economic, and social considerations integral to scientific and technological developments.
- Engineering design courses, particularly at the upper level, should move beyond requiring significant individual effort to requiring collaboration among teams of students formed to work on problems that are not artificially isolated from their social context. And a part of the team effort should bear on the exploration of social consequences and of the problems that arise when technology is used for different purposes. While such projects are inevitably constrained, we believe it is important to require engineering students to begin to work as engineers in ways that reflect to some degree the way actual engineering work is and should be done.
- Students should be prepared for active leadership in the definition and resolution of the issues that arise at the intersection of technology and society. Neither we nor they can afford to sit back and expect other professions to imagine, create, and implement the kinds of solutions that are both socially responsible and firmly grounded in technical realities. Engineers do not hold the sole responsibility here, but the profession must consciously prepare and train itself to do its part: Effective leadership must be learned.

These educational components do not arise only from the challenges facing engineers practicing in the energy field; they are, of course, broadly applicable to all engineering disciplines.

A key element in achieving these educational reforms is to stimulate faculty interest and support in order to cultivate the development of curricula that deal with these interconnected technical and social issues, ideally in the context of the student's major discipline. Without enthusiastic faculty and administrative support, these reforms will not be implemented.

All of these educational objectives aim at the development of engineering graduates who are better prepared and more highly motivated to comprehend this work and their responsibilities in the social and technical threads composing the fabric of responsible, ethical technology development and application.

Energy technology is replete with examples of the need for such perspective:

- Technological approaches to the reduction of greenhouse gas production raise difficult issues of international and intergenerational equity. Who pays? Whose development is constrained? Who benefits?
- The development of consumer products that are more durable, energy efficient, and amenable to eventual recycling raises issues of corporate and consumer cost sharing and incentives that require an appreciation of different values and that will lead to changes in purchasing patterns. How are costs, including the costs of significant externalities, appropriately apportioned? Once again, how does the technology affect who pays and who benefits?

Conclusions

As we have tried to point out by way of examples, there are enormous opportunities for technological innovation in how we supply and use energy. Engineers with vision can provide the means to develop viable strategies in our economically, culturally, and ecologically intertwined world. We must work together to forge a new concept of how engineers work and view the world. As we hope this paper demonstrates, the development of energy technology in a socially and environmentally responsible manner is feasible. Making all this happen with discovery, innovation, and reduction to practice is perhaps one of the greatest challenges we face as engineers, as engineering institutions, and as a society. We believe this conference is starting us off in the right direction by promoting a serious interdisciplinary discussion of energy and environmental technology and policy issues.

Acknowledgments

We would like to thank our colleagues at the Institute for providing many useful suggestions and ideas for this paper. We are particularly indebted to David White, Lawrence Lidsky, Kent Hansen, Richard Lester, and Marvin Miller for their reviews and input. Howard Herzog provided much needed assistance in the preparation of the figures. The editorial advice of Kathryn Lombardi and the word processing assistance of Anne Carbone are also gratefully acknowledged.

References

Alterman, J., *A Historical Perspective on Changes in U.S. Energy-Output Ratios* (EPRI EA-3997), Electric Power Research Institute, Palo Alto, CA (June 1985).

Alterman, J., "Measures of Energy Consumption, Expenditures, and Prices," *Monthly Energy Review,* Energy Information Administration, Washington, DC (May 1988).

Armstead, H. C. H., and J. W. Tester, *Heat Mining,* E. F. Spon, London (1987).

BP Statistical Review of World Energy, The British Petroleum Company, London EC2Y9BU (July, 1988 and 1989).

Carlsmith, R. S., W. U. Chandler, J. E. McMahon, and D. J. Santini, *Energy Efficiency: How Far Can We Go?* (ORNL/TM-11441), Oak Ridge National Laboratory, Oak Ridge, TN (January 1990).

1986 Demographic Yearbook, 38th Edition, Department of International Economic and Social Affairs, Statistical Office, United Nations, New York, NY (1988).

Economic Report of the President, January 1990, US Government Printing Office, Washington, DC (1990).

Energy Information Administration, *Annual Energy Review,* US-DOE, DOE/E1A 0384(88), Washington, DC (May 1989).

Fulkerson, W., *Energy Technology R & D: What Could Make a Difference?* (ORNL-654/V1-4), Oak Ridge National Laboratory, Oak Ridge, TN (May 1989).

Goldemberg, J., T. B. Johansson, A. K. N. Reddy, and R. N. Williams, *Energy for a Sustainable World,* Wiley Eastern Ltd., New Delhi, India (1988).

Johansson, T. B., Virgit Bodlund, and Robert H. Williams (eds.), *Electricity: Efficient End Use and New Generation Technologies, and Their Planning Implications,* Lund University Press, Lund, Sweden (1989).

Jorgenson, Dale W., and Peter J. Wilcoxen, "Global Change, Energy Prices, and U.S. Economic Growth," Energy and Environmental Policy Center, Harvard University, Cambridge, MA (June 13, 1990).

Kaya, Y., contribution to "Policy Strategies for Managing the Global Environment," *Energy and the Environment in the 21st Century,* MIT Press, Cambridge, MA (1990).

Kaya, Y., Kenjl Yamaji, and R. Matsuhoaki, "A Grand Strategy for Global Warming," paper presented at *Tokyo Conference on the Global Environment and Human Response toward Sustainable Development* (September 11–13, 1989).

Longwell, J. et al., *Fuels to Drive Our Future,* National Academy Press, Washington, DC (1990).

Menken, J. (ed.), *World Population and U.S. Policy: The Choices Ahead,* Norton and Company, New York, NY (1986).

Statistical Abstract of the United States 1987, 107th edition, US Government Printing Office, Washington, DC (1986).

Tester, J. W., D. W. Brown, and R. M. Potter, *Hot Dry Rock Geothermal Energy—A New Energy Agenda for the 21st Century,* Los Alamos National Laboratory report #LA-11514-MS, Los Alamos, NM (July 1989).

United Nations, *UN Yearbook of World Energy Statistics,* New York, NY (1980–1982).

US Department of Energy (USDOE), *Energy Conservation Trends,* Offices of Policy, Planning and Analysis, and of Conservation of Renewable Energy, DOE/PE-0092, Washington, DC (September 1989).

US Energy Data Book, General Electric Company (1984).

White, D.C., and D.S. Golomb, *Closing the Energy Cycle: A Challenge for Energy Intensive Industries,* Kohle-Stahl Kolloquium, Berlin, FRG (February 1989).

World Almanac, 1990, Pharos Books, New York, NY (1989).

World Energy Conference, *Survey of Energy Resources* (1974, 1987) and *World Energy Resources 1985–2020* (1978, 1980).

Plenary Session II

Policy Strategies for Managing the Global Environment: A Panel Discussion

John M. Deutch
David O. Wood
Co-Moderators
Massachusetts Institute of Technology
Cambridge, MA

Editor's note: *Evaluating policy strategies for the efficient and equitable management of climate resources is complex and increasingly controversial. Forum members offered their differing perspectives on this subject, followed by questions and discussion with the audience. This excerpt from the Wednesday afternoon Plenary Session II presents the panelists' views.*

Yoichi Kaya, The University of Tokyo: Today I'd like to talk about three things. First, our basic attitude toward global warming. Second, the strategies available to mitigate global warming. And third, and most important, which of these strategies should we choose. Recently, I received a draft report of the Intergovernmental Panel on Climate Change (IPCC) Working Group I, on scientific knowledge about global warming. For example, it is known that we may experience a temperature rise of 1.5–4.5°C in the 21st century. And there may be a surface rise of between 20 and 100 meters in the oceans. There are huge uncertainties about the values predicted in this report. Last September in Tokyo, we had a government meeting on the global environment, and one of the main conclusions at the time was that there are so many uncertainties about global warming matters that we have to work to reduce these uncertainties as soon as possible. Some say that, first of all, we should do research to reduce uncertainties and then implement measures to mitigate global warming. However, I don't think this is a good idea. We know that concentrations of greenhouse gases in the air have been increasing. Today the concentration of CO_2 in the air is around 350 ppm, whereas its value in the middle of the 19th century is thought to have been about 280 ppm. That is about a 25 percent increase. So this is an example of the impact we have had on the environment. I think it's too optimistic to think of this impact as negligible or positive. My basic stance on this problem is that we should take strong measures to mitigate global warming.

The second point is how to determine the strategies we should implement to minimize global warming. I would classify these strategies in three categories. One is the "desperate" strategy—setting a severe greenhouse gas limitation target like that of the Toronto Conference statement, with some sanctions. The second is a "well-moderated" strategy, where we set a moderate greenhouse gas limitation target, probably resulting in some sanctions. And the third is the "conservative" strategy. In this case, we work in terms of energy conservation, but we don't set any CO_2 limitation targets. To determine the most useful approach, you must make a cost/benefit analysis for the three strategies. In this conference, we had presentations from many economists about these matters. But now I'd like to introduce a rather simplified analysis, using simple equations.

Carbon dioxide emissions comprise the following three factors: carbon intensity of energy, energy efficiency, and gross domestic product. The growth rate of carbon dioxide concentrations is equal to the sum of the rate of change of all three other factors. So, if we set the target on this rate of change of carbon dioxide emissions, then we can calculate what must be done in terms of the rates of change of these other factors.

In order to see the details of this kind of analysis we should examine some data. From 1973 to 1986, Japan attained almost 3 percent improvement (reduction) in energy intensity. And in the United States it was about 2.5 percent during this time period. But after 1986, the rate of change slowed to almost zero in most developed countries. The long-time trend for the United States shows the rate of change in energy intensity to be just 1 percent per year. In other words, energy intensity has been decreasing by 1 percent per year over the long-run. This is rather small when compared to the improvement between 1973 and 1986. We should note that conservation efforts were most successful during that period. For Japan, however, energy intensity is almost constant after 1986. Today, people say that energy conservation efforts have virtually come to a standstill. This might be due to lower energy prices, reduced industrial investment in energy conservation, etc. It could be said that in the future we can conserve energy only to a limited degree. The United States long-term data of 1 percent per year improvement suggests that the past trends will continue for the foreseeable future without "desperate" policies. This forecast can be supported by other data. Recently, various countries submitted their energy forecasts to the Energy and Industry Subgroup (EIS) of the IPCC. According to my calculations, the average change in energy intensities is around 1.1 percent, quite close to my observations about past trends.

Another factor we have to consider is the carbon

intensity. The values of carbon intensity after 1973 have been almost constant in most countries with the obvious exception of three: France, Sweden, and Canada. These countries are heavily dependent on nuclear power. However, I don't think it would be easy for other countries to introduce nuclear power very quickly. On the other hand, the base case studies of the IPCC show that the average improvement rate in carbon intensity is only 0.2 percent, quite small. So in this sense, we can say that if we rely upon the past data, it is very probable that carbon intensities of most countries will change at a rate of around 0.2 percent per year, and energy intensities around 1 to 1.1 percent per year. If this is true, then the growth rate of the economy must be reduced if the CO_2 limitation target is a severe one. Of course, this is just the balancing of a mathematical equation. It doesn't say anything about causality among these three factors which are not independent.

I made some simple calculations about the famous targets of Noordwijk and Toronto, as applied to the case of Japan. The Noordwijk target was established in November of 1989 and requires stabilization of carbon dioxide emissions as soon as possible. Applying that condition to Japan, the change in carbon intensity will be at most only a year improvement, which coincides with the government forecast. But in that case, the possible growth rate of the economy is only 2 to 2.5 percent. That is considerably lower than the government target, which is set at 4 percent by the year 2000, and 3 percent by the year 2010. So in order to attain the GNP growth target, we have to improve the energy conservation rate by 2.5 percent, and the carbon intensity changing rate by 1 percent. The problem is that it is not that simple to change the energy supply structure and thus carbon intensity in a short time. For example, the introduction of a nuclear power plant requires at least 15 years. Anyway, the result of this calculation shows that a lot of effort is needed to attain the target, and in the case of the Toronto target, the situation is much, much worse. A decrease of about 1.3 percent per year in carbon dioxide emissions is required, and in this case, assuming that we keep the same values for carbon intensity changing rate and energy conservation rate, the growth rate of the economy will be only between 0.7 to 1.2 percent. Quite low. I don't think government would accept this kind of low-GNP oriented policy. If we try to attain the government GNP growth target, we have to attain 3.3 percent improvement in energy conservation every year. That is higher than that achieved between 1973 and 1986, the most successful period. In that sense, frankly, I don't think it is possible to reach this very ambitious target, even if we introduce drastic measures. Also, this kind of severe target may not be introduced in the United States, or might not be accepted by the US government.

Now I would like to express my recommendation. As I said, we have three strategies and I would actually recommend the moderate strategy, rather than the desperate strategy. One of the reasons is because the total cost of severe measures is very uncertain. It is unlikely that most people could accept a severe target, given the uncertainties about global warming. We are aware that major industrialized countries contribute a greater part of global CO_2 emissions. These countries—including the Soviet Union, China, the United States, and Japan—quite unfortunately, are hesitant to accept severe targets. Even if we try to establish severe targets, these countries will not accept them, eventually doing almost nothing at all. Therefore, if we really want to get agreement from major countries, from a practical point of view, a moderate strategy might be much better than a desperate strategy. The former is weaker than the latter in terms of mitigating global warming, but we can actually invite major countries to join in a more moderate type of agreement. So, in fact, the moderate approach may have greater actual impact on global warming. This is the most important reason for my recommendation of the moderate strategy. Thank you.

Lester Lave, Carnegie-Mellon University: When Professor Kaya put three strategies on the screen, somehow I knew that the middle one was going to turn out to be the best one. Aristotle's "golden mean" seems to be just as compelling in Japan as in the United States.

My first point is to observe that there is a large amount of controversy about the uncertainties of the greenhouse effect. Some scientists get angry and even feel betrayed when other scientists point out the uncertainty. Global climate change depends on emissions of greenhouse gases, on what happens in the atmosphere, and the resulting changes in the oceans, glaciers, forests, and the rest of the managed and less managed environment. The models to date are relatively simple ones that don't incorporate all of the feedbacks, even all of the feedbacks of potential importance within the atmosphere. No one should be surprised that current models don't estimate with precision effects on atmospheric temperature and precipitation, regional effects, or even begin to address what will happen to the ecology.

Nothing is gained by pretending that these uncertainties are small. The climate modeling is uncertain and these uncertainties are further compounded by the interactions between climate change and human reactions. For example, the models use scenarios that have GNP and world population growing exponentially and smoothly. I challenge you to find historical periods when smooth growth took place. The real world has depressions, wars, and every sort of human action to confound prediction. Modeling is a wonderful occupation, unless your clients insist that you predict the future accurately.

The second point is that there are many impediments to action—system lags—as Professor Kaya has mentioned. A first impediment is the delay before scientific consensus is achieved on the size of the climate change and how important these effects will be. Perhaps consensus could be achieved by the end of the century.

After scientific consensus is achieved, actions will be

delayed by the need for public understanding and acceptance. Both scientific consensus and public agreement are required before nations will sign costly international agreements and implement them. Meetings in Toronto and the Hague seem to indicate that international agreement has already been achieved. In fact, these meetings were attended by those who believe the problem is real and important; they are not representative of the world more generally. Even subsequent statements by politicians must be viewed skeptically. Politicians can be expected to express concern when standing in front of the television cameras, and even to pledge action, but there is a long step from these expressions of concern to taking costly actions.

Even after everyone agrees to do something, it takes an enormously long time to actually accomplish it. Several decades would elapse before electricity generation plants could be replaced and perhaps half a century before buildings could be replaced.

Recognizing these lags leads to a simple conclusion: If the climate modelers are correct, there is almost nothing we can do to prevent substantial climate change. I doubt that emissions will decline enough to stop adding to the warming before the middle of the next century. I don't mean to approve these emissions or sound sanguine about them, but climate change seems inevitable.

A further difficulty is something called the "free rider" problem. From the standpoint of all but a half dozen nations, a nation's actions are essentially irrelevant to preventing global climate change. Thailand, or even Norway, can do little by their own action to prevent global climate change. For the vast majority of countries, ending their emissions of greenhouse gases would have little effect on global climate. Thus, if each alone, or even as a group, abated emissions, the effects would be small. Even more importantly, if the most important nations went ahead with substantial abatement, these other nations could continue their emissions without great damage. Under these circumstances, how can you persuade these nations to make painful, expensive sacrifices?

A similar problem is the conflict between rich and poor nations. It is the rich nations that are burning large quantities of fossil fuels and emitting CFCs. They are the overwhelming contributors to the greenhouse gases. If large costs are to be borne, how much should be borne by the rich nations? By the poor ones?

The world greenhouse meetings have announced formulae such as: A 20 percent reduction of emissions from the 1988 level by 2010. That is a difficult goal to achieve for the US, Japan, and Germany. It is an impossible goal for China or India. Insisting that China and India accomplish this amounts to insisting that they shift their efforts from economic development to abating greenhouse gases. Not only would this end any hope of economic development, it would lead to lower standards of living.

Thus, the developed world has been formulating policy as if the world consisted only of rich countries. But countries like China and the USSR have major coal reserves, which they plan to use. It is unrealistic to expect a decline in their carbon dioxide emissions and impossible to exempt them from control, since the growth in their emissions will be substantial.

Abating greenhouse gas emissions will lead to conflict. If the rich nations attempt to compel reductions in emissions, the conflicts will be real and are likely to be bloody. International peace and security will be the victims of such a policy.

I doubt that all of these uncertainties will be resolved—ever. No one can predict the future accurately, especially the future in 110 to 200 years. In 1800, Malthus theorized that population increases would keep people living on the edge of starvation. A glance at the well fed and well clothed people in this auditorium, as well as my difficulty in getting safely across Massachusetts Avenue shows that Malthus was a failure as a futurologist; our real income is many times higher than the level enjoyed by Malthus.

I am not here to ridicule Malthus. Rather, the point is that Malthus wasn't very good at foreseeing the future, even in so fundamental a way as to determine whether people would be on the borderline of starvation or enjoying increasing standards of living.

The purpose of these models is not to make accurate predictions; no one can do that well. Rather, they are designed to trace out the implications of proposed policy and actions; for decisions such as choosing a career or what size dam to build, our current choices will constrain the future. Before we make these choices, we need more information about their implications. To repeat, the reasons for running these projections forward is not because we're making a prediction; the models are pitifully inadequate for that. Instead, the projections are run to inform us about what appear to be sensitive decisions and actions, particularly in cases where failure to act now will preclude sensible decisions in the future. The critical uncertainties are likely to be resolved gradually over time, rather than in a burst of insight.

That is enough about impossibilities and gloom and doom. There are paths, sensible actions that appeal to sensible people—meaning those who agree with me. We all share a concern for the well-being of all people now and in the future and for the environment, now and in the future. We agree that there are pressing needs in many areas other than the environment. As Ronald Reagan learned, every government program has its supporters. Although we all rejoice at the end of the Cold War, there are some people in this very auditorium who do not think that some aspects of military research should be cut.

Few of us are horrified at the prospect of tax cuts and most of us would approve of tax increases only after a concerted effort were made to make current programs more efficient and effective. We would probably concede tax increases to benefit National Science Foundation and National Institutes of Health, and would probably go along with more taxes to support higher education. Perhaps a majority of us would approve higher taxes to help less fortunate individuals and nations, but I suspect that we would put stringent conditions on how the money was spent and who got it. There are many things that currently need our attention.

The problems of research funding, funding for higher education and less fortunate individuals are real and immediate. In contrast, global climate change is highly uncertain and it is in the future, and might not turn out to be a problem. The contrast suggests that we might attend to current pressing issues and leave the greenhouse for our children.

Much of the public, and of Congress, seem to shuttle from regarding global climate change as the most important problem of our time to regarding it as a nonissue. Neither of these viewpoints leads to sensible policy.

More sensible policy requires an analysis of the social cost of essentially ending the emissions of greenhouse gases. To those who regard our current lifestyles as wasteful, or even sinful, the global greenhouse provides another reason for returning to virtue. Eliminating these emissions is not costly, it is altogether a good thing since it banishes sinful lifestyles. Even for those who take a less extreme view, if eliminating these emissions could be done at "trivial cost," it should be done despite the uncertainties about benefits. Some of the papers at this conference show that the emissions could be all but banished at a cost of 1 to 3 percent of GNP. This cost is trivial to some and substantial to others.

If the cost of preventing greenhouse effects is regarded as being nontrivial, analysts must go on to estimate the benefits of abatement, the benefits of avoiding the social cost of climate change. This statement is offensive to many. Any change in the environment induced by humans is viewed as unacceptable. Modern environmental research finds major natural changes in the environment from day to day and year to year. This background variation provides a benchmark against which to judge climate change induced by people.

In my judgment, change per se is not a sufficient criterion. Scientists must examine the environment of concern to humans and estimate the amount of future change and the effects on things that people care about. Sensible policy requires that we try to avoid actions that are ineffective, not cost-effective, or even harmful.

Let me remind you of some of the less illustrious parts of past science policy. For example, there was a lot of concern in the late 1960s that a fleet of commercial supersonic transports would destroy stratospheric ozone with their emissions; it turned out that the atmospheric chemists had overestimated the magnitude and gotten the sign wrong. In the early 1970s, utilities began building dozens of power reactors based on the expectation that the demand for electricity would rise at 7 to 10 percent per year. Those reactors resulted in billions of dollars being wasted and vast excess capacity. In the mid-1970s, concern that we were running out of oil led to billion dollar demonstration projects for shale oil and synthetic fuels, technologies that were unattractive at oil prices below $40 per barrel.

In each of these cases, "far sighted" planners were attempting to head off a future debacle. Unfortunately they were wrong and contributed to a current debacle. People who believe in "prudence" and being "conservative" find no solace in the greenhouse debate; there is no conservative solution.

Ending the emissions of greenhouse gases will pull resources away from other programs that are highly valued, such as current environmental programs, investment in plants and equipment, and in education and research. There is no conservative action here. Rather, society must carefully weigh the benefits and costs of alternative actions and policies.

The two extreme policies seem to be ruled out by such study. It doesn't make sense either to eliminate emissions of all greenhouse gases or to do nothing today. Taking no action today amounts to an affirmative conclusion that business as usual is the best policy. That is ludicrous.

Current uncertainties preclude confident policies. Thus, the first task is to lessen those uncertainties. Those of us who are not currently employed in doing that hope fervently to be employed in this effort soon.

The second task is a little less obvious: The world needs research and development that will increase flexibility and facilitate adjustment. For example, society would benefit from a wider range of energy and energy use technologies; as uncertainties get resolved, having this range of technologies will facilitate quick, inexpensive actions.

Changed climate will require new cultivars and even new crops; the capabilities of plant genetics and biotechnologies must be enhanced. Having a wider range of future plants available puts a premium on preserving current genetic materials; the extinction of species could make adjustments more expensive and difficult.

A third task is to learn the most cost-effective way of abating greenhouse gases. Energy efficiency improvement is a major element of the solution. This improvement isn't limited to improving the efficiency of current equipment, such as refrigerators. The greatest improvements will come from changing systems. For example, aseptic packages could all but eliminate the requirement for refrigerators.

In almost all of the sessions I have attended at this conference, I've heard a discussion of the "engineer's burden." MIT seems populated with technologists who believe that they are guilty, at least in part, for creating the greenhouse situation; they say that it is their responsibility to find new technologies that will eliminate greenhouse gas emissions, that will in addition be cheaper, more reliable, and more attractive than existing technologies in every other dimension.

If there are such technologies, and the second law of thermodynamics leads me to be pessimistic, they will solve the problem. The engineers seem unwilling to trust economists, business leaders, politicians, and the general public. They must do it alone, finding technologies so attractive that people will want them quite apart from their greenhouse benefits. These technologists are misguided, are wasting their time, and are misleading talented students.

The second most common speech has asked: "Why are you people sitting there? We agree on the problem; now get going on a solution. To the barricades!" Earlier I mentioned some of the problems that resulted from rushing to the barricades. I show my gray hair in cautioning that these issues are more complicated than you think. You better think a little bit longer, to find the real problem and desirable solution, rather than rushing off to do something. For example, if we had managed to build a whole industry to produce synthetic liquids and synthetic gas out of coal, it would be producing vastly more carbon dioxide, per unit of useful energy, than current technologies. I think that it is Pogo who remarked, "The main source of problems is solutions."

Irving Mintzer, Center for Global Change, University of Maryland: I'm going to be fairly brief today. The theme of what I want to talk about is not a dream, nor an exhortation, but a suggestion that as we begin to consider the risks of rapid climate change, we look to the situation as one which bears both challenges and opportunities. Although the recent treatment of global warming and the greenhouse effect in the popular press in the United States would lead you to believe that a rampant fear of climate catastrophe has overtaken the country, I'm here to tell you that the sky is not falling. In fact, however, the rate of change in both the composition and behavior of the atmosphere is exceptionally high. It is occurring at rates that are absolutely unprecedented in historical terms, and we are dealing with a complex set of coupled nonlinear systems whose thresholds of rapid response are currently unknown to us. There is danger that this system might respond to some small change with a disproportionately large response, and if we continue heading rapidly toward whatever threshold is ahead, we will quite likely see an accelerated response to greenhouse gas buildup.

I think it is worth noting, in the face of the uncertainty that we have about how the system is working, the current indicators of how it will change in the future, and what its sensitivities are. There are some key signals we might want to pay attention to. One of the earliest signals of the effect of climate change may well be a change in the frequency and duration of extreme weather events—which is not to say that the 80 degree weather in Washington or the snow that occurred last week is an indication that the world's on the edge of rapid climate change—we are seeing signals that have two important implications. One, they give us some foreshadowing of the kinds of events we may see with increasing frequency in the world ahead and some at least anecdotal information about how current institutions might respond. But the point I want to make to you is that despite our attention here, and much of the attention in Washington these days, and in some of the international fora, discussion of energy and environment has focused on climate change as an issue sine qua non. Climate change is just one stress that societies are going to face in the next decade—one stress among many of the demanding challenges and pressing concerns that our allies and friends in the Third World will concurrently face.

Although it's seductively alluring, in my village of Washington, to treat issues like climate change, or ozone depletion, or acid deposition and tropospheric pollution as though they occurred in isolated compartments of the atmosphere—compartments that we could work our way into and work our way out of through intense meetings like this one—you know and I know that such a vision is foolish in the extreme, that at least these problems, global climate change, ozone depletion, acid deposition are tightly linked. They're linked on three important levels. They're linked economically, because it's the same economic activities that generate the emissions that are causing all three problems. They're linked chemically, because the pollutants themselves, once released in the atmosphere, interact in complex and synergistic ways, and they're linked especially at the policy level. So the policy choices we make today to deal with each of these problems, global warming this week, or ozone depletion next week, will unavoidably affect the timing and severity of the remaining threats.

Well, what does this mean for us when what we really want to talk about is policy strategies for efficient and equitable management of climate resources? I think it means essentially two things. It means first of all that we have to recognize the role of averting climate change in the context of the full ensemble of national economic goals, policy strategies, and other needs that national governments and individual corporate leaders have to face each day. And secondly, it encourages us to try to identify strategies that can make one or more of these problems less severe, without making the rest more painful and more extreme. I view this as a challenge to industry as well as the government.

Unfortunately, one of the sidebar costs of the structure of this complex and extremely entertaining meeting has been that in most cases those of us who are most prepared to deal with individual sections of this problem have not taken advantage of the opportunity to talk together as extensively as we might. So we have engineers facing the engineer's burden, economists facing the imponderable difficulties of estimating GDP growth rates for individual countries or whole regions 100 years hence, but not much cross fertilization that would provide a better understanding of how economic changes may affect technical innovation, or how changes in our understanding of technical systems may affect the things that constrain the rate of economic growth. What this leads me to believe is that if we're not clever, we will continue to see environmental issues as a burden on economic growth, as a cost to be borne, and as a painful early payment we make for the heritage we leave our grandchildren. If we're a little bit clever, however, if we're a little bit disciplined and a little bit thoughtful, we may be able to recognize in the current situation an opportunity to use national and international concern about our changing environment as a fulcrum for mobilizing toward sustainable economic growth while dealing with some of the pressing equity issues that threaten to undermine our current period of international stability. That all sounds fine but it's almost impossible to visualize how it might be made operational.

I'll just close by talking a little bit about strategies that might lead us toward reaching these seemingly disparate and contradictory goals. I would suggest to you from my own peculiar prejudice about the importance of market mechanisms that critical to reaching these linked goals will be efforts made today and in the near future to get the prices right for energy and other critical resources, to stop fooling ourselves and fooling our children by systematically excluding from the price of energy the environmental and other costs that must be borne by societies. I believe that, because if we can indeed structure our energy prices so that they represent the full cost of energy supply and use, we can encourage thoughtful consumers and corporate investors to make decisions that represent a path toward efficient use of our scarce resources. In order to do that I think we have to structure both our taxes and other fiscal incentives—physical structures to provide incentives—to meet the joint goals that are important to society, not one single goal that is the most popular theme of the week.

As Professor Lave has pointed out, I think we're going to have to encourage research and development on a wide array of advanced technologies, both those like the development of drought-resistant cultivars that can make societies more resistant to the unavoidable effects of climate change, and those like advanced gas turbines and other low polluting energy conversion devices that can reduce the rate at which the concentration of these troubling gases build up in the atmosphere. But as Professor Kaya pointed out, I think the only way to get there from here is to avoid the allure of flashy desperate actions. In fact, what I would suggest to you is that what we really need is to do the things that make sense, do the things that make sense whether or not we're on the edge of a climate catastrophe. Do the things that make sense to make American business more profitable, more competitive globally, and more enduring over the long term. To forego the attractive allure of our current concentration on the quarterly balance sheet and begin to learn from some of our Japanese colleagues the importance of building for long-term economic growth. But we must develop the elements of such an insurance policy. In my view, since the principal political challenges we're going to face in the next few years involve dealing with the joint problems of the budget deficit and the trade deficit, we have to look for strategies that can address these pressing problems at the same time that we implement our concerns about the environment. In the US I think that means devoting increased attention to measures that can stimulate investments to improve the efficiency of energy use, both the economic efficiency of energy use and the engineering efficiency of energy use. It shouldn't shock us that Japan is out-competing us in many fields, when in fact it takes only one-half as much energy to generate a dollar of real GNP in Japan as it does in the United States. But the difference gives Japanese products in many cases a visible cost advantage, even when competing in the American market. The societies, in fact, that have been most successful in the last decade are those that have captured the costs of energy in the price and have increased their energy efficiency.

In the near term, the biggest bite we can take out of the global warming problem is to deal directly with the problem of ozone depletion. That is, to encourage the rapid phase-out of both the most dangerous fully halogenated chlorofluorocarbon compounds and to encourage the development and orderly phase-out—say 30 years hence—of the somewhat less dangerous hydrogenated CFCs that are being brought on now to replace them. We're going to have to follow the leadership of the President and begin to think about ways that we can sequester more of the carbon we're putting in the air, in the soil, and in the biota, by doing things that are traditionally American—planting trees, maintaining our soil heritage, trying to get back to some of the basic American values that led us to what success we've found—values of thrift, individual responsibility, and protection of our heritage.

We're also going to have to look to the concerns of our brothers in a more open way, I think, in order to increase our own economic prospects. We're going to have to look for opportunities to encourage rapid economic growth in the developing world. I would suggest that the greatest promise lies in opportunities for technology transfer into the Third World during a period of industrialization—the best of the technologies we have available to us today, not just the cheapest, those that can convert and use energy with minimal emissions and make the products that the Third World has a comparative edge to produce more extensive and available in our own markets. I would suggest that the most rapid way to success in that domain is to encourage private-sector cooperation through joint ventures, through co-production and licensing agreements, to bring the powers of our market to work in theirs.

Finally, I think we have to think about ways that we

can encourage an improved market position—here in our own industrialized country markets, especially the US—for less polluting technologies. Think carefully about the way we regulate electric utilities, so that they see their profitability and the incentives for increased profitability in the terms that will meet the joint incentives of the government. But clearly the only role, the only path to success that I can identify, at least that's clear to me, is one that can bring together the forces of the private sector with the concerns of the public sector—and in the academic world—about the environment and the needs of the government to provide a stable, safe place for us all to live.

Stephen Peck, Electric Power Research Institute: Thank you very much. I'm going to talk briefly this afternoon about the greenhouse effect and the role of the electricity industry. I will illustrate my talk with examples taken from the US, but many of my points should be interpreted as applying more generally. I will express my own opinion, not that of EPRI or EPRI's members.

There are four points that I'll make this afternoon. First, electric utility operations are likely to be affected by the climate change issue. Second, electricity will help society respond to that issue. Then I'll show how the industry's likely evolution will facilitate the role of electricity in helping society respond. And finally, I'll recommend a policy framework for the climate issue.

There are, of course, difficulties in knowing how aggressively we should act now, and I'll deal with this when I talk about the recommended policy framework. The uncertainty associated with this issue has been referred to by a number of speakers today. I thought it was amusing when I read that the Australian research organization CSIRO writes a disclaimer, something like the following, on papers reporting global warming projections—"Model projections are imprecise. CSIRO will accept no responsibility for decisions based on these projections."

My first point, then, is that electricity industry operations are likely to be affected. 1) If climate change does occur, then examples of operational considerations are a) stream flow may change, and if it does, then hydroelectric energy and cooling water availability will be altered, b) more frequent and intense storms, like tornadoes and hurricanes, could affect the transmission and distribution grid's reliability, and c) power plant operations in some coastal regions could be affected by a rising sea level. 2) If climate change does occur, then air-conditioning loads will rise in the summer and heating loads will fall in the winter. ICF, in an analysis conducted for the Environmental Protection Agency, estimated that the requirements for new generating capacity could increase by 20 to 30 percent in the southwest, southeast, and southern Plains states by 2055. 3) Finally, if a carbon limit is, in fact, imposed, then it's likely that the electric energy growth rate will be reduced in the early part of the next century due to the consequent increase in electricity price; but the growth rate will be augmented in the latter part of the century because it is projected to be cheaper to get carbon out of the electric sector than the nonelectric sector.

The next point I want to make is that electricity will help society to respond to the climate change issue. 1) New generation technologies are being developed over the long term—and I emphasize the long term—that can help reduce emissions of greenhouse gases from power plant operations. These include more efficient ways of burning fossil fuels, development of renewable sources such as wind and solar energy, and the design of smaller, advanced nuclear plants with improved safety features. 2) The utility industry, partly through the efforts of EPRI, is cosponsoring research on increasing the efficiency of electricity use in the residential, commercial, and industrial sectors. Electric vehicles, for instance, charged overnight using off-peak power, could significantly reduce urban pollution and, depending on the use of nonfossil power plants, could also potentially reduce overall emissions of greenhouse gases. And 3) many utilities have long-standing customer focused conservation programs in place such as helping finance home insulation or the purchase of efficient end-use appliances.

I want to deal briefly with how the structure of the US electric utility industry is changing and whether or not that change in structure is likely to accelerate and facilitate electricity's role in helping society respond to the climate change issue. The outline of the industry's future structure is now becoming apparent. Many utilities are starting to view themselves not as one monolithic company, as in the past, but as three linked activities: generation, transmission, and distribution. On the generation side, because of the passage of the Public Utility Regulatory Policy Act (PURPA) in 1978, and because most utilities face a disincentive to invest in their own service territories, new generation investments are being made by cogenerators and by other independent companies (many of those independent companies are utility companies investing in their own non-native service territories). The choice of the winner for a particular generation project is usually being made by a bidding process which is increasingly incorporating environmental, including climate, considerations. Many new players are thus being encouraged to enter the industry, each with his own idea of how to reduce emissions, including emissions of CO_2.

On the distribution side, many utilities will see themselves as providing what I call "high value energy services," energy conservation, where it's appropriate, and increased services from greater electricity use, where that's appropriate. This focus on increasing the efficiency of distribution activities is also likely to reduce greenhouse gases.

On the transmission side, during the long period over which we will be adjusting to the greenhouse issue, more open access to the transmission system is likely to be achieved. This will inevitably reduce the influence of public utility commissions on generation technology choice, and emphasize the importance of economic rather than environmental considerations in technology choice. At that time the issue of how external costs are internalized will become paramount for the electricity industry. The question will be whether those external costs are internalized by command and control strategies, an approach that has been common in the United States, or whether carbon taxes or other taxes will be used to provide incentives for new investment. That issue, of course, is not specific to the electric utility industry, but applies more generally to other industries.

Now I want to expand my focus beyond the electric utility industry to a national, indeed international, focus. We know that global climate is a high stakes issue. We've seen that there are high costs for reducing and removing greenhouse gases. And we've also seen that there are significant environmental effects that could be caused by climate change. Global climate is also an issue on which there is considerable disagreement, as exemplified by what you've heard in the past several days. Current policy discussions center around a 20 percent reduction from current levels of CO_2, which translates into something like a 50 or 60 percent reduction from levels of CO_2 in the middle of the next century. Bill Nordhaus has calculated that to induce such a large reduction, a tax of more than $100 per ton of carbon would be required, and yet he has also estimated that the likely damage done by a ton of greenhouse gases would justify a tax approximately one order of magnitude less than that. So there's considerable disagreement between what some analysis is suggesting and what the focus of current policy initiatives looks like it's going toward.

Given that disagreement, it would seem that we need a framework for conducting analysis and facilitating debate. I believe that the key insights for building that framework is that global climate is a problem whose solution will take at least a century and that over that time period we will learn much about the science and develop many new options. In that sort of situation, a contingent decisionmaking framework is appropriate.

What I mean by a contingent decisionmaking framework is shown and outlined in the figure. In the year 2000, we'll make a set of decisions based on what we know up to that point. We should recognize when we make decisions in 2000 that there's going to be learning taking place between the year 2000 and 2020. In 2020 we'll make a contingent set of decisions, depending on what we've learned during the period 2000–2020; some uncertainties will be resolved after that and so on.

Up to this point, I've been relatively vague about what decisions and uncertainties I mean; this is also laid out in the figure. According to the work of Dr. Schelling, the decisions that we may make have to do with a set of incentives a) to reduce greenhouse gases, for instance by reducing energy use, b) to remove greenhouse gases, for instance by removing CO_2 from power plant emissions, c) to mitigate the effects of climate change, for instance by cloud seeding, and d) to adapt to climate change, for instance by heating and cooling of buildings. There's also a set of decisions which we can make about research. On science, we know that the value of information is very high, to learn more about the greenhouse effect, if subsequent decisions are in fact driven by what we learn. And we also know, on the technology side, that the payoff to successful research and development is high. Manne and Richels, in a 1989 paper, showed that successful demand and supply R&D could reduce the cost to the US of a stringent CO_2 emissions constraint by approximately 75 percent.

Now the uncertainties that will be resolved have to do with the science of the greenhouse effect, and with technological options which are not now available, but will become available over the next 20, 40, or 60 years. Also the knowledge that we will gain concerns multinational cooperation. It's clear that the climate problem is global and multinational agreement will be required. We will learn if indeed that agreement is feasible and if it is sustained.

If we were to develop such an analytical framework, that would help us decide whether we could afford to take low cost actions now, and to wait for the greater scientific certainty about the greenhouse effect and its likely impact, and also to wait for new options to deal with the global climate problem. Calculations by EPA last year indicate that if action to alter current trends were delayed 20 years by developed countries, and 35 years by developing nations, we would probably have to deal with an additional warming commitment of about 40 percent. Now, significantly more certainty could be gained on a range of issues in a far shorter time period. The first order of business I suggest therefore, for the policy community, is to develop a better understanding of both the consequences and the likelihood of an additional commitment to warming of the magnitude estimated by EPA. Such an understanding would tell us what's at stake by waiting.

To summarize my brief remarks this afternoon, I've talked about the impact of the climate issue on utilities, I've talked about how electricity is likely to play a role in helping society adjust to the climate issue, and finally I've made some recommendations for a framework for policy analysis. Thank you.

Thomas Schelling, Harvard University: I want to go back to something that Bill Nordhaus mentioned. The point that I find important, and perplexing, is this: It is exceedingly difficult to demonstrate that the kind of climate change that is being contemplated for the next hundred years will do any harm to anybody in the United States. It's very hard to demonstrate that there will be reductions in any kind of productivity in any sector of the economy. And it's hard to demonstrate that, in terms of health, comfort, recreation, or anything

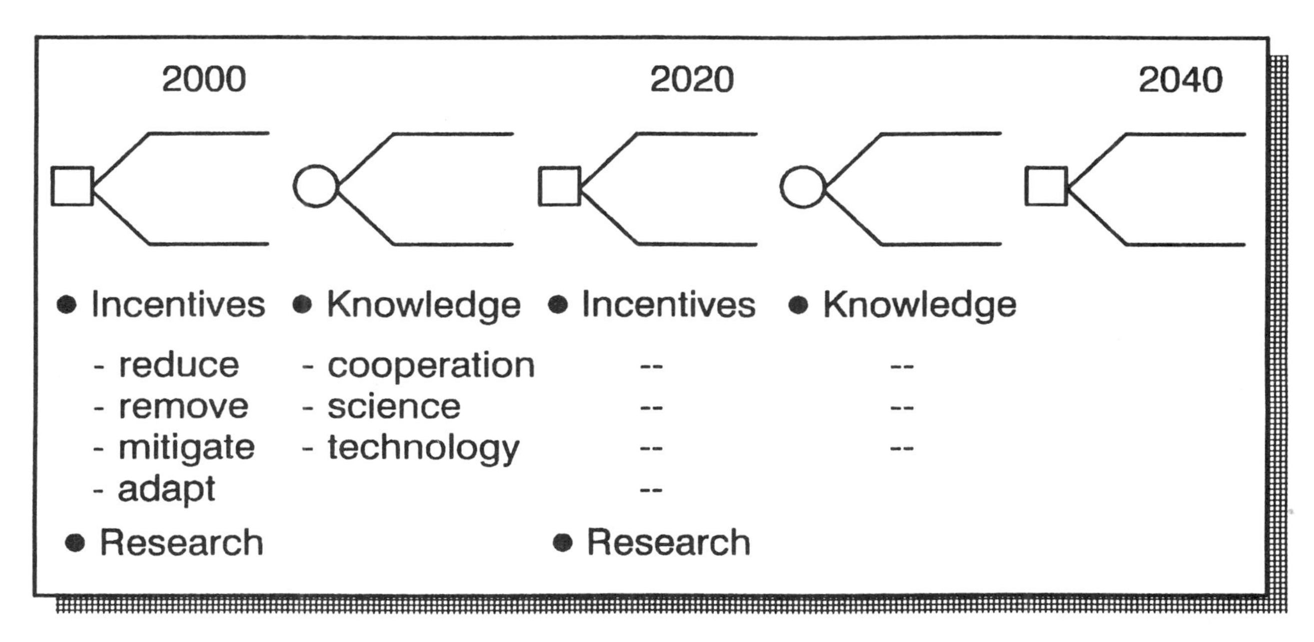

Stephen Peck's recommended policy framework.

else, the kinds of climate change we're talking about will even be noticed 75 or a 100 years from now.

Keep in mind that any of the climate change that is being talked about is less than the climate change I underwent yesterday afternoon in flying here from Washington. Keep in mind, too, that as far as the simulation models are concerned, climates are going to migrate incrementally. Oklahoma may become a little more like north Texas, Kansas will be become a little more like Oklahoma, Nebraska will become a little more like Kansas, and South Dakota will become a little more like Nebraska, but none of them are going to become like Oregon or Massachusetts. Indeed people have been undergoing climate change through all of history with their animals and their crops and their technologies, and surviving perfectly well, and adapting. And if you think of a farmer having to move from South Dakota all the way down to Oklahoma, it is still likely that he wouldn't consider it agriculturally disastrous if he had to use his agronomy in a somewhat more southern climate with perhaps less rainfall, more sunshine, stronger winds, later fall frost, earlier spring thaw, and so on.

The main effects of the global warming will probably not be experienced as warming; it will be changes in rainfall, snowfall, sunshine, frequency of storms, cloud cover, and things of that sort, but if you do think of it as warming, it's worth noticing that in this country, and most of western Europe, nearly everybody goes south for vacation in both summer and winter, and nearly everybody who moves upon retirement goes south, not north. So it's very hard to demonstrate that warming, per se, if that's what you want to focus on, is something that is unwelcome. Indeed if you imagine at the turn of the century, if somebody had said that the average global surface temperature might increase by 3 degrees Celsius by the year 2000, it would have been good news: in those days cold and frost and snow were what mattered, and these days, in our high income climates, it's much easier to get warm in winter than cool in summer, and as a result, nearly all the discussions of warming that you read anywhere, whether it's in your newspaper or even in a journal like *Science,* have to do not with what's going to happen in the wintertime, but what's going to happen in the summertime. It does happen to be good news that the simulations suggest that the warming will occur much more in winter than in summer; and the warming will occur much more in the polar regions than in the equatorial regions.

Now, what I don't know is whether it's good news or bad news that it's hard to demonstrate economic or other damage in the United States, because it may simply be that no administration will be able to persuade the American public to accept measures like a 50 cent gasoline tax per gallon at the federal level. It's a little like the person who is sick but not sick enough to go to bed and take care of himself, and it may be that the inability to demonstrate harm to Americans in the foreseeable future is simply going to mean that we will be unable to participate in any worldwide effort to hold down greenhouse gas emissions, an effort that might be worthwhile. We possibly may, but when I remember that the truck-drivers of America parked their rigs on all highways going in and out of Washington a few years ago because of a rise in the price of diesel fuel, measured in nickels rather than dollars, it seems to me that it's going to be hard to persuade the transportation industry and the farmers that they should cut down on their use of fuel and their use of fertilizer. So I am pessimistic about US participation in any global effort to hold down greenhouse gas emissions. I admit the possibility, though, that within 10 years we and the Canadians and the West Europeans and the Japanese may indeed get something started, and I hope we do. I don't think it will go beyond that in the next 10 years.

What about the Soviet Union, or all of Eastern Europe? I think in the Soviet Union, and in Eastern Europe, and in China so much more is at stake in terms of their economic success that they won't, and they shouldn't, divert any of their energies toward reducing greenhouse gas emissions. I would much rather see Gorbachev succeed in moving toward a successful market economy with at least slightly increasing levels of consumer welfare, and the same in all of Eastern Europe, than see them participate at any expense in a campaign to reduce greenhouse gas emissions, at least for the next 10, 20, or 30 years. The likelihood of catastrophe now or in our children's lifetimes depends much more on whether Eastern Europe and the Soviet Union succeed in what they are attempting now, and I would say the same for China, because I think China will be moving back, before the decade is over, toward a more market-oriented economy. I think it makes more difference to us and our children and our grandchildren that they succeed economically than that they adopt costly restraints on greenhouse emissions. I'd go somewhat further. I don't know whether you saw the *Boston Globe* this morning, about lead poisoning in Romania; there have been stories from Poland and East Germany that are the same. If I were to propose that the East Europeans, including the Soviet Union, do something about noxious and toxic emission for the next 10, 20, maybe 40 years, I'd say you can wait to worry about greenhouse gases. You've got children being poisoned by what they breathe and what they drink, and that's where your priorities should go. So I wouldn't even recommend to them that they participate in something that we and Western Europe and Japan may manage to get started.

I do think that if we ever succeed in getting India, Brazil, China, and very likely Eastern Europe engaged in an effort to cut down greenhouse emissions, we're going to be engaged in the largest trans-boundary financial transfers that the world has ever known. Because it's going to have to be financed by the countries that can afford it and want it. And whether this results in a system of blackmail whereby the Chinese participate in reducing emissions only because we pay the full cost of what they do, I don't know. But if we ask India, China, even the Soviet Union, to take drastic steps to reduce CO_2 emissions, somebody's going to have to pay for it, and it's going to be North America and Western Europe that will pay for it, whether they do it bilaterally or multilaterally or through the World Bank. One of the reasons why any multinational worldwide effort to do something about greenhouse gases is going to be complicated is that we're going to need financial mechanisms; the optimal distribution of costs to reduce greenhouse emissions will not match any optimal distribution of the sharing of the burden.

What do we owe future generations? I notice that Bill Nordhaus's paper talked about the discount rate. I think a crucial issue is whether or not people 50 or 100 years from now are going to be much, much better off than people are now. I don't see any reason to suppose that progress in living standards will not continue around most of the world. I cannot imagine asking my great-grandfather to make economic sacrifices to enhance my welfare in the second half of the 20th century. He lived at the turn of the century in rural America. He lived in such a way that he counts as poor by our standards and he should not have transferred resources forward by making sacrifices. I think it is worth considering whether the people who will be living 50 to 100 years ago will have any claim that we should have made serious sacrifices to enhance their living standards. I can't guarantee you that agricultural productivity will continue to grow boundlessly and that scientific progress will continue to add to productivity all around the world, but I don't see any good reason why that wouldn't go on, and therefore I hesitate to argue that we have this ethical obligation to the wealthy of the future, for whom we should make sacrifices now.

My final point is one that makes me universally unwelcome and unpopular. I'm going to let you in on what I think is likely to be the ultimate solution to the greenhouse problem. When I first began to talk about this, I was the only one I knew who was willing to contemplate it. I wasn't the first to start thinking about it but maybe the first one to go public with it. And that is this: We know how to change the global radiation balance, we know a lot of ways to do it—with CO_2, with methane, with chlorofluorocarbons, with oxides of nitrogen. It just happens that the things that are ineluctable byproducts of what we do anyhow, like driving automobiles, happen to warm the planet rather than cool it. Their molecular structure happens to interfere with the wavelengths of infrared rather than ultraviolet or any other parts of the spectrum. But we do know that there are ways of injecting gases or particulates into the stratosphere that will block incoming radiation. Some volcanos have done it, rather spectacularly. I think it's the volcanos that produce lots of sulfur, not those that produce merely ash. The nuclear winter argument may not yet be resolved but at least it suggests the possibility that if we absolutely had to get something in the stratosphere that would block out incoming radiation there are some drastic ways of doing it. I'd be willing to bet—I might not be around to collect the bet, but I could leave it to my heirs—that within 50 or 75 years we will know how to put things either into orbit or into the stratosphere that can block incoming radiation sufficiently to offset a good part of what we now call the greenhouse effect. Nobody likes people who say there will be a technological fix, don't worry about it. I'm not proposing this is the technological fix, I'm simply predicting that whether you want to think about it and talk about it or not, people are going to.

It also strikes me that this is likely to be the least controversial way to take care of the greenhouse problem. I read a paper the other day of somebody who dismissed this notion in one paragraph by saying that anyhow, he couldn't imagine the countries of the world agreeing on anything as drastic as manipulating the radiation balance. Well, compared with trying to control emissions of methane from the bellies of cows and rice paddies and getting rid of CO_2 from the way the Chinese burn coal to heat their houses and cook their suppers, if we knew of gases or particulates that we could loft into the stratosphere, or plastics that we could put into orbit, it has a huge simplicity. First it's one dimensional in what it's doing. There are only two questions. One is how much to do it? And maybe the easy answer is just enough to offset the greenhouse effect. And the only other question is how do you share the costs? But the costs have to be shared whatever we do. And therefore cost-sharing turns out to be part of any international scheme. What you cannot do is say that this is excessively complicated. Let me make a conjecture. Suppose that methane were not a greenhouse gas but the opposite. And suppose people were saying the whole trouble with this greenhouse warming business is that the CO_2/methane balance is getting shifted. There's not enough methane relative to CO_2. I think people would be out there now, discovering how to release methane in order to offset the CO_2. It just happens that the gases that we are good at releasing as byproducts of what we do, turn out to be warming gases rather than cooling gases. So I will predict to the younger people in the audience that after I have disappeared from this warming earth, you will be reading about what will sometimes come to be probably known as lofting venetian blinds into orbit so that we can turn off the incoming radiation whenever that proves cheaper than facilitating the outgoing radiation. Thank you.

John Deutch, MIT: As we get close to the final moments of this afternoon's session, I'd like to mention a few observations that occurred to me as I contemplated the exchange. We are here to discuss policy strategies for managing the global environment, that is our topic today. We've heard a lot of very interesting and I think important points that testify to the complexity and I would say the robustness of this issue for the future. I would like to make a few personal observations on policy strategies.

First, I have been surprised not to hear my colleagues on the panel, or those of you in the audience, discuss the government's organization, the government's capability for constructing and sustaining the kind of research program that is needed. It may have come up in prior sessions, but I would suggest to you that in talking about policies for strategies to manage the environment, we'd have to have some confidence in starting with the United States government, because as has been mentioned several times, I think it is incumbent on the developed countries to take a leadership role. Is the United States government well organized to design and implement sensible research programs through the whole spectrum from basic research all the way to, perhaps, more applied technology development subjects? Now I'm not asking about the substance of that agenda, but rather whether our government is sensibly organized to do that. And I would just share with you, without going into any great detail, some skepticism on that point. There are just too many government agencies in the pot doing too many things to each other rather than thinking about the problem.

The second point that I would offer is that I believe that we will face problems internationally in incorporating lesser-developed countries. This has been referred to several times by Professor Kaya and by Professor Schelling. The problems include getting both intellectual, and later on, practical cooperation on these global environmental issues, not only in terms of the greenhouse effect, but also in other global issues. This is going to be a really formidable problem. Not only are the Soviet Union, the People's Republic of China, and Eastern Europe going to face more important issues in the short run, but Brazil, India, Pakistan, and many, many other countries of the world are going to find it very, very difficult to really get involved in this issue. And the procedures, the process for achieving international cooperation, is, in my mind, going to be very difficult and quite a great strain on political relationships throughout the globe. I think this deserves much greater attention.

The final observation that I would offer to you is really an intellectual concern. Maybe we're not supposed to have intellectual concerns in public, but I will share mine with you. I have learned to despise exponentials. I used to like them because they were the only things I could integrate, but here we are, it seems to me, stuck using a mode of thinking that really has got as its foundation comparing discounted present value, if you like, discounted benefits and discounted costs. It seems to me that an essential part of this global warming issue is in fact drawing that into question because you are dealing with systems whose uncertainty is enormous, because they are chaotic in their essential physical essence, and because there are difficulties in ever obtaining even the shreds of empirical justifications for parameters that you might use in a cost/benefit analysis. This permits individuals to have a much wider range of differences on these subjects than one would like. I think a central question here is what might replace the traditional cost/

benefit analysis for these global model issues, where the uncertainties are enormous—not because of randomness but because of the inherent chaotic nature of the underlying physical phenomenon itself and because the time scales are enormously long. So that anybody who believes in an exponential will say, let the next person worry about it, let it be done on the next watch. On the other hand there are other people who worry about the irreversibility of these effects, and in my judgment, the balance would point quite properly to the desire to do something now if only we knew what. But this is an issue to which I draw your attention and tell you that I suffer with it quite a bit. The traditional kind of cost/benefit analysis is simply not going to work on this problem.

Professor Lave said that conferences of this sort are important because they allow communities of interested people to look into the next century. I think that certainly has been an important purpose of this conference, "Energy and the Environment in the 21st Century," and I think it's important that all of you took the time to participate in what was surely a very comprehensive look at many of the issues, although not all of the issues that are of concern. I'm pleased that MIT has been the host of this important conference and with that I will declare this plenary session adjourned. I thank you very much for your courtesy and your attention, and also thank the panel members for their participation.

Session A
Transportation Systems

Summary
Session A—Transportation Systems

John B. Heywood
Department of Mechanical Engineering
Malcolm A. Weiss
Energy Laboratory
Session Chairs
Massachusetts Institute of Technology
Cambridge, MA

Effective transportation systems are essential to the economic and social health of all modern societies. Operating those systems requires a large consumption of energy, and that consumption has important impacts on the environment. Transportation systems around the world will inevitably expand and therefore transportation will continue to be a major factor, possibly *the* major factor, affecting energy and the environment in the 21st century.

The world now consumes about 38 percent of its commercial end-use energy to provide transportation services. (In the US, the figure is about 36 percent, including about 62 percent of total petroleum consumption.) Those transportation services move people and goods, by road or rail or air or water or pipe, by sophisticated vehicles or by primitive vehicles, with tight government regulation or with little restraint, for market purposes or for other purposes, with little environmental impact or with overwhelming impact (at least locally).

The agenda of the conference's transportation sessions attempted to reflect at least some of that great diversity of topics; evidence of that attempt is displayed in the following brief descriptions of the papers presented. That diversity also precludes a brief summary of the results in this introduction. The sessions dealt with technological and policy issues, with near-term and long-term problems, and with the needs of both developed and developing countries. Readers are urged to consult the papers of particular interest to them.

Specifically, the program on transportation systems consisted of five sessions, each represented by one to five papers. Session A-1 included four papers that give a broad overview of the transportation system and its environmental impacts.

- Sussman et al addresses the fundamental role of transportation in society and the wide range of alternatives for carrying out that role. He describes how both innovative technology and innovative institutions can improve two of the elements of the system: carrying urban passengers and carrying freight.
- Longwell reviews a recent study by the National Research Council on production technologies for liquid transportation fuels. He estimates US oil and gas resources and cites costs for converting other resources (like coal, tar, and shale) to transportation fuels. He discusses the environmental impacts of fuel alternatives and lists priorities for government funding of research and development (R&D).
- Atkinson et al concentrates on the role of the automobile in urban air pollution. He provides an overview of the recent regulatory history of the automobile, new regulatory developments, and market-based mechanisms that could improve regulatory programs.
- Amann considers the prospects for improving light-duty highway vehicles, considering not only energy-environmental characteristics but also other vehicle characteristics that a consumer seeks. He examines both new types and improved conventional types of power trains and fuels.

Session A-2 consisted of a panel of five speakers expressing their views on self-selected key issues in US transportation energy policy. Most of the speakers focused on the potential for improving the fuel efficiency of automobiles.

- Bleviss argues that improving the fuel economy of light vehicles must be the focus of attempts to reduce environmental impacts of transportation. The government's role should include incentives to vehicle manufacturers, incentives to consumers, and stimulation of R&D.
- Plotkin concludes that the new light duty fleet could average 38 mpg by 2001 if technology diffusion were accelerated, if a few new low-risk technologies were introduced, and if vehicle performance and size were stabilized. The current federal approach to improving fuel economy is limited in effectiveness and distorts the market.
- Bussmann believes that the opportunities for cost-effective improvements in new-car fuel economy are more limited than some other analysts have claimed. His conclusion results from a regression analysis of the fuel economy and technical characteristics of actual 1988 and 1989 model cars.
- Eck focuses on fuels. He opposes the mandating of

any particular fuel, since economics and the free market will arrive at optimal fuels most efficiently. Thus, "costly and inappropriate regulation" of our fuel supply system should be avoided.

- Lave focuses on demand. He believes that the use of personal vehicles in the US will continue to grow but only at about one third of the rate we have become used to. Changing demographics account for that change in growth rate.

Session A-3 shifted from US concerns to those of developing countries. Four members of a panel addressed the problems of expanding transportation services in developing countries and dealing with the energy and environmental effects of those expansions.

- Prud'homme argues that petroleum-fueled motor vehicles will and should be used to provide most transportation growth in developing countries. However, major improvements in energy use and pollution can and should be realized through government pricing actions and through better vehicle choice and usage.
- Polenske considers the constraints on economic growth in China caused by inadequate transportation. More efficient use and transportation of coal (a large burden on the rail network now) would diminish some of the existing energy and transportation bottlenecks.
- Luthra reviews the outlook for expanding the transport system in India during the next 10 to 15 years. Rail capacity will double and shift to diesel and electrification, trucks and buses will grow up to threefold, and personal vehicles will grow even faster. The consequences for urban air pollution loom large.
- Harral discusses the growing problems of pollution and congestion in major urban centers of developing countries. Environmental impacts can be reduced by driving restrictions, taxes, improved bus service, and improved vehicle technologies, but it is difficult to be sanguine about the likely overall environmental outcomes.

Session A-4 focused on case studies. Three speakers were invited to describe how different types of governmental bodies have been dealing with specific problems of transportation energy and the environment.

- Sperling's paper outlines how a particular state, California, is trying to deal with urban air pollution (most importantly, tropospheric ozone) caused primarily by highway vehicles. He gives major attention to fuel choice and discusses the criteria to be considered in making that choice, both in California and, more broadly, in the US.
- Metz summarizes the problems encountered when a group of countries, in this case the European Economic Community, attempts to set standards for car emissions. He goes on to review in more depth the policies of one of the EEC countries (The Netherlands) in dealing with transportation and the environment.
- Trindade describes how a single country, Brazil, has carried out the world's largest experiment in alternative fuels—ethanol from sugar cane. He emphasizes how the experiment's success has waxed and waned depending on the coincidence or conflict of interests of the stakeholders: the government, the state oil monopoly, the sugar growers and alcohol producers, the fuel distributors, the auto manufacturers, and the consumers.

Session A-5, the last session of the transportation program, was devoted to research needs in transportation with emphasis on opportunities to increase energy efficiency and decrease environmental impacts. This session began with a paper by Weiss and Heywood proposing a framework for evaluating research needs in the transportation sector. Their main point is that there is a broad range of strategies for moving toward an environmentally more benign transportation system, but most research agendas do not reflect the full breadth of opportunities.

That paper and other research priorities were then discussed by six panelists: John Brogan of the US Department of Energy, Norman Gjostein of Ford Motor Company, Richard John of the US Department of Transportation, Daniel Sperling (Moderator) of the University of California at Davis, Brian Taylor of Chevron Research Company, and Sergio Trindade of the United Nations Centre for Science and Technology for Development. Following are some of the key points made during the discussion by the panelists or audience members that were directly related to research needs.

- An important research priority in the short term is to learn how to improve the emissions performance of the in-use fleet. In the longer term we need research on the knowledge base required to make good public policy and on the transportation sector as a system. (Taylor)
- We lack the understanding needed for good economic evaluations of transportation options. The economics to be used should be "economics as if the environment mattered." (Trindade)
- There is value in an automotive power plant that can function on diverse existing or new fuels. Research on gas turbines or other versatile power plants, therefore, deserves some priority. (Brogan)
- The Weiss-Heywood framework needs to be developed in more detail in order to be useful. It also needs to recognize that research must yield vehicles with merits beyond those of improved energy and environmental performance. Individual company research should be leveraged by various means such as consortia, university programs, and government programs planned jointly with industry. (Gjostein)
- Federal support for civilian R&D has shifted from technology push and large demonstrations toward knowledge generation, awareness of issues and trends,

and cooperation with nonfederal government and industry. Important research needs include data on the performance of our transportation systems, and improved technology, with cost- and agenda-sharing with industry. (John)

- Only the federal government—not other government subdivisions or industry—can maintain a stable long-term role in transportation research, and that can be done only if the federal role can be insulated from abrupt changes due to new administrations. (Audience, various)
- Data are lacking on transportation energy use and air quality in most developing countries. Even in the US there have been no sophisticated models of the impacts of alternative fuels on urban ozone. (Sperling)

If there is an overall conclusion to be drawn from our transportation program, it is, unsurprisingly, that there is no panacea. There was widespread agreement that 1) there are and will continue to be serious problems associated with energy use and environmental impacts of transportation systems, including transportation's overwhelming dependence on petroleum, 2) easing those problems means increasing the importance given to energy and environmental objectives relative to other social objectives, and 3) a wide range of both technological and policy measures can and should be used to achieve those objectives; portfolios of measures tailored to the needs of particular societies for particular future time frames must be designed and implemented.

The Transportation System: Issues and Challenges

Joseph M. Sussman
Center for Transportation Studies, and
Department of Civil Engineering
Massachusetts Institute of Technology
Cambridge, MA

Carl D. Martland
Nigel H. M. Wilson
Department of Civil Engineering
Massachusetts Institute of Technology
Cambridge, MA

Introduction

Transportation is a fundamental element of society. The manner in which transportation systems are planned and implemented has a profound effect upon political and economic systems and on the quality of life around the world. Decisions made about transportation in antiquity (e.g., the development of the Roman roads systems) and in the present day (e.g., decisions in the developing world about investment in rail vs. highway systems and in the developed world about high speed ground transportation systems vs. future development of air systems) invariably have important impacts upon the ways our society 1) develops economically, socially, and politically, 2) uses scarce energy resources, and 3) impacts the environment.

As an illustration of the critical role that transportation plays in today's society, consider a recent initiative by the United States Department of Transportation (US DOT) to develop a national transportation policy. Under the leadership of US Secretary of Transportation Samuel Skinner, US DOT has embarked on the most substantial effort to develop such a policy in recent US history. The motivation for such an enterprise is clear. To quote the initial documents leading toward this national policy:

> America's unity and vitality are inextricably entwined with the growth of transportation. Throughout our history, transportation has linked farms to markets, oil wells to refineries, factories to consumers, homes to workplaces, and people to academic, cultural, and recreational activities.
>
> **Transportation has been, and continues to be, a powerful engine for America's economic growth.** Some of mankind's greatest advances have been possible only because developments in transportation opened a wider range of resources and economic opportunities. For over 200 years, the ever-increasing expansion of our transportation network of roads, canals, railroads, and airways has stimulated the specialization of economic activity and tied the country together.
>
> Almost one of every five dollars spent on goods and services in our economy—equivalent to 18 percent of America's gross national product—is spent on transportation products and services. Our transportation system handles 3.5 trillion passenger-miles and 3.4 trillion ton-miles of freight annually. Transportation and transportation-related businesses employ one-tenth of America's work force. Annual expenditures for transportation products and services in the United States total nearly $800 billion.[1]

The linkage between transportation and economic growth remains a critical issue. Clearly, growth in a nation's economy will drive growth in the use of transportation services. For example, US data are shown below:

GNP and Transportation Trends
Indexed to 1980

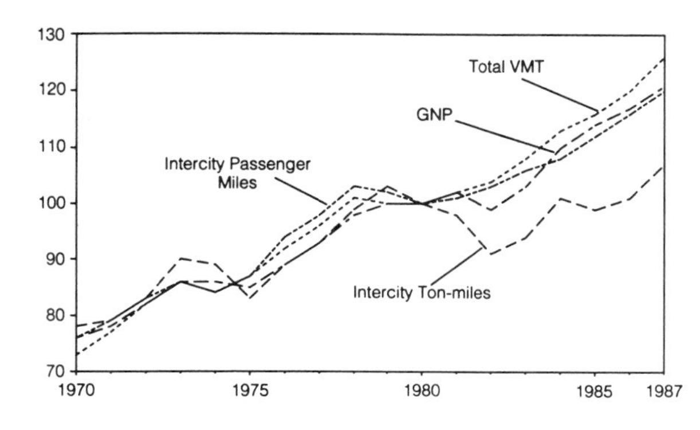

Source: Eno Foundation, *Transportation in America*, Federal Highway Administration, "Highway Statistics," 1970–1987

Historically, transportation has been an important consumer of energy resources. For example, in the

United States about 25 percent of energy consumption is for transportation purposes.[2]

The current modal distribution of energy consumption in the US is as shown below:

Modal Distribution of Transportation's Energy Consumption

Mode	% of Total Transportation Energy Consumption
Auto	50%
Truck	24
Air	7
Water	5
Military	5
Pipeline	4
Rail Freight	3
Bus and Rail	2
	100%

Source: Data supplied by US Department of Transportation, Transportation Systems Center.

Robert Lieb in his text[3] notes that:

> Despite short-term improvements, the automobile, due to the magnitude of its continuing energy requirements, will continue to be viewed as the major area for potential energy conservation in transportation.

The impact of automobiles on the environment is probably equally important.

Transportation as a Complex System—Some Examples

With the above as a brief introduction, let us now consider some illustrative examples of the complex interaction between transportation, energy, and environmental issues. To illustrate anecdotally, consider the recent flurry of activity dealing with Los Angeles' proposal to ban heavy trucks from Los Angeles streets during rush hour. The notion is to ameliorate congestion and improve air quality.

> The measure proposed by Mayor Tom Bradley two years ago quickly earned the support of big political players such as the California League of Cities and the South Coast Air Quality Management District.

But there are opposing viewpoints.

> But as it heads toward the floor of the Los Angeles City Council this month, a state senator from Hacienda Heights and the California Trucking Association are attempting to preempt Bradley in Sacramento by proposing a bill that would bar any municipality from regulating truckers.
>
> . . . allies in Southern California argue that industry would be capsized by the truck ban, forcing truckers, shippers, warehouses and retail outlets to begin delivering goods almost exclusively at night, angering sleeping neighbors, compromising drivers' safety and jacking up retail prices because of higher manpower, employee and utility costs.
>
> Jerry Bakke, president of the Waterfront/Rail Truckers Union, said Bradley's plan would force drivers to work an extra three hours a day to accomplish the same amount of work, a move that would only put more tired and unsafe truckers on the road.
>
> Bakke is not alone in thinking so. Ron Lamb, head of governmental relations for the Los Angeles Area Chamber of Commerce, and Councilman Nate Holden both say the cost of staying open longer at night might force businesses out of the area.[4]

The transportation, environmental, and commercial implications of such a move are profound, and the subtleties of the interactions are complex. Without discussing the pros and cons of this particular proposal, the debate does illustrate the three levels which must be considered when the issues are examined. These are:

1. **Technology**—What are the technical elements that can be brought to bear in dealing with transportation issues (e.g., engine technology)?
2. **System/Economics**—What are the system and economic issues that will impact upon the effectiveness and efficiencies of transportation implementations (e.g., supply/demand equilibrium)?
3. **Institutional/Management**—What are the institutional issues that must be addressed (e.g., deregulation) and the management issues that need to be considered (labor-management relations) in developing transportation systems?

To effectively deal with and understand transportation systems, the subtlety of dealing with each of these levels needs to be addressed. When one includes equally complex systems such as energy and environmental systems, certainly our analyses need to be carried out with extraordinary care.

As our society is transformed by many economic, technologic, and political factors, the transportation implications of this transformation, to say nothing of the environmental and energy implications, are difficult to foresee. We are dealing with a highly complex, dynamic, and interactive system. As we make projections, predictions, and pronouncements, humbleness is the appropriate mind set.

Challenges and Opportunities

We argue that we find ourselves at a very challenging and exciting time in transportation. We are faced with a number of critical transportation issues around the world. Among these issues:

- Congestion
  - Urban
  - Suburban
  - Air
- An increasingly international environment for trade
- Focus on productivity (in both the developed and developing world) with implications for a tighter logistics cycle on an international scale
- Changing face of the transportation industry (i.e., deregulation, changing organization structure)
- Infrastructure deterioration

At the same time, we have the opportunity to take advantage of extraordinary advances in various areas of technology and methodology. Among these are:

- Information sciences
- Communication
- Mathematical methods/operations research
- Materials/structures
- Robotics, automation, and remote sensing
- Organization and economic theory

Our challenge in transportation is to utilize these technological advances to address critical transportation issues. A few examples:

Intelligent Vehicle Highway Systems—The thrust of the argument is that advanced technologies in microelectronics, communications, sensing, controls, artificial intelligence, and other areas will enable the development of a new generation of more automated, much more efficient highway systems. These systems have the potential for greatly ameliorating congestion without substantial investment in conventional infrastructure.

Network Optimization Methods—These methods are highly utilized in complex transportation logistic systems to optimize operations and produce more efficient transportation and manufacturing systems (e.g., just-in-time inventory control). Recent major advances in these methodologies allow us to optimize larger, more complex systems than have been possible previously.

Economic Methods—The development of advanced economic tools to allow more effective analysis of transportation systems has proceeded rapidly in recent years. For example, such methods can be utilized to better understand the linkage between productivity and the decay of transportation infrastructure.[5] This, in turn, leads to better transportation infrastructure, capital, and maintenance decisions.

Materials—There have been revolutionary advances in the materials area over the past decade. Implications for transportation are many. Examples include low weight, high strength vehicles, and innovative materials for infrastructure repair.

We can go on, but the point is clear. The transportation world faces important challenges. That we have the opportunity to improve the system with concurrent improvements in quality of life, energy consumption, and environmental impact is also clear. How to accomplish this is less obvious. We need to consider new technologies and ideas, while also considering the subtle, complex environment in which we operate.

With this daunting background, we discuss urban passenger transportation and freight transportation.

Urban Passenger Transportation

The urban passenger transportation system varies greatly along all significant dimensions between countries in the developed and developing worlds, as well as within each sector. To illustrate this diversity, consider just the United States, in which less than 3 percent of motorized urban passenger miles of travel nationwide is by transit, with the remaining 97 percent by auto. By contrast many cities in developing countries have less than 20 percent of their urban trips by auto, with transit providing the lion's share of urban mobility. Yet even in such an auto dominated society as the US there is the case of New York City in which 72 percent of the trips are by rail transit, and only 14 percent by auto. With such diversity even within a single country, it is obviously very risky to attempt to summarize the state of urban passenger transportation worldwide in just a few pages. It must be recognized at the outset that the existing urban transportation system, and its future evolution, is a function of many influences, including socio-economic, demographic, technological, behavioral, and, perhaps above all, political. The current great diversity in urban transportation systems is a tribute to the diversity in these underlying factors, and in discussing possible system futures, the evolution of all these factors is critical.

Dealing first with economic factors, several influences are clear. First, as per capita income increases so does urban passenger miles of travel per urban resident, independent of the characteristics of the urban transport system. While there is presumably some limit to this relationship, it is clear that most of the developed world, as well as all the developing world, will be facing significant continuing growth in per capita urban trip-making for many years to come. This growth in urban travel is exacerbated by high rates of population growth in most large cities in developing countries. For developing countries, this compounded growth forecast points directly to the second economic factor: the need to make major investments in urban transportation infrastructure. The future urban transportation systems in developing countries will be critically dependent on how the inevitable population and economic growth is accommodated, and what decisions are made on urban transportation infrastructure investment. For developed countries facing slower growth rates and with major sunk investments in urban transportation infrastructure, change in the urban transportation system is likely to be less radical unless major technological innovation occurs.

In terms of technological options, the current array of urban transportation modes—rail, bus, and auto—has existed, without major change in their characteristics, for more than half a century. Each dominates the urban transport system in specific cities, based on the urban geography and governmental policies complementing the inherent strengths of the mode. Urban rail dominates in cities like New York and Tokyo in which extremely high employment and population densities in the central area have resulted in concentrated high volume passenger flows which only rail can serve effectively. In these cities rail is the only existing technology which supports the current urban structure—auto or bus dominated systems would simply destroy them because of the large increase in land demands which would be imposed by these systems. In these cities a large rail network has existed for many years, and because of these past investments, it has been possible largely to avoid major highway investments.

There is no comparable example of a city which has begun constructing a rail network in the past 50 years in which rail now dominates. Buses dominate the great majority of urban transportation systems in medium and large cities in developing countries. Many of these cities are experiencing annual population growth rates of 4 percent or more with increasing per capita trips. Urban population and employment densities are also high enough, given the low auto ownership and per capita incomes, to generate the high passenger flows typically associated with urban rail systems. Indeed, the wave of new urban rail construction activity within the past 20 years has been focused on the major cities in developing countries. However, because of the cost of these projects and the long time between project conception and operation, rail networks to date have been limited in size, and even in the most aggressive rail cities, the rail mode serves only a small percentage of the total public transportation market. In these cities, bus service expansion is relatively inexpensive and is the only viable way to keep up with demand growth, at least in the short run.

Finally, the auto dominates the urban transport in much of the developed world, particularly in North America and Australia. The auto brings significant level of service advantages over public transportation, including convenience and comfort, but at a high price in terms of energy inefficiency, environmental damage, and infrastructure investment. For the auto to enjoy its full service amenity, a relatively low population density and dispersed employment pattern, giving rise to widely distributed person movements at fairly low volumes, is ideal. Most cities which have an auto-dominated urban transportation system tend to evolve to a lower density, suburban form for which it is extremely difficult to offer any efficient public transportation alternative for many urban trips. Thus once an urban transportation system reaches the point of auto domination, it is not clear whether any further system evolution is possible, short of new technology.

Intelligent Vehicle Highway Systems—An Example of New Technology

In terms of new urban transportation technology, there are a number of activities in developed countries on intelligent vehicle highway systems (IVHS) as a strategy to build upon the advantages of the auto-highway system while attempting to overcome some of its deficiencies. To give a sense of what is going on in this important technological area, we quote extensively from a recent IVHS report:[6]

> Advanced technology involving communications, computers, displays, and control processes are the available ingredients for developing IVHS. For purposes of discussion and program development, IVHS have been grouped into four generic elements:
> - Advanced Transportation Management Systems (ATMS)
> - Advanced Driver Information Systems (ADIS)
> - Automatic Vehicle Control (AVC)
> - Commercial Operations
>
> ATMS involve the management of a transportation network including freeways and surface streets. ATMS must be adaptive and responsive in real-time to changes in traffic flow. ATMS must include accurate and reliable surveillance and detection systems, manage both freeway and alternate routes on arterials, provide for interconnection of the systems in adjacent communities, and include effective incident management control strategies.
>
> ADIS are located in an individual's vehicle and interconnected through a communication link with traffic management centers. These systems will provide the motorist with a broad

spectrum of trip and traffic related information. ADIS include a navigation system which provides vehicle location and a route planning assistance system that receives information from the traffic center on congestion, incidents, and other traffic information. An on-board database will contain detailed maps and specific locations of services, hospitals, and tourist related information, such as are found in the "Yellow Pages." Advanced versions will be able to transmit requests for emergency services and information on vehicle location, destination, and speed to the traffic management center.

AVC can lead, eventually, to a fully autonomous vehicle. Such a vehicle would be capable of operating in any traffic environment and traveling portal to portal without driver intervention. Lateral guidance would probably utilize machine vision and pattern recognition. A number of other highly sophisticated sensors and communication devices would provide navigation, collision avoidance, and route optimization capability. Some experiments have begun using collision avoidance systems and adaptive speed controls. On a much less sophisticated scale, cruise control takes over the throttle function from the driver under open road conditions. On a much more sophisticated scale, research on autonomously guided vehicles is under way.

Commercial users comprise a special class of highway users. Their concerns and needs differ from the general public. They require ATT elements that go beyond the needs of the individual motorist. The commercial users operate heavy trucks and other commercial vehicles including taxis. While not 'commercial' in an economic sense, this group also includes public safety vehicles (police, fire, ambulance), and the whole range of vehicles that are not operated by the individual motorist.

It is still much too soon to speculate on what the impact of new IVHS technology might be, although it is clear that any impact will be on cities in developed countries and will not resolve the mobility and investment problems facing the developing world.

Behavioral and Policy Issues

Individual behavior may be a common thread linking the urban transportation systems in different parts of the world. Although tastes and preferences vary among individuals in the same city, let alone among different countries and cultures, to be effective, any urban transportation mode must appeal to enough urban travelers to secure a place in the system. It is also apparent that in much of the world if an urban trip is to be made, the critical issue of mode choice is simply based on answering the question "Is an auto available?" If the answer is yes, then in most cases the auto is used. Experience in the US in particular is very discouraging in terms of the difficulty of persuading an auto-owning commuter to pass up using the car for the trip to work. It is clear that merely offering a heavily subsidized public transportation alternative is not a sufficient inducement, although surely high quality public transportation is a necessary element in doing the job. To be effective public transportation must have clear advantages over the auto in terms of travel time and/or cost to counterbalance the privacy, comfort, and convenience of the auto. In cities with substantial sunk investment in highway infrastructure, with high auto ownership levels and low to medium demand density, this is likely to be simply impossible. While this describes much of the developed world, it does not yet apply to most developing countries; for these countries, at least, real choices still exist.

In developing countries, governmental policies may play a pivotal role in the future evolution of the urban transportation system. A combination of policies which encourage auto ownership and highway construction, and avoid land use planning and control will surely lead in the direction of the current US urban transport system. Let us examine these three elements in turn. Rising levels of auto ownership seem to be an inevitable attribute of economic development, but the unresolved question is whether it is possible successfully to discourage auto usage in the face of rising ownership. Highway construction is also a seemingly inevitable accompaniment to economic development, but here careful planning of grade separated public transportation facilities is a necessary adjunct to highway planning and construction.

Finally, the type and effectiveness of land use planning and control is crucial in determining the urban structure, which in turn is an essential element governing the future viability of public transportation systems. Clearly, there are tremendous stakes in the future evolution of major cities in the developing countries. Energy consumption, environmental impact, and investment could all vary greatly, worldwide, depending on the policies adopted by national governments and by the lending agencies.

Freight Transportation

Introduction

Freight transportation makes possible the specialization that leads to economic development. Neither goods nor resources are distributed uniformly, which inevitably leads to trade among neighboring peoples. The extent of trade is limited by the cost and capacity of freight transportation. Likewise, economies of scale in production enable a single producer to serve a wide market, but only if the costs of transportation do not outweigh the benefits of large-scale production. Similarly, people tend to aggregate in urban areas, but only if food and other resources can be obtained from a wide hinterland. In short, localities and regions can exploit their resources, skills, and location (i.e., their competitive advantages) only if the freight transportation system enables them to trade for the goods and services that they need or desire.

Elements of the Freight Transport System

The basic structure of a freight transport system is simple: freight is consolidated in terminals for movement in vehicles over rights-of-way. However, there is an enormous range of options for each element of the system. The "terminal" could simply be a small grain silo or it could be a highly automated container terminal. The "vehicle" could be a man with a pack animal or it could be a Boeing 747 cargo carrier. The "right-of-way" could be a meandering stream or a rail line designed for heavy axle loads with sophisticated traffic control systems.

Carriers traditionally focus on a single mode of transport—water, rail, highway, or air—and target specific markets where that mode has a competitive advantage in terms of speed or price. Rail and water typically compete for the movement of bulk commodities; rail and truck compete for the movement of agricultural products, forest products, and general merchandise that must move long distances; truck and air compete for very high valued freight. Within any mode, systems can be designed for general freight or they can be specialized for specific types of commodities, such as petroleum, coal, grain, automobiles, or frozen foods. Carriers may focus on serving particular regions or types of commodities, using specific types of equipment, or handling certain size shipments.

There is an important distinction between public and private carriage. Common carriers offer service for specified commodities moving over specified routes at published rates. Common carriers typically are licensed by a regulatory body, which is concerned with such things as the provision of safe service at a fair price, protection of shippers from monopolistic abuses, and protection of carriers from price wars. Private carriers handle a firm's own products, generally with minimal regulatory interference. Somewhere between the two extremes are contract carriers, who negotiate transportation terms with particular shippers. Private and contract carriage offer competitive alternatives to common carriage, which in effect places a limit on common carrier rates.

Long-Term Trends

The intent of this section is to identify the major trends evident in freight transportation, trends that can be expected to continue well into the 21st century. This section is necessarily qualitative. For a more quantitative assessment of the current state of freight transportation, see the statistical reports issued by various modal or governmental agencies[7] or the trade press.[8]

Technological Development. One of the clearest trends in freight transportation is technological development. Over time, vehicles and terminals become faster, larger, more reliable, more efficient, and more complex. Advances in information technology enable better control over far-flung systems, greater capacity on congested systems, and better service for customers. For bulk commodities, technology leads to cheaper means of transloading and transporting very high volumes of bulk commodities. For merchandise transport, technology often leads to increases in speed and reliability, with higher costs of transport justified by savings in other aspects of logistics expense or by the ability to service wider markets.

Commodity Specialization. As an economy develops, the nature of the freight to be transported changes. There are more types of commodities—commodities with higher value/pound, commodities with increasingly specialized transport requirements, and lighter weight commodities. The demands on the freight transport system become more and more intense as shippers seek better ways to move their products. Taken together, these trends mean that the transport market becomes more diverse, with a greater range of service requirements and more opportunities for carrier specialization. UPS found a way to make money moving the small packages that the US Post Office and trucking firms could not handle well and did not want to handle at all. Federal Express pioneered the movement of small packages overnight, which quickly spawned a new, multibillion dollar industry. Railroads and trucking firms battle to find the best equipment for transporting new automobiles cheaply, quickly, and without any damage.

Institutional Complexity. Technological advances reduce the costs of freight transport and thereby reduce the barriers to trade. Producers can seek better inputs from distant locations and sell to a wider market. Transport companies become larger and more complex as they try to meet the demand for transportation. Information technology allows global marketing and global coordination, which paradoxically leads both to global transport companies and to the disintegration of the all-inclusive transport company. Railroads purchase barge lines and ocean carriers, even though they no longer build or maintain their own cars and locomotives. Ocean carriers control port, highway, and rail operations and even become major players in the movement of domestic freight.

Intermodalism and Containerization. Intermodalism has always been an important part of transportation, especially in international trade and in developing countries where water transport is important. The basic idea is to use each mode where it has competitive advantage: truck for pickup and delivery, rail for high volume line haul, water for trans-oceanic, coastwise, and river transportation, and air for rapid transport between cities. Containerization began as a way to minimize the costs of moving cargo through ports, but it has grown rapidly for air and rail transport as well. By using containers, it is easier to move freight between modes, it is easier to prevent loss and damage, and it is possible to simplify rate structures. All of these factors are accelerating the development of containerized systems.

Information Technology. With the need to operate over vast networks, transportation carriers have always had a great concern for communications and information concerning their operations. Railroads, because of the need to maintain safe spacing of trains, have always been particularly concerned with communications and control. With globalization of trade, intermodalism, and increasing institutional complexity of the carriers, information and control are even more critical to carriers. At the same time, shippers and receivers demand better information concerning their shipments as well as more efficient means of communicating with shippers. Also, in international trade, information technology helps to break the log jam at customs. As a result, transportation companies are leaders in the use of electronic data interchange (EDI), computer data bases, operations research techniques, and artificial intelligence applications.

Major Issues

Road/Rail/Intermodal Competition. The boundaries between road, rail, and intermodal competition are not clear, and competition among these modes can be fierce. In general, trucks have a clear economic advantage for short hauls and for small shipment sizes. They can also compete effectively at distances of 1,000 miles or more if trip times and reliability are important to shippers. Railroads have attempted to attract traffic from trucks by using TOFC or COFC (trailer or container on flat car) or other intermodal services. However, shippers generally rate trucks as providing better service than TOFC/COFC and much better service than other rail. As a result, intermodal rates must be lower than truckload rates to attract traffic, and the long-term role and financial prospects of intermodal transport is uncertain. An extreme view would be that railroads will eventually be limited to the transport only of bulk and highly specialized commodities.

Railroad Rationalization. Experience throughout the world shows that construction of a good road network results in trucks capturing a large proportion of the intercity freight that once went by railway. This poses serious problems for the railways, especially those that were constructed when the only competition was the horse and buggy and the riverboat. In the US, for example, the rail system reached its maximum size in the 1920s, and much of the history of the industry since that time has involved the problem of rationalizing the network to reflect the existence of highway competition. Where the railways are state-owned, railway deficits can be enormous. In Japan, for example, the cumulative debt of the Japanese National Railway exceeded $70 billion by the early 1980s, which was one factor leading to the privatization of major portions of the network.

In the less developed countries, the situation is somewhat different. In places like China, where few highways have been constructed, the railway remains the dominant mode for intercity transport for both freight and passengers, and it is very heavily utilized. There the issue is not rationalization so much as expansion. To what extent should the rail network be expanded, given that new highways might alleviate the demand for rail? The future of the railway therefore depends to a large extent upon government decisions concerning both highway and railway development.

In general, it is very difficult to determine the optimal role for the various modes, in large part because of the extensive costs sunk into the existing networks and vehicles and in part because of the peculiarities of past government actions. Government actions affecting freight traffic volumes and mode splits include infrastructure development and financing, economic regulation and safety regulation, and size/weight limitations. In the US and Canada, the governments have been reducing the federal restrictions on railroads and on trucking on the theory that intermodal competition will prevent abuses of market power. In Europe, reductions in transport regulation are very much a part of the changes that will take place in 1992. There are continuing debates concerning the proper size/weight limits and user fees for heavy trucks.

Labor Costs. The transportation industry is one of the most highly unionized industries. Restrictive work rules and high wage rates, often based upon 19th century operating practices, hamper the ability of transportation carriers to take advantage of new technology that allows reductions in crew size, improvements in terminal handling, and more efficient maintenance practices. The problems have been especially acute for ports and for railroads. For a port, labor disputes and high costs can cause carriers to shift to other ports. For a railroad, high costs and poor service can cause traffic to shift to other modes. While railroads have the ability to operate trains with a one-person crew, they are generally required, in the US, to use three-, four-, or even five-person crews. Thus far, rail labor has been insulated from the dramatic changes in pay rates and work rules that followed deregulation in the airline and motor carrier industries.

Multi-modal Transportation Companies. A number of multi-modal transportation companies exist. Many of these are primarily railroads, such as Canadian Pacific and Canadian National in Canada, West Rail in Australia, and CSX in the US. In fact, most large railroads have trucking subsidiaries. However, it is unclear to what extent these companies can actually benefit by providing integrated transportation services. One school of thought is that the modal companies must be allowed to operate independently in order to be effective, in which case multi-modal ownership would have little effect on transport service (although it might be a good financial decision). Another school of thought is that transportation companies must take responsibility for door-to-door service, so that a truly integrated multimodal company can provide better service to shippers. This argument is most compelling to the ocean carriers who must of necessity coordinate with ground transport. American President Lines, a major ocean carrier serving the Pacific rim, has established contracts with US railroads for moving containers inland and acquired a freight forwarder to find domestic freight for the containers that would otherwise return empty to the West Coast.

Logistics Companies. Some carriers already provide logistics services to some shippers. A typical example would be a railroad that owns (and serves) a lumber warehouse, which holds inventory for a variety of shippers who pick up lumber by truck. By handling more of the logistics activities, a carrier may be able to provide more value-added services, thereby becoming more profitable.

Urban Goods Delivery. With increasingly congested urban street networks, deliveries to downtown locations are becoming time-consuming, disruptive, and expensive. This is true both for the small package carriers, which may have several trucks serving a single building, and the truckload carriers, who want to maneuver their 45– or 48–foot trailers into position over a street network designed for, at the most, 40–foot trailers. Newer buildings are designed to permit more efficient pickups and deliveries. At the extreme, the freight traffic can take place on a lower level, thereby segregating freight deliveries from automobile traffic and pedestrians. If the trend toward larger buildings and urban complexes continues, then there could be and will need to be significant improvements in the freight flows in and out of the buildings.

One area where there has been a tremendous amount of change is in small package delivery. Door-to-door and office-to-office deliveries are now commonplace, and people are willing to pay premiums for this service. The ton-miles of such shipments may be very small, but the total annual revenues approach that of the railways. The availability of such service is changing people's buying habits, with catalog shopping becoming very popular, which gives specialty producers a very wide market area. In the future, especially in large residential complexes, might not door-to-door services involve more products than those offered by the Book-of-the-Month Club, Domino's Pizza, and Boylan's Dairy?

Energy and Environmental Implications

Energy Consumption. Energy consumption is a function of the traffic volume, the mode split, modal technology, terrain, and modal operations (load factor, speed, axle loads, empty mileage, etc.). If circuitry and load factors are reasonable, then water is the most energy efficient mode, followed by rail and then air. However, rail unit trains may be more efficient than travel along a much longer water route, and trucks may be more efficient than a local train carrying only a few cars up a poorly maintained branchline. For a particular mode, larger, heavier vehicles tend to be more energy efficient, which explains in part the trend toward larger and heavier trucks and rail cars.

In recent years, traffic has been shifting to more energy intensive modes that provide better service (e.g., rail to truck, boxcar to intermodal, ocean via Panama Canal to minibridge and landbridge rail services from West Coast ports). Clearly, total transportation costs plus the related logistics costs are much more important than energy costs alone in determining mode share. During the 1980s in the US, the real price of diesel fuel declined substantially, truck size/weight limits increased, interest rates increased, and drivers' wages declined: all of these trends made truckload freight more attractive than rail and accelerated the shift of traffic away from rail. Nevertheless, energy consumption per net ton-mile has probably improved because of significant gains in both truck and rail fuel economy, resulting both from more efficient engines and from the use of larger vehicles.

Railroad electrification offers the possibility of reducing the dependence on fossil fuels or at least of concentrating the points where fossil fuels are burned. However, extensive capital investment is required for electrification, which limits its widespread use. In the US, there is very little track that is electrified, and there has recently been a decline in electrified mileage. In Europe, where there is much more passenger traffic, there is much more electrification.

Land Use. Freight terminals occupy large amounts of land, often in prime urban or waterfront locations. In some locations, expansion of freight facilities has been limited to specific areas by local governments. Many of the rail facilities in urban areas are now obsolete or underutilized and offer potential sites for further urban development.

Air Pollution. Freight traffic, in most cases, contributes much less to air pollution than does passenger traffic, simply because there are so many more automobiles in most cities. However, truck traffic can add to the congestion in and near cities, which will increase emissions from all vehicles. Restricting truck traffic to certain streets or to certain times of day is being viewed as a possible way to alleviate congestion by many cities. Los Angeles and Long Beach, for example, are pursuing major plans to improve the access through the city to the ports for both rail and truck in order to reduce congestion and air pollution (as discussed above).

Outlook for the Future

The trends discussed above are all likely to continue into the future. The challenge for governments will be to provide the balanced, rational evolution of all of the available modes. In developed countries, this likely means the rationalization of the rail freight network, improved systems for urban freight movements, and rational limits and prices for the usage of highways by heavy trucks. In less developed countries, this likely means careful analysis of investment options, taking into account the overlapping capabilities of the various modes of transport. Governments must also recognize the institutional complexity of the transport marketplace and the globalization of markets and transportation networks. Carriers will continue to compete aggressively according to whatever rules they are given, using whatever technologies become available. The carriers that provide better service at a better price are the ones that will survive and prosper.

With respect to energy consumption, the demand for freight transportation will increase with general economic growth. If the price of diesel fuel rises very substantially, then there will be some shift from truck back to rail (or to intermodal). However, simply doubling the price of fuel will not be enough to outweigh the great service advantages offered by trucks for high-valued commodities. Much greater increases would be needed to force very substantial amounts of freight back to railroads. Hence, better opportunities for energy conservation are likely to come from such things as improvements in modal efficiency, shifts in economic geography (such as the relocation of auto parts suppliers near assembly plants), and the development and use of lighter materials (plastic and paper products instead of steel).

Extreme changes in energy prices, of course, could have a different effect. A tenfold increase in oil prices would put great pressure on the trucking and air industries. Likewise, a breakthrough in fusion or in superconductivity could vastly change the economics of electrification or even of magnetic levitation.

Conclusion

In this paper, we have tried to address some of the critical issues and challenges in the transportation system. Necessarily, we have had to focus our remarks and have emphasized urban transportation and freight as major sectors. Clearly, there are other issues one could consider. For example:

- **The Automobile of the Future**
  Small, safe, energy efficient or fast, high-tech, energy intensive?
  When will electric propulsion become practical?
- **Intercity Rail Passenger Service**
  Is there really an economical market niche for rail somewhere between short-distance air travel and long-distance automobile travel? How will government price rail service, and how will they finance road construction? Will public transport within metropolitan areas improve and provide better links to train stations?
- **Privatization of Highways**
  Can private enterprise build, maintain, operate, and finance highway capacity more effectively than governments?
- **Airport Congestion**
  Will airport congestion become so great as to force more limits on plane size or restrictions on short-distance trips? Will airport access become an even greater problem that eventually promotes greater use of train, bus, or car for short-distance trips?

One could write an extensive treatise on each of these or on the many transportation system issues noted in this paper.

This paper illustrates the:

- predominant impact of transportation decisionmaking on the very fabric of society,
- tremendous range of transportation alternatives available,
- role of innovation in technology systems and institutions in enhancing transportation service, and
- subtlety of the interaction between the transportation system and other complex systems such as the energy and environmental system.

Clearly, a deep understanding of the transportation system is central to an effective policy and management for "Energy and the Environment in the 21st Century," the theme of this conference. Hopefully, this paper provides a framework for considering these issues.

Notes

1. *Moving America: New Directions, New Opportunities: Volume 1, Building the National Transportation Policy,* US Department of Transportation (July 1989).
2. US Department of Transportation, Transportation Systems Center, *National Transportation Statistics.*
3. Lieb, Robert, *Transportation,* 3rd Edition, Boston Publishing Company (1985).
4. *Traffic World,* pp. 12–13 (August 28, 1989).
5. *Engineering News Record,* pp. 6–7 (September 1, 1988).

6. *Proceedings of a Workshop on Intelligent Vehicle Highway Systems, Mobility 2000,* published by Texas Transportation Institute (February 1989).
7. *Statistical Trends in Transport 1965–1986,* European Conference of Ministers of Transport (ECMT), OECD Publications Service, Paris, France (1988).
8. *Focus: A Decade of Deregulation,* Traffic World Executive Portfolio, pp. 11–23, (November 7, 1988).

US Production of Liquid Transportation Fuels: Costs, Issues, and Research and Development Directions

John P. Longwell
Department of Chemical Engineering and Center for Environmental Health Sciences Massachusetts Institute of Technology Cambridge, MA

Introduction

This paper is based on the recently issued National Research Council report with the above title, which presents a study carried out by their Committee on Production Technologies for Liquid Transportation Fuels during 1989. A major goal of the study was to suggest strategic directions for a five-year research and development program, to be implemented by the Department of Energy (DOE), for production of liquid transportation fuels based on plentiful supplies of domestic resources. A membership list for this committee is attached in Appendix 1.

While this paper concentrates on research and development important to fuels production from domestic resources with priority given to lowest cost, it recognizes that the choice of fuels, the feedstock for their manufacture, and fuel composition will be strongly influenced by additional considerations. Thus, the R&D program should be flexible enough to anticipate and accommodate changes that may be required for environmental and other reasons.

In addition to conventional gasoline, diesel, and aviation fuels, there is a growing interest in alternatives such as methanol and compressed natural gas. Concerted efforts are under way to understand better the consequences to human health, air pollution, and the greenhouse effect from the use of these fuels.

Some anticipation of future conditions is required to plan an R&D program. In particular, it is important to keep in mind the time period at which the R&D program is targeted. The decade 10 to 20 years from now (2000–2010) was chosen since it is believed that this decade will probably be one of major change with major requirements and opportunities for advances in production technologies and for increased use of resources other than oil and gas.

A number of scenarios aid in identifying R&D opportunities. These are:

Economic

I. Oil prices stay at $20/bbl for next 20 years.
II. Oil prices rise to about $30/bbl between 10 and 20 years from now.
III. Oil prices rise to $40/bbl or greater between 10 and 20 years from now.
(All prices in 1988 dollars; substantial price volatility for all scenarios)

Environmental

IV. Increasingly stringent general emissions and waste disposal regulations are established during the next 20 years.
V. Major worldwide effort to control greenhouse gas emissions during the next 20 years.

Government Policy

VI. Neutral.
VII. Encourages domestic production.

For R&D planning, the committee chose Scenario II as the most probable economic scenario, while Scenarios I and III were judged less probable but still likely to occur. Price volatility, however, is expected to be high under any scenario.

Two basic environmental scenarios were considered: (IV) increasingly stringent general emissions, waste disposal, and fuel composition regulations are established during the next 20 years; and (V) because of worldwide concerns about climatic changes, policies to control US greenhouse gas emissions are implemented. These two scenarios are not mutually exclusive.

Consideration was also given to government policies that either encouraged domestic production or were neutral. Even though a trend of increasing international oil prices is expected, price volatility and uncertainty can result in substantial delay in development of domestic resources and in application of new technologies in the absence of government support. Government encouragement in this study is assumed in the form of continued R&D programs with probable cost sharing in large scale demonstrations and pioneer plants.

Several issues were considered in establishing the nature and size of a DOE R&D program for producing liquid fuels:

- expected timing of commercial application;
- potential size of the application;
- potential for cost reduction, improvements in reliability, and diminished environmental impacts;
- the need for DOE participation.

Timing of Commercial Applications

In addition to oil price, the timing of commercial application of new technology depends critically on production costs and environmental impacts. These costs depend on the technology but are also strongly influenced by environmental considerations and by state and federal taxes and tax credits.

While, for research planning, the most probable economic scenario was believed to be for future oil prices to reach $30/bbl (in 1988 dollars) within 20 years (Scenario II); the likelihood that prices would either remain under $30/bbl or exceed $30/bbl appears high enough to necessitate program recommendations that are reasonable given any of the three economic scenarios.

Potential Size of the Applications

Potential size of the applications depends on the size and geographical distribution of the resource. A geographically dispersed resource offers more widespread commercial and employment opportunities and is less vulnerable to local disruption, regulations, and restrictions.

Potential for Cost Reduction

The potential for cost reduction is generally least for mature and technically advanced operations. However, for large-scale activities, such as oil and gas production, even small percentage improvements can justify extensive research.

Need for US Department of Energy Participation

US R&D in transportation fuels production is the sum of industry-supported and government-supported activities. The role of DOE is to help ensure that the major national needs for liquid fuel production technology are met. Where there is substantial and continuing industrial R&D, the role of DOE is generally to support long-range and relevant basic research and in some cases to participate in large demonstration programs. In areas where industrial R&D is low and commercial projects are far in the future, but where continued technological advances are in the national interest, it is logical for DOE to take a lead role.

Resources

At the present time, the US transportation sector depends almost entirely on liquid fuels manufactured from petroleum. It consumed 63 percent of the total US petroleum in 1988. US production of petroleum is declining and is currently approximately 50 percent of the total. This decline results from the high cost of finding and developing domestic resources relative to current international oil prices.

A study of economically producible US petroleum and natural gases for two price levels and two levels of technology was carried out for the committee by the consulting group ICF Resources Inc. (Kuuskraa et al., 1988). Results of this study are summarized in Table 1.

The remaining producible crude oil resources approximately double when the price increases from $25/bbl to $40/bbl and when advanced technology is used. For the high cost level an approximately 50 percent increase in economically producible petroleum is possible with the use of advanced technology. These advanced technologies include the extended recovery techniques of thermal recovery; CO_2 and other miscible flooding; and improved water flooding via polymers, micelles, or pH adjustment. Also included are advances in reservoir modeling to allow effective application of the above advanced extraction technologies.

The ratio of economically recoverable resources to the current US production of approximately 3 billion bbl/yr provides a measure of the time scale involved in using this resource. It indicates that current oil production could be maintained for some decades with a substantial incentive for advanced technology.

Scenario I (with future oil prices less than or equal to about $20/bbl) would result in a continued decline in US oil production, while in Scenario II (prices reach $30/bbl within 20 years) or in Scenario III (prices reach $40/bbl within 20 years), US oil production decline could be reduced for at least the 20-year period of the scenario.

Scenario IV (imposition of more stringent environment controls) seems quite probable. In general, greater environmental controls will increase the costs of exploration and production and will delay the application of

Table 1. Estimated remaining economically producible US crude oil resources—1989.

	Current Practice		Advanced Technology	
	Mod. Cost	High Cost	Mod. Cost	High Cost
Billion bbl oil	75	95	115	140
Resource base to current annual production ratio	25	32	35	47

Moderate cost: $25/bbl
High cost: $40/bbl

advanced oil recovery techniques. Closing frontier areas to exploration and production also reduces the amount of oil available at a given price and shortens the time over which domestic oil could supply a major fraction of US transportation fuels. These trends would increase the need for imports. Energy efficiency improvements can be very important and can also help reduce imports. Scenario V (greater greenhouse gas controls) would tend to limit the effects of thermal-enhanced oil recovery and CO_2-enhanced oil recovery.

For Scenario VI (no government encouragement of domestic oil production), US oil production decline will continue for oil prices below $20/bbl. Even under the price increases of Scenario II and Scenario III, the stabilization of production would require years. Thus, if the US government wanted to retard domestic oil production declines, some form of government encouragement would be required.

Not only is US petroleum production declining, but industry emphasis is changing. The major oil companies are increasingly investing abroad where costs are lower, where the potential for successful large oil fields is higher, and where developing countries are offering special incentives to encourage development of their petroleum resources. In addition, the number and financial health of small independent companies have decreased.

While the traditional form of industry encouragement is through tax incentives, improved technology and its transfer through cooperative efforts will be of increasing importance, especially for the independent operators who, in general, do not have significant R&D programs. To the extent that licensing of technology and use of expert consultants do not facilitate technology transfer to the independent sector, significant advice may be necessary to develop and make available advanced technology to this segment of the domestic oil-producing industry.

Oil cannot be produced to exhaustion at a constant rate but generally declines slowly over time. Even if constant fuel consumption could be maintained, it seems reasonable to expect that significant supplemental sources (either domestic or imported) of transportation fuels will be needed.

Natural Gas

When natural gas is produced, condensible liquids in the gas that are suitable for use in gasoline are recovered. Assuming future yields of liquids from conventional natural gas on the order of those historically extracted, remaining volumes of natural gas liquids are estimated to range from about 12 billion bbl at moderate costs to about 20 billion bbl at gas wellhead prices up to $5.00 per thousand cubic feet. These volumes constitute additions to the liquids potential from remaining crude oil resources. These estimates may be optimistic, since the gas from deeper and unconventional sources is not as rich in liquids.

Economically producible resources of natural gas for two price levels and different levels of technologies are summarized in Table 2. An expansion in the resource base of more than 100 percent is projected at the high price, given use of advanced technology. Significant amounts of gas could therefore be made available as an alternative source of transportation fuels. There are several approaches for exploiting this resource:

- use compressed gas directly for transportation fuel;
- displace fuel oil from power generation and industrial fuel use, making it available for conversion to transportation fuels;
- manufacture hydrogen and carbon monoxide for production of transportation fuels; and
- possibly use advanced, low-cost processes for direct conversion to liquid transportation fuels.

Compressed natural gas vehicles, while not expected to be a significant part of the market because of short vehicle range and onboard storage constraints, have recently attracted much interest as a relatively low polluting alternative for urban fleet use.

A rise in oil price to the range of $25 to $40/bbl would make conversion of heavy fuel oil and heavy oil to transportation fuels more economically attractive. This use is expected to grow. Tar sands bitumen could also be upgraded. Natural gas could be used as the hydrogen source for hydroconversion (which increases liquid yields) of these heavy fuels. Natural gas consumption would also increase from the replacement of the heavy fuel oil that might otherwise be used in power generation and industrial boilers and heaters.

Coal liquefaction and methanol and Fischer-Tropsch (F-T) liquid synthesis from coal are also potentially very large consumers of hydrogen or synthesis gas. For example, the production of the equivalent of 1 billion bbl/yr (2.74 million bbl/day) of crude oil (30 percent of current US production) would require about 40 percent of the gas now produced. While such an increase could come from domestic resources, it would require greatly accelerated exploration and production and would increase the cost of natural gas.

Methane would likely be used for hydrogen manufacture in the initial stages of fuels manufacture from coal and shale because gas price increases will probably lag behind oil price increases, and gas prices may be initially lower than those of the base case used in the economic studies. For the longer term, however, coal gasification to produce hydrogen or synthesis gas will be more economical than use of natural gas to supply hydrogen for coal and shale liquefaction.

Here again advanced technologies can greatly increase the economically producible resource. Many of the same considerations discussed for petroleum also

Table 2. Estimated remaining economically producible natural gas resources.

	Current Technology		Advanced Technology	
	Price 1	Price 2	Price 1	Price 2
Wellhead Gas Price, dollars/Mcf	3	5	3	5
Tcf Gas (billion bbl oil equivalent)	595(107)	770(140)	880(160)	1420(256)
Ratio of Resource Base to Current Production	33	43	50	80

apply to natural gas, with the future health of the industry strongly dependent on government actions. Natural gas imports currently account for a much smaller fraction of US consumption than petroleum imports. An increase in gas price to $5/Mcf would make importation of liquified natural gas much more competitive.

Alternative US Resources

Resources other than oil and gas that have major potential for conversion to transportation fuels are shown in Table 3.

The US geological resource of coal is immense and widely distributed, and the reserve base that is technically recoverable is about 490 billion tons (1.9 trillion bbl of oil in terms of energy equivalent). Full development of underground gasification could extend this resource base.

The largest and richest oil shale deposit (an impure marlstone consisting of silicate and carbonate rock and the organic constituent kerogen) is in the Piceance Creek Basin (Colorado), a part of the Green River formation that occurs in Colorado, Utah, and Wyoming. The bulk of the resource is on government land. Thinner and lower-grade shales occur in the eastern United States. Theoretical conversion estimates of in-place western oil shale resources range from 560 billion to 720 billion bbl; eastern shale resources are estimated at around 65 billion bbl (Lewis, 1980; Riva, 1987). The largest tar sands deposits (a bitumen deposit with greater than 10,000 centi poise in situ viscosity) occur in Utah and Alaska, with smaller deposits in Alabama, Texas, California, and Kentucky. Measured and speculative resources are estimated at about 22 billion and 41 billion bbl, respectively, for a total of about 63 billion bbl. (About 54 billion bbl represent deposits of 100 MMbbl or more.)

Table 3. Alternative US resources.

Resource	Recoverable Amount
Coal	$290–490 \times 10^9$ tons ($1100–1900 \times 10^9$ bbl oil energy equivalent)
Western Shale	$560–720 \times 10^9$ bbl oil
Eastern Shale	65×10^9 bbl oil
Tar Sands	60×10^9 bbl oil
Biomass—methanol	$\sim 1 \times 10^9$ bbl/yr oil equivalent

There is considerable uncertainty regarding the biomass resource base in the US. One estimate is that biomass resources might yield, as a maximum, assuming minimal disruption of the agricultural and silvicultural industries, 1 billion bbl/yr of hydrocarbon fuel equivalents made up of methanol, ethanol, vegetable oils, and others (Sperling, 1988). Emerging estimates suggest greater potential in the resource base. Information available on the practical limits to the production of biomass fuels (i.e., without undue stress on the environment and on agriculture) indicates that biomass could supply a substantial but limited fraction of the total requirements for liquid fuels. Fossil fuels will therefore continue to be dominant in the transportation sector for some time to come.

Production Costs

Conversion of most nontraditional domestic resources into liquid transportation fuels is not economic at current world crude oil prices of about $20/bbl (average refiner acquisition costs in 1988 dollars) but could become competitive as oil prices rise.

The costs of several fuel production technologies were estimated to help compare the economic attractiveness of different feedstock, process, and fuel combinations. Conversion of tar sands, oil shale, coal, natural gas, and wood into transportation fuels as well as the production of ethanol from corn and the use of compressed natural gas (CNG) in vehicles were included. Cost estimates for electric and hydrogen-powered vehicles are not included. Economic estimates were based on US resources, but, since there is currently so much interest in methanol, methanol produced abroad, using natural gas supplies less costly than US natural gas, was also addressed.

Using economic assumptions specified by the committee and summarized in Appendix 2, literature-estimated economic parameters of the processes were initially compiled by SFA Pacific, Inc. (Schulman and Biasca, 1989). These estimates were further refined and updated by various committee members. The final results were used to estimate production costs, on a consistent basis across all feedstocks and technologies, based on current technical understanding. However, many of the technologies have not been commercially implemented, and there remains a high degree of uncertainty about many of the cost elements as well as the total costs.

Annual (constant dollar) discount rates of 10 and 15 percent were used. The costs of natural gas, electricity, and corn depended on crude oil price while wood and coal delivered to the Gulf Coast were assumed to be independent of crude price. Facilities for natural gas conversion, oil shale and tar sand mining, and ethanol production were considered to be located at the resource site. Estimates for natural gas for processing in the Gulf Coast or for use as a transportation fuel included delivery and compression cost.

The results are expressed as the price of crude oil that would produce gasoline at the same price delivered to the consumer as would conversion of these alternative fuels to automotive fuel of the same energy content. Equivalent crude costs for the major processes are shown in Table 4.

Heavy oil and tar sands conversion become competitive at $25 and $26/bbl crude cost to the refiner. These processes produce gasoline by hydrogenation of the tars produced. The substantial amount of hydrogen used is manufactured by steam reforming of methane which, at that petroleum price, costs $3.70 per thousand cubic feet (Mcf) delivered to the refinery. A few hydroconversion plants have been built or planned for special situations where low cost tar or government support is available. Because of the large worldwide resources of tar and tar sands, it is expected that this conversion will grow rapidly in importance, and it is expected that industrial research will continue at a high level.

Table 4. Equivalent crude cost of alternative fuels (in 1988 dollars/bbl, at 10% discounted cash flow).

Process	Cost Estimates for Current Published Technology	Cost Targets for Improved Technology
Heavy Oil Conversion	25	—
Tar Sands Extraction	26	—
Coal Liquefaction	38	30
Western Shale Oil	43	30
Methanol		
Coal Gasification	53	—
Natural gas at		
$4.89/Mcf	45	—
$3.00/Mcf	37	—
$1.00/Mcf	24	—
Compressed Natural Gas	34	—

Coal liquefaction has made substantial progress in the last decade. While published studies resulted in $38/bbl equivalent cost, it is believed that leads exist for reaching $30/bbl or lower. Since, even at this improved cost, our scenario does not call for unsubsidized commercial production for, perhaps, 20 years, it is believed that time is available for discovery and establishment of improved technology before large pilot plant and demonstration programs are required.

The somewhat higher price shown for western shale oil ($43/bbl versus $38/bbl) can be attributed to the relatively small government supported program in recent years and to the absence of published results from the last decade's major industrial programs. It is believed that with a vigorous R&D program, the cost of production from this resource will not be higher than for coal liquefaction and that the same cost target of $30/bbl is appropriate.

Methanol, as a substitute for gasoline, is of growing interest because of possible environmental advantages. Manufacture from coal using current technology is estimated at $35 to $40/bbl equivalent crude price for the same useful transportation energy. Technology for this process is much more mature than for gasoline production by coal liquefaction, and less, but still significant, improvement via continued R&D is expected. Both methanol and Fischer-Tropsch gasoline start with synthesis gas, but since the efficiency of Fischer-Tropsch is somewhat lower than that for methanol manufacture, it is expected that costs for gasoline from this process will be somewhat higher than for methanol. For production of diesel and jet fuel the non-aromatic composition of Fischer-Tropsch liquids is an advantage over conventional coal liquids. However, it is expected that hydrogenation of coal liquids to convert aromatics to napthenes would produce a lower-cost aromatic free product than the Fischer-Tropsch process based on coal-produced synthesis gas. Methanol can also be converted to hydrocarbons by the MTG process at an estimated equivalent crude cost of $62/bbl.

The sensitivity of methanol cost to natural gas cost is also illustrated in Table 4. At $3.00/Mcf it is approximately competitive with methanol from coal; however, in an era where crude cost is greater than $30/bbl it is expected that domestic natural gas would not be available at that price and that methanol would be manufactured at foreign locations where lack of a local market could result in low gas prices and, by this reasoning, methanol is expected to be a high value added import.

Methanol can also be produced from wood and other forms of biomass that can be converted by partial oxidation to synthesis gas. For a 3500 bbl per day (oil equivalent) plant at a wood cost per ton somewhat less than for coal, methanol cost is estimated to be $75/bbl equivalent crude cost. Scaling up to the 50,000 bbl per day size for the coal plant estimate could reduce the cost to about the same as for coal except that, at the fuel prices used, the wood cost would add approximately $10/bbl. Improved technology can be expected to lower methanol cost from wood and coal to about the same extent since the conversion of synthesis gas to product is independent of the source of the synthesis gas, and it is not expected that there will be a significant difference in the cost of producing synthesis gas from the two feedstocks. It therefore appears that, except for special situations where low cost domestic gas may be available, the use of domestically produced methanol would not be justified on purely economic grounds and that special regulations based on other considerations are required if it is to be a major domestically produced transportation fuel.

Environmental Considerations

Environmental considerations can have a major influence on the use and cost of our domestic resources for transportation fuel production and on the composition of the fuels produced. This discussion is divided into consideration of local-regional problems (Scenario IV) and the global problem of the greenhouse effect (Scenario V).

It is reasonable to expect increasingly strict regulation of the many local and regional environmental effects of transportation fuel production and use. The contribution of gasoline vehicle emissions to urban air pollution has generated increased interest in alternative-fueled vehicles using, for example, natural gas or methanol. Another alternative may be reformulating gasoline to facilitate design of improved vehicle emissions control systems. Future vehicle emissions constraints may well affect fuel composition and therefore the choice of conversion processes and related research programs.

Although environmental restrictions may influence automotive fuel composition, the economic, environmental, and health effects of fuel components (paraffins, aromatics, methanol, formaldehyde, and other oxygenates) and optimal control technologies and engine designs are far from well established. The DOE should participate in quantifying these effects and variables to help ensure that production technologies for liquid transportation fuels from domestic resources are properly developed to meet future regulations on vehicle emissions.

Manufacturing processes such as oil refining, methanol production, coal liquefaction, etc., can be controlled to meet current environmental standards with available technology. As regulations become increasingly strict, the added manufacturing cost grows and efficient, lower cost technologies are needed. The situation for oil and gas production is similar, with the major exception of land (and ocean) use environmental considerations. In some cases these considerations are critical, as in offshore drilling and exploration and resource development in wilderness areas and in the Arctic. These areas are the remaining frontiers for the possible discovery of major, low cost reserve additions. According to Kuuskraa et al. (1989), these undiscovered reserves vary between 16 percent of the total resources for $25/bbl current technology, and 22 percent for the high cost advanced technology case—an important, but not dominant, fraction of our future domestic petroleum supplies.

Mining of solids (coal, oil shale, and tar sands) can cause significant local problems and is an especially difficult problem in the use of western oil shale, where a large fraction of the resource is concentrated in the Piceance basin in Colorado, currently a fairly pristine area. Here, existing "prevention of significant deterioration" regulations would, with present technology, severely limit the total production. Water supply and spent stone disposal could also limit operations. Coal, in contrast, is widely dispersed throughout the US, with already established operations and a political constituency in support of existing employment and economic advantages.

The greenhouse effect is of increasing concern. Worldwide, 20 percent of current greenhouse emissions are estimated to come from transportation. The production and use of transportation fuels could be an increasing source of CO_2 and other greenhouse gases. Table 5 shows estimates of relative greenhouse gas emissions for the manufacture and use of transportation fuels from several sources.

Because of coal's low hydrogen content and impurities, the manufacture and use of liquid fuels from coal produce almost twice the CO_2 as use of gasoline from petroleum (see Table 5). Greenhouse gas emissions from the manufacture and use of liquid fuels such as methanol or Fischer-Tropsch gasoline from methane are, however, approximately equal to those from petroleum-based gasoline. Gasoline from oil shale produces less CO_2 than coal, the amount depending on the amount of decomposition of carbonates during retorting. Carbon dioxide emissions can be reduced in all cases by increasing end-use efficiency and by reducing process heat requirements.

The heat necessary to drive processes is conventionally derived from combustion of fossil fuel, with liberation of CO_2. In addition, hydrogen needed for processing is derived from water, where oxygen is eliminated by CO_2 generation. This is a major CO_2 source beyond the CO_2 generated by fuel end use. Nuclear or solar energy and biomass are alternative sources of heat and hydrogen. Water splitting by heat, electrolysis, or photolysis using noncombustion sources of energy is substantially more expensive than use of carbon as an oxygen acceptor (NRC, 1979). However, a long-range

Table 5. Approximate greenhouse gas emissions per mile relative to petroleum-powered internal combustion engines.

Fuel and Feedstock	Percent Change
Current Technology	
CNG, gasoline, diesel, or methanol from biomass[a]	−100
Gasoline and diesel from crude oil[b]	0
CNG from natural gas[c]	−19
Methanol from natural gas[c]	−3
Gasoline from oil shale[d]	+27 to +80
Methanol from coal (baseline)[c]	+98
Potential Advanced Technology	
Gasoline from coal or shale using nonfossil sources for process heat and hydrogen[e]	0

[a] Percent change is for CO_2 only. This is true only for biomass processes that do not use fossil fuel, that do not displace land from forest that would otherwise sequester carbon in its biomass, and that use crops grown every year so that CO_2 from fuel use is taken up by the crops.
[b] Should be increased by 25 to 33 percent for thermally enhanced oil recovery.
[c] Except as noted, the analysis considered emissions of CH_4, N_2O, and CO_2 from the production and transportation of the primary resource (coal, natural gas, or crude oil); conversion of the primary resource to transportation energy (e.g., natural gas to methanol); distribution of the fuel to retail outlets; and combustion of the fuel in engines. N_2O emissions from vehicle engines were not included. Emissions of ozone precursors, chlorofluorocarbons (CFCs) from air-conditioning systems, and water (H_2O) were not considered (available data and models do not allow estimation of the greenhouse effect of emissions of ozone precursors; CFC emissions are independent of fuel use; and H_2O emissions from fossil fuel use worldwide are a negligible percentage of global evaporation). The composite greenhouse gas is actual mass emissions of CO_2 plus CH_4 and N_2O emissions converted to mass amount of CO_2 emissions with the same temperature effect.
[d] Considers only CO_2.
[e] Nonfossil sources could be biomass, nuclear, or solar energy devices.

[Source: Adapted from DeLuchi et al. (1988) and DeLuchi (1989).]

exploratory and basic research program on water splitting is justified. If alternative sources of heat and hydrogen are used in coal liquefaction, coal based transportation fuels become approximately equivalent to those produced from petroleum or natural gas.

Use of biomass to supply heat and hydrogen to fossil fuel processes (if use of fossil fuels in biomass production and processing is minimized) can eliminate or reduce these sources of CO_2. Comparison of the use of methanol generated from biomass via synthesis gas to the conversion of this synthesis gas to hydrogen and its use for coal liquefaction indicates that, for a limited supply of biomass, a greater reduction of fossil carbon-generated CO_2 is obtained by combining biomass gasification with coal liquefaction. System studies and research relevant to this combination are recommended.

Methane is also a greenhouse gas. In Table 5, it is assumed to be 12 times as potent as CO_2 (on a molar basis). The equivalency factor is uncertain and could vary from 5 to 30. Increases in production and use of methane can also be a source of greenhouse gas which deserves more study and evaluation.

Conclusions and Recommendations—R&D Program

A federally funded R&D program on liquid transportation fuels can provide future options for dealing with transportation fuel supply in the face of uncertainties in oil prices and in R&D decisions by the private sector. Domestic resources of oil and gas are sufficient to maintain significant US production for a few decades; however, replacement by use of other resources is required if domestic production is to be stabilized. If oil prices do rise to $30/bbl in 20 years, advances in coal, oil shale, and tar sands technologies can result in economically attractive production from these resources within this time frame. Use of advanced extended oil recovery will become economically attractive and will extend the life of domestic oil and gas supplies. For production from solid fuel resources, this timing does not generally call for immediate large-scale pilot plant and demonstration programs. It is believed that the focus for the next five years should be on establishing further improvements in efficiency, cost, and reduced environmental impact, with the goal of a definite plan for proceeding on large-scale development and demonstration at the end of this five-year period. Our oil and gas resources provide an important buffer for maintaining production, but their optimum use calls for an R&D program that augments a well-established, but ailing, industry. Here demonstration and technology transfer should be a larger fraction of the DOE program.

The recommended areas of R&D have been grouped according to funding levels: major, medium, and modest, as summarized below.

Major Funding

- Participation in R&D and technology transfer for oil and gas production
- Production from coal and western oil shales
- Environmental and end use studies

Medium Funding

- Coal-oil coprocessing
- Tar sands extraction

- Petroleum residuum, heavy oil, and tar conversion processes
- Biomass utilization
- Coal pyrolysis

Modest Funding

- Processes for producing methanol or Fischer-Tropsch liquids from synthesis gas
- Direct methane conversion
- Eastern oil shale

Major Funding Areas

Participation in R&D Technology for Oil and Gas Production. In recent years, the DOE research program in oil and natural gas has been substantially reduced. Industry activity in R&D for domestic oil is also declining. Important opportunities for both cost reduction and improved resource utilization exist, and the degree of DOE participation here should be commensurate with its participation in other energy research areas. The program should focus on those parts of the resource base whose exploitation depends on more comprehensive understanding of geologically complex reservoirs and on technologies yet to be fully developed. The program should be pursued in coordination with industry (both independent oil producers and major oil companies), preferably with direct industry participation. Finally, an effective program of information and technology dissemination is needed.

Production from Coal and Western Oil Shales. Coal and western oil shales both represent very large resources compared to domestic petroleum and natural gas. Estimated costs with current technology require oil prices greater than $38 to $43/bbl, but recent advances suggest that their cost may be reduced to the equivalent crude oil price of $30/bbl or less. Because the cost of producing domestic oil may rise to this level in the next several decades and this is also the time frame required to bring new technology to commercial status, the DOE should establish the goal of reducing the cost of these alternatives to below $30/bbl. The DOE should also take the lead in establishing a demonstration program when pilot plant and engineering studies indicate that this goal can be achieved.

Over the next 5 years, exploratory research on coal should stress new catalysts and processes based on fundamental coal science understanding. The opportunity to reduce costs by integrating hydrogen manufacture should be explored. The program should be guided partly by economic and technical evaluations by engineering firms, petroleum industry operating companies, and qualified consultants. The program should have a five-year objective of reducing the cost of direct liquefaction to $30/bbl or less. If this objective is achieved, preparations for a larger pilot plant (500 to 1000 bbl per day) would begin.

The current shale oil program is considered too small compared to the coal liquefaction program and should therefore be increased. Over the next five years a field pilot facility with a capacity of about 100 bbl per day should be built to further develop surface retorting technologies. These technologies must clearly have the potential for meeting anticipated environmental requirements and for production costs of $30/bbl or less.

Because manufacture of transportation fuels from both of these resources produces more CO_2 byproduct than processes based on oil, gas, or biomass, a special effort should be made to identify and pursue opportunities for reduction in emissions of this greenhouse gas from these resources. Study of nonfossil fuel sources of heat and hydrogen should be included.

Environmental and End-use Studies. There are a number of uncertainties about the implications of alternative fuels use for health, safety, and air quality. With industry and other federal agencies, such as the US Environmental Protection Agency and the National Institutes of Health, the DOE should continue R&D to develop a better data base on these potential impacts. In particular, health effects and different fuel-engine emissions controls combinations should be investigated to identify the safest and most cost-effective combinations and to provide guidance on fuel composition effects for DOE R&D programs for fuels production. This will help ensure that future regulations are balanced and based on a firm technical understanding and that the technologies for liquid transportation fuels production are properly chosen to meet these regulations.

Medium Funding Areas

Coal-oil Coprocessing. Coprocessing of heavy oils or residuum with coal may permit the introduction of coal as a refinery feedstock. It is expected to have rather limited application unless important synergism between oil and coal occurs. Funding of bench-scale research should be continued over the next five years to define the extent of synergy between coal and oil for coprocessing coal-residuum combinations, and to determine its economic impact. If favorable, the results should be confirmed in the Wilsonville, Alabama, test facility to define optimal processing conditions.

Tar Sands Extraction. Domestic tar sands resources are small compared to those of coal and oil shale. However, they are significant compared to proven domestic crude oil reserves, and much of the relevant land area is owned by the federal government. Liquid fuels can potentially be produced from some US tar sands at about $25 to $30/bbl equivalent crude oil price with a hydrocarbon solvent extraction process. Furthermore, there is little industry activity in this area. Therefore, a modest DOE R&D program on tar sands is appropriate if there are sufficient leads toward cost reduction or if costs are low enough to justify development and demonstration of the best technology.

Over the next five years all potential processes and mining techniques applicable to US tar sands should be evaluated both technically and economically. The DOE should sponsor preliminary evaluations by engineering firms, petroleum operating companies, and qualified consultants. The best process should be selected for further development and demonstration in a field pilot plant.

Petroleum Residuum, Heavy Oil, and Tar Conversion Processes. Conversion processes for petroleum residuum, heavy oil, and tar have been under intensive development in both domestic and foreign petroleum industries. Increasing crude oil prices will tend to favor hydroconversion processes over carbon rejection processes because of the higher liquid product yield from hydroconversion. This continuing industrial process development should be supplemented by basic research on the molecular structures of metals, sulfur, and nitrogen binding sites and coke precursor species in heavy oil feeds and upgraded products. Results of this research would help the private sector improve existing carbon removal and hydrogen addition processes. The DOE should involve the private sector in the design of this research program to ensure good technology transfer. This R&D area is considered medium priority because there is considerable activity in the private sector.

Biomass Utilization. Use of biomass resources for the production of liquid transportation fuels is one pathway that can result in less net release of greenhouse gases. Biomass supply constraints and costs will probably require continued use of fossil fuel resources. Use of biomass to produce liquid fuels directly is of continuing interest; however, by integration of the processing of biomass and fossil resources (e.g., by generating process hydrogen from biomass instead of coal), a greater reduction in CO_2 from the combined processes may be achievable. There is little industry activity in this area. Hence, it is recommended that research and systems studies be conducted on the optimum integration of biomass with fossil fuel conversion processes as well as for stand alone biomass conversion.

Coal Pyrolysis. The current DOE program is aimed at production of pyrolysis liquids and metallurgical coke and does not have a high priority for liquid transportation fuels. There is little privately funded R&D in this area. The chemistry and mechanisms of pyrolysis are not well understood, and therefore DOE should place medium priority on a program of basic pyrolysis research, including research in catalytic hydropyrolysis. Systems studies should be carried out over the next five years to evaluate integrating pyrolysis with direct coal liquefaction as well as with gasification or combustion.

Modest Funding Areas

Processes for Producing Methanol or Fischer-Tropsch (F-T) Liquids from Synthesis Gas. Industry is vigorously studying the production of methanol and F-T liquids. While there may be applications in the United States, production is expected primarily outside the United States where low-cost natural gas is available. These factors discourage DOE work in this area beyond fundamental and exploratory research.

Direct Methane Conversion. Direct methane conversion to liquid hydrocarbons or methanol is being studied at the bench scale by various companies, government agencies, and universities. These processes theoretically have the potential for being more energy efficient and less expensive than indirect conversion since they bypass the formation of syngas, an energy-intensive and expensive step. However, potentially significant cost reductions have not yet been achieved.

Even if liquid fuels from direct conversion of natural gas become economically viable, the sources would be predominately based on low-cost natural gas in foreign locations. US government-sponsored research on direct methane conversion technology should be limited to continuing fundamental and exploratory research.

Eastern Oil Shale. Although widespread, most eastern oil shale is low grade, occurs in thin seams, and has a high stripping ratio for mining. Its processing is also inherently more expensive than that of western shale because of its low grade, low hydrogen, and high sulfur content. These disadvantages are expected to outweigh the infrastructure advantages of the eastern location. This resource will be economical only after exploitation of coal or western oil shale. No development is recommended at this time.

References

De Luchi, M., Johnston, R.A., et al., "Transportation Fuels and the Greenhouse Effect," *Transportation Research Record,* 11175, pp. 33–44 (1988).

De Luchi, M., Johnson, R.A., et al., "Methanol vs. Natural Gas Vehicles: A Comparison of Resource Supply, Performance, Emissions, Fuel Storage, Safety, Costs and Transitions," *SAE Technical Paper Series,* 881656, Warrensville, PA, Society of Automotive Engineers (1988).

ICF Resources Inc., "U.S. Petroleum Resources and Natural Gas Reserves," report prepared for the Committee on Production Technologies for Liquid Transportation Fuels, Fairfax, VA (1989).

Kuuskraa, V.A., McFall, K.S., et al., *US Petroleum and Natural Gas Resources: Reservoirs and Extraction Costs,* ICF Resources, Inc. (1990).

Lewis, W.D., "Oil from Shale: The Potential, the Problems and a Plan for Development," *Energy* Vol. 5, pp. 373–387 (1979).

National Research Council, "Hydrogen as a Fuel," Energy Engineering Board, Washington, DC: National Academy of Sciences (1979).

Riva, J.P., Jr., "Fossil Fuels," *Encyclopedia Britannica,* vol. 19, pp. 588–612 (1987).

Schulman, B., and Biasca, F., "Liquid Transportation Fuels from Natural Gas, Heavy Oil, Coal, Oil Shale and Tar Sands: Economics and Technology," report prepared for the Committee on Production Technologies for Liquid Transportation Fuels, SFA Pacific, Mountain View, CA (1989).

Sperling, D., *New Transportation Fuels: A Strategic Approach to Technological Change,* University of California (Berkeley) Press (1988).

Appendix 1. Committee on Production Technologies for Liquid Transportation Fuels.

Oil and Gas Production	
William Fisher (Co-Chairman)	University of Texas—Austin
Robert L. Hirsch	ARCO Oil and Gas Company
Roy Knapp	University of Oklahoma—Norman
Coal Production and Use	
Flynt Kennedy	Consolidated Coal Company
John M. Wootten	Peabody Holding Company
Liquid Fuel Manufacture	
Bruce Beyaert	Chevron Corporation—USA
Robert Hall	Amoco Corporation
Paul A. Kasten	Oak Ridge National Laboratory
Irving Leibson	Bechtel Group Incorporated
Arthur E. Lewis	Lawrence Livermore National Laboratory
John P. Longwell (Chairman)	Massachusetts Institute of Technology
Ronald A. Sills	Mobil Research and Development Corporation
James L. Sweeny	Stanford University
End Use and Environmental	
Seymor Alpert	Electric Power Research Institute
Phillip S. Myers	University of Wisconsin—Madison
Daniel Sperling	University of California—Davis
NRC Staff	
Archie L. Wood	
Mahadevan Mani	
James Zuchetto	

Appendix 2. Economic Assumptions Used in the Cost Estimates.

Natural Gas Price ($/thousand cubic feet)[a]:	$3.91 + 0.05857 × (Poil − 28)
Up to a maximum of:	$5.00/thousand cubic feet
Corn price ($/bushel):	$2.50 + 0.01786 × (Poil − 28)
Corn byproduct price ($/bushel):	$1.20 + 0.01136 × (Poil − 28)
Electricity price (bought) ($/kwh)[b]:	$0.049 + 0.00020 × (Poil − 28)
Gasoline refining (credit) ($/barrel):	$7.00 + 0.18182 × (Poil − 28)
Gasoline distribution/ marketing (Credit) ($/barrel):	$4.00 + 0.02 × (Poil − 28)
Distribution/ marketing for product ($/barrel):	$4 × EQ + 0.02 (Peq − 28)
Coal price ($/ton):	$38.00
Oil shale feedstock ($/ton):	$6.02
Tar sands feedstock ($/ton):	$8.00
Wood price ($/dry ton):	$32.40
Oxygen ($/ton):	$57.05
Formcoke ($/ton):	$100.00
Capital charge rate (%/year):	16.0% and 24%
Consumer discount rate:	10.0% and 15%
Automobile efficiency:	27 mpg
Natural gas delivery cost:	$0.94/thousand cubic feet

Notes to Appendix 2:

Peq = cost per equivalent barrel of oil, net of the additional vehicle cost
Poil = price of petroleum in 1988 dollars per bbl
EQ = 1.8 for methanol; 1.5 for ethanol

[a] EIA Base Case forecast for 2000 has $28/bbl petroleum and $3.91/thousand cubic feet for natural gas; High World Oil Price Case has $35/bbl petroleum and $4.32/thousand cubic feet natural gas. Hence, ($4.32 − $3.91)/7 = 0.05857 is the slope in the equation for projecting gas price as a function of petroleum price.

[b] EIA Base Case forecast of $0.04896/kilowatt-hour for 2000 and High World Oil Price Case of $0.05038/kilowatt-hour for electricity. Hence, the slope in the equation is (0.05038 − 0.04896)/7 = 0.0002.

Role of the Automobile in Urban Air Pollution

D. Atkinson
A. Cristofaro
Office of Policy Analysis
US Environmental Protection Agency
Washington, DC

J. Kolb
Economist, Sobotka, & Co., Inc.
Washington, DC

Abstract

Automobiles are a major source of emissions of volatile organic compounds (VOCs), NO_x, and CO to urban areas. They have been the target of increasingly stringent federal regulations that have significantly reduced the emissions of new automobiles relative to those of the uncontrolled automobiles of the 1960s. Unfortunately, population growth, increases in driving, and a relative increase in the percentage of older cars in the fleet partially have offset the effects of improved emission characteristics of new automobiles. Emissions of VOCs and NO_x (ozone precursors) from stationary sources, which in aggregate account for the majority of these emissions, have remained fairly constant over the past two decades. Thus, many urban areas remain out of compliance with air quality standards. Even with regulatory controls that are likely to be implemented in the near future, most of these areas are expected to remain in violation of air quality standards. This paper will provide an overview of the recent regulatory history of the automobile, new regulatory developments, and market-based mechanisms that could improve regulatory programs.

Introduction

In the Clean Air Act of 1970, Congress established an ambitious schedule of limits for automobile exhaust emissions. Its goal was to obtain a tenfold reduction. Automobiles were a major target for regulation at that time because they were linked to the growing problem of urban air pollution and because automobile manufacturers had no incentive to voluntarily improve the emissions characteristics of cars. In fact, manufacturers adamantly opposed efforts to reduce automobile emissions. Automobiles emit hydrocarbons (HCs) and nitrogen oxides (NO_x), which are precursors to ozone formation, and carbon monoxide (CO). Air pollution problems from ozone, which is the most pervasive pollutant in urban areas across the US, occur mostly during the summer months because the chemical reactions involved depend on sunlight and high temperatures. On the other hand, air pollution problems from CO occur mostly during the winter months.

Although the schedule of emission standards for automobiles was often delayed, the exhaust emission standards envisioned by Congress were finally realized beginning with the 1981 model year. As a result, today's automobile has markedly better emissions characteristics than its uncontrolled 1960s counterpart. Even so, light duty cars and trucks still remain as the largest single source of HCs and NO_x, contributing over 30 percent of national emissions. They are also the major source of CO, contributing over 60 percent of national emissions.

Meeting the nation's air quality goals, as represented by the national ambient air quality standards, has proven to be a difficult undertaking. Notwithstanding the strict regulatory controls that have been placed on automobiles, other mobile sources, and stationary sources over the past two decades, there still are about 100 urban areas that violate the ozone standard and about 60 areas that are out of compliance with the CO standard. Over 100 million people live in these areas (Rosenberg, 1989; EPA, 1988a; OTA, 1989). Unfortunately, most of these urban non-attainment areas are expected to remain out of compliance with air quality standards if the regulatory status quo is maintained. This is what is fueling part of the current debate surrounding the reauthorization of the Clean Air Act.

The imposition of further regulatory controls on automobiles is a central feature of all the strategies that have been proposed to deal with the nation's air quality problems because they represent such a large source of emissions. But the level of such controls, the form of the controls, and the role that alternative fuels might play are issues that have not yet been resolved and are at the heart of the public debate over how best to attain air quality improvements.

In this paper, we will review the regulatory standards that have been imposed on automobiles to develop a sense of how far we have progressed in controlling automobile emissions, and we will examine a number of factors that have partially offset the accomplishments of regulatory programs. We will review projections of emissions and air quality in urban areas assuming that current regulatory requirements are maintained. We will discuss the regulatory approaches that have been suggested to deal with automobile emissions and, finally, we will discuss several market-based mechanisms for controlling automobile emissions that are alternatives to or that could supplement the technology-forcing, standard-setting approach usually favored by Congress and regulatory agencies.

Overview of Automobile Regulations and Their Effects

Congress set a schedule of standards in the Clean Air Act of 1970 that was designed to force automobile manufacturers to develop improved emission control technologies. Emissions limits for other sources of emissions affected by the Act generally were left to the discretion of the EPA. The reasons for the difference in approaches were that automobile emissions constituted a large part of the pollution problem and that automobile manufacturers showed no signs of developing on their own the technologies necessary to deal with air pollution problems (Crandall, 1986; Walsh, 1989). The ambitious schedule established in the 1970 Act was often delayed, either by the EPA or by Congress (as in the Clean Air Act of 1975), but the goal of a tenfold reduction in standards for exhaust (tailpipe) emissions was finally realized for both HC and CO by 1981 (White, 1982).

The exhaust standards that automobiles actually were subject to are shown in Table 1, along with estimates of average emissions of in-use vehicles at 50,000 miles. Although in-use exhaust emissions for HCs and CO exceed the standards at 50,000 miles, the data shows that the in-use exhaust emissions of new vintage automobiles are substantially lower than those of uncontrolled automobiles of the 1960s. Exhaust emissions have been reduced by a factor of about 12 for HCs, 4 for NO_x, and 12 for CO. In addition, evaporative and running loss HC emissions of modern cars are substantially lower than those of older vintage cars (EPA, 1990a). One would think, given the large improvement in the emission characteristics of modern automobiles, that there would have been correspondingly large reductions in aggregate emissions from automobiles and a significant improvement in urban air quality. Unfortunately, that is not the case.

Emission Trends

The bar charts in Figure 1 show estimates of total annual emissions of volatile organic compounds (VOCs), of which HCs are a subset; NO_x; and CO, and the relative contributions of different classes of sources. The data shows that VOC emissions from all sources have declined over the period 1970 to 1987 by almost 30 percent, from about 27 million metric tons (mmt) to just under 20 mmt. Emissions from mobile sources, of which emissions from light duty cars and trucks contribute about 94 percent, declined by about 50 percent, but emissions from stationary sources declined by less than 10 percent. Although emissions from light duty cars and trucks represent about 30 percent of national, manmade VOC emissions, they constitute about 40 percent of sum-

Table 1. Comparison of automobile exhaust emission standards to in-use emissions for HC, CO, and NO_x.

	HC EMISSIONS		CO EMISSIONS		NOx EMISSIONS	
Model Year	Standard	In-use	Standard	In-use	Standard	In-use
Uncontrolled (pre-1968)	-	8.2	-	89.5	-	3.4
1968-1969	4.1	5.6	34.0	69.3	-	4.4
1970-1971	4.1	4.9	34.0	58.0	-	4.4
1972	3.0	4.2	28.0	53.1	-	4.4
1973-1974	3.0	4.2	28.0	54.0	3.1	3.3
1975-1976	1.5	3.0	15.0	35.1	3.1	3.1
1977-1979	1.5	3.0	15.0	35.1	2.0	2.8
1980	0.41	1.38	7.0	14.9	2.0	2.3
1981	0.41	0.81	3.4	10.1	1.0	1.1
1982-1984	0.41	0.70	3.4	7.8	1.0	1.0
1985-1989	0.41	0.66	3.4	7.3	1.0	0.9

Note: Standards and emissions in gpm.
Source: EPA, 1990a.

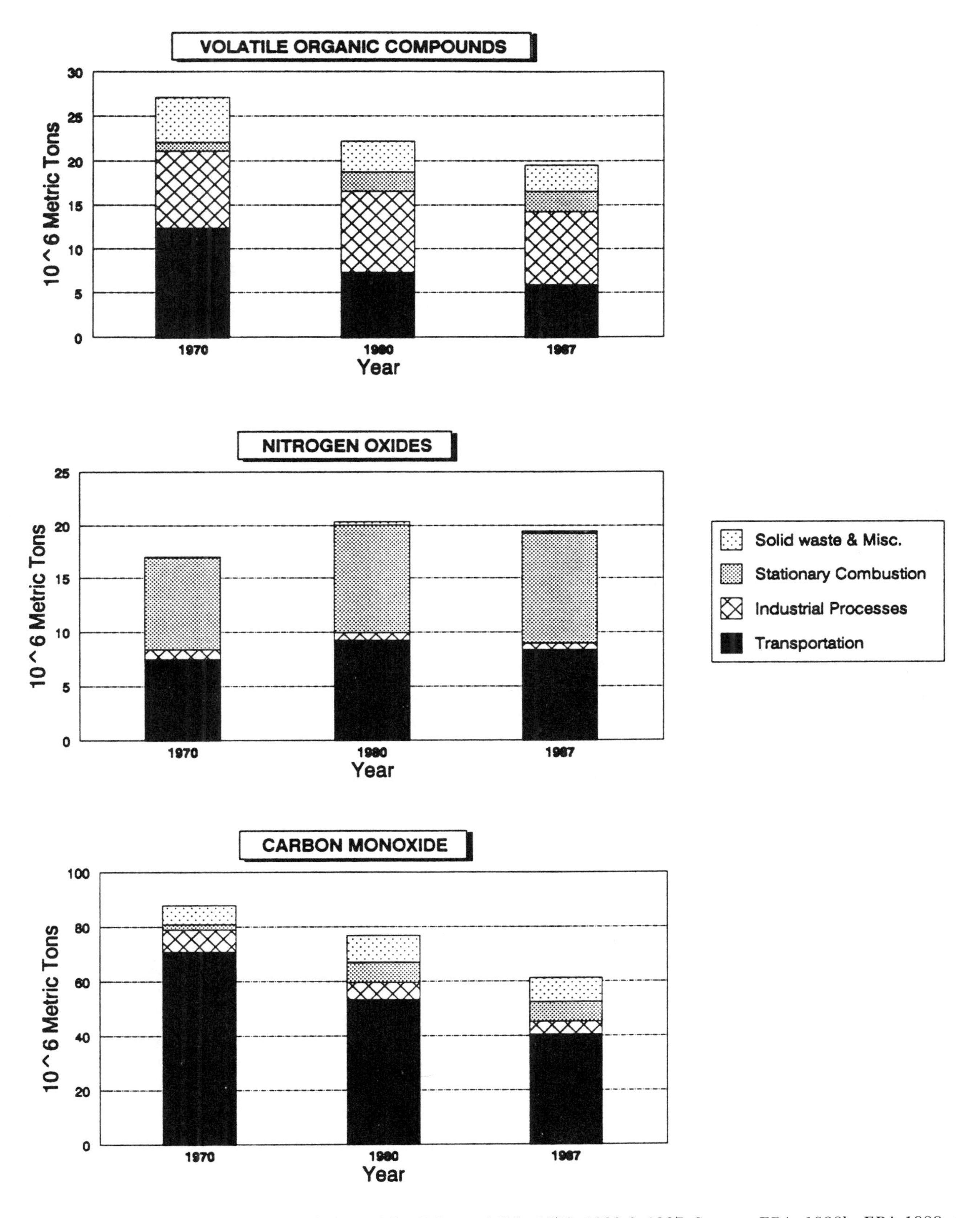

Figure 1. National emissions trends for VOCs, NO_x, and CO; 1970, 1980 & 1987. Source: EPA, 1988b; EPA 1989a.

mertime, manmade VOC emissions in urban non-attainment areas, which are the primary areas of concern (OTA, 1989). (VOCs also are emitted by vegetation, but they are only a small fraction of VOC emissions in most urban non-attainment areas (OTA, 1989).)

Emissions of NO_x, the other precursor to ozone, increased, rather than declined over this period. Emissions of NO_x from all mobile sources (of which cars and light duty trucks account for about 67 percent) increased by more than 10 percent, even though the emissions performance of modern cars has improved significantly. NO_x emissions from light duty cars and trucks now account for about 30 percent of national emissions. Emissions of NO_x from all stationary sources increased by more than 15 percent and now account for about 57 percent of national emissions.

Emissions of CO from all sources declined by about 30 percent over this period. Emissions from mobile sources, of which emissions from cars and light duty trucks contribute about 94 percent, declined by over 40 percent. But this was offset to some extent by an increase in emissions from stationary sources of more than 20 percent. CO emissions from light duty cars and trucks now account for about 60 percent of national emissions.

One pollutant that we have not yet mentioned is lead. The use of tetraethyl lead in gasoline, once the major source of lead in ambient air, gradually has been reduced because of the incompatibility of lead with the catalysts in catalytic converters that are used to meet exhaust standards and because of the lead phasedowns that have been initiated by the EPA. Total lead emissions have declined over the period from 1978 to 1987 by a factor of about 16, and lead emissions from automobiles declined by a factor of almost 40 (EPA, 1989a). Mobile sources accounted for about 90 percent of lead emissions in 1978, whereas today they account for less than 40 percent. This share should decline further as the older cars in the fleet are retired. Additionally, many petroleum refiners no longer market leaded gasoline. Leaded fuel now accounts for less that 13 percent of gasoline sales and its market share is declining by about 25 percent per year (Black, 1989). The large reductions in emissions of lead into the environment and the corresponding decline in the blood lead levels of children in urban areas has been a major regulatory success story.

Trends in Air Quality

The significant, but not overwhelmingly large, reductions in aggregate emissions of VOCs and CO are reflected in changes in air quality over the past two decades. The national ambient air quality standard for ozone specifies a one-hour standard of 0.12 ppm, not to be exceeded more than once per year. Figure 2 shows the trend of the composite average of ozone concentrations at 278 sites in urban areas for which monitoring data is available over the period 1978 to 1987. (This is calculated as the average of the second highest maximum one-hour concentration at each site.) This data indicates that the trend in ozone concentrations is generally downward, but that the reduction in average concentrations has not been large. Because the formation of ozone is affected significantly by meteorological conditions, interpreting trend data is difficult. For example, the summer of 1987 (and 1988) was hotter than normal, so ozone concentrations increased. The average number of exceedances of the ozone standard for the same set of sites has shown more improvement. The average number of exceedances declined from about 12 per year in 1978 to less than seven per year in recent years, a decline of more than 40 percent. (An urban area is considered to exceed the air quality standard for ozone if any of several monitors records ozone levels in excess of 0.12 ppm during any one-hour interval over a 24-hour period.)

Figure 3 provides information on the distribution of ozone concentrations for this set of monitored sites. For example, in 1987, 50 percent of the sites had ozone concentrations between about 0.10 ppm and 0.15 ppm, 15 percent of the sites had ozone concentrations between about 0.15 ppm and 0.17 ppm, and 5 percent of the sites had ozone concentrations between about 0.17 ppm and 0.20 ppm. Not included are the 5 percent of sites above and below the highest and lowest concentrations indicated by the bar graph. This data indicates that, even though there has been a downward trend in the composite average of ozone concentrations, more than 50 percent of the monitored sites still exceed the air quality standards for ozone. More recent information developed by the EPA indicates that over 100 urban areas currently exceed the ozone standard. Additionally, a new study conducted by the Sierra Club of ozone levels in 12 cities suggests that the EPA understates the severity of ozone problems. According to recent studies, health effects from exposure to ozone may occur at levels as low as 0.08 ppm if the exposure is long term. Using an eight-hour average ozone concentration of 0.08 ppm as an indicator of the potential for health problems, rather than the EPA's one-hour average, ozone exceedances were found to occur about 65 percent more frequently than reported by the EPA (Envir. Rep., 1990).

The national ambient air quality standards for CO specify a one-hour standard of 35 ppm and an eight-hour standard of 9 ppm that are not to be exceeded more than once per year. There are very few exceedances of the one-hour standard, so we provide information on trends for the eight-hour standard. Figure 4 provides information on trends in CO levels at 198 sites in urban areas for which data is available over the ten-year period from 1978 to 1987. This data indicates that there has been a significant reduction in the composite

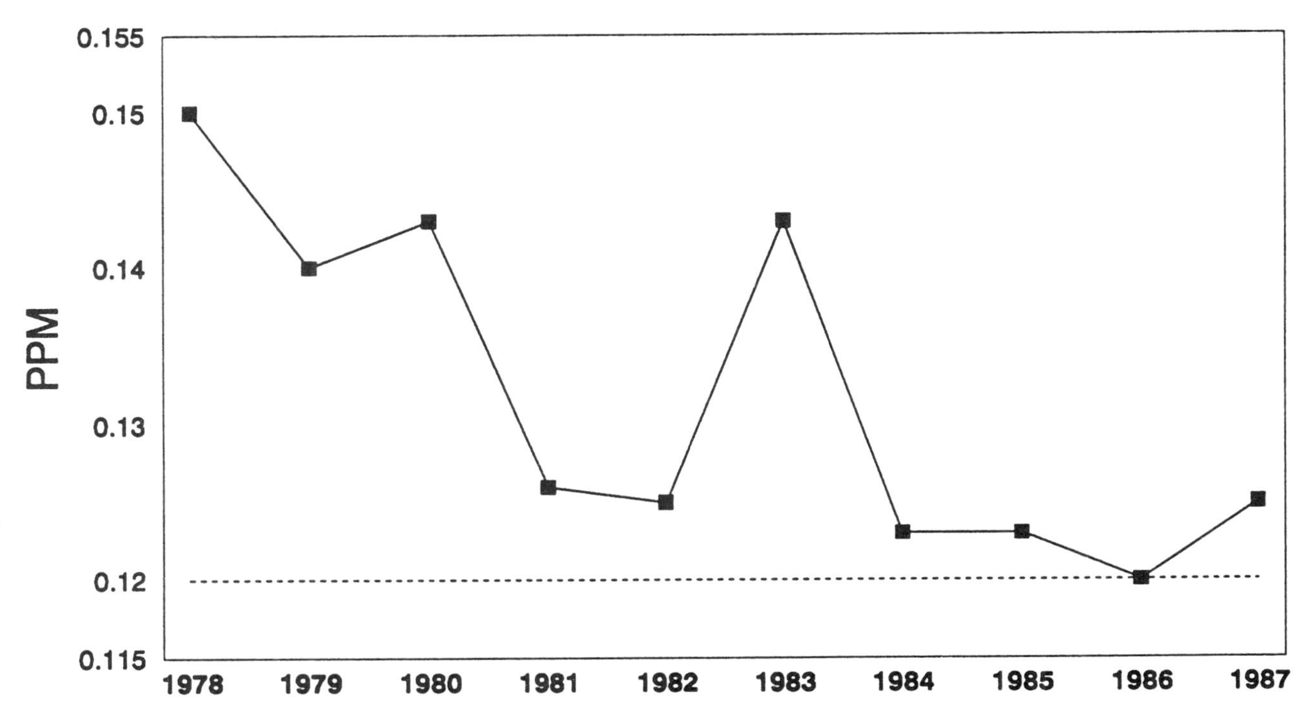

Figure 2. Composite average of ozone concentrations. Source: EPA, 1989a.

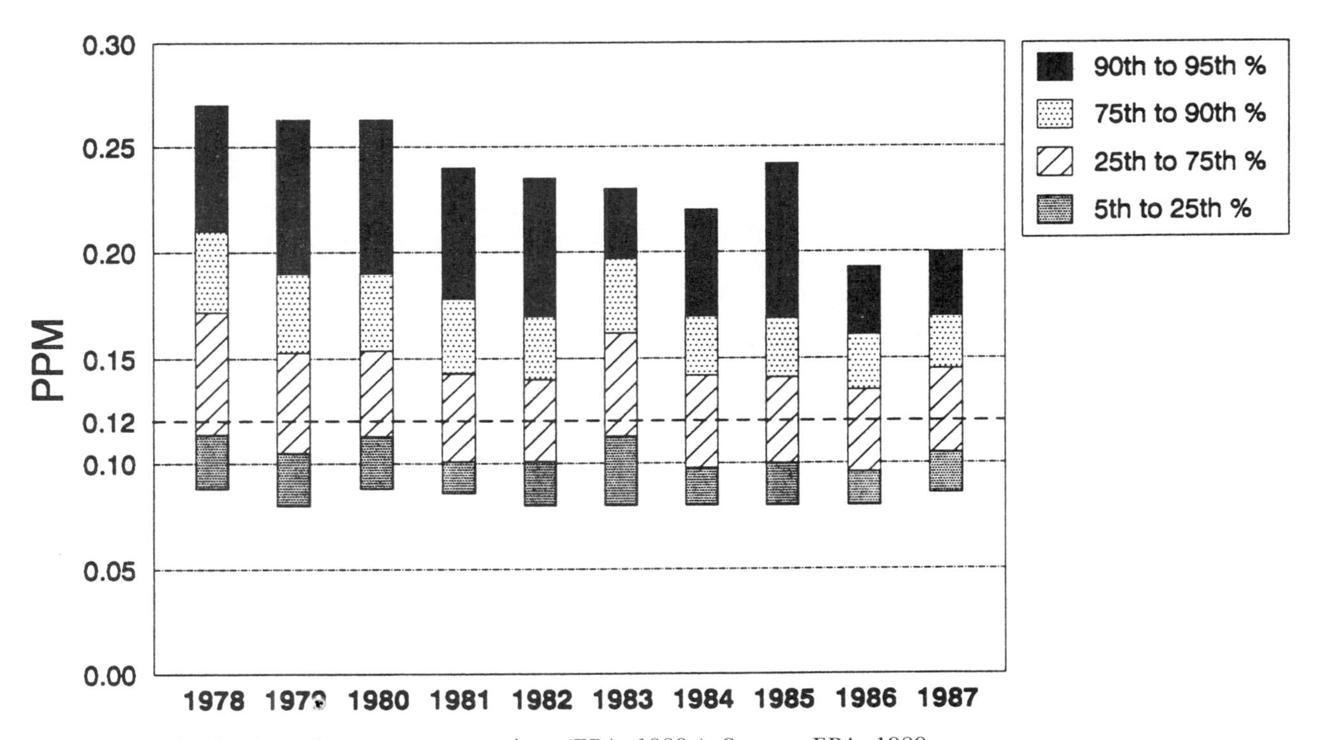

Figure 3. Distribution of ozone concentrations (EPA, 1989a). Source: EPA, 1989a.

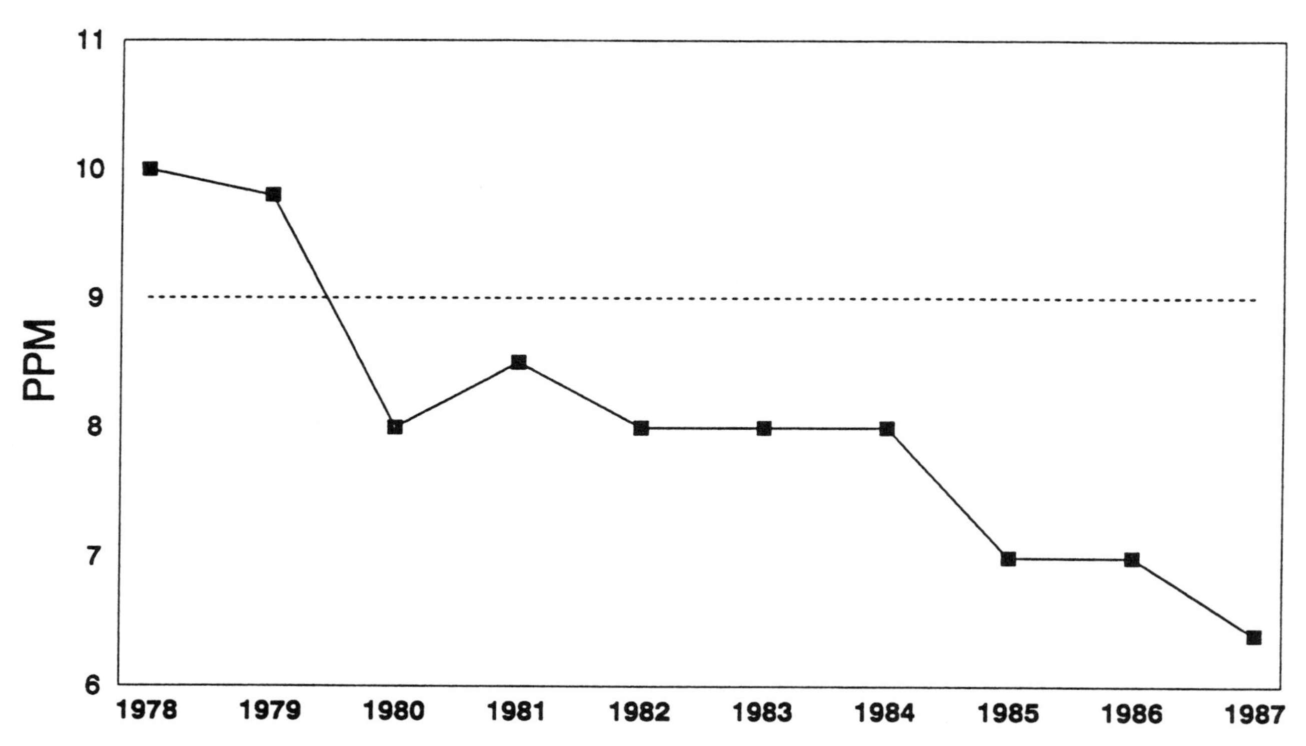

Figure 4. Composite average of CO concentrations. Source: EPA, 1989a.

of average CO concentrations. In addition, the average number of exceedances per site declined from about 14 per year in 1978 to about 1 per year in 1987. Figure 5 shows the distribution of annual CO concentrations. The data indicates that, on average, about 25 percent of the sites were over the CO standard in the late 1980s.

Despite these improvements, CO still remains a substantial concern in many urban areas. Currently, about 60 urban areas exceed the standard and some exceed the CO standard many times per year. For example, the average number of annual exceedances from 1986 to 1987 was 63 for New York, 47 for Los Angeles, 55 for Spokane, and 28 for Denver (Walsh, 1988).

At one point, the EPA projected that most urban areas would be in compliance with the CO standards by 1987. Efforts were even begun in the early 1980s to relax the CO emission standards for automobiles. Unfortunately, the CO problem has turned out to be less tractable than it once was thought to be. The EPA is in the process of collecting additional data to improve its understanding of the CO problem. Available information suggests that excess emissions during cold starts are a major factor in causing CO problems, though auto makers contend that such emissions are a minor factor and that exceedances correlate with atmospheric inversions (EPA, 1988a).

Factors That Offset Improvements in Emission Controls

Given the improved emissions performance of the modern automobile, why have we not seen larger improvements in emissions and air quality? First, reductions in emissions from stationary sources have not kept pace with reductions in emissions from mobile sources. Much of the reduction in emissions from large stationary sources has been offset by growth in emissions from small stationary sources (those that emit less than 50 tons per year), such as paint shops, furniture makers, printers, and households.

Second, total emissions from automobiles reflect emissions not only from new cars, but from old cars that still remain in the fleet. At any time the fleet will consist of a mix of newer cars with low emissions and older cars with higher emissions. Time must pass to allow the fleet to turn over and reach a steady state to obtain the full benefit of the current emissions standards. Clearly, we still have many vehicles of pre-1981 vintage on the road, and they account for a disproportionate share of emissions, even though they are driven fewer miles per year than newer cars. For example, although pre-1981 vintage vehicles accounted for only about 35 percent of

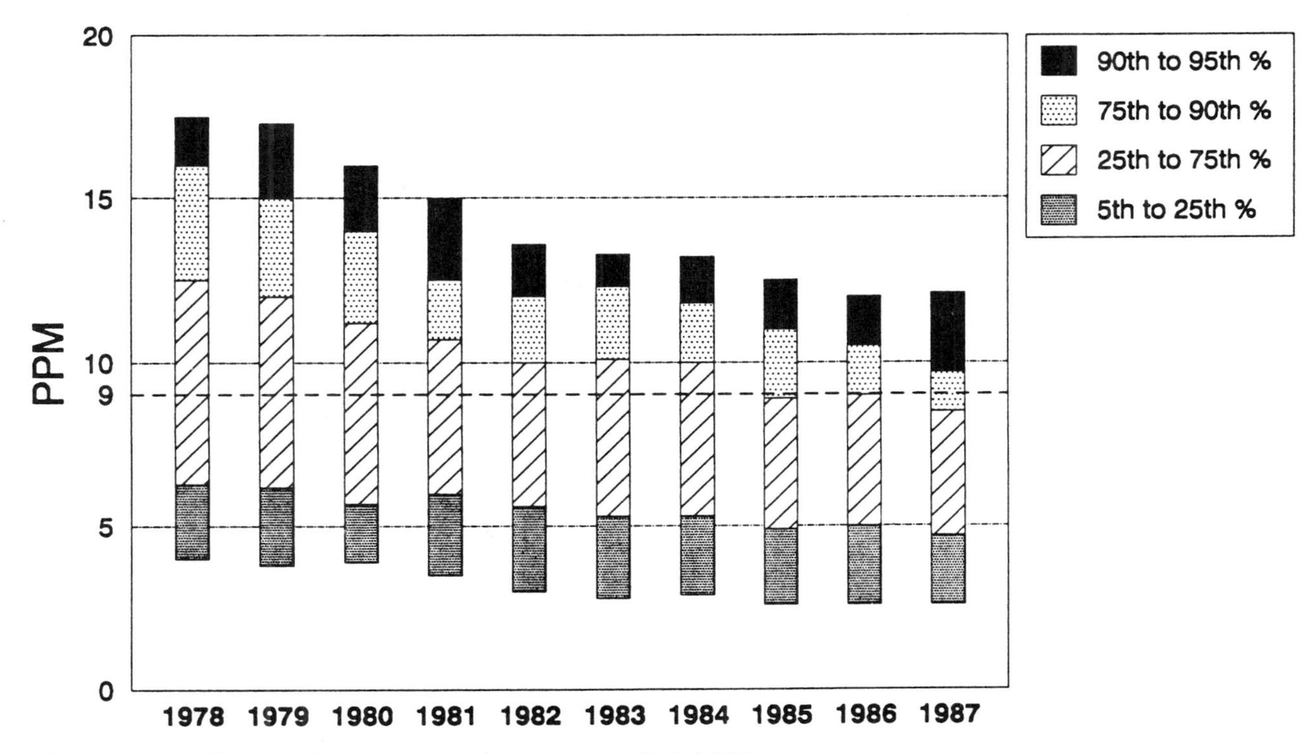

Figure 5. Distribution of CO concentrations. Source: EPA, 1989a.

vehicle miles traveled (VMT) in 1988, they accounted for about 70 percent of HC emissions, 75 percent of CO emissions, and 68 percent of NO_x emissions. The disparity in emissions increases with vehicle age. Automobiles at least 15 years old accounted for less than 8 percent of VMT in 1988, yet they accounted for about 23 percent of HC emissions, 22 percent of CO emissions, and 18 percent of NO_x emissions (EPA, 1990a). This keeps the average of fleet emissions from declining as fast as emissions of new vehicles. Nonetheless, the average emission of HCs of the current fleet is less than a third of that of the 1970 fleet.

Beyond this inventory effect, a number of developments over the past two decades have combined to significantly offset the per vehicle emission reductions gained through the provisions of the Clean Air Act.

Table 2. Vehicle miles traveled (VMT) by passenger cars.

Year	Miles traveled (trillions)	Registered vehicles (millions)	Miles per Car	VMT Per Capita
1987	1.36	137.3	9,883	5,576
1986	1.30	135.4	9,608	5,397
1985	1.26	131.9	9,560	5,280
1980	1.11	121.6	9,141	4,891
1975	1.03	106.7	9,690	4,799
1970	0.92	89.2	10,272	4,494

Source: MVMA, 1989.

Vehicle Miles Traveled

The most significant factor that has offset per vehicle emission reductions is the large increase over the past two decades in the number of automobiles in use and in the total number of miles these vehicles travel (VMT), as is shown in Table 2. If recent trends continue, total VMT in 1990 will be almost 60 percent higher than in 1970. The annual growth rate in VMT averaged about

2.3 percent per year over this period, more than double that of the population growth rate. Vehicle registrations grew about three times faster than population.

The growth in vehicles and in VMT apparently resulted from a number of factors that may be unique to this period: 1) The population eligible to drive grew faster than the total population, due to the maturation of baby boomers; 2) Women entered the workforce in increasing numbers, thereby increasing the number of cars per capita and the number of miles driven per capita; and 3) Incomes increased, allowing more potential drivers to own cars (Lave, 1989). There is some evidence that the ownership market for automobiles has effectively been saturated. This could result in automobile ownership and VMT growing in the future at a rate closer to that of the general population and income (Lave, 1989).

Age of the Fleet

Another development affecting aggregate emissions is the increase in the relative percentage of older cars in the fleet. Figure 6 shows the composition and the average age of the fleet over the past two decades. The average age of the fleet increased from about 5.8 years in the 1970s to 7.6 years in 1985, and since has remained at about that level. Cars nine years of age or older now compose almost 37 percent of the fleet, whereas in the early 1970s they represented only about 19 percent of the fleet. The net effect of the increase in age of the fleet, taking account of the mileage differences between older and newer cars, is to increase emissions from the current fleet by about 20 to 25 percent.

There is some evidence that the additional costs imposed by environmental and safety regulations may have indirectly increased the age of the fleet. Price increases in new cars tend to increase the value of older cars, making it attractive to operate older vehicles rather than scrapping them (Crandall, 1986). Additionally, the large increase in demand for automobiles over this period, the growth in multiple car households, and improvements in the quality of automobiles would lead to an increase in the age of the fleet. The average age of the fleet appears to have stabilized over the past five years, so future emissions may not be adversely affected by continued increases in age. Economic recessions or large increases in the price of new cars, however, could cause the fleet to again increase in age.

Volatility of Gasoline

Another factor affecting automobile emissions is the significant upward trend over the past two decades in the volatility of gasoline, as measured by its Reid vapor pressure (RVP). Figure 7 shows that the average RVP during the summertime (the period during which ozone presents a problem) increased from about 9 psi in 1970 to almost 10.5 psi in the 1980s. The rise in the volatility of gasoline has increased the evaporative emissions from automobiles and has led to an increase in automobile related VOC emissions of more than 30 percent (EPA, 1987a). Evaporative emissions result from: 1) the continued heating of fuel in the fuel line, carburetor, or fuel injection system by the engine after it is shut off (hot-soak emissions), 2) the daily heating of the fuel tank by the outside air (diurnal emissions), and 3) heating of the fuel tank while the car is running (running losses). VOC emissions also result from the displacement of

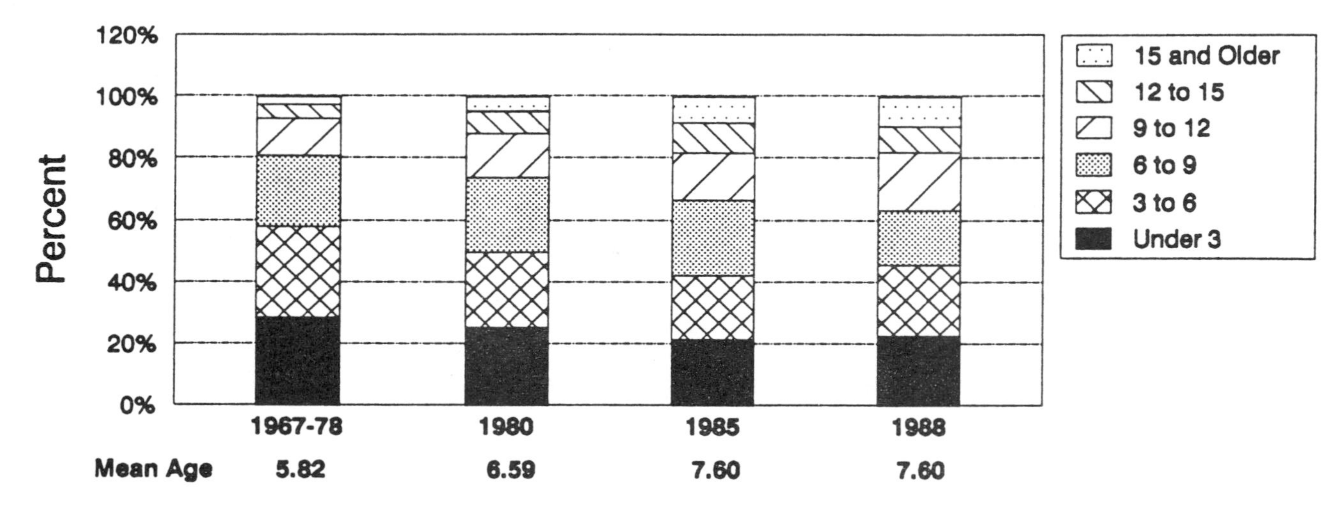

Figure 6. Composition of the vehicle fleet as a percentage of total registrations by age group in various years. Source: MVMA, 1989.

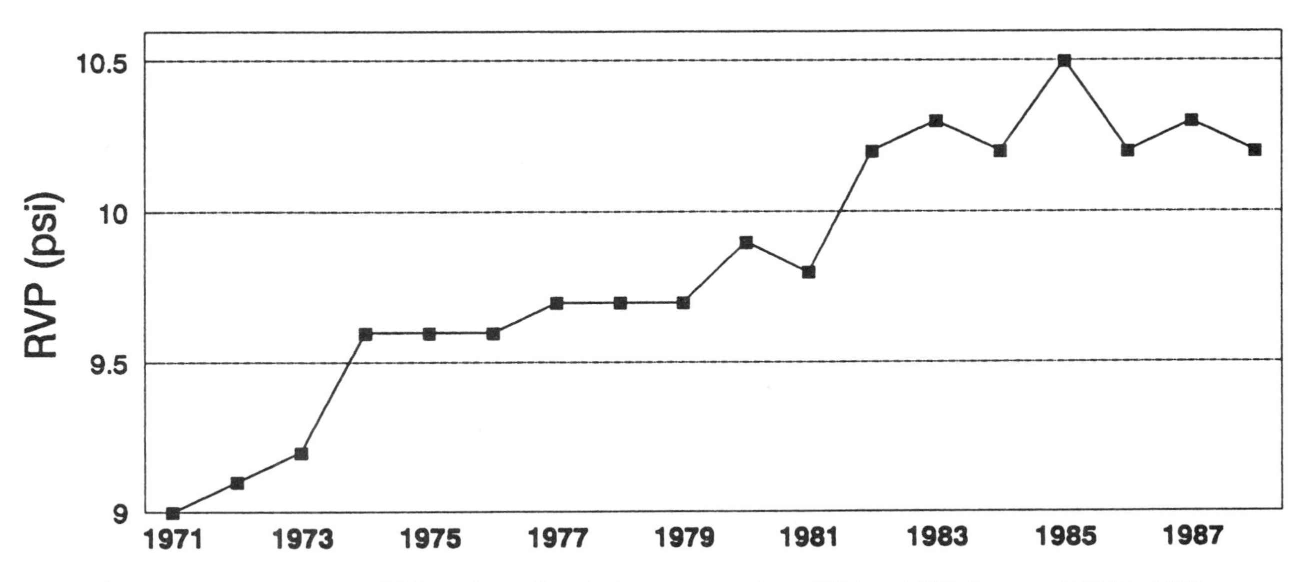

Figure 7. Reid vapor pressure (RVP) of gasoline during summertime, 1971 to 1988. Source: NIPER, 1989.

vapors and some spillage during refueling. The EPA's current certification procedure for automobile evaporative emission control systems uses a nine RVP test fuel. Because the fuel actually used in automobiles during the summer months has a higher RVP and generates significantly more vapors, the evaporative control systems in cars can be overwhelmed. Evaporative and refueling emissions account for about 70 percent of the summertime HC emissions from 1981 and newer vintage automobiles and about half to two thirds of the summertime HC emissions from the entire fleet (EPA, 1990a; OTA, 1989).

The volatility of gasoline increased because of a number of factors. The use of catalytic converters by auto makers to meet exhaust standards required the use of unleaded fuel. Removing lead from gasoline reduces its octane, and to replace the lost octane, refiners blended in low-cost, high-octane/high-RVP butane and also increased the amount of aromatics and olefins. This raises the average volatility of gasoline and increases its reactivity. The EPA's recent lead phasedown of leaded fuel to 0.1 grams per gallon further contributed to this trend. The use of fuel injection systems in automobiles of recent vintage has made them more tolerant of high RVP fuels (EPA, 1987a). Finally, in recent years there has been an increase in demand for higher octane fuels. Over the past five years, the share of premium gasoline nationally has grown from about 14 to 25 percent. This, in turn, has lead to increases in aromatics in gasoline because they are high-octane blending components. Aromatics now represent about 32 percent of the components in gasoline, whereas in the 1970s they represented on the order of 25 percent (Black, 1989; Unzelman, 1988).

The EPA has already promulgated interim standards to reduce the volatility of gasoline by about one psi and has proposed final standards requiring that the RVP of gasoline be further reduced (to nine psi in Class C areas) in the early 1990s (EPA, 1989b). This should result in large reductions in emissions of HCs, on the order of from 30 to 40 percent of HC emissions from automobiles (EPA, 1987a; EPA,1990b).

Net Effects

The net impact of the increase in VMTs and in the age of the fleet was to offset the aggregate emission reductions expected to be gained from controlling new vehicles by a factor of about two. The increase in the volatility of gasoline raises this factor to in excess of 2.5 for HCs during the summer months. This accounts for much of the divergence between the significant improvements in the emission performance of new automobiles and the trend of aggregate emissions from automobiles over the past two decades.

Projections of Future Emissions and Air Quality

Aggregate emissions of VOCs, CO, and NO_x, under the regulatory status quo, are projected to decline in the 1990s, primarily due to regulations affecting automobile emissions, and then to begin to increase after the year 2000 due to the general effects of population growth

(OTA, 1989; EPA, 1987a; Walsh, 1989). We focus on VOC emissions because ozone is the most pervasive urban air pollution problem and because the EPA's strategy for dealing with ozone problems has been to reduce emissions of VOCs.

Figure 8 shows projected emissions from mobile sources and stationary sources through the year 2004. (These projections are based on analyses conducted by the EPA and the OTA. Because there is considerable uncertainty regarding the amount of VOC emissions from various sources, the projections should be viewed as providing only a general indication of the likely trend in VOC emissions and the relative contribution of various sources.) Emissions from mobile sources are projected to decline during the summer months to less than half their current levels as older, higher polluting vehicles are removed from the fleet and if the RVP of gasoline is reduced to nine psi in the early 1990s. (California and some northeastern states already have required such reductions in RVP.) The decline in emissions from mobile sources, however, is expected to be partially offset by growth in emissions from other sources, primarily small stationary sources. Thus, automobile emissions are likely to generally play a smaller role in contributing to urban ozone problems in the future than they have in the past, even if emission standards are not further tightened.

Even with the reductions in VOC emissions attained by lowering the RVP of gasoline, the majority of current non-attainment areas are expected to remain in violation of the ambient air quality standard for ozone (OTA, 1989; EPA, 1987a; Rosenberg, 1989). Many of the urban non-attainment areas that are close to being in compliance (i.e., they have ozone concentrations in the 0.13 ppm to 0.14 ppm range) may be able to come into compliance, but about 60 percent of the urban non-attainment areas, with a combined population of about 90 million, have ozone problems severe enough to require additional control measures to bring them into compliance (OTA, 1989). Reductions in total VOC emissions on the order of from 30 to 50 percent are thought to be necessary to bring most non-attainment areas into compliance (OTA, 1989; EPA, 1987a). However, even this large reduction would still leave some urban areas with high populations and unfavorable meteorological conditions, such as the Los Angeles basin, Houston, Chicago, and the northeastern corridor, in violation of ozone standards.

Potential Regulatory Strategies for Controlling Automobile Emissions

A whole host of regulatory strategies and options for further controlling automobile emissions is now under consideration by the Administration, Congress, the states, and the EPA, as shown in Table 3. Some focus on the automobile owner, but most are aimed at the auto makers and petroleum refiners. Some are short-run strategies and are narrowly focused. Others are longer-run strategies and would require significant technological improvements to implement and massive changes to be made to the fuel supply infrastructure. We provide a brief overview of the major strategies and control options that have entered the public debate.

Near-Term Strategies

Most of the options for further reducing automobile emissions in the near term have already been imple-

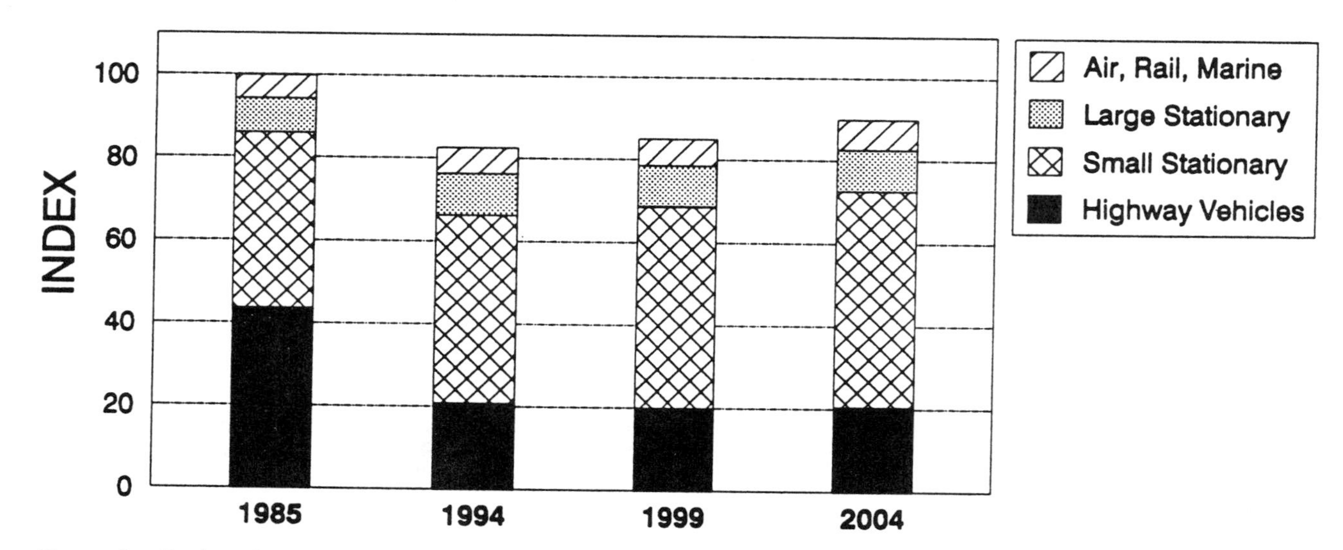

Figure 8. Projected VOC emissions, assuming a reduction in RVP. Source: OTA, 1989, and EPA, 1987.

Table 3. Options for controlling automobile emissions nationally.

NEAR TERM
o RVP of Gasoline
o Evaporative Emissions Controls
o Wintertime Oxygenated Fuel Program
o Reformulated Gasoline (replace unleaded regular)
MID - 1990s
o Lower Exhaust Emission Standards (available technology--first tier)
o Stage II Refueling Controls/On-board controls
o Enhanced I & M
o Cold Start CO Standards
o Reformulated Gasoline (benzene & aromatics)
o Alternative Fueled Fleets in Selected Urban Non-Attainment areas
o Phase-in of Low Polluting/Alternative Fueled Vehicles in Selected Urban Non-Attainment Areas
POST 2000
o Technology Forcing Exhaust Emission Standards (second tier)
o Phase-in of Advanced Clean Fueled Vehicles in Selected Urban Non-Attainment Areas
o Limits on Carbon Dioxide Emissions

mented in part or are likely to be implemented soon. The EPA has already promulgated interim standards to reduce the RVP of gasoline to 10.5 psi in Class C areas and has proposed to further reduce gasoline RVP to 9.0 psi. In addition, the EPA recently proposed testing procedures for evaporative emissions and running losses that probably will cause auto makers to install larger charcoal canisters in cars and to improve the systems for channeling vapors to the engine (EPA, 1990b). General Motors also has proposed a new testing procedure for evaporative emissions. This, along with RVP controls, should reduce automobile emissions of HCs by more than half, and will have the single largest impact on VOC emissions of any of the regulatory strategies under consideration. These controls also are the most cost-effective of any of those under consideration (OTA,1989; Chilton, 1990).

Winter oxygenated fuels programs to address CO problems are already in place in Colorado, New Mexico, and Arizona. These programs will probably be gradually phased out as older cars are retired, because they are relatively ineffective in reducing CO emissions from newer vehicles equipped with adaptive-learning emission control systems.

Although regulatory programs mandating reformulated gasoline are not being considered as near-term strategies (except in California), some refiners have voluntarily introduced reformulated fuels as a replacement for leaded fuel used by older vehicles. HC emissions from such vehicles have reportedly been reduced by over 20 percent. Reformulated fuel has already been introduced in the Los Angeles basin market on a limited basis by Arco and in other markets by Amoco and Diamond Shamrock (ARCO, 1989; OGJ, 1989). It appears that reformulated gasoline will have fewer aromatics (benzene in particular) and olefins; may have a reduced boiling point range; and will have oxygenates, such as MTBE or TAME, added to make up for lost octane (Piel, 1989). Reformulated gasoline would address both the ozone and the air toxics problems. Gasoline-powered automobiles and trucks now contribute a large share of air toxics, including more than 50 percent of the emissions of benzene, a known human carcinogen. Recently, an amendment was proposed to the Senate's Clean Air bill that would define reformulated gasoline along the lines of Arco's product and would require refiners to market gasoline with "equivalent" effects on automobile emissions.

Longer-Term Strategies

There are major differences in the long-run strategies now under consideration to address automobile emissions and urban ozone problems. The Administration's proposed approach is to introduce clean-fueled vehicles into urban areas with severe ozone problems beginning in the late 1990s. These vehicles would run on fuels such as methanol, ethanol, compressed natural gas, or liquified petroleum gas. Initially, such vehicles would likely be flexibly fueled, that is, they could run on gasoline or a blend of gasoline and up to 85 percent methanol (M-85) or ethanol (E-85). Later, vehicles dedicated to operate on a single alternative fuel would be introduced. The Senate bill would require a second tier of technology forcing exhaust emission standards that would take effect in the year 2000. They would go beyond the exhaust emission standards that would be required in 1995 and that are compatible with existing control technologies. The Senate bill would also impose standards for carbon dioxide emissions from automobiles as a method of reducing the generation of greenhouse gases. A third approach, favored by the auto and petroleum refining industries, is to reduce emissions through reformulating gasoline, but not to mandate the use of alternative fuels or a second tier of exhaust emissions standards (Behrens and Ketcham-Colwill, 1989).

As of this writing, it is not clear how the debate will be resolved. Proponents of the alternative fuels approach contend that methanol will be priced about the same as gasoline, that dedicated methanol vehicles operating on 100 percent methanol (M-100) can be developed that will cost about the same to own and operate as conventional automobiles, that methanol vehicles will reduce emissions of ozone precursors by more than 80 percent relative to gasoline powered vehicles, and that emissions of toxics will be substantially reduced. Critics point out the uncertainties of forecasting relative fuel prices and that the government typically is a poor prognosticator of market developments. Some studies have suggested that large price differentials between gasoline and methanol are a distinct possibility, which could turn an alternative fuels program into an expensive option. They also point out the potential difficulties in producing M-100 vehicles, uncertainties in the likely emissions performance of such vehicles, difficulties in persuading consumers to purchase such vehicles, the large cost of developing an infrastructure for supplying the fuel, and problems in forcing states to ensure that alternative fuels are priced "competitively" with gasoline (Behrens and Ketcham-Colwill, 1989).

Proponents of a second tier of national reductions in exhaust emissions argue that the technology will become available to meet such standards, that further reductions in emissions are necessary to bring many urban areas into compliance with ozone standards, and that such emission reductions will be less expensive to obtain than reductions from stationary sources. Critics, on the other hand, point out that it might not be possible to develop the technology necessary to meet the standard, that the approach is likely to be very costly (about $7 billion per year), that the potential reductions in exhaust emissions are small relative to costs, and that the standard is not focused on the urban areas with the largest ozone problems.

The role reformulated gasoline might play in an eventual long-run strategy to reduce ozone in urban areas is unclear, but promising. It has the potential for reducing emissions from the entire vehicle fleet, not just from new automobiles.

California's Proposed Programs

Separate from the debates regarding an appropriate national strategy to address ozone problems are the plans now being developed in California to deal with the severe ozone problems in the Los Angeles basin. California has often led the nation in the formulation of programs to regulate automobile emissions, and this time is no exception. The state is proposing an ambitious and costly set of near- and far-term measures for both automobiles and stationary sources that could significantly affect the lifestyle of southern Californians and could cost many billions of dollars (Portney, 1989). We focus only on the measures aimed at mobile sources.

One innovative measure being considered for the near term is to require on-board diagnostic systems for new cars. This attempts to address problems posed by vehicles whose emission control systems malfunction by alerting the driver to both the problem and its origin. California is in the lead in addressing the toxics problems related to automobile exhaust and evaporative emissions. The state proposes to halve the current percentage of benzene in gasoline from about 1.6 percent to 0.8 percent. Also, California is developing reactivity adjustment factors to normalize emissions from alternatively fueled vehicles in terms of their ozone creating potential (CARB, 1990).

Longer-term measures include requirements for the introduction of specified numbers of Low Emitting Vehicles in the late 1990s and Ultra Low Emitting Vehicles after the year 2000 (CARB, 1990). The proposed HC emission standards for these vehicles are about 30 percent and 15 percent, respectively, of the current California standard for gasoline fueled automobiles, which already reflects the limit of available control technologies. Thus, it is likely that alternative-fueled vehicles will have to be introduced to meet these regulatory requirements.

It will be interesting to observe the political debate in California as the nature, cost, and implications of the proposals affecting mobile and stationary sources become more widely understood by its residents. California may become a testing ground for determining how much communities are willing to spend and how far they are willing to go in changing their lifestyles to meet the air quality standards imposed by the Clean Air Act (Portney, 1989; Brownstein, 1989).

Market-Based Mechanisms

Largely absent from the current debate about strategies for reducing mobile source emissions is mention of the use of market-based mechanisms either as an adjunct to or a substitute for conventional command and control regulations. Market-based mechanisms could be used in conjunction with a number of the strategies that have been proposed. They could also be used in place of some of them.

Market-based mechanisms could be used in conjunction with technology-forcing standards to provide for flexibility in meeting the standards and to create economic incentives for improving emission control technologies. Methods for alleviating problems inherent in mandating a schedule of stringent standards that are beyond the ability of current technology to meet include emissions averaging systems, emission fees, and flexible penalty provisions (White, 1982). Economic incentives could also play a role in encouraging the introduction of lower emission vehicles either nationally or in selected urban areas. Instead of requiring a certain percentage of vehicles to meet low emission standards, a system of averaging fleet emissions could be used to obtain the same emission reduction goal. Auto makers would have the option of introducing low emission vehicles or reducing generally the average emissions of the entire fleet. Alternatively, fees could be imposed on vehicles and fuels that reflect their contribution to air pollution problems. Low emission vehicles would then be accorded a price advantage in the market place. This appears to be a less coercive mechanism than a rule that mandates the introduction of a new type of vehicle.

Market-based mechanisms could provide flexibility to auto makers and petroleum refiners in meeting regulatory objectives. Averaging systems could allow auto makers to pursue promising engine and emission control technologies that might otherwise be foreclosed by requirements that all vehicles must meet specific standards. Banking and trading systems have the potential both to ease the transition of refiners to marketing lower-polluting, reformulated gasoline and to yield environmental benefits beyond those that could be obtained through standard setting.

There are precedents for using market-based mechanisms as supplements to conventional regulations. For example, the EPA combined a lead banking and trading system with a tight regulatory schedule to phase down lead in gasoline to the current 0.1 g/gal level. Under that system, refiners could make immediate reductions and bank lead credits to offset stringent requirements imposed at a later date. And refineries that had difficulty in meeting the standards could purchase the rights to use lead from refineries that could use less lead than the standards allowed. Reliance on this market-based system allowed the EPA to accelerate the removal of lead from gasoline and eliminated the need to develop special exemptions for small refiners, as typically are required under conventional regulations to deal with variations in the capabilities of firms. It also reduced the cost to industry of meeting the lead standards. Another example of the application of market-based mechanisms is the EPA's use of an averaging system for emissions from heavy duty diesel engines. This system is likely to be expanded to allow for banking and trading.

Finally, most of the strategies now under consideration would control aggregate emissions by controlling the emissions of individual vehicles. A different approach would focus on the demand side of the equation. A major part of the current problem from automobiles stems from the large increase in VMT over the past two decades. This upward trend is expected to continue in the future and is part of the reason there is so much concern over future automobile emissions. A straightforward, though probably politically unpalatable, approach for dealing with this aspect of the problem would be to impose an additional tax on gasoline. The tax would be a way of reflecting in the price of gasoline the health and environmental damages associated with driving. This should increase the cost of driving relative to other forms of transportation, leading to somewhat lesser amounts of driving or the purchase of more fuel-efficient vehicles.

Concluding Comments

The emission control systems of automobiles have been improved substantially over the past two decades and will continue to be improved during the 1990s. Unfortunately, the per vehicle gains in emissions have been offset by an increase in driving that resulted from the combination of population growth, increases in income, and increases in labor force participation rates. A large number of urban areas currently do not meet ambient air quality standards for ozone and CO and many of them are expected to continue to violate such standards in the 1990s. Long run strategies to deal with continued ozone problems have not yet been implemented. But given the relative decline in the share of total VOC emissions contributed by automobiles and the increasing cost of more stringent standards, it is important to assess the benefits and costs of suggested regulatory programs and to use market-based mechanisms where possible to improve the efficiency of conventional regulatory approaches.

References

Arco, "Arco's New Low-Emission Gasoline Is First Phase in Helping Clean Up Southern California Air Pollution," Arco Public Affairs (August 15, 1989).

Behrens, L. and Ketcham-Colwill, J., "Cars and Ozone Pollution: Congress Weighs New Fuels, Stricter Standards," Environmental and Energy Study Institute, Washington, DC (Dec. 22, 1989).

Black, F., "Background Overview, Alternative and Conventional Fuels," EPA (1989).

Brownstein, R., "Testing the Limits," *National Journal,* pp. 1916–1920 (July 29, 1989).

California Air Resources Board (CARB), "Low-Emission Vehicles/Clean Fuels and New Gasoline Specifications," Progress Report (December 1989).

Chilton, K. and Sholtz, A., "A Primer on Smog Control," *Regulation,* pp. 31–40 (Winter 1990).

Crandall, R. W., et. al., *Regulating the Automobile,* The Brookings Institution, Washington, DC (1986).

EPA, "Cold Temperature CO," Briefing for the Administrator (August 25, 1988a).

EPA, *Compilation of Air Pollutant Emission Factors, Volume II: Mobile Sources,* (Mobile 4), Fifth Edition.

EPA, "Control of Air Pollution from New Motor Vehicles and New Motor Vehicle Engines: Evaporative Emission Regulations . . . ," *Federal Register* 55 (January 19, 1990b).

EPA, *National Air Pollutant Emission Estimates, 1940–1987,* 450/4-88/022 (1988b).

EPA, *National Air Quality and Emission Trends Report, 1987,* 450/4-89-001 (1989a).

EPA, "Regulation of Fuels and Fuel Additives: Volatility Regulations for Gasoline . . . ," *Federal Register* 52 (August 19, 1987).

EPA, "Volatility Regulations for Gasoline and Alcohol Blends Sold in Calendar Years 1989 and Beyond," *Federal Register* 54 (March 22, 1989b).

Environmental Reporter, "Sierra Club Study Looks at 12 Cities, Finds Greater Ozone Threat," The Bureau of National Affairs, pp. 1648–1649 (January 26, 1990).

Lave, C., "Things Won't Get a Lot Worse: The Future of the US Traffic Congestion," Unpublished Paper, Economics Dept.—Univ. of California, Irvine (1989).

Motor Vehicle Manufacturers Association of the United States, Inc., *Motor Vehicle Facts & Figures* (1989).

National Institute for Petroleum and Energy Research, *Motor Gasolines, Summer 1988,* NIPER-158 PPS 8911 (March 1989).

Office of Technology Assessment (OTA), *Catching Our Breath, Next Steps for Reducing Urban Ozone,* OTA-O-412, US Government Printing Office, Washington, DC (1989).

Oil & Gas Journal, "More US Refiners Offer Low Emission Gasoline," *Oil & Gas Journal,* p. 31 (December 4, 1989).

Piel, W. J., "Ethers Will Play a Key Role in 'Clean' Gasoline Blends," *Oil & Gas Journal,* pp. 40–49 (December 4, 1989).

Portney, P. R., et. al., "To Live and Breathe in LA," *Issues in Science and Technology V(4),* pp. 68–73 (Summer 1989).

Rosenberg, W. G., Testimony Before the Subcommittee on Energy and Power, US House of Representatives (October 19, 1989).

Unzelman, G. H., "Octane Improvements in the 1990s," in Proceedings of the 1988 NPRA Annual Meeting, AM-88-25 (March 1988).

Walsh, M. P., *Critical Analysis of the Federal Motor Vehicle Control Program,* Prepared for Northeast States for Coordinated Air Use Management (July 1988).

White, L. J., "US Mobile Source Emission Regulation: The Problems of Implementation," *Policy Studies Journal V(II),* pp. 77–87 (1982).

Technical Options for Energy Conservation and Controlling Environmental Impact in Highway Vehicles

Charles A. Amann
General Motors Research Laboratories
Warren, MI

Abstract

Manufacturers of light-duty highway vehicles are sometimes caught between the consumers' desire for a reasonable-cost conveyance that is a pleasure to operate and the mandates of regulation seeking societal objectives of energy conservation and preservation of air quality. The prospect for improving conventional vehicles in these areas by the year 2000 are considered. Alternative engines and fuels are reviewed for the same time frame. The status of the battery-electric vehicle is assessed. Shifting attention to the mid-21st century, the possibility of global warming is channeling thought toward nonfossil fuels, with hydrogen being added to the list of options.

Introduction

Private-vehicle ownership has become regarded as a fundamental right of US citizenship, and the wheels of national commerce would grind to a halt without trucks. The growth of automotive transportation has led to two concerns—the sources and availability of petroleum and the environmental impact of its use. In this paper, passenger-car fuel economy and emissions are reviewed, followed by consideration of alternative fuels and engines, and electric vehicles.

Fuel Economy

How high the fuel economy of an automotive vehicle can be raised is a recurring and difficult question. The answer depends on a number of key vehicle characteristics, the efficiency of the powertrain (engine + drivetrain), and the nature of the driving to which the vehicle is subjected.

The passenger car is the focus of the following discussion because over half the energy consumed in highway transportation is in the form of gasoline for automobiles (Davis et al., 1989). Gasoline for trucks, many of which use engines based on those in passenger cars, accounts for another 30 percent of highway energy. Diesel engines, often used in trucks, are deferred for later consideration.

Fundamental Factors

Fuel economy on a specified driving schedule depends on two components of fuel consumption: that consumed during the vehicle-powered modes of acceleration, cruise, and powered deceleration—when the accelerator pedal is depressed, and that consumed during the vehicle-unpowered modes of coasting, braking, and idling—when the accelerator pedal is released.

In the United States, fuel economy is normally evaluated on the Environmental Protection Agency (EPA) urban and highway schedules. For emission testing, the urban schedule forms the backbone of the FTP (Federal Test Procedure), which includes a cold (room-temperature) start. To calculate CAFE (Corporate Average Fuel Economy) for regulatory purposes, fuel economies from the FTP and highway tests are combined by assuming 55 percent of the distance driven is on the FTP and the balance on the highway.

In following either schedule on a given fuel, the powered-mode fuel consumption depends on two powertrain characteristics, three vehicle characteristics, and the accessory load. The relevant powertrain characteristics are the average brake thermal efficiency of the engine and the average efficiency of the drivetrain. The relevant vehicle characteristics are the vehicle mass, the tire rolling-resistance coefficient, and vehicle aerodynamic drag. The added contribution to fuel consumption from the unpowered modes on a specified driving schedule depends on the average unpowered fuel rate. When significant future fuel-economy improvements are projected for the automobile as we know it today, they can come only through improvements in one or more of these factors.

The average effects of these factors on fuel consumption (with a typical accessory load) are ranked in Figure 1 for a three-vehicle sample comprising a compact car, a large car, and a passenger van—all equipped with automatic transmissions. The powertrain factors

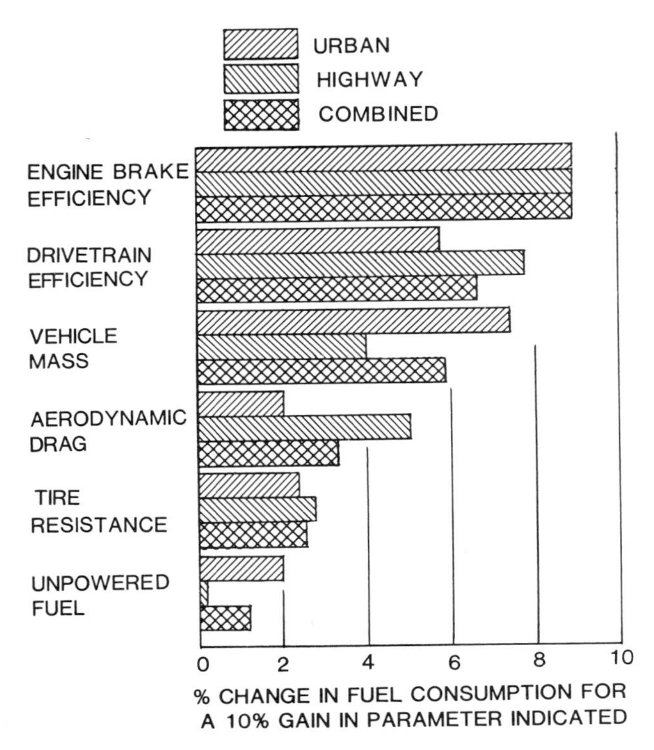

Figure 1. Independent effects of factors affecting fuel consumption on the EPA urban and highway schedules, and the combined value, for typical 1990 light-duty vehicles.

are seen to be the most influential, with vehicle mass being the most significant of the vehicle factors. (Note that fuel consumption is not a linear function of the powertrain parameters in Figure 1, so interpolation or extrapolation from the values shown is inappropriate.)

For many decades, the passenger car has been judged by the product of its mass and fuel economy. A compact car achieving a given fuel economy may actually have a less efficient powertrain than a heavier car that offers more interior room and travels a shorter distance on a gallon of gasoline. The fuel economy index (FEI), obtained here by multiplying vehicle weight (test-weight class) by combined fuel economy, is a parameter that helps to compensate for this difference in vehicle mass and provides a better indicator of powertrain efficiency than fuel economy alone.

For a given level of powertrain technology, FEI can be increased by sacrificing performance. The consumer demands performance for freeway merging, highway passing, hill climbing, etc. In the US the most common performance index has remained the time to accelerate from a standstill to 60 miles per hour (mph)—96.5 kilometers per hour.

Fleet Characteristics

In Figure 2, FEI is plotted against this performance index. Inversion of the abscissa scale places the best-performing cars on the right. The parallelogram essentially encompasses data points for 50 different 1985–86 cars, all equipped with automatic transmissions, which are used in 85 percent of the US fleet (Amann, 1989). The negative slope of this parallelogram indicates the inherent tradeoff between fuel economy and performance for a car of given mass and level of powertrain technology.

The bubbles on Figure 2 illustrate the last decade of progress in two five-year increments. The individual points represent average data for six test-weight classes in each year (Heavenrich and Murrell, 1989). Response to the public desire for increased performance is evident from the rightward drift of the 1989 bubble. Despite this, improved powertrain technology has continued to increase FEI.

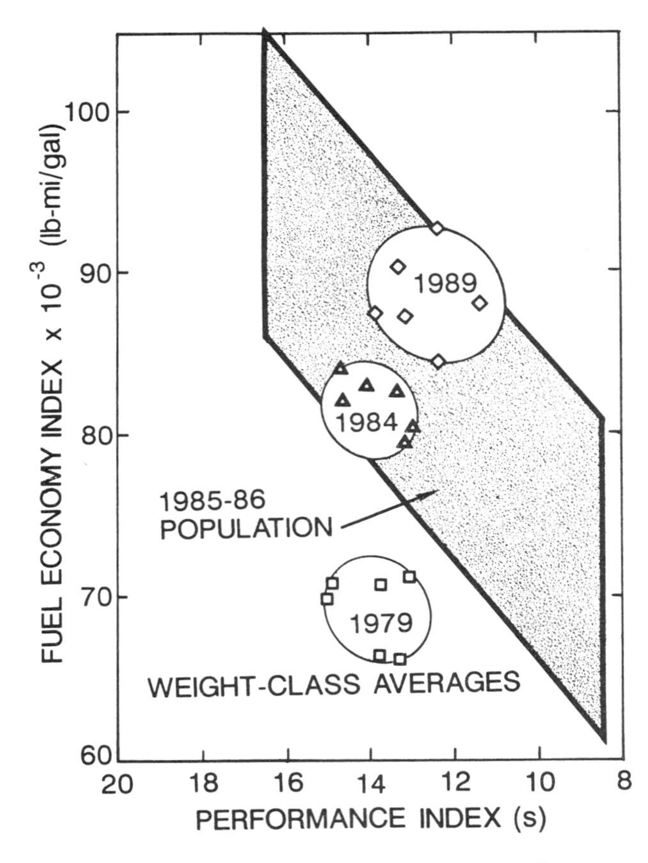

Figure 2. Fuel economy index versus performance index for a 50–car population of 1985–86 cars with automatic transmissions (Amann, 1989) and for the weight-class averages from 1979, 1984 and 1989 (Heavenrich and Murrell, 1989).

Fuel-Economy Projections

Figure 2 depicts history. Of greater significance is what can be expected in the future. It is evident from the parallelogram of Figure 2 that without introducing any unproven technology, the fleet-average fuel economy would improve substantially at constant performance if, at a given performance index, the cars with the lowest FEI were improved to the level of the cars with the best.

This potential has indeed been considered in an EPA study (Heavenrich and Murrell, 1989). First, a hypothetical fleet was constructed from the top-quintile (fifth) fuel-economy cars in each test-weight class, with the distribution of weight classes in the total fleet chosen to match the current new-car population. The result was a 9.6 percent increase in average fuel economy, from 28.2 to 30.9 miles per gallon (mpg). However, the new fleet had an 0.8 second poorer performance index, increasing from 12.5 seconds to 13.3 seconds. This change counters current consumer preference.

Because there is no assurance that this fleet composition would preserve the mix of car models the public has chosen in the marketplace, a second fleet was hypothesized using only the vehicles in each EPA type classification made by the single manufacturer with the best-in-class fuel economy. This resulted in a 15.2 percent increase in fleet-average fuel economy, from 28.2 to 32.5 mpg. However, the average performance index deteriorated further, from 12.5 to 13.6 seconds. Whether this approach accomplished its objective of preserving the consumer-chosen fleet composition is debatable. In the subcompact grouping, for instance, it retained the top-fuel-economy Geo Metro but eliminated the sports-oriented Camaro. Although both cars are classified as subcompacts, each appeals to a different market segment.

Others have attempted to estimate future fuel-economy potential using what are considered to be known technologies. For example, in a Department of Energy (DOE)-sponsored study of domestic cars done from a 1987 base, an average fuel economy of 33.1 mpg was deemed reasonable for the year 2001 (EEA, 1989). This is a 24 percent gain from the stated domestic average of 26.7 mpg in the base year.

In Figure 3, the historic trend line for the US fleet average is shown, along with the above EPA fleet-average projections—arbitrarily assigned an execution date of 2000—and the DOE point for the domestic-fleet potential in 2001. (The steeper slope of the historic 1975–82 trend line coincides with a 1000-pound drop in average car mass.) Parallel projections by the domestic industry are more in line with those of EPA rather than DOE. Discussions are being held to resolve this DOE-industry discrepancy.

Despite their difference over the magnitude of the improvement to be expected, both DOE and industry anticipated the greatest gains from the powertrain. In the engine, the technologies included intake valve control, overhead camshafts, roller followers, lower friction, fuel-system upgrades to throttle-body injection (TBI) and multi-point fuel injection (MPFI), and downsized four-valve engines. In the transmission, they included electronic control of shifts, torque-converter lockup, an increased number of gear ratios, and the continuously variable transmission (CVT). Other contributions were projected from front-wheel drive, lower mass (by 10 percent) through materials selection, lower drag coefficient (from 0.37 to 0.28), improvements in lubricants and tires, and more efficient accessories.

The fuel economy of the *in-use* fleet will increase even if standards are left where they are, solely due to fleet turnover. In 1987, the average in-use fleet fuel economy was 19.2 mpg (MVMA, 1989). Assuming this in-use number is 15 percent below the EPA combined level, that it takes 15 years to turn over the fleet, and that the combined fuel economy for all post–1987 cars is fixed at the 1987 fleet average of 28.5 mpg, it can be projected that fleet average fuel economy will increase 26 percent by the year 2002.

The Concept Car

A number of concept cars have been built to demonstrate high fuel economy, mostly in Europe and Japan (Bleviss, 1988). The majority have employed high-efficiency diesel engines, despite rejection of the passenger-car diesel by the US consumer. Typically, these small, lightweight, hand-built cars lack the air conditioning and automatic transmissions preferred by most US consumers. They do not necessarily meet US emissions and safety standards. Even if safety standards are met, analysis of accident statistics indicates twice the rate of deaths per mile-year in a 2000-pound subcompact as in a 4,500-pound large car (Partyka, 1989). Concept cars lack the ride, handling, freedom from objectionable noise and vibration, roominess, and freeway-merging and passing performance found in larger cars, but they *do* achieve outstanding fuel economy—60 to 80 mpg and more.

Unfortunately, this has led some to question the projections of fuel-economy potential that fall short of 40 mpg. Overlooked is the fact that the 1989 Geo Metro achieved a combined fuel economy of 65 mpg while meeting current US emissions and safety standards. However, the balance between fuel economy and the other characteristics sought by the consumer was such that the Metro accounted for only about 1 percent of General Motors's sales.

Unorthodox Approaches

Beyond those listed above, opportunities to improve engine efficiency are limited. Increasing compression ratio

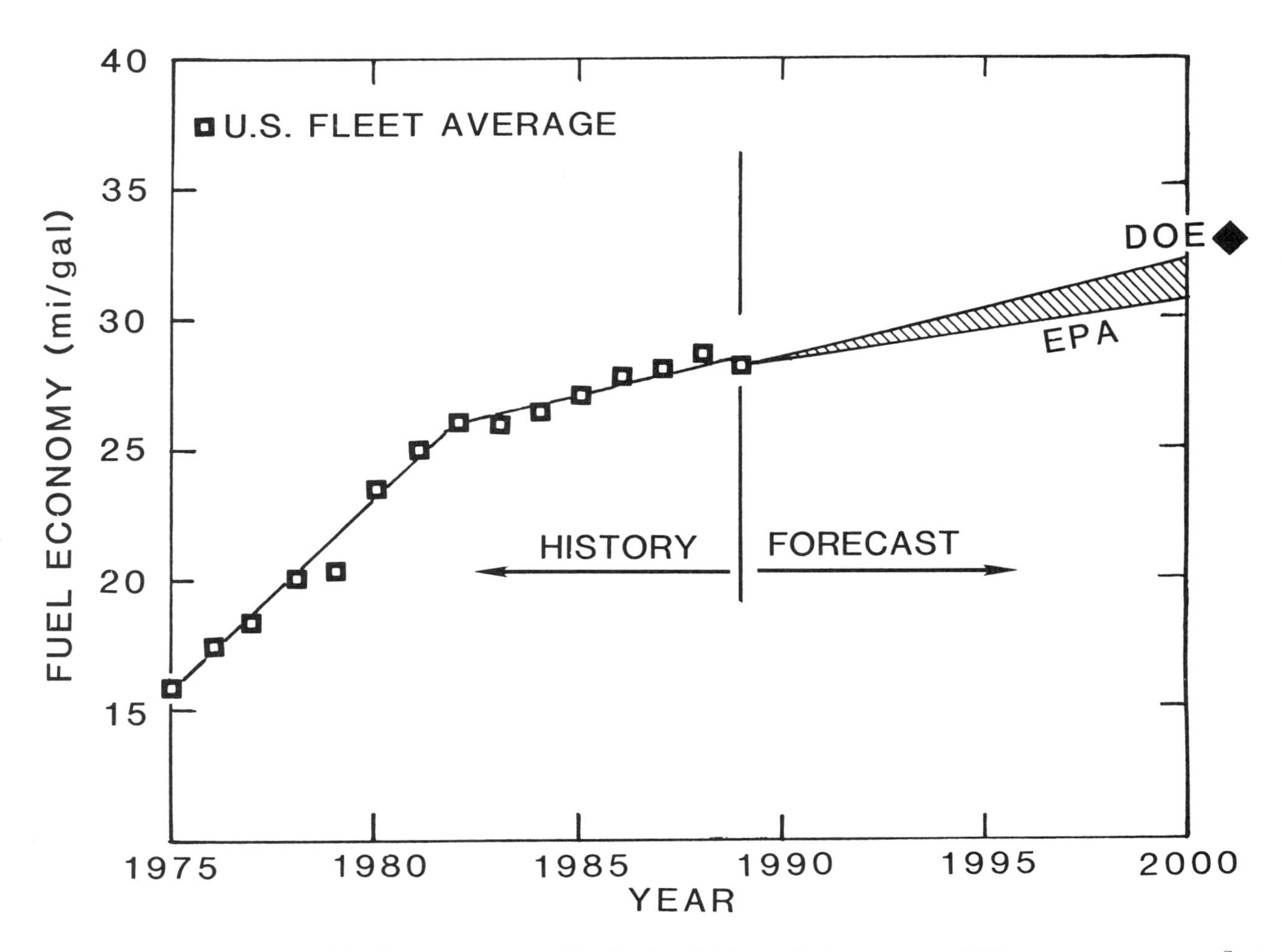

Figure 3. Historical trend in fleet-average combined urban/highway fuel economy of US new passenger-car fleet (Heavenrich and Murrell, 1989) and some future forecasts (Heavenrich and Murrell, 1989; EEA, 1989).

is blocked by combustion knock. Emission regulations have blocked both the burning of leaner mixtures at part load and the elimination of intake pumping loss by replacing throttle control with mixture control. Re-educating the public to accept manual transmissions has been suggested, but a review of 1989 EPA fuel-economy data from 62 pairs of passenger cars offered with both 5-speed manual and 4-speed lockup automatic transmissions showed an average gain for the manual of only 3 percent in fuel economy. Rendering some cylinders inoperative at part load so the remainder will operate more efficiently has been tried several times without acceptance. With this approach, pleasurable driver feel becomes especially difficult in high-economy four-cylinder cars. Fuel cutoff during all unpowered modes has been tried but not accepted, although some current cars with MPFI use limited deceleration-fuel cutoff. Idle cutoff is not suitable when air conditioning is used.

Examination of the energy distribution during urban driving, characterized by frequent stops, prompts interest in recovery of braking energy in an energy-storage system. The flywheel is one approach, but requires a continuously variable transmission with broad ratio range. Hydraulic energy storage is another option, but is hampered by the noise of suitable hydraulic pumps and motors. Electric storage is a third possibility, but faces difficulty with efficient transfer of energy in and out of batteries. The mass and volume of energy-storage systems detract from the gains they offer. The search continues for an acceptable system.

Any of these storage options can be employed in a load-sharing hybrid. This allows use of a less powerful engine forced to operate nearer full load, at higher efficiency. It sacrifices utility, however, because in prolonged high-power operation, such as climbing a lengthy mountain grade, performance capability is sacrificed when the stored energy is depleted.

Regulated Emissions

Air quality is out of compliance with federal standards in many US urban areas, and vehicle emissions are among the many contributors. The first air-quality problem to gain widespread attention was photochemical smog in the Los Angeles basin. Photochemical smog results from the action of sunlight in warm climates on nitrogen oxides (NO_x) and hydrocarbons (HC), which includes most organic compounds but not unreactive methane.

Atmospheric ozone concentration is the normal measure of smog. In Figure 4, isopleths of ozone formed are shown as functions of the initial concentrations of its precursors, NO_x and NMHC (Kelly and Gunst, 1989). The field is divided into three regions. In the HC-saturation region ($HC/NO_x > \sim 13$), decreasing NO_x reduces ozone concentration, but decreasing HC has almost no effect. In the NO_x-inhibition region ($HC/NO_x < \sim 5$), decreasing HC lowers ozone concentration, but decreasing NO_x can actually increase ozone. Within the intervening "knee region," ozone concentration is reduced by decreasing either HC or NO_x. The typical atmospheric HC/NO_x ratio is different in different urban areas, so an emission-control strategy that benefits one area may be ineffective in another.

Nitrogen oxide is formed during combustion and emitted from the vehicle tailpipe. The federal NO_x standard for passenger cars is now 1.0 gram per mile (gpm), compared to 4.1 gpm before emission control. California is already working toward a 0.4 gpm standard.

Unburned hydrocarbons are emitted in several ways: from the tailpipe, as a result of evaporation when the vehicle is inoperative, during refueling, and from the fuel system when the car is running. The federal tailpipe hydrocarbon standard for passenger cars is now 0.41 gpm. Evaporative emissions, also regulated, are absorbed in an underhood charcoal-containing canister and subsequently burned in the engine. Current standards are such that it takes 25 1990–model cars to emit as much tailpipe and evaporative HC as did a single car in pre-control days.

Hydrocarbon emissions during refueling are now effectively captured in St. Louis, MO, Washington, DC, and most of California by using a specially designed service-station refueling system that collects fuel-tank vapor as it is displaced. Collecting this HC on-board the car instead would require a much larger canister that has been cited as a safety concern in case of accidents, and total implementation would be delayed by fleet turnover.

Running losses during engine operation, e.g., fuel vapor escaping through the fuel-tank pressure-relief valve, are most immediately decreased by reducing gasoline vapor pressure. The vapor pressure of gasoline is now regulated in some states.

Estimated breakdowns of HC emissions from the 1989 in-use fleet, making use of EPA data and models, are shown in Figure 5 for different Reid vapor pressures (Rvp) and ambient-temperature ranges (Halberstadt, 1989). These include older cars, some having been poorly maintained and/or misfueled. It is seen that the vapor displaced from the tank when refueling is the smallest contributor. For an Rvp of 11.5 psi, representative of some recent summer-grade gasolines, running losses exceed the tailpipe HC at high ambient temperature.

The third regulated gaseous tailpipe emission is carbon monoxide (CO). The current federal standard of 3.4 g/mi compares to pre-control level of 84 g/mi, indi-

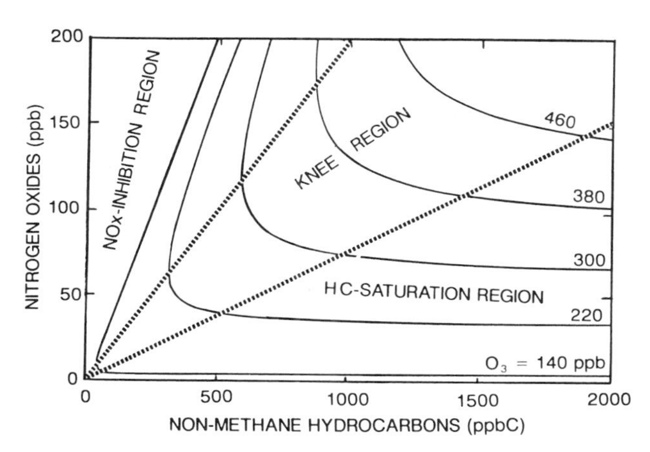

Figure 4. Isopleths of ozone concentration formed as a function of initial concentrations of NO_x and non-methane HC (Kelly and Gunst, 1989).

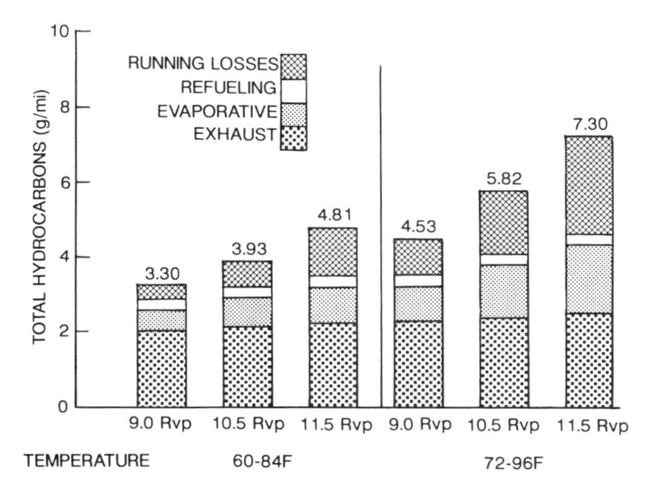

Figure 5. Effects of ambient temperature and Reid vapor pressure on sources of passenger-car HC emissions for the 1987 in-use fleet (Halberstadt, 1989).

cating it would take 25 modern cars to emit as much CO as a single pre-control car. CO is a product of rich-mixture combustion, most likely during cold starts and during wintertime operation. CO emissions tend to be higher at high altitudes, and to be lower for modern cars with closed-loop control of air-fuel ratio than for older open-loop carbureted cars. Denver and other metropolitan areas have mandated addition of an oxygenated-fuel component such as MTBE (methyl tertiary butyl ether) or ethanol to their gasoline during the winter months. In Denver, a resultant decrease in 1988–89 average wintertime atmospheric CO concentration of 12 percent has been estimated (Livo et al., 1989).

Tailpipe emission standards are now so low that the gasoline engine in passenger cars demands an exhaust catalyst. The catalyst is ineffective until it reaches its lightoff temperature, though, somewhere in the neighborhood of 600°F (315°C). Because the FTP begins with a start from room temperature, and because the catalyst can remove on the order of 90 percent of the entering regulated emissions once it is lit off, most of the tailpipe emissions from compliant cars are exhausted during the first driving cycle. (The EPA urban schedule used for emissions measurement consists of 18 different consecutive cycles separated by points of zero vehicle velocity.)

To illustrate this point, cumulative mass emissions are plotted versus distance on the urban schedule in Figure 6. The slope is an indicator of the grams per mile emitted. Each point marks the end of a cycle. Because of the cold start, in this particular car the emissions on the first cycle amount to 92 percent, 98 percent, and 41 percent of the 18–cycle total for HC, CO, and NO_x, respectively. The FTP weights emissions such that these contributions to the final total are diminished, but the cold start is nonetheless a major contributor.

The incentive for faster catalyst lightoff is obvious. Using catalyst substrates with less thermal capacitance, decreasing heat loss in pipes between the engine exhaust port and the converter, and locating the converter closer to the engine have all proven effective. These options have to be exercised with care, however, because they generally increase the likelihood of catalyst exposure to excessive temperature during unusual operating conditions, and too high a temperature causes premature catalyst deterioration. Electrically heating the catalyst on a cold start seems an attractive option but requires significant battery power.

Decreasing emissions from new cars is of little consequence unless those reductions are maintained in the field. From 1982–86, one domestic manufacturer monitored tailpipe emissions from consumer-operated cars as mileage was accumulated (Haskew and Gumbleton, 1988). A significant finding was the reduced deterioration of emission-control systems following introduction of closed-loop control in 1981. As new low-emission cars replace old higher-emission cars, emissions from the in-use fleet decrease. In fact, it was projected that in the year 2000, in-use fleet-average FTP emissions would decrease by two thirds for HC and CO, and over one third for NO_x, if emission standards were fixed at present levels. This is a big improvement over projections made with an earlier EPA model based on 1981 cars that excludes this improved in-use performance.

Alternative Fuels

Federal and state governments are poised to regulate alternative fuels into the marketplace because of concern about photochemical smog in urban areas. Prominent candidates are M85 (85 percent methanol/15 percent gasoline), natural gas (90+ percent methane), M100 (neat methanol), ethanol, and LPG (liquefied petroleum gas, mostly propane). Methanol is currently made from natural gas. In the near term, substantial expansion of its use would encourage its importation from foreign sources, where natural gas is least expensive. In the near term, natural gas for transport use would be sourced in North America. US ethanol is now made principally from corn in insufficient quantities to become a primary transportation fuel. It is used as an additive in gasoline to boost octane rating, however. Neither will LPG—a byproduct of petroleum refining and natural gas production—be available in large quantities. That leaves M85, natural gas, and M100.

The toxicity of methanol is worrisome for both M100 and M85. In addition, the invisibility of a neat methanol flame is a safety concern. So is the fact that the air-vapor mixture in a tank of methanol is more readily ignitable than gasoline at normal ambient temperatures. The safety concern with natural gas is leakage in an enclosed space like a garage.

M85, natural gas, and M100 all burn satisfactorily in a suitably modified spark-ignition engine. M100 has a major starting problem at moderately cool ambient temperatures, which can be overcome with M85 by controlling the volatility of the gasoline added. Being in the gaseous state, natural gas normally avoids this problem but has experienced starting trouble at sub-zero temperatures because of its high ignition temperature.

M85, natural gas, and M100 all have octane ratings superior to those of commercial gasolines, allowing an increased compression ratio for higher thermal efficiency and power. Methanol also offers increased power at a given compression ratio because its high latent heat cools the incoming mixture, increasing charge density. In contrast, the space occupied by natural gas in the intake manifold displaces air, so maximum power decreases significantly at a given compression ratio.

Because of current distribution limitations, M100 and natural gas are presently reasonable only for fleets with central refueling. To avoid this problem, industry is developing cars adapted to both gasoline and metha-

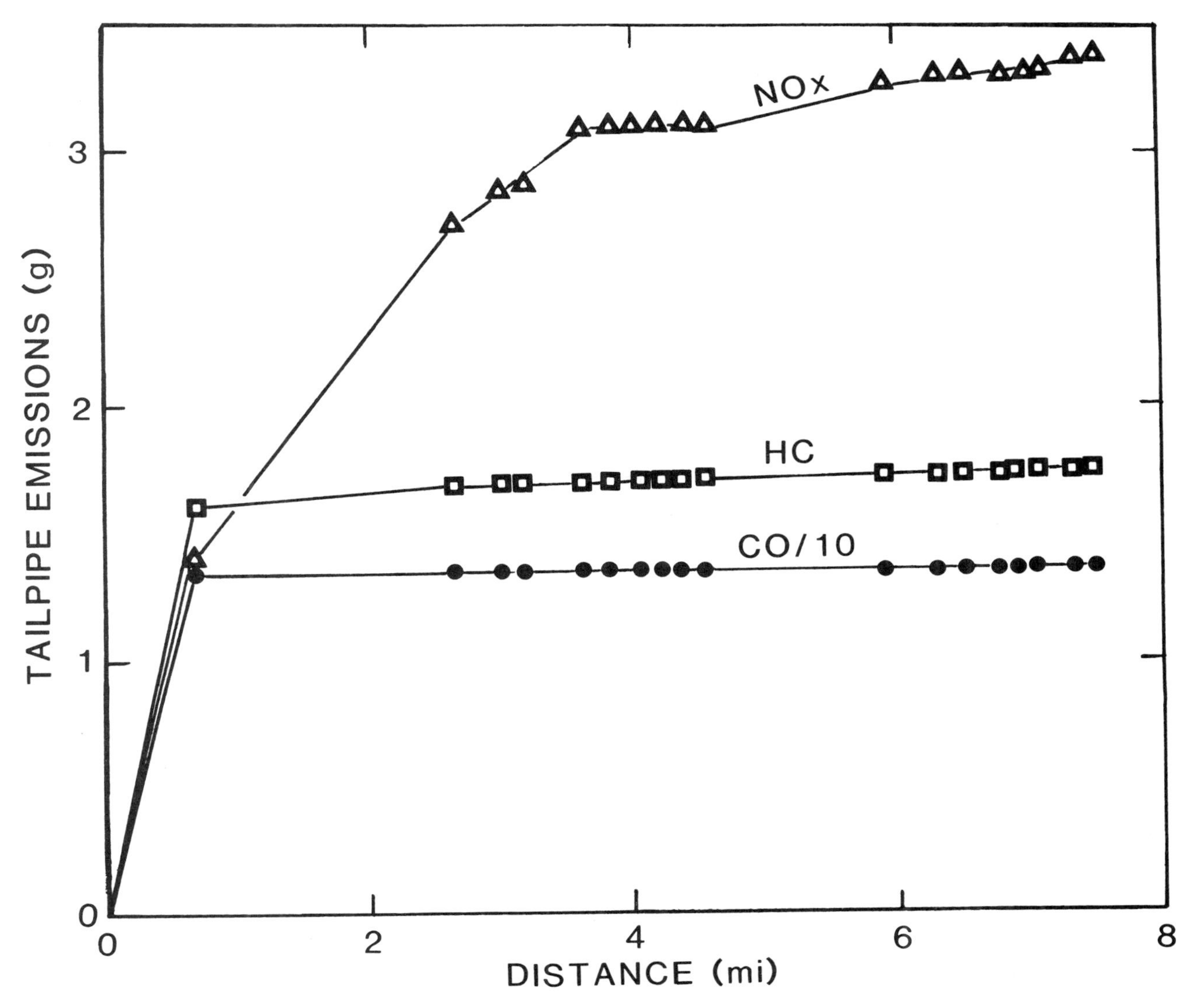

Figure 6. Cumulative mass of regulated tailpipe emissions from a production car versus distance driven on the EPA urban schedule with a cold start.

nol (Chrysler Gasoline-Tolerant Methanol Vehicle (GTMV); Ford Flexible Fuel Vehicle (FFV); General Motors Variable Fuel Vehicle (VFV)). These vehicles operate on any mix of fuels from M85 to neat gasoline, so when the consumer needs fuel in an area not offering M85, gasoline can be substituted. This approach solves the fuel-availability problem but restricts allowable compression ratio to that acceptable to commercial gasoline, thus limiting the feasible increase in thermal efficiency.

The decreased energy density of alternative fuels compared to gasoline reduces range and/or cargo capacity. This is especially true of Compressed Natural Gas (CNG), normally stored on board in gas bottles at 3000 psi (20.7 MPa). The effects on range for equal cargo-carrying capability and on the allowable mass of cargo for equal range are illustrated for a sample of production vehicles in Table 1.

HC reactivity is important in forming smog. With M85 and M100, most of the unburned fuel is methanol, which is considerably less reactive than most of the HC in gasoline exhaust. However, formaldehyde (HCHO), a very reactive compound, is typically five to ten times as abundant as in gasoline exhaust. Past atmospheric-modeling results indicate that substituting methanol for gasoline would reduce peak ozone levels by 5 to 20 percent (DeLuchi et al., 1988). More recent results suggest the ozone-forming potential with M85 is 20 percent

Table 1. Comparative Utility of Various Fuels

	Gasoline	*M85*	*CNG*
Relative Range for Equal Cargo Capability			
Midsize car	1.00	0.60	0.20
Pickup truck	1.00	0.58	0.20
Commercial van	1.00	0.58	0.20
Medium-duty truck	1.00	0.20	0.23
Relative Cargo Capability for Equal Range			
Midsize car	1.00	0.91	0.52
Pickup truck	1.00	0.89	0.30
Commercial van	1.00	0.95	0.77
Medium-duty truck	1.00	0.97	0.84

greater than with the most recent gasoline-fueled cars in cities with HC/NO_x ratios like Los Angeles (Dunker, 1989). In such cities, *if* HCHO from M85 could be adequately controlled, those models suggest 25 percent and 40 percent reductions for M85 and M100, respectively, compared to gasoline. In cities with the high HC/NO_x ratios of Houston, M85 and M100 appear to effect no ozone reductions, however. This work emphasizes the need for better control of HCHO emissions when burning methanol. A suitable long-life HCHO catalyst is still being sought.

Comparable studies on natural gas are lacking. Because methane, the principal HC component of natural gas combustion, has an extremely low reactivity, and because formaldehyde production is much less than with methanol, ozone from natural gas should be low.

An alternative fuel recently added to the list is reformulated gasoline. Past experiments have indicated that exhaust emissions are affected by the aromatic and sulfur contents of gasoline, as well as its 90 percent distillation temperature. Such results have fostered a cooperative research program within the US auto and oil industries that will explore the effects of fuel parameters on exhaust, evaporative, and running-loss emissions. Various levels of aromatics, olefins, and such octane enhancers as ethanol, ETBE, and MBTE in gasoline will be evaluated, along with M85. The first phase is scheduled for completion during 1990.

Alternative Engines

The search continues for an alternative powerplant having greater energy efficiency within anticipated emission constraints. Poor fuel economy has eliminated the steam engine. A passenger-car Stirling engine is unlikely because of its size, weight, and lack of a demonstrated fuel-economy advantage. The fuel-efficient Direct-Injection Stratified-Charge (DISC) engine seems doomed by future NO_x standards. Hope remains for the diesel, the two-stroke engine, and the gas turbine.

Diesel Engine

Despite its unsurpassed fuel economy, the popularity of the passenger-car diesel has waxed and waned in the heart of the US consumer, although diesels for light-duty trucks remain a firm niche market. The diesel is well entrenched in the heavy-duty field, however. Its difficulty there is with future emission standards, especially NO_x and particulates.

This problem is illustrated in Figure 7 (Needham et al., 1989). The particulates-NO_x point for an engine satisfying heavy-duty standards must fall within the appropriate dotted regulatory box. The dashed curves representing the best 1988 diesels of various types dramatize the problem. Techniques expected to move the 1988 curves downward and to the left include refined turbocharging and aftercooling, increased injection pressure, better control of air-fuel mixing, flexible injection timing, decreased oil consumption, decreased fuel sulfur and aromatics, and exhaust aftertreatment. Whether these will suffice to meet 1994 standards is unknown. Particulate trapping is also being pursued, but the traps soon become filled and must be regenerated by particulate burnoff. This approach has proven very expensive, and trap durability remains an issue.

Methanol combustion has the capability of lowering both NO_x and particulates, but methanol is difficult to ignite by compression. Methanol use in diesels is being researched by employing ignition-improving fuel additives, pilot injection of diesel fuel, glow plugs, and spark plugs. In the special case of the two-stroke diesel, methanol can be compression ignited by bypassing the scavenging blower to increase compression temperature. Diesel engines are also being adapted to natural gas. The alternative-fuels issue in diesels remains unsettled.

The Department of Energy is sponsoring research on the Low-Heat-Rejection (LHR) diesel (sometimes called the adiabatic diesel). The objective is to eliminate the conventional liquid-cooling system and employ ceramics, either as structural monoliths or as coatings on metal substrates, to decrease heat rejection. The once-promised 30 percent improvement in fuel economy appears thwarted by the laws of thermodynamics. Reliable ceramics are not yet available, and a lubricant capable of coping with the increased temperatures of the LHR diesel remains to be found. The prospect of meeting future NO_x standards in an engine running at much higher temperatures is not good. Spinoffs from this program may find application in advanced diesel engines of more conventional design, however.

Returning to passenger cars, the diesel is of greater interest in Europe, where emission regulations are less

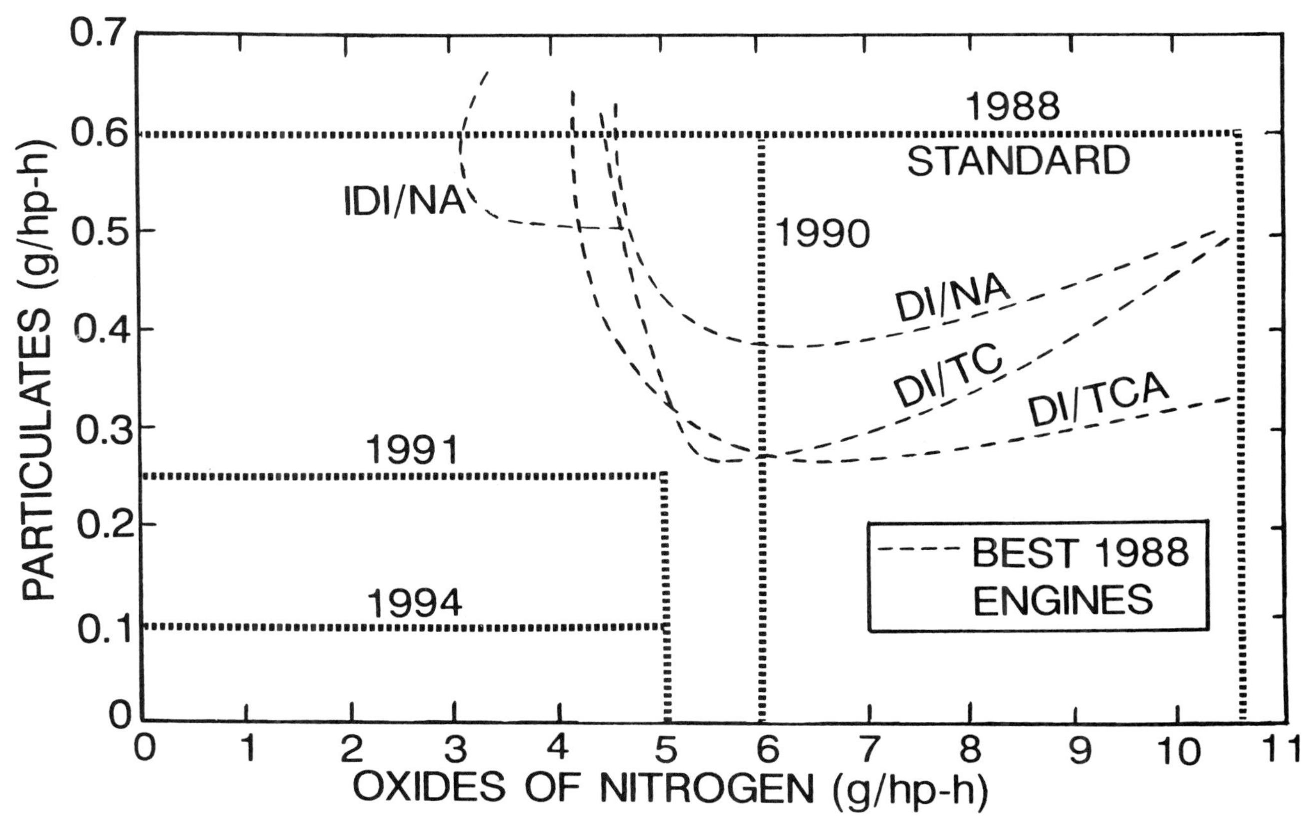

Figure 7. Regulatory particulate/NO_x boxes for heavy-duty diesels and engine status for 1988: IDI = indirect injection, DI = direct injection, NA = naturally aspirated, TC = turbocharged, A = aftercooled (Needham et al., 1989).

stringent and some countries have tax policies that price diesel fuel very attractively compared to gasoline. Given the current US situation, though, widespread use of passenger-car diesels in this country seems unlikely.

Two-Stroke Engine

The two-stroke spark-ignition engine comes in many configurations. A leading candidate is one that replaces intake and exhaust valves with ports in the cylinder wall, uses the underside of the piston to compress the air charge enough to scavenge burned gas from the cylinder, and introduces the fuel into the air charge only after both ports have been closed by the piston. This version, which requires a dry sump and rolling-element bearings, offers packaging advantages because of reduced size and mass. This follows both because it provides a power stroke every revolution of the crankshaft and because the valvetrain atop the cylinder has been eliminated. The engine promises lower fuel consumption because of charge dilution, reduced pumping loss, and less friction. Finally, its doubled firing frequency compared to the four-stroke cycle is more pleasing to the driver.

An oxidizing catalyst is required for control of HC and CO. Because the tailpipe mixture is lean overall, the engine cannot use an NO_x reducing catalyst, but must accomplish NO_x control within the cylinder. The presence of scavenge air in the exhaust lowers exhaust temperature, making it more difficult for the oxidizing catalyst to do its job. However, use of intake and exhaust ports eliminates the need for a valvetrain anti-wear additive in the engine oil that is a catalyst poison. Long-term emissions performance has yet to be demonstrated, and future emissions standards may relegate the engine to small vehicles only. Engine durability questions also remain to be evaluated.

Another version of the two-stroke engine trades the dry sump for a more conventional oil-filled sump and relies on an external blower to facilitate cylinder scavenging. This approach permits use of conventional journal bearings and allows some supercharging for even higher specific output. Some two-stroke engines use valves in the head. Such variations sacrifice some of the fuel-saving and packaging virtues of the crankcase-scavenged type, but may have other advantages. The future of the two-stroke passenger-car engine remains uncertain, but the engine is being pursued aggressively by industry.

Gas Turbine

The automotive gas turbine has attracted industry attention for over four decades. It is now generally conceded that in automobiles, the all-metal gas turbine cannot compete with the piston engine on fuel-economy grounds. However, DOE is sponsoring two contractors to develop ceramics technology for this engine. That would allow significantly higher turbine inlet temperature, a long-recognized key to improved efficiency in this engine. A byproduct of higher temperature is lower specific air consumption, which means smaller turbomachinery. Turbomachinery efficiency, also important to engine efficiency, deteriorates with decreasing size. It is possible for the loss in component efficiency to counter the otherwise beneficial effect of higher temperature on fuel economy (Murphy, 1985).

The automotive turbine, if it succeeds, is still a decade away. Three major barriers must be surmounted before it becomes a serious contender. First, reliable structural ceramics of affordable cost must be proven. Second, satisfactory component efficiencies and low parasitic losses must be demonstrated in what, by gas turbine standards, are very small sizes. Third, a high-temperature low-NO_x combustor must be developed.

Electric Vehicles

Because of energy consumption and emissions difficulties facing engines, the electric vehicle (EV) continues to arouse interest. A popular myth is that the EV is pollution-free. The emission source is merely moved to a remote site. That can be a distinct advantage in meeting *urban* air-quality standards, but is insignificant for more global concerns like acid rain and climate warming.

The DOE breakdown of primary energy sources for generating electricity in the year 2000 is shown in Figure 8 (DOE, 1988). An estimate of emissions from generation of electricity using year-2000 energy sources is presented in Table 2 (Sperling and DeLuchi, 1989). Near elimination of HC and CO is traded for increases in sulfur oxides and particulates, with essentially no effect on NO_x. These estimates assume that half the coal plants will have controls to reduce SO_x by 95 percent and particulates by 99 percent, and that about a quarter of the plants will have NO_x controls.

It is difficult to compare the energy consumption of the EV to that of the engine-powered car because they are seldom operated on the same driving schedule.

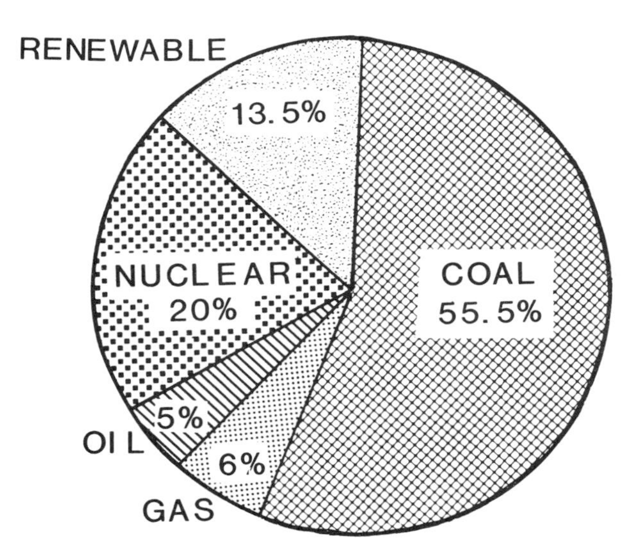

Figure 8. Projected distribution of primary energy sources for US electricity production in 2000 (DOE, 1988).

Table 2. Ratio of Vehicle Emissions, Electric/Gasoline, Year 2000

Reactive hydrocarbons	0.01
Carbon monoxide	0.02
Oxides of nitrogen	1.03
Oxides of sulfur	22.0
Particulate matter	35.0

A decade ago, comprehensive tests were run on an electric adaptation of the Volkswagen Rabbit, using three different battery types (Conover et al., 1980). Battery selection has a minimal effect on the energy consumed by the vehicle. Average battery energy consumption for this car is given in the third column of Table 3 for constant-speed driving and for two standard EV transient schedules with peak speeds of 20 and 45 miles per hour, respectively. Energy requirement is seen to increase with speed. For a given peak speed, more energy is consumed in transient than in constant-speed driving. This is true despite the fact that this car had regenerative braking. Regeneration never recovered more than 4 percent of the propulsion energy on the J227a schedules.

The 55 mph energy in the last column of Table 3 is for the Impact, a recently announced concept car that accelerates to 60 mph in eight seconds with fresh batteries (GM, 1990). This is outstanding performance for an EV but involves no magic. Car mass/power ratio being the most important determinant of acceleration capability, this performance results from combining an extremely light EV (2,200-pound curb weight, including

Table 3. Energy Consumption for Various Driving Schedules (Values are kW-h/mi extracted from battery)

Schedule	Max mph	Car: Rabbit EV	Car: Impact
Constant speed	35	0.17	—
Constant speed	55	0.25	0.11
J227aB	20	0.33	—
J227aD	45	0.34	—

870 pounds of batteries) with motors capable of high output (114 horsepower total) for short periods. This light weight combines with special 65-psi tires having half the normal rolling resistance and an aerodynamic two-passenger body of small frontal area having half the normal drag to give a 55 mph energy requirement about half that of the Rabbit EV.

To determine the primary energy requirement of the EV, the energy leaving the battery for the motor and its controls must be divided by the charge/discharge efficiency of the battery, the efficiency of the battery charger, the efficiency of power distribution from the central power station to the input side of the battery charger, and the thermal efficiency of the central station itself. On that basis, past EVs have consumed more energy than comparable gasoline-fueled cars, but claims for the Impact suggest that on the urban schedule, its primary energy consumption is somewhat less than for the 1989 Geo Metro (at 59 mpg). On a fresh battery, the Impact also beats the Metro on acceleration, but the Metro offers greater cargo capacity, lower cost, greater range, and is suitable for cross-country trips.

The 55 mph range of the Impact is 120 miles. In Figure 9, the energy capacity of the battery in 55 mph tests is plotted against range. For the Rabbit EV, range is affected by battery selection, but all points fall along the 0.245 kW-h/mi line. The 120-mile range of the Impact far exceeds that of the Rabbit not because of its battery capacity, but because the design of the vehicle lowers its kW-h/mi requirement.

Range is very important to the driver not only because it is comparatively short in an EV, but because hours are required to recharge the batteries, rather than the minutes needed to refill a fuel tank. As a result, EVs are special-purpose vehicles that cannot provide the utility of the engine-powered car. This also explains the interest in hybrid vehicles, which include both a battery-electric system and a small engine capable of charging the batteries as they are depleted.

EV performance specifications normally apply to reasonably fresh, well maintained batteries under favorable operating conditions. In one EV, the 52-mile range on an 80°F day fell to 49 miles when a passenger was added, then to 41 miles when tire pressure was reduced

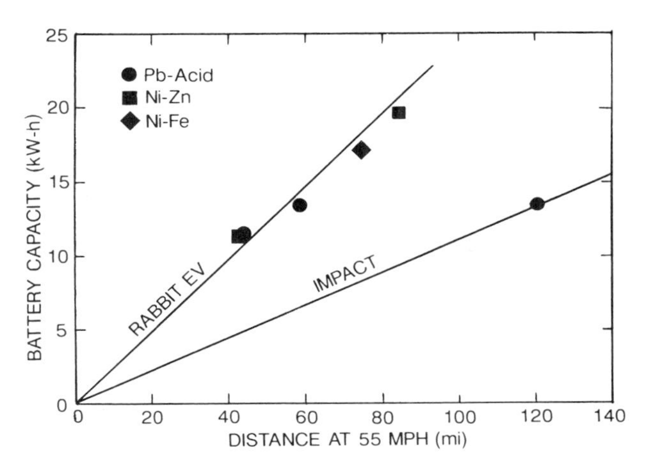

Figure 9. Battery capacity versus range at 55 mi/h for an electric car based on the VW Rabbit (Conover et al., 1980) and for the Impact electric car (GM, 1990).

for a more comfortable ride, and finally to 26 miles when the ambient temperature dropped to 40°F (Gumbleton et al., 1969). In another, when the battery was two thirds discharged, the acceleration time to 30 mph increased by 25 percent, and to 55 mph by 33 percent (DOE, 1989). Continued driving to a state of 89 percent discharge prolonged acceleration to 30 mph by 47 percent and to 55 mph 100 percent.

Batteries also deteriorate with use. To obtain reasonable life from the battery, depth of discharge is normally limited to 30 to 50 percent, which means that an EV with a maximum range of 120 miles has a useful daily range of 35 to 60 miles. Repeated use of maximum range necessitates early battery replacement.

A long list of alternative battery types is being researched to improve on those presently available. One of the more interesting ones, sodium-sulfur, operates at rather high temperature with both elements in the molten state. As high-temperature batteries are perfected, such issues as safety in accidents and restarting after long idle periods, after the elements have frozen, call for attention.

Hydrogen Fuel

The newest environmental concern is the possibility of global warming from altered composition of the greenhouse gases surrounding the earth that regulate its temperature. Carbon dioxide is one of the greenhouse gases, and US highway vehicles are responsible for about a quarter of the national fossil-fuel contribution. Whether warming is actually occurring is controversial. If it is, there is uncertainty over whether it is caused by humans or is the result of natural phenomena not under human control. For example, it has been deduced from an Antarctic ice core that temperature varied over a range of about 11°C over the past 160,000 years, obviously without significant inputs from human activity (Jouzel et al., 1987).

Should it some day prove necessary to eliminate fossil fuel from the transportation system because of global warming, hydrogen fuel is one of the options. It can be made by hydrolyzing water with either nuclear power or solar cells. It has been burned successfully in spark-ignition engines, but the problem is storage. Other nonfossil-fuel options include EVs with batteries charged from nuclear- or solar-energy sources and engine-powered vehicles operated on alcohol from biomass.

Gaseous hydrogen stored at 2400 psi in pressurized cylinders occupies about 18 times the volume of an equivalent amount of energy stored as gasoline and weighs three times as much (Ogden and Williams, 1989). Assuming, for illustration, that the hydrogen engine gains enough efficiency to cancel its adverse mass effect, a gasoline-fueled car with a range of 350 miles would travel only 20 miles on gaseous hydrogen occupying the same vehicle space.

Liquefied hydrogen entails a similar mass penalty but occupies only 3.8 times the volume of gasoline for the same stored energy (Ogden and Williams, 1989). However, it must be kept at -423°F to avoid boiloff. When stored in an automotive dewar, boiloff losses amount to 0.5 to 3 percent per day and would create safety concerns in such enclosed spaces as garages. Moreover, 10 to 25 percent of the fuel boils off during refueling (Ogden and Williams, 1989). Liquid hydrogen cannot be transmitted through pipelines economically for any great distance, and liquefying at a service station is very expensive.

Hydrogen can be stored in a metal hydride, where it is bound in the metal and released for use upon heating, typically by the engine exhaust. Hydrides weigh about as much as high-pressure gas cylinders but occupy only about 25 percent more space than a gasoline tank containing the same energy (Ogden and Williams, 1989). The search for improved hydrides continues.

Conclusions

The fleet-average fuel economy of passenger cars has doubled in the past 15 years. Further improvement will be more difficult as theoretical and practical limits are approached. Projected potential for new cars by the turn of the century from implementing known technologies range from EPA's lower-level US-fleet average of 30.9 mpg to DOE's domestic-fleet average of 33.1 mpg. Industry estimates are nearer EPA's. Discussions are in progress to resolve the DOE-industry discrepancy. Without further fuel-economy gains, the economy of the in-use fleet will rise about 25 percent by the early 2000s because of fleet turnover.

From federally prescribed emission tests on new light-duty vehicles, it takes 25 1990–model passenger cars to exhaust as much HC and CO as did a single precontrol car, and NO_x has been concurrently decreased by 76 percent. In-use deterioration of emission-control systems has decreased since the advent of closed-loop control. Average emissions per car will decrease significantly as the fleet turns over. Despite all this, photochemical smog remains a problem today in some urban areas. Because the automotive vehicle is not the only contributor, tighter controls on other less-regulated sources deserve consideration.

Concern over urban environments is promoting interest in methanol and natural gas, both of which can be burned in spark-ignition engines. A reduction in urban ozone has been estimated for switching from gasoline to methanol in some urban environments provided aldehydes are controlled better than presently possible. In others, that fuel switch is predicted to be ineffective. Natural gas is expected to be cleaner, but its range limitations suit it better for special-purpose fleets than for private passenger cars. Meanwhile, US industry is embarking on a program to learn what can be accomplished with reformulated gasoline.

No automotive engine matches the fuel efficiency of the diesel engine, but the diesel outlook for passenger cars is dim in the US because of emission regulations and consumer preferences. It remains entrenched in heavy-duty trucks, where it is challenged by future NO_x and particulate regulations.

The two-stroke engine has near-term potential for improving fuel economy. Its emissions compliance is uncertain and is being evaluated by industry. It may prove usable in small cars only.

The automotive gas turbine demands ceramics, which are presently too unreliable and too expensive. Proof of acceptable component efficiencies and development of a high-temperature low-NO_x combustor are other hurdles facing this engine.

The battery-electric car (EV) trades lower HC and CO emissions for increased sulfur oxide and particulate emissions, with no significant effect on NO_x. This may be a good trade for local air quality but not global air quality. The EV is hampered by the state of battery technology, which leads to short range. That, coupled with inherently slow recharging rates, limits the EV to special-purpose applications.

Global warming is a controversial issue that may or may not become a future problem. If it does, the ultimate approach is elimination of fossil fuel. Then options would include electric vehicles powered from nuclear or solar-cell sources and engine-powered vehicles using either alcohol made from biomass or hydrogen extracted from water by using nuclear or solar-cell electricity. Hydrogen fuel is suitable in spark-ignition engines but faces a monumental on-board storage problem.

References

Amann, C.A., "The Automotive Engine—A Future Perspective," SAE Paper 891666 (1989).

Bleviss, D.L., *New Oil Crisis and Fuel Economy Technologies—Preparing the Light Transportation Industry for the 1990s,* Quorum, New York, NY (1988).

Conover, R.C., Hardy, K.S., et al., "Vehicle Testing of Near-Term Batteries," SAE Paper 800201 (1980).

Davis, S.C., Shonka, D.B., et al., "Transportation Energy Data Book: Edition 10," Oak Ridge National Laboratory ORNL–6565 (September 1989).

DeLuchi, M.A., Johnston, R.A., et al., "Methanol vs. Natural Gas Vehicles: A Comparison of Resource Supply, Emissions, Fuel Storage, Safety, Costs, and Transitions," SAE Paper 881656 (1988).

Department of Energy, "Energy Technologies & the Environment," DOE/EH–0077 (October 1988).

Department of Energy, "Electric and Hybrid Vehicle Program, Twelfth Annual Report to Congress for Fiscal Year 1988," (February 1989).

Dunker, A.M., "The Relative Reactivity of Emissions from Methanol-Fueled and Gasoline-Fueled Vehicles in Forming Ozone," Paper 89–7.6, Air and Waste Management Association (June 25–30, 1989).

Energy and Environmental Analysis, "Domestic Manufacturers' Fuel Economy Capability to 2001—an Update," Energy and Environmental Analysis, Inc., Report to Martin Marietta Energy Systems (October 1989).

General Motors, "Impact," press release (January 3, 1990).

Gumbleton, J.J., Frank, D.L., et al., "Special Purpose Urban Cars," *SAE Transactions* 78, pp. 1659–85 (1969).

Halberstadt, M.L., "Effective Control of Highway Vehicle Hydrocarbon Emissions," JAPCA 39(6), p. 862 (June 1989).

Haskew, H.H. and Gumbleton, J.J., "GM's In-Use Emission Performance—Past, Present, Future," SAE Paper 881682 (1988).

Heavenrich, R.M. and Murrell, J.D. "Light-Duty Automotive Technology and Fuel Economy Trends Through 1989," EPA Report EPA/AA/CTA/89–04 (May 1989).

Jouzel, J., Lorius, C., et al., "Vostok Ice Core: A Continuous Isotope Temperature Record Over the Last Climatic Cycle (160,000 Years)," *Nature* 329, pp. 403–408 (October 1, 1987).

Kelly, N.A. and Gunst, R.F., "Response of Ozone to Changes in Hydrocarbon and Nitrogen Oxide Concentrations in Outdoor Smog Chambers Filled with Los Angeles Air," Paper 89–152.6, Air and Waste Management Association (June 1989).

Livo, K.M., Miron, W., et al., "1988–89 Oxygenated Fuel Program," Air Pollution Control Division Report to the Colorado Air Quality Control Commission (April 15, 1989).

Motor Vehicle Manufacturers Association, "MVMA Facts & Figures '89," p. 53 (1989).

Murphy, T.E., "Power System Optimization for Passenger Cars," SAE Paper 850030 (1985).

Needham, J.R., Doyle, D.M., et al., "Technology for 1994," SAE Paper 891949 (1989).

Ogden, J.M. and Williams, R.H., *Solar Hydrogen: Moving Beyond Fossil Fuels,* World Resources Institute, Washington, DC (October 1989).

Partyka, S.C., "Registration-Based Fatality Rates by Car Size from 1978 through 1987," National Center for Statistics and Analysis (May 1989).

Sperling, D. and DeLuchi, M.A., "Transportation Energy Futures," *Annual Review of Energy* 14, pp. 375–424 (1989).

The Role for Improved Vehicle Fuel Economy

Deborah Bleviss
International Institute for Energy Conservation
Washington, DC

Abstract

Because light vehicles account for nearly 60 percent of the energy used for transportation, they must be the focus of any strategy to mitigate the undesired environmental impacts of transportation energy use. Improving the fuel economy of these vehicles is a key element of such a strategy. Other options such as increased use of mass transit and conversion to alternative fuels have substantial limitations, particularly in the near term. A government role is needed, however, to stimulate improved fuel economy. That role should consist of a three-pronged approach of offering incentives to manufacturers to produce more efficient vehicles, offering incentives to consumers to buy more efficient vehicles, and stimulating the research and development of fuel efficient technologies.

Introduction

Improving the fuel economy of light vehicles will need to be a major strategy in mitigating the undesired environmental side effects of our transportation energy dependence. Light vehicles account for nearly 60 percent of the energy used in the US for transportation (Davis, 1989). Hence, efforts to contain the adverse environmental effects of transportation energy use must necessarily focus on the use of these vehicles.

There are a limited number of options available to reduce the environmental impacts of these vehicles, however, particularly in the near term. One frequently mentioned option is encouraging greater use of mass transit. Certainly, this strategy should be pursued to its fullest possible extent, but it has its limitations. In the US, if the size of mass transit systems were to triple—a major financial investment—and to be filled to capacity, the energy use for road transportation would decrease by only 10 percent.

Another option that is often mentioned is switching to alternative, non-petroleum based fuels. This, too, has its limitations. At present, there are essentially only three alternative fuels that are commercially viable: methanol produced from natural gas, compressed natural gas, and ethanol produced from agricultural feedstocks. All have problems that would inhibit their widespread use in the near future.

Methanol is attractive because it produces fewer noxious regulated emissions than either gasoline- or diesel-fueled vehicles. However, its emissions of carbon dioxide (CO_2)—a greenhouse gas—are comparable to that of gasoline (per unit of energy burned). Compressed natural gas does offer lower carbon dioxide emissions than gasoline, but problems with storage of the fuel and the range of the vehicles will necessarily limit its penetration. Moreover, there are some concerns about the nitrogen oxide emissions of this fuel. Finally, ethanol, if produced from renewably grown biomass, does offer the advantage of producing no net increase in CO_2 emissions, but it is very expensive. Further, there is insufficient biomass available to convert the nation's, much less the world's, light vehicles to this fuel.

Today's alternatives to petroleum fuels, therefore, can only play a minor role in the near term in helping to meet the nation's transportation needs while minimizing environmental problems. Until cleaner fuels can be developed, a bridging strategy is required. The only option left to provide this bridge is improving the efficiency of today's vehicles. Not only does improved fuel economy reduce CO_2 emissions, but it also generally reduces the emissions of other noxious pollutants, since less pollutant-emitting fuel is burned per mile driven. Moreover, enhanced fuel economy offers the additional benefit of expanding the potential of the resource base from which any future alternative fuel is drawn; a biomass energy plantation will double the number of ethanol-fueled cars it can feed if they achieve 60 miles per gallon (mpg), compared with the 30 mpg vehicle typical of today.

Options to Improve Fuel Economy

The potential exists to achieve fuel economies in vehicles far greater than those on the road today. Perhaps the best illustration of this potential can be seen in the plethora of high efficiency "concept" cars that have been introduced during the 1980s. Most of these achieve a city fuel economy of at least 60 mpg and a highway fuel economy of at least 75 mpg. Fuel efficiencies can be improved in the future by increasing the efficiencies of engines and transmissions, improving vehicle aerodynamics, and substituting lightweight materials for the steel that predominates in today's vehicles. Clearly, as these technologies are developed, care must be taken to ensure that they preserve the safety of vehicle occupants and do not increase noxious emissions.

While the potential is great, the likelihood that these technologies will be tapped any time soon is small. Oil prices are low and a decade has elapsed since the last oil crisis, leading industry, consumers, and government alike to forget the need to pursue higher fuel economy. The US government played a major role in encouraging fuel economy improvements during the 1970s. Among the programs it initiated were fuel economy regulations for new cars and light trucks, information programs, and a gas guzzler tax on the least efficient vehicles sold in the US. Most of these programs, however, are either outdated today or so weakened as to be of reduced effectiveness.

If fuel economy improvements are to be encouraged with all due speed, the US government must step forward again. What is needed is a three-pronged program of offering incentives to manufacturers to develop fuel efficient vehicles, offering incentives to consumers to buy such vehicles, and stimulating research and development of new fuel efficient technologies.

The US already has considerable experience in providing incentives to automakers to improve the fuel economy of their fleets. The fuel economy regulations enacted in 1975 were the single most effective conservation program ever implemented in the US. Between 1979 and 1983, 20 percent of the oil savings achieved by all the signatories of the International Energy Agency was the result of fuel economy improvements in this country (Williams, 1984). Establishing a new set of fuel economy regulations through the end of the century therefore makes good sense; these should strive to achieve a new car fleet average of 45 mpg and a new light truck average of 35 mpg by the turn of the century.

A few refinements, however, are needed in the structure of the old standards, in recognition of changing market conditions. For example, the previous standards were based on a corporate average fuel economy (known as CAFE) standard that was the same for every manufacturer selling vehicles in the US. The structure of this standard made it much easier for manufacturers selling predominantly small vehicles (namely, the Japanese carmakers) to meet the standard than those selling both small and large vehicles (namely, the American carmakers). To make the standard more equitable in the future, many have called for it to be set on a percentage improvement basis, which would require the same degree of effort by all manufacturers.

In addition, it is important that future regulations set the standards for light trucks in law rather than allowing them to be set at the discretion of the Department of Transportation, which has in the past been all too easily influenced by the industry to weaken the standards it proposes. Light trucks are an ever growing portion of light vehicle sales in the US, accounting for a third of such sales today compared with less than 20 percent in 1976 (DOE, 1987). Since these trucks are less efficient than cars and their lifetimes are longer, they account for a disproportionate share—50 percent—of the oil consumed over the lifetime of light vehicles (DOE, 1987).

To provide the market demand for the fuel efficient cars that manufacturers would be producing under a program of new standards, incentives for consumers are needed. As already noted, low fuel prices have encouraged consumers and manufacturers to lose interest in fuel economy. Efforts, therefore, need to be made to shore up fuel prices. It is important to realize that at today's level of fuel economy for new vehicles, increased fuel prices will do little to encourage consumers to demand vehicles with significantly higher fuel economy; after all, fuel costs are a decreasing fraction of the cost of owning and operating a vehicle today. Nevertheless, higher prices will discourage the purchase of inefficient vehicles.

Another option for encouraging consumers to demand fuel economy is through the price of the vehicle itself. A financial penalty can be assessed on the purchase of gas "guzzlers" and a financial reward given for the purchase of gas "sippers." Such a tactic is one way of reflecting the lifetime cost of fuel consumption of the vehicle. As previously noted, the US already has a gas guzzler tax in effect; it is progressive in nature—the lower a vehicle's fuel economy is below a defined threshold, the greater the tax. The threshold below which the tax is assessed, however, has remained constant since 1986 despite improvement in the fuel economy of the country's new light vehicle fleet. Hence, strong consideration should be given to strengthening the existing tax, especially since evidence suggests that domestic automakers have worked hard to avoid having the tax assessed to their vehicles.

To encourage consumers to purchase vehicles at the top end of the efficiency spectrum, many have suggested a rebate or tax credit be given. Bills in the US Congress presently offer a range of ways to structure such an incentive. It has been suggested the incentive be given to the vehicles achieving high fuel economy within each size class, so as to promote the development not only of fuel efficient small cars, but large cars as well.

Finally, there is a need to stimulate research and development in fuel efficient technologies so that the demand for greater fuel economy can be met. Generally speaking, American automakers lag behind their competition, particularly the Japanese, in the development of fuel efficient technologies. European and Japanese automakers have a ready mechanism available to them for being stimulated in the R&D arena because they have had a long history of working together on joint research agendas with their governments. The US has not enjoyed the same characteristic. In fact, when an effort was made by the Carter Administration to establish a cooperative automotive research program, it was strongly resisted by the industry. Nevertheless, now is the time to reexamine the efficacy of a jointly funded research program between the government and industry. Of particular importance is designing a joint program that includes the small innovative US businesses that are developing new efficiency technologies. In recent years, these businesses have been behind much of the innovation in the fuel economy arena, but they have repeatedly found little interest in their products among US automakers. Instead, they have had to go overseas to pursue interest in their products.

Conclusions

To minimize the growing environmental problems from energy use for transportation, we need to begin now to devise mitigative strategies. Such strategies must include enhancing the fuel economy of light vehicles; however, this development will not occur without the active and prompt involvement of the US government.

References

Davis, Stacy, et al., *Transportation Energy Data Book: Edition 10*, Oak Ridge National Laboratory, pp. 2–12 (September 1989).

Department of Energy, *Periodic Energy Report*, Number 2 (December 1987).

Williams, Robert H., testimony at the Hearing on Automobile and Light Truck Fuel Economy of the Subcommittee on Energy Conservation and Power of the US House of Representatives (July 31, 1984).

The Future Fuel Economy of the United States Light Duty Fleet—A Policy Dilemma

Steven E. Plotkin
Office of Technology Assessment
United States Congress
Washington, DC

Abstract

Low gasoline prices and changing consumer preferences have reduced the momentum toward increasing fuel economy in the light-duty highway vehicle fleet; new car fuel economy has remained basically level since 1986. With growing concern about oil imports and projected greenhouse warming, United States policymakers are examining the possibility of raising fuel economy standards beyond the 27.5 miles per gallon (mpg) corporate average fuel economy (CAFE) required today.

Although the solidification of manufacturer plans will make it difficult to accelerate fuel economy gains in the short term, that is, to the 1994 or 1995 model years, prospects for the longer term are good. The fleet could average 38 mpg by the 2001 model year if technology diffusion were accelerated, a few low-risk technologies (e.g., electric power steering, intake valve control) were introduced into the general fleet, and average vehicle performance and size were stabilized after 1995. Still higher fuel economy levels are possible, though more difficult to achieve.

The current federal approach to improving fuel economy is limited in effectiveness and distorts the market. Policymakers should examine alternatives and/or additions to regulatory approaches to achieving higher fuel economy, as well as alternatives to the "one standard for all companies" CAFE standard used today.

Introduction

The United States' fleet of light duty highway vehicles—automobiles and light trucks, including vans—represents a microcosm of the major energy and environmental problems facing the nation. Because this fleet today uses approximately 16 percent of total US energy, it represents an important part of the large US contribution to rising world levels of greenhouse gases. The fleet burns nearly four tenths of the United States annual oil use, at a time when rising demand for oil and falling domestic oil production have pushed US oil imports up to 46 percent in 1989 (American Petroleum Institute, 1990), with higher import levels expected in 1990. The fleet is the major source of emissions of ozone precursors in cities—in 1985, it was responsible for about four tenths of total emissions of volatile organic compounds in cities not attaining the national ambient air quality standard for ozone (Office of Technology Assessment, 1989). All of the above problems are being aggravated by growing demand for travel and by urban congestion, which reduces fuel efficiency and increases emissions levels. Current energy forecasts project light duty vehicle mileage to grow by 2 to 3 percent per year for the next few decades, and recent Department of Energy projections predict that a growing severity of congestion and an increase in the area subject to congestion will cause the efficiency of the highway fleet to decrease by over 15 percent beyond what it would have been with no change from today's level of congestion (Westbrook and Patterson, 1989).

In addressing these problems, the Congress has focused its attention on two primary initiatives: reauthorization of the Clean Air Act, and modification of automobile fuel economy (Corporate Average Fuel Economy, or CAFE) standards. Within the Clean Air Act initiative, separate sections address energy and environmental problems of the light duty fleet: programs encouraging the use of alternative, cleaner fuels replacing gasoline in cities with ozone and carbon monoxide problems; tighter standards for automotive exhaust and evaporative emission controls; and a new automotive standard for carbon dioxide emissions, which is a "back door" way of setting a fuel economy standard. As of January 1990, the fuel economy proposals have incorporated but two options for new standards: a straightforward increase in the current 27.5 mpg standard for all auto manufacturers (from Senator Metzenbaum), and a standard that demands each company to increase their current CAFE by a uniform percentage (from Senator Bryan).

Reauthorizing the Clean Air Act may be the more politically "dangerous" of the two initiatives and, with the multitude of separate types of pollutants and different emission source categories to be controlled, the more technically complicated. Nevertheless, the lack of information available to the Congress about the likely consequences of a new fuel economy standard, and the ambiguity associated with the value to the nation of improving automobile fuel economy, may make the argument over new CAFE standards more of a policy dilemma for the Congress. This argument is the subject of the discussion that follows.

What Is Happening to the Light Duty Fleet?

In every year from 1974 to 1982, the US new car fleet achieved sharp gains in fuel economy over the last year, with a remarkable four mpg gain from 1979 to 1980, a year strongly influenced by the effect of the Iran/Iraq war on gasoline price and supply. With the lower oil and gasoline prices that began in the early 1980s, however, progress in fuel economy has slowed dramatically, and current market forces are making further fuel economy improvements markedly more difficult. Signs of this market difficulty include:

- a remarkable drop in the number of consumers seeking fuel economy as the primary attribute of a new car, from about a third in 1980 to *3 percent* in 1987 (Power and Associates, 1988),
- a dramatic rise in market share of relatively fuel inefficient light trucks, from 17 percent in 1972 to about one third today (Davis et al., 1989),
- a steady increase in vehicle performance (e.g., reduced 0–60 miles per hour (mph) acceleration time, higher horsepower/weight ratio) beginning in 1982 and continuing today (Heavenrich, 1989),
- a recent trend towards increasing vehicle weight and size (Heavenrich, 1989), and
- increasing levels of efficiency-lowering equipment such as four wheel drive, power windows and locks, and so forth.

Coupled with the potential for new standards in other regulatory areas (e.g., emissions and safety) that could conflict with efficiency goals, these trends make it clear that new car automobile fuel economy is unlikely to resume a sharp upward slope without a real change in direction in manufacturer *and consumer* behavior. And without large changes in gasoline price or availability, such a change in direction is likely to come only from government intervention.

If Government Intervenes, What Is at Stake?

It is likely that Congress today views an increase in CAFE standards as a relatively low risk option to slow the growth of oil imports and counter the greenhouse effect. Although there will be *political* costs to reckon with, given strong opposition from some quarters, there does not seem to be much active public opposition to regulating the automobile industry. On the contrary, the perceived success of the previous CAFE standards presumably should make a new round of standards more palatable to the public. Nevertheless, aside from the political costs (and benefits), there are important values at stake in the new CAFE debate.

Just what are the benefits of a stronger CAFE standard? The obvious, sought-after benefit of a new standard is the acceleration of technology development and encouragement of the introduction of available technologies that otherwise wouldn't make the industry's investment "cut." The assumption here is that the industry isn't adding all the technology that is cost effective, and certainly isn't taking into account in its investment strategy any benefits to society that would be obtained from reducing US oil consumption. In general, I conclude that this assumption is correct *if* we buy the idea of "externalities" associated with oil use and also interpret "cost effectiveness" in a particular way.

Although there are sharp disagreements about the costs of fuel economy improvements, past industry investments in fuel economy technology seem to mesh well with an investment criterion that selects technologies that will allow the consumer to recoup his initial investment in about four years. In other words, fuel savings over the first four years of ownership recover the added costs of the fuel economy technologies, using net present value calculations and about a 10 percent discount rate (Duleep, 1990). Although this seems a modest enough criterion, since cars last much longer than four years, it appears unlikely that manufacturers' future decisions will swing more towards emphasizing fuel economy. If anything, these decisions may tend to swing *away*, because past conditions probably were more favorable towards fuel economy than future conditions are likely to be. The companies made their prior investment decisions in the face of a fuel economy standard that, in recent years at least, they were having trouble meeting. Also, during much of this time, their investment decisions were driven by expectations of strong future increases in gasoline prices and strong consumer demand for fuel efficient vehicles—in my view, an unlikely scenario for the immediate future. The implication here is that, without stronger fuel economy regulations and with continued low consumer interest in fuel economy, the companies are likely to pursue only those technologies with a very short payback (or very strong auxiliary benefits) and ignore many technically-viable fuel economy technologies that make sense from the standpoint of vehicles' *total* lifecycle costs. Of course, a strong jump in oil and gasoline prices within the next few years, coupled with consumer fears of further increases in the future, could change this argument radically.

The above argument could be the starting point for hundreds of pages of economic discussion about costs and benefits, and I demur. However, I believe that the following points should be addressed in public discussion:

- Consumer surveys imply that the auto manufacturers, in pursuing fuel economy at a relatively moderate rate, are well in line with consumer desires. Although more fuel-efficient vehicles that have higher first costs may be good for society, it is possible—even likely—that consumers will view them as less attractive than slightly less efficient but cheaper vehicles. On the other hand, the same is true for more stringent emission standards, which yield *no* long-term monetary savings to buyers, yet these have strong public support because of their societal benefits.
- From the standpoint of the average carbuyer, it appears quite rational to be willing to pay only for fuel efficiency measures that will allow cost recovery in four years. Four years is the average length of time that the first purchaser keeps a new car. Unless the used car market gives a significant premium for fuel efficiency, which seems unlikely in today's market, the average buyer will not recover any investment in fuel economy past the four year recovery value.
- The United States' oil imports are rising sharply, and this has provoked sharp concerns about US energy security. There are, however, important differences between the oil market during the first energy crisis, in the 1970s, and today's market. Among these differences: the existence of the Strategic Petroleum Reserve; a much stronger OPEC economic stake in the well-being of the economies of the importing nations; a strong futures market and spot market in oil trade; and so forth. Many of these changes lessen the chances of a crisis, and perhaps weaken the argument that current oil prices do not adequately reflect actual costs to the US economy.
- Despite a body of legislation regulating the environmental impacts of oil use, there remain uncontrolled impacts that would be lessened with improved fuel economy. New concerns about global warming increase the value of higher fuel economy to society. This value is *not* reflected in gasoline prices.

Aside from the question of the benefits to be gained, policymakers should consider the potential costs of new standards. In particular, they should take care to refrain from setting standards that would push vehicle characteristics to the point where a loss of new car sales—and the slower turnover of the fleet, delaying the elimination of older, less efficient vehicles—might counteract the energy efficiency effect of the increased new vehicle efficiency. Further, they should consider carefully how a new standard should be structured. A simple increase in the current standard will perpetuate an important problem—applying the same standard to each manufacturer ignores differences in the market segments they are selling to, and puts little pressure on companies selling primarily small cars. These companies will retain the freedom to satisfy market demands for increased performance and other vehicle attributes at odds with fuel economy, thereby putting pressure on other manufacturers to follow suit or lose market share. Standards putting differential pressures on manufacturers according to the attributes of their fleets or, as in current legislative proposal, on their past CAFE performance can overcome this problem. However, *all* regulations introduce inefficiency into the market. If policymakers wish to pursue a regulatory course towards increased fuel economy, they should take much care in building a regulatory structure that maximizes economic efficiency, minimizes inequities among the different manufacturers, and sends the appropriate signals to the industry about their future behavior.

What Is the Technical Potential?

Most of the recent debate about setting new CAFE standards has focused on how high to make them. In numerous hearings before the Committees of jurisdiction, Congress has heard estimates of the realistic technical potential of the US auto fleet ranging from *at best* about one or two mpg above current standards for the next several years (from the domestic auto manufacturers) to 45 mpg or higher by the year 2000 (from representatives of the conservation community).

Sources of disagreement about technical potential include differing views about:

- percentage improvement in fuel economy possible from specific technologies,
- commercial availability of advanced technologies,
- effects of technologies on other consumer attributes such as driveability,
- consumer costs, and
- the extent to which fuel economy goals should be allowed to interfere with other consumer attributes, especially performance and vehicle size.

Many of the available estimates of technical potential rely on old data and projections or on sketchy analyses that are roughly based on the performance of vehicle prototypes and/or laboratory tests of new engines and other fuel economy technologies. Given the magnitude of investment that would be required if the federal government mandated significant increases in CAFE standards, the level of analysis that is available and that has been supported by the government is surprisingly low.

To my knowledge, the only longstanding, extensive analytical effort in automobile fuel economy is that conducted by Energy and Environmental Analysis, Inc. (EEA), of Arlington, Virginia, and supported primarily by the Department of Energy (DOE). Within the past year, DOE has been joined by the Office of Technology Assessment and others in supporting EEA's efforts.

EEA's estimates have been attacked recently by the three domestic manufacturers, who claim that EEA has consistently overestimated the fuel economy gains obtained from adding specific technologies. It is my view, based upon evaluation of the EEA analysis and attendance at a meeting between EEA and industry analysts where the methodologies of each were explained and critiqued, that EEA's estimates are sound. Much of the disagreement between EEA and the industry appears to stem from misunderstandings about the technology definitions applied by EEA and differences of opinion about what constitutes an appropriate "before and after" case upon which to base estimated fuel economy improvements.

EEA's fuel economy projections are based on a limited set of "scenarios" that should not be interpreted necessarily as appropriate levels for a fuel economy standard. They do, however, provide an excellent technical perspective to help policymakers set such a standard. Some of the projections follow:

1. The total US fleet (domestics plus imports) could reach about 32.5 mpg by 1995 if it returned to 1987 levels of size and horsepower, proceeded with planned dissemination of already-existing-and-in-use fuel economy technologies into new models, and improved the performance of these technologies to levels already achieved by the most successful examples of these technologies in the 1987 fleet.

   Interpretation: Manufacturers have already solidified product plans for 1995, and have begun to order equipment from suppliers. Also, vehicle size and performance have shifted since 1987 in ways that hurt fuel economy: average weight increased from 3032 to 3116 pounds, horsepower increased from 113 to 121, horsepower per pound increased from 0.037 to 0.039, 0 to 60 mph acceleration decreased from 13.0 to 12.5 seconds, large car sales increased from 12.1 percent to 15.2 percent (Heavenrich, 1989). Consequently, unless policymakers want to force a substantial shift in consumer buying patterns, the 32.5 mpg value probably represents an optimistic upper bound for fleet fuel economy for 1995. If current trends of increasing size and performance continue, the fleet will lose about two mpg from this value. It is also important to note that the 32.5 value represents an average. Some companies can do better, some worse, depending on the size, performance, and other characteristics of their fleets. Looking at the three domestic manufacturers alone, for example, because of their larger market shares in larger vehicles, they are likely to achieve only about 31 mpg at 1987 size/power levels, or 29 mpg at expected 1995 levels; the import manufacturers will achieve a correspondingly higher level because of their predominance in smaller cars.

2. The total US fleet could reach about 37 mpg (35 mpg for the domestics) by 2001 if it returned to 1987 levels of size and performance, continued to diffuse currently available fuel economy technology throughout the fleet, and applied moderate levels of low-risk new technologies (advanced weight reduction; advanced drag reduction; intake valve control; 5–speed automatic transmissions; continuously variable transmissions on small cars; and electric power steering). If manufacturers chose to use fuel economy technologies at a considerably higher level (example: 70 percent of vehicles built by the domestic manufacturers use intake valve control in 2001, versus 30 percent in the "moderate" case), the fleet could reach about 40 mpg (39 mpg for the domestics) by 2001. In both cases, if the 2001 fleet remained at projected size and performance levels for 1995 rather than 1987 levels, the resulting fuel economy would be about 2 mpg lower than shown, that is, 35 mpg and 38 mpg, respectively.

   Interpretation: The year 2001 is enough removed from today to allow substantial changes in company product plans. Presuming that a leveling of size and performance trends after 1995 is realistic—and many policymakers probably would consider such an assumption as being too wasteful of fuel economy potential—a fuel economy standard that would achieve a fleet average of 35 mpg by 2001 appears quite modest. A standard that would achieve 38 mpg seems realistic as well, though it will be more expensive. Higher values may be achieved if reductions (from expected 1995 levels) in average vehicle size and power can be attained, or if other technologies—two stroke engines or advanced diesels, perhaps—can be perfected and begin to make significant penetrations into the fleet.

Conclusions

The US automobile fleet has lost the momentum it previously had towards increasing levels of fuel economy. The primary reason for this loss is the shift in consumer preferences toward better performance and larger size vehicles and away from fuel economy.

The stagnation in fleet fuel economy has alarmed policymakers concerned about the greenhouse effect and growing US oil imports, and Congress has been debating the worth of raising the Corporate Average Fuel Economy (CAFE) standards. I conclude that the fleet is capable of raising its average fuel economy in the future, by moderate amounts in 1995 and quite significant amounts by 2001. However, policymakers should examine carefully the costs and benefits of regulating further fuel economy increases. Just how important are rising oil imports to US energy security, given the changing geopolitical conditions and the shifts in the oil market that have occurred since the 1970s? How many years would increased fuel economy "purchase" in terms of delaying global warming, and is this significant? What are the likely costs to the three domestic manufacturers, and to the US economy, of distortions caused by regulating against the grain of existing market trends? Might it not be preferable to attempt to achieve the same ends with economic incentives, e.g., gas taxes, rebates for high mileage vehicles, and so forth?

Further, if policymakers decide to regulate, they should carefully consider alternatives to the current CAFE structure. By treating all manufacturers alike, despite their substantial differences in fleet characteristics, the current regulatory structure creates substantial distortions that might be lessened significantly by alternative structures.

References

American Petroleum Institute, *Monthly Statistical Report for December, 1989* (January 17, 1990).

Davis S.C., et al. *Transportation Energy Data Book: Edition 10*, Oak Ridge National Laboratory, ORNL–6565 (1989).

Duleep, K.G., personal communication, Energy and Environmental Analysis, Inc. Mr. Duleep manages the primary contracting effort in fuel economy analysis for the Department of Energy (1990).

Heavenrich, R.M., et al. *Light Duty Automotive Technology and Fuel Trends Through 1989*, Environmental Protection Agency, EPA/AA/CTAB/89–04 (May1989).

Office of Technology Assessment, *Catching Our Breath: Next Steps for Reducing Urban Ozone*, OTA–O–412 (July 1989).

Power, J.D. and Associates, *The Power Newsletter*, Westlake, CA, reported in Patterson, P.D., *Periodic Transportation Energy Report* Number 4 (September 8, 1988).

Westbrook, F. and Patterson P., "Changing Driving Patterns and Their Effect on Fuel Economy," SAE Government/Industry Meeting, Washington, DC (May 1989).

Potential Gains in Fuel Economy: A Statistical Analysis of Technologies Embodied in Model Year 1988 and 1989 Cars

W. V. Bussmann
Chrysler Corporation
Highland Park, MI

Introduction

In the past year, legislation has been introduced to the Congress that proposes raising the corporate average fuel economy (CAFE) requirements for cars sold in the United States. Fears of global warming as well as increasing dependence on imported oil and desires for cleaner air are the principal forces behind efforts to raise the fuel efficiency of new cars.

One bill, on which much discussion has focused, is S1224 (the "Bryan Bill"). S1224 mandates that each manufacturer selling cars in the United States be required to raise the average fuel economy of cars sold in 1995 by 20 percent from the 1988 level and by 40 percent by the year 2000. These percentages are based on the judgment of the Office of Technology Assessment and the Department of Energy that "increased fuel efficiency is possible utilizing currently available technology and without significant changes in the size, mix, or performance of the fleet" (p.4 of staff working draft of S1224).

The principal work from which the above judgment is derived is a paper written by Carmen Difiglio (US Department of Energy), K. G. Duleep (Energy and Environmental Analysis, Inc.), and David L. Greene (Oak Ridge National Laboratory) entitled, "Cost Effectiveness of Future Fuel Economy Improvements," submitted to *The Energy Journal* (January, 1989, revised August, 1989). That paper contains a table listing the potential gains in fuel economy that can be achieved by more extensive use of currently existing technologies. Using 1987 as a base year and making assumptions about the increased market penetration of each technology, the authors conclude that a 17.1 percent increase in fuel economy is achievable by 1995. In addition, three different scenarios are presented for the year 2000: product plan, cost effective, and maximum technology.

The purpose of this paper is to estimate the fuel efficiency benefits of the technologies embodied in the 1988 and 1989 model year cars. Those estimates are then used to calculate the increase in fuel economy that would be achieved by 1995 if the increases in the use of the technologies assumed by Difiglio et al. actually occur, the mix of cars sold (i.e., large cars vs. small cars) remains at 1987 levels, and no new safety or emission mandates occur to change the weight or efficiency of cars.

No estimate is provided of the fuel efficiency gains possible by 2000 because the basis for the 40 percent number in the Bryan Bill is the maximum technology scenario of Difiglio et al. This scenario involves the application of technologies such as continuously variable transmissions and variable-valve timing that were not used in any 1988 or 1989 cars. Thus, the use of statistical analysis on recent historical data is to no avail.

Brief History

When the Bryan Bill was introduced last summer, representatives of General Motors, Ford, and Chrysler (hereinafter referred to as the Big 3) met and instructed their engineers to assess the accuracy of the results in the Difiglio et al. paper. (As it turns out, most of the technical work on the benefits of available technologies was done by K. G. Duleep of Energy and Environmental Analysis, Inc. As a shorthand, those results will hereafter be referred to as EEA's.)

The engineers' estimates of the possible improvements in fuel economy, using the increased market penetration numbers assumed in the EEA work, ranged between 7.1 percent and 8.6 percent, with a sales-weighted average of 7.6 percent—less than half of the EEA estimate.

Several explanations can be offered for the wide discrepancy between the engineers' estimates and EEA's. First, there is the often-made argument that the Big 3 engineers' estimates are biased on the low side because they work for auto companies. After all, didn't the representatives of the auto companies strongly oppose the CAFE law in the first instance and argue that the 27.5 mpg target could not be met as the 1975 law required? The credibility of those arguments was nil when, in fact, the domestic auto industry did double fuel economy by reducing a typical car's weight by more than 1,000 pounds, improving aerodynamics, reducing engine displacement and horsepower, lowering friction, improving transmissions, and making other engine improvements such as substituting fuel injection for carburetors. Furthermore, the targets were made more difficult to achieve because there was no incentive to the consumer to buy fuel-efficient cars once the real price of gasoline hit a 30-year low in the mid 1980s. In spite of this and other obstacles, such as more stringent emissions and safety requirements, the Big 3 have met the standard in 1989 (after a brief lapse in 1986–1988 when the standard was temporarily rolled back to 26.0 mpg). So why should the engineers now be believed?

Second, the engineers of the Big 3 auto companies could be underestimating the fuel economy benefits because they might be less aware of the relevant technologies than their European or Asian counterparts. Third, the EEA estimates could be overstated because of the lack of access to all the information that the Big 3 engineers have. Fourth, the EEA estimates could be too high because various technologies are not recognized, resulting in interactive effects, double-counting benefits. And fifth, there could be a flaw in the EEA methodology.

The first explanation of the discrepancy between the Big 3 engineers and EEA is little more than a name-calling contest. The rehashing of old arguments about the auto companies arguing against any regulation versus policy makers ignoring market forces is beyond the scope of this paper. The desire to avoid such disputes is the principal reason for applying statistical analysis to a large body of data.

The second explanation—that the Big 3 engineers are not as aware of the potential benefits of various technologies—might be valid. The process of investigating the subject of fuel economy has uncovered some indirect evidence on this topic.

Figure 1 depicts the fuel efficiency of groups of 1988 model year cars within typical marketing segments. The cars are grouped within each segment according to the origin of their design and engineering. Fuel efficiencies of each model are weighted by sales.

Asian-sourced cars are clearly the most efficient subcompacts. The only significant subcompact cars sold by the Big 3 in 1988 are the Escort/Lynx and the Omni/Horizon, both of which were first introduced in the 1970s and which held less than 20 percent of the subcompact market in 1988.

In the small specialty and compact segments, where there are more offerings by the Big 3, the advantage in fuel efficiency shown by Asian-designed cars is less than 0.5 mpg. In the basic middle segment, where the Asian share is only 9 percent, US-designed cars achieve significantly higher fuel economy.

In Figure 2, data for 1990 cars was analyzed and grouped by weight classes. The average mpg for each weight class is not weighted by sales of each car in the class because sales data for the 1990 model year are not yet available.

The conclusion is that for the cars actually sold in the 1988 model year, there is no clear-cut advantage in fuel efficiency shown by Asian-designed cars except in the subcompact segment. In Figure 2, it is clear that the higher fuel economy achieved by Asian-designed cars derives not from their greater efficiency but rather from their lower weight.

In all segments except the subcompact segment, cars of European origin are the least fuel-efficient. This result is not surprising because many European cars sold in the United States are designed for performance rather than for fuel efficiency.

Table 1 depicts the change in fuel efficiency between model years 1985 and 1988 of cars within each segment by region of origin. It is clear from perusing Table 1 that the fuel efficiency of imported cars has dropped between 1985 and 1988. If fuel-efficient technologies were being developed overseas during this time, then either they were used to achieve higher performance (highly likely in view of the drop in real gasoline prices) or else they were introduced after 1988.

The above evidence is illustrative only, but it does not support the notion that Big 3 engineers or car designers or builders are less aware of fuel efficiency technologies than their foreign counterparts. Indeed, from discussions with Asian car company engineers, it is clear that their estimates of the benefits of the relevant technologies are in the ballpark of the Big 3 engineers' estimates and are not in agreement with EEA's.

The third reason for the discrepancy—that EEA does not have access to all of the relevant information that the Big 3 engineers do—is certainly true, but it does not, by itself, invalidate EEA's results. However, it is the fourth and fifth reasons for the discrepancy—the inadequate account taken of interaction effects and the overall methodology—that will be the focus of the rest of this paper.

Regression Model

The methodology used by EEA garnered information from "engineering journals and trade publications, and by means of comparative analyses of similar makes and models with and without the technology in question. Engineering analysis was used to verify the estimated efficiency improvement potentials" (p. 2 of Difiglio et al.).

An obvious statistical problem arises when one or two observations (comparisons of similar makes and models) are used to draw inferences about a larger population. The statistical validity of such results is shaky at best. Chance alone could account for most of the estimated effect. Another problem arises when an investigator estimates the effects on a response series of variables independently of one another and then sums those effects to arrive at a total. Such a methodology ignores interactions among the explanatory variables, and over-

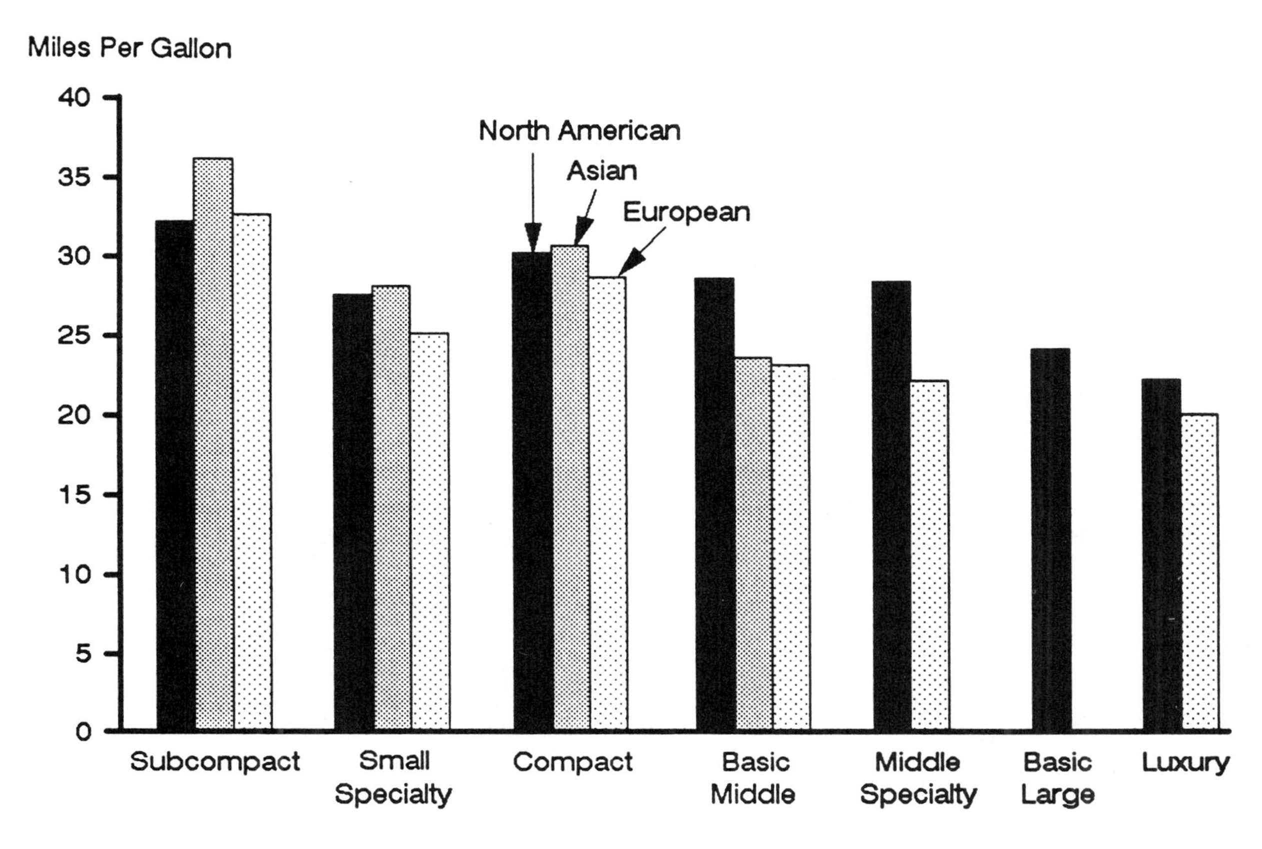

Figure 1. Fuel efficiency comparisons, 1988 MY cars (by region of origin). Note: 1988 EPA data were used (Source: *Automotive News Yearbook*); 1988 model year retail sales data for each model were used as weights within segments (Source: *Wards Automotive*); models were assigned to regions according to the origin of their design and engineering (for example, Japanese transplant cars assembled in the United States are listed as Asian in origin, as are the "captive imports"—cars that are imported from overseas but sold with a Big 3 nameplate); the harmonic mean is used to compute segment averages.

estimates the effect of each explanatory variable to the extent it is positively correlated with other explanatory variables.

There are several possible solutions to these problems. The one that was chosen is a multiple regression model in which miles per gallon is the response series (dependent variable) and the EPA test factors, supplemented by data on engine technologies such as number of valves per cylinder, camshaft arrangement, etc., are the explanatory series. Cars with diesel or rotary engines were excluded, as were Rolls-Royces and the Lamborghini Countach. Less than 1 percent of the data points were excluded, resulting in a total of 1,400 observations for the two model years. Of the 1,400 cars analyzed, 696 were of Asian design, 376 were European in origin, and 328 were designed by the Big 3 US manufacturers. Separate models were estimated for the 1988 and 1989 model years, but since the results were not significantly different the two years' data were pooled into one data base.

The benefit of using a statistical technique compared with engineering analyses, model-by-model comparisons, and quotations from *Automotive News* and other publications is the statistical reliability of the results. The benefit of using a technique like multiple regression is that it accounts for interactive effects among the explanatory series. Therefore its results can be added together without overestimating the total impact of all the possibly correlated explanatory series on the response series. In addition, the results of a multiple regression model estimated on the entire fleet of 1988 and 1989 model year cars show what benefits were indeed attained, on average, by a large number of cars that were actually sold in the marketplace. Thus, it avoids problems such as comparing the worst example of an old technology with the best example of a new technology—the technique often used by EEA. In addition, it avoids the "cherry-picking" approach in which a hypothetical car incorporating all of the best technologies is set up as a standard for the entire

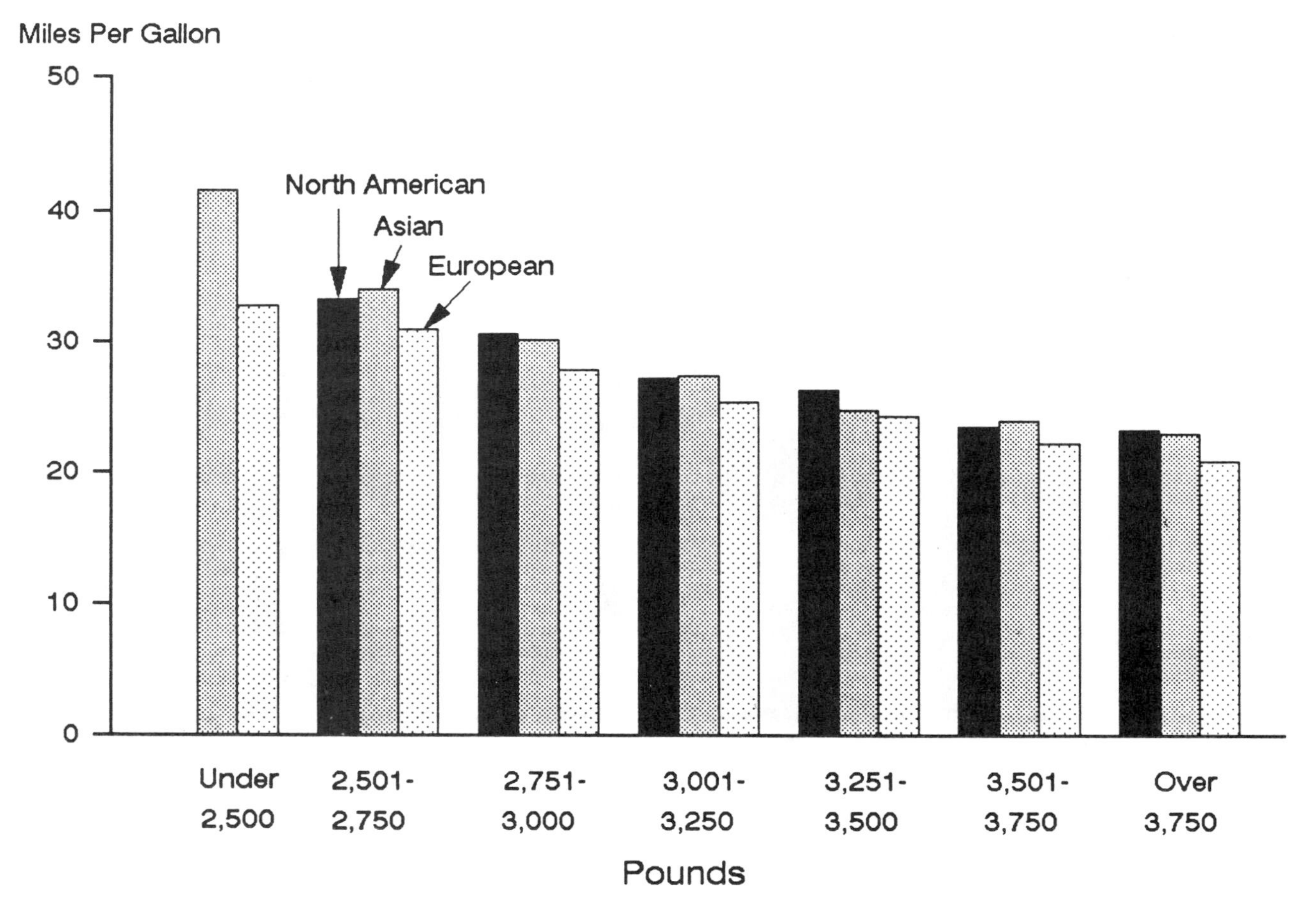

Figure 2. Fuel efficiency comparisons, 1990 MY cars by region of origin, adjusted for weight. Source: 1990 EPA Test Data.

fleet. Such a car may not be very marketable or practical to make.

The estimation of the mpg benefits of individual engine technologies is extremely difficult because many technologies interact with one another in either a positive or negative way. Furthermore, technological improvements, especially in engines, are rarely implemented one at a time. For instance, in designing a new engine, a manufacturer will not incorporate more than two valves per cylinder without also designing overhead camshafts instead of pushrods. In addition, the impact on fuel efficiency of a particular technology may be highly variable, depending on the purpose for which the technology is used. For example, increasing the number of valves per cylinder may be used with other design features either to improve performance or to improve fuel efficiency.

For the above and other reasons, engines were classified into four generations described in the following paragraph. Earlier versions of the regression model were estimated with individual engine technologies as separate variables. However, quantitative measurements are not available for all of the relevant engine technologies (use of low-friction internals, for example), so that the estimates of the impact on fuel efficiency of the factors for which data were available were biased upward. Also, the results were not as statistically reliable as those obtained from the model that incorporates engine generations as separate variables.

The four generations of engines are as follows:

- Old: pushrod or early generation multivalve overhead cam (OHC); no fuel injection (carburetor only).
- Modernized: pushrod or early generation 2-valve overhead cam; some throttle-body fuel injection; some multivalve OHC with carburetor; some conversion to low-friction internals.
- Modern: pushrod with all low-friction internals; no multipoint fuel injection with multivalve cylinders;

Table 1. Miles per gallon comparisons by segment.

	Model Year 1985	Model Year 1988	Change 1985–1988
Subcompact	34.8	34.7	−.1
Asian	36.6	36.1	−.5
Domestic	33.2	30.7	−2.5
European	33.5	32.4	−1.1
Small Special	29.0	27.1	−1.9
Asian	29.5	27.9	−1.6
Domestic	28.7	26.4	−2.3
European	31.4	25.9	−5.5
Compact	29.5	30.6	1.1
Asian	34.2	31.6	−2.6
Domestic	27.9	30.0	2.1
European	32.8	29.4	−3.4
Basic Middle	26.1	27.6	1.5
Asian	25.5	24.7	−.8
Domestic	26.3	28.4	2.1
European	25.1	24.1	−1.0
Middle Special	24.8	27.7	2.9
Asian	—	—	—
Domestic	24.8	27.7	2.9
European	24.9	23.2	−1.7
Basic Large	23.3	24.7	1.4
Asian	—	—	—
Domestic	23.3	24.7	1.4
European	—	—	—
Luxury	23.3	22.3	−1.0
Asian	—	—	—
Domestic	23.1	22.7	−.4
European	23.8	21.6	−2.2
Total Industry	27.6	28.7	1.1
Asian	33.3	32.8	−.5
Domestic	26.3	27.3	1.0
European	27.2	25.1	−2.1

Note: Numbers in the table are harmonic means weighted by model year sales.
Source: EPA data, Wards Automotive

OHC with at least 50 percent low friction internals. No carburetors.
- High: Multipoint fuel injection and multivalve or supercharged; high horsepower/liter; all low-friction internals.

Results

Both linear and logarithmic forms of the regression model were tested. In the linear models, engine displacement, weight, performance, and number of cylinders were tested in nonlinear forms (as their inverses). Theoretical reasons dictate a nonlinear relationship between miles per gallon and these explanatory factors. The final form of the model, chosen because of its superior overall statistical qualities, is logarithmic.

Factors whose significance levels were below 95 percent were excluded from the final model. The least significant variable included in the model (front-wheel drive) has a significance level exceeding 99.6 percent.

Variables of potential importance but which were statistically insignificant in explaining variation in fuel efficiency among cars were EPA interior volume, whether or not the engine was turbocharged, the number of cylinders in the engine, and whether or not the transmission was electronically controlled. The latter variable's insignificance may be attributed to the small number (48) of observations in the entire sample or to the choice of shift points for performance (to satisfy market demand) rather than for economy.

The 15 factors found to be statistically significant "explain" 92 percent of the variation in miles per gallon of the 1400 cars in the '88 and '89 MY data base.

Table 2 lists the 15 factors used in the regression model along with their average values and the effect on miles per gallon of increasing each factor by 10 percent.

The ranking of the factors is by their beta coefficients. A factor's beta coefficient is a measure of the size of the impact on miles per gallon of a 1-percent change in the factor together with the likelihood that the factor will change by 1 percent; it is a measure of the importance of the factor in determining mpg. Thus, a 1-percent change in a given factor may produce a large change in miles per gallon. But if the factor rarely changes by more than, say, 0.5 percent, then its importance in producing changes in miles per gallon is reduced compared with another factor that has a smaller impact but which may typically vary by, say, 10 percent or more.

The statistical insignificance of the number of cylinders is explained by the argument that it is total engine displacement that is important in determining fuel economy, not the number of cylinders. The high degree of correlation ($r = 0.9$) between the number of cylinders and displacement argues for only one variable being included in the regression equation. However, when displacement was excluded in favor of including the number of cylinders, the explanatory power (R-squared) and overall statistical quality (e.g., t-statistics) of the model dropped significantly.

Table 3 lists the regression coefficients and their t-statistics.

The results of the regression model, when combined with the assumed increases in market penetration used by EEA, are that increased use of existing technologies should result in a 5.4 percent increase in CAFE for the entire fleet of cars by 1995. This result is lower than the 7.6 percent weighted average of the Big 3 engineers for at least two reasons.

First, the engineers were asked to report what ben-

Table 2. Impact of EPA test factors on mpg 1988 and 1989 model year cars.

Rank	Factor	Average Value	Percent Change in MPG for 10% Increase in Factor
1	Weight	3097 lbs.	−5.1
2	Engine displacement	144 cu. in.	−2.8
3	N/V (engine revs per mile travelled)	42.5	−3.7
4	Performance (inverse)	26.3 lbs./hp.	+2.1
5	Manual transmission vs. 3-spd. auto*	694 cars	+8.2
6	4-spd auto trans vs. 3-spd auto**	599 cars	+5.2
7	Dynamometer setting	7.96	−1.3
8	High-tech engine vs. old-tech.	266 cars	+6.2
9	Modern engine vs. old-tech.	797 cars	+3.8
10	Compression ratio	9.03	+1.8
11	Modernized engine vs. old-tech.	249 cars	+2.7
12	Axle ratio	3.63	−.5
13	Front-wheel drive	885 cars	−1.2
14	4-Wheel drive	93 cars	−2.8
15	Air conditioning vs. no A/C	1218 cars	−2.2
Factors Found to Be Statistically Insignificant			
	Number of cylinders		
	EPA interior volume		
	Turbocharging		
	Electronically-controlled transmission		

* Manual transmission includes both 5-spd and 4-spd.
** 4-spd auto transmission includes 3-spd with lock-up.

Table 3. Regression model 1988 and 1989 model year cars.

Rank	Explanatory factor	Regression Coefficient	t-Statistic
	Constant	11.12	56.9
1	Weight (log)	−.720	−26.9
2	CID (log)	−.279	−17.1
3	N/V (log)	−.371	−22.6
4	ETW/HP (log)	.211	17.1
5	Manual trans. vs. 3-spd. automatic	.082	11.4
6	4-spd. Auto trans. vs. 3-spd.	.052	7.2
7	Dynamometer setting (log)	−.135	−10.4
8	Old-tech engine vs. high-tech	−.062	−6.2
9	Modern engine vs. high-tech	−.024	−4.8
10	Compression ratio (log)	.184	6.6
11	Modernized engine vs. high-tech	−.036	−5.2
12	Axle ratio	−.053	−3.2
13	Air conditioning vs. no A/C	−.022	−3.9
14	4-Weel drive vs. RWD	−.028	−3.6
15	Front-wheel drive vs. RWD	−.012	−2.2

Adjusted R-Squared: 0.917; Number of Observations: 1400

efits *could* be achieved through the use of given technologies. The model estimates the benefits actually achieved in the model year 1988 and 1989 cars on the road. The reason these results should differ is that engineers admit that one rarely achieves in practice and in mass production what one thinks will be achieved in either laboratory tests or with prototype vehicles.

Second, the engineers were thinking in terms of using the technologies for higher fuel efficiency. Buyers of 1988 and 1989 model year cars demanded increased performance. Indeed, the weight in pounds per horsepower of cars was significantly lower in the 1988 and 1989 model year fleets (26.31) than it was in the 1986 and 1987 model years fleets (28.32), indicating a significant increase in performance in the later years' cars.

The above results emphasize the need to balance the incentives in the marketplace between supply—requiring producers to achieve higher fuel efficiency—and demand—structuring incentives so that consumers want to purchase what producers are required to produce.

As a final check on the validity of the results of the regression model, a comparison was made between the 1988 Chrysler New Yorker and the Fifth Avenue it replaced. Use of a smaller, more efficient engine, better aerodynamics, and, most importantly, elimination of 500 pounds from the vehicle plus other changes resulted in a 16.7 percent improvement in fuel economy for the New Yorker (25.1 mpg versus, 21.5 mpg for the Fifth Avenue). When we inserted the values for each variable of the New Yorker into the regression model, the model predicted a 15.5 percent improvement in fuel economy compared with the Fifth Avenue. (Use of the EEA numbers implies that a 33 percent improvement should have been achieved.)

Conclusion

Statistical analysis of the EPA test data for the 1,400 cars in the 1988 and 1989 model year fleet has resulted in highly significant and very important results. Cleaner air and reduced usage of fossil fuels, which would mitigate any potential greenhouse problems and reduce our dependence on imported petroleum, are the desires of most citizens in the United States. If these goals can be achieved at little or no cost, then most would say they should be adopted. If, on the other hand, attainment of these goals is quite costly, then society must be made aware of the costs if it is to make a rational choice. Basing public policy decisions on faulty analysis or poor data may put unnecessary burdens on some segments of society or may impose such large costs on society in general that eventually those decisions have to be changed. Alternatively, poor data or analysis may lead to a suboptimal improvement in the environment.

The purpose of this paper is to attempt to raise the level of discussion about the fuel efficiency of cars above the "we say—you say" plane and provide a statistically and quantitatively reliable basis for making policy decisions.

The regression model approach overcomes two significant shortcomings of the EEA analysis: the inability to account for interactions among independent variables and the questionable statistical properties of the EEA method. The quest for a sound analytical basis for making public policy has cast serious doubt on the work upon which the Bryan Bill's percentage increases are based. The results of EEA—based as they are on case-by-case engineering results (where available), comparisons of one model with another, and information gleaned from the popular press—stand in stark contrast to the Big 3 engineers' estimates, the statistically reliable regression model results, and the individual predictions by the regression model (only one of which was cited) that agree closely with actual outcomes.

It may be desirable to set as a national goal the improvement of car fuel efficiency by 40 percent. But if we as a society decide to set that goal, let us at least not be fooled into thinking it can be attained with existing technologies and without requiring consumers to drive significantly smaller, probably less safe, and poorer performing cars. (Also, let us not ignore a carbon fee, which is a more economically sound method of improving the efficiency with which we use fossil fuels and of protecting the environment.) In short, the statistical analysis described in this paper lends credibility to the Big 3 engineers' estimates of the fuel efficiency benefits that can be attained with existing technologies—if cars are designed with fuel efficiency and not performance in mind.

Mandating increases in efficiency beyond the engineers' estimates will impose increasing costs on society. Those costs will ultimately be borne by the consumer in terms of restricted choice, lower employment, and higher prices.

Positioning for the 1990s—The Amoco Outlook

Theodore R. Eck
Amoco Corporation
Chicago, IL

Relative to policy issues in transportation energy in the 1990s, my message is a simple one. In this country we have a production and delivery system for transportation energy that is second to none in the entire world. Transportation energy is inexpensive and it is available everywhere. Great progress has been made in cleaning up automobile emissions, and more progress will be made. Automobiles are efficient and will become more efficient. We have the finest system in the world—let's not mess with it.

US gasoline demand increased at an average growth rate of almost 3 percent per year from about 5.8 million barrels per day in 1970 to about 7.4 million barrels per day in 1978. Demand then decreased for four years at about 3 percent per year, due to higher world oil prices and economic recession. Since 1982, demand has been growing again at slightly under 2 percent per year, reaching in 1988 the same level as the previous peak in 1978. In 1989, demand held at about the 1988 level.

Looking out for the next five years, we project that gasoline demand will grow at only half a percent per year. The same holds true if we expand our time horizon to the year 2000. We project that annual vehicle miles traveled will increase at about 2 percent per year over the next decade, but that vehicle miles-per-gallon efficiencies will increase almost that fast, offsetting most of the growth in vehicle miles traveled.

Fuel economies of new US automobiles have risen much faster than those of German and Japanese cars, although they started from a lower base. In 1973, the average fuel economy of new cars was about 14 miles per gallon (mpg) in the United States, but was about 23 mpg in West Germany and Japan. In 1986, US and Japanese new cars were about 28 mpg, while German cars were about 32 mpg. The fuel economies of cars imported into the United States have been consistently higher than those of the domestic fleet, but from 1974 through 1989, the import fleet fuel economy had grown only about 38 percent compared to 104 percent for the domestic fleet.

The amendments to the Clean Air Act being discussed in Congress will likely impose additional tough standards on automobile tailpipe emissions and running losses. The amendments proposed by President Bush were intended to reduce total emissions of volatile organic compounds (VOCs) by about 30 percent with two thirds of the reductions attributed to the tougher automobile standards. Out of the 30 percent, 2 to 5 percent VOC reduction was attributed to the President's proposed alternative fuels program, and that is unquestionably overstated.

On the subject of alternative fuels, Amoco believes that compressed natural gas (CNG) has excellent potential. We are against the mandating of any particular alternative fuel because we believe that economics and the free market can arrive at the optimal fuel much more efficiently than mandates. Compressed natural gas makes economic sense because it is a fuel already proven in fleet use, is environmentally very clean, and has reasonable costs. Mileage between refuelings is limited unless tank size is larger than conventional gasoline tanks or higher pressures (more than 3000 pounds per square inch) are used.

Methanol, frequently proposed as an alternative fuel, is less suitable. The cost of methanol would be higher than gasoline, the fuel probably would be imported, possibly from Middle Eastern countries, it would offer little, if any, ozone advantage over future generation gasoline cars, and would actually increase emissions of greenhouse gases, considering the carbon dioxide released in methanol manufacture.

Reconstituted gasoline also has potential as an alternative fuel. The lowered vapor pressure of reformulated gasoline reduces evaporation and running losses substantially, and the lowered aromatics content cuts benzene emissions. Car design changes can help minimize running and refueling losses.

We expect on-highway diesel demand to grow more rapidly than the demand for gasoline. Diesel fuel demand grew at nearly 5 percent per year between 1983 and 1988, or at about the same rate as industrial production. In 1989, demand grew at about 1.7 percent. Over the next five years, we expect growth of about 2 percent per year. Desulfurization of diesel fuel from the current level of 0.3 percent sulfur to a maximum of 0.05 percent sulfur will increase refining costs by about 3 cents per gallon.

Jet fuel demand is expected to grow at about 1.6

percent per year over the next ten years, considerably less than the growth rate of 6.6 percent from 1983 to 1988 when airline deregulation and cheaper fares were stimulating the growth of passenger miles.

Returning to my initial admonition, let's not rock the boat too much with costly and inappropriate regulation of our motor fuel delivery system or the fuels we rely on. We have the finest production and delivery system for transportation fuels in the entire world. Let's keep it that way!

Future Growth of Auto Travel in the US: A Non-Problem

Charles A. Lave
Economics Department
University of California
Irvine, CA

Introduction

In a conference dedicated to examining various kinds of potential disasters, I will proclaim some good news: auto travel will grow much less in the future. This prediction will surprise most people. We have seen an incredible increase in auto travel during our lifetimes, and we have seen the consequent increases in pollution, fuel consumption, and congestion. I do not deny these observations. My point is that the period of rapid growth is over because the vehicle population is essentially saturated. We now have about one vehicle for every potential driver.

Figure 1 shows the overall story. There has been a strongly disproportionate growth of the vehicle population. The upper curve shows the size of the driving-age population. The lower curve shows the size of the personal-use vehicle fleet. Vehicles have been increasing at a much faster rate than the population of potential drivers—2.9 times faster since 1960.

(I define driving age as 15–64. This range includes too many young people and too few old ones, but it is a reasonable compromise given the data limitations. People in this cohort accounted for 91 percent of all vehicle miles traveled (VMT) in 1983. For vehicles, I focus on the proportion of the total fleet used by individuals for personal travel and commuting. This is the sum of autos plus vans plus 57 percent of light trucks. These personal-use vehicles amount to 90 percent of the entire vehicle fleet.)

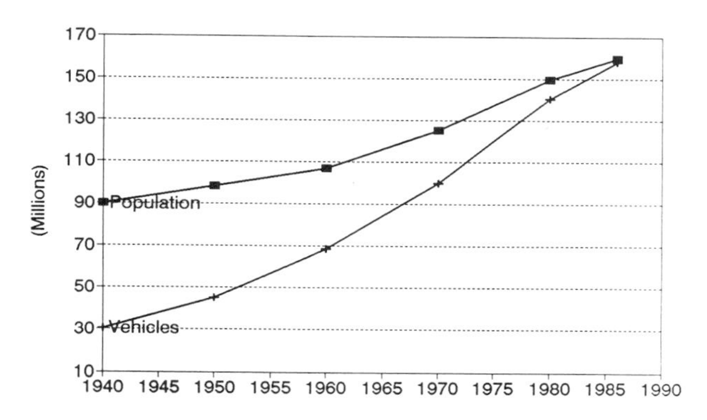

Figure 1. Driving-age population, and personal-use vehicles.

The Transition to a New Demographic Structure

The disproportionate increase in the number of vehicles was the result of some simple demographic changes, but the transition to a new structure is essentially completed.

Changing Age Structure. Drivers' licenses increased much faster than the population during the 1960s and 1970s because a major fraction of the population, the baby-boomers, suddenly became licensable. This transition is over.

Changing Gender Structure of the Labor Force. One reason for the disproportionate growth in women drivers was the disproportionate growth in women workers. In 1947, women were only 27 percent of the total labor force; by 1988, women were 45 percent of the total labor force. Almost all the growth in this ratio occurred in the early period—the ratio grew by 20 percent during the 1970s, but it has only grown by 5 percent during the current decade. The US Bureau of Labor Statistics predicts it will only grow by two percentage points over the next decade (Fullerton, 1989). Women's labor force participation seems to be about at its peak.

The Saturation of Vehicle Demand

What is the evidence for saturation of vehicle demand? We now have about 1.1 vehicles per *licensed driver,* but that ratio does not prove saturation. It has always been high (e.g., there were 0.8 vehicles per driver in 1950), because people don't bother to get licensed until they have the vehicle access to go with it.

To measure vehicle saturation, we must look at the ratio of vehicles to *potential* drivers, that is, all persons of driving age. In Figure 2, the curve labeled "VEH/

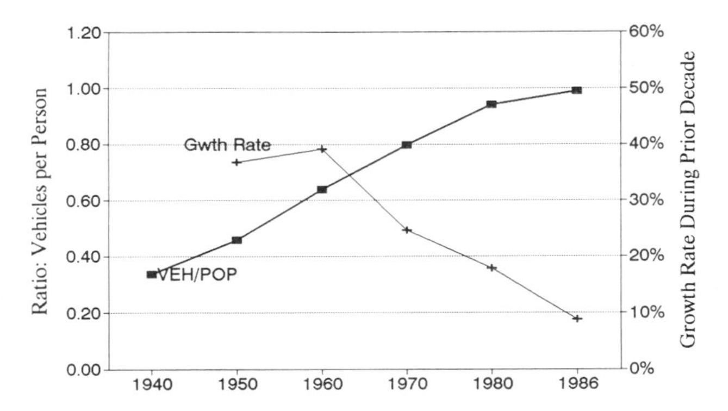

Figure 2. Vehicles/driving-age person and its growth rate by decades.

POP" shows the time path of the ratio: personal-use vehicles per driving-age person. This ratio grows from 0.34 in 1940 to 0.99 in 1986. Will it continue to grow? The second curve, labeled "Growth Rate" shows the percentage change in the ratio, decade by decade. In the decade of the 1950s, the ratio grew by 39 percent; in the decade of the 1960s, it grew by 24 percent; and the growth rate for the 1980s is about 8 percent. The period of disproportionate growth is over; vehicles and population are moving toward equal growth rates.

Analysis by US Regions. Figure 2 is the aggregate pattern for the entire United States. I have plotted similar graphs for 13 typical states across the main regions of the country and gotten essentially the same results.

Why the Implications May Be Incorrect

So vehicle growth will be much slower in the future. But it is travel that creates congestion, not the simple fact of vehicle ownership. What can we say about the likely growth of vehicle *use?*

Historical Patterns. The growth of total VMT has closely tracked the growth of total vehicles over time. This is because average VMT per vehicle has been relatively constant: 11,600 miles per vehicle in 1969; 10,678 miles per vehicle in 1979; 10,315 miles per vehicle in 1983.

Can we depend on the continuation of this relationship? There are two events which might uncouple it: 1) people might decide to drive more than they now do; and in particular, 2) women might begin to drive more. The former is unlikely given the expected increase in congestion. Congested highways make driving less attractive, and a good deal of literature (see *Transportation Research,* January 1981) has given credence to the notion of a limited daily time budget for travel.

But Might Women Drive More? After all, the average female driver now drives only about half as many miles per year as the average male. To evaluate the plausibility of increased driving by women, we must first ask: Why does yearly VMT differ between men and women? The likely explanation is social custom—in general, when a man and a woman travel together, the man drives. Since VMT only accrue to the driver of the vehicle, men will get credit for more VMT.

This theory is supported by the data on person-miles traveled (PMT). You get credit for a person-mile whether you are driving or not, and daily PMTs are almost equal for men and women. The theory is also consistent with overall travel patterns: work trips, when a woman is likely to be driving alone, only amount to about a third of total VMT. The other two thirds of travel, family and social travel, average close to two persons per vehicle, and presumably are more likely to have both a man and a woman on board.

Thus it looks like most of the yearly VMT gap between men and women is nothing more than an accounting quirk. If social norms were to change, and women were to do more of the driving when a man is present, this would only reallocate total VMT between men and women.

A Possible Cautionary Note. Although my conclusion of vehicle-saturation, and hence relatively slow VMT growth seems well supported, there is one piece of contradictory evidence. Starting in 1983, VMT per vehicle began to increase. This increase is so much at variance with past experience that we must ask whether it is a real change or just a statistical error. The VMT-growth data all come from the yearly VMT tables published by the Department of Transportation, and are based on state-supplied estimates. These state estimates were not highly reliable in the past, though a substantial program to improve their accuracy was completed about five years ago, and they ought to be relatively accurate now. The most accurate VMT measurements come from the Nationwide Personal Transportation Surveys (NPTS). The next NPTS will be run this year, and the new data ought to settle this issue.

Distribution Effects. Does the US average of one vehicle per driving age person indicate vehicle saturation? Perhaps some families have more vehicles than drivers, while other families have none. I cannot make a direct evaluation of this possibility with the published data, but I can show that the unequal distribution effect is not large. Of households that have one worker, 90 percent have one or more vehicles available. Of households that have two workers, 82 percent have two or more vehicles available. (This is 1983 data; current data would show even higher percentages.) Clearly, almost all households have at least as many cars as workers, and the zero-vehicle households tend to be very small households located in larger central cities—the New York area alone has 20 percent of the total (Pisarski, 1987). That is, we do not expect much additional vehicle growth due to vehicle purchases from zero-vehicle households.

Conclusion

Yes, vehicle use will continue to rise, but the growth rate should slow to about one third of what we have come to expect. This means that two thirds of the previously expected increase in vehicle-caused problems will go away, all by itself, without any intervention by us with respect to new fuel-efficiency or emission standards.

Acknowledgments

I wish to thank C. Fleet, S. Liss, J. Mauldin, A. Pisarski, M. Wachs, and E. Wiener for help with data, unpublished reports, and interpretations. They are, of course, totally innocent of what I did with their information. S. Handy provided invaluable research assistance. The research was supported by the University of California Transportation Center under US Department of Transportation grant DTO-G-009.

References

Fullerton, Howard N. Jr., "New Labor Force Projections, Spanning 1988–2000," *Monthly Labor Review,* V. 112, no. 11, pp. 3–12, November (1989).

Pisarski, Alan E., *Commuting in America: A National Report on Commuting Patterns and Trends,* The Eno Foundation for Transportation, Inc., Westport, CT (1987).

Tanner, J.C., "Expenditure of Time and Money on Travel," *Transportation Research, Part A,* V. 15A, no. 1, pp. 25–38, January (1981).

US Department of Transportation, *Personal Travel in the US, 1983–84 Nationwide Personal Transportation Study. Volumes I and II, and Survey Data Tabulations,* Washington, DC, November (1985).

Transportation Aspirations of Developing Countries Will Be Met by Oil-Fueled Motor Vehicles

Rémy Prud'homme
University of Paris XII
Paris, France

Introduction

Strictly speaking, there are no "transportation aspirations" in developing countries. These countries aspire to development, not to transportation per se. It just happens that development will mean more transportation of people as well as goods, within cities and between cities. The reasons for this were very clearly stated by Adam Smith more than two centuries ago. Division of labor and economies of scale are the key to improved productivity and growth. They imply trade, and hence transportation. We have only added economies of agglomeration, which can only be reaped if and when large cities function reasonably well, particularly in terms of transportation.

More transportation in developing countries is therefore desirable—and unavoidable. The idea that better planning could mean less transportation and could be a substitute for increased transportation is but a dream. It assumes that planning is costless and effective, which it is not. As a matter of fact, the experience of planned economies, where the ratio of ton-miles per dollar of its output is particularly high, suggests that the more planning, the more unnecessary transportation.

The issue is therefore: What kinds of transportation should or will be utilized in developing countries? At the risk of sounding reactionary, I will answer: mostly oil-fueled motor vehicles. Just as there is an anti-urban bias, and probably for the same reasons, there is a widespread anti-motor vehicle bias, and many people are trying hard to demonstrate that oil-fueled motor vehicles are bad, particularly in terms of energy and environment, that alternative modes of transportation are better, and therefore that developing countries should not follow our patterns of transportation. Let me examine briefly the social cost associated with motor vehicles transportation, the availability and costs of alternative modes, and the policy options that are open to developing countries.

Externalities

Oil-fueled motor vehicles are usually criticized and condemned because of the social costs, or more precisely the externalities, associated with their usage. Three types of social costs are cited: pollution, congestion, and oil consumption.

There is no doubt that noise and air pollution are very serious issues in many large cities of the developing world. It is by now well known that air pollution is much worse in Mexico City, São Paulo, or Cairo than in New York, Tokyo, or Paris. It is also true that oil-fueled motor vehicles are the main culprits for at least three important pollutants: CO, hydrocarbons (HC), and lead. There is no point denying massive traffic congestion in very much the same cities. And it is true that transportation accounts for large shares of oil and energy consumption in developing countries; as a matter of fact these shares are significantly larger in developing countries than in developed countries.

Should these externalities lead us to condemn oil-fueled motor vehicles? No. The importance of pollution and congestion are easily exaggerated, and oil consumption is not truly an externality.

Three points can be made about air pollution.

First, it is obvious that air pollution is only a problem in the very large cities. There is no air pollution in rural areas, nor in smaller urban centers, where the bulk of the population of developing countries still lives. The air is indeed polluted in Calcutta, but hardly more than 1 percent of Indians live in Calcutta. Moreover, intercity transportation and goods transportation do not create any pollution worth speaking of.

Second, motor vehicles are not responsible for all air pollutants. SO_2 and particulates, often considered in developing countries as the worst pollutants, are emitted by industry. For NO_x, the share of industry pollution is about as large as that of motor vehicles. In certain cities such as Seoul and Ankara, domestic heating also contributes significantly to carbon monoxide.

Third, in the cities of developing countries, the bulk of air pollution is contributed by commercial vehicles, not by private automobiles. This is because the share of commercial vehicles relative to total vehicles is very large in these countries and because a bus or a truck is several times more polluting than an automobile.

Similar points can be made about traffic congestion. It is only an issue in certain cities, certain streets, and at certain times. It may be a serious problem for the foreign businessman or professor who has a plane to take, but it is of little concern for the masses of developing countries who live in rural areas or in small cities.

Energy consumption and more precisely oil consumption is often presented as a social cost or an externality associated with motor vehicles usage, but it is not. An externality is an unpriced effect. Oil is priced. People and enterprises that consume gasoline or diesel oil pay for it, just as they pay for any other input or production factor. There is no a priori reason to believe that energy prices are too low, relative to what they should be. And if they were, the thing to do would be to raise them, not to try to restrict motor vehicle usage.

Neither can it be argued that motor vehicles do not pay for the costs they cause in terms of road construction and maintenance. They do by means of taxes, mostly fuel taxes, that they pay. In developing countries (just as in most developed countries) motor vehicles as a whole contribute more to public finance than they get from it. This may not be true for every type of motor vehicle. In many countries, for good and strong social reasons, private automobiles are much more heavily taxed than buses and trucks, largely because gasoline is much more heavily taxed than diesel oil. But it is trucks and buses, not private automobiles, that deteriorate roads. As a result, in some countries, private automobiles subsidize commercial vehicles.

It seems difficult to condemn motor vehicle transportation on the basis of the externalities associated with it: These externalities are either exaggerated or nonexistent.

Alternative Fuels and Modes of Transportation

Critics of motor vehicle transportation like to point out alternative modes and fuels that could be used.

In terms of modes, the main point is that alternatives are more energy efficient. For freight transportation, rail or river transportation requires five to ten times less energy per ton-km. For inter-urban passenger transportation, differences in energy consumed per passenger-km are not so large: Rail is not much more efficient than buses but both are more efficient than private automobiles. The case of urban transportation is rather similar: Subways and trolleys consume as much energy as buses, always on a per passenger-km basis and much less (about three times) than private cars. In addition, it is argued, the energy required for train or subway transportation need not take the form of oil.

The conclusion that is drawn from these figures is simple. A shift from high- to low-intensive modes will save a lot of energy. Since saving energy is desirable, such shifts are also desirable and should be encouraged: fewer trucks and more trains for freight; fewer cars, even fewer buses, and more trains, more subways, more trolleys for people.

Such a conclusion may sound logical, but it is economically stupid. Indeed, energy is scarce and should be saved. But so are labor and capital. Any first year student of economics understands that optimizing relative to just one factor of production makes no sense. The best mode is the one that economizes all factors of production, taking into account their relative scarcity, that is their prices. In simpler terms, the best mode is the cheapest. In many cases, it turns out to be oil-fueled motor vehicles. Furthermore, cost is not the only consideration. Obviously, the railroad cannot offer door-to-door transportation. Motor vehicle transportation is often more flexible and convenient.

The shrinking share of non-motor vehicle transportation in virtually every country, developing as well as developed, is not the result of a world conspiracy engineered by automobile manufacturers. It is simply the result of technological superiority, and also of what could be called managerial superiority. Trucking, and a fortiori automobile transportation, is done by many small enterprises often operating in a competitive environment. Rail transportation, by contrast, is operated by gigantic state monopolies (Indian Railways employs more than two million people!). No wonder the former is more efficient than the latter. For ideological or political reasons, many developing countries ignored these realities and tried to reverse the tide. They all failed, often at great costs. Even in India and China, the share of motor vehicle transportation in total transportation kept increasing.

If we can't escape motor vehicles, can we escape oil-fueled motor vehicles? Are there alternative fuels? No one knows for sure what technology will bring in the coming decades: the water-fueled car might be for tomorrow! But there is one real-life experiment that suggests pessimism: the Brazilian Proalcool plan. In 1976 and again in 1980, the Brazilian government decided to promote alcohol-fueled cars. Technically, the plan was a great success: within years, all the cars produced in Brazil were alcohol-operated, enough sugar cane was produced and the appropriate number of distilleries created. Economically, the plan was a complete failure. The cost of the alcohol produced was and is two or three times higher than the cost of the gasoline it replaces. Yet, it is hard to think of a country more appropriate for sugar cane production and of an experiment conducted on a larger scale.

Policies

To say that the transportation aspirations of developing countries will be mostly met by oil-fueled motor vehicles is not to say that nothing can and should be done about it. Just the opposite. The associated pollution, congestion, and energy consumption problems call for actions and policies.

Take the air pollution externalities. They can and should be reduced. The scope for such reductions is indicated by the experience of developed countries' motorization: Rates are about 10 times higher, but pollution is similar or lower. The reason is that motor vehicles in use are about 10 times less polluting (at least in terms of CO and HC). Such a reduction was achieved at a cost, although not a very high cost. Let the cost be borne by motor vehicles users: If it leads some of them to shift to trains, trolleys or bicycles, so much the better for the environment.

The same applies to congestion. It is indeed an externality: When I take my car on an already crowded street, I slow down traffic even further and inflict a cost upon my fellow street users. I should be made to pay for this cost. This basic economic principle holds for developing countries as well as for developed countries. The best example of its application comes from a developing country, Singapore (but is it still a developing country?), where a fee is required to use the downtown area at peak morning and evening hours. Congestion is limited in space and time: It calls for actions limited in space and time.

Reducing energy consumption is indeed very desirable and should be encouraged, although it is permanently and efficiently encouraged by fuel prices. Changing modes is only one of at least five ways to achieve such a goal. Equally important are:

- Changes in the types of motor vehicles, from high fuel consumption to low fuel consumption vehicles.
- Changes in the weight of motor vehicles, from light to heavy vehicles (although heavy vehicles damage the roads they use more).
- Changes in the usage of motor vehicles, from half-empty to full trucks.
- Changes in road quality, from bad to good roads.

Space does not allow me to elaborate on these shifts. Let me only mention two points. The first is that most of them can only occur as a result of policy changes, be they in taxation, regulation, or expenditure patterns. The second is that each of these shifts can save as much energy as the shift away from motor vehicles often advocated, and can do so at much lower costs.

Interrelationships among Energy, Environmental, and Transportation Policies in China

Karen R. Polenske
Department of Urban Studies and Planning
Massachusetts Institute of Technology
Cambridge, MA

Political events during the summer of 1989 riveted the attention of people around the world on the People's Republic of China. China has over one fifth of the world's population. What happens in China will therefore remain of critical concern to all of us, even though at the moment attention has been shifted to other parts of the world as rapid political changes occur in the Union of Soviet Socialist Republics, Eastern Europe, and South Africa.

Because of China's immense population and its emerging economic importance in the world economy, small shifts in the Chinese economy can have important repercussions on overall energy and environmental concerns. Numbers in the millions are quoted every day for China, from the marchers we saw last summer on television to the large numbers of people who jam the rail systems. A number of momentous issues are confronting Chinese leaders. On the one hand, students and many others have asked for improved human rights, greater roles on the part of the people in making decisions, and less corruption; on the other hand, economic planners in China are confronting severe challenges with energy, the environment, housing, price and wage policies, and transportation problems. These social, economic, and political conditions are closely intertwined.

In this paper, we will focus on only a subset of these issues, namely, transportation and its environmental and energy impacts. Economic policymakers and planners in China realize that transportation is part of the backbone of the Chinese economy and that deficiencies in the transportation system can hinder development and growth in regions throughout the country. The impact that changes in transportation requirements have on energy and the environment and vice versa are of special concern for this conference.

Economic analysts have long disagreed about the relationships between transportation and national, state, and local development. Because of severe budget constraints, policymakers in China are becoming increasingly concerned about how "adequate" transportation can be provided to fulfill the demand for alternative types of service at high quality. A few are now considering a new issue, namely, how changes in the provision of energy and other goods and services can be used to slow down increases in the demand for transportation, thus decreasing the projected quantities of environmental pollution.

The concerns of policymakers range across a broad spectrum of issues. Passenger and freight transportation capacities, for example, are woefully inadequate in China. Long-distance passenger trains are seriously overloaded; in the summer of 1989, for example, more than 700,000 people had to stand each day, with many standing for several days at a time. The lack of freight transport capacity adversely affects both the industrial and household sectors of the economy. Several times each month, for instance, major electric generating plants cannot produce sufficient electricity because of a shortage of coal that results from the lack of coal transport. The end result is that factories in the counties where this occurs must close, and households do not have sufficient coal to use for cooking fuel.

China is the largest consumer of coal in the world. We maintain that if China can drastically reduce the projected increases in the use of coal, some of the major energy and transportation bottlenecks in China will be greatly diminished. By reducing these increases in the use of coal, China will also be taking significant steps to reduce the production of carbon emissions, thus helping to alleviate current global warming trends.

Planning for Chinese Transportation Facilities

According to many Chinese and foreign analysts, one of the major needs in China is for more transportation capacity to move coal from the coal-producing regions to other regions. Xu Yuanchao (1989, p. 2), a reporter for *China Daily,* says that the shortage of energy has created a bottleneck that holds back the development of the Chinese economy. The major source of energy in China is coal, which accounts for 75 percent of the power supply of the country. One of the key coal-producing regions is Shanxi. According to Zhang Quimin (1985), Deputy Director and Engineer of the Research Institute of Agricultural Development in Shanxi province, the volume of coal transported from Shanxi province increased 20-fold within the past 30 years, but the capacity of rails increased only fivefold. Because of the lack of transportation, the inventory of coal in that province increased from 14 million tons in 1980 to 30 million tons in 1983, and the authorities in Shanxi province have had to limit its production. Thus, the production of coal in Shanxi province cannot reach its full potential. The annual shortage of coal for cities along the coastal areas, meanwhile, is about 14 to 18 million tons; therefore, according to Zhang (1985), the energy shortage is due not to the lack of natural resources, but to the inadequacy of transportation.

Wan (1985) and Yang (1987) give two different reasons for the shortage of transportation in China. First, Wan says that China does not have sufficient funds to maintain and replace the fully depreciated transportation facilities. Second, Yang claims that Chinese officials have neglected the importance of an integrated national transportation system because of past development strategies that emphasize regional self-sufficiency in agricultural production.

Looked at from one perspective, Chinese planners should respond by investing in additional transportation, but when the environmental consequences of additional transportation facilities are taken into account, the answer is not so simple. Coal, for example, is one of the major commodities moved on the rail network. More transport would enable more coal to be shipped and used and would lead to even greater carbon emissions. Many of those who advocate increased transport investment in China view transportation in a restricted perspective, however, without taking account of its interrelationship with energy and environmental concerns.

Chinese planners and scholars have suggested numerous ways to solve the transportation, energy, and environmental problems. Wu Mingyu, member of the Center for Research on Economic Technology of the State Council, postulates that planners and officials of China should employ both "soft science" (qualitative analysis) and "hard science" (quantitative analysis) to study the transportation issues in China (Ma, 1987). Besides technical analysis, Wu argues that Chinese planners should also emphasize the design of an integrated intermodal transport system employing techniques from various disciplines, namely, transportation planning, economics, and modern management science, and that they should develop alternative pricing methods for the transport system and for commodities. Li Quohao, professor of the Chinese Science Technology University, proposed that analysts should prepare several areas of study in transportation, including the effects of transportation on development, the management of transportation, and investment and pricing decisions in transportation (Ma, 1987). Finally, using the technique of linear regression, Zhang (1987) studied the relationship between the level of economic development and the volume of freight traffic, using data from 29 regions in China. All of these analysts, however, focus mainly on transport and on means of increasing the transport capacity rather than looking at possible ways to reduce the rate of increase in the demand for transport.

Transportation, Energy, and Environmental Interrelationships

In addition to the proposals just reviewed, which relate primarily to transportation, the Chinese are also studying how the demand for energy and transportation can be reduced through alternative industrial locations and alternative technologies. The Shanghai Statistical Bureau staff, for example, has conducted extensive studies in connection with the Seventh and Eighth Five-Year plans using their Shanghai Economic Planning Model, which incorporates an input-output model (Li Chongxin, forthcoming). For the Seventh Five-Year plan, they used the model to analyze the structure of industries in Shanghai, with emphasis on ways to: reduce the demand for energy and transportation; determine how the consumption of raw materials, such as coal, can be minimized; study the flow of energy into and out of the city; and analyze pollution abatement, using the Netherlands pollution-abatement model. The Shanghai study and other uses of the 30 provincial and national 1987 input-output models for analyses of transportation, energy, and environmental issues are discussed in the book by Polenske and Chen.

In China, many transportation, energy, and environmental analyses have been only partial studies: one mode, rather than all modes; one commodity, rather than all commodities; transportation, energy, and environmental issues in a single region of the country, rather than for many regions. Exceptions are several efforts currently being undertaken at the World Bank. The Bank staff is working in collaboration with Chinese officials to determine which transportation investments are needed in the Yangtze economic zone, with initial focus on the region encompassing Shanghai and the provinces of Anhui, Jiangsu, and Zhejiang. In addition, they are conducting two studies in conjunction with the Ministry of Rail. One is a transport investment study and the other is a coal transport study. For each study, they are training five Ministry of Rail staff in transportation and economic analysis, emphasizing some of the latest computer modeling techniques. Transportation is a derived demand; therefore, it is knowledge about both the direct and indirect interrelationship of freight transportation with the production of all goods and services in the economy that is important for designing and implementing transportation policies both at the national and regional level.

Some of the current research in China is already

focusing on means of reducing the increase in use of energy and transportation. A closely related issue is how such reductions will affect the current global warming trends. China, the United States, and the Union of Soviet Socialist Republics are the three largest coal consumers in the world. In 1988, China consumed 581 million metric tons (24 percent), the United States 480 million metric tons (20 percent), and the USSR 310 million metric tons (13 percent), measured in oil equivalent. Together, they consumed 57 percent of the total world supply. In China, coal accounts for a staggering 80 percent of total energy consumption, and coal consumption is not limited by depletions in the resource base. In fact, the country has sufficient coal reserves to last for the next 100 years, with the reserves rather broadly distributed among the 30 provinces of China.

Given those large reserves and the current low prices of coal, the consumption of coal in China is expected to continue to increase rapidly. It has already increased 53 percent in the last 10 years. It follows therefore that efforts to reduce the increase in the use of coal in China may be a means of greatly improving overall global warming trends by decreasing the rate of release of global greenhouse gases, especially carbon emissions. China is already making some gains in its energy efficiency with the energy use per unit of output decreasing annually by 3.4 percent between 1979 and 1985. Even so, its population growth in aggregate is very large, and its economy is one of the least energy efficient in the world. In order to study these issues systematically, at least three important elements need to be investigated: alternatives to coal as sources of energy, industrial location and regional planning policies, and coal transportation policies. Needless to say, these issues are interrelated.

Future Planning Needs in China

To study the interrelationships among transportation, energy, and environmental policies, Chinese economic planners need to have access to a tool kit of different models, techniques, and computer programs that they can use to help solve different policy problems. Economic analysts use five major types of methodologies for economic planning of transportation: 1) regional and multiregional economic impact analyses and forecasting models, 2) industrial-location models, 3) project-appraisal techniques, 4) network-flow models, and 5) microcomputer transportation analysis techniques (Polenske, 1989). These cover both microeconomic and macroeconomic approaches to the study of transportation, energy, and environmental issues and economic development.

Regional and Multiregional Economic Impact Studies

For these studies, analysts can use many different types of models, including input-output, econometric, economic base, and eclectic models. A vast number of analysts have conducted impact studies. Reviews of some of the literature are contained in Batten and Martellato (1988), Bendavid-Val (1983), Bulmer-Thomas (1982), Ciaschini (1988), Jordan and Polenske (1988), Miller and Blair (1985), Miller, Polenske, and Rose (1989), Pleeter (1980), Polenske (1980), Polenske and Levy (1975), and Round (1988). Mourouzi-Sivitanidou and Polenske (1988) review some of the microcomputer programs available for such economic impact evaluations.

China is very fortunate in regard to input-output tables. Chinese planners have national input-output tables for 1973, 1979, 1981, and 1983, and they have just obtained a 1987 table. The latter was assembled by staff at the State Statistical Bureau using the UN System of National Accounts methodology. In addition, 1987 input-output tables, which are for at least 118 sectors, were assembled in each region except Tibet. Assembly of most of these tables was completed by the fall of 1989. In China, many analysts have been using previously available input-output tables to conduct economic and environmental impact analyses and to forecast the economy to the year 2000, and they have been employing regression, linear programming, and other economic models in their economic analyses. Some of these many uses are documented in a book edited by Polenske and Chen Xikang.

Industrial Location Models

Many formal industrial location theories are so abstracted from reality as to be inapplicable to most real-world problems. This is despite the fact that transportation cost and access are major variables in all location theories. Some analysts, such as Markusen (1987), Massey (1984), and Storper and Walker (1989), are proposing new approaches to location theories in which economic restructuring is often a key factor.

There are applied models that do a reasonably good job of reflecting theory while providing the decision-maker or analyst real guidance concerning what might or should be likely locations for new industrial development. Some of these may be applicable for China. Econometric, simulation, and other models have some potential for use in making accurate predictions of the economic development consequences of transportation investments, but a series of problems may be encountered in trying to apply them to China because of the relative lack of markets. The increasing importance of the market in China would seem to suggest that central place patterns and their growth may provide a means for predicting regional economic development following changes in the transportation system and the ensuing changes in patterns of market centers. Location models do seem to have some potential for contributing significant empirical results to analyze potential economic development effects of changes in transportation in China; however, we have not yet located any studies of this type for China.

Project Appraisal Techniques

These include the use of benefit-cost, rate-of-return, opportunity-cost, and accounting-price concepts and are widely used by international agencies to assess the economic viability of a given transportation project. Most project appraisals include economic, financial, environmental, and engineering components. Here, we define economic project appraisals as those partial equilibrium analyses that are limited in scope to determining whether or not a project is acceptable and the best available alternative. We classify those general equilibrium studies that include analyses of the many direct and indirect effects the project will have upon the economy in the earlier category of regional and multiregional economic impact methodologies. The remainder of this section closely parallels a description contained in Han, Hong, and Polenske (forthcoming). Although project appraisals can be conducted using any of the multiregional and regional methodologies discussed above, many analysts use only the narrowly defined benefit-cost calculations. We will limit our discussion to the economic component.

Considerable research exists on project evaluations. Much of the work by international organizations, such as the World Bank, is not published. Adler (1988) has just updated his general discussion of project evaluations for developing countries, and Bell, Hazell, and Slade (1982) discuss the use of project evaluation for regions. Analysts use two different approaches for economic project appraisals. One is called the UNIDO approach because it is based upon the *Guide to Practical Project Appraisal* written by staff at the United Nations Industrial Development Organization (1978). The other is called the Little-Mirrlees/Squire-Van der Tak approach (Squire and van der Tak, 1975). The end objectives of the two approaches are very different. Those who use the UNIDO approach are interested in what the contribution of the investment project is to the stream of consumption expenditures, i.e., the general welfare of the region or country. On the other hand, those who use the Little-Mirrlees approach are interested in what the project contributes to free uncommitted foreign exchange. In China, project-appraisal analysts are mainly interested in the second objective, while in the United States and other industrialized countries, they usually pursue the first objective.

To determine benefits and costs accurately for the project appraisals, the analyst must make a crucial decision as to the appropriate price to use. Market prices often do not accurately represent the economic price because of subsidies, administered prices, external economies, or other market distortions. Analysts therefore estimate accounting (shadow) prices. For China, Wood (1984) has written an especially helpful summary depicting the major issues in determining accounting prices. Although most analysts use a partial equilibrium means of estimating accounting prices, several have begun to use the dual version of the input-output model for their calculation. These include studies by Han, Hong, and Polenske (forthcoming) for China, Lal (1980) for India, Powers (1981) for Latin American countries, and Schohl for Colombia (1979).

These shadow prices can be calculated for factors of production (land, labor, or capital), intermediate goods, and/or final goods. In most project appraisal work to date, accounting prices have been calculated with the use of a partial equilibrium approach, i.e., the accounting price is determined for only one factor or commodity at a time, or at least little or no effort is made to ensure that the different accounting prices used are internally consistent. Because many input-output tables are not published on a timely basis, only a few analysts have used them either for the determination of shadow prices or for the estimation of direct and indirect benefits and costs. Thus, the benefit and cost measures generally represent only the direct, rather than the direct and indirect, effects unless the analyst makes special efforts to determine the second- and higher-round effects. The accessibility of microcomputers, however, is increasing the use of many different types of data including input-output data. Chinese national and regional planners, for example, will have an advantage over those in most other countries in that they will be able to conduct project appraisals with the use of the 1987 national and regional tables for each of the 30 provinces, except Tibet (Polenske and Chen Xikang, forthcoming).

Network Flow Models

Transportation network flow models are analytical frameworks used to generate a feasible set of hypothetical decisions by shippers (i.e., producers and consumers) and carriers consistent with minimizing transportation costs for goods (and services, in the case of telecommunications). They belong to the general class of spatial price equilibrium models, which determine solutions to cost-minimization problems for spatially separated production points and markets. Fundamental theoretical work on spatial price equilibrium is found in Samuelson (1952) and Takayama and Judge (1971). For policy purposes, the network flow models merit further research for use in China.

Several transportation flow models have been developed during the last 10 years. These include the transportation network model developed by CACI (1980), the rail freight model by Landsdowne (1981), the Princeton railroad network model at Princeton by Kornhauser, Hornung, Harzony, and Lutin (1979), and the Freight Network Equilibrium Model (FNEM), developed jointly by the University of Pennsylvania and the Argonne National Laboratory (Friesz, Gottfried, Brooks, Zielen, Tobin, and Meleski, 1981). Friesz is currently working with World Bank staff in applying these types of models in the People's Republic of China. Additional work has been done on the problem of simultaneity by Harker (1981) and Friesz (1982), and Friesz, Viton, and Tobin (1985) have proposed a model with variational equalities that can generate (or converge to) a unique equilibrium. Recent work on finding efficient algorithms for solving the models includes Dafermos (1979), Dafermos and Nagurney (1987), Fisk and Boyce (1983), Friesz, Tobin, and Harker (1981), Friesz, Tobin, Smith, and Harker (1983), and Nagurney (1987).

Microcomputer Analysis

Within the last few years, analysts have begun to use microcomputers for many of the tedious calculations of economic analyses. In order to conduct any of the studies we have reviewed earlier, analysts must have good quality data and appropriate computer packages. Those who want to study the interrelationships among transportation, energy, and the environment in China will find that microcomputers may be the main type of computer used. Reviews that have been conducted of some of the most recent computer packages in relation to their use for transportation planning include those on impact assessment (Mourouzi-Sivitanidou and Polenske, 1988; Cambridge Systematics, 1989), geographical information systems (Fletcher, 1989; Replogle, 1989; Shaw, 1989), land-use transportation (LUTS) models (Cook, Lewis, and Minc, 1989; Dehghani, Talvitie, and Morris, 1989; Replogle, 1989), and other transportation network analysis tools, such as EMME/2, MICROTRIPS, MINUTP, TMODEL 2, and TRANSCAD (DUSP, 1989). A series of reviews on microcomputer packages is appearing in the *American Planning Association Journal.* All four types of computer packages are useful to planners in China; however, the flexibility of the packages varies a lot in terms of the collection, storage, retrieval, and analysis of transportation and related data. Most Chinese government staff and academic personnel have access to microcomputers and have been using them extensively for several years.

Conclusions

The number of techniques available for use in planning is indeed large. Policymakers in China will have to select carefully those techniques that are of the greatest use for their practical problems. In order to understand the intricate set of interrelationships in China among transportation, energy, and the environment, Chinese analysts will need to be skillful in combining the above techniques in a systematic way. We anticipate that, as in many fields, they will move forward very rapidly, in some cases adopting directly, but, more often, adapting ingeniously to suit their own needs.

Acknowledgment

I thank Yuhung Hong for translating the articles referred to in the section "Planning for Chinese Transportation Facilities" and for helping me in the writing of this section of the paper.

References

Adler, Hans A. 1988. *Economic Analysis of Transport Projects.* Baltimore, MD: The Johns Hopkins University Press.

Batten, David F., and Dino Martellato. 1988. "Interregional Input-Output Models." *Ricerche Economiche,* pp. 204–221.

Bell, C., P. Hazell, and R. Slade. 1982. *Project Evaluation in Regional Perspective.* Baltimore, MD: Johns Hopkins University Press.

Bendavid-Val, Avrom. 1983. *Regional and Local Economic Analysis for Practitioners.* New York: Praeger Publishers.

Bulmer-Thomas, Victor. 1982. *Input-Output Analysis in Developing Countries: Sources, Methods, and Applications.* New York: John Wiley & Sons, Ltd.

CACI, Inc. 1980. "Transportation Flow Analysis: The National Energy Transportation Study (NETS)." Report number DOT-OST-P10(29-32). Washington, DC: US Department of Transportation.

Cambridge Systematics, Inc. 1989. "Highway 29/45/10 Corridor Study: Economic Development Benefits and Cost-Benefit Evaluation." Final Report. Report prepared for Wisconsin Department of Transportation by Cambridge Systematics, Inc. with Donohue and Associates, Inc. and Regional Economic Models, Inc. (March).

Ciaschini, Maurizio, ed. 1988. *Input-Output Analysis: Current Developments.* London: Chapman and Hall.

Cook, Peter, Simon Lewis, and Marcelo Minc. 1989. "Comprehensive Transportation Models: Developmental Overview and Justification of Use." *ITE Journal* (June).

Dehghani, Youssef, Antti Talvitie, and Michael Morris. 1989. "An Integrated Land Use-Transportation Model System for Corridors Analysis and Its Applications." Paper presented at the Second Conference on Application of Transportation Planning Methods, Orlando, FL, April 24–28.

Dafermos, S. 1979. "Traffic Equilibrium and Variational Inequalities," *Transportation Science,* Vol. 14, No. 1, pp. 42–54.

Department of Urban Studies and Planning. 1989. "Transportation Network Analysis Tools: A Workshop on Information Systems and Analysis Tools for Transportation Planning."

Workshop given at the Department of Urban Studies and Planning, Massachusetts Institute of Technology, June 5–9.

Fisk, C.W., and David E. Boyce. 1983. "Optimal Transportation Systems Planning with Integrated Supply and Demand Models." Publication No. 16. Transportation Planning Group, Department of Civil Engineering, University of Illinois, Champaign-Urbana, IL.

Fletcher, David. 1989. "Modeling GIS Transportation Networks." Paper presented at the Second Conference on Application of Transportation Planning Methods, Orlando, FL, April 24–28.

Friesz, T.L., J. Gottfried, R.E. Brooks, A.J. Zielen, R. Tobin, and S.A. Meleski. 1981. "The Northeast Regional Environmental Impact Study: Theory, Validation and Application of a Freight Network Equilibrium Model." Report ANL/ES-120, Argonne National Laboratory, Argonne, IL.

Friesz, T.L., R. Tobin, and P.T. Harker. 1981. "Variational Inequalities and Convergence of Diagonalization Methods for Derived Demand Network Equilibrium Problems." Report number CUE-FNEM-1981-10-1, Department of Civil Engineering, University of Pennsylvania, Philadelphia, PA.

Friesz, T.L., R. Tobin, T. Smith, and P.T. Harker. 1983. "A Nonlinear Complementarity Formulation and Solution Procedure for the General Derived Demand Network Equilibrium Problem," *Journal of Regional Science,* Vol. 23, No. 3, pp. 337–359.

Friesz, T.L., P. Viton, and R. Tobin. 1985. "Economic and Computational Aspects of Freight Network Equilibrium Models: A Synthesis." *Journal of Regional Science,* Vol. 25, No. 1, pp. 29–49.

Han, Jun, Yuhung Hong, and Karen R. Polenske. Forthcoming. "Accounting Price Ratio Model." In *Spreadsheet Models for Urban and Regional Analysis,* edited by Richard E. Klosterman and Richard K. Brail.

Harker, P.T., and T.L. Friesz. 1982. "A Simultaneous Freight Network Equilibrium Model." *Congressus Numerantium,* Vol. 36, pp. 365–402.

Jordan, Peter G., and Karen R. Polenske. 1988. "Multiplier Impacts of Fishing Activities in New England and Nova Scotia." In *Input-Output Analysis: Current Developments,* edited by Maurizio Ciaschini. London: Chapman and Hall.

Kornhauser, A.L., M. Hornung, Y. Harzony, and J. Lutin. 1979. "The Princeton Railroad Network Model: Application of Computer Graphic Analysis of a Changing Industry." Presented at the 1979 Harvard Graphics Conference. Transportation Program, Princeton University, Princeton, NJ.

Lal, Deepak. 1980. *Prices for Planning: Towards the Reform of Indian Planning.* London: Heinemann.

Landsdowne, Z.F. 1981. "Rail Freight Traffic Assignment." *Transportation Research,* Vol. 15A, pp. 183–190.

Li Chongxin. Forthcoming. "Input-Output Model and Application in Shanghai." In *Chinese Economic Planning and Input-Output Analysis,* edited by Karen R. Polenske and Chen Xikang. Hong Kong: Oxford University Press.

Ma Weidi. 1987. "Research on Strategies and Policies of Transportation Development." *Journal of Scientific Proof.* No. 10, p. 4.

Markusen, Ann R. 1987. *Regions: The Economics and Politics of Territories.* Totowa, NJ: Rowman & Littlefield.

Massey, Doreen. 1984. *Spatial Divisions of Labor: Social Structures and the Geography of Production.* London: Macmillan.

Miller, Ronald E., and Peter D. Blair. 1985. *Input-Output Analysis: Foundation and Extensions.* Englewood Cliffs, NJ: Prentice-Hall, Inc.

Miller, Ronald E., Karen R. Polenske, and Adam Z. Rose. 1989. *Frontiers of Input-Output Analysis.* New York: Oxford University Press.

Mourouzi-Sivitanidou, Rena M., and Karen R. Polenske. 1988. "Assessing Regional Economic Impacts with Microcomputers." In *A Planner's Review of PC Software and Technology,* edited by Richard E. Klosterman. Chicago, IL: American Planning Association, pp. 83–90.

Nagurney, A. 1987. "Competitive Equilibrium Problems, Variational Inequalities, and Regional Science." *Journal of Regional Science,* Vol. 27, No. 4, pp. 503–517.

Pleeter, Saul. 1980. "Methodologies of Economic Impact Analysis: An Overview." In *Economic Impact Analysis: Methodology and Applications,* edited by Saul Pleeter. Boston, MA: Martinus Nijhoff Publishing.

Polenske, Karen R. 1980. *The U.S. Multiregional Input-Output Accounts and Model.* Lexington, MA: D.C. Heath and Company, Lexington Books.

Polenske, Karen R. 1989. "Transportation Planning in China." Paper presented at international seminar on "Transportation and Regional Development," May 23–26, in Leningrad, Union of Soviet Socialist Republics.

Polenske, Karen R., and Chen Xikang, eds. Forthcoming. *Chinese Economic Planning and Input-Output Analysis.* Hong Kong: Oxford University Press.

Polenske, Karen R., and Paul Levy. 1975. "Multiregional Economic Impacts of Energy and Transport Policies." US Department of Transportation, Office of Transport Planning Analysis.

Powers, Terry A., ed. 1981. *Estimating Accounting Prices for Project Appraisal: Case Studies in the Little-Mirrlees/Squire-van der Tak Method.* Washington, DC: Inter-American Development Bank.

Replogle, Michael. 1989. "Integration of a Geographic Information System with Computer Transportation Models for Land Use and Transportation Planning." Paper presented at the Second Conference on Application of Transportation Planning Methods, Orlando, FL, April 24–28.

Round, Jeffery I. 1988. "Multipliers and Feedback Effects in Interregional Input-Output Models." *Ricerche Economiche,* Vol. 42, No. 2, pp. 311–324.

Samuelson, Paul A. 1952. "Spatial Price Equilibrium and Linear Programming," *American Economic Review,* Vol. 42, pp. 283–303.

Schohl, Wolfgang W. "Estimating Shadow Prices for Colombia in an Input-Output Table Framework." World Bank Staff Working Paper No. 357. Washington, DC: The World Bank.

Shaw, Shih-Lung. 1989. "GIS as Decision Support Tools in Transportation Analysis." *Proceedings of the 1989 Environmental Systems Research Institute (ESRI) User Conference.*

Squire, Lyn, and Herman G. van der Tak. 1975. *Economic Analysis of Projects.* Baltimore, MD: The Johns Hopkins University Press.

Storper, Michael, and Richard Walker. 1989. *The Capitalist Imperative: Territory, Technology, and Industrial Growth.* Oxford: Basil Blackwell, Ltd.

Takayama, T. and G.G. Judge. 1971. *Spatial and Temporal Price and Allocation Models.* Amsterdam, The Netherlands: North Holland.

United Nations Industrial Development Organization. 1972. *Guidelines for Project Evaluation.* New York: United Nations.

Wan Xiangqin. 1985. "An Urgent Need to Revalue the Underpriced Rail Fixed Assets." *Tie Dao Yun Shu Yu Jingji (Rail Transport and Economics).* No. 11, pp. 3–6.

Wood, Adrian. 1984. *Economic Evaluation of Investment Projects: Possibilities and Problems of Applying Western Methods in China.* World Bank Staff Working Papers No. 631. Washington, DC: The World Bank.

Xu Yuanchao. 1989. *China Daily (Overseas Edition).* January 30. p. 2.

Yang Qingbin. 1987. "Rationalization of the Chinese Transportation System." *Jingji Xue Zhou Bao (Economics Weekly).* No. 11, p. 29.

Zhang Fengbo. 1987. *Economic Analysis of Transportation in China.* Beijing: People's Press.

Zhang Quimin. 1985. "Eliminating the Development Bottleneck of the Shanxi Province." *Shehui Jingji Bao (Journal of Social Economics).* No. 4, pp. 5–11.

Prospective Transportation Developments in India and Their Impact on the Environment

K. L. Luthra
Steering Committee
for Transport Planning
Planning Commission
New Delhi, India

Summary

The transportation system in India needs a quantum jump in capacity if it is to meet the emerging demands that accompany growth in the national economy. The Indian Railways plan to double their capacity by the end of the century through improved higher-capacity rolling stock and upgraded traction with the introduction of higher horsepower locomotives, diesel, and electric. Steam locomotives will be phased out and electrified track doubled. This should help the railways keep their oil consumption in check.

The major user of petroleum—a scarce resource in India—is road transport. The trucking fleet in the country is expected to expand threefold while the number of buses will increase two and one-half times. A much steeper increase is expected in the number of personalized vehicles including cars and two wheelers. The energy consumption of motor vehicles and the corresponding emission of pollutants is excessive due to the obsolete technology of vehicles, higher percentage of overaged vehicles on the road, and poor maintenance of roads and vehicles. The rapidly growing number of two wheelers equipped with two-stroke engines is a major factor accounting for emissions of pollutants. The measures to control air pollution under consideration include upgrading technology of vehicle manufacture, setting standards of emission levels, and monitoring enforcement of these standards.

It will take time for these measures to produce any tangible results. A more effective way to control pollution in the long run will be to develop vehicles using alternative fuels and to introduce grade-separated mass transit systems in the metropolitan cities. This presents a substantial challenge for India, particularly in view of her present scarce-resource constraints.

Introduction

In the four decades since India embarked upon transportation planning in 1950, the country's transportation system has undergone a phenomenal expansion. Road length has increased fourfold and the number of trucks and buses ninefold, from a little over 100,000 to about 1 million. Railroad freight traffic has increased nearly fivefold, while rail passenger traffic has tripled. Nevertheless, the system remains deficient in several respects. Even the basic road network does not link up with all the villages. As many as one third of the villages have no road link, while two thirds lack all-weather motorable road connections. The country has just a little over one million commercial vehicles and 1.5 million cars; that is one car for over 500 people or one vehicle (including cars and commercial vehicles) for about 300 people. The corresponding figures for the United States are one car for 1.7 persons and one vehicle for 1.3 persons. Despite this, emissions of air pollutants from motor vehicles in India have reached high levels in major cities and may assume alarming proportions by the turn of the century.

This paper describes the structural changes in the country's transport system as projected for the next 10 to 15 years and their likely impact in terms of energy consumption and air pollution. In the absence of any monitoring of pollutant emissions from vehicles, the available data are very sketchy, based on just one survey conducted in five large cities in 1985 by the Indian Institute of Petroleum. The discussion draws on this data and some recent official reports, which again use the same data. The scope of this paper is limited to air pollution effects of transportation developments—land abuse and water pollution are not covered. This should not give the impression, however, that there is no cause for concern about conservation of these resources. Land is already a scarce resource and rail and road networks use up 4.6 percent of all usable land in the country. Marine pollution is becoming serious in coastal waters close to major ports, primarily due to industrial and domestic sewage discharge and not oil spillage from ships.

Emerging Transport Demands

With the growth of the national economy, transport demands have increased at a somewhat faster rate than the country's gross national product (GNP). Over the period 1950 to 1987, railway freight as well as passenger traffic increased at a little over 4 percent per year, compared to a 3.5 percent per year increase in GNP. The rate of growth of road traffic during the same period was much higher at 9 to 10 percent per year.

The diversification of the economy, with a shift from agriculture to manufacturing and services sectors, and the growth of urbanization had an important impact on the structure of transport services in the country. The railways' share in the total traffic declined, while demands for road transport increased rapidly. These trends are expected to continue in the years ahead. The country's population is projected to increase from 761 million in 1986 to 986 million in 2001, with the share of urban population increasing from 25 percent to about 33 percent. In absolute numbers, this means an increase of over 200 million people, 20 percent of which will occur in towns with populations of 100,000 or more. The share of agriculture in GNP is expected to decline from 33 percent in 1989 to 25 percent in 2000, while the share of mining and manufacturing may increase from 20 percent to 24 percent and services from 33 percent to 36 percent. The freight and passenger traffic on the railways is projected to double by the year 2000, while road traffic would increase threefold to fourfold.

Strategies For Capacity Increases

There is obviously a need for a quantum jump in the country's transport system. The Indian Railways plan to double their capacity by improving rolling stock, e.g., wagons with higher axle loads and carrying capacity, and upgrading traction, signaling and telecommunication. Steam locomotives are to be phased out completely and electrified broad-gauge network will be doubled from 7,275 km to 15,000 km. At the same time, higher horsepower diesel and electric locomotives will be introduced.

Regarding capacity increases for road transport, the trucking fleet may expand threefold from about 800,000 to 2.1 million and the number of buses will increase from about 200,000 to over 500,000. At the same time, the number of personal vehicles, cars, jeeps, and two wheelers, is likely to increase more steeply—cars from 1.54 million to 3.3 million and two wheelers from 8 million to 35 million or more.

Energy Conservation in the Transport Sector

Transportation is the second largest consumer of energy, next only to industry, and accounts for nearly one-third of the total energy consumption in the country. About 90 percent of its energy consumption is scarce petroleum. Road transport consumes 11 million tonnes of petroleum annually or 77.5 percent of the total petroleum consumption in the country. Civil aviation and railways come next, consuming about 1.40 million tonnes (or a little over 10 percent) and 1.3 million tonnes (or about nine percent of the total) respectively. The country's domestic production of petroleum falls considerably short of the total consumption. The total production of crude is about 32 million tonnes per year, while imports of crude come to about 18 million tonnes per year. In addition, over 6 million tonnes of products are imported annually.

The oil deficit of the country is expected to increase significantly by the turn of the century, hence the need for conservation. If the railways succeed in doubling the electrified network, their consumption of high-speed diesel will be kept in check and, in fact, may decline. Curbing oil consumption in road transportation, on the other hand, presents a real challenge. Several factors account for the present excessive oil consumption by motor vehicles, the more important among these being the obsolete technology of vehicles, a large number of overaged vehicles on the road, poor maintenance of roads—as well as of vehicles—and lack of traffic planning leading to frequent traffic jams on the city roads and crowded highways.

Impact on the Environment—Air Pollution

Railway locomotives and aircraft emit a much smaller volume of air pollutants than motor vehicles. The prospective developments on the railways will help conserve the use of oil and the corresponding emissions of pollutants. Civil aviation is a fast growing sector, yet its effects on air pollution remain minimal. Motor vehicles, the major user of hydrocarbons, are the main contributors to air pollution.

While no nationwide estimates of pollutants have been attempted, the estimates for five metropolitan cities, namely, Delhi, Bombay, Calcutta, Madras and Bangalore, give considerable cause for concern. The annual emissions from motor vehicles in these five cities on current trends will increase enormously between the years 1985 and 2000—carbon monoxide (CO) from 210,000 tonnes to 633,000 tonnes, a threefold increase; hydrocarbons (HC) from 76,000 tonnes to 294,000 tonnes, a fourfold increase; and nitrogen oxides (NO_x) from 65,000 tonnes to 243,000 tonnes, nearly a fourfold increase.

The heavy pollutants from motor vehicles correspond to their higher energy consumption. According to one estimate, the current average rate of pollutant emissions per vehicle in India is about 33 percent higher than that in the United States in the late 1970s. Petrol vehicles are the principal source of carbon monoxide and lead emission, and two wheelers powered by two-stroke engines are the principal emitters of hydrocarbons. Control of pollutants requires modifications in design of vehicles (improved carburetor and ignition, and exhaust gas treatment) and proper maintenance and use of vehicles. For control of lead emissions it will also be necessary to reduce the lead content in gasoline.

The bulk of vehicles belong to the design vintages dating from the late 1940s or early 1950s. While the new models of cars and commercial light vehicles—produced under collaborative arrangements with foreign manufacturers in recent years—have advanced technology, their fuel consumption continues to be excessive compared with some of the modern energy-efficient vehicles produced in advanced countries.

The technology of medium and heavy trucks, on the other hand, has undergone little change since the designs were introduced several decades ago. The derivatives of truck chassis are being used as bus chassis, while rigid two-axle vehicles constitute the bulk of the trucking fleet. These invariably carry heavy loads—much heavier than permissible capacity—consuming excessive fuel and emitting smoke, in addition to causing much damage to roads. The imperative need is to develop technologically advanced buses and trucks and to introduce multi-axle vehicles and, wherever possible, truck-trailer combinations. The sudden spurt in the use of two wheelers in recent years, particularly in large cities, has created serious problems. The gravity of the situation can be realized from the fact that, by the year 2000, nearly one fifth of carbon monoxide and one half of hydrocarbon emissions in five large cities will be accounted for by two wheelers. A switch from the existing two-stroke engines to four-stroke engines in two wheelers might reduce pollutants, but the total impact will be marginal at best.

Recommended Actions

The following measures are presently under consideration by the government; no action plans have, however, been prepared for their implementation.

1. Develop national standards for acceptable emission/smoke levels for gasoline and diesel engines and make them progressively stringent; certain standards have been prescribed under the Motor Vehicle Rules, 1989.
2. Modernize manufacturing technologies to produce more energy-efficient vehicles.
3. Encourage motor vehicle owners to improve maintenance. Weed out overaged vehicles.
4. Develop testing facilities at maintenance workshops to monitor and control pollutants.
5. Introduce legal provisions to ensure that vehicle owners conform to the prescribed emission standards.
6. Explore the commercial viability of vehicles using alternative fuel; field trials are being carried out with the use of methanol and compressed gas as partial substitutes for diesel.
7. Improve traffic management in cities and along the highways.

Conclusion

By the standards of developed countries, road transportation in India is still in its infancy, yet the volume of air pollutants from motor vehicles has already reached high levels in major cities. It will be some years before effective control mechanisms are introduced to check pollutants from vehicles. The growing urbanization, which is difficult to decelerate, gives rise to an unprecedented increase in demands for movements of goods and passengers. With the projected growth of passenger traffic demands, it is an unavoidable necessity to develop adequate public transport services in cities, and the metropolitan cities need grade-separated rapid transit systems. This indeed presents a great challenge to the government, particularly in view of the present severe resource constraints. The sooner the government addresses this challenge, the better. Time is running out while the growing middle-class city dwellers are anxious to buy as many scooters and cars as the country can afford to build and chaos is mounting on the city roads.

Acknowledgments

The paper draws heavily on the work done for the Steering Committee for Transport Planning of the Planning Commission. The valuable contribution made by the members and staff of the Committee is gratefully acknowledged.

Meeting the Transportation Aspirations of Developing Countries: Energy and Environmental Effects

Clell G. Harral
China Department
The World Bank
Washington, DC

In developing countries road transport typically accounts for between two thirds and four fifths of all energy used for transport—not that different from the developed countries—but the personal automobile generally accounts for a much smaller share of the road total, and buses generally do not compensate, reflecting the relatively lower levels of human mobility. Thus, trucks account for 50 to 75 percent of energy consumed on the road, compared to only 30 to 35 percent in the developed countries. Excluding eastern Europe, in only two developing countries—China and India—do railways today play a very significant role in inter-city transport, although there are now at least 21 cities in the Third World where urban railways (metros) are either operating, under construction, or planned. Aviation is not a significant element in the energy equation for any of the developing countries, although a much greater role for this mode may be envisaged for the future in India and particularly China, as well as an enhancement of its already considerable role in Latin America. Water transport remains an important mode for transport in maritime nations such as the Philippines, Indonesia, and China, but its interface with energy and the environment is also limited (apart from the problem of oil spillages).

It is in the cities where the use of energy in transportation raises primary concerns for the environment. Although the overall levels of personal mobility in the developing world are lower, the concentration of populations in very large cities with inadequate road space and inadequate incentives for environmental protection has led to some of the world's worst congestion, noise, and air pollution—and no doubt astronomic fuel consumption coefficients. Examples include Bangkok, São Paulo, and Mexico City. The problem is sure to grow worse: while in 1985 eight of the world's 12 urban agglomerations with populations of 10 million or more were in developing countries, their number will more than double—to 17 out of 23—by the year 2000, while an additional 18 developing country cities will have populations between 5 million and 10 million. Massive urbanization will combine with income growth and high elasticities for passenger travel to spur demand to ever higher levels. Projections of emission levels from passenger transport for 14 large cities by Sinha et al. (1990) indicate doubling or more of emissions by the year 2000 in most cases, even on the assumption that the modal share of the automobile would not increase. But historical evidence from many countries shows rapidly growing levels of automobile ownership as incomes per capita rise except where ownership is restrained (as in Hong Kong, or, until recently, in Korea).

Although a radical tightening of emission standards for individual vehicles to match those of the industrialized countries—as was done in Korea in 1987—can help, some form of restraint in the ownership or use of the automobile is likely to prove unavoidable: the acquisition and use of private automobiles cause a large increase in per capita use of energy for transport in addition to congestion effects. The Area License Scheme operated by Singapore since 1975 has restrained vehicle use, thereby limiting congestion and pollution in the Central Business District (CBD), as well as raising substantial revenues, while the heavy taxes placed on vehicle ownership by Hong Kong from 1982 onwards have had somewhat similar effects. Mexico City—with the world's worst smog problem—has now begun to experiment with limits on auto use as well as enhanced emission standards.

Improvements in public bus services (which already account for over half the urban vehicle kilometers in developing countries) will be of even greater importance, not only to provide a more fuel-efficient alternative to the automobile, but also to provide for greater efficiency in bus services themselves. Over-aged, poorly maintained bus fleets are a major source of pollution and excess fuel consumption in many cities. Improved maintenance of diesel-fueled buses can make a major contribution to improved air quality, as demonstrated in Santiago, Chile. Even greater gains can be achieved by the substitution of compressed natural gas (CNG) for diesel in bus fleets. São Paulo, Brazil, has also successfully operated buses fueled by bio-gas from sewage treatment plants.

Unfortunately, rail mass transit, although it may

provide significant travel time savings and major capacity enhancements (where the high costs can be afforded), is unlikely to yield major improvements in urban congestion, energy consumption, or the environment, judging from a recent major study including case studies in 10 cities (Thomson et al., 1989). That is because new traffic tends to be generated to take the place of any that may be diverted to the metro, inexorably driving street congestion toward an equilibrium condition termed "the threshold of the intolerable."

Bicycles have attracted renewed interest in several countries, partly because they provide a substantial improvement in personal mobility beyond walking for those who cannot afford to ride, and partly because they can provide better door-to-door travel times than buses in heavily congested cities. In China they have also been encouraged by government subsidies (most workers are eligible for subsidies of roughly 50 cents to one dollar per month) which aggregated in 1985 to almost $400 million. Thus it should not be surprising that there has been a rapid growth in bicycles (estimates range upward from 160 million), and a stagnation or even decline in bus usage in many Chinese cities. While the energy and environmental effects, as well as personal mobility aspects, appear quite positive for the individual cyclist, there are problems related to the relatively inefficient use of road space and congestion imposed on motorized traffic where road space is shared. Cities in Latin America, Africa, and particularly Asia are experimenting with traffic segregation on entirely separate rights of ways for bicycles—but in many other cities cycles are viewed primarily as a hindrance to motor traffic, and thus should be banished. In Jakarta, Indonesia, 100,000 tricycle rickshaws have reportedly been destroyed for that reason over the past five years (*The Urban Edge,* March 1990).

Motorization of cycles is the next step up in personal mobility, and this intermediate technology has been particularly popular in Asian countries. It is just beginning in China, but is likely to grow rapidly unless restrained, judging from earlier experience in India, Korea, and Malaysia. Interestingly, motorcycle ownership first grew rapidly and then declined over time in Taiwan, where automobiles were ultimately substituted for motorcycles as income levels rose, but this does not appear to have been the case in Malaysia, where higher levels of ownership of both motorcycles and autos are observed (Sathaye and Meyers, 1989). It should be noted from the environmental perspective that emissions per passenger kilometer of carbon monoxide are about the same or perhaps a bit worse than from the auto, while hydrocarbon emissions may be ten times higher, although emissions of oxides of sulfur (SO_x) and nitrogen (NO_x) may be substantially less (Sinha et al., 1990). Thus, the attractiveness of the motorcycle as a cheap means to an intermediate level of personal mobility is diminished by environmental (as well as safety) considerations.

The availability of plentiful supplies of cheap oil prior to the 1970s provided little stimulus for fuel efficiency. The petroleum price increases of the 1970s did much to encourage attention to more fuel-efficient vehicle technologies, and brought renewed attention to the prospects for inter-modal substitutions (i.e., revival of railways and water transport). The substitution of diesel for gasoline was also encouraged in most countries, as was the search for petroleum alternatives—most notably ethanol in Brazil—although it should be noted that the use of heavily leaded gasoline is still extensive in developing countries. Fuel price subsidies have also been reduced in oil producing countries (Ecuador, Mexico, and Nigeria). Some important examples of energy price distortions still remain, for example in China, where coal and electricity are severely underpriced relative to petroleum, thus encouraging the railways to use steam and electric traction rather than diesel traction. Steam locomotives of vintage design still provide almost 50 percent of the total tractive effort, and consume approximately 20 million tons of coal per annum on the Chinese railways, but are now being phased out.

Outside the cities there is little practical scope for modal substitutions in most developing countries, since highways are the only existing or viable mode. Coastal shipping in maritime countries is the principal exception, but this mode normally can compete only for time-insensitive traffic, which is likely already moving by water where that is feasible. Thus, the prospects for inter-modal shifts to achieve improvements in either fuel efficiency or environmental impacts would not seem very high.

There is substantial room for improvement in the area of vehicle technology and fuel substitution where the gap between what has already been achieved in the developed countries in recent years and the situation in the developing countries is quite large. Most vehicles in developing countries are not equipped with emission control devices, and the standards of vehicle maintenance are very low. Achieving improvements in emission controls, however, is likely to prove problematic. Where the availability and pricing of leaded fuel provide an economic incentive to substitute it for unleaded fuel, control programs based on catalytic converters are unlikely to suceed without regular vehicle inspection and maintenance (VIM). However, similar VIM programs to promote road safety, where the penalties of non-compliance are far more severe (all too manifest risks of injury and death), have generally not worked, so it is difficult to be sanguine in this area.

Policy instruments which operate on fuel consumption and substitution are easier to identify, although not necessarily easier to grasp. In developing countries most motorized transport is provided through sellers of transport services, and, unless cushioned by government subsidies, the profits of the operators are sensitive to fuel efficiency and prices. Appropriately designed fuel taxes, or other types of taxes such as vehicle purchase or annual license fees, by internalizing environmental externalities into the carriers' profit functions, could contribute significantly to improving vehicles' environmental attributes as well as fuel efficiency. Taxes on leaded gasoline can be particularly appropriate, encouraging refineries to shift their product mix and encouraging owners of vehicles to transition to better vehicle designs. Incentives to bus companies to encourage substitution of CNG or bio-gas for diesel-fueled vehicles is another area that merits close attention, especially in poorer countries where bus transit is the dominant mode of motorized transport.

Establishing the economically and socially appropriate level of taxes and other controls is likely to prove problematic, both scientifically and politically. How much should the world's poor countries pay to sustain or improve the environment, and where will they achieve the greatest environmental gain for each dollar spent? How are the benefits and costs of different measures distributed across different income and social groups? Which approaches work well and which do not? Scientific knowledge of the physical processes is still limited, and it is not possible to place a value on the reduction of one air pollutant relative to another, nor is the knowledge of the cost of different control measures adequate. Research in these areas should be encouraged. But ultimately a public policy choice is required to determine the value to be placed on protection of the environment: research can at best merely illuminate the tradeoffs.

Bibliography

"Gridlock Weary, Some Turn to Pedal Power," *The Urban Edge* 14(2), pp. 1–6 (March 1990). [A World Bank Publication]

Gutman, J.S., and Scurfield, R.G., "Towards a More Realistic Assessment of Urban Mass Transit Systems" in *Rail Mass Transit,* Thomas Telford, London (1989).

International Energy Agency, *Substitute Fuels for Road Transport: A Technology Assessment,* Paris (1990).

Saricks, Christopher L., "Technological and Policy Options for Mitigating Greenhouse Gas Emissions from Mobile Sources," paper presented to the 69th Annual Meeting of the Transportation Research Board (Washington, DC, January 7–11, 1990).

Sathaye, J.A., and Meyers, S., "Transport Energy Use in the Developing Countries," paper presented at New Energy Technologies, Transportation, and Development Workshop (Ottawa, Canada, September 1989).

Sinha, K.C., Varma, A. and Walsh, M.P., "Land Transport and Air Pollution in Developing Countries," Discussion Paper INU 60, Infrastructure and Urban Development Department, The World Bank (draft, January 31, 1990).

Sinha, K., Varma, A., et al. "Environmental and Ecological Considerations in Land Transport: A Resource Guide," Technical Paper Report INU 41, Infrastructure and Urban Development Department, The World Bank (March 1989).

Thomson, J.M., Allport, R., and Fouracre, P., "Rail Mass Transit in Developing Cities—The Transport and Road Research Laboratory Study" in *Rail Mass Transit,* Thomas Telford, London (1989).

The World Bank, Transportation and Water Department, "Energy and Transport in Developing Countries: Towards Achieving Greater Energy Efficiency" (February 22, 1983).

An Incentive-Based Transition to Alternative Transportation Fuels

Daniel Sperling
Institute of Transportation Studies
Civil Engineering and Environmental Studies
University of California, Davis

Introduction

Transportation energy issues are moving to the forefront of the public consciousness in the US and particularly California, and are gaining increasing attention from legislators and regulators. The three principal concerns motivating this interest in transportation energy are urban air quality, energy security, and global warming. Transportation fuels are a principal contributor to each of these. The transportation sector, mostly motor vehicles, contributes roughly half the urban air pollutants, about 30 percent of the carbon dioxide in the US, and consumes almost two thirds of the petroleum used in the country (and over three fourths in California).

From a rational technical perspective, the most compelling strategies for responding to these concerns are reduced motor vehicle use and increased vehicular efficiency. Politically, though, these strategies face large barriers.

While the efficiency of light duty vehicles has roughly doubled in the past 15 years—a major technological advance—further efficiency improvements of that magnitude are expected to be considerably more expensive (Difiglio et al., 1990). As a result, vehicle manufacturers have strongly resisted strengthening automobile efficiency standards much beyond the current 27.5 miles per gallon.

At the same time, consumers strongly resist additional fuel and vehicle taxes that are intended to discourage travel and encourage the purchase of more efficient vehicles. Morevoer, they are not very responsive to price inducements. An analysis of a court-ordered program in the San Francisco Bay Area to reduce vehicular emissions indicates that providing free mass transit to riders with incomes of less than $25,000, doubling transit service outside center cities, imposing $1 daily parking surcharges in central cities, increasing bridge tolls by $2.00, and charging a 2-cents-per-mile surcharge on vehicles would each reduce hydrocarbon emissions by only 1 to 2.5 percent (Harvey, 1990). It appears that a large fuel or use tax, much larger than anything under serious consideration, would be needed to gain even a modest change in behavior, an initiative few politicians are willing to propose.

Another strategy to reduce energy use and reduce vehicular emissions would be to encourage ride-sharing, to manage land use and land development in a more efficient and rational manner, and to coordinate expanded transit service with these land use changes. Alas, these changes are painfully slow.

A Technical Fix Solution

Politicians with great relief have found a more palatable strategy for resolving pollution and energy problems: the use of clean-burning alternative fuels. Alternative fuels are the quintessential technical fix. They require minimal changes in personal behavior and in the behavior and organization of local governments. People would still have the mobility and independence of their private motor vehicles, and local governments would not be obligated to surrender their authority to a regional body in order to coordinate and manage growth on a regional level. Indeed, alternative fuels provide the promise of never having to restrict motor vehicle use, because the vehicles would be so environmentally benign. Thus, alternative fuels are attractive because they are less disruptive (though not necessarily less expensive) and politically and institutionally easier to implement than other strategies.

Moreover, using practically any set of conceivable assumptions, one can argue that alternative fuels are inevitable. They are clearly an important part of any long term solution to urban air pollution, global warming, and energy security.

But are alternative fuels also a short term solution? How urgent and how critical are these problems and how appropriate are alternative fuels as a near-term response? Should government intervene now in support of alternative fuels? If so, which fuels and when? And what form should this intervention take?

As we shall see, alternative fuels, indeed, are seen as an important and immediate clean air strategy in California, one that requires a strong government role. I believe this is a reasonable view, but the larger question—how to achieve the transition to clean fuels in an efficient and effective manner—remains.

The first part of this paper evaluates the most attractive alternative fuels; several options are shown to provide major benefits, but none emerges as clearly superior to all others. The second part of the paper is a response to the impending introduction of cleaner fuels (including reformulated gasoline); it is a proposal for an incentive-based framework for regulating fuels and air quality, a regulatory framework designed to fit the changing environmental and fuels situation of the 1990s and beyond. The focus is on California, the region in the US where alternative fuels are likely to be first introduced on a large scale.

Background

Alternative fuels are not new to California, the US or the world; what is new in the US and especially California is that the motivation for introducing them is primarily concern for urban air quality, mostly ozone, not energy security or better trade balances.

Interest in alternative fuels escalated in the late 1980s as a result of several government initiatives: the most prominent was a proposal by President Bush in mid-1989, part of a package of Clean Air Act amendments, to mandate the sale of 9.25 million alternative fuel vehicles during the period of 1995 to 2004 in the nine most polluted cities of the country, and to require, beginning in 1994, that all new buses in areas with one million population operate on clean fuels.

In another initiative, the regional air quality district in the Los Angeles area adopted a plan that will require all vehicles in that region to switch to electricity or other very low-emitting fuels by 2007. The most innovative initiative with the most far-reaching implications is a proposal made in late 1989 by the California Air Resources Board to dramatically tighten vehicle emission standards; it is far-reaching in that, in effect, it would lead to a substantial portion of all new autos and light duty trucks in the state operating on clean-burning alternative fuels by 2000. It is innovative, as addressed later, in that it begins a shift from a command-and-control regulatory approach toward an incentive-based approach.

The motivation for these bold initiatives (as well as more modest initiatives affecting fleets in Texas, and oxygenated blends nationwide) is the continued failure of most urban areas, including all the major areas in California, to meet ambient air quality standards for ozone. First established as part of the 1970 Clean Air Act amendments, with an initial target date of 1975, the standards continue to be violated in all the major metropolitan areas of the country. Some areas are not even close to meeting the ambient standard, even though great improvements have been made in reducing emissions, especially from automobiles. Since mounting evidence suggests that, to protect human health, these difficult-to-attain ambient standards should be even more stringent, there is growing pressure for dramatic reductions in vehicle emissions.

Ozone reduction is not, however, the only reason to introduce alternative transportation fuels in the US; in fact, in the not-so-far-off future, it may not even be the most important.

A Comparative Analysis of Energy Options

Design of a transportation fuel strategy should be predicated upon an understanding of the full range of both private market and social nonmarket costs: private market costs because they are the criterion that industry and individuals use in deciding whether to invest in and purchase alternative fuels, and social (external) costs because they are the justification for government intervention.

In the following paragraphs, the private and social costs and other fuel-specific attributes that determine consumer purchase behavior are compared. Considerable space is devoted to this comparative analysis because of the widely held view, premature I believe, that methanol is the obvious "fuel of the future," and to demonstrate the large nonmarket benefits of alternative fuels.

The transportation energy options analyzed here are electricity, compressed natural gas, and methanol made from natural gas. These are the most attractive near- and medium-term options. Other less important near-term fuel options not included in this list and not considered further are ethanol, liquefied petroleum gases (LPG), solar electricity, and hydrogen. Hydrogen made from water using solar cells and burned directly or in fuel cells, or solar electricity used directly in electric vehicles, are the most environmentally attractive options and may prove to be dominant by the middle or end of the next century, but because they presently are much more expensive than the other fuels, they are not likely to be implemented on a large scale any time soon (DeLuchi, 1989; Ogden and Williams, 1989). For that reason hydrogen and solar-electric options are not included in the comparative analysis, though I strongly believe they merit much greater research funds.

LPG, the light part of crude oil and the heavy part of natural gas, represents a small proportion of oil and gas reserves. It is attractive now because of its low price, but if demand increased in the transportation or other fuels markets, this price advantage would disappear. LPG should not be considered as anything more than a niche fuel.

Ethanol made from farm products, mostly corn, is very expensive and survives in the US only because of the political strength of the farm lobby. It receives a federal subsidy of $0.60 per gallon, plus additional subsidies granted by some states.

Because of lack of reliable data, reformulated gasoline—gasoline that is modified to have lower hydrocarbon (and other) emissions—is also not analyzed here. While it is known that gasoline can be reformulated to provide small emission reductions at small cost, it is uncertain what the costs would be for achieving large emission reductions, reductions of a magnitude similar to those attainable with alternative fuels.

Note that gasoline is routinely reformulated. Gasoline is a non-homogeneous mix of a large number of different molecular compounds, ranging from very light near-gaseous hydrocarbon molecules to heavy complex molecules. In practice no two quantities of gasoline are identical; in fact, refiners purposefully create different gasolines for summer and winter conditions, and in response to properties of available crude stock and available refinery equipment. What is different now is that refiners and regulators are exploring opportunities to reformulate gasoline to reduce hydrocarbon, toxic, and other emissions.

Reformulated gasoline was first proposed as an alternative fuel in summer 1989 in response to the growing pressure for cleaner-burning fuels, in particular the July proposal by President Bush to require the sale of alternative fuel vehicles in the nine most polluted cities of the country. In fall 1989 in southern California, ARCO became the first oil supplier to market a gasoline reformulated for lower emissions. They reformulated leaded gasoline, in part by blending it with MTBE, an oxygenated derivative of methanol. ARCO's self-reported cost differential was 2 cents per gallon; tests indicate that hydrocarbon emissions from the tailpipe were reduced 4 percent and evaporative emissions (which account for much less than half the total hydrocarbon emissions) about 21 percent. Carbon monoxide emissions were reduced 9 percent and NO_x emissions about 5 percent. Other refiners have since introduced their own version of reformulated gasoline, also on a limited scale, in some cases as substitutes for unleaded gasoline, with similarly modest reported emission impacts.

The cost and difficulty of reformulating *un*leaded gasoline is greater because unleaded gasoline already undergoes more severe refining than leaded gasoline to maintain the same octane level and already has had some lower-value light hydrocarbons removed. To remove additional (inexpensive) light hydrocarbons and replace them with less volatile (and more expensive) hydrocarbons will require major changes in refinery design; the cost for producing significantly lower-emitting gasoline fuel is expected to be high. Although it cannot be proven because of lack of data and testing, it is likely that several alternative fuels would be much more cost-effective strategies for reducing emissions.

At present, a joint auto-oil industry coalition is investigating the costs and emissions benefits of reformulated gasoline, but it is uncertain when definitive answers will be attained. In any case, while reformulated gasoline is an inevitable gasoline-refining innovation that eases the urgency of alternative fuels in some areas, in most nonattainment areas the emission reductions associated with reformulated gasoline are not sufficiently large to justify additional delays in pursuing other emission-reducing actions, including alternative fuels.

Air Quality

As mentioned above, the promise of reduced (tropospheric) ozone is currently the primary attraction of methanol, natural gas, and electric vehicles. Ozone is formed from reactions between hydrocarbons and nitrogen oxides that take place in the atmosphere in the presence of sunlight. Other major air quality problems associated with motor vehicles are carbon monoxide and particulates.

Roughly 60 to 100 metropolitan areas (representing 80 to 130 million people) do not meet the statutory ambient air quality standards of the Clean Air Act for ozone. In 1988, the State of California, responding to evidence that the health effects of ozone were even more severe than had previously been thought, established more stringent ambient ozone standards than the federal government (0.09 versus 0.12 ppm over a one hour period, with no exceedances allowed versus three exceedances per three years allowed in the federal rules).

As shown in Table 1 most of the metropolitan areas in California are so far above the ozone standard, and are growing so fast, that they have little hope of attaining the standards in the foreseeable future. These same areas are also in severe violation of the particulate standard. These high pollution levels have dire implications

Table 1. Percent of Days over California State Standard, 1987 Summer and Winter Seasons.[1]

	O_3 1-hr, summer	CO 8-hr, winter	PM10[2] 24-hr
South Coast	90%	42%	78%
SF Bay Area	22%	1%	37%
Sacramento	35%	4%	23%
San Diego	56%	1%	19%
Fresno	59%	3%	59%
Ventura	54%	0%	25%
Kern	61%	0%	66%

[1] Winter season includes December through February and summer season includes May through September.

[2] Particulate matter less than 10 microns in diameter.

Source: California Air Resources Board (1988), *California Air Quality Data: Summary of 1987 Air Quality Data*. Sacramento, CA, Vol. XIX.

for human health and create the risk of federal and state sanctions.

The external (nonmarket) costs of this air pollution are huge: estimates for the US range from $11 to $187 billion per year, the large range depending mostly on uncertainty of the number of deaths and illnesses due to pollution and the monetary value assigned to deaths and illnesses (DeLuchi et al., 1987). As an indication of how large the costs and benefits are, it is estimated that implementation of the Los Angeles area (South Coast) air quality plan will generate benefits of $1.5 to $7.4 billion per year in that region (Portnoy et al., 1989).

Motor vehicles are a principal cause of urban air pollution. The California Air Reources Board (CARB) estimates that cars and trucks contributed 43 percent of the hydrocarbons (reactive organic gases), 57 percent of the nitrogen oxides, and 82 percent of the carbon monoxide emitted in the major urban areas of California in 1987. (Motor vehicles emit relatively little particulates from their exhaust, but airborne particulates (PM10) are composed of up to 35 percent aerosols which are largely the result of atmospheric chemical reactions of the NO_x and hydrocarbons mainly emitted by motor vehicles. CARB estimates that, in addition, over half the PM10 that is directly emitted from anthropogenic sources is dust kicked up by motor vehicle activity on roadways.)

All three energy options are superior to gasoline; initial evidence indicates that electric vehicles would generally provide the greatest ozone reduction, regardless of the source of electricity, followed by natural gas and then methanol.

Methanol combustion will tend to reduce ozone because unburned methanol emissions from methanol vehicles are about five times less reactive than the unburned hydrocarbon emissions from gasoline vehicles. Methanol also produces less carbon monoxide (CO) or nitrogen oxide (NO_x) (but probably not both) than gasoline vehicles.

Carnegie-Mellon researchers estimate that the use of 100 percent methanol (M100) in advanced technology engines would reduce ozone peaks in Los Angeles in a 2010 scenario by 33 percent of the maximum ozone reduction attainable from motor vehicles (Harris, Russell, and Milford, 1988; Russell et al., 1990). (Because of the diminishing importance of motor vehicle emissions in total urban emissions, the nonlinear relationship between emissions and ozone formation, and the existence of natural hydrocarbon emissions, the actual reduction in peak ozone levels due to the use of methanol vehicles would be less.) The use of M85 (85 percent methanol, 15 percent gasoline), the likely specification because it allows easier starting in cold weather and improves the visibility of flames in case of fire, would result in reduced but still significant ozone benefits. In general the substitution of methanol for gasoline in all motor vehicles would potentially reduce the ozone-producing potential of vehicles by about one half. These reductions are not being attained with current technology "fuel-flexible" vehicles that can operate on both methanol and gasoline.

Natural gas vehicles (NGVs) are even more attractive than methanol from an air quality perspective. NGVs emit virtually no carbon monoxide and, because natural gas emissions are highly unreactive, natural gas vehicles would cause less ozone pollution than methanol.

Electric vehicles (EVs) are still better in most areas, especially in California, where little coal is used in powerplants and emission controls on powerplants are stringent (Wang, DeLuchi, and Sperling, 1990; Hempel et al., 1989). Although the use of EVs would generally result in some increases in particulate and sulfur oxide emissions relative to gasoline vehicles, the absolute increases will be small since gasoline vehicles are minor contributors: in California, autos account for only 6.5 percent of total sulfur oxide emissions and only 1.7 percent of particulate emissions (CARB, 1986); elsewhere in the US the share is even less. More significantly, the use of EVs would result in the virtual elimination of hydrocarbon and carbon monoxide emissions, relative to gasoline vehicles, and would reduce nitrogen oxides significantly. Because hydrocarbon emissions are greatly reduced and NO_x emissions are not large, the use of EVs would dramatically lower ozone pollution in most areas.

Two cautionary notes: Ozone air quality models are subject to considerable uncertainty because of inadequate input data, especially outside Los Angeles, and optimized single-fuel engines burn much cleaner than multifuel engines.

This second point is critical because the preceding assessment of emission impacts of alternative fuels was based on the assumption that the engines were designed specifically for those fuels. Commercial versions of such optimized single-fuel engines do not yet exist. Indeed, there is relatively little experience and little current R&D with optimized alternative-fuel engines and catalyst technology. If a serious sustained effort were made to reduce emissions, similar to the 25–year history with gasoline engines, greater emission reductions would be likely. For instance, the ozone reductions with methanol would be even greater than the 33 percent estimated by Carnegie-Mellon.

Automakers apparently do not intend to supply advanced single-fuel vehicles for some time, however. They intend to sell multifuel vehicles designed to operate on both gasoline and the alternative fuel, but optimized for neither, thus not becoming dependent on the establishment of networks of retail stations that sell the alternative fuels. These initial multifuel vehicles will be more costly, less fuel efficient, and more polluting than an optimized single-fuel vehicle: they will not fully capture the potential benefits of natural gas and methanol fuels. As indicated later, regulatory incentives could be used to expedite the transition to the more fuel-efficient and lower-emitting single-fuel vehicles.

Greenhouse Impact

Concern is growing that increasing emissions of carbon dioxide and other greenhouse gases will cause global climate change. The magnitude and geographical distribution of climate change impacts are still uncertain, but there is an emerging consensus that greenhouse gas emissions should be curtailed. Again, methanol does not rate very highly according to this criterion.

Recent studies that take into account carbon dioxide (CO_2) and trace greenhouse gases emitted during fuel production, transport, and combustion are summarized in Table 2.

Electric and natural gas vehicles are the most attractive near-term options for reducing greenhouse gas emissions. Methanol made from natural gas provides little or no benefits; when made from coal it greatly increases greenhouse gas emissions. EV use would nearly eliminate greenhouse gas emissions if the electricity were made from hydro, nuclear, or solar power.

Other attractive options, not listed in the table, are hydrogen made from clean sources, for instance by electrolysis of water with photovoltaic solar cells, and conversion of cellulosic biomass into transportation fuels (ethanol or methanol). Greenhouse gases would be virtually eliminated if biomass fuels were used as the energy input through the entire chain of production and fuel distribution activities (for instance, for farming and operating the fuel production plant). Ethanol made from corn, as currently produced, does *not* reduce greenhouse gases because corn farming consumes large amounts of petroleum and natural gas and because most distilleries use coal as a boiler fuel.

Table 2. CO_2-equivalent Greenhouse Gas Emissions of Alternative Fuels Relative to Comparable Gasoline-Powered Vehicles, Entire Fuel Cycle.

	Greenhouse Gas Emissions Relative to Gasoline
Natural gas vehicle	−25 to −5%
Electric vehicle (natural gas)	−50 to −25%
Methanol from natural gas	−10 to +10%
Electric vehicle (current energy mix)	−5 to 0%
Electric vehicle (coal)	0 to +10%
Methanol from coal	+30 to +80%

Source: Updated from Mark A. DeLuchi, Robert A. Johnston, and Daniel Sperling (1988) and unpublished reports by Acurex Corp. (Mountain View, CA).

National Energy Security

Energy security was the primary energy goal in the 1970s when the Arab OPEC countries embargoed oil deliveries, but concern for energy security has waned since then. Energy security is not considered a high priority by the nation currently, but the unexpectedly rapid increase in oil imports during the past few years—approaching 50 percent of domestic consumption in 1990, higher than levels reached during the 1970s—is rekindling energy security concerns.

If energy security were the only criterion, then the US would be advised to invest in electric and natural gas vehicles, and to a lesser extent in biomass fuels (Sperling, 1988).

Electric vehicles are a highly attractive option from an energy security perspective because electricity is generated almost exclusively from North American resources; less than 5 percent comes from oil in both the US and California. Likewise, biomass fuels—ethanol made from farm products, and methanol and ethanol made from cellulosic wood—would be based on indigenous resources and therefore provide for energy security.

On the other hand, methanol made from natural gas will provide little benefit. Methanol can be made from abundant domestic resources of coal and biomass, but will be made from natural gas into the foreseeable future, because of much lower costs. It is true that proven natural gas reserves in the US are fairly large—over 30 percent greater (in energy content) than proven oil reserves—and that there are enough economically recoverable conventional reserves of domestic natural gas to last for 40 to 60 years at present rates of consumption and possibly over 100 years if economically recoverable unconventional reserves are included. But it is unlikely that this domestic gas would be used to manufacture methanol fuel in the foreseeable future.

Methanol can be made much more cheaply from "remote" natural gas in other countries, where it is available in large quantities at very low cost. This remote natural gas is so abundant and so inexpensive (generally less than $1 per million Btu compared to $2 to $3 in the US) that all current and prospective methanol fuel suppliers intend to import methanol to the US. If the US mounted an ambitious methanol fuel program, it would rely on imports. Although shifting to methanol diversifies our energy supply, this diversification provides relatively minor energy security benefits.

Natural gas vehicles are superior to methanol vehicles from an energy security perspective because natural gas vehicles would not rely on energy imports in the foreseeable future; they would rely on domestic (and Canadian) gas, unlike methanol, because it is less costly to use expensive domestic gas than to incur the high cost of liquefying and transporting the remote foreign natural gas across oceans. Methanol imports are practical, in contrast, because unlike natural gas methanol can be shipped inexpensively across oceans by conventional tankers.

Safety and Toxicity

One of the primary arguments used against methanol has been its toxicity and safety. Methanol can cause blindness or death if drunk, burns with an invisible flame (making it difficult to detect fires), and is highly soluble in water (making it difficult to contain a spill).

The first two of these problems are solved by adding 10 to 15 percent gasoline (or some other combustible denaturant) to the methanol, making the flame visible and giving the liquid a very unpalatable smell and taste. The third issue, solubility of methanol in water, is not necessarily a disadvantage; the greater solubility causes the methanol to quickly dissolve, thus not causing the long-lasting destruction typical of large oil spills. Overall, gasoline is a more threatening fuel than methanol: it is far more flammable and contains many carcinogens.

NGVs and EVs are even safer and contribute no toxic emissions. In New Zealand, for instance, with over 100,000 vehicles in operation for almost 10 years, there has been only one explosion and no fire involving a natural gas vehicle tank, and no one was hurt. The principal danger is the accidental leakage of gas from a natural gas vehicle in an enclosed space—in an open space the gas dissipates quickly upward causing no problems—but again the safety record of these vehicles has been virtually unblemished.

Costs and Cost-Effectiveness

Cost comparisons must be made on a lifecycle basis, whereby differences in vehicle life, fuel and maintenance costs, and initial vehicle cost are translated into a cost per mile basis. Electric vehicles, for instance, have higher initial cost (including batteries) but much longer life than gasoline or methanol vehicles, while natural gas vehicles have much cheaper fuel but higher vehicle cost due to the high pressure tanks. Methanol cars would cost about the same as gasoline cars under mass production conditions, but the fuel price is expected to be somewhat higher. Taking all these differences into account and using expected 1990s prices and technologies, one finds that methanol vehicles will tend to be slightly more expensive to own and operate than natural gas and gasoline vehicles, and probably more so than electric vehicles as well (DeLuchi et al., 1988, 1989).

If, as a means of linking this cost analysis to current regulatory debates, we assume that alternative fuels are being introduced exclusively for ozone control, then the cost analysis may be extended into a cost-effectiveness analysis and alternative fuels may be compared to other ozone control strategies. The only published calculations of the cost-effectiveness of alternative fuels as ozone control strategies that I am aware of were conducted by the US Congress Office of Technology Assessment (1989), the Advisory Board on Air Quality and Fuels to the California Legislature (1989), and Fraas and McGartland (1990). The Advisory Board addressed only methanol, while OTA and Fraas and McGartland addressed methanol and CNG.

The OTA study calculated that using methanol (M85) would cost $8700 to $66,000 to eliminate one ton of "ozone-equivalent" hydrocarbon emissions, and if M100 were used, assuming favorable ozone-reduction parameters, the cost would be $3200 to $22,000 per ton. The comparable estimates for CNG, using dedicated vehicles, would be $1600 to $14,000 per ton. The Fraas/McGartland analysis, using a broader range of methanol fuel prices, estimated a range of −$3900 to $7800 for M100, $3300 to $29,000 for M85, and −$10,293 to $1411 for CNG. The California Advisory Board estimated the cost-effectiveness of M85 at $8000 to $40,000 per ton. (A cost-effectiveness analysis of methanol conducted by Resources for the Future will be finished in mid-1990; its estimates are in the upper range of values reported here.) The various assumptions made in these studies are not consistent, but not unreasonable.

Most other ozone-reduction strategies studied by OTA, those not involving alternative fuels, had cost-effectiveness reductions of $500 to $6000 per ton; however, these other strategies do not have the potential to generate the same magnitude of ozone reductions as methanol and the other alternative transportation fuels. Also, those cost-effectiveness estimates are for the US. In California, the cost-effectiveness of ozone control strategies is considerably higher, because most of the less expensive strategies have already been implemented; additional ozone reduction has a higher marginal cost.

Nonetheless, the OTA estimates suggest that *multi-fuel* methanol cars, which would be inferior to dedicated M85 vehicles, are probably not a cost-effective ozone control strategy, even for California. Given the range of uncertainty in costs and emission reductions, a similarly definitive conclusion regarding optimized dedicated methanol cars is premature. Indeed, if methanol fuel and vehicle prices are not too much higher than those for their gasoline counterpart, and continued advances are made in emission controls of methanol vehicles—two fairly likely assumptions—then dedicated methanol vehicles would be a cost-effective strategy for reducing ozone. The case for dedicated CNG vehicles is even stronger, and for EVs still stronger, if EV cost and performance improvements continue.

Since alternative fuels provide other benefits besides ozone control—with respect to greenhouse warming, safety, toxicity, energy security and indirect economic impacts, as well as carbon monoxide and particulates—the general case for near-term introduction of alternative fuels is strong, especially in California.

Why Does Methanol Dominate the Debate?

If natural gas and electric vehicles are likely to be less expensive and have larger social benefits than methanol, then why does methanol dominate the alternative fuels debate? The answer is simple: the auto industry prefers methanol because it is physically and chemically more similar to gasoline than electricity and natural gas, and is more compatible with gasoline in multifuel engines. Switching to methanol would require less modification to current gasoline vehicles than would gaseous or electric-powered vehicles, and less change in driver behavior. There would be less cost and less market risk. Government regulators, concerned with quick impacts, have accepted auto industry thinking and concerns. The absence of strong lobbying by other non-alcohol alternative fuel interests has allowed the debate to remain focussed on methanol.

The principal fear of the auto industry is that consumers will be unwilling to accept the shorter driving range of natural gas and, especially, electric vehicles. Current EVs travel only about 60 miles per charge. Advanced EVs likely to be available in the late 1990s will probably have a range of 100 to 200 miles, but even this is much less than for gasoline vehicles. Natural gas vehicles have a less severe range problem; assuming that future vehicles will be somewhat more energy efficient than today's vehicles and that auto engineers would slightly redesign an NGV in order to fit more tank capacity into a vehicle, then a future NGV is likely to have a range of about 200 to 250 miles, still somewhat less than today's gasoline vehicles.

A broader, longer term view would suggest that EVs and NGVs may be successful in the future, possibly more so than methanol, as utilities continue to be deregulated and slowly emerge from their lethargy—becoming more aggressive marketers and lobbyists for EVs and NGVs. They are likely to find, based on research by myself and others, that concerns about consumer perceptions may be overblown, that a large number of people would be willing to accept shorter range and even less power, if they were convinced that government and industry were fully behind these technologies and that the vehicles truly were much cleaner-burning (Sperling, Hungerford, and Kurani, 1990). Vehicle and fuel suppliers might even find that the possibility of refueling an NGV and recharging an EV at home proves to be a marketing advantage with those many individuals who dislike refueling at retail stations.

The reality is that industry is conservative and risk averse and, all else being equal, would naturally prefer the least risky path. While this risk-averseness favors methanol in the case of the auto and, to some extent, oil industry, the attractions of NGVs and EVs, the growing aggressiveness of gas and electric utilities, and the absence of a domestic economic constituency for methanol may eventually lead to the emergence of NGVs and EVs as leading transportation energy options (Sperling and DeLuchi, 1989).

Methanol will undoubtedly be an important fuel in the future, but not necessarily the dominant transportation fuel.

In summary, the problem associated with continued reliance on petroleum fuels is not necessarily long run supply, but rather ignored social costs (especially air pollution and global warming) and economic losses resulting from unpredictable oil prices, inflexible responses to oil price changes, and absence of substitute fuels. If market mechanisms were operating efficiently, then optimal consumption and production of oil would follow. But that is not the case. Efficiency improvements and alternative fuels are delayed beyond the time when they would otherwise be economically attractive by uncertain and low oil prices that do not reflect oil's true cost to society.

Moreover, there are also large start-up barriers to alternative fuels. Because of the start-up barriers and a flawed market, new fuels will only be introduced if they receive strong support from government. Significant government intervention will be premised upon the public-good concerns listed above: the greenhouse effect, dependency on foreign oil supplies, economic benefits of lower energy prices, and urban air pollution.

Diesel Fuels

Because of space limitations, I only address light duty spark-ignition vehicles in this paper, even though roughly one fifth of the motor vehicle fuels consumed in the US are diesel fuels, mostly in heavy duty trucks. As an aside, note that equal if not greater pressure is being placed on diesel vehicles to reduce their emissions—mostly particulates and nitrogen oxides—via more stringent EPA emission standards that take effect for transit buses in 1991 and for trucks in 1994, and continuing proposals by the South Coast Air Quality Management District to ban diesel fuels. Natural gas and methanol are likely to be used as substitutes for diesel fuel, although that outcome depends in large part on whether engine manufacturers can produce a reliable diesel engine that meets the new emission standards without incurring large additional costs. In California, the Air Resources Board and the Advisory Board on Air Quality and Fuels have both focussed on gasoline-powered vehicles and deferred actions and recommendations regarding diesel-powered vehicles.

The Impossible Analysis

To determine which fuel or fuels government should promote and to what extent, analysts ideally would calculate the cost-effectiveness of each fuel option in reducing air pollution and greenhouse gases, and enhancing energy security and safety, and compare this rating to other strategies. In other words, they would synthesize all the information presented to this point in a single measure. Unfortunately such an analysis would be impossible to conduct with accuracy and precision at this time. One reason is uncertainty about costs: uncertainty about vehicle and infrastructure costs, engine life, maintenance costs, and future energy prices.

Still greater uncertainty exists on the benefit side of the equation regarding emission characteristics, relationships between emissions and air quality, emissions and global warming, magnitude of safety and toxicity impacts, and impact on energy security. Recall the calculations presented earlier for just one component of this overall measure: the cost-effectiveness of methanol and CNG as ozone reduction strategies. They included only two important factors: fuel/vehicle cost and ozone impact. But even those two factors include considerable uncertainty and unverifiable assumptions, and produced results with a very broad range.

Even more daunting than calculating cost-effectiveness measures for particular impacts, is the issue of how to weight the relative values of improved air quality and safety, reduced global warming, and greater energy security. How much is a 10 percent reduction in greenhouse gases worth? Is it worth more than a 10 percent reduction in hydrocarbon gases?

Facts, Beliefs, and Values

How, when, and where should we initiate a transition to alternative transportation fuels? There is no obvious answer and no consensus. The price of petroleum cannot be predicted, and many of the costs and benefits of alternative fuels are difficult to quantify. Different groups place different values on the important (nonmarket) concerns: energy security, air quality, global warming, and the ease and convenience of a transition. In short, different beliefs and different values lead individuals and organizations to different conclusions about the most desirable path.

For instance, if concerns for self-sufficiency and energy independence were to dominate, then the US should prefer energy options based on abundant domestic resources: fuels from coal, biomass, and oil shale, domestic natural gas, and domestic electricity. Remote natural gas, imported as liquefied natural gas or methanol, would be deemphasized.

If economic efficiency, measured by conventional market indicators, is the dominating value, then hydrogen would be discarded as an option. If environmental quality, avoidance of greenhouse warming, and sustainability takes precedence, then hydrogen and electric vehicles, using clean and renewable energy (probably solar power), would be preferred. If the abiding objective is to make the transition with as little disruption as possible, then petroleum fuels would be retained as long as possible, by increasing oil imports and by reformulating gasoline and diesel fuel to be more environmentally acceptable.

Thus, the choice of transportation energy paths should focus on values and goals, not just projections of market costs, especially when those projected costs are uncertain and not too different from costs for the existing petroleum-based transportation system. Current and projected market prices can be poor criteria for long-term energy choices. Shifting societal goals, values, and preferences will result in redirected government initiatives that will change relative energy prices, while the long-term replacement of today's sunk investments will also cause a shift in long-term energy prices. We should therefore take care not to allow current and extrapolated energy prices to overly influence transition strategies. In the words of Herman Daly, "the choice between . . . energy futures is price determining, not price-determined" (Daly, 1976).

The choice of transportation energy paths should also be open-minded and flexible. There is no one optimal choice for everyone, or every region; the era of one (or two) uniform transportation fuels may be over. This prospective multiplicity of fuel options presents a challenge for business and government. Because many of the benefits resulting from initial alternative-fuel investments do not accrue to the private sector supplier of the fuel, government must take much of the initiative.

The challenge for government is to create a framework for introducing alternative fuels that is flexible in responding to changing economic and technological conditions and shifting priorities, and acknowledges regional differences and the multiplicity of social goals. Pre-selecting one fuel for any particular region, much less all of California or all of the US, would probably be a catastrophically expensive mistake.

Moving Toward a More Efficient and Flexible Regulatory System

Current regulatory initiatives to introduce alternative fuels are based exclusively on vehicle emission standards and mandated use of "low-emitting" fuels. They are part of a first generation regulatory framework that is not suited to the changing circumstances of the future: they do not reflect other social goals, are not flexible in responding to changing economic and technological conditions, are insensitive to regional differences, and do not acknowledge the likelihood of shifting social priorities. There are good reasons for the simplistic inflexibility of past and current approaches—mostly associated with ease of implementation—but they are becoming increasingly inefficient and inappropriate.

Current government initiatives in the US to reduce vehicular emissions and introduce alternative fuels are a continuation of the 1960s command-and-control style of social regulation, an innovation of lawyers. Automotive emissions are currently regulated by requiring every vehicle to meet the same uniform standard, regardless of whether it costs less to reduce the emissions in some vehicles than others and regardless of whether there is an air pollution problem where the vehicle is sold and used.

This current regulatory approach, in which every vehicle is required to meet the same uniform emission standard, provides manufacturers with no incentive to do better than the standard. For instance, if an auto exceeds the standard, the company removes the valuable excess catalyst metals from the catalytic converters, reducing their costs and allowing emissions to increase. This illustrates the flaw of uniform standards: it is not sensitive to differences in the cost of reducing emissions from one vehicle to another. Uniform emission standards are not only an economically inefficient method for reducing emissions, but they provide no incentive to reduce emissions below the standard and therefore no incentive to introduce cleaner-burning alternative fuels.

Continued reliance on the current command-and-control approach implies the use of specific standards, mandates, and direct subsidies to accomplish social goals. This approach requires that government administrators have the foresight to be able to orchestrate which fuels and vehicles should be introduced where and when. Given the uncertainty about the relative attractions of alternative fuels and the best way to introduce new fuels and vehicles, and given the uncertainty about the future, a more efficient and resilient approach would be to offer incentives to industry and consumers that push them in the correct direction—toward lower air pollution, reduced greenhouse gases, and perhaps even domestic resources. This approach is fundamentally more efficient. It does not rely on omniscient government bureaucrats to prognosticate; rather, it relies on bureaucrats to do what they do best—administrate.

An Incentive-Based Approach

An initial element of an incentive-based program might be to provide vehicle manufacturers with the flexibility to average emissions across their fleet of vehicles. An emission standard would be established, as currently done, but in this case it would be an average. Vehicle suppliers could reduce emissions to a lower level in those vehicles where the cost of reducing emissions is less and not reduce emissions as much in other vehicles where the cost would be greater, as long as the average for all vehicles was below the standard. As the emission standard is lowered, resulting in an increasing cost for emission reductions, there will be an incentive for automakers to market vehicles that operate on cleaner-burning alternative fuels.

The new averaging standard would need to be lower than an unaveraged uniform standard in order to gain the same net reduction; because the unaveraged standard is a ceiling and thus all vehicles emit under the standard, the resulting average emission rate is actually considerably lower than the standard.

This averaging approach is not revolutionary. The same concept is used to regulate automotive fuel efficiency; it is not required that every vehicle meet the 27.5 mile per gallon CAFE (Corporate Average Fuel Economy) standard, only that the average for each vehicle manufacturer be 27.5 or greater.

A more sophisticated scheme would be to allow companies to trade emission reduction credits. Thus, those manufacturers who prefer to focus on large engines, jeeps, and other types of vehicles that tend to produce more emissions, could continue to do so, but they would have to buy credits from manufacturers who sell low-emitting vehicles.

An additional procedure would be to allow manufacturers to bank emissions from years when they outperform the average for use in years when they fall short. Emission banking provides flexibility, increasing the efficiency of trading, by allowing emission reduction credits to be traded in a time period after they accrue. Banking also provides an incentive to introduce new technologies and products sooner in anticipation of continuing tightening of emission standards.

Through averaging, banking, and trading, emission reductions would be achieved in the most inexpensive manner possible since industry would have the flexibility and incentive to reduce emissions in the most cost-effective manner.

A study conducted for the US Environmental Protection Agency in 1984 made preliminary estimates of the cost savings associated with emission averaging and trading, but not banking (McElroy, Hayes, and Olson, 1984). They calculated that the differences in emission control costs to automakers between a regime of uniform emission standards and a regime allowing emission averaging and trading between companies, for equivalent reductions in total emissions, was 25 percent. That is, the costs to automakers for reducing emissions would have been 25 percent less if they had been allowed to use emission averaging and trading.

The calculations were made using 1981 emission standards, forecasted vehicle sales for 1984 to 1990, the vehicle and market mix prevailing in 1981, and a set of control cost functions derived from a statistical analysis of 1979 to 1982 certification data and unreported estimated cost functions. The analysis is simplistic, perhaps out-of-date, relies on a poor data base, and uses aggregated data in a manner that underestimates the cost savings. With the vehicle technology, tighter standards, and higher marginal costs of the 1990s, the estimated cost savings would probably be much greater.

Nonetheless, even if one accepts the 25 percent estimate, and estimates that the current marginal cost for emission control is about $500 per vehicle, then emission averaging and trading would generate cost savings of well over $100 million per year in California and over a billion dollars for the US.

Unfortunately, the incentive-based concept was tarnished by the handling of an emissions averaging provision in the mid-1989 Bush Administration proposal. The Administration proposed an averaging standard that was not low enough to gain a net reduction in emissions compared to a nonaveraged uniform standard. Environmentalists objected vociferously, and appropriately so. This averaging provision apparently had been part of a compromise in which automakers had accepted the alternative fuel mandates of the overall proposal in return for this softened averaging standard.

Emissions averaging, banking, and trading constitute the rudiments of a framework for guiding the transition to alternative fuels. Emissions averaging and trading provide the incentive to automakers to develop and market very clean-burning vehicles, whenever the additional cost for doing so is less than the additional cost of reducing emissions from their gasoline vehicles. As emission standards continue to be ratcheted down, the marginal cost of marketing alternative fuel vehicles will eventually drop below that of gasoline vehicles. Emissions averaging would provide the incentive for automakers to gradually phase in clean-burning alternative fuel vehicles by manufacturing them or, through emissions trading, to buy credits or vehicles from an electric, natural gas, or methanol vehicle supplier so that it could continue selling mostly gasoline vehicles. The result is an incentive-driven transition to clean-burning alternative fuels.

Averaging, trading, and banking of vehicle emissions are only one component of an incentive-based approach. By themselves, they provide greater efficiency and are fuel-neutral, but they do not allow for the design of region-specific strategies, because vehicles can be easily moved from one region to another, and do not incorporate other social goals. A more sophisticated and expanded framework is needed.

Fuel Supply Regulation

To incorporate a geographical element into the system it is necessary to involve fuel suppliers—to allow fuel suppliers also to average, bank, and trade credits. Region-specific strategies are desirable because the magnitude and nature of the pollution problem varies greatly from one region to another. For instance, some regions have major pollution problems while others do not. In some cities, the most serious air pollution problem is high carbon monoxide concentrations, while in others the more critical problem is ozone. Even for those cities with ozone problems, the controlling constituent in some is hydrocarbons, while in others it is nitrogen oxides.

Region-specific strategies are possible with fuels regulation because the fuel purchased within a region is consumed within that same region. Vehicles purchased within a region, in contrast, can be readily sold or transferred to another region, a right that government is unlikely to restrict. Thus fuels-based regulations are amenable to region-specific strategies, whereas vehicle-based regulations are not.

The administration of a fuels regulation program would probably be more difficult than a comparable program for automotive emissions, principally because there are many more fuel suppliers than vehicle suppliers and because of the multiple fuel supply industries. There is also less experience with fuels regulation. The only current regulation of fuels is through spot checks of vapor pressure, lead content, and the use of oxygenated blends in some areas (e.g., in Denver to reduce wintertime carbon monoxide).

Fuel regulation would involve hundreds of fuel marketers, and include not only petroleum marketers (who probably would also market methanol, ethanol, and possibly LPG), but also distributors of natural gas and electricity. Fuel regulation would presumably occur at the bulk distribution terminals in the case of liquid fuels, which is the point at which excise and sales taxes on gasoline and diesel fuel are currently collected. Natural gas and electricity regulation would be much simpler since only one supplier operates in any geographical region (they are regulated monopolies) and because the activities of these companies are already heavily regulated and closely monitored.

While fuels regulation is not as familiar as vehicle emissions regulation, and will be more complex administratively, there is no reason to believe it is unworkable. Indeed, as documented by Hahn and Hester (1989), the very successful experience with trading and banking of lead rights for use in gasoline in the 1980s—at one point, over 50 percent of all lead rights were being traded—suggests that fuels regulation would not be onerous or exceptionally difficult.

Since each type of fuel emits differring quantities and types of pollutants, the regulation of fuels as well as vehicle emissions would require that ratings be developed for each fuel indicating the relative harm of pollutants associated with that fuel. Emission equivalency values would be assigned to the different fuels. Ozone reactivity ratings have already been developed for comparing the relative contribution of each type of fuel to ozone formation. In the case of fuels, a rating would be assigned to each carefully specified fuel: for instance, gasoline might be rated 1.0, "reformulated" gasoline 0.9, methanol 0.6, natural gas 0.4, and electricity 0.2. Each fuel supplier would be required physically, or via purchased or banked credits, to supply a slate of fuels that on average meets a rating established by the regional or state environmental regulator.

This regulation of fuels creates the opportunity to develop region-specific strategies in two ways: the equivalency values can be adjusted to reflect the unique aspects of pollution in that area, and the average rating required of each fuel supplier could be raised or lowered depending upon the severity of the problem in that area.

Thus, in Los Angeles the average rating imposed on each fuel supplier might be 0.5, while in San Francisco it might be 0.9. Similarly, because methanol is relatively more effective at reducing ozone in NO_x-rich atmospheres than in hydrocarbon-rich atmospheres, the rating assigned to methanol might be lowered to 0.5 in NO_x-rich Los Angeles and raised to 0.7 in hydrocarbon-rich regions such as Houston. Or, because natural gas vehicles emit very low levels of carbon monoxide, natural gas might be given a low rating of 0.2 in regions such as Denver that have serious carbon monoxide problems.

In fuels-based regulation, each fuel supplier would determine the most cost-effective manner for meeting the specified *average* rating. If it is expensive for an oil refiner to reformulate gasoline to reduce its emissions—because of the design of its refineries—or the average rating is set lower than that achievable with reformulated gasoline, then credits could be purchased from another company which can meet the required rating at less cost. Or the oil refiner might choose to sell natural gas or even electricity itself at its own stations.

Over time, the standards would be gradually tightened on a predetermined schedule (with periodic mid-course adjustments). Fuel suppliers could plan their investments with this schedule in mind; smaller refiners less willing or able to invest in refinery modifications might move more quickly toward alternative fuels, and sell their emission reduction credits to larger refiners who might prefer to focus on reformulated gasoline. Likewise, some automakers might prefer to stick with improving gasoline engine technology, including multi-fuel engines; they would buy emission reduction credits from other companies that sell much lower-emitting EVs and single-fuel natural gas and methanol vehicles.

One last but important refinement would be to design the fuel rating to incorporate other social goals such as reduced emissions of greenhouse gases and toxic gases, and greater energy security. This could be accomplished by converting the emission rating for each fuel into a social index; for instance, the rating for domestically supplied natural gas would be set at 0.3 instead of 0.4 because natural gas vehicles emit fewer greenhouse gases and the gas is domestically produced.

The incentive-based regulatory concept presented here is not new or unknown to government or economists. Economists have long argued for the use of market-like structures in reducing emissions. Slowly, their arguments have been accepted over the past decade by regulators of stationary source emitters (Cook, 1988; Hahn, 1989). But the use of incentives in regulating transportation fuels and motor vehicle emissions has been virtually ignored, by researchers and regulators, with the exception of the brief lead trading experience mentioned earlier. What has changed that compels a reexamination of incentive-based regulation is the difficult and unavoidable issue of determining which alternative fuels should be introduced and when.

The concept of market-like incentives is gaining increasing acceptance not only with researchers but also with policymakers. President Bush, for instance, in his mid-1989 Clean Air Act proposals, though vague in details, endorsed the concept of an incentive-based approach for regulating vehiclular emissions. Also, the previously mentioned Advisory Board to the California legislature, composed of high level government and industry representatives, recommended in its October 1989 final report that a fuel regulation program be established similar to that described above. Labeled "fuel-pool averaging," the intent was to propose a program that was fuel-neutral. Details were not provided.

CARB's Bold Fuel and Vehicle Proposal

Most important of all, in terms of shifting from a command-and-control to incentive approach, is a bold proposal by the California Air Resources Board (CARB), scheduled for adoption in September 1990, to provide limited trading and banking of clean fuel *and* vehicle emission rights for light and medium duty vehicles, as well as averaging of vehicle emissions. It would affect all vehicle marketers and gasoline refiners in California. The proposal establishes equivalency factors between different fuels and vehicles.

The proposal is in flux as this paper is being written. Originally, the proposal created a two-tier system of standards, requiring, in 1994 through 1996, that up to 10 percent of new vehicles sold by each manufacturer would have to meet a more stringent standard of 0.125 grams of hydrocarbons per mile (or the equivalent for nongasoline fuels as determined by CARB), compared to today's standard of 0.41, and the 0.25 standard to take effect in 1994. Many gasoline vehicles would be able to meet this lower standard. In 1997, 25 percent of the vehicles would have to meet a tightened standard of 0.075 grams per mile or equivalent. The percentage would increase thereafter, reaching 100 percent in 2000. Since many gasoline-powered vehicles could not easily meet this 0.075 standard (currently, none can), automakers will have an incentive to use alternative fuels.

In addition, beginning in 1995, some unspecified proportion of the low-emitting vehicles sold would have to meet an even more stringent standard of 0.04 grams per mile or the equivalent, a standard unlikely to be met by gasoline-powered vehicles in the foreseeable future (although CARB claims the contrary). It is proposed that these ultra-low emitting vehicles account for 2 percent of vehicle sales in 2000, 15 percent in 2003, and 100 percent at some future date to be established in future rulemaking.

In this proposal, CARB would allow manufacturers to bank emission reduction credits, and to trade them with other vehicle suppliers—and, in a late reversal of position as this paper goes to press, to allow limited emissions averaging across vehicle fleets. Tightened uniform standards would also be set for nitrogen oxide and formaldehyde emissions, but these would not be subject to trading, banking, or averaging.

On the fuel side, CARB proposed, beginning in 1997, that each gasoline refiner in the state be required to supply a specified amount of *liquid* clean fuel, "liquid" defined to include alcohols and LPG, based on their total fuel sales. These mandated sales would be determined so as to match the total fuel demanded by the low-emitting vehicles on the road. Refiners would be allowed to sell and buy clean liquid fuels that exceed the minimum requirements. Refiners would specifically be exempted from requirements to sell gaseous fuels and electricity, on the premise that enough incentive will already exist to encourage natural gas and electric vehicles, that refiners' facilities are not suited to selling nonliquid fuels, and that natural gas and electricity suppliers are monopolies.

The CARB initiative is bold and innovative, but falls far short of the comprehensive incentive-based approach outlined above. The CARB proposal gives oil refiners clean-fuel credit only for LPG and methanol, not natural gas and electricity, does not establish a mechanism that allows the development of region-specific strategies or the incorporation of non-ozone goals, wavers on the issue of emission averaging, and does not provide a mechanism for trading between vehicle and fuel suppliers. Nonetheless, the CARB proposal is a first step at moving away from command-and-control rules toward an incentive-based approach.

Industry Concerns

For an incentive-based program to be implemented successfully, it must balance the concerns and interests of fuel and vehicle suppliers with the overall societal good. Some of those concerns and interests, and possible responses, are discussed below.

Automakers are concerned that consumers will not buy a vehicle that differs from a conventional gasoline vehicle; as a result they prefer liquid fuel vehicles. They hope to meet their alternative-fuel responsibilities relatively easily and inexpensively by building multifuel vehicles that operate on methanol and gasoline. Once purchased, these vehicles may be used as conventional gasoline vehicles, thereby presenting no marketing risk to automakers. Incentives must therefore be designed to allow the use of multifuel vehicles, but should expedite the transition to cleaner, more energy-efficient and less costly single-fuel vehicles by heavily favoring single-fuel vehicles in the equivalency ratings and indices.

The oil companies face a far greater potential risk than vehicle suppliers. They must invest in methanol fuel at a cost of over a billion dollars per plant, with no assurance that consumers will buy the fuel, or allow natural gas and electricity to enter the transportation fuels market, thereby threatening the oil companies' market share. As a result, oil companies have campaigned hard against alternative fuels, even as they hedge their bets by participating in government methanol programs in California and elsewhere. Mobil Oil, historically the most outspoken of the oil companies, mounted a national media campaign in August and September 1989 with large ads in *Time* magazine, the *New York Times*, and other influential publications opposing and even ridiculing the proposal to mandate alternative fuels. It argued that methanol was toxic, did not improve air quality, worsened the trade deficit, and was expensive.

The oil industry quickly moved beyond this initial tirade to embrace reformulated gasoline, first introduced commercially by ARCO. ARCO, with its prime market in southern California where the pressure to reduce air pollution is strongest, was more subtle and more effective in its opposition. It argued, again in a national media blitz, that alternative fuels were unnecessary since gasoline could be reformulated to emit fewer pollutants. In September 1989 it introduced a reformulated gasoline fuel, but only in southern California and only as a replacement for leaded gasoline. Indeed, ARCO and the oil industry in general have argued on behalf of reformulated gasoline as a superior "alternative" fuel, but have been evasive about the emission characteristics and costs of reformulated unleaded gasoline, saying they are studying the question. (One is left to muse why it took them 20 years to initiate such a study.) This attempt to characterize reformulated gasoline as an alternative fuel must be acknowledged in the development of ratings and indices, but again care must be taken to assure that the fuel ratings and indices appropriately reward the greater social benefits of alternative fuels.

The central industry concern, underlying the auto industry preference for liquid fuels and the oil industry preference for reformulated gasoline, is the very real lack of coordination between the two industries. Would there be fuel available if the auto industry were to sell natural gas and methanol vehicles? Would there be vehicles available to consume the natural gas or methanol if investments were made to sell those fuels? This uncertainty about the other's marketing plans creates huge risks.

In theory, an incentive-based regulatory program would resolve this uncertainty via the workings of the artificial markets. For instance, if the market value of permits for natural gas fuels drops because natural gas vehicles are not being manufactured and there is no market for the fuel, then there would be an incentive for an entrepreneurial oil (or automotive) company to buy up those credits, and to subsidize the manufacture of natural gas vehicles. In practice, it may be necessary initially for the administrative agency temporarily to use command-and-control rules to assure adequate matching supplies of fuels and vehicles. This, indeed, as described above, is what CARB is proposing to do.

The initially high level of uncertainty associated with alternative fuels for both fuel and vehicle suppliers will undoubtedly cause auto and oil companies to focus initially on improving existing engines and fuels. But gradually, following the most cost-effective path, which will differ for each company and perhaps for each region, they will move toward cleaner-burning, more socially desirable nonpetroleum fuels.

Conclusions

There is no analytical basis for definitively determining which fuel is superior and when it should be introduced. The choice depends upon one's values, forecasts of future energy prices, political events, technological advances, and increased knowledge about the greenhouse effect. Nonetheless, choices must be made with incomplete knowledge and limited foresight.

What is needed, and what will best serve us in the long run, is the establishment of an institutional framework that is flexible in responding to new information, shifting values, and beliefs; that incorporates multiple social goals; and that is amenable to region-specific initiatives.

Since no one can claim to know what fuels will be best in the next century and at what rate we should move toward those fuels, the focus of government efforts should be on providing incentives to push industry and consumers in the correct direction—the production and use of vehicles and fuels that are more environmentally benign, safer, and less threatening to our national security; later, mid-course corrections can be made.

Since the US dramatically expanded its commitment to social-style regulation in the 1960s, especially in terms of regulating pollution and safety, we have become much more sophisticated and experienced at how to best regulate the manufacture and use of goods. The time has come to make use of that knowledge and experience, to move beyond simplistic, fragmented efforts at regulation.

Creation of an incentive-based regulatory system does not impose a new layer of regulations on business. Emissions are already regulated and government is obligated one way or another to make certain that the external benefits of alternative fuels are taken into consideration. Rather than imposing highly specific regulations on vehicles and fuels that severely constrain industry initiatives, as occurred historically, an incentive-based regulatory program should be seen as creating a market for social goods, providing industry with the flexibility to unleash its resources.

With a flexible regulatory framework in place, legislators and regulators could make modifications over time: for instance, to weight greenhouse gases more heavily, to drop incentives for energy security if free trade becomes more popular, or perhaps to discourage biomass fuels if soil erosion becomes more critical. The specification of fuel indices and emission standards will not be straightforward, and will be the focus of considerable debate. The advantage of this approach, however, is that the debate is highly focussed and directly addresses specific tradeoffs. Without this structure, working only with the current system of uniform emissions standards along with a potpourri of policy instruments that might influence the introduction of alternative fuels, the debate undoubtedly will continue to degenerate into a cacophony of self-serving interest group arguments.

In conclusion, an incentive-based approach is not a panacea, nor will it be simple to administer. It probably would result in a prolongation of the gasoline era, by stimulating oil refiners to reformulate gasoline so that it is "cleaner," and by stimulating auto companies to use advanced catalytic controls to build cleaner-burning gasoline engines. But this incentive-based framework might result in even greater air pollution benefits in the short term, and almost certainly at lower cost. More importantly, it will provide the structure for guiding the long term transition to cleaner and more socially desirable fuels and vehicles. Another way of thinking about the advantages of an incentive-based approach is that this regulatory approach will prevent government from mandating which fuels should be introduced and when, thereby eliminating the likelihood of expensive mistakes.

In a larger sense, we have come to a crossroad in dealing with pollution and energy use in the transportation sector. We must acknowledge the shortcomings of the command-and-control approach of social regulation that emerged in the late 1960s and 1970s and has dominated pollution control efforts since. Industry, government regulators, and even environmental groups have become accustomed to the certainty that that system provides. It is an anachronism not suited to the needs of the 1990s and beyond. It would be ironic indeed if, as the Soviet Union and Eastern Europe turn away from command-and-control techniques, we were to strengthen our embrace. Although the move away from uniform emission standards will be unnerving to those involved, it should be made—gently, but decisively.

Acknowledgments

This paper is based on research funded by the University of California's Policy Seminar, Transportation Center, and Energy Research Group.

References

California Advisory Board on Air Quality and Fuels, Report to the California Legislature, Vol. 1, Executive Summary (1989).

California Air Resources Board, *Emission Inventory, 1983*, Sacramento, CA (1986).

Cook, B., *Bureaucratic Politics and Regulatory Reform: The EPA and Emissions Trading*, Westport, CT, Greenwood Press (1988).

Daly, H. (1976), cited in A. Lovins, *Soft Energy Paths: Toward a Durable Peace*, New York, Harper & Row, 1977, p. 68.

DeLuchi, M.A., "Hydrogen Vehicles: An Evaluation of Fuel Storage, Performance, Safety, Environmental Impacts, and Cost," *Int. J. Hydrogen Energy* 14, pp. 81–130 (1989).

DeLuchi, M.A., Sperling, D., and Johnston, R.A. *A Comparative Analysis of Future Transportation Fuels*, Univ. of Calif. Inst. of Transportation Studies, Berkeley, CA, UCB-ITS-RR-87-13 (1987).

DeLuchi, M.A., Johnston, R.A., and Sperling, D. "Transportation Fuels and the Greenhouse Effect," *Transportation Research Record* 1175, pp. 33–44 (1988).

DeLuchi, M.A., Johnston, R.A., and Sperling, D., "Methanol vs. Natural Gas Vehicles: A Comparison of Resource Supply, Performance, Emissions, Fuel Storage, Safety, Costs, and Transitions," *SAE* 881656 (1988).

DeLuchi, M. A., Wang, Q., and Sperling, D., "Electric Vehicles: Performance, Lifecycle Costs, Emissions, and Recharging Requirements," *Transportation Research* 23A: 3, pp. 255–278 (1989).

Difiglio, C., Duleep, K.G., and Greene, D.L., "Cost-effectiveness of Future Fuel Economy Improvements," *The Energy Journal* 11:1, 65–86 (1990).

Fraas, A., and McGartland, A., "Alternative Fuels for Pollution control: An Empirical Evaluation of Benefits and Costs," *Contemporary Policy Issues* Vol. VIII (1990).

Gray, C.L. and Alson, J.A., "The Case for Methanol," *Scientific American*, pp. 108–114 (1989).

Hahn, R.W., "Economic Prescriptions for Environmental Problems: How the Patient Followed the Doctor's Orders," *Journal of Economic Perspectives* 3:2, 95–114 (1989).

Hahn, R.W. and Hester, G.L., "Marketable Permits: Lessons for Theory and Practice," *Ecology Law Quarterly*, 16, pp. 361–406 (1989).

Harris, J.N., Russell, A.G., and Milford, J.B. "Air Quality Implications of Methanol Fuel Utilization," *SAE* 881198 (1988).

Harvey, G., Unpublished analysis for Metropolitan Transportation Commission, Oakland, CA (March 1990).

Hempel, L.C., Press, D. et al., *Curbing Air Pollution in Southern California: the Role of Electric Vehicles*, Claremont Graduate School, Claremont, CA (1989).

McElroy, K., Hayes, R., and Olson, A. (1984), "Cost Savings from Emission Averaging for Automobiles," in *Proceedings of APCA Specialty Conference on Mobile Source Issues in the 1980s*, pp. 73–84 (February 1984).

Ogden, J.M., and Williams, R.H., "Solar Hydrogen: Moving Beyond Fossil Fuels." World Resources Institute, Washington, DC (1989).

Portney, P. R., Harrision, D., Krupnick, A., and Dowlatabadi, H. "To Live and Breathe in L.A.," *Issues in Science and Technology*, p. 70, Summer (1989).

Russell, A.G., St. Pierre, D., and Milford, J.B., "Ozone Control and Methanol Fuel Use," *Science* 247: 201–205 (1990).

Sperling, D. *New Transportation Fuels: A Strategic Approach to Technological Change*, University of California Press, Berkeley, CA (1988).

Sperling, D., ed. *Alternative Transportation Fuels: An Environmental and Energy Solution*, Quorum Books/Greenwood Press, Westport, CT (1989).

Sperling, D., Hungerford, D., and Kurani, K. "Consumer Demand for Methanol," in Will Lohl (ed.), *Methanol as an Alternative Fuel Choice: An Assessment*, Johns Hopkins Univ. Press, Baltimore, MD (1990).

Sperling, D. and DeLuchi, M.A., "Is Methanol the Transportation Fuel of the Future?" *Energy*, pp. 469–489 (1989).

US Congress, Office of Technology Assessment, *Catch Our Breath: Next Steps for Reducing Urban Ozone*, OTA-O-412, Government Printing Office, Washington, DC (1989).

Wang, Q., DeLuchi, M.A., and Sperling, D. "Emission Impacts of Electric Vehicles," *JAPCA*, forthcoming (1990).

Transportation and the Environment in the European Economic Community

Bert Metz
Royal Netherlands Embassy
Washington, DC

Abstract

The role of the European Economic Community (EEC) in addressing the environmental problems created by our increasing transportation needs is described. A brief overview is given of the EEC institutions and legal instruments. The specific conditions with respect to transportation and the environment in the EEC are compared with those in the United States. Two cases are discussed to illustrate the European approach. First, the case of the EEC decisionmaking about car emission standards is presented, which illustrates the slow response of the EEC. Second, an example is given of a comprehensive policy response in one of the Member States, the Netherlands, that might indicate the future direction of the EEC.

Introduction

The European Economic Community has gradually gained influence on the environmental policies in the Member States. With the upcoming unification of the internal market in 1992, new challenges are emerging to protect the environment in view of the expected economic boom and the accompanying growth in transportation of goods and people. This paper will look at some of the present European policies in the field of transportation and environment, including the European struggle with the introduction of strict auto emission standards. It will also explore some of the trends for the future, mainly by discussing the policies now emerging in one of the Member States, the Netherlands, where pressure on the environment from transportation has risen to such an extent that major changes in policy were required. First, a brief description will be given of the EEC institutions, procedures, and policies with respect to the environment.

What Is the European Economic Community?

The European Economic Community consists of 12 West European countries (see Figure 1), with a total population of about 325 million and a combined gross domestic product (GDP) of US$3.5 trillion (US GDP = $4.2 trillion) (US Department of Commerce, 1989). The foundation of the EEC was laid by six countries with the treaty of Rome, signed in 1957. The other six joined at a later stage. In 1987, this treaty was replaced with the Single European Act. At that moment, the decision was taken to realize the unified internal market at the end of 1992 by removing all obstacles for movement of goods and people. The Single European Act formalized the responsibilities of the Community with respect to the protection of the environment, although since 1973 those aspects had already been included in the Community policies using the flexibility of the old treaty.

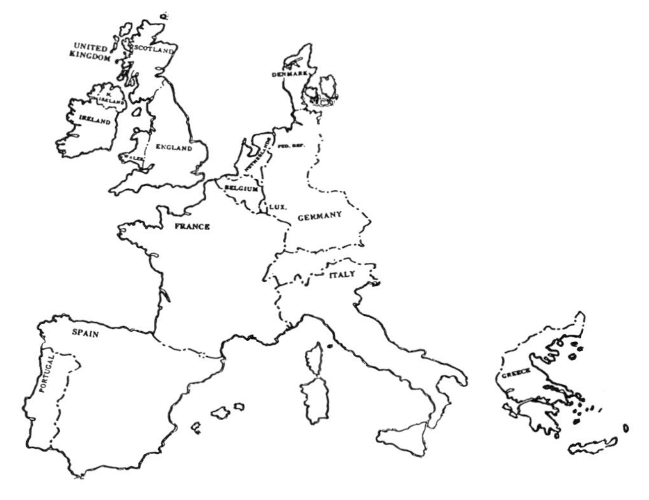

Figure 1. The Member States of the EEC.

The year 1992, therefore, is no more than one of the milestones on the road to a developing EEC environmental policy.

The EEC is more than a group of independent nations bound by their membership in an international organization. It is not a federation, like the United States, either. At this moment the situation is somewhere in between, with a gradual but persistent development towards more federal structures. There are specific tasks allocated to the Community and institutions are created to carry out those tasks (Haig, 1989).

With respect to environmental policies, the European Commission, the executive board of the Community consisting of 17 individuals appointed by their national governments, has authority to propose legislation and has the responsibility to implement Community decisions. The Commission has an administrative apparatus divided into 22 Directorates-General (DG). Separate DGs exist for environment, energy, and transportation.

Decisions are taken, as in all other policy areas where the Community has authority, by the Council of Ministers, where all member countries are represented with one Minister. For environmental matters these are the respective Ministers of Environment and similar Councils are held on transportation and energy matters. Decisions are prepared by a Committee of Permanent Representatives (COREPER), consisting of Ambassadors of the Member States to the EEC, in consultation with their national governments and the Commission. Decisions in the Council take place either by so called "qualified majority" or by unanimity, depending on the legal basis for the decision.

Decisions based on article 100 A usually are related to matters of interstate commerce. This is, for instance, the case with product standards, such as car emission standards, that will be discussed in more detail later. This article specifies that the Commission has to base its proposals on a so called high level of environmental protection. The majority rule applies in this case, which means that 54 out of the total of 76 votes are required for a decision to be adopted. Each Member State has a number of votes, in principle in proportion to its population, but the smaller countries having proportionally more in order to prevent a few big countries overruling the others (see Table 1). Inherent to decisions based on article 100 A is the idea that individual Member States cannot usually enact stricter laws. In other words these decisions preempt national legislation, unless unacceptable environmental damage in a Member State can be demonstrated, in which case unilateral action can be justified.

Article 130 R, S, and T is reserved for other environmental matters, where internal market aspects are not as prominent, such as emission standards for industrial processes, ambient air quality standards, wildlife protection, etc. Decisions based on these articles have to be taken by unanimity, but individual Member States can enact stricter legislation, unless unacceptable distortions in the internal market occur. The procedures are in fact the mirror image of the ones based on article 100 A.

Table 1. Votes for each Member State; 54 votes for qualified majority.

Country	Votes
Belgium	5
Denmark	3
West Germany	10
Greece	5
Spain	8
France	10
Ireland	3
Italy	10
Luxemburg	2
The Netherlands	5
Portugal	5
United Kingdom	10
Total	76

The European Parliament (EP), the members of which are elected by direct voting in each Member State, has only limited powers at this point. Before the Single European Act came into force, only an advisory role could be played. Proposed legislation by the Commission was sent to the EP for comment. The Commission would look at the comments, could amend its proposal or not and send it with the EP opinion to the Council for decisionmaking. The Council then decided, modifying the Commission proposal if need be. Since the European Single Act came into force, the EP can amend or reject—by absolute majority—a Council decision based on article 100 A, after which the Commission has to review the decision and put new proposals before the Council of Ministers. The Council can only change a Commission proposal then with unanimity. For decisions based on article 130 the EP still has the more limited advisory powers. Figure 2 tries to summarize this complicated scheme.

There are three types of legal instruments in the EEC at present:

- **Directives,** which require Member States to achieve the prescribed results via implementation in their national legislation. This is the major instrument used for environmental purposes.
- **Regulations,** which are directly applicable law in the Member States. This instrument has so far been used sparsely for environmental matters. The only examples are the regulation on the trade in endangered species and regulations regarding the assessment of forest damage due to air pollution.
- **Decisions,** which are binding on those to whom the

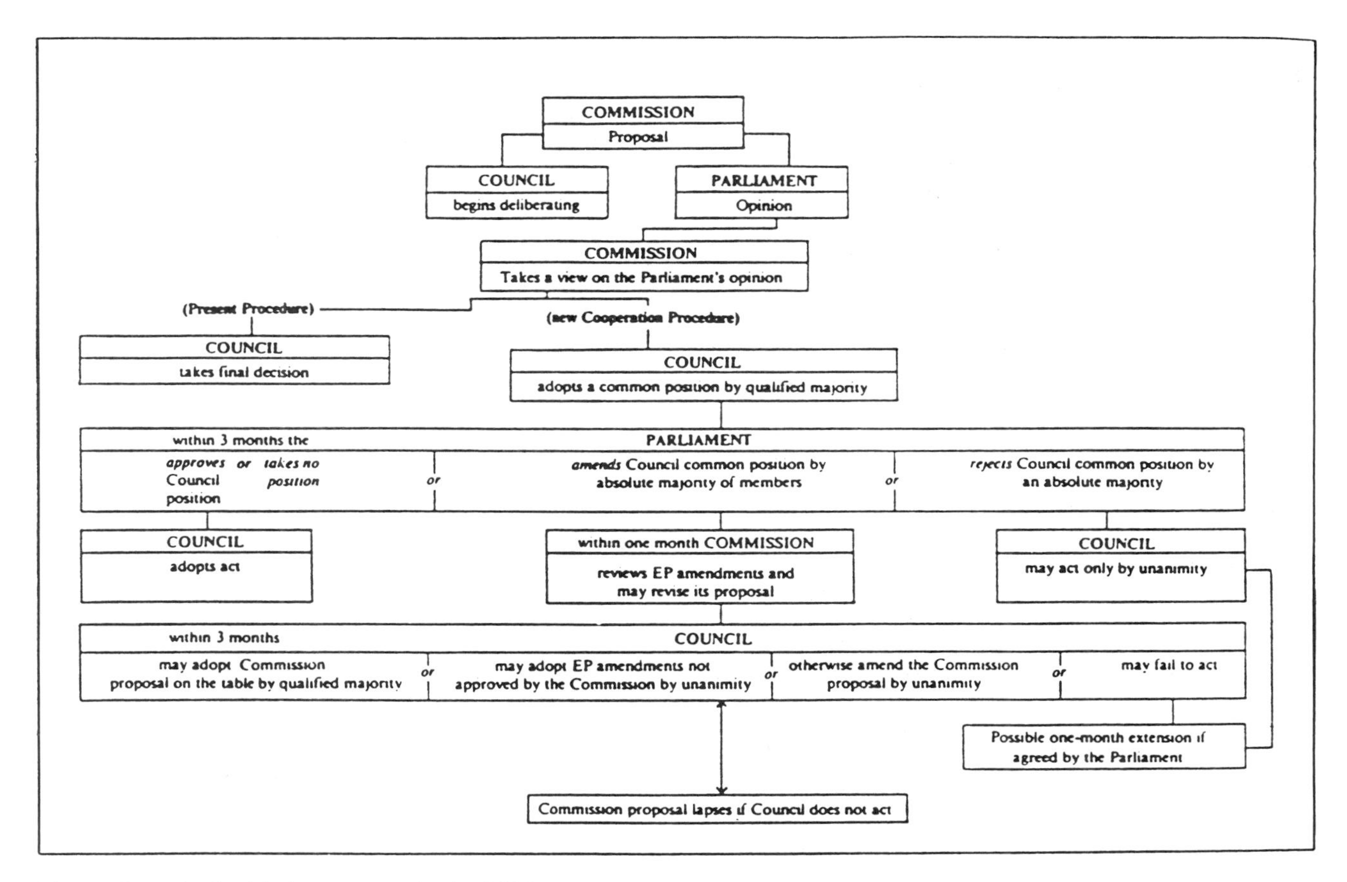

Figure 2. The legislative process in the EEC.

decision is addressed. This instrument has been used predominantly in procedural matters.

Besides these formal instruments, the Council can make recommendations, which have only an advisory character and are not legally binding.

So far more than 150 directives, decisions, and regulations in the environmental area have been published, covering almost all aspects of environmental policies.

Enforcement powers of the Commission are limited. The primary instrument is overseeing the implementation of directives in the national legislation of the Member States and to file suit before the European Court of Justice in case implementation has not been effected in time. The effectiveness of the implementation and the compliance with the national legislation is not being looked at actively by the Commission so far. The Commission is not equipped for that. An inspectorate or compliance monitoring organization does not exist. In case of complaints of citizens, however, the Commission does investigate cases where Member States allegedly are not enforcing the practical implementation of directives.

Since 1973, the EEC has adopted Environmental Action Programs that describe the priority areas for Community Environmental Policy. The present Environmental Action Program covers the period 1987–1992. The main principles of the European environmental policies as indicated in this plan are:

- integration of environmental policy with other areas of Community policy, such as industry, agriculture, energy, trade, and transport;
- environment and economy can go together; a balanced approach can effectively protect the environment and guarantee economic growth in the long term;
- the polluter pays—the use of economic instruments for protecting the environment is essential;
- information and education are vital.

The Fourth Action Program focuses on:

- Pollution prevention
- Source reduction
- Multimedia approaches
- Protection of environmental resources
- International cooperation

The United States and the European Economic Community Compared

To give an idea of the situation with respect to transportation, energy, and the environment in the EEC compared to the US, some data might provide a useful illustration. The transportation energy use per capita, the number of vehicles per person, and the vehicle miles driven per person are much higher in the US than in the EEC. Even expressed per unit of GDP this is still the case. The density (per square kilometer) of energy use, transportation, and vehicles is, however, much higher in the EEC. The same applies to the pollutants associated with transportation, such as NO_x and Volatile Organic Compounds (VOC) (see Tables 2 and 3). Fifty-five percent of NO_x emissions in the EEC is from traffic versus 45 percent for the US. Together with SO_2, NO_x is responsible for acid deposition, one of the most serious environmental problems in Europe. This gives a good indication of the relative efficiency of transportation energy use on the one hand and the high pressure on the environment due to transportation in the EEC on the other.

As far as fuel efficiency of vehicles is concerned, the EEC is in better shape than the US (see Table 4). The average North American car uses twice as much fuel each year due to this effect and more driving (International Energy Agency, 1989). The reason for this difference in fuel efficiency is the traditional considerable difference in fuel prices due to state taxes, up to four times those in the US (see Table 5). Average efficiency is boosted by a substantial share of diesel in the fuel mix in Europe due to extensive use in the passenger vehicle fleet and the virtual absence of light trucks which are so popular in the US (Renner, 1989).

Public transport is more developed in the EEC. The modal split (fraction of passenger-kilometers served by respective transportation mode) for rail and bus transport in the EEC is several times higher than the 0.02 in the US (US Department of Commerce, 1989). Rail and

Table 2. Comparison between EEC and USA, 1987 data (OECD, 1989, IEA, 1989).

Indicator	EEC	USA
Population (Million)	322	239
Area (Million km^2)	2.25	9.37
Population Density (#/km^2)	143	26
Road Network (Million km)	2.59	6.65
Highways (Thousand km)	30.8	82.3
Passenger Cars (Million)	117	138
Pass. Car Density (#/km^2)	52	15
Trucks (Million)	12.9	41.2
Total Vehicle Kilometers (Billion)	1695	3001
Total Veh. km per Capita (Thousand)	5.3	12.6
Total Veh. km per km^2 (Thousand)	753	320
Total Pass. Veh. km (Billion)	1399	2133
Total Truck Veh. km (Billion)	275	859
Transportation Energy Use (Mtoe)	211	465
Transportation Energy per Unit GDP (Mtoe/$1000)	60	111
Transportation Energy per Capita (Toe/capita)	0.66	1.95
Transportation Energy Use Density (Mtoe/km^2)	94	50

Table 3. Pollution load from transportation, 1986 data (OECD, 1989).

Indicator	EEC	USA
Transport NO_x Emissions (10^6 T/yr)	6.1	8.5
Transport NO_x Em. per Capita (T/*cap*/yr)	0.02	0.04
Transport NO_x Em. per km^2 (T/km^2)	2.7	0.9
Transport VOC Emissions (10^6 T/yr)[1]	4.2	8.2
Transport VOC Em. per Capita (T/*cap*/yr)[1]	0.01	0.03
Transport VOC Em. per km^2 (T/km^2)[1]	1.9	0.9

[1] 1980 data

Table 4. Urban fuel efficiency of new passenger cars (Flavin and Durning, 1988).

Country	Fuel Efficiency (miles/gallon) 1973	1985
Denmark	26	33
West Germany	23	31
Italy	28	30
United Kingdom	21	31
United States	13	25

Table 5. Gasoline prices and taxes, 1987 (Flavin and Durning, 1988).

Country	(US$/Gallon) Price (+ tax)	Tax
United States	0.94	0.29
West Germany	2.31	1.34
United Kingdom	2.39	1.53
France	3.06	2.32
Denmark	3.76	2.93
Italy	3.76	2.78

shipping have an important share in the freight transport market.

Present EEC Policies and Trends

Although Western Europe still has an advantage over the United States in terms of its transportation infrastructure and its transportation patterns, the trend in the past decades has been very clearly in the direction of a stronger dependency on private vehicles, a strong growth in and a shift towards freight transport by trucks, a reduced role of the public transportation systems, and an enormous increase in the pressure on the environment from the transportation sector. Just recently, Austria, the preferred passage from West Germany to Italy, closed its roads for heavy and noisy trucks during the night, because the noise levels in the residential areas close to the main highways had become unbearable.

Only in the last five years have serious attempts been made to turn this trend around. Car emissions are being addressed, more than 10 years after the US took serious action. The latest struggle about the emission standards for small vehicles—which will be discussed in more detail later—finally ended with the adoption of US equivalent standards for the 1993 model year about six months ago. Standards for noise levels of cars and trucks have been set by the Community as well.

In France and West Germany, rapid train systems were developed, greatly increasing the speed of long distance travel by train. The French TGV ("Train à Grande Vitesse"), a high speed electric system using traditional rail, became operational about five years ago. Decisions have been made by the EEC Member States for a European rapid train network. The opening of the "chunnel" (train-tunnel from France to Great Britain under the Channel) will be an important contribution to the strengthening of the rail system in Europe. The European Commission has recently completed a proposal for a European Railway Policy (European Commission, 1989b) that aims at harmonizing the railway infrastructure in the various Member States, encouraging the development of a high-speed rail network, and an increased role of freight transport by rail via an integrated road-rail approach for long distance transport.

The European internal market that will come into effect at the end of 1992 will lead to a strong growth in transportation of goods and people. An EEC Task Force recently reported that the pressures on the environment in the EEC will greatly increase and that gains being made via the introduction of cleaner vehicles will likely be more than compensated by the increase of transportation volumes (European Commission, 1989a). The EEC has no comprehensive answer to these disturbing developments as yet. The burden is on the individual Member States to develop their own policies.

The Car Emissions Standards Case

In order to illustrate the way the EEC is handling the problems of addressing environmental issues in a multinational setting, the case of the European car emission standards will be discussed in some more detail. As indicated earlier, this was in fact a rearguard action, since the US, Japan, and even a number of non-EEC European countries adopted strict emission standards long ago.

Discussions in the EEC on car emissions have been ongoing for a long time. In 1970, the first EEC Directive, based on recommendations of the United Nations Economic Commission for Europe (UNECE), was adopted. While the US gradually adopted stricter emission standards for cars during the seventies, nothing happened in Europe. In 1983, the EEC directive was revised, but standards were way above the US values. The use of catalytic converters was not required.

During 1984 and 1985, the serious problems of acid rain and photochemical air pollution led to an intense debate between the European Commission and the Member States about stricter standards. Most European auto manufacturers, especially those in Great Britain, France, and Italy predominantly making small cars, had invested considerable research funds in lean-burn technology, which they considered to be a technically superior method to reduce emissions. No wonder that these countries resisted the adoption of standards that would require catalytic converters. The West German industry tended to favor the use of catalysts. So West Germany, together with Denmark, Greece, and the Netherlands—countries without substantial car manufacturing—pushed for strict standards. Along the same lines, there was the health and ecological argument about the damage done by the car emissions versus the economic argument that the European manufacturers of small cars would weaken their competitive position versus the already powerful Japanese. The addition of catalytic converters would have had a relatively strong effect on the cost of the small vehicles, without the economies of scale that benefit the Japanese. Another argument that was used by the Commission was that the situation in Europe was so different from that in the US with respect to vehicle size, vehicle use, and driving conditions, that the comparison should not be made on the basis of emissions per vehicle, but on overall emissions from mobile sources. A complicated reasoning based on this approach led to the conclusion that Europe could do without catalytic converters.

A compromise was reached based on a division of cars into three engine categories: large cars, over 2 liters engine volume; small cars, under 1.4 liters; and medium cars in between. The Commission proposal was to regulate the large cars (market share about 3 percent) fairly strictly (but not as strictly as the US standards) as of the 1989 model year, requiring the use of three-way catalysts. Medium size cars (market share 31 percent) were to be regulated more loosely, requiring only the use of cheaper oxidative catalysts, to be introduced in the 1992 model year. Small cars, with about 50 percent of the market, were regulated only minimally as of the 1989 model year, with the intention to reduce the standards at a later stage, leaving open the option of lean-burn technology. Diesel fueled passenger vehicles (market share 13 percent) were also subject to the same standards. Due to the required unanimity at that time, the proposal—the so called "Luxemburg Compromise"—did not make it because Denmark refused to agree.

After the Single European Act had come into effect, the Luxemburg Compromise was adopted after all in 1987 by the Council by qualified majority (see Table 6). Denmark objected again. Immediately the fight about the second phase of the small car standards started. The same division as before took place among the Member States. West Germany, Denmark, Greece, and the Netherlands were pushing for US-83 equivalent standards (five grams per test for NO_x + HC). Great Britain, France, and Italy wanted to keep the lean-burn option open (12 grams per test). Finally the Commission proposal (eight grams per test) was adopted, after West Germany had reluctantly given up its resistance. The so called "blocking minority" disappeared.

Strange things happened then. The Netherlands and Greece vowed to keep their tax credit programs for "clean" (US equivalent) cars, which were introduced in 1986 to stimulate the purchase of those cars by making them about as expensive as the "dirty" ones. France objected strongly, accusing those countries of threatening the internal market and thereby damaging the European auto manufacturing industry. It pulled out of the agreement, thereby in fact, but not formally, nullifying the Council decision. The Commission threatened to file suit against the Netherlands before the European Court of Justice if the tax credit program was continued. The Netherlands refused, arguing that article 100 A allows for unilateral action if there is a serious threat to the environment. Besides, the European Court had earlier decided in favor of Denmark in a similar case, where Denmark had introduced a bottle bill applying also to imported beer and soft drinks. This encouraged the Dutch government to stick to its position (Ministry of Housing, Physical Planning and Environment, 1988).

At about the same time that the Commission was deciding on the law suit against the Netherlands, the European Parliament, in its second reading of the earlier Council decision, amended that decision by an absolute majority to the five grams per test value. The Commission decided not to file suit against the Netherlands and to support this amendment and put it again before the Council, which then could only change this with unanimity. Given the upcoming elections for the European Parliament and the "green" tide rising in many Member States, the Council agreed. As of the 1993 model year, US-83 equivalent standards would be introduced (Haig, 1989). In December 1989, the last episode was written, when the Commission proposed to adjust the standards for the medium sized and large cars to US-83 equivalent levels as well (Europe: Environment, 1989). The three-way catalyst has finally been accepted in Europe.

Looking back at this story of epic proportions, a few remarks could be made about the importance of certain forces. Definitely the external pressure from the stricter standards elsewhere (US, later also Japan, Sweden, Switzerland, Austria, and Norway) played an important role. Public opinion and the influence of the European Parliament have also been major factors. Last but not least, the creative pressure brought about by the introduction of fiscal incentives in Greece and the Netherlands helped to change the EEC policy. It will be interesting to see what the EEC reaction will be when a revised Clean Air Act with drastic reductions in emission standards will be enacted this year in the US.

The Case of the Dutch Policy on Transportation and Environment

Within the EEC, the Netherlands is the country with the highest density of population and cars, higher even than

Table 6. EEC emission standards according to "Luxemburg compromise" (Ministry of Housing, Physical Planning and Environment, 1988).

		Standard (grams/test)		
Engine Volume	Date	CO	HC + NO_x	NO_x
>2000 cm^3	1988/1989	25	6.5	3.5
1400< <2000 cm^3	1991/1993	30	8	—
<1400 cm^3, 1st phase	1992/1993	45	15	6

Japan. Although the road network is excellent, the public transportation system is the most extensive of the EEC (share: 13 percent of passenger kilometers), bicycle use is widespread (share: 9 percent of passenger kilometers and in many cities more than 40 percent of the daily trips; see Table 7) (Kram and Okken, 1989), and road traffic is very congested. Predictions for passenger vehicle use in the future show a 70 percent increase by the year 2010 (Ministry of Transportation, 1988). Freight transport is expected to grow 70 to 80 percent in that period. At this moment traffic contributes 18 percent to the alarming levels of acid deposition, 45 percent to emissions of hydrocarbons involved in (excessive) ozone formation, and 15 percent to CO_2 emissions that are too high already, and traffic is responsible for most of the serious noise problems in residential areas and natural habitats (Langeweg, 1989). If present trends continue, clearly the Dutch environment will not be able to cope with the expected amount of road traffic.

During the past few years, a drastic reorientation took place in the policy of the Dutch government with respect to transportation and the environment. In March 1988, a new long-term Policy Plan on Land Use Planning was adopted (Ministry of Housing, Physical Planning and Environment, 1988b). This plan calls for a drastic reduction in commuting by adjusting commercial and residential building locations and improving public transportation systems in order to guarantee a high quality environment for living and doing business and at the same time maintain the traditionally strong transportation and distribution function of the country. In November 1988, a new Policy Plan for Traffic and Transportation was issued (Ministry of Transportation, 1988). For the first time, this plan explicitly makes the environment one of its specific boundary conditions. It defines targets for drastic reductions in passenger vehicle use, improvements in public transportation systems, and shifts towards rail and water transport of freight. In May 1989, a long-term National Environmental Policy Plan (NEPP) was published, covering the period up to 2010 and based on principles of sustainable development (Ministry of Housing, Physical Planning and Environment, 1989). This plan defines targets for massive reductions of vehicle emissions and presents a multi-sectoral set of strategies to reach those targets. Elements of the land-use planning and transportation plans were updated and integrated in this environmental plan. After a political crisis around one element of the financial budget for this environmental plan, which led to the fall of the government (the first government ever to fall on an environmental issue!), new elections took place and a new government was formed in November 1989. The new government promised to strengthen the NEPP even further, because the latest data on both passenger and commercial vehicle use showed an unprecedented growth rate of 6 to 7 percent per year and targets for both acid deposition and greenhouse gas emissions were considered to be insufficient.

Table 7. Cycling as percentage of daily passenger trips (Lowe, 1989).

City	Percent of Daily Trips
Shengyang, China	65
Groningen, Netherlands	50
Beijing, China	48
Delft, Netherlands	43
Erlangen, West Germany	26
Odense, Denmark	25
Copenhagen, Denmark	20
Manhattan, USA	8

What are the policies on transportation and the environment in the Netherlands for the coming 20 years and how are they going to be implemented? Based on the overall targets for emission reductions necessary to achieve an acceptable environmental quality in the year 2010, a series of targets for the transportation sector was derived, involving emission reductions of about 75 percent for most pollutants and a drastically reduced growth of total vehicle miles traveled (see Table 8).

How are these targets to be met? A number of interrelated strategies were developed for that purpose.

Reducing Emissions Per Vehicle

Stricter Emission Standards. The passenger vehicles are covered under the EEC standards and the tax reduction schedule for clean cars as described above will stay in place until all new vehicles are covered under the revised standards. For trucks, a 50 percent reduction in emission standards will be pushed in the EEC. By the year 2010, a 75 percent reduction should be reached. Subsidies are available for development of cleaner trucks and buses and to encourage investment in cleaner vehicles. This is financed via an increase in the diesel fuel tax. In the year 2000, the total NO_x emissions from trucks must be 40 percent less than in 1986 and for passenger vehicles the reduction target is 75 percent.

Reducing Passenger Vehicle Commuting

Shifting Fixed Costs to Variable Costs. Fuel taxes will be raised, among others in the form of a CO_2 tax. Increases are, however, restricted by the harmonization requirements of the EEC. More effective will be a system of "road pricing" that will be introduced and that will enable flexible and variable toll rates via electronic payment systems. With this instrument, costs can be increased selectively in certain metropolitan areas. Re-

Table 8. Policy targets for traffic and transportation, The Netherlands (Ministry of Housing, Physical Planning and Environment, 1990).

	1986	2000[2]	2010[2]
NO_x passenger traffic[1]	163	40 (−75%)	40 (−75%)
NO_x goods traffic[1]	122	72 (−35%)	25 (−75%)
Hydrocarbons[1] passenger traffic	136	35 (−75%)	35 (−75%)
Hydrocarbons[1] goods traffic	46	30 (−35%)	12 (−75%)
CO_2[1]	24,000	24,000 (0)	21,600 (−10%)
Noise passenger cars[3]	80	74	70
lorries/buses[3]	81–88	75–80	70
number of houses[4]	260,000	130,000 (−50%)	
noise nuisance of any extent[5]	2,000,000	1,800,000 (−10%)	1,000,000 (−50%)

[1] NO_x, hydrocarbons and CO_2 in kilotonnes per year
[2] percentages of NO_x and hydrocarbons in relation to 1980
[3] target values for the maximum noise production of vehicles in dB (A)
[4] number of houses exposed to an unacceptably high noise level, reduced by 50 percent in 2000 through measures at source and in the transmission zone
[5] houses subject to noise loading of more than 55 dB (A)

stricting parking space and increasing costs of parking will also contribute. In addition, the current tax breaks for commuters—proportional to the commuting distance—which in fact encourage commuting, will be gradually abolished. Tax-free transportation allowances traditionally paid to employees have already been reduced through a revision in the income tax system.

Carpooling and Shuttle Services. The government is encouraging the introduction of "kilometer-reduction plans" (including carpooling programs) and shuttle services by businesses and institutions on a voluntary basis. Special van pools could, if subsidized like the public transport systems, reduce the overall passenger vehicle use about 5 percent (*Telegraaf,* 1989).

Improving the Public Transportation System

New Investments. An ambitious program of new investments in the infrastructure for bus, tram, and train systems has been laid out. In the next 20 years, about $6 billion will be invested. A high-speed rail system, connecting the major Dutch cities to a European network, will be built. These measures will have to increase the quality of the public transportation system to such a level that it becomes an attractive alternative to the use of the private vehicle. The railway systems "output" in the form of passenger kilometers—now 7 percent of the total (Netherlands Central Bureau of Statistics, 1989)—is scheduled to grow about 50 percent by the year 2005. In 1989, rail passenger kilometers were up 5.5 percent compared to the year before.

Integration of Ticketing System. For bus, tram, and metro services, a uniform ticketing system already exists. Further integration with the railway ticket system will take place. Recently in a number of cities special tickets went on sale that cover connecting taxi services to the final destination.

Subsidies. Government subsidies will remain substantial. At this moment, the average cost per kilometer for the public transportation system is about 17 $c, of which 10 $c is covered by government subsidies (*Telegraaf,* 1989).

Shift to Less Energy Intensive Transportation

Increase Bicycle Use. The already extensive provisions for bicycles (special bike paths, rent-a-bike systems at railway stations, take-your-bike train tickets, bike-parking facilities) will be improved. In this way, the bicycle could partially replace the car for short trips (see Figure 3).

High-speed Rail System. As indicated above, the European high-speed rail system can become a good alternative for air transport (see also Figure 3).

Avoiding New Developments That Generate Passenger Vehicle Use

Residential Developments. Existing policies of coordinating residential developments with public transportation infrastructure will be intensified.

Office and Recreational Locations. The national zoning laws will be used to prevent the construction of new office buildings at locations that are not close to major public transportation facilities. "High-visibility" locations along highways will no longer be allowed. Existing development plans will have to be reviewed as well (Ministry of Housing, Physical Planning and Environment, 1990). New recreation centers ("theme parks" etc.)

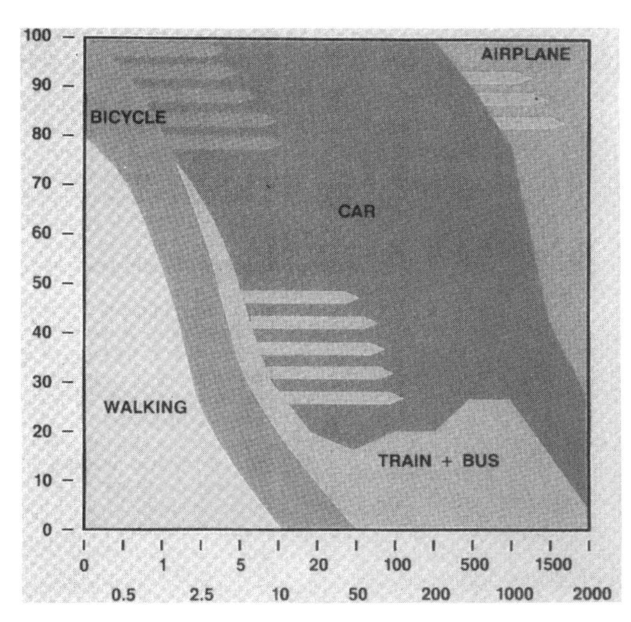

Figure 3. Modal split of passenger transport in the Netherlands (Ministry of Housing, Physical Planning and Environment, 1989).

will also be limited to locations with adequate public transportation facilities. With respect to major events, conditions will be attached to the required permits to provide transportation facilities.

This is just a sample of the major elements of the adopted strategies. There are scores of other supporting programs and projects, such as development of telecommuting, improving information systems for users of public transport, introduction of "traffic impact assessment" for urban planning purposes, education and information programs, etc. Figures 4, 5, and 6 show how the various policy elements contribute to the overall policy targets.

The financial implications of these plans are substantial. In the period 1990–1994, the additional costs for society (individuals, the business community, and government) will grow to about Dfl.1.5 billion per year (about US$0.8 billion). This means that between now and the end of 1994, about Dfl.400 (US$200) per person will be spent additionally. This is only 20 percent of the additional expenditures for all environmental programs together. By 1994, the total environmental expenditures will have doubled compared to 1989 and another doubling will be required to meet the 2010 targets.

It is obvious from these ambitious plans that substantial changes in human behavior will be required. Freedom to use the individual car will no longer be guaranteed, mainly due to greatly increased costs and

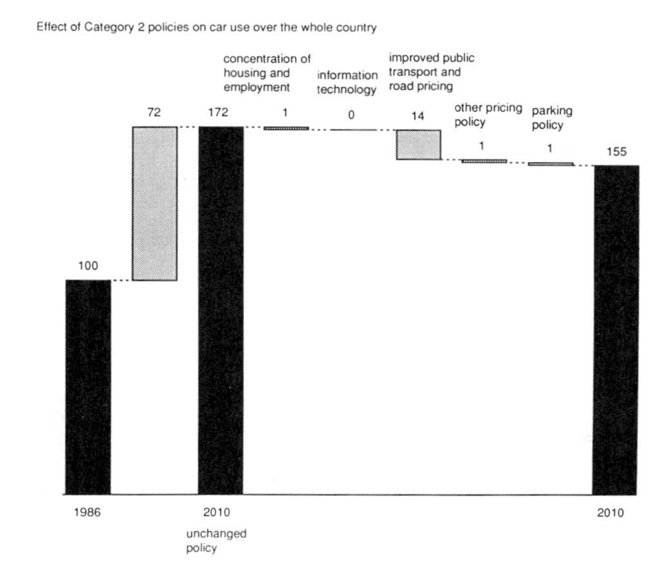

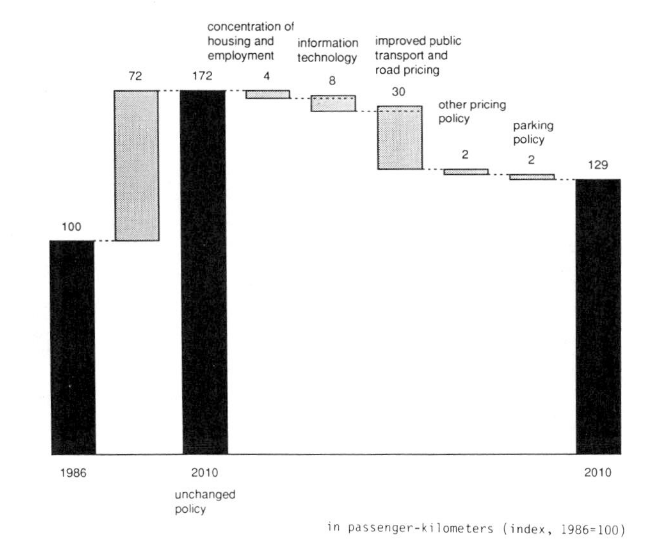

Figure 4. Effects of policies on car use in the Netherlands (Ministry of Transportation and Waterways, 1988).

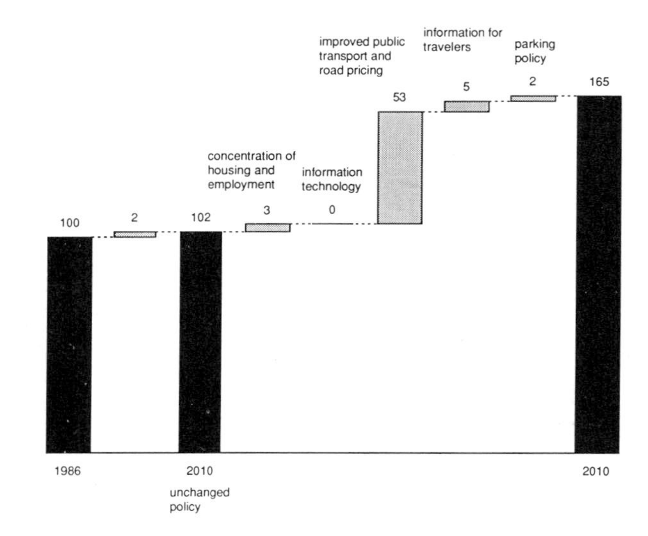

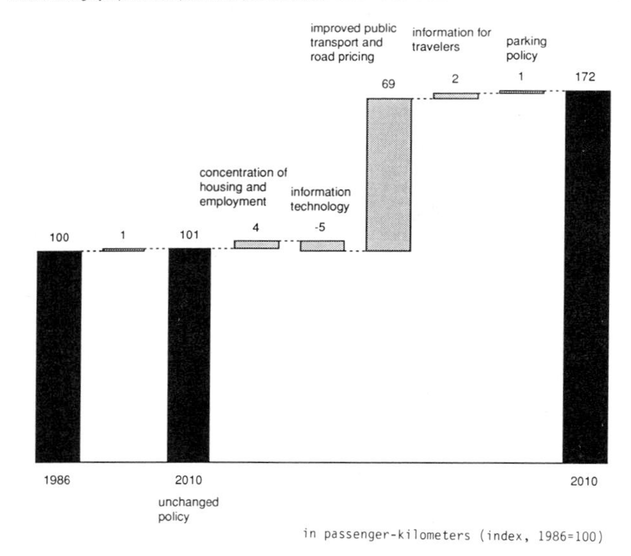

Figure 5. Effects of policies on use of public transport in the Netherlands (Ministry of Transportation and Waterways, 1988).

limiting of parking. The main emphasis, however, will be on providing attractive alternatives to induce people to choose a more environmentally sound means of transportation. The Netherlands will be a different country in 20 years time, and many other countries in the EEC and the rest of Europe will probably head in the same direction in the near future.

Conclusions

So far, the EEC has not found a comprehensive answer to the challenge of balancing the growing needs for transportation with effective protection of the environment. The density of the population and the accompanying economic activity in Europe put severe pressure on the environment. The struggle with adoption of strict car emission standards reflects the complexity of the EEC system and the growing pains of a federal system in its early developmental stage. However, in certain Member States, such as the Netherlands, the problem is being addressed with a multi-sectoral integrated policy approach. This may indicate the direction the EEC will follow in the future.

References

European Commission, *Europe 1992 and the Environment,* Report of a Task Force, Brussels (1989a).

European Commission, RAIL 1992, Press release, Brussels (November 22, 1989b).

Europe, Environment: The European Commission has drawn up proposals for more effective limitations on car exhaust pollutants (December 21 1989).

Flavin, C., Durning, A., *Building on Success: The Age of Energy Efficiency,* Worldwatch Paper 82, Worldwatch Institute, Washington, DC (1988).

Haig, Nigel, "The Environmental Policy of the European Community and 1992," *International Environmental Reporter* 12 (12), pp. 617–623 (1989).

International Energy Agency, *Energy Policies and Programmes of IEA Countries, 1988 Review,* OECD, Paris (1989).

Kram, T., Okken, P.A., Integrated Assessment of Energy Options for CO_2 Reduction, in: *Climate and Energy,* Okken, P.A., Swart, R.J., Zwerver, S., (eds.), Kluwer Academic Publishers, Dordrecht (1989).

Langeweg, K. (ed.), *Concern for Tomorrow—A National Environmental Survey 1985- 2010,* National Institute of Public Health and Environmental Protection, the Netherlands, Bilthoven (1989).

Lowe, Marcia D., *The Bicycle: Vehicle for a Small Planet,* Worldwatch Paper 90, Worldwatch Institute, Washington, DC (1989).

Ministry of Housing, Physical Planning and Environment, *Memorandum on Traffic and the Environment,* The Hague (1988a).

Effect of policy on emissions of nitrogen oxides from private cars

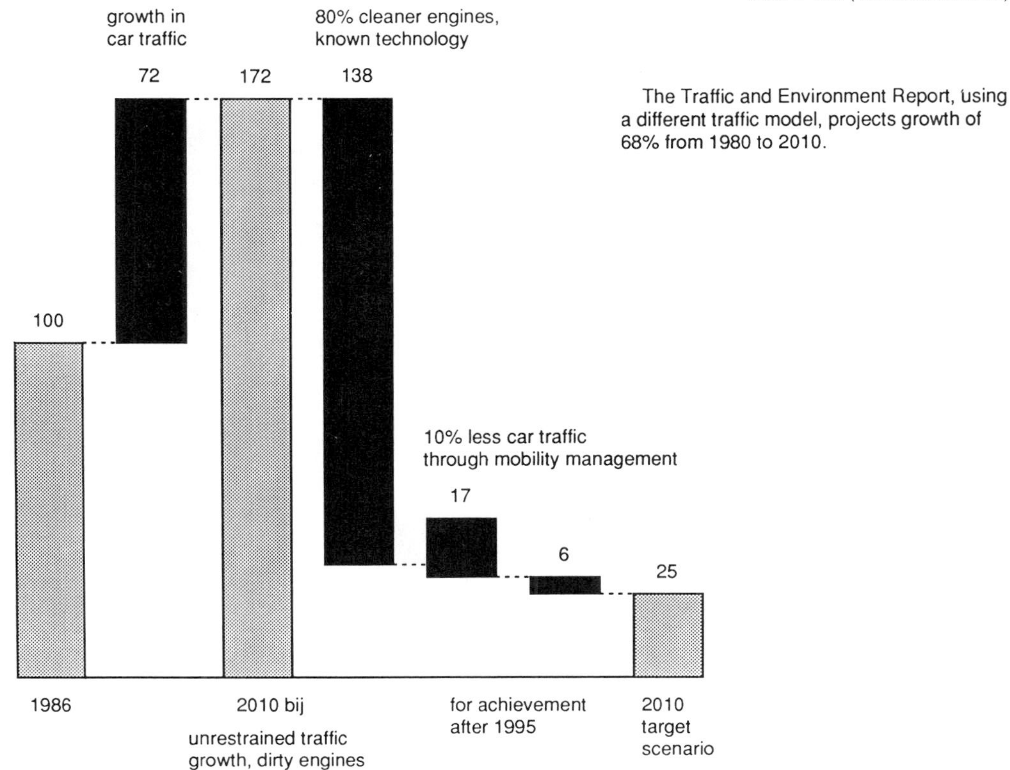

Effect of policy on emissions of nitrogen oxides from heavy goods vehicles

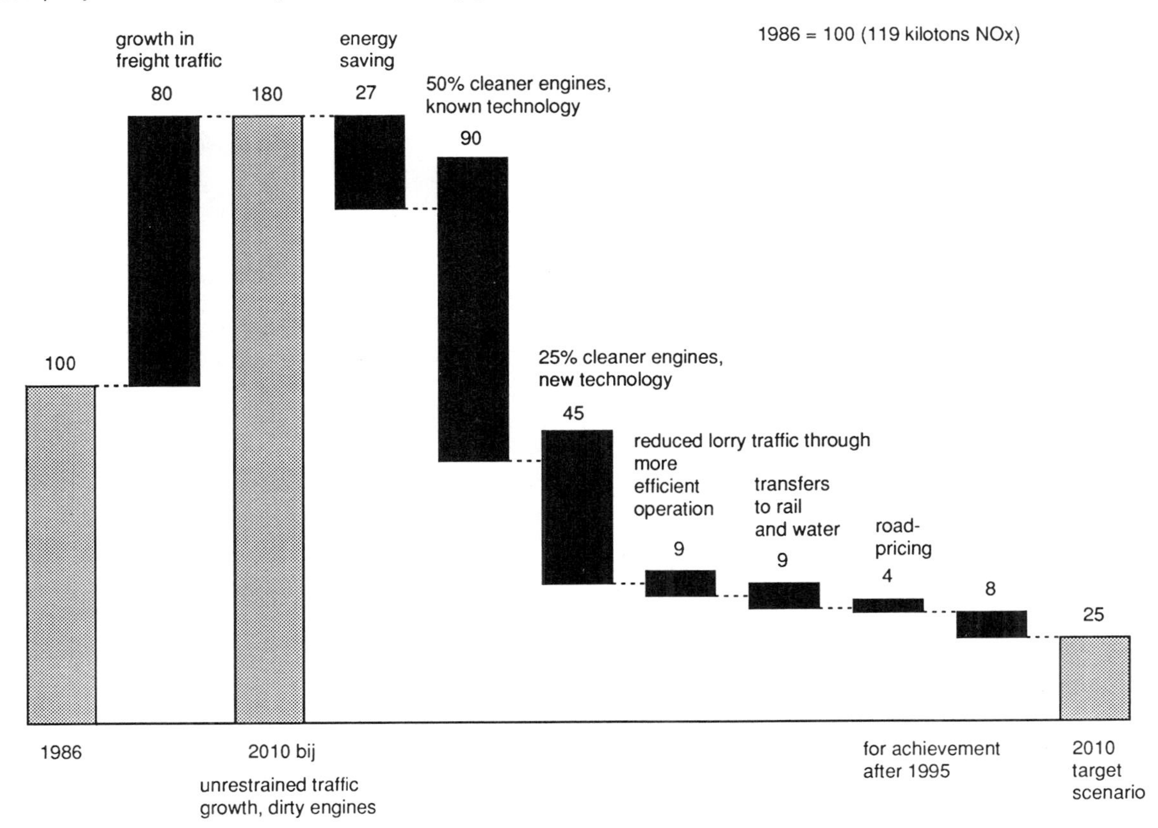

Figure 6. Effects of policies on emissions of nitrogen oxides in the Netherlands (Ministry of Transportation and Waterways, 1988).

Ministry of Housing, Physical Planning and Environment, *On the Road to 2015, Comprehensive Summary of the Fourth Report on Physical Planning in the Netherlands,* The Hague (1988b).
Ministry of Housing, Physical Planning and Environment, *To Choose or to Lose—National Environmental Policy Plan of the Netherlands,* The Hague (1989).
Ministry of Housing, Physical Planning and Environment, *White Paper on the Fourth Report on Physical Planning in the Netherlands,* The Hague (January 1990) (in Dutch).
Ministry of Transportation and Waterways, *Second Transport Structure Plan,* The Hague (1988).
Netherlands Central Bureau of Statistics, *Statistical Yearbook of the Netherlands 1988,* SDU/Publishers, The Hague, p. 231 (1989).
OECD, *Environmental Data Compendium 1989,* Paris (1989).
Renner, Michael, Rethinking transportation, in: *State of the World 1989,* Worldwatch Institute, Washington, DC, p. 103 (1989).
Telegraaf (December 7, 1989) (newspaper article in Dutch).
US Department of Commerce, *Statistical Abstract of the United States,* Washington, DC (1989).

Nonfossil Transportation Fuels: The Brazilian Sugar Cane Ethanol Experience

Sergio C. Trindade*
United Nations Centre for Science and Technology for Development
New York, NY

Introduction

The largest alternative transportation fuels program in the world today is Brazil's Proalcool Program (pronounced approximately "proalcohol"). About 6.3 million metric tons of oil equivalent (MTOE) of sugar-cane-derived ethanol were consumed as transportation fuels in 1988 (equivalent to 133,000 barrels of crude oil per day). Total primary energy consumed by the Brazilian economy in 1988 was 185.1 MTOE, and approximately 4.2 million vehicles—some 25 percent of the total automobile fleet-run on hydrous or "neat" ethanol at the azeotropic composition (96 percent ethanol by volume). Additional transportation fuels available in the country are diesel and gasoline, the latter of which is defined by three grades. Gasoline A (regular, leaded gas) has virtually been replaced by gasoline C, a blend of gasoline and up to 22 percent anhydrous ethanol by volume, and gasoline B (premium gasoline) has been discontinued as a result of neat ethanol market penetration.

The Proalcool Program

There have historically been a number of stakeholders in the Brazilian transportation/fuel industry—the government, ethanol and automobile manufacturers and distributors, the petroleum/gas and sugar cane industries, and Brazilian and international consumers. The economic interests of these stakeholders are often intertwined both in the national and international marketplaces.

For example, ethanol has been used as an automotive fuel in Brazil since the 1920s. Through 1975, it was used occasionally in blends of varying composition with gasoline to protect the important sugar industry from the vagaries of the international market. The production of ethanol absorbed part of the sugar production that was not economically feasible for export, and the government thereby guaranteed a market for ethanol.

By 1957, transnational corporations had begun local car production, and Brazil's automobile industry has since grown, contributing about 10 percent of industrial GNP and taking its place as one of the 10 largest national industries in the world.

Stakeholders and Government Policy

The introduction of the Proalcool program in 1975 coincided with a convergence of the interests of the key stakeholders in Brazil's transport fuel economy. The program was originally conceived to implement the new policy of promoting the market penetration of ethanol as a transportation fuel via gasoline C, thus displacing gasoline A. The rationale behind the policy was the concern over supply prices and the availability of liquid fuels for transportation in connection with the first oil crisis of 1973.

These concerns were exacerbated in 1979 with the onset of the Iranian oil crisis, which led to policies supporting a more radical approach to promote ethanol, in the form of neat ethanol, as a transportation fuel—in addition to gasoline C. Ethanol moved from its earlier role of gasoline extender to being a fuel in its own right.

At the root of the Proalcool initiative were government concerns with the security of the supply of transportation fuels refined in Brazil from predominantly imported crude oil. Furthermore, the oil price hikes of 1973 and 1980 put considerable pressure on the country's balance of payments. Oil imports' share of total imports jumped from 9.8 percent in 1973 to 40.8 percent in 1980. (Tables 1 and 2 provide a more comprehensive picture of the impact of oil imports on Brazil's international finances.) Ethanol is produced from domestic resources paid for in local currency. Its use therefore saves precious foreign exchange.

In founding Proalcool, the government was convinced that the expansion of the ethanol industry could also help Brazil's economic growth. It would provide the automobile industry with a more reliable source of domestic fuel based on the well-established sugar cane agriculture. Direct and indirect job creation connected

* The views expressed in this article are the sole responsibility of the author.

Table 1. Brazilian petroleum imports, 1973–1989.

Year	Avg. Oil Price US $/Barrel	Volume 10^3 bbl/day	Value 10^6 US $	Share of Total Imports (percent)
1973	2.54	652.8	605.2	9.8
1975	10.53	703.5	2,704.1	22.2
1979	16.83	1,019.7	6,263.5	34.6
1980	28.98	886.0	9,372.4	40.8
1985	29.70	545.1	5,749.3	43.7
1987	16.62	676.0	4,100.0	27.3
1988	15.10	642.2	3,545.8	17.4
1989 (estimate)	17.50	501.9	3,205.9	17.7

Sources: SOPRAL, "National Energy Balance"/MME; "Brazil—Programa Econômico"/Brazilian Central Bank

Table 2. Trade balance and foreign debt, 1973–1989.

Year	Exports	Imports	Trade Balance	Net Foreign Debt
1973	6.2	6.2	0	6.2
1975	8.7	12.2	−3.5	17.1
1979	15.2	18.0	−2.7	40.3
1980	20.1	23.0	−2.8	46.9
1985	25.6	12.2	+13.4	84.2
1987	26.2	15.1	+11.1	121.3
1988	33.8	14.6	+19.2	N.A.
1989 (estimate)	34.0	18.2	+15.8	N.A.

N.A.: not available
Source: FGV Conjuntura Econômica, "Brazil—Programa Econômico"/Brazilian Central Bank

with ethanol production would offer growth opportunities for the rural areas and allow the reduction of economic imbalances between the diverse regions of Brazil.

Transportation Fuel Capacity—Expectation and Performance

Currently, Brazil has some 600 distilleries capable of producing about 16 million m^3 (4.2 billion gallons) of ethanol. Ethanol's share of liquid fuels used for ground transportation (which employs both Diesel and Otto engines) increased from 0.5 percent in 1975 to 22 percent in 1988. The pace of market penetration of ethanol fuels is better illustrated, however, by the displacement of gasoline from the Brazilian domestic market, as both ethanol and gasoline are used primarily in Otto engines. On an energy basis, the share of ethanol in this market increased from 0.9 percent to 52 percent during the period from 1975 to 1988 while the market for diesel fuel grew from 8.1 to 15.9 MTOE. Meanwhile, total liquid fuels (ethanol and gasoline) consumed in Otto engines grew by only 19.8 percent to 11.3 MTOE in 1987 (see Table 3).

The large market expansion of ethanol as a transportation fuel stimulated the development of technologies aimed at improving sugar cane yields, more efficient conversion from sugar cane juice and molasses into ethanol, and higher engine efficiency. After all, the ethanol production technology available at the inception of Proalcool was geared to the low volume, high value-added, and traditional potable ethanol market. Otto engine technology, although originally developed with alcohol in mind, was later optimized for hydrocarbon fuels.

To expedite the launching of Proalcool, the government resorted to subsidies for the development of sugar cane agribusiness and ethanol production. These subsidies have practically been removed, and meanwhile, over the past decade, Brazil has undergone acute inflation. While inflation has not impeded economic growth, it has severely disorganized both public and private finances and has distorted relative prices in the economy.

As a result, Petrobras—the main ethanol distributor to all marketers and a company with a former tradition of profitability in its 36 years of existence—has suffered substantial losses in the distribution of ethanol.

Under inflationary conditions and with prices for ethanol ex-distillery and ethanol CIF-retailers set by dif-

Table 3. Transportation fuels consumption in Brazil, 1973–1988.

Year	MTOE				
	Diesel Fuel	Gasoline	Ethanol	Total Otto Fuels	Ratio Diesel/Otto (percent)
1973	6.4	10.4	0.2	10.6	60.4
1975	8.1	11.0	0.1	11.1	73.0
1979	11.9	10.0	1.2	11.2	106.3
1980	12.4	8.7	1.4	10.1	122.8
1985	13.5	5.9	4.1	10.0	135.0
1987	15.4	5.8	5.5	11.3	136.3
1988	15.9	5.4	5.9	11.3	140.7

Source: National Energy Balance, 1989, MME—Ministry of Mines and Energy

ferent government agencies at different times, the distributor often ends up buying ethanol from producers at a higher price than it can sell to the retailers. The slow adjustment of oil product prices under the inflationary regime of the past few years has also compounded Petrobras' losses.

On the other hand, the sugar and ethanol industry has grown extraordinarily with Proalcool and has a large stake in the program's further growth. Yet, the industry has suffered from the price-setting and commercial mechanisms of ethanol sales, and often finds itself at odds with Petrobras, the main ethanol buyer. And more recently, with the increase in international sugar prices, the industry is switching to a considerable extent to more sugar and less ethanol output. Some ethanol producers who do not make sugar are shutting down production due to lack of economic incentive.

Currently, the automobile industry is relatively indifferent to the type of fuel available—ethanol or gasoline—as it is able to supply vehicles for both fuels. Vehicle owners, however, are very sensitive to retail fuel price differences between neat ethanol and gasoline, and consider total vehicle operation costs. Consumer perception of fuel availability is therefore crucial in determining vehicle purchase.

The above issues have led to a situation that challenges the continuation of Proalcool. The reintroduction of gasoline to the market has alleviated the excess of refinery output relative to domestic demand. The prospect of short domestic supply of ethanol has led Petrobras to import methanol and ethanol to fill the gap. A new transportation fuel—an ethanol/methanol/gasoline blend (60, 33, and 7 percent by volume, respectively)—has appeared on the Brazilian market in early 1990.

As of January 1990, retail prices for automotive fuels in Brazil at the official exchange rate were US $0.87 per liter (US $3.29 per gallon) for gasoline C, US $0.66 per liter (US $2.50 per gallon) for neat ethanol, and US $0.42 per liter (US $1.60 per gallon) for diesel fuel. If a more realistic exchange rate is used to convert local currency into US dollars, the above retail prices would be reduced by at least 50 percent.

The Role of Sugar Cane Growers, Sugar Millers, Ethanol Distilleries, and Distillery Manufacturers

Sugar cane agriculture began in Brazil with the arrival of the first Portuguese colonists in the early 1500s and has continued to be a traditional crop. Brazil plays a key role in the international sugar and molasses markets and in the newly created world ethanol market, initially as seller and now as buyer. In all these markets, Brazil has the potential tonnage to significantly affect the prices of these commodities. The size of the country and the recent geographic spread of sugar cane and ethanol production promotes regional development, helps to check migration to urban areas, and cuts down on ethanol distribution costs.

Sugar millers and ethanol distributors in Brazil own the bulk of the sugar cane area required. These stakeholders have responded well to the incentives resulting from government policies to promote growth in ethanol output.

Ethanol producers in both by-product (molasses-based) and independent (sugar cane-based) distilleries and sugar cane growers have generally been well rewarded by Proalcool, except perhaps during the last four to five years as a result of the accelerated pace of inflation when prices lagged behind costs. In addition, ethanol distilleries have increased in number, have introduced new technologies, and are not only supplying an expanded domestic market, but have exported to other countries.

The main concerns of ethanol producers are pricing policy, handling of purchases by Petrobras, and the cost of ethanol storage. Under the highly inflationary economy of Brazil, ethanol producers risk becoming quickly insolvent if ethanol prices are not adjusted in a timely manner to prevent losses. Government pricing policies are therefore crucial to Petrobras—the key commercialization agent—and can have a marked impact on the cash flow of ethanol producers.

Petrobras as an Integral Entity

Petrobras' motivations and rewards do not necessarily coincide with those of the government (see Tables 4, 5, and 6). The distributors of liquid fuels—shippers, marketers and retailers—include the large transnational oil companies and Petrobras, the only marketer with its own sources of oil products, which it sells to all other companies. The rewards of these private distributors are focused on potential profit margins—margins that are established by the government and are the same for all distributors. The performance of these distributors will hence depend primarily on their financial management, particularly in light of Brazil's inflationary economy. Without Proalcool, their total income during the period 1979 to 1985 would have been much smaller as a result of strong pressure on the gasoline supply.

The rapid market penetration of ethanol as a transportation fuel has squeezed gasoline out of the market, while the demand for diesel fuel has remained high. Under a long-standing policy of self-sufficiency in refining, this situation forced Petrobras to invest considerably in adjusting, to the extent possible, to the refining scheme to match crude slate demand. The net result was the need to export gasoline, often at unsatisfactory prices, at the rate of some 100 thousand barrels per day. On the other hand, diesel fuel and particularly liquefied petroleum gas, the main cooking fuel in Brazil, are often imported to supplement domestic refinery supplies.

Throughout the duration of the Proalcool program, the exploration and exploitation efforts of Petrobras have added considerably to oil reserves and oil output, thus toning down the concern over foreign sources of supply. Brazil now produces about 60 percent of its crude requirements, up from 15 percent at the inception of Proalcool.

Automakers and Autodealers

The vehicle dealer network in Brazil reached 4,200 in 1985. Autodealers are in a situation similar to that of fuel distributors (except Petrobras): both are intermediaries selling products they do not manufacture in a market where most prices are directly or indirectly controlled by the government. Hence, they are supportive of Proalcool as long as their market benefits from ethanol fuels.

The powerful automobile industry in Brazil is almost entirely transnational. It was naturally reluctant in the late 1970s to enter into the new market of neat ethanol-fueled vehicles. But, in the end, it went along with the government's programs to improve the performance of the neat ethanol engines. Advances in neat ethanol engine technology have also benefitted the gasoline engine industry in Brazil in the form of increased efficiency, and vehicles made in Brazil can handle all three grades of gasoline: A, B, and C, although vehicles using the neat ethanol mixture have a higher compression ratio and cannot handle gasoline efficiently.

The introduction of neat ethanol engines into the market was not at all smooth (see Figure 1). During the period from 1980 to 1982, widely fluctuating demand severely stressed the industry, but since 1983, the industry has stabilized as a result of improved engine reliability and the security of the neat ethanol supply at a price not higher than 65 percent of gasoline C.

Tables 7 and 8 indicate that the share of neat ethanol vehicles sold in the domestic market in recent years is too high to be maintained over the long term. In fact, the trend is toward a return of gasoline-fueled vehicles to the marketplace. On the other hand, the automobile industry was helped in the domestic market

Table 4. Production and consumption of oil derivatives, 1986–1991.

	Percent			
	1986		1991 Projected	
Oil Product	Refinery Crude Slate	Demand Slate	Refinery Crude Slate	Demand Slate
Diesel Fuel	32	34	37	40
Gasoline	16	13	10	8
Fuel Oil	18	17	18	18
Liquefied Petroleum Gas—LPG	7	11	8	12
Naphtha	10	11	10	10
Other	17	14	17	12

Source: CNE—Brazilian National Energy Council

Table 5. Production and demand annual growth rates for oil products through 1995.

Oil Product	Percent	
	Refinery Crude Slate	Demand Slate
LPG	10.9	12.6
Gasoline	13.3	5.3
Naphtha	10.4	11.2
Diesel Fuel	38.8	45.1
Fuel Oil	16.7	15.0
Other	9.9	10.8

Obs.: Total demand for oil prices in 1995 estimated at 1.56 million barrels/day.
Source: CNE

by the sale of neat ethanol engines in the early 1980s when the sale of gasoline cars began to falter.

It is interesting to note that because automakers have increased exports of gasoline cars since 1985 (by 1987 gasoline car production equaled ethanol car production) the Brazilian-based automakers are not as vulnerable to domestic market shifts as they were 10 years ago.

Public acceptance is crucial to the successful market penetration of ethanol fuels, and the segment of the public most concerned are owners of private cars, and taxi drivers, who constitute a minority of the car-owning population. (Only one in 12 Brazilians owns a car.)

Taxi drivers are an important target group in the dissemination of information about fuel and engine performance. After the initial fiasco of the neat ethanol-fueled cars in 1980 and 1981, persuading taxi drivers to buy new cars at a 50 percent discount (with all excise taxes waived) was crucial to the success of the program. The improvement of the engine quality was also key to regaining public confidence.

As the retail price difference between neat ethanol and gasoline continues to decrease, the demand for gasoline-fueled cars increases. The demand for diesel-fueled vehicles (in Brazil, only commercial vehicles are diesel-fueled) remains unabated as the retail price of diesel fuel has remained relatively low in relation to gasoline.

Government Projections

Since 1975, Proalcool has generally met government objectives—although not necessarily by its own devices. Concerns over the security of supply and high prices have abated due to the increase in domestic crude production, which nearly quadrupled from 1975 to 1987 and has resulted in 60 percent oil self-sufficiency.

The ensuing decrease in imported oil prices and the overall export drive of Brazil has reversed the trade balance. In the final analysis, Proalcool has had a highly positive impact on Brazil's trade balance, but unfortunately, the parallel increase in foreign debt and debt service has offset most trade surpluses. The growth in oil prices since 1973 is one of the causes of the outstanding Brazilian debt.

The rate of economic growth has fluctuated since 1975. Nevertheless, most analyses indicate that agricultural employment and industrial activities connected with ethanol fuel production, distribution, and utilization benefitted substantially from Proalcool.

Economics as if the Environment Mattered

Despite significant technological developments in ethanol production, by the end of 1985 the best Brazilian ethanol producer was barely cost-competitive with gasoline at a crude price of US $28 per barrel. On the other hand, straight economic comparisons do not take into consideration 1) the value of ethanol in reducing oil imports in a country as highly indebted as Brazil, 2) the higher value of ethanol as an octane booster, and 3) ethanol's superior environmental characteristics. As the long-term marginal cost of oil is likely to increase, even in market economic terms, ethanol can become a rational option for some countries.

Table 6. Ethanol and gasoline exports, 1980–1988.

	Thousand barrels/day								
	'80	'81	'82	'83	'84	'85	'86	'87	'88
Gasoline Exports	6	25	27	35	72	81	58	90	88
Domestic Gasoline Consumption	192	184	180	151	136	128	147	133	126
Alcohol Consumption	46	14	64	59	112	139	184	198	209
Combined Domestic Consumption of Gasoline and Alcohol	238	228	244	240	248	287	331	331	335

Source: CNE

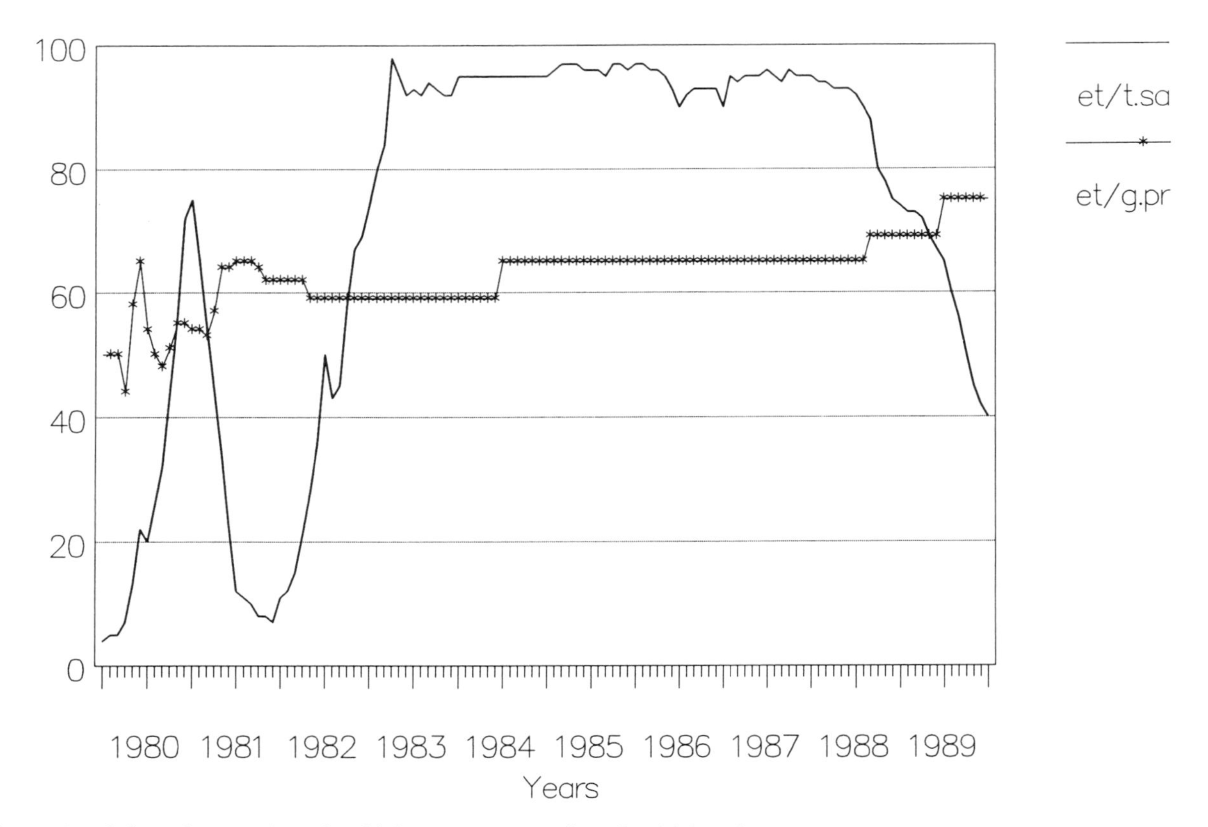

Figure 1. Sales of neat ethanol vehicles as percent of total vehicle sales.

Table 7. Brazilian vehicle sales in the domestic market, 1973–1989.

	Percent by fuel type								
	Cars			Light Commercial			Heavy Commercial		
Year	Gasoline	Ethanol	Diesel	Gasoline	Ethanol	Diesel	Gasoline	Ethanol	Diesel
1973	100.0	—	—	99.5	—	0.5	36.4	—	63.6
1975	100.0	—	—	99.4	—	0.6	20.9	—	79.1
1979	99.7	0.3	—	82.6	0.9	16.5	1.9	0.0	98.1
1980	71.5	28.5	—	63.8	15.2	21.0	0.6	—	99.4
1981	71.3	28.7	—	37.5	11.1	51.4	0.1	1.6	98.3
1982	61.9	38.1	—	24.4	24.3	51.3	0.2	1.9	97.9
1985	4.0	96.0	—	4.8	68.5	26.7	0.1	3.0	96.9
1987	5.9	94.1	—	6.5	66.3	27.2	0.1	0.8	99.1
1989	39.0	61.0	—	29.0	39.3	31.8	0.0	—	100.0

Source: ANFAVEA—Brazilian National Automakers Association

Table 8. Brazilian vehicle sales in the domestic market, 1973–1989.

	Thousands by fuel type								
	Cars			Light Commercial			Heavy Commercial		
Year	Gasoline	Ethanol	Diesel	Gasoline	Ethanol	Diesel	Gasoline	Ethanol	Diesel
1973	558	—	—	106	—	1.	26	—	45.
1975	661	—	—	118	—	1.	17	—	62.
1979	826	2.	—	79	1.	16.	1.	0.	89.
1980	567	226.	—	60	14.	20.	1.	—	93.
1981	319	129.	—	25	8.	35.	0.	1.	64.
1982	344	212.	—	21	21.	44.	0.	1.	48.
1985	24	578.	—	5	67.	26.	0.	2.	61.
1987	23	387.	—	8	72.	24.	0.	0.5	66.
1989	221	346.	—	40	54.	44	0.	—	58.

Source: ANFAVEA

The environment was *not* an explicit concern in the formulation of Proalcool, but as the program evolved, the costs and benefits to the environment became evident. The distillation of ethanol produced from the fermentation of sugar cane juice and/or molasses yields a large volume of liquid effluent (stillage) with a high biological oxygen demand/chemical oxygen demand (BOD/COD) content. Each volume of ethanol yields 13 to 15 volumes of stillage. At the outset of Proalcool, it was traditional to dispose of raw stillage in bodies of water. This process became untenable as ethanol output expanded. The dominant approach today is to return stillage to the sugar cane fields for fertilization and irrigation purposes. In this application, stillage is valued for its potassium content, which is usually sufficient to pay back the required investment within a commercially acceptable time.

Ethanol burning in Otto engines has a generally beneficial effect on emissions. The actual effect depends on a variety of factors including compression ratio, tuning, ethanol content in the blend, and driving cycle. The major negative effect is the increase in aldehyde emissions in relation to gasoline.

The recent concerns over the net contribution of fossil fuels to the level of carbon dioxide in the atmosphere and the ensuing greenhouse effect on climatic change have given sugar cane ethanol an unexpected environmental value. Sugar cane ethanol is the *only* commercial transportation fuel that has a *zero* net contribution of carbon dioxide to the atmosphere.

Conclusions

Government policymakers in Brazil are introducing new policies with the express purpose of arbitrating conflicts among the main stakeholders in the Proalcool program while addressing consumer concerns. Stakeholders and their positions include: Petrobras, which commercializes ethanol and petroleum refining and marketing; ethanol producers, which take most of the added value generated by the Proalcool program, but are exposed to sudden bankruptcy as a result of uneven pricing policies or severe inflation; and automakers, who depend upon the availability of various fuels.

The reintroduction of gasoline into the domestic market reflects an attempt to avoid expensive investment in refining to match market demand for crude slate under a policy of self-sufficiency in refining. The automobile industry output of gasoline-fueled vehicles is likely to increase rapidly in response to this change in policy. Ethanol capacity is likely to remain at the current level in the foreseeable future, although actual ethanol output will depend on the reaction of producers to the pricing policies of the government and the practices of Petrobras in commercializing ethanol. Ethanol and methanol importation is already filling the gap left by domestic producers who are unwilling to sell at prices they do not consider remunerative—particularly those who have the option of switching to other products, such as sugar.

A new consensus among key stakeholders is emerging to replace the understanding achieved during the period 1975 to 1980: There is a new transportation fuels market emerging in Brazil, and the gasoline share is fast increasing. However, this consensus is subject to dynamic change, and it remains to be seen how much value can be attributed to the carbon dioxide recycling feature of sugar cane ethanol—or biomass-derived ethanol in general.

Other countries can learn from the Brazilian Proalcool experience that a delicate balance between key stakeholders is germane to the successful implementation of alternative transportation fuels programs. Other lessons include:

- Decisions on alternative transportation fuels must be seen in long-term perspective;
- Domestic production of alternative fuels cannot be entirely dependent on uncertain and unpredictable oil prices. (It may be better and safer to spend two dollars in local currency at home than one dollar on imports, given heavy indebtedness and soft currency.)
- Consumers respond most favorably to: economic incentives that mirror concern with cost of motor vehicle ownership and operation; consistent and farsighted policies; and the initial reputation of vehicles fueled by alternative fuels;
- Stakeholder consensus is critical for the stable implementation of an alternative fuels program. Consensus achieved at program launch must also be checked and renegotiated periodically throughout the life of the program.
- Air quality can be improved and a total recycling of carbon dioxide emissions can be achieved.

Under current conditions, very few countries in the world should actually embark on an ethanol fuel program. The optimal candidates are countries with biomass surpluses and energy deficits that are land-locked and without their own oil resources. This scenario offers a more viable economic prospect for the penetration of ethanol in the transportation fuel market. The total carbon dioxide recycling feature of biomass-derived ethanol adds a new dimension to the future prospect of this nonfossil transportation fuel.

Acknowledgments

The author wishes to thank Arnaldo Vieira de Carvalho Jr. and Gilson G. Krause of Promon, Rio de Janeiro, Brazil for providing the data contained in the Tables and in Figure 1. Thanks are also due to Julio Borges (Copersucar, São Paulo), Jacy Mendonça (ANFAVEA, São Paulo), Alberto Mortara (São Paulo), Roberto Villa (Petrobrás, Rio de Janeiro), and Juo Alves da Cunha (Usina Guarany, São Paulo), who read an earlier draft and provided useful comments.

References

1. "Politica de Combustiveis Liquidos Automotivos" ("Automotive Liquid Fuels Policy"). CNE—Brazilian National Energy Council, Brasilia, Brazil (1988).
2. Sperling, Daniel. "Brazil, Ethanol and the Process of System Change," *Energy* 12, no. 1 (1987): 11–23.
3. Trindade, Sergio C. and Vieira de Carvalho, Jr., Arnaldo. "Transportation Fuels Policy Issues and Options: The Case of Ethanol Fuels in Brazil" in Sperling, Daniel, ed., *Alternative Transportation Fuels—An Environmental and Energy Solution*, Quorum, New York, NY (1989): 163–186.
4. Trindade, Sergio C. *Oxygenated Transport Liquid Fuels: The Total System*, World Energy Council Monograph, London, UK (1989).
5. Trindade, Sergio C. "Brazilian Alcohol Fuels: An International Multisponsored Program," Rio de Janeiro, Brazil (1984).

Energy and the Environment: A Framework for Evaluating Research Needs in the Transportation Sector

Malcolm A. Weiss
Energy Laboratory
Massachusetts Institute of Technology
Cambridge, MA

John B. Heywood
Department of Mechanical Engineering
Massachusetts Institute of Technology
Cambridge, MA

The Context

The roles of transportation in our 20th-century world are several, important, and generally recognized. An effective transportation system is essential to the economic and social health of all modern societies; individuals place a high value on mobility of both people and goods. Several key ingredients are required to establish and sustain an effective transportation system in any society. One of them, and the one with which we are concerned here, is providing the energy required to operate that transportation system and dealing with the environmental consequences of consuming that energy. In this paper, we examine the issue of deciding how and where research can contribute to solving the energy and environmental problems associated with transportation.

Most transportation systems are regarded as "overloaded" and efforts to expand their capacities do not seem to be able to keep up with demand. In the United States, the current overload is conspicuous in both our highway and air transport systems. The obvious and immediate symptoms of overload are congestion, delay, unreliability, local and regional air pollution, noise, highway deterioration, and land use. Less obviously, but importantly, the transportation system is a major user of petroleum with no currently viable substitutes and thus is a significant contributor to global CO_2 emissions. (In the US, the transportation sector alone consumes more petroleum than the US produces.) Uncontrolled extrapolation of current trends will exacerbate these serious problems.

If we focus on energy for transportation, we can state a broad social objective. That objective is to provide energy sufficient to power transportation services of the scope and quality desired at minimum, stable, and equitable economic and environmental costs during an uncertain future. The question addressed in this paper is: What criteria should be used to decide how and where research can promote this objective either directly or by providing guidance to policymakers? We have chosen not to attempt to define a specific research agenda; in our judgment such a large, important, and necessary task first requires the development of a comprehensive organizing framework. We recognize, and are not overly concerned about, the occasional difficulty of separating energy-environmental issues from the broader issues of transportation systems.

Organizing the Strategic Options

We can categorize the strategic options for achieving the social objective stated above in three broad groups:

1. Options that manage the demand for transportation services;
2. Options that improve the operation of the transportation system;
3. Options that improve the inherent efficiency or "cleanliness" of the vehicles and fuels used.

The strategic options describe what we hope to accomplish with a given measure, technological or nontechnological. All three types of options are important if major reductions in environmental impact are to be achieved. To date, the last option, which ordinarily exploits improved technology, has been viewed as the primary option for change. That reflects a supply-side bias common to most of US energy priorities. We must acknowledge that a broader and more balanced assessment of the strategic options available is essential if significant reductions in environmental impacts are to be achieved. We must also acknowledge that measures now appropriate to the US are not necessarily appropriate to other societies, such as developing countries, or in other time frames; research agendas have to reflect those differences.

The three categories of strategic options are shown in Table 1 with a brief taxonomy; obviously the taxonomy must be carried to further levels of detail before it can be applied in practice. The taxonomy shown provides a framework for considering the potential impact of an option relative to alternative options which would have similar impact. It displays the breadth of strategic options available which could reduce the impact of transportation energy on the environment. It suggests that substantial reductions in environmental impact will come only from implementing many of these options because the impact that can be realistically achieved with most individual options is relatively modest compared to the scope of the environmental problems being addressed. Our research agenda should make full use of that breadth.

We believe that the value of this framework is that it i) broadens our view of options available for changing the environmental impact of our transportation systems, ii) helps us identify potential linkages for positive reinforcement (and also for negative feedback) between individual measures, and iii) allows us to compare the relative potential benefits of research activities designed to further individual options or combinations of options. A deliberate omission in this taxonomy is neglect of the embodied energy (and environmental effects) of vehicles or the transportation infrastructure, since these are usually small. Should that change, however, such effects must then be included.

Criteria for Evaluating Research Areas

The implementation of any of the options (or implied sub-options) of Table 1 requires the adoption of specific measures, since the options describe the objectives to be achieved rather than the means of achieving them. Our next concern then becomes the relevance of research to any proposed specific measure. Academics have an understandable bias toward placing high value on research on many issues. In some cases that bias is well founded, but not in others. Therefore, the first question is whether the introduction and implementation of a measure could in fact be improved substantially (or even, in some cases, made possible) by research.

That question heads a list of evaluative questions, shown in Table 2, that we regard as important for establishing priorities for research proposed to address any particular measure. The other questions deal with impact, timing, ranking, implementation, and, ultimately, prioritization of research in the many different areas that could contribute to achieving our broad social objective.

In an attempt to test the framework of Tables 1 and 2, and to provide some concreteness to it, we held meetings with our colleagues and reviewed much of the relevant literature (see bibliography for examples). One important general conclusion from those meetings is that while some of the theory of linkages between transportation and the economy and public welfare is well developed, the useful application of that theory to real situations and choices has been much more limited—possibly because the theory is sometimes incomplete or invalid. As a consequence, we lack an adequate base of understanding for making informed decisions about our options both for improving the effectiveness of our transportation systems and for decreasing their impact on the environment. Thus, an important overriding research need is to investigate the practical relationships between economic growth, competitiveness, social welfare, and transportation in both industrialized and developing countries. Society is currently making (or avoiding making!) decisions in the absence of that information.

In these meetings with our colleagues, we discussed many strategic options related to the environmental impact of transportation where lack of knowledge prevents an adequate assessment of the potential for reduced impact, or prevents or hinders implementation of that option. The section that follows describes some patterns and examples that we have identified to date.

Examples of Important Types of Research Activities

Several major research themes emerged from our review of the first broad category of options, "demand management." One was the lack of relevant "case studies"; i.e., the careful analysis of the information available on the various impacts of major transportation changes that have the effect of changing demand. Examples are: In developing countries, studies of the extent to which national government or international lending policies have intended to or been able to influence infrastructure to give more efficient "cleaner" transport systems; in the US, a study of how the interstate highway system program has impacted the US economy; in Europe, studies of how changes in operations and policy have modified existing urban/suburban transportation systems to identify their effectiveness and impacts. Also, our understanding of people's transportation expectations is insufficient for us to assess the extent to which these expectations could be influenced in directions that would reduce environmental impacts. An example here is whether the increased "computer literacy" of younger people in the US will increase the potentially wider substitutability of telecommunications in the communications/transportation tradeoff.

When we looked at options in the category of "system operations," we concluded that the extent of the environmental and health impacts of all forms/modes of transportation energy use are not adequately quantified; the issue here is both uncertainty about emissions and uncertainty about the effects of those emissions on health and the environment. This lack of information varies significantly from example to example; however,

Table 1. Strategic options for meeting the broad social objective.

Objective:

An effective transportation system is essential to the economic and social health of every modern society. Our social objective here is to provide energy sufficient to power that system at minimum, stable, and equitable economic and environmental costs during an uncertain future.

Options:

1. Modify or manage demand for transportation
   - 1.1 Encourage land-use planning to reduce transportation needs
     - 1.1.1 Shorten commuting distances
     - 1.1.2 Increase provision of local services
     - 1.1.3 Plan work/residential/service clusters that justify public transport
   - 1.2 Substitute communications services for transportation services
     - 1.2.1 Encourage remote work (telecommuting)
     - 1.2.2 Encourage remote services (telebanking, teleshopping, teleteaching, etc.)
     - 1.2.3 Encourage remote business activity (teleconferencing)
   - 1.3 Reduce the desire for transportation services
     - 1.3.1 Make selected transportation alternatives expensive and/or inconvenient
     - 1.3.2 Encourage a travel-conservation ethic and human-powered options
2. Provide a larger share of transportation services using modes or practices or both with higher efficiency and/or lower emissions
   - 2.1 Shift transportation services to modes with higher efficiency or lower emissions
     - 2.1.1 Shift services up the efficiency/emissions hierarchy: water, rail, highway, air
     - 2.1.2 Shift services into nonfossil modes
   - 2.2 Improve operating practices for given modes
     - 2.2.1 Increase load factors and backhauling
     - 2.2.2 Establish shortest routes and optimum speeds
     - 2.2.3 Implement procedures which promote improved in-use vehicle performance (e.g., inspection/maintenance programs, retirement of older vehicles)
3. Increase the inherent system energy efficiency, and/or decrease the inherent system energy-related emissions for all modes of transportation services
   - 3.1 Improve energy supply systems (from resource to vehicle tank or motor)
     - 3.1.1 Recover, refine, and transport fuels with lower consumption of fossil resources
     - 3.1.2 Provide fuels that burn with lower emissions
     - 3.1.3 Provide a larger share of fuels from lower-carbon fossil resources
     - 3.1.4 Provide transportation energy from nonfossil resources
   - 3.2 Improve vehicle technology (reduced vehicle emissions and increased conversion of consumed energy to motion)
     - 3.2.1 More effective and durable emissions control systems
     - 3.2.2 Increase efficiency of conventional types of propulsion systems, petroleum-based and alternative fuels
     - 3.2.3 Reduce vehicle weight, improve vehicle aerodynamics and rolling friction
     - 3.2.4 Introduce alternative propulsion systems (e.g., electric, fuel-cell-based) which reduce environmental impact

even in some better quantified areas such as urban air pollution there is a serious shortage of data on actual in-use emission characteristics of individual vehicles, modes, and the total system. That lack weakens public policymaking. There are indications that in-use emissions exceed design expectations; if the excess is significant and confirmed, reducing the excess is an objective deserving research attention.

Significant opportunities are thought to exist in the areas of use of information (telecommunications and computerization) to improve the "performance" of the transportation system, and policies which attempt to influence the replacement of inefficient technology and practices with more efficient technology and practices. An example here is the railroads in their competition with trucks for freight haulage. The inherent potential advantages of the railroads in energy efficiency and emissions could be better exploited through improved technology and operations, if the institutional barriers could be overcome.

Especially important in this area, as well as in "demand management," is the analysis of the potential ben-

Table 2. Criteria for evaluating proposed research areas.

1. Research Relevance:	What are the important issues associated with this measure that research would hope to illuminate?
2. Impact:	How large (relatively speaking) is the potential impact on energy use or emissions if the research is successful and the results effectively implemented? What other impacts are significant (for example, on customer satisfaction)?
3. Timing:	How long will it take before a significant impact on energy use or emissions can realistically be expected from implementation of the results of this research?
4. Context:	How important is this measure compared with other measures which could achieve similar objectives?
5. Implementation:	What policies and practices should be adopted by whom to be confident that the research and development will be planned, funded, executed, and transferred effectively?
6. Priorities:	What priority should be given this research and development relative to other R&D on other measures?

efits of combinations of strategic options which have synergistic linkages, where each individual action on its own might have modest impact at best, or where some disadvantage of an otherwise attractive measure can be offset by adopting an additional measure.

The category of "inherent efficiency and cleanliness" seemed easier for us to classify and specify. However, this does not make the category more important than the others; our biases and preoccupations are merely more evident. There were two recurring themes in our discussions. One was that new technologies or significant changes in technologies are seldom adequately evaluated in a systems context. Consider, for example, the very long-term issue of transportation based on nonfossil energy. Broadly based systems studies of widespread use of biomass-based or hydrogen-based transportation energy are only just starting. Research on such total systems is admittedly speculative, but is essential if we are to identify those parts of the system that represent major obstacles and most deserve research attention.

The second theme is that there can be large cumulative changes through the evolutionary improvement of current transportation technologies; that improvement is always helped by expanding the relevant knowledge base and often hindered or prevented by a lack of that knowledge—knowledge that appropriate research can provide. Research aimed at evolutionary improvement is rarely glamorous or attention-getting; but the change in performance of automobiles during the past dozen or so years illustrates how large the effects of that research can be and therefore it should not be undervalued.

There is another broad area of research that does not arise out of measures designed to address specific strategic options. Here we are concerned with developing both data and understanding that illuminate the total costs, benefits, and performance of transportation systems. Those data and understanding are needed to make rational comparisons and choices among broad options. Options for transport of freight or passengers by whatever mode(s) ideally should be compared on a door-to-door basis with the transport alternatives, a comparison that should include all social (including environmental) as well as market costs and benefits. There is no complete methodology to make those comparisons in practice, and we do not have all the data we would need even if we had the methodology, e.g., the health and environmental consequences of emissions, or good indicators of the performance (micro and macro) of our existing transportation systems.

In the preceding discussion of the three categories of strategic options, we have given some examples of transportation research topics that we believe deserve attention. However, we are not proposing here a comprehensive list of such topics, nor do we have a back-pocket list, although we and our colleagues have individual research interests that have been suggested. Preparing a prioritized list is a research task in itself, a task that requires a wide-ranging examination of plausible options and alternatives and that uses the framework developed here to analyze and justify the resulting list.

Implementation Issues

Even widespread agreement on the priorities for research carries no assurance that the research will be done at all, much less done effectively. Means for effective implementation of a research agenda depend crucially on the time scales involved (whether years or decades) and on the nature of the research (technological or nontechnological).

For example, our judgment is that near-term technical research on present types of highway vehicles and fuels should be largely funded and performed by the private sector, sometimes motivated by government incentives or fiat (such as meeting efficiency and emissions standards or eliminating leaded gasoline). That work should be supplemented by basic research that provides the understanding that can lead to future advances. Some of that basic research is and should be undertaken by industry, since it falls within a time frame comfortable to some companies. However, other components of that basic research, and longer-term research, need support from public sources, and will involve universities and other research organizations.

In addition, certain other types of important research, regardless of time frame, require public support (and performers in the nonprofit sector) to avoid either the appearance or reality of conflicts of interest. Conspicuous examples are research on the health and environmental effects of emissions, or on the social costs and benefits of transportation systems alternatives.

To generalize, a serious evaluation of research portfolios should, for each research issue and time frame, consider the questions: Who sets the agenda? Who pays? and Who does the work? In considering those questions, we would like to see more attention paid to new and better mechanisms and policies for planning, funding, executing, and transferring research. For example, there must be more effective mechanisms that can be devised for partnerships—multisectoral or multinational—to plan and conduct research on transportation energy and the environment.

The Next Step

We have argued here that there is a broad range of strategies for moving toward more environmentally benign transportation systems around the world. Research can contribute importantly to moving forward, but a research agenda that reflects the full breadth of opportunities has yet to be developed. The next step in developing that research agenda should be an expansion of the framework proposed here to a much greater level of detail. We would then like to see a reduction to practice of that expanded framework. What specific measures are of potential interest for each of the options identified? What research activities could play an important role in promoting those measures? How should each be evaluated using the criteria listed in Table 2? Accomplishing this task will require participation by a broad range of experts. It will also require strong integrative leadership to be sure the results target the available research resources into areas of greatest potential impact and are not simply compilations of each expert's favorite topics.

Acknowledgments

In preparing this paper we have profited from hearing the views of R. R. John of the US Department of Transportation, and of some of our MIT colleagues: M. E. Ben-Akiva, A. W. Epstein, E. M. Greitzer, J. P. Longwell, R. W. Simpson, J. M. Sussman, D. G. Wilson, N. H. M. Wilson, J. P. Womack, and D. N. Wormley. We have also benefited from the comments of individuals who reviewed a draft of this paper. All those opinions, and those of the authors, reflect personal, and not necessarily organizational, opinions.

Selected Bibliography

National Research Council, *Fuels to Drive Our Future,* National Academy Press, Washington, DC (1990).

Organisation for Economic Co-operation and Development, *Transport and the Environment,* Paris (1988).

Shiller, J. W., "The Automobile and the Atmosphere" in *Energy: Production, Consumption, and Consequences,* National Academy Press, Washington, DC (1990).

Sperling, D., *New Transportation Fuels,* University of California Press, Berkeley, CA (1988).

US Department of Energy, "A Compendium of Options for Government Policy to Encourage Private Sector Responses to Potential Climate Change," Executive Summary (DOE/EH–0102), Volumes 1 and 2 (DOE/EH–0103) (October 1989).

US Department of Transportation, "Moving America: A Look Ahead to the 21st Century," Washington, DC (1989).

US Environmental Protection Agency, "Policy Options for Stabilizing Global Climate, Draft Report to Congress" (February 1989).

Session B
Industrial Processes

Summary
Session B—Industrial Processes

Jefferson W. Tester
Howard J. Herzog
Session Organizers
Energy Laboratory
Massachusetts Institute of Technology
Cambridge, MA

Session B on Industrial Processes was divided into four parts to cover the diverse nature of the different manufacturing industries and the wide range of issues relating to energy consumption in this end-use sector. The papers in Session B-1 provided an overview of major energy-consuming industrial processes and offered a framework for evaluating opportunities to reduce energy use through equipment replacement, process modification, materials substitution, and recycling. The five papers in Session B-2 treated energy consumption and efficiency opportunities in the manufacture of primary materials: iron and steel, aluminum, minerals extraction, cement and concrete, and glass. Session B-3 dealt with three energy-intensive manufacturing processes: chemicals, petroleum, and pulp and paper manufacture. Session B-4 was an open forum where several speakers from the earlier sessions led a discussion that reviewed critical technical, and regulatory issues that could significantly reduce energy consumption. Attilio Bisio chaired that session and prepared a summary paper, "Perspectives on Energy Efficiency and Conservation in the Industrial Sector," that is included in this volume.

Because of the breadth and complex infrastructure of the industrial end-use sector, it was not possible to treat energy consumption and conservation opportunities comprehensively at this conference. Our topics were selected to provide a structure and methodology for evaluating and implementing energy efficient technologies and to illustrate the quantitative level of conservation improvements possible in certain high-level energy-consuming industries. Separate summaries follow for sessions B-1, B-2, and B-3.

Summary
Session B-1—Industrial Processes and Materials Manufacture

James Wei
Jefferson W. Tester
Session Chairs
Department of Chemical Engineering, and Energy Laboratory
Massachusetts Institute of Technology
Cambridge, MA

The first paper, "Modeling the Energy Intensity and Carbon Dioxide Emissions in US Manufacturing," by Marc Ross, provided a foundation for the entire session. Dr. Ross reviewed historical trends in energy consumption and carbon dioxide (CO_2) emissions associated with the manufacture of primary materials and chemicals. He cited mechanisms for reducing CO_2 emissions in the US. Specific opportunities were discussed for conservation, lowering consumption by better energy management and control systems, fundamental process change, fuel switching, cogeneration, materials substitution and recycling, and sectoral shifts in the type and quantity of manufactured products. He concluded his paper with a discussion of a scenario for a 20 percent reduction in CO_2 emissions by 2010, using quantitative modeling components and introducing important policy issues.

The paper by Joan Ogden, Robert Williams, and Mark Fulmer, "Cogeneration Applications of Biomass Gasifer/Gas Turbine Technologies in the Cane Sugar and Alcohol Industries," dealt with a specific case study carried out by the authors at Princeton's Center for Energy and Environmental Studies. It is one of the few papers presented at the conference that treated biomass as a primary energy source for producing electricity and process heat. In developing countries, industries that produce sugar and/or fuel ethanol can efficiently utilize their biomass residues in an integrated gasifer/gas turbine energy conversion system. The paper covers the technology and projected costs for this particular biomass application in considerable detail.

Heat and energy integration methods applicable to chemicals and materials manufacture were treated in the paper by H. Dennis Spriggs, R. Smith, and E. A. Petela, titled "Pinch Technology: Evaluate the Energy/Environmental Economic Trade-Offs in Industrial Processes." A generalized approach to optimizing energy transport (supply and utilization) within a complex process can be developed by employing pinch technology to identify opportunities for improving energy efficiency via heat integration methods. The methodology can also be applied to combined electricity and process heat utilization (cogeneration), as well as to minimization of effluents targeted for environmental reasons.

The final paper of Session B-1, "Energy Conservation in the Industrial Sector," was presented by Robert Ayres. The paper introduced both "first- and second-law" efficiencies to evaluate a broad range of manufactured products and processes. It reviewed worldwide historical trends in energy consumption improvements and their limitations for manufacturing major materials and chemicals. Economic issues for implementing conservation were then discussed in depth to establish marginal costs and rates of return as functions of fractional energy use reduction, using both a general framework and specific examples. The presentation concluded with a brief treatment of policy and regulatory issues.

Modeling the Energy Intensity and Carbon Dioxide Emissions in US Manufacturing

Marc Ross
Physics Department
The University of Michigan
Ann Arbor, MI
Environmental Assessment and Information Sciences Division
Argonne National Laboratory
Argonne, IL

Introduction/Overview

Organizations, such as government agencies and utilities, which make and must justify detailed long-term plans, need to forecast. For many, long-term forecasting has a bad name because forecasts are not accurate. However, when thoroughly grounded in a detailed understanding of past trends, forecasts, or better, scenarios can be useful for planning. This is especially true, if the mathematical process is simple and transparent—at the level of a simple spreadsheet—rather than being so complex as to constitute a black box.

Future energy consumption is a critical area for long-term planning because: 1) energy supply is the most capital intensive of all areas and requires major research and development efforts and 2) energy supply and consumption is the major source of most pollution, including greenhouse gases.

A simple spreadsheet model is developed in this paper that creates scenarios for future energy use and CO_2 emissions by the manufacturing sector, as they depend on production and on measures taken to reduce energy costs and CO_2 emissions.

The United States Congress has requested the Department of Energy to analyze two goals for reduction of US emissions of greenhouse gases: 20 percent absolute reduction in 5 to 10 years and a 50 percent absolute reduction in 15 to 20 years. That request inspired this article. A 20 percent reduction in the carbon dioxide emitted by combustion in the manufacturing sector is studied, with the goal year of 2010. The longer time interval is deemed essential for a realistic attempt to achieve the ambitious goal.

The History of Energy Use and CO_2 Emissions

Energy use by manufacturers in the US constitutes 29 percent of all energy use.[1] It is responsible for a substantial part of the air pollution burden. In particular, the CO_2 emissions from combustion of fossil fuels for heat and power in manufacturing (including the fuel burned to generate the electricity used) in 1985 was about 310 million tonnes of carbon—about 24 percent of CO_2 releases from all fuel combustion in the US.

US manufacturers accounted for 17.4 quadrillion Btu (or 17.4 quads) of fuel consumption for heat and power purposes in 1985, including both fuel used directly and fuel burned remotely to generate electricity used. In addition, they used 3.9 quads of oil and gas for nonenergy purposes, such as feedstock materials in petrochemicals, solvents, lubes, and waxes. In addition to being generated in combustion, CO_2 is generated in calcining of carbonates (especially in manufacture of cement, iron, and glass) and in the oxidation of some feedstocks. (For example, solvents which evaporate are fairly quickly oxidized in the atmosphere.) These sources have not yet been completely evaluated, but their total contribution is roughly 10 percent or less of the CO_2 contribution by fuel combustion in the sector. They will not be considered further in this report.

The major energy form used was electricity (43 percent of all fuel used for heat and power going toward generation of purchased electricity), with direct use of natural gas also substantial (27 percent). (See Table 1.1.)

Table 1.1. Fuel used for heat and power and CO_2 released in manufacturing, 1985. [Sources: Refs. 1, 3, 4]

	Fuel Use (quadrillion Btu)	CO_2 Emitted (million tonnes C)
Coal	2.68	67
Oil	2.80	54
Natural Gas	4.66	68
Purchased Electricity	7.26[a]	119
Total	17.40	308

[a] The average input energy to the electrical utility sector is 3.34 × electrical energy delivered. Total input to the electrical utilities is 26.48 quads. CO_2 emissions are allocated to the manufacturing sector in proportion to its purchased electricity (nationally).

Direct use of petroleum products and coal for heat and power was 15 percent each. The aggregate energy intensity, that is, the ratio of the total energy used by all manufacturers to their total production,[2] was steady in the period from 1958 to 1972 and then fell by more than one third from 1972 to 1985 (Figure 1.1).

The different sources of energy had quite different histories, however. From 1958 to 1985 the aggregate coal and oil intensities fell 70 percent and 60 percent, respectively.[3] During this period, coal was being eliminated, except at large facilities and in very heavy industries. (Quite aside from environmental problems, coal has not been a convenient fuel except at large facilities with special equipment and expertise.) Petroleum was gradually losing share to natural gas, and then, with the second oil shock, its use was drastically curtailed. Between 1958 and 1971, the aggregate natural gas intensity rose 30 percent and then by 1985 dropped by more than 50 percent (Fig. 1.2). Natural gas was favored for its low price and convenience until shortages began in 1971, shortly before the first oil shock.

The combined result of these developments was that the aggregate fossil fuel intensity fell 15 percent from 1958 to 1971 because there was fuel-efficiency improvement in energy-intensive sectors, even though real (inflation corrected) fuel prices were low and falling.[5] Beginning in 1971, the decline quickened, and by 1985 the aggregate fossil fuel intensity declined by another 50 percent. Part of the reason for this accelerated decline was a relative shift away from energy-intensive production.

The pattern for electricity consumption is similar, except that continuing electrification (new uses of electricity) is combined with the other developments. Thus, aggregate electricity intensity grew rapidly between 1958 and 1970 (Figure 1.3), even though the efficiency of electricity-intensive processes, such as the electrolysis of brine to produce chlorine and the smelting of aluminum, was being substantially improved. This was a period of falling real electricity prices. Since 1970, electricity prices have been rising and the aggregate electricity intensity has gradually declined. The two forces for decline, efficiency improvement and the relative decline in electricity-intensive production, have slightly outweighed ongoing electrification.

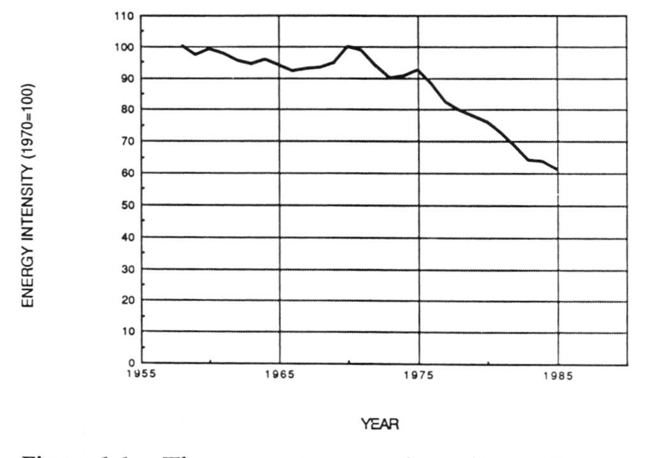

Figure 1.1. The aggregate energy intensity of US manufacturing (relative to 1970).

In 1958, more than 20 percent of total electricity used was generated and used onsite by manufacturers.[6] Onsite generation was falling, however, dropping to less than 8 percent of the total by 1981. With the Public Utilities Regulatory Policy act of 1978, onsite generation began a comeback, rising to 10 percent of the manufacturing total in 1986. This means that, until 1981, utility sales of electricity to manufacturers were growing about 0.7 percent per year faster than total electricity use. Utility sales growth is now slower than the growth of total electricity use. The changing patterns of fuel use are of great importance for CO_2 analysis because the CO_2 emissions per unit of energy vary with energy form (Table 1.2). For example, the average emission characteristic (gC/kWh) for utility electricity declined from 195 to 186 from 1973 to 1985.

Another critical element in understanding trends in energy use is that over 70 percent of CO_2 emissions arise from the manufacture of bulk materials—including pulp and paper, industrial chemicals, petroleum refinery products, glass, cement and clay products, and metals (not including fabrication). This materials manufacturing is roughly 10 times as energy-intensive (defined as the ratio of energy to labor and capital expenditures) as the rest of manufacturing.[7] Moreover, the production and use of these materials in the US have been declining relative to all manufacturing since the early 1970s. It is not useful, therefore, to analyze CO_2 emissions on the basis of the aggregate energy intensity—the use of all fuels in all sectors. Instead one needs to examine three kinds of underlying changes which affect the intensity of CO_2 emissions from fossil fuel combustion in manufacturing: 1) changes in the real energy intensity, i.e., in the energy intensities within each sector of manufacturing; 2) shifts in the composition of production; and 3) relative changes in the energy forms used.

Mechanisms for Reducing CO_2

The Reduction Mechanisms

The 11 major mechanisms for reducing CO_2 emissions in combustion are indicated at the points of their application in Figure 1.4. Three of the mechanisms can reduce energy intensity in the separate process steps. Here, conservation is defined in the narrow sense of

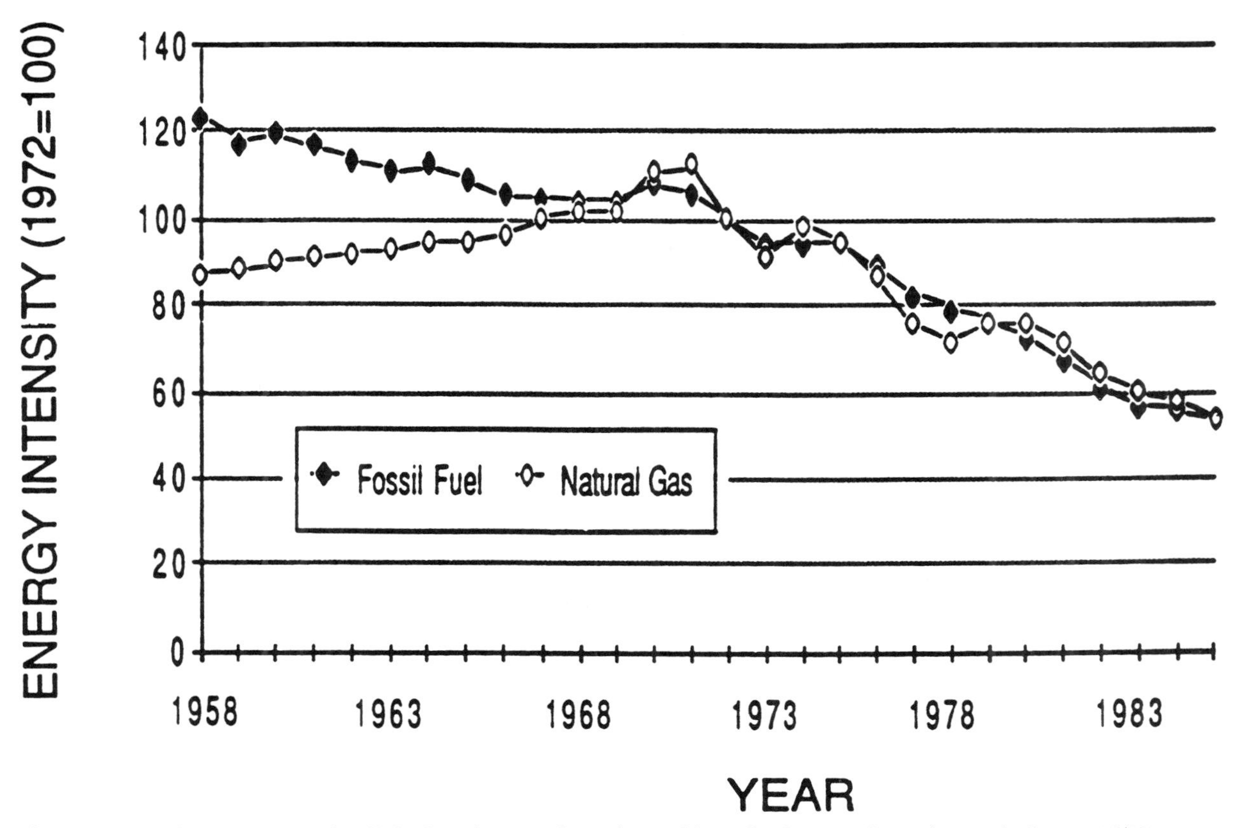

Figure 1.2. The aggregate fossil fuel and natural gas intensities of US manufacturing (relative to 1972).

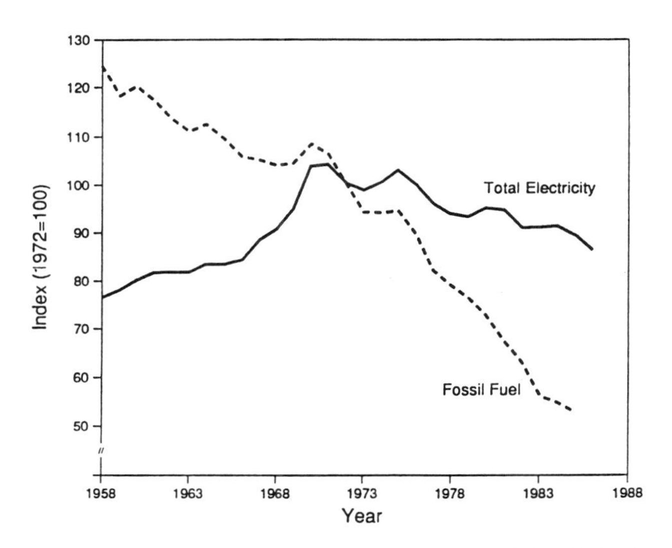

Figure 1.3. The aggregate electricity and fossil fuel intensities of US manufacturing (relative to 1972).

Table 1.2. Carbon emitted to the atmosphere in the form of CO_2 associated with energy use for heat and power.

Natural gas	14.5 kg/million Btu
Distillate	19.0 kg/million Btu
Residual Oil	20.3 kg/million Btu
Coal	25.1 kg/million Btu
Electricity[a]	186 g/kWh
	54.5 kg/million Btu[b]

[a] Characteristics of 1985 utility electricity delivered to final customers
[b] @3412 Btu/kWh.

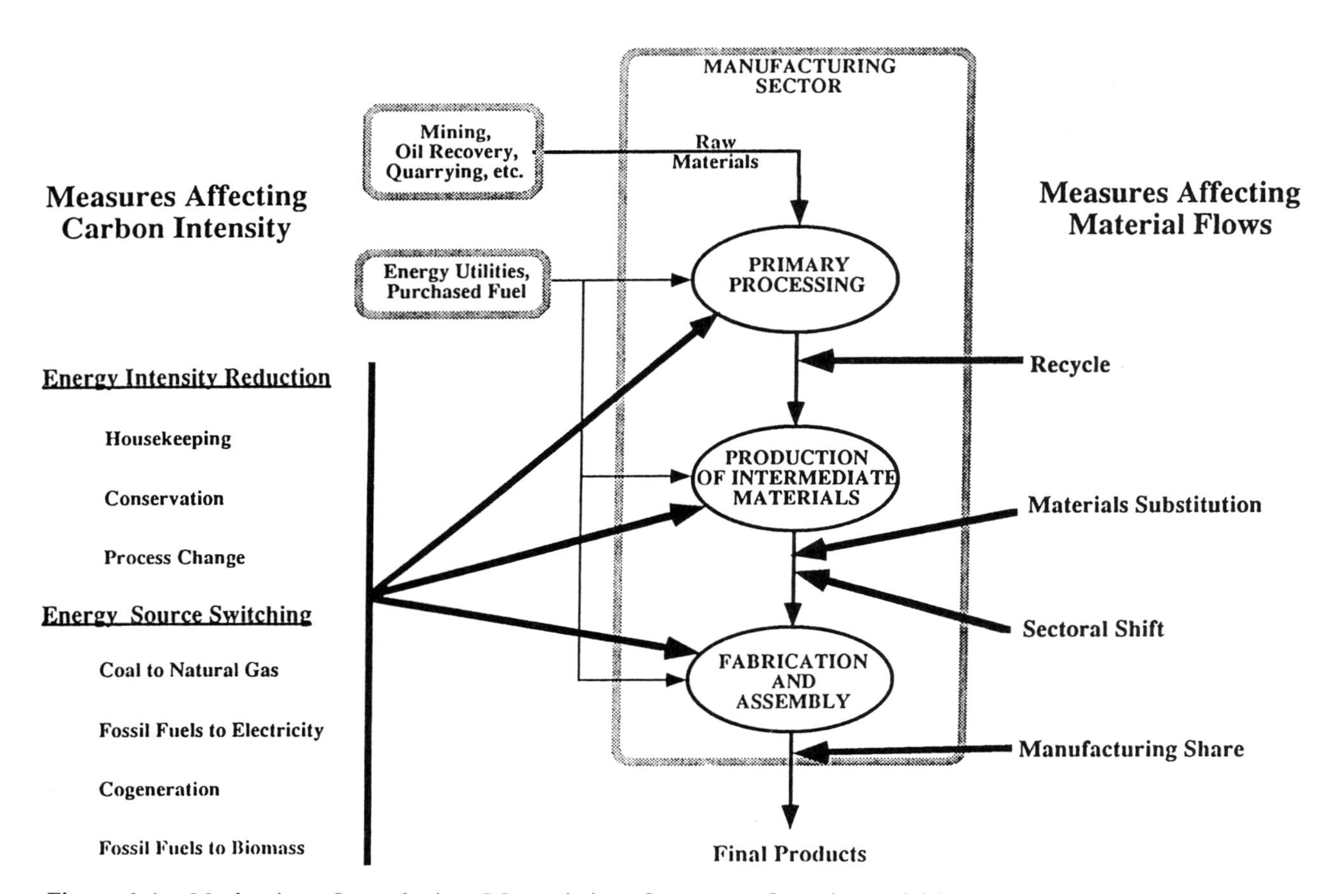

Figure 1.4. Mechanisms for reducing CO_2 emissions from manufacturing activities.

investments for which a major rationale is energy savings. Housekeeping refers to no- or low-investment operational changes. Process change refers to fundamental change in production processes, not primarily motivated by energy savings, although it usually has associated energy savings.

Energy source switching refers to: 1) the potential for substituting natural gas for coal, at least in the short and medium term; 2) the potential for substituting electricity for fossil fuels, if the electricity is nonfossil based, or perhaps if it is natural-gas based; 3) the potential for further growth in cogeneration (including substituting high electricity-to-steam technologies for existing modes); and 4) substituting biomass fuels (or other nonfossil primary energy) for fossil fuels.

The remaining four mechanisms influence intermediate or final product flows. Recycle refers to the use of processed bulk materials which have been recovered from later stages of manufacturing or at the post-consumer stage. Sectoral shift is the effect of the relative shift of production activity from the energy-intensive (materials manufacturing) sectors to downstream fabrication and assembly sectors. (The export of energy-intensive industries will not be considered.) Materials substitution involves substitution among the materials used in the final products, in association with improvements in materials properties and redesign of the final products. Manufacturing share refers to the growth or decline of manufacturing activity relative to that of GNP.

While the 11 reduction mechanisms are distinct, they interfere with one another. For example, the reduction directly associated with process change is less if conservation has been applied to the original process. The model developed below combines the effects of the reduction mechanisms.

In this model, each reduction mechanism is specified algebraically in terms of one or two parameters (such as a reduction factor and a penetration factor). The parameters are estimated in part on the basis of historical trends and conservation supply curves. In addition, they depend on physical limitations to implementation and constraints on effective policymaking. The model and a scenario derived using it will be discussed later.

Feasibility of the Reduction Mechanisms

The 11 reduction mechanisms are assessed in terms of their potential to reduce CO_2 emissions, and their technology-related limitations, as suggested in Table 2.1.

Conservation. Conservation supply curves show that with today's processes and conservation technologies, the expected reductions in energy intensity—associated with energy prices two or three times those today—are about 25 percent.[8] (An example is shown in Figure 2) The conservation potential is higher (but likely to be less rapidly implemented) in non-energy-intensive sectors than it is in the energy-intensive sectors. Housekeeping and cogeneration are not included in the conservation category; they are discussed below. Thus conservation does not include the entire response to energy price increases, perhaps a little more than one half.

A conservation supply curve (CSC) is analogous to an oil or gas supply curve: it shows the cumulative capacity to provide energy, e.g., barrels per day, as it depends on the price of energy. The capacity to provide energy shown by a CSC is actually the energy supply capacity released by saving. The savings can be expressed in energy capacity units or as a percentage of base-year energy use as in Figure 2.

In order to create CSCs, one needs to be able to convert the first costs into effective operating costs in order to relate capital expenditure to energy prices. The price at which an investor breaks even depends on the annualized, or amortized, capital costs. For a capital cost K,

$$\text{annualized capital cost} = \text{CRR} \times \text{K},$$

where CRR is the captial recovery rate. If the CRR is empirically based on consumers' (i.e., manufacturers') behavior, then the CRR may be as large as 0.5 or higher for minor investments.[9] High rates reflect the fact that many consumers do not invest unless the payback is quick. (The payback is essentially 1/CRR in years.) If the CRR is based on the society's cost of capital (a real discount rate typically assumed to be in the range 3 percent to 10 percent), then the CRR may be in the range 0.07 to 0.16. Low rates reflect the world as the society would (probably) desire it to be: After being sure to include all transaction costs in the expenditure K, the CRR used should be based on the actual cost of capital to a well qualified borrower/equity seller. The CSC shown in Figure 2 is based on average manufacturing *behavior* with CRR = 0.33 as shown.

There are two reasons why industry often has short simple-payback requirements like two to three years, or discount rates of 30 percent to 50 percent, while banks charge only 10 percent (and less in real terms). 1) Apparent discount rates appear to be high because there are costs that are not usually included in the simple economic analysis of small projects. Internal management and engineering costs may be relatively large. There is also some risk that a project will interfere with production or that production requirements may change in such a way as to reduce project benefits. It must be stressed that the high discount rates apply to projects deemed to have low risk, but on the other hand, for small projects firms cannot afford to calculate and adequately consider the sensitivity of project benefits to possible variations in the context. 2) More important, extra cash is not normally available to firms at bank interest rates. There is a balance among different sources of cash, and other sources like equity require

Table 2.1. CO_2 reduction mechanisms in manufacturing: technological opportunity and constraints.

Mechanisms for CO_2 Reduction	Overall Reduction Potential*	Physical Limitations	Capital Limitations	Need for New Technology
Conservation	High	Mod	Mod	Mod
Housekeeping	Low	Imp	—	—
Process Change	High?	Mod	Imp	Imp
Energy Switching				
Other fuels to nat. gas	High	Imp	—	—
Fuels to electricity	High	Imp	Imp	Imp
Cogeneration	High?	Imp	Mod	Mod
Fossil to Biomass	Low	Imp	Mod	Mod
Recycle	High	Mod	Mod	Imp
Materials Substitution	Low?	Imp	Mod	Mod
Sectoral Shift	Low	Imp	Mod	Mod
Manufacturing Share	Low	Imp	Imp	Mod

* Judged by size of opportunity *and* potential degree of public policy impact
Imp: Important
Mod: Moderately important
—: Relatively little importance

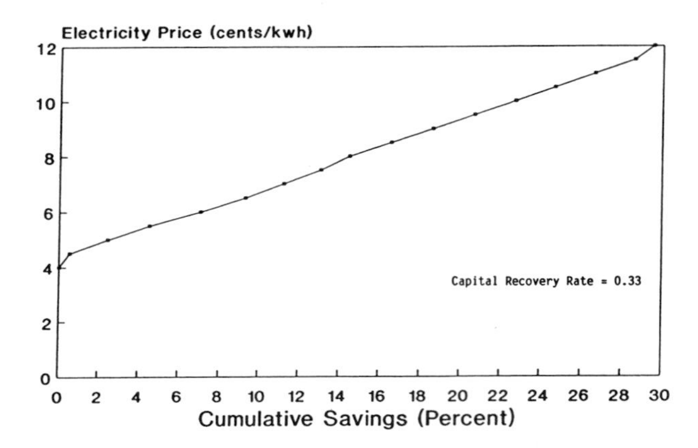

Figure 2. Electricity conservation supply curve—manufacturing.

much higher returns than bank loans. Moreover, and most important, if the flow of earnings is not large enough to cover the projects in question, the management will probably not be willing to go outside the firm for funds. Management risks its position when it seeks outside funds, and making favorable arrangements is extremely time consuming if the firm does not have strong earnings. As a result, outside funds are sought for strategic reasons, like entering a new market, but not for projects which would merely reduce costs. In these situations, capital for such projects is rationed. In today's malfunctioning capital markets such capital rationing also occurs at some firms with good earnings and good investment opportunities. Management fears that large investments in improved processes and products may create a situation in which their stock is undervalued; instead of investing they use stock purchases to increase stock price.

If policies were introduced to reduce the implicit discount rate, the energy-price increase required to achieve a given level of savings would be correspondingly smaller. Thus, if time horizons were lengthened so that the typical CRR = 0.16 (e.g., a 10 percent real discount rate with a 10 year project life), then a much smaller price increase (Figure 2) would be needed to achieve 25 percent savings. The marginal price required would be about 5.5¢/kwh instead of almost 11¢/kwh. Note that the cost of the conservation measures does not involve this uncertainty; the uncertainty concerns financial decisionmaking, given the project costs.

The capital cost of retrofitting today's manufacturing facilities to achieve a 25 percent reduction (in energy intensity associated with heat and power) would be roughly $80 billion (in 1988 dollars). The relation of the capital cost of conservation to energy price and CRR is approximately K = S E p/CRR, where S is the fractional savings (0.25), E is the annual energy consumption, p is the *average* energy price at which this much conservation takes place, and CRR is the implicit capital recovery rate governing the investments. Based on 1985 data, CRR ≈ 1/3; for electricity E = 640 billion kWh, p≃8¢/kWh, and K≃$38 billion; for fuels E = 11 quads, p≃$5/(million Btu), and K≃$42 billion.

For comparison, the total *annual* plant and equipment investment in manufacturing is also typically about $80 billion. In actual practice, in this conservation program not all existing production capacity would be retrofit, but over the time period (to the year 2010), considerable new production capacity would be created with conservation improvements integrated into the design at substantially lower cost. The overall cost would thus be somewhat less than $80 billion. The cost of conservation, to the 25 percent reduction level, is estimated to be about $60 billion, or an average of about $2.5 billion per 1 percent reduction. These conservation investments would be carried out over a period of time—in the model below over a period of 20 years—so in that model the annual conservation investment would be about 4 percent of the current rate of investment in plant and equipment.

For comparison purposes, let us ignore CO_2 issues for the moment and consider the capital costs of replacement supply systems in the early 21st century. If a new fuel supply system could be brought in (in today's dollars) at $50,000 per barrel per day energy equivalent, then the cost would be about $2 billion for 0.1 quad/year, about a 1 percent contribution to fuel supply for manufacturing at today's consumption rates. If a new electricity supply system could be brought in at $2000/kw, or (at 65 percent duty factor, $350 per thousand kWh/year) the cost would be about $2.3 billion for 22 trillion Btu electrical energy delivered per year, about a 1 percent contribution to supply. Substantial operating costs would have to be added to these capital costs for supply. Thus, granting the roughness of the estimates, the cost of a 25 percent reduction through conservation investments would be roughly compensated by reduced needs for new energy supply systems. (Much lower levels of conservation would be significantly cheaper than corresponding supply investments, but much higher levels of conservation could be significantly more expensive.)

The conservation supply curves referred to involve current technology. If rapid development of conservation technology continues, then in 10 to 15 years many important new technologies will begin to contribute to conservation. An impression of the importance of new technology can be obtained from examining the technologies which provided the CSC (Figure 2) for electricity in fabrication and assembly industries. (See Table 2.2.) Often, decades pass between the time an invention is made and the time it achieves commercial success, but in the 1970s development and marketing of energy and electronic technology accelerated. Moreover, the gestation time for small-scale technologies is often much shorter than for the better known large-scale technolo-

Table 2.2. Timing of the introduction of technology for electricity conservation (automotive manufacturing CSC).[10]

New Technology (first marketed after 1970)
Microprocessor-controlled energy management.
Variable speed drive based on inexpensive power semiconductors.
Die cushioning using rubber air bags.
High-efficiency lamps and lighting systems.
Large microprocessor-controlled heating and ventilation systems.
Aircraft-derivative gas turbine cogeneration.
Old Technology
Isolation valving for segmenting compressed-air systems.
Small compressors for satellite compressed-air system.
High-efficiency motors.
Cog belts.

Table 2.3. Examples of potential fundamental process change with important energy and CO_2 consequences.

Chemicals, Paper, and Food Processes
Improved separation processes based on membranes, adsorbing surfaces, critical solvents, freeze concentration, etc.
Ethylene chemistry based on natural gas feedstocks
Waste reduction using closed systems
New and improved catalysts for chemical processing
Recycling paper and plastics (i.e., into new material and product areas)
Continued improvement in the processing and forming of materials
Biological processing of organic materials
Biomass feedstocks for organics
Dry papermaking
Metals Processes
Recycling post-consumer scrap (i.e., into new product areas)
Near net shape casting
Surface treatment with electromagnetic beams
Direct and continuous steelmaking
Coal-based aluminum smelting

gies. In conclusion, there is every reason to expect current CSCs to evolve in the direction of much greater savings potentials (relative to base year energy intensities) in the next decades. Effective technology policies and energy-price signals would accelerate this evolution.

Housekeeping. While of great importance for secondary forms of energy like steam and compressed air, and for lighting and HVAC, housekeeping only has a further overall reduction potential of perhaps 5 percent. The installation of automated energy-management systems, typically requiring expensive rewiring in an existing plant, is included in the conservation category.

Fundamental Process Change. Fundamental process change has a largely unknown potential. It is known that radical new process technology cannot overcome the large minimum thermodynamic requirements for reduction of metal ores or decomposition of brines. (Blast furnace operations, primary aluminum, and chlor-alkali manufacture are the most important examples.) Relatively small but substantial thermodynamic requirements also apply to certain refinery and other organic chemical reformations.[11] But most industrial processes have no or negligible thermodynamic requirements in principle. For all these, radical process change could, in principle, radically reduce energy use. One can, in fact, imagine many processes which require almost no energy use (where now a great deal of energy is used). Four examples of fundamental process changes from recent decades are the float-glass process, basic oxygen steelmaking, induction heating of metals, and surface heating with electromagnetic beams. Examples of fundamental process changes which may be achieved in the near future are listed in Table 2.3. (Where examples of these technologies are already well established, it is quite new applications which are at issue.) In practice, a great deal of research, invention, development, and major investment would be needed to achieve such changes. Dramatic process changes tend to occur slowly.

Longitudinal analysis of energy intensities in separate manufacturing sectors (summarized as the "real intensity" trend) for the period before the early 1970s shows that fuel intensities were declining about 1.2 percent per year and electricity intensities increasing about 1.8 percent per year,[12,13] even though energy prices were low and often declining. (Electricity intensities declined in sectors where electricity use is intensive.) Further analysis shows that these developments are part of a long history of technical change. Fundamental process changes reduced most factors of production, including fuel, but increased electricity use because of its special role in new technology.[14] After the early 70s, real fuel intensity declines quickened and real electricity-intensity growth ceased. Presumably, in addition to the historical trends just mentioned, price- and policy-induced intensity reductions also occurred in this period. In the analysis below these historical trends are assumed to continue (at moderated rates), underlying the policy-induced energy-intensity reductions which are being modeled.

Switching from Coal and Oil to Natural Gas. Several forms of energy source switching can reduce CO_2 emissions depending on specifics. As shown in Table 1.2, natural gas combustion results in less carbon loading of the atmosphere than other fossil fuels. The potential for natural gas substitution for oil or coal is, however, limited (not only because there may be gas supply scarcity). Most combustion of oil products in manufacturing (as in the mining sector) is now at remote sites (e.g., in forest products sectors) or is combustion of by-products (especially still gas and coke) for which there are no other markets at this time. (See Table 2.4.) On this basis, the potential for CO_2 reduction from switching from oil to gas in manufacturing is deemed small and will be ignored in the upcoming scenario. With coal, some uses are also not substitutable at reasonable cost—coking coal for blast furnaces and coal use in remote locations, especially in forest products industries. However, much of the coal use that began around 1973 in response to high gas and oil prices and national priorities is substitutable.

Switching from Fossil Fuels to Electricity. The CO_2 reduction potential of substitutions of electricity for fossil fuel depends on three coefficients:

1. the carbon loading per kWh associated with generating the incremental electricity;
2. the relative efficiency, in the substitution application, of electricity compared to fuel; and
3. the carbon loading per unit of fuel value of the replaced fuel.

In the analysis below, it is assumed that these substitutions are attractive only in end uses where the relative efficiency of electricity is about three (or more) times higher than that of the replaced fuel, because the 1985 average price of fossil fuels is $2.50 to $3 per million Btu, while that of electricity is about $14 per million Btu (electrical). There are quite a few technologies with such high efficiency.[15] It is also assumed that electricity should be preferentially substituted for fuels with high carbon loading. In the analysis below, it is assumed that the carbon loading per kWh of the *incremental* electricity involved in substitution is likely to be the same as today's average. The average efficiency of electricity use in the applications in question is assumed to be three times that of the fuel replaced. Under these assumptions, CO_2 reductions occur with substitution of electricity for coal and oil, but not natural gas. (See Table 1.2.)

Table 2.4. The use of petroleum in manufacturing.

Application	1985 Consumption in Quads
Heat and Power (except refineries)*	
Resid	0.4
Distillate, LPG, gasoline	0.3
Petrochemical Feedstocks	3.4
Refinery Processes‡	1.8
Lubes	0.1
Petroleum coke (aluminum smelting)	0.2
Total	6.2

* Approximately 0.2 of the 0.7 quads is used by forest products industries.
‡ Mostly still gas, resid and coke, fuels with low market value.
Source: Ref. 1.

A critical issue for this mechanism is capital cost. If, for example, a 20 percent reduction in coal and oil use were achieved by substituting electricity at one third of the coal and oil intensity, then in terms of current energy usage: 1) A reduction of 24 million tonnes of carbon (about 8 percent of the total from manufacturing activity) would be achieved if the incremental electricity involved no CO_2 emissions; 2) The capacity of the incremental utility system (not counting the in-factory process equipment) would be about 110 billion kWh/year or about 20 gigawatts (at a 65 percent utilization rate). The corresponding system capital cost would be, very roughly, $40 billion, if low-cost nonfossil fuel systems were successfully developed. To this, add the cost of the new process equipment. Thus, the capital cost of this approach is well over $5 billion per percentage point of CO_2 reduction from manufacturing. (For comparison, the estimate above for conservation was $2 billion to $3 billion per percentage point.) The key consideration is that this high capital cost only applies to substitutions where electrical energy is used three times more efficiently than the replaced fuels. For lower-efficiency applications of electricity, its substitution would be more costly.

Cogeneration. Cogeneration can substantially improve efficiency in the joint production of heat and shaft power, usually steam and electricity. Economical application of cogeneration requires fairly steady heat loads at the site, such as the steam loads in papermaking. The first part of the cogeneration opportunity is simply to exploit these heat loads more systematically in all sectors. The second is to use technology that provides a high ratio of electricity to steam, providing a better match to demands (than outdated steam turbine topping-cycle cogeneration) and larger overall fuel savings.[16] One extraordinary potential in development[17] is the replacement of boilers and steam-turbine generation in black-liquor recovery systems at kraft pulp mills with gas turbines. This substitution will produce about three times more electricity from a given input of black liquor. This black liquor application would increase generation of electricity without fossil fuel input.

Switching from Fossil Fuels to Biomass. The final shift in energy source considered is an increase in combustion of biomass fuels—waste or by-product fuels in forest-products and food-processing industries—to replace fossil fuels. These opportunities have been heavily exploited in the past 15 years. While further such opportunities remain, they are relatively small. There are two major opportunities to substantially increase the biomass contribution: 1) biomass can be used more efficiently and 2) crops can be grown for fuel use (or residues now left in the forest or field may be aggressively pursued).

Recycling. Recycling offers an important opportunity to reduce energy use in manufacturing.[18] The principal sectors affected are primary metals, pulp and paper, organic chemicals/petroleum refining and, to a lesser extent, glass. Where recycled material substitutes for ores in metal manufacture, major energy reductions are achieved. Where they substitute for wood pulp, there can be substantial systems benefits, although the conclusion is complicated by several options in the use of the material. For example, instead of being recycled, waste paper can be used as a fuel, substituting for fossil fuel. Where recycled materials substitute for organic feedstocks, there can be significant CO_2 reductions associated with process energy savings. Glass recycling, unlike glass bottle reuse, has only marginal energy benefits. The major issues for the recycling potential are, in order of importance: 1) the creation of markets for post-consumer recycled material in the manufacture of higher-quality products than heretofore; and 2) the reliable and clean collection of a majority of selected post-consumer materials.

Relatively little attention has been paid to the first issue. The aluminum beverage can, made from about 50 percent recycled aluminum, is the outstanding example of a technologically demanding product for which a system for the supply and use of recycled material has been developed. Other post-consumer recycled materials are typically used in undemanding products: waste paper in cellulose insulation and pasteboard for items like cereal boxes, and scrapped automobiles in concrete reinforcing bars. The challenge would be to convert newspapers into newsprint (as at Garden State Paper Corp.) and autos into sheet metal for autos, etc. Research and development could forward these goals.

Materials Substitution. A wild card in CO_2 reduction policies would be a systematic effort to substitute materials in final products. In many products—construction components, packaging, white goods, automobiles, etc.—different materials are now in head-to-head competition. The CO_2 implications of policy induced substitutions at this margin are complex and have not been studied.

Sectoral Shift. Sectoral shift has been extensively studied at Argonne National Laboratory.[12,13] The analysis shows that since the early 1970s, manufacturing activity has been moving (relatively) downstream, from manufacture of materials (as measured in tons) to fabrication and assembly. For some materials (steel and cement) this relative decline has been apparent for decades; now it is clear that bulk materials overall are in long-term relative decline.[5,19] This sectoral shift has been accelerated by the well-known increases in imports of materials like steel and aluminum. These increased imports have not been the major source of sectoral shift (except in the early 1980s) and will not be considered in the analysis below.

Manufacturing Share of GNP. The final mechanism for reducing CO_2 emissions from the manufacturing sector is a lower share in overall economic activity. The trend in the share of manufacturing in GNP is in dispute due to difficulties in defining price trends for products like computers (and also for many services), but the impression that value added in services is increasing relative to manufacturing is probably correct. The dispute is not critical for this analysis, however, since the poorly known numerator in manufacturing share appears as the denominator in sectoral shift, and so drops out of the overall analysis.

Modeling CO_2 Reduction

The Model

The model represents manufacturing in terms of three stages of processing (Figure 3). The first involves materials processing which can be directly affected by recycling, such as wood pulping and reduction of metal ores. The second involves other materials processing such as manufacture of fertilizers and cement, primary shaping and treating of metals, etc. These two stages account for 72 percent of energy use at present. The third stage is fabrication and assembly comprising all major energy using sectors which are not energy intensive (including food processing and textiles).

The model compares product flows, f_i, and energy intensities EI_{ij} in the base year 0 (1985) with those of a target year T (taken to be 2010). Here i = 1, 2, 3 runs over the stage of processing and j = 1, 2, 3, 4 runs over the energy form, coal, oil, gas, and purchased electricity, respectively (Figure 3). Total use of each energy form

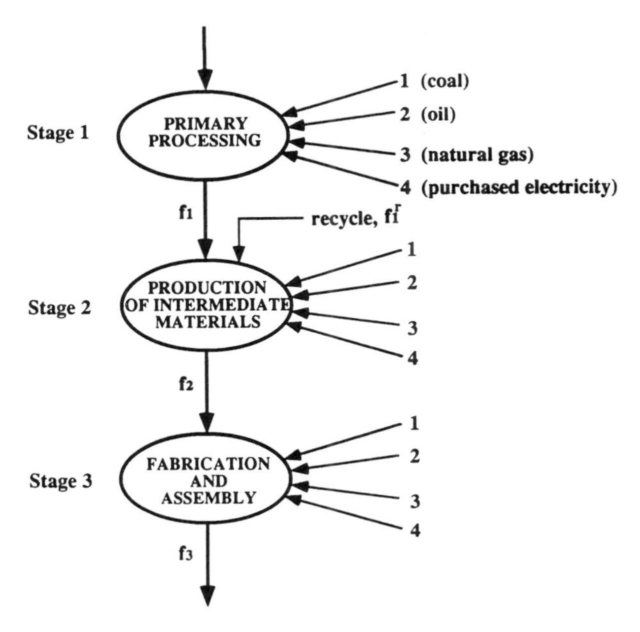

Figure 3. Three stages of production within the manufacturing sector.

at a given process stage is $f_i \times EI_{ij}$. The product flow unit is arbitrary. Here, in the base year

$$f_3 = f_2 = f_1/(1 - f_1^r) = 1$$

The relationships between the flows and intensities in year T and 0 are defined in terms of parameters as shown in Table 3.1.

A 20 Percent CO_2 Reduction Scenario

To illustrate the use of this model and to study the possibilities for CO_2 reduction I will construct a scenario representing what I feel would be possible with strong, well-implemented policies,[20,21,22] but without major net cost. There would be a major trade off in investment activity: from new energy supply to efficient use of energy and materials. These costs and savings would be roughly in balance. Moreover, the pace of these investments would be governed by normal retirement schedules. (As will be clear, however, the selection of parameters in this illustrative scenario is not based on careful quantitative analysis.)

The scenario thus suggests caution about economic commitments to greenhouse gas control. There are two reasons for great caution. First, energy is a commodity that already commands close to 10 percent of the value of total production and a much larger fraction of total capital expenditures. A major increase in the cost of services that energy provides would have serious implications for the entire economy. Second, the issue is climate, a phenomenon subject to great fluctuations and uncertainty. It is global in scope, with no single country in a position to solve the problem. If the US is to undertake ambitious greenhouse gas reduction policies in the foreseeable future, it is essential that the reduction be achieved at low net cost. We face the question: Is it possible that an ambitious CO_2 reduction policy can be carried out in the United States without committing large net resources and endangering the economy? While it is clear that major reductions cannot be achieved without substantial sacrifices or costs in certain subsectors, the net cost could be low if investments in reducing energy requirements are compensated by cost savings in energy supply.

Table 3.1. Relationships and parameters in the model.

Manufacturing production (manufacturing share effect):
$f_3(T) = (\text{GNP growth factor}) \times r_9 \times f_3(0)$

Materials production (sectoral shift effect):
$f_2(T) = r_8 \times f_3(T)$

Initial materials production (recycle effect):
$f_1(T) = f_2(T)\ (1 - f_1^r(T))$

Conservation, housekeeping and historical trend:
$EI_{ij}(T) = \exp(T \times \ln 0.99) \times r_1 \times EI_{ij}(0)$
for j = 1,2,3 (fuels), and for j = 4 (electricity):
$EI_{i4}(T) = \exp(T \times \ln 1.01) \times r_1 \times EI_{i4}(0)$

Process change:
$EI_{ij}(T)' = p_3 \times r_3\ EI_{ij}(T)/r_1 + (1 - p_3)EI_{ij}(T)$

Switch from coal to natural gas (stages i = 2,3 only):
$EI_{i1}(T)'' = (1 - p_4)\ EI_{i1}(T)'$
$EI_{i2}(T)'' = EI_{i2}(T)'$
$EI_{i3}(T)'' = EI_{i3}(T)' + p_4 \times EI_{ij}(T)'$

Cogeneration/biomass:
$EI_{i4}(T)'' = r_6 \times EI_{i4}(T)'$

Switch from coal and oil to electricity:
$EI_{ij}(T)''' = (1 - p_5)\ EI_{ij}(T)''$
for j = 1,2 (coal, & oil), and for j = 4 (electricity):
$EI_{i4}(T)''' = EI_{i4}(T)'' + p_5 \times r_5\ [EI_{i1}(T)'' + EI_{i2}(T)'']$

A 20 Percent CO_2 Reduction Scenario: Parameter Selection

Manufacturing Production. The GNP growth rate drives the model. For this scenario 2.5 percent per year growth is assumed for the period (1985-2010) corresponding to a factor of 1.85. The manufacturing share is assumed to decline 0.5 percent per year for the period, so that $r_9 = 0.88$. This is the historical trend for the Bureau of Labor Statistics real output series. This is the series used in analyzing sectoral shifts. As a result of these two factors, manufacturing production grows by a factor of 1.64.

Materials Production (Sectoral Shift). The sectoral shift analysis discussed above showed a decline in all fuel intensities starting in the early 1970s due to a relative reduction in materials production compared with manufacturing production. The acceleration of this trend in the 1980s associated with increasing net imports will not be considered. The trend in energy intensity due to the relative reduction in material ouput was about −0.5 percent per year. This translates into a trend in f_2/f_3 of about −1.3 percent per year. For the period 1985 to 2010 it is assumed that this trend moderates somewhat to −0.9 percent per year or a reduction factor $r_8 = 0.80$.

Initial Materials Production (Recycle Effect). The energy-of-manufacture-weighted level of recycling of paper, plastics, glass, and metals was about 0.20 in 1985, with roughly 40 percent of inputs to steelmaking being scrap from sources downstream from steel mills,[23] 20 percent of aluminum production being post-mill scrap, and 25 percent of pulp input to papermaking being recycled fiber from sources external to the mill. For the scenario it is assumed that, as a result of strong new CO_2 reduction policies and strong moves to reduce municipal solid waste, the level of recycling increases to a high $f_1^r = 50$ percent. The level of recycling enters through f_1^r by reducing production of initial materials from virgin sources. (See Table 3.1.)

Conservation. The energy intensities for the various energy forms and stages of production are successively modified (starting with base-year values) by conservation, process change, and three energy-source-switching mechanisms. The conservation equation (Table 3.1) incorporates several different effects: underlying historical trends, conservation as such, and a little housekeeping improvement. The historical trends, largely determined in the period before the early 1970s, are for fuel-efficiency improvements and electrification (electricity-intensity increase). These trends are associated with new process and conservation technologies; they are not energy-price driven. There is evidence that both these trends, which were strong in the 1950s and 1960s, are continuing but moderating. For example, electrification now largely involves highly efficient new technologies. The historical trend toward electrification involving *intensive* electrical processes substituting for fuel has moderated.[7] It is assumed that the continuing historical trends are 1 percent per year for electricity intensities and −1 percent per year for fuel intensities. In addition there are changes due to new policies and price changes.

Conservation investments are assumed to be driven by energy price increases and supporting policies. It is assumed that 25 percent energy-intensity reductions are achieved. This would be substantial progress. (See the sample conservation supply curve, Figure 2.) The combined impact of conservation and housekeeping improvement is represented by $r_1 = 0.73$. Without effective support policies and new technologies, energy price increases of about a factor of 2.5, or a 150 percent increase, would be needed.

Process Change. There is not a sound basis for quantitative modeling of policy-induced fundamental process change. To illustrate the possibilities, a penetration factor $p_3 = 25$ percent of new production processes and a 50 percent decrease ($r_3 = 0.5$) in the intensity for all energy forms (within the 25 percent of production at issue) is assumed. (See Table 3.1.)

Switch from Coal to Natural Gas. Coal use, which is replaced fairly easily in process stages two and three, is assumed to be reduced in intensity by $p_4 = 50$ percent. The corresponding absolute increase in gas intensity is taken to be equal.

Advanced Gas-Turbine Cogeneration/Biomass. In the absence of major modeling effort, a 10 percent reduction in purchased electricity requirements is assumed, $r_6 = 0.90$, with no direct fossil fuel consequences. This will be seen in the scenario to correspond to an increment of 69 billion kWh in 2010. For comparison, in 1985 paper industry self-generation was 32 billion kWh and growing rapidly.

Switch from Coal and Oil to Electricity. Absent a quantitative analysis, it is assumed for purposes of illustration that 20 percent of all coal and oil use is displaced by electricity use that is three times as efficient ($p_5 = 0.30$, $r_5 = 0.33$). No displacement of gas is assumed because little, or negative, CO_2 reductions might result.

A 20 Percent CO_2 Reduction Scenario: Results

The scenario results for the year 2010 are developed in a spreadsheet. The first step is to generate a base case scenario for 2010, involving the growth in production and the continuation historical energy-intensity trends discussed above. This analysis is shown in Table 3.2.

The generation of the CO_2 reduction scenario is shown in Table 3.3 with results summarized in Table 3.4. If the utility emissions factor (gC/kWh) is unchanged, there is a 20 percent absolute reduction relative to 1985. If one assumes also that the gC/kWh characteristic is reduced, then one can achieve more than the 20 percent reduction, as indicated by the examples in the tables.

It is seen that conservation investment is the largest single reduction mechanism. Moreover, the parameters chosen, although representing strong policy initiatives, do not overstate the potential from this mechanism. (Note, however, that the ordering of the mechanisms in the calculation affects their stated contributions.) Other important mechanisms are: fundamental process change, reduction in the gC/kWh characteristic of electrical generation, and recycling. Also potentially important are advanced gas-turbine cogeneration, especially if applied with biomass fuels, and a switch from fossil fuels to electricity. Unlike the conservation mechanism, the parameters selected for these five mechanisms may well stretch their potential for the 2010 time period.

The switch from fossil fuel to electricity needs to be put in perspective, because it attracts a great deal of attention given the well-established constituencies and federal programs for increased electricity use. The current high gC/kWh characteristic is likely to apply to any *incremental* electricity use in the period 1985 to 2010. The key policies in the electricity area for CO_2 reductions in this relatively short timescale are increasing the efficiency of existing electricity uses and, if feasible, using low-emissions generating technology in new and replacement power plants. A third policy to increase the electricity load on this time scale would almost certainly mean increasing the continued use of existing coal-fired capacity. In a longer time period, essentially complete substitution of nonfossil-based for fossil-based generation might be possible. After that, substitution of electricity for fossil fuel use in manufacturing could yield substantial CO_2 reduction.

A 50 percent reduction in CO_2 would require, in this scenario, roughly a two-thirds reduction in the gC/kWh characteristic of purchased electricity. To do this by 2010 would involve accelerated replacement of coal-

Table 3.2. Base case scenario: energy use and CO_2 emissions, in manufacturing, 2010

	Fossil fuel		Electricity[h]		CO_2
1985 Energy Use	quads	MtC[a]	quads	MtC	MtC
Intermediate mtls.[b]	7.95	153	1.21	66	219
Fabrication & assembly[c]	2.19	37	0.96	52	89
Total	10.14	190	2.17	118	308
Forecast Parameters		growth rate (% per year)			growth factor (1985–2010)
Manufacturing production		2.0			1.64
Fossil fuel intensity		−1.0			0.78
Electricity intensity		1.0			1.28
Sectoral shift		−0.9			0.80
	Fossil fuel		Electricity		CO_2
2010 Energy Use	quads	MtC	quads	MtC[g]	MtC
Intermediate mtls.	8.14[d]	157	2.04	111	268
Fabrication & assembly	2.80[e]	46	2.02[f]	110	156
Total	10.94	203	4.06	221	424

[a] million metric tonnes of carbon in CO_2 emitted in combustion. Emission factors (kgC/million Btu): 54.5, 25.1, 19.8, 14.5 for purchased electricity, coal, oil and natural gas, respectively
[b] Almost all SICs 26, 28, 29, 32, 33
[c] all manufacturing not included in SICs listed in (b)
[d] 7.95 × 1.64 × .8 × .78 = 8.14
[e] 2.19 × 1.64 × .78 = 2.80
[f] 0.96 × 1.64 × 1.28 = 2.02
[g] The same emission factor (gC/kwh) is used as for 1985; 1 MtC = 1 million tonnes carbon.
[h] @3412 Btu/kWh

fired generating capacity, and would require major technologies or resources not now available. Let us nevertheless consider a cost estimate. Assuming in the base case the same ratios for generation of electricity by coal and nonfossil sources as in 1985, then replacement of about 58 GW of coal fired capacity by 2010 would enable the two-thirds reduction in gC/kwh for the manufacturing sector. If these nonfossil technologies could be brought in at a cost of $2000/kW of system capacity, the capital cost would be $115 billion (in 1988 dollars). Some of this investment would be required in any case to replace retired capacity. If one deliberately replaced the coal-fired capacity at retirement then this cost problem could be mitigated. This would, of course, delay achievement of the 50 percent reduction.

Conclusion

The scenario based on the model constructed here highlights important policy issues. That is its purpose, rather than pretending to be an accurate forecast. Aspects of the scenario important to policy analysis are:

1. The most surprising result is that an absolute reduction in CO_2 emissions associated with manufacturing energy use may be practical.
2. The not surprising result is that, on this kind of time scale, there appears to be little chance for a 50 percent reduction in CO_2 emissions without a massive and rapid scrapping and rebuilding of the electric-power generation sector, using technology not now available.
3. In the two-decade time scale considered and with low net-cost assumptions, fuel switching may not be an important tool for CO_2 reduction.
4. Conservation (in the narrow sense used here) is indicated to be critical to a CO_2 reduction program, with new production-process technology, advanced cogeneration, and recycling also potentially important.

Of the policies to support conservation investment (as defined here), only a massive increase in energy prices (e.g., through a large carbon tax) would clearly be difficult to implement. Conservation and most other reduction mechanisms could be partially supported with technology-push policies that might be implemented without great difficulty.[22] The political difficulties facing major energy taxation are of two kinds: impacted energy industries (especially coal) and their employees would

Table 3.3. Calculation of a 20% reduction scenario for manufacturing (first six columns in million tonnes of carbon emitted).

	Coal	Oil	Natural Gas	Fossil Total	Purchased Electricity[i]	Grand Total	Ratio to 1985	Energy (Quads) fossil fuel	electricity[l]
1985	67	54	68	189	119	308		10.14	2.17
2010									
Historical trends[a]	70	58	75	203	221	424	1.37	10.90	4.06
Conservation and housekeeping[b]	52	42	54	148	162	310		7.95	2.97
Process change[c]	47	39	50	136	149	285		7.33	2.73
Coal to gas[d]	33	39	59	131	149	280		" "	" "
Cogen/biomass[e]	33	39	59	131	134	265		" "	2.46
Fuels to electricity[f]	27	31	59	117	146	263		6.70	2.67
Recycle[g]	21	31	58	110	138	248	0.80	6.37	2.53
Reduced utility emissions[h]	21	31	58	110	106	216	0.70		
Reduced utility emissions[k]	21	31	58	110	44	154	0.50		

[a] See Table 3.2
[b] Reduction factor, $r_1 = 0.73$
[c] Penetration factor, $p_3 = 0.25$, reduction factor, $r_3 = 0.50$
[d] Penetration factor, $p_4 = 0.50$ process stages 2 and 3 only
[e] Reduction factor, $r_6 = 0.90$
[f] Penetration factor, $p_5 = 0.20$, reduction factor, $r_5 = 0.33$, oil and gas only
[g] New recycle fraction 0.50
[h] CO_2 emissions per kWh reduced by factor 0.77 compared to 1985
[i] All except last two rows based on 186 gC/kWh. See Table 1.2
[k] CO_2 emissions per kWh reduced by a factor 0.32 compared to 1985
[l] @3412 Btu/kWh

be singled out for sacrifices and energy-intensive manufacturing sectors would incur cost increases that would damage them competitively, especially with respect to foreign competitors. The second problem might be mitigated by securing international agreement or by correcting for incremental energy costs in imports and exports of energy-intensive products. (If such tariffs and rebates were restricted to the few energy-intensive industries, it might be administratively manageable.)

Ambitious fuel switching policies face several difficulties. The scope of fuel switching is limited unless the major expenditures are made to increase the supply of of primary energy resources (principally, energy to generate electricity that is then used in inefficient non-cost effective substitutions for fuel). Major new technologies for nonfossil electricity would, moreover, need to be developed. In addition, massive fuel switching suffers from the same political barriers as high energy taxation.

Although the scenario suggests that a 20 percent reduction in absolute emissions of CO_2 associated with manufacturing activity might be achievable by 2010, it would require a strong, well-designed policy package involving extraordinary political will. A better understanding of the parameters is needed to build a more convincing scenario. There is a lack of good information on conservation supply curves,[8] on opportunities for process change, and on opportunities for recycling materials not now recycled. Improving this information will enable a more definitive analysis of the paramaters thereby strengthening the scenario.

Acknowledgments

I would like to thank my colleagues at Argonne, Cary Bloyd, Gale Boyd, Ron Fisher, Don Hanson, Ed Kokkelenberg, and John Molburg for their help.

References

1. *Manufacturing Energy Consumption Survey: Consumption of Energy 1985,* Energy Information Administration, US Department of Energy, Washington, DC (1988). Energy used for heat and power and as feedstock. The accounting convention adopted here includes the fuel used at power plants to generate electricity consumed: excludes the use of biomass by-products; includes, within heat and power,

Table 3.4. A 20% CO_2 reduction scenario: summary.

Business-as-usual emissions factor in 2010 relative to 1985		1.37
Policy-Driven Reduction Mechanism	**Individual Reduction (%)**	**Individual Reduction Factor**
Without CO_2 policies	0	1.00
Conservation and housekeeping	27	0.73
Fundamental process change	8	0.92
Switch from coal to natural gas	2	0.98
Advanced gas-turbine cogeneration	5	0.95
Switch from fossil fuels to electricity	1[a]	0.99[a]
Recycle	5	0.94
Cumulative reduction	42[a,b] (55)[c]	0.58[a] (0.45)[c]
Reduction-scenario emissions factor in 2010 relative to 1985		0.80[a] (0.70)[c]

[a] If electrical generation has the same CO_2 implications per kWh as in 1985.
[b] The percent reductions cannot be added; the reduction factors must be multiplied.
[c] The results in parentheses show the cumulative effect of all mechanisms if the carbon loading per kWh is reduced 23% from 1985 to 2010 for the average electricity used in the manufacturing sector.

metallurgical coal used to make coke; and excludes asphalt and road oil. Unless otherwise indicated, energy use will refer to heat and power applications, only.

2. *Productivity Measures for Selected Industrial and Government Sectors,* Bull. 2322, and unpublished supplementary data, Office of Productivity and Technology, Bureau of Labor Statistics. Department of Labor, Washington, DC (1989).
3. Jack Faucett Associates, *National Energy Accounts 1958–1981,* US Department of Commerce, Washington, DC (1984); and current data set from the Office of Business Analysis, Dept. of Commerce.
4. *Monthly Energy Review,* Energy Information Administration, US Department of Energy.
5. Williams, Robert, Eric Larson and Marc Ross, "Materials, Affluence and Industrial Energy Use," *Annual Review of Energy,* Vol. 12, pp. 99–144 (1987).
6. Annual Survey of Manufactures and Census of Manufactures, Bureau of the Census, Washington, DC, various years.
7. Ross, Marc, "Improving the Efficiency of Electricity Use in Manufacturing," *Science,* v. 12, pp. 311–317 (Apr. 21, 1989).
8. "End-Use Technologies in Industrial Applications," Chapter 6, Vol. I of *CO_2 Inventory and Policy Study, a Report to Congress by the Department of Energy,* (Draft 1989); and Marc Ross, "Conservation Supply Curves for Manufacturing," 25th Intersociety Energy Conversion Engineering Conference," American Institute of Chemical Engineering, New York, NY (1990).
9. Ross, Marc, "Capital Budgeting Practices of Twelve Large Manufacturers," *Financial Management,* pp. 15–22 (Winter 1986).
10. Price, Alan, and Marc Ross, "Reducing Industrial Electricity Costs—An Automotive Case Study," *Electricity Journal,* Vol. 2, No. 6, pp. 40–52 (July 1989).
11. Ross, Marc, "Industrial Energy Conservation and the Steel Industry of the US," *Energy* Vol. 12, pp. 1135–1152 (1987).
12. Boyd, Gale, John MacDonald, Marc Ross and Donald Hanson, "Separating the Changing Composition of US Manufacturing Production from Energy Efficiency Improvements: A Divisia Index Approach," *Energy Journal,* Vol. 8, No. 2, pp. 77–96 (Apr. 1987).
13. Boyd, Gale, and Marc Ross, "The Role of Sectoral Shifts in Trends in Electricity Use in United States and Swedish Manufacturing and in Comparing Forecasts," *Electricity,* T.B. Johansson, B. Borlund, R.H. Williams, Eds., Lund University Press (1989), Lund, Sweden.
14. Berg, Charles, *Science,* v. 199, p. 608 (Feb. 10, 1978); and "The Use of Electric Power and Growth of Productivity - One Engineer's View," unpublished, Northeastern Univ. (Apr. 1989).
15. Schmidt, Philip, *Electricity and Industrial Productivity,* Pergamon, New York (1984).
16. Williams, Robert, and Eric Larson, "Expanding Roles for Gas Turbines in Power Generation," see ref. 13.
17. Kelleher, E.G., "Conceptual Design of a Black Liquor Gasification Pilot Plant," a report to the Office of Industrial Programs, US Dept. of Energy (1987).
18. Potter, Richard and Tim Roberts, Eds., *Energy Savings in Wastes Recycling,* Elsevier Applied Science (1985).
19. Larson, Eric, Marc Ross and Robert Williams, "Beyond the Era of Materials," *Scientific American,* v. 254 No. 6 pp. 34–41 (June 1986).
20. *Policy Options for Stabilizing Global Climate, Executive Summary,* Daniel A. Lashof and Dennis A. Tirpak, Eds., a report to the Congress of the United States, US Environmental Protection Agency (Draft 1989).
21. *A Compendium of Options for Government Policy to Encourage Private Sector Responses to Potential Climate Changes,* Richard A. Bradley and Edward R. Williams, Eds., a report to the Congress of the United States by the Office of Environmental Analysis, US Department of Energy (Oct. 1989).
22. Marc Ross, "The Potential for Reducing The Energy Intensity and Carbon Dioxide Emissions in US Manufacturing" in *Energy: The Problem That Didn't Go Away* (working title), Ruth Howes and Anthony Fainberg, eds., American Institute of Physics, scheduled for publication 1990.
23. Marc Ross and Feng Liu, "The Energy Efficiency of the Steel Industry of China," (1990), submitted to *Energy, the International Journal.*

Cogeneration Applications of Biomass Gasifier/Gas Turbine Technologies in the Cane Sugar and Alcohol Industries

Joan M. Ogden
Robert H. Williams
Mark E. Fulmer
Center for Energy and Environmental Studies
Princeton University
Princeton, NJ

Abstract

Biomass integrated gasifier/gas turbine (BIG/GT) technologies for cogeneration or stand-alone power applications hold the promise of producing electricity at lower cost in many instances than most alternatives. BIG/GT technologies offer environmental benefits as well, including the potential for zero net carbon dioxide emissions, if the biomass feedstock is grown renewably. The gas turbine, in various power cycle configurations, is emerging as the technology of choice for thermal power generation, since the low unit capital cost of the gas turbine can now be complemented by high thermodynamic efficiency, owing largely to substantial technological improvements that have been made as a result of support for jet engine research and development for military aircraft applications.

The marriage of coal to the gas turbine through the use of coal-integrated gasifier/gas turbine (CIG/GT) technologies has been demonstrated. Some CIG/GT technologies could be be readily transferred to biomass with modest incremental R&D effort. In fact, biomass is inherently easier to gasify than coal. Moreover, BIG/GT technologies could probably be commercialized more quickly than corresponding CIG/GT versions, because most biomass contains negligible quantities of sulfur (the efficient and cost-effective removal of sulfur is the major technological hurdle to commercialization of cost-competitive CIG/GT technologies).

Eventually biomass grown on plantations dedicated to energy production could be the fuel for BIG/GT systems that generate power only. However, initial applications of these technologies will likely be for the cogeneration of electricity and process heat in industrial and agricultural industries where biomass residues are readily available as fuel.

For developing countries, the sugar cane industries that produce sugar and fuel alcohol are promising targets for near-term applications of BIG/GT technologies. In these industries, bagasse (the residue left after crushing the cane) could be used as BIG/GT fuel to provide the steam and electricity needs of the sugar factory or alcohol distillery during the milling season and to generate excess electricity for "export" to the utility grid. The barbojo (the tops and leaves of the sugar cane plant) could also be harvested and stored for use as fuel to produce more electricity during the off-season.

Depending on the choice of gas turbine technology and the extent to which barbojo can be used to produce electricity, the amount of electricity that can be produced from cane residues could be up to 44 times the onsite needs of the sugar factory or alcohol distillery. Electricity coproduction could help improve the economics of sugar and alcohol production. Under Brazilian conditions coproduction could make fuel alcohol highly competitive, even at the present depressed world oil price.

If sugar cane production were to continue to grow at the historical rate of 3 percent per year, and if BIG/GT technologies were fully deployed in the cane industries in 40 years time, then at the end of this period, potential electricity production from cane in the 80 sugar cane producing developing countries would be up to 2800 TWh/yr, about 70 percent more than total electricity production in these countries from all sources in 1987. Assuming 45 percent of all cane at that time would be associated with producing alcohol, cane alcohol production at the end of this period would be equivalent to 3.3 EJ/yr of crude oil, about 9 percent of total oil use in all developing countries in 1987. Assuming that "cane electricity" displaces electricity that would otherwise be produced with coal and that the fuel alcohol displaces petroleum, total carbon dioxide emissions in these 80 developing countries after 40 years would be 0.75 gigatonnes of carbon per year less than they would be otherwise. For comparison, total carbon dioxide emissions from all fossil fuel use in all developing countries totalled 1.55 gigatonnes of carbon in 1986.

Efficient and cost-competitive BIG/GT technologies suitable for widespread use in the sugar cane industries could be commercialized in less than five years' time at modest cost. Commercialization of these technologies would serve many development objectives while addressing the problem of global warming.

Introduction

The use of biomass as an energy source is one promising strategy for coping with the greenhouse problem. The combustion of biomass grown renewably leads to no net buildup of carbon dioxide in the atmosphere.[1] Moreover, if the biomass feedstock is grown in "energy plantations" with multi-year growing cycles on deforested or previously unforested land, the carbon extracted from the atmosphere in photosynthesis and sequestered in the steady state inventory of such plantations can be significant.

The US Environmental Protection Agency (EPA) recognized the importance of bioenergy in its exploration of global energy strategies for reducing CO_2 emissions (2). The EPA projected that while under business-as-usual conditions global CO_2 emissions would double by 2025 and triple by 2050, emissions could be roughly stabilized by pursuing appropriate energy demand and supply side strategies—with bioenergy strategies accounting for roughly three fifths of the projected potential reduction in CO_2 emissions (Table 1).

Biomass is already a significant global energy source—accounting for an estimated 48 EJ per year in developing countries and 7 EJ per year in industrialized countries at present (3)—in developing countries roughly equivalent to oil and coal use combined (Table 2). Despite its magnitude, bioenergy does not usually show up in the energy production and use statistics of developing countries, because there most of it is used as noncommercial energy—fuelwood, crop residues, and dung. Moreover, it is typically used inefficiently, mostly

Table 1. Role of Bioenergy in Reducing CO_2 Emissions in the US EPA Global Scenarios.[a]

	Global Fossil Fuel-Derived Carbon Dioxide Emissions (billion tonnes of carbon per year)		
Scenario	1985	2025	2050
RCWP	5.1	5.5	5.1
RCWP w/o Bioenergy[b]	5.1	8.6	11.1
RCW	5.1	10.3	15.3
RCW w/o Bioenergy[c]	5.1	10.5	16.5

[a] In 1989, the US Environmental Protection Agency carried out a major study to assess long-term energy strategies for reducing greenhouse gas emissions [US Environmental Protection Agency, "Policy Options for Stabilizing the Global Climate," February 1989 (draft)]. Alternative global energy scenarios were constructed as part of this exercise. Indicated here are variations on the EPA scenarios involving rapid economic growth. The reference case is the Rapidly Changing World (RCW) scenario. For the same economic growth conditions a Rapidly Changing World with Policy (RCWP) scenario was also constructed. The RCWP scenario differs from the RCW scenario in that it involves promoting (via public policy measures) various energy demand side and supply side measures to reduce carbon dioxide emissions.
[b] This shows what carbon dioxide emissions would be if coal were used instead of bioenergy in the RCWP scenario.
[c] This shows what carbon dioxide emissions would be if coal were used instead of bioenergy in the RCW scenario.

Table 2. Energy Use in and Fossil Fuel CO_2 Emissions from Developing Countries.[a]

	1980	1986	Growth Rate (%/yr, 1980–86)
Electricity[b] (TWh)			
Coal	357.8	647.5	10.39
Oil	376.6	415.7	1.66
Natural Gas	80.1	151.8	11.24
Fossil Subtotal	814.5	1215.0	6.89
Hydro	444.9	603.2	5.20
Nuclear	16.2	72.6	28.40
Totals	1275.6	1890.8	6.78
Fuels (EJ)			
Coal	18.69	21.97	2.72
Oil	25.04	28.26	2.03
Natural Gas	4.90	7.25	6.76
Totals	48.64	57.48	2.85
Primary Energy (EJ)			
Coal	23.46	30.61	4.53
Oil	30.07	33.80	1.97
Natural Gas	5.96	9.27	7.64
Hydro	4.87	6.63	5.20
Nuclear	0.18	0.80	28.40
Totals	64.54	81.11	3.88
CO_2 Emissions[c] (Megatonnes C)			
Electricity			
Coal	117.3	212.6	10.39
Oil	99.9	110.3	1.66
Natural Gas	14.4	27.4	11.24
Totals	231.6	350.3	7.14
Fuels			
Coal	459.9	540.3	2.72
Oil	498.4	562.4	2.03
Natural Gas	66.1	97.8	6.76
Totals	1024.9	1200.5	2.67
Primary Energy			
Coal	577.2	752.9	4.53
Oil	598.3	672.7	1.97
Natural Gas	80.5	125.2	7.64
Totals	1256.0	1550.8	3.58

[a] *Source*: Energy Information Administration, "International Energy Annual 1988," DOE/EIA-0219(88), Washington, DC, November 1989.
[b] The distribution of thermal power generation by fuel is from S. Meyers and C. Campbell, "Regional Electricity Supply and Consumption in Developing Countries, 1980–1986," Database Report, Lawrence Berkeley Laboratory, March 1989.
[c] Emission rates are assumed to be 24.6, 19.9, and 13.5 Megatonnes C per EJ for coal, oil, and natural gas, respectively.

in crude cooking stoves. To play a major role as an energy source, bioenergy must be elevated from its present status as the "poor man's oil" into a modern energy source by using advanced techniques to produce it renewably and to convert it efficiently into electricity and gaseous, liquid, and processed solid fuels (4).

Among the advanced technologies for modernizing bioenergy that could be commercialized in the near term, biomass integrated gasifier/gas turbine (BIG/GT) power generating technologies stand out as especially promising. The gas turbine, in various power cycle configurations, is emerging as the technology of choice for thermal power generation in the decades ahead, since the low unit capital cost of the gas turbine can now be complemented by high thermodynamic efficiency, owing largely to substantial technological improvements that have been made as a result of Department of Defense (DOD) support for jet engine research and development for military aircraft applications (5). The marriage of coal to the gas turbine through the use of coal-integrated gasifier/gas turbine (CIG/GT) technology has been demonstrated at Cool Water, California (6). While the particular technology demonstrated there is not likely to be adopted by utilities—because the commercial version would be no less costly than coal steam-electric power generating technology (5)—there are various possibilities for reducing the unit capital cost and improving efficiency (7).

Some CIG/GT technologies could be readily transferred to biomass with modest incremental R&D effort. In fact, biomass is inherently easier to gasify than coal. Moreover, BIG/GT technologies could probably be commercialized more quickly than corresponding CIG/GT versions because most biomass contains negligible quantities of sulfur, the efficient and cost-effective removal of which is the major technological hurdle that must be overcome before cost-competitive CIG/GT technologies can be commercialized (8).

Particularly promising for biomass applications is the combination of an airblown biomass gasifier (fixed bed or fluidized bed) with various aeroderivative turbine technologies (8). Gasification in air, instead of oxygen (used with Cool Water technology), makes it possible to eliminate the oxygen plant, whose capital cost is very scale-sensitive. Using airblown gasifiers with aeroderivative turbine technologies such as the steam-injected gas turbine (STIG—Figure 1a) or the intercooled steam-injected gas turbine (ISTIG—Figure 1b) instead of the gas turbine–steam turbine combined cycle (GT/ST CC—Figure 1c) used at Cool Water makes it possible to achieve both high efficiency and low unit capital costs at the modest scales needed for biomass applications. Because of the dispersed nature of the biomass resource (arising from the low efficiency of photosynthesis), biomass power plants must be relatively small in scale (less than 100 MW) to avoid high fuel transport costs. At these scales the unit capital cost of the GT/ST CC is relatively high, because of the strong scale economy associated with the steam turbine bottoming cycle. But the scale sensitivity of STIG and ISTIG cycles is much more modest, because these cycles do not require steam turbine bottoming cycles to achieve high efficiency (5).

Eventually biomass grown on plantations dedicated to energy production could be the fuel for BIG/STIG or BIG/ISTIG systems that generate power only (9). However, initial applications of these technologies will likely be for the cogeneration of electricity and process heat in industrial and agricultural industries where biomass residues are readily available as fuel.

Suitability of Sugar Cane Industries for Initial Applications of BIG/GTs

Sugar cane is a well-established, photosynthetically efficient crop. Global sugar cane production has been growing since the early 1960s at an average rate of about 3 percent per year and now totals nearly a billion tonnes per year (Table 3). The global average cane productivity is about 58 tonnes per hectare per year (10)[2]—equivalent as an energy feedstock to woody biomass production at a rate of 38 dry tonnes/ha/yr.[3] For comparison, yields of only about 10 dry tonnes/ha/yr have been achieved to date in US short rotation intensive culture demonstrations (13). While yields as high as 40 dry tonnes/ha/yr have been achieved for some plantations in developing countries, such yields have not yet been achieved on a large scale. Someday such productivities may be routinely achievable for wood plantations, but with sugar cane such high yields are already possible. Thus, sugar cane seems to be a good initial target for beginning a transition to an energy economy in which biomass plays a major role.

One economic motivation for cogenerating electricity as a byproduct of sugar production is to help stabilize revenue flows over time to sugar producers. Sugar is a commodity whose world market price has fluctuated widely (Figure 2). In contrast, it should be feasible, in principle, to sell electricity cogenerated at sugar factories to electric utilities for stable prices under long-term contracts. Sales of cogenerated electricity by sugar producers could be an important economic stabilizing activity, since the revenues from power sales would often be comparable to the revenues from sugar sales.

The prospect of shrinking markets for sugar is also stimulating interest in alternative products from cane, such as electricity. The demand for sugar is expected to continue growing in developing countries, owing to population growth and rising living standards, but future growth in the demand for sugar is expected to be much slower than in the past, largely because of the growing importance (in hard currency markets) of grain-based

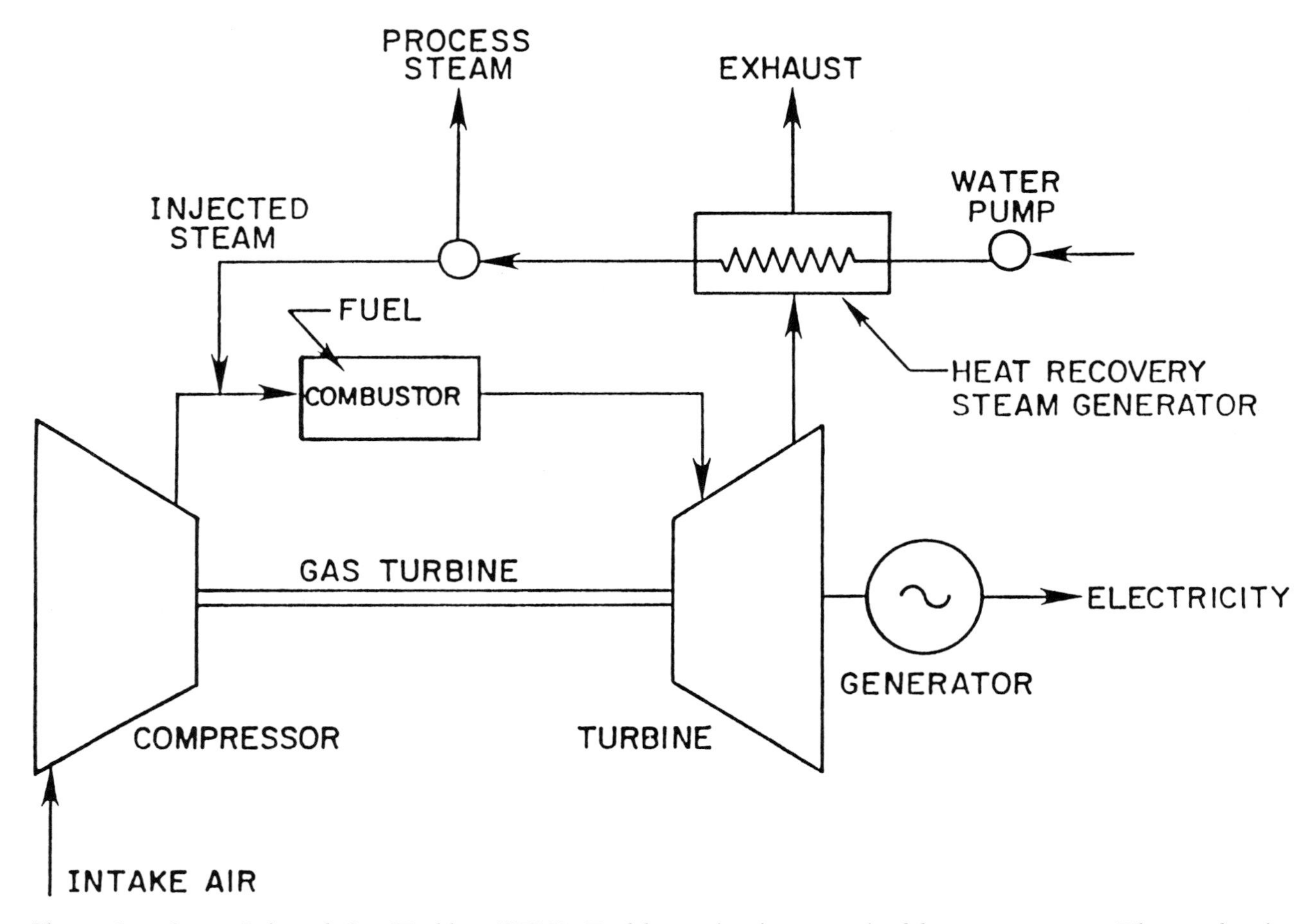

Figure 1a. Steam-Injected Gas Turbine (STIG). Fuel burns in air pressurized by a compressor. The combustion products drive a turbine. The turbine exhaust gases are used to raise steam in a heat recovery steam generator (HRSG). The steam not needed for process is injected into the combustor and at points further down the flow path for increased power output and electrical efficiency.

sweeteners, particularly high fructose corn syrups (HFCS). As a result, the World Bank has projected that world demand for sugar will increase at an average rate of only 1.5 percent per year through 1995 (14). Moreover, two potential developments may further erode sugar demand growth in the period near the turn of the century and beyond: the introduction of crystallized HFCS and the progressive introduction of new low-calorie sweeteners with physical and taste properties closer to those of sugar (14).

One alternative is ethanol produced from the fermentation of sugar juice. The technologies for producing fuel ethanol from sugar cane and using it as an automotive fuel have been developed on a large scale in Brazil (15–17). Alcohol production there increased from 3.4 billion l/yr in 1979/80 to 11.5 billion l/yr in 1987/88, accounting for 4.5 percent of all energy consumed in Brazil then. Several other developing countries are producing fuel ethanol from cane molasses (14,18-27) (Table 4), and in some of these countries national alcohol fuel programs have been implemented or are being considered (14–19).

Unlike the situation for ethanol derived from corn in the US, for which net energy balances can be unfavorable (1), energy balances for producing ethanol from cane are highly favorable. For Brazil, the agricultural energy required to produce a tonne of cane amounts to 305 MJ/tc, while the energy required to manufacture, operate, and maintain the distilleries requires another 143 MJ/tc (excluding the bagasse used as distillery fuel) (28,29). Since each liter of hydrous alcohol is equivalent as a "neat" (unblended) automotive fuel to 0.84 liters of gasoline (11), the 70 liters of alcohol that can be produced from a tonne of cane is equivalent to 4.2 times as much gasoline energy as the energy required to produce the ethanol. In Louisiana, where cane yields are comparable to those in Brazil but cane production is much more mechanized, the energy output/input ratio is

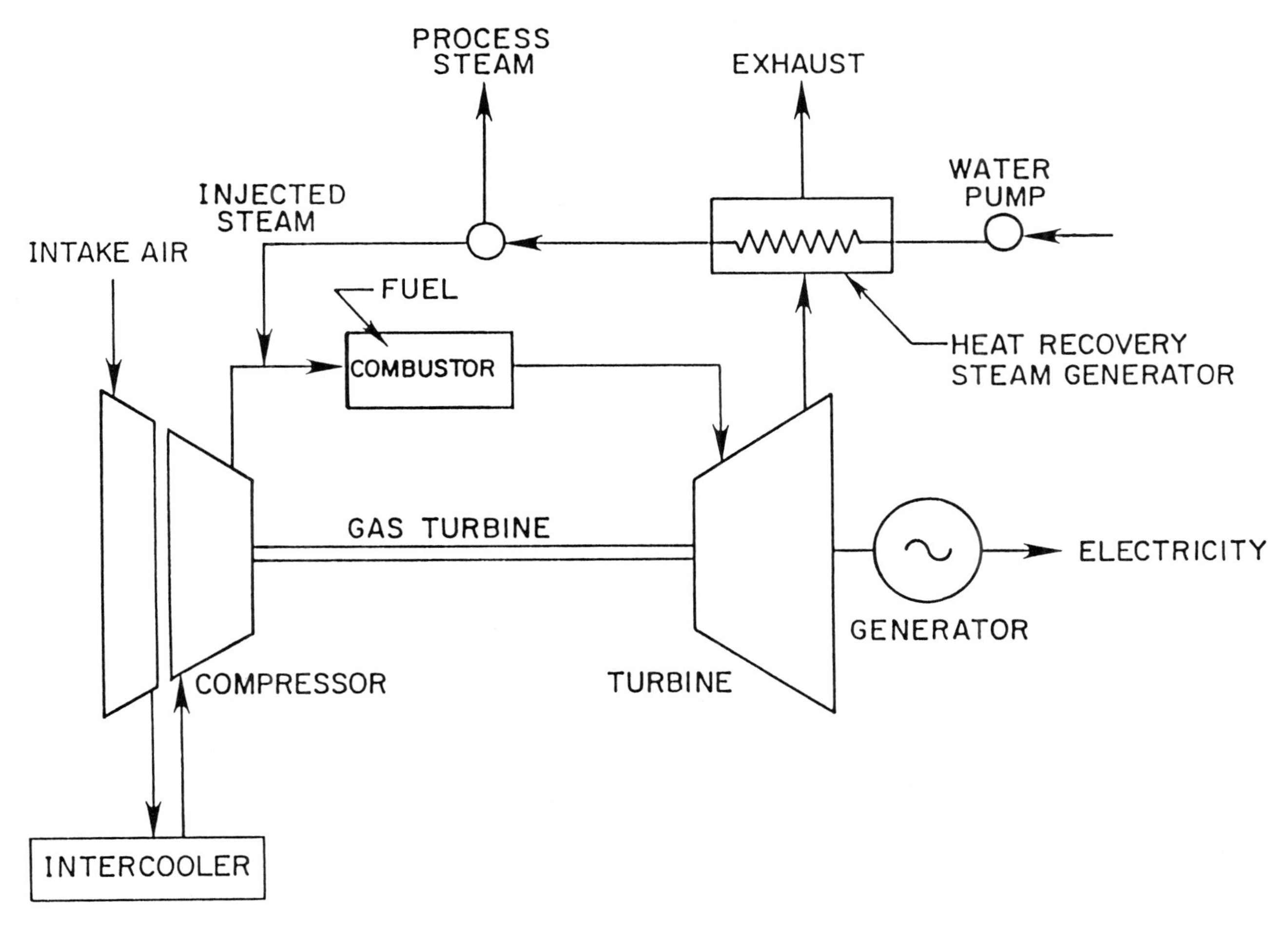

Figure 1b. Intercooled Steam-Injected Gas Turbine (ISTIG). Like STIG except that an intercooler between compressor stages reduces compressor work requirements and allows for operation at much higher turbine inlet temperatures, due to improved air cooling of turbine blades.

lower—2.9:1 (30,31)—but still very favorable. In Hawaii, where cane production is also heavily mechanized but annual cane yields are twice as high as in Brazil or Louisiana, the energy output/input ratio is comparable to that for Brazil (31). The useful energy output/input ratios for cane would be substantially improved if electricity were produced as a byproduct of alcohol, since the energy content of cane residues (bagasse and barbojo) is over five times the energy content of the alcohol.

Despite the impressive advances that have been made in the productivity of alcohol in Brazil,[4] cane-derived ethanol is not competitive as a neat fuel at today's low world oil price, although it is approximately competitive as a gasoline extender (gasohol).[5] As we shall show, the coproduction of alcohol and electricity using BIG/GTs could make ethanol competitive as a neat fuel in Brazil, even at today's low world oil price. Thus sugar cane industries producing sugar or alcohol are good candidates for initial applications of BIG/GT technologies.

Alternative Technologies for Cogeneration in the Cane Industries

In most sugar factories and alcohol distilleries today, small bagasse-fired steam turbine systems supplied with steam at 1.5-2.5 MPa provide just enough steam and electricity to meet onsite factory needs, typically about 350-500 kg of steam, and 15-25 kWh of electricity per tonne of cane milled (34,35). Typically, factories are designed to be somewhat energy inefficient, consuming all the available bagasse while just meeting factory energy demands, so that excess bagasse does not accu-

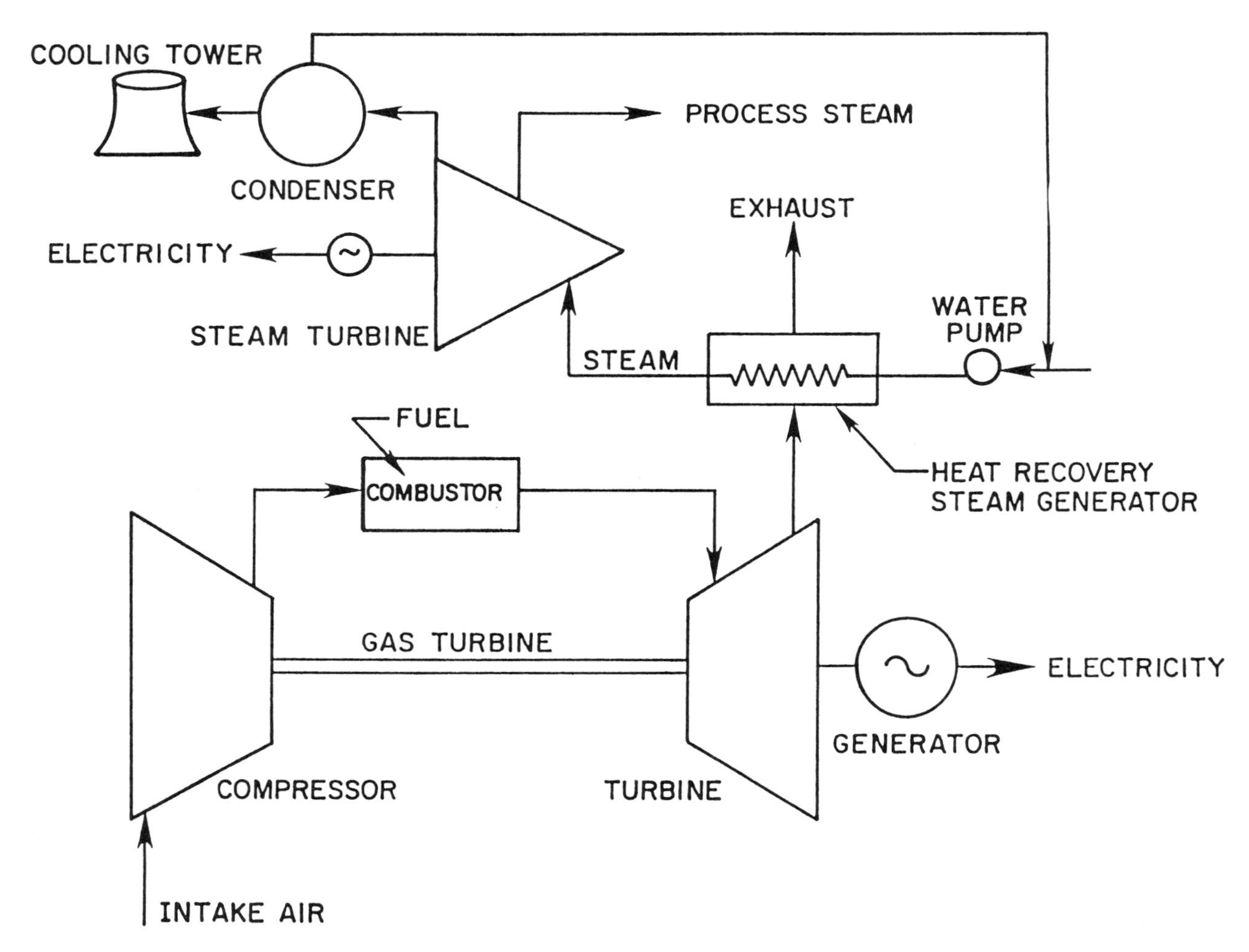

Figure 1c. Combined Cycle. Steam from the HRSG is used to produce extra power in a condensing steam turbine, from which some steam might be bled for process applications.

mulate and become a disposal problem. (In southeastern Brazil, where bagasse is sold as a boiler fuel or as a component of cattle feed [36,37], some factories have been made more energy efficient so as to free up surplus bagasse for these markets [38–41].)

In a few sugar factories and alcohol distilleries, modern condensing-extraction steam turbine cogeneration systems (CEST) operating at turbine inlet pressures of 4.0-8.0 MPa have been installed (42,43,38). With these systems, it is possible to produce enough steam to run a typical factory (350-500 kg/tc), plus 70-120 kWh/tc of electricity, or about 50-100 kWh/tc in excess of onsite needs. The extra electricity can be made available to other users by interconnecting the cogenerator with the utility grid. During the milling season the CEST cogeneration system is fueled with 50 percent wet bagasse, as it comes from the mill. In the off-season, CEST units can be operated in the condensing mode producing power only, fired with barbojo (cane tops and leaves), wood, or heavy fuel oil.

While the introduction of modern CEST units can improve the performance of sugar or alcohol factories, Rankine cycles are relatively inefficient for the steam conditions that can be used at the modest scales needed for biomass applications,[6] and the strong scale sensitivity of the capital cost (Figure 3) makes CEST technology only marginally attractive for many applications.

Much better thermodynamic and economic performance could be achieved at the modest scales needed for biomass-fired applications if BIG/GT systems could be used instead. The first generation BIG/GT technology may well be BIG/STIG (8), combining a commercially available pressurized fixed bed gasifier with a steam-injected gas turbine (Figure 1a), which is commercially available for natural gas applications (5).

In a BIG/STIG cogeneration system, biomass would

Table 3. Historical Data on World Sugar Cane Production.[a]

	Harvested Area (million ha)	Yield[b,c] (tc/ha/yr)	Production[b] (million tc)
1987	16.56	58.44	967.9
1986	15.82	58.77	929.7
1985	15.87	58.44	927.3
1984	15.90	58.87	935.8
1983	15.43	58.12	896.9
1982	15.09	58.82	887.7
1981	13.78	57.24	788.8
1980	13.24	54.54	721.9
1979	13.58	55.55	754.5
1978	13.30	57.31	762.1
1977	13.34	55.13	735.6
1976	12.71	54.43	691.6
1975	12.35	53.47	660.5
1974	12.15	53.88	654.9
1961–65	9.52	49.39	470.1

[a] Source: FAO, *Production Yearbook*, various years.
[b] This is the harvested wet weight, which does not include the weight of the barbojo (tops and leaves).
[c] There is considerable variation in average yield from country to country. The average yield for some countries (Egypt, Malawi, Swaziland, Zambia, Zimbabwe, Peru) is about twice the global average.

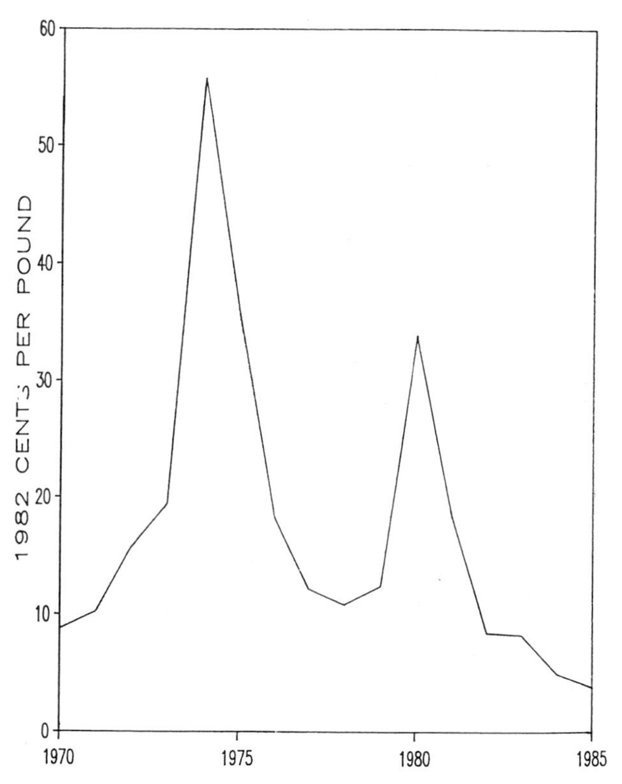

Figure 2. The world market price for sugar, 1970–1985. Current dollar prices (from International Sugar Organization, *Sugar Yearbook*, London, annual) have been converted to 1982 dollars using the GNP deflator.

be gasified in air to form a gas of low energy content (having perhaps 15 to 20 percent of the heating value of natural gas), which fuels a gas turbine. Steam is raised for the mills, for the gasifier, and for process needs in a heat recovery steam generator (HRSG), which utilizes the hot exhaust gases exiting the turbine. Any steam not needed for the factory or the gasifier could be injected into the combustor or the turbine, boosting the electrical output and efficiency of the system.

As in the case of CEST technology, the BIG/STIG system would be fueled with bagasse during the milling season. However, the bagasse would probably have to be densified (either briquetted or pelletized) if used in a fixed bed gasifier originally designed for use with coal (8). In the off-season the system could be fueled with densified barbojo, wood, or distillate oil. Because of their much higher electrical efficiencies (Table 5) and lower unit capital costs (Table 6 and Figure 3), BIG/STIG systems would be able to provide more than twice as much electricity per tonne of cane at lower costs than CEST systems.

BIG/STIG could be the first in a series of increasingly attractive gas turbine systems that could be designed for biomass applications. Even higher efficiencies, more electricity production per tonne of cane, and lower unit capital costs could be achieved by coupling an intercooled steam-injected gas turbine (ISTIG), instead of a STIG unit, to the biomass gasifier (Tables 5 and 6). By adding an intercooler to the compressor (Figure 1b), the output of the turbine would more than double and the turbine efficiency would be significantly increased (Table 5). These performance improvements arise because with intercooling less compressor work is required and it is possible to operate at a significantly higher turbine inlet temperature (TIT). By bleeding air from the compressor to cool the turbine blades it is possible for the turbine to operate at a higher TIT without exceeding turbine blade temperature limits (8). As in the case of the BIG/STIG, the BIG/ISTIG would be fueled with densified bagasse during the milling season and with densified barbojo, wood, or distillate oil during the off-season.

While BIG/STIG and BIG/ISTIG systems are not commercially available at present, they could probably be commercialized much more quickly than the corresponding coal-fired systems because there would be no need for sulfur removal technology with most biomass feedstocks, including cane residues. Not having to remove sulfur means that the unit capital costs for biomass systems would probably be about 20 percent less than for the corresponding coal-fired versions of these technologies (Table 7).

Table 4. Fuel Ethanol Production from Sugar Cane.

	Million liters/yr
South America	
Brazil[a]	11,700
Argentina[b]	380
Paraguay[b]	26
Colombia[b]	38
Central America and Caribbean	
Costa Rica[b]	31
El Salvador[b]	15
Guatemala[b]	0.2
Jamaica[d]	15
Africa	
Kenya[c,e]	18
Malawi[c]	11
Zimbabwe[c,e]	42
Mali[b]	2
Asia and Oceania	
Thailand[b]	203
Phillipines[b]	10
New Zealand[b]	15

[a] For the 1988/1989 season. "Agroindustria Canavieira: Um Perfil," Copersucar, Sao Paulo, Brazil, 1989.
[b] 1984 ethanol production capacity installed. "Worldwide Review of Biomass Based Ethanol Activities," Meridian Corporation Report, Contract No. MC-JC-85FB-008, 1985.
[c] "Electricity and Ethanol Options in Southern Africa," USAID Office of Energy Report, September 1988.
[d] A 180,000 liter/day distillery was installed in 1988, and the planned production is 15 million liters/year. To date, only about one million liters have been produced due to uncertain market conditions. (M.G. Hylton, Jamaican Sugar Industry Research Institute, Bernard Lodge, Jamaica, private communications, 1990).
[e] Capacity installed as of 1984. "Power Alcohol in Kenya and Zimbabwe—A Case Study of the Transfer of a Renewable Energy Technology," United Nations Trade and Development Board Report GE.84-55979, June 5, 1984.

It is estimated that BIG/STIG technology could be commercialized in less than five years (44). The R&D effort required to commercialize the BIG/STIG system would be relatively modest, because much of the development work already carried out for coal integrated gasifier/steam-injected gas turbines (7) would be applicable to biomass (45). ISTIG technology would probably be commercialized first using natural gas as fuel, which would require a four- to five-year development effort (44). If this technology were developed at the same time as BIG/STIG technology, BIG/ISTIG technology could become commercially available shortly thereafter. More detailed discussions relating to the commercialization of these technologies are presented elsewhere (5,8,45).

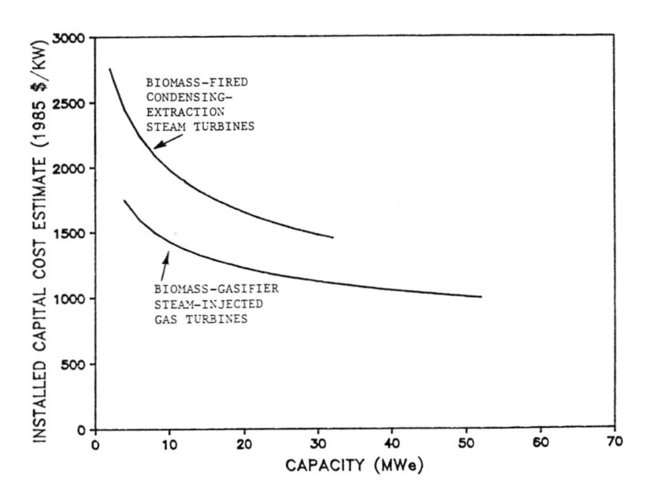

Figure 3. Estimated unit installed capital costs for Biomass-Fired Condensing/Extraction Steam Turbine (CEST) and Biomass-Integrated Gasifier/Steam-Injected Gas Turbine (BIG/STIG) units for cogeneration applications. (Source: E.D. Larson and R.H. Williams, "Biomass Gasifier Steam-Injected Gas Turbine Cogeneration," *Journal of Engineering for Gas Turbines and Power,* vol. 112, pp. 157–163, April 1990.)

Integrating Cogeneration Systems with Sugar Factories or Alcohol Distilleries

Figure 4 shows schematically how various cogeneration systems could be integrated with a sugar factory or alcohol distillery. Figure 4a shows a typical existing sugar factory steam turbine cogeneration system, producing just enough steam and electricity for factory needs. Steam is produced at "moderate" pressure (1.5-2.0 MPa) for use in back-pressure cane mill and turbo-alternator turbines. The "low pressure" (0.15-0.25 MPa) exhaust steam from these turbines is then used to provide heat for processing the cane juice into sugar or alcohol. In Figure 4b, a CEST cogeneration system is shown. Steam is extracted at 1.5-2.0 MPa for use in the factory as before. About 50-100 kWh/tc of electricity in excess of onsite needs can be produced during the milling season. A biomass gasifier/gas turbine cogeneration system is shown in Figure 4c. With the BIG/STIG system about 240 kWh/tc of excess electricity is produced during the milling season. With a BIG/ISTIG system (not shown) excess electricity production would increase to about 285 kWh/tc in season.

In order to use a particular cogeneration system at a sugar factory or alcohol distillery, the system must meet the factory's steam and electricity needs. Figure 5 shows the steam and electricity production for CEST, BIG/STIG, and BIG/ISTIG cogeneration systems, operated on biomass fuel. For each technology, a range of

Table 5. Estimated Performance of Biomass-Fired Cogeneration Systems.

	Cogeneration						Power Only			
	Electricity		Steam		Fuel[d]	Cane	Electricity		Fuel[c]	Cane
	(MW)	(% of fuel[d])	(T/H)	(% of fuel[d])	(T/H)	(T/H)	(MW)	(% of fuel[d])	(T/H)	(T/H)
CEST[a,b]										
Generic	17.5	13.0	65.6	35.9	50.8	169	27.0	20.3	50.2	167
Generic	6.1	11.4	26.4	36.4	20.2	67	10.0	17.8	21.2	71
Generic	1.8	10.1	9.0	37.2	6.73	22	3.0	15.7	7.22	24
BIG/STIG[b]										
LM-5000	38.8	31.3	47.7	30.0	27.6	157	53.0	35.6	33.0	188
LM-1600	15.0	29.8	21.8	33.8	11.2	65	20.0	33.0	13.2	75
GE-38	4.0	29.1	5.7	32.4	3.06	17	5.4	33.1	3.63	21
BIG/ISTIG[d]										
LM-8000	97	37.9	76.2	25.4	57.7	328	111.2	42.9	57.3	325

[a] Adapted from E.D. Larson and R.H. Williams, "Biomass Gasifier Steam-Injected Gas Turbine Cogeneration," *Journal of Engineering for Gas Turbines and Power,* vol. 112, pp. 157–163, April 1990, except that here the gasifier efficiency for biomass is assumed to be the same as for coal.
[b] For 6.3 MPa, 482°C steam at the turbine inlet.
[c] It is assumed that the BIG/STIG and BIG/ISTIG use briquetted bagasse or barbojo (15% moisture), with a higher heating value of 16,166 kJ/kg, and that the CEST uses bagasse (50% moisture) with a higher heating value of 9530 kJ/kg. It is further assumed that 300 kg of bagasse (50% moisture) are produced per tonne of cane milled. If briquetting is required the corresponding quantity would be 176 kg (15% moisture).
[d] Output, in energy units, as a percentage of the higher heating value of the fuel input.
[e] Preliminary estimate of steam and electricity production, based on performance with coal.

Table 6. Estimated Capital and Operating Costs for Biomass-Fired Cogeneration Systems.

	Capacity (MW)	Installed Cost ($/kW)	Maintenance[b] Fixed (1000$/yr)	Maintenance[b] Variable ($/kWh)	Labor[b,c] (1000$/yr)
CEST[a]					
Generic	27.0	1556	664	0.003	129.6
Generic	10.0	2096	246	0.003	97.2
Generic	3.0	3008	73.8	0.003	97.2
BIG/STIG[a,d]					
LM-5000	53.0	990	1304	0.001	297.0
LM-1600	20.0	1230	492	0.001	108.0
GE-38	5.4	1650	133	0.001	97.2
BIG/ISTIG					
LM-8000	111.2[e]	770[e]	2736	0.001	405.0

[a] Adapted from E.D. Larson and R.H. Williams, "Biomass Gasifier Steam-Injected Gas Turbine Cogeneration," *Journal of Engineering for Gas Turbines and Power,* vol. 112, pp. 157–163, April 1990.
[b] Appendix D in E.D. Larson, J.M. Ogden, R.H. Williams, "Steam-Injected Gas Turbine Cogeneration for the Cane Sugar Industry," Princeton University, Center for Energy and Environmental Studies Report No. 217, September 1987, and E.D. Larson, private communication, 1990.
[c] For Jamaican labor conditions.
[d] In general, the estimated unit cost is $2371 * MW^{-0.22}$ where MW is the installed capacity in MW. (See E.D. Larson and R.H. Williams, "Biomass Gasifier Steam-Injected Gas Turbine Cogeneration.")
[e] See Table 7. In general, the estimated unit cost is $2167 * MW^{-0.22}$ assuming the same scaling law applies to both BIG/STIG and BIG/ISTIG technologies.

Table 7. Estimated Installed Capital Cost for IG/STIG and IG/ISTIG Power Plants Fueled with Coal and Biomass (in 1986$/kW).

	CIG/STIG[a]	BIG/STIG[b,c]	CIG/ISTIG[a]	BIG/ISTIG[b,d]
I. Process Capital Cost				
Fuel Handling	39.6	39.6	36.7	36.7
Blast Air System	13.5	13.5	9.6	9.6
Gasification Plant	160.9	160.9	83.1	83.1
Raw Gas Physical Clean-up	8.8	8.8	7.7	7.7
Raw Gas Chemical Clean-up	175.9	0.0	150.9	0.0
Gas turbine/HRSG	294.4	294.4	256.4	256.4
Balance of Plant				
Mechanical	40.2	40.2	22.0	22.0
Electrical	65.0	65.0	48.4	48.4
Civil	65.5	65.5	60.7	60.7
Subtotal	862.9	687.0	686.5	535.6
II. Total Plant Cost				
Process Plant Cost	862.9	687.0	686.5	535.6
Engineering Home Office (10%)	86.3	68.7	68.6	53.6
Process Contingency (6.2%)	53.6	42.6	42.5	33.2
Project Contingency (17.4%)	150.4	119.5	119.6	93.2
Subtotal	1153.2	917.8	917.2	715.6
III. Total Plant Investment				
Total Plant Cost	1153.2	917.8	917.2	715.6
AFDC (1.8%, 2 yr construction)	20.8	16.5	16.5	12.9
Subtotal	1174.0	934.3	933.7	728.5
IV. Total Capital Requirement				
Total Plant Investment	1174.0	934.3	933.7	728.5
Preproduction Costs (2.8%)	32.3	26.2	26.2	20.4
Inventory Capital (2.8%)	32.3	26.2	26.2	20.4
Initial Chemicals, Catalysts	2.5	0.0	2.3	0.0
Land	1.3	1.3	1.3	1.3
Total	1242.0	988.0	990.0	771.0

[a] J.C. Corman, "System Analysis of Simplified IGCC Plants," report prepared for the US Dept. of Energy by the Corporate R&D Center, the General Electric Company, Schenectady, NY, September, 1986.
[b] It is assumed that BIG/STIG (BIG/ISTIG) costs are the same as CIG/STIG (CIG/ISTIG) costs, except that the raw gas chemical clean-up phase required for coal would not be needed for biomass, because of its lower sulfur content.
[c] For a 53 MW unit.
[d] For a 111 MW unit.

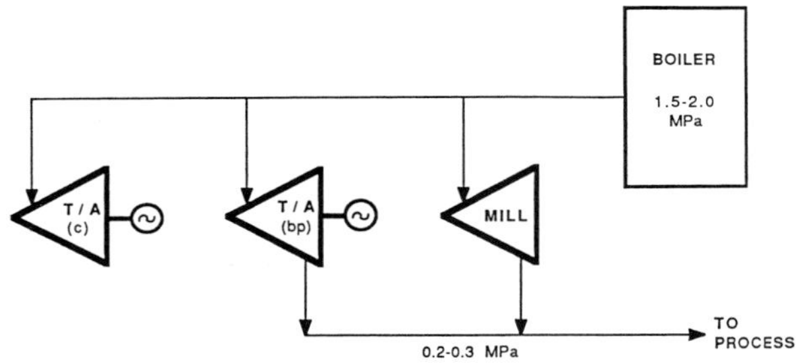

Figure 4a. Cogeneration system for cane sugar factory or alcohol distillery using conventional steam turbines typical of today's factories.

operating values is possible, depending on how much steam is produced for process. For each system, the lower the process steam demand, the higher the electricity production. When the process steam demand is zero, as in off-season operation, electricity production is maximized. The maximum steam production possible with the system is given at the right hand endpoint of each line. Also shown are the steam and electricity demands for typical existing sugar factories or alcohol distilleries.

The more electrically efficient gas turbine cogeneration systems have lower steam production than the CEST system. With BIG/STIG, steam can be produced at a maximum rate of about 300 kilograms (@ 2.0 MPa, 316°C) per tonne of cane milled (kg/tc). With BIG/ISTIG, maximum steam production is about 230 kg/tc.

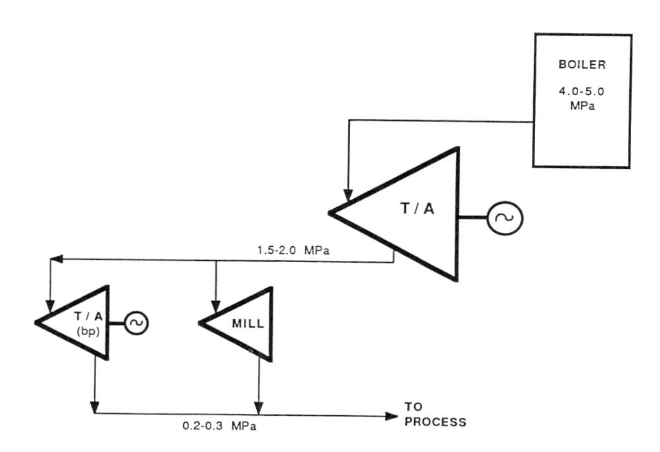

Figure 4b. Cogeneration system for cane sugar factory or alcohol distillery using high-pressure, condensing/extraction steam turbine (CEST).

Since a typical existing sugar factory or alcohol distillery requires about 350-500 kg/tc of process steam, factory steam economy improvement measures must be carried out before gas turbine cogeneration systems can be installed.

It has been shown that by adapting to the cane sugar industry steam economy measures widely used in other process industries, it would be feasible to integrate gas turbine cogeneration technologies into cane sugar factories (46-48). In this paper this analysis is extended to autonomous alcohol distilleries, in which ethanol is produced directly from sugar juice. A more general treatment, also showing how to integrate gas turbine cogeneration technologies into distilleries annexed to sugar factories that coproduce sugar, alcohol, and electricity, is given elsewhere (49).

When sugar cane is delivered to an autonomous distillery, it is washed, chopped, and then milled in a series of roller mills to extract "raw" cane juice, which typically contains over 90 percent of the sucrose in the cane. The raw cane juice is filtered and heated. In some Brazilian autonomous distilleries, the juice is then cooled and sent directly to the fermentation stage. Alternatively, the juice can be limed and clarified, and the clear juice is often concentrated in an evaporator from typical values of 13° Brix (13 percent solids in juice) to about 18-20° Brix, which is preferable for fermentation. The concentrated juice is then fermented to produce a "beer," a water/ethanol mixture that is typically 8 to 10 percent ethanol. The beer is distilled to produce about 70 liters of ethanol per tonne of cane milled. In addition, methane can be produced via anaerobic digestion from the stillage (a waste product of distillation), typically at a yield of about 0.33 GJ/tc (12), about one fifth of the energy content of the alcohol.

Alcohol distilleries would be modest in scale and capital cost compared to fossil synfuel facilities. A "large" distillery would process 4000 tonnes of cane and produce 280,000 liters of ethanol per day—equivalent on a contained energy basis to about 1,200 barrels of gasoline per day. While published estimates of the installed capital cost of autonomous distillery equipment vary widely (14,16,23,24,27,50,51), such a distillery, if built in Brazil, where virtually all distilleries have been built, would cost about $18 million (Table 8). By way of contrast, a plant producing methanol from coal would typically produce methanol at a gasoline energy-equivalent rate of 20,000 barrels per day and cost $1.5 billion (52).

The process steam, mechanical work, and electricity requirements of the distillery are provided by using bagasse as fuel. Steam and electricity use in a typical Brazilian autonomous distillery producing hydrous ethanol are shown in Table 9 (first column) and Figure 6a (38). A conventional distillation system requiring 3.3 kilograms of low pressure (0.15-0.25 MPa) steam per liter (kg/l) of hydrous ethanol is used. The overall factory steam demand for this case is 5.8 kg/l, or 466 kg/tc. (A recent survey of autonomous distilleries in the São Paulo area indicated steam use of 420-550 kg/tc [41].)

Process Technologies for Autonomous Distilleries

There are many possible design options for the coproduction of alcohol and electricity from cane (Table 10). For each cogeneration option, a set of end-use technologies is needed that will make it possible to serve the steam needs of the distillery with the heat output of the cogeneration system, while making as much electricity as can be justified cost-effectively. As already noted, this implies that steam economy measures must be implemented before gas turbine cogeneration systems can be employed.

The main uses of moderate pressure steam in a conventional autonomous distillery (first column, Table 9) are for running turbines to provide the mechanical power needed for cane milling (226 kg/tc) and for generating electricity (175 kg/tc). The low pressure exhaust steam from these turbines (plus some 65 kg/tc of steam which is throttled to low pressure) is then used for distillation (256 kg/tc), evaporation (61 kg/tc), and juice heating (130 kg/tc). Here the various technologies that could be used for each stage of the process are described, noting opportunities for steam conservation. Details are presented elsewhere (49).

Cane Preparation/Milling. To extract the sugar juice, the cane is typically chopped, shredded, and crushed in a series of roller mills. In virtually all autonomous distilleries and sugar factories, the mechanical power requirements for these cane preparation and milling op-

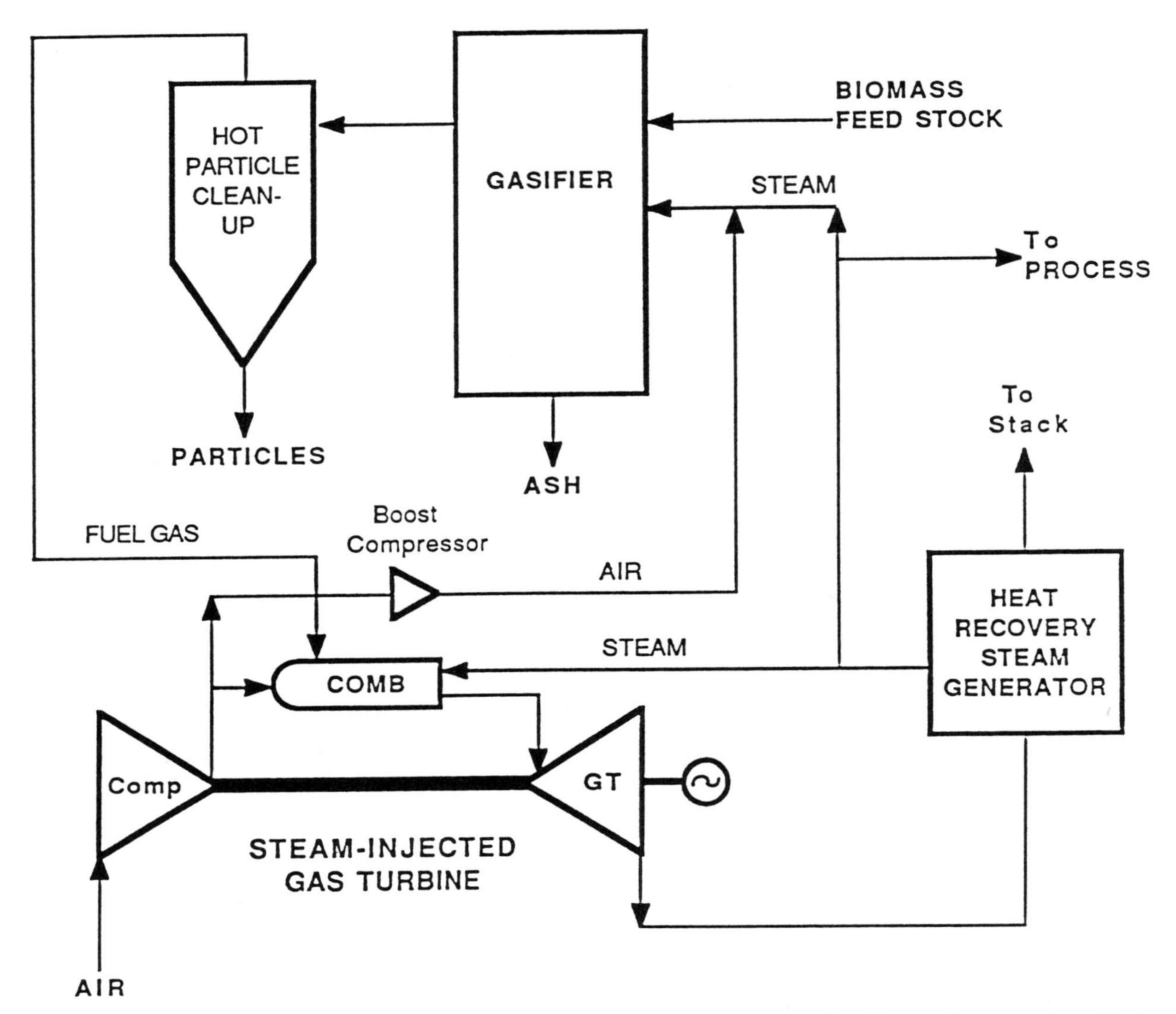

Figure 4c. Cogeneration system for cane sugar factory or alcohol distillery using biomass-integrated gasifier/steam-injected gas turbine (BIG/STIG).

erations are provided by back-pressure steam turbines. Typically, these turbines require 200-250 kg steam/tc at 1.5- 2.0 MPa and exhaust steam at 0.15-0.25 MPa. These operations could also be driven by electric motors. Alternatively, a diffuser could be used to extract the sugar from the cane. In a diffuser the cane is chopped, shredded, and immersed in water to extract the sugar, and the fibers are then dewatered in a single set of roller mills.

The low-pressure (0.15-0.25 MPa) exhaust steam from mill turbines is utilized for process heat (juice heating, evaporation, distillation). Unless the demand for low-pressure steam is less than the amount of this exhaust (e.g., less than about 200 kg/tc), there is little reason to reduce the steam use in cane preparation and milling. However, if the demand for low-pressure steam were reduced, it would be feasible to reduce steam consumption in these operations by using more efficient, higher pressure turbines or by using diffusers or electric mills.

Juice Heating. In autonomous distilleries, juice is generally heated directly with exhaust steam or with vapor bled from the evaporator. Depending on the configuration, bleeding vapor from the evaporator can reduce steam use. In some cases the hot condensates from the evaporator can be utilized, saving some steam.

Evaporation. Evaporation is typically done in multi-effect, short-tube, rising-film evaporators. Increasing

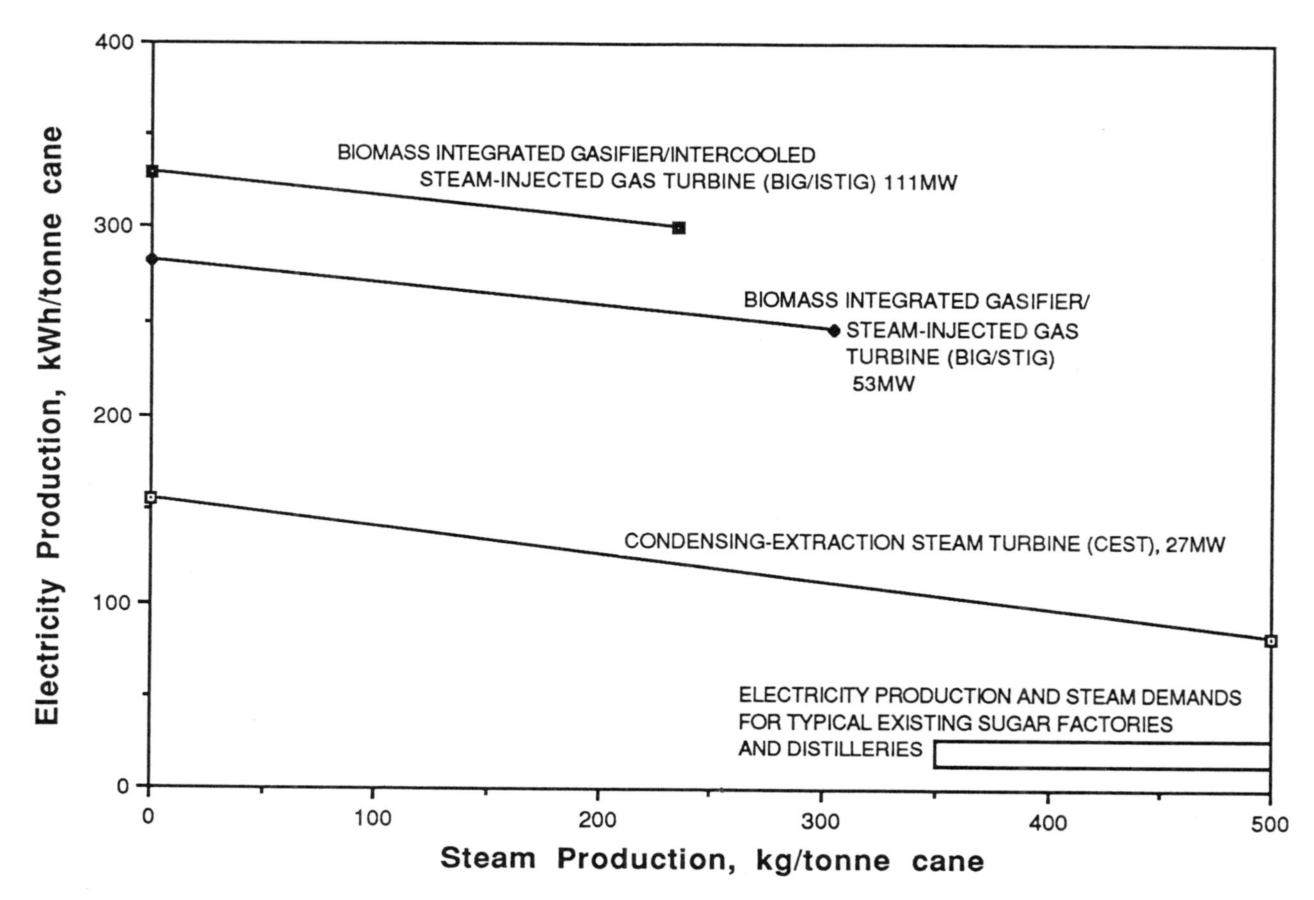

Figure 5. Electricity and steam production estimates for alternative bagasse-fired cogeneration systems. Performance estimates are shown for the condensing-extraction steam turbine (CEST), the biomass integrated gasifier/steam-injected gas turbine (BIG/STIG), and the biomass integrated gasifier/intercooled steam-injected gas turbine (BIG/ISTIG) operating at sugar factories or alcohol distilleries during the milling season. Steam production is given in kilograms of steam produced per tonne of cane milled (kg/tc) and electricity production is in kilowatt hours per tonne of cane (kWh/tc). Also shown are the steam and electricity production from a typical sugar factory or distillery cogeneration system today.

the number of effects can reduce the steam needed for evaporation. If rising-film evaporators are replaced with falling-film evaporators, which can operate at higher temperatures and thus make better use of multiple-effect evaporator configurations, overall steam use can be reduced. Falling-film evaporators are widely used in the beet sugar, dairy, and some other process industries and are being used experimentally in the cane sugar industry (21,46).

Distillation. After fermentation, the ethanol must be separated from the other components of the "beer." (Beer is about 8 to 10 percent ethanol. Most of the rest is water, with traces of fusel oils and other impurities.) Distillation is by far the most common ethanol separation process, although a number of other techniques have been developed (53-55). (See Table 11 for a summary of ethanol separation technologies.) The purity of the alcohol produced determines the type of distillation equipment required. Hydrous alcohol (96 percent to 96.5 percent) is used for unblended motor fuel; absolute or anhydrous alcohol (99.9 percent), is often used for blending with gasoline. The purer the final ethanol, the more complex the distillation process. In this paper emphasis is given to the production of hydrous ethanol for use as a neat motor fuel.

Over one half of the steam use in a typical autonomous distillery goes to distillation (Table 9). Until recently, relatively little attention was given to energy efficiency in cane alcohol distilleries. However, with the large-scale development of a cane-based fuel alcohol industry in Brazil and competitive pressures on this industry as a result of the low world oil price, there is growing interest in more efficient distillation technology. Brazilian manufacturers have developed energy-saving

Table 8. Estimated Capital and Operating Costs for an Autonomous Distillery Milling (excluding the costs for boilers and generating equipment), based on Brazilian Experience.[a,b]

Total Installed Capital Cost (million $)	$18.083
Fixed Operating Costs (thousand $/yr)	
Labor	560
Maintenance	362
Supply	36
Insurance	90
Total	1,048
Cane Cost ($/tc)	8.07
Other Variable Costs ($/tc)	0.176

[a] For a distillery processing 4000 tonnes of cane per day and producing 73 liters of ethanol per tonne of cane.
[b] Source: J. Goldemberg, J.R. Moreira, P.U.M. Dos Santos, and G.E. Serra, "Ethanol Fuel: A Use of Biomass Energy in Brazil," *Ambio*, vol. 14, no. 4–5, 1985; G. Serra and J.R. Moreira, University of São Paulo, Brazil, private communications, 1989.

conserving distillation systems for use in autonomous distilleries. With these commercially available Brazilian systems the demand for steam has been reduced from 3.0-3.5 kg/liter of hydrous ethanol to 1.5-2.0 kg/liter (15,16,36).

One way to reduce steam requirements for distillation is through the use of heat integration techniques that use hot and cold streams within the factory as efficiently as possible to accomplish the required heating with a minimum of external steam input (53,56). With optimized heat integrated ethanol distillation systems, it is feasible to reduce distillation steam use to 1.2 kg/liter for motor grade hydrous ethanol and to 2.2 kg/liter for anhydrous ethanol (56) or less (53) (Table 11). Through the use of mechanical vapor recompression systems, which substitute electricity for steam, overall heat requirements for distillation could be reduced further. These systems have been used for ethanol production but not in the cane sugar industry (53,56).

Dramatic energy savings are theoretically possible if non-distillation separation techniques are used (53,54). Several of these methods have been demonstrated in the laboratory (Table 11), but none are in commercial use.

Design of Autonomous Distilleries for Cogeneration Applications

The alternative distillery designs that can be considered for the coproduction of alcohol and electricity with CEST, BIG/STIG, or BIG/ISTIG cogeneration systems include both conventional and steam-conserving options.

Conventional Distillery Designs. Figure 6a shows a factory flow diagram for a conventional Brazilian autonomous distillery, which mills 125 tonnes of cane per hour to produce 10,000 liters of hydrous ethanol per hour (80 liters of ethanol per tonne of cane) (38). The juice is heated, clarified, concentrated in a conventional short-tube, rising-film, three-effect evaporator, fermented, and distilled in a conventional two-column distillation system requiring 3.2 kg steam per liter of hydrous ethanol produced. The overall steam demand for the factory is 5.8 kg/liter of alcohol or 466 kg/tc—comparable to the amount of steam required in a typical sugar factory (34,35). A CEST cogeneration system could meet the process steam demand for this type of factory, producing 90-95 kWh/tc of excess electricity during the milling season (Table 9, first column).

Steam-Conserving Distillery Designs. The largest single use of steam in the conventional factory is for distillation, about 3.2 kg steam per liter of hydrous ethanol or 256 kg/tc (see Table 9). The next largest use of steam in the factory is for juice processing (juice heating and evaporation to 18° Brix), some 130 kg/tc. There are various opportunities for steam conservation in autonomous distilleries.

Consider first a design requiring 1.9 liters of steam per liter of hydrous alcohol (155 kg/tc) for distillation—consistent with designs demonstrated in Brazil for which steam used for distillation is reduced from 3.0-3.5 kg/liter to about 1.5-2.0 kg/liter by using a higher pressure distillation system with heat integration (16). The total amount of steam used for juice heating plus evaporation can be reduced by using a five-effect evaporator and by bleeding vapor from the evaporator for juice heating. The factory flows for this steam-conserving factory with a low steam use distillery and an energy efficient juice processing system are shown in Figure 6b (38). The total steam demand for this case is 3.2 kg/liter (258 kg/tc), low enough so that a BIG/STIG cogeneration system could be used (Table 9, middle column).

Further reductions in the distillery steam use could be achieved by reducing the distillery steam demand to 1.5 kg/liter via heat integration. The overall factory steam demand would then be about 2.8 kg/liter (223 kg/tc) (Table 9, last column), and the BIG/ISTIG cogeneration system could be used. A flow diagram for this type of factory is shown in Figure 6c. The cost of autonomous distillery equipment would be about the same for each of the cases shown in Figures 6a, 6b, and 6c.

It may be feasible to design distillation systems with even lower energy use. Heat integrated distillery designs using 1.2 kg/liter have been reported (53,56). If electricity is substituted for steam via distillery vapor recompression systems, even lower steam use may be possible (53,55,57,58).

Table 9. Steam and Electricity Demands in Conventional and Steam-Conserving Autonomous Distilleries with Cogeneration.[a]

	Conventional[b]	Steam-Conserving I[b]	Steam-Conserving II
Moderate-Pressure Steam (2.1 MPa, 300°C)			
Total steam used	466 kg/tc	258 kg/tc	223 kg/tc
Cane mills	226 kg/tc	226 kg/tc	223 kg/tc
Back-pressure steam turbines[c]	175 kg/tc	32 kg/tc	—
Throttled to low pressure	65 kg/tc	—	—
Total exhaust steam available	466 kg/tc	258 kg/tc	223 kg/tc
Low-Pressure Steam (mill and turbine exhaust @ 0.25 MPa, 127°C, saturated)			
Total steam used	454 kg/tc	258 kg/tc	223 kg/tc
Evaporator	61 kg/tc	97 kg/tc	97 kg/tc
Direct to Juice Heaters	130 kg/tc	—	—
Distillation	256 kg/tc	155 kg/tc	120 kg/tc
De-Aerator	8 kg/tc	6 kg/tc	6 kg/tc
Electricity Demand (pumps, fans)	12.5 kWh/tc	12.5 kWh/tc	12.5 kWh/tc
Electricity Production			
Factory steam turbines	12.5 kWh/tc	2.3 kWh/tc	—
w/CEST Cogeneration system	92 kWh/tc	123 kWh/tc	129 kWh/tc
w/BIG/STIG Cogeneration system	—	252 kWh/tc	256 kWh/tc
w/BIG/ISTIG Cogeneration system	—	—	298 kWh/tc
Maximum Electricity for Export	92 kWh/tc (CEST)	242 kWh/tc (BIG/STIG)	286 kWh/tc (BIG/ISTIG)

[a] For a distillery processing 125 tonnes cane/hour and producing 80 liters of alcohol per tonne of cane.
[b] Steam use is based on J.L. Oliverio, J.D. Neto and J.F.P. de Miranda, "Energy Optimization and Electricity Production in Sugar Mills and Alcohol Distilleries," presented at the 20th Congress of the International Society of Sugar Cane Technologists, São Paulo, Brazil, October 12–21, 1989.
[c] Assuming 14 kg steam/kWh.

Prospective Economics for the Coproduction of Alcohol and Electricity

For each of the cogeneration technologies under consideration (CEST, BIG/STIG, and BIG/ISTIG), the prospective economics are assessed here for the setup indicated in Figure 7: the distiller markets alcohol and sells cane residues to the cogenerator; and the cogenerator sells steam and electricity to the distiller and the excess electricity to the electric utility. It is assumed that during the milling season the distiller sells bagasse to the cogenerator and pays for process steam and electricity. For the off-season, it is assumed that the cogenerator uses barbojo as fuel, purchased either from the distiller (if the distiller owns the cane) or from independent cane growers. (Of course, if barbojo recovery is not feasible or practical, an alternative fuel such as wood or oil could be used as the off-season fuel.)

Some of the alternative electricity production scenarios considered for the economic analysis are displayed in Figure 8, alongside the situation in typical existing distilleries. With CEST, BIG/STIG, and BIG/ISTIG systems operated year-round, annual electricity production would be 298 kWh/tc, 672 kWh/tc, and 733 kWh/tc, respectively, for a 160-day milling system. Since typically only about three fifths of the barbojo is needed in the off-season, consideration might be given to shortening the milling season, thereby making it feasible to use more of the barbojo during the longer off-season. For BIG/ISTIG systems, reducing the growing season to 133 days would make it possible to use 80 percent of the barbojo and thereby increase electricity production to 897 kWh/tc.

Prospective costs for alcohol and cogenerated electricity[7] are summarized in Figure 9a for all the technological options considered and a 160-day milling season—the norm for Brazil. For each cogeneration option, a range of ethanol and electricity costs are shown, corresponding to different prices received by the distiller from sales of cane residues to the cogenerator. As the price of the cane residues increases (moving from left to right along each line), the cost of electricity increases and the corresponding cost of ethanol decreases, because it is assumed that the distiller takes the excess revenues as a credit against the cost of alcohol produc-

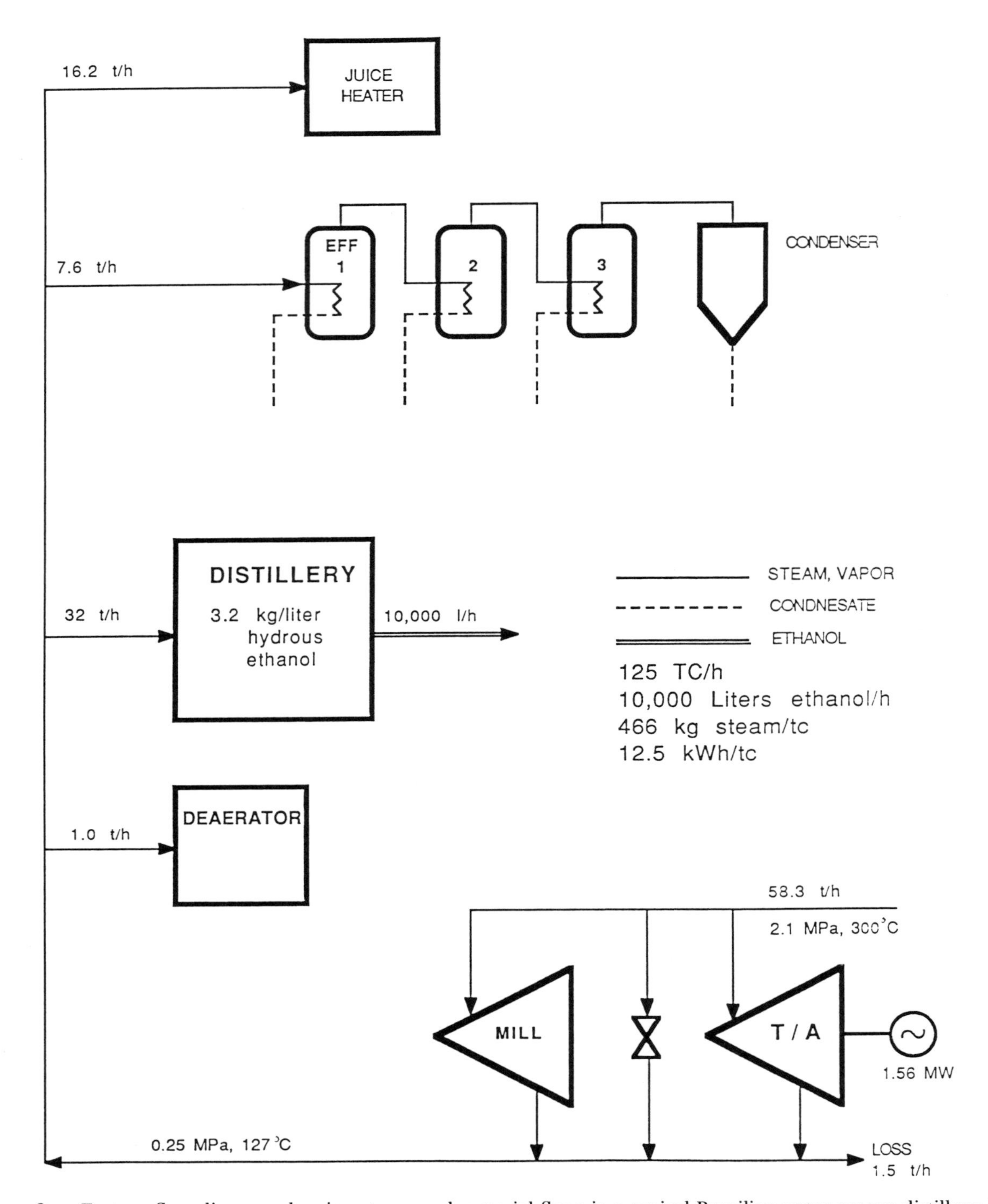

Figure 6a. Factory flow diagram showing steam and material flows in a typical Brazilian autonomous distillery. This distillery mills 125 tonnes of cane and produces 10,000 liters of hydrous ethanol per hour. A conventional distillation system is used, requiring 3.2 liters of steam per liter of hydrous ethanol. The estimated factory steam demand is 466 kg/tc. The electricity demand is 12.5 kWh/tc. (Source: J.L. Oliverio, J.D. Neto, and J.F.P. de Miranda, "Energy Optimization and Electricity Production in Sugar Mills and Alcohol Distilleries," Proceedings of the 20th Congress of the International Society of Sugar Cane Technologists, São Paulo, Brazil, October 12–21, 1989.)

Table 10. Design Options for the Co-production of Electricity and Alcohol from Sugar Cane at Autonomous Distilleries.

Cogeneration System	Condensing Extraction Steam Turbine (CEST) Steam-Injected Gas Turbine (BIG/STIG) Intercooled Steam-Injected Gas Turbine (BIG/ISTIG)	
Fuel		
In-season:	Bagasse Unprocessed (50% moisture) for CEST Briquetted (15% moisture) for BIG/STIG and BIG/ISTIG	
Off-season:	Barbojo, wood or oil	
Cane Milling	Steam turbine drive Electric drive Diffuser	
Juice Processing	*Conventional*	*Steam-Conserving*
Juice Heaters:	Steam/vapor	Condensate
Evaporators:	Short Tube Rising Film	Falling Film, or Falling Film/w Mechanical Vapor Recompression
Fermentation	Batch Continuous	
Distillation	Conventional Low Energy Use (heat integrated design) Mechanical Vapor Recompression	

Table 11a. Energy Use for Ethanol Separation from Water: Anhydrous Ethanol from Dilute Solutions.

Ethanol Concentration (% wt)					
Initial	Final	Process	Energy Use (kJ/l product ethanol)	Process Steam Use[a] (kg/l)	Status
8–10	99.9	Conventional two column distillation in typical cane alcohol distillery + azeotropic distillation w/benzene	9900–11,000	4.5–5.0	Com[b]
8–10	99.9	Heat integrated distillation in innovated cane alcohol distillery + azeotropic distillation w/benzene	6600–7700	3.0–3.5	Com[b]
6.4–10	99.9	Conventional two column distillation + azeotropic distillation with benzene	7630–9650	3.5–4.4	Com[c]
6.3–10	99.9	Conventional distillation with vapor re-use + azeotropic distillation with benzene	5000	2.3	Com[c]
10	99.9	Conventional distillation with vapor recompression + azeotropic distillation with benzene[e]	4400		Com[c]
10	99.9	Conventional distillation with vapor recompression + azeotropic distillation with benzene with vapor re-use[e]	4230		Com[c]
10	99.9	Conventional distillation + water adsorption in cornmeal	3340	1.5	Com[c]
10	99.9	Conventional distillation with vapor recompression + water adsorption in cornmeal[e]	2170		Com[c]
10	99.9	IHOSR distillation + extractive distillation with KAc salts	1700	0.8	Lab[c]
10	99.9	Extraction with CO_2	2232–2791		Lab[d]
10	99.9	Solvent extraction	1005		Lab[d]
10	99.9	Vacuum distillation	10,330		Lab[d]

Table 11b. Energy Use for Ethanol Separation from Water: Hydrous (Azeotropic) Ethanol from Dilute Solutions.

Ethanol Concentration (% wt)					
Initial	Final	Process	Energy Use (kJ/l product ethanol)	Process Steam Use[a] (kg/l)	Status
8–10	95	Conventional two column distillation in typical cane alcohol distillery	6600–7700	3.0–3.5	Com[b]
8–10	95	Heat integrated distillation in innovated cane alcohol distillery	3300–4400	1.5–2.0	Com[b]
6–10	95	Conventional two column distillation	4730–5850	2.1–2.7	Com[c]
6–10	95	Conventional distillation with vapor re-use	1950–3340	0.9–1.5	Com[c]
10	95	Conventional distillation with vapor recompression[e]	1610–1780		Com[c]
10	95	Three column distillation with vapor re-use	4730–5850	2.1–2.7	Com[c]
10	95	Four column distillation	8080	3.7	Com[c]
10	95	Three effect vacuum distillation	2010	0.9	Lab[d]

[a] The process steam is assumed to be saturated at 120°C, with enthalpy of vaporization of 2202 kJ/kg.
[b] G. Serra, Univerity of São Paulo, Campinas, Brazil, private communications, 1989.
[c] A. Serra, M. Poch and C. Sola, "A Survey of Separation Systems for Fermentation Ethanol Recovery," *Process Biochemistry,* pp.154–158, October 1987.
[d] G. Parkinson, "Batelle Maps Ways to Pare Ethanol Costs," *Chemical Engineering,* June 1, 1981.
[e] For vapor recompression, it is assumed that heat is converted into electricity at 33% efficiency.

tion. Two lines are shown for each cogeneration option. The top line represents the case where only bagasse revenues are credited against the alcohol production cost. This would be the case if the off-season barbojo fuel were purchased not from the distiller but from independent cane growers. The steeper line is for the case where the distiller owns the cane fields and sells barbojo as well as bagasse to the cogenerator. The point where these lines meet indicates the ethanol and electricity costs when no net payments are exchanged between the cogenerator and the distillery: bagasse is given to the cogenerator in exchange for the steam and electricity needed to run the plant.[8] The costs of the conventional energy alternatives to cane alcohol and electricity are also indicated in Figure 9a.

For fuel ethanol to be economically attractive fuel it must be competitive with other fuels. At the present world oil price, ethanol would have to be priced at 15 cents/liter (line A1) or less to be competitive with gasoline as a neat fuel and 21 cents/liter or less to be competitive in gasohol (line A2). Similarly, the cogenerated electricity must be competitive with alternative electricity supplies. If the cogeneration facility were competing against an existing oil-fired thermal power plant, the cost of the cogenerated electricity would have to be less than the *operating cost* of this plant, some 3.9 cents/kWh (line E1), based on the current residual oil price. If instead the electrical competition were a new hydropower plant costing $1500/kW, the cost of the cogenerated electricity would have to be less than or equal to 4.3 cents/kWh (line E2). A third possibility is that the competition would be a new coal-fired steam-electric plant with flue-gas desulfurization; for an installed cost of $1400/kW and a coal price of $1.7/GJ, electricity from such a plant would cost about 5.3 cents/kWh (line E3). Still another possibility is that the cogenerator also has access to woodchips as an alternative biomass feedstock. In this case the maximum cost of cane residues to the cogenerator (purchase price plus the cost for processing the biomass into a gasifiable form) would be no greater than about $3/GJ (indicated by black dots), the estimated cost of delivered, air-dried woodchips in Brazil (59).

Figure 9a shows that under certain conditions it would be possible to co-produce ethanol competitively with gasoline and electricity at costs competitive with other supplies:

- The low-cost conditions in Brazil for cane production and processing are key factors underlying the favorable prospective economics for alcohol indicated here. The two most important factors in reaching low ethanol costs are low cane costs[9] and low distillery capital costs.[10] If Brazilian cost conditions are met, it would be possible, without cane residue fuel credits, to produce ethanol at about 21 cents/liter—a cost low enough to make ethanol roughly competitive as an additive to gasoline. The corresponding cost of electricity production would be 5 cents/kWh or less with all the cogeneration systems considered, and less than 3 cents/kWh with the BIG/ISTIG system.
- With cane residue sales credits, it would be possible to produce hydrous ethanol at a net cost of less than 15 cents/liter, making it competitive with gasoline even at the current oil price. At these ethanol costs, gas tur-

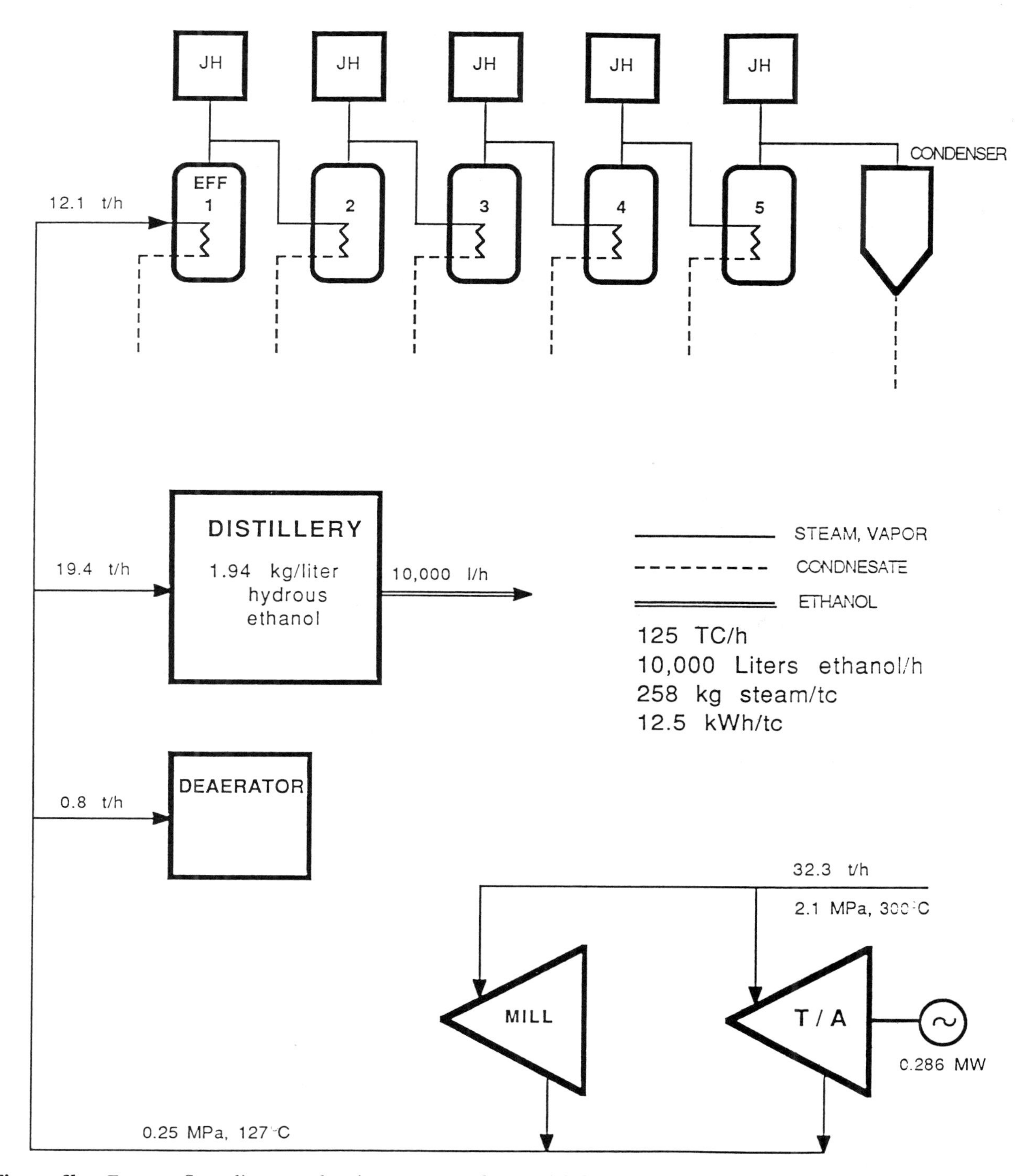

Figure 6b. Factory flow diagram showing steam and material flows in a steam-conserving Brazilian autonomous distillery. This distillery mills 125 tonnes of cane and produces 10,000 liters of hydrous ethanol per hour. Steam use is reduced by using vapors bled from the evaporator for juice heating and a more energy-efficient distillation system with heat integration, requiring 1.9 liters of steam per liter of hydrous ethanol. The estimated factory steam demand is 258 kg/tc. The electricity demand is 12.5 kWh/tc. (Source: J.L. Oliverio, J.D. Neto, and J.F.P. de Miranda, "Energy Optimization and Electricity Production in Sugar Mills and Alcohol Distilleries," Proceedings of the 20th Congress of the International Society of Sugar Cane Technologists, São Paulo, Brazil, October 12–21, 1989.)

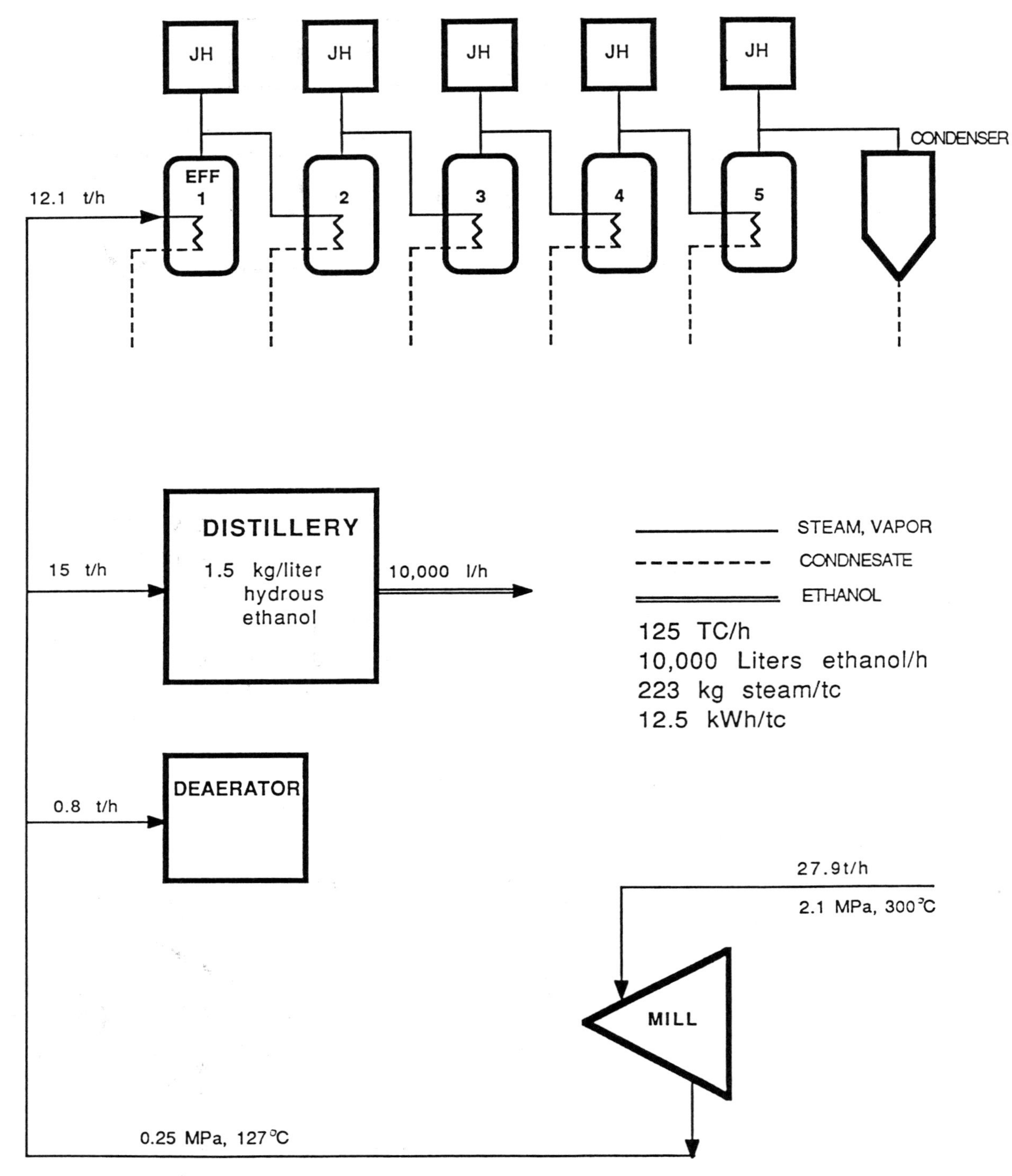

Figure 6c. Factory flow diagram showing steam and material flows in a steam-conserving Brazilian autonomous distillery. This distillery mills 125 tonnes of cane and produces 10,000 liters of hydrous ethanol per hour. Steam use is reduced by using vapors bled from the evaporator for juice heating and a more energy-efficient distillation system with heat integration, requiring 1.5 liters of steam per liter of hydrous ethanol. The estimated factory steam demand is 225 kg/tc. The electricity demand is 12.5 kWh/tc. (Sources: J.L. Oliverio, J.D. Neto, and J.F.P. de Miranda, "Energy Optimization and Electricity Production in Sugar Mills and Alcohol Distilleries," Proceedings of the 20th Congress of the International Society of Sugar Cane Technologists, São Paulo, Brazil, October 12–21, 1989; J. Goldemberg, J.R. Moreira, P.U.M. Dos Santos, and G.E. Serra, "Ethanol Fuel: A Use of Biomass Energy in Brazil," *Ambio,* Vol. 14, No. 4–5, 1985; G. Serra and J.R. Moreira, University of São Paulo, Brazil, private communications, 1989.)

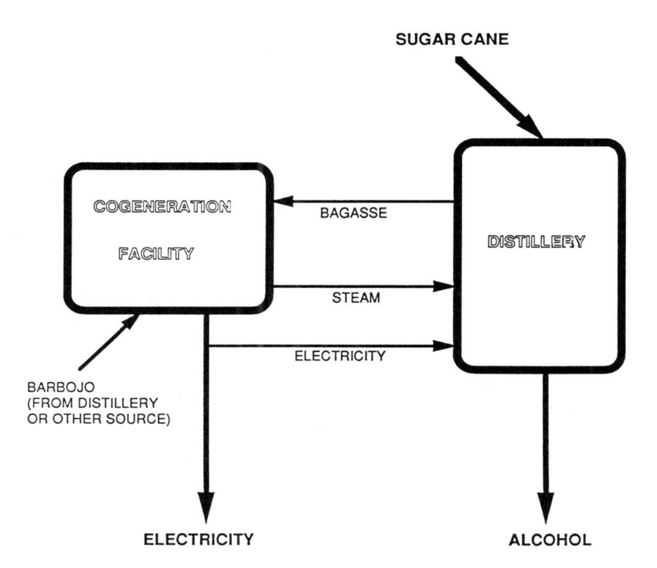

Figure 7. Material and energy flows for an autonomous distillery and cogeneration facility.

bine cogeneration systems (BIG/STIG or BIG/ISTIG) would be required to reach the low production costs needed for the cogenerated electricity to be competitive with other options in most places. The BIG/ISTIG system appears to be particularly promising. With BIG/ISTIG it would be possible to provide ethanol at a cost competitive with wholesale gasoline. At the same time the electricity cost would be less than $0.045/kWh, even for an average cane residue price of more than $3/GJ. Electricity costs with the CEST system would probably be too high to allow this system to compete if the residue credits were high enough to make neat ethanol competitive with gasoline.

- The costs of ethanol and electricity are sensitive to the cane residue price. The lowest net alcohol costs are reached when the distiller can sell barbojo for off-season fuel in addition to bagasse during the milling season. In practice the price of cane residues would probably be determined by their value for nonenergy uses, by the price of competing cogeneration fuels (fuel wood or oil) or by the price of electricity from competing sources (e.g., hydropower), when they are less costly than cogenerated electricity.
- While it is likely that Brazilian alcohol distillery capital costs could be replicated in most other parts of the cane-producing world, the prospects for reducing cane costs to Brazilian levels are limited by regional climate, labor cost, and other conditions. However, with BIG/ISTIG technology, even if cane costs are as high as $20/tc, the ethanol would be competitive with gasoline at the present world oil price and the electricity would be competitive with coal-based electricity, if the distillery capital cost were the same as in Brazil and if revenues from the sale of both bagasse and barbojo could be credited against the cost of alcohol (49).

Figure 9b illustrates the economics of coproduction for a milling season shortened to 133 days to make possible greater use of the barbojo. With a shorter milling season the capacity factor for the distillery would be shortened, which would tend to increase the cost of alcohol production. But if the distiller owned the barbojo, the greater barbojo sales would tend to reduce the cost of alcohol production. A comparison of Figures 9a and 9b shows that for both the BIG/STIG and BIG/ISTIG cases the cogenerated electricity would be competitive with the alternatives at lower net alcohol costs for the shorter milling season than for the normal milling season, indicating that it would be desirable to explore the shortened milling season option for situations where the distiller owns the cane. The major unanswered question about this option is whether it is practical to recover so much of the barbojo.[11]

Global Potential for Cane Energy and Greenhouse Gas Emissions Reduction

To give an indication of the potential role of cane energy in the global energy economy, consider a scenario where cane production grows at 3 percent per year (the historical rate since 1960—Table 3) over the next 40 years, of which 1.5 percent per year growth (the sugar demand growth rate projected by the World Bank for the period to 1995 [14]) is committed to sugar production and the rest to alcohol production (assuming the favorable economic conditions for the coproduction of alcohol and electricity identified for Brazil are generally applicable in the developing world). Further, assume that all the growth in cane production takes place in developing countries. At the end of this period some 3140 million tonnes of cane would be produced annually, of which 1430 million tonnes would be committed to alcohol production.

Suppose that by the end of this period modern cogeneration technologies are fully deployed in developing countries throughout the cane sugar and alcohol industries. Thus, for an average milling season of 160 days, total electricity production in the cane industries in 2027 would be 940 TWh, 2080 TWh, or 2270 TWh for CEST, BIG/STIG, or BIG/ISTIG technologies, respectively, assuming all electricity is provided by one of these technologies. If instead the average milling season were shortened to 133 days, total electricity production would be greater—up to 2780 TWh in the case of BIG/ISTIG—a level that is 70 percent higher than total electricity production from all sources in the 80 sugar cane producing developing countries in 1987 (Table 12).

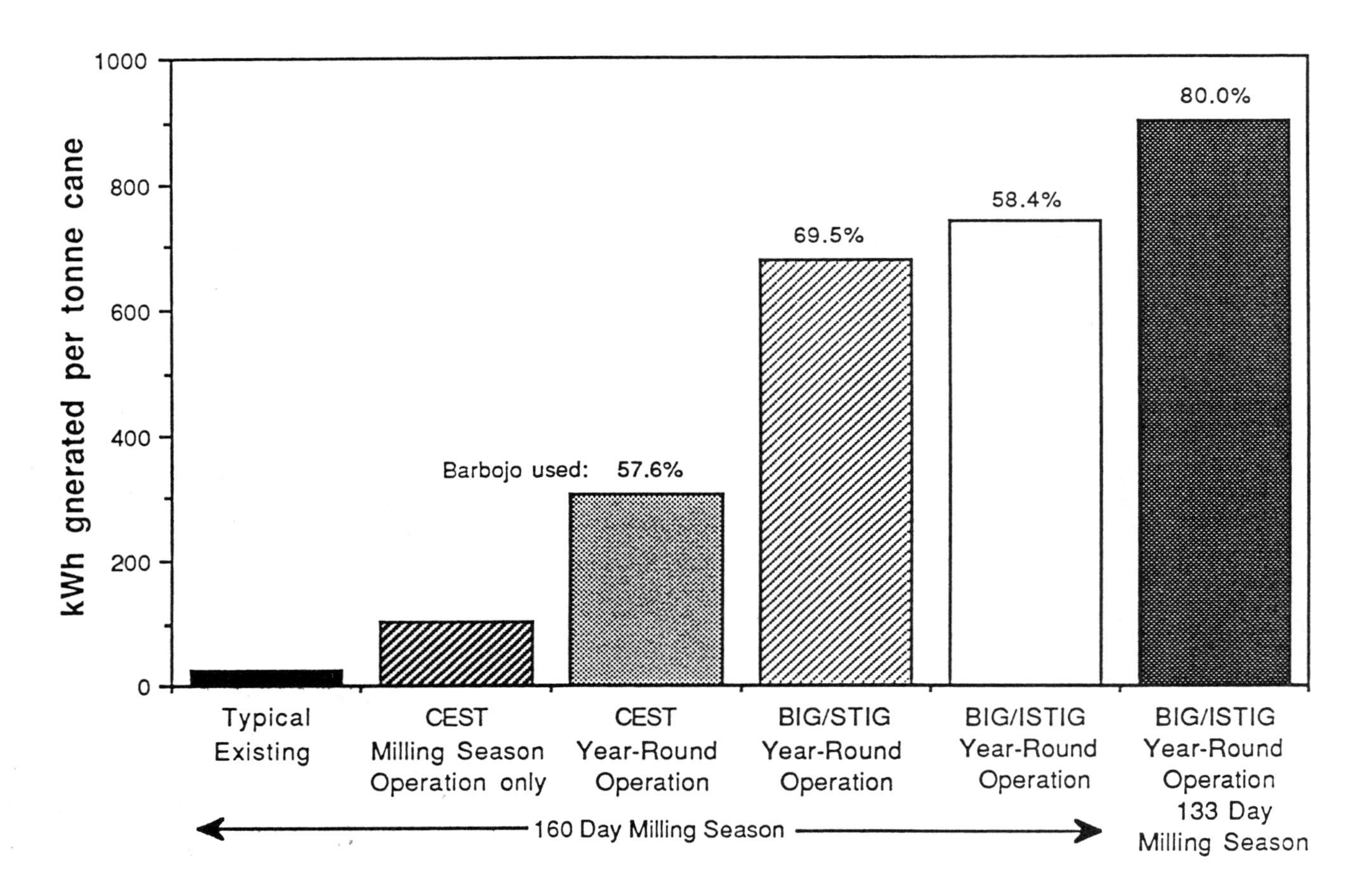

Figure 8. The potential for electricity generated via cogeneration (using sugar cane residues). The first bar is for a typical existing situation at a sugar factory or alcohol distillery during the the milling season. The next bar is for a CEST system operating during the milling season at a factory where steam-saving retrofits have been made. The next three bars are for year-round operation for a plant at which steam-saving retrofits have been made and where the milling season is 160 days: the third, fourth, and fifth bars are for CEST, BIG/STIG, and BIG/ISTIG systems, respectively. The sixth bar is for a BIG/ISTIG unit operated at a steam-efficient plant operated with a 133-day milling season. The number at the tops of the four bars to the right is the percentage of the barbojo used for power generation during the off-season.

Suppose also that alcohol is produced at autonomous distilleries at an average rate of 70 l/tc and that methane is produced via anaerobic digestion from stillage at an average rate of 0.33 GJ/tc. Further suppose that 1.19 liters of ethanol are worth 1 liter of gasoline as a neat fuel (taking into account the modest octane-enhancing benefit of using ethanol in automotive engines). Then annual alcohol and methane production at the end of this 40-year period would be 3.3 EJ/yr (crude oil equivalent) and 0.5 EJ/yr, respectively. Ethanol and methane production in 40 years would be equivalent to 9 percent and 5 percent, respectively, of oil and gas used in all developing countries in 1987.

In estimating the potential reduction in fossil fuel CO_2 emissions arising from cane energy, suppose that the electricity produced from cane displaces electricity that would otherwise be produced from coal and that both alcohol and methane produced at distilleries displace gasoline in the transport sector. Taking into account the fossil CO_2 emissions associated with the growing, harvesting, transporting, and processing of cane for energy purposes under Brazilian conditions (Tables 13 and 14), net CO_2 emissions reductions are estimated to be 52.6 kg C/tc (82 percent of gross emissions reductions) for traditional cane alcohol technology; 73-220 kg C/tc (97 percent of gross emissions reductions) for all sugar/electricity coproduction technologies; and 116-263 kg C/tc (90 to 94 percent of gross emissions reductions) for the alcohol/electricity coproduction options (Table 13). The aggregate net emissions reduction potential for the cane industry in 2027 ranges from 0.062GT C/yr for the traditional alcohol industry, to

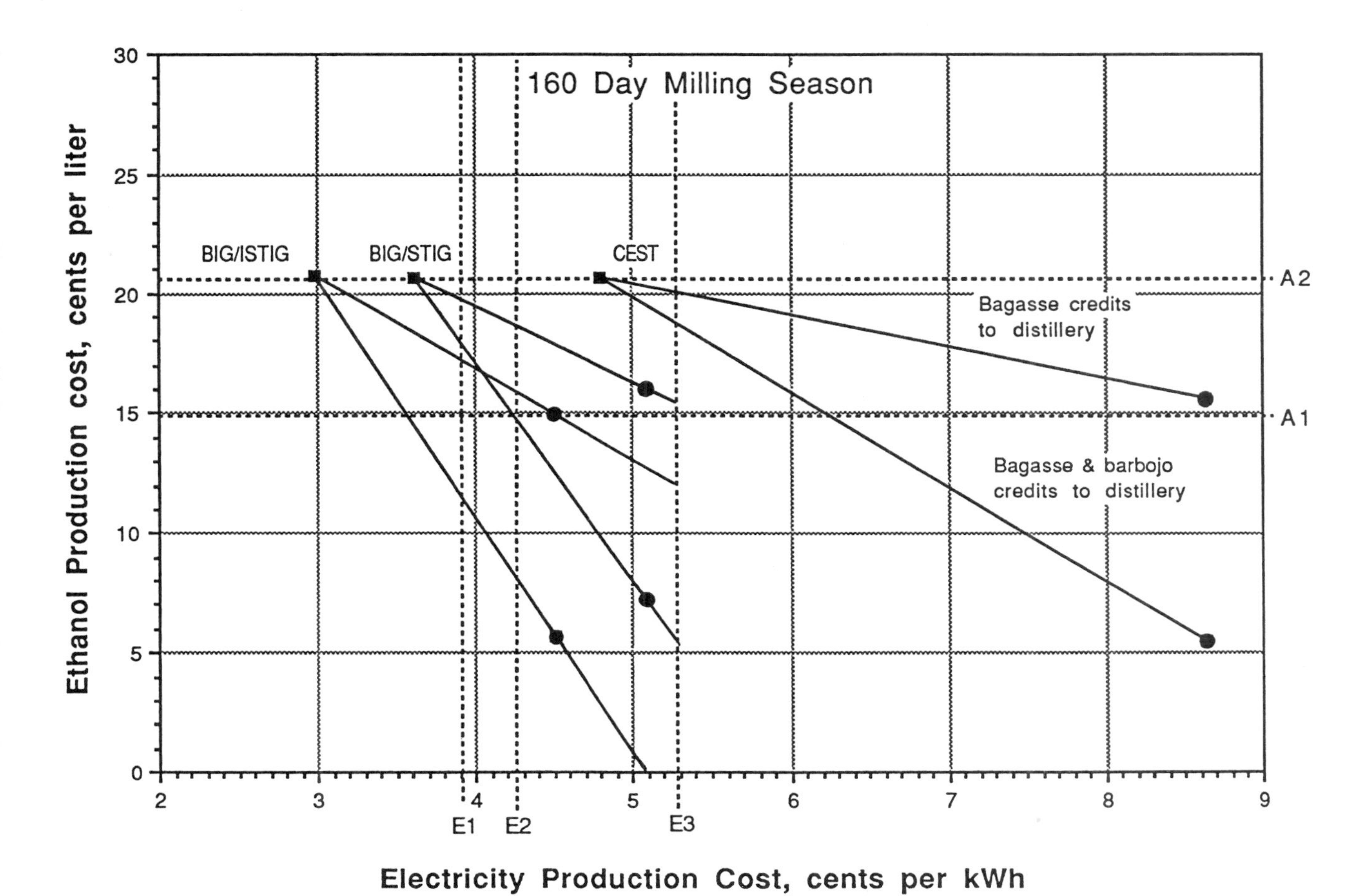

Figure 9a. The cost of ethanol and electricity using alternative cogeneration technologies for Brazilian autonomous distilleries operated 160 days per year. For cogeneration based upon using, as fuel, bagasse during the milling season and barbojo during the off-season. For each technology, a range of ethanol and electricity costs is shown, corresponding to different prices for the cane residues. As the residue price increases (moving from left to right along each line), the cost of electricity increases and the corresponding cost of ethanol decreases, because it is assumed that the distiller takes the increased revenues as a credit against the cost of producing alcohol. Two lines are shown for each technology. The top line represents the case where only bagasse revenues are credited against the alcohol cost. This would be the case if the barbojo were purchased from independent cane growers. The steeper line is for the case where the distiller owns the cane fields and sells barbojo as well as bagasse to the cogenerator. The point where these lines meet indicates the ethanol and electricity costs when no net payments are exchanged between the cogenerator and the distiller: bagasse is given to the cogenerator in exchange for the steam and electricity needed to run the distillery. The black dot on each line is for a maximum residue cost to the cogenerator (purchase price plus the cost for processing the residues into a gasifiable form) of $3/GJ, the estimated cost of delivered, air-dried woodchips in Brazil, for the case where the cogenerator has access to woodchips as an alternative biomass feedstock during the off-season. Also shown are: the prices at which ethanol would be competitive at the present world oil price as a neat fuel (A1), assuming a liter of alcohol is worth 0.84 liters of gasoline, and as an octane-enhancing additive (A2), assuming a liter of alcohol is worth 1.16 liters of gasoline; and the prices at which the cogenerated electricity would be equal to the operating cost of an oil-fired power plant at the present world oil price (E1) (assuming a heat rate of 13,120 kJ/kWh and a fuel oil price of $2.63/GJ), the busbar cost of a new hydroelectric plant (E2) (assuming a capital cost of $1500/kW), and the busbar cost of a new coal plant (E3), assuming a capital cost of $1400/kW and a coal price of $1.7/GJ.

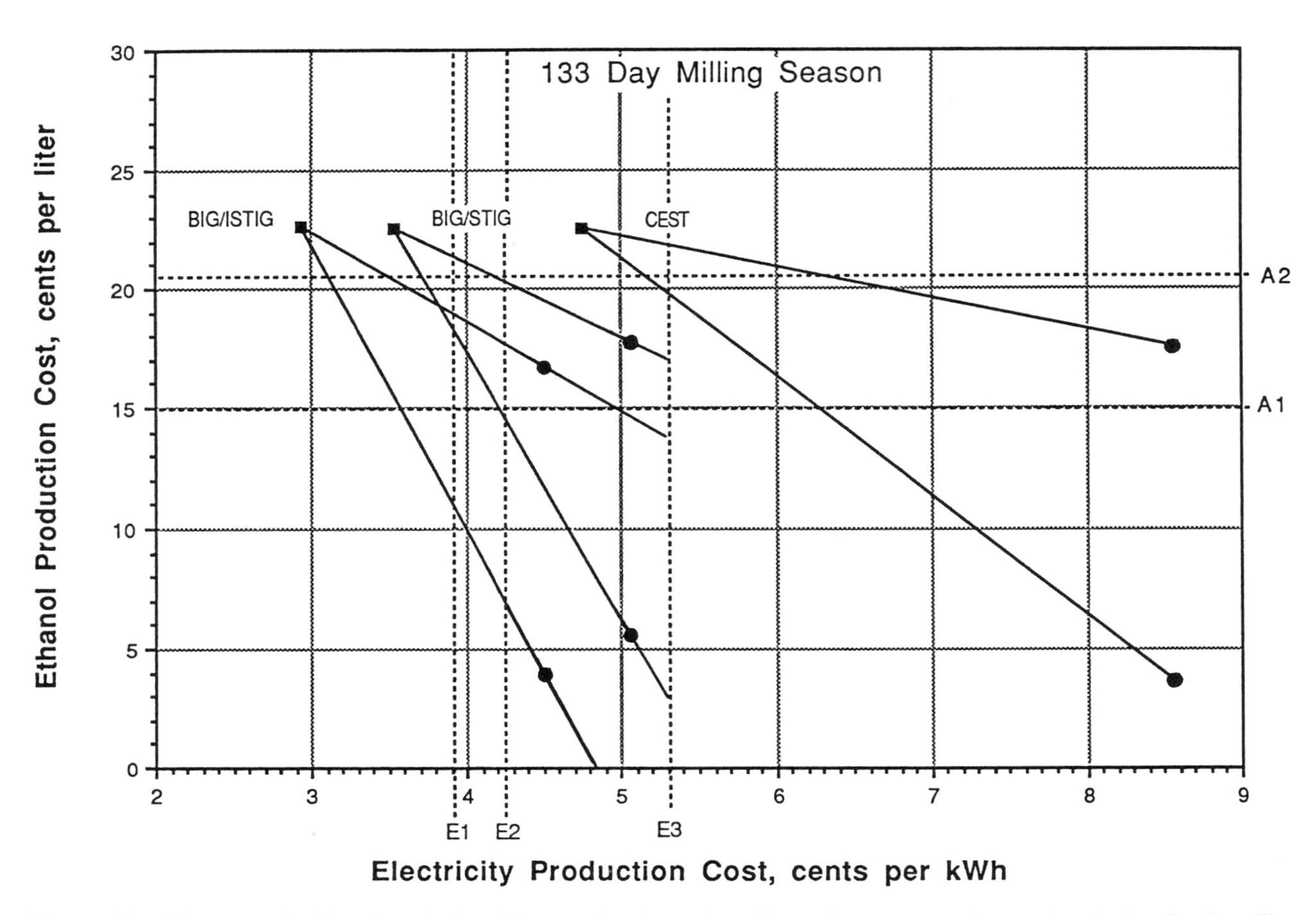

Figure 9b. The cost of ethanol and electricity production using alternative cogeneration technologies for Brazilian autonomous distilleries operated 133 days per year. With a 160-day milling season (Figure 9a) only about 60–70% of the barbojo can be used for power generation in the off-season (Figure 8). A larger percentage of the barbojo could be used if the milling season were shorter. For a 133-day milling season, 80% of the barbojo could be used with BIG/ISTIG technology (Figure 8). The shorter milling season leads to a lower capacity utilization factor for the distillery, which would tend to drive up the cost of alcohol. But if the distiller owns the cane fields, the extra revenues he could get from barbojo sales would more than compensate for the lower distillery capacity factor, making the short milling season option economically attractive.

0.75GT/yr for BIG/ISTIG technology with a shortened growing season (Table 13). The latter is equivalent to nearly one half of total emissions from fossil fuel combustion in developing countries in 1986 (Table 2).

Most of the energy and CO_2 emissions reductions arising from alcohol/electricity coproduction is associated with electricity (Table 13). However, the coproduction of alcohol and electricity would make it possible for the cane industry to expand beyond the limits set by future sugar demand growth, making possible far greater overall emissions reductions than if sugar were the only coproduct of electricity from cane. Moreover, even though the direct impacts of cane alcohol on CO_2 emissions would be small compared to the impact made by cane electricity, the produced alcohol could still play a significant role in the local energy economy as a high quality liquid fuel carrier, as has been demonstrated in Brazil, where in 1988 ethanol consumption was actually 12 percent greater than gasoline consumption (64).

The major uncertainty concerning the alcohol/electricity scenario presented in Table 13 is the extent to which the low-cost conditions that characterize the Brazilian alcohol industry can be replicated in other regions of the developing world. However, the overall potential for electricity production in the cane industry is probably not too sensitive to the global outlook for alcohol. If all the 3080 million tonnes per year of sugar cane production targeted for the year 2027 were used to support sugar markets instead of sugar/alcohol markets, and if all the produced sugar were marketed in developing countries, per capita sugar consumption in developing countries in 2027 would be just 45 kg/capita (assuming

Table 12. Electricity Production Potential with BIG/ISTIG in the Sugar Cane Industries of Developing Countries.

	A	B	C	D	E = C/D
	Cane Production million tc/yr		Electricity Production TWh per year		
	1987[a]	2027[b]	Potential from Cane in 2027[c]	Actual from All Sources in 1987[d]	
Africa					
South Africa	20.00	69.00	61.06	122.30[e]	0.50
Egypt	9.50	32.78	29.01	32.50	0.89
Mauritius	6.23	21.50	19.02	0.49	39.0
Sudan	5.00	17.25	15.27	1.06	14.5
Swaziland	4.00	13.80	12.21	—	—
Kenya	4.00	13.80	12.21	2.63	4.7
Zimbabwe	3.80	13.11	11.60	7.01	1.7
Reunion	2.11	7.29	6.45	—	—
Madagascar	1.80	6.21	5.50	0.50	10.9
Ivory Coast	1.75	6.04	5.34	2.20	2.4
Ethiopia	1.65	5.69	5.04	0.81	6.2
Malawi	1.60	5.52	4.89	0.58	8.5
Nigeria	1.55	5.35	4.73	9.91	0.48
Cameroon	1.29	4.45	3.94	2.39	1.7
Zambia	1.25	4.31	3.82	8.48	0.45
Zaire	1.09	3.75	3.32	5.30	0.63
Tanzania	1.08	3.71	3.28	0.87	3.8
Morocco	0.80	2.76	2.44	8.32	0.29
Senegal	0.70	2.42	2.14	0.75	2.8
Mozambique	0.67	2.31	2.05	0.50	4.1
Uganda	0.60	2.07	1.83	0.66	2.8
Congo	0.51	1.76	1.56	0.24	6.6
Somalia	0.37	1.28	1.13	0.26	4.4
Burkina Faso	0.33	1.14	1.01	0.13	7.8
Angola	0.32	1.10	0.98	0.81[f]	1.2
Chad	0.29	1.00	0.89	—	—
Mali	0.22	0.76	0.67	0.20	3.3
Guinea	0.20	0.69	0.61	0.50	1.2
Liberia	0.16	0.54	0.47	0.83	0.57
Gabon	0.14	0.48	0.43	0.88	0.49
Niger	0.11	0.38	0.34	0.16	2.1
Ghana	0.11	0.38	0.34	4.71	0.07
Sierra Leone	0.07	0.24	0.21	0.20	1.1
Rwanda	0.032	0.11	0.097	0.17	0.56
Subtotals	73.3	253.0	223.9	216.4	1.03
Oceania					
Fiji	3.49	12.05	10.67	0.43	25.0
Papua N. Guinea	0.23	0.80	0.71	1.80	0.39
Subtotals	3.7	12.9	11.4	2.2	5.1
Central America					
Cuba	65.60	226.32	200.29	13.20[e]	15.0
Mexico	42.56	146.83	129.95	104.79	1.2
Dominican Republic	8.60	29.67	26.26	5.00	5.3
Guatemala	6.90	23.80	21.07	2.08	10.2
El Salvador	3.18	10.97	9.71	1.89	5.1
Honduras	3.00	10.35	9.16	1.81	5.1
Costa Rica	3.00	10.35	9.16	3.13	2.9
Haiti	3.00	10.35	9.16	0.45	20.4
Nicaragua	2.58	8.88	7.86	1.24	6.3
Jamaica	2.01	6.93	6.14	2.39	2.6
Panama	1.60	5.52	4.89	2.85	1.7

(continued)

Table 12. *(continued)*

	A	B	C	D	E = C/D
	Cane Production million tc/yr		Electricity Production TWh per year		
	1987[a]	2027[b]	Potential from Cane in 2027[c]	Actual from All Sources in 1987[d]	
Trinidad & Tobago	1.24	4.26	3.77	3.30[e]	1.1
Belize	0.86	2.98	2.64	0.075	35.0
Guadaloupe	0.75	2.57	2.28	—	—
Barbados	0.73	2.52	2.23	0.43	5.2
St. Kitts Nev.	0.25	0.86	0.76	—	—
Martinique	0.25	0.86	0.76	—	—
Bahamas	0.24	0.83	0.73	—	—
Subtotals	146.4	504.9	446.8	142.6	3.1
South America					
Brazil	273.86	944.79	836.14	202.29	4.1
Colombia	24.97	86.15	76.22	35.37	2.2
Argentina	14.00	48.30	42.75	52.17	0.82
Peru	5.95	20.53	18.17	14.20	1.3
Venezuela	7.00	24.15	21.37	50.21	0.43
Ecuador	5.20	17.94	15.88	5.67	2.8
Guyana	3.30	11.38	10.08	—	—
Paraguay	3.19	11.00	9.73	2.83	3.5
Bolivia	2.73	9.42	8.34	1.52	5.5
Uruguay	0.65	2.24	1.99	4.53	0.44
Suriname	0.11	0.38	0.34	—	—
Subtotals	341.0	1176.3	1041.0	368.8	2.8
Asia					
India	182.48	629.55	557.15	217.50	2.6
China	52.55	181.30	160.45	497.30	0.32
Pakistan	31.70	109.37	96.80	28.40	3.4
Thailand	24.45	84.35	74.65	29.99	2.5
Indonesia	21.76	75.09	66.45	34.81	1.9
Philippines	13.33	45.97	40.68	23.85	1.7
Bangaldesh	6.90	23.79	21.06	5.90	3.6
Vietnam	6.60	22.77	20.15	5.20[e]	3.9
Burma	3.28	11.32	10.02	2.28	4.4
Iran	1.15	3.97	3.51	36.80[e]	0.10
Malaysia	1.15	3.97	3.51	16.22	0.22
Sri Lanka	0.80	2.76	2.44	2.71	0.90
Nepal	0.62	2.13	1.88	0.54	3.5
Kampuchea	0.21	0.71	0.63	—	—
Laos	0.11	0.36	0.32	0.88	0.37
Subtotals	347.1	1197.4	1059.7	902.4	1.2
Grand Totals	911.5	3144.4	2782.8	1632.4	1.7

[a] Food and Agriculture Organization, *FAO Production Yearbook*, FAO Statistical Series No. 82, vol. 21, 1987.
[b] Assuming sugar cane grows at a rate of 3.1%/yr.
[c] For sugar factories or alcohol distilleries operated 133 days/yr, with 286 kWh of electricity produced from bagasse/tc during the milling season plus 599 kWh produced from barbojo/tc during the off-season, with BIG/ISTIG units.
[d] Except where otherwise indicated, from J.R. Escay, IENED, "Summary Data Sheets of 1987 Power and Commercial Energy Statistics for 100 Developing Countries," Industry and Energy Department Working Paper, Energy Series Paper No. 23, World Bank, March 1990.
[e] For 1986, from Bureau of the Census, US Dept. of Commerce, *Statistical Abstract of the United States 1989*, 109th Edition, US Government Printing Office, Washington, DC, January 1989.
[f] Public electricity production in 1989. From E.A. Moore (IENED) and G. Smith, "Capital Expenditures for Electric Power in the Developing Countries in the 1990s," Industry and Energy Department Working Paper, Energy Series Paper No. 21, World Bank, February 1990.

Table 13a. Net Potential CO_2 Emissions Reduction for Alternative Energy Systems at Sugar Factories and Alcohol Distilleries (kg C/tc).

Sugar Factories	Conventional Power System	CEST	BIG/STIG	BIG/ISTIG	BIG/ISTIG w/Shortened Milling Season
Gross CO_2 Reduction					
1. If Extra Cogenerated Electricity Displaces Coal-Based Electricity[a]	—	76.2	169.4	184.5	226.4
CO_2 Penalties					
1. Barbojo Recovery[b]	—	1.4	1.6	1.4	1.9
2. Power Plant Construction[c]	—	0.5	1.1	1.1	1.3
3. Power Plant O&M[d]	—	1.3	3.0	3.0	3.6
4. Subtotals	—	3.2	5.7	5.5	6.8
Net CO_2 Reduction	—	73.0	163.7	179.0	219.6
Alcohol Distilleries					
Gross CO_2 Reduction					
1. If Extra Cogenerated Electricity Displaces Coal-Based Electricity[a]	—	76.2	169.4	184.5	226.4
2. If EthOH Displaces Oil[e]	45.3	45.3	45.3	45.3	45.3
3. If CH_4 Displaces Oil[f]	7.3	7.3	7.3	7.3	7.3
4. Subtotals	52.6	128.8	222.0	237.1	279.0
CO_2 Penalties					
1. Cane Production, Harvesting, and Transport[g]	6.1	7.5	7.7	7.5	8.0
2. Distillery Manufacture[h]	1.3	1.3	1.3	1.3	1.3
3. Distillery O&M[h]	2.2	2.2	2.2	2.2	2.2
4. Power Plant Construction[c]	—	0.5	1.1	1.1	1.3
5. Power Plant O&M[d]	—	1.3	3.0	3.0	3.6
6. Subtotals	9.6	12.8	15.3	15.1	16.4
Net CO_2 Reduction	43.0	116.1	206.7	222.0	262.6

[a] For a 160-day milling season, the electricity generated in excess of onsite needs in the milling season (off-season) is 92 kWh/tc (206 kWh/tc) for CEST, 242 kWh/tc (420 kWh/tc) for BIG/STIG, and 286 kWh/tc (435 kWh/tc) for BIG/ISTIG. For a 133-day milling season, offseason production is 599 kWh/tc for BIG/ISTIG. It is assumed that this electricity displaces electricity that would otherwise be produced at coal power plants @ 10.4 MJ/kWh and that 24.6 kg C is released (as CO_2) per GJ of coal burned.

[b] When barbojo is recovered for offseason fuel, it is assumed that the agricultural fuel use increases in proportion to the mass of the recovered barbojo. Assuming that the total amount of barbojo is 660 kg per tonne of cane, the energy penalty for barbojo recovery is FB × 0.66 × (0.18 GJ/tc) (Table 14), where FB is the fraction of barbojo recovered. For a 160-day milling season, FB = 0.58, 0.68, and 0.58, for CEST, BIG/STIG, and BIG/ISTIG, respectively. For a 133-day growing season, FB = 0.80 for BIG/ISTIG. Assuming the fuel used to recover barbojo is oil, 19.9 kg of carbon would be released (as CO_2) per GJ of oil burned.

[c] The energy required to build a power plant is assumed to be 14.8 GJ/kWe (A.M. Perry, W.D. Devine, D.B. Reister, "The Energy Cost of Energy—Guidelines for Net Energy Analysis of Energy Supply Systems," Institute for Energy Analysis, Oak Ridge Associated Universities, August 1977), or 0.493 GJ/yr/kWe for a 30-year plant life. For a 160-day milling season, the extra electrical generating capacity needed for cogeneration equipment is 0.039 kW, 0.088 kW, and 0.092 kW per tc/yr of cane processing capacity, for CEST, BIG/STIG, and BIG/ISTIG, respectively. For a 133-day milling season, the needed extra electrical capacity is 0.110 kW per tc/yr for BIG/ISTIG. It is assumed that the fuel required to build power plants is coal.

[d] As for the Brazilian distillery analysis (Table 14) it is assumed that the maintenance and operation fossil energy penalties are 5.0% and 3.9% of the energy required to construct the power plant, respectively. Thus the O&M penalty is 30 × 0.089 = 2.67 times the construction energy penalty.

[e] It is assumed that 70 liters of fuel alcohol is produced per tc and that 1 liter of alcohol is equivalent as a motor fuel to 0.84 liters of gasoline. Thus, taking into account 10% refinery losses in producing gasoline, the produced alcohol would displace 2.28 GJ of crude oil per tc.

[f] Assuming 0.33 GJ/tc of methane is produced via anaerobic digestion of stillage and used to displace gasoline, or 0.37 GJ of crude oil/tc.

[g] The agricultural energy (assumed to be oil) required for cane production for conventional distilleries is from Table 14. The incremental agricultural energy required for barbojo recovery is estimated in note b.

[h] Energy (assumed to be coal) for distillery manufacture is from Table 14.

Table 13b. Potential CO_2 Emissions Reduction in Developing Countries in 2027 for Alternative Energy Systems Installed at Sugar Factories and Alcohol Distilleries (megatonnes C/yr).

	Conventional Power System	CEST	BIG/STIG	BIG/ISTIG	BIG/ISTIG w/Shortened Milling Season
At Sugar Factories[i]	—	125.0	284.9	306.1	375.7
At Alcohol Distilleries[i]	61.5	166.0	299.6	317.5	375.7
Totals	61.5	291.0	584.5	623.6	751.4

[i] Assuming 1710 (1430) million tonnes of sugar cane are associated with sugar (alcohol) production in developing countries in 2027.

Table 14. Fossil Energy Requirements Associated with Alcohol Production from Sugar Cane at Conventional Plants in Brazil.[a]

	GJ/ha/yr	MJ/tc[b]
Agricultural Energy Requirements		
Manufacture of tractors, trucks, other agricultural equipment	1.68	32.3
Fuel for tractors, harvestors, trucks, etc.	9.37	180.2
N-fertilizer	2.88	55.4
P-fertilizer	0.37	7.1
K-fertilizer	0.40	7.7
Lime	0.15	2.9
Seeds	0.79	15.2
Insecticides	0.01	0.2
Herbicides	0.23	4.4
Subtotal	15.88	305.4
Distillery Energy Requirements		
Manufacture of Distillery[c]		51.5
Distillery Operation[d]		40.2
Distillery Maintenance[e]		51.5
Subtotal		143.2
Total Energy Requirements for Alcohol Production		448.6

[a] *Source:* J.R. Moreira, V.R. Vanin, and J. Goldemberg (Institute of Physics, University of São Paulo, São Paulo, Brazil), "Energy Balance for the Production of Ethyl and Methyl Alcohol," presented at the Workshop on Fermentation Alcohol for Use as Fuel and Chemical Feedstock in Developing Countries, Vienna, Austria, March 26–30, 1979.
[b] The average harvested yield of sugar cane is 52 tonnes/ha/yr.
[c] The energy required to build a 120,000 l/day distillery is estimated to be 254.4 TJ. Assuming the distillery operates 160 days per year @ 90% availability and lasts for 20 years, the manufacturing energy amounts to 736.1 kJ/liter of produced alcohol or 51.5 MJ/tc, assuming 70 liters of alcohol are produced per tc.
[d] Estimated to be 3.9% of the energy of distillery manufacture per year.
[e] Estimated to be 5.0% of the energy of distillery manufacture per year.

a developing country population of 6.85 billion—still only three fourths of per capita consumption of calorific sweeteners in the US). Developing country demand for sugar tends to rise rapidly with income (14), and demand growth could be accelerated by lower sugar prices made possible through sugar/electricity coproduction strategies, just as alcohol/electricity coproduction strategies can be expected to make lower alcohol prices possible. Moreover, reduced sugar prices might even make it possible for cane sugar producers to win back some of the calorific sweetener market that has been lost to high fructose corn syrups.[12]

Public Policy Issues

Deployment of BIG/GT technologies in the sugar cane industries would be a promising initial strategy for modernizing biomass as an energy source in support of broad development goals, while simultaneously helping to reduce CO_2 emissions from the burning of fossil fuels.

A Developing Country Lead in Deploying BIG/GT Technologies? Exploiting energy from cane in the more than 70 developing countries that grow it could help these countries reduce their energy import dependency. Moreover, the unit capital costs for BIG/GT technologies would often be considerably less than the unit capital costs of conventional, large, central-station, fossil-fuel, nuclear, or hydroelectric plants (compare Tables 6 and 15). This is especially important given that sharply rising capital costs for electricity based on conventional technologies (Table 16) are raising the specter of limited capital availability as a constraint on the development of these countries (66,67). The deployment of BIG/GT technologies in rural areas would also help cope with sharply rising transmission and distribution (T&D) costs (Table 15). While most cities and towns are already electrified, many rural areas are not. It is very costly to provide initial electricity supplies to rural areas from large central-station sources, because of the low load factors on the power lines in the early stages of electrification. Considerable T&D savings could be realized by deploying decentralized BIG/GT technologies to serve

Table 15. Investment Requirements for the Power Sector in Developing Countries, as Estimated by the World Energy Conference (WEC).[a]

	1980	2000L[b]	2000H[b]
Unit Cost for New Generating Capacity (1986 $/kW)			
Hydroelectric	2660	3260	3990
Nuclear	2000	2460	3000
Fossil Fuel, Thermal	1000	1200	1460
Average Unit Capital Cost (1986$/kW)			
Generation	1640	2010	2410
T & D	790	1650	2410
Total Investment Requirements (billion 1986 $)			
Generation	28	79	185
Transmission & Distribution	15	65	185
Total	43	144	370
(as % of GDP)	(1.5)	(2.6)	(5.5)

[a] *Source:* H.K. Schneider and W. Schulz, *Investment Requirements of the World Energy Industries 1980–2000*, report prepared for the WEC Study on Long-Term Investment Requirements, Needs, Constraints, and Proposals, World Energy Conference, London, September, 1987.
[b] "L" ("H") refers to the WEC low (high) growth scenario, for which GDP grows at an average rate of 3.5%/yr (4.5%/yr), 1980–2000.

rural local markets during the early stages of T&D development. Moreover, BIG/GT technologies deployed at sugar cane factories could help provide a basis for rural industrialization and employment, helping stem urban migration and reducing rural poverty. Such considerations alone indicate that sugar cane producing developing countries should be powerfully motivated to demonstrate and deploy BIG/GT technologies without taking into account greenhouse concerns.

However, many developing countries are unable to consider seriously major innovations relating to energy. Acting alone, they do not have the financial resources to risk on commercial demonstrations of energy technologies. Moreover, countries capable of carrying out commercial demonstrations are usually given scant encouragement by the bilateral and multilateral assistance agencies. Rapidly industrializing countries like Brazil, for example, are too "developed" to qualify for US AID assistance, while countries that do qualify typically do not have strong enough technological infrastructures to support commercial demonstrations of technologies like the BIG/GT. Also, policies of the multilateral financial assistance agencies do not encourage technological innovations relating to energy. For example, in a 1987 speech on World Bank energy policy, a Bank official stated (68):

> . . . In the world of energy, as in many other areas, [the Bank's] role is not to create innovative technical solutions, or to help countries to gamble on new processes, but to identify the best practices in a developing country situation, and encourage their wider adoption where merited by circumstances . . .

Even if BIG/GT technologies were commercially available, it would be difficult to deploy them in many developing country situations without new public policies. These technologies are unfamiliar, the cane sugar and alcohol industries are not accustomed to embracing major new technologies, and the required investments would be large compared to the investments in the original sugar factories or alcohol distilleries[13]. Moreover, even if the cane industries could be persuaded to deploy BIG/GT technologies, they would often face major problems in trying to market the produced electricity. In developing countries most electric power plants are central-station facilities owned by utilities. Efforts to introduce decentralized BIG/GT cogeneration units would often be met by the same kind of resistance from utilities that cogenerators in the US faced before the adoption of the Public Utility Regulatory Policies Act of 1978 (PURPA). This legislation, which has created a cogeneration boom in the US (5), requires utilities to purchase electricity produced by qualifying cogenerators at an "avoided cost" price. Policy changes along these lines or other approaches that would promote cogeneration and decentralized power generation are needed to facilitate the introduction of BIG/GT technologies in developing country applications.

An Industrialized Country Lead in Deploying BIG/GT Technologies? If developing countries are not likely to take the lead in developing and deploying BIG/GT technologies, what about the already industrialized countries?

There is growing interest throughout the industrialized world in the use of high-efficiency gas turbines for power generation using natural gas as fuel, for both central station and cogeneration applications (5). Most of this activity involves the use of heavy-duty industrial gas turbines and combined cycles. While combined cycles are well suited for power generation at scales of hundreds of megawatts, they do not offer particularly attractive economics at the more modest scales needed for biomass applications where various steam-injected cycles based on the use of aeroderivative turbines would be more attractive.

Because it is generally thought that natural gas may be no more abundant than oil (69) and must be regarded as a transition fuel, there is also considerable interest on the part of industrialized countries in developing technologies that will make it possible to marry the gas turbine to more abundant, lower quality fuels. Because coal is abundant and relatively cheap, ongoing R&D efforts are aimed at marrying the gas turbine to coal through coal gasification. These efforts too are focused on combined cycle applications. There are no demonstration

Table 16a. Present/Future[a] Distribution of Fossil Fuel CO_2 Emissions (GT/yr).

	1985	2025 RCW	2025 SCW	2050 RCW	2050 SCW
Industrialized Countries					
Market	2.6	2.8	2.4	3.2	2.3
Centrally Planned	1.5	2.1	2.1	3.0	2.0
Subtotal	4.1	4.9	4.5	6.2	4.3
Developing Countries					
Market	0.8	3.4	1.9	5.6	2.3
Centrally Planned	0.6	2.0	1.0	3.5	1.3
Subtotal	1.4	5.4	2.9	9.1	3.6
Global	5.5	10.3	7.4	15.3	7.9

Table 16b. Present/Future[a] Distribution of Population (millions).

	1985	2025 RCW	2025 SCW	2050 RCW	2050 SCW
Industrialized Countries					
Market	813	938	943	928	923
Centrally Planned	416	500	514	521	533
Subtotal	1229	1438	1457	1449	1456
Developing Countries					
Market	2500	5024	5522	6212	7543
Centrally Planned	1140	1728	1672	1866	1805
Subtotal	3640	6752	7194	8078	9348
Global	4869	8190	8651	9527	10804

Table 16c. Present/Future[a] Distribution of Per Capita Fossil Fuel CO_2 Emissions (tonnes/yr).

	1985	2025 RCW	2025 SCW	2050 RCW	2050 SCW
Industrialized Countries					
Market	3.20	2.99	2.55	3.45	2.49
Centrally Planned	3.62	4.20	4.09	5.76	3.75
Subtotal	3.34	3.41	3.09	4.28	2.95
Developing Countries					
Market	0.32	0.68	0.34	0.90	0.30
Centrally Planned	0.53	1.16	0.60	1.88	0.72
Subtotal	0.39	0.80	0.40	1.13	0.39
Global	1.13	1.26	0.86	1.61	0.73

[a] The projections were made by the US Environmental Protection Agency for its Rapidly Changing World (RCW) Scenario and its Slowly Changing World (SCW) Scenario in its 1989 study on policy options for coping with global warming [US Environmental Protection Agency, "Policy Options for Stabilizing the Global Climate," February 1989 (draft)].

projects under way aimed at marrying the gas turbine to biomass through biomass gasification.

Why is the private sector not rushing to commercialize technologies like ISTIG? For natural gas based central station power applications ISTIG might offer an advantage of a couple of percentage points in efficiency over the best combined cycle systems on the market, but this advantage is probably not adequate to warrant bringing it to market under present conditions. Its major advantage—high efficiency and low unit cost at modest scale—though important for biomass applications, is not so important for natural gas applications. The US manufacturer that has advanced ISTIG designs has indicated a willingness to commercialize the technology without government support, provided there is sufficient commercial interest (70). But potential users are not likely to be interested in this technology unless natural gas prices are considerably higher than at present. In any case, they seem unwilling to commit financial resources to an undemonstrated technology.

Acting alone, private companies may also not be willing at this time to develop BIG/GT technologies for markets in developing countries. Since many developing countries have abundant, low-cost natural gas resources that are largely unexploited (5), near-term private-sector efforts to expand the use of gas turbines for power generation in developing countries are likely to focus on these lucrative natural gas applications instead. Also, debt problems, conservative attitudes toward new technologies, restrictions on foreign investment, and other financial risks make developing country markets less than attractive.

There are also significant potential near-term industrialized country markets for BIG/GT technologies where there are substantial unused or inefficiently used residues—most notably in the pulp and paper industry (71). But again, private-sector interest in these potential markets is weak at present, owing to low natural gas prices and abundant coal resources.

Toward North/South Cooperation in Deploying BIG/GT Technologies. The emergence of global warming as a paramount environmental concern could radically improve the outlook for commercializing and deploying BIG/GT technologies. The prospect that most future emissions of CO_2 will come from developing countries (Table 16) gives industrialized countries a powerful incentive to help developing countries find ways to meet their development goals with lower CO_2 emissions.

It would be neither realistic nor just to expect developing countries to put greenhouse concerns high on their public policy agenda, in light of far more pressing development challenges. However, they should be willing to pursue low greenhouse gas-emitting energy paths as long as doing so is consistent with meeting sustainable development goals and would not cost them much more than conventional energy.

Could the industrialized countries be expected to shoulder the burden of the extra costs of pursuing low CO_2-emitting energy strategies in developing countries? The extent to which they might be expected to do so depends on how large the costs are expected to be and how they compare to the expected benefits.

Unfortunately, the benefits of reducing greenhouse emissions are extraordinarily difficult (and perhaps impossible) to estimate in satisfactory ways. Some studies of the economics of reducing global CO_2 emissions suggest that the costs might be quite high. Manne and Richels estimate that a long-run equilibrium carbon tax of about $250 per tonne would be needed to constrain CO_2 emissions globally so that emissions from industrialized countries fall to 80 percent of the 1990 level by 2020, emissions from developing countries are constrained to double the present level, and global emissions stabilize at a level 15 percent higher than at present by 2030 (72). In order for developing countries to stay within these carbon emissions limits without incurring extra costs, industrialized countries would have to transfer over $900 billion annually to developing countries by the middle of the next century, under the assumptions of the Manne-Richels modeling exercise.

In their modeling exercise, Manne and Richels assumed that zero CO_2 emissions could be achieved in electric power generation without increasing the cost of electricity beyond that for coal-based power generation—most likely through the use of advanced nuclear power technology. They also assumed that zero CO_2-emitting nonelectric energy supplies would cost twice as much ($19/GJ) as synthetic fuels from coal ($9.5/GJ) that are characterized by a CO_2 emissions rate of 0.0387 tonnes C/GJ. It is the latter assumption that led to their estimated equilibrium carbon tax:

$$(\$19 - \$9.5)/0.0387 = \$250/\text{tonne of carbon.}$$

The present study indicates that sugar cane based bioenergy strategies would reduce CO_2 emissions at much lower marginal costs than is indicated by the Manne-Richels study.

Even without the economic benefits of coproduction, the production cost for alcohol in Brazil ($0.21/liter, or $8.9/GJ [Figure 9a]) would be slightly less than the cost of synfuels from coal. Since net fossil fuel based carbon emissions associated with alcohol production are 9.6 kg C/tonne of cane (Table 13) or 0.0058 tonnes C/GJ of alcohol (assuming no credit for the production of methane from stillage), the cost of carbon removal associated with producing ethanol from cane as an alternative to producing coal-derived synfuels would be negative:

$$(\$8.9 - \$9.5)/(0.0387 - 0.0058) = -\$18/\text{tonne of carbon.}$$

With coproduction strategies the cost of carbon emissions reduction would be even more strongly negative. Consider the alcohol/electricity coproduction case (with a 160-day milling season) where the distiller sells both bagasse and barbojo to the cogenerator, at a price sufficient to reduce the net alcohol cost to 15 cents/liter—the cost needed for alcohol to compete as a neat fuel with gasoline at the present world oil price. At this level, the estimated cost for the 721 kWh/tc of electricity cogenerated with BIG/ISTIG (Table 13) would be 3.6 cents/kWh, compared to 5.3 cents/kWh for a coal plant (Figure 9a), so that the cost of reducing carbon emissions by 0.222 tonnes per tonne of cane processed (Table 14) for this coproduction strategy would be

(721 kWh/tonne of cane)
× [($0.036 - $0.053)/kWh]/
(0.222 tonnes C/tonne of cane)
= −$55/tonne of carbon.

Since these carbon emissions reduction costs are negative, achieving large emissions reductions by deploying BIG/GT technologies in the sugar cane industries would not require large transfer payments to developing countries. Rather, industrialized country commitments are needed to bring to commercial readiness the gas turbine technologies involved and to work with developing countries to facilitate the transfer of these technologies to the marketplace. It is estimated that a 20 MW commercial demonstration project for BIG/STIG would cost about $50 million—twice the estimated cost of a commercial unit (44). The estimated cost of a 110 MW demonstration project for ISTIG (with natural gas firing) is about $100 million (7).

Successful demonstration of these technologies could lead to widespread deployment of BIG/ISTIG technology in the sugar cane industries beginning near the turn of the century and subsequently evolving along the lines depicted for the BIG/ISTIG scenarios described in Table 12. The present (1990) value of the direct economic benefits of such a course, over the period from 1997 to 2027, would be more than $8 billion, corresponding to a benefit/cost (B/C) ratio of more than 50:1 for this $150 million investment, without taking into account any greenhouse benefits.[14]

Such a high B/C ratio suggests that there would be major advantages to both sides from north/south cooperation aimed at facilitating the introduction of BIG/GT technologies for cogeneration applications in the sugar cane industries. A new industry involving north/south joint ventures could be spawned for marketing BIG/GT technologies, with aircraft engine manufacturers providing the gas turbines and a wide range of other companies providing the remaining needed equipment.[15]

To the extent that Brazilian conditions for alcohol production prove to be relevant for other parts of the world, the already large global market for sugar/electricity coproduction could be greatly expanded to an even larger global market that includes alcohol/electricity coproduction as well. Thus Brazilian producers of cane alcohol technology might wish to team with BIG/GT manufacturers in joint ventures to serve such markets. Brazil could thus become a leading marketer of "greenhouse-safe" energy technologies.

Because of the low efficiency of photosynthesis and the high water requirements for the growing of biomass, bioenergy strategies will eventually be limited by land and water availability and by themselves do not offer "the" solution to global warming. But by using high efficiency conversion technologies like the BIG/GT technologies considered here, these constraints can be pushed far into the future, and much can be accomplished by deploying these technologies in sugar cane and other residue markets (8,71) and eventually in markets served by dedicated biomass plantations (9).

Joint north/south initiatives relating to BIG/GT technologies would be consistent with the Bush Administration's "no regrets" policy on climate change, which means, according to Secretary of State Baker (73):

> . . . we are prepared to take actions that are fully justified in their own right and which have the added advantage of coping with greenhouse gases. They're precisely the policies we will never have cause to regret.

Acknowledgments

For research support the authors thank the Office of Energy of the US Agency for International Development, the US Environmental Protection Agency, the New Land Foundation, and the Merck Fund.

For useful discussions the authors thank Pedro Assis, Luiz Paulo de Biase, Harold Birkett, Carlos de Camargo, Stephen Clarke, Stefano Consonni, Michael Hylton, William Keenliside, Eric Larson, Alistair Lloyd, Isaias Macedo, José Roberto Moreira, José Campanari Neto, José Luis Oliverio, Alice de Ribeiro, José Valdir Sartori, Francis Schaeffer, Gil Serra, Walter Siebold, Robert Socolow, Charles Weiss, and Florenal Zarpelon.

Notes

1. If fossil fuels are required to produce biomass and convert it into useful energy carriers, then this is not strictly true. For some high-cost biofuels (e.g., ethanol produced from corn), net energy gains in biofuel production can be small (1), so that net CO_2 reductions achievable with a shift to such biofuels would be small (and might even be negative). However, as will be shown, these fossil fuel energy penalties are small for the sugar cane applications considered here.
2. The mass at harvest of the sugar cane stalks, which does not include the mass of the barbojo (the Latin American word for the tops and leaves of the sugar cane plant). The

barbojo is typically burned off just before harvest, to facilitate harvesting the cane stalks.

3. Alcohol can be produced from sugar cane with a yield of about 70 liters per tonne of cane (tc) or 1.68 GJ/tc. (In this paper higher heating values are used for fuels.) If this much alcohol were produced from wood instead, some 3.36 GJ of woody feedstock would be required, assuming a 50 percent conversion efficiency (11). In addition, methane can be recovered with a yield of about 0.33 GJ/tc from the stillage at cane distilleries via anaerobic digestion (12). To recover this much methane via wood gasification would require 0.49 GJ of woody feedstock, assuming a gasification efficiency of 67 percent. In addition, the bagasse residue left after extracting the sugar juice from cane amounts to 300 kg/tc (50 percent moisture), or 2.85 GJ/tc, and the barbojo is another potentially usable residue amounting to some 660 kg/tc (50 percent moisture), or 6.27 GJ/tc. Thus the total woody biomass equivalent of one tonne of harvested cane is

$$3.36 + 0.49 + 2.85 + 6.27 = 12.97 \text{ GJ or } 0.65 \text{ tonnes dry weight,}$$

for wood at 20 GJ/tonne.
4. Between 1979/80 and 1987/88 the cost of producing alcohol in Brazil declined 4 percent per year, on average. Also, the average productivity of cane fields increased from 2663 l/ha/yr in 1977/78 to 3811 l/ha/yr in 1985/86 (32).
5. The present production cost of ethanol in Brazil is 21 cents per liter. One liter of ethanol is worth 0.84 liters of gasoline as a neat fuel (11) and 1.16 liters of gasoline as "gasohol," an octane-enhancing additive to gasoline (33). At the present world oil price ($20/barrel), the wholesale gasoline price is 17.7 cents/liter, so that the present value of ethanol is 14.9 cents/liter as a neat fuel and 20.5 cents/liter as gasohol. (The wholesale gasoline price Pgas (in cents per liter) can be expressed in terms of Pcrude, the refiner acquisition cost of imported crude oil in the US (taken to be the world oil price), in dollars per barrel, as: Pgas = 6.26 + 0.572*Pcrude, r = 0.99067, obtained by regression of data for the period 1980 to 1989.)
6. A 27 MW steam turbine unit supplied with steam at 6 MPa would have an efficiency of only 20 percent in producing power only (Table 5). Of course higher efficiencies can be achieved with the steam Rankine cycle at the larger scales of central station power plants, for which it is economical to operate at much higher quality steam conditions. A modern 500 MW central station steam turbine unit with 16.5 MPa steam has an efficiency of about 35 percent.
7. For this analysis a real discount rate of 15 percent (10 percent) and a 20-year (30-year) plant life are assumed for distillery (cogeneration) operations, and the insurance cost is assumed to be 0.5 percent per year of the initial capital cost.
8. For BIG/GT systems, the cost of processing the bagasse is estimated to be $1.25/GJ for drying and briquetting (47), which is probably needed to make the bagasse a suitable feedstock for the assumed fixed-bed gasifier; the corresponding cost for recovering, transporting, and briquetting field-dried barbojo is estimated to be $1.35/GJ (47). For CEST systems, no drying or processing would be required for bagasse, but the cost for harvesting, baling, and drying the barbojo is estimated to be $0.97/GJ (47).
9. The cost of cane delivered to the factory depends upon many factors, including cane yield, the type of land, climate, the harvesting system, labor costs, and transportation costs to the factory. Not surprisingly, cane growing and harvesting practices are site specific and vary considerably around the world (60,61). Accordingly, the cost of cane varies from $8/tonne in Brazil to over $20/tonne in Louisiana. In Brazil the large scale of cane production, as well as cane varieties and cultivation practices optimized for high yield, may be a factor in reaching low costs (36). The prospects for reaching these low cane costs in other regions need to be better understood.
10. Brazilian autonomous distillery costs are typically only one half to one third those quoted by US engineering firms. Several factors could account for the difference in autonomous distillery equipment costs (50,62):

    - Autonomous distillery equipment is manufactured in Brazil on a much larger scale than elsewhere, so that manufacturers achieve lower costs by taking advantage of economies of scale in production volume.
    - Because of the experience gained in the PROALCOOL program, the Brazilian alcohol industry may be further along the technological "learning curve."
    - Labor and/or material costs may be lower in Brazil.
    - Engineering standards or codes for process equipment may be different.

    Understanding these issues in detail could shed light on whether the Brazilian experience with autonomous distillery costs could be replicated elsewhere.
11. Technologies are being developed to harvest and recover barbojo (63). However, the extent to which it is practical and cost-effective to recover barbojo may be site-dependent. Studies are needed to understand the prospects for barbojo recovery under the variety of conditions that characterize the cane industry throughout the world.
12. In the US, per capita consumption of high fructose corn syrup sweeteners rose from 0.45 kg/capita in 1970 (out of a total calorific sweetener consumption of 56 kg/capita) to 29 kg/capita in 1987 (out of a total calorific sweetener consumption of 60 kg/capita). In this same period sugar consumption declined from 46 to 28 kg/capita (65).
13. A typical distillery in Brazil producing 120,000 liters of ethanol per day costs about $7 million installed (16). This size distillery could support a 23 MW BIG/ISTIG unit that would cost about $25 million.
14. Consider first the situation where the average cane processing facility at which BIG/ISTIG plants would be sited has a throughput of 325 tc/hour, which could support a 111 MW BIG/ISTIG unit. Assuming that half of the cane is associated with BIG/ISTIG cogeneration technologies deployed at sugar factories and alcohol distilleries operated on 160-day milling seasons (with the alcohol producers selling only bagasse to the cogenerator) and half deployed at sugar factories and alcohol distilleries operated on 133-day milling seasons (with the alcohol producers selling both bagasse and barbojo to the cogenerator), the average cost of electricity (with alcohol selling for 15 cents/liter) would be 3.4 cents/kWh, compared to 5.3 cents/kWh for the coal-based electricity displaced. If total BIG/ISTIG capacity in the cane industries expands from 1 GW in 1997 (producing 7.34 TWh) to 337 GW in 2027 (producing 2473 TWh), the

value of the discounted electricity cost savings (1997–2027) would be $27 billion in 1997 or $14 billion in 1990, assuming a 10 percent discount rate.

This assessment would apply to a situation where there would be considerable centralization of cane processing compared to the present situation—which may well take place, in order to capture the scale economies associated with coproduction. While there are many cane processing facilities in various parts of the world with throughputs well in excess of 325 tc/hour, the average throughput per factory today is instead about 100 tc/hour, which could support a BIG/ISTIG capacity of about 35 MW. At this average capacity, the cost of BIG/ISTIG electricity would be about 4.1 cents/kWh, so that the present value of the benefits would be $8.8 billion instead of $14 billion.

15. Most of the cost associated with BIG/STIG and BIG/ISTIG systems (Table 7) is associated with "low-technology" components (gasifier, electrical generator, heat recovery steam generator, power turbine, etc.) that could eventually be manufactured by many firms in rapidly industrializing countries. The only "high-technology" component (which would have to be imported by developing country users) is the jet engine from which the gas turbine is derived. The jet engine upon which the 53 MW BIG/STIG and 111 MW BIG/ISTIG units are based (Table 6) sells for $6 million (5). It thus represents only 11 percent (7 percent) of the total installed cost of the BIG/STIG (BIG/ISTIG). By choosing high performance systems, developing countries that get involved in joint ventures to produce BIG/GT technologies could reduce foreign exchange requirements to very low levels.

References

1. Office of Technology Assessment, *Energy from Biological Processes,* Washington, DC, 1980.
2. US Environmental Protection Agency, "Policy Options for Stabilizing the Global Climate," February 1989 (draft).
3. David O. Hall, private communication. See also Hall, Barnard, and Moss, *Biomass for Energy in Developing Countries,* Pergamon Press, 1982.
4. Robert H. Williams, "Potential Roles for Bioenergy in an Energy-Efficient World," *Ambio,* vol. 14, nos. 4–5, pp. 201–209, 1985.
5. R.H. Williams and E.D. Larson, "Expanding Roles for Gas Turbines in Power Generation," in *Electricity,* T.B. Johansson, B. Bodlund, and R.H. Williams, eds., University of Lund Press, Lund, Sweden, 1989.
6. S.B. Alpert and M.J. Gluckman, "Coal Gasification Systems for Power Generation," Electric Power Research Institute, Palo Alto, CA, 1986.
7. J.C. Corman, "System Analysis of Simplified IGCC Plants," report prepared for the US Dept. of Energy by the Corporate Research and Development Center, the General Electric Company, Schenectady, NY, September, 1986.
8. E.D. Larson, P. Svenningsson, and I. Bjerle, "Biomass Gasification for Gas Turbine Power Generation," in *Electricity,* T.B. Johansson, B. Bodlund, and R.H. Williams, eds., University of Lund Press, Lund, Sweden, 1989.
9. R.H. Williams, "Biomass Gasifier/Gas Turbine Power and the Greenhouse Warming," in *Energy Technologies for Reducing Emissions of Greenhouse Gases,* vol. 2, pp. 197–248, Proceedings of an Experts' Seminar, OECD/IEA, Paris, April 12–14, 1989.
10. Food and Agriculture Organization of the United Nations, *FAO Production Yearbook,* vol. 41, FAO Statistical Series No. 82, 1987.
11. C.E. Wyman and N.D. Hinman, "Ethanol: Fundamentals of Production from Renewable Feedstocks and Use as a Transport Fuel," Solar Energy Research Institute, Golden, CO, 1989.
12. S. Hochgreb, "Methane Production from Stillage," COMGAS, São Paulo, 1985.
13. Robert D. Perlack and J. Warren Ranney, "Economics of Short-Rotation Intensive Culture for the Production of Wood Energy Feedstocks," *Energy,* vol. 12, no. 12, pp. 1217–1226, 1987.
14. James G. Brown, "The International Sugar Industry: Developments and Prospects," World Bank Staff Commodity Working Paper No. 18, March 1987.
15. "Agroindustria Canavieira: Um Perfil," Copersucar, São Paulo, Brazil, 1989; "PROALCOOL," Copersucar, São Paulo, Brazil, 1989.
16. J. Goldemberg, J.R. Moreira, P.U.M. Dos Santos, and G.E. Serra, "Ethanol Fuel: A Use of Biomass Energy in Brazil," *Ambio,* vol. 14, no. 4–5, 1985; G. Serra and J.R. Moreira, University of São Paulo, Brazil, private communications, 1989.
17. H. Geller, "Ethanol Fuel from Sugar Cane in Brazil," *Annual Review of Energy,* vol. 10, pp. 134–164, 1985.
18. H. Steingass, K. Wentzel, M. Kappaz, F. Schaeffer, R. Schwandt and P. Bailey, "Electricity and Ethanol Options in Southern Africa," Office of Energy, Bureau for Science and Technology, USAID, Report No. 88–21, September 1988.
19. "Worldwide Review of Biomass Based Ethanol Activities," Meridian Corporation Report, Contract No. MC-JC-85FB-008, 1985.
20. "Power Alcohol in Kenya and Zimbabwe—A Case Study of the Transfer of a Renewable Energy Technology," United Nations Trade and Development Board Report, GE.84-55979, June 5, 1984.
21. M.G. Hylton, Director, Factory Technology Division, Sugar Industry Research Institute, Kingston, Jamaica, private communications, 1989.
22. W.V. Saunders, "Petroleum Company of Jamaica's Ethanol Project—An Update," Proceedings of the Jamaican Society of Sugar Technologists, Ocho Rios, Jamaica, November 5, 1987.
23. "Alcohol Production from Biomass in the Developing Countries," World Bank Report, September 1980.
24. A. Jacobs, F. Tugwell, H. Steingass, M. Alvarez, F. Arechaga, W. Klausmeier, R. Richman, "Fuel Alcohol Production in Honduras: A Technical and Economic Analysis," USAID Cane Energy Program, April 1986.
25. "The Sugar Industry in the Philippines," USAID Cane Energy Program, December 1986.
26. "The AID Approach: Using Agricultural and Forestry Wastes for the Production of Energy in Support of Rural Development," USAID Bioenergy Systems Report, April 1989.
27. F.C. Schaeffer and Associates, "Evaluation and Technical

and Financial Assessment, Central Azucerera Tempisque, S.A. (CATSA)," Report to CODESA/USAID, Volume 1, July 1987.
28. J.R. Moreira, V.R. Vanin, and J. Goldemberg (Institute of Physics, University of São Paulo, São Paulo, Brazil), "Energy Balance for the Production of Ethyl and Methyl Alcohol," presented at the Workshop on Fermentation Alcohol for Use as Fuel and Chemical Feedstock in Developing Countries, Vienna, Austria, March 26–30, 1979.
29. J.G. da Silva, G.E. Serra, J.R. Moreira, J.C. Concalves, and J. Goldemberg, "Energy Balance for Ethyl Alcohol Production from Crops," *Science,* vol. 201, pp. 903–906, September 8, 1978.
30. C.S. Hopkinson and J.W. Day, "Net Energy Analysis of Alcohol Production from Sugarcane," *Science,* vol. 207, pp. 302–304, January 18, 1990.
31. J.A. Polack, H.S. Birkett, and M.D. West, "Sugar Cane: Positive Energy Source for Alcohol," *Chemical Engineering Progress,* June 1981.
32. Copersucar, *Proalcool: Fundamentos e Perspectivas,* São Paulo, 2nd Edition, September, 1989.
33. Office of Technology Assessment, "Gasohol," A Technical Memorandum, Washington, DC, September 1979.
34. E. Hugot, *Handbook of Cane Sugar Engineering,* Elsevier Science Publishing Company, New York, NY, 1986.
35. C.A. de Camargo, "Balanco de Vapor e Conservacao de Energia em Usinas e Destilarias," Brazil Acucareiro, Rio de Janeiro, v. 105, pp.44–53, 1987.
36. G. Serra, University of São Paulo, Campinas, Brazil, private communications, 1989.
37. J.C. Neto, Copersucar Technical Center, Piracicaba, Brazil, private communications, 1989.
38. J.L. Oliverio, J.D. Neto, and J.F.P. de Miranda, "Energy Optimization and Electricity Production in Sugar Mills and Alcohol Distilleries," Proceedings of the 20th Congress of the International Society of Sugar Cane Technologists, São Paulo, Brazil, October 12–21, 1989.
39. J.L. Oliverio and R.J. Ordine, "Novas Tecnologias e Processos que Permitam Elevar o Excedente de Bagaco das Usinas E Destilarias," Brazil Acucareiro, Rio de Janeiro, v. 105, pp. 54–88, 1987.
40. "Tecnicas para Producao de Excendentes de Bagacao," Copersucar, São Paulo, 1986.
41. C.A. Camargo, A.M. de Melo Ribeiro, "Manual of Energy Conservation in the Cane Sugar and Alcohol Industry," IPT Report, Instituto de Pesquisas Tecnologias de Etsado de São Paulo, São Paulo, Brazil, 1989.
42. C. Kinoshita, Hawaiian Sugar Planters Association, private communications, 1987.
43. Commission of the European Communities, "24.65 MW Bagasse Fired Steam Power Demonstration Plant," Report No. EUR 10390 EN/FR, 1986.
44. J.C. Corman, "Integrated Gasification Steam-Injected Gas Turbine (IG-STIG)," presentation at the Workshop on Biomass-Gasifier Steam Injected Gas Turbines for the Cane Sugar Industry, Washington, DC, organized by the Center for Energy and Environmental Studies, Princeton University, June 12, 1987.
45. E.D. Larson and R.H. Williams, "Biomass Gasifier Steam-Injected Gas Turbine Cogeneration," *Journal of Engineering for Gas Turbines and Power,* vol. 112, pp. 157–163, April, 1990.
46. J.M. Ogden, S. Hochgreb and M. Hylton, "Process Energy Efficiency and Cogeneration in Cane Sugar Factories," Proceedings of the 20th Congress of the International Society of Sugar Cane Technologists, São Paulo, Brazil, October 12–21, 1989.
47. E.D. Larson, J.M. Ogden, R.H. Williams, "Steam-Injected Gas Turbine Cogeneration for the Cane Sugar Industry," Princeton University, Center for Energy and Environmental Studies Report No. 217, September 1987.
48. E.D. Larson, R.H. Williams, J.M. Ogden and M.G. Hylton, "Biomass-Fired Steam-Injected Gas Turbine Cogeneration for the Cane Sugar Industry," Proceedings of the 20th Congress of the International Society of Sugar Cane Technologists, São Paulo, Brazil, October 12–21, 1989.
49. J.M. Ogden and M.E. Fulmer, "Assessment of New Technologies for Co-Production of Alcohol, Sugar, and Electricity from Sugar Cane," Center for Energy and Environmental Studies, Princeton University, draft report, February 1990.
50. H.S. Birkett, F.C. Schaeffer and Associates, Baton Rouge, LA, private communications, 1989.
51. J.C. Teixeira da Silva, Manager, Department of Product and Process Development, Dedini, S.A., São Paulo, Brazil, private communications, 1987; Quotes from Brazilian manufacturers Dedini, Zanini, private communications, 1988, 1989.
52. US Dept. of Energy, "Assessment of Costs and Benefits of Flexible and Alternative Fuel Use in the US Transportation Sector; Technical Report Three: Methanol Production and Transportation Costs," draft, August 29, 1989.
53. A. Serra, M. Poch and C. Sola, "A Survey of Separation Systems for Fermentation Ethanol Recovery," *Process Biochemistry,* October 1987, p. 154-158.
54. G. Parkinson, "Batelle Maps Ways to Pare Ethanol Costs," *Chemical Engineering,* June 1, 1981.
55. L.R. Lynd and H.E. Grethlein, "IHOSR/Extractive Distillation for Ethanol Separation," *Chemical Engineering Progress,* November 1984, pp. 59–62.
56. R. Katzen, et al., "Low Energy Distillation Systems," in *Bioenergy '84,* vol. 1, p. 309.
57. "Heat Pumps in Distillation," EPRI Report, 1986.
58. Perry's *Chemical Engineers' Handbook,* McGraw Hill Co., New York, NY, Chapter 13, 1984.
59. V.I. Suchek, Director, Division of Development, Jaako Poyry Engineering, São Paulo, Brazil, private communications, 1989.
60. D.H. West, "A Cost Analysis Programme for Mechanical Harvesting of Sugar Cane," 23rd West Indies Sugar Technologists Conference, Bridgetown, Barbados, April 17–22, 1988.
61. C. Hudson, Symposium on "Mechanical Harvesting Systems for Cane," presented at the 20th Congress of the International Society of Sugar Cane Technologists, São Paulo, Brazil, October 1989.
62. J.M. Paturau, "Is Ethanol the Fuel of the Future for Sugar Cane Producing Territories?" Proceedings of the 18th Congress of the International Society of Sugar Cane Technologists, 1980.
63. Ronco Consulting Company and Bechtel National, Inc., "Jamaican Cane Energy Project Feasibility Study," a report

of the Office of Energy, US Agency for International Development and the Trade and Development Program, Washington, DC, 1986.

64. Ministerio das Minas Energia, *Boletim do Balanco Energetico Nacional 1989,* Brasilia, Brazil, 1989.
65. Bureau of the Census, *Statistical Abstract of the United States,* US GPO, Washington, DC, 1989.
66. R.H. Williams, "Are Runaway Energy Capital Costs a Constraint on Development? A Demand Analysis of the Power Sector Capital Crisis in Developing Countries," paper presented at the International Seminar on the New Era in the World Economy, The Fernand Braudel Institute of World Economics, São Paulo, Brazil, August 31–September 2, 1988.
67. US Agency for International Development, *Power Shortages in Developing Countries: Magnitudes, Impacts, Solutions, and the Role of the Private Sector,* a report to the US Congress, Washington, DC, 1986.
68. E.S. Daffern, "World Bank Policy in the Promotion of Energy-Producing Projects in Developing Countries," paper presented in Reading, England, April 7, 1987.
69. C.D. Masters et al., "World Resources of Crude Oil, Natural Gas, Natural Bitumen, and Shale Oil," presented at the Twelfth World Petroleum Congress, Houston, TX, 1987.
70. "New NO_x and CO Limits Spark Utility Interest in I-STIG," *Gas Turbine World,* vol. 18, no. 6, December 1988.
71. E.D. Larson, "Prospects for Biomass-Gasifier Gas Turbine Cogeneration in the Forest Products Industry: A Scoping Study," PU/CEES Working Paper No. 113, February 1990.
72. A.S. Manne (Stanford University) and R.G. Richels (Electric Power Research Institute), "Global CO_2 Emission Reductions—the Impacts of Rising Energy Costs," preliminary draft, February 1990.
73. Secretary William Baker, "Diplomacy for the Environment," Current Policy No. 1254, Bureau of Public Affairs, United States Department of State, Washington, DC, February 26, 1990.

Pinch Technology: Evaluate the Energy/Environmental Economic Trade-Offs in Industrial Processes

H. D. Spriggs
Linnhoff March, Inc.
Leesburg, VA

R. Smith
UMIST
Manchester
England

E. A. Petela
Linnhoff March, Ltd.
Knutsford, Cheshire
England

Introduction

Pinch technology has established itself as a highly versatile tool for process design. Originally pioneered as a technique for reducing the capital and energy cost of new plant, pinch technology was shown to be readily adapted to the task of identifying the potential for energy savings in existing process plants during the period of rapidly escalating oil prices. More recently it has also become established as a tool for debottlenecking, yield improvement, capital cost reduction, and flexibility enhancement. Given the growing concern for the environment, we should now consider whether the power of pinch technology can be used to solve environmental problems.

This paper will address three areas in which pinch technology has already been identified as having an important role to play:

1. flue gas emissions
2. waste minimization
3. evaluation of waste treatment options.

In each of these areas pinch technology has a unique role to play in that it can be used to evaluate the relevant economic trade-offs. However, before this can be considered a brief review of pinch technology is necessary.

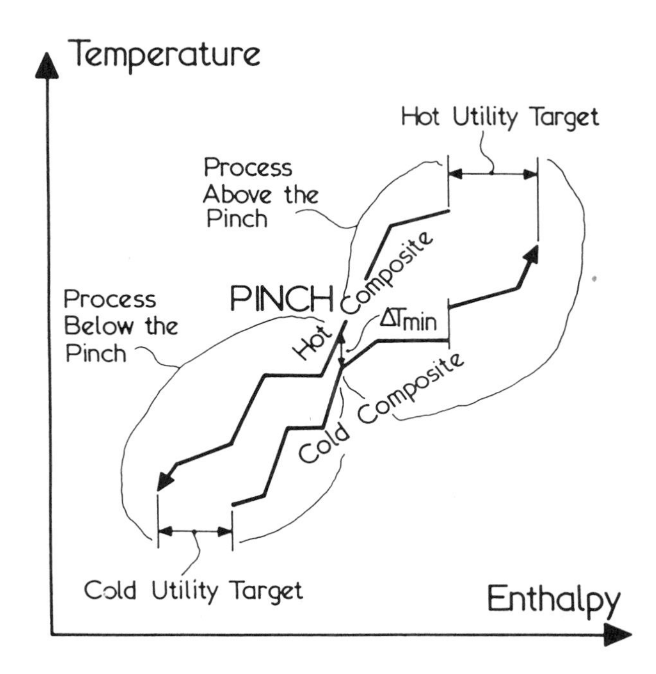

Figure 1. The composite curves allow minimum hot and cold utility to be determined before design. The curves also provide the location of the pinch which is of fundamental importance in design.

Pinch Technology—A Review

Pinch technology first identifies sources of heat (termed hot streams) and sinks (termed cold streams) from the material and energy balance of a process. By combining hot streams in given temperature ranges we can produce the composite curve of all hot streams. Similarly, by combining cold streams in given temperature ranges we produce the composite curve of all cold streams. If we plot composite curves on a common temperature-enthalpy axis we can define the energy target for the process as shown in Figure 1. The relative position of the two curves, and hence the energy target, is fixed by the chosen minimum temperature driving force, ΔT_{min}.

The composite curves are used to determine the energy target for a given value of ΔT_{min}. This minimum driving force will normally be observed at one point only between the hot and the cold composite, called the "pinch" (Figure 1). The pinch point has a special significance.

To understand this significance we divide the process at the pinch, as in Figure 2. Above the pinch (in temperature terms) the process is in heat balance with the minimum hot utility, Q_{Hmin}. Heat is received from the hot utility and no heat is rejected. The process acts as a heat sink. Below the pinch (in temperature terms)

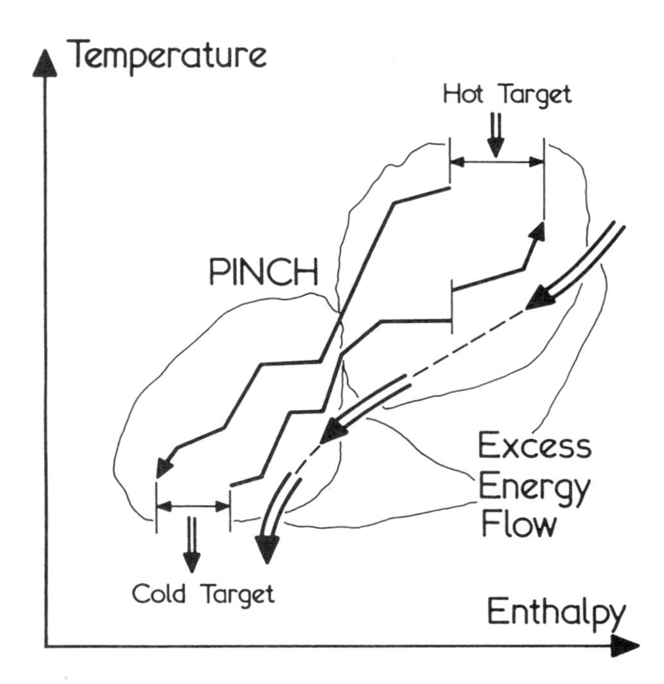

Figure 2. Heat flow across the pinch equals excess heat flow in and out.

the process is in heat balance with the minimum cold utility, Q_{Cmin}. No heat is received but heat is rejected to the cold utility. The process acts as a heat source.

If we choose to transfer heat, 'XP,' from the system above the pinch to the system below, the pinch equation presents the result quantitatively:

A	=	T	+	XP
Actual Energy Consumption		Target		Cross Pinch Heat Flow

In other words, to achieve the energy target set by the composite curves the designer must not transfer heat across the pinch in any form. The following rules, if observed in design, are enough to ensure that the energy target is achieved in practice.[1,2] Don't:

1. transfer heat across the pinch in process to process exchangers
2. use cold utility above the pinch
3. use hot utility below the pinch

The above is based on a fixed relative position of the composite curves, a fixed ΔT_{min}. The correct relative location of the composite curves (and hence the correct ΔT_{min}) is given by an economic trade-off between capital and energy.

Let us now consider how the composite curves are used to bring capital costs into consideration. The capital cost of a heat recovery system depends mainly on three factors:

1. the number of separate heat exchange matches (or units)
2. the total heat exchange area
3. the cost of the utility system necessary to provide the outstanding hot and cold utilities (boiler plant, cooling water system, refrigeration plant, etc.)

The minimum number of heat exchange matches (or units) depends fundamentally on the number of streams involved and on the pinch location. It can be calculated, prior to design, from equations based on Euler's graph theorem.[1,2]

The minimum total heat exchange area can also be calculated prior to any network design.[3,4] The approach is an extension of the standard model for countercurrent heat exchange. When we apply the model to the composite curves for the whole process, the equivalent countercurrent condition requires heat to be transferred vertically from the hot composite curve into the cold composite.

Given the relevant algorithms we can now evaluate the trade-off between capital cost and energy recovery explicitly.[5] The procedure is indicated in Figure 3. For various values of ΔT_{min} (obtained by changing the relative location of the composite curves) we determine the energy target (and hence the hot and cold utility requirements), the overall area required, and the number of units. The larger ΔT_{min}, the larger will be the energy target and the lower the overall heat transfer area. This information is combined and is used to predict ahead of design the overall cost for each ΔT_{min}. The lowest possible overall cost is clearly established (Figure 3).

The composite curves show us the scope for energy recovery and the hot and cold utility targets. Generally, a number of different utilities at different temperature levels and of different cost are available to the designer. Another pinch technology tool, the grand composite curve, helps the designer select the best individual utility or utility mix. The grand composite curve presents the profile of the horizontal (enthalpy) separation between the composite curves with a built-in allowance of ΔT_{min}. As illustrated in Figure 4 its construction involves bringing the composite curves together vertically (to allow for ΔT_{min}) and then plotting the horizontal separation (α in Figure 4).

Figure 5 shows how the grand composite curve reveals where heat is to be transferred between utilities and process and where the process can satisfy its own heat demand. How can this be applied to environmental problems?

Flue Gas Emissions

The relationship between energy efficiency and flue gas emissions is clear. The more inefficient we are in our

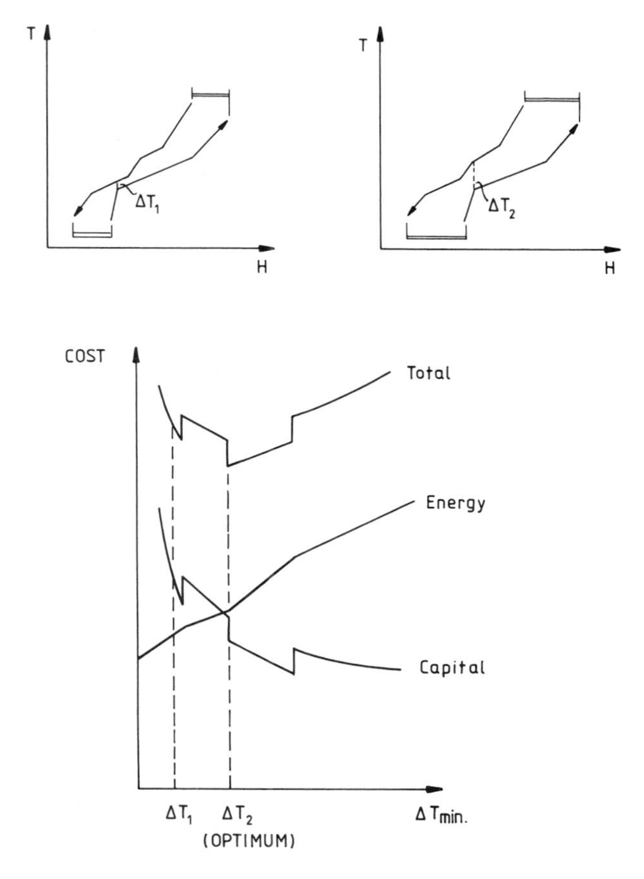

Figure 3. Capital and energy cost targets allow the capital-energy trade-off to be established ahead of design.

use of energy, the more fuel we burn and hence the greater are the flue gas emissions. Pinch technology can be used to improve energy efficiency through better heat integration and hence reduce flue gas emissions. However, with recent developments in pinch technology, we can be sure that this is done at the correct economic level of energy recovery.

Also, basic modifications to the process can be systematically directed to reduce flue gas emissions. Consider the process grand composite shown in Figure 6a. Because the process requires high temperatures, a furnace is required. Figure 6a shows the steepest flue gas line which can be drawn against the existing process. This corresponds with the smallest flue gas flowrate, the smallest fuel consumption, and hence the smallest flue gas emissions.

By contrast, Figure 6b shows the grand composite curve of the same process which has been modified specifically to open up temperature driving forces in the high temperature part of the process. The overall process duty is unchanged. However, the systematic modification of the process as shown in Figure 6b allows a steeper flue gas line to be drawn, leading to reduced flue gas emissions.

As an example of just how powerful pinch technology can be, consider the experience of BASF. BASF carried out an energy efficiency campaign at their Ludwigshafen factory.[6] Pinch technology was used to increase energy efficiency of the individual chemical processes on the site. Individual projects were justified on the basis of their energy saving potential only, with a required simple payback of one year or less. The total energy saved was 500 MW.[6] Figure 7 shows this change graphically and how it was achieved against a background of increased production levels. However, the following environmental relief resulted from reduced power station emissions on the site (where te = 1 metric tonne):[6]

Carbon dioxide	218 te/hour
Sulfur dioxide	1.4 te/hour
Nitrogen oxides	0.7 te/hour
Ash	21 kg/hour
Carbon monoxide	7 kg/hour
Waste water from water treatment	70 te/hour

Remarkable though these reduced emissions are, they are even more remarkable when we consider that they are associated with improved profitability.

Pinch technology can be used to establish the appropriate economic level of energy recovery given the trade-off between energy and capital. How then do we include environmental relief in the trade-offs?

At the moment there is no cost associated with the emission of greenhouse gases such as carbon dioxide and those which cause more direct environmental damage such as sulfur dioxide. If the regulatory bodies responsible for environmental emissions are to impose restrictions on sulfur dioxide then this requires switching to an alternative (more expensive) fuel or the installation of flue gas desulfurization plant. Here pinch technology can be used to evaluate the trade-offs between reduced energy consumption, investment in heat recovery systems for improved energy efficiency, and investment in flue gas clean-up.

Similar trade-offs for carbon dioxide emissions cannot be carried out unless a "carbon tax" is introduced to avert the greenhouse effect. However, the regulatory bodies responsible for environmental emissions could use pinch technology to target emissions rather than target energy consumption.

For a given set of process heating and cooling demands, there is an energy target for a given setting of the capital/energy trade-off. This energy target together with a given fuel and type of combustion equipment results in a minimum flue gas emission. Thus, once the type of fuel and combustion equipment has been set the minimum flue gas emission is set by the process heating and cooling demands and by the capital/energy trade-

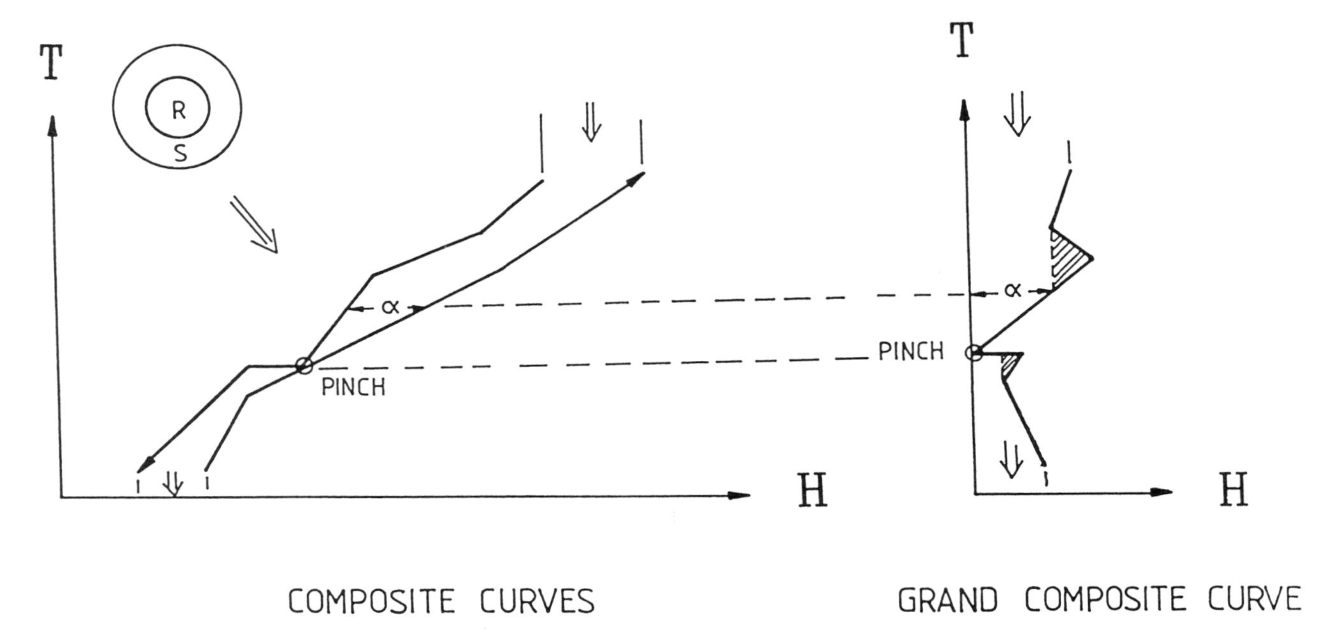

Figure 4. The grand composite curve presents the profile of the horizontal separation between the composite curves with a built in allowance for ΔT_{min}.

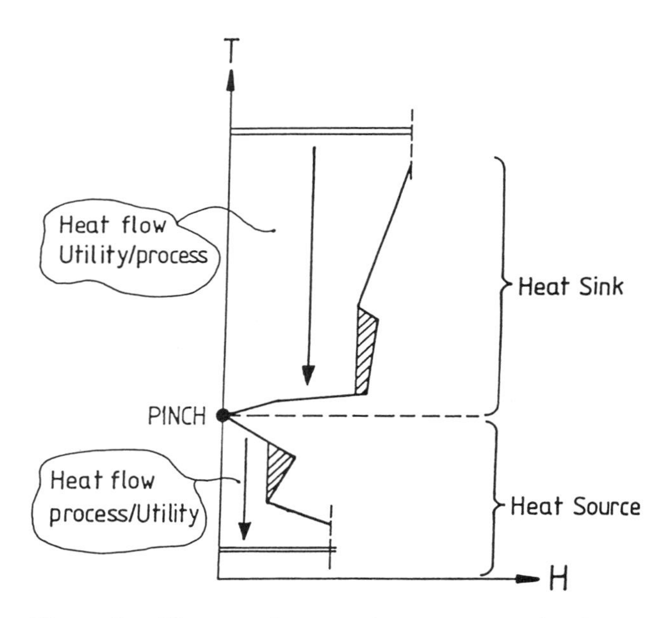

Figure 5. The grand composite curve reveals where heat is transferred between utilities and the process.

off for heat recovery. Pinch technology could thus be used to set practical and economically based targets for flue gas emissions.

Such targets could then be used, for example, to screen alternative process options, to compare centralized power generation with local cogeneration, or to impose a "carbon tax" on those companies which did not meet their targets.

Waste Minimization

The problem of process waste, whether gas, liquid, or solid, is best dealt with by making the process more efficient in its use of raw materials—that is, by waste minimization. Here, both the reaction and separation technology have a key role to play. Sometimes it is possible to reduce the formation of unwanted byproducts in the reactor, which must ultimately be disposed of, by modifying the reactor technology. More often it is possible to minimize waste by changes to the separation system. If the separation systems can be made more efficient such that useful materials can be separated and recycled more effectively, both waste disposal costs and raw materials costs would be reduced.

Many separation processes are driven by the input of heat such as distillation and evaporation. Others may be driven by the removal of heat such as condensation and some crystallization operations. If the separation

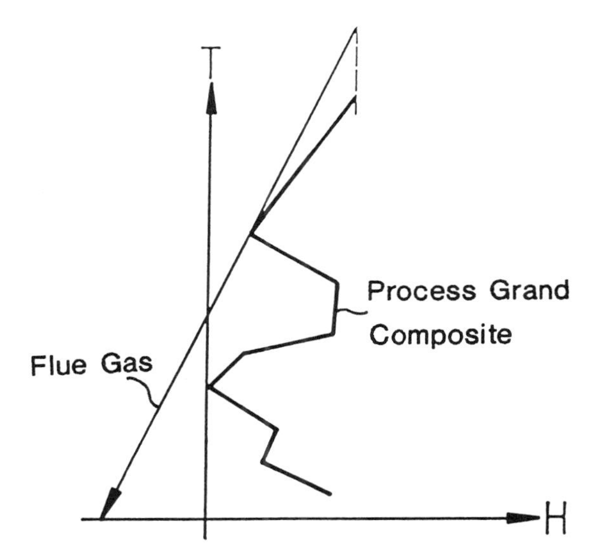

(a) Flue Gas Matched Against the Process is Limited By the Process Above the Pinch

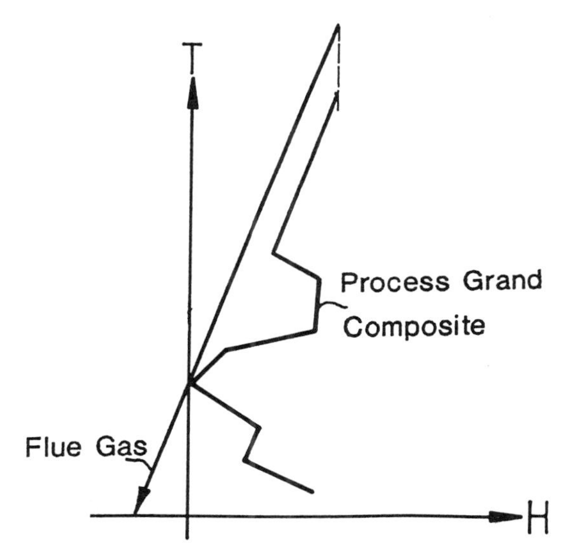

(b) The Modified Process Allows a Steeper Flue Gas Line Which Gives a Reduction in Fuel and Emissions Even Though the Process Duty Has Not Changed.

Figure 6. The grand composite curve allows the minimum flue gas to be established.

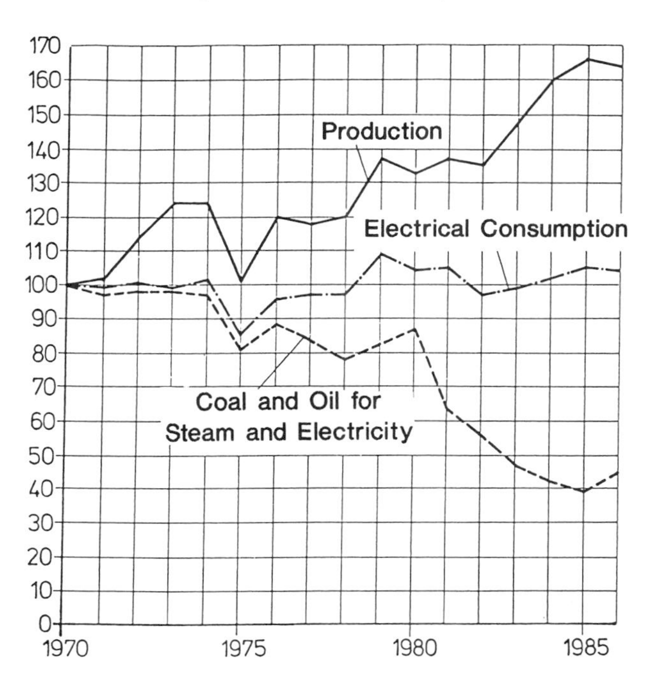

Figure 7. Pinch technology achieved considerable energy savings on the BASF site initiated in 1980 at Ludwigshafen against the background of increased levels of production.

system as a whole requires the addition or removal of heat, then pinch technology has an important role to play. Now we are not just dealing with a trade-off between capital and energy, but a trade-off between capital, energy, raw materials costs, and waste disposal costs. Without the benefit of pinch technology, such trade-offs are difficult or impossible to perform.

Figure 8 shows part of a fine chemicals process. Two feeds (F1 and F2) are reacted together in a stirred tank. The reactor contents are cooled to cooling water temperature, causing the product to crystallize out of solution. The crystalline product is then separated by filtration. The liquor remaining after filtration is disposed of to effluent. The treatment costs associated with the effluent in Figure 8 were considerable. Two alternative means were suggested to increase the efficiency of the separation system in order to decrease the amount of product lost to effluent and thereby reduce the effluent treatment costs:

1. Cooling below cooling water temperature using a refrigerated cooler after the cooling water to cause a greater proportion of the product to crystallize. This involved the installation of new refrigeration

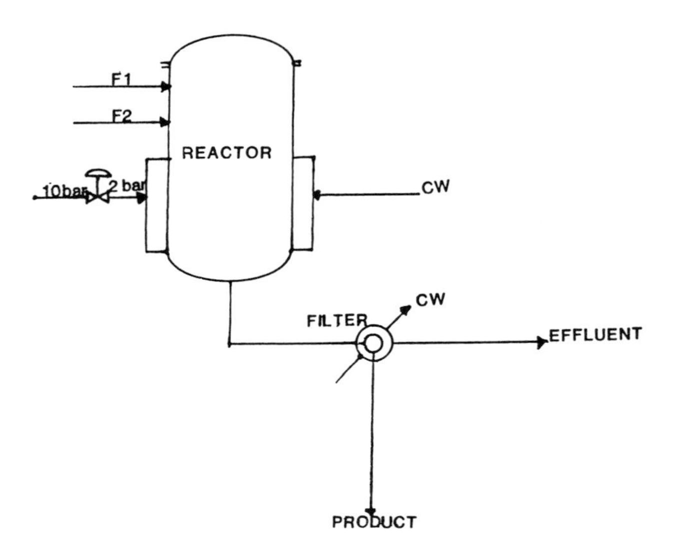

Figure 8. A fine chemicals process.

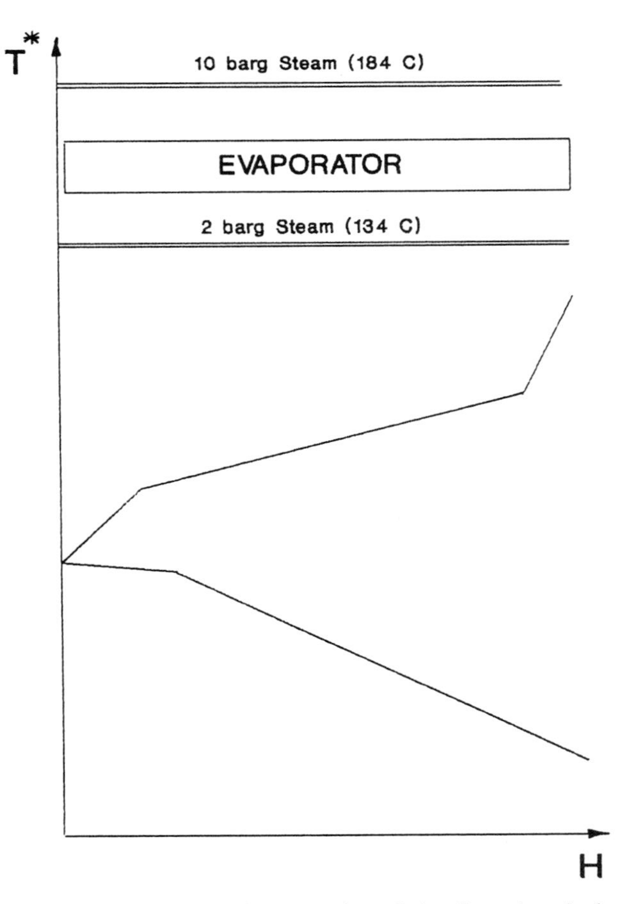

Figure 9. The grand composite of the fine chemicals process indicates that an evaporator can be used to reduce effluent treatment costs without an increase in energy consumption.

plant together with the high electricity cost of running the refrigeration plant.

2. Taking the liquor from the filter in Figure 8 and feeding it to an evaporator. Evaporating part of the liquor again causes a greater proportion of the product to crystallize. Further filtration is then required. In this case, the existing filter could be used because the process was batch in operation. In addition to the capital cost of the evaporator, the scheme requires considerable extra steam to run the evaporator. Unfortunately, both of these schemes would cause an increase in energy costs in exchange for a decrease in effluent and raw materials costs. There seems to be no way out of this effluent treatment problem without an increase in production costs. Pinch technology was then used to analyze this problem.

Steam at 10 bar pressure was originally used to heat the reactor in Figure 8 even though a lower pressure (e.g., 2 bar) would have been adequate. This is clearly shown by the grand composite curve (Figure 9) which plots actual required processing temperatures rather than "existing practice." Instead, in Figure 10 the 10 bar steam has been used in the evaporator, which operates at 2 bar. The steam from the evaporator at 2 bar pressure is used to heat the reactor. Thus the evaporator has been used to increase the product recovery and decrease effluent treatment costs without any increase in energy costs. Energy costs remain the same, while the costs of raw materials and effluent treatment are reduced.

Waste Treatment

Waste minimization at the source is clearly the best way to deal with effluent problems. However, there comes a point where further waste minimization becomes uneconomical, or simply not possible, and we must resort to waste treatment.

At this stage there will often be more than a single option to treat the effluent. The decision as to which option to pursue would be based on several factors such as:

1. Whether the scheme recovers material for recycle (or re-use), or whether the original pollutant is simply destroyed or converted to less harmful material.
2. The capital cost of the effluent treatment plant.
3. The running cost of the effluent plant.

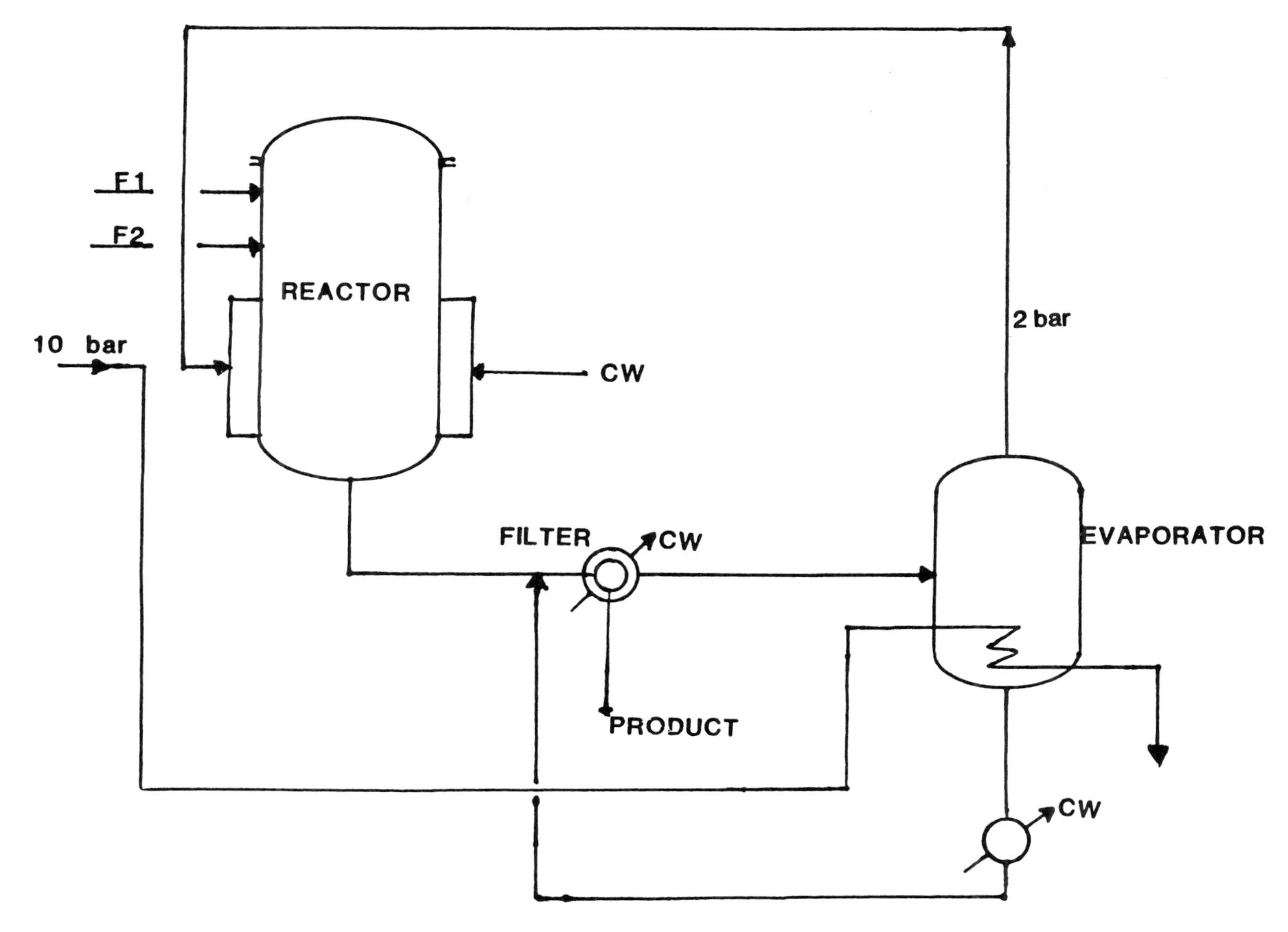

Figure 10. Final flowsheet for the fine chemicals process.

The conventional way of assessing the best method for effluent treatment is to consider the effluent plant in isolation from the existing process plant and to carry out a cost/benefit analysis for each alternative scheme.

During the 10 years or so that process integration techniques have been practiced, if any one lesson has been learned it is that individual unit operations should not be considered in isolation. By drawing boundaries around unit operations many potential benefits of integration can be lost. The lesson of process integration is to draw the boundaries as wide as possible and look for integration potential. Should we not, therefore, apply this lesson to effluent treatment plants in order to discover latent benefits hidden from us by considering effluent treatment in isolation?

Take the simple situation where we have an effluent in the form of an air stream laden with, say, a solvent. An initial market survey reveals that there are two particularly strong candidates for effluent treatment (others are available but much less likely to be worthwhile). The two strong candidates are: 1) water scrubbing followed by distillation to recover the solvent, and 2) incineration. Incineration, in addition to destroying the solvent, could also supply steam from a waste heat recovery unit. Which option is best? On the face of it we might consider that we have sufficient information given the cost of each plant, the value of recovered solvent, and the cost of steam. But are we missing potentially valuable information?

Consider the same problem through pinch technology. Since pinch technology has taught us not to consider systems in isolation, we stand back and consider, at a minimum, the process plant to which the effluent treatment facility is to be added. Grand composite curves are used to give a graphic display of the energy cascade through the process. Figure 11a shows the grand composite curve for the scrubbing/distillation option. The incineration option is shown in Figure 11b. Figure 12 shows the grand composite curve for the process.

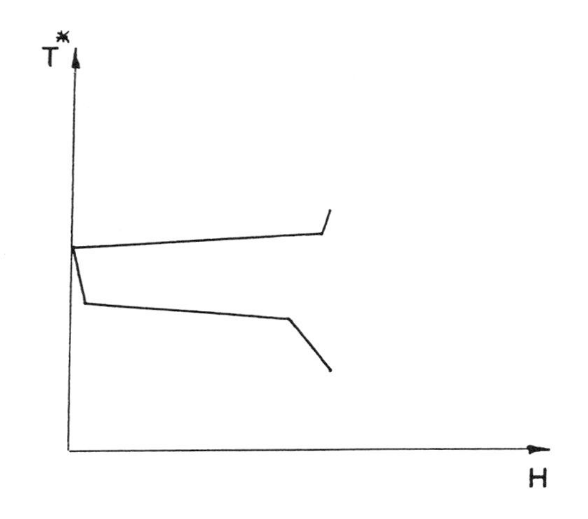

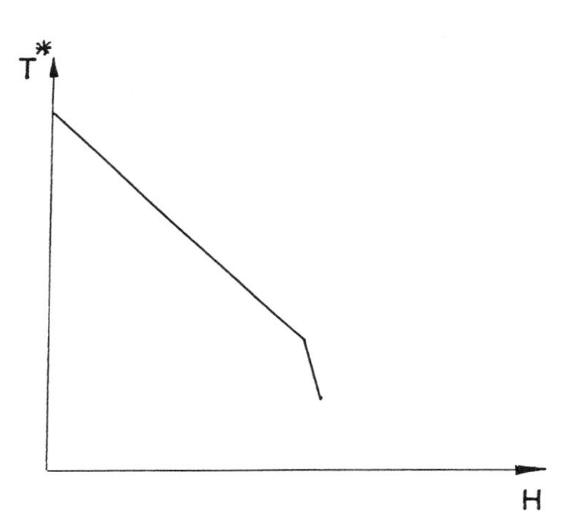

Figure 11. The grand composite curves for the two alternative waste treatment processes.

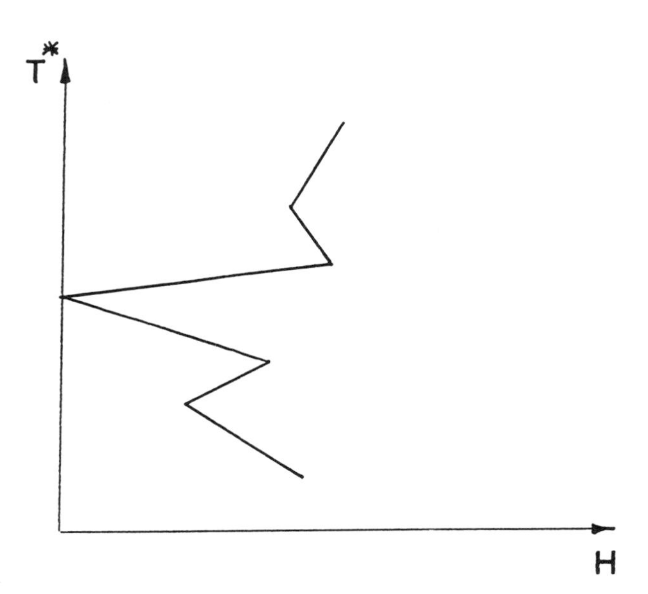

Figure 12. Grand composite curve for the process. An air steam laden with solvent requires treatment.

Superimposing the grand composite curves of the treatment options on the grand composite curve of the process allows the scope for integration to be quantified (Figure 13a). Figure 13a shows the potential for integration between the process and the incineration plant should that option be pursued. The shaded area in Figure 13a shows where waste heat from the process is useful heat for the process plant. Assuming that direct heat transfer is not desirable (or even possible) then line XX defines the optimum steam level for maximum useful heat recovery. This level is not likely to occur where existing steam mains are. Figure 13b shows what often happens in practice. Line YY now represents the site steam header pressure. Clearly much useful heat is being lost due to the constraint of the existing steam distribution system, and the operating economics are different in these two cases shown in Figure 13a and 13b.

Consider now the option of integrating the process plant with the scrubbing/distillation plant. Superimposing these grand composites in Figure 14a again shows the potential for integration. However, another lesson of process integration is that process modifications can be used to provide increased potential benefits.[7] Inspection of the grand composite curve for the scrubbing/distillation process in Figure 11a shows a typical system dominated by a single distillation operation. The operating conditions of the distillation process had, in this case, been set so as to use cooling water in the overhead condenser. An increase in the operating pressure of the distillation process would essentially allow the entire grand composite curve in Figure 11a to rise in the temperature scale. Figure 14b shows the grand composite of the process plant with a high pressure distillation process. The shaded area now shows much greater potential for integration.

In both of the above cases substantial reductions in operating costs were shown to be possible by correctly integrating the effluent treatment plant into the process plant which it serves. In many instances this type of analysis will clearly demonstrate that one of the effluent treatment options can be made far more cost effective than the other options. In some cases the option so selected may not have been the one chosen if this analysis had not been carried out.

While this example is simple, significant benefits

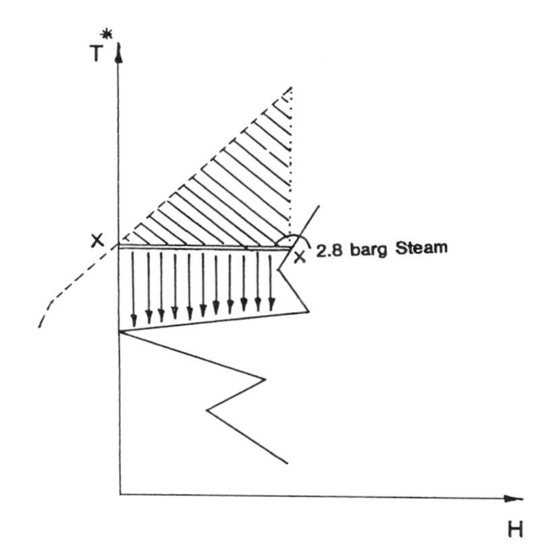

(a) Integration Between the Incinerator and the Process Via Steam Mains at 2.8 barg.

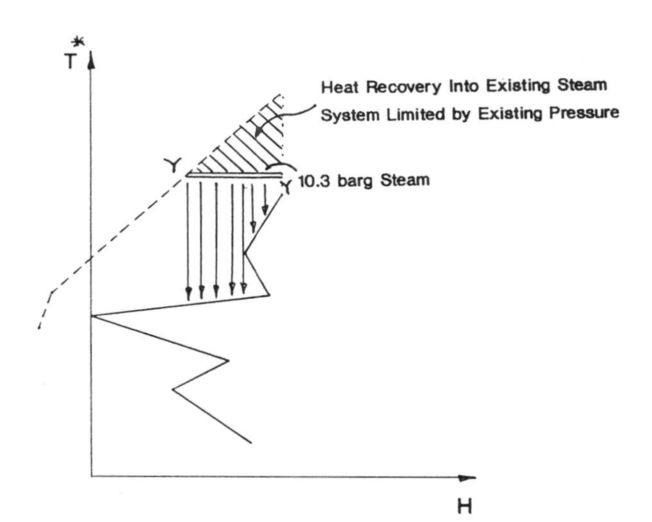

(b) Integration Between the Incinerator and the Process Via the Existing Steam Mains at 10.3 barg.

Figure 13. Integration between the incinerator and the process.

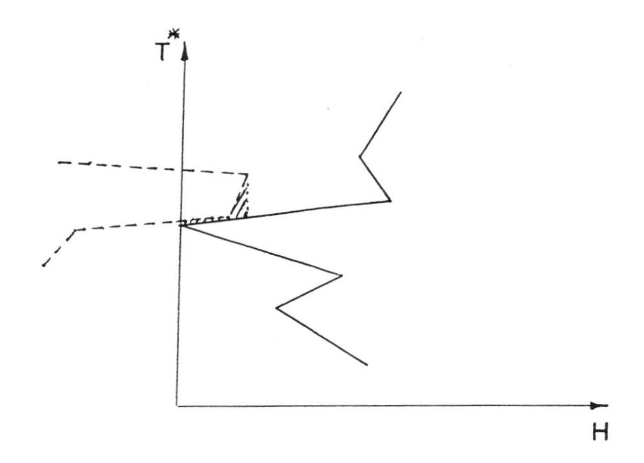

(a) Integration Between the Scrubbing and Distillation Effluent Treatment and the Process.

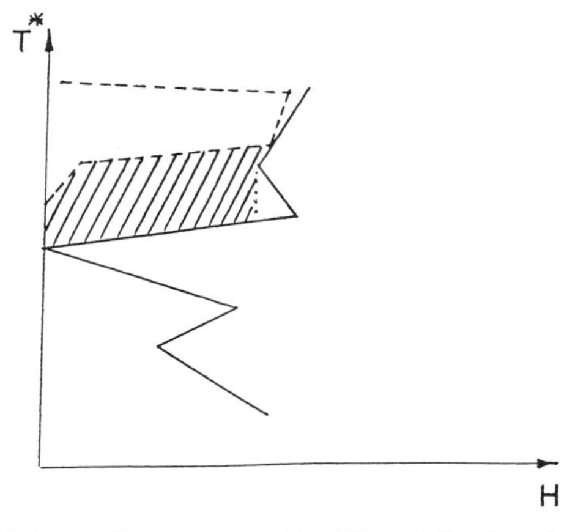

(b) Integration Between the Effluent Treatment and the Process With the Treatment Plant Operating at Higher Pressure.

Figure 14. Integration between the scrubbing and distillation effluent treatment plant and the process.

could be realized from a study of the process and the effluent treatment options using the analytical power of pinch technology. In reality, however, the problem is rarely so simple. There may be many plants on a site and many treatment options. The real power of pinch technology is in its ability to reduce even the most complex series of operations to a simple set of easily interpreted curves.

Many examples are available where this analytical procedure was not carried out and as a result the correct option was not chosen. For example, an operator of a large petrochemical plant chose to incinerate the gaseous effluents because the plant steam balance showed

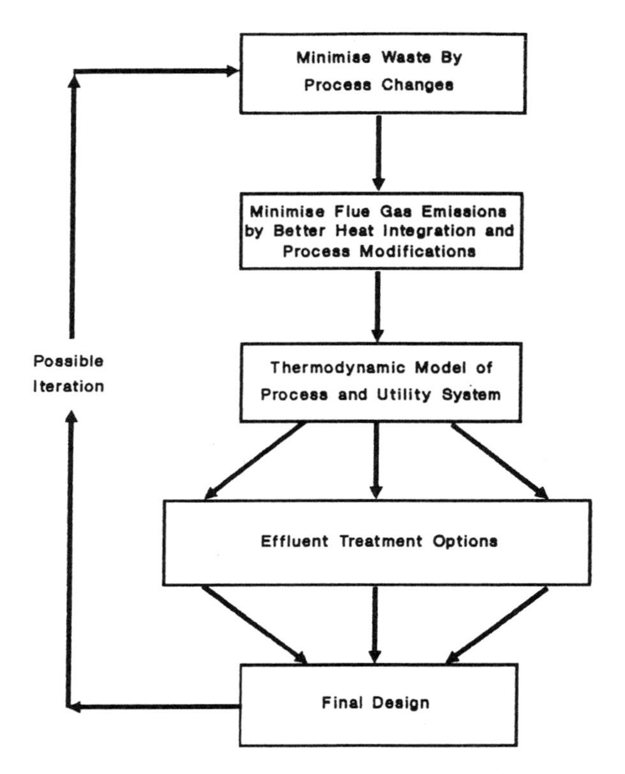

Figure 15. A suggested approach for applying pinch technology to environmental problems.

an import of 20 tons/h of steam. Incineration would provide the necessary steam. An alternative would have been a refrigeration plant which would have recovered much of the effluent for recycle. A process integration study carried out a few years later showed that the majority of this original shortfall in steam could have been removed by simple heat recovery projects having very short paybacks.

A Suggested Approach

Figure 15 summarizes a suggested approach for applying pinch technology to environmental problems. Clearly the place to start is waste minimization. It might not always be simplest to solve environmental problems at source by waste minimization but it is usually the most satisfactory solution in the long term. Reducing the problem at source by modifications to the process reaction and separation technology has the dual benefit of reducing raw material and effluent treatment costs. Once the possibilities for waste minimization have been exhausted by process modifications the minimum flue gas emission can be established (Figure 15). Further process modifications might be required to achieve this as indicated in Figure 6. This establishes the thermodynamic model for the process and utility system.

We are then in a position to assess alternative waste treatment options. It is crucial, however, that these options be assessed by considering the process and its waste treatment system together and any possibilities for integration between them. Figure 15 shows also that it might be necessary to readdress some of the early decisions after the waste treatment options have been considered. At each stage in Figure 15, pinch technology is needed to establish the economic trade-offs.

References

1. Linnhoff, B., Mason, D.R. and Wardle, I., "Understanding Heat Exchanger Networks," *Computers Chem. Engng.* 3, 295–302 (1979).
2. Linnhoff, B., Townsend, D.W., Boland, D., Hewitt, G.F., Thomas, B.E.A., Guy, A.R. and Marsland, R.H., "User Guide on Process Integration for the Efficient Use of Energy," Institution of Chemical Engineers, Rugby, UK (1982).
3. Townsend, D.W. and Linnhoff, B., "Surface Area Targets for Heat Exchanger Networks," I. Chem. E. Annual Research Meeting, Bath (April 1984).
4. Ahmad, S., "Heat Exchanger Networks: Cost Tradeoffs in Energy and Capital," PhD Thesis, UMIST (1985).
5. Linnhoff, B. and Ahmad, S., "SUPERTARGETING: Optimal Synthesis of Energy Management Systems," Tr. ASME, *Journal of Energy Resources Technology,* 111, 121–1300 (1989).
6. Korner, H., "Optimaler Energieeinsatz in der Chemischen Industrie," *Chem. Ing. Tech.* 60, 511–518 (1988).
7. Linnhoff, B. and Vredeveld, D.R., "Pinch Technology Has Come of Age," *Chemical Engineering Process,* 33–40 (July 1984).

Energy Conservation in the Industrial Sector

Robert U. Ayres
Carnegie-Mellon University
Pittsburgh, PA
International Institute for Systems Analysis
Laxenburg, Austria

Abstract

It is estimated that US industry produces materials and material goods at a second-law efficiency of the order of 14 percent. Taking into account the low rate of recycling of energy-intensive materials, especially of plastics, the overall efficiency at present may be closer to 9 percent. According to the "standard" view that is embedded (without serious discussion) in most large-scale economic models, energy savings are only achievable by means that will sharply increase the costs of production. This paper argues that, on the contrary, there are actually a great many opportunities for saving energy and reducing total costs at the same time. One "least cost" estimate suggested that a 10 percent saving in industrial energy use would also reduce net production costs, while a 20 percent saving could be achieved at no increase in total costs of production. The author believes this to be an underestimate of the potential.

Energy Use Efficiency in Industry

The term *efficiency* is commonly used in at least two different ways, which have been denoted "first-law" efficiency and "second-law" efficiency. The former, which merely reflects the standard laws of energy conservation, is defined as the ratio of "useful" energy output to total energy output, for any process. For chains of processes such that the output of one is the input of the next, first-law efficiencies are multiplicative.

"Second-law" efficiency incorporates the notion of increasing thermodynamic unavailability, as reflected by increasing entropy, in a process. However, it can be defined consistently as a ratio. The numerator is the minimum amount of available work theoretically required to accomplish the desired result by the most efficient possible process. The denominator is the actual quantity of available work consumed. Second-law efficiencies can be *defined* (but not necessarily computed very accurately) for all economic activities, subject only to a consistent set of conventions regarding the treatment of energy-containing co-inputs and byproducts.

To illustrate the difference between first- and second-law efficiency, consider the refining of petroleum. From a first-law point of view, the process is roughly 90 percent efficient, in the sense that the "useful" output stream contains 90 percent of the energy that was contained in the input stream. On the other hand, from a second-law perspective—which is adopted hereafter—the process may be only 10 percent efficient, meaning that the most efficient refinery that is theoretically possible would waste only 10 percent as much energy as the existing ones.

Industry, including agriculture, mining, manufacturing, and construction, is the biggest energy consumer among the three major sectors of the US economy (40 percent of the total). Within the sector, 50 percent of the energy is consumed as fuel for process (steam) heat, 17 percent is consumed in nonfuel uses (including coke, asphalt, and chemical feedstocks), 9 percent is used as fuel for internal combustion engines, and the remainder (21 percent) is fuel for electricity generation, for electric furnaces, electrolytic processes, electric motor drive, and illumination. Industrial use of energy in 1979 is broken out in detail in Table 1.

Chemical feedstocks account for about 8.5 percent of industrial energy, or 2.6 quads. About a third of the energy content of chemical feedstocks (inputs) is embodied in "final" chemicals, such as plastics (0.9 quads). About 22 percent of the energy consumed in primary iron ore smelting is embodied in iron and steel, while 13 percent of the energy used in primary aluminum smelting is embodied in aluminum itself. Primary metals embody around 1.1 quads of energy. The remaining direct consumption of fossil fuels (and some electricity) provided process heat for chemical, mineral, and metallurgical processes. At present, most intermediate temperature process heat is delivered in the form of steam (around 12.5 quads in 1978). An efficiency of 25 percent is usually assumed [e.g., Ross *et al.* 75, USOTA 83]. Olivier *et al.* have estimated an average efficiency of 14 percent for process steam in the United Kingdom [Olivier *et al.* 83]. I suspect the latter figure is more realistic.[1]

Table 1. US Industrial Energy and Electricity Consumption in 1979 (in quads).

	Total energy	Inputs to electric power	Inputs to motive power	Inputs to nonfuel uses	Inputs to direct heat	Electric power consumed
Industry, total	32.512	8.698	2.962	5.651	15.201	2.740
Agriculture, fish, forestry	1.352	0.329	0.796		0.227	0.106
Mining, total	2.818	0.615	1.664		0.539	0.165
Oil and gas	1.982	0.244	1.518		0.220	0.046
Other	0.836	0.371	0.146		0.319	0.119
Manufacturing, total	28.162	7.754	0.322	5.651	14.435	2.469
Food and kindred	1.233	0.442	0.047		0.744	0.142
Paper and allied	2.543	0.642	0.020		1.881	0.207
Inorganic chemicals	1.186	0.855	0.009		0.322	0.255
Organic chemicals	1.606	0.404	0.014		1.188	0.130
Products of oil, coal	3.529	0.393	0.015		3.121	0.126
Stone, clay, and glass	1.503	0.350	0.044		1.109	0.113
Primary metals	5.673	1.994	0.045		3.634	0.642
Metalworking	2.227	1.250	0.052		0.925	0.396
Other manufacturing	3.011	1.424	0.076		1.511	0.458
Nonfuel (feeds, asphalt)	5.651			5.651		

The biggest users of process heat are steel, petroleum, chemicals, and pulp and paper. A comparison of actual energy use in these industries in 1968 with theoretical minima has been carried out by Gyftopoulos [Gyftopoulos *et al.* 74] and Hall [Hall *et al.* 75]. The results are expressed below in terms of percentage second-law efficiency.

Integrated iron and steel manufacturing[2]	22.6%
Petroleum refining	9.1%
Primary aluminum	13.3%
Cement production	10.1%

The pulp and paper industry is more difficult to analyze in second-law terms, since most of the energy is used for "digestion" (to separate the lignin from the cellulose) and dehydration. The theoretical minimum energy required for this purpose may not even be precisely known (and would depend on the type of pulp). The industry currently uses a significant amount of purchased fuel although, in principle, it should be able to supply essentially all its own energy from cellulose and lignin wastes and even to sell the surplus. (Indeed, the most advanced pulp and paper mills do so.) A second-law efficiency computation is therefore both uncertain and ambiguous, in the sense that one must decide whether lignin is an output or an input. However, given the fact that purchased energy is a relatively small component of total costs, it is unlikely that the pulp and paper industry is currently as efficient in thermodynamic terms as the four industries noted above.

Rising energy prices have induced significant savings in most of these industries. Energy input per unit manufacturing output (based on the Federal Reserve Bank (FRB) Index of industrial production) declined from 100 in 1970 to 63.4 in 1984 [Doblin 87]. Of this decline, about half was due to structural change and about half of the remainder to changes in output mix [Doblin 87]. Focusing on purchased energy for heat and power, the contribution of technological change was about 33 percent [Doblin 87]. In effect, one may conclude that energy consumption per unit of physical output declined by about 12 percent during the period 1970 to 1984.

However, the overall rate of efficiency improvement in the US steel industry has been slow because of low investment in the so-called integrated sector and a rapid buildup in the share of "mini-mills," which use electric arc furnaces (EAFs) to melt scrap. The scrap remelting process is considerably less efficient, in the second-law sense, than smelting iron from ore. On the other hand, remelting scrap requires only 10 percent as much energy as smelting virgin iron ore. While the current blast furnace smelting technology is old and has changed very little over many decades, dramatic improvements are still possible (Table 2).

In 1983 the US Office of Technology Assessment (OTA) projected further declines in energy consumption to the year 2000 of 39 percent in the unit energy requirements for steel, 10 percent for petroleum refining (despite some increased energy requirements due to the elimination of tetraethyl lead), 9 percent for the chemical industry, and 25 percent for paper-making [USOTA 83].

These incremental improvements do not nearly exhaust the potential savings. They also assume no increase in the efficiency of electricity generation (now around 34 percent) or process-steam generation (14 to 25 percent). A great deal of process heat is still used

Table 2. Iron/Steel Smelting.[a]

Process	GJ/tonne	Second-law efficiency (%)
US average (1980)	27.0	14.4
Swedish average (1976)	22.3	17.5
Swedish average (1983)	16.3	23.9
Swedish, planned (1983)	15.1	25.8
Swedish, possible (1976)	13.0	30.0
Elred Process[b]	11.9	32.7
Plasmasmelt[c]	8.7	44.8
Theoretical minimum	3.9	100.0

[a] Based on 50% scrap.
[b] New process under development by Stora Kopparberg
[c] New process under development by SKF Steel
Source: Goldemberg et al. 87, Table 11.

simply to remove water where it is not wanted, e.g., from brines, alkalis, etc. In principle, this could be done by using solar energy [Kreith & Meyer 83] or waste heat from higher temperature processes such as coking, iron-smelting, and steel-making. The integration of high, medium, and low temperature processes (cascading) has been emphasized in recent years by a number of authors [e.g., Linnhoff *et al.* 82, van Gool 87, Groscurth *et al.* 89, Spriggs *et al.* in this volume].

The problem is partly a practical one of efficient heat transfer from locations where heat is available to locations where it is needed. One approach, increasing in popularity, is for firms to generate their own electricity onsite, using the low temperature waste heat for process purposes [Gyftopoulos *et al.* 74, Diamant 70]. Another source of very large potential savings would be to utilize combustible process wastes, especially carbon-monoxide, for fuel [Rohrmann *et al.* 77].

A great deal of process heat is still lost at relatively high temperatures. Potential savings from waste heat recovery in the industrial sector by using commercially available technology such as heat exchangers (including recuperators and regenerators), heat pumps, and co-generation have been estimated in detail for the Netherlands, West Germany, and Japan [Groscurth *et al.* 89]. Potential savings range from 54 to 64 percent for the Netherlands, 36 to 50 percent for West Germany, and 41 to 67 percent for Japan, depending on how much heat recovery is assumed from electrical devices. The biggest source of such potential savings arises from the use of heat exchangers. This alone accounts for 72 percent of the "low" range of savings for the Netherlands,[3] 94 percent for West Germany, and 92 percent for Japan. Figures for the US are likely to be reasonably similar to the West German case.

The chemical industry is difficult to analyze because its inputs are not clearly segregated between fuel and feedstock. A detailed energy analysis for about 80 major chemical processes was carried out by Ayres [Ayres *et al.* 83]. In this analysis, ratios of the available energy in the product to the available energy of all inputs were computed. (Thus high-energy products like acetylene tend to have high ratios.) Typical "best available technology" output/input ratios (c. 1980) are given in Table 3, with energy losses as a fraction of total inputs in the second column.

Of course most chemical products are intermediates used in the production of other chemicals. Final products are made by "chains," or sequences, of processes with an overall conversion ratio that is the product of the conversion ratios at each stage. The "chain" analogy is an oversimplification, of course, because many processes yield more than one useful product (the chlor-alkali industry is an obvious example), while many products also require two or more inputs. The system as a whole, then, is more like a network, as shown in Figure 1.

If the typical chain has three steps, each of which has a conversion ratio of 0.7, the overall conversion ratio of the chain is around 0.34. A four-step chain would have a conversion ratio of around 0.24. That is, the available energy of the final product might be somewhere between 24 and 34 percent of the available energy of the original feedstocks. The figure one third seems likely to be a conservative estimate. This is essentially the first-law efficiency of the process chain.

The second-law energy conversion efficiency (in terms of theoretical minimum energy use divided by actual use) cannot be computed directly from the numbers in Table 3. However, it is reasonable to suppose that the second-law efficiency of chemical processes in

Table 3. Energy Conversion in Chemical Processes.

Product	First-law (conversion) efficiency	Loss fraction
acetaldehyde	0.73	(0.27)
acetic acid	0.62	(0.38)
acetylene	0.75	(0.25)
ammonia	0.61	(0.39)
butadiene	0.75	(0.25)
chlorine/caustic soda	0.63	(0.37)
chloromethanes ($CHCl_3$)	0.50	(0.50)
cumene	0.82	(0.18)
ethanol	0.70	(0.30)
ethylene	0.85	(0.15)
ethylene glycol	0.67	(0.33)
ethylene oxide	0.62	(0.38)
formaldehyde	0.72	(0.28)
methanol	0.69	(0.31)
polyethylene	0.88	(0.12)
phosphoric acid	0.65	(0.35)
sulfuric acid	0.28	(0.72)

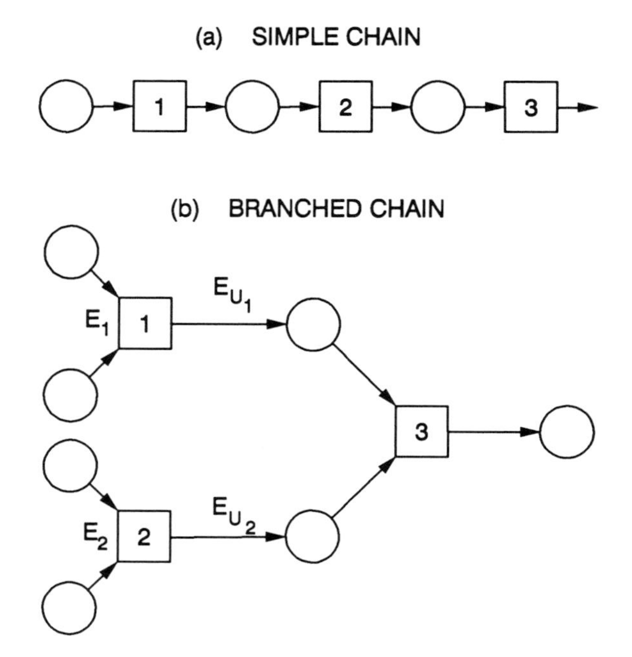

Figure 1. Process chains.

general is lower than the efficiency of petroleum refining, which was about 9 percent in 1968 [Gyftopoulos *et al.* 74] and may be 12 to 14 percent today. This is simply because large petroleum refineries are able to take advantage of scale economies and coproduction savings (economies of scope) not feasible for smaller operations. For most chemical processes the second-law efficiency is probably no better than 10 percent. In effect, this means that a 70 percent feedstock conversion efficiency achieved in practice (30 percent loss) should be compared to a theoretical maximum of 97 percent (3 percent loss). For a three-step chain, then, the minimum possible loss $(1 - (0.97)^3) = 0.087$ (9%) should be compared to actual losses of $(1 - (0.7)^3) = 0.657$. This corresponds to an overall second-law efficiency of 0.087/0.657 or 0.132 (13 percent). For a four-step process, by the same logic, the overall efficiency is about 15 percent.[4] Assuming an average of three steps and a second-law efficiency of 10 percent at each step, the chemical industry as a whole is presumably operating now at something like 0.3 (30 percent) feedstock efficiency and 0.13 (13 percent) second-law efficiency.

The "nonfuel use" category is potentially misleading, since it includes both metallurgical coke (2.1 quads) and natural gas (and natural gas liquids) classed as feedstocks by the chemical industry (2.6 quads). In both cases, energy content of the fuel is utilized in the process and must therefore be counted as an energy input. Thus, the chemical energy content of coke used in steelmaking is explicitly included in Table 2, while the energy content of feedstocks is similarly included in Table 3. Only in the case of asphalt (0.9 quads) is the heat content of the fuel not utilized.

The last major category of industrial energy use is electric drive of machine tools and other equipment. This accounts for roughly three fourths of electricity purchases by industry. (In fact, half of all electricity is used to drive motors.) Electric motors are generally regarded as very efficient, with first-law efficiencies of the order of 90 percent.[5] Yet, in the aggregate, losses are considerable and large gains are theoretically possible. For instance, motor control improvements—especially variable frequency drives for induction motors—could cut internal losses (the major component of power consumption) by a factor of 2, at least.

However, the greatest opportunities for improvement lie in the realm of motor utilization. Metal-cutting is among the biggest uses. The efficiency of the metal-cutting process has increased spectacularly, even since 1960, due to the introduction of harder cutting materials (e.g., alumina-coated carbides). It is not known what the theoretical limits might be for cutting by means of conventional hard-edged cutting tools, but technical progress in this field continues unabated (Figure 2). Harder tools permit much higher cutting speeds with no greater power consumption. Assuming motor and drive-train losses can be cut by at least a factor of 2, and that cutting speeds can be increased by at least a factor of 30 without running into any physical limits,[6] one would have to assume that the overall efficiency of electrical energy use for metal-cutting is no better than 1.5 percent. As a matter of fact, the ultimate lower limit of energy requirements for metal-cutting is probably quite close to zero, by current standards. For other applications of drive motors (except pumps) I would expect similar results.

Using the above data, one can estimate the overall second-law efficiency of the US industrial sector by adding up actual losses. In summary, it appears that chemical feedstocks are currently used with an overall efficiency of the order of 13 percent; iron and steel smelting (with coke) operates at about 25 percent; primary aluminum operates at around 15 percent; petroleum refining is probably now somewhere around 12 percent; and cement is a bit less, while pulp and paper and scrap-melting are somewhat lower. However, precisely because of their energy-intensity, the incentives to conserve have been greatest in these process industries. Elsewhere, where little energy is embodied in the product itself and energy costs have been comparatively insignificant as a fraction of total costs, energy efficiencies tend to be extremely low (as in metal-working).

With categories reorganized in functional terms, data for industrial energy use in 1978 are shown in Table 4.[7] Imputed efficiencies by process are given in the second column. An overall efficiency for the sector can be estimated by calculating imputed minimum energy requirements for the function and summing (col-

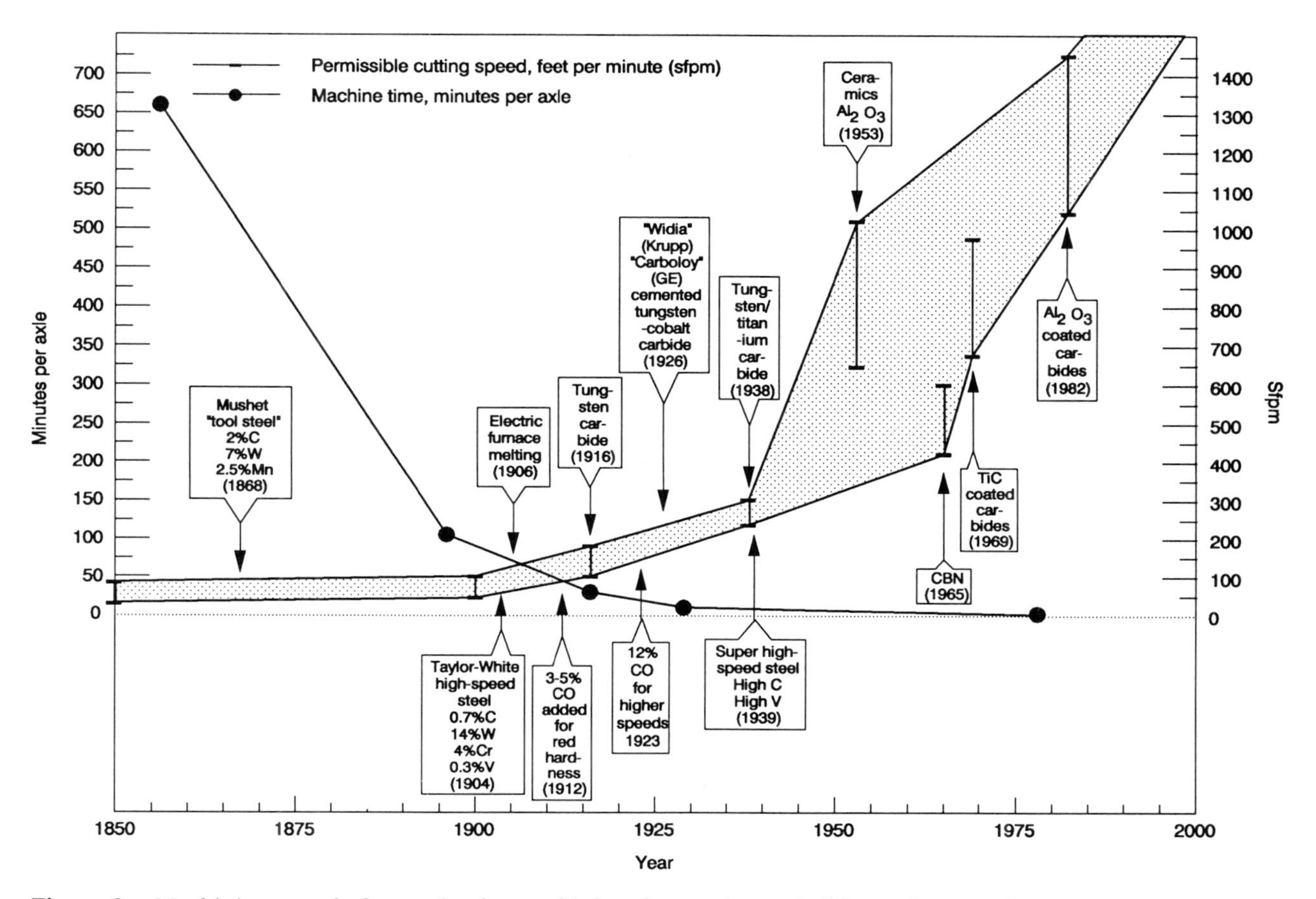

Figure 2. Machining speeds for steel axle; machining time and permissible cutting speed.

Table 4. US Industrial Energy Use by Function (1978).

Function	Consumption (quads)	Imputed efficiency	Imputed minimum
Space heat	0.5	0.01	0.005
Process heat (steam)	12.1	0.14	1.694
Electric furnace melting[a]	0.5	0.15	0.075
Electrolysis	1.5	0.15	0.225
Electric motor drive	8.1	0.015	0.122
Metallurgical coke	2.1	0.25	0.525
Chemical feedstocks (gas)	2.6	0.13	0.338
Gas engines (for pipelines)	1.1	0.30	0.330
Diesel engines (nontransport)	1.0	0.15	0.150
Asphalt	0.9	0.90	0.810
Total	30.4	0.14	4.274

[a] Efficiency of heat use is assumed to be the same as for steam. This is almost certainly an overestimate.

umn 3). Based on this method, energy-use efficiency in the industrial sector comes out to 14 percent. This is an upper limit, of course, since it does not reflect, for instance, the possibility that the quantity of process steam required might be reduced.

But production per se is not the whole story. Bearing in mind that steel, aluminum, paper, plastic, and asphalt all embody significant amounts of energy (at least 3 quads, not including paper and wood products), the question is: How efficiently are final materials subsequently used, and do we need to process as much material as we do? Metals that are recycled require much less energy to re-refine than do virgin metals. As noted already, the energy needed to remelt steel is only 10 percent of that needed to smelt virgin ore. Much the same is true of textiles, paper, and even lubricating oils, although the numbers vary. In principle, the energy embodied in metals, paper, and plastics is not really lost until the metals themselves are dispersed in chemical form or corroded beyond recoverability. If energy-containing materials were not dispersed and lost, we would not need to produce replacements.

The percentage of a metal once produced that is recycled depends on its economic value. It is very high (around 70 percent) for gold; comparatively high for silver, copper, and lead; somewhat lower (around 50 percent) for iron and steel; and less for aluminum, zinc, and paper. Roughly 35 percent of the embodied energy in metals (taken in the aggregate) is currently saved by recycling. This figure is up considerably since 1970, though far less than it could be [Chandler 83]. On the other hand, the recycling rate for paper was only 26 percent in 1980, and the rate for plastics in the US is currently no more than 1 percent. The recycling rate for lubricants, solvents, and other energy-intensive chemicals is also very low.

Note that the imputed minimum of 4.274 quads includes roughly 0.9 quads of final chemical products (mainly plastics and fertilizers) and roughly 1.1 quads of primary metals, of which only the latter is significantly recycled. The effective recycling rate for energy embodied in metals is not more than 40 percent; the remainder is lost in use. Thus, subtracting the 1.5 quads of energy embodied in materials that become unavailable in consumption, but which could be recovered in principle (thus reducing input requirements), yields a corrected energy efficiency for the industrial sector as a whole of about 2.77/30.4, or 9 percent.

The industrial sector (including electricity generation) is, arguably, a more efficient energy user than the transport sector or the buildings sector. If this is so, it would be explained, presumably, by the fact that it is easier to be efficient when operating on a large scale, especially when energy is an explicit and nontrivial element of operating cost. Managers are presumably forced by competition to analyze costs carefully (although, as will be noted later, it is not at all clear that they do so except in industries—such as petrochemicals—where energy costs are a significant fraction of total costs). In large-scale industry it is rare for "best practice" technology to be more than 50 percent or so better than average. (Among small establishments and private households the range is at least an order of magnitude greater.)

Costs of Energy Conservation

Economists tend to assume that managers act rationally. This implies that all possible means of energy production (including conservation) are analyzed and costed by engineers, then ranked in order of cost-effectiveness. If all technological choices were made on the basis of rational analysis by profit maximizers, then the lowest cost alternatives would always be selected first. The result would be a monotonically increasing cost curve such as A(z) in Figure 3.

This set of assumptions is exemplified by the work of Nordhaus using an activity type of model [Nordhaus 73, 75]. Assuming the energy supply/conversion and industrial component of the economy "optimizes" more or less instantaneously to adjust to changing prices (hence, is always in equilibrium), such models can be used to estimate cost curves for various policy assumptions. Nordhaus has, for example, estimated the costs of achieving a given energy output with successively lower amounts of CO_2 production. He found that shadow costs could be expressed by a quadratic function of percentage emissions reductions. A similar result was later obtained econometrically [Nordhaus and Yohe 83]. Nordhaus used this function (updated to 1989 prices) in his recent work [Nordhaus 89].

Another example is found in the well-known ETA-

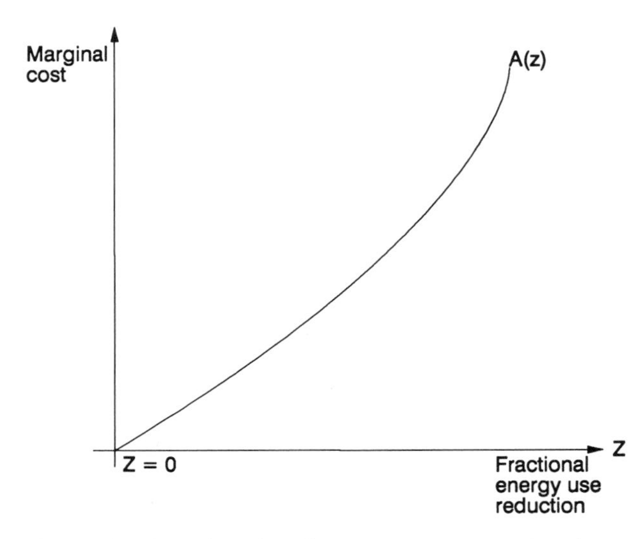

Figure 3. Cost function for energy conservation in an equilibrium economy.

macro model [Manne *et al.* 79, Manne 81, Manne and Richels 89]. This is a quasi general-equilibrium modeling effort linking a macroeconomic model and an energy supply-conversion optimization model of the activity type. There are three underlying assumptions: 1) that the economy is always in an equilibrium state, 2) that it rapidly finds the optimum supply mix for a given demand, and 3) that energy consumption is both an input (factor) and a cost of production and a claim on resources. The last assumption means that energy appears as an input in a production function. When the production function is econometrically fitted to past data on energy consumption, energy prices, and total output of goods and services, it is possible to estimate the reduction in output. This can be interpreted (somewhat loosely) as the economic cost of reducing energy inputs by a given amount.

The interpretation of cost (in the aggregate sense) as "lost gross output" is justified, for most economists, by the notion that GNP is a measure of social welfare.[8] However, it should be noted that costs (in the sense of Figure 3) are more commonly interpreted in another way, viz., as the annualized net *additional* capital and operating costs of investing in and using an alternative technology, as compared to the technology currently in use. It can happen, of course, that little or no new investment is needed or that the substitution results in a *net saving*, rather than a *net cost*. This is a crucial point.

For a business or a householder, a net saving is equivalent to an above-normal profit, or an above-normal return on investment (ROI). The usual standard of comparison is money invested in high quality government bonds or, simply, "money in the bank." In other words, if a given investment produces a greater net return (after taxes) than money safely invested at the current rate of interest, it is profitable in the above sense. If the rate of return is less than the normal interest rate, the investment is unprofitable.

The standard real (inflation-adjusted) interest rate for tax exempt bonds is typically around 3 to 4 percent, which corresponds to a payback time of 25 to 30 years. By contrast, the usual target real rate of return on investment (ROI) for business investments—which tend to be fairly risky, and which must allow for taxes on the profits—is at least 20 percent per annum or higher. This corresponds to a payback time of 5 years or less.[9] If the best return that can be realistically expected from a risky project is only 15 percent, a prudent businessman will normally not make the investment. On the other hand, for a unit of government (which does not have to pay taxes and can borrow money at lower rates than a private business) an 8 percent or 10 percent expected real rate of return is often regarded as adequate justification. (This is often equated roughly with the social discount rate.) It corresponds to a payback time of about 10 years.

Given that capital is scarce, it is rational to invest in the most profitable ventures first. Thus, a business run by a profit maximizer should try to rank-order the various proposals for capital spending (in order of expected ROI) and go down the list until either the available money for investment runs out or the minimum target rate-of-return threshold is reached. In principle, government should do the same. In an equilibrium economy, there should be just enough capital to fund all of the most promising projects, i.e., all the projects with expected ROI above the appropriate threshold level. It follows that the really promising, i.e., profitable, projects should be funded as soon as they are identified. In an equilibrium economy there should be very few opportunities capable of yielding returns far above the average. By the same token, capital should not be available at all for projects with below-threshold ROIs. The existence of a class of underfunded projects with high ROIs, while another class of overfunded projects consistently yields below-norm ROI is an indication of significant departure from economic equilibrium.

In this context, it is relevant to note that most large-scale energy *supply* projects (e.g., hydroelectric or steam-electric plants) yield a long-term real net rate of return between 5 percent and 10 percent (*Economist*, January 6, 1990, p. 59). Since this is below the threshold level for a rational tax-paying profit maximizer, it is difficult not to suspect that noneconomic factors are involved in diverting capital into such investments.

On the other hand, there is ample evidence of underutilization of profitable opportunities for conserving energy. In a major study carried out by the Italian energy research institute ENEA it was shown that technological "fixes" exist with payback times of one to three years—well below the typical threshold for most firms and several times faster than investments in new supplies [e.g., d'Errico *et al.* 84]. The Appendix, taken from the ENEA study, lists numerous examples of commercially available equipment, together with typical applications and payback times.

Even more convincing evidence comes from the experience of the Louisiana Division of Dow Chemical Co. in the US. In 1981 an "energy contest" was initiated, with a simple objective: to identify capital projects costing less than $200,000 with payback times of less than one year [Nelson 89]. In its first full year (1982), 38 projects were submitted, of which 27 were selected for funding. Total investment was $1.7 million and the 27 projects yielded an average ROI of 173 percent. (That is, the payback time was only about seven months.) Since 1982, the contest has continued, with an increased number of projects funded each year. The ROI cutoff was reduced year by year to 30 percent in 1987, and the maximum capital investment was gradually increased. Nevertheless, in the year 1988, 95 projects were funded, for a total capital outlay of $21.9 million and—surprisingly—an average ROI of 190 percent! The average submitted ROI for 167 audited projects over the entire seven years was 189 percent, while the actual (post-audit) average was 198 percent. Tables 5 and 6 summarize the results of the Dow experience.

Table 5. Summary of Louisiana Division Contest Results—All Projects.

	1982	1983	1984	1985	1986	1987	1988
Winning Projects	27	32	38	60	60	92	95
Capital, $Million	1.7	2.2	4.0	9.1	7.1	21.8	21.9
Average ROI (%)	173	340	208	124	106	77	190
ROI Cut-Off (%)	100	100	100	50	40	30	30

Source: Nelson 89, Table 1.

Table 6. Summary of Louisiana Division Contest Results—All Projects.

	1982	1983	1984	1985	1986	1987	1988
Winning Projects	27	32	38	59	60	90	94
Capital, $Million	1.7	2.2	4.0	7.1	7.1	10.6	9.3
Savings, $Thousand/yr							
Fuel Gas[1]	83	−63	1506	2498	798	2550	10790
Capacity						1197	2578
Maintenance	10	45	−59	187	357	2206	583
Miscellaneous						19	−98
Total Savings	1590	3838	5341	7353	6894	11944	18023

[1] All fuel gas savings are based on 1988 incremental fuel gas value.
Source: Nelson 89, Table 1.

It is important to note that, although the number of funded projects increased each year, there is (through 1988) no evidence of saturation. Numerous profitable opportunities for saving energy, with payback times well below one year, apparently still exist at Dow even after the program has been in existence for seven years. One would have to suspect that the program could still be expanded many-fold before reaching the 30 percent ROI threshold. Furthermore, it is important to emphasize that these opportunities exist even at relatively low US energy prices. Should taxes or a new energy crisis force US prices higher (i.e., toward world levels), the number of such opportunities would be multiplied further.

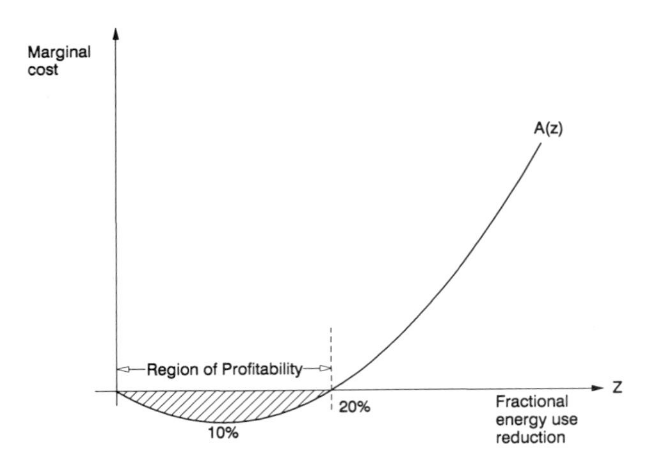

Figure 4. Cost function for energy conservation in a nonequilibrium economy.

Examples like the above suggest that the picture in Figure 4 is more realistic than the conventional picture in Figure 3, despite its apparent heterodoxy. The conclusions of several studies at the macro-level have also suggested that industry may be able to save money by saving energy. For instance, it has been argued persuasively by a study carried out by the Mellon Institute that a least cost strategy for providing energy services for the US in 1978 would have utilized much less primary energy, and in a very different manner, than that which was actually observed [Sant 79].

In economic terms, the Mellon study concluded that the least cost strategy would have saved $800 per family (17 percent), or $43 billion in that year alone.[10] Taking the year 1973 as a standard for comparison, a least cost strategy in 1978 would have involved a sharp reduction in the use of centrally generated electricity (from 30 to 17 percent) and a reduction in petroleum use from 36 to 26 percent. The only primary fuel to increase its share would have been natural gas (from 17 to 19 percent). Interestingly, the Mellon study suggested that *conservation services* would have increased their share from 10 to 32 percent in the optimal case. See Figures 5 and 6.

In particular, the study found significant opportunities for profitable energy conservation in industry, notably by means of regeneration (heat-cascading), heat pumps, and cogeneration (of electricity and process heat). The study concluded that a 10 percent saving in industrial energy use would also reduce net costs. Assuming (for convenience) that the cost curve (Figure 4) is roughly symmetrical around its minimum, it follows that a 20 percent saving could be achieved at little or no increase in total costs of production.

Some Economic and Policy Implications

A question deserving some comment is this: How is it that engineers and other "bottom-up" analysts persistently see profitable opportunities for energy conservation, whereas economists (and econometricians, especially) invariably conclude that conservation will inevitably be costly in proportion to degree ("no free lunch")?

I cannot answer this question authoritatively. There

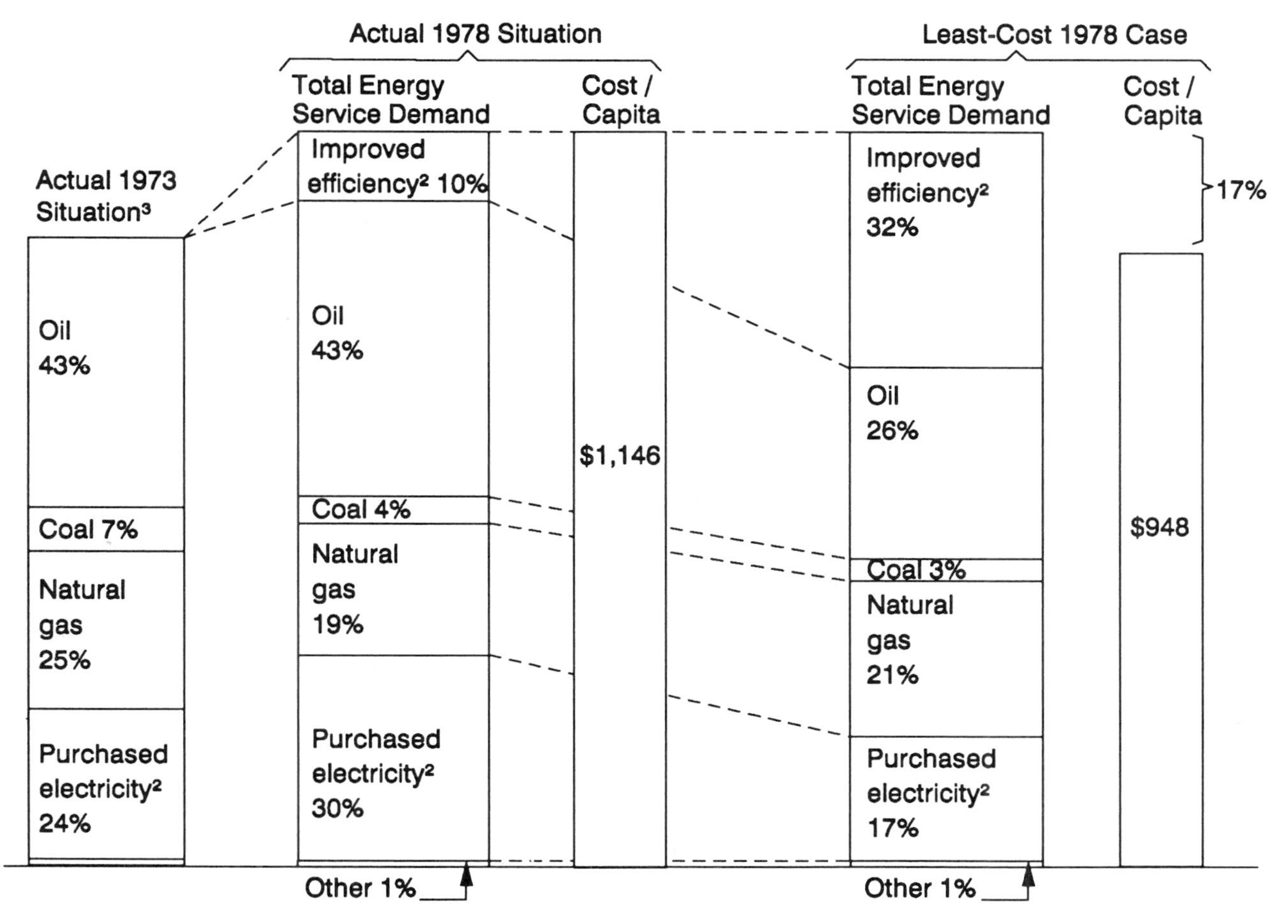

Figure 5. US energy service market shares of various technologies (in terms of primary fuel). Notes: 1) The primary fuel equivalent of service demand in 1978 was 79.0 quads, plus 9.2 quads of improved efficiency (calculated against a base stock and equipment in place in 1973), or a total of 88.2 quads. Actual service demand depends on the conversion efficiency of fuels and equipment utilized. 1 quad = a quadrillion = 10^{15} Btu. Another means of visualizing a quad is 1 million barrels per day of oil equivalent = 10^{15} Btu (quads) per year. 2) Primary fuels demand in 1973 was 74.6 quads.

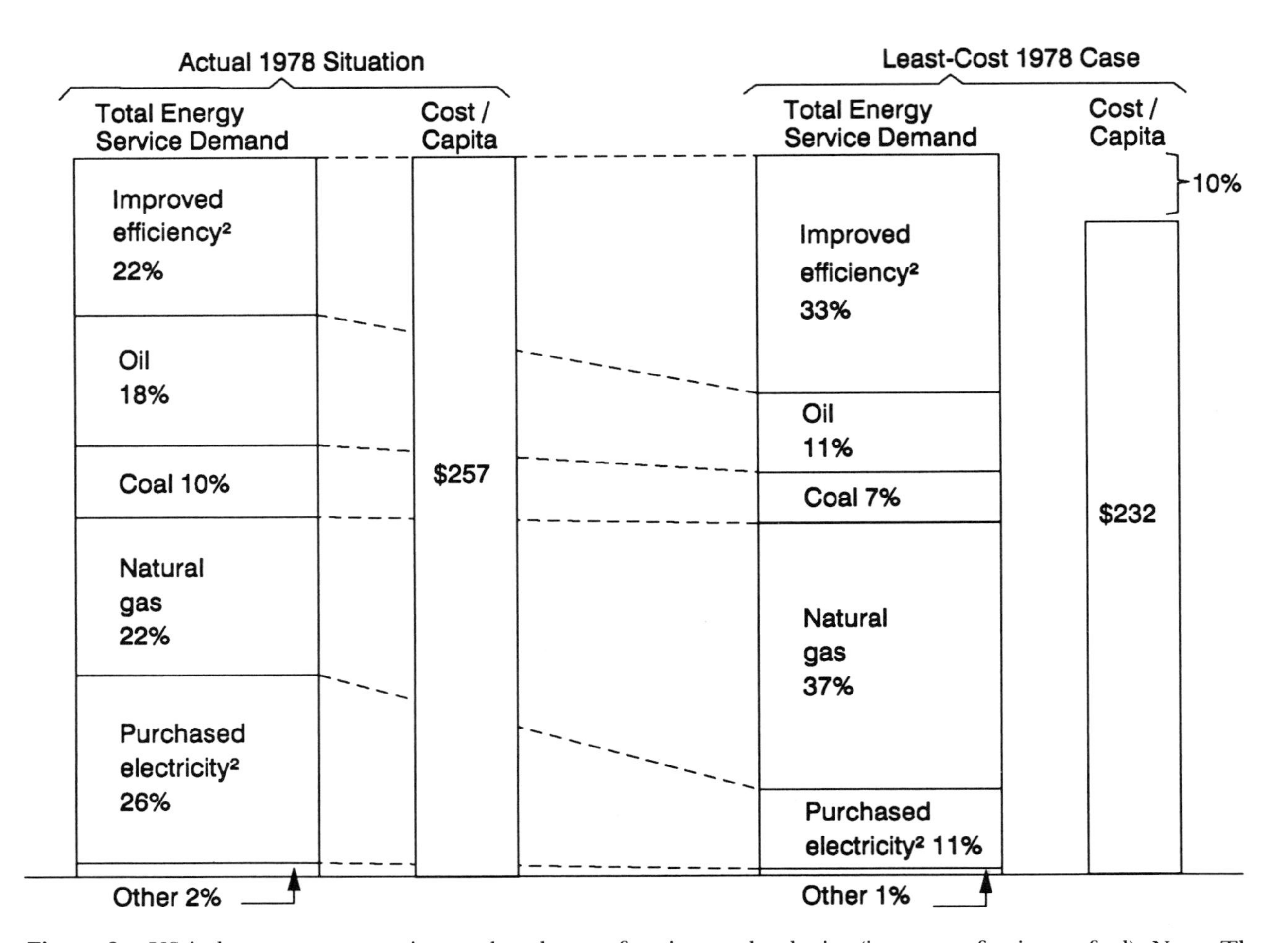

Figure 6. US industry energy service market shares of various technologies (in terms of primary fuel). Note: The primary fuel equivalent of service demand in 1978 was 28.2 quads, plus 7.9 quads of improved efficiency and 0.8 quads of biomass (calculated against a base of stock and equipment in place in 1973), or a total of 36.9 quads.

are several possible answers. One is that most economists start by assuming—as a matter of dogma—that the economy is always in (or near) a state of equilibrium. If this were true, any reduction in energy consumption would necessarily entail substitution of either capital or labor, i.e., a shift in the production function. Such a shift would be economically optimal if and only if it were prompted by price changes occurring naturally (e.g., due to resource exhaustion). According to this view, government intervention would necessarily depart from the optimum and result in a reduced total output of goods and services—hence a "cost."

Econometricians have confirmed this picture (to their own satisfaction) by observing that sharp price rises, as in 1973–74, were followed by recession. A few economists [e.g., Jorgenson 88, Olson 88] even attribute the global slowdown in productivity growth since the early 1970s to the rise in energy prices. However, it is fair to say that most economists now ascribe the slowdown primarily to other causes, notably the "catch-up" (by Europe and Japan) to the US in the decades after World War II.

However, while the tools of econometrics can identify and quantify correlations between hypothetical causes and consequences, they cannot confirm that the economy was (or is) in equilibrium. The most likely explanation for a persistent nonequilibrium condition is the existence of market-distorting institutions and subsidies. That many such distortions exist in energy markets is hardly in doubt. The oil and gas depletion allowance in the US was a direct subsidy to the petroleum and gas producers, and an indirect subsidy to consumers. So was (and is) public funding of roads, dams, and nuclear materials processing and disposal. Britain imposes a value-added tax on labor and materials needed for investments in energy conservation, but there is no value-added tax on fuel.

It is easy to find numerous other examples of policies that favor new energy supplies over conservation. Many such policies are far from obvious to the general public. The first step toward reducing excessive energy consumption, not only in the industrial sector but in the US economy as a whole, should be to identify and eliminate these hidden subsidies. Believers in the competitive virtues of free markets should have no objections to moving away from protectionism of energy inefficiency. Other, more aggressive, policy approaches such as the proposed "carbon tax" clearly have an impact far beyond the industrial sector per se and need not be discussed here.

References

[Ayres, Ayres et al 83] Ayres, Robert U., Leslie W. Ayres et al. "Future Energy Consumption in the Industrial Chemical Industry," in *The Chemical Industry DOE/CS/40151-1* (Series: Industrial Energy Productivity Project, 5), United States Department of Energy, Washington, DC, February 1983.

[Berndt et al 81] Berndt, E. R., M. Manove, and D. O. Wood, *A Review of the Energy Productivity Center's "Least-Cost Energy Strategy" Study,* Massachusetts Institute of Technology, Cambridge, MA, 1981.

[Chandler 83] Chandler, William U., *Materials Recycling: The Virtue of Necessity,* WorldWatch Paper (56), WorldWatch Institute, Washington, DC, 1983.

[Diamant 70] Diamant, R. M. E., *Total Energy,* Pergamon Press, Oxford, England, 1970.

[Doblin 87] Doblin, Claire P., *The Impact on Energy Consumption of Changes in the Structure of U.S. Manufacturing: Part I: Overall Survey,* Working Paper (WP-87-04), International Institute for Applied Systems Analysis, Laxenburg, Austria, February 1987.

[D'Errico et al 84] D'Errico, Emilio, Pierluigi Martini, and Pietro Tarquini, *Interventi di risparmio energetico nell'industria,* Report, ENEA, Italy, 1984.

[Goldemberg et al 87] Goldemberg, Jose et al., *Energy for Development,* Research Report, World Resources Institute, Washington, DC, September 1987.

[Groscurth et al 89] Groscurth, H.-M., R. Kümmel, and W. van Gool, "Thermodynamic Limits to Energy Optimization," *Energy,* 14(5), 1989: 241–258.

[Gyftopoulos et al 74] Gyftopoulos, E. P., L. J. Lazaridis, and T. F. Widmer, *Potential Fuel Effectiveness in Industry* (Series: Ford Foundation Energy Policy Project), Ballinger Publishing Company, Cambridge, MA, 1974.

[Hall et al 75] Hall, E. H. et al., *Evaluation of the Theoretical Potential for Energy Conservation in Seven Basic Industries,* Technical Report (PB-244,772), Battelle Columbus Laboratories, Columbus, OH, 1975.

[Jorgenson 88] Jorgenson, Dale W., "Productivity and Postwar U.S. Economic Growth," *Journal of Economic Perspectives,* 2(4), 1988.

[Kreith & Meyer 83] Kreith, Frank, and Richard T. Meyer, "Large-Scale Use of Solar Energy with Central Receivers," *American Scientist,* 71, November/December 1983: 598–605.

[Linnhoff et al 82] Linnhoff, B. et al., *User Guide on Process Integration for the Efficient Use of Energy,* Institution of Chemical Engineers, Rugby, UK, 1982.

[Manne 81] Manne, Alan S., *ETA-MACRO: A User's Guide,* Report (EA-1724), Electric Power Research Institute, Palo Alto, CA, February 1981.

[Manne & Richels 89] Manne, Alan S. and R. G. Richels, *CO_2 Emission Limits: An Economic Analysis for the USA,* Research Report, Electric Power Research Institute, Palo Alto, CA, June 1989.

[Manne et al 79] Manne, Alan S., R. G. Richels, and J. P. Weyant, "Energy Policy Modeling: A Survey, Operations Research Society of America," January/February 1979.

[Nelson 89] Nelson, Kenneth E., "Are There Any Energy Savings Left?" *Chemical Processing,* January 1989.

[Nordhaus 73] Nordhaus, William D., "The Allocation of Energy Resources," *Brookings Papers on Economic Activity,* 3, Washington, DC, 1973.

[Nordhaus 75] Nordhaus, William D. "The Demand for Energy: An International Perspective," in Nordhaus, ed., *Workshop on Energy Demand,* International Institute for Applied Systems Analysis, Laxenburg, Austria, 1975.

[Nordhaus 89] Nordhaus, William D., "The Economics of the Greenhouse Effect," in *International Energy Workshop,* International Institute for Applied Systems Analysis, Laxenburg, Austria, June 1989.

[Nordhaus & Yohe 83] Nordhaus, William D., and Gary Yohe, "Future Carbon Dioxide Emissions from Fossil Fuels," in *Changing Climate,* National Academy Press (National Research Council-National Academy of Sciences), 1983.

[Olivier et al 83] Olivier, D. et al., *Energy Efficient Futures: Opening the Solar Option,* Earth Resources Research, London, 1983.

[Olson 88] Olson, Mancur, "The Productivity Slowdown, the Oil Shocks, and the Real Cycle," *Journal of Economic Perspectives,* 2(4), 1988: 43–69.

[Rohrmann et al 77] Rohrmann, C. A. et al., *Chemical Production from Waste Carbon Monoxide,* Technical Report, Battelle NW Laboratories, Richland, WA, November 1977.

[Ross 81] Ross, R. M., "Energy Consumption by Industry," *Annual Review of Energy,* 6, 1981: 379–416.

[Ross et al 75] Ross, M. et al., *Effective Use of Energy: A Physics Perspective,* Technical Report: Summer Study on Technical Aspects of Efficient Energy Utilization, American Physical Society, January 1975.

[Sant 79] Sant, R. W., *The Least-Cost Energy Strategy: Minimizing Consumer Costs through Competition,* Report (55), Mellon Institute Energy Productivity Center, VA, 1979.

[Spriggs et al 90] Spriggs, H., E. Petela, and B. Linnhoff. "Pinch Technology: Evaluate the Energy/Environmental Economic Trade-Offs in Industrial Processes," in *Energy and the Environment in the 21st Century,* MIT Press, Cambridge, MA, 1990.

[van Gool 87] van Gool, W., "The Value of Energy Carriers," *Energy,* 12, 1987: 509.

[Yohe & Nikiopoulos 89] Yohe, G. D. Howardh and P. Nikiopoulos, *On the Ability of Carbon Taxes to Fend Off the Greenhouse Warming,* Department of Economics, Wesleyan University Middletown, CT, June 1989.

[USOTA 83] United States Congress Office of Technology Assessment, *Industrial Energy Use,* Report, United States Congress Office of Technology Assessment, Washington, DC, 1983.

Appendix

N	Area of Application	Type	Input	Output	Cost K US$	Saving K US$	Payback years
1	boilers	recuperator	gas waste LE 400° C	hot water GT 70° C	5.7	4.8	1–3
2	boilers	recuperator	gas waste LE 400° C	diathermic oil heating	28.4	20.5	1–3
3	incinerators	recuperator	gas waste LE 1200° C	steam 1–5 ATE	19.3	8.2	1–3
4	boilers	recuperator	gas waste	process water heat	9.2	5.1	2–3
5	boilers	recuperator	gas waste LE 400° C	process water heat LE 80° C	62.5	62.5	1–2
6	boilers	radiation recuperator	gas waste 900–1350° C	process water heat LE 600° C	44.3	42	1–2
7	air condit.	rotary exchanger	air LE 50° C	air cooling GE 15° C	12.5	5.7	2–3
8	boilers	rotary exchanger	gas waste LE 150° C	process air heat LE 120° C	15.9	26.1	0–1
9	boilers	rotary exchanger	gas waste LE 500° C	process air heat LE 400° C	45.5	64.2	0–1
10	air condit.	rotary recuperator	gas waste LE 200° C	external air heat/cooling	9.1	2.7	2–4
11	boilers	gravity recuperator	gas waste LE 170° C	process water heat	41.5	22.7	1–2
12	boilers	tubular recuperator	gas waste LE 440° C	process water heat	28.4	9.1	2–4
13	boilers	inclined tubes recup.	gas waste LE 170° C	process water heat	27.3	11.8	1–3
14	boilers	heating cube recup.	gas waste LE 250° C	process water heat	19.9	18.8	1–2
15	ovens	recuperator	gas waste LE 250° C	process water heat LE 150° C	6	1.7	2–4
16	boilers	recuperator	gas waste	room air heating	1	0.6	1–2
17	boilers	modular recuperator	gas waste LE 140° C	process air heat	4.5	8.5	0–2
18	boilers	recuperator notes	gas waste	process	14.2	6.1	2–3
19	boilers	plate exchanger	liquid waste LE 250° C	process water heat	54.5	33	1–2
20	boilers	rotary recuperator	liquid waste 40–95° C	process water heat	9.7	35.8	1–2
21	boilers	automatic recuperator	liquid waste LE 95° C	process water heat	55.7	37.8	1–2

(continued)

Appendix *(continued)*

N	Area of Application	Type	Input	Output	Cost K US$	Saving K US$	Payback years
22	boilers	re-evaporator	steam power	low pressure steam	7.4	13.9	0–1
23	tunnel exchanger	room air ovens	gas waste	room air heating	11.4	4.9	2–3
24	steel press	room air exchanger	gas waste	room air heating	22.7	15.9	1–2
25	refriger. groups	heat pump	gas waste	water heat LE 55° C	18.2	6.4	2–3
26	refriger. groups	heat pump	gas waste	water heat LE 75° C	198.9	94.3	2–3
27	refriger. groups	refr. heat recuperator	gas waste	water heat LE 55° C	244.3	142	1–2
28	refriger. groups	refr. heat recuperator	gas waste	water heat LE 55° C	8	8.4	0–2
29	refriger. groups	heat pump	gas waste	water heat LE 60° C	28.4	13.8	2–3
30	compressors transform.	heat pump	gas waste 25–35° C	water heat	0.9	0.4	2–3
31	refriger. groups		gas waste 25–35° C	water cooling	1.7	1.9	0–1
32	dryer	refrigerat	gas waste	water heat	1.7	1.9	0–1
33	air compressors	heating recuperat.	liquid waste	room air heating	15.3	11.1	1–2
34	air compressors	energy recuperat.	gas waste	water heat LE 80° C	2.6	1.6	1–3
35	computer center	air condition.	gas waste	water heat LE 70° C	4	21	0–2
36	refriger. groups	heating pump	gas waste	water heat LE 75° C	409.1	154.5	2–3
37	refriger. groups	plate exchanger	gas waste	water heat	25	50.6	0–1
38	refriger. groups	cool accumulat.	gas waste	cooling		14.2	
39	boilers	pyrolitic system	solid waste	fuel substitut.	113.6	49.4	2–3
40	boilers	grape boilers	solid waste	fuel substitut.			0–1
41	wood	chip boilers	solid waste	fuel substitut.	11.9	47.2	0–1
42	boilers	underwater combustion	liquid waste LE 60° C	process heating	23.9	7.8	2–4
43	boilers	radiant tubes	gas waste GT 300° C	room air heating	28.4	10.2	2–3
44	boilers	recuperator	gas waste	process heating			1–4
45	gas boilers	modular condenser	gas waste	process heating	11.4	5.4	2–3
46	room air exchange	air purifier	air waste	fuel substitut.	5	2.7	1–2
47	Industr. building		air waste	fuel substitut.	1.8	1.5	1–2
48	Industr. building		air waste	room air heating (floor)	48.3	22.7	1–3

Notes

1. At a temperature of 200°C only about 40 percent of the heat is thermodynamically available to do work. At 300°C the available fraction is less than 50 percent. Assuming furnace efficiency of 70 percent and distribution losses of 50 percent or so in transport would account for a 14 percent overall second-law efficiency.
2. This figure is based on a typical integrated mill and does not take into account the energy embodied in scrap. When the latter is considered, the overall efficiency of the industry was (and is) much lower. See Table 2.
3. The Netherlands has a relatively small high-temperature metallurgical sector.
4. The fact that multi-stage processes are more efficient in second-law terms than single-stage processes, ceteris paribus, is somewhat counterintuitive. However, it is consistent with the fact that biochemical processes typically involve many stages, each of which is "nearly" reversible.
5. To compute a second-law efficiency for motors one would have to determine a theoretical minimum loss; since a superconducting motor with zero loss can be envisaged, and it is not known even in principle whether room temperature superconductors are physically possible, there is no basis for making such a calculation.
6. The current state-of-the-art machine tools cut at around 3000 surface feet per minute (sfpm). The average for all machine tools on shop floors (most of which are fairly old) is probably between 500 and 1000 sfpm. (Accurate data are not available.) However, high speed prototype cutting machines have already achieved well beyond 30,000 sfpm with coated carbide tools. Given the recent development of technologies for diamond-coating, and their recent application to cutting tools, much higher speeds appear to be potentially achievable.
7. Data for all sectors except agriculture is taken from Ross 81 (Table 9). Data for agriculture for 1979 was derived from the Appendix.
8. Yohe and his colleagues used a similar approach (Yohe and Nikiopoulos 89). They investigated the potential economic costs for a 100 percent carbon tax (phased in over the 20 years from 2000 to 2020) or an equivalent consumption restriction. Their assumptions are based on work suggesting that the sudden increases in petroleum prices that occurred in 1973 and 1979 caused drops in productivity growth (Olson 88). They assumed that any tax on fossil fuels would have the same economically depressing effect. Assuming productivity drops ranging from 0.05 percent p.a. to 0.7 percent p.a. for a carbon tax amounting to 100 percent of current prices, cumulative lost productivity of $75 to $550 billion would be experienced up to the year 2010.

   This argument is questionable for two reasons. First, the essence of the "oil shock" was that money was taken from Western producers and consumers and deposited in bank accounts belonging to (predominantly) OPEC members. They, in turn, increased their spending for consumer goods, military goods, and long-term infra-structure projects, none of which contributed to increased productivity in the West. On the other hand, the impact of a tax collected within the Western economies would, of course, depend on how the money was spent. However, if the new tax were offset, dollar for dollar, by a reduction in other taxes, there is no reason to expect an automatic depressing effect on the economy. Second, we note that the observed productivity drop since 1973–74 has many other possible explanations, of which the most widely accepted seems to be that it reflects the inevitable European and Japanese "catch-up" to the US.
9. Many projects are evaluated in terms of *payback time* rather than *return*. The two concepts are closely related. A project with a payback time of one year corresponds to 100 percent return on investment. A project that pays for itself in six months has an annual return of 200 percent, and so on.
10. It should be noted that the Mellon study was thoroughly critiqued by a group at the Massachusetts Institute of Technology, at the request of the Department of Energy (Berndt *et al.* 81). The critique was extensive and detailed, and a number of significant substantive and methodological criticisms were offered. One criticism was directed at the study's implication that a least cost solution would be achieved automatically if the economy were truly competitive. The authors of the critique asserted that competition does not necessarily yield an optimal result and that regulation might be more effective. (This criticism might be applied even more cogently to the models of Nordhaus and Manne *et al.* mentioned earlier.) The critique also noted that some of the projected savings were "imposed" on the study, rather than being derived endogenously. The examples cited in this regard included projected savings by the use of variable-speed electric motors, cogeneration of electric power and industrial process heat, and dieselization of the bus fleet. In retrospect, the benefits of variable-speed motors were indeed exaggerated somewhat. However, a large number of other specific but minor opportunities for saving energy via the use of available technology were necessarily overlooked, simply because the authors had limited time and resources available to them. Thus, it is more likely that the extent of the conservation opportunities were underestimated than conversely.

Summary
Session B-2—Metals, Minerals, and Ceramics Manufacture

Donald R. Sadoway
Session Chair
Department of Materials Science and Engineering
Massachusetts Institute of Technology
Cambridge, MA

It is estimated that 8 percent of the nation's total energy consumption is in metals production, with the steel industry alone accounting for 5 percent. The aluminum industry consumes in excess of 2 percent of the total generated electrical energy in this country. In the production of both steel and aluminum, vast quantities of carbon are consumed in converting ore into metal: for both metals the ratio is approximately 1:2, i.e., one-half pound of carbon per pound of metal. This carbon reacts to form carbon dioxide which is vented to the atmosphere. Metallurgical grade carbon contains sulfur, typically at concentrations on the order of 1 percent. This sulfur reacts to form sulfur dioxide which also is vented to the atmosphere. The production of portland cement evolves enormous amounts of carbon dioxide during the calcination of limestone. This carbon dioxide is also vented to the atmosphere. Clearly, a discussion of energy and the environment in the 21st century and the role of industrial processes would not be complete without consideration of the materials industry.

A broad segment of the materials industry was addressed in this session. There were five presentations. The first two treated in turn the two most abundant structural metals, namely steel and aluminum. This was followed by a presentation devoted to the more general topic of minerals processing. The last two talks turned attention away from metallurgy with their expositions on cement and glass.

In his paper on iron and steel technology, John F. Elliott reviewed some of the early history as well as current technology. He addressed energy utilization and cited current efforts at improving efficiency. As for environmental problems, carbon dioxide emissions were the prime focus. The efforts of the industry to develop *in-bath smelting* in an attempt to reduce emissions by eliminating coke ovens were described. This stimulated considerable discussion. Questions were raised concerning the motivation for pushing ahead with *in-bath smelting*, the future demand for steel, the advantages and disadvantages of the use of form coke, and the accessibility of such new technologies to developing countries. As regards the last question, Elliott pointed out that any improvement will be adopted only if it is cost effective, and therefore, this should make it less attractive for new players to continue to use technology that is less efficient from the standpoints of energy utilization and pollution.

The next talk, given by Patrick Atkins, Director of Environmental Control at Alcoa, followed much the same format as that of the previous presentation. The technologies for making primary aluminum as well as for recycling were reviewed with attention given to energy requirements, carbon dioxide emissions, and trends in reducing both. The point was made with respect to energy usage that one should look at the product life cycle to appreciate energy savings that accrue from the use of weight reducing materials such as aluminum. The same argument was put forth with respect to carbon dioxide emissions. This was followed by a good deal of discussion about various aspects of recycling aluminum: how successful is aluminum recycling, what are the incentives, metallurgical issues, the role of government, etc. The audience was rather well-informed about environmental problems facing the aluminum industry and asked well-posed questions about the amount of carbon dioxide generated during extraction and about the disposal of spent potliner. In response to the latter, Atkins responded that the industry is prevented by regulation from employing certain methods of treating spent potliner material. At present there are only two options: burial and incineration, which itself produces a somewhat less offensive form of hazardous waste that in turn is buried. Finally, questions were raised about the prospects for implementing inert anode/wettable cathode technology. Atkins expressed skepticism at the success of current research and felt that carbon based technology would be with us for the next decade.

In the context of the minerals industry Jay Agarwal surveyed energy consumption patterns, outlined a strategy for conservation, and argued for a greater appreciation of process economics in decisionmaking. He conceded that we have made many mistakes in the area of metals production and minerals processing; it is an energy intensive, labor intensive, and capital intensive business, but we have to cope. However, maintaining a high standard of industrial health and safety, recycling materials, and protecting the environment are smart business practices. In the question period that followed, two issues were raised. First, how do the economics of mineral production depend on ore grade and how effectively can we identify bodies of high grade ores beneath the earth's surface? Agarwal responded that it is approximately three times more expensive to work underground than at the earth's surface; thus it is unlikely that one could justify shaft mining. However, high grade ores do improve the energy efficiency of a mining operation. As an example, presently in the United States we move 800 tons of earth per ton of copper. Obviously, the higher the ore grade the less rock we must move. Secondly, what are the long term effects of recycling, particularly with respect to the accumulation of impurities? Agarwal's reply was that while costs will go up as scrap contamination rises, the economic incentives are still so large in favor of recycling over mining and minerals processing that recycling will still be practiced. Research is the answer, he stated.

On the subject of portland cement, Stewart Tresouthick reviewed current technology with attention to energy use patterns as well as opportunities for improvement. The problem of carbon dioxide emissions was acknowledged along with the need for further research to reduce them through the discovery of displacive technology. In the ensuing discussion the following questions were raised. What is the motivation to reduce the amount of lime in the mix? The answer: government mandated reduction of carbon dioxide emissions and the desire to decrease the energy requirement of the pyroprocess. What about imports? Imports accounted for only 13 percent of the cement in the United States in 1989, and this figure has been decreasing in recent years. What happens if the industry is forbidden to vent carbon dioxide? Is there a technology to reform lime? Not yet but work is proceeding towards this goal. Finally, what about the use of the rotary kiln to pyrolyze hazardous waste? This has been increasing, but ultimately the industry would like to move away from the use of rotary kilns altogether.

Glass manufacturing technology was reviewed by Greg Ridderbusch, who pointed out that, in spite of the fact that glassmaking is very energy intensive, environmental factors will figure most prominently in decisions about future technology. The audience asked about the status of the advanced glass melter program at the Gas Research Institute. Based upon the concept of entrained solids combustion, this process has been employed at one tenth of full scale to produce fiberglass and the work continues.

Although this session dealt with rather specific sectors of the materials industry, some common themes emerged. First, universally the need for continued research was recognized. We are fast approaching the limits of the technologies available to meet rising environmental standards. Furthermore, major improvements in energy efficiency have been made for the most part. It is clear that without major discoveries there will be no substantial changes in these industries. This does not auger well for the 21st century. Secondly, the capital intensity of the materials industry was made evident. The corollary to this is that adoption of technological change will not be easy to effect. Thirdly, in suggesting new directions the speakers either refused to venture beyond improvements to existing technologies or assessed as rather remote the chances of success of known attempts at radical innovation. Furthermore, the audience did not seem troubled by this state of affairs. It is the personal opinion of this session chair that in view of the minimal attention devoted to the exploration of carbon-free metals extraction technologies there is little evidence to support this pessimism. For example, the capacity of high temperature electrolytic extraction conducted in molten oxides at temperatures of 1000–2000°C has hardly been investigated. Electrolysis should figure prominently in a strategy to develop environmentally acceptable extraction technologies owing to the fact that the electron is a very "clean" reducing agent. Perhaps this unwillingness to look beyond the horizons of today's technologies is symptomatic of some sort of malaise in the materials industry. From an academic perspective, it is clear that materials processing must be infused with a sense of adventure if we are not to squander our most precious resource, the human resource. Unless we continue to attract bright young people who will be inspired to work on the important problems identified above and who will derive satisfaction from their accomplishments, our industries will atrophy and die.

Energy, the Environment, and Iron and Steel Technology

John F. Elliott
Department of Materials Science and Engineering
Massachusetts Institute of Technology
Cambridge, MA

Introduction

The production of iron and steel has always involved high consumption of fuels of relatively high quality. Throughout the history of the iron and steel industry, technological advances have permitted major reductions in energy requirements by introducing measures that directly economize on the use of fuels and improve the yield of product relative to the feed in various steps in the manufacturing process. Low-sulfur coke has served as a source of heat and as a reducing agent in the reduction of iron ores. Autogenous refining of pig iron has all but eliminated energy required for converting pig iron into steel, and the electric arc furnace is an economical means for remelting recycled steel scrap of all kinds. Carbon plays an important role in many of the steps by which steel is manufactured. Eliminating it from the production flow sheet will require major changes in the technological base of the industry.

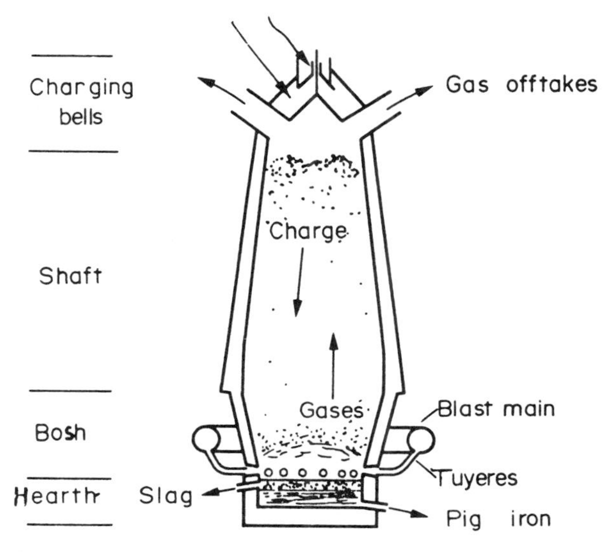

Figure 1. Schematic diagram of a modern blast furnace. Source: *Fundamentals of Metallurgical Processes*[1]

History and Current Status

Since antiquity, the advancement of the technology used in producing pig iron, wrought iron, and steel has been linked with the availability of cheap, good quality fuel and with advances in methods for improving the energy economy of smelting and refining processes. Iron and steel production has always required large quantities of good quality fuels. Charcoal, which is low in sulfur and ash constituents, was the primary fuel employed from prehistoric times to the late nineteenth century for reducing iron ores, smelting crude iron, and producing wrought iron. Pig iron was, and is, produced in a tall shaft furnace, the blast furnace (Figure 1), which is economical in its use of fuels because it is a countercurrent reactor. There was an energy crisis of sorts in England in the late 17th and early 18th centuries because much of the good timber had been converted to charcoal to feed ironmaking blast furnaces. Since coals could not supplant charcoal in ironmaking, and so much forest and timber was lost to build ships for the Royal Navy, a ban was declared on domestic ironmaking. This ban led to increased importation of iron into England from Eastern Europe and to the transfer of English ironmaking operations to the East Coast of North America, where wood was plentiful. After the first commercial use of coke in blast furnaces in 1740, the iron industry flourished once again in England. (Today, charcoal is again used on a relatively large scale to produce pig iron in Brazil.) The stronger structure of coke compared to charcoal has permitted larger and taller furnaces to be built. Today's largest furnaces have production capacities in the range of 7,000 to 12,000 metric tons of pig iron per day.

A major technological advance in the industry was the invention by W. Kelley in Kentucky in 1846, and H. Bessemer in England in 1856, of a method for autogenous refining of liquid crude iron in a pear-shaped vessel to produce liquid steel (Figure 2).[2] The principle of the process is to blow cold air through a mass of molten crude or pig iron to oxidize out the carbon, manganese, and silicon that are present in undesirable concentrations in the liquid. The result was the Bessemer process,

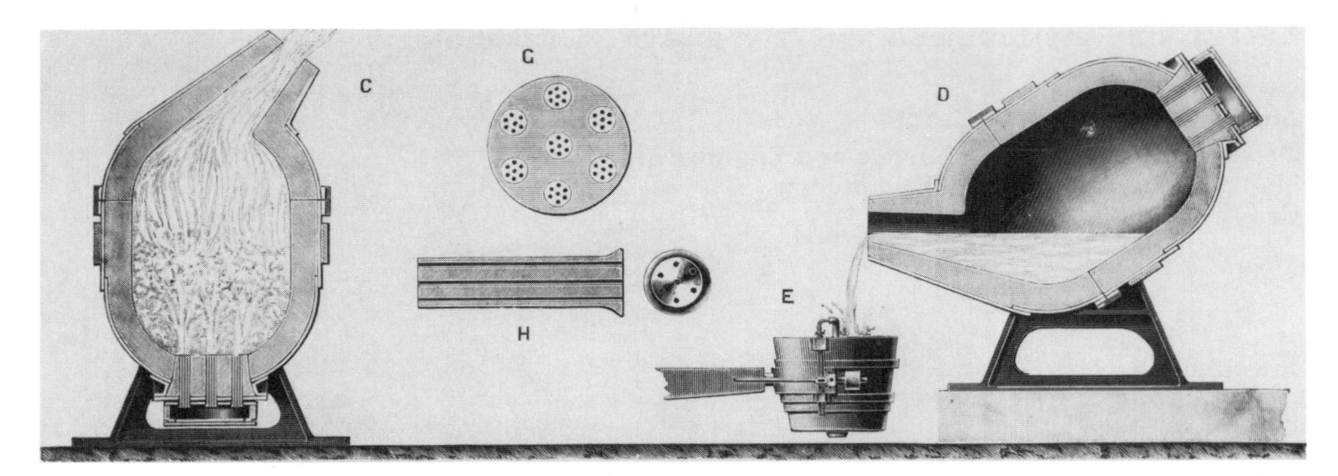

Figure 2. Henry Bessemer's Converter. C—Cross-section of blowing vessel; D—Cross-section of vessel being tapped; E—Ladle for steel; G and H—Transverse and longitudinal cross-sections of tuyere. Source: *Sir Henry Bessemer*[2]

in which the heat required is obtained completely from the oxidation of the impurities in the hot metal. Bessemer's primary motivation in developing the technique was to produce liquid steel in 10 to 15 ton batches, from which large machine parts, gun tubes, etc., could be made. Previous to this invention, the largest quantity of steel or wrought iron that could be made in a single piece was limited by the weight of hot material that could be manipulated by one or two men. The invention resulted in an essentially new material, the ability to produce large steel objects that previously could only be made of cast iron, and a major improvement in the energy economy in the production of steel.

Important steps forward in energy economy in smelting and heating operations came with the introduction of recuperators and regenerators to preheat the combustion air in furnaces with the thermal and chemical energy in the off-gases. The use of regenerators began in 1828 when Neilson utilized the energy in the off-gases from the blast furnace to preheat the air blast to the furnace to 325°C. There resulted a 50 percent reduction in the amount of coke required to smelt one ton of iron (i.e., from 6 G tons to 3 G tons in one specific case[3]). The equivalent fuel requirement for a modern blast furnace ranges between 450 and 550 kg of coke per metric ton of liquid iron. The net energy requirement is approximately 3.2 gigacalories per ton of product after one makes an appropriate correction for the energy in the recovered off-gases. To reduce the coke requirement, oil, tar, coke gas, and natural gas may be injected into the furnace with the air blast. Today, approximately 500 million tons of pig iron are produced annually in blast furnaces;[4] this is equivalent to approximately 95 percent of the iron ores reduced in the world.

The regenerator system of Siemens for preheating combustion air to very high temperatures with the enthalpy in the off-gases from the furnace was invented in 1858. In 1864, Martin applied the method to preheat air and producer gas for reverberatory smelting furnaces.[5] With preheated gas and air, flame temperatures of more than 1700°C —high enough to melt steel—were possible, and consequently steel could be produced in hearth type furnaces. The combination of regenerators and the hearth furnace resulted in the Siemens-Martin, or openhearth, furnace. An important advantage of this new furnace was that charges ranging from cold scrap and pig iron to almost 100 percent liquid pig iron could be refined into steel. The regenerative principle was also applied to furnaces for heating steel ingots and castings where working temperatures in the range of 1200 to 1400°C are required. Today, virtually all furnaces for other purposes in the steel industry are equipped with recuperators.

The beehive coke oven that was originally employed for producing coke wasted energy and polluted the environment to an unbelievable extent. The multiple-slot Koppers-type coke battery has finally replaced the beehive ovens. The newer types of coke-oven batteries also create serious workplace and environmental pollution problems, but very costly modifications to the coking system have eliminated most of them. Difficulties with the environment, and the high costs of building, operating, and maintaining coke ovens remain. It is predicted by many that the use of coke ovens in the US will end within the next 15 years as the coke-oven plants now in operation are retired and no new ones are built. Because an adequate supply of good quality coke is a critical requirement for efficient operation of the blast furnace, the future of the blast furnace-coke oven sequence is in serious question.

The oxygen-blown converter for producing steel first appeared in 1949. It is known by various names, including the basic oxygen furnace (BOF), and the basic oxygen process (BOP). From a historical perspective, the process was made possible by the development in the 1930s of economical methods for producing pure oxygen on a large scale. The converter is a descendant of the Bessemer process in that it is autogenous. The heat for the process is generated by impingement of a supersonic jet of oxygen from a vertical lance onto the furnace charge to generate heat and remove impurities in liquid crude iron. In the BOF there is sufficient excess heat generated that approximately 30 percent of the charge to the furnace can be cold steel scrap. The hot gas leaving the furnace averages approximately 90 percent CO and 10 percent CO_2. These gases are burned, cooled, and cleaned before being discharged into the atmosphere in many plants throughout the world. In Japan, most furnaces are equipped with gas collections systems, and with the recovery of the sensible and chemical heat in the gases leaving the furnace there is a slight overall excess of heat of approximately 0.05 gigacalories per ton of raw steel. Figure 3 is a sketch of a modern oxygen steelmaking furnace[6] in which oxygen can be introduced into the bottom of the vessel through tuyeres as well as from the top through the vertical lance. Large vessels are approximately 10 meters high and 7.5 to 8 meters in diameter and can produce a heat of 300 metric tons of steel in 35 to 40 minutes total time. A steelmaking shop with two vessels can produce from 2.5 to 3.5 million metric tons of steel per annum.

Another important advance that improved the energy economy of the steel industry was the development of the electric arc furnace (EAF) for remelting steel scrap. This furnace was first utilized in 1904 for melting and refining alloyed and stainless steels, and that use continues today. In recent years, the EAF has emerged as the principal method for melting steel scrap to produce ordinary carbon steels. Furnaces with capacities as high as 225 metric tons of steel are in operation, and modern "ultra-high-powered" furnaces have transformer capacities in the range of 0.4 to 0.5 MW per ton of steel capacity. For example, a modern 90 ton furnace would have a 40 MW transformer. Many improvements are being incorporated into electric furnace steelmaking such as the replacement of refractories with water-cooled panels, the use of oxyfuel burners to facilitate scrap melting, the introduction of ladle-processing methods for treating the steel after it leaves the furnace, etc., all of which help to reduce the electrical energy consumption, and some of which improve the quality of the steel. The overall energy requirement for melting scrap in the EAF is approximately 1.5 gigacalories per ton of raw steel, and the overall average energy required for producing a metric ton of raw steel ready for casting via the blast furnace-BOF sequence is approximately 2.6 gigacalories. This figure is based on a charge to the furnace of 70 percent liquid pig iron and 30 percent scrap, and a yield of steel from the charge of 85 percent. The overall average charge for oxygen furnaces in the US today is 26 percent steel scrap and 74 percent liquid pig iron.[7]

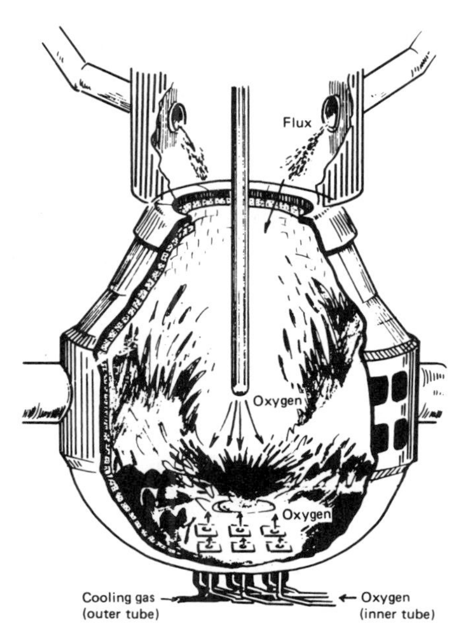

Figure 3. Sketch of a modern basic vessel showing lance and bottom tuyeres for introduction of oxygen, and hood for collection of off-gases. Source: *Steelmaking Plant, Kimitsu Works,* Nippon Steel Corp.[6]

The replacement of traditional ingot casting with methods for continuously casting liquid steel has increased the yield of good steel from the raw liquid steel by 10 to 15 percent, i.e., from 80 to 85 percent to as high as 95 percent. This results in an almost equivalent net savings in the energy required to produce a ton of raw steel. Approximately 0.25 gigacalories of heating fuels are consumed in continuously casting one metric ton of steel. The change to continuous casting also eliminates one step in the rolling sequence required in the conversion of the cast product to a salable product. Bessemer patented a continuous casting method in 1854, but engineering satisfactory casting machines involved many difficulties, and use of the method advanced slowly between 1950 and 1970. Today, a very large fraction of the steel made in Japan and Europe, and approximately 65 percent of that made in the US, is cast continuously.[8] The latter value is rising rapidly. The casters (a sketch of one is shown in Figure 4) are usually installed in pairs, and a large modern shop for casting slabs from which hot and cold rolled sheets and large plates are rolled will cost in the range of $150 to $200 million.

Direct Reduction

Direct reduction of iron ores has become an alternative to blast furnace smelting operations. The principal direct reduction method utilizes a countercurrent shaft

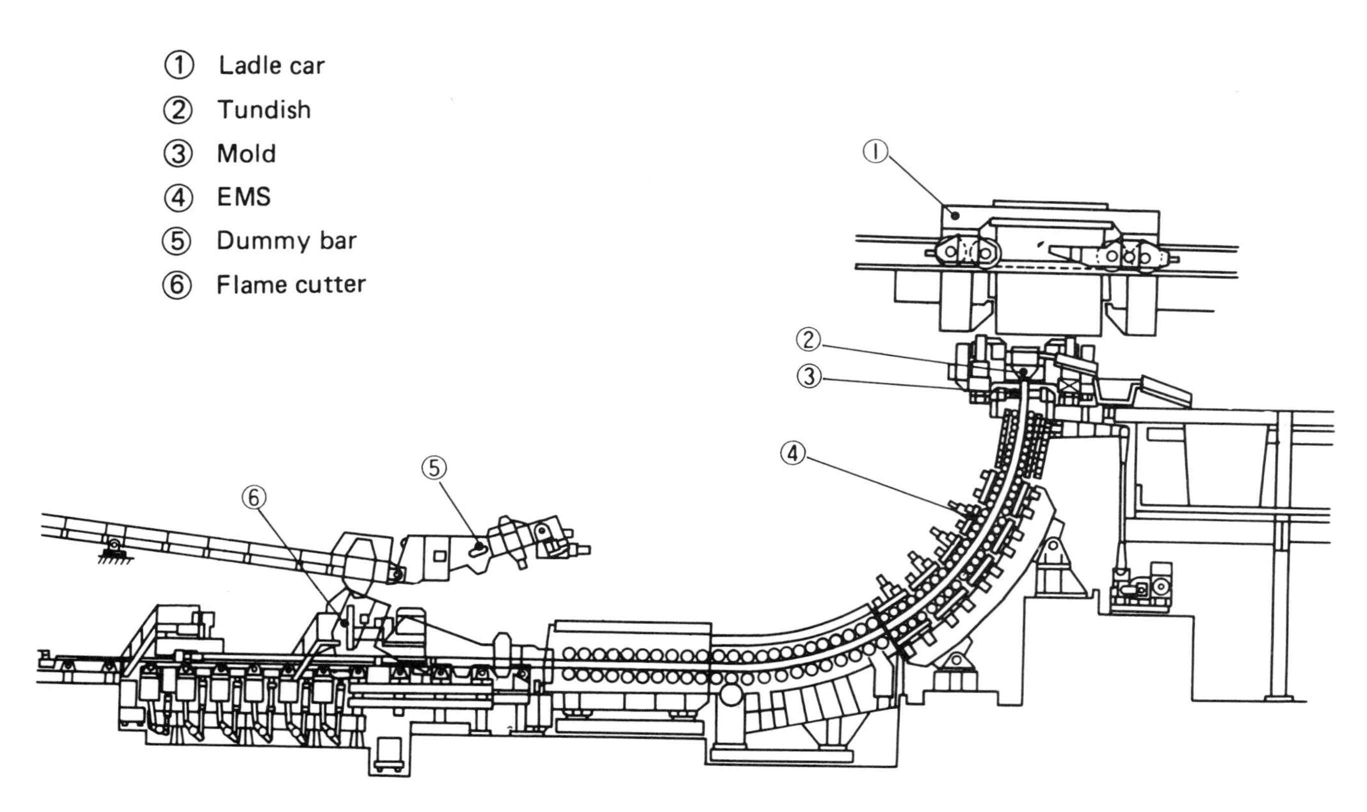

Figure 4. Cross-section through a continuous caster for slabs showing ladle, tundish, casting mold, roller train for support and cooling of slab, electromagnetic Stirrer (EMS) for stirring liquid metal in center of slab, and flame cutter for cutting slab into sections. Source: *Steelmaking Plant, Kimitsu Works,* Nippon Steel Corp.[6]

furnace in which the fuel is reformed natural gas and the ore feed is either coarse ore or pelletized concentrates. Although this type of process was developed in the United States, the technology is employed almost exclusively in regions where natural gas is very inexpensive. Figure 5 is a schematic diagram of a reduction system that shows the gas reformer, the various gas streams, and the furnace shaft in which the ore is heated, reduced and then cooled. Various types of rotary kilns are also employed in which the fuel is usually coal char or relatively reactive coals. This method of producing direct reduced iron is used most extensively in South Africa. The product from both types of operations consists of metallized ore particles or pellets that become the feed for steelmaking operations. The total energy requirement for the production of direct reduced iron in shaft furnaces is in the range of 2.5 to 2.8 gigacalories per metric ton of product. The energy use for the rotary kiln systems varies with the type of kiln and fuel, but is somewhat higher than that for the shaft furnace.

The electric arc furnace is the principal melting furnace for direct reduced iron. The metallized ore carries with it the gangue constituents of the ore so that the steelmaking operation is burdened with a relatively large slag volume. In turn, the yield of steel from the charge is several percent below that obtained when melting good quality steel scrap, and the energy needed in the melting operation per ton of product may be 10 to 15 percent higher than that when melting scrap. The overall energy requirement for the production of finished steel via the shaft-type direct reduction furnace and electric furnace melting is approximately 5.4 gigacalories per metric ton of *raw* liquid steel. Approximately 43 percent of that energy is for the production of electricity.

Energy Use and Prospects

Flow diagrams for five processing sequences for converting iron ore and scrap into finished steel are shown in Figure 6, Coke Ovens-Blast Furnace; Figure 7, Scrap Melting with EAF; Figure 8, Bath Smelting; and Figure 9, Direct Reduction-EAF. The total energy required in the production of a unit of steel in the market-economy, industrialized nations has declined approximately 20 percent in the past decade. Improvement in yields of

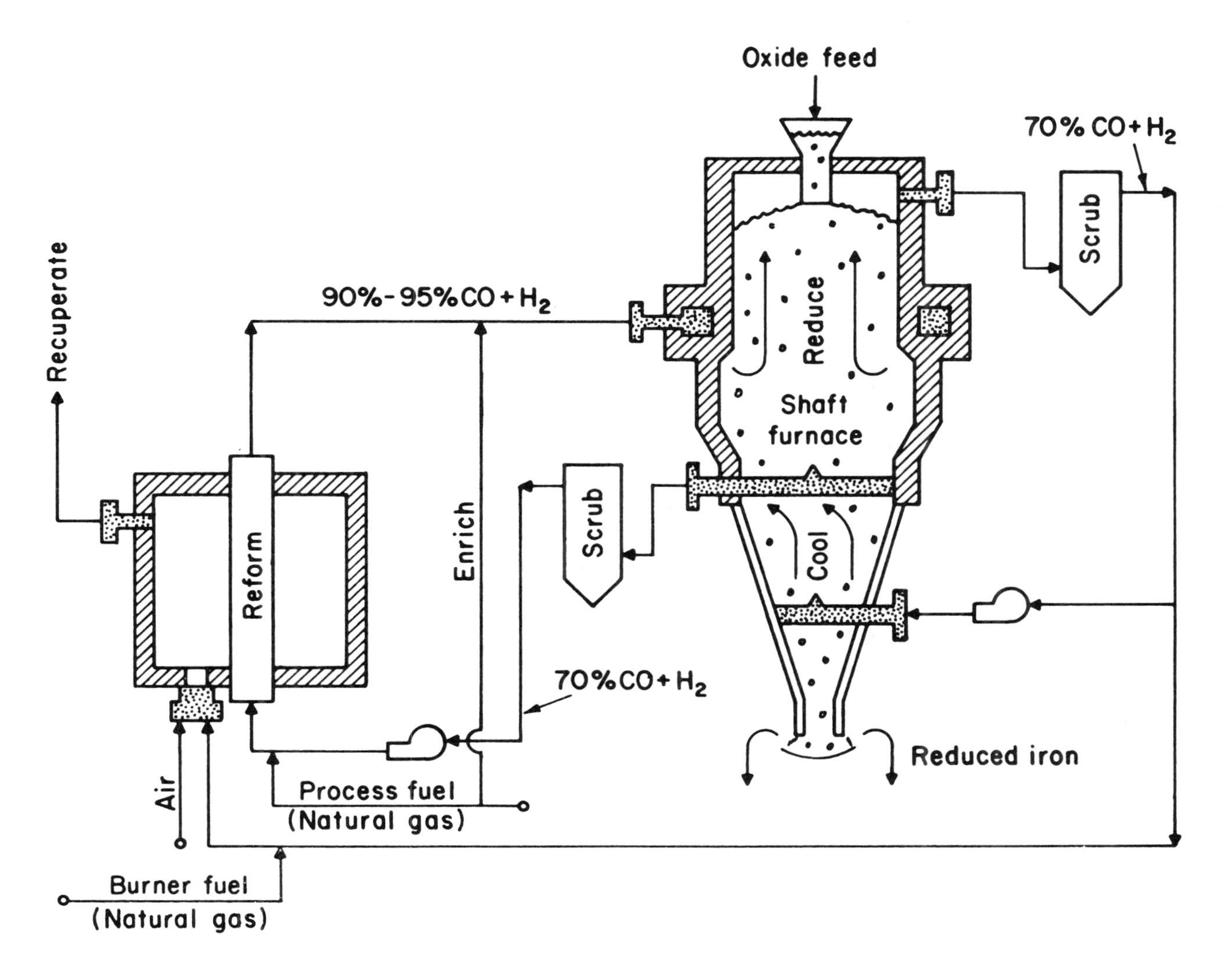

Figure 5. Schematic diagram of a shaft-type direct reduction furnace. Reformed natural gas is fuel and reductant. Source: *Direct Reduced Iron*[9]

good quality material in the various steps in the steelmaking operation, greater economy in the use of fuels, and better furnace designs are primarily responsible for this improvement. The current technology for producing steel in integrated steel plants includes coke ovens, blast furnaces, oxygen blown steel converters, continuous casting, and rolling and finishing operations. The so-called ministeel plants employ electric arc furnaces for melting scrap, continuous casting, and rolling and finishing operations. Some integrated companies also have electric arc steel furnaces for melting home and purchased scrap. The total annual production of raw steel in the world is approximately 700 million metric tons; that of the free market economies is approximately 400 million tons; and in the US, the annual production is approximately 100 million tons.[4] In the US, approximately 35 percent of the raw steel is produced in arc furnaces, 5 percent in the openhearth, and the balance in the basic oxygen furnace. The average overall energy consumption for production of finished steel in the western industrialized countries is 5 gigacalories per ton of *raw* steel, and in Japan it is 4.2 gigacalories (see Table 1). The lower relative use of the electric furnace in Canada as compared to that in Japan and the US is the principal reason for the lower energy consumption in the latter two countries. The average for integrated plants in the US is approximately 5.5 gigacalories, and the overall energy consumption in the ministeel sector is somewhat variable, but averages approximately 2.2 gigacalories per ton of raw steel. The much lower figure for ministeel plants arises for a number of reasons: they melt scrap in electric furnaces; their product mix is

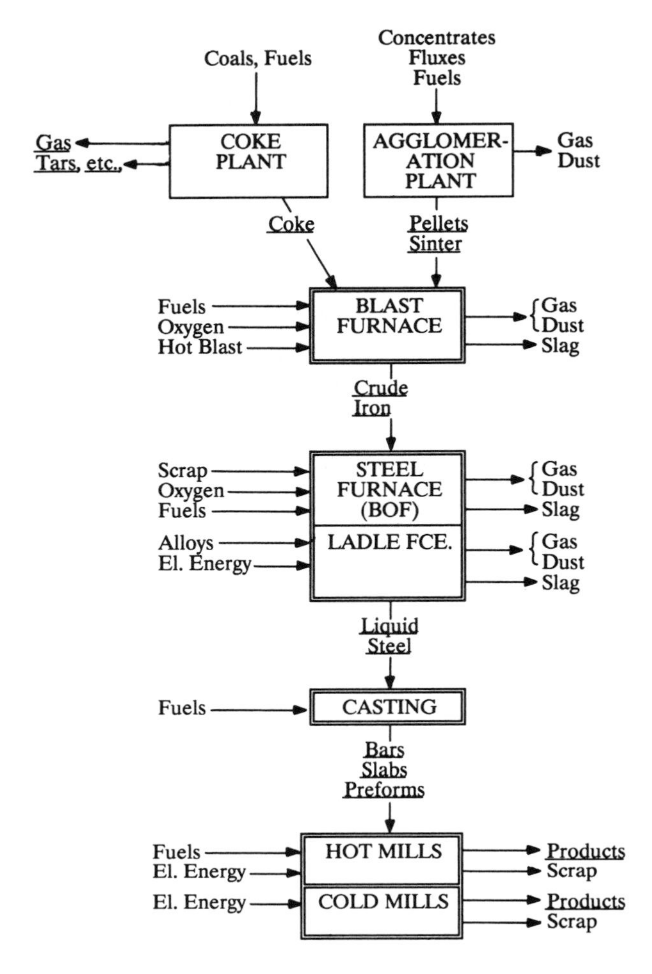

Figure 6. Flow sheet for coke plant-blast furnace based steel production

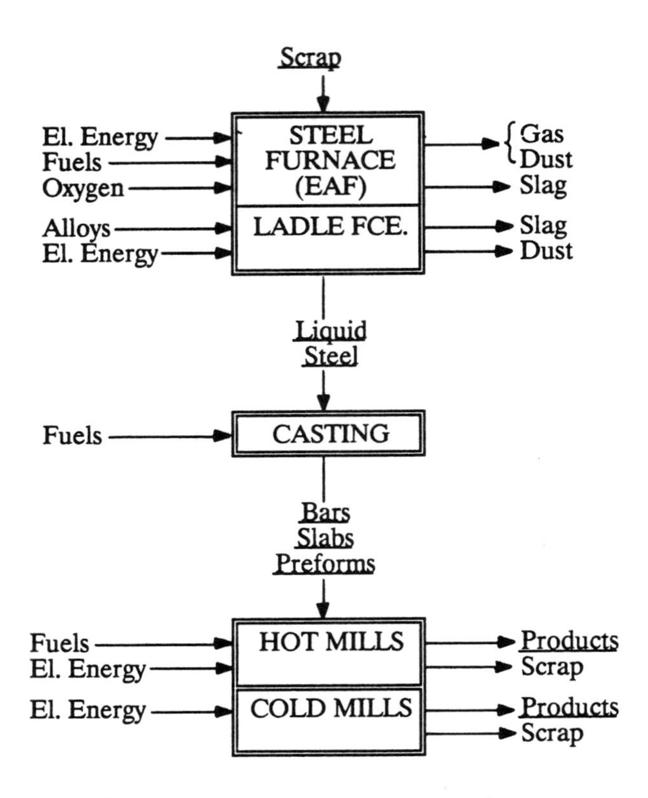

Figure 7. Flow sheet for scrap-based steel plant

relatively simple; each plant includes only a modest number of process steps; and the flat-rolled products that generally require more energy in their production are made predominantly by the integrated sector.

Gases from the blast furnace and coke ovens have sufficient calorific power so that they can be utilized for electric power generation and in various heating and processing steps in a steel plant. During a portion of the blow of a basic oxygen converter, the off-gases have a relatively high concentration of CO. These gases are recovered in Japanese and German steel plants, but this practice is not followed in US and Canadian plants because of the cost of gas recovery systems and because of the relatively low cost of natural gas in those countries. Before being discharged into the atmosphere, the excess waste gases are oxidized to convert CO to CO_2 and any H_2 to H_2O.

Worldwide, the problems with coke ovens have directed attention to an alternative technology by which partially reduced iron ore, oxygen, and coal are reacted in a molten bath of high-carbon, iron-carbon alloy to produce more metal. This method is called "bath smelting," or "in-bath smelting." Development work on processes based on this method is in progress in the US, Japan, Canada, and Australia. The use of coal as the fuel in the smelting step has economic advantages, and if successful, bath smelting could eliminate the need for both coke ovens and blast furnaces in integrated steel plants. It is probable that the overall energy requirements for the production of steel in which the bath smelting operation replaces the coke oven, the blast furnace, and at least part of the steelmaking step, will not differ appreciably from that for a modern integrated steel plant.

There are good prospects for commercial development of further advances in methods for solidifying and casting of steels that may make it possible to eliminate an additional number of the rolling and processing steps now necessary for producing many steel products. The advances are in the area of "near net-shape forming" by which the steel is cast in a form close to the final shape of the steel product. One of the methods is spray casting, by which a product such as a tube or plate is made by spraying layers of liquid metal on a form, so that the spray solidifies as it impacts the surface of the previously cast structure. There are also the so-called

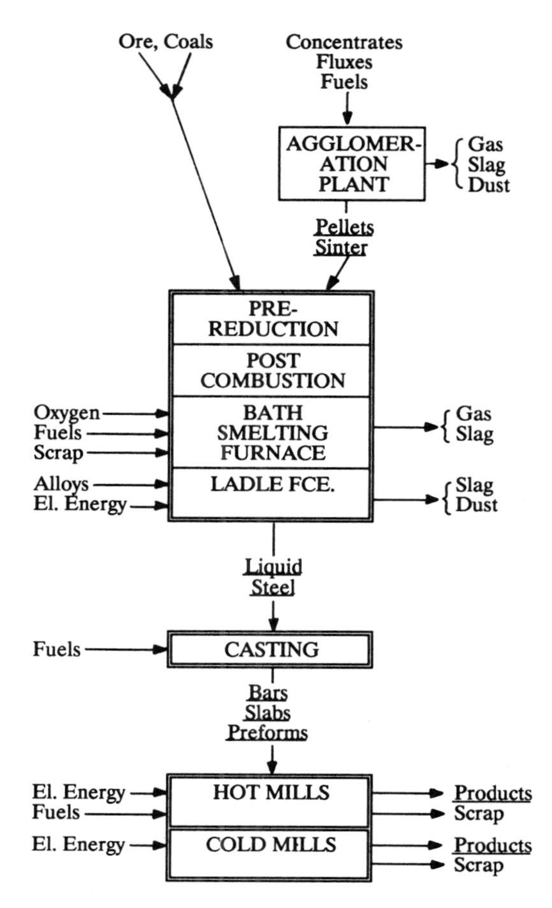

Figure 8. Flow sheet for steel production based on bath smelting

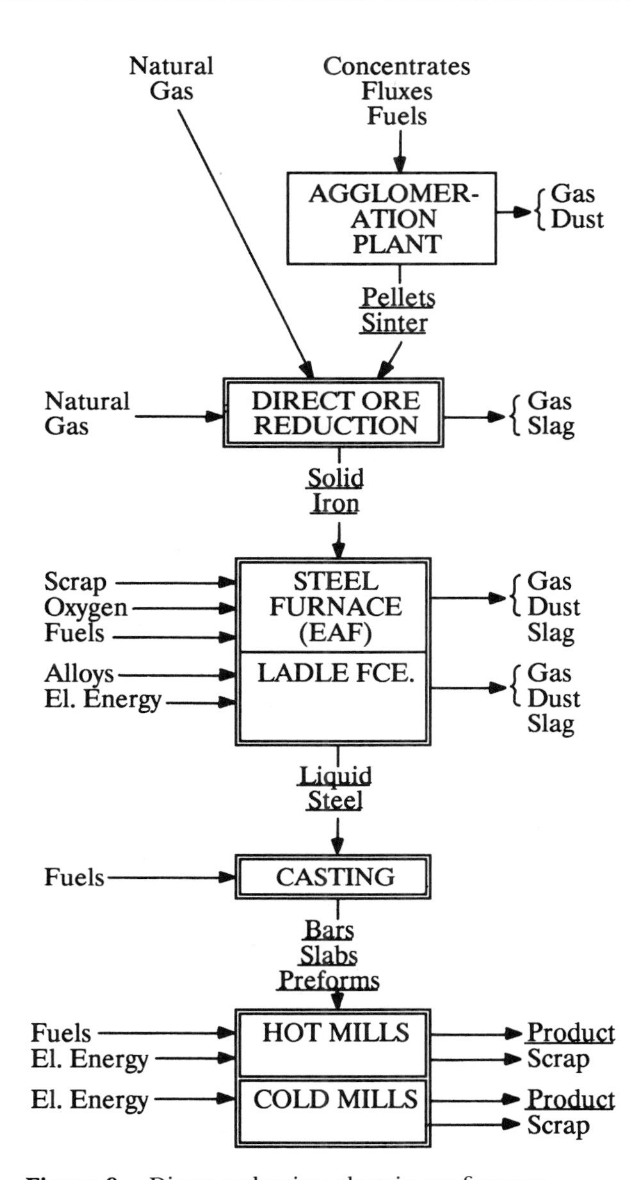

Figure 9. Direct reduction-electric arc furnace flow sheet

thin-strip and thin-slab casting methods, by which liquid steel is converted directly into thin slabs (25 to 35 mm thick) and thin sheets (1 to 3 mm thick) so that relatively few subsequent processing stages are required to produce plates and sheets. In a large integrated steel plant, approximately 40 percent of the total energy required to produce a final product, starting with ores and concentrates, is consumed in processing the steel after it has left the steelmaking furnace. Thus development of these casting methods will probably result in additional reductions in the energy required to produce a ton of steel overall, primarily because of further increases in the yield of finished steel from the raw liquid steel.

Increased reliance on scrap as a raw material for steelmaking in the US and Canada is to be expected in the future because of the relatively abundant supply of scrap. With the use of steel scrap in electric arc furnaces, it is probable that the need for coke ovens and blast furnaces will be reduced, which will result in an overall reduction in energy consumption by the steel industry. Metal obtained from iron ores is generally free of cop-

Table 1. Overall average energy consumption per metric ton of finished steel

Country	Gcal	GJ	% EAF*
Japan	4.2	17.5	30
US	4.9	20.5	36
Canada	5.8	24.2	14

* Electric arc furnace

per, nickel, and tin, but scrap recycled from the general community is contaminated with these elements. Carefully selected grades of scrap including prompt industrial scrap from fabrication plants are generally more free of contaminants, but this advantage will diminish with greater use of anti-corrosion and other types of protective coatings on sheet steels, increased use of alloying elements at low concentrations for many applications (such as auto bodies), and the likely development of steel-based composite materials. Thus far, no practical means has been found to remove these contaminants in steel scrap, and the elements are not eliminated in the steelmaking furnaces. With the greater expected use of scrap in steelmaking, larger tonnages of the higher grades of carbon steel will have to be made in the electric furnace. Accordingly, it will be necessary to keep these contaminants from the scrap at low levels in the steel products. The only practical method now available to accomplish this is to dilute the scrap with virgin metal from ores in the charge to the furnace. This practice of dilution is successful because virtually all commercially available iron ores have very low levels of these contaminants. A future alternative for plants that utilize scrap as their principal source of metallics may be the use of bath-smelting methods to produce liquid metal from ores to act as a dilutent in their steelmaking operations.

The quantity of coal required to produce a metric ton of pig iron in a modern coke oven-blast furnace plant ranges between 720 and 800 kilograms. Complete oxidation of the carbon in the coke in the blast furnace; in the fuels in coke making; and of oil, natural gas, and other products of the coking operation that are used as fuels in other furnaces in the steel plant results in between 1,820 and 2,060 kg of CO_2 being discharged into the atmosphere per ton of pig iron produced. Bath smelting of pig iron could essentially eliminate the need for coking coals in iron- and steelmaking, but the total requirement for coal and the emissions of CO_2 would probably not change very much (See Table 2). An exception to that statement would arise in the CO_2 if the off-gases from the smelter were removed before they were utilized in the partial reduction of iron ore, the generation of oxygen, and the generation of heat in other operations in the plant. In this case, it may be possible to reduce the emission of CO_2 by up to 30 percent.

Table 2. Coal consumption and CO_2 production in smelting ore, US, 1989

Process	Coal,* kg	CO_2 Prod., kg
Blast furnace	720–800	1,820–2,040
Bath smelting: (1) w/o intermed. removal of CO_2.	740–850	1,760–2,060
(2) with intermed. removal of CO_2.	700–800	1,660–1,940
[Scrap melting]	[1.5 Gcal, elec.]	[Eliminate with nonfossil fuel power gen.]

* 80% F. C., 25–28% Vol.

The CO_2 Problem

Carbon figures prominently in virtually all of the steps in the steelmaking flow sheets, and eliminating its use in the production of steels poses some very serious problems. It is the reductant of choice for iron ores, primarily in the form of coke. It is the principal fuel for ironmaking, and it is a major constituent in oil and gases used as fuels in heating, reheating, and annealing furnaces. Carbon acts as an agent to assist refining the steel in steelmaking furnaces and in ladle-refining steps, and it is carried off as CO and CO_2 in the off-gases from such operations. It also is a constituent in steels, ranging from 0.01 wt. pct. to up to 1.0 wt. pct. in a few grades of steels. The carbon steels make up approximately 90 percent of the tonnage of steel produced in the US, and the concentrations of carbon in the bulk of that tonnage are in the range of 0.05 to 0.30 wt. pct.

Carbonaceous fuels could be replaced by electrical heat in most of the heating and reheating furnaces. They are an important agent in steel-refining operations, where they may be difficult to replace because of chemical requirements. However, the quantities of carbon involved are in the range of 5 to 10 kilograms per ton of raw steel. Recycling and recovery of carbon in the off-gases from such operations appear feasible. An agent to carry off the oxygen associated with the carbonaceous gases would be required. Hydrogen would be satisfactory for this purpose, providing there was a ready and inexpensive source of it. Using other elements and thermomechanical treatment methods to eliminate carbon from steels would probably not be required, particularly if recycling of carbon from steel-refining operations is practical. A summary of these matters is shown in Table 3.

Replacement of the coke oven-blast furnace sequence by direct reduction with reformed natural gas and electric furnace melting of the metallized ore has the potential to reduce the output of CO_2 in the ore reduction and steelmaking steps because natural gas has a ratio of H_2/C of two to one as compared to virtually zero in the blast furnace fuels. To make this possible, the electric power for the steel furnace would have to be generated in nonfossil fuel power stations. Also, the evolution of CO_2 could be reduced by greater dependence on recycled steel scrap as the principal source material for steelmaking operations with the attendant use of the electric furnace to melt the scrap. Again, a

Table 3. Means to avoid dependence of steel industry on carbon

(1) Reduction of iron ore
Intermediate term—use $CH_4(g)$ to produce DRI.
Long term—replacement of reductant, that is, with $H_2(g)$, and use electric power as heat source. Electrolysis of aqueous Fe-solutions may be an alternative to electrochemical production of hydrogen.

(2) Steelmaking
Reclaim and recycle carbon used in refining.
Melt scrap and DRI with electric power.

(3) Vacuum and ladle processing
Reclaim and recycle carbon.
Greater use of inert gases.

(4) Hot processing
Alternate fuels, hydrogen and electricity (nuclear).

further reduction would be realized if the electricity for melting came from power stations that did not utilize fossil fuels.

The fundamental problem regarding elimination of carbon from operations in which iron ores are reduced and smelted is that the oxygen in the ores must be discharged into the environment in some form. The principal constituents of these ores are the iron-containing minerals: hematite (Fe_2O_3), magnetite (Fe_3O_4), and silica (the gangue). The majority of steels that are produced are the carbon steels, which are generally in excess of 99 wt. pct. iron. Very broadly, the process of producing steel from ores consists of separating oxygen and the gangue from the ore to leave behind almost pure solid iron. With the present technology, the oxygen is discharged into the atmosphere, i.e., the sink, as $CO(g)$, $CO_2(g)$, and $H_2O(g)$, and the silica is discharged as finely divided tailings and as slag. The overall reactions then are:

$$Fe_2O_3(s) \rightarrow 2\ Fe(s) + 3\ O\ (\text{as } CO, CO_2 \text{ and } H_2O) \quad (1)$$

$$SiO_2(s) \rightarrow SiO_2\ (\text{as } SiO_2(s) \text{ or as } SiO_2 \text{ in slag}). \quad (2)$$

In the actual manufacturing plant, the oxygen contained in effluent gases also includes oxygen from air and oxygen gas that is required for the combustion of CO from the reduction process and for other fuels for the generation of heat needed in various processing steps. Similarly, ash in the fuels and some reoxidized iron are added to the streams carrying the gangue constituents from the process.

The minimum amounts of energy (the reversible work) that is required to separate iron from oxygen in Fe_2O_3 and FeO are shown in Table 4. Also shown is the reversible work required for reduction of the oxides with hydrogen gas. For comparison, the equivalent requirements for several other metals are also shown. The

Table 4. Minimum energy required to separate metal from oxide

(A) $Me_xO_y(s) = x\ Me(s) + y/2\ O_2(g)$
(B) $Me_xO_y(s) + y\ H_2(g) = x\ Me(s) + y\ H_2O(g)$

	Free Energy		
Oxide	KJ/Mole Me Equation A	KWh/MT Me Equation A	KWh/MT Me H_2 Red. Equation B*
Al_2O_3	791.1	8,143	4,481
Cr_2O_3	526.5	2,812	912
Fe_2O_3	371.8	1,848	0
FeO	251.4	1,250	71
H_2O	118.6	32,665	—
MgO	568.9	6,498	3,790
MnO	402.8	2,036	837
SiO_2	866.5	8,468	3,778
TiO_2	889.4	5,156	2,407

* Assuming complete recycle of unused $H_2(g)$

Table 5. Theoretical minimum weight of reduction reagent required to separate one MT of metal from oxide

(A) $Me_xO_y(s) + y\ C(s) = x\ Me(s) + y\ CO(g)$
(B) $Me_xO_y(s) + y/2\ C(s) = x\ Me(s) + y/2\ CO_2(g)$
(C) $Me_xO_y(s) + y\ H_2(g) = x\ Me(s) + y\ H_2O(g)$

	Kg of Reductant, C or H_2		
Oxide	C(CO) Eq. (1)	$C(CO_2)$ Eq. (2)	$H_2(H_2O)$ Eq. (3)
Al_2O_3	667	333	111
Cr_2O_3	346	173	58
Fe_2O_3	322	161	54
FeO	215	107	36
MgO	494	247	82
MnO	218	109	36
SiO_2	854	427	142

stoichiometric amounts of carbon and hydrogen required for the reduction of the iron oxides, and oxides of several other metals, are shown in Table 5. The annual discharge of oxygen directly associated with iron ores smelted in the world for the annual production of approximately 500 million metric tons of crude iron[4] is estimated to be 200 million metric tons. If it is assumed that the gas resulting from the smelting operation has an equilibrium ratio of CO/CO_2 of two, the equivalent amount of carbon associated with this oxygen is approximately 115 million metric tons. This constitutes approximately 1.3 percent of the total annual emissions of carbon into the atmosphere.[11] A more precise calculation would require a correction for the oxygen that combines with hydrogen in the fuel, and for oxygen from the reduction of oxides of metals in the ore such as manganese and silicon; but these corrections would approximately offset each other. It is to be expected that the CO in the off-gases from the smelter would be subsequently converted to CO_2. Accordingly, the minimum theoretical amount of CO_2 resulting from just the chemical removal of oxygen from iron ores is approximately 350 million metric tons. The actual amount of CO_2 generated in iron and steelmaking operations in the world is probably in excess of 700 million tons per year. This does not include the carbonaceous gases generated from mining, beneficiation, agglomeration, and shipping of iron ores and concentrates. That issue is discussed by Agarwal.[12]

An apparently simple means of avoiding the generation of carbonaceous gases in the reduction of ores would be to use electrochemical methods in which the products are solid iron and pure oxygen. Methods for accomplishing this are in use to produce very pure iron, but the cost is very high. Even on a large scale, these methods of electro-winning iron would involve facilities that would be extremely costly to build, operate, and maintain. The energy required would be electricity, again produced by nonfossil-fueled power plants. The prospects for large scale commercial production of iron by this method are not promising. Similarly, the use of high temperature electro-winning of iron in which the electrolyte is a fused salt or a liquid oxide solution is not promising. On the other hand, if hydrogen is to be generated on a very large scale in the future, electro-winning of hydrogen followed by direct reduction of iron may not be as practical a sequence as carrying out electro-winning of iron directly.

Summary

Improvements and changes in the technology for producing iron and steel show promise of improving the energy economy of the steel industry. There are a number of advances by which this may be accomplished, both near term and in the longer term. The major changes that might be realized will come from the elimination of processing stages by techniques such as "near net shape" processing, increased recycling of steel scrap, and advanced smelting methods. Significant reductions in the energy required in the production of steels are to be realized only by elimination of processing steps from the traditional flow sheet. Existing direct reduction technologies with the use of reformed natural gas as the fuel could replace conventional smelting methods and reduce the carbon dioxide emissions from ore reduction by approximately one half. Eliminating essentially all carbon dioxide emissions from steel production plants involves formidable technological obstacles, and serious problems with increased capital and operating costs.

References

1. L. Coudurier, D.W. Hopkins, and I.W. Wilkomirsky, *Fundamentals of Metallurgical Processes,* Pergammon Press, Oxford, 1878, p. 218.
2. H. Bessemer, *Sir Henry Bessemer, F.R.S.,* I.S.I., Gr. Br., 1905.
3. I.L. Bell, *The Manufacture of Iron and Steel,* G. Routledge and Sons, London, 1884, p. 16.
4. F. Schottman, "Iron and Steel," *Minerals Yearbook,* Vol. I, Metals and Minerals, US Department of Interior, 1986–88.
5. J.F. Elliott, "Steel Production," *Encyclopaedia Britannica,* Fifteenth Edition, pp. 637–63, 1974.
6. Anon., *Steelmaking Plant, Kimitsu Works,* Nippon Steel Co., Tokyo, Japan, 1990.
7. R.E. Brown, "Iron and Steel Scrap," *Minerals Yearbook,* Vol. I, *Ibid.*
8. Anon., *Statistical Reports,* American Iron and Steel Institute, Washington, DC (Annual).
9. B.C. Cunningham, and J.G. Stephenson, "Direct Reduction Processes," *Direct Reduced Iron,* R.L. Stephenson, Ed., Iron and Steel Society, AIME, Warrendale, PA, 1980, p. 82.
10. Anon., *Statistical Tables,* International Iron and Steel Institute, Brussels (Annual).
11. J.H. Gibbons, "The Interface of Environmental Science and Policy." *Energy and the Environment in the 21st Century,* MIT Press, Cambridge, MA, 1990.
12. J. Agarwal, "Minerals, Energy, and the Environment." *Energy and the Environment in the 21st Century,* MIT Press, Cambridge, MA, 1990.

Some Energy and Environmental Impacts of Aluminum Usage

Patrick R. Atkins
Aluminum Company of America
Pittsburgh, PA

Herman J. Hittner
Aluminum Company of America
Pittsburgh, PA

Don Willoughby
Alcoa of Australia
Point Henry Works
Geelong, Australia

Abstract

Three major components should be considered when the energy and environmental impacts of a material's use are evaluated. First, the initial energy requirements and associated environmental impacts of converting the raw material and manufacturing the final product must be fully addressed. Second, it is equally important to determine the energy and environmental impacts that occur or are avoided by the reclamation of the material after use to make new products (reuse) or to remake the same product (recycle). A third factor which contributes to the true picture of the energy and environmental impacts is the reduction in energy usage that occurs as a result of the substitution of one material for another over the lifetime of the product.

In the case of aluminum, the energy requirement to recycle discarded products into molten metal suitable for new product manufacture is less than 5 percent of the energy required to produce primary metal from bauxite ore. Similar reductions in environmental impacts also occur. Therefore, the reuse and recycle rate has a major impact on the total energy budget for a product made from aluminum and the overall environmental impacts associated with that product's manufacture.

Since aluminum is lighter, more easily recycled, and more corrosion resistant than many competing materials, its use in applications such as transportation, construction, packaging, and high voltage electrical conductors provides a net energy payback when compared to alternative materials. The use of aluminum in these applications can also produce a net reduction in environmental impacts such as carbon dioxide release when compared to the use of other materials.

Introduction

Energy usage and environmental impacts are often closely linked. Effective energy management programs can also be successful environmental management efforts. Such is the case with the use of aluminum.

Life cycle energy requirements and environmental impacts should be considerations upon which to base the choice of materials. Life cycle requirements include the energy requirements and environmental impacts associated with the manufacture, end use, and recycling of the material. The analysis should include requirements and credits, since, in some applications, the substitution of one material for another produces a net energy savings and a coincidental reduction in environmental impacts.

As consideration is given to the impacts of aluminum on energy use and the environment in the 21st century, the full life cycle impacts should be evaluated. Since there is a growing interest in global pollution issues and concern about the emission of greenhouse gases, carbon dioxide will be used in this analysis as a surrogate measure of environmental impacts. A number of studies on energy impacts of aluminum production and use have been reported by others (Cochran, 1981; Berk, 1982), so no attempt will be made to add significant new information on energy use, except to emphasize the importance of life cycle analyses. The carbon dioxide budget of aluminum use is the primary issue that will be discussed.

Primary Production

Aluminum is the third most abundant element in the earth's crust, present in a variety of hydroxide forms. Most aluminum is produced from bauxite ore in a two-step process. The first step, the Bayer Process, separates impurities, and converts the aluminum hydrates and hydroxides to aluminum oxide. The second step, the Hall-Heroult Process, electrolytically reduces the aluminum oxide to aluminum metal, using a consumable carbon anode. Electricity is the major energy requirement for the aluminum production process, accounting for up to 70 percent of the energy input. If this electricity is produced by fossil fuel fired power plants, carbon dioxide emissions result.

Other significant energy requirements to produce aluminum include the fuel to operate boilers and calciners in the Bayer Process, the carbon required to produce the anodes used in the reduction process, and the fuel necessary to bake the carbon anodes in those plants that use prebaked anode technology. These fuel and carbon uses are also sources of carbon dioxide.

Major strides have been made by the industry to reduce the energy requirements for each of these steps in the process of making primary metal. Figure 1 displays the progress that has been made in energy reduction for alumina calcination. The practical limit using currently available technology and the theoretical limit for removing the excess water from the hydrate are shown. Additional energy reduction is anticipated as processing technology improves, but only incremental progress is likely.

Figure 2 displays the remarkable progress which has been made in electrical requirements for reduction of aluminum oxide since the Hall-Heroult Process was first commercialized. The present generation of cell technologies is approaching the practical limit for electrical requirements. A major technology breakthrough will be required to significantly alter the energy requirements for electrolysis. Efforts are underway to develop these new technologies. However, if a breakthrough technology is developed, commercialization and implementation of significantly different production processes will occur slowly because the current infrastructure will be difficult and expensive to replace. Therefore, the Hall-Heroult Process will continue to be the primary method of aluminum production well into the 21st century (Grjotheim, 1989). This analysis will be based on currently available technologies and the small

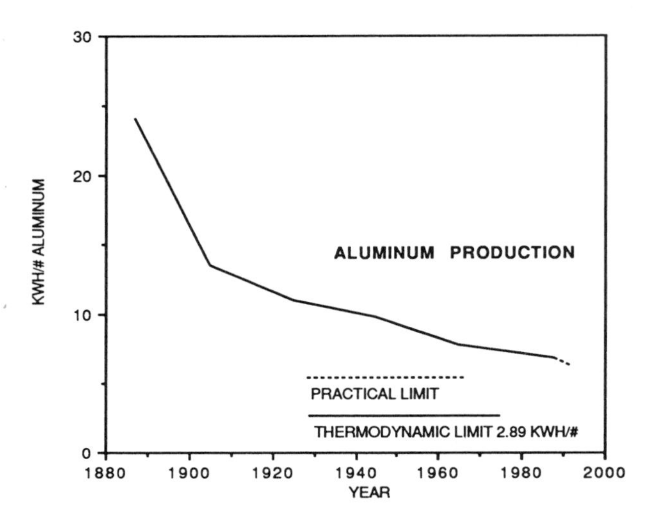

Figure 2. Electric energy consumption. Source: Alcoa Data.

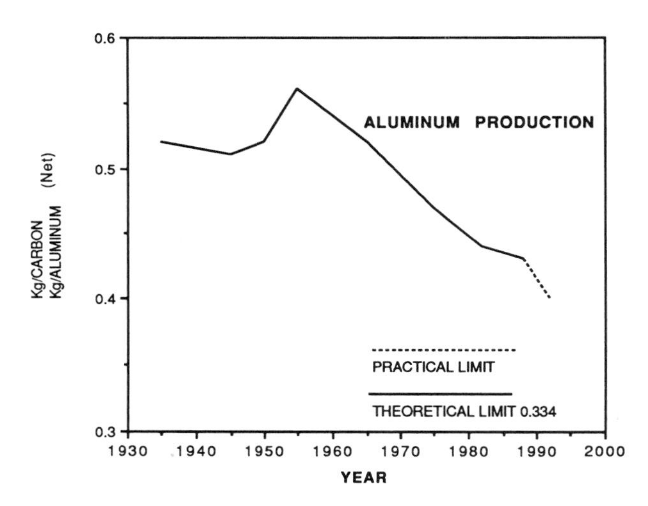

Figure 3. Net carbon consumption. Source: Alcoa Data.

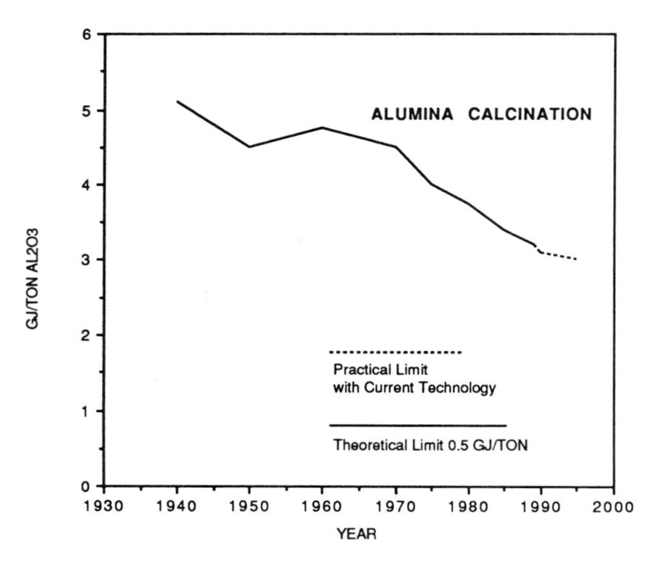

Figure 1. Calcination energy. Source: Alcoa Data.

incremental improvements in energy reduction that are anticipated.

Figure 3 displays the reduction that has occurred in carbon requirements for the Hall-Heroult Process. Some additional improvement in carbon utilization will occur, but the industry is approaching the practical limit. This analysis will assume that carbon utilization levels will continue into the 21st century at essentially the current levels.

Modern plants using these technologies are capable of producing aluminum from bauxite at an electrical energy cost of approximately 13,000 kWh per tonne plus 50 million Btu of fossil fuel energy in the form of coal, gas, oil, coke, and pitch. The US industry average for production of metal using plants that range in age from 10 years to 50 years is approximately 15,000 kWh per tonne for electrolysis and 55 million to 60 million Btu of fossil energy (Atkins, 1973). This includes the energy to produce aluminum oxide from bauxite and molten aluminum from the alumina. The carbon dioxide emissions resulting from aluminum production are directly related to these energy requirements.

A major source of carbon dioxide emissions associated with aluminum production is the power plant used to generate the electricity required for electrolysis. Since primary aluminum production is electricity intensive, smelters are usually located in areas where electric power can be economically generated in large quantities. Initially, aluminum plants were located near hydroelectric facilities, but now a number of smelters obtain power from fossil-fuel fired power stations. In the US between 40 percent and 50 percent of the primary metal is produced by hydroelectric energy. Worldwide, over 50 percent of the aluminum plants use hydroelectric power. In addition, European smelters may utilize some nuclear power sources.

This power mix is important in this analysis because the electric power produced by hydroelectric plants and modern facilities does not result in any carbon dioxide generation, but electric power from a coal-fired power plant results in the production of approximately 1.0 kilogram of carbon dioxide per kWh. Thus, the "average" tonne of aluminum produced by a 50 percent hydropower and 50 percent fossil fuel power mix results in the release of

$$1.0/2\text{kg/kWh} \times 15{,}000\text{kWh/tonne} = 7.5\text{tonnes } CO_2 \quad (1)$$

In addition, the carbon and fuel required in the Bayer and Hall-Heroult Processes (0.44 tonnes of anode carbon plus 1.25 equivalent tonnes of fossil fuel) (Willoughby, 1990) will produce an additional 5.6 tonnes of carbon dioxide. The total carbon dioxide release for the "average" tonne of primary aluminum is thus

$$7.5 + 5.6 = 13.1 \text{ tonnes} \quad (2)$$

Aluminum produced totally from electricity generated by a coal-fired power plant results in 20.6 tonnes of carbon dioxide per tonne. A tonne of aluminum produced totally by hydroelectric power results in 5.6 tonnes of carbon dioxide.

Recycled Aluminum

Once aluminum is produced by the Hall-Heroult Process, the metal is protected from reoxidation by a strong oxide film that forms on all exposed surfaces. However, the metal is available for recycling through a simple melting process, with only small losses. The relatively low melting point of aluminum makes it very cost effective to reprocess the metal.

Aluminum Company of America data show that aluminum scrap can be processed and remelted with 81 kWh of electric power and 4.0 million Btu of fossil energy per tonne of molten metal produced. Thus, recycled (remelted) aluminum requires less than 1 percent of the electric energy and less than 10 percent of the fossil fuel needed to convert bauxite ore to the metal (Atkins, 1973).

Using the same emission factor for carbon dioxide generation from power plants and assuming that the fossil fuel for remelting comes from oil and natural gas, recycled aluminum results in 0.081 tonnes of carbon dioxide from the electric power used and approximately 0.40 tonnes of carbon dioxide from the fossil fuel, or a total of 0.48 tonnes of carbon dioxide per tonne of recycled aluminum. This is less than 4 percent of the carbon dioxide produced when primary aluminum is made from bauxite.

Aluminum recycling obviously has a major impact on the energy use and environmental emissions. For example, in 1989, 61 percent of the aluminum beverage cans produced in the US were recycled (Aluminum Association, 1990). Using the model developed by Atkins (Atkins, 1973), it can be demonstrated that this usage of recycled metal to replace primary aluminum resulted in a 55 percent reduction in electric power requirements and a 33 percent reduction in fossil fuel requirements to maintain the beverage can system, with a similar reduction in carbon dioxide generation. As shown in Table 1, over 1.6 billion pounds of metal in aluminum cans

Table 1. US aluminum can recycling, 1972 to 1989.

Year	Billions of Cans Collected	Pounds Recycled (in millions)
1972	1.2	53
1973	1.5	68
1974	2.3	103
1975	4.1	180
1976	4.9	212
1977	6.6	280
1978	8.0	340
1979	8.5	360
1980	14.8	609
1981	24.9	1017
1982	28.3	1124
1983	29.4	1144
1984	31.9	1226
1985	33.1	1245
1986	33.3	1233
1987	36.6	1335
1988	42.5	1505
1989	49.4	1688

Source: Aluminum Association

was recovered and reused (Aluminum Association, 1990). This is 25 percent more metal than can be produced by the two largest primary production plants in the US. The growth in recycling is expected to continue, contributing to further reductions in energy requirements and carbon dioxide emissions from the "average" tonne of aluminum produced from primary and recycled resources.

Product Life Cycle

In order to determine the full impact of the use of aluminum in a product, the life cycle of the product must be included in the analysis, with appropriate credits for energy savings during use, and recycling after use. A good example where energy credits are important is the use of aluminum in automobiles.

When aluminum is substituted for steel in automobiles, a weight reduction is achieved because of the direct substitution of less dense material. Besides the direct effect of weight reduction on fuel consumption, an additional fuel saving results because the drive train and suspension capabilities can also be reduced without impacting performance as weight is reduced (Cochran, 1986). The energy savings associated with the use of the lighter metal go beyond the savings for the direct substitutions. A previous publication (Aluminum Association, 1980) reported that a typical fuel savings of 290 gallons would occur over the life of a 1979 model automobile which contained 119 pounds of aluminum (typical for that year). This is equivalent to 24.38 liters of fuel saved per kilogram of aluminum (Willoughby, 1990).

Willoughby assumed a specific gravity of 0.85 for gasoline and a carbon content of 84 percent and determined that the equivalent carbon dioxide savings per kilogram of aluminum used is

$$24.38 \times 0.85 \times 0.84 \times 44/12 = 63.8 \text{ kg } CO_2 \quad (3)$$

If the additional assumption is made that the aluminum used in the automobile was produced with the "typical" electric power mix of 50 percent fossil fuel power, 13.1 kilograms of carbon dioxide would have been produced for each kilogram of aluminum used. The savings ratio (kilograms of carbon dioxide saved per kilogram produced) is

$$63.8/13.1 = 4.87 \quad (4)$$

Thus, the use of aluminum in automobiles prevents far more carbon dioxide emissions than it causes.

If a further assumption is made that 25 percent of the aluminum used in automobiles was recycled, the carbon dioxide savings ratio would approach 6. This example becomes more important as aluminum usage in vehicles increases. The aluminum content in automobiles reached 157 pounds per US automobile in 1989 and is expected to grow to 160 pounds in 1991. As the usages of aluminum in automobiles grow, the recycle rate of the metal will also grow, making the 25 percent recycle estimate cited above very conservative.

A similar analysis can be made for aluminum used in railroad cars, aircraft, and high voltage electrical conductor cable (Tribendis, 1984; Aluminum Association, 1980). Table 2, which lists the uses of aluminum in the US market, shows that a significant amount of the aluminum used is in those applications where recycling, fuel savings, weight reduction, and improved conductivity can result in significant energy and environmental benefits over the product life cycle. Complete life cycle analyses should be conducted on all of these major applications to determine the carbon dioxide savings ratio for each use of aluminum and to provide information for those who are considering policy issues associated with material usage in consumer products.

Conclusion

The energy and environmental impacts of aluminum usage in the 21st century can be determined if a full life cycle analysis is conducted. The initial requirements and impacts to produce the primary metal must be integrated with the inputs of recycling and the credits associated with product use. When the full analysis is conducted, the results show that aluminum usage can produce a net energy and carbon dioxide reduction in a variety of applications. The carbon dioxide savings ratio for aluminum usage in automobiles is shown to be 4.9, making aluminum in that application a net reducer of greenhouse gas emissions.

All uses of aluminum benefit from its ease of recycling and the low energy requirements to remelt the metal. This recycling advantage can reduce the carbon dioxide production rate to less than 4 percent of the rate of carbon dioxide production for primary metal.

Full life cycle analyses should be conducted on products to determine their energy use requirements and potential environmental impacts. Anything less than a full life cycle review will produce incomplete infor-

Table 2. Uses of aluminum in the US.

37%	Building and Construction
26%	Packaging
16%	Transportation
12%	Electrical Conductor
5%	Consumer Durables
4%	Industrial Applications

Source: Aluminum Association

mation that may lead decisionmakers to erroneous conclusions.

References

Aluminum Association, *Recycling,* Washington, DC (1990).

Aluminum Association, *Aluminum Recycling, Your Next Assignment,* Washington, DC (1988).

Aluminum Association, *Energy and the Aluminum Industries,* Washington, DC (April 1980).

Aluminum Association, *Aluminum for More Effective Railroad Cars,* Bulletin 66 (January 1980).

Atkins, Patrick R., "Recycling Can Cut Energy Demand Dramatically," *Engineering and Mining Journal* (May 1973).

Berk, Rhea, Howard Lax, William Prast, and Jack Scott, "Energy, Economics and the Environment," in *Aluminum: Profile of the Industry,* published by *Metals Week,* McGraw-Hill, New York, NY (1982).

Cochran, C. N., "Life Cycle Energy Analysis," in *Encyclopedia of Materials Science and Engineering,* Michael B. Bever, editor (1986), Pargon Press, Ltd., Oxford, England, pp. 2543–2545.

Cochran, C. N., "Energy Balance of Aluminum—From Production to Application," *Journal of Metals,* 33: 7 (July 1981), pp. 45–48.

Grjotheim, K., and B. Welch, "Technological Developments for Aluminum Smelting as the Industry Enters the 21st Century," *Journal of Metals,* 41: 11, (November 1989) pp. 12–16.

Tribendis, Joseph J., and Larry E. Carlson, "Light-Weight Coal Car Design: Its Economic Implications," Proceedings of the Spring Conference of Amercian Society for Mechanical Engineers, Chicago, IL (April 1984).

Willoughby, D., "Carbon Dioxide Production and Reduction through the Manufacture and Use of Aluminum," forthcoming.

Minerals, Energy, and the Environment

J. C. Agarwal
Charles River Associates
Boston, MA

Introduction

In 1900, the population of the world was about 1 billion. Today, we have a world population of over 5.5 billion. The ability of the world to support such a large population increase over the last 90 years depends, in great measure, on the increase in mineral wealth and the ability to feed, house, and move people around the world. In addition to the population growth, there has been a significant increase in the global standard of living. In the future, the demand for minerals will continue to escalate as the population grows and the standard of living is more fully equalized throughout the world.

Anything that is extracted, recovered, or mined from the earth and that is mostly inorganic is classified as a mineral. The minerals industries can be separated into four different sectors. First, there are minerals that are not really consumed after extraction and chemical processing, but merely placed in the system for the benefit of mankind. These include most of the metallic ores, such as iron ore, copper ore, and bauxite, which are used for making steel, copper, and aluminum, respectively. These are the major metals that form the backbone of industrial society. Eventually, most of them come back after the products that are made from them, such as ships, bridges, beer cans, electric wiring, cables, and the like, have served their useful life and are recycled. Scrap recycling economics and environmental benefits will be discussed later.

The second class of minerals includes those that are consumed and never recycled. These are normally categorized as industrial minerals. Some of them, such as potash and phosphate, are used to make fertilizers and are essentially lost after their use. Other minerals, such as limestone, are normally made into useful products such as cement, but rarely recycled.

The third category encompasses the energy minerals. The dominant mineral in this category is coal, which is consumed to fulfill various energy requirements around the world. Although uranium is not "consumed" the way coal is consumed through combustion to create carbon dioxide and water vapor, nevertheless minerals containing uranium are also categorized as energy minerals.

The fourth category involves the minor minerals and ores that are required for various applications in the industrial world. Some of them, such as rare earths used in electronics, are useful. Others, such as gold, are not so useful, except as investment vehicles.

When we mine or extract these minerals from the earth, invariably one portion of the solids, ore, is moved, while the rest is waste, or overburden. The earth is moved from the top of the ore so that the ore body can be accessed. The lower-grade ores, which at the time of mining are considered to be uneconomical, are considered waste. Throughout the history of the mining world, the definition of low-grade waste ore has changed. New technology has made possible the recovery of valuable mineral entities in the waste ore, and the demand has increased to support the new way of accessing the mineral entity. A good example of this practice is the copper industry, where the cutoff grade, which is the amount of copper in the ore that would be considered part of the primary ore, has steadily dropped. Today, many mines (especially open-pit mines) process ore that has less than 0.3 percent copper; the same ore would have been considered totally worthless not many years ago. Increasingly, even these wastes are piled into valleys and leached to extract additional amounts of copper by using solution-mining technologies that were unknown just 30 years ago.

The magnitude of the earth and ore moving task is so large that it is interesting to review the whole mining industry. Of nearly 13 billion tonnes of ores and waste mined in the free world (see Table 1), about half is waste material. The task of finding an environmentally safe home for this amount of waste is monumental.

As the overburden is moved away and put into another place to access the ore and transform it into usable products, invariably a void is created by the removal and disturbance of the earth. The management of these voids and of overburden constitutes a challenge that must be met aesthetically and economically so that mineral resources can be exploited without causing the kind of damage that has been done in the past. Exploitation of minerals without *any* insult to the earth is es-

Table 1. Ore and Waste Production by Metal/Mineral (figures exclude Eastern Europe, the Soviet Union, and China).

	Million Tonnes		
Metal/Mineral	Ore	Waste	Ore and Waste
Bauxite	80	25	105
Chromite	9	6	15
Coal	2,300	4,200	6,500
Copper	877	1,381	2,258
Gold	286	388	624
Gypsum	100	30	130
Iron Ore	896	710	1,606
Lead/Zinc	125	75	200
Lime	225	120	345
Manganese	30	15	45
Molybdenum	28	15	43
Nickel	32	16	48
Phosphate	254	365	619
Platinum	30	5	35
Potash	50	10	60
Salt	140	14	154
Silver	25	37	62
Uranium	25	75	100

Source: Charles River Associates, 1990.

sentially impossible. Therefore, the primary requirement when such a disturbance of the earth occurs and huge quantities of solids are moved from one area to another is to manage these solids and the space left behind in an aesthetically acceptable manner. The good news is that the technology exists to make that possible, and mining companies are managing their waste generation more prudently. One of the most promising developments is the use of biological stabilization of the tailing ponds by using plants that are at home in the prevailing environment.

The contaminated runoff water from mining operations should not be allowed to contaminate the other parts of the river and lake systems. In addition, large quantities of water are used to move solids in the form of slurry and to pump the tailings into recovery ponds. This water is recovered, recycled, and stabilized. Most mining companies throughout the world are now decontaminating the process and runoff water by neutralization and other chemical treatment. The use of biological systems to help remove the contamination by taking up these substances in plants shows enormous promise.

All mining has an impact on air quality. It starts with the dusts that are inevitable if the mining is above ground. The control of dust to prevent particulate pollution requires careful planning. Since we are scarring the earth it cannot be completely avoided, but it can be controlled. Next, in the processing of minerals, toxic gases such as sulfur dioxide, NO_x, and carbon monoxide, as well as carbon dioxide, are generated throughout the flowsheet from the mining beneficiation and in the high-temperature processing to make metals. Therefore, it is necessary to treat the offgases to capture as much of these polluting gases as possible and still process the minerals economically into useful products. All of these minerals take energy to extract and use in the final form, and energy consumption always results in some pollution. The question is how much pollution is tolerable concurrent with the growth in industrial activity.

The common pitfall in trying to solve environmental problems in the mineral industry is trying to clean up the effluents while leaving the processes as involved as they are. A better approach, especially when a new operation is being considered, is to design the total system to minimize or eliminate the pollution by careful plant design.

Higher global standards of living will result in higher energy consumption. The energy is consumed not only to produce electricity or move people around or create food, but also in housing people and in creating the infrastructure for modern society. All of these activities need minerals and metals.

Since we need minerals for societal reasons, we must, equally for societal reasons, control the environment in which the production and use of these minerals and metals occur. We should examine both short-term and long-term effects. Societal pressure has changed the way we judge the value and economics of these so-called nonproductive investments in processing minerals and metals. It has been necessary to change that view because we must make use of nature's bounty in a manner that does not destroy lakes, forests, or rivers while we are extracting minerals. Both the easily visible short-term effects, such as smoke or noxious sulfur dioxide fumes, and the long-term effects, which are cumulative and damage the environment for a long time to come, should be addressed.

High-temperature processing of minerals into metals generates some dust, slag, and other solid volatile matter, such as mercury, arsenic, and selenium. It is also possible to release some radioactive material in a more concentrated form than existed before.

Designing benign processes that produce concentrated streams of polluting gases makes it possible to recover and use these gases. Not many years ago, most of the sulfur dioxide from copper production in the western states was discharged without treatment into the atmosphere. Now, most of these copper smelters are recovering this gas and producing sulfuric acid for making fertilizer and for other uses.

Many metals are considered to be injurious and harmful, and indeed they are. Lead is a prime example. Even with the enormous amount of research that has gone on for nearly 100 years, we are still using lead batteries in most cars, buses, and trucks because we have not been able to find a superior chemical substitute.

Energy Consumption in the Minerals Industry

Minerals occur in nature in complex chemical form and are always contaminated with other undesirable minerals. It takes an enormous amount of energy to concentrate or treat the minerals to produce an intermediate product for further chemical processing, which takes additional energy. Once the material has been put into at least a semi-usable form, such as copper, metal, or steel, additional energy is required to fabricate these metals by thermomechanical means into useful products that are then introduced into society. So the consumption of energy in the production of minerals and metals is pervasive, and these operations are generally very energy-intensive.

In the minerals industry, we use energy in three forms: heat, electricity, and chemical energy. The same source, such as coal, can provide all three forms of energy. Coal can provide heat; it can be burned to generate electricity; and it can be used for reduction as a chemical. In any given minerals processing plant, there is always some choice of energy form for producing the end result or product. The choice is generally made in order to save either capital or operating costs. Sometimes these evaluations are systematic, and sometimes they are based on tradition, without much thought for the current economic and environmental requirements.

Table 2 describes various unit operations commonly used in the processing of minerals, the kinds of energy they use, and the interchangeability of these forms of energy. In an examination of this table, two major elements stand out:

- In certain cases, different forms of energy are not interchangeable. For example, electrolysis to recover metals requires electric power, which cannot be replaced.
- Chemical reactions or atmospheres to liberate desired metals or minerals require chemicals and process conditions that will not permit the use of combustion heat to accomplish the desired results. Electric furnaces are used to melt stainless steel, for example, so as not to contaminate and oxidize the charge.

Table 2. Forms of Energy Used in Minerals Processing Unit Operations.

Unit Operation	Energy Form(s) Used
A. Mining	
1. Drilling	Electric or gasoline (diesel)
2. Explosives	Chemicals from oil, natural gas,and coal
3. Movement of ore by trucks, railroad, conveyors, etc.	Gasoline, diesel oil, electric power
B. Crushing and Grinding	Electric power from fossil fuels, nuclear energy, or hydroelectric power
C. Beneficiation	Electric power and chemicals
D. Chemical Processing	
1. Heating	Fossil fuel, coke, electricity
2. Reduction	Carbon, hydrogen, electricity
3. Oxidation	Oxygen, chlorine, sulfur, etc.
4. Neutralization	Chemicals
5. Metal recovery	Electricity, chemicals

Source: Charles River Associates, 1990.

Table 3. Energy Consumption Profile.

Industry	Energy Requirement* 10^6 Btu/ton
Copper	80–100
Nickel (sulfide ore)	200
Zinc	60
Lead	30
Steel	27–30
Nickel (laterite ore)	600
Aluminum	280
Glass	7.4
Lime	6–8
Cement	7.6

*Includes electrical energy at 10,600 Btu/kWh.
Source: Charles River Associates, 1990.

Energy Consumption

Energy consumption in metal production is a major cost item and in some cases, such as aluminum, it is the most important operating cost item. Table 3 gives an energy consumption profile of some of the major industries.

Energy consumption constitutes a major portion of the total cost in these minerals and metals industries. Since the minerals industry is energy-intensive, it is also likely to be capital-intensive.

Energy Loss: Conservation Strategy

In general, the theoretical efficiency of the energy use in these processes ranges from less than 5 percent to 50 percent. The remaining energy must therefore exit the process plants in the following forms:

- Sensible heat in all process and combustion gases leaving the process

- Sensible heat or unrecovered latent heat in cooling water
- Heat losses from process vessels and piping
- Discarded chemical or latent heat in waste streams

Energy in Offgases. Over the years, engineers have succeeded in recovering large quantities of energy from metallurgical plant offgases. Good examples of recovery of fuel value from these gases include:

- Recovery of coke oven gas and its use to underfire coke ovens
- Recovery of blast furnace top gas and its use to fire stoves to preheat blast air
- Recovery of carbon monoxide from electric furnace smelting processes

Despite these successful recoveries, however, large quantities of heat are lost in the form of the sensible heat of flue gases. When the exit temperature of the flue gases approaches approximately 200°C, it is difficult to recover usable energy because the cost of recovery equipment becomes extremely high in proportion to the savings in energy. Therefore, the best ways to conserve energy in this segment of energy losses are the following:

- *Exhaust the minimum quantity of flue gases from the overall system.* This can be achieved by using designs and practices that do not allow unnecessary infiltration of air into the system.
- *Use oxygen-enriched air to minimize the flue gas volume.*

A reduction in offgas volume also decreases the pollution abatement costs for cleaning these dirty gases.

Unrecovered Heat in Cooling Water. In most minerals processing plants, this kind of heat loss is the largest loss and is usually unrecoverable because of low-grade energy contained in the cooling water.

Until recently, large quantities of cooling water have been used for containing high-temperature processing streams and for controlling process operations. But with water pollution abatement regulations, it will become increasingly expensive to continue to use cooling water in this manner because it is expensive to treat the water after use before it can be discharged. New ways of minimizing water usage will therefore have to be devised.

For example, a typical steel plant uses approximately 60,000 gallons of water per ton of steel, but with the design of a cooling system that involves evaporative cooling, it is possible to cut the use to 700 to 800 gallons per ton, resulting in both energy savings and savings in water pollution costs.

Another example is the use of water in the wet grinding of ore, where a large fraction of energy heats the water by a few degrees—a totally unrecoverable transfer of energy to water. If different ways of grinding could be developed that would use less energy, there could be significant savings. The use of water is no longer inexpensive and is extremely wasteful of energy.

Heat Loss from Process Vessels and Piping. Many minerals processing and metallurgical operations are conducted at a high temperature for thermodynamic and chemical reasons. In spite of insulation, heat losses remain high because of

- Indifferent quality of insulation material and techniques
- Insufficient economic incentives

Minimizing these heat losses will require designing systems that have a high production intensity because as production per unit volume goes up, heat losses per unit of production decrease. For example, when oxygen enrichment in normal air from 21 percent to approximately 30 percent was adopted in ironmaking blast furnaces, production increased by nearly 40 to 50 percent and heat losses and the volume of dirty gases to be cleaned up decreased.

Energy Loss in Discarded Process Streams. The minerals processing industry has long realized the merit of beneficiating minerals as early in the flowsheet as possible. There is little benefit in carrying an unnecessarily large amount of gangue material and then discarding it as molten slag. Until now it has not always been possible to increase the grade of a concentrate without sacrificing yield. In the future, we will need to design beneficiation processes that will give high-grade concentrates and high yields. When sulfide mineral concentrates are oxidized to recover metallic values, sulfur dioxide can be easily converted to sulfuric acid if the concentration of sulfur dioxide in the process stream is about 2 percent.

Future Efforts

It is imperative that technical efforts be directed toward developing processes that challenge our long-held beliefs and decrease waste stream volumes to minimize environmental damage.

It is possible both to achieve a lower pollution abatement cost in a given plant and to consume less energy by decreasing the quantity of offgases, cooling water, and waste solids. To realize benefits from both energy conservation and pollution abatement will require an integrated approach to plant design. In some cases, retrofitting may not achieve the desired results at all, but merely result in transferring the problem from one part of the plant to another.

The first part of this paper focused on the production of virgin minerals and metals from four primary sources. However, most metals are not consumed as such, but only placed in the industrial system. For example, once oil is burned, it is gone forever. Metals, on the other hand, eventually come back in the form of scrap when the original use is no longer needed.

In the past, when production of primary metals was dominated by discovery of "good mines," the producers of primary metals did not feel that it was in their economic interest to become major players in the recycling industry because they felt that they would be competing against their own primary production. However, in their business analysis they overlooked the basic economic and engineering fact that scrap recycling is generally more economical than the production of virgin metals, from both a capital and an operating cost point of view. For example, energy consumption for recycling aluminum is only 5 percent of that required for the production of aluminum from bauxite ore.

The savings in other metals, such as steel and copper, are also economically important even though they are not as dramatic. The energy consumption and, by inference, capital requirements are about one third of those for producing these metals from virgin ore sources. This topic will be discussed in more detail later.

Economics of Recycling in the Metals Industry

It is useful to describe the types of scrap and how scrap type affects recycling ability. There are three major categories of scrap:

	Economic Value
• Concentrated Scrap	High
• Diluted Streams	Low
• Contaminated or Mixed Streams	Low

For these different classes of streams, there are two different reasons for recycling. Low-value streams are those whose economic worth is less than $5 per ton or $5 per thousand gallons in 1990 dollars. These low-value streams are processed essentially for environmental reasons, because it is difficult to make a profit on these streams. High-value streams, on the other hand, are processed for profit. It is easy to make a good investment in processing these streams and still make adequate returns. High-value streams could be generated by large quantities of low-price metals, or they could be generated by concentrated, but small quantities of high-priced metal.

Concentrated Scrap

In the following four major categories of scrap, the metallic components are in a concentrated form:

- Obsolete scrap
- Home scrap, secondaries
- Prompt scrap—from fabricators
- Discarded products, i.e., automobiles, batteries, cans, and electrical wiring

These four categories of concentrated scrap have different time horizons. Obsolete scrap can be purchased immediately. Home scrap is also more or less immediately available. Prompt scrap takes a little longer, and discarded products take the longest.

Diluted Streams

In this category of scrap or secondary processing, metals are in the diluted form. Different and difficult processing technologies must be used to recover the metals. In the alternative, these streams must be made environmentally safe. These streams are

- Electronic scrap
- Slag
- Plating solutions
- Ash
- Spent catalysts
- Tailings
- Most dusts
- Pickle liquors

For many of these streams, the best solution is to minimize the amounts by proper process design and operation.

Contaminated or Mixed Streams

The streams in this form are

- Slimes from electrolytic operations
- Catalysts containing two or more metals
- Chemical process solutions and waste streams containing two or more metals or contaminated with other materials

The processing of these streams can be economical if proper process technologies are employed. Generally it is necessary to combine several sources to create an economical operation.

Process Economics of Metals Recycling

The US metallic scrap industry is large. Table 4 shows various commodities that are now recovered and recycled.

Steel scrap is a multibillion-dollar industry, whereas minor metals like titanium and gallium are only multimillion-dollar industries on a national basis. Steel and copper, which are rather mature metals, generate a

Table 4. Recovered and Recycled Commodities in the United States.

Commodity	Annual Amount (in Millions)	Growth (% per Year)	Revenue ($ × 10^6)	Added Value (R × 10^6)
Steel	30–40 tons	2.0	9–12,000	6–8,000
Gold	2.6–3.0 troy oz.	9.0	1,200	200
Aluminum	1.5–2 tons	2.5	1,200	200
Copper	1.4–1.5 tons	1.5	700	150
Silver	55–65 troy oz.	5.0	670	100
Superalloys	20–25 lb.	15.0	200	30
Germanium	0.19–0.2 lb.	6.0	96	10
Titanium	20–22 lb.	10.0	55	10
Gallium	0.018–0.02 lb.	50.0	6	3

Source: Charles River Associates, 1990.

higher amount of total scrap per ton of metal shipped than aluminum, which is a relatively new metal, and the amount of old scrap is relatively small. On the other hand, the amount of prompt scrap for aluminum is much greater because a greater portion of aluminum has gone into the production of beverage cans, which come back into the production cycle within a few months.

Table 5 shows the capital costs for producing metals from scrap and virgin ores for three major metal industries: steel, aluminum, and copper. Steel has the lowest capital cost, for the following reasons:

- The economies of scale because of the high tonnage of steel production
- The high concentration of iron in the iron ores

The scrap processor must acquire the scrap at about 45 to 70 percent of the prevailing price of the slabs or ingots. However, the capital requirements are low, and therefore, even though total operating costs may be the same or even higher, the profitability is generally higher. For example, in the steel industry, the capital cost difference of $0.10 per pound or $200 per ton can be translated into a lower capital recovery charge of about $0.03 per pound or $60 per ton. In addition, a small *amount* of capital is at risk. When these capital charges are taken into account, the total costs become $0.16 to $0.18 per pound of steel from virgin ores, versus $0.13 to $0.15 per pound of steel from scrap. Therefore it is easy to see why a scrap recycling operation is in a better competitive position than the integrated producer on a return-on-investment basis. There are situations in which the primary producers have advantages, such as high-grade ore (for copper) or low-cost electric power (for aluminum). However, the overall economic thrust is obvious when the capital at risk is taken into account.

Table 5. Capital and Operating Costs for Producing Metal from Scrap and Virgin Ores.

	Capital Costs		
	Scrap ($/lb)	Virgin ($/lb)	Ratio
Steel	0.10–0.12	0.20–0.25	2:1
Aluminum	0.2–0.3	2–3	10:1
Copper	0.2–0.3	2–4	12:1

Source: Charles River Associates, 1990.

Capital and operating costs are much lower for recycling scrap than those for production from virgin ore (see Table 6). Since the metals have not been consumed as such, it is important for recycling to play a major role in any industrial society, where a great deal of metal is put into the system and eventually must return to the system.

Industry has made many mistakes in the past about the processing of minerals and metals. It does not have a good public image. People think about mining, minerals extraction, and metals production as dirty businesses. Indeed they are, but polluting the whole environment around the mine and plant is unnecessary. The primary purpose of using these minerals and metals in the service of society can be fulfilled without assaulting other parts of the biosystem. Mining and minerals and metals companies in the past did not understand this to be a requirement for doing business. The cost associated with proper control of the environment is a necessary one.

Protecting the environment is similar to protecting the workers in any given industrial plant. Not long ago, personnel safety in an industrial plant was not a priority. However, industry then learned that a safe plant is a better plant. It is also a more economical plant, from an overall corporate and societal point of view. Pollution control too has been misunderstood in the past. If you want a safe plant to manufacture goods, you also want an environmentally benign plant. The right of human

Table 6. Direct Operating Costs for Producing Metals as First Solid Product.

	From Virgin Ores (¢/lb)	From Scrap (1990 Prices)(¢/lb)		
		Cost of Scrap	Processing Costs	Total
Steel	9–11	5–6	5–6	10–12
Aluminum	40–60	50–60	18–22	68–82
Copper	45–60	70–90	16–20	86–110

Source: Charles River Associates, 1990.

beings to a safe plant and the ability of a plant to be a good neighbor are fundamental to industrial activity.

We still are ignorant about a number of long-term health effects of exposure to pollution, but nobody wants to breathe polluted air. Forty years ago, it was commonplace for people in movies to be smoking. It was a socially acceptable and desirable habit. But that has changed. We now know that there are some long-term health effects, and the American population has responded by changing its habits. If we can change such a deep-seated social habit by reducing or even eliminating smoking in public places, and by cutting down on smoking, it is equally possible to change habits in the industrial sector.

Finally, it is not possible to have high standards of living throughout the world, especially an increasing standard of living in the underdeveloped world, without the use of minerals and metals. Most of the countries in Africa, Latin America, and Asia, unlike those in North America, do not yet have strong infrastructures. Electrical wiring requires metals such as copper, aluminum, tungsten, and steel to support the wires. Great quantities of these metals and minerals will be required to raise the global standard of living. What we have to do is make it a requirement, just as safe plants are a requirement, to build and operate in the minerals and metals industry with the mark of good citizenship. Engineering schools like MIT should teach the fundamentals of designing such plants. Society, through its elected representatives, must demand nonpolluting mines and plants.

Energy and Environmental Considerations for the Cement Industry

Stewart W. Tresouthick
Alex Mishulovich
Chemical/Physical Research Department
Construction Technology Laboratories, Inc.
Skokie, IL

Abstract

Portland cement is the critical ingredient in concrete, the material most used in construction throughout the world. Although the portland cement manufacturing process is one of the most energy intensive, considerable progress has been made in the last 10 to 15 years in decreasing energy requirements. Carbon dioxide is one of the main effluents of cement making that may impact the environment. Carbon dioxide emissions and energy requirements are rather high per ton of cement produced, but are low per ton of concrete, the general end use for portland cement. The use of cement and concrete products should be encouraged to help reduce world energy consumption and environmental stress.

Introduction

Portland cement is a product that everyone uses but few know anything about. Most people confuse cement and concrete, and many believe that portland cement is a brand name. Even the media insist on saying, "The cement dam collapsed in the flood" or "The cement bridge buckled during the earthquake." If dams and bridges were made only of cement, a dry powder, of course they would collapse, but when cement is added to fine and coarse aggregate and water to make concrete, we have a different story. Concrete dams, highways, foundations, pipes, blocks—everywhere we look, we see concrete as the construction material of the modern world, and portland cement is one of its ingredients. Portland cement is by far the most used of the family of hydraulic cements. A hydraulic cement is a chemical powder which, upon addition of water, becomes a hard, strong mass and acts as the "glue" that holds concrete together. Portland cement is a kind of hydraulic cement, not a brand name. People have complained about the giant "Portland Cement Company," which has a stranglehold on the cement business worldwide and should be stopped! The truth is that no individual cement company holds more than about 10 percent of the business in the United States.

The cement industry is very capital intensive. A new plant costs about $250 per annual ton of capacity, with many plants having yearly production of over a million tons. The manufacturing process requires more energy input than almost any major basic industry, and the amount of product that the industry turns out is enormous. In the US, for example, the industry capacity is about 85 million tons of cement per year made in about 130 plants (1). This is some 7 percent of the world's production (2).

Cement is used mostly in concrete. Figure 1 shows a breakdown of areas of usage while Figure 2 details usage by general structure type (3).

The Portland Cement Manufacturing Process

To make portland cement, at least four chemical elements are needed: calcium, silicon, aluminum, and iron. These are derived from naturally occurring rocks and minerals such as limestone, clays, shales, and iron ores. Also, secondary materials can be used, such as blast

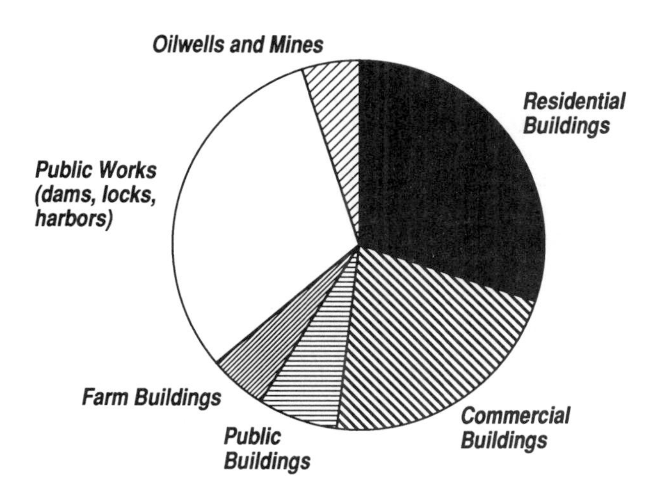

Figure 1. Cement use by structure type.

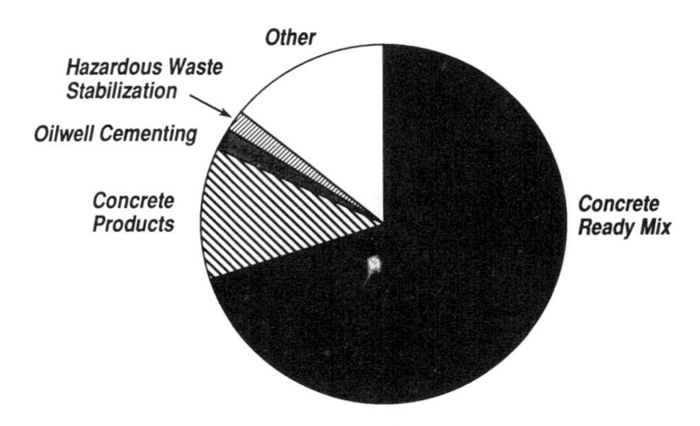

Figure 2. Portland cement usage.

furnace slags, fly ash, aluminum dross, and certain mine tailings. Figure 3 depicts a flowsheet of a generic dry process, preheater cement plant and shows the processing of the raw materials to the finished product. These materials are carefully chemically balanced and proportioned, ground to a very fine powder, blended and homogenized, and fed into a pyroprocessing system. The pyroprocessing system is concentrated on the rotary kiln, which can be arranged in a number of configurations depending on the age of the plant, the nature of the raw materials, and the choice of fuel. The rotary kiln is the high temperature reactor that transforms the starting materials into complex silicates and aluminates at temperatures up to 1500°C. The product of the kiln is an intermediate product called portland cement clinker. The process relies on close control of temperature rate increases and decreases to produce not only the proper chemical compounds or phases, but also their proper crystal structures. The occurrence of several minor elements in the raw materials and fuel greatly complicates the product mineralogy and the process and must be dealt with and controlled.

Clinker produced by the pyroprocessing system is ground with about 5 weight percent of gypsum to a powder, most of which passes a 44 micrometer sieve—a sieve that will hold water. Particle size distribution control is exercised by use of classifiers and complex grinding circuits. Chemical additions made to the cement when grinding include grinding aids, flowability agents, air entraining compounds, and other organic and inorganic materials as required by the customer, specifications, and end use. Almost all cement produced in the US conforms to American Society for Testing and Materials Specification for Portland Cement (ASTM C 150). Most other countries around the world make cement to some standard specification that assures a measure of uniformity and performance that can be relied upon by the user. Cement is shipped mostly in bulk quantities in the US by railroad hopper car, truck bulk carrier, river barge, and oceangoing vessel or barge. Less than 10 percent is shipped in bags, generally going to local building suppliers and hardware stores.

Energy Usage in the Cement Process

In 1985, the US cement industry consumed about 0.38 quads (including electrical losses), accounting for about 1.3 percent of the energy consumed by all US industry and 0.5 percent consumed by the whole nation (4). This is considerably less than the industry used to consume. Figure 4 shows how total energy use and fossil fuel energy has decreased over the past years. However, Figure 4 also shows an increase in electrical energy use. This is due to the installation of more environmental controls and the increasing use of the preheater pyroprocessing system. While the preheater system can reduce specific fossil fuel consumption by almost half over older kiln systems, it nevertheless requires very large draft fans working at high pressure drops. Another reason for a growing use of electrical energy in the cement industry is that the marketplace has continued to demand higher strength cements, putting an extra energy burden on clinker grinding to produce finer cements.

Since cement manufacturing is a multiple step process, it will be helpful to see the extent of energy usage for each major part of the process. Table 1 outlines this for the three major types of cement making processes. As may be seen, pyroprocessing uses the most energy; grinding uses the second most. Table 2 shows the breakdown of the use of different types of energy in the cement manufacturing process, which can be split roughly between 30 percent electrical power and 70 percent fossil or secondary fuels, all on a common Btu basis (5).

Since grinding and clinker production are the two major energy consumers in cement manufacture, it is important to say a few words about the efficiencies obtained by these operations. Grinding is usually carried out in ball mills, although vertical roller mills are now important for raw material grinding in the dry process. Regardless of which type of grinding unit is used, however, grinding efficiency is very low, probably between 2 and 5 percent. This means that only a small amount of the energy put into the system actually accomplishes production of new surface area; the rest of the energy goes into heat, noise, and vibration. This is not unique to the cement industry. All industry relying on comminution to produce fine particles uses the same or similar types of equipment having the same range of efficiencies. Comminution in the cement industry accounts for about 0.06 quads annually, most of it going to produce only heat, noise, and vibration. Efficiencies for pyropro-

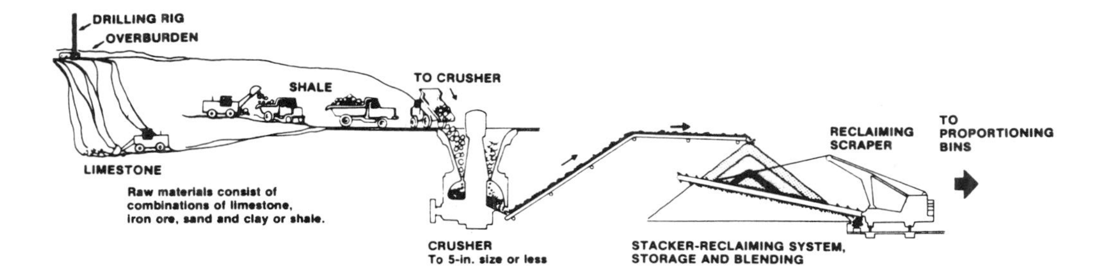

1. Quarrying and blending of raw materials.

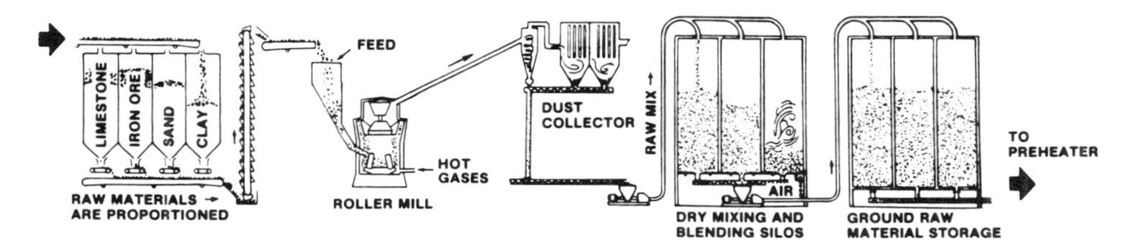

2. Proportioning and fine grinding of raw materials.

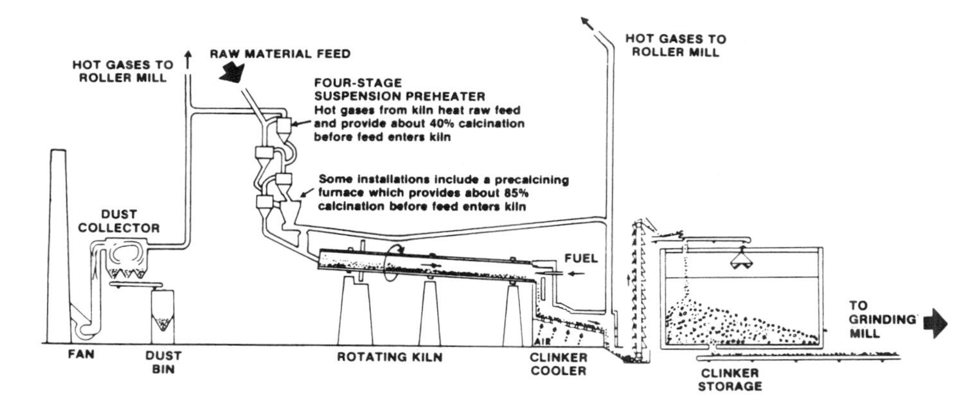

3. Kiln system. Preheating; burning; cooling and clinker storage.

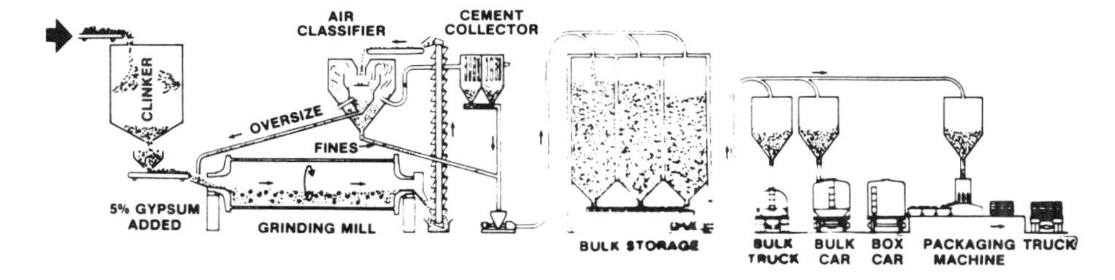

4. Finish grinding and shipping.

Figure 3. Steps in the manufacture of portland cement by the dry process using a preheater.

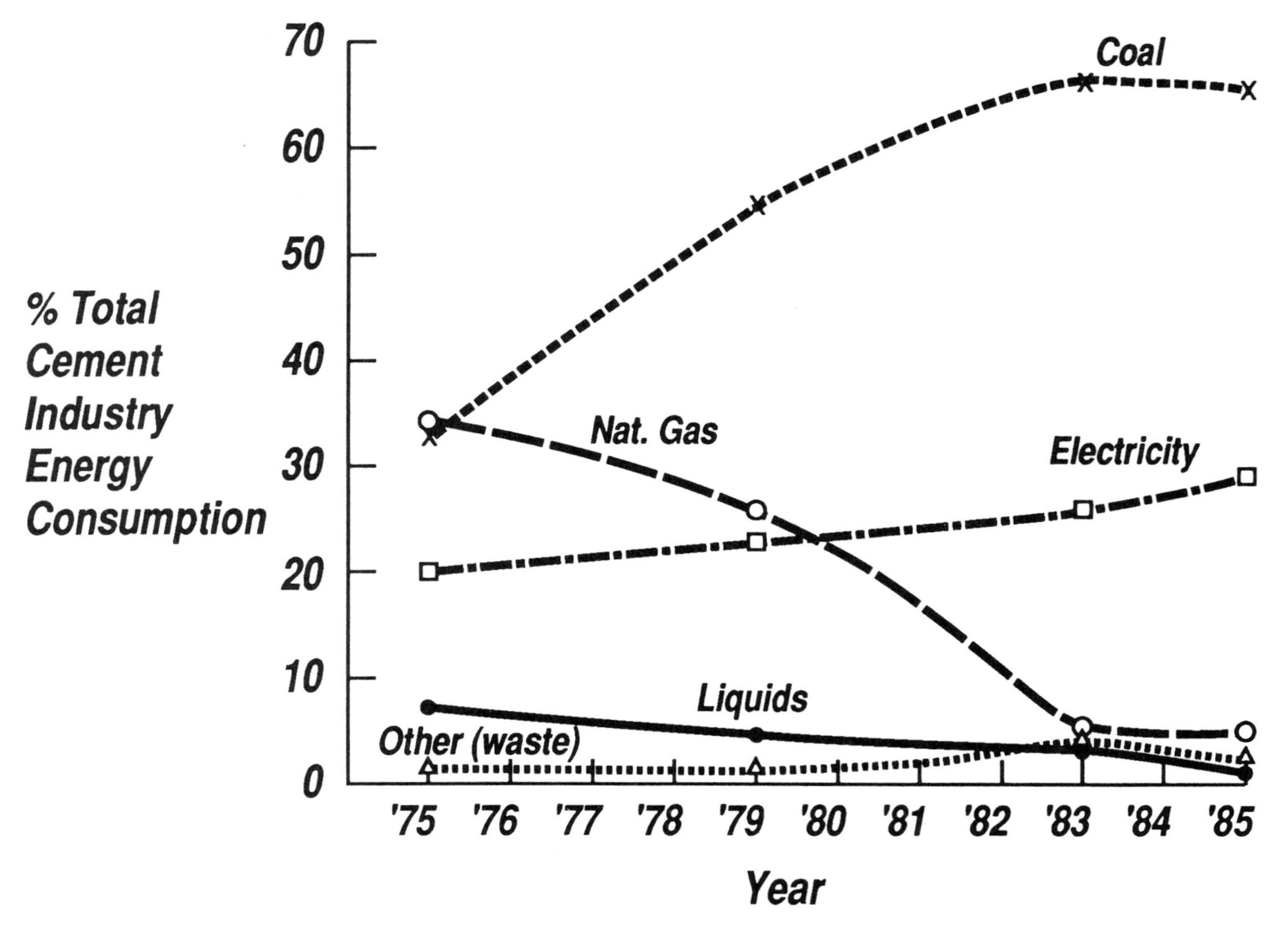

Figure 4. Energy consumption by the cement industry.

Table 1. Specific energy consumption.

Process Steps	Wet Process, %	Dry Process, %	Preheater Process, %
Crushing	0.9	1.0	1.3
Proportioning	0.3	0.4	0.5
Drying	—	5.1	—
Raw Grinding	4.3	5.3	6.8
Blending	0.1	0.1	0.1
Pyroprocessing	86.5	79.6	80.4
Finish Grinding	7.9	8.5	10.9
	100.0	100.0	100.0

Table 2. Cement energy source.

Energy Source	%
Electricity	29
Natural Gas	3
Liquids	1
Coal	64
Other	3
	100

cessing are not as bad as for comminution, but they are not good. The net theoretical heat requirement to produce a short ton of clinker is about 1.5 million Btu. The range of actual specific heat consumption is anywhere from close to 6 million Btu to about 2.8 million Btu per ton. These figures do not include electrical energy but refer only to fuel firing. Therefore, the efficiencies range from about 54 percent to 25 percent. Unless there is a real breakthrough in cement clinker pyroprocessing technology, we seem to have reached a maximum efficiency of 50 to 55 percent.

The US cement industry has made some significant progress, however. Figure 5 shows the gains in overall efficiencies for the three major stages of cement manufacture between 1974 and 1985, representing a 25 percent reduction in overall energy use. However, this required shutting down 43 plants and building almost the equivalent capacity at existing or new locations as new or modernized facilities. The number of kilns decreased from 414 to 253, and the capacity of the average kiln rose from 220,000 short tons per year to 336,000 short tons per year (6). Many kilns today produce one million or more tons annually, and these kilns and new plants are the efficient operations.

Opportunities for Increased Efficiencies

A number of developments in the cement industry in the last few years have proven themselves as methods for increasing efficiencies in comminution and pyroprocessing. These innovations are being installed and operated by a growing number of cement plants both here and abroad.

In the field of comminution in cement plants, several new pieces of equipment and strategies have achieved considerable decreases in specific power consumption for clinker grinding. Table 3 lists some of these and their potential energy savings. The high efficiency classifier has been installed in a growing number of plants, and the high pressure roller press is now working in several plants on clinker grinding. When used together, power savings of up to 40 percent have been reported. Work is being done on developing similar strategies for raw grinding as well, and savings in energy will no doubt be available in this area in the near future. Unfortunately, while a few plants have taken steps to install as much energy saving equipment as can be had at this time, many cement plants have not been

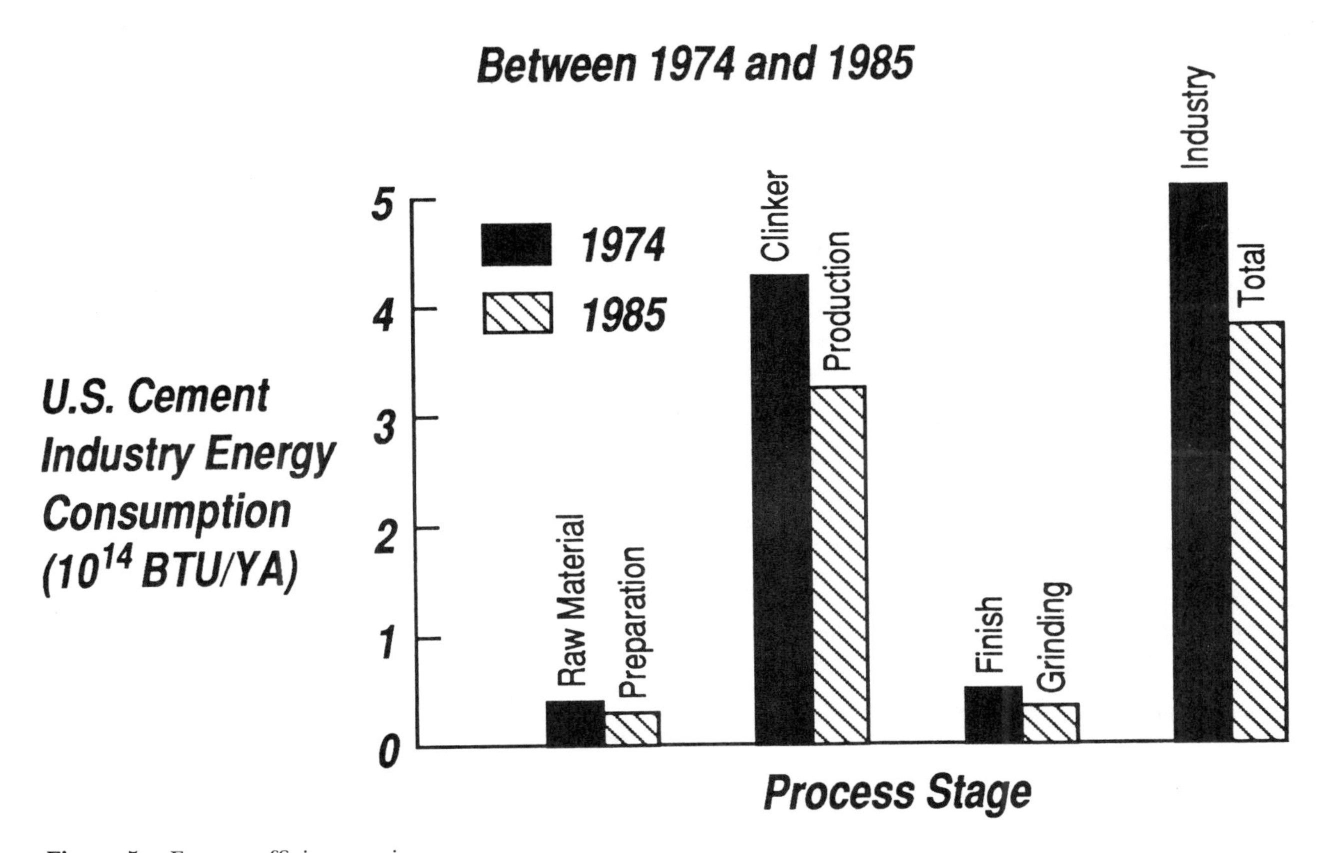

Figure 5. Energy efficiency gains.

Table 3. Opportunities for grinding.

	Average Potential Energy Savings, %
High Efficiency Classifiers	15
Roll Press	20
Controlled Particle Size Distribution	27
Advanced Mill Internals	5
Separate Grinding of Components	5–10
Computer Control	15

able to do so because of high capital requirements. However, as electrical energy costs continue to rise, the economic payback may look better.

Other improvements in grinding in the cement industry have been making themselves felt, although these have not produced large increases in efficiencies. Improvements in grinding aids, grinding media, mill liners, and the use of more complex grinding circuits have made smaller but significant gains. Computer control of the grinding circuits, which has the double effect of increasing efficiency and improving uniformity of product, is also being used in some plants.

There has been a drastic increase in required fineness of cement in the last 10 years. The fineness, measured in Blaine specific surface area, has increased by about 30 percent, requiring close to 50 percent more specific grinding power input. The new energy-saving approaches to grinding mentioned above have helped to partially contain this increase.

Construction Technology Laboratories, working under a Department of Energy (DOE) contract, successfully completed research that developed an optimum particle size distribution for cement that required the least amount of energy to grind and at the same time produced a superior cement (7). This technology has been partly put in place with the advent of high efficiency classifiers, but there seem to be more savings available by using the full technology. This will probably come about slowly since equipment and operating changes must be carried out. Work is going on to develop the use of the roller press as the main grinding apparatus for producing portland cement, with the ball mill being used only as a deagglomerator. Similar development is being pursued for raw material grinding. At this time, totally new concepts in comminution do not appear to be available.

In the pyroprocessing section of the cement plant, the principal energy efficiency gains have been made with preheaters and precalciners. This technology carries out heating and calcining of raw feed using highly efficient countercurrent gas-to-solids heat exchangers rather than the very inefficient rotary kiln. While up to 50 percent reduction of fossil energy input has been attained, capital cost has been enormous and electrical energy usage has increased.

Other new and innovative technologies are being used, studied, or contemplated to decrease energy usage in the pyroprocessing system. Table 4 lists potential energy savings from several current and future developments. For instance, many cement plants are burning waste-derived fuel, which not only conserves energy reserves, but also helps clean up the environment. Computers are now widely used to control and optimize the process, with mixed success. New firing systems have shown improvement and new preheater cyclone designs are lowering the pressure drop across the system, giving hope that electrical energy requirements can be somewhat lowered.

Further improvement in pyroprocessing can come about in two major ways. One is to eliminate the rotary kiln and use a stationary reactor to produce clinker. A number of designs and patents have been put forward over the years; as a matter of fact, clinkering was originally carried out in shaft kilns. A dynamic reactor was tried as early as the 1920s, with limited success, and at least two innovative designs are now being studied that have the potential to lower fuel and electrical power usage by 10 to 30 percent. At this time, however, none of these is developed.

Another approach to saving energy in making clinker is to change its mineral structure and makeup. There are hydraulic compounds that require lower processing temperatures. Ordinarily, these minerals do not have early strength characteristics, but some progress is being made in the US and around the world to solve this problem. We can also add more secondary materials, such as blast furnace slags and fly ashes, to the cement. Many countries outside the US are doing this as a matter of course. However, in this country there are certain problems associated with this, not the least of which is marketing practice. In the US, practically no fly ash or slag is added to cement at the cement plant; instead, it

Table 4. Opportunities for pyroprocess.

	Average Potential Energy Savings, %
Computer Control	3–10
Fluidized-Bed Reactor	10–30
Relaxed Alkali Specification	2–4
Low Pressure-Drop Preheaters	5
Advanced Sensors	2–5
Advanced Preheater/Precalciner Kilns	5–10
Use of Waste Combustibles Fuels	5–50
New Mineralogical Content of Clinker	5–30

is added at the concrete ready mix plant to about 20 to 25 percent of the concrete produced. In Europe, on the other hand, almost all of the cement produced has mineral additives interground with the cement at the cement plant, so that almost all concrete used in Europe contains these materials. In the US, a free competitive market dictates who puts fly ash or slag into concrete, and this pull between cement producer, concrete producer, and user probably does not produce the highest energy savings obtainable by intergrinding mineral additives directly with the cement.

There is also the possibility that certain naturally occurring minerals and rocks or secondary materials can be treated chemically and thermally to provide hydraulic cements that may need very little fossil energy. A number of these ideas have been and continue to be investigated. Not only is it possible that naturally occurring materials can be treated to make them hydraulic, but it is also likely that certain waste materials can be combined and processed to produce hydraulic cements with specialized uses. Such technology could lead to very low energy input materials.

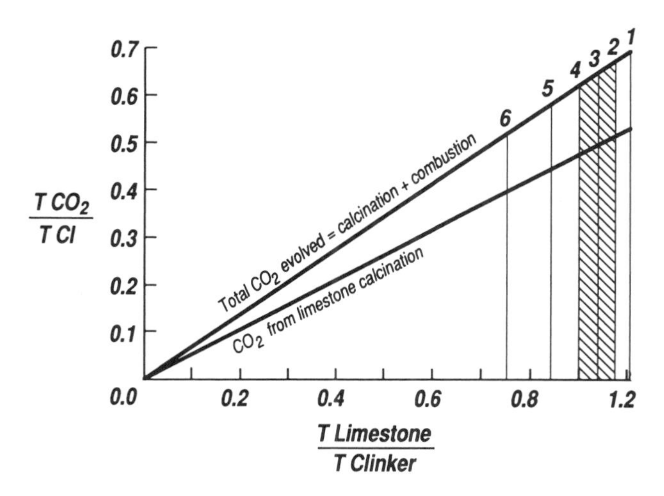

Figure 6. Carbon dioxide produced by the kiln process, expressed in tons of CO_2 per ton of clinker as a function of the limestone to clinker ratio.

Environmental Considerations

The production of portland cement contributes carbon dioxide to the atmosphere in a number of ways. Use of electricity requires the burning of fossil fuel unless electricity is supplied from nuclear or hydroelectric power plants. In the pyroprocess, the coal that is burned produces carbon dioxide. Also, about 80 percent of the feed to the kiln is limestone, which contains carbon dioxide tied up as calcium carbonate. The high temperature used to calcine the limestone liberates this carbon dioxide. Both sources of carbon dioxide are combined as part of the stack effluent. The other constituents of stack gas are very small amounts of particulates (mostly limestone and clay), water vapor, a small amount of NO_x and SO_x, nitrogen, and perhaps a very small amount of carbon monoxide.

The amount of carbon dioxide produced by the kiln process from a given amount of limestone in a selection of different raw mixes has been calculated. Figure 6 shows the amounts produced from coal burning and from calcination of limestone. Using figures for a typical portland cement, the total amount of carbon dioxide emitted by the US cement industry is about 50 million tons per year. This figure can be extrapolated to about 700 million tons of carbon dioxide for the world cement industry.

We must, however, keep this figure in perspective. The amount of concrete and concrete products used in the United States is about 700 million tons per year, which accounts for only about 0.08 tons of process carbon dioxide per ton of in-place concrete, excluding carbon dioxide emitted in producing steel for reinforcing bars, truck transportation, etc. This is a very low figure. Furthermore, cement plants have emissions and byproducts that are generally judged to have relatively low impact on the environment.

Figure 6 also shows other approaches to the manufacture of clinker or substitutes for portland clinker. For instance, while points 2, 3, and 4 represent a normal range for portland clinker, points 5 and 6 represent experimental low lime raw mix designs. Also at point 6 is a raw mix design based on using specific slags as part of the raw material design. Point 7 [not shown] indicates no carbon dioxide production, is based on a yet-to-be developed technology involving naturally occurring minerals. It is clear that production of carbon dioxide, as well as other stack gas constituents, could be lowered as more research and development takes place. Assuming the cement industry could achieve and put in place a technology equivalent to point 5 on the figure, carbon dioxide effluents might be decreased by 50 million tons worldwide. Such a move could also represent a significant decrease in energy usage. Research along these lines has been going on for some time and some progress is currently being made.

It must be remembered, however, that cement is very seldom used by itself. Almost all cement is used in concrete or concrete products that are the end-use materials of the construction world. Concrete has the lowest total energy requirements of all generally used materials of construction other than stone. Table 5 shows relative energy requirements for some construction materials (8). In general, the higher the energy needed to produce a material, the greater the production of carbon dioxide and pollutants. While cement energy usage is high, it

Table 5. Relative energy requirements.

Material	Relative energy
Primary Extruded Aluminum	184
Plastics	92
Copper	37
Ceramics, Vitrified Clay	17
Steel	12
Advanced Cement Products	12
Glass	7
Concrete	1

makes up only about 10 to 13 percent of concrete by weight, and the other ingredients of concrete have very low energy requirements.

Conclusions

For construction materials production, emphasis on the use of concrete, concrete products, and portland cement-based products could lead to lower world energy usage and less environmental stress, particularly if energy requirements for portland cement can be reduced even further than they have been in the last several years. In-place concrete requires a relatively small amount of energy and produces relatively small amounts of carbon dioxide and pollutants. Continued research and development probably can reduce these further as new cement-making technology is incorporated and new cement and concrete products are conceived and brought to market. Civilization started out in the Stone Age and now, with new approaches in cement and concrete, we are entering a New Stone Age to the benefit of us all.

References

1. Portland Cement Association (PCA), *US and Canadian Portland Cement Industry: Plant Information Summary* (December 31, 1988).
2. Energetics, Inc., *The US Cement Industry: An Energy Perspective* (March 1988).
3. Portland Cement Association (PCA), *United States Cement Industry Fact Sheet,* 7th ed. (December 1988).
4. Energetics, Inc., *The US Cement Industry: An Energy Perspective* (March 1988).
5. Weiss, S. J., and Gartner, E. M., *An Expanded Energy Survey of the U. S. Cement Industry,* PCA Report (May 1984).
6. Portland Cement Association (PCA), *US and Canadian Portland Cement Industry: Plant Information Summary* (December 31, 1988).
7. Weiss, S. J., and Tresouthick, S. W., *Energy Savings by Improved Control of Finish Grinding Process in Cement Manufacture,* Final Report to US DOE, DE-FC07-81 CS40419.
8. Gartner, E. M., and Tresouthick, S. W., *High Tensile Cement Pastes as a Low Energy Substitute for Metals, Plastics, Ceramics, and Wood;* Final Report to US DOE, DE-FC07-81 CS40419.A002.

Glass Manufacturing—
Status, Trends, and Process Technology Development

G.L. Ridderbusch
Process Research
Gas Research Institute
Chicago, IL

Abstract

Glass manufacturing is one of the most energy intensive industries in the United States, consuming about 250 trillion BTU per year. The pace and effects of change during the last decade have been dramatic. New technology has permitted modifications to the methods of glass manufacturing and the resulting products have greatly expanded the number of uses for glass. The industry has restructured and consolidated, leaving half as many independent manufacturers today as there were 15 years ago. All sectors of the glass industry have successfully improved process thermal efficiencies and have lowered overall energy consumption. Aggressive change can be expected to continue. Technology is currently under development to make further improvements to glass furnace technology and meet the challenge of lowering process pollutant emissions.

Introduction

The glass industry has changed dramatically during the 20th century. New process and materials technologies have permitted modifications to the methods of glass manufacturing that have not only responded to increased demand, but have yielded a variety of new applications for glass products. Until the 1980s the major portion of glass research and development (R&D) was directed at advancing forming methods and machines, improving quality control techniques, automating process control, and developing new glass types and applications (Tooley, 1989). During this time period less R&D was devoted to improving the heart of the process, the glass melting furnace. Since the fossil-fired regenerative glass melting furnace was introduced in the late 1800s, improvements to this technology have been for the most part evolutionary. However, during the 1980s, with competitive pressure from alternative packaging materials, industry restructuring and consolidation, and increasingly stringent environmental regulations, intensified attention has been placed on advancing glass melting technology. Glass manufacturers, engineering and construction companies, the Gas Research Institute, and other R&D sponsors, including the Department of Energy, are all playing important roles in advancing glass melting furnace technology. This paper presents an overview of the glass industry and glass melting technologies. After reviewing glass industry market conditions that are driving technology development, and quickly discussing conventional furnace technology, several recent development projects that promise to benefit the glass manufacturing sector will be summarized.

Industry Structure

The glass industry is a mature and relatively stable industry. Annual sales of primary glass products exceed $15 billion and annual production exceeds 20 million tons. The glass industry encompasses a wide variety of process technologies and end products. For the purposes of data collection and analysis, the United States Census Bureau classifies the glass industry into four major sectors. The first sector, flat glass (SIC 3211), includes glass for buildings and automobiles, and produced 3.9 million tons in 1986. Container glass (SIC 3221) is the second and largest sector, 11.8 million tons in 1986, whose major products are bottles and jars for food and beverage products. The third sector, pressed and blown glass (SIC 3229), contains the greatest variety of products such as housewares, light bulbs, television picture tubes, glass tubing, and optical fibers, but only accounted for 13 percent of 1986 domestic production, or 2.6 million tons. The fourth sector, glass fiber insulation, also known as mineral wool (SIC 3296), produced 1.7 million tons in 1986 (Darrow, 1989). Due to recent industry consolidation, there are currently only three or four major manufacturers in each sector that account for over 75 percent of production (Figure 1).

Manufacturing facilities are regionally dispersed throughout most of the United States (Figure 2). Site selection for glass manufacturing plants is a complex decision based on many factors. Energy intensive flat glass manufacturing plants were located principally near sources of energy. In the container industry, where transportation costs strongly affect competitive pricing,

SIC	SEGMENT	MAJOR MANUFACTURERS	PRODUCTS
3211	Flat Glass	AFG, Ford, LOF, PPG, Guardian	Automotive Architectural
3221	Containers	Anchor, Ball-Incon, Owens-Brockway, Triangle	Beverage, food
3229	Pressed & Blown	Corning, Libbey, GTE many small manufacturers	Lighting, novelty Electronics
3296	Fiberglass	CertainTeed, Manville, Owens-Corning, PPG	Insulation fiber Textile fiber

Figure 1. US manufacturers and products.

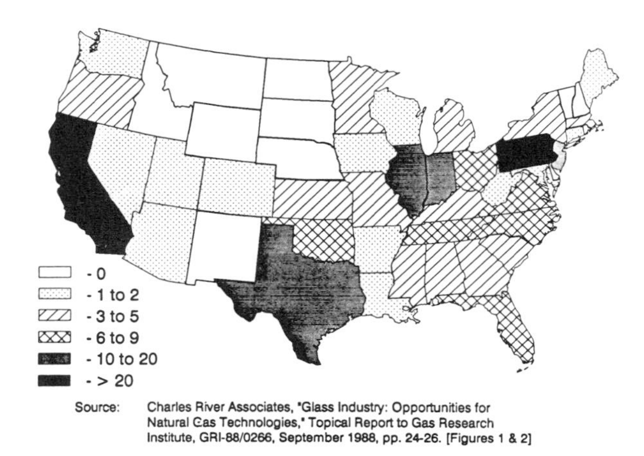

Figure 2. Location and concentration of plants.

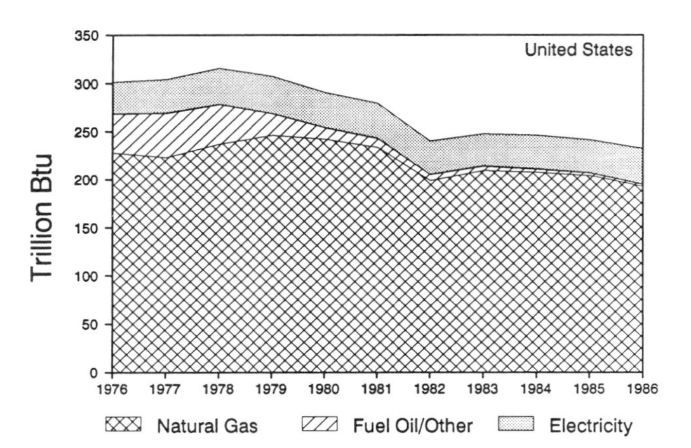

Figure 3. Glass industry energy consumption.

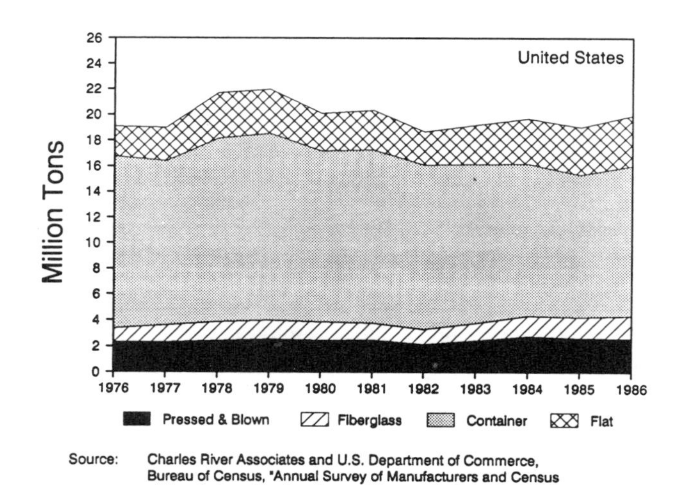

Source: Charles River Associates and U.S. Department of Commerce, Bureau of Census, "Annual Survey of Manufacturers and Census of Manufacturers", various issues. [Figures 3 & 4]

Figure 4. Glass industry production.

the plants were mainly located near areas with population density. However, some container plants were built 50 years ago in the Pennsylvania-West Virginia area because of the availability of coal from which producer gas was derived (Charles, 1988a). In all cases, available cost competitive raw material supplies are essential.

Energy Use and Production Trends

Glass manufacturing is one of the most energy intensive industries in the United States, consuming about 250 trillion BTU (TBTU) per year. Figures 3 and 4 depict the historical trend in energy consumption and total production. In 1986, natural gas supplied 84 percent of the total energy demand, electricity 15 percent, and fuel oil 1 percent. Fuel oil now serves as a backup fuel since natural gas allows operation within SO_x emission limits, and without the added expense of downstream scrubbers needed for oil-fired furnaces. Electricity consumption converted to BTUs in Figure 3 is counted at point of use without adjusting for generation and transmission efficiency. While total energy use has been declining, production has been relatively constant, and thus, as shown in Figure 5, all sectors of the glass industry have been successful in lowering energy consumption per ton of production (Charles,1988a).

Glass production levels can be linked to the general strength of the economy. More specifically, flat and fiberglass production levels correspond with construction industry performance. The early 1980s recession, and the resulting decline in building construction, contracted production in these two sectors (Figure 4). Due to the intense competition with plastic and aluminum packaging materials, glass container production declined from 1979 to 1985 even though total packaging sales increased. As a result, for all sectors, overcapacity and decreasing margins drove industry restructuring and consolidation. In the container glass, flat glass, and fi-

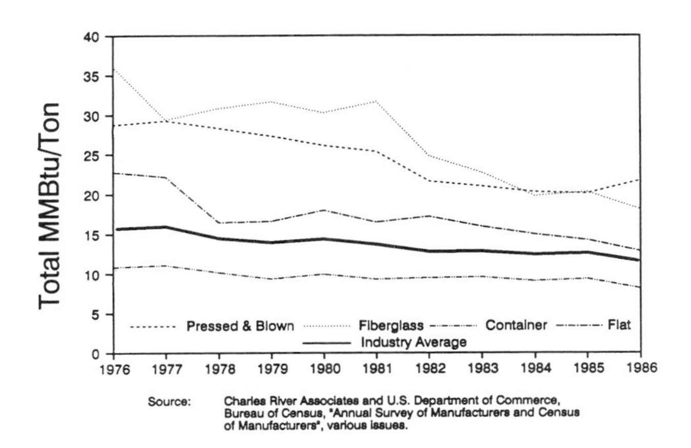

Figure 5. Unit energy consumption.

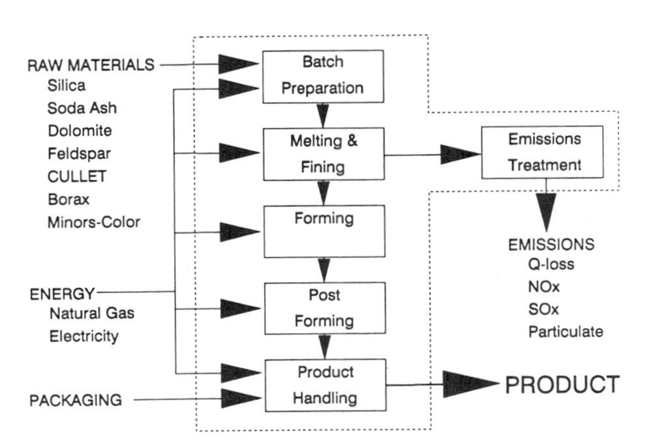

Figure 6. The glass making process.

berglass sectors, there are half as many independent manufacturers today as there were 15 years ago. In the last five years production levels and demand have stabilized. Glass containers now have consumer appeal as a recyclable product compared to plastics (Copperthite, 1990). The general health of the construction industry has increased the demand for fiberglass and window glass products.

Although demand for glass products has stabilized and may even be increasing, new plant construction in the United States has virtually stopped. Due to the capital intensity of glass manufacturing facilities, there has not been sufficient growth in demand to warrant the additional investment to build new plants. Instead, retrofit technologies to boost the production of existing furnaces have been successfully developed. When needed, manufacturers can boost the production of existing facilities for incremental production demand. The most widespread option is "electric boosting," which will be discussed later in this paper. The first new glass container plant to be built in the US for many years will be constructed by Anheuser-Busch. The plant, planned for many years, was first reported on in December 1989. An advanced forming technology developed by Heye-Glas, a West German company, will be installed.

The Glass Making Process

Figure 6 depicts a generic six step process common to all glass manufacturing. In practice, there is obviously a great diversity in the process machinery and production methods to make a specific glass product. *Batch preparation* delivers mixed raw materials to the glass furnace. In the furnace the raw materials are *melted and fined* (bubble removal) to the required quality. During *forming*, molten glass from the furnace is handled and manipulated to make a product. *Post forming* operations include steps like tempering and bending glass windshields for automobiles. *Product handling* includes preparation for shipping. Based on the product, energy use for each of these operations varies in proportion to the total required (Figure 7). In all sectors glass melting furnaces consume the greatest percentage of the total manufacturing energy requirements (up to 85 percent in flat glass), and correspondingly emit virtually all gaseous pollutant emissions attributed to glass manufacturing. *Emissions treatment* equipment is used to clean up the furnace exhaust gas stream to meet regulatory requirements before discharge to the atmosphere (Darrow, 1989).

Conventional Melting Technology

In the container and flat glass industry, there are two principal types of melting furnaces: "side-port" and "end-port." They differ according to where the firing ports are located (on the sides or at the end of the furnace). Figure 8 represents a side view of a typical end-port furnace. The raw material is added continuously from one of two "dog houses" located on either side of the furnace close to the two end firing ports. The firing ports are located at the end of the furnace as shown in the diagram. They fire down the length of the furnace along one side and return exhaust gases back the other side. Waste heat in the exhaust gases is recovered by regenerators. The firing sequence is on an approximate 20 minute cycle. When the left port is firing, the right port exhausts the heated gases through its regenerator heating the refractory lining. During this cycle air is preheated by being drawn through the left regenerator. The firing port and exhausting regenerator switch cyclically. End-port furnaces can be as large as

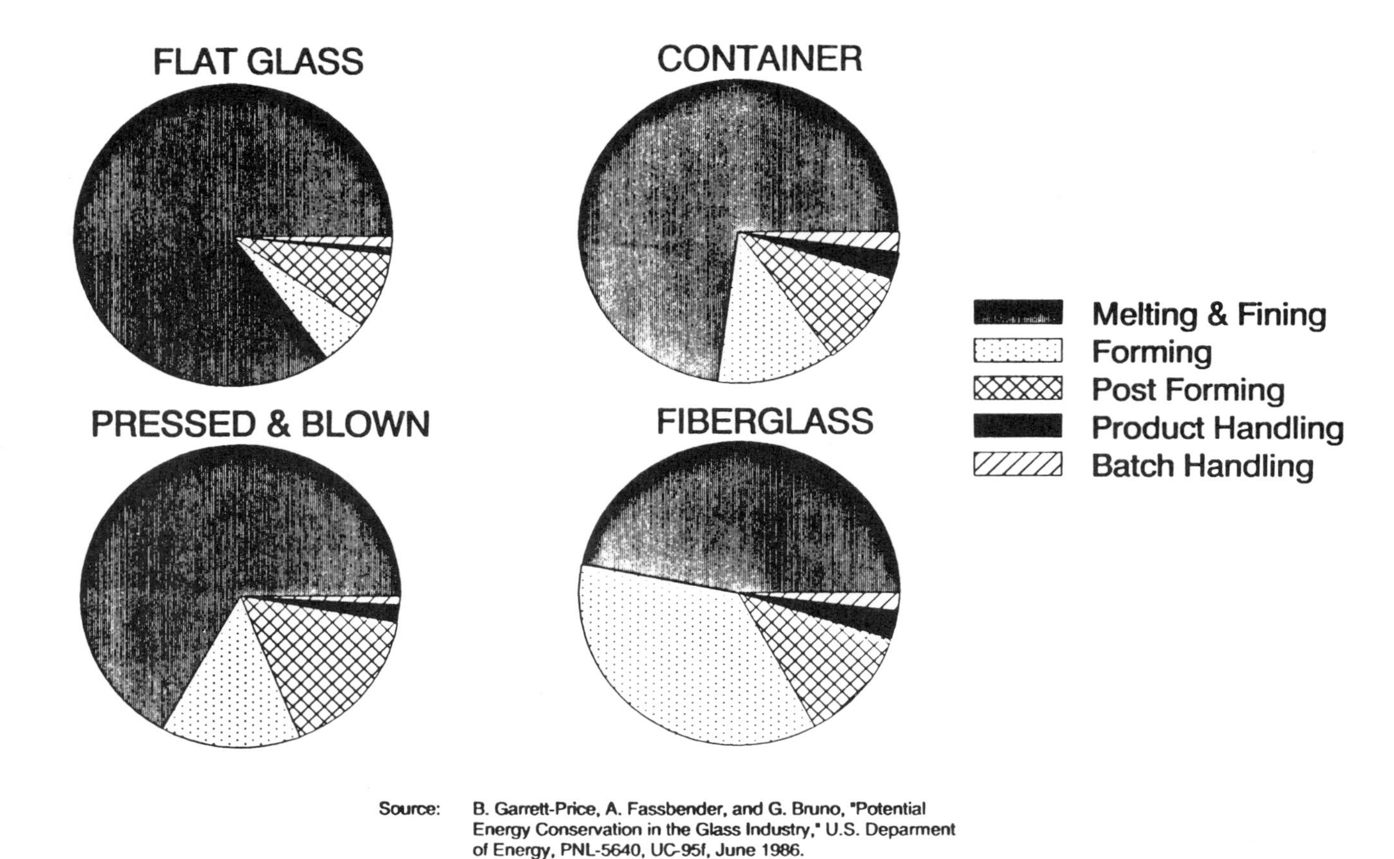

Figure 7. Process energy allocation.

200 tons per day in capacity. Side-port furnaces can be as large as 1000 tons per day. Glass batch is fed from the rear of a side-port furnace and the fuel is fed through three to five burner pairs spaced along the opposing side walls (Charles, 1988b). Many furnaces use supplemental electric resistance heating wherein submerged electrodes are installed through the bottom or side walls of the furnace (Steitz, 1979).

All-electric melters are in operation mainly in the production of insulation fiberglass. Because only electric resistance heating is used for melting, regenerators and exhaust gas cleanup equipment are not required. Figure 9 presents a side view of an electric melter with immersed vertical electrodes. The furnace operates with a layer of batch materials floating on top of the molten pool which allows the furnace to operate at high efficiency since radiation heat loss from the molten glass surface to the refractory enclosure is minimized. This configuration is also known as a "cold top" electric melter (Hibscher, 1982).

"Glass batch" is the term used for the raw materials fed to the melter. Most glass batch components are naturally occurring and reasonably inexpensive. Among these are sand, dolomite, feldspar, soda ash, et al. Silica (sand) is the principal component in most commercial glasses varying from 55 to 75 percent depending on the product (Tooley, 1984). Recycled glass (cullet) from plant rejects or solid waste recycling is also added to the batch. Cullet percentages vary widely across the United States based on supply and quality. To maintain glass quality while using cullet as a raw material, it is vital that the cullet be color sorted and free of contaminants including ceramics, metals, lead crystal, etc. The infrastructure is still developing to effectively recycle glass waste. Cullet percentages average about 20 percent in the United States, although producers in California are targeting 40 percent. In contrast, European producers use close to 70 percent recycled glass in the glass batch.

Furnace peak temperatures vary based on production rate, furnace design, and glass type, with 2800°F being an average value. Heat transfer to the molten glass is principally due to radiation from the refractory crown and side walls, and from the flame; all-electric melters are resistance heated. Preheated air temperatures as high as 2400°F are in use but an average of 1800–2000°F is typical. Higher preheat temperatures produce unde-

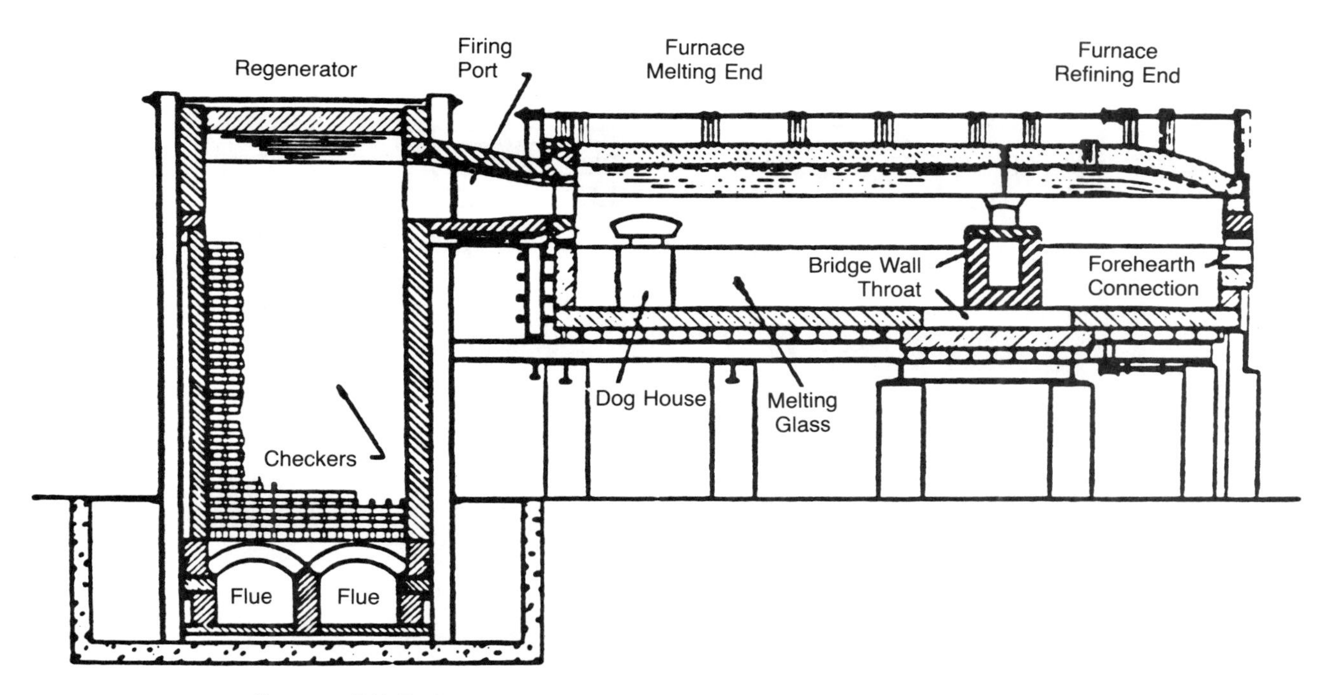

Source: F.V. Tooley, "Handbook of Glass Manufacturing", 3rd ed., New York, N.Y.: Ashlee Publishing Co., Inc., Books for the Glass Industry Division, 1984, p.263.

Figure 8. End port fired furnace.

sirably high levels of NO_x. Regenerator exhaust temperatures can be as high as 1200°F before downstream emissions treatment equipment (Charles, 1988b). Melters operating under these steady state conditions can have a serviceable life (campaign length) exceeding eight years.

Fossil-fired regenerative furnaces with or without electric boost melt about 80 percent of the glass produced in the United States. Figure 10 lists other melting technologies in use. Furnace thermal efficiency varies widely based on furnace configuration, furnace size, glass type, fuel mix, cullet percentage, and furnace age. Figure 11 lists a thermal efficiency range for different melter technologies. As an example, container glass batch theoretically requires approximately 2.8 million BTU per ton (MMBTU/ton) to melt and fine glass from raw materials. Comparatively, an efficient fossil-fired regenerative glass furnace will use about 5.5 MMBTU/ton, and an electric melter will use about 3.1 MMBTU/ton. Fossil-fired furnaces lose heat through the furnace structure and with the exhaust gases. While the electric melter intrinsically has no exhaust gas heat losses, it still loses heat through the furnace structure (Center, 1990). The cullet percentage in the glass batch also affects furnace efficiency. Glass cullet theoretically requires 2.2 MMBTU/ton, or about 25 percent less energy than from virgin batch. Since glass cullet requires less energy to melt, a higher cullet percentage in the batch improves the furnace thermal efficiency (Cole, 1989). For additional information on glass melting technology the reader is directed to some excellent books on the subject listed in the references.

Gaseous Emissions Regulation

Increasingly stringent gaseous pollutant emission standards are causing the glass industry to evaluate, develop, and improve glass melting furnaces. Pollutant emissions from glass manufacturing are regulated on a pound of pollutant per pound of production basis. The principal regulated pollutant emissions are NO_x, SO_x, and particulates. California, with the largest concentration of glass manufacturing plants, has the strictest standards in the nation. The California standards are currently set at 5.5 lb-NO_x, 1.0 lb-SO_x, and 0.2 lb-particulates, all per ton of glass produced. California will lower the standard for NO_x in 1993 to 4.0 lb-NO_x per ton for all furnaces, and is currently enforcing this standard for any new furnace construction (Ross, 1989).

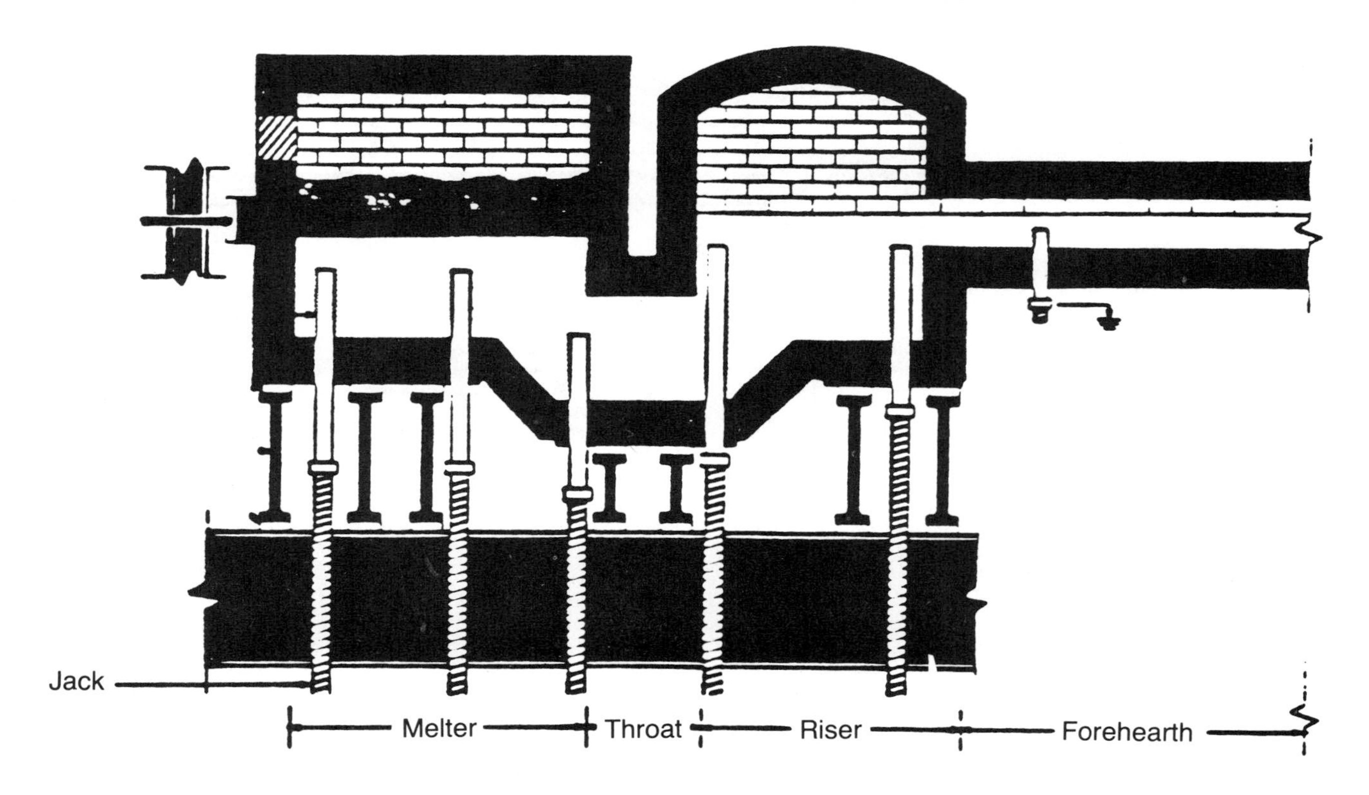

Source: Steitz, W. and Hibacher, C., "Design Considerations For All-Electric Melters" in Proceedings of the 40th Annual Glass Problems Conference, University of Illinois, (November 1979).

Figure 9. COLD TOP electric meter.

SEGMENT	FURNACES	FURNACE SIZE	MELTING TECHNOLOGY
Flat Glass	~ 75	200-1000	Regenerative furnace W/W-out electric boost
Containers	~ 200	150-350	Regenerative furnace W/W-out electric boost
Pressed & Blown	>300	2-250	Unit melter, day tank Electric melter
Fiberglass	~ 90	30-225	Recuperative furnace Cold top electric melter

Figure 10. Melting technology in use.

Melter Configuration	% Efficiency *
Pot Furnace	< 10
Unit Melter	15-18
Recuperative	25-40
Regenerative	20-65
Electric-Semi Cold Top	60-65
Electric-Cold Top	65-90

* Efficiency increases with furnace size

Source: Center for Materials Production, "Electricity in Glassmaking," TechCommentary CMP-047, Carnegie Mellon Research Institute, March 1990.

Figure 11. Furnace efficiencies.

SO_x Emissions and Control

Sulfur oxides can be generated from sulfur contaminants in the fuel or batch, but primarily are released during glass fining. Sulfate compounds, also called fining agents, are purposely added to the glass batch to generate sulfur oxide bubbles that help accelerate molten glass fining. As the sulfur oxide bubble rises to the surface of the melt, other trapped gases are entrained into the bubble (Tooley, 1984). Using this technique, sulfur oxide levels are still generally lower than the regulated limits, and since fuel oil use has virtually disappeared in the United States, the glass industry has not had problems cleaning exhaust gas streams of this pollutant. In Japan, where oil is used for glass melting, wet or dry scrubbers are an acceptable solution for SO_x cleanup.

Particulate Emissions and Control

Particulates can be emitted from a fossil-fired furnace in two ways. Glass constituents like borax (fiberglass compositions) partially volatilize at high temperature in the melter and are carried in the exhaust gas stream. The volatilized constituents subsequently condense during waste heat recovery or exhaust gas cleanup. Carryover is a second particulate source. Glass batch that is fed into the melter requires time to dissolve into solution. Some unmelted raw batch materials on the surface of the melt are entrained into the burner jet firing over the molten surface. Electrostatic precipitators and baghouses are common technology to remove particulates and have been implemented to meet regulated emission levels set by the new source performance standards (NSPS).

NO_x Emissions and Control

The greatest challenge today is economically controlling NO_x emissions from glass melting furnaces. As previously mentioned, NO_x is created during fossil-firing because of the high preheat and melter temperatures. While high preheat temperatures improve furnace efficiency, NO_x creation increases. Meeting the 1993 California standard will be a challenge with current technology. Several techniques are in use to lower NO_x levels, but unavoidably add to the unit production cost of melting glass.

In addition to adding electric resistance heating (boosting) to increase production from an existing furnace, the increased use of electricity provides an environmental benefit to the manufacturer. If electric resistance heating is used to partially supplant fossil-firing, absolute levels of NO_x emissions correspondingly drop. Since the production level of the glass furnace can be maintained with the electric input, the regulated level of NO_x drops. Alternatively, if supplemental electric resistance heating is used to increase the production of a furnace, absolute levels of NO_x emissions remain the same, but the regulated level of NO_x drops; i.e., NO_x per unit of production. This is also true for SO_x and particulate emissions (Center, 1990). A penalty for the increased electric use is the 3.5 to 4.0 cost ratio on a BTU basis for electricity compared to natural gas.

Another technique to minimize NO_x emissions is the use of pure oxygen firing with natural gas. The use of oxy-gas firing is not a new concept but has recently become more economically viable as oxygen costs have continued to decrease with advances in separation technology. Although oxy-gas flame temperatures are higher, it has been demonstrated that refractory temperatures can be equal or lower than comparable preheated air firing. Without the nitrogen in the exhaust stream, the volume of combustion products drops significantly for the same energy input into the melter. Gas velocities drop and therefore particulate entrainment is also minimized. The economics of this technology can also be favorable. In one example, Corning Inc. implemented oxy-gas firing on a small test furnace and demonstrated a 60 percent natural gas savings. This resulted in a furnace energy cost reduction after accounting for both natural gas and oxygen costs (Klingensmith, 1986). It must be recognized that furnace economics are very site and situation specific and thus every application must be carefully evaluated for the best solution.

Trends in Technology Development

Recent technology development to improve glass melting furnaces targets several goals. Lowering pollutant emissions, improving thermal efficiency, lowering unit production costs, and improving operational flexibility are R&D priorities that will enable manufacturers to meet future emissions standards and remain competitive. Lowering pollutant emissions from glass melting furnaces is the highest priority, and virtually all glass furnace R&D currently in progress, or recently available furnace technology, has lower emissions as a key process advantage. The following discussion highlights only a few current examples of melting technology in development.

$LoNO_x$™ Melter

The literature has recently reported on the commercial installation of the $LoNO_x$™ melter at Weigand Glas in Steinbach, West Germany (Figure 12). The furnace concept is a development of Nikolaus Sorg GmbH and Company. Breaking away from the conventional regen-

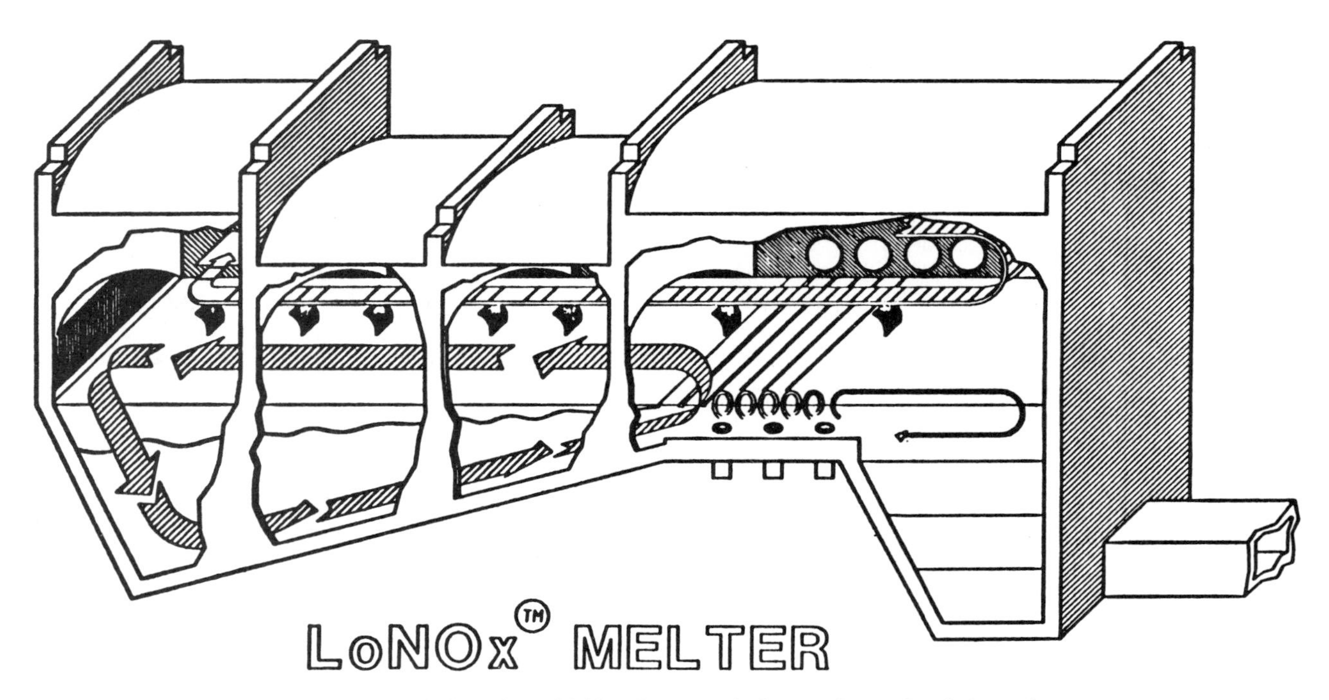

Source: R. Moore, "LoNox Glass Melting Furnace," in Proceedings of 50th Annual Glass Problems Conference, University of Illinois, (November 1989), with permission from Teichmann/Sorg Group, Ltd.

Figure 12. LoNO$_x$™ melter.

erative furnace, the LoNO$_x$™ melter uses radiation shell recuperators that provide lower preheated air temperatures. High thermal efficiency is maintained by using waste heat in the exhaust gases to preheat the batch, cullet, and fuel. With air preheat temperatures less than 1300°F, NO_x minimization is accomplished. Compared to conventional furnaces, fuel costs are actually lower due to the high thermal efficiency. Data from Germany has verified that the melter can meet future US NO_x emissions standards (1.4 lb-NO_x/ton) while operating with 70 percent cullet in the batch. A second commercial installation is planned in Europe with cullet percentages to vary from 20 to 70 percent. The data from this installation will help verify furnace performance with cullet levels typical of United States manufacturing practice (Moore, 1989).

SEG-MELT™ Melter

Another furnace, although not yet proven in operation, is the SEG-MELT™ melter being developed by K.T.G. Glassworks Technology Inc. (Figure 13). The segmented melter takes advantage of the different melting requirements of glass batch and cullet and introduces them separately into the melter. Since cullet can be melted and fined to commercial quality standards much more quickly than glass melted from batch, the segmented melter takes advantage of this difference and introduces the cullet downstream of the batch. In one configuration, an all-electric premelter is used to vitrify the glass batch before it is introduced into the main furnace and oxy-gas burners are used for low NO_x emissions. As in the LoNO$_x$™ melter, cullet preheaters are used to reclaim waste heat from the exhaust gas stream (Argent, 1989).

AGM Melter

A revolutionary departure from conventional furnace designs is the Gas Research Institute sponsored Advanced Glass Melter (AGM) being developed by Vortec Corporation with technical support from Avco Research Laboratory (Figure 14). The AGM uses a natural gas fired combustor which rapidly heats batch materials in suspension that have been injected into the reaction zone of the flame. The products of combustion and the heated batch materials exit the combustor through a high velocity nozzle and are discharged into a melt chamber where the partially reacted glass and exhaust gas are separated. Glass forming reactions are completed in the melt chamber. Because the flame temperature is quenched by the batch in the flame, the AGM

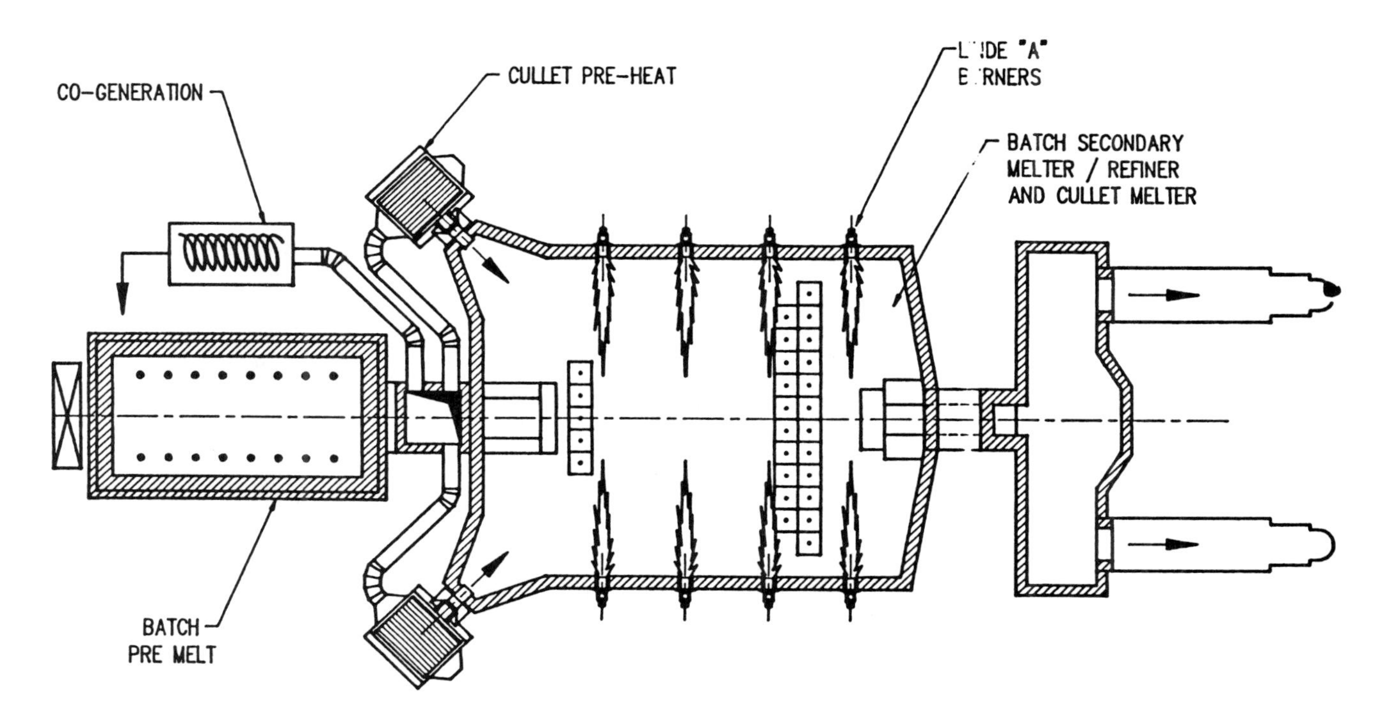

Source: R. Argent, "SEG-MELT," Presentation to American Ceramic Society 42nd Pacific Coast Regional Meeting, Anaheim, California, 10/31 to 11/3, 1989 with permission from K.T.G. Glassworks Technology Inc.

Figure 13. SEG-MELT™ melter.

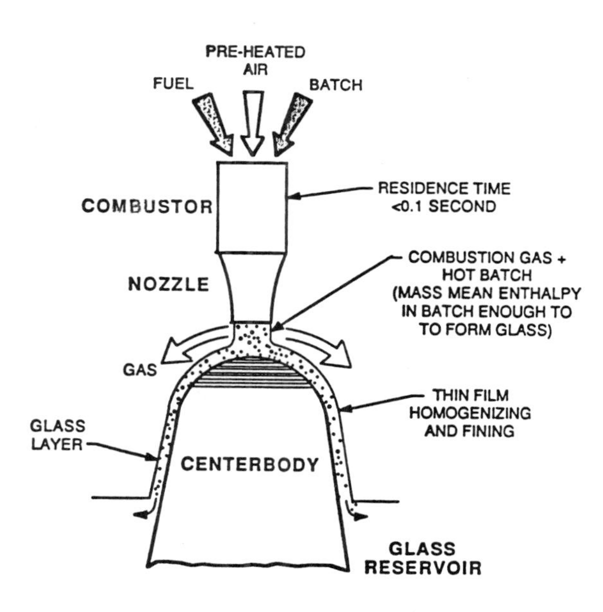

Source: D. Bender, J. Hnat, A. Litka, et al., "Pilot Scale Testing and Commercial System Design of a Gas-Fired Advanced Glass Melting Furnace," in Proceedings of 50th Annual Glass Problems Conference, University of Illinois, (Nov 1989).

Figure 14. Advanced glass melter process.

has intrinsically low NO_x emissions (1.3–1.6 lb-NO_x/ton). AGM process development at the laboratory scale for fiberglass applications is projected to be complete by early 1991. Based on the continued success of the development, scaleup and a field experiment installation would follow in 1991–1993. Test results predict that a commercial scale AGM will meet current and future emissions standards while providing a compact, cost competitive, thermally efficient, and operationally flexible melting system (Bender, 1989).

Conclusion

Glass manufacturers, engineering and construction companies, the Gas Research Institute, and other R&D sponsors, including the Department of Energy, are playing active roles to develop furnace technology and meet the challenges of the 1990s. The industry has successfully improved the thermal efficiency of the melting process over the last decade. It is now responding to the challenge of further improving glass furnace operation to lower emissions. Many examples exist of technology either recently available or in development that will address the industry's needs. In the future, process technology decisionmaking will be increasingly driven by environmental concerns. The domestic glass industry has weathered a past decade of significant change and can anticipate the future decade to be equally eventful.

Acknowledgments

The kind assistance from several gentlemen in developing this paper is appreciated: Mr. Gary Walzer—Center for Materials Production, Mr. Ron Argent—K.T.G. Glassworks Technology Inc., Messrs. Ron Moore & James Aker—Teichmann/Sorg Group Ltd., and Mr. Carl Hibscher—Toledo Engineering Co.

References

Argent, R., *SEG-MELT,* in Presentation to American Ceramic Society 42nd Pacific Coast Regional Meeting, Anaheim, CA (November 1989).

Bender, D., Hnat, J., Litka, A., et al., "Pilot-Testing and Preliminary Commercial Design of a Gas-Fired Advanced Glass Melting Furnace," in *Proceedings from the 50th Annual Conference on Glass Problems,* University of Illinois at Urbana-Champaign, pp. 102–124 (November 1989).

Center for Materials Production, *Electricity in Glassmaking,* CMP TechCommentary CMP-047, Carnegie Mellon Research Institute, Pittsburgh, PA (March 1990).

Charles Rivers Associates, Inc., *Glass Industry: Opportunities for Natural Gas Technologies,* Topical Report to Gas Research Institute, GRI-88/0266, pp. 7–23 (September 1988).

Charles Rivers Associates, Inc., *Advanced Glass Melter Technology Assessment,* Topical Report to Gas Research Institute, GRI-88/0278, pp. 5–28 (October 1988).

Cole, W., Becker, F., et al., "Operation of a Cullet Preheater," in *Proceedings from the 50th Annual Conference on Glass Problems,* University of Illinois at Urbana-Champaign, pp. 53–68 (November 1989).

Copperthite, K., "Glass Container Outlook: Shipments to Rise Slightly in 1990," *Glass Industry Magazine,* 71(1), pp. 8–22 (January 1990).

Darrow, K. and Hyman, B., *Guide to Natural Gas Industrial Processes: Glass Industry,* Topical Report to American Gas Association, Report No. 661R139, Energy International Inc., Bellevue, WA (February 1989).

Hibscher, C., Carle, R., "First Campaign Results of a Cold Top Electric Furnace," in *Proceedings from the 43rd Annual Conference on Glass Problems,* Ohio State University (November 1982).

Klingensmith, L., "Direct-Fired Melter Performance Improved by Gas/Oxygen Firing," *Glass Industry Magazine,* pp.14–18 (March 1986).

Moore, R., "LoNO$_x$™ Glass Melting Furnace," in *Proceedings from the 50th Annual Conference on Glass Problems,* University of Illinois at Urbana-Champaign, pp. 89–101 (November 1989).

Ross, C., Walzer, G., et al., "Panel Discussion on Control of Emissions and Hazardous Waste," in *Proceedings from the 50th Annual Conference on Glass Problems,* University of Illinois at Urbana-Champaign (November 1989).

Steitz, W. and Hibscher, C., "Design Considerations For All-Electric Melters," in *Proceedings from the 40th Annual Conference on Glass Problems,* University of Illinois at Urbana-Champaign (November 1979).

Tooley F., *The Handbook of Glass Manufacture,* 3rd ed, New York, N.Y.: Ashlee Publishing Co., Inc., Books for the Glass Industry Division (1984).

Tooley, F., "The United States Glass Industry—Then and Now," in *Proceedings from the 50th Annual Conference on Glass Problems,* University of Illinois at Urbana-Champaign, pp. 1–14 (November 1989).

Summary
Session B-3—Chemicals, Petroleum, and Pulp and Paper Manufacture

Kenneth A. Smith
Session Chair
Department of Chemical Engineering
Massachusetts Institute of Technology
Cambridge, MA

Three papers were presented in this session: "Energy Efficiency in Petroleum Refining—Accomplishments, Applications, and Environmental Interfaces" by Jerry L. Robertson of Exxon Research and Engineering, "Energy Consumption Spirals Downward in the Polyolefins Industry" by William Joyce of Union Carbide, and "Energy Management and Conservation in the Pulp and Paper Industry" by Howard Herzog and Jeff Tester of MIT.

Each of these papers deals with an enormous industry that is highly energy intensive. Each industry now appears to be technologically mature, but a 20-year perspective includes some of the boisterous youth once enjoyed by the polyolefins industry. Thus, there are both similarities and differences in any comparison of the three industries. Robertson indicates that from the early 1970s to the late 1980s Exxon was able to reduce the energy intensity of its refining operations by about 35 percent. Herzog and Tester indicate that the pulp and paper industry was, for the same period, able to reduce its energy intensity by about 12 percent for total energy and by about 35 percent for purchased energy. Joyce indicates that the energy intensity of polyethylene production dropped by about 70 percent during this period. Each author details the manner in which these impressive improvements were achieved.

In each case, it is probably fair to say that the gains in efficiency were driven largely by economics. It is also important to note that these industries enjoy a scale of operation that permits very large investments. In these businesses, size is a prerequisite for efficiency. The most intriguing element of the comparisons, however, lies in the differences. During this period, the basic technology remained unchanged for the petroleum refining industry and for the pulp and paper industry. For polyethylene, the process technology changed dramatically with the introduction of catalysts, which permitted low-pressure, gas-phase polymerization. It is this type of change that accounts for the considerably greater improvements achieved by the polyethylene industry. Such fundamental changes are certainly more likely in highly profitable, rapidly growing industries as in the polyethylene case. But such changes are also badly needed in certain more mature industries and we must learn to encourage those changes.

Energy Efficiency in Petroleum Refining—Accomplishments, Applications, and Environmental Interfaces

Jerry L. Robertson
Exxon Research and Engineering Company
Florham Park, NJ

Abstract

Refinery energy conservation yields a clear environmental benefit in lower carbon dioxide and pollutant emissions. Improved energy efficiency in all sectors of the economy is critical to balancing economic growth and environmental protection. As both producers of energy in usable form and significant users of energy, Exxon and the petroleum refining industry need to be particularly sensitive to efficient energy use and its related impact on the environment.

The petroleum refining industry has made significant progress in improving refining energy efficiency. For example, by the end of 1988, Exxon had reduced refinery energy consumption, accounting for increases in processing severity, by about 35 percent compared to 1973. This reduction resulted from management commitment and stewardship for improvement. Essential elements of this activity include an energy measurement technique, implementation of employee training programs, deployment of effective communication systems, and the development and application of energy-efficient technology.

Over the past several years, many changes in petroleum refinery processing have responded to the need for continuing improvements in manufacturing higher performance and cleaner burning fuels. Improvements in environmental effects outside the refinery have in turn resulted in increased energy consumption by refineries since the processes required to make the better products demand additional energy.

Accomplishments

In 1969 Exxon senior management instituted a program to reduce refinery energy consumption. By 1981 this program had improved refinery energy efficiency by about 25 percent over 1973 figures, and by 1988 it had improved efficiency by about 35 percent over 1973. These worldwide improvements have increased steadily and have cumulatively resulted in a savings of about 425 million barrels of oil. This energy conservation accomplishment has thus allowed 425 million barrels of oil to stay in the ground for future use.

In addition to the identification and application of appropriate energy technology, there are several other elements necessary to a successful energy conservation program. Energy efficiency measurement and management attention to communications and education are essential to converting theory about energy efficiency to measurable energy savings.

Energy Measurement Technique

A credible energy efficiency measurement method is required to monitor improvements and establish goals for energy use. Exxon developed the Energy Guideline Factor (EGF) method in 1974 and has since used it as a standard to measure refinery energy performance.

Initial efforts to develop criteria that correlated energy requirements with investment, crude throughput, or individual process throughput were unsuccessful. However, when individual process throughput was corrected for processing intensity or severity, a satisfactory energy requirement correlation was achieved. Examples of higher process intensity or severity include operation at higher reflux ratios in distillation towers to achieve higher purities; additional processing to reduce sulfur or nitrogen content of product; and operation of a process at higher temperature or pressure or at lower space velocity to achieve higher conversion. The selection of one or two parameters for each process, such as product Octane Number or distillation reflux, that best predicted energy requirements and the development of the correlations was the subject of an extensive engineering study. These energy/process relationships were based on correcting the required energy of individual process units, e.g., atmospheric pipestills, catalytic reformers, fluid catalytic crackers, etc., to stringent, consistent energy efficiency criteria. Efficient criteria were established for temperatures of existing streams prior to final cooling by water or air, for furnace flue gas temperatures and oxygen content, for stripping steam requirement, and so on. Factors were also developed for converting different forms of energy, such as steam and power, to a common energy measurement unit—the Fuel Oil Equivalent Bbl. Thus, for each process a single efficient energy requirement—the Energy Guideline Factor (EGF)—was established for a particular throughput and operating condition. Energy Guideline Factors for all processes are summed to establish the Guideline energy consumption for the entire refinery. Actual energy consumption, including fuel, steam, and power, is measured and converted to the Fuel Oil Equivalent Bbl to arrive at the refinery total energy consumption. The total energy consumption divided by the Guideline energy consumption represents the refinery Percent of Guideline which is monitored as the overall measure of energy performance.

In summary, the EGF system is theoretically sound, based on real plant designs, yet is simple to use. As a bonus, it provides a method for identifying and quantifying individual process inefficiencies by comparison of the actual process operating conditions with the stringent energy efficiency criteria mentioned previously. It was placed in use by all Exxon refineries in 1975, and plant energy performance was calculated for 1973 and 1974 for comparison. The EGF system is the yardstick by which Exxon refineries measure their energy performance to monitor improvement and is the basis on which the energy savings mentioned previously were calculated.

An overview of the Energy Guideline Factor development and use was given by Lockett (1980), and the technology has been licensed for use by other refiners. Recently, Siegell (1990) has given an overview of the energy loss analysis technique.

Energy Communications and Education

Effective communication of both successful and unsuccessful applications of energy technology among refinery technical personnel is considered important by Exxon management. The first Exxon Energy Conservation Symposium was planned in 1973 and held in February 1974 (before the oil embargo). It was attended by both management and technical representatives from Exxon's worldwide refining operations. Annual meetings, workshops, and awareness seminars aimed specifically at energy conservation and energy efficiency continue to be held. Liaison visits by energy technology experts also facilitate the transfer of technology from theory to specific applications that are tailored to the needs and characteristics of individual refineries.

Another major factor in maintaining and improving energy performance is education and training. Exxon's first energy conservation manual, the "Plant Energy Conservation Guide," was published within Exxon in 1974. This collection of knowledge about current technology and applications has been used as a basis for workshops aimed at training refinery personnel. An overview on saving refinery-furnace fuel, as covered in the manual, was given by Cherrington and Michelson (1974). Many other manuals and reports have been produced and included in workshops aimed at technical and operational methods to improve energy efficiency.

Applications

Petroleum Refining: An Overview

Petroleum refineries consist of crude tankage, a system of separation and conversion processes, individual product tanks, interconnecting lines among the processes and tankage, and a system of utilities that provide and distribute the required supply of steam, power, and cooling. Overlying this equipment are process control systems that assure proper flows, temperatures, and pressures; safety systems that assure that equipment design pressure cannot be exceeded and that discharges are flared in a controlled manner; and environmental systems that assure clean refinery effluents. Figure 1 shows the simplified flow plan of a high conversion refinery of today and includes the environmentally required hydrotreating units.

Crude petroleum is the feed to refineries. Crudes come in many types, ranging from light crude, which contains higher fractions of gasoline and jet fuel, to heavy crudes containing more heavy oil and asphalt. Sour crudes contain more nitrogen and sulfur compounds than sweet crudes. Over the past several years the trend in crude availability has been from light sweet crudes to heavier sour crudes.

The objective of a refinery is to manufacture products ranging from the lightest propane, through gasolines, jet fuels, heating oils, and lubricating oils to the heaviest products, asphalt and coke. The volumetric market demands for these products are seasonal, with higher gasoline demands in the summer and higher heating oil demands in the winter. The variability in crudes, product characteristics, and demands requires that the equipment be flexible enough to operate over a wide range of conditions. Furthermore, over the past several years, the worldwide demand for lighter products such as gasoline, diesel fuel, and jet fuels has been increasing in relation to the demand for heavy fuels. This leads to increased crude conversion.

Refineries are highly energy intensive, consuming 6 to 10 percent of the energy in the crude. Energy needs depend on the processing intensity necessary to convert the crude to the required product mix. For example, more energy is required to convert a heavy sour crude to light products than for converting a light sweet crude to a full range of products. Presently, in high conversion refineries, the energy cost can represent over 40 percent of the operating cost excluding crude costs. In the early 1980s, when crude costs were higher, energy costs rep-

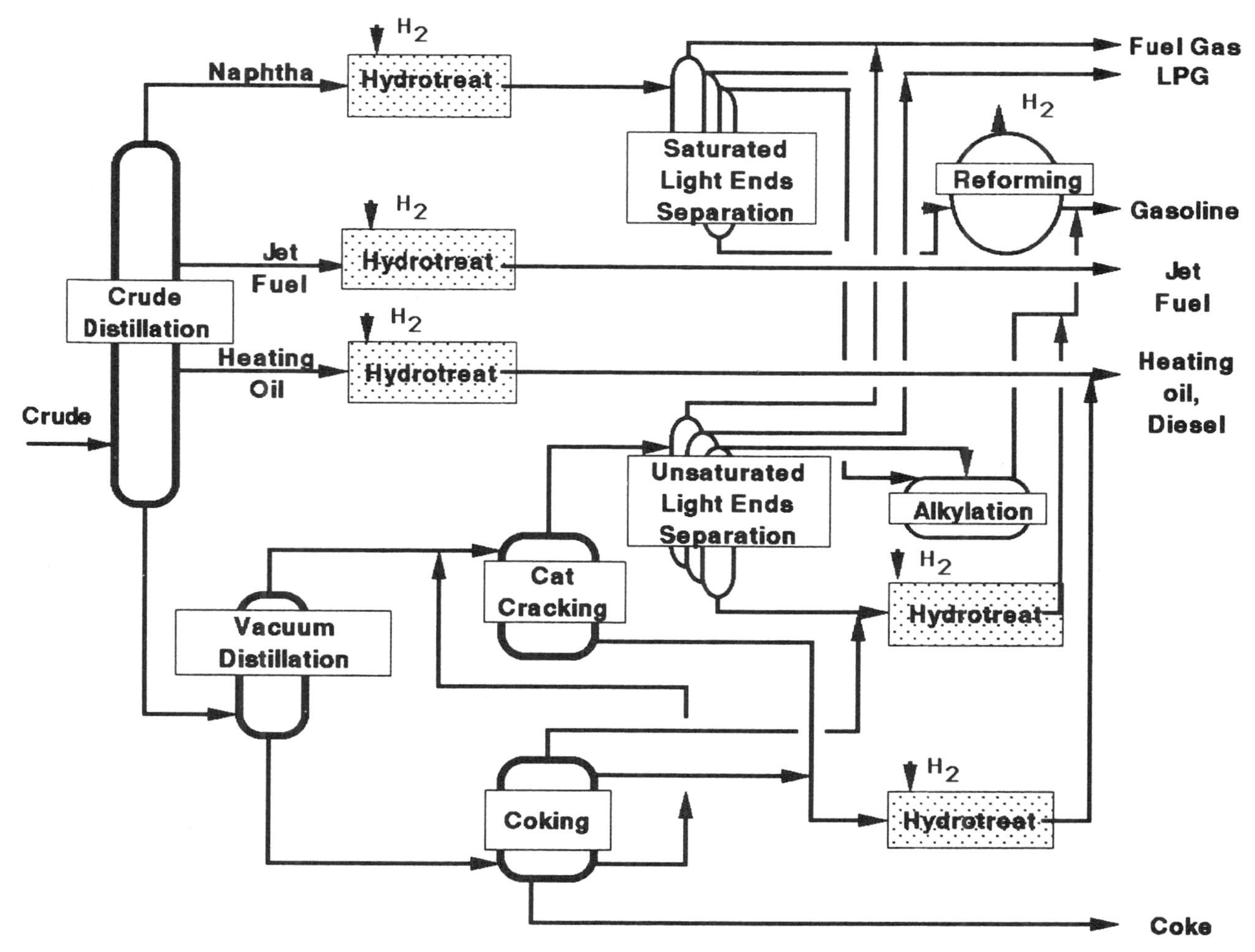

Figure 1. Simplified flow plan of refinery.

resented about 50 percent of the average Exxon refinery operating cost (Eidt, 1981).

High energy costs and changing refinery processing requirements create the opportunity to improve refinery energy efficiency. High energy costs result in energy projects aimed at the singular goal of reducing energy consumption. Important energy efficiency steps are also taken in the design of logistical projects to meet the changing refinery requirements caused by crude supply changes, product mix changes, and product specification changes. Fitting new equipment into existing refinery infrastructures—retrofit projects—offers significantly greater challenges to operation, design, and construction than entirely new refinery construction. Systematic methods of attack are only beginning to evolve for retrofits (Gunderson, 1990). Operations must first be shut down and cleared of hydrocarbons. An optimum set of new equipment compatible with the existing equipment must be designed and installed. Dismantling and removal of obsolete equipment and constructability of the new equipment must be considered to minimize shutdown time.

Energy Efficiency Applications

Applications of energy efficiency technology are highly varied, ranging from the simple and obvious, such as repairing steam leaks, to complex heat, power, and steam system integration. As mentioned previously, petroleum refineries consist of complex interconnections of many separation and reaction processes. A single molecule (or its fragments) may pass through as few as two or three to as many as ten or more of these processes while moving from crude tankage to product tankage. Process requirements often necessitate the heating or cooling of streams between the different processes. Energy efficiency improvements require an understanding of the entire system, including current heat transfer technology, efficient combustion, integration of the heat and power system, process improvements, and implementation of skillful operations and maintenance procedures.

Site Energy Surveys

In 1978, a major energy efficiency research and development effort was initiated. The goal was to consolidate the separate improvements in the various energy related technologies. An early conclusion of this study was that it is necessary to consider the refinery as a whole, rather than as a group of individual process units, in order to accomplish major reductions in plant energy consumption. This is due to the energy interlinking of the process units through the steam and power system as well as interprocess heat integration.

Site Energy Survey Teams were formed that combined knowledge of refining, operations, process design, and steam-power systems. Critical engineering and operating data were gathered from each of the process units. The survey team reviewed these data with plant operating personnel to facilitate brainstorming sessions about potential improvements to each process. Typical improvements included recovery of additional process heat that had previously been rejected to cooling water, switching an energy input from high pressure steam to low pressure steam, switching a low efficiency steam turbine to electric power, and switching large motors to highly efficient back pressure turbines. Lists of these ideas were kept to determine the overall effect on the plant steam/power balance. Often several competing ideas were developed. The study team then screened these ideas on the basis of safety, operating feasibility, and economic attractiveness to develop a site plan for energy utilization.

Since 1976, over 60 of these surveys have been completed. Methodologies have been improved and approaches have been tailored to site requirements. Several refineries have been surveyed more than once because of improvements in survey methods and changes in refining objectives. Energy projects that were developed as a result of these surveys made major contributions to Exxon's success in improving energy efficiency.

Heat Transfer Technology

Application of appropriate heat transfer technology is critical to refinery energy efficiency. Most of a refinery's energy is used to heat the feeds to the various separators and reactors to the temperature required to carry out the separation or reaction. The products from these processes are at high temperature containing high-level heat. Recovery of the heat from the hot products to preheat the incoming colder feeds is accomplished with heat exchangers. Much of the heat required to reboil light end towers can be recovered from these hot products as well. In practice this is often done with a steam "hot-belt"—making steam from "waste heat" in one process and using it in another.

Heat recovery has been practiced for many years, resulting in the development of different kinds of heat transfer equipment to meet specific needs. Research in heat transfer technology and the ongoing development of heat exchanger equipment has produced a myriad of options for selection by the design engineer. Materials of construction, enhanced heat transfer surfaces, turbulence promoters, and configurations of the exchanger internals are among the choices available. Developing the expertise to make the appropriate selection and expanding this experience in applying new heat exchange technology are both essential to a successful energy conservation program.

A recent example of the application of new heat transfer technology is the plate heat exchanger. This type of exchanger is characterized by its compactness, high heat transfer rate, and the relative ease of cleaning fouled surfaces. Initially, there was concern about its use in hydrocarbon service because of the potential for failure of the extensive gaskets during a fire. A protective system was designed and the exchanger was successfully fire tested by a vendor. This test was witnessed by industry and safety experts in mid-1987. In 1989, this technology was installed in one of Exxon's refineries during a retrofit project.

Heat Exchange Networks

There are many streams in a refinery that require heating and cooling. Equal in importance to the selection of appropriate heat exchanger technology is the selection of which hot stream will be used to provide heat to which cold stream. The resulting set of heat exchangers is referred to as the heat exchanger network. In a typical refinery, 75 to 80 percent of the energy required to heat cold feeds to reactor or separation temperatures is recovered in heat exchanger networks. Analysis of the performance of these networks has been greatly enhanced by the development and use of two technologies in the last decade. These are "pinch" technology, a method of establishing targets for maximum potential recovery for an entire refinery (see Spriggs, Petela and Linnhoff, 1990), and HEXTRAN™, a suite of computer applications developed by Simulation Sciences Inc. that facilitates the synthesis, analysis, and optimization of heat exchanger networks. Prior to their development, heat exchange networks were designed based on case study approaches with no assurance of optimization. In the early 1980s Exxon incorporated these technologies into the site survey and retrofit design methods.

The importance of these technologies to refinery energy efficiency cannot be overemphasized. "Pinch" technology provides a new method for visualizing, analyzing, and communicating energy efficiency studies and projects. HEXTRAN™ provides a method for quantifying the effects of adding heat exchangers to the overall efficiency of the entire network. This is important because, depending upon the specific location of the new exchanger in the network, more or less heat can be recovered by the entire network. Furthermore, these new technologies enable the analysis of existing networks for retrofit, including the addition of new heat exchangers, rearrangement of existing networks by moving exchangers, and repiping of networks to allow heat recovery from different streams. These systematic methods and powerful computer tools, along with expertise in their use, have contributed greatly to successful discovery and implementation of refinery energy projects.

Combustion

Combustion provides the required additional process heat that cannot be recovered efficiently in heat exchanger networks. Combustion in furnaces and boilers supplies the highest temperature heat. Their efficiency, unlike that of heat exchangers, is directly related to overall refinery efficiency. Application of currently available combustion technology such as high efficiency–low NO_x burners, combustion air preheaters, stack gas analyzers, and combustion air controls provides the potential for efficient combustion.

The energy demands for a refinery are continuously changing because of changes in crude mix, product mix, and throughput. Furthermore, adding exchangers in heat recovery networks reduces combustion requirements, which can modify the operation of a furnace to an off-design, less efficient mode. Exxon has developed methods for maximizing furnace and boiler efficiency even in these modes. Manuals have been prepared and workshops held to train refinery personnel to recognize opportunities and implement changes to maintain high combustion efficiency.

Cogeneration

Cogeneration is the simultaneous production of work and heat from a single fuel source. One form of cogeneration utilizes gas turbines. The turbine produces the work that can be used to drive large compressors or to generate electricity. The turbine exhaust gas provides heat that can be used either to produce steam that is used to provide process heat or to provide process heat directly. Gas turbine exhaust contains enough oxygen to support combustion. Thus, fuel supplied to the exhaust gas can produce incremental combustion, resulting in additional steam or process heat. Gas turbine exhaust has also been used to provide a preheated, substitute air to large process furnaces.

Another form of cogeneration utilizes very high pressure steam boilers. The high pressure steam is used to drive steam turbines to produce work. Lower pressure steam is extracted from the turbines and utilized to provide process heat.

Both types of cogeneration have been practiced in Exxon refineries for decades. Recent economic and regulatory changes have renewed interest in this technology (Zwicker, 1989). Key to successful implementation is matching the work production and heat or steam production of the particular cogeneration system to the demands of the site. For this reason and because of economy of scale, large sites with many process units favor the installation of cogeneration, which results in energy savings of 10 to 30 percent compared to generating power and steam separately. Many refineries also participate in cogeneration projects with utilities or other third parties, which facilitates the matching of the power and steam demands of the refinery to the capabilities of the generating station.

Steam Systems

Steam, in addition to providing process heat and mechanical work, performs other services for a refinery. Stripping steam removes lighter molecules from heavier sidestreams of distillation towers, as required to meet flash point specifications; "mixing steam" makes flares, when required, burn cleaner; "snuffing steam" clears furnaces of air in case of emergency; "feed steam" is used in reforming plants that produce hydrogen, which is used to remove sulfur compounds from various products. Furthermore, steam is used as a "hot belt" between processes. A process that has an excess of high-temperature heat makes "process steam" for use in another process. All of these types of steam are intermingled in the steam distribution network. Furthermore, there are often three different pressure (and therefore temperature) levels of steam. Proper design and management of the steam system can have a significant impact on overall refinery energy efficiency. For example, projects that "unmake" process steam by using additional heat integration to cool the process stream and projects that substitute cogenerated steam for the "process steam" are often more energy efficient.

Process Improvements

Major improvements in catalyst activity and selectivity and reductions in the tendency for coking have enabled refineries to operate more efficiently. More active hydroprocessing catalysts have allowed sulfur removal processes to operate at lower temperatures while accomplishing the same goal. Improvements in catalytic reforming processes that make high octane gasoline have allowed higher yields and operation at lower pressures. More active fluid catalytic-cracking catalysts have allowed higher yield and once-through rather than recycle operation while processing less coke.

Separations improvements have also led to energy savings. Two examples are the use of structured packing to replace less efficient distillation trays and FLEXSORB, an improved gas treating absorption system that requires 40 percent less energy than previous processes and removes hydrogen sulfide from gas streams down to 0.001 percent.

Operations and Maintenance

While the installation of efficient energy technology provides the capability for energy savings, the system must be properly operated and maintained in order to achieve continuous energy efficiency improvement. Operations must be controlled to assure that proper specification products are made without wasting energy on unnecessary recycles, excessive reflux in distillation towers, or inefficient boilers and furnaces. Best results are obtained when the entire refinery is considered as a complete system, taking into account the current capabilities of the individual processes.

Maintenance of heat transfer systems includes not only monitoring and cleaning of fouled heat exchangers, but also the monitoring of performance requirements to assure that replacement exchangers are the most appropriate choice for the current operation. Steam system maintenance includes monitoring and replacement of leaking or damaged steam traps and replacement of leaking steam system control valves. Combustion systems require maintenance to clean fouled furnace tubes and fouled burners. Energy control systems, including combustion control and stack gas analyzers, also require maintenance to assure high energy efficiency.

Overall Energy Efficiency Improvements

Assignment of savings to individual activities is virtually impossible. What is important is that all of the steps should be evaluated and that the synthesis of their implementation should be the measure of their effectiveness. This is of significance not only to petroleum refining but also to the end uses of the products manufactured by refineries. The singular measure of the effectiveness of these activities is the *overall* reduction in energy consumption.

Environmental Interfaces: A Case Study

Another interface between petroleum refining and the environment, besides energy conservation, involves the processes needed to manufacture cleaner burning products. Early refineries processed relatively sweet, low-sulfur crudes. The processes used then chiefly involved separation of the crude into the various boiling ranges required to make products such as gasoline, kerosene, heating oil, and heavy fuel oil. These early refineries consumed as little as 2 percent of the energy content of the crude. As the marketplace changed, requiring higher octane, lead-free gasolines and cleaner burning, low sulfur, low nitrogen fuels, additional processes were added to the refinery processing complex. Alkylation, catalytic reforming, catalytic cracking, and naphtha isomerization all continue to be instrumental in producing higher octane gasolines. Hydroprocessing reactions are used to lower the sulfur and nitrogen content of fuels. Catalytic cracking and coking reactions are used to convert heavier streams to lighter fuel streams and, in the process, also facilitate desulfurization. Exxon's FLEXICOKING™ process converts heavy residual fuels to light products while it also converts most of the byproduct coke to a low Btu, low sulfur fuel gas that can be burned in refinery furnaces.

Desulfurization Case Study

As a simple example of the environmental effects of product quality improvement on energy consumption, a case study was developed. It was based on desulfurization of diesel fuel from present levels of 0.4 percent sulfur to future levels of 0.05 percent sulfur. Several simplifying assumptions were made. It was assumed that an existing unit would be retrofit to make the new product and that the reactor inlet temperature could only be increased by 35°F because of material limitations. It was further assumed that the remaining improvement in depth of desulfurization would be achieved by increasing hydrogen recycle and that all of the hydrogen for both the base case and higher product quality case was produced by a steam reforming unit. Recycle hydrogen was assumed to be 100 percent pure. These are stringent assumptions and actual process modifications to achieve the product quality goal of 0.05 percent sulfur may differ. However, this case allows the demonstration of the interplay among various process units that will need to be considered for product upgrade retrofits. Also, two alternative heat recovery cases were developed—one for maintaining constant heat transfer area for heat recovery and one for increasing the heat recovery area sufficiently to achieve the same marginal energy economics as the base case.

Figure 2 shows a simplified flow plan of a desulfurization unit. Feed, which contains sulfur, is mixed with recycle hydrogen and heated to reaction temperature. The desulfurization reactor converts sulfur compounds to hydrogen sulfide gas. (The hydrogen recycle requirement is established by the reactor size and the hydrogen partial pressure in the reactor outlet. Lower sulfur products [more desulfurization] require lower hydrogen partial pressure.) Reactor product is condensed by preheating incoming feed and the hot liquid is separated from the hydrogen–hydrogen sulfide gas mixture in the hot separator. High temperature vapor is further cooled by incoming feed. High temperature liquid from the hot separator is used to provide heat to another process unit. Vapor from the cold separator contains primarily hydrogen and hydrogen sulfide. Hydrogen sulfide is removed by absorption and energy is required to remove the hydrogen sulfide from the absorbent. The hydrogen sulfide is converted to sulfur in an exothermic reaction that provides heat (usually through steam production) to other process units. Purified hydrogen is compressed, which requires significant energy input, and recycled to the feed.

Table 1 shows the results of this simplified case study. A typical diesel feedstock containing 1.5 percent sulfur was assumed. Desulfurization was increased by a factor of 8, going from 0.4 percent weight in the base case to 0.05 percent weight in the high desulfurization case. The reactor temperature was increased by 35°F and the hydrogen recycle rate was increased by a factor of 8. The pressure and pressure drop throughout the system were assumed constant.

Figure 3 shows the pinch diagram for this system. The composite heat required versus temperature is plotted for all streams that need to be heated (the lower line) and all streams that need to be cooled. Note that

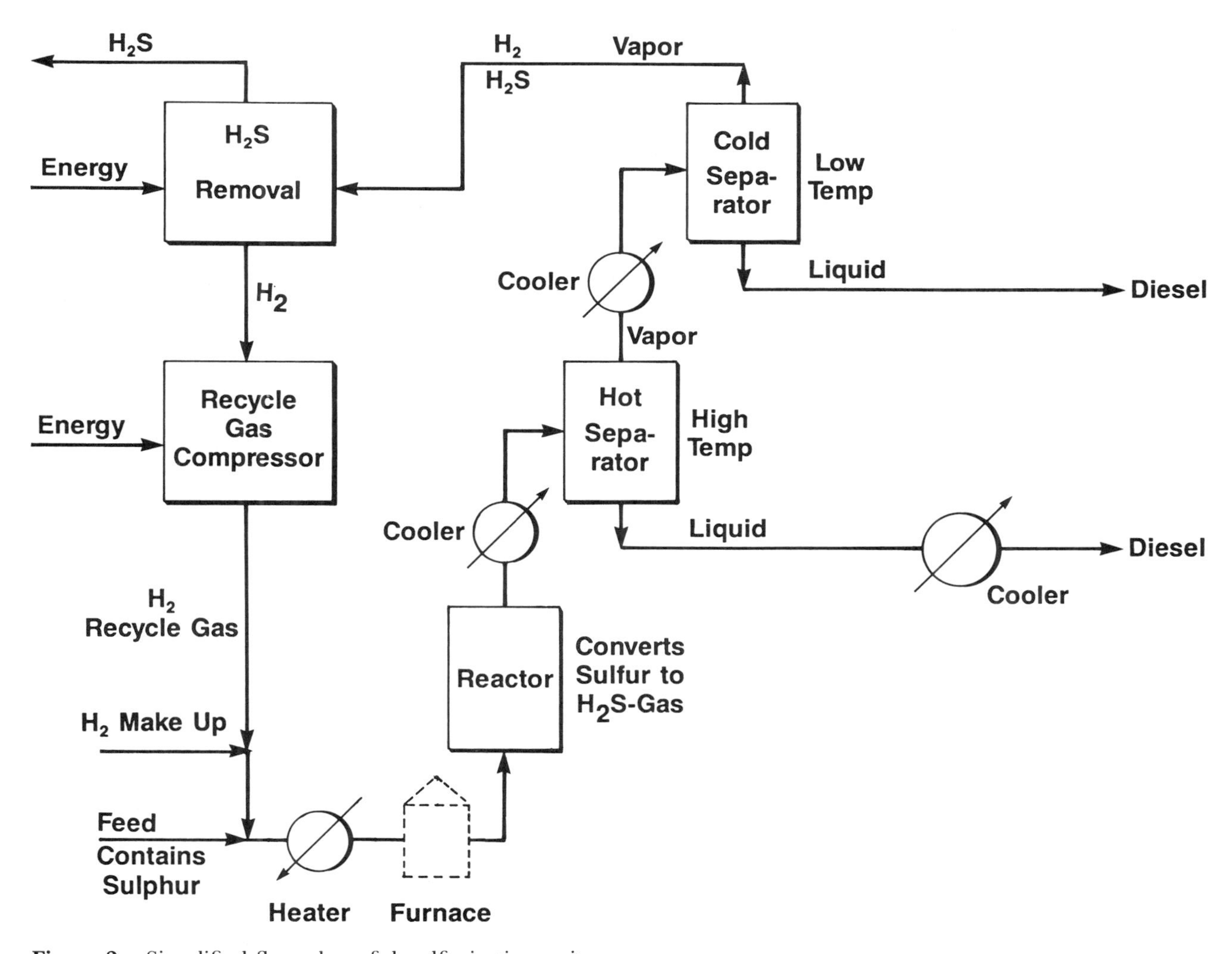

Figure 2. Simplified flow plan of desulfurization unit.

Table 1. Desulfurization Case Study.

Case	Base	Constant Area	Constant IHF
Product Sulfur wt% *	0.4	0.05	0.05
Hydrogen in Reactor Feed	Base	Base × 8	Base × 8
Reactor Inlet Temp.—°F	Base	Base + 35	Base + 35
Furnace Duty-kBTU/bbl Feed	0	28.0	9.8
Cooling Load-kBTU/bbl Feed	26.9	65.6	47.3
Heat Recovery-kBTU/bbl Feed	74.6	120.0	138.3
Heat Recovery Exchanger Area-ft^2	Base	Base	1.85 × Base
Pinch ΔT—°F	Base	Base + 57	Base + 6
Hot Pinch Temp—°F	Base	Base + 154	Base + 154
Total Energy Inputs-kBTU/bbl Feed			
Recycle Gas Compressor	6.3	50.2	50.2
H_2S Removal	7.3	9.8	9.8
Hydrogen Plant	35.3	48.7	48.7
H_2S to S	(9.7)	(13.0)	(13.0)
Furnace Duty	0	28	9.8
Total Energy Input	39.2	123.7	105.5
	Base	3.2 × Base	2.7 × Base

* Feed Sulfur = 1.5 wt%

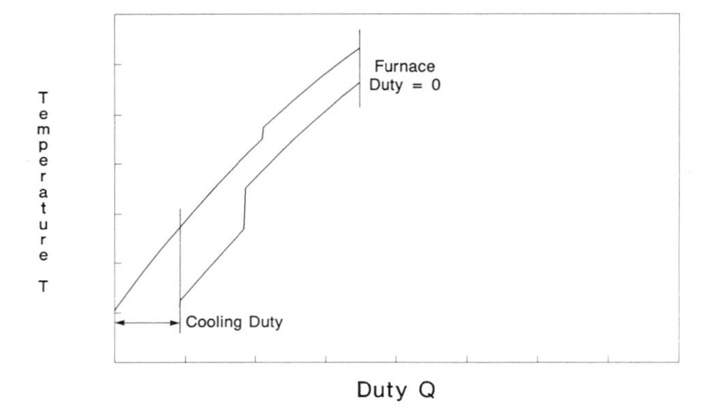

Figure 3. Present desulfurization unit composite streams.

the reaction is exothermic, the reactor outlet temperature being higher than the inlet. This is primarily due to the saturation of olefins in the feed by the hydrogen. Both the desulfurization and the olefin saturation reactions consume hydrogen, and higher desulfurization is accompanied by higher olefin saturation. During base case operation no furnace is required. The lower section of the lower line represents export heat to another process unit while the upper section represents feed preheat. For this study the desulfurization process was assumed to be fed "hot" from upstream separation units. The upper line represents the hot stream from the reactor being cooled prior to the hot separator. This is followed by further cooling of the hot product prior to tankage and the further cooling of the hot vapor prior to the desulfurization and hydrogen recycle. The juxtaposition of the two curves is determined by the amount of heat transfer area in the heat exchanger network that is used to recover heat from hot products to cold feeds. Increasing heat transfer area draws the two curves closer together in the duty (horizontal) direction. In the base case, sufficient heat transfer area was justified to reduce the furnace duty to zero. In general, area is added to the network until the cost of the last square foot of area recovers the present value of the energy saved in the furnace. This is the economic Incremental Heat Flux (IHF) (Ghamarian et al., 1985).

Figure 4 is the pinch diagram for the deep desulfurization case with the same total heat exchanger area as in the base case. Note that both the hot and cold composite curves have changed from the base case. The cold stream duty has increased because of increased hydrogen recycle and higher reactor inlet temperature, and the hot stream duty increased because of a higher outlet temperature and increased hydrogen flow. The reactor outlet is now hot enough that the diesel fuel is partially vaporized. A furnace is now required to supply heat during normal operation. Figure 5 is the "pinch" diagram for adding heat transfer area to the network such that the IHF is the same as the base case (constant energy economics). This moves the cold stream's line horizontally and reduces both the furnace duty and the cooling load.

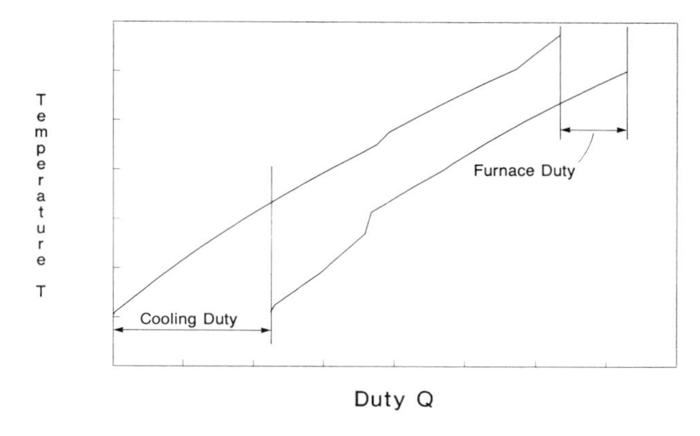

Figure 4. High desulfurization composite streams (constant area).

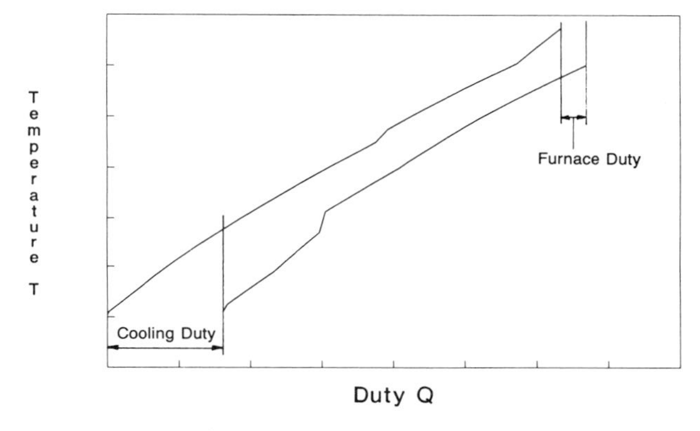

Figure 5. High desulfurization composite streams (constant heat flux).

In the Table 1 results, the furnace duty and cooling loads are given for all three cases. Note that the overall heat recovery for the constant area–high desulfurization case has increased. This is because the delta temperature (T) between the hot and cold streams has increased. One way to characterize this effect is to observe that the pinch delta T has increased by 57°F. For the constant IHF case (which represents constant energy price–heat exchanger cost economics) the pinch delta T has also increased by 6°F. There are two contributors to this increase: the hot pinch temperature has increased by 154°F and the hot streams inlet temperature minus outlet temperature has increased because of the increased reactor temperature. Thus, constant economics does not represent constant delta T at the pinch in this case.

The lower portion of Table 1 summarizes the total energy inputs required for this hypothetical case study. The energy requirement for the recycle gas compressor has increased by a factor of 8. The energy requirements for hydrogen sulfide removal and hydrogen production have also increased as shown. There is some additional recovery of energy from sulfur conversion. Overall energy consumption has increased by a factor of 3.2 for constant area and 2.7 for constant economics. Thus, improved environmental product quality has resulted in increased energy consumption.

As mentioned previously, this study may not represent the energy–environmental process effects for all processes or even all diesel hydrodesulfurization processes. However, it has been the general experience that there is an energy price for environmental improvements. Future research in improved catalysts, lower cost heat exchangers, and lower pressure drop systems are expected to help in offsetting additional energy consumption.

References

Cherrington, D.C., and Michelson, H.D., "How to Save Refinery-Furnace Fuel," *Oil and Gas Journal,* p. 59 (Sept. 2, 1974).

Eidt, C.M., Jr., "Energy Efficient Refineries: Making More from Less," *Proceedings of the Exxon Energy R&D Symposium,* p. 85 (May 1981).

Ghamarian, A., Thomas, W.R.L., Sideropoulos, T., and Robertson, J.L., "Incremental Heat Flux Method for Heat Exchanger Optimization" in *Analysis of Energy Systems—Design and Operation.* Vol. 1, ASME Book No. G 00322 (1985).

Gunderson, T., "Retrofit Process Design—Research and Application of Systematic Methods," *Proceedings—Foundations of Computer Aided Design—89,* Cache, Elsevier, p. 231 (1990).

Locket, W., Jr., "Guidelines for Energy Management," *Chemical Engineering Progress,* p. 57 (Aug. 1980).

Siegell, J.H., "Refinery Energy Conservation," *Hydrocarbon Processing* (forthcoming, 1990).

Spriggs, H.D., Petela, E., and Linnhoff, B., "Pinch Technology: Evaluate the Energy/Environmental Economic Trade-Offs in Industrial Processes," *Energy and the Environment in the 21st Century,* MIT Press, Cambridge MA (1990).

Zwicker, D.A., "Cogeneration Gains Momentum," *The Lamp,* p. 14 (Fall 1989).

Energy Consumption Spirals Downward in the Polyolefins Industry

William H. Joyce
Union Carbide Chemicals and Plastics Company Inc.
Danbury, CT

Introduction

There are few subjects more important than energy and the environment as we head toward the 21st century. Examination of the practices of individual industries should prove useful in the study of energy conservation. Let us focus on the polyethylene industry because it offers a particularly good example of how technological developments can drastically reduce the amount of energy required to produce a product and, at the same time, minimize a product's environmental impact.

Before tracing how energy consumption has declined in the polyethylene industry, it is important to put polyethylene in its proper perspective. First, it is made from by-products of refining crude oil into lubricating and fuel oil and gasoline, or from by-products of natural gas production. These by-products are first heated so that they break down into a two-carbon molecule called ethylene (see Figure 1). Next, the ethylene molecules are joined to form the long chain polymer that is polyethylene (see Figure 2).

Polyethylene is worth examining because it is an important part of the plastics industry and plays an important role in the daily lives of everyone. This has been true for many years, and it will continue to be true well into the 21st century.

How important is polyethylene? Last year, nearly 60 billion pounds of various plastics were sold in the United States. Of that, almost one third of the total, or 18.7 billion pounds, was polyethylene. The second largest volume resin, polyvinyl chloride, was less than half the volume of polyethylene (see Figure 3).

Where does all that polyethylene go? Its uses range from the most sophisticated to the most commonplace, making polyethylene not only the world's largest volume resin, but also the most versatile. It is found in industrial applications, in agriculture, and most definitely around the home.

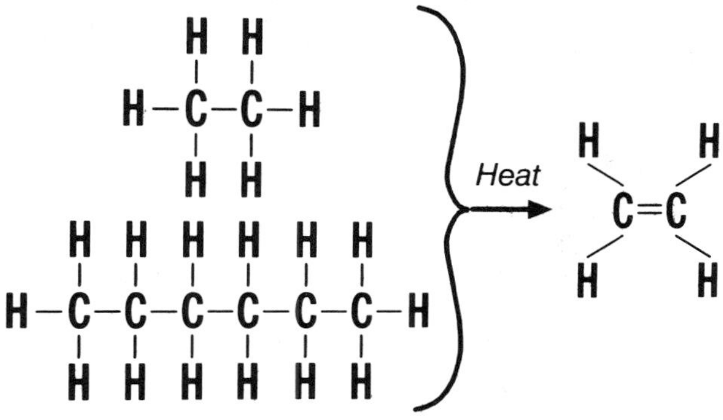

Figure 1. Chemical pathways for ethylene production.

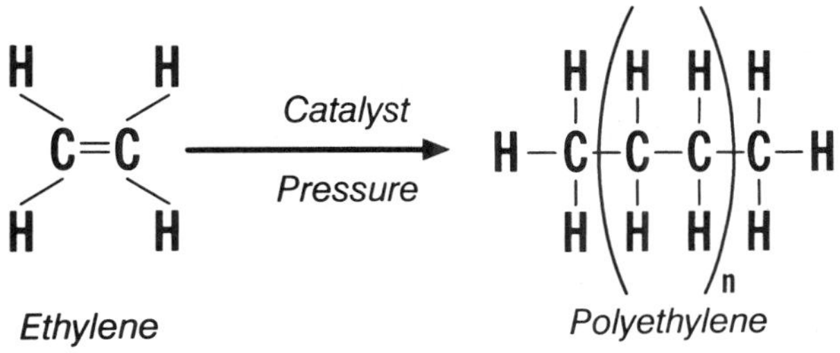

Figure 2. Chemical pathways for polyethylene production.

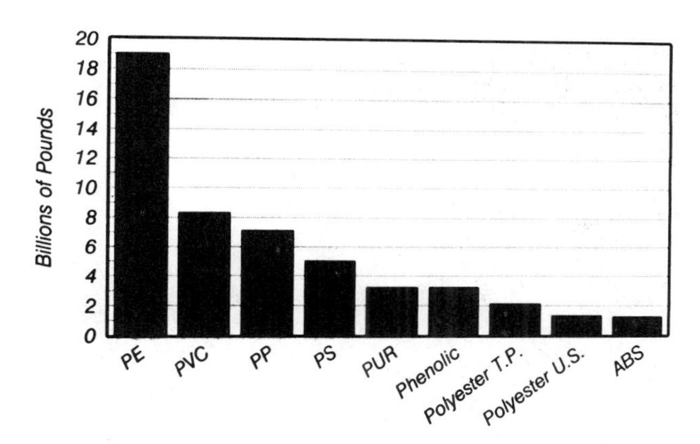

Figure 3. 1989 US plastics consumption. (PE = polyethylene, PVC = polyvinyl chloride, PP = polypropylene, PS = polystyrene, PUR = polyurethane, ABS = acrylonitrile butadiene styrene)

One of the earliest and most sophisticated uses of polyethylene was as an insulation and jacketing material for communications and power cable. Polyethylene is used to produce containers of all sizes and shapes. There are, for instance, polyethylene drums that range from five to 55 gallons in capacity. Polyethylene goes into a variety of plastic bottles, plastic closures, and lids. Polyethylene toys and a wide assortment of polyethylene housewares are common. It is used for hose, tubing, and pipe.

But by far the largest volume use of polyethylene is in plastic film and sheet. These films are used to wrap everything from fruits and vegetables to pallets of industrial products, such as flour, fertilizers, and even other plastics. Polyethylene bags range from plain garbage sacks to elaborately printed bags used in many of the country's finest stores (see Figure 4). It is also used for mulch film around plants to prevent weed growth and conserve water.

So much for the huge size and variety of the polyethylene market. Obviously, all of this did not happen overnight. It was in the 1930s that researchers discovered that ethylene could be polymerized to a high molecular weight resin at high pressure and in the presence of trace amounts of oxygen. But it was not until the 1950s that groups in the United States and in Europe, working independently, discovered that polyethylene could be made at low pressure with heterogeneous catalysts. Polyethylene offers an excellent case study of how processes evolve—slowly but surely over time—getting better with age and experience.

Evolutionary Process Improvement

Evolutionary improvement over time is certainly not unique to polyethylene. It is common in many industries and products. Studies have examined the change in cost over time and have found that there is about a 20 percent reduction in cost to produce the same product every time the cumulative production volume of that product doubles. This concept can be traced back to 1936 when T. P. Wright, participating in the development of the aircraft industry, first described the phenomenon.[1]

Figure 4. Linear low density polyethylene (LLDPE) merchandise bags.

More recently, The Boston Consulting Group made the concept more widely known by expanding upon it in *Perspectives on Experience*.[2] This generalization certainly holds true for polyethylene.

The rate of change of cost for the high pressure production process, as demonstrated by new plants built between 1943 and 1981 is a 19.6 percent cost reduction with each doubling of industry cumulative production volume (see Figure 5). Big gains were made in investment, staffing, overhead, and raw material efficiency from 1943 to 1981. During this period, raw materials efficiency increased from 75.3 percent to 98.9 percent, while raw material losses decreased from 24.7 percent to 1.1 percent. Plant capacity increased from one million pounds to 550 million pounds. Plant investment increased from $4 million for the 1943 plant to $112 million for a plant built in 1981. Since output increased dramatically, the cost per pound decreased from $27.78 to $0.20 cents per pound.[3]

The number of people involved in production has remained surprisingly constant. Even the relative mix of each skill is unchanged. But the number of people per million pounds of production drops from 84 to 0.12. Overhead—that is all the people costs for research and development, marketing, sales, and administration—drops from $0.81 per pound to $0.02.

Utilities are a major cost factor in making polyethylene. Massive savings on utilities are apparent. One-eighth the electrical energy and one-fortieth the low pressure steam were needed in 1981 compared to 1943. Much less cooling water is required today to carry away the waste heat. Losses of raw material are only one twenty-fifth of what they were.

Let us look at the plants themselves to see what changes have occurred over almost 40 years. Figure 6 shows the first polyethylene plant. The scale was relatively small, on the order of one million pounds per year. The major problems in building and operating this plant related to the high pressure used in the process. Pressures up to 50,000 psi and special catalysts were used to get ethylene to form polyethylene.

A modern high-pressure plant is shown in Figure 7. This is the largest, newest high-pressure plant in the United States. This plant, with two lines, is capable of making 500 million pounds of low-density polyethylene (LDPE) per year. High pressure and the danger of fire and explosion are still the biggest concerns.

New techniques of pumping, new catalysts to speed polymerization, new heat transfer techniques, better ways to separate the finished polymer, and, of course, greater scale of operation were the principal technological changes from 1943 to 1981.

The reader should take note of the relative size of the cooling fan shown in Figure 7. It allows comparison

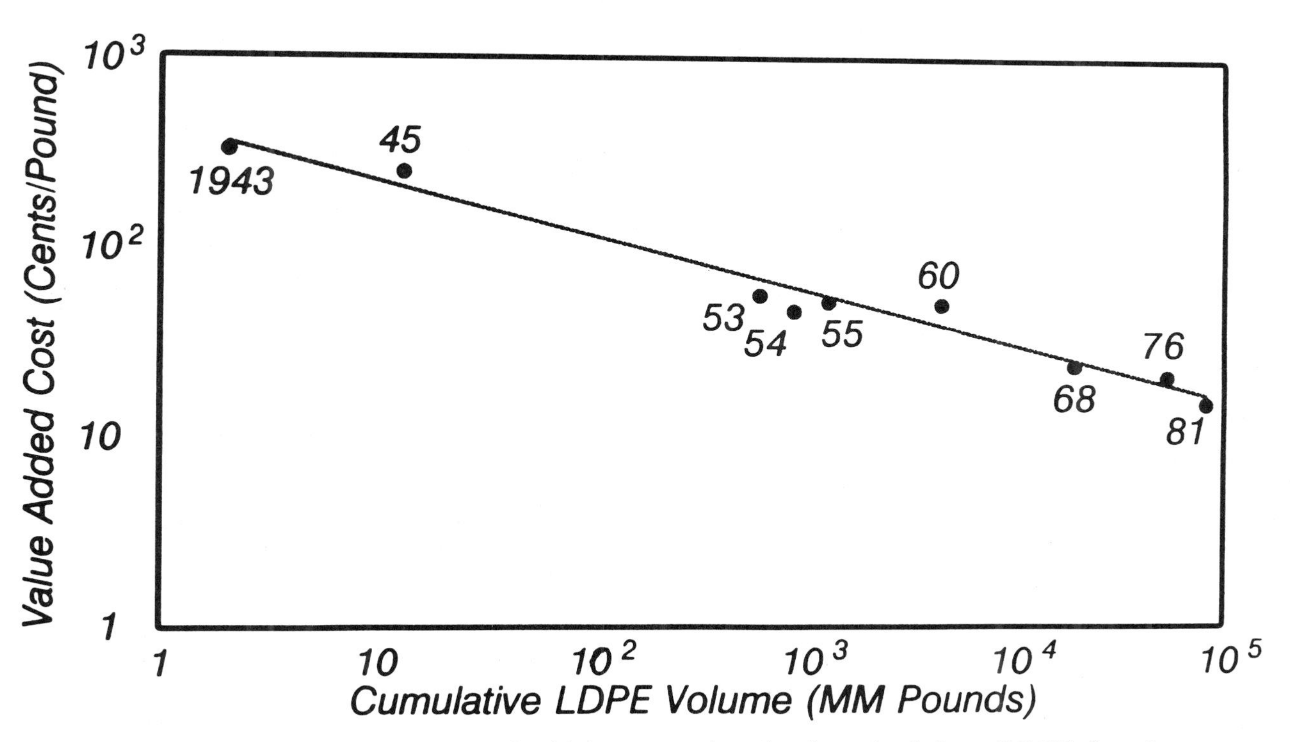

Figure 5. 1943 to 1981 experience curve for high-pressure low density polyethylene (LDPE) (based on constant dollar, value added obtained with 1981 factor prices).

Figure 6. One of the first high-pressure polyethylene plants.

of the size of the new low-pressure reactors to these older, high-pressure plants.

Revolutionary Process Improvement

Sometimes a technological breakthrough comes along that revolutionizes an industry. These breakthroughs seem to happen instantaneously, although, in truth, many years of hard work go into their development. A case in point is the use of light to replace electrical signals in telephones. Bell Labs was developing this technology in the late 1950s and early 1960s, but it was not put into practice until 20 years later.

In 1968, a similar dramatic development revolutionized the polyethylene industry. That development was the first commercial production of polyethylene using Union Carbide's gas-phase, low-pressure Unipol™ process. When Unipol was introduced, the benefits of going to a gas-phase, low-pressure process and eliminating the disadvantages of high pressure and solvent-based processes seemed all too obvious. However, as with Bell Labs' experience with light and telephones, making a gas-phase, low-pressure process work correctly and economically in a world-scale commercial plant took many years of detailed research and development. Based upon what the Unipol process has done to and for polyethylene production, it certainly seems to qualify as a revolutionary development. Today, there are 102 Unipol reactors in operation or under construction in 25 countries throughout the world with a capacity of 19.4 billion pounds per year of polyolefins.

Improvement in Energy Efficiency

One way to measure the amount of energy used in making polyethylene, and the energy conservation that has been realized over the years, is to use what we at Union Carbide call the energy usage index. This is the net energy input to each production output unit, no matter what form the energy input takes—electric power, steam, cooling water, etc. It is based on all energy used as though each input started with natural gas as the sole source of energy. It includes all energy losses along the way. For example, the losses to make electricity are about one-half the starting energy. The energy usage

Figure 7. High-pressure LDPE plant with cooling fans highlighted.

index is expressed in fuel equivalent British Thermal Units (or Btu) per pound of polyethylene.

The low-density polyethylene plants serving the industry in the 1940s—the ones with high-pressure pumping units and stirred autoclaves or tubular reactors—have an energy usage index of 8,400 Btu per pound. While these figures are based on experience at Union Carbide, they are typical of the entire industry.

The early high-density polyethylene plants used slurry loop reactors. Although these reactors operated at much lower pressures than the high-pressure autoclave and tubular reactors, they required a large amount of liquid hydrocarbon diluent. They also required separation and recycle operations, and thus a tremendous demand for utility input. This, of course, translates into energy. As a result, the energy usage index of these plants, although lower, was typically about 5,500 Btu per pound.

By the mid-1970s, Union Carbide had incorporated new technology into its new high-pressure tubular bulk reactors. This technological advance lowered the energy usage index to 4,400 Btu per pound—about half of that in the earlier traditional high-pressure plant.

Let us now take a look at the revolutionary development in polyethylene technology that I mentioned earlier—the development of gas-phase, low-pressure technology, such as the Union Carbide Unipol process. The first commercial Unipol high-density polyethylene (HDPE) plant came onstream in 1968 at Union Carbide's petrochemical complex in Seadrift, Texas. The outstanding feature of the Unipol polyethylene process and other similar processes is their simplicity. They take ethylene, which is a gas, expose it to highly sophisticated catalysts, and convert it directly into solid polyethylene. The ethylene is cooled and recycled, removing the heat of reaction (see Figure 8).

The older polyethylene plants were huge and bulky. By comparison, a Unipol process reactor is simple and compact (see Figure 9). It is easy to build, easy to operate, and easy to maintain. One of these reactors would fit into the cooling fan for the high-pressure process mentioned earlier, but will produce twice as much product.

The Unipol process is also extremely versatile. Each of the older processes produces a different kind of polyethylene. One process produces low-density polyethylene, the flexible resin that goes primarily into plastic film. Another produces high-density polyethylene, the

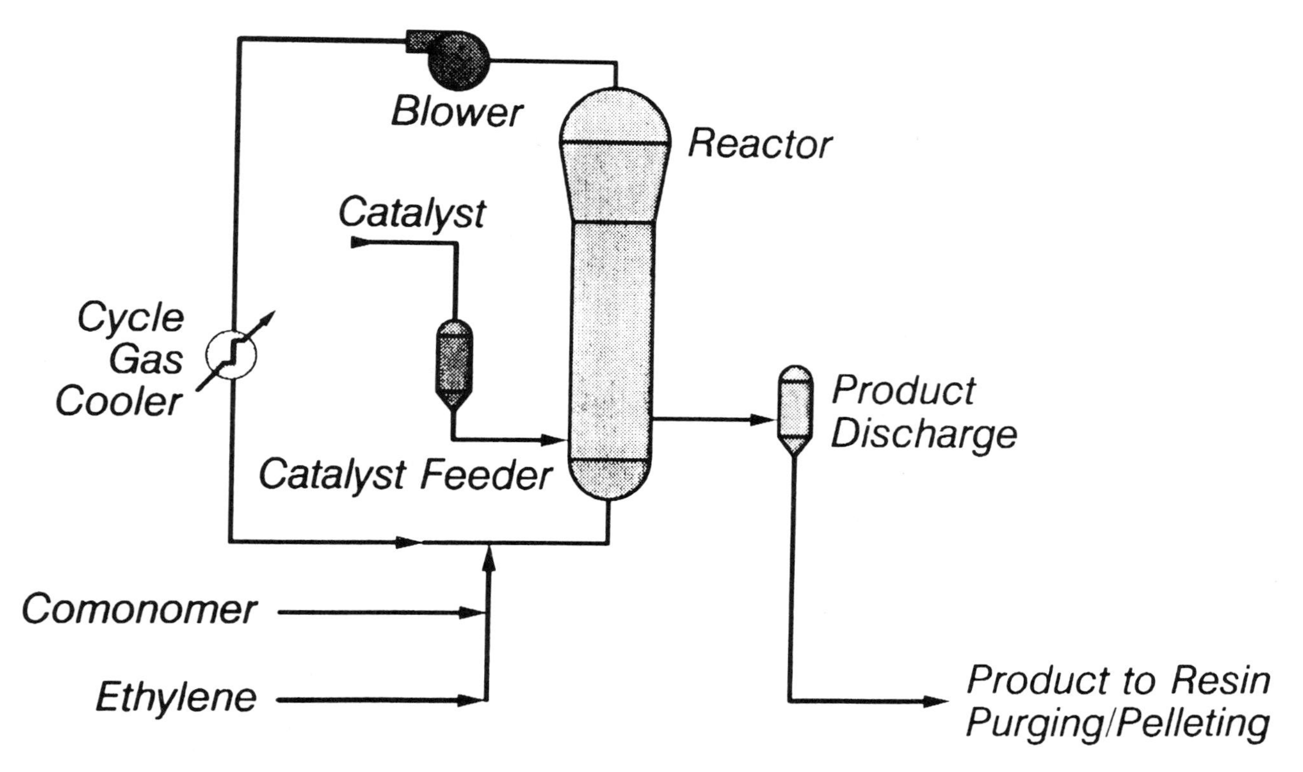

Figure 8. Unipol process flow diagram.

more rigid type that finds its way into molded bottles and drums.

The Unipol process is a single process that can make both varieties—both high-density and low-density material. In addition, the low-density material produced from this process is a new type of polyethylene that itself has revolutionized the plastics industry. We know this new material as linear low-density polyethylene.

As a rule of thumb, the Unipol process reduces capital investment in polyethylene plants by more than 50 percent, compared to plants based on high pressure technology. It even has a 20 percent to 35 percent capital investment advantage over some of the new solution or slurry type plants and other gas-phase linear low-density polyethylene plants.

Comparing the energy conservation of the various processes, we can see the changes that have happened to the energy usage indices since gas-phase, low-pressure polyethylene processes became available. The first Unipol high-density polyethylene unit that started up in 1968 had an energy usage index of 3,800 Btu per pound—about 70 percent of that of other low-pressure processes.

In 1977, the first world scale linear low-density polyethylene plant came onstream. Among other things, it featured higher production capacity per unit volume of reactor space. This increase in production was the result of improvements in the catalyst and the use of streamlined recycle systems. Together, these improvements enabled the Energy Usage Index to drop to 2,200 Btu per pound.

During the 1980s, improvements in gas-phase technology continued at a rapid pace. Energy consumption was further reduced by operating the reaction system in the condensing mode. This simply meant that the cooling gases entering the bed contained a mist of condensed gas. Because the new technology was more efficient at removing the heat generated by the reaction, higher production rates could be attained. Thus, for a given reactor size, higher production rates were possible, thereby reducing energy consumption per pound of polyethylene produced.

In addition, improved systems for the reactor-to-discharge product reduced the energy used to recover gases that leave the reactor with the product. Close coupling of the pelletizing step to the reactor itself eliminated double handling of the resin and captured the heat of reaction for a useful purpose. Melt pumps replaced extruders for more efficient pumping of resin to the pelletizer.

The net result of all these improvements was another huge drop in the energy needed to make polyethylene. The energy usage index of a new Unipol unit dropped to 1,500 Btu per pound of polyethylene.

Figure 9. Union Carbide's Star Plant, Taft, Louisiana, has an annual polyethylene capacity of 1 billion pounds.

Therefore, as a result of technology improvements the 8,400 Btu per pound of polyethylene in the old, high-pressure units was reduced to only 1,500 Btu per pound in the new gas-phase, low-pressure process plants. That is an 85 percent reduction in the energy required for the process. The improvement is particularly significant since it occurred in the process used to make the world's largest-volume plastic. (See Figure 10.)

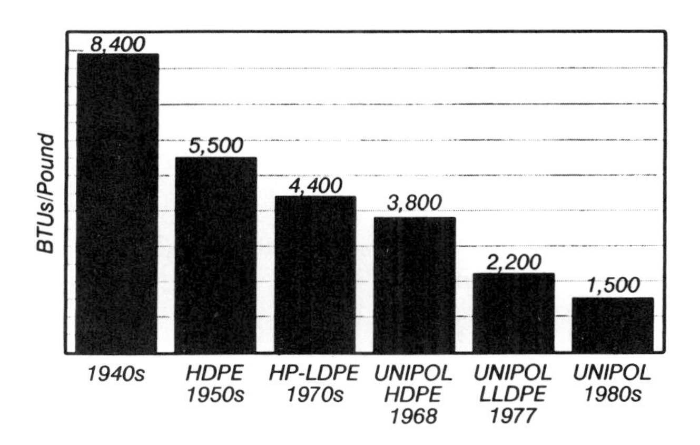

Figure 10. Polyethylene energy usage index.

Related Energy Savings

The savings discussed are just the obvious process savings. Not included are the savings in the energy "inputs" into the production plant itself, for example, energy to produce the material from which the plant is built—bricks, mortar, steel, and so on—and the energy to move that material to the plant construction site and build the plant. If a plant requires less material to build—and a plant using the Unipol process certainly does—there is a net energy savings in materials of construction.

Another item not considered is raw material efficiency. If a plant uses less feedstock—in this case, ethylene—to make a pound of polyethylene than other plants, there is a net energy savings in raw material. Compared to the new gas-phase plants, the original polyethylene plants "wasted" roughly one third of a pound of ethylene per pound of polyethylene produced. That represents a lot of hydrocarbons that could have been used to heat homes or to make other petrochemicals.

Even the number of people required to operate a plant can affect energy consumption. The number of people required to operate a polyethylene plant has not changed radically over the years. But while it took 84.4 workers to produce a million pounds of polyethylene in 1943, it took only 0.12 workers to produce the same volume in 1981. These workers are energy users as well, even though they don't show up "on the meter." For instance, they all use gasoline to drive to work.

Finally, let's consider "down-gauging," a term applied by the industry to the practice of using less of a better material to replace more of an inferior material. Linear low-density polyethylene is a unique material. It makes stronger, more tear-resistant film. It makes tougher, more impact-resistant molded products.

Because they are so much stronger, products made from linear low-density polyethylene can be made thinner. Less of this material—from 15 percent to 50 percent less—will do the same job as larger quantities of conventional high-pressure polyethylene. This, too, results in tremendous energy savings all along the product chain—in the fabrication of the product, the polyethylene needed to make it, and the ethylene needed to make the polyethylene.

With these advantages, the demand for linear low-density polyethylene has been growing dramatically, while demand for conventional low-density polyethylene is stagnant. Last year, an estimated 3.4 billion pounds of linear low-density material was consumed in the United States, with most of that going into film markets.

The Effect of Plastics on Municipal Solid Waste Problems

The other subject of this conference is the environment. Few issues have commanded the attention of the chemical industry in recent years as much as the environment. Despite some scare headlines, the fact is that chemical companies have taken some giant steps toward cleaning up their plants, their emissions, and their waste products. This is true of polyethylene producers. The polyethylene industry has seen a dramatic reduction in emissions to the environment over the past two decades. And it is no exaggeration to say that gas-phase, low-pressure processes, such as Unipol are the cleanest processes and, therefore, have been a major factor in achieving this improvement.

In safety as well, the new units are greatly improved. With the old high-pressure units, a serious incident occurs on an average of once in 20 to 30 reactor years of operation. The new gas-phase Unipol reactors have operated 450 reactor years without a serious incident.

But, in making plastic more energy and cost efficient, are we in fact compounding our nation's solid waste problems? This question can be answered by providing a little insight into the "plastics in solid waste" issue. Dr. William L. Rathje, an anthropologist from the University of Arizona, is widely considered one of the nation's leading authorities on the subject. For 15 years, he has scientifically studied what Americans throw out by digging up and examining landfills around the country. What he found is that plastic on average is only 7.3 percent by weight and 18 percent by volume of the municipal waste stream—compared to paper and paper board, which are 36 percent by weight and 38 percent by volume.[4]

According to Dr. Rathje, plastics are not increasing as a percentage of waste, even though the pounds of plastics used have increased. The reason for this seeming anomaly is that plastic items, like bottles, are now made with thinner walls and, therefore crush and compact better in landfills and take up about the same volume as they did 20 years ago.[5]

Plastics also help shrink food waste via modern packaging techniques that reduce spoilage. The commercial preparation of food removes most of the unusable parts in advance. This leaves paper as the component that is steadily growing as a percent of the waste stream.[6] With new programs under way for using recycled plastics, these products will be more cost and environmentally effective.

In addition to being recycled, waste polyethylene can be used as a fuel supplement. Polyethylene, for example, has a higher heat content than natural gas and burns cleaner than coal or oil. The industry is working to make discarded plastics usable again through recycling and as fuel supplements.

Conclusion

Through examining how technological change can reduce the amount of energy required to produce a product, four key points appear to apply generally to the plastics industry.

1. Every process, no matter how mature, has the potential to save energy.
2. This potential reaches well beyond the process utilities.
3. Plastics are not the sole villain in the solid waste problem. They are a small part of a larger societal problem.

4. There are major advances in processes that will eventually recycle waste plastics to useful products and/or to fuel supplements.

References

1. T. P. Wright, "Factors Affecting the Cost of Airplanes," *Journal of Aeronautical Sciences,* Vol. 3, February 1936.
2. Boston Consulting Group, *Perspectives on Experience* (Boston, Massachusetts, 1968).
3. Union Carbide production records.
4. "Solid Waste Solutions," Dr. William Rathje, Director, Le Project de Garbage, University of Arizona, Speech presented October 4, 1989, to Union Carbide, Polyolefins Division Business Meeting, Orlando, FL.
5. *Ibid.*
6. *Ibid.*

Unipol is a trademark of Union Carbide Chemicals & Plastics Technology Corporation

Energy Management and Conservation in the Pulp and Paper Industry

Howard J. Herzog
Jefferson W. Tester
Energy Laboratory
Massachusetts Institute of Technology
Cambridge, MA

Abstract

From 1972 to 1988, the overall energy intensity of pulp and paper manufacture in the United States has declined 13 percent, with a 32 percent reduction in the intensity of purchased energy. The technology exists to reduce the total energy intensity to about 50 percent of its value today by the year 2010. However, economics will determine the real magnitude of future energy savings. Extrapolating the current energy efficiency trends to the year 2010 results in an additional reduction of total energy intensity for pulp and paper manufacture of 20 percent. Key areas for significant energy savings are in Kraft pulping, chemical recovery, paper drying, and overall heat integration of the process. Not only is the pulp and paper industry the leader in cogeneration, but the 1980s has seen an additional 20 percent growth in cogeneration of electricity relative to pulp and paper production. Recycling is important for the industry, but materials issues drive its use more than energy issues. Finally, in addition to the standard concerns associated with the combustion of fuels for energy, environmental effects are strongly linked to sulfur-based pulping and chlorine-based bleaching.

Introduction

In the US, the pulp and paper industry (Standard Industrial Classification (SIC) code 26) uses about 12 percent of the primary energy consumed by the manufacturing sector. This amounts to about 2.5 quads (1 quad $= 10^{15}$ BTU $= 1.06 \times 10^{18}$ J). It ranks fourth behind petroleum (SIC code 29), chemicals (SIC code 28), and primary metals (SIC code 33). It is a highly energy intensive process, with fuel costs accounting for about 20 percent of the value-added (value of finished goods minus value of raw materials).

In addition to energy, the economics of pulp and paper manufacture have other important considerations. The cost of raw materials (mainly roundwood) is about four times the cost of energy. Potential environmentally harmful effluents are associated with chemical pulping (sulfur-based chemicals), bleaching (chlorine-based chemicals), and recovery furnace operations (particulates and SO_x emissions). Finally, pulp and paper manufacturing is highly capital intensive and competitive, resulting in a relatively small profit margin.

Within this economic and environmental context, the object of this paper is to review the status of energy management and conservation in the pulp and paper industry. This paper presents historical and projected energy use data, identifies opportunities for increasing energy efficiency, and discusses key trends concerning energy use and environmental effects.

The MIT Energy Laboratory is currently conducting a research project on energy management of pulp and paper mills in Maine, under the sponsorship of the Central Maine Power Company. As part of our research, we reviewed studies on energy use in the pulp and paper industry (Hersh, 1981; Elaahi and Lowitt, 1988; Giese, 1988). Data from these studies, the American Paper Institute (API), and the United States Energy Information Administration form the foundation of this paper. Unless stated otherwise, all statistics reported in this paper are based on US data only.

Pulp and Paper Manufacture

The manufacture of paper starts with a source of fiber. Wood is the prime source of fiber, with wastepaper providing 23 to 24 percent and other sources, such as cotton, providing just a few percent. The wood is first debarked and then cut up into chips. In the pulping process, the chips are then broken down into their basic constituents (i.e., lignin, cellulose, etc.) by chemical and/or mechanical means. About 81 percent of the wood is pulped by chemical processes, 10 percent by mechanical processes, and 7 percent by a combination of the two (Elaahi and Lowitt, 1988).

About 35 to 40 percent of the pulp produced in the US is bleached to specified levels of whiteness and brightness. Bleaching is an area of significant environmental concern because most of the bleaching processes use chlorine or chlorine-based chemicals. The bleached or unbleached pulp is made into a furnish, a mixture of pulp (less than 1 percent) and water. The furnish is formed into sheets of paper on large paper machines, then the sheets are pressed and dried to remove the water.

Worldwide, the fiber sources for papermaking are shown in Figure 1. A major difference between the US and the rest of the world is in the use of recycled fiber. The US uses only 23 to 24 percent recycled fiber, compared to about 35 percent for western Europe and almost 50 percent for Japan. Implications of using recycled fibers are discussed later in this paper.

Chemical pulp is produced almost exclusively by the sulfate or Kraft process. The process accepts all kinds of wood and the resulting pulp is of high strength (Kraft is the German word for strong). An important component of Kraft pulping is chemical recovery, where the pulping chemicals are recovered for reuse and the lignin and other unwanted constituents of the wood are burned for energy. The process stream sent to chemical recovery is termed black or spent liquor. The mechanical pulping processes have about twice the yield of the Kraft process because mechanical pulp contains all the wood constituents, not just the cellulose. Mechanical pulping processes are less capital intensive, but more energy intensive, than the Kraft process.

Primary uses of Kraft pulp are for printing and writing papers and for paperboard (paper over 0.012 inches thick). Mechanical pulps are used in lower grade papers, such as newsprint and tissue paper. Table 1 shows US paper and paperboard production figures for 1985 (Elaahi and Lowitt, 1988).

Energy flows in the Kraft process are shown in

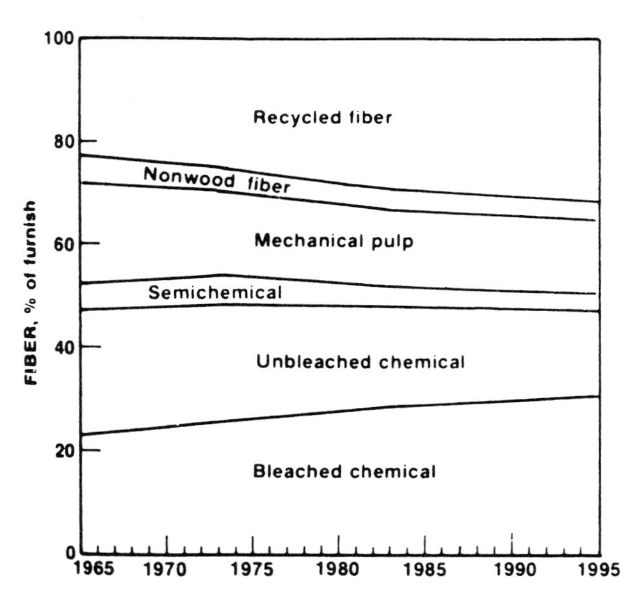

Figure 1. Sources of fiber for pulp and paper manufacture worldwide (Pöyry, 1986).

Table 1. 1985 US Paper and Paperboard Production by Grade. Data from the American Paper Institute.

Paper/Paperboard Grade	Thousands of ADMT*
Paper	
Newsprint	4,940
Printing/Writing Paper	16,790
Industrial Paper	4,730
Tissue Paper	4,490
Subtotal	30,940
Paperboard	
Kraft Paperboard	14,880
Semichemical Paperboard	4,630
Solid Bleached Paperboard	3,590
Recycled Paperboard	6,940
Subtotal	30,040
Construction Paper	1,530
Total	62,510

* ADMT = Air Dried Metric Ton = 1000 kg

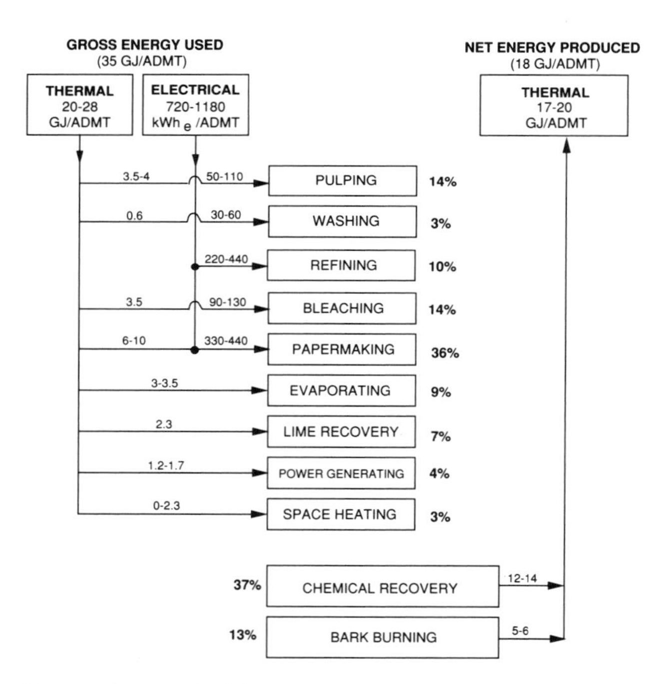

Figure 2. Typical integrated Kraft mill energy flows (Hersh, 1981).

Figure 2. Over one third of the energy goes to papermaking, primarily to press and dry the paper product. Other large users of energy include the digesters (pulping), the bleach plant, and the evaporators (to concentrate the black liquor during chemical recovery). An energy balance for a typical US Kraft mill is shown in Figure 3. Over 50 percent of the energy consumed is produced by self-generated fuels. However, as a result of lower energy intensities and even greater use of self-generated fuels, some integrated Kraft mills require no purchased fossil fuels to generate process steam.

Energy Consumption

Data on energy consumption in the pulp and paper industry has been gathered and analyzed by the American Paper Institute (API) from 1972 to the present, and serves as the source of historical data used in this paper. Because 1972 was the last full year before the first "oil shock," it is used by the industry as a baseline. For easy comparison, the energy consumption from all sources is converted to equivalent thermal energy units. We use API heating values/conversion factors for all energy sources except purchased electricity. API reports purchased electricity without accounting for energy losses associated with electric power generation. For this paper, these energy losses are included by using a conversion factor of 11.13 x 10^{-3} GJ/kWh_e (10,500 BTU/kWh_e) for thermal energy consumed per unit of purchased electricity. This corresponds to an overall thermal cycle efficiency of 32.5 percent.

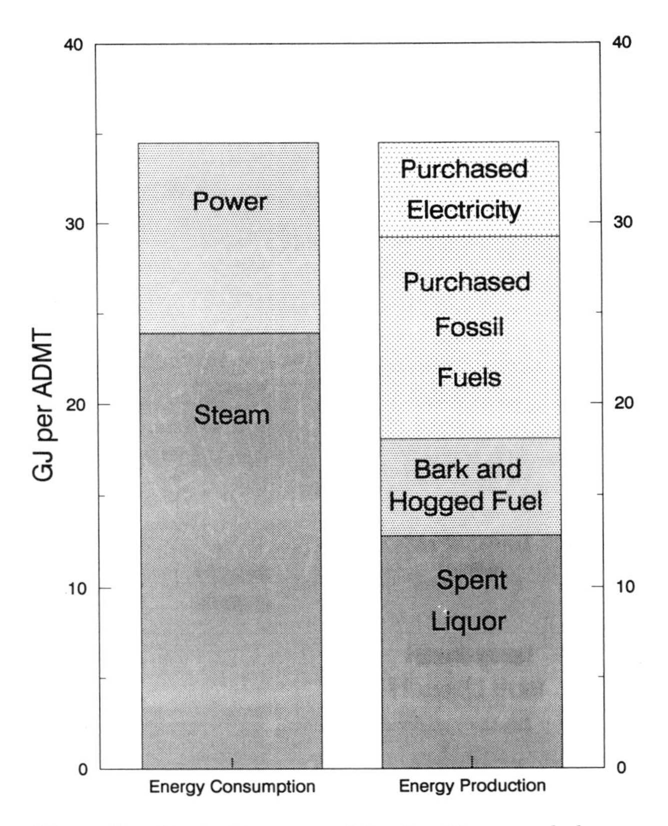

Figure 3. Typical integrated Kraft mill energy balance. Purchased electricity includes energy losses (1 GJ = 90 kWh_e).

Figure 4 presents energy consumption data for the pulp and paper industry plotted as energy intensities (the total amount of GJ of thermal energy consumed, including losses, divided by the tonnage of production). The energy sources are divided into two categories: purchased fossil fuels and electricity and self-generated energy from spent liquor, bark, and other wood wastes. Additionally, the purchased electricity component is plotted as kWh_e/ADMT (Air Dried Metric Ton). Analysis of Figure 4 leads to the following observations:

1. From 1972 to 1988, the overall energy intensity of pulp and paper manufacture decreased by about one eighth (41.1 to 35.6 GJ/ADMT).
2. From 1972 to 1988, the purchased energy intensity of pulp and paper manufacture decreased by about one third (25.9 to 17.7 GJ/ADMT). The majority of this decrease was caused by replacing purchased energy with self-generated energy.
3. Currently, self-generated energy supplies over 50 percent of the energy requirements for pulp and paper manufacture, up from 37 percent in 1972.
4. The energy intensity of purchased electricity has remained relatively flat over the past 10 years.

According to the American Paper Institute, purchased electricity has made up between 51 and 54 percent of the total electricity requirement for each year since 1982. This means that the total electricity energy intensity, as well as the purchased electricity energy intensity, has remained essentially constant over this period, which corresponds to the trend of the entire industrial sector. Gains in more efficient use of electrical energy are offset by new uses of electrical energy, keeping the electrical energy intensity essentially unchanged (Ross, 1989). The trend toward more power-intensive processes and equipment is fueled by goals of higher quality, higher yield, and better environmental performance, subject to the usual economic constraints. Examples of electrification in pulp and paper manufacture include replacing steam drives with electric motor drives, using vapor recompression for black liquor concentration, and increasing water removal in the press section of paper machines. These last two examples are discussed in the next section.

Figure 5 compares the improvement in energy efficiency of pulp and paper manufacture to the industrial sector as a whole. The industrial sector energy intensities

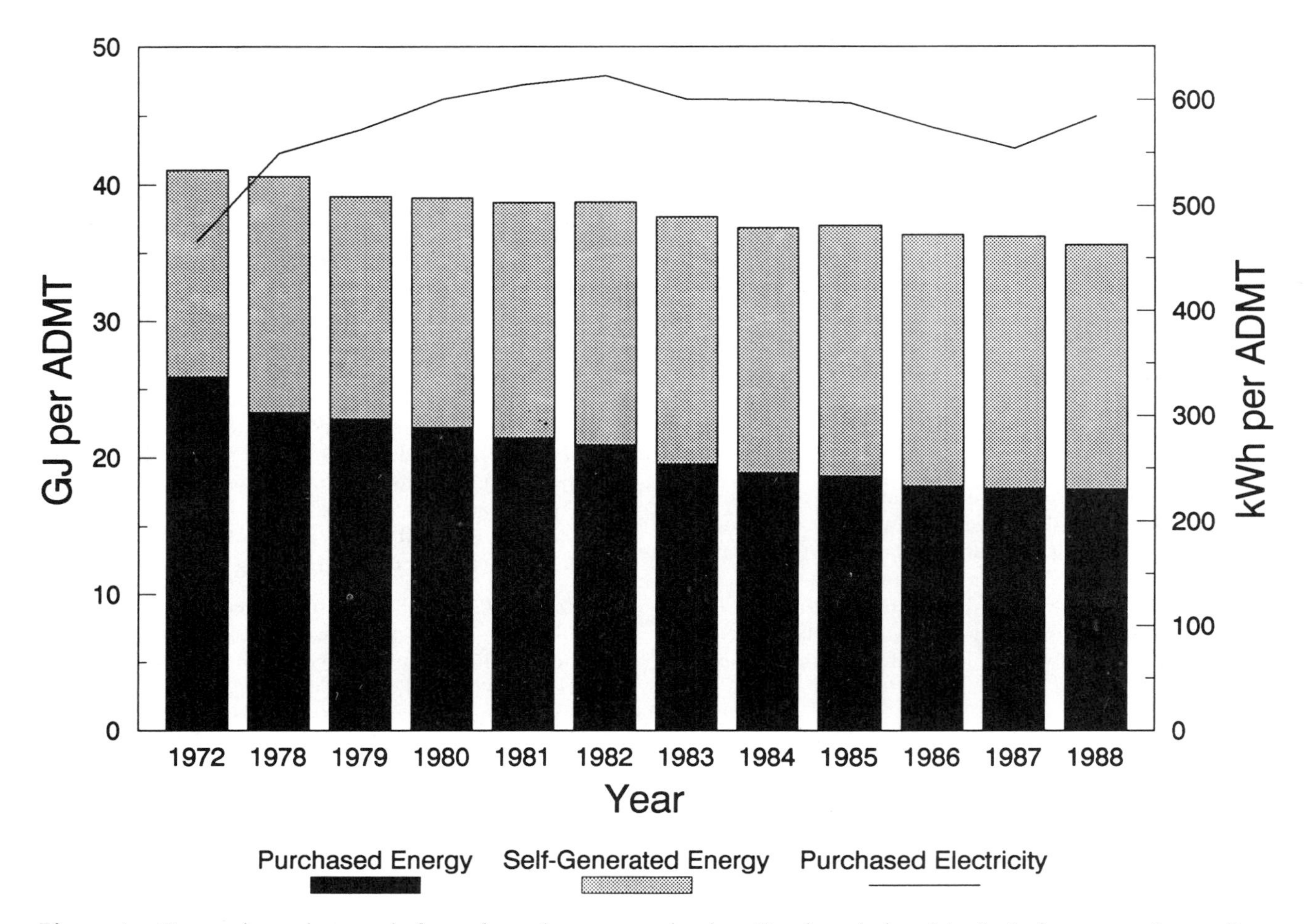

Figure 4. Energy intensity trends for pulp and paper production. Purchased electricity includes energy losses. Data from the American Paper Institute.

are calculated by dividing the total energy consumption (including losses from electricity generation) by the energy-weighted industrial output in constant dollars. Using an energy-weighted industrial output corrects for the shift from energy-intensive industries to less energy-intensive manufacture. All data in Figure 5 are normalized to their 1972 values.

How one appraises the pulp and paper industry's performance on implementing energy efficiency improvements depends partly on whether total energy intensity or purchased energy intensity is used as an evaluation criterion. API has successfully argued with the US government that purchased energy intensity should be the only criterion for measuring the energy efficiency of the industry (Slinn, 1981). However, the authors of this paper feel that total energy intensity is also an important and useful measure of overall process efficiency. Nonetheless, US energy policy since 1972 has focused on reducing dependence on oil and gas, and the pulp and paper industry receives a very good grade in this category (see Figure 6).

Figure 7 presents data on energy efficiency targets for the pulp and paper industry in the year 2010 (Elaahi and Lowitt, 1988). Four scenarios are considered: 1) current energy practices, 2) extrapolation of energy intensity data from 1972–88 to 2010, 3) implementation of proven state-of-the-art technologies available today, and 4) implementation of advanced technologies projected to be commercially available by the year 2010. For comparison, the energy consumption levels assuming 1972 intensities are included. Based on these projections, one may infer that the potential for energy efficiency improvements during the next 20 years is significantly larger than the energy efficiency improvements implemented since 1972. Overall, the energy intensity of pulp and paper manufacture can be cut approximately in one half by the year 2010.

From a technical viewpoint, the potential for energy

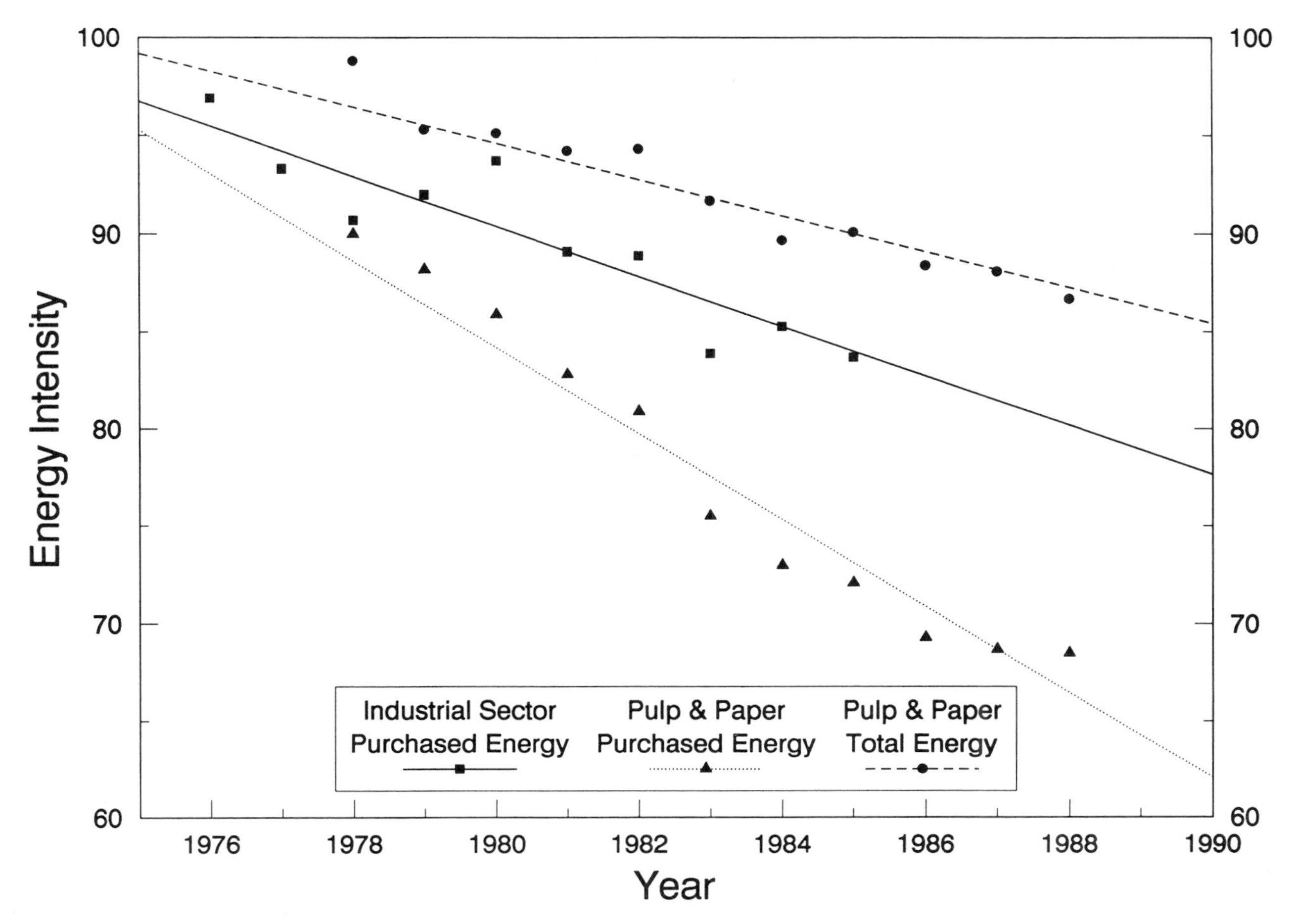

Figure 5. Comparison of energy intensity trends for pulp and paper production to the industrial sector as a whole. All intensities normalized to 100 for 1972. Data from the Energy Information Administration and the American Paper Institute.

efficiency improvements in pulp and paper manufacture appears large. However, to achieve these savings in practice, economic, environmental, and regulatory issues must also be considered.

Examples of Improving Process Energy Efficiency

In this paper, it is not possible to review all the approaches possible to achieve better energy efficiency in the pulp and paper industry. Detailed discussions are provided elsewhere (Hersh, 1981; Elaahi and Lowitt, 1988; Giese, 1988). In this section, four examples are presented to illustrate how the energy intensity of pulp and paper process can be reduced.

Digestion

As seen in Figure 2, about one seventh of the energy required to make Kraft paper is consumed in pulping, with the digester being the main energy consumer. The digester energy consumption can range from 1.5 GJ/ADMT to over 9 GJ/ADMT. Consumption is minimized by using a state-of-the-art continuous digester. In addition to being energy efficient, continuous digesters offer the advantages of being compact in size and automated (leading to lower labor costs). The alternative, less efficient batch digestion, offers lower capital costs and allows for more flexible operations. Still, the present trend in the US is toward the use of continuous digesters.

The majority of batch digesters in existence today in the US still use direct steam heating. By using indirect steam heating, batch digesters can save 5 to 15 percent of the energy consumption. Furthermore, heat recovery through a process called displacement heating can make

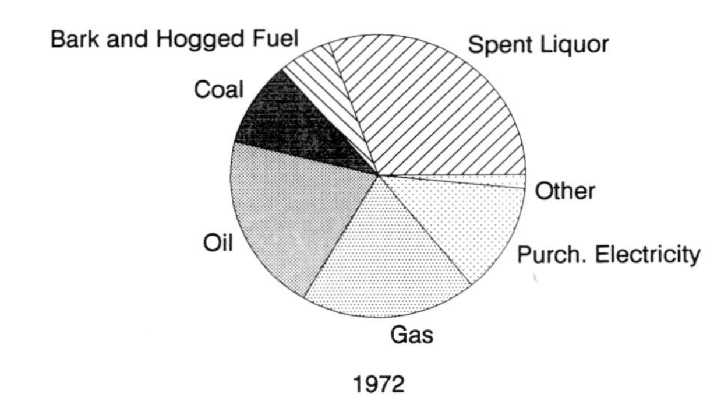

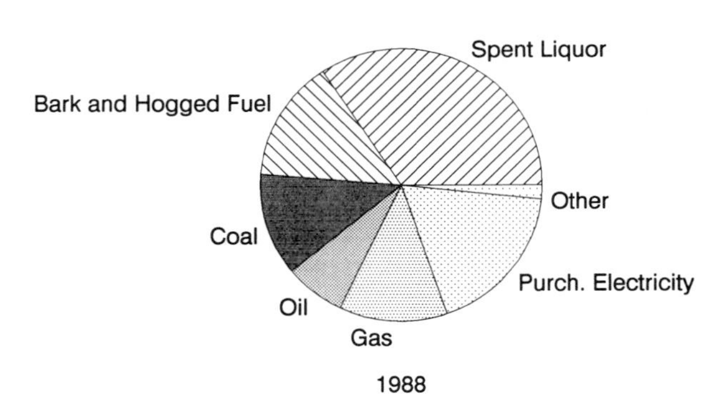

Figure 6. Energy source mix for pulp and paper production. Purchased electricity includes energy losses. Data from the American Paper Institute.

batch digestion almost as energy efficient as continuous digestion (Grant, 1983). The main economic drawback to increasing energy efficiency through indirect steam heating and displacement heating is higher capital costs.

The implementation of energy efficient digesters is both an economic and operational decision. First, the payback on the capital expense must be favorable, generally less than two years. Second, the operator must be willing to accept the risk (or perceived risk) of implementing new technology. If all the digesters were replaced with state-of-the-art continuous or batch digesters, about 4 GJ/ADMT could be saved. Multiplying by total Kraft production in the US, this represents 0.16 quads or over 6 percent of the current energy consumption of the entire US pulp and paper industry (Elaahi and Lowitt, 1988).

Evaporation

A major use of process steam in a Kraft mill is for the concentration of black liquor during the chemical recovery process. In general, this is accomplished using multiple effect evaporation (MEE). Recent design advances have allowed for the use of more evaporation stages. More stages mean higher capital costs, but lower energy requirements (see Figure 8).

Another option to increase energy efficiency is to use vapor recompression evaporation (VRE). The economics of deciding between MEE and VRE depends on the relative price of fuel (for steam) and electricity (for power), as well as a capital cost assessment. State-of-the-art evaporation systems represent a potential energy savings of 40 percent over currently installed evaporation systems. This translates to about 0.06 quads or 2.5 percent of the current US energy consumption in pulp and paper manufacture (Elaahi and Lowitt, 1988).

Papermaking and Drying

The papermaking process is the largest single energy use component in an integrated Kraft mill, requiring about one third of the total energy consumed by the mill. A major use of the energy is to remove water from the paper. Figure 9 shows how water is removed and the relative costs as the paper proceeds through the paper machine. It costs approximately five times as much to remove a ton of water in the dryer section as it does in the press section. Therefore, if more water can be removed in the press section, significant energy savings can be obtained.

A problem in pressing out more water is that at a sheet consistency of 45 to 50 wt. percent pulp further dewatering does not occur from pressing alone. Impulse drying is an advanced technology currently under development which would overcome this theoretical limitation and press the sheet to a 75 wt. percent pulp consistency. Impulse drying uses a long press nip with one hot roll. The resulting high temperatures (175–400°C) and pressures (40–50 atm) are the key to obtaining higher sheet consistencies (Sprague, 1987).

In addition to saving energy, impulse drying will reduce capital costs of new paper machines by replacing a large part of the press section and about one half the dryer section. Also, laboratory tests show impulse drying has positive effects on paper quality, such as increased sheet density and tensile strength. Over half the energy of drying paper will be saved—a savings of over 2.5 GJ/ADMT or about 7 percent of the current energy intensity of pulp and paper manufacture (Giese, 1988).

Heat Integration

When designing or evaluating industrial processes, the engineer must analyze the process in terms of unit operations as well as with the integration and interlinking of these elements to produce the overall system. The

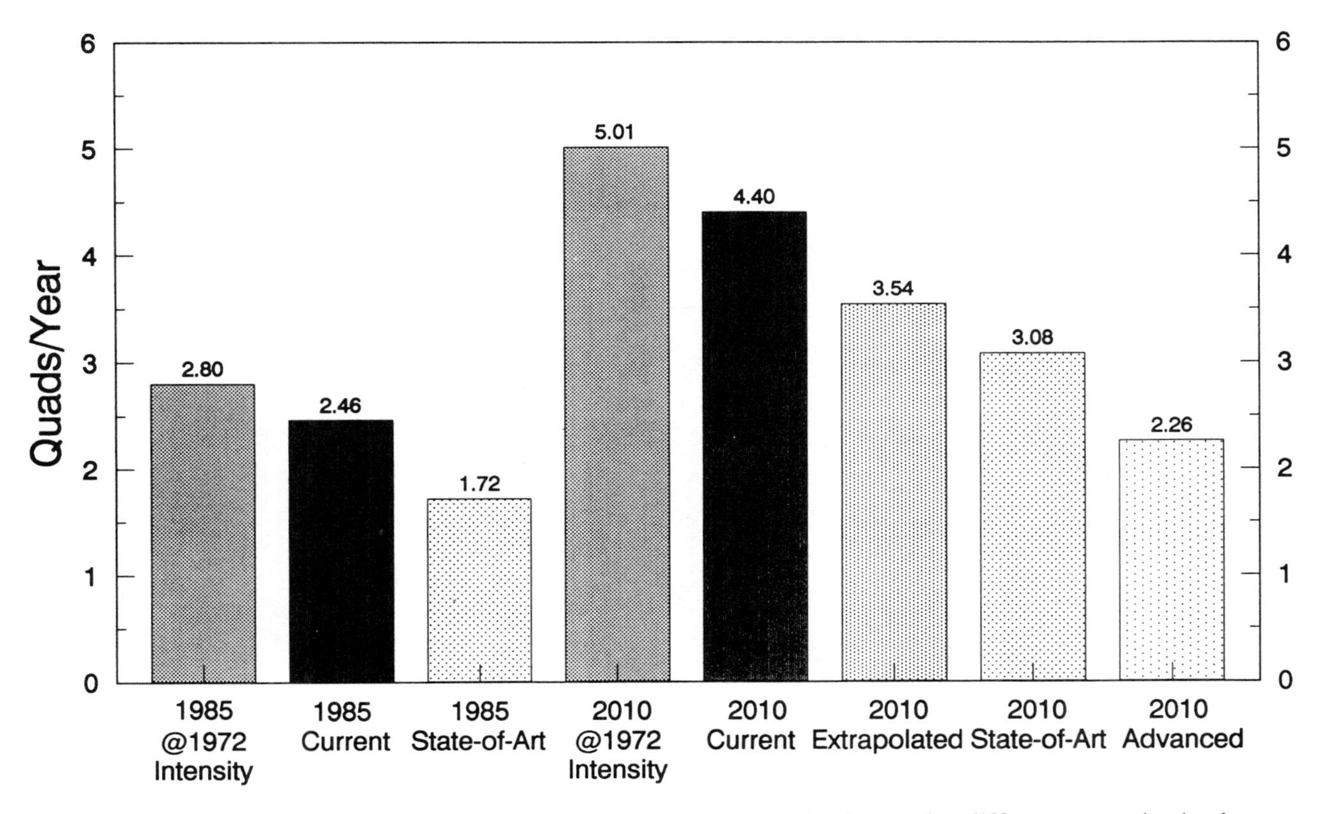

Figure 7. Comparison of energy intensities for pulp and paper production under different scenarios in the year 2010 (Elaahi and Lowitt, 1988).

three examples presented above all deal with individual unit operations. This final example falls under the system integration category.

Heat integration can drastically reduce the energy intensity of pulp and paper manufacture. From a theoretical analysis, a Kraft mill should require no purchased energy. Mills in Scandinavia are coming close to putting this theory into practice (Pöyry, 1986). However, as seen in Figure 3, almost half of US mill energy requirements are still met with purchased fuel or electricity.

A new engineering technique called "pinch technology" (Linnhoff, et al., 1982) has been developed to aid in integrating industrial processes. Another paper in the Industrial Processes section of this conference focuses on this methodology (Spriggs, et al., 1990). Based on thermodynamic principles, the technique has been applied to all major industrial processes, including petroleum, chemicals, pulp and paper, steel, and cement. Typically, new designs show energy savings of 15 to 90 percent and capital savings up to 25 percent. For retrofits, payback periods usually range from six months to two years. Ashton, et al. (1986) applied pinch technology to an existing Kraft mill and reduced steam consumption by 36 percent.

Cogeneration

Cogeneration of steam and electricity is energy efficient because it reduces the energy losses associated with electrical generation. Typically, cogeneration of power and steam will reduce fuel consumption by 25 percent compared to generation in separate cycles (Burwell, 1984). Because pulp and paper manufacture requires much steam and electricity, it is an ideal candidate for cogeneration facilities. As shown in Figure 10, the pulp and paper industry is the leader in cogeneration of electricity.

The amount of cogenerated electricity per ton of production has increased about 20 percent from 1982 to 1988 (Figure 11). This growth came after a long period of decline in self-generated electricity production in pulp and paper mills. In 1962, 55 percent of all electricity used was from cogeneration, but in 1977 only 40 percent of the electricity consumed was cogenerated (Pendergrass and Hu, 1983). This decline was caused chiefly by reliable and inexpensive electricity from public utilities. Higher electricity prices and the Public Utility Regulatory Policies Act of 1978 (PURPA) have changed the economics to make cogeneration not only an energy

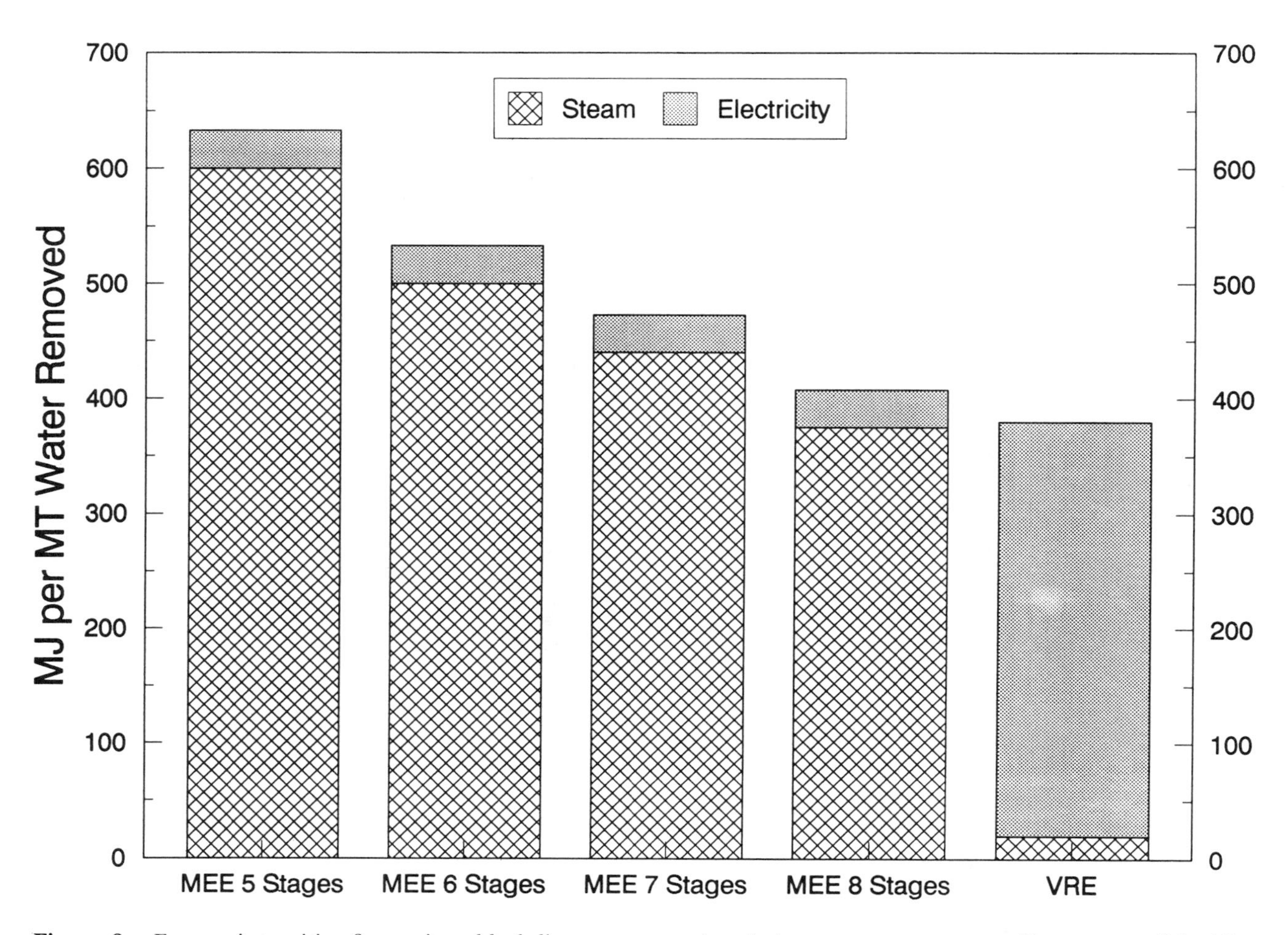

Figure 8. Energy intensities for various black liquor evaporation designs to concentrate to 65 percent solids. Electricity includes energy losses (Elaahi and Lowitt, 1988).

efficient option, but economically attractive as well. Today, almost one half of the 1100–1200 kWh_e/ADMT of electricity used by the pulp and paper industry is cogenerated.

The pulp and paper industry still has room to expand its use of cogeneration, as over one half of its power requirements are still met with purchased electricity. One barrier is decreasing thermal energy requirements. In practice, steam is generally produced in high-pressure boilers at 60–100 atm and reduced to typical process pressures of 5–12 atm using steam turbines to generate electric power. Therefore, the steam requirement is a prime determinant of the amount of electricity cogenerated. As steam demand decreases, so does the amount of cogenerated electricity. One strategy to overcome this limitation is to use cogeneration schemes with higher power to steam ratios than steam turbines, such as gas turbines (see Research and Development section). An alternate strategy, using condensing steam turbines to increase the power to steam ratio, has the adverse effect of lowering the overall thermal cycle efficiency. Another barrier to expanding cogeneration in pulp and paper manufacture is economics. The amortized cost of the cogeneration facility plus the cost of fuel minus the energy saved must be less than the cost of electricity from the power plant. With PURPA allowing the "avoided" cost of electricity to be used, cogeneration facilities have become more attractive and will most likely continue to be built.

Recycling

For the pulp and paper industry, recycling is more of a material issue (i.e., source of fiber) than an energy issue. For most paper products, recycling does not reduce purchased energy requirements (Hersh, 1981). This is because Kraft mills, which internally generate over one half of their total energy requirements, process about

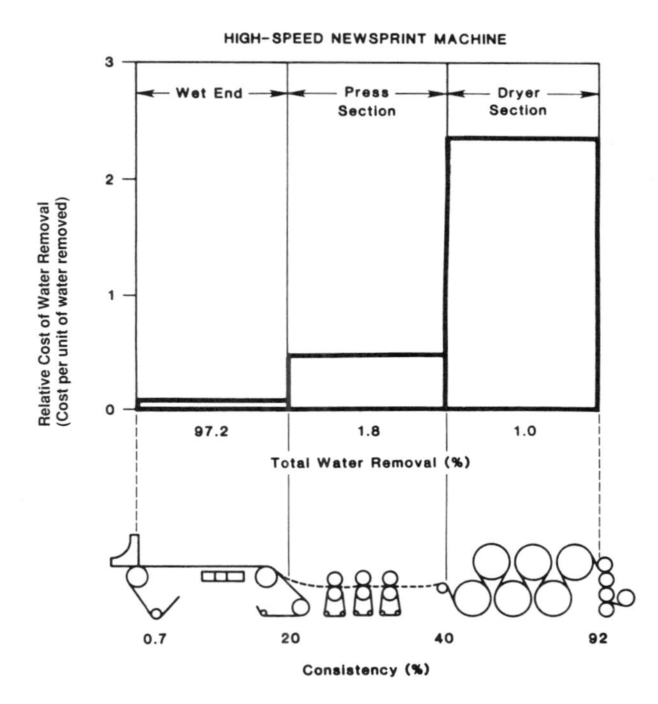

Figure 9. Water removal costs (Hersh, 1981).

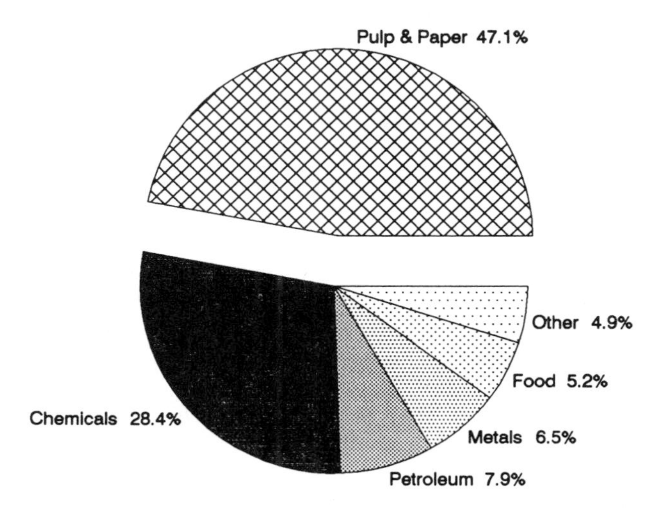

Figure 10. Industrial cogeneration by industry for 1985. Total = 70 Billion kWh_e. Data from the Energy Information Administration.

80 percent of the wood fibers. So while 12–18 GJ/ADMT of total energy can be saved by using wastepaper in place of Kraft wood fiber, over 18 GJ/ADMT of self-generated fuels will be lost. Therefore, more purchased energy is required. On the other hand, by replacing mechanically pulped wood fibers used for newsprint with wastepaper fibers, on the order of 12 GJ/ADMT of purchased energy (mostly electricity) will be saved.

Despite increasing purchased energy costs, replacing Kraft pulp with recycled wastepaper pulp is still economical, as seen by the fact that about two thirds of the wastepaper pulp goes into the manufacture of paperboard. Fiber costs are an important part of the economics of a paper mill, and wastepaper fiber is much less expensive than wood fiber. Also, far less capital equipment is required to make wastepaper pulp than to make Kraft pulp.

Material issues are important to consider when discussing recycling. Roundwood can be used for lumber, fiber, or fuel. By having alternative sources of fiber, the demand on our timber resources is reduced and they can be more effectively managed and renewed. If wastepaper is not recycled, it usually becomes part of a municipal waste stream. Solid waste disposal is a major problem facing the US. Recycling of wastepaper helps reduce the magnitude of this environmental problem as well.

Research and Development

As with most capital intensive industries, the United States paper industry is fiscally conservative. Research and development are geared toward process evolution (improving the current process) rather than revolution (radical process change). In this section, we highlight four areas of research and development that can potentially have a major effect on energy use in the pulp and paper industry and its environmental impact.

Black liquor gasification can potentially have a revolutionary impact on the energy balance of pulp and paper mills. As an alternative to the Tomlinson recovery furnace, black liquor is gasified and its products used to fire a gas turbine combined cycle cogeneration plant. Since recovery furnaces have low thermal efficiencies (about 65 percent versus 80 to 85 percent for fossil-fuel fired furnaces), significant energy savings are possible. In addition, compared to steam turbines, gas turbine combined cycle cogeneration offers the desirable attributes of high power to steam ratios (see Cogeneration section) and high thermal cycle efficiencies. The US Department of Energy has discontinued its gasification program through Champion International (Kelleher, 1985), but is sponsoring work led by MTCI, Inc. (Durai-Swamy, et al., 1989). Also, SKF Plasma Technologies AB of Sweden has a gasification project (Stigsson, 1989).

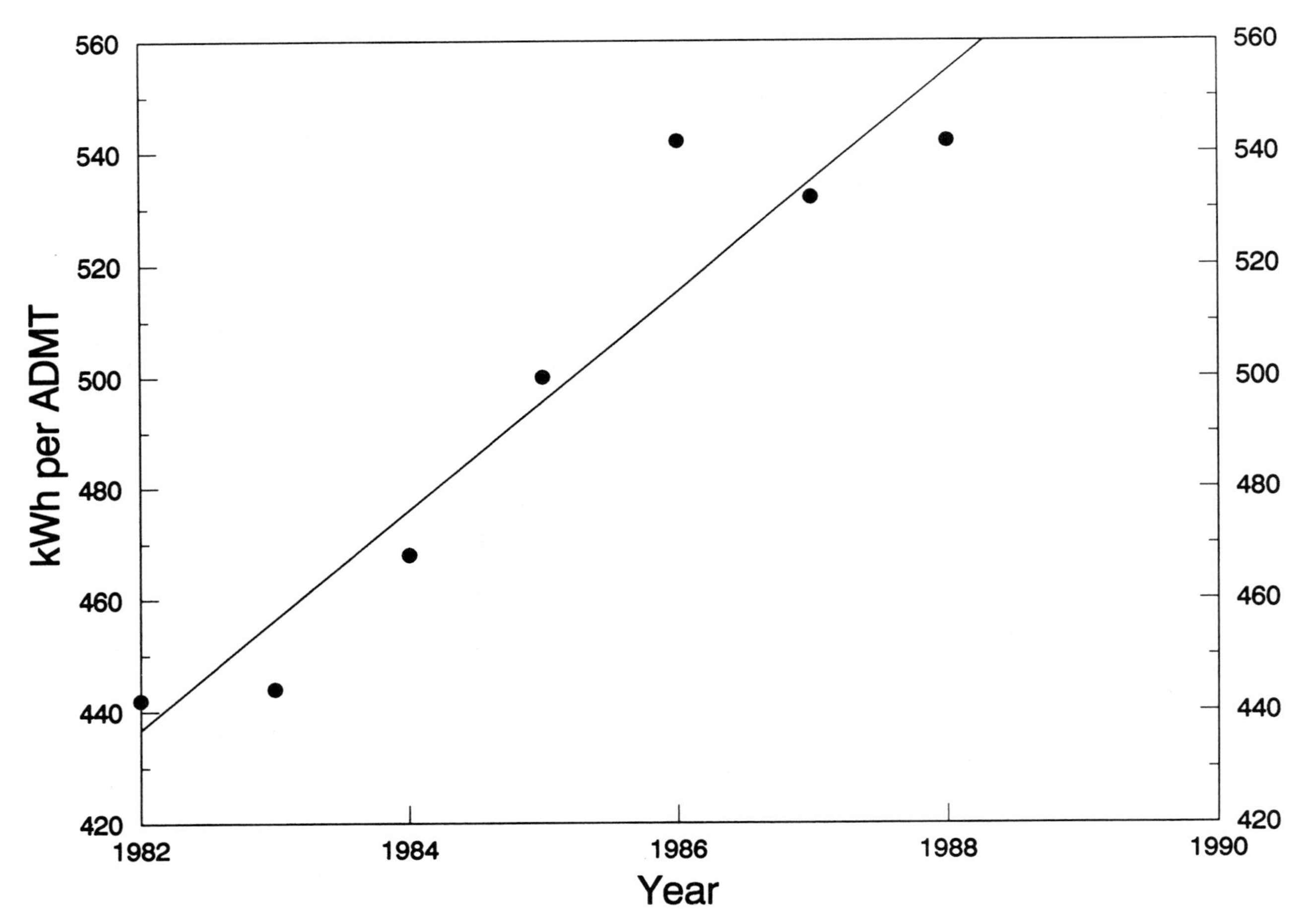

Figure 11. Electricity cogeneration intensity trends for pulp and paper production. Data from the American Paper Institute.

Black liquor gasification technology may result in Kraft mills becoming self-sufficient in both thermal and electrical energy. Gasification systems should save capital costs, improve safety (by reducing the chances of explosion), and reduce adverse environmental impacts (e.g., odor problems).

The Institute of Paper Chemistry is currently developing the impulse drying process. This process, discussed earlier in this paper, is an example of the new technologies under development that address the high energy requirement of drying paper sheets. This is an area where the potential for large energy savings exist.

The bleach plant is an area of environmental concern because of the use of chlorine. Besides being a hazardous chemical, chlorine reacts with the phenols in wood to produce dioxins. The dioxins can appear in effluent streams or, worse yet, in final products, such as coffee filters (and in your coffee!). However, there is no consensus on whether the trace levels of dioxins constitute a health risk, and the residual chlorinated organic byproducts can be treated in secondary treatment lagoons (Hall, 1986). Nonetheless, research on non-chlorine bleaching processes is an active area. Oxygen delignification before bleaching leads to significant reductions in the use of bleaching chemicals. This lessens the environmental impact and lowers operating costs, while capital costs remain about the same. In addition to research on the use of oxygen, other potential approaches are also under investigation, including NO_2/O_2 bleaching, ozone bleaching, and biobleaching.

Another area of research is sensing and control. By implementing sensing and control, product yield and quality are optimized relative to process inputs, including energy. Furthermore, it allows introduction of advanced technologies which require a high degree of control. In general, sensing and control developments are not driven by energy considerations, but energy efficiency usually improves as a byproduct.

Conclusions

This paper has highlighted some of the key issues and trends relating to the energy use and environmental impact of pulp and paper manufacture. It is by no means a complete treatment of the technology. For example, the whole issue of advanced pulping processes was not covered. Through the examples presented, we tried to convey the scope of current practices in the industry and the potential for continued energy efficiency improvements.

From a purely technical viewpoint, the pulp and paper industry can reduce the overall manufacturing energy intensity by about 50 percent by the year 2010, if currently available technology and new advances now under development are implemented. However, the actual reduction will be determined by economics and by the fact that adoption of new technologies is slow in the pulp and paper industry (in part due to a large "sunk capital" base). One of the primary mechanisms for improving energy efficiency is the continual turnover of capital stock. As old equipment is retired, it is replaced by modern equipment, which is generally more energy efficient. Our prognosis is that without any new energy shocks or government regulations the pulp and paper industry will continue to reduce its energy intensity at about its current rate, on the order of 1 percent per year, for the foreseeable future.

References

American Paper Institute, various reports including "US Pulp, Paper, and Paperboard Industry Estimated Fuel and Energy Use" (various years) and "Annual Report of In-Plant Electric Generation, Electricity Sales to Utilities and Cogenerated Electricity by the US Pulp and Paper Industry" (1989).

Ashton, G.J., Cripps, H.R., et al., "Application of Pinch Technology to the Pulp and Paper Industry" in TAPPI Proceedings 1986 Engineering Conference, pp. 489–498 (1986).

Burwell, C.C., *Electrification Trends in the Pulp and Paper Industry,* Oak Ridge Associated Universities Research Memorandum ORAU/IEA–84–11(M), Institute for Energy Analysis, Oak Ridge, TN (November, 1984).

Durai-Swamy, K., Warren, D.W., et al., "Pulse-Enhanced Indirect Gasification for Black Liquor Recovery," in TAPPI Proceedings 1989 International Chemical Recovery Conference (1989).

Elaahi, A. and Lowitt, H.E., *The U.S. Pulp and Paper Industry: An Energy Perspective,* DOE Report DOE/RL/01830–T57, Energetics, Inc., Columbia, MD (April, 1988).

Energy Information Administration, various reports including *Manufacturing Energy Consumption Survey: Consumption of Energy, 1985,* DOE/EIA/–0512(85), (1988) and *Annual Energy Review 1988,* DOE/EIA–0384(88), (1989).

Giese, R.D., *Scoping Study Pulp and Paper Industry,* EPRI Report RP2782-3, Giese & Associates, Kent, WA (December, 1988).

Grant, R.S., "Displacement Heating Trials with a New Process to Reduce Steam," *Tappi Journal,* pp. 120–123 (March 1983).

Hall, F.K., "Bleaching—Opportunities and Issues" in TAPPI Proceedings 1986 Annual Meeting, pp. 31–35 (1986).

Hersh, H.N., *Energy and Materials Flows in the Production of Pulp and Paper,* Argonne National Laboratory Report ANL/CNSV–16 (May 1981).

Kelleher, E.G., *Gasification of Kraft Black Liquor and Use of the Products in Combined Cycle Cogeneration,* DOE Report DOE/CS/40341—T5, Champion International Corporation, West Nyack, NY (July 1985).

Linnhoff, B., Townsend, D.W., et al., *User Guide on Process Integration for the Efficient Use of Energy,* Institution of Chemical Engineers, UK (1982).

Pendergrass, B.B. and Hu, S.D., "Cogeneration for a Bleached Kraft Pulp Mill at a Thousand Tons per Day," *Tappi Journal,* pp. 77–82 (March 1983).

Pöyry, J., "Prospects and Problems in the Growth of the Pulp and Paper Industry," *Tappi Journal,* pp. 30–36 (August 1986).

Ross, M., "Improving the Efficiency of Electricity Use in Manufacturing," *Science* 244, pp. 311–317 (April 1989).

Slinn, R.J., "US Paper Industry Nears 1980 Target for 23% Energy Reduction" in M. Coleman (ed.) *Energy Management and Conservation in Pulp and Paper Mills,* Miller Freeman Publications, Inc., San Francisco, pp. 19–20 (1981).

Sprague, C.H., "Impulse Drying and Press Drying: A Critical Comparison," *Tappi Journal,* pp. 79–84 (April 1987).

Spriggs, H.D., Smith, R., et al., "Pinch Technology: Evaluate the Energy/Environmental/Economic Trade-Offs in Industrial Processes," *Energy and the Environment in the 21st Century,* MIT Press, Cambridge, MA, 1990.

Stigsson, L., "A New Concept for Kraft Recovery," in TAPPI Proceedings 1989 International Chemical Recovery Conference (1989).

Open Forum/Panel Discussion on Industrial Processes Energy Efficiency: Can't We Go Further and Faster?

Attilio Bisio
Editor, *Chemical Engineering News*
Consultant, Atro Associates
Mountainview, NJ

Editor's Note: *Attilio Bisio served as panel moderator of an open forum/panel discussion that concluded the Industrial Processes sessions. His commentary is included here as a summary of the discussion that took place. Panelists included Patrick Atkins of Alcoa, Robert Ayres of Carnegie-Mellon University, William Joyce of Union Carbide, Jerry Robertson of Exxon, and Marc Ross of the University of Michigan and Argonne National Laboratory. The major objectives of this forum were to review and analyze the material presented on energy efficiency and conservation opportunities for the industrial sector and to identify incentives and policy instruments needed to further reduce the environmental impacts of industrial processes.*

Since the 1973 oil crisis, considerable progress has been made in reducing the use of energy in the United States. Total demand for energy in the industrial sector (as defined by the Department of Energy) dropped from 32 quads in 1973 to 29 quads in 1988.[1] If we exclude from the total demand nonmanufacturing activities such as agriculture, mining, and construction, industrial energy usage for 1988 was 22 quads.

This change in energy use by the US manufacturing industry has been driven by improvements in energy use. However, changes in demand for some industrial products and the shifting of some production overseas were also important, and appear to have reduced energy usage by about 5 quads and 3 quads, respectively. Between 1973 and 1985, there appears to have been an increase in the efficiency with which energy is used on the order of 2 to 3 percent per year (some analysts, such as Professor Robert Ayres of Carnegie-Mellon University, have estimated that increase to be as low as 1 to 2 percent per year). This reduction in consumption of energy (from all sources) has more than offset an annual growth rate in industrial product output of almost 2 percent per year between 1972 and 1985.

Recently, improvements in energy efficiency in the industrial sector are apparently slowing down or possibly even being reversed.[2] This is the most likely reason for the 2-3 quads increase in the demand for energy in the industrial sector from 1982 to 1988. The slowdown in capturing improvements in energy efficiency is not believed to be a result of diminishing technical opportunities. There may be an interplay between increased consumption of energy in the industrial sector and the requirements of environmental policies developed during the past decade. Some of the changes desirable from an environmental point of view, such as the reduction of specific emissions, may lead to changes in existing processes that will increase energy requirements. Desirable new products such as reformulated gasoline also appear to require more energy for their production than the fuels currently being used.

The potential for continued improvement in the use of energy in many industrial sectors has been demonstrated in a number of studies. For example, Patrick Atkins and his coworkers at Alcoa have shown the importance of recycling in reducing overall energy consumption in the manufacture of aluminum. Similarly, Howard Herzog and Jefferson Tester of the MIT Energy Laboratory have shown that technology exists in the pulp and paper industry that could cut its energy utilization in half by the year 2010.

If the potential for continued improvement exists, why is there pessimism in many quarters about realizing that potential? Is it simply due to, as some commentators have suggested, a combination of factors such as the decline in energy prices experienced in recent years and the curtailment of government energy conservation programs?

Many of the papers presented at the conference, Energy and the Environment in the 21st Century, stress the importance given to improving energy efficiency over the past 20 years. Companies in all these industries have, to varying degrees, reduced their energy consumption through both "smart engineering" and the

[1] The total energy requirement of 29 quads in 1988 includes consumption in nonmanufacturing sectors like agricultural and construction and electric generation losses. Excluding the nonmanufacturing sectors, the total for 1988 was 21–22 quads. US Department of Energy and Energy Information administration sources give inconsistent data for the industrial sector. The totals reported in this paper are from the *Monthly Energy Review* of February 1989.

[2] A precise evaluation of recent trends is virtually impossible due to the diminishing attention given to this issue during the 1980s and expiration of the Industrial Energy Conservation Reporting Program.

development of new processes that inherently require less energy.

The Exxon group, for example, instituted in 1969 a program to reduce energy consumption in its petroleum refining operation. This program increased energy efficiency in its refineries by 25 percent in 1981 and 35 percent in 1988, compared to 1973. These improvements involved largely "smart engineering" of existing processes, rather than the development of new processes. The cumulative energy savings over this period are judged to be equivalent to about 425 million barrels of oil.

New processes that have inherently lower energy requirements have also been developed by some companies. For example, the Plastics Division of Union Carbide developed and commercialized UNIPOL polyethylene (a fluidized-bed process) and revolutionized that industry. The energy requirement for this new process is only 1,500 BTU per pound of polyethylene produced, as compared to 5,000–8,000 BTU per pound for alternative processes that are still being used throughout the world.

In most instances achieving improvements in the utilization of energy whether through "smart engineering" or the utilization of new processes requires some investments in new facilities or equipment. Often the investment is substantial. This raises many questions of an accounting, financial, and even a philosophical nature.

There are indications that the cost accounting methods currently used in industry to establish the "price" or "value" of energy may inhibit the needed capital investments. For example, steam is probably one of the least understood process flows in a refinery. It is used for many different purposes: driving small steam turbines, as a heat belt to transfer heat from one process to another, as a stripping medium, as a reactant for making flares smokeless to decoke furnaces and to snuff out fires.

Should steam have the same value in all of these examples? Does steam going up a flare once every six days or so have the same value as steam being used in a heat belt? A simple analysis should indicate that it does not have the same value, but it always has the same cost. There is a belief that if one could obtain the "right" value accounting for energy, the amount of energy consumed in many processes would be reduced.

There is no doubt that energy projects have been financially attractive in many cases. Fifteen percent discounted cash flow returns (a typical minimum desirable return for many companies) have not been uncommon. Some projects have had returns as high as 80 percent. However, the returns actually achieved (as compared to predicted) have been dependent upon the location of the plant, accuracy of the prediction of energy price, and the actual cost of the required changes. This last factor, the actual cost of the project, is critical. Many energy improvement projects involve the modification of existing plants and processes. Estimating and controlling the cost of these modifications is extremely difficult.

Moreover, an energy improvement project practiced in one plant or one country may not be attractive or practical in another. For example, a project that is attractive in the United States may not be attractive for a similar plant located in Argentina. The price of energy in Argentina has been regulated by the government for many years. The cost of new heat exchangers there is about three times the price of heat exchangers in the rest of the world.

In a given industry or company in the US, the plants may be reasonably similar. However, there are almost always some differences in the specifics of the processes being used, the products being made, and the age of the plant. These differences make projects more difficult to accomplish (and, therefore, more costly) in some plants as compared to others. In general, the shorter the distance the energy has to be transported, the more attractive a heat integration project will be. Similarly, if one can transfer hot streams from one step of a process to another, the overall process will be more efficient. Unfortunately, some of these concepts can only be employed in relatively new plants.

While considerable progress has been made in reducing the consumption of energy for some processes, we cannot count on reducing the consumption infinitely. There will always be a minimum practical level of energy consumption needed. For example, for petroleum refineries the best judgment—economics aside (that is, buying the needed equipment at "zero cost")—is that "smart engineering" may be able to achieve another 20 to 25 percent reduction in the total energy requirements. However, there could be concerns about the operability of such a refinery since it would have limited flexibility.

There are many behavioral and political forces that limit the nature of the changes in a plant that managers are willing to consider. Many, if not most, managers feel more comfortable making a series of small steps (through smart engineering) that lead to immediate benefits, rather than considering a major new process with significant capital requirements. Moreover, the development of a revolutionary new process such as UNIPOL is the result of a long-term program over many years. When a new process (or product) is commercialized often has little or no relationship to the peaks and valleys of concerns over energy reduction.

There are many forces in US industry today that lead an entire industry or a company to resist the development of revolutionary processes. In the US metals and minerals industries, for example, the perceived risks for a given company to build the first plant based on a new technology are large. Major breakthroughs have come from other countries. Only when commercialized elsewhere will a "revolutionary" process be used in the US.

US managers have become conservative in their outlook. Often it appears that development of a new technology is undertaken by a company when it is perceived to be the only option for survival. Unfortunately, by then the company may not be able to marshall the needed funds for the commercialization.

Often, development of a company culture that avoids risk is the result of a "bad experience." Alcoa, for example, in 1973 announced a smelting process that was the culmination of 10 years of intensive research and development. That new process was expected to reduce the electrical requirements for making aluminum from 6.8 kilowatt hours per pound to 4.2 kilowatt hours per pound. Alcoa built and operated a commercial plant for about 10 years. In 1984, the plant was abandoned with the admission that the process was not viable due to technical limitations that could not be overcome. Over $250 million was spent between 1973 and 1984, and, before that, a large amount had been spent on research and development. That experience had a fairly significant impact in developing risk aversion at Alcoa.

There is some agreement that the risk-reward syndrome of many companies, which limits the willingness to undertake energy projects, could be changed by some combination of taxes and regulations. For example, in the aluminum industry a carbon tax would give additional incentive for the development of inert (nonconsumable) anodes. Research and development programs to develop inert anodes would reduce an aluminum company's dependence on carbon significantly. They would also reduce the electrical power requirement and, therefore, the consumption of fossil fuel. A carbon tax would certainly change an aluminum company's view of the risks associated with such a project. Whether it would make an inert anode project more viable is not clear.

Alcoa, for example, has worked on the development of inert anodes for several years. Other companies are also working along similar lines. A carbon tax might lead Alcoa and some other aluminum companies to double or triple their current efforts to solve existing technical and economic problems. However, a carbon tax alone may not solve the problems or make a risky technology work. The tax would more likely force the companies to make minimum incremental changes as they tried to become more efficient. Whether or not a carbon tax would force revolutionary changes within the aluminum industry as a whole is not clear.

For the metal industries, a carbon tax might reduce carbon emissions from most metal production in the US by about 25 percent. For example, if the tax were high enough, the steel industry might be forced to consider making significant changes in the production of iron. A tax would certainly accelerate the use of natural gas for the reduction of iron ore and accelerate development of technology to reduce intermediate cooling steps.

While the tax would have a limited impact on the most efficient producers, it could result in major shifts within the industry. However, it would not have a significant impact on how steel is used by consumers. Whether or not the elimination of significant process steps in the steel processing industry would occur is again not clear. However, it is possible that interest in eliminating the liquid metal stages in steelmaking would grow.

Not all companies in a given industry are utilizing the most efficient processes. For example, half of the worldwide polyethylene industry is still using processes that require more energy than the newer processes. There are a variety of reasons why. While some companies would make more money running the new processes, they do not have the cash flow to take that step. Moreover, a company may not be prepared to take the risk of losing some of its present markets.

In addition, some industries and many US companies have lost the ability to make technological changes. In some cases making significant technological improvements in processes and products will require cooperation between companies. This is not easy to achieve, although this process has had limited success in areas other than energy efficiency.

Management in the United States has, in recent years, paid increased attention to environmental and safety issues. In part this is simply because not paying attention to them could significantly increase the cost of doing business. Regulations have forced a review of long-standing business practices. Some of these changes have been quite minor. For example, refineries now sweep the streets more than they did years ago. This is because any dirt that gets into the sewerage system becomes part of the hazardous waste that must be disposed of by that refinery. As the price of disposing of hazardous waste goes up, there is an incentive to sweep the street.

Just the requirement to measure and report, if done in a sound manner, is motivating to companies. Making measurements leads to the setting of goals and programs. The more attention such programs get, the more changes occur. It is essential that the statistics be consistent, informative, and not political in nature. For example, ignoring the losses inherent in generating electricity (which is the case with some of the statistics being collected in the US today) can lead some companies to consider purchase of electricity rather than generation.

Companies—like individuals—appear to make *some* changes only when and if they have to. If "one's back is to the wall," as may be the case with global change, companies will make changes that would not otherwise be contemplated. We can be optimistic about future improvements in energy efficiency, as the "backs" of US industry may well be approaching the wall.

Session C
Building Systems

Summary
Session C—Building Systems

Leon R. Glicksman
Session Chair
Department of Architecture
Massachusetts Institute of Technology
Cambridge, MA

Introduction

The buildings area, including both residential and commercial buildings, has special importance because of the large amount of energy consumed and the unique environmental problems that exist within interior living and work spaces. Currently over one third of the total energy consumed in the US is used to heat, cool, and light buildings or to operate equipment, such as computers, used by the occupants of buildings. Over one third of the electricity in the US is consumed in buildings, and peak loads on electric power systems are primarily caused by cooling and lighting demand fluctuations within buildings.

Not only does the buildings sector have a substantial global environmental impact because of CO_2 emissions from fossil fuel consumption, but there are also more immediate environmental problems within buildings themselves. These range from excessive radon levels in homes to increased concentrations of potentially dangerous substances in the work place. Paradoxically, increased concern about indoor air quality stems in part from measures taken to improve the energy efficiency of buildings. For example, buildings have been made tighter and ventilation rates have been more closely controlled: many new office buildings, shopping centers, and hotels are now made with permanently sealed window areas.

The phasing out of chlorofluorocarbons (CFCs) has had a major impact on buildings. Chlorofluorocarbons are used in refrigeration and air conditioning systems. Foam insulations containing CFCs are the most efficient thermal insulations used in building envelopes and appliances. The imposition of less energy efficient substitutes therefore raises the issue of the tradeoff between reduced stratospheric ozone depletion and greater CO_2 production.

Although the energy efficiency of buildings in the US and the West has increased substantially in the last decade, much of this progress has been achieved by applying technologies already at hand. Additional gains will require a concerted effort by industry, government, and researchers to develop innovative solutions and promote widespread adoption. Progress will be hindered by the fragmented nature of the building industry, aversion to new building practices that have an element of risk, and the very modest budgets for research and development, both in industry and the government.

The goals of the Building Systems Session were to identify key problem areas; highlight promising new solutions, particularly those incorporating advanced technology; and discuss the roles of industry, government, and academia in achieving a unified vision of future buildings that implements an integrated, system-wide approach. Accordingly, the Building Systems Session was organized into three sections. In the first, speakers characterized the buildings sector: the nature of the industry, historical energy use and efficiencies, and prospects for future improvements. The second group of papers concentrated on several specific areas and associated technologies. The final section discussed means of encouraging the development and widespread use of innovations and the possible roles of the major parties—government, industry, and academia—involved in this task.

The Buildings Sector

Rosenfeld pointed out that in the "efficiency boom years" there was a dramatic increase in energy efficiency in the buildings sector, as in other US sectors, because of energy price increases. There are still many technologies for conservation that are cost effective when compared to conventional energy production rates. What is required are revenue-neutral incentives to encourage the adoption of efficient technologies in the present era of low energy prices.

Schipper distinguished measures of the intensity of energy use, e.g., energy per degree day per unit of floor area to heat a house, from the level of energy use, which depends on the size of the building. The former is a measure of technical progress toward efficiency while the latter is an issue of lifestyle. The intensity of energy use in the buildings sector of the US is comparable to that of most other wealthy nations, but the average building area per occupant is much larger in this country.

Grimsrud reviewed the increasingly important issue of indoor air quality. From a public health perspective, air quality is primarily a buildings problem. Increased energy efficiency in buildings will require a better understanding of ventilation. Key problems of the 1990s include health effects due to long-term exposure to low pollutant concentrations, understanding physical and chemical properties of pollutant sources within buildings, and policy responses to scientific findings.

Carlson examined the buildings industry and discussed the barriers to the introduction of new concepts and technology. Typically, new products in the buildings area have required one or more decades to achieve significant market penetration. This phenomenon is due to industry fragmentation, liability concerns in products with long lifetimes, complex codes and standards, and the short-term focus of the industry. There is a need for industry leadership and comprehensive research projects.

Pellish advocated joint collaboration of industry and government as a means to promote the development of new technologies for buildings. While keeping the building form unchanged, innovative building envelopes, for example, could utilize a large fraction of incident solar energy for lighting and heating needs. A fresh look at the overall function of building components yields other equally innovative technologies.

Important Research Areas

Kohonen reviewed the steps taken in Finland to carry out a policy of increased energy efficiency, including several comprehensive research programs related to buildings. One important factor has been coordination of conservation measures with the buildings as they are designed and constructed. McElroy gave an overview of the research program on building materials carried out at Oak Ridge National Laboratory for the Department of Energy. Important research areas include cooperative work with industry on large scale building components and tests of existing materials. Advanced materials work has emphasized alternative technologies for insulations containing CFCs. A promising new insulation is an evacuated powder filled insulation that is an order of magnitude more efficient than present insulations.

Weizeorick pointed out that the energy efficiency of refrigerators increased by a factor of 2 over the past 16 years (this is the energy consumed per unit volume of the refrigerator). This progress may be due in part to the concentration of the industry into five major producers. The industry currently faces the challenge of meeting stricter energy standards while eliminating the use of CFCs, formerly a major factor in efficiency improvement, in insulation and refrigeration systems. McMahon presented data to show that more efficient appliances with a simple payback of one to three years have been available but are seldom purchased. He made the case for mandatory appliance efficiency standards to correct market failures. Other economic incentives must be developed, and research on energy efficient improvements is needed. Future technologies must include acceptable replacements for CFCs; the next technical option for insulation appears to be vacuum panels.

In building controls, discussed by Bohn, considerable research has focused on physical means of providing cooling and heating storage for buildings. However, the control technology for these systems has received scant attention. Proper use of thermal storage systems requires control strategies that include algorithms dealing with weather and load forecasts. To optimize response, a control system has been designed on a simple computer: it balances hourly spot pricing from electric utilities, which reflects social costs, with the private needs of the individual building. The yearly savings are almost equal to the initial cost of the control system.

Two trends of the 1980s that have stalled conservation in commercial buildings, as reported by Nall, were the increased intensity of energy use by computers and other new equipment and a market structure in which building developers had little incentive to invest in energy efficient designs. Remedies include spot pricing to reduce peak loads and regulations to impose increased systems efficiencies.

Encouraging New Solutions

Millhone and Teichman discussed the role of the federal government in promoting improved building design and accelerating the development and commercialization of new, efficient technologies that will simultaneously improve energy efficiency and indoor air quality. Actions range from labeling and test procedures to financial incentives and research and demonstration activities. There are positive signs of increased efforts in the buildings area from both the Department of Energy and the Environmental Protection Agency related to buildings.

Groth and Chafee described new building initiatives in plastics and in the window industry, respectively. In the plastics industry an innovative house incorporating many new materials and building systems has been constructed through a partnership of 54 companies. This house has already heightened the interest of the public and other members of the building industry in the potential of advanced technologies. In the windows industry, new energy saving concepts, such as argon filled multiple glazing and thin film coatings, have gained increasingly wide acceptance.

Senator Lieberman described two new legislative initiatives in the area of building conservation and renewable energy. A proposed federal-private sector cooperative would make federal buildings the test bed for new technological construction initiatives. This would lessen many of the liability concerns that hold back innovations in the buildings field.

Socolow recapitulated the themes of this session. Analysis of US energy policy reveals the priority of investment in energy efficiency through technological and policy innovations rather than through investments in energy supply options. End use research should focus on 1) the overall system, i.e., the building, its internal operations, and the surrounding urban area, 2) experiments for measuring the real performance of innovations and better understanding them, and 3) high technology that contributes to major innovation in the field. Three recent considerations in building efficiency have been public health, ecology, and a global application of solutions.

Conclusions

The papers and discussions in this session revolved around several general themes. A large portion of the energy consumed in the US and worldwide is used to provide building comfort, services, and appliance needs. The buildings industry is fragmented, and there is little international integration. The industry relies on proven materials and components; research is minimal and, aside from evolutionary improvements, few new concepts are developed. Government support for energy efficiency in the buildings area has been meager.

Precisely because of the slow evolution of building practices, new concepts from disciplines such as materials science, automation, control, and systems analysis can be applied to the buildings field to make a major impact on global energy use and environmental stress. Many of these applications will require a level of scientific and technical expertise equal to that applied to innovative energy supply technologies. The rewards will be substantial; for example, if the efficiency of refrigerators were doubled, the savings in operating energy for the refrigerators produced in one year would be about 300 MWe, which is equivalent to one third of the generating capacity of a large power plant.

The stratification of the industry, separating builders from users, has distorted the marketplace and held back the adoption of energy efficiency measures. For the next century, there is a need for further financial and regulatory incentives by utilities and the government, accompanied by a new vision of buildings from a global perspective.

The planning and conduct of the Building Systems Session was facilitated by the efforts of many colleagues who organized and chaired sections and helped in the review of papers. I am particularly grateful to James Axley, Lesley Norford, and Richard Tabors, all of MIT; David Pellish and John Millhone, both of the US Department of Energy; Kevin Teichman of the US Environmental Protection Agency; Ross Bisplinghoff of Raytheon Corp.; Arthur Rosenfeld of Lawrence Berkeley Laboratories; and Erv Bales of the New Jersey Institute of Technology.

Energy-Efficient Buildings in a Warming World

Arthur H. Rosenfeld
Center for Building Science
Lawrence Berkeley Laboratory
Department of Physics
University of California, Berkeley
Berkeley, CA

Abstract

Our quality of life depends in large part on the heating, lighting, cooking, transport, and communications services that are provided by energy. Unfortunately, the same fossil fuels that have met so many of our energy demands to date have also added billions of tonnes of CO_2 to the atmosphere. The menace of global climate change, spurred by greenhouse gases, has naturally disrupted conventional notions of "quality of life." Increased energy efficiency, however, can provide more services with less fossil fuel, less money, and less environmental impact.

To a great extent, energy prices motivated the efficiency gains of the "boom" period. Today with low prices, we need non-price leadership in the form of policy that will repeat the efficiency gains of 1973 to 1986 and save the US a total of 30 EJ by the year 2005. See Figure 1. Since buildings use 40 percent of US energy and two thirds of all electricity consumed in the US, they are clearly key players in an efficient future.

Our proposals fall into the categories of revenue-neutral market-based incentives; efficiency standards and labeling for buildings, appliances, and lighting; utility fees, offsets, and rebates; and urban infrastructure improvements to battle summer heat islands and greenhouse gases.

Introduction

US residential and commercial buildings consume 40 percent of all US energy and 66 percent of all electricity and are therefore key players in an efficient future. In fact, since US electricity contributes 36 percent of US CO_2, electric efficiency in buildings is a prime target for policies aimed at the mitigation of greenhouse gases. For example, electricity contributes 500 million tonnes of carbon emissions to the atmosphere yearly, nearly twice the 280 million tonnes from gasoline use.

From 1973 to 1986, improvements in US energy efficiency captured dramatic pollutant and cost savings. However, since oil prices collapsed in 1986, our focus on efficiency has blurred and growth in energy use once again parallels growth in the gross national product (GNP). Even though low energy prices and lack of political commitment have delayed the adoption of a comprehensive national strategy for reestablishing earlier gains in energy efficiency, targeted policy proposals can revive a national commitment to efficient building systems.

Consumers and the energy industry must pursue the environmental and economic benefits of an efficient buildings sector through efficiency labeling and standards for structures, appliances, and lighting; carbon fees or offsets for utilities; simple urban infrastructure improvements; and the newest idea, revenue-neutral market-based incentives.

We cannot afford to continue manufacturing suboptimal buildings with life-cycle energy requirements that must be met for the next 50 years, when energy supply will be costly and imported. If we do choose to saddle our children and grandchildren with an inefficient infrastructure, they will find it hard to understand how we could be so short-sighted.

Efficiency Boom Years[1]

Prior to 1973, oil use and US GNP were inexorably linked. Figure 1 shows the spectacular, and by now familiar, decoupling of US GNP and energy consumption that occurred from 1973 to 1986. During this period, US GNP grew 35 percent while total energy use remained constant and oil and gas use decreased 1.2 percent annually (Figure 2). Twenty-five annual exajoules[2] (EJ) of anticipated energy usage, worth $165 billion, never materialized. When offset by initial investments in efficiency, net annual energy savings gained from efficiency improvements equal $100 billion.

In addition to the 35 percent dollar savings, a

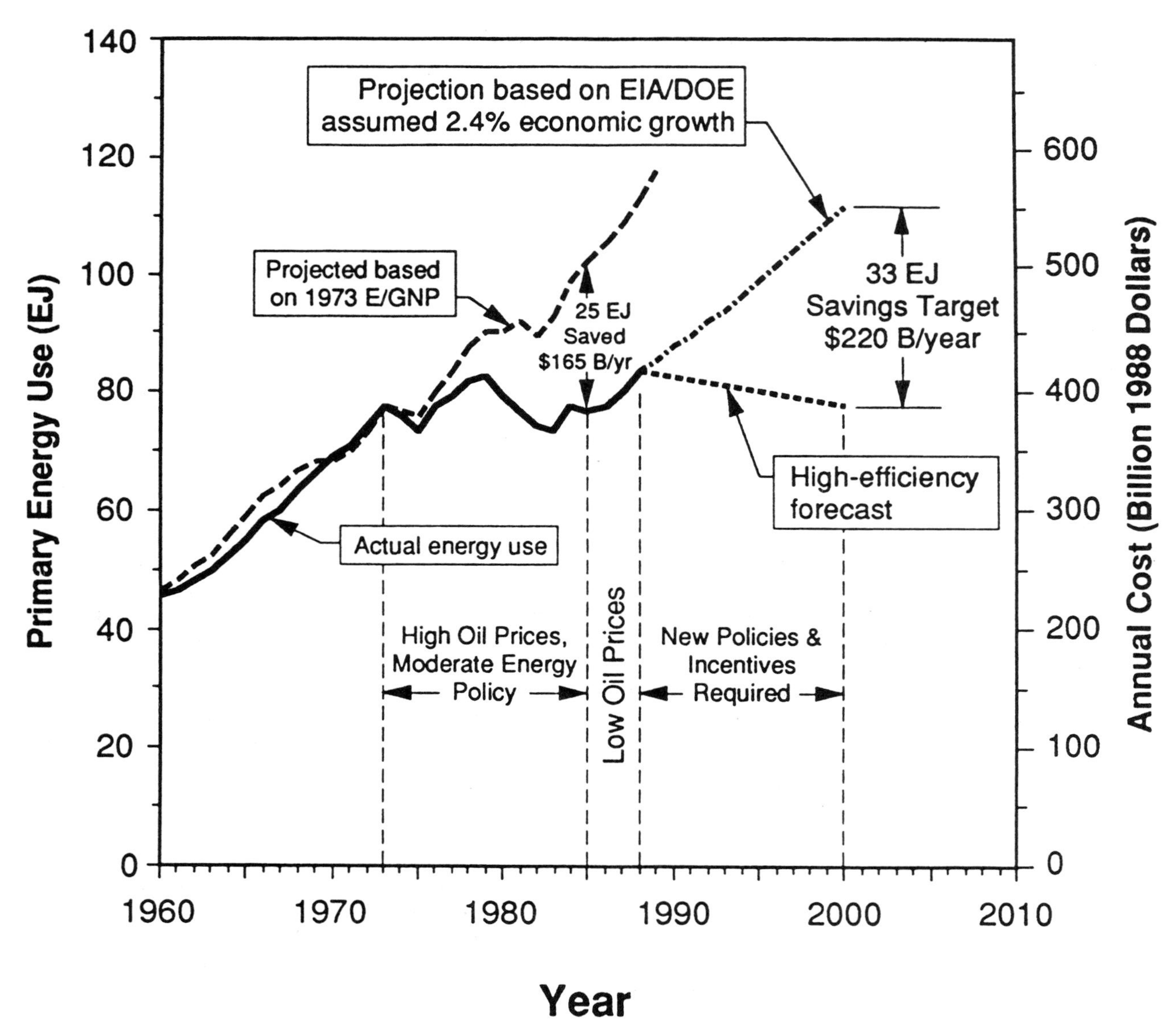

Figure 1. Energy efficiency temporarily breaks the lockstep of frozen efficiency and GNP. Before the first oil crisis, in 1973, energy use was in lockstep with GNP. During the 13 year period of high oil prices, from 1973 to 1986, improved energy efficiency and conservation allowed a 35% increase in GNP while total energy use stayed almost constant (around one). By 1986, this translated into savings of $165 billion per year with oil and gas savings amounting to 14 million barrels of oil per day or one-half the current capacity of OPEC. Shortly after oil prices collapsed in late 1985, pre-OPEC frozen efficiency reappeared and now energy use is once again in lockstep with GNP.

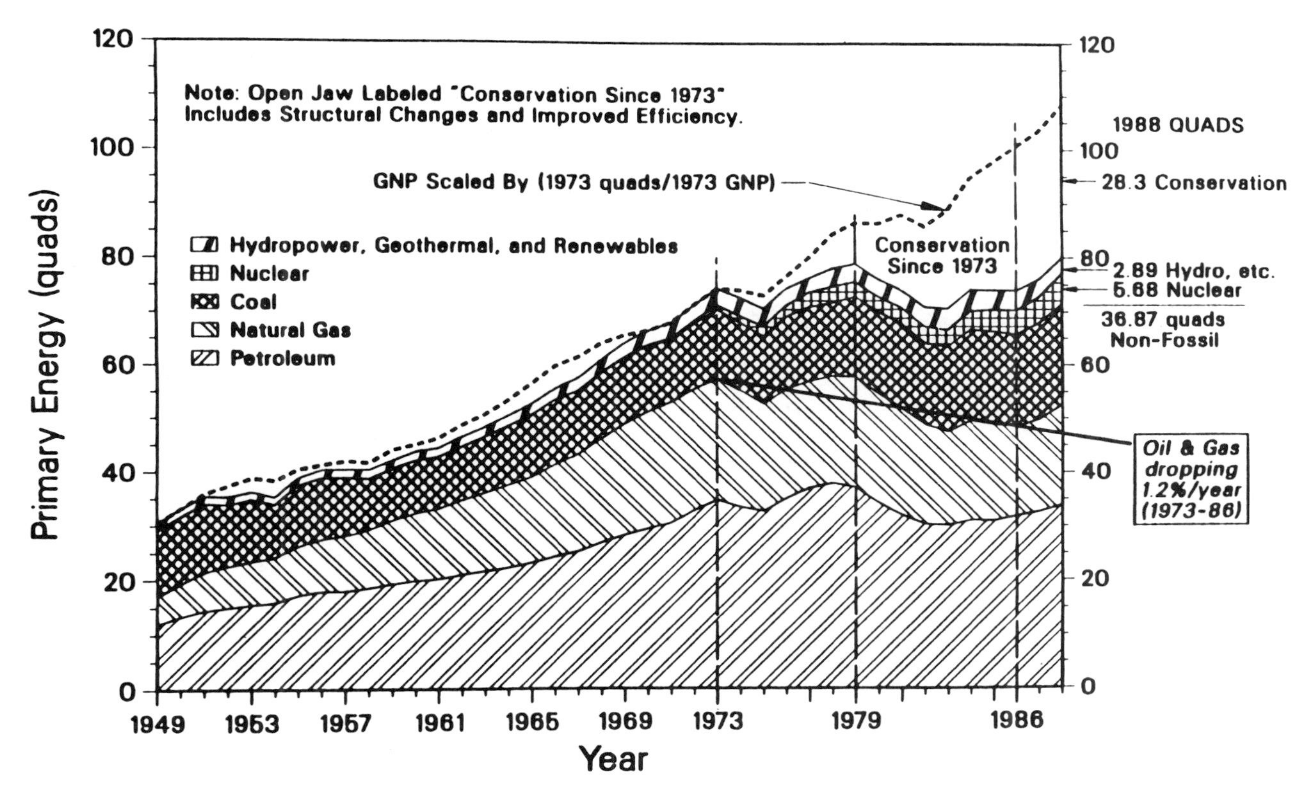

Figure 2. US primary energy use: actual vs. predicted by GNP (1949–1988). The figure shows energy consumption by source. Oil and gas use dropped an average of 1.2% per year from 1973 to 1986 explaining why OPEC lost control of world oil prices. The upper dashed line is GNP, in 1982 dollars, scaled to go through 74.3 quads in 1973, and illustrates how GNP and energy use were in lock-step before 1973. Energy conservation and improved efficiency have made a significant contribution to the US economy since 1973 and currently provide annual savings of more than 24 quads—worth $165 billion. The figure does not include about 4 quads of wood waste energy, but adding the contribution from conservation to the other nonfossil sources gives almost 37 quads—equal to one third of the energy we would have used if we were still operating at 1973 efficiencies. (Source: US Energy Information Agency. Note: 1 quad = 1.055 EJ = 10^{15} Btu.)

roughly 50 percent increase in CO_2 was avoided during this period. Without conservation, in fact, coal use during these years would have doubled. If the US were still operating at 1973 efficiency levels we would have dumped 50 percent more carbon dioxide, sulfur dioxide, and nitrogen dioxide into the atmosphere in 1986.

Improvements in US electricity consumption were more impressive than improvements in total energy consumption. And since buildings consume two thirds of total US electricity (three quarters of all dollars spent on electricity), improvements in this sector contributed significantly to total electricity savings. Until 1973, total electricity use was growing at a rate of 7 percent per year. After OPEC, electricity use grew only 2 percent a year, for an annual savings of 5 percent. See Figure 3. Compared to pre-OPEC trends, 55 percent of projected electrical usage was avoided between 1973 and 1986, compared to 35 percent of all energy use. This savings is equivalent to almost $80 billion per year, or the annual output of 250 baseload power plants.

However, from 1973 to 1986, total energy efficiency gains in the buildings sector were more modest than savings realized throughout the economy as a whole. Still, primary energy use per unit of floor space in commercial buildings fell 12 percent, and per household fell 20 percent. Incredibly, the energy required for heating buildings decreased by 1.2 million barrels of oil per day (an amount equal to two-thirds the daily output of the Alaskan pipeline) despite the fact that 20 million new homes were built nationwide, and commercial floor space increased by 40 percent. The buildings sector avoided $45 billion in annual energy bills and probably averted a 20 percent increase in annual CO_2 emissions through technological improvements in its heating and air conditioning systems, appliances, and building "envelopes."

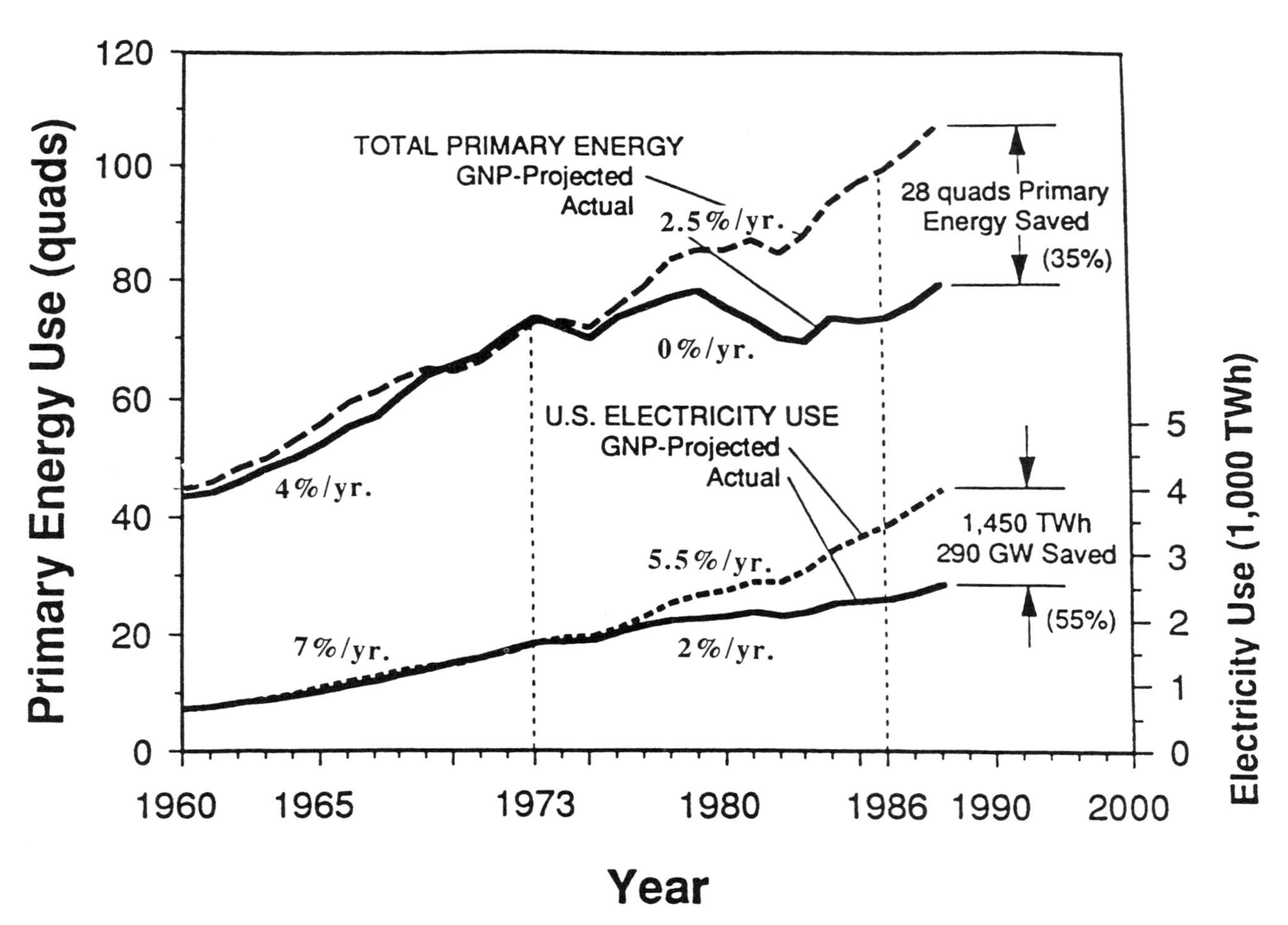

Figure 3. Total US primary energy and electricity use: actual vs. GNP projected (1960–1988). The dashed lines are GNP-projected energy and electricity use. GNP-projected energy values are based on 1973 efficiency and GNP. The electricity projections include an additional 3%/year that accounts for increasing electrification. Before the first oil embargo in 1973, total primary energy use was growing at the same rate as GNP, and electricity use was growing 3% faster than GNP. Electricity use is given in terms of total equivalent primary energy input (quadrillion Btu—left-hand scale), and net consumption (1,000 trillion Wh or TWh—right-hand scale). Compared to pre-1973 trends, in 1988 we saved 35% in primary energy and 55% in electricity. (Source: Energy Information Agency, US Department of Energy.)

Conservation "Supply Curves" for Electricity[3]

Conservation "supply curves" relate energy savings achieved by implementing a given efficiency measure, to that measure's "cost of conserved energy" (CCE).

Cost of Conserved Energy (CCE) =

$$\frac{\text{Annualized Investment (\$ per year)}}{\text{Annual Energy Saved (kWh per year)}}$$

The initial investment in an efficient-technology or program is annualized by multiplying it by the "capital recovery rate" (CRR).

$$\text{CRR} = \frac{d}{[1 - (1 + d)^{-n}]}$$

where "d" is the real discount rate and "n"is the number of years over which the investment is written off, or amortized.

Conserved energy is liberated to be "supply" for other energy demands, and therefore it may be thought of as a resource and plotted on a supply curve. On a conservation "supply curve," the y axis shows the CCE values for different efficiency measures (all our measures cost less than the current retail price of 7.5 cents per kWh), and the x axis aggregates all the measures' individual contributions to conserved supplies. The curve is upward-sloping because the measures are ranked in order of ascending CCE.

There are two different sorts of conservation "supply curves." One shows *technical potential*, based on engineering calculations without concern for the probability of implementation or possible failure to achieve calculated savings in the real world. This is analogous to ignoring the fact that a "27.5 mpg" car actually gets about 23.5 mpg on the road. The second type of curve shows *achievable scenarios* that are based on actual experience.

Figure 4a is a compilation of six potential conservation "supply curves" depicting the technical potential for electricity savings in the United States. These curves were calculated by different groups and compiled at Lawrence Berkeley Laboratory (Rosenfeld, 1990). They show nearly the same shape and a modest spread (split roughly into two ranges) regarding the potential electricity savings available in the US by the year 2000 and beyond. The curves mainly depict the potential to conserve electricity in buildings which consumed 64 percent (1677 BkWh) of the 2630 BkWh (BkWh = 10^9 kWh) sold in 1989. The discrete steps composing each curve show the average cost of conserved electricity (CCE) from improving a given end-use plotted against potential electricity savings (expressed as a percentage of total building sector electricity use). For example, the CCE and potential electricity savings for refrigerators would constitute a step. The dashed curve is an "eyeball fit" which reaches 50 percent savings at a CCE of 7.5 cents/kWh; the current US average price of electricity for buildings.

One of the curves in Figure 4a, labeled MEOS (Michigan Electric Options Study, LBL-23025, April 1988), shows that it is technically possible to conserve 60 percent of electricity demand at a CCE of only 3.5 cents/kWh, the marginal short-term cost of operating a Michigan coal plant. In this case, efficiency measures are not only cheaper than building new power plants but cheaper than operating existing plants.

Assessments of technical potential, however, ignore the logistical difficulties and costs of implementing efficiency measures. In Figure 4b we differentiate between a calculated potential and an estimated "achievable scenario," depicting actual savings as typically just half of potential savings. Recent modest utility conservation programs (generally amounting to 1 to 2 percent of utility revenues) have normally saved one-half to two-thirds of the technical potential. The MEOS conservation potential of 60 percent, for example, shrank to 28 percent in actual savings from utility conservation programs. Since Michigan utilities forecasted a modest 10 percent drop in sales from market-induced gains in efficiency, the right-hand curve of Figure 4b has been scaled down by 10 percent, from 60 percent to 54 percent. In addition, MEOS reduced each step according to measurements of conserved electricity for each utility program. The average down-scaling was by about one half, causing 54 percent to shrink to 28 percent.

This sort of shrinkage is typical. Conservation programs generally achieve about 50 percent of their technical potential. An aggressive program, such as Bonneville Power Administration's (BPA) Residential Construction Demonstration Project can achieve substantially more.

An Integrated Supply Curve: New Supply Compared to Electric Efficiency

Barriers to exploiting the full technical efficiency potential of electric end-uses are subject to change and even removal depending on regulatory and other influences. Accordingly, Figure 5 compares the costs and technical (rather than achievable) potential of demand-side electric efficiency to new electric supply. The curve in Figure 5 illustrates electric savings from efficiency measures corresponding to 1315 BkWh, 50 percent of total US electric sales in 1989. Half of this amount (660 BkWh) should be readily achievable. The remaining 1315 BkWh is assumed to be supplied by combined-cycle natural gas-fired generation. Efficiency improvements in industry are assumed to mirror those in the buildings sector.

The shaded area in Figure 5 corresponds to annual net savings in billions of dollars. The potential gross savings in electric bills is 7.5 cents/kWh multiplied by 1315 BkWh, which is equivalent to about $110 billion. The potential net savings is about two thirds of that, or $67 billion/year. The $33 billion difference represents the annualized cost of the efficiency improvements (the more expensive refrigerators, lights, etc.).

From Curves of Conserved Electricity to Curves of Avoided Carbon

Fossil fuel combustion produces carbon pollution as well as electricity. Figure 6 shows the costs and magnitude of potential conservation savings of carbon pollution (as CO_2) wrought by the efficiency and supply options shown in Figure 5. Net costs were calculated from the reference value of 7.5 cents/kWh for electricity in Figure 5. Using this base case, the costs and quantities of conserved tonnes of carbon in Figure 6 were derived from the costs and quantities of conserved kWhs shown in Figure 5. Table 1 shows the relative costs and effectiveness of alternative supply options and efficiency mea-

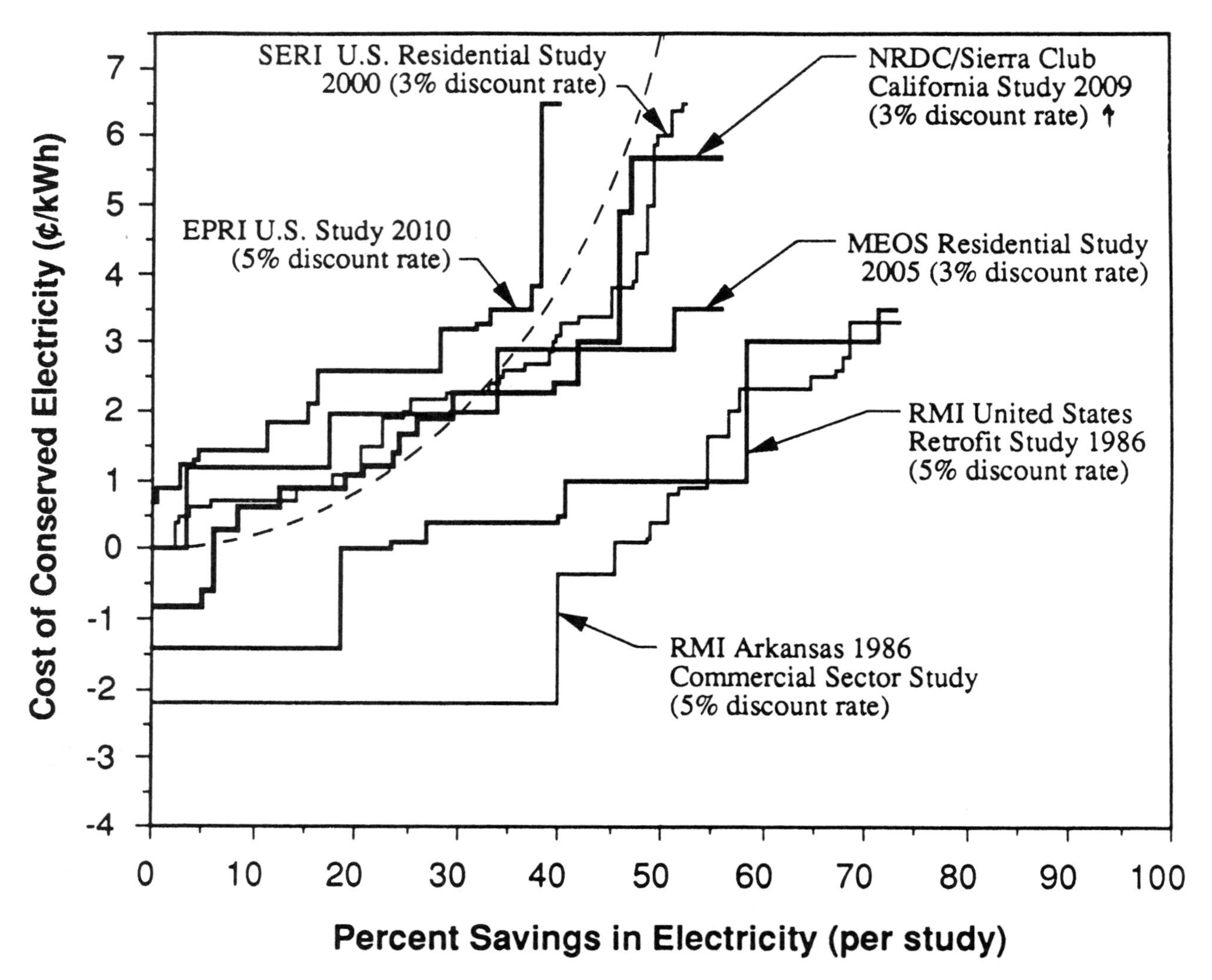

Figure 4a. A collection of efficiency supply curves indicating the technical potential for electricity savings in the United States. These curves imply that for an average cost of less than 2¢/kWh it is technically possible to save 40–75% of annual electricity use within the United States. The Electric Power Research Institute (EPRI) study examined all sectors of the US in the year 2010. The Natural Resources Defense Council/Sierra Club study examined projected commercial lighting and residential lighting and appliances in California in year 2009. The MEOS study examined residential lighting and appliances. The Solar Energy Research Institute study examined all sectors of the US for the year 2000, but only the residential electricity efficiency supply curve is reproduced here. Rocky Mountain Institute examined retrofitting the entire US and the Arkansas commercial sector using 1986 as a base year. (Source: *Efficient Electricity Use: Estimates of Maximum Energy Savings,* prepared by Barakat & Chamberlin, Inc., EPRI CU-6746, March 1990 (and personal communication A. Rosenfeld with A. Faruqui, Barakat & Chamberlin, Inc.). *Initiating Least-Cost Planning in California: Preliminary Methodology and Analysis,* D. Goldstein, R. Mowris, B. Davis, K. Dolan, Natural Resources Defense Council and The Sierra Club, prepared for the California Energy Commission, Docket No. 88-ER-8, February 1990, *A New Prosperity, Building a Sustainable Future,* The Solar Energy Research Institute (SERI) Solar/Conservation Study, Brick House Publishing, Andover MA, 1981, and *Competitek,* A. B. Lovins, R. Sardinsky, P. Kiernan, T. Flanigan, B. Bancroft, M. Shepard, Rocky Mountain Institute (RMI), 1986, Analysis of Michigan's Demand-Side Electricity Resources in the Residential Sector, F. Krause, A. Rosenfeld, M. Levine *et al,* Lawrence Berkeley Laboratory, LBL-23026, 1987.)

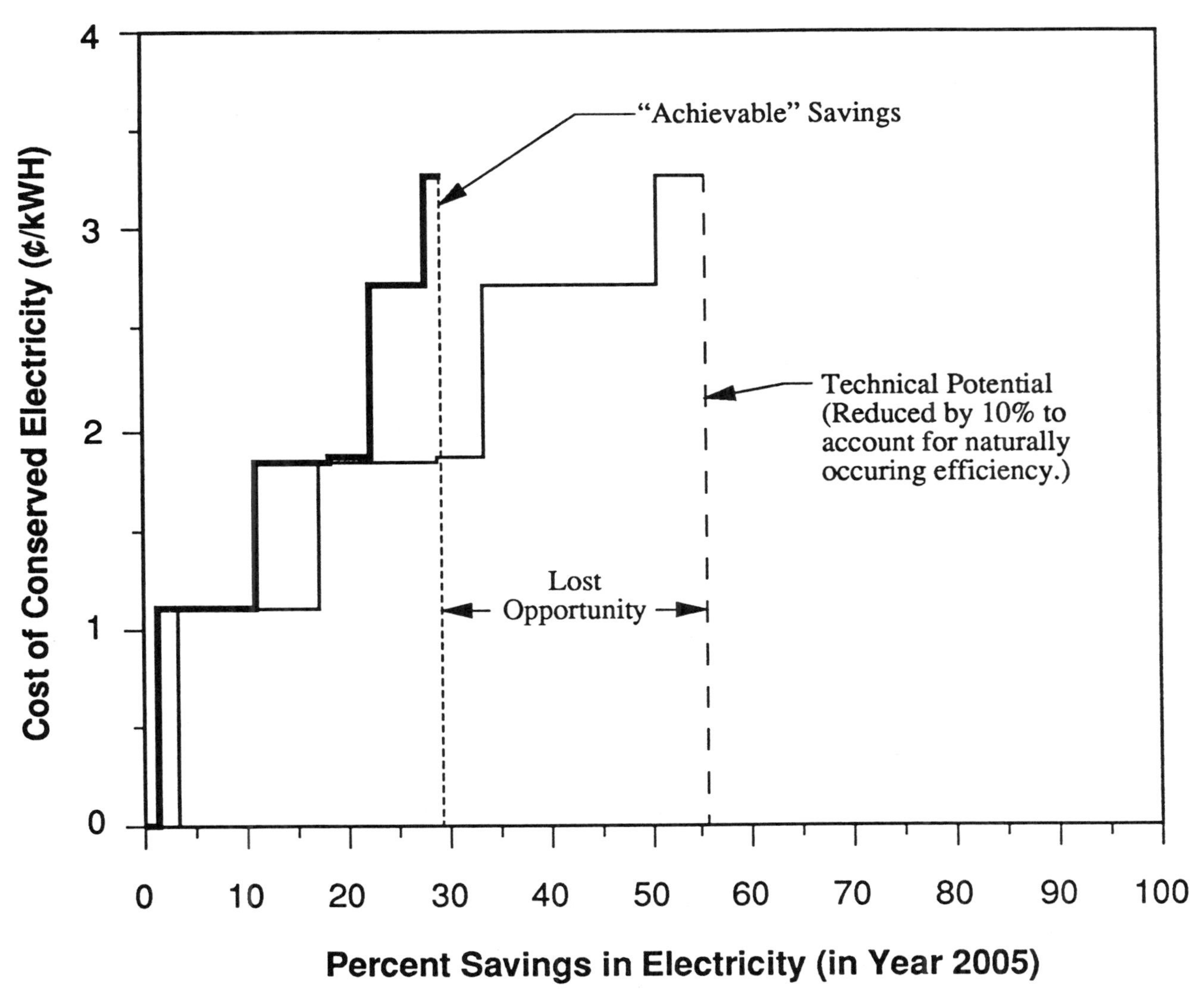

Figure 4b. Comparison of estimated "achievable" vs. technical potential savings. The estimated achievable savings are about two times less than the technical potential savings. The difference between the two curves represents *loss opportunity* that must be made up by purchasing expensive new supply resources. (Source: *Analysis of Michigan's Demand-Side Electricity Resources in the Residential Sector,* F. Krause, A. Rosenfeld, M. Leving *et al,* Lawrence Berkeley Laboratory, LBL-23025, February 1987.)

sures. The national mix of US generation options typically produces 0.2 kg of carbon as CO_2 for each kWh of electricity. Therefore, conserving a kWh is tantamount to saving 0.2 kg of carbon. Similarly, generating a kWh by an alternative method such as wind or biomass, which produces almost no CO_2 per kWh, avoids 60 percent of the 0.2 kg of carbon associated with conventional generation.

The shaded area in Figure 6 corresponding to the cost of conserved carbon (CCC) and magnitude of CO_2 savings from combined-cycle natural gas (C-C gas) generation, reflects the assumption that the remaining 50 percent (1315 BkWh) of electricity demand not accounted for by efficiency is supplied solely by C-C gas.[4]

Making a Market for Efficient Products

Renewable Supply Options

Any comprehensive strategy to improve US energy efficiency includes the development and implementation of alternative supply options such as biomass, photovoltaic, and wind technologies. Energy-efficient equipment

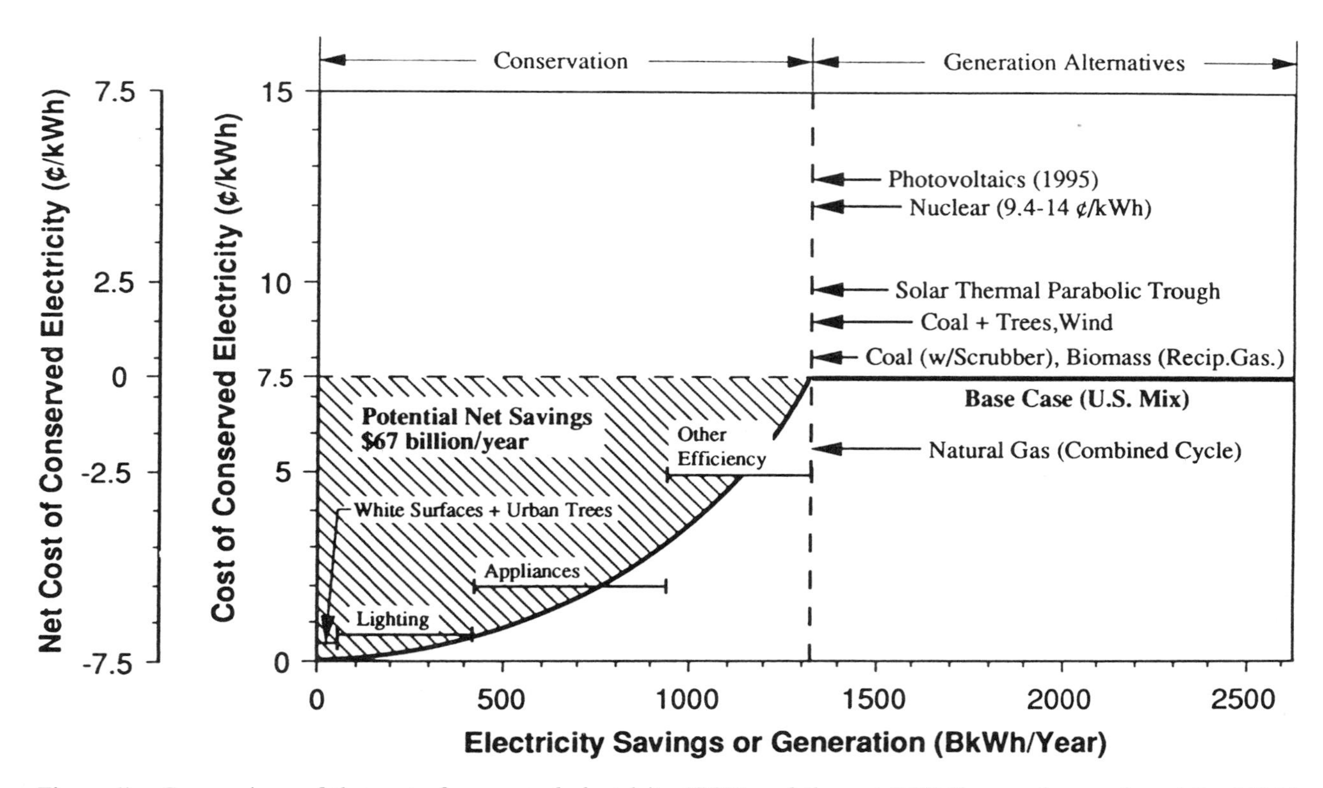

Figure 5. Comparison of the cost of conserved electricity (CCE) and the net CCE (from columns A and B of Table 1) to the cost of generation alternatives. The net CCE is measured from a base of 7.5¢/kWh for the US mix at the customer's meter. Thus, the net CCEs for conservation measures are negative corresponding to savings. The length of the x axis corresponds to current US annual electricity use of 2630 BkWh in 1989. The technical potential for conservation is drawn as a curve going out to 1315 BkWh (50% of 2630 BkWh) based on existing efficiency supply curves (see Figure 4a). The shaded area represents the estimated net savings of $67 billion/year assuming 100% penetration of technically available efficiency measures. Some of the estimated technical potential savings are based upon products already proven, but not yet commercially available. Similarly, some of the generation technologies, such as 12.7¢/kWh photovoltaics, will not be commercially available for a few years.

can enhance the attractiveness of these renewables. By reducing energy demand and the pollution associated with it, improved efficiency will give us the environmental stability and available capital to pursue alternative supply options that may require 20 years to perfect. See Figures 5 and 6, and Table 1 to compare efficiency measures to alternative supply for costs and carbon savings.

Utility Programs, Research & Development, Payback Periods, Efficiency Standards, and Revenue-Neutral Incentives

A strategy designed to capture available energy savings will call for a variety of approaches including:

- Reforming utility profit-structures to reward utilities that switch to cleaner sources of energy and sell "least cost" and "least carbon" energy services, rather than raw energy.
- Leveling the "tilted playing field" where investors accept 10 year paybacks for energy supply, but consumers require 3 year paybacks for efficient technologies.
- Increasing R&D for energy efficiency, which is currently minuscule at 0.1 percent of our national energy bill, especially compared to an average of 2 percent for all non-military industries and 1 percent for most mature industries.
- Strengthening building and appliance efficiency standards that are currently stuck at about 3 year paybacks.
- Implementing revenue-neutral consumer-based financial incentives that strengthen the market for efficient technologies by rewarding energy-efficient purchases and penalizing inefficient buying. These incentives can also factor the life-cycle costs of pollution "externalities," resulting from energy use, into the first cost of the product. The sliding-scale building hookup fees

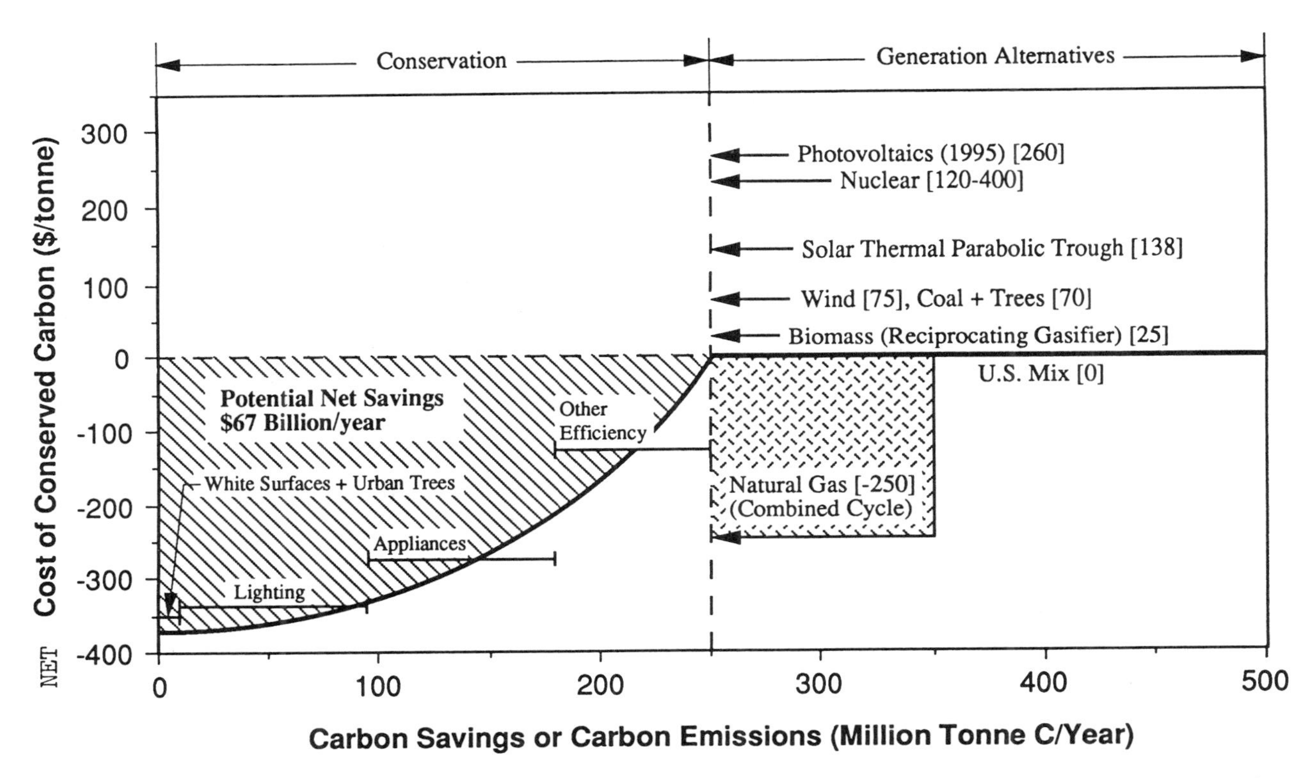

Figure 6. Comparison of the cost of conserved carbon (CCC) for efficiency and generation alternatives (from column E of Table 1). Conservation measures could potentially reduce carbon emissions by 50% at an average CCC of −$290 per tonne. The length of the x axis corresponds to total 1989 US emissions from electricity generation of 500 million tonnes.

discussed below are an example of a revenue-neutral incentive in the buildings sector.

Two Policy Initiatives

The two approaches described below are examples of targeted governmental policies that strengthen the market for efficiency.

"All-Sources" Bidding

The function of US utilities is slowly being redefined to emphasize the delivery of energy services to consumers, rather than the delivery of raw energy. Consequently, the provision of cheap and clean energy services is a priority for a growing fraction of utilities. Conservation frees electricity capacity by improving the energy efficiency of existing commercial and industrial buildings, thereby obviating, or at least minimizing, increases in demand and associated investments in new generating capacity. Throughout the nation, several Public Utility Commissions (PUCs) are rewriting profit-making rules to allow utilities to receive profits equivalent to 10 to 30 percent of avoided energy bills resulting from demand-side management programs. Several PUCs have also adopted (or are in the process of adopting) a competitive process called "all-sources bidding" that pits new supply against greater efficiency on the customers' side of the meter.

States that have approved all-sources bidding systems include California, Colorado, Connecticut, Maine, Massachusetts, New York, and Virginia. Regulations are also being developed in Michigan, New Jersey, and Vermont. Utilities in Maine, Massachusetts, and Virginia have already conducted all-sources bidding auctions, and have awarded contracts to Energy Service Companies (ESCOs) to install retrofits.

Typically ESCOs contract to retrofit lighting, heating, ventilating, and air conditioning systems, improve the energy efficiency of industrial processes, make building shell improvements, and/or operate and maintain building systems in a fully integrated manner.[5]

Table 1. The cost of saving electricity and carbon through conservation compared to generation alternatives. Generation options are calculated at the customer's meter, not at the busbar. Column A (Cost of Conserved Electricity—CCE) is plotted in Figure 5 and Column E (Cost of Conserved Carbon—CCC) is plotted in Figure 6.

Measure	(A) CCE ¢/kWh	(B=A -7.5[a]) Net CCE ¢/kWh	(C) C Emission kgC/kWh	(D=0.2[a] - C) Net C Saved kgC/kWh	(E=10 x B/D) CCC $/tonne	Potential U.S. Savings based on 2,630 TWh/yr BkWh
Conservation[b]						
White Surfaces + Urban Trees[c]	0.5	-7.0	0	0.2	-350	45
Lighting	0.7	-6.8	0	0.2	-340	420
Appliances	2	-5.5	0	0.2	-275	480
Conservation (50% savings)	3	-4.5	0	0.2	-225	1315
Generation[d]						
Base Case (U.S. Mix)	**7.5**	**0**	**0.2**[e]	**0**	**0**	**2630**
Natural Gas (Combined Cycle)[f]	5.5	-2.0	0.12	0.08	-250	See text
Baseload Coal (With Scrubber)[f]	7.9	0.4	0.27	n/a	n/a	
Biomass (Reciprocating Gasifier)	8.0	0.5	0	0.2	25	Limited Land
Coal (from above) + Tree Planting	8.9	1.4	0	0.2	70	Limited Land
Wind	9.0	1.5	0	0.2	75	
Solar Thermal Parabolic Trough (Hybrid with Natural Gas)	9.8	1.8	0.07	0.13	138	
Nuclear (EPRI)[g]	11.7	4.2	0.04	0.16	260	
Photovoltaics (1995)	12.7	5.2	0	0.2	260	

[a] Base Case, Average retail price for U.S. Mix, *Monthly Energy Review*, EIA, February 1990.
[b] White Surfaces + Urban Trees, H. Akbari, LBL, 1988, all other estimates by R. Mowris, LBL, March 1990 (savings for lighting include cooling bonus and are based on currently available technology, other estimates based on studies referred to in Figure 4a).
[c] Tree planting costs: $15/tonne of carbon sequestered, H. Akbari, A. Rosenfeld, *Conservation Supply Curves for Reducing CO2 Emissions*, January, 1990.
[d] Unless otherwise noted, costs are busbar plus 3 ¢/kWh based on average difference between retail and wholesale, *Monthly Energy Review*, EIA, February 1990. Busbar costs from *Energy Technology Status Report*, Gerald Bemis *et al*, California Energy Commission, Draft, March 1990.
[e] Carbon emission from R. Richels, EPRI TR045; DOE/NBB-0085, 1987 (The precise number is 0.19—rounded here to 0.2).
[f] Cost from J. Koomey, LBL-28313, April 1990, based on EPRI TAG—Tehnical Assistance Guide, 1986, adjusted for capacity, reserve, $1990, including T&D.
[g] Based on range of 9.4-14 ¢/kWh assuming 63% capacity, 15% fixed charge, 20% reserve, incl. T&D (from Koomey).

Sliding-scale Hookup Fees and Rebates in New Residential and Commercial Buildings

This initiative begins with setting target energy consumption and peak demand values for different categories of new buildings. Buildings that use more watts per square foot than the target values would be charged a fee that would then be rebated to buildings that use less. (We propose the value of $1000 dollars per kW because that is roughly the cost of avoided peak capacity.) The target would be adjusted annually to keep the account revenue-neutral. A portion of the fees would be allocated to cover administrative costs for the utility or state agency running the program, while some other portion could be shared with the utility as a reward for promoting conservation. A bill proposing this policy has been introduced in the Massachusetts legislature by Representative Lawrence R. Alexander, and Senator Nicholas J. Costello.[6]

Two Buildings-Related Efficiency Measures

In order to show the power of two particularly attractive efficiency measures, we explore the environmental and economic impact of compact fluorescent lamps (CFLs), and strategies for mitigating summer urban heat islands below.

Compact Fluorescent Lamps (CFLs)

The potential for improving the lighting systems in US buildings is enormous. CFLs alone have made, and will continue to make, huge contributions to efficiency. Each CFL is four times as efficient as an incandescent bulb and lasts ten times longer.

Assuming that a CFL's life is 10,000 hours, replacing a 60 W incandescent bulb with a 16 W CFL saves

440 kWh at the meter or 478.3 kWh at the busbar of the power plant. One CFL, kept on all year, will obviate the need for a sequence of twelve 60 W incandescents. The simple payback time for one CFL ranges from one-ninth to one-fifth of the lamp's life. In all, the electricity saved by one CFL is equivalent to 450 lbs. of coal or 40 gallons of gasoline. The gasoline equivalent spared by one CFL would drive a family car 1000 miles, from San Francisco to Denver. In fact, the energy saved by 12 lamps burning continuously (24 hours a day) is equivalent to enough gasoline to run a new car 10,000 miles a year indefinitely.

Just one CFL factory costing about $7.5 million and producing 1.8 million CFLs per year represents energy savings equivalent to: the output of a 4500 barrels of oil per day (bod) off-shore oil platform, taking 188,000 cars off the road, or fueling six Boeing 757 jets flying continuously around the globe.

In a national context incandescent lighting represents 10 percent of US electricity use, i.e., 300 BkWh per year. At a total cost of $1.5 billion, the output of 200 CFL factories ($7.5 million each) would conserve 50 percent of this demand or 150 BkWh, which represents the full-time output of 25 power plants. These avoided power plants cost $2 billion each for a total cost of $50 billion.

Reducing Temperatures and Air Conditioning in Urban Summer Heat Islands: Urban Shade Trees and Light-Colored Surfaces

Summer temperatures in cities are typically about 5°F higher than temperatures in the surrounding countryside, and the size and intensity of urban "heat islands" is growing. On average, summer temperatures in cities throughout the country are increasing by 0.25 to 0.5°F each decade. Temperatures in Los Angeles are increasing by 1°F per decade and air conditioning demand is up about 1 GW.[7] See Figure 7. Avoidable air conditioning now costs about $100,000 per hour on a hot afternoon in LA. And an increase of approximately 20 percent in smog episodes can also be attributed to LA's heat island warming. See Figures 8 and 9.

Adding the effects of global warming to the temperatures of heat islands compounds the warming trend. Even if CO_2 emissions were to grow rather modestly at 1.5 percent/year, CO_2 concentrations would double by the year 2050. The present global circulation models predict warming of about 3.5°C or 6°F, with a slope of about 1°F per decade. If we do nothing to mitigate heat islands, their temperatures will double by the year 2050 (an increase of roughly 10°F). Adding this 10°F rise to the 6°F increase from CO_2 emissions would cause LA's downtown temperature to rise a total of 16°F, making LA hotter than any city today in South or Central America.

Cheap urban landscape improvements can mitigate summer heat islands and reduce smog and the cooling energy demands of buildings. Mitigation measures include shade trees, park and street trees, and white surfaced roofs, streets, and parking lots. Shade trees and white roofs reduce the energy needs of buildings by blocking and reflecting solar radiation directly, and by improving buildings' microclimates (the wind and temperature around them) indirectly.

Such cost effective energy efficiency measures could reverse the trend toward summer urban heat islands by the year 2010. Preliminary analyses suggest that city-wide increases in tree coverage and surface albedo (reflectance) can reduce cooling energy needs in many American cities by as much as 50 percent. These targeted improvements of the man-made microclimate in urban areas would save over 0.50 quads per year[8] in air conditioning worth over $3 billion per year to electric rate-payers, and reduce peak cooling demand by 20 GW worth $20 billion in avoided generation costs and building heating, ventilation, and air conditioning (HVAC) equipment. Concomitant quality of life improvements are guaranteed as emissions of CO_2 and other pollutants from power plants are avoided and smog is reduced.

Conclusion

Improving energy efficiency in buildings can help cap world fossil fuel use and buy the time required to make today's expensive solar, biomass, wind, and storage alternatives economically competitive. We can reduce greenhouse gases with affordable technology currently available throughout the buildings sector at about twice today's efficiency—and at such a low cost that investors can get their money back from avoided energy bills in only three years. Available efficient products can give us the 20 year reprieve that we need to make renewable nonfossil energy supplies reliable, safe, and cost-effective.

Potential energy, pollutant, and cost savings in buildings can only be captured by a comprehensive strategy that relies on consumer-based financial incentives, utility regulations and programs, and infrastructure and urban planning improvements. It is time to revive a national commitment to efficient building systems without waiting for a resurrection of OPEC's pressure of exorbitant energy prices or the exigencies of global warming.

Notes

1. Rosenfeld and Hafemeister, "Energy-efficient Buildings," *Scientific American*, Vol. 258, No. 4, pp. 78–85 (April 1988).

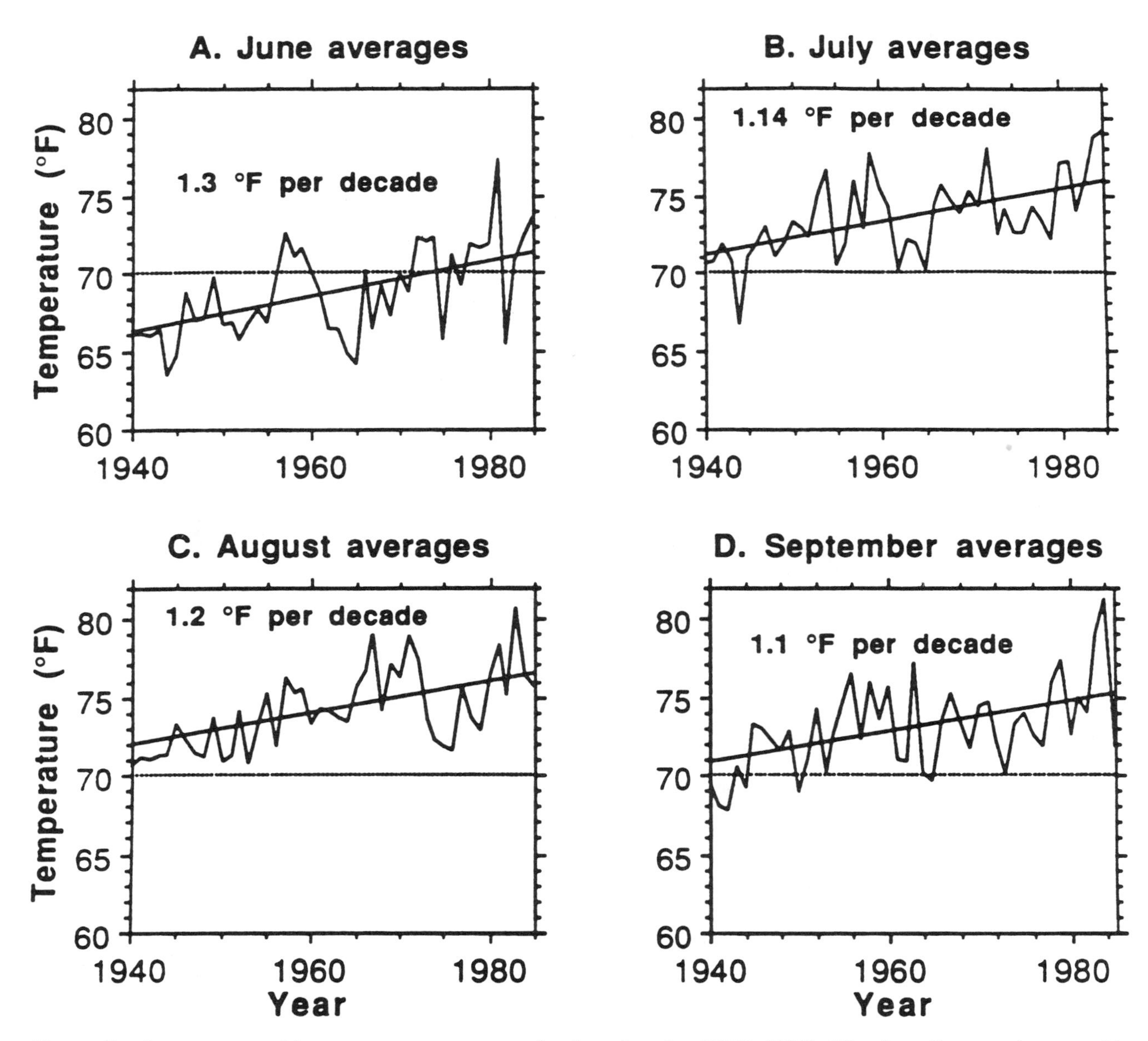

Figure 7. Summer monthly temperature averages for Los Angeles (1940–1985). The four figures give monthly averages for June, July, August, and September. The figures have remarkably similar slopes of about 1.2°F per decade or 0.12°F per year. The 45 years of data show a warming trend of about 5°F, and this will hit 10°F in another 30 years. Note: 1°F = 5/9°C.

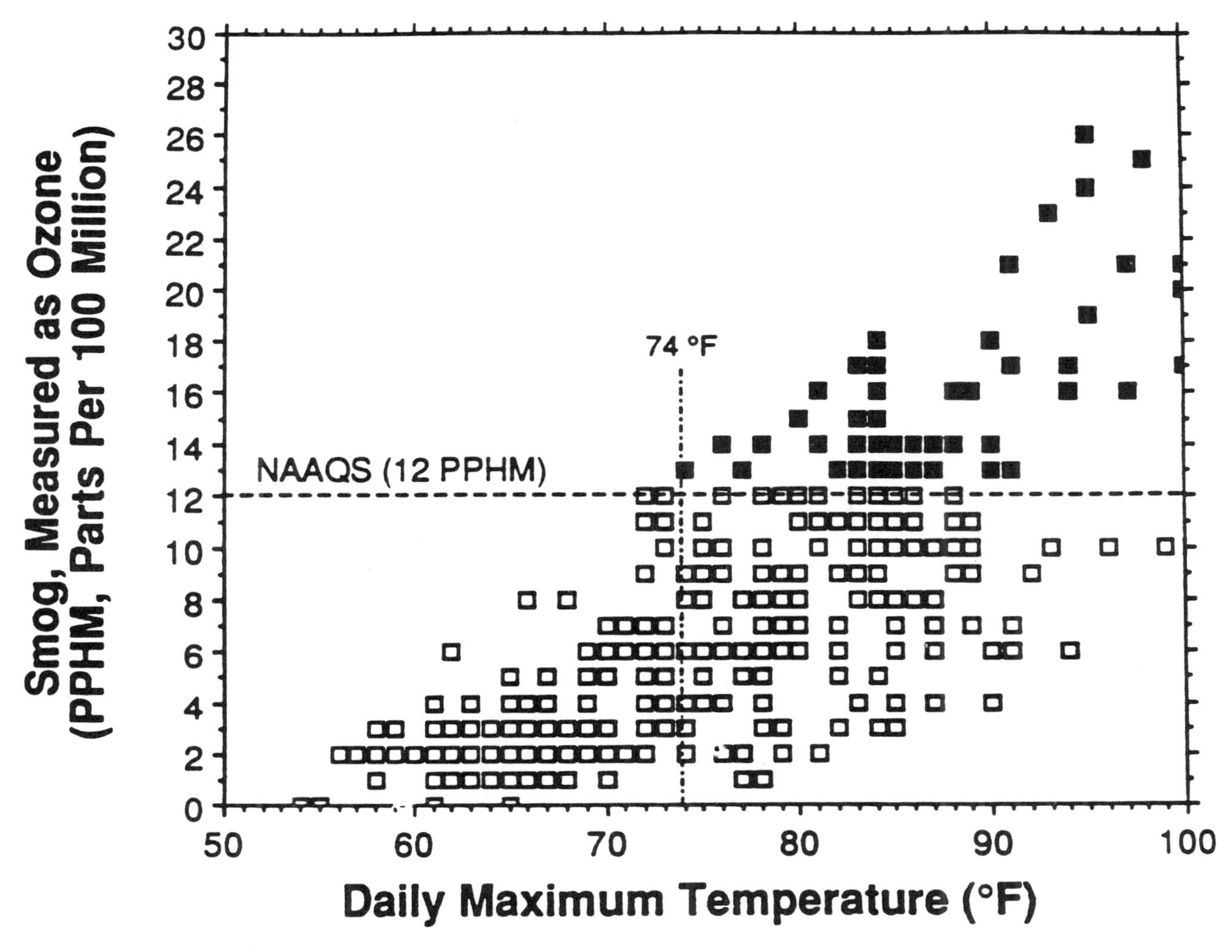

Figure 8. Los Angeles smog concentration measured as ozone in PPHM (parts per hundred million) at North Main Street (1985 Data). The horizontal line marks the National Ambient Air Quality Standard (NAAQS) for ozone of 12 PPHM, and shows that Los Angeles never experiences smog episodes when the daily maximum temperature is below 74°F. However, when the temperature is greater than 90°F smog incidents increase dramatically. (Source: Laura Wilson, University of California, Berkeley, from data provided by the California Air Resources Board, Technical Support Division, *California Quality Data,* Volume 17, No. 1–4, 1985, and the US Dept. of Commerce, National Atmospheric and Oceanic Administration, *Climatological Data, California,* 1985.)

2. One exajoule (10^{18} joule) equals 1/1.054 quadrillion (10^{15}) Btu.
3. A. Meier, J. Wright, and A.H. Rosenfeld, *Supplying Energy through Greater Efficiency: The Potential for Conservation in California's Residential Sector,* 1983, pp. 15–21.
4. Replacing the remaining 1315 BkWh (250 million tonnes of carbon [MtC]) with C-C gas-fired generation will save 40 percent of the base case carbon emissions equivalent to 100 MtC, because C-C gas generation produces 60 percent (0.12 kgC/kWh) of the carbon produced by the US mix (0.2 kgC/kWh). The shaded area is correct because the CCC has been adjusted by a factor of 2.5 (to $250/tonne) as shown in Table 1.
5. All-sources bidding is similar in spirit to the competitive procedures resulting from the Public Utility Regulatory Policies Act (PURPA) of 1978, that required utilities to purchase power from small and independent power producers at the "avoided cost" a given utility would have to pay to produce the same unit of power on its own.
6. House Bill 5277, "An Act Reducing the Greenhouse Effect by Promoting Clean and Efficient Energy Resources," is expected to be approved in 1990.
7. The temperature in the LA heat island has increased 5°F in downtown and probably an average of 3.5°F throughout the LA Basin. Approximately 300 MW of demand is invoked with each 1°F increase. Consequently, 3.5°F multiplied by 300 MW per °F equals a 1 GW increase in demand out of the total 18 GW peak demand in the LA Basin.

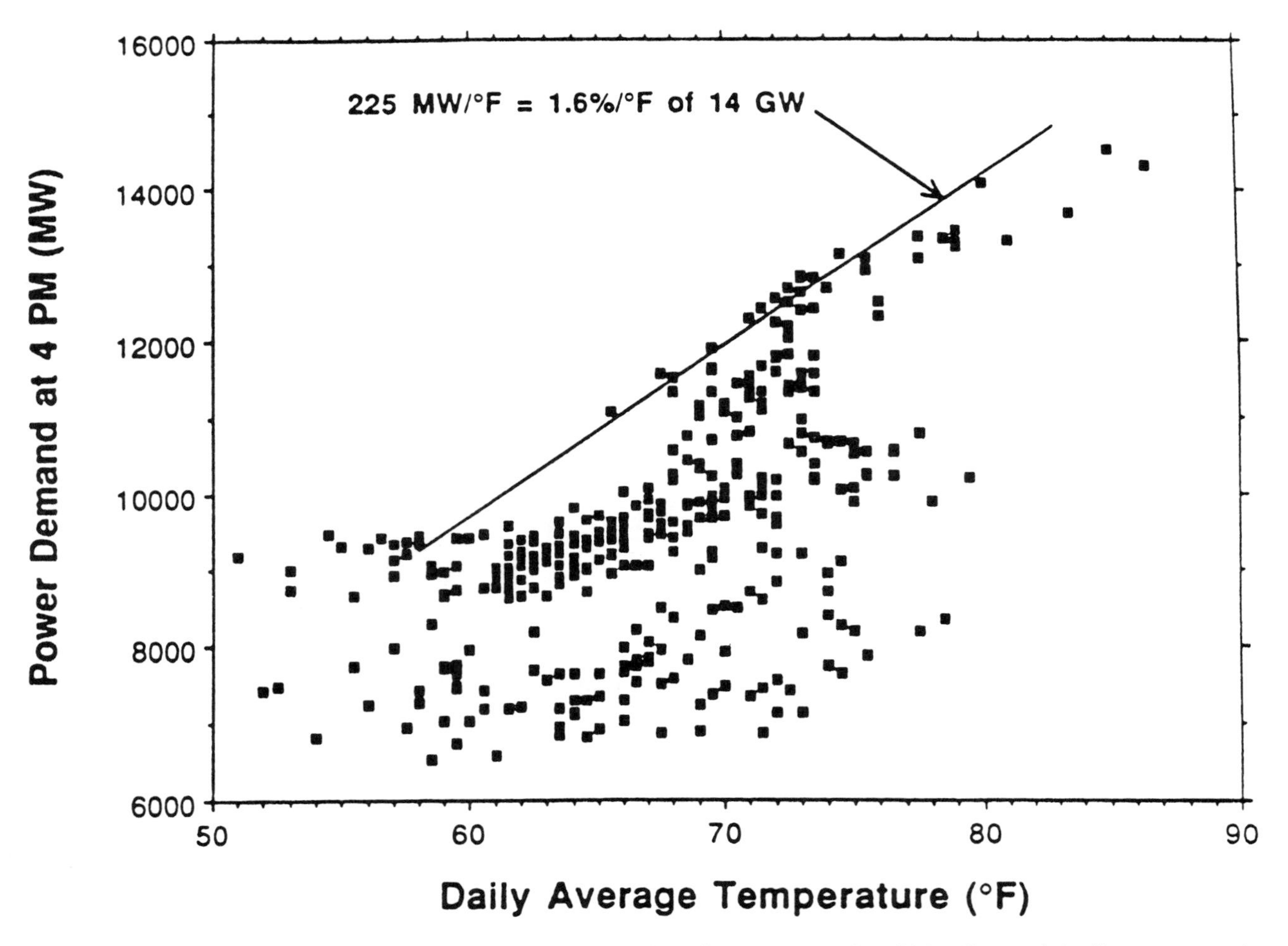

Figure 9. Electric power demand for Southern California Edison at 4 PM in 1986. The straight line representing the envelope of the scatter of data has a slope of 225 MW per °F, corresponding to 1.6% of the peak demand of 14 GW. Adding in Los Angeles Department of Water and Power, which covers the rest of the LA basin, the total is about 300 MW per °F. Los Angeles is already 5°F hotter than it was in 1940, and will be an additional 3°F hotter in another 30 years, if present trends continue. In addition, predicted global warming will add another 1°F/decade. If peak electricity is worth 10¢/kWh, then the rise of 5°F increases demand by 1.5 GW or $150,000 per hour. (Source: Akbari, *et al*, 1989.)

8. Akbari, H., Rosenfeld, A.H., Taha, H., "Summer Heat Islands, Urban Trees, and White Surfaces" in proceedings of ASHRAE January 1990 Meeting, Table 4.

Acknowledgments

We thank Ellen Ward for her substantial writing contributions to this paper, and Robert Mowris for creating many of the figures and tables. The work described in this paper was funded by the Assistant Secretary for Conservation and Renewable Energy, Office of Buildings and Community Systems, Building Systems Division of the US Department of Energy, under Contract No. DE-AC03-76SF00098.

References

Akbari, H., Huang, H., Martien, P., Rainer, L., Rosenfeld, A., Taha, H., *The Impact of Summer Heat Islands on Cooling Energy Consumption and Global CO_2 Concentration*, Proceedings of the American Council for an Energy-Efficient Economy 1988 Summer Study on Energy Efficiency in Buildings, Vol. 5, pp. 11–23, Asilomar, CA, August, 1988.

Akbari, H., Rosenfeld, A., Taha, H., Summer Heat Islands,

Urban Trees, and White Surfaces, Proceedings of the ASHRAE Meeting, Atlanta, GA, January 1990.

Chandler, W., Geller H., Ledbetter, M., *Energy Efficiency: A New Agenda,* American Council for an Energy Efficient Economy, 1001 Connecticut Avenue, N.W., Washington, DC, July 1988.

Difiglio, C., Duleep, K., Greene, D., "Cost Effectiveness of Future Fuel Economy Improvements," Draft, Submitted to *The Energy Journal,* January 1989.

EIA, Energy Information Administration, *Monthly Energy Review, January 1989,* DOE/EIA-0035(89/01), 1989.

Ernst, M.D., Update from Research Director, Energy Committee on revenue-neutral incentive bill in Massachusetts legislature (H. 5058), titled *An Act Reducing the Greenhouse Effect by Promoting Clean and Efficient Energy Resources,* April 1989.

Geller H., *Financial Incentives for Reducing Automobile Emissions: A Proposal for Clean Air Act Ammendments,* American Council for an Energy-Efficient Economy, July 1989.

Geller, H., Harris, J., Levine, M., and Rosenfeld, A., "The Role of Federal Research and Development in Advancing Energy Efficiency: A $50 Billion Contribution to the U.S. Economy," 12:357–95, *Annual Energy Review,* Annual Reviews Inc., 1987.

Gordon, D., Levenson, L, *Drive+: A Proposal for California to Use Consumer Fees and Rebates to Reduce New Motor Vehicle Emissions and Fuel Consumption,* Final Report prepared for the US Environmental Protection Agency, July, 1989.

Laitner, S., *Designing Energy Strategies to Incorporate External Costs into Public Policy: Where LES Is More,* Economic Research Associates, Lincoln, Nebraska, 1989.

Michigan Electric Options Study, LBL-23025, April 1988.

Moscovitz, D., "Cutting the Nation's Electricity Bill," pp. 88–93, Vol. 5 No. 3, *Issues in Science and Technology,* Spring 1989.

Rosenfeld, A., Mowris, R., Koomey, J., *Energy Efficiency and Least-Cost Planning,* Draft LBL Report, revised 1989.

Sierra Research, Inc., *The Feasibility and Costs of More Stringent Mobile Source Emission Controls,* prepared for the US Congress Office of Technology Assessment, January 1988, Sierra Research, Inc., Sacramento, CA (916)

SCAQMD, South Coast Air Quality Management District, *Air Quality Management Plan,* September 1988.

Wiel, S., "Making Electric Efficiency Profitable," in *Public Utilities Fortnightly,* Arlington, VA, July 6, 1989.

Energy Saving in the US and Other Wealthy Countries: Can the Momentum Be Maintained?

Lee Schipper
International Energy Studies, Energy Analysis Program
Applied Science Division
Lawrence Berkeley Laboratory
Berkeley, CA

Introduction

Recently many scientists have voiced concern that increased human-made production of so-called greenhouse gases, such as CO_2, methane, or chlorofluorocarbons (CFCs), may have lasting and possibly costly effects on the earth's climate (1). This concern has led to interest in reducing the emissions of these gases, particularly through more efficient use of fossil fuels, because more efficient fuel use also saves money.

Indeed, between 1972 and 1985, the wealthy industrialized countries reduced their energy use through greater efficiency by nearly 32 exajoules (EJ), or 16 million barrels per day of oil equivalent (2). These savings played a crucial role in forcing down world oil prices. But the pace of these improvements in energy use has begun to slow (3). This paper examines trends in key energy use sectors, commenting on recent changes that may signal a slowdown in energy efficiency.

Measuring Energy Efficiency

How can progress toward more efficient energy use be measured? The ratio of energy to GDP is often used as a measure of the differences in energy efficiency among economies (4). But Schipper and Lichtenberg (5) found that this comparison was misleading as an indicator of energy efficiency performance. Further, changes over time in that indicator were only vaguely related to changes in individual energy intensities. And Schipper, Howarth, and Wilson (6) found that while the energy/GDP ratio in Norway fell 30 percent between 1973 and 1986, sectoral energy intensities (energy use per unit of residential heating, passenger or freight transport, services GDP, or manufacturing output) was constant or increased! Clearly, use of the energy/GDP ratio can give very misleading signals about changes in energy efficiency. Additionally, one striking finding is that differences among various countries' sectoral energy intensities are almost always significantly smaller than the differences among countries' energy/GDP ratios. Therefore, we reject this indicator for the present analysis, focusing instead on the measures of energy intensity from each sector or subsector of each economy.

Summary Findings from National Studies: The United States and Other Countries

We recently reviewed changes in US energy use intensity since the early 1970s (7). The principal findings of that review may be summarized as follows:

- Aggregate energy intensities—the ratio of energy use to activity—in the US residential, services, manufacturing, freight, and passenger transportation sectors, adjusted for changes in the level and structure of sectoral activity, fell by an average of 23 percent between 1973 and 1987. Adjusted primary energy intensities fell by an average of 19 percent. Since the US energy/GNP ratio fell by 31.8 percent for delivered energy and 26.3 percent for primary energy over this period, this analysis suggests that about three quarters of the decline in the energy/GNP ratio was induced by improved energy efficiency, while the remainder was caused by structural change and fuel substitution.
- Actual energy use for the five sectors surveyed in detail was 51.8 EJ in 1987, or 71.4 EJ including electricity generation and transmission losses. Taking into account changes in the level and structure of energy-using activities, the efficiency improvements described above translate into savings of 15.6 EJ of delivered energy, or 17.7 EJ of primary energy.
- The largest reductions in energy intensities occurred for automobile and air travel, home heating, and fuel use in the manufacturing and service sectors. The energy intensity of truck freight, in contrast, actually increased. A decline in load factors and a rise in the importance of light trucks for personal transportation together limited the decline in the system intensity of private vehicles, measured in energy per passenger-km, to only 13 percent.
- Overall, structural change within sectors increased US energy delivered use by only 1.2 percent and increased primary energy use by 5 percent. Thus, in most sectors the impact of structural change was small.

A recent slowdown in the improvement of US energy efficiency has manifested itself in almost every sector, with the possible exception of manufacturing. This slowdown represents a market plateau, not the confrontation with thermodynamic or technological limits. Public policies could restore some of the interest in raising the efficiency of energy use.

How do these global results compare with those from other countries? Unfortunately, there have been few studies of this scope performed, but in analyses of the Federal Republic of Germany (8) and, more recently, of Norway (9) we found that the results for the US fall between the results from these two countries. For example, manufacturing energy intensity declined more in Germany, but less in Norway, compared with that of the US. Transportation energy intensity increased in both of these European countries, but decreased in the US. Residential heating intensity fell more in the US than in Germany, because the low penetration of central heating in Germany in 1973 (less than 55 percent) increased to over 70 percent by 1987, pushing up heating use significantly.

These three country studies showed that only part of the change in the ratio of energy use to gross national product measures improvements in energy efficiency. Indeed, the increase in energy-intensive activities in both Germany and Norway, particularly driving and heating, was far greater than in the US, and offset energy savings in other sectors in the European countries. Moreover, some of the differences in efficiency among these countries decreased because automobiles in the US improved more rapidly than in other countries. In all three countries, however, a significant energy saving potential remains.

Is energy use in the US less efficient than in other countries? Using the other countries we have analyzed as guides, as well as the individual sectoral studies we describe below, we can make some important estimates of how US energy efficiency compares with that of other countries. American cars, homes, appliances, and service buildings still use 20 to 33 percent more energy per unit of activity than do those in other industrialized nations. American industries use 10 to 25 percent more energy per unit of activity. Thus, while improvements in the US have kept pace with improvements elsewhere, and even narrowed the gap considerably for space heating and driving, significant differences between energy efficiency in the US and other countries still exist. However, these differences are much smaller than the differences in energy/GDP ratios imply.

Brief Sectoral Comparisons

Presently, International Energy Studies is analyzing the intensity of energy use in the major consuming sectors (passenger transport, freight, manufacturing, households, and the services sector) of major countries in Europe, as well as in Japan, the USSR, and important energy consuming developing countries. In this summary we focus on the most important activities in the major developed countries, using the United States as our reference country.

Transportation

The transportation sector accounts for 20 to 35 percent of final energy use in OECD countries. In spite of oil price shocks, transportation energy use in most OECD countries was considerably higher in 1987 than in 1972. The main reason was more cars. But trends in the US were different from those in other countries. Car ownership increased more rapidly in Japan and Europe than in the US (a consequence of higher US ownership prior to 1973), as did total travel. While the level of travel relative to GDP fell in the US between 1973 and 1987, this indicator increased in most other countries, as shown in Figure 1. Air travel increased its share of total travel, and rail and bus transit lost some of their shares to air travel and the automobile in almost every country. As a result of these shifts toward auto and air travel, the aggregate energy intensity of passenger travel increased more in other countries than in the US. Still, the US travel structure is some 33 to 50 percent more energy intensive than it is in other countries: Americans have more large cars and travel more in them (and in the air) than do Europeans or Japanese. One important finding often overlooked, however, is that automobiles dominate passenger transportation and energy use in Europe as well as in North America, and their share is growing rapidly in Japan. Thus, future energy savings in this sector must focus on automobiles.

Freight patterns in the US are different than those in other countries. The volume of freight, relative to GDP, is higher in the US than in virtually every other industrialized country (the USSR is an important exception). [One reason is the sheer size of the US, and the importance of shipments of bulk materials and energy over long distances. Another is that much of the trade that could be considered international (by sea) between Central Europe, Scandinavia, or Japan is not counted in domestic freight activity. The same activities take place by domestic routes within the US.] Freight volume, relative to GDP, fell in almost all industrialized countries. But the share of energy-intensive truck freight in the US—20 percent—is lower than in most other countries, and remained steady while increasing elsewhere. This means that improving the efficiency of trucks will be an important step in reducing future energy use for freight.

Overall, the level of passenger travel, relative to GDP, declined in the US, but increased in most other

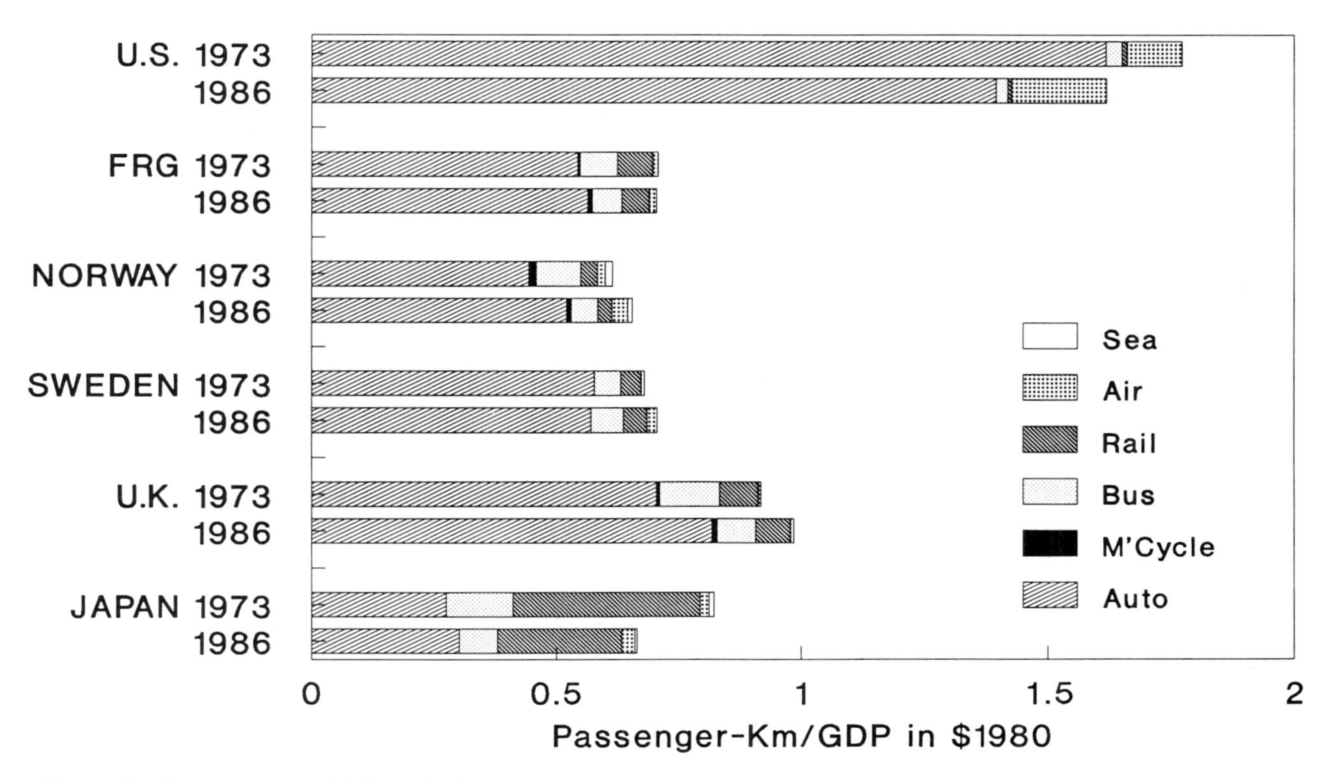

Figure 1. Passenger travel (all modes).

countries. The level of freight activity fell in all countries. These structural changes narrowed the gap between the US and other countries, but the US still remains one of the most transportation-intensive countries.

The energy intensities in transportation behaved in a mixed way. The intensity of air travel fell everywhere, principally because the same new, fuel-efficient aircraft, manufactured by Boeing, McDonnell Douglas, and Airbus Industries, entered almost all the fleets of the industrial nations. The fuel intensities of automobiles behaved in a different way. The sales-weighted fuel intensity of new US autos declined dramatically, as noted above; that for light trucks also declined but from a higher initial level. Although the share of light trucks increased to over 20 percent of personal vehicles in the US, the figure for combined new-vehicle fuel intensity still fell more than 45 percent. The same measure for new autos in most other countries showed very little change, declining by 20 percent at most, as shown in Figure 2 (10). When the fleet fuel intensities for these countries are displayed (Figure 3), the results are different: Intensity in the US fell by 30 percent, in the other countries by 10 percent at most. The disparity between new car test economy and on-the-road performance in Japan and Europe was greater than in the US, both because of the inaccuracy of the sales weighted new car intensity figures in Japan and Europe, and because the increase in congestion and city driving was so great. By 1987, the fuel intensity of new cars in the US was only slightly higher than that of the wealthiest countries in Europe, but the US car/light truck fleet fuel intensity was still 30 to 40 percent greater than intensities in Europe and Japan. Thus, other countries narrowed the structural gap with the US level of travel, while the US narrowed the differences with European or Japanese fuel intensity. In this way one major difference between US energy use and that in other countries shrank considerably.

What are the prospects for continued energy savings in transportation? The worlds' airlines are still adding newer, more efficient aircraft, which will reduce average energy use per km of travel. But both Boeing and McDonnell Douglas have postponed plans to introduce a very fuel-efficient prop-fan engine as a consequence of low oil prices and high interest rates, according to experts at one of these companies. The slowdown in improvements in new-car efficiency, coupled with the small improvements in fleets (outside of North America) is more worrisome as the role of the automobile in-

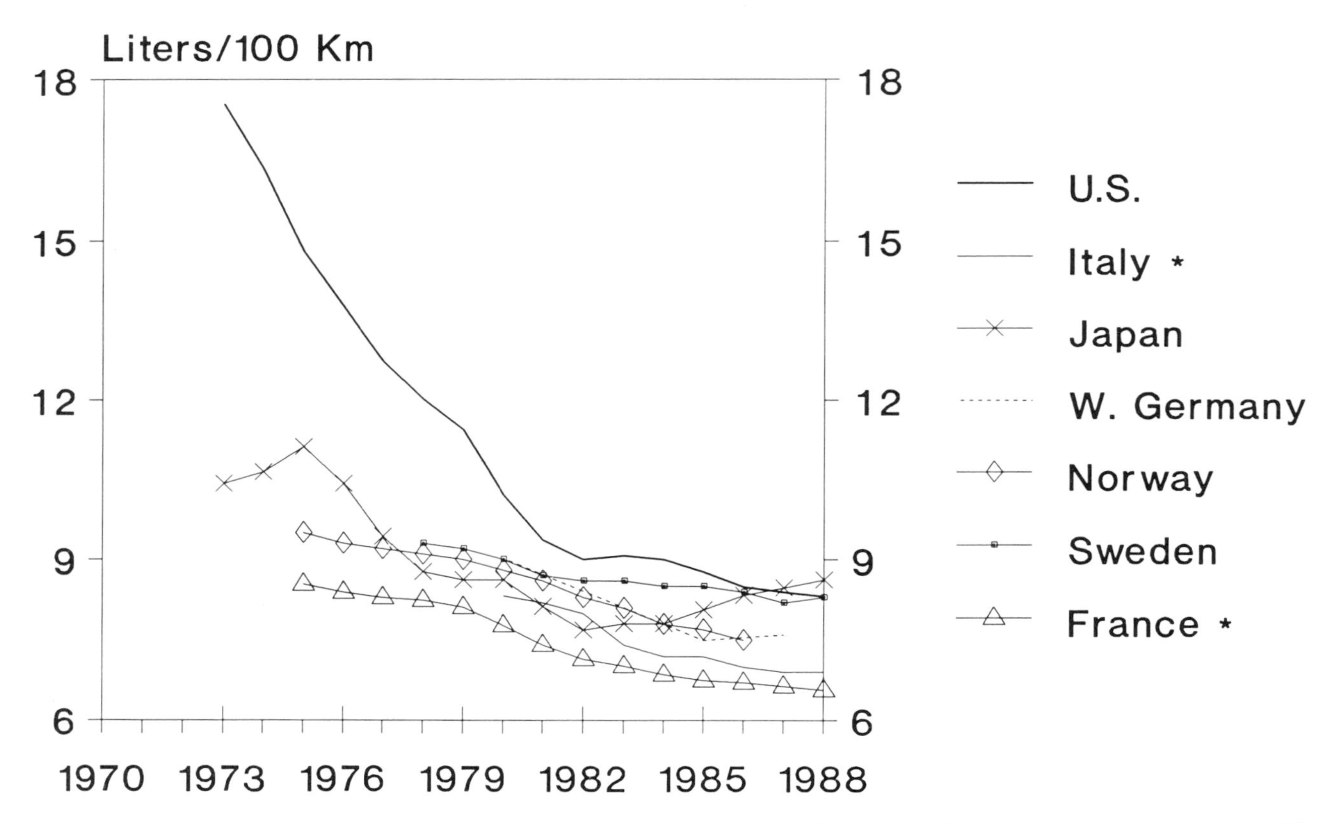

Figure 2. Automobile fuel economy—new car fleet averages (tests). Figures weighted approximately by sales. US figures (1989 est.) include light trucks. *Including diesel cars.

creases. While many high MPG prototypes exist, the challenge of the next decade appears to be the improvement of more conventional, i.e., family-type, cars that dominate present markets by improving motors and drive trains, and reducing drag and weight through design and new materials (11). The present fleets in Europe average about 9 liters/100 km (26 MPG), those in North America about 12 liters/100 km (19 to 20 MPG). These could be improved to about 5.5 liters/100 km (about 40 MPG) at less than a 10 percent increase in vehicle cost (Ross 1989). Equally important, however, is the reduction of emissions from all vehicles, as well as the improvement of traffic, which plays a key role in future fuel economy.

Manufacturing

Manufacturing accounts for between 25 and 35 percent of final energy use in most countries. Important changes took place in the role of manufacturing in overall energy consumption in most other OECD countries (12). The most transparent structural change in manufacturing energy use is the reduction in raw materials production per unit of GDP. Figure 4 shows the situation for steel in 1973 and 1986, for example, which is indicative of that for cement as well. Note that the US produces somewhat less steel per unit of GDP than do other countries, particularly Japan, Sweden, and West Germany. Indeed, the share of US manufacturing value-added in six major energy-intensive industries [Iron and Steel (ISIC 371), Non-Ferrous Metals (342), Basic Chemicals (351), Petroleum Refining (354,4), Stone-Glass-Clay (36), and Paper and Pulp (341)] (16.1 percent in 1985, excluding refining) is lower than the shares of most other countries (23 percent in 1985–87 in Japan; 20 percent in the EU-5 [France, West Germany, Norway, Sweden, and the United Kingdom]) (see Figure 5). Thus, if the US produced the same mix of output as do these other countries, manufacturing energy use in the US would be higher than it actually is.

Changes in the role of these industries since 1973 had a measurable impact on manufacturing energy use in all countries. Overall, the shift away from production of raw materials (measured in value added) reduced energy use for manufacturing by 7 percent in five European countries (EU-5), by 12 percent in Japan, and by 15 percent in the US, as the first three indices in

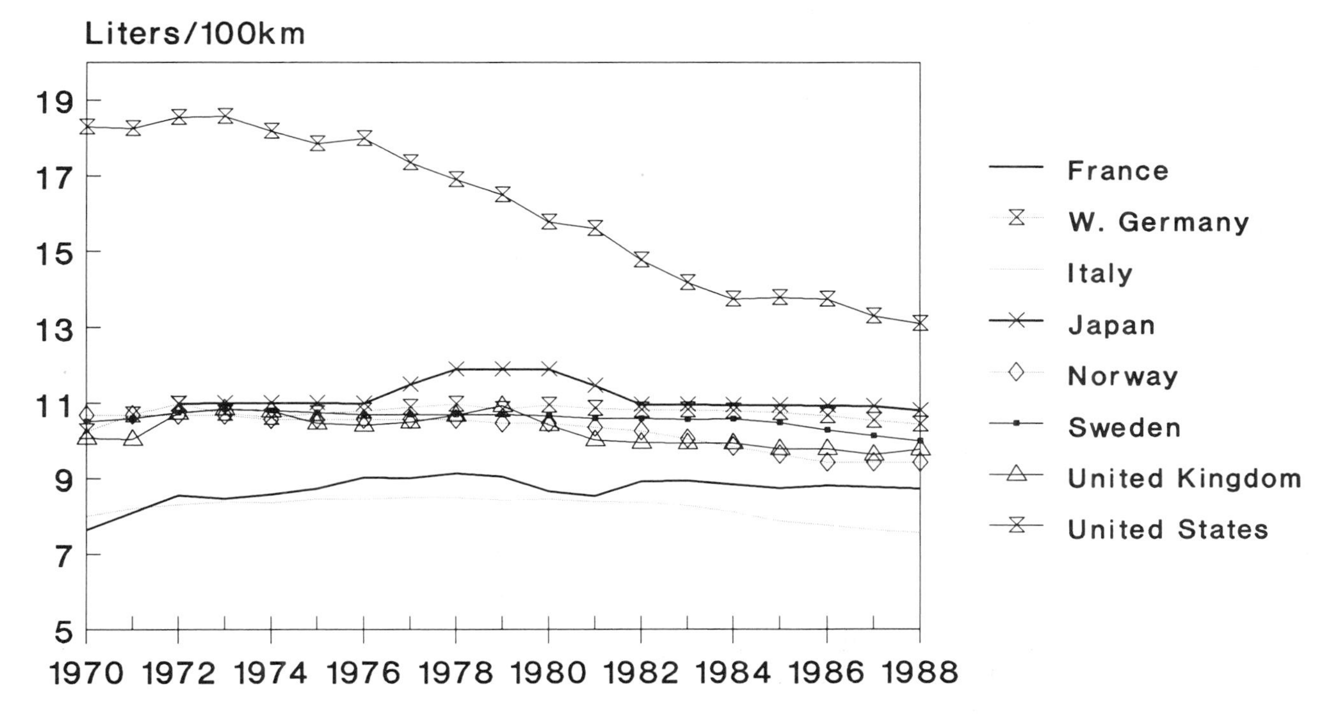

Figure 3. Automobile fuel economy—on the road fleet averages (includes diesels, personal light trucks). (Sources: US, DOE; W. Germany, DIW; Norway, Esso France, AFME; Japan, MITI; Sweden, TR; Finland, KTM.)

Figure 5 show. Structural change had a larger impact on manufacturing energy use in the US than in the other countries. Corrected for this structural change (i.e., holding the 1973 mix of industries constant), energy use per unit of value added fell by 30 percent in the US, by 38 percent in Japan, and by 27 percent in five European countries, as the last three lines in Figure 5 show. Note that the intensity indices fell significantly more than the structural indices, i.e., intensity changes caused more energy saving than structural changes. The share of oil in all countries fell markedly, and the intensity of oil use fell by more than 66 percent everywhere. The overall US record, measured as reduced energy intensity of manufacturing, lies between that of Japan (greatest decline) and Europe (least decline). While energy use per unit of monetary output is not always a reliable indicator of changes in efficiency, available data show that energy use per unit of physical output for most of the key raw materials fell in all countries.

The improvements in energy efficiency in manufacturing since 1973 represent only an acceleration of a long-term trend. Indeed, few observers find any slowdown in the improvement of efficiency occurring after the oil price crash, because the increase in output from basic industries that followed the recovery in the mid-1980s stimulated turnover of old, inefficient plants and innovation to save all resources. Thus, it is widely believed that industry will continue to save energy.

Residential Sector

The residential sector accounts for nearly 25 percent of energy use in most OECD countries (13). In spite of two sharp hikes in energy prices, standards of comfort and convenience increased in all OECD countries. However, differences in standards also narrowed because house area per capita and appliance ownership grew more rapidly in Europe and Japan than in the US, while the penetration of central heating increased to over 70 percent of homes in Central Europe, and water heating reached nearly every home. Still, Americans enjoy 50 to 80 percent more house area per person than Europeans and more (or larger) major appliances as well.

Fuel choices and fuel shares in the US evolved in line with the pattern in Europe: The choice of oil as heating fuel fell drastically and that of LPG declined as well (14). Substitutes were principally gas and wood in existing homes in the US, and gas, district heat, and electricity in Europe. In new homes, gas and electricity filled the gap left by oil. Electricity made important gains in water heating and cooking in both the US and most European countries. The only countries where oil maintained an important share of final consumption were West Germany (45 percent) and Japan (65 percent).

The US space heating intensity is about average, with those of Scandinavian countries being lower, and those of the United Kingdom, Holland, and Germany

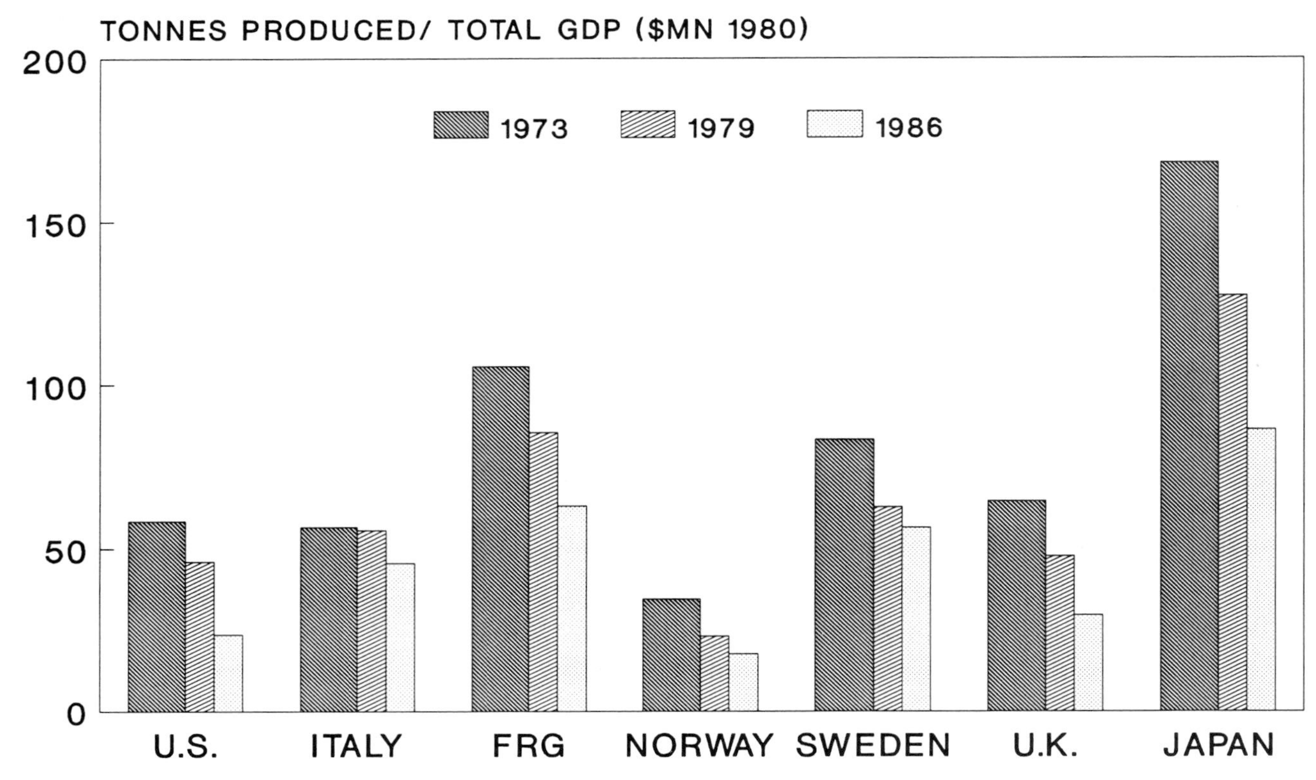

Figure 4. Energy intensive materials production—raw steel. (GDP at 1980 purchasing power parities.)

higher, as shown in Figure 6 (15). [The intensity for Japan (not shown) lies far below those displayed because heating is only intermittent and by room. Note that Japan has only 1975 degree-days (base 18°C); the US has 2599 (Meyers 1987), while the European countries we consider range from 2450 (France), 2800 (Britain), 3017 (Holland), 3113 (West Germany), and 3316 (Denmark), to over 4000 for Sweden and Norway. The average for the European countries, weighted by population, is about 15 percent greater than the US population weighted average.] The energy intensity of heating (in kJ/sqm/deg day of useful energy, i.e., the energy delivered to the heating system, minus losses in combustion of liquid, gaseous, and solid fuels) fell somewhat more in the US than in most other countries. The decline in the United Kingdom, West Germany, France, and Holland understates the real savings in space heating energy since 1973 in these countries, however. This is because the penetration of central heating rose from under 50 percent to over 70 percent in these countries in the period considered, a change that would have boosted space heating energy use per home by some 33 percent in these countries. Norway and Japan, by contrast, have little central heating (circulating hot water or hot air systems). Nevertheless, heating intensity grew steadily in these countries as more heat was supplied to more rooms.

There were several components of these improvements (16). Rapid reduction in comfort in oil-heated homes in most countries, including the US, led the decline. Occupants of gas-heated and electrically heated homes also cut back on comfort. Gradually some or much of this short-term savings was supplemented by efficiency improvements throughout. Heating equipment efficiency improved; in the US, the penetration of heat pumps rose higher than in every country except Japan, but there they are now used as complements with kerosene heaters. Improvements in the thermal integrity of new homes in most European countries, as implied by stiffer building code requirements (17) or actual insulation values (18), contributed to a significant decline in heat requirements for new homes. The thermal resistance of attics virtually doubled in all countries, but the improvements in insulation in walls were uneven. The insulation levels now required in walls of new homes built in Europe are twice the pre–1973 values, but improvements in the US were much less (19). In all, space heating intensity in the most recent homes in the US and Europe dropped by 30 to 50 percent compared with that in representative homes built before 1975. The intensity of US residential space heating fell far enough between 1973 and 1987 to be nearly comparable to the average for European countries; the significantly large size of US homes would permit economies of scale, so

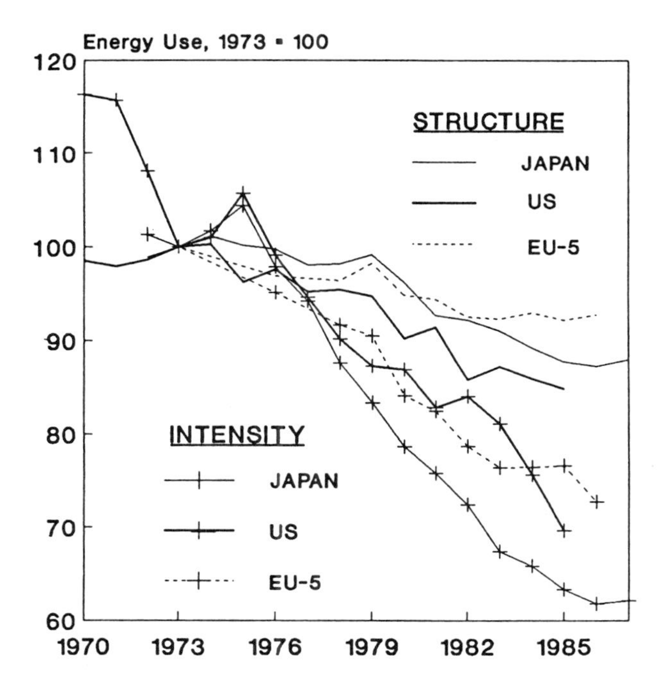

Figure 5. OECD manufacturing energy use—intensity and structure effects. (Energy use if only structure, or only intensity, changed relative to 1973.)

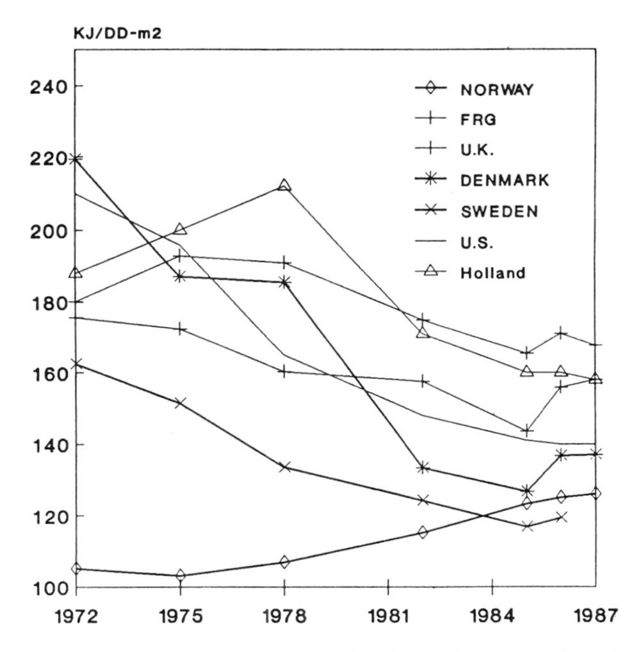

Figure 6. Home space heating intensity—Scandinavia and OECD countries. (Useful energy per gross floor area (m^2) per degree-day (DD), degree-days base 18°C.)

US values should be lower than those in Europe for a given thermal integrity. Considering all factors, US homes are about 20 to 25 percent more energy intensive for space heating than those in Europe.

We examined electric appliances and other home electricity uses in 12 OECD countries (20). We found that appliance efficiency improved in all countries, with the US improvements about average. Figure 7 summarizes the developments for the US. We have displayed electricity use per year for a variety of appliances in the 1973 and 1987 stock, as well as average values for those sold today. Note the decline in average consumption, as well as the lower consumption of new models. (Appliances sold in other countries show similar progress.) More efficient appliances have reduced household electricity use for appliances (40 to 66 percent of total household electricity use) by 10 to 20 percent in almost every country. After these improvements are counted, however, new US appliances still consume about 20 to 30 percent more electricity per unit of service than do those in Europe.

In conclusion, structural differences between the US and Europe narrowed significantly between 1973 and 1987. Still, if the residential sectors of European countries (or Japan) had US characteristics, residential energy use would be nearly 50 percent (100 percent in Japan) larger than it is. The intensity of heating and electric appliances in the US fell significantly between 1973 and 1987, but US homes and appliances still require roughly 20 to 25 percent more energy per unit of service than do those in Europe.

The outlook for improvements in energy use in the residential sector is mixed. There is still a large fraction of each country's housing stock with poorly insulated walls and leaky windows, and even in Sweden and Norway improvements to existing homes make sense. The insulation levels in new homes could be improved somewhat as well. New appliances could be improved significantly through adoption of the energy-saving features

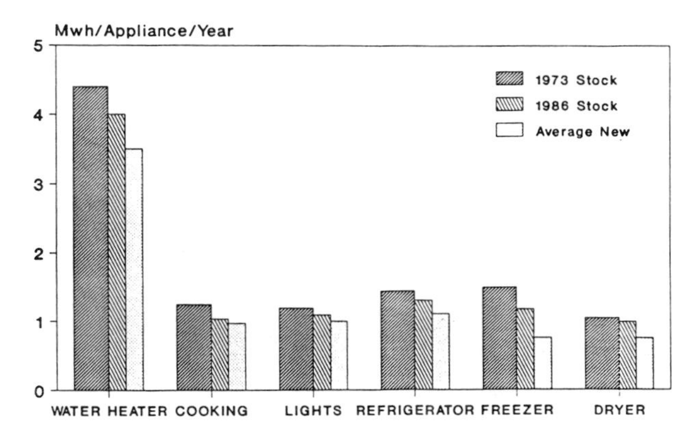

Figure 7. Appliance electricity use in the US—unit consumption: stock and new.

already found in some models of every appliance sold (21). All of these improvements pay off.

Figure 8, for example, illustrates the potential for electricity saving in key appliances sold in Sweden. The first three levels of consumption for each appliance reflect the average for each appliance in stock in the year shown. Next are shown models with the highest and lowest yearly consumption of those typical in size and features among models sold in 1987. Finally, we show values for a Low Energy Refrigerator sold in Denmark, designed by a team lead by Professor Joergen Noergaard of the Technical University of Lyngby (see the review in Schipper and Hawk, 1990), as well as Lawrence Berkeley Laboratory projections for refrigerator-freezers (combis) and freezers. If consumers pay no attention to electricity use, they will on average buy models that use about 25 percent less electricity than the average of the stock, the rough mean of the highest and lowest models shown. If they focus on models using the lowest level of electricity, usage per appliance will fall by 50 percent. And if consumers send a strong signal over electricity efficiency to manufacturers, then 66 to 75 percent of present-day use per appliance could be cut from household electricity bills as manufacturers make the "possible" practical.

Unfortunately, present-day markets provide little stimulus to homeowners or occupants to make these improvements, because of a variety of market and non-market barriers. Put simply, consumers have very short time horizons and ignore efficiency except where the payoff is very rapid, i.e., in two years or less. A combination of standards and other incentives is probably necessary to accelerate the technical progress that will lower the energy needs of comfort and convenience.

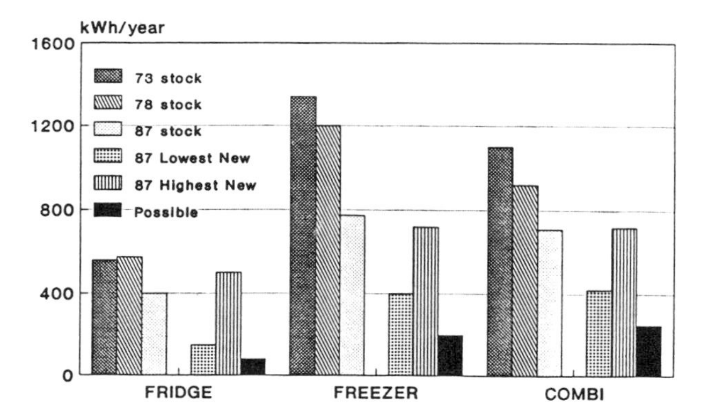

Figure 8. Refrigeration in Sweden—Appliance-specific electricity use (Sources: Konsumentverket, Vattenfall. Size ref:150–200; frz:250–300; comb:350).

Services Sector

The services sector accounts for only 10 percent of delivered energy in most countries, but a higher (and growing) share of electricity. International comparisons of energy use in the services sector are difficult. Nevertheless, available data do make some partial comparisons meaningful (22). The share of services GDP increased slightly in almost all countries, and total service-sector area increased slightly faster. The total area of buildings in the service sector grew to more than 24 sqm per capita in the US, versus 15 to 19 sqm per capita in Europe or Japan. As with the residential and transportation sectors, then, US activity level is significantly higher than that of any other country. But the gap between the US and Europe narrowed in the 1970s and 1980s.

The balance between electricity and fossil fuels or district heat in the US is different from that of Central Europe, and close to that of Sweden (over 35 percent of delivered energy as electricity). The high share of electricity in the US is a result of high cooling and electric heat penetration, as well as high electricity use for lighting and other building services. Only Norway and Canada, with higher penetration of electric heating, have a higher share of electricity in delivered energy. After 1973, oil heat lost share to gas, district heat, or electricity in every country, and the share of oil in overall delivered services energy fell everywhere.

Schipper *et al.* (1986) found that US space heating intensity was considerably higher than that in other countries and judged uses of electricity somewhat less efficient than in other countries, particularly because of high lighting levels. Fuel intensity, measured in GJ/sqm, fell in most countries in the 1970s and 1980s, but the US intensity still appears to be somewhat higher than those of colder, northern European countries. Electricity intensity increased in most countries, even where the penetration of electric heating did not. Computers and other information technology have been important factors in raising electricity use. By contrast, electricity intensity in the US did not rise significantly after 1979.

The outlook for energy savings in the services sector is bright. This is because increasing use of electricity is putting pressure on local and national supply authorities to work with building owners or operators to reduce peak and even average loads. Improvements in lighting technologies, optical coatings on windows to reduce undesirable heating gains or losses, and more efficient motors and compressors all promise to reduce electricity intensity or at least reduce growth. Unlike residential buildings, commercial buildings represent an opportunity where a small intervention by the user and supplier together can bring about an enormous absolute saving in electricity (23). The same is true for reducing oil and gas use for space and water heating.

Conclusions: International Improvements in Energy Efficiency

This brief comparison reveals that wealthy countries reduced many important energy intensities and thus saved energy. In the case of manufacturing, these savings represented an acceleration of historical trends. In the residential and service sectors, the savings broke with past trends. In transportation the record was mixed; indeed, only in the air travel sector were there clear energy savings. In the personal vehicle sector outside of North America the overall efficiency of producing passenger-kilometers increased marginally. In the truck freight sector in every country, the overall efficiency of tonne-kilometers has improved very slightly or gotten worse. But on balance, OECD countries used about 32 EJ of delivered energy less than they would have in 1985 if 1972–73 energy intensities had not fallen. About 90 percent of this decline could be attributed to energy conservation, the rest to structural changes that also reduced energy demands.

The brief comparison presented above also shows that, relative to GDP, larger homes and appliances raise US residential energy use compared to that in Europe significantly. Larger cars, more driving, more flying, and more freight increase US energy intensity even more compared to that of Europe or Japan. Greater service sector area also increases US energy intensity. Output of energy intensive raw materials in the US, by contrast, is somewhat lower than in Japan or Western Europe. On balance, these *structural* differences between the US and other countries account for more than 50 percent of the difference in the energy/GDP ratio. But increases in travel, and increases in built space for homes and services have kept significant pressure on electricity and oil use in almost every country.

Since 1973 differences in energy use among wealthy countries have become smaller, particularly as other nations have approached US levels of activity, but considerable gaps remain. Additionally, some of the change in the US energy/GDP ratio was caused by structural change: manufacturing output, home floor area, passenger travel, and freight all grew less than or equal to the rate of growth in GDP in the US. These considerations illustrate why it is dangerous to use changes in the energy/GDP ratio as indicators of energy savings.

References

1. Schneider, S., "The Greenhouse Effect: Science and Policy," *Science*, Vol. 243, 10 Feb 1989, pp. 771–781. See also Tirpak, D., and Lashof, D., *Policies to Stabilize the Climate*. Washington, DC: US Environmental Protection Agency, 1990.
2. Schipper, L., "Energy Saving Policies of OECD Countries: Did They Make a Difference?" *Energy Policy*, Vol. 15, No. 6, December 1987, pp. 538–548.
3. Schipper, L., and Ketoff, A., "Energy Efficiency: The Perils of a Plateau," *Energy Policy*, December 1989.
4. See, for example, International Energy Agency, *Electricity Conservation*. Paris: Organization for Economic Cooperation and Development, 1989.
5. Schipper, L., and Lichtenberg, A., "Efficient Energy Use and Well Being: The Swedish Example," *Science*, 197, 6 December 1976, p. 1013.
6. Schipper, L., Howarth, R., and Wilson, D., *A Long Term Perspective on Norwegian Energy Use*. Prepared for the Royal Norwegian Energy and Oil Ministry. Berkeley, CA: Lawrence Berkeley Laboratory (in press).
7. Schipper, L., Howarth, R., and Geller, H., "United States Energy Use from 1973 to 1987: The Impacts of Improved Efficiency," *Annual Review of Energy*, 16 October 1990 (preprint), Palo Alto, CA: *Annual Reviews, Inc.*
8. Schipper, L., *Energy Use in Germany in the Long Term*. Berkeley, CA: Lawrence Berkeley Laboratory, 1988.
9. Schipper, L., Howarth, R., and Wilson, D., 1990. *A Long Term Perspective on Norwegian Energy Use* (6, above).
10. These data are taken from national sources: The Agence pour La Matrise d'energie (France); *Verkehr in Zahlen* (German Ministry of Transport, Bonn, West Germany); Transportraadet and National Energy Board (Sweden); Transport Oekonomisk Institute (Norway); Ministry of International Trade and Industry (Japan); Ministry of Trade and Industry (Finland).
11. Ross, M., "Energy in Transportation," *Annual Review of Energy*, 1989, Vol. 14, Palo Alto, CA: Annual Reviews, Inc. See also the article by Sperling and DeLuchi ("Transportation Energy Futures") and that by Mellde, Maasing, and Johansson ("Advanced Automobile Engines") in the same volume.
12. Howarth, R., Schipper, L., and Duerr, P., *Manufacturing Oil and Energy Use in Seven OECD Countries*, LBL-27887. Berkeley, CA: Lawrence Berkeley Laboratory, 1990.
13. Schipper, L., Ketoff, A., and Kahane, A., "Explaining Residential Energy Use by International, Bottom-up Comparisons," *Annual Review of Energy*. Palo Alto, CA: Annual Reviews, Inc., Vol. 10, 1985, pp. 341–405.
14. Schipper, L., and Ketoff, A., "Household Oil Savings in the OECD: Permanent or Reversible?" *Science*, Vol. 230, 6 December 1985, pp. 1118–1125.
15. Calculated from I.E.S. data for each country. See Schipper, L., Ketoff, A., and Kahane, A., 1985 (13, above).
16. Schipper, L., Ketoff, A., and Kahane, A., 1985 (13, above).
17. Wilson, D., Schipper, L., Tyler, S., and Bartlett, S., *Policies and Programs for Promoting Energy Conservation in the Residential Sector: Lessons from Five OECD Countries*, LBL-27289. Berkeley, CA: Lawrence Berkeley Laboratory, 1989.
18. Schipper, L., Meyers, S. and Kelly, H., *Coming in from the Cold: Energy-Wise Housing from Sweden*. Cabin John, MD: Seven Locks Press, 1985.
19. Meyers, S., "Energy Consumption and Structure of the US Residential Sector: Changes between 1970 and 1985," *Annual Review of Energy*, 1987, Vol. 12, pp. 81–97.
20. Schipper, L., Ketoff, A., Hawk, D., and Meyers, S., "Residential Electricity Consumption in Industrialized Countries: Changes since 1973," *Energy*, Vol. 12, No. 12, 1987,

pp. 1197–1208. See also Schipper, L., and Hawk, D., *More Efficient Household Electricity Use: An International Perspective*, LBL-27277. Berkeley, CA: Lawrence Berkeley Laboratory, 1990.

21. Geller, H., *Update on Electricity Use in the Residential Sector*. Report to the US Office of Technology Assessment. Washington, DC: American Council for an Energy-Efficient Economy, 1988. See also Schipper and Hawk, 1990 (20, above).
22. Schipper, L., Ketoff, A., and Meyers, S., "Energy Use in the Service Sector: An International Perspective," *Energy Policy*, Vol. 14, No. 3, June 1986, pp. 201–218.
23. Geller, H., *Update on Electricity Use in the Service Sector*. Report to the US Office of Technology Assessment. Washington, DC: American Council for an Energy-Efficient Economy, 1988. Or Krause, F., *et al.*, *Analysis of Michigan's Demand-Side Electricity Resources in the Residential Sector*. LBL-23025 (3 vols.). Berkeley, CA: Lawrence Berkeley Laboratory, 1988.

Indoor Air Quality—Toward the Year 2000

David T. Grimsrud
Minnesota Building Research Center
University of Minnesota
Minneapolis, MN

Abstract

This paper describes the development of indoor air quality over the past 20 years and discusses major scientific problems of the next decade. Policy questions are becoming more important as the field matures, and the responsiveness of policy to new scientific information is an ongoing concern.

Introduction

The study of indoor air quality, the comfort and health characteristics of air within buildings, has undergone several strange twists in the past two decades. If we learn from our history we must expect that similar twists will occur in the future as well. Since this area of study is likely to encounter these same changes, any projections made today with longer than five-year time horizons have the same validity as long-range weather forecasts, that is, very little or none.

History of the 1970s

The study of indoor air quality began in the early 1970s when a few outdoor pollution scientists such as John Yocum at The Research Corporation, Craig Hollowell at Lawrence Berkeley Laboratory, and John Spengler at Harvard began to look at pollutant sources that existed inside buildings (Yocum et al., 1974; Hollowell et al., 1977; Spengler et al., 1979). They were surprised to find that pollutant concentrations were, in some situations, substantially larger than outdoor levels. The researchers quickly realized that even though the sources were small, the air masses affected (defined by the building volume) were also small, so that the concentrations of these pollutants within buildings reach substantial values. This observation, together with the realization that people spend most of their time within buildings, caused this small group of researchers to argue that air quality, from a public-health perspective, was primarily a buildings problem.

This cry, though repeated often, was largely ignored during the latter half of the 1970s. Indoor air quality research that was done during this time was sustained from an unusual source—not, as one might assume, from the Environmental Protection Agency (EPA) with the creation of "Earth Day One," but from the Department of Energy (DOE) with the creation of "Oil Embargo One." Energy conservation became the by-word in the late 1970s. Building energy conservation included tampering with a building's ventilation system to reduce energy use. The goal of reducing energy use implied reducing ventilation, which meant potential problems if air quality in buildings were compromised as a result. Thus, DOE supported indoor air quality research to stay ahead of those who would attack conservation in buildings because of its potential negative impact on air quality and thus on building occupants. Despite and perhaps because of this support, the study of air pollution in buildings remained a minor, fringe activity within the air pollution field.

The comments above, and those that follow, reflect a personal view of the history of the field. While significant work has been done in Europe (Bierksteker et al., 1965; Molhave, 1979; Jonassen and McLaughlin, 1980) and indeed, major international indoor air quality and climate conferences in Copenhagen (1978), Stockholm (1984), and W. Berlin (1987) have helped shape and focus research in the field, this paper discusses indoor air quality in the US context. The interaction between federal funding sources and researchers is known best in the US context and is described in the material that follows.

History of the 1980s

The official change of stance within the EPA began when Stanley Watras began to trigger radiation alarms as he entered a Pennsylvania nuclear power plant then under construction. The high levels of radon found in residences in Pennsylvania and New Jersey following this event caused the EPA to become interested in air quality in buildings as an issue that should be considered as part of their mandate (Henschel and Scott, 1986). Unfortunately, there was little money for this work. The federal government was effectively paralyzed by the debt incurred by the supply-side economics of the 1980s (although it could find the will to act in unusual circumstances, such as rescuing medical students in Granada and bailing out unregulated savings and loan corporations). Consequently, even though indoor air quality had emerged as a significant research issue for both the EPA and DOE, few resources to date have been devoted to it.

With the understanding, finally, that indoor air quality is a substantial public health problem (enough to attract the attention of Senate majority leader Mitchell, Representatives Schneider and Kennedy; cf. The Indoor Air Quality Act, H.R. 1530 and S. 657, US Congress 1990), its progress, again, is taking a strange turn. Environmental issues have changed in the last few years from concern about air pollution as a public health issue to concern about air pollution and the end of nature as we know it. The threat that global warming may fundamentally change the climate of the planet—at a rate beyond the possible response rate of an ecosystem—is accepted by most major newspapers of the land. With that issue as a major part of the environmental agenda of this country and other developed nations, what lies ahead for the study of indoor air quality?

Global warming and other environmental concerns call out for an increase in energy efficiency. This, in turn, implies tampering with the ventilation in a building once again to reduce energy use. If doing so compromises the air quality of the building much will be lost, little gained, and the resources provided by the Department of Energy to support this research field in the 1970s will have been wasted. However, I am optimistic. We *can* learn from the past. Many studies have shown the compatibility of energy efficiency in buildings and good indoor air quality. Several are reviewed in the recent book of duPont and Morrill (1989).

Major Issues of the 1990s

At this point it is useful to step back from this personal sense of political history and ask about the important scientific problems that must be attacked to assure building occupants that the quality of the air within buildings will not compromise their health. The important tasks have been and continue to be: 1) studies of the health effects of long-term exposure to low pollutant concentrations, 2) characterizing the chemical and physical behaviors of the sources of these pollutants, their transport and fate in buildings, and 3) developing procedures and policies to reduce harmful pollutant concentrations that exist within buildings. That is a short list. However, its brevity does not imply that the goals will be reached quickly. Studies of indoor air quality have often been organized by pollutant (e.g., radon), by pollutant groupings (volatile organic compounds or VOC), or by source (unvented combustion appliances, for example). The major questions cut across all categories. Some progress has been made in several areas but a substantial amount remains to be done before a coherent policy regarding indoor air quality in buildings can be declared.

Health Effects. The dominant problems in the field continue to revolve around health effects. What are the health effects of long-term exposures to low pollutant concentrations? The issue is phrased in this way because health effects of even short-term exposures to high concentrations can be dramatic. Each winter in northern regions of the country there will be deaths due to carbon monoxide poisoning from malfunctioning heating systems. Carbon monoxide poisoning also occurs at lower concentrations but the symptoms are less dramatic and the effects may be misdiagnosed. The health effects issues, because they are difficult to investigate, take substantial time and resources. Without the results from these studies, concentration guidelines in buildings cannot be justified and the building industry is left with historical tradition alone as the basis for design standards and operating procedures.

Health effects of long-term exposure to low levels of radon are better understood than those for any other indoor pollutant. However, even in this case there is no direct evidence that radon in buildings has ever resulted in a lung cancer. Indeed, the epidemiological problem of demonstrating lung cancers from radon alone when confounded by the large "noise signal" provided by tobacco smoking may mean that the relationship between lung cancer and radon cannot be proved. This is an important problem and substantial work in this area continues (Radon, 1989).

Source Characterization. In parallel with the health-effects studies that are under way are investigations of physical and chemical properties of pollutant sources that occur within buildings. Why is this important? Ultimately, control of pollutant concentrations within buildings to levels consistent with the guidelines determined by health studies depends on a) knowing what pollutant sources are present and b) understanding their properties so that if the sources remain in the space their pollutant emissions can be controlled. Since eliminating a pollution source from a space is the most effective control technique, even a simple catalog of emissions from sources typically used in buildings would be of substantial aid to a designer or architect as a new building goes through its various design stages.

Policy Issues

Indoor air quality is a young scientific field. Progress has been made in understanding many specific pollutants. Substantial investigative work remains to be done in most health-effect areas and in characterizing the sources of volatile organic compounds—compounds that change dramatically as new building materials and consumer products are introduced into buildings. Adequate characterization of particles, both as entities in their own right and as carriers of adsorbed gases, remains a major issue for the field. Reducing harmful concentrations of indoor pollutants in the nation's building stock requires policies that are sensitive to the needs of the public and are responsive to new scientific information. The policy questions are important for they decide the implementation of results of the scientific work currently under way. How are indoor air quality policies to be implemented? In particular, *when* should these policies be implemented? Can the scientific results affect a policy once it is implemented? In this section I will examine asbestos, radon, and ventilation standards.

Two major indoor air issues have been treated by federal policy. These are asbestos and radon. The policies are administered by different groups within the EPA, each with a specific charge from Congress. The policies have some similarities and some important differences. Both are under attack by respected scientists. If we expect the public to listen to scientists discuss global issues facing our planet on the occasion of "Earth Day Two," we must be sensitive to the image of an agency grappling with a policy that is under attack from the scientific community.

Asbestos. The EPA asbestos policy is based on the assumption that if there is a source of asbestos present within a building it should be removed. EPA training has helped develop a rapidly growing industry to remove the asbestos sources. The procedures are costly for building owners. In particular, a 1986 interpretation of the Asbestos Hazard Emergency Response Act required public and private schools to inspect for asbestos and inform parents if asbestos-containing materials were present. This policy neglected direct measurements of airborne concentrations and thus any estimate of the exposure to an occupant of these buildings.

Growing data show that concentrations in buildings are comparable to outdoor concentrations. Thus, assumptions that risks are high indoors, high enough to justify the enormous cost of removal, appear unjustified.

A recent review of the asbestos issue by Mossman et al. (1990) in *Science* concludes that

> . . . available data and comparative risk assessments indicate that chrysotile asbestos, the type of fiber found predominantly in U.S. schools and buildings, is not a health risk in the nonoccupational environment.

The authors later conclude with the statement that

> Prevention (especially in adolescents) of tobacco smoking, the principal cause of lung cancer in the general population, is both a more promising and rational approach to eliminating lung tumors than asbestos abatement.

A subsequent editorial by Philip Abelson in the same journal uses a stronger voice to argue that present policy with respect to asbestos abatement should be changed.

> . . . Unless policies are modified, the sums wasted in abatement and litigation will proliferate. Regulations should be modified to take into account the greatly differing hazards of the various asbestiform materials. Standards for indoor air should be based on actual measurements of types and amounts of fibers.

Radon. EPA radon policy has developed along similar lines. Since radon exposure occurs primarily in individual homes, i.e., in private not public air space, EPA chose to educate the public rather than regulating the air within the 80 million houses in the United States. In this model the EPA would warn the public, the public would measure concentrations, and if necessary, find a contractor to mitigate the problems identified. This policy requires the existence of a group of contractors trained to mitigate those problems. While it is true that techniques for mitigation have been developed and are accessible to those familiar with the published research literature, builders and contractors do not learn new techniques by reading those journals in university research libraries. Rather, the EPA established a group of radon training centers at five universities around the country to provide information about radon to government officials and the general public and to establish training courses in radon measurement and abatement for building professionals.

Thus, the pieces are in place to effect a national indoor air quality policy related to a major indoor pollutant. Is it the correct policy? Are there other models? How should decisions be made among alternatives? Can existing policy be changed? What investment does EPA have in existing policies? Will these policies be models for future policies related to other indoor pollutants?

A critic of the present policy, who has contributed substantially to our present understanding of the radon problem, estimates that the cost of the present policy is \$8.8 billion plus \$0.8 billion annually for maintenance (Nazaroff, 1990). Furthermore, he estimates that the present EPA recommendations about radon will, if completely successful, reduce the lung cancer mortality in the general population by the same amount as a permanent reduction in the smoking rate of 3 percent.

He does not disparage the good intentions of those who have created the present policy but points out the major differences between the present policy that will reduce the risk of many people slightly and a less expensive alternative that would focus on the seventy thousand or so houses in the United States having radon concentrations above 800 Bq/m^3 ($\sim$ 20 pCi/l). Long-term exposure to radon at these concentrations is comparable to occupational limits for uranium miners, an exposure regime where health effects are not in dispute.

How can a federal agency modify policy once it is in place? One model is the standard-setting process where regular revisions occur.

Ventilation Standards. The dominant mechanism used to control air quality in buildings is ventilation with outdoor air. Thus the role of ventilation standards, the documents that provide guidance to the engineering and design community, is crucial for the maintenance and improvement of indoor air quality.

In North America the most important ventilation standard is Standard 62, "Ventilation for Acceptable Indoor Air Quality," published by the American Society of Heating, Refrigerating and Air Conditioning Engineers (ASHRAE). This standard has recently been revised and published as Standard 62-1989 (ASHRAE, 1989). This improved standard, based on new information and new experience since the publication of Standard 62-1981, resulted from a consensus process involving representatives of many disciplines and a public review of the draft standard. This process and the compromises inherent to such an activity have been described in other papers and will not be reviewed here (Janssen, 1989).

It is important to recognize that a standard is a transition document. Revision is mandatory, particularly in a field that is changing as rapidly as indoor air quality. The present scientific basis for the Standard 62, although based on the best available research, is not strong (Grimsrud and Teichman, 1989). There are important data necessary to establish a solid foundation for the Standard that do not yet exist. These include the health effects and source characterization issues mentioned above. In the absence of appropriate data, assumptions have been made that must be tested experimentally and verified. Much more should be known from measurements in actual buildings before the Standard can be justified rigorously on its scientific merits.

It is useful to spend some time imagining a standard that could be written if unequivocal experimental data about pollutant sources, health effects, and ventilation rates existed. This is another way to see where this field is headed and what questions must be investigated.

The purpose of Standard 62-1989 is (and that of any future standard is assumed to be) "To specify minimum ventilation rates and indoor air quality that will be acceptable to human occupants and are intended to avoid adverse health effects" (ASHRAE, 1989).

The ideal standard is constructed from knowledge of the health effects of all the pollutants that are found within the building. Once these are known, relative risks of exposures to these pollutants can be determined and acceptable concentration limits can be set. At this point a performance procedure (called the Indoor Air Quality Procedure in Standard 62-1989) could be used. The designer of the building would be free to choose any technology available to achieve the air quality specified by the set of concentration guidelines.

The ideal standard has a prescriptive path as well as a performance path to give the designer a well-understood and explicit procedure to satisfy the standard. For each pollutant found in the building an upper bound on source strength and a lower bound on ventilation rate will assure that the concentration limit for that pollutant is not exceeded. These bounds would be determined using verified indoor-air-quality models that would accurately simulate pollutant emission and transport, ventilation parameters (e.g., ventilation effectiveness), and occupant exposure.

The importance of limiting the source strength cannot be overemphasized. Since the goal is to limit pollutant concentrations (the ratio between a source strength and a ventilation rate), source strengths must be less than some value while ventilation rates must be larger than some related value. Field measurements demonstrate that most problems in buildings occur because pollutant source strengths are too large rather than from insufficient ventilation (Nero et al., 1983; Traynor et al., 1987; Turk et al.,1989). This simply means that a wider variation is seen in source strengths than in ventilation rates as one examines groups of similar buildings. Excessive concentrations are found in buildings with both low and high ventilation rates (measured against a norm). Similarly, concentrations within recognized bounds are also found within buildings having high and low ventilation rates.

Setting source strength upper limits and minimum ventilation rates for our ideal standard will be an iterative process, since source characteristics for some materials depend upon concentrations in the surrounding air and the dynamics of the adsorption-desorption process (Matthews et al., 1985). Typical source terms for materials found in buildings of a similar design and recognized concentration limits will be used to produce minimum ventilation rates. A realistic ventilation rate for the building type will then be chosen and maximum source terms adjusted accordingly.

The prescriptive portion of Standard 62-1989, the Ventilation Rate Procedure, largely ignores source strengths. The author views this as its major failing—one of the primary areas that must be improved in any future revision of the Standard. The concentration guidelines of the Indoor Air Quality Procedure are not part of the Ventilation Rate Procedure. The only source term considered in determining the ventilation rates of Table 2 of Standard 62-1989 is the CO_2 generation rate of the building occupants. This value, coupled with the observation from many studies that CO_2 concentrations larger than 1000 parts per million (ppm) are associated with an increase in occupant complaints in buildings, leads to the minimum ventilation rate of 15 cfm/occupant in the Standard. Standard 62-1989 clearly notes that the CO_2 concentration limit is not, in itself, a physiological limit. Rather it substitutes for many pollutants associated with occupancy that may cause discomfort in the space.

This lack of coupling between the criteria adopted for the performance procedure in Standard 62-1989 (admittedly an incomplete list) and the prescriptive procedure forces the nature of Standard 62-1989 to change from one based on avoiding adverse health effects to one based on acceptability of the air within a space, that is, from health to comfort. This change means that the key element in the purpose of Standard 62-1989, ". . . avoid adverse health effects . . . ," cannot be achieved directly. Scientific data to establish concentration limits for pollutants in buildings do not exist. Only modest information about source strengths of common pollutants is available.

Summary

Major scientific questions remain in the field of air quality in buildings. These questions are reflected in current policies and standards that translate knowledge about the issue to public action. The lack of health effect and source characterization information that persists in this field weakens attempts to initiate policy and write standards for buildings. Two important issues that must be addressed repeatedly are flexibility and accountability. How well do national policies change in response to new scientific information? Standards have a fixed revision schedule; federal policies could well adopt a similar procedure.

References

Abelson, P.H., "The Asbestos Removal Fiasco," *Science* 247, p. 1017 (1990).

ASHRAE Standard 62, *Ventilation for acceptable indoor air quality,* American Society of Heating, Refrigerating, and Air Conditioning Engineers, Atlanta (1989).

Bierksteker, K.G., DeGraff, H., Nass, C.A.G., "Indoor Air Pollution in Rotterdam House," *International Journal of Air and Water Pollution* 9, p. 343 (1965).

duPont, P. and Morrill, J, *Residential Indoor Air Quality & Energy Efficiency,* American Council for an Energy-Efficient Economy, Washington, DC (1989).

Grimsrud, D.T., Teichman, K.Y., "The Scientific Basis of Standard 62-1989," *ASHRAE Journal* 31 (10), pp. 51–54 (1989).

Henschel, D.B., Scott, A.G., "The EPA Program to Demonstrate Mitigation Measures for Indoor Radon: Initial Results," *Indoor Radon,* Air Pollution Control Association, (1986).

Hollowell, C.D., Budnitz, R.J., Traynor, G.W., "Combustion-Generated Indoor Air Pollution," *Proceedings of the 4th International Clean Air Congress,* Tokyo, Japan, (May 1977).

Janssen, J.E., "Ventilation for Acceptable Air Quality," *ASHRAE Journal* 31(10), pp. 40–48 (1989).

Jonassen, N., McLaughlin, J.P., "Exhalation of Radon-222 from Building Materials and Walls," in T.F. Gesell, W.M. Lowder, eds.. Natural Radiation Environment III, Vol. 2, pp. 1211–1224. US Department of Energy Technical Information Center, 1980.

Matthews, T.G., Reed, T.J., Tromberg, B.J., Fung, K.W., Thompson, C.V., Simpson, J.O., Hawthorne, A.R., "Modeling and Testing of Formaldehyde Emission Characteristics of Pressed-Wood Products: Report XVIII to the US Consumer Product Safety Commission 1985." Oak Ridge National Laboratory Report No. TM-9867, Oak Ridge, TN, 1985.

Molhave, L., "Indoor Air Pollution Due to Building Materials," in P.O. Fanger, O. Valbjorn, eds. *Indoor Climate.* Copenhagen : Danish Building Research Institute, 1979, pp. 89–110.

Mossman, B.T., Bignon, J., Corn, M., Seaton, A., and Gee, J.B.L., "Asbestos: Scientific Developments and Implications for Public Policy," *Science* 247, pp. 294–301 (1990).

Nazaroff, W.W., "Indoor Radon: Exploring US Federal Policy for Controlling Human Exposures," to be published in *Environmental Science and Technology,* 24 (1990).

Nero, A.V., Boegel, M.L., Hollowell, C.D., Ingersoll, J.G., Nazaroff, W.W., "Radon Concentrations and Infiltration Rates Measured in Conventional and Energy-efficient Houses," *Health Physics* 45, pp. 401–406 (1983).

"Radon," International Workshop on Residential Radon Epidemiology, Conf-8907178, US Department of Energy (1989).

Spengler, J.D., Ferris, B.G., Jr., Dockery, D.W., "Sulfur Dioxide and Nitrogen Dioxide Levels Inside and Outside Homes and the Implications on Health Effects," *Environmental Science and Technology* 13, p. 1276 (1979)

Traynor, G.W., Apte, M.G., Carruthers, A.R., Dillworth, J.F., Grimsrud, D.T., Gundel, L.A., "Indoor Air Pollution Due to Emissions from Wood-burning Stoves," *Environmental Science and Technology,* 21, pp. 691–697 (1987).

Turk, B.H., Grimsrud, D.T., Brown, J.T., Geisling-Sobotka, K.L., Harrison, J., Prill, R.J., "Commercial Building Ventilation Rates and Particle Concentrations," *ASHRAE Transactions,* 95 (I), pp. 422–433 (1989).

Yocum, J., Cote, W., Clink, W., "A Study of Indoor-Outdoor Air Pollutant Relationships," Summary Report, Washington DC: National Air Pollution Control Administration (1974).

Change in the Building Industry

Tage C. G. Carlson
Construction Technology Laboratory
USG Corporation Research and Development
Libertyville, IL

Abstract

This paper examines the general theme of change in the building industry with a specific focus on the time for introduction of new products and systems. A brief overview of the building industry is given. In the residential building sector, contemporary building practices are examined with focus on material and labor intensities and overall costs. In order to examine the time scales and barriers in changing contemporary building practices, a review of several systems, processes, and material changes which have occurred, or are occurring in the industry, are studied. Specifically these are:

- Gypsum wallboard
- Factory-built housing
- Rigid foam insulation
- Pressure treated wood foundations (PWF)

From an industry perspective, the explicit and implicit barriers to change in the industry are also discussed. Specifically:

- The initial cost versus life-cycle cost issue will be reviewed with emphasis on how and when it is used.
- In the past 15 years, liability issues have risen to the forefront in many industries. Their impact in risk aversion on building practices is discussed.
- Codes and standards are a key element in any change in the industry. The national and local code influences are discussed.
- End-user types and their views are critical to change. These views are discussed from a functional (e.g., does it work?) and a perceptive point of view (e.g., "I just don't like it!").
- The general business philosophy in the United States and its impact on research and development in the building industry are considered.

In conclusion, several opportunities are discussed relative to rate of change in the building industry.

Introduction

For many who work in the building industry, there is the feeling that the rate of change relative to technology is slow when compared to other major industries. To begin a review of this perception, a general overview of the building industry is in order. The total building construction industry today is in excess of $400 billion and represents approximately 8 percent of the Gross National Product. The industry reached a peak growth of nearly 12 percent right after World War II. Figure 1 illustrates the change in the total construction industry during the period from 1984 to 1989 in constant 1982 dollars. The construction industry has an anticipated growth somewhat slower than the economy throughout the next several years. An interesting side note of Figure 1 is that the rate of change from 1984 to 1987 was a direct result of mortgage rates dropping from the high teens in late 1983 to 9 and 10 percent in 1984.[1] The construction industry is divided into three major seg-

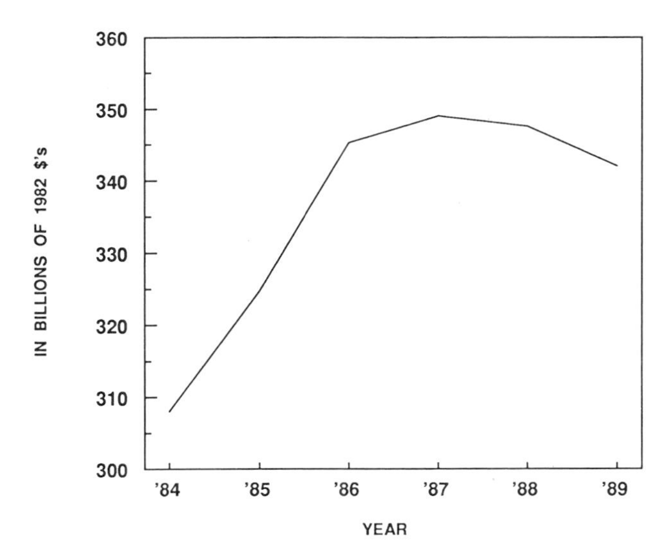

Figure 1. Total new construction, 1984–1989. Source: *US Industrial Outlook* 1989.[1]

ments as shown in Figure 2. The largest segment is residential construction, which is essentially privately owned and rental housing. The second segment is nonresidential, which is privately and publicly owned commercial buildings such as high-rise offices, hotels, and light commercial spaces. The segment labeled "other" is buildings associated with factories, farms, churches, etc.[1]

For the purpose of this paper, the residential building segment will be used to draw examples to illustrate the major points to be made. The residential building industry has on the order of 90,000 residential builders in it. The top 20 builders in the industry build approximately 10 percent of the annual housing starts. The top 400 builders account for 30 to 40 percent of the housing starts. One can see that the industry is highly fragmented with no one in a dominant position. The composite product output for the residential segment is the detached single or attached multifamily home.

When one looks at the cost of housing, labor and materials are still the major components of the total house price. In 1988, labor and materials, on average, represented approximately half the total cost of a new home.[2] In 1949, this figure was 70 percent. The reason for the decrease in percentage size is not so much due to efficiencies in labor and material, but that land costs have increased by almost 400 percent since the mid-1970s and now account for a larger portion of the total cost. On average in 1988, land accounted for approximately 27 percent of the total cost of the home. Figure 3 shows the historical material and labor trend since 1975 in residential housing.[3] Two points can be made from this figure. The first is that the price of material and labor has grown over time at an annual growth rate of about 5.5 percent per year. Secondly, and perhaps more important, is that the ratio of labor to material has not significantly changed. This trend suggests that perhaps there are still greater opportunities for more efficient use of materials and labor within the housing construction process. An interesting question to ask might be how other major industries (e.g., automobiles) have evolved in terms of the labor/material ratio over the years. The answer most likely would be a much higher material to labor ratio and a more efficient use of both elements.

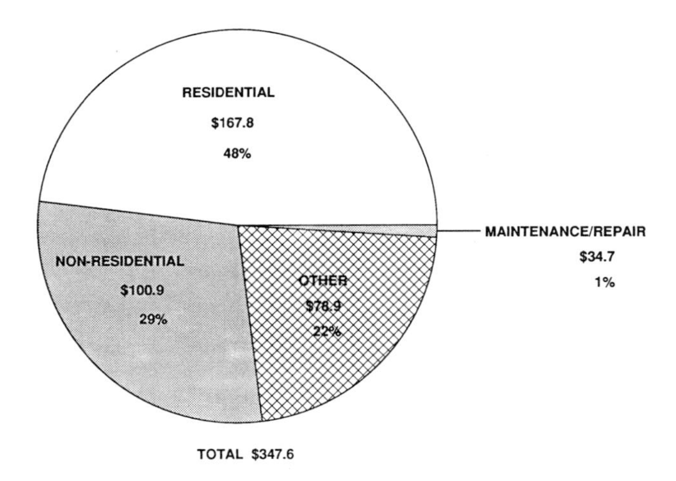

Figure 2. 1988 new construction by type (in billions of 1982 dollars). Source: *US Industrial Outlook* 1989.[1]

The rising cost of home ownership is a major issue for the building industry. Figure 4, from a recent professional builder survey[4] illustrates the four major obstacles to home ownership. All four obstacles have to do with cost. The first deals with the absolute cost of the house. The second is not being able to afford the house with the money that is available to the potential buyer. The third is raising the down payment and the fourth is the high cost of financing. In 1988, the median house price was $112,000 and in recent years, the median price has grown at an average annual rate of 8.4 percent. Currently, the average down payment needed to finance a new home is approximately $36,000 and since 1980, the down payment needed has increased at an annual compound growth rate slightly more than 5 percent.[5]

Changing Contemporary Building Practices

In order to examine the time scales and barriers in changing contemporary building practices, a short review of several systems, processes, or material changes which have been introduced over the years will be made. Specifically these are: gypsum wallboard, factory-built housing, cellular plastic foam insulations, and pressure-treated wood foundations.

Gypsum Wallboard

The major constituent in gypsum wallboard is calcium sulfate ($CaSO_4 + 2H_2O$), commonly referred to as gypsum. Gypsum had its first known use as a plaster and decorative material in 3000 BC, with contemporary use beginning in the early 1900s.

The gypsum mineral is mined from the earth, crushed, screened, heated to a point at which one and a half molecules of the chemically bonded water are driven off, and ground. The result is a powdered material referred to as plaster. When water is reintroduced to the plaster, rehydration occurs and forms a solid material. The earliest contemporary uses for gypsum were in traditional plaster type applications for walls and ceilings. In 1917, the US Gypsum Company invented gypsum wallboard and it was called SHEETROCK®.

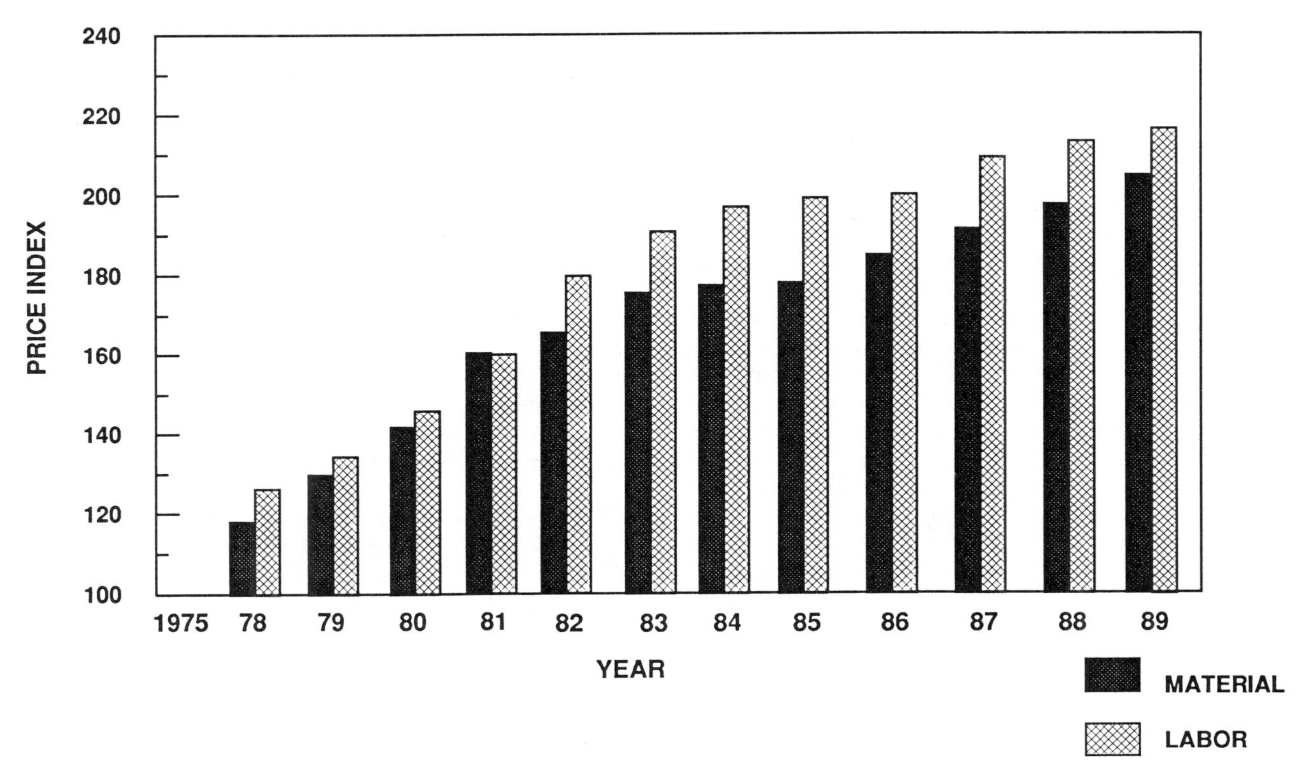

Figure 3. Historical material and labor trends. Source: *Building Design and Construction* 11/89.[3]

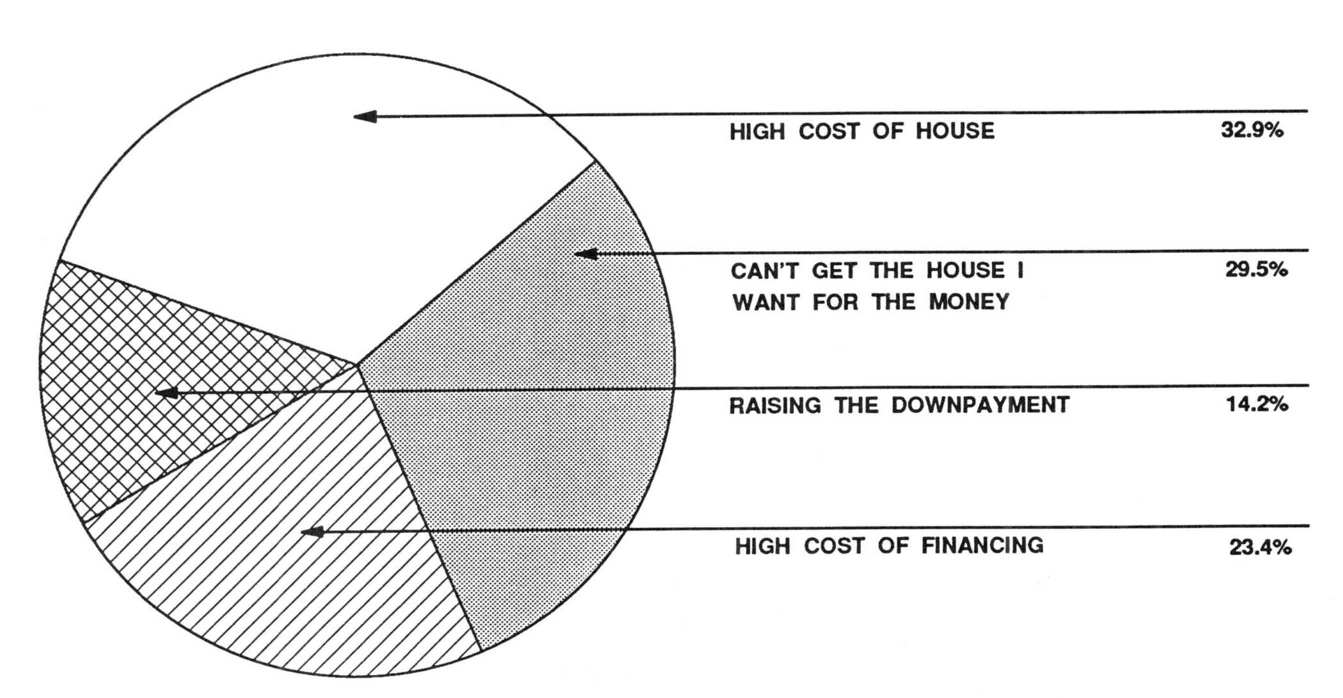

Figure 4. Obstacles to home ownership—buying barriers. Source: *Professional Builder* 12/89.[4]

Wallboard is made by adding gypsum plaster, water, and several additives to form a slurry. The slurry is placed between two sheets of paper, dried, and formed into boards ranging in various width lengths and thicknesses.

Early US Gypsum Co. net sales, starting at the turn of the century through approximately 1952, are shown in Figure 5.[6] Initial use for the SHEETROCK® material occurred right after World War I in erecting military barracks and headquarters where speed of building was important. The interesting feature in Figure 5 is the rapid increase in use of this material began around the 1930s. The success factors that led to continued rapid growth were due in most part to the post World War II rapid housing expansion and demands created by demographic shifts. Specifically, the new SHEETROCK® system offered speed, less labor sensitivity, and economy in erecting the new housing stock. The sheer demand of housing and demographic shifts created shortages in skilled trades, such as plasterers, that further drove the use of this more efficient building material. It is estimated that it took nearly 40 years from wallboard invention to the point where it was considered a standard interior sheathing material in the industry. Today, nearly 80 years later, there are currently 20 billion square feet of wallboard sold annually in this country.

Factory-Built Housing

Factory-built housing had its inception in the early 1930s when Wilbur and Schulp founded the Sportsman Trailer Company, employing 20 carpenters that produced a small two-wheel unit.[7] This early beginning was rather modest, with approximately 1,300 units produced in 1930. Not until the late 1950s and through the mid-1970s did the basic idea of building a liveable unit in the factory begin to infuse itself into the primary residential market. It was not until 1956 that the public image of factory-built housing began to shift from recreational trailers to the mobile homes that are being manufactured today.

There are several distinct advantages to factory-built housing. These are as follows:

- Today, factory-built modular homes can be completed and placed on-site within 30 to 70 days versus stick-built or site-built homes that take from 120 to 150 days when major factory components are not used.
- Consistency and quality are more easily controlled in a factory environment.
- In the factory, there is literally a 365 day per year building season where inclement weather is not an issue.

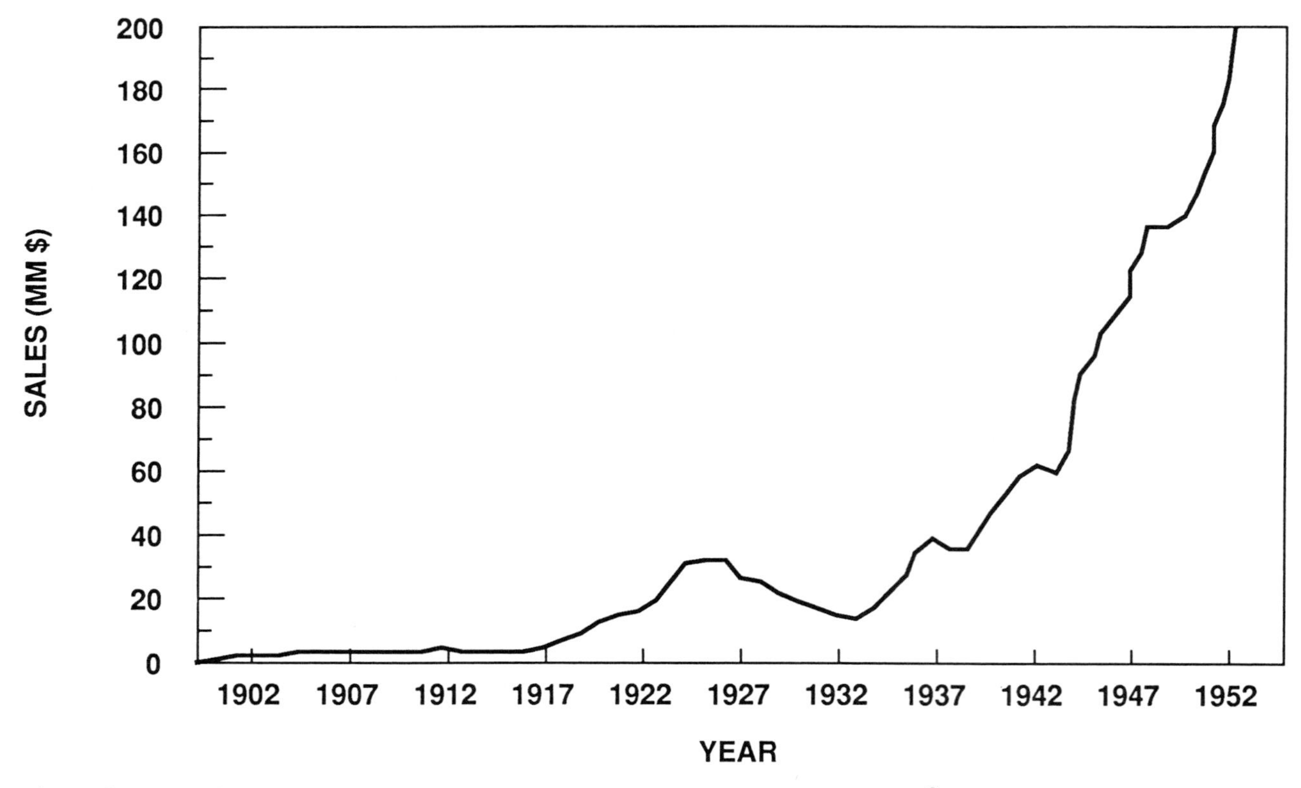

Figure 5. United States Gypsum Co. net sales. Source: United States Gypsum Co.[6]

- Building inside a factory has less dependency on skilled labor due to the assembly line techniques and fixtures used.
- Factory-built housing can also have a significant impact on financing costs for builders who are involved in multi-unit projects.

The question is "How well has this type of construction infused itself into the residential building marketplace?" Figure 6 is a breakdown of the type of building that has gone on from 1980 to 1988.[8] The four groups are:

- Site builders. These are builders who construct the major portion of the house on-site with site labor. Some use of factory components such as roof trusses and prehung doors would be included.
- Panelized Builders. These builders essentially erect the outer shell, and even inner walls, with factory made panels that have been delivered to the job site.
- Modular Manufacturer. Those builders who assemble two or more complete box-like modules at the home site. These modules have been made in a factory and are almost completely finished inside and out.
- HUD. Represents the mobile home industries that produce approximately 220,000 units yearly.

It is interesting to note that even though the panelized builders are growing, there is still much field labor that occurs in this type of construction. Modular growth, which perhaps better approximates true factory housing, really has not seen a significant growth during the 1980s. Nearly 40 years have passed since the idea of factory-built housing became a reality. Depending upon whether you use modular or panelized construction figures, only 15 to 50 percent penetration has occurred.

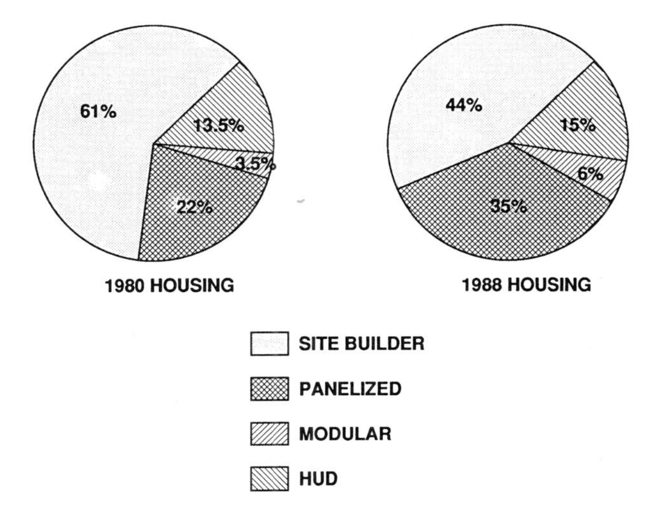

Figure 6. Housing production by industry segment. Source: *Automated Builder* 2/89.[8]

Here is a building process with proven rational benefits, that still has not reached its full potential.

It's the author's opinion that there are several hurdles that have not yet been overcome or are still in the process of being overcome in manufactured housing. The first is customer perception, which is driven by tradition. The idea of houses that are built in the field with traditional materials and practice being "better" is still very strong. Early perceptions of factory housing in the 1950s still leave lingering doubts to the home buyers of today, but in recent years these perceptions have begun to erode. A second hurdle relates to the fragmented builder base. As stated earlier, there are over 90,000 builders in this country and many of them, as can be seen from earlier graphs, are still site building. These builders are not just going to go away. The labor dynamics of this industry are such that conversion to factory systems will not occur as quickly as one might expect. In many cases, large factory producers sell a complete home as well. Many small builders see factory construction as a competitive force (i.e., another builder) and not as a potential advantage to their future business health. One final example is that custom builders still do not see the flexibility in factory-built housing applied to their product. Custom builders are those builders who build in the trade-up 2, trade-up 3 and above category. These builders typically build to specific designs and construct under 200 homes per year. In these homes, the foot print and roof designs vary significantly. Many have not found traditional panelizers to be of value to them.

Cellular Plastic Foam Insulation

The first cellular plastic foam was a phenol formaldehyde resin developed in 1909.[9] The first commercial cellular foam was a sponge rubber that came out shortly thereafter. The definition of a cellular plastic foam is one in which the apparent density of the material is decreased substantially by the presence of many cells that are disposed throughout its mass.[9] Rigid foams, which are used primarily in insulation applications, fall into two major categories. The first, a polyisocyanurate, is the most broadly used insulation material in construction, refrigeration, and transportation. This type of foam is characterized by a closed cell structure and a relatively high compressive strength. It is chlorofluorocarbon (CFC) blown and has a thermal resistance value on the order of seven R-values per inch. The second category of rigid foams is polystyrenes. The major forms are either extruded or expanded polystyrene boards. The extruded board is CFC blown, has good compressive strength, and a thermal resistance value of between four to five R-values per inch. The expanded polystyrene board is formed from expanded polystyrene beads typically impregnated with 5 to 8 percent pentane for flame retardancy.[9] The beads are pre-expanded in the presence of steam and a vacuum, and are finally molded in steam heated block molds. Pentane is the blowing agent for this material. The foam is lighter in density and has a thermal resistance value of three to four R-values per inch.

Between 1967 and 1987, the compounded growth rate for polymer foam used was on the order of 12 percent per year. Between 1967 and 1982, the growth was nearly 20 percent.[9] Unlike the past examples, here is a basic new material that seems to have experienced a rapid acceptance in the 5-to-10-year time frame. It appears the success factors that drove this rapid increase were due, for the most part, to the 1970s national energy crisis. Specifically:

1. Utilities were highly driven towards conservation. Highly insulated buildings offered opportunities for shaving peak demands and reducing the need for new generating capacity.
2. Homes that had higher insulation values had lower annual energy bills. This reduced cash flow qualified more buyers for new homes during this period.
3. Due to the higher thermal resistance of rigid foams, higher wall and ceiling R-values could be achieved without significantly changing the structural dimensions or building process.

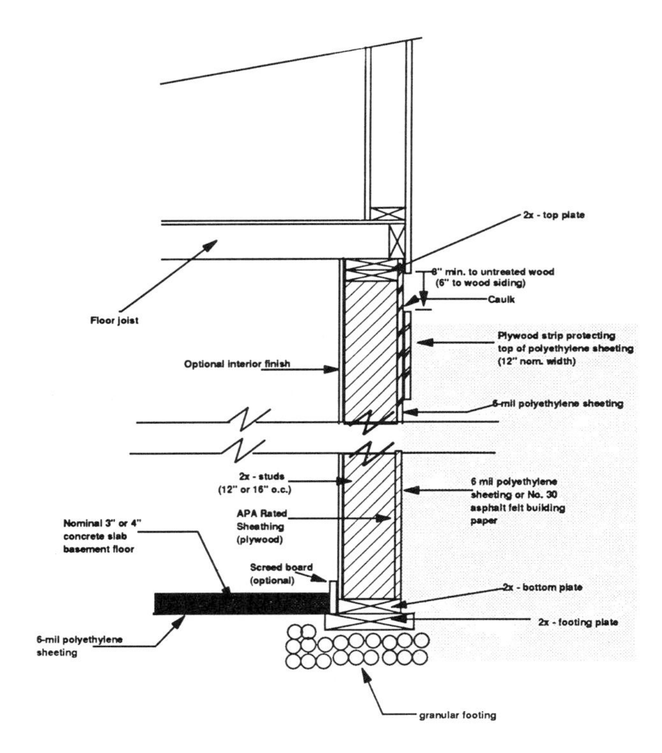

Figure 7. Pressure-treated wood foundation. Source: *Permanent Wood Foundation—Guide to Design and Construction*, 1/90.[10]

Pressure-Treated Wood Foundations

The Pressure-Treated Wood Foundation (PWF) as illustrated in Figure 7 is an example of a "rational man" argument that still seems to be searching for a place in the industry.[10] In discussions with the American Plywood Association, the idea of panelizing a foundation system was first initiated in Canada in the early 1960s. The basic system involved a pressure-treated wood footer system and panel system that could be placed on-site quickly and without the need for using the traditional concrete or block materials. In the late 1960s, due to a general problem of leaks in basements, Housing and Urban Development (HUD) and the National Association of Home Builders (NAHB) showed interest in this system. The system evolved here in the United States until approximately 1980, when code acceptance for the PWF was given in many states. The major advantages of the PWF are:

1. Faster Construction. Typically, poured or block walls can take as long as four weeks to construct when one considers weather delays, availability of labor, forms, etc. The PWF systems can be installed in two to three days.
2. Design Flexibility. PWF can be used in simple and multistory construction. It can be factory or field fabricated. Remodeling contractors find the PWF ideal for room additions, especially where access is an issue.
3. Ease of Interior Finishing. Since studs are exposed on the interior side, standard plumbing, wiring, and finishing techniques used in aboveground frame construction can be used with PWF systems as well.

Since the early work with PWF, nearly 250,000 wood foundations have been installed in North America. Most of these foundations can be found in Wisconsin, Minnesota, Alaska, and North Dakota. Currently, wood foundations are being installed at the rate of 10,000 to 15,000 per year. In comparison, there are approximately 375,000 traditional full-concrete or block basements that were installed in 1989—therefore, PWF accounts for only 2 to 3 percent of the total foundation market. Twenty years have passed since this concept was first introduced, yet very little market penetration has occurred for what appears to be a more rational approach to a major hurdle in the construction process. Eight or 10 years ago, significant efforts were placed on introducing this system by several large builders in the United States with very limited success. In discussions with several of these builders, some with extensive background in PWF, there are three major reasons why PWF has not grown faster:

1. There is a high degree of builder skepticism. Their perception deals directly with the suitability of the basic material used, specifically wood being the major structural element under a home and lasting over a period of time.
2. There is a high degree of difficulty for the homeowner to recognize the value of this system. Compounding this is the perception, similar to that of the builder, of wariness concerning the use of non-traditional material for this application.
3. Many builders have stated that the labor availability of people sufficiently knowledgeable about the system and how to install it is limited.

Table 1 summarizes the approximate times for implementation of the four examples given. One conclusion that can be drawn is that change occurs very slowly, even under accelerated circumstances as with rigid foams.

Impediments to Building Industry Change

The reasons why things in this industry do not change quickly are not trivial and are many in number. The following will be a brief review of several implicit and explicit hurdles that constrain the process and should be considered as areas for opportunity in the future.

First Cost

The sensitivity to equivalent first cost arguments is extremely high, and therefore, presents a major hurdle to value based systems on product introductions (e.g., lower maintenance, higher material/labor ratio, faster trade eliminations, etc.). In most cases the residential homebuyer will also be its resident; therefore, the value of a new product must be easily observable throughout the period of tenure in the home or simply be on the buyer's "want list" (e.g., Jacuzzis, larger bathrooms, microwave ovens, etc.). Since the average new homeowner lives in the house for only approximately five to seven years, life-cycle costing arguments tend not to stand up with the initial buyer. Value arguments to the builder are also difficult if this product or process cannot be eventually sold to the homeowner or does not have an immediate impact on lowering first cost. Because the residential housing market is so cyclical in nature, many of the 90,000 builders in this country tend to be risk averse and will not venture too far from a first-cost short-term focus. In the commercial arena, many times the typical tenant of the building is not the owner. In these cases, first cost truly is a major driver to the developer and life cycle costing arguments are difficult as well.

Table 1. Time for Change.

	% of Potential	Time
Factory-Built Housing	30–40	40 Yrs.
Gypsum Wallboard	95	40 Yrs.
All Weather Wood	—	20 Yrs.
Foam Insulation	50	5–10 Yrs.

Liability

Discussions with several architects, architectural and engineering firms, and members of the American Institute of Architects reveal that one major constraint on profitability in recent years has been liability insurance. There has been approximately a 23 percent increase in liability premiums in 1988 and liability insurance costs have doubled in the last five years. In addition, there are soft costs associated with liability that show up in the increased manpower needed for the specification review process. The liability question for suppliers of materials and products to the industry has become a major element and a small gate through which a new product must pass. Recent statements by people in National Institute of Building Science also indicate that the current liability situation in this country has reached a point where we are beginning to withhold new products from the marketplace.[11]

Codes

Currently, there are three model building codes that cover construction throughout the United States. These are:

- BOCA (Building Officials and Code Administrators International), the basic building code.
- SBCCI (Southern Building Code Congress International), the standard building code.
- ICBO (International Conference of Building Officials) the uniform building code.

The jurisdictions of these codes are approximately as shown in Figure 8.[12] In principle, the idea of model building codes applying uniformity to the construction process is a good one, but is still evolving. In addition to the model codes, there are four distinctive state codes and different city codes for Chicago, New York, Boston, Los Angeles, Detroit, Baltimore, and Atlanta. Thirty-four states have adopted preemptive state codes relative to modular housing and 16 states have no codes at all for factory housing. Enforcement systems vary depending on the state. Over 30 states administer these codes through state officials, eight states use county officials,

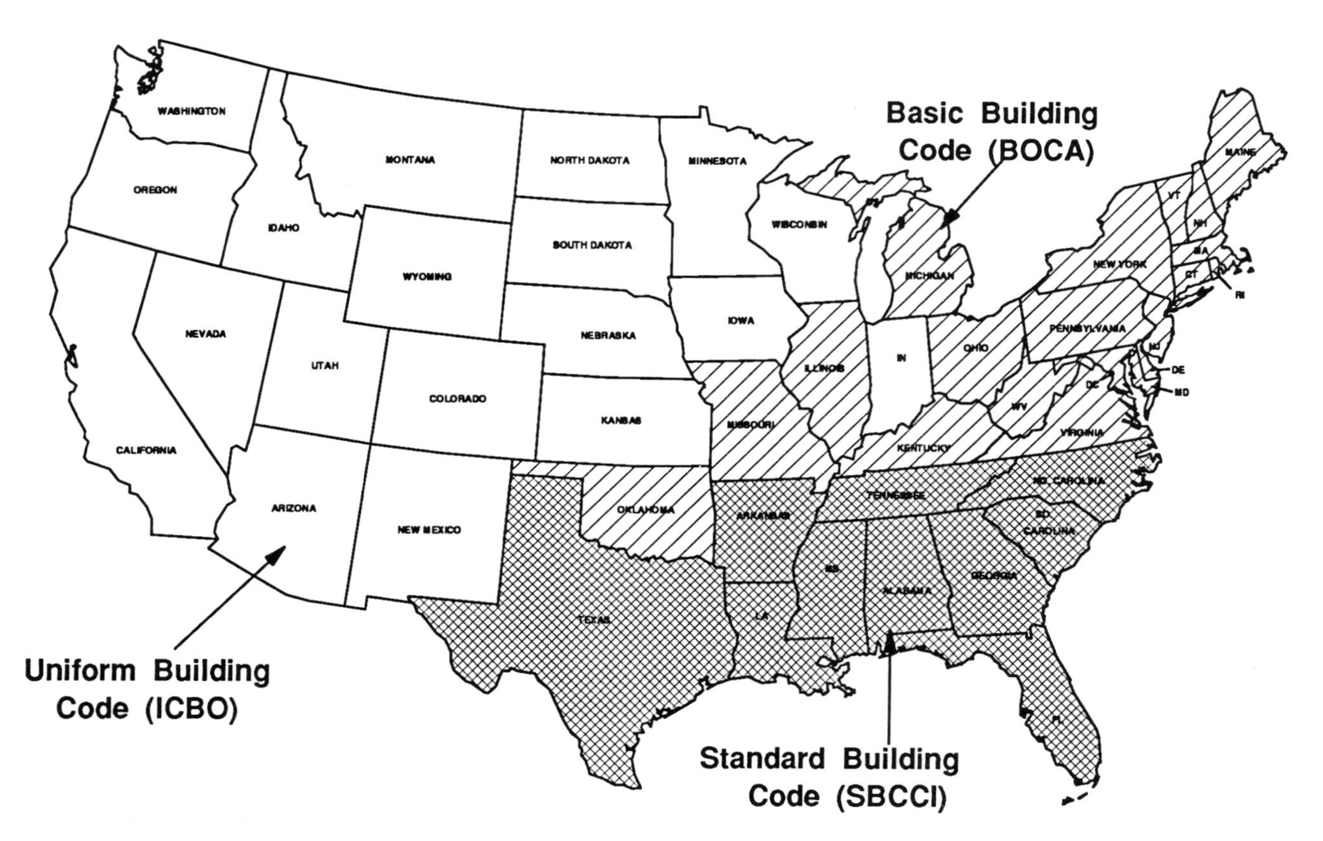

Figure 8. Approximate areas of code influence. Source: *Southern Building*, June/July 1982.[12]

and 13 states use municipalities. Code adoption happens through a consensus process. One could argue that as the number of decisionmakers increases, the likelihood for consensus is reduced. For a national home builder, the number of code inspections varies significantly across the United States. There is no question, when talking to builders, that code variations and the current system continue to be a major source of irritation. When one considers the perturbations with the model codes that are in force and the number of decisions needed to reach agreement on acceptance of new materials and new products, time to introduction can be long.

Short-term Focus

In the last eight to ten years, a more recent phenomenon continues to compound the hurdles for new opportunities. Many companies today have felt the influence of the short-term investment view that has been demanded by the marketplace. Witness to this has been the high number of mergers, recaps, and leveraged buyouts that have occurred in the last five years. In many cases, a side effect of this activity is a reduction in size and scope of research and development efforts, and focus is placed on short-term issues where returns can be seen quickly, at the expense of implementing new technology. Unlike Japan, where the major construction firms have upwards of 100 people in their own research and development departments, contractors and building firms in this country rely solely on suppliers for innovation. In light of the long lead time for the industry, as argued earlier in this paper, care must be taken that R&D be given the resources needed over time to be able to effect change.

Conclusion

It would be naive to believe that there are simple solutions to increasing the implementation rate of new technology into the industry. Following, however, are several general thoughts on opportunities that might address several of the issues that have been raised earlier.

In the area of codes, there still appears to be more room for uniformity in the administration of the code system in this country. Perhaps the process that the Europeans will begin in 1992 is worth watching. The ultimate goal there is a uniform code system where reciprocity exists between the various countries. In our country, perhaps incentives by the government could be given to cities and municipalities for adhering more to the uniform codes without the local perturbations. This of course would require close cooperation with local trade factors and preferential building practices.

In the area of liability, perhaps a cap on the liability exposure on new products over a period of time should be considered. This would be a variation on the HUD insurance system that currently exists, but would exclude the HUD financing criteria. To ensure that the new technology being introduced is functional, perhaps a national laboratory (e.g., National Institute of Standards and Technology (NIST)) could be used as the review vehicle prior to allowing the liability cap.

Currently, the industry has no leader identity. When you think of the automotive industry, General Motors, Chrysler, and Ford immediately come to mind. Having critical mass will be a major factor for implementing new technology into the construction industry in the future. This could be accomplished through major company alliances. Perhaps the acronym FHA in the future will stand for "Future Housing Alliance," where manufacturers, suppliers, builders, and developers will joint venture either new building systems or new products to produce the next generation of housing. Signs of these efforts can be seen in General Electric Plastics Living Environment Program and the MIT Innovative Housing Construction Consortium. Heightening end-user awareness of new products and technology, and providing the convincing value arguments is a must. Through the mass media, government sponsored general education messages could begin to make the public aware of benefits associated with the available new technologies and their impact on housing. As alliances form, sponsored television programs, such as "This New House," could be generated and buttressed by traditional advertisement. Backward integration of this information to the design curriculum at the college level and even at the high school level must occur. Academia can play a critical and proactive role in accelerating the rate of change for the building industry.

It is evident that change in the building industry occurs slowly. Opportunities to accelerate change exist, but vary in degree of difficulty. Several elements of change will be in new products, systems, and processes. Clearly, many solutions for more efficient building systems exist and are waiting in the wings. The challenge for the industry will be to implement them. Without the industry's best technology and systems in hand, the industry's dominant position in this country's market—as well as opportunities in other parts of the world—will be at risk.

References

1. *1989 US Industrial Outlook,* US Congress, 1989.
2. *NAHB Housing Backgrounder,* January 1990, p. 20.
3. "Historical Material and Labor Trends," *Building Design and Construction,* November 1989, p. 46.
4. Benderoff, Eric L., "What 1990 Buyers Want in Housing," *Professional Builder,* December 1989, pp. 100–117.
5. Gilmore, Lisa, "Trends and Statistics," *Building Design and Construction,* August 1989, p. 13.
6. *Building a Better Tomorrow,* US Gypsum Co. publication, 1952, p. 51.
7. Graham, Charles, "Manufactured Housing," *The Texas Planner,* 1989, pp. 11–14.
8. Carlson, Dan O., "A Look Ahead after 25 Years," *Automated Builder,* February 1989, pp. 8–10.
9. *Encyclopedia of Polymer Science and Engineering,* Volume 3, Cellular Materials, pp. 38–45.
10. *Permanent Wood Foundation—Guide to Design and Construction,* Wood Products Promotion Council, January 1990.
11. *Professional Builder,* July 1987, p. 28.
12. *Southern Building,* June/July 1982.

Buildings in the Next Century

David M. Pellish
Department of Energy
Washington, DC

Abstract

Radically improved building components, particularly those incorporated in walls and roofs, will enable buildings in the 21st century to benefit from the abundant, environmentally benign contributions of solar energy to heating, cooling, daylighting, moisture control, and indoor air quality improvements required for buildings. These revolutionary changes will evolve from such currently emerging developments as electrochromic optically switching "smart" windows and phase change thermal storage wallboards. They will provide architects, engineers, and others in interdisciplinary teams the maximum design flexibility required for the widest variety of new and existing buildings while minimizing requirements for non-renewable energy. Instead of considering the building envelope as the equivalent of an overcoat, requiring more layers to minimize heat gains and losses, future walls and roofs will be designed as "controllable membranes." They will consist of new or improved conventional materials capable of transmitting heat and daylight deep into building interiors in the winter, rejecting that heat in the summer without reducing transmission of natural light, and treating moisture as a thermal transport or storage medium instead of considering it an energy-consuming hindrance to human comfort.

Introduction

More than 150 years ago, the American historian H. H. Prescott wrote the following in his classic study on Peru:

> The surest test of the civilization of a people . . . is to be found in their architecture, which presents so noble a field for the display of the grand and the beautiful, and which, at the same time, is so intimately connected with the essential comforts of life.[1]

As we look toward the 21st century, we must ask ourselves: Will we have the wisdom to meet the challenges to civilization by using our knowledge to enrich and elevate our built environment? Can we achieve the noble objective of providing the next generation with the essential comforts of life without destroying the legacy we have received from previous generations in the form of a healthy and irreplaceable environment? My reply to both questions is a strong affirmative. Yes, our buildings can, and will, provide those essential comforts of life by using the beneficial contributions of the abundant renewable natural resources now available.

Before describing how those future developments may be achieved, let us first look to the past for guidance. In his discourse on the "Culture of Cities,"[2] the late Lewis Mumford reminded us of a question posed by Nathaniel Hawthorne in 1851. "Why," he inquired, "should each generation go on living in the quarters that were built by its ancestors? These quarters, even if not soiled and battered, were planned for other uses, other habits, other modes of living; often they were mere makeshifts for the very purpose they were supposed to serve in their own day; the best under existing limitations that now no longer hold."

It is significant that 50 years after Mumford's reminder of Hawthorne's comments, one must now again pose the same questions when we consider buildings in the next century. We are currently living through an historic period of dynamic social, political, and economic change that will have profound impact on the future. Just as major changes in building design and construction of our buildings and urban centers reflected great historical periods, I suggest that the built environment of the 21st century will also produce revolutionary changes.

Adapting New Technologies

It would be presumptuous to discuss aesthetics and architectural styles that may be preferred by a future generation. Such considerations should be deferred to the fully integrated teams of architects and engineers who are expected to inherit the design responsibilities and leadership in the next century.

On the other hand, it is now appropriate to discuss the development and nurturing of innovative concepts that could benefit future generations. A long lead time is required to introduce major changes in every sector of our society. For the buildings sector, at least a decade of intensive research, development, and promotion is absolutely critical before a major change receives general acceptance. More often, it takes 15 to 20 years to achieve market penetration.

More than any other sector of the economy, the building industry has serious institutional barriers to the introduction of technological innovations. It is unnecessary to repeat the usual litany of roadblocks, such as building codes, construction practices, and the fragmented nature of the industry. Nevertheless, it has been this writer's experience, as Assistant Director of the President's National Commission on Urban Problems[3] in the 1960s, as the Technology Officer of the New York State Urban Development Corporation in the 1970s,[4–6] as the US Department of Energy's program manager for the Model Energy Code and for solar energy building code requirements,[7–8] that these institutional problems can be resolved equitably, to the satisfaction of both the building industry and the public.

When adequate attention was given to creating and maintaining an environment receptive to new technologies, this author found that the contentious issues were settled and the differences within the industry were reconciled fairly and reasonably. Admittedly, there is no easy fix in this subject area, but there have been recent successful precedents.

The real challenge is the search for simple, manageable improvements in energy efficiencies for the many different types of buildings in different regions of the nation. If we are to improve our existing building stock and to succeed in introducing energy-saving technologies in new structures, the most direct and effective course is to adapt innovative building technologies to generally accepted building design and construction practices. There can be no single energy-saving panacea that would be applicable to all of the different types of buildings now, or in the next century.

As much as some would like to see dramatic changes in the design of the whole building, regardless of its functions or specific site conditions, both the industry and the public are more receptive to improvements that can be easily assimilated in contemporary buildings. The recent history of advanced building technologies has shown that new building materials, components, and systems are more welcome than revolutionary building designs.

Those of us who were fortunate to study or work with such visionaries as Buckminster Fuller know that radical changes in building design can be exciting and technically meritorious. But we also are aware that "Bucky" faced insurmountable barriers, and that his ingenious ideas were hardly ever accepted, except as buildings in world fairs and defense-listening outposts in Greenland and the Arctic wastelands.

The most effective strategy lies in changing the physical characteristics and properties of building components, such as those that comprise the building envelope. This approach has proven to be very powerful in at least two current instances. First, the "smart" window optical switching films, now being developed by Tufts University and EIC Labs in the Boston area and at Lawrence Berkeley National Laboratory in California will ultimately be applied to windows in both buildings and automobiles. It is now projected that this technology, when combined with automated-dimming controls, could provide up to 78 percent of the lighting requirements for perimeter spaces while reducing perimeter space cooling loads by 50 to 60 percent in a typical office building.[9] Second, the storage of thermal energy in conventional gypsum wallboards is now possible as a result of work conducted at the Dayton University Research Institute.[10–12] These innovations will be readily adapted to any building because the emphasis has been on improving components produced by those enterprising companies that have traditionally been in the forefront of investing in research and development and in promoting innovative technologies.

Before discussing anticipated advancements in future buildings, I shall posit two fundamental assumptions. The first is that there will be improved conditions at our universities and in industry conducive to widespread research in building technologies from a fresh, uninhibited point of view. Second, the transfer of new and innovative technologies to the marketplace will be facilitated by a bolder partnership between industry and government, firmly established before the beginning of the next century.

Solar Innovations

With these assumptions in place, I shall now focus on future buildings. My discussion will center on innovative solar energy concepts. These new technologies will supply a significant portion of building energy requirements—without causing adverse environmental impacts. The ideas suggested are intended to stimulate creative research and development, but should not be considered complete or all-inclusive. Certainly, it should be understood that such developments will also be accompanied by energy conservation measures, such as improved insulation materials and more energy efficient mechanical systems.

To begin with, we must accept the truism that the 84 million existing buildings and the millions of new structures anticipated in the future are now, and will be, quite different from each other in their functions, configurations and size. Nevertheless, all of them will have common requirements: they must provide for the essential comfort of their occupants. To meet those requirements energy will be required for heating, cooling, and lighting.

Up to now, meeting these requirements has been the function of different engineers in this era of specialization. Notwithstanding the theory of interdisciplinary teams that has been discussed for decades, building design may still be perceived as an arcane process in great need of improvement. If any change is to be accommodated, it is essential that all professionals engaged in the design of buildings have an equal voice around the drafting table in the development of creative ways to produce energy-efficient buildings.

To better explain this problem, let us review the relatively simple process of designing the exterior wall of an office building. First, the architect determines the wall's shape and general assembly of different materials. After that, the structural engineer proposes ways to support that wall assembly. The mechanical engineer may suggest appropriate insulation materials and the required thermal properties of the windows. Unlike the symphony orchestra, which must respond to the conductor's baton in absolute unison, the building design team more often acts like a relay running team, where the baton is handed over from one runner to another.

Now, let us examine the implications of this obsolete design process insofar as the occupants' comfort and the environment are concerned. To use a metaphor, the building exterior wall separates the occupants from the outside weather in the form of an overcoat with holes, through which they can see out. In colder climates, the architect or mechanical engineer seeks additional layers for that overcoat by providing insulation to prevent excessive heat loss. In southern regions, tinted glass in windows is popular because it reduces transmission of the sun's heat and thus lowers the building's cooling load.

If buildings of the 21st century are to meet the challenge, these concepts must change radically. Instead of perceiving the exterior wall as the equivalent of an overcoat, the wall must be treated as a "controllable membrane" separating building interiors from the outdoors. There is abundant natural energy available in the environment that could, and should, be used to heat, cool, and light buildings. Exterior walls and roofs need not be constructed so as to completely shut off the beneficial interface between the building and the environment.

A symbiotic relationship between buildings and the natural elements can be achieved by employing some of the latest scientific advances in solid-state physics, optics, and electrochemistry. An excellent example of this proposed approach is the recent development of electrochromic films for windows, which will be discussed later in more detail. Following this route will not only reduce current building requirements—which account for more than one third of the total energy consumed in the US. It will provide a renewable energy source of supply that is both environmentally benign and will reduce the energy requirements from sources that now pollute the environment.

"Smart Windows" and Beyond

Referring to natural solar energy in this discussion should not imply that our focus is on the commonly used solar collector jutting out of a roof for heating domestic hot water. Returning to the metaphor of the exterior wall as a "controllable membrane," let us first examine a few facts. As indicated in Figure 1, in the most northerly states of the US, the incident solar energy on a building exterior provides a house with at least four times the heat it requires over the winter.[13] In that region, more than 20 times the light required to light building interiors is available from natural sunlight. Needless to say, those ratios improve dramatically as one proceeds to more southerly regions. It is therefore self-evident that new, creative technologies are needed to capture and use that natural heat and light, when and as required by building occupants. On the other hand, the exterior wall should also be able to reflect that natural heat in the summer before it enters a building, so that energy for cooling purposes is not required. Let us also not forget that moisture transported indoors in hot, humid states in the southeast now presents a challenge, but could be put to useful purposes. Figure 2 summarizes the status of windows and US energy consumption. Figure 3 shows the relationship between solar radiation and heat emission.

We now have scientific evidence that these concepts are valid. The marketplace will soon witness the emergence of new "smart windows" containing solid-state electrochromic optically switching films (see Figures 4 and 5). With a momentary electrical charge of low voltage, it will be possible to transmit the sun's heat into a building in the winter and to shut out the unwanted heat in the summer, without adversely affecting the occupants' ability to see and enjoy the external environment. Because auto windows have less rigorous requirements than architectural windows, and because autos in the US sunbelt are no longer able to keep passengers comfortable, it is expected that this new technology will emerge in new cars as early as two to three years from now. (See Figure 6 for the role of windows in US energy consumption.)

Having proven that this concept works for windows, one may well ask, why stop there? If the passage of heat may be deliberately controlled in 25 percent of the building skin, why is it not possible to expand the use of these scientific advances to the opaque wall? If heating buildings remains our greatest energy challenge after the best conservation measures are employed, we must reexamine the commonly accepted dogma of building

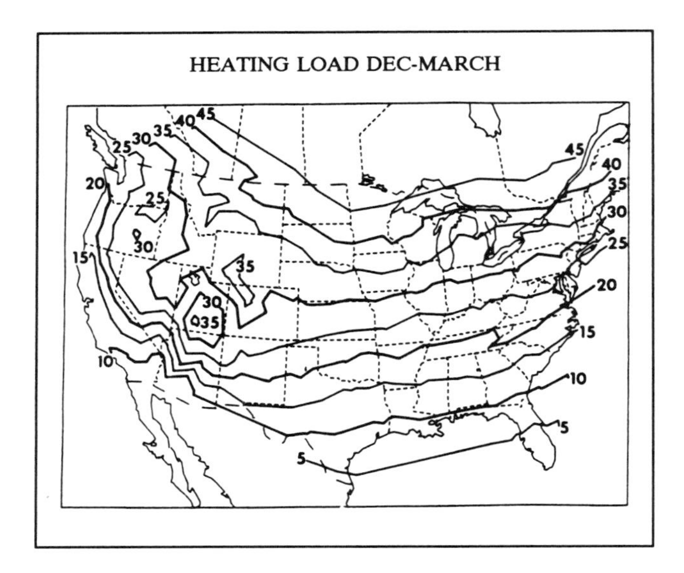

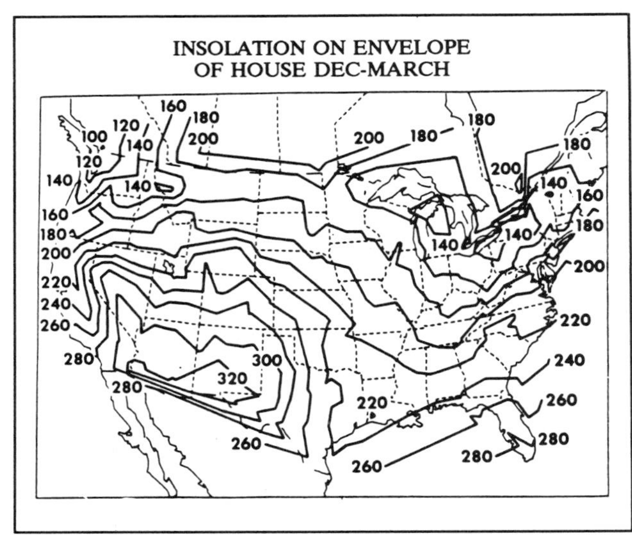

Units: MMBtu

SOURCE: Neeper, D.A., and McFarland, R.D. Some Potential Benefits of Fundamental Research for the Passive Solar Heating and Cooling of Buildings, Los Alamos National Laboratory (LA-9425-MS), August 1982.

Figure 1. Comparison of heating energy requirements and available insulation on envelope of 1200 ft^2 house.

HEAT GAINS AND LOSSES THROUGH WINDOWS

- o WINTER HEAT LOSSES
  - 25% OF HEAT LOSSES FROM RESIDENTIAL BUILDINGS
  - 4% TOTAL U.S. ENERGY CONSUMPTION THROUGH WINDOWS
  - LOSS EQUIVALENT TO 1.4 MILLION BARRELS OF OIL PER DAY
- o SUMMER UNWANTED HEAT GAINS
  - 1 TO 1.5% OF TOTAL ENERGY CONSUMPTION FOR COOLING
  - EQUIVALENT TO 350-525 THOUSAND BARRELS OF OIL PER DAY

LIGHTING ENERGY REQUIREMENTS EXCLUDING DAYLIGHTING

- o ELECTRIC LIGHTING ACCOUNTS FOR 5% OF TOTAL ENERGY CONSUMPTION

IMPACTS ON TOTAL U.S. ENERGY CONSUMPTION

- o TOTAL ENERGY REQUIRED 10 TO 10.5%
  - EQUIVALENT TO 3.5 TO 3.7 MILLIONS OF BARRELS OF OIL PER DAY
- o TOTAL NATIONAL EXPENDITURES: $30 BILLION

Figure 2. Windows and US energy consumption—current status.

design so as to use the sun's heat in a more sophisticated way. The building standing free on the landscape can, and should, be considered a heat transfer mechanism. The current conventional wisdom applied to designing exterior walls and roofs will be replaced in the future by new interdisciplinary teams consisting of architects, engineers, physicists, chemists, and material scientists, among others. Wall materials and assembly techniques will change radically in the 21st century.

Let us not forget the remarkable advances that have been made in transporting light. If the photon can be transported thousands of miles along telephone lines via fiber optic materials, and if the surgeon can literally see man's internal organs with the same technology, why is it not possible to transport natural sunlight into the interior of buildings without gouging out energy-guzzling building sections, now commonly called atria? New laser technology applied to holographic films for windows has shown that natural sunlight striking a window from many different angles in one day, and from one season to another, can be programmed to diffract effectively up to 30 feet deep into building interiors.

A study was conducted under the direction of Professor Anand of the University of Maryland to examine opportunities for continuing research that could further advance such ideas. Leading experts at such institutions as MIT, IIT, Tufts University, the Universities of Colorado, Wisconsin, Arizona, and South Carolina, as well as national laboratories at Los Alamos, Lawrence Berke-

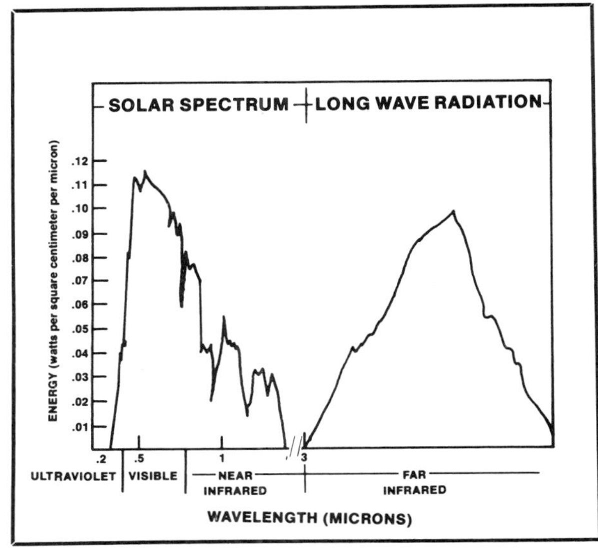

• **Most of the energy available in sunlight is in the short wavelength region-- the visible and near infrared portions of the Solar Spectrum.**

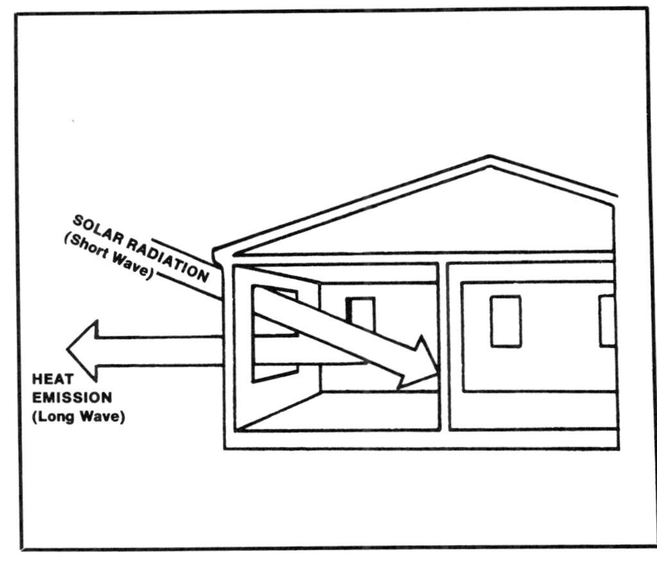

• **Thermal energy that is emitted falls in the long wavelengths-- the far infrared portion of the spectrum**

Figure 3. Relationship between solar radiation and heat emission.

ley, and the Solar Energy Research Institute contributed new, far-sighted ideas to this study. Their comments indicate that we have only crossed the threshold in the development of these concepts.

These scientists concluded that there is a wide variety of opportunities for increasing the contributions of natural solar energy to buildings in the future. They suggested that moisture can be viewed as a beneficial resource, not a necessary evil, that could be used as an energy storage or transport medium by means of new desiccant materials. In addition, the extension of electrochromic films to the opaque parts of the building skin with transparent convection suppression materials could lead to "smart" walls and roofs, which could reflect heat during the summer and transmit heat into the building in the winter. They also considered use of the "so-called" "atmospheric window" in the far-infrared portion of the spectrum, which could permit transmission of unwanted heat from a building in the summer, while reflecting unwanted heat away from the building.

Among other suggestions from these experts was the further development of compound parabolic concentrators with light guides for collecting and transporting heat or light striking the exterior building skin. Additionally, there is a proposed new approach that anticipates the development of advanced membranes through which vapor will pass in only one direction, permitting the removal of moisture at room temperatures, instead of requiring the current practice of cooling air to about 40°F with electric air conditioners. Problems of mitigating indoor air pollution could be resolved with increased fresh air intake at minimal energy costs, using solar energy assisted and air-to-air heat exchangers. Finally new uses of heat pipes and phase change transport loops could permit exterior walls to act as thermal diodes.

We must accept the fact that, left to its own devices, the building industry lacks the necessary research resources. It consists primarily of hundreds of thousands of small businesses that are confronted with overwhelming financial uncertainty. Under such conditions, it is not difficult to understand why the average homebuilder, who only produces about ten homes a year, is unable to support research.

Joint Enterprise of Industry and Government

Given that the building construction industry has historically accounted for about 10 percent of total GNP, that buildings consume more than 37 percent of total US energy and 62 percent of the electricity supplied by utilities, it must be assumed that the nation has much at stake in the future of America's buildings. From any reasonable point of view, we cannot ignore the opportunities for improving the energy effectiveness of buildings in the 21st century while protecting the environment.

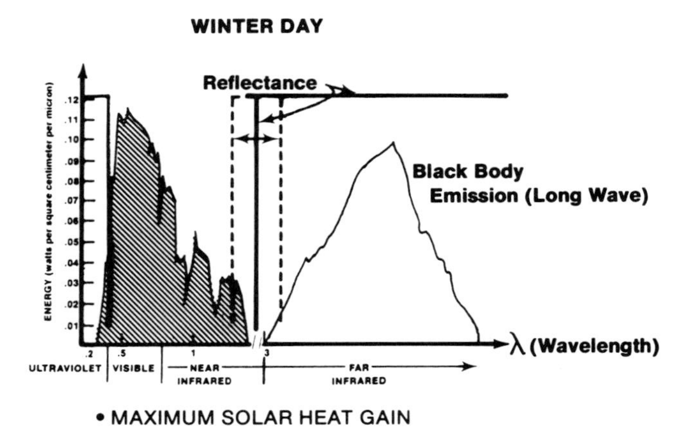

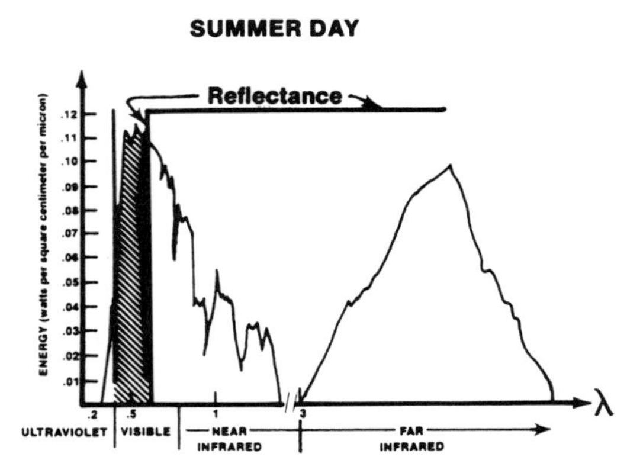

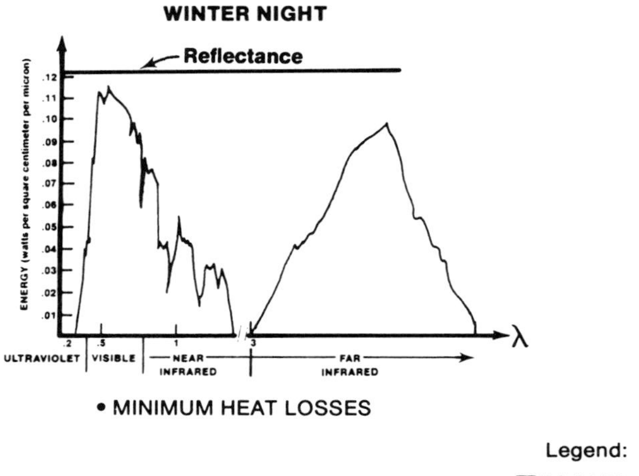

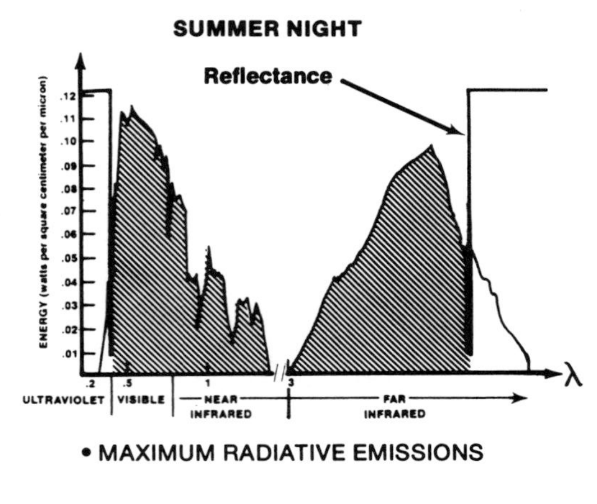

Legend:
Light/Heat Transmitted
Light/Heat Reflected

Figure 4. Modulating solar heat gains and losses.

The development of new technologies such as those I have suggested can only be undertaken as a joint enterprise of industry and government. Evidence that this approach has proven successful may be found in the work of the New York State Urban Development Corporation (UDC) in the early 1970s. Under the dynamic leadership of Ed Logue, the UDC made it possible to introduce technological innovations in a statewide program that was the nation's largest producer of subsidized housing. It is time to think of major new strategies for buildings in the next century. Just as most other major industrialized nations have already adopted institutional changes, a new partnership between industry and the US government will be organized to carry out the necessary research, development, field testing, and final applications to buildings across the nation.

There is reason to be optimistic that these ideas will be implemented within the next decade, and that such environmentally benign energy sources as solar energy will be able to supply the major portion of building energy requirements. Renewed interest recently emerged in the idea that the long, expensive process required to gain market acceptance of new building technologies can be reduced with the assistance of public agencies. To meet that challenge, Senator Joseph Lieberman of Connecticut recently introduced new Congressional legislation to facilitate the introduction of innovative technologies in government-owned buildings by means of a working partnership between government and industry.

It is no longer valid to assume a government-supported research report, when tossed over the wall to the

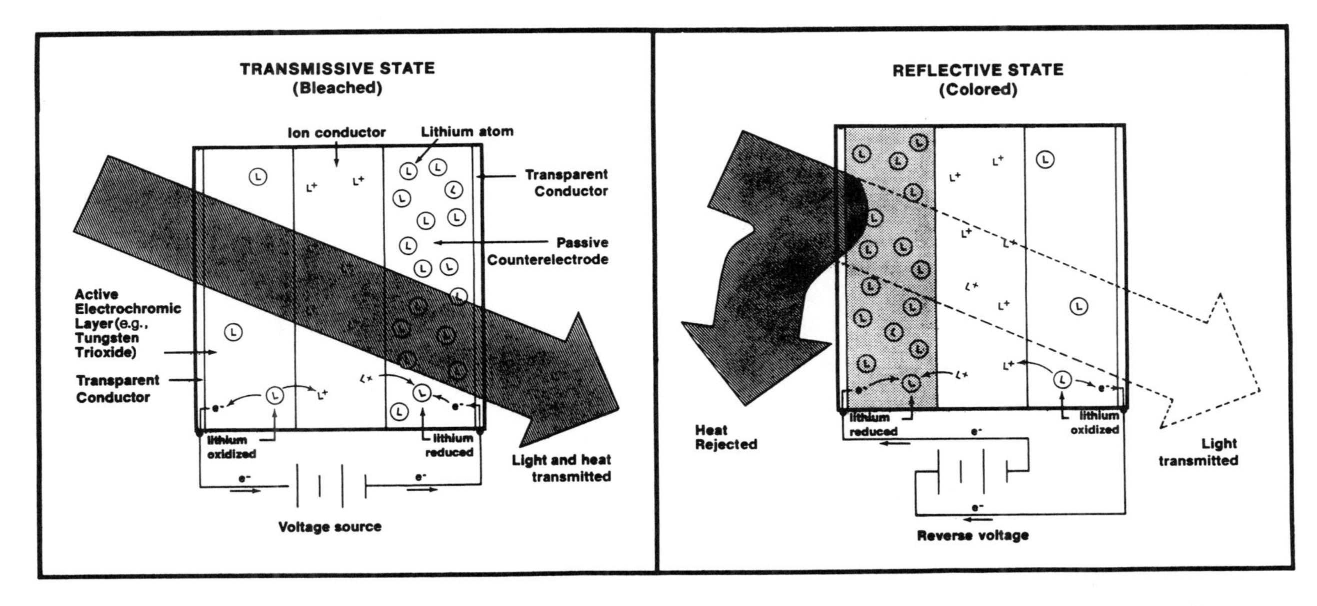

- **Electrons supplied to passive counterelectrode**
- **Window becomes increasingly transparent as lithium atoms are shuttled from left to right**

- **Direction of electron flow has been reversed**
- **As lithium atoms are shuttled from right to left, the window becomes less transparent**

Figure 5. Operation of electrochromic "smart window" (example using tungsten trioxide and lithium atoms).

ENERGY SAVINGS OPPORTUNITIES

- o SUPPLY HEAT IN THE WINTER
- o REDUCE HEAT GAINS IN THE SUMMER
- o REDUCE ELECTRICAL LIGHTING REQUIREMENTS
- o REDUCE AIR CONDITIONING LOADS

EXAMPLE OF APPLICATION: A TYPICAL OFFICE BUILDING

- o REDUCE HVAC EQUIPMENT SIZE BY 30%
- o REDUCE PEAK ELECTRIC DEMANDS BY AS MUCH AS 30%
- o REDUCE TOTAL BUILDING ENERGY OPERATING COSTS BY 40%
- o INCREASED WORKER PRODUCTIVITY

POTENTIAL CONTRIBUTIONS OF SOLAR BUILDINGS APERTURE RESEARCH:

- o REDUCE TOTAL U.S. ENERGY CONSUMPTION BY 3% - 5%
- o DISPLACE 1 TO 1.75 MILLIONS OF BARRELS OF OIL PER DAY
- o SAVE $9 - $15 BILLION IN ANNUAL ENERGY EXPENDITURES

Figure 6. Windows and US energy consumption—potential contributions from windows.

private sector, will be picked up and then rapidly assimilated into the building industry. Senator Lieberman's bold ideas are symbolic of the emerging changes that will help the next generation to meet the challenges of the 21st century.

References

1. Prescott, W.H., *Conquest of Peru,* 1847.
2. Mumford, L., *Culture of Cities,* 1938.
3. National Commission on Urban Problems to the Congress and to the President of the US, (Report), *Building the American City,* 1968.
4. Pellish, D.M., "A New Approach to Code Problems," *Journal of the American Institute of Architects,* January 1969.
5. Pellish, D.M., "A New Government and Industry Partnership for Building More Housing," *Architectural Forum,* July/August 1970.
6. Pellish, D.M., "New York State Moves Ahead with Industrialized Building Systems," *Journal of American Society for Testing and Materials,* November 1972.
7. Council of American Building Officials, *Recommended Requirements for Code Officials for Solar Heating, Cooling and Heat Water Systems,* June 1980.
8. Council of American Building Officials, *Model Energy Code,* 1989.
9. E.A. Mueller, and Science Applications International

Corp., *Estimates of the Performance and Economic Value of Advanced Solar Buildings Technologies,* 1990 (A Report to DOE Solar Buildings Technology Program).

10. Salyar, I.O. and Sircar, A.K., "Development of PCM Wallboard for Heating and Cooling of Residential Buildings," *DOE Thermal Energy Storage Research Activities Review,* 1989 Proceedings.
11. Salyar, I.O. and Sircar, A.K., "Phase Change Materials for Heating and Cooling of Residential Buildings and other Applications," *25th Intersociety Energy Conversion Engineering Conference Proceedings,* August 1990.
12. Tomlinson, J., "Government-Sponsored Research in Thermal Energy Storage," *Proceedings of the ETS/TES 1990 Meeting,* April 1990.
13. Neeper, D.A. and McFarland, R.D., *Some Potential Benefits of Fundamental Research for the Passive Solar Heating and Cooling of Buildings,* Los Alamos National Laboratory Report, 1982.
14. Anand, K.K., et al. *Innovative Solar Energy Concepts to Meet Building Energy Requirements,* April 1990 (A report to DOE Solar Buildings Technology Program).

Building Construction in Finland: Energy Considerations and Research

Reijo Kohonen
Technical Research Centre of Finland
Laboratory of Heating and Ventilation
Espoo, Finland

Introduction

The development and implementation of Finland's energy supply is guided by national energy policy programs. The Council of the State approved the first of these programs in 1979, and the second in 1983. The third energy strategy is under preparation. The need for energy program adjustments stems from legislative amendments, changes in the availability and price of energy, the introduction of energy and pollution taxes, and, above all, the pursuit of new policies by committees and advisory bodies concerned with environmental and energy issues. Preparatory work on the national energy strategy should be completed in the latter part of 1991.

Efforts are being made to produce diverse material to assist in the preparation of the next energy strategy. The most notable of the policy committees engaged in this work are the Energy Committee, the Committee for Sustained Long-Term Development in Finland, and the Committee for Environmental Economy. The Ministry of Trade and Industry is drafting alternative energy supply scenarios for the period up to the year 2025. In addition, work on the Energy Conservation Project is under way in the Ministry of Trade and Industry. The objectives of this project are to study already existing means of energy conservation, to compile international comparisons, and to evaluate energy conservation opportunities and potential.

The 10 national energy research programs initiated by the Ministry of Trade and Industry in 1988 will yield material for the preparation of the next energy strategy. Two of these programs are concerned with energy use in buildings, LVIS-2000 (Future Building Services) and ETRR (Energy Efficient Buildings and Building Components).

The Production and Use of Energy in Finland

In 1988, the total end-use energy consumption of Finland was 24 Mtoe. The population of Finland is about 5 million and the GNP about 320 billion Finnish marks.

Finland's energy system is diverse, involving a variety of energy sources and forms of energy production (Figure 1). Oil is the most important of these energy sources. However, its role has declined rapidly, and now its share is 32 percent of Finland's total energy resources. The use of oil is concentrated in the sphere of transportation, where it is difficult to replace. Domestic fuels like wood, peat, and forest-industry waste are the second most important sources of energy. Their share is about 18 percent. Forest-industry waste is the most used domestic fuel. The share of peat is only 3 percent. The share of nuclear power in energy supply is 16 percent, and in electricity supply 32 percent. Coal accounts for 32 percent of energy use. The share of hydropower in total energy is 11 percent, while that of imported electricity is 6 percent. In electricity supply, the corresponding shares are 23 percent and 13 percent. The share of natural gas is 4 percent.

The heating of buildings accounts for about 30 percent of total energy consumption in Finland (Figure 2). Energy conservation measures implemented over the past 10 years have halted the growth of heating energy consumption, even though the building stock has grown by 100 percent since 1960. The contribution of domestic

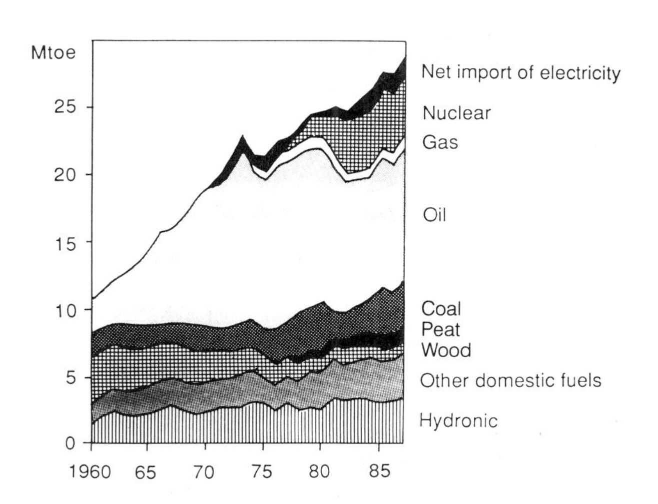

Figure 1. Energy production in Finland.

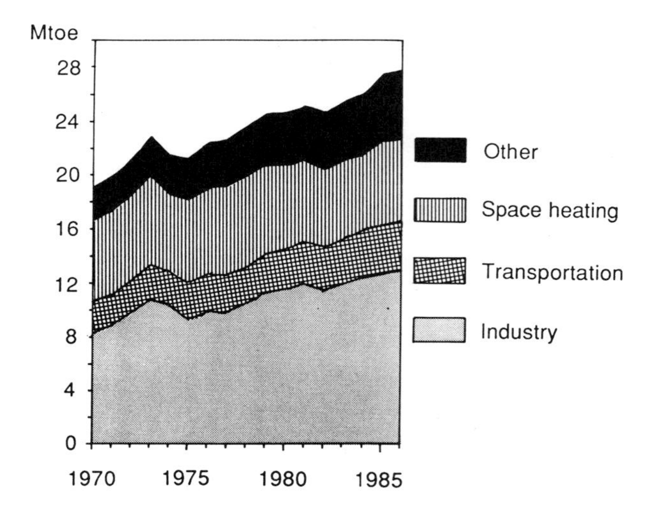

Figure 2. End use sectors of the total energy consumption in Finland.

fuels in heating has also started to grow, having been only about 20 percent at the end of the 1970s. The share of oil in heating has declined faster than had been planned, primarily due to the steady growth of district heating and the popularity of electric heating during the 1980s.

The specific consumption of useful heat energy has declined in residential, commercial, and public buildings by more than 30 percent since the beginning of the 1970s, when it was over 80 kWh/m^3a. Since 1981, specific consumption has fallen to below 60 kWh/m^3a. Average specific consumption has been reduced by energy conservation measures aimed at old buildings, the construction of new buildings that use little heating energy (less than 40 kWh/m^3a), and changes that have taken place in usage habits. However, the trend of development has changed, and unit consumption embarked on an upward track during the second half of the 1980s.

The fundamental question as far as Finland's energy supply is concerned is the choice between two policies. The first policy would be to continue the present line of development. It would be based upon continued economic growth and the need to safeguard the operating requirements of an industrial base that uses large amounts of energy. The other energy policy would be to strive with all haste toward a low-energy economy, in which the importance of energy to production would be substantially less than it is now.

The energy system is changing, but only slowly. For some time to come, Finland's energy supply will continue to be based on the present energy sources. The need for a change of energy sources only exists in some areas of use. Wind and solar energy will continue to be of minor importance in the decades ahead. Future living conditions and changes in lifestyles may well be reflected by the changes in energy use taking place in buildings.

Energy Strategy and Its Effects on the Use of Energy in Buildings

Finland's energy policy is and will continue to be oriented toward ensuring the continuity of present economic growth. The alternative, rapid movement toward a low-energy economy would require government action in the form of price intervention and statutory usage controls. As far as the use of energy in buildings is concerned, this might mean that restrictions will not be placed on energy consumption; instead, efforts would be made to bring economical energy solutions into use through the promotion of public awareness and education. Furthermore, it is likely that restrictions on different forms of heating will be loosened; for example, the electrical heating of residential buildings would be allowed.

The subsidization of energy conservation investments was stopped in the 1980s, due mainly to changes in the political climate in Finland. These investments have been beneficial from the standpoint of both the economy as a whole and individual consumers. In order to control the energy consumption and make energy consumers more enlightened, studies are being carried out to determine how to promote consumer awareness of environmental and energy issues. These studies also address the coordination of consumer and environmental policies and the modeling of a voluntary, positively discriminating marking system for environmentally conscious products. The possible introduction of "Energy Labeling" for appliances and even whole buildings has also been debated in Finland.

Mirroring proposals made by a world commission set up by the General Assembly of the United Nations, the Finnish Committee for Sustained Long-Term Development proposed that the energy strategy be amended to hold the consumption of primary energy at the 1989 level until the year 2000, and then reduce it by 10 percent by the year 2010. This, however, contradicts the present policy of maintaining the energy platform for economic growth. The committee's proposal would mean a level of consumption that would be about 30 percent lower than the level allowed by the present policy. This energy savings could be obtained with high efficiency heat recovery, boiler and appliance stock, and with super insulation and windows.

Most of the simple and inexpensive energy conservation solutions have already been implemented. As a consequence, future energy conservation will involve larger-scale measures that, in most cases, will not be possible without basic repair and renovation of buildings and installations.

Potential savings in the consumption of useful heat (net heating energy) in buildings are estimated to be 6 TWh/a, about 12 percent of present consumption. When potential energy savings in building-specific heat production are taken into consideration, the estimated potential heat conservation rises to 8 TWh/a. The use of electricity as a means of heating new detached and semi-detached family houses has grown rapidly. It is estimated that electricity used for heating purposes will account for about 12 percent of Finland's electricity consumption in 1995; the potential for energy savings is estimated at 10 percent. Figure 3 illustrates the estimated (according to the third energy strategy) energy consumption of space heating until 2030.

Development Needs of Heating, Ventilation and Air Conditioning (HVAC) Systems

HVAC systems and equipment are crucially important in the realization of potential energy savings. Efficient energy use in buildings will require the ability to use a number of alternative energy sources.

According to the National Energy Research Committee, research and product development in the field of HVAC systems should be oriented toward the more efficient systems. The main development objectives regarding heating systems are: practical and precise control over individual room temperature, flexibility in the use of different energy sources, and the applicability of installations to industrialized construction. The efficiency of individual devices and systems should also be improved.

Conserving electricity in buildings is a prime objective. Flexible installation systems should be developed for electrical heating. It will be possible to reduce the specific consumption of electricity by using equipment that can function in a distributed mode only when necessary, by automating the control of equipment and lighting, by improving the efficiency of equipment, and other similar means. The development of equipment technology will facilitate the combined production of electricity and heat on a building-specific basis.

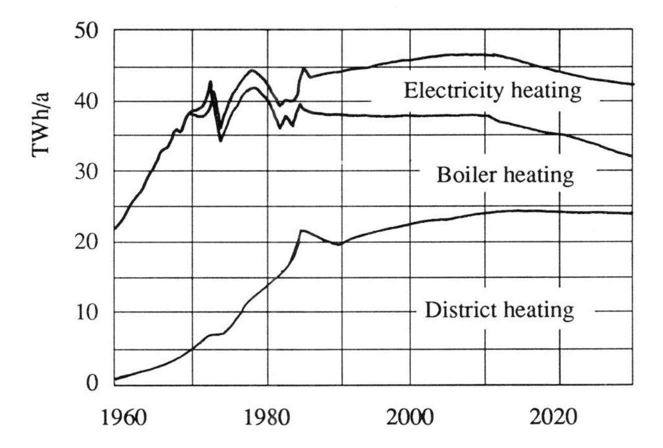

Figure 3. Heating energy consumption of buildings until 2030.

The control of HVAC systems will be improved by integrating the operation of HVAC devices and controllers. More and more attention will be paid to the controllability of HVAC systems. Digital controls, for example, will open the door to improved interior climate and energy economy.

The Finance and Organization of Energy Research Activities

The contents of the two research programs addressing energy use in buildings, which are described in the remainder of this report, largely correspond to the research needs specified by the Energy Committee. Although the financial support provided for energy research in Finland has, by international standards, been relatively high (about 100 million Finnish marks), the research and development input should be increased if conservation targets are to be reached. It has been proposed that energy research funding be increased by 9 percent a year.

Research Administration

Scientific and technological research in Finland is done mainly at the Technical Research Centre of Finland (VTT) and the universities of technology. With the exception of research work in the fields of petrochemicals and pulp and paper, technological research in Finland is concentrated at VTT, which operates under the Ministry of Trade and Industry. Correspondingly, the principal financiers of research are the Ministry of Trade and Industry's Energy Department (energy research) and Technology Development Centre, TEKES, Finland (product development finance, e.g., the industrial building technology program). Research in the construction sector is also financed to a modest extent by the Finnish National Fund for Research and Development (SITRA) and the Academy of Finland. Most of industry's research and development work is carried out internally, and only a part of their R&D is contracted out to VTT. However, product testing at VTT is mainly financed by industry. Research in the field of environmental protection is to some extent financed by the same administrative sectors as in energy research; so opportunities for coordination do exist. Figure 4 illustrates the organization of research administration in Finland.

Energy and environmental research in Finland is characterized by its programmed structure. There are currently 10 national research programs in progress in the field of energy research, and the Technology Development Fund is financing 11 technology programs. In the field of environmental protection there are two to three research programs. Research establishments, universities of technology, and private companies are all involved in the research.

In principle, the research is well organized and coordinated. The exploitation of results and the penetration of new technical solutions largely depends on the willingness of the private sector to invest in technology that will conserve energy and protect the environment, as well as on government subsidies and energy pricing policies (i.e., taxes). Recent political stances point to a tightening of energy taxation, which will probably take the form of a pollution tax. Revenues thus generated would be used to finance environmental protection investments and the development and introduction of new equipment solutions into industrial companies.

Another way to steer the introduction of energy and environmental conservation technology is through the development of building codes, which are issued by the Ministry of the Environment. At the present time, the only requirements relating to energy use take the form of U-values for building components. The general attitude is, however, that regulations concerning construction should be reduced.

Technical Research Centre of Finland (VTT)

The Technical Research Centre of Finland (VTT) is a modern research establishment that carries out a wide range of technical and techno-economic research requiring a high level of expertise. VTT consists of 34 modern research laboratories covering a diverse range of technologies. It employs some 2,700 people, half of whom are university graduates. The purpose of VTT is to maintain and raise the level of technology and to meet the research and testing needs of the private and public sectors.

VTT is the only major research establishment in Finland. Its activities are organized into a number of research divisions, that are further subdivided into laboratories, such as the Laboratory of Heating and Ventilation, the Building Material Laboratory, the Laboratory of Structural Engineering, etc. Cooperation among the various laboratories occurs both spontaneously and on a formal project basis, and the state of affairs at VTT in this respect is getting better all the time. The application of management-by-objectives principles is also having beneficial effects on the orientation and fruitfulness of the research.

Energy technology research is regarded as important at VTT, and its various laboratories are involved in many of the national energy and technology programs. There is also one internally financed energy research program in progress at VTT (Energy and the Environment), and preparations are in hand for the start of a second (Energy-Saving Equipment). However, VTT's opportunities to initiate its own research programs are limited, as only about 20 percent of its research costs are financed directly by appropriations from the government budget. The remainder has to be generated by contract research for the public and private sector.

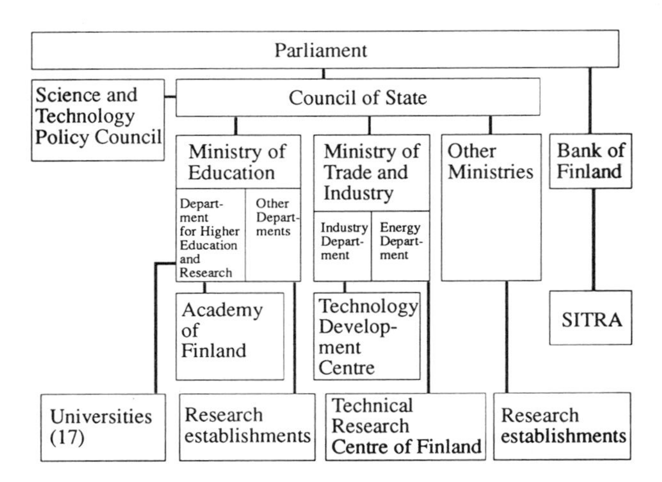

Figure 4. Organization of research administration in Finland.

Training and Education

The introduction of new construction and design methods will require changes and improvements in both the basic and supplementary training of engineers. Studies should start from building system theory and progress through various solution models to an examination of equipment engineering. Energy and environmental matters should be seen as built-in characteristics of systems.

In Finland, HVAC engineering is taught in mechanical engineering departments of the universities of technology. We therefore have a good opportunity to lean on the contribution of special courses on heat transfer and fluid mechanics. The technologies involved in HVAC equipment and conservation are, generally speaking, applications of thermodynamics.

The advantage of the system engineering approach is that "theory" (the general system engineering model and the operational functions of systems) is permanent, whereas application solutions are developed and equipment technology updated. Moreover, altering the design practice is possible with this approach to the teaching.

Another important development need in the teaching of HVAC engineering is conversance with whole entities, i.e., an understanding of the interaction between and integration of different systems. This will require good abilities in the application of computer and simulation techniques on the part of both teachers and students.

In my view, considerable resources will have to be devoted to the development of teaching in the HVAC field (both basic and supplementary training) in the near future, so that the results of the industrialized building technology projects and the energy research programs can be exploited according to the planned schedule.

Technical universities are also carrying out some energy-related research in the building sectors. For example, the HVAC laboratory of the Technical University of Helsinki is concentrating on indoor air quality studies. At the Technical University of Tampere there are some activities on refrigeration techniques and energy techniques of buildings.

National Energy Research Programs Related to Construction

At present in Finland there are a number of ongoing building technology development projects that are aiming at the achievement of better quality, lower costs, and industrialized production in the building industry (Figure 5). The energy research programs ETRR (Energy Efficient Buildings and Building Components) and LVIS-2000 (Future Building Services) started in 1988 will, from the standpoint of energy technology, yield alternative solutions for the structural, heating and cooling, electrical and automation systems that will be needed in buildings of the future. In accordance with the energy strategy, the objective is efficiency in the use of energy—i.e., the achievement of a reduction in specific energy consumption. This research is being carried out in the form of "energy conservation studies," a term which aptly describes the role of the energy research programs in the national energy strategy: They must produce the basic technological capabilities and the "conservational" alternatives that the consumer will hopefully select as a result of energy conservation campaigns and changes in taxation practices. It would appear that energy economy must be seen as a "built-in" characteristic of future buildings in Finland and elsewhere.

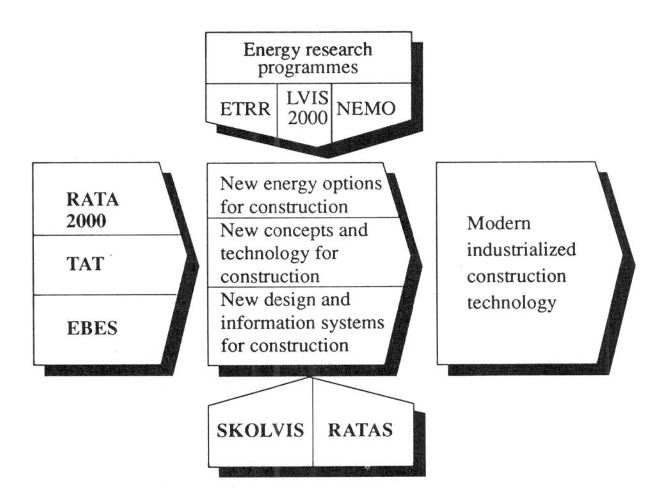

Figure 5. Research and technology programs aimed at industrialized construction.

Common to all the building technology research programs is the development work being done from the starting point of system engineering. A building must be seen as a complete functioning entity that is composed of various systems. The systems produce an outcome, that is, a product or service. Correspondingly, certain performance characteristics as well as compatibility are required of the subsystems and components. RATAS and SKOLVIS projects are developing new building information (database) and design systems and methods.

The intention is that the RATA-2000 model of the construction process and the TAT building system will form the basis of a new generation of industrialized building technology in Finland. The term "TAT" originally referred to product development of apartments and office buildings. It could now be understood as standing for "Totally Adaptable Technology." The building system is defined as a hierarchical, modulated system containing functional and spatial systems, a structural system, systems for ventilation, water supply, electricity, information technology, and waste management as well as connecting, dimensioning, and tolerance systems. There are five hierarchical levels in the system: the building level, the sub-building level, the module level, the component level, and the basic element level. Guidelines for design, production, service and product and method development are defined. Although TAT is in a conceptual stage a couple of pilot construction projects have already been initiated.

The ETRR and LVIS-2000 research programs will produce energy options and lay the foundations for energy techniques and interior environment quality. Interpreted broadly, the quality is seen as an intermediate means of improving the efficient energy use.

LVIS-2000: Future Building Services

The research program LVIS-2000: Future Building Services is concerned with the control of the indoor climate and energy use in buildings of the future. The objective is to present alternative technical solutions at the system level and to lay the groundwork for appropriate product development. The research is primarily oriented toward new construction, although applicability to repair and restoration has also been included as one of the starting points of the systems development work. The studies encompass residential, commercial, and public buildings. The objective of experimental research is to apply results to the product development of HVAC systems and to the development of installation and operational techniques.

The LVIS-2000 research program encompasses three areas of research as well as a number of demonstration projects:

1. Energy systems of buildings
2. Building energy management and control systems (BEMCS)
3. Indoor climate and energy
4. Joint product development and demonstration projects.

The objective of research on the energy supply systems of buildings is to develop heating and cooling techniques (free and indirect humidification cooling), to improve control over electricity consumption in buildings, and to better the performance and operational reliability of HVAC systems. Real-time diagnostics for HVAC processes are also being studied.

The objective of the BEMCS research is to optimize energy use in buildings through the development of control and energy management programs, equipment, information technology, and man-machine interfaces. The development of testing methods for building energy management and control systems also contributes toward effective energy use.

Research on the indoor climate is primarily intended to support studies aimed at determining future system requirements. This research is concerned with questions relating to the hygiene, health, and thermal comfort of the indoor climate as well as new ventilation systems.

The results of the LVIS-2000 research program find practical application within joint product development and demonstration projects. Initially, these projects are started on the basis of present knowledge concerning the performance requirements and development needs of the systems. Furthermore, feedback from these product development and demonstration projects is used to guide the direction of the research program. At a later stage, new technology based on results obtained from the research program will be applied in the joint development projects. Typically, over 50 percent of the research funds necessary for the joint development projects are supplied by the companies involved.

The application of results from the LVIS-2000 program takes place within jointly financed product development and demonstration projects. Development projects started in 1988 and 1989 include:

- Future water radiator heating
- A Finnish air conditioning system
- A new electrical installation system
- Conservation of electrical heating energy
- Demonstration of man-machine interface (MMI) of a BEMC-system

A number of new joint projects based on research results are planned to be started in 1990. They include demonstration of absorption heat pumps in heating and cooling of buildings; testing of the general concept of the "LVIS-2000 BEMC" system; design of a "CFC-free" office building; a warm air-heated house of flats with demand controlled ventilation; and condensing boilers. Figure 6 summarizes the research components of the LVIS-2000 program.

Getting a clear idea of what living and working life will be like in the year 2000 is an important starting point for the systems development work. The objective of studies in this area is to identify future impulses and tendencies. Tests are also being conducted to determine consumer attitudes toward different heating and air conditioning solutions and systems employing information technology.

ETRR: Energy Efficient Buildings and Building Components

The objective of the research program ETRR (Energy Efficient Buildings and Building Components) is to present building technical solutions that reduce and improve energy use in both new and existing buildings while taking account of other building usage requirements such as room climate. In order to attain this goal, the research program focuses on the study and evaluation of applications of new building methods and materials technology.

In new building production, the research emphasis is initially being placed on the development of small low-energy houses. The objective is to present house concepts in which solutions satisfying the requirements of the users (e.g., good room climate) can be attained with the minimal possible energy use in heating, cooling, ventilation, lighting, and other electrical consumption. Concepts will be introduced for houses of 75, 50, and 25 percent consumption levels.

The main objectives in the small house project are the study and development of building design, new material technologies and building techniques, and methods of simulating energy consumption. The studies include different house concepts, new glass and window solutions (light openings), new heat insulation and the structures that are based on them, and the integration of structures and heating and ventilation equipment.

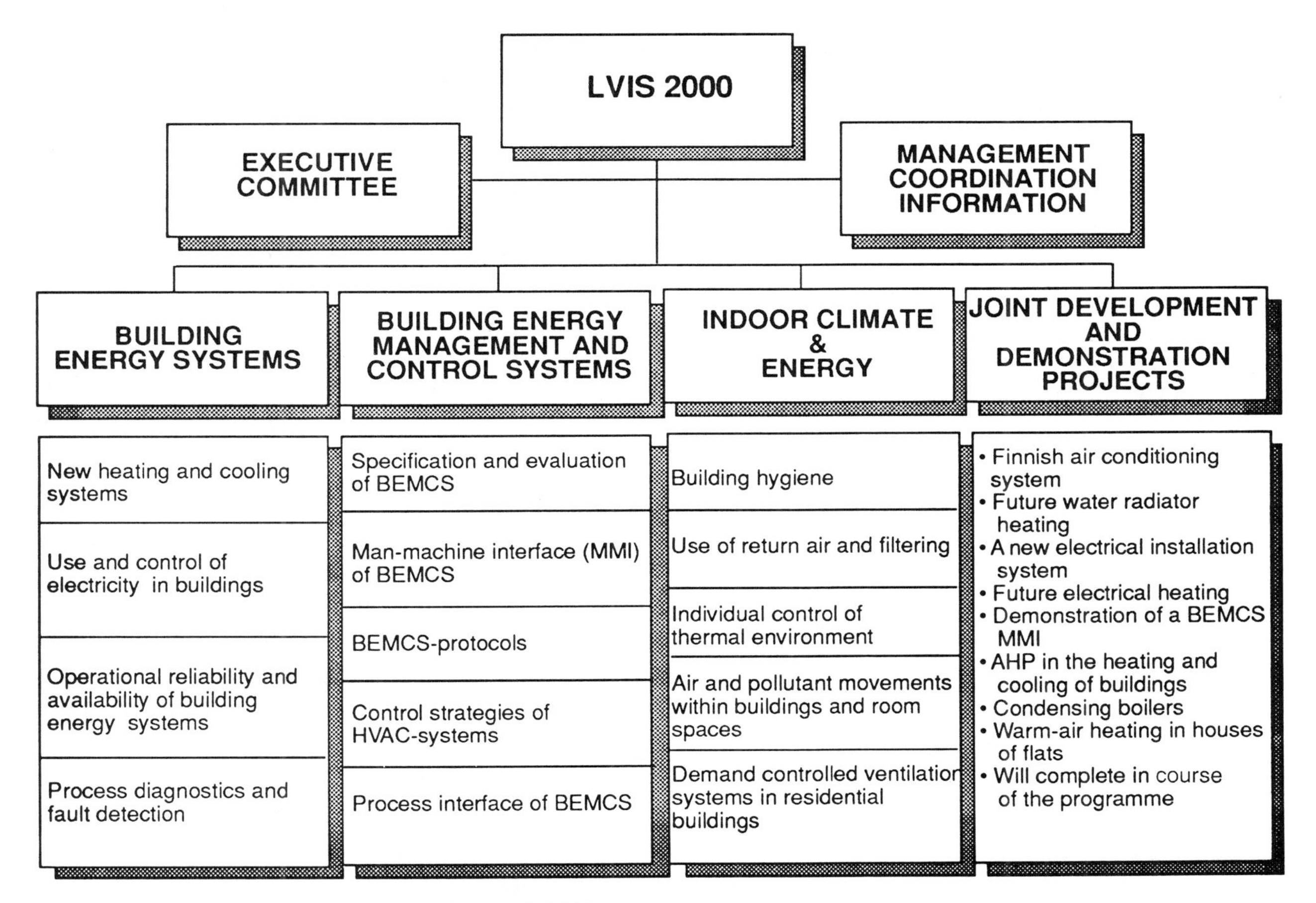

Figure 6. Research components of the LVIS-2000 program.

The same type of research also concerns high-rise buildings.

In rehabilitation, the objective is to achieve an essential reduction in the unit energy consumption in the existing building stock. During the early phases of the research program, the need for repair of the building stock and the measures that it requires are being studied in collaboration with the Ministry of the Environment. The focus of the research is on the development of flexible rehabilitation methods.

One key objective of the research program is experimental building, which tests the feasibility of new building techniques from the standpoints of user needs and the demands of new building production.

The research areas are:

1. New building techniques
2. Retrofitting of existing buildings
3. Energy calculation and design methods
4. Joint product development and demonstration projects

NEMO Research Program

The NEMO program seeks to assess the impact of advanced energy technologies on Finnish energy production systems.

Basically, the program will evaluate various new technologies that show promise in the development of more efficient energy systems. Areas of interest are divided as follows:

1. Energy system analyses
2. Energy storage and transmission
3. Renewable energy sources
4. New nonfossil fuels

Most of the results derived from energy system analyses will be incorporated directly into other NEMO projects, although the possibility for wider application exists. Innovative storage systems will be studied, and emphasis will be placed on design evaluation and improvement of system performance. This segment of the program also covers battery design and superconductivity applications.

Renewable energy sources, i.e., wind and solar power, are covered in the third segment. Wind-driven generators will be studied, especially from the standpoint of construction materials. The goals are to eventually create a full-scale advanced prototype system and conduct a wind farm experiment.

Finally, the new fuels group will study fuel cell design, such that industrial-scale applications will be possible in the near future. Hydrogen storage systems constitute a second area of investigation.

Summary

The development and implementation of Finland's energy supply is guided by national energy policy programs. The Council of the State approved the first of these programs in 1979, and the second in 1983. The third energy strategy is currently under preparation and is expected to be completed in the latter part of 1991.

It is clear that Finland's energy policy is and will continue to be oriented toward ensuring the continuity of present economic growth. The alternative (rapid movement toward a low-energy economy) would require government action in the form of price intervention and statutory usage controls. As far as the use of energy in buildings is concerned, this might mean that restrictions will not be placed on energy consumption.

Energy conservation in building energy use plays a key role and will continue to do so in the future. It would appear that energy economy must be seen as a "built-in" characteristic of future buildings in Finland and elsewhere. HVAC systems and equipment are crucially important to achieving energy savings. Efficient energy use in buildings will require the possibility of using a number of alternative energy sources. Research and product development in the field of HVAC systems should be oriented toward the systems that consume energy more efficiently than those existing at present. The main development objectives with regard to heating systems are: practical and precise control over individual room temperatures, flexibility in the use of different energy sources, and the applicability of installations to industrialized construction. The efficiency of individual devices and systems will also be improved.

Energy and environmental research in Finland is characterized by its programmed structure. There are currently 10 national research programs in progress in the field of energy research, and the Technology Development Fund is financing 11 technology programs. Two of these programs are concerned with energy use in buildings, namely LVIS-2000: Future Building Services, and ETRR (Energy Efficient Buildings and Building Components).

The exploitation of results and the market penetration of new technical solutions largely depends on the willingness of the private sector to invest in technology that will conserve energy and protect the environment as well as on government subsidies and energy pricing policy (specifically, taxation). Recent political stances point to a tightening of energy taxation, which will probably take the form of a pollution tax. Revenues thus generated would finance environmental protection investments as well as the development and introduction of new equipment solutions particularly into industrial companies.

Considerable resources will have to be devoted to the development of teaching in the HVAC field (in both basic and supplementary training) in the near future, so that the results of the industrialized building technology projects and the energy research programs can be implemented according to the planned schedule. Important developments require familiarity with whole entities, i.e., an understanding of the interaction between and integration of different systems. This will require skillful application of automation, computer, and simulation techniques.

Building Materials Research Program

D. L. McElroy
Metals and Ceramics Division
Oak Ridge National Laboratory
Oak Ridge, TN

Abstract

This paper describes selected aspects of the Building Materials Research Program conducted by Oak Ridge National Laboratory (ORNL) for the Building Systems Division of the Office of Buildings and Community Systems (OBCS) of the US Department of Energy (DOE) in three areas: energy conservation measures to reduce the energy use in buildings; test procedures and properties of currently used materials; and advanced materials (alternative materials for CFC-based insulation and high-resistance powder-filled panels). This paper reports recent progress in these three areas.

Introduction

Energy use in buildings has risen from 33 percent of national energy consumption in 1979 to 36 percent in 1985. Improvement of the energy performance of building envelopes constitutes a major opportunity to conserve energy. This paper is about materials research supported by the US Department of Energy (DOE) to pursue this opportunity.

ORNL staff provides management for the DOE Office of Buildings and Community Systems (Division of Building Systems) project on Building Thermal Envelope Systems and Materials (BTESM) in three areas: roofs and attic systems, walls and foundations, and building materials. The research at ORNL on roofs and attic systems is performed in a National User Facility that includes a Roof Thermal Research Apparatus (RTRA) and a Large Scale Climate Simulator for field and laboratory tests of roof and attic assemblies in cooperation with private industry and universities. Research on walls and foundations focuses on field tests and development of models to describe the behavior of these subsystems. The Building Materials Research Program conducts research on advanced materials and currently used materials, and recommends energy conserving measures. Results from projects in these areas are described in the BTESM Monthly Report.[1] The information in this paper is provided for use by codes, standards, and handbooks.

Energy Conservation Measures

Proper application of thermal insulation to residences can be an effective energy-conserving measure. The Building Materials Research Program produced the DOE Insulation Fact Sheet, which recommends use of insulation in residences as an energy-conserving measure.[2] Over 50,000 copies of the Fact Sheet have been printed and are being distributed by ORNL staff, the Conservation and Renewable Energy Inquiry and Referral Service (DOE hot line 800-523-2929), state energy offices, and industry. The Insulation Fact Sheet is one means to transfer technology to users. It provides recommended thermal resistance (R-value) levels keyed to the first three digits of the Zip Code of a residence. The computer program to provide the recommendations, ZIP 1.0, is available on a floppy disk for customized estimates by users.[3] The recommendations are based on an analysis of cost-effectiveness, using average local energy prices, insulation costs, equipment efficiencies, climate factors, and energy savings for both the heating and cooling seasons. A basic assumption is that no structural modifications are needed to accommodate the added insulation. The recommendations are based on a 30 year lifecycle cost for new houses and a 20 year lifecycle cost for retrofit. Table 1 lists the recommended total R-values for existing houses in eight insulation zones. The ZIP program calculates economic levels of insulation for attic floors, exterior wood frame and masonry walls, floors over unheated areas, slab floors, and basement and crawlspace walls. The economic analysis can be conducted for either new or existing houses. Climate parameters are contained in a file on the ZIP disk and are automatically retrieved when the program is run. Regional energy and insulation price data are also retrieved from the ZIP disk, but these can be modified by the user to reflect local prices.

Research is in progress to develop recommendations for cathedral ceilings, water heaters, and ducts. These results will be incorporated into a new fact sheet

Table 1. Recommended total R-values for existing houses in eight insulation zones.[a]

Component	Ceilings Below Ventilated Attics		Floors over Unheated Crawlspaces, Basements		Exterior Walls[b] (Wood Frame)		Crawlspace Walls[c]	
Insulation Zone	Oil, Gas, Heat Pump	Electric Resistance	Oil, Gas, Heat Pump	Electric Resistance	Oil, Gas, Heat Pump	Electric Resistance	Oil, Gas Heat Pump	Electric Resistance
1	19	30	0	0	0	11	11	11
2	30	30	0	0	11	11	19	19
3	30	38	0	19	11	11	19	19
4	30	38	19	19	11	11	19	19
5	38	38	19	19	11	11	19	19
6	38	38	19	19	11	11	19	19
7	38	49	19	19	11	11	19	19
8	49	49	19	19	11	11	19	19

a. These recommendations are based on the assumption that no structural modifications are needed to accommodate the added insulation.

b. R-value of full wall insulation, which is 3½ inches thick, will depend on material used. Range is R-11 to R-13. For new construction R-19 is recommended for exterior walls. Jamming an R-19 batt in a 3½ inch cavity will not yield R-19.

c. Insulate crawl space walls only if the crawl space is dry all year, the floor above is not insulated, and all ventilation to the crawl space is blocked. A vapor barrier (e.g., 4- or 6-mil polyethylene film) should be installed on the ground to reduce moisture migration into the crawl space.

and Version 2.0 of the ZIP computer program. In addition, a consumer-oriented handbook on reflective insulation has been produced.[4]

A technical assessment of the energy-conserving potential of active systems (variable thermal resistance and switchable emittance) for residential building thermal envelopes and fenestration showed that annual loads can be decreased by up to 30 percent.[5] The potential for energy savings was established using the EPRI Simplified Program for Residential Energy (ESPRE) to determine the hourly annual load for heating and cooling for a 1600 ft^2 lightweight structure located in Lexington, Kentucky; Minneapolis, Minnesota; or Phoenix, Arizona. Calculations were performed for 430 passive cases which have fixed levels of resistance and absorptance/transmittance levels. A passive insulation system was considered to be a building envelope with fixed roof, wall, and floor thermal resistance; fixed roof and wall absorptance; and fixed window transmittance. An active system was considered to be a building envelope with variable resistances, absorptances, and/or transmittances that could be changed with time to minimize the building load. The load for active systems was derived from these passive cases by minimizing the hourly loads for selected switchable levels. The annual loads for recommended insulation levels were Phoenix: 58, Lexington: 62, and Minneapolis: 90 MBtu/yr. (Note: 1 MBtu = 1,000 Btu.)

Table 2 shows the decrease in annual load from 183 to 55 MBtu that was calculated for the structure in Lexington as the passive thermal resistance levels were increased for the attic, floor, and walls. If the attic, floor, and wall resistances were made active, the annual load would decrease to 51 MBtu, a reduction of 4 MBtu below the case with R-38 resistance levels and 11 MBtu below the recommended levels. Table 3 shows that decreasing the passive absorptance of roofs and walls and passive transmittance of windows for fixed resistance levels reduces the annual load by 5 MBtu. An active system reduces the annual load to 48 MBtu and over half of the annual reduction is due to active window transmittance. Clearly, active fenestration systems deserve further study.

Research on Currently Used Materials

The goal of research in this area is to establish improved and valid measurement techniques and to use these to determine the performance of currently used, commercially available materials, particularly insulations.

Table 2. Annual loads, MBtu/year, with passive and active resistance at absorptance/transmittance levels of 0.8/0.5/0.5 for roof/walls/windows.

Resistance Levels Attic	Floor	Walls	Lexington Annual Load
0.5	0.5	0.5	183
19	19	19	66
38	19	19	62
38	38	38	55
Active A&F&W between 38 or 0.5			51

Table 3. Annual loads, MBtu/year, for passive and active absorptance/transmittance systems at resistance levels of 38/19/19.

Roof Absorptance/ Wall Absorptance/ Window Transmission	Annual Load MBtu
0.8/0.8/0.8	65
0.8/0.5/0.5	62
0.1/0.1/0.1	60
Active 0.8 or 0.1	48

Test Methods

Research to improve test procedures resulted in a new standard test method, ASTM C1114,[6] that covers the ORNL unguarded thin heater technique.[7] The technique yields steady-state thermal resistance results with an uncertainty of less than 2 percent. In addition, a computer-controlled transient test procedure was developed for this apparatus.

The unguarded thin heater apparatus is shown in Figure 1. The apparatus is an absolute, longitudinal heat flow measurement method and consists of an unguarded, electrically heated, flat, large area nichrome screen wire heat source sandwiched between two horizontal layers of insulation with flat isothermal bounding surfaces. The screen wire heat source has a low thermal conductance which reduces lateral heat flow and minimizes the need for edge guarding. The heat source provides vertical heat flow in its central region across the insulation to two temperature-controlled, water-cooled copper plates. The screen area is large, 0.9 by 1.6 m, and is instrumented with 11 thermocouples for temperature measurement and voltage taps for power

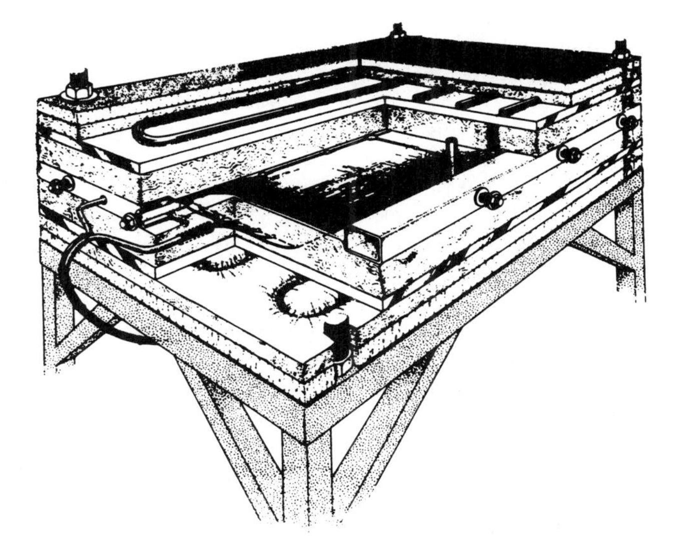

Figure 1. Unguarded thin heater apparatus.

measurements. A measured direct current passes through the screen, and the heat generated passes through the two layers of insulation. When steady-state is reached, standard potentiometric equipment is used to measure the thermocouple outputs, the current, and the voltage. These quantities and the known spacing between the plates allows the apparent thermal conductivity to be calculated for two-sided heat flow from the measured temperature difference (ΔT). This equipment can also be used for one-sided heat flow tests.

The measurement errors of the thin heater apparatus have been assessed. An error analysis predicts a maximum uncertainty of 1.7 percent if ΔT is 5 K and 0.7 percent if ΔT is 30 K. The most probable uncertainty is 1.2 percent and 0.4 percent, respectively, for these ΔT values.

Transient tests can be performed with the unguarded thin heater apparatus. A step-change in heat flux to the thin heater yields an initial heater temperature increase that is a linear function of the square root of time, and the slope is proportional to the product of thermal conductivity, density, and specific heat of the specimen (see Fig. 2). This procedure was used to determine the thermophysical properties of gypsum boards containing 0, 15, and 30 weight percent wax ($C_{18}H_{38}$) in the temperature range 20–50°C. Since this wax melts near 28°C, this combination provides a phase-change material to increase the thermal storage capacity of the gypsum.[8]

Standard Reference Materials

In 1978, a working group of the American Society for Testing and Materials (ASTM) Committee C16 addressed the issue: Adequate standard materials are not available for thermal conductance measurements in the field of thermal insulations. The working group developed a plan for selecting materials for Standard Reference Materials (SRMs) of low thermal conductivity (approximately 0.023 W/mK).[9]

In 1989, the thermal conductivity of fumed-silica insulation board was measured using the one-meter guarded hot plate at the National Institute of Standards and Technology [NIST].[10] Measurements were conducted for the following ranges: bulk density, 304.5–325.4 kg/m^3; and chamber air pressure, 97.51–103.43 kPa. The effect of moisture content on room-temperature measurements was minimized by prior conditioning of the specimen at 100°C for 24 hours. A linear equation having two independent parameters—bulk density and ambient air pressure—was fit to 35 data points. Certified values of thermal resistance at room temperature were calculated for the following ranges of bulk density and ambient air pressure: 300–330 kg/m^3 and 97–102 kPa, respectively. Seventy-five samples, each 600 by 600 by 25.4 mm, were transferred to the Office of Standard Reference Materials of NIST in Gaithersburg, Maryland, and designated as Standard Reference Material-1449 having a low thermal conductivity at room temperature.

ASTM Round Robins

Two ASTM C-16 round robins are in process with the goal of strengthening the precision and bias statements in ASTM standards for testing materials. ASTM C687-88 describes the standard practice for determining the thermal resistance of loose-fill building insulations.[11] The existing precision and bias statement is based on a round robin conducted in 1987 that yielded the results shown in Table 4. The imprecision (two standard deviations) determined for the measurement equipment was 3 percent using the glass fiber blanket, but unexpectedly large values of imprecision were found for the loose-fill materials. This large imprecision was primarily associated with specimen preparation procedures. The preparation procedures have been rewritten and a new round robin will be conducted in 1990 using the materials listed in Table 4.

An ASTM round robin to measure the corrosiveness of insulations is in progress. This round robin is targeted to develop a standard test procedure that uses electrochemical tests on insulation leachates, rather than wet immersion tests of metal coupons. The latter are time consuming, whereas the former can be completed in less than two hours. A novel potentiostat (the Corrater) that provides direct readings of corrosion rates (mils per year) has simplified the round robin.[12]

Research on Advanced Materials

Research on advanced materials has emphasized alternative technologies to replace insulations containing chlorofluorocarbons (CFCs) that harm the environment. The CFC issue is enormous. US industry produces over 400,000 metric tons of rigid foam-board insulation annually, and consumes over 60,000 metric tons of CFC-11 and CFC-12 in doing so. If environmentally acceptable alternative blowing gases are not available, the estimated energy impact for buildings applications alone is between 0.65 and 1.5 quads/year (1 quad = 10^{15} Btu).[13] Research on alternative materials for CFC-based insulations focused on rigid foam boards for roofing and high resistance powder-filled evacuated panels.

A cooperative industry/government research project for CFC alternatives is in progress to evaluate the thermal performance of rigid foam board blown with CFC-11 (control), HCFC-123, and HCFC-141b. This project evolved from two public/private workshops on CFC alternatives organized by ORNL.[14] The project is

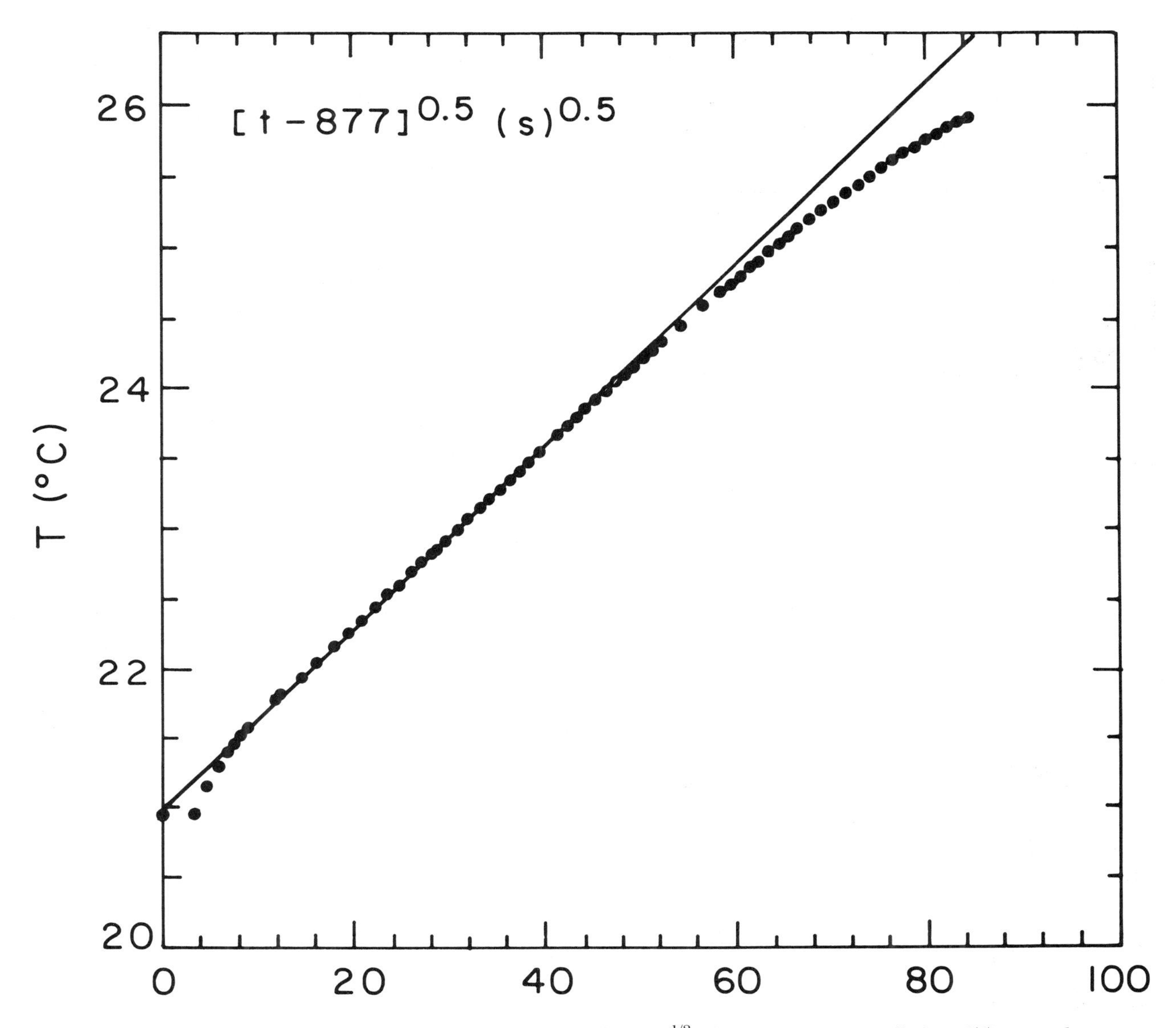

Figure 2. Temperature of the thin heater as a function of $(\text{time})^{1/2}$ after power was applied at 877 seconds.

directed by a steering committee with representatives from the cooperating organizations: Society of Plastics Industry, Polyisocyanurate Manufacturers Association, National Roofing Contractors Association, Department of Energy, Environmental Protection Agency, and ORNL. The prototypical boards were produced by industry and are being evaluated by industry and ORNL. Laboratory and field tests are being conducted on the thermal performance of rigid foam boards foamed with five blowing agents (CFC-11, HCFC-123, HCFC-141b, and two blends: 50/50 and 65/35 of HCFC-123/HCFC-141b). The thermal resistances of the boards were measured prior to exposure in the ORNL Roof Thermal Research Apparatus (RTRA) using the unguarded thin heater apparatus. The exposed boards will be removed, tested, and replaced at six month intervals to obtain laboratory thermal resistance tests. The ORNL-RTRA yields the thermal performance of membrane-covered panels exposed to field conditions from continuous measurements of temperature and heat flux. A comparison of laboratory and field tests will be made. The thermal resistance of sliced foam boards that are aging at room temperature is being measured using a heat flow meter apparatus. The R-value change evaluated by slicing is found to be proportional to $[\text{time}/(\text{thickness})^2]$.

Research at MIT demonstrated reduction of radiative heat transport in foams by adding small carbon particles to the foam, and measured effective diffusion

Table 4. C-687 loose-fill insulation thermal resistance imprecision.

1987 Materials/Results		1990 Materials
	Two-Sigma %	
Glass Fiber, Blanket	3	Fiberglass Blanket
Cellulose	19	Perlite, Pouring
Bonded Glass Fiber	10	Rockwool
Rock/Slag Wool	8	InsulSafe III
Unbonded Glass Fiber	10	Cellulose

coefficients for O_2, N_2, and blowing agents to provide data for models to predict long-term R-value.[15]

Figure 3 compares results of modeling and testing of thin specimens of rigid board foamed with CFC-11 and aged at 75°F. The MIT model considered 50.8 mm thick specimens aged for 5,400 days and 5.08 mm thick

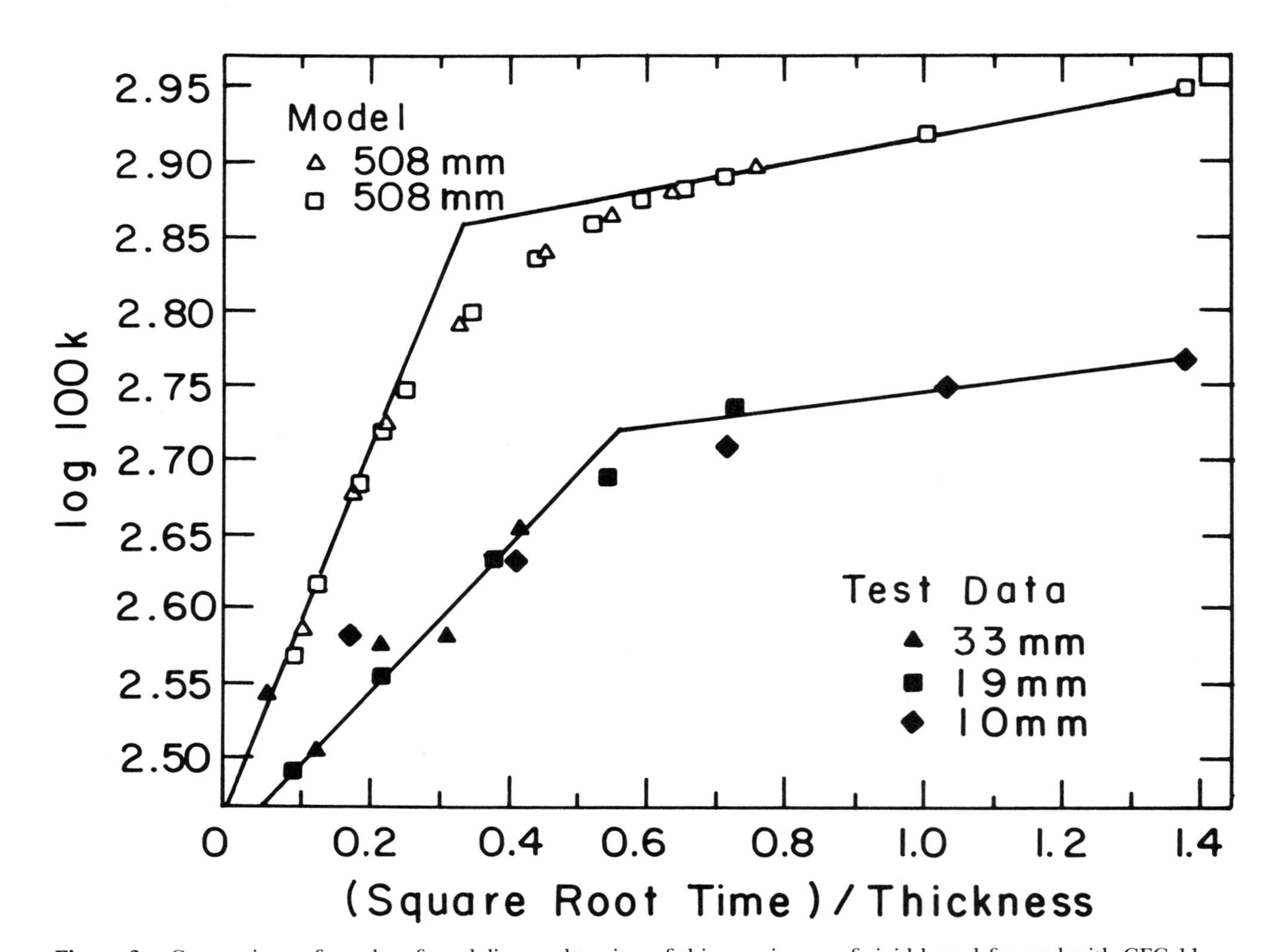

Figure 3. Comparison of results of modeling and testing of thin specimens of rigid board foamed with CFC-11.

specimens aged for 15 days. The results yield similar behavior when log (100 x k) is plotted versus $(\text{time})^{1/2}$/thickness. Two distinct linear regions are noted: a rapid initial increase due to the influx of air components, and a slower steady increase due to loss of CFC-11. The results for five tests up to 190 days on three thicknesses are similar to the modeling results. This result suggests that slicing is an attractive accelerated-aging protocol.

Significant changes in refrigerator/freezers (R/F) are needed because of the CFC issue and new appliance energy-use standards.[16] Insulation with thermal resistivities of 20 h · ft^2 · °F/(Btu · in.) could reduce the annual energy consumption of an 18 cubic-foot R/F from 1,100 kWh to below 600 kWh and reduce annual national energy consumption by one quad. Three advanced high resistance systems are being studied. The Solar Energy Research Institute[17] is investigating Compact Vacuum Insulation, which is a thin, high-vacuum panel with thin stainless steel walls held apart by spacers. This concept is similar to the vacuum thermos bottle. The Lawrence Berkeley Laboratory and Thermalux[18] are studying silica aerogel tiles that achieve R-20/in. at 100 mm of Hg. The tiles provide a silica structure to reduce heat flow. At ORNL, we are studying powder-filled evacuated panels that obtain R-20/in. below 10 mm of Hg.[19,20] These panels include a compacted silica powder in a porous pouch and a sealed plastic gas barrier. The barrier materials are heat sealable multilayer plastic films with low gas permeability. These panels were produced by industry and the barriers are proprietary information. Figure 4 shows R/in. values obtained on panels as old as seven years. The R-value decreases with age due to gas diffusion through the plastic barrier envelope, which raises the internal pressure and thermal conduction of the panel. Problems to be solved include reliability, aging to identify lifetimes, automated production, and cost reduction. The effectiveness of panels foamed into the walls of portable coolers was demonstrated.[21] A reproducible ice-melting-rate test was developed and applied to portable coolers with and without evacuated panels. The thermal performance of coolers with panels was not as good as predicted by one-dimensional models but was in excellent agreement with three-dimensional models.

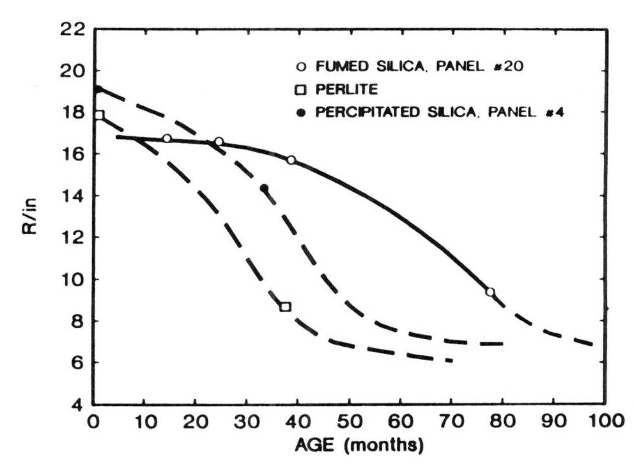

Figure 4. The thermal resistivity of powder-filled evacuated panels as a function of panel age.

Conclusions

The ORNL Building Materials Research Program being conducted for DOE (OBCS) is pursuing advanced and currently available materials. A cooperative project with industry is providing field and laboratory results on boards foamed with alternatives to CFC-11 and CFC-12. The performance of powder-filled evacuated panels is being established and appears promising for near-term use in R/F. Improved test procedures are being defined for currently available materials. Cost-optimized energy-conserving measures are being recommended to consumers.

Acknowledgments

This research was sponsored by the Office of Buildings and Community Systems, Building Systems Division and Building Equipment Division, US Department of Energy, under contract DE-AC05-84OR21400 with Martin Marietta Energy Systems, Inc. The author is grateful for reviews by R. S. Graves, T. G. Kollie, J. E. Christian, and D. F. Craig. Special thanks are extended to G. L. Burn for preparing the manuscript.

References

1. *Monthly Progress Report,* Building Thermal Envelope Systems and Materials (BTESM) and Research Utilization/Technology Transfer Progress Report for DOE Office of Buildings and Community Systems, December 1989 and January 1990, ORNL/M-1069 (March 1990).
2. DOE Insulation Fact Sheet, DOE/CE-0180 (January 1988).
3. Stephen R. Petersen, *ZIP - The ZIP Code Insulation Program* (Version 1.0) Economic Insulation Levels for New and Existing Houses by Three Digit ZIP Code, Users Guide and Reference Manual, ORNL/TM-11009, NISTIR 88-3801 (January 1989).
4. A.O. Desjarlais and R.P. Tye, Research and Development Data to Define the Thermal Performance of Reflective Materials Used to Conserve Energy in Building Applications, ORNL/Sub/88-SA835/1 (April 1990).
5. H.A. Fine and D.L. McElroy, Assessment of the Energy Conservation Potential of Active Systems for Residential Building Thermal Envelopes and Fenestration, pp. 639–

654 in *Thermal Performance of the Exterior Envelopes of Buildings IV,* ASHRAE (December 1989).

6. ASTM C1114-89, Standard Test Method for Steady-State Thermal Transmission Properties by Means of the Thin-Heater Apparatus, *1989 Annual Book of ASTM Standards,* Vol. 04.06; pp. 601–606 (1989).
7. D.L. McElroy, R.S. Graves, D.W. Yarbrough, and J.P. Moore, "A Flat Insulation Tester that Uses an Unguarded Nichrome Screen Wire Heater," *Guarded Hot Plate and Heat Flow Meter Methodology,* ASTM STP 879, pp. 121–139 (1985).
8. R.S. Graves, D.L. McElroy, D.W. Yarbrough, and H.A. Fine, "The Thermophysical Properties of Gypsum Boards Containing Wax," *Thermal Conductivity 21,* Lexington, KY, Proceedings (1989).
9. ASTM Subcommittee C16.30, "Reference Materials for Insulation Measurement Comparisons," Thermal Transmission Measurements of Insulation, ASTM STP 660, R. P. Tye, Ed., American Society for Testing and Materials, pp. 7–29 (1978).
10. R.R. Zarr and T.A. Somers, "Thermal Conductivity of Fumed Silica Board, SRM 1449 at Room Temperature," The 21st International Thermal Conductivity Conference, Lexington, KY (October 15–18, 1989).
11. ASTM C687-88, Standard Practice for Determination of the Thermal Resistance of Loose-Fill Building Insulation, *1989 Annual Book of ASTM Standards,* Vol. 04.06, pp. 327–332 (1989).
12. E.E. Stansbury, Overview of the Applicability of Electrochemical Methods to Evaluation of the Corrosiveness of Residential Building Thermal Insulations with Proposed Cooperative Test Program, ORNL/Sub/88-SB716/1 (October 1989).
13. D.L. McElroy and M.P. Scofield, CFC Technologies Review, Foamed-Board Insulation for Buildings, ORNL/TM-11291 (May 1990).
14. J.E. Christian and D.L. McElroy, Results of Workshop to Develop Alternatives for Insulations Containing CFCs—Research Project Menu, ORNL/CON-269 (December 1988).
15. L.R. Glicksman and M.R. Torpey, A Study of Radiative Heat Transfer Through Foam Insulation, ORNL/Sub/86-09099/3 (October 1988).
16. National Appliance Energy Conservation Act, 1987, Public Law 100-12, March 17, 1987 and amendments.
17. David K. Benson, The High-Vacuum Insulation Option: Benefits and Problems, Seminar at 1989 ASHRAE Annual Meeting, June 24–28, Vancouver.
18. Thermalux 1989 Silica Aerogel Refrigerator Insulation Presentation at DOE Public Heating (January 12, 1989).
19. D.L. McElroy, F.J. Weaver, D.W. Yarbrough, and R.S. Graves, Thermal Resistance of Fine Powders at Atmospheric Pressure and under Vacuum, *Insulation Materials Testing and Applications,* ASTM STP 1030, D.L. McElroy and J.F. Kimpflen, Eds., American Society for Testing and Materials, Philadelphia, pp. 52–65 (1990).
20. R.S. Graves, D.W. Yarbrough, and D.L. McElroy, The Thermal Resistance of Flat Powder-Filled Evacuated Panels, Proceedings of Int. Conf. of Thermal Insulation, Vol. 5, 20–34 (February 27–28, 1989).
21. R.W. Barito, T.G. Kollie, D.L. McElroy, F.J. Weaver, and S.H. Werst, Fabrication, Evaluation, and Application of Evacuated Panel Insulation (EPI) to Portable Coolers, ORNL/Sub/88-SD 731/1 (March 1990).

A "Place on Which to Stand" for the Appliance Industry
A Commentary on Appliance Engineering: Energy, Environment, and Ethics

Jack Weizeorick
Engineering and Technical Services
The Association of Home Appliance Manufacturers
Chicago, IL

The futurists tell us that the most accurate way to forecast the future, even a decade out, is to see the present as it really is. And the most effective way to begin to do that is to glance at the past.

The appliance industry has had profound impact on 20th century consumer life and lifestyles. The wringer washer brought us a long way from the washboard, the automatic was a far cry from the wringer. The dryer was a leap from the clothesline. The refrigerator was a giant step from the iceman. The dishwasher a country mile from the dishpan. And so on. Now the microwave oven is a far cry into the future. Today's lifestyles are different from, and better than, they were before the major appliance industry off-loaded much of the drudgery of household chores.

In fact, major appliance industry history reminds me of the words of one of my favorite engineers, Archimedes, who invented the lever and pulley in the second century before Christ. He said: "Give me a place on which to stand, and I will move the earth." That could have been the battlecry for the major appliance industry. The place on which we stood throughout the 20th century was that if we dreamed big enough, and worked hard enough, and systematically enough, gradually we could "move the earth" enough to transform American home life.

We did that, consistently, effectively, and with gratifying consumer response. We managed to provide the marketplace with an array of appealing new products, year after year. At the same time, we set a standard for product durability and affordability that the consumer now has come to expect, and that other industries envy.

In fact, the March 12th issue of *Fortune* said that appliance makers have managed production economies so effectively that, if we had been making the Chevrolet Caprice since 1980, its price would have inched from $7,209 to $9,500, instead of the actual 1990 sticker of $17,370.

This production effectiveness has not come easily. For example, some 98 percent of this industry headcount active in the 1950s have folded their appliance tents entirely. Today, the US marketplace is served chiefly by five producers, General Electric, Maytag, Raytheon, White Consolidated Industries and Whirlpool—and their nearly 20 brands of major appliances. These firms were able to capitalize on the economies of scale, and to dedicate enough engineering, manufacturing, and marketing resources to serve the vast volume needs of a price-conscious marketplace that consistently called for "more-bigger-better" products.

In the early 1970s, the appliance industry joined the rest of the country in developing energy consciousness and conscience. "More-bigger-better" no longer was enough. We added "energy-stingy" to our list of adjectives.

We took the initiative to shift much of our research and development focus to higher efficiency, leaving only enough features–innovation work to provide market appeal. We fine-tuned engineering and production facilities to maximize the potential of the available technology. And we have been making steady gains in that area ever since. In fact, our industry has been able to realize a 96 percent efficiency improvement in refrigerators (our most energy-intensive product) when 1988 is compared with 1972. We also made notable gains in dishwashers, clothes washers, and room air conditioners.

Part of the challenge was to achieve these gains while producing appliances that consumers were willing to buy in volumes sufficient to aid national energy objectives, and to keep the market alive.

As we all know, the energy picture has stayed on the minds of the industry, the citizenry, the media, and legislatures, every day since. Obviously, major appliances in the homes of each of these people are constant reminders every day of their lives.

While we were evolving more efficient ways to use what we had, in terms of technology, facilities, and tooling, some state legislatures began to enact laws that set a whole spectrum of efficiency standards for appliances. With the volume requirements on which our manufacturing and distribution systems were based, the prospect of a potential 50 different sets of state standards for eight or nine different products each was chilling. We knew we could not sustain that without collapse.

That is one reason that in 1987, the Association of Home Appliance Manufacturers (AHAM) welcomed the prospect of a national standard for appliance efficiency. We got the first olive out of the jar by volunteering to spearhead development of a "National Appliance Energy Conservation Act of 1987" (NAECA, for short).

From AHAM perspective, NAECA's major promise was to offer our industry another specific toe-hold, a "place on which to stand" for all the industry's efficiency efforts. This 1987 federal law spelled out specific maximum energy consumption standards for specific appliance product categories manufactured after January first of 1990. The statute also empowered the Department of Energy to review standards regularly and to decide when, or if, more stringent energy requirements would be appropriate. At the time, we believed NAECA's results were tough and ambitious. We knew they would require dedicated, accelerated engineering achievement from our industry, but we thought they were fair.

While we were embroiled in Washington, hammering out NAECA with a coalition of industry, government, and environmentalists, 25 developed nations were meeting in Quebec to discuss the state of the stratosphere. The outcome of their meeting was the Montreal Protocol of 1987, which identified chlorofluorocarbons (CFCs) as major contributors to the depletion of the ozone layer. The Protocol detailed a program to cut CFC production and use in two by 1998. At about the same time, the US Environmental Protection Agency developed similar regulations.

Now, the US Congress is debating amendments to the Clean Air Act, which would eliminate CFCs entirely by the year 2000. And the Montreal Protocol group will reconvene this summer, probably to reinforce this decision from an international perspective. The reality is that CFCs will not even be available by the end of the decade.

This creates a major dilemma for refrigeration products. Our remarkable record in energy improvements since the 1970s was predicated on using more, not less, CFC technology. In fact, refrigeration companies have increased their use of CFC-11 by more than 55 percent in the past 15 years, in the interest of energy conservation. Until three years ago, we were confident that CFCs were a big part of the energy solution, certainly not a part of the environmental problem . . . much less a flashpoint for the future of life on earth as we know it.

It is really immaterial now whether Chicken Little is right regarding CFCs. The threat is perceived as true. And it will be treated as though it were true. The thought that appliance products might conceivably threaten our fragile stratosphere alarms us, personally and professionally. Ethically, we will respond as though the most outrageous claim were fact. We will correct the CFC situation because not to do so would be immoral, but it is a tricky balancing act.

We know that American lifestyles are not about to surrender the standard of living that CFC-based products have provided, in terms of food supply and preservation. So it is in everyone's best interests if we discover an alternative that allows the advantages without the threat. We need to manage the problem, not react to it irresponsibly, and we know the best way to deal with the future is to gang up on it.

To do that, we have established an unprecedented technological partnership, the Appliance Industry-Government CFC Replacement Consortium, Inc., an AHAM subsidiary made up of refrigerator and freezer manufacturers, the Department of Energy, and the Environmental Protection Agency, along with compressor manufacturers and chemical suppliers. Its charter: to find acceptable replacements for CFC-11 and CFC-12 within the next two to three years, at the outside limit. Its promise: to get this monumental job done, effectively, cooperatively, and on time.

That is a quick profile of environmental and energy challenges to the major appliance industry, especially refrigerators and freezers, on the brink of the 1990s. Except for one little thing: Even before NAECA-specified products began to fill the 1990 pipelines, DOE published energy standards for refrigerators and freezers, due to go into effect on January 1, 1993. These standards are staggeringly stringent. The appliance industry was stunned. Almost all models that achieved NAECA goals for January 1 of this year will be required to show energy-consumption reductions of another 20 to 30 percent within another three years.

How are we going to achieve two simultaneous engineering miracles? I simply do not know. Our industry is called upon to ride off in two directions at once: 1) to meet DOE's interim 1993 mandate for stepped-up CFC use; 2) to meet our promise to Montreal, the EPA, and the 21st century, by learning to step down, and finally to step out of, CFCs. We certainly promise to try, and if we cannot succeed, we'll tell DOE promptly.

But now we are truly daunted. If DOE has chosen to clobber refrigerators and freezers (products they know to be under the technological and circumstantial gun), what might be their plans for clothes washers and dryers, ranges, dishwashers, or even the microwave oven (a real anorexic in energy consumption)?

It is true that the appliance industry has been accused of crying wolf when difficult demands have been made. Sometimes our members have raised a chorus of "We can't do that!" then went off to their labs and did it. This time though, it looks like the wolf truly might be at the door.

The saddest part (and the hardest part) in a situation like this is that it's easy to forget that we are all on the same team. It often feels like its "us-versus-them." But self-pity is not productive and blaming does not move the ball. They are the real energy-wasters. The reality is that we have only one planet to walk on, only one sky to look up to, and only a little bit of time. We share the responsibility to be their stewards. I have a personal stake in the future; so do you.

This is not a boardroom squabble nor an academic argument. We have a critical and common goal. And it is so overwhelming that I think that we sometimes forget that partnership is essential to its solution. If we are to fulfill our responsibilities to ourselves, one another, and to the future of the planet, we need the best that each of us can muster. We must be able to trust each other's work and words.

I never want to paint a picture of federal government agencies as arbitrary and punitive. That's far too easy and naive. The government has a host of special interest "publics" to which they must respond, as does every industry, including our own. It is absolutely essential that our partnership be based on reality: the critical reality of the government's situation and the genuine reality of appliance industry's strengths, capacities, and resources. With our blinders off, we can work together effectively, productively, cooperatively, and communicatively.

With that kind of mission in mind, our program planners have asked me to talk about ways to improve joint government and appliance industry efforts. I believe we are seeing the first signs of a wholesome turn in our mutual work. But there is room for more candor, balance, and mutual respect.

For example, when DOE and its bretheren announce a publication date for an ANPR or NPR, our industry schedules around it, scrambles around it, and prepares to devote our best resources to it. When the agency overshoots its own published deadline, generally it shows the statutory 60-day comment period, regardless of the complexity of the issue at hand or the seemingly haphazard scheduling.

Often agencies dismiss our objections by saying, "You have got to understand how the government works." Our partnership with the government will truly be enhanced when we can say to them, "You have got to understand how our industry works." When we have scheduled for a time crunch that does not materialize, everything backs up. It is not that our engineers have nothing to do while they wait; our industry plates are full. But the resources that were primed for agency assignment now need to be reassigned to other essential obligations, like engineering research, production innovation, reliability, and manufacturing productivity. When the agency lobs a document over the net, everything else on our side must come to a screeching halt, in order to provide the required fast detail and accuracy. Pragmatically, it clobbers our calendars and affects every other item on our industry agendas. Psychologically, it feels unfair.

In the proposed rulemaking research process, our industry is called upon to provide painstaking detail to document the validity of every claim and decimal point. We do not object to that. But our experience shows that sometimes an agency gives equal weight to the options of special interest groups with more decibels than details in their presentations. We *are* griping about that. As of this afternoon, we are waiting for a final ruling on clothes washers, clothes dryers and dishwashers. We are in the middle of rulemaking activities on ranges, room air conditioners, and even microwave ovens. Will the government give us time to respond, responsibly? I certainly hope so.

Our industry's engineering ranks are spread dangerously thin, and it is not simply a matter of hiring more. We need appliance professionals, people who truly understand our industry and how it works, to do an effective job for the government, the consumer, and the stratosphere. Our industry's historical success in flexing with the marketplace has been due to long-range planning for a foreseeable future (a luxury we no longer can claim), and on evolutionary, not revolutionary, changes in design and production.

When appliance producers initiate product changes, they plan for maximum use of existing facilities and gradual evolution for tooling, building, testing, and re-testing. For example, when CFC-12 was introduced as refrigerant, it was applied only to a few models at first, then a few more, until all the new tooling was in place and 100 percent of the line utilized the new technology.

In fact, the CFC Consortium is organized around this reality. Those experts are looking first for replacement candidates that are as close as possible to drop-ins. Why? Because that is the effective way to make changes swiftly enough for maximum impact with less change to the product, less retooling, lower cost, and, most importantly, less uncertainty about reliability.

Our critics suggest that our pursuit of a close alternative to CFC means we are too closed-minded to see the potential of other refrigeration technologies, for example, evacuated panels. As a matter of fact, we are pretty optimistic about the panels—at some point down the road. First they must be researched properly so that we can come up with a design we can trust. Even so, we do not have the acumen to mass-produce them now. It would be ludicrous to revamp our facilities for vacuum panels in the roughly six million refrigerators consumers take into the hearts of their homes every year . . . only to discover reliability or other problems we missed in our rush to the market.

There is another issue on the candor front, that concerns only refrigerators and freezers. It would be an enormous aid to our industry if the DOE and EPA could share with us a coherent policy regarding ozone depletion and greenhouse warming. Does one outweigh the other in long-term priority? Yesterday's rule commanded absolute dedication to one priority, and we made appropriate pronouncements and promises; today's rule undermines the urgency of the other, possibly sacrificing the first. Our resources and efforts get cornered and baffled.

The place on which the appliance industry stands has turned into two places which we must straddle: on the one side, the environmental concerns of energy,

CFCs, and the government; and on the other side, the marketing issues of innovation, reliability, consumer appeal, and price. It is pretty difficult to "move the earth" from this position.

Still, I'm optimistic—and what is the evidence for my new optimism?

I hear the EPA talking about the need for a hybrid approach to resolving environmental problems, a team, like our Consortium, made up of government, industry, and environmentalists.

I have observed DOE's new willingness to listen and reconsider, when an objection from industry makes genuine sense to them. DOE does not specialize in arbitrary demonstrations of power. DOE seems now to acknowledge the appliance industry's commitment to realistic problem solving—and they seem newly willing to help us succeed.

DOE seems now to believe the appliance industry wants to give this decade our best shot, and that we can not do that if they keep "taking the shell out of the chamber."

Archimedes had also another famous quote, and if my guarded optimism is warranted, the government and industry will repeat it in chorus as we trudge into the next millenium. The words: "Eureka! [We] have found it!"

National Appliance Efficiency Regulations and Their Impacts*

James E. McMahon
Energy Analysis Program
Lawrence Berkeley Laboratory
Berkeley, CA

Household appliances consume 35 percent of the electricity and 25 percent of the natural gas in the US. In the context of global warming, the importance of the residential and commercial sectors may be indicated by their electricity consumption, which in 1988 was equivalent in primary energy to total US coal consumption (18.7 quadrillion Btu), or 62 percent of US electricity sales (Figure 1) (USDOE, 1990).

This paper focuses on residential sector mandatory efficiency standards, using refrigerators as an example. The paper discusses the following: market failures that provide an opportunity for regulatory remedies; history of energy performance standards affecting household appliances; projected changes in energy performance of new refrigerators; components of a technical analysis of proposed efficiency standards; and paths for additional efficiency increases.

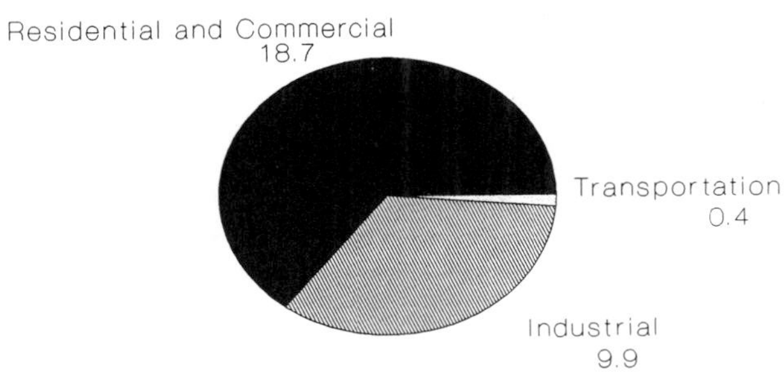

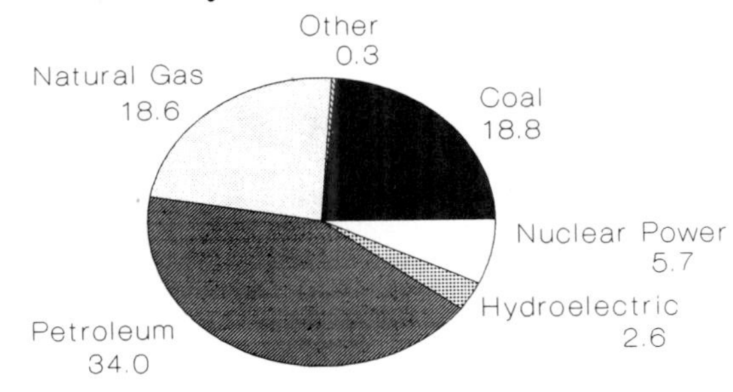

Figure 1. Electricity consumption (including losses) in the residential and commercial sectors is 18.7 quadrillion BTU, equivalent to total US coal consumption. (Source: *Monthly Energy Review* 12/88.)

Market Failures

From the perspective of energy efficiency, market failures are common for most residential appliances (Ruderman, 1987). The reasons for this include: 1) lack of information to purchasers about the choices available, and the costs and savings of those alternatives; 2) purchases made by builders or landlords, who do not consider the operating costs of the appliances; 3) operating cost savings may be small (a few dollars a year), and therefore are ignored; 4) capital constraints may prevent some purchasers from buying more efficient appliances; 5) efficient models may not be available in all local markets; 6) marketing strategies on the part of manufacturers and retailers may promote other qualities than energy efficiency; and 7) manufacturer's strategic planning may be indifferent to, or contrary to, improvements in energy efficiency.

The result of these market failures is that in the 1970s economic investments which should have been attractive, but which took the form of energy-efficient appliance designs, seldom occurred in the marketplace. A comparison of actual average purchase behavior for new energy-using equipment against the range of efficiencies and engineering costs available showed that simple paybacks from less than one year up to three years were available in the form of more efficient appliances, but were seldom purchased (Table 1, Ruderman et al., 1987). Table 2 shows that these corresponded to real returns on investment of 18 percent to over 825 percent for different products, in the period 1972 to 1980. The

* This work was supported by the Assistant Secretary for Conservation and Renewable Energy, Office of Buildings and Community Systems, Buildings Equipment Division of the US Department of Energy, under Contract No. DE-AC03-76SF0098.

Table 1. Payback period in years for appliances 1972–1980.

Appliance	1972	1978	1980
Gas Central Space Heater	2.98	2.38	2.21
Oil Central Space Heater	2.33	1.70	1.18
Room Air Conditioner	5.11	4.77	5.25
Central Air Conditioner	4.96	4.16	5.18
Electric Water Heater	.48	.41	.41
Gas Water Heater	1.50	1.07	.98
Refrigerator	1.35	1.45	1.69
Freezer	.60	.67	.72

Table 2. Aggregate market discount rates in percents for appliances 1972–1980. (Source: Ruderman, H., M.D. Levine, and J.E. McMahon, "The Behavior of the market for energy efficiency in residential appliances and heating and cooling equipment," in *The Energy Journal,* v.8, 1987.)

Appliance	1972	1978	1980
Gas Central Space Heater	39	51	56
Oil Central Space Heater	52	78	127
Room Air Conditioner	20	22	19
Central Air Conditioner	19	25	18
Electric Water Heater	587	825	816
Gas Water Heater	91	146	166
Refrigerator	105	96	78
Freezer	379	307	270

market generally failed to capture these returns even during a period of increased interest in energy. States and the federal government enacted mandatory efficiency regulations in an attempt to correct some of the market failures.

Refrigerator Regulations

In the case of refrigerators, the State of California enacted minimum efficiency regulations affecting new models beginning in 1978 (California Energy Commission, 1983). Those regulations were periodically updated. In 1987, the National Appliance Energy Conservation Act (NAECA) was passed (see McMahon, 1987, for brief background history), setting minimum efficiency (or maximum energy consumption) standards for most residential appliances, including: furnaces, room and central air conditioners, heat pumps, water heater, refrigerators, freezers, and cooking appliances. In 1988, fluorescent light ballasts were added to NAECA.

Figure 2 shows that refrigerator efficiencies improved nearly 100 percent from 1972 to 1987. In addition to state regulations, the transition to automated assembly lines, using polyurethane foam in place of hand-installed fiberglass as insulation in the walls, resulted in improved energy efficiency. However, it would be a mistake to view this accomplishment as exhausting the technological potential for efficiency improvement. In 1988, US Department of Energy (DOE) studies identified additional design changes that could achieve substantial additional improvements in energy efficiency using cost-effective current technologies (USDOE, 1988). Among the designs analyzed (Table 3) were improved compressors, additional insulation, more efficient fans, and adaptive defrost. Each of these was available on some models in 1989, but no one model contained all of them.

Table 3. Design options for refrigerators and freezers.

Foam Insulation Substitution in Cabinet
Foam Insulation Substitution in Doors
Increased Cabinet Insulation Thickness
Increased Door Insulation Thickness
Reduce Heat Load of Through-the-Door Feature*
Double Door Gasket
Improved Foam Insulation
Evacuated Insulation Panels
High-Efficiency Compressor Substitution
Adaptive Defrost
Fan and Fan Motor Improvement
Anti-Sweat Heater Switch
Increased Evaporator Surface Area
Hybrid Evaporator*
Enhanced Heat Transfer Surfaces
Mixed Refrigerants
Improved Expansion Valve
Fluid Control Valves
Two Compressor System*
Variable Speed Compressor
Two-Stage Two-Evaporator System*
Use of Natural Convection Currents
Location of Compressor, Condenser, and Evaporator Fan Motor

* Design options for refrigerator-freezers only.

Future Technologies

In 1989, a similar analysis was performed (USDOE, 1989), but it assumed the chlorofluorocarbons (CFCs) would be phased out in the mid-1990s. Since CFCs only represent about $2 (of $240) cost to the manufacturer, the analysis concluded that replacement with alternative refrigerant and foaming agents with about 5 percent of the ozone-depleting potential would not represent a large burden on either manufacturers or consumers. However, work is needed to complete toxicity testing of the CFC substitutes, and to find solutions compatible with other components (solvent action of some alternatives must be addressed). The availability of some of the more efficient designs is brought into question with the elimination of CFCs, such as certain compressors. These may have to be redesigned, at unknown cost, to perform well with the new refrigerants. Those designs were elim-

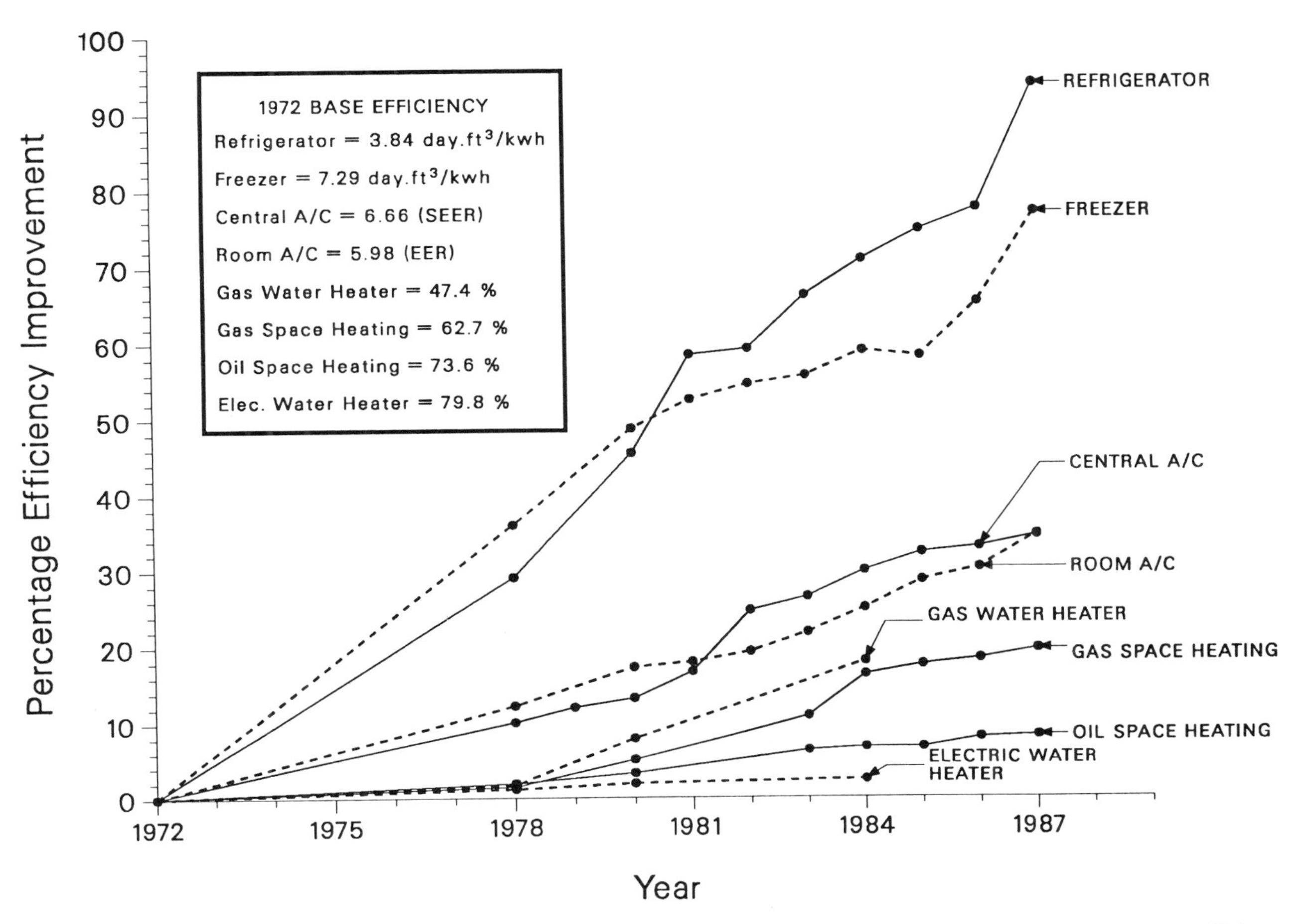

Figure 2. Percent improvement in efficiency in major residential appliances. (SEER = seasonal energy efficiency ratio; EER = energy efficiency ratio.)

inated from the list of options considered in the update to the efficiency regulations.

Even with the elimination of CFCs, DOE found a 25 percent additional reduction in unit energy consumption (for the typical 18 cubic foot, top-mount, automotive defrost class) of refrigerators/freezers technologically possible and economically justified (Table 4). Consequently, this reduction is mandated for new refrigerators manufactured beginning in 1993.

The next technological option for refrigerators appears to be advanced insulation, namely, vacuum panels. There are three alternative technologies under development that may become commercially available at competitive costs in this decade. The significant issues for commercialization are the ability to produce panels that are large enough and durable enough to maintain vacuum for a refrigerator life of 20 years. As research is performed on the performance characteristics of new refrigerants, additional design changes to improve energy efficiency can be imagined for the future.

Analyzing Standards

The process of analyzing possible efficiency standards involves:

1. listing designs that are technologically possible;
2. defining classes and baseline units as starting points from which changes can be made and analyzed;
3. simulating the energy consumption of alternative designs;
4. estimating the production costs associated with those designs, and ranking the designs by cost-effectiveness;
5. modeling the manufacturers' and retailers' markups to determine consumer prices for each design;
6. analyzing the economics from the consumer perspective;
7. projecting the demand for appliances in the future,

Table 4. Manufacturer cost and unit energy consumption top-mount auto-defrost refrigerator-freezer (no CFCs). (Column 4 lists *annual* energy use; AV = adjusted volume.) (Source: US Department of Energy.)

Design	Branch	Design Option	Energy Use(kWh)	Manufacturer Cost ($87)
0		Baseline	955	224
1		0 + Enhanced Evap. Heat Transfer	936	224.1
2		1 + Foam Refrigerator Door	878	225.55
3		2 + 5.05 Compressor	787	228.95
4		3 + 2 inch Doors	763	232.65
5		4 + Efficient Fans	732	241.65
6	#1	5 + 2.6"/2.3" Side and 2.6" Back Insulation	706	249.10
7		5+ 3.0"/2.7" Side and 3.0" Back Insulation	690	253.95
8	#2	5 + Evacuated Panels*	577+	287.65
9		8 + Two-Compressor System	508	337.65
10		9 + Adaptive Defrost	490	353.65

Volume = 18.0 cubic ft., AV = 20.8 cubic ft., Baseline; 4.5 EER compressor.
Side wall insulation: 2.2" foam in freezer; 1.9" foam in refrigerator.
Door insulation: 1.5" foam in freezer; 1.5" fiberglass in refrigerator.

* Evacuated panels could be powder-filled, aerogel, or steel. Manufacturer costs will vary for each technology. The cost estimated above is for powder-filled panels.
+ For steel evacuated panels, the energy use could differ depending upon the R-value achieved.

including shipments, energy consumptions, and usage patterns;

8. selecting trial mandatory standard levels;
9. projecting the impacts of each trial standard level on shipments, energy consumption, fuel switching (where applicable), consumer economics, and usage patterns; calculating net present value to consumers;
10. projecting the impacts of each trial standard level on manufacturers' finances, and on competitiveness within the industry;
11. quantifying impacts on electric utilities, including net revenue losses (gross revenue losses less decreased operating costs) and capacity deferrals;
12. estimating environmental impacts (CO_2, SO_x, NO_x, CFCs).

These analyses are described in DOE publications with each rulemaking or update (e.g., USDOE, 1989). Appliance efficiency standards enacted to date are projected to save from 17 to 21 quadrillion Btus (primary), equivalent to 1 to 1.3 years of residential energy over the next 25 years. Consumer spending for equipment will increase, but will be more than compensated by decreased operating expenses (by about 2 to 1). The need for power plants in 2015 will be reduced by more than 15,000 to 21,000 MW. Annual emissions of CO_2, NO_2, and SO_2 will be reduced by 1 to 2 percent in 2015 by these regulations alone, largely through reduced electricity consumption. The present value of the net savings is about $34 billion dollars (1987 dollars, discounted to 1987 at 7 percent real), without placing any value on the deferred generating capacity or the emissions reductions.

Future Directions

Increased concern over global climate change is focusing attention on policies which might decrease emissions of CO_2, along with other pollutants. The analysis of possible appliance efficiency standards offers both a database of technological options and their costs at the end-use level, and a methodology for assessing other technological options. The timetable already enacted into law involves updates to appliance efficiency requirements in the US through the year 2012 (Table 5). A number of promising technologies have already been identified which can provide energy, and carbon savings into the next century.

Additional options can be developed if we:

- Innovate. Research, development, and demonstration are needed to explore the additional cost-effective potential for energy-efficiency improvements.
- Educate. Consumers (including builders and landlords) need to be informed of the technological possibilities available to them in the marketplace when they purchase energy-using equipment, and the economic opportunities for operating-cost savings associated with energy-efficient alternatives. Manufacturers need to be informed of the possibility of changing designs in a profitable manner. Mortgage grantors

Table 5. Effective dates for appliance standards.

On the books (end 1989):	
1990	Furnaces, water heaters, refrigerators, freezers, room A/C Fluorescent light ballasts
1992	Central air conditioners, heat pumps
1993	Refrigerator, freezer (Update)
Under development (1990-91):	
1993	Clothes dryer, clothes washer, dishwasher
1994	Mobile home furnaces
1995	Water heaters (Update), pool heaters, direct heating Room A/C (Update), ranges and ovens (Update), Fluorescent light ballasts (Update)
Coming attractions:	
1998	Refrigerator, freezer (Update) Clothes dryer, clothes washer, dishwasher (Update)
1999	Central air conditioners, heat pumps (Update)
2000	Room air conditioners (Update) Ranges and ovens (Update)
2002	Heat pumps (HSPF) Furnaces (Update)
2005	Water heaters, pool heaters, direct heating (Update)
2006	Central air conditioners, heat pumps (Update)
2012	Furnaces (Update)

should consider energy expenses in determining eligibility for loans.

- Motivate. The key decisionmakers may need some incentives to move toward increased energy efficiency. Electric utilities have been successful in a variety of programs in motivating customers (of all sectors: residential, commercial, and industrial) to conserve energy. Rebates, tax incentives, pricing strategies, and other economic approaches should be explored, experimented with in pilot programs, analyzed, and implemented on a larger scale. Such programs may be applicable to manufacturers, retailers, builders, landlords, other purchasers, and electric and gas utilities.
- Mandate. Depending upon the rate of progress from the measures listed above, additional mandatory elimination of less efficient designs may be necessary. Each building or piece of energy-using equipment put in place now that consumes more energy than necessary to provide a service will continue to contribute to global climate change over its useful life. Limiting potential future damage can be accomplished by hastening the transition to more energy-efficient products.

References

California Energy Commission. "California's Appliance Standards: An Historical Review, Analysis, and Recommendations," Staff Report P400-83-020, July 1983.

McMahon, J.E. "Enactment of Federal Energy Efficiency Standards for Residential Appliances," Lawrence Berkeley Laboratory Report LBL-24064, Berkeley, CA, 1983.

Ruderman, H., M.D. Levine, and J.E. McMahon. "The Behavior of the Market for Energy Efficiency in Residential Appliances and Heating and Cooling Equipment," in *The Energy Journal* 8 (1) 1987.

US Department of Energy. "Monthly Energy Review, December, 1989." (DOE/EIA-0035 (89/12)), 1990.

US Department of Energy. "Technical Support Document: Energy Conservation Standards for Consumer Products: Refrigerators and Furnaces," (DOE/CE-0277), November 1989.

US Department of Energy. "Technical Support Document: Energy Conservation Standards for Consumer Products: Refrigerators, Furnaces, and Television Sets," (DOE/CE-0239), November 1988.

Socially Responsive Buildings: Real-Time Control of Storage HVAC Systems

Roger E. Bohn
Center for Technology, Policy, and Industrial Development
Massachusetts Institute of Technology
Cambridge, MA

Introduction

A microprocessor revolution is making it possible to add digital control to many energy uses that previously were controlled with relatively simple mechanical or pneumatic analog systems. The resulting intelligence allows greater efficiency in the use of electricity and other costly inputs, and also greater effectiveness in meeting the needs of customers. In particular, it is quite common to have systems with sophisticated scheduling and closed loop control for building heating, ventilating, and air conditioning (HVAC) systems. A single integrated building control system can also provide for fire alarms, security, and other functions which are easy to conceive and implement in a digital world, particularly with distributed computation.

Another generation of capabilities is developing for buildings, as well as larger electricity users such as factories. This is the shift from purely internal feedback loops toward outward-looking buildings that communicate and respond to the physical and societal environment around the building. In order to make such systems work effectively, several requirements must be met. First is the use of systems which give the building managers and occupants incentives to respond to overall social good, not just to their own needs. Second is the availability of real time data on price and weather conditions. Third is the use of forecasts of future conditions, so that the building can respond intelligently to anticipated events, not just reactively. All of these requirements are achievable today for a specific application, the storage heating and storage cooling of buildings.

This paper discusses one experiment involving storage heating of three buildings and storage cooling of another building, in response to real-time prices of electricity. This system is designed to work in a completely unattended fashion. Price and weather information is gathered electronically, the building status is sensed, and cost-minimizing storage plans for the building are calculated and put into effect, all under computer control. The additional computer technology required beyond the existing building control system is an IBM-AT class computer, and a modem for communications. A few additional sensors are required to measure the status and performance of the HVAC system in more detail than was formerly required. Total hardware cost of the system is on the order of $1,000.

Electricity-based storage cooling and, to a lesser extent, storage heating are becoming well established technologies in the United States and Europe. The basic idea is that although electricity is expensive to store in its initial form, it is often quite inexpensive to store *after* conversion to the final energy form. Thus large amounts of electricity can be used during the night to heat or cool a volume of water (or ice). This water can then be used during the day to condition the building spaces. Since electricity is more expensive during the day, the result is considerable reduction in cost, corresponding to a reduction in the societal fuel inputs needed to generate the electricity (Science Applications International Corp., 1989).

Considerable research and commercial development are being done on alternate physical technologies for storage cooling (e.g., ice versus water) and heating. In contrast, the control technology (i.e., the algorithms used to decide when and how to charge and discharge the storage systems) have received much less attention. Existing control systems for storage HVAC are based on two drastic simplifications. First, a forecast of tomorrow's building load is needed in order to decide whether to fully charge the storage system or only partially charge it. This is provided in existing systems by very simple mechanisms such as looking at current outdoor air temperature, which is then used as a crude forecast of tomorrow's building thermal load. The second requirement is a response rule to minimize the cost of electricity consumption. Existing control systems use the existing utility rate structures, which are optimal for the building owner, but as we will see is not socially optimal since these rate structures do not reflect the true day-to-day and hour-to-hour variations in the cost of electricity.

Our goal was to retrofit a much more sophisticated control system onto an existing hardware system. Preliminary results suggest that it may be possible to achieve a 50 percent improvement in the benefits of storage heating by using a socially responsive control system as will be described. We expect that results for storage cooling will be at least this good. Because the storage heating experiment has been completed and data analysis is in progress, the rest of this article will deal primarily with storage heating. However, storage cooling is a more economically promising technology since air conditioning is almost always electrically driven.

The next section of this paper will discuss the concept of electricity spot prices, which are a social signaling system. Basically the hour-by-hour cost of electricity reflects the true opportunity cost to society (marginal cost of generation to the utility) for that hour. This leads to highly variable prices with numerous opportunities to save money through time shifting of electrical loads. The following section describes the system design used in the storage heating experiment. The fourth section will discuss results to date, and the final section will discuss extensions and conclusions. We will discuss both the use of storage heating and cooling for air conditioning, and the possibility of extending the concept of socially responsive buildings to other applications besides electricity use.

Spot Pricing of Electricity for Control of Storage Devices

The concept of spot pricing was developed to reflect the true opportunity costs of electricity generation and use. The incremental cost of generation varies in real time over a day, a week, and a year based on a number of factors, many of which change randomly over time. The short-run marginal cost of the next kilowatt hour of electricity generated (and consumed) depends on what generating unit is on the margin, its fuel type and efficiency, and on transmission losses. The underlying principles of spot pricing are documented in Schweppe et al. (1988). Basic calculation of spot prices is comparatively straightforward for utilities that are members of pools that trade within the pool on an hourly basis. Thus the hourly price of inter-utility pool transactions (often referred to as "economy A") can be used as an estimate of the hourly spot price.[1]

Our experiment was conducted for customers of New York State Electric & Gas Company (NYSEG). NYSEG is a member of the New York power pool, and we used the pool-to-NYSEG price on economy interchanges of electricity as the starting point for our spot price. This is a true spot price in that it changes in real time (it is a "real-time price") and it is based on a marginal cost calculation.

The resulting prices are driven by changes in demand and generator availability, both within NYSEG and for other members of the New York power pool. Figures 1-1 and 1-2 show four weeks of the prices we calculated.[2] Although some days follow a regular and somewhat predictable pattern, other days do not. Thus weekdays are not always cheaper than weekends, and even in the middle of the night it is occasionally possible to have high spot prices. Ratios of 2:1 between highest and lowest prices in a day are normal.

We will not justify the particular way in which these prices were calculated (see the reference for further discussion). The point is simply that there does exist a set of prices that change in real time and condense and internalize all the relevant social costs of electricity used in that hour. These relevant costs include the fuel choice of generators, transmission losses, curtailment premia if some customers are about to be browned out or blacked out, and, in principle, even pollution effects. For example, on a polluted day, if electrical demand rises so high that a utility has to bring on some generators which are heavy emitters, this could be reflected in a price premium for those generators leading to a higher spot price during those hours.

Under conventional control of storage heating or cooling, customers are not even aware of these hourly spot prices, and if they were aware, they would have no incentive to respond accordingly. Instead, the usual rate structure for storage systems is a time-of-day rate structure with an unvarying price pattern. For NYSEG, the rate structure consisted of high prices from 7:00 am until 10:00 pm every weekday, and low prices on weekends and at night during the week. Thus the cost minimizing response of the building operator would be to fill up storage by using electricity after 10:00 pm, and on weekends.[3] In contrast, the socially optimal response, and the privately optimal response under spot pricing, would be a more sophisticated attempt to exploit the lowest cost hours shown on Figures 1-1 and 1-2. The ideal strategy is to charge (fill) storage as fast as possible during those hours, and not to use electricity at other times. This leads to storage from one day to the next, and different amounts of charging on different nights even if the weather forecast is the same.

The basic principle for socially optimal response is that the building is responding to two sets of forces: social costs as reflected in the spot prices charged for electricity, and private needs/opportunities as they affect the building load. The utility cannot easily control the storage for the customer because daily customer requirements can vary depending on occupancy, scheduling of holidays, physical performance of the storage system, and other building specific and not predetermined factors. Thus the computational approach we took was a distributed one, where the utility calculates the spot price, and broadcasts that price to the customers, whose computers then sense and optimize the storage response of their individual buildings. For convenience in the experiment we actually implemented the optimization for three buildings on a single PC located in Cambridge, Massachusetts.[4]

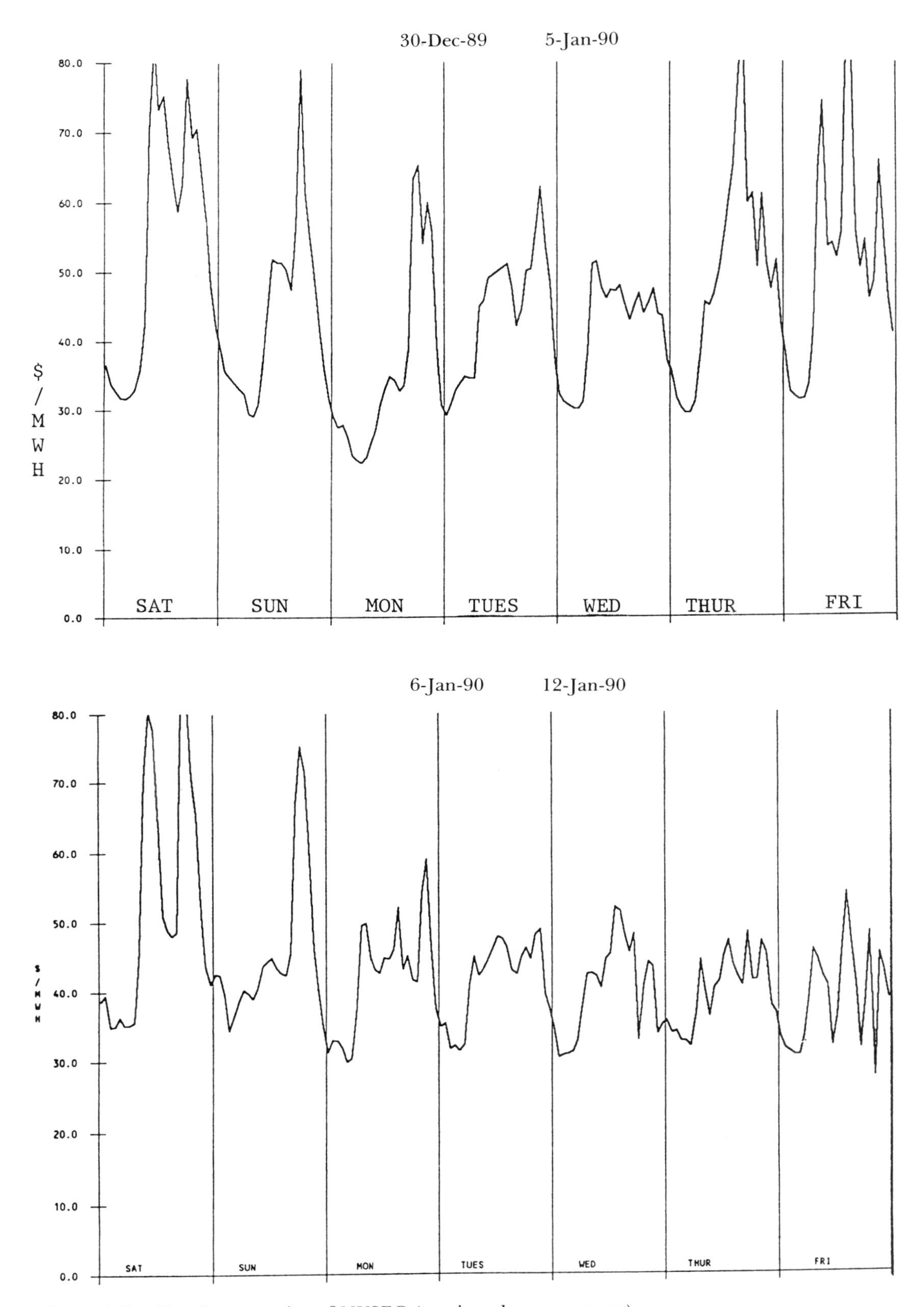

Figure 1-1. Hourly spot price of NYSEG (continued on next page).

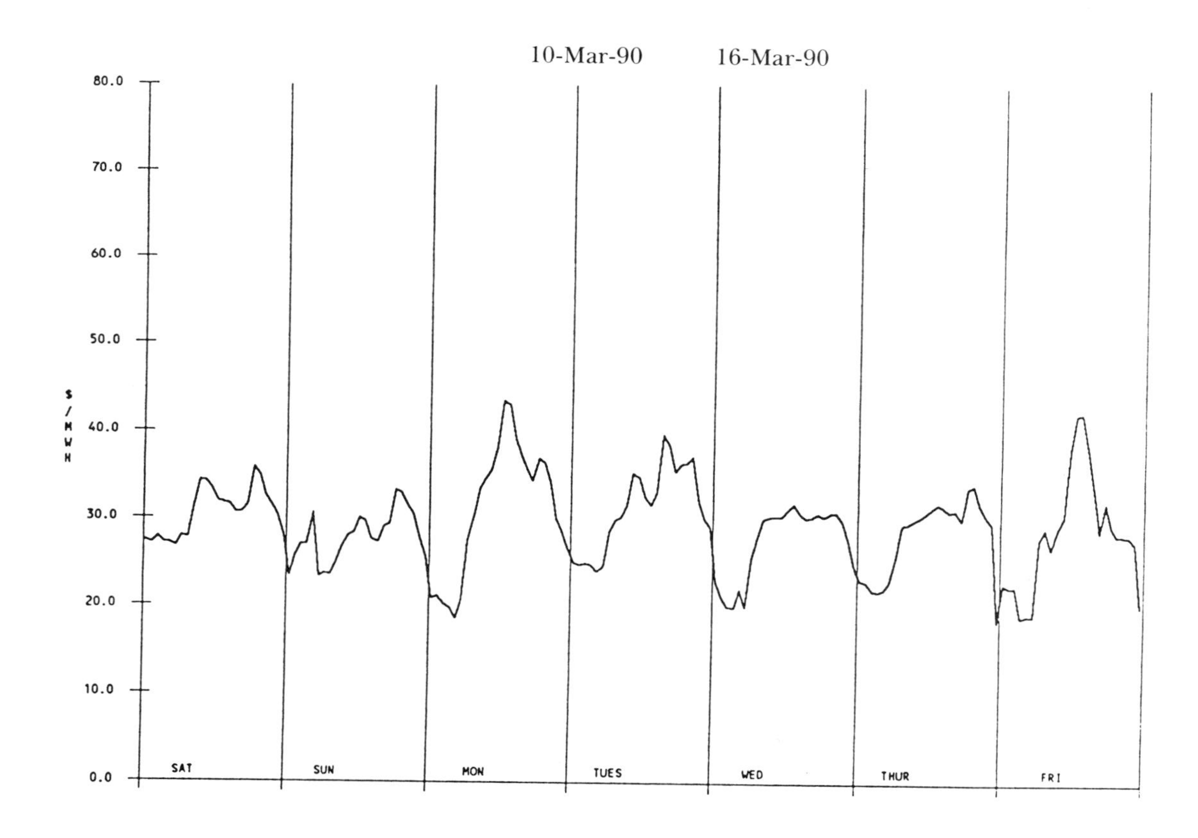

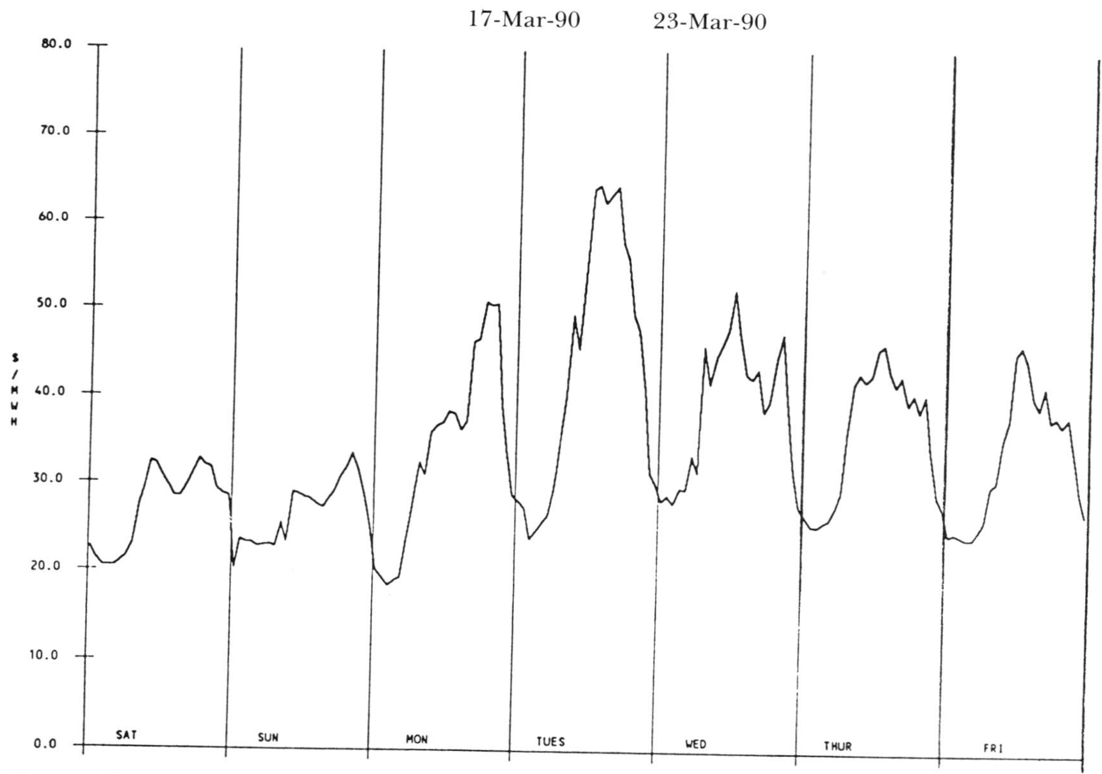

Figure 1-2. Hourly spot price of NYSEG.

System Design

The system architecture is shown in Figure 2. The spot price response system is designed as a retrofit add-on to the existing building control system (shown in the lower right area). The only function that changes is the timing of the storage decisions. The space conditioning temperature profile and demands made by the occupants are sensed by the system but are not altered in any way. Instead, an hourly schedule of storage actions is downloaded into the building control system.[5] This hourly schedule tells the HVAC system when to activate the storage charging (i.e., the consumption of electricity). The existing conventional control system decides on its own when to use the stored energy to heat the building, in response to conventional thermostatic control or time clock control of a night setback thermostat.

Calculation of the hourly schedule is done by a personal computer running several software programs. This computer receives building weather data, weather forecasts, current prices, and price forecasts via modem. These come from a commercial service for the weather forecasts, and from the NYSEG utility control room for the price information. In addition, the central computer receives historical data from the building about what actually happened over the previous day. This is combined with the weather information to calculate a building load forecast for the next three days.

The load forecast and the price forecast are then used in a two-step procedure to calculate storage timing. First, a deterministic linear program is solved to calculate an optimal schedule, assuming the price forecast and load forecast are completely accurate. The nonsimplex algorithm takes advantage of special features of the problem and is much faster than a standard linear programming (Daryanian, Bohn, et al., 1989). However, the load forecast and especially the price forecasts can

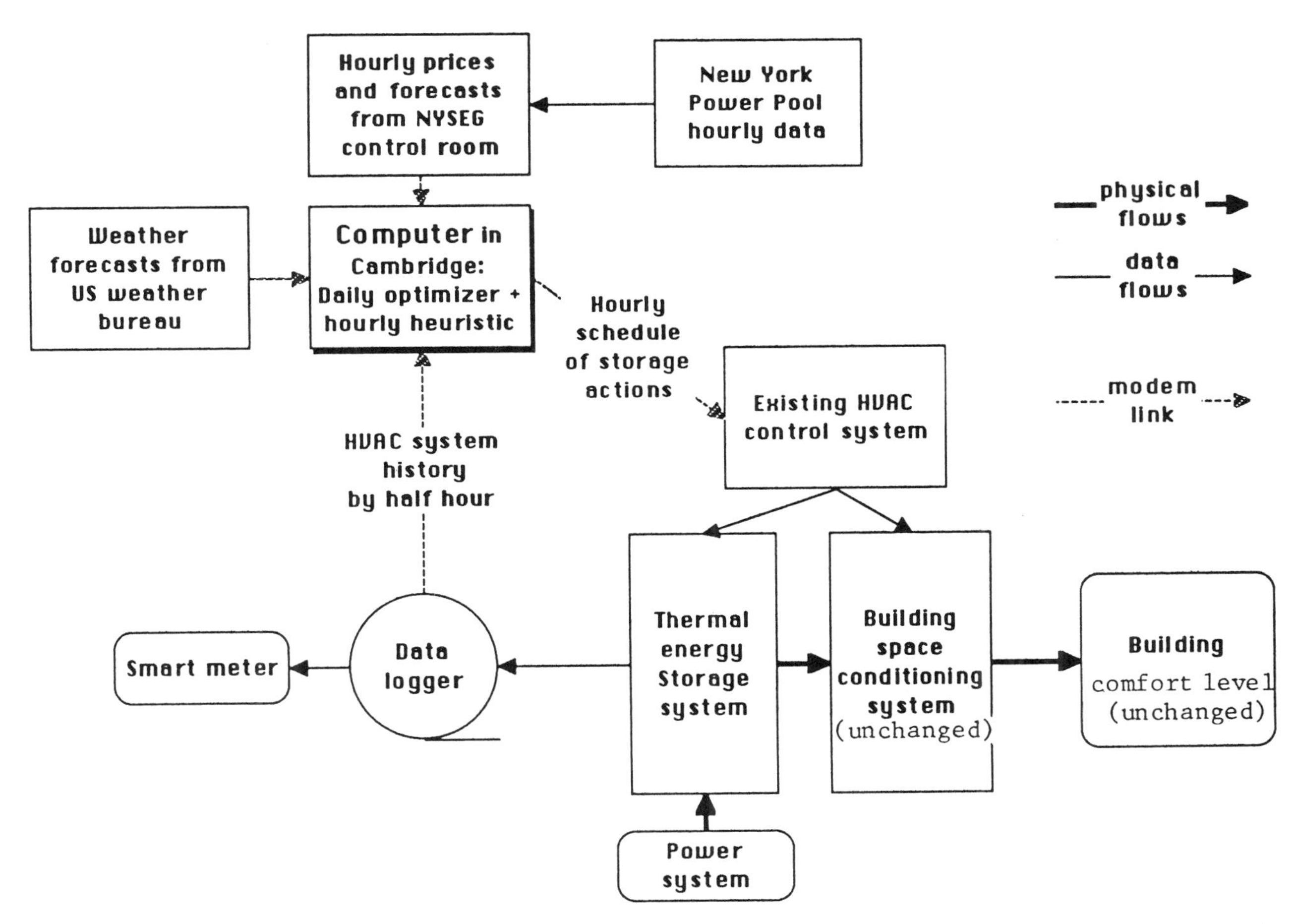

Figure 2. Spot pricing responsive storage: control system architecture (responsive storage architecture, Wednesday, April 25, 1990, 11:12).

never be completely accurate. Therefore, a second stage heuristic procedure is used to determine, hour by hour, when charging should be done. If the forecasts are perfect, then the heuristic simply gives the same results as the deterministic optimization. More often, it reacts to unexpected changes (either increases or decreases) in the prices and loads. It does this by calculating a threshold price and comparing it with the actual hourly price.

The system is designed to be fail-safe in several respects. For example, if telephone communication with the building is lost, the building has preprogrammed instructions covering the next 18 hours. Thereafter it reverts to conventional (time-of-day-based) control. If price and weather forecasts are not available, crude forecasts are done based on average historical information. We found that a 48 hour forecast, updated at noon each day, was adequate.

Behavior and Results

This architecture was applied to three heat storage systems during the winter of 1989-90. Two of them used hot water tanks and were located in Brewster, New York. One used earth storage and was located in Plattsburg, New York. One of the water storage systems had a sophisticated existing energy management system to control the building, while the other was a simple thermostat-based unit. The earth storage system has limited degrees of freedom, since the storage medium is the ground. Once energy is stored in the ground, it inexorably moves upwards into the building. It will not be discussed further.

The two water storage systems had independent storage and heating loops, so that energy could be put in under the spot pricing system control, while hot water was drawn under the control of the building's pre-existing heating system. For this experiment, the physical characteristics of the buildings and their storage systems were not altered. Only the control of storage was changed.

Figure 3 shows the behavior of the larger (pressurized) water storage system over a one-week period. The diamond line on the top is the temperature of the water in the storage tank, shown on the right axis. The first day shown is Saturday. The highest temperature was reached on Sunday night, at approximately 245°F. The lowest temperature was reached Wednesday night at approximately 170°F. During each night, the system was recharged using electricity, giving the temperature rise. Thus the tank temperature responds visibly to the amount of charging done the night before.

Actual charging rates each hour are shown by the thin lines. How were the decisions made about charging each night? The dark triangles show the actual spot prices as they unfolded during the week. This information was available only after the fact; the system had to operate based on one-hour-ahead price forecasts, rather than actual price. Despite this, the performance was good. Basically, the system was trying to find the cheapest hours each night and charge during those hours. For example, on Saturday and Sunday it waited until the early morning hours for the lowest prices, rather than charging immediately at 10:00 pm as it would have done under time of day control. The system also acted to optimize as much as possible across days. Thus prices on Sunday night were several mills higher than on Saturday night, and the number of hours of charging was notably less on Sunday night. Similarly there were four low-priced hours Tuesday night, but the other hours were much higher in price. Therefore the system only charged for a short period on Tuesday night, although it did not pick exactly the lowest hours. This led to the decline in tank temperature to its lowest level on Wednesday, followed by a build-up back to 220°F by early Thursday morning.

Careful examination of the diagram will reveal that the actual charging schedule, although quite good, was not precisely optimal. This resulted because the hour-ahead price forecasts, and especially the day-ahead price forecasts, were not completely accurate. Therefore the system occasionally charged during hours which in fact were slightly more expensive than necessary. However, the discrepancies were generally only one to two mills, and as we will see had only a moderate effect on overall economic performance.

How would a conventional time-of-use-based (TOU) control system have done for the same week? Although the data are not shown on the graph, we simulated such a system for each week of the experiment. The logic in use would have turned on the storage charging at 10:00 pm each week night. On Friday night, the system would have begun charging and would have charged through Saturday until the tank temperature reached near its maximum. A very simple system called "outdoor temperature reset" was used to determine when the tank was charged enough under conventional controls. Judging from the times of low prices in Figure 3, this system would have performed reasonably well, except for the early morning price spike on Wednesday. At that time the price went to 10 cents per kilowatt hour (100 mills) at 6 am. This hour might have been used under conventional TOU control. Another impact of TOU control would be the loss of the ability to make tradeoffs from one day to the next. As we have seen, these tradeoffs were not perfect because forecasts are not perfect, but any level of look-ahead is better than no look-ahead.

The experiment ran in January, February, and March 1990. During that interval, the large Brewster

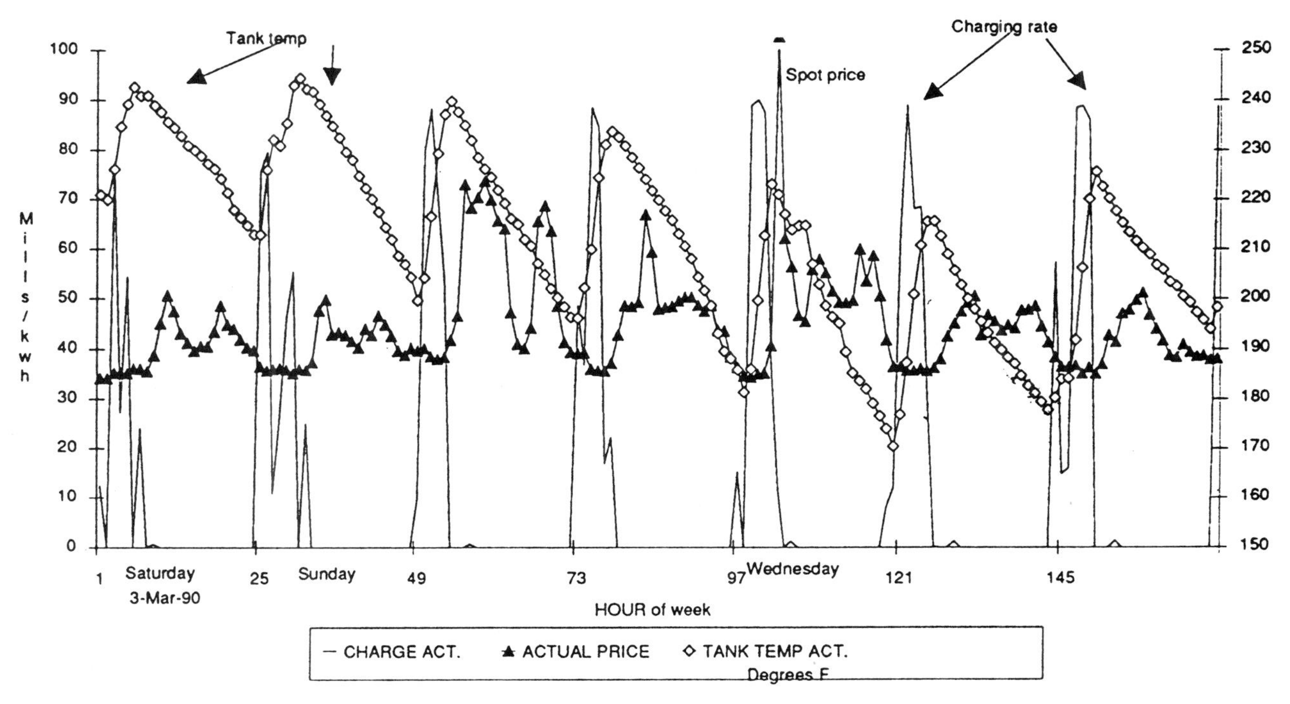

Figure 3. Performance of Brewster large site, week 9.

building used 112,000 kilowatt hours of heating. We simulated what would have happened if there had been no storage system, or if the existing storage system had been under time-of-use control. With no storage system, the electricity would have been used at the same moment as demand. The total social costs in the no-storage case would have been $3,979.[6] With conventional TOU-based control of the existing storage, the cost would have fallen to $3,636, an 8.6 percent savings. Under the existing spot-price-based control system which we used, the cost fell an additional 6.9 percent, to $3,360. Thus spot-price-based control led to an 80 percent increase in the savings (from $343 to $619). Considering that the cost of the control system is $1,000 to $2,000, while the cost of the storage system hardware itself was several tens of thousands of dollars, this is a very favorable ratio. We estimate that if perfect one-hour-ahead price forecasts had been available, the savings would have risen by an additional 3 percent, bringing the total cost down to about $3,140.

The results for the smaller water storage building were not as impressive. The storage system itself under conventional TOU-based control would have reduced costs by about 20 percent of a much smaller base. The addition of spot-price-based responsive control saved approximately an additional 3 percent, for a total savings of 23 percent of the original cost. Notice the importance of scale here; the cost of communication and control is approximately independent of the building size, and only slightly dependent on the building complexity. On the other hand, the *savings* are directly proportional to the building size. Therefore, this kind of system makes more sense for large buildings than small commercial buildings. Implementation in the residential sector raises a number of cost benefit questions, since to be cost effective the control system would have to be mass produced and designed right into the storage system from the beginning.

We are now beginning implementation of a similar system to control an ice storage cooling system in an office building. Cooling will require a more sophisticated algorithm and control system, because decisions must be made about the time at which to draw down the stored ice, and when to cool the building using the existing chillers. Thus, the building has a four-state control system: idle; charging the storage system during the night; cooling during the day with chiller priority; and cooling during the day with storage priority. Our system will calculate the optimal times to be in each of these four modes, and download the schedule to the building. For storage cooling applications we expect larger savings than for storage heating, because there are additional ways in which to save. The first is to optimally time the storage charging (ice making) to "fill in the valleys" of the price curves in Figure 1. In addition there is opportunity to save by timing the optimal discharge of the ice storage so as to "shave off the peaks" of the price curves. In a lightly loaded building (or one with an oversized storage system) the timing of such peak shaving may be critical only on very hot days. In a more heavily loaded building or one with a smaller storage system, there will be considerably more benefits from spot-price-based control, since the optimal timing will be an issue on many days of the summer.

Conclusions

We have seen that it is possible to control an existing building using a more socially beneficial response technology, based on continuous hourly communication between the building and the environment. When the application is storage heating and cooling, the relevant "environmental" variables are electricity cost and weather. In this application it is possible to design and operate a system that is completely transparent to the building occupant so that they do not know when the system is operating, yet it still operates in a both socially and privately efficient manner.

Spot electricity prices are a critical link between the social goals and private cost minimization. Such a system has several advantages over conventional control of storage based on fixed price schedules. Not only does it have operating cost advantages, which we have emphasized here, but it also gives building owners incentives to maintain and upgrade their system over time, in order to increase the benefits of the system. Some other demand management incentives, in contrast, motivate customers to skip maintenance.

Other incentive schemes for controlling storage cooling and heating may also give building owners incentives to perform expensive changes to the system to reduce their costs, without much affecting the social welfare. For example, rates with heavy demand charges lead to the installation of sophisticated systems for demand management of an individual building. At the level of the overall utility there may be little value in such demand management because load diversity randomizes the unsystematic demands of the buildings, while the coincidence of weather in a region ensures that the utility will still see the full air conditioning load on the hottest days. Under spot pricing there is no incentive for the individual to make an investment unless it is also socially beneficial, and there is no private benefit unless the system continues to operate effectively.

It is interesting to speculate on the use of socially responsive control systems for products other than electricity. Natural gas prices and oil/gas/electric fuel switching are obvious and to an extent ongoing areas for this concept. Many industrial loads and industrial cogeneration can benefit from responding to spot electricity prices. Going farther afield, it is conceivable that pollution indices could be used to provide daily or even hourly signals to alter building and occupant behavior. For example parking prices could have a $20 premium on days with temperature inversions. Again, the key issue is whether it is socially useful to affect behavior on average over long periods, or better to influence behavior in response to actual conditions.

Once such systems are in operation for a few customers, there is no reason to limit them. The cost of creating the hourly prices and price forecasts and of communicating them are basically fixed costs. The cost of developing response algorithms can be spread over a large number of customers. If customers, or a class of customers, are willing to pay the increased cost of metering and communications hardware, there is no obvious reason to avoid encouraging them to use spot prices and to respond in a socially beneficial manner.

Notes

1. This bypasses a number of issues such as revenue reconciliation requirements, the difference between pool and utility marginal cost, and so on. However, it does pick up the dominant hourly and stochastic determinants of "true" spot prices and is based on a well developed and accepted arrangement among the respective utilities. Thus although it might require modification before use as the basis of a widespread rate structure, it is quite appropriate for initial experimentation with spot prices.
2. The major adjustment to the price was to allow for distribution system losses. These add approximately 9 to 16 percent to the price, and tend to be higher at times of higher spot prices.
3. The conventional price usually includes both energy and a demand charge. The demand charge is paid only for use during the peak period, further encouraging the customer into a rigid time pattern of response. For an economic projection of a one million dollar California cool storage system, see Bryant (1990).
4. In principle the utility could sell computational services to building owners, even though the building owners retain autonomy and the final decisionmaking on how and whether they choose to respond. The building owners would tell the central utility their criteria and set up real-time sensing to communicate the necessary information. This will become clearer as we look at the specific architecture.
5. This is an oversimplification. For the purposes of this experiment, a stand-alone energy management unit was put in place to turn on and off the storage charging electrical elements.
6. This storage system was well insulated, so we have ignored storage losses in this calculation.

References

Bryant, Frank, "Medical Center Installs $1M Eutectic Salt TES," *Energy User News,* p. 14 (March 1990).

Daryanian, B., Bohn, R., et al., "Optimal Demand-Side Response to Electricity Spot Prices for Storage-Type Customers," *IEEE Transactions on Power Systems,* 4(3), pp. 897–903 (August 1989).

Schweppe, F., Caramanis, M. et al. *Spot Pricing of Electricity,* Kluwer Academic Publishers, Hingham, MA (1988).

Science Applications International Corp., "Operation and Performance of Commercial Cool Storage Systems," EPRI Report CU-6561 (1989).

Building Energy Conservation in the 1990s: Barriers and Remedies

Daniel H. Nall
Jones, Nall & Davis
Atlanta, GA

Abstract

Energy conservation progress in the buildings sector begun during the last energy crisis has stalled during the 1980s. The two major reasons for this lack of progress are more energy-intensive occupancy activities and a breakdown of market mechanisms for recapturing capital expenditures for energy conservation. Two remedies for the above situations offer some hope for avoiding major electric generation capacity shortages in the next decade. The first of these is regulation to impose building and building system efficiencies that cannot be realized through the marketplace. The second is variable spot pricing of electricity that will cause the marketplace to reschedule energy intensive activities to less expensive off-peak periods. Because energy consumption in the buildings sector is growing rapidly, these two remedies will help avoid new electrical generation capacity and the environmental problems that accompany it.

Introduction

It is now almost 10 years since the change in federal administration began to dismantle the massive 1970s energy conservation initiative. In that time, oil imports have climbed back to 50 percent of US domestic consumption, and brownouts have once again become news items. Although oil prices per barrel have remained low, electricity prices, spurred by the immense capital cost of new capacity, have risen in current if not in real dollars since the beginning of the decade. The costs of environmental protection and cleanup threaten to drive electricity prices even higher.

With the prospect of higher energy prices, one would assume that energy conservation would again arise as a major concern in the building industry. Building energy consumption, however, is not particularly sensitive to price. Consumption, therefore, is on the rise and can be expected to continue to rise until halted by the curtailments that will probably be imposed in the next energy crunch.

The reasons for this price insensitivity and the increasing consumption are diverse, but can be categorized into two general areas:

1. The changing characteristics of occupancy, and
2. Institutional barriers to marketplace-driven conservation.

Barriers to Conservation

The first area incorporates a number of technological and economic trends that have intensified energy consumption in most building-housed activities. Homes contain many more energy-consuming appliances than previously. Retail establishments feature much more energy intensive lighting and display techniques. Engineering designers have been forced by the marketplace to increase by 50 percent their design lighting power allowances for retail rental space in the last decade. The office automation trend, particularly, has dramatically increased the energy intensity of offices. The personal computer, the laser printer, the fax machine, lower priced/higher output copiers, and the digital telephone switch have all contributed to office energy consumption. According to the Landmarks Group, an Atlanta based southeastern developer, the average 1980 office tenant had an average peak receptacle load of approximately 0.75 watts per usable square foot and an average sustained consumption of about 0.5 per square foot. These numbers have more than doubled in the past 10 years. The current standard allowance for Class A office buildings in Atlanta is for a sustained receptacle load of two watts per usable square foot. Some of this load, furthermore, has become 24-hour load. Cooling tower water loops for auxiliary cooling are now a requirement for Class A office buildings whereas they were unheard of in 1980. Larry Spielvogel, a prominent energy consultant, has spoken of an air conditioning capacity crisis as today's office buildings begin to experience 1990s occupancies. Current standards for air conditioning capacity will be inadequate for the increasing power intensity of office equipment during the next decade, he says.

Institutional occupancies have experienced the same sort of increase in energy intensity. Energy-consuming medical equipment has proliferated and environmental standards have become more exacting. Libraries and schools now feature personal computers and videocassette players in addition to microfiche readers. Institutional and even industrial facilities are often more elaborate. The swimming pool has become the natatorium and features full environmental conditioning and dehumidification. The heated and ventilated workshop has become the Class 10 clean room.

In the face of the increased energy intensity of occupancies, the marketplace has not driven buildings and their installed systems to higher efficiencies for one major reason: utility bill payers are not making capital decisions affecting energy efficiency. This trend is pervasive throughout commercial and residential development. The typical apartment development, for example, utilizes energy code minimum equipment and insulation levels because the tenant pays the energy bills. The developer is driven to moving costs from areas that can't be seen to those that can in order to attract more tenants. Leasability is the primary consideration for rental property and a dollar spent for landscaping or kitchen cabinetry has a much more positive impact on leasability than a dollar spent for energy conservation.

Residential HVAC (heating, ventilation, and air conditioning) and water heater manufacturers typically market three or four efficiency levels for their equipment. The first level meets the prevailing code minimums. The second level is only very slightly better and meets typical utility energy conservation standards such as "Good Cents" or "Energy Wise" programs. These two levels represent the preponderance of equipment sold. Efficiency levels higher than these are typically sold only to special interest builders or to homeowners who are upgrading as they replace existing equipment.

Even in houses and condominiums built for sale, a builder measures the marketability of energy conservation measures against that of richer levels of finish or a whirlpool tub for incorporation within his limited construction budget. Even luxury speculative housing typically features "Good Cents" or "Energy Wise" minimum standards of efficiency, a level of efficiency much less than current state of the art and much less than the demonstrated lowest life cycle cost level of efficiency. Energy conservation measures beyond those required for the utility standards programs mentioned above are simply not cost effective from a marketing point of view when compared with more visible improvements. Acquisition of this "energy conserving" label from the utility gives the maximum marketing benefit for a low first-cost premium. Unfortunately, it represents only a minor efficiency improvement over standard practice.

Similarly in office buildings, most tenant leases are designed so that energy costs above an initial set maximum are directly transferred to the tenant. The increased energy intensity of office occupancy and its greater variability among tenants recently has made almost impossible the design of leasing instruments that can recapture the developer's increased capital costs for energy conservation. At least one southeastern developer, in the early 1980s, experimented with leasing instruments which returned a share of the energy savings estimated from his conservation investments. Unfortunately, the tenants in his first building utilizing these instruments were companies in the vanguard of office automation. As a result, this owner found these leasing arrangements were transferring the increased energy costs of the tenant's high tech office equipment back to him rather than transferring energy conservation savings.

Retail tenants, furthermore, are usually individually metered for utilities, even while decisions about their building space are made by developers. Most retail developments are single story, utilizing packaged rooftop HVAC equipment, a condition which would seem ideal for the utilization of airside economy systems to reduce cooling energy during cool weather. In fact, most smaller retail developments do not include this measure despite the fact that its capital cost can be as low as 10¢ per square foot with a payback of less than two years. The reason that this conservation opportunity is lost is that the developer who provides the equipment cannot recapture his capital because all of the savings accrue to the tenant who pays the utility bills.

Institutional owners, on the other hand, are much more conscious of energy costs because they both control their investments in their physical plant and pay their own operating costs. Design team selections for state and municipal buildings are often very sensitive to demonstrated energy conservation capabilities. Because many institutions, both educational and health care, perform their own maintenance, they are very sophisticated consumers and are often more knowledgeable than designers about the long-term efficacy of conservation measures. The Institutional Conservation Program funded by the US Department of Energy with oil overcharge monies has been very effective in inducing institutions to start or to improve conservation programs. For the institutional owner, energy waste is generated as often by inoperative or malfunctioning equipment as it is by purely inefficient equipment. Through this program many institutional owners have been able to increase their energy efficiency at the same time they are ridding themselves of maintenance headaches. The effectiveness of this program is based on its amplification of existing functional market mechanisms that allow recapture of capital expenditures for reduction in operating costs. Unfortunately, this mechanism is not operative throughout most of the buildings sector.

Remedies to Conservation Barriers

Because of this breakdown in the marketplace, regulation has become a much more important stimulus for conservation than cost savings. The single most effective instrument for increasing conservation in buildings during the 1980s has been the energy code. Even though the currently most widely enacted model energy code is not sufficiently stringent to prevent a floor-to-ceiling tinted glass office building, it has placed limits on what can be designed and constructed. Energy code compliance has become an issue in building design. The most effective limitations regard envelope standards. Insulation is pervasive and floor-to-ceiling clear glass is never seen unless protected by shading devices. The current trend for energy standards is toward more flexible formats, which, while more demanding of efficiency, are more flexible in the means by which the building may comply. If ASHRAE (American Society of Heat, Refrigeration, and Air Conditioning Engineers) Standard 90.1 becomes widely recognized by building code authorities, it will have a dramatic effect on new building energy efficiency. Increasing rigor in codes is particularly unpopular in the development community because of the increased capital cost entailed by the mandated efficiency improvements. The business community generally forecasts doom in the face of more restrictive codes. The issue that should be considered, however, is the quality of our capital stock of buildings. Will our country be better off with 20 energy inefficient buildings constructed for the same cost as 19 highly efficient ones? In the absence of market mechanisms to insure that efficient buildings are built, energy codes are our best bet for controlling the quality of our capital investment for buildings.

The argument that regulation is a barrier to innovation and to new technologies has been disproven by the long-term record of conservation techniques. The techniques utilized to achieve enhanced levels of efficiency mandated by codes or demanded by sophisticated institutional owners have not been those that achieved the limelight of the 1970s. Conservation has mainly been achieved by evolutionary means rather than revolutionary means. Solar projects are very few today. The southeast's most prominent solar project, in the Georgia Power corporate headquarters, has been decommissioned. This project failed because a number of compounding optimistic assumptions forecast a performance that could never be realized. Even though each assumption was only slightly optimistic—atmospheric turbidity, pipe fouling, long-term reflectivity of the concentrator, etc.—the combined effect was disastrous. In fact, the collectors could rarely generate water hot enough to run the absorption chiller, and the maintenance and operations requirements were too complex for the average maintenance staff. Similarly, most daylight-sensitive dimmers have been deactivated as have most active shading devices because of maintenance problems and perceived tenant inconvenience. Energy efficiency has instead come in the form of the tri-phosphor T-8 fluorescent lamp, the solid state ballast, the fluorescent downlight, the solid state variable speed motor controller, the digital control system with solid state sensors and sophisticated logic, low emittance glass coatings, closed cell foam insulations, and other extensions of existing technology. One of the most important energy conservation measures for office buildings is the unheralded and now ubiquitous fan-powered terminal box. This terminal allows a zone to obtain heating and ventilation without reheating previously cooled air by utilizing return air from other zones. Current technologies to be improved include refrigeration cycles utilizing non-CFC refrigerants, multi-state glazing films, vacuum insulation, low cost occupancy-sensing light switching, and more sophisticated controls. These technologies can be easily incorporated into building codes.

Unfortunately, these technologies do not offer solutions for the changing occupancy characteristics that will likely cause electric generating capacity shortages in some areas during the 1990s. How, then, will buildings, businesses, and residences cope with this peak power limitation? The precedent is already being set throughout the East Coast with electric utilities offering incentives for peak demand reduction, either through automatic disengagement of devices during peak demand periods or by utilization of alternative sources during declared demand events. Georgia Power Company, for example, offers what is effectively an interruptible rate, which reduces the billing demand of a client by the amount of load they can shed when called upon by the utility. In the face of increasingly energy intensive occupancies and the resulting increase in cooling requirements, the only alternatives to increased electrical generating capacity will be load displacement, achieved either through mechanical systems, local generation, or revision of the daily business schedule.

One prototype of such a load displacement mechanism is the spot-pricing rate developed by the Georgia Power Company for use in its Transtext System. In this system, demonstrated in several hundred residences near Atlanta, the homeowner inputs his comfort elasticities to a processor in the form of desired temperatures at various costs of power. The utility sets hourly electricity rates from a previously approved schedule of rates based upon its marginal costs of generation. These hourly rates are forecast by the utility 24 hours in advance, with the exception that a system peak-demand crisis rate may be scheduled with only one hour's notice. Algorithms in the processor not only reset thermostat settings based on the current and upcoming rates, but also anticipate rate changes by preheating or pre-cooling to minimize energy consumption during high rate periods. Enhancement to the basic Transtext processor would allow the user to schedule certain discretionary energy consuming activities, such as electric clothes drying, during periods of low rates. This system offers a much more user-friendly alternative to the rebated duty cyclers for residential air conditioners and water heaters offered by some utilities.

While the system peak-demand impact of a residential variable spot-pricing control system might be small, the extension of such a system to commercial and industrial customers could be much more dramatic. Controllers on intermittently used devices could lock out such devices or make them available only on a priority basis during peak load hours. Lighting in perimeter offices could be automatically dimmed and the thermal intertia of spaces could be automatically exploited by referencing and anticipating this broadcast spot-pricing rate schedule. Through the use of this scheme, the user is made to pay for its impact on the utilities' peak demand. In effect the user pays for the increased generating capacity it is requiring.

Conclusion

The energy crisis of the 1990s will probably affect the electric utilities far more than the crisis of the 1970s. Little increased generation capacity is in the pipeline and our appetite for electrical enhancements to business, industry, and domestic life has dramatically increased. New electrical generating capacity, furthermore, has significant environmental risks, whether the fuel source is fossil or nuclear. Although energy codes and regulations have been and will be very effective at increasing the efficiency of buildings and building systems, they have little or no effect on energy intensive activities in the buildings. Occupancy loads, furthermore, will increase far more than can be offset by improvements to envelope and system performance. Coordination of individual electric demands and relocation of demands to off-peak demands to reduce system demand is the only feasible short-term answer to the generation capacity crisis. Communication between the utility and the user and a system of incentives is the most available system for coordinating our usage to avoid increased system demand.

Government Incentives for Improving Building Performance and Developing New Building Technologies

John P. Millhone
Office of Buildings and Community Systems
US Department of Energy
Washington, DC

Kevin Y. Teichman
Office of Research and Development
US Environmental Protection Agency
Washington, DC

Abstract

Government actions that can promote improved building design and performance and accelerate the development and commercialization of new technologies fall into seven strategies. These include: developing test procedures or consensus protocols; labeling products, building components or buildings; establishing standards; providing financial incentives; disseminating information and promoting education; exercising taxation, import, and other options; and conducting research, development, and demonstration activities. This paper provides examples of how the US Department of Energy and the Environmental Protection Agency use these strategies to promote both energy efficiency and indoor air quality.

Introduction

Governments encourage improved building design and performance and the commercial development of new building technologies by using their broad and pervasive authority to create an environment in which these enhancements are recognized, encouraged, and diffused rapidly into the domestic and international market.

The US Department of Energy (DOE), through new incentives and the National Energy Strategy, is seeking to accelerate the commercial development of new energy-efficient building technologies. The government recognizes that these technologies contribute to five major national objectives: saving finite energy supplies, reducing the cost of energy services, improving national security, improving US competitiveness, and reducing outdoor air pollution, including reducing acid rain, global climate change, and stratospheric ozone depletion.

At the same time, DOE and the US Environmental Protection Agency (EPA), in coordination with other Federal agencies, conduct research programs devoted to indoor air pollution. The objectives of these efforts include studying the relationship between energy conservation and indoor air quality and promoting the development of new technologies that optimize these building needs.

The government actions that can promote improved building design and performance and accelerate the development and commercialization of new technologies fall into seven strategies. These include:

1. Developing test procedures or consensus protocols;
2. Labeling products, building components or buildings;
3. Establishing standards;
4. Providing financial incentives;
5. Disseminating information and promoting education;
6. Exercising taxation, import, and other options; and
7. Conducting research, development, and demonstration activities.

In some cases, as described below, the private sector has voluntarily chosen to implement one, or a combination, of these strategies. This is especially true for strategies that promote indoor air quality.

The Seven Strategies

Test Procedures and Consensus Protocols

The development of test procedures or consensus protocols is an essential first step toward measuring the energy efficiency of products, components, or buildings themselves. Without some acceptable yardstick for measuring energy efficiency, there would be uncertainty about the claimed superiority of new technologies. A major achievement of the DOE Appliance Standards program—which usually goes unnoticed—is the test procedures that have become accepted as a reliable way of measuring the energy efficiency of major household appliances.

Similarly, researchers have developed testing protocols for measuring indoor pollutant concentrations and the emissions from building materials and products. Because there are many different indoor pollutants and acceptable levels of each can vary greatly, however, it is difficult to establish a simple yardstick for quantifying indoor air quality performance. Surrogate yardsticks that have been used include a minimum ventilation rate expressed in cubic feet per minute per occupant, or a maximum concentration of carbon dioxide (e.g., 1,000 ppm).

The government can play an important role in both conducting the basic research and balancing the contending interests that are required to establish uniform test procedures. For example, over 1,000 firms participate in the EPA's Radon Measurement Proficiency (RMP) program. The primary objective of the RMP program is to provide both the public and the States with information about radon measurement companies and their capabilities. A second, more long-term objective is to promote standard radon measurement and quality assurance procedures. The RMP program is a voluntary proficiency program, not a federal certification program. Nevertheless, many manufacturers of radon detectors have chosen to label their products as having successfully participated in the EPA RMP program.

Labeling

After test procedures are established, the next step is frequently to label products, building components or buildings so buyers and users will be able to obtain this information. Here, again, the DOE Appliance Standards labeling program is an example of a government strategy applied to energy efficiency. The developers of new technologies are assured that prospective buyers will be able to compare their product with the others being offered for sale. In addition, the many efforts to develop and implement home energy rating systems recognize the influence these systems can have in encouraging home buyers to purchase energy-efficient housing.

Product labeling by the private sector has been used to promote indoor air quality, even in the absence of federal requirements to do so. For example, manufacturers of pressed wood products have chosen to label products that minimize formaldehyde emissions. In addition, building upon the home energy-rating systems concept, discussions are currently underway with homebuilders to evaluate "indoor air quality upgrades" that would promote indoor air quality.

Standards

The establishment by government of minimum energy efficiency standards promotes greater energy efficiency in buildings and in the products from which they are constructed and operated. The degree of pressure for new building technologies depends upon the stringency of the standards, i.e., whether they are intended to eliminate the least efficient products or to force the development of new technology. The Department of Energy has responsibility for two building energy-regulatory programs—Building Guidelines and Standards, and Appliance Standards (DOE, 1976a; DOE, 1976b).

For example, the DOE recently promulgated its energy standard for new federal nonresidential buildings. This standard, which also serves as a guideline for the construction of nonfederal buildings, is very similar to the American Society of Heating, Refrigerating, and Air-Conditioning Engineers' (ASHRAE) voluntary industry consensus standard 90.1-1989, "Energy-Efficient Design of New Buildings Except New Low-Rise Residential Buildings" (ASHRAE, 1989a). Both of these standards set minimum requirements for the energy-efficient design of new buildings so they may be constructed, operated, and maintained in a manner that minimizes the use of energy without constraining the building function or the comfort and productivity of the occupants.

Similarly, ASHRAE Standard 62-1989, "Ventilation for Acceptable Indoor Air Quality," is a voluntary industry consensus standard for promoting indoor air quality (ASHRAE, 1989b). Standard 62-1989 provides two alternative procedures for achieving acceptable indoor air quality: the ventilation rate procedure, which specifies the quality and quantity of ventilation air for different applications, and the indoor air quality procedure, which specifies acceptable limits for indoor contaminants. There is no analogous federal indoor air quality standard, since the federal indoor air quality program is primarily devoted to research and information dissemination. (It is interesting to note, however, that legislation currently before Congress would require the EPA to conduct a program to analyze the adequacy of existing ventilation standards and guidelines to protect the public from indoor contaminants.)

Financial Incentives

The use of financial incentives can provide an additional stimulus for the improvement of building performance and the development and selection of advanced technologies. Financial incentives—like many of these strategies—may be beneficial or detrimental, depending upon how they are applied. For example, if incentives stimulate the introduction of a new technology, the result is beneficial. If, however, incentives prompt the premature expansion of an advanced technology before adequate field testing or supporting service capability is in place, the experience may be disappointing.

Examples of financial incentives include the retrofit and solar tax credits the federal government provided in the late 1970s and early 1980s to promote energy efficiency. Similarly, federal lawmakers have proposed legislation that would provide a tax credit for the installation of radon mitigation systems. States and utilities have also offered similar financial incentives.

Information and Education

Government information and education programs can also encourage the adoption and use of advanced technologies. The building sector is composed of a great number of decision makers—homeowners, home builders, commercial and industrial building owners and managers, architects, engineers, construction contractors, and building product suppliers. Because of this decentralization and diversity, government support for technology transfer can be particularly beneficial in encouraging improved building performance and the adoption of advanced technologies. Governments can also be influential through the example they set in their own building and purchase decisions.

Examples of EPA publications that draw upon the results of federal and other research efforts include: *The Inside Story: A Guide to Indoor Air Quality, A Citizen's Guide to Radon,* and a series of indoor-air fact sheets that address, among other topics, ventilation, sick buildings, environmental tobacco smoke, and air cleaners (EPA, 1988; EPA, 1986). In addition, the development of guidance documents for building owners and managers is a top EPA priority. Publications currently under preparation include a technical manual for design professionals on preventing indoor air quality problems in new buildings, a guide for building owners and managers on managing existing buildings to assist in both preventing and addressing indoor air quality problems, a new home construction guide for developers of residential housing, and a model school district indoor air quality management plan.

Taxation and Import

Taxation, import, and other government decisions that affect the relative price of energy and therefore the attractiveness of efficiency investments are other ways to influence the adoption of energy-efficient technologies. These decisions affect the levelness of the playing field between investments in new energy supplies and improved efficiency. They may therefore either encourage or discourage new technologies. In the past—and still today—when utility stockholder earnings have been tied to kilowatt hours of sales, the field has favored new supply investments and discouraged the introduction of new, efficient technologies. The DOE, working with utilities and state public utility commissions, is seeking to reform this regulatory climate.

Research, Development, and Demonstration

The final strategy for achieving change is support for research, development, and demonstration (RD&D) activities that accelerate the development and commercialization of new technologies. An RD&D program in energy efficiency was started immediately after the 1973 OPEC embargo—before there was a DOE—and has continued, with ups and downs, to the present. Similarly, research explicitly devoted to indoor air quality was initiated in the late 1970s as concern for increasing indoor pollutant levels associated with decreasing ventilation rates grew. Whereas the other six strategies influence the nation's receptivity to improved building performance and new technologies, RD&D has this increased receptivity as its primary purpose. The results of RD&D produce the new design and operating concepts and the products and components that extend the limits of energy efficiency and enhanced indoor air quality choices available to us. The long-term progress toward a more efficient and less damaging environment—both indoors and out—depends upon the success of RD&D activities.

The seven strategies discussed above have been, and continue to be, combined by governments in myriad combinations. The US approach to energy efficiency in the late 1970s, for example, was to try something in all of the strategies. The approach during most of the 1980s was to reduce the overall effort and concentrate on generic, high risk, but potentially high payoff research. Today mix of strategies is again changing. Energy efficiency is once more gaining support in the form of a budget request in the buildings area in FY 1991 that is three times higher than the administration's request a year ago and nearly 20 percent above last year's Congressional appropriation. Similarly, the FY 1990 budget for the EPA's indoor air program office activities is four times its FY 1989 budget, and the president's FY 1991 budget request for indoor air research is nearly double the FY 1990 budget.

Future DOE Directions

The new DOE initiatives announced earlier this year show a willingness to use the regulatory strategy with an expanded demonstration of the recently issued building guidelines and standards and the suggestion that standards be considered for commercial lighting. A major increase in support for integrated resource planning recognizes the gains that can be achieved by leveling the supply-demand playing field of utilities. The initiative to relight federal buildings shows a desire to set an example through the federal government's own activities. These initiatives and other changes in the DOE budget request for fiscal year 1991 signal a shift from a singular focus on long-term research to a cluster of strategies designed to move new and emerging technologies toward accelerated commercialization.

The DOE is also considering other new directions that should accelerate the commercialization of new energy efficient technologies. These include:

- The combination of energy efficiency and solar energy programs into a coordinated buildings program. The two programs have complementary objectives, the first to reduce energy loads and the second to increase the portion of those loads provided by renewable sources. A synergistic strategy that seeks to optimize the contribution of each of these technologies should stimulate new areas of investigation.
- The identification of increased efficiency and the use of renewables as a value-added feature in buildings. Whereas the concern for energy efficiency waxes and wanes depending upon prices and policy priorities, there is an enduring public search for high value in the products that are acquired. Energy-efficient products have a higher quality of design, engineering, and function that represents increased value. By stressing these qualities, we hope to achieve a more sustained appreciation for environmentally sensitive, energy-efficient products.
- The recognition that regional patterns in energy sources and uses provide an opportunity for federal, state, utility, private sector, and consumer environmental collaborations. Energy supply systems and energy use patterns have distinctive differences in different parts of the United States. The Northeast, for example, is characterized by oil dependence, an aging building infrastructure, and, in New York State, a high percentage of utilities with mixed electricity and natural gas operations. These patterns offer unique opportunities for particular kinds of new technologies. The identification of these regional features and the technologies they encourage offer the potential for new government-industry collaborations.

The most significant activity by the Department is the pursuit of a new National Energy Strategy (NES). Secretary Watkins has made it clear that the future direction of the Department will be shaped by the recommendations that come from the NES. A conscientious effort is being made to base the NES on the inputs obtained from hearings held throughout the country. An interim report on the NES was published in April 1990. The interim report outlines the opportunities we face in all areas of energy policy, the obstacles to these opportunities, and the options available to us as a nation. During the summer, DOE will encourage comments on this report. Based upon these comments a final NES is scheduled to be completed by December. Active participation in this process is encouraged as a way to influence the direction of future energy programs.

Future EPA Directions

In August 1989, the EPA submitted its "Report to Congress on Indoor Air Quality" (EPA, 1989). The report concluded:

> Sufficient evidence exists to conclude that indoor air pollution represents a major portion of the public's exposure to air pollution and may pose serious acute and chronic health risks. This evidence warrants an expanded effort to characterize and mitigate this exposure.

In addition, the report presented six recommendations for future directions for both the public and private sector's indoor air quality activities. These included:

1. Expanding research devoted to better characterizing exposure and health effects of chemical contaminants and pollutant mixtures commonly found indoors.

Although EPA is beginning to devote greater attention to characterizing noncancer health effects from various exposure routes, information on exposure in homes and buildings is limited to a very few pollutants and groups of pollutants. In addition, very little is known about cancer and noncancer health effects due to low-level respiratory exposures to multiple chemical contaminants. An expanded research program in this field is needed to help characterize causes and solutions to the "sick building syndrome" and to investigate emerging health issues such as multiple chemical sensitivity.

2. Developing a research program to characterize and develop mitigation strategies for biological contaminants in indoor air.

The EPA's historical experience in addressing environmental hazards has been predominantly focused on chemical contaminants. However, biological contaminants in indoor air are predominantly responsible for known building-related illnesses that include Legionnaire's disease and hypersensitivity pneumonitis, and have been increasingly associated with poor hygienic and maintenance practices in buildings. Reliable and inexpensive measurement techniques are needed to both quantify total and speciate individual indoor biological concentrations and to characterize baseline indoor levels of these contaminants.

3. Expanding research devoted to identifying and characterizing significant indoor air pollution sources and evaluating appropriate mitigation strategies.

Source control is the most effective control option when major sources can be identified and characterized, and it may be the only viable option in some situations. However, significant resources must be devoted to identifying and characterizing sources to enable the EPA and other federal agencies to take appropriate control actions under existing authorities or to advise the public of the health risks from specific sources and other actions they can take to reduce risk. Furthermore, research into innovative control technologies and evaluation of technologies developed by the private sector should be significantly enhanced.

4. Developing and promoting, in conjunction with appropriate private sector organizations, guidelines covering ventilation, as well as other building design, operation, and maintenance practices to ensure that indoor air quality is protective of public health.

An effective national program to control indoor air pollution will require the application of generic strategies involving provisions for adequate ventilation, and provisions to avoid problems through proper building design, operation, and maintenance. This approach, combined with programs targeted to specific individual high risk sources and pollutants, would provide a comprehensive, feasible, and cost-effective control strategy. A pollutant-by-pollutant approach alone, encompassing target levels for individual pollutants, is not adequate to resolve many indoor air quality problems.

5. Expanding current efforts to provide technical assistance and information dissemination to inform the public about risks and mitigation strategies, and to assist state and local governments and the private sector in solving indoor air quality problems. Such a program should include an indoor air quality clearinghouse.

While the EPA has joined the ongoing Federal and private sector efforts to disseminate information on indoor air quality, as our experience with radon has demonstrated, a program is needed that can keep pace with the needs of state and local governments, architects, building owners and managers, researchers, the medical and health communities, building occupants, and the general public. A program to transfer both technical and nontechnical information and develop capabilities in the public and private sectors would include a variety of technical assistance and information dissemination activities comparable to those developed to address the radon problem. An indoor air information clearinghouse is needed to enhance coordination and access to such information.

6. Undertaking an effort to characterize the nature and pervasiveness of the health impacts associated with indoor air quality problems in commercial and public buildings, schools, health care facilities, and residences, and to develop and promote recommended guidelines for diagnosing and controlling such problems.

The available indoor air quality literature suggests that indoor air quality problems are pervasive in a wide spectrum of buildings, but the prevalence of such problems, the nature of their sources, and the amount of human exposure attributable to these sources is not well understood. However, an increasing number of complaints are being registered to government agencies, and a growing number of private sector firms are attempting to respond to a rapidly emerging market for diagnostic and mitigation services. A major study is needed to determine the scope and character of such problems, and to develop recommendations to guide and control the quality of diagnostic and mitigation services in the private sector.

Conclusions

Governments can encourage improved building design and performance and the commercial development of new building technologies by using their broad and pervasive authority to create an environment in which these enhancements are recognized, encouraged, and diffused rapidly into the domestic and international market. Any combination of seven strategies can be used to accomplish this goal. The application of these strategies in the future activities of both the DOE and the EPA will hopefully provide the appropriate environment for both increased energy efficiency and reduced pollution, both indoors and out.

Disclaimer

The opinions expressed in this paper are those of the authors, and do not necessarily reflect those of the US Department of Energy or the Environmental Protection Agency. No official endorsement is implied.

References

ASHRAE, 1989a. American Society of Heating, Refrigerating, and Air-Conditioning Engineers, "Standard 90.1-1989: Energy-Efficient Design of New Buildings Except New Low-Rise Residential Buildings," 1989.

ASHRAE, 1989b. American Society of Heating, Refrigerating, and Air-Conditioning Engineers, "Standard 62-1989: Ventilation for Acceptable Indoor Air Quality," 1989.

DOE, 1976a. Department of Energy, "Energy Conservation Standard for New Buildings Act of 1976," Public Law 94-385, 1976, as amended.

DOE, 1976b. Department of Energy, "Energy Conservation Program for Consumer Products Other Than Automobiles, Energy Policy and Conservation Act, Part B," Public Law 94-163, 1976, as amended.

EPA, 1986. Environmental Protection Agency, *A Citizen's Guide to Radon,* OPA-86-004, August, 1986.

EPA, 1988. Environmental Protection Agency, *The Inside Story: A Guide to Indoor Air Quality,* EPA/400/1-88/004, September, 1988.

EPA, 1989. Environmental Protection Agency, "Report to Congress on Indoor Air Quality," EPA/400/1-89/001, August, 1989.

Consortium Planning: A New Approach to Building Innovation

Constance J. Groth
General Electric Plastics
Pittsfield, MA

Abstract

Through the initiative of General Electric Plastics, a multimillion dollar cooperative program has begun to speed the development and commercialization of beneficial technologies for the building industry, specifically those that utilize engineering thermoplastics.

The focal point of this program is the "Living Environments" concept house located in Pittsfield, Massachusetts. The house provides an excellent illustration of the following strategies:

Partnerships with other companies. This strategy utilizes the expertise of industry partners to develop the best possible technologies and drive these innovations more quickly to market. The Living Environments house, for example, was built in alliance with 53 other companies, providing a strong consortium for future development.

The systems approach. Viewing the house as a whole, instead of a series of separate components, allows the potential for synergies that can lower costs, save energy, and make the home a better place to live. This is beneficial to the homeowner, and will also make the participating companies more competitive in the world market.

Figure 1. The 3,000-square-foot Living Environments concept house, shown here in its front view, showcases advanced design and building methods, processes, and materials, and will serve as a laboratory to explore the feasibility of widespread use of engineering polymers in construction markets.

The Consortium Approach

Recognizing the wide diversity of decisionmakers within the building industry, the consortium approach assembles the talents and expertise of players ranging from architects and homebuilders to designers, building-product manufacturers, and distributors. The consortium combines these forces in a major, unified effort to develop innovative technologies designed to lower costs, save energy, and improve performance of building systems and components.

The focal point of this approach is the recently constructed concept house in Pittsfield, Massachusetts (Figs. 1 and 2), which provides a showcase to industry and the general public for emerging building technologies. Fifty-four separate companies have pooled their expertise and innovation in this joint effort.

In line with the program's major goals and objectives, the house has been named "Living Environments." Its construction was the first step in a carefully conceived plan for future cooperative efforts. The initial results of this joint venture provide living proof that the new concept can work very effectively. In the highly competitive world market, building systems and component manufacturers that wish to compete must provide the homeowner with what we call the "three dreams" of homeownership: a house that is affordable, inexpensive to maintain, and better to live in. The goal is to provide these "three dreams" to the ultimate homeowner.

The consortium approach involves working in partnership with other companies, architects, home builders, and researchers who have special insights in their par-

Figure 2. More than 50 partner companies from leading building and construction suppliers, as well as component manufacturers and members of academia, are involved in the Living Environments concept house, shown here from a back view. The basement contains a 2,500-square-foot business center, offices, and a display area for prototype models.

Figure 3. Three companies are providing roofing materials for Living Environments. Nailite International and Masonite Corp.'s roof coverings using engineering thermoplastics are showcased on the house; Carlisle SynTec Systems is working with GE Plastics on the design of another plastic panelized roof system, a prototype of which is displayed inside.

ticular portions of the building industry. For example, GE Plastics is a raw materials supplier, providing only pellets and sheets of engineering thermoplastics, not finished parts or components. But this knowledge of engineering thermoplastics when combined with the expertise of a company involved in roofing systems and possessing special experience in the roofing industry's codes, distribution methods, costs, and installation methods, can produce improved new products and systems for the market and drive those innovations toward quicker market acceptance (Fig. 3).

New Technologies

The house as it stands right now documents some of the many possibilities for using engineering polymers in building and construction today. But more importantly, Living Environments is intended to serve as a living laboratory that will continue research into improved housing systems and components for many years to come.

The Living Environments' technologies are classified into three broad areas, corresponding with how long they will take to reach the market. For example, short-term applications are those that already exist in engineering polymers or will in the very near future. A mid-term application has a clear material and technology fit, but a development program with a key industry partner is needed to make it happen. The more challenging developments are long term—possibly five years or more out. Here we need a committed industry research and development partner and a major mutual development effort. The goal is to continually initiate long-term projects, which will become mid-term, then short-term applications, so there is a constant stream of technology being commercialized in the marketplace.

The home-building industry is an obvious strategic fit for engineering thermoplastics—with their ability to be easily molded or extruded into complex shapes that are impossible or expensive for traditional materials; their integral color; their impact resistance; their imperviousness to rot, rust, and corrosion; their light weight, which saves both shipping costs and back-breaking effort at the construction site; and their inherent recyclability. The house also represents a comprehensive "material strategy." One important part of that strategy is resource recovery. For example, the sidewalk in front of the house and portions of the basement are made from a special "engineering concrete." Nearly 60 percent of the mix is plastic, composed of re-ground fenders and other automotive parts.

It was never the intention to build an "all plastics" house. The idea was to demonstrate the tremendous performance options of polymers used in harmony with traditional materials such as wood and stone (Fig. 4). In fact, many people, upon entering Living Environments, ask "Where's the plastic?" We used natural materials where they made the most sense, polymers where they best fit, and combinations of the two where that was most appropriate—as in the polymer concrete just mentioned. There were approximately 45,000 pounds of polymers used in the house construction; 30 percent of the components use plastics. In designing the house, we based it on an imaginary family of four. Throughout the building of Living Environments and into the future, it is of primary importance to bear in mind that we're building homes for real people, trying to make them both less costly *and* better to live in.

Figure 4. The interior of the Living Environments concept house reflects the modern lifestyle and daily living requirements of a family of four, with traditional materials used in harmony with engineering polymers. The living room features a home entertainment center with a conceptual 43-inch flat-screen TV, developed in conjunction with Sony Corporation of America. Ceiling-mounted audio speakers are part of the Bose Acoustimass® built-in music system installed throughout.

The Systems Approach

This full-scale home spotlights the systems approach to home building. It includes both prototypes and commercially available components, sub-systems, and systems. The systems concept is absolutely critical to the success of Living Environments and we believe it will be critical for global competition. Some examples of systems use in the house are:

- The Total Environment Control System or TEC (Fig. 5). This is a whole-house concept that integrates all the heating, cooling, and water conditioning needs in the home instead of having a separate furnace, air-conditioning compressor, and water-heater. A water-collection system collects gutter run-off and could be used for solar-collected heat storage . . . or for watering your lawn.
- The Central Processing Unit. Many functions in the building—including security, energy, and lighting—are controlled by the Central Processing Unit, or CPU. While the CPU performs many functions now, down the road its uses will multiply—for example, it will shave energy loads during peak periods to both save money and conserve.
- Integrated Baseboard Raceways (Fig. 6). The entire house is tied together with an extruded baseboard system that is really an electronic raceway linking power, telecommunications, temperature, and other systems to the CPU.
- Panelized Construction. There are many concepts in the house showing how a modular or panelized approach to both horizontal and vertical panels can make construction quicker and maintenance easier. For example, one wall system concept that has raised much interest consists of corrugated plastic foam panels that bond together using factory-installed Velcro (Fig. 7). Together they form integrated channels that can be snapped open to route electrical wiring. Air is blown directly into the channels, virtually eliminating the need for conventional ductwork. Other panels include: a modular flooring system (Fig. 8) through which flexible pipes can be threaded, doing away with the need for a common plumbing wall; a roof panel combining the roof covering with insulation; and a basement panel that could ultimately eliminate poured concrete foundations.

The basement walls, in fact, are a good example of the development strategy mentioned earlier. For example, a near-term application is a recently developed form that was used to pour the concrete basement walls for Living Environments. The form is molded using a glass-filled polymer for strength and rigidity. It is light and easy to handle, leaves an excellent finish on the

Figure 5. The Total Environment Control (TEC) system, a joint project of Carrier Corp. and GE Plastics, is the heart of the Living Environments concept house. The conceptual unit combines the functions of five home comfort appliances, and uses thermoplastics to provide aesthetics, heat and corrosion resistance, light weight, easy serviceability and installation, and design-for-manufacturability.

Figure 6. The wire management system in the house is made by Molex Inc. The raceway is a wire channel that runs along the wall, much like a baseboard, and contains electrical outlets, receptacles, telephone plugs, and cable connections.

Figure 7. A developmental wall system combines an experimental corrugated engineered strand product from Weyerhaeuser Co. with low-density NORYL® foamable resin and reclaimed material from hamburger packages.

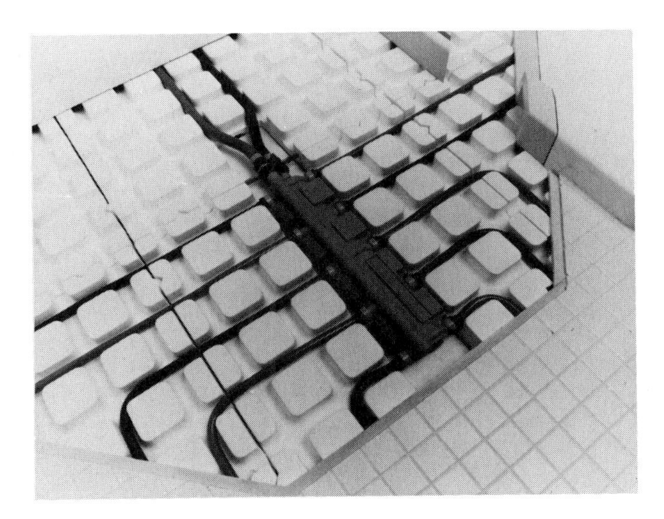

Figure 8. The interior of the Living Environments concept house features a GECET™ resin substrate as part of a distribution floor developed by Infill Systems, B.V., the Netherlands, through which wiring and piping can be routed.

concrete, and does not have to be oiled like wood forms. A mid-term application would be a plastic form that is left in place as an insulation barrier after the concrete is poured (Fig. 9). For the long term, it may be possible to avoid pouring concrete basement walls altogether by using a pre-fabricated modular wall system.

The Living Environments program will create significant growth opportunities for GE Plastics' partners in the building, construction, and related industries. It will be the proving ground for resource recovery strategies, reclaiming engineering thermoplastics from packaging and automotive applications for use in high-tech building components.

Most important is the project's goal of designing more attractive and functional homes affordable to broad segments of the population. With our partners, we believe we can make this vision a reality.

Figure 9. One key strategy is to have a blend of short-, mid-, and long-term applications under development so there is a constant stream of technology being commercialized in the marketplace. These leave-behind concrete forms, for example, illustrate a mid-term application. Not only do the forms provide insulation for the basement, but their design allows modules, such as this workbench or shelving, to be easily hung.

Encouraging the Search for New Solutions

Glenn Chafee
Research and Development
Rolscreen Company
Pella, IA

Abstract

Approximately 5.5 percent of all the energy consumed in the United States is lost through windows. New products, including double- and triple-paned windows, special low-emissivity glazings, and argon- and krypton-filled multi-glazed windows, can increase energy efficiency by as much as 70 percent over the single-glazed windows found in most homes today. Additionally, older technologies, such as double-glazed windows, could save the equivalent of 160 million kilowatts annually if retrofitted.

While industry continually seeks ways to manufacture high-quality, energy-efficient products, homeowners and builders do not always have sufficient incentive to install them. Recognizing the need for a cohesive conservation policy is the first step toward utilizing the technological solutions that are currently on the market or will be introduced during the next decade. Many state and local governments have developed innovative programs to address this problem. There are enormous savings possible by applying new and existing solutions to energy conservation. Government and utility incentives can act as a catalyst to speed change, especially in the retrofit market. It is only through education and cooperation between government, industry, and the consumer that the promise of an energy-secure future can be achieved.

Introduction

This paper addresses the challenge of energy conservation by looking at solutions that deal directly with inadequate windows which dissipate heat in the winter and increase cooling loads in the summer. These are major problems in the building sector of the economy, whose total energy consumption accounts for more than one third of the nation's energy requirements. It will discuss the fenestration industry's most sophisticated energy conservation technology available, and will report the findings of the Research and Development Group of the Rolscreen Company, which is actively involved in the search for new energy-efficient windows and doors.

The search for a complete solution to energy conservation, however, must go beyond technological development. Toward this end, government and industry must work together to create the incentives necessary for those products and methods to be produced and accepted on a mass scale.

Recognizing the Need for a Strong Conservation Policy

While interest in the depletion and pollution of our natural resources has been increasing for decades, the so-called energy crisis of the 1970s was the event that confronted the public with the impact of the problem. Taxpayers were alarmed by skyrocketing energy costs and long lines at filling stations. In response, political leaders sought policies to reduce energy use and reliance upon foreign energy sources.

As we enter the 1990s, nearly two decades after energy policy became a leading item on the national agenda, little has been resolved. Energy policy remains a problem in need of a solution, or, more accurately, a set of solutions. The fact of the matter is that energy policy is an extremely complex and tangled issue; much too complex to be addressed by a single solution. Many people must become involved at a multitude of levels in order to address the issue effectively.

The first step is to recognize the magnitude of the problem. Although the national intent has been to wean ourselves away from our dependence upon overseas sources, the reality is just the opposite:

- The United States currently obtains nearly one half of its oil requirement from foreign sources. This figure may reach 60 percent by the end of the decade. According to the Commerce Department, in just the first month of the new decade alone, oil imports cost this country $5.86 billion dollars as volume hit a record 291.3 million barrels.
- Political and social concerns have brought the nuclear power industry to a halt. Since 1973, not a single US

nuclear power plant has been ordered or put into construction. Utility companies have cancelled plans for some 65 plants since the 1979 Three Mile Island incident. Plants that were operational or nearly operational have been abandoned, with investors and utility companies absorbing billions of dollars in losses.
- The nation's electricity-generating industry has warned that it will not be able to supply enough electricity to support the country's economic growth over the next decade if current barriers to power plant construction are not eased.
- Pending legislation may curtail the use and increase the cost of coal, our most abundant fuel.
- Hydropower's role as a major energy supplier is in jeopardy as a result of government actions.[1]

In short, declining capacity is the disease, and we are beginning to see the symptoms, which were particularly apparent during the past winter's cold snap. In Florida, the demand for electricity was so high utilities instituted "rolling blackouts," or regional shutoffs, to keep the entire system from going down. For example, the Tampa Electric Company, with a capacity of 2800 megawatts, saw demand reach a peak 3230 megawatts during the season's hottest weather.

In the summer of 1988, New England experienced "voluntary brownouts" when heat waves boosted electricity consumption past capacity. Customers who agreed to service interruptions during these "voluntary brownout" periods received reduced rates. These and other innovative programs across the country may help to manage a crisis situation, but is compensating the consumer for an inability to meet service demand the best we can hope for in the coming years?

Energy out the Window

Studies at the Lawrence Berkeley Laboratory have indicated that 5.5 percent of all the energy consumed in the United States literally goes through windows as a result of winter heat losses and summer heat gains. Even though today's state-of-the-art window technology has advanced dramatically, more than half of the 70 million homes built prior to 1980 have windows that are single-glazed. In fact, a large percentage of homes being built today still feature single-glazed windows. Retrofitting double glazing into just one million of these homes would save about 150 million therms of energy per year. This is equivalent to 160 million kilowatts annually. This does not involve the industry's newest technology—just new applications of proven existing technology.

The choices available to a homeowner today, however, are dramatic improvements over yesterday's products. New technologies can produce windows that are actually more energy-efficient than the walls in which they are placed. These products, including double- and triple-paned windows, special low-emissivity glazings, and argon- and krypton-filled multi-glazed windows, increase energy efficiency by as much as 70 percent over single-glazed windows. A reference point on the differences in energy efficiency follows:

- Single-glazed windows have an R-value less than R-1.
- Typical double-glazed windows double that value to almost R-2.
- Low emission glaze windows are slightly under R-3.
- Argon-filled windows are approximately R-4.

Better windows mean greater energy conservation. On a scale the size of all available housing in the US, that conservation of energy can be a significant factor in reducing overseas export of oil. An $8 million investment in a sputtering plant to produce low-emissivity coatings saves as much energy—36 million barrels of oil over a 10 year period—as a $300 million investment in an offshore oil platform might be expected to produce.

But building a better window takes more than just technology. It has to appeal to the homeowner and the contractor. It has to sell and be implemented on a mass scale for the energy conservation properties to have a significant effect. To achieve this end, building a better window has to become an art, as well as a business.

The Art of Designing Windows

Fenestration, the art of designing windows and doors, must fulfill three requirements demanded by customers—aesthetics, control, and light.

Whether for commercial or residential use, windows and doors must be attractive. The second requirement is the ability to control the inside environment, regardless of what is happening outdoors. And finally, light is the third requirement of fenestration. "Let there be light"—but make sure it is modified to control glare and heat.

These needs present a variety of challenges to the building trade. Although there are many programs under way to improve the thermal performance of windows that meet the above criteria, the areas that still demand attention are:

- Improving the thermal properties of the glass edge and frame. A window with R-8 glass might actually be an R-4 wall element because of heat loss through the glass edges and frame. Until now, most of the work done has been to improve the center glass insulating value. It remains to improve edge and frame construction so that the R-value of the center pane is consistent throughout the wall unit.
- Improving shading coefficient without sacrificing aesthetics or visible light. This is particularly important

for southern climates and for construction that specifies overhead glazing.
- Providing both individual and computer control of glare and heat gain. The most common method is through indoor window coverings (i.e., shades, blinds, and drapes). Reflective glass has been effectively used to reduce heat gain in commercial buildings, but it is not aesthetically acceptable in residential use.

Given the demands of the market, and the particular concerns of the fenestration industry to improve product performance, what then are some of the answers research and development has identified in our search for solutions?

Evolving Technologies of the Nineties

Superwindows

Superwindows are windows with overall R-values of 6 to 8 that will outperform the very walls that surround them. Glazing systems with an R-8 value at center pane are already available. Because they are made with today's insulating glass technology, however, significant heat is still lost along the edges. Research is focusing on how to reduce the heat loss along the perimeter of the glass.

A typical residential window has a glass area of about 70 percent. The remainder is sash and frame. Work done by Rolscreen and confirmed by other American Society of Heating, Refrigerating and Air Conditioning Engineers (ASHRAE) data shows the insulating value of the wood components to be about R-2.5. Since the overall window R-value is proportional to the R-values of the individual components, it quickly becomes apparent how important it is to reduce glass-edge and frame-edge heat loss to successfully produce a superwindow.

A joint research project between Canada and the United States was initiated in 1988 for the purpose of coordinating window performance research in both countries. The goal of the project is to develop standardized, validated calculation procedures to determine the thermal properties of windows, including the center of the glass, glass edge, frame, and overall window assembly. The research coordinators recently issued their first report, with much more planned for the future.[2] These findings will help make superwindows widely available by the end of the decade.

Smartwindows

Smartwindows have properties that cause their surfaces to change from clear to opaque in order to curb solar heat gain and glare. Current smartwindow research is concentrated on photochromic, thermochromic, electrochromic, and electric crystal switching systems.

All of these windows change transparency in response to photons, temperature, or electricity. The first two technologies, photochromic and thermochromic windows, provide no control override for the user. Thus, their use in residential markets is limited; and, because they are expensive, they will probably remain a small part of the commercial market.

Electrochromic windows do have override control potential and this technology is being pursued by a number of laboratories. Here again, however, customer demand and market dynamics play a major role in product development. The key concern in our current research is to produce a color in the switched stage that is aesthetically pleasing to the consumer. The functionality of the window may be perfect, but the end result may not have mass appeal due to cosmetic qualities. A small market demand would render these windows prohibitively expensive.

Another area being addressed by research and development is the improvement of privacy in the switched stage. Given the progress our industry has made with electrochromic windows, it is highly probable that they will be introduced to the public sometime during the new decade.

Finally, another form of electrically controllable opacity employs use of liquid crystals. When electric voltage is applied to the panel of these windows, the crystals align in such a manner as to make them transparent. When the charge is removed, the material becomes opaque.

Problems that must be solved regarding electric crystal windows are the slight hazing effect caused by light scattering from the crystals in the transparent stage, and the need to reduce the cost of these windows to a price acceptable to the customer.

Currently, the material costs for smartwindows are at least $75 per square foot—250 times that of clear double strength glass. With high volume this can be reduced, but even a premium of moderate proportions will limit the usefulness of this technology on a mass scale. Superwindows and smartwindows show great promise for the future, but, while searching for solutions, we must be careful not to overlook the technology developed and proven in the past.

Merging Existing Techniques with Cutting-Edge Technologies

If we are to become more energy-efficient as we head toward the 21st century, we must merge yesterday's proven successes with tomorrow's cutting-edge technology. In addition, the manufacturing expenses of future energy-efficient products must be such that these products can be mass produced.

The building trades will ignore a product that adds initial cost they feel is not justified. Insulating glass suffered this initial rejection, as did low emissive coatings for glass. Only the energy crisis of the 1970s allowed insulated glass to gain widespread acceptance. Low-e glass has benefitted from massive advertising and public relations campaigns, but both of these innovations are still often omitted from use in new residential construction in order to cut costs.

While society strongly encourages the search for new solutions, economics too often encourages the depletion of natural resources rather than investments to develop energy-efficient products. While research and development departments in industry continue to seek ways to manufacture the highest quality energy-efficient products possible, homeowners and builders have not had sufficient incentive to install them.

Experiences and Future Prospects

The experiences of Rolscreen, which produces the "Pella Window," have confirmed that much progress in energy savings can be made by improving the properties of glazing materials and the physical characteristics of windows. The following data, applicable to the "Pella Windows," illustrates the energy-related achievements and prospects of windows:

1. In 1980, 90 percent of windows sold by Rolscreen consisted of a double glazed panel, composed of clear single or double strength exterior and interior lights, separated by a seven-eighths inch thick air space. The U-value was 0.47 BTU/hrft^2°F.
2. In 1990, clear double-glazing still preliminates Rolscreen's current product. Other options that we now produce which were not available 10 years ago are: 1) Low-e pyrolite coating (emissivity - 0.2) on the double-glazing panel to yield U=0.37; 2) Argon gas-filled sputtered low-e coating (emissivity = 0.1) insulating glass to yield U=0.34; and 3) Low-emissivity slimshades (mini-blinds) between two clear glass lights to yield U-0.29 in the closed position.
3. The industry currently has future options to yield U-values below 0.25. The designs consist of three lights of glass or two lights of glass used in combination with plastic films. Various combinations of gas fills and location of the low-e coating or coatings have been suggested. These designs are technically feasible but may not be cost-effective depending upon the geographic location and the cost of fuel.

Government and Utility Incentives

Many of the best solutions do not require new technologies, just a new mindset and financing. Government and utility incentives could be a catalyst in these changes—especially in the retrofit market. This is becoming increasingly popular as many states begin to adopt strong conservation policies.

During the energy crisis two decades ago, the federal government became actively involved in promoting conservation as a means to preserving natural resources. The National Energy Conservation Policy was adopted during this era. Under this plan, the states assured that utilities within their jurisdictions would establish various residential conservation service programs, including energy audits, weatherization projects, and curtailment procedures. Even as the energy crisis became less an issue, it became apparent that the Residential Conservation Program was not working.

Congress amended the policy with the Conservation Reform Act of 1986. This paved the way for state agencies to revisit the issue of conservation. Not surprisingly, this trend is most apparent in the northern states, with Washington, Iowa, Illinois, Wisconsin, Michigan, and Massachusetts instituting conservation policies. Even southern California has adopted a policy, however, as have some municipalities.

A review of some of this legislation gives an indication of its scope and creativity:

- The residential conservation service program in the state of Washington includes virtually all residential buildings. In Tacoma, for example, residents are eligible for rebate incentives if they have:
  submitted an energy audit;
  installed city approved conservation devices through city-authorized contracts;
  passed a city-conducted inspection of the conservation devices.
- The Iowa Utilities Board has long shown an interest in encouraging conservation, and has been a leader in initiating short-term pilot projects for weatherization assistance, commercial refrigeration efficiency, assistance to low-income groups, and furnace and boiler replacement. Last year, the board approved a pilot project called "Make Cents." The program, aimed at senior citizens as well as low-income ratepayers, provides loans for installing improved heating systems that closely monitor energy consumption.
- The Illinois Commerce Commission emphasizes the use of solar energy. It identifies specific insulation standards for windows, ceilings, walls, floors, ducts, pipes, and water heaters. In 1986, the Illinois commission authorized Commonwealth Edison to offer rebate programs for those customers choosing to replace or install high efficiency appliances, including

water heaters, refrigerators, air conditioning systems, or lighting systems. In addition to the rebate program, the utility company was allowed to provide zero- or low-interest loans for weatherization projects, including energy-efficient windows.

- The Wisconsin Public Service Commission has issued orders for almost every utility in its jurisdiction. In the past two years, the commission has directed several utilities to initiate sales promotion and advertising programs for upgraded, more efficient appliances. The program included significant loan and rebate programs.

  One particularly creative program provides gift certificates that can be purchased for other customers to finance weatherization or appliance refurbishment.

- The Michigan Public Service Commission emphasizes interest-free financing for conservation and the installation of usage-reduction devices. It also has endorsed the installation of clock thermostats, high-efficiency furnaces, and floor, ceiling, and sidewall insulation.
- In Massachusetts, 15 utilities joined to found Mass-Save, Inc., an organization devoted to promoting conservation and providing assistance and advice to participants in conservation programs. Mass-Save has gone beyond simple residential energy audits to more affirmative efforts such as load management, household energy efficiency improvement loans, and rebate programs.
- And, in southern California, Edison offers a rebate of one dollar per square foot to homeowners who retrofit windows with dual glazing, and three dollars per square foot for retrofitting with energy efficient coatings.[3]

Conclusions

Utility incentives will be an important agent of changes to improve energy conservation, particularly in retrofit markets. As we move through this decade, more and more states will join the trend toward applying new solutions to the conservation problem.

Fenestration savings will become increasingly important in the search for new energy solutions. The American Architectural Manufacturers Association predicts an increase in sales of windows and doors for 1990 and 1991 due to new construction and continued growth in the replacement and remodeling market.[4] The potential energy savings in this area are so great, the National Fenestration Research Council is developing a rating system to quantify the energy efficiency of each window system.

There are many other solutions that can be used in retrofit markets, as well as new housing construction, that would substantially reduce energy demands. The challenge is to make these "solutions" general practice. All of the professionals allied with the building industry—including manufacturers, contractors, state and federal governments, and technical researchers—have a critical role in meeting this challenge.

Working together to apply new solutions to conservation problems, we can create a sizable national energy dividend—which translates to significant savings for the individual. Some of these solutions are technical, but others simply require a new mindset that fosters creative use of the products and technologies already in existence. The potential of energy conservation can be realized if we work as a team, not just as individual researchers, but as a nation searching for new solutions.

References

1. Bacon, Donald C., "A New Energy Crisis," *Nation's Business,* pp. 20–26 (February 1990).
2. Elmahdy, A. H., "Joint Canadian/US Research Project on Window Performance: Project Outline and Preliminary Results," *ASHRAE-Transactions* V.96, Pt. 1 (1990).
3. Sponseller, Diane, "The Return to Conservation Principles," *Public Utilities Fortnightly,* pp. 50–54 (February 18, 1988).
4. "Rebound in Residential Demand Projected by AAMR Report," *Fenestration* pp. 27–28 (January/February 1990).

The Search for New Solutions: Federally Supported Innovation in Construction and the Federal Role in Supporting Conservation and Renewables

United States Senator Joseph I. Lieberman (CT)
Washington, DC

Introduction

A number of us in the United States Senate recognize that the solution to many of America's environmental problems lies in energy policy. We will not solve acid rain or ozone smog or global warming until we fully integrate energy policy and environmental policy. We now have the technology, or will have it soon, to address our energy needs in an environmentally friendly way. Moving these technologies into common use in the construction industry, for example, will require that the federal government play a new role. Most of the solutions do not demand exotic technologies. Rather, they require new uses for existing technologies. The new federal role is not that complicated. The cooperative role played by the government in the introduction of new medicines provides, as I will discuss, a useful model for how to bring these technologies into the marketplace.

Two pieces of legislation that I recently introduced in Congress illustrate the role the government can play in integrating energy and environmental policy and in encouraging the energy industry to participate in promoting a clean environment.

The Advanced Building Consortium

Last year I introduced a bill in the Senate (S. 1479) that supports the search for new ways to reduce building costs and increase energy efficiency. The bill would create a federal-private sector cooperative program: the Advanced Building Consortium, which would remove the uncertainties presently associated with introducing and employing technological construction innovations. In a nutshell, the industry consortium would have the chance to test and monitor the performance of new cost- and energy-saving products in the laboratory of a federal building.

The Fragmented Nature of the Construction Industry

The unique nature of the building industry serves as the greatest disincentive to innovation in construction. The industry is highly fragmented; no company accounts for more than 1 percent of new construction nationally. The business is dominated by small, thinly capitalized firms with little ability to develop or finance new products. It frequently takes 15 to 20 years for a new building product to gain general market acceptance, even a product proven safe and reliable. The simple smoke detector waited 14 years before reaching a level of general acceptance.

Compounding this problem is the reluctance of architects, builders, and engineers to use new products. This reluctance is fueled by fear of high liability insurance costs and the prohibitive price of any new product not yet in full production. Our thinly capitalized contractors are profit maximizers and the risk of new products is difficult for them to manage.

An Incentive for Innovation

This institutional aversion to change has grown more troublesome in the recent past. Energy costs have soared, yet building technologies have remained, for the most part, at the level they were immediately after the Second World War. The building sector consumes 62 percent of the nation's $150 billion annual electric bill. Conservation of resources and protection of our environment will require a more efficient use of energy in the nation's buildings. Unless we update our technology in construction, greater efficiency is an impossibility. An impetus to development is needed now.

The bill I introduced last year would provide a spur to initiation of housing construction technology. The bill calls on the leaders of the construction industry—product manufacturers and contractors, engineers, and architects—who have the strongest interest in innovative products to form a consortium to develop new ideas.

This independent consortium will recommend the most promising innovations to the government for use in one of its 400,000 buildings. If a product is found to be workable and cost saving, that information will be provided to the public and all of the industry. Using federal buildings as a test ground would sharply reduce the amount of development time. This experience in the federal building will serve as testimony to the product's usefulness. The government and taxpayers will be beneficiaries of the development. They will utilize the new products without bearing the development costs, and will enjoy lower operating costs when the products prove successful.

Previous Efforts

There is historical proof that a program such as the one the bill would create can succeed. The New York State Urban Development Corporation worked along lines similar to those of the Advanced Building Consortium. In a three-year period, during the early 1970s, the Urban Development program constructed 33,000 housing units using new developments in construction. Like the Advanced Building Consortium bill, the Urban Development Corporation required all participants to guarantee their products for a five-year period.

The Army Corps of Engineers' Construction Productivity Advancement Research Program offers another example of a similar program. Both programs aim at introducing innovative building technologies in federally owned facilities. They differ in the fact that the Army Corps program is a cost-sharing program, whereas the Advanced Building Consortium bears the expenses and the responsibility of development. The Army Corps of Engineers is currently cooperating with the solar power industry and the Department of Energy under a joint research program.

The federal government is no stranger to research and development projects for building technology. It has spent tens of millions on such testing without industry assistance. Sadly, the fruits of these tests—paper reports—have resulted in only a modest level of innovation. The people who would use the innovations want actual results, not a paper report produced in a laboratory. The Advanced Building Consortium would improve this situation with industry's cooperation. The result would be a proven product rather than a report.

A Parallel in the Introduction of Medicine

The average physician faces problems similar to those before the typical builder. Each attempts to stay abreast of developments and innovation in his field, but is unable to implement them on his own because of the threat of personal liability and prohibitive prices. The architect or the engineer has no protection if he makes a decision to use an untried product in an effort to assist the client. The doctor, however, has a legal shield to protect him when a new drug is prescribed. This shield is the approval process conducted by the federal government.

Both the medical industry and the government spend millions in the development of new drugs. Cooperation between industry experts and government agencies serves as an effective mechanism which removes the threat of liability from the doctor. The term "approved by the FDA" is magic to the world of medicine; it means a product is proven safe and effective.

No such words of approval exist for building products. The creation of a similar cooperative system would allow the same assurance. As a result, new effective building technologies can be freely "prescribed" and the outdated building materials and equipment replaced with more cost-effective substitutes.

If the government were to recognize this proposed consortium, new cost-saving and energy efficient technologies could be implemented in government buildings. Within a short period of time, the successful results of these pilot programs would be disseminated to all other areas of the building and construction industry. Use of the technologies in private construction would have the approval of experience in federal buildings.

Proposed Clean Air Act Amendments

The second piece of legislation is an amendment I have placed in the Clean Air Act Amendments (Section 403 of S. 1630) currently before the Senate.

This amendment requires the federal government to take energy conservation steps in federally owned or leased facilities in areas where acid rain originates. It also requires federal agencies to install renewable and alternative non-polluting energy technologies in these facilities. It would enable federal facility managers to enter into shared-savings, energy conservation contracts.

The Federal Government Is Largest Electric Consumer

The federal government is the largest single consumer of electricity in this country. We spend about $2.4 billion on electricity for federal facilities each year. Electricity represents about two thirds of the energy costs in operating federal buildings.

Because the production of electricity usually involves the burning of fossil fuels, which result in emissions which cause acid rain, the demand for and the consumption of electricity is at the heart of the acid rain problem. As the largest single consumer of electricity, the federal government is a large part of the problem.

Energy conservation is the most cost-efficient technology available to prevent acid rain and protect natural resources. Conservation reduces electric demand. Reduced electric demand leads directly to reduced electric generation. Simply stated, reduced electric generation leads directly to fewer emissions.

Ideas that cost money are always a concern in Washington and rightly so. But this amendment will save money as well as reduce pollution. We are asking each federal agency with facilities in a service area of 107 utilities targeted by the acid rain provisions of the Clean Air Act Amendments to install energy conservation and efficiency improvements, and alternative and renewable clean energy technologies that have a payback time of less than 10 years. We are also providing federal procurement officials with access to "shared-savings" contracts. This is a type of contract that finances energy conservation investments with no capital costs for the property owner. With these contracts, private capital will finance conservation improvements in public buildings, with no requirement for increased federal budget commitment.

The impact of energy conservation in the federal sector is illustrated by a recent program proposed by the Department of Energy. On January 26, 1990, DOE announced a series of energy efficiency and renewable energy initiatives. According to Deputy Secretary of Energy Henson Moore "Abundant testimony [regarding the National Strategy] . . . has documented the substantial potential of efficiency and renewables in our nation's future energy mix." Mr. Moore noted, "Using energy more efficiently under these initiatives will also avoid damage to the environment by an estimated 800 million tons of carbon dioxide, 2.3 million tons of sulphur oxides and 2.1 million tons of nitrogen oxides annually . . ."[1]

We are going to impose stringent new emissions regulations on electric utilities to control acid rain. The least we can do is insure that the federal government is a model consumer.

Nonpolluting Technologies

Fuel cells, solar, wind, geothermal, and biomass are energy producing technologies that offer tremendous environmental benefits. They produce energy with virtually no emissions. In many cases they are renewable.

We often perceive these technologies as having little application in the real world. We are continually told that these technologies are unproven, prohibitively expensive and, if they're ever going to be available, it's in the far distant future.

In reality, geothermal and biomass already competitively supply baseload capacity. Since 1982 the costs of wind and solar power have decreased dramatically. Competitively priced fuel cells will be available in the next few years.

Most sources agree that electricity costs about 6 to 8 cents per kilowatt hour in the United States.[2] Renewable and clean energy sources are competitive right now. Wind turbines generate electricity for 7 to 9 cents per kilowatt hour. That's down from 25 cents per kilowatt hour in 1981. In the same period, the capital costs of wind turbines have dropped from $3,100 per kilowatt to $1,100 per kilowatt.[3,4] Solar energy includes photovoltaics and solar thermal technologies. During the 1980s, costs per kilowatt hour fell 50 to 70 percent.

Phosphoric acid fuel cells are capable of producing electricity for 13 cents per kilowatt hour. They will be available in two to three years, at a capital cost of around $1,500 per kilowatt. Molten carbonate fuel cells will be available a few years later at around $1,000 per kilowatt. In Japan, regulations allow environmental costs to be factored into utility rates. Based on this rate structure, Tokyo Gas recently purchased two American-made fuel cells.

These technologies exist today. My legislation recognizes these promising, nonpolluting technologies in two ways. First, it directs federal agencies in the affected areas to install them, if they meet a cost-effectiveness test that requires a 10-year payback on the investment under a formula that would include environmental costs. Second, it establishes a goal to have federal facilities meet 5 percent of their electric energy needs with alternative and renewable technologies by January 1, 1997.

What will it cost the United States as a competitor in the world energy production market if we fail to support these technologies? Other countries recognize that the world needs new sources of energy; sources of energy that do not pollute. Right now, if federal agencies go out to the marketplace to purchase solar, wind, biomass, and fuel cell technologies, they will most likely be purchasing US products. But if we fail to support the domestic industries that produce this equipment now, they may not be there in the future.

There are major social and environmental costs we will all pay if we fail to broaden our technological base to include nonpolluting energy generating technologies. The damage to the health of our citizens is an environmental cost. Damage to our lakes and forests is an environmental cost. It is hard to put a dollar value on a small lake in rural Maine made lifeless by acid rain. It is even more difficult to put a dollar figure on the pain of a person struggling with asthma, emphysema, or heart disease.

Various sources have suggested that the pollution and social costs of burning fossil fuels range from two to six cents per kilowatt hour. When the federal government determines the cost effectiveness of energy conservation improvements at federal facilities, the environmental costs must be part of the calculation. This

Amendment directs the administrator of the Environmental Protection Agency to set cost estimates for environmental emissions. These estimates would be incorporated into the cost effectiveness analysis methodology developed by the National Institute of Standards and Technology.

Alternative, renewable, and clean energy technologies must be cost effective to find customers in the federal government. But the cost effectiveness test must include all the costs. In time, these technologies will be competitive with traditional energy sources and I think that day will come soon.

We do not anticipate that this amendment, on its own, will cause utilities to develop alternative energy sources. Rather, for their own reasons utilities will, over time, grow more receptive to alternative sources. Some far-sighted utility companies have already accepted alternative sources as a necessity for the future. The amendment will, however, require federal facility managers to install new technologies to meet their direct electric and cogeneration needs where such independent installations are feasible.

Alternative Financing

Will energy conservation improvements require federal dollars? Fortunately, that is not necessarily the case. For the last 15 years, many state and local governments have been retrofitting government buildings with energy conservation improvements without any capital investment, taking advantage of alternative funding arrangements.

The federal government now has only limited involvement in these beneficial public-private partnerships. This Amendment contains a mechanism that will allow all federal agencies to participate in alternative financing arrangements.

These shared savings contracts, as many of you are aware, operate as follows: The energy services company pays for the conservation measures; it does not require an expenditure by the owner. The contractor then shares in the building owner's reduced cost. This immediately reduces the amount the building owner has to pay for electricity.

Conclusion

The two bills I have described would result in clear and immediate benefits for the federal government and the public it serves. The nation at large would receive the indirect benefits of more energy efficient federal facilities: less pollution and lower utility costs. Through the use of innovative technologies in federal buildings, products could be demonstrated to the public and the building industry as safe, efficient, and effective means of reducing energy and construction costs.

The Advanced Building Consortium bill would lead the way for use of government buildings as a test bed for untried developments in construction. The federal facilities amendment to the Clean Air Act would place the government in the lead by requiring the use of alternative and clean energy sources in buildings powered by utilities that produce acid rain. Because of its size as a utility customer, the federal government has a duty to search for more efficient and environmentally friendly technologies. These two bills would give the government the responsibility and the ability to serve as an example and an innovator. That leadership is necessary if cost-effective construction and energy use are to be the rule in the coming years.

References

1. Henson Moore, United States DOE Press release dated January 26, 1990.
2. The 1990 *Annual Energy Outlook* published by the Energy Information Administration, United States DOE, pp. 45.
3. Statement by the American Wind Energy Association at the National Energy Strategy hearing conducted by the Dept. of Energy and Treasury, February 2, 1990.
4. Table 2 of The Public by the Safe Energy Communication Council, Winter 1990, MYTH-Busters, No. 5, Table 2.

Themes of the Building Systems Sessions

Robert H. Socolow
Center for Engineering and Environmental Studies
Princeton University
Princeton, NJ

New Questions, Old and New Answers

Throughout the two days of Building Sessions, I have been asking myself which of the answers we were proffering were new ones, and which were ones that have been on our lists for the 20 years that I have participated in this enterprise. The celebration of the twentieth anniversary of Earth Day in less than a month may have something to do with this retrospective frame of mind.

Professor Glicksman in his opening remarks correctly observed that in many instances the answers we were proffering were the same, but the questions had changed over time. Answers of this kind are the best sort; they have been called spherically sensible—they make sense no matter how you look at them. (The converse are proposals that are spherically senseless—senseless no matter how you look at them!)

The most important spherically sensible answer we have been giving for 20 years is that it is productive to disaggregate energy demand in many different ways (regionally, by economic sector, by physical component—for example, window vs. roof), in order to answer in considerable detail the question: "Where is the energy going?" Indeed, the recognition that little was actually understood about patterns of energy use was the single most important insight of the energy research community during the first period of heightened environmental consciousness 20 years ago, and it has served as a touchstone for analysis ever since.

Specifically, answers that illuminate how energy is used have been helpful in answering at least three recurring policy questions: 1) How does one maximize the productivity of the economy? 2) How does one minimize the vulnerability of the national economy to foreign pressures made potent by large imports of oil? 3) How does one minimize the environmental impacts (local, regional, and global) of energy use? Although the weightings of various supply technologies in optimal responses are different if one is primarily concerned with Joules, with oil, or with carbon atoms, nonetheless in each case a crucial component of the analysis is end-use energy efficiency. Such analysis time and again reveals the priority of investments in further energy efficiency through technological and policy innovations, relative to investments in any of the supply options.

Three Older Perspectives on End-Use Research

The recurrently correct answer that it is immensely productive to disaggregate total energy use promotes three perspectives that were explicit in the original formulations of nearly 20 years ago: a systems perspective, an experimental perspective, and a high-technology perspective.

The systems perspective is a fundamental element of end-use analysis, in the sense that the object of inquiry is not vehicles but transportation, not buildings but housing (or, an even more encompassing word, *habitat*). During one of our early sessions we were engaged in an argument (related to this distinction), which revealed some confusion of vocabulary that I would like to remedy. The orthodox strategy of end-use energy analysis is to decompose total energy use into a sum of products:

$$\text{Total energy use} = \text{Sum}\,[(\text{level of use}) \times (\text{intensity})],$$

where the sum extends over the component sectors, and within each sector one strives to load onto "level of use" as much as possible of lifestyle and onto "intensity" as much as possible of technology.

The best known example is probably transportation, where the sum might be over modes (air, automobile, etc.), the level of use would be vehicle-kilometers flown or driven, and the intensity would be the liters of fuel per vehicle-kilometer. Flying and driving less or improving load factors (passengers per vehicle) is addressed in parallel with improving the fuel economy of engines and reducing vehicle drag. No one should be surprised when time-series data reveal improvements in intensity without improvements in total energy use; indeed, improvements in technology have done much to limit energy growth for transportation, in the face of great expansions of travel by airplane and automobile.

The same holds true in the buildings sector, where floor space per capita has increased and buildings have become more heavily equipped with appliances. At the same time building thermal integrity has improved and appliances have become more energy-efficient. New objectives for indoor environments are becoming widespread; for example, the enclosed shopping mall and the high school indoor swimming pool are both now frequently encountered in United States suburbs. It is important, accordingly, that the energy analyst keeps abreast of every new demand made on the indoor environment, searching for both technological and policy innovations that can lessen the impact of its associated resource use on the environment.

The buildings research community is continually challenged to confront the nested structure of the energy system—to make sure, for example, that the reference system is large enough to include not only issues driven by the characteristics of the exterior of buildings but also issues driven by the technologies at play within buildings. Architects in particular have little training along these lines. Thus, there was a time when people were uncritically enthusiastic about British commercial buildings that were heated without boilers or furnaces, just by their lighting—an accomplishment which may or may not be economically and ecologically optimal. Around 1980, as Dan Nall reminded us, architects of commercial office buildings were blind-sided by the large increases in cooling loads that came with the electronic office; today, we correctly understand the personal computer to be one of the leading energy-using appliances, rivaling the refrigerator. Recent field studies of energy use by computers are clarifying the technological and policy issues associated with providing the services of computers at economically and environmentally optimal levels of energy use.[1]

Along with a concern for what goes on within and at the boundaries of buildings, a systems view of habitat includes a concern for what goes on around buildings—issues of city and regional planning. When one disaggregates end use into "buildings," "transportation," and "industry," land use issues can easily fall into the crack between the first two, and that is what happened at this conference. We have to remember that being thoughtful about where buildings are built and, in some instances, incorporating community-wide technologies like district heating and cooling (often in conjunction with cogeneration) are important ways to meet environmental objectives.

The experimental perspective affirms that one cannot model energy use in the absence of measurement. One of the most important achievements on the part of the end-use research community over the past two decades is to have gained acceptance for measured data, an achievement requiring new software and methodologies that took considerable time to develop (and that are still imperfect). Cases in point include PRISM (software for evaluating energy efficiency in space heating and cooling by means of a well researched weather-correction methodology) and the EPA urban and highway driving cycles.

When measured and predicted performance improvement are compared, the predictions (as in other fields) are generally found to be overestimates. However, the crucial and sometimes forgotten point is that in the course of measurement the innovation becomes more deeply understood. Often, measurement reveals unanticipated linkages. One class of examples, stressed in Peter Brothers's presentation, addresses the role of the human operator, which, until it is carefully incorporated, undermines the full potential of an energy management system in a building.

Another class of examples concerns the interaction among pieces of hardware. In the research of Les Norford and Scott Englander the performance of variable-speed-drive motor controllers on the supply and return fans of large buildings was found to fall short of prediction, as a result of operating rules for duct pressure that derive from pre-computer-age needs to reset control parameters as infrequently as possible. Conversely, rethinking variable-speed-drive control in conjunction with electronically controlled room-level ventilation systems reveals opportunities for achieving greater energy savings (and greater comfort) than would be predicted from independent evaluations of the components.[2]

Generalizing, I would state the following rule: Every innovation *underperforms* relative to claims when deployed exactly as described, but *overperforms* when fully understood in context.

The high-technology perspective has been crucial to end-use research in its struggle for respectability—against the expectation that first-class scientists and engineers should confine their efforts to issues related to energy supply. The struggle continues to this day. At the National Science Foundation, the Engineering Directorate is taking a fresh look at how their sponsored research relates to the environmental sciences. At the US Department of Energy (as part of the reorganization of the Division of Conservation and Renewables) I hope that the Division of Energy Research is assessing its role in fostering the basic science elements of the ongoing programs. At MIT, we have this conference, remarkable in many ways, but especially for giving such prominence to the end-use perspective.

Those of us who engage in end-use research are continually confronted with the interplay of technological options from low-tech through high-tech. Examples at the high-tech end described in the Building Sessions at this conference include the extraordinary progress made in selective window coatings, involving state-of-the-art thin film technology and semiconductor physics. Our sessions also gave a glimpse of forthcoming sophisticated software embodying revolutionary concepts in computer-based building management, notably the concept, expounded by Roger Bohn, that building electricity consumption (and the associated management of storage systems) will soon be optimized with reference to service-area-wide objectives rather than being restricted to building-specific objectives. (The two objectives will coincide ultimately, when continuous spot-market charges for electricity use in buildings are in place.)

One important reason to keep the high-technology aspects of end-use efficiency in full view at this time is that there is about to be a peace dividend in the form of scientists and engineers previously engaged in military research who are now looking for new research areas. As always, the most self-confident of our colleagues will move first, and they are bound to rejuvenate our field.

Three Newer Perspectives on End-Use Research

Although the systems perspective, the experimental perspective, and the high-technology perspective have been explicit in end-use analysis for most of the past 20 years, three other perspectives have been more recently acquired. I have in mind the incorporation of a public health perspective, an ecological perspective, and a planetary perspective. Each was explored in our sessions.

The public health perspective places what happens in houses within a new and powerful context, where time-integrated exposure to pollution is the unifying measure. Arguably "what is the total human exposure, indoors and outdoors?" will become as revolutionary a question for environmental analysis as the question "what is the total societal cost to provide services, from supply through end use?" has been for energy analysis. This approach links indoor radon and nuclear power, cigarettes and vehicle exhaust, and (in a slightly different context) natural foods and additives.

This question could hardly have been formulated 20 years ago, when almost all environmental measurement focused on the workplace (not the home) and outdoor (not indoor) air. The most important byproduct of end-use research in buildings has been the creation of a new research field addressing indoor air quality. Many lives will be saved and much illness avoided because of new approaches to public health inherent in this work. To a large extent, the scientists and engineers who brought instruments into houses to document the characteristics of energy use were the same ones who opened up this adjacent field. Conceptually, the key step was the realization that houses were worth studying with the methods of modern science.

The ecological perspective focuses on how energy efficiency technologies interact with the natural environment. The chlorofluorocarbon issue is a case in point, where the ecological perspective highlights the need to reconcile incongruent objectives. On the one hand, the chlorofluorocarbons are devastating the stratospheric ozone layer. On the other hand, the chlorofluorocarbons are of great importance for energy efficiency, for at least two reasons: they are currently the most economical agents for the production of the foams of low thermal conductivity, and they are currently the most economical thermodynamic working fluids for the cooling cycles of refrigerators and air conditioners. Our sessions gave insights into how the multi-objective pursuit of alternatives to the chlorofluorocarbons has stimulated a wide range of innovative technologies based on low-cost, low thermal conductivity materials, including vacuum powders and aerogel.

From an ecological perspective one must feel hopeful about the current restructuring of the Division of Conservation and Renewables of the US Department of Energy, which has as one of its objectives the integration of the buildings programs on energy efficiency and renewable energy. As David Pellish reminded us, window research increasingly incorporates both the management of sunlight and the management of thermal radiation. The full scope of this integration may take the Department of Energy beyond the union of its current programs, however.

An integrated research program should include consideration of the sun as a source of photons, not only of heat and light. Photovoltaic systems have already become interesting at the building level, and an applied research program addressing photovoltaic roof shingles, for example, could accelerate the rate of innovation and deployment of this technology by design professionals. The electricity available from a building roof (20 or more watts per square meter on average in most locations, and peaking when cooling needs are greatest) can be a significant factor in building design and operation, and will have additional implications for grid-connected systems.

As a second example, considering the ecological aspects of building technology, the choices among building materials (including choices between plastics and wood, between plastics and metals, and between one metal and another) raise many new research issues. Such materials choices, for example, involve questions of resource availability, recycling potential, and toxicity of waste streams in the environment.

As a third example, the integration of biomass fuels with building and community energy systems presents intriguing research opportunities. New technological approaches to the conversion of biomass to higher-value energy forms, through gasification and gas-turbine-based electricity generation, suggest the possibility of integrating—in entirely new ways—forest and crop management, decentralized industrial production, community design, freight and personal transport, and building energy management.

The planetary perspective moves emphatically beyond the residential and commercial building stock of the United States to the building stock of the world. Global issues, such as the greenhouse effect and large-scale regional issues like acid rain, enlarge our understanding of the community to whom end-use energy research is addressed. For energy policy in the United States, the planetary perspective appears in two complementary guises.

First, the planetary significance of US energy policy derives from the goal of establishing the level of national credibility necessary to secure the cooperation of other countries in the implementation of effective global environmental strategies. No international consensus will ever be reached unless the United States, in particular, is persuasive about the depth of its own commitment. Our level of resource use is too conspicuous and too well known abroad (it may even be overestimated) for us to expect others to join in any constraining policy initiative related to energy supply, unless it is part of a package of initiatives that includes vigorous measures designed to enhance the objective of energy efficiency.

Second, the planetary significance of US energy policy derives from its immense leverage, because the United States is so widely imitated. Our cars, our refrigerators, our house designs, the organization of our commercial space, the layout of our cities—these are widely perceived to define "the good life." It is hard to see this changing. As a result, to the extent that energy efficiency abroad is a help to environmental objectives in the United States, the *benefits* to the United States inherent in discovering new technologies and market strategies enhancing energy efficiency are larger than would be estimated on the basis of national impact alone—while the *costs* of doing at least the pioneering steps of discovery and trial-and-error implementation are nearly unchanged. It follows that estimates of benefit-cost ratios for domestic energy efficiency research and development will be underestimated, if they are based on the assumption that new innovations will have their only impacts within the United States.

I would not be surprised if the conference record reveals that the Building Sessions have been the most parochial and the most US-focused of the entire conference, for our buildings research community has often felt stretched to the limit interacting with the notoriously diffuse and disconnected domestic buildings industry. Those currently responsible for developing the new academic programs in technology and architecture, at MIT and elsewhere in the United States, will have to swim upstream to become truly international centers.

The choices among non-US-based research problems is wonderfully varied. Problems related to energy use in the less developed countries loom particularly large. New ideas related to building-level and community-scale energy use for cooking and heating in rural villages, for example, have immense potential. So do inventive strategies to accelerate the incorporation of advanced building technologies into the large office buildings in the urban cores of new cities in developing countries, In both cases, the underlying concept is called "technological leapfrogging," in which new technologies and policies are deployed *first* in developing countries. Indeed, one of the great stimuli to innovation over the next decade will come from confronting the problems of the "south" as new problems. Some of the solutions proposed will apply in the north. Others that will not apply in the north will still be welcomed for alleviating the stress on our common environment.

Also of global significance is the overall energy efficiency of the formerly planned economies of the former Soviet bloc. A few of us at several US institutions have been working with Soviet counterparts with the joint objective of accelerating the incorporation of the energy efficiency concepts and buildings technologies into that exceptionally energy-wasteful economy. In one strategy, the Natural Resources Defense Council is shipping a pair of low-emissivity double-glazed window systems to the Soviet Union, one destined for a research laboratory and one destined for deployment in a research apartment in one of the immense residential apartment blocks. The expectation is that to conduct simultaneously both field trials and laboratory investigations related to buildings standards will ease the adoption of this new technology. Such coordinated strategies may have their place in promoting innovation in the United States as well; gas-fired heat pumps may be an apt example.

It is hard to overestimate the importance of making rapid progress in thinking through the technologies and policies of the post-industrial, ecologically constrained world. Our small planet cannot sustain world-wide deployment of the current technologies that characterize the built environment and the transportation systems of the United States. It is incumbent on Americans to take the lead in inventing a more ecologically responsible future.

The research community does not yet have most of the answers. In fact, we are sure to discover new problems and constraints as we understand this planet better, and these will lead to new solutions. What the research community does have is history's most productive methodology: the scientific method. It is critical is to keep alive the spirit of innovation, of trying and being willing to fail, of evaluating carefully the extent of our successes, and fully informing our colleagues and the public worldwide.

The Spotlight Returns to Energy Efficiency. Are We Ready?

Politics is a searchlight, moving from topic to topic, rarely concentrating on any single topic for more than a year or two, yet returning to each topic after a time. When favored solutions work imperfectly, the public becomes tired and the searchlight moves on. But the issues (drugs, health care, care of the elderly, illiteracy, nuclear weapons, the environment) persist, and the professionals stay at work. My message to fellow professionals working on the efficiency of energy use is that the searchlight is heading our way.

We have not seen the searchlight for nearly a decade. This time around, the energy efficiency agenda should be less distressing to society as a whole, as compared to the decade of the seventies when the arguments were shrill and some of the professionals associated with the energy supply industries saw themselves as the losers of a nasty argument. Indeed, the Reagan administration approached energy efficiency and solar energy as if it had a score to settle.

In the intervening years the case for energy efficiency, like the case for cogeneration, has picked up important allies in industry. Enterprises large and small have recognized the huge markets (domestic and international) for efficient lighting, motor control, high-performance thermal insulation. advanced compressors, and, to be sure, devices not yet unveiled.

During the 1980s, our ranks thinned. As a manifestation of the real peace dividend, in talent more than in dollars, our ranks are about to swell to record numbers. Indeed, it will take considerable imagination, drive, and even grace, on the part of those of us who have stayed the course (adequately represented by the participants in these sessions) to retain leadership as the newcomers arrive. It is entirely possible that the new arrivals will shoulder us aside, much as we shouldered aside the buildings energy establishment 20 years ago.

References

1. Norford, L., A. Hatcher, J. Harris, J. Roturier, O. Yu, "Electricity Use in Information Technologies," *Annual Review of Energy* (1990) 15: 423–53.
2. Englander, S., "Ventilation Control for Energy Conservation: Digitally Controlled Terminal Boxes and Variable Speed Drives," *Center for Energy and Environmental Studies Report No. 248,* Princeton University, Princeton, NJ, 1990.

Session D
Electric Power Systems

Summary
Session D—Electric Power Systems

David White
Session Chair
Energy Laboratory
Massachusetts Institute of Technology
Cambridge, MA

Electric power, economic development, and improved quality of life are closely intertwined in the world's industrialized economies. There is strong evidence that electric power is also a critical ingredient for the effective and healthy growth of productive economies in the developing nations.

The continued expansion of new uses for electrical power—communications, computations and data processing, robotics, industrial process control, building space conditioning, etc.—has created a steady growth in demand for electricity in developed economies. The ability to perform both physical and intellectual tasks is enhanced by electricity. Its value is even greater because it can be delivered and controlled by simple configurations of electric conductors and components. The limit of what electricity can do for humankind in meeting its many needs and desires has not yet been reached. The potential for further diversification and growth remains. It is therefore nearly axiomatic that electric power will supply an ever growing fraction of the need for energy services. This growth in electric power demand will be amplified as the economies of developing nations expand and mature.

The world now consumes about 30 percent of its primary energy resources to produce electricity. Typical examples of primary energy to electricity are: US, 34 percent; OECD, 37 percent; USSR, 27 percent; and developing nations, 16 percent. This electric power is derived 69 percent from fossil fuels and 42 percent from coal. Because of the enormous consumption of primary energy (332 quads) and particularly the large consumption of coal (102 quads or 3.6 million tons), electric power is a major factor to consider in controlling regional and global environments. The coupling between electric power production and environmental effects becomes even more important when considering energy use in the 21st century.

There are electricity generating technologies that do not emit gaseous pollutants to the atmosphere and there are advanced versions of fossil-fuel based electric generating systems whose atmospheric emissions per unit of electric power generated are significantly lower than current designs. How these technologies will improve and which will be installed in future electric power systems is a critical issue for electric power development in the 21st century. It is also possible to increase the productivity of electric end use and thereby reduce the demand for primary energy resources. The installation of improved end-use devices is often controlled by socioeconomic factors that do not optimize the total energy/environment equation. Thus, both technological and socioeconomic innovation are important in determining the size and structure of the electric power systems in the next century.

To address this diverse set of issues and evaluate the potential for more efficient and environmentally acceptable electric systems, the electric power topics were divided into the following sessions: Electric Power for Developing Nations (D-1), Dietmar Winje, chair; Fossil-Fuel Based Electric Power Technologies (D-2), János Beér, chair; Nuclear Based Electric Power Technologies (D-3), Kent Hansen, chair; Alternative-Energy Based Electric Power Systems (D-4), Jon McGowan, chair; Electric Power End-Use and Network Efficiency (D-5), David White, chair. Summaries of these sessions and the respective papers follow. Jon McGowan's paper provides an overview of alternative energy options and serves as the summary for Session D-4.

Summary
Session D-1—Electric Power for Developing Nations

Dietmar Winje
Session Chair
Energy Engineering and Resource Economics
Technical University of Berlin
Berlin, Federal Republic of Germany

The electric power sector in most developing countries will play an important role in economic development and the improvement of living standards. Presently the developing countries, excluding the People's Republic of China, account for approximately 16 percent of the 7.6 billion toe of global commercial primary energy consumption. About 30 percent of the primary energy consumed in developing countries is used for electricity generation. For the future, it is expected that in the developing countries almost 50 percent of primary energy input will be needed for electricity generation. The expected growth of electricity generation confronts the developing countries with a variety of technological, economic, and environmental problems. In industrialized countries the electric power sectors are concerned with environmental problems related mainly to generation, while so far in many developing countries environmental issues are not considered a high priority. The problems of the electricity sectors in the developing world differ among regions and countries.

In many countries of Africa, for example, indigenous commercial energy resources play a small role as a base for electric power generation. As a result, oil must be imported, leading to major difficulties in the debt situation in these countries. The societies' needs for electricity exceed the supply. Existing electricity systems are in most cases isolated, with small generation units that operate at low efficiency and reliability. Considering these basic problems, environmental issues in the power sector remain of relatively small importance compared with the necessity of improving and expanding the existing systems. Especially in Africa, the need for cooperation with industrialized countries in the development of national and regional electricity master plans must be emphasized.

In Latin America, where electricity generation is about evenly distributed between hydroelectric and thermo-electric plants, the development of the power sector is overshadowed by the general debt problem. However, the environmental impact of electricity supply has become more important in the last 10 years. Some countries have begun to consider the environmental impacts of mining fossil fuels and constructing large hydropower and thermal power plants. In some areas reforestation projects are now required by law. In Latin America there seems to be a need for concerted international action to ensure the development of an efficient electricity system in a sound environment.

In terms of energy intensity the Eastern European countries—where energy resources are not generally scarce—consume a larger amount of primary energy and electricity than industrialized countries. The electricity intensity is two to three times, and the energy intensity three to four times, higher than in industrialized countries. Besides the inefficiency of energy generation and industrial and agricultural production, the poor quality of energy carriers such as coal and lignite and the large portion of primary industries contribute to these high ratios. They also explain the high level of pollution containing SO_2, NO_x, and fly ash. To improve the situation, the restructuring of the fuel mix in the power sector and the implementation of stack gas cleaning equipment seem to be important; both measures should be undertaken in cooperation with industrialized countries.

The quantitative analysis of the electric power sector reveals that in 1987 around 36 percent of the world's primary energy was used for electricity generation; the respective share in the developing countries is 31 percent. It is expected that the primary energy input for electricity generation in developing countries will increase to 46 percent by 2010, making the electric power sector dominant. This is caused by the high annual electricity growth rate of 5.6 percent for the developing countries, compared to a 3.2 percent increase in the world.

In looking at the potential for reducing pollutants in the electricity sector of developing countries, the size of the sector should be kept in mind. The electricity sector in these countries contributes about 4 percent of the world's CO_2 emissions, 3 percent of SO_x emissions, and 3 percent of NO_x emissions, whereas the figures for the industrialized countries are much higher. Considering energy consumption and emissions per capita, the huge differences between the developed and developing world become more obvious. In the area of energy conservation, and electricity conservation in particular, as well as the CO_2, SO_x, and NO_x problems, the industrialized countries and the centrally planned economies need to take action first. However, developing countries must also make efforts to reduce environmental pollution; therefore, some means of reducing emissions and some technology and policy instruments have been suggested.

Electric Power in Africa—Issues and Responses to Environmental Concerns

John Gindi Boutros
National Electricity Corporation
Khartoum, Republic of Sudan

Energy Resources and Use in Africa

For centuries, wood and charcoal fuels have been the primary sources of household energy in Africa in urban and rural areas, as a majority of the poor depend on these traditionally accessible fuels for cooking, heating, baking, and brickmaking. In large areas of Africa, commercial energy ("modern energy") has not yet replaced the old energy sources such as wood and charcoal fuels, human and animal muscle, and agricultural wastes. Deforestation is occurring rapidly in African areas because new agricultural schemes are replacing the forest land (in addition to the still common use of wood fuel). At the current deforestation rate, it is estimated that by the year 2020, accessible forests will disappear altogether, seriously damaging the environment. This continued reliance on traditional fuels leads to desertification and drought. As a result of this inefficiency and serious environmental degradation, many African countries face political instability, financial problems, and environmental collapse.

In some areas of Africa, oil has become the main source of commercial energy used for transportation, industry, agriculture, electricity generation, and household needs. The oil price increases in 1973 and 1980 caused many African countries difficulties in supplying and financing oil. Many borrowed heavily to finance oil imports. As a result, national deficits grew and development programs were delayed. Some African countries had their own crude oil and natural gas resources during the energy crises of the 1970s and 1980s. Unlike Algeria, Angola, Congo, Egypt, Gabon, Libya, Nigeria, and Tunisia, they were not particularly affected by those crises.

Thermal generation plants in African countries vary in type and size. Both large- and medium-scale power plants exist in South Africa, Zimbabwe, Nigeria, Egypt, Algeria, Morocco, and Zambia. The majority of these units burn oil or coal. Small steam, diesel, and gas turbines are installed in many towns and villages in Africa. Some of them are connected, working in parallel as a limited national grid system. Others work as isolated systems, or directly supply areas or consumers, or serve as standby units. Small plants cost more to operate compared with larger plants, especially high- and medium-speed diesel units and gas turbines that use light fuels. This type of power generation is becoming more common in Africa as an expedient way to introduce electric power to new areas, or to provide additional energy during shortages and outages of the main electricity supply.

Electric Power Generation and Use in Africa

Electric power provides the most efficient and economical means of motive power for industry, agriculture, domestic use, and those services needed to improve the present living and environmental conditions for many African citizens and regions. Therefore, a large-scale electrification program is needed. Existing electric facilities in African countries supply primarily the growing industrial sector, mining, agriculture, and large towns. Still, the majority of land and activities are not yet electrified. Average generation and consumption of electric power per capita in Africa varies from one country to another. Generally, it is very low compared with other countries. Table 1 compares Africa's per capita energy consumption with other countries for the year 1986 (in kilowatt hours per year). From these figures, it is clear

Table 1. African per capita energy consumption as compared with other countries for 1986, in kilowatt hours per year. (Source: Reference World Energy Statistics and Balances, 1971/1987.)

Africa		Others	
Country	KW.HR./Annum	Country	KW.HR./Annum
Algeria	580	Pakistan	300
Cameroon	235	Iraq	1500
Egypt	730	Kuwait	8500
Ethiopia	45	Oman	3600
South Africa	4200	Bolivia	230
Sudan	75	Peru	630
Tanzania	50	Hungary	2350

that major action, including medium- and long-term programs must be adopted to shift Africa from the use of primitive and inefficient energy sources to modern and sound energy sources.

Many systems should be immediately modernized with electric power to improve output and save human and animal power, wood fuel, and oil. Two such systems common in Africa today are SAGIYA and SHADOOF. SAGIYA is a simple continuous circular conveyor elevator used to drag water from a river or lake using animals as the driving power. SHADOOF uses a simple lifting beam, fixed at its middle point with a bucket attached to a long rope at one end and a counter weight at the other. Human power is used to work on the empty bucket, pulling it downwards to water level. Both processes are used to pump water from rivers and lakes for irrigation and could be replaced with simple and efficient electric pumps. Electricity could also be used for lighting, cooking, grain and oil mills, refrigeration, etc., replacing kerosene and both human and animal power.

Essential improvements await African citizens in areas that now use no electricity. A progressive program to introduce electricity to these areas would improve standards of living, productivity, health, education, transportation, and communication. Such a large project needs a high level of planning to select the most appropriate technologies to substitute for locally available energy and to recommend the least-cost solution to achieve its objectives step by step given the national and regional resources available. The investment required would be arranged through international funding. The full participation of the citizens of Africa in the planning, completion, and operation of new facilities is very important.

Problems and Issues with Electric Power Systems in Africa

In all African countries there are existing power systems, as well as isolated small systems or individual generating units. The generation facilities in many of these countries do not satisfy basic electric power requirements and operate at low levels of productivity with high systems losses. Many countries face shortages and unreliable power supplies. As previously mentioned, in some African countries the majority of the population is living without any electricity services at all.

The main problems and issues in supplying electric power include:

- inadequate investment in generation capacities, particularly hydropower,
- inadequate use of the local primary energy resources on a national or regional level,
- lack of applying new technologies to improve or replace existing inefficient systems,
- delay in utilizing renewable energy technologies,
- delay in construction of power projects,
- inadequate maintenance for existing power facilities,
- lack of spare parts,
- inadequate training of local staff,
- inefficient management and organization of most of the African National Utilities,
- manpower brain drain,
- uneconomical and inefficient electricity pricing,
- problems of billing and collection for electricity consumption,
- uncontrolled demand (growth without priorities),
- lack of demand management,
- loss reduction and conservation policy.

Much can be done to resolve these problems if they are approached in scientific and systematic ways for the benefit of the African economy and African citizens. The main goals are:

- to modernize and improve the efficiency of existing power systems,
- to finance power projects,
- to develop and use local and regional primary energy for large-scale generation,
- to develop national grid systems and encourage regional interconnections,
- to obtain the materials and equipment so that local renewable energy can replace wood fuel in remote areas,
- to recognize the need for a national and regional master plan for electrification
- to put into place energy policy, standards, and regulations to improve environmental aspects of energy use and to protect nature.

Electricity in Africa in the 21st Century

A master plan for electrification of Africa is essential. Such a plan must have clear objectives and include the participation of all African countries. This endeavor would need the attention of international organizations and the financial support of the World Bank, the African Development Bank, the European Investment Bank, and other international and regional financial institutions, in addition to support from industrialized nations.

One of the major goals of the master plan is to recommend national power projects. These projects should seek to:

- improve the existing facilities in each country,
- concentrate on national and regional primary energy resources,

- recommend large, economical generation programs (especially hydropower) to produce cheap and reliable power,
- replace "primitive" energy sources (human and animal power, and wood) with electric power,
- achieve these electrification goals in all areas of Africa,
- introduce renewable energy sources such as solar, wind, mini-hydro, biomass and geothermal,
- develop the management and organization of national utilities in terms of both administration and facilities, and
- ensure the participation of African citizens in the study of the plan—as well as its execution, operation, and maintenance—through widespread training facilities and programs.

Conclusion

I hope this conference will initiate further discussion of these concerns, and that one of its recommendations will be for a special conference to be held in one of the African countries. This special conference should present the issues discussed in this paper for evaluation and study by all African countries. Their acceptance of these ideas will be the essential first step toward a formal study of an African master plan for electrification. Such a plan will directly improve the standard of living for millions of African citizens; will open new opportunities for jobs and industries, both inside and outside Africa; and, mainly, will help control the environmental impacts of energy use in Africa.

Appendix A. Hard coal production by region and country (million MT).

COUNTRY	1971	1972	1973	1974	1975	1976	1977	1978	1979	1980	1981	1982	1983	1984	1985	1986	1987
Morocco	0.5	0.5	0.6	0.6	0.7	0.7	0.7	0.7	0.7	0.7	0.7	0.7	0.8	0.8	0.8	0.8	0.9
Mozambique	0.3	0.3	0.4	0.4	0.6	0.6	0.3	0.1	0.2	0.2	0.3	0.2	0.1	-	-	-	-
Nigeria	0.2	0.3	0.3	0.3	0.2	0.1	0.2	0.2	0.2	0.2	0.1	0.1	0.1	0.1	0.1	0.1	0.1
South Africa	58.7	58.5	62.1	66.0	68.7	76.5	85.2	89.2	103.7	115.1	130.4	144.2	142.7	159.1	169.8	172.5	173.1
Zambia	0.8	1.0	0.9	0.8	0.8	0.8	0.7	0.6	0.6	0.6	0.5	0.6	0.5	0.5	0.5	0.6	0.6
Zimbabwe	2.8	2.6	2.8	2.7	2.7	2.8	2.6	2.5	2.6	2.7	2.4	2.4	2.6	2.7	3.0	3.0	3.4
Others	3.1	2.8	3.1	3.1	2.8	3.9	3.6	3.8	3.8	3.7	3.7	3.5	3.9	3.9	4.0	4.8	5.0

Reference world energy statistics and balances 1971/1987

Appendix B. Crude oil and LNG production by region and country (mtoe).

COUNTRY	1971	1972	1973	1974	1975	1976	1977	1978	1979	1980	1981	1982	1983	1984	1985	1986	1987
Algeria	37.1	50.1	50.9	47.4	45.7	49.7	54.0	58.9	62.1	52.1	46.4	46.9	45.9	45.8	45.7	47.5	47.9
Angola/Cabinda	5.7	7.1	8.2	8.5	7.8	7.6	7.1	6.5	7.2	7.4	7.2	6.4	8.8	10.1	11.5	13.9	17.3
Benin	-	-	-	-	-	-	-	-	-	-	0.2	0.2	0.2	0.2	0.3	0.4	0.2
Cameron	-	-	-	-	-	-	-	0.5	2.0	2.7	4.3	5.3	5.7	7.4	9.2	9.0	8.8
Congo	-	0.3	2.1	2.5	1.8	2.0	1.8	2.6	2.8	3.3	4.3	4.5	5.4	6.0	5.9	5.9	6.0
Egypt	15.0	10.7	8.5	7.5	11.8	16.6	20.9	24.4	26.5	29.3	32.1	33.2	36.3	41.7	44.9	40.9	45.9
Gabon	5.8	6.3	-	-	11.3	11.3	11.3	-	9.8	8.9	7.7	7.7	7.9	7.9	8.6	8.1	7.8
Ivory Coast	-	-	-	-	-	-	-	-	-	0.1	0.4	0.7	1.0	1.1	1.3	1.3	0.7
Libya	132.5	108.3	104.8	73.4	71.2	93.7	99.3	95.5	100.7	88.4	58.7	54.7	54.7	48.6	50.2	50.7	48.1
Nigeria	75.5	89.9	101.4	111.3	88.0	102.4	103.5	93.6	113.7	101.8	71.1	63.7	61.0	68.7	74.0	72.4	63.1
Tunisia	4.1	4.0	3.9	4.1	4.6	3.7	4.3	4.9	5.5	5.6	5.4	5.1	5.5	5.5	5.4	5.2	5.0
Others	0.2	0.2	0.2	0.2	0.2	1.4	1.3	1.4	1.3	1.7	2.4	3.4	4.5	4.9	5.1	5.4	5.6

Reference world energy statistics and balances 1971/1987

Appendix C. Marketed production of natural gas (,000 toe).

COUNTRY	1971	1972	1973	1974	1975	1976	1977	1978	1979	1980	1981	1982	1983	1984	1985	1986	1987
Algeria	2448	3102	4120	4763	5866	7387	7071	10965	17973	12300	14337	18379	34271	33111	34479	37043	-
Angola/Cabinda	37	49	56	58	56	48	60	60	60	66	77	77	88	99	99	109	-
Congo	62	71	13	15	14	7	5	2	2	1	-	2	2	2	3	2	2
Egypt	72	60	74	37	33	104	353	583	863	1616	1844	2023	2376	3046	3733	4306	-
Gabun	83	92	138	151	147	144	151	141	128	136	122	128	101	124	140	141	-
Libya	412	1943	2625	2316	2879	3254	3336	4027	4012	2943	2308	1858	2513	3284	3417	3439	-
Morocco	44	48	60	68	61	68	74	70	64	57	73	67	71	71	74	73	-
Nigeria	176	235	361	337	309	492	663	868	1168	1395	2046	2202	2452	2410	2662	2720	-
Tunisia	1	20	115	203	212	216	232	289	333	358	395	427	430	430	412	429	804
Others	-	-	-	-	-	-	-	-	-	-	-	-	-	-	-	-	-

Reference world energy statistics and balances 1971/1987

Appendix D. Electricity production in Africa: I - thermal electricity; II - hydro/geothermal electricity; III - nuclear electricity; IV - total electricity production (in Gigawatt hours).

	1971	1972	1973	1974	1975	1976	1977	1978	1979	1980	1981	1982	1983	1984	1985	1986	1987
I	69251	73560	81031	86695	94707	101056	104666	113290	123555	138286	157913	168277	179923	195664	211750	220702	-
II	32356	35798	37911	41619	43142	47672	51230	53278	59816	65948	57536	56948	55250	94671	51547	5315	-
*III	-	-	-	-	-	-	-	-	-	-	-	-	-	3925	5315	5315	-
IV	101672	109423	119007	128379	137912	148798	155965	166641	183444	203858	215609	223433	235409	249513	268858	278769	-

* South Africa

Reference world energy statistics and balances 1971/1987

Policymaking Pertaining to the Environmental Impact of Energy Use in Latin American and Caribbean Countries

Eduardo Del Hierro
Ministry of Mines and Energy and
Former Senator of the Republic of Colombia
Bogotá, Colombia

Abstract

The careless use of natural resources in Latin American and Caribbean countries to tend to the livelihoods of a growing population and to earn sorely needed foreign currencies has brought widespread destruction and contamination. Common occurrences in this region are: deforestation, emission of noxious carbon and nitrogen oxides from forest and grass fires, water and air pollution due to all sorts of waste products, uncontrolled hunting and fishing, and construction of public works regardless of their environmental impact.

Weak national governments, inefficient or unconcerned, are preoccupied only by difficult economic and political problems; therefore environmental protection gets a low priority in governmental actions.

Much technical work at the local level is required to ensure effective and positive future actions. There is a need for reliable data collection on the environmental impact of public works and on the extent of the damage caused to river basins. Engineering designs should be improved to minimize the negative impact on the environment.

The case of Colombia shows that even when the national governments enact legislation in order to ensure the rational use of natural resources and to protect the environment, these laws are ineffectual. Colombian authorities have tried time and again to protect the environment with rather mediocre results.

In Latin America and the Caribbean there is an urgent need for concerted international action in all the economic and political aspects of policymaking and law enforcement in relation to environmental protection. National governments in this region need international cooperation to avoid further indiscriminate destruction of natural resources because isolated local actions are ineffectual.

Introduction

The per capita energy consumption of this region is far lower than that of developed countries. The yearly per capita consumption of gasoline is around 40 gallons on the average; even oil exporting countries such as Mexico, Venezuela, Colombia, and Ecuador have per capita consumptions of only about 60 gallons of gasoline. These data can be compared to an estimate of 300 gallons in the United States.

The yearly per capita consumption of electric energy is about 1300 KWH on the average for the whole region. Venezuela and Puerto Rico have a consumption higher than average: Mexico, Argentina, Brazil, Chile, Cuba, Costa Rica, and Colombia have average consumption and the rest are much lower.

The generation of electric energy is about evenly distributed between hydroelectric plants and thermoelectric plants. Argentina, Mexico, Venezuela, Ecuador, and the Caribbean islands rely mainly on thermoelectric power, while Brazil, Colombia, Peru, Chile, Paraguay, and Central America depend mostly on hydroelectric power. The contribution of nuclear plants is small.

The yearly average consumption of coal is about 100 kilograms per capita. The highest consumption corresponds to Colombia with 200 kilograms per capita. Mexico and Brazil have average consumptions, while some Central American countries, Paraguay, and Ecuador do not use any coal.

As a consequence of poverty and low levels of industrialization, the region is not a large consumer of electric energy or petroleum products. However, due to the great diversity of natural resources, some countries are net energy importers as buyers of petroleum products while other countries of the region export crude oil, petroleum products, or coal.

Since the governments of these countries are directly involved in electric energy, petroleum, or coal production, data for these items are available although some years late. On the other hand, data for the use or misuse of biomass are not very reliable. A fair estimate can be made of the production and use of sugarcane bagasse and of other agricultural by-products, but it is difficult to ascertain the amounts of wood used or wasted.

Active deforestation proceeding in most countries of the region, inefficient logging, and slash and burn agriculture waste a large fraction of the wood cut. The burning of vegetation and the rotting of deadwood produce large amounts of carbon dioxide and other undesirable substances.

The environmental impact of energy use and energy projects of the Latin American and Caribbean countries has to be analyzed, not only in relation to the direct local impact of each energy development, but also in relation to deforestation, water pollution, and the energy exports which bear on the global environmental conditions. It appears that the emission of carbon and nitrogen oxides from the burning of fossil fuels in the region is not commensurate with its population—estimated at 410 million people for 1990—because of the low per capita consumption indexes. On the other hand, the effects of deforestation as a consequence of the relentless cutting of tropical and subtropical forests, is a cause of deep concern because of the large emissions of carbon and nitrogen oxides, the destruction of the habitat of vegetable and animal species, and the subsequent destruction of soils by erosion. The deforestation increases the amount of silt and dissolved chemicals carried by the rivers affecting waterways, harbors, and hydroelectric plants. Furthermore, deforestation causes torrential water run-off during the rainy seasons leading to floods in the plains and lack of water from drying mountain springs in the dry seasons.

The building of roads and other works of physical infrastructure, without due regard to environmental impact, wreaks havoc with natural resources in many Latin American countries. A typical case is the opening of new roads in the foothills of the Andes mountain chain toward the Amazon river basin in Colombia, Peru, Ecuador, and Bolivia with the purpose of opening new land for farming or for oil field developments.

At first the roads are used for logging or to move in the drilling rigs, then the most valuable trees are felled in the heterogeneous tropical forest. Next, farmers move in to cut down the remaining trees to clear the land for subsistence agriculture using slash and burn techniques. Most of the forest is lost to burning and rotting since the wood is too far from any sizable markets. Finally comes the establishment of cattle ranches for extensive grazing or the planting of small plots of rice, corn, plantains, cassava roots, etc. Only a few areas are fertile land; most of the land in the Amazon basin in those countries has very poor soil that cannot sustain intensive agricultural use. Soil authorities in Colombia give the following description for this soil: "The soils in the Amazon region have very low fertility, low pH, very low basic saturation, very low levels of interchangeable calcium, magnesium, and potassium, low available phosphorus contents, and very high contents of interchangeable aluminum which turn them highly toxic for many crops." Moreover, the hot and humid weather prevailing in the Amazon basin promotes the rapid deterioration of the thin soil when the forests are cut down.

The foregoing remarks indicate the need for an integrated approach to set the correct policies for energy projects development and for the rational use of natural resources in every country in the region, to satisfy the socioeconomic need of the present and of future generations.

There is urgent need to make national and multinational policies for rational use and safeguard of the natural resources within the restrictions imposed by the present political and economic conditions existing in this region. Due enforcement of environmental protection laws is direly needed.

Policymaking and Law Enforcement for Environmental Protection in Latin America under Economic and Political Crises

Many electric power plants, oil fields and refineries, and large coal fields in Latin America are government owned; therefore, management and policymaking for the energy sector are closely related to the ever changing economic and political conditions. Unfortunately, in the last two decades, the development of the energy sector in the Latin American and the Caribbean nations has taken place under unfavorable economic and political circumstances.

The governments of the region have followed prevailing economic theories and have tried at all costs to increase the supply of electric energy and hydrocarbon fuels. In the seventies, the ready availability of petrodollars offered as loans by the private foreign banks and also by the multi-national banks made it possible to push ahead with many hydroelectric and thermoelectric projects for the region. At the same time, the high prices of crude oil brought increasing investments in oil and natural gas. The governments of the region were eager to satisfy the needs of their people for electric power and fuels, and the international banks were equally eager to lend out readily available funds. Neither the banks proffering the loans nor the industries of developed nations that sold machinery gave much thought to ecology or to environmental considerations. Very few people in Latin America or in the banking community were concerned or worried about the environmental impact of the energy projects on the local natural resources. These decisions were all market driven either by the forces of the money markets or by the demands of the people of Latin American and Caribbean nations.

The consequences of these events were to last for many years because of the long time needed to design, finance, build, and commission large energy projects. Moreover, delays and cost overruns were quite frequent. Many projects started in the 1970s were carried through only in the 1980s without many changes in design to take into account environmental impact.

The 1980s brought dramatic changes in the economic circumstances affecting Latin America and the Caribbean. The prices of petroleum and of many commodities plunged, interest rates went up and money markets became tight. The ensuing debt crises are well known. A very detailed study made in July 1989 by Guillermo Perry and Mauricio Cabrera for OLADE (Organización Latino-Americana de Energía) on the financial situation of the electric companies of Latin America shows a very sorry sight. The Chilean government had to bale out the electric energy companies in order to sell them to the private sector; the governments of other countries also had to use scarce budgetary funds to avoid the economic collapse of the electric sector. The foreign debts are in hard currencies such as dollars, yen, marks, or francs but the income from electric energy sales is in devalued local currencies. The companies have lost most of their working capital and have severe difficulties investing in additional plants. Since electric energy tariffs are a sensitive political issue facing weak or demagogic governments, the economic squeeze is very hard to avoid.

In many countries, the debt owed by the electric sector is a large fraction of the total public foreign debt. Ongoing negotiations of these debts will undoubtedly affect the electric sector. The International Monetary Fund and the international banks will have a lot to say in future financial decisions of the electric energy sector in Latin America.

The petroleum and natural gas sector has also had difficult times in the last 20 years. Income from crude oil exports has been much less than expected, making it hard for some countries such as Mexico, Venezuela, and Ecuador to meet foreign debt payments.

Political events in Peru have scared away investments in the oil fields turning this country from an oil-exporting country into an oil-importing country. Guerrilla fighters in Colombia have frequently blown up the pipeline from the Caño Limón oil field to the coast of the Caribbean Sea. About 510,000 barrels of crude oil have been spilled, thus polluting farms and rivers in Colombia and Venezuela. ECOPETROL, the Colombian state-owned oil company has had to learn to cope with these catastrophic spills.

To make matters worse, the lack of financial funds has delayed the development of projects to clean the smoke stacks of old refineries that send up large quantities of sulfur containing flue gases.

The development of the open-pit coal mine of Cerrejón in the Colombian peninsula of Guajira brings environmental problems. The $3 billion project to produce 15 million tons of low-sulfur steam coal for export — a joint venture between the Colombian government and Exxon — is already in full production. The mine operator claims to have taken actions to minimize the impact of the mining operation on the desertic land but some government authorities believe that more should be done.

Underground mining of coal in Colombia, the region's largest coal producer, is carried out by very many small miners under very poor conditions with very little government control. Six to seven million tons of steam and coking coal per year are produced by these small mines for local consumption and for export.

Road building to improve the transportation network, to open new land for farmers, or to develop mining prospects is going on throughout the whole area. These new roads are financed using foreign loans and local resources but the environmental impact is of little concern to public works authorities. The destruction of the Amazonian forest in Brazil, Peru, Bolivia, Colombia, and Ecuador is a clear example of economic market forces at work without much restraint from legal regulation.

The present political condition of Latin America is one of unstable and weak democracies fighting a clearly deteriorating economic condition, while the rest of the world seems bent on creating strong trading blocks.

The task of policymaking and law enforcement in relation to environmental issues becomes very difficult for governments facing awesome state problems. The oil exporting nations, Mexico, Venezuela, and Ecuador, are burdened with heavy foreign debts and strong inflationary pressures which have stifled economic growth. Central America is immersed in guerrilla fighting in Nicaragua, El Salvador, and Honduras. Panama has deep problems recovering political and economic sanity. Chile, Bolivia, and Guatemala are healing their economies from past problems. Argentina and Brazil have terrible economic crises with astonishingly high internal inflation rates. Peru is slowly sinking into chaos. Colombia, which has made valiant efforts to maintain monetary discipline and to honor all its foreign debts, is in the midst of murderous violence caused by guerrilla warfare, cocaine trade, and common crime. Furthermore, the falling price of coffee, Colombia's main export, creates additional economic problems.

To rule properly under such conditions becomes extremely difficult and due enforcement of environmental protection laws gets a low priority. In the best of cases, well meaning government officials in faraway capital cities are hampered by budgetary deficits and bureaucratic inefficiency. Some areas in the Andes mountains and in Central America are not controlled by the national governments but by guerrilla fighters or cocaine traders. Sometimes placer gold miners and subsistence farmers will open footpaths which later become public roads for local political reasons, destroying large tracts of forest.

There is evident need for concerted international action involving not only the national governments of this region but also those of the main developed nations. National and international public opinion has to be clearly informed about the rational use of natural resources and the urgency of environmental protection.

The communications media, especially television and radio, are powerful agents to promote public action. The World Bank and the other international banks should require compulsory environmental protection actions in all the works of infrastructure financed by them in this area such as: waterworks, hydroelectric plants, thermoelectric plants, public roads and bridges, refineries, oil fields and pipelines, etc. It is time to reserve and protect certain areas of forest by precluding human settlements and stopping the building of any roads and bridges, in order to shelter animal and vegetable species, to avoid forest fires and to protect the soils.

It is clear that these environmental protection actions have high costs which must be paid by all concerned. Therefore, the involvement of governments and international institutions is absolutely essential; otherwise studies may be made and reports written but the destruction of natural resources will go on unchecked. Already the United Nations and other entities such as OLADE (Organización Latino-Americana de Energía) emphasize environmental protection, but their actions must be strengthened.

Soft loans and grants, economic pressures, and trade conditions could be used to advantage in a framework of mutual international cooperation with the purpose of achieving quick action for environmental protection. In a shrinking world, the response to global problems has to be global in order to be effective.

Environmental Protection in Colombia — 1970–1990

An energy rich country such as Colombia with its wide variety of energy sources is a good case to study the environmental impact of energy use and of energy projects. The country has oil and natural gas fields, coal mines, mountains with a large hydroelectric potential, extensive tropical forests and prairies, and two long seashores in the Atlantic and the Pacific oceans. All kinds of environmental problems may arise. Besides these considerations, the author finds it proper and expedient to describe the problems occurring and the mistakes made in his own country.

In 1968, the Colombian government created INDERENA (National Institute for Natural Resources and Environment) to handle all matters pertaining to the proper use of natural resources and the preservation of the environment. The law intended this institution to control and oversee public entities, private companies, and individuals.

Early in the 1970s it became evident to the Colombian authorities that there was an urgent need for an integrated approach to environmental protection and to the rational use of natural resources. The enactment of the Natural Resources Code (Decree 2811 of 1974) set up a very detailed legal framework granting to the executive branch of government all the powers required to enforce the proper use of natural resources. This code is a very comprehensive collection of rules and regulations thoroughly covering the use of air, water, soil, and subsoil; the preservation of endangered vegetable and animal species; the control of the environmental impact of oil fields, mines, and electric energy plants; the studies to be submitted before power plants and dams are built; the legal conditions for the use of public lands and public forests; the prohibitions and penalties to be imposed to prevent grass and forest fires, etc.

Unfortunately, this code proved ineffective in many cases. The fiscal and economic conditions made it difficult for the national government to spend enough money in environmental protection — the funds available were meager for the task to be accomplished.

The Ministry of Public Works in charge of roads and railroads paid very little attention to the environmental impact of road building. The agrarian reform authorities paid attention only to the pressures of poor farmers and demagogic politicians and granted squatter's rights to anybody clearing public lands. Logging, hunting, and fishing continued without due respect to the law.

The environmental impact of electric energy projects moved the National Congress in 1981 to enact a law imposing a 4 percent surcharge on the wholesale price of electric energy sold, to finance environmental protection and rural electrification in the river basins and the regions affected by hydroelectric plants, thermal plants, and the mining of fossil fuels for energy generation (Law 56 of 1981). This law requires companies generating electric energy to carry out reforestation projects and to take measures to protect river basins and reservoirs. Reforestation projects were also needed to fulfill the needs for timber in underground coal mining. Rural electrification reduces the use of kindling wood by small farmers. Compliance with this law by the electric energy companies, all of them government owned, has not been totally satisfactory because, quite often, cash flow problems prevent the expenditure of these funds as ordered by the law.

In 1986 and 1989, two Colombian economists, Hernando Roa-Suárez and Astrid Blanco-Alarcon, studied the environmental impact of four large hydroelectric projects — Betania, Guavio, San Carlos, and Salvajina — with a total generating capacity of 2.310 MW. They show that even though environmental impact was taken into consideration from the early stages of these projects as ordered by the above-mentioned laws, the actual environmental effects have been considerable. These projects were carried out after 1974, and three of them are finished — the Guavio project will start generating electricity in 1992. Efforts are being made to protect the environment and to avoid silting up the water reservoirs. The river water levels downstream from these plants have been affected, sometimes causing floods in the rainy season as in the case of Betania.

The environmental impact of the country's four coal-fired thermoelectric plants in Yumbo, Zipaquirá, Paipa, and Tasajero (with a total capacity of 768 MW) was studied by Carlos Martínez-Mora in 1989. There is clear evidence that little is done to protect the environment in these old-fashioned steam plants. However, these thermoelectric plants do not operate often because of the availability of low-cost hydropower.

Political and economic factors have influenced investment decisions in energy projects without due consideration given to environmental impact. A typical case is the assignment made by law in 1978 to the Soviet Union to build two hydroelectric plants in the upper San Jorge River. Environmental impact studies for these two projects, URRA I and URRA II, show large negative consequences. In spite of this, local demagogic politicians and the Soviet government — interested in the sale of the machinery — keep insisting on these projects. The fact that thousands of hectares will be flooded, submerging trees and vegetation and destroying the habitat of many species of animals is of no consideration for the advocates of these projects. The electric energy to be generated could easily be generated by other hydroelectric plants elsewhere in the country.

For many years, ecological considerations had no weight for the financial institutions of the developed nations interested in selling capital goods for electric energy generation. In the seventies, the World Bank and the rest of the Western and Japanese banks did not have any environmental protection requirements in their loans for physical infrastructure.

Only in 1984 did the World Bank set any guidelines for environmental protection in its loans. It was too late for many projects started and financed in previous years. Therefore, roads, bridges, hydroelectric plants, thermoelectric plants, waterworks, etc., were built without due consideration to environmental protection — in spite of Colombian legal restrictions. Some of these projects are still in the process of construction, following the original design specifications.

In 1982, Colombian consultants, with the aid of West Germany, made a complete study of the energy sector for the Ministry of Mines and Energy and for the Department of Economic Planning. This study has a short chapter on the environmental impact of energy projects and energy usage. The study indicates that although the Colombian laws are clear on environmental protection, compliance with the law leaves a lot to be desired. This study is especially significant because it considers that environmental protection must be necessarily included in general economic planning.

Lately, the Ministry of Mines and Energy has had to act again regarding new legislation to strengthen the already existing laws. Present and future mining operations must take measures to protect the environment as ordered by the new Mining Code (Decree 2655 of 1988) enacted by dictate of Law 57 of 1987. In 1989, the Ministry of Mines and Energy imposed compulsory contractual obligations for all new petroleum joint ventures signed with foreign oil companies in order to protect the environment of the areas assigned to them for oil exploration and exploitation (Decree 2782 of 1989). These obligations are over and above those imposed by previous laws.

The Colombian case shows that even when national governments enact legislation to ensure the rational use of natural resources and to protect the environment, these laws are ineffective. A national government needs international cooperation and the help of worldwide public opinion to achieve positive results.

Recommendations for Action

To realize progress in environmental protection and the rational use of natural resources, prompt and concerted action in the economic and political spheres is required. The leadership in all nations concerned should be exerted to make use of existing institutions and to create needed organizations to achieve these goals.

Useful instruments at hand are regional agreements such as: the Andean Pact (Venezuela, Colombia, Ecuador, Peru, and Bolivia); the Treaty of Amazonian Cooperation (Bolivia, Brazil, Colombia, Ecuador, Guyana, Peru, Surinam, and Venezuela); the Caribbean Free Trade Association involving independent Caribbean nations; the Permanent Commission for the South Pacific (Colombia, Ecuador, Peru, and Chile); the treaties between Central American countries; and bilateral agreements in North, Central, and South America. All these treaties and the corresponding organizations could become actively involved in environmental programs at the national and international levels.

The United Nations, as well as the Organization of American States, should implement measures on environmental protection and on the proper use of natural resources through the corresponding agencies and such entities as SELA (Latin American Economic System) and OLADE (Latin American Organization for Energy). Great Britain, Holland, and France should be invited to joint efforts with all the American nations, given the traditional presence of those countries in the Caribbean region. In the economic sphere, the multilateral banks and the governments of the United States and other developed countries could exert pressure to get results and provide the needed financial resources for these programs.

In 1992, there will be a meeting on environmental issues planned by the United Nations. This could be an excellent opportunity for a wide multilateral agreement. Let us hope that effective action can be taken to get concrete results for the mutual benefit of all nations.

Conclusion

Experience in Latin America shows a strong need for concerted national and international action, both in the economic and political aspects of policymaking and law enforcement, to get concrete results in environmental protection. Not only must the Latin American and Caribbean nations bear the high cost of the region's environmental protection, but developed countries should also pay some of the cost. These countries buy goods and raw materials and will eventually be affected by global environmental changes.

International action is required in many areas, for instance: reliable data collection using modern technologies, credit policies requiring debtor nations to use environmentally sound engineering, financial aid through soft loans or grants to supplement local efforts, bans on the trade of endangered vegetable and animal species, etc. The ecological problems described here must be part of the negotiations between developed nations and Latin American and Caribbean countries. Trade, foreign debt, and aid negotiations should bear on the actions needed for environmental protection.

References

1. Blanco-Alarcon Astrid, Editor "Colombia gestión ambiental para el desarrollo" Editora Guadalupe, Bogotá, 1989.
2. Colombian Government, Decree 2811 of 1984, Código de Recursos Naturales Renovables y del Medio Ambiente.
3. Colombian Government, Law 1 of 1978.
4. Colombian Government, Law 56 of 1981.
5. Colombian Government, Departamento Nacional de Planeación y Ministerio de Minas y Energía, Estudio Nacional de Energía, 1982.
6. Colombian Government, Ministerio de Minas y Energía, Decree 2655 of 1988, Código Minero.
7. Colombian Government, Ministerio de Minas y Energía, Decree 2782 of 1989.
8. Hecht, Susana, Editor, "Amazonía. Investigación sobre Agricultura y Usos de Suelos," CIAT, Cali-Colombia, 1982.
9. OLADE, "Lineamientos para un Plan de Acción Latinoamericano sobre Energía y Medio Ambiente," Document prepared by Szekely, F. and Galan, A., July 15, 1988. Quito-Ecuador.
10. OLADE, "Financiamiento del Sector Eléctrico en América Latina," Document prepared by Perry, G. and Cabrera, M., July 1989. Quito.
11. Roa-Suárez, H. and Blanco-Alarcon, A. "Hidro-electricas en Colombia. Impactos Ambientales y Alternativas" FEN, Bogotá 1986.
12. UNITED NATIONS-CEPAL, ILPES, and PNUMA, "La dimensión Ambiental en la Planificación del Desarrollo," Volúmenes I y II, Grupo Latinoamericano, Buenos Aires, Argentina, 1986, 1988.

Present and Future Electric Power Systems in Eastern Europe: The Possibilities of a Broader Cooperation

A. Lévai
T. Jászay
Technical University of Budapest
Budapest, Hungary

Introduction

The political events of 1989 resulted in extraordinary political and economic changes in Eastern Europe. Treating these countries as a single unit no longer seems appropriate. From an energy perspective, however, these countries (Bulgaria, Czechoslovakia, Hungary, the German Democratic Republic, Poland, Romania, the Soviet Union, and Yugoslavia) still form a unique entity, interconnected by a network of electricity transmission lines and gas and oil pipelines. In addition, these countries look to the same source to meet the majority of their hydrocarbon demand. This source is the Soviet Union. It is important to realize, however, that although the Soviet Union provides the energy resources, the nations utilizing this supply contribute substantially to the construction of energy supply systems—often investing large sums in these projects. In spite of the latter fact, reliability of energy deliveries has recently deteriorated, largely as a consequence of problems emerging in the Soviet economy.

Energy in Eastern Europe [1, 2]

It is not useful to address electric energy separately from the overall energy situation. We, therefore, first review the general energy picture in the region and then compare it with that of the world. In the interest of a realistic analysis, comprehensive data are given on energy reserves, making a distinction between proved recoverable reserves and the estimated resource base of the most important energy components (such as coal, oil, natural gas, uranium, hydropower, and biomass). Based on 1987 consumption figures, comparative virtual rates of exhaustion are calculated and figures for per capita reserves and production are given.

Coal

Roughly 21 percent of the world's proved recoverable coal reserves can be found in Eastern Europe (of these, 13 percent is bituminous coal, 30 percent is brown coal, and 40 percent is lignite). Sixty-four percent of these reserves are found in the Soviet Union. The estimated additional resources are considerably higher than the proved recoverable reserves—primarily due to the large Soviet coal reserves. Eastern Europe holds approximately 74 percent of the world's total coal reserves (including 72 percent of all bituminous coal, 87 percent of the world's brown coal, and 64 percent of its lignite).

A comparison of the proved recoverable reserves and the rate of primary production shows that, in 1987, Eastern European countries accounted for about 34 percent of world coal production (including 24 percent of the world's bituminous coal, 30 percent of its brown coal, and 65 percent of its lignite). The annual rate of coal production in the region equalled 0.49 percent of proved recoverable reserves (including 0.58 percent of bituminous coal reserves, 0.24 percent of brown coal reserves, and 0.48 percent of lignite reserves). Comparing this specific rate of consumption with world figures, the rate of consumption of proved recoverable reserves in Eastern Europe is 1.7 times the world average. (Bituminous coal is consumed at 1.9 times the world average, lignite is consumed at 1.6 times, while the regional consumption rate of brown coal is about equal to the world average.) The situation appears more favorable when additional resources are considered. Taking these into account, the consumption of both bituminous and brown coal is essentially only one third the world average, while the rate of lignite consumption is equal to the world average. In these terms, the overall rate of coal consumption is about one-half the world average. Comparing regional per capita and average world per capita figures, Eastern Europe's proved recoverable coal reserves are about two and one half times the world average, estimated coal resources are nine times the world average, and coal consumption is four times the world average.

Oil

Only about 7 percent of the world average oil reserves are found in Eastern Europe. Ninety-seven percent or more of these are in the Soviet Union. (In the literature, data regarding possible and hypothetical oil resources varies significantly depending on the reservoir type, costs of extraction, nonconventional sources, etc., that are considered in the estimate.) In 1987, Eastern European oil production was about 23 percent of global production. The ratio of annual oil production and proved recoverable resources is 7.8 percent for Eastern European countries as a whole, while the average ratio for the world is 2.3 percent. This means that Eastern European oil reserves are consumed at 3.3 times the average world rate of consumption. The Eastern European per capita proved recoverable oil reserves are close to the world average, while per capita production is 2.5 times the world average.

Natural Gas

Approximately 38 percent of the proved recoverable natural gas reserves and—due to the rich reserves of the Soviet Union—74 percent of the world's estimated natural gas resources is found in Eastern Europe. (Ninety-eight percent of the region's natural gas reserves are in the Soviet Union.) In 1987, gas production equalled about 14 percent of world production, and thus consumption of proved recoverable reserves was practically identical with the world average. Regional consumption of estimated additional resources was lower—about one-half the world rate. Eastern European per capita recoverable natural gas is 4.6 times the world average, per capita estimated resources 9 times the world average, and per capita annual production about 5 times the world average.

Nuclear Fission (Uranium)

The availability of uranium ore—the basic fuel of nuclear power plants—is uncertain both in Eastern Europe and the world. Military and safety considerations are partially responsible for this uncertainty; however, the extent of additional resources is not clear. One way to classify world uranium resources is to establish a relationship between the amount of uranium recoverable as a function of the cost of extraction. The above uncertainties notwithstanding, Eastern Europe (primarily the Soviet Union) holds about 4 to 5 percent of the world's uranium reserves recoverable at costs below $130/kg (in US dollars). Yet, in 1987, uranium production in the region was only 33 percent of world production. Accordingly, the rate of consumption of proved (or admitted) reserves is only 0.73 times the world average, although per capita uranium production is four times the world average.

Hydropower

Given the economically available hydropower resources, Eastern Europe is in a good position to utilize this energy source. Electricity produced by hydroelectric power stations in the region may increase to 30 percent of world hydropower production (compared with 13 percent effective production in 1987). The per capita magnitude of hydropower resources is 3.5 times the world average, while actual electricity production is only 1.6 times the world average. Presently only about 6 percent of the considerable Soviet hydropower resource is being utilized.

Biomass

There is little reliable information about the present or potential quantity of biomass available for energy production. In Eastern Europe, the use of biomass for energy production is estimated at only 4 to 5 percent of the world figure. Given the immense territory of the Soviet Union and the strength of agriculture in the rest of Eastern Europe, the energy potential of biomass deserves closer attention.

Energy Intensity [3, 4]

It is common practice to use the ratio of per capita energy consumption per year to per capita gross domestic product (GDP) to link energy consumption and economic development. This relationship is by no means straightforward, even theoretically, because it is affected by numerous external factors (e.g., climate and lifestyle) as well as the way in which primary energy consumption is calculated (e.g., using a thermal equivalent for kWh in the case of hydro or nuclear power). The economic indicator used by Eastern European countries (with the exception of Yugoslavia) is per capita national income; thus it is generally difficult to make comparisons. Hungary has been using both economic indicators for years, and ordinarily GDP-based figures are 22 to 24 percent higher than those based on per capita national income. To achieve an objective comparison, national incomes described in US dollars have been increased by 23 percent. This adjustment to a normalized GDP allows a more realistic comparison between Eastern European countries and capitalist countries.

Table 1 lists primary and electric energy consumption, GDP, and primary energy and electric energy intensities for several countries. Note that the values for the different capitalist countries were converted by United Nations experts into US dollars (at the 1985 rates). It is important to note that for the Eastern European countries, only Hungary and Yugoslavia use a

Table 1. Primary energy and electricity consumption, GDP, and energy intensity for 1986 (1 kWh = 10 MJ). (Note: only Hungary and Yugoslavia use a US dollar exchange rate to reflect the value of their currency.)

Country	Primary Energy Consumption GJ/capita/year	Electric Energy Consumption MWh/capita/year	GDP $/capita, year	Primary Energy Intensity MJ/$	Electric Energy Intensity KWh/$
Bulgaria	188	5.11	3,716	50.5	1.38
Czechoslovakia	197	5.54	3,801	51.8	1.46
GDR	242	7.00	6,066	39.9	1.15
Hungary	124	3.63	1,991	62.3	1.82
Poland	140	3.74	2,166	64.6	1.73
Romania	140	3.30	3,140	44.4	1.05
USSR	202	5.60	3,204	63.1	1.75
Yugoslavia	85	3.37	1,964	43.5	1.71
EECPE	186	5.18	3,154	58.9	1.64
Greece	76	2.96	3,394	22.5	0.87
Spain	79	3.31	4,376	18.1	0.76
Austria	154	5.71	8,804	17.4	0.65
Denmark	157	6.01	11,714	13.4	0.51
FRG	185	6.77	10,427	17.7	0.65

US dollar exchange rate to reflect changes in the value of their national currencies. For the others, the officially declared, and therefore not quite reliable, values of the national currency in US dollars is indicated. Thus, the energy intensity data on the degree of economic development versus energy consumption for some countries is also not reliable and could alter the ranking of the countries as well. A slightly more realistic picture is possible if the corrected GDP values mentioned above are converted into real income, with the actual internal purchasing capacity taken into consideration. These factors are accounted for in the European Comparison Program (ECP) and the International Comparison Program (ICP). It was not possible to include these corrections in this paper. In order to get a broader picture, data for some semi-industrialized countries (e.g., Greece and Spain) and heavily industrialized countries (e.g., Austria, Denmark, Federal Republic of Germany) are also included in Table 1.

As can be seen, the energy intensity of most Eastern European countries is three to four times that of industrialized capitalist countries. The electricity intensity is somewhat lower than energy intensity, but is still two to three times greater than in industrialized countries. These differences can be explained in the following ways:

- Energy intensity is affected not only by factors such as geographic location, climate, and consumer behavior, but also by the structure of the available primary energy sources, technical and economic standards, the structure of industrial and agricultural production, and the technical and economic standards of energy consumption.
- The structure of primary energy sources differs greatly between Eastern European countries and industrialized European countries. Coal is the predominant form of energy consumed in Eastern Europe, including brown coal, which is of low heating value. Industrialized countries, on the other hand, mainly use liquid and gaseous hydrocarbons—if coal is used at all, it is high grade bituminous coal.
- The difference in technical and economic standards of production between East and West is self-evident. Most production plants and equipment in Eastern Europe are so obsolete that they would be considered unacceptable in a capitalist environment. Still, the structure of production includes a large quantity of basic materials or semi-finished products of high energy input.

Given the above, it is obvious that differences in per capita primary energy and/or electricity consumption, which are considered to measure standards of living, are not nearly as significant as the differences in energy intensity. For example, the relatively high per capita energy and electricity consumption in the Soviet Union and the German Democratic Republic is associated with their rather unfavorable energy intensity. This can be attributed primarily to outmoded production technology. Consider also the significant difference in per capita GDP between Hungary and Yugoslavia, on the one hand, and Bulgaria and Romania on the other. This difference results from the artificial national currency/US dollar exchange rates used by Bulgaria and Romania, rather than from actual industrial conditions. These factors are clearly reflected in the real living standards of these countries.

Electric Power Systems [5]

Those countries in Eastern Europe that have been in close cooperation with the Soviet Union since the 1950s expressed early interest in modernization and safety improvements in their electricity supplies. Increased unit capacities, utilization of large power stations, and significant transportation and exchange of electricity were primary objectives. Complex and branching networks were needed for transmission and distribution of the generated electricity and for the economic improvement of reserve operations.

This cooperation among the countries resulted in what is known today as the "Interconnection of Power Systems" (IPS). It began in the 1960s with the construction of 400 kV transmission lines within and between the countries. (Prior to that, only 110 kV and 220 kV lines were operated as a bilateral link between countries and for electricity exchange and reserves.) In 1974, a 750 kV international transmission line was constructed between Hungary and the Soviet Union—the first of its kind on the European continent. Since then, IPS development has continued through a uniform, long-term plan approved by each country involved. The first plan covered the period 1976 to 1990 and the second, 1982 to 2000.

Since 1967, development plans for the national systems have been coordinated by the Central Dispatching Organization (CDO) of IPS, located in Prague. Currently, the members of the CDO are Bulgaria, Czechoslovakia, Hungary, German Democratic Republic, Poland, Romania, and the South Power System of the USSR. Their total installed capacity of 170.7 GWe includes 21.7 GWe of nuclear power, 18.5 GWe of hydropower, and 15.8 GWe coming from industrial power stations. All power sources other than hydro and nuclear are fired with organic fuels. [December, 1987] The IPS covers 1,628,000 square kilometers and serves a population of nearly 170 million. The annual peak utilization of the IPS is very high—more than 6,700 h/yr. The energy exchanges between the countries are established via four 750 kV and twelve 400 kV and 220 kV transmission lines.

The estimated annual growth rate in electricity demand (from 1980 to 2000) for Eastern European countries is about 3 percent. The majority of these countries plan to construct nuclear power plants utilizing Soviet 1000 MWe pressurized water reactor (PWR) designs. The exceptions are Hungary and Poland, whose programs also include conventional fuel power stations. Between 1980 and 2000, the percentage of conventional thermal power stations in the region is expected to drop from 74 percent to 47 percent.

Up to now, except in the Soviet Union, only Soviet PWR nuclear power stations have been constructed in these countries. Developing and improving the safety systems of these nuclear plants has become a key issue: specifically, better control of nuclear power stations; updating and organizing maintenance, control, and protection of the technological processes; more efficient utilization of nuclear fuel; employing modern construction techniques; staff education; and radioactive waste disposal concerns. From the point of view of Eastern Europe as a whole—perhaps for the entire European continent—recycling spent fuel from identical or nearly identical reactors and/or the development of an equilibrium system of spent fuel and recovered fissionable materials may prove important.

As the number of base-load nuclear power plants increases, construction of peaking plants is considered especially important, with pumped storage concepts given priority. (A total capacity of 17 GWe will be needed in these countries before the year 2000.) The contribution of pumped storage plants is expected to be 7 percent in the GDR and 6 percent in Czechoslovakia. In addition, it is likely that gas turbines will be integrated in some development plans. It would be most economical to introduce them at sites with a steam-cycle to cogenerate heat and power. Within the IPS, hydropower may reach 63 to 67 percent of the total installed peaking capacity by the year 2000. At this point, hydroelectric power plants and pumped storage plants would represent about 14 percent of the total installed generating capacity.

Despite these developments, there remain input capacity deficiencies and unreliable equipment. Parallel operation of systems in the integrated Eastern European energy network does not meet today's safety and reliability standards. For this reason, significant efforts are being made to enable connection to the 330,000 MWe Western European grid (UCPTE). This proposal has been studied by experts in both the East and the West for some time. There are a number of advantages to an East-West grid, which are summarized below. [6, 7]

- A mutually advantageous energy exchange through improved electricity generation and utilization will be financially beneficial for both East and West. For example, the peak of an "all European" load can be reduced by 1.5 to 2.5 percent (or even 5 to 8 percent) due to the time zone differences between regions. Maximum consumption occurs at a different time in Eastern and Western European countries. Seasonal and daily fluctuations in energy demand may also increase availability of energy resources for both regions.
- An integrated grid would permit savings of 10 to 15 percent in spinning reserves, and thus in generating

capacity, compared with the total reserve available when the two systems operate independently.
- It would be easier to provide mutual assistance in emergencies.

Potential savings of an East-West integration are expected to be $15 billion to $20 billion (US dollars), for a generating capacity of 15,000 to 20,000 MWe and assuming a specific investment of $1,000/kWe (US dollars).

While the idea of an integrated European energy grid is appealing, it can only be realized step by step. The strong interconnection of the two systems by high capacity links requires multinational investment, which to date has not been possible. And for both technical and economic reasons, it is not likely to come into being in the foreseeable future. Some of the more significant problems are listed below.

- Both systems are very large and have high capacity. Expensive ultra high-voltage (>1,050 kV) alternating current (UHVAC) links are required to connect them. At the same time, uniform frequency control techniques would have to be introduced in both the East and West. Alternatively, ultra high voltage (1,500–2,000 kV) direct current (UHVDC) links would be required if each system were to maintain its own frequency control techniques. Although either of the two connected systems could then be operated and controlled independently, UHVDC links would be as expensive as in the first case.
- The demand for safe and reliable power continues to increase everywhere. To meet this demand, both regions are attempting to guarantee their own reserves on the basis of their own resources.
- Political and economic problems (arising from shared investments, distribution and accounting of transported energy, etc.) can be expected in a multinational agreement involving so many countries.

An economical and feasible approach to integrating the European energy system is through bilateral contracts for construction of high-voltage direct current (HVDC) back-to-back links, without any change in frequency control techniques. This way, interconnection occurs incrementally, eventually resulting in a totally interconnected European grid.

According to these bilateral contracts, the two countries involved would construct HVAC transmission lines (preferably 400 kV) at the most appropriate places near their borders. Converter-inverter equipment at a substation would allow direct current (dc) interconnection between the two alternating current (ac) systems. By using the converter-inverter to change the dc voltages, control of energy transport in both directions is flexible. Although such a station is rather costly, this scheme is advantageous for both countries—provided cooperation is contractually assured on mutually agreeable terms over a reasonably long duration. HVDC back-to-back links are in operation between the Soviet Union and Finland, as well as between Czechoslovakia and Austria [8]. The rated capacities of these links are, respectively, 3 x 350 MWe and 550 MWe. The Czechoslovakia-Austria link has been designed for electricity transport from Poland to Austria and for seasonal electricity exchange between the Soviet and Austrian electric power systems.

The CSSR-Austria link was put into operation in 1985 and a bilateral contract for construction of an HVDC back-to-back link between Hungary and Austria was signed in 1986. Similar contracts are being considered in a number of European countries to prepare the way for future "all European" cooperation.

Air Pollution and Environmental Protection

Emissions of air pollutants, especially sulfur dioxides, are significant in Eastern Europe—the region has a bad reputation with environmentalists. These high emissions rates result from the undesirably high content of oil and low-grade high-sulfur coal in the fuel mix. As discussed earlier, low-grade coal is by far the largest fuel reserve of the region. Moreover, capital is limited and cutting costs in energy projects usually involves eliminating environmental protection efforts (e.g., omitting flue-gas desulfurization equipment).

Table 2 shows sulfur (SO_2) emissions figures for Europe in 1980 and 1987 [9]. As can be seen, three Eastern European countries (the Soviet Union, GDR, and Poland) emit the most SO_2, while Czechoslovakia ranks sixth, Hungary tenth, Bulgaria twelfth, and Romania twenty-first. This ranking remained essentially unchanged between 1980 and 1987. Although emissions in Poland increased 10 percent and in Bulgaria 9 percent, the other Eastern European countries reduced emissions. This reduction can be attributed to the following factors: 1) the increased role of nuclear energy in electricity production, 2) the increased share of natural gas in the fuel structure, and 3) reduced energy intensity in some of the countries. The improvement in emissions resulted from changes in energy structure and production that affected the absolute values of sulfur emissions, rather than from a direct commitment to environmental protection. Carbon dioxide emissions, which have also become a concern, are also favorably affected by these structural changes.

Given the rapid and fundamental political and economic changes occurring in Eastern Europe, it would be venturesome to predict future economic changes in the region. Still, it is reasonable to expect continued modernization and, as a result, reduced energy intensities. It is also likely that economic stagnation will be replaced by more efficient utilization of assets. As more

Table 2. Sulfur emissions in Europe (kilotons of sulfur per year).

Country	1980 Emissions	1987 Emissions	Percentage Difference	Promised Reduction
Soviet Union	6,400	5,100	- 20	30% by 1993
GDR	2,500	2,500	0	30% by 1993
Poland	2,050	2,270	+ 10	—
United Kingdom	2,335	1,840	- 21	30% by 1993
Spain	1,625	1,581	- 3	—
Czechoslovakia	1,550	1,450	- 6	30% by 1993
Italy	1,900	1,252	- 34	30% by 1993
FRG	1,600	1,022	- 36	65% by 1993
France	1,779	923	- 48	50% by 1990
Hungary	817	710	- 13	30% by 1993
Yugoslavia	588	588	0	—
Bulgaria	517	570	+ 9	30% by 1993
Belgium	400	244	- 39	50% by 1995
Greece	200	180	- 10	—
Turkey	138	177	+ 22	—
Finland	292	162	- 44	50% by 1995
Denmark	219	155	- 29	50% by 1995
Netherlands	244	141	- 42	50% by 1995
Portugal	133	116	- 13	—
Sweden	232	116	- 50	68% by 1995
Romania	100	100	0	—
Ireland	110	84	- 24	—
Austria	177	75	- 58	70% by 1995
Norway	70	50	- 29	50% by 1994
Switzerland	63	31	- 51	57% by 1995
Albania	25	25	0	—
Luxembourg	11	6	- 45	58% by 1990
Iceland	3	3	0	—
Total	26,078	21,471	- 8	

capital becomes available, the door will open for investments in environmental protection.

Extended Cooperation

Eastern Europe, especially the Soviet Union, has significant energy sources and a close cooperation among the countries has developed based upon an extensive network of electric transmission lines and oil and gas pipe lines. While links for electricity are in operation between Eastern Europe and some of the Central and Northern European countries, the current volume of transport or exchange has been insignificant. The transport of gas is more widespread in terms of both volume and geographical distribution.

At the same time, the countries of East Central Europe are trying to reduce their dependence on energy imported from the Soviet Union and to diversify the sources of energy imports. The anticipated shift from exchange of goods through interstate contracts to trade against convertible currency will help in these efforts.

A unique form of diversification has been presented in which Western electricity companies would construct a nuclear power plant in Hungary and export part of the electricity generated by the plant to Western European markets. This idea will understandably encounter initial resistance. However, given Hungary's serious shortage of capital, and the fact that 25 percent of its electricity is imported, a Western-financed nuclear plant may eventually be considered seriously.

Assuming this idea can become reality, the Hungarian electric energy system should be coordinated with the West European grid. The dc links necessary to deal with the frequency difference between East and West should be built near the eastern border of Hungary rather than the western border. Although this idea requires further development, it represents the potential range of greater cooperation.

In the long term, the energy supply and demand conditions in Europe will accelerate this cooperation. The Soviet Union, with its ample reserves, is strongly interested in energy supply and resource export because the technical and economic requirements are possible. Because production of hydrocarbon reservoirs in the North Sea has been reduced and the demand for more environmentally benign fuels has increased, Western Europe is very interested in importing natural gas and electricity. Both Eastern and Western Europe are interested in technology transfer, including energy conversion, transport, and environmental protection technologies. The quality and reliability of energy supply is expected to improve in the small countries of the eastern region of Central Europe, as the result of the developing interface between the Soviet and Western European energy systems.

Both Eastern and Western Europe will benefit through the coordination of their energy systems. The authors hope the present political changes will translate into broader cooperation in these regions and that energy will become an international priority item for the benefit of all.

References

1. *Survey of Energy Resources.* World Energy Council, 1989.
2. *Britannica Atlas.* Encyclopedia Britannica, 1987.
3. United Nations Statistical Data, 1987.
4. *International Comparison of Primary and Electrical Energy Consumption and Intensity in Hungary.* MVMT Hungarian Electricity Works, Budapest, 1989 (in Hungarian)
5. *Twenty-five Years Central Dispatching Organization of the Interconnection of Power Systems (IPS),* Prague, 1988.
6. A. Lévai, *Energetical Links between Eastern and Western Europe with Fixed Transport Lines: Participation in the World's International Energy Trade.* Acta Technica. Acad. Sci. Hung., 100/1–2.
7. Kogal, Y., et al. "Deepening of Electrification and Possibilities for Electric Energy Exchange," 13th Congress of WEC, Cannes, 1986.
8. Dwek, M.G., et al., "Use of HVDC Links to Interconnect European AC Systems of UNIPEDE Countries," UNIPEDE Athen Congress paper, 1985.
9. Alm, H., "Emissions Are Falling, But Is It Enough?" *Acid Magazine,* No. 8, Sept. 1989.

Electricity and the Environment in Developing Countries with Special Reference to Asia

Mohan Munasinghe*
Environmental Policy Division
The World Bank
Washington, DC

Introduction

The experience of the industrialized countries emphasizes that a reliable supply of electric power is a vital prerequisite for economic growth and development. Thus, the observed trends relating to electricity demand in developing countries (which indicate annual growth rates in the region of 6 to 12 percent) are consistent with the development objectives that these countries all share. Up to the present time, many developing countries have been struggling with the formidable difficulties of meeting these demands for electricity services at acceptable levels of reliability. If these needs cannot be met, economic growth will almost surely slow down and the quality of life will fall.

Given these already existing handicaps, the growing additional concerns about the environmental consequences of power generation considerably complicate the policy dilemma facing the developing countries. Historically, industrial countries that faced a tradeoff between economic growth and environmental preservation invariably gave higher priority to the former. The countries that are presently developed have only recently awakened to the environmental consequences of their economic progress, and only after a broad spectrum of economic objectives have been reached. This model of economic and social development has been adopted by many third world regions (especially in Asia). Indeed, at the present time, this appears to be the model of development that holds the widest appeal. Consequently, there is a real likelihood that even sincere Western concerns about environmental degradation in the developing countries, may be perceived as both unrealistic and inconsistent.

The concept of fairness or equity is very relevant in the context of the current global environmental concerns. Of the present emissions of man-made carbon dioxide in the atmosphere, as much as 70 percent can be attributed to the developed countries where 25 percent of the world's population reside (Flavin, 1989). This is the result of massive disparities in access to and consumption of energy. The low-income economies (accounting for half of the world's population) had an average 1987 per capita gross national product (GNP) of US $290, compared with a per capita GNP of $18,530 in the US. In the two largest developing countries, India and China, per capita GNP was $300 and $290 respectively. Per capita energy consumption in the US was 7265 kilograms-oil-equivalent (kgoe) in 1987, compared to 208 and 525 kgoe in India and China respectively (World Bank, 1989). Consequently, in the period from 1950 to 1986, North America alone has contributed over 40 billion tons of carbon resulting from fossil based CO_2 emissions and the countries of Western and Eastern Europe account for another 25 and 32 billion tons respectively. On the other hand, all the developing countries accounted for only 24 billion tons, representing 18 percent of the cumulative world total. On a per capita basis, this comparison is even more dramatic—over the same 1950 to 1986 period, North America emitted almost 20 times as much fossil-based CO_2 as did the developing countries.

Ironically, the predicted consequences of global change are likely to affect the developing countries more severely, since these countries lack the flexibility to respond quickly to environmental changes. The drastic impact of droughts, storms, and flooding in the poorer countries is well known historically and an increase in the occurrence of these events will exacerbate the burden on the less-developed countries.

The recent report of the World Commission on Environment and Development (WCED, 1987), which has been widely circulated, analyzed the concept of sustainable development, based on the interaction of two aspects: human needs, especially those of the poorer segments of the world's population, and resource limitations, which are imposed by the ability of the environment to meet those needs. These resources, broadly defined, include both the assets that are used in productive activity (such as oil, gas, and minerals) as well as the ability of the global ecosystem to absorb the byproducts of this activity. The development of the pres-

---

* Until recently, the author served as Senior Energy Advisor to the President of Sri Lanka. The opinions expressed in this paper are those of the author, and do not necessarily represent the views of any institution or government. The invaluable assistance provided by Chitru Fernando is acknowledged, in preparing this paper.

ently industrialized countries took place in a setting which emphasized needs and neglected limitations. The development of these societies has effectively exhausted a disproportionate share of global resources. Indeed, this resource-intensive development path implies that the developed countries owe a significant "environmental debt" to the larger global community.

The division of responsibility in any future global efforts is clear from the above arguments. The existence of this environmental debt should provide an equitable and sustainable basis on which the developed and developing countries can work together to share the resources that remain. The developing countries can help in the global effort to the extent that this participation is fully consistent with and complementary to their immediate economic and social development objectives. The developed countries, on the other hand, have already attained most reasonable goals of development and can afford to increasingly substitute environmental objectives for further economic expansion.

In the context of the foregoing, this paper goes on to identify some specific issues that are critical to the current environmental concerns about power sector growth in developing countries (with special focus on the Asian region). It also explores some policy implications of such issues at the national and global level for both the developing countries as well as the wider international community, and examines the role of emerging mechanisms such as the global environmental fund, in the mobilization and allocation of resources for addressing global environmental problems.

Power Sector Role in Developing Countries

Despite some anomalies, the link between energy demand and gross domestic product (GDP) is well established. Electric power, in particular, has a vital role to play in the development process, with future prospects for economic growth in the third world being closely linked to the provision of adequate and reliable electricity supplies. Figure 1 indicates the relationship between electricity use and income for both developed and developing countries. A more systematic analysis of World Bank and United Nations (UN) data over the past two or three decades indicates that the historical ratio of percentage growth rates (or elasticity) of power system capacity to GDP is about 1.4 in the developing countries.

Assuming no drastic changes in past trends with respect to demand management constraint, the World Bank's most recent projections indicate that the demand for electricity in less-developed countries (LDCs) will grow at an average annual rate of 6.6 percent during the period 1989 to 1999 (World Bank, 1990). This compares with actual growth rates of 10 percent and 7 percent in the seventies and eighties, respectively. These rates of growth indicate the need for total capacity additions of 384 gigawatts (GW) in the 1990s and annual energy growth of 2135 kilowatt hours (kWh) by 1993 (see Figure 2). As indicated in Figure 3, the Asia region requirements dominate, accounting for almost two thirds of the total, while coal and hydro are the main primary sources—both of which have specific environmental problems associated with their use.

The investment needs corresponding to these indicative projections are also very large. Table 1 shows the projected breakdown of LDC power sector capital expenditure in the 1990s. Of a total of $745 billion (constant 1989 US$), Asia, which includes both India and China, accounts for $455 billion or over $45 billion annually. In comparison with the total projected annual requirement for LDCs of $75 billion, the present annual rate of investment in developing countries is only around $50 billion. Even this present rate is proving difficult to maintain. Developing country debt, which averaged 23 percent of GNP in 1981, increased dramatically to 42 percent in 1987. In low-income Asian countries, outstanding debt doubled, from 8 percent in 1981 to 16 percent in 1987. Capital intensive power sector investments have played a significant role in this observed increase.

If the developing countries follow this projected expansion path, the environmental consequences are also likely to increase in a corresponding fashion. There is already a growing concern about the environmental impacts of energy use at the national level in developing countries. At a recent workshop on acid rain in Asia, participants reported on a wide range of environmental effects of the growing use of fossil fuels, especially coal, in the region (Foell, 1989). For example, even in 1985 total sulfur dioxide emissions in Asia were estimated at around 22 million tons, and these levels, coupled with high local densities, have led to acid deposition in many parts of Asia.

The developing countries feel that any attempts to mitigate these environmental effects, however, cannot jeopardize the critical role played by the power sector in economic development. Similarly, the allocation of resources to environmental programs in developing countries cannot diminish the resources needed to fund this projected expansion. Energy and environmental policymakers in both developing countries and the global community are, indeed, confronted with a formidable dilemma.

Framework for Addressing Energy-Environmental Issues in the LDC Power Sector

The foregoing discussion has helped to establish a rational and equitable basis for addressing the problems

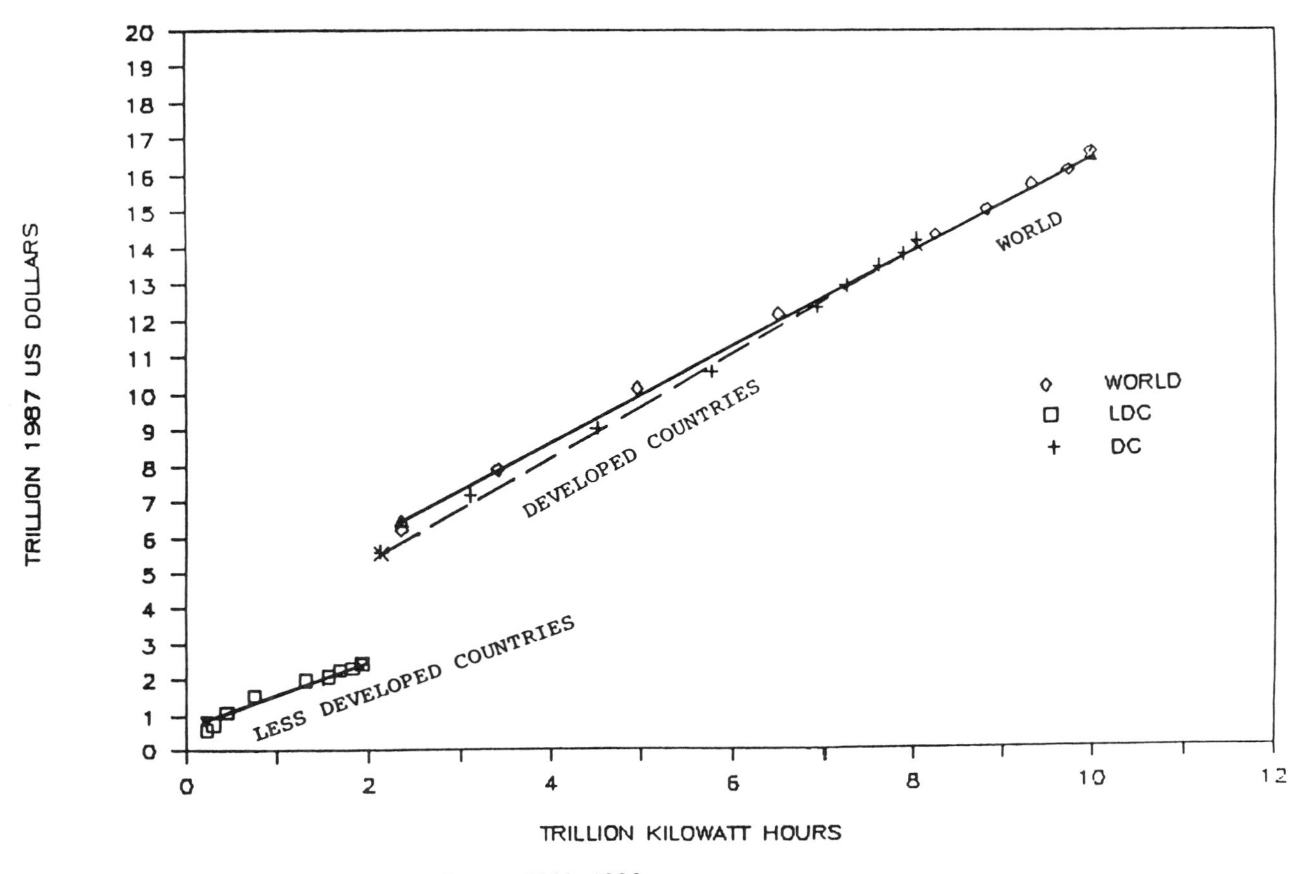

Figure 1. Gross product vs. kilowatt hours, 1960–1986.

of global environmental mitigation. In this section we present an energy-economic analytic framework which ties together the issue of environmental protection with the existing energy sector goals of efficiency and economic growth.

It is convenient to recall here that the specific prerequisites for economic efficiency have included both efficient consumption of energy, by providing efficient price signals that ensure optimal energy use and resource allocation; and efficient production of energy, by ensuring the least-cost supply mix through the optimization of investment planning and energy system operation.

A new issue, which has emerged in recent decades as an area of particular concern, is the efficient and optimal use of our global natural resource base, including air, land, and water. Since there has been much discussion also about the key role that energy efficiency and energy conservation might play in mitigating environmental costs, it is useful first to examine how these topics relate to economic efficiency. Specific issues dealing with the formulation and implementation of economically efficient energy policies are presented in the next section.

Major environmental issues vary widely, particularly in terms of scale or magnitude of impact, but most are linked to energy use. First, there are the truly global problems such as the potential worldwide warming due to increasing accumulation of greenhouse gases like carbon dioxide and methane in the atmosphere, high altitude ozone depletion because of excessive release of chlorofluorocarbons used mainly in refrigeration devices, pollution of the oceanic and marine environment by oil spills and other wastes, and over-depletion of certain animal and mineral resources. Second in scale are the transnational issues like acid rain or radioactive fallout in one European country due to fossil-fuel or nuclear emissions in a neighboring nation, and excessive downstream siltation of river water in Bangladesh due to deforestation of watersheds and soil erosion in nearby Nepal. Third, one might identify national and regional effects, for example those involving the Amazon basin in Brazil, or the Mahaweli basin in Sri Lanka. Finally, there are more localized and project specific problems

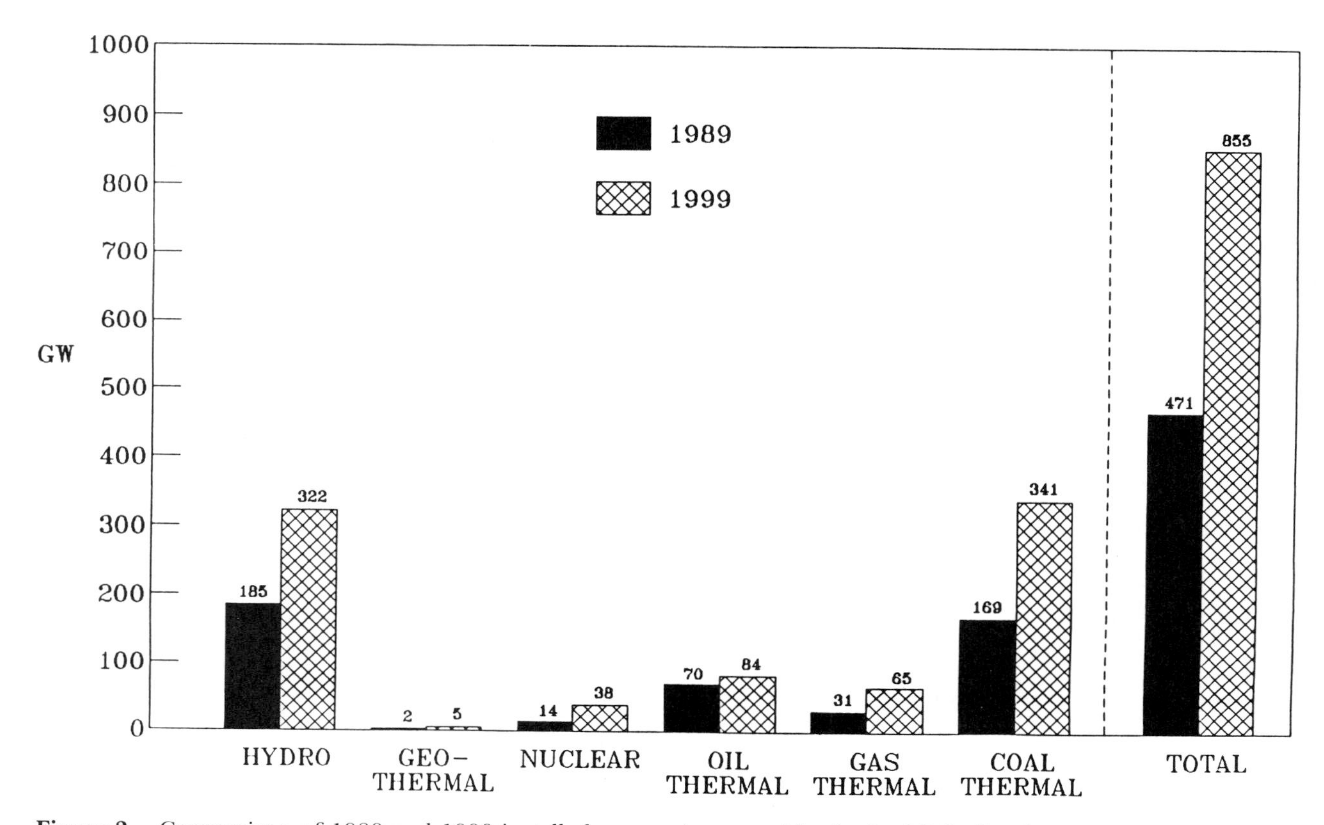

Figure 2. Comparison of 1989 and 1999 installed generating capacities in the LDCs (in gigawatts).

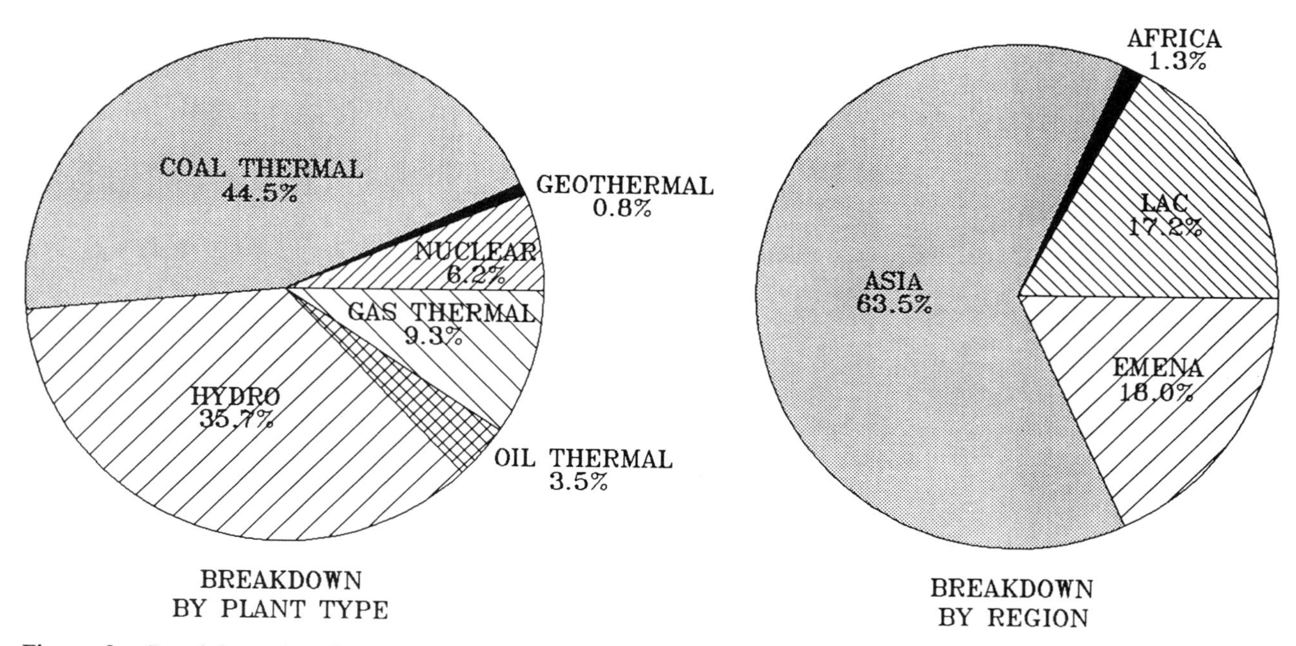

Figure 3. Breakdown by plant type and region of capacity expected to be added in the LDCs in the 1990s (384 gigawatt total).

Table 1. Regional Breakdown of LDC Power Capital Expenditures in the 1990s (billions of 1989 US$).

	Asia	EMENA*	LAC**	Africa	Total
Generation	277	82	83	6	448
Transmission	39	8	32	2	81
Distribution	100	23	27	2	152
General	39	11	13	1	64
Total	455	124	155	11	745
%	61.1	16.6	20.8	1.5	100

* Europe, Middle East and North Africa (Mediterranean region)
** Latin America and the Caribbean
Source: The World Bank

like the complex environmental and social impacts of a specific hydroelectric or multipurpose dam.

While environmental and natural resource problems of any kind are a matter for serious concern, those that fall within the national boundaries of a given country are inherently easier to deal with from the viewpoint of policy implementation. Such issues that fall within the energy sector must be addressed within the national policymaking framework. Meanwhile, driven by strong pressures arising from far-reaching potential consequences of global issues like atmospheric greenhouse gas accumulation, significant efforts are being made in the areas of not only scientific analysis, but also international cooperation mechanisms to implement mitigatory measures.

Given this background, we discuss next some of the principal points concerning energy use and economic efficiency (Munasinghe, 1990a). In many countries, especially those in the developing world, inappropriate policies have encouraged wasteful and unproductive uses of some forms of energy. In such cases, better energy management could lead to improvements in economic efficiency (higher value of net output produced), energy efficiency (higher value of net output per unit of energy used), energy conservation (reduced absolute amount of energy used), and environmental protection (reduced energy-related environmental costs). While such a case may fortuitously satisfy all four goals, the latter are not always mutually consistent. For example, in some developing countries where the existing levels of per capita energy consumption are very low and certain types of energy use are uneconomically constrained, it may become necessary to promote more energy consumption in order to raise net output (thereby increasing economic efficiency). There are also instances where it may be possible to increase energy efficiency while decreasing energy conservation.

Despite the above complications, our basic conclusion remains valid—that the economic efficiency criterion which helps us maximize the value of net output from all available scarce resources in the economy (especially energy and environmental resources in the present context), should effectively subsume purely energy-oriented objectives such as energy efficiency and energy conservation. Furthermore, the costs arising from energy-related adverse environmental impacts may be included (to the extent possible) in the energy economics analytical framework, to determine how much energy use and net output that society should be willing to forego, in order to abate or mitigate environmental damage. The existence of the many other national policy objectives—including social goals that are particularly relevant in the case of low-income populations—will complicate the decisionmaking process even further.

The foregoing discussion may be reinforced by the use of a simplified static analysis of the trade-off between resource use and net output of an economic activity. In Figure 4, Y represents the net output of productive economic activity as a function of some resource input (say energy)—without accounting for environmental impacts. Due to policy distortions (for example, subsidized prices), the point of operation in many developing coun-

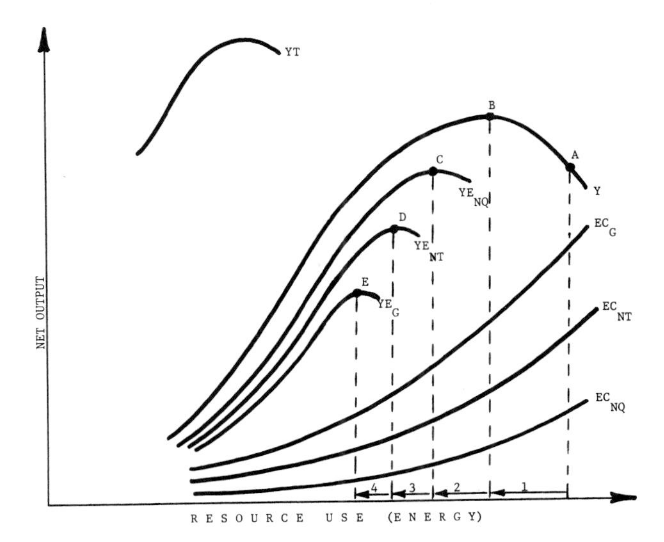

Figure 4. Net output, resource use and environmental cost. $YE_{NQ} = Y - EC_{NQ}$; $YE_{NT} = Y - EC_{NT}$; and $YE_G = Y - EG_G$.

tries might be A, where the resource is being used wastefully. Therefore, without invoking any environmental considerations, economic use and resource use efficiency (i.e., energy efficiency) could be improved by moving from A to B. A typical example might be improving energy end-use efficiency or reducing energy supply system losses.

Now consider the curve EC_{NQ} which represents economically quantifiable national environmental costs associated with resource use. These might include air-pollution-related health costs of a coal power plant or the costs of environmental protection equipment (like scrubbers and electrostatic precipitators to reduce noxious gas and particulate emissions) installed at such a plant, or the costs of resettlement at a hydropower dam site. A corrected net output curve is defined by : $YE_{NQ} = Y - EC_{NQ}$. The maximum of this curve at C lies to the left of B, implying lower use of (more costly) energy.

Next, we define the curve EC_{NT} which shows total national environmental costs (both quantifiable and non-quantifiable). With this further correction, the corresponding net output curve is $YE_{NT} = Y - EC_{NT}$. The maximum has shifted to D and the optimal energy consumption has also declined.

Finally, EC_G represents global environmental costs of energy use (including national impacts), and $YE_G = Y - EC_G$ is the correspondingly corrected net output which implies an even lower level of optimal energy use. The curve YT shows net output for a technologically advanced future society that has achieved a much lower resource intensive production.

This framework illustrates the crucial dilemma for LDCs. In Figure 4, all countries (including the poorest) would readily adopt measures that will lead to the shift (1) which simultaneously and unambiguously provides both economic efficiency and environmental gains. Most developing countries are indicating increasing willingness to undertake the shift (2). However, implementing shift (3) will definitely involve crossing a "pain threshold" for many developing countries, as other pressing socioeconomic needs compete against the costs of mitigating nonquantifiable adverse environmental impacts. Clearly, the shift (4) which implies optimization of a global value function would hardly appeal to resource constrained developing countries, unless external financing was made available.

Therefore, we may conclude briefly that, while the energy required for economic development will continue to grow in the developing countries, in the short to medium run there is generally considerable scope for most of them to practice better energy management, thereby both increasing net output and using their energy resources more efficiently. In the medium to long run, it will become possible for developing countries to adopt newer and more advanced (energy efficient) technologies that are now emerging in the industrialized world, thus enabling their transformed economies to produce even more output using less energy, as indicated by the curve YT in Figure 4.

Energy-Environmental Policy Issues and Options at the National Level

The last section has introduced a rational framework for addressing environmental issues. It is clear that the developing countries could be expected to cooperate in global environmental programs only to the extent that such cooperation is consistent with their national growth objectives. The challenge for both national planners and the international community is to find areas in which such consistency exists. The role of the developed countries, on the other hand, is to incur the risks inherent in developing innovative technological measures which are the prerequisites for the next level in environmental protection and the mitigation of consequences. These risks include the possibility that these measures may turn out to be unnecessary or inapplicable after all, given the prevailing uncertainty about the future impact of current environmental developments. The following sections review some policy options at the national level in the above hierarchy.

Technological Options

In addition to improving traditional tasks such as demand forecasting, least cost system planning, and optimal system operation, there is a spectrum of technological options which the developing countries could potentially utilize in order to improve energy efficiency and thereby reduce environmental effects arising from power sector activity (Munasinghe, 1990b). These range from simple infrastructural retrofits to the use of advanced generation technologies.

Among the short term technological options for the LDC power sector, reducing transmission and distribution losses, and improving generation plant efficiencies appear to be the most attractive. While estimates of LDC power system losses vary, they all point to levels which are far in excess of accepted norms: 6 to 8 percent in transmission and distribution as a percentage of gross energy generation, and a further 1 to 7 percent in station use. In contrast, total losses in LDC power systems are estimated to average in the 16 to 18 percent range. The average system losses in South Asia have been estimated at 17 percent and in East Asia at 13 percent. Table 2 presents data for some Asian LDCs in comparison with industrialized countries. The consequences of reducing these losses can be quite remarkable. On the basis of our previous estimates of capacity requirements, a one percentage point reduction in losses now would

Table 2. Electrical Transmission and Distribution Losses (percent of generation).*

Country	Losses
Pakistan	28%
India	22%
Bangladesh	31%
Sri Lanka	18%
Thailand	18%
Philippines	18%
South Korea	12%
Japan	7%
US	8%

* These loss estimates include non-technical losses (i.e., due to deficient metering and theft).
Sources: The World Bank and USAID

reduce required capacity by over 1 percent or over 2.2 GW for Asian LDCs. The estimated saving in capital investment would be around $4 billion. Over time the savings would, obviously, be even greater. The Agency for International Development (USAID, 1988) has estimated that the average heat rate of LDC power plants is around 13,000 Btu/kWh, compared to 9,000 to 11,000 Btu/kWh if these plants were operated efficiently. The energy savings (and positive environmental consequences) implied in these figures are considerable.

Similar gains are possible by conservation on the demand side. Johansson et al. (1989) provide an insightful review of the developments that have been taking place in end-use technologies which can have a major impact on power efficiency. These technologies (which developed in the industrialized countries as a response to the oil price escalation in the seventies) can be easily applied towards more efficient lighting, heating, refrigeration and air conditioning around the developing world.

Substitution of primary energy sources in power generation is another potential means of achieving dual benefits. In the developing world, natural gas is the most likely candidate for fossil fuel substitution. The economic benefit of natural gas substitution comes from either import substitution for petroleum products or releasing these products for export. On the environmental front, natural gas firing typically achieves reduction in carbon emissions of 30 to 50 percent. Many Asian countries are endowed with significant resources of natural gas, including Malaysia, Indonesia, and Thailand.

In the longer term, the developing countries will need to rely on more advanced technological options which are currently being developed in the industrialized countries. As we have discussed above, power generation capacity in developing countries is expected to nearly double by the turn of the century, and will increase further thereafter. This provides opportunities to add state-of-the-art technologies which have been designed with regard to both economic and environmental criteria. Clean coal technologies, cogeneration, gas turbine combined cycles, steam-injected gas turbines, and other advances are all part of this menu of technologies which have important potential in developing countries. Similar applications will become available for emission control technologies. However, as we have argued previously, the developing countries will look to the industrialized nations to provide the leadership in refining and proving these technologies before they are implemented in the developing world.

Economic/Financial Options

Providing the correct economic signals, through the use of marginal cost pricing, offers the most attractive demand-side option for improving power sector efficiency. While the economic principles of electricity pricing are now well understood, pricing policy in developing countries is guided by a tradeoff between economic efficiency on the one hand and a series of socio-economic considerations on the other. The recognition that electric power is fundamental to economic growth and that access to electricity improves living standards of the people has driven a policy in which affordability has often replaced economic efficiency as the principal criterion. Furthermore, in practice it has been difficult to separate social and economic criteria in a single pricing structure, leading to the dominance of the social theme.

A recent study of the electric power sector in developing countries (Munasinghe et al., 1988) indicates that electricity tariffs have not kept up with cost escalation. The operating ratio (defined as the ratio of operating costs before debt service, depreciation and other financing charges, to operating revenue) for the almost 400 power utilities studied, deteriorated from 0.68 in the 1966 to 1973 period to 0.80 between 1980 and 1985. In some countries these deviations are significantly greater.

This study and other available evidence indicates that a significant shift towards marginal-cost-based electricity pricing would be possible without creating undue hardship to the poorer segments of the population especially if lifelike rate tariff structures are carefully crafted. While extensive information (in the form of price elasticities) is not available for most developing countries, many recent studies provide a reasonable basis for projections. Assuming a price elasticity for electric power to be -0.3, a 20 percent real increase in electricity prices (which would restore the above operating ratio to its level in the 1960s and 1970s) would result in a 6 percent reduction in electricity demand.

Apart from price rationalization, there is also scope for applying other incentive schemes aimed at improving the efficiency of power sector energy use. These would include taxes and subsidies based on fuel type, technology, research and development, retrofits, conservation programs, etc. In most instances, programs involving coordinated use of both price and non-price policy instruments are likely to achieve desirable effects on both energy use and environmental impacts. Fiscal instruments such as emission fees and carbon-based user fees can be used to control environmental impacts more directly. In many LDC applications, however, problems of implementing and monitoring such mechanisms are significant.

Organizational/Institutional Options

The power sector in developing countries is typically owned and controlled by the government, and is characterized by large monolithic organizations. While there is some rationale for this centralization, it also remains a critical barrier in the path of greater efficiency and improved flexibility. The desperate circumstances of many developing country power utilities have generated pressures for new approaches to organizing the power sector. In particular, there appears to be considerable interest in the scope for more decentralization and greater private participation. Developing countries in the Asian region have been very active in studying this option, and some countries have already prepared the necessary legislative and institutional groundwork for this transition. India plans to install as much as 5000 megawatts (MW) of private power capacity over the 1990 to 1995 period, and similar plans are underway in Indonesia, Malaysia, Thailand, Philippines, and Pakistan. In Sri Lanka, a publicly owned but independent company has been distributing power since the early eighties, and significant efficiency and service improvements have been observed during this period.

Despite these trends, enhanced private participation in the power sector is likely to be more successful when it is one element in a broader economic package involving policy reforms in other parts of the economy. Market forces confined to the power sector in a highly distorted economy may not necessarily improve the power sector situation since private participants will try to maximize financial rather than economic costs. Thus, private sector participants would make full use of cheaper generation inputs such as coal or natural gas even when this is potentially detrimental to the economy. Even in a reasonably market-oriented economy, the introduction of private participation in the power sector is unlikely to lead to environmental benefits, unless the costs of pollutants can be fully captured (i.e., internalized) in the financial cost to the participant. Thus, while private participation is likely to bring significant gains by the infusion of new capital and innovative management methods, it is likely to remain only one of several methods aimed at restructuring the power sector.

With regard to environmental issues, the national framework in which the utility functions is likely to play an equally important role. Actions of the utility need to be backed up by a set of consistent national policies and legislative support. The development of environmental standards and regulations is likely to (and should) take place outside the utility, while the public needs to understand the importance of a commitment to a program of environmental mitigation.

Global Environmental Issues in the LDC Context

The discussion in previous sections has clearly established the context within which the developing countries are capable of participating in environmental mitigation efforts at the global level. It is quite obvious that LDCs do not have the ability to contribute financially to global environmental cleanup efforts where the measurable benefits to the national economy are too low to trigger investment. Indeed, this paper has argued that many LDC projects which do have positive measurable benefits at the national level are being bypassed on account of capital constraints.

The principle of assistance to developing countries for environmental mitigation efforts, in terms of technology transfer, financial support and other means, is already well established. The Montreal Protocol, which was adopted in 1987 as a framework within which reduction in the consumption and production of certain types of chlorofluorocarbons (CFCs) is to be achieved, recognized the need for global cooperation and assistance to the developing countries. More specifically, a subsequent Ministerial Conference on Atmospheric Pollution and Climate Change, held in November 1989 and attended by representatives of 67 countries, declared the need to:

> Urge industrialized countries to use financial and other means to assist developing countries in phasing out their production and consumption of controlled substances as soon as possible, by providing them with sufficient means to enable them to meet their target date. The development of alternative technologies and products in developing countries should be promoted.

Currently, discussions are underway among world bodies and governments to define effective criteria and mechanisms for both generating and disbursing funds from a global environmental fund. While a broad workable agreement will not be easy to reach, global financing issues might be analyzed and resolved through a tradeoff involving several criteria: affordability/additionality, fairness/equity, and economic efficiency (Munasinghe, 1990c).

First, since LDCs cannot afford to finance even their present energy supply development, they will need financial assistance on concessionary terms in addition to existing conventional aid, in order to address global environmental concerns. Second, as noted earlier, past growth in the industrialized countries has exhausted a disproportionately high share of global resources, suggesting that the developed countries owe an "environmental debt" to the larger global community. This approach could help to determine how the remaining finite global resources may be fairly shared and used sustainably. Finally, the economic efficiency criterion indicates that the "polluter pays" principle may be applied to generate revenues, to the extent that global environmental costs of human activity can be quantified. If total emission limits are established (eg., for CO_2), then trading in emission permits among nations and other market mechanisms could be harnessed to increase efficiency.

Based on these principles, a core multilateral fund—the Global Environmental Fund (GEF)—has been proposed to be set up as a pilot over the next two or three years. This fund would finance investment, technical assistance and institutional development activities and will be managed under a collaborative arrangement between the United Nations Development Program, United Nations Environmental Program, and the World Bank. The disbursements from this fund might be guided by the principles previously discussed in this paper. In particular, the GEF would fund those investment activities that would provide cost-effective benefits to the global environment, but would, however, not be undertaken by individual countries without concessions. Thus, the fund is specifically designed to fill the void which is created by the lack of individual national incentives for those activities which would, nonetheless, benefit us all.

Conclusions

International pressures to implement environmentally mitigatory measures place a severe burden on developing countries, including those of Asia. The crucial dilemma this poses to LDCs is how to reconcile development goals and the elimination of poverty—which will require increased use of energy and raw materials—with responsible stewardship of the environment, and without overburdening economies that are already weak. This paper has argued that in view of the severe financial constraints that developing countries already face, the response of these countries in relation to environmental preservation cannot extend beyond the realm of measures that are consistent with near-term economic development goals. More specifically, the environmental policy response of LDCs in the coming decade will be limited to conventional technologies in efficiency improvement, conservation, and resource development.

The developed countries are ready to substitute environmental preservation for further economic expansion and should, therefore, be ready to cross the "pain threshold," providing the financial resources that the LDCs need today and developing the technological innovations and knowledge base to be used in the 21st century by all nations. The Global Environmental Fund, which is presently being established, will be in a position to facilitate the participation of LDCs at the global level.

References

Asian Development Bank. "Private Sector Participation in Power Development." Manila, Philippines, November 1988.

Economic and Social Commission for Asia and the Pacific (ESCAP). *Structural Change and Energy Policy.* United Nations, May 1987.

Flavin, C. *Slowing Global Warming: A Worldwide Strategy,* World Watch Institute Paper No. 91. Washington, DC, October 1989.

Foell, W. "Report on the Workshop on Acid Rain in Asia." Asian Institute of Technology, Bangkok, Thailand, November 1989.

Johansson, et al. *Electricity.* Lund University Press, Sweden, 1989.

Krause, F., Bach, W., and Koomey, J. *Energy Policy in the Greenhouse, Vol. 1.* International Project for Sustainable Energy Paths (IPSEP), El Cerrito, CA.

Munasinghe, M., Gilling, J. and Mason, M. "A Review of World Bank Lending for Electric Power," Industry and Energy Department, Energy Series Paper No. 2. The World Bank, Washington, DC, March 1988.

Munasinghe, M. *Energy Analysis and Policy.* Butterworths Press, London UK, 1990a.

Munasinghe, M. *Electric Power Economics.* Butterworths Press, London UK, 1990b.

Munasinghe, M. "The Challenge Facing the Developing World." *EPA Journal* March/April 1990c.

Razavi, H. and Fesharaki, F. "Electricity Generation in Asia and the Pacific: Power Sector Demand for Coal, Oil and Natural Gas." mimeo, February 1989.

Siddiqi, T. "Implications of Climate Change for Energy Policies in Asia," Workshop on "Responding to the Threat of Global Warming: Options for Asia and the Pacific." Honolulu, HI, June 1989.

Starr, C. and Searl, M. "Global Projections of Energy and Electricity," Paper presented at American Power Conference Annual Meeting. Chicago, IL, April 1989.

The World Bank, "Capital Expenditures for Electric Power in the Developing Countries in the 1990s," Industry and Energy Department Working Paper No. 21. Washington, DC, February 1990.

The World Bank. *World Development Report 1989.* Oxford University Press, 1989.

USAID. "Power Shortages in Developing Countries." Washington DC, March 1988.

Wilbanks, T. and Butcher, D. "Implementing Environmentally Sound Power Sector Strategies in Developing Countries," Paper prepared for the *Annual Review of Energy*. January 1990.

Winje, D. "Electric Power and the Developing Countries," *Energy and the Environment in the 21st Century*, MIT Press, Cambridge, MA, 1990.

World Commission on Environment and Development. *Our Common Future*. Oxford University Press, London UK, 1987.

Electric Power and the Developing Economies

Dietmar Winje
Energy Engineering and Resource Economics
Technical University of Berlin
Berlin, Federal Republic of Germany

Abstract

This paper provides an overview of the energy demand pattern in developing countries in general and, in particular, looks at the electric power sector and its impacts on environmental pollution. Presently the developing countries, excluding the People's Republic of China, account for approximately 16 percent of the 7.6 billion tons of oil equivalent (toe) of global commercial primary energy consumption. About 30 percent of the primary energy consumed in developing countries is used for electricity generation. It is expected that in the future almost 50 percent of the primary energy input in the developing countries will be needed for electricity generation. At present the electric power sector in the developing countries contributes 4 percent to world emissions of CO_2 and 3 percent each to SO_x and NO_x emissions. For the future it can be expected that, due to higher electricity demand growth rates, the share of the developing countries will increase, though the level of emissions will remain low compared to developed countries. Developing countries must evaluate means to reduce environmental pollution. Technology and policy means for a reduction of SO_x, NO_x, and CO_2 emissions in the electricity sector of developing countries and supporting recommendations on policy instruments are discussed here.

Introduction

In the past 20 years the environmental problems related to electric power generation have become obvious in many countries. The heavy environmental damage caused by the emission of sulfur oxides and nitrogen oxides was first realized in industrialized nations. Besides these emissions, the greenhouse effect may be a challenge for humanity to cope with if the current rate of carbon dioxide emissions cannot be controlled. The consequences of a CO_2 accumulation in the earth's atmosphere cannot easily be quantified, and global climate is a dynamic system that has not yet been adequately modeled, but a rise in global temperatures and the effects thereof cannot be ignored (Enquête-Kommission, 1988). The electric power sector in the world already has had a major impact on CO_2 emissions. Although these problems have been recognized for some years, the importance of pollution control measures varies greatly from country to country.

Electricity Generation in Developing Countries

The following set of graphs shows the contribution of electric power generation in developing countries to global primary energy consumption and the proportions of the energy carriers used to generate electricity in these countries. Through them we can ascertain the possible contribution of developing countries to a global reduction of pollutants and the efforts that developing countries will need to make in order to reduce emissions while meeting the future electricity demands of their growing populations.

In the year 1987 developing countries, excluding the People's Republic of China, consumed approximately 16 percent of the global primary energy consumption of 7,615 billion toe, as Figure 1 shows.

About 30 percent of the primary energy consumed in developing countries is used for electricity generation. Thus about 5 percent of global primary energy consumption is used by the electricity sector in developing countries. The contribution of these countries to world energy consumption and pollutant emissions is shown to be relatively small.

If world population by region is compared with primary energy consumption and electricity production as shown in Figure 2, it can be seen that the industrialized countries account for only 12 percent of the world's population but for more than 50 percent of primary energy consumption and 59 percent of electricity production.

The developing countries, excluding the People's Republic of China, make up 56 percent of the world's population and account for 16 percent of primary energy consumption and 15 percent of electricity produc-

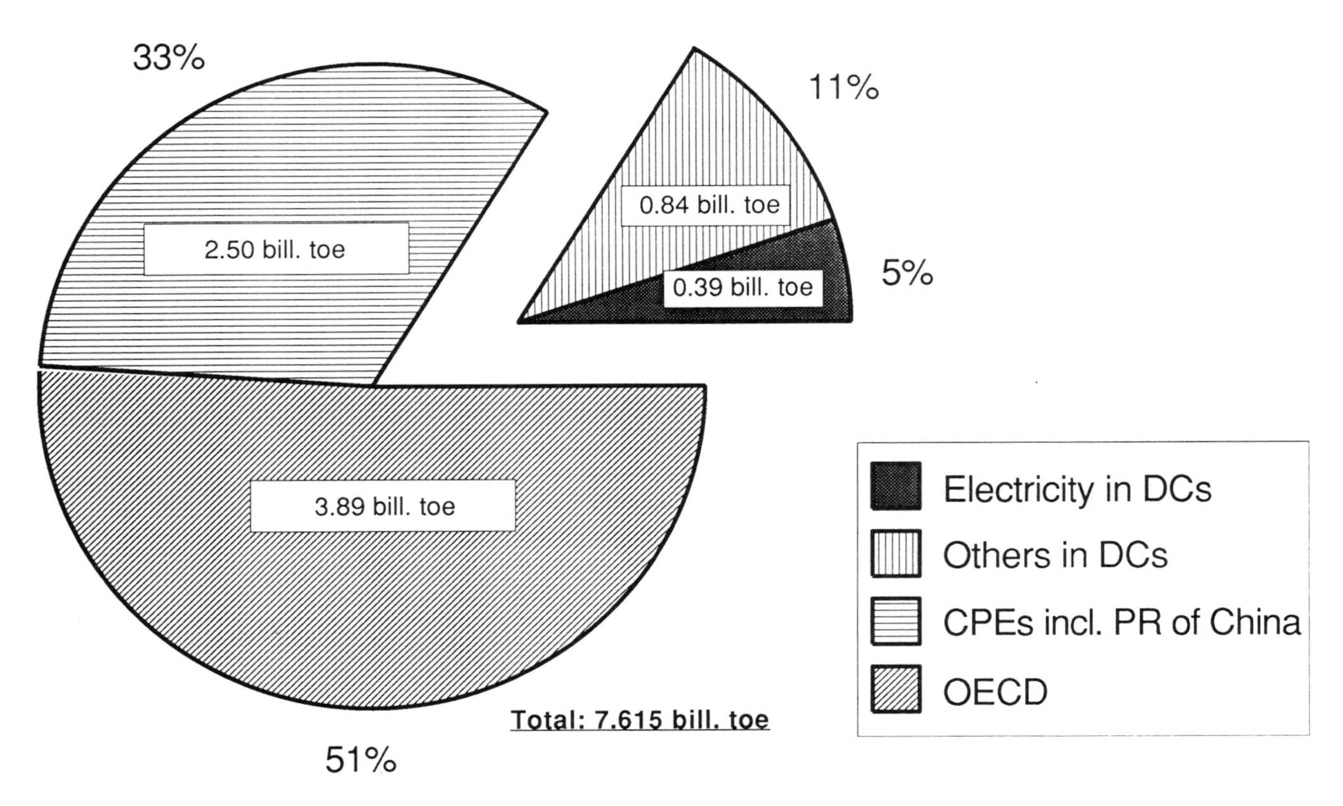

Figure 1. Global primary energy consumption and share of electricity consumption in the developing countries in 1987. Source: IEA, 1989.

tion. This indicates a relatively low level of energy consumption and the potential for further growth.

Considering the development of electricity consumption from 1950 to 1988, as shown in Figure 3, two major issues become obvious. First, there is a high increase of 8 to 9 percent per year from 1950 to 1970, approximately 5 percent from 1970 to 1980, and around 3 percent in the 1980s, resulting in about a tenfold increase over the 1950 level (Voss, 1990). Second, the Figure shows the growing share of the developing countries in the world's electricity consumption, though this remained small compared with the population level.

Figure 4 shows a proportional survey of the energy carriers used to generate electricity. The energy carrier mix in developing countries differs from that in the rest of the world. The proportion of electricity generated from hydroelectric sources in developing countries is 29 percent, compared with a global figure of 17 percent. Nuclear energy accounts for around 5 percent of electricity generation in developing countries.

The most conspicuous differences lie in the proportions of oil and coal used: developing countries generate 22 percent of their electricity from oil-fired plants and 30 percent from coal-fired plants, while the figures in member nations of the Organization of Economic Cooperation and Development (OECD) are 8 percent and 43 percent, respectively. Hydroelectric power and nuclear power do not contribute to air pollution.

It is generally assumed that, among other factors such as economic development, electricity demand depends largely on the size of the population. According to forecasts, the global population will grow from the present 5 billion to 6.1 billion by the year 2000 and 7.8 billion by the year 2020; that is a growth rate of almost 2 percent (WMO, 1986). In developing countries, where large sections of the population are not yet supplied with electricity, pent-up demand as well as population growth will contribute to increased demand for electricity. Forecasts of the future development of electricity generation in developing countries differ to a great extent.

An investigation by the Directorate-General for Energy of the Commission of the European Communities predicted an annual growth rate of the electricity sector in developing countries of almost 6 percent, as shown in Figure 5. The per capita electricity consumption in these countries is more than 20 times lower than in the OECD.

Increases in electricity demand are expected to be significantly lower in centrally planned economies (3.2

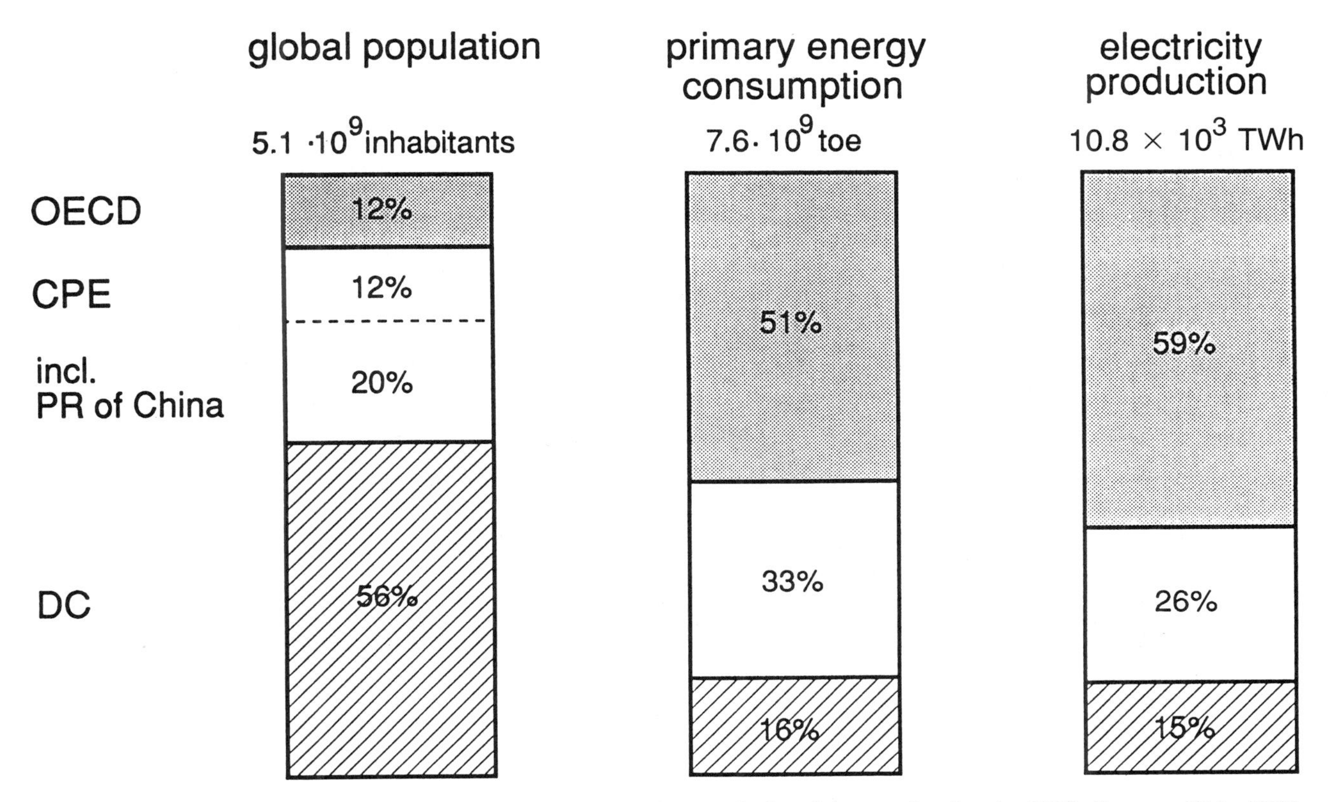

Figure 2. Global population, primary energy consumption, and electricity production in 1987. Source: IEA, 1989; *Statistisches Jahrbuch*, 1989.

percent) and in the OECD (2.3 percent). Increases projected for the different country groups average 3.2 percent.

Figure 6 shows primary energy consumption in developing countries as a proportion of global primary consumption in the year 2010, isolating the proportion of global primary energy expected to be used for electricity generation. The developing countries are expected to account for a larger proportion of the global primary energy consumption in the future.

Whereas in the year 1987 the developing countries consumed 16 percent of global primary energy, in 2010 they will consume 24 percent. In 2010 the primary energy needed for electricity generation will be almost four times higher than in 1987 (see Figure 5). In developing countries primary energy input for electricity will account for almost 50 percent of the total primary energy consumed, showing the growing importance of electric power.

These figures suggest that the most important investments that developing countries will have to make in the future will be to cover their growing electricity demands. As shown in Figure 5, the growth of electricity demand in TWh in the developing countries will be almost as high as in the OECD countries. The investments for new power plants could amount to around $50 billion a year—excluding investments for the replacement of old power plants, air pollution reduction equipment, and the extension of the electricity grid (VIK, 1988).

Emissions Due to Electricity Generation

Data from the above forecasts for electricity demand can be used to quantify increases in pollutant emissions from electricity generation in developing countries, assuming all other factors remained unchanged.

First we will look at carbon dioxide emissions (Figure 7). In 1987 total CO_2 emissions due to energy consumption were 22.7 billion tons worldwide. The CO_2 emissions from electricity generation amounted to 6.6 billion tons globally, while the corresponding figure for developing countries was around 0.9 billion tons. In developing countries per capita CO_2 emissions due to electricity consumption were 0.3 tons, while in the developed countries they were 4.4 tons.

Figure 7 also shows that CO_2 emissions in developing countries will increase from the 1987 level of 0.9

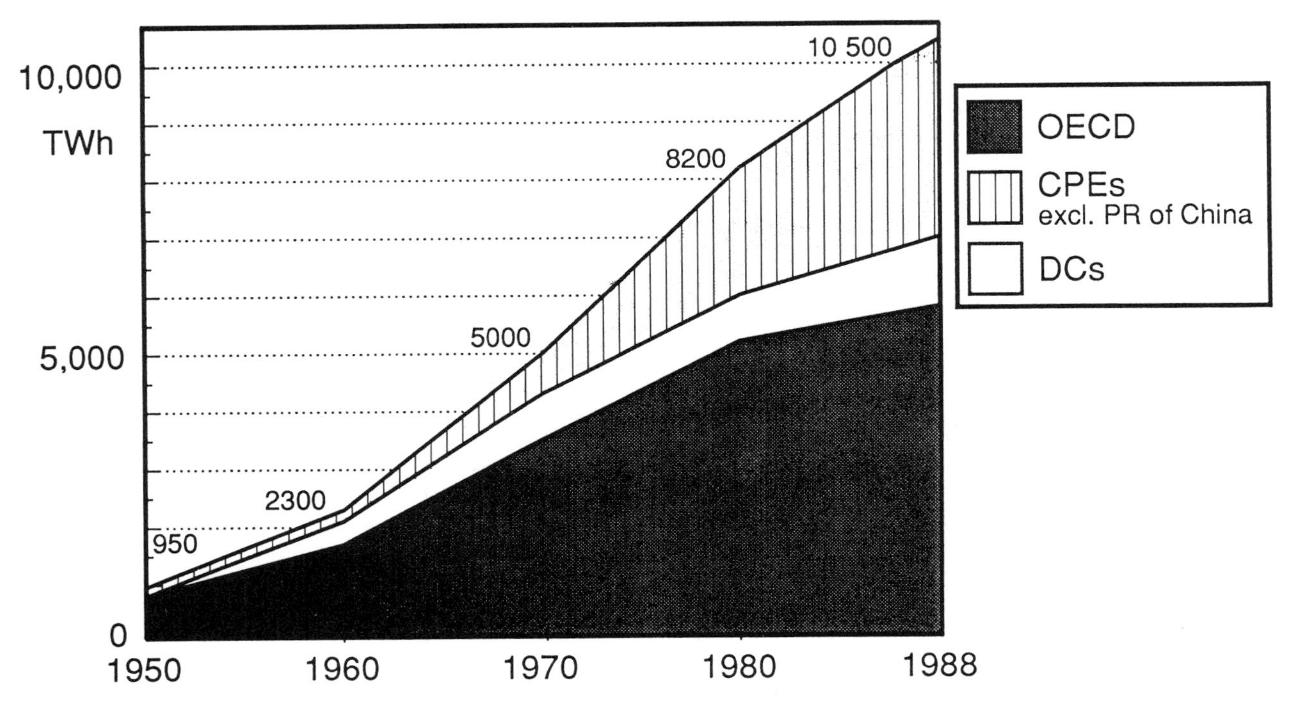

Figure 3. Development of electricity consumption in the world from 1950 to 1988. Source: Voss, 1990.

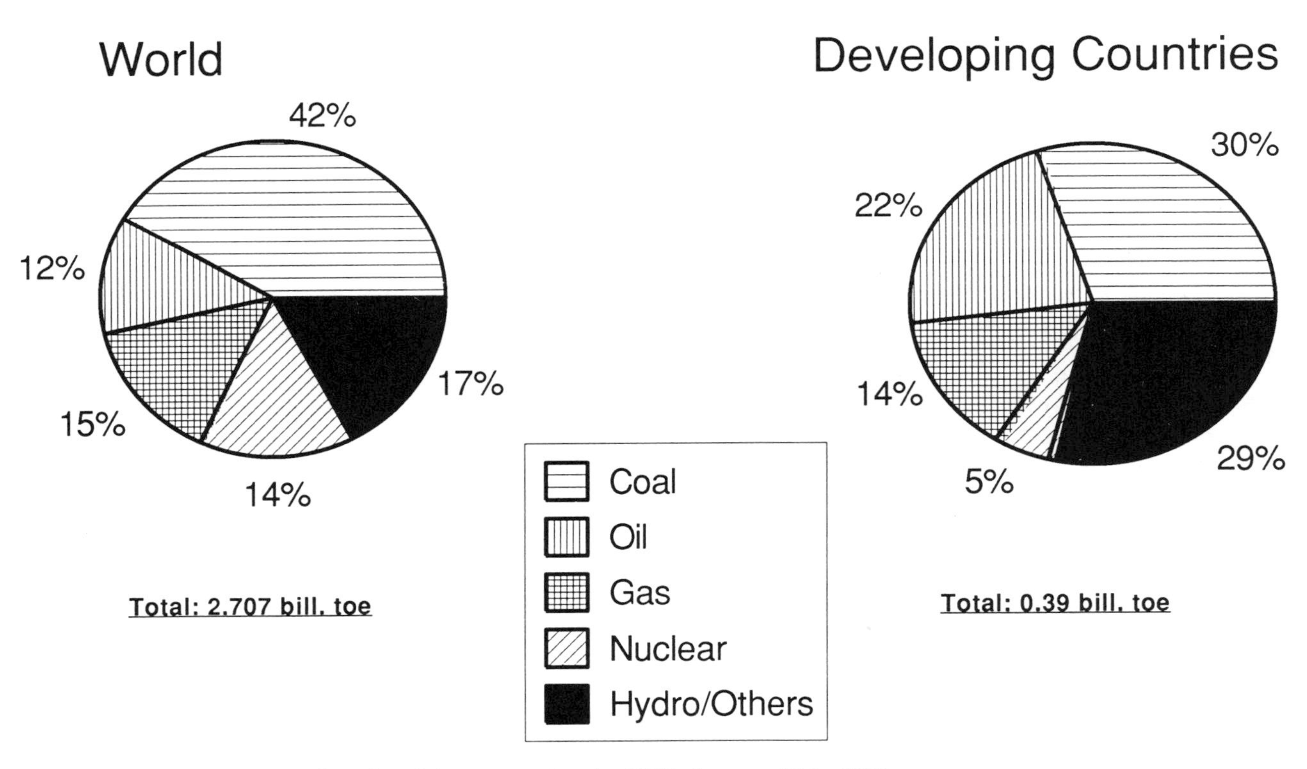

Figure 4. Energy sources for electricity generation in 1987. Source: IEA, 1989.

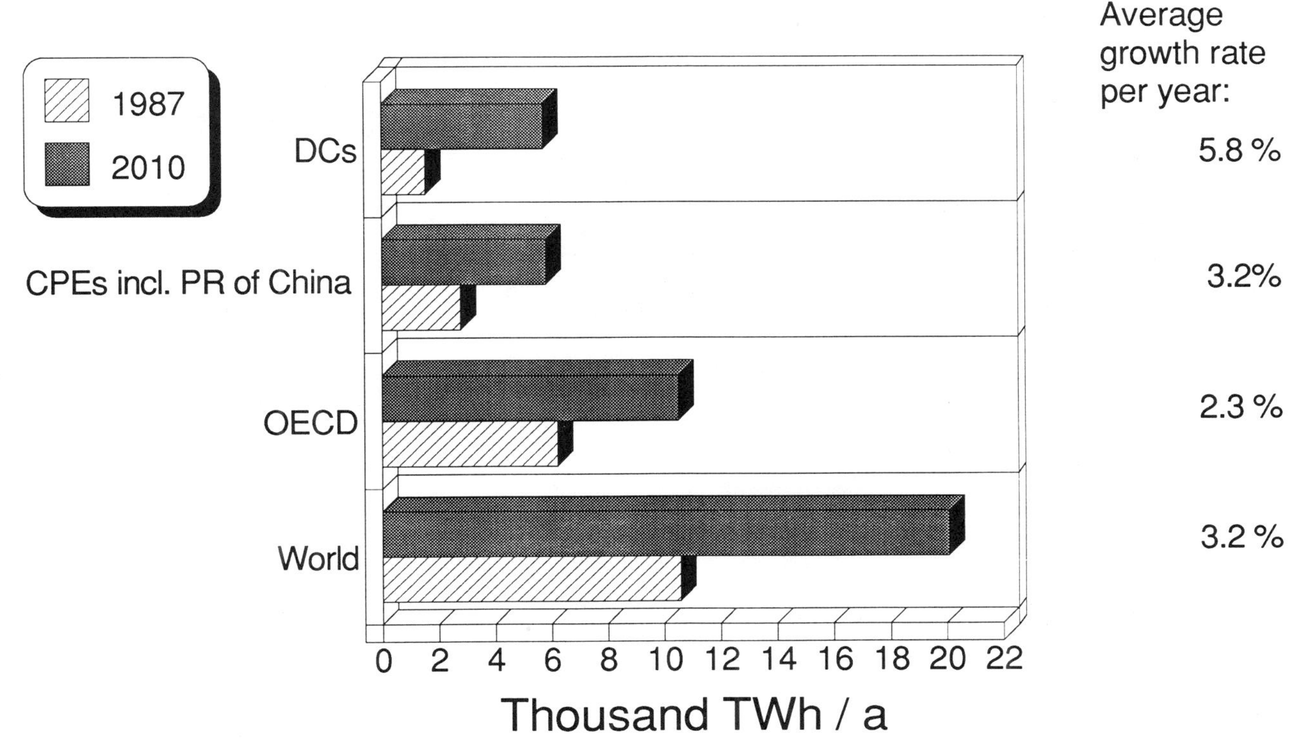

Figure 5. Forecast of growth rate of electricity generation. Source: CEC, 1989.

billion tons to around 3.4 billion tons in 2010. Still, it can be concluded that, ceteris paribus, the developing world's share of CO_2 will remain relatively low.

Estimated SO_x emissions in 1980 are about 110 million tons worldwide (Figure 8). The electric power sector in the developing world contributed approximately 3 million tons; the corresponding figure for the OECD was 16 million tons.

As shown in Figure 8, SO_x emissions in developing countries will, ceteris paribus, increase to 15 million tons by the year 2010 if no measures are taken. This figure would still be smaller than the level of SO_x emissions in the OECD in 1980.

The proportion of world NO_x emissions from developing countries is smaller than the proportion of SO_x emissions because of the relatively low level of motorized transport. Total NO_x emissions worldwide amounted to 69 million tons in the year 1980 (Figure 9). The portion contributed by the electric power sector in the developing countries was about 3 percent of this, while the OECD electric power sector contributed 9 percent. The reduction of NO_x emissions will be of major concern for the transportation sector rather than the electric power sector.

The figures describing CO_2, SO_x, and NO_x emissions show that the developing countries in general and their electric power sector in particular contribute rather small portions of the world's emissions. But differences exist not only between developing and developed countries; a look at the developing countries alone reveals that five nations, namely India, South Africa, Mexico, South Korea, and Taiwan, produce almost 40 percent of all the electricity in the developing world and are therefore responsible for a great amount of the pollution.

Means to Reduce Emissions

The following technical measures deserve attention as reactions to the environmental consequences of energy consumption, particularly electricity consumption.

Electricity Conservation. A reduction of emissions in developing countries by electrical conservation is a necessity. The end use of electricity is highly inefficient in most developing countries, leaving great potential for energy conservation in all sectors. The main reasons for high inefficiency by the end user are that appliances and plants are technically antiquated and electricity prices are too low to offer any incentive to save energy (World Bank, 1983). Furthermore, most consumers are not conscious of the environmental problems caused by wasteful uses of energy.

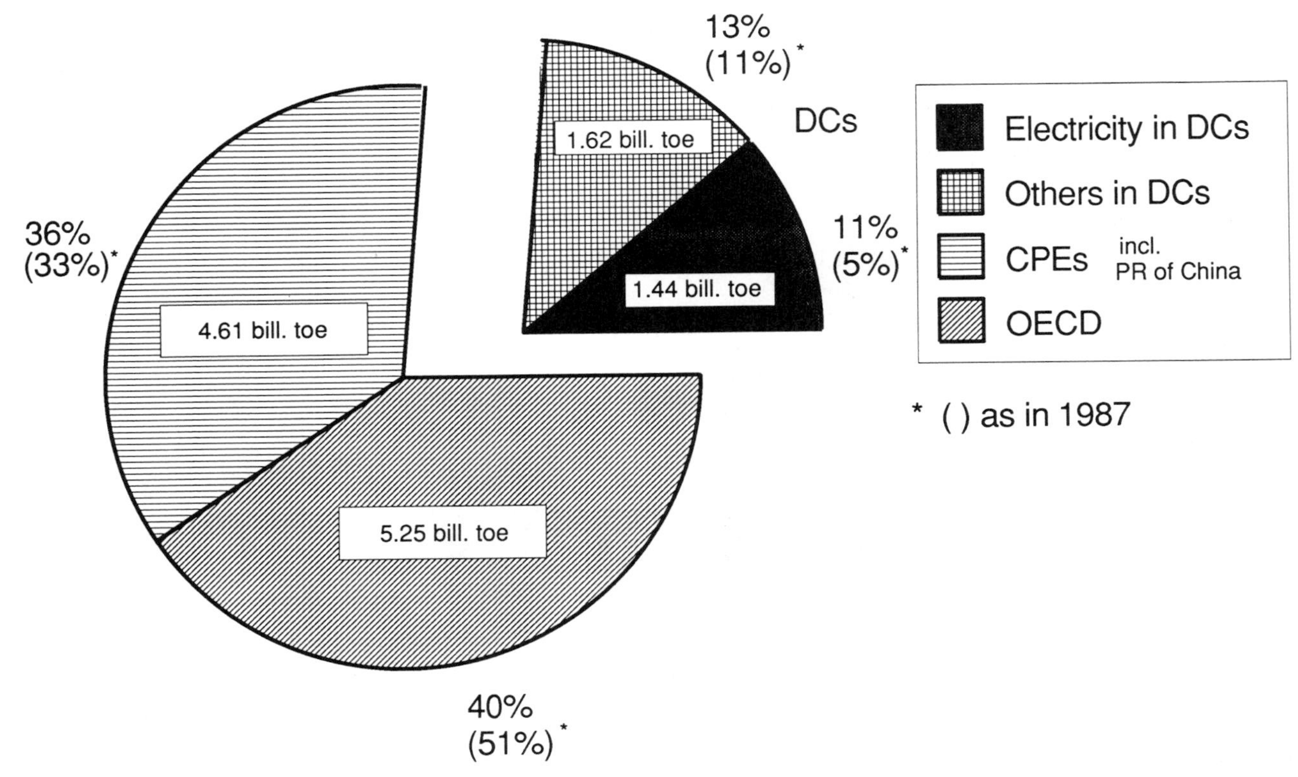

Figure 6. Global primary energy demand and share of electricity consumption in developing countries in 2010 (without noncommercial and renewables other than hydropower).

Efficiency Improvement of Generating and Distribution Systems. Substandard generating equipment and distribution networks are also a source of losses. The electricity production in developing countries, compared to that of OECD countries, exhibits low efficiency in terms of gross electricity production versus primary energy input. For example, the grid losses for the transport and distribution networks are 17 percent in South Asia and 21 percent in East and West Africa (Grawe, 1989). The use of modern electricity generating equipment with high technical efficiencies, such as combined-cycle power stations, is beneficial to the environment and resource conservation, but requires high investment and technical expertise.

Fuel Switches from High to Low Polluters. The fuel used for electricity generation often has high sulfur content, causing high emissions. A noticeable reduction of the specific pollutant emissions can be achieved by substituting quality coal or other fuels for low-grade coal. Investing in gas-fired and oil-fired power plants would also reduce CO_2 emissions. However, for most developing countries, this choice would mean increasing their dependence on imported oil or gas. Because of the low volume of export in most developing countries, this option requires special consideration. In 1980, after the second oil price shock, the non-OPEC developing countries spent 35 percent of their export earnings for oil imports (Grawe, 1989). Additionally, increasing the use of gas would mean making major investments in a gas transportation network.

Use of Renewable Energy Sources. The use of renewable energy sources such as hydro and solar power offers many options, though in some regions it is restricted to the electricity supply. Although hydro power resources are available, further development of hydroelectric power requires, in many cases, exploiting uneconomic hydro power potentials or, in other cases, the use of large-scale hydro power with impacts on the environment. In many developing countries electricity consumption is concentrated in a few densely populated urban centers, where about one third of the developing world's population lives. The electricity distribution network is often restricted to these areas. Solar energy sources generally require large land areas and cannot be used in these urban centers. On the other hand, the potential use of solar energy as a source of electricity for decentralized energy supply is attractive. Currently solar generators and wind energy generators require

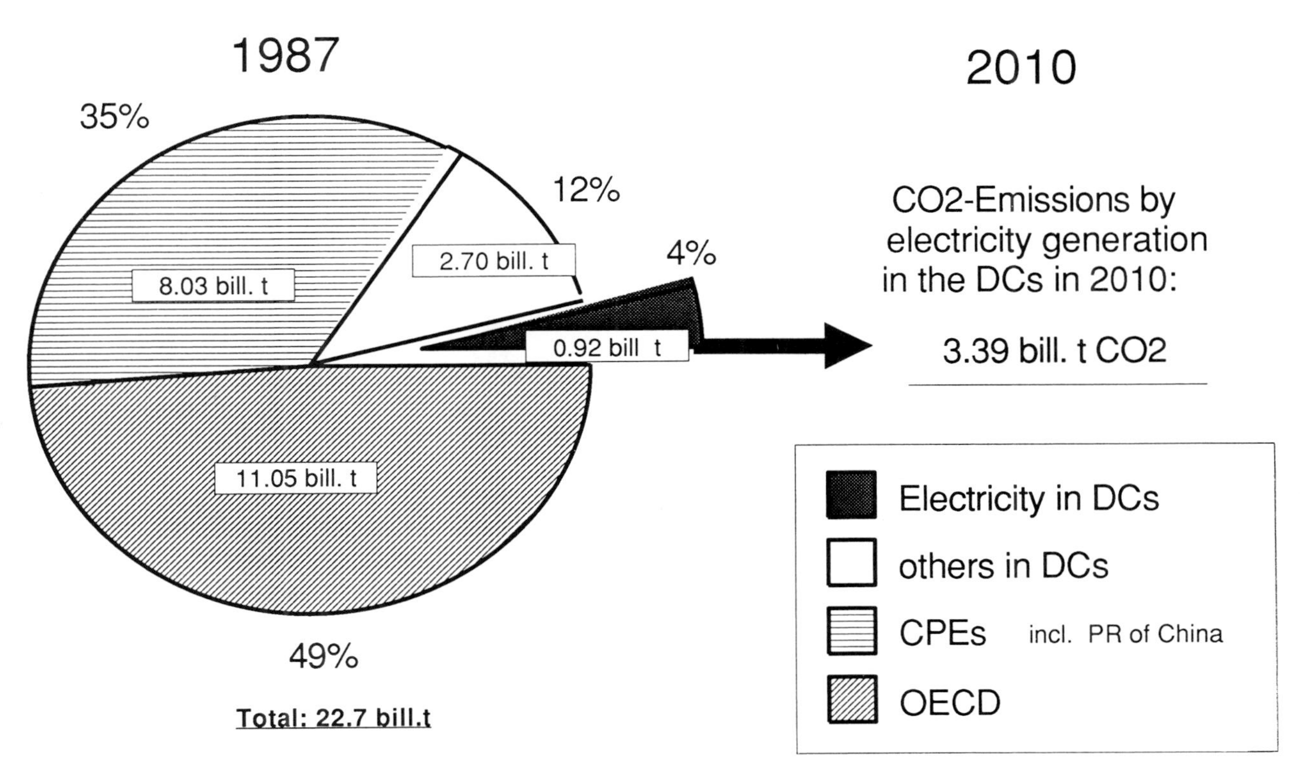

Figure 7. CO_2 emissions through energy consumption.

high investments. This technology is restricted to rural electricity supply in developing countries. However, the future for electricity generated from renewables seems to be promising.

Re-evaluation of Nuclear Power. In several mostly developed countries discussion of the social acceptability of nuclear power has led to delays in construction and to a reduction of nuclear power programs. Increasing evidence of the influence of carbon dioxide emissions on global warming may lead to a re-evaluation and revitalization of the nuclear industry. In light of the debt burdens of developing countries and the relatively high investment and knowledge requirements for nuclear power, this technology may prove its value more in industrialized than in developing countries.

Flue-gas Desulfurization and NO_x Reduction Systems. Flue-gas desulfurization and NO_x reduction systems reduce emissions of SO_x and NO_x in large power stations but require higher investments than can be made by developing countries, considering the absolute potential for reduction of emissions in these countries. Fitting all existing coal- and oil-fired power plants with flue-gas desulfurization and NO_x reduction systems would increase electricity production costs in developing countries by $10 billion per year. This figure assumes that all existing power plants can be fitted with collectors. To date, no technology exists that can economically clean CO_2 emissions from exhaust.

Recommended Action

Options for reducing pollutant emissions require policy instruments in order to be fulfilled. These must be tailored to the specific problems of individual developing countries (Winje, 1988).

Pricing Policy. Pricing is a central element of efficient resource allocation and, in particular, of energy demand management. This instrument can be used to influence consumer behavior and the market for electric appliances and plants indirectly. Electricity prices are subsidized in many developing countries and do not reflect the true generation cost. This leads to inefficient allocation and wasteful use of electricity and steers investments in the wrong direction.

Developing countries must reduce subsidies for electricity. The strain this will put on low-income households can be lessened through direct transfer payments to these households instead of through a system of differential electricity prices. Electricity supply side measures should begin with increasing the efficiency of

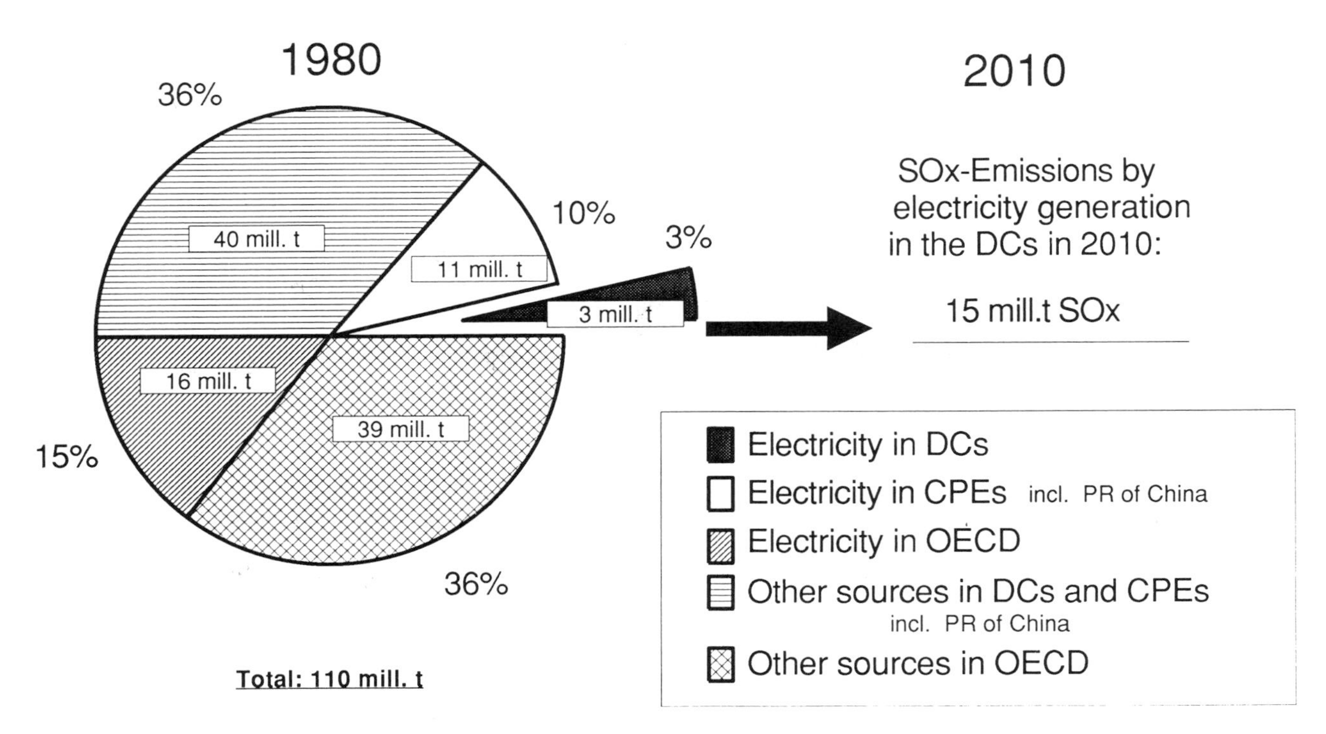

Figure 8. SO_x emissions from energy consumption. Source: OECD, 1989.

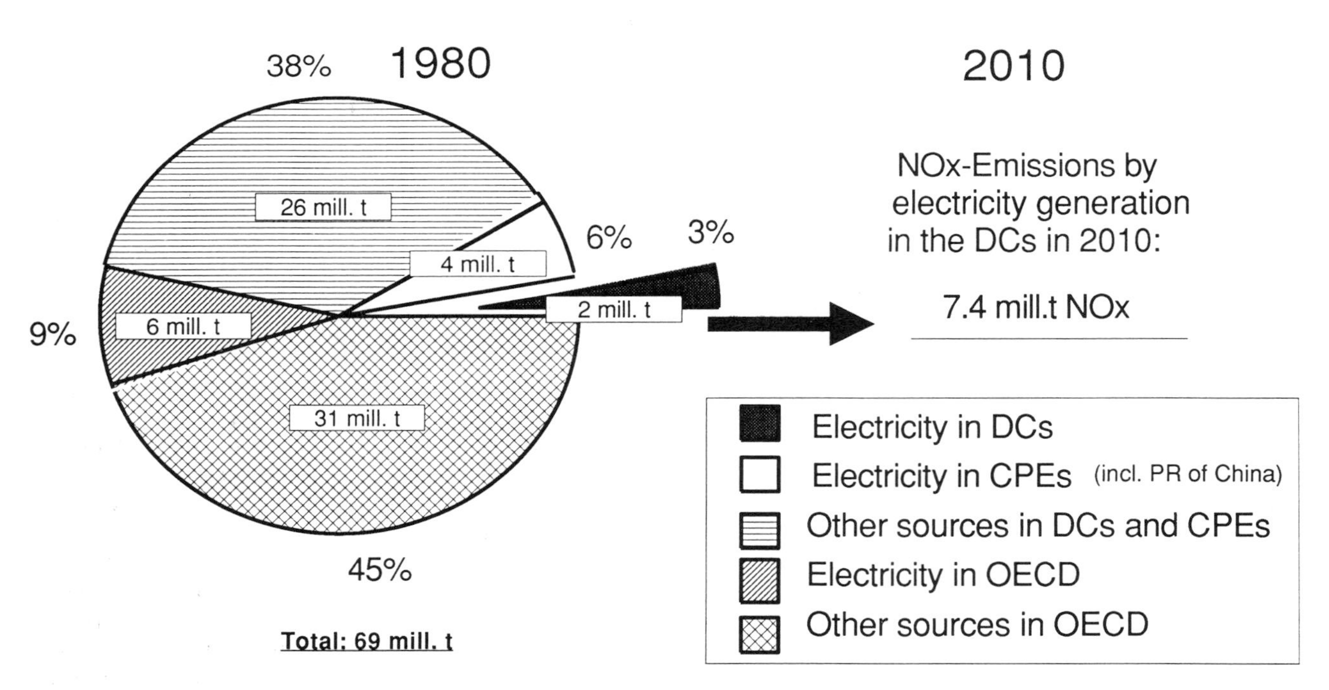

Figure 9. NO_x emissions from energy consumption. Source: OECD, 1989.

existing plants, substituting energy carriers, and implementing flue-gas cleaning systems.

Standards and Regulations. The electricity sector must be committed to improving efficiency and the environmental situation through minimum standards, at least for conversion systems. Improvement in efficiency is the least we can expect of developing countries in order to reduce emissions.

The pollution of the environment with SO_x and NO_x can be limited by the establishment of pollution control standards such as those already introduced in some countries. However, the capital requirements for desulfurization and NO_x reduction systems may prove too high for developing countries.

Financial Support. The developing countries can hardly succeed independently if they must both contribute to the solution of the world's environmental problems and expand their electric power sector. Financial assistance from the developed world may be necessary to support developing countries. Besides direct financial aid, knowledge transfer and engineering support may help to ease the considerable problems. Within individual developing countries energy and environmental efficiency can be influenced through financial support and taxation frameworks that aim to direct electricity generation and use in a more meaningful way. For example, the use of renewable energy sources and the substitution of energy carriers can be accelerated by other policy instruments.

Training and Education of Management. Management often has insufficient means at its disposal and must deal with problems through a poor infrastructure. Frequently management concentrates on permanent problems, neglecting other activities such as training employees, constructing effective control structures, and developing strategies for the future. Management of the electric power sector in developing countries can be improved through intensive education and training programs, for which the assistance of the industrialized nations is required.

Global Cooperation. The expansion of electric power systems and the reduction of SO_x and NO_x emissions are tasks that can be regionally solved. However, developing countries will not be in a position to achieve noticeable successes in reducing their pollutant emissions on their own. In order to develop their electricity supply systems and at the same time reduce environmental pollution they need the assistance of the industrialized nations. To improve the technical efficiency of their power plants and achieve their energy conservation potentials, they must depend on financial and technical assistance from the developed countries.

The reduction of CO_2 emissions is a task that must be solved on the global level. Though most efforts should be undertaken in the developed world, the developing countries must also make their contributions. This global task needs global approaches. For example, the technology transfer to developing countries must be accelerated through joint ventures. This reduces the need for hard currencies in the developing countries and is the best way to improve their electricity supply.

Research projects must take the specific needs of developing countries into account, and efforts in this direction should be supported. This includes building research and development centers in developing countries so that these nations can learn to help themselves.

Concluding Remarks

This paper has surveyed the energy demand pattern in developing countries in general, focusing in particular on the electric power sector and its impacts on environmental pollution.

The quantitative analysis of the electric power sector revealed that in 1987 around 36 percent of the primary energy in the world was used for electricity generation; the respective share in the developing world is 31 percent. For the future it can be expected that the primary energy input for electricity generation in developing countries will increase to 46 percent by 2010, making the electric power sector dominant. This is caused by the high annual electricity growth rate of 5.8 percent for the developing countries, compared to a 3.2 percent increase in the world.

The size of the electricity sector in developing countries should be kept in mind when looking at their potential for reducing pollutants in this sector. The electricity sector in developing countries contributes about 4 percent to the world's CO_2 emissions, 3 percent to SO_x emissions, and 3 percent to NO_x emissions, whereas the figures for the industrialized countries are much higher. Considering the energy consumption and emissions per capita, the huge differences between the developed and developing world become more obvious. With respect to energy conservation, electricity conservation, and the CO_2, SO_x, and NO_x problems, the industrialized countries and the centrally planned economies need to take action first. However, developing countries must also make efforts to reduce environmental pollution. To this end, some technology and policy instruments have been suggested.

References

Commission of the European Communities, Directorate-General for Energy, "Major Themes in Energy," *Energy in Europe, Special Issue*, pp. 51–54 (Sept. 1989).

Enquête-Kommission des 11. Deutschen Bundestages, *Schutz der Erdatmosphäre. Eine internationale Herausforderung*, Interim Report: Vorsorge zum Schutz der Erdatmosphäre, Deutscher Bundestag, Bonn (1988).

Grawe, J., *Energie für Entwicklungsländer,* VDEW, Frankfurt (1989).

International Energy Agency, *World Energy Statistics and Balances 1971–1987,* OECD, Paris (1989).

Organisation for Economic Co-operation and Development, *OECD Environmental Data, Compendium 1989,* OECD, Paris (1989).

Statistisches Bundesamt, *Statistisches Jahrbuch 1989 für die Bundesrepublik Deutschland,* W. Kohlhammer GmbH, Stuttgart (1988).

Vereinigung Industrielle Kraftwirtschaft, *Statistik der Energiewirtschaft 1986/87,* VIK, Essen (1988).

Voss, E., "Weltweite Energieversorgung—globale und regionale Aspekte des Kraftwerksmarktes," Brennstoff-Wärme-Kraft 42 (1/2), pp. 15–25 (1990).

Winje, D., "A Framework for Formulating Policies for Fossil Fuels in Oil-Importing Developing Countries," *Petroleum for the Future,* Hilal A. Raza and Arshad M. Sheikh, Islamabad, pp. 353–362 (1988).

World Bank, *The Energy Transition in Developing Countries,* The World Bank, Washington DC (1983).

World Meteorological Organization, *Report of the International Conference of the Assessment of the Role of Carbon Dioxide and of Other Greenhouse Gases in Climate Variations and Associated Impacts,* WMO Publ. No. 661, Geneva (1986).

Summary
Session D-2—Fossil-Fuel Based Electric Power Technologies

János M. Beér
Session Chair
Department of Chemical Engineering, and
Combustion Research Facilities
Massachusetts Institute of Technology
Cambridge, MA

Presentations at this session of the conference (D-2) were made by Kurt E. Yeager (EPRI), Frank T. Princiotta (EPA), Jack S. Siegel and Jerome R. Temchin (DOE), Steven J. Freedman (GRI), and Christian Bolta (ABB-CE). Written texts of the first four of these presentations are included here. The authors of these papers shed light on problems of electric power generation and potential solutions of these problems from the viewpoints of, respectively, the power generating industry, environmental protection, national energy policy, and the gas industry. The following summary highlights salient points developed in the discussion and draws attention to areas of general consensus as well as issues of disagreement.

Electricity Growth. There is broad agreement among the authors on continued electricity growth in the US, with electrical energy reaching nearly 50 percent of all energy consumption by the turn of the century. Due to improvements in efficiency, a reduction is expected in the rate of increase of primary energy to one half that of GNP growth. Internationally, future economic growth, especially in developing regions, can be expected to be associated with very large increases in electricity growth.

The Role of Coal. There is consensus also on the continued dominant role of coal in power generation during the first half of the 21st century. Coal accounts for 70 percent of worldwide and 95 percent of US fossil fuel reserves. Coal use, however, is contingent on the penetration of Clean Coal Technology, i.e., early application of new advanced coal utilization systems, combustion process modifications, and postcombustion cleanup capable of reducing pollutant emissions through increased energy efficiency and specific reduction of individual pollutants. There is general agreement that coal use must be joined by a nuclear renaissance and, from the global perspective, promising renewable energy opportunities.

Given this level of global dependence on coal, it is noteworthy that most of the world's coal resources are controlled by the US, China, and the USSR, and that US energy consumption will strongly diminish in terms of global energy consumption because of increased electrification and improved efficiency (Yeager).

Emissions of Acid Rain Precursors (SO_x, NO_x). The authors agree that pollutant reduction processes demonstrated under the aegis of the Clean Coal Technology Program can provide methods for SO_x and NO_x emissions reductions to satisfy the Bush Administration's proposed Clean Air Amendments. They differ, however, in their estimate of the costs and benefits of retrofitting existing power plants with commercially available flue gas desulfurization and low NO_x control technology.

Siegel and Temchin (DOE) present scenarios both with and without retrofit of existing coal fired units until 2004. Beginning in 2005 they consider repowering by Integrated Gasification Combined Cycle (IGCC) in 42 percent cycle efficiency and very low SO_x and NO_x emissions to meet incremental demand. They assume that Fuel Cell Technology will achieve 55 percent cycle efficiency beyond 2015. They do not give cost estimates for their scenarios except to mention that, while their "total control" scheme including retrofit of existing coal fired power plants would not be found economically feasible by the utility industry, it illustrates the potential of Clean Coal Technology to stay well below the cap of SO_2 emissions of 8.9MM (million) tons SO_x required by the year 2000 in the Bush Administration's proposal.

Princiotta (EPA) presents results of a retrofit cost estimate developed for 516 boilers in 188 plants using a number of Clean Coal Technologies, including fuel switching, combustion process modification, and post-combustion flue gas clean up. He estimates that achievement of the President's proposal to reduce SO_2 emissions by 10MM tons per annum by the year 2000, with a reduction of 5MM tons of SO_2 per annum by 1995, will cost $700 million per annum for the first five years and $3.8 billion per annum from 1995 to 2000. This cost would add 2 percent to the US electricity bill of $160 billion per annum by the year 2000.

Yeager (EPRI) does not find justification for the massive application of emissions control technology on existing power plants (retrofit strategy) and emphasizes the cost effectiveness of replacement strategy instead. He argues that if, instead of retrofit, accelerated deployment of more productive Clean Coal Technology for new and repowered plants were encouraged, $200 billion to $400 billion could be saved during the next 60 years. Also, some undesirable side effects of retrofit technologies, including higher volume of solid wastes, lower energy efficiency, and increased CO_2 emissions, could be avoided. In Figure 16 (page 662) of his paper, Yeager compares the performances of the retrofit and replacement strategies over the period 1980 to 2050.

Global Warming (CO_2 Emissions). Princiotta discusses options for reducing CO_2 emissions and presents the results of calculations using his Utility CO_2 Emission Model (UCEM) for estimating the impact of different energy options on CO_2 emissions. The options include energy conservation, the use of nuclear energy, renewable solar and biomass energy, the application of Clean Coal Technology, the use of new supplies of natural gas, and combinations of these. In his model, Princiotta makes judgments about the MW penetrations assumed for nuclear, solar, biomass, and clean coal, and the amount of natural gas in quads that may be available to meet expected increase in electrical demand. Energy conservation is assumed in terms of electricity growth rates (1.5 percent annual growth was assumed for the "conservation case," and 2.5 percent for the "base case") (see Princiotta, Table 3, page 643).

Even with uncertainties about the assumptions on the penetration of the technologies over the period 1990–2020, Princiotta's analysis clearly illustrates that:

- Conservation will have to be an indispensable part of any strategy for CO_2 emissions reductions.
- Application of Clean Coal Technology, even in combination with a nuclear renaissance and renewable (solar, biomass) energy use, is insufficient for holding the rise of the CO_2 emissions from the present level of 2.2×10^9 tons per annum to below 3.5×10^9 tons per annum by the year 2020.
- New natural gas has a key role in combination with conservation, Clean Coal Technology, nuclear and renewable energy in maintaining CO_2 emissions at near present levels (2.5×10^9 tons of CO_2 per annum by 2020) (see Princiotta, Figures 15 and 17, pages 646 and 647). Gas is attractive as a fuel because of its lower carbon to hydrogen ratio and because it lends itself readily to use in energy efficient combined cycles.

Efficiency as a Basic Requirement—Natural Gas Use. Higher efficiency requirements are driven primarily by environmental considerations, whether they concern acid rain or global warming. Internationally, there is great interest in the improvement of energy efficiency in coal fired power generation. It can be expected also that natural gas, which is capable of fueling gas turbine–steam combined cycle plants of high energy efficiency and of relatively low first cost, will play an increasingly important role, particularly in the next two decades, in power generation. While the resource base of natural gas may be sufficient to support some 100GW of new power generation capacity (Freedman), the deliverability of gas varies regionally and the unavailability of long-term natural gas supply contracts hinders this expansion. Nevertheless, natural gas use in power generation can be expected to more than double by the end of this decade, with a price approaching $4–$5/MMBtu, the cost of producing synthetic gas from coal.

The Power Generation Challenge. In the year 2010 one half of the US electric power generating capacity (about 300GW) will be more than 30 years old; yet less than 40GW of regulated and 34GW of unregulated new generating capacity is scheduled for the 1990s. To avoid power shortages by the end of the decade, 20GW of new capacity would have to be added annually. Yeager analyzes the causes of this problem into three categories: investment disincentives due to the regulated nature of the utility industry, uncertainties arising from tightening environmental regulations, and the regrouping of nationally oriented power equipment suppliers into several integrated multinational corporations. The latter are forcing electric utilities to rethink the processes of designing, purchasing, and constructing power plants. There is a trend away from individually customized plants toward more standardized, shop-fabricated modular plants—a trend dictated by international competition and a drive toward improved quality of construction. Yeager's criteria for cost effective power plants are 90 percent or greater reliability, 6500–8500 Btu/kWh heat rate, $500–$1000/kW capital cost, and capability for compliance with environmental regulations. These constitute major challenges to US equipment manufacturers and require acceptance of greater risk in research and development for sustaining leadership in power generation innovation.

Role of Clean Coal Technology in Electric Power Generation in the 21st Century

Jack S. Siegel
Jerome R. Temchin
Office of Fossil Energy
US Department of Energy
Washington, DC

Abstract

Deployment of high efficiency, low polluting clean coal technologies allows widespread use of coal energy resources for electric power generation while minimizing environmental effects. By mid-century, high efficiency coal-based systems, such as integrated gasification combined cycle, fuel cells, and magnetohydrodynamics, can reduce worldwide CO_2 emissions by 37 percent compared to conventional technology. For acid rain control, advanced clean coal technologies achieve significantly lower SO_2 and NO_x emissions than currently available technology and can provide the long-term reduction of acid rain gases required by pending legislation. Even higher efficiencies (80 to 90 percent) are possible through the use of combined heat and power systems.

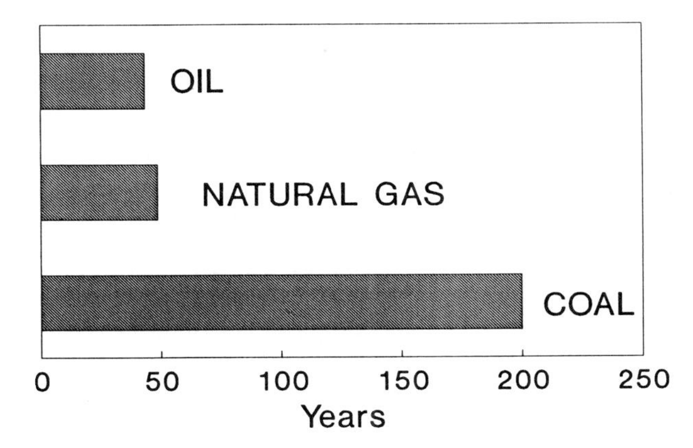

Figure 1. Years' supply of recoverable reserves based on 1988 production rates. Source: USDOE/EIA, *International Energy Annual,* 1988.

Introduction

Coal is the world's most abundant fossil fuel. It constitutes 70 and 95 percent, respectively, of the heat content of worldwide and US fossil fuel reserves. It represents 28 percent of the world's energy consumption. At current production rates, existing coal reserves will last 200 years (see Figure 1).

Over half of US electricity generation is powered by coal. Coal consumption for electricity generation is expected to grow at the same rate as electricity demand—2 to 3 percent per year. Worldwide, electricity demand has been growing even faster, especially in less developed countries, where it has been growing at a rate of 7.5 percent per year since 1980. Coal is expected to remain the fuel of choice for electricity generation in the foreseeable future.

However, the use of coal carries with it several environmental concerns. The purpose of this paper is to present the potential environmental benefits that could be achieved through the widespread deployment of advanced coal utilization technologies in the electric utility sector.

The biggest concern today regarding coal use is global warming. Carbon dioxide, a major "greenhouse" gas, is produced when any fossil fuel is burned. Carbon dioxide is responsible for half of the increases in the greenhouse effect, and 73 percent of the CO_2 emissions are generated by fossil fuels. However, the worldwide fossil energy resource base, economic development policies, and energy/economic growth relationships indicate continued reliance on these fuels.

If it is determined that greenhouse gas emissions should be reduced, more efficient fossil energy technologies can play a role. Efficiency improvements are much preferred over other methods, such as CO_2 capture, which significantly derate the power unit and thus cause an increase in other undesirable emissions such as SO_2 and NO_x. In addition, efficiency improvements reduce the production of CO_2 and therefore eliminate the difficult problem of CO_2 disposal. Efficiency improvements in countries currently using fossil energy technologies, together with reliance on advanced fossil technologies in developing countries, could achieve long-term reductions in greenhouse gas emissions. Acid rain is also of great concern in many parts of the world, including the US. A major cause of acid rain is thought to be SO_2 and NO_x emissions from coal burning electric power plants. As will be shown, deployment of clean coal technologies can have a significant impact on the reduction of SO_2

and NO_x emissions. Some of these technologies are also highly efficient and can be useful for reducing both the greenhouse and acid causing gases.

Description of Advanced Technologies

Through its research, development, and demonstration programs, the US Department of Energy is developing advanced coal technologies for use not only in electric power plants but also for all other market applications. In addition to a $200 million research and development program that is researching technologies that could greatly improve performance of coal-based systems, the Department is also engaged in a $5 billion Clean Coal Technology (CCT) Demonstration Program. The CCT Program is jointly funded by the government and private industry. In this program, the most promising of the advanced coal-based technologies are being moved to the marketplace through demonstration. The demonstration effort is at a scale large enough to generate all of the data needed by the public sector to judge the commercial potential of the processes being developed. The goal of the program is to make available to the US energy marketplace a number of advanced, more efficient, and environmentally responsive coal utilization technologies. These technologies will reduce or eliminate the environmental impediments that limit the full use of coal. This activity and the resulting processes that will be commercialized recognize the strategic importance of coal to the US economy and the international marketplace. These efforts aim to resolve the conflict between the increasing use of coal and the growing concern about the environmental impact of such use.

The technologies being developed under the CCT Program include advanced coal cleaning, combustion, conversion, and post-combustion cleaning. These technologies may be utilized in new as well as existing plants and are therefore an effective way of reducing emissions in the nation's aging inventory of coal-fired units. Several of these systems are not only very effective in reducing SO_2 and NO_x emissions, but also emit lower amounts of CO_2 per unit of power because of their higher efficiencies. These technologies include integrated gasification combined cycle (IGCC), coal-fired fuel cells, and magnetohydrodynamics (MHD) and will be discussed in more detail in this paper. Other high efficiency technologies include pressurized fluidized combustion and several advanced concepts being considered in the R & D program. These technologies can be used for repowering existing power units as well as greenfield applications.

In the repowering of an existing boiler with IGCC (see Figure 2), the existing boiler is typically supplemented or replaced by gasifiers which, after passing through a gas cleanup system, supply fuel gas (partially combusted coal gas) to a combustor that drives a gas turbine, creating additional electric power capacity. Waste heat from the combustion turbine can be used to produce high quality steam to run the existing steam cycle. An IGCC can be designed to achieve a more efficient heat cycle than a pulverized coal (PC) unit, resulting in a significant savings in the cost of electricity. One of the advantages of IGCC systems is that they can be built in 100-150 MWe modules. This flexibility in capacity expansion means that power can be increased as demand warrants, thus reducing the problem of overbuilding capacity. Also, this modular design allows a significant portion of the IGCC unit to be shop-fabricated and thus enables standardized design, better manufacturing control, and lower labor rates, resulting in lower unit costs than on-site fabrication.

Large SO_2 emissions reductions (over 99 percent) are achieved with an IGCC with use of sorbents (e.g., limestone) and/or the gas cleanup system at the exit of the gasifier. Very low NO_x levels (0.1 pound per million Btu or better) are also achieved with an IGCC because of high removal of the fuel-bound nitrogen in the coal and lower combustion temperatures that prevent the formation of thermal NO_x. Partial combustion in the gasifier takes place in a reducing atmosphere where intermediate combustion products of the fuel-bound nitrogen—hydrogen cyanide and ammonia—are produced. (No formation of thermal NO_x takes place since it requires an oxidizing atmosphere.) These intermediate products are removed in the gas cleanup system downstream of the gasifier. Formation of thermal NO_x is minimized in the combustor by combining the fuel gas with water to reduce combustion temperature.

Another advanced coal based technology currently under development is the fuel cell (Figure 2). Fuel cells do not rely on combustion. Gasified coal is used as a source of hydrogen to supply the fuel cell for the electrochemical reaction that releases chemical energy when oxygen atoms and hydrogen from the hydrocarbon fuel are combined to form water. The hydrogen is fed down one of two electrodes in the cell's "power section," where it gives up electrons. Oxygen is fed down the other electrode, where it picks up electrons. The flow of electrons through an electrolyte from one electrode to the other creates a current of electricity. The fuel cell is extremely clean and highly efficient. In a CCT configuration, the fuel cell is fueled by hydrogen extracted from the coal gas made by a coal gasifier. The impurities in the coal gases (e.g., SO_2 and NO_x) are removed by a gas cleanup system. The principal waste product from the fuel cell is water. The most mature fuel cell is the phosphoric acid fuel cell. It has been used in buildings and is now being developed for utility use. The phosphoric acid fuel cell is powered by natural gas, is low temperature (400°F), and has shown operating thermal efficiencies of over 40 percent. Other concepts are also being developed, such as the molten carbonate fuel cell,

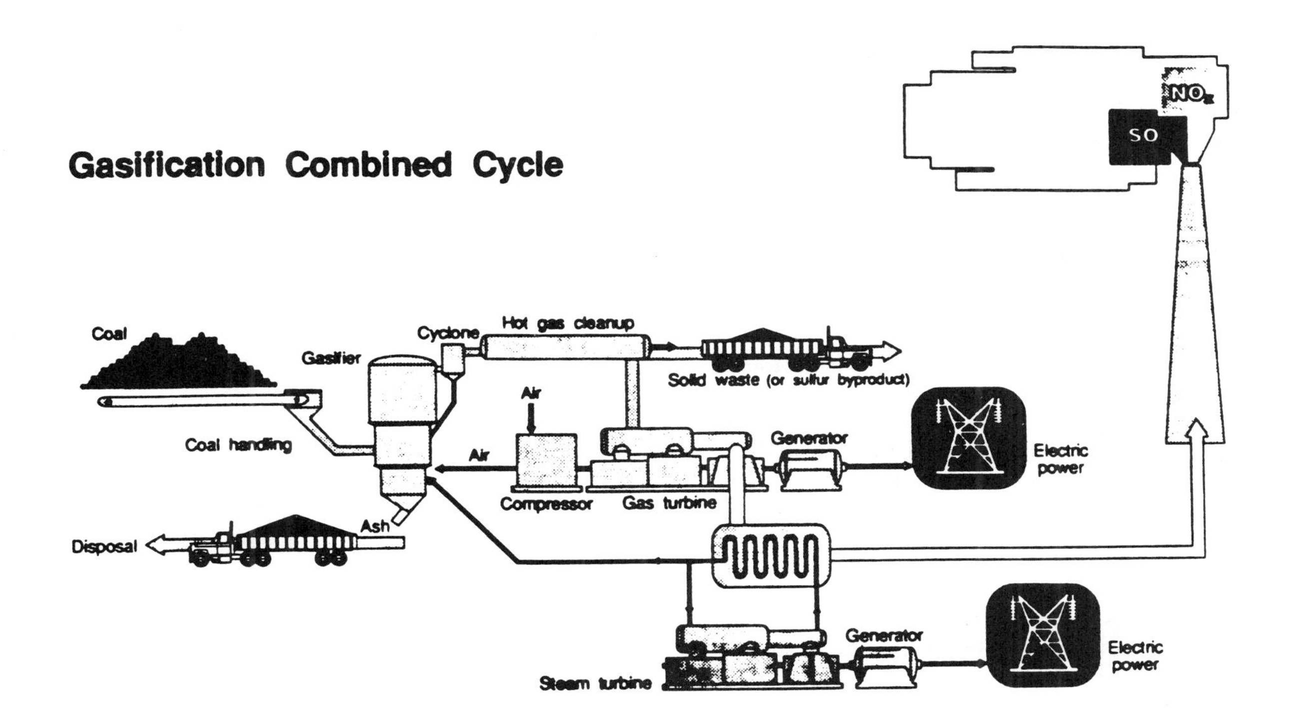

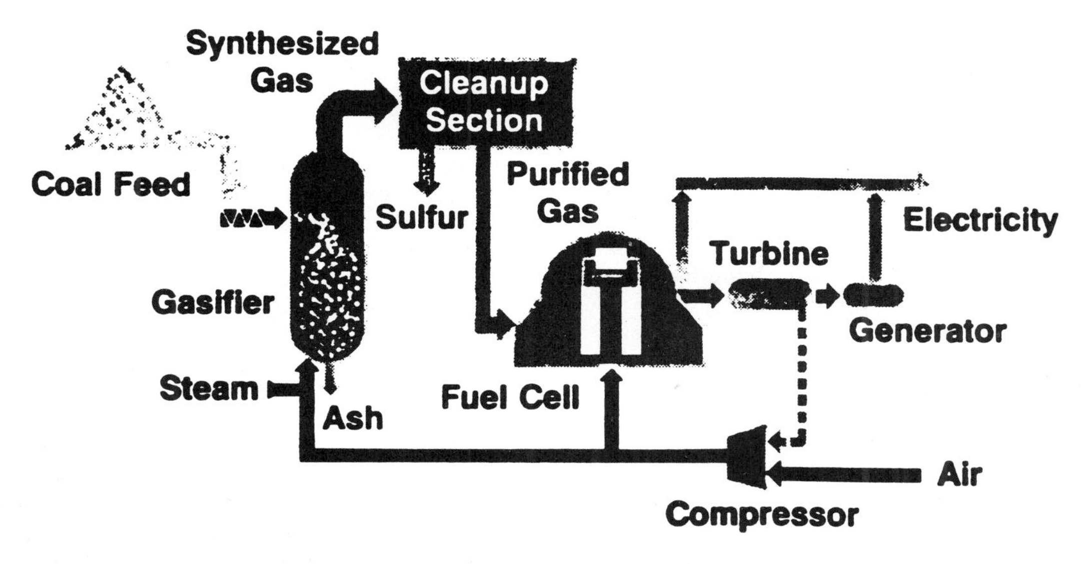

Figure 2. High efficiency clean coal technologies.

which uses a hot mixture of lithium and potassium carbonate as the electrolyte. The most advanced type is the solid oxide fuel cell, which uses a hard ceramic material instead of a liquid electrolyte. Both molten carbonate and solid oxide fuel cells can use coal gas as well as natural gas, have higher operating temperatures (1200°F and 1800°F, respectively), and have potential efficiencies of 50 to 60 percent. The higher operating temperatures allow these technologies to be used in a combined cycle configuration to achieve additional power production.

Magnetohydrodynamics (MHD) is based on burning coal at high temperature (5,000°F) to release a plasma of highly charged particles that is channeled through an intense magnetic field that generates electricity. Salts are added to the plasma to increase conductivity. The electricity is tapped by electrodes in the channel walls. The exhaust gases leaving the channel are sufficient to run a steam cycle, as in a conventional plant. The salts chemically combine with sulfur in the coal, thus greatly reducing SO_2 emissions. Nitrogen oxide formation is minimized by staging the burning with the appropriate stoichiometry of fuel and oxygen. MHD systems can also be used in a combined cycle mode because of the high exhaust temperature (3700°F) flowing from the MHD channel. This technology is capable of achieving efficiencies of 50 to 60 percent. Although these technologies are currently in varying stages of development and demonstration, through an accelerated commercial demonstration program they are likely to be available by the beginning of the next century. Initially, it is anticipated that IGCC technology will be commercialized on a broad scale through the repowering or replacement of existing units. In the near term, CCT conversion efficiencies will be in the 40 to 45 percent range, compared with conventional coal-fired power plants with scrubbers, which have thermal efficiencies of 30 to 35 percent. For each 5 percent improvement in efficiency, CO_2 emissions are reduced by approximately 15 percent. In the more distant future (2010–2020), coal-fired fuel cells and/or MHD systems will become commercially available with thermodynamic efficiencies in the 50 to 60 percent range.

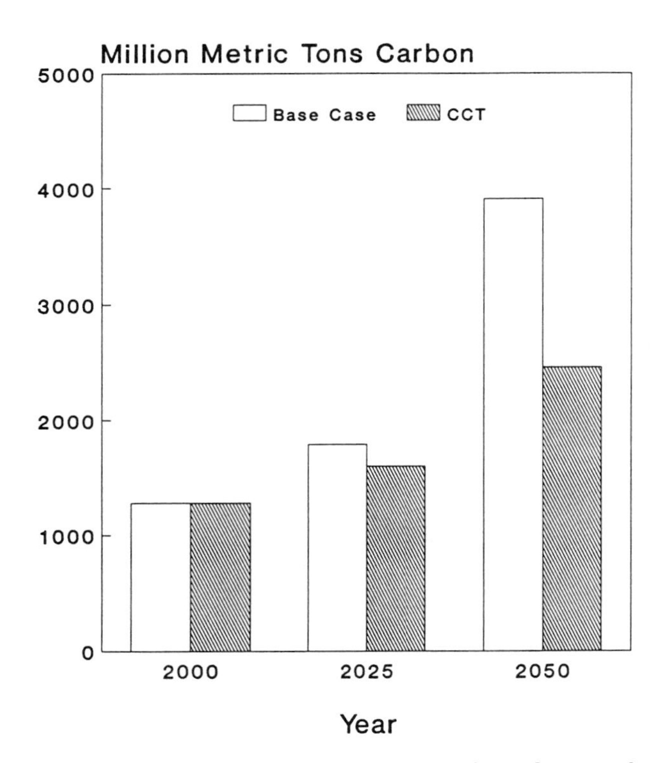

Figure 3. World CO_2 emissions (as carbon) from coal-fired electric utility generating units. Notes: a) Incremental demand for the years 2000 to 2025 assumed to be satisfied by IGCC repowering (42 percent thermal efficiency). b) Incremental demand for the years 2025 to 2050 assumed to be satisfied by (green-field) coal-fired fuel cells (55 percent efficiency). [Derived from Institute for Energy Analysis, Oak Ridge Associated Universities, Global Energy-CO_2 Model: PC Version 3.05 (Edmonds-Reilly Model)]

Worldwide CO_2 Emission Reduction

Carbon dioxide is responsible for half of the potential increase in the greenhouse effect. About 38 percent of world CO_2 emissions comes from coal use and is therefore responsible for about 19 percent of the potential global climate change.

Figure 3 shows the impact on worldwide CO_2 emissions (as carbon) if CCT is deployed extensively throughout the world beginning in 2000. Between 2000 and 2025, it is assumed that existing coal-fired electric utility units are repowered with IGCC technology with a 42 percent thermal efficiency. Although this introduction of advanced combustion technology represents a significant improvement in efficiency—a 23 percent improvement over conventional units—this initial response would reduce CO_2 emissions by a little over 10 percent, compared to conventional technology, because of the large percentage of existing conventional power units assumed to be remaining in operation during this period.

After 2025, it is postulated that all the new and most of the existing utility coal demand (except the demand satisfied by IGCC repowered units) would be satisfied by fuel cells and/or MHD systems with a 55 percent thermal efficiency. Nearly all of the coal demand is satisfied by fuel cells because it is assumed that all of the conventional units will have been retired or repowered (with IGCC or fuel cell technology) by this time.

Based on a projected 150 percent increase in utility coal demand between 2025 and 2050, CO_2 emissions would increase on the average about 3.2 percent per year during the same time period if conventional coal burning technology is utilized (base case). In contrast, if advanced coal based technology is used by utilities during this period to satisfy additional demand, CO_2 emissions will increase only 2 percent per year.

US Electric Utility CO_2 Emission Reduction

US coal use contributes approximately 6 percent of global CO_2 emissions or about 3 percent to global climate change. The potential for CO_2 reduction in coal-fired electric utility units in the US shows a pattern similar to what can be expected on a worldwide basis with widespread deployment of advanced coal-based technologies. As Figure 4 indicates, by 2030, advanced technologies can help reduce CO_2 emissions by 24.2 percent compared to exclusive use of conventional combustion systems. In the base case, which only considers conventional pulverized coal (PC) units, CO_2 emissions show an average annual increase of 2.6 percent between 2000 and 2030, compared to 1.7 percent per year with deployment of advanced technology.[1]

Although US energy projections are not available beyond 2030, it can be expected, as with worldwide deployment, that as advanced technology replaces existing units, CO_2 emissions levels from electric utility units will be substantially below the base case.

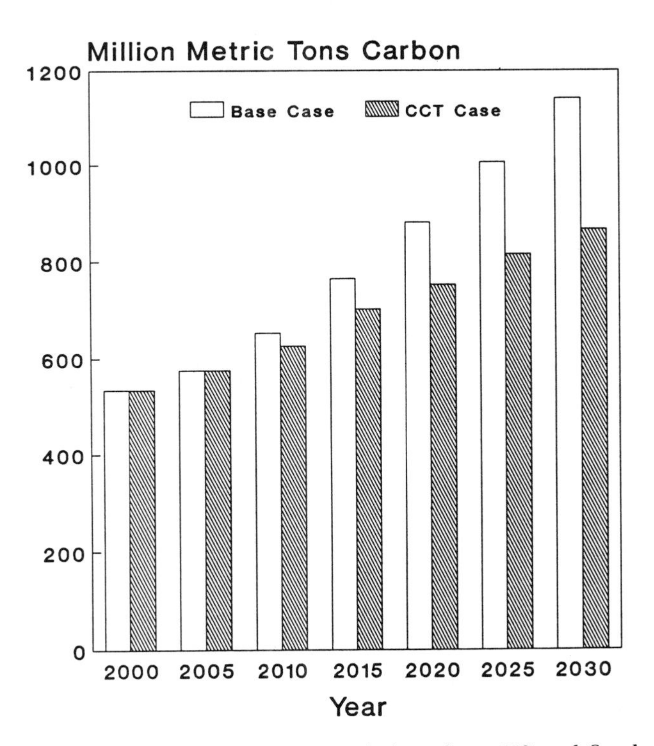

Figure 4. CO_2 (as carbon) emissions from US coal-fired electric utility units. Notes: a) Incremental demand for the years 2000 to 2005 assumed to be satisfied by PC/FGD (NSPS) units. b) Incremental demand for the years 2005 to 2015 assumed to be satisfied by IGCC repowering (42 percent efficiency). c) Incremental demand for the years 2015 to 2030 assumed to be satisfied by repowering with coal-fired fuel cells. [Derived from US DOE Global Warming Study, Draft Reference Case, 12/15/89]

US Electric Utility SO_2/NO_x Emission Reduction

To reduce the suspected causes of acid rain, SO_2 and NO_x emissions, the Administration has proposed to the Congress amendments to strengthen existing provisions of the Clean Air Act. Title V of the Administration's bill includes a required 10 million tons per year reduction in SO_2 emissions from electric utility power plants compared to 1980 levels by the year 2000, with an intermediate reduction of 5 million tons per year by 1995. Total SO_2 emissions are to be limited to year 2000 levels after that time. Required reductions in NO_x emissions are also included.

As with the production of greenhouse gases, deployment of advanced fossil technologies can also play a significant role in reducing acid gas emissions from electric utility power plants. Potential SO_2 and NO_x emissions reductions using CCT are presented for two scenarios:

Business-as-Usual (BAU)

Base Case. Existing units not required to meet New Source Performance Standards (NSPS) are life-extended for a total service life of 60 years. New PC units meeting NSPS are built through 2030 to satisfy incremental demand. New PC units are fitted with commercially available flue gas desulfurization (FGD) units and low NO_x control (LNC).

CCT Case. IGCC technology is commercially deployed in 2005 through the repowering of existing units to satisfy incremental demand. Beginning in 2015, repowering is assumed to utilize fuel cell technology. Otherwise, existing units[2] are life-extended for a total service life of 60 years without pollution controls added. Fuel cell technology begins to penetrate in 2015.

Total Control

Base Case. Existing non-NSPS units are life-extended and retrofitted with commercially available FGD units and LNC. All other units are PC units originally designed to meet NSPS with commercially available control technology.

CCT Case. New PC units meeting NSPS are built until 2004 to satisfy new demand. New PC units are fitted with commercially available FGD units and LNC. Starting in 2005, existing plants are repowered with IGCC technology to satisfy incremental demand. Beginning in 2015 repowering is assumed to utilize fuel cell technology. Existing non-NSPS units not repowered with CCT are life-extended and retrofitted with CCT consisting of advanced flue gas cleanup systems for SO_2 and NO_x control.

Assuming business-as-usual, Figure 5 shows that use of conventional control technology results in SO_2 emission levels steadily rising through the year 2010, then reaching a plateau and beginning to decrease in 2020 as significant capacity from older, uncontrolled units begins to be retired and replaced by capacity from new units meeting NSPS. In contrast, use of CCT results in a steady decline of SO_2 emission levels beginning in 2005—the year CCT begins to penetrate the market. By 2030, the introduction of CCT yields SO_2 emission levels that are nearly 70 percent lower than the base case.

Nitrogen oxide emission levels steadily increase through 2030 in the base case. However, the Clean Coal Case shows an opposite trend (Figure 6) because of the retirement of older units and the very low NO_x emission levels obtainable with advanced CCT systems.

When all units are assumed to be controlled with conventional technology, both SO_2 and NO_x emission levels show a significant upward trend (Figures 7 and 8). Even with increasing demand, however, the introduction of CCT on all units maintains a steady level for

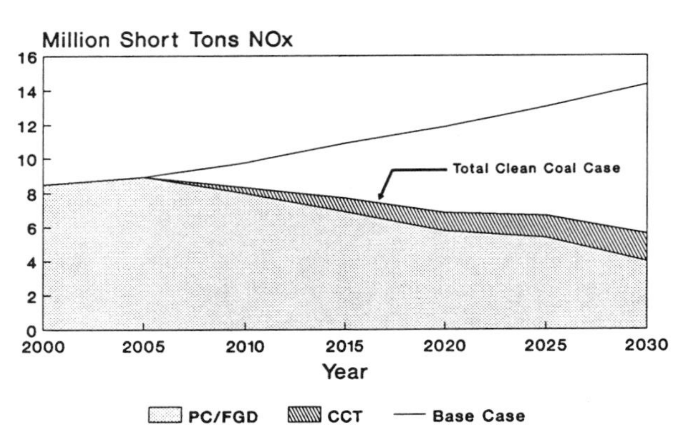

Figure 6. NO_x emissions from coal-fired electric utility units (business-as-usual).

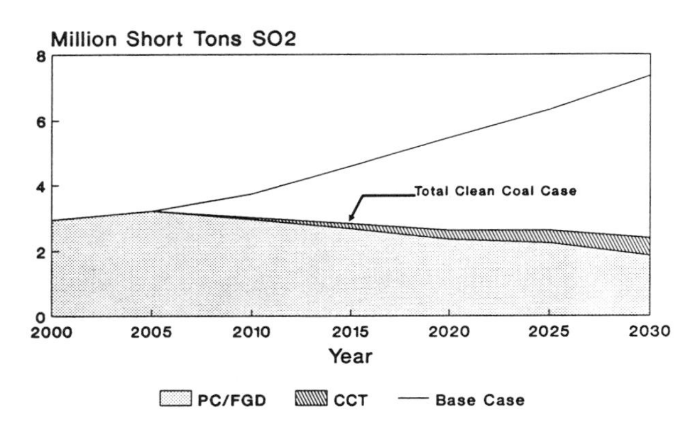

Figure 7. SO_2 emissions from coal-fired electric utility units (total control case).

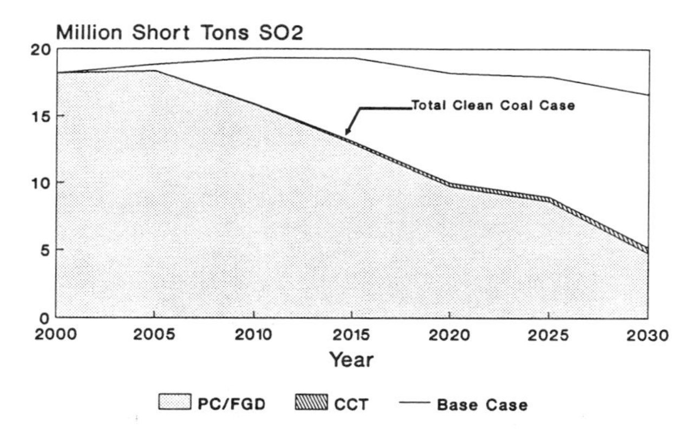

Figure 5. SO_2 emissions from coal-fired electric utility units (business-as-usual).

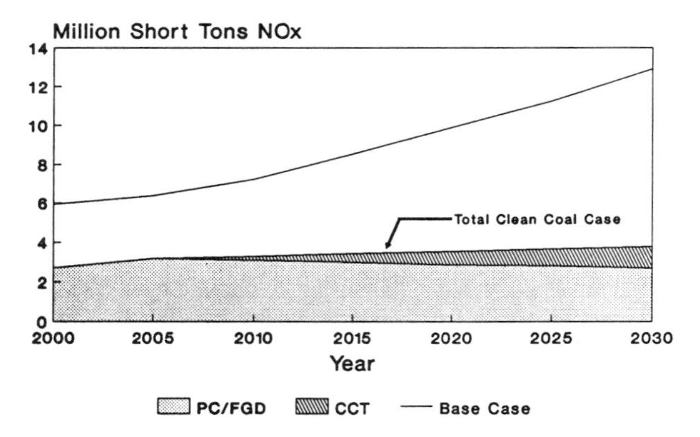

Figure 8. NO_x emissions from coal-fired electric utility units (total control case).

both these pollutants. By the year 2030, widespread use of CCT yields SO_2 and NO_x emission levels that are about 70 percent lower than the base case.

Although the Total Control scenario is not termed an acid rain control scenario because it deploys more control technology than utilities would find economically feasible, it does illustrate the potential for using CCT to stay well below the cap on SO_2 emissions required by the year 2000. It is estimated that the cap on SO_2 emissions will be about 8.9 million tons in the year 2000. At the rate that SO_2 emissions are increasing in the base case, that level will be reached by the year 2034. Extensive use of CCT, however, pushes that target out beyond the 21st century.

Combined Heat and Power Systems

Even greater thermal efficiencies can be obtained through the use of the waste heat from electric power plants. As the diagrams in Figure 9 show, combined heating and power can increase thermal efficiencies to 80 percent or more. Moreover, the waste heat from power plants contributes to thermal pollution in lakes and rivers with possible ecological consequences. This concept, which includes cogeneration and district heating and cooling and process heat for industry, has been used in various places throughout the US and Europe

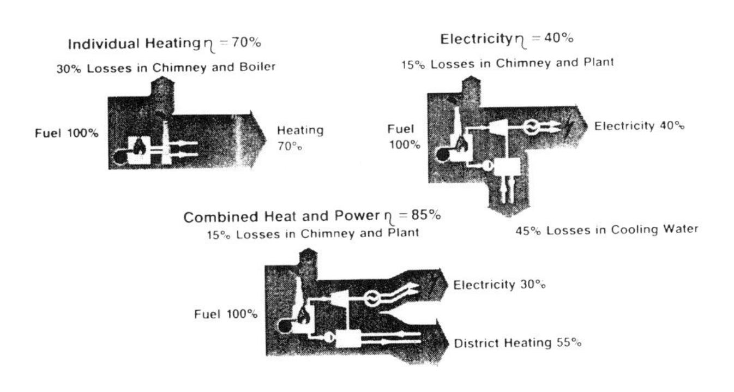

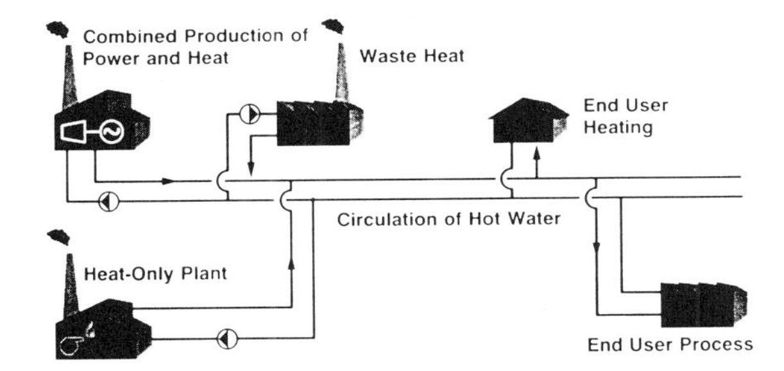

Figure 9. Combined heat and power systems. (Courtesy of B. Hojlund Rasmussen, Consulting Engineers and Planners A/S.)

for the past 100 years. Cogeneration is starting to play a larger role in electricity production in the US, and non-utility growth in electricity generation—much of which is cogenerated—is expected to grow at a faster rate than electricity supplied by utilities.

The fuel savings and consequent environmental benefits of widespread cogeneration, especially from use of the waste heat from electric power plants, presents an enormous potential. If it is assumed that 60 percent of the heat input that flows out as waste heat to the environment can be captured for other energy needs and that 50 percent of the existing plants could be converted to cogeneration, CO_2 (as carbon) could be reduced by 67 million metric tons each year through the replacement of natural gas heating with predominately coal-fired combined plants, based on current electric energy generation.[3] This represents a reduction of 13 percent below 1987 CO_2 emissions from US electric power plants.

Summary and Conclusions

Widespread deployment of high efficiency Clean Coal Technology can significantly reduce the increase in CO_2 emissions from electric power generation. Clean Coal Technology is even more effective in reducing gases that cause acid rain. The technologies sponsored by the Department of Energy and others are only the beginning of the challenge to develop and demonstrate new and more efficient coal-based energy systems. Efficiencies as high as 65 percent are considered possible with fuel cell technology. When used alone or in combined heat and power configurations, these technologies can make a major contribution to minimizing emissions of polluting gases. Future coal-based power systems can play a major role in supplying the world's electricity needs without contributing to the deterioration of the environment.

Notes

1. The use of advanced technology in the US obtains more reduction over conventional technology than in the worldwide deployment scenario discussed previously because of the assumption that fuel cell technology is deployed earlier in the US (2015) than in the world at large (2025). Also, the US calculations take into account retirements of existing coal-fired units, while the world calculations do not, due to lack of data.
2. Existing units are defined in this paper as coal-fired utility boilers in service in 1985 that were not subject to the 1979 New Source Performance Standards.
3. This information is based on a preliminary estimate by the Office of Community Planning and Development, US Department of Housing and Urban Development.

References

Assistant Secretary for Fossil Energy, *Clean Coal Technology: The New Coal Era,* US Department of Energy, Washington, DC, DOE/FE-0149 (November 1989).

Committee on District Heating and Cooling, *District Heating and Cooling: Prospects and Issues,* Energy Engineering Board and Building Research Board, Commission on Engineering and Technical Systems, National Research Council, National Academy Press (1985).

Miller, C.L., and Pell, J. "New Clean Coal Technologies for Control of Acid Rain Emissions Move to Demonstration," presented at the 1989 Joint Power Generation Conference and Exposition, Dallas, TX (October 25, 1989).

Randlov, Peter, "Tomorrow's Solution: The Environment and Combined Production of Power, Heating and Cooling," fourth draft, B. Hojlund Rasmussen, Consulting Engineers and Planners A/S (January 3, 1990).

Temchin, J., and Feibus, H., "Clean Coal II: Low NO_X Technologies," *Proceedings of the 1989 Symposium on Stationary Combustion Nitrogen Oxide Control,* vol. 1 San Francisco, CA (March 6–9, 1989).

Pollution Control for Utility Power Generation, 1990 to 2020

Frank T. Princiotta
Air and Energy Engineering Research Laboratory
Office of Research and Development
US Environmental Protection Agency
Research Triangle Park, NC

Abstract

The major environmental challenges facing the utility industry in the 1990 to 2020 time frame are acid deposition control in the near term and potential global warming mitigation in the longer term. President Bush has proposed an ambitious acid rain control program requiring reduction of 10 million tons of sulfur dioxide (SO_2) by the year 2000. Options available to the utility industry include coal switching and flue gas desulfurization, as well as such emerging lower cost technologies as limestone injection multistage burners (LIMB) and advanced silicate (ADVACATE), both developed by the Environmental Protection Agency (EPA); selective use of gas to reduce NO_x and SO_2 in coal-fired boilers; and the use of natural gas combined cycle plants which may later be converted to integrated coal gasification combined cycle (IGCC). Since utility boilers, especially coal boilers, are major emitters of carbon dioxide (CO_2), they are important targets in the effort to mitigate global warming. Utility options to deal with this problem include user conservation, increased use of nuclear power, renewable generators (biomass and solar), clean coal technologies, and increased use of natural gas. Model analysis suggests that in the 1990 to 2020 time frame conservation is critical. Nonfossil fuel technologies can also make significant contributions to reducing CO_2 emissions, but their role will be limited by low penetration in this time frame. Clean coal technologies (e.g., integrated coal gasification combined cycle) appear to provide only marginal CO_2 emissions reduction benefits.

Introduction

The US electric utility industry has shown extraordinary growth since the early part of this century. Figure 1 shows US electrical production from 1930 to 1985. This figure also illustrates the dominance of coal as the major source of the derived electrical generation. Although oil and gas were significant contributors to electrical production in the 1960s and 1970s and nuclear power was a significant contributor in the 1980s, coal continues to be the dominant fuel source. Electrical use has grown from only 1 quad (10^{15}Btu) of energy utilized in the 1930s to 26 quads used in 1985. This 26 quads represents over one third of all the energy used in the US, which was 74 quads in 1985.

Unfortunately, the utilization of coal involves major environmental and safety problems throughout the fuel cycle. From mine safety and the ecological damage of strip mining to air pollution associated with coal combustion to the ultimate disposal of fly ash and sulfur-bearing sludge, coal use presents major challenges. Perhaps the most important of these challenges is controlling gaseous emissions of important air pollutants associated with coal combustion. Table 1 describes the major regulated (criteria) and CO_2 pollutants for which coal combustion is an important source. Sulfur dioxide (SO_2), nitrogen oxides (NO_x), and particulates are all currently regulated for both new and existing sources under the US Clean Air Act. Carbon dioxide is unregulated but is a major contributor to potential global warming. Table 1 also summarizes the major health and environmental concerns and the role of electrical production, particularly coal combustion, compared to other sources for each of these pollutants.

Major Environmental Challenges for the Future—Acid Rain and Global Warming

The major environmental challenges for the electric utility industry appear to be associated with acid deposition and global warming. Response to the acid rain challenge appears imminent in light of the near term prospect of US legislation. Global warming is a longer term challenge that the international community is struggling to understand and deal with.

Acid Rain—The Near Term Challenge

Acid rain, the popular term for acid deposition, is most closely associated with damage to aquatic systems. It may also contribute to forest damage and damage to build-

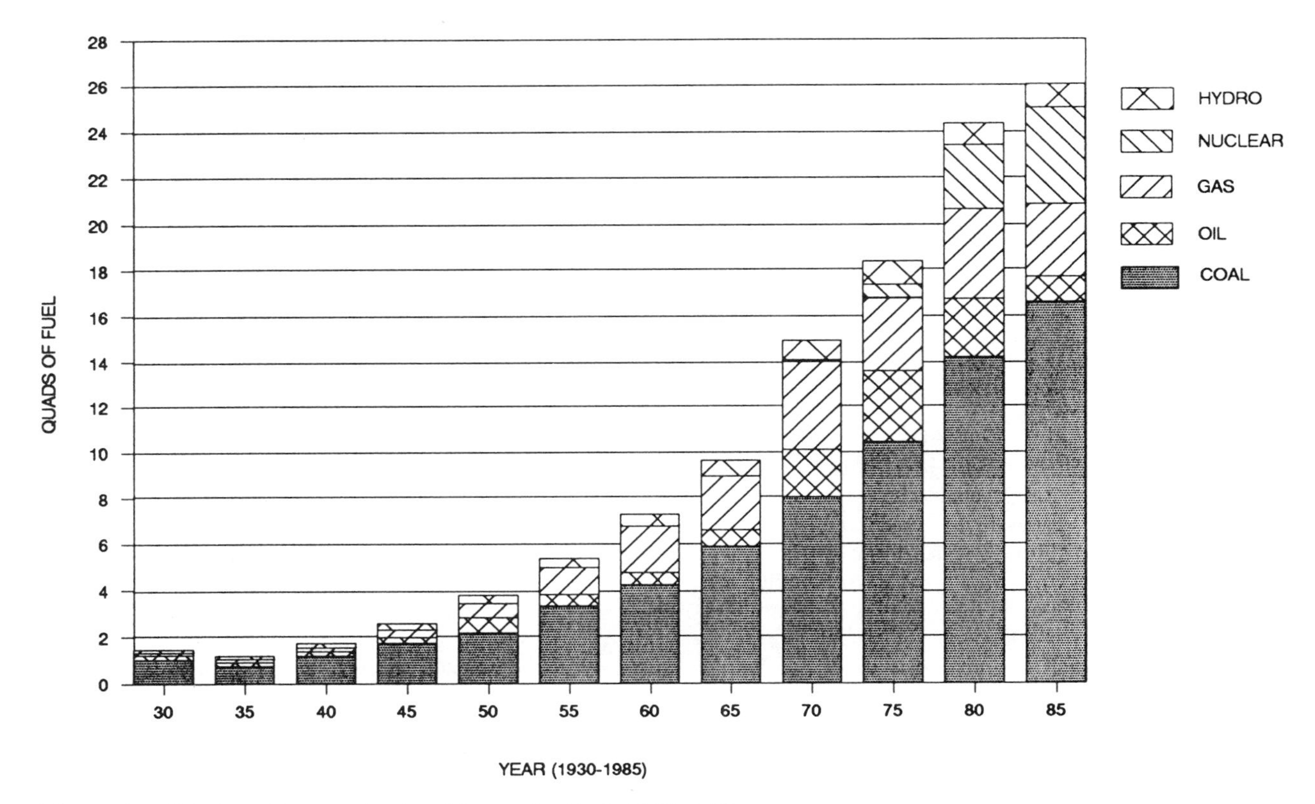

Figure 1. US electricity production history, by fuel, 1930–1985.

Table 1. Important electrical power plant pollutants.

Pollutant	Health Concern	Environmental Concern	Coal Role
Sulfur Dioxide	• Respiratory tract problems. • Permanent harm to lung tissue.	• Acid deposition damage to lakes, streams, and forests. • Visibility.	Coal combustion is dominant source.
Nitrogen Oxides	• Respiratory illness and lung damage.	• Acid deposition. • Increased ozone: forest damage.	Coal combustion and autos are major sources.
Airborne Particulates	• Eye and throat irritation, bronchitis, lung damage.	• Visibility.	Power plants have diminished in importance.
Carbon Dioxide	• Potential major impacts of global climate warming. • Indirect health impacts unknown.	• Destruction of sensitive eco-systems by global warming. • Increased air pollution.	Coal most important fuel due to large use and high carbon/hydrogen ratio.

ings, monuments, and other structures. Acid rain occurs when SO_2 (emitted primarily by coal-fired electric utilities) and NO_x (primarily from transportation sources and utilities) are chemically transformed, transported into the atmosphere, and deposited on the earth in wet or dry form. Based on this potential environmental damage, the Bush Administration has recently proposed an aggressive amendment to the Clean Air Act that would control both SO_2 and NO_x from existing electric utility plants. Specifically, the President's proposal calls for

- A reduction of 10 million tons annually of SO_2 by the end of the year 2000 using a base line year of 1980 for tons of SO_2 emitted, primarily from coal-fired power plants.
- A two-phase program in order to ensure early reductions. A reduction of 5 million tons is required during Phase 1, by the end of 1995.
- A 2 million ton reduction of NO_x in Phase 2. The plan would also allow utilities to trade reductions of NO_x for reductions of SO_2 or vice versa and thus represents a call for a total annual reduction of 12 million tons in pollutants that cause acid rain.
- A three-year extension of the Phase 2 deadline for plants adopting "clean coal" repowering technologies combined with regulatory incentives designed to smooth their transition into the marketplace.
- Freedom of choice in cutting pollution. The plan requires all plants of a certain size to meet the same emissions standard, but does not dictate to plant managers how the standard should be met.
- Maximum flexibility in obtaining reductions. The plan would allow utilities to trade required reductions so that they would be achieved in the least costly fashion. Full interstate trading would be allowed in Phase 2, although Phase 1 trading would be limited to electric plants within a state or within a utility system.
- The EPA has estimated the cost of the president's proposal to be $3.8 billion annually in Phase 2 and approximately $700 million per year in Phase 1. This is estimated to represent an increase of over 2 percent by the year 2000 in the nation's $160 billion a year electricity bill.

Dealing with the Likelihood of New Acid Rain Legislation

At the time of this conference, it appears likely that acid rain legislation with provisions similar to the Administration's proposals will pass the Congress this session. If so, the US electric utility industry will have to make many cost intensive decisions to comply with provisions of the legislation. For SO_2 control, the industry will have the choice of locating an adequate supply of low sulfur coal, selecting a control technology, or selectively burning natural gas. The utility will probably look for available low sulfur coal supplies from both eastern and western mines to determine the most economical fuel for a particular utility system. The utility will likely compare the coal switching option to the control technology options available. Table 2 describes current and emerging technologies for SO_2 and NO_x control. Included are control technologies available for current pulverized-coal boilers as well as those that could be applied as new boiler technology with inherent SO_2/NO_x control capabilities. These new SO_2/NO_x technologies in Table 2 reflect modified combustion where both SO_2 and NO_x are reduced in the process of fuel combustion. The table briefly describes the technology, the estimated level of control of SO_2 and NO_x, and the projected commercial availability, including comments primarily related to capability. Note that the overwhelming current choice of US utilities for SO_2 control technology has been lime and limestone wet scrubbers.

The EPA's Air and Energy Engineering Research Laboratory (AEERL) has been actively developing acid rain technology capable of low cost retrofit in anticipation of acid rain legislation. Three of the technologies listed in Table 2 have been actively developed by AEERL: LIMB, E-SO_x, and ADVACATE. LIMB technology, which was recently successfully demonstrated at a 100 MW boiler in Ohio, can be a cost effective choice for certain applications. LIMB appears particularly cost effective for older plants burning low to moderate sulfur coal with lower load factors. Another low cost option capable of 50 to 60 percent SO_2 removal is E-SO_x. It is being pilot tested in Ohio. We are particularly excited about the ADVACATE technology, a novel concept of duct sorbent injection using a calcium silicate sorbent that is easily prepared from waste fly ash and lime and that has unique properties of rapid SO_2 absorption and moisture absorption and adsorption. The ADVACATE technology appears to have substantially lower costs than wet scrubbers for almost all retrofit applications. We hope to demonstrate this technology successfully so that it can play a role in controlling SO_2 cost effectively in the time frame associated with the likely acid rain legislation.

Figure 2 describes the major cost elements in terms of dollars per ton of SO_2 removed for LIMB, ADVACATE, and conventional limestone FGD technologies. FGD is dominated by capital and labor material costs, whereas LIMB is dominated by sorbent costs. The LIMB process utilizes a hydrate material that typically costs $70 per ton, whereas ADVACATE and FGD utilize lower cost limestone reagents. Note that ADVACATE has an advantage over FGD in all major cost categories, allowing it to cost approximately half as much as the wet scrubber. Also note that ADVACATE is one of the few low cost technologies capable of achieving 90 percent SO_2 control.

Table 2. SO_2 and NO_x control technologies for coal-fired boilers.

Technology	Description	Control %*		Estimated Commercial** Availability	Comments
		SO_2	NO_x		
Wet flue gas desulfurization (FGD)	Limestone or lime in water removes SO_2 in a scrubber vessel. Additives may be used to enhance SO_2 removal. A wet waste or gypsum is produced.	70–97	0	Current for new boilers and retrofit.	State-of-the-art for higher S coal and FGD. Certain retrofits difficult.
Dry FGD	Lime in water removes SO_2 in a spray dryer, which evaporates the water prior to the vessel exit. Produces a dry waste.	70–95	0	Current for low to moderate S coal for new boilers. High S coal retrofit, 5 yrs.	Demonstration for high S coal retrofit is necessary, but may be limited to 90% SO_2 removal.
E-SO_x/in-duct injection	Lime and water are injected in a boiler duct and/or ESP (E-SO_x) and react with SO_2 similar to a spray dryer.	50–70	0	Pilot scale only. Demonstrations required, 3–7 yrs.	Potentially low cost retrofits. May be site-specific limits.
Limestone injection multistage burners (LIMB)	Low NO_x burners and upper furnace sorbent injection. May use humidification to improve SO_2 capture and ESP performance.	50–70	40–60	Wall, current; T-fired, 3–4 yrs.	T-fired wall-fired demo complete. Applicable to $\leq$ 3% S coal retrofits.
Advanced silicate (ADVACATE)	Several variations. Most attractive: adding limestone to boiler, generating lime. Lime/fly ash collected in cylone and reacted to generate highly reactive silicate sorbent. Moist sorbent added to downstream duct.	Up to 90	0	Pilot scale only. Demo required, 3–7 yrs.	Most promising emerging retrofit technology. Capable of 90% removal with costs 50% of wet scrubber.
Low NO_x burners, overfire air modifications	Burner/boiler design controls coal/air mixing to reduce NO_x formation.	0	40–60	Now, new boilers and retrofit.	Additional retrofit demos desirable.
Natural gas reburning	Boiler fired with 80–90% coal. Remaining fuel (natural gas) is injected higher in boiler to reduce NO_x. Air added to complete burnout. Sorbent may be injected to capture SO_2.	Without sorbent, 10–20; with sorbent 50–60	50–85	Demos starting, available in 3 yrs.	May be only combustion NO_x control for cyclones. Sensitive to natural gas price. New or retrofit.
Selective catalytic reduction (SCR)	Reacts NO with NH_3 over a catalyst at 500–700°F (260–370°C).	0	80–90	Pilot plant only in US, 4 yrs.	Catalyst cost and life main issues. Retrofit or new, if demos in US.

Table 2. *(continued)*

Technology	Description	Control %* SO$_2$	Control %* NO$_x$	Estimated Commercial** Availability	Comments
		SO$_2$/NO$_x$ New Boiler Systems			
Atmospheric fluidized-bed combustion (AFBC)	Coal and air are burned in fluidized bed of sorbent that captures SO$_2$ and limits NO$_x$ formation.	70–90	50–70	Now, for industrial boilers. Utility demo in progress (2–3 yrs.).	Primarily for new boilers. Repowering (retrofit) applicability limited.
Pressurized fluidized-bed combined cycle (PFBCC)	Coal burned in a fluidized bed at elevated pressure to capture SO$_2$ and limit NO$_x$ formation. Gas cleaned and run through gas turbine that generates electricity. Waste heat converted to steam to generate electricity.	90–95	70–80	Pilot plant data. DOE demo initiated, 5–10 yrs.	Primarily for new boilers. Repowering involves replacing the entire boiler.
Integrated gasification combined cycle (IGCC)	Gasifier converts coal/air to fuel gas, sulfur removed, and fuel gas burned in a combustion turbine. Waste heat used to generate steam. Electricity generated by both combustion and steam turbines.	95–99	40–95	Cool water demo conducted, economics questionable. Two DOE demos initiated, 5–10 yrs.	See PFBCC comments. Also, economic operation depends on hot fuel gas cleanup and high temperature turbines which are not demonstrated.

*Control efficiency is % reduction from emission levels for uncontrolled coal-fired power plants.
**Estimated commercialization for some technologies is strongly dependent on successful demonstrations.

Cost Estimates

To elaborate on the choices facing the utility industry, it is worthwhile to summarize the results of a recent study (Emmel and Maibodi, 1989) sponsored by AEERL. The objective of this study was to improve significantly the accuracy of engineering cost estimates used to evaluate the economic effects of retrofitting SO$_2$ and NO$_x$ controls to the top 200 SO$_2$-emitting coal-fired utility boilers. This project was conducted in several phases. In Phase 1, detailed, site-specific procedures were developed and used to evaluate retrofit costs at 12 actual plants. In Phase 2, simplified procedures were developed to evaluate the site-specific costs and to evaluate retrofit costs at 50 plants. In Phase 3, costs were evaluated at all remaining 138 plants. This recently published report presents the cost estimates developed for 576 boilers in 188 plants using the simplified procedures. The study evaluated retrofit costs for the following technologies:

- limestone FGD
- additive-enhanced limestone FGD
- lime spray drying FGD
- physical coal cleaning
- coal switching and blending
- low NO$_x$ combustion
- furnace sorbent injection with humidification (LIMB)
- duct spray drying
- natural gas reburning
- selective catalytic reduction
- fluidized-bed combustion or coal gasification retrofit

To generate retrofit costs for each plant, a boiler profile was completed using sources of public information. Additionally, boiler design data were obtained from power plants, from a data base maintained by *Power* magazine (Elliot), and aerial photographs obtained from state and federal agencies. The plant and boiler profile information is used to develop the input data for the performance and costs models. To enhance the credibility of cost information, which is almost always controversial, the performance and cost results incorporate recommendations from utility companies and a technical advisory group. This group included the utility industry, FGD vendors, and government agency representatives.

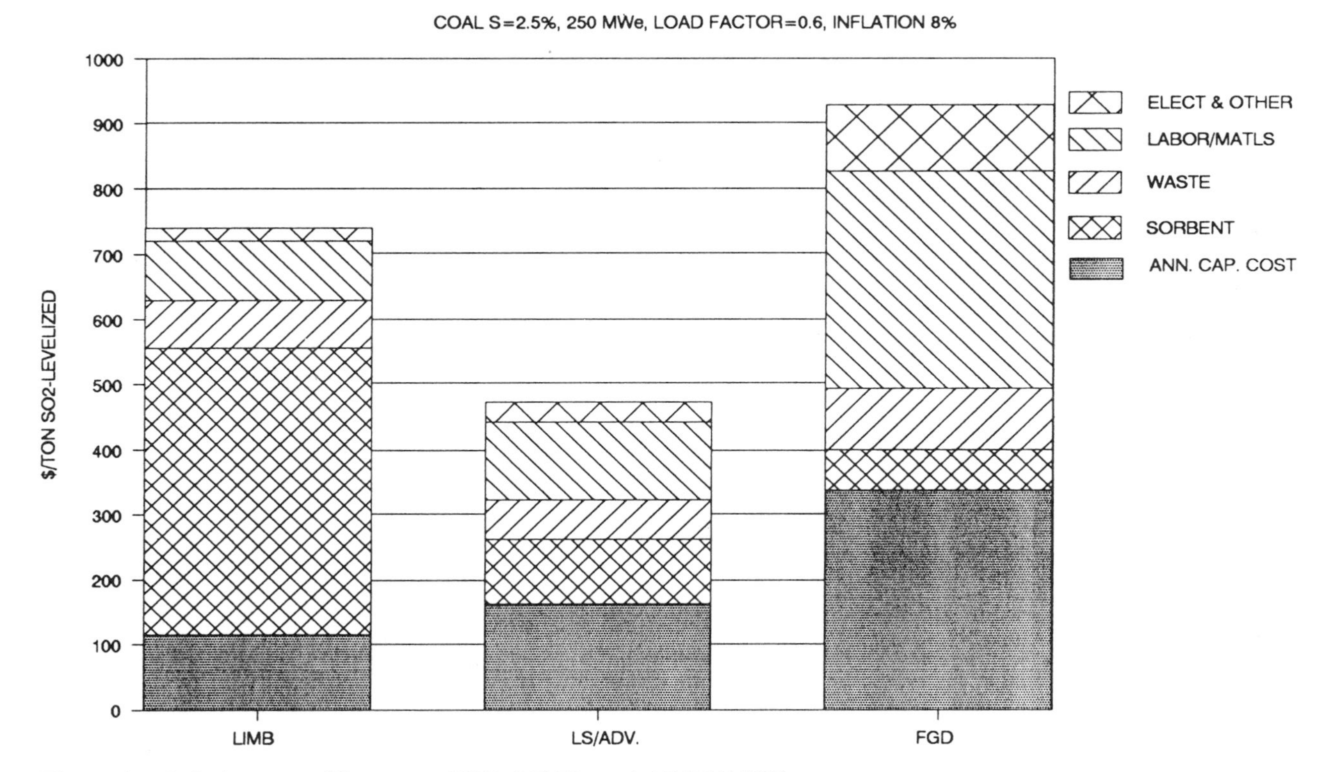

Figure 2. Relative cost of limestone FGD, LIMB, and ADVACATE processes.

All the cost estimates were developed using the integrated air pollution control systems (IAPCS) cost model. The IAPCS model was upgraded to include the technologies being evaluated in this program.

The results of this study confirm that costs of various acid rain retrofit options vary considerably from plant to plant. What might be an economical approach at one plant could be prohibitively expensive at another plant due to unique local conditions, such as lack of space or other site-specific factors. Figures 3 through 6 summarize some of the results of this study. They describe the costs of retrofit control for coal switching, lime/limestone desulfurization, LIMB (for SO_2 control), and three combustion technologies for NO_x control. Figure 3 summarizes the cost per ton of SO_2 removed for coal switching and blending. Price differentials of both \$5 and \$15 per ton of coal were assumed in this cost analysis since they bracket the likely differential for many existing boilers in the eastern US. Note that, for about 50 percent of the applicable boilers for a \$5 price differential, the levelized cost of control will be substantially less than \$1,000 per ton of sulfur removed. (All costs were calculated on a levelized basis; i.e., they were increased over first year costs to take into account likely inflation over the control's lifetime.) However, for boilers already burning relatively low sulfur coal, even this relatively small coal price differential can yield substantially higher cost of controls per ton of sulfur removed. For the higher priced differential, typically for plants far from sources of low sulfur coal, only 25 percent of the boilers can be controlled at less than \$1,000 per ton. Utilities will likely look very closely at the low sulfur coal option, which in many cases will be the least expensive option.

Figure 4 summarizes the cost per ton of SO_2 removed for lime or limestone FGD technology. As shown, certain plants can be controlled for less than \$1,000 a ton; but for about 75 percent of the plants, costs will be higher than that. For the most expensive 25 percent of boilers, costs will be quite high due primarily to difficulty of retrofit.

Figure 5 summarizes the cost per ton of SO_2 removal for LIMB technology. Two cases are studied corresponding to 50 or 70 percent SO_2 removal from the LIMB/humidification technology. For most cases, this technology is less expensive per ton of SO_2 removed, especially if 70 percent SO_2 removal is achievable for a given plant.

The last figure in this series, Figure 6, summarizes costs per ton of NO_x removed utilizing three low NO_x combustion technologies: low NO_x burners (LNB), natural gas reburning (NGR), and overfire air (OFA). As

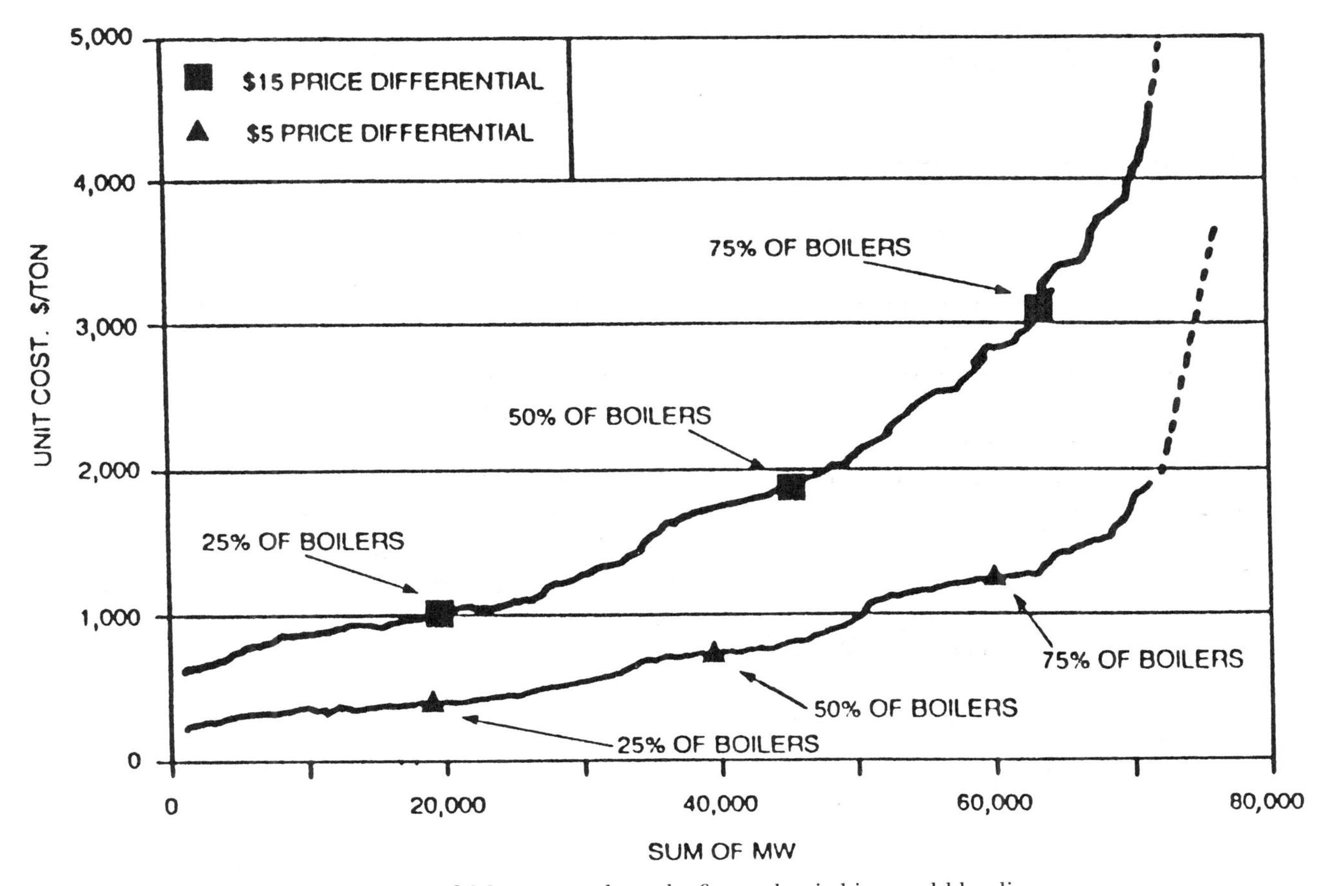

Figure 3. Summary of cost per ton of SO_2 removed results for coal switching and blending.

shown, the combustion technologies LNB and OFA are considerably less expensive than NGR. However, for certain classes of boilers, such as cyclones, reburning may be the only feasible option. Also note that 75 percent of the boilers can be controlled with a low NO_x burner or overfire air system for costs below $500 per ton of NO_x removed.

Results of this study should be useful to utilities, states, and others who will likely be making or monitoring the difficult choices of control mandated by the expected acid rain legislation. In January 1990, the authors of the retrofit study (Emmel and Maibodi) were asked to apply their results to a hypothetical 10 million ton per year SO_2 reduction program (from 1980 emission levels). The objective was to estimate the maximum potential benefit of emerging technologies (i.e., LIMB and ADVACATE) to an acid rain retrofit program. The methodology involved selecting the lowest cost option for a particular plant, ultimately achieving the required 10 million ton reduction by retrofitting the top 200 SO_2-emitting plants.

For this analysis, the following limited sets of available control options were assumed:

Cases 1 and 2
coal switching/blending
limestone FGD
LIMB (50% removal)

Cases 3 and 4
coal switching/blending
limestone FGD
LIMB (50% removal)
ADVACATE (limestone, 90% removal)

Cases 1 and 3 assumed a low sulfur coal incremental cost of $5 per ton; whereas Cases 2 and 4 assumed a $15 per ton differential. Cases 3 and 4 included the ADVACATE process to estimate the impact of such a technology, assuming costs at half those of wet FGD and retrofittability similar to that of wet FGD. Note that this is not a demonstrated technology; cost savings should only be considered an upper limit of what might be achievable if successfully demonstrated and freely selected by the utility industry, despite lack of extensive field operation experience.

Figure 7 shows the results of this analysis. For Cases 1 and 2, coal switching, FGD, and LIMB would all play

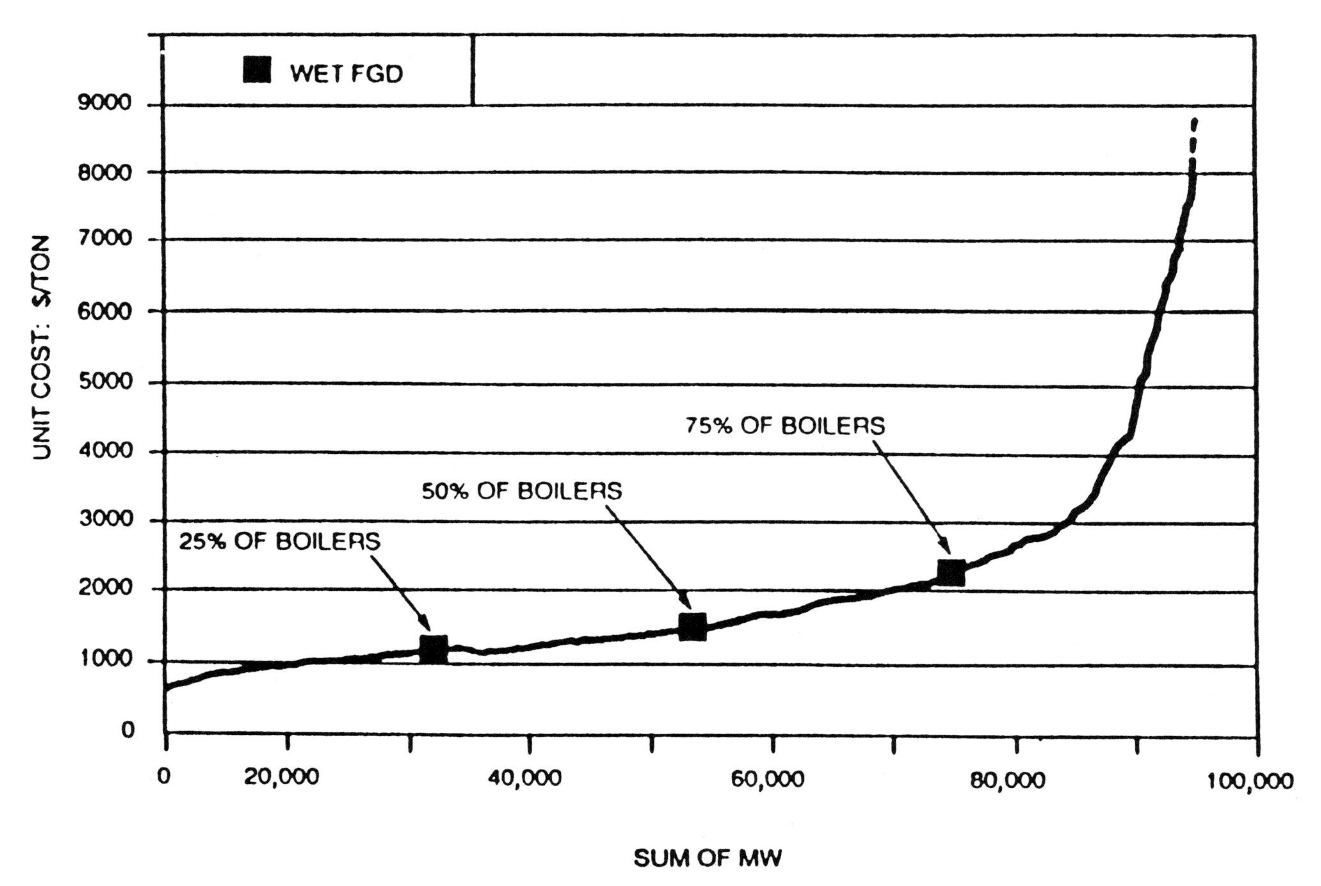

Figure 4. Summary of cost per ton of SO_2 removed results for lime/limestone flue gas desulfurization.

major roles, with coal switching particularly important at the low ($5 per ton) coal price differential (Case 1). For Cases 3 and 4, ADVACATE would play the major role, essentially displacing all other options for the high ($15 per ton) coal price differential (Case 4). Maximum possible annual cost savings associated with ADVACATE technology availability are in the order of $2 billion.

Global Warming—The Longer-Term Utility Challenge

The most important long-term challenge in environmental protection for the electric utility industry appears to be global warming. Major global warming is projected to occur in the early to mid 21st century. Although the sources of gases that contribute to global warming are ubiquitous, electric utilities represent one of the more significant of these sources. All fossil fuels generate CO_2 during combustion. However, coal is a particularly important contributor since it generates about 220 pounds of CO_2 per million Btu, compared to half that amount for natural gas and about two thirds that amount for oil. Figure 8 (Lashoff and Tirpak, 1989) illustrates that the global warming problem is clearly an international one, and shows that the US currently represents about 20 percent of the problem. The Soviet Union, Europe, China, Brazil, and India are also major contributors. Figure 9 (Smith [IEA], 1988; Smith and Tirpak, 1989; Mintzer, 1987; Ramanathan et al., 1985; Lashof and Tirpak, 1989) illustrates that, although CO_2 is the major global warming trace gas, chlorofluorocarbons (CFCs), methane, nitrous oxides, and other anthropogenic gases contribute significantly to the anticipated global warming. Figure 9 also estimates the relative importance of the various sources that will contribute to global warming in the 2030 time frame. The utility industry is clearly a major source. Also important are other anthropogenic activities, including industrial, transportation, and residential combustion, as well as industrial noncombustion activities that generate CFC emissions. Other sources include deforestation that generates CO_2, and livestock production, agriculture, biomass burning, and other sources that generate methane and disperse it into the atmosphere.

If the international community is to deal with this

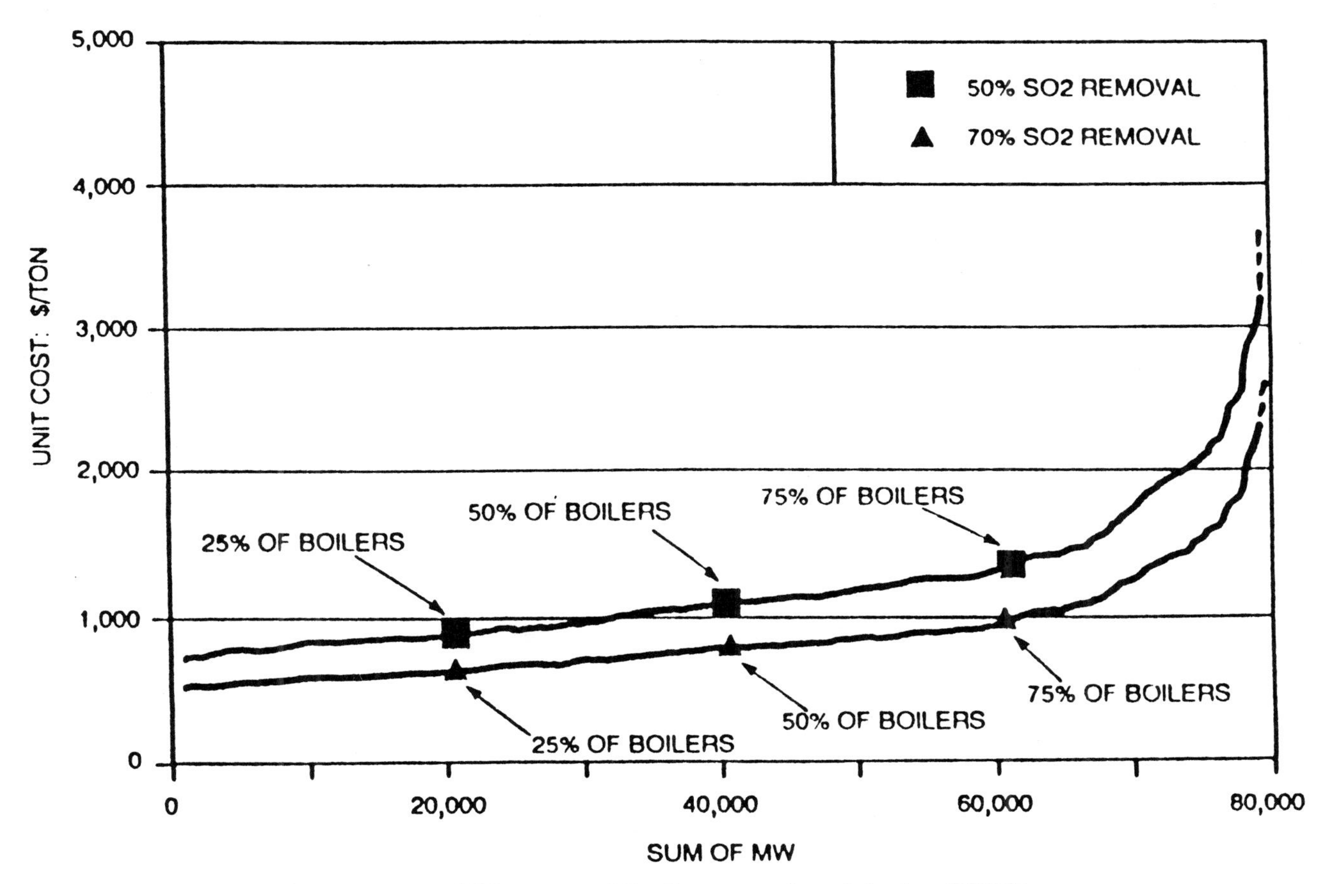

Figure 5. Summary of cost per ton of SO_2 removed for furnace sorbent injection (LIMB).

problem seriously, the electric utility industry may be among those sources targeted for modification.

Options for Decreasing Carbon Dioxide Emissions

Discussed briefly here are some of the more obvious approaches that could be used by the utility industry between now and 2020 to decrease CO_2 emissions. Following this discussion a simple emissions allocation model will be used to quantify the effectiveness of these options. The most straightforward approach is user conservation. By decreasing the demand for electricity or, more accurately, by decreasing the growth in demand over historical patterns, major reductions over "business as usual" can be realized.

Another option is to utilize nuclear power to offset emissions from fossil fuel units. However, the prospects for further capacity additions are uncertain. According to the DOE, "No additional orders for nuclear powerplants are likely in this country until the conditions of the past few years are changed" (US DOE, 1987). The primary reasons for this bleak outlook are concern over safety, waste disposal, nuclear proliferation, and high capital costs. Several of these concerns may be alleviated with some of the advanced nuclear fission designs. One of the promising technologies is the modular high temperature gas reactor (HTGR). This technology is compartmentalized and is immune to the uncontrollable runaway reaction potential of conventional technology. The HTGR concept has not yet been demonstrated, however. Current estimates are that the technology could be made commercially available in 1995. With an implementation time of nine years, power generation using HTGR could begin in 2004. Unfortunately, nuclear fusion, which utilizes water as the nuclear fuel and potentially avoids the safety and waste problems of nuclear fission, is not likely to be commercially available before 2020.

A third option would be large-scale application of solar electric technologies, again displacing fossil fuel emissions. There are two general methods for producing electricity from sunlight: photovoltaics (PV) and solar thermal systems. Estimates of the potential contributions to the nation's future electrical energy requirements from these two technologies have large

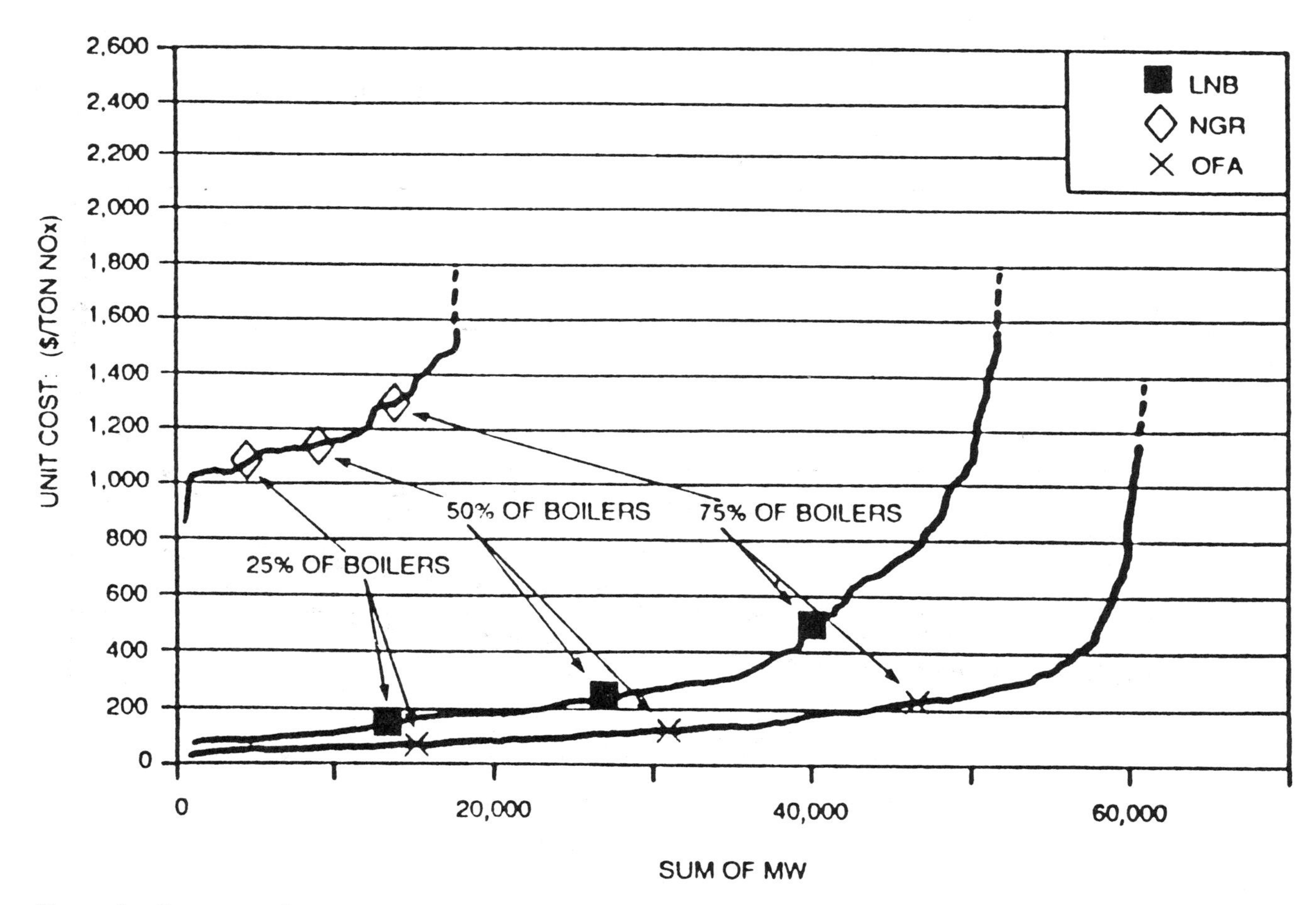

Figure 6. Summary of cost per ton of NO_x removed for low NO_x combustion.

uncertainties. For either of these methods to be commercially significant for power generation, major improvements must be made in the area of capital cost, electricity storage, and conversion efficiency. Currently, such systems are generally not competitive with coal-fired power plants, although costs have been decreasing in recent years.

The use of biomass electrical generating capacity also presents a potentially attractive approach. By burning biomass, it is possible to generate electricity without any net emissions of CO_2 to the atmosphere. This is accomplished by utilizing photosynthesis for atmospheric removal via fixation of the same quantity of CO_2 generated during the combustion process. Potential fuel sources include forest wood, unused wood residues, crop residues, herbaceous energy crops, livestock and poultry wastes, and municipal solid wastes.

Forest wood energy research is focusing on biotechnological engineering for yield improvement and stress resistance. Research on methods for efficient biomass processing is encouraging. Cofiring with various combinations of woody biomass, nonhazardous solid waste, herbaceous biomass, coal, and natural gas is also promising and is currently implemented commercially on a small scale.

Another option is the development and implementation of so-called clean coal technology. This technology could include integrated coal gasification combined cycle (IGCC) processes, as well as pressurized fluidized-bed combustion. Table 2 describes these processes. The objective is to generate electricity from coal at higher efficiency, thereby decreasing CO_2 emissions. For IGCC, the primary reason for integrating the gasification system with the combined cycle plant is that this design configuration substantially improves the overall system energy efficiency or heat rate. Although all components (i.e., gasifiers, gas coolers, acid gas removal systems, combined cycles) included in an IGCC configuration have been demonstrated to operate at full commercial scale, they have only recently been operated in unison in a complete system to generate electric power. Integrated control and operation of such plants in a commercial environment must be demonstrated on a large scale before the electric utility industry will seriously

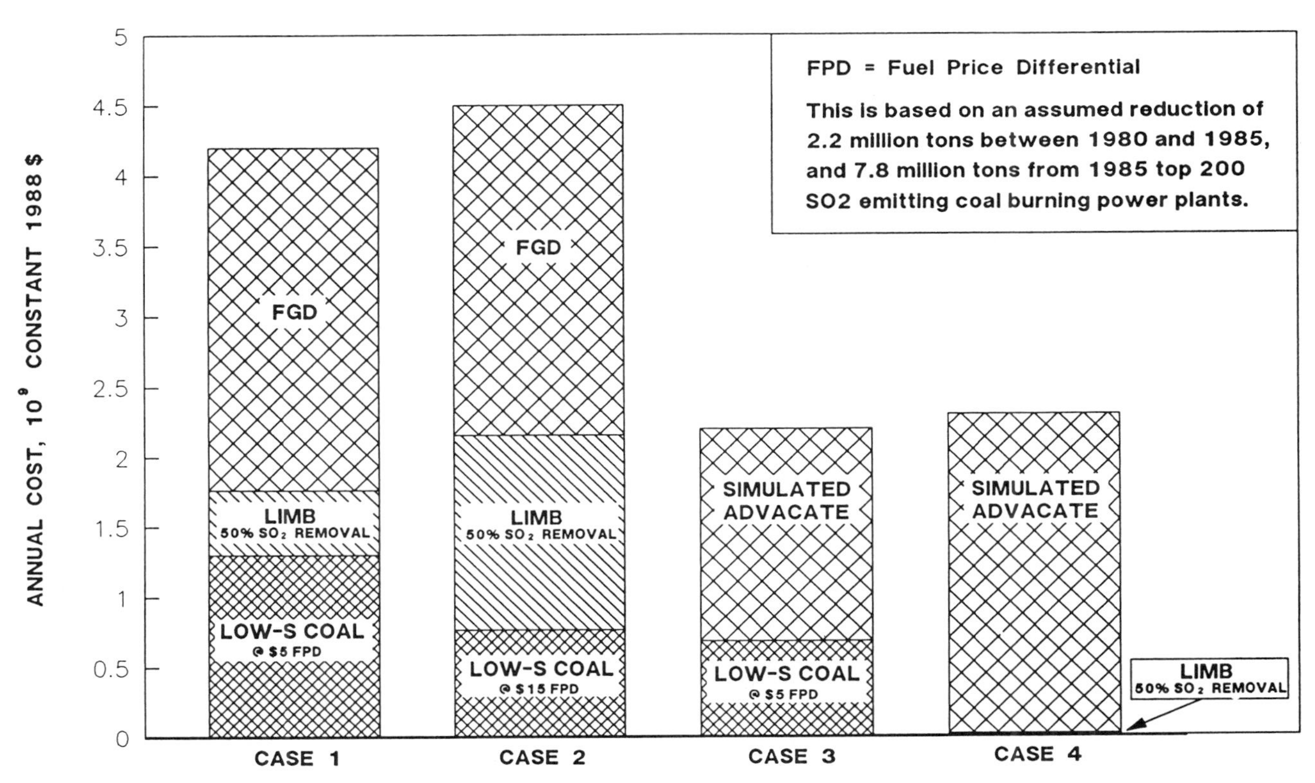

Figure 7. Annual cost of achieving a 10 million ton reduction of SO_2 per year from 1980 emission levels in the eastern region of the United States.

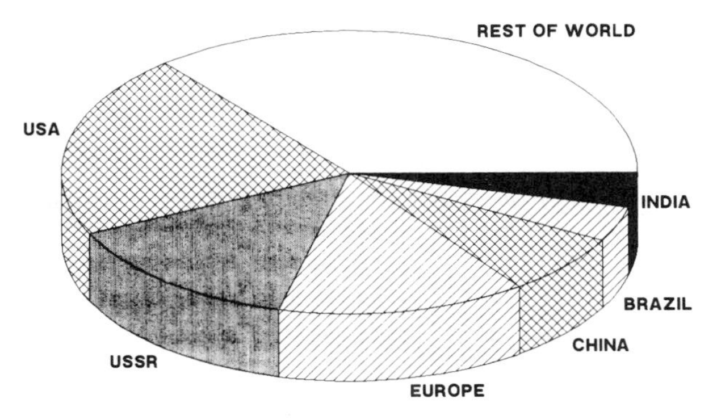

Figure 8. Projected regional contributions to global warming.

consider adopting IGCC systems for electric power generation. Since this technology is still under development, it will probably be unavailable for commercial application before 1996. Assuming a seven-year implementation time for the technology, the first units could not begin electrical production before 2003.

Another potentially promising option is to substitute natural gas for coal, resulting in lower CO_2 emissions due to the higher hydrogen to carbon ratio of natural gas. Using natural gas as fuel in conventional boilers for electric power production would substantially decrease emissions of CO_2. Natural gas firing of utility boilers could begin almost immediately since existing coal-fired boilers can be easily retrofitted to burn gas. The primary limitations on use of this resource are fuel cost and limited availability. In 1985, gas consumption by US utilities was about 3.2 quads. Total natural gas consumption in 1985 was 17.0 quads. US reserves are estimated to last about 70 years at that rate of consumption.

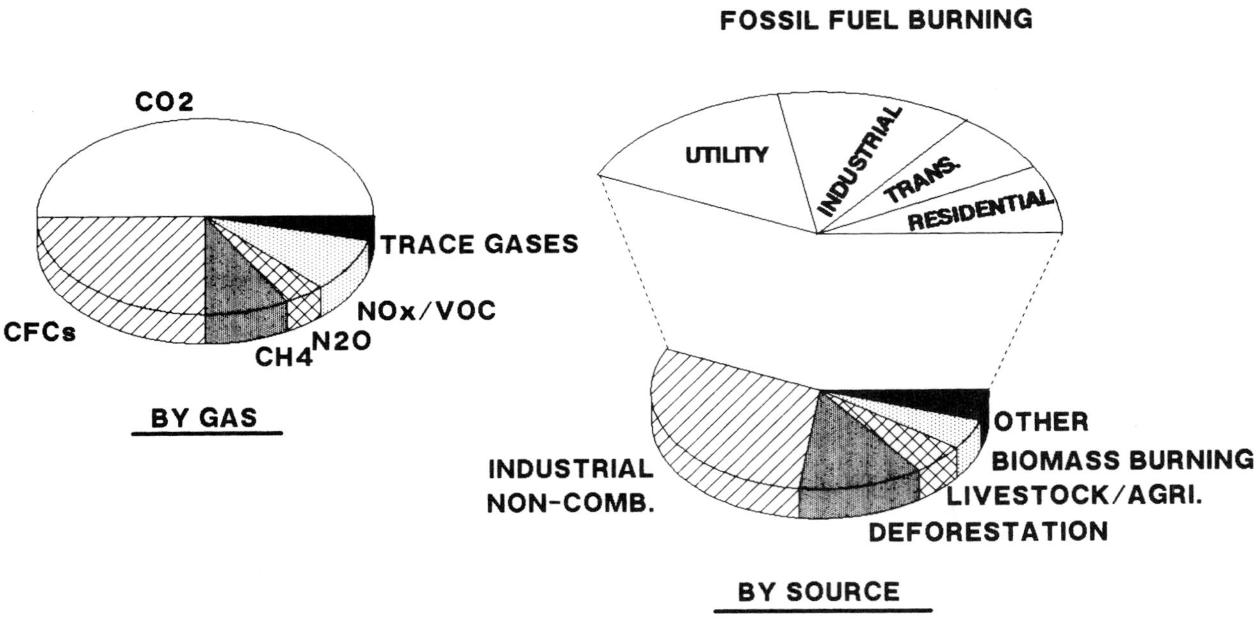

Figure 9. Projected contributions to global warming in 2030 by gas and source.

Utility Carbon Dioxide Emission Model

In order to estimate the impact of some of these options on utility industry generation of CO_2, I have developed a simple emissions allocation model which I call the utility CO_2 emissions model (UCEM). This model assumes a certain annual growth rate in electrical demand. It also estimates retirement of existing capacity by retiring units once they reach an assumed retirement age. It then fulfills new and replacement demand by adding capacity via one or more combinations of new technologies whose capacities are input into the model. The UCEM utilizes all the assumed new technology and then allocates any remaining needed capacity via conventional pulverized coal technology.

Table 3 describes the 10 options evaluated and the input data for the projected penetration of these technologies and approaches from the year 2000 to the year 2020. Note that of these 10 options only the base case (1) assumes a "conventional wisdom" annual electric growth rate of 2.5 percent. This base case assumes all new and replacement capacity will be conventional pulverized coal generators. All the other options assume user conservation based on a 1.5 percent growth rate. Also note the assumed conversion efficiency for the technologies. This is important since higher efficiencies lead to lower CO_2 emissions for a given electrical output. The megawatt penetration I have assumed for nuclear, solar, biomass, clean coal, and the amount of natural gas that may be available in quads to meet this expected increase in electrical demand is based on my judgment as to what might be reasonable upper limits to such penetration in this time frame. Clearly, among persons knowledgeable in this field technology penetration predictions will differ.

Figures 10 through 13 summarize the results of this model analysis for 10 options. Figure 10 projects CO_2 emissions for the base, conservation, nuclear (plus conservation), and new gas (plus conservation) options. Conservation is effective with gas, adding a significant near term benefit, with nuclear providing some longer term benefit (assuming, of course, the penetration assumptions listed in Table 3).

Figure 11 compares the same base case and conservation options with a clean coal (plus conservation) scenario. The clean coal approach offers only a modest advantage over conservation alone. This is because clean coal technologies are still major contributors of CO_2 since their higher efficiency only modestly reduces CO_2 emissions compared to conventional coal units. This approach will probably have only a marginal benefit over the next 30 years.

Figure 12 compares the base and conservation options with the solar (plus conservation) and biomass (plus conservation) options. Again, some significant but modest improvements over conservation alone are realized. Of course, higher utilization of these technologies could allow these technologies to make a more significant impact.

The last figure in this series, Figure 13, compares

Table 3. Ten scenarios analyzed by Utility CO_2 Emission Model (UCEM). (*Represents an average of about 36% for baseline gas steam plants and 44% for new combined cycle plants.)

Option Abbrev.	Option No.	Option	Assumed Penetration All in MWe, Except No. 9: quads					Conv. Eff.	Annual Electric Growth Rate, %
			2000	2005	2010	2015	2020		
Base	1	Base Case - No New Technology						0.36	2.5
Conserv.	2	Conservation - No New Technology						0.36	1.5
Nuclear	3	Nuclear (& Conservation)	0	10,000	20,000	40,000	80,000	0.33	1.5
Solar	4	Solar (& Conservation)	0	5,000	10,000	20,000	40,000	----	1.5
Biomass	5	Biomass (& Conservation)	0	5,000	15,000	30,000	60,000	----	1.5
Clean Coal	6	Clean Coal (& Conservation)		10,000	40,000	80,000	100,000	0.42	1.5
Renewable	7	Solar + Biomass + Conservation (Renewable Option)	See Solar (4) and Biomass (5) Above						1.5
Max. Tech	8	Option 7 + Nuclear	See Solar, Biomass, and Nuclear (3) Above						1.5
New Gas	9	Gas (& Conservation)	4.0	6.0	7.0	7.0	7.0	0.40 *	1.5
Max. Tech & Gas	10	Option 8 + Gas (9)	See Solar, Biomass, Nuclear, and Gas (9) Above					----	1.5

the same two base and conservation options with two combination options. The renewable option shows the effect of biomass and solar penetration, whereas the so-called maximum technology plus gas option shows the effect of all of these technologies plus natural gas applied in tandem. Only this option (of those evaluated) yields stabilization of CO_2 emissions in this time period. I could come up with no credible combination that would show a major decrease of emissions from current levels. Only more drastic conservation measures appear to allow for such a possibility.

Figures 14 through 17 present additional model outputs for two of the options. Figures 14 and 15 show electric and CO_2 generation by source, for the maximum technology plus gas option (Option 10) that allows a possible approach to stabilizing CO_2 emissions. Similarly, Figures 16 and 17 present electrical production and CO_2 projections by source, for the clean coal option.

Since conservation appears to be such an important component of any electric utility CO_2 stabilization program, it is informative to look at historical patterns of electrical production. Figure 18 shows the historical pattern of annual electric growth rate in the US from 1935 to 1988. For the period 1935 through 1985, the bars represent averages for the preceding five years, while for the period 1985 through 1988, bars represent single-year growth rates. Historical growth rates, except for the Depression-impacted 1930 to 1935 and the recession-impacted 1986 time frames, have never been below 3 percent. Most experts project a 2.5 percent annual growth rate between now and the early 21st century (DOE, 1987). Therefore, my conservation options (Table 3), assuming only a 1.5 percent growth rate, clearly suggest that fundamental changes will have to be made, most likely in the way we price energy if we are serious about decreasing the growth rate substantially below historical levels.

Conclusions

The utility industry is likely to face two major waves of environmental challenges in the next 20 years. The first is associated with the requirement to limit emissions of both SO_2 and NO_x from existing coal-fired boilers. It appears that acid rain legislation mandating such a program will pass the US Congress during 1990. Utilities will probably be able to select the most cost-efficient option from available alternatives: coal switching, wet FGD, and emerging technologies such as LIMB and possibly ADVACATE.

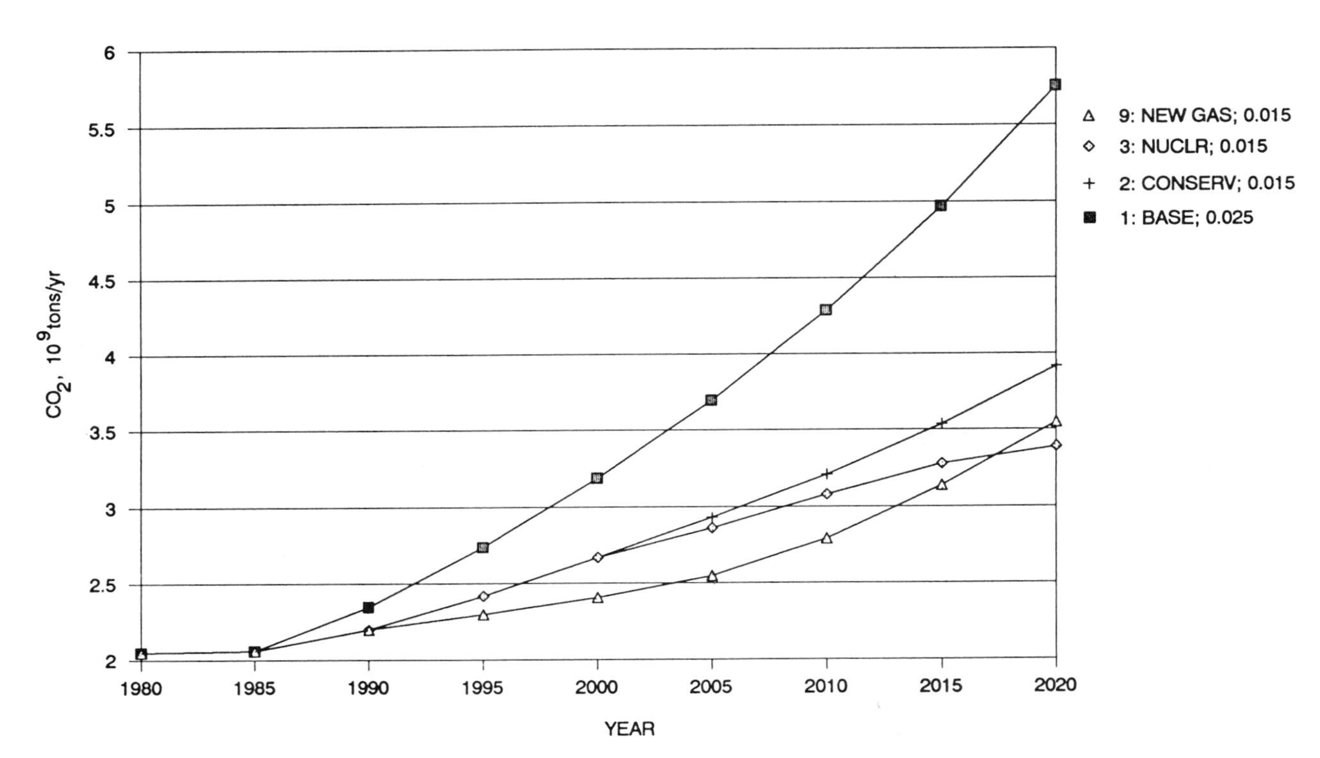

Figure 10. Projected utility CO_2 emissions for base, conservation, nuclear, and new gas options.

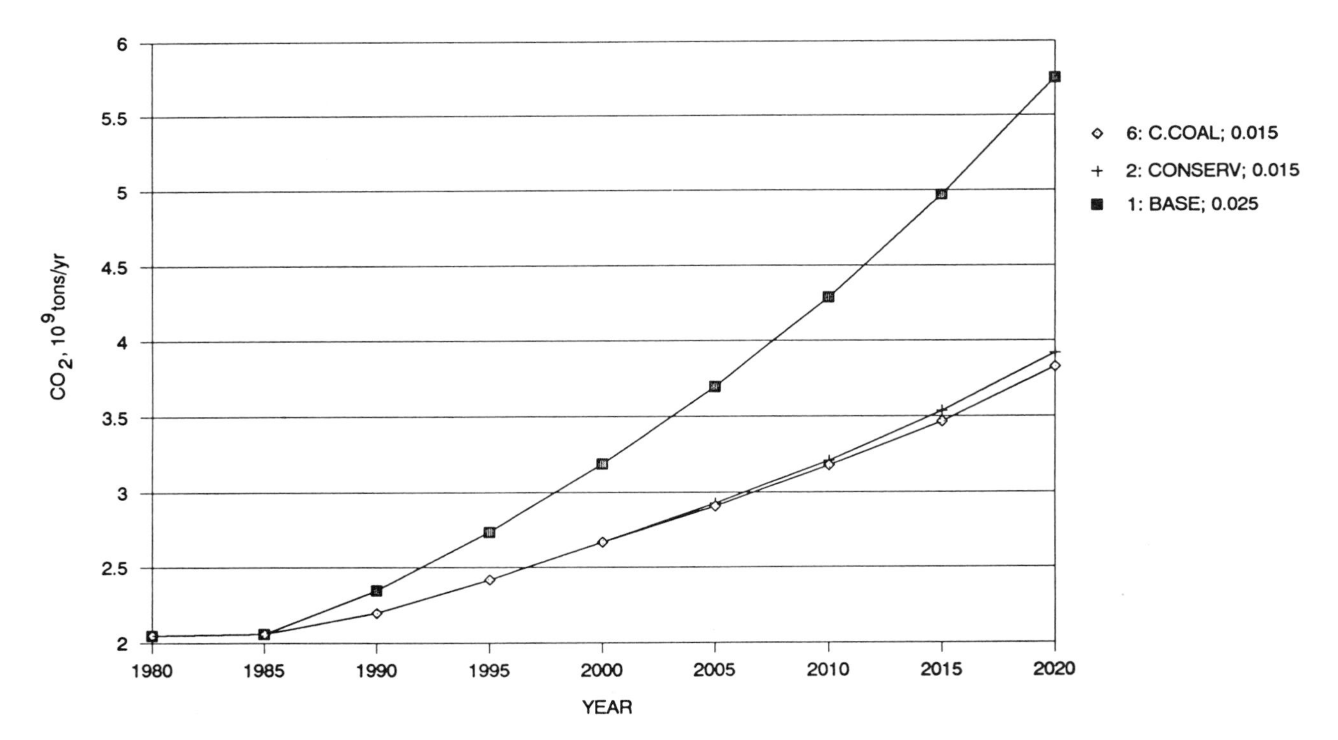

Figure 11. Projected utility CO_2 emissions for base, conservation, and clean coal options.

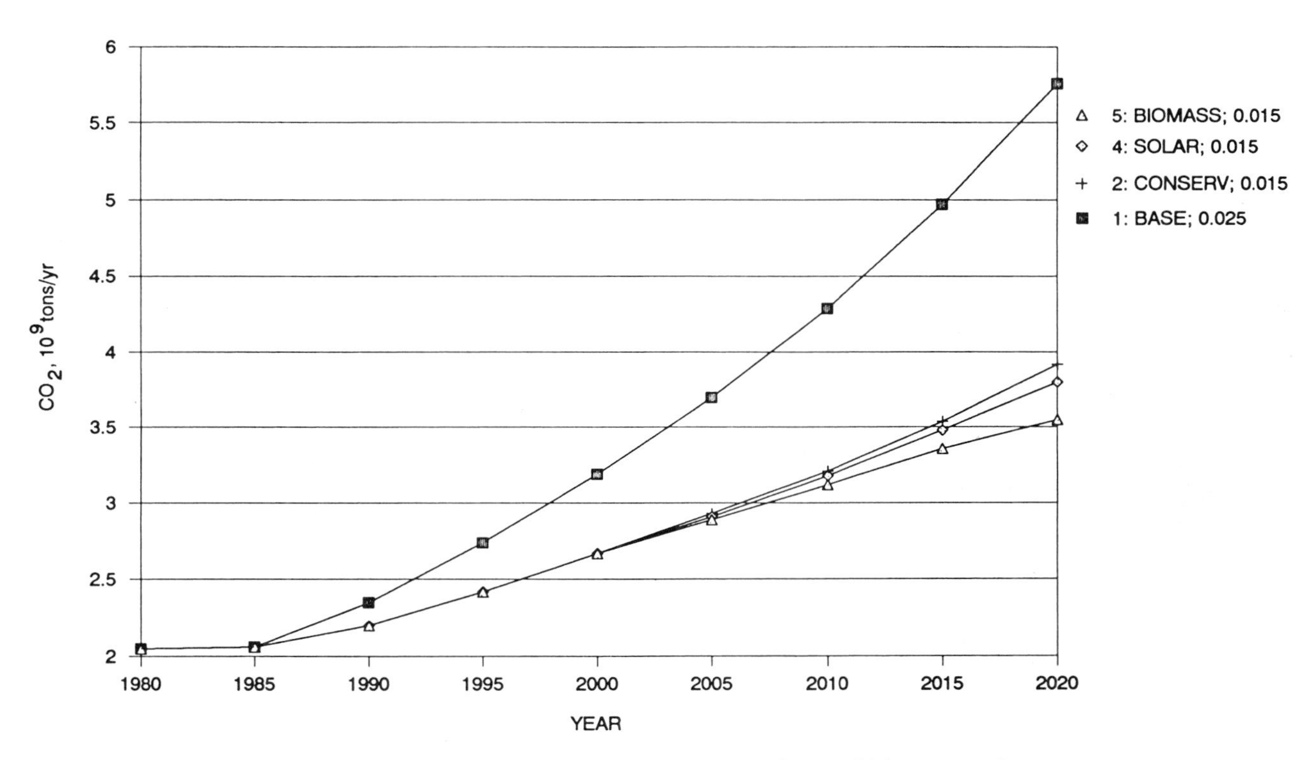

Figure 12. Projected utility CO_2 emissions of base, conservation, solar, and biomass options.

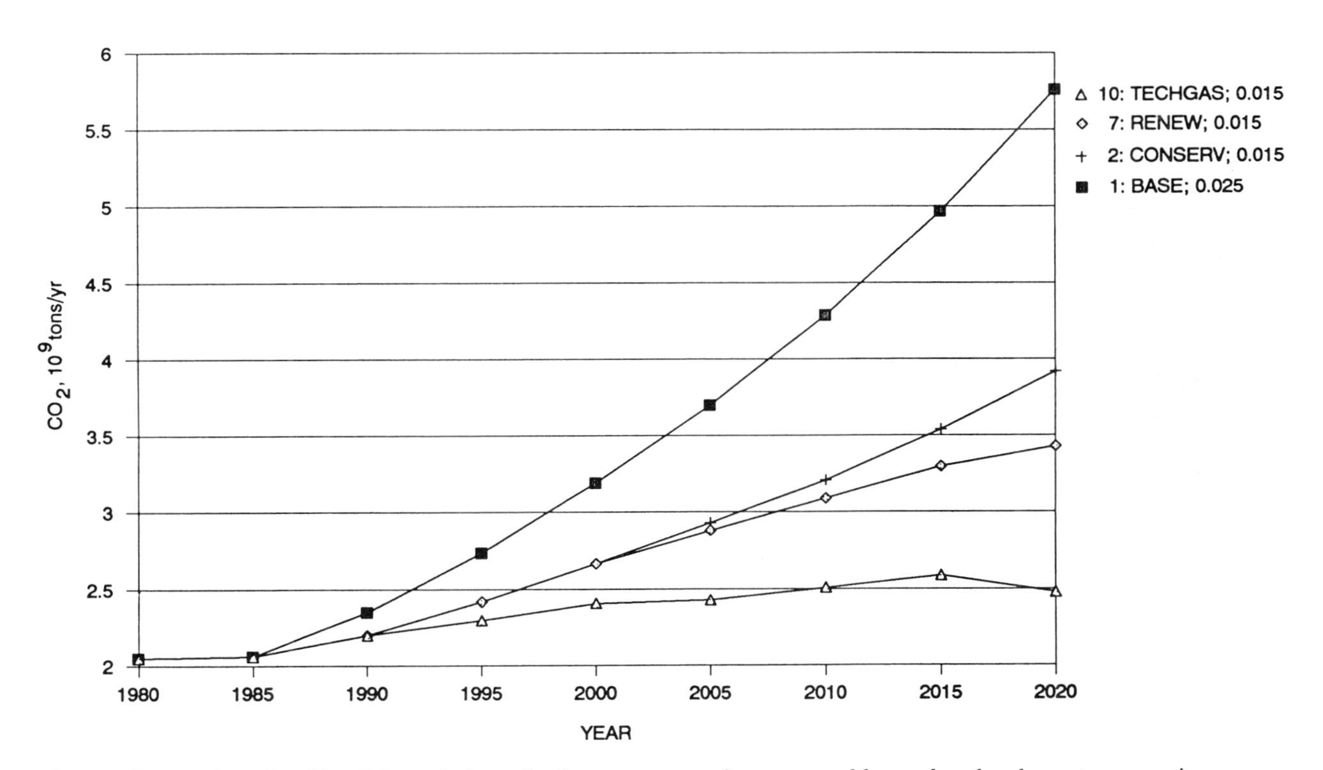

Figure 13. Projected utility CO_2 emissions for base, conservation, renewable, and technology + gas options.

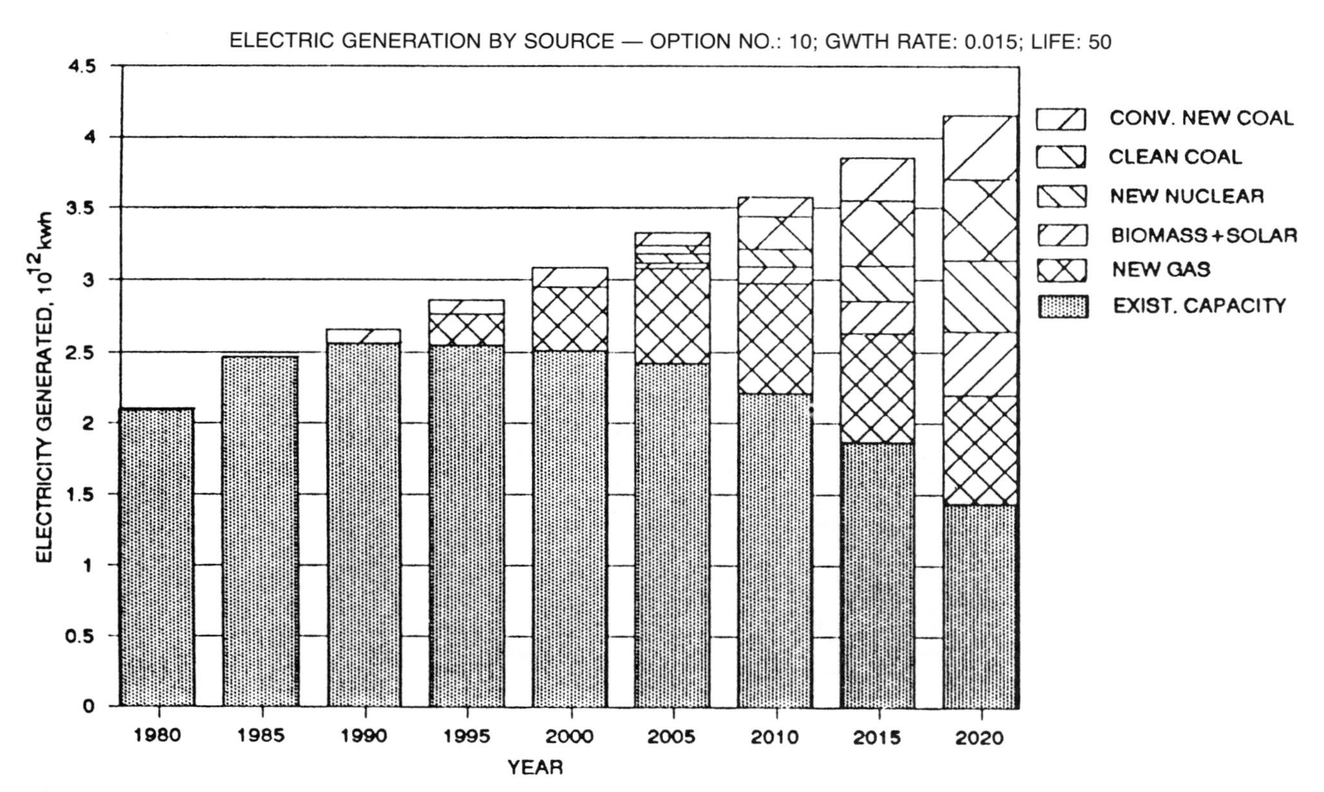

Figure 14. Electric generation by fuel/technology—max. technology + new gas option.

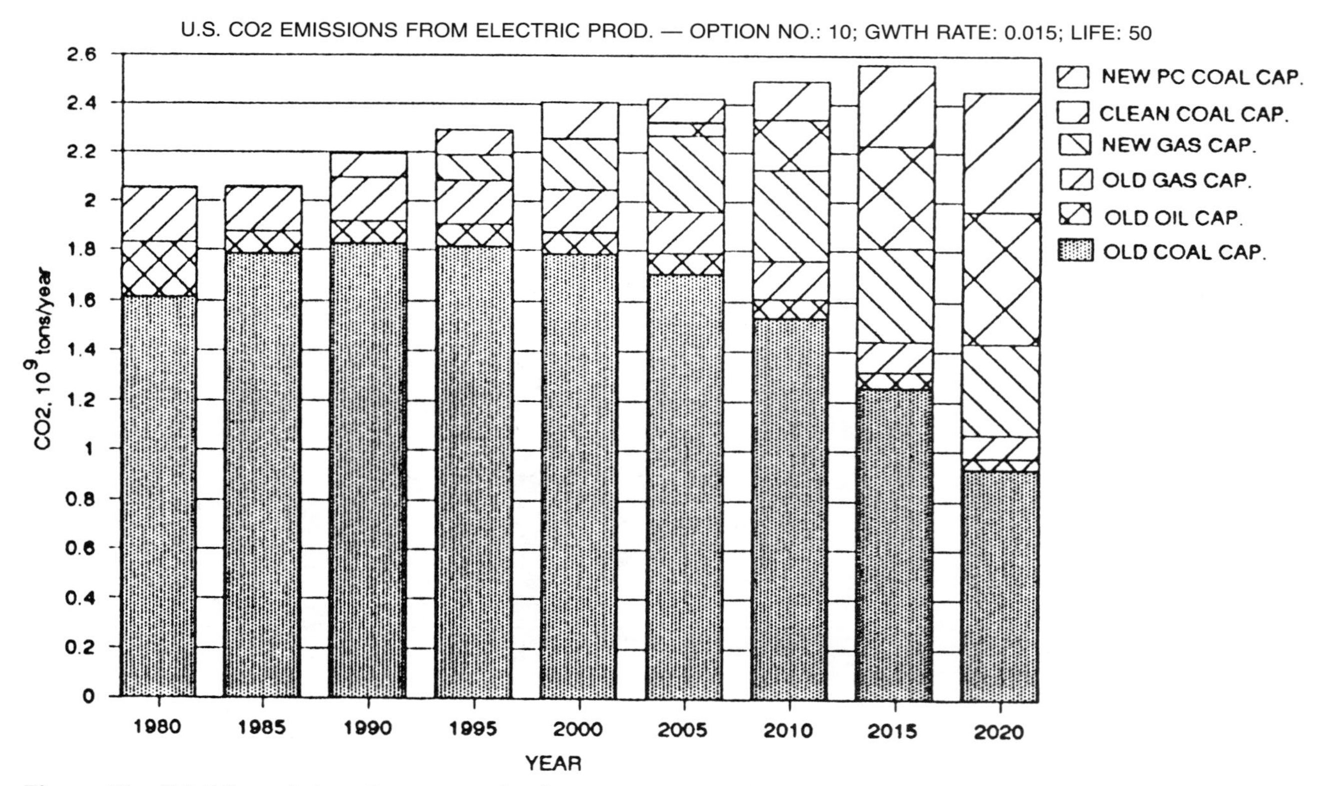

Figure 15. US CO_2 emissions for max. technology + new gas option.

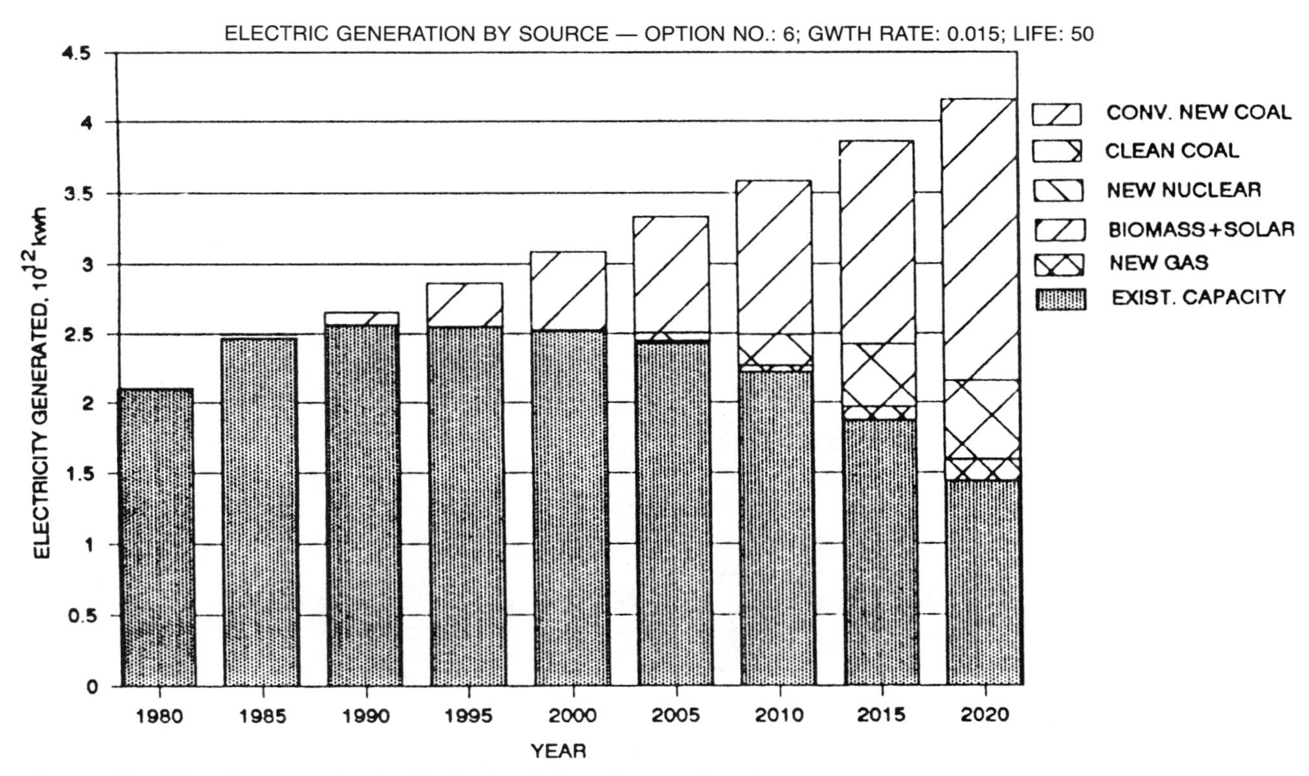

Figure 16. Electric generation by fuel/technology—clean coal option.

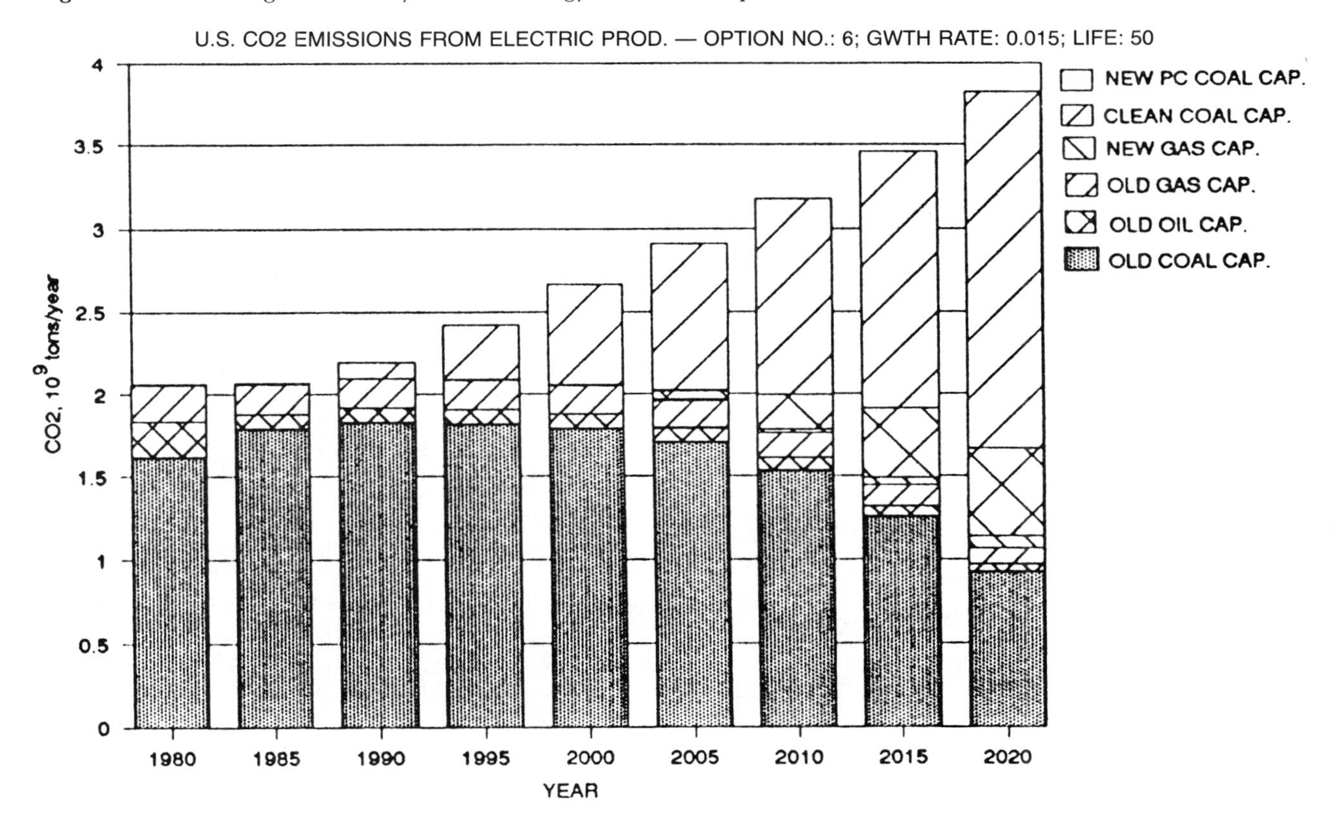

Figure 17. US CO_2 emissions for clean coal option.

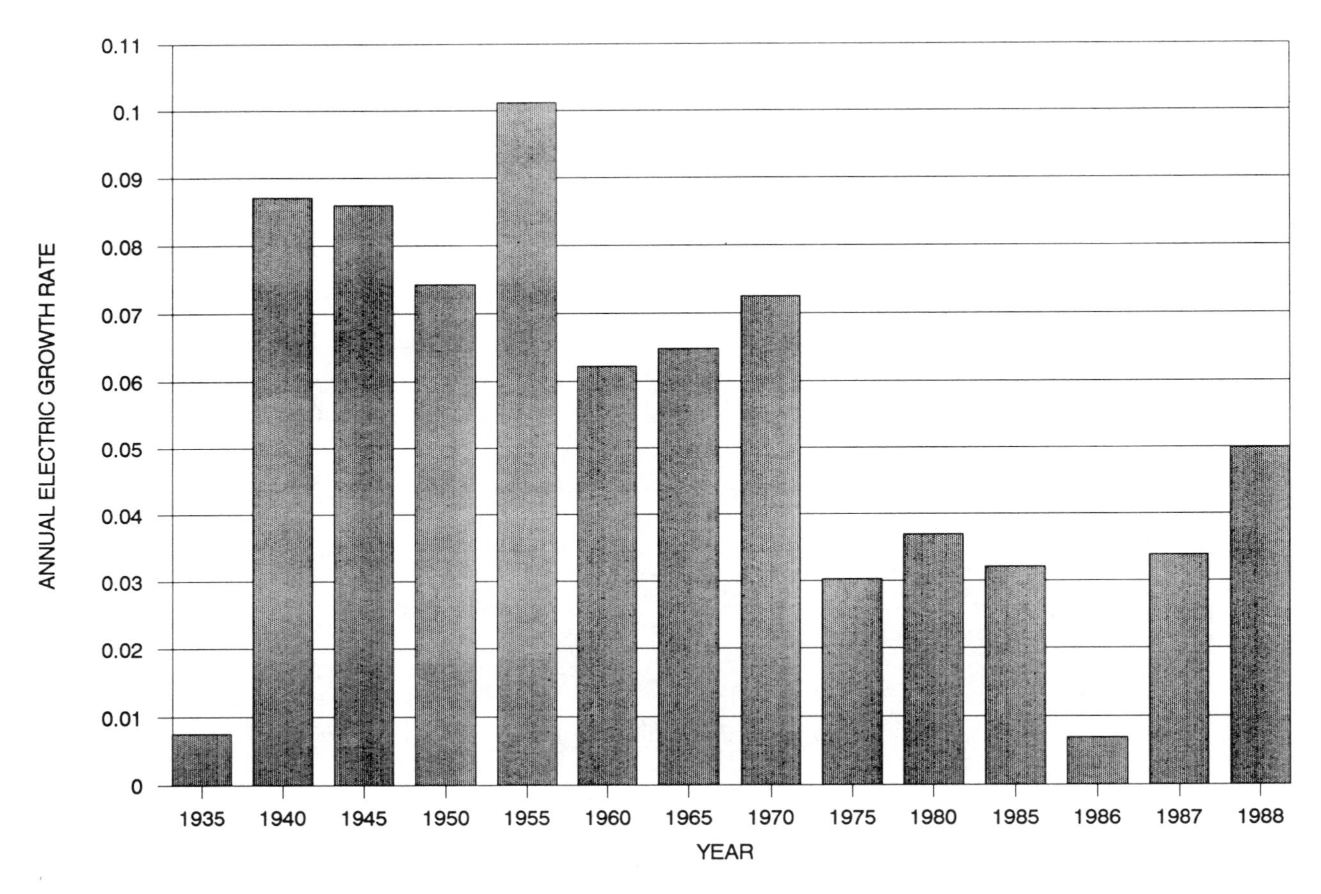

Figure 18. US annual electric growth rate, 1935–1988.

The longer-term and more formidable challenge is associated with potential global warming, to which utility power generation is an important contributor. Dealing with this problem will require fundamental changes in how we use and generate electricity. Only by combining end-use conservation and massive quantities of nonfossil fuel generating capacity can CO_2 emissions be stabilized.

References

Elliot, T.C., ed., *Powerplants Database, Details of Equipment and Systems in Utility and Industrial Power Plants, 1950–1984,* McGraw-Hill, Inc., New York, NY, 1985.

Emmel, T., and M. Maibodi, "Retrofit Costs for SO_2 and NO_x Control Options at 200 Coal-Fired Plants," Draft Report, EPA Contract No. 68-02-4286, December 1989.

Lashof, D.A., and D.A. Tirpak, "Policy Options for Stabilizing Global Climate," Draft Report to Congress, US Environmental Protection Agency, Washington DC, 1989.

Mintzer, I.M. (World Resources Institute), "A Matter of Degrees: The Potential for Controlling the Greenhouse Effect," Research Report No. 5, Holmes, PA, April 1987, 60 pp.

Ramanathan, V., R.J. Cicerone, H.B. Singh, and J. T. Kiehl, "Trace Gas Trends and Their Potential Role in Climate Change," *Journal of Geophysical Research,* 90 (D3):5547–5566, June 20, 1985.

Smith, I.M., "CO_2 and Climatic Change," IEACR/07, IEA Coal Research, International Energy Agency, 1988.

Smith, J.B., and D.A. Tirpak, "The Potential Effects of Global Climate Change on the United States," Report to Congress, US Environmental Protection Agency, Washington, DC, December 1989.

US Department of Energy, "Energy Security, A Report to the President of the United States," Washington, DC, April 1987.

Additional Reading:

Jozewicz, W., and G.T. Rochelle, "Flyash Recycle in Dry Scrubbing," *Environ. Progress,* 5, 219, 1986.

Jozewicz, W., and J.C.S. Chang, C.B. Sedman, and T.G. Brna, "Characterization of Advanced Sorbents for Dry SO_2 Control," *Reactivity of Solids,* 6, 243, 1988.

Jozewicz, W., and J.C.S. Chang, "Evaluation of FGD Dry Injection Sorbents and Additives, Vol. 1, Development of High Reactivity Sorbents," EPA-600/7-89-006a, May 1989.

Jozewicz, W., C. Jorgensen, J.C.S. Chang, C.B. Sedman, and T.G. Brna, "Development and Pilot Plant Evaluation of Silica-Enhanced Lime Sorbents for Dry Flue Gas Desulfurization," JAPCA, 38, 6, 796, June 1988a.

Jozewicz, W., J.C.S. Chang, C.B. Sedman, and T.G. Brna, "Silica-Enhanced Sorbents for Dry Injection Removal of SO_2 from Flue Gas," JAPCA, 38, 8, 1027, August 1988b.

Jozewicz, W., J.C.S. Chang, T.G. Brna, and C.B. Sedman, "Reactivation of Solids from Furnace Injection of Limestone for SO_2 Control," *Environ. Sci. Technology,* 21, 664, 1987.

Chang, J.C.S., and C. Jorgensen, "Evaluation of FGD Dry Injection Sorbents and Additives, Vol. 2, Pilot Plant Evaluation of High Reactivity Sorbents," EPA-600/7-89-006b, May 1989.

Blythe, G., V. Bland, C. Martin, M. McElroy, and R. Rhudy, "Pilot-scale Studies of SO_2 Removal by the Addition of Calcium-Based Sorbents Upstream of a Particulate Control Device," in *Proceedings of Tenth Symposium on Flue Gas Desulfurization,* Vol. 2, EPA-600/9-87-004b, February 1987.

Chang, J.C.S., and C.B. Sedman, "Scale-up Testing of the ADVACATE Damp Solids Injection Process," in *Proceedings of First Combined FGD and Dry SO_2 Control Symposium,* Vol. 3, EPA-600/9-89-036c (PB89-172167), March 1989.

Jorgensen, C., J.C.S. Chang, C.B. Sedman, and D.C. Drehmel, "Pilot Plant Evaluation of Post-Combustion LIMB SO_2 Capture," in *Proceedings of First Combined FGD and Dry SO_2 Control Symposium,* Vol. 2, EPA-600/9-89-036b (PB89-172167), March 1989.

Powering the Second Electrical Century

Kurt E. Yeager
Electric Power Research Institute
Palo Alto, CA

Foreword

In preparing this paper, I have been particularly indebted to concepts developed by Dr. Chauncey Starr in his essays *Current Issues in Energy.* They are collected in a Pergamon Press volume published in 1979.

> The world has always known resource limits—the predominant historical limits have been food and water. These limits have been overcome in two ways—geographic expansion and technical innovation. The possibility of geographic expansion has been largely exploited, but a wide range of technical solutions to resource pressures remains. Technical development represents the only truly unlimited resource. The concerns associated with projected resource shortfalls have always been a result of a failure to recognize the important contribution of technical development. The global problems that we now face should not be viewed from a "doomsday" perspective, but rather as challenges and opportunities for stimulating technical innovation.

Chauncey Starr, "Technical Innovation: The Answer to Resource Depletion," October 1977.

Introduction

In forecasting the technology future for electricity generation, the vision required depends principally upon acceptance of that which is already visible. Therefore, the vision offered here is limited to an arguable set of possibilities focusing on coal as the industry's dominant fuel source for the dawn of the Second Electrical Century. Although this vision makes no pretense of being all-encompassing, its logic is applicable to other dimensions of power supply, as well as to delivery and end use.

One can also argue that anomalous events we cannot yet identify will be the largest determinants of the future. It is quite likely that in coming years we will have to deal with new examples of energy supply interruptions, environmental politics, Chernobyl-like occurrences, and similar discontinuities. While we cannot predict the specifics, or when and where they will happen, history suggests the certainty of their occurrence. Such events, however, can be expected to accelerate the least cost, innovative approach to coal-based power generation outlined here. In the broader context, this approach is offered as an essential step to balanced resolution of our national energy security, economic competitiveness, and environmental protection goals during the first half of the 21st century.

Fundamental to this vision is a more proactive American response to global competition among relatively equal economic competitors. In evaluating this preeminent national issue, the conclusion is the same as Pogo's—"we have met the enemy and he is us." The greatest threat to our competitive position is not foreign but our own national preoccupation with short-term, innovation-averse policies, as opposed to those of our more strategic and globally oriented competition. This attitude, long encouraged by the popular perception that post–World War II US global economic advantage could be taken for granted, is a luxury that the US cannot afford if it is to sustain national economic prosperity and ecological well-being into the next century. What follows assumes that the US reawakens—either through proactive leadership or in response to self-induced crises—and that a greater sense of national responsibility and destiny is encouraged.

Such a transformation is essential if we are to overcome our historically reactive posture and marshall national strengths that will enable us to leap-frog the global competition technologically, commercially, and, as a result, environmentally. The recent progress in East-West relations provides a particularly timely opportunity to address the North-South relationship, which is likely to become at least equally important in the Second Electrical Century. This will require a change in public perception to one that mandates a sustained commitment to the unmatched US "soft" power advantage in resources and technology equal to the commitment that has supported our military leadership over the past 50 years. This commitment must be reinforced by public policies recognizing that national economic and environmental security are as dependent on industrial technology as military security. The result can be both the foundation for rebuilding America, and the arsenal of electricity-based energy innovation powering the new century.

Energy and Environment—A Transition in Perception

This transformation seems particularly germane to the interconnected themes driving the Secretary of Energy's National Energy Strategy initiative. These themes, as reported by the Department of Energy (DOE) from public testimony, are increased energy efficiency; secure future energy supplies; environmental protection; and fortifying of the technological infrastructure, including scientific research, education, and facilities. A strategic element linking these themes is electricity and the technology for its global development. In spite of such comprehensive strategic intentions, official US energy policy has historically remained a political creature reflecting popular perceptions of the moment. It has been made up of numerous narrow policies intended to stimulate specific actions for explicit goals, usually established in differing times and circumstances. It is not surprising that these disconnected policies are often in conflict and unable to maintain a progressive balance among numerous competing interests.[1] Achieving such a balance would require a perspective of 30 years or longer and a, to date, politically unrewarding concern for the energy structure available to future generations. For example, only immediately after the oil supply interruptions of 1973 and 1978 was there sufficient temporary popular perception of the precarious nature of our dependence on foreign fuel sources to encourage even a fleeting reflection of oil supply reality in public policy.

Similar policy confusion, but on an even broader scale, compounds national technological progress. Because public policy has not addressed the issue of the commercial application of technology, government often tends to impede, rather than assist, industrial efforts to bring new technology rapidly to the world's market. Federal support alone is dissipated among a dozen Executive branch agencies responsible for various aspects of science and technology and an equivalent number of Congressional appropriation subcommittees. In contrast to our international competitors, we still lack government mechanisms for strategic coordination of priorities and for sustaining the support vectors necessary to technological, to say nothing of commercial, success.

The one enduring factor underlying US energy policy has been the societal pressure to eliminate environmental impacts and risks associated with energy systems. Unfortunately, because of the popular perception of unconstrained energy security and unaffected global competitiveness, the policy response to risk minimization has focused on short-range clean-up expediencies rather than providing incentives for innovative technologies and practices that reduce pollutant formation. This undermines the goal of a sustainable national and global balance between energy use and the environment. The result has been ever larger energy and environmental stresses and diminished economic competitiveness. This situation underscores the fact that most of the expedient, "band-aid" responses have been made to balance energy security, environmental protection, and economic competitiveness. Progress will require a much more systematic balancing of costs and benefits plus development of politically viable means to "compensate the losers."[2]

Resolution of this tension among energy, economics, and the environment is dependent on the cultural process of internalizing environmental costs, i.e., including them in the price of products and services. This promises to be a major global-scale societal challenge continuing well into the new century. It represents the latest step in an on-going social process which, historically, has confronted a series of so-called "external costs" beginning with raw materials, and extending, for example, through labor, education, and industrial safety.[3] This evolution, summarized in Figure 1, also reflects a correspondence between the pressure to internalize more costs, and the pressures of exponentially increasing population density on finite resources. Just as in the past, the corporations and nations that prosper in the Second Electrical Century are likely to be those that most effectively harness technological innovation and new business opportunities to facilitate this relentless cost internalization process.

Historically, this process of change has been difficult and often unjust because the adversarial methods used by society tend to accentuate, rather than resolve, the

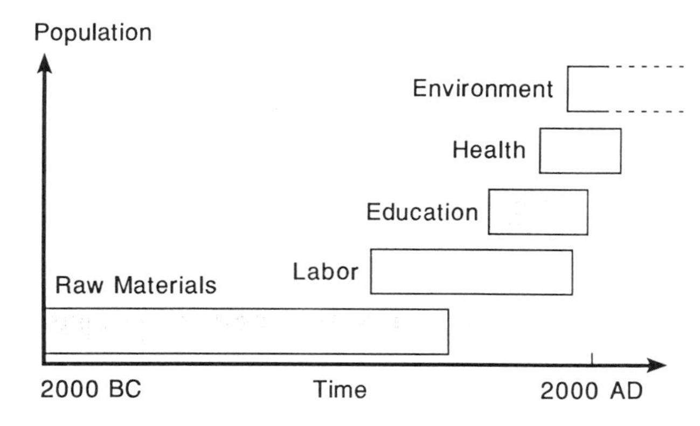

Figure 1. The cost integration process.

differences between perceived winners and losers. In the case of the environment today, promoters of change seem intent on insuring that such change occurs only at the expense of those with vested interests in the status quo. It is also obvious from historical precedent that resolution of cost conflicts, on both the personal and global level, comes more easily to the rich than to the poor, and is discouraged by moralistic judgments.

In this context, the value of science and technology depends on its ability to recognize that perceptions, environmental and otherwise, are manifestations of underlying social realities. Science and technology serve best as a resource to reconcile these often flawed perceptions with their underlying reality, thus reducing the dislocation and injustice involved. The value of science and technology is dissipated when it loses its focus on resolution of root causes and becomes trapped in adversarial debate over perceptional effects. This logic supports the theme and interconnected examples presented in the following sections.

The more we study the major energy and environmental problems of our time, the more we come to realize that they cannot be understood, let alone solved, in isolation. They are systemic problems, interconnected and interdependent, and their solution suffers from a "crisis of perception" reinforced by political expediency.[4] Science has shown that a holistic view—seeing the world as an integrated system rather than separate parts—is essential if we are to sustain progress in environmental protection while developing the resources needed for global economic development. Technology is providing the means to achieve that sustainable future, if we sustain our commitment to innovation.

Within this reality, national leadership will continue to depend on the ability of the private sector to recognize the critical trends in a rapidly changing world, to guide public perceptions accordingly, and to reconcile energy security, environmental protection, and economic competitiveness within the rules of what is likely to remain a perception-driven public policy umbrella. The key question that we face is: What constitutes a sustainable energy future—corporately, nationally, and globally?

Global energy use is likely to increase sharply in the next half century because of the combination of increasing world population and economic growth. This higher demand for energy will significantly affect US fuel availability and cost. In spite of this, we continue to encourage an inefficient hunter-gatherer mode for energy supply, wasting nature's stored resources.[1] Our goal should be to phase into a managed energy supply without limits. This is analogous to the prehistoric conversion to a cultivation basis for agriculture, which was also a cost internalization process responding to increasing population density and resource constraints. A more efficient refining approach to coal utilization, as developed in this paper, is an essential element in the transition to that globally secure energy and environmental future.

The Second Electrical Century

Electricity—The Engine of Progress

It can be confidently predicted that providing for future growth in electricity supply and assuring that electricity is used efficiently will continue to be inseparable from achievement of the nation's economic prosperity in a globally competitive market. Figure 2 shows the US historical trend of increasing electrification while decreasing total energy per unit of economic output. During the economic expansion since 1982, electricity growth has paralleled GNP growth while real electricity price has declined 16 percent and overall energy efficiency has improved 11 percent. Continued productive improvement in energy efficiency through electrification means that the rate of increase in primary energy consumption is expected to shrink, by 2010, to less than half the annual rate of US GNP growth.[5] As these figures indicate, economic progress, energy efficiency, and increased electrification have proceeded hand-in-hand, and continued economic growth is expected to raise electricity use to nearly 50 percent of US energy consumption by the turn of the century.

It is also likely that the role of electricity in the US (and world) energy economy will continue to grow as dramatically during the next 50 years as it has during the past 50 years. This prediction is based on three emerging, environmentally related market opportunities where electricity offers unique advantages. All three provide commercial breakthrough opportunities for those who most effectively harness innovation to electricity production and use. These opportunities are:

- The use of electricity to reduce the current dependence of the transportation sector on petroleum. This opportunity is inevitably driven by the increasing scarcity and cost of oil combined with the inability of vehicles powered by relatively inefficient, individual,

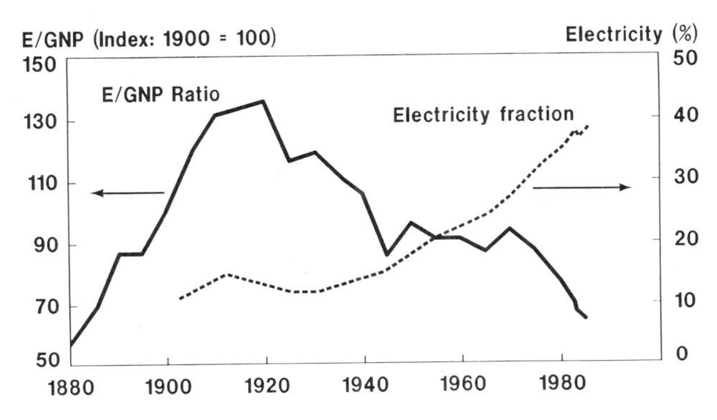

Figure 2. US energy/GNP ratio and electricity fraction, 1880–1985.

internal combustion engines to resolve the inexorable environmental pressures facing urban centers.

- The ability to transform urban wastes, through electric power production, into usable energy and refined materials of commerce. This will transform an otherwise increasingly unmanageable social and environmental debit into economic credits while protecting the viability of the cities and the quality of life for their inhabitants.
- The essential importance of electric power production in biomass management and the utilization of other renewable energy sources intended to protect global ecology while supporting economic progress in many currently underdeveloped regions.

Electrification has become a global economic imperative because electricity-based technologies have served to raise the net economic productivity from all factor inputs, including energy.[6] Thus, growth in electricity generation is a major factor in global energy and environmental projections. Its ability to improve the efficiency of resource utilization, as suggested by the above examples, is essential if these conflicting requirements are to be balanced.

Coal—A Bridge to the Future

As we enter the Second Electrical Century, the growth of electricity will, however, continue to depend, as it did in the first, on low-cost energy from coal. Coal use must be aggressively joined by a nuclear renaissance and promising new resource opportunities essential to a sustainable energy future. Even then, it is likely that coal will continue to produce at least half the nation's electric power through the middle of the next century and, in so doing, act as the bridge to that sustainable future. This forecast is based on the importance of coal as the primary domestic energy source during this period (Figure 3).

Based on examinations of US energy use patterns that consider alternative economic, social, energy availability, environmental, and technological scenarios, we should be preparing for at least a doubling of coal use and an increase in its share of total US energy consumption from today's 26 percent to about 40 percent over this period.[7] This is mirrored by the expected growth in electricity and the role of coal in its generation, as shown in Figure 4. Notably, 1990 will likely mark the year in which the US first reaches an annual coal production rate of one billion tons. Thus, the challenge is not whether coal will be used but how to assure its most efficient utilization and the resulting lessened impact on the environment. This will be essential to the concept of sustainable global economic and environmental development in the Second Electrical Century.

The importance of the technological link between

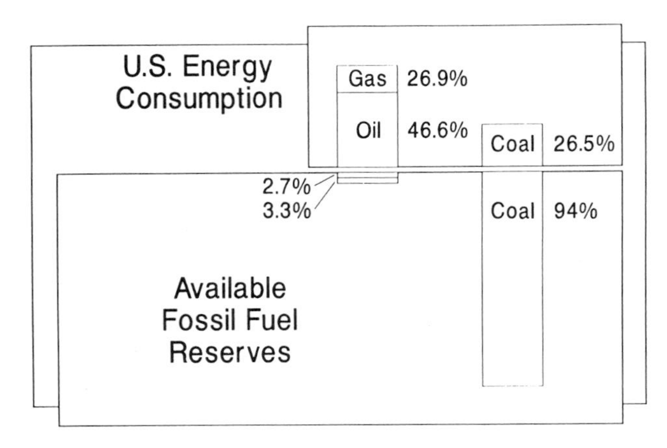

Figure 3. The domestic energy imbalance. (Source: US DOE, 1988)

electricity and coal is underscored by the Conservation Commission of the World Energy Conference,[8] which projects that:

- Future global economic growth, particularly in the developing regions, will be associated with very large increases in electricity growth.
- Coal, which provides over 80 percent of the world's nonrenewable energy resources (Figure 5), will become increasingly prominent in the coming decades. Worldwide, coal usage is expected to exceed the use of petroleum during the first decade of the next century and to continue its preeminence during at least the following 50 years.
- Technological innovation to support this expansion in coal use will be essential for the world's economy and its environment.
- The geopolitical importance of coal is also likely to increase since the world's coal resources are predominantly controlled by three nations: the US, China, and the USSR.

Figure 6 compares the world's nonrenewable energy resource base from Figure 5 with the projected global energy consumption range.[8] This indicates that proven global reserves other than coal, i.e., resources that can be recovered economically under local conditions using existing technologies, are likely to be consumed by the end of this decade. That reality will place escalating cost pressure on the use of additional noncoal energy. Escalation is likely to be capped, in the case of oil, by the $40 to $50 per barrel (1990 dollars) cost of producing synthetic fuel oil from tar sands and coal. This comparison of energy consumption versus resource also underscores the need for accelerated, safeguarded breeder reactor development that could extend the limited uranium resource by a factor of 60. Similarly, the urgency of pursuing renewable energy technology is

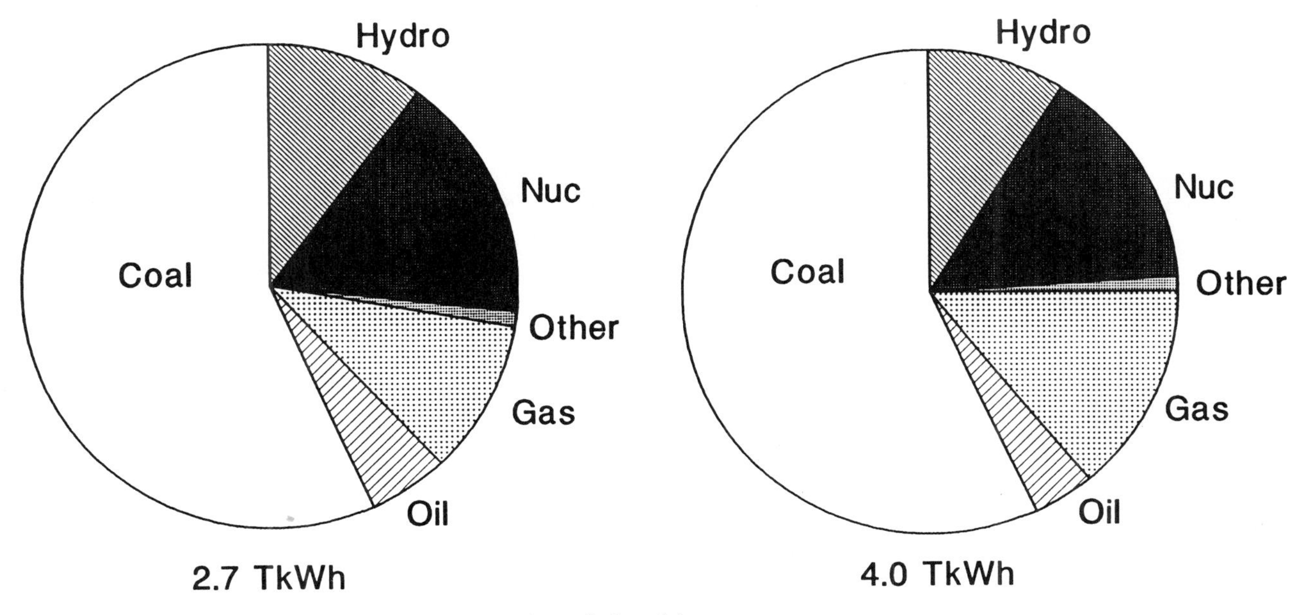

Figure 4. Current and projected US generation of electricity.

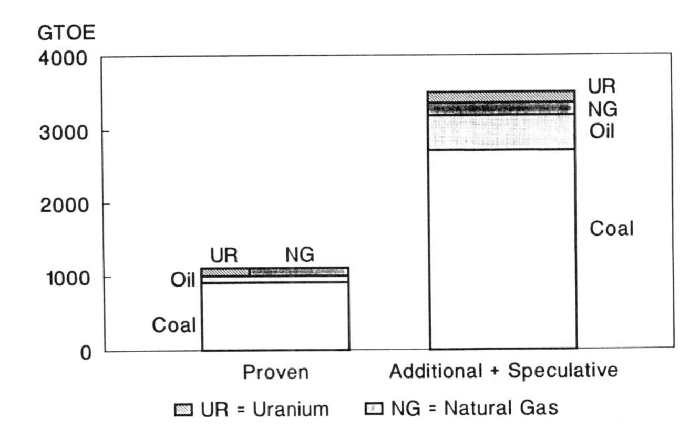

Figure 5. World nonrenewable energy resources (GTOE = gigatonnes of oil equivalent).

clear if the developing world, in particular, is to move beyond what is likely to become an overwhelming dependence on coal during the first half of the 21st century, as reflected in Figure 7. Figure 8, in turn, displays the diminishing role of the US in world energy consumption, principally through its relatively advanced state of electrification and related energy efficiency improvements.

The Power Generation Challenge

The US electric power business faces an even more immediate challenge, however, viz., the need to keep up with rising demand for its product even as many existing facilities reach the end of their planned useful lives and investment by regulated utilities is discouraged. In the next 20 years no less than one half of the electric utility industry's generating capacity, or about 300,000 MWe, will reach 30 years of age and thus be eligible for retirement.

In spite of this, investment disincentives mean that less than 40,000 MWe of new capacity is scheduled to enter commercial operation during the entire 1990s by regulated utilities, plus an additional 34,000 MWe by nonutility generators.[9] By far the largest single category in utility planning to meet future electricity demand requirements is purchased power. Who the seller will be on the scale required is typically not yet determined. The challenge facing the electricity power industry mirrors that facing the nation, i.e., the need to reverse short-term, innovation-averse policies in the face of competing market pressures where ultimately only the lowest cost/highest quality provider (and producer) can be assured of prosperity or even survival.

As a result, the challenge of meeting power requirements in the 1990s will be a technological prologue for the even larger issues facing power generation in the 21st century. The 1990s will see the end of excess generating capacity in the US and the need to cope again with exponential demand growth, probably in the range of 2 percent to 3 percent per year.[10,11,12] Although this sounds modest compared to the 7 percent per year growth of the 1960s, it implies the annual addition of 20,000 MWe of new capacity by the turn of the century.[13,14] This rate of addition is equivalent to the capac-

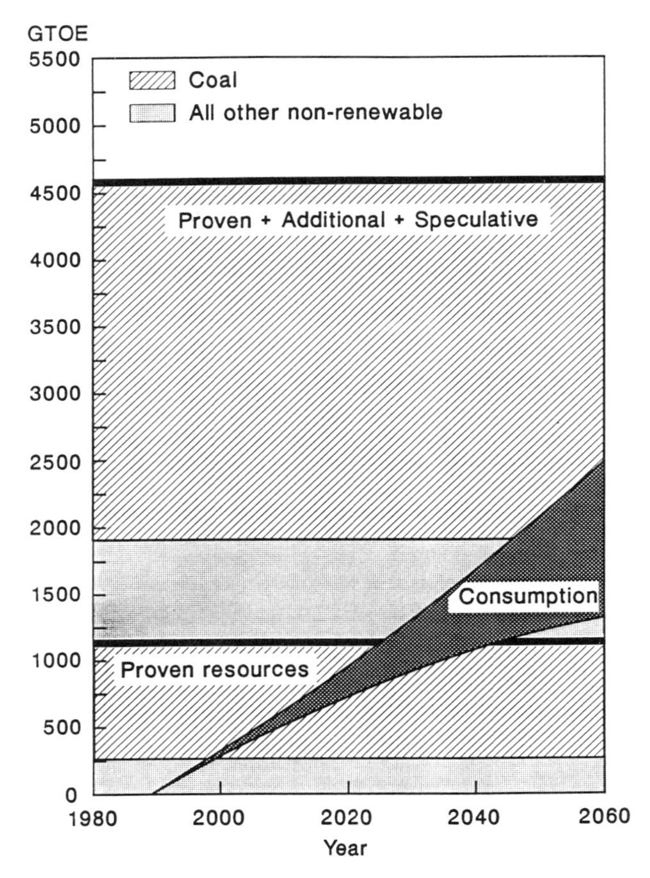

Figure 6. Accumulated world energy consumption vs. resources.

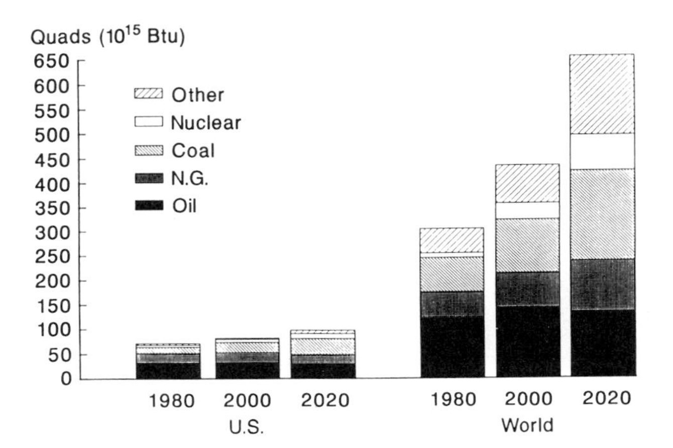

Figure 8. Evolution of world energy supply/demand.

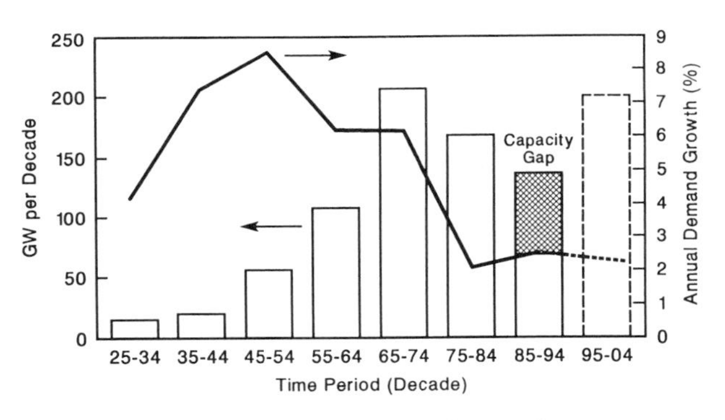

Figure 9. US capacity vs. demand growth.

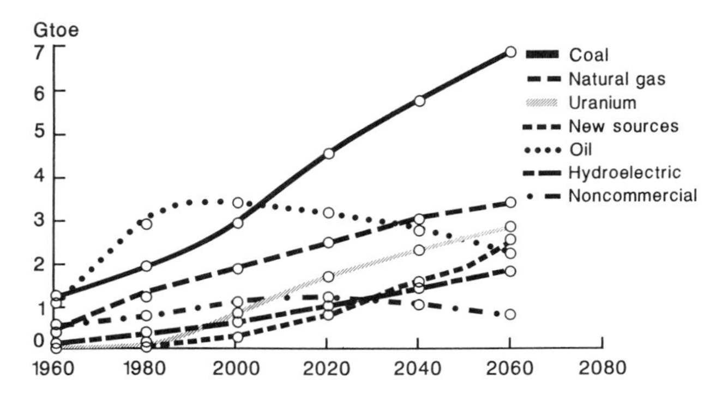

Figure 7. Evolution of world energy utilization. (Source: World Energy Conference—Conservation Commission, 1986)

ity addition rate seen in the 1960s and 1970s. It results from the much larger existing generation base now and the need to offset the capacity "gap" created during the 1980s and early 1990s, when investment disincentives led to the deferral of new capacity commitments, as shown in Figure 9. It is this exponential nature of electricity demand growth, versus limited resources, that inevitably presses for ever higher levels of efficiency on both the supply and demand sides of the consumption equation.

Let us briefly examine the changes that are likely to herald a more competitive, entrepreneurial era in the business of power production as we enter the Second Electrical Century. In many ways we are returning to conditions that parallel those facing the electric power pioneers at the beginning of the first electrical century. Now, as then, those who prosper in the power generation business are most likely to harness new technological opportunities and anticipate the organizational and cultural changes necessary to capture the competitive advantage of being the low-cost producer. This ability underscores the industrial definition of innovation—introduction of a method that is inherently less costly than its predecessor because it cuts, or eliminates outright, certain costs which are intrinsic to its predecessor.[15]

The root causes that seem to be simultaneously creating, and responding to, this change are the investment disincentives affecting regulated electric power production, the pressure of increasingly stringent environmental control requirements, and the internationalization of the power equipment supply industry (Figure 10). The following will discuss technological implications of these root causes.

Investment Disincentives

Harnessing the Microprocessor

Investment disincentives are leading electric utilities to place increasing emphasis on the productivity of their existing plants in order to defer the need for major new capital investment. While this is, at best, a stop gap measure with respect to the industry's rapidly expanding capacity needs, it nevertheless is resulting in an important strategic technology trend. Specifically, the industry is harnessing the electronic microprocessor revolution to achieve higher availability, lower heat rate, and extended operation through improved diagnostics, plant control, and expert advisory systems guiding the man/machine interface. This trend is driven by the realization that in a competitive environment, nonfuel operations and maintenance (O&M) costs become a critical factor in determining the low-cost producers. It is further amplified by the need to plan for reliability in an increasingly decentralized, competitive power supply system.[16]

Early electronic applications in the power plant emphasized the tedious analytical tasks that utility personnel were previously called upon to handle. These included calculations for stresses, temperature, vibration, and the processing of raw signals for further diagnosis.

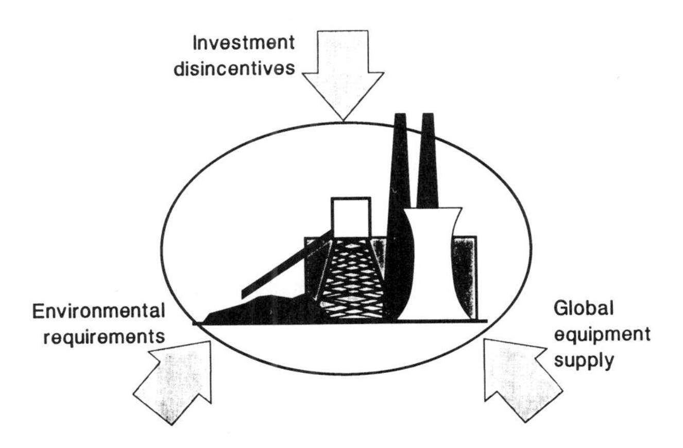

Figure 10. Messengers of technological change.

The next step has been to include diagnostic functions intended to detect incipient damage in advance of failure. If equipment damage can be forestalled by normal maintenance, the huge costs of a forced outage can be avoided and utilities can plan maintenance and upgrade activities based on a day-to-day knowledge of machinery condition.

This leads to the third step—prognosis. Whereas diagnostic systems analyze past data, knowledge-based "expert systems" are able to look into the future and forecast what might happen as well as the possible consequences and needed actions. As a result, the industry is now preparing to take the next step where the integrated knowledge and intuition of specialists can be electronically focused to augment the diagnostic analyses and guide plant operations as well as maintenance. The ultimate goal is the means to operate power plants economically without forced outages or age limits.

This revolutionary change in O&M approach is being accelerated by the attrition of an aging work force whose experience and judgment will be increasingly difficult and expensive to replace. It also sets the stage for the equally revolutionary change in power generation technology and operating demands that will follow. As one of my utility colleagues succinctly put it, "80 percent availability, 60 percent capacity factor, and 30 percent efficiency are not the figures of merit you would choose to succeed in a competitive environment."

In the longer term, however, the current disincentives to invest will inevitably have a debilitating effect on productivity. For example, although investor-owned utility operating income increased about 40 percent, on average, over the past six years, average regulated earnings have dropped by over 50 percent during the same period. This, regrettably, encourages sacrificing investments in innovation, and its supporting technology development, in favor of maintaining immediate shareholder return. This is just one example of how national policies, in effect, encourage liquidation of capital assets for the short term. The end result is to put corporate, and ultimately national, competitiveness at risk by denying investment. Productivity improvement and competitive survival depend on innovative technology for new plant that fundamentally increases the efficiency of energy conversion and power supply.

Initially, as increased capacity is needed, it is being added marginally through peaking duty combustion turbines using natural gas, plus plant repowering and small-scale nonutility generation, both often utilizing fluidized bed combustion (FBC) to take advantage of low-cost or waste fuels. The technological innovation of FBC marks the first important change in coal-based, commercial power production in over 60 years. It also marks a fundamental engineering step toward more effectively internalizing environmental costs through a process methodology that reduces pollutant formation. Another dimension of technological change is emphasis on energy storage as a means to increase the capacity factor of existing power plants.[17] New storage technologies, such as compressed air energy storage and superconducting magnetic energy storage (SMES), in addition to the more classic methods of pumped and battery storage, are being encouraged by this trend. These not only can increase the effective capacity of both existing generation and transmission systems, but also can reduce power plant cycling requirements with an attendant reduction in operating stresses.

All these steps are tactically intended to forestall the financial exposure of the regulated utilities but, as an unintended effect, they are also ushering in a new diversity of electricity supply technology encouraged by immediate marginal power requirements, the small unit sizes being purchased, and the correspondingly lower risk involved.

The Rush to Natural Gas

Of particular interest is the growing short-term reliance of electric power production on natural gas, and the resulting impact on the expected availability and delivered price as shown in Figures 11 and 12. This situation results from the commitment to at least 40,000 MWe of new combustion turbine capacity during the 1990s. These turbines represent the bulk of firmly planned new capacity by both regulated and nonregulated generators because of their low capital cost, short lead-time, and peaking duty advantages compared to alternative generating methods. Technologically, this combustion turbine commitment sets the stage for combined cycle power plants, capable of moving beyond the efficiency limits of the cycle, and the coal-derived synthetic gas production ultimately necessary for their operation. Growth in natural gas usage also reflects the anticipated co-burning of natural gas with coal for expedient emissions control, based on the present relatively low price

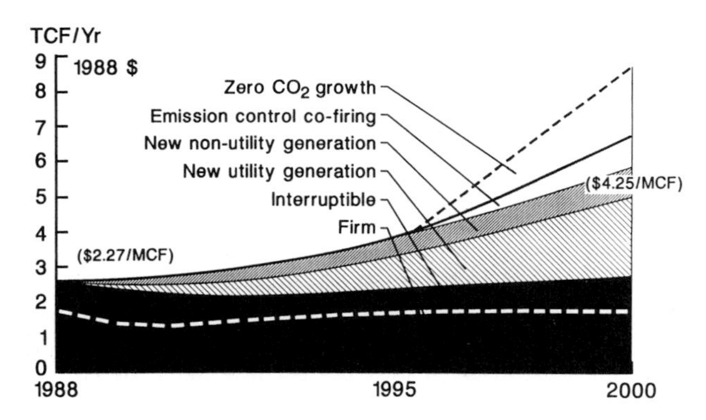

Figure 11. Natural gas consumption for power generation. (MCF = 1,000 cubic feet and TCF = 1 trillion cubic feet) (Source: NERC/EIA)

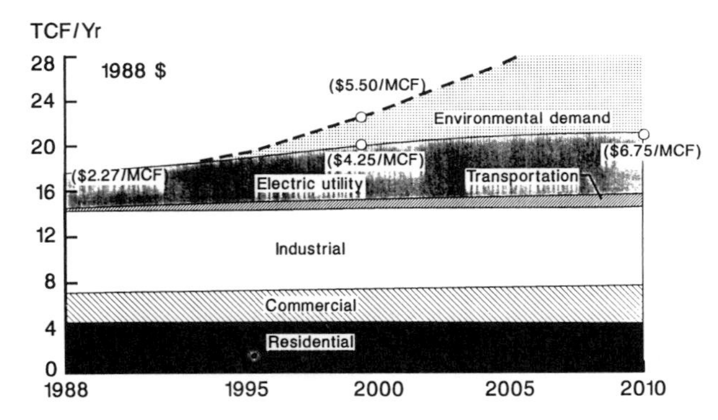

Figure 12. Natural gas consumption. (Source: GRI, Campbell & Martin)

of natural gas. This environmental demand factor is shown both in terms of the probable response to pending Clean Air Act modifications and the much larger requirements under consideration to control CO_2 emissions growth.

The result is that electricity generation requirements for natural gas are expected at least to double by the year 2000.[5,11,12] Only a fraction, however, will be supplied on a firm, noninterruptible basis since the gas industry relies on shedding load for peak demand control rather than on building reserve capacity as in the electric utility industry. If the gas industry can meet the total need by discovering sufficient gas or increasing productivity at the wellhead, the regional differences are still substantial enough to affect deliverability. Some areas are already experiencing pipeline capacity constraints. Although studies indicate that the natural gas resource base is sufficient to meet expected demand during at least the coming decade, a precarious supply/demand balance is forecast and nearly all uncertainties deteriorate this balance.[18] Under these conditions, price is expected to increase to ensure reasonable deliverability and to achieve the $4 to $5 per thousand cubic feet (MCF) break-even point with coal-derived synthetic gas during this decade. This is anticipated even with modest consideration of the expected emissions control demand component.

Environmental Control Requirements

Impact on Power Plant Design

The changes in technology for power production, particularly from coal, are driven primarily by the need to reduce the cost of complying with ever-increasing environmental requirements. As a result, the environmental performance criterion in power plant design is being raised to a level equivalent to the classic engineering criteria of thermal efficiency and reliability, and is being integrated into power generation technology accordingly. This trend, reflected in the "Clean Coal Technology Initiative" made possible through the technological leadership efforts of the electric utility industry, is bringing a variety of innovative technology options for coal-based power generation into serious commercial consideration after decades of relative technological stagnation.[19]

The new options are intended to provide opportunities to resolve the historical conflicts between coal use and the environment. These conflicts are reflected in the fact that environmental protection costs have typically risen to more than 35 percent of the investment for a conventionally designed and constructed, coal-fired power plant (Figure 13).

As a result of innovative technology options, the electric power industry stands at a threshold of change in its technological base for coal-fired power generation that can better meet the environmental performance criterion. The US has pursued two primary objectives, viz.,

- To provide an array of retrofittable emissions control options that, at reduced cost, fill the gap between present physical coal cleaning capabilities and flue gas scrubbing for existing power plants.
- To provide improved coal utilization alternatives for both repowering and new plant application. Of particular significance are fluidized bed combustion and coal gasification–combined cycles, which combine high levels of emissions and effluent control with reduced cost and improved efficiency.

The latter objective, in particular, provides the nation with the opportunity to stabilize, or even reduce, the levelized cost of electric power. Guiding this innovative clean coal technology development is the principle that sustained environmental improvement can only be achieved when emissions and cost reduction are achieved concurrently, rather than pitted against each other. The results provide a unique opportunity to satisfy two national goals at once: a) to continue the substantial progress made in emissions control and b) to meet the rapidly approaching electricity generating capacity challenge with reduced cost and increased productivity.

The importance of this technological change can be further appreciated by examining historical trends in coal plant design, viz., thermal efficiency and capacity. Figure 14 shows the rise in thermal efficiency of coal-fired steam generation from 5 percent in Edison's first plants in the 1880s to nearly 40 percent by the late 1960s. This resulted in an 85 percent reduction in fuel consumption per kilowatt hour (kWh) of power produced. During the same period boiler size increased from 50 kWe to 1200 MWe. Consequently, the cost of new generating capacity dropped from $350 per kWe in 1920, for example, to $130 per kWe in 1967 (constant 1967 dollars), and average residential service cost dropped from $0.25 per kWh to about $0.02 per kWh.

By the 1970s, however, the diminishing performance returns from Rankine Cycle power plants were joined by new societal and institutional demands on power production, particularly environmental control requirements, with their impact on efficiency and reliability. The result was an abrupt end to the historical trend of declining real electricity cost and a corresponding increase in regulatory attention. This institutional impact also extended siting, licensing, and construction time and cost. This constraint has effectively overwhelmed the economy of scale laws that had previously guided power generation. The result is that the economic advances in coal-based power generation achieved through technology up to 1967 have, in effect, been reversed during the past 23 years.

It is precisely this challenge that new technology for coal-based power production must meet in the Second Electrical Century: controlling the upward trend in real

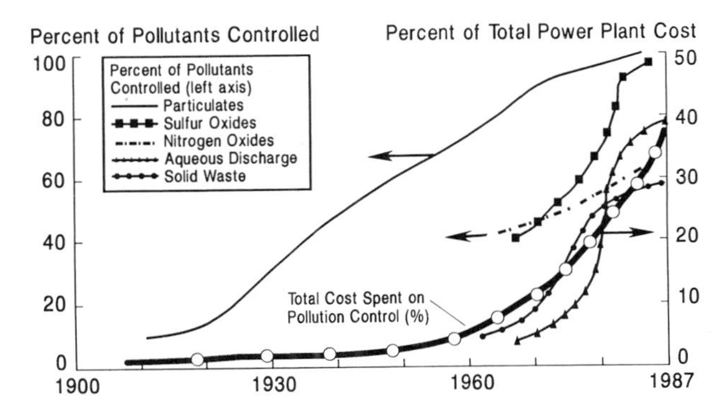

Figure 13. Percentage of power plant cost spent on pollution control as percentage of pollutants controlled increases. (Source: "Coal-fired Power Plants of the Future," *Scientific American*, Sept. 1987)

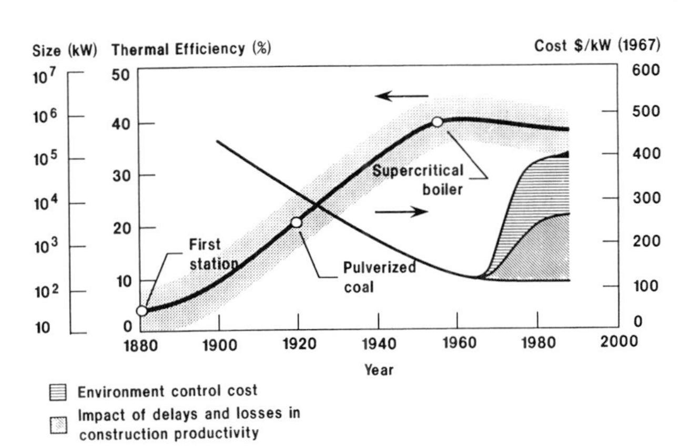

Figure 14. Coal-fired power plant evolution.

electricity generation cost by more effectively addressing the new demands on power production. This requires, first, that auxiliary emissions and effluent controls be eliminated wherever possible, either by avoiding formation of the pollutants or through more efficient process integration with power production. Second, plants must be designed and built in a manner that minimizes construction time and best matches plant size with electric power demand growth rates under a wide range of local conditions.

A major result of this new attention to the economies of time and efficiency is recognition that coal-based power plants are rapidly evolving away from the boiler-turbine-generator configuration toward integrated resource processing facilities, whether the prime technology is pulverized coal, fluidized combustion, or coal gasification–combined cycle. The electric utility industry and the Electric Power Research Institute (EPRI) have invested over $2 billion to date on this new Clean Coal Technology base and will spend more than $3 billion additionally in what is now a national joint venture, with federal and state governments sharing the cost to accelerate confident commercial deployment.[19] By the end of the decade, these technological advances addressing the economies of time and efficiency should be mature enough to support the return to an economy of scale basis for the base load power generation additions required by then. Impending environmental legislation will determine the pacing of these changes.

Figure 15 highlights the international nature of the Clean Coal Technology Initiative in terms of the nearly 100 demonstrations under way both in the US and overseas.[20] This international attention is driven by a growing worldwide clean coal technology market that will be valued at about $80 billion per year by 2000. This market is expected to remain relatively stable in value well into the 21st century. Table 1 summarizes the basis for this market resulting from the expected global expansion in coal utilization for electric power production.

Total coal-fired generating capacity worldwide is expected nearly to double over the next 30 years. Essentially, all this new capacity will be a candidate for innovative clean coal power generation technology. The US is not likely to play a main role in this market, however, unless our short-term, risk-averse national environmental, technology, trade, and investment policies are changed to encourage innovative leadership nationally and globally. The impending Clean Air Act amendments are a highly visible example of this counterproductive policy pattern.

Clean Air Act Amendments

With respect to the impending Clean Air Act amendments, Clean Coal Technology provides additional technical options that can reduce the compliance costs for power production.[21] This economic and environmental advantage led to the initial political consideration of compliance schedules under the Clean Air Act necessary to encourage Clean Coal Technology deployment. Near-term cost advantages depend, however, on freedom of choice and incentives for using innovative technology that reduces pollutant formation but has higher risk for the initial users.

Regrettably, this principle is overlooked by legislative proposals that deal with emissions reductions as absolute issues independent of their cost or the benefits to be achieved. This counterproductive tunnel vision on environmental issues is a painfully graphic example of the observation made by Henry Adams over a century ago: "Practical politics consists in ignoring facts." The result is a continuing focus on after-the-fact pollution cleanup through regulatory disincentives, rather than incentives to utilize innovative, new and repowered plant technology that reduces pollutant formation while improving productivity.

In spite of the lack of scientific justification for a crisis response, the proposed Clean Air legislation for controlling acid rain precursors demands the immediate, massive application of emissions control technology on existing power plants (the retrofit strategy). This is in contrast to encouraging accelerated deployment of more productive and pollutant preventive technology for new and repowered plants (the replacement strategy).[22] The replacement strategy assumes reinforcement of the existing Clean Air Act by requiring 95 percent SO_2 control on new power plants and 90 percent SO_2 control on repowered plants. Existing plants, unless repowered, would not require additional controls. The objective of these assumptions is to provide a distinct contrast in the rate of SO_2 reduction between the two strategies so that any resulting differences in SO_2 effects can be clearly identified. No attempt has been made, however, to optimize efficiency or holistic environmental benefit through technology.

Comparison between the retrofit and replacement strategies indicates that both will have essentially the same impact on SO_2 effects. However, the retrofit strategy will result in both higher cost and greater environmental loading of SO_2 by-products, i.e., in solid wastes and carbon dioxide (CO_2).[23] This conclusion results from the most current acid precipitation scientific knowledge established during the past decade of research sponsored by National Acid Precipitation Assessment Program (NAPAP), the Environmental Protection Agency (EPA), DOE, and Electric Power Research Institute (EPRI).[23] The results of this comparison are summarized in Table 2.

Emissions and other projections for each strategy are based on a range of uniform and reasonable assumptions with regard to electricity demand growth rates, capacity reserve margins, plant retirement ages, fuel use patterns, and rates of deployment for advanced

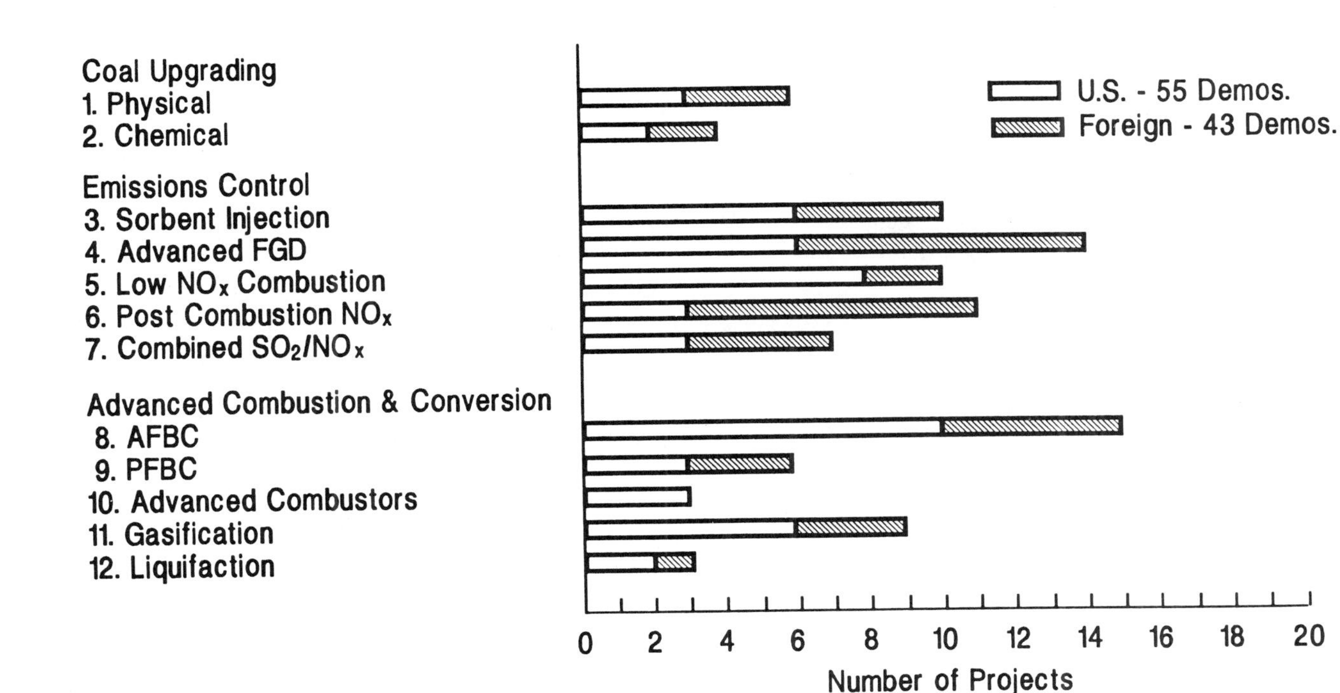

Figure 15. Clean coal demonstrations.

Table 1. Coal-fired generating capacity growth.

	1990 Installed Coal Capacity (GW)	Expected Coal Growth 1990–2000		Expected Coal Growth 2000–2020	
		%/Yr	New GW	%/Yr	New GW
N. America	340	1.0	30	2.0	180
W. Europe, Japan, Australia	220	2.0	40	1.5	100
E. Europe, USSR	250	1.1	30	1.5	100
China, India, Africa	200	1.3	30	2.0	110
Pacific Rim, S. America	60	7.5	60	4.7	190
Total	1040		190		680

technologies.[22] As indicated in Figure 16, both strategies result in a nominal 10 million tons per year (TPY) reduction in SO_2. The retrofit strategy is projected, however, to be more costly than the replacement strategy under any set of assumptions. This additional cost is likely to be in the range of $200 billion to $400 billion over 60 years, even under the very conservative technological assumptions used. By comparison, the relative environmental benefits resulting from the retrofit strategy are negligible, whether measured in terms of lake impacts, forest impacts, crop impacts, or materials damage. Only in the case of visibility is a temporary advantage expected prior to 2025.

The more immediate impact is indicated by Figure 17. The proposed retrofit strategy legislation, if implemented as scheduled, will be likely to reduce the generating capacity reserve margin in the eastern portion of the US alone from a marginally adequate 17 percent to a completely inadequate 10 percent during the latter half of this decade, when the actual power plant retrofits are occurring. Under these conditions, it appears that either the implementation schedule must slip or an additional 30,000MW_e of gas turbine–combined cycle capacity beyond that now contemplated must be installed, for example, in order to avoid an unprecedented impact on electric service reliability and cost.

Also of technological and economic significance are 1) the concept of a cap on SO_2 emissions, to assure that emissions do not rise again after the new emissions restrictions have taken effect, and 2) an emissions credit

Table 2. Tradeoffs between SO_2 control alternatives (retrofit vs. replacement).

Factor	Probable Impact
SO_2 Emission Rate Reduction	Same—10 MTPY relative to 1980
Date at Which Emission Rate Reduction Is Achieved	Retrofit—2000 Replacement—2025
Electricity Cost	Retrofit—\$200–\$400 Billion higher
Cumulative SO_2 Emitted	Retrofit—7 to 11% less (45–90 million tons)
Visibility Impact	Retrofit—marginally better before 2025—no difference after
Other SO_2 Effects Impact	No discernable difference
By-product Wastes	Retrofit—11 to 17% more (2.5–4.2 Billion Tons)
CO_2 Production	Retrofit—3 to 4 % more (10–14 Billion Tons)
Coal Consumption	Retrofit—3 to 4% more (4–5.5 Billion Tons)

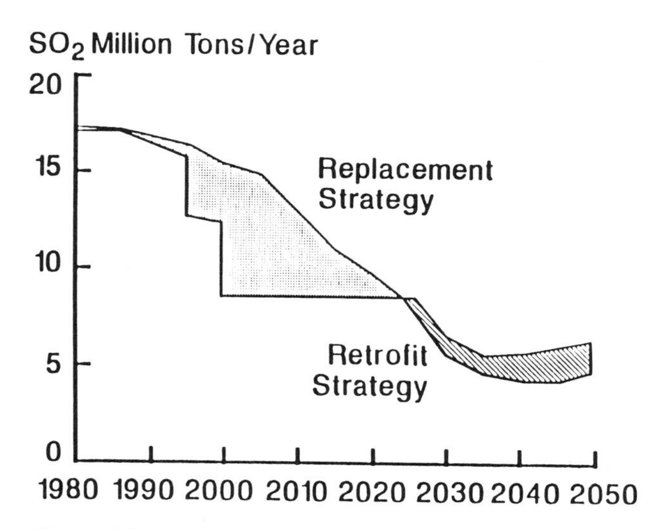

Figure 16. Utility SO_2 emissions—Contiguous US.

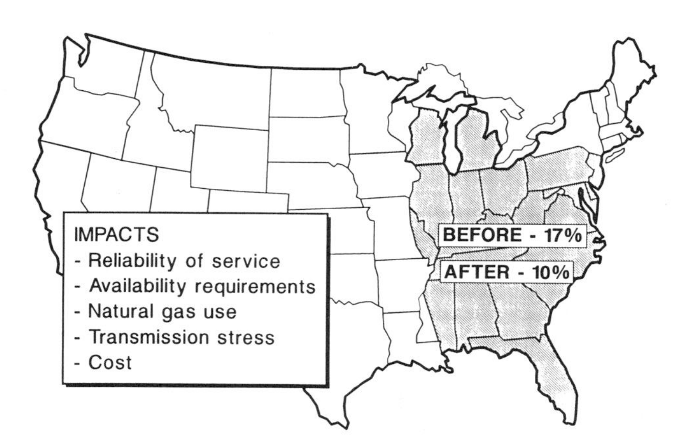

Figure 17. Impact of acid rain legislation on NERC reserve margin forecast (1998).

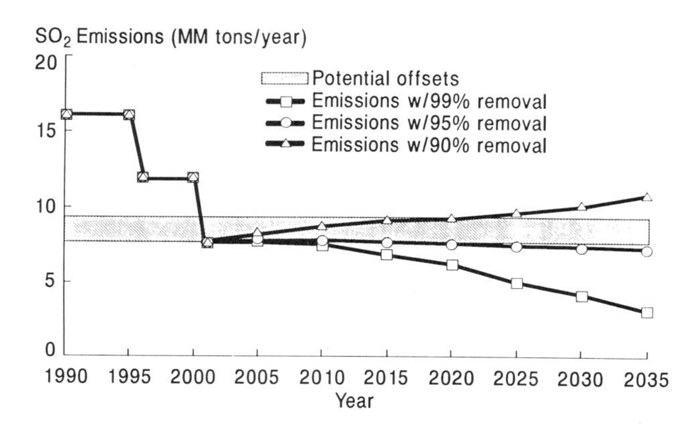

Figure 18. National capacity expansion—SO_2 control implications (2.4% growth, with retirements).

"market" to encourage emissions reductions exceeding those required for compliance by individual, controlled power plants. The first means that all new capacity built must provide an emissions offset equal to its net emissions rate. The second, although politically dubious, is intended to motivate existing plants to control more intensively where it is either cost-effective to market the resulting credits or to bank them as offsets for future capacity growth. If implemented, these requirements will result in ratcheted levels of future emissions control, exceeding the current New Source Performance Standards (NSPS) as shown in Figure 18. As currently constructed, these requirements will be based on locally capped emissions rates. This, in effect, means that the lower the emissions rate of a plant at the time of the cap, the lower its emissions must be maintained independent of environmental impact, and the earlier the plant is likely to become environmentally obsolete.

The choice in strategies reflects the need to look at both the short- and longer-term future in a manner that optimizes all benefits. In the transition to a new century, we have a major opportunity to incorporate our growing knowledge of environmental science and technology into a more efficient framework for using energy while conserving resources and protecting the prospects of future generations. This analysis encourages national policies that embrace a systematic, holistic approach if we are to achieve a sustainable framework for global development rather than diminish the prospects of future generations. Our choice is clear: either continue to focus on "retro" cleanup that only moves the environmental impact around and reduces national productivity, or focus on innovative power production technology that reduces the formation of pollutants while improving national productivity.

Figure 19 indicates another dimension of the fragmented approach to environmental policy development under the Clean Air Act that may ultimately have even greater impact on power production costs. Here the issue again results from the lack of a risk assessment framework for evaluating the relative costs and benefits of control strategies. For example, current legislative proposals for controlling air toxics, in effect, propose to eliminate *all* risk at *any* cost. Not only are such absolutes unattainable in a literal sense, but they also divert immense capital resources that could better serve the environment by investment in more productive technology that eliminates the formation and release of these species. The challenge for public policy is to avoid pressures for simple solutions and, instead, to craft informed and balanced decisions that serve the public interest. This interest seems better served by collaborative information sharing and risk assessments that determine shared values than by the aggregation of selfish interests through today's adversarial processes.

Efficiency as a Strategic Requirement

Historically, emissions control has evolved from local issues typically perceived as and characterized by smoke, to regional issues popularly perceived as and characterized by acid rain, to global issues perceived as and characterized today by potential climate change. Local issues could be effectively resolved by pollutant cleanup technology. As the issues expand geographically, however, the cost-effectiveness of after-the-fact cleanup becomes increasingly unsupportable. The previous discussion on acid rain, for example, underscored the much greater value of innovative power generation technology that reduces pollutant formation.

As the focus of public perception moves to the global stage, harnessing efficiency improvements to prevent pollution becomes the only practical means of addressing interactive, global, core resource issues. This process becomes increasingly urgent under the pressures of economic development, urbanization, and population growth. Economic development has proven historically to be the most effective, nonapocalyptic means of population control.

The effects of continuous growth in global energy use on fuel resources, energy economics, geopolitics, and, most recently, the environment have created international concern. In particular, the prediction of a greenhouse effect on climate has stimulated intensive discussion of the various means of mitigating such potential climate change.[4] Figure 20 summarizes the predominant role of energy in the expected contribution by sector to the global greenhouse gas inventory.

In the greenhouse question, the economic development in China, for example, could have a greater effect on atmospheric accumulation of CO_2 than that of any other nation and could far outweigh any control measures taken by the US. China's critical role stems from its large and growing population, poor energy

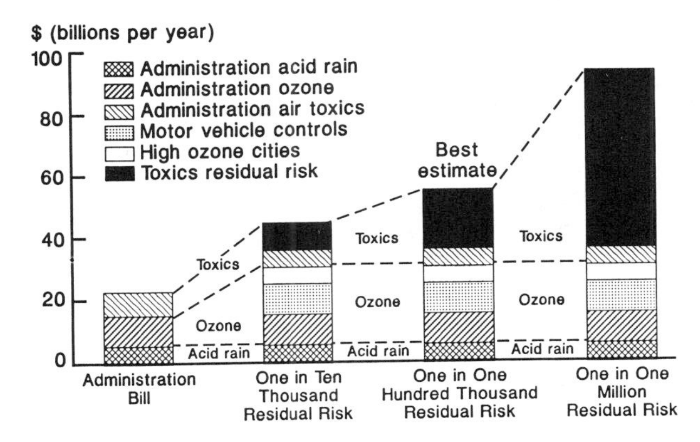

Figure 19. Costs: incremental to the Administration bill as introduced. (Source: The Business Roundtable, January 1990, Washington, DC)

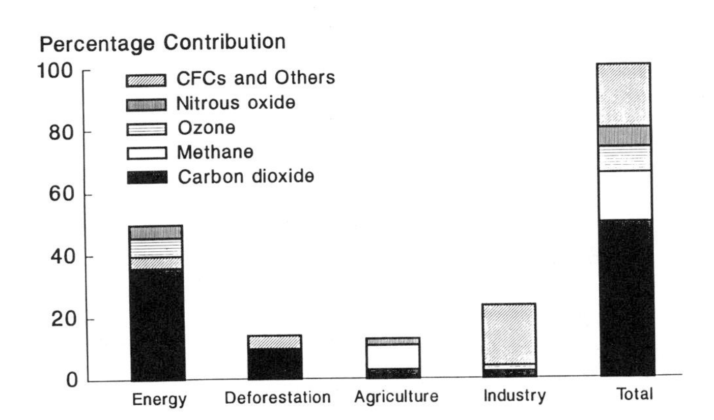

Figure 20. Greenhouse gases, 1980–2030 (projected). (Source: UNEP/GEMS, 1989)

efficiency, and massive reliance on coal. The question again is not *if* coal will be used in the developing world but *how* it will be used.[24]

Developing countries face hard choices because of the close relationship between levels of economic development and energy use, as indicated in Figure 21. This linkage is likely to increase the political pressure for greater energy efficiency in both the developing and developed world.[25] It is interesting to note that Western Europe and Japan, both at economic levels comparable to North America and well beyond Eastern Europe and the USSR, prosper on about one half of US per capita energy intensity. This is only partially explainable by geographic scale differences. At the other extreme, countries at the bottom of the global economic ladder subsist on less than 10 percent of US per capita energy consumption. Thus, while efforts to improve overall energy efficiency in the US are justifiable on the basis of economic competitiveness, their impact on energy availability and the economies of the developing world will be negligible. It is also likely that most US efficiency improvement will be reflected in reduced petroleum consumption. Petroleum, however, is not an affordable energy resource for economic development by nations at the bottom of the economic ladder.

Figure 22 indicates that an even more consistent and dramatic relationship exists between economic development and electrification. Here, Japan, Western Europe, and North America are all contained in the highest block, in terms of generating capacity per capita. On the other hand, the subsistence countries in the lowest block, already representing over half the global population and growing rapidly, exist on about 1 percent as much electricity generation per capita. Comparison of Figures 21 and 22 suggests that the ratio of electrification to total energy consumption is a more confident predictor of relative energy efficiency and prosperity, and should be the focus of global economic strategies intended to conserve resources.

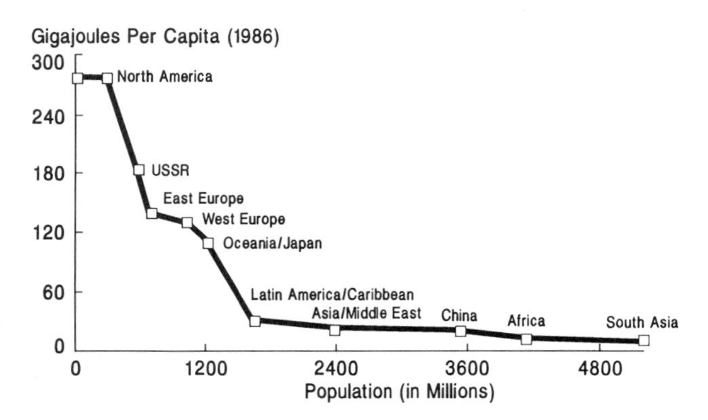

Figure 21. Per capita energy consumption by region. (Source: World Resources, 1989; RFF Analysis)

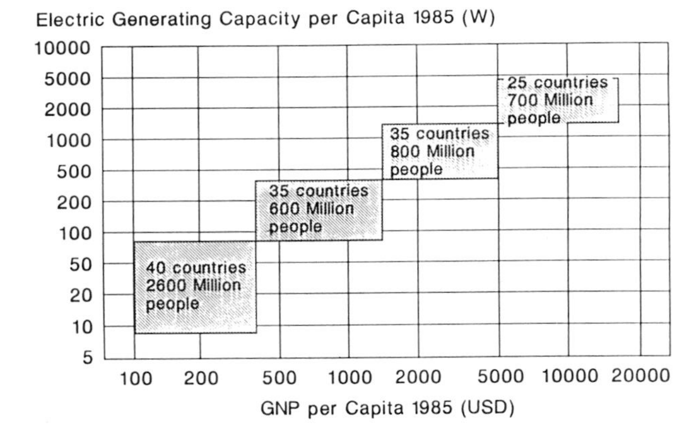

Figure 22. Prosperity and electric energy. (Source: ASEA Brown Boveri)

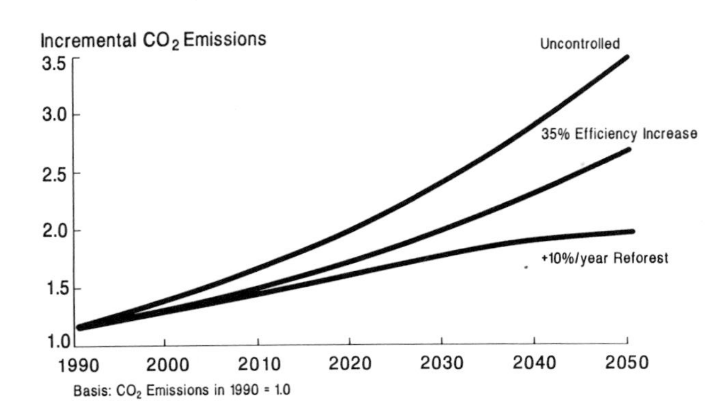

Figure 23. Incremental CO_2 emissions—illustrative situation.

Fortunately, innovative technology can substantially reduce the amount of energy required to provide a given level of goods and services while simultaneously reducing environmental impact. On the supply side, chemical feed stocks and a refining approach to coal utilization for electricity transport fuel would be important elements of such a technological strategy. This would process the total value of a nation's indigenous coal as well as its urban and biological waste resources into an array of products, including electricity, and thus provide the foundation for both continued economic development and environmental protection.

Figure 23 illustrates, as a preliminary example, the ability to control CO_2 emissions globally over the next 60 years based on a realistic assessment of technological opportunities. This suggests that global emissions could be stabilized at about twice the current rate by mid-century. Implicit in this illustration is recognition that the world will remain dependent on a carbon-based energy cycle throughout this period. As a result, the two primary, and complementary, paths to CO_2 control in this time frame involve improving the efficiency of fossil-based energy production and use, and reforestation to provide a CO_2 storage capability. Two basic questions are 1) how much incremental capital investment would be required? and 2) can this investment be compensated by energy cost savings?

Achieving a global 35 percent energy efficiency improvement by 2050 is likely to require on the order of $25 trillion (1990 dollars). This expected annual global investment increases from $90 billion in 2000 to $1.25 trillion in 2050 (all 1990 dollars). On the other hand, the corresponding total undiscounted energy cost savings from that efficiency improvement would amount to some $50 trillion (1990 dollars) over that same time period. Figure 24 indicates that at a 6 percent real annual discount rate these estimates result in a probable payback period of 5 to 20 years after the initial investment.[26] Stated another way, these estimates indicate a broad range in rates of return depending on the specific circumstance. Some efficiency improvements are extremely cost-effective (particularly in certain end-use applications) and already are being financed privately all over the world. Other improvements necessary to the global efficiency objective are at the low end of the profitability scale and will require a change in public policy to encourage private investment with longer-term paybacks and in less developed regions where economic risks are greater.

This analysis also indicates that the efficiency of capital investment, defined as the ratio between energy cost savings and capital investment requirements, is higher for energy efficiency improvements in both electric supply and end-uses than for nonelectric energy consumption. Furthermore, the less developed countries are likely to have higher capital investment efficiencies than the developed countries. For example, in the less developed countries, the efficiency of capital use is forecast to be greater than 7 to 1 for the conversion of

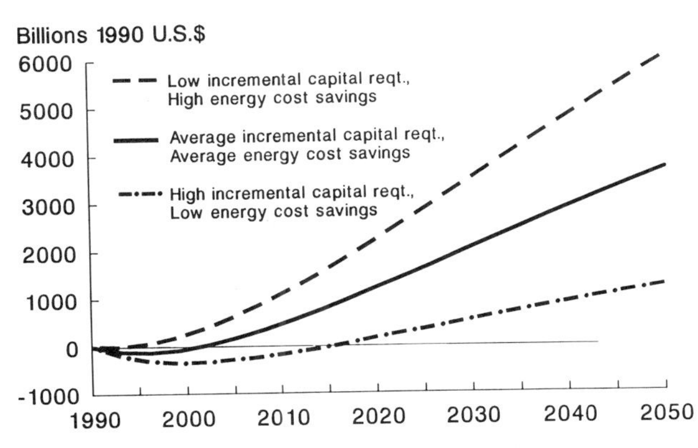

Figure 24. Cumulative discounted net cash flow, 1990–2050.

certain industrial processes to electric energy, and in excess of 4 to 1 for improving the efficiency of electric power production.[26] This result is reasonable since capital shortages have caused less developed countries to lag behind developed countries in adopting capital-intensive technologies, even when these technologies are more energy- and cost-effective. As their growth accelerates, this capital availability issue will be the prime determinant of global economic and environmental progress.

The complementary path, carrying a reforestation program capable of absorbing an incremental 10 percent of the new CO_2 generated each decade through to the year 2050, would require a total land area for reforestation of less than 4 billion acres, representing approximately 10 percent of the earth's surface. The total investment over the intervening 60-year period, including land procurement, is estimated to be some $200 to $400 billion (1990 dollars) depending on the regions involved.[27] Since only growing plants absorb CO_2, this biomass "proto-coal" resource would be harvested and replaced at maturity, increasing to provide some 10 trillion kWh per year by 2050—about the same as the world's electricity demand today.

This preliminary analysis indicates that global efficiency improvement in both energy supply and end use can be financially viable as well as essential to the conservation of global economic and environmental resources. While the financial sums involved are staggering at first glance, they reflect a long-term, manageable investment more fundamental to global quality of life than merely to deterring a possible greenhouse effect. The investment is necessary both to internalize the environmental costs of energy on a global scale and to move toward a managed energy supply without limits.

It is also useful to put this investment in perspective, for example, with the escalating cost of imported oil, which now exceeds $60 billion per year in foreign trade deficit and supply line defense cost to the US alone. Even more dramatic is the realization that the global commitment to military defense already exceeds $1 trillion per year, or about 6 percent of the global GNP. By comparison, the necessary annual global financial commitment to energy efficiency improvement in the Second Electrical Century would be about 1 percent of expected global GNP, and with a far greater assurance of investment return for most nations.

The harder issue implicit in this illustration, however, will be developing the global cooperation and sustained political will to ensure the necessary business incentives for investment. This is particularly true in the developing world, where the greatest efficiency gains can be made. In the final analysis, such policies and investments will only be justified if the time gained by the effort is used to expedite technologically a post-carbon global energy economy by the second half of the 21st century. This is the minimum time frame likely to be consistent with such an unprecedented technological transformation, and it underscores the urgency of arriving at the necessary global consensus.

In spite of these global realities and today's limited scientific understanding of the climate modification issue, the political process is rushing to fill the policy vacuum by supporting special interests that tend to ignore the long-term international character of the issue in favor of unilateral national actions giving the impression of progress. For example, these fragmented policy proposals have suggested the following steps to maintain zero US CO_2 emissions growth indefinitely: a $50 per ton carbon tax on coal; producing 70 percent of electricity from nuclear power by 2010; a 2.5 percent compounded annual energy efficiency improvement; a 50 percent compounded increase in natural gas use each decade; or flue gas scrubbing of CO_2 followed by sequestering in the ocean. While each of these proposals may have some political value, their unrealistic quantification and lack of global context renders them ineffective, if not counterproductive.

In considering such policy suggestions, it is important to note the declining percentage share of the US and other developed countries in human-made global CO_2 mitigation strategy[28,29] (see Figure 25). Also note the rapidly growing role of the developing countries and their resulting importance in any strategic CO_2 mitigation strategy. This parallels the energy supply/demand patterns shown earlier in Figure 6.

Thus, these fragmented proposals, even if implemented, would only delay the date of a doubling of the global atmospheric CO_2 concentration from pre-industrial levels by a fraction of a decade. Paradoxically, such unilateral restrictions could actually increase global CO_2 emissions by driving up the domestic price of energy and forcing more industry to move offshore, where the efficiency of energy use is generally lower.

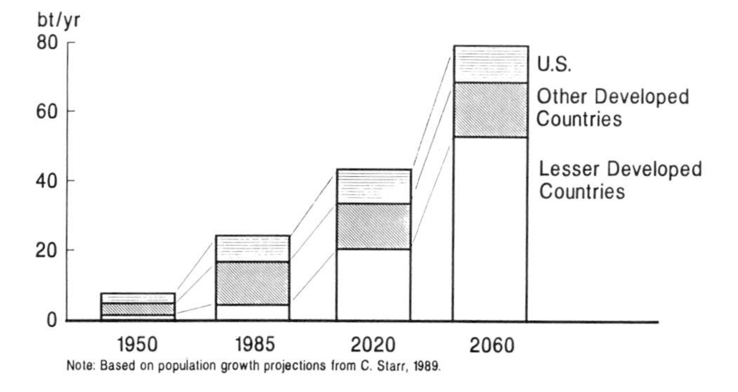

Figure 25. Contribution to global human-made CO_2 emissions.

Globalization of Equipment Supply

Standardization

The regrouping of nationally oriented boiler, turbine/generator, and emissions control suppliers into several integrated multinational corporations seeking competitive advantage in a global economy is forcing electric utilities and the supporting engineering/construction community to rethink the whole process of designing, specifying, purchasing, and constructing power plants. The economic feasibility of individually customized power plants is being superseded by ever-larger and more standardized shop fabricated modules, including prefabricated power plants, that take advantage of internationally competitive supply opportunities. This trend is likely to be reinforced by the classic global marketing strategy of achieving economies of manufacturing scale and time through globally standardized "world-class" equipment designs.[30] There is concern that this may not result in responsive market differentiation, but instead encourage a lower standard of technology and performance than designs targeted at a US level of technical and economic development. US electric utilities may be particularly vulnerable because even the largest are relatively small entities in a global scale market, and their regulated incentives today encourage purchasing the lowest first cost equipment rather than selecting on the basis of life cycle costs.

Thus, collaboration should be encouraged among the US utilities to establish common plant design and performance criteria, in this way helping to maintain their technical advantage in productivity, and more effectively leveraging the global equipment supply market. Electric utilities should also emphasize their inherent economy of scale and quality advantages in what will again be an expanding generation capacity market. On the more positive side, standardization should encourage a reduction in construction time that will overcome the investment disincentives that delay new capacity commitments.

This emphasis on reduced construction time will be critical in a market driven by deferred commitments and increasingly controlled by competitive bidding for construction of new power generation facilities. Failure to compete effectively in this expanding market for new capacity, on the other hand, will ultimately cause utility assets to shrink and, in effect, lead to their liquidation over time. Remaining competitive under these circumstances may, however, require new alliances for the construction of generation equipment among utilities, and between utilities and others. We see the beginning of that trend today in the formation of creative industrial joint ventures in which utilities share risk and cost.

If the utility industry collaboratively maintains control of equipment design and performance criteria, the results should also include improved quality control and advantageous integration of the innovative coal technologies now entering the commercial arena. Under these circumstances, the competitive, value-added opportunities for the equipment supplier and constructor/engineer will result less from the bulk steel or manhours involved and more from exploiting accelerated power generation/environmental control process design and fabrication, new material selection, and the electronic "smart parts" that increasingly differentiate construction costs as well as the economics of plant operability and maintainability. We see that trend today, for example, in the introduction of pressurized fluidized bed combustion (PFBC) as a potentially more efficient approach to new and repowered capacity, under competitive conditions, than either atmospheric FBC or conventional pulverized coal plants.

The Role of the United States

A major concern during this period of technological change is the ability of the US to sustain domestic leadership in power generation innovation when the nation's laws and policies have not yet caught up with the realities of global competition among relative economic equals. Figure 26 displays the recent comparative history of US productivity growth and capital formation compared to other economically developed nations. In part as a result of these policies, US corporations have lacked the long-term investment and trade financing incentives necessary to remain power equipment innovators in a global market. Rather, they have increasingly focused on short-term profit production under the pressure of achieving maximum, immediate shareholder return in what has been in recent years a shrinking domestic market.

The 20-year depressed domestic power plant equipment market, shown previously in Figure 9, combined with the lack of national investment and trade policy incentives necessary to compete globally, has resulted in the depression in US supplied power plant sales summarized in Figure 27.[31] Since 1970, US heavy power plant equipment industry sales have diminished by 70 percent. This spectacular decline in market share by US producers exceeds even the fall in the domestic market and includes the loss of the few export markets (mainly Pacific Rim) once held by the US. By contrast, the European response to its more recent decline in market share has been to consolidate and establish an aggressive global strategy through multinational corporations. The Rest of the World category is dominated by Japan, which has made remarkable inroads into world markets, mirroring in reverse the US decline. In the cases of both Europe and Japan, national trade and tax policies have encouraged their more strategic commercial approach.

This comparison demonstrates that the sum of short-term optimized corporate financial decisions does not assure long-term corporate or national economic prosperity, i.e., does not assure a steady stream of innovation leading to sustained national competitive advantage and import replacement. As a result, the US faces a major competitive challenge. The success of foreign products in our market, the relative inability of US companies to compete overseas, the decline of major industries, and the trade deficit indicate the magnitude of this challenge. We are already experiencing changes in the power equipment industry, where corporate headquarters and their innovative nerve centers are migrating globally, just as production has already done, to achieve maximum corporate advantage and minimum risk. Innovation today must also compete on a global scale for corporate investment.

In many ways, the discussion of short-term versus long-term is a discussion about the increasing tension between the US financial and manufacturing sectors. Financial markets, faced with a broad array of international investment opportunities, are increasingly unwill-

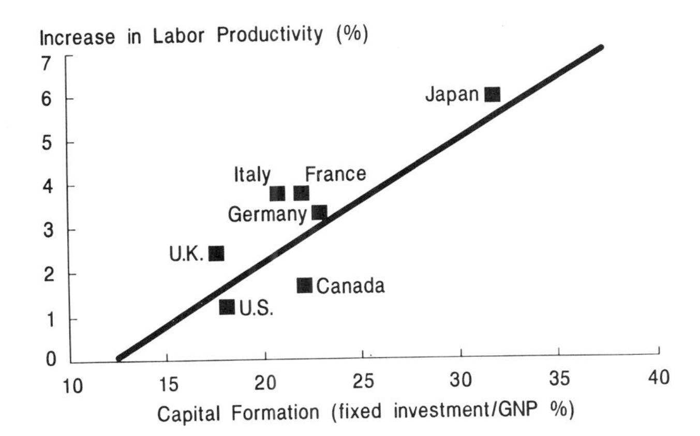

Figure 26. Productivity growth and capital formation international comparison, 1960–1983.

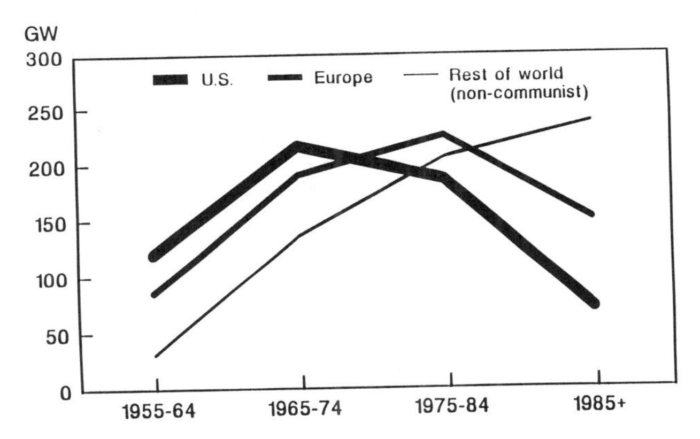

Figure 27. Sales of power plant by source.

ing to make long-term investments in US manufacturing when they can earn a higher return in a shorter time period by investing in other areas. In this regard, there seems to be a fundamental lag between the rate of technological and commercial change in a global economy and the rate of adjustment to these changes by US policy makers who remain preoccupied with *intra*national rather than *inter*national trade issues. The resulting policies, such as the 1986 Tax Reform Act, fail to recognize the establishment of a highly efficient world market to match lender and borrower and to pool resources and share risk on an international scale without regard to national boundaries. Although the structure of the competitive market is changing internationally, the basic rules that determine who wins and who loses still apply. Risk elimination and innovation remain incompatible.

The result is that US leadership in power generation technology, including environmental controls, is being eclipsed by both Europe and Japan. There the strategic value of energy innovation coupled with expedited production techniques has been nurtured through strategic investment and trade financing policies that are attracting the corporate nerve centers of the rapidly emerging multinational power equipment supply industry. Thus, short-term revitalization of the US conventional coal combustion/environmental control technology foundation for power production has become dependent on foreign technology and investment.

A New Arsenal of Energy Innovation

The US is best served by focusing on innovative technological opportunities to regain international industrial leadership and competitive advantage. Success will depend on transforming technological advancements—wherever they may be discovered around the globe—into high-quality US products, coupled with more strategic trade and investment policies that level the international market playing field, particularly in the developing world.

US international competitiveness is central to our national strategy and influence in the decades ahead. In the future, foreign policy will have to give more serious consideration to strategic industrial issues, such as the strength of the US manufacturing base and the leverage of key technologies across industries. This shift in focus will require a reassessment of both the priorities that govern foreign policy and the government institutions that are involved in the policy making process. National security can no longer be viewed exclusively in military terms; economic security must be taken into account. This will necessarily have an impact on the management of international relations, since important political allies are often hard-nosed economic competitors in commercial technology.[32]

At one time US manufacturing was viewed as synonymous with moribund industries that had more to do with America's industrial past than its high-tech future. It is now generally recognized that a strong manufacturing base is essential to leading-edge industries as well as to mature industries, and that technological and commercial "food chains" connect so-called low-tech and high-tech industries. Moreover, mastery of the manufacturing process is increasingly viewed as an essential part of the technical competence that is necessary to advance existing technologies and create new ones.

Fortunately, the US still remains a technological and commercial leader for the higher value-added gas turbine, coal conversion, and silicon microprocessor technologies, all essential to a least cost, refining approach to power production and a coal-based national energy economy. These technologies provide incentives for US corporate innovative and commercial leadership because their advantages in the global marketplace have, so far, been sufficient to offset the deficiencies of US trade, tax, and commercial collaboration policies. For example, gas turbine and other post–Rankine cycle energy conversion technology leadership reflects US leadership in defense and aeronautics technology. Coal conversion technology results from strategic research and development (R&D) investments by the petroleum and chemical industries, which control the bulk of US coal resources and are culturally and technically familiar with a chemical processing approach to resource utilization. Finally, microprocessor technology is a powerful and rapidly expanding result of the $400 billion per year US electronic industry, where 20 percent of all US industrial research is conducted.

Paradoxically, these three relatively robust technological foundations, not dependent on the power industry, will become increasingly important to US power generation in assuring both its competitive future and that of the nation to whose prosperity its own is so closely tied. When combined, they provide the means to break the Rankine cycle efficiency barrier for coal-based electric power production with an impact similar to that of jet aircraft on the sound barrier and air travel 40 years ago. Figure 28 indicates the resulting resumption in both heat rate and emissions control efficiency progress that can be achieved after a 30-year period of relative technological stagnation. The ultimate result can be the transformation of coal-fired power plants from relatively resource inefficient, mechanical combustion units into fully integrated coal refineries. This can be the basis for the "world class" power plant capable of best addressing the interdependent economic and environmental challenges facing the Second Electrical Century. Through its combination of technology and coal resources, the US is poised to provide world leadership for this transformation. The payoff is not just commercial advantage but resource conservation and environmental protection on a global scale.

Invention will not be enough, however. Sustained

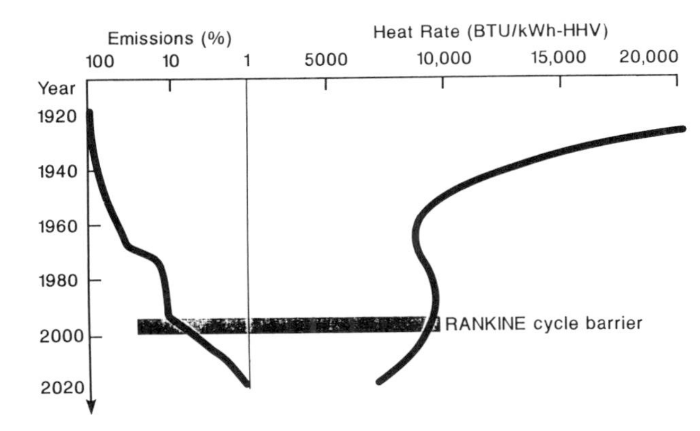

Figure 28. Progress in thermal power production.

competitiveness through innovation will also require a better linkage of government and private R&D funding to the productivity of commercial products; strengthening the manufacturing base; the establishment of, and commitment to, higher technical quality standards; investment in the technological training of workers; and a better basic education for all citizens.[33] Policies will have to be altered to encourage sustained long-term investment in improved skills and facilities, to support long-term global market development, and to promote innovative commercial leadership, increasingly dependent on collaborative R&D.

The current environment of increased international corporate flexibility in the quest for economic advantage encourages collaborative national enterprises that focus public and private resources and risk-sharing for advantage in attracting and building the corporate economic base. This strategy of establishing national "arsenals of innovation" is already being exploited successfully by other nations. Their importance to the US may be as much in building the new relationships among businesses, universities, and the government necessary to compete internationally, as in actually creating technological innovations.[32] Underpinning all of these steps is the need for a greater sense of personal responsibility for the future and a more cooperative, less adversarial approach by both the public and private sectors to marshalling our unmatched national capabilities.

Synthesis

The synthesis of these three effects—investment disincentives, environmental requirements, and globalization of equipment supply—signals a fundamental transition in the power generation business as we enter the new century. This transition is likely to be marked by the following overriding trends:

- An increasingly competitive electricity supply market where the low-cost producer is the only assured survivor. The message of the last two decades is that electricity demand *is* dependent on price, and this will remain the case for the foreseeable future.
- Increasing emphasis on coal, not only as the primary fuel for electric power production, but also as the only domestically secure hydrocarbon base for transport fuel.

The first of these trends has already become sufficiently clear to have gained a measure of political orthodoxy. The second, in an era of seemingly abundant low-cost oil and gas, unfortunately remains beyond the horizon of popular perception. As a result, the former is likely to proceed with relatively greater predictability and fewer crises than the equally assured latter trend which will, ultimately, also serve to accelerate this process of change.

Figure 29 indicates the cost and difficulty of sustaining the current projection for US petroleum supply.[34,35] Even though today the US imports more oil than it did at the time of the 1978 oil embargo, business-as-usual dependence on foreign oil continues, adding about $50 billion per year to the foreign trade deficit, plus associated defense costs. This policy, unless changed to emphasize domestic resources, is likely to result in the importation of two thirds of the US oil supply by 2000, with a $150 billion per year negative impact on the nation's foreign trade balance and thus a threat to national competitiveness and security. Under these conditions, domestic oil replacement resources based on coal-derived synthetic fuels, can phase down oil imports and ease the transition from direct internal combustion powered vehicles to electrified urban personal and public transportation.

Conversion to domestic resources is encouraged by the linkage among three trends that appear likely to reach their climax as we enter the new century: First,

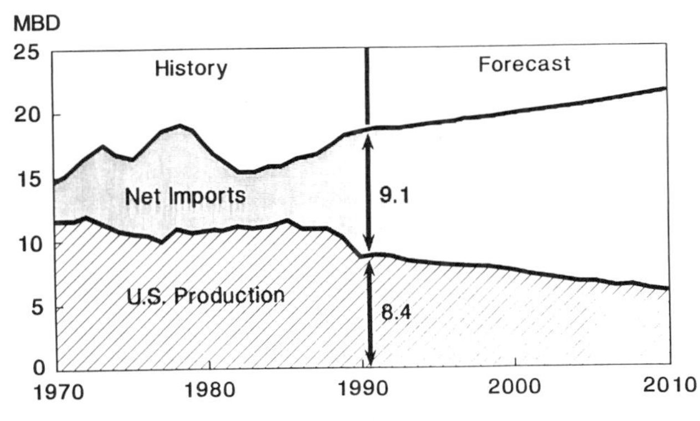

Figure 29. US oil supply and consumption. (References: EIA/Annual Energy Outlook, 1990; American Petroleum Institute, 1990)

the probability that oil prices will rise rapidly as demand exceeds proven reserves and Middle East "preferred" production capacity is exceeded, even under politically stable regional conditions; second, political pressure in the urban centers of the US, Europe, and Japan to phase out internal combustion-powered vehicles; and third, the escalating negative impact of imported oil dependence on economic competitiveness and the resulting political stresses among the US, Europe, and Japan.

The coincidence of such climactic trends, all driven by scarce resources, has historically forced economic crises triggering rapid political reaction. Fortunately, among the three major Western economic blocs, the US has the advantage in being able to phase out its imported oil dependence in favor of indigenous resources. This is likely to be an exceptionally important international business advantage while protecting political cooperation from the stress of competition for dwindling resources. As a result, US coal may prove to be as important to protecting the Western alliance during the next 50 years as the US military umbrella has been during the past 50 years.

We estimate the average annual investment necessary to reduce oil imports by these means to no more than 25 percent of national consumption within the next 20 years to be less than $25 billion per year. Regrettably, neither the private nor the public sector is encouraged to make such an investment under the current US short-range policy umbrella until oil prices actually reach the threshold of immediate profitability. This will build in at least an additional five- to ten-year delay during which energy costs, competitiveness, and security will likely deteriorate at an escalating rate.

In summary, an increasingly likely chain of events on this path to a lowest-cost, coal-refining based power generation industry includes the following:

- Continued electric efficiency improvement and conservation measures driven by a combination of regulatory, environmental, economic, and supply issues shortage.
- Saturation of the peaking power market by at least 40,000 MW_e of new simple cycle gas turbines firing natural gas.
- Increasingly stringent environmental restrictions on coal use that emphasize "refining" coal as the most environmentally and economically acceptable approach for new power generation.
- A growing shortage of base and intermediate load capacity caused by continued load growth, deferred capital commitments, and new environmental requirements that reduce productivity and encourage capacity retirement. This will also reflect the inability of performance improvements to existing capacity, plus incremental nonregulated power production, to keep pace.
- The resulting need for rapid installation of large blocks of new capacity (approximately 20,000 MW_e per year) to overcome this shortage, coupled with regulatory changes that encourage utilities and/or deregulated entities to make the necessary investments.
- Naturally or artificially induced discontinuities in the supply and cost of natural gas resulting from the large commitment of the power industry to new gas turbines, plus the coburning of gas with coal as a short-term emissions control expedient.
- Oil supply interruption and/or price increases that force a cap on oil imports, coupled with policies to encourage increased production from domestic energy sources.

The organizations best equipped to survive such a transition, in terms of staying in the power generation business, are those that come out the other end best able to maximize the value-added element of their invested capital and purchased fuel. These organizations are likely to be characterized by their successful adaptation, directly or through alliances, to a process methodology for coal utilization.[24,36] Figures of merit achievable through the application of technology now available or in advanced development are outlined in Table 3.

These criteria are most likely to be achieved through integrated processing facilities capable of translating coal and other carbonaceous material into a variety of products including, but not limited to, electricity. Thus, as conceptually shown in Figure 30, the Clean Coal Technology building blocks with coal gasification at the core can evolve coal-fired power plants from relatively resource-inefficient and environmentally constrained combustion units into fully integrated coal refineries. Power production from these facilities will be one product of a broader resource processing capability that can also competitively produce transport fuel and chemical feedstocks and recover marketable mineral credits.[37] Controlling the commercial high ground in this coal-based energy revolution will likely be one of the most important economic determinants facing the utility industry, and this nation, over the next 20 years.

The economic importance of such a coproduction approach is underscored by its potential to reduce electricity generating costs by 25 percent or more. Figure

Table 3. Criteria for future fossil power plants to be cost competitive.

Availability	90% +
Heat rate	6500 Btu/kWh – 8500 Btu/kWh
Capital cost (1990$)	$500/kW – $1000/kW
Technology choice	Determined by environmental requirements and new business opportunities

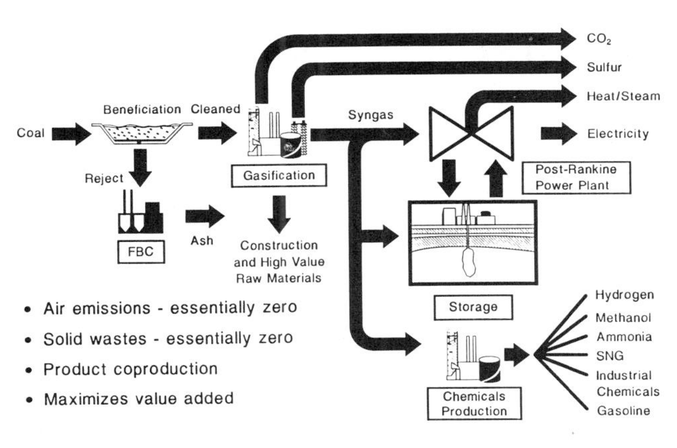

Figure 30. Integrated coal refinery concept.

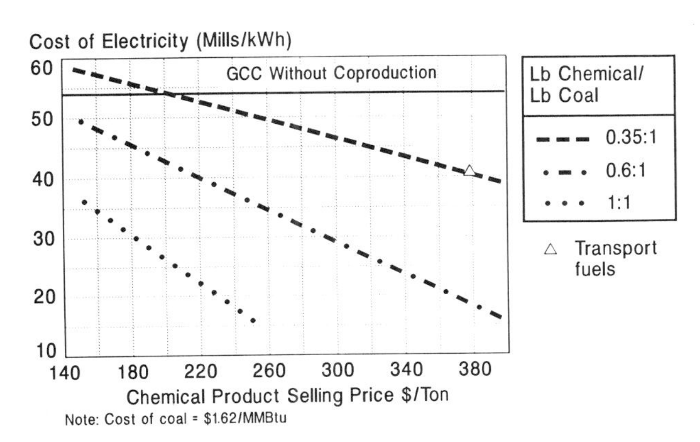

Figure 31. Cost of electricity vs. dollars per ton of chemical byproducts from a 400 MW GCC.

31 indicates this economic potential when coproducing, for example, transportation fuel from excess syngas at a 25 percent capacity factor and at an FOB price of $1.20 per gallon for the product fuel.[38] Figure 32 summarizes, from the perspective of the Shell Oil Company, the time frame in which they expect coal gasification to become competitive with conventional US sources for various refined products.[39] A band is shown for each application to reflect the uncertainty in the cost of the coal-derived product as well as the sensitivity of the end product price forecast to small changes in energy prices. Confidence in coal refining technology is reflected by the commitment of the SEP electric utility in the Netherlands to build a 2,000 ton per day coal gasification combined cycle power plant generating 250MW$_e$ of electricity for startup in 1993. Similar projects are also under consideration for the 1990s here in the US, including the Commonwealth Energy Project in Massachusetts, if public perceptions do not prevent their timely imple-

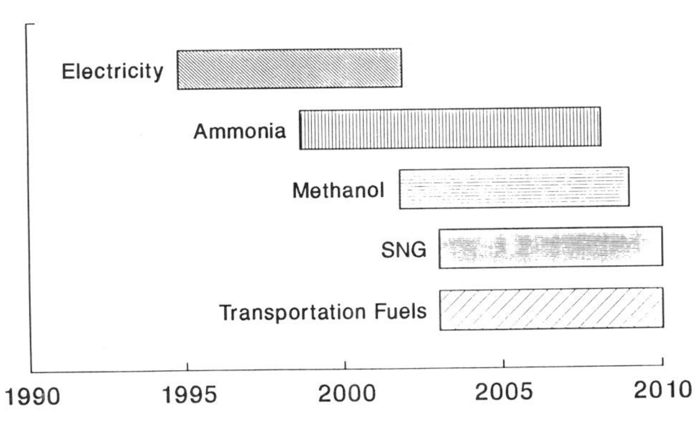

Figure 32. Timing for earliest US commercial applications of coal gasification. (Source: Shell Oil Company)

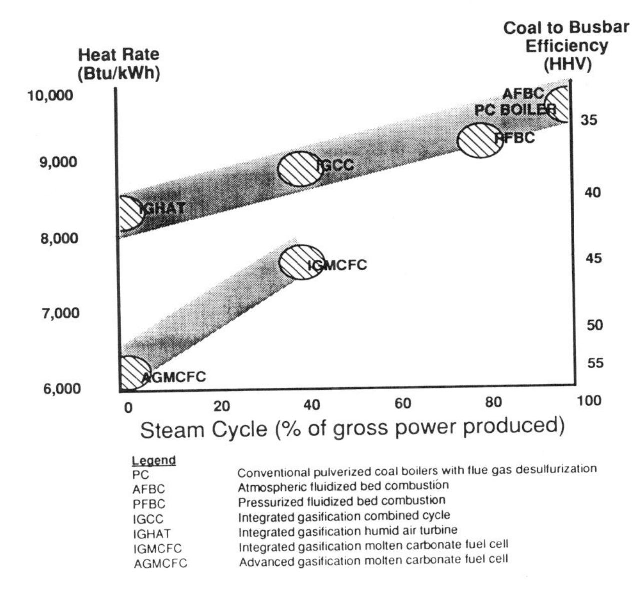

Figure 33. Coal fired power cycles (heat rate vs. steam cycle contribution).

mentation. The progression of technological opportunities, already within our engineering grasp, that mark the expected advance in post-Rankine cycle power production are visualized in Figure 33.

In short, the result of this technological transformation will be an integrated energy and resource production foundation based on indigenous fuel which can serve not only as a transition energy capability for the US and other developed nations but as a fundamental economic base for the less developed world as well. The mortar that cements, or integrates, these coal utilization building blocks and makes them an operationally practical reality is the same microprocessor revolution that is now being technically and culturally assimilated to enhance the productivity of existing power plants. This "smart" instrumentation will enable knowledge-based diagnostic and control systems to manage plant operations as well as robotic systems for maintenance and inspection.

The national goals for this progress require little vision. They are 1) providing 200 GW_e to 300 GW_e of new electrical generating capacity and 2) replacing 10 million barrels per day (MBD) of oil imports.

Achievement of these goals by 2010 is not simply desirable but appears essential to our national economy, security, and environment. The policy steps needed to accomplish these goals are incentives to immediately phase down oil imports in favor of domestic sources and to reduce pollutant formation through the harnessing of real technological innovation to domestic energy production.

The direct commercial benefits to the nation through lower cost electricity, increased power equipment exports, and reduced oil imports alone are expected to escalate, over the next 20 years, to at least $150 billion per year, or six times the required $25 billion per year investment. To this should be added the broad fabric of economic benefits resulting from a more competitive economic posture in the world market. Finally, the results will have a fundamentally positive impact on resolution of the global dilemma concerning energy, economic development, and the environment. Within the required investment, less than 1 percent would be necessary to complete development of the already well-defined technology base. As attractive as this opportunity seems, its initial payback period is simply too long and too subject to political manipulation to attract investment under current short-term, risk-averse national policies. As pointed out repeatedly through this paper, these policies, and the perceptions on which they are based, must be changed if this inevitable technological transition is to occur with the greatest commercial and environmental benefit to this nation.

The performance of US technology-intensive sectors, the strength of the manufacturing base, the leverage of key technologies across different industries, the state of the nation's technical infrastructure—all of these issues should increasingly influence US policies in the future. As these issues grow in importance, they will also exercise increasing influence over our relationships with other nations as well.[32]

We have unparalleled means to continue to prosper as a nation while being the global leader in productive resource stewardship. The issue remaining is whether, also as a nation, we have the patience and proactive will to invest in this sustainable future or simply continue short-sighted expediencies. The bottom line is whether we care enough to guide our own destiny or will leave it to others by default.

Conclusion

Whatever else may be said of the concepts advanced here, they cannot be characterized as novel. Rather, they reflect developments already accelerating throughout the electric power industry. An increased emphasis on harnessing technological and market opportunity to new business strategies, either directly or through alliances, is likely to encourage integrated energy coproduction capabilities in which electricity is one essential product of the most efficient use of coal. Uncertain regulatory economics, increasingly stringent environmental controls, and a global economy are effective harbingers of this change. At issue is not whether change will occur, but who will determine the resulting high ground, corporately and globally.

Thus, the last decade of the 20th century and the first decades of the 21st will be a period of major technical, economic, and organizational change for power production—even dismissing the possibility of further energy supply discontinuities such as those experienced in the 1970s. While these changes pose many challenges, they also bring even greater opportunities. Whether it will be the best of times or the worst is, to a large extent, up to the power industry, its regulators, and their combined effectiveness in adapting to these changing circumstances. We can be assured, however, that the decisions made concerning electricity and the environment in the 1990s will determine the pattern of new generation far beyond the turn of the century and, in doing so, will impact a major portion of our current power generation capability as well.

Success is likely to depend on a growing national collaboration between innovative technological industries and electric power to meet the real competitive challenges and opportunities of the global economy in the 21st century. Maintaining control of our technical and commercial destiny through creation of a national arsenal of energy innovation will be essential in a competitive global market where the US cannot afford to squander either competitive advantages in technology or the resources necessary for their commercial success. This will be encouraged by a less adversarial public policy framework that focuses on industrial competitiveness and technology. The results will be a win-win opportunity for our nation and the world and a necessary step in meeting the challenge of a sustainable future.

In the broader context, coal represents one link in the chain of resources on which the world must rely to meet its energy requirements; and it too shall pass. By the end of the Second Electrical Century, the legacy of coal will be measured not only in terms of its contribution to global economic development over the preceding 300 years, but also by its role in the formation of global policy governing stewardship of all resources.

References

1. Starr, C., "National Energy Policy: A Retrospective," *Energy Systems and Policy,* Vol. 13, pp. 51–61, Pergamon Press, London (1988).
2. Fri, R. W., "Energy and Environment: A Coming Collision?" *Resources* No. 48, Resources for the Future, Washington, DC (1990).
3. Hardin, G., *Exploring New Ethics for Survival,* Chapter 10, Penguin Books, Inc., New York, NY (1972).
4. Capra, F., *The Futurist,* p. 64, World Future Society, Bethesda, MD (Jan-Feb 1990).
5. Starr, C., and Searl, M.F., *Global Electricity Futures: Demand and Supply Alternatives,* presented at American Nuclear Society Annual Meeting, San Francisco, CA (Nov. 1989).
6. Schurr, S. H., et al., *Electricity in the American Economy: Agent of Technological Progress* (to be published in 1990).
7. Carlton, D. M., Liebson, I., et al., *The Long-Range Role of Coal in the Future Energy Strategy of the United States,* Draft Report to the Secretary of Energy, National Coal Council, Arlington, VA (Jan. 1990).
8. Frisch, J. R., *Future Stresses for Energy Resources—Energy Abundance: Myth or Reality?* World Energy Conference—Conservation Commission, Graham and Trotman Ltd., London (1986).
9. *New Capacity Update: Utilities 40 GWe, Cogen/IPPs 34 GWe,* Utility Data Institute, Inc., Washington, DC (March 1990).
10. *1988 Electricity Supply Demand,* North American Electric Reliability Council, Princeton, NJ (1988).
11. *Outlook for US Electric Power through 2000,* Energy Information Administration/Annual Outlook for US Electric Power 1989, Washington, DC (1989).
12. *Annual Energy Outlook 1985 with Projections to 1995,* Energy Information Administration, DOE/EIA-0383 (85), Washington, DC (1985).
13. "1989 Annual Statistical Report," *Electrical World,* pp. 61–68, McGraw-Hill, New York, NY (April 1989).
14. "Annual Industry Forecast," *Electrical World,* pp. 43–48, McGraw-Hill, New York, NY (Oct. 1989).
15. Berg, C., "A Suggestion Regarding the Nature of Innovation," *Electric Power Research Institute Workshop on Energy, Productivity, and Economic Growth,* Palo Alto, CA (Jan. 1981).
16. Kahn, E. P., "Structural Evolution in the Electric Utility Industry," *Public Utilities Fortnightly,* Vol. 125, No. 1, pp. 9–17, Arlington, VA (Jan. 1990).
17. "Energy Strategies for Energy Storage," *EPRI Journal,* Palo Alto, CA (July/Aug. 1989).
18. *Natural Gas for Electric Power Generation: Strategic Issues, Risks, and Opportunities,* Strategic Decisions Group, EPRI, P-6820 (1989).
19. Yeager, K. E., Liebson, I., et al., *Innovative Clean Coal Technology Deployment,* Report to the Secretary of Energy, The National Coal Council, Arlington, VA (Nov. 1988).
20. *Global Prospects for U.S. Coal and Coal Technologies,* Report of Hearings (no. 40), US House of Representatives, Committee on Science, Space, and Technology, US Government Printing Office, Washington, DC (1987).
21. *America's Clean Coal Commitment,* DOE FE-0083, US Dept. of Energy, Office of Fossil Energy, Washington, DC (1987).
22. *Analysis of Alternative SO_2 Reduction Strategies,* Draft Summary, Electric Power Research Institute, Palo Alto, CA (March 1990).
23. *State of Science and Technology Reports,* Draft, National Acid Precipitation Assessment Program (NAPAP), Washington, DC (Jan. 1990).
24. Yeager, K. E., "Coal in the 21st Century," The 1989 Robens Coal Science Lecture, BCURA, The Royal Institution, London (Oct. 2, 1989).
25. Gibbons, J. H., Blair, P. D., and Gwin, H. L., "Strategies for Energy Use," *Scientific American,* Vol. 261, No. 3, pp. 136–143 (Sept. 1989).
26. Yu, O., and Kinderman, E., *Estimation of Incremental Capital Requirements and Energy Cost Savings for Energy Efficiency Improvements,* Stanford Research Institute draft report for EPRI (March 1990).
27. Sedjo, R. A., "Forests to Offset the Greenhouse Effects," *Journal of Forestry,* pp. 12–15 (July 1989).
28. Starr, C., *Global Climate Change and the Electric Power Industry,* Presentation to the National Climate Program Office Strategic Planning Seminar, Washington, DC (Jan. 1988).
29. "The Politics of Climate," *EPRI Journal,* Palo Alto, CA (June 1988).
30. Yip, G. S., "Global Strategy—In a World of Nations?" *Sloan Management Review,* pp. 29–41, Massachusetts Institute of Technology, Cambridge, MA (Fall 1989).
31. "Market Share and Development in the Heavy Electrical Industry," *Power in Europe,* London (Jan. 18, 1990).
32. Inman, B. R., Burton, D. F., Jr., "Technology of Competitiveness: The New Policy Frontier," *Foreign Affairs,* Vol. 69, No. 2, pp. 116–134. The Council on Foreign Relations, Inc., New York, NY (1990).
33. Reich, R. B., "The Quiet Path to Technological Preeminence," *Scientific American,* Vol. 261, No. 4, pp. 41–47 (Oct. 1989).
34. Campbell, S., and Martin, W., *Natural Gas: A Strategic Resource for the Future,* Washington Policy Analysis Group, Washington, DC (Nov. 1988).
35. *Energy Security White Paper: U.S. Decisions and Global Trends,* American Petroleum Institute, Washington, DC (1988).
36. Linden, H.R., "Electrification of Energy Supply—An Essay on Its History, Economic Impact and Likely Future," *Energy Sources,* Vol. 10, No. 2, pp. 127–149 (1988).
37. Gluckman, M. J., "The Long-Term Incentives for Coal Gasification in Electric Power Generation," *Proceedings of the XVI Energy Technology Conference,* pp. 557–569, Washington, DC (March 1989).
38. Electric Power Research Institute, *TAG—Technical Assessment Guide, Vol. I: Electricity Supply-1989,* EPRI P-6587-L, Vol. I, Rev. 6, Palo Alto, CA (Sept. 1989).
39. Cremer, G. A., Hauser, N., and Bayens, C.A., "Application of the Shell Coal Gasification Process," Presented at AIChE Spring National Meeting, AIChE, New York, NY (March 1990).

The Role of Natural Gas in Electric Power Generation, 1990 to 2020

Steven I. Freedman
Technology Evaluation and Coordination
Gas Research Institute
Chicago, IL

Abstract

The demand for electricity is continuing to grow. New generating plants will have to be built in the mid-1990s. Natural-gas-fueled power plants are the most efficient, least costly, and emit the least pollutants and greenhouse gases of all the fossil-fueled power plant technologies available. Hesitancy to use natural gas is based on the shortages of the 1970s; presently we are in the seventh year of a natural gas surplus of deliverability over consumption with relatively low prices for natural gas supplies and proven reserves. Because of the relationship between oil and gas prices and the lack of confidence in the stability of future oil prices, long-term contracts for gas are difficult to obtain. Many options exist for natural-gas-fueled cogeneration in which byproduct heat from power generation is used for space and water heating and industrial steam production so as to materially increase the overall efficiency of fuel use. Natural gas is also being proven as an NO_x reducing agent for use in pollution reduction in coal-fired power plants. Given all of these technologies, the nation can look forward to a period when a significant portion of electric power generation comes from the use of natural gas in a clean and economic manner.

Background

The ability to use energy to make life better has been the driving force behind the progress of civilization from the actions of nomadic tribes 5,000 years ago, through the agricultural and industrial revolutions to today's energy-intensive society. Without vast supplies of energy and a diversity of engines to convert raw energy into meaningful services, we would not be able to enjoy the types of conveniences and comforts to which we have grown accustomed. The major energy use services can be classified as space conditioning (including heating, cooling, and humidity control), transportation, industrial processing, lighting, mechanical power, and information transmission and processing. The use of electricity to provide these energy services has been increasing.

Traditionally, energy economists categorize energy delivery from a supply side perspective, that of fuel type and business sector, rather than from the perspective of energy service delivered. However, it is the energy service that the consumer wants. As our society advances technologically, more uses of electricity are discovered, as witnessed by the recent rise in the use of VCRs and computers. Convenience, ease of control, and unobtrusiveness at the point of use have resulted in a continued growth in demand for electricity. While it may be surprising to hear about the high value of electricity to the user and the continuing growth in the demand for electricity from someone from the gas industry, the historical data regarding electricity use speaks for itself. While forecasts of the future have implicit uncertainties, growth in electricity use is projected in general, notwithstanding the possibility that the introduction of more efficient equipment to deliver energy services from electricity could mitigate such growth. Figure 1 shows the historic and forecasted consumption of electrical energy for a range of growth rates considered reasonable by economic forecasters.[1]

Benefits of Gas Use in Electric Power Generation

Where will the added electricity supply come from? Institutionally it will come from a mix of utility-owned coal- and gas-fueled central station power plants; from large plants extremely similar to utility plants owned by independent power producers;[2] from large and modest-sized industrial cogeneration plants; from some additional hydroelectric plants; and, in lesser amounts, from small commercial cogeneration plants and alternative energy sources. An Electric Power Research Institute/Edison Electric Institute conference on "Utility Opportunities for New Generation" was held in June 1989 in Boston.[3] At that conference, it was noted that there will be numerous forms of organizations that will own and operate new power generation facilities and that natural

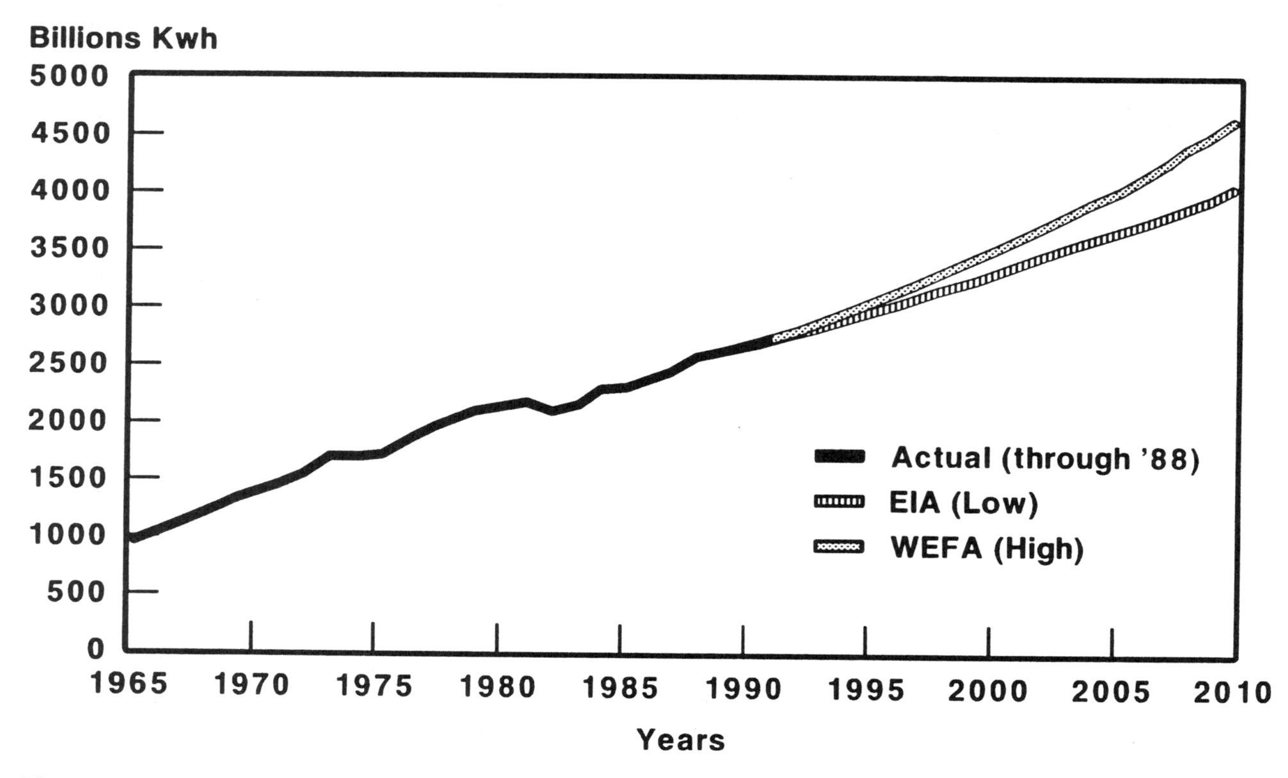

Figure 1. Electricity sales—actual and projected. Growth in electrical energy consumption, 1965–2010.

gas will play an important role in providing the additional electricity needed in the near future.

Many of these new generation facilities will be fueled by natural gas. Why? The virtues of natural gas are 1) it is principally clean, contains virtually no ash or sulfur, and produces essentially no SO_2 or particulates and much less NO_x and CO_2 when burned than other fossil fuels, 2) it is used in equipment that is both higher in efficiency and less expensive than alternatively fueled equipment, 3) it is supplied by a system that is extremely secure and independent of transportation problems, and 4), of substantial importance for security considerations, it is principally domestic. In today's climate of responsibility for our stewardship of the quality of our air and other natural resources, many consider natural gas to be the fuel of choice. Table 1 shows the allowable emissions from new large fossil-fueled power plants. Coal- and residual-oil-fired plants operate at or close to these limits. Natural-gas-fueled power plants have only traces of SO_2 (typically only between 0.001 and 0.002 pounds of SO_2 per MMBtu (million Btu) when firing gas that has been odorized)[4], emit a negligible amount of particulates, and can achieve current NO_x standards. Many siting requirements are demanding even lower NO_x emissions from natural-gas-fueled large combined-cycle power plants, and the plants are achieving such lower emissions.

Economics of Gas Supply

Indeed, why are other fuels used at all? Even though the price of natural gas is not high, it is not the cheapest fuel. Coal and uranium are less expensive per Btu. However, the plants necessary to convert these solid fuels to electricity are much more expensive than those fueled by natural gas; are much more difficult to site and permit; entail substantial operational, maintenance, and decommissioning costs; and require very lengthy and unpredictable construction times, resulting in substantially increased interest cost during construction. Consequently, at today's prices, gas-generated electricity is the lowest cost for many, if not most, applications. The attractiveness of natural gas is tempered only by the uncertainty in future price and supply availability. If utility management were sure of future gas supplies at today's prices, the economics would result in gas-fueled electricity generation being the least expensive power generation expansion option with the possible exception of

Table 1. Federally allowable emissions from large fossil fuel power plants.*

	New Source Performance Standard (NSPS)
SO_2	
Coal	1.20 lb/MMBtu** & 90% reduction (when over 0.60 lb/MMBtu) 70% reduction (when under 0.60 lb/MMBtu)
Oil and Gas	0.80 lb/MMBtu & 90% (unless under 0.2 lb/MMBtu)
NO_x	
Coal	0.50 lb/MMBtu Subbituminous Coal 0.80 lb/MMBtu (if more than 25% NGP lignite) 0.60 lb/MMBtu Other Coal
Oil	0.30 lb/MMBtu
Gas	0.20 lb/MMBtu
TSP (particulates)	
All Fuel	0.03 lb/MMBtu

* Sec. 111, 301(a) of the Clean Air Act as amended (42 U.S.C. 7411, 7602[a]) and additional authority.
** MMBtu = 10^6 Btu

some additional hydroelectric plants in a few selected localities.

The principal impediment to wider use of gas for power generation, viz., the uncertainty of future price and supply, is due mainly to the short supplies of gas in the 1970s and the consequential high prices of the late 1970s and early 1980s. While much hindsight analysis can be offered regarding what happened in the 1970s and why circumstances are different now, the fact of the matter is that the price of gas has been quite low for the last five years. During this period we have had a protracted surplus of gas deliverability in excess of demand, referred to as the gas bubble.[5] Figure 2 shows the extent and duration of this surplus deliverability. The extent to which the change from shortage to surplus is related to deregulation, to improved technology and efficiencies in exploration and production, or to decreased gas use due to improved conservation practices and the use of more efficient equipment is a separate topic. The fact is that more gas is being produced from smaller reservoirs than previously expected[6] according to the techniques and data used by M. King Hubbert[7] in the 1960s that led to a forecast of a continuing decline in gas supply. Figures 3 and 4 show the size of the ultimate recovery of total hydrocarbons in terms of barrels of oil equivalent per wildcat well based on five-year averages through 1985. The data for the early period clearly shows a steeply falling discovery rate; however, the data for the

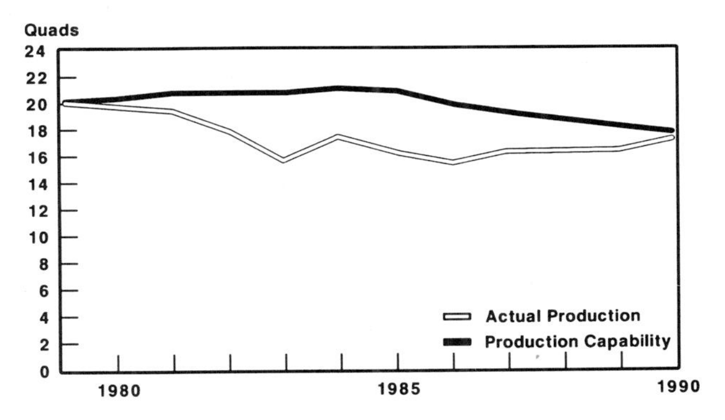

Figure 2. Production and production capability. Excess of natural gas deliverability over consumption. Source: AGA.

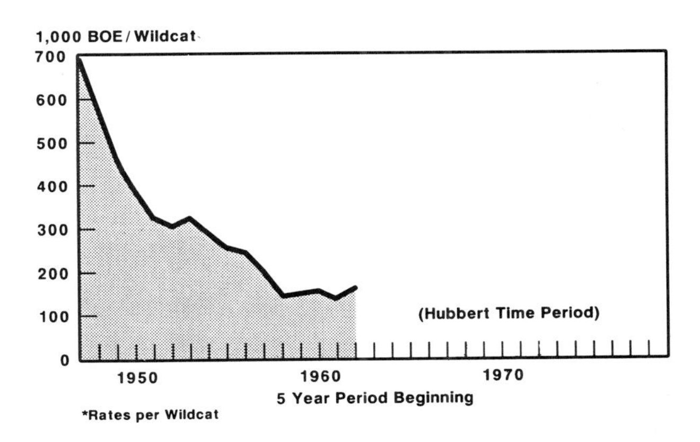

Figure 3. Ultimate recovery of BOE per wildcat well, data through 1967.

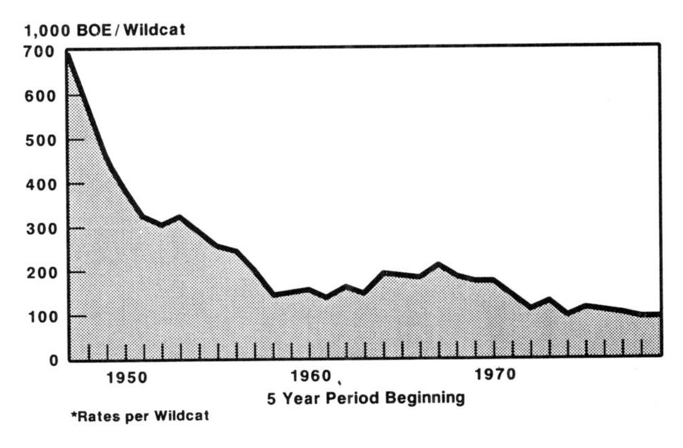

Figure 4. Ultimate recovery of BOE per wildcat well, data through 1984.

later period shows how the recovery rate has stabilized. Production of new gas in the 1980s includes a substantial amount of gas from new wells in established fields in which numerous small deposits are being tapped, as well as new resources such as coal bed methane coming online. It is believed that improvements in the computer analysis of seismic information have substantially contributed to these increases in discoveries of gas and oil. In spite of the Cassandras of the 1970s, gas discoveries have continued even though drilling rates have recently been extremely low. The extent to which the gas bubble has continued is a testament to the resourcefulness of the gas production industry. Important variables in gas production are the technology of discovery and the efficiency of drilling, and, fortunately, these technologies have been improving.

Natural Gas Resource Base

Gas not delivered is described in terms of reserves and resources. Gas reserves are either proven producing reserves that are from wells drilled, tested, and producing, or proven undeveloped reserves that are considered likely as a result of production from nearby wells. Proven reserves are acceptable financial assets and are recognized by the Securities and Exchange Commission on business balance sheets, while resources do not have such a degree of certainty or financial recognition. Oil and gas leases on reserves are acceptable by banks as loan collateral. Gas resources are less certain than reserves. Speculative resources are in geological formations in which there is some likelihood that producible gas will exist but which are not yet proven by producing wells. In order to create a reserve from a resource, exploratory wells must be drilled, and drilling is expensive. Gas producers only drill wells when they can expect to make a profit from such activity. For such economic reasons, it is only prudent to invest money in exploration to create an inventory backlog of producible gas when the return on investment (ROI) obtainable over the production period is greater than the ROI obtainable from alternative investments. Thus, the reserve/production ratio is a function of economic conditions—the higher the interest rate, the quicker the inventory must be turned over—rather than an indication of how much gas is indeed in the ground. Briefly, that is why the reserve/production ratio has hovered in the neighborhood of 12 years for the last 6 decades. The discounted present value of gas that takes 13 years to deplete is quite low and is not worth drilling for. Drilling for scientific research purposes by the United States Geological Survey is negligible in comparison to commercial exploratory drilling. Gas resources are less certain than reserves in exact occurrence but are a more meaningful indicator of how much gas exists that is producible economically with known technology.

Key conditions in describing gas resources are the price at which and the technology with which they are producible. Improvements in drilling technology, formation fracturing and other well operating procedures, building offshore structures, and, especially, computerized analysis of seismographic and downwell data have progressed, and will continue to advance, thereby helping to maintain cost competition of new gas and increasing our resource base of economically producible gas. The Gas Research Institute (GRI) and gas producers engage in a substantial program of natural gas supply research and development that is improving the efficacy of gas production, thereby increasing the economic resource of natural gas. Indeed, the GRI R&D program has already made a significant contribution in the field of coal bed methane that is now being produced commercially. Here both the resource base and the reserves have been increased.

With this background, let us look at what our current reserves and, more important, our gas resources are. According to the US Potential Gas Committee, an authoritative group of gas geological experts, we have probable resources of 154 Tcf (trillions of cubic feet), possible resources of 269 Tcf, and speculative resources of 197 Tcf in the lower 48 states,[8] as shown in Table 2. With the inclusion of Alaska, the resource potential is 739 Tcf, based on current and foreseeable technology. In a separate process, the National Petroleum Council has estimated the natural gas resource contained in tight gas sands alone to be 924 Tcf.[9] New technology in drilling, seismic technology and analysis, and operations in deeper offshore locations in the next 50 years can be expected to increase this resource base further. To these resources must be added those of Canada and Mexico that those nations are willing to market in the United States. Liquified natural gas (LNG) is once again being imported into the United States, thereby further increasing the economic gas resource available to the United States and the duration of gas availability. With all this knowledge, one can be certain that there is enough additional gas beyond today's consumption for at least 100 Gw of new combined-cycle plants for at least 25 years. This additional 100 Tcf will be enough to take care of our incremental electric power generation needs for the next few decades, at which time we can reappraise our resources to see how much longer our supply of moderate cost natural gas will last. While scientific evidence of additional resources does not yet exist, the record of continuing improvements in exploration and production technology encourages the belief that such advances will continue. After all, until the 1950s, the North Sea was not considered to be an oil or gas resource.

Table 2. Estimated potential natural gas resources (billion cubic feet).

	Mostly Likely Estimates			
	Probable	Possible	Speculative	Total
Lower 48 States				
Onshore	128,183	200,046	156,169	484,398
Offshore	25,462	69,148	40,768	135,378
Total	153,645	269,194	196,937	619,776
Alaska				
Onshore	6,590	16,400	27,690	50,680
Offshore	2,400	12,700	53,000	68,100
Total Alaska	8,990	29,100	80,690	118,780
Total United States	162,635	298,294	277,627	738,556

Difficulties in Long-Term Contracting

Why, then, is it difficult to obtain long-term contracts? It is because the specter of the Organization of Petroleum Exporting Countries and the enticing opportunities for gas sales that would be created by an oil price increase hang over the market. Producers will sell their gas at today's prices only for a modest period as long as the long-term price of oil is believed to be substantially uncertain. In addition to future oil price increases, many believe that the dearth of drilling since the 1985 oil price collapse has resulted in a shortage of wells capable of future production and that a gas price increase will occur in a few years.[10] While such reasoning has existed for several years, no sign of a major price increase has been seen. Indeed, recent statistics from the Department of Energy's (DOE's) Energy Information Administration stated that in 1988 gas reserve additions were 130 percent of consumption.

The oil and gas exploration and production business is noted for having some of the biggest risk takers of the century. These fiercely independent and inherently optimistic capitalists are well aware of the recent increase in US oil imports and the consequential opportunity for cartel-controlled oil price increases. One must accept the fact that oil and gas prices are more uncertain than those of most other commodities. Those desiring long-term supplies are welcome to go out and buy reserves. Indeed, USX announced in October 1989 that they were selling 1.2 Tcf of reserves of their Texas Oil and Gas subsidiary because of low gas prices.[11] Anyone desiring to buy a 20-year gas supply for a 1000 Mw plant (or any part thereof) was invited to Pittsburgh to enter a bid. Newspaper reports claimed that the value of such gas reserves was in the neighborhood of $1 per Mcf (thousand cubic feet), or $1.2 billion for 1.2 Tcf. The supply of gas is there; all we are now doing is negotiating the price.

While natural gas is abundant and cost competitive, concern exists regarding how long such circumstances will continue. Because of the seasonal nature of the demand for gas for space heating, gas is more expensive in the winter than in the summer. A substantial part of the industrial and electric utility load is switchable between gas and oil, and most gas turbines and gas-fired boilers can burn gas or oil. Consequently, price competition exists between gas and oil for this substantial market, placing the gas business appreciably at the mercy of OPEC. Because of such uncertainties and differences in opinion on the future price of oil, gas producers are reluctant to enter into long-term supply contracts. Such long-term fuel supply contracts are, however, needed by the electric utilities to assure the regulatory commissions that it will be in the electric ratepayers' economic interest to build a new plant whose principal fuel is gas. Additionally, changes in the regulation of transportation of gas and of the roles of pipelines and distribution companies make this already complex issue still more complicated. Producers are learning the economic facts of life in doing business with electric utilities and vice versa, and long-term contracts with complex indexing and escalation clauses are being negotiated. The progress required for increased use of gas for electric power generation is less the advancement of technology than the advancement of the art of writing long-term gas supply contract conditions. This is not to state that technical advances are not worthwhile—indeed they are, but their influence on the use of gas for electric power generation has less significance than the ability to negotiate conditions for long-term gas supply contracts.

Forecast of Gas Use in Electric Power Generation

In returning to the topic of the role of natural gas in electric power generation, a number of factors should be emphasized. Because of its low emissions characteristics, natural gas is the fuel of choice where the emission of acid rain precursors is of concern. Also, because the equipment to use it is so low in cost, it is the fuel of choice for incremental peaking and intermediate duty service and for industrial and commercial cogeneration. It is forecasted that these factors will create an addition of 50 percent in the use of gas for electric power generation by 2010.[12] While we can use forecasts of the elasticity of natural gas supply and demand to project the future price of natural gas, such price forecasts have greater uncertainty than forecasts of the quantities of energy use. The latest forecast of the price of gas to electric utilities shows an increase, in terms of constant 1988 dollars, from $2.30 per MMBtu in 1988 to $5.64 per MMBtu in 2010, an average increase of 4.2 percent per year. In considering the contribution of increases in efficiency and decreases in fuel cost to reductions in cost of electricity, it is worthwhile to note that for a combined-cycle power plant having 1990 technology and burning fuel that costs $4.50 per MMBtu, an improvement in efficiency of one percentage point (requiring five to ten years of effort and hundreds of millions of dollars of R&D) would result in a production cost decrease of 0.8 mills per kwh. The change in fuel price necessary to achieve such a savings in only 10¢/MMBtu. The uncertainty in the price of gas in 2010 is at least this amount. This is why there is so much concern for long-term supply contracts. A plan that enables electric utility management to hedge their investment is to first build a gas-fired plant with sufficient land and transportation access for coal and then to add an integrated coal gasification plant if the gas price rises far enough to warrant conversion to coal.[13] Several utilities appear to be following such a strategy of incremental plant construction.

Equipment and fuel to supply electric power is generally characterized as providing power for baseload, peaking, or intermediate duty loads. Baseload power is generated by plants with the lowest incremental operating costs that include fuel, labor, and maintenance. The investment in baseload plants must not be too high in order to assure the regulatory commissions that the annual cost of running and owning these plants is indeed less than alternatives. Peaking plants are gas- and oil-fired gas turbines that run only a few hundred to 15 hundred hours per year and are selected on the basis of lowest installed cost per kilowatt of capacity, provided the efficiency is not too low or the fuel price too high. Intermediate duty plants were historically older, moderate-efficiency baseload plants that were substantially depreciated and replaced by new, higher efficiency baseload plants. These were supplemented by a few low-cost, modest efficiency bare-bones steam plants built for intermediate duty when sufficient old baseload plants were not available. Coal and nuclear energy are used to fuel the large steam plants presently used for baseload service. This traditional picture has changed because of the maturing of the gas turbine combined-cycle plant and because of the financial incentive to keep in baseload service the old pre–New Source Performance Standards (NSPS) coal-fired plants that were "grandfathered" by the Clean Air Act and which can therefore operate without the economic and efficiency penalties imposed by the use of scrubbers. Acid-rain legislation currently under consideration by Congress may substantially change such incentives. Natural-gas-fired combined-cycle plants are substantially higher in efficiency than oil- or gas-fired intermediate duty steam plants—45 percent versus 37 percent—so that combined-cycle plants are now the plant of choice whan a new plant has to be built for intermediate duty operation. On a lower heating value basis (used for engines instead of the higher heating value basis used for power plants), new combined-cycle plants are over 52 percent efficient. At gas prices below $4.50 per Mcf, combined-cycle plants are most economical even for baseload use, as shown in Figure 5.[14,15] Should gas prices rise to higher levels, then integrated coal gasification and fuel gas cleanup plants can be added to the natural gas combined-cycle plants at an incremental cost of about $1000 per kw, and the plant can be continued in baseload service.

The above analysis addresses national average gas and coal prices when, in actuality, such fuel prices have a considerable geographic variation. Inexpensive coal can be obtained in the vicinity of coal seams, and inexpensive natural gas can be found in areas with geologic formations containing large amounts of natural gas. Figure 5 shows annual revenue requirements for peaking, intermediate duty, and baseload power plants for a representative levelized constant dollar gas price of $4.50 per MMBtu. The economic attractiveness of natural gas is apparent.

Examination of the candidate technologies and possible areas of economic improvement—reducing plant cost, reducing fuel cost, and increasing plant efficiency—shows that the influence of lowering life-cycle fuel costs for combined-cycle plants can offer consumers benefits comparable to or greater than those that could result from more impressive technical advances in equipment performance. Concern of electric utilities for their fuel availability and cost is not without precedent. Electric utilities have performed R&D on plants to produce synthetic oil for use in residual oil-fired steam plants and have built and operated the 100 Mw coolwater integrated gasification combined-cycle plant to ensure future fuel availability options. Of course, a prime concern

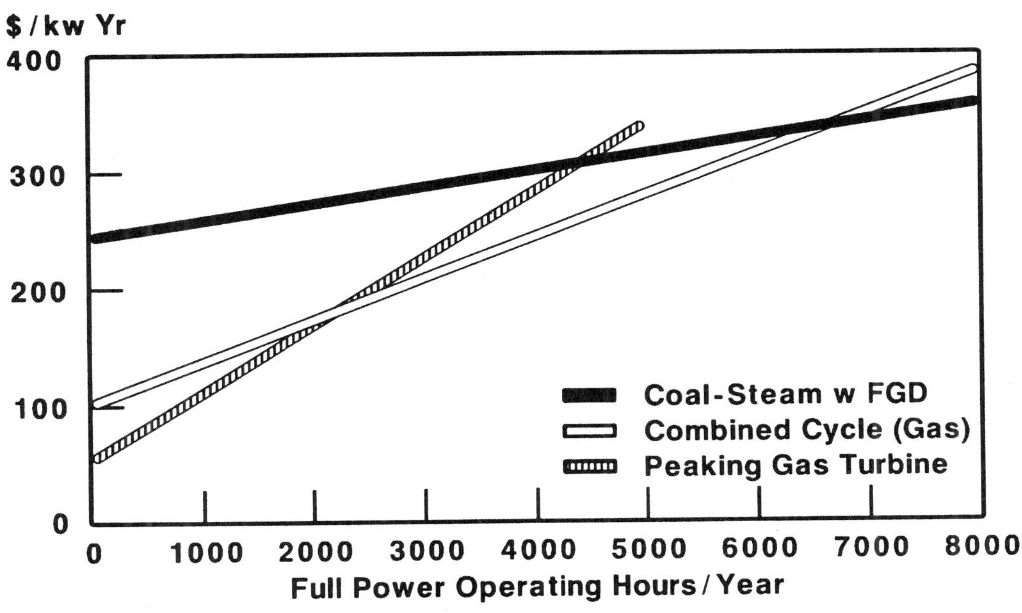

Peaking Gas Turbine $350/kw, Fixed O&M $1/kwYr, Heat Rate 12000 BTU/kwh
Combined Cycle $600/kw, Fixed O&M $7/kwYr, Heat Rate 8000 BTU/kwh
Coal-Steam w FGD $1500/kw, Fixed O&M $20/kwYr, Heat Rate 9700 BTU/kwh

Coal at $1.50/MMBTO
Natural Gas at $4.50/MMBTU
Capital Charge Factor, 16%

Figure 5. Revenue requirements for electric power generation plants.

of the gas industry is the price and availability of gas. Indeed, R&D expenditures continue to be made by GRI and members of the gas industry in the field of lowering production costs and thereby increasing the economic gas supply. It is worthwhile pointing out that the beneficiaries of such gas industry R&D include the electric utilities and their ratepayers.

Technology of Gas-Fueled Electric Power Generation Equipment

Gas Turbines

The largest gas turbines are now approaching 150 Mw in capacity by themselves and over 250 Mw in capacity when used in a combined cycle. These machines have grown in size, efficiency, temperature, and reliability in the 40 years since their first introduction. At present their engineering is quite mature and sophisticated even though incremental improvements are continuing. Efficiency improvements have only been about two to three percentage points per decade, and the R&D to achieve this has cost hundreds of millions of dollars. The high technology parts of a gas turbine are the complex internal cooling passage design in the first stage turbine blades, as shown in Figure 6, and the combustor design to give low emissions of NO_x at progressively higher firing temperatures. The challenge of increasing efficiency through increasing firing temperature while simultaneously reducing NO_x emissions is extremely demanding on the skills of the gas turbine development engineers. Continuing advances in heat transfer design enable gas turbines to run reliably with cooled blades when the temperature of the high-pressure combustion products in the first stage turbine section is at the melting point of stainless steel. These machines can be sited, purchased, manufactured, delivered, installed, and producing power for revenue in only 18 months. The addition of the heat recovery boilers, steam turbines, and condensers to convert the simple cycle plant to a high-efficiency combined-cycle unit takes an additional 6 to 12 months to complete. Should power shortages develop in various regions in the mid-1990s, as many expect, then such equipment will be installed in short order

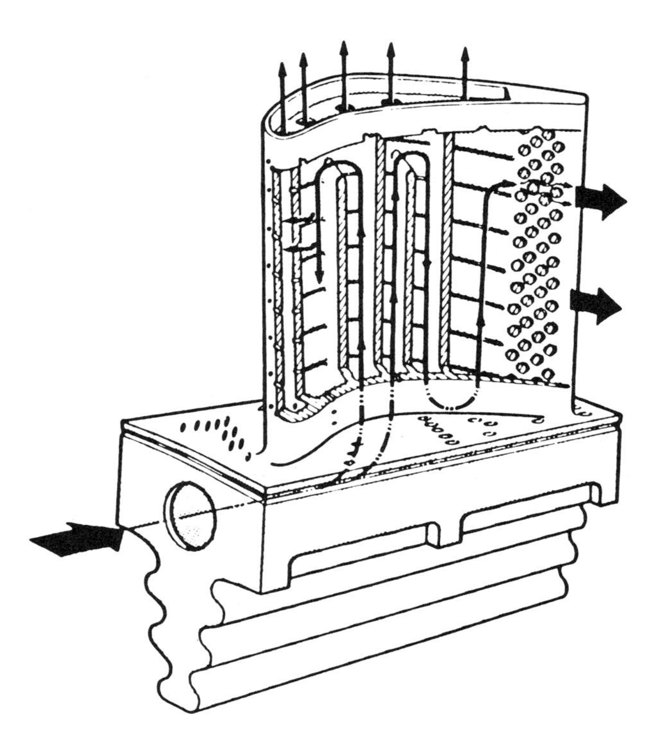

Figure 6. HPF multipass blade. Gas turbine blade internal cooling flowpath.

whether owned by an electric utility, an independent power producer, or an industrial organization.

Next in size are the large aircraft derivative gas turbines for utility power and industrial cogeneration. For the most part, the government pays for the R&D for advancing the technology of such engines. These gas turbines can be used as simple-cycle machines, can be steam injected with steam produced by an add-on boiler, or can be used with the steam driving a separate extraction or condensing steam turbine in a combined-cycle mode of operation. Steam and water injection lowers NO_x as well as inexpensively raising the efficiency and increasing the power of gas turbines.

Reciprocating Engines

In smaller sizes, reciprocating engines and small gas turbines are used in cogeneration facilities in the size range from about five kilowatts to a few megawatts. Automotive engines are used in the small end of this size range, and at somewhat larger sizes, truck engines and converted stationary diesels are used.

Fuel Cells

The newest competitor in the small cogeneration market is not an engine at all in the thermodynamic sense. It is the fuel cell that converts the chemical energy of hydrogen, produced by the steam reforming of natural gas, into electricity electrochemically. Such fuel cells emit the least NO_x per kwh produced of any available power generation technology and, consequently, are expected to be in demand in urban areas that are unable to meet ambient air quality standards. Perhaps we will eventually see the emergence of this new product as the source of ultraclean high-efficiency electric power and cogenerated heat.

Economy of Scale in Production

Until now, electric power generation has benefited from an economy of scale in progressing to larger and larger units. This economy of scale traditionally applied to custom-built plants in circumstances when interest during construction was not a major part of plant cost. Starting in 1968 and becoming more prevalent during the 1970s, large nuclear and coal steam plants took so long to build in an era of high interest rates that the contribution of interest during construction to the total cost when completed made electricity from these plants cost more than electricity from existing plants. Delays in construction due to permitting and construction management difficulties in many cases exacerbated the interest cost expense. Then, for the first time since the beginning of the century, electric utilities required rate increases, rather than decreases, as the result of the addition of new plants. Such economic handicaps continue today for very large plants.

An alternative strategy is to examine the opposite end of the power plant size spectrum—the tiny plant of only a few kilowatts, perhaps as small as 1 kilowatt or as large as 150 kilowatts. A different economy of scale exists for these plants—the economy of scale of mass production. As an example, loose engines can be bought from the automobile manufacturers for under $2,000 for an engine that can produce 50 to 75 kw at 1800 rpm and over 100 to 150 kw at 3600 rpm. This is $40 to $13 per kw, enormously lower than even the price of peaking gas turbines. Of course, these engines must be combined with electric generators, heat rejection radiators, switch gear, and controls. Complete engine-driven generator sets can be bought for $250 per kw in quantities of one. Lower prices would result if such complete systems were mass produced and sold in large numbers like the engines. While the life of such automotive derivative engines is not as long as desired, the engines can be inexpensively replaced when in need of overhaul. Experience with truck engines operating at 1800 rpm shows their life to be in excess of 20,000 hours before overhaul is required. A unique opportunity also exists to use the byproduct heat from these engines for space and water heating and possibly to run thermally driven air-conditioning systems such as conventional absorption or advanced sorption systems.

Electric utilities are aghast when presented with the concept of a system with a very large number (millions) of tiny generators in it. However, they have accepted as normal business the operation of a system with millions of tiny users of electricity, each one in turn composed of dozens of individual loads and each one totally out of the realm of control of the electric utility. The reason why they accept the large number of small, uncontrolled loads is that they have adequate experience and data to handle such loads on a statistical basis and the interconnecting grid serves to transform the numerous small loads into a few large loads of acceptably predictable time-varying behavior. In order for large numbers of small generators (or cogenerators) to be placed in service, a generic procedure for permitting (such as the Environmental Protection Agency [EPA] certification of specific automobile models) will have to be developed. Electric utilities may have to depend on such dispersed power generation as the difficulty in building additional transmission lines in crowded areas continues to grow.

Cogeneration

Statistical techniques can be applied to electricity generation as well as electricity consumption, provided the statistical data can be developed. If electricity can be generated economically by large numbers of small dispersed cogenerators, there will be many benefits. There will be a greater efficiency of primary energy use, approaching 80 percent in many cases, as engine byproduct heat is used economically for space and water heating at the engine site with concomitant reduction in the emission of pollutants and greenhouse gases. The voltage drop in areas of high power use can also be materially reduced. Capacity additions will be made in small increments so that load growth can be matched by generation growth without overbuilding and with an extremely short time period between capacity purchase and capacity operation in revenue service. Thus, the risk of exposure of capital invested in new plant is minimized. Such small cogeneration systems would be sized to the thermal load rather than the electric load of the building to facilitate integration and enhance overall economics. If built with heat dump capability, cogeneration could be dispatched by the utility during times of peak demand and so have a substantial capacity credit value to a utility with need for additional peaking capacity.

Much progress has been made in the field of cogeneration interconnection with the grid in terms of protective relaying and microprocessor-controlled load switching. With the possibility that future electric utility planning may accept the concept of numerous small generation modules, the economies of scale of high volume production can be used to economic advantage where before utilities benefited from the economies of scale of size. Perhaps the first step towards serious consideration of such systems is not to think of them as power plant engines whose waste heat is used for space and water heating, but as space and water heaters that produce electric power as a byproduct.

The technology of all of these gas turbines and converted gasoline and diesel engines is quite mature. Advances in performance are small. Due to the pressure of environmental needs, much work is being done to reduce NO_x emissions without increasing CO emissions or compromising performance. Most notably, NO_x has been reduced by using lean-burn technology. Gas engines are now available whose NO_x emissions are below 2g per horsepower per hour. Add-on tail gas treatment for further NO_x reduction is possible, but expensive. The question of how much more expensive power generation will become in the search for an acceptable tradeoff between clean air and inexpensive power has not yet been answered. Indeed, we have not yet reached the end of a protracted experiment to determine the tradeoff between air quality and the price of power.

Use of Natural Gas for Pollution Reduction in Coal-Fired Utility Boilers

Besides being used as the primary fuel for power generation in new equipment, natural gas is being developed as a supplementary fuel to be partially used in coal-fired boilers to reduce emissions of SO_2 and NO_x. Two different technologies are now under development for such combustion modification—cofiring and gas reburn. In cofiring, gas is substituted for coal with conventional firing so that SO_2 is reduced due to the essentially zero sulfur content of natural gas. Since there is no fuel-bound nitrogen in natural gas, NO_x emissions are often lower. Such emission reductions can be used in existing power plants—those built before the EPA NSPS went into effect—to meet State Implementation Plan Standards (SIPS) emission requirements and, if they become law, new acid rain precursor reduction standards. The advantages of gas cofiring were initially believed to be only the ability to trim emissions modestly at minimum capital investment. However, cofired gas and coal put less mineral matter into the combustion chamber, thereby reducing the propensity of slag to form on furnace walls. Interest exists in determining the extent to which the reduction in combustion zone fuel ash resulting from cofiring can reduce maintenance costs, outages, and capacity derates. Additionally, recent data has shown that SO_2 emission reductions with gas cofiring are slightly better than expected and that stack opacity is reduced with natural gas cofiring. Such experience may enable coal-fired plants that are derated due to furnace slagging, convective pass deposit conditions, or plume opacity to increase their rating. Additionally, plants operating at full rating could possibly burn less expensive higher ash coal when cofiring, thus realizing a net economic benefit to the operator and ratepayer.

With gas reburn technology, coal firing is reduced in the top part of the coal combustion zone, and a layer of natural gas is placed in this region. A high temperature reducing zone is thereby produced in which NO_x is reacted to form molecular nitrogen. After the combustion products pass through the reducing zone, they are reacted with overfire air that burns out the remaining reducing gas and obtains heat from the fuel value of the injected gas. With gas reburn, it is expected that NO_x may be reduced by 50 percent from straight coal-firing levels at the same power output. The injected gas or the overfire air may be used to bring pulverized lime, limestone, or other sorbent into the furnace for reaction with the SO_2, forming gypsum (calcium sulfate). In this manner, appreciable SO_2 reduction, perhaps as large as 60 percent, can be achieved at economic levels of sorbent consumption. Such use of natural gas for pollution control in the operation of coal-fired electric utility plants is expected to be a low-cost, moderate-performance option to reduce NO_x and SO_2 from old coal-fired power plants, thereby contributing to the use of natural gas for pollution control in the electric utility sector.[16]

Conclusion

The compelling appeal of natural gas as an electric utility fuel is the combination of its extremely low emission of pollutants and greenhouse gases, the low cost and high efficiency of the equipment needed to use it, the large resource base, and the secure pipeline system to deliver it. In order that this promising potential be realized, it will be necessary for the gas producers and electric power plant owners to arrive at acceptable long-term supply contract terms. Electric power will be generated from natural gas in myriad types of power plants—large electric utility and independent power producer-owned, moderate-sized industrial cogeneration, and small-sized commercial and possibly residential cogeneration equipment. With new pollution-reducing gas combustion technologies in addition to the traditional technologies for gas turbines, combined cycles, steam injected gas turbines, and gas-converted diesel and automotive engines, natural gas will be playing a substantial role in producing clean electric power for the growing needs of the United States.

References

1. Spencer, Dwain F., *Electricity and Related Issues,* Atlantic Council, Knoa, HI, February 4–6, 1990.
2. Hilt, Richard H., "Independent Power Production: How Big is Big?" International Association for Energy Economics, 11th Annual North American Conference, Los Angeles, CA, October 16–18, 1989.
3. "Utility Opportunities for New Generation," Edison Electric Institute and Electric Power Research Institute, Boston, MA, June 28–30, 1989.
4. Attari, Amir, Institute of Gas Technology, private communication, February 9, 1990.
5. *Natural Gas Production Capability 1990,* Issue Brief 1990–2, American Gas Association, Arlington, VA, January 2, 1990.
6. Woods, Thomas J., "Resource Depletion and Lower-48 Oil and Gas Discovery Rates," *Oil and Gas Journal,* October 28, 1985.
7. Hubbert, M. King, "Energy Resources of the Earth," *Scientific American,* September 1971.
8. Potential Gas Committee, Potential Gas Agency, *Potential Supply of Natural Gas in the United States,* Colorado School of Mines, Golden, CO, December 1986.
9. *NPC, Unconventional Gas Resources, Volume V; Tight Gas Resources,* National Petroleum Council Federal Advisory Committee Report to the Secretary of Energy, Department of Energy, 1980, Washington, DC.
10. Pitts, L. Frank, "A Gas Producer's Perspective of the Gas Supply and Contracting Situation," GRI/EPRI Workshop on Natural Gas as a Fuel for Electric Generation, October 27, 1987.
11. *The Wall Street Journal,* October 3, 1989.
12. Holtberg, Paul D., Woods, Thomas J., Lihn, Marie L., and McCabe, Nancy D., *Baseline Projection Data Book, 1989 GRI Baseline Projection of US Energy Supply and Demand to 2010,* GRI, Washington, DC.
13. Linden, Henry R., "The Case for Increasing Use of Natural Gas in Power Generation," *Public Utilities Fortnightly,* February 18, 1988, Arlington, VA.
14. Hedman, Bruce, and Oates, George K., "Meeting Future Electric Power Needs with Natural Gas," *Public Utilities Fortnightly,* January 5, 1984, Arlington, VA.
15. Hedman, Bruce A., Oates, George K., Willke, Theodore L., Linden, Henry R., and Goins, Dennis, *Meeting Future Electric Power Needs with Natural Gas,* Iowa State Regulatory Conference, Ames, IA, May 18, 1983.
16. Ashby, Anne B., and Holtberg, Paul D., "The Effects of Acid Rain Legislation on Electric Utility Consumption of Natural Gas," *Gas Research Insights,* Gas Research Institute, Chicago, IL, October 1986.

Summary
Session D-3—Nuclear Based Electric Power Technologies

Kent Hansen
Session Chair
Department of Nuclear Engineering
Energy Laboratory
Massachusetts Institute of Technology
Cambridge, MA

There is a consensus among the authors in this session that nuclear power offers very great environmental benefits over conventional means of electric energy production. The papers deal largely with the key issues involved in capturing these potential benefits. The issues include: nuclear safety, nuclear power economics, and the regulation of the industry.

Research and development for the future will concentrate on enhanced safety through simplified designs that reduce hardware, while taking advantage of natural processes for heat removal. Cost reductions will be achieved via simplification and improved plant capacity. Long-duration generation cycles are anticipated for all reactor designs. An unresolved challenge for the future is the role of nuclear power in lesser developed countries. Current plants are too large and complex for use in small grids. It is unclear whether or not small, easy-to-operate, and ultrasafe systems can be made available. Until the industry regains commercial success, efforts in these directions are unlikely.

A brief description of this session's papers follows.

"Nuclear Power Development to Enhance Environmental Quality: An Overview" (John J. Taylor). The priorities of the nation's energy R&D programs have shifted from a search for technologies to supply abundant energy sources to technologies compatible with environmental and public health standards, states John Taylor. What is the role of nuclear energy within the context of these new priorities? This paper discusses the obstacles, disincentives, and R&D issues related to reviving options for nuclear power expansion in the US. A review of the generic technical features of advanced reactors is included along with specific discussions.

"The Advanced Pressurized Water Reactor—Meeting the Energy Needs of Today and Tomorrow" (Richard J. Slember). If nuclear energy is to be reintroduced as a prevalent energy source, new reactor designs must address safety and reliability concerns. In addition, these designs should be standardized, with predictable costs and construction schedules. Richard Slember presents the Westinghouse AP600, the reactor design that meets these criteria. Slember discusses the benefits of advanced pressurized water reactors and how the AP600 exemplifies those advantages.

"The Advanced Boiling Water Reactor and the Road to Revival of Nuclear Power" (Bertram Wolfe). This paper describes the history, technical characteristics, and licensing of the advanced boiling water reactor (ABWR). The ABWR (an advanced large evolutionary light water reactor) may be the key to revitalizing the nuclear energy option in the US, says Bertram Wolfe. As the lead plant under review by the Nuclear Regulatory Commission (NRC) for certification under its new standard licensing guidelines, the ABWR could bolster public confidence by demonstrating the effectiveness of the US licensing and plant construction process. In doing so, this reactor design could overcome the current institutional impediments to a return to nuclear power.

"Environmental Impacts of the Modular High-Temperature Gas-Cooled Reactor (MHTGR)" (David D. Lanning and Scott W. Pappano). David Lanning and Scott Pappano argue that the Modular High Temperature Gas-Cooled Reactor (MHTGR) offers the most promise for an environmentally sound form of electric energy production and process heat. The design for MHTGRs is described and the environmental impact of both routine operation and potential accidents is evaluated within the context of currently permissible environmental release levels. Lanning and Pappano assess the research and development issues that must be resolved in order to realize the potential of this technology.

"Advanced Reactor Development: The Liquid Metal Integral Fast Reactor Program at Argonne" (Charles E. Till). Technological ground rules used in designing the reactors of the 1950s and 1960s are no longer appropriate, and the reactors once embraced by the public are now met with extreme skepticism, maintains Charles Till. He contends that building a better reactor begins with the selection of basic materials. This paper recounts the development of the Integral Fast Reactor at Argonne National Laboratory and discusses how this technology addresses technological concerns, environmental issues, and problems with public acceptance of nuclear power.

Nuclear Power Development to Enhance Environmental Quality: An Overview

John J. Taylor
Electric Power Research Institute
Palo Alto, CA

Introduction

As the energy research and development (R&D) programs are formulated for the 1990s to meet the nation's energy needs in the next century, a substantial shift in emphasis and priority can be observed. The major R&D efforts of the past three decades searched out technologies that would provide the most abundant sources of energy. Coal, fission and fusion power, syn-fuel and renewable energy sources, such as solar, are dominant examples. This emphasis was spurred by a perception that there was a limited availability of fuel resources and that national security and economy required significant national energy independence.

Now there is a focus on satisfying stringent environment and public health standards first while striving for economic competitiveness compatible with those standards. This shift has resulted from a recognition that the people and their political representatives in the developed countries have much greater concern for public health and environmental effects, and comparatively little concern as to the availability of fuel resources and energy production capacity. In addition, experience in energy development in this century has made the industry and the nation's R&D organizations realize that there is no lack of abundant energy resources. What creates the limitation is the ability to provide those resources in sufficiently economic form while meeting environmental and safety requirements.

A corollary to this economic issue is the cost of R&D itself. Vast sums of money and the marshalling of many skilled people and dedicated facilities are required to achieve success in energy development, particularly with the increased emphasis on assurance of public health and environmental compatibility. These large costs of R&D have led to a much greater emphasis on setting priorities, selecting a few of the most promising candidates for development, and sharing costs through international cooperation. It has also been recognized that environmental and public health issues are global in scope and require international cooperation for their resolution. For these reasons, the desire to achieve national energy independence has shifted to the desire to establish international energy R&D cooperation and interdependence. In no element of energy R&D are these shifts more apparent and perhaps more advanced than in the development of nuclear power. This overview will summarize these new emphases in nuclear power development in the 1990s which are setting the stage for the expanded utilization of nuclear power in the next century.

US Utility Consensus

A report prepared by the Advanced Reactor Corporation—a group comprising primarily chief executive officers of US nuclear utilities—has been submitted to the Department of Energy (DOE) and key Congressional committees (ARC Report, 1990) and describes the US utilities' consensus on the need for renewed expansion of nuclear power and the related R&D program priorities and direction. A summary of that consensus follows.

As things stand, most utility capacity planning is based on coal, oil, and gas because of the uncertainty and high cost ascribed to nuclear capacity additions and subsequent operations in the current regulatory environment. However, reliance on these primary resources to meet future needs is not in the best national interest, because of the volatility of oil and gas prices, the uncertain availability of oil, and the increasing emission controls on coal. The US electric utilities remain convinced that nuclear power will be needed to assure the nation of an adequate supply of electric power at reasonable prices. They urge (Edison Electric Institute, 1985, and O'Connor, 1989) that the United States should begin now to take steps that will again make nuclear power an option that utilities can prudently consider to meet future needs for electric generating capacity. Changes in the present institutional arrangements governing nuclear power are judged to be necessary to restore nuclear power as a viable option. The US electric utilities also advocate continued research and development, funded both by industry and government, to define the technical characteristics of advanced reactor systems which will best serve that option in the future.

Prerequisites for the Nuclear Option

US electric utilities face several obstacles and disincentives to building new nuclear power plants. A reformed licensing process that permits true one-step licensing and rate regulatory reform are prerequisites, as is timely implementation of the technical and political solutions of high-level waste disposal by the federal government and low-level waste disposal by state governments. The lack of substantive progress on the federal nuclear waste disposal program appears to be a principal reason for lack of public confidence in nuclear energy. It is, therefore, recommended that DOE place high priority on establishing real progress on high-level waste implementation and continue to support the one-step licensing process.

Although these remedial steps and regulatory, financial, and legislative reforms are essential, improved nuclear plant performance is also a prerequisite to expanded use of nuclear power in the United States (O'Connor, 1989). An estimated $500 million annually is being spent by the utility industry to increase availability and improve the effectiveness of operations and maintenance of all existing US nuclear power plants to the levels of performance already being achieved in the best performing nuclear plants in the United States. These actions emphasize safety and dedication to excellence in all elements of nuclear power plant operations.

Finally, a more tangible acceptance of the need, benefits, and residual risk of nuclear power by concerned institutions is an essential impetus for the return of the nuclear option. The industry recognizes the importance of regaining the confidence of the public in nuclear power as a key prerequisite to expanding the nuclear option. The industry accepts its responsibility to reestablish that confidence, working with the cognizant government agencies, through the "Drive for Excellence" of operation of the present nuclear power plants, the resolution of radioactive waste storage, the development of improved nuclear power for the future, and the communication of these results to the public.

The Next Increment of Nuclear Capacity

Achievement of the above prerequisites is expected to increase confidence in the safety, reliability, and economic superiority of existing US nuclear power plants to where the nation's electric utilities can prudently include nuclear power in their capacity planning. Once the institutional reforms are implemented and their effectiveness generally accepted, confidence will be restored in the processes to build and operate additional nuclear capacity. The question then becomes, what reactor concept will be embodied in the next increment of nuclear power plants?

It is the present judgment of the majority in the nuclear power industry that the light water reactor (LWR) will remain the dominant nuclear power technology for the next several decades (O'Connor, 1989; Testimony to Senate, 1988). Over 25 years of generally favorable operating experience worldwide, along with extensive development and testing programs, support the LWR and provide a sound basis for confidence in this concept. Moreover, the industrial infrastructure to support the LWR is well established. Thus, an improved version of the LWR is expected to be the leading candidate for the next increment of nuclear capacity ordered in the United States.

Other advanced nuclear alternatives should be pursued. Based on results of the conceptual design and preapplication licensing review, the modular high-temperature, gas-cooled reactor (MHTGR) offers a diverse approach to containing the business risks of nuclear plant operation and is applicable to process heat applications. Based on pilot plant testing, the advanced liquid metal cooled reactor (ALMR) system also offers a diverse business approach that will ultimately assure abundant nuclear fuel supplies in the next century. Thus, it would be prudent to develop the MHTGR and ALMR as potential nuclear options for the next century.

Stability and Cooperation in Research and Development Funding

There is concern as to the adequacy and stability of DOE funding of nuclear power R&D. A more adequate overall level of DOE funding of the program in the range of $200 million annually, excluding facilities, is needed if the high-priority goals of the program are to be met successfully. The industry is sharing the funding burden primarily in LWR technology where it carries the lion's share of the R&D programs on the present plants and essentially 50–50 sharing on future advanced light water reactor (ALWR) R&D.

Evaluation of past successes in industrial development has clearly shown that a long, persistent, and stable development effort is a necessary element of success. Energy competitiveness in price and technology is at stake. If the United States is to remain a major supplier of nuclear reactor systems to the global markets, it must continue to be a major contributor to the advancement of the technology. This requires a long-term commitment to advanced reactor research and a consistent pattern of multiyear funding. Only in this way can requisite talent be attracted into the program and encouraged to remain involved over the long term. A fundamental related concern is erosion of the supply of new technically skilled people. It is difficult to induce young people into a technical field without a sense that there are new challenging technical and business opportunities in its future. New talent is needed not only to pursue advanced nuclear power development but to replenish the professional work force in existing nuclear power generation activities to assure their continued safety and excellence in performance.

Industrial competitiveness in the global market does not preclude international cooperation in advanced reactor development. A prerequisite for a stable market for nuclear power is an international consensus on the technical characteristics and standards required to assure reactor safety and environmental compatibility. In-depth research, development, and demonstration that result in technically sound and reliable nuclear power plants are also essential, and there is merit in sharing the costs, facilities, and human resources among the supplier nations to implement that R&D. Such international cooperation should, therefore, be fostered. It is encouraging that the level of international cooperation in R&D is increasing in all three developments: ALWR, MHTGR, and ALMR.

Generic Technical Features in the Advanced Reactors

There is a substantial degree of commonality in the technical improvements and innovations in the three advanced systems—light water, gas-cooled, and liquid-metal—that are directed first at assuring public safety; second, minimizing environmental impact; and third, achieving substantial economies in capital and operating costs (Taylor, 1989). Subsequent papers in this session will describe these features, and I would simply like to summarize common, generic characteristics which make each of them promising in future deployment of nuclear power capability.

Each of the advanced systems is clearly emphasizing safety, environment, and economy. All have set goals for reactor safety to at least the level of probability of 10^{-5} per reactor year for severe core degradation and 10^{-6} per reactor year for a significant release of radiation to the atmosphere, a level of safety about ten times better than present reactors. These goals are being met by greater utilization of natural processes for core and containment cooling, by increasing the margins for system and component operation, and by major advances in the human engineering of the plant. These safety goals are also environmental goals, because if there is assurance that a severe accident does not occur or will be contained, the environmental benefit of nuclear power will be achieved: an energy-generating system with essentially no atmospheric pollution.

The economic goals are being met by the substantial simplification that is being sought by substituting natural processes for electric-powered machinery and trading off high performance (and its concomitant high demands on equipment and operators) for simpler, more rugged components and systems. A move to smaller unit power output and modular design has also been made to achieve lower capital cost, shorter construction times, and a greater level of factory, as contrasted to field, construction.

I would like now to summarize the recommendations (ARC Report, 1990) of the utilities on the course of development of the three advanced reactor types.

Advanced Light Water Reactor (ALWR)

High priority should be given to the ALWR program as the leading technology to reopen initially the nuclear option in the United States. The advanced LWR program, presently sponsored jointly by the utility industry and DOE, offers the most promising approach to developing the next increment of nuclear power plants. This program has three objectives. The first is the development of utility requirements for future LWRs to achieve a substantial increase in safety, increased lifetime reliability, reduced capital costs, shorter construction time, and lower operating and maintenance costs. A substantially simpler plant design is also called for to help achieve the safety, reliability, and cost goals. The second objective is to obtain Nuclear Regulatory Commission (NRC) certification of the designs of large (1300 MWe) evolutionary Boiling Water Reactor (BWR) and Pressurized Water Reactor (PWR) plants already developed by the reactor manufacturers and based on these utility requirements as well as on the detailed design data and testing base from which the evolutionary designs have evolved. The third objective is to develop the option for an economical smaller and simpler unit power output (600 MWe) PWR and BWR and to obtain NRC certification based on the utility requirements, the extensive LWR operating experience base, and greater use of natural processes for core and containment cooling.

The limited resources and competition for federal government funding have caused some suggestions for an "either/or" approach to large, "evolutionary" plants and smaller, "passive" plants. Both systems have a place among the options for future expansion by US utilities and provide a needed flexibility in unit power output to match individual utilities' growth, network capacity, financial requirements, and construction timing. Both systems have improved safety, reliability, and economic potential. The evolutionary plants will be available as an option sooner than the passive plants and provide the most immediate potential for export sales by US manufacturers/architect-engineers. The passive plant, in addition to filling a niche at lower power output without entailing an economic penalty in capital cost per kilowatt, offers the potential of simplification and reduced burden on the operator because natural cooling processes have been substituted for electric-powered cooling systems. The passive plant also provides the potential to achieve a much higher level of standardization than has been achieved to date on US LWRs because its design will be based rigorously on standardized utility requirements. The greater degree of simplification and reliance on natural processes lends itself more easily to detailed design standardization, and the detailed design is being implemented with a well defined set of industrial, operational, quality, and regulatory standards. If the promises materialize, the design might well emerge as a "winner" not only in the United States but in international markets where the smaller size has traditionally had strong market appeal.

Standardization is a key objective in the ALWR program both to improve licensability and to achieve greater economy. The foundation for standardization is being developed through the definition of detailed utility requirements for future light water reactors. This three-volume Utility Requirements Document stipulates safety, economic, and reliability goals, key performance parameters, safety and power margins, and human factor attributes. The major thrust of the requirements is to elicit the simplest, practical plant to build, operate, and maintain. To assure consistency with NRC's standardization policy and conformance with regulatory requirements, the Requirements Documents are being submitted for review and approval to NRC to provide a reference base for the design certification process. These reviews are being carried out in parallel with evolutionary plant certification and in series with the passive plant certification.

The participation of NRC in reviewing and approving the utility requirements documentation and completing the design certifications is essential to the success of the program. The fundamental concept in this effort is that NRC, by review of the requirements, will establish its position on safety and licensing generically and then, on the basis of that position, will complete the certification processes for individual designs.

Although the establishment of uniform and stable utility and regulatory requirements are essential to achieve standardization, significant additional effort is needed by the industry to achieve standardization in design, construction, operation, and maintenance. This level of standardization should permit the application of one set of engineering, fabrication, and construction planning, and operational and maintenance procedures for each of the PWR and BWR series rather than for each plant.

Modular High-Temperature Gas-Cooled Reactor (MHTGR) Development

The development of high-temperature gas-cooled reactor (HTGR) technology in the United States occurred primarily within the private sector and focused on plant deployment until the collapse of the commercial nuclear market in the mid-1970s. Over the past decade, however, DOE has been sustaining design engineering development at an annual level of approximately $25 million in recognition of HTGR's potential as an advanced alternative for electric power generation as well as its long-range potential as a source of high-temperature process heat. The goal of the MHTGR program is to achieve economical, safe, reliable power through the use of an innovative modular design with a reactor module power output of 140 MWe. The present DOE plan is to pursue final design approval of a standard four-module nuclear island by 1996. Utility/user support is currently organized through Gas-Cooled Reactor Associates (GCRA), with supplemental technical support by Electric Power Research Institute. This private-sector interest and potential to build a lead plant is required to complement the DOE program such that certification and commercialization can be realized.

Through the years of development, testing, and operation, the HTGR has shown advantages that are derived from the inherent material properties associated with ceramic-coated fuel particles, graphite core structure, and helium coolant. The design, material properties, and operating characteristics of the MHTGR result in enhanced safety margins without reliance on AC-powered safeguard systems and/or operator actions. These margins provide the technical basis for fulfilling the GCRA requirements to eliminate the need for off-site public evacuation and sheltering plans. As a result, the MHTGR has the potential to provide eased siting and licensing.

In August 1988, DOE recommended that a variant of the civilian MHTGR design be deployed as a half-size new production reactor (NPR) at the Idaho National Engineering Laboratory as part of a dual-technology, dual-site program to assure an adequate supply of materials for the nation's nuclear defense requirements. The planned schedule projects start up in the late 1990s. For fiscal year 1990, approximately $120 million has been appropriated for the MHTGR-NPR. Utility interest for the MHTGR-NPR project has developed based on the expected benefits this project will bring to the civilian MHTGR program from the common design and technology development programs.

To the fullest degree practical, the synergism between the MHTGR civilian and defense production (NPR) reactor programs should be fostered, consistent with nonproliferation objectives. A key requirement to achieving this potential is that the nuclear island and all elements of its safety systems, accident prevention, and mitigation, be as similar as practical so that the NPR can provide prototypical experience for the commercial MHTGR nuclear power plant. At present, there are significant design differences between the two systems involving both power production and safety systems. Prominent among these differences are coolant circulator bearings, quality control requirements, and containment/confinement provisions. Prompt decisions are needed to make changes in either the commercial or military versions (or both) to remove as many of these differences as practicable.

The MHTGR-NPR will be reviewed and approved by the new Defense Nuclear Facilities Safety Board. This process should closely parallel the NRC licensing process whose approval has strong prospects of being tantamount to a commercial license. Because of its potential role as a prototype power plant, the MHTGR-NPR should also be expected to meet utility (owner-operator) requirements for design to assure effective operational/maintenance characteristics. The DOE has already started the process of developing such requirements and is seeking utility input to that process. This process should be completed in an in-depth manner.

Whether or not the MHTGR-NPR project is deployed, the civilian MHTGR program must establish a lead project that will design, license, construct, and operate a full-scale, prototypical plant in a civilian environment and thereby demonstrate that the MHTGR is safe, reliable, economical, and acceptable to the investors and public.

Advanced Liquid Metal Reactor (ALMR) Development

The DOE has sponsored the development of the LMR since the early 1950s and has amply demonstrated its technical feasibility and excellent passive safety characteristics. Although the LMR has not yet been demonstrated to be economically competitive, the innovative efforts of the past several years have given promise that economic competitiveness can be achieved. The DOE has, therefore, established economic competitiveness as a major goal for its ALMR development program. If this goal is achieved, this system could be deployed commercially in the next century. Then, as increased reliance on lower-grade uranium resources forces prices upward, the ALMR would become, in the 2030–50 time range, increasingly attractive as a power producer because of its vastly superior fuel utilization efficiency, especially when used as a breeder.

Several major developments have emerged recently from the DOE's ongoing R&D program activities. The new conceptual designs of the LMR incorporate additional passive safety characteristics which reduce and, in some cases eliminate, the need for engineered safeguards equipment or emergency operator response, greatly reducing chances of a severe accident. Significant progress has also been achieved toward improving the economics of the ALMR through such innovative engineering approaches as: modularized construction, improved materials, natural circulation cooldown, greater automation, and shop fabrication of the nuclear steam supply system.

The ALMR program has been reoriented to be effective and focused and strongly innovative as befits its goals of safety and economy as well as its timing. There are several areas in which emphasis on the ALMR should be increased. The budget reductions have primarily been applied to reduce the industrial contractors' participation in the program. This will weaken attention to the plant design, fabrication, and construction aspects as well as lose design know-how contained in the reactor manufacturers and architect-engineer organizations. Utility experience should be solicited through design reviews and the definition of owner-operator requirements, and evaluation of economic potential. A start has been made on this, but continued in-depth pursuit of this experience is needed. Ultimately, an NRC license will be required when the time comes to build a prototype plant. It is important that NRC continue to review the technical features of the ALMR designs with the purpose of reconciling any significant differences of opinion bearing on the licensability of the ALMR.

Conclusion

The US utilities have formally expressed for many years the need to reopen the option for nuclear power expansion in the United States. That expression has increased in urgency recently for several reasons:

- The projected shortfall in baseload electric generation capacity in the mid-1990s is becoming more certain.
- New requirements for air pollution controls on coal plants, such as those identified in the Administration's Clean Air Bill, will increase the cost and regulatory uncertainty of coal generation.
- Increased concern regarding the long-term effect of combustion processes on global warming calls for greater priority in developing and utilizing noncombustion processes for electric generation.

Although institutional reforms in regulation and financing are essential to reopening the option, the utilities advocate increased priority on vigorous R&D programs, both to enhance the capability of the present nuclear power plants and to define future plants that are greatly improved. They are fostering the development of systems that achieve the full potential of nuclear power to assure public safety, environmental quality, and cost competitiveness. The utilities recognize that this full potential, and the responsibilities it entails, must be realized globally to be effective. Thus, international cooperation in R&D is considered to be essential.

That assurance will be provided by more complete utilization of the intrinsic technical characteristics of these systems to achieve safe operation with essentially no atmospheric pollution: self-regulating neutronic processes in the core to assure stability, passive cooling to assure decay heat removal, and containment cooling in the event of an accident. To provide further assurance, the systems are being developed to be more forgiving of equipment failure and human error: They are more simple and rugged, have more safety and performance margins, and have been human-engineered in all aspects. The simplicity and lifetime reliability, as well as reduced maintenance costs, will more than compensate for the added cost to provide these margins.

The future is hazy as to when and how the institutional barriers to reopening the option for expanded nuclear power will be removed. But when they are, these R&D programs will have provided the basis for nuclear power plants which meet the environmental and economic needs of society in the 21st century.

References

"Report of the ARC Ad Hoc Committee on US Department of Energy Advanced Reactor Development Plan," Advanced Reactor Corporation, January 10, 1990.

Edison Electric Institute, "Report of the CEO Task Force on Nuclear Power," February 1985.

O'Connor, James J., Chairman, Nuclear Power Oversight Committee Report, "Nuclear Energy for the Future: What We Must Do," forwarded to Secretary of Energy Watkins by letter, January 27, 1989.

Testimony on May 24, 1988, to the Senate Subcommittee on Energy Research and Development by James J. O'Connor, Chairman and CEO of Commonwealth Edison Company, John J. Taylor, VP, Electric Power Research Institute, and John Landis, Sr. VP, Stone & Webster Engineering Company, as well as to the Nuclear Regulatory Commission by Sherwood Smith, Chairman, President and CEO of Carolina Power & Light Company.

Taylor, John J., "Improved and Safety Nuclear Power," *Science*, Vol. 244, April 21, 1989.

The Advanced Pressurized Water Reactor—Meeting the Energy Needs of Today and Tomorrow

Richard J. Slember
Energy Systems Business Unit
Westinghouse Electric Corporation
Pittsburgh, PA

Abstract

In today's world, increased demands for energy, coupled with concern for the environment, provide an opportunity for the further development of nuclear energy. Advanced pressurized water reactors (PWR) like the Westinghouse AP600 can help provide the world with secure, reliable electricity in an environmentally acceptable manner today, and well into the 21st century. With design simplification, emphasis on economics, modular construction, and enhanced safety, the advanced pressurized water reactor represents a prime energy source for both developed and developing nations through international cooperation. Continued improvements of the advanced PWR design will yield additional simplification. The solution to the safe disposal of radioactive waste exists and will eventually be implemented. The outstanding performance of today's light water reactors, along with the deployment of advanced designs, promises to provide our nation and the world with clean, safe, nuclear energy.

As we enter the last decade of the 20th century, we find the world in a constant state of change. The sweeping changes in the political structure of world governments, the increased global awareness of the earth's environmental vulnerability, the continued advancement of industrialized and developing nations, and the increased demand for energy—especially electricity—are vital issues in today's world. Many of these factors are interrelated, so that a change in one will have an impact on the others. This paper will look at the role of nuclear energy and the advanced PWR in meeting our energy needs in this changing world.

Like the rest of the world, America needs more electric generating capacity. A strong economy, able to welcome new members to the labor force and to produce world-class products at competitive prices, demands reliable and low-cost electric power. To remain strong, we need more generating capacity, and we need it now.

After a long period of excess capacity nationally, and despite continued adequate reserve margins in most western states, we face power shortages in the northeast and along the Atlantic coast. The Pennsylvania–New Jersey–Maryland interconnection was forced to impose voltage reductions four times last summer; the Long Island Lighting Company once. A massive pre-Christmas cold wave forced utilities to institute rolling blackouts for the first time ever in Texas and the first time since 1983 in Florida.

We will need significantly more new power capacity in the decade just ahead; 115 gigawatts is the most widely accepted estimate of new power needed in the 1990s. Beyond that, we see estimates of 168 gigawatts of new capacity being required between 2001 and 2010. This includes 38 gigawatts of replacement capacity and 100 gigawatts of new baseload capacity. These factors place the nation at the beginning of a period of power expansion that is absolutely necessary. Where should we look for this new power? We should look everywhere. Energy security should be sought in energy diversity—conservation, clean coal, the renewables, and nuclear.

Nuclear power passes all the tests we should apply to a primary energy source. It provides one fifth of the electricity in the United States, 17 percent of the world's electricity, with none of the greenhouse gases or acid rain emissions. Since 1973, it has replaced $114 billion worth of oil imports. And despite the country's inability to agree on a site for the final storage of high-level radioactive wastes, the relatively small amount of wastes is being stored safely. It is clear that nuclear power is, and will continue into the 21st century to be, essential to our energy security, essential to our economy, and essential to protecting our environment. Nuclear energy is not the total answer, but it is a vital component of our energy mix.

What kind of nuclear reactor is best suited to meet our need for power? It must be a reactor that can be standardized and provides utilities certainty in costs and construction schedules. And it must be a reactor that is more than safe. It must satisfy the public's perception of safety.

By most standards, the present generation of light water reactors has been a remarkable success. They have an enviable safety record. They can be reliable and economical. The Wolf Creek plant produced more electricity in 1989 than any other US generating station—nuclear or fossil. Between 1985 and 1989, the Callaway plant produced 16 percent more electricity than any other US nuclear plant. Three Mile Island (TMI) 1 had a capacity factor of 100 percent in 1989. A combination of conditions, however, has made utilities reluctant to build nuclear energy plants. The cost of more recent units was escalated by construction and licensing delays. In addition, some state regulatory commissions have held utilities responsible for changes in electrical-use patterns in the aftermath of the oil embargoes, making it difficult for utilities to earn adequate returns on their later units.

If utilities are to resume building nuclear energy plants, new reactor designs must address these concerns. Utilities must be certain of construction costs and schedules. Despite the safety record of present reactors, new designs must offer some segments of the public even greater confidence. To allow utilities to follow load growth more closely, the plants should be smaller than recent units. The plant design should be standardized, so that once it is approved by the Nuclear Regulatory Commission (NRC), it can be reproduced with only site specific changes. Finally, because of our immediate need for power and our already dangerously high dependence on foreign oil, new reactors must be based on proven technology so we can begin building them in the next decade. The advanced light water reactors being developed under the combined leadership of the US Department of Energy, the Electric Power Research Institute, and the nuclear power industry meet all of these criteria and more.

One such reactor is the AP600, the advanced light water reactor Westinghouse is developing in the Department of Energy (DOE)/Electric Power Research Institute (EPRI) program. The AP600 is an advanced, simplified 600 megawatt pressurized water reactor (PWR) with automatic safety systems. It is conservatively based on proven, licensed technology with a new emphasis on safety features that rely on natural forces. In comparison to other advanced modular nuclear energy plant concepts, the AP600 achieves the benefits of natural safety systems while it simultaneously retains maximum lineage to the pressurized water reactors operating in the US and around the world. The AP600 has the unique advantage of being both new and proven at the same time. The AP600 plant arrangement is shown in Figure 1.

The design incorporates 3,000 reactor-years of PWR operating experience worldwide; improvements and lessons learned from Three Mile Island; and input from utilities, regulators, and the public. The major components and systems are the same as those in over 200 operating PWRs in the western world, but they are configured in a simplified design to meet new criteria and pass both regulatory and public scrutiny.

Safety, simplicity, and reliability dominate the design process. For example, the simplified primary loop employs inverted canned motor pumps, which have consistently demonstrated their reliability in the field. As shown in Figure 2, the pumps are mounted in a close-coupled configuration with the steam generator. The design eliminates the loop seal between components and simplifies the support system. Here a simpler design brings tangible improvements in operations and reliability. In the primary loop, long radius bends in the reactor coolant piping reduce pressure drop and improve efficiency. The lower power density of the core results in improved safety margins.

The AP600 design results in major reductions in equipment and buildings. The plant will have roughly half as many valves, large pumps, pipes, heat exchangers, ducting, cable, and building volume. Such simplification contributes much toward a lower capital cost and a shorter construction schedule.

The AP600 safety systems are simple on a universal scale. There are no safety-grade diesels, no evaporators, no safety-grade pumps, no chillers or coolers, and no safety-related cooling units or service water. Each component that is not there cannot fail. Simplicity is also achieved through automatic safety systems which rely on gravity, convection, conduction, and evaporation to cool the reactor in the event of an accident.

When cooling water is needed for emergency use, gravity and pressure bring it to the reactor core from several large tanks inside of the containment, positioned above the reactor vessel. When the water storage tanks are emptied, the containment is flooded to a level above the reactor coolant loop, and a natural long-term cooling cycle begins.

Convection brings air to cool the outside of the containment building and moves it across the surface to naturally remove heat as shown in Figure 3. If the pressure and temperature in the containment exceed set points, water storage tanks located above the containment automatically release water at a controlled rate. The water flows naturally downward, evaporating under the influence of the surface heat and convective air flow. Convection, conduction, and evaporation work together to cool the containment. Tests have confirmed that this system of natural circulation and heat removal could continue safely, with no operator action required, until the plant has cooled and returned to a stable condition.

The simplicity of the design leads to simple means for assuring safety. No active systems are required to maintain safe operation under accident conditions. In the worst kind of assumed loss-of-coolant-accident (LOCA) natural processes provide adequate cooling to prevent core melt or even core damage.

The plant is significantly more tolerant of human errors should they occur. Margin is designed into the AP600 to give operators additional time to act when necessary. Safety margins of passive reactors, aided by a strong containment, are extraordinary.

The AP600 is designed with ease of construction in mind. The plant will be largely prefabricated, using factory-built, factory-inspected construction modules.

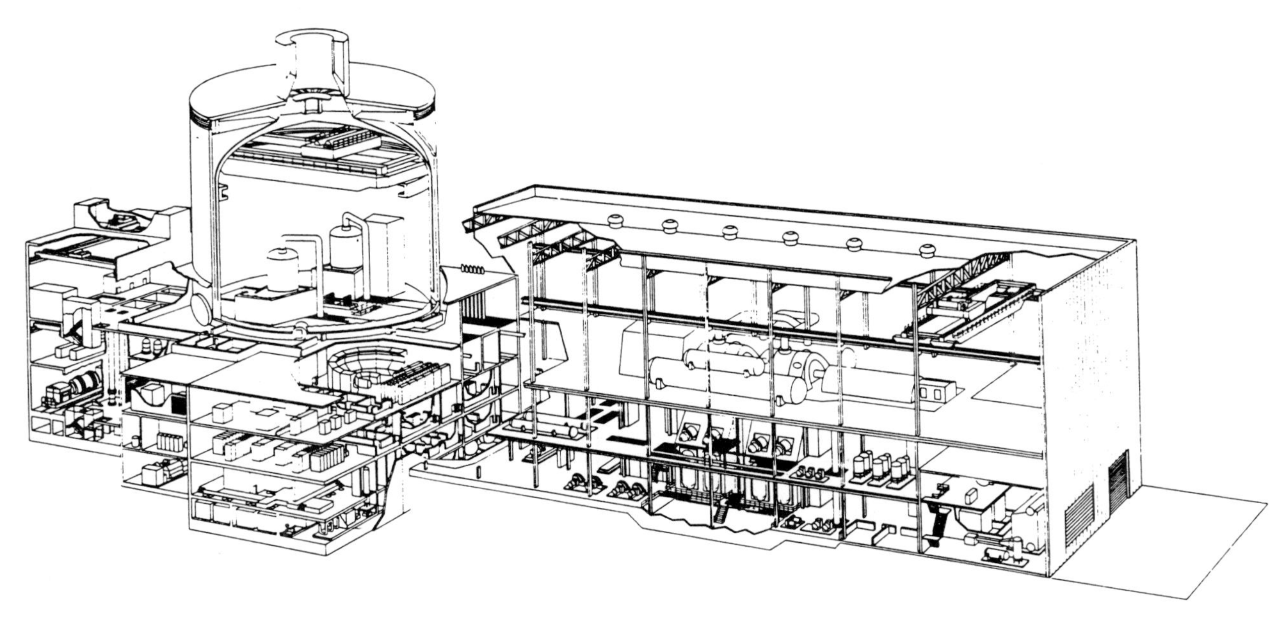

Figure 1. AP600 plant arrangement.

Used in modern shipbuilding, factory prefabrication enhances quality and increases worker productivity. Our partner in this part of the design is Avondale Shipyards, the leader in modular construction. The AP600 modules will travel by barge or railroad car to the site where they will be assembled. Fast modular construction of this type can put a nuclear plant on line in just three years, greatly reducing interest on capital.

Besides a simplified safety system that should create greater public confidence, the AP600 addresses utility concerns for low-cost electricity. The AP600 is the optimized progression from the Westinghouse two-loop, 600 megawatt PWR. Six of these plants in the United States consistently perform 10 percent above the national average and are among the top performing plants in the world. Simplification will further enhance plant inspection and maintenance.

Commercializing the AP600 will also be cost effective. Because the AP600 makes use of existing PWR technology, no prototype plant will be required. A prototype plant could take a decade to build and cost several billion dollars.

The US Department of Energy strongly endorsed the AP600 in 1989 when it awarded Westinghouse a $50 million contract to continue to develop the technology through the design certification stage. The effort also includes $70 million from program participants and EPRI. Westinghouse expects to receive design certification from the Nuclear Regulatory Commission in 1994.

In order for the first advanced PWR to operate in the 1990s, we must overcome institutional obstacles that have kept America from gaining the full benefit of its nuclear technology. The NRC has taken major steps to avoid unnecessary license delays. We believe their rule combining construction and operating licenses must, however, become federal law. We must educate the public on the benefits of nuclear energy and inform them of the higher safety margins and tolerances for human error of the simplified reactors being developed. Furthermore, if the existing nuclear power plants continue to operate efficiently and safely, public acceptance will increase. These accomplishments will help when utilities begin seeking permits for new nuclear units.

The next milestone in breaking down institutional barriers to nuclear energy is to build an advanced PWR. Westinghouse believes a "lead plant" will be required to clear the path. To this end, Westinghouse and a team of its industrial partners are proposing a lead plant program using the Westinghouse AP600 as the model plant for stepped-up development. Under this plan, an AP600 plant could be in operation by the end of the decade.

In the lead plant strategy, several key activities would be undertaken in parallel, rather than in sequence. These would include site selection and certification, and the completion of first-of-a-kind lead plant engineering. By conducting these activities in parallel with reactor design and certification, with the efforts to achieve the necessary institutional improvements, the lead plant approach would expedite the operation of the first advanced PWR plants by three years or more. To make this plan a reality, Westinghouse is fully committed to innovative approaches and working with all parties and the federal government.

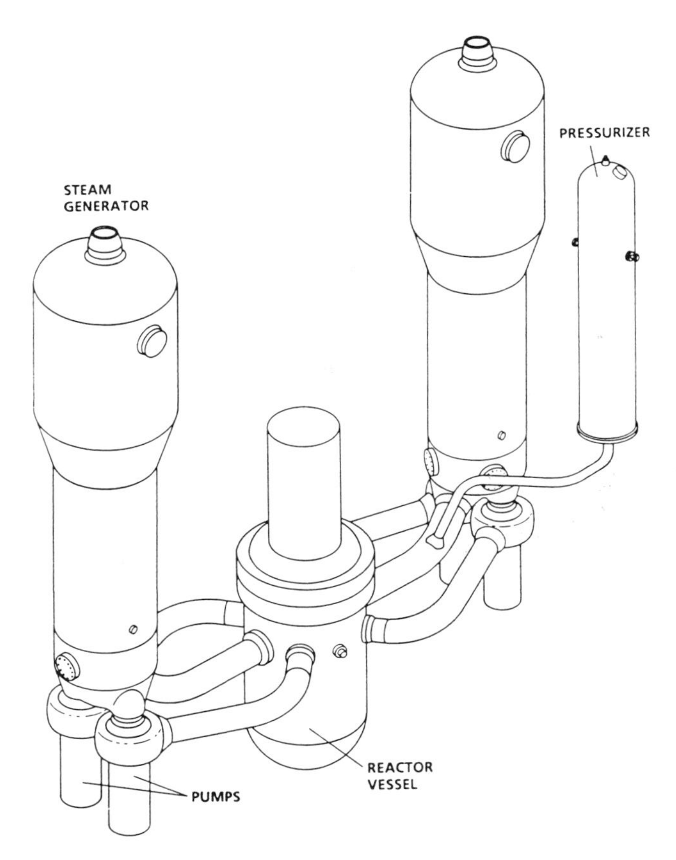

Figure 2. AP600 nuclear steam supply system.

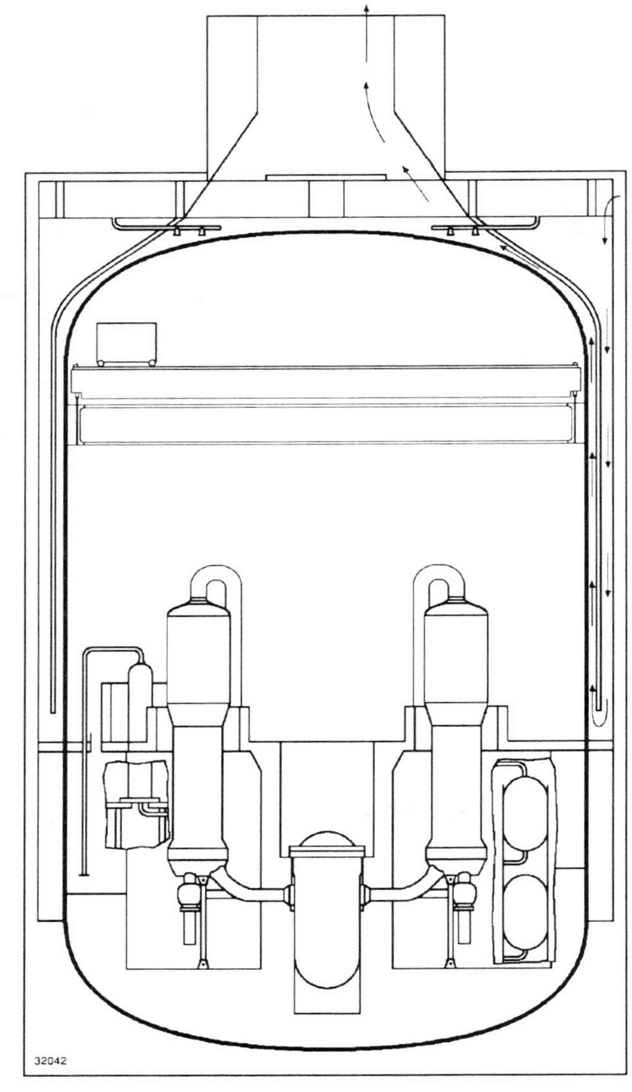

Figure 3. AP600 natural containment cooling system.

Today's electrical generating stations in the United States are an average of 21 years old. The older plants will soon need to be replaced as they reach the end of their operating life. The old fossil plants could be replaced with advanced light water reactors which would help to alleviate the environmental concerns over global warming and acid rain. The older nuclear plants could have their original 40–year design life objective extended into the 21st century through the process known as plant life extension. Using the previous operating history, a re-analysis of the plant is performed to determine the actual unused structural fatigue portion of critical components such as the reactor vessel. The utility would then apply to the NRC for an extension of their operating license. Having completed a feasibility study, a lead plant license renewal project is currently under way for both a pressurized water and a boiling water reactor plant. If successful, it will pave the way for extending the generating capacity and environmental benefits of existing nuclear plants into the next century.

When a nuclear plant has finally reached the end of its economic life, it will be shut down and decommissioned. Studies are being conducted with various scenarios for reusing the site for a new plant. This would be advantageous to a utility in terms of owning a qualified nuclear site and the local acceptance of the plant gained through years of safe and reliable operation in the community. The site would then be available for the construction of a new advanced light water reactor such as the AP600.

Of the estimated 100 gigawatts of new baseload capacity required between 2001 and 2010, 60 percent of it could be nuclear. With a design life objective of 60 years, advanced PWR plants built in the 1990s and the early part of the next century will be able to supply the United States with electrical energy well into the 21st century. This would be in line with estimates of available uranium supplies to power the light water reactors.

Beyond this time, hard spectrum PWR convertor reactors and breeder reactors, which would be able to utilize the spent fuel from today's light water reactors, would come on line to meet our energy needs for hundreds of years. In addition, advances in fusion and solar energy may make their unlimited sources available on a commercial basis by the middle to late 21st century.

As far as future PWR design innovations are concerned, we can look ahead to evolutionary process. Advanced control rooms will be developed incorporating intelligent decisionmaking features through the application of adaptive learning algorithms to the plant instrumentation and control systems. One proposed control room design would use the concept of a "wall" of graphical information that surrounds the operator. The surface of the wall will present, simultaneously, all information needed for the operation of the plant. The operator need only look at the right area of the wall to see required information. Thus, all the information that is needed is available in parallel. The operator interacts with the wall using a cursor or verbal commands to operate the plant. A three-dimensional plant database and a combination of conventional and holographic displays will be used to assist in operation and maintenance as necessary. This advanced instrumentation and control system will reduce the probability of human error and control system failures, thereby reducing risk. They will also reduce the operation and maintenance costs of the plant.

With design simplification, emphasis on safety, and modular construction, the advanced PWR will become ever more available to developing nations as we move into the next century. Under cooperative international guidance, advanced PWRs based on AP600 technology could be made available to meet the specific requirements of those nations. This has worked in the past for industrialized nations with the transfer of pressurized water reactor technology from the United States. The United Kingdom, for example, has been able to access America's 30 years of PWR experience in developing their Sizewell B project.

The lack of sufficient electrical capacity in developing countries is one of their major problems. Today, the Third World has just 120 watts of electric generating capacity per person. The United States alone has 2,900 watts per person. In the Third World, the demand for electricity has been growing at more than 7 percent per year over the past 20 years. Developing countries already spend an average of 25 percent of their public development budgets on electric power.

Following the patterns of the past will result in building two types of power plants, fossil and hydroelectric, both accompanied by their existing problems. Increased burning of fossil fuels will accelerate the greenhouse effect worldwide. Building many hydroelectric projects would displace people, communities, and cultures on a large scale.

Advanced PWRs can provide the reliable electric power supplies that developing nations need to industrialize without the problems associated with the other sources. Their automatic safety features simplify both the plant and the challenge of operating it. The advanced PWR is a plant that is within the safe operating capabilities of a developing nation. Innovative financing and international cooperation can turn this idea into a reality. Developing nations could then share in the benefits of advanced nuclear technology as they industrialize and grow in the world community of the 21st century.

Whenever nuclear energy is mentioned with regard to protecting the environment, the question of radioactive waste will always be raised. As a means of comparison, between now and the year 2000, all the high- and low-level wastes from America's nuclear plants could be stored in a single landfill the size of a baseball field and 50 feet deep. Ash and scrubber sludge from America's coal plants require 1,000 landfills that size every year.

The nuclear industry as a whole supports the eventual long-term burial of high-level radioactive waste. Nuclear waste is not a technological problem. The French derive more than 70 percent of their electricity from nuclear, and they manage their wastes safely and calmly. In the US, the waste is a political and emotional issue. People have become so alarmed by misinformation that the government has a problem with any locale accepting a waste repository. Public education and public involvement would help to assuage the concerns over nuclear waste. In the meantime, management of wastes from commercial nuclear plants poses no environmental threat. Currently, we safely store high-level wastes at the power plant sites. Each facility is licensed by the federal government, and there have been no releases to the environment.

An interim solution to the waste disposal dilemma in the United States could be Monitored Retrievable Storage (MRS). MRS would provide an acceptable alternative as the delays to construct a permanent, underground repository seem to continue. The MRS would not require any new technology to store the spent fuel assemblies for a limited time. It would also provide excellent experience in waste management techniques and the spent fuel assemblies would be readily accessible if a change in the current fuel cycle philosophy should occur.

Regarding radioactive waste generation by the advanced PWRs, the AP600 produces substantially less liquid, gaseous, and solid radioactive waste in normal operation compared to current PWRs. In addition, spent fuel reprocessing, which may be of future environmental importance, is available technology for the AP600.

As we move toward the 21st century with energy sources like the advanced PWR, we have the opportunity to help satisfy humankind's great need for energy and avoid an ever-growing threat to our atmosphere and environment. However, we will not achieve our goals for tomorrow without our best and most concerted efforts today and without the understanding and support of the public. The challenge is a pressing one, but one we must accept in order to provide the world with clean, safe, and reliable energy.

The Advanced Boiling Water Reactor and the Road to Revival of Nuclear Power

Bertram Wolfe
GE Nuclear Energy
General Electric Company
San Jose, CA

Abstract

The United States today does not have a viable nuclear option. A 15-year hiatus in US nuclear orders has been accompanied by the development of a number of institutional impediments which must be removed before a return to nuclear power can occur. These impediments include an unpredictable regulatory process at the federal level and the lack of effective design standardization that has characterized the US nuclear program. There have been encouraging recent developments in these interrelated areas.

In 1988, the Nuclear Regulatory Commission (NRC) began the process of revising their rules to streamline the regulatory process. In May 1989, the NRC's new 10 CFR 52 regulations went into effect. Part 52 represents a new regulatory approach to licensing standardized nuclear plants in the United States.

US manufacturers have also taken important initiatives toward the development of future standard advanced light water reactor (ALWR) designs. In General Electric's case, our leading effort is focused on the development of the advanced boiling water reactor (ABWR)—an advanced 1350 MWe evolutionary LWR, followed by the exciting new 600 MWe passively safe BWR, the SBWR. The ABWR, undergoing standard licensing in the US today and ready for start of construction in Japan next year is described in this paper.

Status of Nuclear Power

Since the Arab oil embargo of 1973 when energy prices rose, and growth of energy use declined, there has been no need for new nuclear plants or, for that matter, new baseload electrical generating plants of any kind in the US (see Figure 1). Utility purchases of new plants through 1973 produced electrical capacity well in excess of that necessary to meet demand until recently. In fact, many nuclear as well as coal units were suspended or cancelled in the middle and late 1970s, while still leaving a surplus of capacity. Since new base load plants weren't needed, it was not necessary to confront the difficult choices among greater reliance on conservation, increased burning of fossil fuels, or construction of more nuclear plants, or combinations of such new capacity alternatives.

The excess of electric generating capacity that has existed for the past 16 years, however, is coming to an end. Brownouts have occurred in the eastern US during the past two summers, and the nation as a whole will face electrical shortages unless new baseload plants are constructed in the 1990s (see Figure 1). The difficult choices must now be faced.

The environmental impact of burning fossil fuels is becoming increasingly of national and international concern. The "greenhouse effect," acid rain, air pollution, and oil spills are receiving public attention, and are the subject of major legislative actions by the Bush Administration and Congress. In addition, the United States is becoming increasingly dependent on Mideast oil, today importing more than 40 percent of its oil.

Perhaps the problems, or apparent problems, with fossil fuels will be resolved, but under the circumstances it would appear irresponsible not to revive a viable nuclear option. Nuclear power, while not risk-free, is today producing 20 percent of our electricity safely and economically without polluting the air, creating acid rain, contributing to the "greenhouse effect," or subjecting our energy supply to the vagaries of Mideast governments. It could play an even greater role in meeting our future energy and environmental challenges.

Unfortunately, the United States today does not have a viable nuclear option. The hiatus of US nuclear orders since 1973 has been accompanied by the development of a number of institutional impediments which must now be removed before any return to nuclear power can occur.

Institutional Impediments

One of the impediments is an unpredictable regulatory process at the federal level. Today, no one—not even the Chairman of the Nuclear Regulatory Commission (NRC)—can predict with any certainty the licensing schedule, or even the eventual licensability, of a new

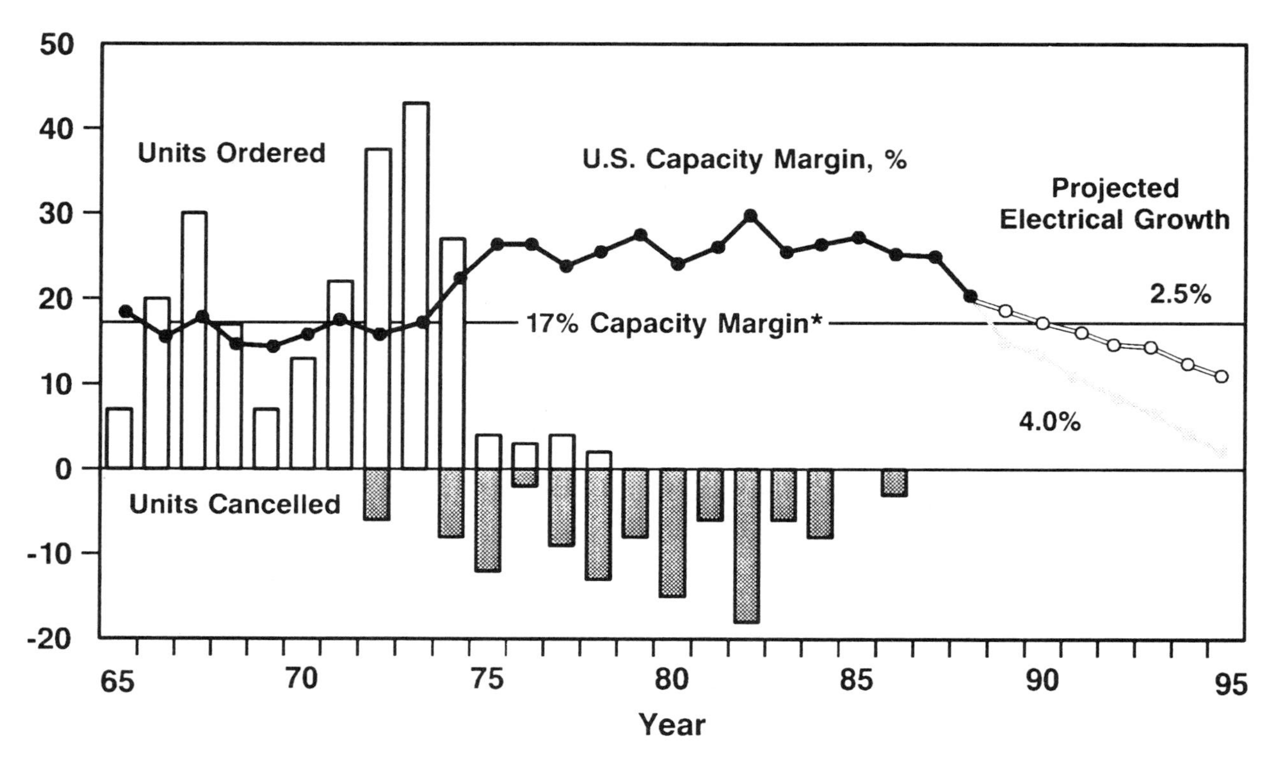

Figure 1. US capacity margin and nuclear orders. (*Minimum required for reliability.) (Source: US Council for Energy Awareness, June 1989).

plant identical to one recently licensed. The current process is characterized by continuous and unpredictable regulatory changes, and opportunities for obstructionist intervention, which make it impossible for a utility to establish construction costs and schedules with any confidence. The resulting financial and planning risks discourage commitment to a new nuclear plant under current conditions.

Another institutional impediment is the lack of effective design standardization that has characterized the US nuclear program. The 113 licensed nuclear plants in the US are, with few exceptions, all different. The lack of standardization is the result of many factors including unique utility requirements, continuously changing regulatory requirements, and different design choices by the many combinations of manufacturers and architect engineers who designed existing plants. The plant unique approach to design was understandable during the light water reactor's (LWR) evolutionary development. Today, however, the LWR has matured and the plant unique approach to design and licensing carries an overhead of inefficiencies and costs which can no longer be sustained by a competitive nuclear option.

Revival of the Nuclear Option

Although several reactor concepts are being promoted for the future, near-term revival of the nuclear option will depend upon the light water reactor. The LWR, in contrast to the more developmental liquid metal reactor (LMR) and gas-cooled reactor (GCR) concepts, is backed by 35 years of experience with hundreds of plants and an infrastructure fully knowledgeable of LWR design, licensing, and operation. Experience with the LWR has taught that, regardless of initial promise, problems can and will occur during the 40+ year product life cycle of a nuclear system. In the case of the LWR, the unforeseen technical problems have been met, regulatory criteria are in place for new plants, a demonstrated technical base exists for developing optimized standard plant designs, and organizations experienced in the technology are in place and ready to serve.

The LWR has demonstrated its capability to generate safe, economical electricity. The US, as pioneer of LWR technology, has, to some extent, carried the development and learning burden of LWR technology while other countries, building on US technology, have avoided the pitfalls and reaped great benefits. The potential of this technology is amply demonstrated by looking not only at the United States but at worldwide experience.

France and Japan are frequently cited as examples where the LWR technology has been exploited to the fullest. Both countries have achieved enviable results and continue to expand their nuclear base. From a US vantage point, France is somewhat unique due to its reliance on a single national utility, single LWR technology, and single manufacturer. Japan, on the other hand, is more like the United States. Japan has nine utilities, three domestic manufacturers, and employs both boiling water reactors (53 percent) and pressurized water reactors (PWR) (47 percent) in its 48-unit LWR fleet. LWR plants in Japan and France are routinely constructed in four to five years, operate at 70+ percent capacity factors and achieve generation costs superior to all other alternatives. This ongoing overseas experience is based on US LWR technology as provided by General Electric and Westinghouse. Indeed as noted below, the advanced BWR (ABWR) is being developed by General Electric and its two Japanese partners, Hitachi and Toshiba, and is scheduled to start construction in Japan next year.

In the US, programs are in place within the government and private sector to address the institutional impediments discussed above. Hopefully this will permit us to meet the cost and performance achieved abroad. Earlier this year, the NRC promulgated a new regulatory approach to licensing standard plants in the US. These regulations, contained in 10 CFR Part 52, provide a preapproval of nuclear plant sites, certification of standard plant designs, and issuance of a combined Construction Permit/Operating License to the prospective owner of a nuclear plant. The NRC is to be commended for its forward-looking approach to standard plant licensing. The new regulations provide the vehicle to assure full public participation in the process while eliminating much of the regulatory uncertainty which has plagued past nuclear projects.

The new standard plant licensing regulations, however, may not be sufficient to restore investor confidence in the regulatory process. The Part 52 regulations still provide the opportunity for a preoperational hearing, albeit of restricted scope, following plant construction and prior to operation. This maintains the exposure of fully constructed plants to the vagaries and uncertainties of a preoperational hearing, notwithstanding the thorough reviews, public hearings, and regulatory approvals of plant site and design before construction began. Although these regulations represent the NRC's intention today, they could be modified in the future if the makeup of the NRC changes in a new political environment. Legislative action to put a Congressional mandate behind 10 CFR Part 52 and eliminate the unwarranted risk of a preoperational hearing would greatly increase industry confidence in this important NRC initiative and the prospects that its practical use will be realized.

The United States utilities have taken a complementary initiative toward standardization by commissioning the Electric Power Research Institute (EPRI) to develop an Advanced Light Water Reactor (ALWR) Requirements document—in effect a standard utility bid specification—for future ALWRs. This effort, which has the support and participation of utilities, manufacturers, architect engineers, Department of Energy and the NRC, represents a key step toward future plant standardization. The ALWR Requirements Document for large, evolutionary standard plants is scheduled for completion in 1990 and GE, working with EPRI, expects that its ABWR will conform to EPRI requirements.

The ABWR is the lead LWR in the US Standardized licensing program today. The Combustion Engineering System 80+ pressurized water reactor follows the ABWR by about a year. Recently, both General Electric and Westinghouse, after competitive bidding, received DOE and EPRI shared cost awards to develop new standardized 600 MWe passively safe LWRs. These exciting plants, the GE Simplified BWR (SBWR) and the Westinghouse AP600, are scheduled for NRC Certification within the 1995 time frame.

The Advanced Boiling Water Reactor

US manufacturers as noted have taken important initiatives toward the development of future standard ALWR designs. In GE's case, our leading effort is focused on the development of the ABWR—an advanced large evolutionary LWR. The ABWR was conceived by an international team of BWR manufacturers under GE leadership. The team's mission was to combine the best features of BWRs from the United States, Europe, and Japan into a single, proven design. The ABWR development effort has been sponsored by the Tokyo Electric Power Company (TEPCO), the world's largest private utility with 17 BWRs in its nuclear program, and largely performed by GE and its technical associates in Japan—Hitachi and Toshiba. Some $250 million has been invested in its development with more than another $100 million in design definition.

The ABWR represents the leading edge in advanced LWR technology. It meets or exceeds NRC and EPRI ALWR requirements, resolves all of the NRC's Unresolved and Generic Safety Issues, and has been adopted as the next generation standard BWR in Japan. In 1987 TEPCO committed to the ABWR for its next two plants. These will be supplied by a joint venture of GE, Hitachi, and Toshiba, with GE supplying the nuclear steam supply systems, fuel, turbines, and generators for both units.

Plant Size/Core

The reactor thermal output is 3926 MWt, which provides for a turbine-generator gross output in excess of 1356 MWe. The reactor core consists of 872 fuel bundles operating at a power density of 50 kW/liter. The ABWR, like all BWRs, is capable of using the most current advanced fuel/core design features. Examples of recent fuel improvements are fuel rods with a zirconium barrier liner, axial variation of enrichment and gadolinia, high fuel exposure, minimal control cells, no shallow control rods, and no rod pattern exchanges.

Reactor Assembly

The reactor assembly (Figure 2) utilizes a BWR reactor pressure vessel configuration equipped with 10 internal pumps for recirculation flow and with electrical-hydraulic control rod drives for fine motion rod control. The reactor pressure vessel has a single forged ring for the internal pump mounting nozzles and the conical support skirt. Forged rings are also utilized for the core region of the vessel shell sections. The elimination of external recirculation piping and the use of the vessel forged rings has resulted in over a 50 percent reduction in the weld requirement for the primary system pressure boundary. This inherently increases the integrity of the system and, of course, results in a sizable capital cost and in-service inspection savings.

The reactor assembly design permits the following design and operating improvements:

- Elimination of external recirculation piping.
- Reduction of containment radiation levels by over one-half compared to current plants.
- Over 10 percent excess flow capability at rated power.
- Lower recirculation flow pumping power.
- Elimination of large reactor pressure vessel nozzles below the reactor core top elevation.
- Reduced control rod drive maintenance.
- Electrically ganged control rod drives.
- Diverse control rod drive insertion capability.

The reactor pressure vessel (RPV) is approximately seven meters in diameter and 21 meters in height. The vessel is of standard BWR vessel design except for two items. The annular space between the RPV shroud and the vessel wall is increased to permit the positioning of the 10 internal pumps used for recirculation flow. Also, the standard cylindrical vessel support skirt has been changed to a conical skirt—again to permit the use of the internal recirculation pumps. The internal pumps are of the wet-motor glandless type. Significant plant operation experience with these pumps is now being accumulated in a number of European BWR plants. In addition, this type of pump has been the favored design for high-pressure fossil boiler circulation pumps for many years.

Emergency Core Cooling

The emergency core cooling system (ECCS) and the residual heat removal (RHR) system have been designed on a three-division basis. Two divisions each provide high-pressure and low-pressure coolant injection capability. A third division combines a reactor steam-driven turbine pump for high-pressure coolant injection and a low-pressure coolant injection system. A triple redundant water delivery/decay heat removal system provides the RHR function.

The elimination of large nozzles on the reactor vessel below the core permits the design of an ECCS system with no core uncovery during the design basis loss-of-coolant accident (LOCA), and at the same time a reduction in total ECCS pump capacity to approximately one-half of that required for an equivalent-size external recirculation loop BWR plant.

Control and Instrumentation

The control and instrumentation systems feature digital/solid state equipment. This equipment permits a design that increases the system redundancy, and provides fault-tolerant operation and the inclusion of self-diagnostics while the system is in operation. Coupled with the digital controls is a multiplexing arrangement for control and instrumentation (C&I) signal transmission. The multiplexing system complements the digital control design. It includes fiber optic isolation capability. The C&I changes defined are most dramatic and result in improved operability and availability, and save considerable construction time with a resultant decrease in capital cost.

Reactor Building/Containment

The reactor building structural design is of reinforced concrete with a seismic design basis of 300 gal. for the S1 or Operating Basis Earthquake (OBE) and 450 gal. for the S2 or Safe Shutdown Earthquake (SSE). The containment design is of the pressure suppression type with a covered suppression pool. From a structural design standpoint, it is a lined, reinforced concrete structure. The containment design features a unique horizontal vent system for the LOCA venting of the drywell to the suppression pool. From an arrangement standpoint (Figure 3), the elimination of the external recirculation piping system permits significant improvements in drywell space utilization—particularly in access for inspection and maintenance activity. The containment

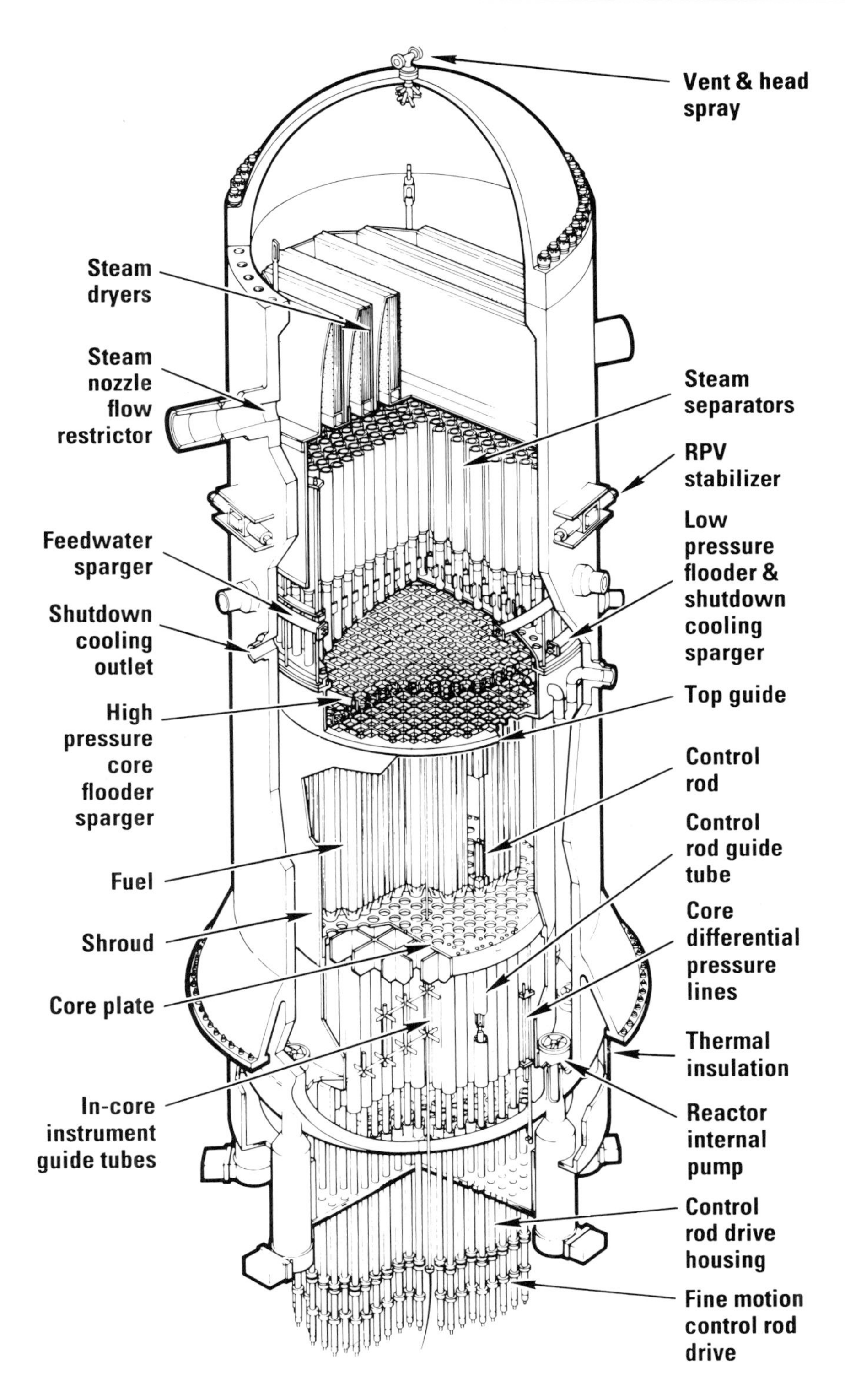

Figure 2. Advanced boiling water reactor assembly.

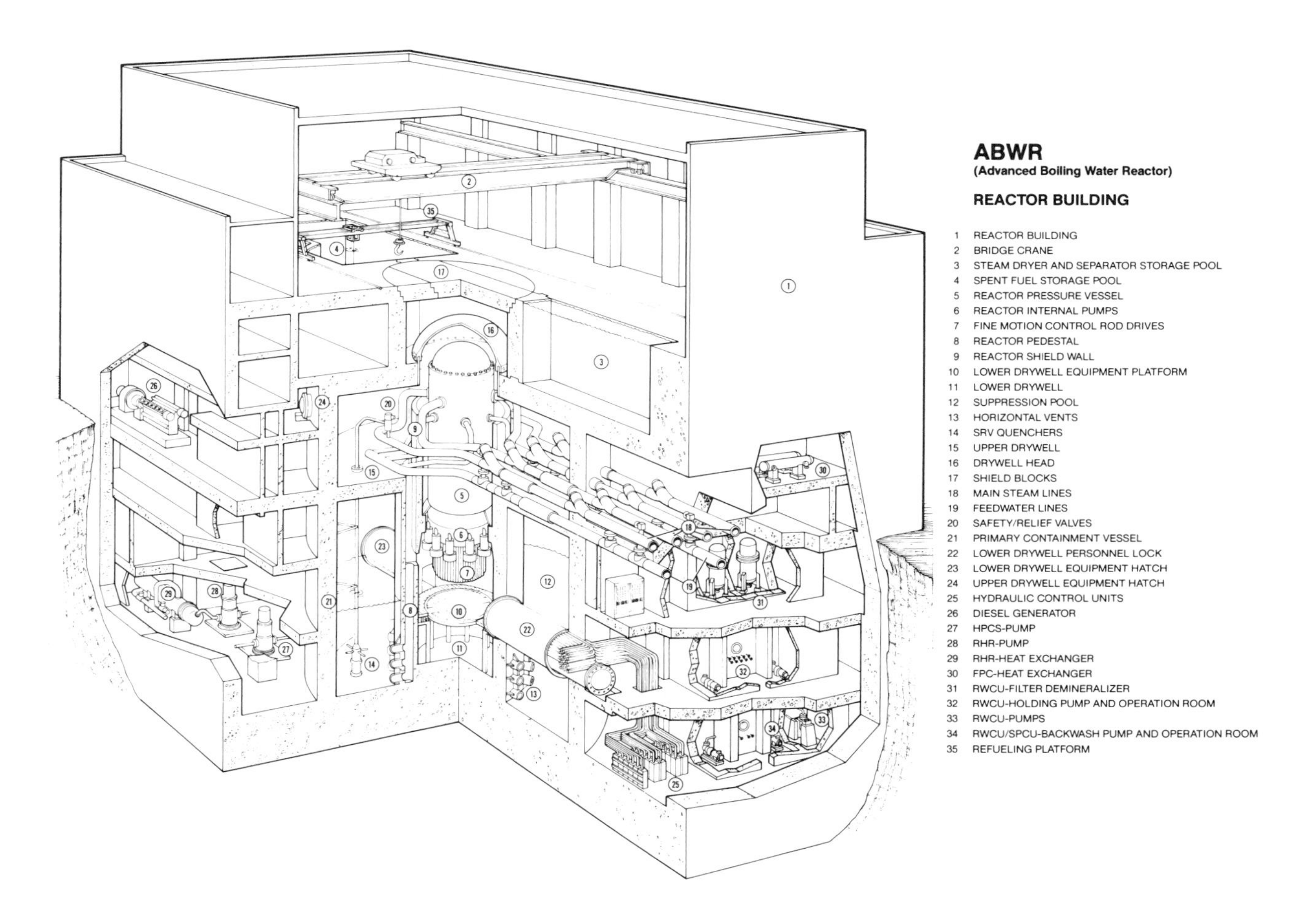

Figure 3. Advanced boiling water reactor building.

area beneath the reactor pressure vessel has also been optimized to provide operating/maintenance space for the internal recirculation pumps and fine motion control rod drives (FMCRD). The containment structure and the reactor building structure have been fully integrated from a structural design standpoint. This has permitted an improved structural capability and has also enhanced the layout of the equipment and structure. The ABWR reactor building total enclosed volume compares very favorably with past BWR plant reactor building designs.

The ABWR performance characteristics are summarized below:

Electrical output	1350 MWe
Construction schedule	48 months
Capacity factor	86%
Daily load following range	50–100% rated power
Core damage probability	$<10^{-6}$ per year
Occupational exposure	100 mrem/year
Solid radwaste	100 drums/year

Severe Accident Capability

The ABWR capability to prevent severe reactor accidents from occurring, and the capability to withstand a severe accident in the extremely unlikely event that one should occur were evaluated with a probabilistic risk assessment (PRA). This evaluation indicates that events resulting in damage to the reactor core are extremely unlikely, but if such events were postulated to occur, passive accident mitigation features would limit the offsite dose so that the effect on the public would be insignificant.

In addition to normally operating systems, the primary ABWR features which prevent core damage are the three divisions of ECCS and RHR. As noted previously, the former includes a reactor steam-driven turbine pump that does not rely on alternating current (AC) electrical power. These features are supplemented by a gas turbine-generator that can be used as an alternate AC power source for electrically powered ECCS and a means of adding fire protection water to the reactor vessel with a pump directly driven by a diesel engine or a fire truck. The fire water addition system is very resistant to earthquakes well beyond the plant design basis. The frequency of events which damage the reactor core was calculated to be about four times in 10 million reactor-years of operation.

Several ABWR passive plant features serve to mitigate the consequences of a severe accident in the extremely unlikely event that one should occur. The containment atmosphere is inerted with nitrogen gas to prevent the evolution of hydrogen from causing combustion or detonation and threatening containment integrity. A system that floods the drywell with water to prevent noncondensible gas generation from a reaction of core debris with concrete reduces the threat of containment integrity loss from excessive pressures. This system is initiated by high temperatures in the lower drywell which result from the molten core debris. The suppression pool previously discussed also serves to trap most fission products in water so that they cannot be released to the environment, even if the containment should be overpressurized. The containment has an overpressure relief device in which a rupture disk opens to allow the transfer of decay heat from the containment while retaining most fission products in the suppression pool. In addition to these passive accident mitigation features, fire protection water can also be added to the drywell to prevent excessive drywell atmosphere temperatures from threatening containment integrity.

The effectiveness of the above-mentioned core damage prevention and passive severe accident mitigation features was quantified in several different ways in addition to the core damage frequency noted above. The conditional (i.e., assume core melt) containment failure (loss of function, defined as off-site dose exceeding 25 rem) probability was estimated at 0.4 percent. This provides a measure of the ability of the containment to provide defense-in-depth. The probability of exceeding a 25 rem off-site dose was estimated at 2×10^{-9}. The increase in risk to members of the public was found to be several orders of magnitude below the Nuclear Regulatory Commission's Safety Goal Policy Statement goal of 0.1 percent.

ABWR Design Certification

As noted, the ABWR is currently the lead plant being reviewed by the NRC for certification under the NRC's new standard plant licensing approach. This effort is co-funded on a 50/50 basis by the United States Department of Energy (DOE) and GE, under the DOE's ALWR Evolutionary Plant Certification Program. The objective of this program is to demonstrate implementation of the new 10 CFR Part 52 regulations. The program was initiated in 1987 and is now approximately 70 percent complete. Final NRC design approval of the ABWR is scheduled for 1990, with certification in 1992. DOE and Combustion Engineering are supporting a parallel effort on the System 80+ pressurized water reactor.

The ABWR under review by the NRC consists of an essentially complete plant. The design will be certified to an envelope of site parameters (seismology, meteorology, etc.) encompassing approximately 80 to 90 percent of available US sites. In principle, once certified, the ABWR could be constructed by any utility on a preapproved site with no further NRC review of the design. The NRC would conduct "sign-as-you-go" conformance reviews during construction to ensure that the plant is constructed in accordance with the combined license and the certified design it references.

Conclusion

Successful certification of the advanced boiling water reactor and, ultimately, a demonstration that the complete licensing and construction process works efficiently in the United States should enhance public confidence in nuclear power and should enable utilities to plan with greater predictability to meet future energy needs. With the energy supply problems now confronting the nation, it would seem irresponsible if this program did not move forward expeditiously.

Environmental Impacts of the Modular High Temperature Gas-Cooled Reactor (MHTGR)

David D. Lanning
Scott W. Pappano
Department of Nuclear Engineering
Massachusetts Institute of Technology
Cambridge, MA

Introduction

During the last 15 years, domestic electricity consumption in the United States has risen at a rate of approximately 2.5 percent annually. This rising demand, however, has not been met by a parallel increase in the construction of new electricity production facilities. A continued rise in the growth rate of electrical consumption means the United States will soon have to embark upon a revival of electrical plant construction or face an energy shortage by the end of the decade. Assuming this challenge to increase America's supply of electrical energy will be met, utilities are faced with choosing a specific avenue for new construction based on many diverse factors—one of these factors is the potential impact future electrical power plants may have on the environment. In this paper, we present one of the possible choices, namely, the Modular High Temperature Gas-Cooled Reactor (MHTGR), and describe the basis for our conviction that this choice represents an environmentally benign electricity-production option relative to existing and proposed nuclear and fossil-fuel power plants.

MHTGR Design Description

The MHTGR design has been chosen as the primary thrust of Department of Energy (DOE) and US corporation efforts with HTGR development programs. The MHTGR design utilizes four 350 MWt graphite-moderated, helium-cooled reactor modules which are cross-headered to feed two steam turbine generators connected in parallel in a non-reheat steam cycle. This cycle has been developed to produce a net electrical output of 537.6 MWe at an overall plant efficiency of 38.4 percent. Additionally, the MHTGR design incorporates passive safety features based on: low core power density (5.9 kWt/liter); large core heat capacity; negative temperature effect of reactivity; inherent fuel particle stability; and passive heat removal capability by conduction, convection, and radiation (GCRA, 1987). Figure 1 shows a flow diagram of this system; the helium in the

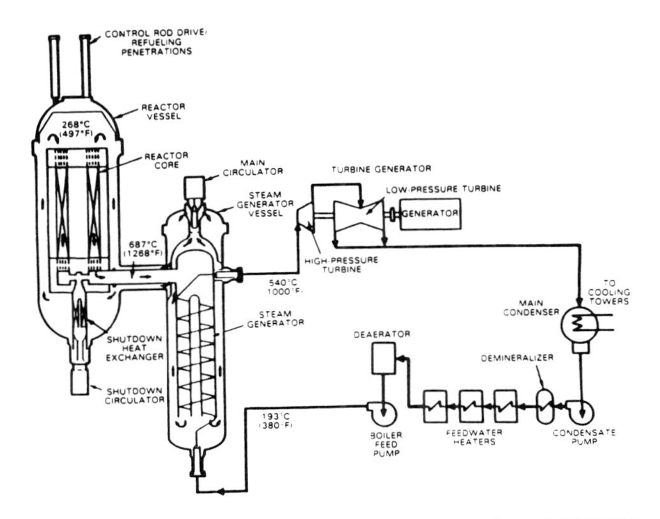

Figure 1. Simplified flow diagram for MHTGR USDOE design.

primary system is circulated to cool the core and carry the heat to the steam generators.

The inherent safety of the MHTGR is primarily a function of its fuel design. The MHTGR utilizes low-enriched uranium/thorium (LEU/TH) fuel in spherical grains that are encapsulated in a porous buffer layer and a silicon carbide (SiC) coating sandwiched between two layers of pyrolytic carbon (PyC). The porous buffer prevents coating damage due to recoiling fission products and provides a free volume for fission gases, while the SiC coating, with a tensile strength of 500-1000 MPa, serves as a high-temperature fission product release barrier; the two PyC layers act primarily as pressure vessel shells which protect the SiC coating from internal and external attack (Lohnert et al., 1988). These coated fuel particles, measuring less than 1.0 mm in diameter, can withstand temperatures in excess of 1600°C without significant fission product release (Lanning, 1989).

To form the "prismatic" fuel elements utilized in the MHTGR core, these fuel particles are embedded in a graphite binder and encased in an outer graphite structure to form fuel sticks. As shown in Figure 2, the fuel sticks are then fitted into holes in large, hexagonal

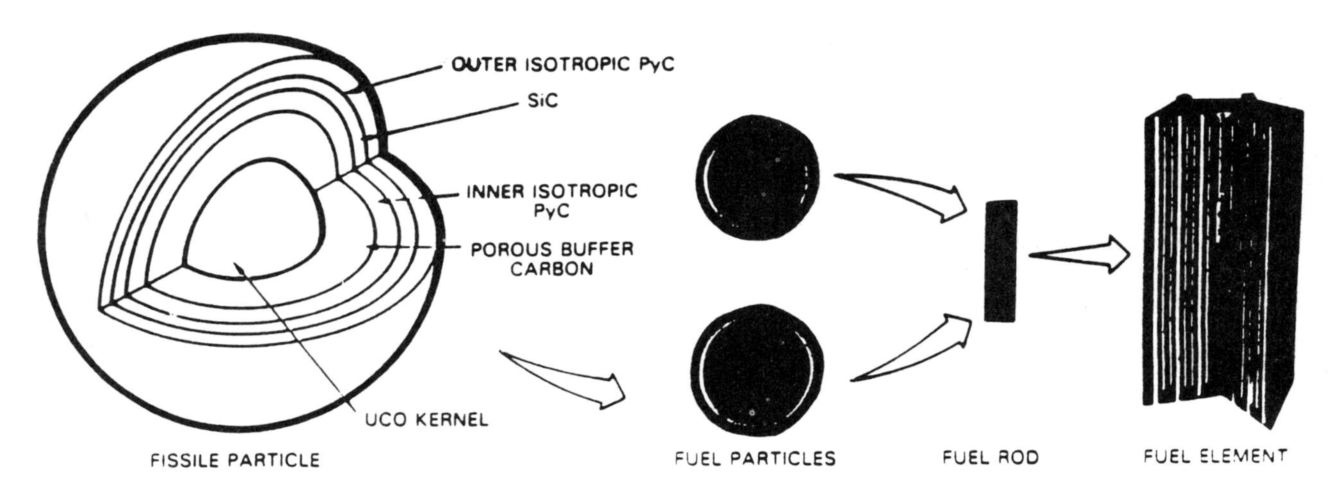

Figure 2. Triso-coated fuel particles and fuel components.

graphite blocks (normally 360 mm across and 793 mm high), which also contain flow passages for the helium coolant. The graphite fuel blocks, called prismatic fuel assemblies, are arranged in an annular core configuration, bordered by an internal central graphite reflector and a surrounding outer graphite reflector (Turner et al., 1988). As shown in Figure 3, the control rods can be inserted in the central or surrounding graphite reflector regions (6 rods and 24 rods, respectively), while boronated spheres of the reserve shutdown system can be dropped directly into the active core region in emergency situations. Equilibrium fuel burnup is estimated to be 82,460 megawatt days per tonne (MWd/T), requiring a refueling schedule of one-half core every 20 months. The large heat capacity of the massive graphite core results in slow heat-up and cool-down times, allowing ample time for operator intervention to correct unlikely malfunctions in the automated control systems. The core design incorporates a temperature effect of reactivity that remains negative under all operating and

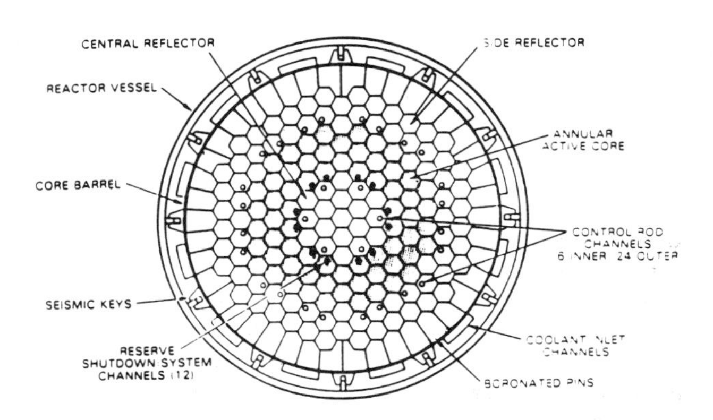

Figure 3. 350 MWt MHTGR core cross-section.

transient conditions, thus shutting down the reactor if the fuel temperature should rise from power increases or coolant flow decreases—even if the control rods or other active safety systems are not engaged (Turner et al., 1988).

The reactor core is contained within an uninsulated steel reactor vessel that is located below grade and connected, via a crossduct, to a helically coiled steam generator bundle in a side-by-side configuration. Both components are housed within a large steel vessel comparable to a Light Water Reactor (LWR) primary vessel, as shown in Figure 4. Each modular reactor assembly is surrounded by a separate reinforced concrete silo that protects the module from seismic amplifications, acts as an initial confinement area, and houses the reactor cavity cooling system (RCCS)—a passive, natural circulation cooling system that circulates air within enclosed panels surrounding the reactor vessel (Neyland et al., 1988). The thermal radiation from the uninsulated reactor vessel to the RCCS allows for passive decay heat removal under loss of forced cooling (depressurization) accidents, maintaining the fuel below coating failure temperatures (1600°C) without requiring any active cooling systems or operator intervention. Should all active forced cooling systems and the RCCS fail simultaneously, decay heat removal would take place via radiation from the vessel and RCCS and then conduction through the concrete silo to the earth, still maintaining fuel temperatures below critical levels and preventing dangerous fission product releases to the environment (albeit with some investment loss due to structural concrete failures) (Lanning, 1989).

All four reactor modules are housed beneath a common reactor service building (RSB), which contains facilities, systems, and components shared by all the reactor modules. Reactor auxiliary and service buildings primarily concerned with helium processing, low-level

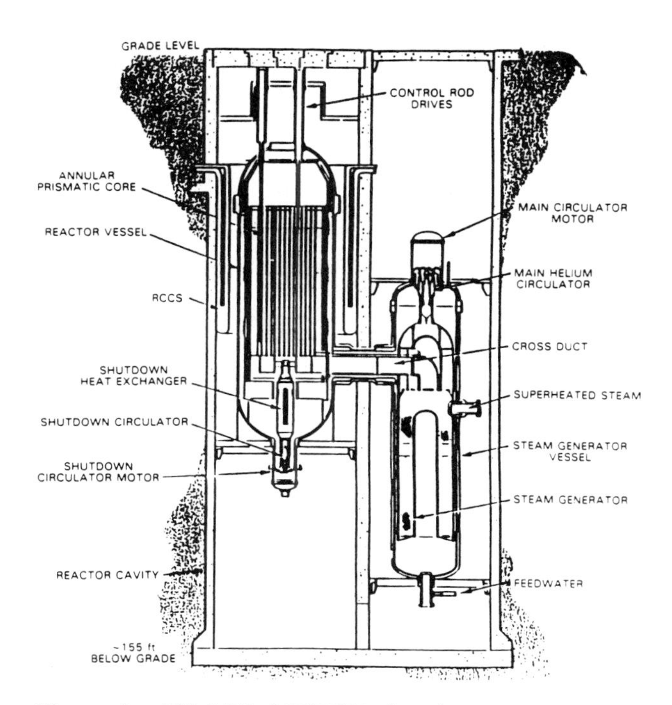

Figure 4. 350 MWt MHTGR elevation.

waste handling and storage, and fuel handling and storage, form the remainder of the nuclear island (NI) portion of the plant layout. The non-nuclear portion of the plant is termed the energy conversion area (ECA) and contains the operations center, the turbine generator building, the mechanical draft cooling towers for condenser heat rejection, and other equipment comparable to fossil-fired steam powerplants (Williams et al., 1989). While the USDOE efforts have centered around an MHTGR design utilizing a steam cycle for electricity production, another option for MHTGR use involves linking the helium-cooled reactor modules to gas turbine generators powered directly by the helium coolant flow. Using today's technology, this option offers an expected single unit power supply of about 100 MW(e) from a module with an estimated overall plant efficiency of better than 45 percent; additionally, each independent module can be coupled with other modules for increased power demands. All of the passive safety features of the current MHTGR design are also included in the HTR gas-turbine design with a reduced potential for water ingress, a simplified balance of plant, and simpler operation and maintenance characteristics (Staudt and Lidsky, 1987). While this innovation may enhance the future acceptability of nuclear power plants, further research and testing is required to fully demonstrate the ability of this unique design to operate as predicted in preliminary analyses.

A final consideration supporting the deployment of either MHTGR design is the fact that the MHTGR is the only current nuclear concept that has the potential to displace fossil fuels for process heat applications in the industrial sector (Northup and Penfield, 1988). From an environmental perspective, the MHTGR also offers the best possible utilization of total heat produced. The relatively high operating temperatures of the MHTGR offer potential process heat applications, such as cogeneration or desalinization, without significantly affecting the current design or electrical output capabilities.

Permissible Environmental Release Levels

The top-level regulatory criteria for radioactive emissions from nuclear power plants are defined by various sections of the Code of Federal Regulations (CFR) and the Environmental Protection Agency (EPA) 520/1-75-001 Protective Action Guidelines (PAG) for sheltering and evacuation. For routine reactor operation, 10CFR20 limits whole body doses to 0.5 rem annually, 0.1 rem in any consecutive seven days, and 0.002 rem in any one hour. Limits for annual doses from the entire fuel cycle are defined by 10CFR190 as 0.025 rem whole body, 0.075 rem thyroid, or 0.025 rem any organ, while 10CFR50 Appendix I limits annual expected doses from gaseous pathways, liquid pathways, and particulates to 0.005 rem whole body (0.015 rem any organ), 0.003 rem whole body (0.010 rem any organ), and 0.015 rem any organ, respectively (Silady et al., 1988).

For accident scenarios, whole body and thyroid doses are limited to 25 rem and 300 rem, respectively, by 10CFR100 guidelines. The limiting criteria for the MHTGR, however, stem from the user requirements that the design should not necessitate offsite sheltering or evacuation under accident scenarios; hence, the EPA PAG limits become the limiting design criteria for the MHTGR. Specifically, maximum offsite design doses are limited by sheltering criteria of 1 rem whole body or 5 rem thyroid doses, evaluated 36 hours after the accident (Silady et al., 1988). These dose limits, however, have been under review by the EPA and will likely be reduced to 0.5 whole body and 2.5 rem thyroid, evaluated four days after the accident, in upcoming revisions to the PAGs (EPA, 1989).

Environmental Impacts from Routine Operation

Routine operation of the MHTGR, like all nuclear power plants, results in periodic radioactive discharges to the environment through gaseous, liquid, and solid pathways—routine electricity production by any type of power plant directly impacts the environment through waste heat rejection, land and water usage, and industrial toxin discharges. The effects of these environmental perturbations, in conjunction with other potentially hazardous activities associated with electricity production (such as fuel mining and transportation), pose health risks both to workers in various fuel cycle stages and to the general public.

Normal MHTGR operation results in radioactive gaseous emissions stemming primarily from the helium purification system, the gaseous radioactive waste system, reactor building ventilation, and activated air in the silo and the RCCS. The emissions from the primary system are monitored, filtered, and released by the gaseous waste system (GWS). Normal filtration consists of high efficiency particulate/charcoal filtration assemblies and redundant waste gas exhaust blowers, but high-activity emissions can be diverted to a waste-gas vacuum tank, compressed by diaphragm compressors, and retained in surge tanks for 30 days to allow for radioactive decay. Maximum annual releases for dose contributing radionuclides from the primary system and GWS are estimated to be 40 Ci/yr for Kr–85 and 10 Ci/yr for both H–3 and Xe–133; additionally, while Ar–41 is not treated by the GWS, release levels are anticipated to remain below 20 Ci/yr (Dunn et al., 1990). All of these radionuclide release levels fall well below 10CFR20 gaseous effluent release limits and, using conservative site and meteorological parameters, lie over an order of magnitude below 10CFR50 Appendix I dose limits (Dunn et al., 1990).

Liquid effluents resulting from normal MHTGR operation consist primarily of low-conductivity (less than 50 μmho/cm) waste, but some high-conductivity waste results from highly tritiated effluents and decontaminating equipment; both waste types are collected in separate receiver tanks for subsequent treatment and handling. Liquid effluents are processed through a filter/demineralizer, collected in a test tank, and discharged to the environment after a cooling tower blowdown. Assuming a 20 to 1 dilution due to the blowdown, annual radioactive discharges are estimated to be 4.9e-6, Ci/yr for I-131, 2.2e-4 Ci/yr for Cs-137, and 3.1e-7 Ci/yr for Ba-140, or five to eight orders of magnitude below 10CFR20 maximum permissible concentration limits (Dunn et al., 1990).

Radioactive solid MHTGR effluents can be classified as low-level or high-level wastes. Low-level radioactive wastes consist of solidified wet wastes, spent resins from liquid waste demineralizers, spent filter cartridges, high-temperature filter units, compressible wastes, high-efficiency particulate absorbers (HEPA), charcoal filtration units, miscellaneous activated solid materials, and spent graphite reflector blocks. The annual low-level waste volume, which is primarily composed of used reflector blocks, is estimated to be 90 cubic meters, resulting in an activity level of approximately 470 Ci/yr (due primarily to chemical solutions from decontamination operations) (Dunn et al., 1990). Comparatively, the Fort Saint Vrain (FSV) HTGR has generated only 134 cubic meters of low-level solid waste containing an estimated 437 Ci over more than 14 years of operation and, on a waste per unit of energy production basis, the FSV gaseous, liquid, and solid wastes were lower than the average levels for the nuclear power industry (Dunn et al., 1990).

High-level waste from the MHTGR fuel cycle is composed of spent prismatic fuel elements. The relatively low power density and short fuel cycle of the MHTGR, in conjunction with the additional volume associated with disposing of the entire prismatic fuel assemblies intact, result in a relatively large volume (approximately 130 cubic meters/GW(e)-year, including the graphite fuel blocks) of high-level waste. The level of activity release from this spent fuel, however, is relatively low due to the presence of multiple radionuclide release barriers, including the fuel kernel coatings and the surrounding graphite block (Dunn et al., 1990). Preliminary calculations indicate that 100 days of onsite storage in six spent fuel storage pools (whose capacity permit spent fuel storage for up to one year) provides ample time for sufficient reductions in decay heat generation from fully exposed fuel elements to allow offsite shipment (Dunn et al., 1990).

Several methods have been proposed to reduce the large volumes of high-level waste generated in the MHTGR, including fuel rod pushout and crush/burn operations (Holloway, 1989). Fuel rod pushout involves removing the fuel compacts from the graphite fuel blocks and treating the vacated blocks as low-level waste, reducing the high-level wastes by up to a factor of four. Crush/burn operations involve further reducing the fuel compacts to their most basic stage—the original coated fuel particles—thus lowering high-level waste levels by a factor of ten from initial volumes. Both these high-level waste reduction methods, however, increase resultant levels of low-level waste and can result in additional detrimental effluents, such as carbon dioxide and radioactive carbon; consequently, extensive research, and development in the field of high-level MHTGR waste reduction and/or treatment needs to be conducted.

Based on the ability of HTGRs to achieve higher efficiencies than existing fossil-fired plants and non-gas-cooled reactors, both water usage levels and waste heat levels rejected to the environment by the MHTGR are generally lower than existing power plants (Dunn et al., 1990). Non-radioactive effluents from nuclear power plants stem primarily from biocides or anti-corrosion agents in water streams; thus, lower water usage requirements for the MHTGR equate to lower chemical effluent levels released to the environment than LWRs or fossil-fired plants. The passive safety of the MHTGR also allows for lower chemical effluent levels than LWRs based on the elimination of active safety heat-removal systems and sinks. Additionally, the MHTGR does not suffer from carbon dioxide, nitrogen oxide, and sulfur oxide emissions—gaseous effluents associated with fossil-fired plants. These noxious gases can contribute to the problems of air pollution, acid rain, and the "greenhouse effect," thus yielding detrimental environmental effects. Finally, coal-fired plants, in particular, are further hampered by the production of solid effluents resulting from flue-gas desulfurization, particulate collection, and ash removal systems, making them inferior to the MHTGR in non-radioactive effluent comparisons (Dunn et al., 1990).

The routine fuel cycle effluents discussed and other potentially hazardous consequences of electricity production present acute (i.e., causing immediate death) and late (i.e., causing disease) health risks to both fuel cycle employees and the public. Based on risk-analysis work performed by Fritzshe (1989), the MHTGR fuel cycle results in occupational mortality risks (in fatalities/GWy(e)) which are over an order of magnitude lower than the coal fuel cycle and slightly less than the LWR fuel cycle for both acute and late risks. The acute public risk of the MHTGR is on the same order as the LWR and up to two orders of magnitude below the coal cycle (due primarily to the risk associated with transporting large fuel and solid effluent quantities); the late public risk associated with the MHTGR, however, can be up to two orders of magnitude below the LWR (due to radioactive emission differences) and up to four orders of magnitude below the coal cycle (due to large volumes of noxious gas emissions). Based on available data, the MHTGR offers the least severe environmental impact and lowest human health risk when compared to existing nuclear and fossil fuel options during routine plant operation.

Environmental Impacts of Potential Accidents

Based on the ability of the passive shutdown features of the MHTGR to maintain fuel particle temperatures below minimum fuel coating failure levels (1600°C) during a large spectrum of postulated accident scenarios, DOE has deviated from LWR and earlier HTGR designs by proposing a mechanistic determination of radionuclide releases to the environment (Williams et al., 1989). Because this mechanistic approach assumes postulated accidents would not result in fuel particle failure with subsequent augmentations of radionuclide inventories in the helium coolant loop, normal operating levels of plated-out and circulating radionuclides in the primary system form the basis of the siting source terms proposed by DOE.

Using the proposed mechanistic source term, upper limits for doses received due to accidents are established in 10CFR100 as 25 rem whole body and 300 rem thyroid; however, limiting criteria for the MHTGR, as imposed by the user/utility group, is based on remaining below current PAG sheltering dose limits of 1 rem whole body and 5 rem thyroid, evaluated at 36 hours. A spectrum of accidental release scenarios, represented by a set of logically chosen, bounding, conservative licensing basis events (LBEs) is used as a basis for determining MHTGR site suitability source terms (Dunn et al., 1990). MHTGR LBEs are further divided into two categories, design basis events (DBEs) and safety-related design condition (SRDC) events. SRDC events represent initiating events for each of the DBEs, which, in turn, represent accident scenarios that are expected to occur in the lifetime of several hundred plants (frequencies of $2e^{-2}$ to $1e^{-4}$ per plant-year) (Williams et al., 1989).

While not all DBEs result in offsite doses at the 425 meter exclusion area boundary (EAB), those that do cause offsite doses all have estimated values that fall below current PAG limits for sheltering and, subsequently, 10CFR100 guidelines, as illustrated in Figure 5. The information in Figure 5 shows the mean whole body doses at the EAB and predicts frequencies (with uncertainties) for various anticipated operational occurrences (AOOs), design basis events (DBEs), and emergency planning basis events (EPBEs). This information was derived by DOE based on conservative LBE requirements for the DBEs and a best estimate method for the EPBEs. The maximum median offsite dose for any postulated DBE is less than 0.002 rem, which results from a depressurized conduction cooldown (DPC) with a moderate primary-coolant leak (DBE–10). Other DBEs involving 1) a primary coolant leak without heat transport system (HTS) and shutdown cooling system (SCS) cooling (DBE–11) and 2) various forms of moisture ingress accident scenarios (DBEs 6 through 9) result in substantially lower postulated offsite doses (Williams et al., 1989). Events labeled DBE–1 to DBE–5, which represent pressurized conduction cooldowns (PCCs) with 1) loss of HTS and SCS cooling, 2) HTS transient without control rod trip, 3) control rod withdrawal without HTS cooling, 4) control rod withdrawal without HTS or SCS cooling, and 5) earthquakes, do not result in any radionuclide release to the environment (Silady et al., 1988). Based on the information presented in Figure 5, all postulated MHTGR accident scenarios with a mean frequency greater than $1e^{-8}$ per plant-year (which extends beyond the design basis region) result in median offsite doses which fall well below PAG requirements for sheltering. Additionally, bounding events (BEs) analyzed by DOE yielded estimated maximum offsite doses of 2.6 rem thyroid and 0.011 rem whole body for accidents having frequencies extending as low as $2e^{-12}$ per plant-year, illustrating the potential ability of the MHTGR's inherent passive safety features to maintain offsite doses resulting from accidental radionuclide releases below

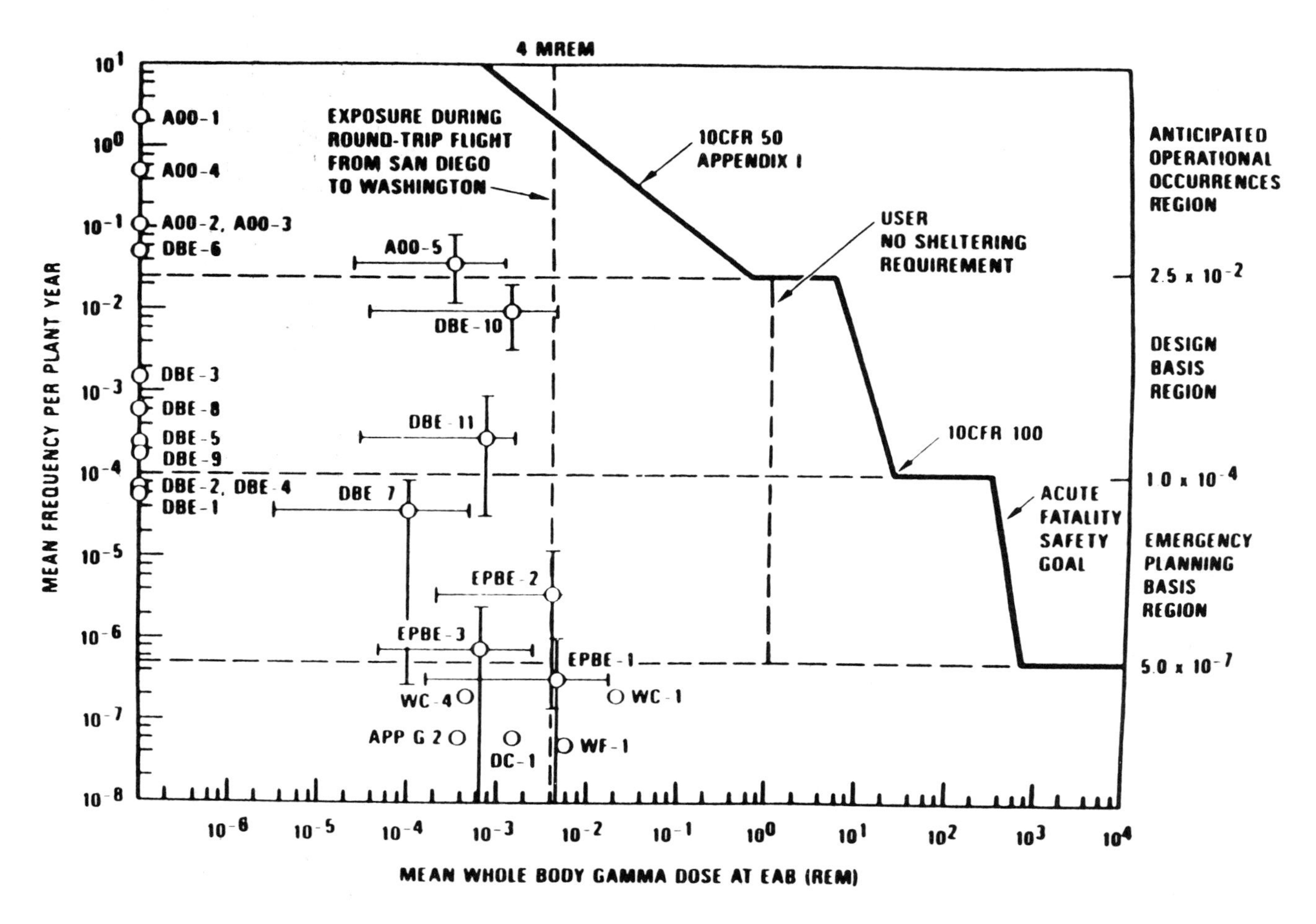

Figure 5. Top-level regulatory criteria and results of safety analysis for the MHTGR (Williams et al., 1989).

PAG sheltering requirements, even for extremely unlikely accident scenarios (Williams et al., 1989). By remaining below PAG sheltering criteria for even highly improbable accidents, the need for active offsite emergency planning drills is also negated, resulting in a reduced negative public impact, thus making the MHTGR a more acceptable nuclear option to surrounding communities.

The small potential for severe accidents (generally defined as causing 10 or more simultaneous fatalities), however, poses occupational and public risks, previously discussed relevant to routine plant operation, which stem from both acute and, only in the case of nuclear power, late health hazards associated with various accident scenarios which occur throughout individual fuel cycles (Fritzsche, 1989). Based on information compiled from 1969 to 1986 by the Swiss Reinsurance Company (Zurich), Fritzsche (1989) compared severe accidents associated with various fuel cycles to determine the relative risks of nuclear power. According to this data, coal usage has accounted for 62 severe accidents associated with mine disasters which have killed an average of over 200 people per year, while oil usage has accounted for an average of over 115 lives per year due to capsizing oil platforms and fires/explosions at refineries and during transport. Even the presumably safe electricity production method of using hydroelectric dams has killed an average of over 200 people per year resulting from the over-topping of dams due to flooding. During this same time span, however, nuclear power has accounted for only one severe accident, the Soviet-Russian RBMK reactor at Chernobyl, which accounted for 31 acute fatalities. The occurrence and severity of the Chernobyl accident was primarily a function of inherent nuclear instability, a combustible moderator, a deficient shutdown system, a lack of effective containment, and bad management and personnel training (Fritzsche, 1989). These drawbacks are heavily regulated in western countries and, thus, the Chernobyl accident bears little relevance to the probability of a similar severe accident in a western country, especially if the inherent safety of the MHTGR is considered as a basis for future design.

Gross numbers of fatalities, however, do not provide a relative comparison between various fuel cycles. To establish a comparative basis, acute fatalities must be referenced to the total electrical power produced by individual fuel cycles. While data was not available for coal and oil, nuclear energy could be readily compared to hydroelectric power. From 1969 to 1986, hydroelectric energy accounted for 3200 GWe y of electricity worldwide, indicating an acute risk of 1.2 fatalities/GWe y. During the same period, worldwide electrical production using nuclear power was 1240 GWe y, resulting in a risk of 0.025 fatalities/GWe y, a risk 50 times lower than risks associated with hydroelectric power (Fritzsche, 1989).

Comparisons of acute nuclear energy risks relative to fossil-fuel options were made by Fritzsche (1989) based on work by MIT Professor Norman Rasmussen in the US and the German Risk Study (DRS) in Europe. This data indicates that an individual severe accident resulting in a specific number of acute fatalities is four orders of magnitude more likely for the coal fuel cycle than if the same electrical energy was produced using nuclear power (LWR). Based on the inherent passive safety and lower potential source term releases of the MHTGR relative to the LWR, even greater incremental safety gains related to severe accidents could be expected by using the MHTGR.

Only nuclear power subjects the public to appreciable late risks resulting from accidental radionuclide releases to the environment, which can cause delayed fatalities through cancer. According to a German study (Fritzsche, 1989), an extremely severe accident might occur during the production of one million GWe y of electricity, leading to about 50,000 cancer deaths over 50 years (up to 17,000 cancer deaths are predicted for Chernobyl). This quantity of electricity production, however, represents the total current electricity production in the US summed over 4,000 years and the resulting cancer deaths represent only 0.1 percent of the total cancer deaths which would statistically occur over the 50 year period following the accident. Fritzsche (1989) concludes that electricity production utilizing nuclear power results in health risks for both routine operation and potential accident scenarios which are lower than any conventional energy option. Furthermore, the inherent passive safety features incorporated in the MHTGR design result in lower environmental radionuclide releases relative to existing LWR designs, making the MHTGR a more attractive nuclear option than the LWR from a risk-analysis viewpoint.

MHTGR Research and Development Needs

The Draft Preapplication Safety Evaluation Report for the MHTGR (NUREG–1338), prepared by the Nuclear Regulatory Commission (NRC) (Williams et al., 1989), provides an initial safety review of the MHTGR and suggests specific research and development needs for the design. While the opinions expressed in NUREG–1338 are subject to change, the document serves as an inspiration to continued MHTGR development because DOE proposals concerning unique licensing criteria for the MHTGR seem to be acceptable to the NRC, conditional on evidence that the design will perform as presented in the preliminary safety assessments. Specifically, the reviewing staff approved of both the use of siting source term based on a mechanistic analysis of fuel failure and the radionuclide inventory in the primary system (both circulating and plated-out) and the potential of the MHTGR to meet present PAG criteria at the site boundary without the use of a containment structure compatible with LWR designs (Williams et al., 1989).

Much of the research required for the MHTGR design involves validating the use of the mechanistic source term. Limiting the magnitude of this source term is based primarily on the fuel working as predicted; consequently, the reviewing staff felt that DOE had to provide more data to prove the fuel performs as predicted and that fuel meeting required quality levels can be consistently delivered by its manufacturers. In pursuit of these needs, the NRC staff called for additional research to provide more experimental data concerning fuel performance at elevated temperatures (both in-pile and out-of-pile); performance over longer-range testing scenarios; performance changes due to internal composition variations over fuel lifetime (i.e., plutonium buildup); and the effects of water, oxygen, or nitrogen environments on fuel failure modes. Once the fuel capabilities have been proven, the licensee must also provide data to prove the fuel manufacturing industry's ability to produce fuel that meets design needs (Williams et al., 1989).

Secondary to proving fuel performance data but also necessary for validating the mechanistic source term criteria is the need to validate resuspension and transport phenomena for circulating and plated-out radionuclides in the primary system. To prove these assumptions, more experimental work is needed to demonstrate local flow velocities due to depressurization; radionuclide augmentation by washoff from steam ingress; evaporation effects from elevated temperature surfaces in the primary circuit; the nature and location of plated-out radionuclides; and the detrimental chemical effects on primary system metals by helium contaminants, including oxidants, hydrogen, hydroxides, nitrogen, chlorides, and carbon dust (Williams et al., 1989).

A final requirement for validating the use of a mechanistic source term is proving the ability of the passive shutdown systems to function without allowing the fuel to reach critical failure temperatures, thus resulting in elevated radionuclide releases. A demonstration of the MHTGR and its passive safety features is suggested as the final step for licensing and for public demonstration of its expected capabilities. Questions concerning the possibility for thermal stratification in the RCCS, graphite thermal conduction characteristics, and vessel emissivity will need to be studied further (Williams et al., 1989).

Based on research and development in the areas listed above, improved data on fuel-failure thresholds, fission product transport, and RCCS unavailability should be substituted into the MHTGR probabilistic risk assessment (PRA) to improve data concerning accident modes and consequences. Additional updated information to be incorporated in the PRA should include improved reactor physics model characteristics which have been validated through experimentation. Most important are features concerned with proving the passive safety of the MHTGR, such as control rod worth measurement variations due to water ingress or core cooldown, reactor transient responses, and the negative temperature effect of reactivity near the end of the fuel cycle (Williams et al., 1989). Again, validating these characteristics and other features of the MHTGR hinges primarily on the design, construction, operation, and testing of an MHTGR prototype, although independent assessments of the MHTGR design performed by the Oak Ridge National Laboratory (ORNL) and the Brookhaven National Laboratory (BNL) to date confirm the predicted safety features of the design (Williams et al., 1989).

While the NRC reviewing staff felt optimistic about the MHTGR's ability to remain below PAG sheltering limits without the use of a containment structure, they felt either a containment or confinement building should be considered to help maintain a thorough defense-in-depth strategy (Williams et al., 1989). Based on the wishes of the staff, confinement and containment strategies relative to original release scenarios should be explored, and research comparing incremental safety gains to incremental cost increases should be conducted to determine the need for further barriers against radionuclide release.

If the proposed PAG modifications are approved, these lower design criteria for the MHTGR might require either further radionuclide release barriers (confinement or containment structures) or lower source terms based on lower levels of circulating and plated-out activity. Based on available fuel performance data, the MHTGR could remain within the present PAG limits at the site boundary (Silady et al., 1988); however, for the new proposed guidelines and with updated data concerning fuel performance or potential accident frequency, the present MHTGR design may exceed the proposed PAG and necessitate the redevelopment of fuel or core designs to remain below the minimum sheltering criteria (0.5 rem whole body or 2.5 rem thyroid over 4 days) without using a containment structure. Research concerning the effects of these proposed PAG changes and the possibility for fuel modification, core modification, or use of a confinement or containment structure must certainly be explored.

Finally, there is one area of research and development outside of the licensing arena that is of prime importance—economic comparison of the costs and benefits for all types of energy sources being considered for future deployment is required on a total and consistent basis. This comparison must include the cost of construction, operation, fuel cycle waste disposal, and the relative costs and benefits of environmental impacts associated with the fuel cycles under consideration. Only after such a comprehensive study has been made, and the uncertainties assessed, will it be reasonable to select specific options as future domestic energy sources.

Conclusions

From an environmental perspective, we are convinced that the MHTGR currently offers the most promising form of electrical energy production and process heat in the United States. Compared to existing nuclear technology, the passively safe MHTGR design provides the advantage of lower radionuclide release levels, lower chemical effluent levels, higher thermal efficiencies, lower accident probabilities and severities, and the potential for waste heat utilization through process heat applications. Additionally, the MHTGR does not pollute the atmosphere with noxious gas emissions or greenhouse gases associated with fossil-fuel power plants.

Based on a mechanistic siting source term (with current data), the inherent passive safety features of the MHTGR limit offsite doses due to accidents to below PAG sheltering limits without the use of a containment structure. Preliminary analyses by the NRC indicate acceptance of this mechanistic source term contingent upon proving: 1) that the MHTGR fuel performs as predicted, 2) that fuel meeting required quality levels can be consistently delivered by its manufacturers, 3) that models for resuspension and transport phenomena of circulating and plated-out radionuclides are valid, and 4) that the passive features of the MHTGR always maintain fuel temperatures below critical limits. While independent analyses by ORNL and BNL confirm the predicted safety of this design, final demonstration of the validity of the MHTGR concept must come through the design, construction, operation, and testing of an MHTGR prototype or demonstration plant. Continued research and development should also be devoted to demonstrating licensing and total costs for further in-depth comparisons with other available systems. The safety, reliability, licensing, and inherent advantages of the direct-cycle gas-turbine MHTGR design should be studied for future applications of both electricity production and process heat.

A prime consideration for decisions concerning potential energy sources utilized in the 21st century will be the economic evaluation comparing available energy options. It is extremely important that such an evaluation be made on a consistent basis and incorporate the total system costs and benefits—including environmental impacts—over the entire expected lifetime of the plant. It should be noted that even if this information is highly accurate and in hand, it may be wise to choose a system that costs a little more but minimizes environmental impacts.

References

Dunn, T., Cardito, J., et al. "Perspectives of Modular High Temperature Gas-Cooled Reactor (MHTGR) on Effluent Management and Siting," *Nuclear Engineering and Design* (to be published) (1990).

EPA/520/1–75–001, "Manual of Protective Action Guides and Protective Actions for Nuclear Incidents (DRAFT)," US Environmental Protection Agency, Washington, DC (1989).

Fritzsche, A.F., "The Health Risks of Energy Production," *Risk Analysis* 9(4), pp. 565–577 (1989).

GCRA 87–011, "A Utility/User Summary Assessment of the Modular High Temperature Gas-Cooled Reactor Conceptual Design," Gas-Cooled Reactor Associates, San Diego, CA (1987).

Hollaway, W.R., "An Assessment of High-Level Radioactive Waste Disposal for Advanced Reactor Fuel Cycles," S.M. Thesis, MIT Nuclear Engineering Department (1989).

Lanning, D.D., "Modularized High Temperature Gas-Cooled Reactor Systems," *Nuclear Technology* 88(2), pp. 139–156 (1989).

Lohnert, G.H., Nabielek, H., et al. "The Fuel Element of the HTR-Module, A Prerequisite of an Inherently Safe Reactor," *Nuclear Engineering and Design* 109, pp. 257–263 (1988).

Neyland, A.J., Graf, D.V., et al. "The Modular High-Temperature Gas-Cooled Reactor (MHTGR) in the US," *Nuclear Engineering and Design* 109, pp. 109–105 (1988).

Northup, T.E., and Penfield, S., "Perspectives on Deployment of Modular High Temperature Gas-Cooled Power Plants," *Nuclear Engineering and Design* 109, pp. 23–30 (1988).

Silady, F.A., Millunzi, A.C., et al., "Safety and Licensing of MHTGR," *Nuclear Engineering and Design* 109, pp. 273–279 (1988).

Staudt, J.E., and Lidsky, L.M., "Design Study of an MGR Direct Brayton Cycle Power Plant," *MITNPI–TR–018*, Massachusetts Institute of Technology (1987).

Turner, R.F., Baxter, A.M., et al., "Annular Core for the Modular High-Temperature Gas-Cooled Reactor (MHTGR)," *Nuclear Engineering and Design* 109, pp. 227–231 (1988).

Williams, P.M., King, T.L., et al., "Draft Preapplication Safety Evaluation Report for the Modular High-Temperature Gas-Cooled Reactor," NUREG–1338, US Nuclear Regulatory Commission, Washington, DC (1989).

Advanced Reactor Development: The Liquid Metal Integral Fast Reactor Program at Argonne

C. E. Till
Engineering Research Division
Argonne National Laboratory
Argonne, IL

Abstract

Reactor technology for the 21st century must develop with characteristics that are quite different from the things important when the fundamental materials and design choices for present reactors were made in the 1950s. It is clear, now, that such characteristics are important for the future.

Since 1984, Argonne National Laboratory has been developing the Integral Fast Reactor (IFR). This paper will describe the way in which this new reactor concept came about; the technical, public acceptance, and environmental issues that are addressed by the IFR; the technical progress that has been made; and our expectations for this program in the near term.

Introduction

I have two points to make before getting into the specifics of our liquid metal reactor (LMR) advanced reactor development program. They are so fundamental to the directions advanced reactor development must take for the 21st century that they need to be made upfront. They are simple points, and really are perfectly obvious, but they have not been as widely appreciated in recent discussions of the directions for advanced reactor development as they should be.

The first point bears on greenhouse effect arguments. In the welter of imperfect information, deductions, and guesses that, taken together, make the case for climate change, it is hard to be sure about anything, much less about what to do.

A few things, however, are clear: With the tremendous growth of the last few decades, human activities that generate energy now produce, and release, amounts of gases each year that are significant fractions of the normal atmospheric content. So the very composition of the atmosphere is changing. It certainly would not be surprising if our environment was fragile in the face of this.

It is only a mild overstatement to say that today all human energy generation is done by burning carbon in the form of oil, coal, or natural gas, to form carbon dioxide—nine tenths of it is, anyway. If nuclear power is to make any real impact on the greenhouse problem, it has to substitute for carbon burning on the same scale that carbon burning is actually taking place. The point is that the scale is tremendous—so large that the inefficient consumption of uranium in today's reactors limits them to only a minor role for this purpose simply because uranium itself is also a limited resource. Even if all other constraints on nuclear power were swept aside—safety, waste, radiation fears, etc.—and only uranium availability was left as a constraint, you could not fuel enough present-day reactors long enough to make any appreciable difference in carbon-dioxide generation.

The simple fact is that nuclear reactors generate electricity without generating gases that pour into the atmosphere—and that the uranium resource constraints on nuclear can be removed if the reactors breed their own fuel. So nuclear reactors could generate all the electricity the world could ever need—even if transportation was also electrified—and do this without altering the atmosphere. If reactors breed, uranium resources are ample. If they do not, uranium is a too-scarce commodity for atmospheric carbon dioxide reduction. It is that simple—reactors for the future must breed. That is my first point.

The point is easy to see: world consumption of coal, oil, and gas is of the order of 270 quads per year; present nuclear plants require about 65,000 tons of uranium oxide over their nominal 30-year lifetime to produce one quad of energy a year. The world total resources of economically recoverable uranium, reasonably assured [that is, already located] and estimated additional [not yet found], are estimated to be about 6.4 million tons of uranium oxide. The world's total uranium resources therefore translate to 100 quads per year, over a 30-year reactor design lifetime, just 40 percent of present world usage, for just 30 years, when used in present-day reactors—and energy use is growing. Obviously, used in this way, nuclear provides no solution for a long-term global concern. Factors of two in resource utilization, or estimated resources, don't change this overall picture. The hundred-fold multiplication allowed by breeding is fundamental.

My second point bears on other characteristics that reactors must have and what this implies for policies now.

The magnitudes required to significantly reduce gas-generating electrical production imply that significant improvements are needed in nuclear safety, nuclear waste, and probably in other areas too, in addition to the necessity for breeding. As we are hearing this afternoon, reactor designs of today—the Light Water Reactor (LWR) and the High Temperature Gas-Cooled Reactor (HTGR)—are being reoptimized to accentuate certain of these characteristics, safety, of course, in particular. These are solid, proven, demonstrated technologies, but in a sense, their strengths are also their weaknesses. Just as their practicality and workability are well established, so are their limitations. Like all reactor systems, their fundamental characteristics, good and bad, are set by the original choices of the basic materials that make up the reactor. To radically change characteristics, you must change the basic materials.

To do this is a research and development proposition—not a reoptimization of present designs and technologies, but whole new reactor systems conceived and developed to yield the characteristics we can now see as necessary for the future. That is my second point. New reactor systems with much improved characteristics are possible, but they require real R&D. Redesign and reoptimization of present technologies are not sufficient for the needs of 21st century. The national advanced reactor program must concentrate a significant effort on real R&D.

The reactor technology developed must be flexible. A key characteristic should be that it is amenable to reoptimization to meet needs as they arise. The precise characteristics desirable or necessary at some point in the future can't be seen today. This is demonstrated by the current events taking place in Eastern Europe and the implications they have for the cold war, and in turn for weapon arsenals, and thus in turn for plutonium stocks, and finally then for the desirability of reactor systems that could burn such stocks—and expand them again as needed for power production. No one could have predicted any of this even a year ago. Our technologies for the next century must be flexible.

Even today we suffer from the implications of our reactor technologies being largely defined by technological assumptions of the 1950s and 1960s. The reactor world has changed so much. The bywords then were "engineered safety"—"we build 'em safe." Waste was not thought to be any problem. Quite the reverse; it was said to be easily buried and it was a point of pride that there is remarkably little of it. Good efficient breeder uranium utilizations were expected to follow naturally from plutonium recycle in LWRs. The public was impressed and supportive. The only problem really faced then was how reactor-produced electricity could actually be made economical. All material choices, design choices, and optimization choices were made to achieve this one goal. Really, only the reactor itself was considered in-depth. Everything else—fuel cycle, waste disposition, any deep thought about safety, safeguards, transportation, etc.—all these were left for down the road.

Few of these assumptions are true today. A long succession of events have changed the characteristics necessary for a reactor system to adequately meet even current conditions, much less conditions projected well into the next century. Three Mile Island, and then Chernobyl, caused a transformation in safety thinking. The perceived connection to weapons, first, but economics too, stopped LWR recycle and reprocessing. The Liquid Metal Fast Breeder Reactor (LMFBR) characteristics came to be seen as inadequate, and Clinch River Breeder Reactor (CRBR) was cancelled. Waste disposal became almost impossible because of the fear of its extremely long-term radiological risk. Today, the public, far from supportive, is skeptical in the extreme.

Reactor systems with better characteristics, aimed at meeting present and future conditions, are required. Although there is a temptation to call nuclear's problems "institutional," technical characteristics that address the public concerns are likely to be essential. Safety, safeguards, transportation, and waste are now critically important reactor system characteristics. In the long term, uranium is a scarce resource too, so resource efficiency for a next-generation system is important as well.

Reactor system characteristics are set fundamentally by the choice of basic materials for coolant and for fuel, primarily, and less so for structural materials. For example, the choice of H_2O as coolant, oxide for fuel, and zircaloy for clad define the LWR. Similarly helium and graphite define the HTGR. Sodium coolant, mixed oxide fuel, and stainless steel clad define the LMFBR. Good design can play up or play down a characteristic but material properties define the possibilities.

Better reactors can be developed now, but if radical improvements are to be made, changes in the basic fuel-coolant choices are necessary to allow new processes and new technology to be developed. The first requirement is that the necessary characteristics be explicitly recognized. The right goals must be set. This, I submit, is not always the case these days. Then, second, materials must be chosen and exploited specifically to give the necessary characteristics.

A better exploitation of the inherent properties of liquid metal cooling can give revolutionary improvements in reactor characteristics. The single most significant property is the atmospheric-pressure primary system. To take full advantage of it requires the complementary selection of a new fuel material. Properly chosen, the new fuel can be completely compatible with the coolant, if both are metal. In turn, a metallic fuel material allows radically different fuel cycle processes. Thus, the introduction of metallic fuel and the metallurgical processing it allows can provide the necessary breakthroughs to a truly economical fuel cycle. Such new fuel processes also open up the possibility of limited radiological-life nuclear waste. If the process is properly chosen, a naturally diversion-resistant fuel cycle with little or no incremental proliferation risk can result. Compactness is a characteristic of metal processes. Metal is very dense. Compactness sufficient to co-locate within a plant, if desired, eliminates the need for fuel transport. Fuel recycle to increase resource utilization a hundredfold over present commercial reactor types follows naturally.

The LMFBR, as personified by the CRBR, did the last of these things, but really only if Purex reprocessing could be made acceptable, both economically and politically. But all these are the exact features of the Integral Fast Reactor development at Argonne.

The key to obtaining new characteristics is the choice of the fuel and the technology of the fuel cycle. The fuel cycle is interpreted here in a broad sense, including the effect of the fuel choice on the reactor behavior itself, particularly in accident situations. It also includes whatever the fuel makes possible in processes for spent fuel and nuclear waste, and also for meeting diversion and proliferation concerns.

At Argonne we had had some rather special experience with metal fuel. Commercially, metal fuel had really never been thoroughly investigated. It had been dropped early, in favor of oxide, when it was found not to sustain reasonable burnups (at only 1 or 2 percent burnup, early uranium metal alloys would swell and burst the clad), but the Experimental Breeder Reactor-II (EBR-II) became the significant exception. EBR-II was fueled with a metallic uranium alloy from the beginning. Through the 1960s and 1970s, development of metal fuel continued at Argonne, because metal had so many other attractive qualities.

By the late 1970s, the burnup problem was solved. Almost any burnup was achievable insofar as the metal fuel itself influences the lifetime. The solution turned out to be simple and is now well known: allow sufficient initial clearance radially for the fuel to expand. With an initial 75 percent or so smear density, the metal is made porous enough by the accumulated fission product gases that when it does swell to the cladding (at 2 percent burnup or less), its porosity prevents the fuel from causing stresses sufficient to challenge the cladding integrity. By the early 1980s, the standard EBR-II fuel ran routinely to 8.5 percent heavy metal burnup, many experimental assemblies had gone beyond 10 percent, and an exploratory assembly had gone past 18 percent burnup.

Meanwhile, the intensive examination of fuel forms and fuel cycle that Argonne had done as part of the International Nuclear Fuel Cycle Evaluation (INFCE) studies of the late 1970s raised real possibilities for both reactor safety and fuel cycle improvements with metal fuel.

The foundations for a greatly simplified fuel cycle based on metal fuel had been laid at Argonne in the late 1960s. In the period 1964–69, a crude form of pyrometallurgical reprocessing and injection casting fabrication of metal fuel had been demonstrated at the EBR-II Fuel Cycle Facility. Successful though it was in demonstrating features of what has come to be known as the pyroprocess, the technology of the late 1960s was inadequate in several respects. First, it dealt only with uranium recovery, plutonium being left to an undemonstrated future process. Second, even the uranium process was incomplete in that the noble metal fission products were not separated significantly from the uranium.

Still, what was demonstrated in the late 1960s was that a simple process could be housed in a very compact facility, and remotely operated, to close the metal-fueled breeder reactor fuel cycle. During the 1970s and early 1980s, some thinking continued on methods of addressing the deficiencies of the early pyroprocess. The fabrication of metal fuels continued to improve because the main EBR-II fuel remained a metal, and there was continuous motivation to make its fabrication as easy (and its burnup as high) as possible. The reprocessing side of the metal-based pyroprocess became feasible—at least in principle—only with the discovery that electrorefining, useful in other applications, could be adapted to a one-step approach to reprocessing.

So by the early 1980s, the stage was set for a detailed look at what kind of reactor system a metallic fuel, now with high burnup, might make possible. It was clear that, to be seen as viable in the conditions of the day, the entire system would have to be brought along at once: reactor, fuel cycle, and waste technology. This then was the background of the Integral Fast Reactor (IFR) concept, so named because all the elements of a complete breeder reactor system would be developed and optimized as a single entity and could in fact, if desired, all be made an integral part of a single plant.

The IFR is based, in one way or another, on the earlier Argonne directions newly relevant for two basic reasons. First, new discoveries had been made within the Argonne program that indicated new possibilities in fuels, safety, and fuel cycle technology. Second, when the new factors affecting nuclear power were recognized, reactor system properties not thought to be important before now seemed very important indeed.

In the summer of 1984 the IFR program was started. The program was governed from its inception by four overriding requirements:

- Passive or inherent safety characteristics
- Economically competitive
- Environmentally sound
- Proliferation- and diversion-resistant

For the concept to be feasible, three basic developments were needed: a specific metal alloy was required, establishing concomitant improvements in reactor safety was essential, and showing the feasibility of the new metal-based fuel cycle was perhaps the most important of all.

Fuel

The fabrication of any metallic fuel alloy promised to be cheap and readily adapted to remote operation. The EBR-II fuel had been made at Argonne for years, with one simple casting operation instantly producing enough fuel for one assembly. As noted above, in the period 1964–69, fuel had been made remotely for EBR-II.

The standard EBR-II metal fuel available at that time would not do—outstanding in-reactor though it was—because it didn't use plutonium. In a closed cycle breeder system, plutonium is the bred material, so any IFR alloy had to include plutonium.

The alloy selected was a (Uranium-Plutonium-)Zirconium alloy that had very limited trials in the late 1960s, but appeared to have the basic characteristics needed. It used plutonium. It had a high melting point and a high eutectic point with stainless steel, even higher than those for EBR-II fuel.

But would the U-Pu-Zr alloy provide adequate burnup? In the fall of 1984, a new plutonium fuel fabrication capability was put in place. The Experimental Fuels Laboratory, or EFL, was created in just four months, once again demonstrating the simplicity of the fabrication process. Early in 1985, three lead assemblies of the new IFR fuel were put into EBR-II.

The fuel development has been a remarkable success. In Figure 1, the improvement in burnup through the years is shown, for the uranium-bearing alloy before ~1988 and more recently for the IFR alloy. Experience with the latter is examined in more detail in Figure 2, where the present database on IFR-alloy metal fuel is depicted (these are intact fuel elements either discharged from EBR-II at the burnups shown, or still in the reactor with present burnups quoted). The database with IFR fuel will now grow at a significant rate, for as of the beginning of 1989, EBR-II was completely fueled with prototypic IFR alloys.

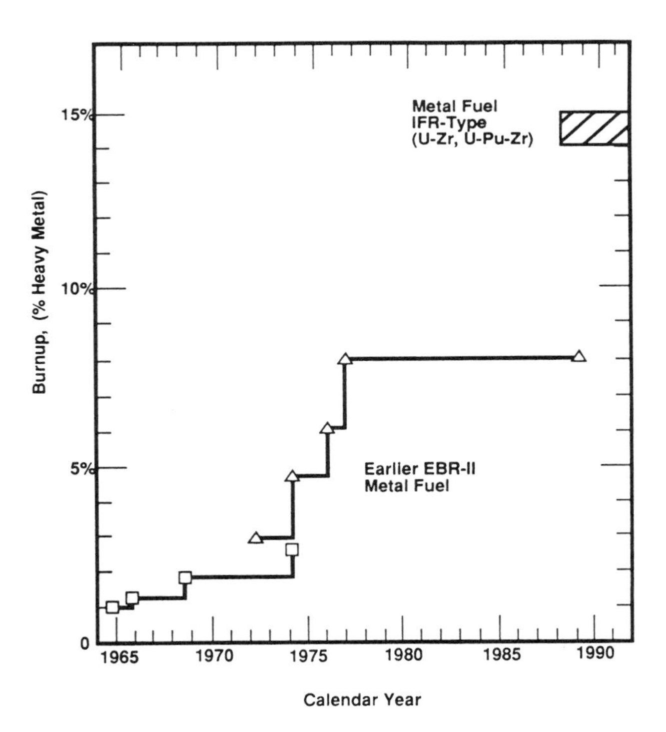

Figure 1. EBR-II fuel burnup capability.

Safety

It was important to demonstrate the unique safety properties made possible in the IFR with the use of metallic fuel. On April 3, 1986, two carefully planned tests were carried out. From full power to EBR-II, with the normal safety systems temporarily bypassed, the power to the primary pumps was shut off, simulating station blackout, or loss-of-flow without scram. The reactor shut itself down without safety-system or operator action, because of the reactivity feedback characteristics of the IFR. No damage occurred either to the fuel or to any of the system structures. Later in the day, the reactor was brought back to full power and loss-of-heat sink without scram test was also carried out. The result again was without harm of any kind. These tests dramatically demonstrated what is possible for incorporating passive safety features in IFR plants. Parenthetically, later in that month, the Chernobyl accident occurred and the stark contrast between the consequences of these two loss-of-flow events in the same month gave much added impetus to IFR development.

EBR-II is an electricity-generating power reactor which, although small (20 MWe), has a power density typical of that in larger fast reactors. The features of EBR-II which allowed it to shut itself down in these two tests are typical of larger IFR plants as well.

The third of the classical fast reactor Anticipated Transient-Without-Scram (ATWS) events, the transient overpower (TOP) accident, is also reduced in consequence because of the metal fuel. The higher core conversion ratio offered by the higher fuel atom densities achievable with metal gives rise to reduced reactivity swings during a cycle. This, in turn, reduces the control requirements, allowing lower-worth control rods. The initiators of TOP events are the control rods and the transient initiated by control rod withdrawal can be

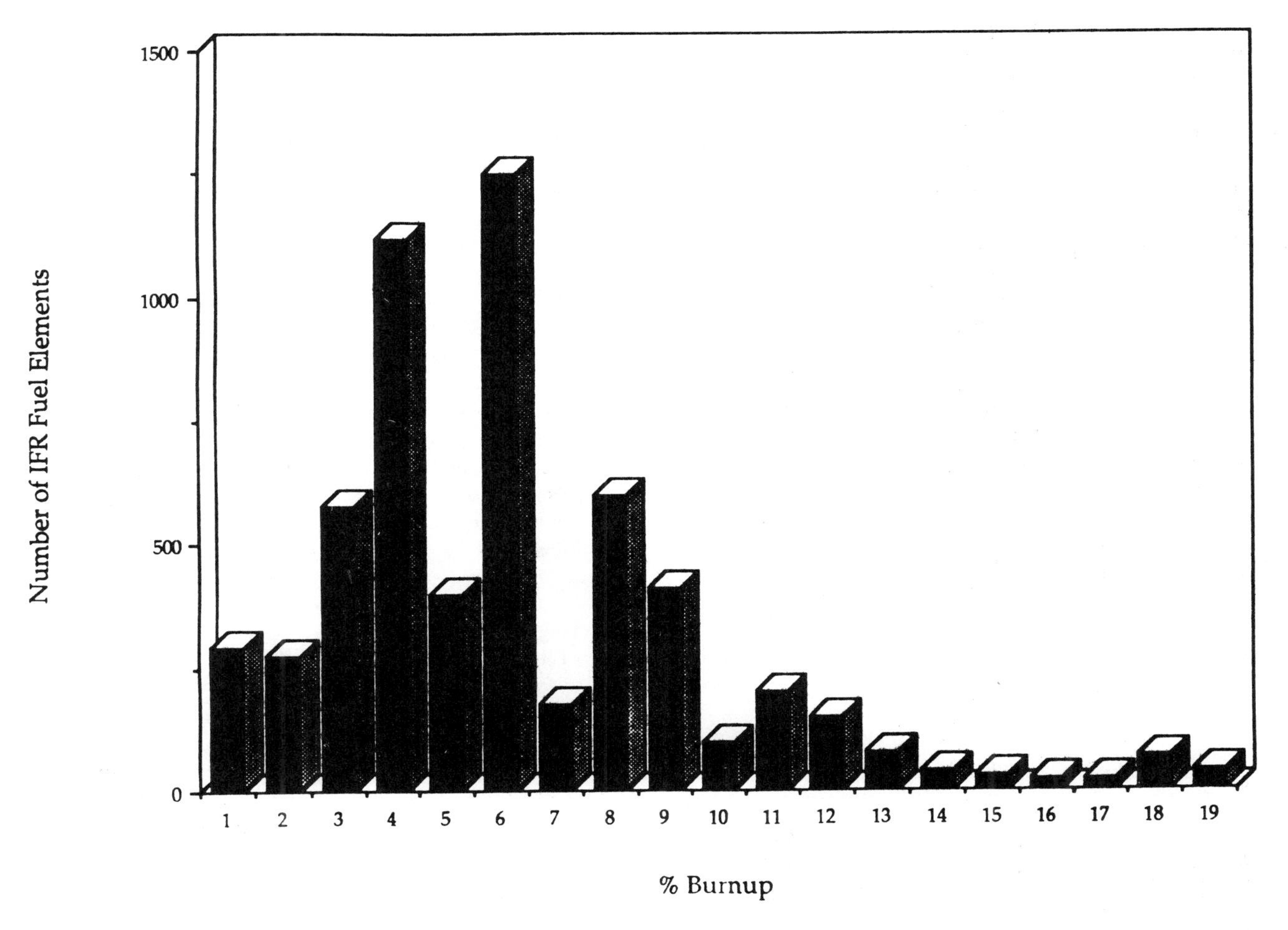

Figure 2. Histogram of IFR fuel element burnup in EBR-II.

made much less severe than would otherwise be the case. The goal is to limit the available excess reactivity contained in control rods to levels which would cause no damage if inadvertent rod runout at power were to occur.

The safety case is further strengthened by the fact that significant margin exists before fuel failure would occur in fast reactivity transients. Tests in the Transient Reactor Test facility (TREAT) reactor have demonstrated that linear power can be increased by about four times above normal before cladding failure occurs, and that during these excursions the metal fuel is expanding axially within intact cladding, driven by entrapped fission gas pressure, which provides a strong negative reactivity feedback. Tests at EBR-II addressing other plant transients, such as overcooling associated with a sudden increase in the speed of coolant pumps or a rapid depressurization of the steam system, have also been done and show that they can also be accommodated without safety system action.

Fuel Cycle

Metal fuel opens up the possibility of using a very different process for reprocessing spent fuel, and this process is described more fully elsewhere. Electrorefining, instead of solvent extraction, can be used. Electrorefining has different properties, some of which are very advantageous.

The basic process is electrochemical. The fuel to be processed forms the anode of an electrolytic cell. The electrolyte is a molten salt and the product heavy metal is collected on a cathode. Proper selection of the voltage draws uranium and plutonium from the spent fuel, leaving the fission product waste behind. The separation is done in this single step, at relatively low temperatures (about 500°C), and the device in which this is done is very small and compact: a 1.5 m diameter module would be sufficient for a CRBR-size plant. The process separates uranium and plutonium from the fission products adequately for fast reactor purposes, but leaves a highly radioactive diversion-resistant product.

Development began on a few-gram scale, and both a uranium-only process (necessary to recycle uranium for the breeder blanket) and a uranium-plutonium process (for the core) are now proven. The two processes differ only in cathode design. The uranium-only process has now been scaled up and operates routinely at plant scale, about 10 kg per cathode (see Figure 3). The U-Pu process scale-up awaits the completion of our Idaho fuel cycle demonstration facility.

In tests conducted recently, fuel segments sheared from a fuel element irradiated to 10 percent burnup have been successfully dissolved in an experimental apparatus. This was done electrolytically, essentially the reverse of electrorefining. We found that the heavy metal could be driven quantitatively from the cladding (only 0.04 percent of the heavy metal remained) in a one-hour period. We thus have evidence that within a single electrorefiner, fuel can be successfully dissolved electrolytically, and electrotransported selectively to cathodes. The key reprocessing steps of dissolution and separation are thereby achievable in a simple, compact, piece of equipment.

Figure 3. Uranium metal cathode from electrorefining experiments.

Importantly, we have found that the transuranics go with the product, so the waste product radiological lifetime is dramatically reduced, since the transuranics can be recycled and burned in the reactor. In an IFR reactor, with the high-energy neutron spectrum unique to a metal-fueled fast reactor, these elements are efficiently fissioned, and essentially provide more fuel, not waste.

Future IFR Program Activities

The basic feasibility of all elements of the IFR has now been proven. The next important step will be to close the fuel cycle at EBR-II. EBR-II is our prototype. It is sodium-cooled; it is a pool-type reactor configuration; it is now completely fueled with IFR fuel. When we have the new processes in operation, we will have the complete prototype—integral cycle and all. It will demonstrate each of the essential features of the IFR: passive safety, ease of operation, fuel performance, reprocessing and recycle, and transuranic burnup to improve the waste form.

Modifications are in progress to fully demonstrate the new fuel cycle at the EBR-II Fuel Cycle Facility. Following its use in the 1960s, this facility was converted to an examination facility (and renamed the Hot Fuel Examination Facility/South). The facility is being modified to bring it up to today's standards and regulations for such a facility. The facility modifications are detailed elsewhere. Briefly, they are: 1) confinement improvements, 2) provision of a new class 1-E emergency power system, 3) installation of a new safety-class exhaust system, and 4) construction of a new area within the facility in which to repair contaminated equipment. All of the work associated with these four areas is scheduled for completion in September 1990.

The process equipment is now in fabrication. There are nine items of main equipment (all quite compact, in all cases able to pass through a 2 m diameter, 2.5 m tall transfer lock):

- assembler/dismantler machine
- element chopper
- electrorefiner
- cathode processor
- injection casting furnace
- pin processor
- element settling furnace
- element welder
- leak-detection module

Together with a small amount of other support equipment, all these equipment items are also scheduled for installation in the hot cells in September 1990, ready for cold operations.

In early 1991, we will start reprocessing and refabricating fuel for EBR-II and the Fuel Cycle Facility (FCF) will have a dual mission: produce all the fuel needed for EBR-II and serve as a test bed for optimization of the process.

There is still basic development to be done, and this will be going on simultaneously with fuel cycle operations and experiments in the FCF. Through the early 1990s, the IFR prototype will be recycling fuel; the recycled fuel will be tested and proven and the whole system (fuel, fuel cycle, and waste process) optimized.

All of this will be done on the Argonne National Laboratory-Idaho site, which has all the necessary facilities, without large expenditures. Our colleagues from General Electric, with the Power Reactor Inherently Safe Module (PRISM) system based on the IFR, will be ready—and perhaps others as well—to proceed with the next step.

And what does the IFR promise?

- In safety, fuel with larger overpower margins, resilience to transients, completely nonreactive with the coolant; a reactor with built-in ability to survive both loss-of-heat sink and loss-of-flow without scram events.
- In breeding, metallic fuel is the best possible. In addition to the obvious resource conservation reasons, this also allows the limited control rod worths that help in Transient Overpower (TOP) situations, adding again to safety.
- A simple closed fuel cycle, with recycle, and reuse of uranium, plutonium, and the other transuranics as well. All transuranics go with the fuel product, and are not left in the waste.
- Recycled plutonium fuel always accompanied by uranium, always carrying the other transuranics, and some small amount of fission products as well, removing diversion concerns, and adding nothing incremental to proliferation risk.
- No transportation of fuel or spent fuel, and, if desired, on-site storage of wastes for the life of the plant.
- A waste product that has all long-lived transuranics removed, such that the carcinogenic risk from the waste has decayed to less than the original ore in about 200 years. The public is asked to accept a change in the kind of risk, not just degree of risk, associated with waste disposal.

Although no one attribute alone may make the case, the whole of its attributes makes the IFR system a revolutionary improvement in fission energy for the 21st century.

Summary
Session D-4—Alternative-Energy Based Electric Power Systems

Jon G. McGowan
Session Chair
Department of Mechanical Engineering
University of Massachusetts
Amherst, MA

Four papers were presented on alternative-energy based systems that utilize renewable energy sources. The first reviews solar and wind energy systems in the context of electric power generation using solar thermal, solar photovoltaic, wind, and ocean thermal energy sources. In addition, the potential for hydrogen as an energy storage and transmission medium is also discussed. The second paper, by Ghazi Darkazalli, focuses on photovoltaic (PV) systems for small electric power plant applications. By emphasizing their positive environmental attributes and modular character, Darkazalli suggests that PV-generated electricity applications will continue to grow at the current rate of 30 percent per year into the foreseeable future. (See also "Perspectives on Renewable Energy and the Environment," by Hartley and Schueler in Session F of this volume, Advanced Energy Supply Systems.)

The third and fourth papers deal with geothermal energy. Ronald DiPippo's paper, "Geothermal Energy: Electricity Production and Environmental Impact, A Worldwide Perspective," provides an excellent review of hydrothermal resource utilization technology on an international scale. DiPippo covers power plant developments in the US and other countries stressing the type of power plant hardware used and its environmental impacts. He emphasizes the minimal impacts of geothermal power systems relative to existing fossil and nuclear technologies. Carel Otte's paper, "Geothermal Energy Opportunities for Developing Countries," completed the session on a positive note by citing the commercial success of geothermal developments in the Philippines and Indonesia. In large part, US-based technology developed for The Geysers and other fields in California was successfully transferred to these developing countries. Readers interested in reviewing the status of advanced hot dry rock geothermal systems should consult the paper by Brown, Potter, and Myers found in Session F of this volume.

Large-Scale Solar/Wind Electrical Production Systems—Predictions for the 21st Century

Jon G. McGowan
Department of Mechanical Engineering
University of Massachusetts
Amherst, MA

Abstract

A review of renewable or solar/wind systems that produce utility-scale electrical power is presented. This includes the solar electric category (photovoltaics, wind, and ocean thermal) and solar thermal systems. The current state-of-the-art of these systems is discussed in the light of recent technology and large-scale electrical power generation applications. Based on the use of these types of renewable energy systems, a discussion of predicted developments for these systems in the 21st century is made. In addition to continued technological performance improvements and cost reduction, these include the expansion of international markets for large-scale renewable energy systems, developments in energy storage and the use of hydrogen as an energy storage medium, and utilization of offshore solar/wind resources.

Introduction

Since the oil crisis of the early 1970s, there has been active worldwide research and development in the area of renewable solar/wind energy systems. During this time, energy conversion systems that were based on renewable technologies appeared most attractive because of the projected high cost of oil and because of most favorable early cost-effectiveness estimates and predicted ease of implementation of such systems. Furthermore, in recent times, it has been realized that renewable energy systems can have a beneficial impact on the following technical, environmental, economic, and political problems of the world:

- Depletion of the world's nonrenewable energy resources
- Increasing energy use in developing countries
- Greenhouse effect and global warming
- Acid rain and air pollution problems
- Environmental degradation

A comprehensive review, or even complete definition, of these global problems is beyond the scope of this paper; however, some of them are addressed in other papers from this conference or in recent publications on the subject, for example, recent Worldwatch Institute publications (Brown, et al., 1988).

As pointed out by a recent International Energy Agency study (IEA, 1987), almost no generalities can be made about renewable energy technologies without a loss of accuracy or usefulness of the statement. That is, there is not only tremendous diversity in the types of energy sources that are considered renewable, but also in: 1) the amount of each resource that different countries possess, 2) the technologies that use these energy sources and deliver useful energy to end users, and 3) the stage of development that each renewable energy technology has achieved.

In an attempt to clarify the last two points, this paper will present a review of renewable or solar/wind driven systems that produce utility-scale electrical power. Utility-scale electrical systems are defined as grid-connected systems with power outputs on the order of megawatts (MW) or greater. [Editor's note: All MW or kW in this paper refer to electric, not thermal, power unless otherwise stated.] It should be pointed out, however, that renewable energy systems have and will continue to make major contributions in many nongrid-connected applications requiring either electrical or thermal energy inputs. For electrical systems, these include small stand-alone or remote-scale power systems (on the order of one kW) or village-scale renewable or hybrid (e.g., wind/diesel) energy systems in the 50 to 200 kW size range. A complete discussion of these smaller scale renewable energy systems is beyond the scope of this paper—a recent review by the American Solar Energy Society (Andrejko, 1989) gives an overview of the most significant renewable technologies for all scales of energy production.

In this paper, the utility-scale electric power system category will include solar electric (photovoltaics, wind, and ocean thermal) and solar thermal systems. As will be noted, under this category significant advances have been made, and numerous large-scale demonstration power plants have been constructed and operated. It is not the objective here to discuss the total resource potential of each of these systems, or technical details, design, and performance of such systems—an excellent summary of these subjects is given in the recent text of Twidell and Weir (1986). Also, comprehensive assessments of the current state-of-the-art of these technologies are given in the recent publications of the American Solar Energy Society (Andrejko, 1989) and the International Energy Agency (IEA, 1987).

Based on the expanded use of these types of renewable energy systems, a discussion of predicted developments for them in the 21st century is made. In addition to continued technological performance improvements and cost reduction, these include the expansion of international markets for large-scale renewable energy systems, developments in energy storage and the use of hydrogen as an energy storage medium, and utilization of offshore solar/wind resources. Although most of the detailed examples in this section are for wind energy conversion systems, future progress is expected in all areas of large-scale solar electric and thermal power systems.

Current Status and State-of-the-Art of Utility-Scale Solar/Wind Electric Systems

In this section the current status or state-of-the-art of solar or wind powered utility-scale electric sytsems will be discussed. In a recent European study (IEA, 1987), utility-scale wind, photovoltaic, and solar thermal systems can be classified as "under development" technologies—that is, technologies which need more development to improve efficiency, reliability, or cost in order to become commercial. According to these researchers, this categorization would include materials and systems development, pilot plant or field experiments to clarify technical problems, and demonstration plants to illustrate performance capabilities and to clarify problems for commercialization. Following the same type of classification of commercial status, ocean thermal energy conversion systems fit in the "future technologies" category—a technology that has not yet been technically proven, even though it is scientifically feasible. This category would include basic research and development on components and bench-scale model development at laboratory levels. However, as other technical experts (such as some of the authors in this conference) may point out, one or more of these technologies may have already reached commercial cost-effectiveness in certain locations of the world. Short reviews of the four technologies follow.

Photovoltaic Systems

The recent research and development progress in photovoltaic (PV) technology can be summarized with respect to increases in cell and module conversion efficiencies, new PV alloy materials and device structures, and improvements in system design (Herwig, 1989; and Palz and Helm, 1988). In general, the technological performance of PV systems is expressed in terms of efficiency and it is in this area that major improvements have been and will continue to be made in the future, directly affecting the PV module cost. PV cells are typically classified as crystalline, "thick-film" cells, thin-film cells, and concentrator cells (used in conjunction with a lens or focusing mirror that concentrates the sunlight on a small area). A description of the current status of each of these technologies is given in the recent American Solar Energy Society (Andrejko, 1989) review: each of them has achieved a high level of efficiency in production (12 to 16 percent) and has shown stable operation in the field. It also should be noted that this recent review concludes that, since all three options are capable of long, economical life, the choice between them will ultimately be decided by a tradeoff between production and cost efficiency. Furthermore, at present, no clear-cut superior technology exists, and each is thought to be capable of reaching the $0.12/kWh near-term utility entry market.

In the area of large-scale or utility sized applications, the United States Department of Energy has an active program in this area (Photovoltaics for Utility Scale Applications, PVUSA) and a number of large plants are already in operation in the US, with more to be constructed in the future. It also has been noted that, in the United States, the most likely place for a huge market growth is considered by the PV industry to be electric utilities. Herwig (1989) states that there are at least six US PV companies in an advanced planning stage for building 10 MW_{peak} (annual) production plants. He notes that until the potential electric utility market appears to be accessible, the business decision and capital for implementing these production plans are not expected to be available.

On a worldwide basis, notably in Europe and Japan, there have been a number of large scale PV systems installed (IEA, 1987). Similar to the United States market, it is generally believed that the large-scale market offers the best opportunity to make PV systems cost-effective via cell and module mass production techniques. The IEA study also concludes that, in addition to continued improvement in module and system lifetimes and operating reliability, other key areas where technological improvement is needed include:

1. Balance of system costs. At present, these represent a major portion of total system costs, and these costs will become more important as module prices are reduced.
2. Optimum system configuration, performance, and operation. Construction of additional large-scale systems will help provide essential data on perform-

ance and operating experience (see Hoffner, 1989, for an example of this type of documentation).

3. Current state of knowledge of the solar resource. In many countries, this must be improved for the optimum design and siting of PV plants. For example, before utilities can integrate PV into large-scale central plants, they would require minute-by-minute understanding of insolation patterns at proposed sites and considerably more data specific to the needs of PV system designers.

Despite these technological problems and the need for a major reduction in system costs, proponents of PV technology predict a growing future for utility-scale PV systems. For example, Herwig (1989) believes that the potential US utility market to be served by PV systems is a minimum of tens of gigawatts of utility systems (even without a sustained consideration of the environmental advantages of renewable energy systems).

Wind Systems

The development of wind energy conversion systems during the last 15 years has undergone significant progress. From its problem-ridden beginnings and a heavily subsidized growth era, the wind industry has attained the present-day status of an increasingly reliable, competitive power source. Through 1988, over three billion dollars worldwide has been invested in the purchase and installation of over 20,000 turbines, with the market reaching a peak of about 600 MW_e installed during 1985 (Lindley, 1988). In the United States, more than 1400 MW of installed capacity, primarily in California and Hawaii, were added in the 1980s. Furthermore, wind power now represents about 3 percent of California's installed electric generation capacity, the highest market penetration of wind energy in the world (AWEA, 1990).

In wind turbine technology, the horizontal axis type of wind turbine has emerged as the predominant type. However, there are still many possible configurations of this type of machine and choices of specific system designs. For example, there is no clear choice between upwind and downwind machines, between fixed and variable pitch, and among different rotor materials. Also, one of the key questions that has been addressed many times over the past 20 years is concerned with the optimum size of wind turbines. Specifically, system analysis studies conducted during the early part of the US wind energy program indicated optimum wind turbine sizes in the 1–3 MW range. Such studies gave some impetus to other countries' national wind programs (e.g., in Sweden, UK, and Germany) to fund the development of MW rated large machines.

In the United States, the majority of the initial wind energy funds went toward a multimegawatt wind turbine development program that produced four series of machines ranging from the 100 kW (38m diameter) NASA MOD-0 machine to the 3.2 MW Boeing MOD-5B machine with its 98m diameter and a variable speed generator. On the international scale, other large wind turbines have been developed by a number of countries. In general, the development of large turbines has been slower, more expensive, and technologically more difficult than that of smaller machines (IEA, 1987). Also, none of the large machines have been built in production quantities such that the economic benefits of volume production could be accurately determined. Despite these problems, at present about 10 multi-Megawatt wind turbine prototypes are being operated or tested, or are in the final stages of construction (Beurskens and Schmid, 1988).

Following recent experience with the development and production of smaller wind turbines for windfarm applications, wind turbines with ratings from approximately 50–350 kW are presently the most economical. Also, based on current information, it appears to be generally accepted that the maximum economic size (taking into account all costs) lies in the 300–750 kW rating size (Lindley, 1988).

Through the use of large groups of medium sized (50–500 MW in rated power output) wind turbines (windfarms) it has been demonstrated in locations such as California, Hawaii, Denmark, and the Netherlands that wind energy conversion systems can generate electricity at costs comparable to conventional utility-scale plants. In the United States, California, which contains approximately 90 percent of the world's installed wind generating capacity, has been the proving ground via the deployment of the windfarm concept. According to Lynette (1989), this rapid development has occurred due to the following favorable economic conditions in California:

- Federal and California tax credits
- Accelerated depreciation
- Favorable power purchase contracts
- Access to investors in need of tax credits
- Excellent wind resources near utility access
- Growth of California economy
- Emphasis shifted away from coal and nuclear

The rapid growth of wind energy in California did not occur without problems. Several wind turbine manufacturers became involved in the industry without an appreciation of the technical challenges associated with the design and maintenance of large wind turbines. During the last few years, the industry has undergone consolidations and major changes, especially with the phase-out of federal and state tax credits. For example, the number of US manufacturers of wind turbines has dropped from several dozen to less than 10 in 1989 (Andrejko, 1989). On the other hand, the remaining US manufacturers and a number of European manufacturers are committed to improving reliability, reducing operation and maintenance costs, and increasing productivity. According to the latest reviews of the subject (Lynette, 1989), they appear to be succeeding.

Solar Thermal Systems

As noted in a recent review of renewable energy sources (IEA, 1987), solar thermal technology has excellent potential because its solar concentrator based systems produce thermal energy at elevated temperatures that match well with conventional power cycles to produce electricity. These systems also may offer further advantages due to relatively rapid construction times and the potential for modularity. It has been concluded that solar thermal technology, in addition to its compatibility with conventional energy systems in existing plants, has the potential to generate electricity at scales from kilowatts to hundreds of megawatts, has established the highest efficiency for conversion of solar radiation to electricity (approximately 32 percent), and lends itself to energy storage systems that employ latent heat.

Most large-scale electrical generating plants using solar thermal technology have been based on the use of parabolic trough or central receiver systems, with some small electrical generation applications based on the use of parabolic dish systems. The US has demonstrated significant progress by bringing solar thermal technology to the marketplace with 200 MWe of electrical capacity on-line and 500 MWe planned for the future (Morse, 1989). As described in a recent technical paper (Kearney, et al., 1989), the most developed technology is the parabolic trough type of system, and currently seven solar electric power plants with a total capacity of 194 MW are in operation in the Mohave desert in southern California. These plants utilize parabolic trough collectors, now in the third generation of design evolution, with improved system design and controls yielding peak noon solar-to-net electric efficiencies on the order of 23 percent. According to a recent paper by Morse (1990), these commercially mature systems are now providing electricity to the Southern California Edison grid at a cost of $0.10 to $0.12/kWh, and the latest plant (Luz SEGS VIII) is predicted to produce electricity at $0.08/kWh.

The initial central receiver demonstration project in the US was accomplished with the Solar One system at Barstow, California—a 10 MW generating plant using a water/steam receiver that met all its design goals and operated successfully on the Southern California Edison grid for six years (from August 1982 through September 1988). This system has been shut down and is currently in "caretaker" status (Whyte, 1990). Following the successful startup of the Solar One system, the US government stressed commercialization of the central receiver concept, and funded several feasibility studies for designs of large central receiver systems. However, after these studies, US government involvement in large-scale commercial demonstrations was reduced, and left to the private or utility sector to pursue. Many researchers believe that the large-sized solar thermal technologies are likely to have difficulty achieving commercial readiness in the current economic environment. For example, most companies in the industry feel that a central receiver technology is all ready for a new system demonstration in the 30–100 MW range—but the capital cost of such a system represents a risk that industry is not prepared to undertake alone (IEA, 1987; Andrejko, 1989).

With the exception of solar thermal parabolic trough technology, recent reviews of the subject (IEA, 1987; Andrejko, 1989) note that, on an international scale, few specific plans exist for demonstration or commercialization projects for utility-scale solar thermal electricity plants. In addition to lower capital costs and higher annual efficiency, it needs to be demonstrated that such systems can perform reliably with high system availability and low maintenance costs. The proponents of central receiver systems believe that these barriers can be overcome and continue to believe in the economic viability of this technology. For the parabolic trough technology, which is the closest to commercial status, the competing economics of conventional energy systems appear to be the only barrier (IEA, 1987).

Ocean Thermal Energy Conversion

Ocean thermal energy conversion (OTEC) systems are based on the use of the oceans as a natural solar collector and operate on the principle that energy can be generated using the temperature difference between the hot and cold regions of the ocean. As pointed out by its proponents, OTEC has the advantage that it can serve as a source of baseload electricity, providing power on a continuous basis. Some researchers (IEA, 1987) consider it ideal for supply power to isolated tropical and subtropical locations, as well as its additional potential for the production of fresh water, and/or the cultivation of kelp, shellfish, and other marine life by using the nutrient rich cold water from the OTEC powerplant.

A renewed study of utility-scale ocean OTEC powerplants began in the early 1970s, as one of the original six solar technologies in the United States energy program. The original research carried out in the US through the mid-1970s (McGowan, 1976) emphasized the technological design of large floating power plants that could supply electrical energy to the mainland, selected islands, or large floating platforms that could produce energy intensive products. Most of these designs were based on the use of closed-cycle OTEC systems, with ammonia usually specified as the working fluid. Furthermore, it was generally concluded that the major problem areas for the floating type designs were the following:

- Heat exchangers
- Cold water pipe
- Power transmission cables
- Ocean mooring system

With these subjects identified, the US sponsored a research and development program addressing key components of these problem areas during the mid-1970s and early 1980s. As a result of this program, significant technological advances have been achieved in the areas of heat exchangers (and corrosion and fouling prevention), cold water pipe design, and power transmission cables.

In the 1980s, research interests and programs in the United States shifted to near-shore or onshore systems. This included a change in interest from vertical cold water pipes to bottom mounted ones, and from floating systems to ones mounted onshore or on artificial islands or near-shore shelves. As a result of some of the previously mentioned problem areas, researchers started a program of open cycle system and component design for OTEC power systems. A privately sponsored project (MINI-OTEC), was the first successful demonstration of a 35 kW (15 kW net) floating OTEC plant. The US government sponsored the construction of OTEC-1, a closed cycle floating OTEC plant that was tested in Hawaiian waters in 1980. Design for a 40 MW OTEC closed cycle, near-shore plant, sponsored by a private corporation, with cost-sharing by the US government and the state of Hawaii, was also initiated in the 1980s.

Internationally, a number of countries have carried out significant research and design projects on OTEC systems during the past 15 years, and the United States and Japan have built prototype sized experimental OTEC systems. A country by country summary (IEA, 1987; Walker, 1988) of some of these recent efforts follows:

France has conducted feasibility studies on both open and closed cycles, and planned a 5 MW demonstration plant for Tahiti.

Japan, through the government sponsored "Sunshine" project, has carried out extensive research and development on heat exchanger technology, ocean engineering for floating plants, materials for OTEC systems, and design studies for floating and submerged platforms and their associated cold water pipes and mooring systems. The government is also sponsoring the construction of a 5 kW floating platform model and is proceeding with a design for a 10 MW floating plant. The Tokyo Electric Company sponsored the construction of a 100 kWe, shore-based, closed-cycle plant on Nauru. Plans are in progress for a 20 MW shore-based plant on the same island.

India carried out several OTEC design studies, and concentrated on the development of a 1 MWe closed cycle pilot scale plant.

The Netherlands government has supported efforts of domestic companies in design and feasibility studies for OTEC systems.

The United Kingdom has numerous government-sponsored technical and economic feasibility studies for OTEC systems and key system components. The Department of Industry and several companies and universities have worked together on the design of a floating 10 MW closed-cycle demonstration plant to be located in the Carribean or Pacific.

As can be seen, due to the potentially large resource, interest in the development of OTEC has been widespread, yet, no OTEC systems are ready for commercialization. The projected costs of most designs are still high compared to other renewable energy technologies. The large funds required for demonstration OTEC power plants are a financial barrier to commercial projects on OTEC systems, and will require government support. Once a small megawatt-sized prototype plant is operated for a significant time period, and commercial viability and cost effectiveness can be demonstrated for a scaled-up design, a worldwide OTEC industry could emerge.

Predicted Developments into the 21st Century

What developments can be expected in the 21st century in the four previously discussed renewable energy systems?

Continued Technological Performance Improvement and Cost Reduction

The requirements and potential for cost reduction are quite different for each of the utility-scale solar/wind systems discussed here, and Table 1 (abstracted from IEA, 1987) presents one summary of some of the probable technological areas where cost reduction can be achieved.

In the area of PV systems, a major area for potential improvement is in the increase of cell efficiency—a comprehensive summary of worldwide research and development efforts on this subject is given by Hamakawa (1988). As previously stated, PV cells are typically classified as thin-film cells, crystalline ("thick-film") cells, and concentrator cells. Herwig (1989) notes that at least six principal semiconductor materials and their alloys are under intensive cooperative investigation by the US PV industry and the federal PV program. He also notes that the efficiencies of today's commercial PV modules ap-

Table 1. Most Important Areas for Improvement of Economic Viability of Utility-Scale Solar/Wind Systems.

System	Research and Development			Mass Production	Scale-up
	Resource Assessment	Configuration*	Other**		
Photovoltaics	X	X	X		
Wind	X	X			
Solar Thermal	X			X	X
OTEC	X	X	X		X

* Including the match between supply and demand
** Including new technological breakthroughs
(Ref.-IEA, 1987)

proximate the 1978 laboratory efficiencies of their respective cells—specifically, about 5 percent for a number of amorphous silicon (thin-film) modules, about 15 percent for some of the best crystalline silicon commercial modules, and more than 20 percent for a few concentrator modules. Since today's best laboratory efficiencies are about 15 percent for thin-film cells, about 23 percent for crystalline cells, and 31 percent for concentrator cells, following past trends, it can be expected that these will be the efficiency values of the different types of PV cells at the start of the 21st century. Along with efficiency, it should also be noted that improvements in PV technology will also be made in the areas of increased system lifetime, reliability, and reduced overall costs (including balance-of-system, operation and maintenance, installation, and manufacturing).

A summary of the potential technological improvements for each system is beyond the scope of this paper—some may be discussed by other authors in this session. However, as a representative example, the types of potential technological improvements for utility-scale wind turbines that might occur in the future will be summarized. These include variable speed operation, optimized airfoils, and optimal control.

Variable Speed Operation of Wind Turbines. Most conventional utility-scale wind turbines have a rotor that turns at a nearly constant speed in order to produce constant frequency power. The variable speed concept offers the possibility of operating the generator at a variable speed such that the turbine always captures maximum energy for any wind speed. In addition to improved energy capture, other benefits of variable speed wind turbines include: enhanced utility system compatibility, structural load alleviation, operational flexibility, and improved site adaptability. This concept is seen by many in the wind energy industry as integral to the third generation of wind turbine development.

Optimized Airfoils for Horizontal Axis Wind Turbines. Today's wind turbines employ airfoils whose designs were originally developed for fixed wing aircraft applications. While these designs are adequate for overall energy production, they are not optimized for the complexities of wind turbine operation. Most conventional wind turbines use fixed pitch, stall-regulated rotors which are designed to control peak rotor power under strong wind conditions. However, the conventional airfoils do so at the expense of optimized energy production at moderate wind speeds. The blades are also subjected to high loads at peak power conditions, leading to potential problems. In addition, today's blades are highly sensitive to roughness of the blade's leading edge (insect buildup or erosion), which can lead to significant performance losses. In order to improve airfoil performance, researchers are developing several new "special-purpose" airfoil designs. This work is directed at developing thick and thin airfoil designs, for rotors of diameter 10–30 m, that can enhance annual energy output by 10 to 15 percent at low to medium wind speeds and provide more consistent operating characteristics with lower fatigue loads at high wind speeds.

Optimal Control. There are three broad areas where machine control is needed: power limitation, power control, and system control. The control system for a wind turbine represents a key component with respect to both power and machine operation. The basic objective of an optimized control system is to allow the wind machine to produce a maximum amount of energy while suffering a minimum of fatigue load. Thus, optimal control strategies should increase the productivity of the wind machine, result in lower fatigue, and minimize loads at winds with high damage potential. Overall, the advantage of optimal control is that it allows a given system configuration to perform at its maximum capacity. It is expected that the following improvements in control systems could increase the cost effectiveness of wind turbines:

- Start up/shut down algorithms for low winds
- High speed wind power limitation for fatigue reduction
- Enhanced system reliability
- Control strategies for variable speed operation.

With these types of technological improvements, it is expected that wind turbine costs will continue to decline, and result in the following predictions (Lynette, 1990) being realized, perhaps as early as the year 2000:

- Wind turbine costs will continue to decline (below $400/kW).
- Wind turbine efficiencies will increase by 10 to 20 percent.
- Wind turbine sizes will increase to 300–600 kW, resulting in lower balance-of-station costs per kW.
- Cost of generating electricity will decrease by at least 40 percent.

In the United States, it is expected that these kinds of technological improvements will result in expansion of the windfarm market to sites other than California. For example, as recently studied by Bailey (1988), this could include a number of large sites in the northeastern US.

Expansion of Worldwide Markets for Renewable Energy

When technological improvements are made to large-scale renewable energy systems, their delivered electricity costs should decrease, opening new markets for such systems, as fuel and operating costs of conventional electrical power systems increase as expected.

As pointed out in recent European studies (Hohmeyer, 1988), another factor that should be included in the economic comparison of renewable and conventional energy conversion systems is the social costs of energy production. It has been concluded that since conventional (fossil or nuclear) powered energy technologies are able to pass on to society a substantial part of their costs (social costs), renewable energy sources, which do so to a far lesser extent, are put at a systematic disadvantage. Specifically, for the Federal Republic of Germany, it has been calculated that if social costs had been included in the market price of conventionally generated electricity, cost effectiveness of wind energy would have been achieved by 1984. It is concluded that, if social costs continue to be excluded, it may take until 1994 for cost effectiveness of wind energy generation to be realized. Furthermore, it is noted that similar considerations applied to photovoltaic based solar electrical energy production may delay cost effectiveness to the year 2000.

Regarding the international market, recent studies (Pourier and Meade, 1990) conclude that a worldwide market for independent utility-scale energy facilities is developing quickly. As given in Table 2, their analysis of the world market shows a potential for 43,500 MW of capacity from independent energy sources to be ordered outside the United States through the year 2000. As shown, they divide the international market opportunities for independent energy into the following five markets:

Table 2. The World Independent Power Market (Orders-1989–2000).

Market Segment	Size (MW)	No. of Projects
Wholesale	11,500	80
PURPA-Overseas	14,200	840
Self-Generation	13,500	1,500
Frontier	3,600	200
Specialty	600	3,570

(Ref.- Poirier and Meade, 1990)

1. Wholesale. The wholesale market includes power plants in the 60–600 MW range that operate as baseload units with 15 to 20 year contracts and sell power in a range from $0.04 to $0.08/kWh. Presently, typical plants included in this category include large hydroelectric, solid-fuel fired boiler steam turbine or gas-fired turbines/combined cycle plants.
2. PURPA—Overseas. This market involves cogeneration or small power projects, generally in the 3–100 MW range that are grid connected and sell electricity at prices ranging between $0.05 to $0.12/kWh. Most PURPA-overseas projects will be subject to seasonal dispatchability requirements.
3. Self-generation. The self-generation market is the most established independent energy market and includes power plants where power is used onsite with no regulation entity involved. These projects can be either on- or off-grid and can typically range between 100 kW to 40 MW. Electricity prices have a wide range between $0.02 to $0.20/kWh, particularly when diesel generators are used. It includes both new and retrofit sites, with large potential existing in new industrial parks, tourism centers, resorts, desalination plants, and new communities.
4. Frontier. This market is found whenever developing countries are opening new territories that call for rural electrification and irrigation. In a typical project, a new region is being developed on a large scale and a national/regional grid is being extended or mini-grids are being created. Electricity prices can range between $0.02 to $0.20/kWh in these applications.
5. Specialty. The specialty market includes projects in sizes below 250 kW in "high-value-added" applications such as remote villages, military facilities, satellite tracking bases, offshore applications, etc. Electricity is a costly commodity in these applications and can range between $0.50 to $1.00/kWh. The type of technology presently used for these systems can range from diesel engines to renewable energy sources such as wind and PV.

As can be seen, the predicted market for independent energy projects is going to be large in the next decade, and from all indications, it will increase rapidly in the 21st century. For grid-connected applications, it is expected that windfarms, utility-scale PV systems, and solar thermal plants can capture some of this potentially large market. Due to recent technological progress in the area of hybrid solar or wind/diesel systems, it is expected that there also will be a large market in the medium nongrid or small-grid applications area. A review of this subject is beyond the scope of this paper; however, recent summaries and reviews of the technological developments in the field of wind/diesel hybrid systems are presented by Chertok, et al. (1989), and McGowan, et al. (1988).

Improvements in Energy Storage/Use of Hydrogen Energy

As noted in a recent Electric Power Research Institute (EPRI) review (Boutacoff, 1989), the ability to store energy on a large scale would have a profoundly liberating effect on the electric utility industry. That is, supply and demand would not have to be balanced instantaneously, so utilities would have much greater flexibility in operating their systems and conducting power transactions. Furthermore, the effect that cost-effective energy storage systems would have on typical solar or wind powered utility systems, with their much higher energy output variations, would be even more dramatic.

For conventional utility-scale power plants, today's available energy storage technology options include compressed-air energy storage, battery storage, and pumped-hydroelectric storage. According to EPRI (Boutacoff, 1989), pumped-hydro systems are most economical in large sizes (over 1000 MW) and are best suited to baseload duty, above eight hours per day. Compressed-air storage systems cost much less per hour of storage and are most cost effective when operated from 5 to 12 hours per day of generation. Battery systems are most appropriate for providing peak power about three hours per day. Regardless of which of these modes of energy storage are used, present-day costs are quite high and add significantly to the capital costs of the total power plant. However, although technical progress is expected to yield cost reductions in most of these systems in the future, a major breakthrough in cost reduction is not expected.

For renewable energy powered sytems, a number of researchers (e.g., Lodhi, 1989) have concluded that hydrogen as an energy carrier is the most efficient means for the collection, storage, and transportation of energy. In general, these scientists have concluded that hydrogen is the most economical product to produce and store energy, as well as being the cleanest and recyclable. Also, using solar or wind energy as an input, hydrogen can be produced either electrolytically or thermochemically.

Right now, there are a multitude of methods for the storage of hydrogen, several of which are currently in use commercially. Others, such as metal hydrides storage are under active investigation. Furthermore, as concluded by Lodhi (1989), the most economical large-scale transportation of hydrogen is by pipeline. As a result of numerous commercial operations in the past, there exists a wealth of experience and information for the safe operation of hydrogen systems.

Numerous researchers, with most current activity occurring in Europe, have pointed out the advantages of combining the utility-scale energy produced from solar and wind systems with a hydrogen production/storage/distribution system. For example, Winter and Nitsch (1989) conclude that solar energy could become a commercial commodity and would, as a component of a prospective hydrogen energy economy, participate in the energy trade of the world, perpetuating it beyond the point in time where the fossil fuel supplies are exhausted. In another recent study, Knoch (1989) compared solar-wind-hydrogen energy systems to conventional power plants, concluding that electric power and hydrogen generation systems, using wind energy systems combined with decentralized power supply systems, could be set up immediately. Solar (PV)–wind–hydrogen power plants could be economically feasible in the near future permitting the exploitation of daytime and nighttime solar energy.

On the other hand, a clear, or most cost-effective methodology for implementation of a hydrogen energy economy on a national or international basis has yet to be formulated. Many researchers believe it should have a high priority and will require concerted support by many governments. If established, it will require one of the major technological programs of the 21st century, and will most probably utilize large-scale solar and wind energy conversion systems.

Utilization of Offshore Solar/Wind Resources

The electrical energy potential of solar or wind systems sited in the oceans of the world is huge, and constitutes a resource that may be hard to ignore in the next century. For example, the following estimates have been made for this resource on the Atlantic Ocean side of the United States (McGowan and Heronemus, 1975):

1. Ocean Thermal Power Plants. The available power from the Gulf Stream is estimated to be about 1.4 trillion kWh per year.
2. Offshore Wind Energy. Offshore wind energy systems sited off New England and the eastern seaboard could produce about 650 billion kWh per year.

Since wind energy generation systems are in a more advanced state of development, the first projected use of this resource will be based on systems using large groups of offshore wind turbines. A short summary of the key points of these potential systems follows.

In this concept, wind machines are installed in offshore locations. For depths less than about 20m, they would be placed on fixed platforms. In greater depths, they could be supported by floating structures. The prime justification for considering offshore windfarms is the generally higher winds. Proposed offshore wind sytems are relatively large to compensate for the expected higher fixed costs. One feature unique to the offshore concept is the need for an undersea cable to transport electrical energy to land (unless a hydrogen based pipeline system is used).

Offshore wind systems could take advantage of higher average winds (for example, it is typically estimated that annual productivity will be 40 percent greater compared with equivalent land-based systems), and land costs are less relevant. Social acceptance problems should be much less, especially in terms of aesthetics, noise, and TV interference. Safety would be enhanced. Fatigue of the machines should be less due to relatively low wind shear and turbulence in offshore winds. Large structures are easier to transport on water than on land. This could help make multi-megawatt machines more cost effective offshore.

In the area of technological problems faced by such systems, the system would need to be large to be cost effective, since there are very large fixed costs. Submarine transmission cables would be needed, and structures would need to withstand wave and ice loadings, as well as possibly higher survival winds. Dynamics of the systems would need to be carefully analyzed, taking into account all of the possible loadings. The dynamic response of floating supports or multi-rotor structures would be particularly problematic. Foundation design and construction would be more costly than with onshore machines. In the oceans, the salt water environment would need to be considered in the design and fabrication—marine engineering of many of the components would be necessary. Access for inspection and routine maintenance would be relatively difficult, especially in bad weather. Particular environmental concerns would involve navigation, fishing, boating, shore facilities, and microwave interference. Availability of large seagoing cranes for installation of wind systems would be limited.

The first design studies of utility-scale offshore wind systems were carried out in the United States at the University of Massachusetts (Heronemus, 1972). This proposed 80 GW offshore wind system was to be located in operation off the New England coast in the relatively shallow waters of the Gulf of Maine and over the Banks. Several horizontal axis wind turbine designs were suggested, with concrete and aluminum structures to carry three or four separate wind turbines. Both floating and seabed mounted structures were proposed, with the produced electricity used to electrolize water, producing hydrogen for transmission to the land. Energy storage was to be provided via the use of submerged pressure balanced storage containers. As originally proposed, this system had the potential to convert all petroleum and coal fired heating processes to electricity or hydrogen combustion, and to also provide a synthetic fuel (hydrogen) to all road, train, and air transport in New England.

Offshore wind is currently the subject of considerable research in Europe, especially Sweden (Hardell and Bjork, 1989) and the UK (Dixon and Swift, 1986 and Swift-Hook, 1989). Numerous feasibility and design studies have been made, but no wind machines have yet been installed in the open sea. Furthermore, some of the latest economic analyses show offshore wind systems to be competitive with onshore wind systems. Specifically, these studies predict cost of energy at about $0.04 to $0.10/kWh in sites with mean wind speeds of 9–10 m/s.

In Europe, the first steps were taken toward offshore development in some shore-edge installations in the outer harbor of Zeebrugge, Belgium (DeWilde, 1986). In addition, 11 large scale (450 kW) wind turbines are to be part of the world's first offshore wind farm to be constructed 1.5 km off the coast of Denmark. Although this plant is not expected to be able to produce electricity as cheaply as a comparable land-based plant in Europe ($0.08/kWh vs. $0.05/kWh), a major goal of the project is to establish the technical feasibility of offshore windfarms.

Conclusions

During the later part of the 20th century, impressive technological advances have been made in the development of utility-scale electricity producing systems powered by the wind or the sun. In the 21st century, the continued development of these systems will rely on technical improvements from a concerted research and development program in the areas of wind, photovoltaics, solar thermal, and ocean thermal energy conversion. In some areas such as windfarms, economic viability and cost effectiveness compared with conventional large-scale power generation systems have already been established. Most of these technologies will require continued government support for large demonstration projects so that system costs can be further reduced via the development of mass production techniques. The benefits of these systems that produce less damage to the environment—and hence lower societal costs—must be reflected in future cost calculations for energy producing systems.

On an international scale, the available market for these renewable energy technologies is predicted to expand at a rapid rate. Currently, despite the pioneering technological efforts of the United States in these areas, the international market, and even the market in the US, is wide open to all the industrial countries of the world. Due to large-scale government support, it is expected that Japan and Europe (especially Germany) will play a major part in the commercialization of these renewable energy technologies.

References

AWEA (American Wind Energy Association), "Wind Energy for a Growing World," Washington, DC (1990).

Andrejko, D. A., "Assessment of Solar Energy Technologies," American Solar Energy Society (1989).

Bailey, B. H., "An Assessment of Wind Energy Feasibility in the Northeastern U.S.," *Proc. 1988 Annual Conference American Solar Energy Society*, Cambridge, MA (1988).

Beurskens, J. and Schmid, J., "Review of Large Wind Turbine Developments," *Proc. of the Euroforum-New Energies Congress*, Saarbrücken, Federal Republic of Germany, October (1988).

Boutacoff, D., "Emerging Strategies for Energy Storage," *EPRI Journal*, July/August (1989).

Brown, L. R., et al., *State of the World 1988*, W. W. Norton, New York (1988).

DeWilde, M., "The Windfarm in the Outer Harbour of Zeebrugge," *Proc. European Wind Energy Conference (EWEC)*, Rome, Italy (1986).

Dixon, J. C. and Swift, R. H. "Offshore Wind Power Systems: A Review of Developments and Comparison of National Studies," *Wind Engineering*, 10, No. 2 (1986).

Hamakawa, Y., "Recent R & D Efforts for Solar Cell Efficiency Improvement," *Proc. of the Euroforum-New Energies Congress*, Saarbrücken, Federal Republic of Germany, October (1988).

Hardell, R. and Bjork, B., "Offshore Based Power Plants in Sweden," *Proc. European Wind Energy Conference (EWEC '89)*, Glasgow, Scotland (1989).

Heronemus, W. H., "Power from the Offshore Winds," *Proc. Marine Technology Society, 8th Annual Congress* (1972).

Herwig, L. O., "Photovoltaic Technology Advances, Industry Progress, and Market Promise," *Proc. 1989 Annual Conference American Solar Energy Society*, Denver, CO (1989).

Hoffner, J. E., "Analysis of the 1988 Performance of Austin's 300-Kilowatt Photovoltaic Plant," *Proc. 1989 Annual Conference American Solar Energy Society*, Denver, CO (1989).

Hohmeyer, O. H., "Macroeconomic View on Renewable Energy Sources," *Proc. of the Euroforum-New Energies Congress*, Saarbrücken, Federal Republic of Germany, October (1988).

IEA, International Energy Agency, *Renewable Sources of Energy*, OECD (1987).

Kearney, D. W., et al., "Design and Operation of the Luz Parabolic Trough Solar Electric Generating Plants," *Proc. 1989 Annual Conference American Solar Energy Society*, Denver, CO (1989).

Knoch, P. H., "Energy without Pollution: Solar-Wind-Hydrogen Systems: Some Consequences on Urban and Regional Structure and Planning," *Int. J. Hydrogen Energy*, 14, No. 12 (1989).

Lindley, D., "The Commercialization of Wind Energy," *Proc. of the Euroforum-New Energies Congress*, Saarbrücken, Federal Republic of Germany, October (1988).

Lodhi, M. A. K., "Collection and Storage of Solar Energy," *Int. J. Hydrogen Energy*, 14, No. 6 (1989).

Lynette, R., "Status of Wind Power Industry," *Proc. 1989 American Wind Energy Association (AWEA) Annual Meeting* (1989).

McGowan, J. G. and Heronemus, W. H., "Ocean Thermal and Wind Power: Alternative Energy Sources Based on Natural Solar Collection," *Environmental Affairs*, Nov. (1975).

McGowan, J. G. "Ocean Thermal Energy Conversion—A Significant Solar Resource," *Solar Energy*, 18 (1976).

Morse, F. H., "Solar Thermal Technology Applications and Opportunities," *Proc. 1989 Annual Conference American Solar Energy Society*, Denver, CO (1989).

Morse, F. H., "Opportunity Knocks for Solar Thermal," *Solar Today*, 4, No. 1, pp. 16–19 (1990).

Palz, W. and Helm, P., "PV Power Generation—An Overview," *Proc. of the Euroforum-New Energies Congress*, Saarbrücken, Federal Republic of Germany, October (1988).

Poirer, J. and Meade, B., "Assessing the World Market," *Independent Energy*, Jan. (1990).

Twidell, J. W. and Weir, A. D., *Renewable Energy Resources*, E. F. Spon, London, 1986.

Walker, J. F., "Ocean Technologies as a Source of Renewable Energy," *Proc. of the Euroforum-New Energies Congress*, Saarbrücken, Federal Republic of Germany, October (1988).

Whyte, M. D., Personal Communication, March (1990).

Winter, C. J. and Nitsch, J., "Hydrogen Energy—A 'Sustainable Development' towards a World Energy Supply System for Future Decades," *Int. J. Hydrogen Energy*, 14, No. 11, pp. 785–798 (1989).

Future Effects and Contributions of Photovoltaic Electricity on Utilities and the Environment

Ghazi Darkazalli
Spire Corporation
Bedford, MA

Abstract

Renewable energy technologies have gone through significant development stages in the past two decades. Now these technologies represent vital sources of energy that are reliable, economical, and safe to the environment. Utilities, industry, and commercial and residential consumers have realized the value of using renewable energies such as hydroelectric power, photovoltaics, geothermal, wind, and solar to help meet their energy demands.

Introduction

In the last decade, we have witnessed the evolution of renewable energy sources, such as photovoltaic and wind, into mature industries. Renewable energy has maintained a healthy growth rate. Its contribution to the energy supply in the United States has grown by close to 30 percent since the early 1980s. Currently, more than 9 percent of US energy demand is met by solar, wind, geothermal, hydro, and other renewable energy sources. As the United States and the world strive to improve public health and preserve the environment, renewable energy sources will play a growing role. The US alone can cut a minimum of 5 percent and a maximum of 19 percent of the projected annual total of 6.4 billion tons of CO_2 by the year 2000 (Rader, 1989).

The Environment

With growing concern about the effects of acid rain and global climate changes, renewable energy sources are attracting great interest.

Acid rain comes mostly from the emission of sulfur dioxides resulting from fossil fuel burning, particularly coal. The damage done by these emissions has become blatantly clear with the death of forests and lakes and the increased corrosion of metals and stone structures. These effects are felt worldwide.

Global climate changes are due to the presence in the air of the "greenhouse" gases released from fossil fuel burning. Carbon dioxide will account for one half of the global warming problem over the next 60 years. Other greenhouse gases such as methane, nitrous oxide, and chlorofluorocarbons will account for the other half of this greenhouse effect.

Nuclear power proponents have pointed to nuclear power as the solution to the greenhouse effect. However, nuclear power's contribution in the United States is limited because no plants are scheduled for construction in the near future due to their high cost and public resistance.

A photovoltaic (PV) cell is a thin layer of semiconductor material, usually silicon, which converts sunlight directly into electricity. No carbon dioxide is emitted during this process. Fuel costs are zero, and maintenance costs are low because PV systems are solid-state and can operate without any moving parts.

We can easily establish the principal attributes that make photovoltaics a socially beneficial technology when compared to other options. These attributes exhibit both the promise of photovoltaics and their barriers. The fundamental promise of PVs lies in their characteristics, including environmental benignity, modularity, universal availability, very low operating costs, and a 30-year life expectancy. To clearly illustrate the environmental comparisons, Table 1 represents air emissions of some conventional technologies and those of PV systems. The values given in the table represent emissions over the life of a facility, including construction, manufacturing, fuel cycle, and decommissioning.

The use of photovoltaic power generating systems will substantially reduce emissions from conventional power plants as shown in Table 2.

Utilities

According to the US Department of Energy (DOE), the new generating capacity needed by electric utilities be-

Table 1. Air Emissions: Electric Generation Systems (Tons per GWh).

	Conventional Plant	AFBC Plant	IGCC Electric Plant	Boiling Water Reactor	PV Central Station
CO_2					
Fuel Extraction	-	-	-	1.642	NA
Construction	1.048	1.048	1.048	1.088	5.890
Operation	1057.143	1055.942	822.945	5.861	NA
Total	1058.191	1057.090	823.993	8.590	5.890
NO_x					
Fuel Extraction	0.066	0.066	0.052	0.022	NA
Construction	0.001	0.001	0.001	0.001	0.008
Operation	2.914	1.484	0.198	0.011	NA
Total	2.986	1.551	0.251	0.034	0.008
SO_x					
Fuel Extraction	0.055	0.055	0.043	0.024	NA
Construction	0.002	0.002	0.002	0.001	0.023
Operation	2.914	2.911	0.291	0.003	NA
Total	2.971	2.968	0.336	0.029	0.023
Particulates					
Fuel Extraction	1.482	1.480	1.173	0.002	NA
Construction	0.001	0.001	0.002	0.001	0.017
Operation	0.143	0.143	0.001	neg	NA
Total	1.626	1.624	1.176	0.003	0.017
CO					
Fuel Extraction	0.061	0.061	0.048	0.002	NA
Construction	0.001	0.001	0.001	0.001	0.003
Operation	0.206	0.205	-	0.016	NA
Total	0.267	0.267	-	0.018	0.003

Mass. PV Center

Table 2. Emission Displacement Using Photovoltaics (Metric Tons/GWh).*

Fuel	CO_2	SO_2
Coal (AFBC)	960	10
Oil (Steam Turbine)	700	5
Gas (Steam Turbine)	480	neg
Nuclear	8	N/A

Source: US Department of Energy
*1GWh is the annual output of 600kW of photovoltaics operating 4.5 hours per day.

tween now and the year 2000 is approximately 111 GW. The major portion of this capacity, 73 GW, has not yet been announced or ordered.

It is estimated that the global primary energy demand will double by the year 2025. To meet this tremendous growth requirement, many developing countries will turn to renewable energy sources. This will be in the form of local stand-alone power plants. Photovoltaics represent a cost-effective, environmentally safe system. Using photovoltaics will eliminate the need for elaborate central power plants and extensive distribution networks.

Utility Experience with Photovoltaics

More than 25 megawatts of photovoltaic systems are interconnected in the United States with the utility grid. Approximately 40 US utilities have had experience with photovoltaic generation. This experience may be classified in three categories; 1) experimental test facilities, 2) stand-alone power plants, and 3) grid-connected photovoltaic powered plants. The utility interactive systems range in size from small residential systems such as those used by New England Electric System (2.2 kW roof mounted) to large central stations such as Pacific Gas and Electric Company Carissa Plains (6.5 MW generator).

In a report by V.V. Risser and H.W. Stokes entitled "Photovoltaic Field Test Performance Assessment, 1987," the authors concluded that notable performance improvements in photovoltaic powered plants have occurred (Risser and Stokes, 1987). These improvements include: 1) higher availability during sunlight hours, 2) efficiencies improved from the earliest studies (Phoenix Sky Harbor, 6.5 percent) to the most recent one (Austin, 9.8 percent), and 3) operation and maintenance costs which are much lower for single-axis tracking systems than for those of two-axis tracking.

Photovoltaic System Contribution

The photovoltaic contribution to world energy demands is estimated at 10 to 30 percent by the year 2050. This contribution will result from the 1) availability of inexpensive cells, 2) economical time-shifting of energy available through storage, or 3) significant time zone transmission. Many Third-World countries will elect to use photovoltaic technology because it can be manufactured locally. It does not require an extensive energy infrastructure and will reduce the country's current dependence on expensive electricity for its development.

The improved photovoltaic cell efficiencies, combined with larger manufacturing plants currently under construction, are expected to produce cells that can generate electricity at approximately one dollar per peak watt. Even with today's technologies, economies of scale employing larger manufacturing facilities could bring the cost of photovoltaic electricity down to 12–15 cents/kWh, which is comparable to the current cost of generating electricity.

The world photovoltaic module shipments grew from 35.2 MW in 1988 to 42.1 MW in 1989. This 20 percent increase represents the second year in a row with growth of 20 percent or more. It is anticipated that photovoltaic module shipments in 1990, 1995, and 2000 will exceed 54, 120, and 145 MW, respectively.

Table 3 shows the photovoltaic module market as a function of price. It is very important to note that at current prices, $4 to $6 per watt, the PV market is mostly concentrated in the remote power applications. It is predicted that improvements in the manufacturing techniques, processing technology, and conversion efficiency will decrease module prices to the $1 to $2/Wp (dollars per peak watt) level. This price reduction will result in an increase in the market size to approximately 500 to 2000 MWp (peak megawatts) per year. The major area of this increase will be in the rural power and suburban power sectors which will represent 35 percent and 50 percent of the total market.

The potential of photovoltaics in energy production, emissions, and land and water use (as compared to conventional and next generation coal and nuclear plants) is very favorable for the following reasons (Sklar, 1989):

- Photovoltaic power plants produce the least emissions of any conventional power plant, including operation, construction, and extraction.
- Photovoltaic power plants utilize the same land area needed by conventional energy plants when operation, construction, and extraction and transportation are considered.
- Photovoltaic power plants use at least one third less water than any conventional energy power plant.

In addition to the environmental advantages of photovoltaic power plants, utilities' objectives can be easily met by using photovoltaics for load shaping and demand-side alternatives, such as peak clipping, valley filling, load shifting and others.

Future Scenario

The Federal Energy Regulatory Commission is considering a number of issues relating to avoided costs, competitive bidding, and power production. This interest on the part of government, coupled with the demonstrated advantages of photovoltaic power plants, makes photovoltaics one of the choice energy applications in the 1990s.

PVUSA is a photovoltaic power project to demonstrate utility scale applications. It is a national cooperative research and development project aimed at assessing possibilities for direct harnessing of the sun to generate commercial quantities of electricity. This project will evaluate and compare the electrical performance and reliability of a number of photovoltaic technology systems for utility use. PVUSA was initiated in 1988; a number of new and emerging PV technologies were chosen and contracts were awarded for 20 kW segments. During 1990 to 1992, larger utility scale segments of 200 kW to 400 kW will be awarded. These larger selections will be based on their best potential for bulk power generation.

PV/2000 is an initiative program, introduced to the US Congress by the Solar Energy Industries Association (SEIA), which provides loan subsidies for utility companies to purchase photovoltaic peak power plants. The plan, if adopted, will in effect add 2000 MW of photovoltaic power plants to the US utility grid.

Conclusions

Over 40 utilities in the United States and many others around the world have had successful experience with small and large scale photovoltaic power systems. The

Table 3. PV Market.

Price ($/Wp)	6	4	3	2	1
Volume (MWp)	20	40	100	500	2,000
Revenue ($M/Yr.)	120	160	300	1,000	2,000
SEGMENTS (%)					
Remote Power	75	40	20	5	1
Rural Power	3	10	30	35	35
Suburban Power	-	-	2	40	50
Motive Power	2	25	28	15	9
Consumer Package Goods	20	25	20	5	5
GEOGRAPHIC DISTRIBUTION (%)					
North America	35	30	25	30	35
Europe	30	35	25	25	25
Japan	25	25	20	15	10
Developing Countries	10	10	30	30	30

* INTERNATIONAL SOLAR ENERGY INTELLIGENCE REPORT

photovoltaic industry is mature and growing at approximately 30 percent per year.

Photovoltaic power systems can be utilized for small electric power plants. These plants are a natural match to localize power distribution systems. Photovoltaic systems are modular, which allow the utilities to add capacity easily, as needed. Photovoltaics may be used for residential, village electrification, and central power stations.

Photovoltaic systems are environmentally safe. They produce the least emissions of any conventional power plant, require the same land area as conventional energy plants, and use one third less water than any conventional energy power plant.

The significant value of photovoltaic electricity—from an energy and environmental point of view—is sufficient to justify widespread use of this technology.

References

N. Rader, "Power Surge, The Status and Near-Term Potential of Renewable Energy Technologies," *Public Citizen,* Washington, DC (May 1989).

W.V. Risser and H.W. Stokes, "Photovoltaic Field Test Performance Assessment," Southwest Technology Development Institute (1987).

S. Sklar, "Solar Energy Regulatory and Policy Issues in the United States: A 1989 Photovoltaic Review" (Proceedings).

Geothermal Energy: Electricity Production and Environmental Impact, A Worldwide Perspective

Ronald DiPippo
Mechanical Engineering Department
Southeastern Massachusetts University
North Dartmouth, MA

Abstract

This paper presents the technology for utilizing natural hydrothermal geothermal energy for the generation of electricity. Emphasis is placed on the potential impact of geothermal power plants on the environment. Comparisons are drawn between geothermal plants and other means of power generation. Overall, geothermal plants emerge as perhaps the most benign of all plants now available with regard to environmental impact. Geothermal energy is a worldwide resource that can make significant local contributions to meeting present and future demands for electricity in an economic and environmentally safe manner.

Worldwide Power Plant Development

Geothermal energy is now an accepted means of generating electricity. There are 20 countries today with operational geothermal power plants. All six non-polar continents are represented: Africa, Asia, Australia, Europe, North America, and South America (including Central America). There are over 35 additional countries with power plants either under construction or in the advanced planning stage, or with exploration programs that appear to hold promise for geothermal development underway or planned.

This paper will deal solely with the form of geothermal energy that has been proven commercially and economically feasible so far, namely, hydrothermal energy. The other forms—geopressured, hot dry rock, direct tapping of shallow magma bodies, and low-temperature radiogenic sources—are the subject of research and development.

Historical Development

Italy began using geothermal energy for electric power generation in 1904 at Larderello. The first commercial plant went on line in Italy in 1913. In 1958 New Zealand commissioned its first plant at Wairakei, and the United States opened its Geysers Unit 1 in 1960. Since then the growth in installed capacity has been impressive, particularly in the United States. On the average, the US has increased its geothermal capacity by 30 percent annually over the period from 1960–1989. During the 1980s, the average annual growth rate in the US was 13 percent. The total worldwide geothermal power capacity grew at an average annual rate of nearly 14 percent from 1960 to 1989, and at about 12 percent for the decade of the 1980s.

Figure 1 shows the growth patterns for the US (lower curve), the rest of the world (middle curve), and the whole world (upper curve). Note that since the ordinate is plotted as $\log_{10}$ of the installed megawatts, the limits of the graph are from 10 to 10,000 megawatts (MW_e) (DiPippo, 1988a).

Table 1 gives a summary of the status of development by country. In each case, the types of geothermal power plants, the total number of power units (i.e., turbine-generator sets), and the total capacity are given. The technology represented by the plants includes dry steam (DS), single flash (1F), double flash (2F), binary (B), and hybrid (H) systems. These will be described in the next section.

The number of units, as defined in this paper, has skyrocketed with the recent widespread use of binary units, which are typically small (0.5–1.5 MW_e) and designed for modular applications. The average size of a geothermal plant, using the totals given in Table 1, turns out to be about 19 MW_e. More typically, geothermal steam units (DS, 1F, 2F) tend to be sized from 30–55 MW_e in contrast to the binary units, which average about 1 MW_e.

The United States operates about 50 percent of the total 316 power units and accounts for about 48 percent of the total 5900 MW_e.

Future Development

The following countries have very good prospects for developing their geothermal resources to the point of electric generation by the year 2000: Bolivia, Canada,

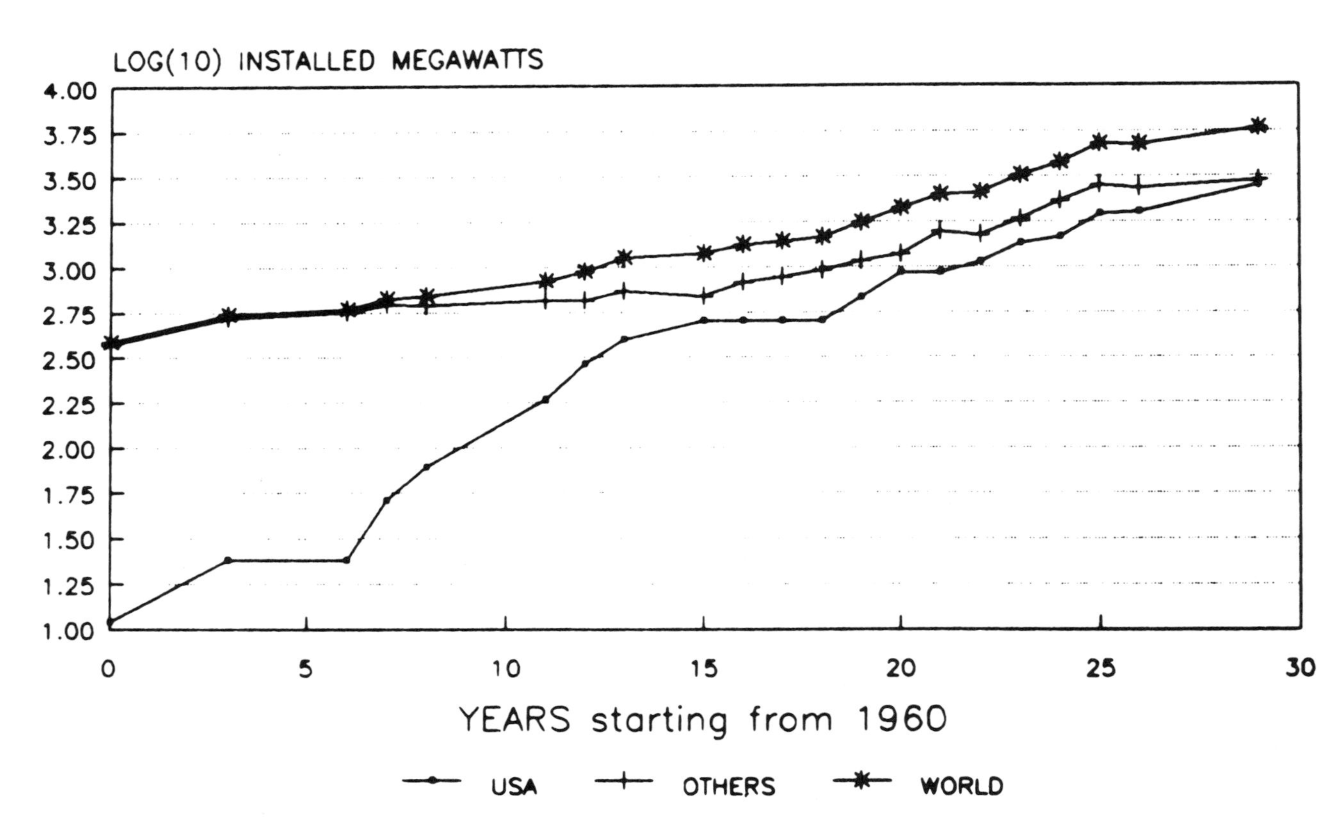

Figure 1. Log_{10} of installed geothermal capacity (MW_e): 1960-1989.

Chile, Costa Rica, Djibouti, Ethiopia, Guatemala, Honduras, India, Saint Lucia, and Thailand.

Although less advanced in geothermal exploration, the following countries have interesting prospects and with further exploration may prove to have commercially valuable geothermal resources (Dickson and Fanelli, 1988; DiPippo, 1980): Algeria, Austria, Brazil, Burma, Colombia, Domenica, the Dominican Republic, Ecuador, Granada, Greece, Haiti, Hungary, Iran, Israel, Jordan, Madagascar, Nepal, Pakistan, Panama, Peru, Portugal (Cape Verde), Spain (Canary Islands), Sri Lanka, Tanzania, Uganda, Venezuela, Viet Nam, and Yugoslavia.

With the emergence of small, modular binary power units, many geothermal reservoirs previously thought to be unsuitable for power generation because of relatively low temperature are now good candidates for development and commercialization.

Geothermal Power Plant Technology

In this section we will describe the means of converting geothermal energy from hydrothermal resources to electricity. Descriptions and simplified schematic flow diagrams will be presented for the major types of plants: Direct (or dry) steam, single flash, double flash, and binary plants.

Our focus will be on the potential impacts of plant operations on the environment. We will point out those places in each plant where the possibility exists for plant/

Table 1. Geothermal power plants worldwide—1990.

Country	Plant Types[1]	No.Units[2]	Total MW_e[3]
United States	DS,1F,2F,B,H	159	2826.49
Philippines	1F	23	894.0
Mexico	DS,1F,2F	21	725.0
Italy	DS,1F	42	504.2
New Zealand	1F,2F,B	16	285.5
Japan	DS,1F,2F	9	215.1
Indonesia	DS,1F	5	142.25
El Salvador	1F,2F	3	95.0
Nicaragua	1F	2	70.0
Kenya	1F	3	45.0
Iceland	1F,2F	5	39.0
Turkey	1F	1	20.6
China	1F,2F,B	17	20.586
Soviet Union	1F	1	11.0
France (Guadeloupe)	2F	1	4.2
Portugal (Azores)	1F	1	3.0
Romania	B	3	1.5
Argentina	B	1	0.6
Zambia	B	2	0.2
Australia	B	1	0.02
20 Countries	DS,1F,2F,B,H	316	5900.246

[1] DS = dry steam; 1F = single flash; 2F = double flash; B = binary; H = hybrid.
[2] A "unit" is defined as a turbine-generator unit.
[3] Total installed capacity.

environment interaction. In the third section, we will deal with the environmental impacts in detail.

Direct Steam Plants

In certain hydrothermal systems, the fluid and reservoir rock characteristics are favorable for the formation of dry (essentially saturated vapor) or slightly superheated steam. Such reservoirs exist, for example, at Larderello (and neighboring areas) in Italy, The Geysers in California, Kawah Kamojang in Indonesia, and Matsukawa in Japan.

It is relatively simple to exploit a dry steam reservoir. Wells drilled into such a formation yield steam (possibly contaminated with certain gases such as carbon dioxide and hydrogen sulfide) but no liquid. Thus, a turbine may be designed to expand the moderate-temperature (say, 180°C or 350°F), moderate-pressure (say, 700 kPa or 100 lbf/in^2, abs) steam and drive a generator to produce electric power.

Figure 2 is a simplified schematic of such a plant. Since a typical well might be capable of generating 5–10 MW_e, several wells are connected to a single turbine-generator set (a unit) to yield, say, 55 MW_e. An in-line, axial, centrifugal separator is fitted at each wellhead to remove entrained particulates such as rock dust which could cause erosion problems. Drain pots (not shown in Figure 2) are often installed at regular intervals along the steam pipelines to purge the steam of any condensate that forms during transmission. Spent steam from the turbine is condensed and recycled via a cooling tower to serve as cooling water for the condenser. There is more than enough geothermal steam condensate to accomplish this task, and the excess water from the cooling tower is available for reinjection into the reservoir. Roughly 20 percent of the steam flow may be returned to the formation, where it will help maintain reservoir pressure. The gases that accompany the steam are noncondensable and must be purged by some type of ejector or compressor. Normally, these gases along with the evaporated condensate from the cooling tower eventually end up in the atmosphere. Depending on the composition of the gases, abatement systems may or may not be required.

The presence of hydrogen chloride in the vapor phase has caused corrosion problems recently at some wells at The Geysers field. Although this problem is being studied intensively, it is speculated that it may be associated with excessive production as manifested by a dramatic drop in steam pressure in the reservoir.

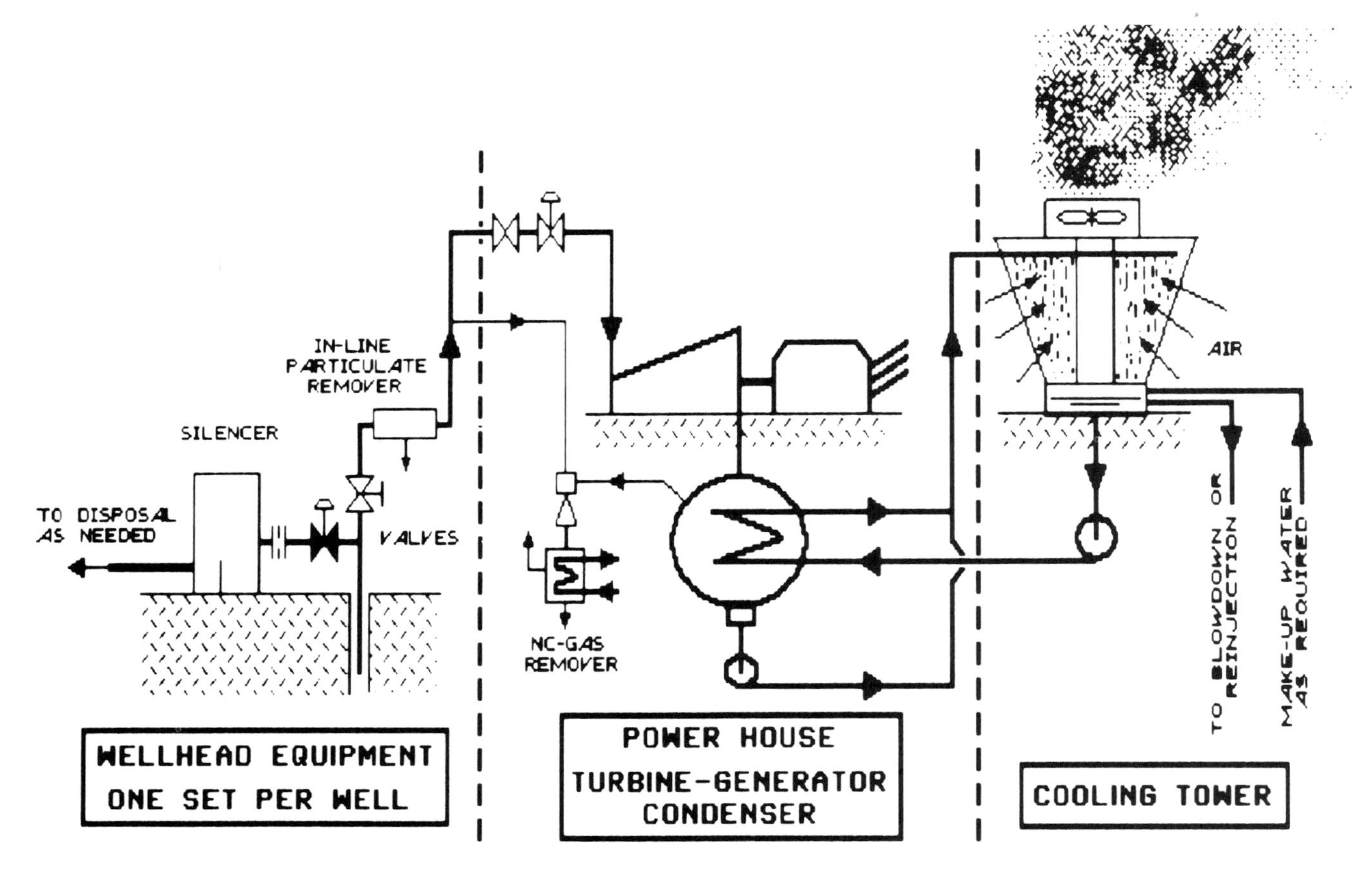

Figure 2. Simplified schematic of direct-steam plant.

Single Flash Plants

Dry steam reservoirs are not common. The conditions in most geothermal hydrothermal reservoirs favor the production of a two-phase, liquid-vapor mixture at the wellhead. The addition of the liquid phase creates several problems for the plant designer. The most critical problems stem from the impurities that are dissolved into solution through fluid-rock reactions and which are carried by the liquid. These dissolved solids may render the geofluid highly corrosive, or may cause scaling or plugging of pipes upon precipitation triggered by temperature or pressure drops. The dissolved solids may include heavy metals or other toxic elements that could cause serious environmental hazards if not handled properly.

The simplest type of plant for use with a two-phase geofluid is the single flash plant. Figure 3 is schematic of such a plant. The major new feature, compared to a dry steam plant, is the cyclone separator. Its purpose is to separate the two-phase geofluid into a vapor stream and a liquid stream. It does this by centrifugal action and can produce very dry steam (say, 99.95 percent vapor) with the aid of a final demister (not shown). The separated steam then follows a series of processes essentially identical to the steam in the dry steam plant. The solids-laden liquid is usually disposed of by means of reinjection wells into remote parts of the reservoir.

Double Flash Plants

The fraction of the total available work in the geofluid at the wellhead that is discarded with the separated liquid in a single flash plant is significant—typically, 20–30 percent. To achieve a greater efficiency of conversion from geothermal energy to electricity, it is important to recover at least a part of this discarded potential.

A double flash plant, such as that schematized in Figure 4, can recover much of that potential loss and can achieve 15- 20 percent higher efficiency than a single flash plant. There are two steam-generating components—a high-pressure separator (similar to that used in single flash plants) and a low-pressure flasher. The liquid from the separator is let down in pressure (via, say, a control valve or an orifice), thereby releasing lower

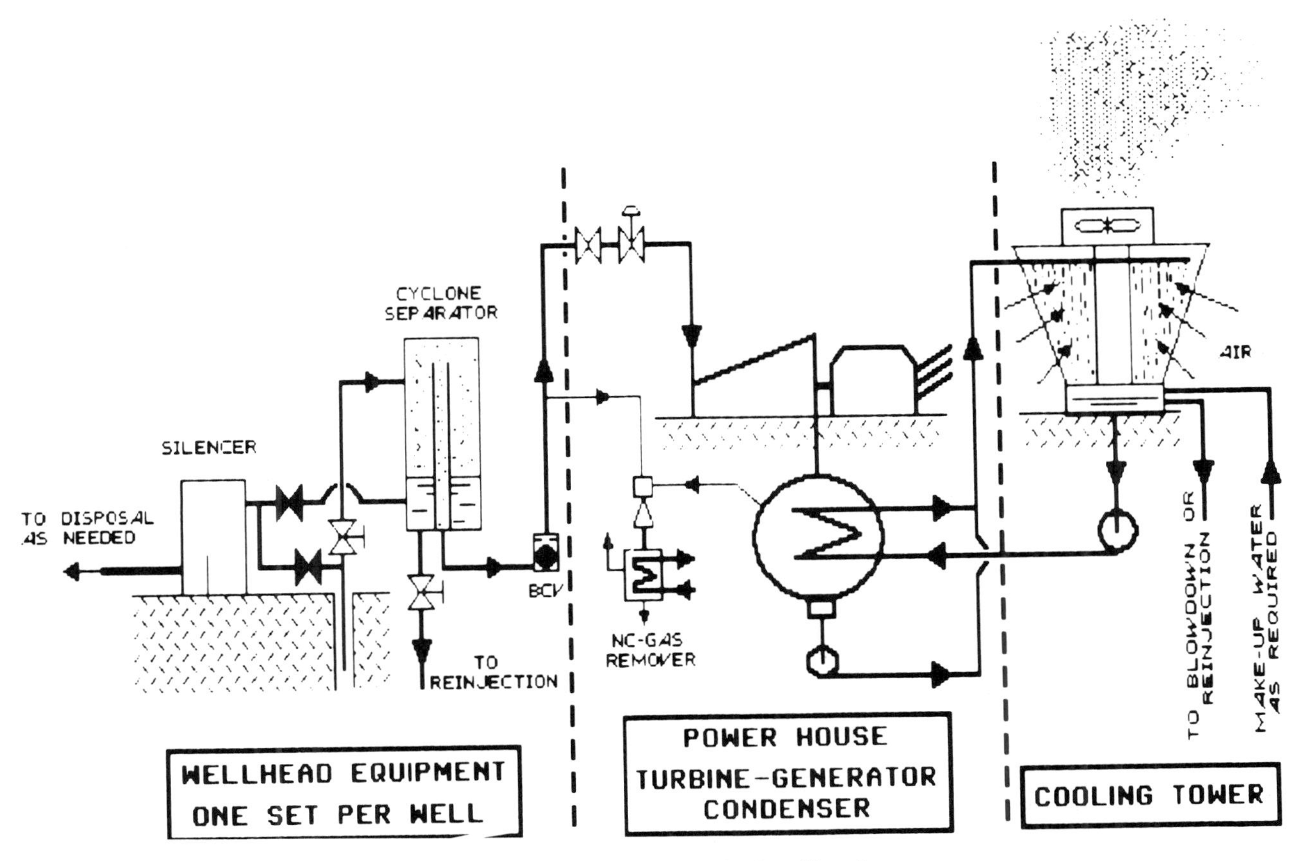

Figure 3. Simplified schematic of single-flash plant with individual wellhead separators.

pressure steam that is admitted to the turbine at an appropriate stage. The rest of the plant is similar to the ones just described.

The liquid from the flasher is now highly concentrated with dissolved solids, having lost perhaps 30–35 percent of its water content via steam separation. Saturation or supersaturation conditions may exist with respect to any of certain minerals. The major cause for concern is silica—amorphous silica—which is highly supersaturated at the temperatures usually encountered in plants of this type. The rate at which the silica will precipitate is a complex function of several variables, including the pH of the solution. Appropriate means must be devised to prevent this precipitation from interfering with the operation of the plant.

The hypersaline, high-temperature brines found at the Salton Sea field in California have been mastered through the use of flash-crystallizer/reactor-clarifier (CRC) technology. Several commercial plants operate with brines having 200,000-300,000 ppm of total dissolved solids. The CRC systems allow for smooth operation by purging the waste brine of the solids prior to reinjection.

Binary Plants

Whenever it is inadvisable to allow the geofluid to flash (i.e., to evaporate or boil) anywhere in the system, starting from the reservoir, then binary plants may be used as the energy conversion system. The geofluid may be prevented from flashing by maintaining it under a pressure that exceeds the saturation pressure for the fluid temperature. A downwell pump is usually employed for this purpose, as shown in Figure 5, a schematic of a basic binary plant. With the geofluid under pressure, it is passed through a heat exchanger where it is cooled while it heats and vaporizes a secondary working fluid (say, a hydrocarbon such as isobutane or isopentane). The working fluid is chosen so that its thermodynamic properties allow a good match with the geofluid properties (which of course are site-specific). The working fluid executes a closed cycle, passing through a turbine, condenser, and feed pump before returning to the vapor generator. The vapor generator and power house, as shown in Figure 5, can be engineered as a compact, transportable unit, factory-built for rapid installation at a geothermal site.

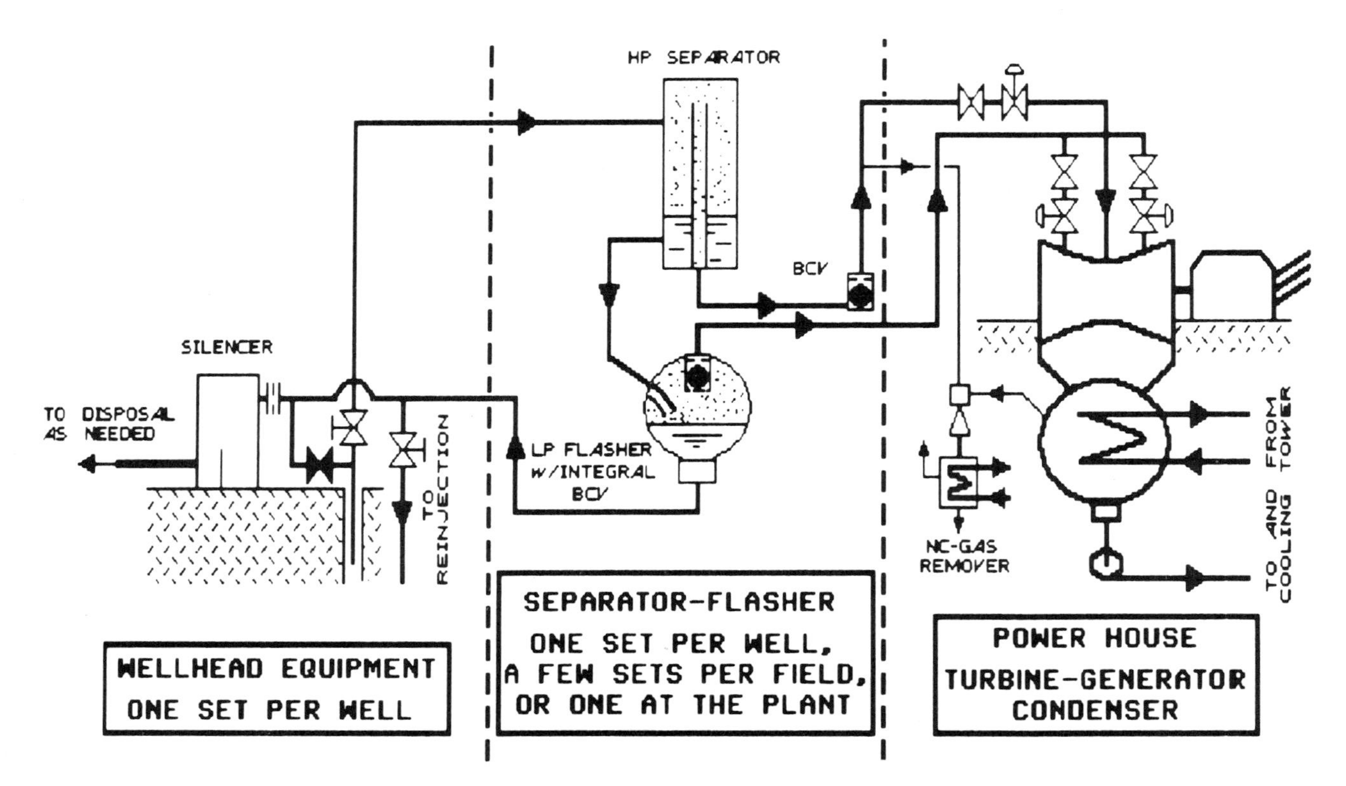

Figure 4. Simplified schematic of double-flash plant.

Note that the geofluid is contained within a closed loop when it is fully reinjected after use. Since the working fluid cycle is also closed, the only place where fluids are discharged to the atmosphere is the cooling tower, where water vapor evaporates. If the condenser is air-cooled, even this vapor discharge can be eliminated.

Emission Points

During normal operation (i.e., excluding emergency conditions), a geothermal power plant may interact with its environment at the following points:

- Vent from noncondensable gas ejector
- Plume from cooling tower
- Silencers at wellheads and at power house
- Drains and traps from geofluid pipelines
- Blowdown of excess condensate from cooling tower.

For certain plant designs, the following emission points may also be important:

- Condenser cooling water outlet (if using once-through cooling and a direct-contact condenser)
- Separated or flashed liquid discharge (if not reinjected)
- Geofluid exhaust from turbine (if using a non-condensing, discharge-to-atmosphere unit)
- Outlet of gas abatement system (if used).

Finally the operation of the plant has a thermal effect on the environment, as required by the laws of thermodynamics, although this is usually not thought of as an emission.

Environmental Impacts and Controls

The environmental impact of geothermal energy use for power production has been the subject of numerous studies and conferences. For example, we draw the reader's attention to the works of Armstead (1976), Axtmann (1975), Bowen (1973), Hartley and DiPippo (1980), Pasqualetti (1980), and Thind (1989), to cite only a few. The categories of impact may be listed as follows:

- Air pollution
- Water pollution
- Thermal pollution
- Noise pollution
- Land usage
- Water usage
- Land subsidence

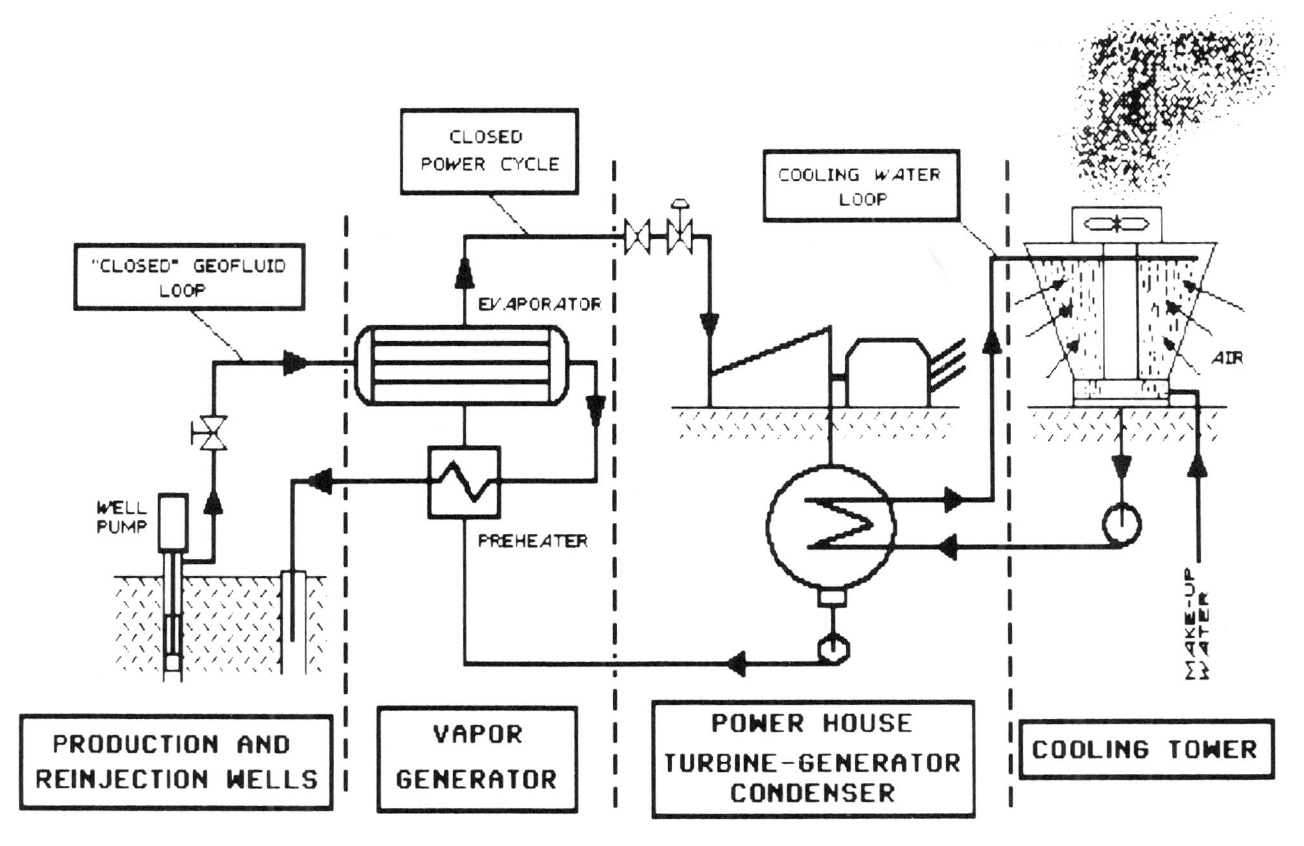

Figure 5. Simplified schematic of basic binary plant.

- Loss of natural beauty and scenic wonders
- Catastrophic events

We shall discuss each potential impact separately.

Air Pollution

The main problem arises from the gaseous elements, carried by the steam, that are discharged to the air. By and large, the noncondensable gases consist of CO_2 and H_2S, plus very small amounts of other gases such as methane, hydrogen, and ammonia. The emissions of H_2S are strictly regulated at all geothermal plants in the US because of the gas's harmful effects on human and plant life in certain concentrations. Its "rotten egg" odor, detectable at 30 parts per billion, is an added nuisance. The emissions of CO_2 are, as yet, unregulated, but CO_2 has been implicated in the so-called "greenhouse effect" and has become a focal point for those concerned about the environment.

The amount of gaseous emissions per unit of electricity generated may be used to rate various methods of power production. DiPippo (1988b) showed that geothermal steam plants have the lowest emissions of CO_2 per kWh of electricity than any other plant type having CO_2 emissions. Geothermal steam plants emit typically only 5 percent of the CO_2 emitted by a coal plant and about 8 percent of the CO_2 from an oil plant, per kilowatt·hour.

The amount of gas emitted per unit of electricity depends on the percent of noncondensable (NC) gas in the steam, the percent of each gaseous constituent, and the efficiency of steam utilization. A typical steam consumption for geothermal plants is nine kg steam/kWh (19.8 lbm/kWh). Based on this value, Figure 6 gives the specific H_2S emissions (if unabated) in kg/MWh as a function of the percent of NC gas in the steam and the percent H_2S in the NC gas. Figure 7 gives the specific CO_2 emissions.

By using appropriate technology, the H_2S emissions may be reduced drastically. At The Geysers plants, Stretford process systems achieve better than 90 percent reduction of total H_2S emissions.

In principle, one could capture all the vent gas from the noncondensable gas ejector, compress it, and reinject it back into the reservoir, providing that the thermal, chemical and physical properties of the formation are

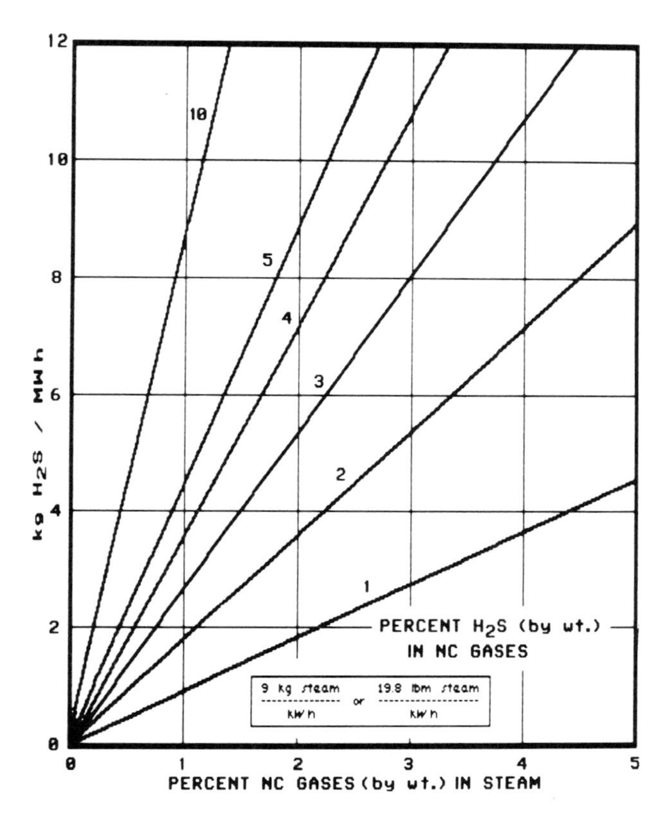

Figure 6. Specific emissions of H_2S as a function of percent noncondensible (NC) gases and H_2S concentration.

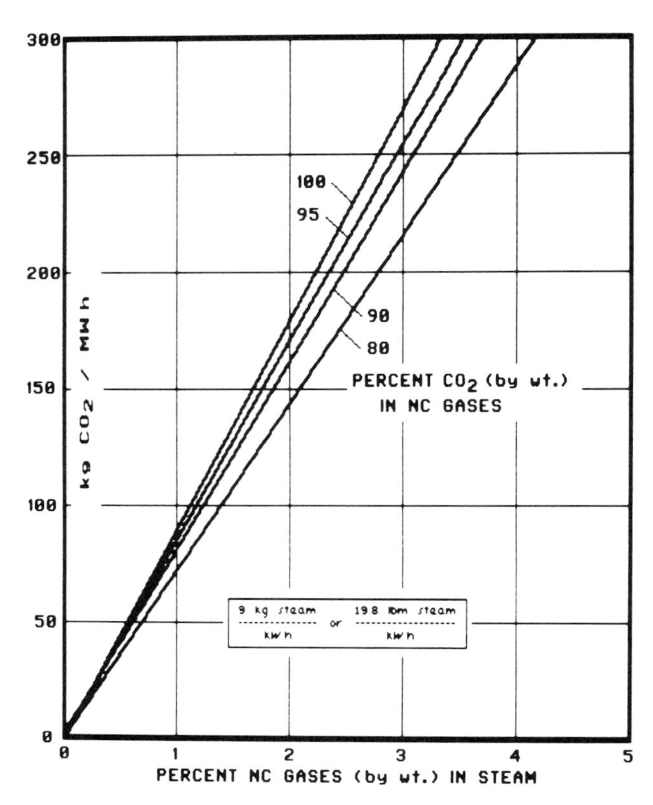

Figure 7. Specific emissions of CO_2 as a function of percent noncondensible (NC) gases and CO_2 concentration.

favorable. The work required for gas compression, however, could be a significant drain on the plant output. Even in this case, if the geosteam condensate is used as cooling tower makeup, a certain amount of the noncondensable gas will be dissolved in the condensate and will be released into the atmosphere with the water vapor plume, unless the condensate is treated chemically. This principle is used in the plants at the Coso field in California. The work of gas compression is minimal due to highly favorable reservoir pressures in the reinjection wells.

It is interesting to note that the only geothermal plants with H_2S abatement systems are in the United States. Finally, binary plants of the type shown in Figure 5 have no gaseous emissions.

Water Pollution

Water pollution encompasses surface and ground waters. Surface waters can become contaminated through run-off of geothermal liquid to streams, creeks, etc. High-temperature reservoirs, say, greater than 230°C(450°F), generally produce liquids containing an extensive menu of dissolved minerals, some of which could poison surface waters. However, lower temperature reservoirs often yield relatively clean fluids with less than 1000 ppm of total dissolved solids. Nevertheless, it is good practice to collect all discharge liquids and dispose of them either by reinjection or in large, impermeable evaporation ponds.

Ground water contamination may occur through failure of the casings in reinjection wells, allowing fluid to leak into shallow aquifers, or if holding ponds are not impermeable. Both of these possibilities can be essentially eliminated by careful design, attention to quality control during drilling and construction, and proper monitoring during operation.

Outside of the United States and Japan, where reinjection of waste fluids is standard practice, surface disposal of waste fluids is common. It is encouraging that the new 116.4 MW_e Ohaaki power plant in New Zealand will reinject the waste liquid instead of discharging it into the adjacent Waikato River, which in fact receives the liquid discharged from the Wairakei geothermal station only a few kilometers away (Brown, 1989). Also, provisions are being made to reinject all spent liquid from the 55 MW_e plant under construction at the Miravalles field in Costa Rica.

Thermal Pollution

Geothermal plants reject much more heat per unit of electrical output than any other comparable plant (Horsak, 1979). Figure 8 illustrates this point. The values for a typical single flash plant were based on a turbine efficiency of 75 percent and a condenser temperature of 52°C(126°F). Shown for comparison are the values for an ideal Carnot cycle operating between the reservoir and condenser temperatures, and typical gas turbine (GT), nuclear (N), coal (C), and gas-turbine combined cycle (GTCC) plants. A geothermal plant is two or three times worse than a nuclear plant with respect to thermal pollution *at the plant site*. However, the situation is less disadvantageous for the geothermal plant when one considers the heat dissipated during the mining, processing, fabrication, transportation, and reprocessing of the nuclear fuel.

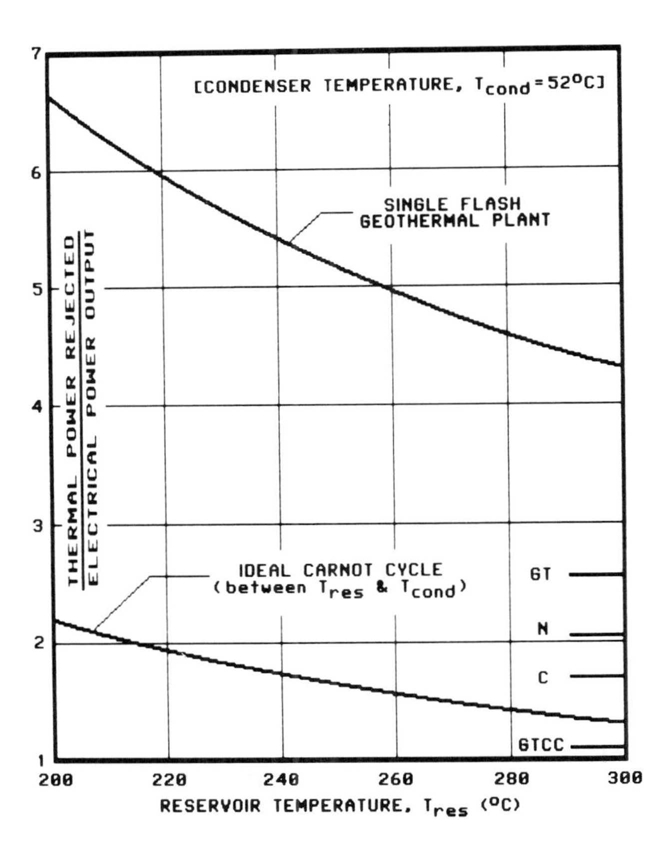

Figure 8. Specific waste heat as a function of reservoir temperature: geothermal single-flash plant compared with other plants (GT = gas turbine, N = nuclear, C = coal, GTCC = gas-turbine combined cycle).

Nevertheless, the waste heat rejection systems for geothermal plants are significantly larger than for comparable sized fossil or nuclear plants (Robertson, 1980). For example, the waste heat system for a 100 MW_e geothermal plant will be about the same size as that for a 500 MW_e gas-turbine combined cycle.

Noise Pollution

The most serious noise pollution is generated during the well drilling phase. During normal plant operation, a geothermal station does not emit objectionable noise. However, under emergency conditions, usually for brief periods, the possibility arises for high noise levels when it may be necessary to vent steam because of equipment outages. Simple mufflers are routinely installed to curtail the velocity (and hence the noise) of the venting steam.

Figure 9 shows a comparison of noise levels from a variety of geothermal operations with noises of everyday life. The data were taken from Hartley and DiPippo (1980) and refer to plants at The Geysers in California. The main point is that even the worst possible noise associated with geothermal operations, when perceived from a distance of about one km (3,000 ft), is lower in level than that associated with a typical noisy urban area. Furthermore, the routine noise of plant operation (excluding well drilling, testing, or venting) is practically indistinguishable from other background noises at about one km.

Land Usage

Geothermal plants are built on the geothermal reservoirs. Long geofluid transmission lines are not practical because of losses in pressure and temperature. Land is required for the power house and its related equipment such as cooling towers and electrical switchyard, for the well pads, and for the geofluid pipelines. Since the latter are normally mounted on stanchions and do not preclude parallel usage (such as cattle grazing), the land taken out of service for pipe routes is minimal. The well field, however, can cover an extensive area. The total area encompassing all the wells serving the 180 MW_e Cerro Prieto I plant in Mexico amounts to roughly 5.4 x 10^6 m^2 or 540 ha (ha = hectares) (Manon, Bermejo, et al, 1987). The power house is included within this area. The area actually occupied by the well pads, however, is far smaller, roughly 0.12 x 10^6 m^2 or 12 ha. In the case of Cerro Prieto, the space between well pads is unusable because it is barren desert, but in many geothermal fields, the open space is free for agricultural or other uses. Since reinjection has not been practiced at Cerro Prieto, a large evaporation lagoon has been created for the waste brine. It covers about 4 x 10^6 m^2, or 400 ha.

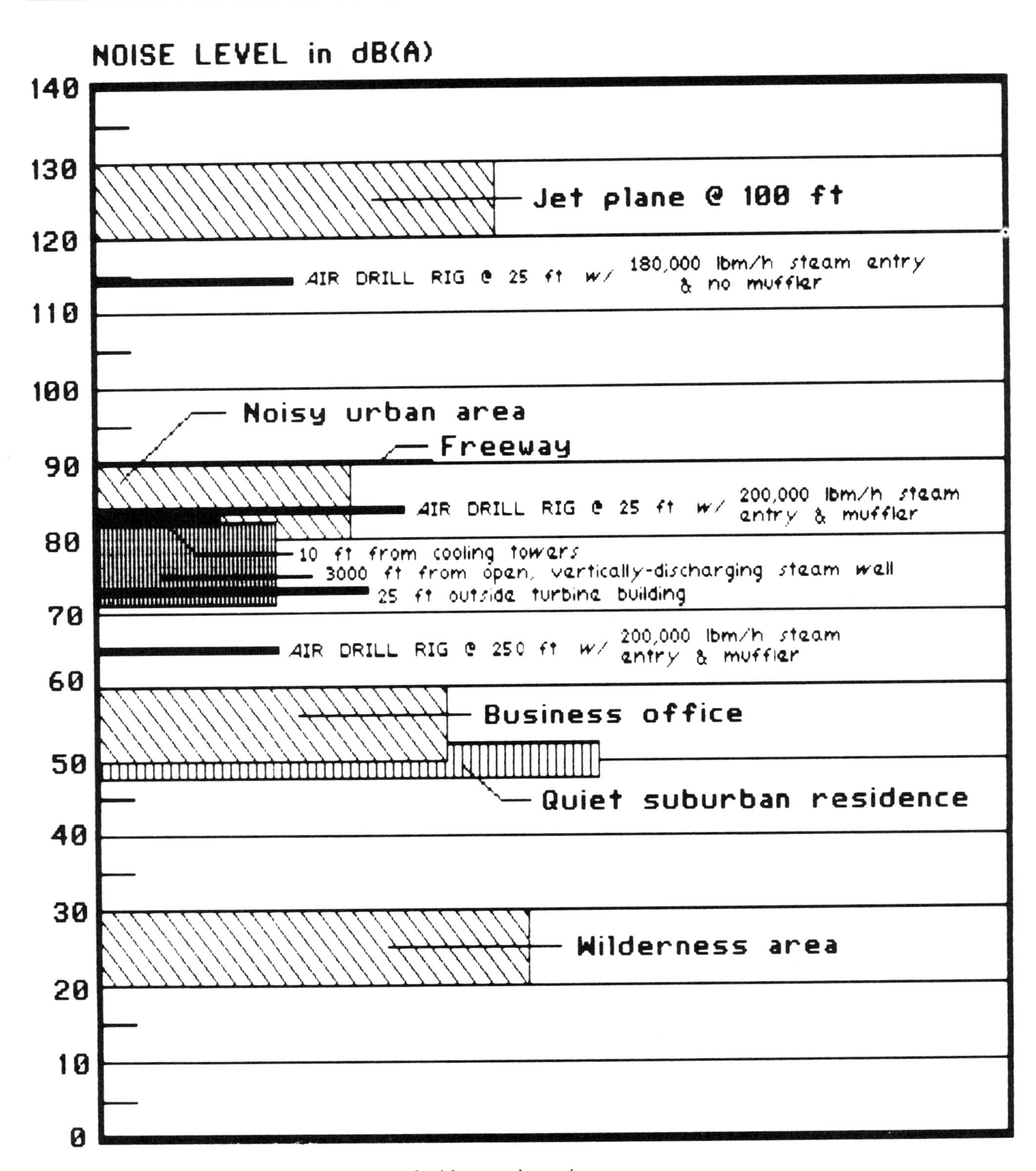

Figure 9. Geothermal noise levels compared with everyday noises.

Where reinjection is carried out, the land usage is far less extensive. Figures 10 and 11 illustrate the cases of a typical flash plant and a large binary plant, respectively. The flash plant is roughly patterned after the 2 x 55 MW_e Miravalles plant under construction in Costa Rica and the binary plant after the 65 MW_e (gross), 45 MW_e (net) Heber Binary Project in California (Nelson, 1986). The flash plant uses 1260 m^2/MW_e; the binary plant uses 3700 m^2/MW_e. If we do not include the well area allotted for future replacement wells in Figure 11, then the binary plant uses 2700 m^2/MW_e. The flash plant uses individual well pads for each well, whereas the binary plant uses multiple wells drilled directionally from "islands." The latter arrangement is more efficient as regards land usage.

Geothermal plants require less land per megawatt than competing power plants. For example, the three-unit, 2258 MW_e total, coal-burning Navajo plant in Arizona requires 40,000 m^2/MW_e, including 30 years of coal strip-mining (Pasqualetti, 1980). A solar-thermal plant soon to be built in the Mojave desert of California will need a total of 1.3 x 10^6 m^2 for an 80 MW_e (peak) plant, or about 16,000 m^2/peak MW_e. Factoring in that the plant operates as a solar plant for only 14 hours/day, the land usage per average MW_e soars to 28,000 m^2/average MW_e (Chiles, 1990). Solar photovoltaic plants are even worse. For plants in the American Southwest where the insolation (averaged over 24 hours) is the highest in the US, namely 0.25 kW/m^2 (Boes, Hall, et al, 1976), the collector area alone turns out to be 33,000 m^2/average MW_e, assuming 12% conversion efficiency between photons and electrons. When one allows for "elbow room" between collectors, the total required land area doubles to 66,000 m^2/average MW_e.

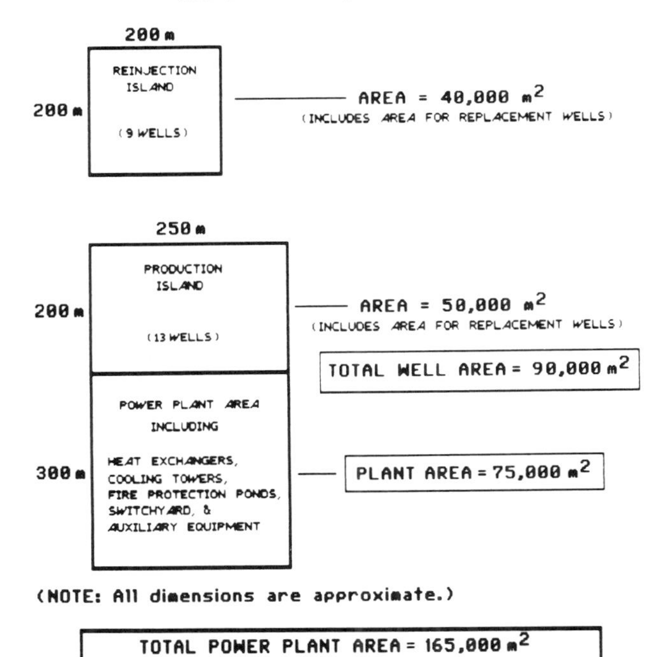

Figure 11. Direct land usage for a large binary plant.

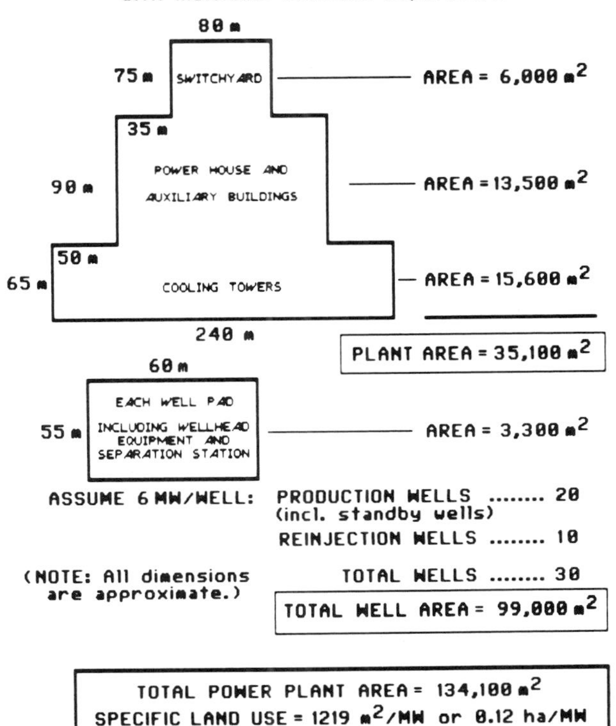

Figure 10. Direct land usage for a typical 110 MW_e single-flash plant.

Water Usage

A single flash plant using a resource at 260°C would require about 108 kg/s of continuous cooling water per MW_e, but once-through cooling systems are rare. The Wairakei station in New Zealand is the prime example (Thain and Stacey, 1984). A coal plant would need only 37 (kg/s)/MW_e. However, since the geothermal steam condensate may be used as the cooling medium after passing through a cooling tower, a geothermal steam plant needs no outside cooling water. Of course, a source of fresh water is necessary to fill the tower basin for start-up and periodic blowdown, but these are not major quantities.

By contrast, a geothermal binary plant has no geofluid condensate and the cooled brine is usually too contaminated for use in a cooling tower without chemical treatment. An attempt to do just that at the Raft River binary test plant in Idaho proved unworkable and uneconomical (Whitbeck, 1982). The Heber Binary Project plant would use about 188 (kg/s)/MW_e under full power conditions. Smaller binary plants can be built with air-cooled condensers, thereby eliminating the need for cooling water. This is the case, for instance, at the 7 MW_e Mammoth-Pacific plant in California (Holt and Campbell, 1984).

Land Subsidence

Wherever large quantities of water are produced from aquifers, there is always the possibility that the overburden may sag as a result. The problem may be worse if the produced fluid is not reinjected. The worst geothermal case of subsidence has occurred at the Wairakei field where the maximum drop in ground level exceeds 7.5 m (25 ft) and is continuing at an annual rate of about 0.4 m (1.3 ft), according to Thain and Stacey (1984). Since this is confined to a small area away from the power house, it has caused no major difficulties. With the exception of the Wairakei field, there are no known cases of significant subsidence at geothermal plants.

Loss of Natural Beauty and Scenic Wonders

Some of the best geothermal resources occur in or near national parks, not only in the United States, but also, in Japan, New Zealand and Costa Rica, for example. To the extent that countries value the natural beauty of their lands, care will be taken to ensure that geothermal development will proceed in a careful manner with full regard for the preservation of scenic wonders for the enjoyment of future generations. Although it is often argued that "beauty is in the eye of the beholder," it is doubtful whether anyone would seriously propose to build a geothermal power plant in, say, Yellowstone Park at the risk of destroying one of the most spectacular and awe-inspiring regions known to humankind.

With regard to the despoiling of natural vistas, geothermal plants can be designed to blend into the natural surroundings (Armstead, 1976). This is not the case for coal plants, which often have stacks as high as 500 feet, nuclear stations with massive containment structures and tall natural draft cooling towers, or wind farms with thousands of wind turbines perched atop 100-foot high supports, arrayed on hillsides and in mountain passes.

Catastrophic Events

Under this category, we consider: (1) well blowouts, (2) pipe ruptures, (3) landslides, and (4) induced seismicity.

Well blowouts have occurred at a number of fields around the world. For example, at The Geysers, well Thermal 4 blew out in September 1957 and has resisted all efforts to completely shut it in (Mogen and Maney, 1985). At Wairakei, Bore 26 blew out in January 1960, but was brought under control in November of the same year by drilling a relief well, Bore 26A (Thain and Stacey, 1984). One blowout, the "Rogue Bore" at Ti Mihi in New Zealand, became a popular tourist attraction until 1973 when it stopped erupting. Such events are dangerous, can cause bodily injury, and damage the environment. The likelihood of such an event is minimized by the proper use of blowout preventer equipment during drilling when most blowouts occur. The drilling of geothermal wells is a highly regulated operation in the United States. A summary of the requirements and procedures for California may be found in Division of Oil and Gas (DOG, 1986).

The rupture of a steam line or hot water line could lead to serious injury or major contamination. This is an unlikely event since pipelines are designed according to safe engineering practice with adequate margins of safety. Furthermore, any spill would be of short duration since control valves would be activated to isolate the broken section. The problem is no worse than at any fossil-fueled steam plant. At some nuclear plants, however, the steam and water are radioactive.

Often geothermal fields are in rugged volcanic mountains prone to landslides. Owing to the normal seismic nature of such regions, landslides are not uncommon. Danger arises from the possibility of a landslide resulting in the shearing of a well casing, the burying or rupture of wellheads or piping, or damage to other plant buildings, equipment or access roads. Careful geologic mapping of the field is necessary to locate ancient landslides so they may be avoided as drill sites, roadways, etc. Areas of potential landslides should be reinforced or buttressed.

It is well known that injection of high-pressure liquid into a reservoir can cause the opening of fractures. The process is called "hydraulic fracturing" (Maurer, 1976) and is useful in stimulating low permeability reservoirs. The question is whether the reinjection of waste liquid (at moderate pressure) might create a new fracture or lubricate an existing one to the extent of causing a significant seismic event. Studies have shown that this does not occur and, at most, microseismic events may be associated with production or injection of geothermal fluids (Pasqualetti, 1980).

Even under the worst plausible accident, geothermal plants fare far better than, say, nuclear stations, where a major accident has global repercussions, as did the calamity at Chernobyl. Hydroelectric dams have broken and inundated towns, killing thousands of people. The effects of the worst imaginable catastrophe at a geothermal plant would be confined to a small area.

As with any industrial facility, geothermal plants are subject to normal accidents, but the safety record at geothermal plants around the world is exemplary, mainly because of the safety precautions that are built in at every stage from initial design to everyday operations. Safety valves, over-speed protectors, air/gas/water monitoring equipment, conservative design factors of safety, fire protection gear, personnel training—all these contribute to the enviable safety performance record of geothermal power plants.

Conclusions

The technology to exploit hydrothermal geothermal energy for the generation of electricity is now at hand. It is proven and economical. Resources as low as 100°C may be tapped using simple binary-plant technology; moderate-temperature resources may be exploited with flash-steam technology. Even hypersaline brines can now be harnessed using flash-crystallizers and reactor/clarifiers together with conventional steam turbines. Many companies around the world have the expertise to design, engineer, construct and operate geothermal plants. Many countries are blessed with natural geothermal resources, which could make a major impact on meeting the present and future local demands for electricity. While not perfectly benign with regard to environmental impact, geothermal on balance is arguably the most benign of the present technologies for power generation.

Acknowledgments

The manuscript was typed by Ms. Joan M. Sincero in her usual meticulous fashion. The final art work was produced by the Photographics Dept. of SMU from original materials prepared by the author.

References

Armstead, H.C.H., "Summary of Section V - Environmental Factors and Waste Disposal," *Proc. Second United Nations Symposium on the Development and Use of Geothermal Resources, vol. 1*, San Francisco, CA, pp. lxxxvii–xciv (1976).

Axtmann, R.C., "Environmental Impact of a Geothermal Power Plant," *Science* 187(4179), p. 795 (1975).

Boes, E.C., Hall, I.J., et al, "Distribution of Direct and Total Solar Radiation Available for the USA," *Proc. 1976 Annual Mtg. Am. Sec. ISES*, pp. 238–263 (1976).

Bowen, R.G., "Environmental Impact of Geothermal Development" in *Geothermal Energy - Resources, Production, Stimulation*, Kruger, P. and Otte, C., eds., Stanford Univ. Press, Stanford, CA, pp. 197–215 (1973).

Brown, D., "The Ohaaki Geothermal Development," *Geothermal Resources Council BULLETIN* 18(10), pp. 3–7 (1989).

Chiles, J.R., "Tomorrow's Energy Today," *Audubon* 92(1), pp. 60–73 (1990).

Dickson, M.H. and Fanelli, M., "Geothermal R&D in Developing Countries: Africa, Asia and The Americas," *Geothermics* 17(5/6), pp. 815–877 (1988).

DiPippo, R., *Geothermal Energy as a Source of Electricity*, U.S. Dept. of Energy, DOE/RA/28320-1, Gov. Printing Office, Washington, DC (1980).

DiPippo, R., "International Developments in Geothermal Power Production," *Geothermal Resources Council BULLETIN* 17(5), pp. 8–20 (1988a).

DiPippo, R., "Geothermal Energy and the Greenhouse Effect," *Geothermal Hot Line* 18(2), pp. 84–85 (1988b).

DOG, "Drilling and Operating Geothermal Wells in California," 4th ed., Dept. of Conservation, Div. of Oil and Gas, Sacramento, CA (1986).

Hartley, R.P. and DiPippo R., ed., "Environmental Considerations" in *Sourcebook on the Production of Electricity from Geothermal Energy*, Kestin, J., ed.-in-chief, US Dept. of Energy, DOE/RA/28320-2, US Gov. Printing Office, Washington, DC, pp. 786–869 (1980).

Holt, B. and Campbell, R.G., "Power Plant Development at Mammoth Project," *Geothermal Resources Council BULLETIN* 13(4), pp. 4–6 (1984).

Horsak, R.D., "Heat Rejection from Geothermal Power Plants," EPRI Rep. No. ER-1216, Electric Power Research Institute, Palo Alto, CA (1979).

Manon, A., Bermejo, F., et al, "Operation of Surface Equipment for Recovery of Geothermal Fluids at Cerro Prieto I," *Proc. Ninth Annual Geothermal and Second IIE-EPRI Geothermal Conference and Workshop, vol. 2*, English Ver., EPRI Rep. No. AP-4259-SR, Electric Power Research Institute, Palo Alto, CA, pp. 9.1–9.38 (1987).

Maurer, W.C., "Geothermal Drilling Technology" in *Proc. Second United Nations Symp. on the Development and Use of Geothermal Resources, vol. 2*, San Francisco, CA, pp. 1509–1521 (1976).

Mogen P. and Maney, J., "The Thermal 15 Relief Well and Production Performance of the Thermal Shallow Reservoir," *Proc. Tenth Workshop on Geothermal Reservoir Engineering*, Rep. No. SGP-TR-84, Stanford University, Stanford, CA, pp. 141–144 (1985).

Nelson, T.T., "Heber Binary-Cycle Geothermal Demonstration Power Plant: Summary of Technical Characteristics," EPRI Rep. No. AP-4612-SR, Electric Power Research Institute, Palo Alto, CA (1986).

Pasqualetti, M.J., "Geothermal Energy and the Environment: The Global Experience," *Energy* 5(2), pp. 111–165 (1980).

Robertson, R.C., "Waste Heat Rejection from Geothermal Power Stations" in *Sourcebook on the Production of Electricity from Geothermal Energy*, Kestin, J., ed.-in-chief, US Dept. of Energy, DOE/RA/28320-2, US Gov. Printing Office, Washington, DC, pp. 541–655 (1980).

Thain I.A. and Stacey, R.E., "Wairakei Geothermal Power Station - 25 Years' Operation," Electricity Div., Min. of Energy, Wellington, New Zealand (1984).

Thind, P.S., "Environmental Characteristics and Occupational Hazards Associated with Abatement of Non-Condensable Gases at The Geysers," *Geothermal Resources Council TRANSACTIONS, vol. 13*, pp. 101–106 (1989).

Whitbeck, J.F., "Raft River 5 MW Plant Startup Experience," *Proc. Sixth Annual Geothermal Conference and Workshop*, EPRI Rep. No. AP-2760, Electric Power Research Institute, Palo Alto, CA, pp. 2.20–2.23 (1982).

Geothermal Energy Opportunities for Developing Countries

Carel Otte
Geothermal Division
Unocal Corporation
Los Angeles, CA

Introduction

Geothermal energy is, in the broadest sense, the "natural heat of the earth," but as a practical matter this heat is not directly available to us for capture. Normally we think of geothermal energy in terms of the fluids that are in contact with hot rocks. It is these heated fluids that can be captured and from which energy can be extracted. This form of geothermal energy, also called hydrothermal energy, is the only form that has proven commercially feasible to date. Geopressured energy, hot dry rock (HDR) geothermal energy, and energy derived from shallow magma bodies are still in the research phases.

The temperature within the earth is considerably higher than at the surface, and this difference causes heat to flow toward the surface—a flow that occurs everywhere, though we are not normally aware of it. The normal geothermal gradient is about 27°C per km (1.5°F per 100 ft), so at a depth of 4570 m (15,000 ft) the temperature is about 135°C (250°F) assuming a surface temperature of 15°C (27°F). This temperature is too low to be useful and too deep to capture economically. However, in certain areas, molten rock or magma, formed at great depths in the crust, succeeds in working itself very close to the surface, causing a sharp steepening of the geothermal gradient, which may be 10 times the normal gradient, or more.

The areas of above-normal heat flow and steepening of the geothermal gradient occur in zones or belts that extend around the world. These are zones of crustal weakness, and they are characterized by such phenomena as high seismicity, evidence of volcanic activity, and geologically young mountains. One such belt extends from the tip of South America through Central and North America, Alaska, and around the western Pacific, through Kamchatka (USSR), Japan, and the Philippines to Indonesia. Another belt extends from Indonesia along southern Asia into southern Europe as shown in Figure 1. According to the theory of plate tectonics, these zones mark the borders of the stable, but moving, continental plates.

In these areas of weakness, magma ascends closer to the surface. If groundwater is adjacent to these magmatic bodies or is mixed with hot gases and steam emanating from the crystallizing molten rock, the water will be heated and begin to rise toward the surface, sometimes causing hot springs, geysers, and fumaroles.

Hot water in a continuous column is subjected to the pressure of its own weight, raising the boiling point of water progressively with depth. For instance, at a depth of 300 m (1,000 ft), water boils at a temperature of about 215°C (450°F) as seen in Figure 2. If a well is drilled deep into a fissure that is bringing thermal fluids to the surface, the hot water can be relieved of its overlying weight; it will begin to boil and then will flash into steam. The higher the temperature, the higher the ratio of steam to water when it comes to the surface. If the original heat content of the rocks is high, or the formation fluid pressure is below normal, the fluid may occur in superheated form and be all in the steam phase, and it can be piped directly from the well into the generating plant. This is called a vapor-dominated system. Normally aquifer systems are near hydrostatic equilibrium and a geothermal system will be liquid-dominated. In case of flashed hot water a steam separator is required. The excess water is then disposed of on the surface or by reinjection. It is thought that the hot water geothermal reservoirs will be by far the most abundant.

It is important to realize that we are attempting to produce heat by means of fluids in the ground—either in the form of steam or hot water—not by means of the direct thermal energy contained in the magma itself. This, in turn, means that the rocks need to contain adequate void space so that they contain hot water or steam and that they are sufficiently permeable to yield these fluids at acceptable rates. In fact, open fracture systems permitting high fluid production rates are a prerequisite for a commercial geothermal operation.

Finally, the fluids should possess relatively suitable chemical composition so that they do not corrode or scale up the casing and piping systems. This is an important aspect of geothermal development; we need to locate a resource with all the suitable attributes.

Utilization of Geothermal Energy

The temperature of geothermal systems is normally in the range of 175-315°C (350-600°F), which is considered

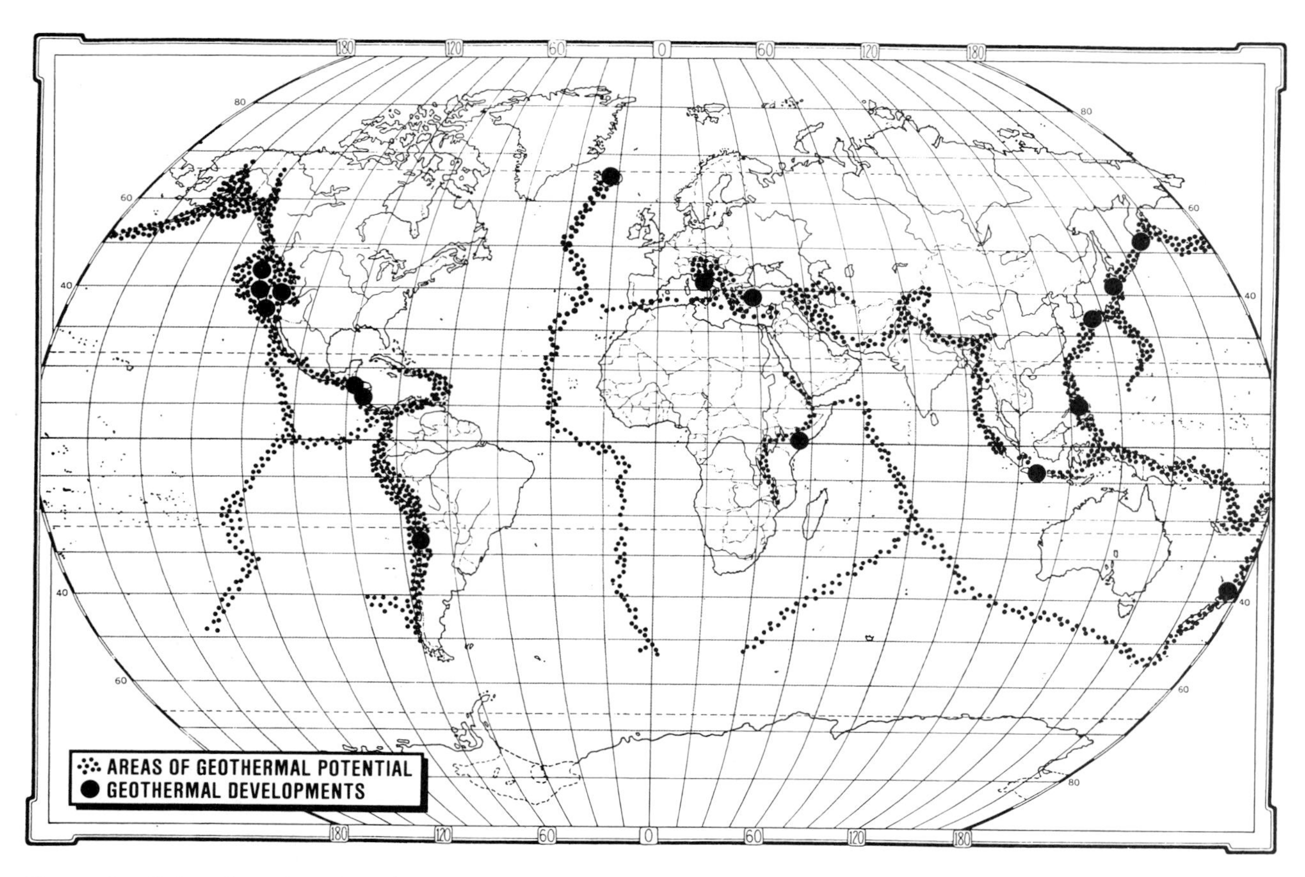

Figure 1. Belts of earthquakes along crustal plates and areas of geothermal development.

low-quality heat by fossil fuel standards. For this reason, the most efficient utilization of geothermal energy would be for the purpose of "process heat" in industrial applications. But the distance over which the energy can be transported economically is very limited—approximately two miles. Beyond this, there is too much heat or pressure loss. Thus, the utilization for process heat calls for a unique set of circumstances bringing the resource and the industrial application together in a single geographic setting. There are only a few such unique developments—for space heating in Iceland and in such towns in the United States as Klamath Falls, Oregon, and Boise, Idaho, and for the paper industry in New Zealand. It is estimated that approximately 10,000 MW_t of thermal energy are being used for direct heating or process purposes around the world.

By far the largest industrial application of geothermal energy today is the generation of electric power. It is discussed here as an alternative to fossil-fueled power generation. A pound of steam coming from a man-made boiler fired by conventional fuel is indistinguishable from a pound of steam coming from the earth's boiler, and the steam turbine does not know the difference. Accepting, then, that electric power can be generated and transported over a transmission system, the development of a geothermal deposit will depend solely upon the available load centers requiring the energy. In the United States a dense power grid now brings every geothermal deposit within reach of such a load center, so good deposits will see ready development. In other areas of the world, however, the transmission cost to take the energy to the market is an additional factor in comparing the feasibility of geothermal power with an alternative source of electric generation such as coal, nuclear, or oil, where the fuel can be transported to its point of utilization at the generating plant.

In the United States, geothermal power is now developed in increments of 50-100 MW_e, which appear to be optimum blocks of electric power for the number of wells required, the pipeline distance, and the size and cost of the turbine. At increments of 100 MW_e, geothermal power is economically competitive with energy produced by fossil fuel and nuclear plants of about 750-1000 MW_e size that enjoy the advantage of economy of scale as shown in Figure 3.

Many developing countries cannot handle incre-

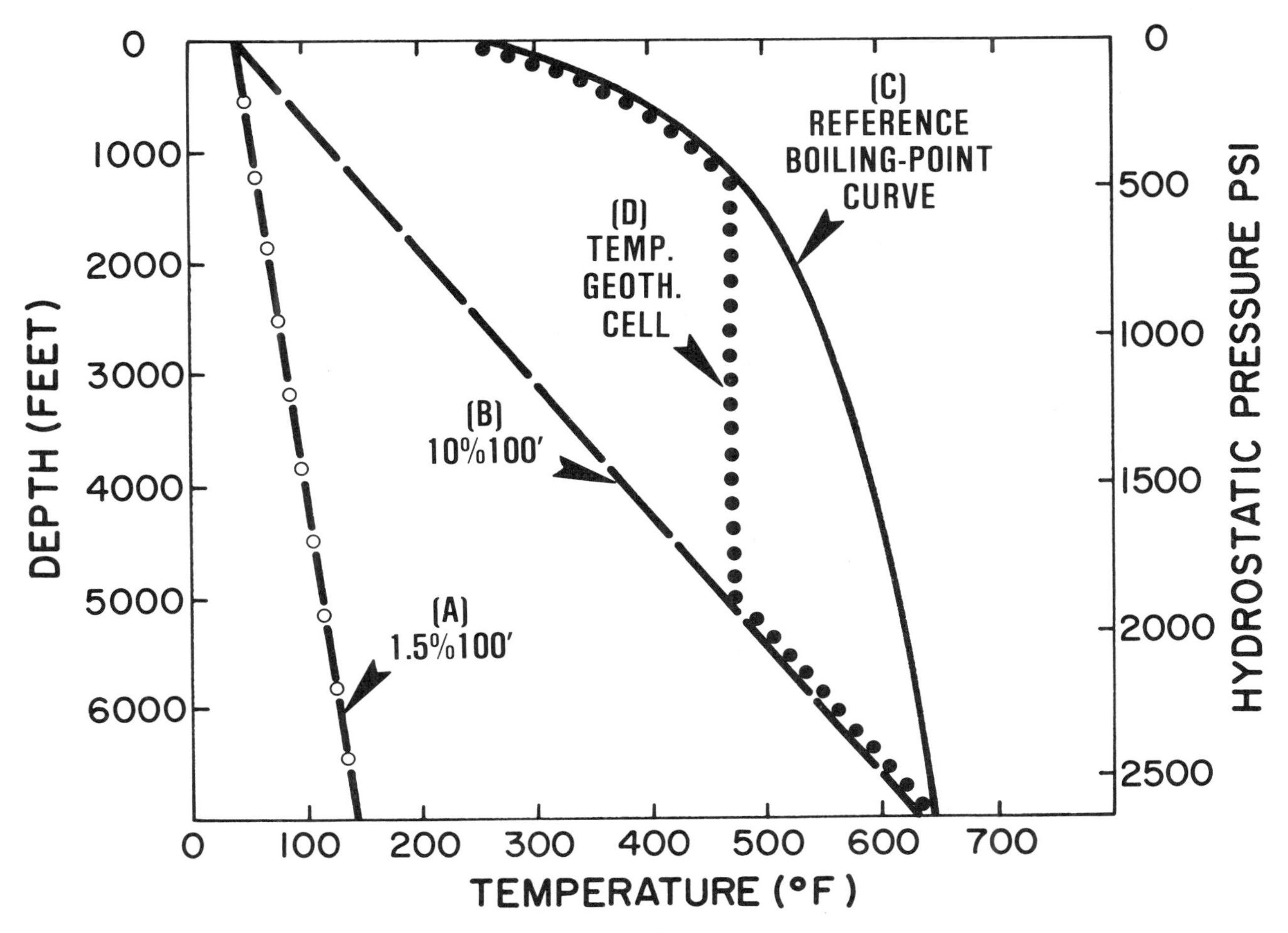

Figure 2. Geothermal gradient.

ments as large as 1000 MW_e to their installed capacity, and smaller blocks of fossil power are uneconomical. This makes geothermal power a very attractive form of power generation in those countries that have the geologic potential. Some 50 countries around the world are now active or interested in geothermal exploration (DiPippo, 1990).

Applications in the United States

In the United States, the best areas for geothermal potential are found in the states west of the Great Plains from Canada to the Mexican border. The Geysers project in Sonoma County, California, is the largest geothermal project in the world. This development began in 1957, and by 1960 it was producing enough steam to power a 12.5 MW_e generating plant. Development of the field has reached maturity with an installed capacity exceeding 2000 MW_e, which is sufficient to supply all the electrical requirements of a typical city of over two million people.

The Geysers field is an example of a steam field with water occurring in the vapor phase underground and ready to be used directly in the turbines when brought to the surface. The fact that this particular resource can be used so directly and efficiently has been a key element in the successful development of The Geysers to its current capacity of over 2000 MW_e.

Another area with the potential for major development is the Imperial Valley of California. Current activities there are a resumption of an effort that began in the early 1960s in the Salton Sea area to develop hot geothermal brines for their potash and heat content, for both electric and process heat application. However, the fluids found there were so highly corrosive they proved very difficult to handle with the equipment and technology available to industry at that time.

The scaling and corrosion problems were finally resolved in the Salton Sea field and a total of 236 MW_e of geothermal power is now installed with much addi-

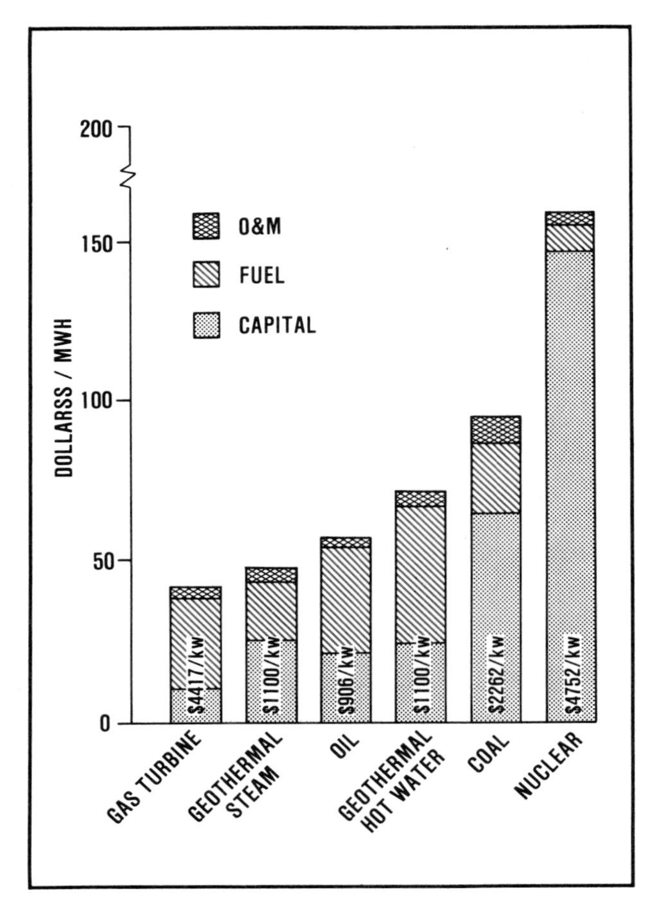

Figure 3. Estimates of electricity costs in dollars ($) per MWh of electricity generated. Oil is assumed at $18.50 per barrel. Capital power plant costs in $/kW installed are also shown. (Source: California Energy Commission, 1985.)

tional potential. The Heber and East Mesa fields are relatively low-temperature deposits of about 175°C (350°F). Heber supports a 50 MW_e double flash steam unit, while East Mesa has a total installed capacity of 135 MW_e, consisting of 92 MW_e generated by many small binary cycle power plants located at the wellhead and a 43 MW_e double flash steam unit.

Another major hot water geothermal development in California exists at Coso Hot Springs, east of the Sierra Nevada, where California Energy Company has installed 240 MW_e of generating capacity. A novel feature of this development is the injection of the noncondensable gases into the reservoir.

All these major developments on a number of resources with wide-ranging conditions, different power plant designs (Figure 4), and problems competing in the marketplace against fossil-fuel fired and nuclear power plants indicate that geothermal energy has come of age and is now a reality. A country possessing geothermal potential can now plan for its future development to broaden its resource base.

Applications in Developing Countries

More than 50 nations around the world today are active or interested in geothermal exploration and development (DiPippo, 1990).

In addition to the major developments in the United States, Mexico, Italy, New Zealand, and Japan, there are also substantial developments in countries such as Nicaragua and El Salvador in Central America, and the Philippines and Indonesia in Southeast Asia. The biggest success story in the developing world is the Philippines, where my former company, Unocal Corporation, had a major role to play in geothermal development. The story is related here as a case history for other countries to study.

The Philippines: A Case Study

The Philippine government, through its Commission of Volcanology and the National Science Development Board, pursued geothermal investigation throughout the archipelago starting in 1964 (Budd and Horton, 1981). These investigations focused eventually on the Tiwi area in southern Luzon about 240 km (150 miles) southeast of Manila, where a few shallow prospecting wells were drilled to 180 m (600 ft).

In 1970 the US Agency for International Development (US AID) referred Philippine government officials to Unocal for possible assistance in developing their geothermal resources. Also at this time the National Power Corporation (NPC), a government-owned utility company, assumed control of the entire Tiwi prospect area to determine the viability of geothermal resource development for generating electrical power on a commercial basis.

To provide the technical expertise, risk capital, and funding necessary for exploration and subsequent development of the resource, NPC entered into a service contract in 1971 with a subsidiary of Unocal, Philippine Geothermal, Inc. (PGI), covering the Tiwi area. Under the terms of the agreement, NPC committed itself to finance, construct, and operate the power plants and transmission facilities to utilize the geothermal energy that Unocal would produce.

The first deep exploratory well was completed in mid-1972, and three additional wells completed by January 1974 verified commercial feasibility over a significant area. In December of 1974, after long-term production and reservoir tests, NPC committed itself to the

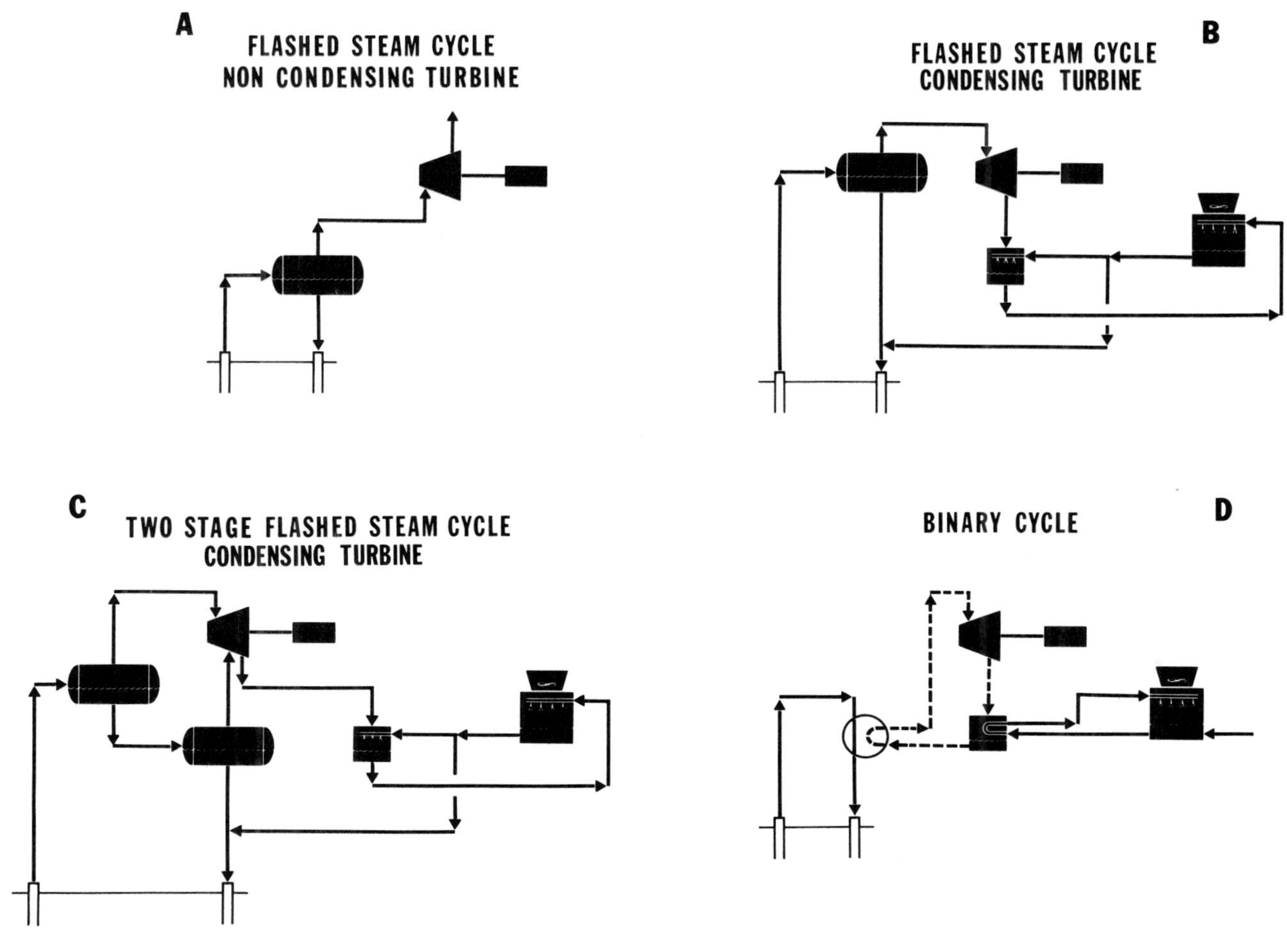

Figure 4. Geothermal power cycles.

first 110 MW_e plant. This plant went on line in mid-1979.

Following this success at Tiwi, the NPC-PGI contract was amended to include a separate area located approximately 56 km (35 miles) southeast of Manila, the Mak-Ban area. Following the necessary geological and geophysical investigations, the first deep test was completed as a discovery in January of 1975. Five additional step-out wells were then drilled over the succeeding 18 months, confirming a commercial resource. NPC then committed itself to the first 110 MW_e power station in the Mak-Ban area; the station was commissioned in 1979.

The shock of the first OPEC oil price increase came in 1973, during early geothermal exploration activities and feasibility determination. Because of a secessionist revolt by the Muslim population on the island of Mindanao, the Philippines suffered greatly from the Arab oil embargo. The Philippines was fully dependent on imported oil for its power generation needs. Besides, nuclear and fossil power plant construction could not keep pace with surging energy demand in the mid-1970s, resulting in power shortages in Luzon. The national energy situation demanded immediate attention. With the discoveries in the Tiwi and Bulalo fields NPC decided to accelerate the development of geothermal power stations and to reduce dependence on the importation of either coal or oil. NPC ordered four additional 55 MW_e units ahead of "reserve" confirmation.

The second 110 MW_e power plant in Tiwi was completed in early 1980 and the third 110 MW_e facility in 1982, for a total capacity of 330 MW_e. The Mak-Ban field's generating capacity was similarly expanded with two additional 110 MW_e power plants, the first plant completed in 1980 and the final in 1984.

A second round of OPEC price increases occurred in 1979, proving the wisdom and timeliness of these bold decisions. Oil acounted for some 13 percent of the country's import bill in 1973, and was up to 32 percent by 1980 (Alcaraz and Barker, 1989). By 1984 the Tiwi–

Mak-Ban geothermal projects constituted 25 percent of NPC's generating capacity on the island of Luzon and provided 30 percent of the generation for the year because of high operating load factor.

While NPC was constructing the power plants and transmission facilities, Unocal (PGI) was engaged in field development with the drilling of wells and the installation of surface pipeline facilities. Today PGI has drilled a total of 132 wells in the Tiwi field and 85 wells in the Mak-Ban field, including the wells for disposal of the geothermal brines and other liquid wastes.

The geothermal projects are a good example of a successful partnership between a private resource developer and a state-owned utility to achieve national objectives; each party was responsible for an area where it had a comparative advantage. Unocal shouldered the exploration risk and guided the steam field development, and NPC, with access to government financing, shouldered the heavy infrastructure investments and the construction and operation of the power plants.

Expeditious development is critical to economic performance of a geothermal power plant. Because the geothermal power plant relies on a single energy source, excessive preproduction investments create a drag on potential profitability (Stefansson, 1987). A financial study done for the Asia Development Bank (ADB) in Manila (ELC, 1987) found that the net cash flows, including avoided oil costs, had paid for the Tiwi project by 1986. A similar early pay-out schedule would apply to the Mak-Ban project.

Technology transfer was another outfall of the close working relationship between Unocal and the Philippine government. The Philippine National Oil Company (PNOC) undertook the exploration and development of other potential geothermal sites in the 1980s, resulting in 118 MW_e of geothermal power installed on the island of Palimpinon in southern Negros, and 112 MW_e on the island of Leyte. These geothermal developments combined with the Tiwi and Mak-Ban projects produce a total of 894 MW_e of geothermal power, making the Philippines the second largest producer of geothermal power in the world after the United States. This development occurred in record time.

Geothermal Energy in Indonesia

Following the success in the Philippines, Unocal entered into a joint operating contract with two state entities of the Indonesian government in 1982. Unocal is a service contractor to Pertamina, the state-owned oil company which has responsibility for geothermal resource development in Indonesia. The installation of its power plants and associated facilities is entrusted to the state electricity authority (PLN).

The contract covers the exploration and development of the Gunung Salak area, located on the western flank of a huge volcanic complex, about 50 miles south of the capital city of Jakarta on the island of Java. Seventy percent of the nation's 165 million people live on Java, and the island's electricity demand has been growing 20 percent annually.

Because the Gunung Salak area is mostly located in a tropical rain forest with average rainfall of 51 cm (20 inches) per month, the task of providing access, while taking great care to minimize environmental disruption, has been difficult. A total of nine exploration and step-out wells have been drilled to evaluate the resource potential of the area. PLN is planning to install three 55 MW_e power plants in the Gunung Salak area and thereby more than double its geothermal power on the island of Java.

Although Indonesia currently produces 1.3 million barrels of oil per day, most of which is being exported, it is simultaneously moving ahead to develop its significant geothermal potential. This will allow Indonesia to export more of its oil and gas to earn valuable foreign currency for the near term. In the long range, Indonesia believes that with further growth of its domestic economy it could become a net importer of oil, and thus geothermal energy will be a better source for central power generation than oil, which is currently the main fuel.

Conclusion

Technological advances coupled with improved economics of both small-scale and large-scale power installations have served to accelerate the growth of geothermal energy and expand its use. There is no doubt that geothermal development can be a significant factor for developing countries with geothermal potential and diverse needs to change their current fuel consumption patterns. By backing oil out of the boiler feed stream for central power generation, these countries reduce oil imports, save foreign currency, and provide for broader-based domestic economic development and greater national security.

References

Alcaraz, A.P., Barker, B.J., et al., "The Tiwi Field: A Case Study of Geothermal Development for the National Interest," Proc. 11th New Zealand Geothermal Workshop, 1989.

Budd, Jr., C.F., Horton, R.M., et al., "Commercial Development of the Tiwi and Mak-Ban Geothermal Fields in the Philippines," Proc. 2nd ASCOPE Conf., Manila, Philippines, 1981.

California Energy Commission, "Relative Cost of Electricity Production," Sacramento, CA, July 1985.

DiPippo, R., "Geothermal Energy: Electricity Production and Environmental Impact, A Worldwide Perspective," *Energy and the Environment in the 21st Century,* MIT Press, Cambridge, MA, 1990.

ELC—Electroconsult, "Geothermal Steam Pricing Study: Report to Asia Development Bank," Manila, Philippines, April 1987.

Stefansson, V., "Success in Geothermal Development," Proc. 9th PNOC-EDC Geoscientific Workshop and Geothermal Conference, 1987.

Summary
Session D-5—Electric Power End-Use and Network Efficiency

David C. White
Session Chair
Energy Laboratory
Massachusetts Institute for Technology
Cambridge, MA

Electric power is a major factor in increasing productivity in supplying society with goods and services. It is also a major potential source of environmental pollution. The way electric power is transmitted and priced has profound implications for how efficiently and cleanly (in terms of the environment) it is utilized.

This session addresses some of those issues under the following broad categories: Networks (in "Networks of the 21st Century," by Lionel O. Barthold); System-Wide Planning for Technical, Economic and Environmental Efficiency (in "System-Wide Evaluation of Efficiency Improvements: Reducing Local, Regional, and Global Environmental Impacts," by Stephen R. Connors and Clinton J. Andrews, and in "Transitional Strategies for Emission Reduction in Electric Power Generation," by Richard Tabors and Burt Monroe); End-Use Services (in "Environmental Protection through Energy Conservation: A 'Free Lunch' at Last?" by Larry E. Ruff, in "Encouraging Electric-Utility Energy-Efficiency and Load-Management Programs," by Eric Hirst, and in "Structural Factors Underlying the Increasing Demand for Electricity in Japan," by Hisao Kibune).

Networks of the 21st Century

Lionel O. Barthold
Power Technologies, Inc.
Schenectady, NY

Abstract

Historically, the drive behind increases in transmission voltages has been growth in power plant size. That drive has reversed due to the high financial risk, regulatory delays, and escalation in cost of conventional plants. Those factors, coupled with technical progress in combustion turbine cycles, and a growing taste for competitive generation market will continue the trend toward a more dispersed generation pattern. This, in turn, will cause greater variations and complexity in dispatch, greater wheeling demands, and more flexibility in use of the inherent thermal capacity of existing circuits—largely through uprating of lines and rights-of-way, and wider application of thyristor controlled devices. Existing networks will increasingly be viewed as an input to planning rather than a consequence of an integrated plan. An increasingly competitive generation market will create pressures for horizontal segregation of the industry and, irrespective of institutional changes, may cause transmission networks to be looked upon as profit centers.

The trend to decentralization of generation is, barring new breakthroughs in technology, also a commitment to fossil fuels, putting the United States on a collision course with the global warming problem. The resolution of this conflict will have a major bearing on the shape of future transmission networks.

Introduction

The last decade's shifts in the role of electrical networks have been driven largely by 1) public concern with the environment and consequent politicization of power supply issues, 2) shifts in economies of scale for generating plants, and 3) swings in fuel economics. While the industry enjoys at least temporary stability in fuel prices, the first and second of these drivers remain hard at work. Public environmental concern has been newly reinforced by fears of electromagnetically induced health effects. Economies of scale in generation, on a total cost basis, have often reversed. Technology, up to now not really a major driver, shows signs of becoming the most important of all. Progress in cogeneration cycles, for example, is reinforcing a trend to decentralized generation. Meanwhile, a growing appetite for deregulation and competitiveness is affecting basic system functions and making their future requirements more difficult to predict.

Most of the comments in this paper apply to the bulk power supply system—in general, lines of 230 kV and above. Trends in subtransmission, discussed briefly, are likely to be quite different.

Finally, any paper which is rich in generalizations can hope, at best, to improve insights into what may lie ahead by identifying basic drivers. US systems are sufficiently diverse that these drivers may apply to most but undoubtedly not to all.

Network Evolution

From the turn of the century to the 1960s, networks were regularly overlaid by higher and higher voltages—each increasing maximum circuit capacity by a factor from four to eight. Research programs demonstrating 345 kV, 500 kV, and 800 kV were not even complete before construction of those voltages began. Eleven hundred kV and even 2000 kV were regarded as further steps in this unceasing trend.

Then it all stopped. Eleven hundred kV research programs grew increasingly irrelevant as plans for that voltage were tabled except in the USSR and Japan. High voltage direct current (HVDC) projects also slowed down, nor were any new precedents set in dc transfer capability. The trend to higher circuit capacities saturated in most industrialized countries, because the principal forces changed:

- Economies of scale in larger generator size ended, both because of availability problems and materials limitations. Disproportionately high financial risks inherent in large stations then strongly reinforced the attractiveness of smaller scale alternatives.
- Better economies in transmission, realizable by further increases in maximum voltage, were thwarted both by

environmental concern and saturation in transmission need.[1] More and more countries realized their highest voltage was high enough and 400 kV remains Europe's basic building block while 500 kV and 800 kV are about as bold as most people think in terms of new construction in the US.

- Growth in electrical load, once steady and reasonably predictable, slowed and became a guessing game.
- The ability to plan in an orderly fashion and to finance new economically justified facilities gave way to tightly constrained alternatives and struggles, on some utilities' part, to maintain solvency.

The 1970s saturation in generation-driven transmission requirements gradually gave way, at least regionally, to new pressures on power transfers—this time not driven by unit size but by imports into high fuel-cost or capacity-limited regions. The matrix of electric energy suppliers began to change too, to some extent due to subsidies given to small producers, but increasingly because the economic, financial, and regulatory climate makes generation an attractive business to industrial companies and smaller independent power producers. They do not run the risk of midstream changes in design requirements or ex post facto rejection of their investment's allowance in the rate base.

The tough question is where will systems go from here? With that question in mind, it is interesting to speculate on six key questions which seem basic to the future of tomorrow's networks:

1. What will happen to electrical energy demand?
2. Who will supply electric energy?
3. How will the function of networks change?
4. How will those changes affect institutional boundaries?
5. How will today's networks accommodate future needs?
6. What surprises might significantly alter system patterns?

All of these questions will be considered in the US context, recognizing that there will be major differences between trends in this country and in developing countries.

Electrical Energy Demand

Electrical energy consumption, now increasing at about 2 percent per year, is still very unpredictable. Major opportunities for conservation remain—witness recent advances in lighting efficiency. Their exploitation would be accelerated by an end to the current honeymoon in oil prices. The US still imports half its needs and remains highly vulnerable to OPEC, political instability in the Middle East, or environmental restrictions on ocean transport.

Yet there are just as many reasons to be braced for a major increase in demand. General Motors's new electric car comes very close to the economic breakthrough needed to compete with internal combustion engines, implying a staggering effect on electricity demand. It is also safe to predict that the last electric appliance has not been invented, nor has the last opportunity for electrification in industry.

Just how long today's growth rate continues or whether it changes by several percentage points in either direction is critical to the market for equipment. That variation is probably less germane to the nature of transmission networks than are changes in the pattern of generation sources that meet the incremental demand for power.

Electric Energy Suppliers

In the short range, US options for new, domestically sited capacity consist mainly of traditional 200–600 MW fossil-fired stations and 25–300 MW cogenerating or industrial supply plants.

While we have yet to see the last import deal from Canada, the difficulty of access, growing Canadian demand, and the fact that there simply are no more Canadas, suggests that hydro imports will not have a major impact on transmission networks of the next century except possibly on a very regional basis.

The next century will certainly see a number of today's experimental options go commercial. Prospects include:

- "Fail-safe" nuclear units of "standard" design in the range of 100–500 MW. The inevitable climb in world oil prices, antagonism toward fossil fuels based on fear of global warming, and the gradual restoration of objectivity on the part of the public, utilities, and government can only demonstrate how environmentally benign nuclear can be made to be. The key limit will not be acceptance of the plants but resolution of the 30–year battle over waste disposal.
- Advanced fossil-fired plants which, while more expensive than today's, will reduce toxic emissions to acceptable standards, though these and other fossil-fired options will draw increasing criticism as evidence mounts supporting the greenhouse effect.
- Combined cycle plants with efficiencies and capital costs which, when evaluated on a total function basis, will compete with larger free-standing power plants.
- Solarvoltaic options, increasingly linked to new end-use devices.
- Fuel cells will emerge as an important part of the

energy supply matrix to the extent that gas reserves are developed and prices are contained.

The actual mix will, as now, be influenced by environmental impacts, technology, regulation, politics, economics, and network capability—probably in that order of importance.

Of all the options that develop, small fossil-fired cogeneration units will undoubtedly see the greatest technical advance through the early part of the 21st century. Innovations involving industrial process heat utilization (steam, air, or water) plus waste-burning, are outpacing progress in larger conventional plants and will serve as one of the major drivers for a more decentralized generation pattern. With a relatively new *external* market for electricity, industrial power producers have adopted far more efficient cycles. The world crisis in waste disposal will keep pressure on those developments, while the relative ease of siting generators in industrial environments will encourage their application.

Those prospects have drawn attention to two very basic questions:

1. Is a decentralized generation pattern, heavily weighted by independent producers, compatible with an adequate and reliable electric energy supply?
2. If that pattern develops, what pressures will it put on today's institutional (corporate) boundaries?

To the first question, it is argued that independent power production:

- Will not allow a utility to prepare an orderly plan to meet its obligation to supply adequate power to its customers.
- Cannot be accommodated by a transmission network planned and built for a specific pattern of centralized plants.
- Cannot be operated independent of the overall network requirements.

Those arguments underestimate the flexibility of a market economy, the variety of contractual options, and evolutions in system technology that will make the whole game easier.

Utilities are already getting used to the same "make-or-buy" decision that manufacturers face in product decisions. In fact, some utilities are becoming partners in offering Independent Power Producers (IPPs) to other utilities. Many are finding that, level playing field or not, there *are* ways to mix internal and third-party capacity reliably.

Contracts with industrial producers will reflect the latter's need for flexibility in electricity production commitments. That constraint differs only in degree to the traditional need for accommodation of forced outages and will be handled much the same way—by sufficient reserves.

While most industrial and municipal entities maintain their own plants, maintenance service will become an attractive market for utilities interested in diversification, particularly since it will enhance unit reliability.

It is ironic that the trend to decentralization, driven directly or indirectly by environmental concerns, could be the worst course of any from an environmental standpoint. While some small industrially based plants achieve high energy efficiency by recovering some waste heat, most do not. Only the latest technologies for non-process-integrated plants approach the efficiencies of large central stations. More importantly, the trend represents an investment in a long-term energy infrastructure that locks the country into dependence on premium fossil fuels—an arguable course from a resource point of view and certain bad news from the standpoint of global warming.

Changing Network Functions

Bulk supply networks can no longer be looked upon as an element tailored to a master "system plan." The plan for which most networks were designed has long since been abandoned. Networks will increasingly be looked on as a "given"—an asset to be used for what it is—or augmented in situ to make it flexible enough to meet a generation dispatch with an extremely high number of variables.

The functions of networks will not change in dimension, but will change significantly in emphasis. Extra High Voltage (EHV) transmission networks will:

1. Continue to be the basic means by which power is reliably delivered from generation to load centers. Both the sources of power and the point of delivery will reflect greater variety and less predictability.
2. Serve as a means to back-up generating capacity—for the utility and, increasingly, for others as well.
3. Provide for economy interchange among areas and users.

To the degree the beneficiaries of networks broaden, so will the perspective of just what it is that a utility has to sell. Network capacity itself will be an important source of income to its owner, and will lead to innovations which stretch transfer capacity to allow serving both one's own customer obligations and the demands of third-party transactions.

The network, once a *consequence* of generation siting, will increasingly be its *determinant.* An independent producer, industrial firm, or municipal plant, bidding to serve utility generating capacity will see its bid severely enhanced or penalized by transmission limitations inherent to its location on the network.

To the extent that regulatory constraints or subsidies, fair or unfair, bias the "free-market" evolution of networks, that bias is likely to further encourage a decentralized pattern. Increasingly stringent pollution limits will fall hardest on the regulated segment of the industry.

A proliferation in the number of generation sources and an increasingly competitive generation market will force transmission networks to accommodate many more transactions. Circuit loading will be more variable because of more rapid shifts in dispatch. Barring breakthroughs in storage, those networks will have overload capability in order to achieve regional reliability in the face of less predictable local generation.

It is important to point out that the degree of dispersion or centralization in power supply will have a far stronger influence on higher voltages than on subtransmission. The latter will, in virtually any future scenario, see steady increases in demand but is, on the other hand, the easiest to expand or uprate.

Pressures on Institutional Boundaries

Mergers among utilities will accelerate and continue well into the next century. They will likely extend to nonutility entities who find financial or operational synergies with the power supply industry. The real advantage of inter-utility mergers will be structural and financial. Most of the advantages realizable at the transmission level are already being identified and exploited by power pools and reliability councils.

It will be more interesting to watch the way in which changing system functions put pressure on institutional boundaries. A recent US Office of Technology Assessment (OTA) report, "Electric Power, Wheeling and Dealing," cites five scenarios ranging from today's structure to horizontally segregated generation, transmission, and distribution businesses. The transmission system, as an independent profit center, might either look to generation companies as its supplier and distribution companies as its customers or simply be a transport company accommodating deals between independent generation and distribution companies or large industrial users. There are two central issues here:

1. Can there be fair competition among alternative generating entities (traditionally the money-making end of the business) when one of them owns access routes to their common market?
2. How much regulation can be displaced by competition without jeopardizing the perceived rights of the public to a reliable electricity supply?

These issues are extremely complex, and germane here only to the extent that they affect network structure and function. Would the transmission system, if operated as an independent profit center, be any different than if it were part of a vertically integrated system? The purpose of regulation, in the latter case, would be to cause the answer to be no. It is difficult, however, to imagine a regulatory framework so effective as to cause the owner of the means of energy transport to make the type of transmission equipment investments which would increase transmission revenues at the expense of far greater (internal) generation income or investment recovery. Thus, the institutional question may well be germane. Pressures for institutional change will be fueled by continued efforts to build a competitive energy supply environment and by the emergence of transmission investment options which may do more to accommodate nonutility (or neighboring utility) generators of power than to serve the utility on whose system those options must be incorporated.

The OTA report, as do so many others addressing this topic, speaks of the need for competition without fully recognizing the extent to which competition and entrepreneurial incentive can displace regulation. A company whose sole source of revenues is the transport of electric power, for example, does not have to be told which generation entities it must serve. In a free market and flexible pricing structure, it will seek out *any* reasonable opportunity to transmit and, if the history of other businesses is any guide, it will become very clever at accommodating new customers within necessary remaining regulatory constraints, i.e., a mandate to maintain reliability of supply to distribution and large industrial power users.

Whether transmission companies are simply implementers (transporters and dispatchers) of contracts between generators and users of power, or whether they serve as broker/deliverer of power is probably less germane to transmission system evolution than the question of whether, in either form, they continue to generate power themselves.

These arguments ultimately led to those major institutional changes being made in the United Kingdom and in New Zealand and those being considered by other countries whose utilities have, up to now, been nationalized. The move to horizontal segregation is to some extent self-reinforcing. A competitive generation market argues for segregation, while it is argued that segregation, once implemented, will attract additional competition at the generation level.

Those arguments also point out that, whether by virtue of horizontal segregation of the industry, revised regulation, or simply changed attitudes, the real impetus to full utilization of the nations' transmission capabilities will be realized *only when the business of transmitting power is looked upon as a profit opportunity in its own right, and is priced and actively sold accordingly.*

Absent the ridiculous scenario of competing, overlapping networks, some degree of transmission system regulation will always be necessary to assure reliable electricity supply. The form of that regulation will, to some extent, affect the evolution of networks themselves. The key question is whether that regulation will encourage or discourage more entrepreneurial approaches to network operation and planning and hence their evolution as profit centers. For example, in a future where reliability performance will be more easily specified in quantitative terms, it is not difficult to imagine incentive-based regulation in which allowable revenues at both transmission and distribution levels, are tied to reliability.

Of the three major system segments, generation holds the only (far distant) prospect of total deregulation, should sources of power become truly competitive and diverse. Generation might then join the long list of products and services to which the public perceives it has a "right," but which are supplied by a free-market economy with a minimum of regulation, e.g., transportation, telecommunication, and even more fundamentally, food!

Accommodation of Future Transmission Demands

It is probably safe to assume that the transmission requirements of the next century will have to be based largely on lines, or at least rights-of-way, in place today. Given that premise, there are three separate, but interrelated aspects to their accommodation of future transmission requirements:

- How can the transport limits of existing lines and rights-of-way, based on either thermal or reactance considerations, be extended?
- How can equipment help systems exploit the full intrinsic thermal capacity of the network?
- Will dispersed sources affect system planning and operations?

Overcoming Limitations in Existing Lines and Rights-of-Way

The bellwether program in improving intrinsic right-of-way capability, started in the early 1970s, targeted increases in capability of existing lines and rights-of-way at 138 kV.[2] It was demonstrated that 138 kV circuits could be built on rights-of-way formerly characteristic of distribution lines. That program brought a broader focus on means for making rights-of-way, at all voltage levels, work harder.

Most of the basic technical parameters inherent in compaction of lines are well understood and documented, yet there remain some very fruitful opportunities for research in practical application of that knowledge. For example, better techniques are needed to bypass sections of line quickly and cheaply as new, higher capacity circuits are built on the same right-of-way. Techniques have also been proposed, and should be developed, for "Hot-Line Restringing" of lines, allowing thermal uprating of lines which cannot be taken out of service.

Existing programs of temperature monitoring also hold more promise, though it is not likely that any system will be able to detect hot-spots and thus take full advantage of overall thermal limits based on ambient conditions on the right-of-way.

One of the most interesting technologies relating to lines themselves is High Phase Order (HPO). It has been shown both theoretically and in test projects, that the transfer capacity of six or 12 phase lines is very significantly higher for the same line-to-ground voltage and the same overall construction dimensions as a three-phase circuit. For example, it is estimated that a 12-phase line with a line-to-ground voltage of 460 kV would have the same transfer capacity as an 1100 kV three-phase circuit. The former could also have lower environmental impact.

For at least 25 years, the public has been waiting for the news that the technology that put a man on the moon can at last make it economical to transmit power underground. Unfortunately this will remain an inherently expensive option in most areas, due in part to the fact that much of the cost is dependent on civil works where technology is slowest to impact. Eventual exploitation of higher-temperature superconductors in practical cables will eventually make superconducting cables economically and technically attractive for reasonable power transfer levels. Underground will see a resurgence in construction—not because of major changes in cost, but because in more and more cases where new capacity is critical, underground will be the only viable option. Underground distribution will continue its inroads, possibly spurred by the need and the ability to use existing overhead distribution rights-of-way for transmission circuits.

High Voltage Direct Current (HVDC) is unique in its ability to use available rights-of-way very efficiently and simultaneously control transfers to best use existing thermal capacity. In a highly dispersed generation pattern, HVDC has little to offer. As in the past, however, there are bound to be large new interregional transfer opportunities or requirements. New or upgraded rights-of-way will be regarded as a precious resource and used to their utmost capability, arguing for either HVDC or, High Phase Order as the synchronous option. Conversion of existing ac lines to HVDC is a "back-to-the-wall" option. Barring drastic restrictions in generation options, it is unlikely that this will ever be cheaper than alternative generation siting.

Equipment to Enhance Transfer Limits

A broad spectrum of possibilities for increasing utilization of transmission rights-of-way was presented in a 1987 Electric Power Research Institute (EPRI) research report entitled "Technical Limitations to Transmission System Operations."[3] Equipment-related techniques to allow higher ac line loading was the subject of EPRI's more recently funded projects on RP3022, "Flexible AC Transmission Systems" (FACTS). Initial programs make it clear that power thyristors can play a very important role in enhancing transfer capacity of ac systems by redirecting flows to take full advantage of latent network thermal capacity, while extending stable transfer limits through modulation of effective system impedance. The primary options visualized, both of which offer control on a time scale short enough to enhance dynamic stability, are: 1) thyristor switching and modulation control of series capacitors; and 2) thyristor controlled tap changing (voltage and/or phase angle).

Fast-valving and braking resistors were also considered. Combinations of these options showed prospective increases of nearly two-to-one in the transfer capability of example systems. Widespread use of thyristor controls on ac systems will seriously weaken the prospect of long HVDC lines within ac systems.

Both the planning and operation of networks will become more challenging as they move closer to the thermal limits of conductors and equipment. Proximity to those limits does not necessarily imply reduced reliability, but will surely call for more precise definition of limits and more carefully engineered margins.

Widespread use of thyristor-controlled devices will raise some new system control questions. For example, would a proliferation of controllable devices, all responding to the same or related system inputs, produce artificial resonances?—and to the extent that is a concern, how would one best model a very large, complex system to predict that danger, recognizing that the response of devices and the system itself may have to be modeled in considerable detail. Concern over the reliability of the controls themselves and the communication systems have been voiced by those in operation.

Relay systems for FACTS applications will be adaptive, receiving inputs from FACTS controllers so the protective functions properly account for the ever-changing impedance of the transmission lines.

There will obviously be evolution in all types of transmission equipment, including a gradual shift to gas-insulated substations. Protective relaying will probably see breakthroughs, many shifting the "trigger" for trips or other responses to the event itself rather than the system's response to that event. For example, a sudden imbalance in available generation and required load, rather than frequency decay, is now used as a basis for load shedding in certain large industrial plants.

Operational Aspects of Dispersed Generation Systems

A system heavily weighted, or even dominated, by relatively small independent, industrial, or waste-burning plants, would have a much more rapidly varying generation mix, due to the dual demands served by many such plants. Its dispatch will still be optimized by cost but heavily biased by constraints in reliability, thermal capability, contractual constraints, and environmental impacts.

Solving that dispatch for both real and reactive power in a rapidly changing system environment will require software that considers thousands of permutations of generation supply, contractual terms, and system control options, explored iteratively to assure specified reliability at minimum evaluated cost. That software will track the existing state of the system and calculate *quantitative* risk of service interruptions—both for the system as currently dispatched and for all reasonable options. Reliability calculations will take advantage of more accurate event and failure probability data to establish outage time and recurrence rates for each class of service, recognizing the probability and consequences of dynamic events, and prospective measures to enhance security, e.g., diversion of transmission loading, and their resulting economic penalty to dispatch.

That task, formidable as it seems, is basically one of simulation and information management—precisely the area where technology is making its most rapid strides. The past year has seen major advances in relational data bases. Within the past six months, commercially available optimal load flows have cut solution times by from one to two orders of magnitude and have yet to be stumped on any system configuration. Contingency analysis methods are also advancing by quantum steps.

Thus the ability to simulate, either in planning or operation, the complex dimensions inherent in a competitive, dispersed generation pattern, will not inhibit trends towards that pattern driven by economic or regulatory factors.

Effect of "Upsets" on Future Systems

It is interesting to speculate over what "upsets" in the present course of events could alter the general pattern for transmission system development, recognizing that most past changes of course were based on events hard to predict in advance. Suppose, for example:

New Sources of Natural Gas Were Discovered. Suppose, despite the more pessimistic view now gaining favor, the "Deep Natural Gas" scenario is validated, i.e., that the earth's core really *is* a tank of methane, simply waiting for advances in deep drilling technology. Abundant reserves of natural gas, regardless of the source, would certainly accelerate decentralization of generation, possibly through fuel cells used all the way down to the consumer level.

An Intensified Greenhouse Effect Alarm. New data on the greenhouse effect would lead to severe restrictions on fossil fuel use. The time constant of switching, to a significant degree, to nuclear or other nonfossil sources is so long as to put tremendous pressure on conservation. This would ease transmission requirements in general, but could severely strain transmission from areas rich in hydro or nuclear, probably encouraging a new generation of long-distance transmission projects.

A Breakthrough in High Density, Medium Scale Energy Storage. Of all the technical breakthroughs, this would surely have the most profound influence on system structure. In the limit, moderate-scale storage could decouple transmission and distribution, allowing the former to be base-loaded. In addition, a dramatic increase in thermally based energy transfer capability, transmission dynamic limits, enhanced by the stabilizing capability of storage devices, would be further extended by the new reliability criteria reflecting the systems ability to survive transmission outages without interruption of load.

A Major Breakthrough in Solar Photovoltaic Efficiency. Conversion efficiencies of commercially manufactured solar photovoltaic devices are now reaching 35 percent. With each gain in efficiency, new applications appear. If the history of technological evolution is any guide, solar photovoltaic's inroad will be driven more by new functions, best performed by freestanding systems, than by dollar-for-dollar competition with other forms of energy.

A "Northeast Blackout" Equivalent. A new major blackout would, in today's world of a power-system conscious public, draw an immense level of political attention to the industry's structure. It would likely intensify political pressures for a greater degree of local power independence.

A Major Nuclear Accident. No amount of confidence rebuilding or innovation in nuclear cycles could survive another Chernobyl, either in the US or elsewhere. A "second Chernobyl" would probably force closing of existing nuclear plants, drastically curtailing generation capacity and imposing extremely severe burdens on system transfer limits. The result would again be pressure for conservation, but also for new sources with short delivery cycles, thus reinforcing decentralization.

New Evidence Linking Electromagnetic (E/M) Fields to Adverse Health Effects. New evidence, *perceived* as constituting a serious health hazard would, at a minimum, bring public pressure for load limitation on both high voltage transmission lines and distribution lines, translating strongly and directly to incentives for locally generated power.

A Sharp Escalation in Oil Prices. A gradual climb in oil prices is inevitable. A precipitous rise would once again plunge the US into economic recession, thus curtailing demand but straining ties from areas having abundant coal, nuclear, or hydro capacity. Such an escalation could result from OPEC action, political turmoil in the Middle East, or severe restrictions in ocean transport resulting, for example, from a few more Valdez-equivalents. Future major increases in oil prices would probably be considered permanent, causing a resurgence in coal-fired construction which, for economic reasons, would reinforce traditional system patterns of centralized generation, or at least slow down any trend towards decentralization. These effects assume that coal liquefaction technology will not have reached the point where costs for locally based generation once again make that option viable.

Resurgence in Load Growth. A rapid resurgence in load growth would work strongly in favor of short delivery generation alternatives and accommodation of existing transmission limits. As with many other "upsets" above, this too would favor the trend to decentralized capacity.

A Serious, Prolonged Recession. A recession would cut growth in electric power demand, at least temporizing pressures for increased transfer capacity. Technical progress tends not to respect economic downturns, however, which suggests that technically driven changes in system structure, while delayed, would likely have a more sudden effect in the recovering economy.

A summary of how various upsets might affect existing load trends on bulk supply systems is suggested in Table 1. Column one in the table estimates whether average loading of EHV transmission circuits will increase or decrease, column two addresses the variability of loading resulting from wheeling agreements or changes in dispatch, and column three addresses the need for major new interregional ties. Of the ten upsets hypothesized, only a sharp rise in load growth, simply incapable of being served by distributed plants, suggests a return to the historical pattern of centralized supply and broad scale increase in new EHV line construction. Three suggest the need for major new interregional ties.

It is not difficult to mix these upsets, or speculate over others, to reach quite different conclusions. For example, suppose pressures to limit fossil-fuel consumption became sufficiently extreme to force a major resurgence of nuclear and a shift to electric cars. That could force construction of new nuclear plants and a resumption of historical trends to higher capacity circuits. A breakthrough in fusion, showing it practical only in very large plant sizes, would produce the same result.

Research Implications

Research in energy conversion will have a greater effect on the nature of future transmission systems than will those in transmission. That observation notwithstanding, it is interesting to speculate over those most important to the transmission sphere. Highest on the list would

Table 1. Probable impact of various "upsets" to existing trends.

UPSET	(1) GENERAL LOAD LEVELS		(2) DEMAND FOR FLEXIBILITY	(3) MAJOR NEW PROJECTS
	+	-		
A. ABUNDANT GAS			X	
B. GREENHOUSE ALARM		X	X	X
C. STORAGE BREAKTHROUGH		X		
D. SOLARVOLTAIC BREAKTHROUGH				
E. NEW BLACKOUT			X	
F. MAJOR NUCLEAR ACCIDENT			X	X
G. E/M FIELD SCARE		X	X	
H. SHARP OIL PRICE RISE			X	X
I. SHARP INCREASE IN LOADS	X		X	X
J. SERIOUS RECESSION			X	

be means for extending the transfer capability of existing circuits, e.g.:

- Means for uprating (current and voltage) existing circuits without their removal from service.
- In situ uprating of substations, i.e., accommodation of uprated circuits within fixed available substation boundaries.
- Thyristor applications for increased flexibility in power control.

There are, within that subject area, a number of options less dependent on research than on prototype installation. Foremost among them is High Phase Order. Other key research opportunities include: continued research in the effects of E/M fields; and monitoring and life extension methods for transmission lines and equipment.

There are obviously countless opportunities for research (and invention) in hardware and software products. Most are "opportunity-driven," and provide sufficient commercial incentive to cause timely development by private industry. Software is probably today's fastest moving opportunity-driven field.

Conclusions

Factors driving system evolution increasingly favor smaller, decentralized supply sources. However, that trend is strongly based on the availability of reasonably priced oil, natural gas, or a synthetic substitute for them—probably a safe assumption for the next several decades, but also a collision course with the global warming problem.

While decentralization of generation sources will, in general, extend the capability of the transmission networks, transmission loading will be far more variable. "Stretches" in transfer capability, achieved through uprating of lines and rights-of-way, as well as through new thyristor applications, will be required to accommodate economy transfers among utilities, contracts with independent producers and cogenerators, and backup to sources with lower reliability than traditional utility plants. System operation will be far more complex, requiring more realism, accuracy, scope, and speed in simulation.

The future role of bulk transmission systems has probably never been harder to forecast. That uncertainty itself—in technical, economic, and political spheres—is an important driver for the "small step" approach, which translates, in the near term at least, to greater dispersion of energy sources. This trend will probably continue for 20 to 30 years, by which time fears of global warming will force a reversal in the use of fossil fuels. Society will then buy time through strong conservation efforts, beyond which nuclear may be the only viable option. Should small scale, fail-safe nuclear plants be developed or (less likely) solar photovoltaic become viable at the 50 to 100 MW level, the dispersed, multisource pattern of generation will be sustained. Absent either, plants will again be highly centralized and transmission will revert to its former climb in capacity levels.

That forecast, too, could be upset by massive reforestation or the development of artificial photosynthesis which, perhaps in combination, could extend the life of fossil-based power sources indefinitely.

Notes

1. Growth in transmission voltage was inherently coupled to generator unit size since plants are usually sized for from two to six units and since planners usually allow about the same number of major lines to provide egress for the power generated. Interconnections also tend to track unit size since reliability criteria tend to treat the loss of either an interconnection or a major generator as equivalent threats.
2. That objective remains central to most of today's important areas of research in networks . . . in transmission lines, substation equipment, control, and system engineering.
3. "Technical Limitations to Transmission System Operations," Power Technologies, Inc., EPRI Research Project Report RP5005–2, Final Draft, August 1987.

System-Wide Evaluation of Efficiency Improvements: Reducing Local, Regional, and Global Environmental Impacts

Stephen R. Connors
Clinton J. Andrews
The Analysis Group for Regional Electricity Alternatives
Energy Laboratory, Massachusetts Institute of Technology
Cambridge, MA

Abstract

Future electric service needs may be met through a variety of possible strategic combinations of demand-side (zero-emissions) and supply-side (positive-emissions) resources. This paper explores the efficacy, in environmental and cost terms, of 60 different long-term strategies for the New England region. We simulate the regulatory status quo by using a reliability-based least-cost planning logic, in which new demand- and supply-side resources compete with one another to fill capacity blocks ensuring target reserve margins. According to our analysis, strategies that focus only on improving end-use efficiency fare poorly in reducing sulfur dioxide, nitrous oxides, and particulate emissions when compared to strategies that balance efficiency improvements on both the supply and demand sides. These results suggest that the current technology-based planning framework, in which new supply options and demand-side options compete at the margin for financing, is not the most effective way to create a "least social cost" electric power system. It appears that a system-wide approach is better, because it teams new options with existing plants to evaluate their combined effectiveness in addressing cost and environmental goals. A group of New England's electric power decision makers are working cooperatively to develop such an integrated strategy for the region.

Introduction

The complexity of choosing and implementing electric power resources has increased dramatically in the last decade. Controversy and uncertainty have hindered planning efforts, while the utilities' mandate has shifted from simply providing electricity to ensuring adequate energy services. The options utility decision makers must now consider encompass both the supply and demand sides of electric service, and may be provided directly by the utilities or through the actions of third parties. Likewise, the uncertainties that utilities, regulators and consumers must consider have increased. The risks associated with load growth, fuel and capital markets, and a host of other possible occurrences must now be considered in the decision-making process.

In addition, the goals that the electric power system must strive to meet in its design and operation have become more diverse. Cost and reliability of electric service are major concerns as competitiveness in a world economy becomes important. Similarly, the environmental impacts associated with the provision and use of electric power have also become central issues. Utilities must seek to simultaneously minimize costs, environmental impacts, and reliability problems.

These goals are often conflicting, both between and within categories. While some tradeoffs between cost, reliability and environmental impacts are intuitively apparent, there are also tradeoffs within each group that can only be evaluated using a rigorous system-wide approach that incorporates the uncertainties of an unknown future. The heterogeneous nature of environmental impacts, be they local site-related problems, changes in regional air quality, or impacts upon our climate and atmosphere, illustrates the need for a method that deals in an integrative manner with these complexities. Carefully crafted, multifaceted strategies need to be developed during the planning process.

Yet finding a suitable course of action is not enough. The implementation of such a strategy requires a level of consensus and understanding between the different decision makers. As environmental impacts have local, regional, and global components, the responsibility for decision making, approval, and implementation requires the coordination of participants at all levels. The method that identifies the most suitable strategies must involve enough of those participants that implementation of those strategies becomes feasible.

This paper describes one such method now being developed at the MIT Energy Laboratory for a group of New England's electric utilities, regulators, and environmental and industry concerns. Called the New England Project: Analyzing Regional Electricity Alternatives, it draws upon the region's electric power and environmental expertise to formulate studies aimed at identifying the tradeoffs between different strategies' ability to meet such diverse goals as cost, reliability, and various environmental concerns. After briefly discussing the structure and scope of the project, results of the latest set of analyses will be presented, followed by a discussion of the environmental tradeoffs and their implications for the development of robust strategies.

The New England Project

The scientific and technical complexity of regional electric power planning demands significant joint efforts at fact finding, identifying options, and evaluating their implications. Several years ago, a group of MIT faculty, staff, and graduate students offered to provide the analytic support for such an effort, and by early 1988, regulators, environmentalists, utility personnel, and electricity customers started meeting at MIT on a regular basis. Funding from a consortium of regional utilities and industrial customers has allowed the group to expand its analytic capabilities, and by 1989 every state in the region was represented on the advisory group for the New England Project.

Figure 1 shows the structure employed by the New England Project to foster the joint fact-finding approach to long-range, integrated electric power system planning. The external participants in the New England Project form an advisory group, which interacts periodically with the MIT analysis team. The interactions have been set up in the form of *public* public-policy analysis exercises. Using this structure, the diverse concerns of the advisory group are channeled into analyzable form by the MIT analysts.

During interactions with the advisory group, the analysis team constructs a set of scenarios—strategies combined with a set of possible futures—which incorporate the options available to, and constraints faced by, the electric power system for meeting its desired goals

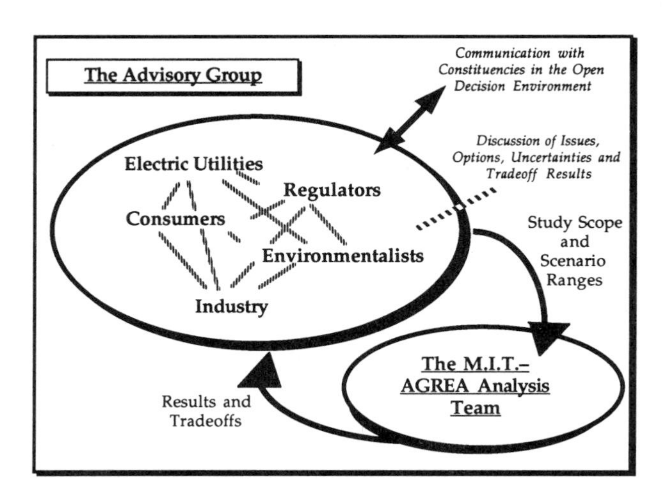

Figure 1. Structure of the New England Project.

of low costs, increased reliability, and reduced environmental impacts. Input data, provided by advisory group members' staffs and other credible sources, is used to piece together the scenarios so that standard utility power system simulation programs, such as Electric Power Research Institute's Electricity Generation Expansion Analysis System production costing model, can be used to perform the analysis.

Using these planning tools, the analysis team evaluates each scenario by calculating a set of attributes that measures how well each scenario meets the overall goals. Scenarios, and therefore strategies, can then be compared with how other strategies perform if load growth, fuel prices, and other major uncertainties behave unexpectedly.

The interactions between the advisory group and analysis team serve not only to identify the primary issues of concern, the options and strategies available, and the uncertainties faced, but also to reduce the "black box" nature of the analysis work. By opening up the analytic process, as well as the components of the analysis itself, controversy over the results themselves is reduced. Through the joint exploration of the tradeoffs identified in the results, it is hoped that the decision makers will achieve a better understanding of the impacts of different strategies, and will draw upon that understanding to invent and implement even better strategies.

Discussions with the advisory group have focused on three primary issues: the environmental impacts of various strategies, the cost of electric service for those scenarios, and the reliability of electric service. Results presented here focus primarily on the tradeoffs between cost and environmental impacts. To measure how well a strategy fared with respect to these issues, the following attributes were used:

- Total Cost of Electric Service—The net present value of all costs associated with the delivery of electric service over the 20 year study period (1989–2008), including capital and operating costs, plus customer and utility investments in conservation and load management.
- New Site Megawatts in 2008—The amount of new capacity built on new sites in New England by the last year of the study, used as a proxy for site-related local environmental impacts.
- Cumulative Sulfur Dioxide Emissions—The total amount of sulfur dioxide emitted to the atmosphere by all units (including power purchases) over the study period (1989–2008), used to gauge the relative impact for a regional environmental issue, acid rain.
- Cumulative Carbon Dioxide Emissions—Similar to the above attribute, the total amount of carbon dioxide emitted to the atmosphere, used to gauge the relative impact for a global environmental issue, global warming.

Many more attributes were calculated than the four above. Many were descriptive in nature, such as fuel consumption and electricity production by fuel type. Other attributes behaved similarly to the ones above. For example, total NO_x and particulate emissions behaved similarly to sulfur dioxide, so reporting trends based on SO_2 results was sufficient to describe the behavior of all three.

In total, over 2,160 scenarios were analyzed, a combination of 60 strategies over 36 futures. Table 1 shows the basic composition of the 60 strategies. Each is a combination of a supply-side and a demand-side option set.

There are 15 demand-side option sets composed of a combination of current utility demand-side management (DSM) programs, residential appliance standards, commercial building standards, commercial lighting, and commercial heating, ventilation, and air-conditioning (HVAC) subsidies. Each is modeled on an end-use basis. Because the modeling system cannot accommodate period-to-period interactions of the price of electricity on its demand, all demand trajectories are bracketed. Optimistic load growth cases are assumed to have low electricity prices and strong economic growth, while pessimistic load growth cases have high electricity prices and weak economic growth.

Four separate supply-side option sets are combined with each demand-side option set. Each supply-side option set employs a different mixture of technologies to meet future generating needs. There are three classes of new generating technologies: new units on new sites, power purchases, and repowered units/new units on old sites. New units on new sites are the first three technology options listed under supply option sets, covering peak, intermediate, and cogeneration technologies. The power purchases refer to contracted capacity from Canada, broken down into hydro and nuclear purchases, and Canadian coal-fired purchases—for which emissions are counted. The repowering options are substantially different. Under the repowering option, a new unit could replace an existing unit as needed, if the existing unit was installed over 30 years ago and is over 40 MW in size. In these cases, new generating technology was added to replace both the old MWs and to meet additional capacity needs.

An important assumption in this modeling framework is that, except for the repowering cases, old units are never retired during the 20 year study period. How-

Table 1. Strategy definitions.

Strategies

Options are combined into Option Sets which are in turn combined into Strategies.

Supply Option Sets	0	1	2	3
Technology Options	Gas Dependent	Power Purchases and Gas	Repowering and Gas	Repowering and Purchases
(New Capacity)	(%-New MW)	(%-New MW)	(%-New MW)	(%-New MW)
Combustion Turbine 80	20%	10%	10%	5%
Combined Cycle 200	70%	45%	-----	-----
Cogeneration 40	10%	10%	10%	10%
Power Purchases	0%	35%	0%	35%
Life-Extend Existing Units	No Retirements	No Retirements	None except Repowerments	
Repower Existing Units	None	None	80%	50%

Demand-Side Option Sets

	Options Utility Programs	*Residential Appliance Eff. Stds.*	*Commercial Building Eff. Stds.*	*Commercial Lighting Subsidies*	*Commercial HVAC Subsidies*	*Combined*	*Strategies*	*(Key)*	
A	—	—	—	—	—	A0	A1	A2	A3
B	—	■	—	—	—	B0	B1	B2	B3
C	—	—	■	—	—	C0	C1	C2	C3
D	—	—	■	■	—	D0	D1	D2	D3
E	—	■	■	—	—	E0	E1	E2	E3
F	—	—	■	■	■	F0	F1	F2	F3
G	—	■	■	■	■	G0	G1	G2	G3
H	■	—	—	—	—	H0	H1	H2	H3
I	■	■	—	—	—	I0	I1	I2	I3
J	■	—	■	—	—	J0	J1	J2	J3
K	■	■	■	—	—	K0	K1	K2	K3
L	■	—	■	■	—	L0	L1	L2	L3
M	■	—	■	■	■	M0	M1	M2	M3
N	■	■	■	■	—	N0	N1	N2	N3
O	■	■	■	■	■	O0	O1	O2	O3

ever, operation and maintenance costs are increased at the end of book life for these "Life Extension" units to account for the increased overhead required to maintain reliability. Except for some Canadian power purchases, and repowered coal units, all new capacity is dual fueled with natural gas and oil 2.

The planning logic used in constructing strategies is based on current definitions of "least-cost planning" in the region. New capacity is assumed only to be developed to meet system reliability requirements. Since existing capacity is typically not retired, then new supply- and demand-side investments are made only to meet the reserve margin requirements imposed by new load growth.

The 15 demand-side option sets combined with the four supply-side option sets yield a total of 60 strategies. These 60 strategies are then evaluated for 36 futures composed of the four types of uncertainties listed in Table 2. Three economic growth uncertainties are the primary determinants of electricity demand to which the demand-side options were then applied. Two fuel price uncertainties help gauge the relative impacts upon system dispatch and cost for an increase in natural gas prices. Two additional types of uncertainties deal with the impacts of discrepancies in the expected versus actual cost and electricity savings of demand-side initiatives.

Study Results

To be comprehensible, the task of interpreting the results of 2,160 scenarios must be systematic. The presentation of the study's results reflects our approach. First, the effects of uncertainties are observed to see how sensitive the strategies are over possible futures, and to gauge the variability of the attributes over the range of futures. Second, because of the complex composition of the individual strategies, the separate supply- and demand-side options are examined against single attributes. With this understanding of how individual components behave, multi-attribute analysis of the different strategies across the 36 futures is performed. Additional steps aimed at addressing the individual concerns of advisory group members can also be performed, but are not part of the analysis presented here.

The Effects of Uncertainties

As seen previously in Table 2, four types of uncertainties were addressed. These four uncertainties behaved rather predictably, that is, the results under each uncertainty moved about uniformly showing little sensitivity to the individual strategies being evaluated.

With increased economic growth (and lower unit prices) from the low to the high load scenarios, overall fuel use and investment and operating costs increased, as did all environmental emissions and land requirements, since more GWhs were required and the resources necessary to produce them increased. However, the increased need for generating capacity increased the average operating efficiency of the region's fossil-fueled generating capacity as a larger number of newer and more efficient units were added than in lower growth scenarios.

The impact of higher natural gas prices also caused costs to rise. Relying more on older, non-natural gas-fired units caused SO_2, NO_x, and particulate emissions to increase more than CO_2 emissions. In addition, this greater reliance on the region's older oil and coal-fired generating capacity decreased the region's operational efficiency for electricity generation.

Increases in the cost of demand-side measures only caused the total and unit costs of electricity to rise. However, the effect of less than expected electricity savings due to the unanticipated overlap of demand-side pro-

Table 2. Uncertainty definitions.

Futures — *A future is a combination of different uncertainties*

Load Key	*Economic Growth*
H	High/Optimistic
L	Low/Pessimistic
M	Medium/Base

DSM Cost Key	*DSM Cost Effectiveness (¢/kWh)*
C	On-Budget DSM Programs and Measures
E	50% Cost Overrun—Programs and Measures
B	50% Cost Underruns—Programs and Measures

Fuel Key	*Fuel Prices*
F	Base Fuel Prices
G	High Nat. Gas Prices

Interactions Key	*DSM Interactions (GWh & Peak)*
D	No Interactions - Utility Progams and Measures
5	50% Reduction in Utility Programs/Interactions

grams had a wider impact, with cost and emissions increasing as the supply-side had to make up the shortfall in conservation measures. These impacts were small, however, when compared with the magnitude of total system costs and emissions.

Single Attribute Evaluation of Option Sets

In this section we explore how individual demand-side and supply-side option sets compared against each other for different attributes. Later, we will examine how these option sets, combined into strategies, perform for multiple attributes.

It is important to note that performance of the strategies was quite consistent, in relative terms, across the ranges of uncertainty explored here. We confirmed this both with the Friedman test statistic, which showed consistent strategy ranks at the 0.999 level across futures, and by visual inspection of the strategies' relative placement in scatter plots for each of the 36 futures. The consistent trends are shown here using the average cost or environmental impacts for each option set (based on 144 scenarios for each DSM option set, and 540 scenarios for each supply-side option set).

Figure 2 shows how the average total cost of electric service varied with increasing conservation. The DSM option sets are shown in order of increasing GWh savings. The background histogram shows the relative GWh savings of each option, starting with no savings for option set "A," and ending with an average total savings of nearly 155,000 GWh for option set "O." As can be seen, there is a definite trend of reduced overall costs, with increasing investments in conservation.

Figure 3 shows how total cost varied for supply-side option sets. The pie charts in the lower portion of the figure give the relative contribution of the technology groups described above and in Table 1. The higher total costs for the option sets with repowering reflect the increased capital expenditures associated with replacing old units as well as building new ones. However, there is only a 0.9 percent increase in average total costs between the repowering only and gas dependent option sets. This is small compared to the 4.8 percent difference between the "A" and "O" demand-side option sets in Figure 2. The small increase in total cost between repowering and nonrepowering strategies is due, in part, to the reduced operating and fuel costs of the new technologies offsetting the increased capital expenditures.

If we turn now to the various environmental attributes, different stories emerge. Focusing first on local impacts, measured by increased land requirements for

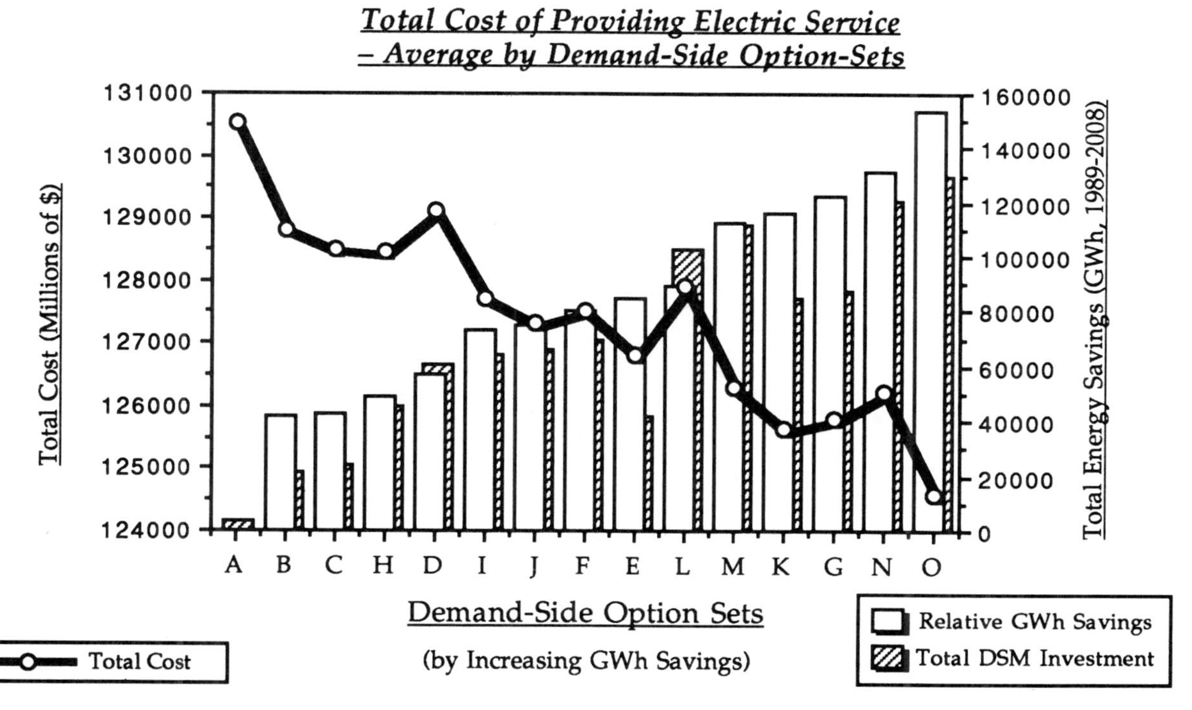

Figure 2. Total cost impacts by DSM option sets.

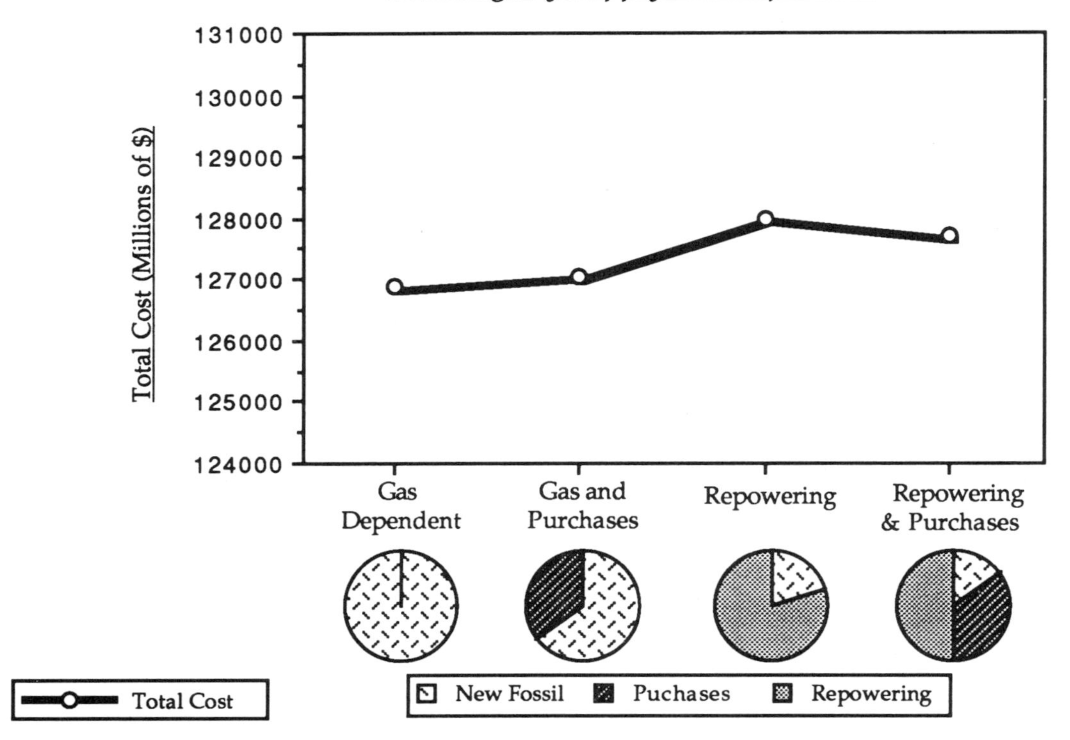

Figure 3. Total cost impacts by supply-side option sets.

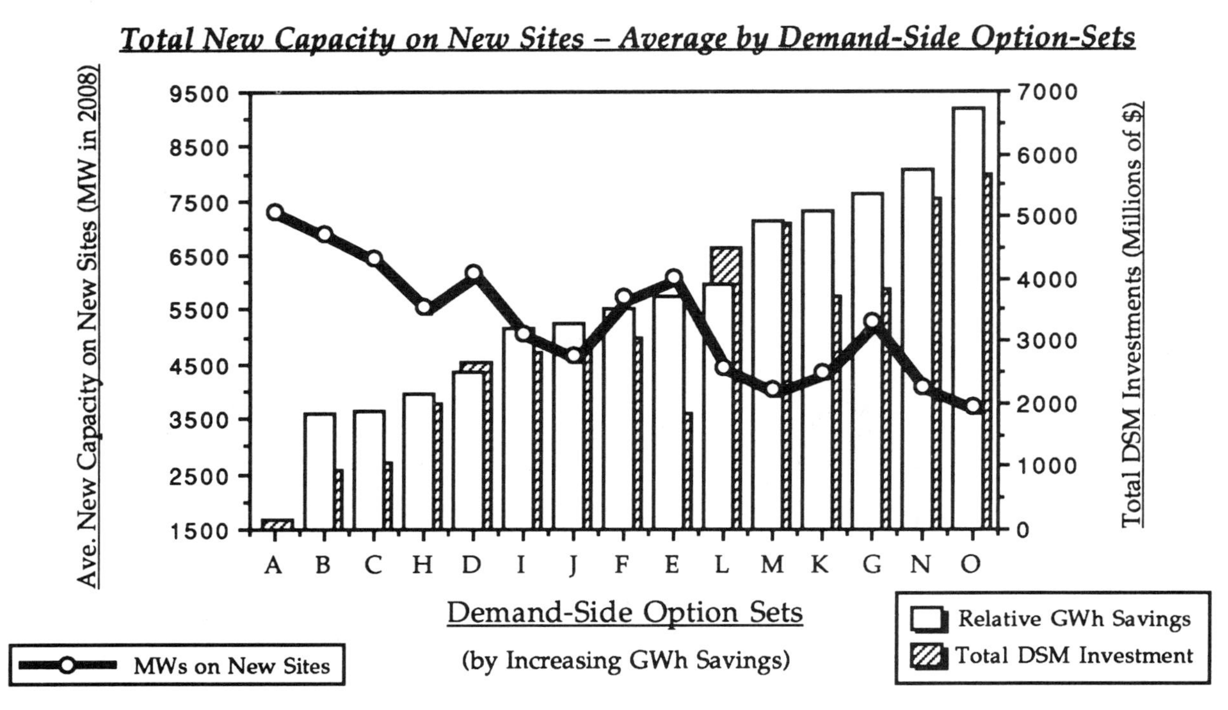

Figure 4. New site megawatts impacts by DSM option sets.

new generators, we see in Figure 4 a downward trend for new sites with increasing DSM activity. The variability in the curve can be attributed to the relative differences in peak reduction versus GWh reduction between DSM option sets. Higher GWh saving option sets "F," "E," and "G" do not include current utility conservation programs, and have significantly higher new site MW needs than their neighbors.

Figure 5 shows the same comparison, but for the supply-side option sets. Here we see that the repowering options displace the need for new generating sites more effectively than power purchases alone. The combination of both repowering and power purchases was most effective at reducing local land use environmental impacts across all the futures evaluated.

Figure 6 shows the average emissions for SO_2 and CO_2 across the DSM option sets. Increased conservation was effective in reducing carbon dioxide emissions by an overall average of 4.2 percent. However, the results show that strategies with increased conservation had higher SO_2, NO_x, and particulate emissions compared to the other strategies. The average SO_2 emissions for "O" are nearly 10 percent higher than the residential-standards-only case "B."

The increases in these emissions are attributed to the retention of older, less efficient, and higher emitting power plants in the generating mix. SO_2, particulate, and NO_x emissions are concentrated in these older units. If the conservation programs fail to displace the electricity generated by these units, and decrease the rate at which new, cleaner technologies enter the mix, more of these emissions are generated compared to other strategies.

Looking again at Figure 4, we see that the moderate conservation DSM option sets of "F," "E," and "G" had higher average new site MWs than their neighbors, reflecting a greater influx of new technologies. In Figure 6, we can see a corresponding decrease in their average SO_2 emissions. As we will see below, SO_2, NO_x, and particulate emissions are affected more by the composition and operation of the installed capacity than by the conservation impacts of any particular demand-side option sets.

Figure 7 shows the same comparison for the supply-side option sets. Here it is clear that the repowering options are the most effective at reducing SO_2 emissions, while power purchases are best for reducing carbon dioxide. Compared with the gas dependent option set, the repowering set exhibits a 21 percent reduction of SO_2, but less than a 1 percent reduction in CO_2. Conversely, the gas and purchases option set shows only a 3 percent reduction in SO_2, but a 6.6 percent reduction in CO_2 compared to the gas dependent set. The sensitivity of sulfur dioxide emissions to the composition of the installed capacity is borne out by the increase in average SO_2 emissions from the repowering to the repowering and purchases supply-side option sets.

Multi-Attribute Evaluation of Strategies

By examining demand-side and supply-side option sets in isolation we have seen that increased conservation decreases costs, the number of new generation sites required, and total CO_2 emissions, but was ineffective at reducing other atmospheric emissions. In comparison, repowering of old existing units was effective in decreasing new site requirements and SO_2 emissions. Repowering, however, was slightly more expensive, and was not as effective in reducing CO_2 emissions as power purchases.

How do the combined strategies of supply- and demand-side option sets perform with the competing goals of cost and environmental emissions reduction? To evaluate the relative impacts of the strategies on competing goals we will use scatter plots that reveal tradeoffs by showing the position of one strategy versus others for different combinations of attributes. Since the relative positions of strategies did not change from future to future, the tradeoff curves show the average attribute values, rather than those for a specific future.

Figure 8 shows the tradeoffs between cost and the global environmental attribute of cumulative carbon dioxide emissions. Since our goals are to reduce both cost and CO_2 emissions, strategies—identified by a single point and label—that lie closest to the lower left hand corner are preferred. As can be seen, the high DSM repowering and purchases strategies are closest to this corner. As expected, the higher the DSM intensity, the lower the cost. In fact, the clusters of similar supply-side strategies—for which frontiers have been drawn—are nearly vertical, demonstrating that the cost savings of increased DSM activity greatly outweighs the cost impacts of the supply-side options.

The shapes of the curves drawn on the figure indicate the competing nature of DSM and power purchases in decreasing carbon dioxide emissions. Without the purchases, the most DSM intensive strategies, "O0" and "O2," are clearly best. But with both DSM and power purchases, DSM strategies "G3" and "K3" move to the frontier.

Figure 9 shows the cost and regional environmental tradeoff between total cost and sulfur dioxide emissions. While exhibiting the same cost trend as the previous figure, the repowering strategies instead of the power purchase strategies have clustered on the left. In fact, the repowering alone supply-side frontier outperforms the repowering and purchases frontier. However, only the repowering and purchases strategy cluster is in the left hand pair of clusters for both figures.

Figure 10 shows the cost versus local siting-impacts tradeoff. Unlike Figure 8, where DSM and purchases competed in their effectiveness at reducing CO_2 emissions, neither interact here, both acting to reduce the need for new sites. Here the repowering and purchases with high DSM strategies fare best.

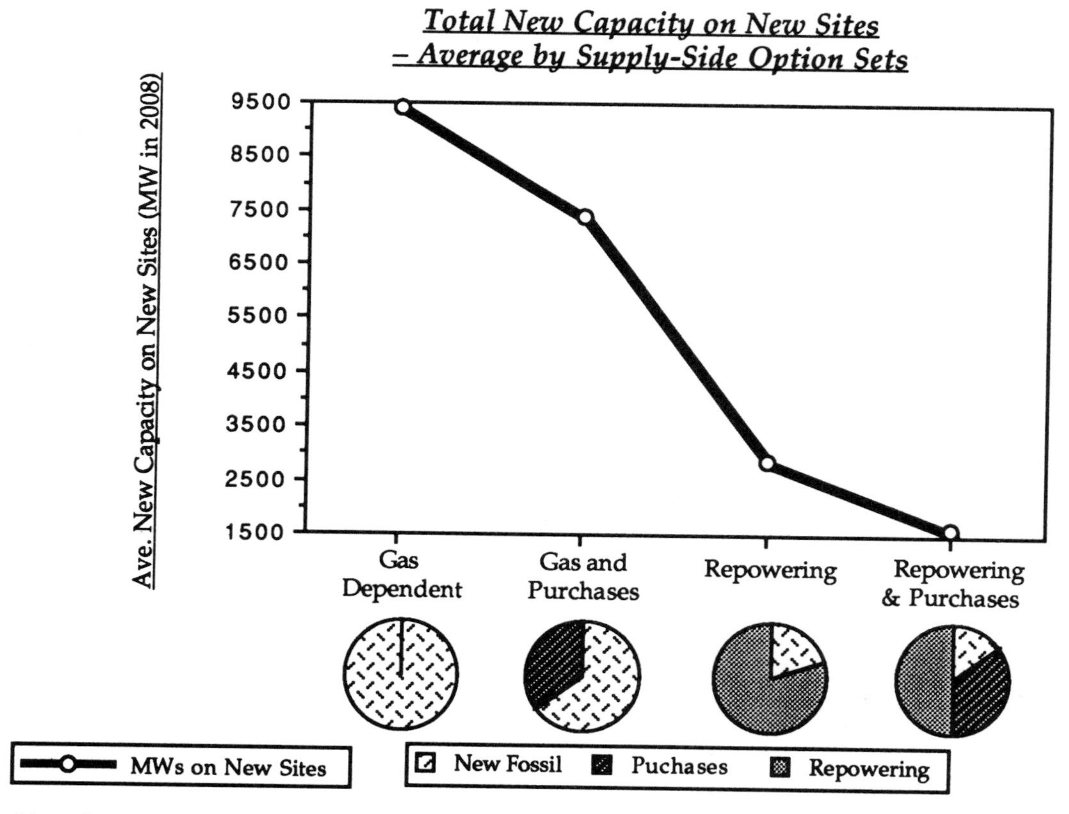

Figure 5. New site megawatts impacts by supply-side option sets.

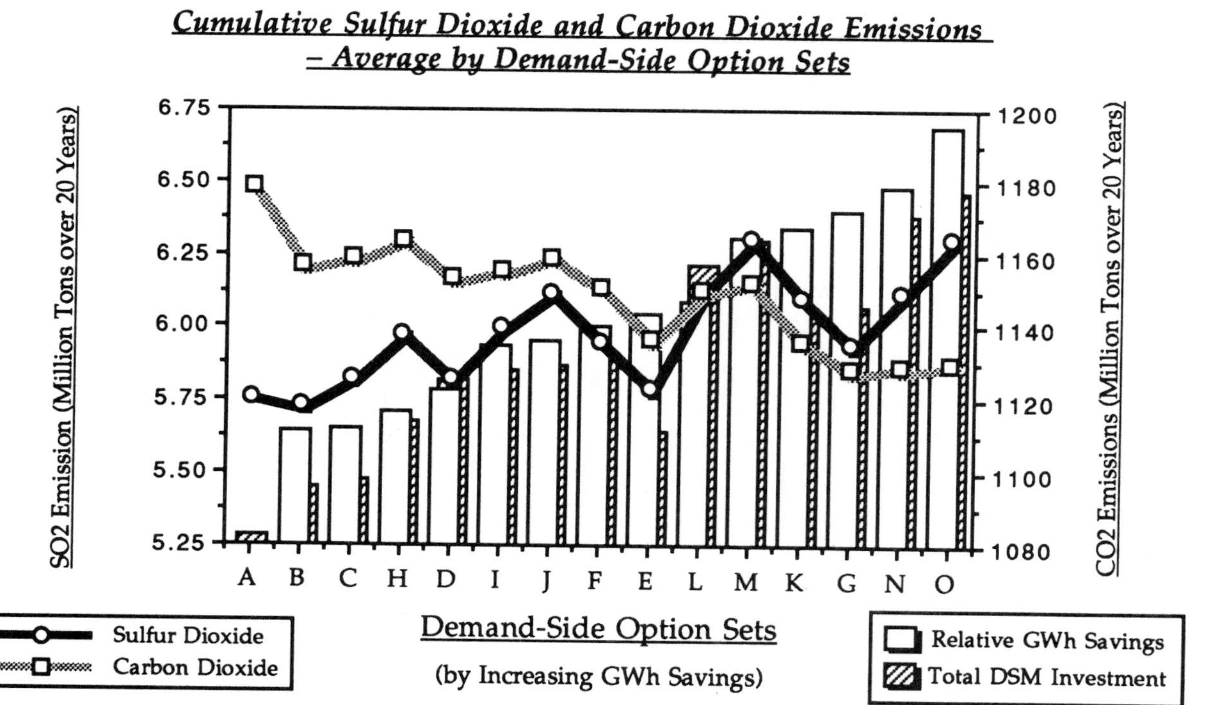

Figure 6. SO_2 and CO_2 impacts by DSM option sets.

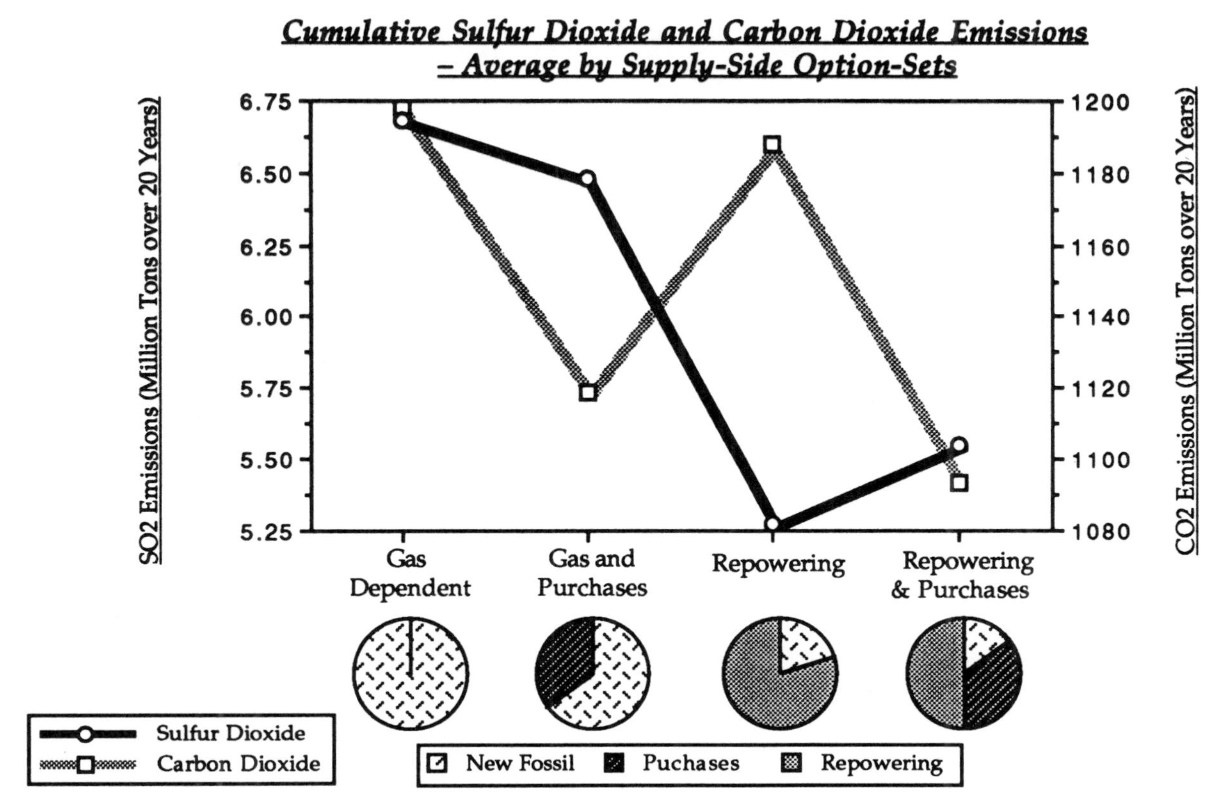

Figure 7. SO_2 and CO_2 impacts by supply-side option sets.

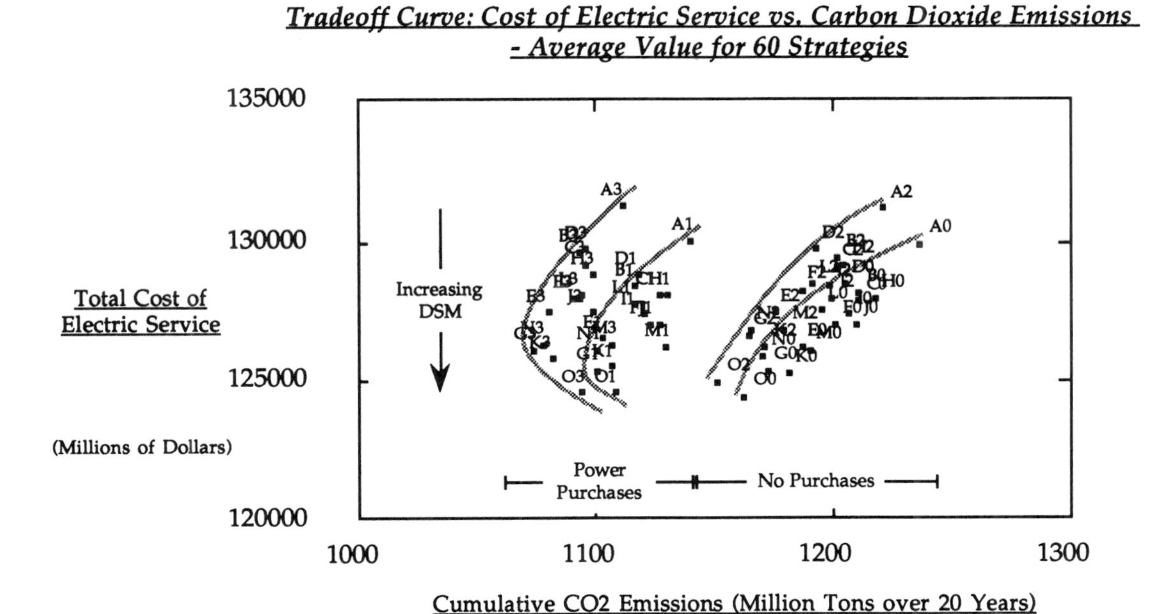

Figure 8. Cost versus carbon dioxide tradeoffs.

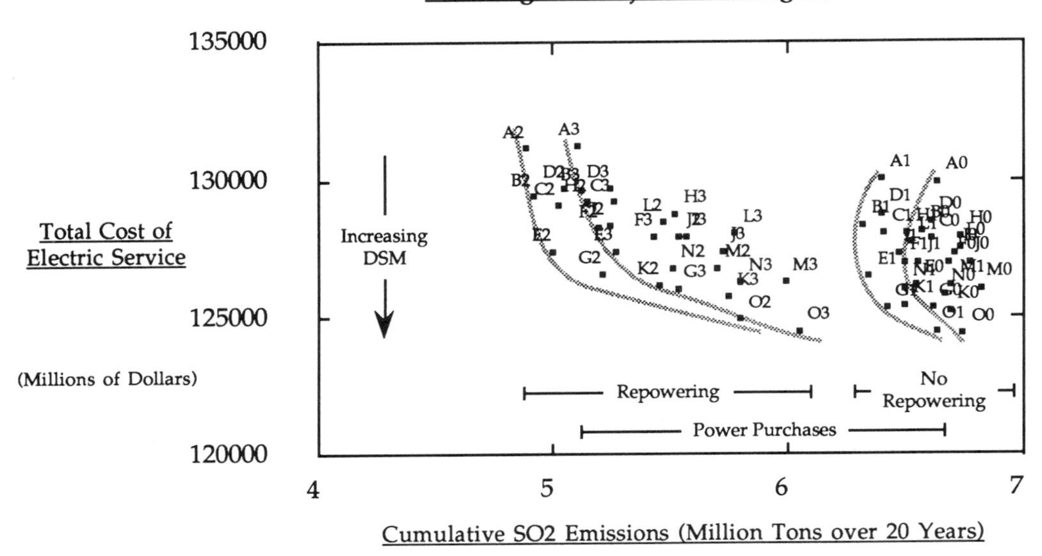

Figure 9. Cost versus sulfur dioxide tradeoffs.

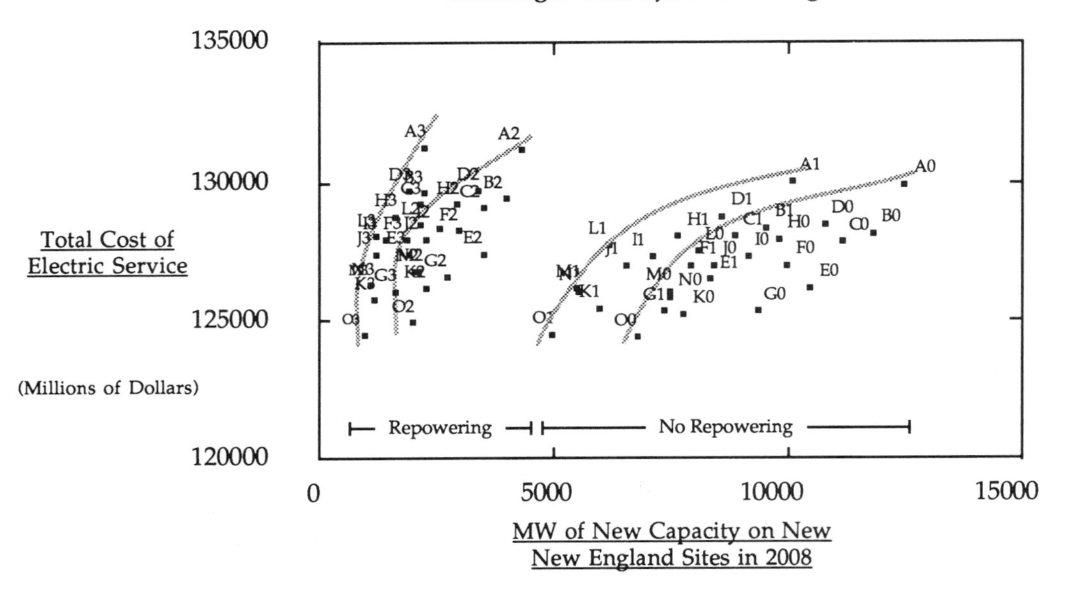

Figure 10. Cost versus new site megawatts tradeoffs.

Since the sensitivity to cost depends predominantly on the level of demand-side activity, we can look at local versus regional versus global environmental tradeoffs, knowing that the higher the level of conservation the lower the total cost of providing the electric service. Looking first at global versus regional environmental tradeoffs, what strategies dominate others for both SO_2 and CO_2?

Figure 11 shows the SO_2/CO_2 tradeoff. It is clear, within this limited set of strategies, that the way to attain the greatest reductions for both SO_2 and CO_2 emissions is to repower old units, and contract for nongreenhouse technologies through power purchases. Within this cluster it appears that "B" and "E" DSM option sets give the best additional reductions. Only the "E" DSM option set has moderately high levels of DSM, and offers an additional cost incentive.

Figure 12 shows the local versus regional tradeoff. Although different in appearance from the previous figure, the strategies that lie closest to the lower left hand corner have not changed. The repowering and purchases with moderate DSM still outperforms the others on emissions alone.

Figure 13 shows the third and final environmental tradeoff—local versus global impacts. Similar to Figure 11, the repowering and purchases cluster of strategies lies in the lower left corner. This is a relatively tight cluster where there is very little difference between DSM option sets. However, the order within the cluster has flipped, with the moderate and high DSM strategies situated along the frontier, rather than the lower DSM strategies, as seen in the CO_2/SO_2 and SO_2/new site megawatts tradeoffs.

General trends between the major components of strategies and the primary issues can be summarized from the tradeoff analysis. Table 3 summarizes these trends, with a plus sign indicating that the presence or increase in that component served to better performance (e.g., to lower cost or emissions). Increased DSM is the primary factor in lowering costs, with repowering slightly increasing costs when present. Power purchases are the biggest factor for reducing CO_2 emissions, with DSM and repowering being additional positive factors. Interactions between power purchases and DSM relative to CO_2 emissions were observed.

Repowering was the single most effective method for both reducing SO_2 emissions and reducing the number of new generating sites. Strategies incorporating higher levels of DSM (thus postponing repowerments) tended to have higher SO_2 emissions for the strategies evaluated, with power purchases (again postponing repowerments) also slightly increasing SO_2 emissions. Because of the reliability block-based planning framework, both increased DSM and power purchases reduced the need for new and repowered generating units.

Discussion

In order to develop even better strategies for meeting local, regional, and global environmental goals as well as cost, the operational characteristics that support or detract from a strategy's performance must be fully understood.

New sites may be avoided with the help of all three of the options explored above. It is clear that the repowering of existing units, in this case constructing an entirely new plant on the existing site, is the most effective method of reducing the need for new sites. But promoting DSM and contracting for extraregional sources of generation could also reduce local land requirements for power generation.

The methods for reducing global environmental impacts are also clear. The nonfossil components of power purchases and GWh reductions due to increased demand-side management both play important roles in reducing carbon dioxide emissions. The difference between the two components is that one eliminates the GWhs of electricity demanded, while the other meets the remaining GWhs with noncarbon generating technologies. The difference between reducing electric demand and meeting that demand with an alternative generating technology is important in understanding strategies that will reduce all environmental impacts.

The greatest reduction in regional environmental impacts, i.e., SO_2 emissions, comes from repowering in this set of scenarios. The repowering option has such a strong showing because it is the only one that displaces existing dirty capacity. A planning logic that invests in new capacity only to meet reliability concerns cannot get old, dirty capacity off line. Table 4 shows the age and efficiency breakdown for New England's existing fossil-fueled generating capacity. Looking at oil 6–fired capacity, by the year 2000 there will be 46 units, representing over 4200 MW of capacity that will have been in operation for over 30 years. These units have an average efficiency of 32.4 percent, and are old enough that many are exempt from existing environmental regulations.

On a MW per MW basis, the combustion efficiency ($\approx$ 45 percent) of new gas-fired combined-cycle units is much better than existing oil 6–fired units. Not only do these new technologies use one quarter less fuel to generate the same GWh, their environmental characteristics per unit of fuel burned are far superior. Compared to the older units, a gas-fired combined-cycle unit emits no sulfur dioxide, NO_x emissions are reduced by a factor of 12, and particulate emissions are nearly eliminated.

Furthermore, since fuel costs are reduced by such a large margin the plant is less expensive to operate. So whether the new technology is added to meet new demand, or to replace an old unit on a megawatt-by-mega-

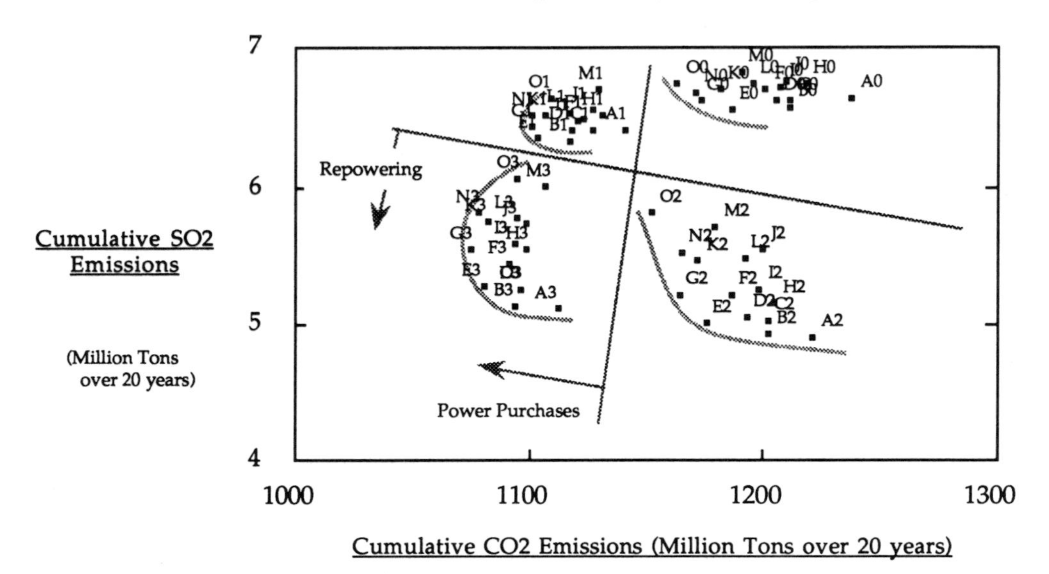

Figure 11. Sulfur dioxide versus carbon dioxide tradeoffs.

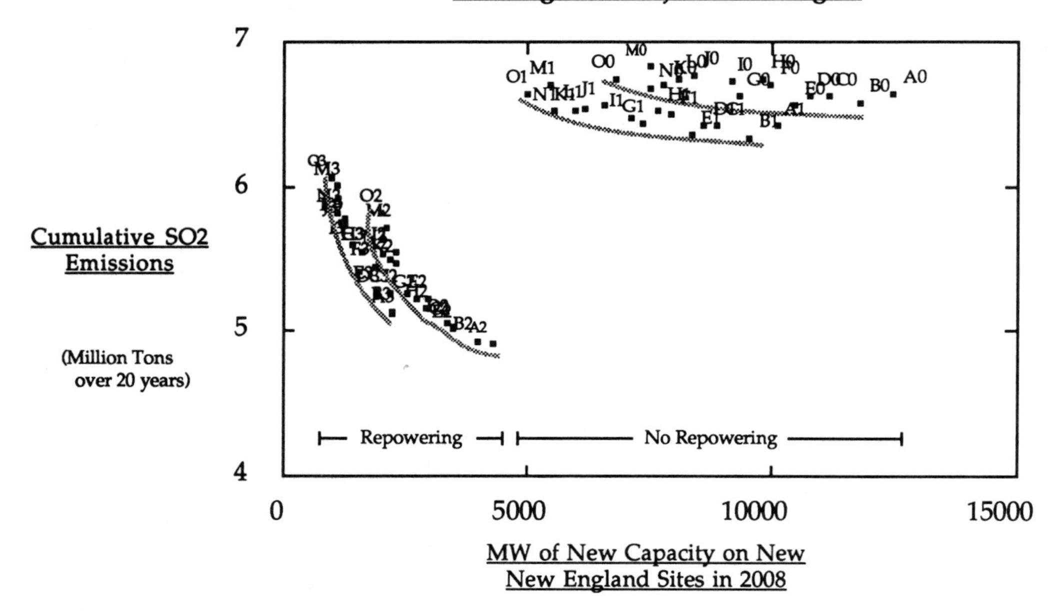

Figure 12. Sulfur dioxide versus new site megawatts tradeoffs.

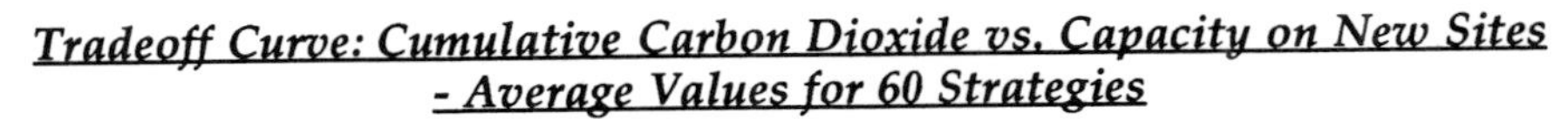

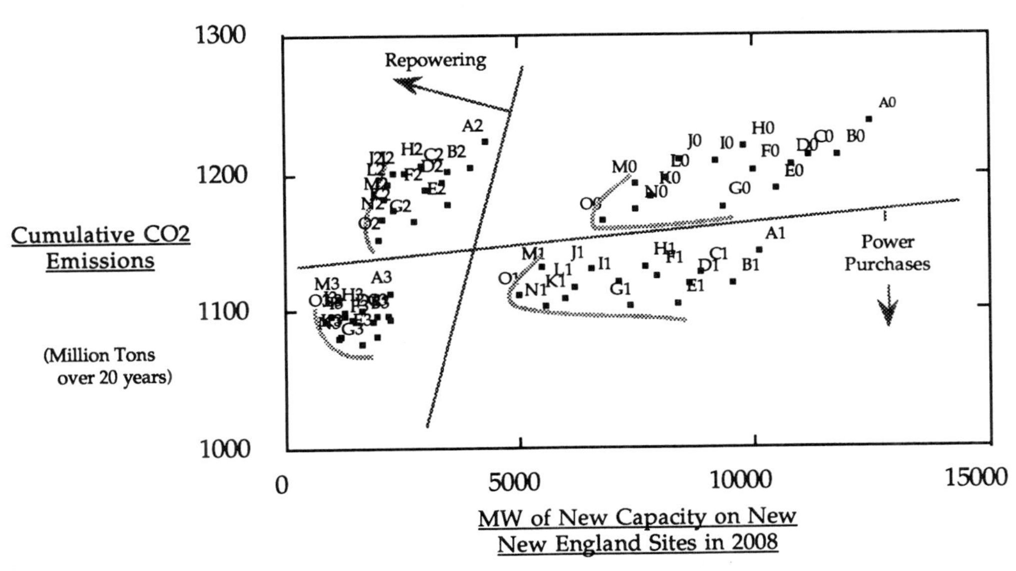

Figure 13. Carbon dioxide versus new site megawatts tradeoffs.

Table 3. Performance characteristics of strategies.

Issue	*Attribute*	*Demand-Side Management*	*Unit Repowering*	*Power Purchases*
Cost	Total Cost	+	≈	≈
Global Environment	CO2	+	≈	+
Regional Environment	SO2	−	+	±
Local Environment	New Sites	+	+	+

Table 4. New England's fossil-fueled generating capacity.

	Number of Units	Rated Capacity (MW)	Forced Outage Rate	Average Heat Rate (Btu/kWh)	Average Efficiency (%)
Coal-Fired	15	3151	0.086	10025	34.2
Oil 2-Fired	96	2079	0.109	11825	29.8
Gas-Fired	8	165	0.095	10491	32.6
Oil 6 Fired Generating Capacity					
Time Period	Number of Units	Rated Capacity	F.O.R.	Average Heat Rate	Average Efficiency
Prior 1951	16	602	0.059	12922	27.6
1951 - 1955	10	420	0.053	12445	28.0
1956 - 1960	11	1120	0.068	10372	33.0
1961 - 1965	6	1211	0.082	9720	35.0
1966 - 1970	3	930	0.084	10533	33.4
1971 - 1975	7	2915	0.080	10351	32.9
1976 - 1980	2	1162	0.082	9748	34.9
1981 - 1985	-	-	-	-	-
1986 - 1990	-	-	-	-	-
All Years	55	8360	0.073	10870	32.1
Total Non-Nuclear Thermal Generating Capacity					
All Units	174	13755	0.093	10956	32.5

watt basis, it will be run more than the older, less efficient unit.

While demand-side programs were effective at reducing costs, carbon dioxide emissions, and the need for new sites where the cost and environmental characteristics were more homogeneous across all GWh, they were ineffective at reducing SO_2, NO_x, and particulate emissions. The inability of the higher DSM strategies to reduce SO_2 emissions is explained not by the zero-emissions characteristics of individual DSM options but by these strategies' ineffectiveness at displacing the dirty oil 6 generating capacity in the heart of the dispatch order. The strategy of increasing end-use efficiency, thus displacing a generic GWh of demand, was unsuccessful in displacing a specific GWh of electricity production by an old plant.

Cumulative sulfur dioxide emissions were higher with high DSM strategies for two reasons: the failure of demand-side programs themselves to displace the dirtiest GWh being produced by the system, and the effectiveness of DSM in reducing the need for additional supply-side resources, technologies that would have displaced the dirtiest GWhs.

What we see is that the best overall environmental strategies employed components that increased fuel-use as well as end-use efficiency. Demand-side options contributed a great deal to reducing costs, carbon dioxide emissions, and the need for new sites. Power purchases contributed predominantly to reducing carbon dioxide emissions and the number of new sites. Repowering, by design, displaced the oldest and least efficient units, utilizing those existing sites to maintain and increase the region's generating capacity, thereby reducing SO_2, NO_x, and particulate emissions as well as siting requirements.

Which characteristics of the strategies we evaluated were detrimental to the reduction of environmental impacts? Clearly, the most detrimental component was the nonrepowering of units, or life extension. Where repowering was not an option, the less efficient, dirtier old units remained in the generation mix. Their fuel consumption and environmental emissions were higher compared to all the other strategies. In addition, these old sites could not be used to meet additional capacity needs, thereby increasing the number of new sites needed to be found. If SO_2, NO_x, and particulate emissions were the only environmental issues of merit, then alternative fueling of these units with low-sulfur oil 6 would be a primary consideration.

The only detrimental aspect of increasing demand-side efforts was that it tended to crowd out the introduction of more efficient and cleaner generating technologies. This impediment to increasing fuel-use efficiency occurred because within the strategies considered, old capacity was only replaced or supplemented in order to increase total capacity requirements. An alternative decision rule, where old units were replaced independent of load growth or the level of DSM, would have eliminated the end-use versus fuel-use efficiency tradeoff observed with increased demand-side management. What impact this would have on costs and reliability remains to be determined.

Conclusions

System-wide evaluation of strategies aimed at meeting competing cost and environmental goals is necessary if robust strategies are to be identified. In New England, the tradeoffs between end-use and fuel-use efficiency must be reconciled if diverse environmental goals are to be effectively met. Examining the technological characteristics of individual options will not reveal these tradeoffs.

Due to the region's aging capacity, significant improvements in fuel-use efficiency can be made. The capital costs of replacing older units are offset by the significant savings in fuel from increased power plant efficiencies. The improved environmental characteristics of the new technologies are an added benefit to upgrading the region's generating capacity.

The environmental and cost-related benefits of demand-side programs make them an important part of any coordinated strategy. However, too great an emphasis on end-use efficiency can inhibit improvements on the supply-side. Increasing coordination, instead of competition, between options is needed to increase end-use and fuel-use efficiencies, to meet cost goals, and to improve environmental quality. Both decreasing the total electricity needs of the region and improving the way it is generated are equally important goals.

Developing and evaluating such coordinated strategies in conjunction with a group of the region's decision makers increases the likelihood that coordinated action can occur. The set of results presented in this paper identifies some of the most likely components of such an integrated strategy. Strategies that focus further on coordinating end-use and fuel-use efficiency improvements are being developed through this process.

References and Acknowledgments

Details of the modeling assumptions used in this analysis are summarized in the February 1990 Scenario Set Background Packet, which is available from the authors. NEPOOL, individual utilities, and other regional organizations provided the raw information upon which these scenarios were based. This work was inspired by the late Fred Schweppe, and has benefitted from the freely shared wisdom of David White, Richard Tabors, and the many members of the advisory group for the New England project. Funding from a consortium of the region's electric utilities and businesses is gratefully acknowledged.

Transitional Strategies for Emission Reduction in Electric Power Generation

Richard D. Tabors
Burt L. Monroe III
Laboratory for Electromagnetic and Electronic Systems
Technology and Policy Program
Massachusetts Institute of Technology
Cambridge, MA

Introduction

The electric power sector in the United States accounts for roughly one third of the country's annual emissions of CO_2. Worldwide, the electric power sector accounts for a lesser but rapidly growing percent of total CO_2 emissions. As a long run strategy, electricity generation will probably need to focus on either non-carbon-based fuels such as nuclear energy or solar energy, or will need to be based on recycled carbon such as that from biomass. The ultimate challenge is to develop the socially acceptable technologies to achieve dramatically lower emissions by the second or third decade of the next century. To do so will require the development of acceptable new technologies, or the redefinition of acceptability, and the restructuring of the electric power system as we know it today along one of several feasible paths.

While the ultimate solution in all likelihood lies in new technologies, there is a need to understand the potential for existing technologies and/or evolved technologies to significantly reduce emissions over the next decade or two.[1] It is important to ask the following questions:

- What are the relative roles of electricity generation technologies and end-use technologies in serving the demand for electricity-related services?
- What current generation technologies could be effectively employed to reduce CO_2 emissions if cost per kWh were not the primary objective?
- What tradeoffs between cost and emission levels are possible with existing technologies?
- What combinations of technological changes (fuel switching and increased efficiency, for instance) could provide the greatest improvements in emission reduction with the lowest total cost?
- What technological alternatives could provide the most robust transitions to what we are now discussing as the long-term, near-zero net emission technologies?

Realities of the Electric Power System

The objective of this paper[2] is to evaluate a range of transitional technologies that could play a major role in the reduction of CO_2 in the near term. The paper focuses on fossil and nuclear-based generation technologies and load reduction (conservation) as likely short-run alternatives. It evaluates a set of strategies available for achieving a change in the operating behavior of the electric industry ranging from a carbon tax to environmental dispatch modification. The units of comparison presented are cost of generation and absolute volumes of greenhouse gases and acid rain emissions.

The study is based on evaluation of the operating characteristics of the electric utility system, not upon simple technology substitution. System operation is simulated to capture the dispatch effects of changes in both technology and input prices. While it is often convenient to think of "changing out" an oil or coal steam unit for a natural gas fired unit, in reality, the old and new coexist on the system with the new fuel efficient (and less environmentally degrading) facility appearing lower in the loading order. The question is how much lower in the order and to what overall effect on the operation of the system?

A significant underlying hypothesis of this effort is that, because of the current structure of the utility system in the United States, there are regions in which it may be possible to reduce the emissions with little if any increase in the average cost of energy delivered to the end user. This occurs because of increased efficiency of power production and an increase in the ratio of hydrogen to carbon in the fuel. In a transitional time frame, countervailing forces are at work. Demand reduction minimizes consumption but reduces the need to change over to newer, more fuel efficient technologies.

This can be best illustrated through an examination of the "pollution equation" often used in the global warming debate.[3] This equation states that carbon emissions from electric power are determined as follows:

$$\text{Carbon} = \left(\frac{C}{GWH}\right)\left(\frac{GWH}{GNP}\right)\left(\frac{GNP}{\text{Population}}\right)(\text{Population})$$

where C/GWH is "marginal carbon emissions,"[4]
GWH/GNP is "energy intensity,"
and GNP/Population is a measure of standard of living.

This equation is often used to point out the futility of emissions abatement strategies given likely increases in population and the desire for increases in standard of living. It is possible, however, that the present value of the energy intensity term is much larger than technologically or economically necessary—implying potential for reductions through end-use efficiency and conservation measures. The marginal carbon emissions term is *infinitely* larger than technologically necessary. Not only can the marginal carbon emissions term be reduced, but it can be lowered to zero through the use of nonfossil fuel sources such as nuclear or solar.[5] A more detailed examination of marginal carbon emissions can provide insights into the effectiveness of various strategies for reducing emissions.

Let us disaggregate the marginal carbon emissions term, examining the carbon emitted over any particular time period in a particular electric power system. The system emissions are the sum of emissions from each individual technology type:[6]

$$\frac{C}{GWH_{sys}} = \sum_{i=1}^{N}\left[\left(\frac{C}{GWH_i}\right)\left(\frac{GWH_i}{GWH_{sys}}\right)\right]$$

where N = Total number of technologies within the system,
GWH_i = Energy generation of technology i,
and GWH_{sys} = Total energy generation of the system.

The importance of the (GWH_i/GWH_{sys}) term, which represents the percentage of the total system energy that is derived from any particular technology, is evident. If the level of utilization of a technology is low, a low (C/GWH_i) term will not have much effect on overall system emissions. The level of utilization of a particular technology is a complex function of the system capacity mix, the system reserve margin, the outage rates of all technologies, the character of system demand, and the relative marginal costs of various technologies. In general, the generating units with the lowest marginal cost are dispatched first (if available), followed by more expensive units, until system demand is met. In this way, the costs of electricity generation (given any particular capacity mix) are minimized. Since such factors as demand, unit availability, and fuel prices vary continually, calculating such outputs as generation costs and environmental emissions is a nontrivial task, even when the values of the input variables are known. Of particular note is the fact that the relationship between capacity and generation is highly nonlinear (for instance, a system with 40 percent coal capacity might generate 70 percent of its energy from coal). More detailed examination of these factors is included within the modeling efforts described below, but is beyond the scope of this discussion.

It can be noted, however, that change of these factors is traditionally driven by shifts in the relative costs of technologies and by load growth or plant retirement, which facilitate the construction of new capacity. Additional changes can be driven by artificially changing the costs of various technologies (taxes, changes in dispatch rules) or by artificially inducing changes in capacity mix (early retirement of existing plants). Load reduction, while reducing the (GWH/GNP) term of the original equation, can suppress changes in the system which might otherwise occur. The balance of these two factors may not be easily predictable.

How low can the (C/GWH_i) term be? As stated previously, for many nonfossil technologies, such as nuclear, hydroelectric, or solar, the marginal carbon emissions are zero. For those technologies that utilize an input fuel, the (C/GWH_i) term can be expressed:

$$\frac{C}{GWH_i} = \left(\frac{C}{MMBTU_i}\right)\left(\frac{MMBTU_i}{GWH_i}\right)$$

where $MMBTU_i$ = the amount of energy contained in fuel i,[7]
and $MMBTU_i/GWH_i$ = the heat rate of technology i.[8]

The C/MMBTU of a fuel is a function of carbon-hydrogen ratios within the fuel, with coal having the highest value (~ 0.03 tons C/MMBTU), followed generally by oil (~ 0.022) and natural gas (~ 0.017).[9] When contemplating a "fuel switching" strategy, however, one must also account for the efficiency (or heat rate) of the technology that utilizes the fuel. In general this leads to carbon efficiencies in the range of 250–400 tons C per GWH for conventional coal technology, the range resulting from a mix of inefficient older plants and more efficient newer plants. Existing oil capacity has typical values in the range of 200–300 tons C per GWH, with natural gas in the range 175–225 tons C per GWH. Relative to existing coal, new combined-cycle fossil fuel plants can have improvements on the order of 10 percent (gasified coal), 30 percent (oil), or 50 percent (natural gas). Note that sulfur scrubber retrofits *raise* this value by approximately 10 percent, due to loss of efficiency.

In general, then, it can be stated that the most effective strategies will be those which strike an optimum balance between the displacement of high-emission generation with low-emission generation and load reduction. Closely related to this argument is its corollary that, given the economic/financial structure of the US electric power system and its present operating rules, some policy options will achieve the desired greenhouse gas emission reduction objectives more cost effectively than others. Given the logic of today's dispatch centers, options that change the relative prices of inputs—fuels, for instance—can easily be incorporated while those that change the basic rules (dispatch according to emissions instead of costs) are far more difficult and costly as, in the short run, they would require the basic reprogramming of the dispatch centers themselves.

The Methodology

Within the US electric utility industry there are models and data bases that have been legitimized by their industry acceptance. By using these accepted tools, discussions can focus on inputs and outputs of the analytic exercise rather than on uncertainties about model structure. This is a nontrivial concern that has limited the usefulness of many previous modeling efforts. These have typically been so large in scope that actual system structure is buried in many levels of assumed aggregation, causing debate to focus on assumptions rather than on results.[10] Given this concern, this study utilized the Electric Power Research Institute's (EPRI) Electric Generation Expansion Analysis System (EGEAS) modeling system, the EPRI Regional Systems (ERS) Database, and the EPRI Technology Assessment Guide (TAG) in order to conduct a structurally detailed analysis of CO_2/electric power interactions in a few geographic regions over a limited time frame.

The Model

MIT and Stone and Webster Engineering Corporation developed the Electric Generation Expansion Analysis System (EGEAS) for EPRI in the early 1980s. It is now in use at over 100 utilities in the US and abroad.[11] The EGEAS user provides the model with information about the individual plants within an electric power system and potential alternatives for future capacity expansion. Among this data are size and age of plants, performance data (heat rates, forced outage rates, etc.), capital costs, operating costs, fuel use, environmental characteristics, and financial data. Data about external factors, such as fuel prices and inflation rates, are also provided. EGEAS can then be used to determine cost-optimal plans for future capacity expansion[12] and to simulate the system operation over time. The modeling outputs—production costs, environmental emissions, fuel use, etc.—are reasonably accurate due to a sophisticated production costing algorithm within EGEAS, which accounts for many power system subtleties without the prohibitive computational requirements of chronological models.

The Database

Because the US utility system is effectively fully interconnected, isolating one utility or even one actual region for analysis is a major task. During the period of the so-called energy crisis, EPRI began the development of a set of synthetic regional utility databases that could be used for technology and policy analyses. The current version of the EPRI Regional Systems (ERS) database[13] is being used in this study in order to present results based on system data that are accepted by those in the industry. This analysis covers the Northeast region of the United States, which includes New England, New York, New Jersey, Delaware, eastern Maryland, eastern Pennsylvania; and the East Central region of the United States, which includes the heavy coal-burning region of Indiana, Ohio, West Virginia, Kentucky, lower Michigan, western Pennsylvania, western Maryland, and western Virginia.

The primary uncertainties considered within the study were fuel price and load growth. An attempt was made to choose a set of possible futures which would provide a reasonable resolution with which to view possible outcomes, without prohibitively increasing computational requirements. Two fuel price trajectories—base and high—and four load growth trajectories—low, base, high, and very high—were selected. Each possible pair of these uncertainties constitutes a "future."[14] The base fuel price was taken from Data Resources Institute (DRI) forecasts[15] and adjusted regionally according to guidelines in the EPRI Technology Assessment Guide (TAG).[16] The high fuel price uncertainty represented significantly higher prices for both oil and gas. The base load growth uncertainties were taken from the ERS.[17] Low growth uncertainties were 1 percent lower than the relevant base growth; high growth 1 percent higher; very high growth 3 percent higher.

Technology Options

The generation technology options were taken from the TAG. These included: coal (28 technologies evaluated); liquid and gas (15 technologies evaluated); and nuclear (3 technologies evaluated).

The analysis did not include evaluation of nondispatchable options such as solar and wind[18] and did not include the addition of storage technologies. Those options that appeared in the optimal pathway (a small subset of those considered) for either region or both were:

	Size MW	Cost $/kW	Heat Rate BTU/kWh	MCE Tons C/GWH
Pulv. Coal w/scrubber	500	1281	9700	300
IGCC	800	1467	9000	257
Advanced CT, oil-fired	140	385	11100	257
Advanced CT, gas-fired	140	373	11500	200
Advanced GTCC, oil	210	531	7360	174
Advanced GTCC, gas	210	518	7514	130
Advanced LWR	1200	1524	10220	0

Strategies

The study defined seven policy strategies for analysis. These are:

Base: For each future an optimal (cost-minimizing) 25-year expansion plan was developed based on the regionally and technically available expansion alternatives. In most cases, the optimal plan involved early construction of natural gas-fired combined-cycle (GTCC) plants with an eventual switch to integrated gasifier combined cycle (IGCC) plants.[19] Oil-fired combined cycle was occasionally superior to natural gas in high fuel price futures. Some high growth futures also led to the construction of combustion turbines fired either by natural gas or oil. Nuclear was not an option in the base strategy. Early plant retirement was also not a possibility.

Nuclear (NUC): The nuclear strategy differed from the base in that nuclear options were offered as the only available baseload capacity option (coal options were removed) in the optimization runs. In the Northeast, the nuclear option was marginally more economical than the IGCC option, while marginally less so in the East Central. Capacity expansion followed similar trajectories with coal options generally substituted by nuclear while the role of the GTCC remained roughly the same.

No Nuclear No Coal (NNNC): In this strategy, the system was forced to choose only natural gas and oil based technologies, thus forcing GTCC and CT technologies.

Dispatch Modifier (DM): The optimal expansion pathway of the Base strategy was used to define the plant additions. The units were then operated according to carbon emissions instead of marginal cost.

Early Retirement (ER): For all coal plants, the existing operating life was reduced by 10 years (generally from 50 to 40). This required a substantial increase in construction, particularly in the early years and specifically in the coal dependent East Central region.

Carbon Tax (CT): A substantial tax on the use of fossil fuels was assumed based on carbon content ($5.70/ GJ for coal; $2.30/GJ for oil; and $1.10/ GJ for natural gas).[20] Capacity expansion patterns similar to the base strategy were observed, with the exception that fuel switches from natural gas to oil or coal occurred later in the planning horizon.

Conservation: No attempt was made to explicitly model conservation efforts. For each region, however, four possible load growth paths were defined and simulated separately in order to determine what benefits, if any, might be obtained through switches from high growth futures to low growth futures.

Optimization and Simulation

Within each study region, each possible combination of a future and strategy constitutes a "scenario." For each scenario, EGEAS defined an optimal expansion path, or timetable of technology choice over a 25-year study period.[21] The simulation was then run, providing various output attributes which could then be used for strategy evaluation. The primary output attributes of interest for this discussion are the costs and environmental emissions in each scenario. These included:

- Total discounted cost
- Annual costs in 1988 dollars
- Annual emissions of carbon, SO_2, NO_x, TSP, methane, and N_2O

Results

The results are divided into two sections and are based on the included figures. The first is a summary of the environmental performance of each strategy relative to total discounted cost. The second is a summary of the general conclusions of the study with regard to the underlying technological characteristics that determine the success of individual strategies.

Strategy Performance

The first set of figures (Figures 1–1, 1–2, and 1–3) present tradeoff data for the Northeast region. Each strategy is graphed for two attributes of primary concern for a single future (one combination of demand growth rate and fuel price increase)—the base case. Each figure represents the tradeoff between two attributes, chosen from carbon emissions, sulfur dioxide emissions, and cost. Emissions numbers represent total emissions over the

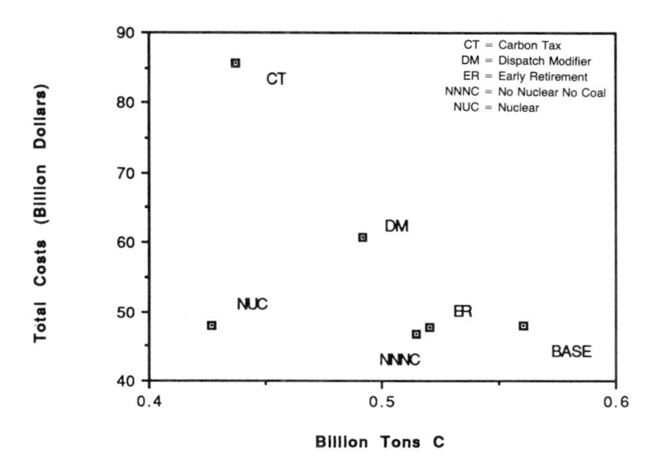

Figure 1-1. Carbon vs. cost (Northeast—base future).

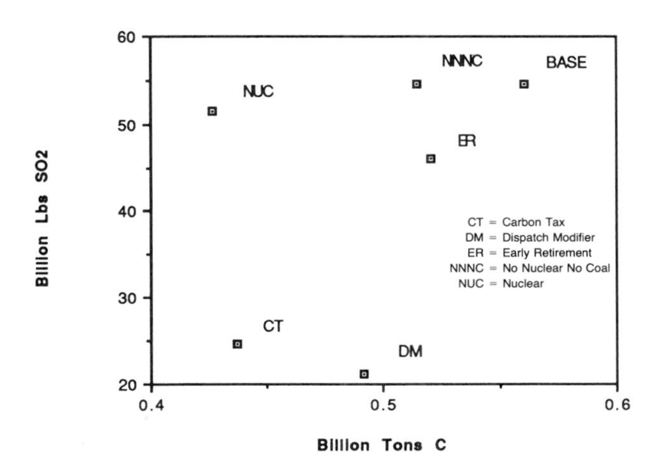

Figure 1-3. Carbon vs. SO_2 (Northeast—base future).

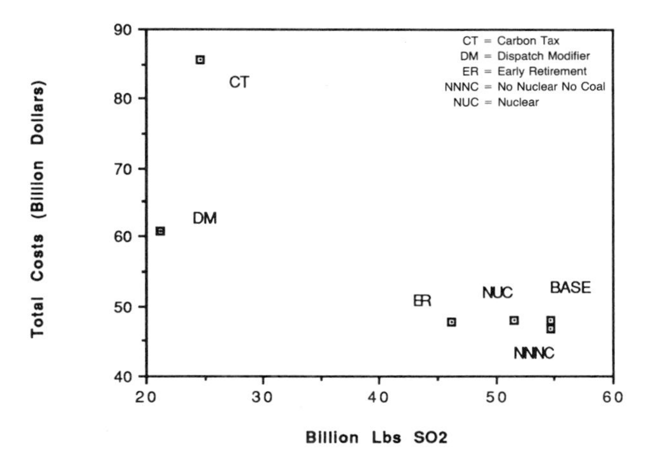

Figure 1-2. SO_2 vs. cost (Northeast—base future).

25-year study period. Costs are total net present value of system costs as faced by the utility.[22] For any particular graph, larger values on either axis are undesirable; the most desirable position for any particular strategy is in the lower left corner (approaching zero emissions and zero cost). Strategies that dominate other strategies—that have no other strategies both to the left of and below them—represent the optimum choices for those two attributes. Figure 1–1 illustrates the tradeoff between carbon emissions and costs. The nuclear strategy is dominant over all other strategies for these two attributes, given the assumptions of the study. Figure 1–2 illustrates the tradeoff between sulfur dioxide and costs. Of note here is that only the carbon tax and base strategies have clearly dominated positions—that is, for each of these two strategies, there is at least one other strategy that is superior in both attributes. Finally, the tradeoff between the two environmental attributes, and the clear dominance over the base strategy by all other strategies, is shown in Figure 1–3.

The second set of figures depicts the trajectories of carbon emissions over time for each strategy, again in the base case. Figure 2–1 shows the Northeast region and Figure 2–2 shows the East Central region. Differences in effectiveness across the two regions are apparent.

Base: The base strategy results in rapid increases in the emissions of carbon dioxide and total suspended particulates (TSP) due to the continued use of coal-fired units. Since new baseload units are IGCCs, however, some reduction in the levels of acid rain emissions, sulfur dioxide, and nitrogen oxides is realized. This is illustrated in Figures 3–1 (Northeast) and 3–2 (East Central), with each environmental emission shown relative to its 1989 level.

Nuclear: Based on the EPRI capital cost numbers used in this study, nuclear has a set of obvious advantages. While maintaining costs at or below the level of the base strategy, nuclear results in significant decreases in all atmospheric emissions. Carbon emissions decrease both relative to the base strategy and in absolute terms, while acid rain emissions (SO_2 and NO_x) are similar to those in the base strategy.[23] Nuclear is the only strategy that reduces all emissions in both regions regardless of the rate of growth in demand. This reduction is shown in Figure 4 and is particularly dramatic in comparison to the base strategy shown in Figure 3–1. The issues of nuclear waste and nuclear safety were not, however, taken into account. Similarly, the capital costs appear to be optimistic given recent experience, and social acceptability is still the major issue.

No Nuclear No Coal: This option results in modest environmental gains at approximately the same cost as the base strategy. This is best illustrated in Figure 1–1, where the no nuclear no coal option is seen to improve

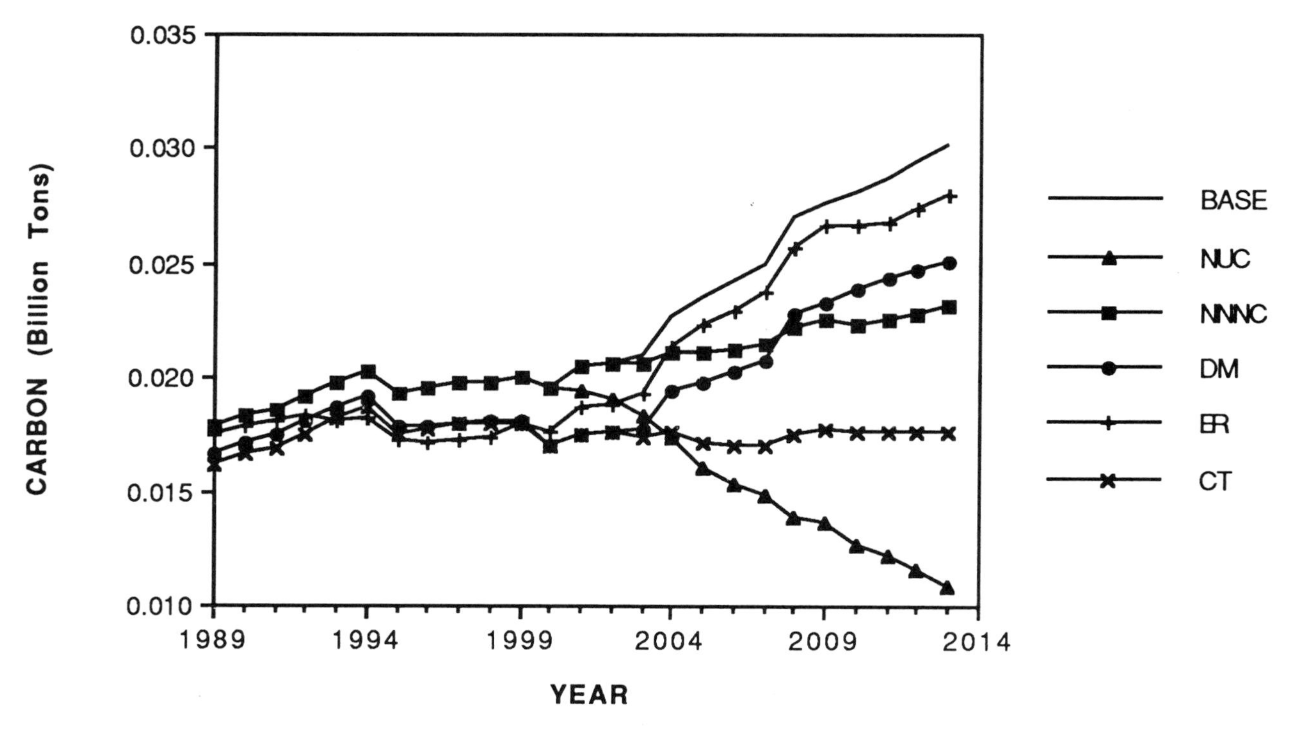

Figure 2-1. Northeast base case carbon emissions by strategy.

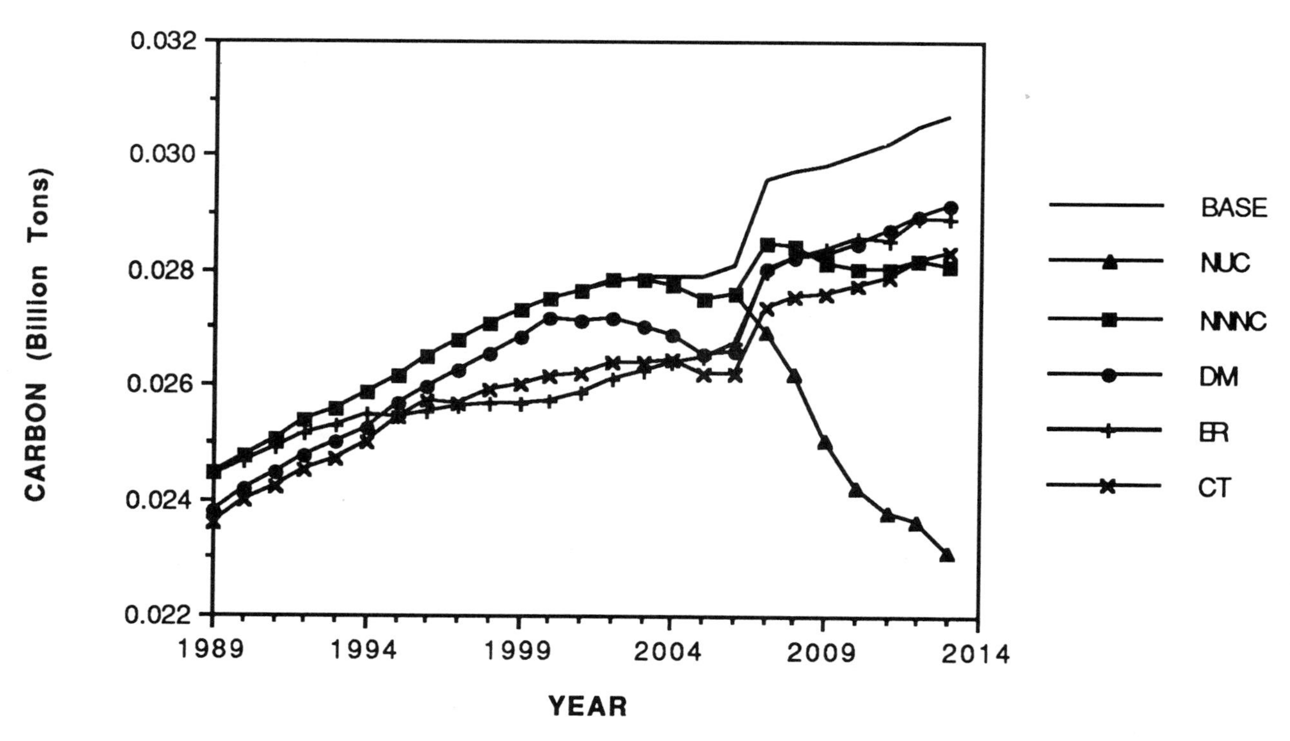

Figure 2-2. East Central base case carbon emissions by strategy.

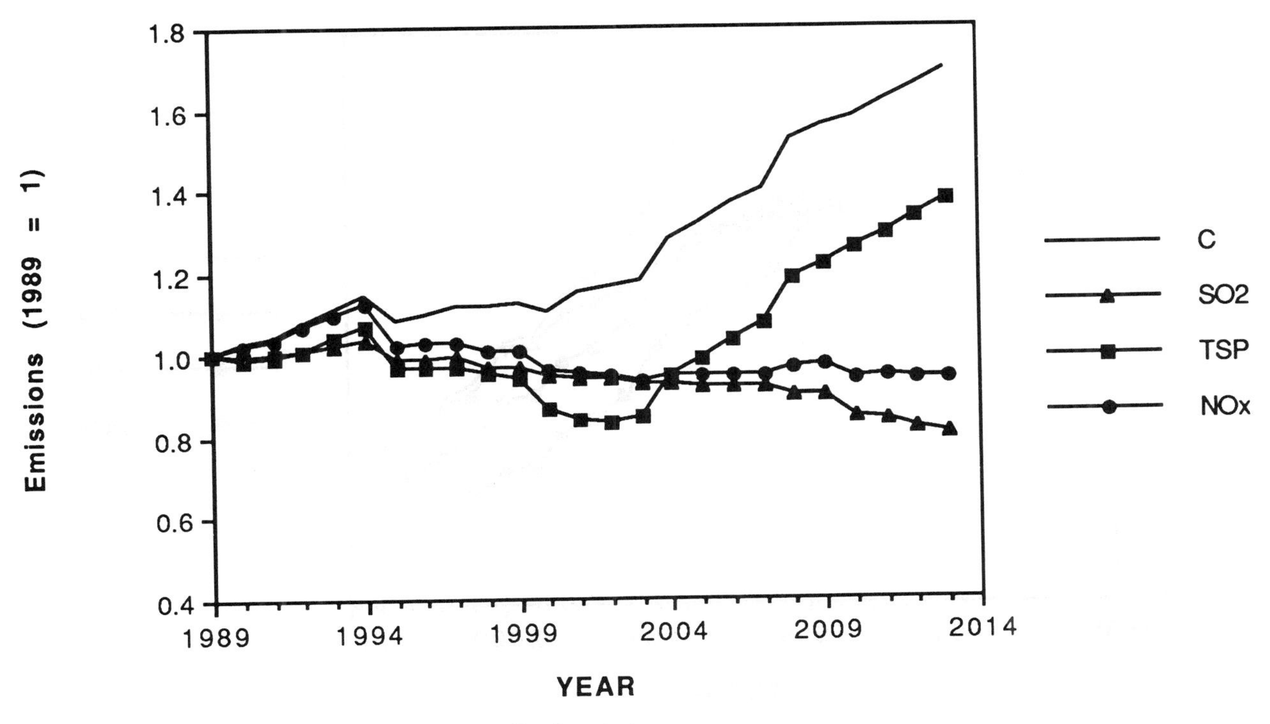

Figure 3-1. Northeast base strategy normalized emissions.

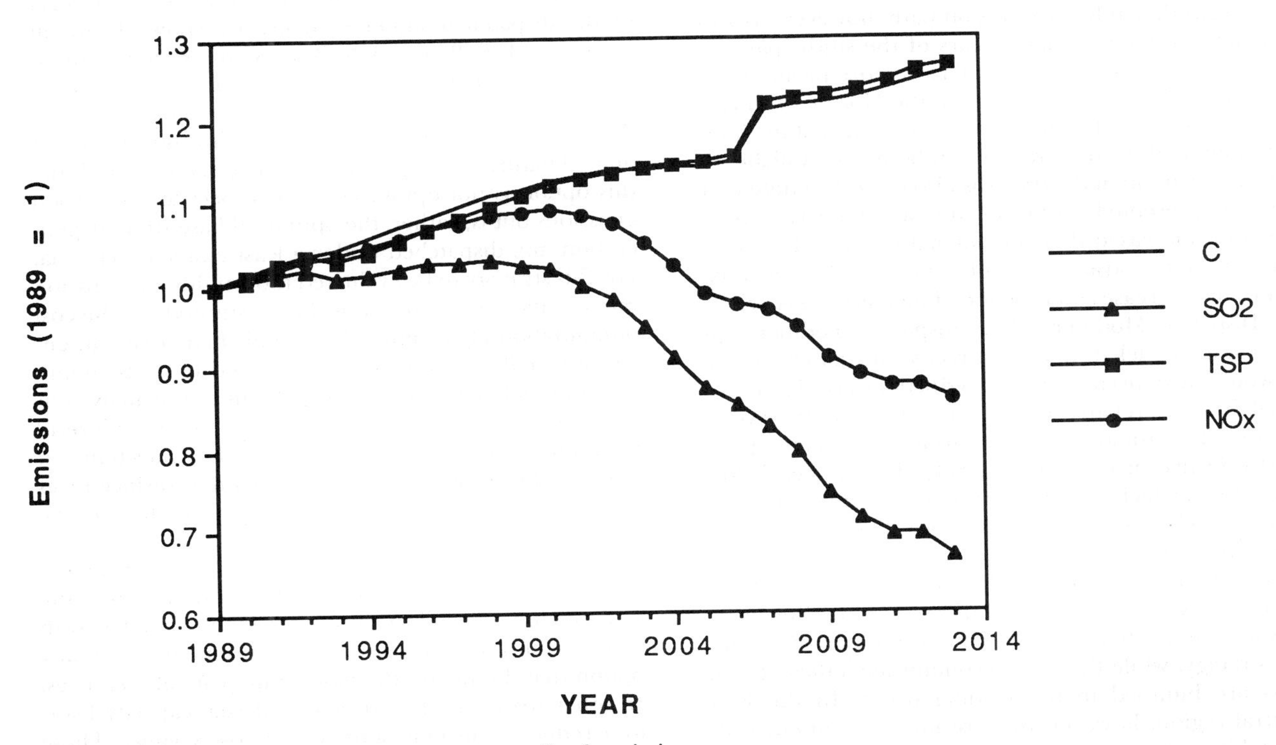

Figure 3-2. East Central base strategy normalized emissions.

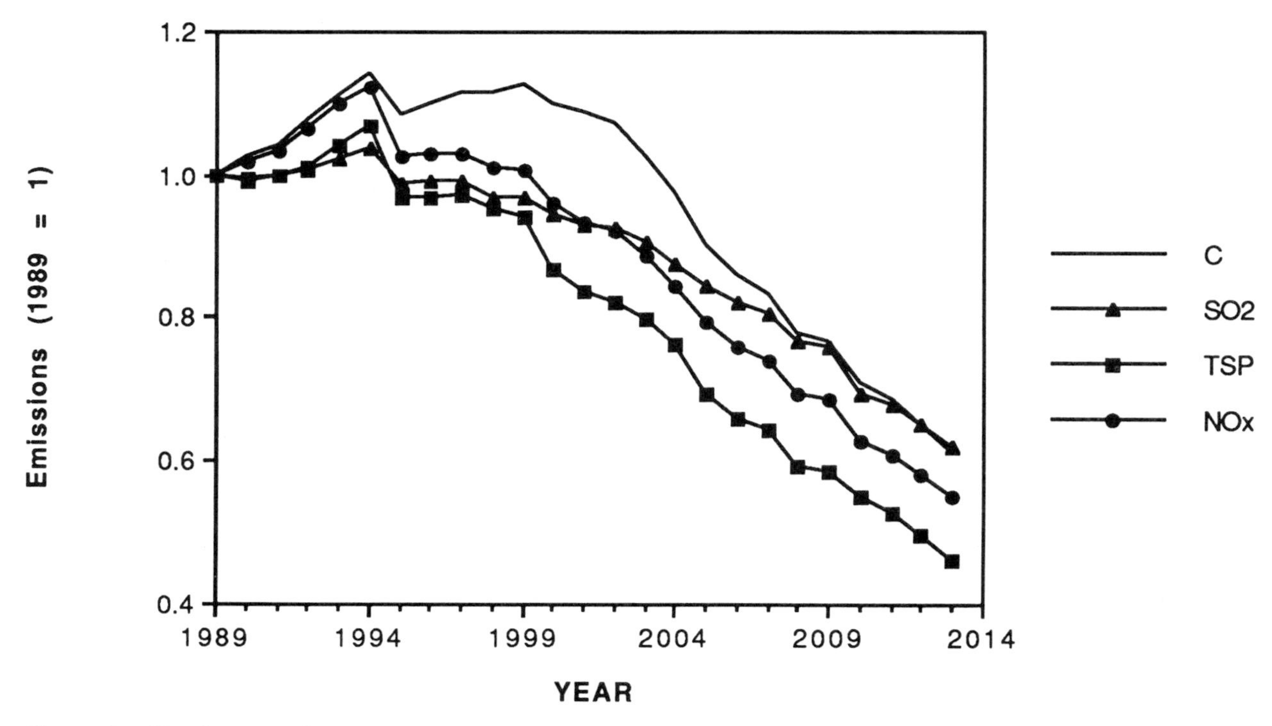

Figure 4. Northeast nuclear strategy normalized emissions.

the level of carbon emissions at essentially no cost. The natural gas dependent expansion path, however, creates reliability problems in later years of the study, particularly at higher growth rates.[24] This is more pronounced in the East Central region than in the Northeast due to the current mix of technology. It is unclear that the level of gas implied in this strategy can be made available to the electric utility industry. It is also critical to note that the costs presented do not assume any increase associated with increased demand for natural gas. The basic structure of the strategy would not change but the total cost would increase as a function of increasing gas prices.

Dispatch Modifier: The dispatch modifier approach, as modeled, is very attractive for all emissions. As would be expected, carbon reductions are significant in all futures, relative to the base strategy. In several futures, the dispatch modifier approach is among the most effective in reduction of SO_2. This is seen below in Figure 5, which depicts Northeast SO_2 emissions over time for each strategy in the base case. These gains come at significant cost in the Northeast region, which can be seen in the tradeoff curves of Figure 1. Because of the capacity mix in the Northeast, there are many relatively expensive units that are run for increased periods under this strategy, while the more economically efficient coal units are bumped in the loading order. In the East Central region, however, the cost impacts and environmental impacts are minimal. The tradeoffs for the East Central region base case, and the minimal cost impact of the dispatch modifier strategy, are shown below in Figures 6–1, 6–2, and 6–3. Because the system is dominated by coal, the dispatch modifier does not cause significant change in the loading order, but does allow for some improvements in emissions through the use of limited natural gas capacity. It must be remembered that this option is not optimized with respect to the carbon emissions but is, rather, the optimized base strategy generation mix dispatched under a least emissions criteria. For this strategy to be evaluated thoroughly, an emission minimization algorithm must be substituted for the cost minimization algorithm. This would then create an environmentally optimal plan as opposed to an economically optimal plan as the starting point of the analysis of any given future. Such a strategy may provide a useful operating rule for the short run in the transition but not for the long run (due to economic disincentives against investment in the construction of high marginal cost, low-emission capacity).

Early Retirement: The value of the early retirement option varies regionally and with the load growth of the system. Under all but the highest of the modeled growth scenarios in the Northeast, early retirement is the only option that dominates the base strategy in all measures. In other words, early retirement of coal capacity leads to a reduction in emissions with a cost savings. These gains are due to the displacement of inefficient, more

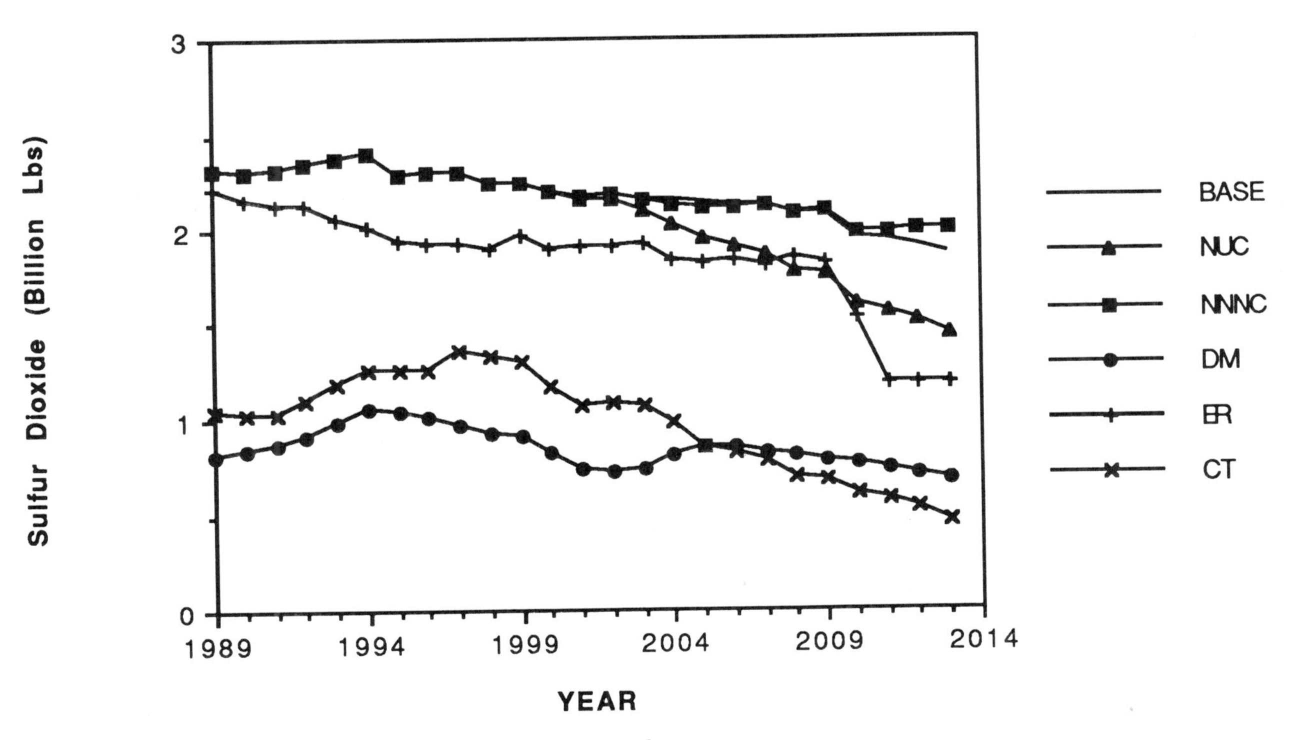

Figure 5. Northeast base case sulfur dioxide emissions by strategy.

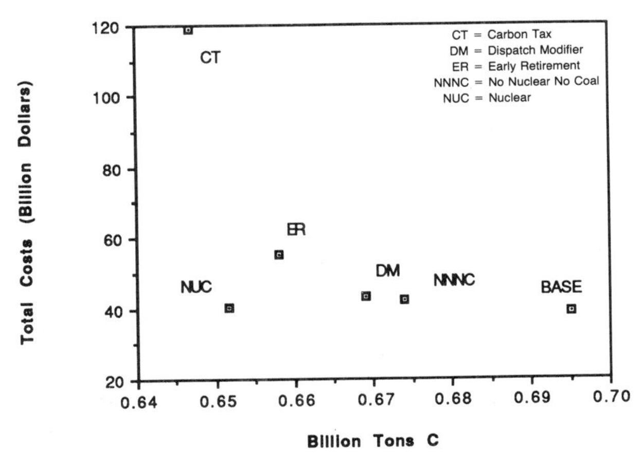

Figure 6-1. Carbon vs. cost (East Central—base future).

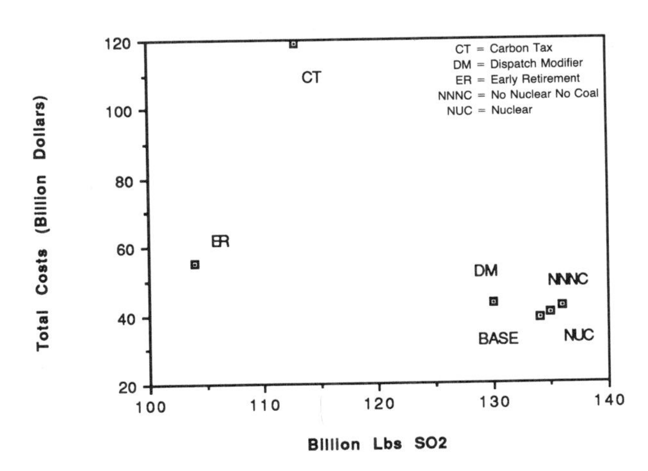

Figure 6-2. SO_2 vs. cost (East Central—base future).

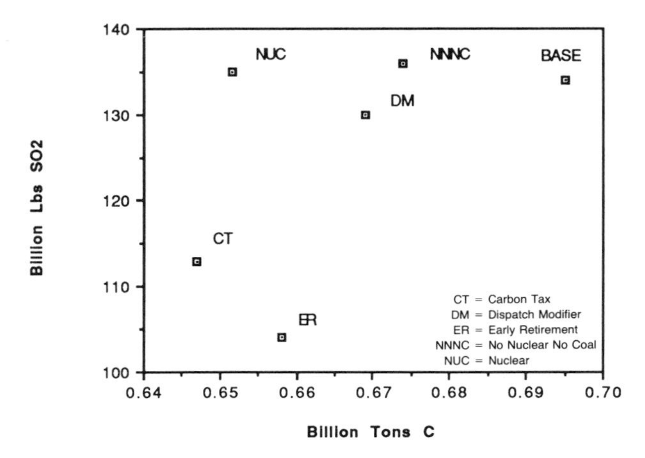

Figure 6-3. Carbon vs. SO_2 (East Central—base future).

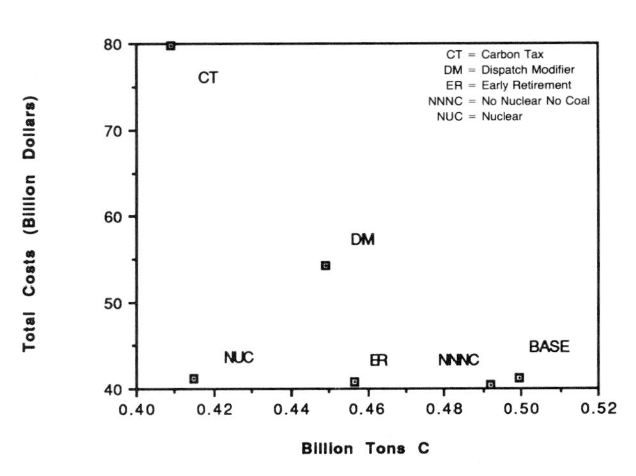

Figure 7-1. Carbon vs. cost (Northeast—low growth).

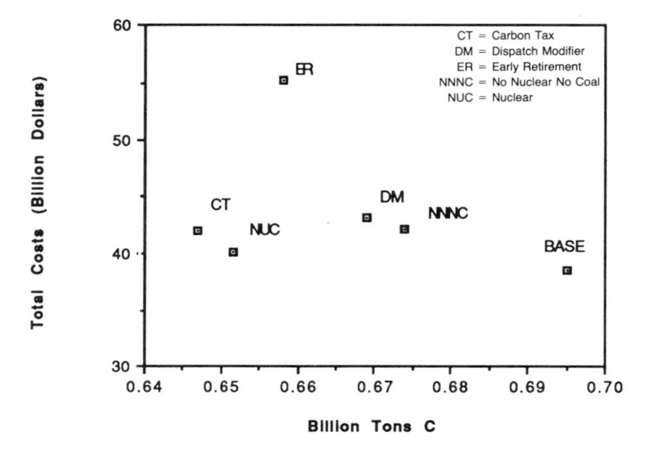

Figure 6-4. Carbon vs. net societal cost (East Central—base future).

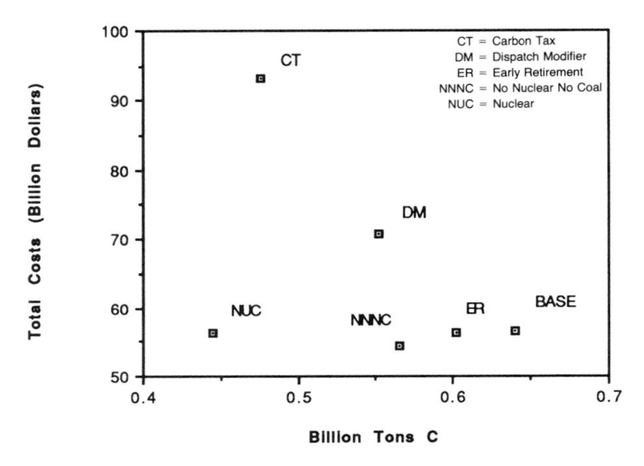

Figure 7-2. Carbon vs. cost (Northeast—high growth).

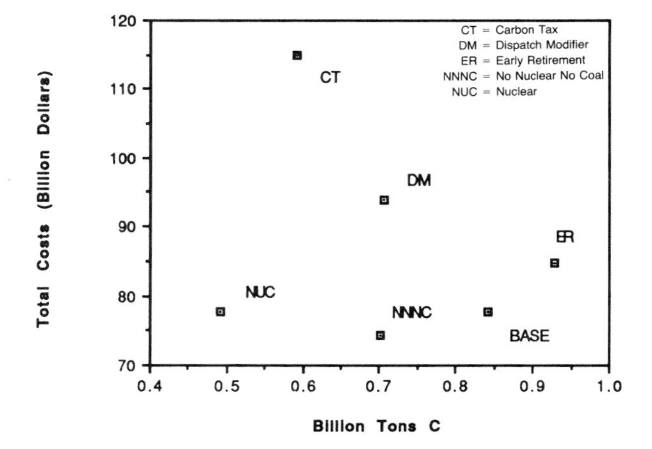

Figure 7-3. Carbon vs. cost (Northeast—very high growth).

polluting capacity with efficient, cleaner technologies. They are felt more in the beginning of the study period than in the end. As load growth increases, the need for new coal capacity in early years becomes greater, until the point at which the potential for carbon gains is no longer present. The migration of the early retirement option can be noted in the carbon-cost tradeoffs of Figures 7–1, 7–2, and 7–3, which show the desirability of the option in the low growth case (7–1) and its ineffectiveness in the very high growth scenario (7–3), where even the base strategy has better attributes. Early retirement is environmentally attractive in the East Central region but carries a greater cost penalty due to the magnitude of the retired capacity in the early years of the study. While in the Northeast base case, early retirement has similar costs to the base strategy, early retirement in the East Central results in an increase in total costs over the base strategy of approximately 20 percent (see Figure 5.1). These results imply that early retirement may be a very attractive option if applied in conjunction with aggressive load management, allowing for an acceleration of the environmental gains offered by the transition.

Carbon Tax: The carbon tax is consistently the highest cost alternative, as would be expected. It should be noted, however, that this is in reality a distributional effect, since this alternative provides significant revenues that are available for other purposes or that might in some way be returned to electricity consumers. For instance, in the base case of the East Central region, the total net present value of the costs of electricity generation under the carbon tax strategy is \$119 billion. Over this time period, however, a total tax revenue of \$77 billion was created. The costs of the strategy *to society as a whole* are actually \$42 billion, less than 10 percent more than the base strategy. Figure 6–4 illustrates the change in the original tradeoff curve (Figure 6–1) that results if net costs to society are plotted instead of utility costs. When viewed from this perspective, the carbon tax is much more attractive.

The environmental effects of the tax are substantial. This strategy allows for very high levels of sulfur dioxide reduction (see Figure 5) and for a stabilization of carbon emissions[25] in many futures (for an example, see Figure 2–1). It is the only nonnuclear case that is fairly robust in this regard. The strategy, like several others, creates a strong incentive for the use of natural gas, while also creating strong incentives for quicker introduction of post-transition nonfossil alternatives. The disadvantages of the strategy clearly lie in the unknown social impacts of such drastic economic measures. It should be noted that this strategy also creates an incentive for demand reduction,[26] an effect unaccounted for within this formulation. The price elasticity of demand is negative, meaning that demand will decrease with the increase in energy cost, but the current formulation is equivalent to an assumption that this elasticity is zero. This demand reduction would certainly cause the costs faced by the utility to be lower than stated and would probably lower emissions as well.

Conservation: This strategy was not modeled explicitly as a policy option as were the previous six. By examining the effects of variations in load growth on various strategies, however, we can identify those strategies in which load reduction is desirable. It is assumed that some load reduction can be obtained at some cost, but attempts to explicitly determine such costs have not been made. The environmental effects of conservation depend strongly on both the characteristics of the existing generation system and the characteristics of expansion alternatives. Clearly, with any given supply system, conservation reduces the emissions of all types from that system. What conservation also does, however, is reduce the need to change the system through the addition of new capacity. Continued or expanded use of existing capacity can, in this scenario, lead to an increase in emissions. It is the relative weighting of these two factors that determines the overall effect of conservation. The effects of a shift from a high-growth future to a low-growth future therefore may or may not be desirable. For example, with the implementation of the base strategy in the Northeast, increases in load growth clearly increase carbon emissions, as shown in Figure 8–1. Figure 8–2 illustrates that the positive effects of conservation are not so clear in the nuclear case, particularly in the later years of the study period.

Underlying Technological Characteristics of Successful Strategies

The somewhat counterintuitive effects of conservation deserve further examination. Due to the differing environmental characteristics of various expansion alternatives, the effects of conservation vary depending on which emissions are of concern. In order for the effect of an expansion alternative to be great enough to offset conservation savings, it must have significantly lower emissions *and* displace a significant portion of the original high-emission capacity. In the case of carbon emissions, only two existing alternatives accomplish this goal: nuclear and gas-fired combined cycle. The nuclear alternative not only has zero emissions, but its ability to operate economically at baseload allows it to displace large amounts of generation from original capacity. (Problems with this option, however, are substantial, as previously discussed.) The gas-fired GTCC option has lower emissions than typical coal capacity (by about 50 percent), but significant operation at baseload is costly and possibly infeasible due to natural gas resource limitations.

In the case of acid rain emissions, SO_2 and NO_x, an additional alternative is available: IGCCs. The ease with which this option displaces original coal capacity and its

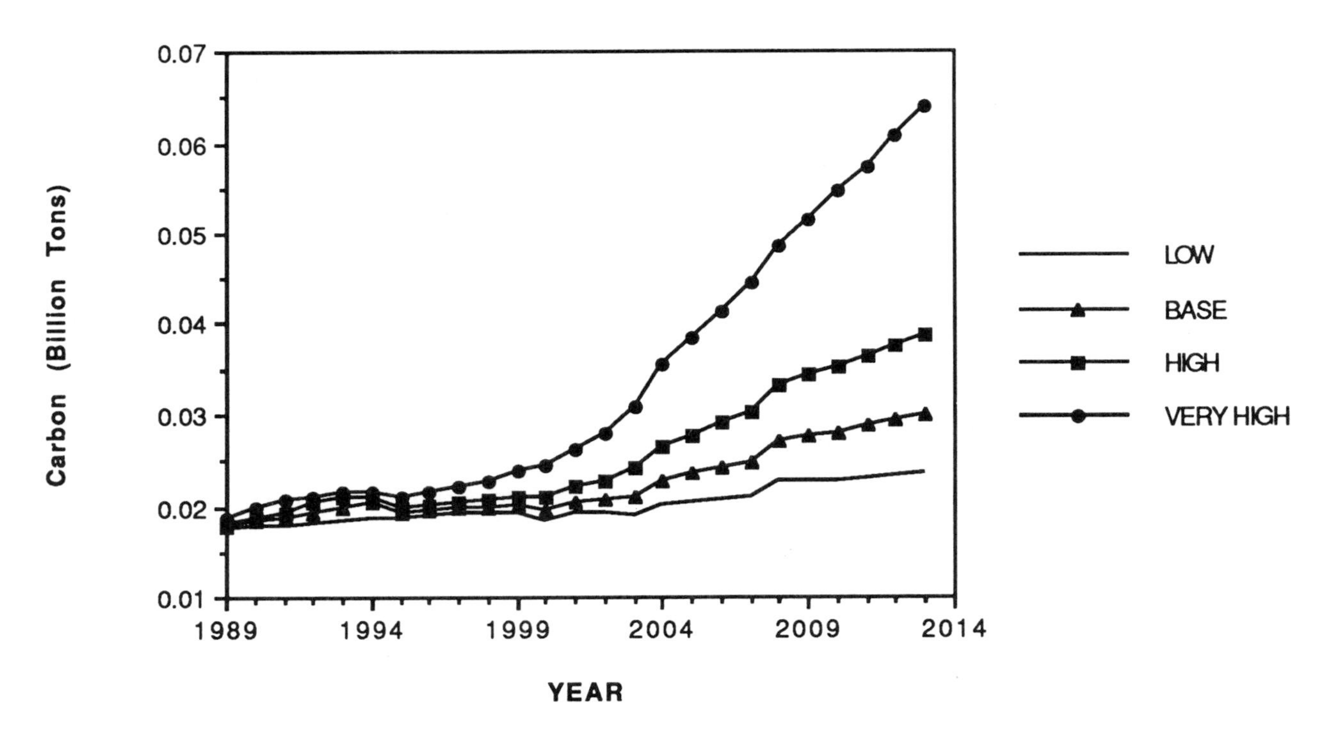

Figure 8-1. Northeast base strategy carbon emissions by growth rate.

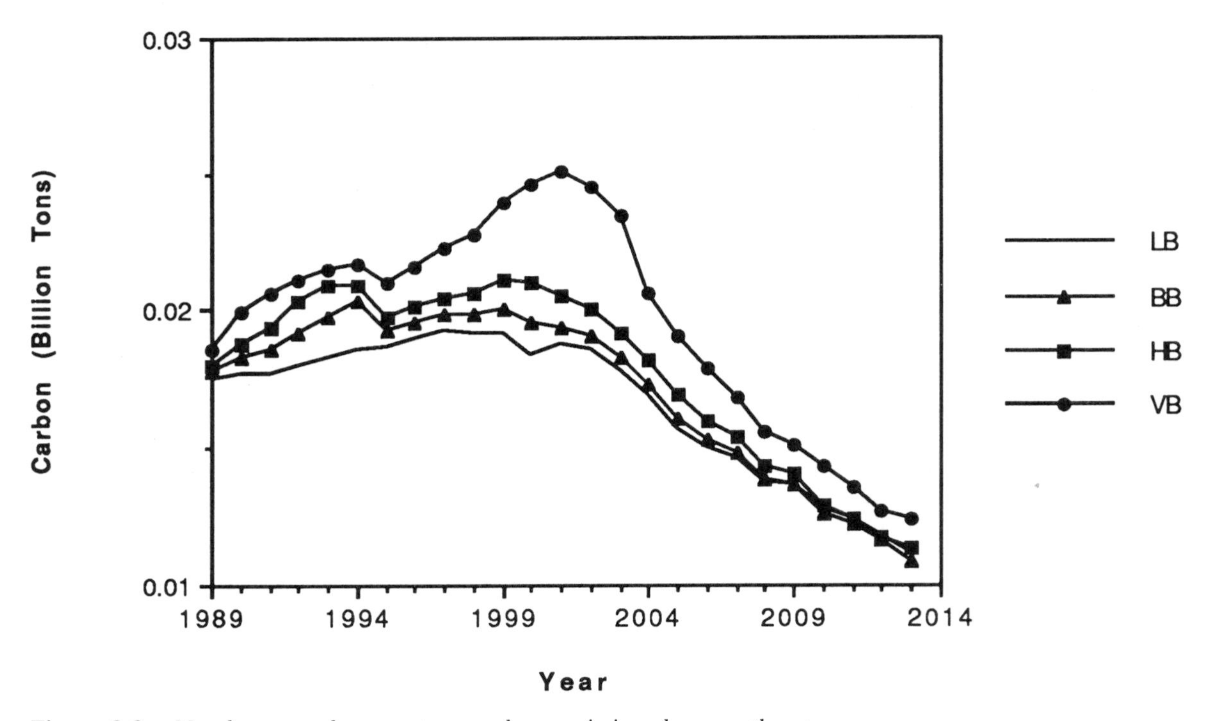

Figure 8-2. Northeast nuclear strategy carbon emissions by growth rate.

favorable environmental characteristics make this a highly economical option for reduction of these emissions. Conservation can delay the construction of new IGCC and GTCC capacity, actually causing SO_2 and NO_x emissions to increase over what might otherwise be possible.

What then, are the implications for technology choice? If your concern is carbon emissions, then the use of coal burning base load and residual oil burning intermediate load must be avoided. The most effective measures, without regard for resource or cost constraints, would then be the following measures, in rough order of importance:

- *Increased use of zero-carbon baseload*
  Nuclear, hydroelectric, and nonfossil renewables.
- *Increased use of low-carbon baseload*
  Gas-fired GTCC (forced baseload).
- *Reduced demand*
  Assumes demand reductions are on the same order as load growth. More substantial reductions could move this higher in list.
- *Increased use of decreased-carbon intermediate/peaking load or higher efficiency baseload*
  Unforced GTCCs or IGCCs. The effects of these are different but both small compared to above actions.

If your concern is acid rain emissions, the following measures are most effective (again in the absence of resource or cost constraints) for avoiding the use of high-emission capacity (uncontrolled coal baseload and uncontrolled residual oil intermediate load):

- *Increased use of zero- or low-sulfur baseload*
  Nuclear, hydroelectric, nonfossil renewables, forced GTCCs, pulverized coal with scrubbers, IGCCs. The natural baseload operation of the "clean coal" technologies makes SO_2 emissions a much more tractable problem than CO_2.
- *Reduced demand*
- *Increased use of low-sulfur intermediate and peaking load*
  Gas- and oil-fired GTCCs.

Note that the cost-effectiveness issue is very important. The technologies best suited for emissions reduction by their technological characteristics may not necessarily provide the most cost-effective means of reduction. The uncertainties about nuclear power and the constraints on widespread natural gas use are clear examples of this. It should also be noted that while conservation may indeed be the most cost-effective method of emissions reduction at the margin, there is a finite limit to the amount of conservation that can be accomplished in the long term.

The various policy strategies that have been modeled accomplish these tasks with varying effectiveness. The nuclear option is particularly effective because it accomplishes the most significant tasks (zero-emissions baseload) at little or no change in total cost. Again, the feasibility of this option is questionable for many reasons. The no nuclear no coal strategy provides some emissions reductions because new demand caused by load growth and by plant retirement allows natural gas combustion to replace some amount of coal combustion. The effects on baseload, however, are slow and minimal with sacrifices of reliability as the baseload demand increases. The dispatch modifier is very effective because low-emissions capacity (gas-fired GTCC) is forced into baseload. Again, the high usage of natural gas is definitely costly and may be infeasible. The early retirement option is highly effective because the most inefficient and highly polluting plants are explicitly removed from the system. When pursued at a reasonable level, substantial emissions benefits can be achieved with little or no cost increases due to the usage of high-efficiency new capacity. Clearly there is an optimal level, however, as costs are seen to increase rapidly with the high demand for new capacity seen in the East Central region or in the high growth cases of the Northeast. Lastly, the carbon tax is highly effective because the carbon emissions are (crudely) internalized within the system, allowing the appropriate levels of each reduction action to be chosen according to an optimization criterion. The feasibility or advisability of such drastic economic measures is not clear, however.

As a final note, it should be stated that the range of policy strategies examined here is not intended to be either comprehensive or exhaustive. Obviously, there are many other possible policy strategies with varying possibilities for success. Even with the strategies considered, there are many possible combinations (i.e., early retirement/conservation or carbon tax/nuclear) or modifications (i.e., phased carbon tax or SO_2–dispatch) that might be considered. It is hoped that the limited window provided by this analysis helps to highlight those strategies that are likely to be successful and worthy of further investigation.

Notes

1. Given the long time lags between increased CO_2 emissions and the possible effects of increased CO_2 concentrations, even small reductions in CO_2 emissions in the near term may have significant long-term impact.
2. The authors wish to acknowledge the support of the MIT Center for Energy Policy Research in carrying out this effort.
3. The general formulation of this equation—Impact = Population * Impact per capita—is attributed to Erlich and Holdren, 1971.
4. As with "marginal costs" (i.e., $/GWH) in an electric power system, this value is marginal to the system and not necessarily to the technology with which the cost is associated. The marginal costs or emissions are the *average* cost or

emissions per GWH of the technology which is loaded at the margin of the system.

5. Carbon dioxide emissions are no more technologically fundamental to electric power than tetraethyl lead was to gasoline use in automobiles. It is economic and political infeasibility that limits nonfossil fuel use.
6. Units are classed within a single technology type if all characteristics of the units are identical, including fuel type used. At the finest grain level, this summation could be across single generating units in order to fully account for differences.
7. This is an industry standard unit which can be confusing. An MMBTU is 10^6 BTUs, not 10^9.
8. Equivalent to BTU/kWh, a more typical unit.
9. Typical values for fuels used in the United States.
10. Most previous efforts have modeled the entire global energy/CO_2 system over time scales on the order of 100 years. Perhaps foremost among these are Edmonds and Reilly, 1986 (used as the basis of many energy/climate studies); Nordhaus and Yohe, 1983; and Manne and Richels, 1990. Detailed evaluations of these and other modeling efforts can be found in Keepin, 1986, and Ausubel and Nordhaus, 1983, among others.
11. EPRI, 1982.
12. Using one of three methods: linear programming, Bender's decomposition, or dynamic programming. Bender's decomposition was the primary method used in this study.
13. EPRI, 1989a.
14. The "base future" or "base case" for any region refers to the combination of the base fuel price and base load growth for that region.
15. DRI, 1989.
16. EPRI, 1989b.
17. Growth rates vary annually, but are approximately 1.5 percent per year in the Northeast and 1.0 percent per year in the East Central.
18. The Technology Assessment Guide deems these technologies to be feasible only in the West region. While clearly the feasibility of these technologies is more limited and longer term in the Northeast and East Central regions, their complete exclusion may not be inherently necessary.
19. The use of IGCC plants in the base strategy implies that the cost estimates for the technology are accurate and that the cost effectiveness of the option will be recognized by utility planners. If this is not the case, a more accurate base case might involve new coal capacity with sulfur scrubbers, an option with higher costs and significantly higher environmental emissions.
20. These taxes, in 1985$, are the same as those used in several EPA studies. In the EPA studies, the taxes were phased in over a period from 1985 to 2050. In the present study, the taxes were implemented immediately (1989) and are, thus, even more extreme.
21. With a 25-year extension period to account for end effects.
22. This is as opposed to social costs, a distinction that is significant in evaluating the carbon tax strategy. This is discussed in greater detail below.
23. The dominant effect in the acid rain emissions from a power system is the percentage of system energy generated by uncontrolled coal and oil capacity. While the SO_2 and NO_x emissions from nuclear power per kWh are well below those of the IGCC capacity constructed in the base case, both are orders of magnitude below the emissions per kWh from existing uncontrolled capacity. Since each displaces a similar amount of generation from this older capacity the overall acid rain emissions are similar between the two scenarios.
24. In later study years where the cost of gas is high, the system may actually choose to let energy go unserved instead of operating a large amount of high cost generation, sacrificing system reliability in order to keep costs down. As a result, the costs of the no nuclear no coal strategy, while similar to those of the base strategy, do not represent costs for similar qualities of service.
25. This should not be confused with a stabilization of atmospheric CO_2 concentrations, which would require significant worldwide cuts (on the order of 50 percent).
26. "Demand reduction" is defined here as a decrease in load in response to price increases. "Conservation" is defined as programmatic efforts to reduce load growth through physical change.

References

Ausubel, J.H., and Nordhaus, W.D., "A Review of Estimates of Future Carbon Dioxide Emissions," in National Research Council (NRC), *Changing Climate,* National Academy Press, Washington, DC, 1983, pp. 153–185.

Data Resources Institute (DRI), *Forecast of Fuel Prices,* DRI, October, 1989.

Edmonds, J., and Reilly, J., "The IES/ORAU Long-Term Global Energy-CO_2 Model: Personal Computer Version A84PC," Publication No. 2797, Institute for Energy Analysis, Oak Ridge Associated Universities, Washington, DC, 1986.

Electric Power Research Institute (EPRI), *Electric Generation Expansion Analysis System (EGEAS),* EPRI EL-2561, EPRI, Palo Alto, CA, 1982.

EPRI, *The EPRI Regional Systems Database: Version 3.0,* EPRI P—6211, EPRI, Palo Alto, CA, 1989a.

EPRI, *TAG™—Technical Assessment Guide, Volume 1: Electric Supply—1986 (Revision 6),* EPRI P-6587s-L, EPRI, Palo Alto, CA, 1989b.

Erlich, P.R., and Holdren, J.P., "Impact of Population Growth," *Science,* Vol. 171, 1971, pp. 1212–17.

Keepin, B., "Review of Global Energy and Carbon Dioxide Projections," *Annual Review of Energy,* Vol. 11, 1986, pp. 357–92.

Manne, A.S., and Richels, R.G., "CO_2 Emission Reductions: A Regional Economic Cost Analysis," *Energy and the Environment in the 21st Century,* MIT Press, Cambridge, MA, 1990.

Nordhaus, W.D., and Yohe, G.W., "Future Paths of Energy and Carbon Dioxide Emissions", in NRC, *Changing Climate,* National Academy Press, Washington, DC, 1983, pp. 87–153.

Environmental Protection through Energy Conservation: A "Free Lunch" at Last?

Larry E. Ruff
Putnam, Hayes & Bartlett
London, England

Energy conservation will surely play a large role in the costly effort necessary to reduce environmental problems. Furthermore, utilities (among others) can facilitate conservation by "reducing market imperfections," i.e., by finding innovative ways to offer conservation services in the market in competition with energy itself. But the current enthusiasm for utility demand-side programs goes beyond the commonsense observation that conservation will thrive if given the opportunity to compete with supply as a way to meet consumer needs at minimum economic and environmental cost. Instead, it is commonly argued that, if only normal market concepts and processes are set aside where conservation is concerned, both energy use and energy costs can be reduced simultaneously, yielding environmental benefits as virtually a "free lunch."

This paper examines the view that utility demand-side programs offer an environmental free lunch on a large scale. It is shown that energy conservation neither requires nor justifies setting aside standard economic and business principles or processes. It is also shown that the more enthusiastic projections of the "cost-effective" conservation potential are based on advocates' hypothetical claims about costs and benefits that are not supported by the available market evidence. This helps explain why calls for aggressive energy conservation efforts are directed almost exclusively at electric utilities: only administrative regulatory processes allow advocates' claims to replace valid market tests; and only a protected monopoly has the power to "tax" its captive customers to cross-subsidize demand-side options that cannot compete in a competitive market because they are not truly cost-effective.

Of course, if energy is underpriced because of uninternalized environmental costs, there is a theoretical argument for also underpricing conservation services as a "second-best" alternative to raising energy prices. However, if environmental costs are reasonably internalized on the supply side, where most of them occur, using emission charges, marketable discharge permits, or economically based regulations, the resulting price increases will stimulate the proper pattern and amount of energy conservation. Simple numerical examples suggest that the appropriate price-correcting subsidy is likely to be small compared to the level of subsidy implicit in many demand-side programs, and that most of the cost-effective environmental benefit is likely to come from supply-side rather than demand-side responses; this may be true even for pollutants such as carbon dioxide for which supply-side controls are most difficult.

The Concept of Bundled End-use Services

Kilowatt-hours (kWh) are never eaten raw, but are combined with other things—insulation, light fixtures, machinery—to produce the end-use energy services consumers ultimately desire—heat, light, mechanical drive. Thus, at least conceptually, a utility should be viewed not simply as a seller of commodity kWh, but as a potential seller of "bundles" of kWh and complementary services, designed to meet consumers' demands for energy services at the lowest possible total cost.

In this view, utility provision of demand-side services is simply a downstream value-adding strategy. Such value-adding or bundling strategies are essential to competitive survival in most markets and have been the key to the growth of utilities in the past; vertically integrated regional utilities bundling generation, transmission, and distribution into a higher-value service called "reliable, delivered electricity" have prospered because consumers found this bundled service a good value relative to nonelectric energy or self-generation.

A utility should constantly reevaluate the service bundles it offers to determine whether different combinations might better serve its customers' end needs. It should be prepared to unbundle its services and even to reduce the scope of its internal activities if this appears competitively advantageous. For example, a utility with a particular advantage in transmission because of experience or location should consider offering transmission as an unbundled service; and a utility that can buy electricity from an independent source more cheaply than it can produce it itself should do so. Eventually, utilities and even consumers may have the option of buying generation, transmission, and distribution from separate vendors and assembling their own bundles of reliable, delivered electricity.

By the same reasoning, a utility should expand its activities to offer bundles including additional services if doing so allows it to offer customers better value for their money. A utility with a cost advantage in mining coal should integrate upstream into this business, providing its customers with a bundle that includes utility-mined coal. A utility with an advantage in bundling its commodity electricity with other services to provide end-use energy services should integrate further downstream into consumer markets. There is nothing sacred about the traditional utility service bundles.

The Ultimate Level Playing Field

As a utility seeks higher-value services to offer its customers, it should not discriminate against service bundles that meet customers' needs with less electricity and more conservation. However, leaving environmental issues aside for the moment, neither should it discriminate *in favor of* conservation-intensive bundles, *just because* such bundles contain less commodity electricity. Any utility activity, whether downstream integration into high-efficiency lighting, upstream integration into coal mining, horizontal integration through a merger with a neighboring utility, or even diversification into an unrelated activity such as baking pizzas, might lower customer costs and enhance utility competitiveness. If it does, there is no reason in principle a utility should not offer this service, whether it results in more or less commodity electricity being included in the utility's bundle.

This is the ultimate "level playing field": all activities that have economic efficiency as their principal rationale should compete on their ability to lower the costs of providing consumers with the end-use services they desire. As discussed presently, some fine-tuning of the playing field at the point of consumption may be appropriate in principle if environmental costs are not reasonably internalized where they occur. However, trying to manage supply-side environmental problems with demand-side programs is relatively ineffective and inefficient, compared to appropriate supply-side policies that may have surprisingly little demand-side effect. Thus, environmental issues can be set aside for the moment, to be revisited in a concluding section.

Not only do all efficiency-seeking activities have the same fundamental objective of lowering consumers' end-use costs, but all such activities pursue this objective with the same fundamental strategy: reducing information and transaction costs. It is common, at least where utility demand-side programs are concerned, to attribute information and transaction costs to something called "market imperfections." But the reduction of such costs, whatever their cause, is the *raison d'être* of any organized economic activity. No activity, whether inside or outside a utility, deserves special consideration just because it seeks to reduce information and transaction costs.

Consumers in all times and places meet their end-use needs by learning about technical options, finding people who will lend them resources, obtaining physical inputs and organizing cooperative efforts to assemble the various pieces—in short, by obtaining information and engaging in transactions, whether to create end-use energy services or pizza services. These universal activities always involve time and inconvenience—i.e., markets are always "imperfect"—and hence the cost of organizing delivery of a service can often be lowered by an enterprise that specializes in that job. Such an enterprise is normally called a business, and its success at reducing information and transaction costs is normally measured by its ability to sell its services to willing buyers for enough to cover its (allegedly lower) costs.

A utility conservation program tries to lower the information and transaction costs involved in delivering end-use services to customers. But precisely the same is true of a utility generation program, a utility coal mining program, a utility pizza baking program, or a non-utility program. Each such economic activity, whether inside or outside a utility, and whether on the demand side or the supply side, should be required to justify its existence, not by pointing to the universal truth that information and transactions costs are *high,* but by demonstrating its ability actually to *reduce* these costs.

Utility Downstream Integration as an Efficiency Measure?

Viewing a utility as a provider of end-use services rather than just commodity electricity provides a logical way to justify demand-side programs on standard economic and business grounds. However, this even-handed view of demand-side programs raises an immediate question: What reason is there to think that utilities have the kind of competitive advantage in end-use markets that would justify the more enthusiastic claims about conservation potential?

On the face of it, it is hard to see why utilities should have much comparative advantage in providing bundled end-use services to consumers. For one thing, downstream integration of utilities into dynamic, service-oriented consumer markets would run directly counter to the unbundling and vertical disintegration movements sweeping modern economies. In telecommunications, transport, finance, computers, even on the supply side of the electricity business itself, the clear trend of modern technology and institutions is toward unbundling and entrepreneurial competition in the market. What reason is there to think that large, monopolistic, regulated utilities have significant cost advantages in providing tailored demand-side services to millions of individual consumers, when so many utilities are having trouble competing even in the large-scale, engineering-oriented generation businesses they have been in for decades, and when so many utility diversification efforts have proven disappointing or worse?

Of course, utilities can learn new businesses and can restructure themselves along more entrepreneurial lines. Some of the most successful firms in dynamic, service-oriented industries are large; AT&T, American Airlines, Citibank, and IBM are surviving, albeit not without difficulty in the face of increased competition and unbundling. But these large companies do not try to do everything, and what they do is done largely through decentralized, specialized profit centers that either pay their own way as separate businesses or are closed down or spun off: Boeing is very good at making airplanes, but lets American Airlines bundle these with other services for millions of consumers; IBM makes good business machines, but sold off its retail stores because other firms were better at bundling this commodity with end-use services for small consumers; and AT&T's computer business may benefit from being associated with AT&T's telecommunications business, but had better pay its own way or it will be spun off.

A utility demand-side business that is serving both consumers and the utility should be able to buy commodity electricity on an arm's-length basis from the generation side of the business or from another supplier, buy complementary goods and services in the market, and sell bundled end-use services in competition with all the other sources of such end-use services. But evidence from energy service markets does not support the view that such activities would accomplish dramatic results. There is no shortage of firms selling high-efficiency lights, insulation, and advice; these firms are constantly looking for new ways to offer bundled services that consumers value. What reason is there to think that these normal, competitive market processes that work so well in providing most other end-use services are barely scratching the surface of the potential end-use energy service market?

The Basis for Claims about Conservation Potential

If there really is a large, untapped potential for cost-effective utility demand-side programs, it must be due to some combination of two factors: either commodity electricity is badly underpriced, so that consumers have too little incentive to buy cost-effective conservation services from utilities or anybody else; or utilities have some under-appreciated cost advantages that allow them to thrive in markets where ABC Conservation Company cannot. If neither of these is true, then serious consideration must be given to the possibility that the definition of the "cost-effective potential" is badly distorted where utility conservation programs are concerned.

The possibility that electricity is underpriced because of environmental costs and the implications for a utility demand-side business are discussed presently. Electricity underpricing for other reasons was a problem following the oil price shocks of the 1970s and persists today in special circumstances, particularly in regions with subsidized hydropower and during peak demand periods in many regions. Of course, even if prices are providing inadequate conservation incentives, the logical solution is to correct the price signals, perhaps using focused subsidies where necessary, to give consumers the right incentives to conserve in everything they do. It is quite illogical to leave electricity underpriced so that only a few conservation services provided by the utility can compete with it, and then only if they, too, are underpriced. But underpriced electricity is not a widespread problem today and is not the most common basis for claims that there is a huge, unexploited potential for cost-effective conservation.

The more common defense of implausible claims about conservation potential is that utilities have large advantages in reducing information and transaction costs caused by "market imperfections." The argument is that a utility demand-side business has more credibility with customers, more ability to study the options and hire the experts, better access to capital, better service personnel, more comprehensive data bases and billing systems, etc., than does ABC Company, and hence can provide conservation services with much lower transaction and information costs. Thus, a utility's demand-side business should be able to deliver conservation services at prices that ABC Company cannot be expected to match.

The only problem with this argument is that few people, and least of all demand-side advocates, really believe it. The ability of a utility to reduce market imperfections is seldom cited to explain why the utility's demand-side business is making money selling better services at lower prices, thereby beating ABC Company at its own game. Instead, market imperfections are invariably cited as excuses for exempting utility demand-side programs from the need to compete on price and service. The typical claim is that the experts know which conservation services are cost-effective for which consumers in which applications at which times, and hence there is no need to define, measure, and price utility demand-side services, or to convince individual customers that the benefits are worth the costs. Instead, the utility demand-side business should do what ABC Company cannot do—give away its services where the experts claim they are cost-effective and raise the price of electricity to pay for them.

This argument is sometimes stated as a research proposition, along the following lines: hypothetical arguments will never determine how large the cost-effective conservation potential really is, so let us run a market experiment; however, because we cannot sell utility conservation services for enough to cover their costs, we should give them away, paying for them with the revenue from higher electricity prices; if we can give away enough of these costly services on an experimental basis, we will expand the giveaway program, on the grounds that we are somehow lowering customer costs and enhancing utility competitiveness—even though customer electricity prices are higher and/or utility profits are lower. Such market research designs and marketing strategies would not last long in the insurance or automobile business, or even on the supply side of the utility business; but they appear to be taken seriously where utility demand-side programs are concerned.

Now, there is no doubt that utility conservation programs can be expanded to virtually any extent if the level playing field is interpreted in this way. The experts and advocates will never run out of services they can "prove" to be cost-effective based on their own definitions and estimates, as long as they do not have to suffer the cost and indignity of dealing with consumers as consenting adults and defining, measuring, and pricing utility services. But using the utility's monopoly power in the commodity electricity market to escape competitive disciplines in downstream markets should never be confused with reducing market imperfections; indeed, it is a sure way to create new market imperfections. As long as demand-side advocates are so unwilling to subject their claims about cost-effectiveness to any reasonable market test—and trying to give away valuable and costly services hardly qualifies as a valid market test—their claims are at best unproven.

Since the more enthusiastic estimates of the cost-effective conservation potential are not based on claims that electricity is underpriced or that utility programs can lower costs enough to compete successfully in the market, one can be excused a heretical thought: *Maybe the alleged potential is not there.* Maybe there is a reason that special concepts, criteria, and jargon have had to be invented to justify demand-side programs. Maybe, if utility lighting and water heater wrapping programs were subjected to the same tests routinely imposed on utility generation, coal mining, or pizza baking programs, the same commonsense conclusion would be reached in each case: a utility with good management and entrepreneurial abilities should be able to compete in any of these markets—but only as one of many competent players, not with such an overwhelming advantage that it is likely to change fundamentally the competitive landscape or general consumption patterns.

The "No-cross-subsidy" Test

The utility demand-side movement has spawned a legion of special concepts and "tests"—the "no-losers test," the "all-ratepayers test," the "non-participants test"—all for the purpose of answering one simple question: how can it be determined whether some utility program will benefit consumers, on balance? But there is nothing new or difficult about this question, for which generations of academics, managers, and regulators have developed a very logical and elegant answer: in a market economy, an efficient activity is one that results in lower average prices—not lower revenues or customer bills of a specific enterprise, but lower *prices*.

There is no more fundamental economic proposition than this: an economic enterprise should be judged by its ability to keep its unit costs, and hence its prices, low. In keeping with demand-side tradition, this proposition can be stated as a "test"—the no-cross-subsidy test. According to the no-cross-subsidy test, a utility should not engage in any activity if doing so results in average prices for other utility services being higher than they would otherwise be. Although this test has appeal on distributional or equity grounds, it is strictly an efficiency test based on very simple logic: if those who benefit from a utility program are not able to pay its incremental costs and still be ahead of the game, then the program must have more costs than benefits and is not cost-effective by any reasonable definition of that term.

The most natural and effective way to satisfy the no-cross-subsidy test is to require the costs of each service to be recovered through the prices charged for the service, so that alleged beneficiaries have the option of declining to pay for the service if its costs exceed its benefits. However, in principle the no-cross-subsidy test can also be met by recovering the costs of a class-wide service through a surcharge on customers in the class. This method of cost recovery might be applicable in special cases where some simple service is so obviously beneficial for all customers in a class that it can simply be delivered and its costs collected through a surcharge or tax on the lucky beneficiaries—an approach that at least gives the class of alleged beneficiaries an incentive, and presumably an opportunity through the regulatory process, to object if they do not think the benefits are worth the costs.

The no-cross-subsidy test, under different names, is routinely applied to utility supply-side programs, despite persistent confusion on this matter. When demand is growing, a utility with an obligation to serve has no choice but to meet that demand, even if doing so increases prices for the "old" customers whose demand is not growing. The no-cross-subsidy test says that the best option in such cases is the one that results in the *least* increase in electricity prices. When only supply-side options are being compared, the demand to be met is the same for all options, and hence the option that minimizes the present value of revenue requirements also minimizes average prices—since REVENUE = PRICE x QUANTITY, and QUANTITY is the same for all supply-side options, minimizing REVENUE is equivalent to minimizing PRICE. In this special case, the no-cross-subsidy test is equivalent (apart from rate-design issues that are not directly relevant here) to the traditional minimize-revenue-requirements test.

For evaluating utility options more generally, the no-cross-subsidy test remains valid while the minimize-revenue-requirements test is clearly inappropriate. If a utility were to propose adding coal mining or pizza baking to its activities, there would be no fundamental reason to object, as long as engaging in these activities reduced average *prices*. If getting into the pizza business allowed the utility to lower its electricity prices, nobody would object that the total utility revenue requirement or utility bill, now covering both electricity and pizza, is higher. But if the only way the utility can pay for mining coal or baking pizza is to charge higher prices for electricity, consumers and regulators—and probably even most demand-side advocates—would protest vehemently. They would say that, while utility integration into such businesses is not objectionable in principle, if it must be cross-subsidized through higher electricity rates it must be inefficient and should not be undertaken by the utility.

Even though there is no objection in principle to a utility getting into coal mining or pizza baking as long as no cross-subsidy is involved, in practice regulators are understandably reluctant to allow utility diversification outside the core electricity business, simply because they are *afraid* of cross-subsidization. They fear that accounting cost allocations and other difficult-to-monitor devices will be used to raise the price of the commodity electricity all customers buy, with the proceeds used to cross-subsidize utility activities that cannot otherwise compete. Nor are they much swayed by the fact that the subsidized services are on the demand side (e.g., appliance sales), are in markets that are highly imperfect, are cost-effective for many customers, or involve "only a transfer payment" from ratepayers who choose not to use the subsidized services to those who do. They know, as surely as they know anything, that even running the risk of such cross-subsidies is bad economics, bad business, and bad regulatory policy.

Despite this deep and proper aversion to cross-subsidizing utility programs in general, for some reason different standards have become accepted where demand-side programs are concerned. Of course, nobody advocating a demand-side program is foolish enough to say that it is cost-effective even though it must be cross-subsidized; instead, they say that it is cost-effective even though it cannot pass a "no-losers test" or a "rate-impact test." But, under this rhetorical smokescreen, they are quite explicitly arguing that utilities should be allowed, even forced, to raise the price of commodity electricity to pay for utility entry into markets where the utility cannot otherwise compete. If this is not a cross-subsidy, what is? And if this is the basis for the claims about the massive potential for "cost-effective" utility demand-side programs, why should these claims be taken seriously?

Why Utility Conservation and Not Utility Pizza?

The widespread view that demand-side programs are somehow different, and hence should be exempt from standard economic and business criteria such as the no-cross-subsidy principle, has absolutely no basis in economic or administrative theory. Indeed, the entire intellectual foundation for this view comes from simple examples purporting to prove that applying a no-cross-subsidy/no-losers/no-rate-increase test discriminates against demand-side programs. Yet, hidden in these examples is a critical, implicit assumption that would be immediately spotted and rejected if applied to, say, utility pizza programs.

A typical example allegedly "proving" the inapplicability of a no-cross-subsidy test for conservation programs goes as follows: a utility is selling electricity for its marginal supply cost of 8 cents/kWh; demand-side advocates say that a high-efficiency light can save a kWh for only five cents at no cost to customers in terms of quality or convenience of service—if the utility "reduces market imperfections" by installing the light itself; however, if the utility provides the light rather than the generated kWh, it will have higher total costs with no increase in kWh sales and will have to increase average electricity prices; thus, applying a no-cross-subsidy test would prevent the utility from implementing an obviously cost-effective demand-side program.

What is wrong with this picture? Most people tend to nod in agreement at each step in this argument; if the conclusion appears unusual, they apparently assume that it has something to do with the special energy-saving or market-improving character of demand-side programs. After all, if a utility pays for a conservation measure that only it can implement, even a very cost-effective one, of course it will have higher costs with smaller electricity sales and will have to raise average electricity prices to pay for it. But, in fact, the conclusion in this example has nothing to do with any special characteristics of utility conservation programs and everything to do with an implicit assumption about utility pricing. What is it?

Give up? As a hint, try applying precisely the same logic to a utility pizza program: pizzas cost $8 in the market; utility pizza advocates say that pizzas just as tasty and nutritious can be made for only $5—if the utility "reduces market imperfections" by making the pizzas itself; however, if the utility produces pizzas, it will have to raise the price of electricity to pay for them; thus, applying a no-cross-subsidy test would prevent the utility from implementing an obviously cost-effective pizza program.

When the utility service involves pizzas instead of lightbulbs, most people will immediately see the logical flaw in this example: Why is it assumed that the utility must give away pizzas? If pizza advocates are correct that a $5 utility pizza is as good as the competing $8 pizza, the utility can easily sell pizzas for enough to prevent electricity prices from increasing; indeed, if the utility gives pizza consumers a $1 benefit by selling pizzas for $7, gives shareolders $1 as a reward and incentive for thinking of such a good idea, and uses the remaining $1 pizza profit to lower electricity prices, there are no losers. If the utility cannot get customers to pay enough for its pizzas to produce such a result, then the claims about cost or quality must be wrong and utility pizzas are not cost-effective after all.

Of course, *some* utility pizzas may be cost-effective for *some* consumers even if the utility does not charge for them. But when consumers queue up for free pizzas, the utility and its regulators will have no effective way to decide which consumers really like pizza enough to deserve it, much less which should have the expensive special-with-anchovies rather than the standard cheese-and-tomato. The pizza experts will claim to know, of course, and will be quite willing to testify on the matter, telling the utility and its regulators how ratepayers' money should be spent; indeed, when hamburger-loving consumers complain about paying for pizzas through higher electricity prices, the experts will argue for raising electricity prices even further to pay for free hamburgers so that everyone will have a give-away program to defend. Happily, at least where pizzas rather than lightbulbs are concerned, it is generally recognized that there are better ways to accomplish truly cost-effective results.

Substitute "high-efficiency lights" for "pizza" in this example and none of the logic changes. A utility can always give away or underprice a costly service that allegedly "reduces market imperfections"; if it does so, it will have to raise the price of some other service to make up its loss, whether kWh sales are higher or lower, and even if some of the underpriced service manages to find its way into cost-effective uses. But this arithmetic tautology hardly invalidates the basic economic and business principle that any utility service that really "reduces market imperfections" enough to lower end-use costs *can* be sold at a price that does not require such a cross-subsidy and *should* be sold at such a price in order to limit waste, fraud, and abuse. And this is strictly an efficiency principle; the fact that it is "fair" for those who benefit from a service to pay (at least) its incremental costs is merely a bonus.

Why is it taken for granted that costly and valuable demand-side services should be given away, or at least cross-subsidized by higher electricity prices? Why do so few people spot or question the critical pricing assumption implicit in literally every example "proving" that cross-subsidies are necessary for cost-effective conservation, when almost everybody would recognize and reject the same assumption if it were applied to utility pizzas? This is not the place to speculate on the history, reasons, or motives that have created this peculiar intellectual blindspot. But there can be no doubt that it is a blindspot or that it affects many analysts and regulators or that, like all blindspots, it is potentially very dangerous.

Leveling the Tilted Playing Field

The assertion that demand-side programs should be exempt from the well-accepted no-cross-subsidy test is really quite extraordinary. It is neither more nor less than the assertion that social resources should be allocated to demand-side programs on the basis of hypothetical claims by interested parties who expect that, if they can just convince regulators that their favorite programs "reduce market imperfections" enough to be "cost-effective," costly services will be given away so that their claims about cost-effectiveness will never have to face a valid market test. This is a strange definition of a level playing field; there must a better one.

If conservation measures are really as incredibly cost-effective as advocates claim, they can easily be implemented with programs that recover enough of their costs from participants to meet a no-cross-subsidy test. Consumers have had no trouble finding insulation, caulking, and plastic storm windows in their local home centers, have found room on their credit cards to finance them, and have spent many a weekend installing them. These same consumers should be able to find, finance and screw in the $20 lightbulb that, according to advocates, will reduce electricity bills by $2 per month for ten years with absolutely no other cost to the consumer.

Even if utility purchasing, installation, endorsement, and/or finance are necessary to "reduce market imperfections" enough to make this miracle lightbulb cost-effective, there is still no justification for a cross-subsidy. The utility can include in its monthly bills information describing the $2 per month saving and inviting consumers to return a pre-stamped postcard accepting the entire bundled service for a $1 monthly surcharge—a price that will give the utility a nice profit to divide between higher shareholder dividends and lower electricity prices. Customers who will gain a net of $1 per month for simply mailing the postcard will do so in droves; even tenants and landlords suffering from the "split incentives" so often cited to justify conservation subsidies should be able to find some way to divvy up the windfall. Consumers will decline the offer only when they know something about their specific situations, habits, and tastes that makes them doubt they will benefit even by $1, i.e., when the light is not, in fact, cost-effective.

If advocates find it too onerous to let consumers participate in the process even to this extent, they can instead demonstrate to the regulators that a class of customers will gain $1 per month from the service, even after paying its incremental costs through the $1 surcharge on their class-specific bills. Regulators will jump at this chance to do ratepayers such a favor, knowing that any consumers who manage to figure out what has happened will thank them for the $1 per month windfall. The utility can then deliver the service, add the surcharge to the bills of the lucky beneficiaries, increase dividends, and decrease rates.

Significantly, this picture of consumers, shareholders, and regulators happily dividing up a free lunch is far from what is actually observed in the market and in regulatory proceedings, suggesting that the realities of conservation costs and benefits are rather different than the claims. Conservation services alleged to be incredibly cost-effective go unsold, often even when heavily subsidized; so demand-side advocates, in the name of maintaining a level playing field, advocate larger subsidies. Even the simple device of recovering easily identified costs from easily identified beneficiaries is opposed by demand-side advocates, presumably because consumers will not want the benefits if they must pay the (allegedly much lower) costs.

Of course, consumers may decline to buy the latest conservation gadget or may object to the utility installing, financing, and charging them for it simply out of ignorance or inertia. But it is also conceivable that consumers know something about their individual preferences and opportunities that the experts have not fully considered. Perhaps consumers do not want the high-efficiency light because of its color balance or flicker, or because it looks silly in their fixtures, or because they do not use any single fixture enough to justify the $20 investment, or just because they do not want to be bothered for a potential and uncertain saving of a few dollars a year. Perhaps the landlord/tenant team knows that the commercial space will be remodeled soon, or that conventional lighting is a cheap way to keep clients and staff happy; anybody who has fought the automated lighting systems in some modern office buildings knows that conservation has costs the designers do not fully consider.

Even if the experts are, in some technical sense in some specific case, right about how other people should spend their money, there is some merit in the old-fashioned idea that those who will ultimately pay for something should have the incentive and the opportunity to compare the costs to the benefits and decide for themselves. Before having a claim on consumers' money, a utility's demand-side business should go through much the same process its supply-side business or its pizza business must go through before having a claim on consumers' money: determine what consumers want; develop services that meet these needs; find ways to deliver and finance these services without raising other prices; and convince consumers, individually or through their regulatory representatives, that the new service is worth having *and paying for.*

Of course, where electricity is underpriced relative to its marginal cost, the competition between electricity and conservation is distorted. If it is impossible to redesign rates to reduce this market imperfection, there is a theoretical case for offering price-correcting subsidies that give consumers the right incentives to conserve—provided the inefficiencies inherent in any such subsidy scheme do not swamp any theoretical benefits. But even here, the no-cross-subsidy test applies: the highest conservation subsidy that can be paid without providing *too much* incentive for conservation is also the highest subsidy that can be paid *without increasing prices* above what they would otherwise be—the difference between marginal cost and price.

Today, the more common market imperfection stemming from mispriced electricity is a price that exceeds marginal supply costs, tilting the playing field *in favor of* conservation. In such cases a utility should reduce electricity prices selectively if necessary to avoid loss of profitable sales and should price any conservation services to earn the same margins earned on electricity sales. This "second-best" solution also meets the no-cross-subsidy test, since cutting prices selectively if necessary to prevent uneconomic conservation will result in lower average prices than letting profitable sales go without a fight.

In summary, requiring a utility demand-side business to cover enough of its costs to keep electricity prices from increasing above what they otherwise would be is perfectly consistent with truly cost-effective conservation, even when electricity is mispriced. To a first approximation, this means running a demand-side program as an independent business or at least as a separate profit center; any subsidies truly justified by electricity mispricing will generally be small and narrowly focused, and should be available to any supplier of demand-side services, not just the utility's own demand-side business. This is the only administratively practical way to maintain a truly level playing field.

Buying "Saved kWh" and Demand-side Bidding

The current interests in energy conservation and competitive purchasing of generation have led to proposals that utilities buy "saved kWh" the same way they buy produced kWh; the "negawatt auction" or all-source bidding process is the clearest expression of this idea. Of course, if such purchasing is done "right" it will get the "right" result; but this is a tautology that begs the real question of whether there is any right way to buy saved kWh that is worth the trouble or the risk that it will be done very wrong.

In the simplest and most distorted approach to buying saved kWh, a utility would invite offers of saved kWh alongside offers of produced kWh, selecting and paying for those kWh with the lowest bid prices per kWh saved or produced. This form of demand-side bidding involves an elementary logical flaw: buying a kWh is fundamentally different in economic and business terms from not-selling a kWh. When a utility buys a produced kWh for, say, four cents, the seller is compensated for the cost of producing the kWh and the utility gets a kWh that it sells on to a willing buyer for something near four cents; it is a simple exchange of value-for-value. But when the utility pays four cents for a saved kWh, the seller is paid for doing nothing except not buying a kWh for a fair price; four cents in value flows from the utility to the saver, but little or no value flows the other way.

Of course, a monopoly utility can, if its regulators allow or even encourage such nonsense, pay four cents for nothing and then recover its loss from its captive customers who have no say in the transaction and get no benefits from it. But this use of monopoly power to tax consumers hardly makes the value-for-nothing payment for a saved kWh equivalent in economic or business terms to the value-for-value payment for a produced kWh. Even if the utility is losing money on incremental sales, the value to the utility of not-selling a kWh is only what it is losing on marginal sales, which (unless the utility is giving away kWh) is nowhere near the full value of the kWh being offered by a supply-side bidder.

In effect, a utility always has a standing offer on the table to "buy back" kWh at the retail price of, e.g., five cents: a consumer is always free to take one less kWh and keep five cents. If the consumer chooses not to sell back some kWh at the going price of five cents, it must be because it would cost at least five cents to get the same end-use service some other way. If the utility offers four cents for saved kWh in addition to the five cent standing offer, the total payment for saved kWh becomes nine cents, which will induce consumers to incur costs up to that level to save kWh. The utility will buy more saved kWh if it offers nine cents rather than five; but that does not make it cost-effective for it to do so.

Charles Cicchetti and William Hogan of Harvard's Energy and Environmental Policy Center have developed a clever approach to demand-side bidding, based on the view that a utility provides end-use services rather than just commodity electricity (*Public Utilities Fortnightly*, June 8, 1989). As a provider of end-use services, a utility should charge for the service provided, not just for the kWh bundled in the service. Since a utility payment for a saved kWh allows the consumer to obtain the same end-use service with fewer kWh, it is only fair and efficient that the consumer should continue paying the same amount to the utility, i.e., should pay not only for the kWh it continues to consume, but also for the kWh it has been able to save as a result of the utility's payment.

Looked at yet another way, the only way a non-generating consumer can obtain any kWh "resources" to sell to the utility is to buy more kWh than it consumes. Thus, the total kWh the saving consumer must pay the utility for is the sum of the kWh it consumes and the saved kWh it is paid for selling back.

So there is a "right" way to include conservation options in an all-source bidding process: the utility selects the winning bidders on the basis of the bid prices, pays all winners the market-clearing price, but adds the saved kWh onto the demand-side bidders' meter readings and charges the retail price (less distribution-related costs) for them. Of course, if the retail price is approximately the same as the marginal supply cost (as it should be), winning a demand-side bid is not worth much because the saving consumer will pay as much to a kWh "resource" as it obtains from selling it back. But this is just the appropriate reflection of the fact that consumers already have a conservation incentive equal to the retail price, and hence this amount should be subtracted from any payment made for conserved kWh in order to give the right price signals.

Demand-side bidding *á la* Cicchetti and Hogan could help correct some pricing distortions and finance conservation investments; the utility would make an up-front payment to a consumer who invests in conservation measures, with the loan/investment paid off through the monthly additions to the utility bill. This is a perfectly logical and economically correct approach to buying saved kWh; the fact that it is so different from the negawatt auction and "qualified conservation provider" ideas being advocated in Congress and elsewhere merely demonstrates the economic flaws in these ideas.

The Level Playing Field for Environmental Benefits

Finally, there are the environmental benefits of energy conservation—more accurately, the environmental costs of energy production. There is no doubt that energy production imposes costs on society through its effects on the environment, or that these costs should be taken into account in comparing generation and conservation alternatives. Nor is there any doubt that conservation on a large scale can be a very cost-effective response to higher energy costs; the experience of the last 20 years has demonstrated conclusively that energy demand will respond dramatically to price signals. The real issues concern how best to take environmental costs into account and, in particular, whether direct conservation subsidies are a cost-effective environmental policy compared to the supply-side and demand-side alternatives.

With insignificant exceptions, the *consumption* of electricity is environmentally benign; it is the *production* of electricity that damages the environment, and then to widely varying degrees, depending on how the electricity is produced. Burning strip-mined high-sulfur coal in an inefficient, uncontrolled power plant causes far more environmental damage per kWh of electricity produced than does burning natural gas in a high-efficiency combined cycle plant with nitrogen oxide controls. Subsidizing the conservation of kWh independent of how the kWh are produced will be far less beneficial for the environment than incurring the same cost to conserve selectively those kWh produced in the most damaging ways.

In general, the way to obtain the most environmental benefit for a given total cost is to shift consumption from relatively "dirty" to relatively "clean" kWh, and then to conserve even the cleaner kWh to the extent justified by their higher cost and remaining environmental damages. The best way to accomplish this is to be sure that each production activity pays its full social costs, including the costs of appropriate environmental controls and any remaining environmental effects. This will encourage producers to seek cost-effective ways to reduce their environmental impacts and to pass through in consumer prices the costs of both environmental controls and remaining environmental damages, encouraging consumers to conserve energy with the appropriate emphasis on those sources that are most difficult to control.

Supply-side pollution taxes and marketable discharge permits are conceptually the best way to internalize environmental costs, although reasonable results can be obtained with other supply-side policies. However, demand-side subsidies for conservation attack the problem from precisely the wrong end; such subsidies provide no price signals or incentives for the decisions that directly affect the environment—decisions about fuels, technologies, and locations used to produce the kWh. Paying a consumer four cents to induce conservation of a random or "average" kWh, if paying one cent to an equipment vendor would convert a dirty kWh into a clean kWh, is bad economics and bad environmental policy; it accomplishes less than it should and reduces society's willingness and ability to solve environmental problems.

The inefficiency of attacking supply-side environmental problems with demand-side programs can be illustrated with the control of sulfur oxide emissions from power plants. An emission charge of 25 cents per pound of sulfur dioxide would almost surely reduce sulfur dioxide emissions by 50 percent or more. The resulting higher consumer prices would also encourage conservation; but the total effect on consumer prices of the control costs plus the charge on remaining emissions would be less than a cent per kWh, even for a utility that was originally relatively "dirty." If this price increase reduced electricity sales for this dirty utility by 10 to 20 percent, it would reduce emissions an additional 5 to 10 percent of the original level, compared to the 50 percent reduction likely to be accomplished on the supply side at the same per-unit cost.

Put another way, 25 cents spent on electricity conservation for which a utility must pay four cents/kWh would save about six kWh. If it were possible to assure that only very dirty kWh were conserved, this could reduce sulfur dioxide emissions by 0.2-0.5 pounds; if only "average," post-supply-side-control kWh are conserved, the reduction in sulfur emissions could be an order of magnitude less. The same 25 cents spent on scrubbing of the remaining dirty sources could remove several pounds of sulfur. Thus, demand-side measures to reduce sulfur emissions could easily be 10 to 20 times less cost-effective than supply-side policies that attack sulfur emissions directly.

Even for reducing carbon dioxide emissions, where conservation is often regarded as the only effective solution, subsidizing conservation directly is likely to be an inefficient policy. Conserving one kWh will save roughly half a pound of carbon if the saved kWh is produced from typical coal, half as much if the kWh is produced from natural gas in a combined cycle plant, none if the kWh is produced from nuclear or hydropower. If a subsidy of four cents per kWh is necessary to induce conservation of one kWh, the implied control cost is about $160 per ton of carbon if only coal-generated kWh are conserved; if an "average" kWh is conserved, the implied carbon control cost could be well above $200 per ton of carbon.

A $200 per ton carbon tax/rebate would increase powerplant coal prices by a factor of three to five and would stimulate a dramatic and costly supply-side response: some combination of massive shifts to less carbon-intensive fuels, application of highly-efficient fossil fuel burning technologies, removal of carbon dioxide from stacks with disposal in the deep oceans or underground, tree planting on a global scale, etc. Whether society should be or will be willing to pay costs of this magnitude to control carbon dioxide is a question well beyond the scope of this analysis; after all, even adjusting to a warmer climate is an alternative with finite costs. But if society is not willing to bear carbon control costs of this magnitude on the supply side, there is no reason it should bear them on the demand side either.

If society is willing to pay carbon control costs at the level suggested by a $200 per ton carbon tax, to what extent is electricity conservation part of the most cost-effective response? There has been little analysis in this area, but some illustrative calculations are instructive. A

$200 per ton carbon tax would increase average electricity prices by (very) roughly four cents per kWh if there were no supply-side response, and less to the extent there are supply-side control measures costing less than $200 per ton; suppose the price increase is two to three cents per kWh, amounting to some 30 to 50 percent of average retail prices. Such a price increase might reduce electricity consumption 20 to 30 percent, reducing carbon emissions by this fraction of the emissions remaining after taking account of the supply-side response; i.e., if the supply-side response reduced emissions per kWh by half, conservation would contribute another 10 to 15 percent. Overall, the supply-side response in this example would be four to five times as large as the demand-side response.

These are illustrative figures only, not intended to prejudge the question of the importance of conservation in an overall environmental control policy. If supply-side measures are much less cost-effective than assumed above, conservation will have to bear the brunt of any emission control society is willing to pay for. But it is likely that supply-side measures will dominate a cost-effective approach to environmental control even where carbon dioxide is concerned, and that focusing on demand-side subsidies may turn out to be very costly and ineffective.

The potential costs of environmental control are far too high for society to tolerate inefficient policies. The emphasis must be on supply-side policies that attack these problems at their sources and cause downstream price signals to reflect upstream environmental costs. These downstream price signals will make utility (and non-utility) demand-side businesses profitable where conservation is a cost-effective response. However, if such direct environmental policies result in less energy conservation than demand-side advocates would like to see, this is not an argument for additional conservation subsidies. Quite the contrary: it is evidence that forcing or bribing consumers to save energy is not a cost-effective approach to environmental improvement.

Conclusions

Utility demand-side programs deserve to be given full and careful consideration as a potential way to give consumers better end-use energy services for their money, utilities an edge in an increasingly competitive world, and society a way to reduce the environmental costs of energy production. But in each of these areas, demand-side programs offer no free lunches and have no inherent advantages over supply-side programs. There are many ways for consumers to meet their needs, for utilities to prosper, and for the environment to be protected; all options, whether inside or outside the utility and whether on the supply-side or the demand-side, should be considered on a truly level playing field that is not tilted and confused by misleading jargon and concepts.

If utility demand-side programs make sense on economic and business grounds, they should be able to meet the standard economic and business tests applied to other utility and non-utility activities. If utility demand-side programs make sense on environmental grounds, they should be able to demonstrate their cost-effectiveness relative to other, primarily supply-side, measures society is willing to undertake to control environmental effects.

If this all sounds simple and obvious, it is. There is nothing about utility demand-side programs that cannot be understood by applying the same economic and business commonsense that works for utility supply-side programs and for non-utility activities.

Encouraging Electric-Utility Energy-Efficiency and Load-Management Programs

Eric Hirst
Energy Division
Oak Ridge National Laboratory*
Oak Ridge, TN

Abstract

During the past few years, electric utilities have begun to view conservation and load-management programs as low-cost alternatives to some power plants. In addition, such programs save money for customers, improve utility relations with customers and regulators, reduce financial risks to utility shareholders, and improve environmental quality. Utilities could play much larger and more active roles in acquiring such demand-side resources. State regulators and the US Department of Energy (DOE) can play important supporting roles.

Introduction

Electric utilities account for almost 40 percent of the nation's primary energy consumption (Figure 1). Further, electricity production is capital intensive and environmentally damaging. Electric power plants account for one third of carbon dioxide emissions, two thirds of sulfur dioxide emissions, and one third of nitrogen oxide emissions in the US. Therefore, improving the efficiency with which customers use electricity not only saves money but also provides large indirect benefits.

During the past several years, utilities have increasingly turned to programs that affect the efficiency and timing of customer electricity use, called demand-side management (DSM) programs. Such programs include promotion of new lighting systems in office buildings; efficient appliances in homes; adjustable-speed drives for industrial motors; and direct control of electricity-intensive equipment, such as air conditioners, at certain critical times. These utilities recognize the benefits of focusing on electric-energy *services* rather than on electricity as a *product.* Such energy-efficiency and load-management efforts can:

- Provide low-cost alternatives to construction of new power plants,
- Save money for customers,
- Improve relations with customers by providing additional services,
- Improve relations with state public utility commissions (PUCs),
- Reduce financial risks to utilities,
- Improve environmental quality, and
- Enhance the economic competitiveness of utilities and their customers.

* Work reported here was sponsored by the Office of Conservation and Renewable Energy, US Department of Energy.

Past Progress and Future Potential

Between 1950 and 1973, US electricity use increased at an average rate of 8 percent per year (Figure 2). Between 1973 and 1988, electricity use grew at only 2.7 percent per year (Energy Information Administration [EIA] 1989). This slower growth was caused by increases in electricity prices, government efficiency standards for appliances and new construction, utility DSM programs, and development of new technologies.

The number and scope of utility DSM programs have increased during the past few years (Cogan and Williams 1987):

> More than 85 percent of the 123 utilities surveyed by IRRC [Investor Responsibility Research Center] have implemented demand-side management programs, including 101 with "formal" conservation programs and 68 with "formal" load management programs. More than half of these programs have been established during the 1980s, indicating that demand-side management is a relatively recent and still-evolving phenomenon.

IRRC estimated that utilities spent more than $1 billion in 1988 on DSM programs and cut peak demand by 21 GW (a 4 percent reduction in peak demand), almost double the savings achieved five years earlier.

Three of the many utilities offering DSM programs are noted here (Hirst 1989). Duke Power, one of the nation's largest utilities, has operated such programs since the mid-1970s. In 1987, these programs cut sum-

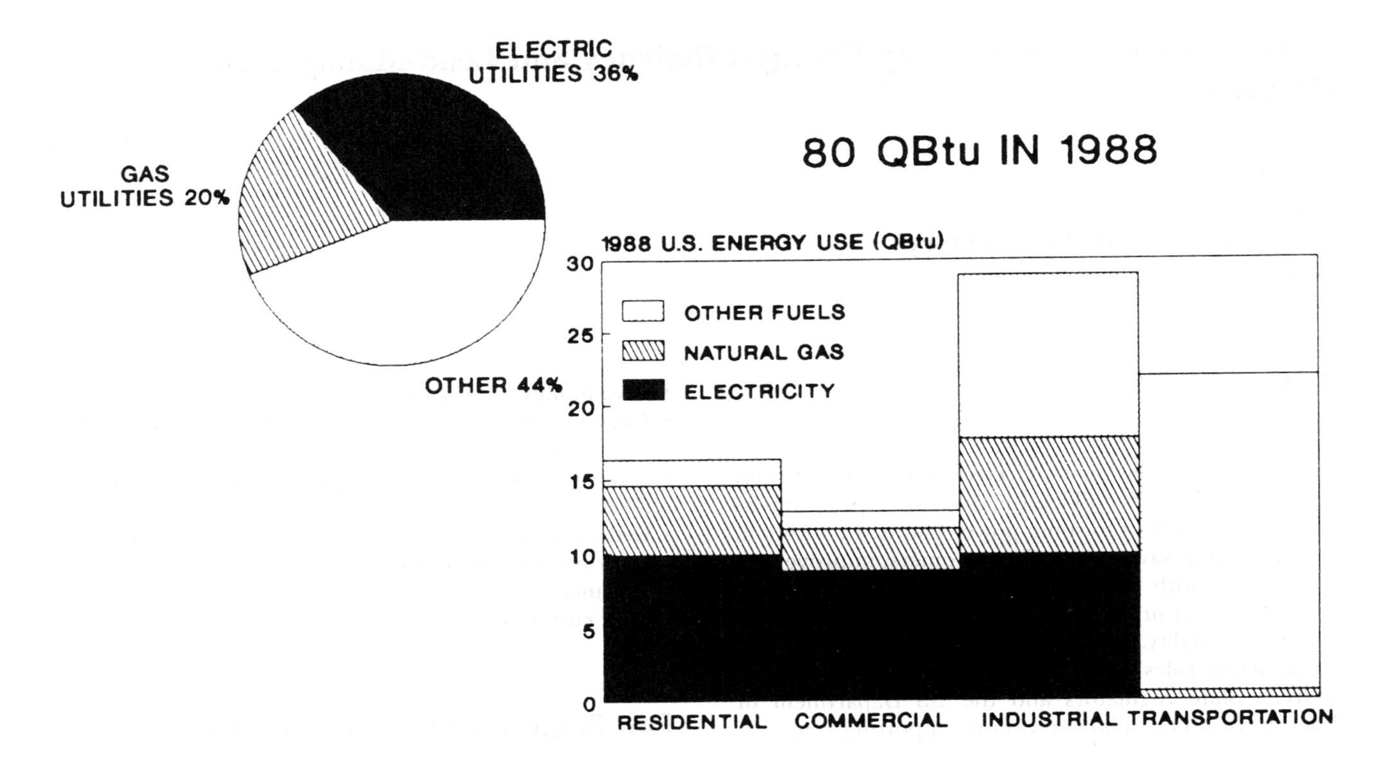

Figure 1. Almost 40% of US primary energy consumption is devoted to electricity production. The residential, commercial, and industrial sectors use roughly equal amounts of electricity.

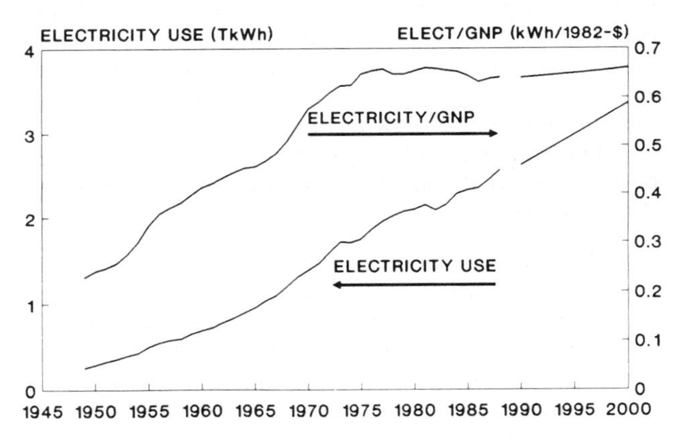

Figure 2. US electricity use and the ratio of electricity use to gross national product, from 1949 through 1988 and the EIA forecast to the year 2000.

mer and winter peaks by 18 percent and 24 percent, respectively. Duke's residential programs promote construction of efficient new homes, use of high-efficiency heat pumps rather than resistance heating, and purchase of energy-efficient appliances. In addition, Duke operates load-control programs that interrupt residential air conditioners and water heaters at critical times. Commercial-sector programs are aimed at lighting, insulation, heating equipment, and interruptible and standby-generator programs. Finally, industrial programs deal with load control, baseloads, heating equipment, time-of-day rates, and interruptible and standby-generator programs.

Central Maine Power, a medium-sized utility, has been aggressively running DSM programs since 1984. Savings from its programs were expected to reach 3 percent of total sales in 1989. Interruptible rates alone cut the 1988 winter peak by 7 percent. The company spends more than $20 million a year on its programs, roughly 3 percent of operating revenues.

Puget Power, a medium-sized utility in Washington State, has been offering rebates and loans for energy-efficiency improvements in existing buildings since 1978. (Because the Pacific Northwest electric system is dominated by large hydro facilities, the system is energy constrained, not demand limited, as are most US electric systems. Therefore, most DSM programs in the Northwest focus on energy efficiency rather than on load management.) Recently, Puget Power expanded its program to include financial incentives for construction of energy-efficient homes and design assistance for new commercial buildings (Table 1). The company spends almost 3 percent of its annual revenues on these energy-efficiency programs.

Although many improvements in electric-energy efficiency have been achieved during the past 15 years, the well is far from dry. The large potential for additional savings comes from many small improvements. These improvements include wider applications of known technologies, prompt commercialization of new technologies, and development of new ideas. Technological opportunities exist for most end use/customer class combinations. These technologies include advanced electric heat pumps; "smart" buildings; high-efficiency motors and controls; industrial cogeneration; and new lamps, fixtures, ballasts, and lighting controls. Utility efforts are especially important in hastening adoption of known and emerging technologies.

Research sponsored by the US Department of Energy (DOE) has helped to create new energy-efficient technologies. For example, DOE's Oak Ridge National Laboratory is developing advanced electric heat pumps to improve efficiency by 40 percent by the year 1995. Computer models to help design new variable-speed, capacity-modulation heat pumps were recently made available to manufacturers. Several other DOE research projects (Brown, Berry, and Goel 1989), including low-emissivity windows, solid-state ballasts for fluorescent lights, dielectric coatings for lighting fixtures, unequal parallel compressors for supermarket refrigeration, heat-pump water heaters, and radiant barriers, are likely to gain significant market penetration during the next several years, improving the efficiency of electricity use in buildings.

Several studies have combined individual improvements, examples of which are noted above, to estimate electric-energy efficiency potentials for New York, Michigan, and the Pacific Northwest. The New York study (Miller, Eto, and Geller 1989) identified a total cost-effective savings, aggregated over all sectors, equal to one third of both annual use and summer-peak demand in 1986 (Table 2). Plans filed by the utilities with the New York Public Service Commission in 1988 show that they intend to reduce peak demand by only one fifth of this potential by the year 2000.

Table 1. Puget Power energy efficiency programs.

Program	Cumulative cost, 1978–1988 (million-$)	Estimated savings (GWh/year)
Residential		
retrofit	122	244
water heating	11	227
new construction	3	3
Commercial/industrial	46	226
Street lighting	4	11
totals	186	711

Source: Swofford (1989).

Suggestions for Future Utility Activities

Understand End-Use Markets

The amount of information on the DSM resources that electric utilities can deploy is much less than the amount available on power plants. This difference exists largely because utilities have been operating power plants for a century, but have been running DSM programs for only a decade.

Utilities could conduct more technology assessments to ensure that they stay abreast of developments in load-management and energy-efficiency technologies. Recent advances in heat pumps, motors and controls, computerized energy-management systems, lighting, and other end-use systems can only be incorporated into DSM programs if the utility knows about and understands these technologies.

Utilities also need to learn more about their customers, market segment by market segment. Utilities need to monitor closely the electricity-use patterns and trends of their customers, conduct periodic surveys and focus groups, and listen closely to their field personnel. Be-

Table 2. Partial results of New York study on electricity-saving potentials.

Commercial-sector measures	Cost of conserved energy (¢/kWh)	Potential savings[a] (GWh/year)
Reset supply air temperature	0.5	1200
New reflectors in lights	1.0	4100
Variable-air-volume conversion	1.3	2800
Variable-speed drive on fans	2.1	3300
Resized chillers	3.8	2300
Daylighting controls	4.6	1700
Other measures costing less than 5¢/kWh		2700

[a] The total savings shown here (18,100 GWh) is 45% of 1986 commercial sector electricity use in New York.

Source: Miller, Eto, and Geller (1989).

cause participation in DSM programs and the operation of DSM systems depend on customers, utilities need to know more about their customers' attitudes toward and interest in different kinds of programs.

Load research data and cost-of-service studies can provide valuable information on the temporal patterns of electricity use and the costs to provide electricity to each customer class. Although collecting end-use load data is expensive, sophisticated sampling methods can reduce the number of facilities that need to be metered (Northeast Utilities 1990). Also, methods can be developed to transfer such data from one location to another, further reducing the cost of such data.

Consider a pilot program run by New York State Electric and Gas Corporation as an example of the types of experiments utilities should run to assess alternative ways to market DSM programs (Xenergy 1989). The utility designed this pilot to assess different approaches to encourage participation in its planned energy-audit program for commercial customers. Potential participants were contacted either by mail or through on-site personal visits, using three different fee structures. Results showed that, in spite of the high cost of personal contacts, the cost per audit "sold" was considerably lower than the cost per audit sold by direct mail marketing (Figure 3).

Finally, utilities need to evaluate the electricity savings and costs of their DSM programs. Utility confidence in the performance, reliability, and cost effectiveness of these programs will increase only as empirical evidence is produced (Hirst 1990; California Collaborative 1990).

Aggressively Acquire Energy and Capacity Resources through DSM Programs

Utilities could build on the technology and market research described above to implement programs that acquire all cost-effective DSM resources (i.e., that cost less than the power plants they displace). The term "acquire" is used to emphasize the importance of treating DSM programs as substitutes for some supply resources.

Various strategies are used to reach customers (Table 3), ranging from information and education approaches to financial incentives and utility installation of DSM measures. Innovative program ideas continue to emerge beyond those summarized in Table 3. Recently, utilities in New England have acquired DSM resources through competitive bidding (Cole, Wolcott, and Weedall 1988). A group of state agencies and utilities in Rhode Island combined forces to develop new lighting programs aimed at commercial customers (Berry and Hirst 1989). And the Conservation Law Foundation established collaborative projects with several New England utilities to design and implement energy-efficiency programs.

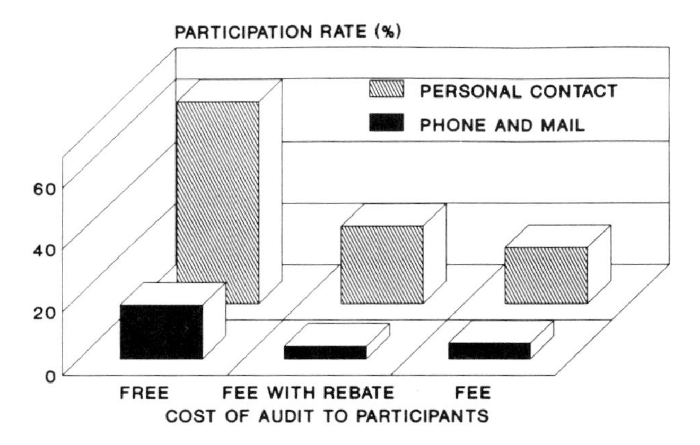

Figure 3. Participation rates in a commercial energy-audit pilot program operated by New York State Electric and Gas Corporation (Xenergy 1989).

Integrate DSM Programs into Long-Term Resource Planning

Integrated resource planning (IRP) is a new process that consistently assesses various demand and supply resources to meet customer energy-service needs at the lowest economic and social costs (Cavanagh 1986; Hirst 1988; National Association of Regulatory Utility Commissioners 1988). IRP involves deliberations among utility planners and executives, PUCs, and customers. These deliberations lead to development of a plan that will ensure reliable and low-cost electric service to customers, financial stability for the utility, a reasonable return on investment for shareholders, and protection of the environment.

Typically, a utility begins its IRP process by developing alternative forecasts of future electric loads. The utility then estimates the costs and remaining lifetimes

Table 3. Types of demand-side management programs.

General information programs (e.g., mass media, bill stuffers, and direct mailings)
Onsite energy audits and technical assistance (e.g., home energy audits, design assistance for architects and engineers, and workshops)
Financial incentives (e.g., rebates, low-interest loans, and rate discounts)
Direct installation (e.g., water-heater wraps and low-flow showerheads)
Cooperation with trade allies (e.g., manufacturers and dealers, architects, engineers, builders, and community organizations)

of its existing resources, and identifies the need for additional resources.

The utility next assesses a broad array of alternatives that could satisfy the need for more electricity, including supply, demand, transmission, distribution, and pricing options. Different combinations of these resources are then analyzed to see how well they meet future electricity needs and how expensive they are. These analyses are repeated time and again to test diverse resource portfolios for their resilience against various uncertainties. Such uncertainty analysis helps to identify a mix of options that meets the growing demand for electricity, is consistent with the utility's corporate goals, avoids exposure to undue risks, and satisfies environmental and social criteria. The long-term resource plan of New England Electric (1989), for example, relies on a diverse mix of resources. DSM programs are expected to provide more than one fourth of the resource additions during the next 20 years (Figure 4).

Then the utility prepares a formal report, which presents the preferred resource plan. After approval by the PUC, the plan is implemented and resources are acquired. While the plan is in force, the utility monitors changes in its environment and its implementation of the plan, and the plan is modified as events and opportunities change over time.

IRP is relevant here because 1) it explicitly includes DSM programs as alternatives to power plants, 2) it involves public participation, 3) it considers environmental factors as well as direct economic costs, and 4) it requires careful analyses of the risks that different resources pose. Public participation is important because both regulators and customers often prefer DSM programs to power plants and transmission lines. Environmental factors are important because of growing concerns about acid rain, global warming, and other forms of pollution; and because of local opposition to construction of power plants, transmission lines, and even substations (the "not-in-my-backyard" syndrome). Uncertainty analysis is important because DSM programs have inherent advantages over power plants in risk reduction.

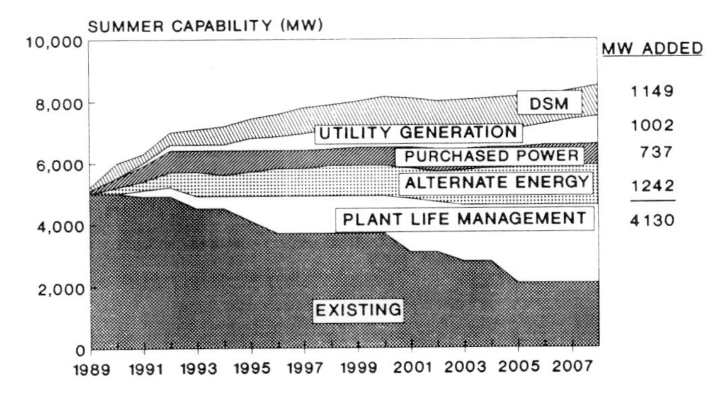

Figure 4. The demand and supply resources that New England Electric (1989) plans to use to meet growing loads between 1989 and 2008.

Work with PUCs to Update Rate Regulation

Utilities are properly concerned that the electricity savings afforded by their DSM programs will increase electricity prices (by spreading fixed costs over fewer kWh sales). The National Association of Regulatory Utility Commissioners (1989) adopted a resolution stating that "a utility's least-cost plan for consumers should be its most profitable plan. However, because incremental energy sales increase profits, traditional rate-of-return calculations generally provide much lower earnings to utilities for demand-side resources than for supply-side resources."

These disincentives to DSM programs have sparked debates over the need to reform ratemaking to align the financial interests of utility shareholders with the interests of customers. Several proposals have been made that incorporate one or more of the following elements:

- Utility recovery of the net lost revenue (difference between revenue forgone because of reduced consumption and reduction in operating costs) caused by DSM programs,
- Recovery of the utility costs to operate DSM programs, and
- Provision of financial incentives to utilities for exemplary delivery of DSM services.

The underlying idea is to make DSM investments profitable for utilities. The National Association of Regulatory Utility Commissioners, with support from the DOE Least-Cost Utility Planning (LCUP) Program, analyzed diverse proposals (Moskovitz 1989). Utilities should work with their PUCs to explore and implement such new regulatory formulas. Already, such methods are being tested in Massachusetts, New York, and Rhode Island.

In addition, utilities and their commissions need to examine regulatory practices related to resource planning, acquisition, and financing (Hirst 1988). In particular, agreement must be reached on the appropriate economic tests to use in assessing DSM programs and on whether DSM program costs should be capitalized or expensed.

Promote Government Energy-Efficiency Standards

Electric utilities are not always the best agents to overcome market barriers that prevent adoption of cost-effective efficiency improvements. Improving efficiency of new buildings, appliances, and equipment may be accomplished most efficiently through state or federal standards that mandate minimum efficiency levels.

Although utilities do not play a primary role in implementing standards, they can play important supporting roles. Utilities are often influential in state legislatures and can use this influence to encourage adoption of meaningful standards. Utility organizations, such as the Edison Electric Institute and the American Public Power Association, can similarly affect federal efficiency standards.

As shown in the Pacific Northwest, utilities can train local code officials and builders and provide financial incentives for energy-efficient construction that meets local codes. The Bonneville Power Administration runs a Super Good Cents program, which provides cash rebates to builders that meet the region's Model Conservation Standards.

Utilities can implement programs that encourage efficiency improvements beyond those required by current standards. The Bonneville Power Administration provides design assistance to architects and engineers to encourage construction of new commercial buildings that are more efficient than required by current codes. And New England Electric offers design assistance and rebates for installation of cool-storage systems in new commercial buildings because such systems cut peak demands, but are not required by state building codes.

The National Appliance Energy Conservation Act of 1987 (P.L. 100-12) was passed with the support of electric utilities. McMahon (1990) estimates that the standards for refrigerators and freezers alone will save 52,000 GWh/year and save consumers almost $3 billion by the year 2015. Electric utilities could encourage manufacturers to produce, dealers to stock, and consumers to purchase appliances whose efficiencies exceed those of the federal standards.

Develop Innovative Pricing Strategies to Modify Loads and to Encourage Participation in DSM Programs

The crucial role of electricity pricing in affecting the level and timing of electricity use is well recognized. Traditionally, rates have been based on average costs. Since passage of the Public Utility Regulatory Policies Act (P.L. 95-617), debates about marginal-cost pricing have occurred in most states. Participants agree that pricing electricity close to its marginal cost is desirable for economic efficiency. Differences occur, however, on how to define marginal costs, on whether to use short-term or long-term marginal costs, and on how to apply such pricing schemes to customers that use little electricity.

Recently, many utilities have offered additional rate options, usually to influence temporal use patterns. For example, Potomac Electric Power Company (1989) planned to have more than half its retail revenues billed under time-of-use rates by 1990. The company already has 300 of its largest commercial and government customers on interruptible rates (which offer rate reductions to customers willing to shed loads when system peaks or emergencies occur).

Utilities could expand their use of pricing as a marketing tool and set prices to encourage participation in DSM programs. For example, utilities in North Carolina offer residential customers a discount if their homes meet certain energy-efficiency criteria.

Finally, advances are rapidly occurring in microcomputers, telecommunications, and other technologies. These advances could affect meter reading, billing, and the costs of time-of-use metering, demand metering for residential customers, and smart meters for real-time pricing and should be assessed for their applicability and cost-effectiveness.

Federal Actions to Encourage Utility DSM Programs

Because electricity production consumes almost 40 percent of the primary energy used in the US, electricity must be a major part of national energy policy. In addition, concerns about environmental quality, economic productivity, international competitiveness, and national security suggest a larger role for the federal government in working with utilities to implement DSM programs. Improving energy efficiency through electric utilities is a particularly effective way to reach energy consumers. Utilities have direct monthly contact with all their customers (i.e., meter reading and billing) and are generally well-respected organizations in their communities. Thus, the federal government can work with a few hundred utilities and, through them, reach tens of millions of households and millions of businesses.

Require Utility Wholesale Contracts to Incorporate DSM Programs

The Federal Energy Regulatory Commission (FERC) approves all wholesale transactions among utilities. Currently, FERC reviews of proposed contracts involve no consideration of energy efficiency. FERC could require the buying utility to demonstrate that it has acquired all the DSM resources in its service area that cost less than the proposed purchased power before FERC approves the contract.

Consider, for example, a proposed 10-year contract for purchase of baseload power at a cost of 6 cents/kWh. In its application to FERC, the purchasing utility would have to include documentation showing its plans to acquire all the end-use efficiency improvements available in its service area at a total cost (including the utility's administrative costs) of less than 6 cents/kWh. Similar requirements for load-management programs would apply to purchases of peaking power. Presentation of an integrated resource plan, approved by the utility's PUC, could satisfy this requirement.

Require Federal Utilities to Expand DSM Programs

The federal Power Marketing Agencies (PMAs, part of DOE) and the Tennessee Valley Authority (TVA, an independent federal corporation) account for almost one tenth of the electricity consumed in the United States. Traditionally, TVA and the Bonneville Power Administration (the largest PMA) have operated large DSM programs, which not only saved energy for their customers but also served as examples for other utilities. Unfortunately, short-term budget considerations forced substantial reductions in these programs. Indeed, TVA recently canceled all its conservation programs.

New legislation could require these federal power authorities to expand their conservation programs and to consider explicitly environmental factors in their benefit/cost analyses of all resource alternatives. Such legislation would be a logical extension of the Pacific Northwest Electric Power Planning and Conservation Act (P.L. 96-501), which explicitly made conservation the electricity resource of choice and gave it a 10 percent bonus in economic analyses of alternative resources. The 10 percent bonus reflects the environmental and social benefits of conservation relative to power plants. The other federal utilities could employ similar factors in their resource assessments.

Expand DOE Technology-Transfer Activities to Utilities

DOE's LCUP Program (Goldman, Hirst, and Krause 1989; Berry and Hirst 1989; DOE 1990) manages a variety of projects aimed at improving the resource-planning processes, data, and analytical methods used by utilities. DOE also sponsored conferences on utility planning in 1988, 1989, and 1990 as well as five regional workshops in 1990. LCUP could be expanded to fund additional cooperative projects with utilities and PUCs. This approach focuses on cost-sharing projects, with DOE assistance provided through the national laboratories and other government contractors.

The underlying rationale for this option is the knowledge that many innovative and successful programs operate throughout the country. However, information on these successes is hard to obtain because the sponsoring utility or state agency has little incentive to publish information on the program. Thus, DOE can play a valuable technology-transfer role by participating in these programs, ensuring that the programs are carefully evaluated, and then funding preparation of reports and conference presentations that effectively and widely disseminate information to other utilities and state agencies. The Northeast Region Demand-Side Management Data Exchange (NORDAX), funded in part by DOE, is a good example of such technology transfer. The initial phase of NORDAX, a consortium of more than 20 utilities, yielded a database with information on 90 DSM programs operated by 17 utilities (Berry and Hirst 1989).

Conduct More Research and Development on Energy-Efficient Technologies

DOE's Office of Conservation sponsors a variety of R&D projects aimed at improving efficiency of energy use in the residential, commercial, and industrial sectors. Many of these projects are co-funded by industrial partners, including the Electric Power Research Institute.

Additional research, focusing on demonstration of new technologies, would hasten commercialization of energy-efficient, electricity-using systems. Commercial lighting, industrial motors, and residential refrigerators and air conditioners are especially promising because they are major electricity end uses and because their potential for efficiency gains is large.

Collect More Information on Electricity Use

The Energy Information Administration is responsible for collecting, evaluating, analyzing, and disseminating information on energy reserves, production, demand, and technologies. The bulk of EIA's efforts are related to the supply of, rather than the demand for, energy. For example, EIA's (1989) *Annual Energy Review* contains separate chapters on fossil-fuel reserves, petroleum, natural gas, coal, electricity, nuclear energy, renewable energy, financial indicators, and only one chapter on energy consumption.

EIA collects detailed information from electric utilities on individual power plants related to their construction cost and capacity; annual operations and maintenance expenses; and monthly fuel consumption, generation, availability, and air emissions. And FERC's Form 1 collects information on construction costs and operations for power plants, and the costs and characteristics of transmission lines, substations, and transformers. Unfortunately, neither EIA nor FERC collects comparable data about utility energy-efficiency and load-management programs.

The forms completed by electric utilities for EIA and FERC should be expanded to include information on utility DSM programs consistent with information reported on power plants (Table 4). FERC's Form 1, for example, requires data on the operating and maintenance costs of power plants, broken down into 63 separate elements. On the other hand, Form 1 includes not a single identifiable element for DSM program costs.

EIA conducts detailed energy end-use surveys of households, commercial buildings, and manufacturers. These onsite surveys are conducted once every three years. (By comparison, EIA collects *monthly* data on production of all fuels and electricity.) The end-use surveys collect valuable information on the energy-related characteristics of the facility and its energy-using equipment and on the occupants. This information is then linked to a year of fuel-consumption data for the building or industrial plant.

Additional information on the patterns, trends, and determinants of energy use would be very helpful in designing and managing energy-efficiency programs. Such information might indicate whether and why consumers participate in various types of conservation programs, what factors influence their purchase of energy-using equipment, and what factors affect their operation and maintenance practices. And EIA could expand the data-collection forms completed by utilities to require information on their DSM programs. These activities would help to redress the substantial imbalance between the supply and demand sides in EIA's data-collection activities. In addition to expanding EIA's collection of data related to energy use and to DSM programs, DOE could devote more effort to analyzing these data to improve our understanding of energy-related behaviors in all sectors of the economy.

Conclusions

Improving the efficiency with which electricity is used can save money for US electricity consumers, reduce the need to build new generation and transmission facilities, improve the financial performance of utilities, reduce emissions of greenhouse gases and other pollutants, improve economic productivity, and improve utilities' relations with their customers and regulators.

Electric utilities, with the support of their public utility commissions, can play a vital role in realizing this large potential to save energy and cut peak demands. Utilities need to learn more about their customers and about energy-efficient technologies. This knowledge must be used to design, test, implement, and evaluate a broad and vigorous array of DSM programs. Information programs, site-specific technical assistance, financial incentives, direct installation of measures, and cooperation with trade allies should be deployed to reach all major end uses in all sectors.

The federal government, working with a few hundred utilities, can influence the energy-efficiency actions of tens of millions of consumers. Useful federal activities include improved regulation of wholesale power transactions, expansion of DSM programs run by federal utilities, expansion of DOE's Least-Cost Utility Planning Program and other technology-transfer activities, research and development support for energy-efficient electric technologies, and collection of more data on electricity use. Fortunately, the federal actions proposed here require little additional funding and will therefore not worsen the federal deficit.

Table 4. Demand-side analogs to data reported by utilities on power plants.

DSM programs	Power plants
Participation rate	Availability and capacity factors
Energy savings	Heat rate
Program costs	Plant costs
Fixed	Construction and fixed O&M
Variable	Variable O&M and fuel costs

References

Berry, L. and Hirst, E., *Recent Accomplishments of the US Department of Energy's Least-Cost Utility Planning Program,* ORNL/CON-288, Oak Ridge National Laboratory, Oak Ridge, TN (August 1989).

Brown, M. A., Berry, L. G., and Goel, R. K., *Commercializing Government-Sponsored Innovations: Twelve Successful Buildings Case Studies,* ORNL/CON-275, Oak Ridge National Laboratory, Oak Ridge, TN (January 1989).

California Collaborative, "Measurement Protocols for DSM Programs Eligible for Shareholder Incentives," Appendix A of *An Energy-Efficiency Blueprint for California,* California Public Utilities Commission, San Francisco, CA (January 1990).

Cavanagh, R. C., "Least-Cost Planning Imperatives for Electric Utilities and Their Regulators," *The Harvard Environmental Law Review* 10(2), 299–344 (1986).

Cogan, D. and Williams, S., *Generating Energy Alternatives,* Investor Responsibility Research Center, Washington, DC (1987).

Cole, W. J., Wolcott, D. R., and Weedall, M., "Competitive Bidding of Demand-Side Management," *Proceedings of the 1988 ACEEE Summer Study on Energy Efficiency in Buildings,* American Council for an Energy-Efficient Economy, Washington, DC (August 1988).

Energy Information Administration, *Annual Outlook for US Electric Power 1988,* DOE/EIA-0474(88), US Department of Energy, Washington, DC (August 1988).

Energy Information Administration, *Annual Energy Review 1988,* DOE/EIA-0384(88), US Department of Energy, Washington, DC (May 1989).

Goldman, C., Hirst, E., and Krause, F., *Least-Cost Planning in the Utility Sector: Progress and Challenges,* LBL-27130 and ORNL/CON-284, Lawrence Berkeley Laboratory, Berkeley,

CA, and Oak Ridge National Laboratory, Oak Ridge, TN (May 1989).

Hirst, E., "Integrated Resource Planning: The Role of Regulatory Commissions," *Public Utilities Fortnightly* 122(6), 34–42 (September 15, 1988).

Hirst, E., *Electric-Utility Energy-Efficiency and Load-Management Programs: Resources for the 1990s,* ORNL/CON-285, Oak Ridge National Laboratory, Oak Ridge, TN (June 1989).

Hirst, E., *Measuring Performance: Key to Successful Utility Demand-Side Management Programs,* Oak Ridge National Laboratory, Oak Ridge, TN (May 1990).

McMahon, J. E., "Technical Basis for DOE Appliance Standards," Lawrence Berkeley Laboratory, Berkeley, CA (April 1990).

Miller, P. M., Eto, J. H., and Geller, H.S., *The Potential for Electricity Conservation in New York State,* American Council for an Energy-Efficient Economy, prepared for the New York State Energy Research and Development Authority, Albany, NY (September 1989).

Moskovitz, D., *Profits & Progress through Least-Cost Planning,* National Association of Regulatory Utility Commissioners, Washington, DC (November 1989).

National Association of Regulatory Utility Commissioners, *Least-Cost Utility Planning, A Handbook for Utility Commissioners,* Washington, DC (October 1988).

National Association of Regulatory Utility Commissioners, *NARUC Energy Conservation Committee Discussion Paper on Conservation Profitability & Financial Incentives,* Working Draft #1, Washington, DC (February 1989).

New England Electric, *Conservation and Load Management Annual Report,* Westborough, MA (May 1989).

Northeast Utilities, *Conservation and Load Management Evaluation Plan, 1990–1994,* Hartford, CT (January 1990).

Potomac Electric Power Company, *1988 Annual Report,* Washington, DC (1989).

Swofford, G. B., Direct Testimony before the Washington Utilities and Transportation Commission, Puget Power, Bellevue, WA (1989).

US Department of Energy, *Least-Cost Utility Planning FY 1990 Program,* Office of Conservation and Renewable Energy, Washington, DC (January 1990).

Xenergy, Inc., *Final Report, Commercial Audit Pilot,* draft report prepared for New York State Electric and Gas Corporation, Binghamton, NY (1989).

Structural Factors Underlying the Increasing Demand for Electricity in Japan

Hisao Kibune
The Institute of Energy Economics
Tokyo, Japan, and
Center for Energy Policy Research
Massachusetts Institute of Technology
Cambridge, MA

Abstract

This paper surveys the characteristics of Japan's electricity demand, then uses this information to discern future trends. Major findings are as follows. First, saving electricity is difficult for structural reasons. These include economic structure, a steady penetration of appliances requiring electricity, and a powering-up of the appliances for longer and longer periods each day. Second, society now and for the future is increasingly supported and maintained by a flowering electronics technology. This technology tends toward electricity as its energy source, and so influences the demand for electricity. Third, the impact from this situation includes the likelihood that the demand for electricity will grow, a signal that it is time to review the reliability of supply, and that the power load gap may become sensitive to seasonal temperature extremes and the load gap between peak and bottom may widen. In this case, the supply side of the power industry should concern itself with overall rate reduction as well as greater diversity in the individual rates that are offered; improving the load factor by increasing demand in the lower demand segment and allocating peak demand across other energy suppliers; measures to insure reliability and backup for the supply, and institutionalizing a system for compensating accidents; and improving the efficiency of the power supply in generation and transportation, through co-generation, waste heat recovery, and on-site generators.

Introduction

In the past two or three years, the demand for electricity in Japan has surged more than 5 percent annually, surpassing the GNP growth rate. When compared with other energy sources, the stable growth in the demand for electricity stands out as remarkable. This relatively higher growth is a consistent trend from the decade of the 70s, the so-called higher-energy-prices era when total energy demand was stagnant. Against this background, energy conservation in fossil fuels has progressed dramatically, but electricity conservation has not done well and may suffer from structural impediments. In addition, the recent growth in electricity demand appears to be encouraged by economic recovery. As a consequence, electrification has developed significantly. Taking conversion losses into account, the primary energy supply has grown faster than final energy demand.

This paper aims to clarify why these trends have appeared and why electricity conservation is difficult, given the background situation that explains these trends, and then to propose which energy and electricity policies should be considered, through analyses of past and current trends in electricity demand. The paper is divided into seven sections, including this introduction. The next section examines past electricity demand and abstracts the structural factors that are at play. The third, fourth and fifth sections analyze in a concrete fashion the factors behind electrification in each demand sector: industrial, commercial, and residential. The sixth section examines the relation between electricity demand and the electronics technologies that strongly encourage electrification. The final section is a summary and conclusion.

New Trends in the Demand for Electricity

The Progress of Electrification

Comparing growth rates among energy sources, it is clear that electricity has a relatively stable growth rate (Fig. 1). The contribution, by sectors, to incremental electricity demand is shown in Figure 2. The drastic reduction of demand for all energy sources in the industrial sector has kept the growth of demand for primary energy stagnant. Nonetheless, the impact of this reduction, on just electricity, has not been so great in the past decade and a half.

Consequently, the structure of demand for energy in Japan has a tendency toward electrification. At the same time, the two oil crises led to a drastic reduction in and a much more efficient energy intensity per GNP. When we compare these two factors, electrification and intensity, we see that the ratio of electrification to the primary energy requirement is increasing continuously,

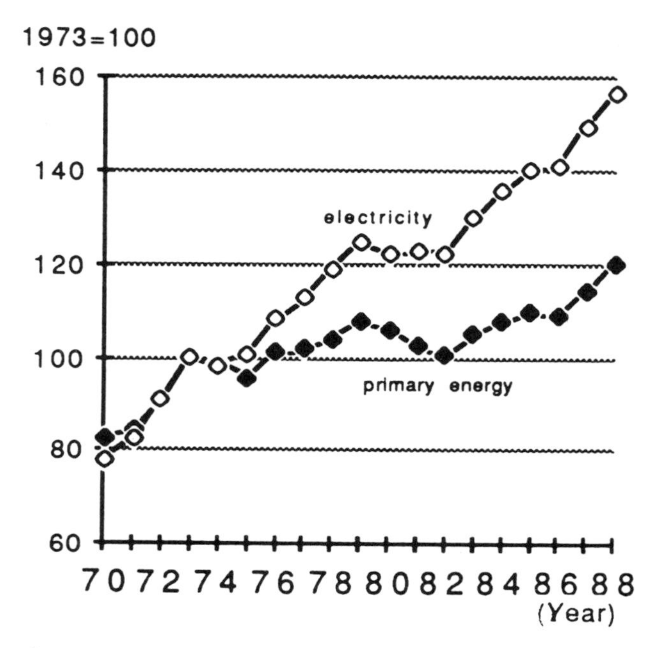

Figure 1. Primary energy and electricity requirements. Source: "Energy Balance Tables," IEE, Japan.

and that electrification looks like an inverse proportion to intensity (Fig. 3). This chart shows that as the energy demand structure changed to become a much more efficient one, the rate of electrification grew, from a macro point of view.

The demand in each sector shows a similar trend when we look at electrification rates by sector (Fig. 4). In 1970, the ratio of electrification in the industrial sector was computed to be 14 percent; by 1988 it had increased to 19 percent. A similar comparison in the residential sector shows 25 percent electrification in

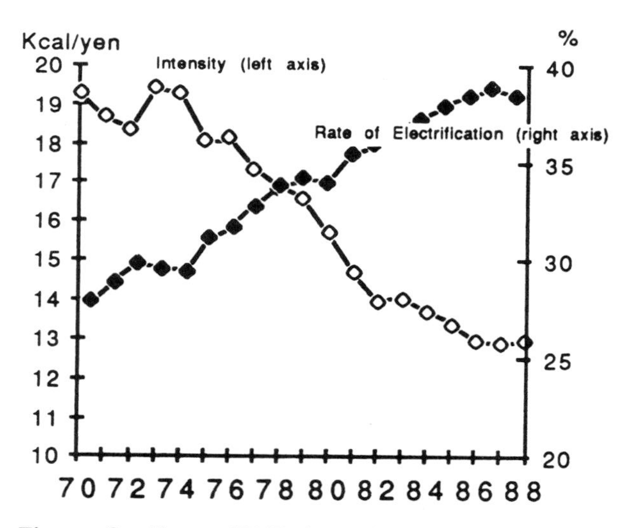

Figure 3. Energy/GNP intensity and electrification. Source: "Energy Balance Tables," IEE, and "National Accounts," Economic Planning Agency.

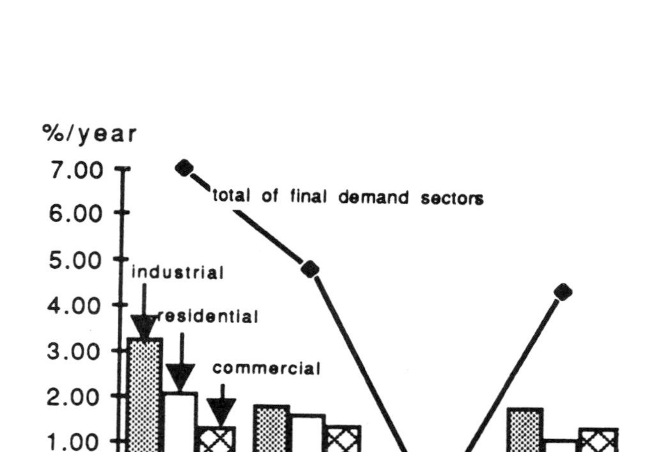

Figure 2. Electricity demand growth rate and its contribution by sector. Source: "Energy Balance Tables," IEE, Japan.

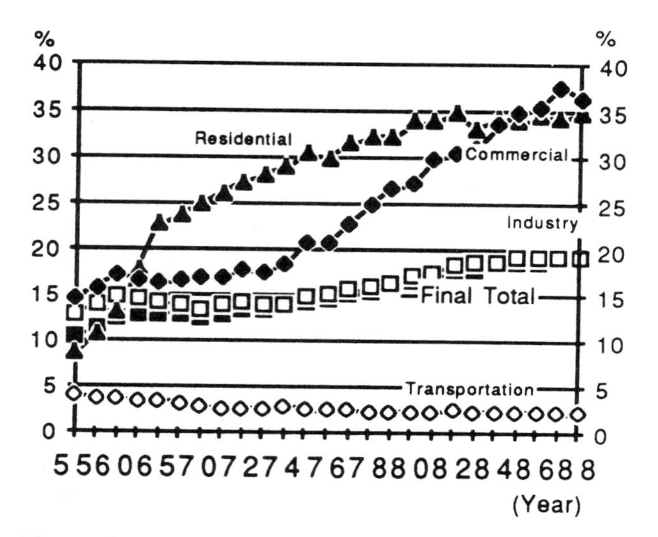

Figure 4. Trends of electrification by sector. Source: "Energy Balance Tables," IEE, Japan.

1970, growing to 35 percent by 1988. In the commercial sector the figures are 27 percent, growing to 38 percent. As a result, electricity's share is 19 percent of the final energy demand in 1988, up 6 points from the figure in 1970.

Factors Supporting Electrification

Four major factors lie behind these trends toward electrification.

The first factor is a shift in demand among sectors. In terms of overall shares, demand has shifted from the industrial sector, which relies mainly on nonelectricity sources, and toward the residential/commercial sectors, where electricity is predominant. Shifts within the industrial sector have also contributed to this result. The materials industries, which rely on nonelectricity sources and are also energy-intensive users, lost share to processing/assembly industries, which use more electricity. We should note, however, that in a period of economic recovery the thrust toward electrification is slowed, because materials industries expand their consumption of nonelectricity energy. Between 1970 and 1988 the share held by nonelectricity energy in the industrial sector fell from 87 percent to 81 percent.

The second factor is that electricity was not a main target for energy conservation. For the past decade and a half, the dominant focus for energy conservation was the industrial sector. And there the materials industries were the main target, especially heating processes. Here, the main energy source was petroleum. It is true that there was some improvement in the efficiency of electric appliances. But the effects of improving burning systems, as well as improving the process whereby oil is consumed, far outweigh improvements in appliances.[1]

The third factor is the way the characteristics of the demand for electricity make the effort to conserve it somewhat difficult. The use of electricity for the necessities of life is taking a larger portion of the economic activities of both manufacturing and households. Beyond the necessities, the needs of consumers are changing, becoming more diverse and sophisticated, and taking forms which require much electricity. For example, more safety and cleaning devices are becoming available in households, and flexible production lines are needed to manufacture these items. As a result, we need additional energy in order to produce higher value-added product, typically at the end of the manufacturing process. Most of the incremental energy required for this is electricity. The energy source for Factory Automation (FA) and Flexible Manufacturing System (FMS), which are both at the forefront of manufacturing systems, is limited to electricity. In basic materials, computer programs that are designed to minimize energy cost greatly contribute to the conservation of nonelectricity energy; the use of extra electricity, on the other hand, continues.

The fourth factor is the proliferation of electric appliances, particularly spurred by the advance of electronics technologies. Though this is a topic of a later section in the paper, let's note here that many appliances, based on electronics technologies, have spread through each of the sectors of the demand for energy. The motivations for introducing these gadgets include: rationalization and the pursuit of a higher value-added mix of products in manufacturing, greater convenience and security in households, and, for the commercial sector, the demands of internationalization and a high-speed era that arises from an information economy. Devices for FA, FMS, Office Automation (OA), Store Automation (SA), Audio Visual (AV), and Home Automation (HA) gradually penetrated in the late 1970s and early 1980s. However, the recent expansion of their use has been terrific. There is, as it were, a flowering of technological innovation; in the background we may even say that an era of electronics has arisen.

Maintaining Competitiveness and the Industrial Demand for Electricity

Issues Facing the Manufacturer

After 1987, Japan experienced a strong demand for energy, reflecting the largest economic boom since the first oil crisis. The upward trend in energy has been caused by investments in housing and by personal consumption expenditures. The increase results from a sharp rise in capital gains from investments in securities and land, rather than from lower energy prices. In turn this caused the basic materials industries to recover production levels and has led to a rise in energy demand in the industrial sector.[2] Figure 5 shows trends of energy demand in the industrial sector.

How long will this recent upward trend in the demand for energy last? It is extremely difficult to answer this question. The environment in which Japan's manufacturing industries operate has not changed substantially as a result of the recent economic recovery. Basically speaking, the problems encountered by manufacturing industries in the 1985–1986 period amidst the *endaka fukyo* (the business recession due to the yen's appreciation) remain unresolved, although Japan found some signs that the domestic market may be larger than had been previously thought.[3]

Those problems include: the yen's continuing strength; friction with trade partners; and the rising competitiveness of newly industrialized economics (NIEs), particularly the Republic of Korea and Taiwan. These are problems in the international area. Domestically, on the other hand, the effort to achieve efficiency in production requires producing a variety of products in small batches. This reflects both the diversification

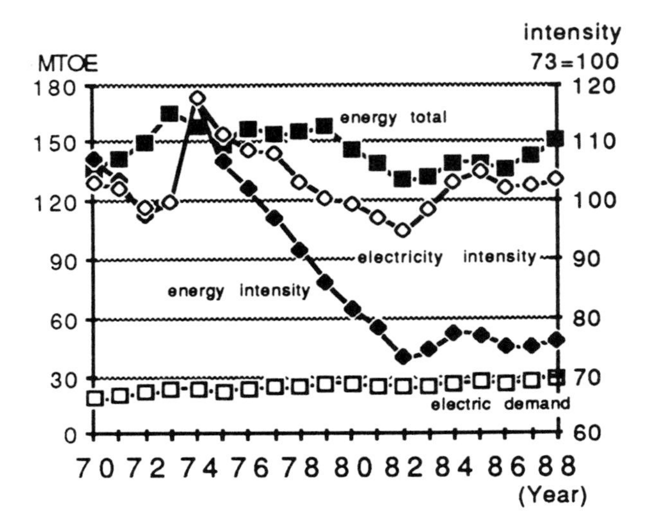

Figure 5. Energy and electricity demand and intensity by IIP in the industrial sector. Source: "Energy Balance Tables," IEE, Japan, and "MITI Statistics," Ministry of International Trade and Industry (MITI). (MTOE = millions of tonnes of oil equivalent)

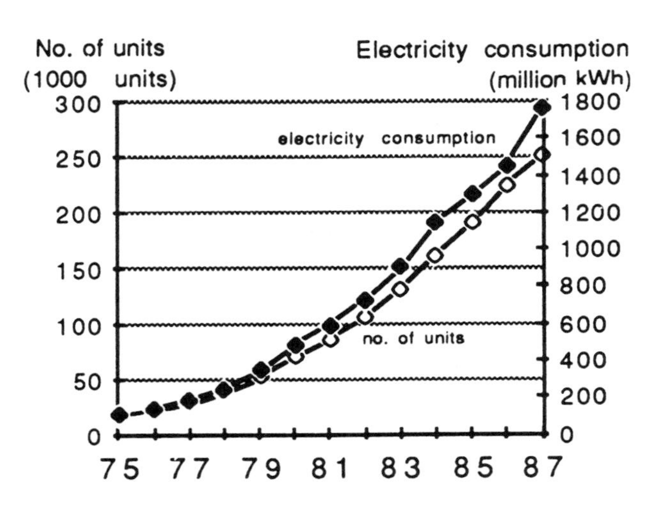

Figure 6. Industrial robot and its electricity consumption. Note: Numbers are in operation. Source: Uekusa (1988).

and the sophistication of consumers' needs. This means it is necessary to rationalize if we are to maintain the entrepreneurs' competitiveness and to move to a higher value-added mix of products. Factories are in the process of responding to these problems. The following subsections present an analysis, from this perspective, of energy and electricity use for industrial purposes.

Electricity and Rationalization

Today, Japan's manufacturing industries are engaged in rationalization; their intent is to improve productivity and reduce the costs of domestic production. At the same time, they are increasing overseas production through Direct Foreign Investment (DFI). By installing their factories in foreign countries, they transfer energy demand from the mother country to the overseas host economy (assuming, of course, that all production is moved overseas, rather than some being retained in Japan, which may well occur). Since production lines in Japan are relatively energy efficient, the technologies of energy conservation are being spread around the world by DFI.[4]

On the other hand, however, major investments undertaken for the rationalization of domestic production include stepped-up introduction of factory automation and electronic devices such as robots and computers. The impact of these investments on energy demand is twofold. One, the electricity consumed (Fig. 6) by the devices themselves is expected to increase as their numbers increase. Two, there is an indirect effect produced by the introduction of such devices. Robots, for example, require less lighting than humans do for the same job. Moreover, the introduction of computers for the purpose of energy management is designed to minimize energy costs. Thus we can say that the introduction of electronic devices has both a hardware effect and a software effect on energy demand.

Taking into account both hardware and software aspects, the impact of the penetration of electronic devices on energy demand is believed to be different between the basic materials industries and the processing/assembly industries. The cost reduction achieved by computerization or robotization is the same for both industries in terms of manpower cost reduction. However, in the basic materials industries energy costs account for a major portion of total costs, and computer programs designed to minimize energy costs can greatly contribute to energy conservation—from a 1 percent to a 6 percent reduction, according to several reports.

Most of the energy conserved in materials is not electricity, however. Between the two sectors then—materials and processing/assembly—the overall impact of increasing use of electronics, such as the computers, is to shift demand share toward electricity.

Electricity and Higher Value-Added Products

Typically, two different measures are used to raise the level of value added in the product mix. One is to increase quality level. That may require greater precision, for instance homogeneity in temperature levels throughout an oven, or homogeneity in the form of purity for materials content; or it may require special processing. The other measure is to increase the degree of discrimination among products flowing through the manufacturing system. This necessitates a variety of products, manufactured in small batches.

The first measure—increasing quality, perhaps with special processing—generally requires the addition of another process. This incremental work tends to be a process that requires precise control of the gas and temperature used for special plating, coating, and processing. Such additional processes tend to use electricity rather than other energy sources. Electric appliances are stable and highly controllable, and they are capable of ultra-high temperature. The second measure—variety and small lot production—depends on FMS, which mainly consumes electricity.

Figure 7 gives an example to illustrate how higher value-added products crave additional energy, especially electricity. We can see that electricity is the main incremental energy source for these processes that are added to increase value-added in the product mix. Hence, electrification continues to be pushed into the economy.

Electricity and Component Changes in Manufacturing

The industrial structure of Japan is shifting to a post-industrial process, or service, economy. Even in manufacturing industries, "heavy-thick-long-big" industries like the materials industry are being replaced by "light-thin-short-small" industries such as processing/assembly and high-tech industries. These latter industries are gradually assuming the dominant position in total value added.

The ratios of electrification are larger for processing/assembly industries compared with the ratios in materials industries (Fig. 8). It is well known that the fabrication of semiconductors and integrated circuits requires clean rooms equipped with efficient electric dust collectors, since those products cannot tolerate even micro dust. Furthermore, the new raw materials expected to be important in the future, such as carbon fiber and new ceramics, are electricity intensive.[5] We can therefore say, "High-tech prefers electricity."

There are at least two crucial reasons why processing/assembly industries tend toward electricity. The first reason has to do with the use of energy in these industries. Fundamentally, these users are oriented to electricity, since they use energy principally to power motors, lighting, and air-conditioning. The second reason derives from energy costs. Energy costs represent a small portion of the total production costs of these industries (again, Fig. 8). As a result, this industry is not so serious about its choice of energy and about energy management, as compared with the materials industries. These users also have a tendency to choose electricity because there is less labor required for maintenance of electricity, compared with other energy sources.

Electricity and the Rise of a Service Economy

Energy Demand Trends in the Commercial Sector

Energy demand in the commercial sector, especially for electricity, is climbing dramatically, driven by the rise of a service economy. In the past, demand increased in proportion to the increase in floor space used by tertiary

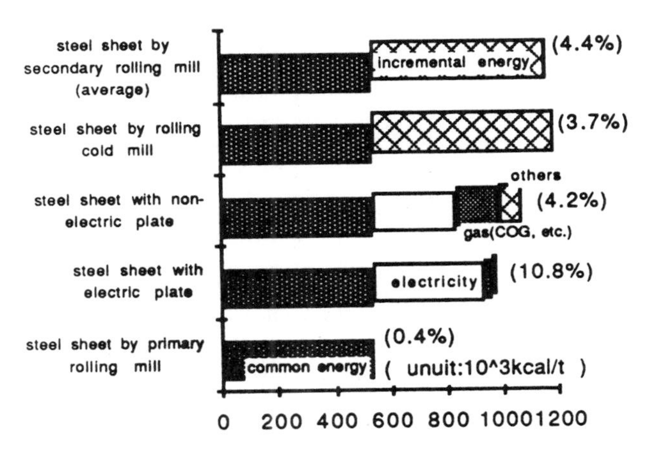

Figure 7. Incremental energy for higher value-added products. Note: Counted energy is only in down-stream processes. Figures in parentheses are annual growth rate of production (1983–1988 FY). Source: Federation of Japan Iron and Steel Industries (1988).

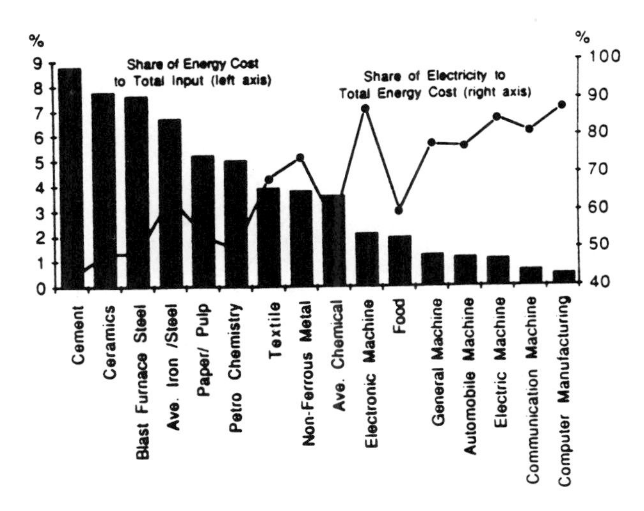

Figure 8. Cost share of energy and electricity. Source: "Industrial Statistics Tables 1987," MITI.

industries. And though energy intensity per unit of area was actually declining until 1982, in recent years that trend reversed and is now upward (Fig. 9).

The sources for this increasing energy use are, first, electricity and, second, city gas. However, the demand for petroleum has been almost flat over these years. The parts of the commercial sector that exhibit particularly strong demand are office buildings, wholesale/retail businesses, service businesses, foods service, hotel, theater, and other amusement businesses. As for applications, major incremental demand grows from cooling, lighting, and other electricity usage including Office Automation devices (Fig. 10).[6]

What is the background that lies behind these trends? At least three factors can be pointed out. The first is the shift to a service economy. In Japan, service industries such as finance and various service businesses are gradually taking a larger part of the industrial structure. This reflects a next stage of economic development. Expansion of the service industries has two aspects. One is that the service industries literally grab a larger stake in the economy. The other is that planning and administration functions become much more important in firms. Thus the strategic management section in the company's office building becomes substantial, even in the manufacturing industries.

The second factor is that energy conservation is at a standstill, as new electric appliances appear and the hours of their operation are prolonged. The tempo of energy saving by actual users has slowed down, although the efficiency of individual energy-consuming devices continues to improve. And there may be the possibility for improving the efficiency of buildings, when there is

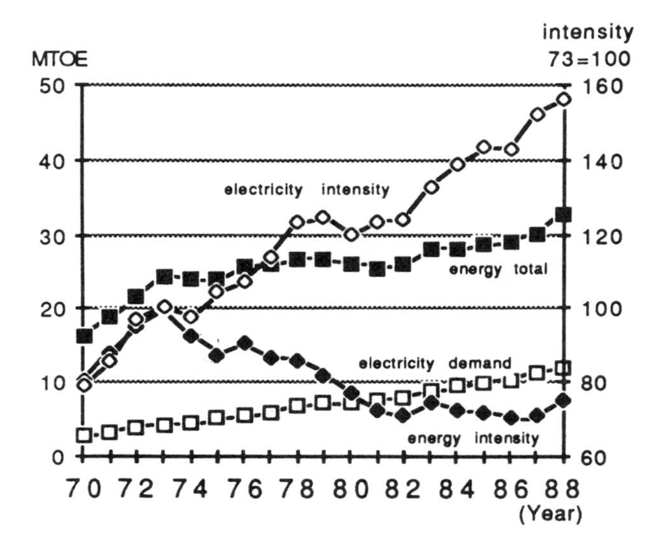

Figure 9. Energy and electricity demand and intensity by floor space in the commercial sector. Source: "Energy Balance Tables," IEE, Japan, and several others.

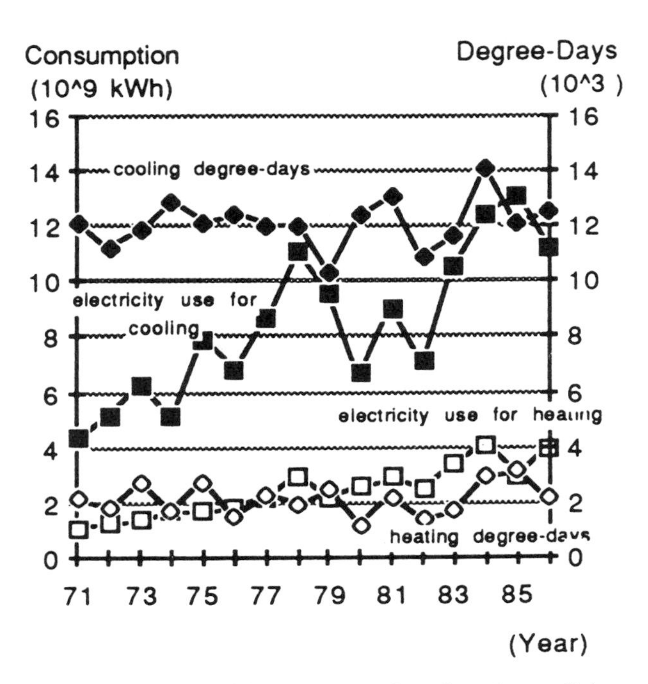

Figure 10. Electricity consumption for air-conditioning. Source: Usuki and Onosaki (1988).

rebuilding. Some of this stagnation in conservation trends may be the result of relatively cheap energy prices, but there are already a few factors imbedded in our economy that worsen intensity of use. Those are the factors surrounding a 24-hour operating system in urban areas. For example, Tokyo, as an international financial center, is watching the world around the clock, for each of the 24 hours. If there are some persons working, even if they are not physically at the office, there are needs for 24 hours of service operation and various machines continue to be in service.

The increasing energy intensity per unit area derives from, and points to, these factors: a prolongation of the hours over which devices operate, and the appearance of new equipment that consumes electricity.

All economic actors need the ability to manage information well and to process data on a timely basis. To do this, there has been widespread utilization of electronic devices and the operation of these devices for longer periods during the day. This corresponds to current trends of internationalization, a high speed society, and diversifying consumers' needs. Twenty-four hour operation has come to pass through the introduction of on-line systems in offices and service businesses. In parallel with the development of 24-hour operations, there has been a change in the terms of employment so that nighttime work hours are included. Also there has been the development of unmanned operations.

The following subsections analyze the two factors mentioned above—the shift to a service economy and factors operating against conservation.

Changes in Electricity Use among Components of the Commercial Sector

In recent years the greatest increment in the demand for electricity has come from office buildings, wholesalers/retailers, and service businesses (Fig. 11). Those three have increased their floor space; meanwhile their electricity intensities per unit area have surged.

Area per office employee rose to 13.5 m^2 by 1986, an increase of 20 percent from 10 years earlier. In these offices, the administration function has grown in importance. New office machines require space as well. Wholesalers/retailers have steadily been integrating, with the result that their operations are larger in scale. Sales of department stores have recorded double digit annual growth, supported by an expansion of domestic demand and growth in household expenditures. And food service, hotel, theater, and amusement businesses are generally expected to continue to grow, since our lifestyles are becoming more Americanized: enjoying an unmarried single life, preferring leisure to work, developing a stronger orientation toward comfort.

Office buildings and wholesalers/retailers (though not the service businesses) are oriented to the use of electricity, which means that electricity takes the share in total energy demand.

Major applications for this electricity include air-conditioning, especially cooling, lighting, and other needs for power. As for cooling, electricity is not the only energy source; city gas may also be used. Mainly in the metropolitan area, cooling systems that use city gas are becoming popular. Since electricity-oriented users are growing more rapidly than other businesses, electricity use in the commercial sector is showing a steady growth.

Electricity Use and the Penetration of Electronic Devices

Various electronic devices, in the form of new appliances, directly and indirectly impact the demand for electricity. Now there is a proliferation of instruments, such as several kinds of computers, copying machines, word processors, etc., for offices; point of sales machines for distributors; and cash dispensers and automatic teller machines in the banking business.

Originally the purpose behind introducing these devices was to simplify and rationalize and, in general, to speed the work across a desktop. However, a subsequent stage has now arrived. Now stress is placed on supporting creative, in addition to routine, work. Creative work on the computer particularly depends upon developing a network that organically integrates computers that are located at different points in the value-added chain. And these devices have turned out to be indispensable tools for planning and executing an organization's strategy.

Figure 12 shows an extraordinary expansion of these appliances. Parallel with this penetration, demand for electricity has risen, as shown in Figures 13 and 14. As mentioned before, the adoption of these tools leads to incremental floor space and yet additional electricity demand for lighting and air-conditioning.[7] Office au-

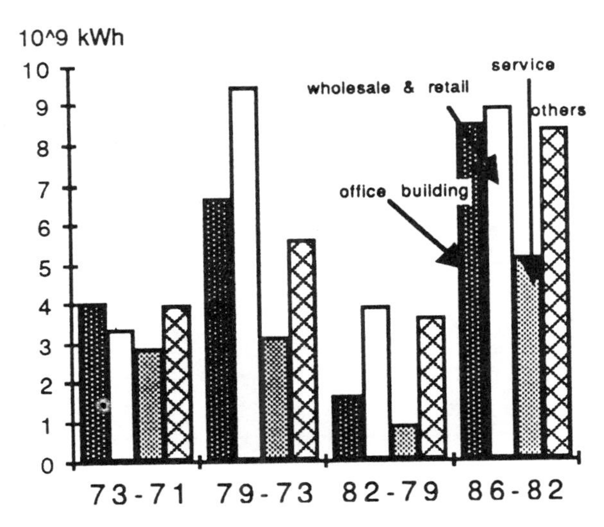

Figure 11. Incremental electricity by users. Note: "Service" includes restaurant, hotel, inn, theater, and other amusement places. "Others" includes school, hospital and others. Source: IEE, Japan.

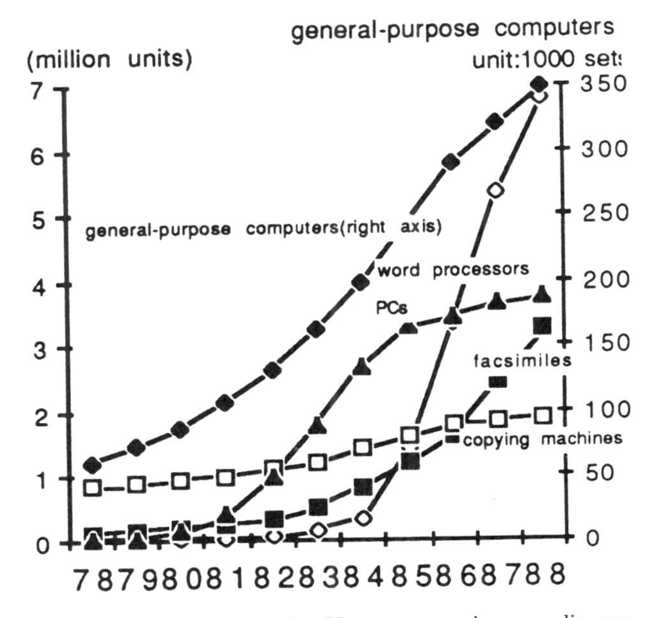

Figure 12. Spread of office automation appliances. Note: Figures are "In operation." Source: "Statistics on Machinery," MITI, and several other sources.

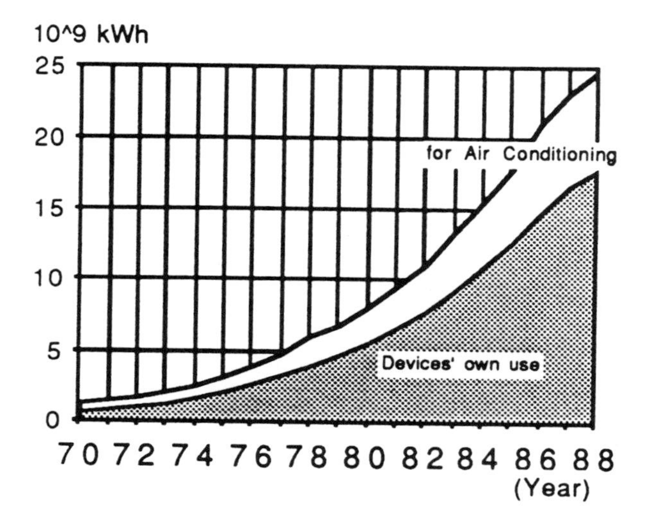

Figure 13. Electricity demand by office automation devices. Source: IEE, Japan.

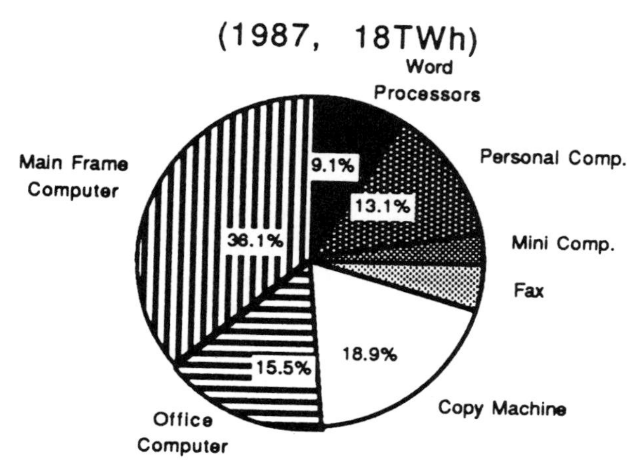

Figure 14. Demand by device. Note: Excludes air-conditioning. Source: IEE, Japan.

tomation devices alone account for fully one fourth of all electricity demand in the Japanese commercial sector in recent years.[8]

Trends that characterize these appliances are that speed of their penetration has not yet slowed down, the performance required from computing increases while electricity consumption per unit of performance is declining, and the hours of operation for these devices are prolonged each day, because they are linked to other terminals for real-time communication. Consequently, the effect of office automation devices on commercial energy use is important and must be well accounted for in the analysis.

The Pursuit of Amenities in Daily Life and the Demand for Electricity

The Background behind Growing Demand for Electricity in the Residential Sector

Over the past two decades, the residential demand for energy has grown steadily. Growth in energy intensity, that is, the demand per household, has contributed much more to this trend than has the increase in the number of households (Fig. 15).

The increase in electricity consumption per household has been relatively stable, though it declined due to climate in the 1970s. Recently there has been a strengthening upward trend. This new stronger demand is led by cooling as well as by another set of factors which we cannot see too well yet. There are several factors behind this strong demand for electricity, despite the fact that individual appliances are becoming more efficient.

The first factor is that electricity consumption for each appliance is larger than the model that has been officially sampled. New models in the market have additional functions, and the size of the model has grown larger. For example, the average size of refrigerators being sold has grown to almost twice that of 15 years ago. And color TV sets now make several accessory functions available, such as receiving multiplex transmissions from a direct broadcast satellite or linkage to stereo sets, and so on. The second factor is that the penetration of durable goods is moving upward. The third factor is the appearance of new electric equipment, along with increases in its hours of operation.

On an income basis, the Japanese have already be-

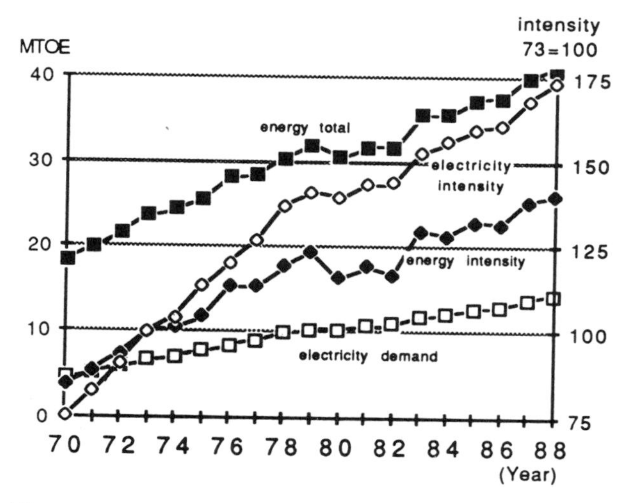

Figure 15. Energy and electricity demand and intensity by household in the residential sector. Source: "Energy Balance Tables," IEE, Japan, and several others.

come rich; now they are climbing to the next level of an affluent society, where durable goods are for personal, rather than family use. We can see the clues to changing consumers' needs in their preference for products that offer higher quality (particularly in terms of security, cleanliness, comfort) and more amenities for their lives. Japanese society is in the process of moving toward: a more aged population; an increase in the number of wives who work; an increase in unmarried "singles" households; greater diversity in views about what is valuable; and an information-oriented society. An individual's sense of value is changing so that, for instance, greater importance is attached to leisure or amenities than to the company that employs the individual (or to being a workaholic) and an individual's life takes precedence over cooperating with others in the society.

With this as background, the following sections analyze 1) the increase in the size of appliances and the number of functions they offer, 2) the increase in their penetration, and 3) the appearance of new devices and the prolongation of their hours of use.

Energy Saving with Appliances and the Increasing Demand for Electricity

According to MITI surveys, the energy efficiency of home electric appliances has improved quite well. For example, refrigerators, color TVs, and air-conditioners have improved their efficiencies 67 percent, 41 percent, and 42 percent, respectively, from 1973 to 1987, on the basis of same model types. More than half the newly purchased air-conditioners do not exclusively cool, but are heat pumps that function both to cool and to heat. Additionally, most refrigerators today have two or three doors, because that is more convenient. As the number of doors increases, the total length of the door seal soars, and so does the possibility for letting out the cold inside air. The color TV models on the increase are those with high quality picture tubes to provide higher resolution images and remote control functions. Also, the number of speakers attached to a TV set is increasing, and the capability of those speakers is growing to support bilingual and stereo broadcasting.

Ways of watching television have changed. Now the behavior of the television audience is proactive; they watch what they like. A decade ago when the choice of amusement was narrow, the audience was more passive. Having a variety of options means our wealth has increased. The appearance of the VCR has certainly contributed to this change. And although its saturation is far from that in the US, commercial cable TV has recently become available in urban areas. The broadcasting satellite, teletext and CAPTAIN, a videotext information network provided by NTT (Nippon Telegraph and Telephone), have each also become available.

These new styles of television viewing add demand for electricity in addition to that for the television set itself. Watching cable TV adds 18w of electricity for the cable connector box. A tuner for satellite broadcasting, a decoder for teletext, and a home terminal for CAPTAIN increment the electricity load by 25w, 15w, and 50w, respectively. High Definition TV (HDTV), expected to be the TV set of the future, is said to need 300w to 500w for operation; the average load for TV sets currently in use is 80w.

When we look at instruments for lighting such as the individual bulb and the fluorescent lamp, we see that they have become more efficient. However, lamps that produce more light have substituted for darker ones, in the bathroom and in corridors. And since lighting has become integral to interior decoration, a chandelier which will use more than one bulb has become more popular than the simple fluorescent lamp for the living room.

As we have already discussed, users adopt electric appliances to gain the amenities embodied in their use. In response, the appliances have become larger in size and added a variety of functions, so that the demand for electricity expands.

The Penetration of New Electric Appliances

The saturation of color TV sets has reached almost two units per family. A decade ago, a single student in an apartment would have thought a television set, a refrigerator, or a vacuum cleaner was a luxury, but today they are necessities in such a household.

In recent years, the electric appliances with dramatic increases in penetration ratios are the VCR, microwave oven, air-conditioner and electrically heated carpet (Fig.16). These are owned by more than half of Japanese households. The CD player, word processor, and personal computer are near 10 percent saturation and on the rise. But it is home-security systems and the videophone which probably give clues to future developments. These devices, along with changes in the use of television, relate to information processing. This is the direction expected for future development of the market, as information media diversify and there is pursuit of greater comfort.

Electricity consumption per Japanese household is a third to a half that of European or American households. Despite the differences in house size and climate, when we account for the recent Westernizing of lifestyles there appears to be a big potential to expand the demand for electricity in households if the price of appliances and of electricity becomes cheaper. Table 1 shows the new appliances that are expected, as listed in a report by the Agency of Science and Technology.[9]

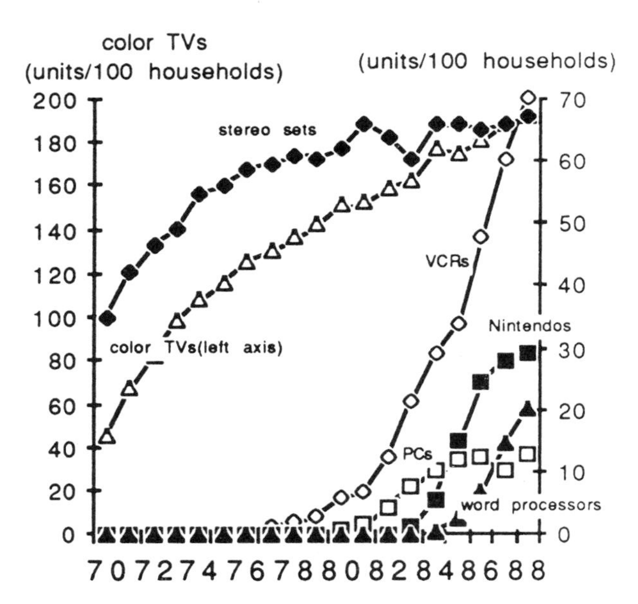

Figure 16. Spread of electric appliances. Source: "Consumption Trend Surveys," Economic Planning Agency.

Table 1. Outlook for the saturation of home electronic appliances.

Appliances	Pattern of Penetration	Ratio % 1995	2010
System for Health Check	semi-choice	2	10
Multi-Function Bed	semi-choice	3	15
Home Elevator	semi-choice	1	8
Home Computer	necessary/choice	20	50
All Automatic Washer/ Dryer Machine	necessary	25	65
All Automatic Dish Washer/Dryer	necessary	25	85
Automatic Cooker	sonii-necessary	15	45
House Keeping Robot	semi-necessary	10	45
Home Control System	semi-necessary	10	30
Home Facsimile	semi-choice	10	50
HDTV	necessary	10	50
General AV System	semi-necessary	20	50
Videophone	choice	25	50
Home Security System	choice	10	25
Environment System	choice	7	25
Multi-Function Heat Pump	choice	7	30

Note: Pattern of penetration is divided by whether goods are necessary or not.
Source: The Agency of Science and Technology (1987)

The Prolongation of Hours of Operation

Along with the impact of the increasing penetration of devices, we should also consider the impact of the hours of operation on the demand for electricity.

Devices for relaying information on a real-time basis, and for quick starts, have become popular. The home security system, to prevent crime and disasters, is a representative case. A remote home control system is another similar case. It controls other home appliances, such as the air-conditioner and the bath, by means of a telephone call from the outside. These kinds of systems for maintaining a comfortable existence are examples of the appliance that works 24 hours, around the clock. Other examples, closer to our daily lives, are quick starts for TV sets and user-programmed recording with VCRs. These need sensors in operation 24 hours a day. Although the load for a sensor is not so big (less than 10w) its influence is nonetheless not so small because of the number of sensors that pile up. As for facsimile, it consumes about a 20w electric load as it sits on stand by.

There is another factor behind the prolongation of devices' operating hours. That is linkage to other devices and *their* added functions. For example, the home control system can turn the air-conditioner on without anyone in the room. In the case of the telephone, a weak electric current through wire cables was enough to communicate in the past. But if we add functions to the telephone, such as an answering machine or a facsimile, we add electric load and also expand the hours in operation. Communication by personal computer is a similar case. Linking the TV to the stereo set, to form AV, makes about 20w work for the amplifier over each of the hours that the TV is in operation.

The Demand for Electricity in the "Electronics Era"

So far, we have investigated the facts behind possibilities for expansion of current and future demand for electricity in each sector. As the economy moves toward a service basis, there is a need to maintain international competitiveness and a need to respond to a people who prefer the enjoyment of comfort and affluent lives. By what means will these needs be realized? That means, I believe, is electronics technology. This section analyzes issues that lie between the flowering of electronics technologies and the demand for electricity.

Technological Innovation Supporting the Trends

The current blooming of electronic technology appears to be one of the most significant technological innovations ever, the basis for a new era. Electronic technology is as fundamental as the steam engine, the automobile, and petrochemistry. These earlier technologies were driving forces to change their times. I think the coming decade or two will be the so-called Electronics Era.

Electronics has supported the solutions to our various problems. In manufacturing, the introduction of electronic devices, as a tool for rationalization, is progressing. It resolves several complicated problems simultaneously. To maintain competitiveness and improve productivity, production must be able to respond to a variety of consumer needs. In the distribution and service businesses, electronic devices were first used to speed work at the desk. Now, electronics turn out to be indispensable for managing strategies, where communications-linked data processing is vital for the job of marketing and planning. And in households, appliances with microcomputers and sensors add amenities to consumers' lives and become the preferred mode.

In the future an information network will link consumers, distributors and producers; consumers will be able to do their economic chores in their homes. Producers will use their information for marketing surveys. Such times will come. Electronic contrivances will usher in this era.

According to G.F. Ray (1983), major technological breakthroughs and the principal energy sources that serve the era of those breakthroughs join together to form pairs.[10] He pairs coal with steam engines, and oil with automobiles and petrochemistry. From this perspective, the principal energy source for the "electronics era" is obviously electricity. This view might be a bit different from Ray's, since electricity is a secondary energy. Needless to say, the choice of energy sources for generation, and of technology and systems for power supply, becomes a key issue.

Impacts on the Demand for Electricity

How does this electronic era impact the supply and demand for electricity? We will think about four items: the volume of power demand; quality of power; power load; and the power supply system (energy policy).

The volume of power demand. In the electronics era, the growth of electricity demand may be accelerated, for two major reasons. These are the penetration of new appliances, and the prolongation of their operating hours. Though the introduction of electronic devices will improve the efficiency and productivity of individual work, it will also lead to added functions and expanded economic activity, as I've already discussed at length.

Quality of power (high reliability and backups). As information is processed by computers and electronics technology spreads widely, the quality of electric power, which supports this technology at work, becomes important. In the case of manufacturers and business users, it is not so difficult to take preventive measures, such as the installation of small generators as a backup against accidents. But it is not so easy for households to keep standby power sources or machines. If households are included widely in a network, a kind of guarantee for the quality of electric power is necessary. The effect on computers of a cut in power supply is shown at Fig. 17.

If a stable supply is not guaranteed, what shall we do? The second best solution is to provide backup. That means backups for machines, and also for the power supply. In the future, we expect that power storage systems will emerge. From this perspective, the development of a cheap and efficient battery is necessary.

Power load. The electronics era may possibly widen the load gap between summer peak and winter bottom. A computer-based society requires energy for air-conditioning, especially cooling. The processing/assembly components of the manufacturing industries gradually increase their shares, and air-conditioning use, responsive to the climate, holds a high share in those industries. As mentioned before, in the commercial sector the use of electricity rises with the increase in office automation devices and air-conditioning. Also in residential, the demand for cooling is expanding.

The question is, how high the mountain and how deep the valley? Already there are, as a matter of course, devices in use which start automatically by sensing the temperature. But the number of such devices installed will be quite different in the future compared with now. Appliances that have sensors and are controlled by computer turn on and off frequently and accelerate to produce a load simultaneously with all other similarly controlled equipment. The load curve shapes so that, the higher the mountain, the deeper the valley. We cannot

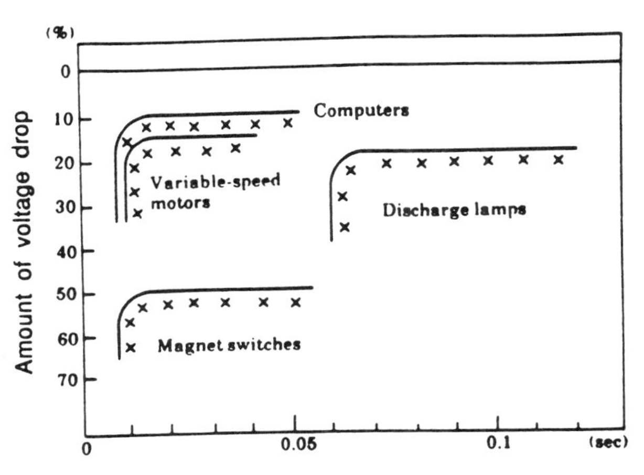

Figure 17. Influences of instantaneous voltage drop. Source: "A New Development of Energy System" by a team headed by Y. Kaya.

neglect the possibility for a speculative load curve, in the way that computer-based trading caused Black Monday on the stock market.

On the other hand, the nighttime load might be increased. Internationalization and an information society lead urban areas to 24-hour operations. And the appearance of flexible terms of employment increases the population who works in the middle of the night or early morning.

The power supply system (energy policy). For the electronics era four issues should be considered by the power supplier.

The first is an across-the-board reduction in power rates, along with a more diversified rate menu. The space issue in small and narrow houses, together with appliances that have higher initial and running costs, impedes the penetration of devices providing amenities.

Along with reducing the price of the devices, the power rate should be reduced because it also represents a variable cost. Since there is a segment of users for whom supply stability is crucial and who therefore require higher power quality, it is natural that we have a rate menu that takes demand priority into account. Scholars point to differences in production cost, depending on time and the development of demand, as the argument for a diversified rate menu. But I'd like to propose a diversified rate menu based on demand priority as an argument equal to that of production cost.

The second issue involves efforts to improve the shape of the load curve. Power utilities have already attempted to level the load gap with water heaters that load at night and with efficient heat-pumps for winter load. It goes without saying that they should continue to press these efforts harder. At the same time, we should pay attention to sharing a portion of demand with other energy sources. For instance, we could consider leaving to city gas the demand for cooling office buildings in urban areas. Cooling equipment that uses city gas has already operated with high efficiency. If we use this kind of cooling and allocate demand between the two energy sources, we can stabilize the demand load in both the electric and the gas utilities. We need to pay attention to total efficiency as a system of social infrastructure.

The third issue is making sure of supply reliability, and the related question of establishing backup systems and compensation programs. Of course, we need stable supply sources; but if that fails, who takes the responsibility for a stable supply and for damage from power outage? The issue of responsibility could become a big matter in the future. Since an institutional system for compensating a power outage has not yet been established, each damage case depends on a negotiation between an injured party and a utility. In the network society, electronic devices that process data even for a household will spread to the general family. If there is a blackout and data files are lost, who compensates the loss? We should build an institutional measure to guarantee that responsibility.

The fourth issue is improving the efficiency of power generation and transportation to reduce the input energy required simply to supply electricity. Since in the present power supply system two thirds of the input energy is lost in the conversion and transportation of output electricity, the incremental demand for electricity accelerates the requirement for primary energy when compared with the increment to other energy sources. Greater efficiency in power generation, using waste heat and co-generation, is necessary to reduce input energy. Also, establishing small generators near the area where there is demand, in other words using an on-site generator, contributes to lowering the transportation loss. We should develop and expand means to supply electricity without increasing primary energy demand. This kind of research and development will become increasingly key.

Conclusion

Saving electricity is difficult for several reasons. The economy is shifting toward the industries and places where electricity is the biggest energy source. Electric appliances, including new devices, are steadily penetrating. The hours of their use also tend to be prolonged through the day.

In the industrial sector the processing/assembly industries, which are oriented to the use of electricity, are going to take a big share. In order to maintain their competitiveness each manufacturer is making efforts to introduce automatic machines that consume electricity for the purpose of rationalization and for enhancing the product mix. In the commercial sector the electricity-oriented user, such as the office building, is taking a larger share of their sector; at the same time a lot of office automation devices are penetrating widely. In the residential sector, families in pursuit of amenities for their lives tend to prefer electric devices that offer safety and cleanliness and can easily be connected to automatic systems. In the society at large, internationalization and diversification are progressing; and as the information-oriented society grows, real-time response and 24-hour service operations are required.

Present and future society is also supported and maintained by a flowering electronics technology. Electronics, as one of the biggest technological innovations ever, can change our times. This technology prefers electricity as its energy source, and influences the demand for electricity.

The impact includes the likelihood of growing demand for electricity (the continuing advance of electrification); the importance of reviewing the quality of electric power (supply reliability), and the effect on power load gap (becoming sensitive to weather temperature, a widening load gap between summer peak and winter bottom).

Given this situation, the supply side of the power industry needs to respond to the following issues: 1) an across-the-board rate reduction and a more diversified rate table; 2) improvement of the load factor (increase demand in the lower demand segment, allocate demand with other energy suppliers); 3) supply reliability and backup measures (also, an institutional compensation system for accidents); and 4) improving the efficiency of the power supply (generation and transportation—co-generation, waste heat recovery, and on-site generator).

References

A. Matsushima, "Changing Life-Style and Residential Energy," *Energy in Japan* (82), pp. 69–93, Nov. 1986

The Agency of Science and Technology, *A Report on Advanced Utilization of Electronics Technologies in the Households,* Dec. 1987

G.F. Ray, "Energy and the Long Cycles," *Energy Economics,* Jan. 1983

EPRI, *Office Productivity Tools for Information Economy,* EPRI P/EM-6008, Oct. 1988

H. Kibune, "Internationalization of Industries and Energy," *Energy in Japan,* pp. 67–94, Feb. 1988

H. Kibune, "Issues Related to Energy Supply and Demand in Japan's Economic Restructuring," in Proceedings of *Energy for Tomorrow,* World Energy Conference 14th Congress, Montreal, 1989

The Institute of Energy Economics (IEE), Japan, *Energy Supply/Demand 1988 and Short Term Outlook,* Research Report, IEE-SR 214, 1989

The Institute of Energy Economics (IEE), Japan, *The Study on Energy Demand Structure by Application in the Industrial Sector,* Research Report, IEE-SR 180, 1986

K. Fujime and T. Onozaki, "Outlook for Growing Energy Demand," *Energy in Japan* (98), pp. 1–25, July 1989

R. Squitieri, O. Yu, and C. Roach, "The Coming Boom in Computer Loads," *Public Utilities Fortnightly,* Dec. 25, 1986

M. Usuki and T. Onozaki, "Changing Energy Demand in the Commercial Sector and Its Outlook to the Year 2000," *Energy in Japan* (91), pp. 17–20, May 1988

M. Watanabe, *New Materials and Energy,* Research Report, IEE-SR 186, 1987

Notes

1. There are voluminous materials on this subject of energy conservation in the Japanese industrial sector. For a typical treatment, see IEE, Japan (1986).
2. The recent trend of strong demand for energy is analyzed in Fujime and Onozaki (1989) and IEE, Japan (1989).
3. Opening the possibility for a larger domestic market could be the turning point for the Japanese economy. However, I don't think economic growth led by domestic demand will continue, because of serious problems with land. In order to realize a society more oriented to domestic demand, we need to restructure the domestic infrastructure. This would involve reviewing land policies and decentralizing factors, instead of centralizing them around Tokyo. At this moment, it is difficult to imagine that such a dramatic review and policy change can happen. As a result I draw the conclusion that Japan's economy will face serious problems in the future.
4. Kibune (1988) provides an analysis of the relation between DFI and domestic energy demand.
5. For instance, we need 12-25kWh electricity to produce 1kg carbon fiber. This is more than ten times for steel (0.6kWh/kg), and for glass (1kWh/kg). See Watanabe (1987).
6. S. Usuki and T. Onozaki (1988) present a recent analysis of Japan's energy demand in the commercial sector.
7. In the United States, a similar impact on electricity demand, by the same rise of these devices in the commercial sector, has been analyzed by EPRI. See EPRI (1988), and R. Squitieri, O. Yu, and C. Roach (1986).
8. Y. Uekusa (1988).
9. See The Agency of Science and Technology (1987).
10. G.F. Ray (1983), p.7

Session E
Economics and Policy

Summary
Session E—Economics and Policy

David O. Wood
Session Chair
Center for Energy Policy Research
Massachusetts Institute of Technology
Cambridge, MA

The purpose of this session was to present and discuss four recent studies of the economic costs and consequences of policies designed to reduce greenhouse gas emissions. Diversity of modeling approach was the distinguishing feature of these presentations, with process simulation models, linear and mathematical programming models, and econometric models all being represented.

The session E papers begin with "An Economic Analysis of Greenhouse Effects," written by Lester Lave. While this paper was not presented at the conference, it is included here because it provides an excellent summary of many of the economic and policy issues confronting environmental concerns. Lave discusses the modeling, climate, and human uncertainties that are inextricably part of the greenhouse problem and consequently part of its solution. An approach to the problem, given these uncertainties, is suggested, and the elements of an "attractive" greenhouse policy are outlined. Additional comments from Lester Lave can be found in "Policy Strategies for Managing the Global Environment: A Panel Discussion," earlier in this volume.

David Montgomery won the "model diversity" award with his report of the Congressional Budget Office study "Effects on Energy Markets and the US Economy of Measures to Reduce CO_2 Emissions from Fossil Fuels." This study distinguishes near-term and long-term possibilities for CO_2 emissions reductions and associated economic costs. It employs three models in the analysis of the short-term effects, including the Energy Information Administration's energy market model, the DRI macroeconomic model, and the Jorgenson dynamic general equilibrium model. Synthesizing the results from these models, Montgomery finds that a $100 per ton carbon tax will at least stabilize US CO_2 emissions at current levels, and could reduce them by as much as 20 percent by the year 2000. The economic costs of this initiative are between 3 to 6 percent loss of US GNP, with the loss depending critically upon assumptions about the uses of the tax revenues and complementary changes in other tax rates.

Montgomery's long-term analysis is based upon the Environmental Protection Agency's adoption of the Edmonds-Reilly long-term global energy–CO_2 model. The basic result of this analysis is that a US tax of $100 per ton would slow doubling of global CO_2 emissions by about 3 years while a similar global tax slows doubling by about 20 years. These latter results are purchased at the expense of about a 1 percent reduction in global GNP.

The Environmental Protection Agency study, "A Least Cost Energy Analysis of US CO_2 Reduction Options," presented by Barry Solomon, is based upon a linear programming model (MARKAL) of the US energy market. A distinguishing feature of this study is its emphasis on analyzing the sensitivity of the study's results to systematic variations in key input parameters and data, including natural gas prices, alternative estimates of exogenous improvements in end-use efficiency, variations in the costs of non-hydro renewables, and nuclear availability. The impact of input variations on results is significant. While the estimated marginal cost of reducing emissions to 80 percent of 1990 levels by the year 2010 is about $40 per ton carbon, under conservative assumptions about nuclear power availability the estimate rises to $130 per ton.

The study by Alan Manne and Richard Richels, entitled "CO_2 Emissions Reductions: A Regional Economic Cost Analysis," is based upon a mathematical programming model of global energy markets and the economy/energy interactions. This model is perhaps the most elegant of those employed in the studies presented in this session, being based upon a relatively few broadly defined energy technologies that are characterized by time profiles of expected costs, efficiencies, and emissions coefficients, and by cost minimization behavioral assumptions. The study focuses on an emissions reduction scenario in which developed countries stabilize their CO_2 emissions by the year 2000 and decrease them by 20 percent by 2020, while China and other developing countries limit emissions to twice 1990 levels.

The most striking results from the analysis of this emissions reduction scenario are the differential effects on economic growth in different regions. By 2030, the US would suffer about a 3 percent loss in GNP, roughly twice the losses suffered by the other OECD countries. The long-term effects on the developing countries seem more serious; by the year 2100 China's GNP would be 10 percent below its unrestricted level, while the GNP in all other developing countries would be reduced by about 5 percent. These results provide arithmetic drama regarding the difficulties in designing and implementing equitable and efficient international policies to reduce greenhouse gas emissions.

Irving Mintzer, in his study "Insurance Against the Heat Trap: Explorations of Strategy to Reduce the Risks of Rapid Climate Change," takes a very different approach from the other studies presented in this session. Instead of analyzing the emissions and cost implications of a specific set of policy initiatives, he examines the effects on emissions and economic growth of a comprehensive strategy to promote sustainable development. His analysis of a global sustainable growth policy is expressed in terms of changes in such economic parameters as labor productivity growth rates, income elasticities, rates of improvement in end-use energy efficiency, renewable technology costs, and other key economic variables. These changes in basic economic parameters are then employed, together with the World Resources Institute model of warming commitment to estimate the impact on emissions and economic growth.

With this approach, Mintzer finds wide scope to reduce overall energy consumption and consequent greenhouse gas emissions. For example, by the year 2075 overall energy consumption between the "business as usual" and the "aggressive response" scenarios is reduced from 1,774 to 740 exajoules, while consumption of solids drops from 1,246 to 55 exajoules. This occurs while GNP growth rates increase from 4.1 to 6.0 percent per year in developing countries, and from 2.1 to 2.5 percent in developed countries. These are very provocative results that, if borne out by other studies, suggest that more comprehensive policy strategies to stimulate sustainable economic growth can produce very different estimates from those of most other studies of the costs of mitigating greenhouse gases; indeed, there are primarily only benefits to be reaped. Not surprisingly, commentary as to whether Mintzer might be right or wrong occupied much of the discussion during this interesting and well-attended session.

An Economic Analysis of Greenhouse Effects

Lester B. Lave
Carnegie-Mellon University
Pittsburgh, PA

Uncertainties in Modeling

Global climate change is but one force that will influence the 21st century. Potential world wars, regional conflicts, or ethnic strife probably will prove more critical. Furthermore, significant improvement in the social welfare of developing countries may more than offset the effects of concomitant global climate change of the magnitudes now being considered.

Discussions of needed actions concerning greenhouse effects are models of miscommunication. One group starts from the position that there is only one earth and that humans have demonstrated their ability to damage not only local environments, but also regional (acid rain, ozone depletion) and global environments (destruction of stratospheric ozone). As population and economic activity increase, our ability to damage the environment increases. Another group starts from the position that doubling atmospheric concentrations of carbon dioxide may not lead to substantial climate change and the climate change may turn out to be beneficial, rather than harmful. Still another group stresses the social costs and disruptions associated with rapid abatement of carbon dioxide emissions; these costs will be painful to poor nations striving to develop as well as to the poor people in the United States.

The sometimes sharp dispute among these three groups need not mean they disagree on the actions that are needed to manage greenhouse effects. They each have valuable insights to contribute to the discussion. But first they need to understand the facts and uncertainties. Greenhouse policies must recognize these uncertainties, but should not be paralyzed by them.

Climate Modeling. Forecasts of greenhouse effects are awash in a sea of uncertainty. One prominent government scientist testified before the US Senate that he had "99 percent confidence that the warming of the 1980s was associated with the greenhouse effect" (Hansen, 1988). This confidence is excessive for several reasons. First, the current models are recognized to be deficient in their representations of clouds, oceans, and other phenomena (Abelson, 1990). Second, there is less than full agreement among the models, even about the extent of mean global warming. Third, the measured temperature increase is about half of what the models predict and well within the "noise" of daily and annual variation. Fourth, a new paper by Spencer and Christy (1990) concludes that there has been no warming of the earth's lower atmosphere since 1979, despite surface measurements showing an increase in global mean temperature. Spencer and Christy's work supports the contention that the measured temperature increase is due to problems with the data, rather than real warming (Nierenberg, 1990).

Scientists don't agree on how much confidence to put in current model runs (Lindzen, 1990; George C. Marshall Institute, 1989; Schneider, 1989; Schneider and Rosenberg, 1989; US Environmental Protection Agency, 1989; Yohe, 1989). In general, scientists tend to be overconfident about their models and abilities to measure events (Fishchhoff and Henrion, 1988). Simon (1990) urges caution in interpreting models as complicated as either the climate or economic models. One indication of both progress and uncertainty is that only within the last decade has the importance of greenhouse gases other than carbon dioxide been appreciated (Schneider, 1989; US Environmental Protection Agency, 1989). While estimates of the rise in sea level during the next century span a wide range, many scientists have come to agree that the rise will be smaller than one meter (Kerr, 1989). Almost nothing is known about climate changes for each region, aside from growing agreement that there may be a warming and drying of the mid-continent (Schneider, 1989). Future storm frequency and severity are largely conjecture; a bit more is known about effects on water resources (Waggoner, 1990). Perhaps the most certain prediction is that our knowledge of climate change and its sequelae will itself change quantitatively and even qualitatively over the next decade.

Economic Modeling. Uncertainties also abound in the models of the costs of abating carbon dioxide emissions (Manne and Richels, 1990; Nordhaus, 1990; Nordhaus and Yohe, 1983; Yohe, 1989) or examining the social costs of the resultant changes (Lave, 1988; Lave and Vickland, 1989; Schelling, 1983). Emissions of carbon dioxide and other greenhouse gases are determined by factors that cannot be forecast confidently for the 21st century, such as future population, economic activity, and new technology. Abelson (1990) notes several reasons why fossil fuel use is unlikely to grow exponentially through the next century, including the price increases an exponential increase would be expected to produce. None of the modeling takes sufficient account of technological advances that could reduce emissions or even reverse climate change.

Effects of Climate Change. The effects of climate change are difficult to predict, both for the managed and less managed biospheres. Knowing global mean increases for temperature and precipitation is not very informative. Estimating the effects of climate change requires knowing temperature and precipitation changes in each region, not only in terms of annual averages but also in terms of the time patterns and the extremes (Cooper, 1982; Schneider and Rosenberg, 1990). Even then, biologists might be able to predict that a particular plant would not fare well, but predicting related ecological effects is virtually impossible. Many varieties could fill the vacated niche, depending on the complex interactions among plants, insects, and animals.

The effect of climate change need not be harmful. One group of experts (Coolfont Workshop, 1989) estimates that greenhouse induced climate change would benefit US forests, agriculture, and water supply, and impose only small costs due to sea level rise. At equilibrium, they estimate that total biomass in forests would increase due to greenhouse effects by perhaps 10 percent in the US and 20 percent in the USSR. They predict that national crop yields would rise in the US by about 15 percent, in the USSR by 40 percent, and in comparable amounts in Australia, China, Brazil, and Europe. A sea level rise of 50 cm would increase the costs of protective coastal structures over half a century by $25 billion for the US, $1 billion for the USSR, $25 billion for Europe, and small amounts for other countries. Bakun (1990) suggests that greenhouse warming could intensify upwelling, increasing the productivity of fisheries.

"Normal" Variation. Temperature and precipitation display considerable variation from day to day and year to year (Schneider and Rosenberg, 1990). A major ice age covered much of North America with ice sheets 18,000 years ago, and a little ice age caused starvation in Europe from 1300 to 1900, with rapid warming since 1700. Given this history, it is easy to understand why the temperature decline between 1940 and 1960 led to fears that another ice age had begun. Periodically El Niño changes world climate. The decade long "dust bowl" in the United States and the Sahel drought are "normal" variations in climate that may go beyond the predicted greenhouse effects of the next century. "Normal" variation is also large with respect to ecology and disease: Dutch elm disease and gypsy moths have devastated trees; smallpox and influenza epidemics have killed millions of people.

Coping with Climate Change. Normal weather variability and a host of other challenges have led farmers to develop ways of coping with "unfriendly" climates, from irrigation and pesticides to developing new plants and cultivars (Cooper, 1982). The existing techniques will aid adjustment to climate change; even more helpful will be the research, development, and dissemination network implemented to cope with current variability. However, uncertainties and current institutions, such as US agricultural support programs, can prevent adjustment. The droughts in the US during the 1930s and in the Sahel show that "normal" variability can lead to environmental damage, as farmers struggle to maintain their livelihoods. These occurrences are exacerbated by pressures from increasing population.

Prevention. Preventing greenhouse effects is more complicated than current discussions assume. Abating the emissions of all greenhouse gases would prevent climate change; however, as currently configured, the US economy cannot function without fossil fuels. People in developed countries won't give up their automobiles and air conditioners easily—and developing countries want these consumer goods also.

Even minimizing emissions proves problematic. One option is substituting natural gas for coal, insofar as possible. This switch would have no effect on carbon dioxide emissions in 2050, but would lower emissions in the next decade or so, giving more time to explore the importance of greenhouse effects and accomplish research and development on energy and the environment. A second option would be to focus on the long term, expediting the replacement of fossil fuels with the current nuclear energy and renewable energy technologies. The first option focuses on the short run, but may provide the time needed to resolve crucial uncertainties. The second option starts work immediately on long-term carbon dioxide reduction, but invests billions of dollars in deploying current generation technologies even though technological advances promise vast improvements during the next decade or so.

Uncertainty enhances the value of short-term activities that decrease greenhouse gas emissions today, without making large investments that could be useless tomorrow. The uncertainty means that we don't know enough today to focus an aggressive campaign on the most important emissions; the campaign might allocate resources to useless, or even harmful, technologies. The best investment might be in improving intellectual and physical capital so that society will be able to react quickly and forcefully when we learn what is needed.

Predicting Human Reactions

Impossibility of Forecasting 2090. As already noted, future greenhouse gas emissions cannot be predicted with confidence. For example, the Montreal Protocol on CFCs could be extended to a complete ban in the next decade, or it could be ignored. The next decade could bring rapid economic growth and a baby boom, like the 1950s, or depression and fertility drop, like the 1930s. The poor nations might adopt small family size, along with the products and lifestyle of the rich nations. The mix of economic activity is also important. An economy that focuses on building roads, bridges, and heavy industry will use more energy than one focused on microelectronics. If world energy consumption grows 4 percent annually, energy use will increase by a factor of 100 before 2100. Supplies of petroleum and natural gas, however, are unlikely to permit such large growth.

Aggressive, worldwide actions to abate emissions of greenhouse gases, assuming the effects are large and harmful, would be delayed by three lags (Lave, 1988b):

Recognition Lag. The first lag is the delay before scientific consensus has been reached that greenhouse effects are large and harmful. More encompassing and sophisticated models are unlikely to compel scientific consensus by themselves. Scientists will await observations that corroborate model predictions, such as a change in the temperature gradient of the atmosphere. Major climate change isn't necessary to demonstrate the validity of the models, but scientists will demand data that support counterintuitive predictions of the models. The scientific models are nonlinear and highly interdependent. In such systems, a small change in one parameter can produce a qualitative change in an outcome. For example, warmer temperatures increase precipitation over Antarctica and Greenland, thickening the ice cap and possibly lowering sea level (Zwally, 1989). Technological change could make the greenhouse problem irrelevant. For example, a technology like cold fusion could end the use of fossil fuels.

The future (beyond a few years) cannot be predicted with confidence. What predictions would have been made about 1990 by experts in 1890? They couldn't have known about the scientific, political, and economic forces that have shaped our world. Even the predictions of such bold futurists as H. G. Wells and Jules Verne now seem quaint and irrelevant.

Agreement-Negotiation Lag. The second lag is the time required to move from scientific consensus to social consensus to world agreement on an enforceable treaty. Even after most scientists agree, some notable scientists will continue to debunk the greenhouse effect. The public currently hears sharp disagreements among scientists. These sharp disagreements will continue even after most scientists come to regard the case as proven. Under these conditions, persuading the public to make major sacrifices will take some time.

Negotiations will prove even more difficult. If rich countries are asked to reduce emissions to the per capita levels of the poor countries, we may have to reduce fossil fuel use by 90 percent. Under these conditions, even the most benevolent nations will be tested. Additionally, such a treaty will determine, literally, the economic hopes of most poor countries, and so they can be expected to bargain hard against restrictions.

Implementation Lag. Even after a treaty is ratified, implementation will take decades. More than a decade has been required to build a nuclear power plant. To increase energy conservation and abate greenhouse gas emissions, nearly all products must be redesigned and produced in rebuilt factories. The substitutes for current CFCs require redesigned refrigerators and air conditioners. Replacing these appliances and capital equipment will take decades.

These processes—achieving scientific consensus, persuading the public, negotiating enforceable treaties, and implementing emissions reduction goals—cannot take place concurrently. Each awaits reasonable completion of the prior step. Decades are likely to elapse before emissions can be curtailed to break-even levels.

Conflict or Cooperation. Climate change could lead to a new era of world peace or to escalating conflict (Lave, 1988a). Changing climate will challenge individuals and societies to adapt (Lave and Vickland, 1989). People could rise to the challenge through world cooperation and an end to international conflict and domestic strife, or the challenge could exacerbate nationalism and ethnic conflicts.

Climate change may require individual farmers to plant new crops, change to irrigated agriculture, or even migrate to land made desirable by the change in climate. Nations may have to build massive irrigation systems. Climate change will require modifications in lifestyles and economic activities, especially for poor countries. Market economies are subject to constant challenge and so both lifestyles and economic activities tend to be flexible. More traditional societies, however, are likely to find it difficult to adjust to the new opportunities and constraints (Lave and Vickland, 1989).

Deciding on Actions in a Sea of Uncertainty

There is no "conservative" decision about what actions to take in managing greenhouse issues. On the one hand, continuing on our current path could lead to disastrous climate change, such as crop failure and massive starvation. On the other hand, mounting a major campaign to curtail emissions throughout the world could impede the development hopes of rich and poor nations, reduce current consumption, decrease resources available for government programs such as education, medical care, and research and development, and even increase current pollution problems. Much of this expenditure could be wasted or prove counterproductive.

Economic Models. The social cost of reducing emissions of greenhouse gases has been modeled extensively (see Manne and Richels, 1990; Montgomery, 1990; Nordhaus, 1990; Nordhaus and Yohe, 1983). These models differ in many ways, but they all find the least cost fuel mix that produces required energy when there is a constraint on carbon dioxide or other emissions. They are subject to all of the uncertainties associated with long-term, large-scale models.

Although the models were developed independently using a range of techniques, the results are similar. For all these models, small (perhaps 10 to 20 percent) reductions in emissions would increase costs little. However, as the abatement increases, the social cost rises more rapidly. Subject to the overwhelming uncertainties mentioned above, the models suggest that even a virtually complete elimination of carbon dioxide emissions would increase costs by less than about 3 percent of GNP for the US. The cost is small because energy represents a small proportion of GNP in the US and other developed countries. These countries have the technology to increase conservation and produce energy from nuclear and renewable energy sources. For poor countries, eliminating greenhouse gas emissions would prove all but impossible.

These models suggest R&D priorities such as developing low cost nuclear reactors that have public confidence, and demonstrating that large quantities of carbon dioxide can be injected into and retained in the deep oceans.

Modeling Human Adjustment. The social cost of climate change has received little study (Schelling, 1983, and Kates et al., 1985, are notable exceptions). Lave and Vickland (1989) noted that only a tiny portion of economic activity in the rich countries is subject to disruption from climate change. Perhaps the greatest problems for the United States and the Netherlands would occur if there were a large rise in sea level. Almost all of the world's largest cities are ports, built at sea level. A sea level increase of several meters would flood these cities or force erection of dikes and other protective structures. Fortunately, sea level is estimated to rise less than one meter during the next century (Kerr, 1989). The Coolfont Workshop (1989) estimated that the costs of dealing with sea level increase would be relatively small in the countries they examined.

Nordhaus (1990) has quantified these effects by estimating that only about 3 percent of US national income is sensitive to climate change (this includes agriculture, forestry, and some coastal activities). Another 10 percent of national income is "modestly sensitive" to climate. He estimates that a doubling of atmospheric carbon dioxide concentrations would reduce national income in the US by "around one fourth of 1 percent." (Nordhaus, 1990, p. 9). These conclusions are subject to the usual assumptions that the model includes the relevant aspects of the economy and that other aspects are unimportant, e.g., problems with pests are handled easily.

In contrast, Lave and Vickland (1989) note that developing countries have a much larger proportion of their economic activity in agriculture, and thus are more at risk from climate change. They assert that these countries would be less able to adjust to climate change because they lack trained agronomists and biotechnologists, as well as the capital for large-scale irrigation, and their traditional societies are likely to adjust more slowly. Thus, the effects on poor countries could be much larger and potentially devastating. Finally, they note that efforts by the developed countries to help poorer nations have produced little in the past 40 years, despite the valiant efforts of many men and women. At a time of immense disruption, the efforts of rich countries might be still less successful in helping these poor nations.

For the rich countries, the models agree that the cost of preventing climate change is likely to be only a few percent of GNP, and the cost of adjusting to a new climate regime similarly would be relatively small. Even so, in absolute terms, these costs are large. For example, Manne and Richels (1990) estimate the present value of a severe carbon dioxide constraint to be $3.6 trillion for the US. Even if that is only a few percent of GNP, it represents a large amount of resources that could be available for many other beneficial purposes. Moreover, since none of the models give extensive treatment to uncertainties and untoward effects, they might underestimate the the cost of abating emissions or of social adjustment by a large margin.

The Dangers of Aggressive Abatement. The initial estimates of the environmental damage from some activities—for example, acid rain and damage to stratospheric ozone from supersonic transport (NAPAP, NAS)—have been overstated. The depletion of natural resources has not proven to be a major problem (Barnett and Morse, 1963; Smith, 1979). In the mid-1970s the US government embarked on an ambitious program to increase nuclear energy, produce liquid and gaseous fuels from coal, and disseminate existing solar technologies. If these programs had not been moderated, the economy might be saddled with many more nuclear reactors abandoned during construction and with synthetic fuel plants emitting more than twice the amount of carbon dioxide for every unit of useful energy from natural gas or gasoline. Most Americans would also know from first-hand experience that solar technologies were unreliable and inadequate. Until renewable energy sources are reliable and economically attractive, embarking on an aggressive program to disseminate these resources could prove as wasteful, and even harmful, as the ill-conceived programs of the past. Thus, an aggressive program to abate carbon dioxide emissions has high potential costs. Analysts must first think through the implications for reductions in government programs and private consumption, as well as the possible adverse implications for the environment.

Moderate programs focused on energy conservation and R&D in specific areas appear to have low social costs and some benefits. "Business as usual," however, is almost certainly not the best policy. A decision not to do anything today is still a decision, with costs and benefits as real as a decision for aggressive abatement. Individuals and governments need to weigh the uncertainties of greenhouse effects carefully in order to formulate policy that is neither overly aggressive in abatement nor mired in inaction.

Elements of a Solution

An attractive greenhouse policy will recognize the following concerns and contain the following elements:

Environmental Concern. The "green" movements in the US and Europe, as expressed in voting and opinion polls, reflect widespread concern for the quality of our environment.

Concern for the Future. People are concerned that the environment we leave to our grandchildren not be despoiled.

Pressing Current Needs. The US has a situation common to many countries. President Bush wants to eliminate the budget deficit by cutting some domestic programs, but he is reluctant to raise taxes. The public agrees on deficit reduction and not raising taxes, but doesn't agree that domestic services, such as medical care, education, and transportation should be cut. The agreement to eliminate the deficit is vitiated by the lack of agreement on which services to cut. Americans want to help the poor in the US and to help poor nations develop. Again, there is no agreement about where these resources would come from. Thus, any successful policy for managing greenhouse effects must have low costs, improve current environmental quality, and offer other benefits such as energy conservation.

Research and Development to Lessen Uncertainty. Recognizing the scale and scope of current uncertainties is a major impediment to developing policy. An aggressive program formulated today is unlikely to be efficient or effective. Research and development must improve estimates of the extent of climate change, both globally and regionally, and the ecological effects of this change.

Research and Development to Facilitate Adjustment. To reduce greenhouse gas emissions, new use technology must be energy efficient and new generating technology must curtail greenhouse gas emissions. For example, a new generation of nuclear reactors and improved renewable energy technology is needed. As climate change occurs, new cultivars and crops will be needed and agriculture will have to cope with drier or wetter, hotter or colder conditions. Unfortunately, climate change will push some plants and animals to extinction. This is not only deplorable in itself, but it also threatens our adjustment by losing genetic material that could help to develop new crops and cultivars in the future. Finally, research is needed on why some cultures adjust easily and others are more rigid. What can be done to facilitate the dissemination of innovations?

Cost-Effective Adjustment. We must know the effects of each greenhouse gas and the costs of abatement. For example, CFCs are much more important, molecule for molecule, than carbon dioxide. Thus, it is worth much more to abate a molecule of CFC than a molecule of CO_2. Decisionmakers need to know what improvements in energy efficiency are possible at what real costs (including individual attention to conservation).

Research and Development to Manage, Given the Uncertainty. At present and for the foreseeable future, the most important questions cannot be answered. Intelligent policy requires acknowledging this uncertainty and working out policies that are robust in the face of these unknowns. The research and development agenda should be structured, in part, to resolve the uncertainties that are most important for policy formulation; analysis is required to determine which are the most important uncertainties.

More attention should go immediately to emphasizing fuels that produce less greenhouse gases per unit of useful work. For example, natural gas produces less carbon dioxide per Btu than does coal. Synthetic liquids and gases from coal lead to much greater emissions of carbon dioxide per unit of useful work than do petroleum or natural gas.

In many facilities, methane and other "waste" gases are flared, since collecting the energy isn't worth the trouble. Once greenhouse effects are taken into account, less of this flaring is justified.

Forests are a huge store of carbon. Preserving and replanting forests, both temperate and tropical, is a relatively inexpensive way to prevent carbon dioxide emissions.

Developing nations, such as China, are emitting large and rapidly growing amounts of carbon dioxide. Energy policy in these countries might be changed by offering other technologies, and possibly subsidies to use them. The nations likely to suffer most from climate change, particularly those that we are least likely to be able to help in the event of climate induced disruption, must be helped now.

Other Elements. Greenhouse management will affect current environmental programs; by absorbing resources, greenhouse programs can curtail expenditures on research and development for other issues, expenditures on education, and expenditures on economic development. These are valuable areas and care must be taken not to curtail resources so much that valuable programs are lost.

Conclusion

The uncertainties put special weight on finding programs that curtail the emissions of greenhouse gases, but also serve other social goals, such as saving fuel, lowering environmental pollution, and preventing the destruction of farmland. In many cases, aggressive programs are justified by these other social goals; helping manage greenhouse problems is simply an additional benefit. Such programs should be emphasized.

An aggressive world program to abate greenhouse gases is likely to prove expensive and disruptive. The press in each country seems to focus on the deficiencies in other countries without recognizing the corrections that need to be made at home, e.g., criticizing tropical deforestation while ignoring the fact that most of the North American forests have been cut.

Using uncertainties as an excuse for continuing current policies is a bankrupt approach. Society needs to recognize and act on the loss of humanity's environmental innocence. What is not needed is an aggressive, good-hearted, soft-headed campaign that could as easily do harm as good.

The focus of current efforts should be reducing critical uncertainties; modeling these uncertainties to determine which are most critical and how to formulate policy; developing technologies that will allow effective and cost-efficient abatement of greenhouse gas emissions; investigating how to facilitate adjustment to current variations in climate, and thus to climate changes; taking action to abate current environmental pollution; and pursuing policies justified by current concerns, such as energy conservation, that also serve to lessen emissions of greenhouse gases.

References

Abelson, P., 1990, "Uncertainties about Global Warming," *Science,* 247:1529.

Bakun, A., 1990, "Global Climate Change and Intensification of Coastal Ocean Upwelling," *Science,* 247:198–201.

Barnett, H. and C. Morse, 1963, *Scarcity and Growth: The Economics of Resource Availability,* Baltimore, MD: Johns Hopkins University Press.

Coolfont Workshop, 1989, National Climate Program, National Oceanic and Atmospheric Administration.

Cooper, C., 1982, "Food and Fiber in a World of Increasing Carbon Dioxide," in W. Clark (ed), *Carbon Dioxide Research 1982,* New York: Oxford University Press.

Hansen, J., 1988, Testimony before the Senate Energy Committee, June 23, 1988.

Kates, R., J. Ausubel, and M. Berberian, 1985, *Climate Impact Assessment: Studies of the Interaction of Climate and Society (SCOPE)* 27, New York, NY: John Wiley & Sons.

Kerr, R., 1989, "Bringing Down the Sea Level Rise," *Science,* 246:1563.

Lave, L., 1988, "The Greenhouse Effect: The Socioeconomic Fallout," in D. Abrahamson and P. Ciborowski (eds), *The Greenhouse Effect,* Minneapolis, MN: University of Minnesota.

Lave, L., 1988, "The Greenhouse Effect: What Government Actions Are Needed?" *Policy Analysis,* 7:460–70.

Lave, L. and K. Vickland, 1989, "Adjusting to Greenhouse Effects: The Demise of Traditional Cultures and the Cost to the USA," *Risk Analysis,* 9:283–91.

Lindzen, R., 1990, "Some Coolness Concerning Global Warming," *Bulletin of the American Meteorological Society,* Vol. 71, No. 3.

Manne, A. and R. Richels, 1990, "CO_2 Emission Reductions: A Regional Economic Cost Analysis," *Energy and the Environment in the 21st Century,* Cambridge, MA: The MIT Press.

George C. Marshall Institute, 1989 (F. Seitz, K. Bendelsen, R. Jastrow, and W. Nierenberg), *Scientific Perspectives on the Greenhouse Problem,* Washington, DC.

Montgomery, W. David, 1990, "Effects on Energy Markets and the US Economy of Measures to Reduce CO_2 Emissions from Fossil Fuels," *Energy and the Environment in the 21st Century,* Cambridge, MA: The MIT Press.

Nierenberg, W., 1990, letter to the editor, *Science,* 247:14.

Nordhaus, W., 1990, "Economic Policy in the Face of Global Warming," *Energy and the Environment in the 21st Century,* Cambridge, MA: The MIT Press.

Nordhaus, W. and G. Yohe, 1983, "Probabilistic Forecasts of Fossil Fuel Consumption," in *Changing Climate,* Washington, DC: National Academy Press.

Schelling, T., 1983, "Climate Change: Implications for Welfare and Policy," in National Research Council, *Changing Climate,* Washington, DC: National Academy Press.

Schneider, S., 1989, "The Greenhouse Effect: Science and Policy," *Science,* 243:771–81.

Schneider, S., 1990, "Prediction of Future Climate Change," *Energy and the Environment in the 21st Century,* MIT Press, Cambridge, MA, 1990.

Schneider, S., N. Rosenberg, W. Easterling, P. Crosson, and J. Darmsteader 1989, "The Greenhouse Effect: Its Causes, Possible Impacts, and Associated Uncertainties," in *Greenhouse Warming: Abatement and Adaptation,* Washington, DC: Resources for the Future.

Simon, H., 1990, "Prediction and Prescription in Systems Modeling," *Operations Research,* 38:7–14.

Smith, V., 1979, *Scarcity and Growth Reconsidered,* Washington, DC: Resources for the Future.

Spencer, R. and J. Christy, 1990, "Precise Monitoring of Global Temperature Trends from Satellites," *Science,* 247:1558–62.

US Environmental Protection Agency, 1989, *The Potential Effects of Global Climate Change on the United States: A Report to Congress,* EPA-230-05-89-050, December 1989.

Waggoner, P., 1990, *Climate Change and US Water Resources,* New York, NY: John Wiley & Sons.

Yohe, G., 1989, "Uncertainty, Global Climate and the Economic Value of Information: An Economic Analysis of Policy Options, Timing and Information under Long Term Uncertainty," working paper, Woods Hole Oceanographic Institution.

Zwally, H., 1989, "Growth of Greenland Ice Sheet: Interpretation," *Science,* 246:1589–91.

Effects on Energy Markets and the US Economy of Measures to Reduce CO_2 Emissions from Fossil Fuels*

W. David Montgomery
Natural Resources and Commerce
US Congressional Budget Office
Washington, DC

Introduction

A properly designed tax on oil, gas, and coal could provide an effective incentive to reduce CO_2 emissions into the atmosphere. Such a tax is commonly called a "carbon charge," because in order to encourage the most cost-effective means of emissions reduction the tax would be made proportional to the carbon content of the various fuels. About twice as much CO_2 is released from coal, per unit of energy produced, as from gas, with oil falling in between; a carbon charge would reflect these differences.

Reducing CO_2 emissions from fossil fuel combustion is likely to be part of any strategy designed to cope with the potential problems of global warming. Increasing concentrations of certain trace gases, including notably CO_2, CFCs, methane, and ozone, may lead to rising average temperatures throughout the world. The rate at which global average temperatures may increase is highly uncertain; estimates of how much temperatures could rise over the next century range from negligible to dramatic. At one extreme, the effects of warming would have only localized effects; at the other, economic, environmental, and geophysical effects could be severe, including rising sea levels, increased intensity of storms, extinction of certain plant and animal species, and disrupted agriculture. A number of means of delaying the onset of global warming, or of minimizing its effects when it occurs, have been considered. These include reducing emissions of the trace gases that contribute to warming, cultivating plants and forests that remove trace gases as part of their growth process, and taking steps to adapt agriculture, protect areas threatened with inundation, and otherwise minimize the harm that rising temperatures will cause.

Reducing CO_2 Emissions

Carbon dioxide emissions from fossil fuel combustion are important potential contributors to global warming. According to an Environmental Protection Agency (EPA) estimate, they now constitute over 50 percent of worldwide greenhouse gas emissions and are likely to account for a larger share over the next decade as the most reactive gases, such as CFCs, come under effective international regulatory controls. Carbon dioxide emissions are also a large share of the US contribution to global greenhouse emissions. Since the US accounts for less than 20 percent of global CO_2 emissions, action taken unilaterally by the US will not significantly slow global warming, but it is also inconceivable that an international solution will be possible without US participation.

Costs and Benefits of Reducing CO_2 Emissions

The benefits of mitigating global warming come from slowing the rate of increase in average global temperatures, thereby delaying the time when harm will begin and reducing the amount of damage that will occur by any given time. The long time scales involved in global warming suggest that the question of *when* action should be taken is as important as the questions of *what* and *how much* should be done. Activities that increase information—such as research on the various geophysical, environmental, and economic processes that contribute to global warming—are particularly suited to such an uncertain situation. Activities that make more alternatives available in the future when more is known about the likely progress and causes of global warming are also worth considering. These could include research and development directed, in the case of CO_2 emissions, at development of technologies for using non–carbon bearing forms of energy and for increasing energy ef-

* This paper is based on preliminary results from a study of carbon charges conducted in the Natural Resources and Commerce Division of the Congressional Budget Office by Thomas Lutton, Patrice Gordon, and Mark Chupka, but statements and conclusions in this presentation are the sole responsibility of the author, and do not necessarily represent the position of the Congressional Budget Office. The final Congressional Budget Office study, "Carbon Charges as a Response to Global Warming: The Effects of Taxing Fossil Fuels," focuses on a somewhat different scenario of phased-in charges, reports results of some models for different years, and corrects some minor inconsistencies in the early results.

ficiency. But one other factor must be kept in mind: accumulation of CO_2 in the atmosphere is a cumulative phenomenon—unless removed by one of the carbon sinks, be it a tree or the ocean, CO_2 remains in the atmosphere forever. Thus there are risks and costs to delaying action to reduce CO_2 emissions.

Information about how much harm rising temperatures will do compared to stable temperatures is irrelevant to the decision; what matters is getting an idea of how much harm could be avoided through feasible measures to control CO_2 emissions. Even with such information, the classic conundrum remains of how to balance uncertain future benefits against much more easily identified current costs. Standard economic analysis involves discounting future benefits for both time and risk; but even astronomical costs 50 years off nearly vanish when normal discount rates are applied.

Economists and philosophers have debated for many years whether moral feelings of obligation to future generations, or commonly held conceptions of justice, require analyzing widespread and irreversible phenomena like global warming without regard for discounting. Such a decision must be part of the development of a national and global program addressing climate issues. Without a consensus on how to weigh future benefits there can be little progress on the question of how much should be done now to mitigate global warming. But it must be kept in mind that such a question needs to be asked. There are likely to be measures that are so costly and disruptive that there would be widespread agreement that the harm they avoid is not worth the cost.

A closely related question, and one which must also be answered, is how any given target for reduction in emissions of greenhouse gases can be achieved most efficiently. To help in understanding where to draw that line it is useful to ask how alternative policies for achieving any given target compare in their effectiveness in reducing emissions and in their impacts on the economy. Choice of inefficient policies for reducing emissions can, by increasing costs, make some levels of mitigation appear undesirable even though more carefully designed policies might bring those mitigation levels in at affordable cost.

Policies for Reducing CO_2 Emissions

This report focuses on CO_2 emissions from fossil energy combustion in the United States. Such emissions form a large share of the US contribution to the greenhouse effect and may become increasingly important as the exceptionally damaging gases like CFCs and ozone come under effective regulatory controls. For some time to come, reducing these CO_2 emissions will require reducing the amount of fossil energy that is burned. We know of no way to remove CO_2 from combustion gases after fuels are burned that is not extremely expensive.

Carbon dioxide emissions come from the burning of coal, oil, and gas. Because of its chemical composition, coal produces roughly twice the CO_2 emissions produced by natural gas, with the emissions level for oil falling between coal and gas. Fossil energy plays a pervasive role in the economy, but there are a number of ways in which its consumption could be reduced. Reducing CO_2 emissions is possible through reducing energy consumption overall or by substituting energy forms that create less CO_2 for those that create more. Means of accomplishing these ends fall in five rough categories: reducing consumption of all goods in order to reduce use of energy, inducing changes in consumption patterns, inducing changes in how goods are produced, encouraging fuel substitution, and stimulating technological advances. The feasibility of these different measures, and how great a contribution each can make, depends in part on how much time is allowed for adjustments to take place. The categories are arranged roughly in the order of how much time it takes for each to become an effective means of reducing consumption.

Reductions in GNP. One way of consuming less energy is by simply doing with less of everything. There is fairly clear evidence that much of the drop in energy consumption that followed the oil price shocks in the 1970s was due to the depressed economy and lower levels of GNP that the price shocks caused. This is probably not the intended consequence of any policy, but to the extent that reductions in GNP occur as a result of policies that are designed to discourage fossil energy use through the other mechanisms described below, there will be an additional "income effect" tending to reduce consumption further. Consumers may also choose to make do with less energy, for example by traveling less or by reducing temperature settings when heating their residences. In this case, the income they do not spend on energy will be available for other purposes, which in turn can lead to changes in consumption patterns.

Changes in Consumption Patterns. Consumers can substitute other consumption activities for those that involve direct purchases and use of energy, like driving. But energy is not consumed only in the form of automotive gasoline and natural gas for heating and cooking. Different goods purchased by consumers embody different amounts of energy in their production. Energy is contributed both directly, as in the case of glassware, which must be heated to melt and shape the glass into useful forms, and textiles, which must be dried, and indirectly, as in the case of manufactured products assembled from components, each of which required energy in its manufacture. Rearranging consumption patterns away from goods with heavy direct and indirect content of energy and toward goods lighter in their requirements for energy can reduce energy consumption. Consumers can also change their direct energy consumption patterns by purchasing more energy efficient appliances, modifying heating systems, and otherwise making investments that improve the efficiency of energy use.

Changes in How Goods Are Produced. The amount of energy used in producing any good can also be changed, by means of changes in process design, installation of more energy-efficient equipment, or through better housekeeping practices designed to reduce waste. These changes often take the form of substituting other inputs—such as capital equipment, labor, or materials—for energy.

Fuel Substitution. Households, businesses, and electric utilities, in particular, can often find ways of substituting one fuel for another. Sometimes this is relatively quick and simple, as in the case of businesses equipped to burn more than one type of fuel in their boilers or electric utilities that have sufficient capacity in different types of generating plants so that they can choose among types of fuel to burn. In other cases fuel substitution involves investing in different types of fuel burning equipment and may require substantial time and expenditures. Fuel substitution can be a means of reducing CO_2 emissions because of the different carbon content of fuels. In particular, substituting oil or gas for coal produces lower carbon emissions for the same amount of energy input. Substituting nonfossil energy for fossil energy is likely to be more limited in scope, at least for some time.

Development of New Technologies. New technologies that allow for improved efficiency of energy use are possible in all sectors, ranging from automobiles and other forms of transportation, consumer durables, manufacturing equipment, and powerplants, to electricity generation. New technologies may also make substitution away from coal into oil or gas or into non–carbon bearing forms of energy easier. New technologies making such noncarbon sources as nuclear and solar energy affordable, acceptable, and usable on a wide scale can not only encourage substitution but allow growth in energy demand to be satisfied from sources that do not add CO_2 to the atmosphere.

Time Scales

The most painful means of reducing energy consumption is the quickest. Reductions in GNP and economic activity translate into less energy use immediately, as do decisions by consumers to use less energy directly. Some changes in the composition of consumption can be made fairly readily, as consumers reallocate their purchases of consumable items and rearrange daily activities; changes that involve purchasing different appliances, insulating houses, etc., may take a longer time. Consumers' habits and preferences may also take some time to change. Improvements in manufacturing energy efficiency obtained through better housekeeping may come quickly, but those that involve substitution of other inputs for energy, or changes in capital equipment, may take longer. Results of research and development into new technologies do not come quickly; and even after new technologies are available, the same process of adoption through gradual replacement of the capital stock may be necessary. For example, it may take decades before a prototype of an "inherently safe" nuclear powerplant is operated, and then more decades until the infrastructure and manufacturing capability exists to meet a large share of the increase in electricity demand with nuclear power.

Alternative Measures

What, then, can be done to bring about some or all of these changes, and how can a decision be made as to how much should be done? The policy options commonly discussed to reduce CO_2 emissions involve extensive interventions in energy markets: regulations, standards, taxes, or technology mandates. Carbon charges affect the broadest range of decisions that determine energy use, because no decision can escape their influence. A carbon charge policy would directly affect the supply and demand for primary energy by changing the prices of fossil fuels. Charges on fossil fuels levied at the point where they enter the economy—at the mine, wellhead, dock, or pipeline—are likely to promote a wide range of economic responses at all market levels, as the charges work their way through the economy.

Carbon Charges

Carbon charges, properly designed, would shift production and consumption decisions away from the economic activities that directly or indirectly produce the most carbon dioxide. The range of potential responses includes all the possible responses outlined above: the mix of energy inputs could change, the energy intensity of manufacturing processes could decrease through substitution and technological change, and consumers might reduce purchases of goods produced with the most fossil fuels. In their efforts to adjust to higher prices, the participants in energy markets will tend to make adjustments that reduce CO_2 emissions as cheaply as possible. A cost-effective mix of emissions reduction activities or technologies would reduce CO_2 emissions with a minimal sacrifice of economic output. Differentiating carbon charges among fuels would encourage the greatest conservation efforts for the fuels that release the greatest amount of carbon per unit of energy, e.g., coal, and induce substitution of forms of energy that release less carbon.

Carbon charges can provide an incentive for reduction in carbon emissions from fossil energy combustion in a cost-effective manner. What they cannot do is correct other market imperfections that prevent those who make decisions about energy consumption from fully recognizing or acting on these incentives. As they affect the list of potential actions described above, carbon charges

- allow consumers to decide if all the energy they are consuming (directly and indirectly) is worth the higher cost, and provide incentive to cut back if they do not find some uses of energy or energy-intensive goods worthwhile at higher prices.
- encourage consumers to change consumption patterns, reducing consumption of goods that embody the most direct and indirect energy, and increasing consumption of those that embody less.
- encourage businesses to adopt more efficient processes and practices, to invest in equipment that allows greater efficiency, and otherwise substitute other resources for energy.
- especially in electric utilities, stimulate switching from the more carbon-intensive fuels like coal toward those creating less CO_2 emissions, like gas, solar, and nuclear energy.
- increase the rewards for developing and bringing into the marketplace innovations that allow greater energy efficiency or make use of non–carbon bearing forms of energy.

Working in this manner, carbon charges are an alternative to regulatory or incentive programs designed to encourage specific actions that would reduce carbon emissions. Instead of restricting energy supply or requiring more efficient end-use technologies in downstream markets, carbon charges provide an incentive to individual choices. In comparing charges with more specific regulatory and incentive approaches, the amount of information that the government can ever have at hand becomes the crucial distinction. Although appliance labeling programs, efficiency standards, and utility-sponsored conservation programs may affect a limited number of decisions, it is difficult to conceive of a strategy that would affect all the consumer decisions that have been described as influencing energy consumption. In supporting such a strategy some analysts have argued that cost-effective technologies appear to be underutilized, and in some cases standards and regulations could increase economic efficiency. On the other hand, calculating the most cost-effective mix of regulations in all end-use energy markets is extremely difficult, in part because only limited information exists on the current stock of energy-using equipment. Thus, an effective end-use strategy from the standpoint of lower energy use might not be the most economically efficient way to reduce energy demands.

In particular, the inducement to move consumption patterns away from goods with high direct and indirect carbon content and toward goods with lower carbon content is impossible to provide except through prices. Certain key industrial processes likewise might be targeted by conservation programs, but the range of decisions involved, from better housekeeping practices to redesign of whole processes, seems more likely to be stimulated by the pursuit of lower costs. Even fuel switching has turned out to be a more complex process than one that can be addressed by simple formulas banning the use of specific fuels in certain uses, as was discovered by those assigned to implement the Power Plant and Industrial Fuel Use Act (PIFUA) which (paradoxically, from the point of view of CO_2 emissions) restricted gas use and required conversion to coal.

Dealing with Other Market Failures

What carbon charges cannot do is offset fully potential impediments to efficient choices. The impediments thought to characterize some portions of energy markets have been described in many places. They include

- Difficulties that consumers may face in obtaining information or financing necessary to make efficient choices among consumer durables, heating systems, etc.
- Situations in which those who make decisions affecting energy consumption do not pay their own energy bills, as may be the case in some buildings where renters pay fixed charges for heating and utilities as part of their rent.
- Regulatory distortions in incentives. Removal of oil and natural gas price controls has ended many distortions, but electric utility regulation may still cause delivered prices to diverge from the marginal cost of generating electricity and may bias the capital investment decisions of electric utilities in inefficient directions.
- Difficulties which an individual innovator or developer may have in capturing the economic benefits that come from what may be expensive processes of developing new technologies.

A number of energy programs, such as utility-sponsored conservation programs and broad government support for research and development (R&D) in energy efficiency and alternative forms of energy, address these market failures. A carbon charge system could work well with these other programs. For example, since alternative forms of energy creating no carbon emissions would pay no carbon charge, the system of carbon charges would make those alternatives much more attractive. An innovator bringing such forms of energy to market would reap increased rewards, and as a result the payoff from successful R&D in alternative forms of energy would be greater. Although they would not correct the basic market failures, carbon charges could help direct into appropriate channels the activity stimulated by other government programs such as R&D tax credits.

How High a Charge?

In the absence of much clearer information than is now available about how the damage from global warming would be reduced through reductions in CO_2 emissions, it is impossible to estimate a charge that would internalize the full social cost of carbon emissions. It is possible, however, to take a step by step approach to setting a level for carbon charges that leads toward such an estimate. This approach involves

- Estimating how high a charge it would take to achieve some specified goal of emissions reduction,
- Asking what economic and other impacts such a charge would have and projecting what difference the charge would make to global warming, and
- In light of that estimate and projection, considering whether the goal should be revised upward or downward.

We began by examining a charge of $100 per ton of carbon imposed in 1990.

What Does It Take to Achieve Reductions by 2000?

Fossil energy consumption and CO_2 emissions are likely to grow significantly over the next decade. The base case projection for emissions thus plays a central role in determining how high a charge is needed to reduce emissions. It is also responsible for some of the uncertainty about how quickly global warming will occur in the absence of new policies.

In this study we used as a baseline for energy demand the 1989 Energy Information Administration (EIA) Annual Energy Outlook, and applied the CO_2 emissions coefficients to generate a CO_2 baseline. Under this baseline, CO_2 emissions would rise to 17 percent above 1988 levels by the year 2000. The goal of 20 percent reduction in emissions by that year is therefore an ambitious one under this baseline, and not likely to be achieved without far-reaching changes in how energy is produced and used. An initial idea of how high a charge it would take to achieve that goal can be obtained by examining some simple relationships between energy use and economic activity. This examination focuses on three of the factors described above as determining energy use: the relation between levels of GNP and levels of energy use, the substitution of other resources for energy in producing goods, and the possibilities for switching from one form of energy to another. A number of studies of these factors have been conducted in an attempt to understand how energy markets and the economy reacted to price changes and supply shocks in the 1970s and 1980s. These studies in many cases tried to quantify the elasticity of response to energy price changes—how much of a change in energy use would be caused by a given change in energy prices. Using middle ground estimates based on this period, a carbon charge of $100 per ton of carbon contained in each type of fuel might produce results in the neighborhood of a 20 percent reduction from 1988 levels by the year 2000. A charge of $25 per ton might achieve stabilization, depressing emissions levels for a time but by the year 2000 allowing emissions to rise back to their 1988 levels.

These are very gross and aggregate results. The Energy Information Administration's Personal Computer Annual Energy Outlook (PCAEO) model was used to generate more detailed estimates of the short-term impacts of a $100 per ton carbon charge. We also used the Data Resources Inc. quarterly macro model, but found its results, although suggestive of some of the risks of rapid implementation of large energy taxes, too unstable to be relied on for precise numerical estimates of macroeconomic impacts. To explore the longer-term possibilities for substitution away from energy-intensive processes and goods, we used the Dynamic General Equilibrium Model developed by Dale Jorgensen and provided to the Congressional Budget Office (CBO) through the courtesy of Federal Emergency Management Administration (FEMA). In the very long term, endowments of energy resources and the availability (or nonavailability) of new energy technologies make a tremendous difference in carbon emissions. To explore these trends through the year 2100, we used, through the courtesy of EPA, the Atmospheric Stabilization Framework (ASF) (Edmonds-Reilly) model being utilized by EPA to study global warming issues. All of these models were modified and run by CBO for this exercise, so that projections and conclusions should not be attributed to their developers or proprietors, who were not directly involved in the model simulations and may not necessarily agree. Finally, results of the Manne-Richels analysis were used to explore the implications of another view of the longer-term implications of policies restricting CO_2 emissions.

Projected Effects of Carbon Charges

Since coal, oil, and gas release different amounts of CO_2 emissions, basing a charge on carbon content would imply different charges for each fuel. If the carbon charge were $100 per ton, the resulting fuel-specific taxes would be $60 per ton of coal, $1.65 per thousand cubic feet of natural gas, and $13 per barrel of oil. Such charges would almost triple the price of coal, but raise prices paid by final consumers for oil and gas by only about 30 percent (see Figure 1). There are two reasons

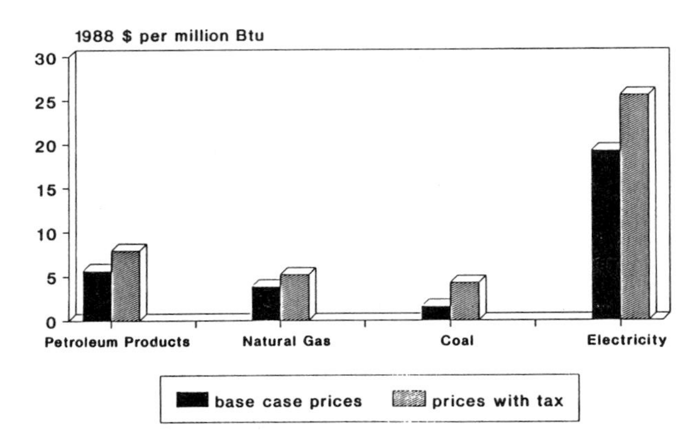

Figure 1. Price impacts for a $100 per ton carbon charge by fuel type: a short-run view.

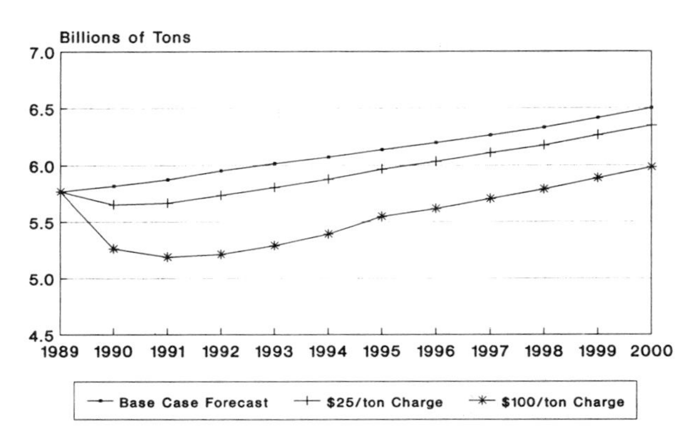

Figure 2. US carbon dioxide emissions (EIA energy model).

for this: even at the point of production, the energy in oil and gas costs considerably more than the same amount of energy in coal, and because of processing and distribution costs there is a much higher margin between the basic cost of the energy in oil and gas and their price delivered to consumers.

To look at these price impacts in more detail, it is necessary also to consider how the supply and demand of each of these fuels will respond to the imposition of such charges, and to estimate thereby how market prices will change and how much of the tax will be absorbed by producers of different energy forms and how much passed on to consumers. The Energy Information Administration's PCAEO model was used for this purpose. This model projected that emissions would be roughly stabilized by this tax (see Figure 2). EIA's projections also underlie the base case projections of emissions described above.

Price Impacts. When a tax is applied to any good, it can be expected to reduce demand for that good and place downward pressure on the prices received after tax by the sellers of that good. When carbon charges are applied to the EIA model, only coal shows a noticeable drop in the price received by sellers, and even in the case of coal the drop is small. The reason for the small drop in coal prices is that moderate increases or decreases in coal production are met by opening or closing mines whose cost conditions are nearly identical. Since the cost of producing such marginal supplies determines the price of coal, some reduction in supply can occur with little drop in price. In the case of oil, the domestic price is set by the world oil price, whose level would be affected infinitesimally by a drop in oil demand of the magnitude projected. In the case of gas, demand is also largely unaffected in this time period by the carbon charge, for reasons described below, and as a result the price changes little.

Electricity prices paid by consumers are also driven up by the carbon charge, for two reasons. Utilities that burn oil, gas, and coal face increased prices for these fuels, which they will pass on to their consumers. In response, electricity demand can be expected to drop. But since over this time period the costs of utilities related to their installed capacity are fixed, spreading those fixed costs over a lesser demand base increases the cost per unit of electricity.

Effects on Demand. Figure 3 illustrates how energy demand will change over time (specific fuel impacts are similar). The EIA model indicates that these reductions would occur within one to two years of the imposition of the charge. In discussing more detailed results, we focus on an intermediate year, 1995, in which many of the effects incorporated in the EIA model have had time to work out. The results are best interpreted as being the immediate results of imposing the tax, before many of the adjustments described above as possible in the longer term occur. As shown in Figure 4, demand in all the end-use sectors (residential, commercial, industrial,

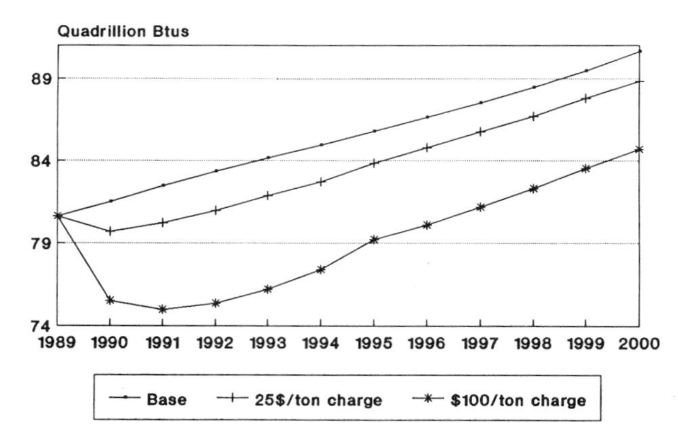

Figure 3. US consumption of energy (EIA energy model).

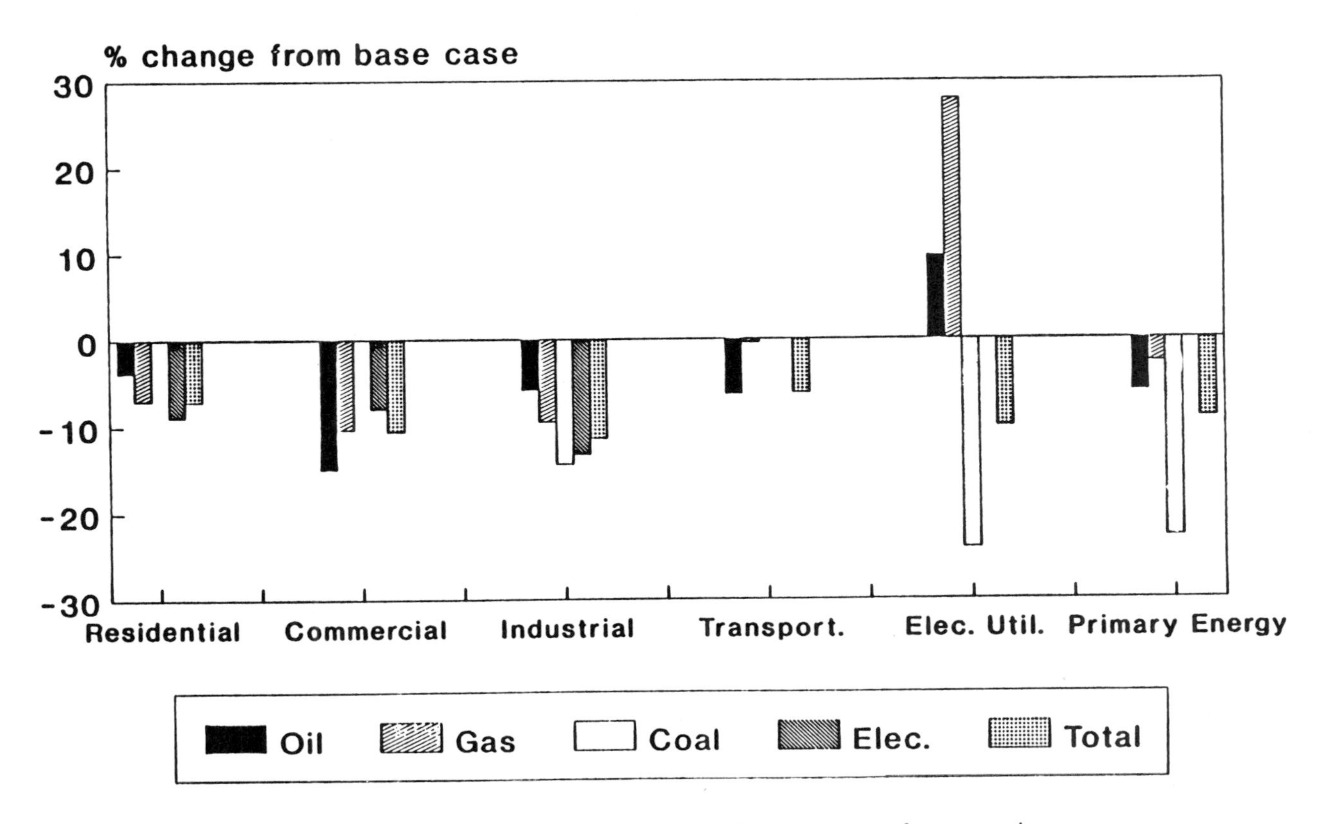

Figure 4. Energy consumption impacts for a $100 per ton carbon charge: a short-run view.

and transportation) falls compared to the base case when carbon charges are applied. Coal demand drops substantially in the industrial sector, the only sector in which noticeable quantities of coal are consumed, while oil, gas, and electricity consumption drops in all sectors.

Electric output also drops with the decrease in electricity demand, and this drop is responsible for some of the reduction in demand for fuels by the electric utility sector. But utilities strongly show another effect to be expected when the price of coal rises so drastically compared to the price of other fuels: they substitute natural gas for coal by cutting back the hours that they run coal-fired powerplants and increasing their utilization of gas-fired powerplants. These dispatching decisions are made on the basis of the relative operating costs of different units (largely fuel costs), and carbon charges make many coal-fired plants more expensive to operate than gas-fired plants within the same power pool. (These results would change by the year 2000, or if the charge were phased in, giving utilities time to cancel new construction—in this case utilities reduce gas as well as coal and oil use.) As a result of this substitution, the reduction in gas demand in the end-use sectors is almost exactly offset by an increase in gas demand by electric utilities.

Effects on Domestic Energy-producing Industries. The coal industry is hit hard by a carbon charge (Figure 5). Even in the very short run, and if only stabilization of emissions were to be accomplished, the coal industry would suffer about a 25 percent drop in output. This

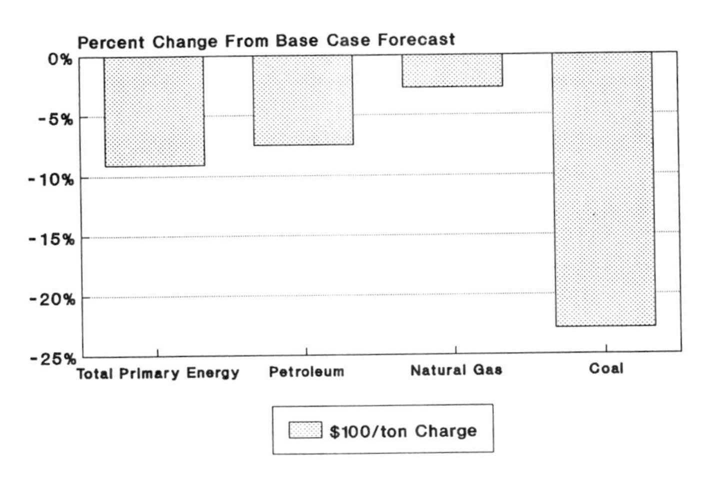

Figure 5. US primary energy consumption impact (EIA energy model: very short run).

kind of effect occurs in every carbon charge scenario examined, for one simple reason. Coal is the most carbon-laden fuel. Given this fact, as well as the cost and difficulty of doing without other fuels in the end-use sectors and the relative ease of substituting gas for coal in electric utilities in the short run, reduced coal consumption is a necessary part of any effective program to control CO_2 emissions in the near term. This is true whether the policy instruments are carbon charges or regulatory restrictions on coal use.

Natural gas supply would be virtually unaffected by carbon charges under this scenario where gas substitutes for coal. Impacts on the domestic oil industry depend largely on whether the reduction in coal demand has a noticeable effect in depressing world oil prices. The reduction in oil demand is only about 3 percent of US demand, or less than 1 percent of world demand. Using another of the EIA's models that examines supply, demand, and pricing in world oil markets, it is estimated that such a drop in demand would affect world oil prices (now about $18 per barrel) by less than 25 cents per barrel. Thus there is likely to be no noticeable effect on domestic oil production; the reduction in demand would all come out of imports. Alternative fuels not subject to the carbon charge are likely to benefit, leaving the coal industry as the largest loser among US energy producers in the short run. This picture changes in the longer term, when demand for all fossil fuels is likely to fall in order to achieve further reductions in CO_2 emissions.

Tax Collections. Gross tax collections are likely to be about $150 billion from the $100 per ton charge, and $50 billion from the $25 per ton charge. The reason for the lack of proportionality between tax rates and receipts is the reduction of energy demand, and substitution of cleaner fuels for coal, induced by the higher charge. These collections might be offset in part by reduced taxes elsewhere in the economy.

Impacts on the Economy. Effects on energy markets are constrained in the short term by the limited flexibility that exists to change energy consumption through measures other than simply doing with less. The EIA model shows stabilization as the most that could be hoped for through a carbon charge for two reasons:

- In the form used for this study, the EIA model allows for little adjustment of the capital stocks of consumers or businesses, and thus embodies none of the types of changes in consumption that might be expected to show up after a few years of higher prices.
- The EIA model also shows limited impacts of the carbon charge on overall economic performance (see Figure 6).

Another attempt to examine the short-run effects of the carbon charge on energy markets and the economy was based on the Data Resources, Inc. (DRI) annual model of the US economy. This model represents energy markets in a more rudimentary way than the EIA model, but it has a much richer structure for examining the effects of such taxes on the economy and the potential of various macroeconomic policies to offset their effects. The DRI model was chosen for one other reason: it has consistently, in past studies, shown far more damage to the economy from taxes like the carbon charge than any of its competitors. Thus it is likely to represent another extreme in the responses that it shows, allowing us to bracket the likely range of possibilities.

The DRI model suggested that imposing a $100 per ton charge without compensatory policy or a phase-in period could create a damaging shock to the economy, including significant inflation. For the worst year of the implementation, the DRI model showed a drop of 6 percent from baseline GNP, and a rise in prices greater than can be explained by energy price increases alone, suggesting a wage-price spiral.

These results point up the serious short-run risk to the economy that rapid implementation of a large tax entails, but more than that should not be read into the results. The DRI projections are extreme, and although the risk they imply cannot be ignored, it should not be looked on as a necessary result. Such a tax is so large that the DRI model actually failed to solve in many of the simulations we ran.

Policies to Offset Impacts. A number of approaches could be taken to mitigate the near-term impacts of a carbon charge while preserving most of the desirable long-term incentives that it creates. We examined three, using the DRI model: lowering the charge, phasing in the charge, and returning some of the tax revenues to the economy through lower income and payroll taxes.

A lower tax ($25 versus $100) does less damage to the economy and accomplishes less in reducing emissions. One reason that might be suspected for the smaller impact of a reduced charge is the lessened drain from the consumer's pocket. The traditional aggregate demand analysis would suggest that suddenly reducing the deficit by over $100 billion would so dry up demand that a recession would surely result, and that these effects would be largely offset if the charge were used for tax reduction rather than deficit reduction. That the phenomenon of "fiscal drag" is not the main reason why a carbon charge reduces GNP is suggested by the results derived from the DRI model when the charge is returned to the economy through reductions in income taxes. Impacts are only slightly lessened, even if other taxes are reduced by the full amount of revenues from the charge.

The imposition of a carbon charge is more a supply shock than a demand shock; by making energy much more expensive, its effects on the economy are akin to that of the oil embargo of the 1970s. Businesses and consumers find that they must make do with less energy,

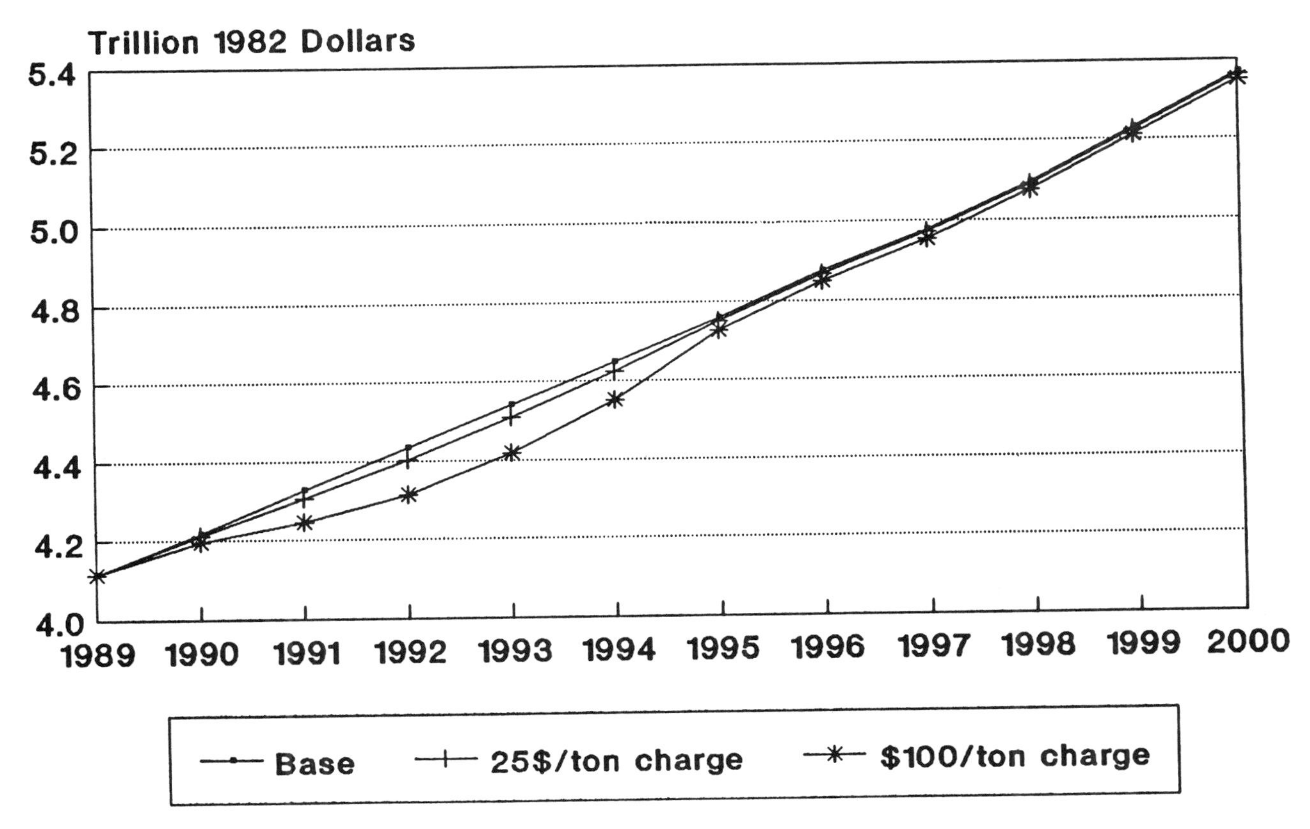

Figure 6. Real GNP (EIA energy model).

and rearranging their purchasing patterns takes some time. The result is that the economy is temporarily jolted out of equilibrium, and it takes some time for markets to get back into balance. [See Bohi and Montgomery, *Oil Prices, Energy Security and Import Policy.*]

Phasing in the charge can allow time for some of these adjustments to take place, while returning some of the revenues to the economy through tax rate reduction can reduce fiscal drag. One way to institute both of these policies would be to begin with a charge of $25 per ton in 1990, rising in equal annual increments to $100 by 2000, and to reduce income tax rates sufficiently to reduce income tax collections by one half of the carbon charge receipts. This combined policy eliminates 80 percent of the damage to the economy while preserving almost half of the CO_2 emissions reductions achieved with the $100 charge. Reducing taxes that show up directly in price levels, such as payroll taxes, might be an even better way to counteract the damage of the carbon charge, by offsetting some of its inflationary pressures directly. With a policy of phasing in and partially offsetting a $100 per ton carbon charge, the loss in GNP would be a steady 1 to 2 percent through the year 2000.

Differences between the Two Short-run Models. It is only to be expected that two models designed to look at very different aspects of energy-economy interactions would differ in their conclusions about as dramatic an event as a high carbon charge. The DRI model projects much larger effects from a given charge on CO_2 emissions, but also much higher risk of great economic harm unless a charge is phased in gradually and its fiscal drag offset. The two models also showed very different near-term effects of a carbon charge on CO_2 emissions. The difference between the two is due in part to the greater loss in GNP predicted by the DRI model, and in part to significant differences between the models in their optimism about the ease of attaining greater energy efficiency or fuel substitution. This difference highlights how large a role "doing with less" may play in forecasts that suggest the possibility of large reductions in CO_2 emissions.

But another difference is in the degree to which the energy use to GNP ratio changes: in the EIA results the ratio declines to about 13,000 million Btu per 1988 dollar of GNP, while DRI takes the ratio down to 11,500 as a result of the charge. The difference is largely in the greater flexibility that the DRI model assumes to exist in adaptation to higher energy prices.

As indicated, the EIA model is not really well suited to looking at the impacts of a carbon charge after about five years, and the DRI model is not well suited to long-term extrapolation either, given the somewhat erratic cyclical behavior it exhibits. Such substitutions can take a much longer time than that to work out, especially if changes in the stock of buildings and capital equipment are important. To explore these possibilities of longer-term substitution, we used a model developed to look explicitly at such relationships.

How Much Substitution?

This model, originated by Dale Jorgensen and called the Dynamic General Equilibrium Model (DGEM), assumes that in the long term there is great flexibility in how the economy can adjust to higher energy prices and that new non–carbon bearing energy technologies will be readily available. CBO used its own version of the DGEM model to examine the consequences of a $100 per ton carbon charge. The DGEM model represents essentially an extrapolation of past experiences with energy price increases into the future. It incorporates all the adjustments in energy markets described above: reduction in GNP, changes in purchasing patterns of consumers, substitution between energy and other factors of production, interfuel substitution, and development of new technologies. Statistical techniques are used to see how the changes in each of these areas were related to changes in energy prices in the past; the results of this historical analysis were that in all these areas there was a very large response to energy price changes. To forecast the future it is assumed that these relations will continue to hold throughout the forecast period.

This is the normal practice in econometric forecasting, and it has much to recommend it. The huge changes in patterns of energy use that occurred in the 1970s and 1980s, and the many years in which the US economy grew while energy consumption fell, are highly instructive. It would be foolish to ignore the lessons of history on how much it is possible to increase energy efficiency and introduce new sources of energy. But it must also be recognized just how strong an assumption it is to expect these patterns to continue. Tremendous amounts of waste were wrung out of the economy as rising energy prices forced many energy users to think for the first time about how they used energy. Moreover, the substitution of new forms of energy for traditional oil, gas, and coal largely took the form of introduction of nuclear power for electricity generation—a trend which few expect to continue.

Thus the results of the DGEM model can be expected to be biased in favor of greater ease of stabilizing carbon emissions than may be real. But the patterns of change which this model shows to be required to achieve stabilization are interesting, and its estimates of economic loss associated with these changes are clearly a lower bound. In particular, this approach shows that changes in energy consumption come about largely through rearrangements by consumers in the types of nonenergy goods that they buy—toward those with less direct and indirect energy content and away from those with more. Whole industries rise and decline by noticeable amounts, and significant increases in non–carbon bearing forms of energy are required. At the same time the coal industry would shrink to one-half its current size.

Overall, the results of these changes would be to reduce CO_2 emissions by 36 percent below the baseline. If this were to occur by the year 2000, a goal of 20 percent reduction in emissions would be more than achieved (see Figures 7 and 8). The cost would be less than 1 percent of the GNP that would otherwise be produced in the year 2000, if the entire adjustment process with whatever costs it entails could be achieved by then (Figure 9). These results would be accomplished through a reduction of almost 40 percent in total fossil

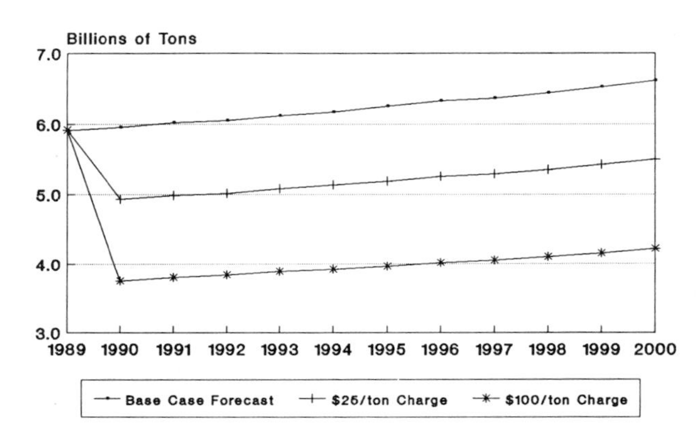

Figure 7. US carbon dioxide emissions (Dynamic General Equilibrium Model).

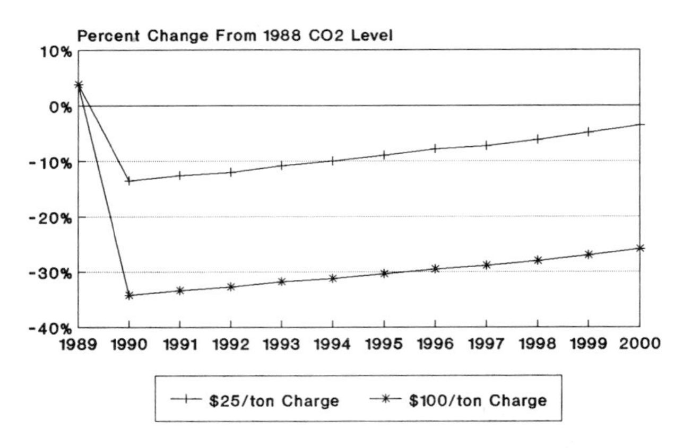

Figure 8. US carbon dioxide emissions (Dynamic General Equilibrium Model).

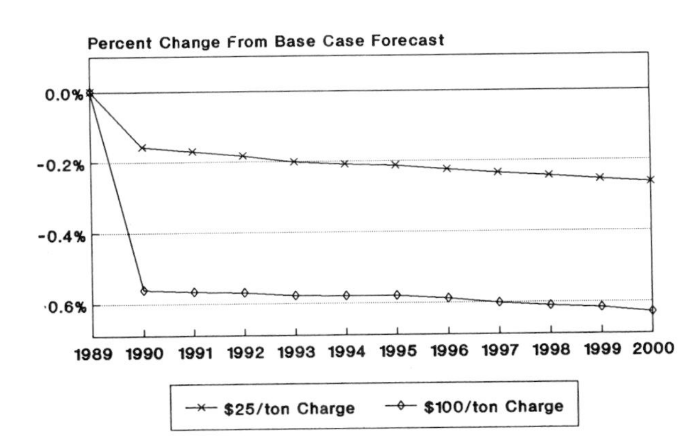

Figure 9. US real GNP (Dynamic General Equilibrium Model).

energy consumption, and an increase of 80 percent in the amount of nonfossil energy utilized. How, then, could these massive changes in energy markets be achieved at so little cost?

Reductions would be required in every form of energy consumption in every end-use sector, with the largest percentage decreases in coal consumption (see Figure 10). As end-use consumption of electricity declines relative to the base case, demand for fossil fuel for electric power generation would also fall, again with the bulk of the reduction concentrated in coal, but with no form of fossil energy benefiting from its ability to substitute for coal. (This result is due to the excessive optimism about the potential for nonfossil energy in this time frame, but results not too different are also obtained if it is assumed that no additional supplies of nonfossil energy are available.)

US coal production would fall by 70 percent below the baseline, to less than half current levels, and even natural gas production would fall by about 20 percent. With a decline in US oil demand of about 20 percent, an effect on world oil prices is likely, so that domestic producers would face lower prices and reduce production accordingly. The EIA model of world oil markets mentioned above projects that such a decline in US energy demand would cause a $2 to $3 per barrel drop in world oil prices in the 2000–2010 time period, and a 300,000 barrel per day drop in US oil production by 2010 (from 7.3 million bd to 7.0 million bd). The carbon charge itself amounted to $13 per barrel of oil, so that world market reactions absorb about 20 percent of the tax.

The residential and transportation sector results show that consumers may be assumed to be willing to give up 20 to 30 percent of their direct energy consumption when faced with prices that are higher by only 30 percent or so for oil, gas, and electricity. But they also change their demands for all other goods they consume. It is only this change in consumer purchasing habits that allows the substantial reductions in commercial/industrial energy use observed. To some extent, makers of goods also substitute components made with less direct and indirect energy content for those made with more. The result is the pattern of changes in output across industrial sectors seen in Figure 11. Some industries expand by more than 10 percent while others decline in like amount. The amount of energy used per unit of GNP falls much lower than in the projections of impacts done with the EIA model (Figures 12 and 13), as should be expected in a model that allows much greater substitution at all levels of the economy from final consumption to factor markets. In addition, GNP is sustained to some extent through greater effort (Figure 14), in that greater labor inputs are substituted for energy. Since full employment is always assumed in DGEM, these increased labor inputs represent an additional loss in welfare over and above that captured in GNP loss.

None of these changes are likely to occur quickly. The immediate impact shown in the DGEM results occurs simply because the model ignores short-term difficulties in adjustment, focusing as it does on long-run responses. Even the year 2000 may be too soon to see all the implicit adjustments work out. Whatever the ultimate ability of the economy to rearrange itself, the kind of economic adjustments projected by DGEM are likely to be accompanied by a difficult and painful adjustment period not considered in this model. Moreover, given the likely timing of increases in demand for energy, exhaustion of gas resources, and availability of new carbon-free energy sources, maintaining stable levels of carbon emissions beyond the year 2000 may also be very difficult.

Effects in the Longer Term

The more time that passes after a carbon charge is implemented, the greater are the opportunities for the capital stock and patterns of energy use to adjust and for new technologies to come into play. But secular changes also occur with the passage of time: the economy, and the economic base for CO_2 emissions, will grow, and some energy resources will become harder to find and more expensive to produce. Moreover, there are limits to how rapidly the development and introduction of new technologies can be accelerated. Depending on how the time scales for these varied changes work out in relation to one another, the period after 2000 could be one in which substantial progress toward stabilization of emissions is achieved or one in which stabilization could be achieved only through tremendous economic sacrifices. These contrasting possibilities can be illustrated through the results of two different models

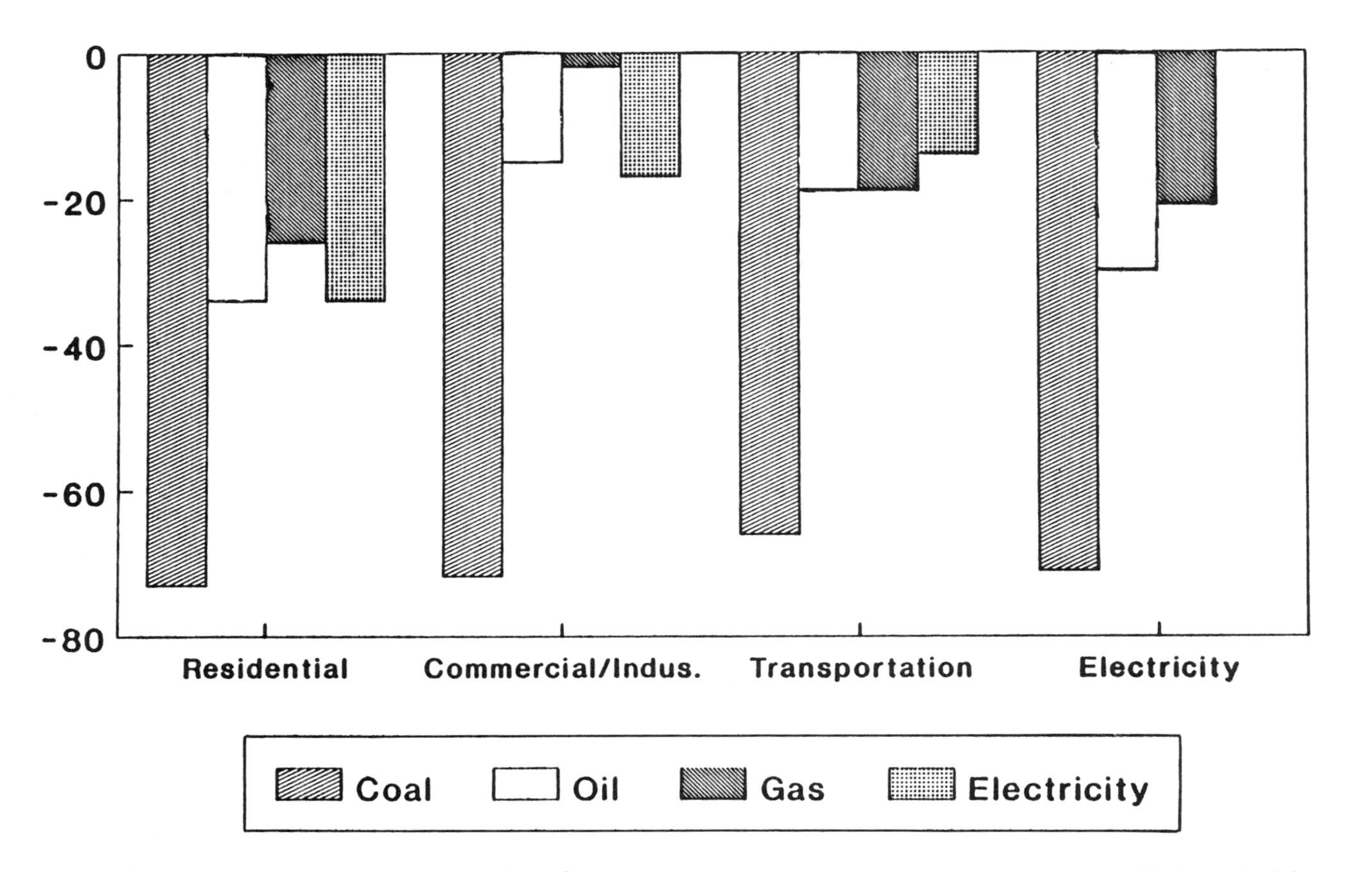

Figure 10. Energy consumption impacts for a $100 per ton carbon charge (a Dynamic General Equilibrium Model).

designed to analyze the longer-term impacts of changes in energy prices.

One approach is based on work done at Oak Ridge National Laboratory by James Edmonds and John Reilly, and has since been modified and expanded significantly by the EPA in their work on global warming. CBO obtained a version of this model from EPA in order to explore these projections independently. The other, developed by Alan Manne and Richard Richels, looks explicitly at some real-world constraints on the supply of relatively carbon free resources and at the time required to deploy new energy technologies.

The two models differ dramatically in the base cases that they envision beyond the year 2000. Two key factors driving these very long-term forecasts are rates of change in energy use per unit of GNP and the availability of nonfossil energy technologies. Energy use per dollar of GNP using ASF assumptions drops from its current level of 12 million Btu per dollar to between 2 and 3 million Btu per dollar by 2100, while in the Manne and Richels work it drops to something more like 5 to 6 million Btu per dollar. ASF shows US demand for nonfossil fuels in the base case rising from 6 to over 25 exajoules, while Manne and Richels place much lower limits on the rate of introduction of these fuels. By 2100, ASF projects US CO_2 emissions of just over 8 billion tons, while Manne and Richels project emissions of about 16 billion tons in 2100. As a result of these optimistic baseline assumptions, ASF shows that a carbon charge imposed multilaterally and rising from $100 per ton in 2000 to $300 per ton by 2100 could keep US CO_2 emissions below their current levels through 2100 and hold world CO_2 emissions to half their base case values by 2100 (Figures 15 and 16).

Negative effects on the economy are also limited. A $100 per ton charge would have a permanent effect of causing a loss of about 1 percent of GNP compared to baseline in every year, and would reduce CO_2 emissions from just over 8 billion tons in 2100 to just under 4.5 billion tons. A charge rising to $300 would have an impact increasing to just over 2 percent of GNP (Figure 17).

The Manne-Richels perspective is very different. They asked what level of carbon charges it would take to achieve a somewhat more modest near-term target—zero growth in CO_2 emissions through 2000, and then a 20 percent reduction by 2010 and zero growth from then on. Through the year 2000, their conclusions are

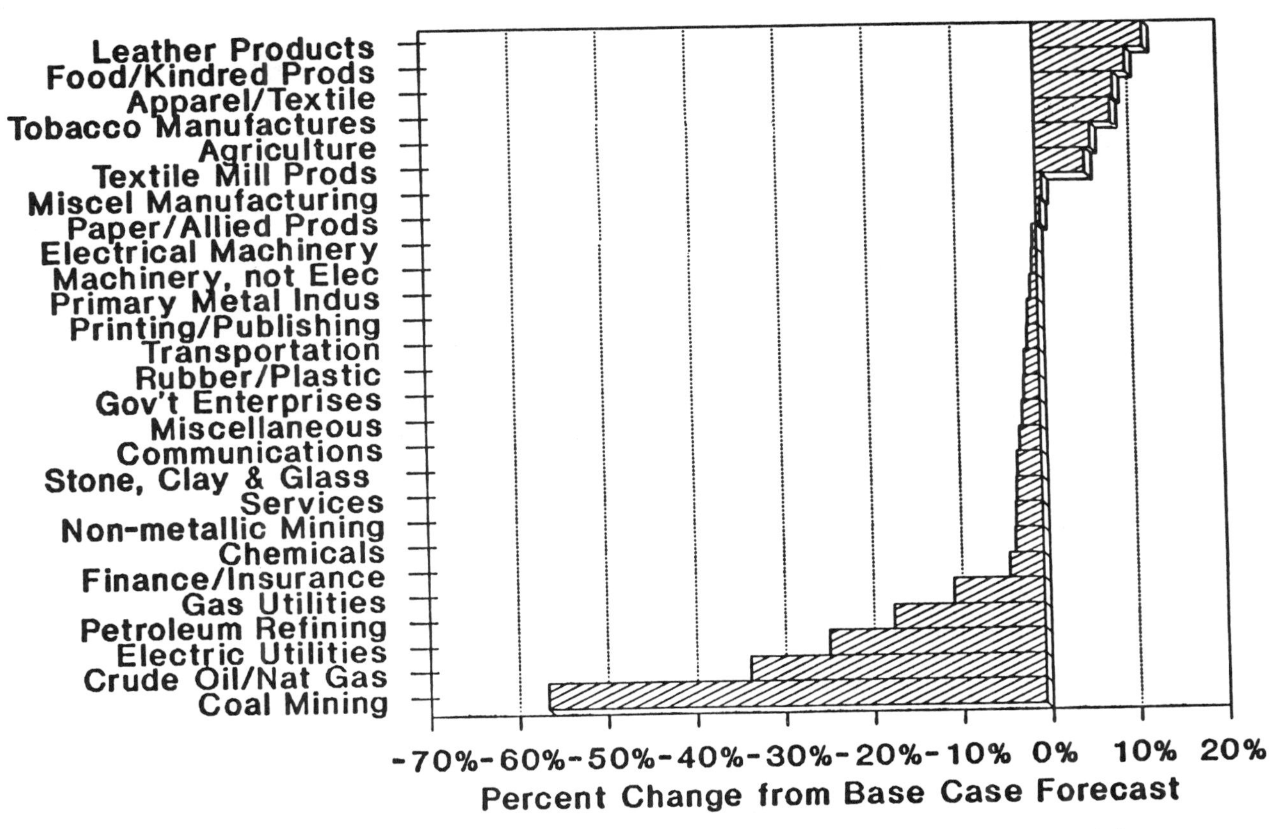

Figure 11. Industry effects for a $100 per ton carbon charge (Dynamic General Equilibrium Model).

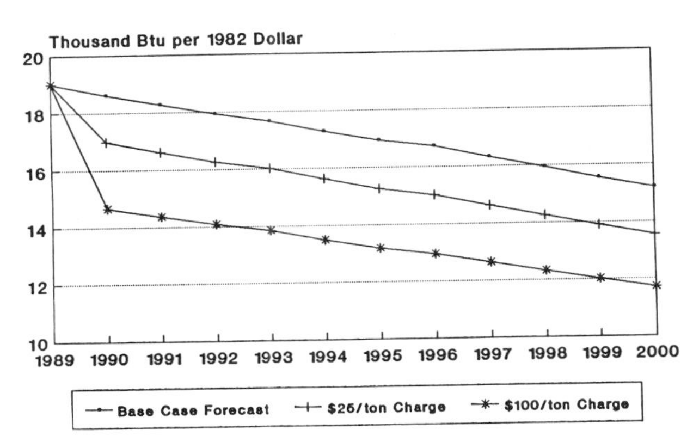

Figure 12. Gross energy use per dollar of GNP (Dynamic General Equilibrium Model).

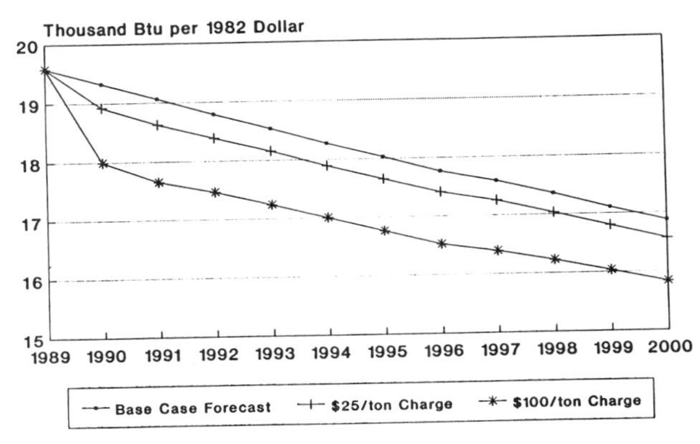

Figure 13. Gross energy use per dollar of GNP (EIA energy model).

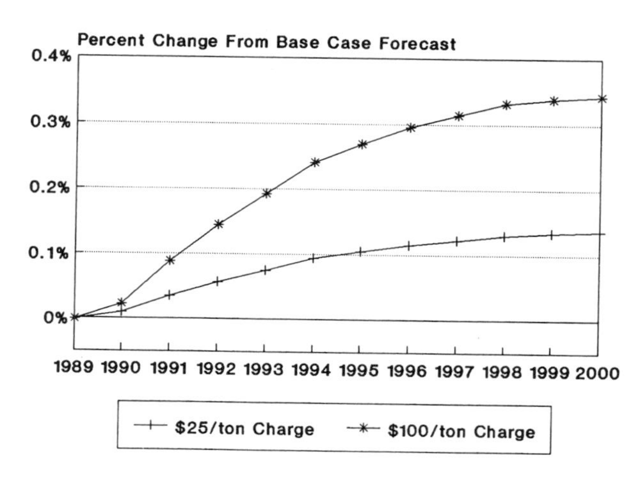

Figure 14. Quantity of labor services supplied (Dynamic General Equilibrium Model).

similar to those of DGEM, viz., that zero growth in CO_2 emissions could be achieved with modest losses in GNP. As in the case of DGEM, the Manne-Richels approach may be overly optimistic about the ease of adjustment, because it does not consider the costs of transition to lower energy use, but concentrates on what happens beyond 2000. They conclude that the US would lose about 3 percent of GNP in every year from 2030 on, in order to reduce its emissions to a level about 20 percent below 1990 levels. This is about the same level of emissions that results from the ASF model with a $100 charge, but it requires a much larger reduction from baseline emissions. As a result, Manne and Richels project a charge that rises to $400 per ton by 2020, and then recedes to about $250 per ton as less expensive non–carbon bearing technologies finally become available on a wide enough scale. Both models appear to agree that a 75 percent reduction in emissions would require a charge of $250 to $300 per ton in 2100 and cost 2 to 3 percent of GNP in perpetuity.

In performing their analysis, Manne and Richels have introduced three key new ideas with implications for the very long term:

- Long-term growth rates for energy demand imply a growing need for forms of energy with little or no carbon content through the next century.
- Domestic natural gas resources may begin to be exhausted in the early 21st century.
- Technologies allowing the commercial use of carbon-free energy sources may not be widely available until after 2030.

Even higher forecasts of energy demand growth than assumed by Manne and Richels may be justified by recent research into relations between energy prices and economic growth, energy efficiency, and technical change being carried out by William Hogan and Dale Jorgensen. This research suggests that in the absence of energy price hikes during the 1970s, the economy would actually have become less energy efficient, i.e., would have used more energy per dollar of GNP. The conventional wisdom in energy forecasting is that the energy-GNP ratio will improve at a rate of about 1 percent per year, while the Hogan and Jorgensen studies suggest that it may be declining slightly, by 0.1 percent per year. Manne and Richels adopt what may now be a conservative assumption that drives demand growth through the 21st century.

Concern about how large the natural gas resource in the United States may be has existed for some time. The gas glut that existed through most of the 1980s muted that concern, and it must be recognized that any estimate of the size of the natural gas resource is highly speculative. Nevertheless, there is some evidence that natural gas will become increasingly difficult to find and produce sometime early in the next century if current rates of extraction continue. If this is the case, it will become increasingly difficult to substitute what is now the best available alternative for coal as a means of reducing CO_2 emissions.

There is not believed to be any way to remove and store CO_2 at reasonable cost after fossil fuel is burned. Finding clean alternatives to coal requires developing means of utilizing energy sources that do not release CO_2 into the atmosphere, and that are capable of widespread enough use to supply the entire increment to demand. Some resources, such as geothermal energy, are currently available but cannot be expanded sufficiently. Others, such as fusion energy, are too far off to help until late in the 21st century. Development of inherently safe and economically feasible fission reactors might provide a sufficiently large source. New technologies for improving energy efficiency may also have a role to play. But the time scales on which such alternative sources of energy may become available drive the projection of how difficult it will be to hold the line on CO_2 emissions.

With substantial growth in energy demand and limitations on the supply of cleaner fossil fuels, preventing growth in CO_2 emissions will require finding non–carbon bearing energy sources to supply the entire increment to demand in every year. If those sources are not available, the carbon charge must be set high enough to choke off growth in demand in order to accomplish stabilization. Given the paucity of alternatives, choking off energy demand means giving up a large portion of the economic growth that would otherwise be possible. This bleak prospect lies behind the conclusions of Manne and Richels about the cost of limiting fossil fuel emissions. If sufficient time is allowed for the introduction of new nonfossil energy sources, the charge required to achieve stabilization of emissions falls. Manne and Richels find that by the latter part of the 21st century a charge around $250 per ton would be sufficient

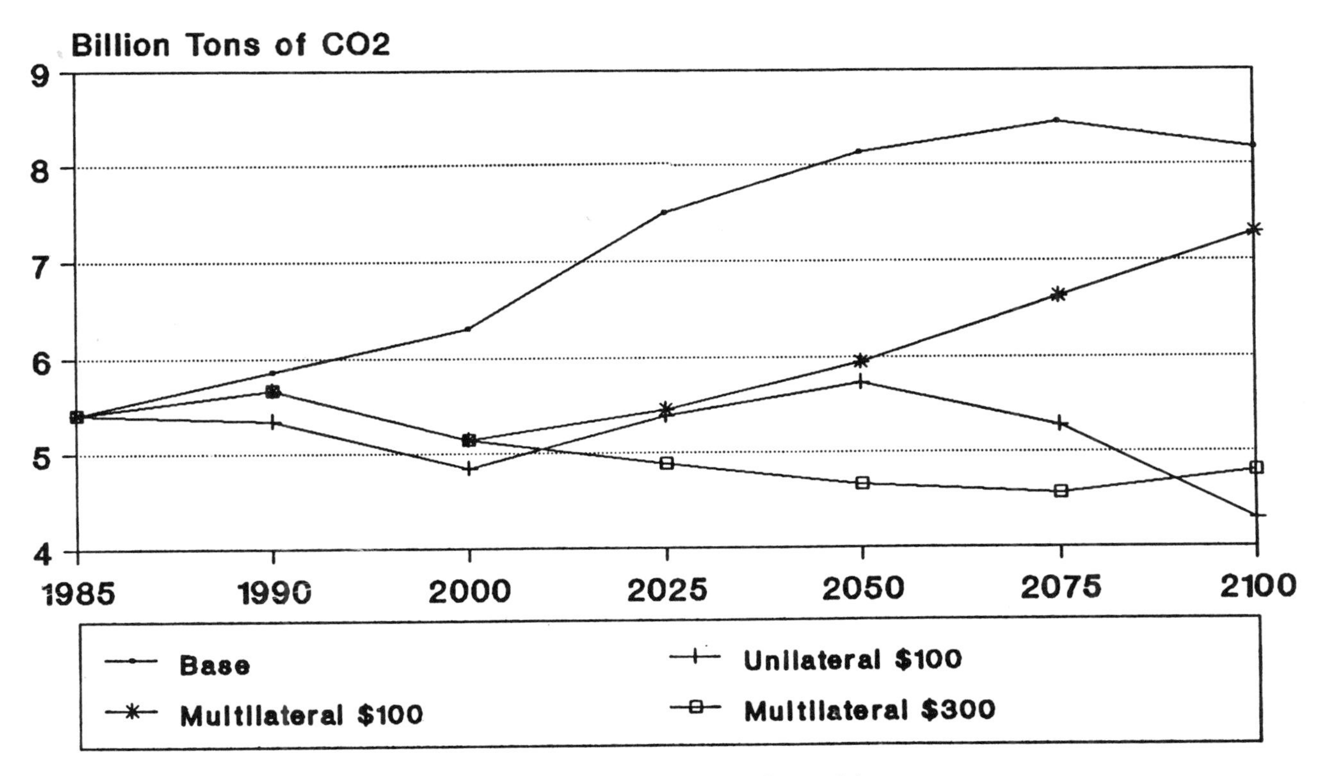

Figure 15. US carbon dioxide emissions (ASF: Edmonds and Reilly Model).

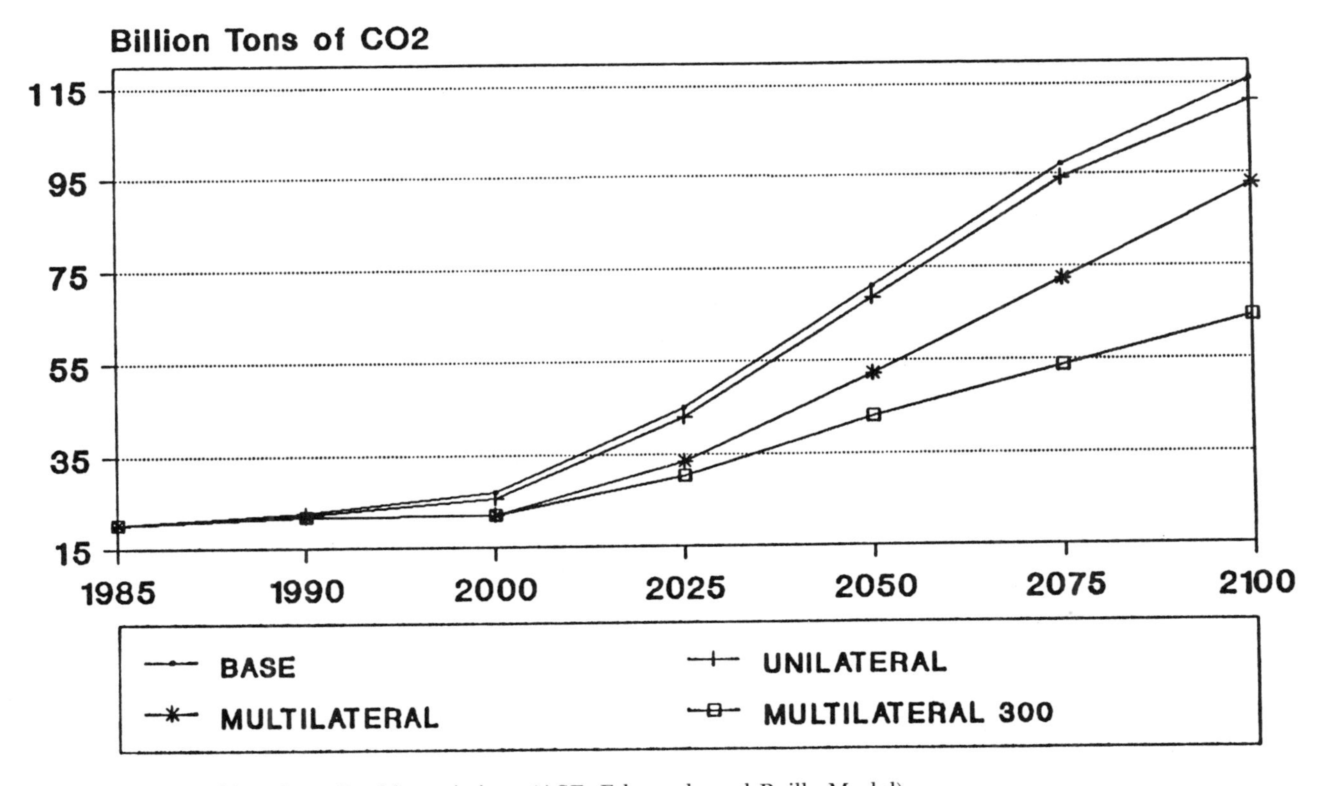

Figure 16. World carbon dioxide emissions (ASF: Edmonds and Reilly Model).

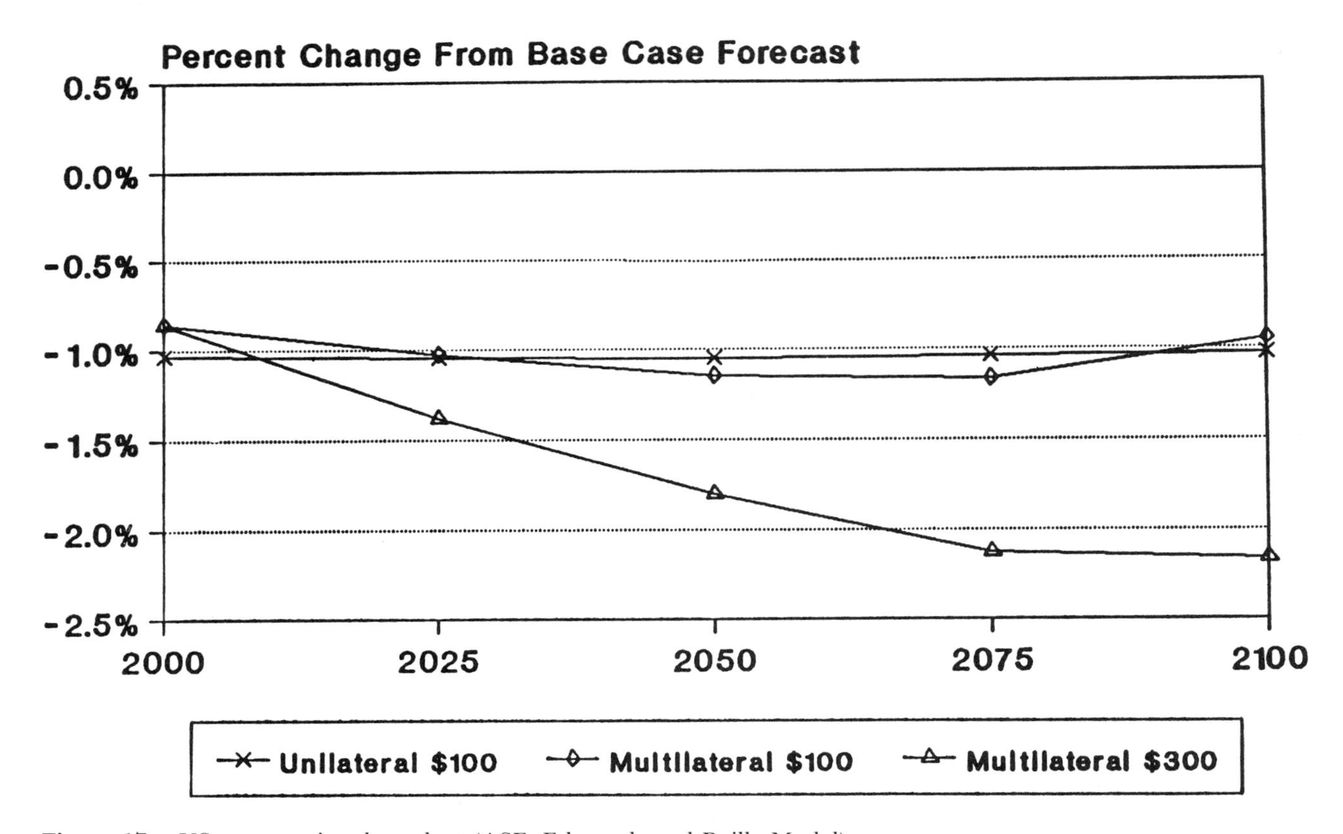

Figure 17. US gross national product (ASF: Edmonds and Reilly Model).

to ensure that increases in energy demand were met from sources with no CO_2 emissions. This conclusion also points out the value of successful R&D directed at technologies that are capable of sufficiently widespread adoption to become the marginal sources of energy for economic growth.

Since the US is responsible for 20 percent or less of global CO_2 emissions, unilateral action is unlikely to achieve a large result. Simulations with the ASF model suggest that unilateral imposition of a $100 per ton charge would only delay the doubling of atmospheric concentrations of CO_2 by three years. The results of multilateral action could be much more dramatic—a delay of about 17 years in the time of doubling. If a rising tariff—to approximately $300 by 2100—were adopted, doubling might be put off until after 2100.

Conclusions

Carbon charges can work to reduce CO_2 emissions from fossil fuel combustion in the US. A charge of $100 per ton of carbon would succeed by the year 2000 in accomplishing something between stabilization and a 20 percent reduction in such emissions. More certain predictions are not possible to make, because achieving these limitations of fossil energy use requires large changes in energy production and use pervading the entire economy. These results would be accomplished at a long-term cost of 1 to 3 percent of GNP, but with much larger costs for a few years during the 1990s as the economy adjusted rapidly to doing with less energy. These transition costs could be reduced substantially by phasing in the charge and returning some of the revenues to the economy by reducing other taxes. Over the longer term, maintaining a carbon charge at $100 per ton would hold GNP at least 1 percent lower than it would be without charges. Even higher charges would be required to prevent growth in CO_2 emissions after 2000. The $100 per ton charge imposed unilaterally by the US might delay the doubling of global CO_2 emissions by three years, while a charge of that amount imposed by all countries might buy almost two decades.

These costs could not be reduced by policies other than economic incentives, such as those embodied in carbon charges, to reduce fossil energy use, and coal use in particular. Charges provide measured incentives for all the different ways energy consumers find to reduce their reliance on fuels that produce CO_2 emissions; no other policy can address as wide a range of possible responses. Programs to stimulate R&D in improved energy efficiency and alternative energy sources may also be needed, as might other programs designed to address market imperfections that hinder adoption of some measures to enhance energy efficiency.

The question then becomes whether the costs imposed by carbon charges, or any other program equally effective in controlling CO_2 emissions, are worth bearing in order to obtain the likely delay in the date when CO_2 concentrations will double. The cost of approaching the global warming problem through reducing fossil energy use also suggests the merit of looking into other ways of dealing with global warming.

There are other greenhouse gases that are more reactive—more effective in causing global warming—and potentially cheaper to control than CO_2. Some analyses have suggested that complete elimination of CFCs should at least be considered before drastic reductions in fossil energy use are undertaken. Methane is another atmospheric trace gas whose control is worth examining. Cultivation of plants and forests whose growth can remove CO_2 from the atmosphere is also a potential alternative or complement to reducing CO_2 emissions. There is clearly a role for R&D in improved energy efficiency and new technologies for using on a large scale forms of energy that do not release CO_2 emissions.

Finally, the time scales before doubling occurs are very long. Over such time periods in the past the location of economic activity has changed dramatically, completely new techniques of farming and crops have come into use, and the capital stock of the nation has changed over many times. The same will happen over the next 70 years regardless of climate change; more attention could be paid to finding ways of designing incentives to ensure that forthcoming changes will recognize the risks of climate change and will exploit opportunities for mitigating the effects of global warming.

A Least Cost Energy Analysis of US CO_2 Reduction Options

Samuel C. Morris
Brookhaven National Laboratory
Upton, NY

Barry D. Solomon
US Environmental Protection Agency
Washington, DC

Douglas Hill
Consulting Engineer
Huntington, NY

John Lee
Brookhaven National Laboratory
Upton, NY

Gary Goldstein
Brookhaven National Laboratory
Upton, NY

Introduction

Public policy debate on global climate change is increasingly focused on the cost of mitigating greenhouse gas emissions. Discussion in the US has centered on national energy policy and the desirability and cost of increased energy conversion efficiency and end-use conservation, and on shifting from high greenhouse gas emitting fuels to natural gas, renewable, and nuclear-based energy sources.

This paper overviews the US MARKAL model, a dynamic linear programming (LP) model of US energy supply and demand. Useful energy projections are specified exogenously to the model, which then determines the optimal energy supply that can meet the demand. We have updated MARKAL with currently available energy technology cost and market penetration data and have applied it to the CO_2 reduction problem for the US. In addition, we have varied some key inputs to the model to test the sensitivity of the energy system to alternative assumptions and to overcome some of the key limitations of the input data.

We view this exercise as one step in the ongoing process of developing and refining useful models for evaluating national energy policy. The usefulness of such an effort is twofold. First, it helps policymakers to further bound plausible estimates of the costs of control of CO_2 emissions, while identifying the sensitivity of the results to assumptions about input data and the model structure. Second, it provides guidance for future modeling efforts. It helps to illustrate the strengths and limitations of our modeling approach, highlights comparisons to alternative modeling frameworks, and identifies improvements that should be made in future efforts. It also helps to identify key parameters and uncertainties, as well as potential interrelationships that may be important and considered in future modeling.

The results of this analysis suggest what some of the key determinants are for estimating the costs of a carbon constraint in the US. MARKAL analyses to date have focused on technological options for reducing emissions and the cost of technologies, but not on specific policy measures required to implement these technologies. Implementation of a carbon reduction program would require major shifts in energy policies. Examples include carbon taxes (possibly with a commensurate reduction in nonenergy taxes), improved energy efficiency standards for appliances, commercial lighting, buildings, and motor vehicles, least cost utility planning, improved access to capital for industrial efficiency investments, and various combinations of these measures. The cost-effectiveness of nonenergy measures to reduce greenhouse gas emissions, such as chlorofluorocarbon (CFC) substitution, tree planting, and reversal of deforestation, are not considered here.

The MARKAL analysis reported herein examines a base case and three CO_2 constrained cases, one with CO_2 emission levels in 2010 equal to those in 1990, one with CO_2 emissions in 2010 limited to 90 percent of those in 1990, and one with CO_2 emissions in 2010 limited to 80 percent of those in 1990. All cases assume that energy investment decisions from 1986 onward are based on MARKAL optimality criteria. This assumption, which is of course violated by actual behavior since 1986, influences the reported results for technical feasibility and cost. Public policy could move future behavior toward that embedded in the optimization model, but there would be decision and implementation lags. Sensitivity analyses were performed for each scenario. In the constrained cases, supplementary emission constraints were made to gradually meet the 2010 targets, and further emission constraints were continued in 2015 and 2020. In the most stringently constrained case, these aim toward reaching a 50 percent reduction by 2050.

As with all modeling efforts, there exist uncertainties. Sources of uncertainty include stochastic variability, input information deficiencies, and modeling errors. Sensitivity tests were conducted to consider how some of the key uncertainties in the model affect our results.

For example, uncertainties exist about the input data assumed for this exercise. These tests indicate that the results of our analysis are highly sensitive to certain input assumptions, as well as to the model structure (e.g., absence of macroeconomic feedback mechanisms). We are, therefore, performing ongoing investigations into the validity of the input assumptions used here, as well as the credibility of alternative data bases.

Our analyses suggest that an important advantage of the MARKAL LP framework is that it is a technology rich energy model (as opposed, say, to a traditional econometric model) that consistently compares energy supply and efficiency options. It should be combined, however, with a more complete demand-side economic model in order to construct an integrated framework that approaches a general equilibrium model. Such an integrated framework could capture important interrelationships, feedback mechanisms, and some key uncertainties. Also, although the MARKAL model leads to a least cost solution, it does not necessarily lead to a "social optimum" in a social welfare sense.

Approaches to Estimating CO_2 Reduction Costs for the US

Alternative Modeling Frameworks

Alternative approaches exist to modeling energy systems. It is, for example, important to understand the distinction between LP and econometric models. The underlying objective of an LP model is to optimize some specified objective function. In the case of MARKAL, some amount of end-use service, e.g., for transportation, space heating, etc., is exogenously specified and the model meets that demand while minimizing total system cost. In contrast, econometric frameworks in their pure form estimate production efficiencies (and costs) using production (and cost) functions derived from historical data. Such production function models assume that future relationships between energy prices and economic activity will be similar to past relationships, rather than assessing future penetration of specific technologies.

Another important distinction is between partial and general equilibrium models. An LP model such as MARKAL typically consists of a partial equilibrium framework, i.e., a particular economic sector is isolated, such as the energy sector, and information on all other sectors is specified exogenously to the model. Since MARKAL is a partial equilibrium framework that looks at the energy sector in isolation and specifies effective levels of energy demand exogenously to the model, important feedback mechanisms are neglected. In contrast, a general equilibrium framework endogenously integrates all key economic sectors and permits all economic agents in all sectors to optimize jointly. Multiple objective functions are simultaneously optimized, feedbacks and interrelationships between sectors are captured, and the model solves for the general equilibrium for the entire system.

Dynamic general equilibrium models that incorporate engineering-based information can be used to capture the effects of interproduct competition, regional differences, changing technology, and government and private supply decisions. These models can be very useful for forecasting, evaluating strategic planning alternatives, and analyzing the impacts of market uncertainties and government policy actions. They can also be adapted to account for imperfect competition.

Partial equilibrium models are often not adequate for assessing the important feedback mechanisms and interactions that characterize complex systems, but they can effectively highlight technical decisions in a single sector. As the system being studied becomes more complex, a general equilibrium approach becomes essential for fully assessing significant direct and indirect effects and costs of energy and environmental policies.

General equilibrium models are, however, much more difficult to develop, and can introduce significant biases or errors if not properly implemented (Zimmerman, 1990; Simon, 1990). For example, the compounding of modeling and input data errors through the model may be more significant than in a partial equilibrium framework unless adequate diagnostics are performed. Many analysts believe that the partial equilibrium approach is necessary to incorporate greater detail in analyses of particular sectors.

Literature Survey

The pioneer scholar of the economics of global climate change is William Nordhaus (1977). Nordhaus's approach is to derive optimal carbon charges from a discounted net present value standpoint. The assumption made is that CO_2 damages (though highly uncertain) would be minimal until sometime after the atmospheric concentration doubled, after which damages would be severe. He showed that at initial levels of CO_2 control, cost would be very low, but would eventually rise to very high levels. This approach has major implications for the timing of CO_2 control policy, since the costs of mitigating emissions may be incurred many years before quantifiable damages from greenhouse gas buildup could be detected (when the equilibrium warming commitment may be unacceptably large). In addition, the benefits of controlling emissions in terms of avoided costs or damages to ecological systems may be incalculable, resulting in insufficient control. Nordhaus follows this approach in a major update of his earlier work (Nordhaus, 1990). He shows that if we assume a high damage function equivalent to 2 percent of national output, and a zero discount rate, roughly a 47 percent reduction in total greenhouse gases would be the most efficient level of control at a cost of $107 per ton of CO_2 equivalent. However, a medium or low damage function with a 1 percent discount rate would only justify 10 to 17 percent control, at a cost of $3 to $13 per ton.

Edmonds and Reilly (1983; 1985) developed a model of long-term global energy supply in order to estimate CO_2 emissions. While their parametric energy model became the state of the art for developing carbon emission scenarios, its partial equilibrium framework (a characteristic shared by MARKAL) and incomplete picture of the economy somewhat limits its value as a costing tool. Even so, Edmonds and Reilly have used this model to estimate the impact of various carbon taxes on emissions (for the US or the world). Edmonds is currently developing a second generation model that introduces a general equilibrium framework.

Kosobud et al. (1983) attempted to estimate an optimal carbon tax by adapting the ETA-MACRO energy-economic model of Alan Manne (Manne et al., 1981), which uses econometric methods and nonlinear programming. Following a modeling philosophy similar to Nordhaus, they concluded that the optimal tax is currently low but would build up rapidly over time. They found that nuclear power, in particular, would be substituted for the taxed fossil fuels, and become a major source of electricity by 2020. Their results were very sensitive to energy supply estimates, cost data, and key parameter values in the economic model, but especially to the specification of the environmental damage function.

More recently, Manne and Richels (1990) adapted ETA-MACRO into GLOBAL 2100, an analytical framework for estimating the costs of reducing carbon emissions from energy use. While detailed critiques of this paper have been provided elsewhere (Williams, 1990; Hogan, 1990), a few critical aspects of their US analysis deserve discussion here. The first issue is the daunting task of forecasting technology costs, as the authors do, over 110 years. This may lead to their result that a carbon tax of almost $600 per ton would be needed by 2020 in order to reduce emissions to the target level, though the tax would eventually settle at $250 per ton and remain there through 2100. They also conclude that the total cost to the economy of lowering CO_2 emissions by 20 percent over current levels, through 2100, would be $0.8 trillion to $3.6 trillion. Although the high tax would be phased in gradually, at some upper price level fossil fuel consumption would presumably stop. Williams (1990) suggests that GLOBAL 2100 may overstate the costs of important alternative energy technologies, such as renewables and energy efficiency improvements.

Moreover, conservation as modeled in GLOBAL 2100 is primarily a result of the high carbon tax. There is considerable evidence on the significant potential for cost-effective energy efficiency that is not being pursued because of a range of market barriers or failures (Hirst, 1989). Manne and Richels also include a factor in their model for non-price-induced conservation, i.e., that due to capital stock turnover and government policies, that strongly influences their cost estimates. This factor equals zero in their base case and 1 percent per year in an extreme case, since they claim that there is no evidence for an autonomous energy efficiency trend in the post–World War II record. This claim, however, is contradicted by the findings of Edmonds and Reilly (1985), Schipper et al. (1990), and Ross (1989) for the industrial sector, and by Greene (1989) for the transportation sector, i.e., Corporate Average Fuel Economy (CAFE) standards. In short, while the correct base case assumption is debatable, evidence exists for a higher extreme case efficiency assumption.

Hogan and Jorgenson (1990) focus on the estimation and use of long-term productivity trends and the associated technical biases in energy-economic models such as GLOBAL 2100. They estimate technical bias of factor inputs in production functions as the negative of the rate of change in total factor productivity with respect to input price of factor i, which equals the rate of change in the expenditure share for that factor. If technical bias, e.g., for energy inputs, is less than zero, a technical energy-saving bias independent of price would decrease the expenditure share on energy inputs over time, and a rise in the relative input price of energy (such as through a large carbon tax) is a cost that would increase productivity and economic welfare. Using a 1958 to 1979 time series of data for 35 sectors of the US economy, they calculate technical bias for electric and nonelectric energy, and aggregate across sectors weighted by value added in 1982. The authors surprisingly find that the average technical bias is positive for energy inputs, capital, and labor, and negative for materials only. With this result, Hogan and Jorgenson suggest that Manne and Richels may have underestimated by half the total economic costs of CO_2 emission reduction. Only the final quarter of the Hogan and Jorgenson sample period, however, captures an environment of high energy prices, probably too small a period to fully account for technical innovation in energy efficiency.

A final analysis of interest is the comparison of a carbon tax to a gasoline tax by Chandler and Nicholls (1990). Using the same price elasticities of demand in simplified analyses and the latest version of the Edmonds-Reilly model, they conclude that a carbon tax would be two to three times more effective than a gasoline tax in reducing CO_2 emissions. The authors find that an $85 per ton carbon tax could hold US CO_2 emissions constant through 2000, a tax level which is three times higher than the $28 per ton tax level found by the Congessional Budget Office (CBO, 1990) to have the same effect. Moreover, they emphasize that the overall economic impact of a carbon tax could be near zero if the revenues were used to offset other taxes. But increased government revenues could decrease government borrowing, thereby decreasing interest rates and stimulating real GNP.

The MARKAL Model

MARKAL was developed in the late 1970s at Brookhaven National Laboratory (BNL) and at Kernforschungsanlage Julich (KFA), in West Germany, as part of an International Energy Agency program. It was based on the earlier BNL models BESOM (Hoffman, 1973) and DESOM (Marcuse et al., 1976). MARKAL was described in detail by Fishbone and Abilock (1981). The last US MARKAL run prior to this analysis was made in 1985. In the interim, however, MARKAL continued to be widely used internationally. The current analysis was performed using PC-MARKAL running on a 386 microcomputer in conjunction with an integrated data base management and analysis support tool (MARKAL User's Support System).

MARKAL is an LP model that optimizes a network representation of an energy system with respect to several possible criteria. The network extends from energy resources through energy-transformation and end-use demand devices to exogenously specified demands for useful energy. Each link in the network is characterized by one or more typical technologies, available for the model to choose among. Many such energy networks, or Reference Energy Systems, could be drawn for each time period. MARKAL creates the "best" energy system network for each time period by selecting a set of typical technologies, optimizing over all time periods.

For a feasible solution to be derived by the model, the demand must be met in each period. Sensitivity analyses show that under certain scenarios, such as one in which conservation improvements are slow, renewables are costly, and no additional nuclear power plants are built, the model is unable to derive a solution, i.e., the problem is overconstrained. This limitation is not surprising given that MARKAL consists of a partial equilibrium, as opposed to a general equilibrium framework. The effective result of minimizing total discounted cost over all time periods is that the model behaves as if it had full knowledge of the past and present for decisions made at each time period; it uses existing facilities most efficiently and can plan ahead when investing in new facilities. Operationally, least cost means the least cost network subject to a set of constraints. Categories of constraints used are given in Table 1.

MARKAL is a technology-oriented model, which has over 200 energy supply, conversion, and demand technologies. US MARKAL's flexibility and technological depth was enhanced by the European influence on its development; for example, energy technologies such as district heating, rare in the US, receive detailed treatment. However, the model does not account for the distribution of new technology costs that depends on the particular application for which the technology is being applied. The policy constraint was on the overall emission of CO_2. MARKAL was run for nine five-year time periods from 1980 to 2020.

Table 1. Constraints in the MARKAL energy model.

Fossil, nuclear, and renewable fuel balances.
Fuel availability bounds and interperiod growth.
Cumulative fuel availability.
Seasonal district-heating balance.
Electricity balance seasonally and diurnally.
District-heating winter peaking characterization.
Interperiod transfer of energy technology capacity and activity.
Process (e.g., refinery) balance with fixed or variable outputs.
Process capacity availability.
Maintenance scheduling seasonally for electric-power and heating plants.
Demand-device capacity sufficiency.
Ratios linking proportional capacity of different technologies (e.g., fossil backup for solar devices).
Upper bound on total system CO_2 emissions by time period.

Source: Fishbone et al. (1983).

MARKAL is driven by a set of demands for useful energy specified exogenously for each time period by end-use category. Demands are listed by category in Table 2. MARKAL meets these demands by selecting evenhandedly among energy supply and end-use conservation technologies in an associated data base, a distinctive feature of the model. Demands are in terms of 1980s technologies, building insulation levels, etc. Future end-use energy conservation is included through a dummy fuel conservation to account for potential fuel savings. This allows the selection of conservation technologies to be internalized in the model and provides a convenient way to summarize the role of conservation in each time period.

The US MARKAL analysis to date has some key limitations related to problems with available input data. These limitations also apply to other currently available national modeling analyses. The quantitative importance of these limitations is uncertain and is the subject of ongoing research. For example, the model uses a supply curve for conservation technologies based upon life cycle costs using national energy prices and average equipment utilization rates. Actual life cycle costs will occur as a distribution determined by variations in prices, utilization rates, and market penetration rates. Thus a segment of the supply curve which this analysis treats as flat may actually be upward or downward sloping. With these inputs the model will overpredict the amount of some energy-efficiency technologies and un-

Table 2. Useful energy demand input to MARKAL model, petajoule (PJ).

	1990	1995	2000	2005	2010	2015	2020
Industrial							
I1 Iron and steel	596.	655.	720.	791.	869.	954.	1049.
IA Aluminum electrolytic	310.	319.	327.	340.	344.	348.	352.
ID Indust. steam heat	3251.	3086.	3005.	3041.	3098.	3088.	3077.
IE Elec. machine drive	1636.	1867.	2129.	2391.	2399.	2642.	2906.
IF Other indust. heat	7958.	7557.	7358.	7444.	7584.	7560.	7532.
IK Oth. ind. use light dist.	785.	857.	916.	980.	1057.	1105.	1155.
IM Elec. for fabrication	1003.	1144.	1305.	1466.	1471.	1620.	1781.
Residential							
R1 Res. space heat	4256.	4295.	4282.	4295.	4328.	4360.	4459.
R5 Res. water heat	958.	988.	1007.	1027.	1037.	1047.	1066.
R9 Air conditioning	3222.	3446.	3804.	4028.	4296.	4610.	4818.
RD Res. misc. appliances	2018.	2042.	2167.	2291.	2416.	2540.	2640.
Commercial							
RX Comm. heat/hot water	3563.	3699.	3876.	3972.	4036.	4229.	4510.
RC Commercial cooling	837.	951.	1112.	1260.	1409.	1547.	1707.
RN Commercial misc.	2505.	2774.	3195.	3598.	4002.	4405.	4842.
Transportation							
T1 Rail	625.	677.	712.	755.	809.	867.	929.
T4 Automobile	12000.	12216.	12610.	13194.	13987.	14828.	15720.
T5 Trucks, heavy	3300.	3500.	3690.	3848.	4047.	4257.	4418.
T6 Bus	205.	212.	221.	232.	246.	262.	278.
T7 Trucks, light	4500.	4773.	5031.	5247.	5519.	5805.	6106.
T9 Air transport	1850.	2003.	2127.	2274.	2439.	2616.	2806.
TC Ship	1595.	1728.	1845.	1978.	2127.	2286.	2457.

derpredict the level of others (especially newer technologies) that would optimally be applied to meet the energy demand or carbon constraint.

A second limitation is caused by the high level of geographic aggregation. For instance, the model treats the electric utility sector as one interlocking system (e.g., power demand in New England can be met from supply throughout the US) when, in fact, it is composed of a set of regional power grids. Costs of many of the electricity supply options are region specific, e.g., natural gas and geothermal. The model may overestimate the levels of utilization or underestimate costs for some technologies because of lack of regional detail.

It should also be kept in mind that the modeling analysis begins in the year 1980 and proceeds in five-year steps. For any given scenario, the model will begin implementing least cost options immediately after the starting year of the analysis. Thus, technology choices made in the model for 1980 to 1990 will not reflect actual historic decisions (although MARKAL was calibrated to match actual fuel use for 1980 and 1985). Moreover, there would be decision and implementation lags associated with policies designed to move private decisionmakers toward the MARKAL concept of optimality. Since energy capital decisions in the recent past and immediate future cannot be made to conform with MARKAL, simulation runs will systematically overestimate feasible reductions and underestimate costs that would actually be achieved at various future dates. Thus, these simulations should not be interpreted as projections. However, the CO_2 reduction targets are based on the assumed 1990 emissions.

A last concern is that the analysis may not fully assess the multi-attribute nature of many of the energy technology options. For example, compressed natural gas use in vehicles may result in performance penalties in terms of acceleration. Nonprice, nonenergy characteristics of many options that affect consumer welfare cannot be analyzed in this type of modeling framework. Behavior deemed optimal by MARKAL may not necessarily be consistent with a broader definition of individual optimality that encompasses these considerations.

MARKAL is a normative model. Unlike a simulation model in which one tries different policy options until one produces a satisfactory CO_2 reduction, in MARKAL we specify the CO_2 reduction desired and the model produces the energy mix that can meet that reduction at least cost. As noted by Fishbone and Abilock (1981), and confirmed by our sensitivity analyses, the exact nature of an optimal solution derived by MARKAL is sensitive to the technological and cost data or estimates supplied by the user to characterize energy technologies. The direct costs estimated by MARKAL predict a lower bound on actual future direct costs given the same demands and technology availability. Other factors in real-world decisions may introduce inefficiencies that will increase costs.

Technology selection decisions in the model were made using a 7 percent real discount rate. One implication of this is that there are many end-use energy conservation technologies that have net cost savings at a 7 percent discount rate, but that are not implemented because individuals or firms lack relevant information, assess adversely other characteristics of the alternative technologies, have below-average utilization patterns, or apply higher discount rates in their investment decisions. In this analysis, most such conservation measures are implemented in the base case. Estimating indirect effects on the economic system resulting from increased expenditures in the energy system and estimating potential feedbacks to the actual market penetration of new technologies would require coupling with an economic model and are not included in this analysis.

Data Sources

For each technology MARKAL requires data on energy inputs and outputs, conversion efficiency, investment cost, operating costs (fixed and variable), the fraction of time during which supply technologies are available and the fraction of that time that is unscheduled outage, the year the technology becomes available (for new technologies not currently in the energy system), the economic life of the technology, and the cost of delivering fuel to the technology. Upper and lower bounds on capacity or activity of the technology can be specified to limit the model to realistic growth or to insure that a technology is included in the energy system at a minimum level. As an example of the latter, a lower bound was set on large automobiles to require their use at existing levels and prevent the model from making the entire automobile fleet small, more efficient cars.

Energy demands were derived from the latest annual projections of the Energy Information Administration (EIA, 1990) and from preliminary results of the Department of Energy's (DOE's) FOSSIL2 model projections. Both sources project fuel use by end-use category, so useful energy demands were calculated by applying end-use efficiencies.

Energy supply and technology characterization data relied heavily on the existing (1985) US MARKAL data base. These data were considerably modified and updated:

- Energy technology data were updated primarily from information obtained from a recent survey of DOE laboratories, to be detailed in a forthcoming DOE report to Congress.
- Substantial improvements were made in the availability of energy conservation technologies within the data base, drawing on several ongoing efforts of the Environmental Protection Agency (EPA) and the DOE.
- Limits on natural gas supplies were modified in a sensitivity analysis to be in accord with the midprice scenario of the American Gas Association (AGA, 1989).
- Technological activities were fixed to assure that use of various energy sources matched reality for 1980 and 1985.

Results

Since MARKAL is an optimization model rather than a simulation model, the base case does not represent a "business-as-usual" case, but the result of least cost decisions throughout the energy system. All conservation measures that are cost-effective on a life cycle basis at a 7 percent discount rate will be implemented in the base case even without any CO_2 constraint. These reduce total system cost and fuel use and achieve a slower increase in CO_2 emissions. In addition to CO_2 reduction from conservation technologies, the base case also includes stricter requirements for SO_2 removal from new coal-fired electric power plants, as required by the acid rain sections of the proposed Clean Air Act Amendments, and anticipated rates of retrofitting and repowering of existing coal-fired plants (ICF, 1989). Since this increases the cost of coal-steam electric plants, it tends to cause a shift to more efficient technologies or to lower carbon fuels, both decreasing CO_2 emissions.

The amount of CO_2 emission savings included in the base case can be estimated by comparing it to the EIA's Annual Energy Outlook (AEO) base projection, 1990 to 2010 (EIA, 1990). This is a business-as-usual forecast that includes some minor increases in efficiency of energy use (cf. Chandler and Nicholls, 1990) and a small increase in use of renewable energy sources, but no new environmental constraints. The MARKAL base case results in lower annual growth rates during this period for all fossil fuels, with a higher growth rate for renewables (Table 3). A base case in which MARKAL optimization began at some future date, however, would show higher emissions growth.

The difference in fossil fuel use makes a corresponding difference in CO_2 emissions. Applying the CO_2 emission coefficients used in the MARKAL analysis to the DOE fossil fuel forecasts yields an annual CO_2 growth rate (1990 to 2010) of 1.26 percent compared to 0.34 percent in the MARKAL base case. Thus, implementation of energy conservation measures and renewable energy sources economically attractive at a 7 percent discount rate and implementation of the 1990

Table 3. Annual Growth Rates Expected, 1990 to 2010 (percentages).

	MARKAL Base Case Results	AEO Base Case Results
Oil	-0.3	0.7
Gas	-1.7	0.9
Coal	1.8	2.1
Renewables	2.5	2.0
Nuclear	0.8	0.5

Source for AEO Results: EIA (1990).

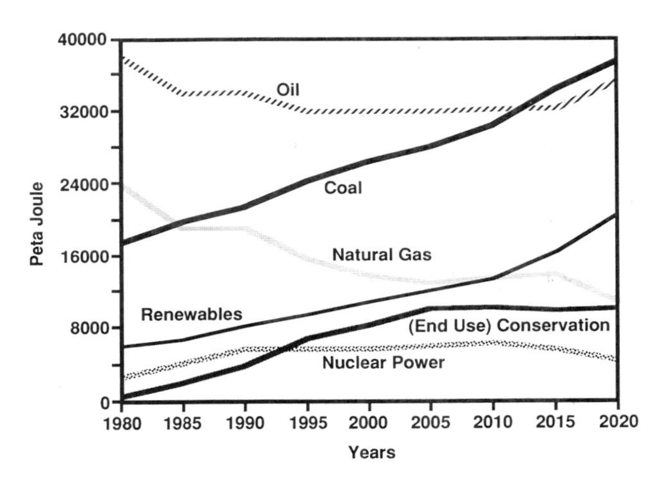

Figure 1. Primary energy supply: unbounded CO_2.

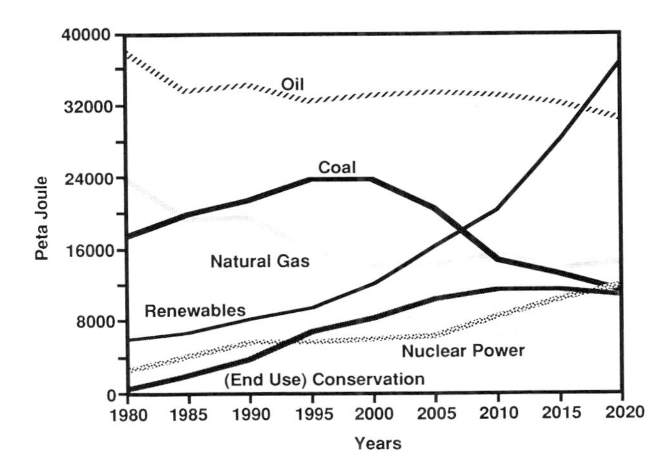

Figure 2. Primary energy supply: 20% CO_2 reduction.

Clean Air Act Amendments, ceteris paribus, are sufficient to produce almost a leveling of CO_2 emissions through 2010 (CO_2 emissions increase sharply after 2010 in the MARKAL base case). Cumulative emissions calculated from AEO fuel usage for 1990 to 2010 was 7500 megatons of carbon, compared to 6370 megatons for the MARKAL base case during the same period. The MARKAL base case thus exhibited 15 percent lower total CO_2 emissions over the 1990 to 2010 period than the DOE base case.

We initially planned to set CO_2 constraints on MARKAL to bracket the recommendations of the 1988 Toronto Climate Conference resolution (Ferguson, 1989). Preliminary MARKAL runs indicated that stabilization by 2000 was technologically feasible (in a simulation which began optimal investment decisions in 1986), but that a 20 percent reduction from 1990 was difficult to obtain by 2005. This led us to apply less stringent constraints, which set the CO_2 emission level in 2010 equal to the 1990 AEO level, and 10 percent and 20 percent lower than the 1990 level. (The feasibility of actually achieving stabilization or reduction goals would depend on how soon and how completely the MARKAL solution, which differs markedly from historic behavior, could actually be implemented.)

MARKAL results for fuels and end-use conservation over time are shown for the base case in Figure 1 and for the 20 percent CO_2 reduction case in Figure 2. The principal change is that the growth of coal in the base case is reversed, to be replaced primarily by renewables. Oil supply in the model is still relatively flat. The differences over time in use of coal, renewables, nuclear power, and end-use conservation are detailed in Figures 3 through 6. (Note that the Figures 5 and 6 are on an expanded scale.) Figures 7 through 9 show the distribution of different fuels over time in the residential and commercial, industrial, and transportation sectors. There is a substantial increase in use of renewables in the residential and commercial sectors, and an increase in the use of district heating and cooling in the residential, commercial, and industrial sectors. Transportation remains mostly petroleum based, but there is a shift to more efficient cars and to diesel cars. From 1990 to 2010, the model supplements petroleum products with compressed natural gas as an automotive fuel, and in 2020 introduces alcohol fuel from wood.

Figure 10 shows CO_2 emissions over time for the four cases. As discussed above, the base case shows very minor growth through 2010, but increases beyond that. The reduction in CO_2 emissions in the post-1995 period in the constrained cases reflects the constraints imposed on the model.

The simplest way to measure the cost of emissions reductions in the constrained cases is to accumulate the total CO_2 emissions and total energy system costs over the 1990 to 2020 study period for the four cases. The first level of CO_2 constraint (2010 = 1990) saves an annual average of 56 megatons at an incremental annual cost of \$220 million, or \$4 per ton (1989 dollars). The second level CO_2 constraint (2010 = 10 percent reduc-

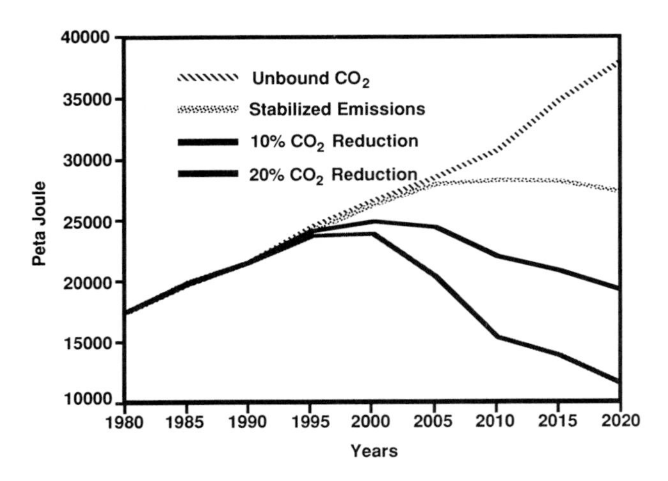

Figure 3. Coal energy supply.

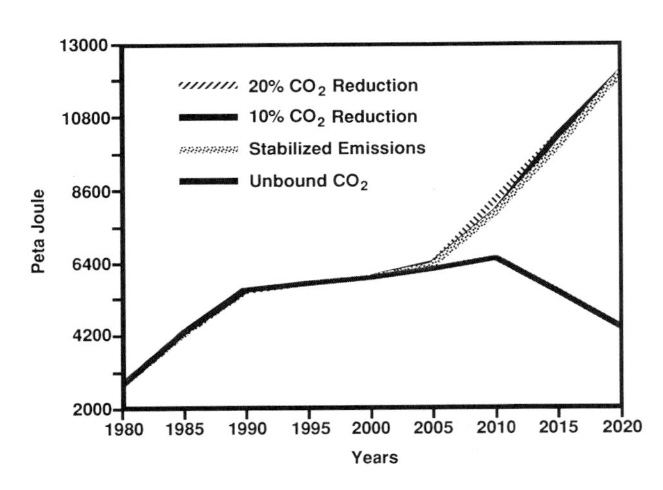

Figure 5. Nuclear energy supply.

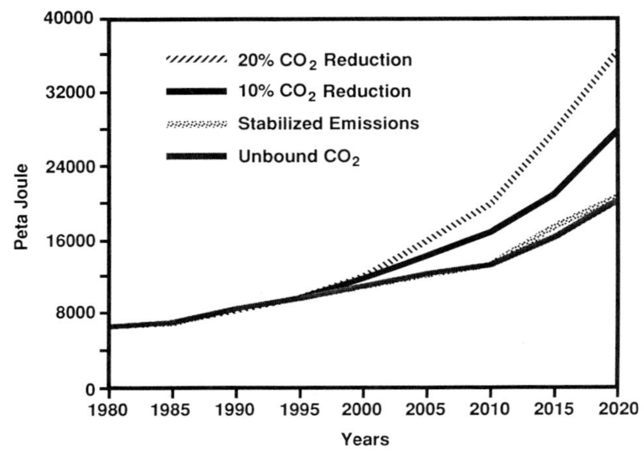

Figure 4. Renewable energy supply.

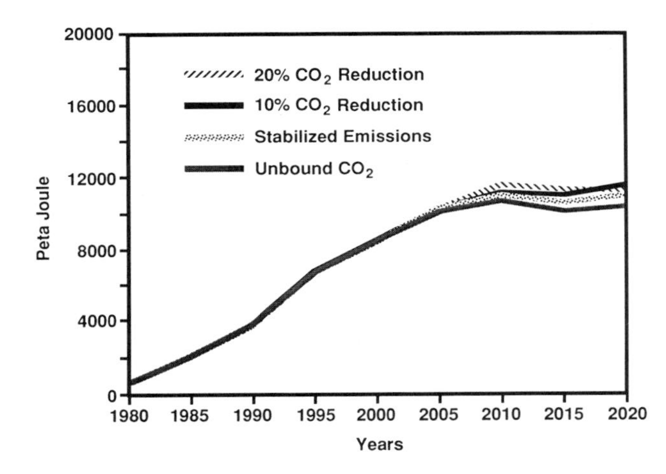

Figure 6. End use conservation.

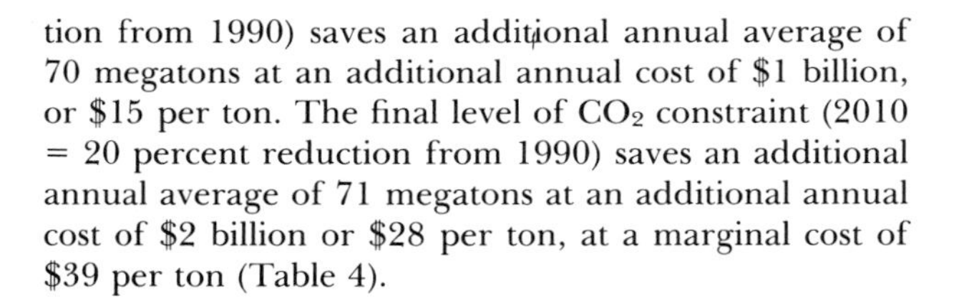

tion from 1990) saves an additional annual average of 70 megatons at an additional annual cost of $1 billion, or $15 per ton. The final level of CO_2 constraint (2010 = 20 percent reduction from 1990) saves an additional annual average of 71 megatons at an additional annual cost of $2 billion or $28 per ton, at a marginal cost of $39 per ton (Table 4).

Sensitivity Analyses with MARKAL

Several sensitivity analyses were conducted with US MARKAL (Table 4). The sensitivity analyses generally used more pessimistic assumptions than did our base case, not because we lack confidence in the base case but rather to consider the implications of "worst case" scenarios. First we investigated the implications of alternative assumptions for natural gas prices and supplies. Use of the American Gas Association's middle price scenario (which is higher than DOE projections) resulted in a higher marginal cost by about 8 percent, while gas use in 2010 declined by about 28 percent. We also found that even if we allow for unlimited gas availability at a price just below the lowest cost oil, gas use as a percentage of total energy supply never rises above its 1985 contribution (24 percent). We then examined a case with about 15 percent less end-use conservation and the price of nonhydro renewables being 50 percent higher than assumed in our base case. In this case, incremental costs in MARKAL rise by about 40 percent. The most sensitive factor, however, was the future of nuclear power. When we do not allow for any new investment in nuclear power plants, the marginal cost rises to $130 per ton of

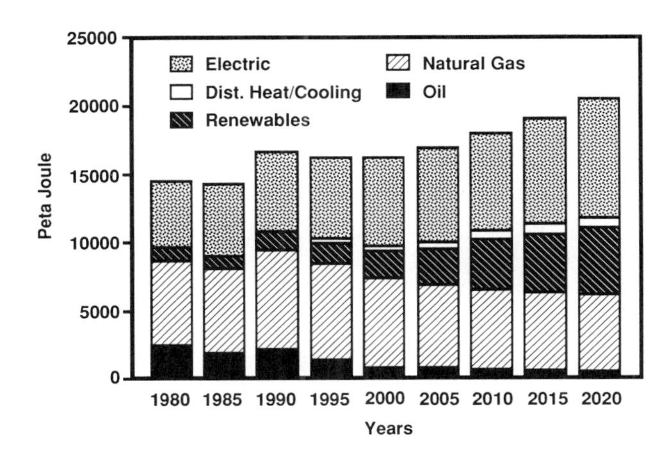

Figure 7. Residential and commercial fuels and electricity—20% CO_2 constraint.

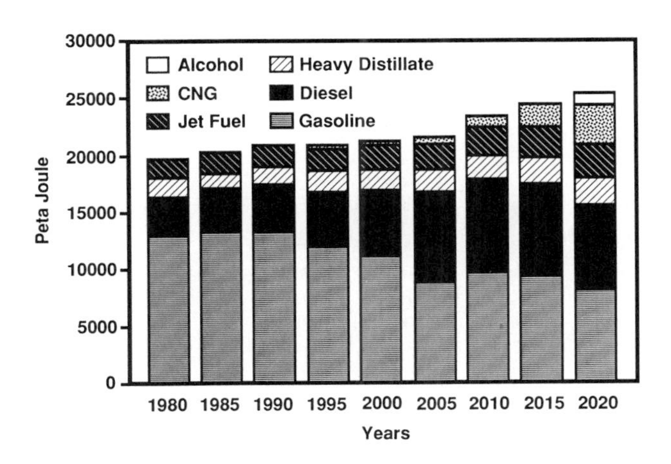

Figure 9. Transportation fuels—20% CO_2 constraint.

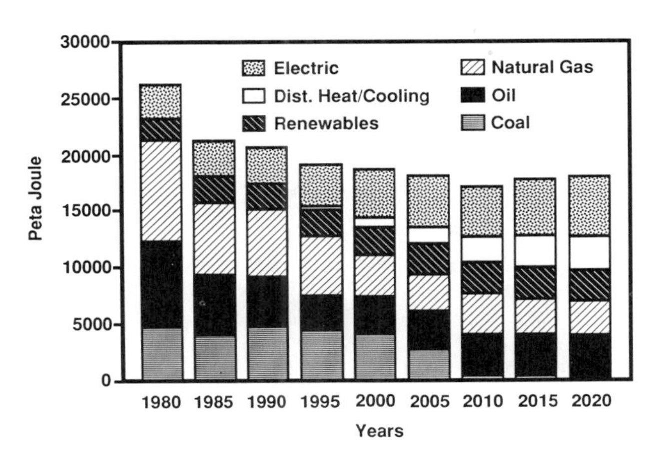

Figure 8. Industrial fuels and electricity—20% CO_2 constraint.

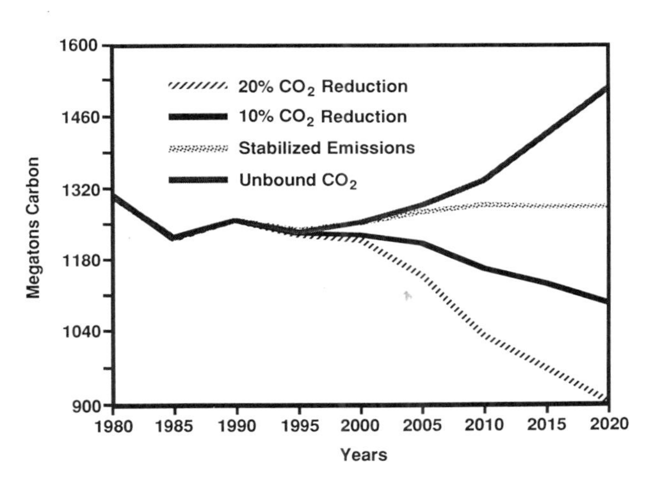

Figure 10. Total CO_2 emissions.

carbon, though lower priced renewables would of course reduce this figure accordingly. In this scenario with a 20 percent CO_2 constraint, gas use rebounds to above its actual level in 1980. Finally, when we assume no new investment in nuclear power, reduced conservation, and higher priced renewables, the incremental cost at the 10 percent CO_2 reduction level surpasses \$800 per ton, and thereafter the solution becomes infeasible.

Discussion

It should be recognized that costs estimated by the model represent choices made with perfect knowledge. These technological costs do not consider the effects of price elasticity on demand or macroeconomic effects such as those produced by large investments in new energy technologies. The effect of these factors can work in different directions. Real-world choices involve nonprice, nonenergy characteristics, institutional constraints, variation across individuals and regions, and uncertainties that may lead to higher system costs in implementing MARKAL solutions. These factors are reflected in private decisions. Price elasticities and other economic effects would tend to reduce demand, reducing overall costs in the energy system but possibly increasing total national costs because of secondary effects. The latter effects may be further clarified in a planned coupling of MARKAL with an economic model.

To understand the changes in the energy system that yield these reductions in CO_2 emissions, one must consider how CO_2 emissions can be reduced in MARKAL. There are five ways: 1) end-use conservation, 2) introduction of more renewables, 3) fossil fuel switching,

Table 4. Summary of Cost/Sensitivity Case Analyses (1989 dollars per ton).

	Base Case	Case 1	Case 2	Case 3	Case 4
Step 1	3.95	9.92	7.46	21.84	—
Step 2	14.90		23.71	46.62	828.79
Step 3	27.91	28.10	39.74	64.48	—
Step 4	38.93	41.90	43.98	129.52	—

Base Case = Base Case assumptions.
Case 1 = AGA Natural Gas Midprice assumptions.
Case 2 = Reduced conservation plus 50 percent higher costs for renewables other than hydropower.
Case 3 = No new nuclear power plants after 2000.
Case 4 = Case 2 and Case 3 assumptions.

Step 1 = No CO_2 constraints to stabilized emissions by 2010.
Step 2 = Stabilized emissions to 10 percent reduction from 1990 by 2010.
Step 3 = 10 percent reduction to 19.5 percent reduction by 2010.
Step 4 = 19.5 percent reduction to 20 percent reduction by 2010.

Note: All simulations for 2010 reflect MARKAL model results from 1986 onward.

4) introduction of more nuclear power, and 5) introduction of more efficient technologies. The role of each of these is discussed below.

Most end-use conservation measures are selected in the base case, leaving little to be added to meet CO_2 constraints. Our 2010 estimate of conservation is about 72 percent of the total cost-effective level found with a 7 percent discount rate by Carlsmith et al. (1990), a team from five of the national energy laboratories. End-use conservation technologies included in the analysis, however, are now available or are expected to be available in the near term. Unlike new supply technologies, there are no future conservation technologies that become available in the post-2010 period, which explains why base case emissions accelerate after 2010. This undoubtedly is due to our current inability to fully characterize or conceive of future conservation technologies.

While CO_2 reductions in MARKAL build on a base of end-use conservation, they are primarily the result of substitution of renewables and to a lesser extent nuclear power for coal. None of the so-called "clean coal" technologies are selected very much in the model. Overall growth of renewables in the four cases is shown in Figure 4 while specific renewables responsible are detailed in Table 5. Within each category, technologies in Table 5 are listed roughly in order of preference to the model, taking into account both shadow price and the size of the contribution.

While attractive to the model, there is little potential to develop additional hydroelectric power. Geothermal energy (especially from hot water reservoirs), wind power, and central solar electricity are found to have greater growth potential. Photovoltaics, herbaceous crop-based gas, and wood-based ethanol, in contrast, are not found by MARKAL to be generally cost competitive until early in the next century. While the attractiveness of these technologies increases with more stringent CO_2 constraints, their rank order does not change.

Table 5. Principal renewable energy technologies, ranked by category.

Electric Generation
Hydroelectric
Small scale hydro
Geothermal
Wind (central)
Wind (local)
Solar central thermal
Solar central photovoltaic
Ocean thermal electric
Biomass*
Municipal solid waste (MSW)*
Wave power (central)*

Demand Devices
Firewood
Solar water heat
MSW for industrial heat
Geothermal for industrial heat
Solar for industrial heat

Processes
Landfill gas production
Agricultural crops for pipeline quality gas*
Wood for liquid fuel*

* appears only in CO_2 constrained cases

Nuclear power grows rapidly after 2010 in the CO_2 restricted cases, contrasting with a decline after 2010 in the base case (Figure 5). All forms of nuclear power available to the model are ultimately brought into the energy system under CO_2 constraints. This includes advanced light water reactors (LWR), high temperature reactors, liquid metal fast breeders, and integrated fast reactors. Despite the increase in nuclear power, electricity production as a whole is curtailed under the increasing CO_2 constraint. The 20 percent CO_2 constraint case has less than a 1 percent annual growth in electricity. Our base case results in a generally steady decline in natural gas use. There is only a modest increase in total gas use as CO_2 constraints are applied, despite introduction of gas heat pumps and compressed natural gas powered automobiles.

There is almost no change in oil use from the base case through the different CO_2 constrained cases. While declining in the residential and commercial sector, and to a lesser extent in the industrial sector, oil continues to be the main fuel for a growing transportation sector through 2020 even under CO_2 constraints. The model results illustrate the pressure on the transportation sector in the sequence of technology changes in automobiles. Higher efficiency gasoline cars are selected immediately and a shift is also made to higher efficiency diesel cars. Compressed natural gas cars are introduced too (about 5 percent of total automotive fuel use in 2000), and are supplemented in 2020 by cars fueled by ethanol produced from wood.

Conclusions

Several preliminary conclusions are suggested by the MARKAL analysis conducted thus far. First, stabilizing or substantially reducing CO_2 emissions from the US energy system by 2010 appears technologically feasible across the range of input assumptions and constraints applied. Aggregate costs of achieving specific targets vary considerably depending on technology cost and availability and on future fuel prices, all of which are inputs to the model. As discussed previously, work on improving many of the detailed technology inputs is currently under way, addressing several limitations and uncertainties which have been identified in existing information. Further review and refinement of these estimates is clearly a high research priority. In addition, the model and data should be refined to allow simulation of optimal investment decisions to begin at a specified future year.

MARKAL results illustrate that major changes in the structure of energy systems will be required to achieve significant CO_2 reductions. Changes of this magnitude stretch the limits of existing energy and economic analysis tools. Much further work is needed to improve our ability to model and evaluate major departures from historic trends. Much more extensive sensitivity analysis using MARKAL and other models could make a significant contribution to this work.

Finally, it is clear that such extensive restructuring of energy systems would require a complex and aggressive mix of policy measures by governments. Analyses to date have tended to focus on a single policy measure, e.g., a carbon tax, or on a range of technical options without regard to specific policy implementation measures (as was the case in the current MARKAL analysis). Major analytic efforts and tools will be needed to evaluate a broad range of specific policy measures alone or in combinations.

Acknowledgments

The authors are grateful to Rick Bradley of the US Department of Energy for providing energy technology cost data and to Leonard Hamilton for his support in re-establishing the US MARKAL model. We thank Howard Gruenspecht, Dan Lashof, Dick Morgenstern, Joel Scheraga, Paul Schwengels, Dennis Tirpak and Mary Beth Zimmerman for their comments on a previous draft of this paper. The research reported in this paper was supported by the Climate Change Division of the EPA. The views, opinions, and conclusions expressed herein are solely those of the authors, and do not necessarily reflect the policies of the EPA, DOE, or any other agency of the US government.

References

AGA. 1989. *The gas energy outlook 1990–2010.* American Gas Association, Arlington, VA.

Carlsmith, R.S., W.U. Chandler, J.E. McMahon, and D.J. Santini. 1990. *Energy efficency: how far can we go?* (ORNL/TM-11441). Oak Ridge National Laboratory, Oak Ridge, TN.

Chandler, W.U., and A.K. Nicholls. 1990. "Assessing carbon emissions control strategies: a carbon tax or a gasoline tax?" American Council for an Energy-Efficient Economy, ACEEE Policy Paper No. 3, Washington, DC.

CBO (Congressional Budget Office). 1990. *Reducing the deficit.* Government Printing Office, Washington, DC, p. 441.

Edmonds, J.A., and J.M. Reilly. 1983. "Global energy and CO_2 to the year 2050." *The Energy Journal* 4: 21–47.

Edmonds, J.A., and J.M. Reilly. 1985. *Global energy: assessing the future.* Oxford University Press, New York.

EIA. 1990. *Annual energy outlook 1990: with projections to 2010* (DOE/EIA-0383(90)). Energy Information Administration, Washington, DC.

Ferguson, H.L. 1989. "The changing atmosphere: implications for global security," in D.E. Abrahamson (ed.), *The Challenge of Global Warming,* Island Press, Washington, DC, pp. 44–67.

Fishbone, L.G., and H. Abilock. 1981. "MARKAL, a linear-programming model for energy systems analysis: technical description of the BNL version." *Energy Research* 5: 353–375.

Fishbone, L.G., G. Giesen, G. Goldstein, H. A. Hymmen, K.J. Stocks, H. Vos, D. Wilde, R. Zölcher, C. Balzer, and H. Abilock. 1983. *User's guide for MARKAL* (BNL/KFA Version 2.0-BNL-51701). Brookhaven National Laboratory, Upton, NY.

Greene, D.L. 1989. *CAFE or price? an analysis of the effects of federal fuel economy regulations and gasoline price on new car MPG, 1978–89.* Oak Ridge National Laboratory, Oak Ridge, TN.

Hirst, E. 1989. *Federal roles to realize national energy-efficiency opportunities in the 1990s* (ORNL/CON-290). Oak Ridge National Laboratory, Oak Ridge, TN.

Hoffman, K.C. 1973. *A linear programming model of the nation's energy system* (BNL-ESAG-4). Brookhaven National Laboratory, Upton, NY.

Hogan, W.W. 1990. "Comments on Manne and Richels: CO_2 emission limits: an economic cost analysis for the USA" (draft). Harvard University, Cambridge, MA.

Hogan, W.W., and D.W. Jorgenson. 1990. "Productivity trends and the cost of reducing CO_2 emissions" *The Energy Journal* (in press).

ICF. 1989. *Economic analysis of Title V (acid rain provisions) of the Administration's proposed Clean Air Act Amendments* (H.R. 3030/S.1490). ICF Resources Inc., Washington, DC.

Kosobud, R.F., T.A. Daly, and Y.I. Chang. 1983. "Decentralized CO_2 abatement policy and energy technology choices in the long run," in *Government and Energy Policy,* International Association of Energy Economists, Westview Press, Alexandria, VA.

Manne, A.S., R.J. Condap, and P.V. Preckel. 1981. *ETA-MACRO: a user's guide.* Electric Power Research Institute, Palo Alto, CA.

Manne, A.S. and R.G. Richels. 1990. "CO_2 emission limits: an economic cost analysis for the USA." *The Energy Journal* (in press).

Marcuse, W., L. Bodin, E.A. Cherniavsky, and Y. Sanborn. 1976. "A dynamic time dependent model for the analysis of alternative energy policies," in K.B. Haley (ed.), *Operational Research 1975,* North Holland, NY.

Nordhaus, W.D. 1977. "Strategies for the control of carbon dioxide." *Cowles Foundation Discussion Paper,* No. 443.

Nordhaus, W.D. 1990. "The cost of reducing greenhouse gas emissions," *The Energy Journal,* (in press).

Ross, M. 1989. "The potential for reducing the energy intensity and carbon dioxide emissions in US manufacturing" (draft). University of Michigan, Ann Arbor, MI.

Schipper, L., R.B. Howarth, and H. Geller. 1990. "United States energy use from 1973 to 1987: the impacts of improved efficiency." *Annual Review of Energy* (forthcoming).

Simon, H.A. 1990. "Prediction and prescription in systems modeling." *Operations Research* 38: 7–14.

Williams, R.H. 1990. "Low-cost strategies for coping with CO_2 emission limits." *The Energy Journal* (in press).

Zimmerman, M.B. 1990. "Assessing the costs of climate change policies: the uses and limits of models." The Alliance to Save Energy, Washington, DC.

CO_2 Emissions Reductions: A Regional Economic Cost Analysis*

Alan S. Manne
Department of Operations Research
Stanford University
Stanford, CA

Richard G. Richels
Environmental Risk Analysis Program
Electric Power Research Institute
Palo Alto, CA

Introduction

In recent years there has been growing concern that the increasing accumulation of greenhouse gases in the earth's atmosphere will lead to undesirable changes in global climate. This concern has resulted in a number of proposals, both in the US and internationally, to set physical targets for reducing greenhouse gas emissions.

With CO_2 believed to be responsible for approximately half the problem, the energy sector plays an important role in strategies to delay climate change. Two bills introduced in the US Congress during 1989 (the National Energy Policy Act of 1989 and the Global Warming Prevention Act) called for a national energy policy to reduce global warming. Each bill would establish national goals of reducing carbon dioxide emissions by 20 percent by the year 2000.

Calls for reducing CO_2 emissions have been a common theme at international conferences on global warming. For example, the final statement from the June 1988 Toronto conference on "The Changing Atmosphere" called for 20 percent worldwide reductions of CO_2 emissions by the year 2005. The Hamburg Conference (November 1988) called for a 30 percent reduction by the year 2000. In each instance, the major share of the reduction would be borne by the industrialized countries.

In the diplomatic arena, perhaps the most notable group is the Intergovernmental Panel on Climate Change (IPCC). This is being sponsored by the World Meteorological Organization and the United Nations Environmental Program. In addition to developing a comprehensive initial assessment of the scientific evidence and impacts of climate change, the IPCC will be exploring strategies for policy responses.

As the global climate debate moves toward the consideration of specific legislative initiative and policy options, international negotiations are likely to take on increasing importance. The greenhouse effect is inherently a global problem. Most countries are taking the position that if significant measures are required to reduce CO_2 emissions, they should only be taken in the context of an international agreement.

The negotiation of such an agreement would be extraordinarily complex. Major reductions in carbon emissions will be expensive. The difficulty in achieving a given target is likely to vary greatly among nations. Some nations might incur enormous costs in order to achieve relatively modest reductions, while the converse might hold for others.

For a set of limits to be broadly accepted, they must be perceived as equitable, enforceable, and based on gradual rather than abrupt changes in the status quo. Economic efficiency is likely to be a secondary criterion, but could be achieved through international markets in carbon emission rights. For projecting the evolution of such markets, it is important to understand how the costs of emissions abatement vary among regions. This requires an explicit analysis of how the abatement costs might be distributed among the potential signatories to any treaty.

In the following analysis, we seek to improve our understanding of how the costs of a CO_2 emissions limit will vary depending on the stage and pattern of economic development, the fuel mix, and the initial endowment of hydrocarbon resources. The analysis is based on Global 2100, an analytical framework for estimating the economy-wide impacts of rising energy costs.

We explore how emissions are likely to evolve in the absence of a carbon limit, and how the regional pattern is likely to shift during the next century. We then examine strategies to limit global emissions, calculate the impact of higher energy costs upon conventionally measured GDP, and indicate the size of the carbon tax that would be required to induce individual consumers to

* The research reported in this paper was funded by the Electric Power Research Institute (EPRI). The views presented here are solely those of the individual authors, and do not necessarily represent the views of EPRI or its members. This paper provides a summary of a more detailed technical report which is forthcoming in the *Energy Journal*. See Manne and Richels (1990).

reduce their dependence on carbon-intensive fuels. Finally, differences in the time path of carbon taxes among regions are analyzed to identify potential opportunities for trade in emission rights.

We have not attempted to estimate the *benefits* of slowing down the rate of climate change through a reduction in worldwide CO_2 emissions. Our analysis is confined to the impacts of carbon emission limits upon the *cost* of energy—and the resulting effects upon the economy as a whole. This measurement of costs is only a part of the story. It is a far more formidable task to estimate the benefits from reduced emissions, and this is well beyond the scope of the present analysis. Clearly, policymakers will need to balance both the benefits and the costs in order to arrive at an overall judgment.

An Unconstrained Carbon Emissions Scenario

The Global 2100 model employs five major geopolitical groupings: the US; other OECD (OOECD) nations (Western Europe, Canada, Japan, Australia, and New Zealand); the USSR and Eastern Europe (SU-EE); China; and ROW (rest of world). In defining these regions, we have attempted to employ the minimal level of disaggregation necessary to provide meaningful insights into how the costs of a carbon constraint may vary among nations.

A key issue for the present analysis is the size and pattern of future emissions in the absence of measures to slow growth. In projecting future emissions, one major factor is the rate of GDP growth. This rate depends both upon population and per capita productivity trends. In parallel with the slowdown of population growth during the 21st century, there will be a diminishing rate of growth of GDP, and hence a slowdown in the demand for energy. Figure 1 shows our assumptions about the rate of growth within each region. Typically, these represent extrapolations of performance during the past two decades.

Figure 2 compares global emissions at two points in time, 1990 and 2100. Although emissions grow considerably over the next century, the average rate of growth slows to 1.4 percent per year. This is low by historical standards. A good deal of the explanation for this slowdown lies in our assumptions about the slowdown in population growth and in its implications for future GDP.

Energy efficiency improvements also play a role in reducing the growth in carbon emissions. As in the case of future GDP growth, we make a series of assumptions on the potential for costless energy efficiency improvements and also for price-induced energy efficiency. Figure 3 shows average annual GDP and total primary energy (TPE) growth rates for all five regions through 2100. The growth rate in energy demand is considerably lower than that for GDP. Together, price- and non-price-induced energy efficiency improvements lead to a significant decoupling of these two growth rates.

Figure 2 also shows a significant shift in the pattern of global contributions to CO_2 emissions. Today, the industrialized nations (US, OOECD, and SU-EE) account for 71 percent of carbon emissions. By the end of the next century, their contribution is projected to drop to less than one half of the total.

The Costs of Limiting Carbon Emissions

There are many possible ways to define a global agreement on carbon emissions. Although there is considerable disagreement over the appropriate level of reduction, there is agreement that the burden must fall disproportionately on the industrialized nations. If global income inequalities are to diminish, developing countries will experience much faster rates of economic expansion and energy demand growth than their currently industrialized counterparts. The developing nations are unlikely to accept any agreement which fails to provide for some increase in their carbon emissions. For them, the issue will be how to limit their *growth* in emissions.

There have been a number of calls to reduce global carbon emissions by 20 percent below current levels. Some of the proposals apply solely to the industrialized countries, while others refer to the world as a whole. Recognizing the need for some emissions growth in the developing countries, we begin by assuming that the 20 percent reduction is confined to the industrialized nations only. Specifically, we assume that the US, OOECD, and SU-EE agree to stabilize carbon emissions at their 1990 level through the year 2000 and then gradually reduce them by 20 percent by the year 2020. In return, the developing nations (China and ROW) would limit their emissions to *twice* their 1990 levels. Although these targets are not as ambitious as those contained in some recent proposals, they nevertheless represent a significant reduction in emissions, especially when compared with a business-as-usual view of the future.

Figure 4 compares carbon emissions with and without the carbon limits. Overall, the proposed limit would lead to a 15 percent increase in global emissions between 1990 and 2030, but no further increase thereafter. By 2100, this leads to a 75 percent reduction in the emissions level that would have been reached in the absence of an international agreement.

The feasibility of any scheme to reduce worldwide carbon emissions will depend on the costs to individual nations. Using Global 2100, we may add together the impacts of rising energy costs in each region and cal-

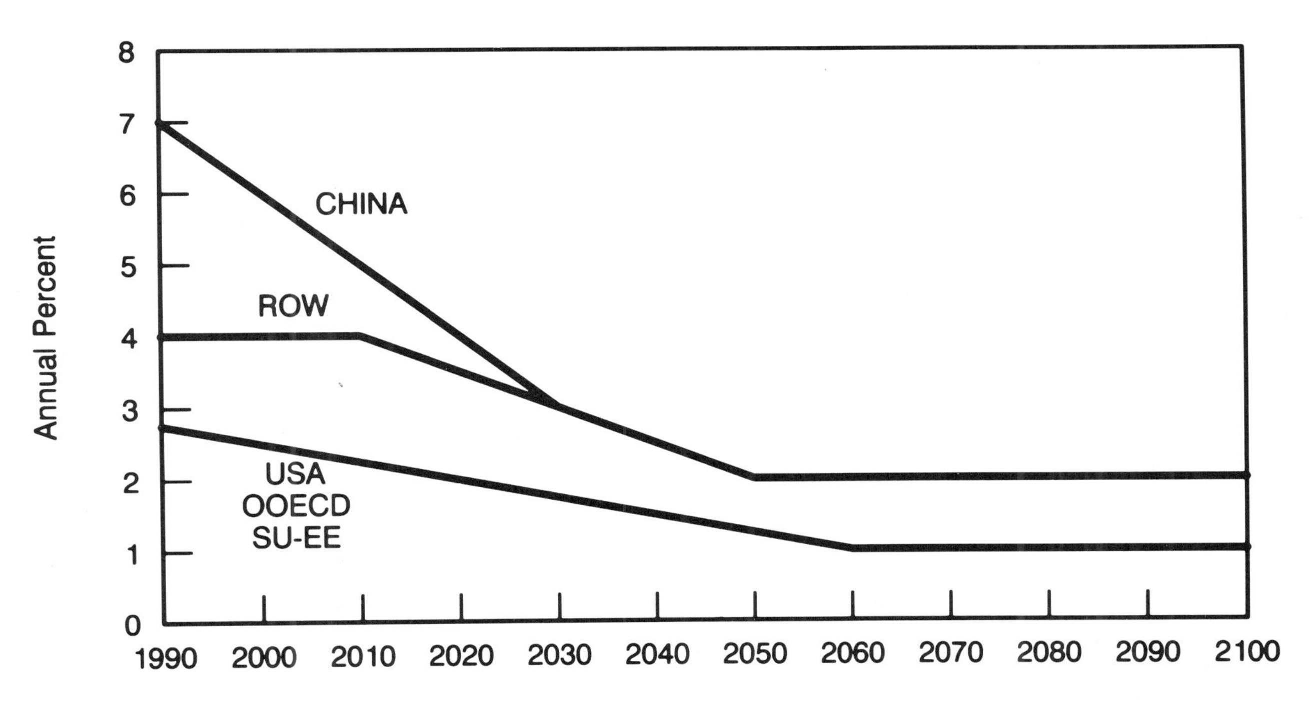

Figure 1. Potential GDP growth rates.

culate the annual losses due to the carbon constraint. Figure 5 shows the annual losses as a percentage of aggregate GDP. For the US, the effects of a carbon constraint do not begin to have measurable macroeconomic consequences until after 2000. At that point the rise in energy prices begins to have a significant effect upon the share of gross output available for current consumption and investment. By 2030, roughly 3 percent of the total annual US GDP is lost as a consequence of the carbon constraint. This percentage remains roughly constant for the remainder of the time horizon. For the other OECD countries, the losses do not begin to accrue until 2010, and they are limited to the 1 to 2 percent range. The costs of the carbon constraint are somewhat lower than those in the US. They have a relatively higher proportion of undiscovered oil and gas resources, and their nuclear power industry is larger. Moreover, these countries currently emit less carbon per unit of GDP than the US. Their economies are less dependent on carbon-based fuels.

The Soviet Union and Eastern Europe will find it more difficult to adjust to a 20 percent emissions cutback than their Western counterparts. By 2030, roughly 4 percent of total macroeconomic consumption is lost as a consequence of the carbon constraint. The higher costs relative to the OECD follow directly from the assumption that it will be more difficult to decouple GDP and energy growth in this region.

Since ROW includes OPEC, Mexico, and other potential oil exporters, the costs of a carbon constraint to that region are negligible until 2020. Gradually, however, their oil and natural gas resources are exhausted. For them too, it will become increasingly difficult to find low-cost carbon-free energy sources, and the macroeconomic consequences will begin to mount. By the end of the 21st century, approximately 5 percent of GDP will be lost as a consequence of the emissions constraint.

Eventually, China would become the region most heavily affected by the international carbon reduction agreement. Figure 5 shows that China's annual GDP losses would exceed 10 percent by the latter half of the 21st century. Their rapid rate of economic development would place enormous upward pressures on energy demands. Since China's fossil fuel resource base is dominated by coal, it will be costly to accept any constraints on carbon emissions. In a carbon constrained energy future, the principal alternatives for China will be conservation along with biomass and other carbon-free substitutes for conventional fuels. If required on a large scale, these will command a high price tag.

According to our calculations, the percentage losses accruing to China would be far greater than those experienced by other regions. The question arises as to how much the carbon limits would need to be relaxed to bring China's losses in line with those of the rest of the world. If international negotiators were to adopt the criterion of equal percentage GDP losses across all regions, we calculate that a quadrupling in emissions (the

4X scenario) will be required for China. This would result in a 37 percent increase in global emissions between 1990 and 2030, but no further increase thereafter. Even with this exception for China, global emissions in 2100 would be only 30 percent of the level that would have been reached in the absence of an international agreement.

The Chinese case highlights the difficulty of achieving a 20 percent reduction in worldwide emissions. Clearly, any increases by developing countries would need to be offset by reductions in industrialized countries. A back-of-envelope calculation shows that if China and ROW are permitted to double their emissions, then the industrialized countries would need to reduce theirs by nearly 70 percent below current levels in order to achieve a worldwide reduction of 20 percent. If China is permitted to quadruple its emissions, the industrialized countries would need to reduce theirs to zero!

Carbon Taxes

There are a variety of policy instruments available for reducing CO_2 to the desired levels. An efficient option would be to impose a tax upon those activities responsible for carbon emissions, and to vary the tax rate according to the carbon content of individual fuels. The purpose would be to discourage those activities with relatively high carbon emissions.

Figure 6 compares the region-by-region time path for the carbon tax that would be required to provide the price signals consistent with the regional carbon limits considered in this paper. It turns out that the long-run equilibrium tax level is determined to raise the price of coal-based synthetic fuels sufficiently so that consumers will find that carbon-free fuels are equally attractive. This works out to be $250 per ton of carbon. A carbon

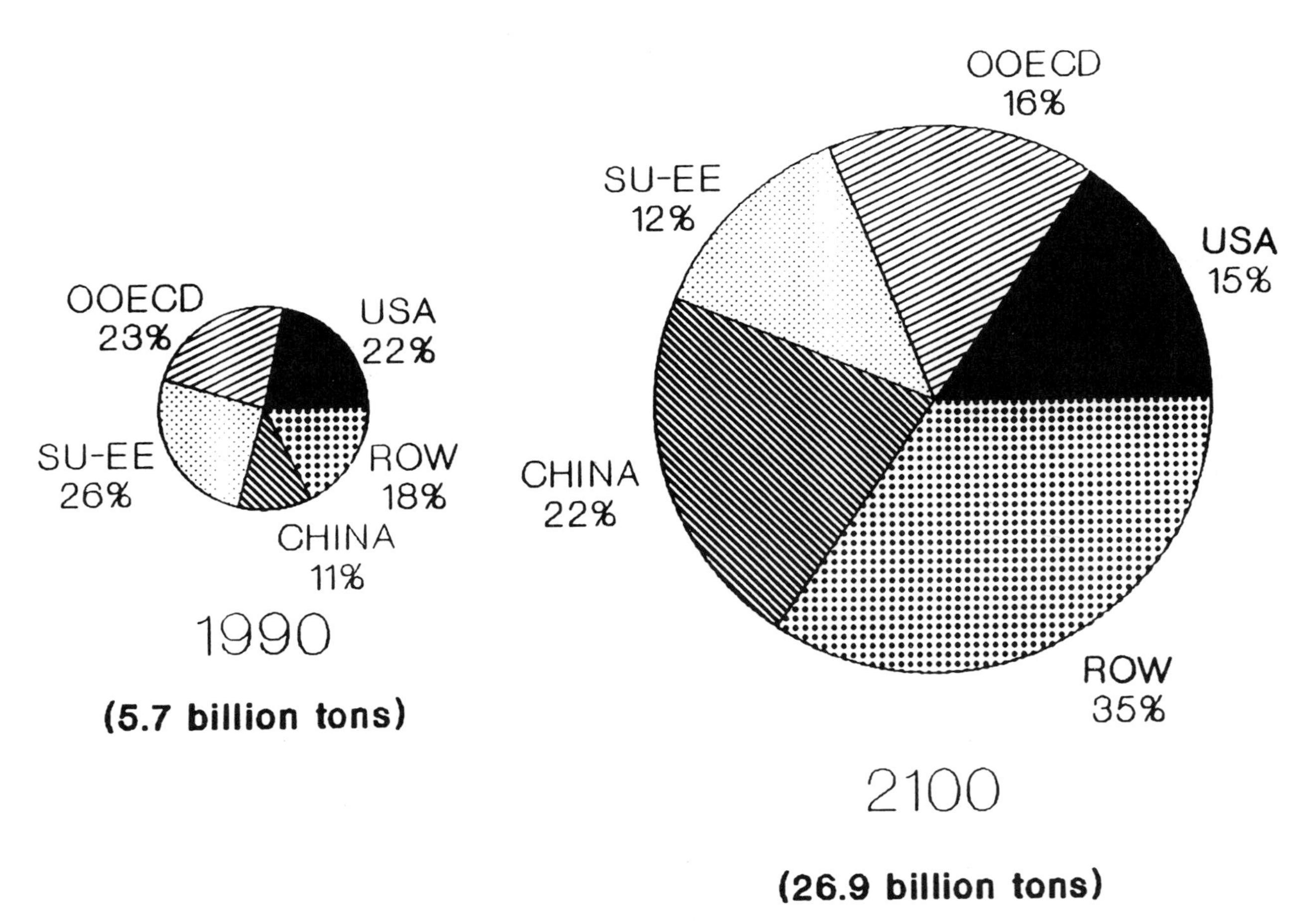

Figure 2. Unconstrained carbon emissions scenario.

tax of $250 would imply a fivefold increase in the US price of coal—or an increase of $0.75 per gallon of refined petroleum products. If one region were to sell just 100 million tons of carbon emission rights to another, this tax rate would imply a financial transfer of $25 billion annually.

The specific paths to the equilibrium level depend on several factors. A tax is required earlier in the industrialized regions because of the agreement to reduce carbon emissions by 20 percent by 2020. In contrast, China and ROW are allowed to increase emissions beyond current levels. Pressures to switch away from carbon intensive fuels do not begin to build until somewhat later.

Although the industrialized countries face identical percentage reductions, the time path for carbon taxes varies considerably. Those which find it less difficult to decouple energy consumption from GDP growth will tend to have an easier and smoother transition. Lacking the demand-side alternatives of its western counterparts, the SU-EE will need to maintain taxes at a higher level to induce consumers to switch away from high carbon fuels.

In the event of an international agreement to limit carbon emissions, region-by-region differences in time paths could be exploited to identify cost-effective strategies for emissions reduction. At a given point in time, regions which find it more difficult to adjust to their emissions limits (those requiring higher taxes) should be willing to purchase emissions rights from regions experiencing less difficulty. For example, based on the carbon taxes reported here, all three industrialized regions would be buyers of emissions rights prior to 2020. Having agreed to reduce emissions by 20 percent by that year and lacking sufficient supply- and demand-side alternatives to achieve such reductions without a sizeable tax, countries in these regions should be willing to pay a great deal for the right to emit more carbon. By contrast, the developing regions do not begin to bump up against their limits until 2020. Prior to that year, countries in these regions should be willing to sell some of their emissions rights to the industrial nations.

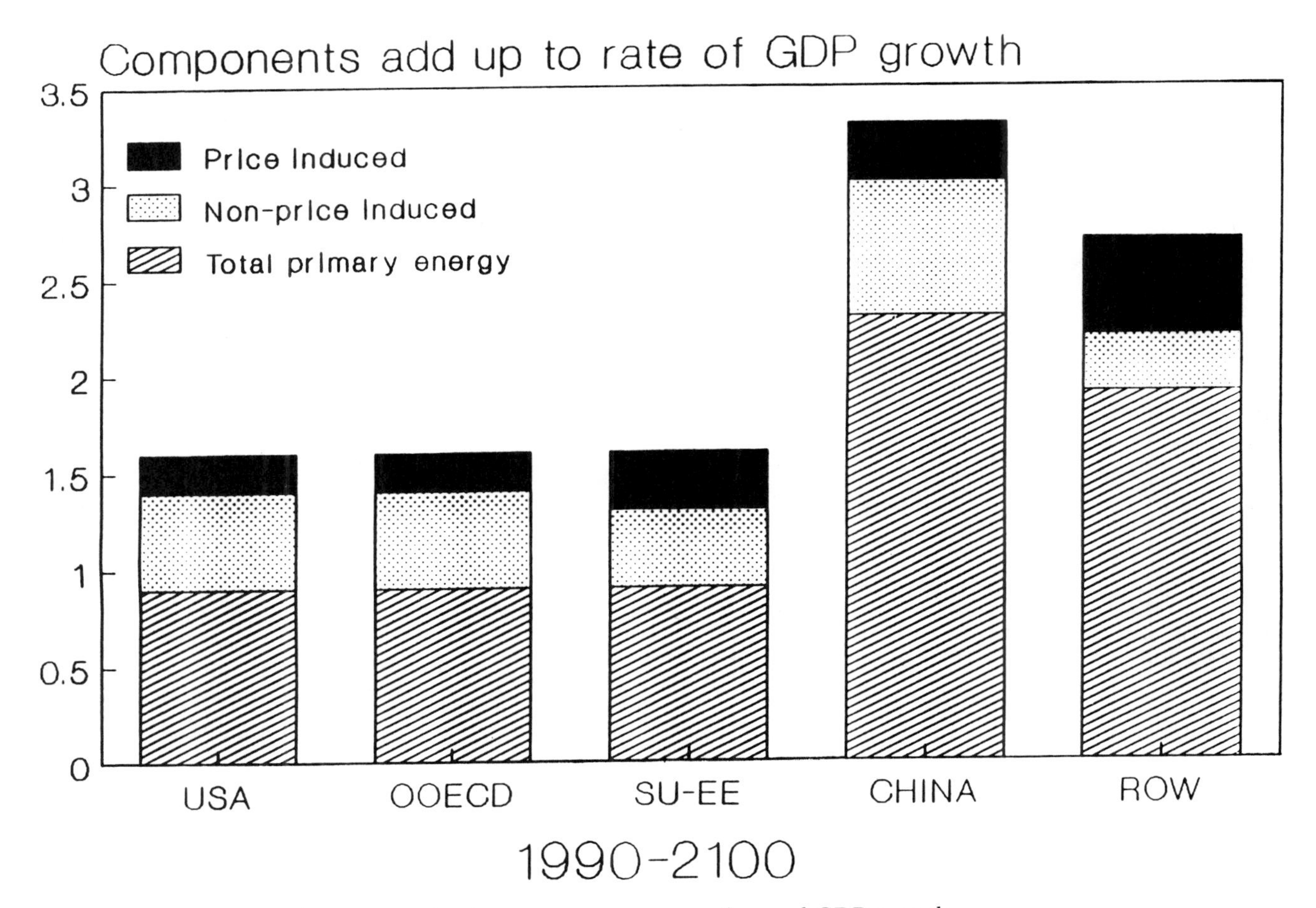

Figure 3. Decoupling between total primary energy consumption and GDP growth.

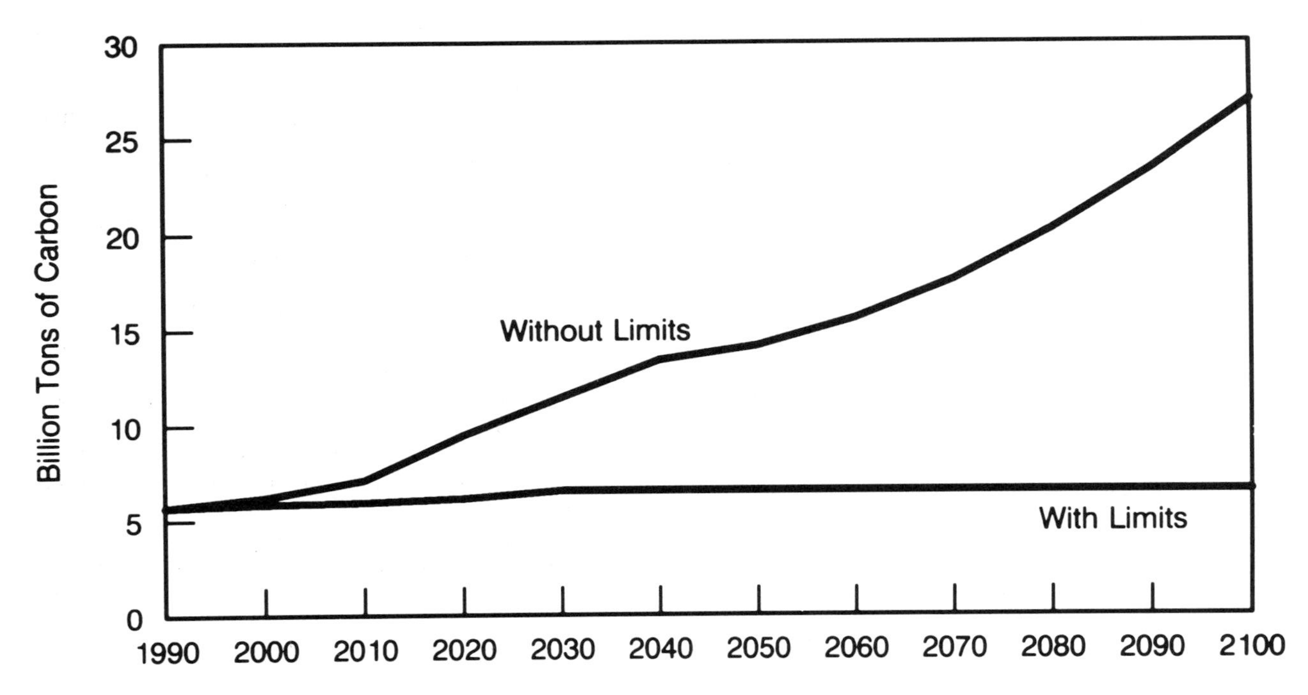

Figure 4. Carbon emissions (with and without carbon limits).

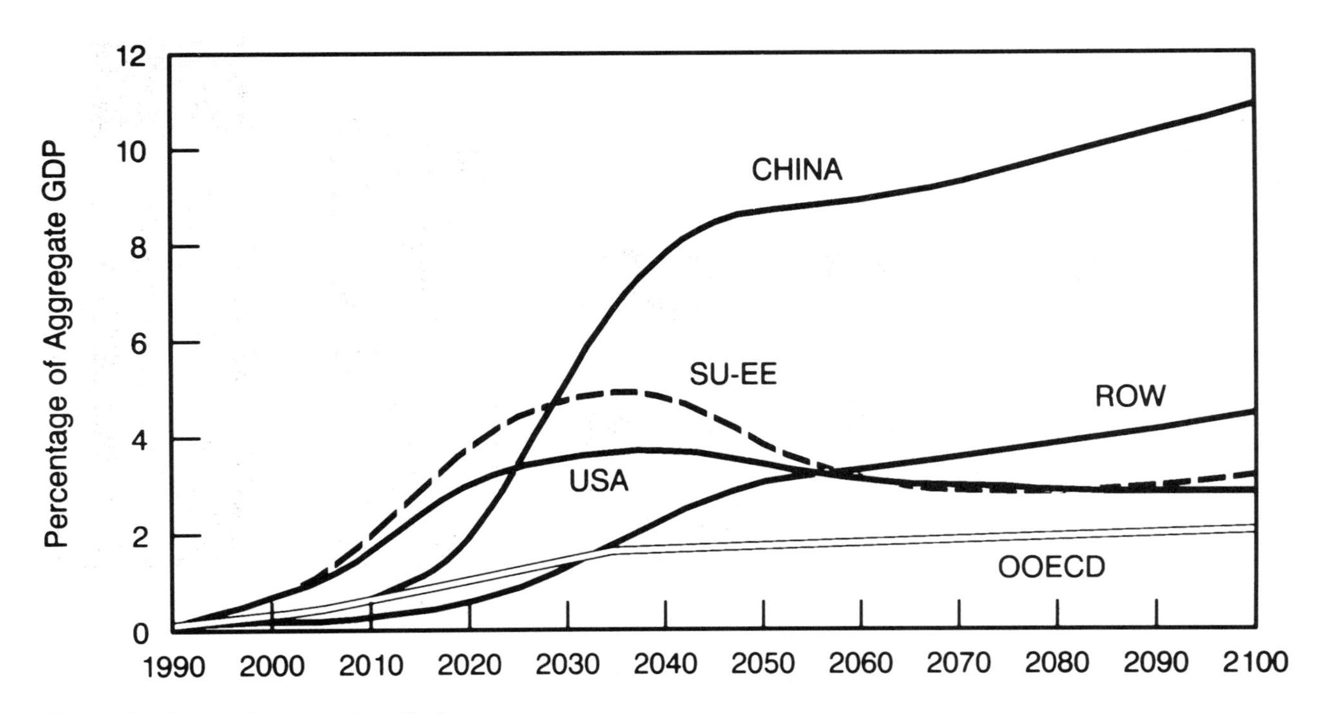

Figure 5. Losses due to carbon limit.

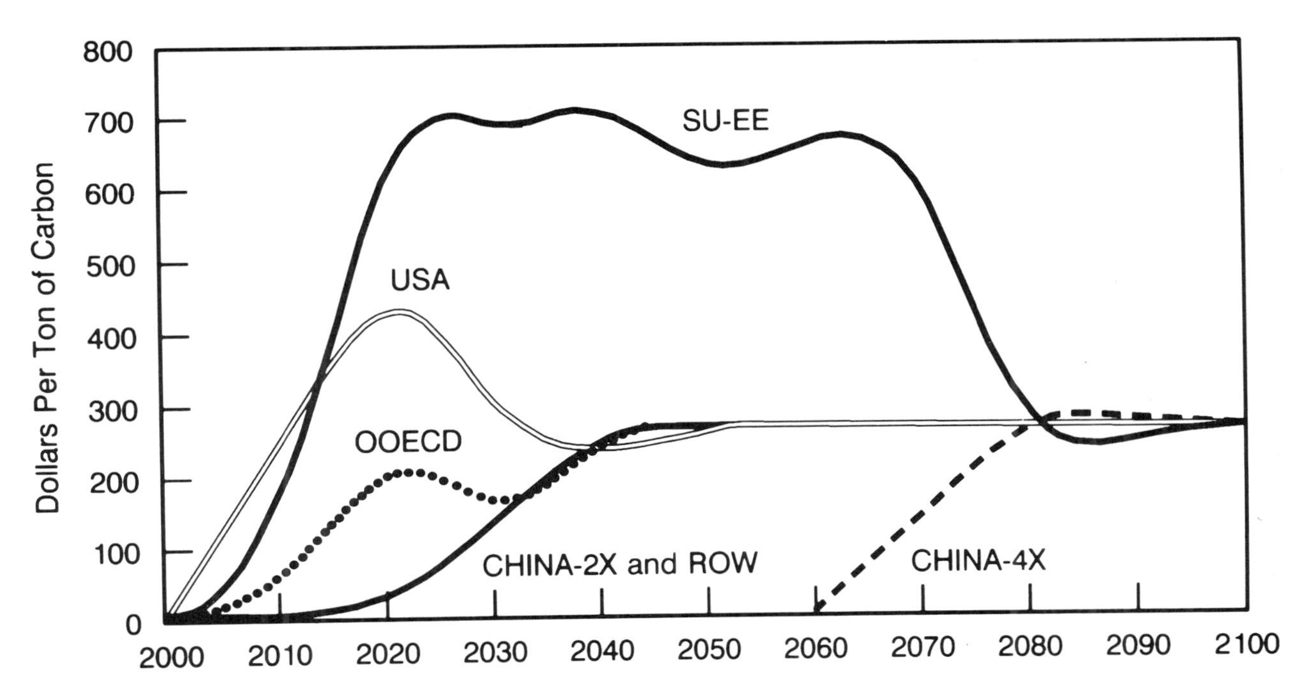

Figure 6. Carbon taxes.

Summary and Conclusions

This paper seeks to improve our understanding of how the costs of limiting carbon emissions are likely to vary among regions. An analysis that encompasses a period exceeding a century is necessarily subject to enormous uncertainty. Nevertheless, we believe that the calculations have provided some useful results.

In the absence of an international agreement to limit growth, carbon emissions are likely to increase considerably, perhaps by a factor of four or more over the next century. During this period, there is apt to be a significant shift in the regional pattern of emissions. In 1990, the industrialized countries accounted for 71 percent of human-made carbon emissions. By 2100, their share is likely to drop below 50 percent.

In recent years, there have been numerous proposals to reduce worldwide emissions by 20 percent or more. Our analysis indicates that this target would be difficult to achieve. If per capita income inequalities are to be significantly reduced, developing countries will have to achieve much higher rates of economic growth than their currently industrialized counterparts. This will place enormous upward pressure on their demands for commercial energy. For developing countries, the issue will not be how far to reduce emissions, but how far to limit their rate of emissions growth.

Against this background, achieving a 20 percent *worldwide* reduction would prove difficult. Suppose, for example, that the developing countries were to agree to limit emissions to a 100 percent increase. The industrialized countries would then need to reduce their emissions by nearly 70 percent below current levels in order to achieve a 20 percent reduction in worldwide emissions.

We have explored less ambitious but perhaps more nearly feasible carbon limits: a 20 percent reduction in emissions by the industrialized countries, a doubling or quadrupling by China, and a doubling in the other developing countries. This would result in a 15 to 37 percent overall increase between 1990 and 2030, but no further increase thereafter. By 2100, global emissions would be only 25 to 30 percent of the level that would have been reached in the absence of an international agreement.

We have also investigated the size of the carbon tax that would be required in each region to induce consumers to reduce their dependence upon carbon-based fuels. Under the assumptions adopted here, it turns out that the long-run equilibrium tax is the same in all regions—$250 per ton of carbon. Such a tax would imply a fivefold increase in the price of coal and an increase of $0.75 per gallon of gasoline. If one region were to sell just 100 million tons of carbon emission rights to another, this tax rate would imply a financial transfer of $25 billion annually.

There are significant regional differences in the

time path to the long-run equilibrium carbon tax level. This would point to opportunities for international trade in emissions rights. At a given point in time, those regions that find it more difficult to adjust to their carbon limits should be willing to purchase emissions rights from other regions experiencing less difficulty.

Our analysis has focused exclusively on the *costs* of reducing carbon emissions. We have not attempted to estimate the *benefits* of slowing down the impacts of global climate change. Without such information, it is unclear whether it makes sense to undertake sizeable reductions in emissions. If it does turn out that significant abatement measures are worthwhile, it is important to understand how the costs of these measures might vary between nations. This could be critical to the negotiation of an acceptable international agreement.

Acknowledgments

The authors are much indebted to Diane Erdmann and to Lawrence Gallant for research assistance. Helpful comments have been provided by Robert Dorfman, George Hidy, William Hogan, Henry Lee, Lu Yingzhong, Stephen Peck, Lee Schipper, Stanley Vejtasa, John Weyant, and Robert Williams.

References

Manne, A.S., and R. G. Richels, 1990. "Global CO_2 Emission Reductions—the Impacts of Rising Energy Costs." *Energy Journal,* forthcoming.

Insurance against the Heat Trap: Explorations of Strategy to Reduce the Risks of Rapid Climate Change

Irving M. Mintzer
Center for Global Change
University of Maryland
College Park, MD

Introduction

A string of unusual weather events during the last three years has focused the attention of politicians and citizens on our changing atmosphere. Much of this attention has been directed toward the effects of a continuing buildup of radiatively active gases, popularly called greenhouse gases. These compounds, including carbon dioxide (CO_2), methane (CH_4), nitrous oxide (N_2O), tropospheric ozone, and various chlorofluorocarbons (CFCs), are transparent to incoming solar radiation but absorb and re-emit the thermal radiation released by the earth's surface. The effect of this re-emission is to increase the average surface temperature of the planet. Many scientists now believe that a continuing buildup of these greenhouse gases could raise the average surface temperature by as much as 4–5° C during the next century. A rise of this magnitude could cause rapid changes in regional climate with potentially severe effects on national economies and natural ecosystems.

During the last century, atmospheric scientists have developed numerical models to simulate the behavior of our planet's atmosphere. These models can be used to estimate the impacts on global and regional climate of future changes in the composition of this thin shield of invisible gases that separates life on our planet from the harsh environment of outer space. In the last decade, these computer-based models of physical processes that take place in the atmosphere have been linked to other, more policy oriented models that simulate the effects of future economic activities on the emissions of radiatively active gases.

The first section of this paper offers an overview of the physical models currently used to explore the impacts of changing atmospheric composition on climate. The second section focuses on one particular model that links policy choices and technological changes to the future rate of greenhouse gas buildup. This policy model, called the Model of Warming Commitment, has been used to explore the possibility of stabilizing future global temperatures during a period of international stability and steady economic growth. This paper summarizes the results of that preliminary exploration.

Global Climate Models

Two types of models are currently used to explore possible future climate conditions. One family of models, principally used in the Soviet Union, involves extensive analysis of historical climate data, going back to paleolithic times. These models analyze empirical data for climate variables believed to be good proxies for concurrent instrumental measurements. These proxy data include the types and distribution of plant pollen, the sizes of tree rings, and the distribution of fossil remains to evaluate the temperature and moisture record of past climates.

The applicability of these models to the analysis of future climates is limited because they cannot recognize the effects of evolutionary change in the climate system. They assume that the same processes and functions that led to the initiation of warm or wet periods in the past and to the historical distribution of regional climates will continue into the future. In the case of global warming due to the greenhouse effect, this is clearly not the case.

By contrast, the global climate models most widely used in the United States and Europe apply numerical simulation methods to describe the physical responses of the atmosphere to changing concentrations of greenhouse gases. These models range from very simple to enormously complex mathematical structures. They are often classified in terms of the number of dimensions used to represent the climate system. One-dimensional models treat the earth as a point source of energy, simulating both radiative and convective heat-transfer processes. Two-dimensional models represent the planet as a plane with longitude and latitude. Three-dimensional models divide the planet into grid-boxes in which longitude, latitude, and altitude can be represented. This last class of models, which includes general circulation models (GCMs), is the most complex and requires the use of large mainframe computers or supercomputers.

Even the most elaborate and sophisticated numerical climate models have serious limitations. These models offer only crude and incomplete representations of the complex physical processes known to occur in the atmosphere. They do not include many known and important feedbacks. In particular, the models do not include accurate representations of ocean-atmosphere interactions or of the dynamic role of living systems.

Despite using the world's fastest supercomputers, the models are crude and very gross in their regional disaggregation. Limited by the costs of computer runs, these models typically divide the world into gridboxes that are far too large to provide the region-specific information demanded by many national policymakers. It is unlikely that these limitations will be overcome.

Recognizing the limitations of both these approaches to forecast or predict the future, a broad consensus has developed in the atmospheric science community on the global consequences of a continuing greenhouse gas buildup. The Scientific Working Group (WG-I) of the Intergovernmental Panel on Climate Change (IPCC) recently articulated this consensus in a report released in May 1990.[1] Assuming that current trends continue, the WG-I report estimates that global mean temperatures will rise to 1.3–2.5° C above the preindustrial level by 2020 and to 2.4–5.1° C above preindustrial by 2070. Precipitation is expected to increase by 3 percent in 2020 and 7 percent in 2070.

The regional effects of this warming cannot now be predicted with certainty, but some broad trends can be discerned. The heating is expected to be most pronounced in high northern latitudes. This is likely to decrease the extent of seasonal sea-ice and snowfall. Despite the rise in global precipitation, substantial (but unidentified) areas of decreased precipitation are anticipated.

The models predict that air temperatures will rise more rapidly over land than over oceans. Thus, the Northern Hemisphere will warm more rapidly than the Southern Hemisphere. The warming is expected to be 50 to 100 percent greater than the global average in the high northern latitudes in winter. Less warming will occur around Antarctica than in the North Atlantic, in part due to the presence of moderating cold currents in the southern ocean. These changes are expected to occur at a rate substantially greater than anything that has been experienced in the last 10,000 years.

Linking Climate to Economics and Policy Concerns

Concerns about the still uncertain risks of such rapid climate change have caused many authors to begin to explore the economic costs of reducing the risks.[2] After accounting for the direct costs (but not the direct and indirect benefits), some have suggested that the expense of reducing the projected concentrations of greenhouse gases may be prohibitively high. Others have suggested that the costs of measures which have reduced the historic rate of emissions may have already substantially penalized at least the US economy.[3]

The study reported here summarizes a modeling effort implemented at the World Resources Institute.[4] It illustrates the characteristics of a scenario leading to a stabilization of global temperatures at levels below the doubled-CO_2 effect by the middle of the next century. The present study suggests that the preliminary analyses indicating prohibitively high costs may not reflect the complete picture. Indeed, by examining the effects of a comprehensive strategy to promote sustainable development, the present study suggests that increases in global greenhouse gas concentration can be minimized where national economic growth continues at historical rates in industrialized countries and at faster than historical rates in developing countries.

Approach

This study utilizes a model, called the Model of Warming Commitment (MWC), developed by the author to test the effects of policy choices on the rate of future economic growth and greenhouse gas emissions.[5] It incorporates a regionally disaggregated partial-equilibrium economic model developed at the Institute for Energy Analysis (the Edmonds-Reilly model) and various sub-models to link future concentrations of the non-CO_2 gases to rates of future economic growth. Emissions of CO_2, N_2O, and CFCs are tracked through to concentrations in the atmosphere. Future growth rates of methane and tropospheric ozone are estimated exogenously.

The following sections describe an exploration of two scenarios of future economic growth developed with the MWC. The first scenario, called Business As Usual, represents a continuation of current trends in the growth of primary energy use with no special effort made to reduce the rate of greenhouse gas buildup. The second scenario, called Aggressive Response, describes a world in which national governments make a concerted effort to promote economic growth, especially in the Third World, while minimizing the rate of buildup of greenhouse gases.

The Business-As-Usual Scenario

This scenario illustrates a world in which recent and current trends continue unabated. This continuation of current trends would mean that lifestyles would not change much until forced to do so by an altered climate. New energy technologies might or might not be introduced, depending on the marketplace. The current gaps between rich and poor, North and South, East and West, would continue to divide the population of our planet into competing and increasingly anxious factions.

Economic growth continues in this scenario at close to historical levels. The income gap between rich and poor countries narrows slightly but persists throughout the period. The ratio of the average per capita income in the rich countries to the average per capita income in the poor countries declines from about 10 to one in 1975 to about seven to one in 2025.

Urbanization continues as a worldwide trend. In the advanced industrial economies, more of the population moves into suburbs and areas just outside the boundaries of today's largest cities. In the US, immigration and internal migration continues well beyond the turn of the century with population continuing to gravitate toward the coasts, and especially toward the southern and western regions of the country. In Western Europe, Japan, and the Soviet Union, the trend is toward ever more extensive industrial centers surrounded by rural areas increasingly dependent on extensive, mechanized agriculture.

In many areas of the Third World, urban cores grow radially, surrounded by a dense ring of poorer suburbs. The central cores become oases of modern technology and convenience for those who can afford to live there. The poor communities on the city's edge are pushed farther and farther out from the "downtown" center, eating into the rich agricultural land that surrounded many cities in 1985. Lack of capital resources at the national and municipal levels inhibits the smooth extension of basic services and public utilities. Pollution, hunger, disease, and poverty strain the social fabric of society.

Growth in personal incomes continues, and oil prices increase less rapidly than per capita GNP. The global fleet of private cars continues to expand. Automobile production grows in all the advanced economies except the US, and increases, although at a slower rate, in developing countries. In the US, the Federal Republic of Germany, France, Great Britain, and several other industrialized countries, the trend is back toward larger, heavier, faster, and more powerful cars.

Developing countries rapidly expand their transportation systems, increasing dependence on cars and light trucks. Most countries evolve vehicle fleets similar to those found today in Western Europe, although some Third World countries expand on the Los Angeles model, with every adult believing he or she needs a large, fast car. Today's global fleet of about 400 million cars and light trucks grows to over 1 billion by 2025. Virtually all automobiles and light trucks are designed to run on gasoline.

Air pollution also mounts in this world with rapid global urbanization, a growing, mostly unregulated global car fleet, and dramatic increases in coal combustion. As temperatures rise and stratospheric ozone is depleted by CFCs, the frequency of photochemical smog alerts and the concentration of ground-level ozone increase in urban areas on six continents. In short, the planet becomes hotter, dirtier, and more crowded (although somewhat more affluent). A declining level of natural amenities is shared unequally among a much larger population.

The Aggressive-Response Scenario

The Aggressive-Response scenario illustrates a quite different world. In this scenario, very high priority is given to policies that can ensure sustainable economic growth while protecting the environment. All nations make a commitment to implement policies designed to slow the buildup of greenhouse gases. Free and open trade policies promote international interdependence, exploiting the comparative advantage of different countries to produce various goods and services.

Compared to the world of Business As Usual, in the Aggressive-Response scenario, significant changes in public values accompany newly earned wealth. Government policies emphasize economic cooperation and the expansion of social services at the expense of military hardware and international adventurism.

In Aggressive Response, governments make a dedicated and systematic effort to promote sustainable economic growth. Public policy instruments are used to incorporate the full economic and environmental costs of energy supply and use into the average price of fuels. Investments in new energy supply options compete for available capital with opportunities to increase the efficiency of energy use on a "level playing field." Public attention and financial resources are redirected from extravagant military adventures and exotic hardware toward cooperative efforts to confront the common enemies of poverty, hunger, disease, and environmental degradation. In this scenario, the gaps in per capita income between North and South, East and West begin to close.

As in the Business-As-Usual scenario, in Aggressive Response economic growth continues throughout the simulation period. GNP grows rapidly in this scenario at slightly less than historical rates in the industrialized countries but at faster than historical rates in the developing world. In the developing countries of the South, more extensive education, better nutrition, and improved health care in rural areas dramatically increase the literacy rate. This in turn, leads to a rapid increase in labor productivity. As a result, real GNP grows much faster than occurs under Business as Usual, at a rate of approximately 6 percent per year. Thus, in Aggressive Response over a 50-year period, the developing country share of global GNP increases from less than 20 percent in 1975 to over 50 percent in 2025. By contrast, in the Business-As-Usual scenario, the developing country share of global GNP is only about 40 percent of the total in 2025.

In Aggressive Response, incomes are more equitably distributed among countries and within countries than they are in Business As Usual. The gap between rich and poor, North and South, closes significantly in this scenario. The ratio of per capita incomes in the industrial countries to those in the developing world falls from 10 to one in 1975 to about four to one in 2025.

The distribution of population is also somewhat different in this scenario, compared to Business As Usual. Urbanization continues as a global trend, but less intensively. In the advanced industrial countries, cityscapes have more the character of southern England or northern Germany than of southern California. The pattern is one of multiple, urban centers in regional clusters, each surrounded by a ring of suburbs, and separated by intensively cultivated agricultural land. This is the alternative to massive Los Angeles-style "slurbs," sprawling into the countryside, swallowing farmland and wetlands indiscriminately and importing water, fuel, and food over ever-increasing distances.

In the Third World, development follows a new pattern. The introduction of small industrial facilities into rural areas spreads economic growth. Rural development policy focuses on the provision of attractive, economically profitable opportunities in traditional agricultural areas. Mass migration to the cities is discouraged.

First in the industrialized economies and later in the developing world, the new patterns of urbanization and rural development encourage new patterns of energy supply and use. Substantial investments are also made to increase efficiency.

Electric utilities treat investments in electricity conservation and efficiency improvements on an equal footing with investments in new supply. Major programs to improve the efficiency of lighting, refrigeration, and electric motor drives are implemented first in the industrialized countries of the Organization for Economic Cooperation and Development (OECD) and then duplicated in the East Bloc and the Third World. Improvements in energy-intensive industrial processes, including the introduction of advanced computer-aided process controls, the retrofit of variable speed motor load controllers on industrial prime movers, and the development of continuous (rather than batch) process lines in basic materials production plants combine to reduce the energy intensity of industrial production.

The process of technology transfers, with new developments often provided on concessionary terms to Third World countries, is mediated both by national governments and, increasingly, by multinational corporations that operate in both industrial and developing countries. Aggressive efforts to promote the early transfer of the best and most efficient new technologies are viewed by the governments of most OECD and Council of Mutual Economic Assistance (CMEA) countries as an essential ingredient in the process of industrialization needed to support sustainable development in the Third World.

The pattern of energy use in transportation is quite different in Aggressive Response from what is seen in Business As Usual. As cities expand and the new regional complexes evolve, provision for mass transit is made from the start. Car use expands less rapidly in this scenario, discouraged by high fuel taxes, heavy parking fees, and high road taxes. Greater emphasis is put on use of local buses and light rail systems in town, following more closely the Dutch model of urban transport in the 1980s than the American model of the 1960s. Fast intercity trains are combined with high-efficiency jet aircraft and barges for moving passengers and freight over long distances.

Electricity plays an increasingly important role in this scenario. Not only does the share of electricity increase as a fraction of total primary energy in Aggressive Response, but the mix of electricity sources shifts as well. As concern about the environmental impacts of combustion increases worldwide in these scenarios, the nonfossil sources become increasingly important. Electricity grids are expanded less by building behemoth new power plants and long transmission lines and more by the incremental development of decentralized local generators feeding regional mini-grids. Distributed solar power systems, small hydro facilities, and small to medium-size cogeneration systems make up an increasing fraction of the new generation in these mini-grids. The mini-grids are redundantly interconnected to insure system reliability with minimal amounts of spinning reserve. In this politically unrealistic and improbable environment, humankind is able to limit the risks of rapid global climate change and stabilize atmospheric temperatures.

Input Assumptions for the MWC Scenarios

The most important input assumptions driving the Model of Warming Commitment are those that represent the human factors in the equations. These include demographic trends, improvements in human capabilities, and behavioral responses to a changing environment. The second most important set of assumptions takes into account the ways that the engineering characteristics of energy technologies change over time. The third set of parameters reflects assumptions affecting greenhouse gases other than carbon dioxide. These shall each be taken up in turn below.

Human Factors

The basic building block in the Model of Warming Commitment is a partial-equilibrium, macroeconomic model developed at the Institute for Energy Analysis.[6] This model (referred to below as the Edmonds-Reilly model) estimates future economic growth, energy use, and CO_2 emissions for nine world regions. Economic activity, represented as projected GNP for each region, is determined principally by the rate of growth in population and the rate of improvement in labor productivity for each region.

In both of the two principal scenarios evaluated in this study, global population follows approximately the UN/World Bank mid-range growth projection, stabilizing in about 2075.[7] In this projection, population growth is much faster in the developing world than it is in the industrialized countries. Population levels off quickly in the advanced industrial economies but continues to grow in developing countries. (See Table 1.) Most of the growth in population occurs in the period between 2025 and 2075. Worldwide, population doubles from the current level of about 5 billion to 10.5 billion people by 2075.

It is not just the number of people in the world that affects the rate of economic growth in this model. The second human factor that strongly influences GNP and the demand for energy is the rate of increase in labor productivity. For this set of experiments, very different assumptions have been made about the regional rates of growth in labor productivity. In the Business-As-Usual scenario, relatively modest rates of growth in labor productivity have been assumed, rates close to historical experience. However, in the Aggressive-Response scenario, an assumed increase in investments by the governments of developing countries in health care, nutrition, education, and other social services is expected to result in rapid increases in literacy and in output per worker hour. This belief is represented in the model by an assumption of higher than historical rates of growth in labor productivity for these regions throughout the simulation period. (See Table 2.)

The presumption of increased investment in education and social services is assumed to have other effects on regional economies as well. In the Business-As-Usual scenario, the pattern of industrialization follows the conventional western model. The Aggressive-Response scenario assumes, however, that the social investment programs described above will simultaneously change cultural values and consumer behavior as well. In particular, this scenario assumes that as a result of increased public education concerning environmental issues, citizens in developing countries will choose to avoid some of the patterns of conspicuous and wasteful consumption that are so evident in the advanced industrialized economies. This emerging consciousness about the global environment is represented in the model by assuming a lower income elasticity of demand for energy for developing countries in the Aggressive-Response scenario than is assumed for these countries in the Business-As-Usual scenario. (See Table 3.) In light of the higher rate of GNP growth experienced by developing countries in the Aggressive-Response scenario, this assumption has a strong impact on the projection of future energy demand and CO_2 emissions.

Another set of assumptions that have a strong impact on the projections of future energy demand and CO_2 emissions is the values assigned to the parameters representing technological change in the model. The most important of these is the parameter used in the

Table 1. Population Growth in the WRI Scenarios (population in billions).

	1975	2025	2075	Annual Growth Rate 1975–2025	Annual Growth Rate 2025–2075
Industrialized Countries	1.142	1.437	1.502	0.5%	0.1%
Developing Countries	2.834	6.782	9.030	1.8%	0.6%
World Total	3.976	8.219	10.530	1.5%	0.5%
US	214	287	292	0.6%	0.0%
W. Europe and Canada	405	522	543	0.5%	0.1%
OECD Pacific	128	155	153	0.4%	−0.0%
USSR and E. Europe	395	473	514	0.4%	0.2%
PRC and C.P. Asia	911	1589	1706	1.1%	0.1%
Middle East	81	276	409	2.5%	0.8%
Africa	399	1593	2654	2.8%	1.0%
Latin America	313	722	896	1.7%	0.4%
South and East Asia	1130	2602	3365	1.7%	0.5%
World Total	3976	8219	10532	1.5%	0.5%

Table 2. Labor Productivity Increase in the WRI Scenarios (annual rate of improvement).

	Business-As-Usual Scenario	Aggressive-Response Scenario
Industrialized Countries	1.5–2.0%	2.0–2.3%
Developing Countries	2.5%	2.5–4.8%

GNP Growth in the WRI Scenarios (GNP in trillions of 1985$)

	1975	2025	2075	Annual Growth Rate 1975–2025	Annual Growth Rate 2025–2075
			Business-As-Usual Scenario		
Industrialized Countries	9.8	33.8	76.6	2.5%	1.7%
Developing Countries	2.3	23.5	128.0	4.7%	3.5%
World Total	12.1	57.2	204.6	3.2%	2.6%
			Aggressive-Response Scenario		
Industrialized Countries	9.8	41.9	116.8	3.0%	2.1%
Developing Countries	2.3	45.1	530.4	6.1%	5.1%
World Total	12.1	87.1	647.3	4.0%	4.1%

Table 3. Long-Run Income Elasticity of Demand for Energy in the WRI Scenarios.

	Business-As-Usual Scenario	Aggressive-Response Scenario
OECD Countries	1.03	0.95
USSR and E. Europe	1.08	1.03
Developing Countries	1.60	1.05

Table 4. End-Use Efficiency Increase in the WRI Scenarios (annual rate of improvement).

	Business-As-Usual Scenario	Aggressive-Response Scenario
OECD Countries	0.80%	1.2–1.6%
USSR and E. Europe	0.80%	1.5%
Developing Countries	0.75–0.85%	2.0–3.0%

Edmonds-Reilly model to represent the rate of improvement in the efficiency of energy use in each region. This parameter aggregates the effects of many individual changes made in the key end-using technologies in the residential, commercial, industrial, and transportation sectors. For the two scenarios studied here, this parameter takes on very different values. In the Business-As-Usual scenario, the parameter takes on a median estimate, close to historical experience. By contrast, in the Aggressive-Response scenario, the presumption of increased investments in the development and use of energy-efficiency improving technologies has led to a much higher assumed value. (See Table 4.) The effect of this assumption is to reduce the projected demand for energy in the Aggressive-Response scenario relative to the energy demand at the same GNP level in the Business-As-Usual scenario.

The Edmonds-Reilly model also incorporates parameters that affect the relative price of energy from different technologies. One group of these parameters represents the rate of efficiency improvement for each energy supply technology. For the two scenarios, generally similar assumptions have been made about the rate at which such future improvements will occur. Two important differences are worth noting, however. The rates of improvement in efficiency for nuclear fission and for coal electric technologies in the Aggressive-Response scenario are slightly more rapid than the rate assumed in Business As Usual.

Another set of parameters determines the rate at which advanced technologies penetrate the energy markets. For solar technologies and for synthetic fuels from coal and shale, the model assumes that prices will fall over time to some minimum level, measured in constant dollar terms. The minimum price level and the time it takes each technology to reach it are assumptions specific to each scenario. In the Aggressive-Response scenario, for example, assumed additional investments in solar energy development reduce the minimum price for solar electricity to a level that is less than half the price assumed in the Business-As-Usual scenario. By contrast, due to relatively more limited investments in synthetic fuels development, the minimum prices for synfuels are assumed to be about 50 percent higher in the Aggressive-Response scenario than in the Business-As-Usual scenario. The effect of these assumptions is to increase the rate at which solar energy captures market share in the Aggressive-Response scenario compared to the role it achieves in Business As Usual. (See Table 5.)

Another set of factors affecting the relative price of fuels (and thus their market share over time) in these scenarios is the cost assigned to the environmental impacts of energy supply and use. In the Aggressive-Response scenario, policies are assumed to try to capture these environmental costs in the price of energy. In the Business-As-Usual scenario, much more limited efforts are assumed. These policies are reflected in values assigned to a tax penalty that is applied to the cost of energy from each technology. In the Aggressive-Response scenario, for example, a carbon-based fuel tax is assumed to be applied to the costs of energy derived from coal, oil, natural gas, and synthetic fuels. No such tax is assumed in the Business-As-Usual scenario. In addition, the environmental charge placed on nuclear electricity is increased in the Aggressive-Response scenario by about 35 percent compared to the charge assumed in the Business-As-Usual scenario. Finally, starting in 2025, a consumption tax is assumed in the Aggressive-Response scenario on the use of oil, gas, and coal in all industrialized countries. (See Table 6.) The effect of these assumed charges is to increase the cost and decrease the competitiveness of fossil fuel and nuclear technologies relative to investments in renewable energy systems and energy-efficiency improving technologies.

Other model assumptions also affect the rate of greenhouse gas buildup in the MWC scenarios. Because the model does not explicitly represent activities associated with deforestation and land use changes, an assumption must be made about the rate of biotic CO_2 emissions in each scenario. For the Business-As-Usual scenario, deforestation is assumed to increase, although at a pace somewhat slower than that seen in the last few years. Thus, biotic emissions of CO_2 are expected to more than double between 1985 and 2075. By contrast, in the Aggressive-Response scenario, an active and vigorous program of worldwide reforestation and soil conservation is assumed to reduce dramatically the biotic contribution to global CO_2 emissions. In this scenario, the biota is assumed to become a declining source over time, reaching a neutral state (relative to net CO_2 emissions) by 2025. By 2075, in the Aggressive-Response scenario, this process has made the terrestrial biota a net sink for about 0.8 Gt of carbon per year.

Many factors affect the rate of buildup of methane and tropospheric ozone in the atmosphere. In the Business-As-Usual scenario, historical rates of growth are assumed for both methane and tropospheric ozone. Methane concentration is assumed to grow by about 1 percent per year throughout the simulation period and tropospheric ozone to increase about 15 percent by 2030. In the Aggressive-Response scenario, by contrast, efforts to limit methane leakage and to control carbon monoxide emissions from automobiles and from biomass burning could reduce the rate of methane buildup to about half the historical rate, or 0.4 percent per year. Tropospheric ozone is assumed to increase by only about 10 percent by 2030.

The final set of assumptions concern the rate of future production and use of chlorofluorocarbons (CFCs). The processes controlling the use and release of these compounds are well understood but the policy choices that will limit their ultimate release have not yet been fully implemented. For the Business-As-Usual scenario, it is assumed that the Montreal Protocol on Substances that Deplete the Ozone Layer will be ratified by all countries and implemented as written, with no cheating. It was also assumed for this case that the HCFCs (safer substitutes that still deplete the ozone layer and add to the greenhouse effect, although at a lower rate than the CFCs they replace) will be phased out of commercial use by industrialized countries in 2030. (A sensitivity case was also run in which no further international agreements beyond the Montreal Protocol to control these dangerous gases was assumed during the simulation period.) The Aggressive-Response scenario assumes that substantial additional agreements to control these gases will be implemented and that the phaseout of both CFCs and HCFCs will thus be accelerated. In particular, it assumes that the use of the most dangerous chlorofluorocarbons will be phased out by the turn of the century in the industrialized countries. It also assumes that the compounds that replace those now in use will be carefully controlled. This scenario, in

Table 5. Minimum Cost of Solar Energy and Synfuels in the WRI Scenarios (costs in 1985 US$/GJ).

	Business-As-Usual Scenario		Aggressive-Response Scenario	
	Min Cost	Yr Achieved	Min Cost	Yr Achieved
Solar Electricity	$24.00	2010	$10.20	2005
Synthetic Oil	$10.50	2025–30	$14.50	2050
Synthetic Gas	$8.30	2025–30	$12.30	2050

Table 6. Energy Taxes in the WRI Scenarios.

	Business-As-Usual Scenario		Aggressive-Response Scenario	
	Tax	Yr Imposed	Tax	Yr Imposed
	Taxes on Energy Supplies (Imposed in All Regions, 1985 US$/GJ)			
Gas	$0.00	2010	$0.80	2010
Oil	$0.00	2010	$1.50	2010
Coal	$0.90	2010	$1.80	2010
Unconv. Oil	$5.00	2010	$9.50	2010
Nuclear	$15.00	2010	$19.00	2010
	Taxes on Energy Consumption (Percentage Increase, Industrial Countries Only)			
Gas	0.00%	NA	30%	2025
Oil	0.00%	NA	50%	2025
Coal	0.00%	NA	65%	2025

effect, assumes that by 2020, all chlorofluorocarbon compounds will be eliminated from use in the industrialized countries (with the exception of some very small-scale medical applications). It assumes that the same compounds will be eliminated from use in the developing countries by 2030. It further assumes that the compounds replacing the current generation of chemicals will, after these dates, be limited to those which have no ozone-depleting potential and no global warming effect.

The assumptions illustrated above are represented explicitly by the mathematical parameters in the Model of Warming Commitment. They are generally consistent with the illustrations of two alternative worlds given above. The model results reflecting the impact of these assumptions on the timing and extent of global warming are described in the paragraphs below.

Model Results: The Business-As-Usual Scenario

The Business-As-Usual scenario roughly corresponds to a continuation of current trends. Economic growth continues in this scenario at slightly faster than historical levels. In the industrialized countries of the North, gross national product (GNP) increases from 1985 to 2025 at a rate of about 2.5 percent per year, and at a somewhat slower rate in the following 50 years. In the developing countries, economic growth is more rapid during the next 40 years, approaching an average annual rate of about 4.7 percent. The income gap between rich and poor countries narrows slightly but persists throughout the period. In 2025, per capita incomes in the industrialized countries average approximately $23,500 (in 1985 US dollars) compared to an average of $3,460 in developing countries.

No specific effort is made to limit growth in primary energy demand or to reduce the environmental risks of economic activity. Global primary energy demand continues to grow between 1985 and 2025 at an annual rate of about 2.2 percent per year, approximately equal to the historical rate from 1970 to 1985.[8] After 2025, demand grows at a more moderate rate, approximately 1.7 percent per year. Thus, in this scenario, energy consumption reaches approximately 445 exajoules (EJ) by the year 2000, compared to approximately 260 EJ in 1975. [One exajoule equals 10^{18} Joules or approximately 0.95 quadrillion BTUs.] Energy supply is approximately 765 EJ in 2025 and 1775 EJ in 2075. (See Figure 1.)

The Business-As-Usual scenario represents a world where more concern is placed on avoiding energy shortages than on the problems created by energy surpluses. Government planners assume that demand will grow and focus investment capital on increasing supplies of energy. Because energy efficiency increases slowly in this scenario, achieving these levels of energy supply requires vigorous and simultaneous efforts to expand energy production of every type in all countries. In the US, primary energy demand increases at an annual rate of 1.2 percent per year from 1975 to 2025. This compares with a historical growth rate of 1.1 percent per year from 1967 to 1987. For the total OECD, Business As Usual implies an annual rate of growth in commercial energy demand of about 1.2 percent per year from 1975 to 2025. For the Soviet Union and the East Bloc, primary energy demand grows in this scenario at an annual rate of 0.5 percent from 1975 to 2025. Commercial energy demand grows especially rapidly in the developing world, approximately 4 percent per year during the next 40 years. As a result of these regional increases, the global growth rate follows the historical trend in this scenario, with primary energy demand increasing an

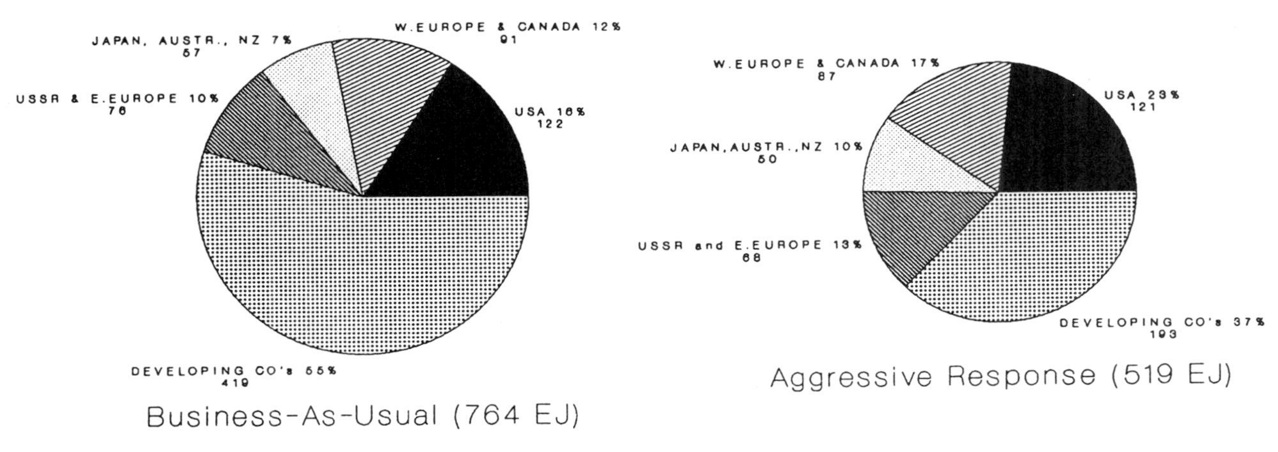

Figure 1. Primary energy use in the WRI scenarios (energy use for 2025 in exajoules).

average annual rate of 2.2 percent for the next four decades. This scenario implicitly assumes that the future will be smooth and surprise-free during this period with no big oil price shocks, no supply disruptions, and no global depressions due to bank failures or to the debt crisis.

Meeting the needs of such a high demand future requires a smoothly managed market for liquid fuels throughout the next 40 years. OPEC and non-OPEC oil producers must cooperate wholeheartedly to meet the growing global appetite for petroleum products. World oil production must increase by 50 percent between 1985 and 2000 in this scenario, then remain largely constant for the next 25 years.

Electrification intensifies in all countries in this scenario, with the most rapid rates of demand growth occurring in developing countries. In the period 2000–2025, electricity growth rates approach 5.6 to 8.4 percent per year for developing countries compared to 1.1 to 3.2 percent per year for the industrialized countries.

In this scenario the petroleum era gives way to a world fueled by coal, nuclear, and solar power. Increasing reliance is placed on coal to fire the engine of world economic growth. The level of coal consumption grows at an annual rate of about 1 percent between 1975 and 2025, and at a rate of about 4.7 percent annually in the following 50 years. Coal demand reaches about 10 times the 1975 level at the end of the simulation period. The supply of electricity from nuclear fission grows at almost 4 percent per year from 1985 to 2025. The output of hydroelectric plants increases by more than a factor of five over this 100-year period.

Energy prices rise steadily in the Business-As-Usual scenario, with oil prices increasing by about a factor of three between 1985 and 2025. The real price of oil increases by an additional 50 percent between 2025 and 2075. The price of natural gas increases at an annual rate of nearly 3 percent for the period from 1985 to 2075. The price of coal more than doubles in constant dollars by 2025 and increases by a factor of three by 2075, relative to the 1985 levels.

As a result of changing relative prices and declining resource availability, the fuel share of conventional oil declines from 40 percent of the global total in 1986 to 23 percent in 2025 and to less than 1 percent in 2075. The share of conventional gas in the fuel mix declines from about 20 percent of the total in 1986 to a little over 10 percent in 2025 and to less than 1 percent in 2075. The coal share increases in this scenario from 28 percent in 1986 to 38 percent in 2025 and to over 65 percent in 2075. Nuclear energy increases by a factor of more than 20 from 1975 to 2075 but continues to supply only 5 percent of the world's primary energy supply. Commercial solar power takes off from nearly zero in 1975 to supply about 6 percent in 2025 and 15 percent in 2050. (See Table 7.)

The impact of this pattern of energy use is a steady increase in the emissions of CO_2 from fossil fuel combustion. At the global level, total emissions increase at an annual rate of about 2 percent per year from 1987 to 2025, doubling from about 5.5 Gt of CO_2 in 1987 to about 11.3 Gt in 2025. The increase accelerates after 2025, growing at a rate of about 1.9 percent per year. By 2075, fossil fuel-related CO_2 emissions reach about 29 Gt per year. (See Table 8.)

Lack of concern about environmental quality leads to increasing biotic emissions of CO_2 as well. Here, a growing rate of deforestation increases biotic emissions. Tropical forests are increasingly converted to pasture lands. Primary growth is burned off, stumps are left to decay, and the land is seeded in grasses. Forest soils are stable and productive in these applications for only three to five years. After that, each recently cleared area must be abandoned and left to lie fallow. Denuded of forest cover, tropical soils rapidly erode. Their carbon content

Table 7. Primary Energy Use in the WRI Scenarios (annual rate of increase, 1975–2025).

	Business-As-Usual Scenario	Aggressive-Response Scenario
OECD Countries	1.52%	1.43%
USSR and E. Europe	0.49%	0.28%
Developing Countries	4.33%	2.73%
World Total	2.37%	1.58%

	Oil	Gas	Solids	Nuclear	Solar	Hydro	Total
			Business-As-Usual Scenario (Exajoules per Year)				
1975	119	41	76	4	0	19	258
2000	178	66	132	10	1	57	444
2025	177	84	319	29	50	104	764
2050	17	117	766	52	161	118	1230
2075	13	33	1246	89	273	119	1774
			Aggressive-Response Scenario (Exajoules per Year)				
1975	119	41	76	4	0	19	258
2000	137	65	71	10	15	49	346
2025	106	53	77	25	171	91	524
2050	32	58	69	22	329	120	628
2075	27	24	55	30	478	126	740

Table 8. Fossil Fuel CO_2 Emissions in the WRI Scenarios (annual rate of increase, 1975–2025).

	Business-As-Usual Scenario	Aggressive-Response Scenario
OECD Countries	1.30%	−0.65%
USSR and E. Europe	0.18%	−1.84%
Developing Countries	3.77%	1.03%
World Total	2.02%	−0.32%

	Conv. Oil	Shale Oil	Synoil	Coal	Syngas	Conv. Gas	Total
			Business-As-Usual Scenario (Gigatons of C per Year)				
1975	2.2	0.0	0.0	1.8	0.0	0.6	4.6
2000	3.3	0.0	0.0	2.8	0.0	0.9	18.2
2025	3.2	0.0	0.5	6.4	0.0	1.1	30.4
2050	0.2	0.1	6.6	10.4	0.2	1.6	48.2
2075	0.1	0.2	9.8	15.9	2.5	0.5	29.0
			Aggressive-Response Scenario (Gigatons of C per Year)				
1975	2.2	0.0	0.0	1.8	0.0	0.6	4.6
2000	2.5	0.0	0.0	1.1	0.0	0.9	4.4
2025	1.9	0.0	0.0	0.7	0.0	0.7	3.4
2050	0.6	0.0	0.0	0.0	0.0	0.8	1.4
2075	0.5	0.0	0.0	0.0	0.0	0.3	0.8

is washed away or picked up by the wind and quickly oxidized to CO_2.

Total CO_2 emissions from the biota increase from about 1.25 Gt of C per year in 1985 to three times that amount by 2075. This estimate does not take account of the releases of CO_2 from soil carbon that result from heat-induced increases in respiration rates by soil bacteria. Even without this potentially strong positive feedback, by 2075, total annual emissions of CO_2 in this scenario are more than 30 Gt of C, more than five times the current level.

As a result of this combination of biotic and fossil fuel-derived CO_2 emissions, the concentration of CO_2 reaches twice the pre-industrial level by about 2050. (See Figure 2.) Absent any other increases in greenhouse gas concentration, this CO_2 buildup alone would commit the planet to a warming of about 1.5–4.5°C. But the concentrations of other greenhouse gases continue to increase in this scenario as well.

The rapid increase in coal use leads to a nearly 15 percent increase in N_2O concentration by about 2025 and close to a doubling by 2075. The rate of growth in N_2O concentration in this scenario, about 0.7 percent per year, is much more rapid than the historical rate of about 0.2 percent per year. (See Figure 3.)

This scenario assumes one further international agreement on CFCs beyond the Montreal Protocol on Substances that Deplete the Ozone Layer. Business As Usual assumes that all industrialized countries reduce their emissions by 50 percent from the 1986 levels and that all developing countries take full advantage of the special provision for low-consuming countries. It further assumes that CFCs are fully phased out in 2000 and their replacements, the HCFCs, are phased out in the industrialized countries by 2030. As a result, the concentration of CFC-11 increases at an annual rate of about 0.5 percent per year, compared to a historical rate of more than 5 percent per year. Thus, CFC-11 concentration increases from 0.2 ppbv in 1985 to about 0.53 ppbv in 2025 and then declines to 0.3 ppbv in 2075. (See Figure 4.)

The concentration of CFC-12 grows slightly less rapidly in this scenario. It increases at a rate of about 0.9 percent per year from 1985 to 2075, compared to the historical rate of over 5 percent. Over this period, the atmospheric concentration grows from about 0.4 ppbv to a peak of 0.9 ppbv in 2025 and then declines to about 0.65 ppbv in 2075. (See Figure 5.) The increases in CFC concentration in this scenario could have significant implications for stratospheric ozone depletion as well as for global warming.

The concentration of methane continues to grow at

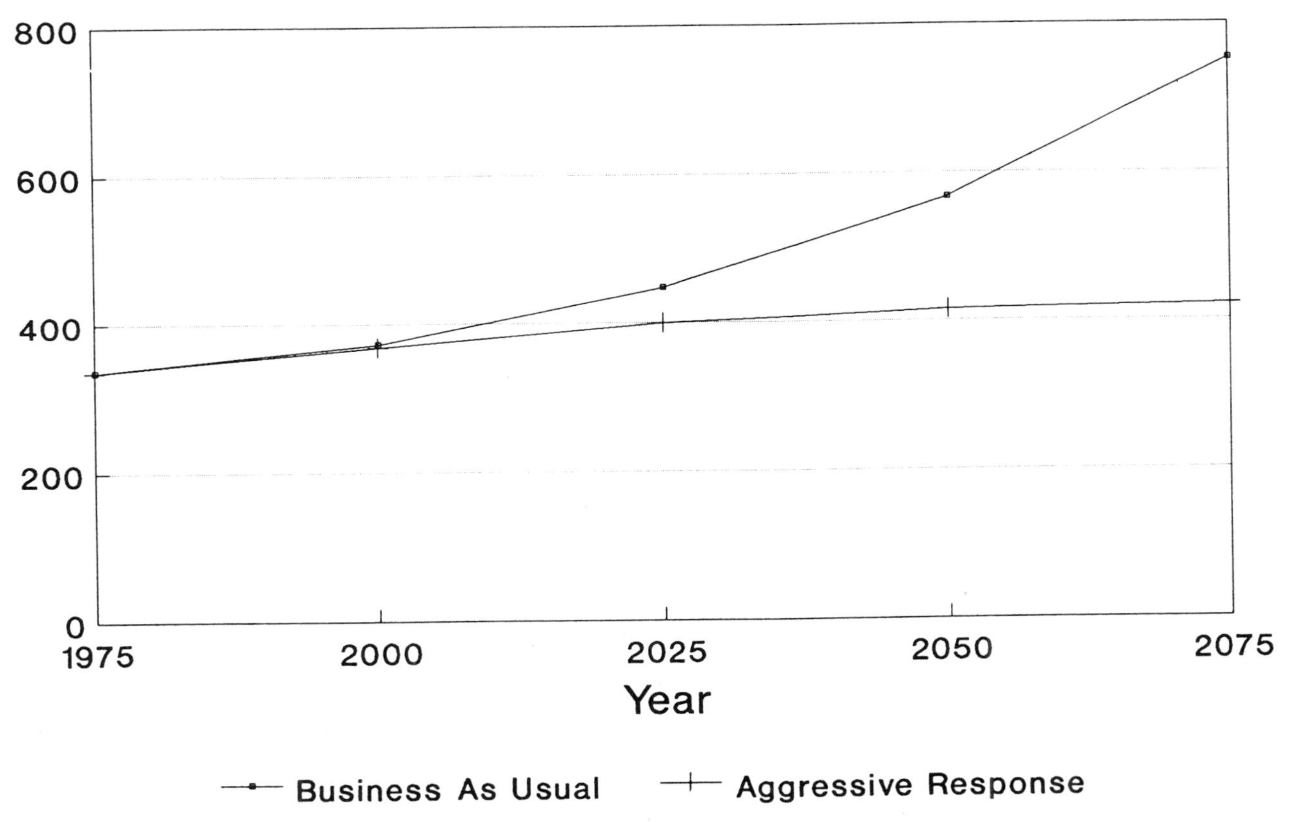

Figure 2. CO_2 concentration in the WRI scenarios (concentration in ppmv).

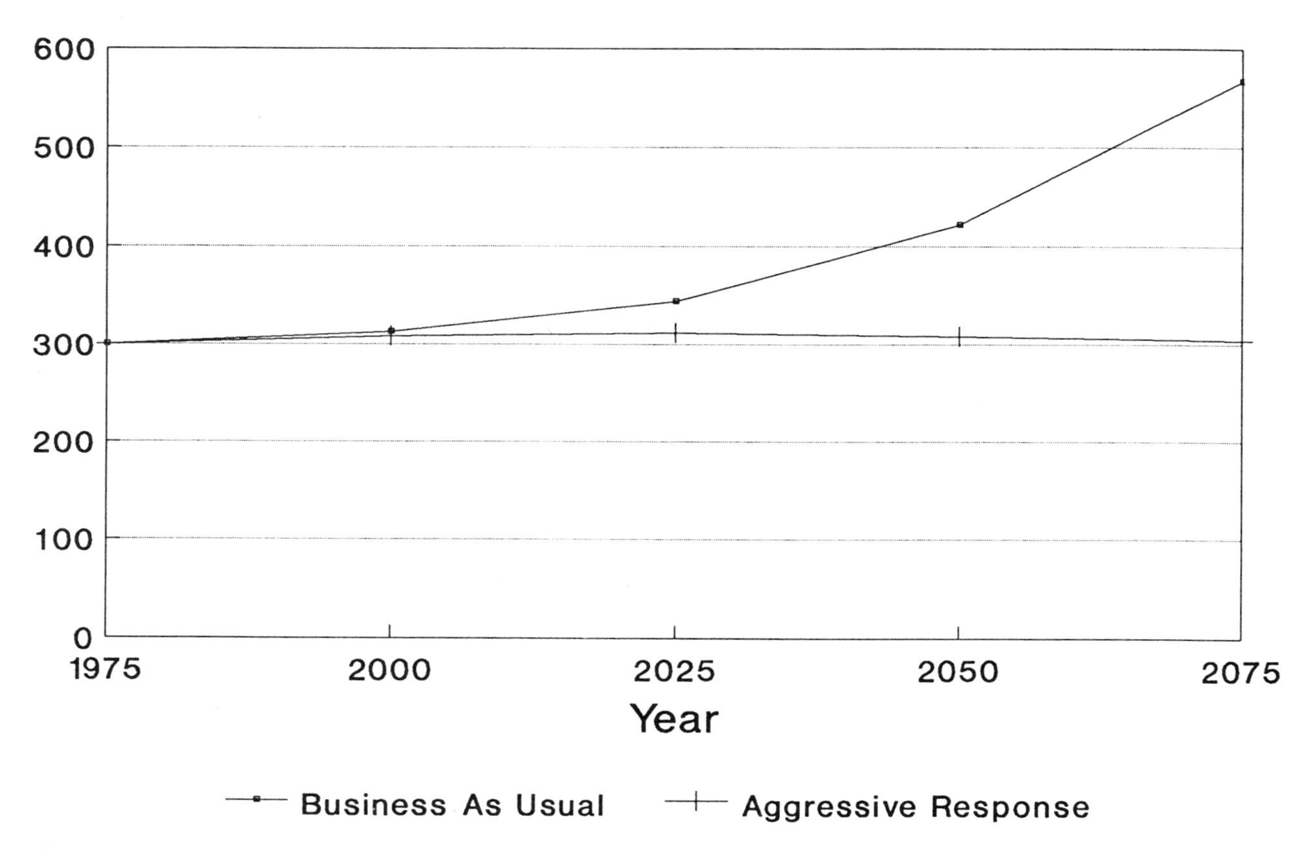

Figure 3. N_2O concentration in the WRI scenarios (concentration in ppbv).

the historical rate of 1 percent per year under Business As Usual. Many factors are assumed to contribute to this rapid methane buildup. The growing human population in developing countries is assumed to increase the demand for paddy rice cultivation. Rising incomes are expected to lead to a growing demand for meat in human diets. Both of these activities result in larger biotic emissions of methane.

The growing global demand for natural gas between 1985 and 2050 is assumed to lead to higher losses from pipeline leakage and releases from local distribution systems. At the same time, the growing global car population, the rising demand for fuelwood, and the inefficient burning of biomass contribute to emissions of carbon monoxide that reduce the atmospheric sink for methane. These factors together combine to stimulate a rise in the atmospheric concentration of methane from about 1.65 ppmv in 1985 to over 4 ppmv only 90 years later. (See Figure 6.)

Air pollution is also assumed to increase, leading to higher levels of ozone pollution. As temperatures rise and stratospheric ozone is depleted by CFCs, the frequency of photochemical smog alerts and the quantity of ground-level ozone is expected to increase in urban areas on six continents. The average global concentration of tropospheric ozone is assumed to increase by about 15 percent in this scenario from 1985 to 2025. (See Figure 7.)

In Business As Usual, the combined effect of greenhouse gas emissions commits the planet to a large and rapid global warming. If the trends described in this scenario continue, the planet will be committed to a warming relative to the pre-industrial atmosphere that is equal to the effect of doubling the pre-industrial concentration of CO_2 alone, a warming of 1.5–4.5°C, by about 2010. [This takes into account an estimated commitment to warming of 1–2.5°C relative to the pre-industrial atmosphere that is already "in the pipeline," due to emissions between 1880 and 1980.] By 2025, the planet will be committed to a warming of 2.0–5.5°C, relative to the pre-industrial atmosphere. By 2075, the commitment to future warming will be approximately 3.5–10°C relative to the pre-industrial atmosphere. The continuing buildup of greenhouse gases throughout this scenario suggests that temperatures would rise quite rapidly. The climate would not be expected to equili-

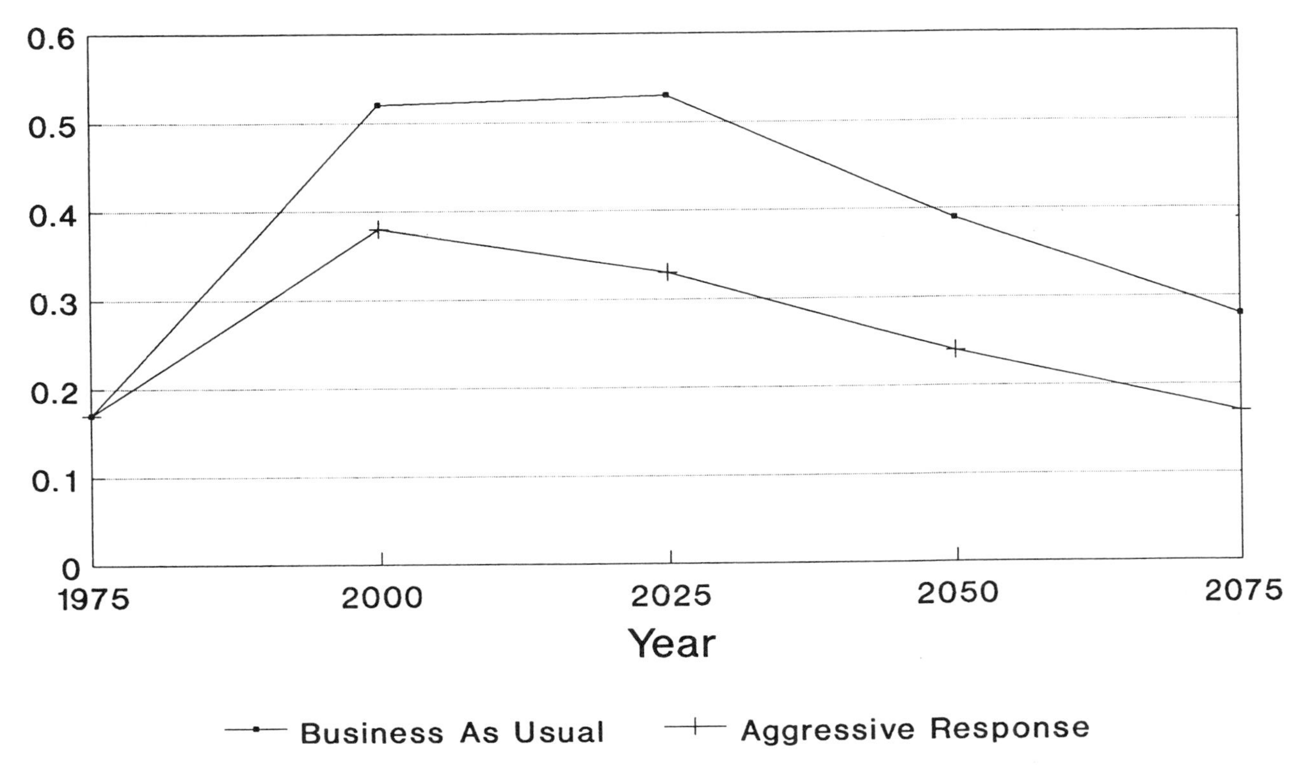

Figure 4. CFC-11 concentration in the WRI scenarios (concentration in ppbv).

brate at the elevated level of 2075 but rather global surface temperatures would continue to increase into the indefinite future. (See Figure 8.)

Model Results: The Aggressive-Response Scenario

The Aggressive-Response scenario shares some important similarities with the Business-As-Usual scenario but represents a very different world. The principal difference is that vigorous, systematic, and widespread efforts are made to ensure sustainable economic growth while minimizing the long-term environmental consequences of energy supply and use. As a result of this multifaceted policy strategy, world growth in primary energy demand is moderated, the trend toward increasing deforestation is reversed, and global climate stabilizes during the third quarter of the 21st century.

The principal similarities between the two futures have to do with population and economic growth. In both scenarios, global population grows significantly between 1985 and 2075, ultimately stabilizing at about 10.5 billion people. (See Table 1.) Here again, most of the population growth takes place in the developing world during the period between 2025 and 2075.

As in the Business-As-Usual scenario, economic growth continues throughout the simulation period. GNP grows rapidly, at slightly higher than historical rates in the industrialized countries, and at faster than historical rates in the developing world. In the industrialized North, real GNP grows at a rate of about 2.9 percent per year between 1975 and 2025. In the developing countries of the South, assumed investments in more extensive education, better nutrition, and improved health care in rural areas dramatically increase the literacy rate. This, in turn, leads to an assumed rapid increase in labor productivity. As a result, the model projects real GNP to grow much faster during this period, at a rate of approximately 6 percent per year. Thus, in Aggressive Response, over a 50-year period, the developing country share of global GNP increases from less than 20 percent in 1975 to over 50 percent in 2025. By contrast, in the Business-As-Usual scenario the developing country share of global GNP is only about 40 percent of the total in 2025.

At first blush, this seems to be an unlikely and inconsistent result. How could GNP grow rapidly in the developing countries if economic growth slows down in the industrialized world? The answer lies in a fundamental change in the pattern of development. In this scenario, economic growth in the developing world is not principally directed toward export markets but in-

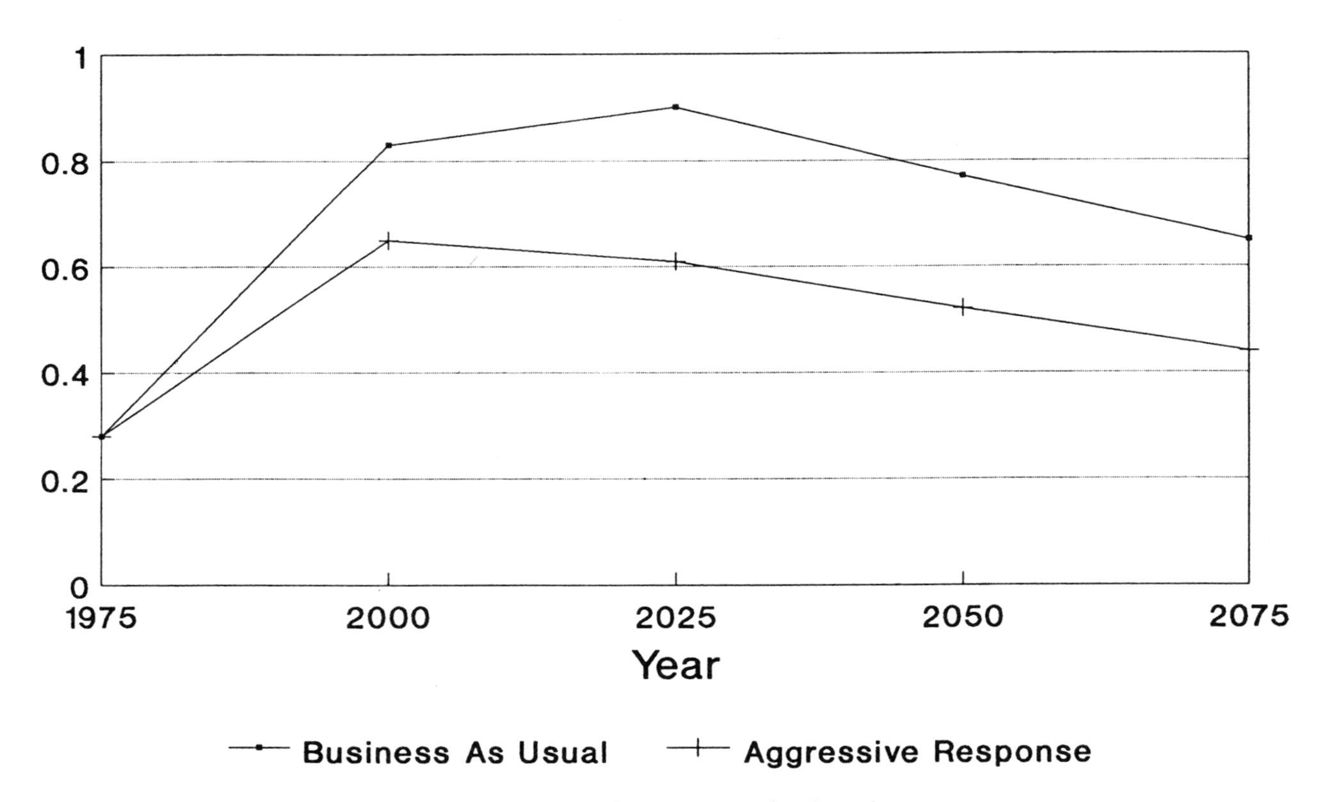

Figure 5. CFC-12 concentration in the WRI scenarios (concentration in ppbv).

stead is aimed at meeting basic human needs and building capital infrastructures in the rural areas of each country. This pattern of development is assumed to be facilitated in this scenario by a program of technology transfer and development assistance from the industrialized world that is designed to promote ecologically sustainable development and to minimize global environmental problems. Motivated by the need to protect the global commons rather than to enrich the metropolitan countries of the First World, this new pattern of industrial development is expected to cause a net flow of capital to the developing world.

Reflecting this shared global commitment to limiting environmental risks, energy demand is projected to grow much less rapidly in Aggressive Response than in Business As Usual, about 1.4 percent per year from 1975–2025 in this scenario compared to 2.2 percent per year for the same period in Business As Usual. Primary energy use grows at an annual average rate of 0.5 percent in the industrialized countries and about 2.3 percent per year in the developing world. At these rates, by 2025, in the Aggressive-Response scenario, global primary energy demand reaches only 525 EJ, about twice the 1975 level compared to more than three times the 1975 level in Business As Usual.

In Aggressive Response, primary demand grows even more slowly after 2025, reaching only 740 EJ in 2075. This represents a rate of about 0.7 percent per year from 2025 to 2075, compared to almost three times that rate in Business As Usual.

This seemingly counterintuitive result—high rates of GNP growth with slow growth of energy demand in the developing world—results from the lower income elasticity of demand for energy that was assumed in the Aggressive-Response scenario. As a result of comprehensive educational and social welfare programs, the income elasticity of demand is assumed to be approximately 1.05 for developing countries in the Aggressive-Response scenario, compared to 1.6 for the same countries in the Business-As-Usual scenario. Thus, even though per capita incomes are substantially higher in the Aggressive-Response scenario, aggregate demand for energy is substantially lower than in Business As Usual. Whereas per capita demand for energy grows by about 1.5 percent per year in industrialized countries in the Business-As-Usual scenario, it grows by only 1 percent per year between 2000 and 2025 in the same countries in Aggressive Response.

Because the demand for energy grows more slowly in this scenario, the world oil market is much easier to

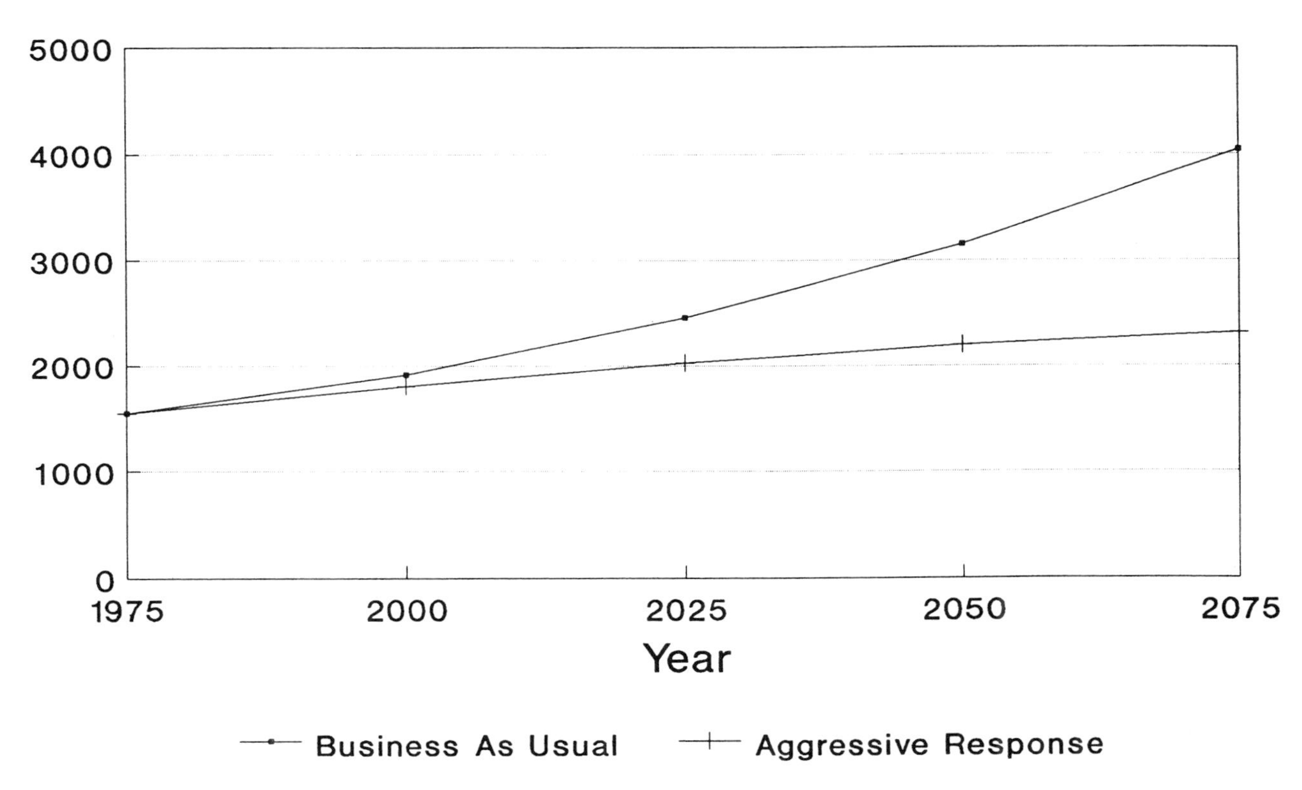

Figure 6. Methane concentration in the WRI scenarios (concentration in ppbv).

manage in a nondisruptive fashion. Largely because the size and activity of the world car fleet is limited in Aggressive Response, by 2000 oil demand grows only slightly (i.e., a little over 10 percent) over the 1980 level. Growth in oil demand levels off, then declines to about 105 EJ per year in 2025 (approximately 53 million barrels per day). After 2025, oil demand falls rapidly, to less than 23 percent of the 1980 level in 2075. In this slowly growing market, the geopolitical pressures associated with assuring steady supplies of oil at increasing volumes decline substantially. As in Business As Usual, the US and Japan are almost completely dependent on imported oil by 2025. The share of global production coming from Canada, Western Europe, and the Soviet Union also increases. Nonetheless, by 2025, the Middle East is supplying over 40 percent of the global supply of conventional oil in Aggressive Response.

Electricity plays an increasingly important role in this scenario. For the US, electricity demand grows from less than 40 percent of primary energy demand in 1975 to about 78 percent in 2025. In Japan, Canada, and Western Europe, electricity increases from about 45 percent of primary energy demand in 1975 to over 80 percent in 2025. For the Soviet Union, the increase is from about 35 percent in 1975 to about 70 percent in 2025. In China, electricity use grows from less than 15 percent of primary energy demand in 1975 to almost 50 percent in 2025. And for South and East Asia, the increase is nearly as dramatic, from less than 35 percent in 1975 to about 70 percent in 2025.

Not only does the share of electricity increase as a fraction of total primary energy in Aggressive Response, but the mix of electricity sources shifts as well. As concern about the environmental impacts of combustion increases worldwide in these scenarios, the nonfossil sources become increasingly important. Solar, nuclear, and hydropower represent 55 percent of global primary energy supply in 2025 compared to less than 10 percent in 1975. Solar alone contributes almost 33 percent of the total, or about 130 EJ in 2025. By contrast, in the Business-As-Usual scenario, the contribution from solar, nuclear, and hydropower together represents less than 25 percent of total primary energy supply in 2025. The solar fraction alone is about 6 percent of the total in Business As Usual, or a little more than 50 EJ for 2025. (See Table 7.)

The demand for gas increases significantly during the next several decades in Aggressive Response. Gas use is increased largely because of its environmental advantages relative to coal and oil consumption. In effect, gas in this scenario is used as a "bridge fuel" to an era of dependence on "smokeless technologies." Global gas demand reaches a peak in 2000 at about 55 EJ per year in this scenario, an increase of about 25 percent over the 1975 level.

Commercial biomass plays a much larger role in

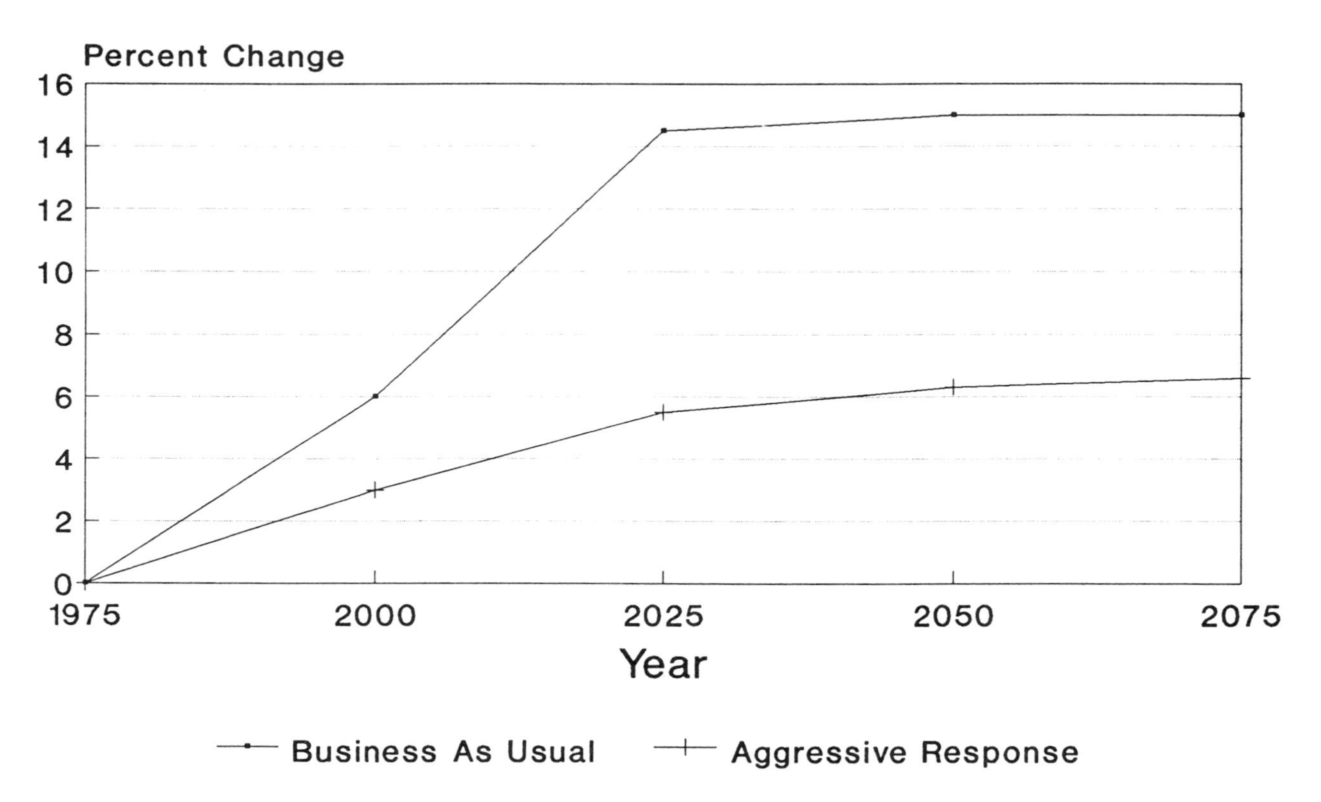

Figure 7. Tropospheric ozone concentration in the WRI scenarios (percent change relative to 1980).

Aggressive Response than it does in Business As Usual. Intensively cultivated on scientifically managed commercial plantations, biomass supplies an increasing share of global demands for solid fuels. On a regional basis, commercial biomass harvesting and cultivation are controlled so that CO_2 uptake in new growth just matches the CO_2 released when the annual harvest is burned. Major programs to develop energy plantations substantially reforest China, South and East Asia, Latin America, and Africa. By 2025, biomass plantations are providing almost four times as much energy annually as are coal mines in Aggressive Response. By 2050, coal production for energy has been phased out in this scenario.

Energy prices increase in Aggressive Response, but because overall demand grows less rapidly, most prices remain somewhat lower than in Business As Usual. By 2025, the real price of oil has nearly doubled from the 1988 levels but remains about 25 percent below the expected price in Business As Usual. The real price of gas in Aggressive Response is slightly more than twice the 1988 level in 2025. This is about 20 percent less than the real price in Business As Usual for 2025.

By contrast, the average price of coal (before taxes) and biomass in Aggressive Response in 2025 is about 15 percent higher than the price in Business As Usual. The share of commercial biomass in Aggressive Response is almost 80 percent of the total supply of solid fuels on a global basis in 2025. The price of commercial biomass for energy is elevated by the competition for land with the enormous global reforestation program assumed in this scenario. Beyond this, the price of coal in the Aggressive-Response scenario is pushed up by the environmental taxes applied to it.

By contrast, in Business As Usual coal, which is substantially cheaper to produce, represents about 90 percent of the total solids supply in 2025. The carbon tax placed on coal supplies is less than one-third the size of the carbon tax applied in the Aggressive-Response scenario.

As the scenarios unfold, the demand for coal grows rapidly in the Business-As-Usual scenario. By 2075, the price comparison for solid fuels in the two scenarios is reversed. Largely because the demand for solid fuels is so high in the Business-As-Usual scenario—approximately 1250 EJ per year or 5 times total world primary energy supply in 1975 versus only 75 EJ per year—the average price of solid fuels is approximately twice as high in constant dollars in the Business-As-Usual scenario as it is in Aggressive Response.

As prices change and new technologies evolve, the mix of energy sources consumed in the Aggressive-Response scenario shifts. The fuel share of conventional

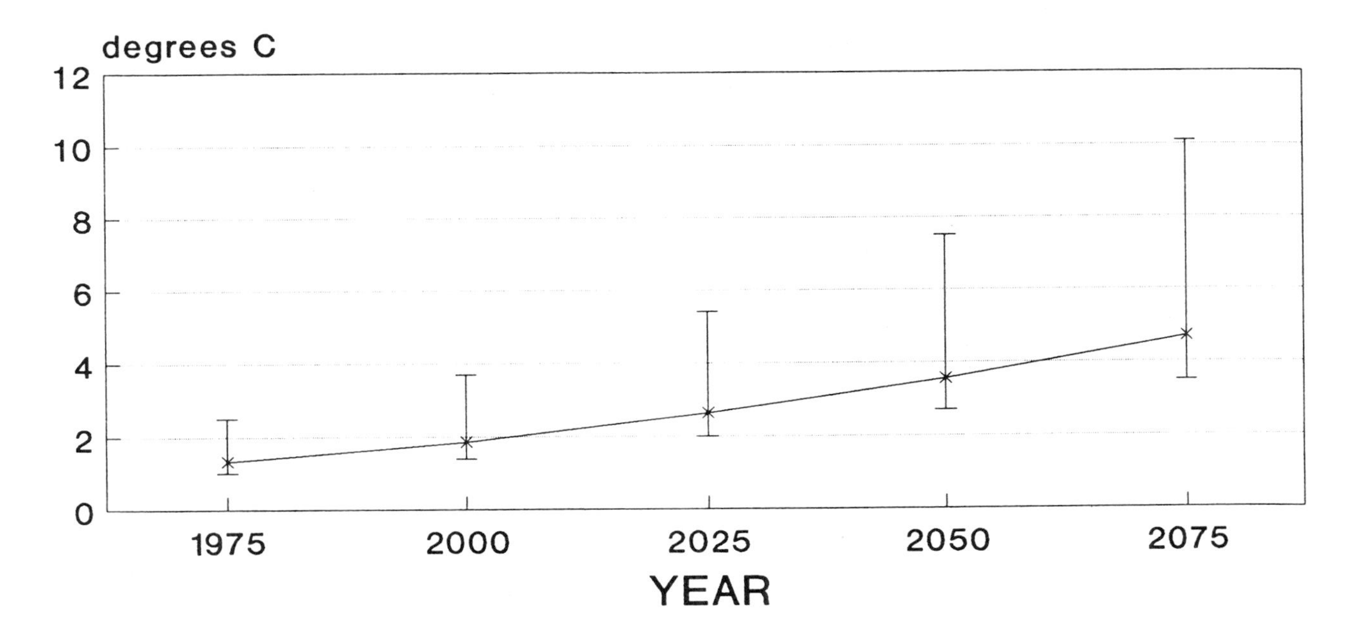

Figure 8. Warming commitment in the Business-As-Usual scenario (assumes 2C sensitivity to 2 × CO_2 + H_2O).

oil declines from 40 percent of the total in 1986 to about 20 percent of the total in 2025 and to about 3.5 percent 50 years later. The share of conventional gas declines from about 20 percent of the global total in 1986 to a little more than 10 percent in 2025 and to about 3 percent in 2075. The coal share declines from 28 percent in 1986 to about 6 percent in 2025 and to zero in 2075. Nuclear energy use increases in absolute terms but remains less than 5 percent of the total primary energy supply through 2075. Solar-derived electric energy, on the other hand, grows from virtually nothing in 1986 to almost 30 percent of world primary energy supply in 2025 and to over 60 percent of the total in 2075.

The consequence of this change in the fuel mix and the slow growth in primary energy demand in the Aggressive-Response scenario is that carbon dioxide emissions from fossil fuel combustion decline dramatically. Emissions decline by 26 percent from 1986 to 2025. By 2075, CO_2 emissions fall by 85 percent from the 1986 level, declining to approximately 0.8 Gt of carbon per year. (See Table 8.)

Growing worldwide concerns about environmental quality combined with the strong international programs to develop scientifically managed, commercial biomass plantations lead to a radical reduction in CO_2 emissions from the biota.

Large areas of the United States, Europe, Asia, and Africa are reforested in this scenario. Multiple-use projects in developing countries provide fuelwood, fodder, and building materials as well as commercial solid fuels from carefully managed woodlots and forests. In the United States and other industrial countries, CO_2 offset programs encourage the planting of new forests to offset the emissions of CO_2 from new industrial facilities. Urban tree-planting programs increase CO_2 uptake, decrease the local albedo, and provide shade and comfort in the summer while reducing building cooling loads. Trees planted as wind-breaks and hedgerows stabilize rural soils, reduce erosion losses, and increase carbon storage.

By 2025, biotic emissions fall from 1.25 Gt of C per year in 1986 to about 0.5 Gt per year in Aggressive Response. By 2075, the biota has become a net sink for atmospheric CO_2, withdrawing 0.8 Gt of C per year from the air. At this point, the enhanced CO_2 uptake capability of the biota just offsets the fossil fuel releases each year, resulting in no net injection of CO_2 into the atmosphere.

Because of these controls on CO_2 emissions, the atmospheric concentration of carbon dioxide grows very slowly in the Aggressive-Response scenario. By 2025, the estimated concentration is 399 ppmv in Aggressive Response, less than 13 percent higher than in 1986. By

2075, the atmospheric concentration of CO_2 is only slightly more than 420 ppmv, an increase of only about 70 ppmv from current levels. Thus, in Aggressive Response, the concentration of CO_2 does not exceed twice the pre-industrial level during the simulation period. This situation contrasts with that in Business As Usual, where the atmospheric concentration of CO_2 reaches twice the pre-industrial level (approximately 550 ppmv) by about 2050. In Business As Usual, the concentration of atmospheric CO_2 by 2075 is about 750 ppmv, more than twice the 1986 level.

Without any other increases in greenhouse gas concentration, the buildup of CO_2 in Aggressive Response would commit the planet to an ultimate warming of only 0.4–1.35°C. But the concentration of other greenhouse gases also increases in this scenario.

Given the limited use of coal in this scenario, and the expectation that fertilizer use will be managed with the goal of limiting emissions of nitrous oxide in many regions, the buildup of N_2O is rather slow in Aggressive Response. The concentration of N_2O peaks in 2025, and by 2075 has declined to approximately the 1986 level. As a consequence, the commitment to future warming from N_2O buildup in Aggressive Response is very small, relative to the 1980 atmosphere.

Aggressive Response assumes substantial further reductions in the production and use of the most dangerous chlorofluorocarbons (CFCs) and halons. In fact, this scenario assumes that new agreements to supplement the Montreal Protocol on Substances that Deplete the Ozone Layer will result in the full and rapid phaseout of all such compounds. Furthermore, it assumes that the alternative compounds that perform the services now provided by CFCs will be limited to those with no global warming impacts and no ozone depleting effects. Aggressive Response assumes that the industrialized countries implement this total phaseout by 2020. By 2030, developing countries also reduce their use of these infrared-absorbing compounds to nearly zero.

As a consequence of these international control measures, the concentration of CFC-11 grows at about 3.4 percent per year until it peaks at 0.39 ppbv in 2005. Following the phaseout, concentrations decline slowly, falling to about 0.17 ppbv in 2075, about equal to the 1980 level. The situation is similar for CFC-12. Concentration increases at about 3.4 percent per year, peaking in 2005 and falling thereafter. By 2075, the concentration of CFC-12 in this scenario is approximately 0.44 ppbv, slightly less than three times the 1980 level.

The concentration of methane continues to grow in this scenario, but at an average rate of only 0.4 percent per year, significantly slower than historical rates. The slower rates of growth in this scenario are due to reductions in the rate of methane emissions from fossil fuel extraction, mobilization, and use; to reductions in emissions from inefficient biomass combustion; and to reductions in the rate at which carbon monoxide emissions destroy the atmospheric sink for methane. Methane concentration thus grows at only 0.5 percent per year from 1985 to 2025, or about half the historic rate of 1 percent per year, reaching an atmospheric level of about 1980 ppbv in 2025. Concentration grows at about 0.2 percent per year in the following 50 years, reaching approximately 2175 ppmv by 2075.

The combined effect of this greenhouse gas buildup is to increase the ultimate level of global surface temperatures. In Aggressive Response, the planet is committed, by 2025, to a warming of 1.5–4.0°C, when the effects of a 1–2.5°C warming already "in the pipeline" are taken into account. The interesting thing about this scenario, however, is that the warming effect levels off not much after the middle of the century. By 2045, the planet is committed to a warming of 1.6–4.3°C. But after this date, the combined effects of future trace gas emissions commit the planet to almost no further temperature increase. That is, sometime after 2050 (and probably before 2075), when the full effects of earlier emissions are observable, in this scenario the climate of the planet will stabilize with respect to the greenhouse effect at a level of about 1.6–4.4°C above the pre-industrial average. (See Figure 9.)

Conclusions

This exercise illustrates the possibility that the net costs of dramatically reducing the risks of rapid climate change may be small and that there could even be net economic benefits. By measuring the overall effects of such strategies on regional GNP it is possible to identify combinations of policies that could sustain the prospects for economic growth while minimizing the buildup of greenhouse gases.

Of course, the fact that such a scenario is technically and economically feasible does not suggest that it is politically practicable. Achieving the kinds of results suggested in the Aggressive Response scenario implies a level of political cooperation not presently visible in the world as we know it. Whether the recognition of the potential dangers of rapid climate change could stimulate such a change in political perceptions remains to be seen.

Notes and References

1. Intergovernmental Panel on Climate Change, 1990. *Scientific Assessment of Climate Change,* Report of Working Group I. World Meteorological Organization and United Nations Environment Programme, London, United Kingdom.
2. See, for example, Nordhaus, W.D., "To Slow or Not to Slow: The Economics of the Greenhouse Effect," paper given at

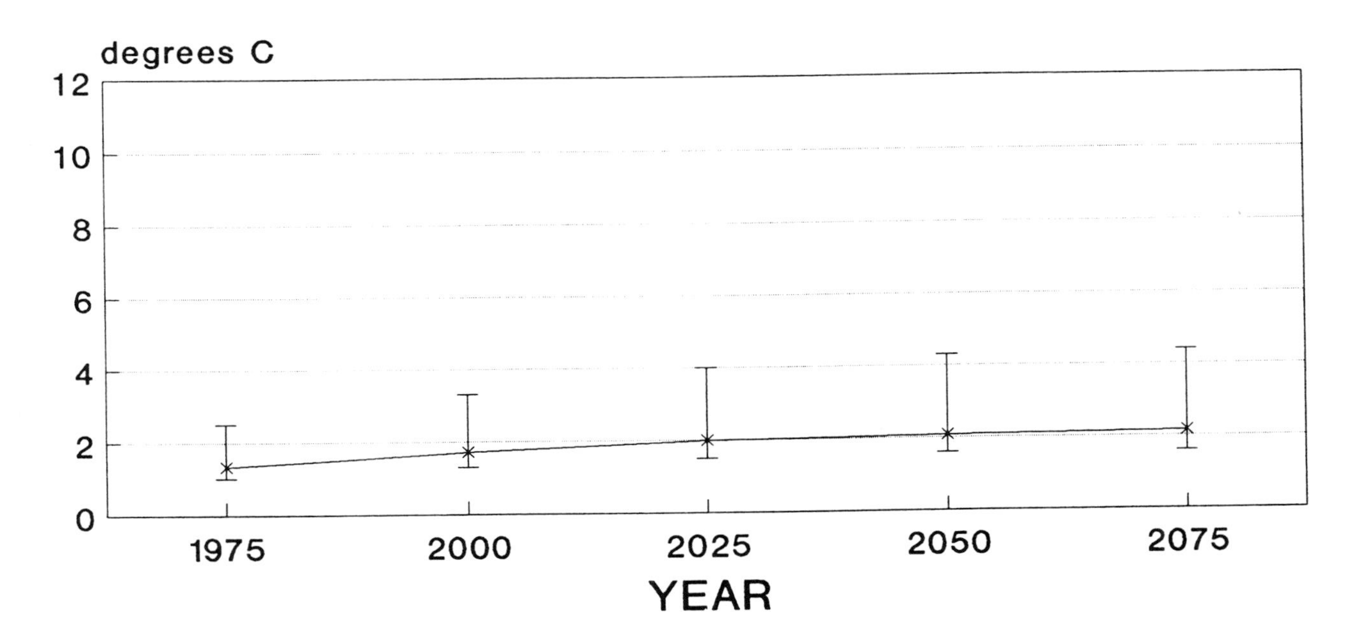

Figure 9. Warming commitment in the Aggressive-Response scenario (assumes 2C sensitivity to 2 × CO_2 + H_2O).

the Annual Meeting of the American Association for the Advancement of Science, Jan. 15–17, 1989, New Orleans, LA; Jorgenson, D.W. and Wilcoxen, P.J., "Energy Prices and US Economic Growth," paper prepared for the Panel on Policy Implications of Global Warming, Committee on Science, Engineering and Public Policy, National Research Council, Washington, DC; and Chandler, W.U. and Nicholls, A.K., "Assessing Carbon Emissions Control Strategies: A Carbon Tax or a Gasoline Tax?" American Council for an Energy Efficient Economy, Washington, DC.

3. Jorgenson, D.W. and Wilcoxen, P.W., Ibid.
4. Mintzer, I. and Moomaw, W., "Escaping the Heat Trap: Policy Options to Stabilize Future Global Temperatures," World Resources Institute, Washington, DC, in press.
5. Mintzer, I., *A Matter of Degrees: The Potential to Control the Greenhouse Effect,* World Resources Institute, Research Report no. 5, Washington, DC.
6. Edmonds, J. and Reilly, J. 1983. "A Long-Term Global Energy-Economic Model of Carbon Dioxide Release from Fossil Fuel Use," in *Energy Economics,* Volume 5, Number 2, pp. 74–88.
7. Vu, My T., 1986. *World Population Projections: 1980–2100,* World Bank, Washington, DC.
8. The historical rate of 2.2 percent per year is calculated from data provided in Table 20.1, p.306, of WRI, *World Resources Report, 1988–89,* Washington, DC, 1989. It is similar to the growth rate for 1975–1985 calculated from pp. 31–32, British Petroleum, *BP Statistical Review of World Energy,* June 1988. Approximately the same result can be obtained for the period 1975–1985 from the data in Table 102, p. 237, of US Department of Energy, *Annual Energy Review: 1987,* National Technical Information Service, Springfield, VA, May 1988.

Session F
Advanced Energy Supply Technologies

Summary
Session F—Advanced Energy Supply Technologies

Robert Duffield
Energy Laboratory
Ronald Davidson
Plasma Fusion Center
Session Chairs
Massachusetts Institute of Technology
Cambridge, MA

This session comprised papers dealing with the present status and future prospects of several technical approaches to delivering energy to the consumer. The approaches differ widely in the type of fuel used, in the method of converting the fuel energy into a useful final form, and in perceived environmental impact.

The substantial technical progress, remaining technical challenges, and arguments in favor of fusion were set forth in the paper "The Prospect for Fusion Energy in the 21st Century." Fusion requires the extraction of the deuterium fuel from water, the reaction of the deuterium in a hot confined plasma, releasing heat energy, and the conversion of that heat energy into electricity via thermal power plant technology, i.e., steam-driven turbine generators. The very large size of the fuel resource is one strong factor indicating that fusion energy should be pursued. The challenge has been and continues to be understanding and controlling the complex phenomena involved in heating and confining the hot plasma to cause the fusion reactions and produce a net quantity of heat. A strong argument in favor of fusion is the fact that the power-producing reaction does not emit any of the undesirable atmospheric pollutants that are produced by the combustion of fossil fuels (carbon dioxide, oxides of nitrogen, oxides of sulfur, etc.).

The utilization of solar energy was the subject of "Perspectives on Renewable Energy and the Environment." Here the supply of fuel exists, and the reactions that provide the source of energy take place in the sun, which is some 90 million miles away. Any noxious products of these reactions, therefore, do not affect the earth. The technical challenges are quite different from those in fusion, viz., how to interpret the diffuse flux of radiation and convert it to an attractive economical form. The present status of such technology was set forth in this paper; utilization of solar energy was also the subject of several papers in Session D.

The layers of the earth's crust become increasingly hot with distance from the surface. The paper "Hot Dry Rock Geothermal Energy—An Emerging Energy Resource with Large Worldwide Potential" argued that technology now available in the drilling industry can penetrate deeply enough into the earth's crust at selected locations so that rock at sufficiently high temperatures is accessible. The stored heat in the rock can then be extracted or mined and made available at the earth's surface. This energy can either be used directly as heat or be converted by developed power plant technology into electricity. The paper presented the results of experiments to accomplish this and made the case that the accessible resource is very large and that the probable environmental impacts are low. The fuel resource in this approach is the energy stored inside the earth, and no combustion process producing noxious effluents is involved in the exploitation of this stored energy. The technological challenge in this method of energy production is to locate, through geological prospecting, areas that are attractive as energy production sites because of their large geothermal gradient, satisfactory rock type, surface topography, etc., and then through drilling and fracturing to create a down-hole heat extraction reservoir so that fluid may be circulated in the system to deliver heat at the earth's surface at a useful temperature. The delivery of geothermal energy for useful purposes was also one of the subjects in Session D-4. There, papers by Ronald DiPippo and Carel Otte described the present success in the production of electricity from hydrothermal sources, both in the US and in developing countries, and the attractive costs of such energy.

The other two papers at this session, "Economic, Environmental, and Engineering Aspects of Magnetohydrodynamic Power Generation" and "Fuel Cells: Power for the Future," presented quite different technologies, with different scientific and engineering challenges. Each of these technologies converts the energy released by oxidation of a fuel into electrical energy by direct means, i.e., by electrical phenomena rather than by means of a heat engine as in a steam-driven turbine generator set. The fuels that have received considerable attention are coal for MHD generation, and natural gas or coal gas for fuel cells. In addition, MHD is envisioned as a topping cycle in conjunction with a heat engine in order to achieve a high overall thermal efficiency. The nature of the processes and the use of appropriate control techniques result in low levels of emission of undesirable effluents. The papers presented the current status of technology development and the opportunities for the future.

A general objective of this conference was to describe the present status of various energy production technologies, particularly with regard to environmental consequences, both good and bad, and to determine whether some degree of consensus might emerge as to what new actions should be taken to move toward lower levels of environmental degradation, while at the same time ensuring that a reliable and adequate economical supply of energy would be available on a worldwide basis. Decisions regarding the emphasis on various technologies should be based on the best scientific and engineering assessments that can be made for each of the competing energy supply systems. One difficulty in making informed technical judgments on the *relative* merits is that the particular scientific disciplines and engineering judgments appropriate to one energy production system, e.g., geothermal energy, are not particularly appropriate when applied to a competing energy production system, e.g., nuclear fusion. Important decisions as to national policy and financial support for research and development are made in the political arena. However, as eloquently stated by Senator Gore at the beginning of the Conference, the elected and appointed, predominantly nontechnical persons involved in the decision-making process should be able to rely on informed, balanced technical assessments by the scientific and technical community.

The Prospect for Fusion Energy in the 21st Century

D. Bruce Montgomery
Plasma Fusion Center
Massachusetts Institute of Technology
Cambridge, MA

Abstract

Magnetic fusion has been pursued as a long range energy option since the 1950s. Experimental devices have now reached the point where substantial neutron energy (10 MW) can be produced for times (10 seconds) limited only by heating and energy supply limitations for the resistive magnets. It is now possible to design an engineering test reactor and to have it in operation early in the next century. Environmental impact studies detailing the potential for reduced risk associated with fusion reactors have been published. Fusion reactors will need to deal with activated structures and with the potential release of tritium, but inventories and public exposure risks can be several orders of magnitude less than possible with fission reactors. A recent report of the National Research Council has recommended that the funding for fusion programs be increased, and that the US join in an international program aimed at providing the insurance of a viable fusion option for the 21st century.

Introduction

We quote from the introduction to the recently published "Report of the Senior Committee on the Environmental, Safety, and Economic Aspects of Fusion Energy," referred to as the ESECOM Report (Holdren, et al., 1989):

> In many ways, fusion is the ultimate energy source: It is the fire that lights the sun and the rest of the stars; it is the source of the elements that constitute the universe; it can wring from a gram of deuterium fuel as much energy as the combustion of 10 tons of coal, enabling the trace of deuterium in each gallon of sea water to provide as much energy as 300 gallons of gasoline. It is also so difficult to harness for meeting civilization's energy needs that more than three decades of intensive scientific and engineering effort have not sufficed to accomplish the task.

Though the task has not been completed, substantial progress has been made. For example, in the leading approach to fusion energy, magnetic confinement, modern tokamaks produce substantial quantities of fusion neutrons for times approaching 1 second. The largest devices, the Tokamak Fusion Test Reactor (TFTR) in the US, and the Joint European Tokamak (JET) in Europe, have achieved conditions close to "scientific break even" (when the plasma would make as much fusion energy as it takes to heat the plasma). Were TFTR and JET operated with a 50–50 mixture of deuterium and tritium fuel rather than deuterium alone, they would have produced ~ 10 MW of fusion power for several seconds.

The task is far from complete, however. It is not enough that fusion systems simply work. Again quoting from the ESECOM introduction:

> The scientists and engineers working to develop fusion energy have long considered the attractions of achieving their goal to be self-evident: the attainment of an energy that would be at once safe, clean, affordable, and inexhaustible. But safety, cleanliness, affordability, and inexhaustibility are all relative attributes, not absolutes. The yardsticks by which these attributes are measured, and society's weighing of their importance relative to one another depends on the context: the time frame of interest, the projected energy demand in the time frame, the characteristics of all energy sources potentially available to meet those demands, and the priorities given to economic vs. environmental and safety concerns.

The ESECOM report, some 350 pages in length, represents a several-year-long examination of the potential for magnetic confinement fusion by a senior group of individuals with fusion and fission backgrounds. The report, from which we will borrow in more detail in later sections, concludes, in brief:

> Our analysis indicates that magnetic fusion energy systems have the potential to achieve costs

of electricity comparable to those of present and future fission systems, coupled with significant safety and environmental advantages.

The most important potential advantages of fusion with respect to safety and the environment are:

1. High demonstrability of adequate public protection from reactor accidents (no early fatalities off site), based entirely or largely on low radioactivity inventories and passive barriers to release rather than on active safety systems and performance of containment buildings;
2. Substantial amelioration of the radioactive waste problem by eliminating or greatly reducing the high-level waste category that requires deep geological disposal;
3. Diminution of some important links with nuclear weaponry (easier safeguards against clandestine use of energy facilities to produce fissile materials and no inherent production or circulation of fissile materials subject to diversion or theft.)

These advantages are potentially large enough to make a difference in public acceptability of magnetic fusion energy, as compared with fission.

Approaches for Achieving Controlled Fusion Energy

Alternate Approaches

There are two approaches to controlled fusion that receive substantial funding in the United States: magnetic confinement and inertial confinement. The former is supported by the Department of Energy with the stated goal of establishing "the scientific and technological base for fusion energy early in the next century, on a schedule which would allow a timely assessment of the role of fusion energy in the nation's energy future (Decker, 1988)." Inertial fusion has been entirely supported by the defense programs within the Department of Energy over the last two decades, on the basis of its potential role in weapons simulation. It is now funded at approximately half the level of magnetic confinement. In Japan and Europe, laser fusion is primarily an energy research program and is funded at less than 10 percent of the level of magnetic confinement.

In the inertial confinement approach to fusion, a frozen pellet of fusion fuel is compressed and heated by bombarding it with laser or particle beams called drivers. The pellet is compressed to a density some 10 billion times the density of the diffuse plasmas in the magnetically confined approach, and the necessary confinement time to reach fusion temperatures is therefore short enough for fusion to take place before the pellet blows apart (OTA-E-338, 1987). In a hypothetical inertial confinement reactor, micro-explosions of pellets would be produced several times per second.

Quoting from the 1987 Office of Technology Assessment study of fusion energy (OTA-E-338, 1987):

> The technical requirements for commercial applications of inertial confinement go considerably beyond the requirements for weapons effect simulation. Due to the relatively low efficiencies (e.g., 10 to 25 percent) at which the drivers operate, each target explosion must generate several times more energy than it is driven with to reach break even. An additional factor of four to 10 is required beyond break even to produce substantial net output. In a commercial reactor, therefore, as much as 100 times as much energy must be released in a pellet explosion as is required to heat and compress the pellet to the point where it can react.
>
> A significant potential advantage of inertial confinement over magnetic confinement is that the complex and expensive driver system can be located some distance away from the reaction chamber, where the effects of the radiation, neutron-induced activation, and micro-explosion thermally induced stresses would be confined.
>
> On the other hand, inertial confinement also has disadvantages compared to magnetic fusion. Inertial confinement is inherently pulsed; the systems needed to recover energy and breed fuel in the reaction chamber have to withstand explosions equivalent to a few hundred pounds of TNT several times a second. Inertial confinement reactors must focus high-power driver beams precisely on moving targets in this environment.

Relatively little study has been given to the potential of inertial fusion for commercial power generation although several reactor studies have been carried out, as typified by the HiBall reactor study by the University of Wisconsin (1984). It is the intention of the Department of Energy to revisit the energy application of inertial fusion, and a reactor study will be undertaken during 1990/91. Inertial fusion reactors as well as magnetic confinement reactors are discussed in a very recent paper by Conn et al., "Fusion Reactor Economic, Safety, and Environmental Prospects," (Conn, 1990). We will not address further the inertial fusion approach in this paper, and will address only magnetic confinement in detail.

The Magnetic Confinement Approach

The world fusion programs presently support three toroidal approaches to magnetic confinement—the tokamak, the stellarator, and the reversed field pinch (RFP). The great majority of funding goes to the tokamaks, with a smaller amount to the stellarators, and a much smaller fraction to the RFPs. These devices are all toroidal, and are referred to as "closed confinement systems." The mirror confinement devices, which are "open systems," are no longer funded at the level they were, but an effort comparable to RFP research is still carried out in Japan and the Soviet Union.

The tokamak devices, exported from the Soviet Union in the late 1960s, dominate confinement research. There are currently more than 12 major tokamaks in the world program, and perhaps twice that number of smaller scale devices. In the tokamak, the confining field is produced by combining the field from the toroidal magnets that surround the torus and the field from the current flowing in the plasma. In most tokamaks the plasma current is induced by transformer action from external coils, resulting in a finite pulse time set by the volt-seconds available in the transformer. It has been demonstrated, however, that the current can be maintained at a steady-state level by the application of external particle beam or phased radio frequency sources. These "current drive" mechanisms, unfortunately, appear limited to relatively low efficiencies unless very high fractions of self-generated "bootstrap" current can be generated.

In present day tokamaks, the plasma is heated by a combination of the resistive dissipation from the current flowing in the plasma and from heat input from external sources, such as neutral particle beams or rf waves. In a tokamak reactor, the principal heating would come from the self-heating of the released fusion alpha particles. External sources would be used only for burn control, or to drive current.

The stellarator differs from the tokamak in that the confining fields are all produced by external windings, eliminating the requirement for current within the plasma. Elimination of the plasma current leads more naturally to a steady-state device, and eliminates the electromagnetic stresses that accompany plasma current disruptions in tokamaks. Stellarators, while an earlier concept, are a generation behind the larger tokamaks. Nevertheless their performance in comparable scale devices approaches that of tokamaks.

The reverse field pinches (RFPs) are tokamak-like in that they utilize induced plasma currents to produce part of the confinement field. In RFPs, however, the induced currents play a more prominent role, allowing the device to have rather modest toroidal fields, and to exhibit strong heating from the relatively larger plasma currents. These attributes may lead to more compact and lower capital cost reactors. Progress in this line is well behind that of the tokamaks and stellarators.

All three confinement devices could in principle provide the "fusion core" of a fusion reactor. The emitted neutrons would pass through the torus first wall and be converted to heat in a surrounding blanket (Conn, 1983). In a reactor operating on deuterium/tritium fuel (DT) (the reaction with the highest nuclear cross section and hence requiring the lowest plasma density and temperature), the blanket would also be required to breed tritium for recycling. The reactor would be fueled by continuous injection of frozen pellets of DT, injected at velocities of a few kilometers/second to assure adequate penetration.

Current Status of Magnetic Confinement Research

The current status of magnetic confinement research can be described by discussing some of the key experimental results from several of the large tokamak experiments—the JET, the Japanese Tokamak Model 60 (JT-60), TFTR, DIII-D [a tokamak fusion device at General Atomic Company in California], and ALCATOR C. This collection of experiments has demonstrated each of the parameters required for a magnetic fusion reactor:

- density $> 3 \times 10^{14}$ cm^{-3} was demonstrated by ALCATOR C;
- confinement time > 1 s demonstrated by JET;
- average ion temperature > 10 keV demonstrated by TFTR;
- beta $\sim 10\%$ demonstrated by DIII-D;
- noninductive current drive at the MA level demonstrated by JT-60

The total of the results from these large tokamaks provided a solid basis for the design of the next generation of tokamaks.

As an example of one of these machines, the major parameters of the JET machine are given in Table 1, and the machine cutaway drawing in Figure 1. The device was commissioned in 1983, and has completed more than 20,000 high performance discharges. The machine has been operated to date only in deuterium gas to limit the nuclear activation of the internal components, thereby allowing personnel access to the chamber interior for component replacement and upgrade. It is planned to operate JET in DT toward the end of the next five-year experimental plan, and the team has already developed and tested a full set of remote maintenance tools for that phase of operation.

In fusion devices, typified by JET, the vessel wall is lined with protective tiles. Among other functions, the tiles allow the designers to choose materials that will

Table 1. JET device parameters.

Major radius	2.96 m
Minor radius (horizontal)	1.25 m
Minor radius (vertical)	2.10 m
Plasma current	7.0 MA
Flattop pulse length	20.0 s
Toroidal field strength	3.45 tesla
Weight of the vacuum vessel	100 tonnes
Weight of the toroidal field coils	385 tonnes
Weight of the transformer core	2800 tonnes
Toroidal field coil power	380 MW
Neutral beam power	20 MW
RF input power	20 MW

minimize the effect of wall impurities on the plasma. Tiles also protect the vacuum wall from the effects of a loss of plasma control, called a disruption. JET has recently improved their plasma performance by a factor of two by coating the existing carbon tiles with beryllium.

The most recent results of TFTR and JET are summarized in Table 2 (Bell, 1988; JET Results, 1989), and are put into the context of previous results in Figure 2 (Fusion Power Associates, 1989). The plasma energy confinement time (the average time over which the energy content of the plasma leaks out and is replaced by the auxiliary heating) exceeded one second for the first time. The plasma is held in a stable condition for 10 seconds, limited by heating in the magnets and the depletion of energy stored in the magnet power supplies. The temperature and confinement time achieved are considered adequate for operating a fusion reactor; however, the density falls short by a factor of 10.

These recent results in JET, coupled with nearly comparable results in the US TFTR tokamak, have led the JET team leader to conclude "that for all practical purposes, break even conditions have been demonstrated and the principal objective of establishing the scientific feasibility of nuclear fusion as an energy source has been demonstrated. We now have to concentrate on reducing impurities and controlling the plasma fueling for long enough for a reactor (JET Press Conference, 1989)."

JET plans an extension of their present deuterium operating phase to explore these reactor related issues. To control impurities they will install a new magnetic coil within the vacuum chamber to create a well-defined magnetic divertor configuration, which will employ steady-state cooled beryllium tiles. At the completion of this next engineering phase, JET will operate with DT fuel starting in 1995. TFTR plans to add additional heating power before operating with DT fuel starting in 1993.

Magnetic Fusion Technology Development

A sense of the technology status can be gained from the success of the JET systems, but additional technologies are required for a magnetic confinement reactor that do not appear on JET.

Prominent among these are the superconducting magnet systems that would be required in a reactor to achieve a net power balance. Several smaller tokamaks have been constructed utilizing superconducting magnets, the largest in operation being the Tore Supra in France (Komarek, 1989). Large coil test arrays as well as the large MFTF-B magnetic mirror device have also been built from superconductors. These devices are approaching the scale necessary for a reactor, but have been designed to operate at magnetic fields somewhat below those now thought to be necessary for a reactor.

Another technology not represented on JET is the blanket elements that would be necessary for power conversion and tritium breeding. Only small-scale experimental work has been carried out in this area utilizing test canisters in fission reactors (Abdou, M., 1985).

Perhaps the most important technology missing from JET in terms of the future of fusion's acceptability is the use of conventional steels rather than low activation structural materials. Low activation materials such as vanadium or silicon-carbide remain largely laboratory-scale materials (DOE/ER-031/6, 1989). The use of fiber-reinforced silicon-carbide, is growing, however, through industrial applications in aerospace and high-temperature heat exchangers.

Remaining Uncertainties

Despite the dramatic progress in achieving plasma performance, as illustrated in Figure 2, the ability to confidently predict the performance of future machines by extrapolation from present machines remains elusive. The trends are obvious, of course, e.g., greater size, higher magnetic fields, and plasma currents, but precisely how much of each, or in what combination, is necessary to meet the requirements for a reactor remains to be determined. The focus of this uncertainty, however, is not over whether reactor conditions can be achieved, but rather how large and potentially expensive such devices might have to be. We will note this impact when we discuss economics.

A major initiative has been undertaken in 1990 to improve the theoretical and experimental database in existing machines in order to improve our ability to extrapolate to the next generation of machines, which already will be prototypical reactors.

Steady-state operation remains an important con-

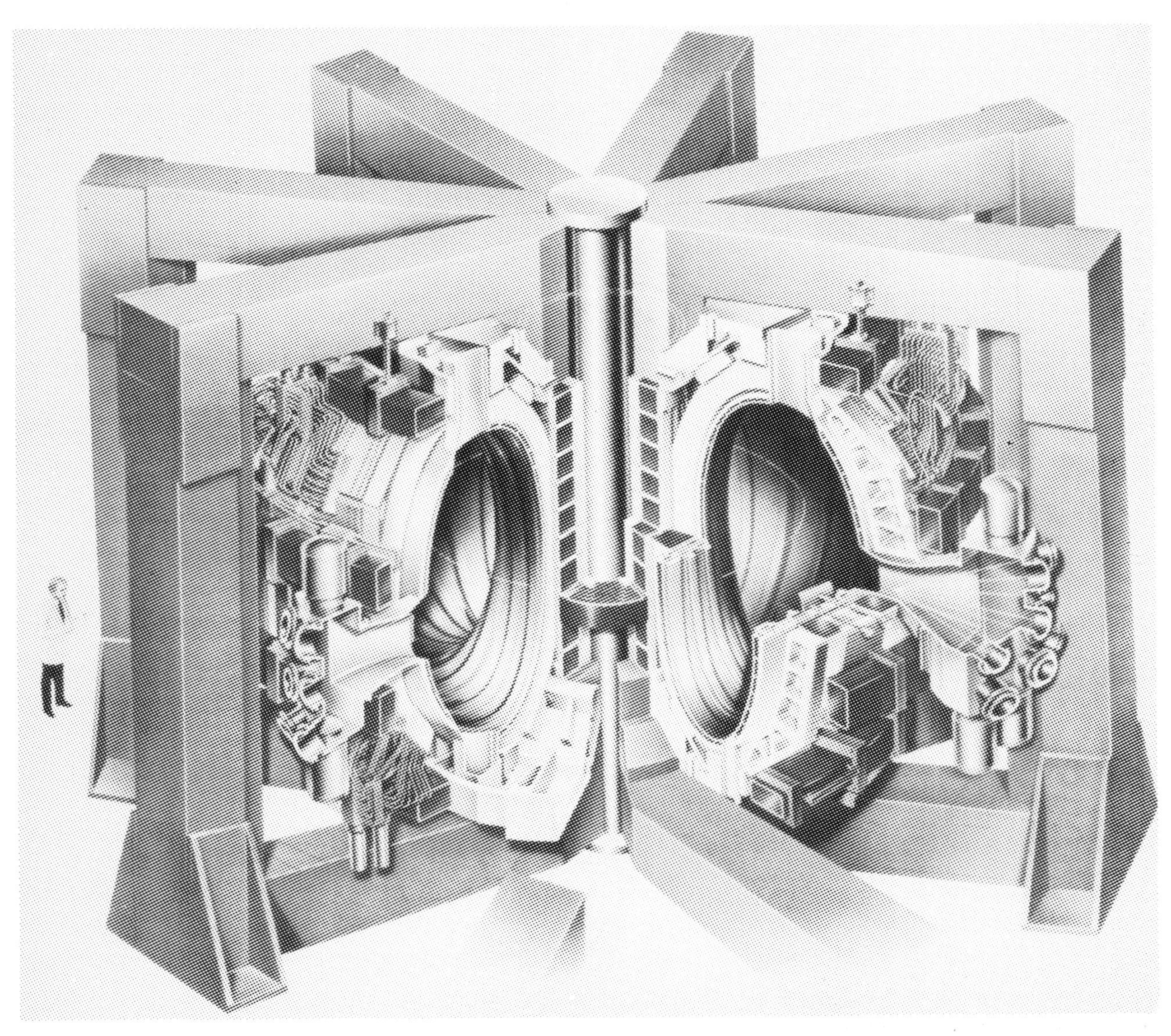

Figure 1. Overview of the joint European torus (JET) tokamak, commissioned in 1983 in the United Kingdom.

Table 2. JET/TFTR performance parameters.

	JET	TFTR
Plasma Temperature (keV)	22	27
Plasma Ion Density (10^{13} cm^{-3})	3.7	6.7
Energy Confinement Time (s)	1.0	0.2
Projected Q_{DT}	0.7–0.8	0.5

Q_{DT} = projected fusion energy release in DT/plasma heat loss

cern. Fusion reactor studies have always shown the desirability of steady-state, as opposed to pulse operation. A pulsed device must include the complication of a thermal storage mechanism, must be designed for a very large number of thermal and stress cycles, and pay a reliability penalty for systems that must repeatedly cycle. A conventional tokamak reactor without external current drive, for example, would need to bring the plasma current to zero for several minutes approximately once per hour, in order to recharge the transformer.

The use of an external current drive has been demonstrated on a number of tokamaks, but the projected efficiency of the process is in the region of 20 percent,

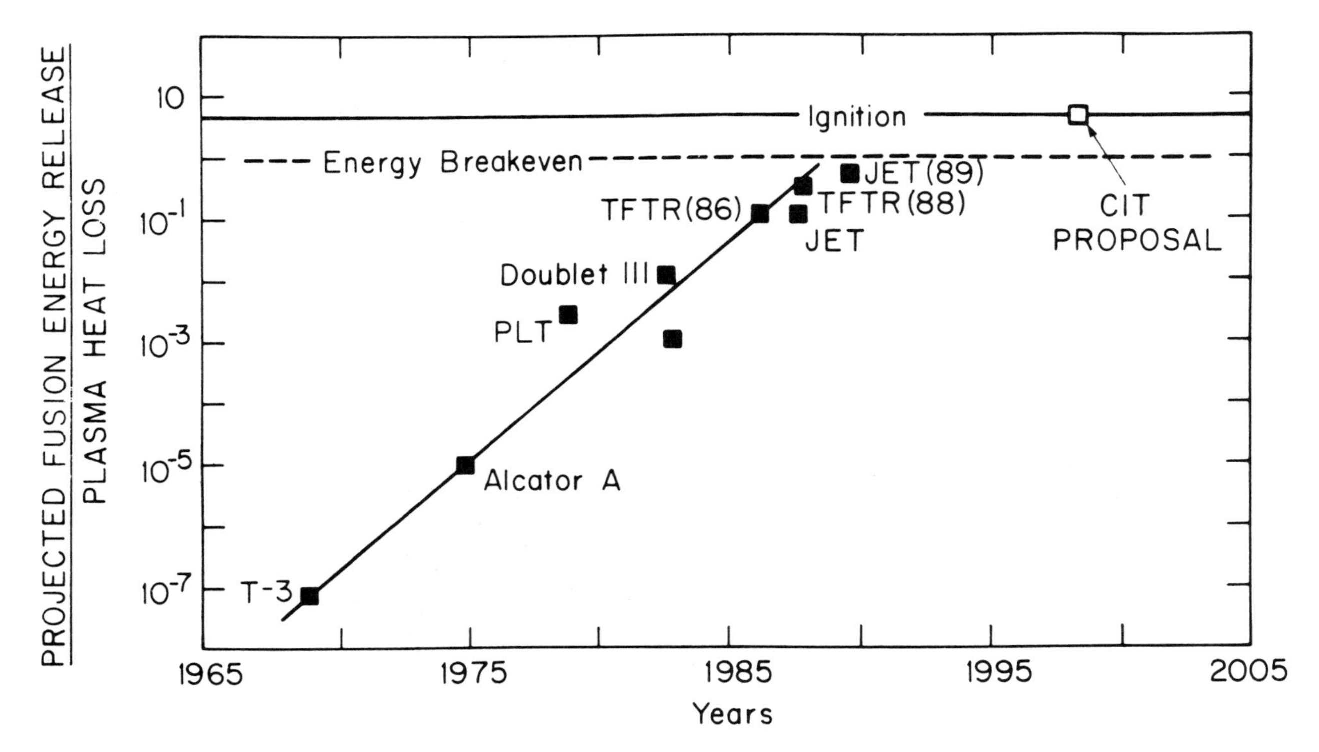

Figure 2. Progress in the tokamak devices in achieving the conditions required for fusion power.

projecting a requirement for several hundred megawatts of recirculating power. The ability to drive current at higher efficiencies remains an important uncertainty for the tokamak.

A stellarator reactor would eliminate the requirement for current drive, but carries its own set of uncertainties. For one, the complex helical windings required in stellarators complicate the reactor design and maintenance. For another, stellarator performance lags well behind the current large tokamaks, making the extrapolation to reactor scale even more uncertain. Nevertheless, stellarator research is supported largely because of the desirability of currentless reactor operation.

Reactor relevant materials performance remains a major uncertainty for the future. At the most fundamental level, materials in the first tens of centimeters of the first wall/blanket assembly must retain their performance under neutron fluxes of several MW/m^2, at a substantially different neutron energy spectrum (14 MeV) than in fission reactors. This difference in spectrum makes performance prediction uncertain, and has given rise to the future research and development requirement for an accelerator-based neutron source for materials testing. The uncertainties are further compounded by the desirability of developing low activation materials, for which only a small database currently exists.

Successful development of low activation materials is required if fusion is to realize a substantial advantage over competing fission options (Holdren et al., 1989).

Materials uncertainties also exist for first wall and divertor elements, where heat fluxes of five to 10 MW/m^2 must be sustained in steady-state cooled tile assemblies subject to erosion and sputtering from the plasma. This uncertainty is driving the next phase of the JET experimental program.

Finally there remains a pervasive and important question, namely: Can machines of the apparent complexity of fusion devices obtain sufficiently reliable operating levels and repairability to achieve the required reactor availability? Substantial improvements in systems availability over that of existing devices, impressive as that has become (Parasells and Howard, 1987), will be required.

Economic Considerations

The ESECOM study (Holdren et al., 1989) represents the most complete look yet taken at potential fusion economics. As stated in the introduction to that report:

> Our analysis indicates that magnetic fusion systems have the potential to achieve costs of electricity comparable to those of present and fu-

ture fission systems, coupled with significant safety and environmental advantages. This conclusion is based on (a) assumptions about plasma performance and engineering characteristics that are optimistic but defensible extrapolations from current experience, and (b) consistent application of an elaborate set of engineering/economic and safety/environmental models to a range of fusion and fission reference cases, with the known characteristics of fission light-water reactors as a bench mark.

Typical illustrations from the ESECOM report are repeated in Figures 3, 4, and 5. Figure 3 illustrates the potential cost impact of various performance assumptions for tokamaks, and includes a base case representing conventional tokamak operation and three advanced "second stability regime" (SSR) tokamaks, a mode of operation yet to be demonstrated. A compact reverse field pinch is also illustrated.

Figure 4 illustrates the impact of several more assumptions, and we repeat it only to illustrate the impact of including advanced engineering features and potential inherent safety credits.

Figure 5 illustrates the dependence of COE on the plant scale for two fusion devices, and compares them with PWR power costs. The two PWR costs represent the best present experience (BPE) and the median experience (ME). By and large, all the fusion extrapolations examined by ESECOM fall between these two PWR cases.

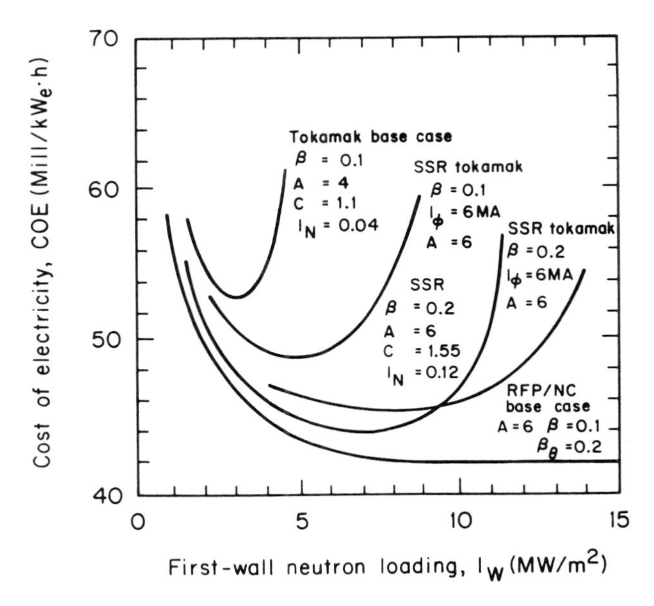

Figure 3. Dependence of the cost of electricity on neutron wall loading for several toroidal approaches: tokamak base case (first stability regime), three advanced SSR tokamaks (second stability regime), and a high power density reverse field pinch (RFP) [from ESECOM, 1989].

Commenting on a preliminary version of the ESECOM report available in 1987, the Office of Technology Assessment stated (OTA-E-338, 1987):

> Given all the uncertainties, OTA finds that the economic evidence to date concerning fusion's cost effectiveness is inconclusive. No factors identified in the fusion research program conclusively demonstrate that fusion will be either much more or much less expensive than possible competitors, including nuclear fission. Fusion appears to have the potential to be economically competitive, but making reliable cost comparisons will require additional technical research and a better understanding of nontechnical factors, such as ease of licensing and construction, that can have a profound influence on the bottom line.

Environmental Considerations

Fusion, like fission, has the obvious environmental benefit of eliminating the emission of hydrocarbons into the atmosphere. The importance of this benefit will be the center of climate-change debate for some time, but the trend in thinking in the developed economies is certainly toward the wisdom of reducing hydrocarbon emissions as an insurance policy. This would favor nuclear technologies, fission, and in the longer term, fusion.

The assessment of relative risk and severity of accident is an environmental issue which can potentially differentiate fusion and fission. The OTA study of fusion energy (OTA-E-338, 1987) comments that assuring public safety with fusion reactors should be easier than with fission because:

> Fusion reactors cannot sustain runaway reactions. Fuel will be continuously injected, and the amount contained inside the reactor vessel at any given time would only operate the reactor for a matter of seconds.
>
> Fusion reactors should require simpler post-shutdown or emergency cooling systems than fission reactors, if such systems are needed at all. With appropriate materials choices, afterheat from fusion reactors (activated components) should be much smaller than from fission reactors (fuel elements).
>
> Potential accidents that could occur in fusion reactors should be less serious than those that could take place in fission reactors. With suitable materials choices, the radioactive inventory of a fusion reactor should be considerably less hazardous than that of a fission re-

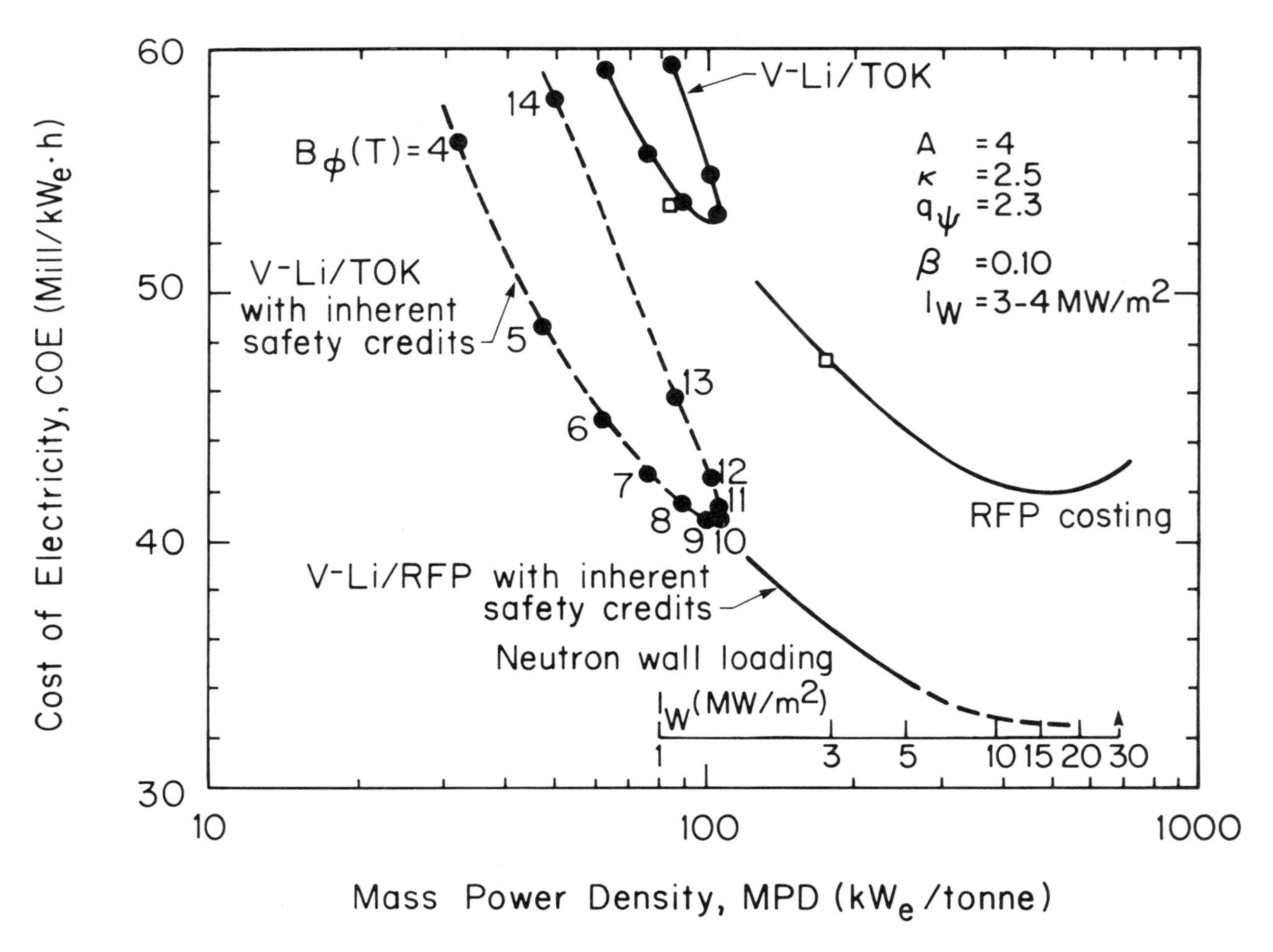

Figure 4. Economic impact of postulated full safety-assurance credits and advanced superconducting magnets [simplified from ESECOM, 1989].

actor. Moreover, the radioactive materials in a fusion reactor would generally be less likely to be released in an accident than would those in a fission reactor, since they would be largely bound in the structural elements. The only volatile or biologically active radioactive component in a fusion reactor would be the active tritium inventory; gaseous and volatile radioactive products in a fission reactor would be present in amounts orders of magnitude greater.

The proper choice of blanket and structural materials in a fusion reactor is important to maintain a significant advantage in this environmental area. Figure 6, reprinted from Conn (1983), illustrates the comparison of two fusion designs with a liquid metal fast breeder reactor (LMFBR). If ferric steels are used, as they were in the STARFIRE reference reactor, the fusion advantages are not dramatic until 10 years after shutdown, when the shorter half-lives of the steels relative to fission fuels show up. On the other hand, if low activation materials are used, the advantages are dramatic and immediate.

Directly related to the above discussion of inventory is the issue of long-term disposal of radioactive waste. The first wall of the fusion reactor might be expected to require replacement every 10 years, and at end of life all the core elements will require disposal. The total volume of activated material will be considerably greater than the accumulated spent fuel and activated reactor components of a fission reactor. However, as stated in the OTA report (OTA-E-338, 1987), "the ESECOM report concluded that although fusion wastes may have greater volume than fission wastes, they will be of shorter half-life and intensity and should be orders of magnitude less hazardous."

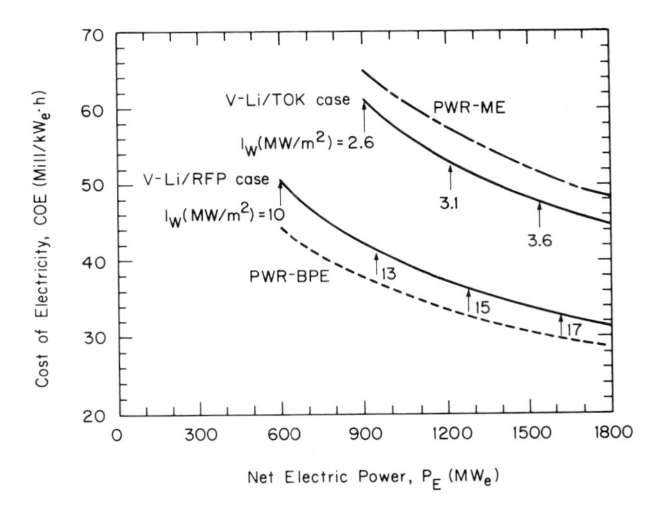

Figure 5. Dependence of cost of electricity on net electric power for a tokamak and an RFP, compared with the best performance experience and median experience of PWRs [from ESECOM, 1989].

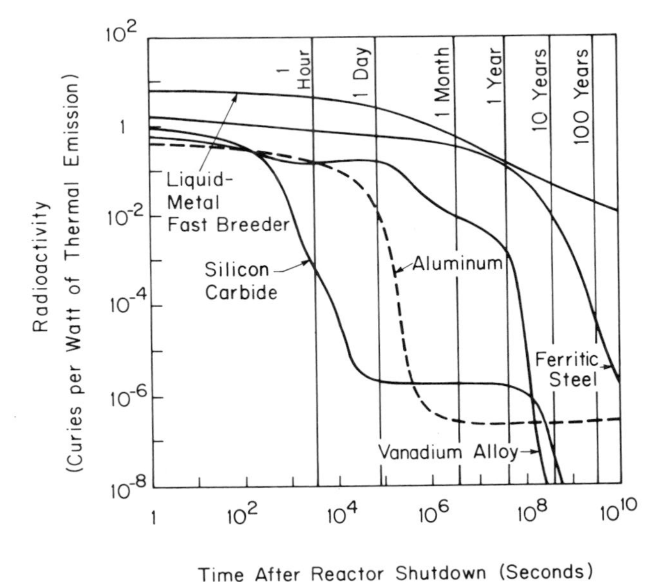

Figure 6. Decline of induced radioactivity in potential materials for a fusion reactor compared with radioactive waste from a liquid-metal fast breeder fission reactor [from Conn, 1983].

Future Research and Development Program Recommendations

The fusion community published a major technical planning activity in 1987 (TPA, 1987) to carry the program through an engineering test reactor stage that would establish the scientific and technological base required to assess the economic and environmental aspects of fusion power. The TPA planning exercise was constrained by logic, but not by budget considerations, and led to an assessment in the year 2005. If that assessment were positive, the program would then move forward into a demonstration plant of the 400–500 MW class, presumably located at an electric utility site.

The 1987 planning exercise has been revisited in 1989 in connection with a draft position paper [Brolin, 1989] prepared for input to a national fusion program planning exercise. Quoting from that position paper:

> The key elements of the proposed program are:
>
> - A multi-purpose technology R&D program sufficient to support:
>   - Reactor materials development, including radiation-resistant and low activation materials;
>   - A theoretical and experimental program adequate to develop a physics-based transport scaling relationship and a database sufficient to project performance to the tokamak ignition demonstration. It is assumed that existing experimental devices would be sufficient for this purpose.
>   - Advanced tokamak and advanced alternate concepts in parallel with the mainline tokamak developments, e.g., follow-on devices to the existing Advanced Tokamak Facility (ATF), Plasma Beam Experiment Modification (PBXM), and ZT-H [the ZT-H is a reverse field pinch fusion device at Los Alamos National Laboratory] devices.
> - A fusion irradiation materials test facility (FITF), necessary because of the extensive development work required to obtain reactor materials that are both technically and environmentally acceptable. This could be an international facility because of the benefit to ITER, DEMO and future world fusion reactors but has been assumed in the proposed program to be funded solely by the US.
> - A burning plasma experiment (CIT), operating before the year 2000. (The device would be expected to generate 100–200 MW of fusion power.) The burning plasma demonstration would provide operational guidance and R&D input to ITER, the next phase in the

world program. It would be considered part of the US ITER contribution.

- An international thermonuclear engineering test reactor (ITER) operating by the year 2005 with the following capabilities: Ignition, Thermal (steam) power generation equivalent to 50–100 MWe, Engineering and Physics R&D.
  - The design and R&D phase of ITER would directly influence design decisions made for DEMO, and ITER operating experience, influence DEMO operations.
  - ITER is considered an international project, with international collaboration desirable because of the large investment both financially and in the scientific and engineering talent and facilities required to ensure success.
- A fusion reactor demonstration plant (DEMO) of 400–500 MWe capability located at a utility site and operating by the year 2014. Commercial fusion reactor licensing requirements would be applied and would be developed in conjunction with the NRC and international bodies. DEMO is assumed to be an international collaboration.

The CIT burning plasma experiment and the ITER (International Thermonuclear Engineering Reactor) are the major centerpieces of the National and International Fusion programs, and both have received concentrated design attention over the last three years. Their parameters are given in Table 3.

The CIT, illustrated in Figure 7, is a relatively short pulse machine (five to 10 seconds), with the pulse duration limited by magnet heating (Jardin, 1989). The device utilizes copper conductor coils, and has been designed to be located at an existing US fusion site, utilizing existing power systems in order to achieve burning plasma conditions at the minimum cost. The burning core would generate 100–200 MW of fusion power during the duration of the pulse.

The ITER, illustrated in Figure 8, is designed as a steady-state machine utilizing external current drive and superconducting magnets (International Atomic Energy Agency, Vienna, 1989). The design work has been carried out by an international team based at Garching, set up under the aegis of IAEA following a joint declaration by Reagan and Gorbachev at the Geneva Summit. The conceptual design phase is nearing completion, and a quadripartite five-year engineering design and research and development phase is proposed.

Toward the end of that period, the four partners (the US, the European Community, Japan, and the Soviet Union) would decide whether to move forward into a construction phase.

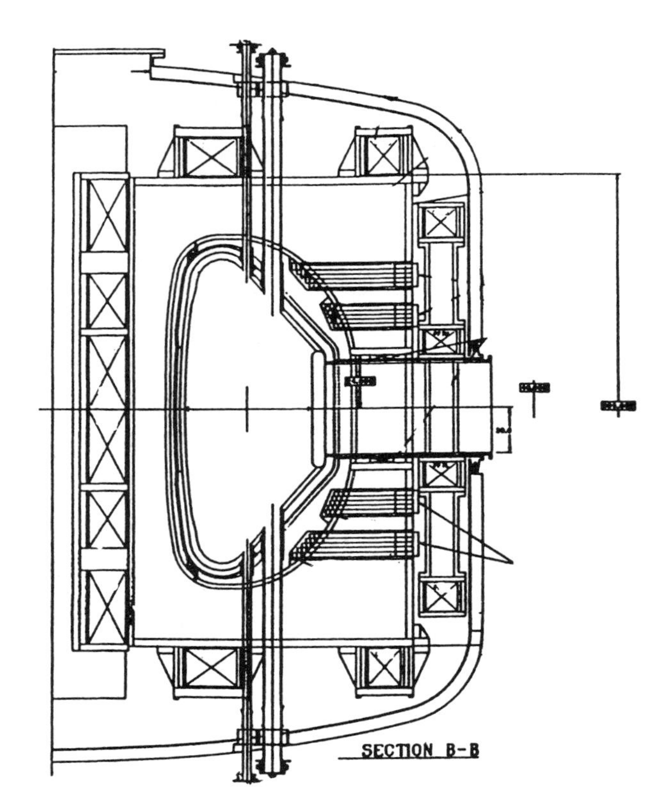

Figure 7. Elevation view of the proposed compact ignition tokamak (CIT). The 2.1 m major radius device is somewhat smaller than current large tokamaks, but achieves higher performance because of the higher magnetic field.

Table 3. Parameters of the CIT and ITER tokamak devices.

	CIT	ITER
Major radius	2.14 m	6.0 m
Minor radius	0.66 m	2.15 m
Plasma current	12.0 MA	22.0 MA
Magnetic field on axis	11.0 T	4.85 T
Peak TF magnetic field	19.2 T	11.1 T
Flattop time	5 s	steady-state
Coil material	copper	superconductor

Pacing the US Magnetic Fusion Program

A possible scheduling of the program elements discussed above is indicated in Figure 9 (Brolin, 1989). As in the 1987 TPA plan, this schedule is logic constrained, and would require a change in the current level of funding. Quoting from the position paper:

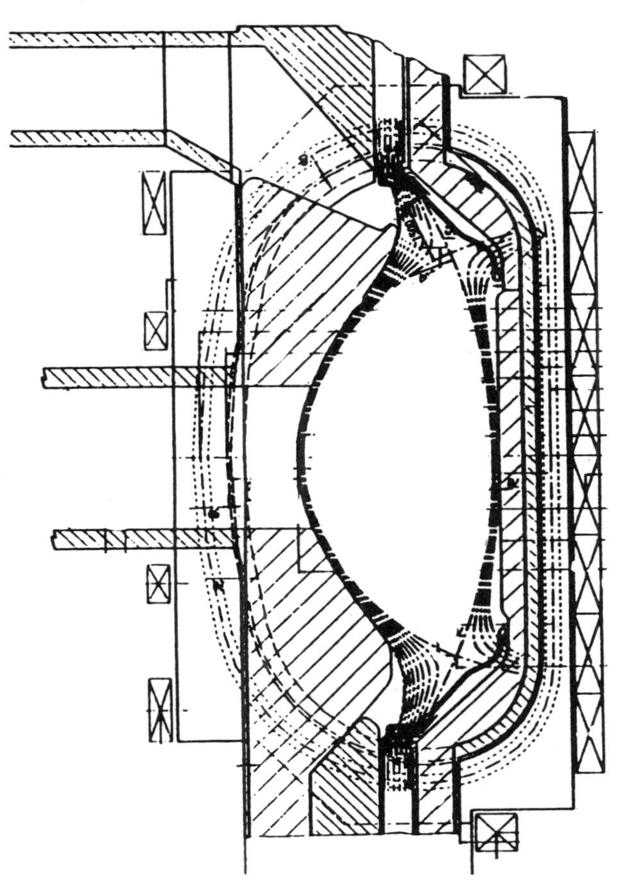

Figure 8. Elevation view of the proposed international thermonuclear engineering reactor (ITER). The 6 m major radius device would use superconducting magnet systems, and be fitted with net energy producing blanket modules.

The program has been structured to determine in a relatively short time whether or not magnetic confinement fusion reactors can contribute to the world energy supply. An accelerated program is considered optimal because of:

- the world need for an unlimited, environmentally benign energy source
- the cost savings that would result from moving into an application phase instead of unnecessarily prolonging basic research
- the need to have a program which moves the intellectual resources from one project to another with minimum disruption and loss of knowledge

The cost to the US of such a program (assuming that ITER and DEMO are both done internationally) would require an increase in the current level of the fusion program by approximately 50 percent.

The rationale for an increase in the current fusion budget has also been laid out by a panel of the National Research Council of the National Academy of Engineering (NRC of NAE) in a report called "Pacing the US Magnetic Fusion Program," published in 1989 (White et al., 1989). That panel was chaired by Irvin L. White, of the New York State Energy Research and Development Authority. We quote from the committee recommendations:

- The committee recommends that the federal government develop a broad, balanced array of technological alternatives as an insurance strategy for meeting US and global long-term energy needs. Decisions about the government's role, which alternatives to support, and funding levels should be based on the results of a comprehensive assessment. In the interim, the committee offers the three recommendations below as to the priority, pace, and direction of the US magnetic fusion program.
- The committee recommends that the United States enter into partnership arrangements for the next major steps in an international fusion program as the most cost-beneficial US approach to fusion over the next decade.
- The committee recommends an increase over current fusion program funding of about 20 percent, held steady for the next five years, followed in the mid- to late-1990s by an additional increment of about 25 percent. These increments are expected to permit construction and operation of the Compact Ignition Tokamak (CIT), resolution by the early 1990s of the central scientific feasibility question, and participation in the construction of an international engineering test reactor beginning in the mid- to late-1990s.
- The committee recommends that the Department of Energy develop a revised plan providing greater participation by US companies to ensure that those companies are not placed at a disadvantage relative to foreign competitors. Such participation could, for example, follow the Japanese practice of assigning industry responsibility for both the design and construction of major systems and subsystems.

The NRC/NAE committee also addressed the consequences of a reduction in the US fusion budget:

> Given the constraints already imposed by the fusion program's current budget, a decrease in funding would result in a major retrenchment in US fusion research and development efforts. CIT and ITER would not be options, nor

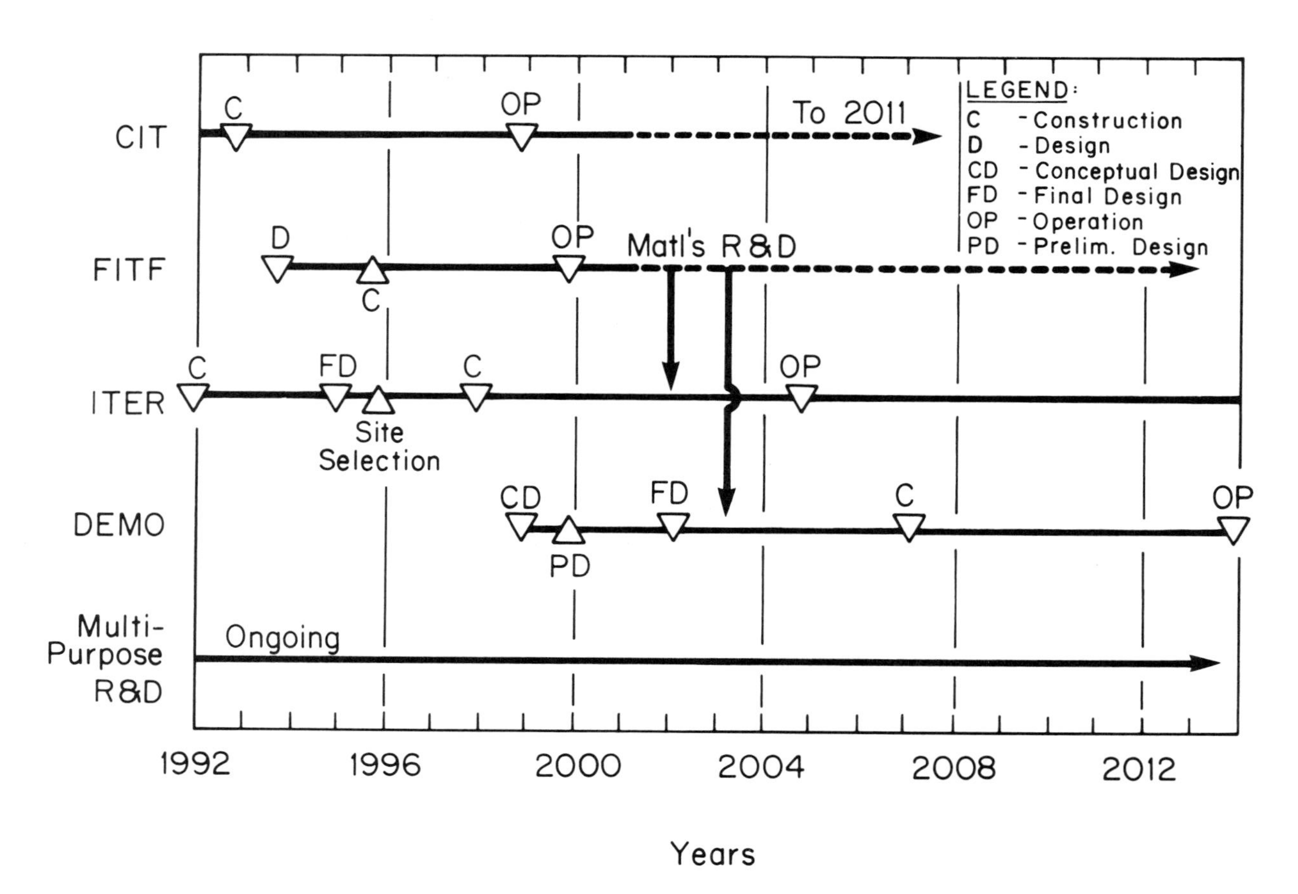

Figure 9. Construction schedule for the major elements in the Magnetic Fusion Program. The compact ignition tokamak (CIT), the fusion irradiation test facility (FITF), the international thermo-nuclear engineering reactor (ITER), and a 500 MW utility-based demonstration reactor (DEMO).

would any major US fusion experiments. Resources could be concentrated on basic research on plasma physics, but the importance would be limited because the values of plasma parameters would not be in reactor-relevant regimes.

The reduced program capability, including the loss of science and engineering personnel, would greatly constrain US opportunities for collaborative research with the world's major fusion programs. Decreased funding for three to five years would dilute the program to the point that restoring research and development activity to its original depth and breadth could take a decade or more. It would take a comparable amount of time to restore confidence in the United States as a partner in international fusion projects.

A historical perspective of the US fusion budget over the last 35 years is shown in Figure 10 (Brolin, 1989). The program increased dramatically following the 1975 oil shock, but has been falling in the decade since, and

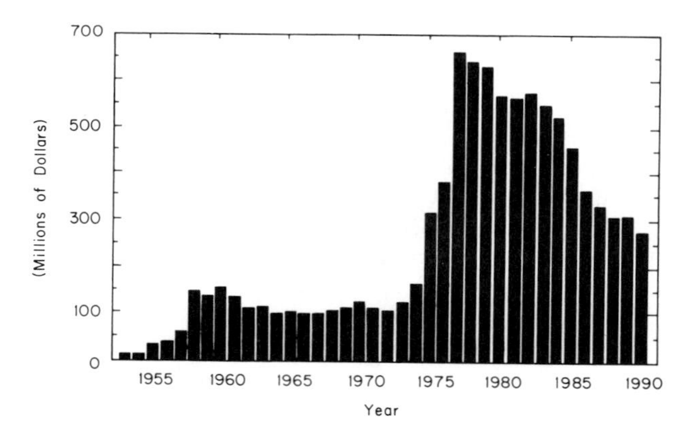

Figure 10. Historical magnetic fusion research and development funding over the 1951–1987 period (in 1986 dollars).

is now down to its pre-1975 levels. The US is rapidly losing its leadership position to the European and Japanese programs, both of which have continued at the typical 1975 level.

Fusion, despite its ultimate potential, is not without its critics, particularly in the present budget climate. One characteristic of the fusion program which surely has led to criticism is the long horizon. The program has been funded for nearly 40 years, and even an aggressive future program stretches out 25 years to the operation of a 500 MWe demonstration reactor. Progress has in fact been dramatic, but has often taken longer than promised.

Secretary Watkins has recently been critical of fusion's pace, recalling that in a visit to Oak Ridge in 1958 he was told that scientists were within seven years of achieving fusion (Davidson, 1989). Perhaps the program has occasionally overpromised, but it is more likely that what is meant by "achieving fusion" keeps changing as progress is made. In 1958 the Oak Ridge scientists probably thought of "achieving fusion" as producing and detecting the first fusion neutrons. Today, when we can generate 10 MW of fusion neutrons for times limited only by the heating in the magnets, achieving fusion means a 100 MWe demonstration plant.

References

Abdou, M., FINESSE Phase 1 Report, "Technical Issues and Requirements of Experiments and Facilities for Fusion Nuclear Technology," University of California, PPG-909, UCLA-ENG-85-39, December 1985.

Bell, M.B., et al., *Plasma Physics and Controlled Nuclear Fusion Research,* Vol. 1, p. 27, 1988.

Brolin, E., Position paper "US Magnetic Fusion Program," Princeton Plasma Physics Laboratory, (1989). Also Linford, R.K., "A Plan for the Development of Magnetic Fusion Energy," Los Alamos National Laboratory, FPAC presentation, April 1990.

Conn, R.W., "The Engineering of Magnetic Fusion Reactors," *Scientific American,* (1983).

Conn, R.W., et al., "Fusion Reactor Economic, Safety, and Environmental Prospects," White paper submitted to the International Fusion Research Council (IFRC) by UCLA, Institute of Plasma and Fusion Research, January 1990.

Davidson, K., interview with Secretary Watkins, *San Francisco Chronicle,* October 29, 1989.

Decker, J.F., Acting Director, Office of Fusion Energy, DOE Statement to Subcommittee on Energy Research and Development, House Science, Space and Technology Committee, March 1988.

Fusion Reactor Materials, Semiannual Progress Report, US Department of Energy (DOE), DOE/ER-0313/6, March 1989.

Holdren, J.P. et al., Report of the Senior Committee on Environmental, Safety, and Economic Aspects of Magnetic Fusion Energy, LLNL report UCRL-53766, September 1989.

International Atomic Energy Agency (IAEA), ITER Concept Definition, ITER Documentation Series, No. 3, Vienna, September 1989.

Jardin, S., "Physics Overview of CIT," Proceedings, 13th Symposium on Fusion Engineering, Knoxville, TN, October 1989 (in press).

JET Results, American Physical Society Division of Plasma Physics, San Diego, CA, November 1989.

Komarek, P., "Present Achievements and Prospects for Superconducting Tokamaks in the World," in Proceedings 11th International Conference on Magnet Technology (MT-11), Tokyo, August 1989 (in press).

Magnetic Fusion Community Position Paper, May 1989. Available from Fusion Power Associates, Gaithersburg, MD.

Office of Technology Assessment (OTA) Starpower, The US and International Quest for Fusion Energy, US Congress OTA-E-338, October 1987.

Pacing the US Magnetic Fusion Program, A report prepared by the Committee on Magnetic Fusion in Energy Policy, Energy Engineering Board Commission on Energy and Technical Systems, National Research Council, National Academy Press, 1989.

Parasells, R.F. and Howard, H.P., "QA Support for TFTR Reliability Improvement Program in Preparation for DT Operation," IEEE CH2507-2/87/0000, (1987).

Rebut, D.P., JET Press Conference, Abingdon, England, November 7, 1989.

Technical Planning Activity (TPA), Final Report, ANL/FPP-87-1, Argonne National Laboratory, January 1987.

Univ. of Wisconsin, HiBall-II, "An Improved Heavy Ion Beam Driven Fusion Reactor Study," UWFDM-625, 1984.

Perspectives on Renewable Energy and the Environment

Dan L. Hartley
Donald G. Schueler
Sandia National Laboratories
Albuquerque, NM

Over the next several decades, it is expected that the use of renewable energy technologies will greatly expand as these technologies mature, as the cost of conventional energy supply increases, and as the environmental impact of fossil fuel usage is better understood. A critical question is whether the impact of renewables, in terms of displaced fossil fuel use and reduced environmental effects, can be significant. Certainly, public policy will have dramatic effects on this question. Given a positive policy environment, renewables have the potential to displace a significant fraction of projected energy use within 30 to 40 years. For the United States, renewables could contribute as much as 25 to 55 exajoules of energy annually by the year 2030, or 15 to 35 percent of the projected total US energy consumption.

Introduction

Renewable energy technologies produce marketable energy by converting natural phenomena into useful energy forms. These technologies utilize the energy inherent in sunlight and its direct and indirect impacts on the earth (photons, wind, falling water, heating effects, and plant growth), gravitational forces (the tides), and the heat of the earth's core (geothermal) as the resources from which they produce energy. These resources represent an energy potential that is incredibly massive, dwarfing that of equivalent fossil resources. Therefore, the magnitude of these resources is not a key constraint on energy production. They are, however, generally diffuse and not fully accessible, some are intermittent, and all have distinct regional variabilities. These aspects of their character give rise to difficult—but solvable—technical, institutional, and economic challenges inherent in development and use of renewable energy resources.

Expanded worldwide research and development on renewables over the past two decades have brought them serious attention as potential alternative energy resources. Today, significant progress continues to be made in improving collection and conversion efficiencies, lowering costs, improving reliability, and understanding where and how these technologies are most useful. Given the current encouraging pace of technological advances in renewable energy technologies and the increasing cost of conventional energy supplies, the use and mix of these resources are expected to greatly expand over the next several decades as these technologies mature.[1–3] Renewable energy technologies generally should be expected to reduce environmental impacts below those of energy systems based on fossil or nuclear sources. The critical question is whether the impact of renewables, in terms of displaced fossil fuel use and reduced environmental effects, can be significant. The projections of energy cost and market size for renewables presented in this paper are based on a recent interlaboratory assessment of the status and potential of renewable energy resources.[1]

Progress of Renewable Technologies

As illustrated in Figure 1, the renewable energy technologies are a mix of older, mature concepts such as hydropower, geothermal, and biomass; technologies that are currently entering the market because of developing technology or preferential tax or rate treatment; and advanced ideas that offer significant potential for future energy supplies. For the newer technologies, the research, development and demonstration efforts of the past two decades have greatly advanced the technologies and brought several of them to the point of competitive market entry. Today, renewable technologies, principally hydropower and biomass, supply approximately 7.1 exajoules (EJ) per year, or about 8 percent, of the energy used by the United States annually. Figure 2 shows the current relative contribution from the various renewable energy technologies.

The following sections present a brief assessment of the status and prospects of the renewable technologies from the viewpoint of the end-use energy they produce.

Electricity

Between 1960 and 1985, worldwide consumption of electricity increased by 267 percent. In the United States, the energy used for generation of electricity ac-

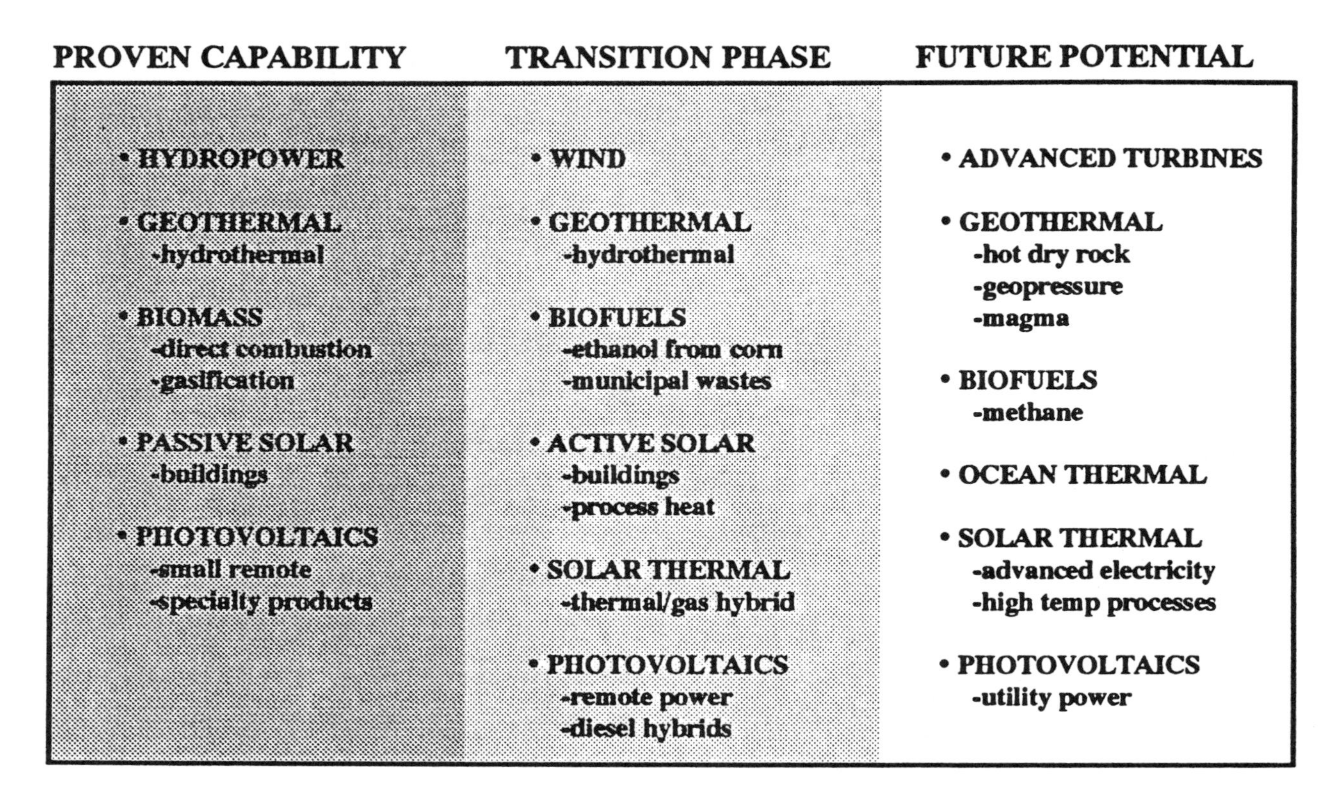

Figure 1. Maturity of renewable energy technologies (after Reference 1).

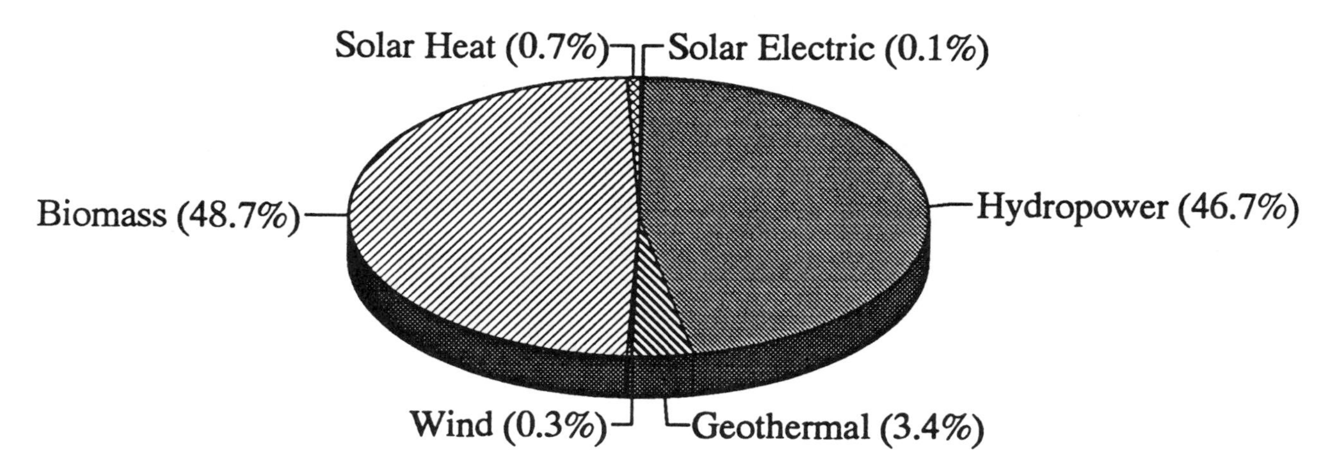

Figure 2. Current contribution to US energy supply by various renewable technologies (Total = 7.1 EJ/Year).

counts for approximately 35 percent of our total primary energy consumption. Projections of new electric generation capacity needed to meet load growth and retirement of aging plants in the United States between now and the year 2030 indicate a cumulative need for about 1100 gigawatts (GW) of new capacity,[4] compared with the current installed capacity of 680 GW. As such, electric utility applications represent a large potential market for renewable energy technologies.

Electricity produced today from renewable energy resources comes mostly from hydropower, with 71 GW of installed generating capacity, and biomass, with 8 GW of installed capacity. Expanded use of hydropower is seriously limited by environmental and siting constraints. The biomass generating capacity in the United States today is primarily in the form of wood and agricultural-waste-burning cogeneration plants. Future growth in the biomass electric technologies is expected to come from expanded use of municipal solid wastes and the development of new high-yield energy-crop feedstocks.

The current geothermal generating capacity in the United States is about 2.9 GW and is derived mostly from high-temperature hydrothermal sources such as the Geysers dry-steam field in northern California. Estimates are that an additional usable resource for electric generation of 95 to 150 GW exists in the western continental United States, Alaska, and Hawaii. Beyond today's technology, advanced concepts such as hot dry rock, geopressured sources, and magma are being researched and could expand the resource base dramatically.

Wind energy systems have experienced significant commercial market development over the past decade, taking advantage of the combination of tax incentives, favorable utility power purchase agreements, and the best wind sites. The current US installed wind-generating capacity is about 1.5 GW and is located primarily in California. After a problematic start due to poor turbine reliability, considerable progress has been made in wind turbine technology over the past several years. The cost of energy from wind turbines is extremely dependent on the wind characteristics of the site. Today, wind turbines can compete with moderate-to-high-cost conventional fuels in the best of wind sites, such as mountain ridges and valleys. In the near future, improvements in turbine technology should allow wind systems to compete in moderate wind resource sites such as those typical of large areas of the Great Plains.

Solar thermal generation of electricity is in the market transition phase, providing energy competitive with moderate-to-high fuel costs in those circumstances that combine a good solar resource site with a favorable tax and financing environment. Currently, about 300 megawatts (MW) of parabolic trough collectors are operating in California. These systems use a hybrid combination of solar energy and natural gas to achieve a high utility capacity credit. A number of significant technology improvements, such as membrane reflectors, direct absorption receivers, and high-performance heat engines should make next-generation solar thermal systems competitive in a larger arena of applications.

Photovoltaics enjoy a substantial and growing worldwide market (over 40 MW of annual sales) for consumer product and specialty remote power applications. Declining costs for photovoltaic systems is making them competitive in expanded remote power and related applications where energy cost is typically greater than 20 cents per kilowatt-hour (kWh). Although numerous residential, industrial, and megawatt-scale utility photovoltaic power systems have been built and successfully operated in the United States, costs must further decline for photovoltaics to be competitive in domestic electric grid-connected applications. Ongoing research on a variety of photovoltaic technologies such as crystalline silicon, thin-film compound semiconductors, and concentrating collectors show great promise for photovoltaics in energy-significant electric utility applications by the 2000 to 2010 timeframe.[5,6]

Figure 3 illustrates the current and projected cost of energy for selected developing solar renewable technologies (wind, solar thermal, and photovoltaics), along with the current and projected range of energy costs for conventional peaking, intermediate, and base load utility generating plants.[1,4] Wind and solar thermal systems, with current energy costs ranging from 8 to 15 cents per kWh, are competitive today with the upper range of peaking power cost. Proven and future technological improvements that can be applied in next-generation hardware should make all three of these technologies broadly competitive with conventional peaking and intermediate generation during the 2000 to 2010 timeframe. In the longer range, and with the anticipated application of advanced electric energy storage technologies such as pumped hydro, batteries, and superconducting magnetic energy storage, these renewables can also become alternatives in a mix with coal-fired and nuclear base load electric generation.

Transportation

Biofuels represent the only current significant use of renewable resources in the US transportation sector, which accounts for over 25 percent of our primary energy use. Ethanol, produced primarily from corn, is used for 10 percent blending with gasoline and now displaces about 1 percent of the US gasoline consumption. Such blends are being increasingly used to improve the oxygen content of transportation fuels used in large urban areas with air quality problems. Future growth of renewables for transportation purposes is expected to come from the development of new cellulosic biomass feedstocks for the production of ethanol, the develop-

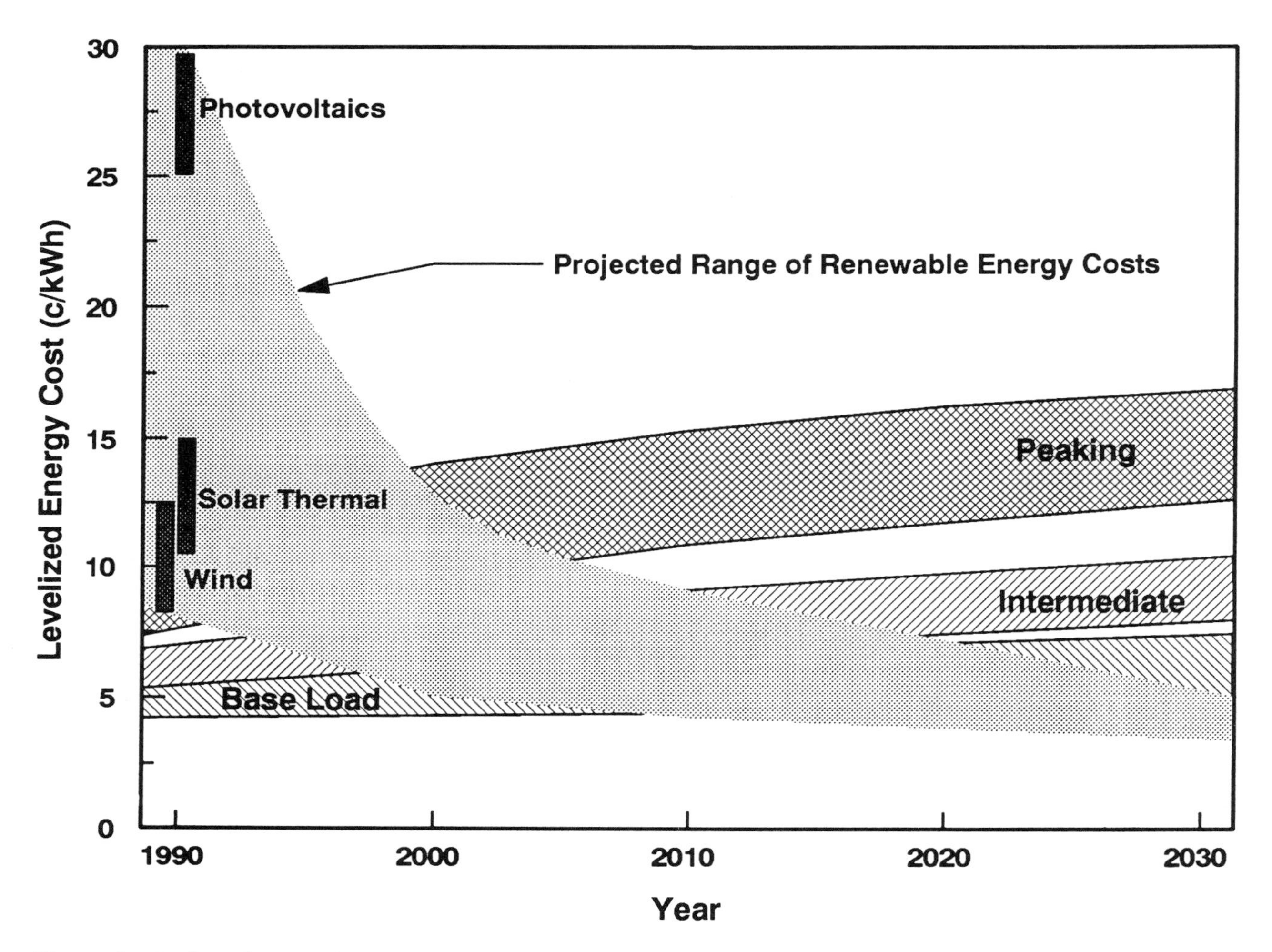

Figure 3. Projected competitive status for selected solar renewables in the electric utility sector.

ment of efficient methods to produce methanol from biomass feedstocks, and the expanded use of electric vehicles that could use electric energy produced by renewable technologies.

Buildings and Industrial Applications

Buildings and industrial applications represent the remaining 40 percent of US primary energy consumption. The burning of wood and wood wastes for space heating and industrial process heat is the largest current renewable contributor, producing about 1.9 EJ of energy annually. The current installed base of active and passive solar energy systems in buildings and industrial applications supplies about 0.05 EJ. Future inroads of renewables into this energy sector are likely to come from continued market penetration by active and passive solar space conditioning systems, day-lighting concepts, solar thermal industrial process heat applications, and innovative industrial applications of highly concentrated sunlight, such as destruction of hazardous wastes.

The Potential of Renewables

Projection of the future contribution of renewable energy technologies is obviously very difficult and imprecise. Factors such as future energy demand, energy cost and availability, environmental issues, public attitudes, tax and financial policies, and the international political climate, to name a few, will affect the eventual rate of growth of existing renewable technologies and the development and introduction of new technologies.

A recent analytical assessment of the market potential of renewable resources considered likely technological improvements in performance and cost, the projected growth in regional energy markets, and various public policy scenarios to construct a range of projected market penetrations for renewables in the United States to the year 2030.[1] Figure 4 presents the results of that study. The range of projected energy contribution from renewables results from a variety of assumptions about research and development (R&D) priorities and environmental concerns over the next 40 years. In the most pessimistic case, renewables are projected to grow slowly from the current 7.1 EJ to nearly 25 EJ by 2030, representing 15 percent of total energy use. This lower bound is based on a "business as usual" scenario of continued modest R&D funding for renewable technology development, but with no significant energy supply or environmental issues to change current tax or regulatory policies for renewables. The most optimistic case for renewables results from assuming either a policy of intensified R&D for renewables (technology push) over the next 20 years, or a policy of national market incentives of 2 cents/kWh or $2/million Btu (market pull) for renewables because of environmental or other national concerns. In this case, the contribution from renewables reaches the previous 25 EJ by 2010, 20 years earlier, and grows to about 55 EJ by the year 2030, or 35 percent of total energy use. Figure 5 shows the projected relative contribution to US energy supply by the various renewable technologies in the year 2030. The predominate contributors at that time are expected to be the solar electric technologies: photovoltaics, solar thermal electric, wind, and biomass.

The above discussion considers only the US energy marketplace. Renewables could also make similar important contributions worldwide. As an example, electrification is a key issue in developing countries. Most of these countries rely heavily on fossil fuel plants and plan expansion around these familiar energy systems. Renewables could offer a viable alternative in these countries, provided an experience base is developed around early, cost-effective niche applications. Many of the renewable technologies could also stimulate greater economic development in rural areas worldwide. Successful deployment of renewable technologies in the developing countries could provide substantial benefits as the result of local economic development, reduced fossil fuel

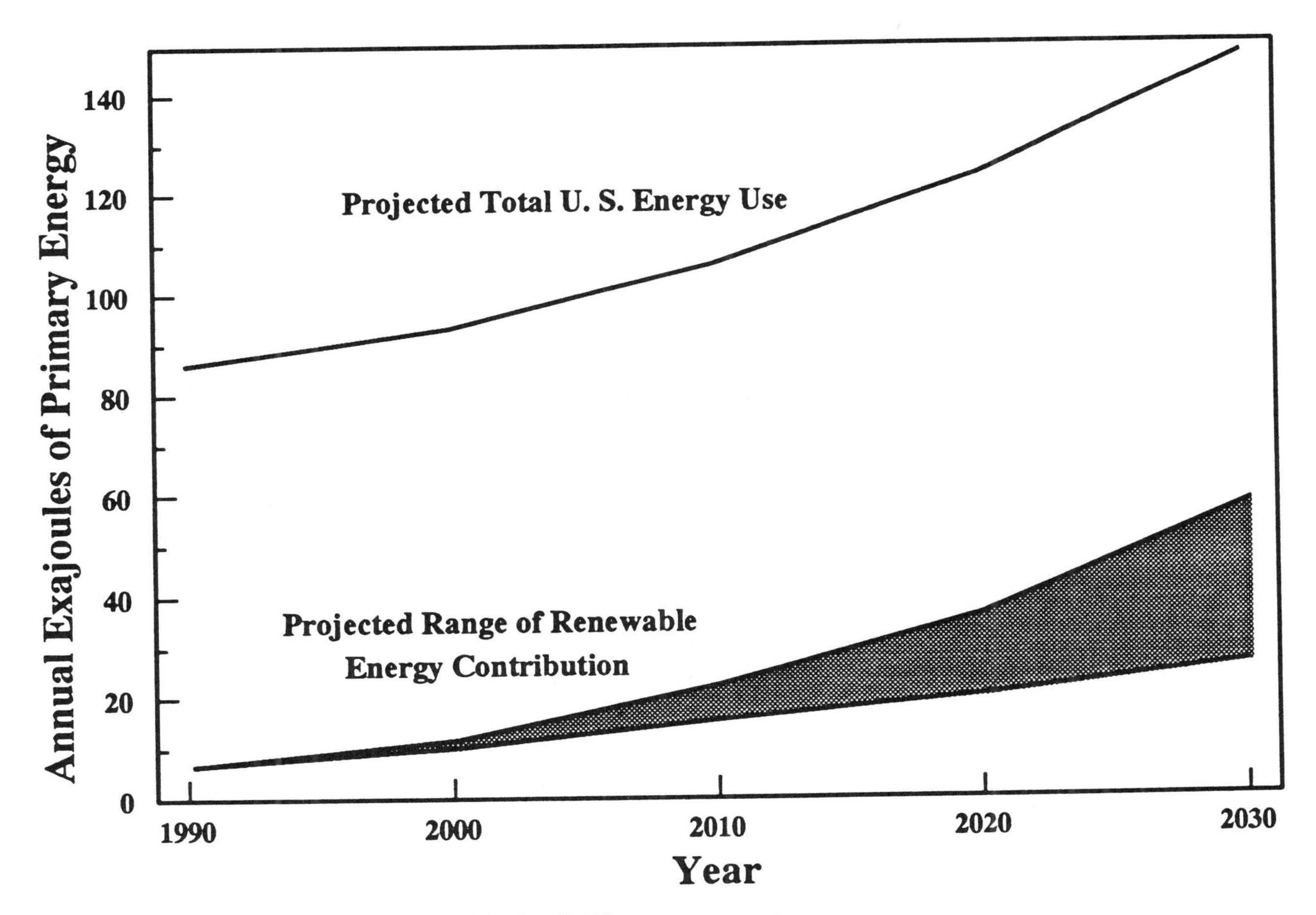

Figure 4. Projected growth of renewables in all US energy use sectors.

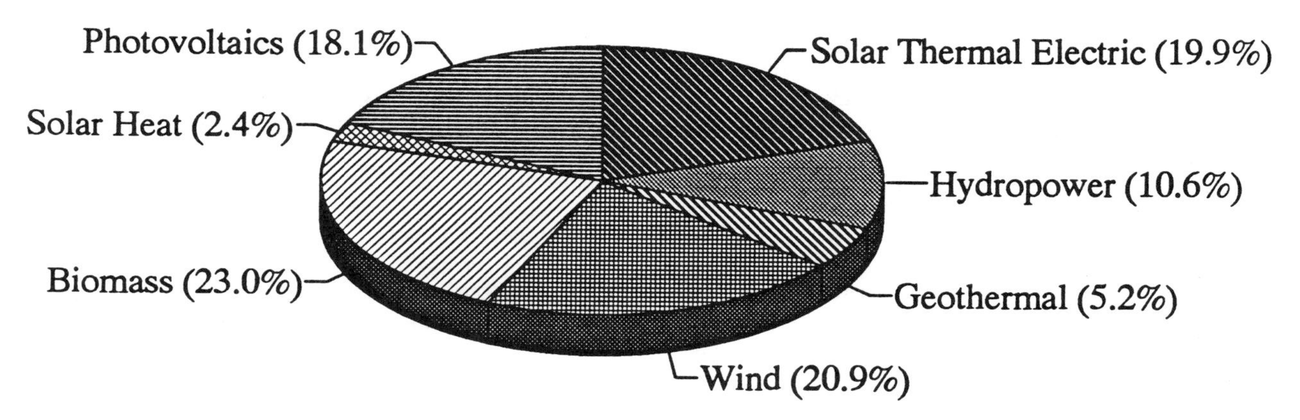

Figure 5. Projected contribution to US energy supply by various renewable technologies in 2030.

usage, and reduced political stress. In addition, perhaps the greatest economic contribution of renewable energy would be in the form of moderated demand of conventional energy resources worldwide.

Environmental Perspectives

The use of renewable energy provides several significant benefits, not all of which are reflected in the commercial energy marketplace. These benefits include reduced environmental stresses, enhancement of energy security, and a variety of direct and indirect economic benefits. In spite of their significance, most of these benefits are difficult to quantify. For example, it is nearly impossible to assess the value of renewables in helping to keep oil demand and prices from reaching levels they might otherwise attain in the absence of competition from renewables. It is encouraging, however, that analysts and policy makers are beginning to think in terms of "total societal cost" and "fuel cycle cost" in an effort to recognize these factors and to compare energy systems on a more comprehensive basis.[7]

Because renewable energy systems extract relatively small quantities of energy from massive, on-going natural processes, their impact on the environment is generally minimal and localized. The environmental effects that do result from constructing and operating renewable energy systems are varied; certainly land and water use are among the most obvious and important. Other environmental impacts result from use of raw materials, manufacture and transport of system hardware, and in some cases direct emissions into the atmosphere (e.g., biofuels and geothermal).

An increase in the world's greenhouse effect and consequent global warming poses a long-term threat to our environment, which is potentially more serious than any other previously encountered environmental issue. Global warming has the potential to significantly alter climate, and thereby cause serious agricultural, economic, political, and social disruptions. It is estimated that worldwide energy production is responsible for 57 percent of the anthropogenic contributions to the greenhouse gases,[8] with CO_2 being the most significant. Any serious attempt to combat the greenhouse effect will require worldwide actions aimed at effecting reduced CO_2 emissions; this makes coal use less attractive. Although not free from environmental consequences, most renewable energy resources eliminate or drastically reduce the release of greenhouse gases into the atmosphere. They therefore offer a partial solution to the problem of CO_2 emission, and also provide a non-nuclear alternative to fossil fuels.[8–11] It is important to note that the atmospheric benefit resulting from renewable energy resources depends greatly on the type of fossil fuel they displace. Emissions of CO_2 per unit energy of natural gas, oil, coal, and synthetic gas are 13.8, 19.7, 26.9, and 40.7 Megatonnes carbon per exajoule, respectively.[12] Worldwide CO_2 emissions from the production of energy are currently about 6.5 Gigatonnes of carbon per year (GtC/year) and are expected to reach nearly 20 GtC/year by 2030.[8] About 80 percent of those emissions are anticipated to come from coal and coal-derived gas and oil consumption. Figure 6 shows the percentage of projected worldwide CO_2 emissions from energy consumption that could be displaced by various amounts of renewable energy resources, depending on which fossil fuel type is displaced.

Conclusions

A limited number of renewable energy technologies have established themselves as viable energy resources in the current mix of energy alternatives. Several additional renewable technologies are in transition to energy-

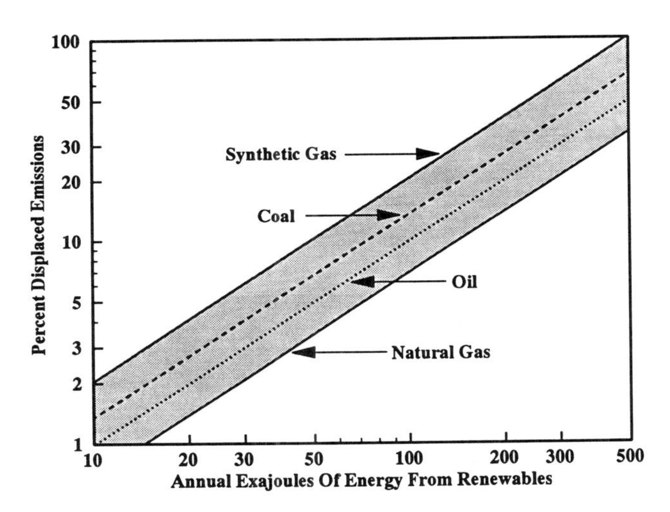

Figure 6. Percentage of projected year 2030 world-wide CO_2 emissions resulting from energy production which could be displaced by increasing amounts of renewables, for various displaced fuel types.

significant market applications and advanced technologies hold promise for ever-wider application. The rate of future growth in the use of renewable resources will depend strongly on many factors such as advances in technology, the world energy supply/demand situation, and worldwide public policy. Depending on the influence of these factors, renewables are projected to grow from their current 8 percent contribution to US energy supply to 15 to 35 percent over the next 40 years. Since electricity generation represents a prime opportunity for renewables, the solar electric technologies—solar thermal electric, photovoltaics, wind, and biomass—are expected to be the major future energy contributors from renewables. If renewables can indeed displace 15 to 35 percent of our future energy consumption, they can lead to a substantial diversification of the world's energy mix and also reduce the environmental impact of ever-increasing energy consumption.

References

1. "The Potential of Renewable Energy—An Interlaboratory Analytic Paper," (prepared for the DOE Office of Policy, Planning, and Analysis by INEL, LANL, ORNL, SNL and SERI), SERI/TP-260-3674, March 1990.
2. "New Electric Power Technologies: Problems and Prospects for the 1990s," US Office of Technology Assessment, OTA-E-246 (July 1985).
3. "Strategic Implications of Alternative Electric Generating Technologies," Economic Division, Edison Electric Institute (April 1984).
4. Based on extrapolation of trends in DOE/EIA's *Annual Energy Outlook 1989, Annual Outlook for U.S. Electric Power 1989*, and the DOE Fossil2 demand projection model.
5. "Five Year Research Plan: 1987–1991," US Department of Energy, DOE/CH10093-7, Washington, DC (May 1987).
6. Hubbard, H. M. "Photovoltaics Today and Tomorrow." *Science* 244, pp. 297–304 (April 1989).
7. Hohmeyer, O. *Social Costs of Energy Consumption.* Springer-Verlag, New York, NY (1988).
8. Aronson, E. A., and Edenburn, M. W., "The Potential Impact of Conservation, Alternative Energy Sources, and Reduced Nonenergy Emissions on Global Warming," SAND89-1380, Albuquerque, NM, Sandia National Laboratories (December 1989).
9. Ogden, J. M. and Williams, R. H. "Solar Hydrogen: Moving Beyond Fossil Fuels." *World Resources Institute* (October 1989).
10. Haraden, J. "An Appraisal of Magma Power Generation and the Greenhouse Effect." *Energy* 14 (6), pp. 333–340 (1989).
11. Carlson, D. E. "Fossil Fuels, the Greenhouse Effect and Photovoltaics." *Proc. Twentieth IEEE Photovoltaic Specialists Conference*, pp. 1–7 (1988).
12. Mintzer, M. "A Matter of Degrees: The Potential for Controlling the Greenhouse Effect," World Resources Institute, Washington, DC, Research Report 5 (April 1987).

Hot Dry Rock Geothermal Energy—An Emerging Energy Resource with Large Worldwide Potential

D. W. Brown
R. M. Potter
C. W. Myers
Earth and Environmental Sciences Division
Los Alamos National Laboratory
Los Alamos, NM

Abstract

Hot dry rock (HDR) geothermal energy, which utilizes the natural heat contained in the earth's crust, is a large and well-distributed resource of nonpolluting energy that is available globally. Its use could help mitigate climatic change and reduce acid rain, two of the major environmental consequences of our ever-increasing use of fossil fuels for heating and power generation. The earth's heat represents an almost unlimited source of energy that can be exploited within the next decade using heat-mining concepts being developed in the United States, Great Britain, Japan, and several other countries. This new energy source can be used for electric power generation or for direct-heat applications, or indirectly in hybrid geothermal/fossil-fuel power plants.

In the HDR concept—which has been demonstrated in two different sites, one in the US and the other in the UK, and flow tested for periods approaching five years—heat is recovered from the earth by pressurized water in a closed-loop circulation system. As a consequence, minimal effluents are released to the atmosphere and no wastes are produced.

This paper describes the nature of the HDR resource, reviews the technology required to implement the heat-mining concept, and assesses the commercial feasibility of HDR.

Motivation and Scope

Continuing studies have shown that 20th century energy production methods—including the use of fossil fuels for electric power generation as well as their direct use for heating, and industrial and transportation needs—are having an adverse effect on the environment. Of primary concern are the deleterious effects of the vast amounts of CO_2 and smaller amounts of other gases such as SO_2 and NO_x that are liberated in the combustion process (Brown et al., 1988; Lovins et al., 1981; Shepard, 1988; and Mathews, 1987). These gases play major roles in the processes that cause global warming (the greenhouse effect) and pH changes in regional precipitation (acid rain).

It is difficult to predict with certainty the direction and effects of global temperature change. Because the consequences are unclear, the tendency is to resist change and maintain the status quo. It appears that the only practical solution to this dilemma is to significantly reduce our use of fossil fuels. Active conservation programs may achieve additional gains (Chandler et al., 1988), but the real solution seems to lie with the application of alternative energy generation technologies. Many studies have proposed greatly increasing the use of nuclear energy and other nonfossil energy options as rapidly as possible. Alternative options mentioned typically include solar, fusion, biomass, and wind turbines. Conspicuously absent (at least to the authors of this paper) is the inclusion of geothermal energy as a potential replacement for a portion of our future fossil fuel needs.

Geothermal energy, in the form of naturally occurring high-temperature water or steam, is presently providing an economical and trouble-free source of electric power in a number of locations worldwide (see also Otte, 1990, and DiPippo, 1990, in this volume). Installed worldwide capacity is about 5000 MWe to 2900 MWe in the US alone. With availability factors ranging from 70 percent to as high as 95 percent, geothermal power has been demonstrated as a reliable source of baseload electricity. The world's largest geothermal field development is located at The Geysers site in northern California, where it presently produces over 2000 MWe, about 7 percent of California's power demand. Furthermore, electric power from The Geysers field competes quite favorably with both fossil and nuclear supplied electricity. Geothermal development activity has been increasing in the western US where non-utility projects have begun to tap high-grade, liquid-dominated hydrothermal resources (e.g., at Coso Hot Springs and the Imperial Valley in California). Although limited to the few regions where these hydrothermal anomalies occur, the magnitude of the US natural geothermal resource base (National Research Council (NRC), 1987) is still substantial at some 44,000 quads of energy. [1 quad = 10^{15} Btu

$= 1.055 \times 10^{18}$ J $= 181.8 \times 10^6$ barrels of oil $= 2.475 \times 10^7$ metric tonnes of oil]. Worldwide, the natural geothermal resource base is estimated to be about 130,000 quads (Armstead and Tester, 1987).

The broader application of geothermal energy, however, will probably depend on the success of one or more of the advanced concepts being investigated: hot dry rock (HDR), geopressured, and magma. Together, these advanced concepts would expand the total geothermal resource base to about 100,000,000 quads worldwide (see later section on Resource Size). Of course, it remains to be seen how much this enormous resource base can be converted to an economically viable reserve. Nonetheless, in comparison to the world's current annual energy consumption of about 320 quads, the potential for geothermal is very large. Furthermore, unlike many renewables, geothermal is inherently capable of providing an uninterruptible supply of energy, regardless of weather conditions, for example.

HDR geothermal energy represents the largest portion of this broader available resource (see Figure 1). In the HDR concept, a fractured reservoir is created by stimulating existing joints in a deep, hot region of the crust by injecting water under pressure—the commonly employed technique of hydraulic fracturing used in the oil industry. Research in the US and Japan is being directed toward engineering such systems in tight crystalline basement rock where joints are well sealed and very little natural permeability exists. In Great Britain, France, and West Germany, other concepts are being pursued to utilize higher permeability systems with partially open, natural fractures providing predominant fluid pathways. In all cases, some form of hydraulic stimulation is used—the actual characteristics and extent of this procedure will, of course, be controlled by prevailing geologic conditions within the reservoir. In future applications, reservoir stimulation and diagnostic techniques developed for HDR could be employed to utilize the significant geothermal resource intermediate between hydrothermal and HDR—at the margins of existing hydrothermal fields, where there is insufficient permeability and/or fluids.

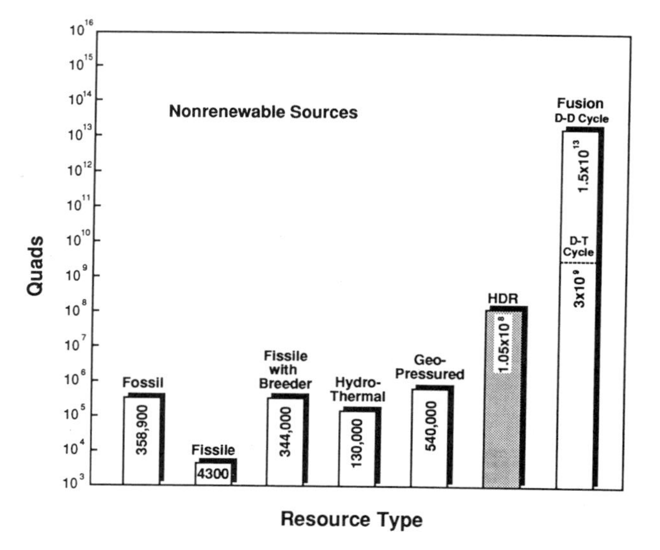

Figure 1. Worldwide resource base estimates for nonrenewable resources. (Adapted from Armstead and Tester, 1987.)

If developed in a timely fashion, HDR geothermal energy could play a substantial role in evolving national policies for reducing fossil fuel use. In addition to providing electricity, this renewable energy resource has the potential for supplanting significant quantities of natural gas and heating oil now used for space heating and for process heating in lower-temperature industrial processes. This application of HDR geothermal energy will help preserve the more limited quantities of high-grade fuels for use in high-temperature industrial processes and transportation systems. In this mode, HDR geothermal energy could significantly extend the lifetime of our domestic reserves of petroleum and natural gas. Immediate direct-heat uses and the potential for significant power generation from the higher-grade HDR resources in the mid-continent and western regions of the United States exist now. In order for the more widely distributed lower-grade HDR resource to have a substantial future role in the generation of electricity, advances in hard-rock drilling technology are needed to lower costs.

During the next several decades, with an active research and development program in place, the economic drilling of HDR boreholes should be possible to depths of 10 km, which is now the maximum depth for oil and gas well drilling. However, oil, gas, or scientific drilling to depths of 10 km has never been undertaken at a well diameter of 8½ (21.6 cm) required for HDR applications. Using current drilling technology, HDR systems are thought to be limited to about 6 km. However, technical limitations for HDR drilling may not be as severe as originally perceived because of the inherent borehole stability of hard, crystalline rock. With this stability, larger diameter holes, in principle, should be drillable to depths of 10 km—without the excessive use of intermediate casing strings now needed for ultra-deep oil and gas well drilling. Casing and hole lining issues still need to be addressed, however.

Advances in drilling technology would ensure that rock temperatures in excess of 200°C—appropriate for HDR-augmented electric power generation—will be accessible almost everywhere in the US. The resulting dispersion of energy production could be of great practical importance, helping to ameliorate current environmental and social problems, such as those associated with the siting of mega-sized power plants near population centers.

Characteristics of an HDR-Based Power System

As depicted in Figure 2, thermal energy can be extracted from artificially fractured HDR reservoirs that emulate natural hydrothermal systems. The primary technique for engineering HDR reservoirs utilizes fluid pressure to open and propagate fractures from an initial well, creating artificial permeability within a fracture network as illustrated in Figure 2. This hydraulically stimulated region of high permeability—consisting of an interconnected array of open joints and fractures—is then connected to a second well to complete the underground system. Heat is extracted by circulating water from the surface, down one well, through the fractured rock network, and up the second well. The heated water then passes through an appropriately designed power plant on the surface where, for instance, electricity or process steam is generated. The cooled fluid is then reinjected to complete the closed-loop cycle. In effect, we would be mining heat in a fashion analogous to the way other resources are mined (Smith et al., 1975; Armstead and Tester, 1987). HDR energy offers the following benefits as a primary energy source:

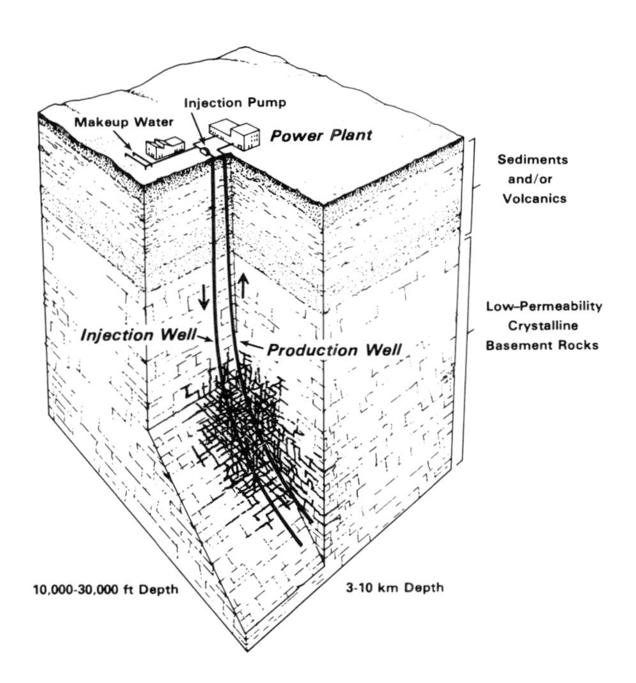

Figure 2. Hot dry rock (HDR) geothermal system concept for low-permeability formations.

Minimal Environmental Impact

No fluids or gases (no CO_2, NO_x, SO_2, etc.) are released to the environment since HDR systems operate in a closed-loop circulation mode. Thus, HDR geothermal energy will not contribute to increased levels of atmospheric greenhouse gases. Further, waste heat rejected from HDR electric power plants will be so widely dispersed that its effect on any local environment will be minimal. In some instances, an external water source will be required for heat rejection from the power plant, for example, for evaporative cooling to condense power cycle working fluids. In addition, seismic risk and subsidence must be dealt with on a case-by-case basis. Field testing to date has shown these effects to be minimal. Finally, drilling access wells and laying pipelines may temporarily disturb the land surface, but such changes should be on a much smaller scale, for instance, than those associated with strip mining coal or building hydroelectric dams.

Available Everywhere

HDR energy is available virtually everywhere on the earth's surface, with the resource temperature inexorably increasing with depth. The HDR resource base is defined as the thermal energy above a certain minimum temperature stored in rock beneath the surface at depths accessible using conventional drilling technology. Therefore, in developing the HDR resource, the primary parameter determining the local grade of the resource is the average temperature gradient or, conversely, the drilling depth required to reach a temperature suitable for a specified application.

End-Use Flexibility

HDR production fluids can be used directly for local space or process heating needs, or for lower-temperature power plant heating requirements. HDR can be used as a single supply of energy to a power plant, or used in conjunction with coal, oil, or natural gas in a hybrid plant to improve efficiency. In the first case, fossil fuel consumption and CO_2 generation rates per unit of electricity produced are entirely eliminated, and in the second case, they are significantly reduced. As the energy supply is in the form of hot water under pressure, HDR-based power plants are inherently as safe or safer than other existing power generating systems, and would be simpler to construct and operate than fossil or nuclear plants. The optimum size for HDR-based electric generating plants is in the 10–100 MWe range (e.g., Bechtel, 1988), in line with current utility plans for incremental power generation increases. This represents a significant change from past large-utility capacity additions in the hundreds of MWe size range.

Each Reservoir Represents a Large Thermal Resource

Field experience has shown that fractured HDR reservoirs can be created economically (but not necessarily fully utilized for energy production) and that their size can be almost arbitrarily large, depending primarily on the volume of injected water. In principle, by increasing reservoir size, one can provide for multiple cogeneration needs by producing both power and process heat, with only a very modest increase in reservoir development costs. However, in practice, the exploitation of these large thermal reservoirs has yet to be accomplished in the field. The economic feasibility of HDR depends on finding a satisfactory solution to this critical problem (see later section on Remaining Research and Development Issues).

Resource Size

Earlier we stated that the world HDR resource base was more than 10^8 quads; now we would like to justify that figure. Because hot dry rock systems do not require contained hot fluids and high permeability, the HDR resource is larger and more widely distributed than the hydrothermal geothermal resource. For example, the US HDR resource is not a localized geothermal resource limited to the western states. It ranges from low-grade regions having normal to near-normal temperature gradients of 20–30°C/km, to high-grade regions with above-normal gradients greater than 50°C/km. The moderate-grade resource (30–50°C/km) is distributed more or less uniformly throughout the mid-continent of the US as shown in Figure 3, while the higher-grade resource is found primarily in the western half of the country.

The US resource estimates have been provided by the US Geological Survey Circulars 726 and 790 (White and Williams, 1975; Muffler, 1979), by the Los Alamos National Laboratory [Hot Dry Rock Assessment Panel (HDRAP), 1977; Rowley, 1982], and by others (NRC, 1987; Armstead and Tester, 1987). Although the actual numerical estimates differ, their underlying methodologies are similar. Basically, an average geothermal temperature gradient—or an actual distribution of gradients across the US land mass—is integrated from the surface to an assumed accessible depth, in order to estimate the total thermal energy contained in the rock relative to some specified lower use temperature, like 85°C. An initial minimum rock temperature is used to define the shallow boundary of the resource, while the rock temperature at an assumed maximum drilling depth defines the deeper, lower boundary. The HDR resource base in the United States can then be estimated by using a gradient of 25°C/km as a reasonable average in rock having an average volumetric heat capacity of 2.2×10^{15} $J/km^3 \cdot °C$. If we assume 150°C for the minimum initial rock temperature and 10 km (33,000 ft) for a maximum drilling depth, the total HDR resource base for the 9.36×10^6 km^2 (3.61×10^6 sq. miles) of the US land area is about 10 million quads. At the current US rate of energy consumption, approaching 80 quads per year, this HDR resource base could conceivably supply a portion of the country's energy needs for many thousands of years.

Estimates of the higher-grade HDR resource base have also been made. For example, recent studies suggest that more than 2 percent, or about 1.8×10^5 km^2 (6.9×10^4 sq. miles), of the US land area should have average gradients in excess of 45°C/km (Kron and Heiken, 1980). If we follow the same approach used for the lower-grade HDR resource, the higher-grade resource base would amount to some 650,000 quads.

Armstead and Tester (1987) have applied this methodology to estimate the worldwide HDR resource base at more than 10^8 quads. Thus, on almost any basis, the amount of potentially usable thermal energy contained in the HDR resource is vast, orders of magnitude larger than the sum total of all fossil and fissionable resources. (See Figure 1 for estimated worldwide distribution of major energy sources.) However, these HDR resource estimates are clearly upper limits. It remains to be shown just how much of this vast resource base can be recovered within existing technical and economic constraints.

Status of the Technology

For the past 19 years, the Los Alamos National Laboratory (LANL) has conducted a major research effort at the Fenton Hill site in New Mexico to develop techniques for creating such reservoirs, sponsored by the US Department of Energy (USDOE). The Federal Republic of Germany and the government of Japan have supported a portion of this work under a collaborative agreement with the United States. In addition to the US effort, the British have been involved in the major development of HDR technology for about 17 years with a large field test being conducted at the Rosemanowes site in Cornwall (Batchelor, 1987; Parker, 1989). The Japanese have been active in the field for about five years at the Hijiori site in Yamagata Prefecture (NEDO, 1989). Other significant work is being carried out by research teams in the Federal Republic of Germany, France, and the USSR.

These research programs have already met many of the crucial technical goals that are required before large-scale HDR geothermal energy development can occur. These are:

- **Accessibility to the resource**—Conventional oil and gas drilling methods and capabilities have been successfully adapted to the harsh environment encoun-

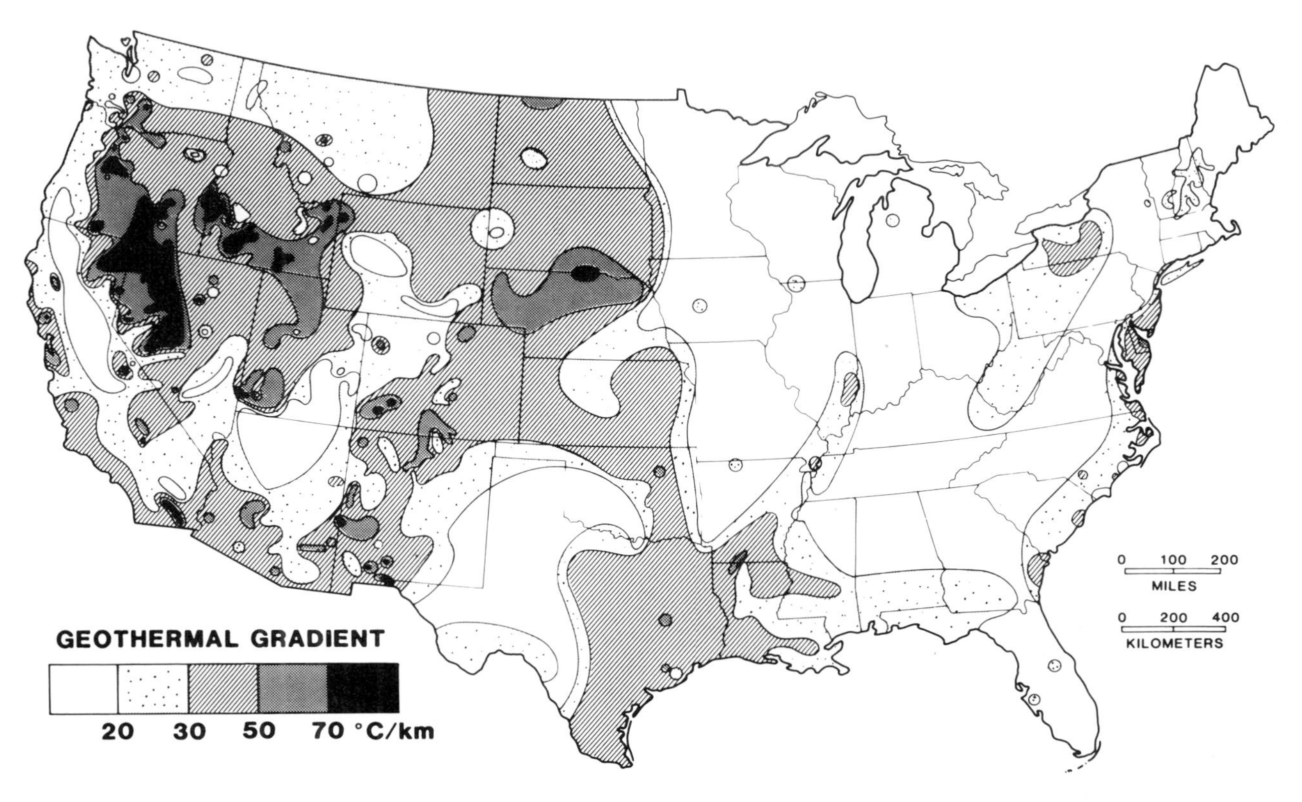

Figure 3. Geothermal gradient map of the United States. (Adapted from A. Kron and J. Stix, 1982, Geothermal Gradient Map of the United States, National Oceanic and Atmospheric Administration, and Los Alamos National Laboratory.)

tered at depth in hard crystalline rock where useful temperatures (about 150–300°C) are found. The US program has demonstrated that we can efficiently drill hard crystalline rock using current technology, to temperatures above 300°C and depths of 5 km. (Figure 4 shows trends in drilling costs for HDR wells compared with oil and gas wells at similar depths.)

- **Creation of usefully large fracture systems**—Again, oil industry methods of hydraulic stimulation have been successfully adapted to this harsh environment, resulting in the opening of numerous interconnected flow paths within very large volumes of naturally jointed hot rock.
- **Extraction of thermal energy**—The heat extraction process in these fracture systems has been shown to be both predictable and suitable for commercial use.
- **Determination of the size and physical structure of the created fracture system**—Microseismic emissions accompanying the creation of these fracture systems have been used successfully to determine the location, size, and orientation of the fractured region (i.e., the HDR reservoir). Fracture interconnectivity and wellbore placement within this microseismically determined region is crucial to optimizing the extraction of the thermal energy.

Two major constraints to commercialization remain: 1) a demonstration that reservoirs of sufficient size and lifetime can be created to maintain economic fluid production rates and 2) a reduction in the relatively high costs of drilling wells in hard crystalline rock. HDR energy supply costs are inherently linked to a combination of these factors. For example, a key economic parameter is the cost of producing a unit of fluid at a specified temperature. This can be expressed in $/kg/s of fluid produced or in $/kW of thermal power. Higher reservoir flow rates per well pair and/or lower individual well-drilling costs reduce the overall cost of supplying HDR energy. On the other hand, higher flow rates through the reservoir increase pressure drops and accelerate thermal drawdown, as thermal energy is removed from the rock at a higher rate.

Economic feasibility of HDR depends on finding satisfactory solutions to these constraints. Data collected during field testing in the US and UK suggest that by vigorously applying and extending existing technology in the area of reservoir development and drilling, tractable solutions are within reach.

Early data from HDR field drilling suggested that well costs strongly depend on depth, but are two to three times higher than those for oil and gas wells (see wells

EE-2 and EE-3 on Figure 4). The LANL drilling programs at Fenton Hill in the mid-1970s represented, at that time, the most extensive drilling effort in deep granite anywhere in the world. However, repeated drilling stops for geophysical measurements, coring, fracturing tests, and diagnostic logging combined with industry inexperience in this hostile environment made overall well costs significantly higher than those for conventional oil and gas drilling. Now, however, many of the problems associated with deep granite drilling have been resolved, as evidenced by the Laboratory's successful redrilling programs in 1985 and 1987. (Refer to the cost figure for the redrilled EE-3A well shown on Figure 4).

The world's first HDR reservoir was completed at Fenton Hill in 1977 at a depth of about 2.6 km (9000 ft), with rock temperatures of 185°C. This system was enlarged in 1979 by additional hydraulic stimulation, and was operated successfully for about one year to test the feasibility of the heat extraction concept and to measure the thermal and hydraulic performance of the reservoir. Overall, the results of this early testing were very positive, clearly demonstrating that heat could be extracted at reasonable rates from a hydraulically stimulated region of low-permeability, hot crystalline rock without serious technical or environmental problems. Although the initial reservoir was too small for commercial use, it provided an excellent test bed for verifying reservoir models designed to simulate thermal performance, and for developing techniques and equipment to characterize reservoir geometry.

To extend HDR technology to the higher temperatures and heat production rates required to support a commercial-sized electric generating plant, the development of a larger and hotter HDR reservoir was initiated in late 1979. This Phase II HDR system was designed to have two inclined wellbores—EE-2 and EE-3—with a vertical separation of approximately 300 m. However, attempts to connect the wells using hydraulic fracturing techniques were unsuccessful despite the substantial volumes of water injected during massive fracturing operations from 1982 to 1983. The last major attempt occurred in December 1983, when almost 21,500 m^3 of water were injected at a depth of approximately 3500 m in EE-2, at surface pressures up to 48 MPa (7000 psi) and at an average injection rate of 108 kg/s. Analyses of the microseismic events induced by water injection during this test showed that the stimulated zone was three-dimensional rather than planar, with dimensions of approximately 800 m high by 900 m wide in the north-south direction by 200 m thick in the east-west direction. Unfortunately, this zone did not intersect the upper EE-3 well as originally predicted, but instead inclined to the east at about the same angle as the directionally drilled boreholes. Subsequent analyses by the Los Alamos team indicated that the microseismically active region represented the stimulation of multiple natural joints rather than a single vertical fracture, and that a shear-slippage mechanism was dominant in the microseismic signals. These findings agreed with results from the British tests, but were inconsistent with conventional tensile failure theories originally thought to be operative.

With no hydraulic connection to the upper well, the Los Alamos team decided to sidetrack and redrill EE-3 to create a workable reservoir. The microseismic event map from the 21,500 m^3 injection test was used to design the EE-3A trajectory for maximum penetration through the seismically active region. Major flow connections to EE-2 were detected during the redrilling and further opened during several stimulation operations from EE-3A. The final result was a well-connected HDR reservoir ready for testing. A 30-day high-pressure reservoir flow test, referred to as the Initial Closed-Loop Flow Test, was conducted in May and June of 1986 (Dash et al., 1989), with very favorable results. Measured tracer-determined residence-time distributions highlight the differences in flow patterns between this and the earlier reservoir at Fenton Hill. A larger reservoir (modal) volume, with a larger portion of the flow distributed in longer residence-time flow paths, was found for the deeper system. During the 30-day test, the modal volume increased from 270 to 350 m^3. With a constant reinjection temperature of 20°C, the production temperature steadily increased to about 190°C, corresponding to a thermal power level of about 10 MW, with no thermal drawdown observed. Furthermore, silica and Na-K-Ca geothermometers indicated *in situ* reservoir temperatures of about 242°C and 222°C, respectively, consistent with an initial measured rock temperature of 232°C at an average reservoir depth of 3550 m. Temperature surveys of the production and injection regions indicated that a multiple network of fractures connects EE-2 and EE-3A.

Present plans call for continued testing of the current reservoir at Fenton Hill, with a long-term (one- to two-year duration) circulation test scheduled to begin in 1991. The objectives are to characterize steady-state power production, flow impedance, and water loss by monitoring injection and production flow rates and temperatures, and to infer the state of reservoir thermal drawdown by using chemically reactive tracers.

One of these long-term reservoir test objectives may already have been met by a pressure test presently under way. Since May 1989, the reservoir has been maintained close to a constant pressure of 15 MPa (2176 psi) above hydrostatic, in order to measure the temporal variation of the reservoir water loss rate due to permeation to the far field (Brown and Robinson, 1990). During this test, the production well has been shut in and thus able to provide a continuous measurement of the reservoir pressure at a position 110 m from the injection well. Figure 5 shows the permeation loss rate plotted versus the logarithm of time, which is appropriate for transient

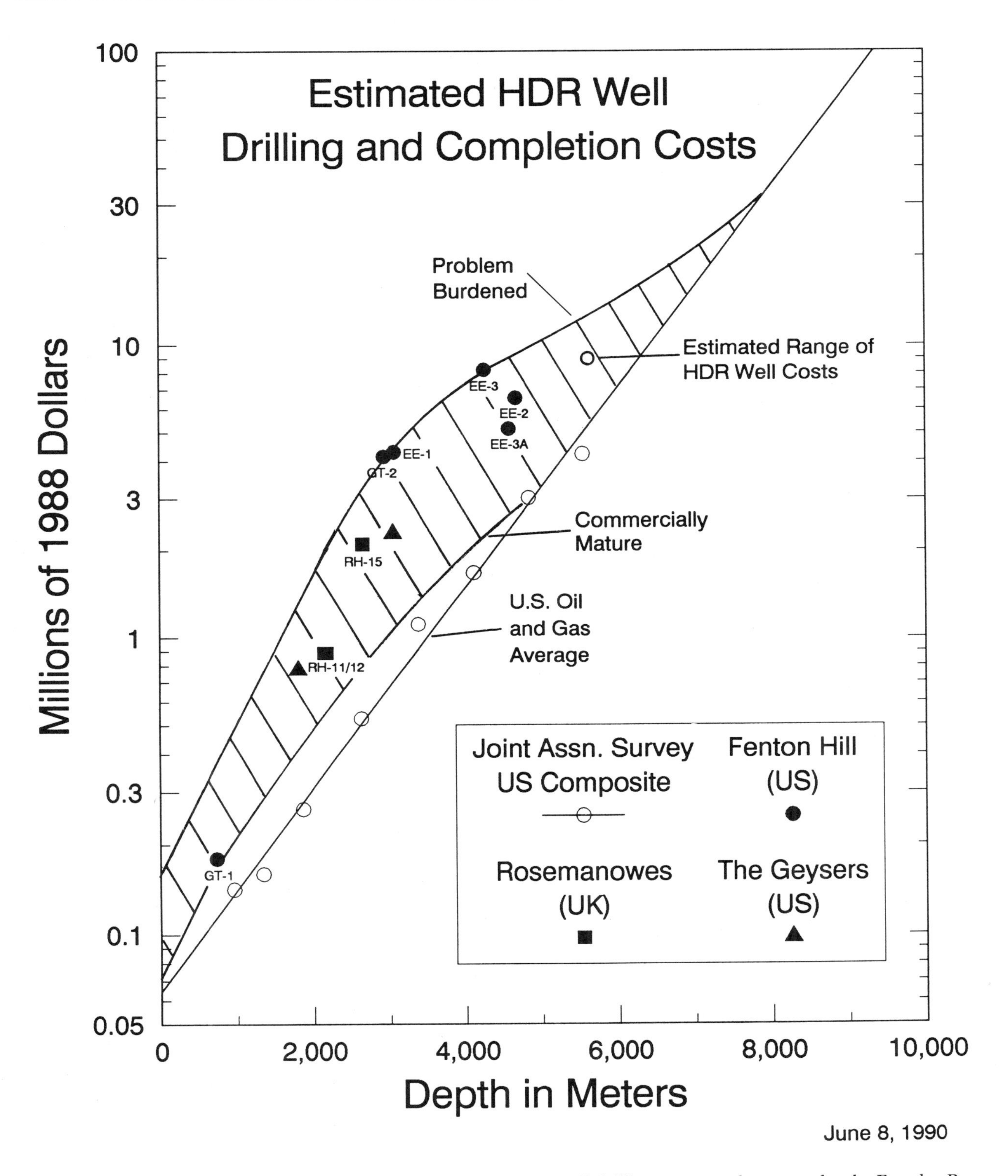

Figure 4. HDR well-drilling costs compared with oil and gas well-drilling costs to the same depth. For the Rosemanowes UK HDR wells, British pounds were assumed equal to dollars to account for the higher relative drilling costs in the UK. Joint Association Survey data (1988) using the methodology of Tester and Herzog (1990) are shown.

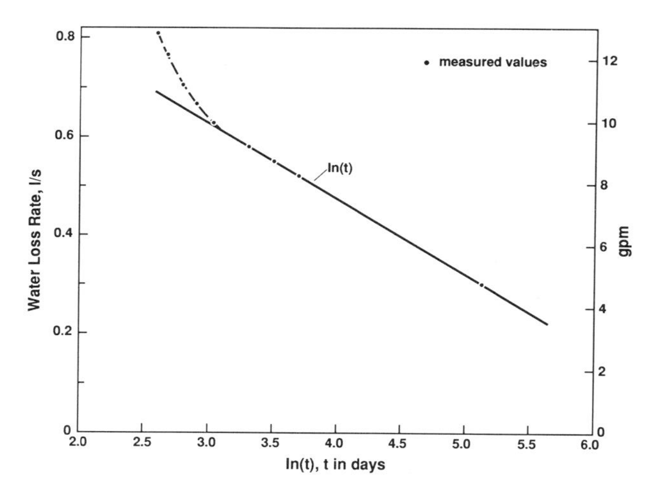

Figure 5. Long-term reservoir permeation loss rate profile at 15 MPa above hydrostatic. (June through October, 1989.)

diffusion from a nonplanar reservoir region. As can be seen, after six months of constant pressure operation [ln(182) = 5.2], the measured permeation loss from this 0.3 km^3 stimulated region has declined to only 0.3 liters/s (4.6 gpm). This result is significant, since the anticipated mean reservoir pressure during the planned long-term flow test will be approximately 15 MPa at a surface injection pressure of 24 MPa (3500 psi).

Utilization Options

One of the axioms of efficient power conversion is to transfer heat to the working fluid of the power cycle at a temperature as close as possible to the heat source temperature. This introduces the concept of availability (or Second Law efficiency) into the design and selection process for power generation systems. Fossil fuels, as "high-grade fuels," would be better utilized by supplying heat to working fluids at high temperature. Conversely, geofluids—that is, "low-grade fuels"—are more appropriate for supplying heat to working fluids at lower temperature. Consequently, the best utilization (in a thermodynamic sense) of the very large HDR resource would be low-temperature heating applications—for direct use to supply process or space heat needs, or for feedwater heating to a conventional fossil-fueled steam Rankine-cycle power plant. As pointed out by Khalifa (1980), effective feedwater heating holds for HDR geofluid production temperatures up to at least 250°C. The corollary is that "high-grade" fossil fuels and obviously electricity itself are poorly utilized in a thermodynamic sense when they are used for low-temperature applications such as space and domestic hot water heating.

Five utilization options for HDR were selected by Tester et al. (1989). These options are listed here to show the breadth of possible applications for the heat produced from HDR.

Hybrid steam cycle with HDR geofluid feedwater preheating. In a conventional fossil-fueled steam power plant, HDR geothermal energy could supply the lower-temperature feedwater heating requirements, typically replacing all but the last stage of regenerative feedwater heating.

HDR-augmented combined gas turbine cycle. Most (or all) of the feedwater heating requirements for the turbine exhaust-driven bottoming steam cycle portion of the combined gas-turbine cycle could be supplied by the geofluid, in a manner analogous to the hybrid steam cycle concept above.

HDR direct heat use. The geofluid would be used, either directly or through a heat exchanger, for process or space heating.

HDR baseload electric power generation. The geofluid would be used to generate power in either a dual-flash or a binary cycle.

HDRALPS: HDR-augmented liquid pumped storage. This power-peaking and load-leveling concept utilizes the inherent mechanical elasticity of a fractured HDR reservoir during cyclic pressurization. Excess off-peak baseload power would be used to inject fluid into the HDR reservoir, with either a cyclic or a continuous geofluid production flow.

In addition to these applications, an HDR system could conveniently be used to sterilize and distill sewer treatment plant effluents, and thereby produce potable water. In the arid southwestern United States, this method of water conservation might become very attractive. For example, the million or so gallons of fluid discharged on any given day from the local Los Alamos County system (at about 700 gpm) is sufficient to pressurize the Fenton Hill reservoir to 15 MPa in the absence of any permeation outflow, or even to provide the primary source of injection flow at a thermal power level of about 25 MW.

Economic Issues

The key economic components for HDR geothermal power systems are similar to those for natural hydrothermal systems, namely, well costs for drilling and completion and power plant costs. Power plant capital costs for HDR-based systems are expected to be comparable with those for existing organic Rankine cycle or multi-stage flash plants designed for low-salinity hydrothermal brines ranging in temperature from 200 to 300°C. The situation for drilling costs is less comparable.

Two primary factors control the cost of producing electricity or heat from HDR. The first is the grade of the resource, which can be quantitatively represented by the effective geothermal gradient. This resource factor strongly influences drilling costs. The second is the productivity of the reservoir, which is defined by several reservoir performance parameters including initial fluid temperature, flow impedance, mass flow rate per pair of wells, and the thermal drawdown rate. The initial fluid temperature for a given HDR resource can be selected by drilling to a certain depth with a defined gradient, whereas the drawdown rate will depend on a single loading parameter, the effective reservoir heat transfer surface divided by the mass flow rate. Minimal drawdown rates are desirable, but they cannot be obtained by merely reducing the mass flow rate, for obvious economic reasons. The alternative is to maximize the effective heat transfer surface to permit commercially acceptable flows and thermal production rates per well pair. Therefore, the key economic component influencing HDR development is the total cost required to drill into a region of hot rock and to create sufficient reservoir area and volume for a commercial system. These constraints contribute to the risk associated with HDR. Because we are in an early developmental stage of a new technology, the element of risk is high. Government incentives, or even full subsidies, for the first generation of plants may be required to lower perceived risks to a level acceptable to private investors. An open issue is the extent to which a new technology such as HDR must demonstrate its viability before becoming commercially sustainable. Until the costs associated with environmental externalities of competing energy supplies from fossil and nuclear sources are accounted for properly, HDR will be somewhat at a disadvantage in an "open market" economy.

Early economic modeling of HDR prototype systems considered the interaction between these two economic factors in trying to establish minimum performance goals for reservoir development (Armstead and Tester, 1987, and Tester and Herzog, 1990). Assuming prevailing costs for drilling and completing wells and for surface power plant facilities, breakeven conditions were estimated and sensitivity analyses carried out to quantify the importance of the various technical and economic factors. With reasonable assumptions regarding resource quality and reservoir performance, the economic picture looks good; that is to say—if HDR reservoirs are created in areas with above-average temperature gradients and if they have productivities comparable with natural geothermal systems already in operation around the world—the estimated costs for generating electricity are competitive with other sources. As expected, these early economic forecasts carried out by EPRI (Cummings and Morris, 1979) and Los Alamos (Murphy et al., 1982) show a strong dependence of breakeven electricity prices on drilling costs, debt interest rates, and anticipated rates of return on equity capital. The base-case condition assumed a 40–60°C/km resource, a reservoir productivity of 75 kg/s per pair of wells with about 20 percent drawdown in 10 years, drilling costs about two and one-half times those of oil and gas wells (see Figure 4) of comparable depth, and real interest and equity rates of 9 and 12 percent. The resulting breakeven price was 5 to 7 cents/kWh of electricity generated (in 1989 dollars).

In previous forecasts of HDR-generated electricity, high-grade HDR resources would be developed first for practical economic reasons, as these resources require much shallower drilling depths and consequently result in significantly lower reservoir development costs. For example, for a 60°C/km resource, one needs to drill only to depths of 4 km (13,000 ft) to reach initial rock temperatures of about 250°C, which is more than sufficient for generating electricity. However, by using HDR in an integrated energy conversion system where the geothermal fluid would be used for feedwater heating, lower-grade HDR resources could be utilized economically now in the generation of electricity. In the eastern US, where the average geothermal gradient is lower and where production temperatures from a resource depth of 4.5 km (15,000 ft) would be in the 140–170°C range, the use of an HDR heat source in a combined-cycle power plant could displace up to 15 percent of the fossil fuel requirements for generating electricity (City of Burbank, 1977). Although the quantitative impact would depend on the source and type of fossil fuel displaced, this application of HDR would have a positive effect, both economically and environmentally.

An industrial group headed by Bechtel National, Inc. completed an independent study of HDR economics in 1987. The Bechtel study (1988) considered the development of an HDR system for generating electricity in a high-grade (~78°C/km) resource located at the Roosevelt Hot Springs site in central Utah. After a careful review of the state of the art in HDR technology, they developed a design for both the underground reservoir and the surface power plant to produce about 50 MWe, at a total capital cost of about $52.6 million for the well/reservoir system and $68.9 million for the power plant. When operating and maintenance costs were added, a levelized revenue requirement of 5.0 cents/kWh resulted, remarkably close to costs predicted by Armstead and Tester (1987) and the other studies cited earlier. Other predictions from a British study (Shock, 1986) and EPRI (Roberts, 1986) show that HDR-produced heat and electricity are competitive with oil priced at about $14/barrel for heat and $24/barrel for electricity. These correspond to an estimated price (in 1986 US dollars) of $2.50 per million Btu for delivered heat or about 6.1 cents/kWh for electricity. Again, these estimates agree quite well with earlier projections.

Most recently, Tester and Herzog (1990) have critically reviewed the above HDR economic studies, as well as those done by the Meridian Corp. (Entingh, 1987) and by the Japanese (Hori et al., 1986). In their review, they normalized these previous HDR economic studies to a common basis, and then presented a reformulated economic model for HDR geothermal energy with revised cost components. Also included in the Tester and Herzog study is a general evaluation of the technical feasibility components of HDR technology, since they are important in establishing drilling and reservoir performance parameters required for any rational economic assessment. Tester and Herzog reached conclusions similar to most of the previous studies, namely that electricity produced from high-grade HDR resources would be competitive at today's energy prices. They further conclude that electricity produced from midgrade (~50°C/km) HDR resources would be only marginally competitive at today's prevailing prices. For low-grade HDR resources (30°C/km or less), significant advances to reduce drilling costs are needed in order for HDR to compete with fossil and nuclear electric systems.

Remaining Research and Development Issues

After successfully flow testing the deeper reservoir at Fenton Hill for one to two years, the primary technical tasks remaining are: 1) the enlargement of the present two-well reservoir system—with an additional production wellbore on the other side of the reservoir—to meet commercial power production requirements and 2) the demonstration of the HDR concept in other geological settings. Based on our present understanding of HDR reservoirs, the next phase of the project would require the logical extension of established HDR reservoir development technology to even larger sizes or different geologic environments. In this context, the HDR heat-mining concept appears amenable to physical extension. If we can build on our success of creating an interconnected network of fractures with the existing pair of wells, then this doublet system can be treated as a modular unit for scale-up, to extract heat from much larger adjacent volumes of hot rock. If HDR reservoirs can be engineered as systematically as envisioned, then this last technical hurdle of creating and testing a commercial-sized system at Fenton Hill is well within reach during the next five years. However, even if this field testing operation is completely successful, several major issues still must be resolved to reduce the risks (both real and perceived) and costs of developing commercial HDR ventures at other sites. These issues deal primarily with the continued engineering of the underground system, and include:

Drilling and completion. Improved conventional and new hard rock drilling techniques that lead to increased penetration rates and reduced costs will encourage commercial HDR development. Further, new or improved methods of wellbore completion and reservoir isolation are needed—particularly in the case of multiple HDR reservoirs created from the same wellbore—that can circumvent most of the problems associated with conventional cemented-in liner and/or casing completions. This is because the formulation and placing of cements suitable for temperatures approaching 300°C at depths up to 5 km is presently very difficult.

Reservoir stimulation. Improved techniques are needed for multiple hydraulic stimulations from the same wellbore, yielding several reservoir regions connecting injection and production wells, with combined swept heat transfer areas large enough to support commercial heat production rates for periods of 10 to 30 years. Improved geophysical diagnostic techniques are required to measure *in situ* stresses and to better understand how natural joint systems open and extend under fluid pressurization. For example, microseismic event maps need to be still more quantitatively related to active reservoir volumes.

Long-term reservoir performance. A commercial-scale HDR reservoir needs to be operated for a significant period of time to demonstrate its thermal-hydraulic performance and to verify models being developed to predict reservoir performance. Because of severely reduced funding, only very limited testing of the Phase II reservoir at Fenton Hill has been accomplished to date (May, 1990). Although the results of testing have so far been very positive, it is too early to predict long-term reservoir performance in terms of the rate of thermal drawdown, impedance changes, induced seismicity, and geochemical behavior.

Work at Los Alamos, at Rosemanowes, and other sites is addressing all of these issues, and at this time we can report considerable progress toward reducing the costs associated with drilling and completion, and with reservoir stimulation. Recent deep hard-rock drilling experience at Fenton Hill has shown that HDR drilling costs can be significantly less than equivalent hydrothermal drilling costs, and closer to conventional oil and gas drilling costs (see Figure 4). In the early 1980s, the Los Alamos National Laboratory directionally drilled two deep holes (EE-2 and EE-3) at Fenton Hill at average completed costs in excess of $16 million in 1988 dollars ($3550/m, or $1080/ft). However, inexperience in drilling highly deviated holes in this high-temperature hard-rock environment greatly added to the costs of these two holes (see Figure 4). More recently (1985 to 1987), the Laboratory has redrilled the deeper, hotter portions of both these (EE-2A and EE-3A) holes at an overall average cost of $500/m ($152/ft), as a result of improved drilling efficiency.

One of the most significant factors that has reduced drilling costs—in combination with more experience—has been the elimination of most of the directional drilling. Analysis of the Phase II reservoir structure (Fehler et al., 1987) indicates that significant amounts of directional drilling may not be required to create and access HDR reservoirs. Typically, one or more of the naturally occurring joint sets will be somewhat inclined from the vertical. Further, most deep boreholes end up being inclined at from 10–20° from the vertical—without any overt directional drilling procedures. The combination of these two factors should allow the initial drilling, reservoir development, and the drilling of the access borehole(s) to be done with a minimum of directional drilling.

In both of the more recent redrilling programs at Fenton Hill, two other significant changes were made in drilling strategy: 1) high-temperature clay-based drilling fluids were used instead of water for lubrication and to remove cuttings, and 2) comprehensive inspections of the drill pipe and drilling assemblies were scheduled, and the drillpipe was rotated on a routine basis. These two drilling procedure changes completely eliminated pipe twist-offs and most stuck drilling assembly problems that plagued earlier drilling programs at Fenton Hill (see Armstead and Tester, 1987, for details). Based upon 1310 m (4300 ft) of straight-hole (nondirectional) drilling during the Fenton Hill redrilling programs in EE-3A and EE-2A over the depth range from 2900 to 4000 m (9,500 to 13,200 ft), the overall average drilling cost was $500/m ($152/ft). For comparison, the associated 350 m (1,150 ft) of directional (downhole motor) drilling averaged $775/m ($236/ft). Completion costs for EE-2A, including a cemented-in liner and a full tieback string of 7-in (178-mm) casing to the surface, were $840,000. These completion costs should be quite representative of the anticipated completion costs for a typical HDR well to a depth of about 3.5 km.

Summary

The challenge remaining before the HDR concept is ready for commercialization in the US is a sufficient demonstration—sufficient from the viewpoint of the energy industry—of efficient and sustained high levels of heat production from deep, hot fractured reservoirs. Demonstrating HDR technology to stimulate private investment will involve the development and extended production testing of several HDR reservoirs in different geological settings and depth regimes, both in the western and eastern United States.

Several technology milestones have been reached for HDR. We have adequately demonstrated that one can drill into regions of hard crystalline rock at temperatures above 300°C, can then create large zones of open, multiply interconnected fractured rock by hydraulic stimulation, and can mine heat for extended periods without dire practical or environmental consequences. We have not seen the failure of any hypothesis involving basic physical principles regarding the effectiveness of the heat-mining process. The pressurization of the natural joints within the reservoir region during normal production flow at elevated pressures appears to sufficiently dilate the reservoir, decreasing the overall impedance to flow, while at the same time accessing a large portion of the previously stimulated reservoir. These volumetric systems will undoubtedly result in improved heat extraction rates over the idealized parallel set of discrete vertical tensile fractures originally envisioned. Finally, the hydraulic stimulation of naturally jointed crystalline rock is usually accompanied by low-level microseismic events that can be located with great precision.

Still, much remains to be done. The demonstration of commercial production rates with acceptable impedances and water losses has yet to be achieved in the United States, Great Britain, or elsewhere. In addition to these required hydraulic characteristics, the major remaining technical objective is to achieve high efficiency in extracting the thermal energy contained in the large stimulated HDR zones defined by microseismic event maps.

We believe that an appropriate national energy policy, regarding alternate energy development programs, would be one that allocates government funding based on two criteria: 1) the potential magnitude of the resource, and 2) the time scale for its implementation. Such a policy implies a continuing strong role for the government during the next phase of HDR development. During this phase, industry would be involved in constructing and testing several HDR demonstration systems at different geological sites, to provide the private sector with sufficient experience and data for large-scale HDR commercialization.

HDR, with its positive environmental features, range of thermal quality, and its inherent simplicity could provide directly, and in concert with other fossil and nonfossil thermal sources, a significant portion of the world's energy demand. Maintaining a sufficiently funded, viable research and development program is the first step.

Acknowledgments

This study was performed under the auspices of the US Department of Energy. We wish to acknowledge the numerous contributions to this paper by Prof. Jefferson W. Tester, Director of the MIT Energy Laboratory, since significant sections have been taken from a previous report authored by J.W. Tester, D.W. Brown, and R.M. Potter (see Tester et al., 1989). This paper has been updated where appropriate, based on new experimental results from Fenton Hill and other recent information.

We would also like to thank Dr. A.S. Batchelor for his critical review and suggestions.

References

Armstead, H.C.H., *Geothermal Energy,* 2nd ed., E.F. Spon, London (1983).

Armstead, H.C.H. and Tester, J.W., *Heat Mining,* E.F. Spon, London (1987).

Batchelor, A.S., "Development of Hot-Dry-Rock Geothermal Systems in the United States," IEE Proc. 134 (pt A, #5), 371–380 (May, 1987).

Bechtel National, Inc., "Hot Dry Rock Venture Risks Investigation," Final report for the USDOE, under contract HDE-AC03-86SF16385, San Francisco, CA (1987).

Brown, D.W. and Robinson, B.A., "The Pressure Dilation of a Deep, Jointed Region of the Earth," to be presented at the ISRM International Conference on Rock Joints, Loen, Norway (June 4–6, 1990).

Brown, L.R., Durning, A., et al., *State of the World, 1988: A Worldwatch Institute Report on Progress Toward a Sustainable Society,* W.W. Norton & Co., New York (1988).

Chandler, W.U., Geller, H.S., et al., *Energy Efficiency: A New Agenda,* American Council for an Energy-Efficient Economy, Washington, DC (1988).

City of Burbank, "Site-Specific Analysis of Hybrid Geothermal/Fossil Power Plants," Prepared for ERDA/DGE, Contract E(0-4-1311), Public Service Dept., Burbank, CA (June 1977).

Cummings, R.G. and Morris, G.E., "Economic Modeling of Electricity Production from Hot Dry Rock Geothermal Reservoirs: Methodology and Analysis," EPRI report EPRI-EA-630, Palo Alto, CA (1979).

Dash, Z.V. (Ed.), "ICFT: An Initial Closed-Loop Flow Test of the Fenton Hill Phase II HDR Reservoir," Los Alamos National Laboratory report LA-11498-HDR (1989).

DiPippo, R., "Geothermal Energy: Electricity Production and Environmental Impact—A Worldwide Perspective" in *Energy and the Environment in the 21st Century,* MIT Press, Cambridge, MA (1990).

Entingh, D. "Historical and Future Cost of Electricity from Hydrothermal Binary and Hot Dry Rock Reservoirs, 1975–2000," Meridian Corp. report 240-GG, Alexandria, VA (Oct., 1987).

Fehler, M., House, L., et al., "Determining Planes Along Which Earthquakes Occur: Method and Application to Earthquakes Accompanying Hydraulic Fracturing," *J. Geophys. Res. 92,* 9407–9414 (1987).

Hari, Y., et al., "On Economics of Hot Dry Rock Geothermal Power Stations," and related documents, Corporate Foundation, Central Research Institute for Electric Power, Hot Dry Rock Geothermal Power Station Cost Study Committee report 385001, Japan (March 1986).

Hot Dry Rock Assessment Panel (HDRAP), "Hot Dry Rock Geothermal Energy-Status of Exploration and Assessment," Report #1 June 1977, Energy Research and Development Administration, ERDA-77-74, Washington, DC (1977).

Khalifa, H.E., "Hybrid Fossil/Geothermal Power Plants," *Sourcebook on the Production of Electricity from Geothermal Energy,* Ch. 4, J. Kestin et al. (Eds.), US Govt. Printing Office, Washington, DC, 471–503 (1980).

Kron, A. and Heiken, G., "Geothermal Gradient Map of the United States—Exclusive of Alaska and Hawaii," Los Alamos Scientific Laboratory map LA-8476-MAP (1980).

Lovins, A.B., Lovins, L.H., et al., *Least-Cost Energy—Solving the CO_2 Problem,* Brick House Publishing Co., Andover, MA (1981).

Mathews, J.T., "Global Climate Change: Toward a Greenhouse Policy," *Issues Sci. Technol.* 3 (3), 57–68 (1987).

Muffler, L.J.P. (Ed.), "Assessment of Geothermal Resources of the United States—1978," US Geological Survey Circular 790 (1979).

Murphy, H.D., Drake, R., Tester, J.W. and Zyvoloski, G., "Economics of a 75-MWe Hot Dry Rock Geothermal Power Station Based upon the Design of the Phase II Reservoir at Fenton Hill." Los Alamos National Laboratory report LA-9241-MS (1982).

National Research Council (NRC), *Geothermal Energy Technology: Issues, R&D Needs, and Cooperative Arrangements,* National Academy Press, Washington, DC (1987).

Otte, C., "Geothermal Opportunities for Developing Countries," in *Energy and the Environment in the 21st Century,* MIT Press, Cambridge, MA (1990).

Parker, R.H., "Hot Dry Rock Geothermal Energy Research at the Camborne School of Mines," GRC Bulletin 18 (9) 3–7 (Oct. 1989).

Roberts, V., "EPRI Economic Study," paper presented at the National Hot Dry Rock Program Development Council meeting on July 15, 1986, in Albuquerque, NM (1986).

Rowley, J.C., "Worldwide Geothermal Resources," *Handbook of Geothermal Energy,* Ch. 2, L.M. Edwards et al. (Eds.), Gulf Publishing, Houston, TX (1982).

Shepard, M., "The Politics of Climate," *EPRI J.* 14 (4), 4–15 (1988).

Shock, R.A.W., "An Economic Assessment of Hot Dry Rock as an Energy Source for the UK," Energy Technology Support Unit report ETSU-R-34, UK DOE, Oxfordshire, UK (1986).

Smith, M.C., Aamodt, R.L., Potter, R.M. and Brown, D.W., "Manmade Geothermal Reservoirs," Proc. 2nd US Symposium on Geothermal Energy, San Francisco, CA, 1781–1787 (1975).

Tester, J.W., Brown, D.W., and Potter, R.M., "Hot Dry Rock Geothermal Energy—A New Energy Agenda for the 21st Century," Los Alamos National Laboratory report LA-11514-MS (1989).

Tester, J.W. and Herzog, H., "Economic Predictions for Heat Mining: A Review and Analysis of Hot Dry Rock (HDR) Geothermal Energy Technology," MIT Energy Laboratory report MIT-EL-90-001, Cambridge, MA (July 1990).

White, D.F. and Williams, D.L. (Eds.), "Assessment of Geothermal Resources of the United States—1975," US Geological Survey Circular 726 (1975).

Economic, Environmental, and Engineering Aspects of Magnetohydrodynamic Power Generation

Robert Kessler
Avco Research Laboratory, Inc.
Everett, MA

Abstract

Coal-burning magnetohydrodynamic (MHD) electric power generation technology is described, and its economic and environmental advantages are discussed. Advanced MHD/steam plants can achieve efficiencies of 55 percent to 60 percent with less environmental intrusion than from conventional coal-burning steam plants. The national program for development of MHD power generation is outlined and the development status of individual components and subsystems is presented. An extensive engineering database exists, including analytical models, design information, and test data from experimental facilities at substantial scale. Significant progress has been made in development of all critical component and subsystem technologies. It is concluded that MHD's potential advantages of higher efficiency, lower generating costs, and less environmental intrusion than other direct coal-burning technologies are of national significance, and that there are no insurmountable problems to prevent its successful development.

Introduction

Magnetohydrodynamic (MHD) power generation is a technology for generating electric power from fossil or nuclear fuels by passing an electrically conducting fluid through a magnetic field without rotating machinery or moving mechanical parts. For applications of interest here, the working fluid is composed of coal combustion products. The absence of moving machinery allows the MHD generator to operate at much higher temperatures and, therefore, higher efficiency than is possible with other power generation technologies. Coal-burning central station MHD power plants promise to generate power at up to 50 percent greater efficiency and with lower cost of electricity than can be achieved with current coal-burning power plants.

Environmental intrusion from MHD plants is far below permissible levels, without requiring expensive exhaust gas cleanup systems. This is because pollutant control is inherent in the basic design of MHD power plants. Emissions of SO_x, NO_x, and particulates are well below the New Source Performance Standards (NSPS) of 1979, and, because of the higher efficiency and consequent lower fuel usage of MHD plants, emissions of CO_2, solid wastes, and waste heat are much lower than for less efficient plants. The environmental benefits of MHD power generation are probably even more important to national interests than its substantial advantages of high efficiency, lower fuel usage, and lower cost of electricity.

A number of issues and problems remain to be resolved. A well-defined program for the development of MHD technology is in place, funded by the US Department of Energy (DOE). A proof-of-concept test program for critical components and subsystems is in progress to establish an engineering database that will allow the private sector to evaluate the risks and benefits of the technology before proceeding with commercial demonstration of MHD power generation. This demonstration is planned as a retrofit of an MHD power system to an existing coal-burning steam plant on a utility site. Conceptual designs of two site-specific MHD retrofits, which could be built by the late 1990s, have been completed. Technological progress in important areas provides confidence that a near-term demonstration plant can be built and operated successfully and that commercialization of the technology will follow.

Key Technical Elements

Central station MHD power generation exploits two fundamental principles:

- the Faraday effect (Faraday,1832), by which electric current is induced in a conductor moving through a magnetic field;
- the Carnot principle (Carnot,1824), which leads to the prediction that the efficiency of thermodynamic cycles increases with the maximum cycle temperature.

The implementation of the Faraday effect to produce MHD power is shown in Figure 1. An electrically conducting fluid flows with velocity U, through a magnetic field B, to induce an electric field E, which is orthogonal

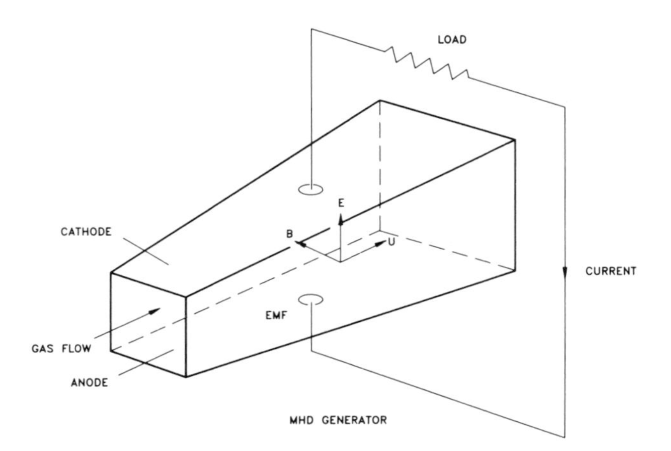

Figure 1. MHD principle of operation.

to both the flow direction and the magnetic field direction. If the flow is contained in a duct, as shown, the two walls perpendicular to the electric field will be at different potentials. If these two walls are connected through an external resistance or load, the field will cause a current to flow through the load, thus generating power.

In principle, any electrically conducting fluid can be used as the working fluid, and power generation has been demonstrated with a number of such fluids, varying from liquid metals to hot ionized gases. For central station power plants of interest here, the working fluid is composed of the products of coal combustion.

In application to electric utility power generation, MHD is combined with steam power generation in a binary cycle. The MHD generator is used as a topping unit to the steam bottoming plant, as illustrated in the simple block diagram in Figure 2. Starting with combustion products at a pressure of 5 to 10 atmospheres and a temperature sufficiently high (about 2900 K) to produce a working fluid of adequate electric conductivity when seeded with an easily ionizable salt such as potassium, the hot ionized gases flow through the MHD generator at approximately sonic velocity. The MHD generator duct, or channel, extracts energy from the gas and the flow is expanded so that it can maintain its velocity against the decelerating forces resulting from its interaction with the magnetic field. The combination of energy extraction and flow expansion causes the gas temperature to drop. Energy is extracted until the gas temperature becomes too low (about 2300 K) to have a useful electric conductivity. The gases exhausting from the generator still contain significant useful heat energy. This energy is used in the bottoming plant to raise steam to drive a turbine and generate additional electricity in the conventional manner of a steam plant, and also to preheat the combustion air. By preheating the combustion air to temperatures of 1650 K to 2000 K, the required high combustion temperatures can be achieved, but this method requires the use of high-temperature refractory heat exchangers. The requisite combustion temperatures can also be attained by enriching the combustion air with oxygen and preheating the oxygen-enriched air to more moderate temperatures, which can be reached in conventional metal tubular heat exchangers. This latter method is satisfactory for use in early commercial MHD plants based on current technology. Development and use of the high-temperature refractory preheaters in future advanced MHD plants would allow realization of the full potential of MHD to reach very high efficiencies.

Those components of a combined MHD/steam power plant which are most directly associated with the MHD process are referred to collectively as the MHD power train, or topping cycle components. The rest of the plant consists of the steam bottoming plant, a cycle compressor, the seed regeneration plant, and the oxygen plant, if necessary. The MHD topping cycle components are the magnet, the coal combustor, nozzle, MHD channel, associated power conditioning equipment, and the diffuser. The magnetic field required in a commercial plant is typically 4.5 to 6 tesla; hence, the magnet is superconducting, as a conventional magnet would require uneconomical amounts of electric power to operate.

Power conditioning is necessary between the MHD channel and the transmission grid. This is because, first, the channel produces DC electric power, as both the magnetic field and the channel flow are in steady state; therefore, an inverter is required to convert the DC MHD output to AC for transmission. Second, a large channel may contain several hundred pairs of electrodes, on opposing walls, to conduct electric current out. As an electrical circuit, the channel consists of a large number of two-terminal outputs, each at a different potential and all coupled internally. Power conditioning is needed to consolidate the outputs from all the terminal pairs for delivery into the main load inverter. The power conditioning circuitry is needed also to control local fields and currents within the channel, to prevent electrical faults, or to clear faults which do occur.

The steam bottoming plant of the combined MHD/steam power plant consists basically of a heat recovery and seed recovery system (HRSR) and a turbine/generator for additional power production. The HRSR is essentially a heat recovery boiler and oxidant preheater that is fired by the exhaust gases from the MHD channel. However, in addition to generating steam, the HRSR system must also perform the functions of NO_x control, slag tapping, seed recovery, and particulate removal.

Control of NO_x is achieved by means of a two-stage combustion process. Primary combustion occurs in the coal combustor under fuel-rich conditions. Secondary combustion takes place in the heat recovery boiler upon cooling the fuel-rich MHD generator exhaust gases in a

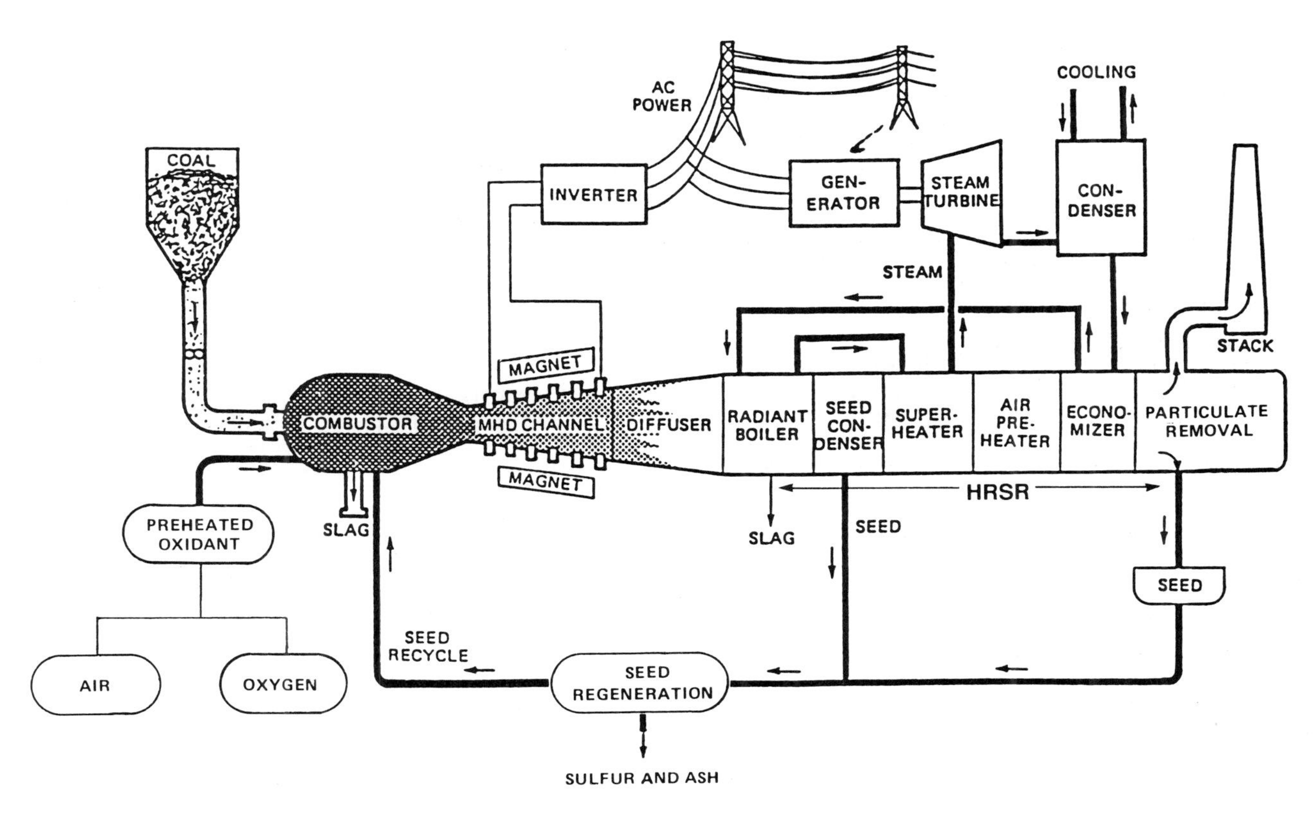

Figure 2. MHD/steam power plant.

radiant furnace that provides a residence time of at least two seconds at a temperature above 1800 K. Conditions in the radiant furnace allow the NO_x to decompose into N_2 and O_2; the cooling rate of the exhaust gas is the key element in this decomposition. The required residence time is provided by appropriate design of the primary radiant chamber.

Control of SO_x is intrinsic to the MHD process. Because of a strong chemical affinity, the potassium seed in the flow combines with sulfur to form potassium sulfate, which condenses in the gas. The condensed potassium sulfate is removed as particulate by the downstream stack gas cleaning equipment, which also removes the fly-ash. The recovered potassium sulfate is then delivered to a seed regeneration unit where the ash and sulfur are removed, and the potassium, in a sulfur-free form such as formate or carbonate, is recycled to the MHD combustor. Thus, SO_x control is combined with recovery and regeneration of seed.

Economic and Environmental Factors

Because the MHD generator has no moving mechanical parts, it can operate at a much higher combustion temperature than other power generating systems. This is the key factor that allows the combined MHD/steam cycle to achieve higher thermal efficiency and lower cost of electricity than other systems. System studies, summarized in Table 1, predict that early commercial MHD/steam power plants will have thermal efficiencies in the 42 percent range, and as the technology matures, thermal efficiencies will increase to 55 percent to 60 percent. This may be compared with 33 percent to 38 percent for modern coal-fired steam plants, with scrubbers.

The early commercial MHD plants shown in Table 1, with 40 percent to 42 percent efficiency, use oxygen-enriched air with moderate (900 K) preheat temperature as the oxidant. The future, most advanced MHD plant shown in the table achieves its very high efficiency of nearly 59 percent mainly because it uses high-temperature (2000 K) air heaters fired directly by the channel exhaust gas instead of oxygen enrichment, and has a more advanced topping cycle, with a higher magnetic field and a channel with higher electrical stresses. Also, the plant operates at higher pressure and has improvements in other parts of the cycle.

Competitive capital costs together with high thermal efficiency yield very attractive cost of electricity (COE) estimates for MHD. Figure 3 shows results from two separate studies by General Electric Co. which compare

Table 1. Comparison of MHD/steam plant with conventional steam plant. Primary fuel for each plant is pulverized Illinois #6 Coal.

Type of Cycle	Conventional Pulverized Coal Steam	First Commercial MHD/Steam	Advanced Direct-Fired MHD/Steam
Net Plant Output, MWe	954	212 - 492	961
Net Plant Efficiency, %HHV	37.4 (a)	40.2 - 41.7	58.5
Combustor Oxidant Mole Concentration of Oxygen	air	air + oxygen 32 pct	air
Type Air Heater	Lungstrom-type regenerative	recuperative metal	direct-fired regenerative refractory
Combustion Air Preheat, °F Combustion Pressure, atm	600 1.0	1200 5.5	3100 14.5
MHD Stress Level	none	POC (b)	advanced
Steam Cycle:	Current	Current	Current
Throttle Pressure, psi	3500	2400	3500
Temperatures:			
Superheater Outlet, °F	1000	1000	1000
1st Reheat Outlet, °F	1050	1000	1050
2nd Reheat Outlet, °F	none	none	none

a) Obtainable only in modern plants; average for all U.S. plants is 32.8% b) Similar to levels planned for POC tests

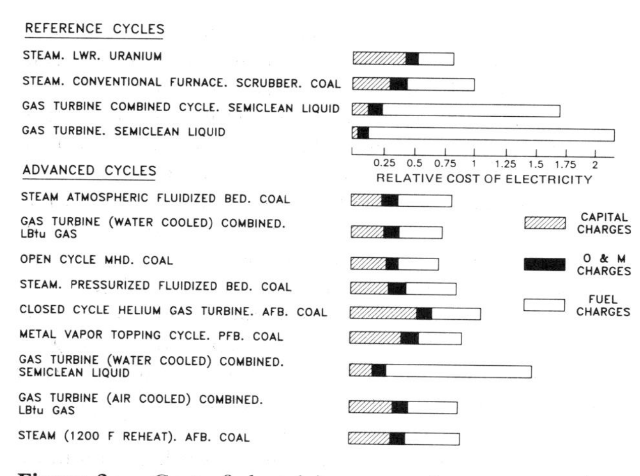

Figure 3a. Cost of electricity comparison—comparison with advanced cycles.

projected COE from various cycles. Figure 3a (Pomeroy, 1978) shows COE comparisons for a number of advanced power cycles, and Figure 3b (Muller, 1983) shows comparisons of MHD plants with conventional steam plants, at various sizes. The cost of electricity from MHD plants is substantially lower than from a modern steam plant. The earlier study, Figure 3a, assumed a coal cost of $0.97/MBtu or $18/ton (1978 dollars; MBtu = 1000 Btu). As the cost of fuel increases, the COE advantage of MHD increases correspondingly because of its higher thermal efficiency.

Basic to the design of the MHD system is the ability to achieve high plant efficiencies while at the same time controlling pollutant emissions at very low levels. Figure 4 shows measured emissions of SO_x, NO_x, and particulates compared to NSPS requirements. The figure shows test results obtained from a subscale experimental HRSR fired by a simulated MHD flow train operating on Illinois No. 6 coal with sulfur content of 3.3 percent to 4 percent (Johanson and Muehlhauser,1989).

About 90 percent of the particles in the exhaust gas stream consist of K_2SO_4. The rest are fly-ash. No toxic metals or compounds have been detected in the effluent from the experimental HRSR plant. Particle emissions are below NSPS standards with either baghouse or ESP (Electrostatic Precipitator) particle removal devices (At-

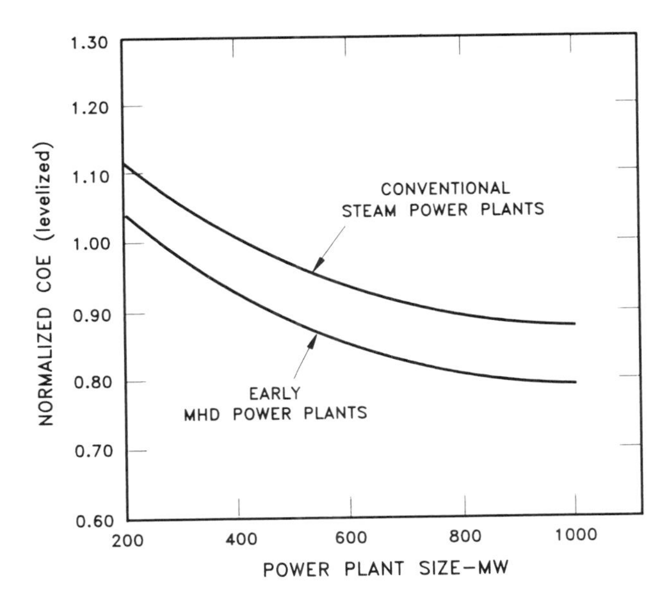

Figure 3b. Cost of electricity comparison—comparison with conventional steam plant.

tig et al., 1989). Other equipment, e.g., wet scrubber, is not necessary.

Table 2 compares carbon dioxide emissions, solid wastes, and heat rejection from advanced MHD/steam power plants with those from conventional plants. These are considerably lower from MHD plants than from less efficient plants because of MHD's lower fuel usage.

National MHD Development Program

The national MHD development program (Wright, 1989a) is currently funded by the US Department of Energy (DOE), through its Pittsburgh Energy Technology Center. The program objectives are to establish, through proof-of-concept testing, an engineering database that will allow the private sector to evaluate the risks and benefits of the technology before proceeding with commercial demonstration of MHD. The Proof-of-Concept (POC) program, shown in Figure 5, is aimed at performance and lifetime of major components and subsystems and contains four major elements:

1. The Conceptual Design program, to prepare conceptual designs of two possible MHD retrofit plants.
2. The Integrated Topping Cycle program, to demonstrate scalability, performance, and lifetime of the MHD topping cycle through long-duration (1000 hours) testing.
3. The Integrated Bottoming Cycle program, to develop technical and environmental data for the bottoming cycle through long-duration (4000 hours) testing.
4. The Seed Regeneration program, to design and construct a system to regenerate a sulfur-free seed compound from the mixture of spent potassium, sulfur, and fly-ash removed from the MHD exhaust gas.

These programs were initiated in 1987–88, and follow a lengthy prior period of component development. The major test programs will be conducted at two DOE test facilities, the integrated topping cycle at the 50 MWt Component Development and Integration Facility (CDIF), in Butte, Montana (Hart et al., 1989), and the integrated bottoming cycle activities at the 28 MWt Coal Fired Flow Facility (CFFF), in Tullahoma, Tennessee (Johanson and Muehlhauser, 1989). The CDIF has a complete coal-fired MHD power train, a 3 tesla iron core magnet, and an inverter that interfaces with the local utility grid. The CFFF has a complete HRSR system fired by a coal-fired flow train, but does not operate with a magnet.

Conceptual Design of MHD Retrofit Plant

The most favorable path to pilot scale demonstration of MHD, the necessary first step in its commercialization, is the repowering of an existing utility plant. This approach is taken also by other nations which are developing coal-fired MHD (Wright et al., 1989b). Repowering of existing plants allows the use of many existing systems at considerable cost savings compared to the cost of building a new plant from the ground up. Existing systems which can be used are, for example, the steam turbine and generator, major parts of the cooling water system, the electrical transmission system, waste handling, auxiliary support systems such as fire protection, heating, ventilation and plant utilities, and, of course, the site itself.

Conceptual designs of two repowered existing coal-fired plants have been performed (LaBrie and Egan, 1989; Van Bibber et al., 1989). These designs are useful in evaluating the practicality of the proposed repowering, in identifying interfaces between the MHD units and the existing plants, and in providing cost and schedule information. They also provide a framework within which to identify important issues and technology gaps to be addressed in the overall proof-of-concept program. The two existing plants studied were the Montana Power Company's Corette plant in Billings, Montana, and Gulf Power's Scholz plant, in Sneads, Florida. The conceptual designs for the two plants were performed by teams led by the MHD Development Corporation, Butte, Montana, and the Westinghouse Electric Corporation, Pittsburgh, Pennsylvania, respectively.

The MHD power train for the Corette plant has a

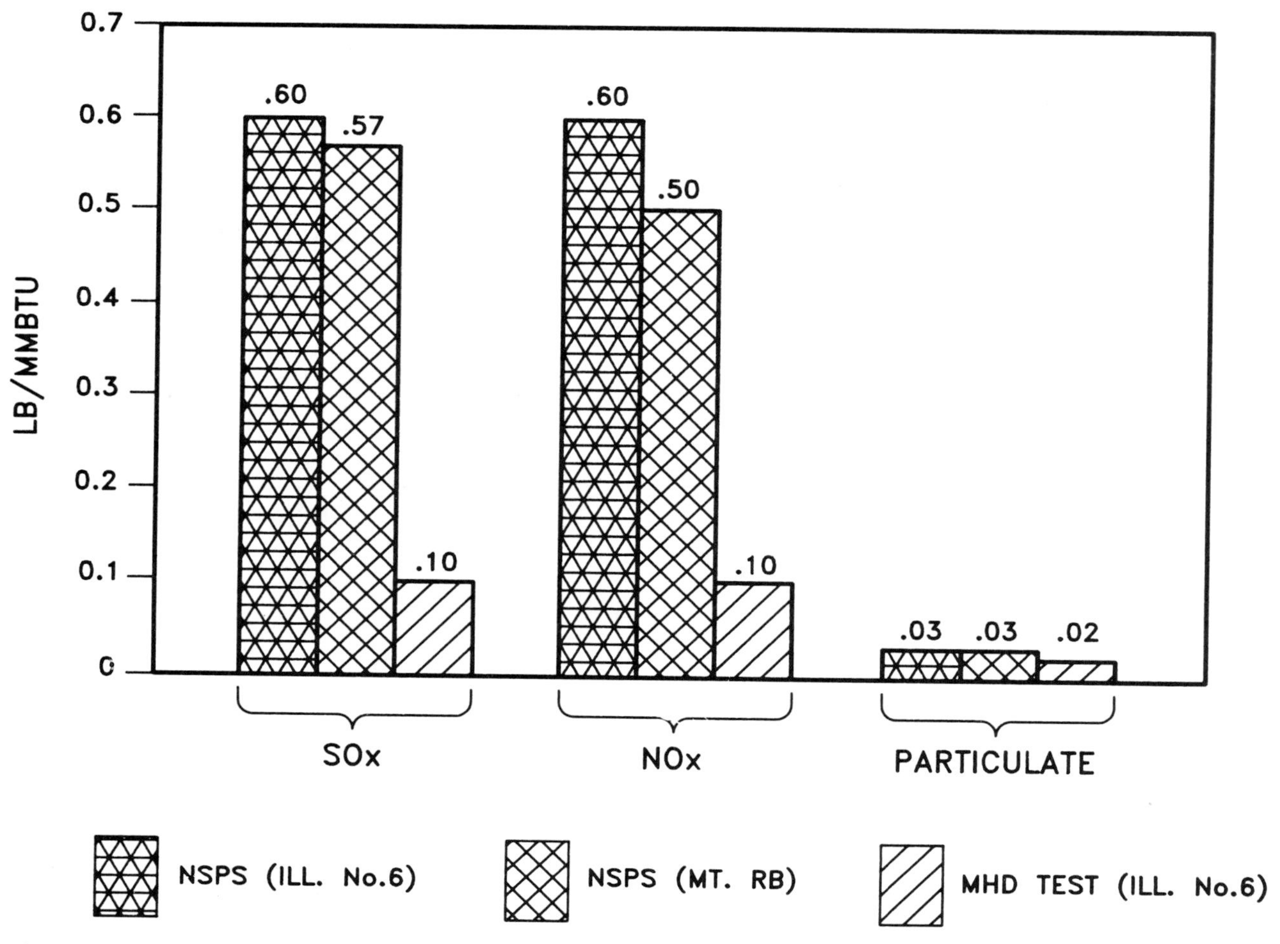

Figure 4. MHD environmental benefits.

thermal input of 250 MWt, which represents a scale-up by a factor of five from existing coal-fired test components, and is about half the size of an early commercial MHD plant. This is considered reasonable for a demonstration plant. The Corette plant operates with Montana Rosebud coal. The type of retrofit designed for the plant is shown in Figure 6. A steam connection is used between the MHD unit and the existing plant components. This means that steam generated in the MHD unit is sent to the existing turbine to generate electricity, together with steam from the existing boiler. The electrical output from the MHD unit is about 98 MWe, of which about 28 MWe comes from the MHD channel and the rest from the MHD bottoming cycle components. The total project schedule, from Title 1 preliminary design through start-up and checkout, is six years.

The Scholz plant burns Illinois No. 6 coal. It would be repowered by means of a steam-side connection, like the Corette plant. The MHD repower unit would produce about 74 MWe, of which 24 MWe would be produced by the MHD generator itself, and 50 MWe would be produced by the MHD bottoming plant.

Integrated Topping Cycle

The objective of the Integrated Topping Cycle program is to build and test, for a total of 1000 hours, an integrated coal-fired MHD flow train consisting of a combustor, nozzle, channel, associated power conditioning equipment, and diffuser. The flow train and the operating conditions are intended to be prototypical of hardware for commercial plants, so that design and operating data can be used to project component performance, lifetime, and reliability in commercial plants. The flow train will operate at 50 thermal megawatts. The integrated topping cycle program follows an earlier program of component testing at the CDIF, and will conclude in 1993. The program is being conducted by: a team consisting of TRW, Inc., which is responsible for

Table 2. Environmental intrusion comparison.

	MHD/STEAM POWER PLANT		CONVENTIONAL STEAM POWER PLANT (a)
	EARLY	ADVANCED (a)	
Carbon Dioxide - lb/MWhr	1550	1151	1861
Solid Wastes - lb/MWhr	220	165	255
Cooling Tower Heat Rejection - Million Btu/MWhr	3.25	1.58	4.28
Cooling Water Consumption - lb/MWhr	3166	1995	4798

(a) Weinstein and Boulay, 1989

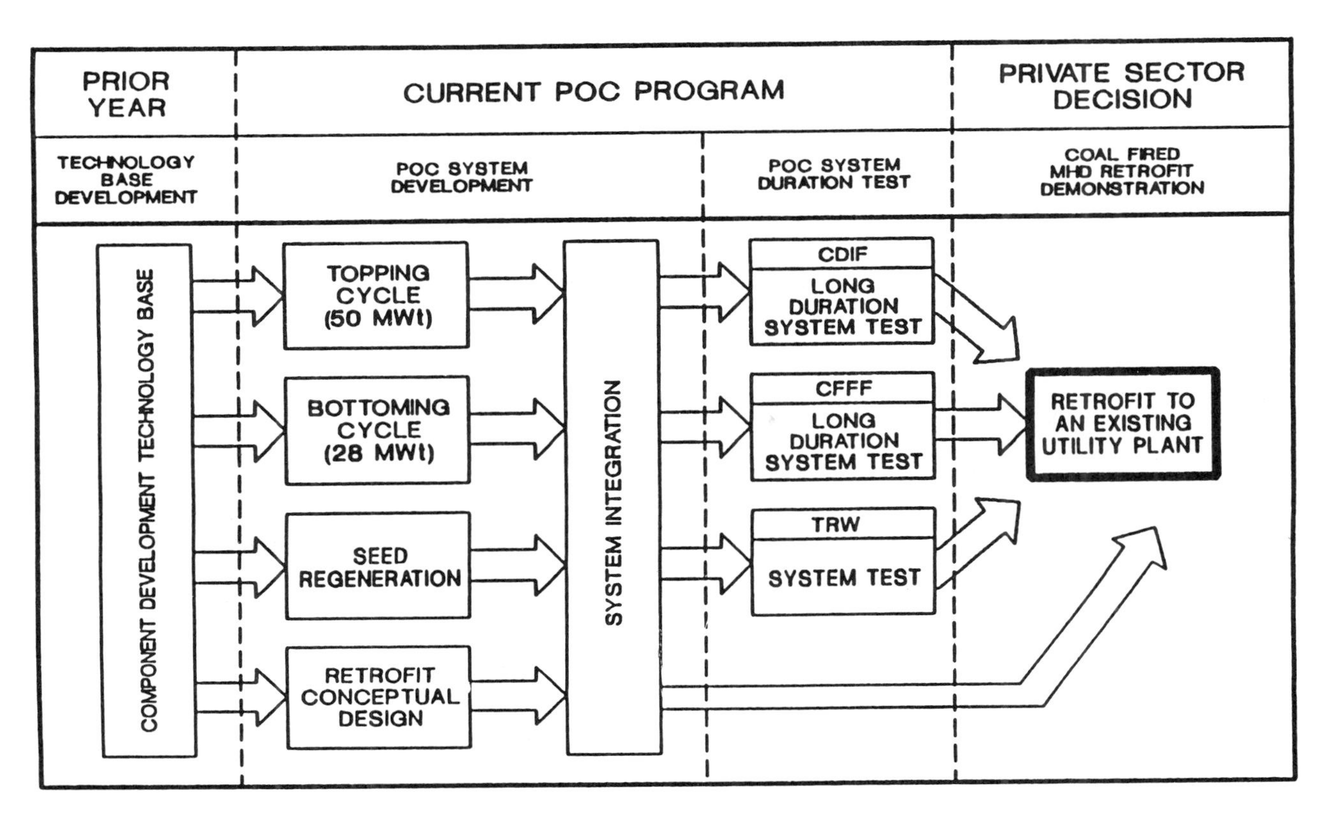

Figure 5. MHD proof of concept program.

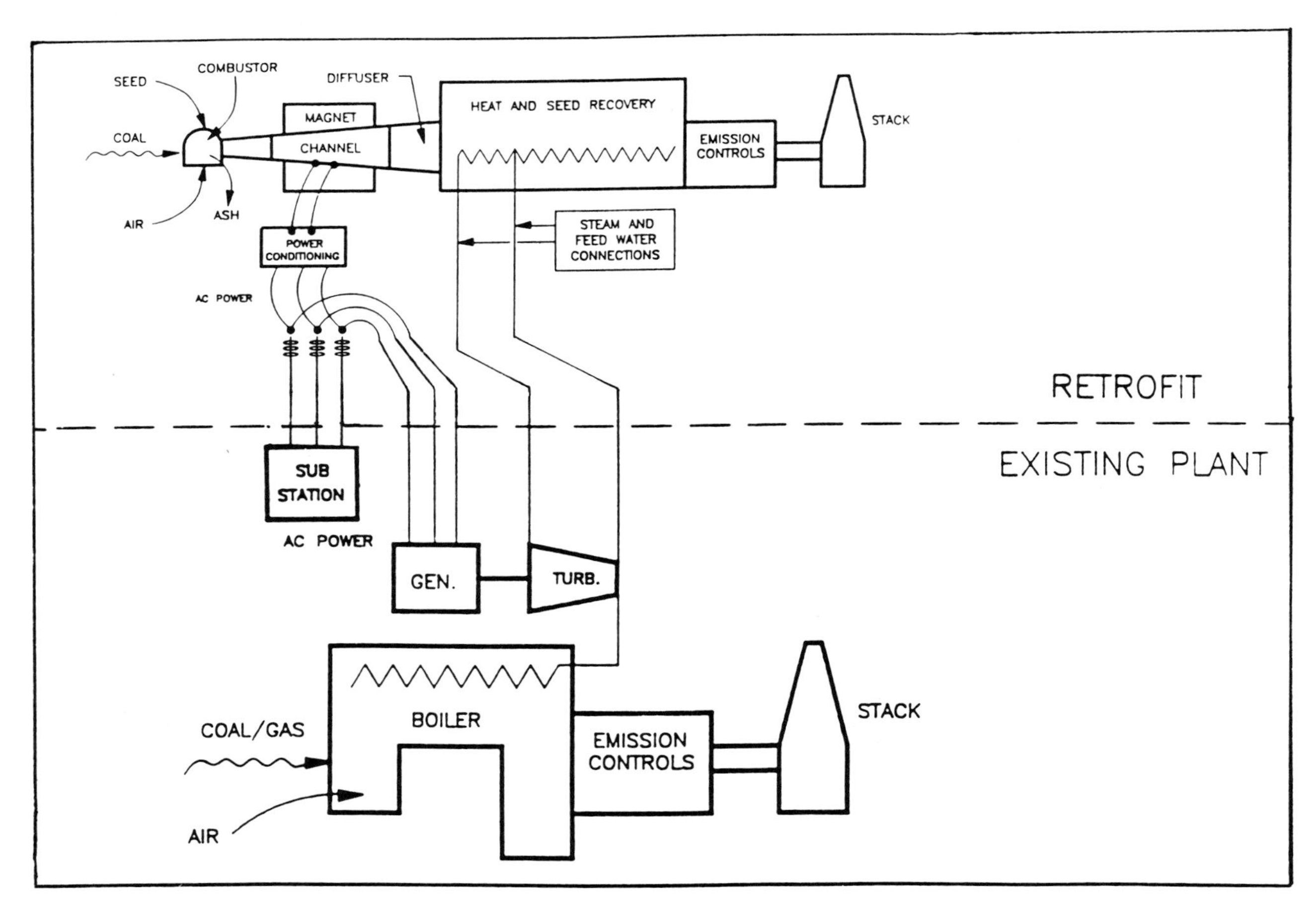

Figure 6. Corette plant retrofit configuration.

the coal-fired combustor, as well as for integration of the overall effort; Avco Research Laboratory, Inc., which is responsible for the nozzle, the channel, the diffuser, and the channel power conditioning equipment; and the Westinghouse Electric Corporation, which will provide current consolidation equipment for the power take-off electrodes at the channel ends.

In order to support the design and fabrication of the prototypical flow train to be tested for the 1000 hours, extensive component testing at the 50 thermal megawatt level is in progress at the CDIF. To date, over 100 hours of topping cycle component testing has been conducted. In addition, testing of electrodes and sidewalls and of coal-fired channels of 20 MWt size is in progress at Avco.

Integrated Bottoming Cycle

The bottoming cycle program is being conducted with a heat recovery and seed recovery system, located at the Coal-Fired Flow Facility in Tullahoma, Tennessee. The system includes a radiant furnace, secondary combustor, air heaters, superheater modules, cyclone scrubbers, a baghouse, and an electrostatic precipitator. It is fired by a coal-fired flow train rated at 28 MWt. Long-term testing is conducted, with the goal of obtaining 2000 hours of operation on each of two types of coal. The two coals of interest are Montana Rosebud, considered to be a representative low sulfur western coal, and Illinois No. 6, a representative high sulfur eastern coal. The goal of the integrated bottoming cycle program is to obtain an engineering database for the heat recovery and seed recovery components of the bottoming cycle, so that operational characteristics, reliability, maintainability and materials performance applicable to commercial systems can be established. The program is being conducted by the University of Tennessee Space Institute (UTSI), with assistance from Babcock and Wilcox Corporation. A small program to provide supplemental materials test data is in place at Argonne National Laboratory.

With the goal of providing data necessary for scale-up to commercial systems in mind, the bottoming cycle components use existing boiler technology and materials wherever possible. Appropriate gas-side conditions are provided by a coal-fired flow train that simulates as closely as possible the composition, temperature, and residence time, among other conditions, in a commercial MHD/steam plant. Specific aspects of system operation investigated include slagging, fouling, erosion, and corrosion of heat transfer surfaces; identification and measurements of gaseous and particulate pollutants; waste management; seed recovery; heat transfer; and system integration and scaling characteristics.

Seed Regeneration

The objectives of the current program are to provide experimental verification of the feasibility of one of the seed regeneration processes which have been selected for early commercial use. Specifically, the program goals are to test a specific process at bench scale, to design a proof-of-concept test plant, and to prepare a conceptual design of a seed regeneration unit for a commercial MHD plant. The specific process under evaluation is the Econoseed process, which is based on the conversion of the recovered potassium sulfate seed to potassium formate, by reacting it with calcium formate (Meyers,1989). The sulfur is removed as calcium sulfate, and the potassium formate is recycled. The work is being performed by TRW, Inc.

State of Development

The current state of development of MHD plant components and subsystems is summarized in Table 3. Requirements against which to compare the current development status are shown for two plants: the retrofit demonstration plant, and a larger, early commercial plant that more realistically represents projected commercial design and operation. The state of development of individual components is described below.

Coal Combustor

The MHD coal combustor must produce a homogeneous uniform product gas with adequate electrical conductivity. Typical MHD power plant conditions require that the combustion gas have a temperature of 2800 K to 2900 K, a pressure of 5 to 10 atmospheres, and an electrical conductivity of about 10 mho/m. The combustor must withstand a molten slag-laden environment, at voltages of 20 kV to 40 kV below ground potential, because it is in contact with the MHD channel. These requirements differ sufficiently from those of conventional coal combustors so as to require essentially new technology for the development of MHD coal combustors.

The combustor must reject a large fraction (50 percent ~ 70 percent) of the ash content of the coal burned, as low slag rejection (high ash carry-over) makes efficient seed recovery more difficult. Operation of the MHD channel itself is relatively insensitive to the amount of slag in the flow, above a level of about 10 percent.

Slag-rejecting coal combustors of 20 MWt scale have been operated at Avco, TRW, and Rocketdyne. A 50 MWt coal combustor built by TRW, Inc. (Braswell et al., 1984) is operated at the CDIF to fire a channel. Satisfactory performance properties for channel operation have been demonstrated. Slag rejection has also been demonstrated, but in a manner not suitable for long duration operation.

Achieving and maintaining electrical isolation of the slag handling system during continuous combustor operation remains yet to be demonstrated. This is planned at the CDIF during 1990.

A feed system suitable for feeding coal into pressurized vessels has been developed successfully at the CFFF, where a 28 MWt non-slag-rejecting coal combustor is in operation. Scaling this coal feed system to commercial size is not considered a problem.

Channel

The MHD channel is the heart of the MHD power generation system. It is the component that produces the MHD power, and its requirements determine the major specifications for other components and subsystems of the MHD power plant. Many types of MHD channels have been designed and tested (Kessler, 1981). Most of these have been linear channels of various electrical and mechanical configurations, although some channels of disk configuration have also been tested.

The basic requirements for channel development are governed by overall plant requirements of high coal-pile to bus-bar efficiency, low cost of electricity and high reliability and availability. To satisfy these plant requirements, three major MHD channel design criteria can be identified (Hals, 1980; Petrick and Shumyatsky, 1978). These are: 1) fraction of thermal energy input extracted from the gas as electric power output (enthalpy extraction ratio); 2) isentropic efficiency; 3) durability and reliability. For the retrofit plant, the required enthalpy extraction is at least 10 percent at an isentropic efficiency of 50 percent. The channel should operate for at least 2000 hours between scheduled maintenance. For early commercial plants, the goals are enthalpy extraction of at least 16 percent at isentropic efficiency of 60 percent, or greater, and operation for several thousand hours between scheduled maintenance.

Figure 7 shows predicted and achieved values of channel enthalpy extraction ratio, as a function of a

Table 3. MHD development status.

Component	Requirements/ Parameter	Retrofit (Corette)	200 MWe early commercial	Technology Status	Achievements/comments
Combustor	thermal input, MWt	250	520	Engineering research	50 MWt combustor operating at CDIF. > 50% slag removal achieved. Electrical and pressure decoupling of slag removal system to be demonstrated.
	oxidant enrichment, %	38	32		
	pressure, atm	5.1	5.2		
	temperature, °C	2550	2550		
	air preheat, °C	650	650		
	slag rejection	>50%	>50%		
Generator	gross MHD power, MWe	28	85	Engineering research	Channel durability demonstrated at low power. 11.8% enthalpy extraction demonstrated for short duration with clean fuel. Electrode life of 5000 to 8000 h predicted from 1000 h tests .
	channel length. m	10	10		
	channel inlet dim. m	0.6 x 0.25	0.6		
	channel exit dim. m	0.6 x 0.75	1.2		
	gas velocity, Mach no	>1.0	>1.0		
	enthalpy extraction, %	11.7	16.3		
	isentropic efficiency, %	45	60		
Magnet	tesla	4.5	4.5	Engineering development	173 ton, 6 tesla magnet with stored energy of 210 MJ built and tested. Stored energy of commercial scale magnet > 2000 MJ.
Inverter	must convert DC output from as many as 500 internally coupled, two terminal outputs to AC			Engineering development	Fully integrated power conditioning and inverter system connected to an electric utility power grid tested at power levels up to 1.5 MWe at CDIF
Heat Recovery/	air preheat, °C	650	650	Design basis available	CFFF testing in progress.
Seed Recovery	seed recovery,%	>90	>90		
Gas cleaning				Design basis available	Control of NOx ,SOx and particulates demonstrated in CFFF. Further design data required
Seed Reprocessing				Engineering design data required	Bench scale demonstration of components.
Steam Power train				Available Technology	
Coal preparation/ injection				Available Technology	High pressure and high density coal injection demonstrated at CFFF

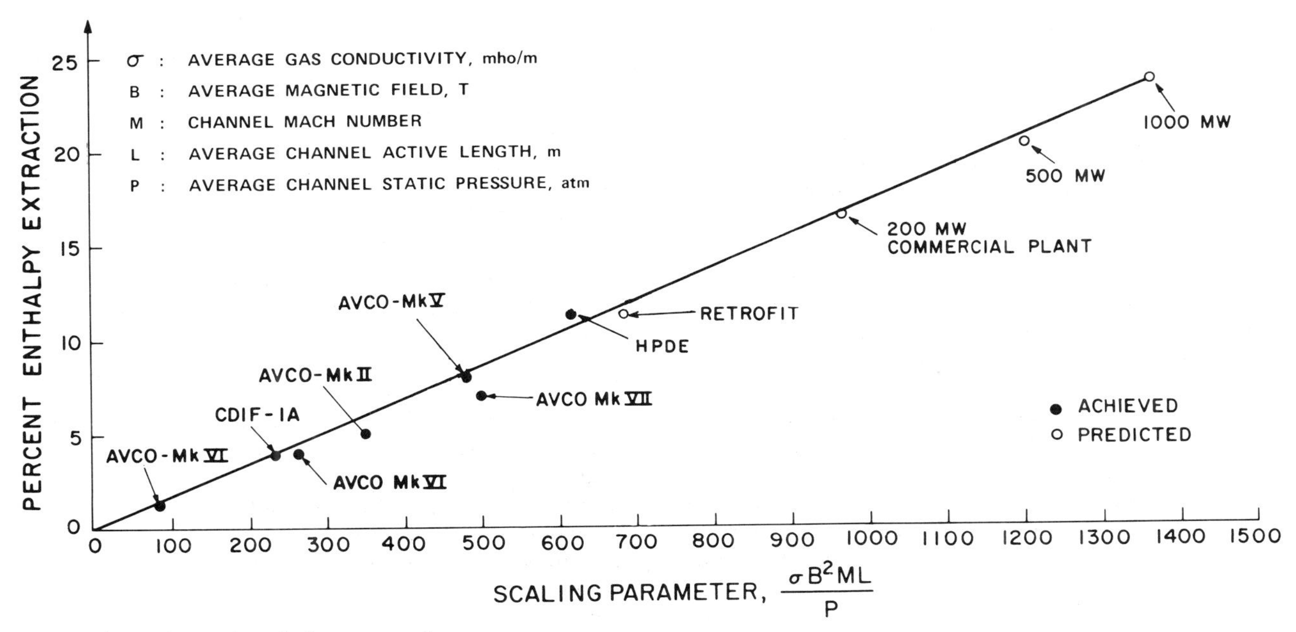

Figure 7. Channel enthalpy extraction.

scaling parameter, for a number of channels operated to date. Enthalpy extraction equal to that required by the retrofit plant has been achieved, in a combustion-driven linear channel of 300 MWt size operated at Arnold Engineering Development Center, although it operated on a clean fuel and only for a short duration (Whitehead et al., 1983). The channel operated at over 50 percent isentropic efficiency. Most channels operated in the US to date were used for duration testing and have been much smaller (20 to 50 MWt). They therefore cannot achieve the performance projected for larger size channels because of the adverse influence of effects which depend on the surface to volume ratio. Although the major value of subscale channel testing is in the development of long-duration designs and operating capability at realistic stress levels, one use of subscale channel testing is to verify predictive capability and scaling laws. Figure 7 shows that channels have, in fact, performed generally in accordance with predictions, and that measured performance does support scaling laws from which performance of full scale channels is projected.

Figure 8 shows the operating durations achieved by various combustion-driven experimental channels. Two significant demonstrations are noted. The first was the operation of an Avco Mk VI channel, at 20 MWt scale, for 500 hours (Demirjian et al.,1979). The electrochemical and thermal stress levels were similar to those expected in commercial coal-burning plants. The test was conducted in two 250 hour segments. Results from this test indicate that durability of properly designed and operated channels can be extrapolated to several thou-

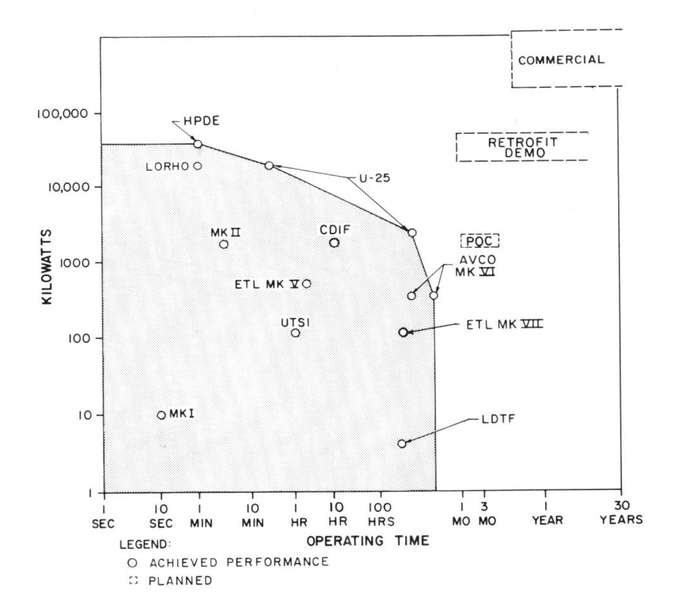

Figure 8. Generator operating durations.

sand hours. The second was a test, also at 20 MWt scale, of anode electrodes for an accumulated duration of 1300 hours (Hruby et al., 1982). Test results indicated a projected lifetime between 5000 hours and 8000 hours for these electrodes. Both of these tests were performed prior to the availability of adequate MHD coal combustors. Coal-burning operating conditions were simulated by injecting ash and sulfur into an oil-fired MHD combustor. Actual coal-fired operation, at 50 MWt scale, is in progress at the CDIF, where tests up to 12 hours long have been achieved to date, at power levels of about 1.5 MWe. Tests up to 100 hours in length are planned, in preparation for the 1000 hour POC test scheduled for 1993.

Figures 7 and 8 show that considerable progress has been made towards achieving acceptable channel performance (power and enthalpy extraction) and durability. However, performance and durability have not yet been demonstrated simultaneously; the next step toward this goal will be the POC tests at the CDIF, followed by the retrofit.

Further work is also required in the area of mechanical design and construction of large MHD channels, especially with respect to construction features which are scalable to commercial size channels. The mechanical design of channels is aimed, first, at achieving electrical and structural integrity of the channel, and, second, at achieving the most efficient use of the magnet bore in order to minimize the required volume of magnetic field, and hence the magnet cost. This is done by compact packaging of channel structure, electrical wiring, water manifolds, hoses, etc. Additional important considerations are channel installation, maintenance, and repair.

Power Conditioning System

Current development goals in this area are to determine optimum combinations of channel electrical configuration and power take-off, effective methods for local field and current control, and reliable and economical techniques for power consolidation. Work has also been done on the integration of MHD plants into electric utility transmission grids (Jackson et al., 1984).

Much early channel testing was done using resistive loads, with the generated power being dissipated. More recently, inverters have been used, in both the US and in the Soviet Union. In the US, inverters have been built and operated with channels at UTSI, at Avco, and at the CDIF. The CDIF inverter is a relatively large 3.5 MWe system built by Westinghouse under Electric Power Research Institute (EPRI) sponsorship (Koester et al., 1983). It is a line-commutated device, using existing inverter technology. No major problems are anticipated in the design, construction, and operation of similar, but larger, units for commercial plants.

Power conditioning devices which perform consolidation and control functions use the same basic circuitry. Such devices have been built by Avco (Hruby, 1984) and tested successfully both at Avco and at the CDIF. At Avco, consolidation and control devices are used on a complete Mk VI channel with 140 electrode pairs, and their effectiveness has been successfully demonstrated at electric power levels of a few hundred kilowatts. At the CDIF, channels have operated partially equipped with such devices, and a channel with full consolidation and control circuitry, capable of handling 1.5 to 2 MWe, will be tested during the POC program. No major difficulties are foreseen in the building and operation of such devices at commercial scale.

The power conditioning circuitry tested to date performs its current control function by forcing the generator electrodes to carry the same average current. An alternative concept, which is in a less advanced stage of development, allows the current from each electrode pair to be controlled independently. A few units of this type have been designed and built by Westinghouse and have undergone preliminary testing.

Superconducting Magnet

Although most MHD channel testing to date has been performed with conventional magnets, commercial-scale MHD power plants will use superconducting magnets, for economic reasons. Magnetic fields of 4.5 tesla to 6 tesla will be required, over warm bore volumes with typical dimensions of 3 to 4 meters diameter and 15 to 20 meters long. Stored energies in such magnets will be 2,000 MJ or greater. The external dimensions will be of the order of 15 m in diameter by 25 m in length. Engineering design of such large structures— which must withstand the high mechanical loads imposed by the magnetic fields while still retaining cooling integrity so that internal temperatures of 4 K can be maintained without excessive heat loss—is a formidable task.

Some technology for large superconducting magnets has been developed, mostly for large bubble chamber and fusion reactor applications. MHD magnets will require complicated saddle-shaped windings, which are more efficient for MHD use. Thus, new winding and fabrication techniques need to be developed. Magnets of commercial size will certainly be too large to transport, and so they will require field assembly with the attendant need to develop suitable fabrication techniques.

MHD superconducting magnets have been built in Japan and in the US. A magnet with a peak central field of 5 tesla and stored energy of about 20 megajoules was built by Argonne National Laboratory for the U-25B facility in Moscow (Kirillin, 1978), as part of a US–USSR cooperative program. The magnet has a tapered warm bore of approximately 50 cm diameter and a length of

2.5 meters. The magnet is about 2 m in overall diameter and 4 m long. The U-25B magnet operated reliably for several years.

The largest MHD superconducting magnet to date was built by Argonne National Laboratory in 1981 (Argonne National Laboratory, 1983). The magnet was designed to be cost-effective and scalable to commercial size. It was successfully tested at its design field of 6 tesla. Its characteristics are given in Table 4.

Studies of MHD superconducting magnets are in progress at the Plasma Fusion Center at MIT to investigate various detailed aspects of magnet design. These include materials studies, which are concerned with the properties of highly stressed structural members operating at temperatures near absolute zero; studies of superconductor configurations and winding techniques; studies of shipping and on-site assembly methods to establish the degree of modularity required; and studies of scaling factors and costs, to enable credible cost estimates to be made.

Heat Recovery and Seed Recovery System (HRSR)

Although much technology developed for conventional steam plants is applicable to HRSR design, several problems peculiar to HRSR design remain to be resolved. These include:

1. material lifetime in the corrosive environment created by the presence of hot combustion gases containing sulfur, seed, and ash. Of particular concern is the effect of condensed seed compounds, mainly K_2SO_4, on the boiler tubes and recuperative air preheater;
2. adequate understanding of the physical processes involved in NO_x and SO_x control. These include the kinetics of NO_x chemistry, radiative heat transfer and gas cooling rates, fluid dynamics and boundary layer effects in the boiler, final combustion of fuel-rich MHD generator exhaust gases, and seed/sulfur chemistry;

Table 4. Argonne 6T superconducting magnet design characteristics.

TYPE	S.C.
PEAK ON-AXIS FIELD (HORIZONTAL)	6T
INLET BORE DIA.	0.8m
OUTLET BORE DIA.	1.0m
ACTIVE FIELD LENGTH	3.0m
OVERALL HEIGHT & WIDTH	4.9m x 4.1m
OVERALL LENGTH	6.4m
TOTAL WEIGHT	173 Tonnes
STORED ENERGY	210 MJ

3. methods for efficient recovery of seed and removal of particulates from the effluent gas, including testing of gas cleaning equipment (ESP, baghouse), investigation of physical and chemical properties of seed/ash particles, and their size distribution and condensation in the gas;
4. techniques for removing slag and seed deposits from heat transfer surfaces, including studies of seed/ash chemistry, and particle condensation, deposition, and fouling of heat transfer surfaces;
5. radiative and convective heat transfer characteristics from seed and ash-laden combustion gases, including effects of seed condensation and particle deposition on heat transfer surfaces,

Extensive testing aimed at resolving these issues is in progress at the CFFF in Tullahoma, Tennessee, where a 28 MWt HRSR, fired by a coal-burning flow train, is being operated. To date, more than 1400 hours of coal-fired testing has been performed, of which 1160 hours have been with eastern coal. Testing with eastern coal is expected to be completed in 1990, and long-duration testing with western coal is expected to begin in late 1990. Measured NO_x and SO_x emissions are well below NSPS standards, and are shown in Figures 9 and 10 (Johanson and Muehlhauser, 1989). Particulate emissions are also well below NSPS standards; either a baghouse or an ESP was used for particulate removal. The concentration of priority pollutant organics in ash/seed samples was about the same as or less than that from

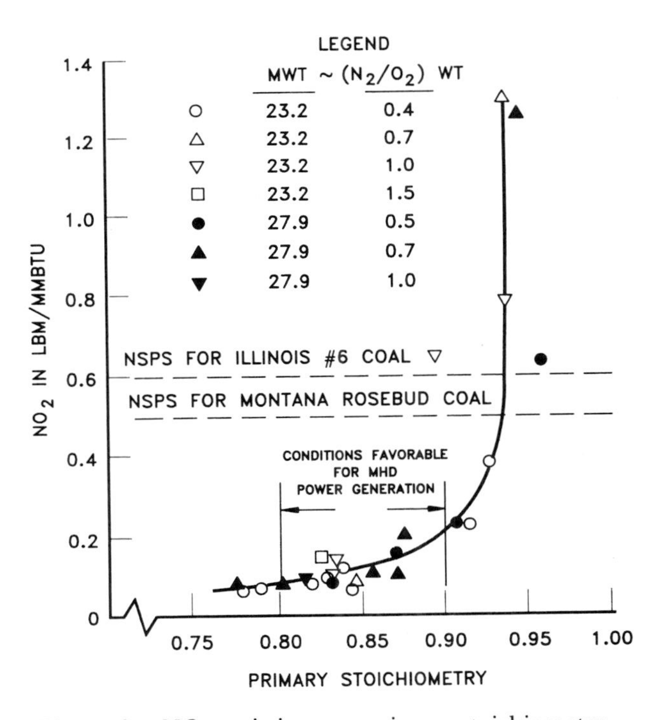

Figure 9. NO_x emissions vs. primary stoichiometry.

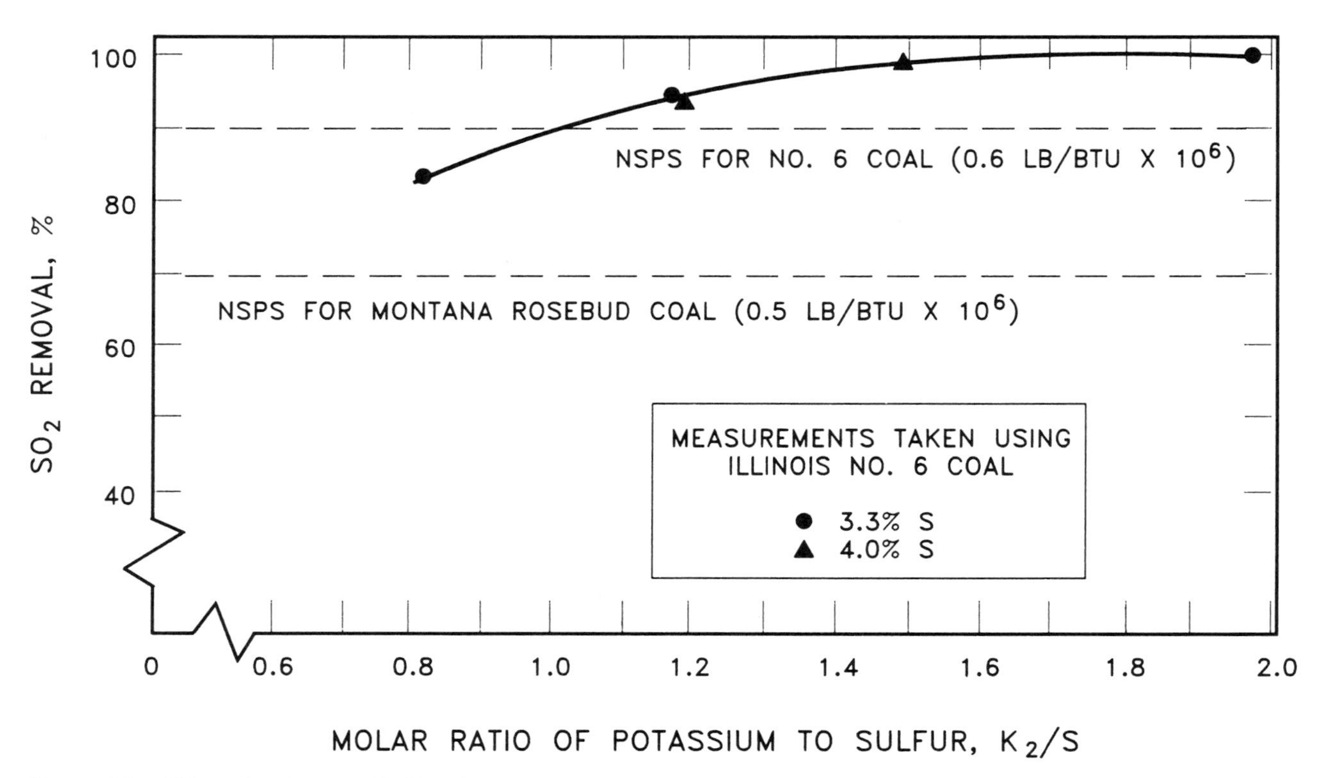

Figure 10. SO_x reductions vs. K_2/S ratio.

conventional coal-burning plants so that no difficulty is expected in disposing of ash or slag from MHD plants.

Seed Regeneration

Although a number of studies of seed regeneration processes have been performed, little experimental work has been done to verify either technical or economic predictions. Current work is focused on the Econoseed process, developed by TRW, Inc., a version of the potassium formate process.

In the Econoseed process, recovered potassium sulfate is converted to potassium formate, by means of reactions with calcium formate, $Ca(COOH)_2$, which is made by reacting hydrated lime, $Ca(OH)_2$, and carbon monoxide. The potassium formate is recycled to the combustor as seed, and sulfur is removed as solid calcium sulfate which is removed by filtration and disposed of. A similar process has been used in Germany for industrial production of potassium formate. The formate process appears practical for MHD application.

The process consists of five sections, shown in Figure 11. Operation of each of the units has been demonstrated at bench scale. Evaluation of the data shows that the process is technically and economically viable and is ready for proof-of-concept testing at a scale of 50 lb/hr to 250 lb/hr.

High-Temperature Air Preheaters

Early commercial plants will use moderate temperature (900 K) oxidant preheat, which can readily be achieved by existing metal recuperative heat exchangers. Future advanced power plants will use higher preheat temperatures (1650 K to 2000 K), which will require refractory-lined heat exchangers fired directly by the exhaust gases from the MHD generator. Heat exchangers able to achieve the required temperatures have been developed, and are used in the glass and steel industries. These are fired with clean fuels. Considerable work on the development of heat exchangers able to operate with the ash and seed-laden MHD generator exhaust gases has been done (DeCoursin, 1983), but more is required. Such equipment, however, is not necessary for early commercial MHD power plants.

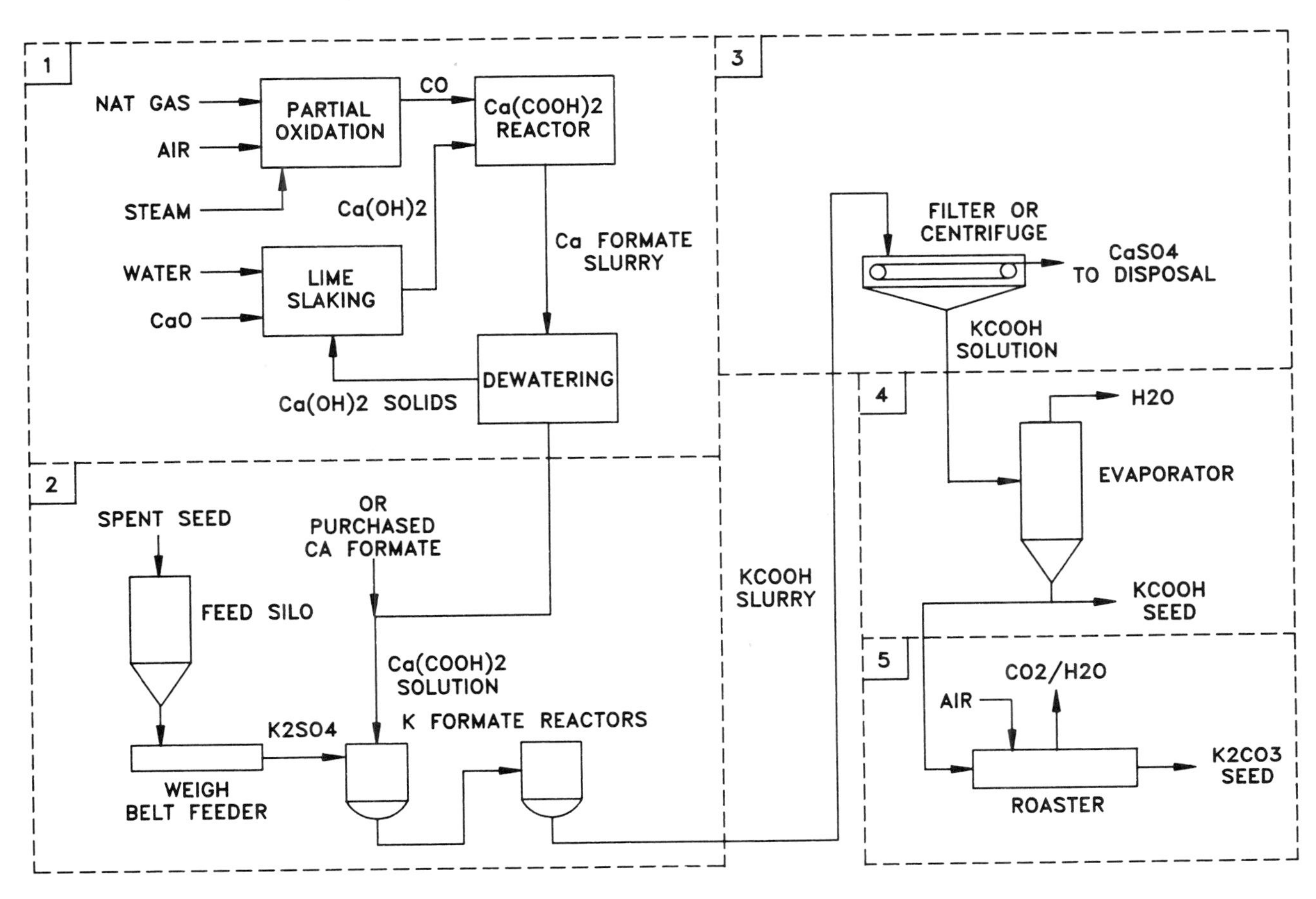

Figure 11. Econoseed process.

Conclusions

Coal-burning MHD/steam power plants promise significant economic and environmental advantages compared to other coal-burning power generation technologies. A well defined proof-of-concept program aimed at performance and lifetime of major components and subsystems is in place so that, by the late 1990s, an MHD demonstration plant can be built. The most feasible path to such a demonstration is a retrofit of an MHD power system to an existing coal-burning utility plant. The demonstration plant will provide evidence that the attractive economic and environmental objectives of commercial MHD power generation can be realized. An extensive engineering database has been developed, including analytical models, design information, and test data from experimental facilities at substantial scale. Significant progress has been made in development of all critical component and subsystem technologies. No insurmountable technological problems are foreseen which will prevent the successful achievement of program goals.

References

Argonne National Laboratory, "Coal-Fired Flow Facility Superconducting Magnet System," Proc. of 21st Symposium on the Engineering Aspects of MHD, (June, 1983)

Attig, R. C., et al. , "Status of POC Testing at the CFFF," Proc. of 27th Symposium on the Engineering Aspects of MHD, (June, 1989)

Boulay, R. B., et al., "Seed Regeneration Integration Options for Commercial MHD/Steam Power Plants," Gilbert/Commonwealth Inc. Report No. 2833 (Draft), (January, 1990)

Braswell, R., et al., "Early Development of the TRW 50 MWt MHD Coal-Fired Combustor," Proc. of 22nd Symposium on the Engineering Aspects of MHD, (June, 1984)

Carnot, S., "Reflections sur la Puissance Motrice du Feu," Bachelier, (1824)

DeCoursin, D., "MHD Air Heater Technology Development," Paper presented at MHD Contractors' Review Meeting, Pittsburgh, (November, 1983)

Demirjian, A., et al., "Long Duration Channel Development and Testing," Proc. of 18th Symposium on the Engineering Aspects of MHD, (June, 1979)

Faraday, M., "Experimental Researches in Electricity," Phil. Trans. R. Society, (1832)

Hals, F. et al., "Results from Study of Potential Early Commercial MHD Power Plants," Proc. 7th International Conference on MHD Electrical Power Generation, (June, 1980)

Hart, A., et al., "Coal-Fired MHD Topping Cycle Hardware and Test Progress at the CDIF," Proc. of 27th Symposium on the Engineering Aspects of MHD, (June, 1989)

Hruby, V. J. et al.,"1000 Hour MHD Anode Test," Proc. of 20th Symposium on the Engineering Aspects of MHD, (June, 1982)

Hruby, V. J., "Current Control and Consolidation Circuit Development at AERL," Proc. of 22nd Symposium on the Engineering Aspects of MHD, (June, 1984)

Jackson, W. D., et al., "Integration of MHD Plants into Electric Utility Systems," Proc. of 22nd Symposium on the Engineering Aspects of MHD, (June, 1984)

Johanson, N. R. and Muehlhauser, J. W., "MHD Bottoming Cycle Operations and Test Results at the CFFF," Paper presented at the 2nd International Workshop on Fossil Fuel Fired MHD, Bologna, Italy, (1989)

Kessler, R., "Open-Cycle MHD Generator Channel Development," *Journal of Energy,* p.178 - p.184, (1981)

Kirillin, V.A., et al., "The U25B Facility for Studies in Strong MHD Interaction," Proc. of 17th Symposium on the Engineering Aspects of MHD, (March, 1978)

Koester, J. K., et al., "Integrated MHD Generator-Inverter System Test Results at CDIF," Proc. of 21st Symposium on the Engineering Aspects of MHD, (June, 1983)

LaBrie, R. and Egan, N., "The Corette Retrofit: Conceptual Design Summary," Proc. of 27th Symposium on the Engineering Aspects of MHD, (June, 1989)

Meyers, R. A., et al., "TRW Econoseed Process for MHD Seed Regeneration," Proc. of 27th Symposium on the Engineering Aspects of MHD, (June, 1989)

Muller, D. J., "Definition of the Development Program for an MHD Advanced Power Train," Proc. 8th International Conference on MHD Electrical Power Generation, (September, 1983)

Petrick, M., and Shumyatsky, B., eds., *Open-Cycle MHD Electrical Power Generation,* Argonne National Laboratory, Argonne, IL, (1978)

Pomeroy, B. D. et al., "Comparative Study and Evaluation of Advanced Cycle MHD Systems," Electric Power Research Institute Report AF-664, v.1, (February, 1978)

Van Bibber, L., et al., "Conceptual Design of the Scholz MHD Retrofit Plant," Proc. of 27th Symposium on the Engineering Aspects of MHD, (June, 1989)

Weinstein, R. E., and Boulay, R. B., "1000 MWe Advanced Coal-Fired MHD/Steam Binary Cycle Power Plant Conceptual Design," Gilbert Commonwealth, Inc. Report No. 2738 (Draft), (June, 1989)

Whitehead, L., et al., "High Performance Demonstration Experiment Test Results," Paper presented at MHD Contractors' Review Meeting, Pittsburgh, (November, 1983)

Wright, R. et al., "Fossil Energy MHD Program in Electrical Power Generation in the United States," Proc. 10th International Conference on MHD Electrical Power Generation, (December, 1989a)

Wright, R. et al., "Open Cycle Coal-Fired MHD Retrofit," Proc. 10th International Conference on MHD Electrical Power Generation, (December, 1989b)

Fuel Cells: Power for the Future

Frank C. Schora
Elias H. Cámara
M-C Power Corporation
Burr Ridge, IL

Abstract

Because of their high efficiency and other favorable operating characteristics, fuel cells should find wide application in the future on a cost-competitive basis with other power conversion techniques.

Three different fuel cell types are currently under development in the US, Japan, and Europe as power generation devices for stationary applications. Fuel cell systems have the capability of attaining high efficiency in the conversion of natural gas or coal gas to electricity with negligible emissions of pollutants such as SO_x and NO_x. They accomplish this as electrochemical devices rather than as heat engines. The phosphoric acid fuel cell is now being offered for commercial application, both in the US and Japan. An advanced concept, the molten carbonate fuel cell, will be ready for commercial application by the mid 1990s, and it is projected that the solid oxide fuel cell will be at this level of development by the year 2000.

Introduction

This discussion is presented as an overview and update on the commercialization of power generation through fuel cells. As such, the paper will necessarily raise more questions than it answers. For a more detailed treatment of the technology, the reader is specifically referred to the recent *Fuel Cell Handbook* by Appleby and Foulkes (1989).[1] Even that excellent effort, however, is limited by space and includes over 2800 references with more detailed analysis.

This paper does *not* present technical detail; no technological breakthroughs are reported (and none are expected, because the current engineering effort only whittles away at the frontiers). Rather, the paper is designed to familiarize its audience with the prospects for the commercialization of this important technology.

This presentation is biased toward the molten carbonate fuel cell because, in the authors' opinion, this technology will emerge as *the* preferred power generation system of the future. Of course, only time can tell if this opinion is justified.

Historical Perspective

Fuel cells are not new. It can be argued that they have existed in nature for millions of years in the form of the organs and muscles that supply the shock of the electric eel and other electric fish.[2] Present day technology, however, did not evolve from a fish, but was the invention of Sir William Robert Grove, a jurist in England, who devised and built the first fuel cell in 1839.[3] In this first cell, he used hydrogen as fuel, oxygen as oxidant, and electrodes made of platinum and dilute sulfuric acid for the electrolyte. A symposium to commemorate the 150th anniversary of this occurrence was held in London this past year.

Today, fuel cells are used regularly in our space program to supply the necessary electric power to the Space Shuttle Orbiter. They were also used in the past in the Gemini and Apollo programs.[4] These systems require high purity hydrogen and oxygen and can be considered a special application, as such purity gases are not practical for the economical terrestrial applications that this paper will address.

The fuel cell has many attributes that, because of abundant low-cost fossil fuel and the lack of environmental concern in the past, have been slow in developing as highly desirable energy conversion devices.

Since the time of Sir William Grove's invention, there have been several attempts at developing a domestic fuel cell. Most notable was a team led by Francis T. Bacon, an engineer, who started in 1932 to develop an alkaline fuel cell.[5] He developed the technology for porous gas diffusion electrodes, which provide the basis of much of our work today. In 1959, Bacon and his co-workers announced they had demonstrated a 5 kWe system.[6] In the same year, Allis Chalmers demonstrated their 20-horsepower fuel cell powered tractor.[7]

Fuel cells, in the form of the alkaline fuel cell (AFC), found application in the space program because they had a distinctive energy-to-weight ratio advantage over other power supplies. For a 10-day mission they are superior to primary batteries by a factor of about eight. Because of this, NASA funded an extensive fuel cell research and development program that resulted in the fuel cell systems used in space today. This technology, although well-developed for aerospace, has little place for practical power generation on earth due to cost and the requirement of very pure hydrogen and oxygen as feed gases.

Meanwhile, in 1967, the American Gas Association (AGA), representing the gas utilities, launched a nine-year program with United Technologies Corporation (UTC) as the prime contractor and the Institute of Gas Technology (IGT) as subcontractor.[8] IGT had been supported in the development of fuel cell technology by AGA and several gas utility companies since about 1960.[9]

The object of the AGA effort, which was formally known as the "TARGET" (Team to Advance Research in Gas Energy Transformation) program, was to utilize natural gas as the basic fuel, developing the phosphoric acid fuel cell (PAFC) as the primary technology with the molten carbonate fuel cell (MCFC) as the back-up technology. Advances in PAFC technology at IGT were considered sufficient (in 1967)[10] to institute a commercialization program, and MCFC technology was expected to follow.

The oil embargo of 1973–74 heightened the interest in fuel cells for power generation because of their high conversion efficiencies. In the late 70s, federal support for terrestrial applications began and has continued at various levels to this date. While federal and private monies were initially spent mostly on the PAFC, an increasing proportion is now being directed to the MCFC.

The PAFC is now considered to be ready for introduction into the marketplace, and efforts in this regard are underway. The molten carbonate fuel cell should see market entry by the mid-1990s. Large amounts of monies are now being expended in Japan for fuel cell development, and serious interest is growing in Europe. Emphasis appears to be shifting toward the development of the MCFC for a number of reasons, which will be covered in this paper.

Current focus is on the development of fuel cell systems that are in the range of 200 kWe and higher. Applications can be identified for commercial, industrial, and utility use, with increasing interest in cogeneration and distributed power generation. Environmental issues, which are becoming an increasing concern in the world, are a further stimulus to developing this technology since it is virtually pollution free (as discussed later). Also, the increasing cost of fuels and their uncertain availability will stimulate the development of technologies with the highest thermal-to-electric efficiencies, especially those having the potential for cogeneration. Fuel cells, as we shall see, meet these criteria.

The Electrochemistry

A fuel cell is an electrochemical device that directly converts the chemical energy of a fuel and an oxidant into DC electricity. Rather than having the reactants as an integral part of the cell, as is the case with primary and secondary batteries, fuel cells are only converters, to which fuel and oxidants are fed, and electricity and reaction products removed. Since nothing is consumed or deposited in the cell, it can operate over long periods of time, and unlike the battery, will not deplete itself of energy. Fuel cells of the types discussed here react gaseous hydrogen and oxygen (usually air). The overall cell reaction is $H_2 + \frac{1}{2}O_2 \rightarrow H_2O$. The fuel cell consists of three active components: a fuel electrode (anode) and an oxidant electrode (cathode), which sandwich an ion-conducting electrolyte. The electrolyte employed identifies the fuel cell type, largely dictating the operating conditions and composition of electrodes for achieving the required catalytic activity.

To illustrate the electrochemistry involved with fuel cell operations, Figure 1 shows the reactions when using an acid electrolyte. The ionic transfer of hydrogen through the electrolyte must be balanced by an electric current within the electron-conducting external circuit. Energy is then extracted by placing a load across the external circuit. In theory, such a cell should produce 1.229 volts when operating reversibly at 1 atm and 25°C.[11] However, in practice, a cell exhibits polarization, which reduces cell voltage to about 0.7 to 0.8 volts at practical current densities.[12] Polarization is of three types: activation, concentration, and ohmic. A major effort in the development of the technology involves finding ways to reduce these polarization effects. The best way to achieve this has been to select sufficiently active catalysts to promote electrode reactions and to engineer component and cell designs that reduce the

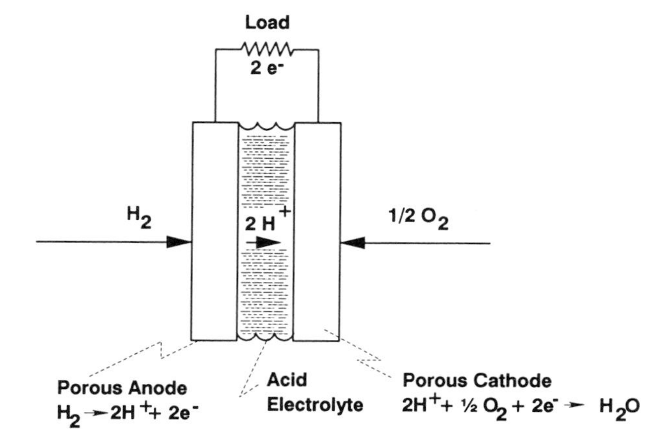

Figure 1. Typical fuel cell acid electrolyte.

active component thicknesses, minimize electrical resistance, and increase gas and ionic diffusion rates.

Another major engineering challenge of the fuel cell results from the low voltage output of a single cell. As it is not practical to drive a load or to invert to AC with such voltages, it is necessary to place the cells in series, in order to achieve practical system voltages (several hundred volts input are required for high efficiency DC/AC inverters).[13] This is accomplished by "stacking" the cells so that the cathode of one cell is in electrical contact with the anode of the next cell while keeping a simple fuel and air feed system. The cleverness of this stacking design will have a major impact on the success and cost of the fuel cell system.

Cell Types

Of the various types of fuel cells developed throughout the evolution of this technology, three appear to be most practical from the standpoint of stationary power sources for large power production. They are:

- phosphoric acid fuel cell (PAFC)
- molten carbonate fuel cell (MCFC)
- solid oxide fuel cell (SOFC)

Each concept is distinctly different from a fabrication and engineering point of view. Electrode reactions and operating conditions are shown in Table 1.

Phosphoric Acid Fuel Cell (PAFC)

Of the three fuel cell systems, the PAFC is the most advanced in development and is being offered for commercial application today. It has been termed "first generation." Although the cell will operate at temperatures as low as 35°C, the technology has evolved at temperatures of about 200°C to minimize polarization and to enable the design of practical cooling.[14] In order to design a stack of multiple cells, the electrolyte (phosphoric acid) is immobilized between the electrodes in a porous silicon carbide matrix. Figure 2 illustrates the sequence of elements that make up the stack configuration.[15]

A graphite bipolar plate backs the anode to allow electrical flow from the anode of one cell to the cathode of the next while keeping the anode gas (fuel) separate from the cathode gas (air). Passages are provided between the bipolar plate and the anode to carry fresh fuel to the face of the porous anode, and to vent spent fuel. The porous anode, made from carbon/graphite and impregnated with platinum, converts hydrogen gas to hydrogen ions and electrons. The electrons flow from the anode to the cathode through an external circuit, while hydrogen ions pass through the phosphoric acid electrolyte, which is held in a porous silicon carbide matrix. Upon reaching the cathode, these ions are converted to water vapor by reacting with oxygen on the platinized carbon/graphite surface within the porous cathode. Again, there are gas passages to supply oxygen to the cathode and to carry away the water vapor produced. The cathode, in turn, abuts the bipolar plate, separating the cell gas passages from the cell below it. It is in this fashion that the stack is constructed. The number of cells per stack determine the ultimate DC voltage produced, which must be sufficient to drive an inverter, if alternating current is to be produced. In actual design, as compared to that shown for illustrative purposes in Figure 2, the fuel passages run at right angles to the air passages (cross flow) in order to facilitate the feeding and removal of gases from the stack sides. End plates, on the top and bottom of the stack, conduct the electrical current from the stack and apply force to maintain good contact between stack elements. Manifolds are clamped to the four sides to supply fresh feed to and conduct spent gases from the passages in the cells as illustrated in Figure 3, a simplistic top view of a PAFC stack. Because of the need to replenish some of the phosphoric acid (H_3PO_4) electrolyte during the cells' operating life, provisions for storage of make-up acid in the anode are provided. Also, to remove heat produced in the stack, cooling plates are periodically located between the cells, with cooling accomplished by circulating either water or air through the plates.

The PAFC is sensitive to poisoning by sulfur gases, as well as carbon monoxide, in the feed hydrogen. If operated on reformed natural gas, both of these constituents must be reduced to low levels. Carbon dioxide acts as a diluent in the feed gas, as does any unconverted methane.

Molten Carbonate Fuel Cells (MCFC)

The molten carbonate fuel cell has been termed the "second generation fuel cell" as it has lagged in development efforts compared to the PAFC. It operates at nominal temperatures of 650°C in order to keep adequate ionic mobility in the lithium carbonate/potassium carbonate eutectic which serves as the electrolyte.[16] This higher temperature has two benefits: first, a less active catalyst is needed to maintain the necessary electrode reactions, and second, the waste heat is generated at sufficiently high temperature (greater than 535°C) to make it a useful co-product (excess heat can be recovered into efficient high-pressure steam cycles with reheat). Unlike the PAFC, the anode is not poisoned by carbon monoxide (CO) but rather promotes the reaction of steam and CO to produce additional hydrogen fuel in the cell. Again, as with the PAFC, cells must be designed so that they can be "stacked" to produce the voltage necessary to drive an inverter efficiently. While pressure operation is very advantageous with the PAFC

Table 1. Fuel cell reactions.

FUEL CELL TYPE	ANODE REACTION	ELECTROLYTE TRANSFER ION	CATHODE REACTION	OPERATION TEMP. & PRES.
PAFC	$H_2 \rightarrow 2H^+ + 2e^-$	$2\ H^+$ --->	$2H^+ + O_2 + 2e^- \rightarrow H_2O$	150-250°C 1-8 atm
MCFC*	$H_2 + CO_3^=$ $\rightarrow H_2O + CO_2 + 2e^-$	$CO_3^=$ <---	$\frac{1}{2}\ O_2 + CO_2 + 2e^- \rightarrow CO_3^=$	650°C 1-10 atm
SOFC	$H_2 + O^=$ $\rightarrow H_2O + 2e^-$	$O^=$ <---	$\frac{1}{2}\ O_2 + 2e^- \rightarrow O^=$	1000°C 1 atm

* Both oxygen and carbon dioxide must be fed to the cathode in order to enable $CO_3^=$ formation. CO_2 is supplied by directing a portion of combusted depleted anode gas to the cathode along with air.

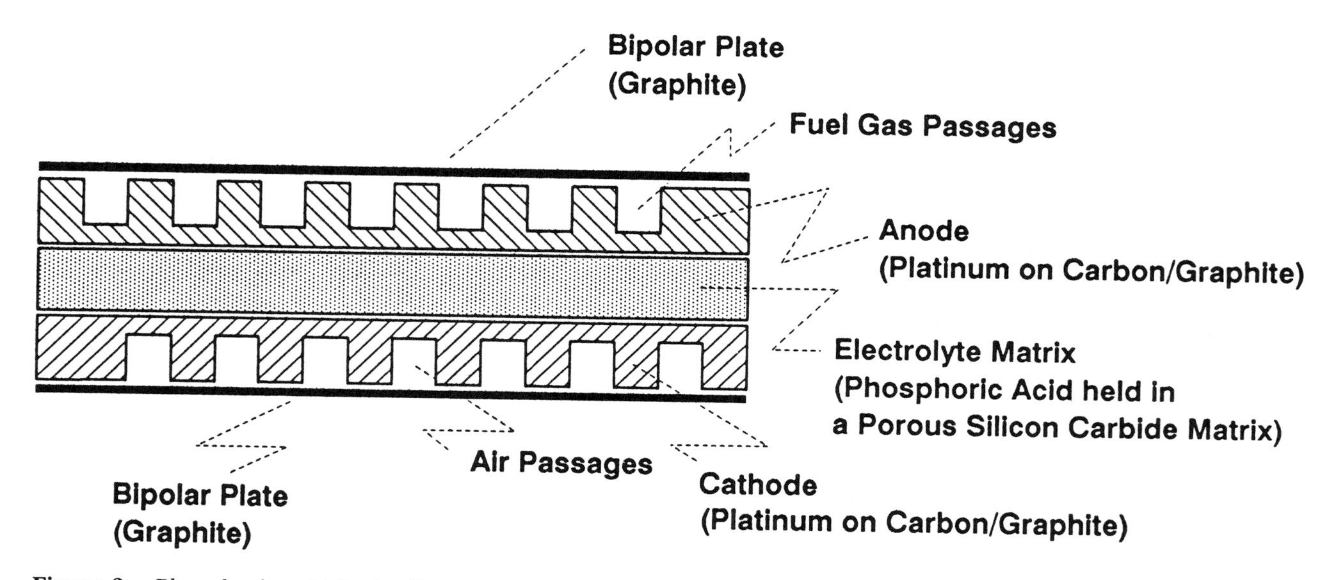

Figure 2. Phosphoric acid fuel cell.

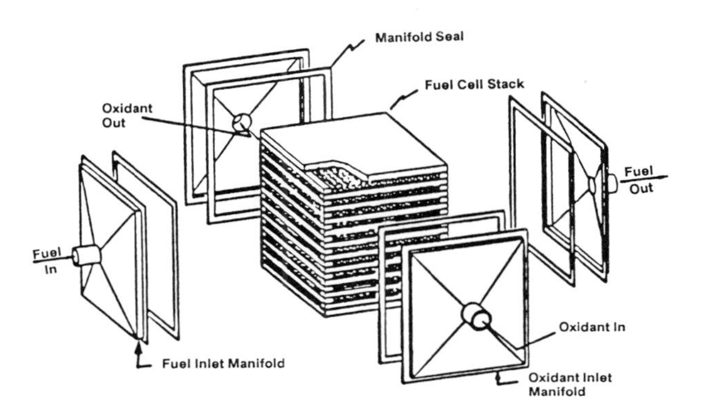

Figure 3. Externally manifolded stack.

to prevent vaporization of the acid electrolyte and to decrease polarization losses at the cathode, a molten carbonate cell does not require pressure for suitable operations. Pressure, however, can be beneficial in providing lower pressure drop through the stack, and will enable higher current densities at equivalent voltages. As with the PAFC, the electrolyte is held between the electrodes in a porous nonreactive matrix.

Figure 4 shows a simplified cross section of an MCFC fuel cell as it is configured in an internally manifolded fuel cell stack. If external manifolding were used, the cell and stack configuration would be similar to that of the PAFC shown in Figures 2 and 3. In Figure 4, the corrugated, stainless steel, bipolar separator plate has three functions: 1) it maintains electrical continuity from the cathode of one cell to the anode of the next, 2) it keeps the anode fuel gas and cathode air separated, and 3) it provides the channels for fuel and air flow. With the internal manifold design, co-current or counter-current flow of the fuel and air are possible, which has advantages over the cross flow pattern dictated by external manifolding. The next element in the MCFC is a flat anode, composed of porous sintered nickel along with some minor additives that inhibit loss of surface area during operation. At the operating temperature of 650°C, the nickel surface is the only catalyst required. The anode is in direct contact with the electrolyte matrix, which is porous lithium aluminate filled with the lithium carbonate/potassium carbonate eutectic. The cathode is porous nickel oxide, which is initially fabricated in the form of porous sintered nickel and is subsequently oxidized during the burn-in or start-up procedure. The cathode is exposed to air passages, in the next lower bipolar plate. Again, the number of plates determines the stack voltage.

In one modification of this design, the concept of internal reforming has been investigated for use with natural gas. This concept involves the addition of a reforming catalyst in the fuel gas passages adjacent to the anode. Methane is added to the fuel gas which, along with steam, reacts to produce hydrogen and carbon monoxide within the cell. The water gas shift reaction also occurs, reacting the carbon monoxide with steam to produce additional hydrogen. The reforming reaction absorbs heat from the stack, thus assisting in stack cooling and eliminating the fuel requirements for external reforming. Fuel to electric conversion efficiencies in excess of 60 percent can be achieved, but poisoning of the reforming catalyst by carbonates remains a major challenge.

Two approaches to stacking are currently being

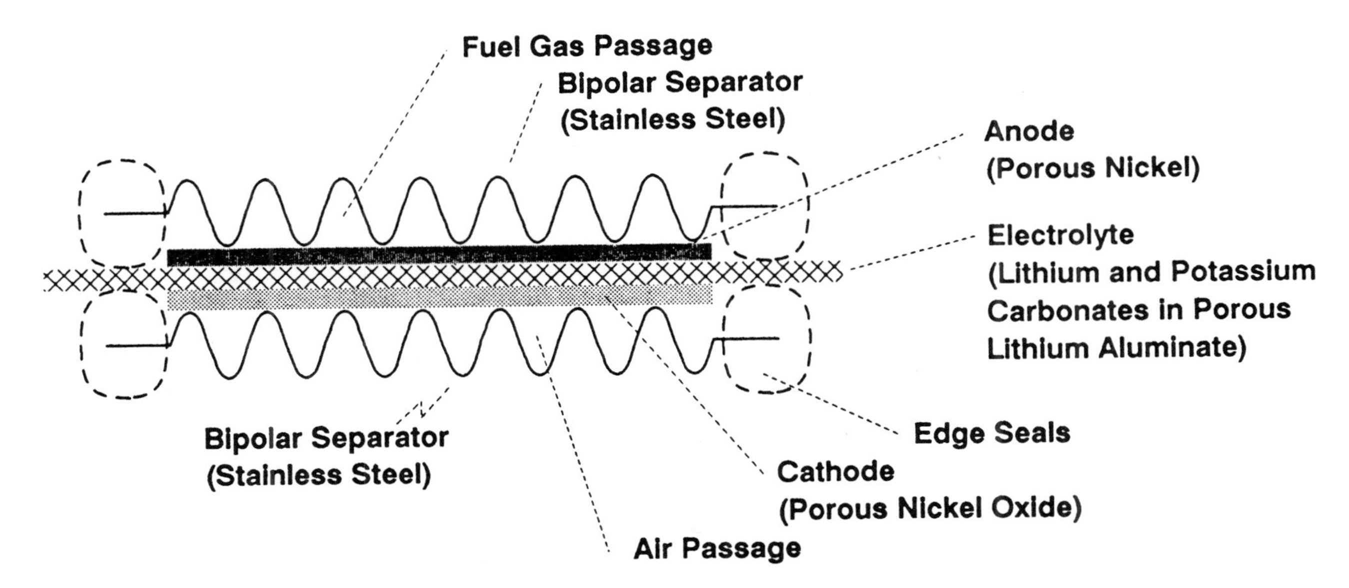

Figure 4. Molten carbonate fuel cell.

used in the US and overseas. Internal manifolding is being developed in the US by M-C Power and overseas by the Japanese and Europeans, while external manifolding is being used by two other US developers.

An MCFC stack designed with external manifolds is similar in configuration to the PAFC, as shown in Figures 2 and 3. Bipolar plates are fabricated from nickel clad stainless steel, and fuel and air passages are either cut in the electrodes or fabricated from metal. Cross flow of gas is required and the manifolds are clamped to the stack sides.

In the internal manifold configuration, however, gas manifolds are created within the stack by providing holes in each plate. This creates gas passages through the stack. Each plate is provided with the necessary porting to direct gas flows from the various manifolds to either the anode or cathode side of the particular bipolar plate. In recent tests, this concept has proven itself to have many advantages over the external manifold designs. Two major effects realized are the reduction of internal "hot-spots" through more uniform current density and the elimination of "carbonate pumping," which impacts stack performance. The details of these effects are beyond the scope of this discussion. A typical plate design is shown in Figure 5.

The molten carbonate fuel cell is believed by many to be the most practical cell type for adaptation and integration into electrical power generation. It has several advantages over the other systems:

- Low cost for materials of construction
- Simple component manufacture
- Ease of assembly
- Highest efficiency of the three cell types
- Recovery of heat at high quality
- Can be used with natural gas, coal gas or other fuels.

Solid Oxide Fuel Cell (SOFC)

The solid oxide fuel cell, often called the "third generation" technology, presents many materials and engineering challenges. These cells, which operate at nominally 1000°C, are built using sophisticated thin film techniques. The electrolyte is an yttria- or calcia-doped zirconia, in which oxygen ions transfer the charge across the electrolyte. In the development of this technology, a number of "stacking" approaches are being used, as dictated by the cell design. The structure of the basic cell is shown in Figure 6. A support tube or a flat sheet is necessary to build on when thin film techniques are used. Most work has been done on the tubular cell, which is shown here for illustrative purposes.[17]

The support tube is made from porous, calcia-stabilized zirconia. Since gases must diffuse through the tube to the cathode, which is formed on the tube's exterior surface, the tube should be as thin as practical. The cathode is a film of porous, strontia-doped lanthanum manganite perovskite deposited on the tube. After this, an electrolyte of yttria- or calcia-stabilized zirconia is deposited. This is then covered by the anode, which is a cermet of metallic nickel or cobalt with yttria- or calcia-stabilized zirconia. An electron-conducting interconnection material, which is used to series-connect one cell to the next, is composed of dense magnesium- or strontium-doped lanthanum chromite.

These tubes can then be "stacked" by contacting the cathode interconnection material with the surface of the anode on the adjacent tube, as indicated in Figure 6.

Another approach used with tubular cells is the construction of a series of cells along the tube, with electrical isolation of cells at a point where the anode of one cell is connected to the cathode of the adjacent cell. Again, all of this is necessary to increase the voltage to

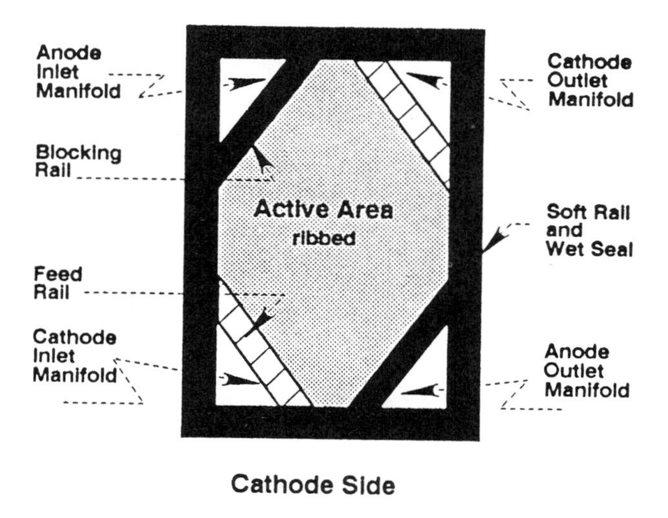

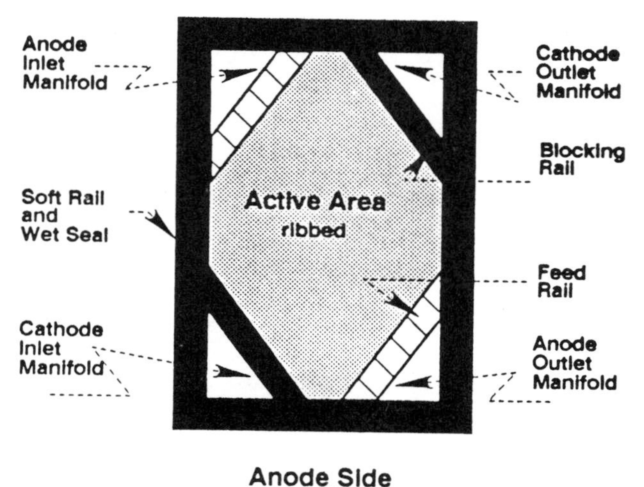

Figure 5. Internal manifold configuration.

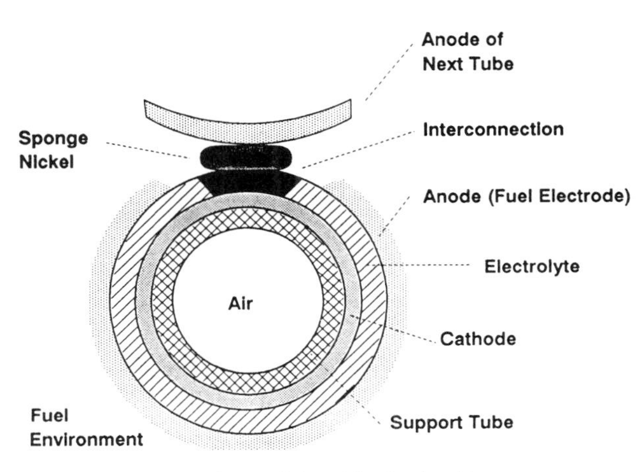

Figure 6. Tubular solid oxide fuel cell.

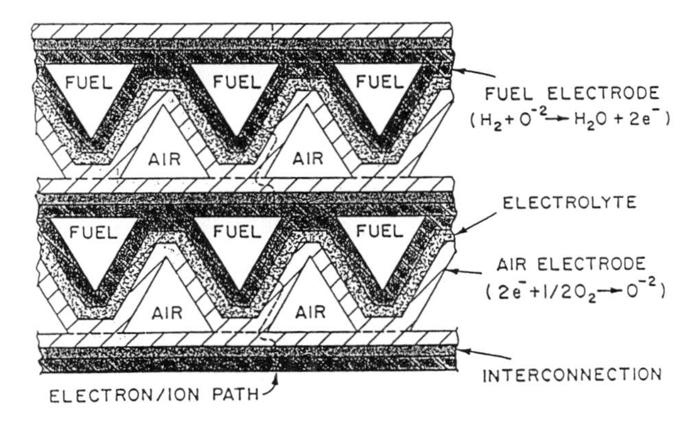

Figure 7. Monolithic fuel cell.

a reasonable level. Various other approaches at "stacking" are being developed. One of the more prominent is the monolithic solid oxide fuel cell, which has a honeycomb arrangement as shown in Figure 7. This configuration has the promise of very high power output on a volumetric basis, and will probably find use in non-stationary applications. Table 2 shows possible construction materials for the various cell types.

Fuel Cell Support Systems

To ensure the proper operation of a fuel cell stack, a well engineered support system must be provided. Two types of systems must be considered: 1) systems which because of their size (i.e., a utility operation) will have operators in attendance at all times, and 2) smaller systems for stand-alone cogeneration applications that must be capable of unattended operation. In both instances, however, the major components of the support systems are similar and will consist of the following:

Fuel Processing. For natural gas or light hydrocarbons, this will principally involve steam reforming (i.e., to convert methane to hydrogen and carbon monoxide fuels) after suitable sulfur removal from the fuel. If fuel cells are not to be base loaded, the reformer systems must be capable of operating at partial load and of load-following with fast response time. For coal-based systems, a coal gasifier is necessary to produce the fuel gas for cell operation.[18]

Fuel Gas Purification. With natural gas as the fuel, no gas purification (except for a preliminary canister to trap odorants) will be necessary, with the exception that for the PAFC, carbon monoxide must be reduced to low levels. In a coal-based operation, particulates, tar oils, sulfur, and chloride compounds must be removed from the gas prior to use. The purification system required will depend on the gasifier used, but satisfactory engineering designs have been developed for most gasifier types.[19]

Air Feed System. All stationary fuel cells will be operated using air as the oxygen source, with filtering, compression, and preheating required for proper stack operation. Also necessary for MCFCs is the blending of combusted anode exhaust gas with the air to supply the carbon dioxide required for carbonate ion formation at the cathode.[20]

Stack Cooling and Heat Recovery. Because heat is generated within the stacks, temperature control and heat removal systems are essential in addition to electricity. In the PAFC system, heat is removed by air or water cooling, while with the MCFC and SOFC, cooling is accomplished by recirculation of excess air and/or fuel gas through the stack. In the support system, this heat is then used to generate steam. It is anticipated that in large coal or natural gas fueled systems, a bottoming cycle will be used to generate additional electric power.[21]

Power Conditioning. Most applications will require the conversion of DC to AC and then a step-up in voltage for distribution. It is important to control frequency so that the systems can be integrated into a power grid.

Control System. Process control of a fuel cell system is very important if good load following is to be achieved. This is especially true in smaller systems, where unattended operation will be required.

Environmental Issues

Engineering analyses of coal-based fuel cell systems have concluded that the environmental releases of criteria pollutants will be orders of magnitude less than Federal New Source Performance Standards for alternative power production from coal.[22] (In fact, cost-effective system designs have been generated that have *no* emis-

Table 2. Materials of construction.

	ANODE	CATHODE	ELECTROLYTE MATRIX	BIPOLAR SEPARATOR PLATE
PAFC	Platinum on Carbon/Graphite Substrate	Platinum on Carbon/Graphite Substrate	H_3PO_4 Silicon Carbide	Graphite
MCFC	Nickel	Nickel Oxide	$LiCO_3 + K_2CO_3$ Lithium Aluminate	Stainless Steel
SOFC	Nickel Zirconia	Lanthanum Manganite	Yttria or Calcia Stabilized Zirconia	Support Material Zirconia with 15% Calcia

sions of SO_2.)[23] Because the electrocatalysts of fuel cell systems are sensitive to sulfur poisoning, the fuel gases must be highly purified. Therefore, because the sulfur must be removed from the fuel gas, it is readily recovered for minimal environmental release. Similar recovery of nitrogen compounds is possible, for minimal NO_x discharge. Systems fired by natural gas are, of course, more readily treated to be benign.

The high efficiency of fuel cell systems also causes the reduced production of CO_2 per kWh, which contributes to the greenhouse effect and may lead to long-term global warming. Compared to conventional power generation from coal with flue gas desulfurization, the release of CO_2 is reduced by a third and thermal emissions are halved as the efficiency is improved from 33 percent to 50 percent.

The reduced environmental cost of power generation from fuel cells is so significant that Electric Power Research Institute (EPRI) assigns a capital cost credit to fuel cell systems when comparing the costs of alternative approaches to power production.[24]

Economic Issues

Engineering analyses have indicated that coal-based, central fuel cell power stations will be competitive with alternative systems when the technology is mature.[25] These analyses do not include the environmental credit suggested by EPRI; furthermore they employ financial factors that strongly penalize efficiency relative to initial capital costs. Similar analyses of natural gas fuel cell systems indicate superiority to alternative advanced power production techniques—for example, steam injected gas turbine (STIG), intercooled steam injected gas turbine (ISTIG), and combined cycle—at sizes to 50 MWe.[26]

Applications

Because of their high efficiency and negligible production of pollutants, fuel cell systems will find broad application. In the early days of development, fuel cells were being considered for residential use in single family homes, with the waste heat being used for water heating.[27] More recently, a number of 40 kWe phosphoric acid systems were placed in commercial applications for evaluation purposes.

Certain characteristics of the fuel cell types should be recognized in identifying applications. The PAFC, which operates at relatively low temperatures, can only produce hot water or low quality steam. Also with PAFCs, small amounts of carbon monoxide have deleterious effects on the electrodes and are therefore generally viewed for use only with natural gas since reducing the large amounts of carbon monoxide to the necessary low levels in a coal based gas is expensive.

MCFCs and SOFCs operate at sufficiently high temperatures to allow for the recovery of high quality waste heat and can tolerate carbon monoxide. Because of these factors, both are candidates for integrated coal gasification fuel cell systems (IGFC).

Three areas of application are: commercial; electric utility; and industrial.

There are many advantages in using fuel cells as compared to other power generating systems.[28] They include:

- High efficiency at full and partial load
- Fast load response
- Ease in siting (quiet, clean, no vibration, extremely low emissions)
- Modularity
- Unattended operation
- Operation without external water supply
- By-product heat recoverable
- Low carbon dioxide emissions
- Cost competitive

Table 3 shows anticipated efficiencies for the various applications.

Commercial Sector

The commercial sector will be the most attractive for near-term application. On-site cogeneration of electricity and thermal energy using natural-gas-fired fuel cells offers great potential. Overall efficiencies of 80 percent (HHV) have been projected for such systems.[29] Other fuels can also be used in this application, such as liquids and bio-gas. Waste heat can be used for heating and cooling by absorption chillers. Units in the range from 50 kWe to 2 MWe can be used in such applications as apartment buildings, office buildings, restaurants, motels, hotels, and hospitals, to mention just a few. Market studies indicate a very substantial demand if capital costs are at or below the $1500/kWe level.[30]

Electricity Utility Sector

The concept of combining fuel cells with coal gasifiers and bottoming cycles is the long-range market for fuel cells in the electric utility market. Most of this market may go to MCFCs because of their operating characteristics, discussed above. Initial applications in the utility sector, we believe, will be for "dispersed" generation in urban and suburban areas using gas-based systems of 10 to 50 MWe in size. One reason for this is the interest utilities have in supplying both steam and electricity. In addition, some public resistance has been encountered in expanding electric transmission systems due to the concerns over the effects of 60-cycle electromagnetic radiation from overhead power lines. It is more economical to distribute natural gas to dispersed fuel cell systems than to utilize underground electric transmission. With the acceptance of siting such systems in urban areas, units as large as 100 MWe may become feasible.

After the year 2000, and with the acceptance of fuel cells by the utility industry, integrated gasification fuel cell (IGFC) systems for base-load central station power generation should become feasible. Also, recent studies made for the molten carbonate system indicate that the

Table 3. Anticipated fuel cell system efficiencies (in percents) based on higher heating value (HHV).

APPLICATION	PHOSPHORIC ACID *(PAFC)*	MOLTEN CARBONATE *(MCFC)*	SOLID OXIDE *(SOFC)*
Electric Utility			
Natural Gas Fueled	40-45	50-60	45-60
Coal Fueled	38-42	50-55	50-55
On-Site			
Natural Gas Fueled	36-45	45-60	50-55

repowering of existing coal power plants shows merit from a siting, environmental, and economic viewpoint.[31]

Industrial Sector

In the industrial sector there are many potential applications, some of which are simple cogeneration situations and others which involve fuel cells integrated into a process such as supplying power for a DC electrolytic cell operation. Waste process gases such as hydrogen and carbon monoxide can also be utilized as fuels. The high efficiency afforded by fuel cells, along with their projected reasonable cost, in the size range of 1–20 MWe, add to this attractiveness. Since industrial applications usually call for higher thermal to electrical energy requirements than commercial cogeneration applications, this market will lean toward either MCFC or SOFC systems. In most instances, such systems are special applications of the fuel cell and may not become practical until after their acceptance in the commercial cogeneration market. Market studies indicate areas of application such as municipal waste treatment plants, breweries, paper making, petroleum, metal refining, and chlor-alkali production.[32]

A projection of the minimum market has been made for the various fuel cell types in the 1990 to 2000 time frame. This study, shown in Table 4, is in total megawatts for each category and does not indicate system sizes.[33] An effort to commercialize the PAFC in the 200 kWe-rated system is now under way in the US, and 50 kWe units are being offered from Japan.[34] Market entry for the MCFC should be in the 200 to 500 kWe size range and should occur in the mid 1990s. SOFCs, it is projected, will be introduced in the 50-100 kWe range.

It is difficult to project the market penetration that will be achieved by fuel cells, as their high efficiency even at a reduced load and their negligible emission of pollutants are both key factors in how they will be adopted. If fuel prices increase, and the enhanced greenhouse effect is found to be real, then these factors will become a very strong motivating force for fuel cell use.

Political factors should not be overlooked. As an example, California's South Coast Air Quality Management District recently announced their intention to implement an air quality management plan, which will entirely eliminate combustion processes for power generation in their region by the year 2000. These regulators believe that fuel cells will allow for the commercial generation of electric power with minimum creation of pollutants.

Status of Development

The three principal areas in the world where fuel cell development is active are the United States, Japan and Europe.

Table 5 is a listing of US activities; only the major participants are listed, as it would be impractical to reference everyone involved. Again, it should be kept in mind that this paper addresses only PAFC, MCFC, and SOFC developments. Where team efforts have been established, members of the teams are listed.

In the United States, principal sources of funding are the US Department of Energy, the Electric Power Research Institute, and the Gas Research Institute. In addition, there is some private industry support of these programs and, in certain instances, there is cost sharing by the developer. Identified funding by the various US organizations is shown in Table 6.

In Japan, there is a large commitment on behalf of the Japanese government and industry to the successful

Table 4. Minimum estimated market in MWs 1990 to 2000.

	COMMERCIAL		INDUSTRIAL		UTILITY		TOTAL	
	Demo Units	Commercial Sales	Demo Units	Commercial Sales	Demo Units	Commercial Sales	Demo Units	Commercial Sales
PAFC	10	100	30	60	60	100	100	260
MCFC	10	10	10	20	20	160	40	190
SOFC	3	0	1	0	1	0	5	0
						Subtotal	145	450

Table 5. US fuel cell programs.

DEVELOPER	PAFC	MCFC	SOFC
Allied Signal Combustion Engineering, Inc. (CE) Argonne National Laboratories (ANL)			Monolithic
Energy Research Corp. (ERC)		External Manifold Internal Reforming	
International Fuel Cells Corp. (IFC)	Selling 200 kW Systems	External Manifold	
M-C Power Corp. (MCP) Combustion Engineering, Inc. (CE) Institute of Gas Technology (IGT)		Internal Manifold	
Westinghouse Electric Corp. (WEC)	Developing Air Cooled Stacks		Tubular Cell Design

Table 6. 1990 US fuel cell program budgets for combined PAFC, MCFC, and SOFC development ($ million).

FUNDING ORGANIZATION	BUDGET
DOE - Fossil Energy	38.5
DOE - Conservation and Renewable Energy	5.6
EPRI	3.5
GRI	2.8

development of fuel cell systems for cogeneration, utility and industrial use. Funding for their programs comes principally through their "Moonlight Project," which is funded by the Ministry of International Trade and Industry (MITI).[35] The programs are, for the most part, contracted through the New Energy Development Organization (NEDO). Funding of fuel cell programs is probably in excess of $100 million annually. The program is well coordinated, with many manufacturers working on its various segments. At present, their major emphasis is on MCFC development. As an example of their teamwork, they have established an MCFC research association, consisting of fourteen organizations working in MCFC development. Table 7 shows the principal participants in the various programs. Work on the solid oxide fuel cells is relatively small at this time, with a budget of about $2 million annually.

In Europe, there is considerable renewed interest in fuel cells. ENEA in Italy has instituted a project "VOLTA," for bringing fuel cells into practical use.[36] Although they are purchasing a PAFC stack for a fuel cell demonstration, their emphasis is on the develop-

Table 7. Japanese fuel cell programs.

COMPANY	PAFC	MCFC
Fuji Electric	Offering a 50 kW System 1 MW Plant Demonstrated	Self Supported Program
Hitachi Ltd.		Segmented Internal Manifold Stack *(Moonlight Program)*
Toshiba	Constructing Multi MW System	Self Supported Program
Ishikawajima-Harima Heavy Industries, Ltd. *(IHI)*		Internal Manifold Stack Development Support System *(Moonlight Program)*
Mitsubishi Electric Corp. *(MELCO)*	200 KW Unit *(Moonlight Program)*	Internal Reforming Stacks *(Moonlight Program)*

ment of MCFCs, and they are trying to arrange for a cooperative program with either a US or European group. In the Netherlands, ECN has been working on MCFC development through joint funding with NOVEM, a Dutch government organization.[37] There is presently an effort to establish a group of Dutch companies to work on commercializing this technology. Several other European companies are active in SOFC development.

Remaining Technical Issues

There are two primary technical issues that are fundamental to all three fuel cell technologies:

- Development of manufacturing procedures to reduce stack cost to acceptable levels.
- Fabrication of fuel cell stacks that will have a productive life of about five years.

The manufacturing techniques for the three types of fuel cells, most of which are proprietary, vary widely. Simplification of component manufacture and stack assembly will be needed to reduce cost. Presently, much of the work going on in the United States and in Japan is geared toward developing these techniques.

It is important for fuel cell stacks to have a reasonable life. A target of five years is generally recognized; however, it will be the cost of electricity that will ultimately dictate acceptable stack-life. With low-cost stacks and a facility design that would allow for easy stack replacement, a shorter life may be acceptable. In this regard, there is a tradeoff that applies to all three cell types. Thinner components reduce stack material costs and improve efficiency by reducing cell polarization. Thinner components, however, are more prone to early failure and place more demands on manufacturing and quality control.

Conclusions

Fuel cells represent a novel approach to electric power generation. These devices have found successful application in the space program and should find a place in terrestrial power-generation in the years ahead because of their high efficiency and low emissions. Operation on natural gas currently appears most attractive, with the fuel of choice changing to coal-gas when the cost of natural gas becomes relatively more expensive.

Phosphoric acid fuel cells are being offered for sale today; advanced molten carbonate types are expected to be available by the mid-1990s, and solid oxide fuel cells should be ready by the year 2000. At this time, projected system costs appear competitive with other advanced power generation technologies.

Continued research and development will be necessary to reduce fuel cell stack and system costs and to extend fuel cell stack operating life to four to five years. The sizeable amounts of research and development funding in Japan, Europe, and the US in fuel cell development, if maintained, should be sufficient to see these systems reach a mature state of development.

References

1. Appleby, A.J. and Foulkes, F.R., *Fuel Cell Handbook*, Van Nostrand Reinhold, NY (1989).
2. Shumov, Y.S., Chakmakhchyan, S.S., et al., *Zh. Fiz. Khim.*, p. 53 (1979).
3. Grove, W.R., *Philosophical Magazine,* p. 14 (1839).
4. Crowe, B.J., "Fuel Cells: A Survey," NASA Report (SP-5115). Washington, DC (1973).
5. Bacon, F.T., *Fuel Cell Symposium*, American Chemical Society Meeting, Atlantic City, NJ (1959), published by the American Chemical Society, Washington, DC.
6. Adams, A.M., Bacon, F.T., et al., *Fuel Cells*, W. Mitchell, Jr., editor, Academic Press, NY (1963), pp. 129–192.
7. Crowe, B. J., *loc. cit.*
8. Burlingame, M.V., in *Fuel Cell Systems II, Advances in Chemical Series 90*, American Chemical Society, Washington, DC (1969), p. 377.
9. Baker, B.S., et al., "Numerous reports to the American Gas Association under IGT Project ZB–56," Institute of Gas Technology Library (1960–1967).
10. Fleming, D.K., personal communication (1990).
11. Appleby, *loc. cit.*, p. 18.
12. *Ibid*, p. 21*ff*.
13. Appleby, A.J., personal communications (1989).
14. Lee, J.M., *National Fuel Cell Seminar Abstracts*, 1983, p. 56.
15. Appleby, A.J., "Phosphoric Acid Fuel Cells (PAFC's)," *Energy: The International Journal*, Vol. 11, No. 1/2, January/February, pp. 13–94 (1986).
16. Ackerman, J.P., *Progress in Batteries and Solar Cells*, Vol 5, p. 13, JEC Press, Cleveland, OH (1984).
17. Isenberg, A.O., *Proceedings Symposium Electrode Materials Processes Energy Conversion Storage*, p. 572 (1977).
18. Borys, S.S., and Ackerman, J.P., *Proceedings of the 14th Intersociety Energy Conversion Engineering Conference*, Vol. 1, p. 563 (1979).
19. Krumpelt, M., Ackerman, J.P., et al., *Fuel Cell Power Plant Designs: A Review*, DOE/CC/49941-1833, US Department of Energy, Washington, DC (1985).
20. Pierce, R.D., Smith, J.L., et al., *Proceedings of the Electrochemical Society Symposium on Molten Carbonate Fuel Cells*, Montreal, May 9–14 (1982), published by the Electrochemical Society, Pennington, NJ.
21. Krumpelt, M., et al., "Molten Carbonate Fuel Cells for Coal and Natural Gas Fuels," *Proceedings, 7th Annual Energy Sources Technology Conference and Exhibit*, ASME, New Orleans, LA, February 12–16 (1984).
22. Degan, L.K., et al., *Molten Carbonate Fuel Cell Power Plants*, GE-PG & E-SCE joint publication, March 1980.
23. *Systems Analysis of 200 MWe Power Plants*, US DOE Contract No. DE-AC21-88MC25026, Combustion Engineering, Inc, (1990).
24. Rigney, D.M., *National Fuel Cell Seminar Abstracts*, p. 1 (1981).
25. *Ibid.*
26. M-C Power Corporation, unpublished study (1989).
27. "Final Report, Project S-233," Institute of Gas Technology, (1968).
28. Appleby, A.J., *Loc. cit.*, p. 28*ff*.
29. Appleby, A.J., *Loc. cit.*, p. 83.
30. Bryant, P., personal communication (1989).
31. *Ibid.*
32. Liu, G., *National Fuel Cell Seminar Abstracts*, p. 106 (1983).

Appendices

Appendix A: Conference Program

Energy and the Environment in the 21st Century
Complete Conference Program

Plenary Session I

Monday, March 26

8:30 am–8:40 am

Welcoming Remarks by Dr. Paul Gray
President, MIT

8:40 am–9:10 am

Introductory Address by Richard Morgenstern
Director, Office of Policy Analysis, US Environmental Protection Agency

9:10 am–9:45 am

Keynote Address by Albert Gore, Jr.
Senator, United States Congress

9:45 am–1:00 pm

Plenary Session I: Environmental Science and the Energy/Environmental Technology and Policy Agenda

1. Environmental Science
Scientific evidence of environmental change from a global perspective (atmospheric chemistry, climate change and the greenhouse effect, stratospheric ozone depletion, etc.) and from a local and regional perspective (e.g., acid rain, tropospheric ozone, visibility). Uncertainties in relevant data and model predictions and how they affect policies relating to technology development.

Ensemble Assessments of Atmospheric Emissions and Impacts
Thomas Graedel, AT&T Bell Laboratories, and Paul Crutzen, Max Planck Institute

Global Atmospheric Chemistry and Global Pollution
Ronald G. Prinn, MIT

Prediction of Future Climate Change
Stephen H. Schneider, National Center for Atmospheric Research

The Interface of Environmental Science, Technology, and Policy
John H. Gibbons, Office of Technology Assessment

1:00 pm–2:00 pm

Luncheon

2:15 pm–5:45 pm

2. Energy and Environmental Policy for Economic Growth and Social Well-Being
Political, economic, and energy/environmental policy strategies for achieving sustainable and equitable world economic growth and social well-being.

Energy and Environment: Strategic Perspectives on Policy Design
William C. Clark, Harvard University

Energy and Environmental Policies in Developed and Developing Countries
José Goldemberg, University of São Paulo, Brazil

Population and the Global Environment
Nazli Choucri, MIT

Economic Policy in the Face of Global Warming
William Nordhaus, Yale University

3. Energy Technology
An overview of energy supply and end-use applications to show how environmental effects such as carbon dioxide buildup and acid rain are influenced by fuel and conversion-process choice and utilization efficiency. Industry's range of response to environmental regulations. Identification of the technical and economic constraints to efficiency improvement and conservation. Discussion of priorities for research and development.

Energy Technology: Problems and Solutions
Paul Gray, Jefferson Tester, and David Wood, MIT

6:30 pm–8:30 pm

Reception for Conference Attendees
Hyatt Regency Hotel

Parallel Sessions A–F

Tuesday, March 27 (8:30 am–5:30 pm) through Wednesday, March 28 (8:30 am–12:00 noon)

Sessions on Energy Services, Economics and Policy, and Advanced Energy Supply Technologies
Four concurrent sessions consider technology and other components specific to: transportation systems (Session A), industrial processes (Session B), building systems (Session C), and electric power systems (Session D). Discussion focuses on current and future uses of conventional and nonconventional energy supplies, including fossil (existing oil, gas, and coal and synthetic fuels), nuclear fission, geothermal, and renewables (solar, biomass, and wind). Session participants evaluate the environmental impacts of technology development possibilities and options for increased efficiency, conservation, and cogeneration. Emphasis is placed on the interaction between technology and political and economic factors and discussion of policy and regulatory issues.

Two special sessions are also included. A session on economics and policy (Session E, on Tuesday afternoon) considers national and global economic studies of strategies for reducing greenhouse gas emissions. A session on advanced energy supply technologies (Session F, on Wednesday morning) reviews advanced concepts such as fusion, fuel cells, photovoltaics, and hot dry rock geothermal and evaluates their potential for the future.

Plenary Session II

Wednesday, March 28

12:00 noon–1:30 pm

Luncheon
The chairpersons of the concurrent sessions present short summaries reporting the results and recommendations of their sessions.

1:30 pm–4:30 pm

Forum: Policy Strategies for Managing the Global Environment
Moderators: John Deutch and David Wood, MIT
Evaluating policy strategies for the efficient and equitable management of climate resources is complex and increasingly controversial. Forum members present and discuss their differing perspectives on this subject, followed by questions and discussion with the audience.

Forum members include:
Yoichi Kaya, The University of Tokyo
Lester Lave, Carnegie-Mellon University
Stephen Peck, Electric Power Research Institute
Irving Mintzer, University of Maryland
Thomas Schelling, Harvard University

Session A: Transportation Systems

Current trends in energy use for land, water, and air transportation. Transportation's current and prospective contributions to regional and global air pollution. Developments in conventional engine and vehicle technology and in conventional fuels. New propulsion system concepts and alternative fuels.

Tuesday, March 27

8:30 am–12:00 noon

Transportation Systems and Their Environmental Impacts: An Overview
Chair: John Heywood, MIT

The Transportation System: Issues and Challenges
Joseph Sussman, Carl Martland, and Nigel H.M. Wilson, MIT

US Production of Liquid Transportation Fuels: Costs, Issues, and Research and Development Directions
John Longwell, MIT

Role of the Automobile in Urban Air Pollution
D. Atkinson and Alex Cristofaro, US Environmental Protection Agency, and J. Kolb, Sobotka & Co.

Technical Options for Energy Conservation and Controlling Environmental Impact in Highway Vehicles
Charles Amann, General Motors

12:00 noon–1:00 pm

Luncheon

1:00 pm–3:00 pm

Panel on Key Issues in US Transportation Energy Policy

Moderator: John Heywood, MIT

Positioning for the 1990s—The Amoco Outlook

Theodore Eck, Amoco Corporation

The Role for Improved Vehicle Fuel Economy

Deborah Bleviss, International Institute for Energy Conservation

The Future Fuel Economy of the United States Light Duty Fleet—A Policy Dilemma

Steven Plotkin, US Congress, Office of Technology Assessment

Future Growth of Auto Travel in the US: A Non-Problem

Charles Lave, University of California, Irvine

Potential Gains in Fuel Economy: A Statistical Analysis of Technologies Embodied in Model Year 1988 and 1989 Cars

Van Bussmann, Chrysler Corporation

3:15 pm–5:30 pm

Panel on Meeting the Transportation Aspirations of Developing Countries: Energy and Environmental Impacts

Moderator: Daniel Roos, MIT

Transportation Aspirations of Developing Countries Will Be Met by Oil-Fueled Motor Vehicles

Remy Prud'homme, University of Paris XII

Interrelationships among Energy, Environmental, and Transportation Policies in China

Karen Polenske, MIT

Prospective Transportation Developments in India and Their Impact on the Environment

K. L. Luthra, Planning Commission, India

Meeting the Transportation Aspirations of Developing Countries: Energy and Environmental Effects

Clell Harral, The World Bank

Wednesday, March 28

8:30 am–10:15 am

Case Studies on Transportation Energy and Environmental Policy

Chair: Malcolm Weiss, MIT

An Incentive-based Transition to Alternative Transportation Fuels

Daniel Sperling, University of California, Davis

Transportation and the Environment in the European Economic Community

Bert Metz, Royal Netherlands Embassy

Nonfossil Transportation Fuels: The Brazilian Sugar Cane Ethanol Experience

Sergio Trindade, United Nations

10:30 am–12:00 noon

Energy and the Environment: A Framework for Evaluating Research Needs in the Transportation Sector

Malcolm Weiss, John Heywood, MIT

Panel on Research Needs

Moderator: Daniel Sperling, University of California, Davis

Richard John, US Department of Transportation
Brian Taylor, Chevron Research Company
Sergio Trindade, United Nations
John Brogan, US Department of Energy
Norman Gjostein, Ford Motor Company

Session B: Industrial Processes

Existing and new processes for the manufacture of materials, their energy intensity, and environmental impact. Potential for improving the energy efficiency of major energy-consuming industrial processes and the costs associated with achieving such improvements. Materials substitution, life cycle, and recycling issues. Process change versus retrofit with more efficient components.

Tuesday, March 27

8:30 am–12:00 noon

Industrial Processes and Materials Manufacture: An Overview

Chair: Jefferson Tester, James Wei, MIT

Modeling the Energy Intensity and Carbon Dioxide Emissions in US Manufacturing

Marc Ross, University of Michigan

Cogeneration Applications of Biomass Gasifier/Gas Turbine Technologies in the Cane Sugar and Alcohol Industries

Joan Ogden, Robert Williams, and Mark Fulmer, Princeton University

Pinch Technology: Evaluate the Energy/Environmental Economic Trade-Offs in Industrial Processes

H. Spriggs, E. Petela, and B. Linnhoff, Linnhoff March

Energy Conservation in the Industrial Sector

Robert Ayres, Carnegie-Mellon University and International Institute of Applied Systems Analysis

12:00 noon–1:30 pm

Luncheon

1:30 pm–5:00 pm

Industrial Processes—Metals, Minerals, and Ceramics Manufacture
Chair: Donald Sadoway, MIT

Energy, the Environment, and Iron and Steel Technology
John Elliott, MIT

Some Energy and Environmental Impacts of Aluminum Usage
Patrick Atkins, Don Willoughby, and Herman J. Hittner, Alcoa

Minerals, Energy, and the Environment
Jay Agarwal, Charles River Associates

Energy and Environmental Considerations for the Cement Industry
Stewart Tresouthick, Alex Mishulovich, Construction Technology Laboratories, Inc.

Glass Manufacturing—Status, Trends, and Process Technology Development
Greg Ridderbusch, Gas Research Institute

Wednesday, March 28

8:30 am–10:15 am

Industrial Processes—Chemicals, Petroleum, and Pulp and Paper Manufacture
Chair: Kenneth Smith, MIT

Energy Efficiency in Petroleum Refining—Accomplishments, Applications, and Environmental Interfaces
Jerry Robertson, Exxon Research and Engineering

Energy Consumption Spirals Downward in the Polyolefins Industry
William Joyce, Union Carbide

Energy Management and Conservation in the Pulp and Paper Industry
Howard Herzog, Jefferson Tester, MIT

10:15 am–10:30 am
Break

10:30 am–12:00 noon

Open Forum/Panel Discussion on Industrial Processes
Moderator: Attilio Bisio, Atro Associates, and Editor, *Chemical Engineering News*

Topics include:
- Process energy efficiency and conservation limits
- Materials life cycle and durability issues
- Recycling of metals and nonmetals
- R&D opportunities for new products and processes
- Critical policy and regulatory issues

Session C: Building Systems

Levels of technical and economic conservation for new and existing buildings. Performance limits and environmental effects of interior space conditioning, building materials, lighting systems and appliances, and control systems. Potential for integrated cogeneration systems.

Tuesday, March 27

8:30 am–12:00 noon

Building Systems Overview Papers
Chair: Leon Glicksman, MIT

1. **Today's Buildings**
   Energy-Efficient Buildings in a Warming World
   Arthur Rosenfeld, Lawrence Berkeley Laboratory
   Energy Saving in the US and Other Wealthy Countries: Can the Momentum Be Maintained?
   Lee Schipper, Lawrence Berkeley Laboratory
2. **Environmental Issues**
   Indoor Air Quality—Toward the Year 2000
   David Grimsrud, University of Minnesota
3. **The Building Industry**
   Change in the Building Industry
   Tage Carlson, USG Corporation, Research and Development
4. **Buildings in the Next Century**
   Buildings in the Next Century
   David Pellish, US Department of Energy

12:00 noon–1:30 pm
Luncheon

1:30 pm–5:30 pm

Energy Use and Environmental Effects in Building Systems
Chair: Leon Glicksman, MIT

Comfort and Health: Interior Conditioning/Quality
(Session Chairs: James Axley and James Woods)
James Woods, Virginia Polytechnic Institute, and Reijo Kohonen, Technical Research Center of Finland

Envelopes: Building Materials
(Session Chair: Erv Bales)
Dennis Ross, Jim Walters Corporation, and David McElroy, Oak Ridge National Laboratory

Services: Appliances
(Session Chair: Ross Bisplinghoff)

Jack Weizeorick, Association of Home Appliance Manufacturers, and Jim McMahon, Lawrence Berkeley Laboratory

Systems Integration: Controls, Communications, and Interfaces
(Session Chairs: Les Norford and Richard Tabors)
Roger Bohn, Harvard Business School; Daniel Nall, Jones, Nall, and Davis; and Peter Brothers, Johnson Controls

Centers of Research and Education
Michael Joroff, MIT

Wednesday, March 28

8:30 am–12:00 noon

Cross-Cutting Technology and Policy Issues for Building Systems
Chair: Leon Glicksman, MIT

Regulations, Incentives, and New Solutions

1. **Regulations, Performance, Planning**
(Session Chair: Kevin Teichman, US Environmental Protection Agency)
2. **Government Incentives for Commercial Development of New Technology**
(Session Chair: John Millhone, US Department of Energy)
3. **Encouraging the Search for New Solutions**
(Session Chair: David Pellish, US Department of Energy)
The Consortium Planning Approach
Constance Groth, General Electric Plastics Division
Encouraging the Search for New Solutions
Glenn Chafee, Rolscreen Corporation
The Search for New Solutions: Federally Supported Innovation in Construction and the Federal Role in Supporting Conservation and Renewables
US Senator Joseph Lieberman (Paper presented by David Pellish)
4. **How to Develop Advanced Technologies for Buildings**
Low-E Windows and Energy Efficient Buildings—Lessons for Future Innovative Building Technology
Stephen Selkowitz, Lawrence Berkeley Laboratory

Recommendations for Critical Actions: Session Observers, Organizers, and Speakers
Robert Socolow, Princeton University

Session D: Electric Power Systems

Quantitative evaluation of present electric power systems in developing and developed nations with particular emphasis on environmental factors. Creation of a structure for evaluating future options that incorporates technical, political, economic, and social factors. Examples of how electric power systems can adapt to changing environmental constraints and societal goals over the short and long term.

Tuesday, March 27

8:00 am–12:30 pm

Plenary Session for Electric Power

Electric Power Systems
Chair: David White, MIT

Electric Generation Technologies
Floyd Culler, Electric Power Research Institute

Networks of the 21st Century
Lionel Barthold, Power Technologies Inc.

Break

Electric Power for Developing Nations
Chair: Dietmar Winje, Technical University of Berlin

Electric Power and the Developing Economies
Dietmar Winje, Technical University of Berlin

Problems, Issues, and Responses to Environmental Concerns of Electric Power Generation and Use in Four Important Continental Regions: A Panel Discussion

Policymaking Pertaining to the Environmental Impact of Energy Use in Latin American and Caribbean Countries
South America—Eduardo Del Hierro, Colombia

Electricity and the Environment in Developing Countries with Special Reference to Asia
Asia—Mohan Munasinghe, World Bank

Electric Power in Africa—Issues and Responses to Environmental Concerns
Africa—John Gindi Boutros, National Electricity Corporation, Republic of Sudan

Present and Future Electric Power Systems in Eastern Europe: The Possibilities of a Broader Cooperation
András Lévai and Tamás Jászay, Technical University of Budapest, Hungary

12:30 pm–1:30 pm

Luncheon

1:30 pm–5:30 pm

Four Parallel Sessions on Electric Power Technology

Session D-I: Fossil Fuel
Chair: János Beér, MIT

Powering the Second Electrical Century
Kurt Yeager, Electric Power Research Institute

Role of Clean Coal Technology in Electric Power Generation in the 21st Century
Jack Siegel and Jerome Temchin, US Department of Energy

Pollution Control for Utility Power Generation, 1990 to 2020
Frank Princiotta, US Environmental Protection Agency

The Role of Natural Gas in Electric Power Generation, 1990 to 2020
Steven Freedman, Gas Research Institute

Manufacturing Technology Challenges in Meeting Demands for Environment-Friendly Power Generation Systems
Jack Sanderson, Combustion Engineering

Session D-II: Nuclear
Chair: Kent Hansen, MIT

Nuclear Power Development to Enhance Environmental Quality: An Overview
John Taylor, Electric Power Research Institute

The Advanced Boiling Water Reactor and the Road to Revival of Nuclear Power
Bertram Wolfe, General Electric Nuclear Energy

The Advanced Pressurized Water Reactor—Meeting the Energy Needs of Today and Tomorrow
Richard Slember, Westinghouse Electric Corporation

Advanced Reactor Development: The Liquid Metal Integral Fast Reactor Program at Argonne
Charles Till, Argonne National Laboratory

Environmental Impacts of the Modular High Temperature Gas-Cooled Reactor (MHTGR)
David Lanning and Scott Pappano, MIT

Session D-III: Alternative Energy
Chairs: Jon McGowan, University of Massachusetts, and Ben Holt, The Ben Holt Company

Photovoltaic Systems, A World Market
Charles Gay, Arco Solar

Future Effects and Contributions of Photovoltaic Electricity on Utilities and the Environment
Ghazi Darkazalli, Spire Corporation

Wind as a Renewable Source in the 21st Century
Jamie Chapman, Consultant

Alternative Energy in Europe
Wolfgang Palz, Commission of the European Communities

Geothermal Energy: Electricity Production and Environmental Impact, A Worldwide Perspective
Ronald DiPippo, Southeastern Massachusetts University

Geothermal Energy Opportunities for Developing Countries
Carel Otte, Unocal Corporation

Large-Scale Solar/Wind Electrical Production Systems—Predictions for the 21st Century
Jon McGowan, University of Massachusetts

Session D-IV: Demand Management Alternatives
Chair: David White, MIT

Environmental Protection through Energy Conservation: A "Free Lunch" at Last?
Larry Ruff, Putnam, Hayes, and Bartlett, UK

Structural Factors Underlying the Increasing Demand for Electricity in Japan
Hisao Kibune, The Institute of Energy Economics, Japan

Encouraging Electric-Utility Energy-Efficiency and Load-Management Programs
Eric Hirst, Oak Ridge National Laboratory

Transitional Strategies for Emission Reduction in Electric Power Generation
Richard Tabors, MIT

System-Wide Evaluation of Efficiency Improvements: Reducing Local, Regional, and Global Environmental Impacts
Stephen Connors, Clinton Andrews, MIT

Wednesday, March 28

10:30 am–12:00 noon

Policy and R&D Issues in Electric Power

Discussion of the major policy issues identified in the four parallel sessions on March 27 (Fossil Fuel, Nuclear, Alternative Energy, and Demand Management Alternatives). Reports by the four session chairs consider the following: factors that control electric energy use, special requirements of the LDC's in meeting electric energy demand, technology options including environmental constraints.

R&D Opportunities

A panel report by the four session chairs including interaction with participants on an appropriate R&D agenda to adjust to regional and global environmental issues.

Session E: Economics and Policy

Presentation and discussion of national and global economic studies of strategies and policy initiatives for reducing greenhouse gas emissions.

Tuesday, March 27
1:00 pm–5:30 pm

Opening Session on Economics and Policy
Chair: David Wood, MIT

Effects on Energy Markets and the US Economy of Measures to Reduce CO_2 Emissions from Fossil Fuels
W. David Montgomery, US Congressional Budget Office

A Least Cost Energy Analysis of US CO_2 Reduction Options
Barry Solomon et al., US Environmental Protection Agency

DOE Estimates of Costs Associated with Alternative Greenhouse Gas Emission Reduction Policies
Edward Williams and Richard Bradley, US Department of Energy

CO_2 Emission Reductions: A Regional Economic Cost Analysis
Alan Manne, Stanford University, and Richard Richels, Electric Power Research Institute

Session F: Advanced Energy Supply Technologies

Special session covering advanced energy supply technologies that could have significant long-term impact on a worldwide scale. Technical and economic factors are reviewed for each technology.

Wednesday, March 28
8:00 am–10:30 am

Special Session on Advanced Energy Supply Technologies: Current Status, Remaining R&D Issues, and Potential for the Future
Chair: Robert Duffield, Salk Institute

The Prospect for Fusion Energy in the 21st Century
D. Bruce Montgomery, MIT

Economic, Environmental, and Engineering Aspects of Magnetohydrodynamic Power Generation
Robert Kessler, Avco Everett Research Laboratory, Inc.

Fuel Cells: Power for the Future
Frank Schora, E.H. Cámara, M-C Power Corporation

Hot Dry Rock Geothermal Energy—An Emerging Energy Resource with Worldwide Potential
C.W. Myers, Donald Brown, and Robert Potter, Los Alamos National Laboratory

Perspectives on Renewable Energy and the Environment
Dan Hartley and D.G. Schueler, Sandia National Laboratories

Appendix B: Unit Conversion Table

	BTUs	quads	calories	kWh	MWy
BTUs	1	10^{-15}	252	2.93×10^{-4}	3.35×10^{-11}
quads	10^{15}	1	2.52×10^{17}	2.93×10^{11}	3.35×10^{4}
calories	3.97×10^{-3}	3.97×10^{-18}	1	1.16×10^{-6}	1.33×10^{-13}
kWh	3413	3.41×10^{-12}	8.60×10^{5}	1	1.14×10^{-7}
MWy	2.99×10^{10}	2.99×10^{-5}	7.53×10^{12}	8.76×10^{6}	1
bbls oil	5.50×10^{6}	5.50×10^{-9}	1.38×10^{9}	1612	1.84×10^{-4}
tonnes oil	4.04×10^{7}	4.04×10^{-8}	1.02×10^{10}	1.18×10^{4}	1.35×10^{-3}
kg coal	2.78×10^{4}	2.78×10^{-11}	7×10^{6}	8.14	9.29×10^{-7}
tonnes coal	2.78×10^{7}	2.78×10^{-8}	7×10^{9}	8139	9.29×10^{-4}
MCF gas	10^{6}	10^{-9}	2.52×10^{8}	293	3.35×10^{-5}
joules	9.48×10^{-4}	9.48×10^{-19}	0.239	2.78×10^{-7}	3.17×10^{-14}
EJ	9.48×10^{14}	0.948	2.39×10^{17}	2.78×10^{11}	3.17×10^{4}

To convert the first column units to other units, multiply by the factors shown, e.g., 1 BTU = 252 calories.
Key: MWy—Megawatt-years; bbls—barrels; tonnes—metric tons = 1000 kg = 2204.6 lb; MCF—thousand cubic feet; EJ—exajoules.
Assumed calorific values: oil—10180 cal/g; coal—7000 cal/g; gas—1000 BTU/ft^3 at standard conditions.

bbls oil equivalent	tonnes oil equivalent	kg coal equivalent	tonnes coal equivalent	MCF gas equivalent	joules	EJ
1.82×10^{-7}	2.48×10^{-8}	3.6×10^{-5}	3.6×10^{-8}	10^{-6}	1055	1.06×10^{-15}
1.82×10^{8}	2.48×10^{7}	3.6×10^{10}	3.6×10^{7}	10^{9}	1.06×10^{18}	1.06
7.21×10^{-10}	9.82×10^{-11}	1.43×10^{-7}	1.43×10^{-10}	3.97×10^{-9}	4.19	4.19×10^{-18}
6.20×10^{-4}	8.45×10^{-5}	0.123	1.23×10^{-4}	3.41×10^{-3}	3.6×10^{6}	3.6×10^{-12}
5435	740	1.08×10^{6}	1076	2.99×10^{4}	3.15×10^{13}	3.15×10^{-5}
1	0.136	198	0.198	5.50	5.80×10^{9}	5.80×10^{-9}
7.35	1	1455	1.45	40.4	4.26×10^{10}	4.26×10^{-8}
5.05×10^{-3}	6.88×10^{-4}	1	0.001	0.0278	2.93×10^{7}	2.93×10^{-11}
5.05	0.688	1000	1	27.8	2.93×10^{10}	2.93×10^{-8}
0.182	0.0248	36	0.036	1	1.06×10^{9}	1.06×10^{-9}
1.72×10^{-10}	2.35×10^{-11}	3.41×10^{-8}	3.41×10^{-11}	9.48×10^{-10}	1	10^{-18}
1.72×10^{8}	2.35×10^{7}	3.41×10^{10}	3.41×10^{7}	9.48×10^{8}	10^{18}	1

Index